W9-CLQ-042

EDITION: 41

Steam

its generation and use

The Babcock & Wilcox Company
a McDermott company

Edited by J.B. Kitto and S.C. Stultz

Disclaimer

The information contained within this book has been obtained by The Babcock & Wilcox Company from sources believed to be reliable. However, neither The Babcock & Wilcox Company nor its authors make any guarantee or warranty, expressed or implied, about the accuracy, completeness or usefulness of the information, product, process or apparatus discussed within this book, nor shall The Babcock & Wilcox Company or any of its authors be liable for error, omission, losses or damages of any kind or nature. This book is published with the understanding that The Babcock & Wilcox Company and its authors are supplying general information and neither attempting to render engineering or professional services nor offering a product for sale. If services are desired, an appropriate professional should be consulted.

Steam/its generation and use. 41st edition.
Editors: John B. Kitto and Steven C. Stultz.
The Babcock & Wilcox Company, Barberton, Ohio, U.S.A.
2005

Includes bibliographic references and index.
Subject areas: 1. Steam boilers.
2. Combustion – Fossil fuels.
3. Nuclear power.

The editors welcome any technical comments, notes on inaccuracies, or thoughts on important omissions. Please direct these to the editors at SteamBook@babcock.com.

ISBN 0-9634570-1-2
Library of Congress Catalog Number: 92-74123
ISSN 1556-5173

Printed in the United States of America.

Steam, Edition: 41

Steam/its generation and use is the longest continuously published engineering text of its kind in the world. It has always been, and continues to be, written and published by The Babcock & Wilcox Company, *the Original*, headquartered in Barberton, Ohio, and incorporated in Delaware, The United States of America.

a McDermott company

The Babcock & Wilcox Company

Preface

Dear Reader:

The founders of our company, George Babcock and Stephen Wilcox, invented the safety water tube boiler. This invention resulted in the commercialization of large-scale utility generating stations. Rapid increases in generation of safe, dependable and economic electricity literally fueled the Industrial Revolution and dramatically increased the standard of living in the United States and industrialized economies worldwide throughout the twentieth century.

Advancements in technology to improve efficiency and reduce environmental emissions have continued for nearly 140 years, creating a unique and valuable body of applied engineering that represents the individual and collective contributions of several generations of employees. As in other areas of science and engineering, our field has continued to evolve, resulting in an extensive amount of new material that has been incorporated into our 41st edition of *Steam/its generation and use*. This edition required an extensive amount of personal time and energy from hundreds of employees and reflects our commitment to both our industry and our future.

Today it is clear that the challenge to generate power more efficiently from fossil fuels, while minimizing impacts to our environment and global climate, will require significant technological advancements. These advances will require creativity, perseverance and ingenuity on the part of our employees and our customers. For inspiration, we can recall the relentless drive and imagination of one of our first customers, Mr. Thomas Alva Edison. For strength, we will continue to embrace our Core Values of Quality, Integrity, Service and People which have served us well over our long history as a company.

I thank our shareholders, our employees, our customers, our partners and our suppliers for their continued dedication, cooperation and support as we move forward into what will prove to be a challenging and rewarding century.

To help guide us all along the way, I am very pleased to present *Edition: 41*.

David L. Keller
President and Chief Operating Officer
The Babcock & Wilcox Company

Table of Contents

Acknowledgments .... viii to ix
System of Units: English and Systême International x
Editors' Foreword xi
Introduction to *Steam* .... Intro-1 to 17
Selected Color Plates, Edition: 41 .... Plates 1 to 8

Section I – Steam Fundamentals

Chapter 1 Steam Generation – An Overview .... 1-1 to 1-17
2 Thermodynamics of Steam .... 2-1 to 2-27
3 Fluid Dynamics .... 3-1 to 3-17
4 Heat Transfer .... 4-1 to 4-33
5 Boiling Heat Transfer, Two-Phase Flow and Circulation .... 5-1 to 5-21
6 Numerical Modeling for Fluid Flow, Heat Transfer, and Combustion .... 6-1 to 6-25
7 Metallurgy, Materials and Mechanical Properties .... 7-1 to 7-25
8 Structural Analysis and Design .... 8-1 to 8-17

Section II – Steam Generation from Chemical Energy

Chapter 9 Sources of Chemical Energy .... 9-1 to 9-19
10 Principles of Combustion .... 10-1 to 10-31
11 Oil and Gas Utilization .... 11-1 to 11-17
12 Solid Fuel Processing and Handling .... 12-1 to 12-19
13 Coal Pulverization .... 13-1 to 13-15
14 Burners and Combustion Systems for Pulverized Coal .... 14-1 to 14-21
15 Cyclone Furnaces .... 15-1 to 15-13
16 Stokers .... 16-1 to 16-11
17 Fluidized-Bed Combustion .... 17-1 to 17-15
18 Coal Gasification .... 18-1 to 18-17
19 Boilers, Superheaters and Reheaters .... 19-1 to 19-21
20 Economizers and Air Heaters .... 20-1 to 20-17
21 Fuel Ash Effects on Boiler Design and Operation .... 21-1 to 21-27
22 Performance Calculations .... 22-1 to 22-21
23 Boiler Enclosures, Casing and Insulation .... 23-1 to 23-9
24 Boiler Cleaning and Ash Handling Systems .... 24-1 to 24-21
25 Boiler Auxiliaries .... 25-1 to 25-23

Section III – Applications of Steam

Chapter 26 Fossil Fuel Boilers for Electric Power .... 26-1 to 26-17
27 Boilers for Industry and Small Power .... 27-1 to 27-21
28 Chemical and Heat Recovery in the Paper Industry .... 28-1 to 28-29
29 Waste-to-Energy Installations .... 29-1 to 29-23
30 Wood and Biomass Installations .... 30-1 to 30-11
31 Marine Applications .... 31-1 to 31-13

Section IV – Environmental Protection

Chapter 32 Environmental Considerations . . . . . 32-1 to 32-17
33 Particulate Control . . . . . 33-1 to 33-13
34 Nitrogen Oxides Control . . . . . 34-1 to 34-15
35 Sulfur Dioxide Control . . . . . 35-1 to 35-19
36 Environmental Measurement . . . . . 36-1 to 36-15

Section V – Specification, Manufacturing and Construction

Chapter 37 Equipment Specification, Economics and Evaluation . . . . . 37-1 to 37-17
38 Manufacturing . . . . . 38-1 to 38-13
39 Construction . . . . . 39-1 to 39-19

Section VI – Operations

Chapter 40 Pressure, Temperature, Quality and Flow Measurement . . . . . 40-1 to 40-25
41 Controls for Fossil Fuel-Fired Steam Generating Plants . . . . . 41-1 to 41-21
42 Water and Steam Chemistry, Deposits and Corrosion . . . . . 42-1 to 42-29
43 Boiler Operations . . . . . 43-1 to 43-17

Section VII – Service and Maintenance

Chapter 44 Maintaining Availability . . . . . 44-1 to 44-21
45 Condition Assessment . . . . . 45-1 to 45-21

Section VIII – Steam Generation from Nuclear Energy

Chapter 46 Steam Generation from Nuclear Energy . . . . . 46-1 to 46-25
47 Fundamentals of Nuclear Energy . . . . . 47-1 to 47-15
48 Nuclear Steam Generators . . . . . 48-1 to 48-15
49 Nuclear Services and Operations . . . . . 49-1 to 49-21
50 Nuclear Equipment Manufacture . . . . . 50-1 to 50-13

Appendices

Appendix 1 Conversion Factors, SI Steam Properties and Useful Tables . . . . . T-1 to T-16
2 Codes and Standards . . . . . C-1 to C-6
Symbols, Acronyms and Abbreviations . . . . . S-1 to S-10
B&W Trademarks in Edition: 41 . . . . . TM-1
Index . . . . . I-1 to I-22

Acknowledgments

Steam/its generation and use is the culmination of the work of hundreds of B&W employees who have contributed directly and indirectly to this edition and to the technology upon which it is based. Particular recognition goes to individuals who formally committed to preparing and completing this expanded 41st edition.

Editor-in-Chief/Project Manager

S.C. Stultz

Technical Editor/Technical Advisor

J.B. Kitto

Art Director/Assistant Editor

G.L. Tomei

Assistant Technical Editors

J.J. Gaidos
M.A. Miklic

Lead Authors

M.J. Albrecht
G.T. Bielawski
K.P. Brolly
P.A. Campanizzi
P.L. Cioffi
R.A. Clocker
P.L. Daniel
R.A. Detzel
J.A. Dickinson
W. Downs
D.D. Dueck
S.J. Elmiger
J.S. Gittinger
J.E. Granger*
G.R. Grant
G.H. Harth
T.C. Heil
D.A. Huston
B.J. Jankura
C.S. Jones
K.L. Jorgensen
J.B. Kitto
D.L. Kraft
A.D. LaRue
M.P. Lefebvre
P. Li
G.J. Maringo
W.N. Martin
E.H. Mayer*
D.K. McDonald
R.M. McNertney Jr.
J.E. Monacelli
T.E. Moskal
N.C. Polosky
E.F. Radke
K.E. Redinger
J.D. Riggs
D.E. Ryan
D.P. Scavuzzo
S.A. Scavuzzo
W.G. Schneider
T.D. Shovlin
T.A. Silva
B.C. Sisler
J.W. Smith
R.E. Snyder
W.R. Stirgwolt
J.R. Strempek
S.C. Stultz
J.M. Tanzosh
G.L. Tomei
D.P. Tonn
S.J. Vecci
P.S. Weitzel
R.A. Wessel
L.C. Westfall
P.J. Williams

* The editors offer special acknowledgment to authors J.E. Granger and E.H. Mayer who passed away during the preparation of Edition: 41.

Primary Support Authors

S.A. Bryk
D.E. Burnham
D.S. Fedock
J.T. Griffin
B.L. Johnson
N. Kettenbauer
T.P. Kors
G.J. Lance
R.C. Lenzer
E.P.B. Mogensen
G.M. Pifer
K.J. Rogers
B.J. Youmans

Executive Steering Committee

B.C. Bethards
E.M. Competti
J.S. Kulig
D.C. Langley
J.W. Malone
M.G. Morash
R.E. Reimels

Production Group

J.L. Basar
L.A. Brower
P.L. Fox
L.M. Shepherd

Outside Support

P.C. Lutjen (Art)
J.R. Grizer (Tables)

System of Units
English and Système International

To recognize the globalization of the power industry, the 41st edition of *Steam* incorporates the Système International d'Unitès (SI) along with the continued use of English or U.S. Customary System (USCS) units. English units continue to be the primary system of units with SI provided as secondary units in parentheses. In some instances, SI units alone have been provided where these units are common usage. In selected figures and tables where dual units could detract from clarity (logarithmic scales, for example) SI conversions are provided within the figure titles or as a table footnote.

Extensive English-SI conversion tables are provided in Appendix 1. This appendix also contains a complete SI set of the Steam Tables, Mollier diagram, pressure-enthalpy diagram and psychrometric chart.

The decision was made to provide exact conversions rounded to an appropriate number of figures. This was done to avoid confusion about the original source values.

Absolute pressure is denoted by *psi* or *kPa / MPa* and gauge pressure by *psig* or *kPa / MPa gauge*. The difference between *absolute pressure* and *pressure difference* is identified by the context. Finally, in Chapters 10 and 22, as well as selected other areas of *Steam* which provide extensive numerical examples, only English units have been provided for clarity.

For reference and clarity, power in British thermal units per hour (Btu/h) has typically been converted to megawatts-thermal and is denoted by MW_t while megawatts-electric in both systems of units has been denoted by MW.

The editors hope that these conversion practices will make *Steam* easily usable by the broadest possible audience.

Editors' Foreword

When we completed the 40th edition of *Steam* in 1992, we had a sense that perhaps our industry was stabilizing. But activity has again accelerated. Today, efficiencies are being driven even higher. Emissions are being driven even lower. Many current technologies are being stretched, and new technologies are being developed, tested and installed. We have once again changed much of *Steam* to reflect our industry's activity and anticipated developments.

Recognizing the rich history of this publication, we previously drew words from an 1883 edition's preface to say that *"we have revised the whole, and added much new and valuable matter."* For this new 41st edition we can draw from the 1885 edition to say *"Having again revised Steam, and enlarged it by the addition of new and useful information, not published heretofore, we shall feel repaid for the labor if it shall prove of value to our customers."*

We hope this new edition is of equal value to our partners and suppliers, government personnel, students and educators, and all present and future employees of The Babcock & Wilcox Company.

Introduction to *Steam*

Throughout history, mankind has reached beyond the acceptable to pursue a challenge, achieving significant accomplishments and developing new technology. This process is both scientific and creative. Entire civilizations, organizations, and most notably, individuals have succeeded by simply doing what has never been done before. A prime example is the safe and efficient use of steam.

One of the most significant series of events shaping today's world is the industrial revolution that began in the late seventeenth century. The desire to generate steam on demand sparked this revolution, and technical advances in steam generation allowed it to continue. Without these developments, the industrial revolution as we know it would not have taken place.

It is therefore appropriate to say that few technologies developed through human ingenuity have done so much to advance mankind as the safe and dependable generation of steam.

Steam as a resource

In 200 B.C., a Greek named Hero designed a simple machine that used steam as a power source (Fig. 1). He began with a cauldron of water, placed above an open fire. As the fire heated the cauldron, the cauldron shell transferred the heat to the water. When the water reached the boiling point of 212F (100C), it changed form and turned into steam. The steam passed through two pipes into a hollow sphere, which was pivoted at both sides. As the steam escaped through two tubes attached to the sphere, each bent at an angle, the sphere moved, rotating on its axis.

Hero, a mathematician and scientist, labeled the device *aeolipile*, meaning rotary steam engine. Although the invention was only a novelty, and Hero made no suggestion for its use, the idea of generating steam to do useful work was born. Even today, the basic idea has remained the same – generate heat, transfer the heat to water, and produce steam.

Intimately related to steam generation is the steam turbine, a device that changes the energy of steam into mechanical work. In the early 1600s, an Italian named Giovanni Branca produced a unique invention (Fig. 2). He first produced steam, based on Hero's aeolipile. By channeling the steam to a wheel that rotated, the steam pressure caused the wheel to turn. Thus began the development of the steam turbine.

The primary use of steam turbines today is for electric power production. In one of the most complex systems ever designed by mankind, superheated high-pressure steam is produced in a boiler and channeled to turbine-generators to produce electricity.

Fig. 1 Hero's aeolipile.

Today's steam plants are a complex and highly sophisticated combination of engineered elements. Heat is obtained either from primary fossil fuels like coal, oil or natural gas, or from nuclear fuel in the form of uranium. Other sources of heat-producing energy include waste heat and exhaust gases, bagasse and biomass, spent chemicals and municipal waste, and geothermal and solar energy.

Each fuel contains potential energy, or a heating value measured in Btu/lb (J/kg). The goal is to release this energy, most often by a controlled combustion process or, with uranium, through fission. The heat is then transferred to water through tube walls and other components or liquids. The heated water then changes form, turning into steam. The steam is normally heated further to specific temperatures and pressures.

Steam is also a vital resource in industry. It drives pumps and valves, helps produce paper and wood products, prepares foods, and heats and cools large buildings and institutions. Steam also propels much of the world's naval fleets and a high percentage of commercial marine transport. In some countries, steam plays a continuing role in railway transportation.

Steam generators, commonly referred to as *boilers*, range in size from those needed to heat a small building to those used individually to produce 1300 megawatts of electricity in a power generating station – enough power for more than one million people. These larger units deliver more than ten million pounds of superheated steam per hour (1260 kg/s) with steam temperatures exceeding 1000F (538C) and pressures exceeding 3800 psi (26.2 MPa).

Today's steam generating systems owe their dependability and safety to the design, fabrication and operation of safe water tube boilers, first patented by George Babcock and Stephen Wilcox in 1867 (Fig. 3).

Because the production of steam power is a tremendous resource, it is our challenge and responsibility to further develop and use this resource safely, efficiently, dependably, and in an environmentally-friendly manner.

The early use of steam

Steam generation as an industry began almost two thousand years after Hero's invention, in the seventeenth century. Many conditions began to stimulate the development of steam use in a power cycle. Mining for ores and minerals had expanded greatly and large quantities of fuel were needed for ore refining.

Fig. 2 Branca's steam turbine.

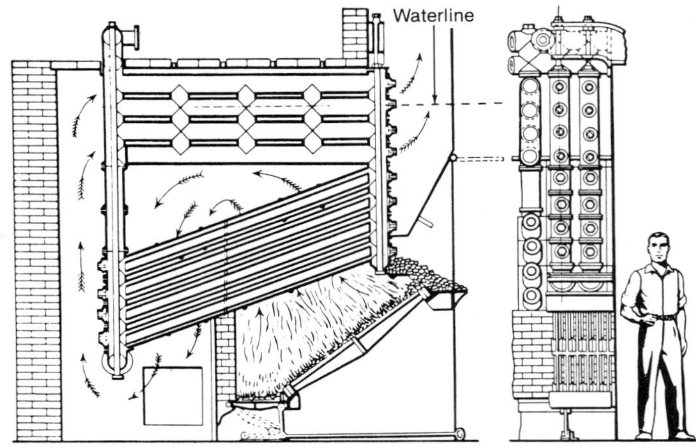

Fig. 3 First Babcock & Wilcox boiler, patented in 1867.

Fuels were needed for space heating and cooking and for general industrial and military growth. Forests were being stripped and coal was becoming an important fuel. Coal mining was emerging as a major industry.

As mines became deeper, they were often flooded with underground water. The English in particular were faced with a very serious curtailment of their industrial growth if they could not find some economical way to pump water from the mines. Many people began working on the problem and numerous patents were issued for machines to pump water from the mines using *the expansive power of steam*. The early machines used wood and charcoal for fuel, but coal eventually became the dominant fuel.

The most common source of steam at the time was a *shell* boiler, little more than a large kettle filled with water and heated at the bottom (Fig. 4).

Not all early developments in steam were directed toward pumps and engines. In 1680, Dr. Denis Papin, a Frenchman, invented a steam digester for food pro-

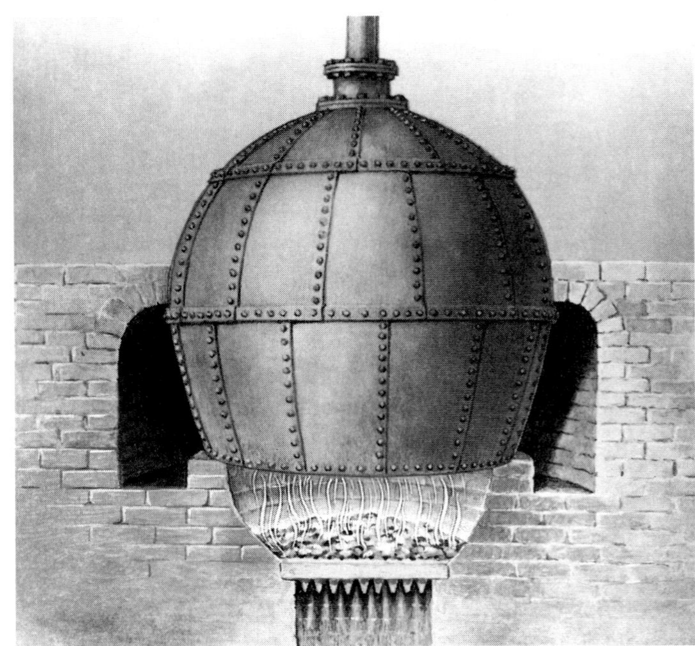

Fig. 4 Haycock shell boiler, 1720.

cessing, using a boiler under heavy pressure. To avoid explosion, Papin added a device which is the first safety valve on record. Papin also invented a boiler with an internal firebox, the earliest record of such construction.

Many experiments concentrated on using steam pressure or atmospheric pressure combined with a vacuum. The result was the first commercially successful steam engine, patented by Thomas Savery in 1698, to pump water by direct displacement (Fig. 5). The patent credits Savery with an engine for raising water by the impellant force of fire, meaning steam. The mining industry needed the invention, but the engine had a limited pumping height set by the pressure the boiler and other vessels could withstand. Before its replacement by Thomas Newcomen's engine (described below), John Desaguliers improved the Savery engine, adding the Papin safety valve and using an internal jet for the condensing part of the cycle.

Steam engine developments continued and the earliest cylinder-and-piston unit was based on Papin's suggestion, in 1690, that the condensation of steam should be used to make a vacuum beneath a piston, after the piston had been raised by expanding steam. Newcomen's atmospheric pressure engine made practical use of this principle.

While Papin neglected his own ideas of a steam engine to develop Savery's invention, Thomas Newcomen and his assistant John Cawley adapted Papin's suggestions in a practical engine. Years of experimentation ended with success in 1711 (Fig. 6). Steam admitted from the boiler to a cylinder raised a piston by expansion and assistance from a counterweight on the other end of a beam, actuated by the piston. The steam valve was then closed and the steam in the cylinder was condensed by a spray of cold water. The vacuum which formed caused the piston to be forced downward by atmospheric pressure, doing work on a pump. Condensed water in the cylinder was expelled through a valve by the entry of steam which was at a pressure slightly above atmospheric. A 25 ft (7.6 m) oak beam, used to transmit power from the cylinder to the water pump, was a dominant feature

Fig. 6 Newcomen's beam engine, 1711.

of what came to be called the *beam engine*. The boiler used by Newcomen, a plain copper brewer's kettle, was known as the Haycock type. (See Fig. 4.)

The key technical challenge remained the need for higher pressures, which meant a more reliable and stronger boiler. Basically, evolution of the steam boiler paralleled evolution of the steam engine.

During the late 1700s, the inventor James Watt pursued developments of the steam engine, now physically separated from the boiler. Evidence indicates that he helped introduce the first *waggon boiler*, so named because of its shape (Fig. 7). Watt concentrated on the engine and developed the separate steam condenser to create the vacuum and also replaced atmospheric pressure with steam pressure, improving the engine's efficiency. He also established the measurement of horsepower, calculating that one horse could raise 550 lb (249 kg) of weight a distance of 1 ft (0.3 m) in one second, the equivalent of 33,000 lb (14,969 kg) a distance of one foot in one minute.

Fig. 5 Savery's engine, circa 1700.

Fig. 7 Waggon boiler, 1769.

Fire tube boilers

The next outstanding inventor and builder was Richard Trevithick, who had observed many pumping stations at his father's mines. He realized that the problem with many pumping systems was the boiler capacity. Whereas copper was the only material previously available, hammered wrought iron plates could now be used, although the maximum length was 2 ft (0.6 m). Rolled iron plates became available in 1875.

In 1804, Trevithick designed a higher pressure engine, made possible by the successful construction of a high pressure boiler (Fig. 8). Trevithick's boiler design featured a cast iron cylindrical shell and dished end.

As demand grew further, it became necessary to either build larger boilers with more capacity or put up with the inconveniences of operating many smaller units. Engineers knew that the longer the hot gases were in contact with the shell and the greater the exposed surface area, the greater the capacity and efficiency.

While a significant advance, Newcomen's engine and boiler were so thermally inefficient that they were frequently only practical at coal mine sites. To make the system more widely applicable, developers of steam engines began to think in terms of fuel economy. Noting that nearly half the heat from the fire was lost because of short contact time between the hot gases and the boiler heating surface, Dr. John Allen may have made the first calculation of boiler efficiency in 1730. To reduce heat loss, Allen developed an internal furnace with a smoke flue winding through the water, like a coil in a still. To prevent a deficiency of combustion air, he suggested the use of bellows to force the gases through the flue. This probably represents the first use of forced draft.

Later developments saw the single pipe flue replaced by many gas tubes, which increased the amount of heating surface. These *fire tube* boilers were essentially the design of about 1870. However, they were limited in capacity and pressure and could not meet the needs that were developing for higher pressures and larger unit sizes. Also, there was the ominous record of explosions and personal injury because of direct heating of the pressure shell, which contained large volumes of water and steam at high temperature and pressure.

The following appeared in the 1898 edition of *Steam*: *That the ordinary forms of boilers* (fire tube boilers) *are liable to explode with disastrous effect is conceded. That they do so explode is witnessed by the sad list of casualties from this cause every year, and almost every day. In the year 1880, there were 170 explosions reported in the United States, with a loss of 259 lives, and 555 persons injured. In 1887 the number of explosions recorded was 198, with 652 persons either killed or badly wounded. The average reported for ten years past has been about the same as the two years given, while doubtless many occur which are not recorded.*

Inventors recognized the need for a new design, one that could increase capacity and limit the consequences of pressure part rupture at high pressure and temperature. *Water tube* boiler development began.

Early water tube design

A patent granted to William Blakey in 1766, covering an improvement in Savery's steam engine, includes a form of steam generator (Fig. 9). This probably was the first step in the development of the water tube boiler. However, the first successful use of a water tube design was by James Rumsey, an American inventor who patented several types of boilers in 1788. Some of these boilers used water tube designs.

At about this time John Stevens, also an American, invented a water tube boiler consisting of a group of small tubes closed at one end and connected at the

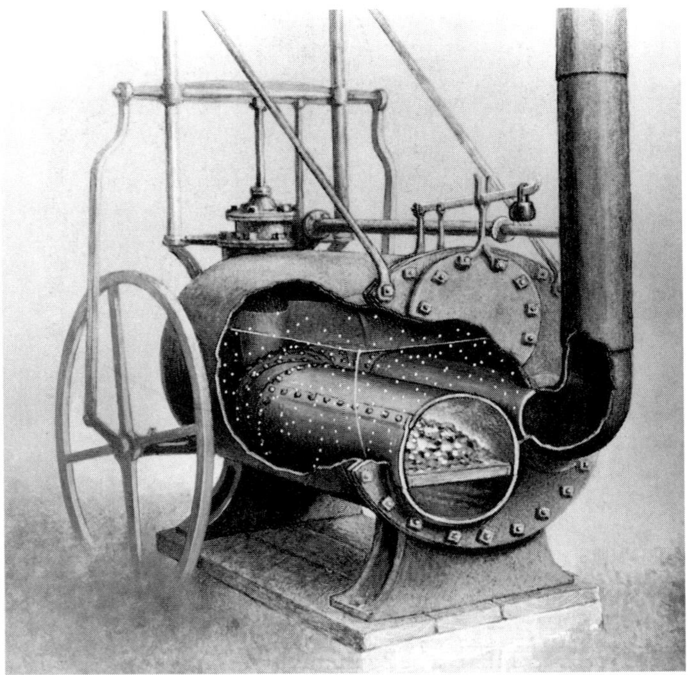

Fig. 8 Trevithick boiler, 1804.

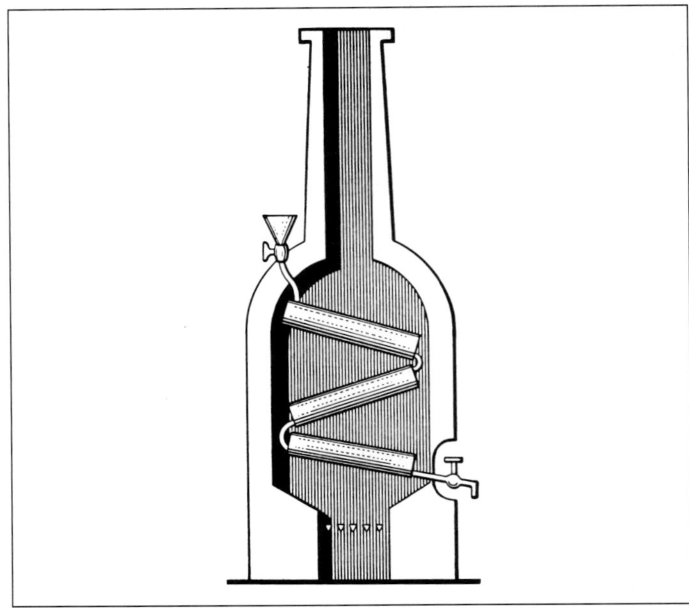

Fig. 9 William Blakey boiler, 1766.

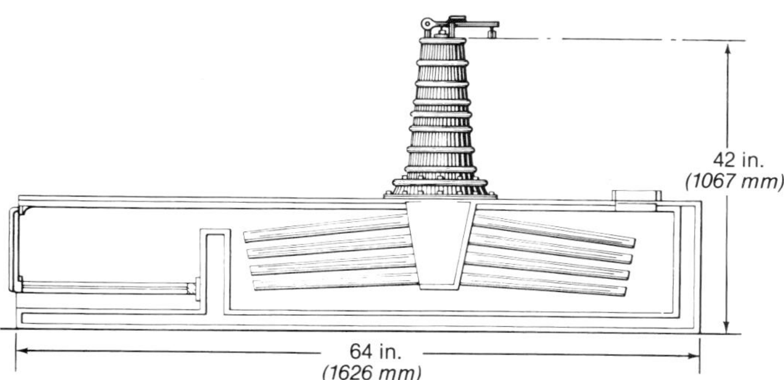

Fig. 10 John Stevens water tube boiler, 1803.

other to a central reservoir (Fig. 10). Patented in the United States (U.S.) in 1803, this boiler was used on a Hudson River steam boat. The design was short lived, however, due to basic engineering problems in construction and operation.

Blakey had gone to England to obtain his patents, as there were no similar laws in North America. Stevens, a lawyer, petitioned the U.S. Congress for a patent law to protect his invention and such a law was enacted in 1790. It may be said that part of the basis of present U.S. patent laws grew out of the need to protect a water tube boiler design. Fig. 11 shows another form of water tube boiler, this one patented by John Cox Stevens in 1805.

In 1822, Jacob Perkins built a water tube boiler that is the predecessor of the once-through steam generator. A number of cast iron bars with longitudinal holes were arranged over the fire in three tiers by connecting the ends outside of the furnace with a series of bent pipes. Water was fed to the top tier by a feed pump and superheated steam was discharged from the lower tier to a collecting chamber.

The Babcock & Wilcox Company

It was not until 1856, however, that a truly successful water tube boiler emerged. In that year, Stephen Wilcox, Jr. introduced his version of the water tube design with improved water circulation and increased heating surface (Fig. 12). Wilcox had designed a boiler

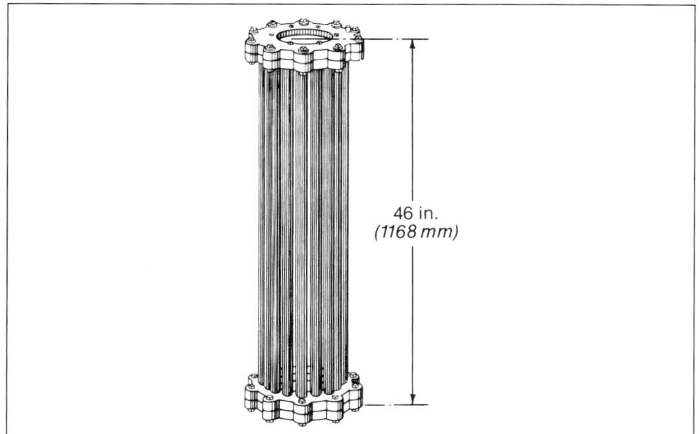

Fig. 11 Water tube boiler with tubes connecting water chamber below and steam chamber above. John Cox Stevens, 1805.

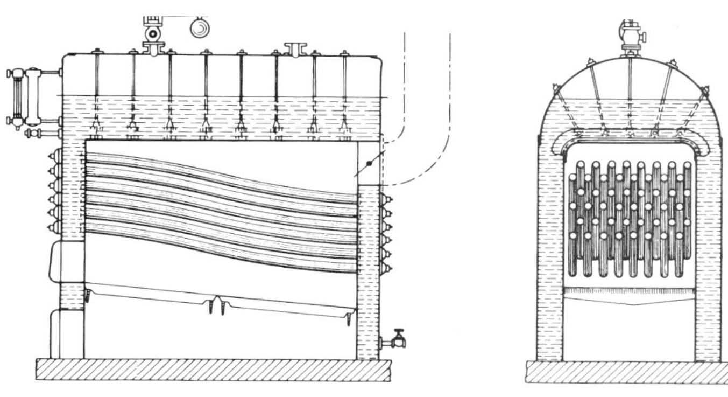

Fig. 12 Inclined water tubes connecting front and rear water spaces, complete with steam space above. Stephen Wilcox, 1856.

with inclined water tubes that connected water spaces at the front and rear, with a steam chamber above. Most important, as a water tube boiler, his unit was inherently *safe*. His design revolutionized the boiler industry.

In 1866, Wilcox partnered with his long-time friend, George H. Babcock. The following year, U.S. Patent No. 65,042 was granted to George H. Babcock and Steven Wilcox, Jr., and the partnership of Babcock, Wilcox and Company was formed. In 1870 or 1871, Babcock and Wilcox became the sole proprietors, dropping *Company* from the name, and the firm was known as Babcock & Wilcox until its incorporation in 1881, when it changed its name to The Babcock & Wilcox Company (B&W). (see Fig. 3).

Industrial progress continued. In 1876, a giant-sized Corliss steam engine, a device invented in Rhode Island in 1849, went on display at the Centennial Ex-

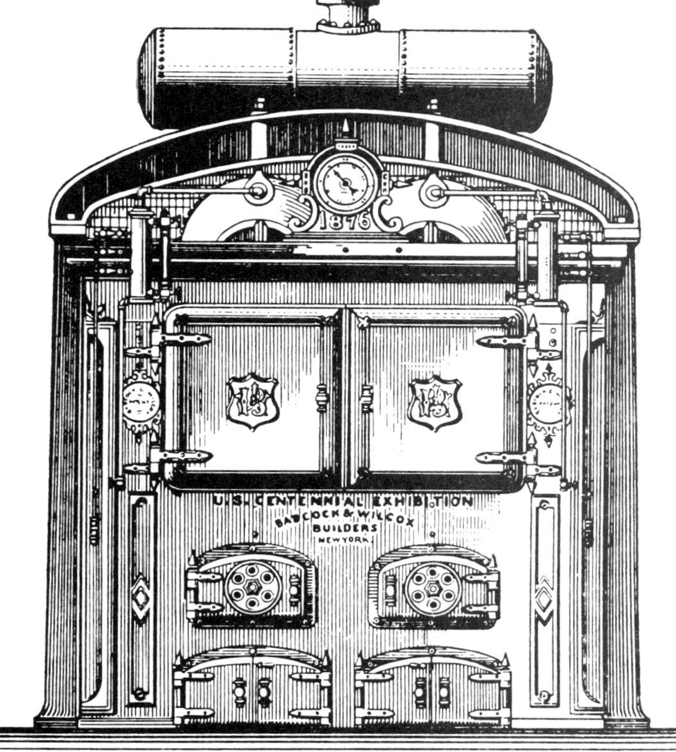

Fig. 13 Babcock & Wilcox Centennial boiler, 1876.

hibition in Philadelphia, Pennsylvania, as a symbol of worldwide industrial development. Also on prominent display was a 150 horsepower water tube boiler (Fig. 13) by George Babcock and Stephen Wilcox, who were by then recognized as engineers of unusual ability. Their professional reputation was high and their names carried prestige. By 1877, the Babcock & Wilcox boiler had been modified and improved by the partners several times (Fig. 14).

At the exhibition, the public was awed by the size of the Corliss engine. It weighed 600 tons and had cylinders 3 ft (0.9 m) in diameter. But this giant size was to also mark the end of the steam engine, in favor of more efficient prime movers, such as the steam turbine. This transition would add momentum to further development of the Babcock & Wilcox water tube boiler. By 1900, the steam turbine gained importance as the major steam powered source of rotary motion, due primarily to its lower maintenance costs, greater overloading tolerance, fewer number of moving parts, and smaller size.

Perhaps the most visible technical accomplishments of the time were in Philadelphia and New York City. In 1881 in Philadelphia, the Brush Electric Light Company began operations with four boilers totaling 292 horsepower. In New York the following year, Thomas Alva Edison threw the switch to open the Pearl Street Central station, ushering in the *age of the cities*. The boilers in Philadelphia and the four used by Thomas Edison in New York were built by B&W, now incorporated. The boilers were heralded as *sturdy, safe and reliable*. When asked in 1888 to comment on one of the units, Edison wrote: *It is the best boiler God has permitted man yet to make*. (Fig. 15).

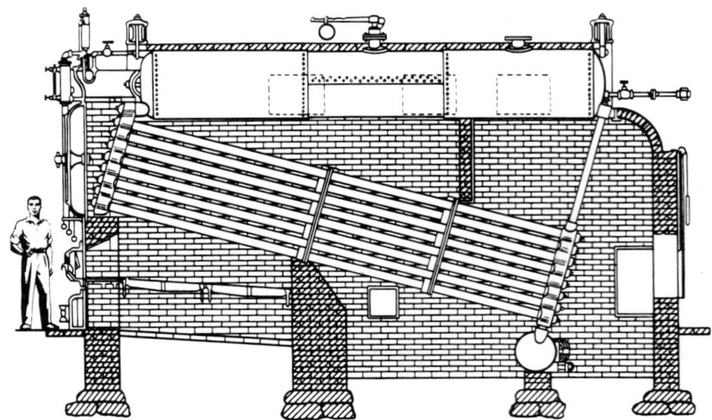

Fig. 14 Babcock & Wilcox boiler developed in 1877.

The historic Pearl Street Central station opened with 59 customers using about 1300 lamps. The B&W boilers consumed 5 tons of coal and 11,500 gal (43,532 l) of water per day.

The B&W boiler of 1881 was a safe and efficient steam generator, ready for the part it would play in worldwide industrial development.

Water tube marine boilers

The first water tube marine boiler built by B&W was for the *Monroe* of the U.S. Army's Quartermaster

George Herman Babcock

George Herman Babcock was born June 17, 1832 near Otsego, New York. His father was a well known inventor and mechanic. When George was 12 years old, his parents moved to Westerly, Rhode Island, where he met Stephen Wilcox, Jr.

At age 19, Babcock started the *Literary Echo*, editing the paper and running a printing business. With his father, he invented the first polychromatic printing press, and he also patented a job press which won a prize at the London Crystal Palace International Exposition in 1855.

In the early 1860s, he was made chief draftsman of the Hope Iron Works at Providence, Rhode Island, where he renewed his acquaintance with Stephen Wilcox and worked with him in developing the first B&W boiler. In 1886, Babcock became the sixth president of the American Society of Mechanical Engineers.

He was the first president of The Babcock & Wilcox Company, a position he held until his death in 1893.

department. A major step in water tube marine boiler design came in 1889, with a unit for the steam yacht *Reverie*. The U.S. Navy then ordered three ships featuring a more improved design that saved about 30% in weight from previous designs. This design was again improved in 1899, for a unit installed in the U.S. cruiser *Alert*, establishing the superiority of the water tube boiler for marine propulsion. In this installation, the firing end of the boiler was reversed, placing the firing door in what had been the rear wall of the boiler. The furnace was thereby enlarged in the direction in which combustion took place, greatly improving combustion conditions.

The development of marine boilers for naval and merchant ship propulsion has paralleled that for land use (see Fig. 16). Throughout the twentieth century and into the twenty-first, dependable water tube marine boilers have contributed greatly to the excellent performance of naval and commercial ships worldwide.

Bent tube design

The success and widespread use of the inclined straight tube B&W boiler stimulated other inventors to explore new ideas. In 1880, Allan Stirling developed a design connecting the steam generating tubes directly to a steam separating drum and featuring low headroom above the furnace. The Stirling Boiler Company was formed to manufacture and market an improved Stirling® design, essentially the same as shown in Fig. 17.

The merits of *bent tubes* for certain applications were soon recognized by George Babcock and Stephen Wilcox, and what had become the Stirling Consolidated Boiler Company in Barberton, Ohio, was purchased by B&W in 1906. After the problems of internal tube cleaning were solved, the bent tube boiler replaced the straight tube design. The continuous and economical production of clean, dry steam, even when using poor quality feedwater, and the ability to meet sudden load swings were features of the new B&W design.

Electric power

Until the late 1800s, steam was used primarily for heat and as a tool for industry. Then, with the advent of practical electric power generation and distribution, utility companies were formed to serve industrial and residential users across wide areas. The pioneer stations in the U.S. were the Brush Electric Light Company and the Commonwealth Edison Company. Both used B&W boilers exclusively.

During the first two decades of the twentieth century, there was an increase in steam pressures and temperatures to 275 psi (1.9 MPa) and 560F (293C), with 146F (81C) superheat. In 1921, the North Tess station of the Newcastle Electric Supply Company in northern England went into operation with steam at 450 psi (3.1 MPa) and a temperature of 650F (343C). The steam was reheated to 500F (260C) and regenerative feedwater heating was used to attain a boiler feedwater temperature of 300F (149C). Three years later, the Crawford Avenue station of the Commonwealth Edison Company and the Philo and Twin

Stephen Wilcox, Jr.

Stephen Wilcox was born February 12, 1830 at Westerly, Rhode Island.

The first definite information concerning his engineering activities locates him in Providence, Rhode Island, about 1849, trying to introduce a caloric engine. In 1853, in association with Amos Taylor of Mystic, Connecticut, he patented a letoff motion for looms. In 1856, a patent for a steam boiler was issued to Stephen Wilcox and O.M. Stillman. While this boiler differed materially from later designs, it is notable as his first recorded step into the field of steam generation.

In 1866 with George Babcock, Wilcox developed the first B&W boiler, which was patented the following year.

In 1869 he went to New York as selling agent for the Hope Iron Works and took an active part in improving the boiler and the building of the business. He was vice president of The Babcock & Wilcox Company from its incorporation in 1881 until his death in 1893.

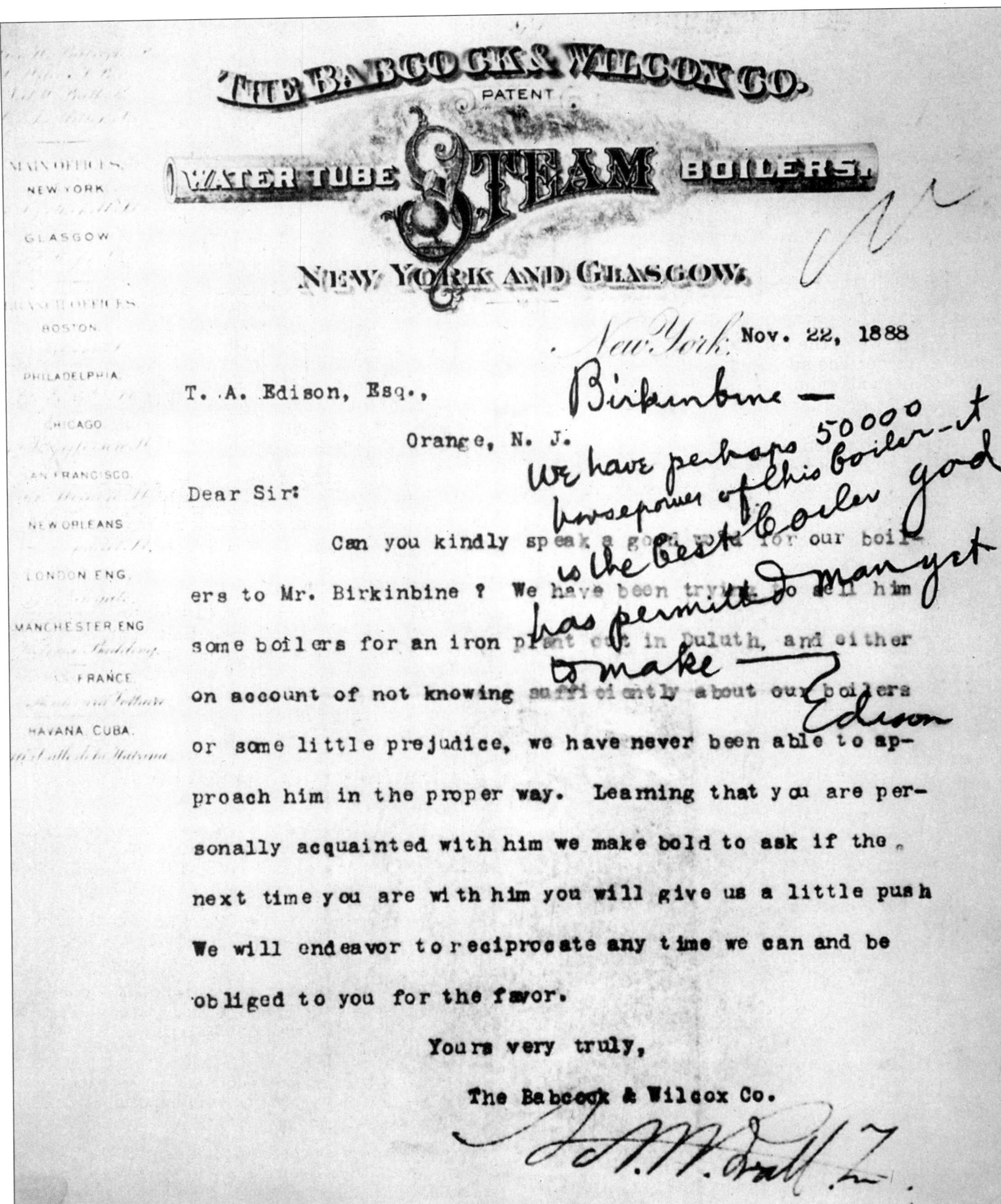
THE BABCOCK & WILCOX CO.
PATENT
WATER TUBE STEAM BOILERS.
NEW YORK AND GLASGOW.

MAIN OFFICES. NEW YORK
GLASGOW
BRANCH OFFICES. BOSTON
PHILADELPHIA
CHICAGO
SAN FRANCISCO
NEW ORLEANS
LONDON ENG.
MANCHESTER ENG.
FRANCE
HAVANA CUBA.

New York. Nov. 22, 1888

T. A. Edison, Esq.,

Orange, N. J.

Dear Sir:

Can you kindly speak a good word for our boilers to Mr. Birkinbine ? We have been trying to sell him some boilers for an iron plant out in Duluth, and either on account of not knowing sufficiently about our boilers or some little prejudice, we have never been able to approach him in the proper way. Learning that you are personally acquainted with him we make bold to ask if the next time you are with him you will give us a little push We will endeavor to reciprocate any time we can and be obliged to you for the favor.

Yours very truly,

The Babcock & Wilcox Co.

Birkinbine — We have perhaps 50000 horsepower of this boiler—it is the best boiler god has permitted man yet to make — Edison

Fig. 15 Thomas Edison's endorsement, 1888.

Branch stations of the present American Electric Power system were placed in service with steam at 550 psi (38 MPa) and 725F (385C) at the turbine throttle. The steam was reheated to 700F (371C).

A station designed for much higher steam pressure, the Weymouth (later named Edgar) station of the Boston Edison Company in Massachusetts, began operation in 1925. The 3150 kW high pressure unit used steam at 1200 psi (8.3 MPa) and 700F (371C), reheated to 700F (371C) for the main turbines (Fig. 18).

Pulverized coal and water-cooled furnaces

Other major changes in boiler design and fabrication occurred in the 1920s. Previously, as power generating stations increased capacity, they increased the number of boilers, but attempts were being made to increase the size of the boilers as well. Soon the size requirement became such that existing furnace designs and methods of burning coal, primarily stokers, were no longer adequate.

Pulverized coal was the answer in achieving higher volumetric combustion rates and increased boiler capacity. This could not have been fully exploited without the use of water-cooled furnaces. Such furnaces eliminated the problem of rapid deterioration of the refractory walls due to slag (molten ash). Also, these designs lowered the temperature of the gases leaving the furnace and thereby reduced fouling (accumulation of ash) of convection pass heating surfaces to manageable levels. The first use of pulverized coal in furnaces of stationary steam boilers had been demonstrated at the Oneida Street plant in Milwaukee, Wisconsin, in 1918.

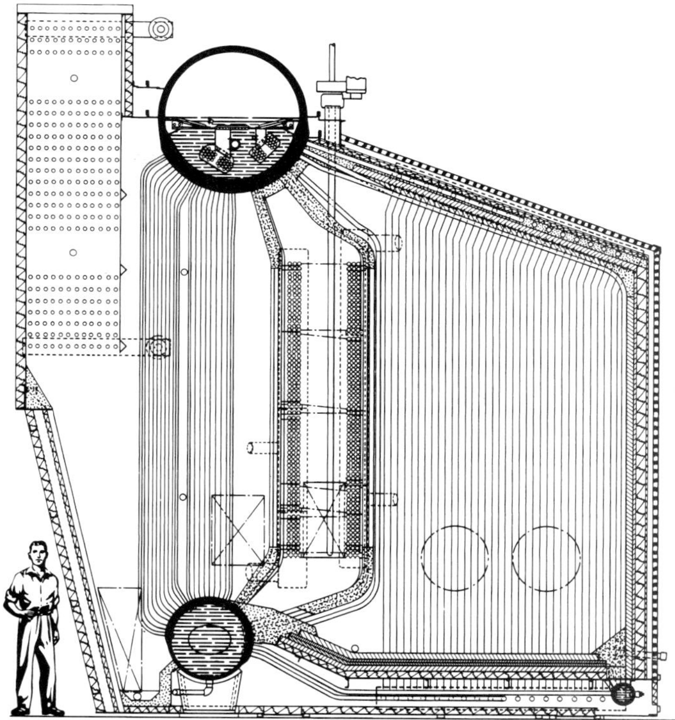

Fig. 16 Two drum Integral Furnace marine boiler.

Integral Furnace boiler

Water cooling was applied to existing boiler designs, with its circulatory system essentially independent of the boiler steam-water circulation. In the early 1930s, however, a new concept was developed that arranged

Requirements of a Perfect Steam Boiler – 1875

In 1875, George Babcock and Stephen Wilcox published their conception of the perfect boiler, listing twelve principles that even today generally represent good design practice:

1st. Proper workmanship and simple construction, using materials which experience has shown to be best, thus avoiding the necessity of early repairs.

2nd. A mud-drum to receive all impurities deposited from the water, and so placed as to be removed from the action of the fire.

3rd. A steam and water capacity sufficient to prevent any fluctuation in steam pressure or water level.

4th. A water surface for the disengagement of the steam from the water, of sufficient extent to prevent foaming.

5th. A constant and thorough circulation of water throughout the boiler, so as to maintain all parts at the same temperature.

6th. The water space divided into sections so arranged that, should any section fail, no general explosion can occur and the destructive effects will be confined to the escape of the contents. Large and free passages between the different sections to equalize the water line and pressure in all.

7th. A great excess of strength over any legitimate strain, the boiler being so constructed as to be free from strains due to unequal expansion, and, if possible, to avoid joints exposed to the direct action of the fire.

8th. A combustion chamber so arranged that the combustion of the gases started in the furnace may be completed before the gases escape to the chimney.

9th. The heating surface as nearly as possible at right angles to the currents of heated gases, so as to break up the currents and extract the entire available heat from the gases.

10th. All parts readily accessible for cleaning and repairs. This is a point of the greatest importance as regards safety and economy.

11th. Proportioned for the work to be done, and capable of working to its full rated capacity with the highest economy.

12th. Equipped with the very best gauges, safety valves and other fixtures.

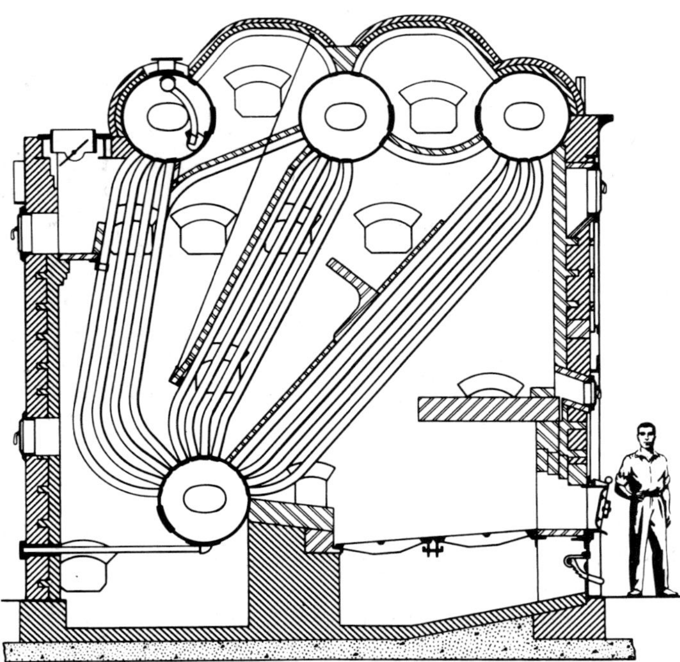

Fig. 17 Early Stirling® boiler arranged for hand firing.

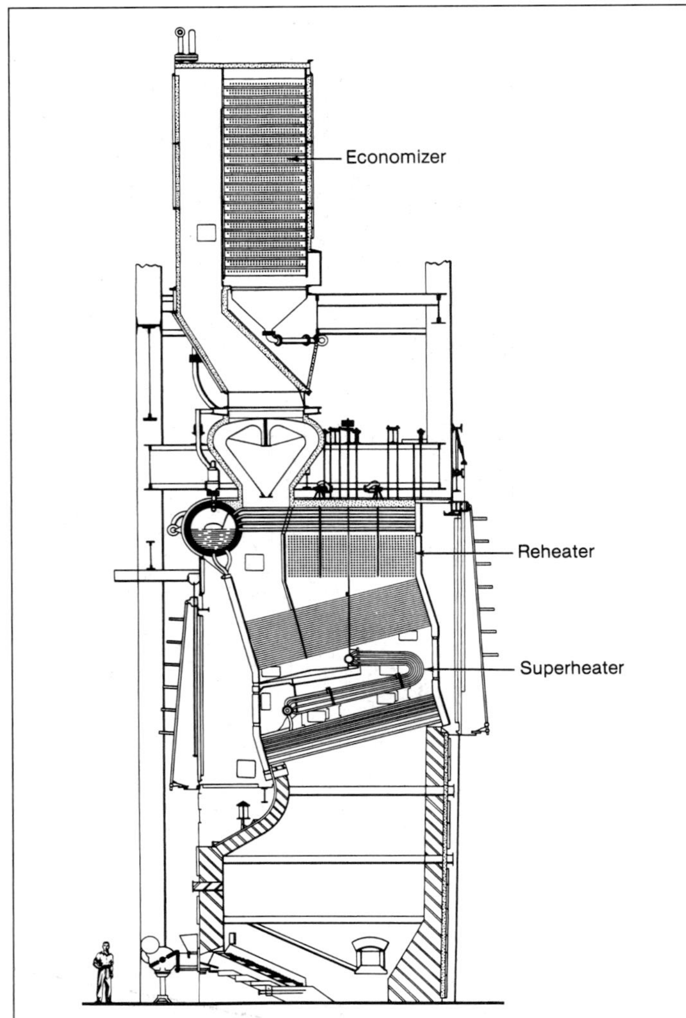

Fig. 18 High pressure reheat boiler, 1925.

the furnace water-cooled surface and the boiler surface together, each as an integral part of the unit (Fig. 19).

Shop-assembled water tube boilers

In the late 1940s, the increasing need for industrial and heating boilers, combined with the increasing costs of field-assembled equipment, led to development of the shop-assembled *package* boiler. These units are now designed in capacities up to 600,000 lb/h (75.6 kg/s) at pressures up to 1800 psi (12.4 MPa) and temperatures to 1000F (538C).

Further developments

In addition to reducing furnace maintenance and the fouling of convection heating surfaces, water cooling also helped to generate more steam. Boiler tube bank surface was reduced because additional steam generating surface was available in the furnace. Increased feedwater and steam temperatures and increased steam pressures, for greater cycle efficiency, further reduced boiler tube bank surface and permitted the use of additional superheater surface.

As a result, Radiant boilers for steam pressures above 1800 psi (12.4 MPa) generally consist of furnace water wall tubes, superheaters, and such heat recovery accessories as economizers and air heaters (Fig. 20). Units for lower pressures, however, have considerable steam generating surface in tube banks (boiler banks) in addition to the water-cooled furnace (Fig. 21).

Universal Pressure boilers

An important milestone in producing electricity at the lowest possible cost took place in 1957. The first

Fig. 19 Integral Furnace boiler, 1933.

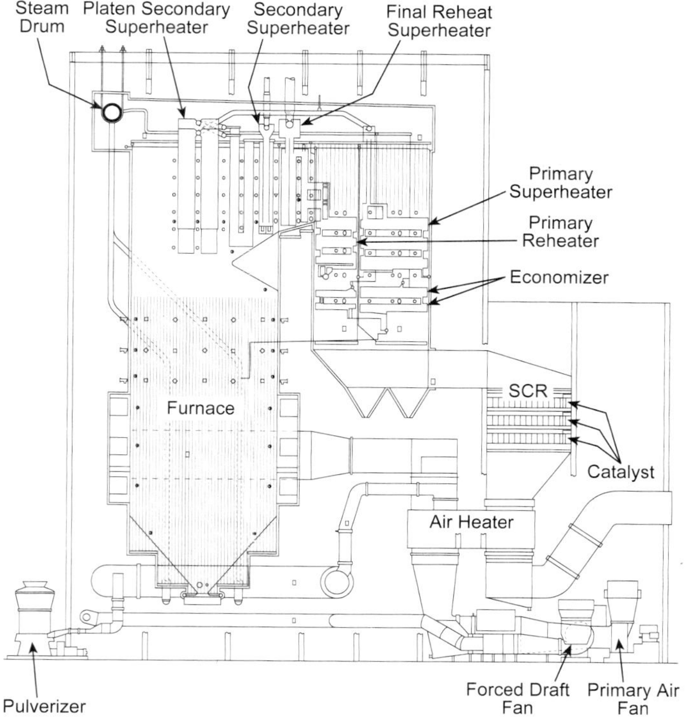

Fig. 20 Typical B&W® Raciant utility boiler.

boiler with steam pressure above the critical value of 3200 psi (22.1 MPa) began commercial operation. This 125 MW B&W Universal Pressure (UP®) steam generator (Fig. 22), located at Ohio Power Company's Philo plant, delivered 675,000 lb/h (85 kg/s) steam at 4550 psi (31.4 MPa); the steam was superheated to 1150F (621C) with two reheats to 1050 and 1000F (566 and 538C).

B&W built and tested its first once-through steam generator for 600 psi (4.1 MPa) in 1916, and built an experimental 5000 psi (34.5 MPa) unit in the late 1920s.

The UP boiler, so named because it can be designed for subcritical or supercritical operation, is capable of rapid load pickup. Increases in load rates up to 5% per minute can be attained.

Fig. 23 shows a typical 1300 MW UP boiler rated at 9,775,000 lb/h (1232 kg/s) steam at 3845 psi (26.5 MPa) and 1010F (543C) with reheat to 1000F (538C). In 1987, one of these B&W units, located in West Virginia, achieved 607 days of continuous operation.

Most recently, UP boilers with spiral wound furnaces (SWUP™ steam generators) have gained wider acceptance for their on/off cycling capabilities and their ability to operate at variable pressure with higher low load power cycle efficiency (see Fig. 24).

Subcritical units, however, remain the dominant design in the existing worldwide boiler fleet. Coal has remained the dominant fuel because of its abundant supply in many countries.

Other fuels and systems

B&W has continued to develop steam generators that can produce power from an ever widening array of fuels in an increasingly clean and environmentally acceptable manner. Landmark developments by B&W include atmospheric fluidized-bed combustion installations, both bubbling and circulating bed, for reduced emissions.

Waste-to-energy systems also became a major effort worldwide. B&W has installed both mass burn and refuse-derived fuel units to meet this growing demand for waste disposal and electric power generation. B&W installed the world's first waste-to-energy boiler in 1972. In 2000, an acquisition by Babcock & Wilcox expanded the company's capabilities in design and construction of waste-to-energy and biomass boilers and other multi-fuel burning plants.

For the paper industry, B&W installed the first chemical recovery boiler in the U.S. in 1940. Since that time, B&W has developed a long tradition of firsts in this industry and has installed one of the largest black liquor chemical recovery units operating in the world today.

Modified steam cycles

High efficiency cycles involve combinations of gas turbines and steam power in cogeneration, and direct thermal to electrical energy conversion. One direct conversion system includes using conventional fuel or char byproduct from coal gasification or liquefaction.

Despite many complex cycles devised to increase overall plant efficiency, the conventional steam cycle

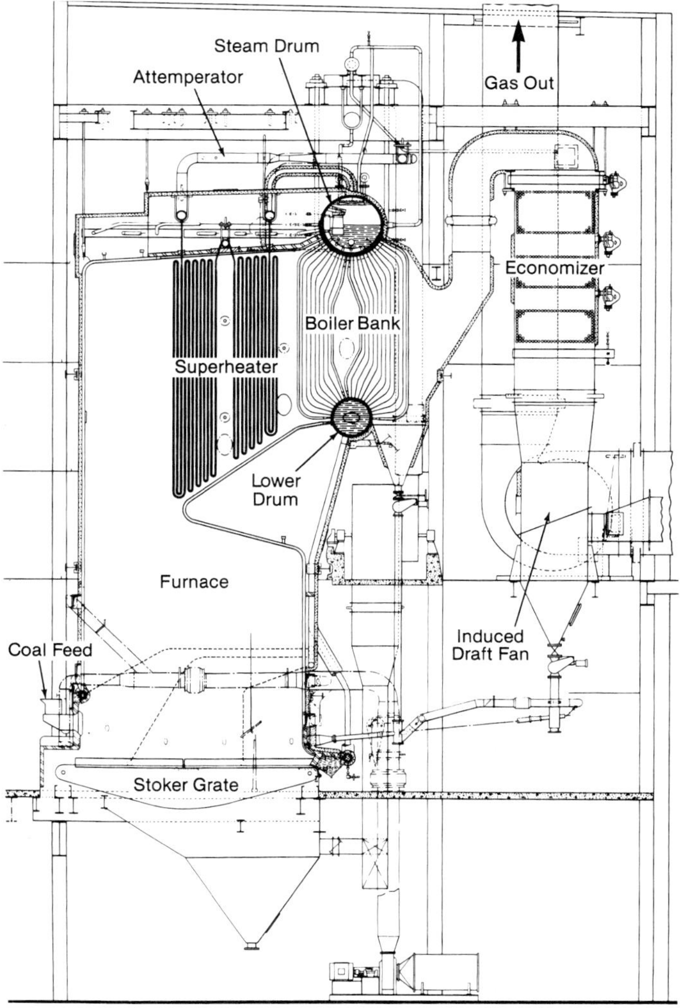

Fig. 21 Lower pressure Stirling® boiler design.

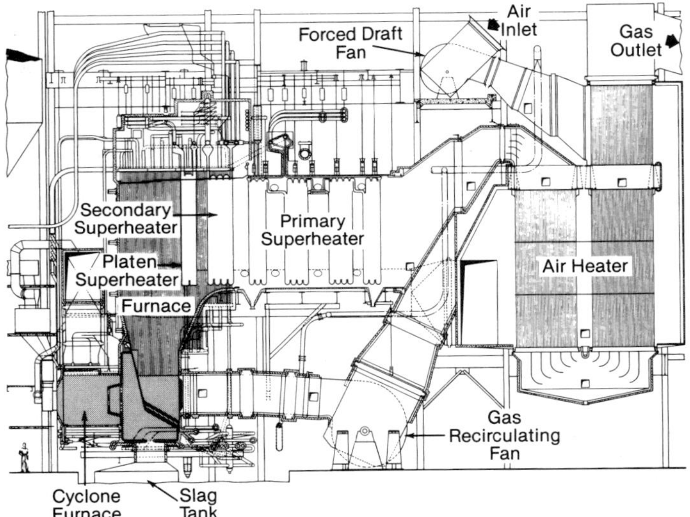

Fig. 22 125 MW B&W® Universal Pressure (UP®) boiler, 1957.

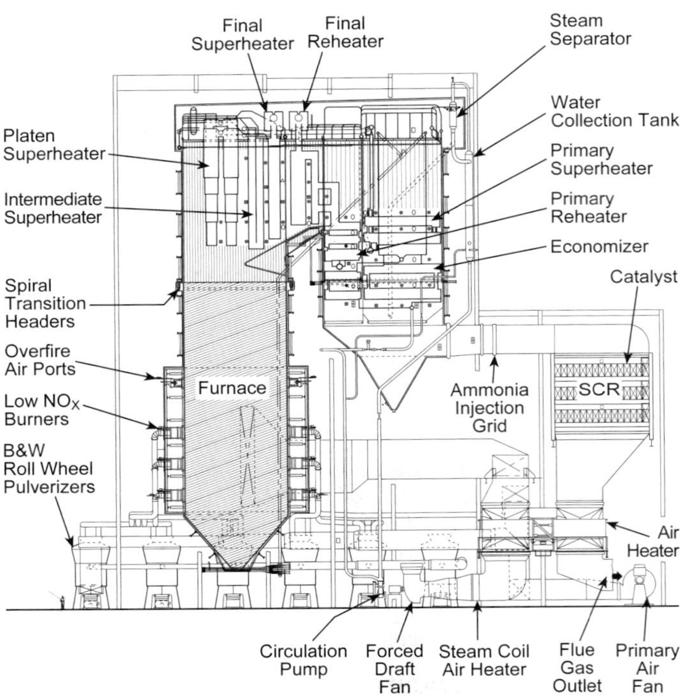

Fig. 24 Boiler with spiral wound universal pressure (SWUP™) furnace.

remains the most economical. The increasing use of high steam pressures and temperatures, reheat superheaters, economizers, and air heaters has led to improved efficiency in the modern steam power cycle.

Nuclear power

Since 1942, when Enrico Fermi demonstrated a controlled self-sustaining reaction, nuclear fission has been recognized as an important source of heat for producing steam for power generation. The first significant application of this new source was the land-based prototype reactor for the U.S.S. *Nautilus* submarine (Fig. 25), operated at the National Reactor Testing Station in Idaho in the early 1950s. This prototype reactor, designed by B&W, was also the basis for land-based pressurized water reactors now being used for electric power generation worldwide. B&W and its affiliates have continued their active involvement in both naval and land-based programs.

The first nuclear electric utility installation was the 90 MW unit at the Shippingport atomic power station in Pennsylvania. This plant, built partly by Duquesne Light Company and partly by the U.S. Atomic Energy Commission, began operations in 1957.

Spurred by the trend toward larger unit capacity, developments in the use of nuclear energy for electric power reached a milestone in 1967 when, in the U.S., nuclear units constituted almost 50% of the 54,000 MW of new capacity ordered that year. Single unit capacity designs have reached 1300 MW. Activity regarding nuclear power was also strong outside the

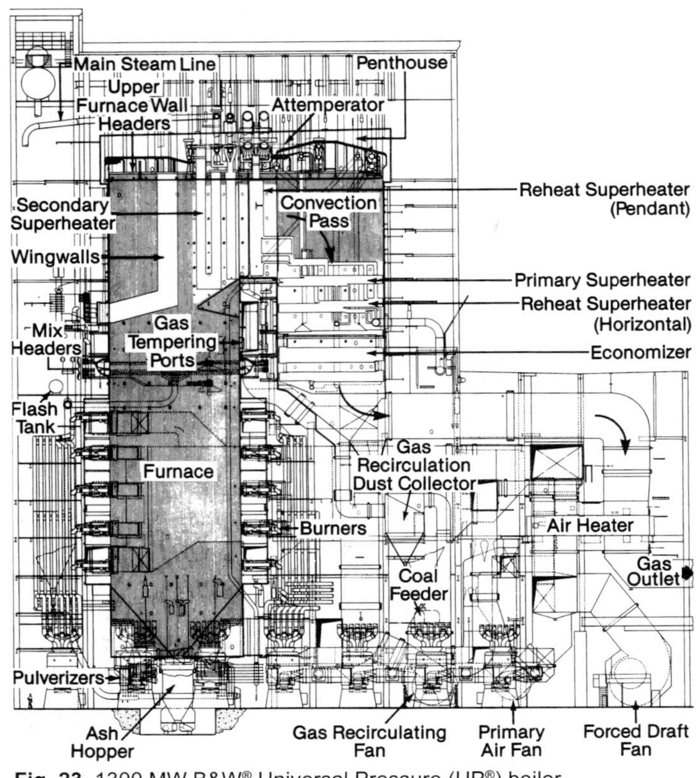

Fig. 23 1300 MW B&W® Universal Pressure (UP®) boiler.

Fig. 25 U.S.S. *Nautilus* – world's first nuclear-powered ship.

U.S., especially in Europe. By 2004, there were 103 reactors licensed to operate in the U.S. Fifty of the operating units had net capacities greater than 1000 MW.

Throughout this period, the nuclear power program in Canada continued to develop based on a design called the Canada Deuterium Uranium (CANDU) reactor system. This system is rated high in both availability and dependability. By 2003, there were 21 units in Canada, all with B&W nuclear steam generators, an additional 11 units operating outside of Canada, and 18 units operating, under construction or planned that are based on CANDU technology.

The B&W recirculating steam generators in these units have continually held excellent performance records and are being ordered to replace aging equipment. (See Fig. 26.)

While the use of nuclear power has remained somewhat steady in the U.S., the future of nuclear power is uncertain as issues of plant operating safety and long-term waste disposal are still being resolved. However, nuclear power continues to offer one of the least polluting forms of large-scale power generation available and may eventually see a resurgence in new construction.

Materials and fabrication

Pressure parts for water tube boilers were originally made of iron and later of steel. Now, steam drums and nuclear pressure vessels are fabricated from heavy steel plates and steel forgings joined by welding. The development of the steam boiler has been necessarily concurrent with advances in metallurgy and progressive improvements in the fabrication and welding of steel and steel alloys.

The cast iron generating tubes used in the first B&W boilers were later superseded by steel tubes. Shortly after 1900, B&W developed a commercial process for the manufacture of hot finished seamless steel boiler tubes, combining strength and reliability with reasonable cost. In the midst of World War II, B&W completed a mill to manufacture tubes by the electric resistance welding (ERW) process. This tubing has now been used in thousands of steam generating units throughout the world.

The cast iron tubes used for steam and water storage in the original B&W boilers were soon replaced by drums. By 1888, drum construction was improved by changing from wrought iron to steel plates rolled into cylinders.

Before 1930, riveting was the standard method of joining boiler drum plates. Drum plate thickness was limited to about 2.75 in. (70 mm) because no satisfactory method was known to secure a tight joint in thicker plates. The only alternative available was to forge and machine a solid ingot of steel into a drum, which was an extremely expensive process. This method was only used on boilers operating at what was then considered high pressure, above 700 psi (4.8 MPa).

The story behind the development of fusion welding was one of intensive research activity beginning in 1926. Welding techniques had to be improved in many respects. Equally, if not more important, an acceptable test procedure had to be found and instituted that would examine the drum without destroying it in the test. After extensive investigation of various testing methods, the medical radiography (x-ray) machine was adapted in 1929 to production examination of welds. By utilizing both x-ray examination and physical tests of samples of the weld material, the soundness of the welds could be determined without affecting the drum.

In 1930, the U.S. Navy adopted a specification for construction of welded boiler drums for naval vessels. In that same year, the first welded drums ever accepted by an engineering authority were part of the B&W boilers installed in several naval cruisers. Also in 1930, the Boiler Code Committee of the American Society of Mechanical Engineers (ASME) issued complete rules and specifications for the fusion welding of drums for power boilers. In 1931, B&W shipped the first welded power boiler drum built under this code.

The x-ray examination of welded drums, the rules declared for the qualification of welders, and the control of welding operations were major first steps in the development of modern methods of quality control in the boiler industry. Quality assurance has received additional stimulus from the naval nuclear propulsion program and from the U.S. Nuclear Regulatory Commission in connection with the licensing of nuclear plants for power generation.

Research and development

Since the founding of the partnership of Babcock, Wilcox and Company in 1867 and continuing to the present day, research and development have played important roles in B&W's continuing service to the power industry. From the initial improvements of Wilcox's original safety water tube boiler to the first supercritical pressure boilers, and from the first privately operated nuclear research reactor to today's advanced environmental systems, innovation and the new ideas of its employees have placed B&W at the forefront of safe, efficient and clean steam generation and energy conversion technology. Today, research and development activities remain an integral part of B&W's focus on tomorrow's product and process requirements.

Fig. 26 B&W replacement recirculating steam generators.

A key to the continued success of B&W is the ability to bring together cross-disciplinary research teams of experts from the many technical specialties in the steam generation field. These are combined with state-of-the-art test facilities and computer systems.

Expert scientists and engineers use equipment designed specifically for research programs in all aspects of fossil power development, nuclear steam systems, materials development and evaluation, and manufacturing technology. Research focuses upon areas of central importance to B&W and steam power generation. However, partners in these research programs have grown to include the U.S. Departments of Energy and Defense, the Environmental Protection Agency, public and private research institutes, state governments, and electric utilities.

Key areas of current research include environmental protection, fuels and combustion technology, heat transfer and fluid mechanics, materials and manufacturing technologies, structural analysis and design, fuels and water chemistry, and measurement and monitoring technology.

Environmental protection

Environmental protection is a key element in all modern steam producing systems where low cost steam and electricity must be produced with minimum impact on the environment. Air pollution control is a key issue for all combustion processes, and B&W has been a leader in this area. Several generations of low nitrogen oxides (NO_x) burners and combustion technology for coal-, oil- and gas-fired systems have been developed, tested and patented by B&W. Post-combustion NO_x reduction has focused on both selective catalytic and non-catalytic reduction systems. Combined with low NO_x burners, these technologies have reduced NO_x levels by up to 95% from historical uncontrolled levels. Ongoing research and testing are being combined with fundamental studies and computer numerical modeling to produce the ultra-low NO_x steam generating systems of tomorrow.

Since the early 1970s, extensive research efforts have been underway to reduce sulfur dioxide (SO_2) emissions. These efforts have included combustion modifications and post-combustion removal. Research during this time aided in the development of B&W's wet SO_2 scrubbing system. This system has helped control emissions from more than 32,000 MW of boiler capacity. Current research focuses on improved removal and operational efficiency, and multi-pollution control technology. B&W has installed more than 9000 MW of boiler capacity using various dry scrubbing technologies. Major pilot facilities have permitted the testing of in-furnace injection, in-duct injection, and dry scrubber systems, as well as atomization, gas conditioning and combined SO_2, NO_x and particulate control. (See Fig. 27.)

Since 1975, B&W has been a leader in fluidized-bed combustion (FBC) technology which offers the ability to simultaneously control SO_2 and NO_x formation as an integral part of the combustion process, as well as burn a variety of waste and other difficult to combust fuels. This work led to the first large scale (20 MW) bubbling-bed system installation in the U.S. B&W's research and development work has focused on process optimization, limestone utilization, and performance characteristics of various fuels and sorbents.

Fig. 27 B&W boiler with SO_2, NO_x, and particulate control systems.

Additional areas of ongoing environmental research include air toxic emissions characterization, efficient removal of mercury, multi-pollutant emissions control, and sulfur trioxide (SO_3) capture, among others (Fig. 28). B&W also continues to review and evaluate processes to characterize, reuse, and if needed, safely dispose of solid waste products.

Fuels and combustion technology

A large number of fuels have been used to generate steam. This is even true today as an ever-widening and varied supply of waste and byproduct fuels such as municipal refuse, coal mine tailings and biomass wastes, join coal, oil and natural gas to meet steam production needs. These fuels must be burned and their combustion products successfully handled while addressing two key trends: 1) declining fuel quality (lower heating value and poorer combustion), and 2) more restrictive emissions limits.

Major strengths of B&W and its work in research and development have been: 1) the characterization of fuels and their ashes, 2) combustion of difficult fuels, and 3) effective heat recovery from the products of combustion. (See Fig. 29.) B&W has earned inter-

Fig. 28 Tests for multi-pollutant emissions control.

national recognition for its fuels analysis capabilities that are based upon generally accepted procedures, as well as specialized B&W procedures. Detailed analyses include, but are not limited to: heating value, chemical constituents, grindability, abrasion resistance, erosiveness, ignition, combustion characteristics, ash composition/viscosity/fusion temperature, and particle size. The results of these tests assist in pulverizer specification and design, internal boiler dimension selection, efficiency calculations, predicted unit availability, ash removal system design, sootblower placement, and precipitator performance evaluation. Thousands of coal and ash samples have been analyzed and catalogued, forming part of the basis for B&W's design methods.

Combustion and fuel preparation facilities are maintained that can test a broad range of fuels at large scale. The 6×10^6 Btu/h (1.8 MW_t) small boiler simulator (Fig. 30) permits a simulation of the time-temperature history of the entire combustion process. The subsystems include a vertical test furnace; fuel subsystem for pulverizing, collecting and firing solid fuels; fuel storage and feeding; emission control modules; gas and stack particulate analyzers for O_2, CO, CO_2 and NO_x; and instrumentation for solids grinding characterization.

Fig. 29 Atomic absorption test for ash composition.

Research continues in the areas of gas-side corrosion, boiler fouling and cleaning characteristics, advanced pulp and paper black liquor combustion, oxygen and oxygen enhanced firing systems, and coal gasification, among others.

Heat transfer and fluid dynamics

Heat transfer is a critical technology in the design of steam generation equipment. For many years, B&W has been conducting heat transfer research from hot gases to tube walls and from the tube walls to enclosed water, steam and air. Early in the 1950s, research in heat transfer and fluid mechanics was initiated in the supercritical pressure region above 3200 psi (22.1 MPa). This work was the technical foundation for the large number of supercritical pressure once-through steam generators currently in service in the electric power industry.

A key advancement in steam-water flow was the invention of the ribbed tube, patented by B&W in 1960. By preventing deterioration of heat transfer under many flow conditions (called critical heat flux or departure from nucleate boiling), the internally ribbed tube made possible the use of natural circulation boilers at virtually all pressures up to the critical point. Extensive experimental studies have provided the critical heat flux data necessary for the design of boilers with both ribbed and smooth bore tubes.

Fig. 30 B&W's small boiler simulator.

Closely related to heat transfer, and of equal importance in steam generating equipment, is fluid mechanics. Both low pressure fluids (air and gas in ducts and flues) and high pressure fluids (water, steam-water mixtures, steam and fuel oil) must be investigated. The theories of single-phase fluid flow are well understood, but the application of theory to the complex, irregular and multiple parallel path geometry of practical situations is often difficult and sometimes impossible. In these cases, analytical procedures must be supplemented or replaced by experimental methods. If reliable extrapolations are possible, economical modeling techniques can be used. Where extrapolation is not feasible, large-scale testing at full pressure, temperature and flow rate is needed.

Advances in numerical modeling technology have made possible the evaluation of the complex three-dimensional flow, heat transfer and combustion processes in coal-fired boiler furnaces. B&W is a leader in the development of numerical computational models to evaluate the combustion of coal, biomass, black liquor and other fuels that have a discrete phase, and the application of these models to full boiler and system analysis (Fig. 31). Continuing development and validation of these models will enhance new boiler designs and expand applications. These models are also valuable tools in the design and evaluation of combustion processes, pollutant formation, and environmental control equipment.

Research, analytical and field test studies in boiling heat transfer, two-phase flow, and stability, among other key areas, continue today by B&W alone and in cooperation with a range of world class organizations.

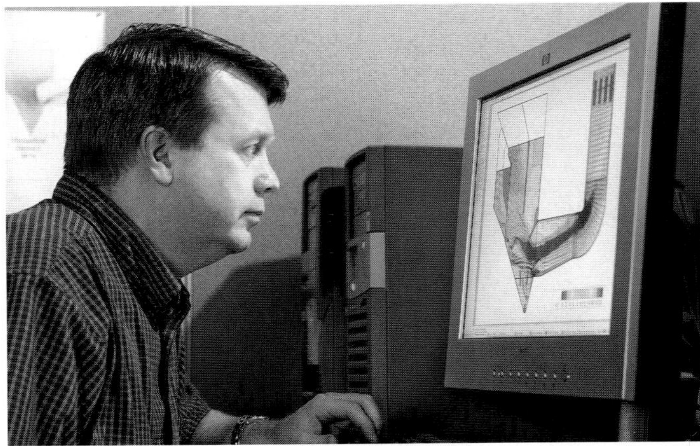

Fig. 31 B&W has developed advanced computational numerical models to evaluate complex flow, heat transfer and combustion processes.

Materials and manufacturing technologies

Because advanced steam producing and energy conversion systems require the application and fabrication of a wide variety of carbon, alloy and stainless steels, nonferrous metals, and nonmetallic materials, it is essential that experienced metallurgical and materials science personnel are equipped with the finest investigative tools. Areas of primary interest in the metallurgical field are fabrication processes such as welding, room temperature and high temperature material properties, resistance to corrosion properties, wear resistance properties, robotic welding, and changes in such material properties under various operating conditions. Development of oxidation-resistant alloys that retain strength at high temperature, and determination of short-term and long-term high temperature properties permitted the increase in steam temperature that has been and continues to be of critical importance in increasing power plant efficiency and reducing the cost of producing electricity.

Advancements in manufacturing have included a process to manufacture large pressure components entirely from weld wire, designing a unique manufacturing process for bi-metallic tubing, using pressure forming to produce metallic heat exchangers, developing air blown ultra-high temperature fibrous insulation, and combining sensor and control capabilities to improve quality and productivity of manufacturing processes.

Research and development activities also include the study of materials processing, joining processes, process metallurgy, analytical and physical metallurgical examination, and mechanical testing. The results are subsequently applied to product improvement.

Structural analysis and design

The complex geometries and high stresses under which metals must serve in many products require careful study to allow prediction of stress distribution and intensity. Applied mechanics, a discipline with highly sophisticated analytical and experimental techniques, can provide designers with calculation methods and other information to assure the safety of structures and reduce costs by eliminating unnecessarily conservative design practices. The analytical techniques involve advanced mathematical procedures and computational tools as well as the use of advanced computers. An array of experimental tools and techniques are used to supplement these powerful analytical techniques.

Computational finite element analysis has largely displaced experimental measurement for establishing detailed local stress relationships. B&W has developed and applied some of the most advanced computer programs in the design of components for the power industry. Advanced techniques permit the evaluation of stresses resulting from component response to thermal and mechanical (including vibratory) loading.

Fracture mechanics, the evaluation of crack formation and growth, is an important area where analytical techniques and new experimental methods permit a better understanding of failure modes and the prediction of remaining component life. This branch of technology has contributed to the feasibility and safety of advanced designs in many types of equipment.

To provide part of the basis for these models, extensive computer-controlled experimental facilities allow the assessment of mechanical properties for materials under environments similar to those in which they will operate. Some of the evaluations include tensile and impact testing, fatigue and corrosion fatigue, fracture toughness, as well as environmentally assisted cracking.

Fuel and water chemistry

Chemistry plays an important role in supporting the effective operation of steam generating systems. Therefore, diversified chemistry capabilities are essential to support research, development and engineering. The design and operation of fuel burning equipment must be supported by expert analysis of a wide variety of solid, liquid and gaseous fuels and their products of combustion, and characterization of their behavior under various conditions. Long-term operation of steam generating equipment requires extensive water programs including high purity water analysis, water treatment and water purification. Equipment must also be chemically cleaned at intervals to remove water-side deposits.

To develop customized programs to meet specific needs, B&W maintains a leadership position in these areas through an expert staff for fuels characterization, water chemistry and chemical cleaning. Studies focus on water treatment, production and measurement of ultra-high purity water (parts per billion), water-side deposit analysis, and corrosion product transport.

B&W was involved in the introduction of oxygen water treatment for U.S. utility applications. Specialized chemical cleaning evaluations are conducted to prepare cleaning programs for utility boilers, industrial boilers and nuclear steam generators. Special analyses are frequently required to develop boiler-specific cleaning solvent solutions that will remove the desired deposits without damaging the equipment.

Measurements and monitoring technology

Development, evaluation and accurate assessment of modern power systems require increasingly precise measurements in difficult to reach locations, often in hostile environments. To meet these demanding needs, B&W continues the investigation of specialized sensors, measurement and nondestructive examination. B&W continues to develop diagnostic methods that lead to advanced systems for burner and combustion systems as well as boiler condition assessment.

These techniques have been used to aid in laboratory research such as void fraction measurements for steam-water flows. They have also been applied to operating steam generating systems. New methods have been introduced by B&W to nondestructively measure oxide thicknesses on the inside of boiler tubes, detect hydrogen damage, and detect and measure corrosion fatigue cracks. Acoustic pyrometry systems have been introduced by B&W to nonintrusively measure high temperature gases in boiler furnaces.

Steam/its generation and use

This updated and expanded edition provides a broad, in-depth look at steam generating technology and equipment, including related auxiliaries that are of interest to engineers and students in the steam power industry. The reader will find discussions of the fundamental technologies such as thermodynamics, fluid mechanics, heat transfer, solid mechanics, numerical and computational methods, materials science and fuels science. The various components of the steam generating equipment, plus their integration and performance evaluation, are covered in depth. Extensive additions and updates have been made to the chapters covering environmental control technologies and numerical modeling. Key elements of the balance of the steam generating system life including operation, condition assessment, maintenance, and retrofits are also discussed.

Selected Color Plates — Edition: 41

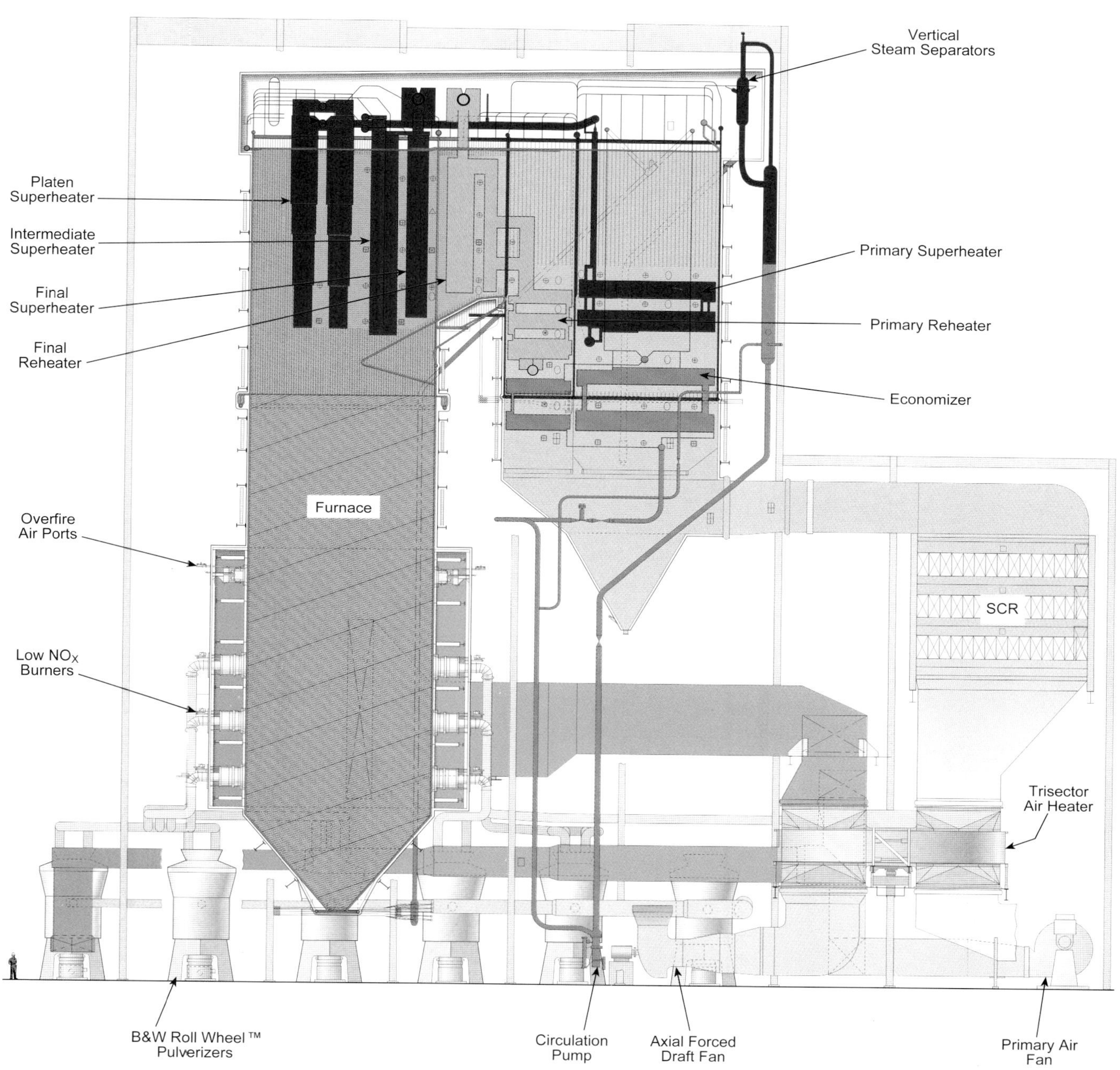

B&W supercritical boiler with spiral wound Universal Pressure (SWUP™) furnace.

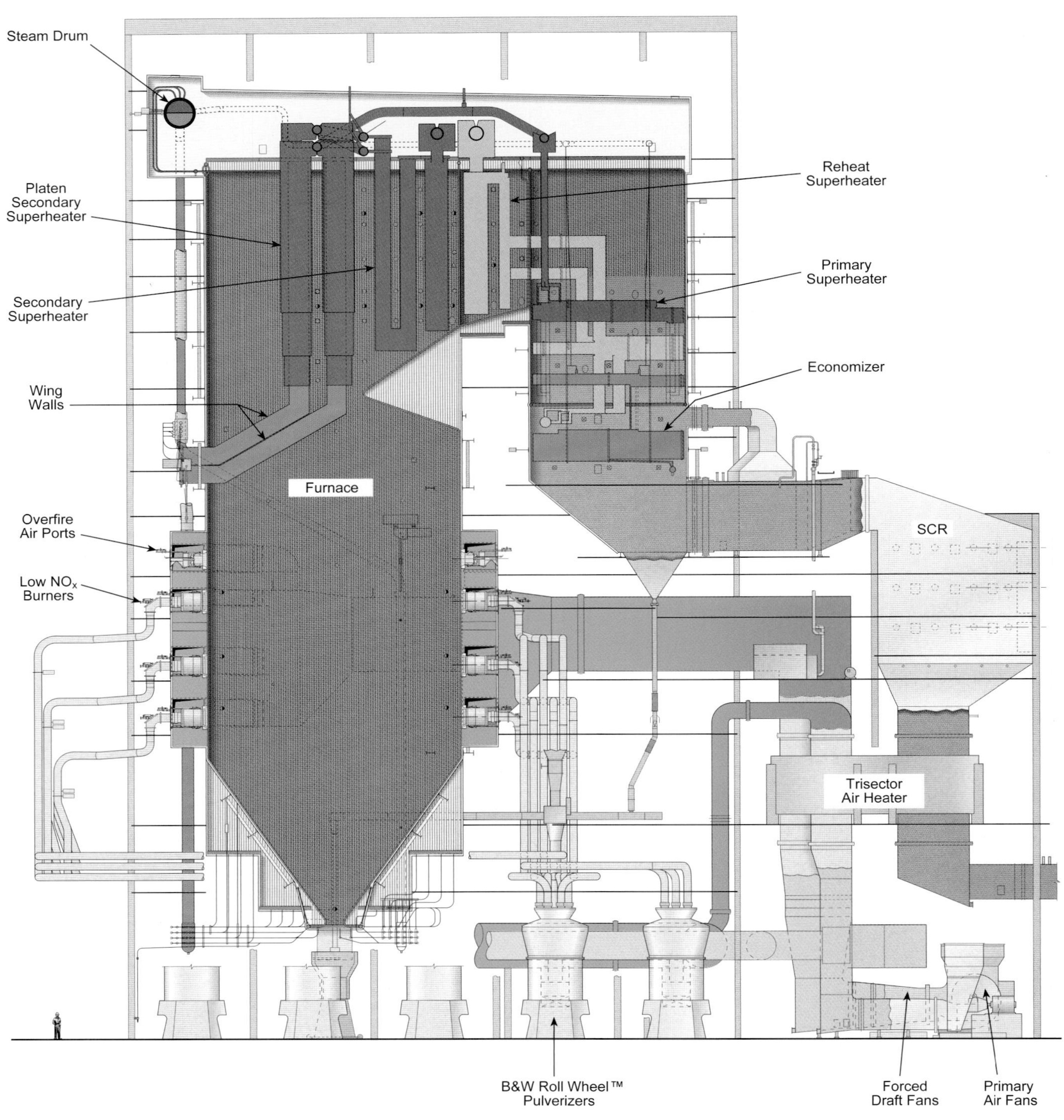

Carolina-type 550 MW Radiant boiler for pulverized coal.

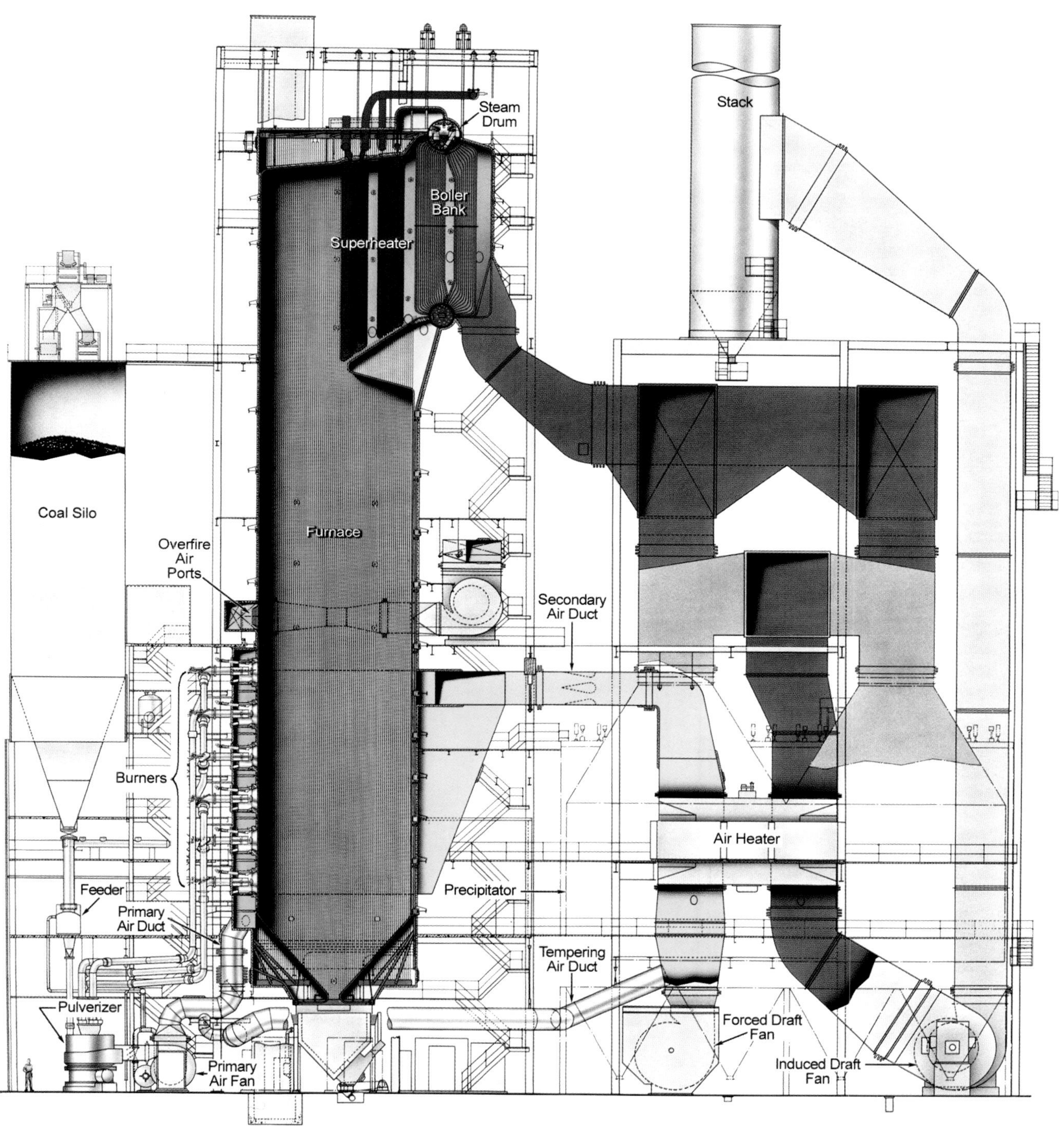

Large coal- and oil-fired two-drum Stirling® power boiler for industry, 885,000 lb/h (112 kg/s) steam flow.

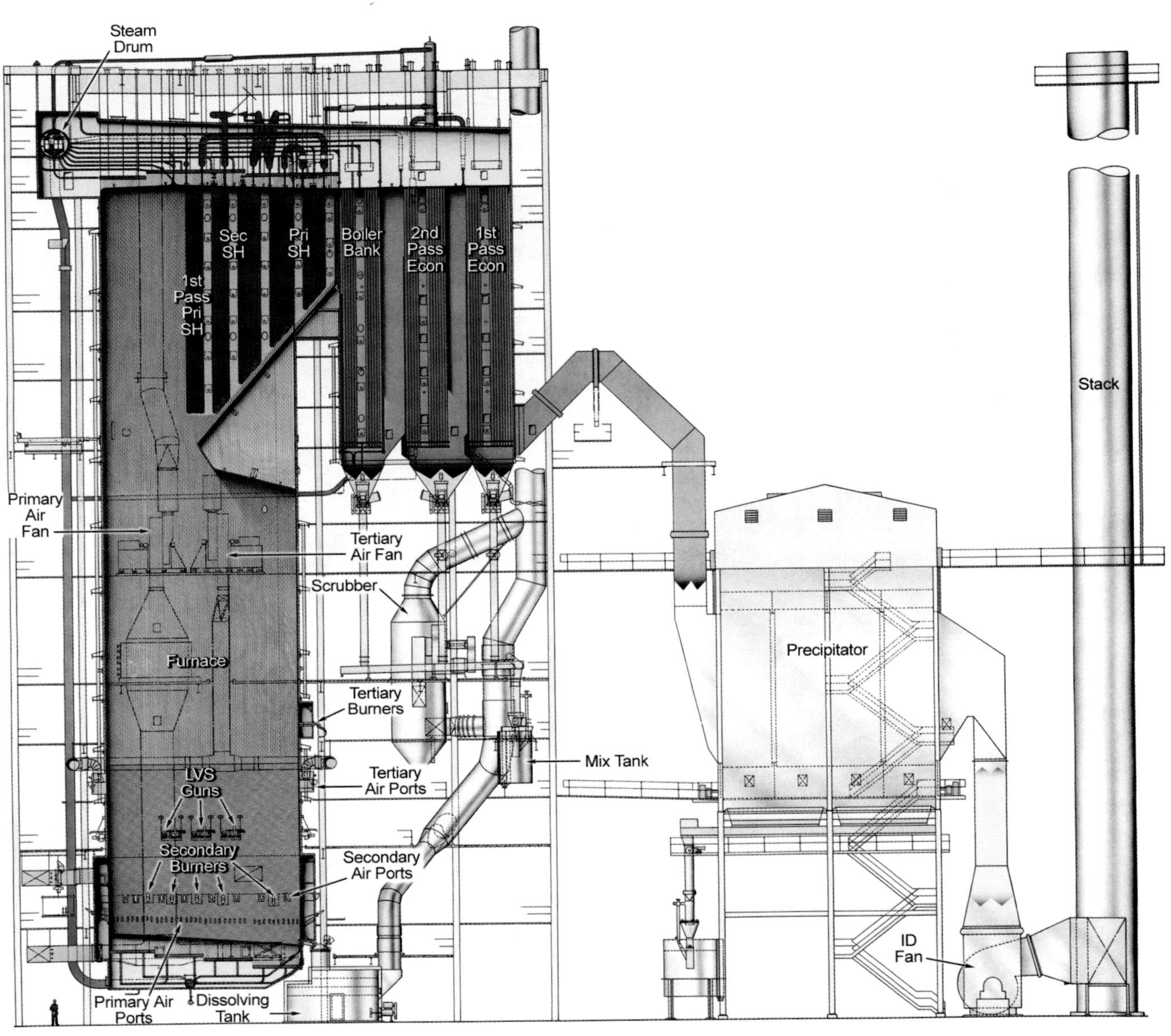

Single-drum chemical recovery boiler for the pulp and paper industry.

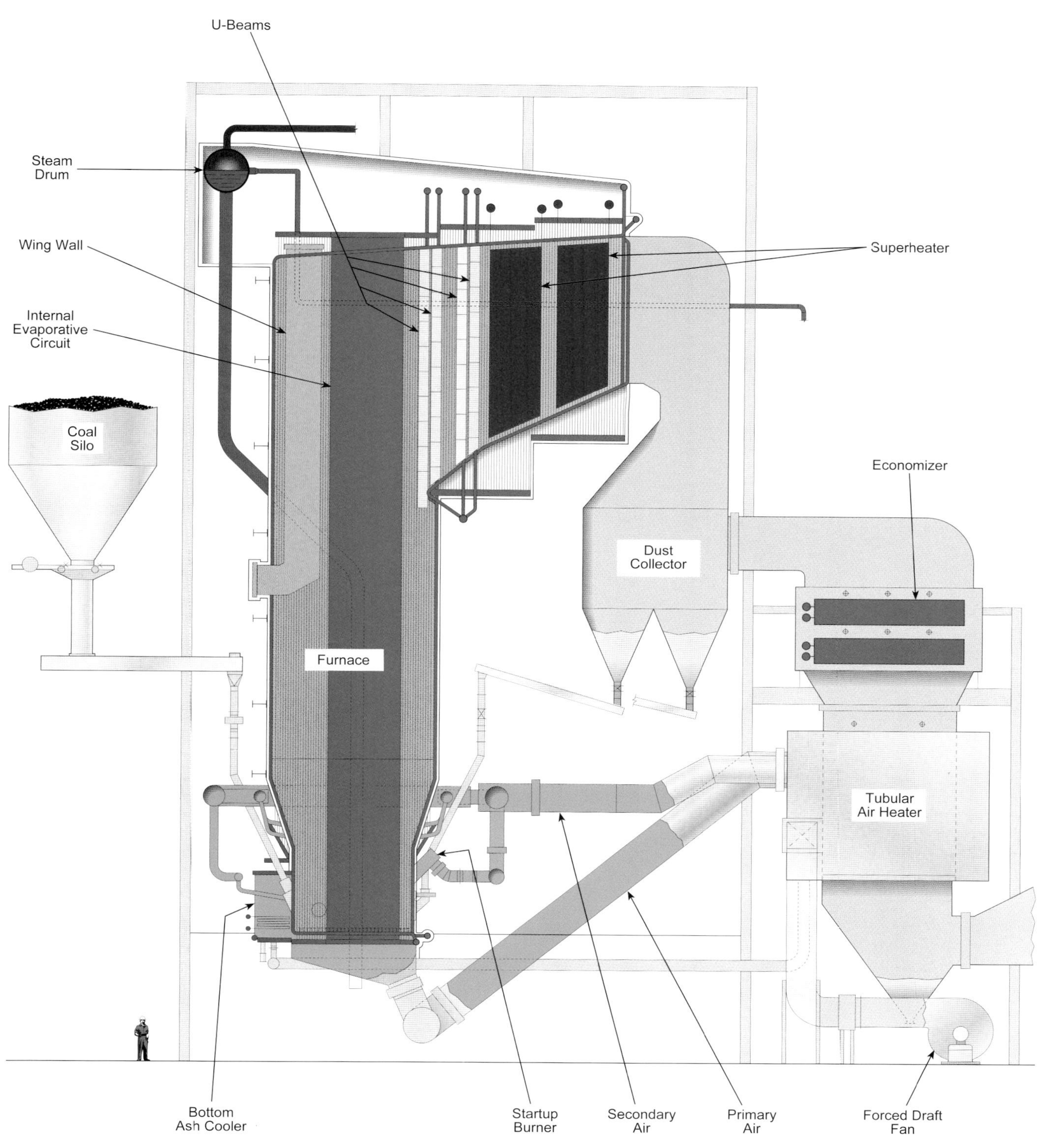

Coal-fired circulating fluidized-bed combustion steam generator.

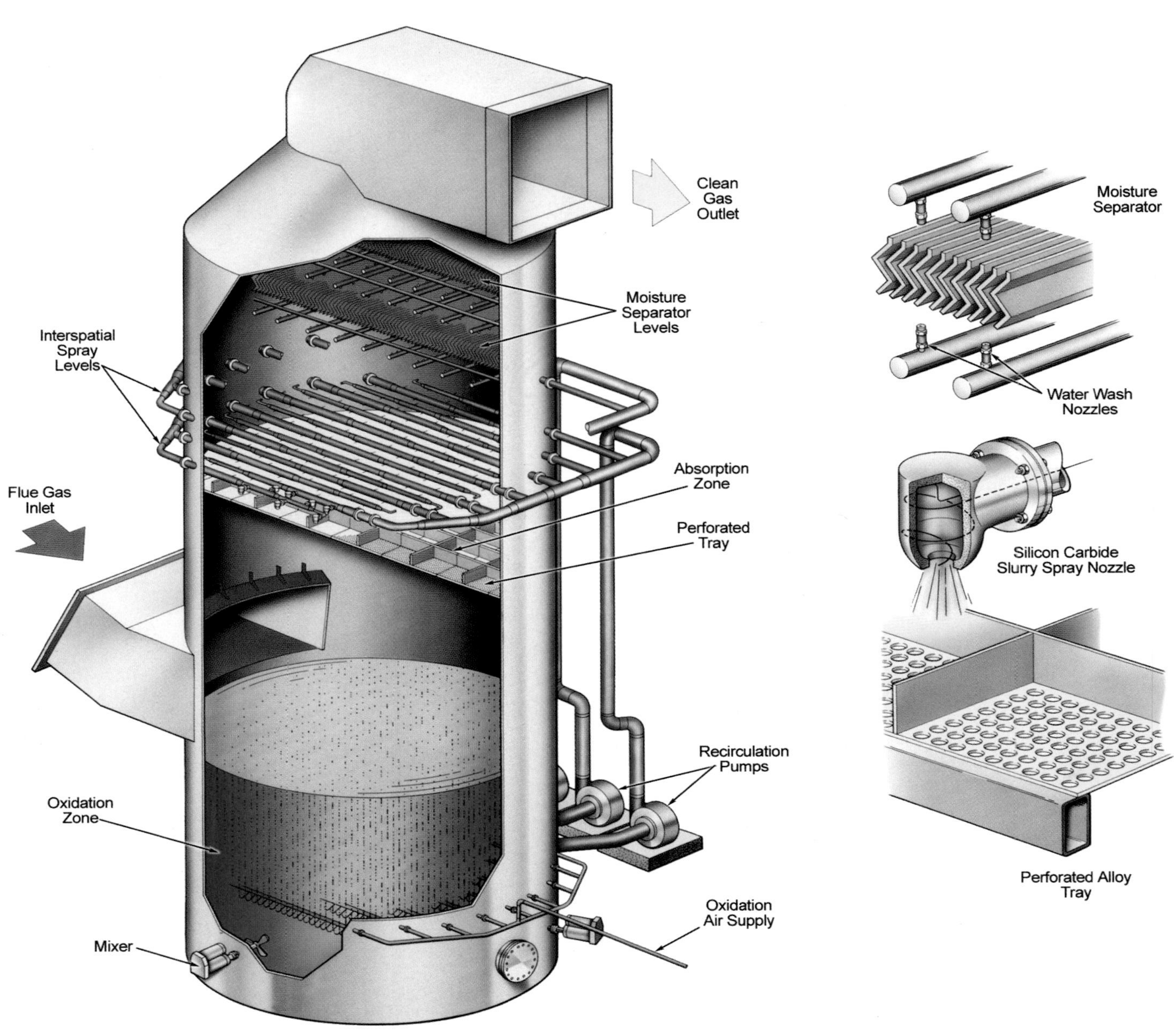

Wet flue gas desulfurization scrubber module for sulfur dioxide control.

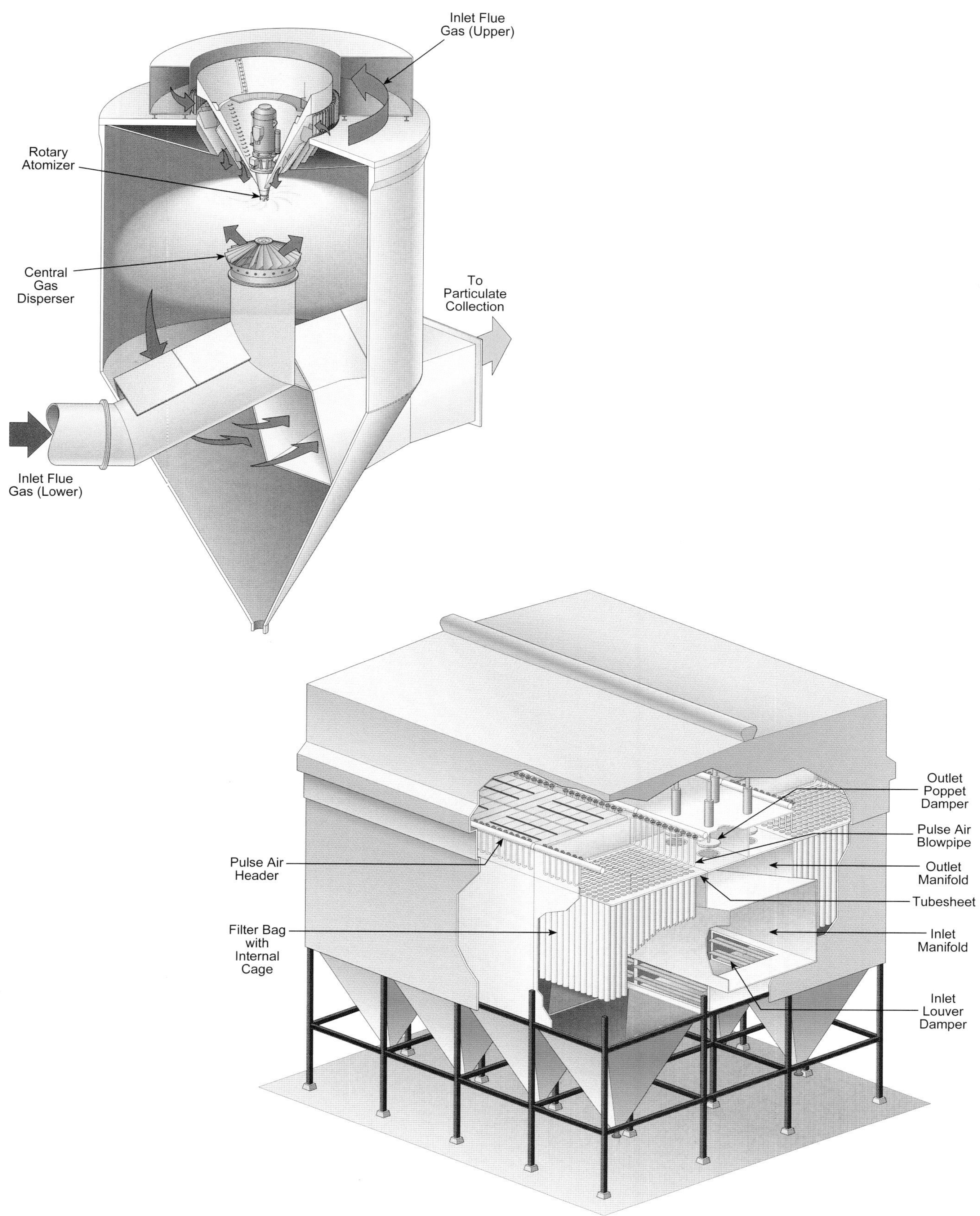

Dry flue gas desulfurization spray dryer absorber for sulfur dioxide control (upper left) and fabric filter baghouse for particulate control (lower right).

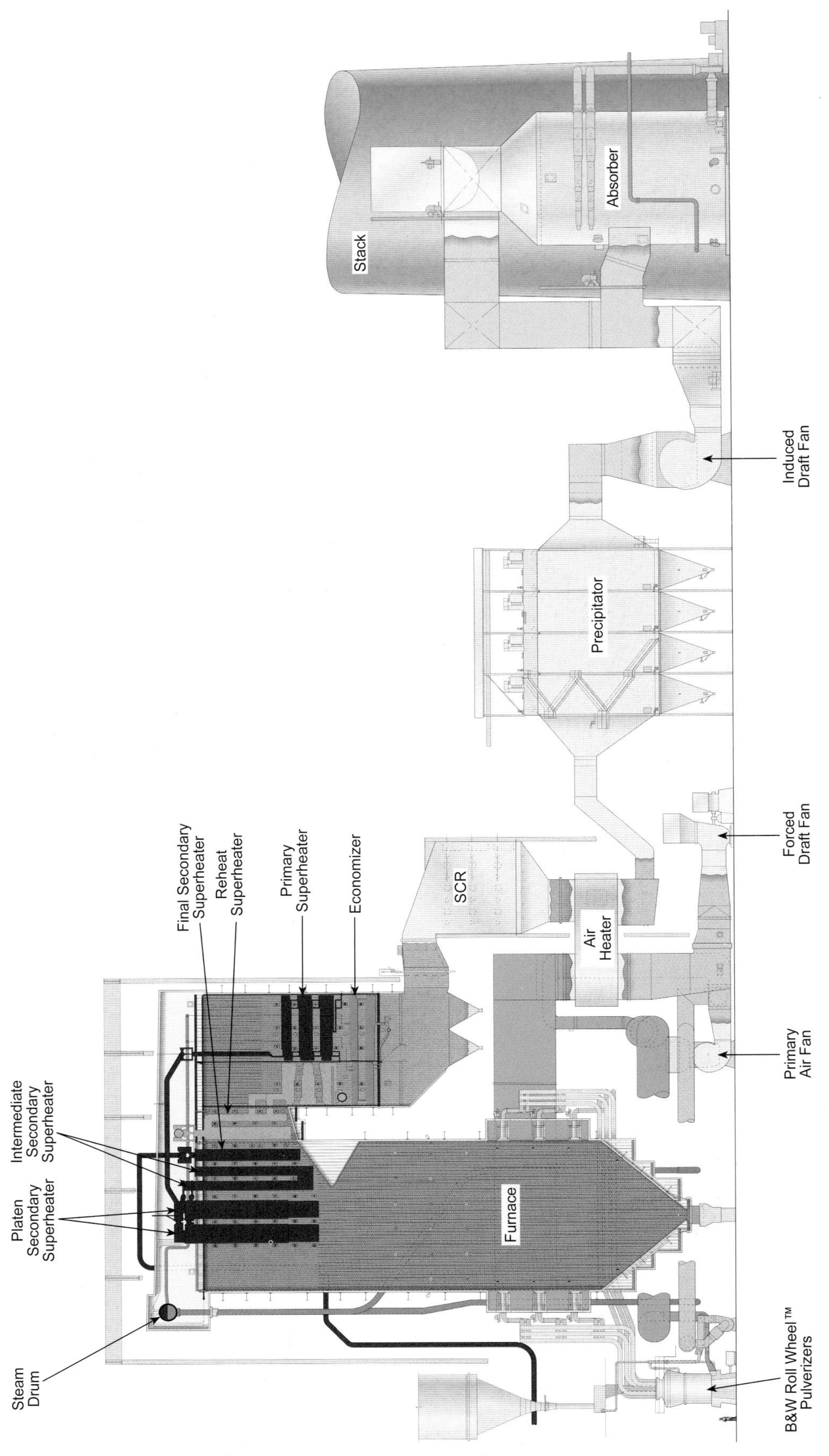

Modern 660 MW coal-fired utility boiler system with environmental control equipment.

Section I
Steam Fundamentals

Steam is uniquely adapted, by its availability and advantageous properties, for use in industrial and heating processes and in power cycles. The fundamentals of the steam generating process and the core technologies upon which performance and equipment design are based are described in this section of eight chapters. Chapter 1 provides an initial overview of the process, equipment and design of steam generating systems, and how they interface with other processes that produce power and use steam. This is followed by fundamental discussions of thermodynamics, fluid dynamics, heat transfer, and the complexities of boiling and steam-water flow in Chapters 2 through 5. New Chapter 6 is dedicated to exploring the dramatic increase in the use of advanced computational numerical analysis in the design of modern steam generators. The section concludes with Chapters 7 and 8 discussing key elements of material science and structural analysis that permit the safe and efficient design of the steam generating units and components.

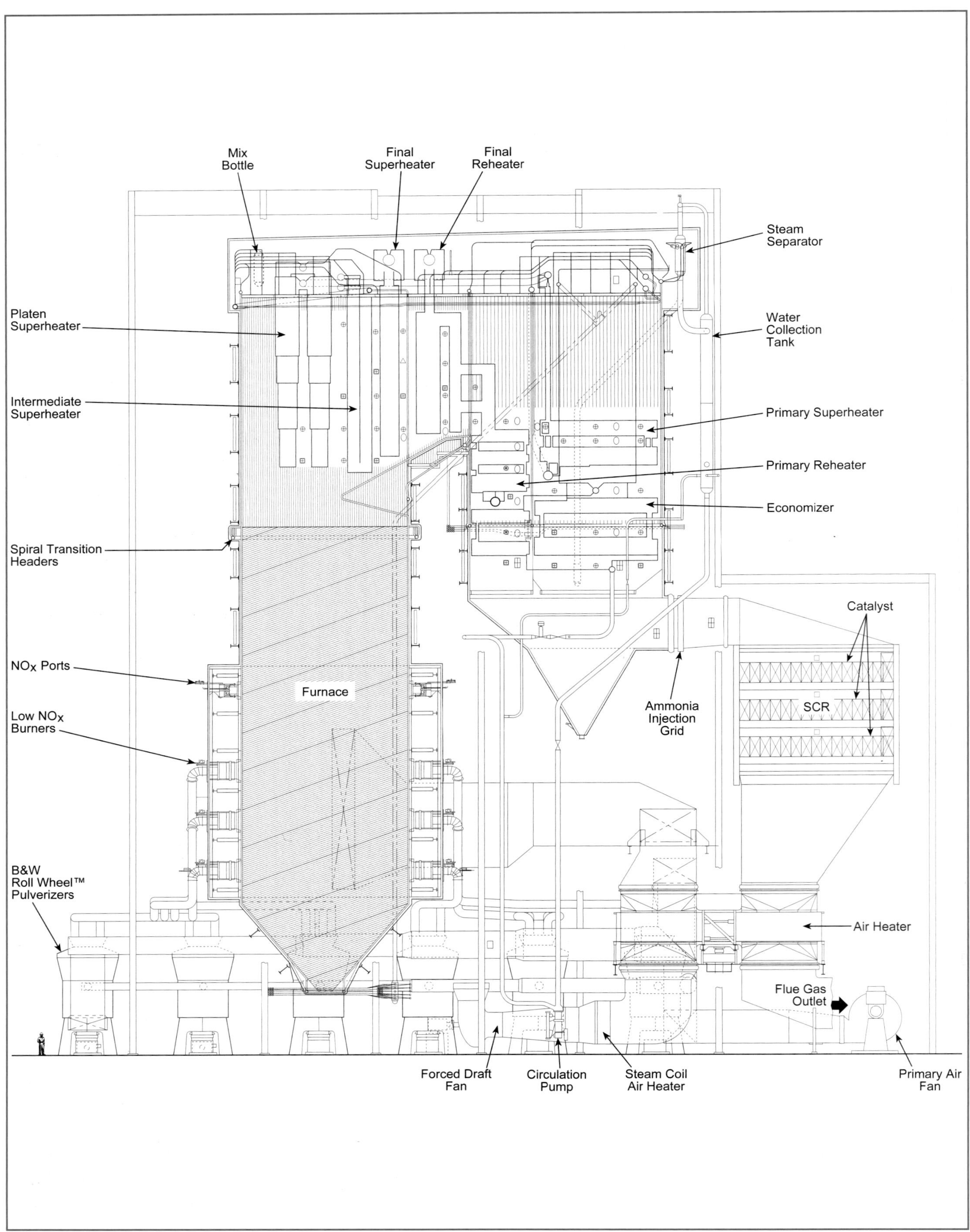

750 MW once-through spiral wound universal pressure (SWUP™) coal-fired utility boiler.

Chapter 1

Steam Generation – An Overview

Steam generators, or boilers, use heat to convert water into steam for a variety of applications. Primary among these are electric power generation and industrial process heating. Steam is a key resource because of its wide availability, advantageous properties and nontoxic nature. Steam flow rates and operating conditions are the principal design considerations for any steam generator and can vary dramatically: from 1000 lb/h (0.1 kg/s) in one process use to more than 10 million lb/h (1260 kg/s) in large electric power plants; from about 14.7 psi (0.1013 MPa) and 212F (100C) in some heating applications to more than 4500 psi (31.03 MPa) and 1100F (593C) in advanced cycle power plants.

Fuel use and handling add to the complexity and variety of steam generating systems. The fuels used in most steam generators are coal, natural gas and oil. However, nuclear energy also plays a major role in at least the electric power generation area. Also, an increasing variety of biomass materials and process byproducts have become heat sources for steam generation. These include peat, wood and wood wastes, bagasse, straw, coffee grounds, corn husks, coal mine wastes (culm), and waste heat from steelmaking furnaces. Even renewable energy sources, e.g., solar, are being used to generate steam. The steam generating process has also been adapted to incorporate functions such as chemical recovery from paper pulping processes, volume reduction for municipal solid waste or trash, and hazardous waste destruction.

Steam generators designed to accomplish these tasks range from a small package boiler (Fig. 1) to large, high capacity utility boilers used to generate 1300 MW of electricity (Fig. 2). The former is a factory-assembled, fully-automated, gas-fired boiler, which can supply saturated steam for a large building, perhaps for a hospital. It arrives at the site with all controls and equipment assembled. The large field-erected utility boiler will produce more than 10 million lb/h (1260 kg/s) steam at 3860 psi (26.62 MPa) and 1010F (543C). Such a unit, or its companion nuclear option (Fig. 3), is part of some of the most complex and demanding engineering systems in operation today. Other examples, illustrating the range of combustion systems, are shown by the 750 t/d (680 tm/d) mass-fired refuse power boiler in Fig. 4 and the circulating fluidized-bed combustion boiler in Fig. 5.

The central job of the boiler designer in any of these applications is to combine fundamental science, technology, empirical data, and practical experience to produce a steam generating system that meets the steam supply requirements in the most economical package. Other factors in the design process include fuel characteristics, environmental protection, thermal efficiency, operations, maintenance and operating costs, regulatory requirements, and local geographic and weather conditions, among others. The design process involves balancing these complex and sometimes competing factors. For example, the reduction of pollutants such as nitrogen oxides (NO_x) may require a larger boiler volume, increasing capital costs and potentially increasing maintenance costs. Such a design activity is firmly based upon the physical and thermal sciences such as solid mechanics, thermodynamics, heat transfer, fluid mechanics and materials science. However, the real world is so complex and variable, and so interrelated, that it is only by applying the art of boiler design to combine science and practice that the most economical and dependable design can be achieved.

Steam generator design must also strive to address in advance the many changes occurring in the world to provide the best possible option. Fuel prices are expected to escalate while fuel supplies become less certain, thereby enforcing the need for continued efficiency improvement and fuel flexibility. Increased environmental protection will drive improvements in combustion to reduce NO_x and in efficiency to reduce carbon dioxide (CO_2) emissions. Demand growth continues in many areas where steam generator load

Fig. 1 Small shop-assembled package boiler.

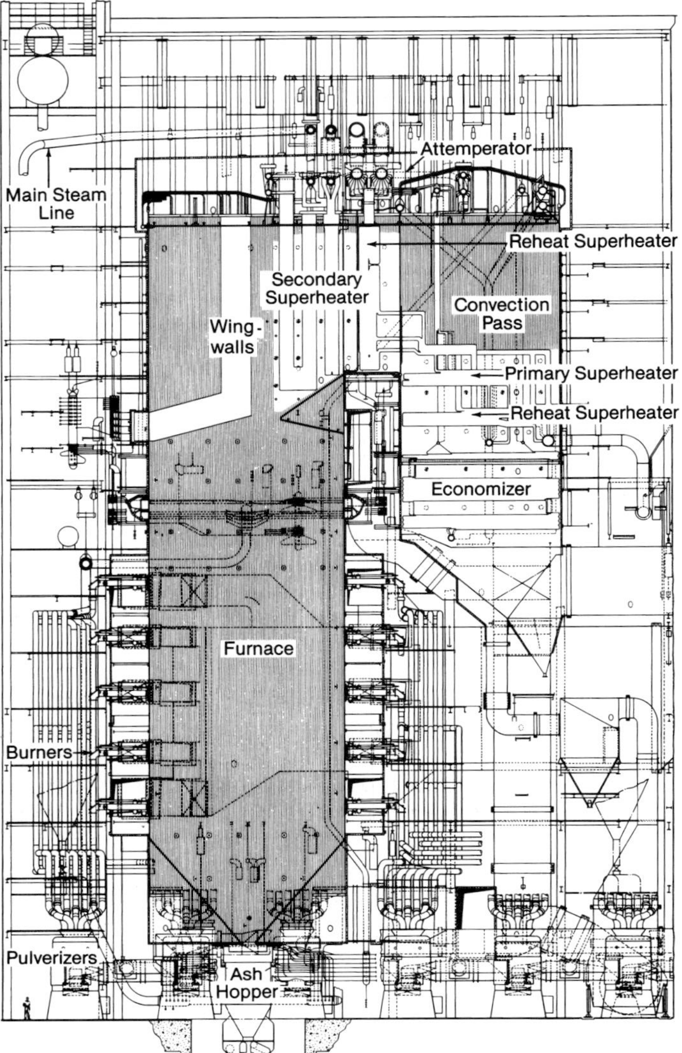

Fig. 2 1300 MW coal-fired utility steam generator.

may have to cycle up and down more frequently and at a faster rate.

There are technologies such as pressurized fluidized-bed combustion and integrated gasification combined cycle systems, plus others, which actually integrate the environmental control with the entire steam generation process to reduce emissions and increase power plant thermal efficiency. Also, modularization and further standardization will help reduce fabrication and erection schedules to meet more dynamic capacity addition needs.

Fig. 3 900 MW nuclear power system.

Steam generation fundamentals

Boiling

The process of boiling water to make steam is a familiar phenomenon. Thermodynamically, instead of increasing the water temperature, the energy used results in a change of phase from a liquid to a gaseous state, i.e., water to steam. A steam generating system should provide a continuous process for this conversion.

The simplest case for such a device is a kettle boiler where a fixed quantity of water is heated (Fig. 6). The applied heat raises the water temperature. Eventually, for the given pressure, the boiling (*saturation*) temperature is reached and bubbles begin to form. As heat continues to be applied, the temperature remains constant, and steam escapes from the water surface. If the steam is continuously removed from the vessel, the temperature will remain constant until all of the water is evaporated. At this point, heat addition would increase the temperature of the kettle and of any steam remaining in the vessel. To provide a continuous process, all that is needed is a regulated supply of water to the vessel to equal the steam being generated and removed.

Technical and economic factors indicate that the most effective way to produce high pressure steam is to heat relatively small diameter tubes containing a continuous flow of water. Regardless of whether the energy source is nuclear or fossil fuel, two distinct boiling systems are

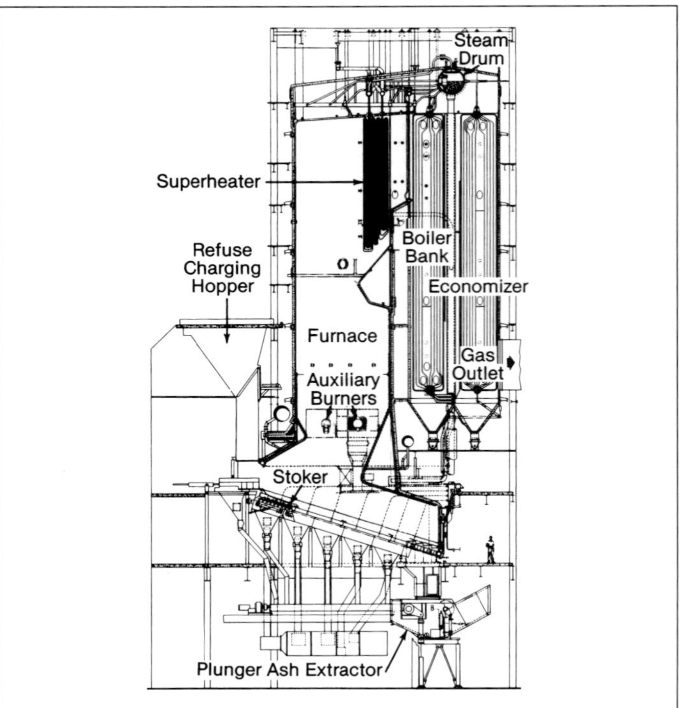

Fig. 4 Babcock & Wilcox 750 ton per day mass-fired refuse power boiler.

used to accomplish this task: those that include a steam drum (see Fig. 7a), or fixed steam-water separation point, and those that do not (see Fig. 7b), termed once-through steam generators (OTSG).

The most common and simplest to control is the steam drum system. In this system, the drum serves as the point of separation of steam from water throughout its boiler's load range. Subcooled water (less than boiling temperature) enters the tube to which heat is applied. As the water flows through the tube, it is heated to the boiling point, bubbles are formed, and *wet* steam is generated. In most boilers, a steam-water mixture leaves the tube and enters the steam drum, where steam is separated from water. The remaining water is then mixed with the replacement water and returned to the heated tube.

Without a steam drum, i.e., for an OTSG system, subcooled water also enters the tube to which heat is applied, but the flowing water turns into steam somewhere along the flow path (length of tube), dependent upon water flow rate (boiler load) and heat input rates. Shown in Fig. 7b, the flow rate and heat input are closely controlled and coordinated so that all of the water is evaporated and only steam leaves the tube. There is no need for the steam drum (fixed steam-water separation point).

Circulation

For both types of boiling systems described above, water must continuously pass through, or circulate through, the tubes for the system to generate steam continuously. For an OTSG, water makes one pass through the boiler's tubes before becoming steam to be sent to the turbine-generator. However, for those boilers with a fixed steam-water separation point or steam drum, a molecule of water can make many passes through a circulation loop before it leaves as steam to the turbine-generator. Options for this latter system are shown in Fig. 8.

Two different approaches to circulation are commonly used: natural or thermal circulation, and forced or pumped circulation. Natural circulation is illustrated in Fig. 8a. In the *downcomer*, unheated tube segment A-B, no steam is present. Heat addition generates a steam-water mixture in segment B-C. Because the steam and steam-water mixture in segment B-C are less dense than the water segment A-B, gravity will cause the water to flow downward in segment A-B and will

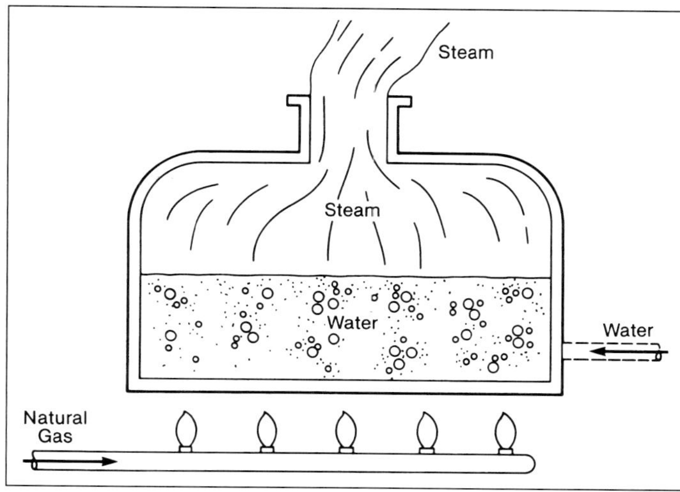

Fig. 6 Simple kettle boiler.

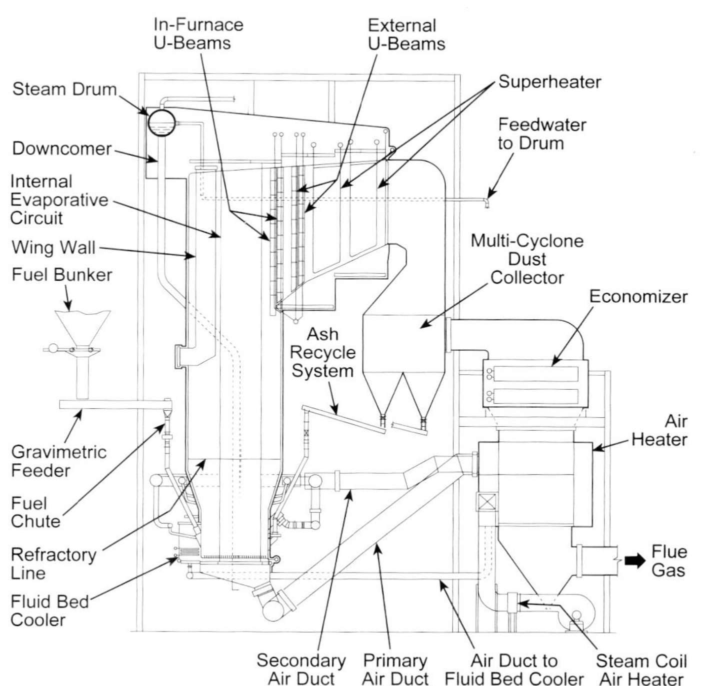

Fig. 5 Coal-fired circulating fluidized-bed combustion steam generator.

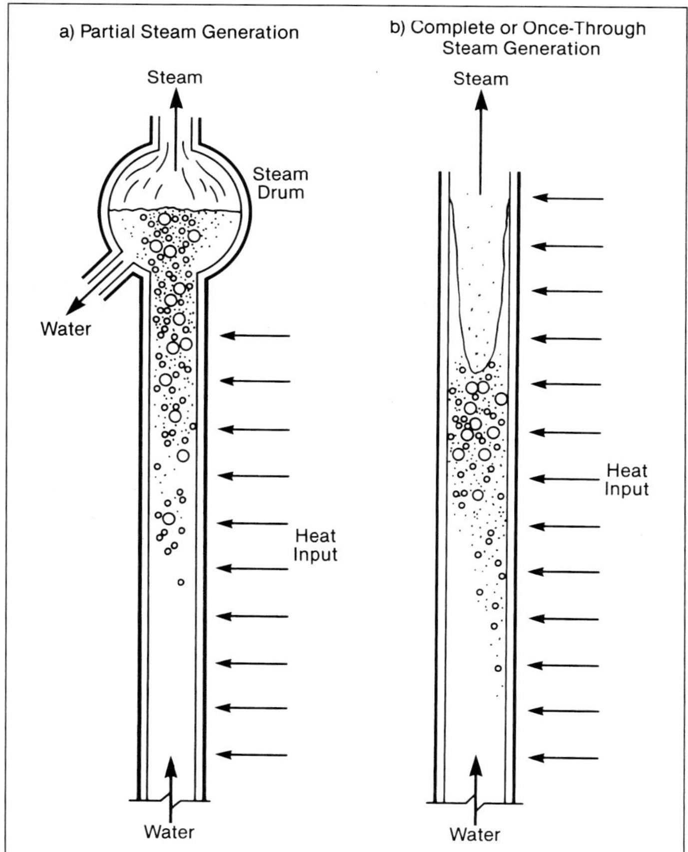

Fig. 7 Boiling process in tubular geometries.

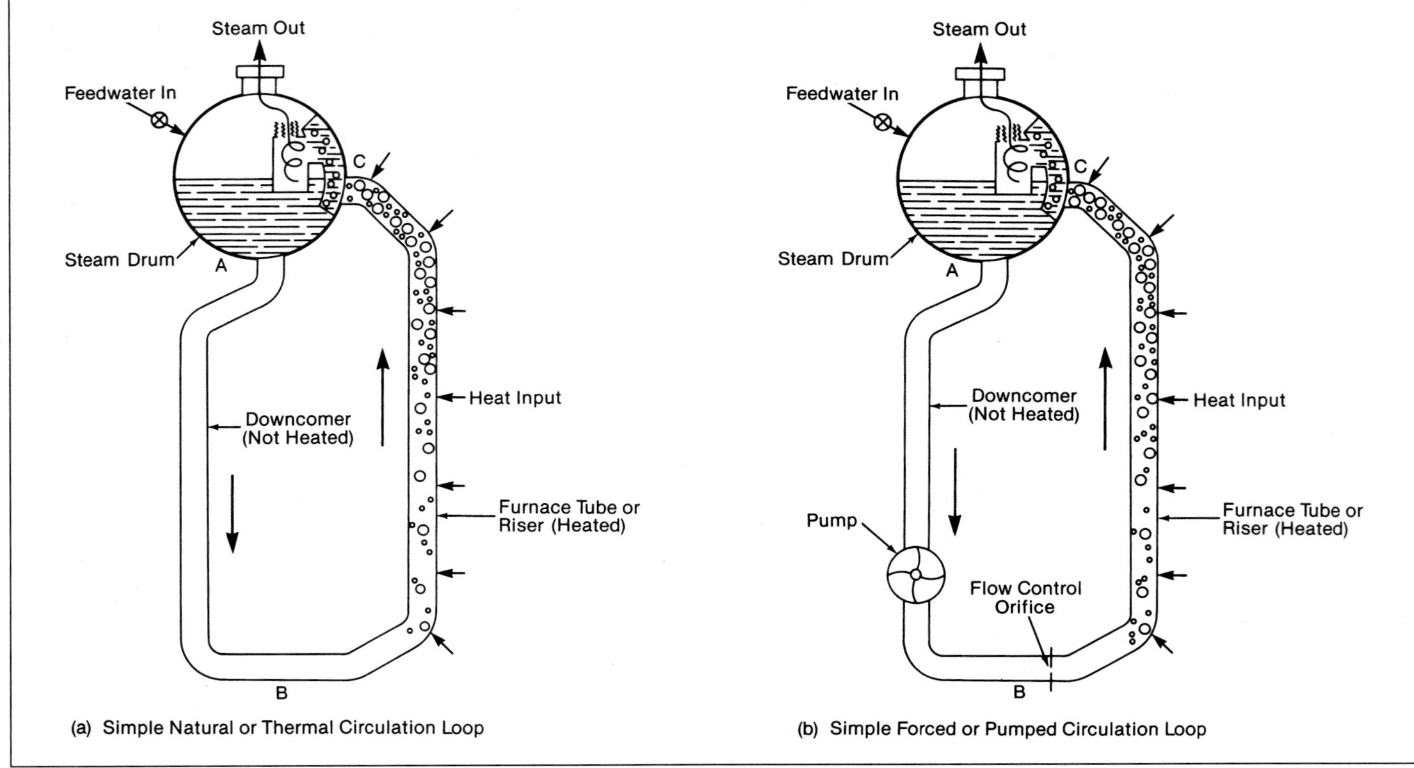

Fig. 8 Simple circulation systems.

cause the steam-water mixture (B-C) to move upward into the steam drum. The rate of water flow or circulation depends upon the difference in average density between the unheated water and the heated steam-water mixture.

The total circulation rate in a natural circulation system depends primarily upon four factors: 1) the height of the boiler, 2) the operating pressure, 3) the heat input rate, and 4) the free flow areas of the components. Taller boilers result in a larger total pressure difference between the heated and unheated legs and therefore can produce larger total flow rates. Higher operating pressures provide higher density steam and higher density steam-water mixtures. This reduces the total weight difference between the heated and unheated segments and tends to reduce flow rate. Higher heat input typically increases the amount of steam in the heated segments and reduces the average density of the steam-water mixture, increasing total flow rate. An increase in the cross-sectional (free flow) areas for the water or steam-water mixtures may increase the circulation rate. For each unit of steam produced, the amount of water entering the tube can vary from 3 to 25 units.

Forced or pumped circulation is illustrated in Fig. 8b. A mechanical pump is added to the simple flow loop and the pressure difference created by the pump controls the water flow rate.

The steam-water separation in the drum requires careful consideration. In small, low pressure boilers, steam-water separation can be easily accomplished with a large drum approximately half full of water. Natural gravity steam-water separation (similar to a kettle) can be sufficient. However, in today's high capacity, high pressure units, mechanical steam-water separators are needed to economically provide moisture-free steam from the drum. With such devices installed in the drum, the vessel diameter and cost can be significantly reduced.

At very high pressures, a point is reached where water no longer exhibits boiling behavior. Above this critical pressure [3200.11 psi (22.1 MPa)], the water temperature continuously increases with heat addition. Steam generators can be designed to operate at pressures above this critical pressure. Drums and steam-water separation are no longer required and the steam generator operates effectively on the once-through principle.

There are a large number of design methods used to evaluate the expected flow rate for a specific steam generator design and set of operating conditions. In addition, there are several criteria which establish the minimum required flow rate and maximum allowable steam content or quality in individual tubes, as well as the maximum allowable flow rates for the steam drum.

System arrangement and key components

Most applications of steam generators involve the production of electricity or the supply of process steam. In some cases, a combination of the two applications, called *cogeneration*, is used. In each application, the steam generator is a major part of a larger system that has many subsystems and components. Fig. 9 shows a modern coal-fired power generating facility; Fig. 10 identifies the major subsystems. Key subsystems include fuel receiving and preparation, steam generator and combustion, environmental protection, turbine-generator, and heat rejection including cooling tower.

First, follow the fuel and products of combustion (flue gas) through the system. The fuel handling system stores the fuel supply (coal in this example), prepares the fuel for combustion and transports it to the steam generator. The associated air system supplies air to the burners through a forced draft fan. The steam generator subsystem, which includes the air heater, burns the fuel-air mixture, recovers the heat, and generates the controlled high pressure and high temperature steam. The flue gas leaves the steam generator subsystem and selective catalytic reduction (SCR) system if supplied, then passes through particulate collection and sulfur dioxide (SO_2) scrubbing systems where pollutants are collected and the ash and solid scrubber residue are removed. The remaining flue gas is then sent to the stack through an induced draft fan.

Next, follow the steam-water path. The steam generator (boiler) evaporates water and supplies high temperature, high pressure steam, under carefully controlled conditions, to a turbine-generator set that produces the electricity. The steam may also be reheated in the steam generator, after passing through part of a multi-stage turbine system, by running the exhaust steam back to the boiler convection pass (reheater not shown). Ultimately, the steam is passed from the turbine to the condenser where the remaining waste heat is rejected. Before the water from the condenser is returned to the boiler, it passes through several pumps and heat exchangers (feedwater heaters) to increase its pressure and temperature. The heat absorbed by the condenser is eventually rejected to the atmosphere by one or more cooling towers. These cooling towers are perhaps the most visible component in the power system (Fig. 9). The natural draft cooling tower shown is basically a hollow cylindrical structure which circulates air and moisture to absorb the heat rejected by the condenser. Such cooling towers exist at most modern power plant sites, both nuclear- and fossil fuel-fired.

Fig. 9 Coal-fired utility power plant.

For an industrial power system, many of the same features are needed. However, the turbine-generator and heat rejection portions are replaced by the process application, such as radiant space heaters or heat exchangers.

In a nuclear power system (Fig. 11), the fossil fuel-fired steam generator is replaced by a nuclear reactor vessel and, typically, two or more steam generators. The coal handling system is replaced by a nuclear reactor fuel bundle handling and storage facility, and the large scale air pollution control equipment is not needed.

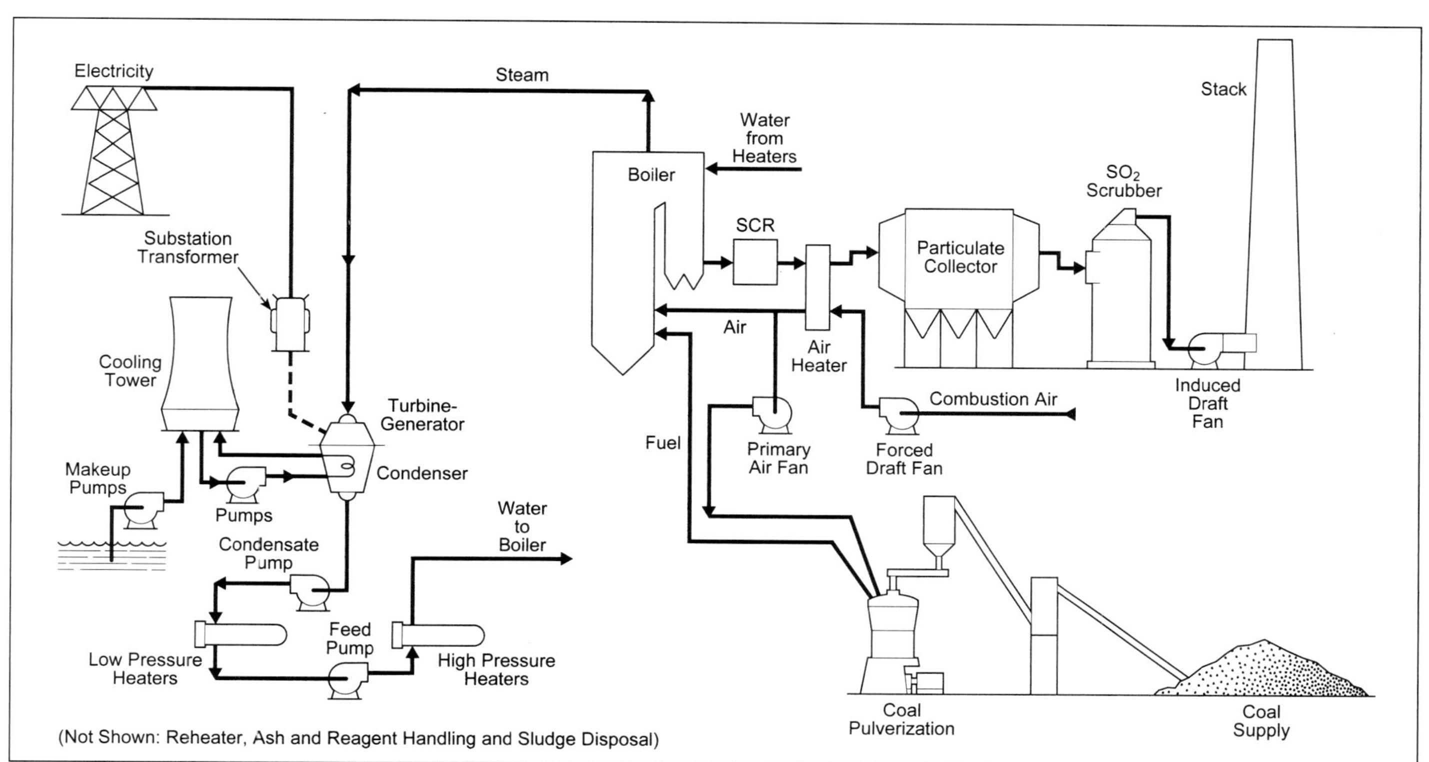

Fig. 10 Coal-fired utility power plant schematic.

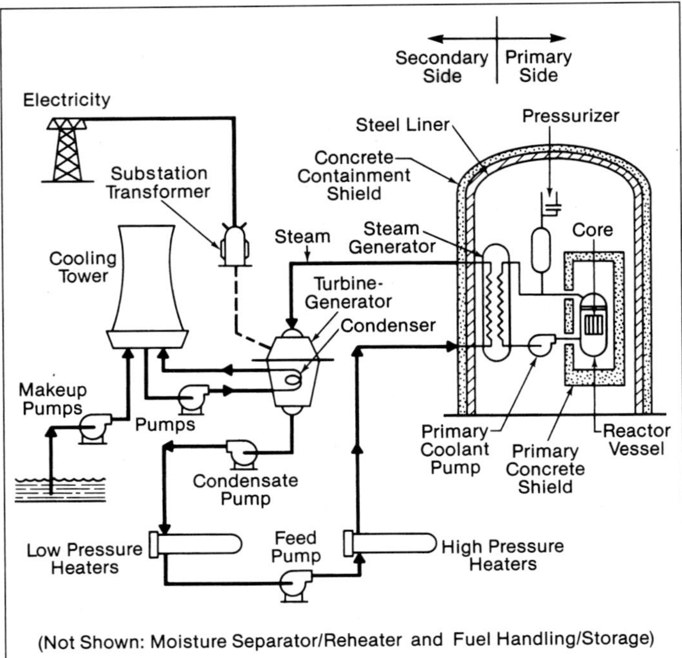

Fig. 11 Nuclear power plant schematic.

Fossil steam generator classifications

Modern steam generating systems can be classified by various criteria. These include end use, firing method, operating pressure, fuel, and circulation method.

Utility steam generators are used primarily to generate electricity in large central power stations. They are designed to optimize overall thermodynamic efficiency at the highest possible availability. New units are typically characterized by large, main steam flow rates with superheated steam outlet pressures from 1800 to 3860 psi (12.41 to 26.62 MPa) with steam temperatures at or above 1050F (566C). A key characteristic of newer units is the use of a reheater section to increase overall cycle efficiency.

Industrial steam generators generally supply steam to processes or manufacturing activities and are designed with particular attention to: 1) process controlled (usually lower) pressures, 2) high reliability with minimum maintenance, 3) use of one or more locally inexpensive fuels, especially process byproducts or wastes, and 4) low initial capital and minimum operating costs. On a capacity basis, the larger users of such industrial units are the pulp and paper industry, municipal solid waste reduction industry, food processing industry, petroleum/petrochemical industry, independent power producers and cogenerators, and some large manufacturing operations. Operating pressures range from 150 to 1800 psi (1.04 to 12.41 MPa) with saturated or superheated steam conditions.

Impact of energy source

The primary fuel selected has perhaps the most significant impact on the steam generator system configuration and design. In the case of nuclear energy, a truly unique system for containing the fuel and the nuclear reaction products has been developed with an intense focus on safety and protecting the public from radiation exposure. Acceptable materials performance in the radiative environment and the long term thermal-hydraulic and mechanical performance are central to system design. When fossil, biomass, or byproduct fuels are burned, widely differing provisions must be made for fuel handling and preparation, fuel combustion, heat recovery, fouling of heat transfer surfaces, corrosion of materials, and emissions control. For example, in a natural gas-fired unit (Fig.12), there is minimal need for fuel storage and handling. Only a small furnace is needed for combustion, and closely spaced heat transfer surfaces may be used because of lack of ash deposits (fouling). The corrosion allowance is relatively small and the emissions control function is primarily for NO_x formed during the combustion process. The result is a relatively small, compact and economical design.

If a solid fuel such as coal (which has a significant level of noncombustible ash) is used, the overall system is much more complex. This system could include extensive fuel handling and preparation facilities, a much larger furnace, and more widely spaced heat transfer surfaces. Additional components could be special cleaning equipment to reduce the impact of fouling and erosion, air preheating to dry the fuel and enhance combustion, more extensive environmental equipment, and equipment to collect and remove solid wastes.

The impact of fuel alone on a utility boiler design is clearly indicated in Fig. 12, where both steam generators produce the same steam flow rate. Further adding to the size and cost difference, but not shown, are the facts that the coal-fired boiler will be wider (dimension not shown) and will require more flue gas cleanup equipment to meet emissions requirements. The particular challenge when burning different solid fuels is indicated in Fig. 13, where provision is made for burning both pulverized (finely ground) coal using the burners and for wood chips and bark which are burned on the moving grate (stoker) at the bottom of the unit.

Impact of steam conditions

The steam temperature and pressure for different boiler applications can have a significant impact on design. Fig. 14 identifies several typical boiler types, as well as the relative amount of heat input needed, for water heating, evaporation (boiling), superheating, and reheating, if required. The relative amount of energy needed for evaporation is dramatically reduced as operating pressure is increased. As a result, the relative amount of physical heat transfer surface (tubes) dedicated to each function can be dramatically different.

Fossil fuel systems

Fossil fuel steam generator components

Modern steam generators are a complex configuration of thermal-hydraulic (steam and water) sections which preheat and evaporate water, and superheat steam. These surfaces are arranged so that: 1) the fuel can be burned completely and efficiently while minimizing emissions, 2) the steam is generated at the required flow rate, pressure and temperature, and 3)

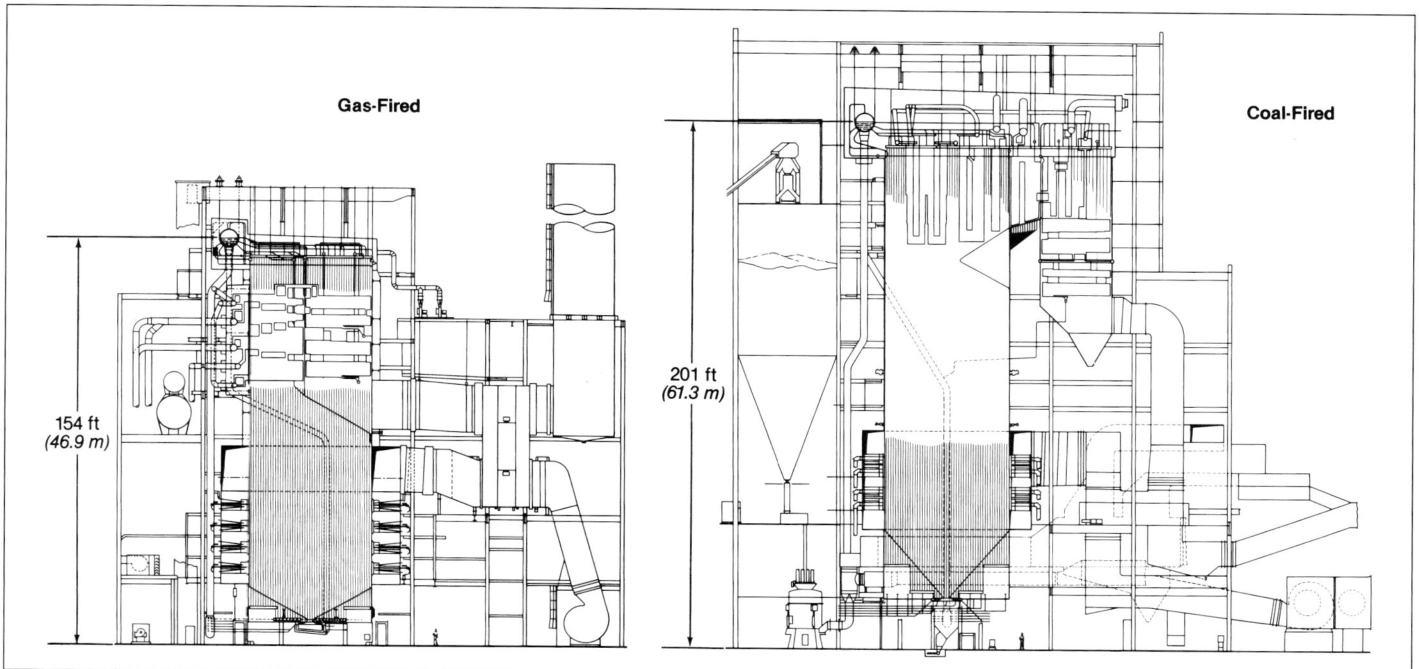

Fig. 12 Comparison of gas- and coal-fired steam generators.

the maximum amount of energy is recovered. A relatively simple coal-fired utility boiler is illustrated in Fig. 15. The major components in the steam generating and heat recovery system include:

1. furnace and convection pass,
2. steam superheaters (primary and secondary),
3. steam reheater,
4. boiler or steam generating bank (industrial units only),
5. economizer,
6. steam drum,
7. attemperator and steam temperature control system, and
8. air heater.

These components are supported by a number of subsystems and pieces of equipment such as coal pulverizers, combustion system, flues, ducts, fans, gas-side cleaning equipment and ash removal equipment.

The *furnace* is a large enclosed open space for fuel combustion and for cooling of the flue gas before it enters the convection pass. Excessive gas temperatures leaving the furnace and entering the tube bundles could cause particle accumulation on the tubes or excessive tube metal temperatures. The specific geometry and dimensions of the furnace are highly influenced by the fuel and type of combustion equipment. In this case, finely ground or pulverized coal is blown into the furnace where it burns in suspension. The products of combustion then rise through the upper furnace. The superheater, reheater and economizer surfaces are typically located in the flue gas horizontal and vertical downflow sections of the boiler enclosure, called the *convection pass*.

In modern steam generators, the furnace and convection pass walls are composed of steam- or water-cooled carbon steel or low alloy tubes to maintain wall metal temperatures within acceptable limits. These tubes are connected at the top and bottom by headers, or manifolds. These headers distribute or collect the water, steam or steam-water mixture. The furnace wall tubes in most modern units also serve as key steam generating components or surfaces. The tubes are welded together with steel bars to provide membrane wall panels which are gas-tight, continuous and rigid. The tubes are usually prefabricated into shippable membrane panels with openings for burners, observation doors, sootblowers (boiler gas-side surface cleaning equipment) and gas injection ports.

Superheaters and *reheaters* are specially designed in-line tube bundles that increase the temperature of saturated steam. In general terms, they are simple single-phase heat exchangers with steam flowing inside the tubes and the flue gas passing outside, generally in crossflow. These critical components are manufactured from steel alloy material because of their high operating temperature. They are typically configured to help control steam outlet temperatures, keep metal temperatures within acceptable limits, and control steam flow pressure loss.

The main difference between superheaters and reheaters is the steam pressure. In a typical drum boiler, the superheater outlet pressure might be 2700 psi (18.62 MPa) while the reheater outlet might be only 580 psi (4.0 MPa). The physical design and location of the surfaces depend upon the desired outlet temperatures, heat absorption, fuel ash characteristics and cleaning equipment. These surfaces can be either horizontal or vertical as shown. The superheater and sometimes reheater are often divided into multiple sections to help control steam temperature and optimize heat recovery.

The heat transfer surface in the furnace may not be sufficient to generate enough saturated steam for

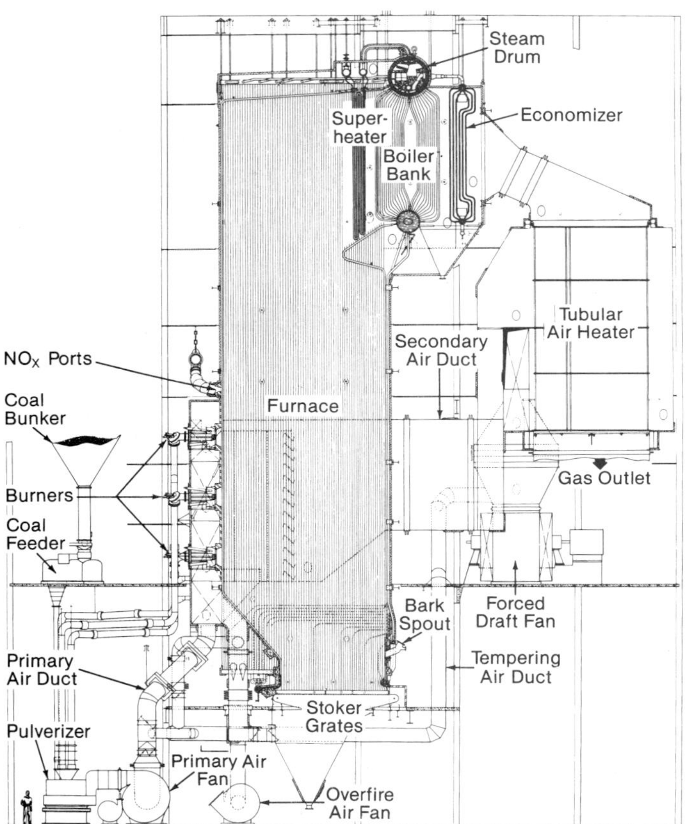

Fig. 13 Large industrial boiler with multiple fuel capability.

the particular end use. If this is the case, an additional bank of heat exchanger tubes called the *boiler bank* or *steam generating bank* is added. (See Fig. 13.) This is needed on many smaller, low pressure industrial boilers, but is not often needed in high pressure utility boilers. This boiler bank is typically composed of the *steam drum* on top, a second drum on the bottom, and a series of bent connecting tubes. The steam drum internals and tube sizes are arranged so that subcooled water travels down the tubes (farthest from the furnace) into the lower drum. The water is then distributed to the other tubes where it is partially converted to steam and returned to the steam drum. The lower drum is often called the *mud drum* because this is where sediments found in the boiler water tend to settle out and collect.

The *economizer* is a counterflow heat exchanger for recovering energy from the flue gas beyond the superheater and, if used, the reheater. It increases the temperature of the water entering the steam drum. The tube bundle is typically an arrangement of parallel horizontal serpentine tubes with the water flowing inside but in the opposite direction (counterflow) to the flue gas. Tube spacing is as tight as possible to promote heat transfer while still permitting adequate tube surface cleaning and limiting flue gas-side pressure loss. By design, steam is usually not generated inside these tubes.

The *steam drum* is a large cylindrical vessel at the top of the boiler, in which saturated steam is separated from the steam-water mixture leaving the boiler tubes. Drums can be quite large with diameters of 3 to 6 ft (0.9 to 1.8 m) and lengths approaching 100 ft (30.5 m). They are fabricated from thick steel plates rolled into cylinders with hemispherical heads. They house the steam-water separation equipment, purify the steam, mix the replacement or feedwater and chemicals, and provide limited water storage to accommodate small changes in unit load. Major connections to the steam drum are provided to receive the steam-water mixture from the boiler tubes, remove saturated steam, add replacement or makeup water, and return the near saturated water back to the inlet of the boiler tubes.

The *steam temperature control* system can be complex and include combinations of recirculating some of the flue gas to the bottom or top of the furnace, providing special gas flow passages at the back end of the steam generator, adjusting the combustion system, and adding water or low temperature steam to the high temperature steam flow (*attemperation*). The component most frequently used for the latter is called a *spray attemperator*. In large utility units, attemperators with direct injection of water or low temperature steam are used for dynamic control because of their rapid response. They are specially designed to resist thermal shock and are frequently located at the inlet of the superheater or between superheater sections to better control the superheater outlet metal temperatures. Positioning of individual superheater sections can also help maintain proper outlet steam temperatures.

The *air heater* is not a portion of the steam-water circuitry, but serves a key role in the steam generator system heat transfer and efficiency. In many cases, especially in high pressure boilers, the temperature of the flue gas leaving the economizer is still quite high. The air heater recovers much of this energy and adds it to the combustion air to reduce fuel use. Designs include tubular, flat plate, and regenerative heat exchangers, among others.

Steam-water flow system

The steam-water components are arranged for the most economical system to provide a continuous supply of steam. The circulation system (excluding reheater) for a natural circulation, subcritical pressure, drum type steam generator is shown in Fig. 16. Feed-

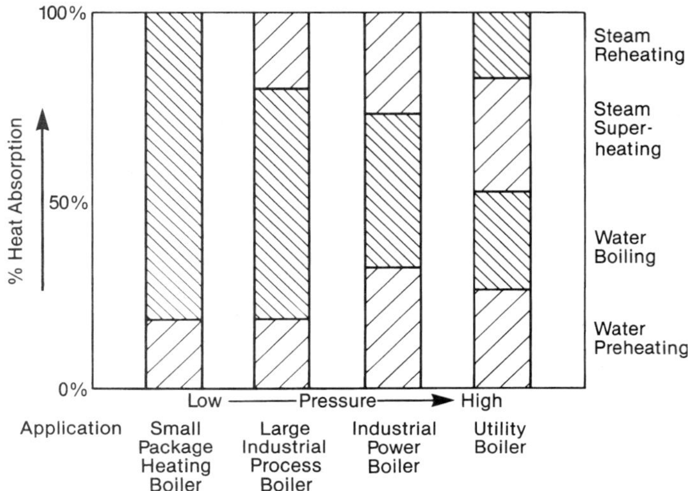

Fig. 14 Steam generator energy absorption by function.

water enters the bottom header (A) of the economizer and passes upward in the opposite direction to the flue gas. It is collected in an outlet header (B), which may also be located in the flue gas stream. The water then flows through a number of pipes which connect the economizer outlet header to the steam drum. It is sometimes appropriate to run these tubes vertically (B-C) through the convection pass to economizer outlet headers located at the top of the boiler. These tubes can then serve as water-cooled supports for the horizontal superheater and reheater when these banks span too great a distance for end support. The feedwater is injected into the steam drum (D) where it mixes with the water discharged from the steam-water separators before entering connections to the *downcomer* pipes (D-E) which exit the steam drum.

The water travels around the furnace water wall circuits to generate steam. The water flows through the downcomer pipes (D-E) to the bottom of the furnace where *supply* tubes (E-F) route the circulating water to the individual lower furnace panel wall headers (F). The water rises through the furnace walls to individual outlet headers (G), absorbing energy to become a steam-water mixture. The mixture leaves the furnace wall outlet headers by means of *riser* tubes (G-D), to be discharged into the drum and steam-water separators. The separation equipment returns essentially steam-free water to the downcomer inlet connections. The residual moisture in the steam that leaves the primary steam separation devices is removed in secondary steam separators, and dry steam is discharged to the superheater through a number of drum outlet connections (H-I and H-J).

The steam circuitry serves dual functions: cooling the convection pass enclosure, and generating the required superheated steam conditions. Steam from the drum passes through multiple connections to a

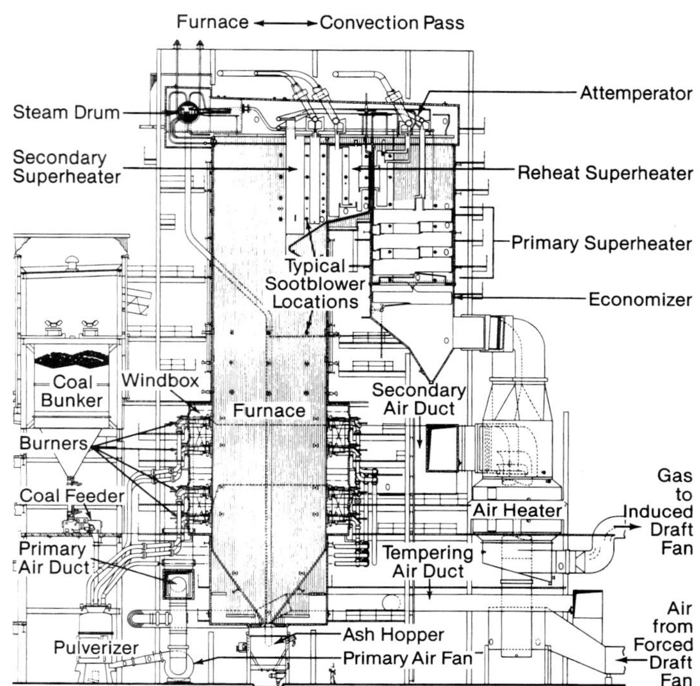

Fig. 15 Coal-fired utility boiler.

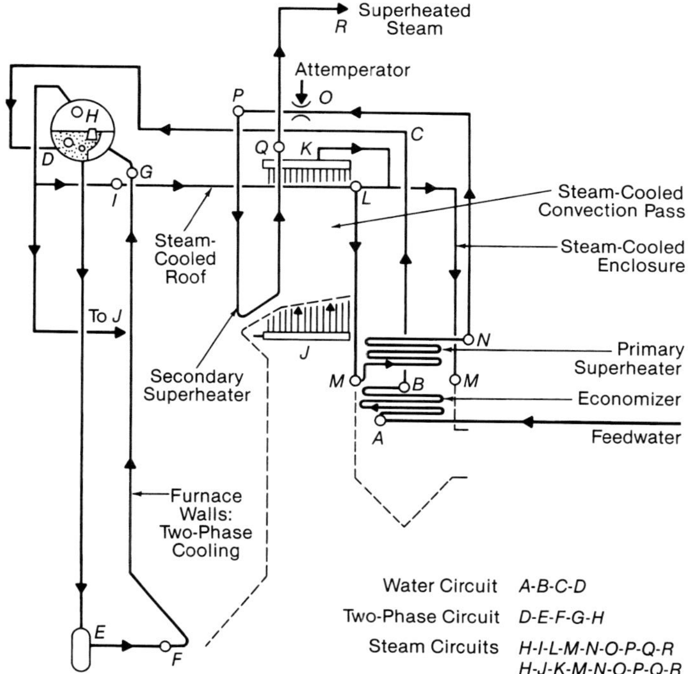

Fig. 16 Coal-fired boiler steam-water circulation system.

header (I) supplying the roof tubes and, separately, to headers (J) supplying the membrane panels in the pendant convection pass (so named because the superheater/reheater vertical tubes are hanging from supports above). The steam flows through these membrane panels to outlet headers (K). Steam from these headers and the roof tube outlet headers (L) then provides the cooling for the horizontal convection pass enclosure (L-M) (so named because the superheater/reheater/economizer tubes are horizontal in this flue gas downpass). Steam flows downward through these panels and is collected in outlet headers (M) just upstream of the economizer bank.

Steam flow then rises through the primary superheater and discharges through the outlet header (N) and connecting piping equipped with a spray attemperator (O). It then enters the secondary superheater inlet header (P), flowing through the superheater sections to an outlet header (Q). A discharge pipe (R) terminates outside of the boiler enclosure where the main steam lines route the steam flow to the control valves and turbine.

Combustion system and auxiliaries

Most of the non-steam generating components and auxiliaries used in coal-fired steam generators are part of the fuel preparation and combustion systems. These include:

1. fuel preparation: feeders and coal pulverizers,
2. combustion system: burners, flame scanners, lighters, controls, windbox,
3. air-gas handling: fans, flues and ducts, dampers, control and measurement systems, silencers, and
4. other components and auxiliaries: sootblowers (heat transfer surface cleaning equipment), ash collection and handling equipment, control and monitoring equipment.

Because of their intimate relationship with the steam generation process, many of these components are supplied with the boiler. If not, careful specification and interaction with the steam generator manufacturer are critical.

The combustion system has a dramatic impact on overall furnace design. Wall mounted burners are shown in Figs. 15 and 17. These are typical for large coal-, oil-, or gas-fired units today. However, a variety of other systems are also used and are continuing to be developed to handle the variety of fuel characteristics and unit sizes. Other combustion systems include stokers (Figs. 4 and 13), Cyclone furnaces, and fluidized-bed combustion units (Fig. 5). All have their strengths and weaknesses for particular applications. Key elements of these systems involve the need to control the formation and emission of pollutants, provide complete efficient combustion, and handle inert material found in the fuel. The fuel characteristics play a central role in how these functions are met and how the equipment is sized and designed.

Air-gas flow system

Many of these auxiliaries are identified in Fig. 17 along with the air-gas flow path in the large coal-fired utility boiler. Air is supplied by the *forced draft* fan (A) to the air heater (B) where it is heated to recover energy and enhance combustion. Most of the hot air (typically 70 to 80%), called secondary air, passes directly to the windboxes (C) where it is distributed to individual burners. The remaining 20 to 30%, called primary air, passes to the booster (or primary air) fan and then to the coal pulverizers (D) where the coal is dried and ground. The hot primary air then pneumatically conveys the pulverized coal to the burners (E) where it is mixed with the secondary air for combustion. The coal and air are rapidly mixed and burned in the furnace (F) and the flue gas then passes up through the furnace, being cooled primarily by radiation until it reaches the furnace exit (G). The gas then progressively passes through the secondary superheater, reheater, primary superheater and economizer before leaving the steam generator enclosure (H). The gas passes through the air heater (B) and then through any pollution control equipment and *induced draft* fan (I) before being exhausted to the atmosphere.

Emissions control

A key element of fossil fuel-fired steam generator system design is environmental protection. A broad range of government regulations sets limits on primary gaseous, liquid and solid waste emissions from the steam generating process. For coal-, oil-, and gas-fired units, the primary air pollutant emissions include sulfur dioxide (SO_2), nitrogen oxides (NO_x) and airborne particulate or flyash. Water discharges include trace chemicals used to control corrosion and fouling as well as waste heat rejected from the condenser. Solid waste primarily involves the residual ash from the fuel and any spent sorbent from the pollution control systems.

The gaseous and solid waste from the fuel and combustion process can be minimized by fuel selection, control of the combustion process, and equipment located downstream of the steam generator. SO_2 emissions may be reduced by using fuels which contain low levels of sulfur, by fluidized-bed combustors, or by using a post-combustion scrubber system. Combustion NO_x emissions are typically controlled by using equipment such as special low NO_x burners or fluidized-bed combustors. Where it is required to significantly reduce NO_x emissions (to levels lower than is practically achieved by combustion techniques alone), back-end or post-combustion techniques, such as selective catalytic reduction (SCR) or selective noncatalytic reduction (SNCR) technologies, are employed. Flyash or airborne particulate is collected by either a fabric filter (baghouse) or electrostatic precipitator (ESP) with removal efficiencies above 99%. The particulate collection equipment and SO_2 scrubbers produce solid byproduct streams which must be safely landfilled or used for some industrial applications.

The water discharges are minimized by installing recirculating cooling systems. An example of this is cooling towers that reject the waste heat from the power cycle to the air, instead of to a water source. These are used on virtually all new fossil and nuclear power plants. Chemical discharges are minimized by specially designed zero discharge systems. A set of emissions rates before and after control for a typical 500 MW power plant is shown in Table 1.

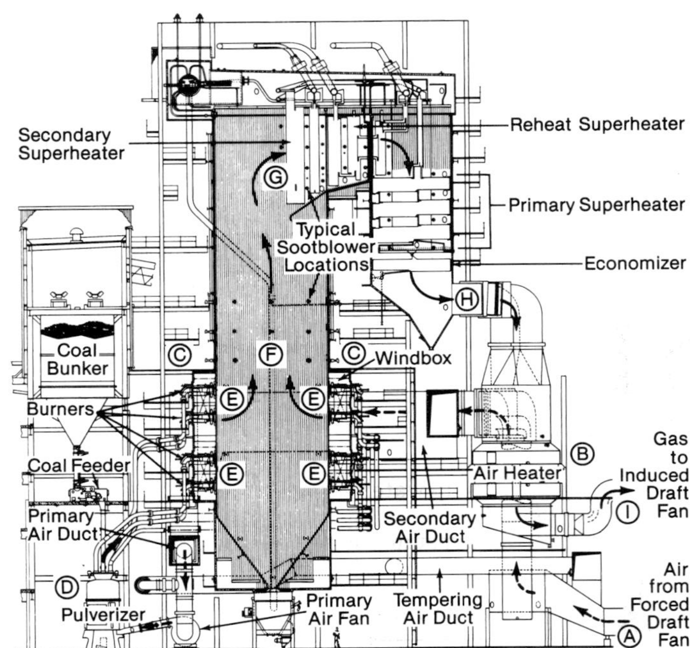

Fig. 17 Coal-fired boiler air/gas flow path.

Nuclear steam generating systems

Overview

Nuclear steam generating systems include a series of highly specialized heat exchangers, pressure vessels, pumps and components which use the heat generated by nuclear fission reactions to efficiently and

Table 1
Typical 500 MW Subcritical Coal-Fired Steam Generator Emissions and Byproducts*

Power System Characteristics
- 500 MW net
- 196 t/h (49.4 kg/s) bituminous coal
  - 2.5% sulfur
  - 16% ash
  - 12,360 Btu/lb (28,749 kJ/kg)
- 65% capacity factor

Emission	Typical Control Equipment	Discharge Rate — t/h (t_m/h) Uncontrolled		Controlled	
SO_x as SO_2	Wet limestone scrubber	9.3	(8.4)	0.3	(0.3)
NO_x as NO_2	Low NO_x burners and SCR	2.9	(2.6)	0.1	(0.1)
CO_2	Not applicable	485	(440)	—	—
Flyash to air**	Electrostatic precipitator or baghouse	22.9	(20.8)	0.05	(0.04)
Thermal discharge to water sources	Natural draft cooling tower	2.8 x 10^9 Btu/h	(821 MW_t)	~0	(0)
Ash to landfill**	Controlled landfill	9.1	(8.3)	32	(29)
Scrubber sludge: gypsum plus water	Controlled landfill or wallboard quality gypsum	0	(0)	27.7	(25)

* See Chapter 32, Table 1, for a modern 615 MW supercritical coal-fired steam generator.
** As flyash emissions to the air decline, ash to landfill increases.

safely generate steam. The system is based upon the energy released when atoms within certain materials, such as uranium, break apart or *fission*. Fission occurs when a fissionable atom nucleus captures a free subatomic particle – a neutron. This upsets the internal forces which hold the atom nucleus together. The nucleus splits apart producing new atoms as well as an average of two to three neutrons, gamma radiation and energy.

The nuclear steam supply system (NSSS) is designed to serve a number of functions: 1) house the nuclear fuel, 2) stimulate the controlled fission of the fuel, 3) control the nuclear reaction rate to produce the required amount of thermal energy, 4) collect the heat and generate steam, 5) safely contain the reaction products, and 6) provide backup systems to prevent release of radioactive material to the environment. Various systems have been developed to accomplish these functions. The main power producing system in commercial operation today is the pressurized water reactor (PWR).

A key difference between the nuclear and chemical energy driven systems is the quantity of fuel. The energy released per unit mass of nuclear fuel is many orders of magnitude greater than that for chemical based fuels. For example, 1 lb (0.454 kg) of 3% enriched uranium fuel produces about the same amount of thermal energy in a commercial nuclear system as 100,000 lb (45,360 kg) of coal in a fossil-fired steam system. While a 500 MW power plant must handle approximately one million tons of coal per year, the nuclear plant will handle only 10 tons of fuel. The fossil fuel plant must be designed for a continuous fuel supply process, while most nuclear plants use a batch fuel process, where about one third of the fuel is replaced during periodic outages. However, once the steam is generated, the balance of the power producing system (turbine, condenser, cooling system, etc.) is similar to that used in the fossil fuel plant.

Nuclear steam system components

A typical nuclear steam system from The Babcock & Wilcox Company (B&W) is shown in Fig. 3 and a simplified schematic is shown in Fig. 18. This nuclear system consists of two coolant loops. The primary loop cools the reactor, transports heat to two or more steam generators (only one shown), and returns coolant to the reactor by four or more primary coolant pumps (only one shown). The coolant is high purity, subcooled, single-phase water flowing at very high rates [350,000 to 450,000 GPM (22,100 to 28,400 l/s)] at around 2200 psi (15.17 MPa) and an average temperature of approximately 580F (304C). The primary loop also contains a pressurizer to maintain the loop pressure at design operating levels.

The secondary loop includes the steam generation and interface with the balance of the power plant. High purity water from the last feedwater heater passes through the steam generator and is converted into steam. From the steam generator outlet, the saturated or superheated steam flows out of the containment building to the high pressure turbine. The operating pressure is typically around 1000 psi (6.9 MPa). The balance of the secondary loop resembles fossil fuel-fired systems. (See Figs. 10 and 11.)

The center of the NSSS is the reactor vessel and reactor core (Fig. 19). The fuel consists of compressed pellets [for example, 0.37 in. (9.4 mm) diameter by 0.7 in. (18 mm) long] of 2.5 to 5% enriched uranium oxide. These pellets are placed in zircaloy tubes which are sealed at both ends to protect the fuel and to contain the nuclear reaction products. The tubes are assembled into bundles with spacer and closure devices. These bundles are then assembled into the nuclear fuel core.

The reactor enclosure (Fig. 19) is a low alloy steel pressure vessel lined with stainless steel for corrosion protection. The rest of the reactor includes flow distribution devices, control rods, core support structures,

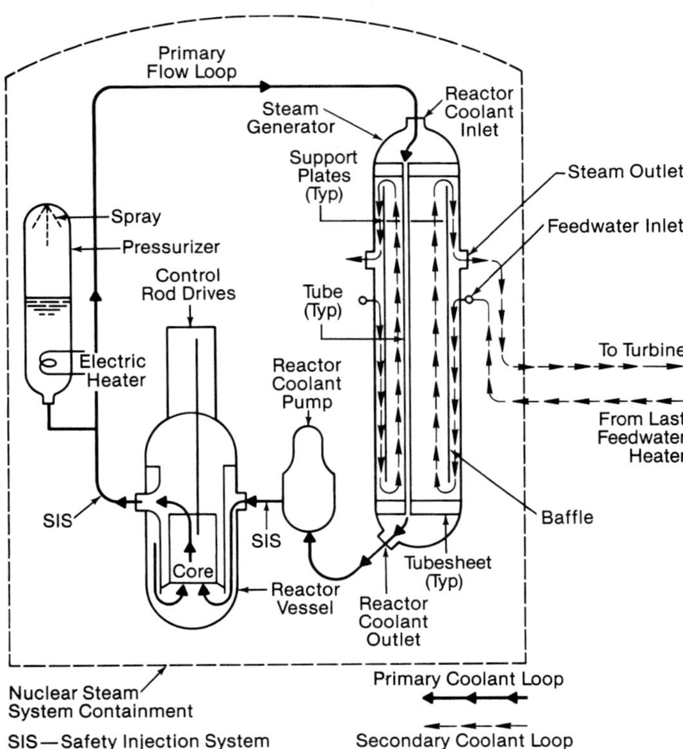

Fig. 18 Nuclear steam system schematic.

thermal shielding and moderator. The moderator in this case is water which serves a dual purpose. It reduces the velocity of the neutrons thereby making the nuclear reactions more likely. It also serves as the coolant to maintain the core materials within acceptable temperature limits and transports thermal energy to the steam generators. The control rods contain neutron absorbing material and are moved into and out of the nuclear core to control the energy output.

The steam generators can be of two types, once-through (Fig. 20) and recirculating (Fig. 21). In both types, the pressure vessel is a large heat exchanger designed to generate steam for the secondary loop from heat contained in the primary coolant. The primary coolant enters a plenum and passes through several thousand small diameter [approximately 0.625 in. (15.9 mm)] Inconel® tubes. The steam generator is a large, carbon steel pressure vessel. Specially designed tubesheets, support plates, shrouds, and baffles provide effective heat transfer, avoid thermal expansion problems, and avoid flow-induced vibration.

In the once-through steam generator (OTSG), Fig. 20, the secondary loop water flows from the bottom to the top of the shell side of the tube bundle and is continuously converted from water to superheated steam. The superheated steam then passes to the high pressure turbine.

In the recirculating steam generator (RSG), Fig. 21, water moves from the bottom to the top of the shell side of the tube bundle being converted partially into steam. The steam-water mixture passes into the upper shell where steam-water separators supply saturated dry steam to the steam generator outlet. The steam is sent to the high pressure turbine. The water leaving the steam generator upper shell is mixed with feedwater and is returned to the bottom of the tube bundle.

The pressurizer is a simple cylindrical pressure vessel which contains both water and steam at equilibrium. Electrical heaters and spray systems maintain the pressure in the pressurizer vessel and the primary loop within set limits. The primary loop circulating pumps maintain high flow rates to the reactor core to control its temperature and transfer heat to the steam generators.

A number of support systems are also provided. These include reactor coolant charging systems, makeup water addition, spent fuel storage cooling, and decay heat removal systems for when the reactor is shut down. Other specialized systems protect the reactor system in the case of a loss of coolant event. The key function of these systems is to keep the fuel bundle temperature within safe limits if the primary coolant flow is interrupted.

Nuclear steam system classifications

A variety of reactor systems have been developed to recover thermal energy from nuclear fuel and to generate steam for power generation. These are usually identified by their coolant and moderator types. The principal systems for power generation include:

1. *Pressurized water reactor (PWR)* This is the system discussed above, using water as both reactor coolant and moderator, and enriched uranium oxide as the fuel.
2. *Boiling water reactor (BWR)* The steam generator is eliminated and steam is generated directly in the reactor core. A steam-water mixture cools and moderates the reactor core. Enriched uranium oxide fuel is used.

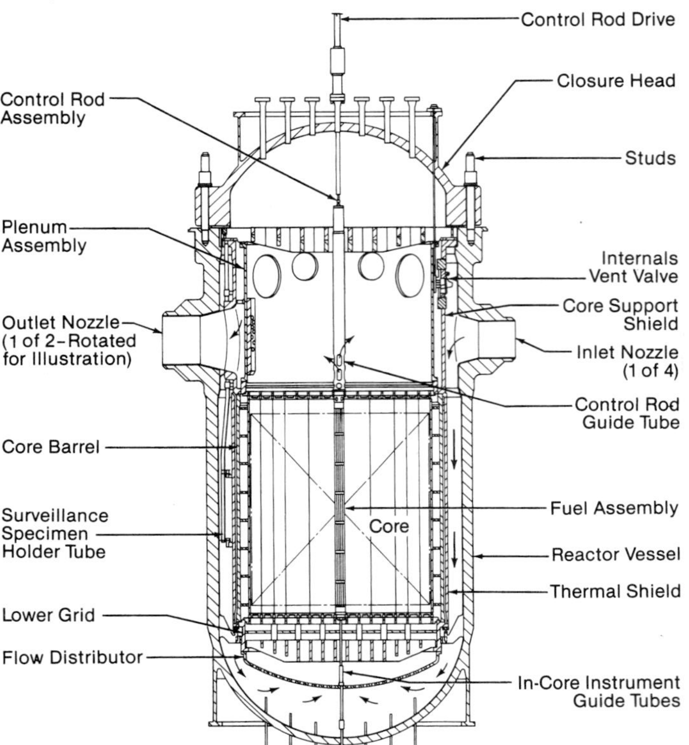

Fig. 19 Reactor vessel and internals.

3. *CANDU (PHWR)* Heavy water (deuterium) is used as the moderator and primary loop coolant. The reactor configuration is unique but the steam generator is similar to the recirculating steam generator for the PWR. Natural (not enriched) uranium oxide is used as the fuel.
4. *Gas-cooled reactors* These are a variety of gas-cooled reactors which are typically moderated by graphite and cooled by helium or carbon dioxide.
5. *Breeder reactors* These are advanced reactor systems using sodium as the reactor coolant with no moderator. These systems are specially designed to produce more fissionable nuclear fuel than they use.

Engineered safety systems

Safety is a major concern in the design, construction and operation of nuclear power generating facilities. The focus of these efforts is to minimize the likelihood of a release of radioactive materials to the environment. Three approaches are used to accomplish this goal. First, the nuclear power industry has developed one of the most extensive and rigorous quality control programs for the design, construction and maintenance of nuclear facilities. Second, reactor systems are designed with multiple barriers to prevent radioactive material release. These include high temperature ceramic fuel pellets, sealed fuel rods, reactor vessel and primary coolant system, and the containment building including both the carbon steel reactor containment vessel and the reinforced concrete shield building. The third approach includes a series of engineered safety systems to address loss of coolant conditions and maintain the integrity of the multiple barriers. These systems include:

1. emergency reactor trip systems including rapid insertion of control rods and the addition of soluble neutron poisons in the primary coolant to shut down the nuclear reaction,
2. high and low pressure emergency core cooling systems to keep the reactor core temperature within acceptable limits and remove heat from the fuel in the event of a major loss of primary coolant or a small pipe break,
3. a heat removal system for the cooling water and containment building, and
4. spray and filtering systems to collect and remove radioactivity from the containment building.

Because of the high power densities and decay heat generation, the reactor integrity depends upon the continuous cooling of the nuclear fuel rods. Multiple independent components and backup power supplies are provided for all critical systems.

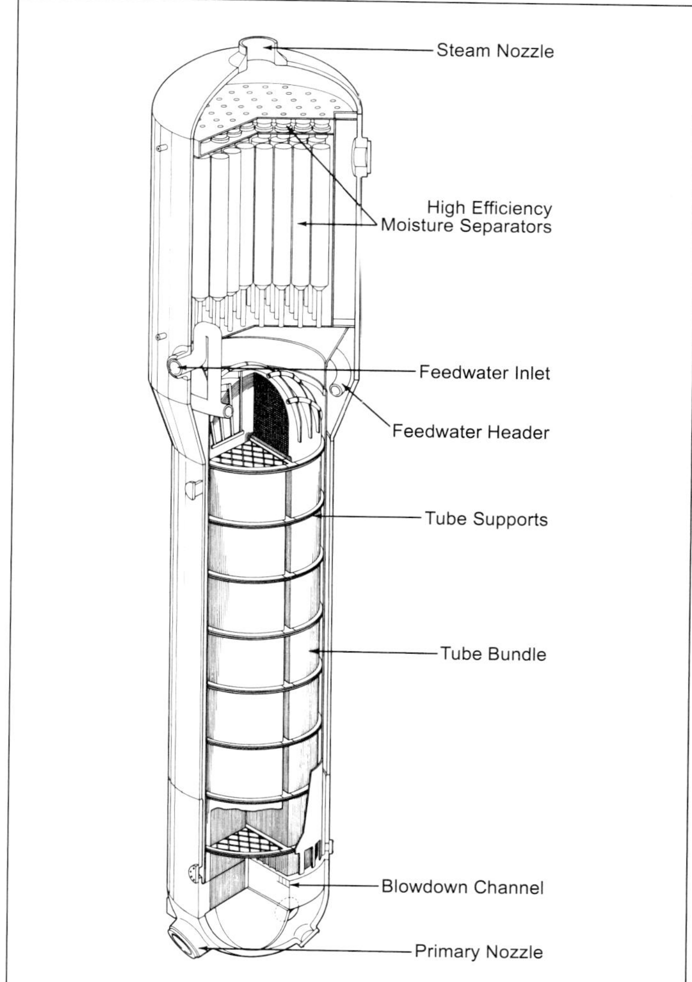

Fig. 21 Recirculating steam generator.

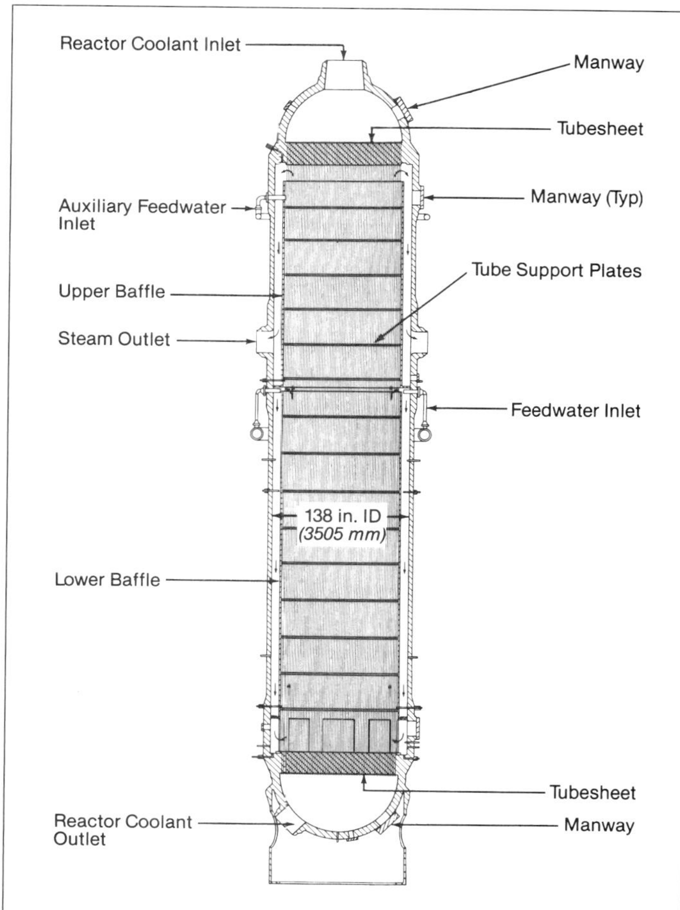

Fig. 20 Once-through steam generator.

Steam system design

Steam generator interfaces

The steam generator system's primary function is to convert chemical or nuclear energy bound in the fuel to heat and produce high temperature, high pressure steam. The variety of fuel sources, the high temperature nature of these processes, and the large number of subsystem interfaces contribute to the challenging nature of the design process. The initial steps in evaluating the steam generating system include establishing key interfaces with other plant systems and with the power cycle. These are typically set by the end user or consulting engineer after an in-depth evaluation indicates: 1) the need for the expanded power supply or steam source, 2) the most economical fuel selection and type of steam producing system, 3) the plant location, and 4) the desired power cycle or process steam conditions. The key requirements fall into six major areas:

1. Steam minimum, nominal, and maximum flow rates; pressure and temperature; need for one or more steam reheat stages; auxiliary equipment steam usage; and future requirements.
2. Source of the steam flow makeup or replacement water supply, water chemistry and inlet temperature.
3. The type and range of fuels considered, including worst case conditions, and the chemical analyses (proximate and ultimate analyses) for each fuel or mixture of fuels.
4. Elevation above sea level, overall climate history and forecast, earthquake potential and space limitations.
5. Emissions control requirements and applicable government regulations and standards.
6. The types of auxiliary equipment; overall plant and boiler efficiency; access needs; evaluation penalties, e.g., power usage; planned operating modes including expected load cycling requirements, e.g., peaking, intermediate or base load; and likely future plant use.

When these interfaces are established, boiler design and evaluation may begin.

Systematic approach

There are a variety of evaluation approaches that can be used to meet the specific steam generator performance requirements. These include the multiple iterations commonly found in thermal design where real world complexities and nonlinear, noncontinuous interactions prevent a straightforward solution. The process begins by understanding the particular application and system to define conditions such as steam flow requirements, fuel source, operating dynamics, and emissions limits, among others. From these, the designer proceeds to assess the steam generator options, interfaces, and equipment needs to achieve performance. Using a coal-fired boiler as an example, a systematic approach would include the following:

1. Specify the steam supply requirements to define the overall inputs of fuel, air and water, and the steam output conditions.
2. Evaluate the heat balances and heat absorption by type of steam generator surface.
3. Perform combustion calculations to define heat input and gas flow requirements.
4. Configure the combustion system to complete the combustion process while minimizing emissions (fuel preparation, combustion and air handling).
5. Configure the furnace and other heat transfer surfaces to satisfy temperature, material, and performance tradeoffs while meeting the system control needs.
6. Size other water-side and steam-side components.
7. Specify the back-end tradeoffs on the final heat recovery devices such as water heaters (economizers) and air heaters.
8. Check the steam generating system performance to ensure that the design criteria are met.
9. Verify overall unit performance.
10. Repeat steps 2 through 9 until the desired steam mass flow and temperature are achieved over the specified range of load conditions.
11. Use American Society of Mechanical Engineers (ASME) Code rules to design pressure parts to meet the anticipated operating conditions and complete detailed mechanical design.
12. Design and integrate environmental protection equipment to achieve prescribed emissions levels.
13. Incorporate auxiliaries as needed, such as tube surface cleaning equipment, fans, instrumentation and controls, to complete the design and assure safe and continuous operation.

The life cycle and daily operation of the steam generator (and the plant in which it will operate) are important elements to be considered from the beginning of the design and throughout the design process. Today, some steam generators will be required to operate efficiently and reliably for up to 60 years or more. During this time, many components will wear out because of the aggressive environment, so routine inspection of pressure parts is needed to assure continued reliability. Unit operating procedures, such as the permitted severity and magnitude of transients, may be monitored to prevent reduced unit life. Operating practices including water treatment, cycling operation procedures, and preventive maintenance programs, among others, can significantly affect steam generator availability and reliability. Key unit components may be upgraded to improve performance. In each case, decisions made during the design phase and subsequent operation can substantially enhance the life and performance of the unit.

System design example

Now that the basic fossil fuel and nuclear steam generating systems have been described, it is appropriate to explore the general design and engineering process. While each of the many systems requires specialized evaluations, they share many common elements. To illustrate how the design process works, a small industrial B&W PFI gas-fired boiler has been selected for discussion. (See Figs. 22 and 23.)

Basically, the customer has one overriding need: when the valve is turned on, steam is expected to be supplied at the desired pressure, temperature and flow rate. In this example, the customer specifies 400,000 lb/h (50.4 kg/s) of superheated steam at 600 psi (4.14 MPa) and 850F (454C). The customer has agreed to supply high purity feedwater at 280F (138C) and to supply natural gas as a fuel source. As with all steam generating systems, there are a number of additional constraints and requirements as discussed in *Steam generator interfaces*, but the major job of the steam generator or boiler is to supply steam.

Combustion of the natural gas produces a stream of combustion products or flue gas at perhaps 3600F (1982C). To maximize the steam generator thermal efficiency, it is important to cool these gases as much as possible while generating the steam. The minimum flue gas outlet temperature is established based upon technical and economic factors (discussed below). For now, a 310F (154C) outlet temperature to the exhaust stack is selected. The approximate steam and flue gas temperature curves are shown in Fig. 24 and define the heat transfer process. The heat transfer surface for the furnace, boiler bank, superheater and air heater is approximately 69,000 ft^2 (6410 m^2).

From a design perspective, the PFI boiler can be viewed as either a steam heater or gas cooler. The latter approach is most often selected for design. The design fuel heat input is calculated by dividing the steam heat output by the target steam generator thermal efficiency. Based upon the resulting fuel flow, combustion calculations define the air flow requirements and combustion products gas weight. The heat transfer surface is then configured in the most economical way to cool the flue gas to the temperature necessary for the target steam generator efficiency. Before proceeding to follow the gas through the cooling process, the amount of heat recovery for each of the different boiler surfaces (superheater and boiler) must be established.

Fig. 25 illustrates the water heating process from an inlet temperature of 280F (138C) to the superheater

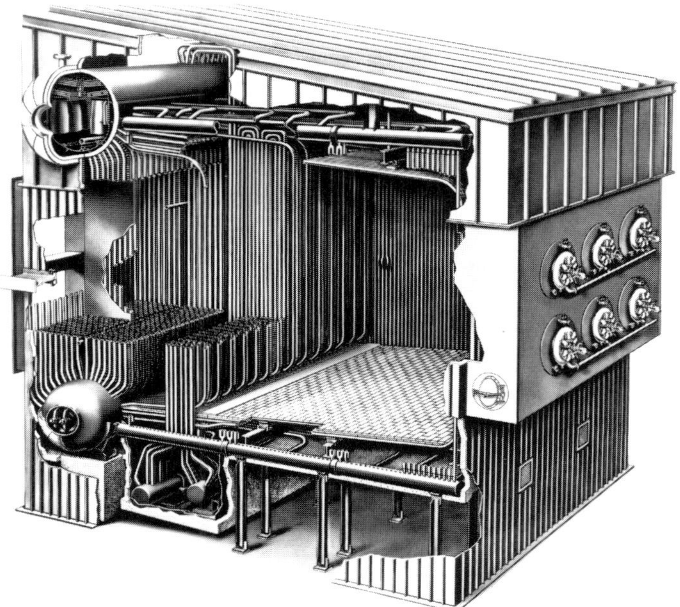

Fig. 22 Small PFI industrial boiler.

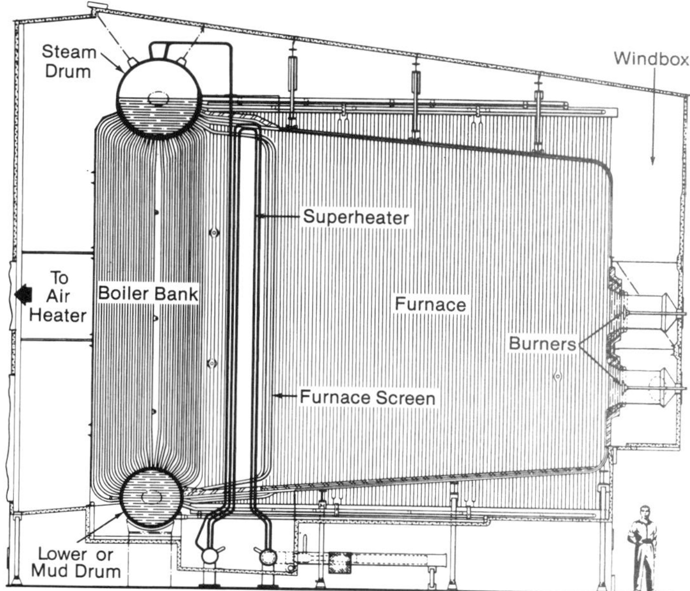

Fig. 23 Small industrial boiler – sectional view.

steam outlet temperature of 850F (454C). This curve indicates that about 20% of the heat absorbed is used to raise the water from its inlet temperature to the saturation temperature of 490F (254C). 60% of the energy is then used to evaporate the water to produce saturated steam. The remaining 20% of the heat input is used to superheat or raise the steam temperature to the desired outlet temperature of 850F (454C).

The fuel and the combustion process selected set the geometry of the furnace. In this case, simple circular burners are used. The objective of the burners is to mix the fuel and air rapidly to produce a stable flame and complete combustion while minimizing the formation of NO_x emissions. Burners are available in several standardized sizes. The specific size and number are selected from past experience to provide the desired heat input rate while permitting the necessary level of load range control. The windbox, which distributes the air to individual burners, is designed to provide a uniform air flow at low enough velocities to permit the burners to function properly.

The furnace volume is then set to allow complete fuel combustion. The distances between burners and between the burners and the floor, roof, and side walls are determined from the known characteristics of the particular burner flame. Adequate clearances are specified to prevent flame impingement on the furnace surfaces, which could overheat the tubes and cause tube failures.

Once the furnace dimensions are set, this volume is enclosed in a water-cooled membrane panel surface. This construction provides a gas-tight, all steel enclosure which minimizes energy loss, generates some steam and minimizes furnace maintenance. As shown in Fig. 23, the roof and floor tubes are inclined slightly to enhance water flow and prevent steam from collecting on the tube surface. Trapped steam could result in overheating of the tubes. Heat transfer from the flame to the furnace enclosure surfaces occurs primarily by thermal radiation. As a result, the heat input rates per

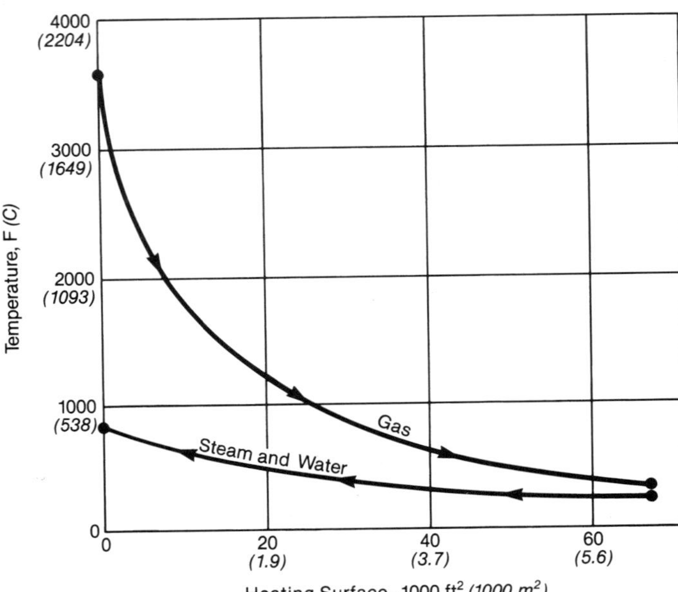

Fig. 24 Industrial boiler – temperature versus heat transfer surface.

unit area of surface are very high and relatively independent of the tubewall temperatures. Boiling water provides an effective means to cool the tubes and keep the tube metal temperatures within acceptable limits as long as the boiling conditions are maintained.

Fig. 26 shows the effect of the furnace on gas temperature. The gas temperature is reduced from 3600 at point A to 2400F at point B (1982 to 1316C), while boiling takes place in the water walls (points 1 to 2). A large amount of heat transfer takes place on a small amount of surface. From the furnace, the gases pass through the furnace screen tubes shown in Fig. 23. The temperature drops a small amount [50F (28C)] from points B to C in Fig. 26, but more importantly, the superheater surface is partially shielded from the furnace thermal radiation. The furnace screen tubes are connected to the drum and contain boiling water. Next, the gas passes through the superheater where the gas temperature drops from 2350 at point C to 1750F at point D (1288 to 954C). Saturated steam from the drum is passed through the superheater tubing to raise its temperature from 490F (254C) saturation temperature to the 850F (454C) desired outlet temperature (points 5 to 4).

The location of the superheater and its configuration are critical in order to keep the steam outlet temperature constant under all load conditions. This involves radiation heat transfer from the furnace with convection heat transfer from the gas passing across the surface. In addition, where dirty gases such as combustion products from coal are used, the spacing of the superheater tubes is also adjusted to accommodate the accumulation of fouling ash deposits and the use of cleaning equipment.

After the superheater, almost half of the energy in the gas stream has been recovered with only a small amount of heat transfer surface [approximately 6400 ft^2 (595 m^2)]. This is possible because of the large temperature difference between the gas and the boiling water or steam. The gas temperature has now been dramatically reduced, requiring much larger heat transfer surfaces to recover incremental amounts of energy.

The balance of the steam is generated by passing the gas through the boiler bank. (See Figs. 22 and 23.) This bank is composed of a large number of water-containing tubes that connect the steam drum to a lower (mud) drum. The temperature of the boiling water is effectively constant (points 5 to 6 in Fig. 26), while the gas temperature drops by almost 1000F (556C) to an outlet temperature of 760F (404C), between points D and E. The tubes are spaced as closely as possible to increase the gas flow heat transfer rate. If a particulate-laden gas stream were present, the spacing would be set to limit erosion of the tubes, reduce the heat transfer degradation due to ash deposits, and permit removal of the ash. Spacing is also controlled by the allowable pressure drop across the bank. In addition, a baffle can be used in the boiler bank bundle to force the gas to travel at higher velocity through the bundle, increase the heat transfer rate, and thereby reduce the bundle size and cost. To recover this additional percentage of the supplied energy, the boiler bank contains more than 32,000 ft^2 (3000 m^2) of surface, or approximately nine times more surface per unit of energy than in the high temperature furnace and superheater. At this point in the process, the temperature difference between the saturated water and gas is only 270F (150C), between points 6 and E in Fig. 26.

Economics and technical limits dictate the type and arrangement of additional heat transfer surfaces. An economizer or water-cooled heat exchanger could be used to heat the makeup or feedwater and cool the gas. The lowest gas exit temperature possible is the inlet temperature of the feedwater [280F (138C)]. However, the economizer would have to be infinitely large to accomplish this goal. Even if the exit gas temperature is 310F (154C), the temperature difference at this point in the heat exchanger would only be 30F (17C), still making the heat exchanger relatively large. Instead of incorporating an economizer, an air preheater could be used to recover the remaining gas energy and preheat the combustion air. This would reduce the

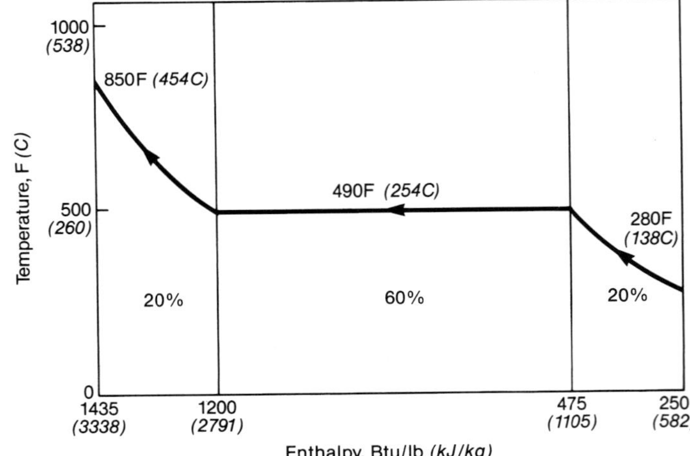

Fig. 25 Steam-water temperature profile.

natural gas needed to heat the steam generator. Air heaters can be very compact. Also, air preheating can enhance the combustion of many difficult to burn fuels such as coal. All of the parameters are reviewed to select the most economical solution that meets the technical requirements.

In this case, the decision has been made to use an air heater and not an economizer. The air heater is designed to take 80F (27C) ambient air (point 9) and increase the temperature to 570F (299C), at point 8. This hot air is then fed to the burners. At the same time, the gas temperature is dropped from 760F (404C) to the desired 310F (154C) outlet temperature (points E to F). If a much lower gas outlet temperature than 310F (154C) is used, the heat exchanger surfaces may become uneconomically large, although this is a case by case decision. In addition, for fuels such as oil or coal which can produce acid constituents in the gas stream (such as sulfur oxides), lower exit gas temperatures may result in condensation of these constituents onto the heat transfer surfaces, and excessive corrosion damage. The gas is then exhausted through the stack to the atmosphere.

Finally, the feedwater temperature increases from 280F (138C) to saturation temperature of 490F (254C). In the absence of an economizer, the feedwater is supplied directly to the drum where it is mixed with the water flowing through the boiler bank tubes and furnace. The flow rate of this circulating water in industrial units is approximately 25 times higher than the feedwater flow rate. Therefore, when the feedwater is mixed in the drum, it quickly approaches the saturation temperature without appreciably lowering the temperature of the recirculating water in the boiler tubes.

Reviewing the water portion of the system, the feedwater is supplied to the drum where it mixes with the recirculating water after the steam is extracted and sent to the superheater. The drum internals are specially designed so that the now slightly subcooled water flows down through a portion of the boiler bank tubes to the lower or mud drum. This water is then distributed to the remainder of the boiler bank tubes (also called risers) and the furnace enclosure tubes where it is partially converted to steam (approximately 4% steam by weight). The steam-water mixture is then returned to the steam drum. Here, the steam-water mixture is passed through *separators* where the steam is separated from the water. The steam is then sent to the superheater, and from there to its end use. The remaining water is mixed with the feedwater and is again distributed to the downcomer tubes.

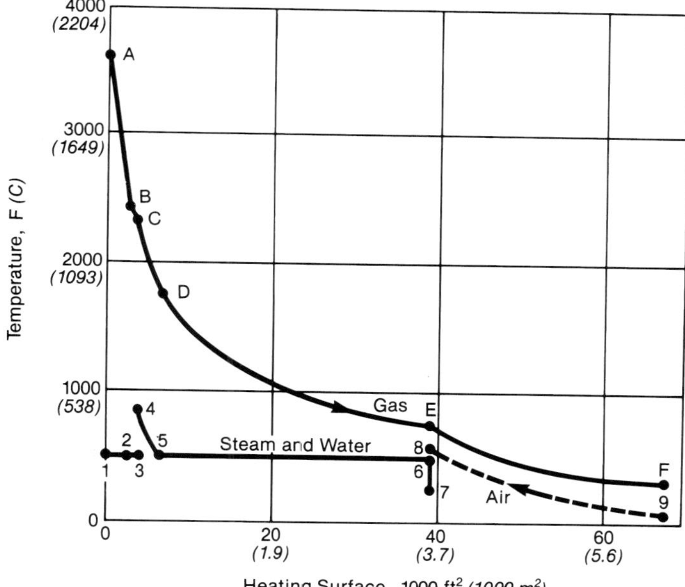

Fig. 26 Gas and steam temperature schematic.

Other steam producing systems

A variety of additional systems also produce steam for power and process applications. These systems usually take advantage of low cost or free fuels, a combination of power cycles and processes, and recovery of waste heat in order to reduce overall costs. Examples of these include:

1. *Gas turbine combined cycle (CC)* Advanced gas turbines with heat recovery steam generators as part of a bottoming cycle to use waste heat recovery and increase thermal efficiency.
2. *Integrated gasification combined cycle (IGCC)* Adds a coal gasifier to the CC to reduce fuel costs and minimize airborne emissions.
3. *Pressurized fluidized-bed combustion (PFBC)* Includes higher pressure combustion with gas cleaning and expansion of the combustion products through a gas turbine.
4. *Blast furnace hood heat recovery* Generates steam using the waste heat from a blast furnace.
5. *Solar steam generator* Uses concentrators to collect and concentrate solar radiation and generate steam.

Common technical elements

The design of steam generating systems involves the combination of scientific and technical fundamentals, empirical data, practical experience and designer insight. While technology has advanced significantly, it is still not possible to base the design of modern systems on fundamentals alone. Instead, the fundamentals provide the basis for combining field data and empirical methods.

Even given the wide variety of shapes, sizes and applications (see Fig. 27), steam generator design involves the application of a common set of technologies. The functional performance of the steam generator is established by combining thermodynamics, heat transfer, fluid mechanics, chemistry, and combustion science or nuclear science with practical knowledge of the fouling of heat transfer surfaces and empirical knowledge of the behavior of boiling water. The design and supply of the hardware are aided by structural design and advanced materials properties research combined with expertise in manufacturing technologies and erection skills to produce a quality, reliable product to meet the highly demanding system requirements.

Fig. 27 This energy complex in the northern U.S. includes four coal-fired steam systems, ranging from 80 to 330 MW, installed over a 25-year period.

The ASME Boiler and Pressure Vessel Code is the firm basis from which steam generator pressure parts can be safely designed. Once built, the operation and maintenance of the steam generator are critical to ensure a long life and reliable service. Water chemistry and chemical cleaning are increasingly recognized as central elements in any ongoing operating program. The impact of fuel and any residual flyash is important in evaluating the corrosion and fouling of heat transfer surfaces. The use of modern techniques to periodically inspect the integrity of the steam generator tubes leads to the ability to extend steam generator life and improve overall performance. These are accomplished by the application of engineered component modification to better meet the changing needs of the steam generating system. Finally, the control systems which monitor and operate many subsystems to optimize unit performance are important to maintain system reliability and efficiency.

All of these functions – functional performance, mechanical design, manufacture, construction, operation, maintenance, and life extension – must be fully combined to provide the best steam generating system. Long term success depends upon a complete life cycle approach to the steam generating system. Steam generator system operators routinely require their equipment to operate continuously and reliably for more than 60 years in increasingly demanding conditions. Therefore, it is important to consider later boiler life, including component replacement, in the initial phases of boiler specification. Changes in design to reduce initial capital cost must be weighed against their possible impact on future operation.

Bibliography

Aschner, F.S., *Planning Fundamentals of Thermal Power Plants*, John Wiley & Sons, New York, New York, 1978.

Axtman, W.H., and American Boiler Manufacturers Association staff, "The American Boiler Industry: A Century of Innovation," American Boiler Manufacturers Association (ABMA), Arlington, Virginia, 1988.

"Boilers and auxiliary equipment," *Power*, Vol. 132, No. 6, pp B-1 to B-138, Platts/McGraw-Hill, New York, New York, June, 1988.

Clapp, R.M., Ed., *Modern Power Station Practice: Boilers and Ancillary Plant,* Third Ed., Pergamon Press, Oxford, England, United Kingdom, April 1, 1993.

Collier, J.G., and Hewitt, G.F., *Introduction to Nuclear Power*, Taylor and Francis Publishers, Washington, D.C., June 1, 2000.

Elliot, T.C., Chen, K., Swanekamp, R.C., Ed., *Standard Handbook of Powerplant Engineering*, McGraw-Hill Company, New York, New York, 1997.

El-Wakil, M.M., *Powerplant Technology*, McGraw-Hill Primis Custom Publishing, New York, New York, 1984.

Foster, A.R., and Wright, R.L., *Basic Nuclear Engineering*, Pearson Allyn and Bacon, Boston, Massachusetts, January, 1983.

Fraas, A.P., *Heat Exchanger Design*, Second Ed., Interscience, New York, New York, March, 1989.

Gunn, D., and Horton, R., *Industrial Boilers*, Longman Scientific and Technical, Longman Science & Technology, London, England, United Kingdom, April, 1989.

Hambling, P., *Modern Power Station Practice: Turbines, Generators and Associated Plant*, Third Ed., Pergamon Press, Oxford, England, United Kingdom, April 1, 1993.

Jackson, A.W., "The How and Why of Boiler Design," Engineers Society of Western Pennsylvania Power Symposium, Pittsburgh, Pennsylvania, February 15, 1967.

Kakaç, S., Ed., *Boilers, Evaporators and Condensers*, Interscience, April 15, 1991. See Chapter 6, "Fossil Fuel-Fired Boilers: Fundamentals and Elements," by J.B. Kitto and M.J. Albrecht.

Li, K.W., and Priddy, A.P., *Power Plant System Design*, Wiley Text Books, New York, New York, February, 1985.

Shields, C.D., *Boilers: Types, Characteristics, and Functions*, McGraw-Hill Company, New York, New York, 1961.

Wiener, M., "The Latest Developments in Natural Circulation Boiler Design," Proceedings of the American Power Conference, Vol. 39, pp. 336-348, 1977.

Inconel is a trademark of the Special Metals Corporation group of companies.

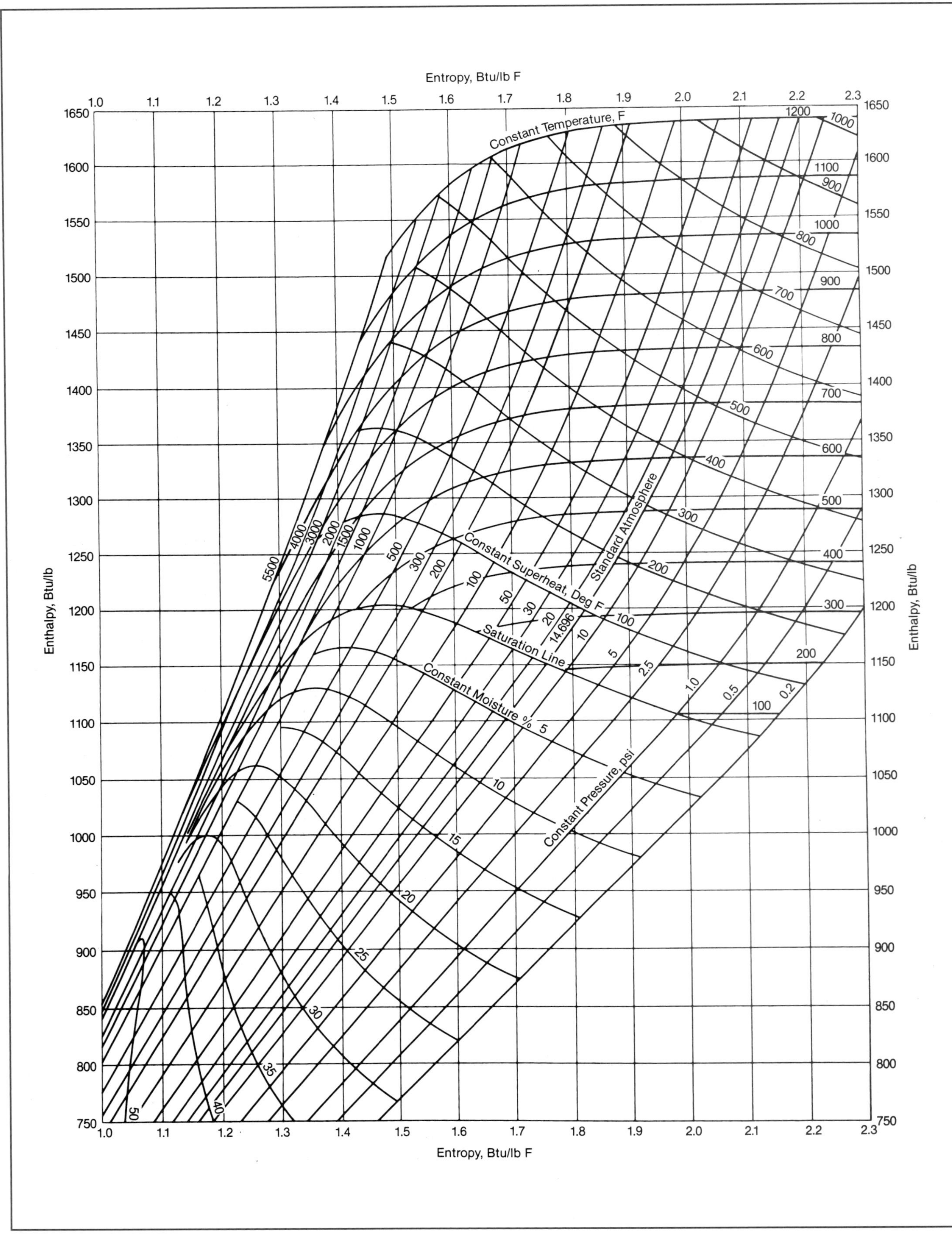

Mollier diagram (*H-s*) for steam.

Chapter 2

Thermodynamics of Steam

Thermodynamics is the science which describes and defines the transformation of one form of energy into another – chemical to thermal, thermal to mechanical, and mechanical to thermal. The basic tenets include: 1) energy in all of its forms must be conserved, and 2) only a portion of *available* energy can be converted to *useful* energy or work. Generally referred to as the first and second laws of thermodynamics, these tenets evolved from the early development of the steam engine and the efforts to formalize the observations of its conversion of heat into mechanical work.

Regardless of the type of work or form of energy under consideration, the terms heat, work, and energy have practical significance only when viewed in terms of systems, processes, cycles, and their surroundings. In the case of expansion work, the *system* is a *fluid* capable of expansion or contraction as a result of pressure, temperature or chemical changes. The way in which these changes take place is referred to as the *process*. A *cycle* is a sequence of processes that is capable of producing net heat flow or work when placed between an energy *source* and an energy *sink*. The *surroundings* represent the sources and sinks which accommodate interchanges of mass, heat and work to or from the system.

Steam may be viewed as a thermodynamic system which is favored for power generation and heat transfer. Its unique combination of high thermal capacity (specific heat), high critical temperature, wide availability, and nontoxic nature has served to maintain this dominant position. High thermal capacity of a working fluid generally results in smaller equipment for a given power output or heat transfer. The useful temperature range of water and its high thermal capacity meet the needs of many industrial processes and the temperature limitations of power conversion equipment.

Properties of steam

Before a process or cycle can be analyzed, reliable properties of the working fluid are needed. Key properties include enthalpy, entropy, and specific volume. While precise definitions are provided later in this chapter, *enthalpy* is a general measure of the internally stored energy per unit mass of a flowing stream, *specific entropy* is a measure of the thermodynamic potential of a system in the units of energy per unit mass and absolute temperature, and *specific volume* is the volume per unit mass.

In the case of steam, a worldwide consensus of these and other thermophysical properties has been reached through the International Association for the Properties of Steam. The most frequently used tabulation of steam properties is the American Society of Mechanical Engineers (ASME) Steam Tables.[1,2] Selected data from this tabulation in English units are summarized in Tables 1, 2 and 3. Corresponding SI tabulations are provided in Appendix 1. These properties are now well described by formulas that have been agreed to by the International Association for the Properties of Water and Steam, and are available from a number of sources on the Internet as add-on functions to spreadsheets and other software products.

The first two columns of Tables 1 and 2 define the unique relationship between pressure and temperature referred to as saturated conditions, where liquid and vapor phases of water can coexist at thermodynamic equilibrium. For a given pressure, steam heated above the saturation temperature is referred to as superheated steam, while water cooled below the saturation temperature is referred to as subcooled or compressed water. Properties for superheated steam and compressed water are provided in Table 3. Reproduced from Reference 1, Fig. 1 shows the values of enthalpy and specific volume for steam and water over a wide range of pressure and temperature.

Under superheated or subcooled conditions, fluid properties, such as enthalpy, entropy and volume per unit mass, are unique functions of temperature and pressure. However, at saturated conditions where mixtures of steam and water coexist, the situation is more complex and requires an additional parameter for definition. For example, the enthalpy of a steam-water mixture will depend upon the relative amounts of steam and water present. This additional parameter is the thermodynamic equilibrium quality or simply quality (x) defined by convention as the mass fraction of steam:

$$x = \frac{m_s}{m_s + m_w} \qquad (1)$$

Note: The following steam tables and Fig. 1 have been abstracted from *ASME International Steam Tables for Industrial Use* (copyright 2000 by The American Society of Mechanical Engineers), based on the IAPWS industrial formulation 1997 for the Thermodynamic Properties of Water and Steam (IAPWS-IF97).

Table 1
Properties of Saturated Steam and Saturated Water (Temperature)[1]

Temp F	Press. psia	Specific Volume, ft³/lb Water v_f	Specific Volume Evap v_{fg}	Specific Volume Steam v_g	Enthalpy,[2] Btu/lb Water H_f	Enthalpy Evap H_{fg}	Enthalpy Steam H_g	Entropy, Btu/lb F Water s_f	Entropy Evap s_{fg}	Entropy Steam s_g	Temp F
32	0.08865	0.01602	3302	3302	-0.02	1075.2	1075.2	-0.00004	2.1869	2.1868	**32**
35	0.09998	0.01602	2946	2946	3.00	1073.5	1076.5	0.0061	2.1701	2.1762	**35**
40	0.12173	0.01602	2443	2443	8.03	1070.7	1078.7	0.0162	2.1427	2.1590	**40**
45	0.14757	0.01602	2035.6	2035.6	13.05	1067.8	1080.9	0.0262	2.1159	2.1421	**45**
50	0.17813	0.01602	1702.9	1702.9	18.07	1065.0	1083.1	0.0361	2.0896	2.1257	**50**
60	0.2564	0.01603	1206.0	1206.1	28.08	1059.4	1087.4	0.0555	2.0385	2.0941	**60**
70	0.3633	0.01605	867.2	867.2	38.08	1053.7	1091.8	0.0746	1.9894	2.0640	**70**
80	0.5074	0.01607	632.4	632.4	48.07	1048.0	1096.1	0.0933	1.9420	2.0353	**80**
90	0.6990	0.01610	467.4	467.4	58.05	1042.4	1100.4	0.1116	1.8964	2.0080	**90**
100	0.9504	0.01613	349.9	349.9	68.04	1036.7	1104.7	0.1296	1.8523	1.9819	**100**
110	1.2766	0.01617	265.0	265.0	78.02	1031.0	1109.0	0.1473	1.8098	1.9570	**110**
120	1.6949	0.01620	202.95	202.96	88.00	1025.2	1113.2	0.1647	1.7686	1.9333	**120**
130	2.2258	0.01625	157.09	157.10	97.99	1019.4	1117.4	0.1817	1.7288	1.9106	**130**
140	2.8929	0.01629	122.81	122.82	107.98	1013.6	1121.6	0.1985	1.6903	1.8888	**140**
150	3.723	0.01634	96.92	96.93	117.97	1007.8	1125.7	0.2151	1.6530	1.8680	**150**
160	4.747	0.01639	77.17	77.19	127.98	1001.9	1129.8	0.2313	1.6168	1.8481	**160**
170	6.000	0.01645	61.97	61.98	137.99	995.9	1133.9	0.2474	1.5816	1.8290	**170**
180	7.520	0.01651	50.15	50.17	148.01	989.9	1137.9	0.2631	1.5475	1.8106	**180**
190	9.350	0.01657	40.90	40.92	158.05	983.8	1141.8	0.2787	1.5143	1.7930	**190**
200	11.538	0.01663	33.59	33.61	168.10	977.6	1145.7	0.2940	1.4820	1.7760	**200**
212	14.709	0.01671	26.76	26.78	180.18	970.1	1150.3	0.3122	1.4443	1.7565	**212**
220	17.201	0.01677	23.12	23.13	188.25	965.0	1153.3	0.3241	1.4198	1.7440	**220**
230	20.795	0.01684	19.356	19.373	198.35	958.6	1157.0	0.3388	1.3899	1.7288	**230**
240	24.985	0.01692	16.299	16.316	208.47	952.1	1160.5	0.3534	1.3607	1.7141	**240**
250	29.843	0.01700	13.799	13.816	218.62	945.4	1164.0	0.3678	1.3322	1.7000	**250**
260	35.445	0.01708	11.743	11.760	228.79	938.6	1167.4	0.3820	1.3043	1.6862	**260**
270	41.874	0.01717	10.042	10.059	238.99	931.7	1170.7	0.3960	1.2769	1.6730	**270**
280	49.218	0.01726	8.627	8.644	249.21	924.7	1173.9	0.4099	1.2502	1.6601	**280**
290	57.567	0.01735	7.444	7.461	259.5	917.5	1177.0	0.4236	1.2239	1.6476	**290**
300	67.021	0.01745	6.449	6.467	269.8	910.2	1180.0	0.4372	1.1982	1.6354	**300**
310	77.68	0.01755	5.609	5.627	280.1	902.7	1182.8	0.4507	1.1728	1.6235	**310**
320	89.65	0.01765	4.897	4.915	290.4	895.0	1185.5	0.4640	1.1480	1.6120	**320**
340	118.00	0.01787	3.771	3.789	311.3	879.2	1190.5	0.4903	1.0994	1.5897	**340**
360	153.00	0.01811	2.940	2.958	332.3	862.5	1194.8	0.5162	1.0522	1.5684	**360**
380	195.71	0.01836	2.318	2.336	353.6	844.9	1198.5	0.5416	1.0062	1.5478	**380**
400	247.22	0.01864	1.8454	1.8640	375.1	826.4	1201.5	0.5667	0.9613	1.5280	**400**
420	308.71	0.01894	1.4818	1.5007	396.9	806.7	1203.6	0.5915	0.9171	1.5086	**420**
440	381.44	0.01926	1.1986	1.2179	419.0	785.9	1204.9	0.6161	0.8735	1.4896	**440**
460	466.7	0.0196	0.9755	0.9952	441.5	763.7	1205.2	0.6405	0.8304	1.4709	**460**
480	565.9	0.0200	0.7980	0.8180	464.4	739.9	1204.4	0.6648	0.7874	1.4522	**480**
500	680.5	0.0204	0.6551	0.6756	487.9	714.5	1202.3	0.6890	0.7445	1.4335	**500**
520	812.1	0.0209	0.5392	0.5601	511.9	687.0	1198.9	0.7133	0.7013	1.4145	**520**
540	962.2	0.0215	0.4441	0.4656	536.7	657.3	1194.0	0.7377	0.6575	1.3952	**540**
560	1132.7	0.0221	0.3654	0.3875	562.3	624.9	1187.2	0.7624	0.6128	1.3752	**560**
580	1325.4	0.0228	0.2995	0.3223	589.0	589.3	1178.2	0.7875	0.5668	1.3543	**580**
600	1542.5	0.0236	0.2438	0.2675	616.9	549.7	1166.6	0.8133	0.5187	1.3320	**600**
620	1786.1	0.0246	0.1961	0.2207	646.6	505.0	1151.6	0.8400	0.4677	1.3077	**620**
640	2059.2	0.0259	0.1543	0.1802	678.7	453.3	1132.0	0.8683	0.4122	1.2804	**640**
660	2364.8	0.0277	0.1167	0.1444	714.5	390.9	1105.3	0.8991	0.3491	1.2482	**660**
680	2707.3	0.0303	0.0809	0.1112	757.3	309.3	1066.6	0.9354	0.2714	1.2068	**680**
700	3092.9	0.0368	0.0378	0.0747	823.6	167.0	990.6	0.9910	0.1440	1.1350	**700**
705.1028	3200.1	0.0497	0	0.04975	897.5	0	897.5	1.0538	0	1.0538	**705.1028**

1. SI steam tables are provided in Appendix 1.
2. In the balance of *Steam*, enthalpy is denoted by H in place of h to avoid confusion with heat transfer coefficient.

Table 2
Properties of Saturated Steam and Saturated Water (Pressure)[1]

Press. psia	Temp F	Volume, ft³/lb			Enthalpy,[2] Btu/lb			Entropy, Btu/lb F			Internal Energy, Btu/lb		Press. psia
		Water v_f	Evap v_{fg}	Steam v_g	Water H_f	Evap H_{fg}	Steam H_g	Water s_f	Evap s_{fg}	Steam s_g	Water u_f	Steam u_g	
0.0886	31.986	0.01602	3303.8	3303.8	-0.03	1075.2	1075.2	0	2.1869	2.1869	0	1021.0	**0.0886**
0.1	35.005	0.01602	2945.0	2945.0	3.01	1073.5	1076.5	0.0061	2.1701	2.1762	3.01	1022.0	**0.1**
0.15	45.429	0.01602	2004.3	2004.3	13.48	1067.6	1081.1	0.0271	2.1136	2.1407	13.48	1025.4	**0.15**
0.2	53.132	0.01603	1525.9	1525.9	21.20	1063.2	1084.4	0.0422	2.0734	2.1156	21.20	1028.0	**0.2**
0.3	64.452	0.01604	1039.4	1039.4	32.53	1056.8	1089.4	0.0641	2.0164	2.0805	32.53	1031.7	**0.3**
0.4	72.834	0.01606	791.8	791.9	40.91	1052.1	1093.0	0.0799	1.9758	2.0557	40.91	1034.4	**0.4**
0.5	79.549	0.01607	641.3	641.3	47.62	1048.3	1095.9	0.0925	1.9441	2.0366	47.62	1036.6	**0.5**
0.6	85.180	0.01609	539.9	539.9	53.24	1045.1	1098.3	0.1028	1.9182	2.0210	53.24	1038.4	**0.6**
0.7	90.05	0.01610	466.80	466.81	58.10	1042.3	1100.4	0.1117	1.8962	2.0079	58.10	1040.0	**0.7**
0.8	94.34	0.01611	411.56	411.57	62.39	1039.9	1102.3	0.1195	1.8770	1.9965	62.39	1041.4	**0.8**
0.9	98.20	0.01613	368.30	368.32	66.24	1037.7	1103.9	0.1264	1.8601	1.9865	66.23	1042.6	**0.9**
1	101.69	0.01614	333.49	333.51	69.73	1035.7	1105.4	0.1326	1.8450	1.9776	69.73	1043.7	**1**
2	126.03	0.01623	173.70	173.72	94.02	1021.7	1115.8	0.1750	1.7445	1.9195	94.01	1051.5	**2**
3	141.42	0.01630	118.69	118.70	109.39	1012.8	1122.2	0.2009	1.6849	1.8858	109.38	1056.3	**3**
4	152.91	0.01636	90.61	90.63	120.89	1006.1	1126.9	0.2198	1.6423	1.8621	120.87	1059.9	**4**
5	162.18	0.01641	73.507	73.52	130.16	1000.6	1130.7	0.2349	1.6090	1.8438	130.15	1062.7	**5**
6	170.00	0.01645	61.963	61.98	137.99	995.9	1133.9	0.2474	1.5816	1.8290	137.97	1065.1	**6**
7	176.79	0.01649	53.632	53.65	144.79	991.8	1136.6	0.2581	1.5583	1.8164	144.77	1067.1	**7**
8	182.81	0.01652	47.328	47.34	150.83	988.2	1139.0	0.2675	1.5381	1.8056	150.80	1068.9	**8**
9	188.22	0.01656	42.387	42.40	156.27	984.9	1141.1	0.2760	1.5201	1.7961	156.24	1070.5	**9**
10	193.16	0.01659	38.406	38.42	161.22	981.8	1143.1	0.2836	1.5040	1.7875	161.19	1072.0	**10**
14.696	211.95	0.01671	26.787	26.80	180.13	970.1	1150.3	0.3121	1.4445	1.7566	180.09	1077.4	**14.696**
15	212.99	0.01672	26.278	26.30	181.18	969.5	1150.7	0.3137	1.4413	1.7549	181.13	1077.7	**15**
20	227.92	0.01683	20.075	20.09	196.25	959.9	1156.2	0.3358	1.3961	1.7319	196.18	1081.8	**20**
30	250.30	0.01700	13.7312	13.748	218.9	945.2	1164.1	0.3682	1.3313	1.6995	218.8	1087.8	**30**
40	267.22	0.01715	10.4832	10.500	236.2	933.7	1169.8	0.3921	1.2845	1.6766	236.0	1092.1	**40**
50	280.99	0.01727	8.4998	8.517	250.2	924.0	1174.2	0.4113	1.2475	1.6588	250.1	1095.4	**50**
60	292.69	0.01738	7.1588	7.176	262.2	915.6	1177.8	0.4273	1.2169	1.6443	262.0	1098.1	**60**
70	302.92	0.01748	6.1896	6.207	272.8	908.0	1180.8	0.4412	1.1907	1.6319	272.5	1100.4	**70**
80	312.03	0.01757	5.4554	5.473	282.2	901.2	1183.3	0.4534	1.1678	1.6212	281.9	1102.3	**80**
90	320.27	0.01766	4.8792	4.897	290.7	894.8	1185.6	0.4644	1.1473	1.6117	290.4	1104.0	**90**
100	327.82	0.01774	4.4146	4.432	298.6	888.9	1187.5	0.4744	1.1288	1.6032	298.2	1105.5	**100**
120	341.26	0.01789	3.7107	3.729	312.6	878.1	1190.7	0.4920	1.0964	1.5883	312.2	1108.0	**120**
140	353.04	0.01802	3.2019	3.220	325.0	868.4	1193.4	0.5072	1.0685	1.5757	324.5	1110.0	**140**
160	363.55	0.01815	2.8163	2.834	336.1	859.4	1195.5	0.5207	1.0440	1.5647	335.6	1111.6	**160**
180	373.08	0.01827	2.5137	2.532	346.2	851.1	1197.3	0.5328	1.0220	1.5549	345.6	1113.0	**180**
200	381.81	0.01839	2.2696	2.288	355.5	843.3	1198.8	0.5439	1.0021	1.5460	354.9	1114.1	**200**
250	400.98	0.01865	1.8252	1.8439	376.2	825.4	1201.6	0.5679	0.9591	1.5270	375.3	1116.3	**250**
300	417.37	0.01890	1.5245	1.5434	394.0	809.4	1203.4	0.5883	0.9229	1.5111	393.0	1117.7	**300**
350	431.75	0.01913	1.3071	1.3262	409.8	794.6	1204.5	0.6060	0.8914	1.4974	408.6	1118.6	**350**
400	444.63	0.0193	1.14225	1.1616	424.2	780.9	1205.0	0.6217	0.8635	1.4853	422.7	1119.1	**400**
450	456.32	0.0196	1.01283	1.0324	437.3	767.9	1205.2	0.6360	0.8383	1.4743	435.7	1119.2	**450**
500	467.05	0.0198	0.90840	0.9282	449.5	755.5	1205.0	0.6490	0.8152	1.4643	447.7	1119.1	**500**
550	476.98	0.0199	0.82229	0.8422	460.9	743.6	1204.6	0.6611	0.7939	1.4550	458.9	1118.8	**550**
600	486.25	0.0201	0.75002	0.7702	471.7	732.2	1203.9	0.6723	0.7740	1.4464	469.5	1118.4	**600**
700	503.14	0.0205	0.63535	0.6559	491.6	710.3	1201.9	0.6928	0.7377	1.4305	489.0	1117.0	**700**
800	518.27	0.0209	0.54830	0.5692	509.8	689.5	1199.3	0.7112	0.7050	1.4162	506.7	1115.0	**800**
900	532.02	0.0212	0.47983	0.5011	526.7	669.4	1196.2	0.7279	0.6751	1.4030	523.2	1112.7	**900**
1000	544.65	0.0216	0.42446	0.4461	542.6	650.0	1192.6	0.7434	0.6472	1.3906	538.6	1110.0	**1000**
1100	556.35	0.0220	0.37869	0.4006	557.6	631.0	1188.6	0.7578	0.6211	1.3789	553.1	1107.0	**1100**
1200	567.26	0.0223	0.34014	0.3625	571.8	612.4	1184.2	0.7714	0.5963	1.3677	566.9	1103.7	**1200**
1300	577.50	0.0227	0.30718	0.3299	585.5	593.9	1179.5	0.7843	0.5727	1.3570	580.1	1100.1	**1300**
1400	587.14	0.0231	0.27861	0.3017	598.8	575.7	1174.4	0.7966	0.5499	1.3465	592.8	1096.3	**1400**
1500	596.27	0.0235	0.25357	0.2770	611.6	557.4	1169.0	0.8084	0.5279	1.3363	605.1	1092.1	**1500**
2000	635.85	0.0256	0.16255	0.1882	671.8	464.7	1136.5	0.8622	0.4242	1.2864	662.3	1066.9	**2000**
2500	668.17	0.0286	0.10208	0.1307	730.8	360.7	1091.5	0.9130	0.3199	1.2329	717.6	1031.1	**2500**
3000	695.41	0.0344	0.05015	0.0845	802.9	213.6	1016.5	0.9736	0.1849	1.1585	783.8	969.5	**3000**
3200.11	705.1028	0.0497	0	0.0498	897.5	0	897.5	1.0538	0	1.0538	868.0	868.0	**3200.11**

1. See Note 1, Table 1.
2. See Note 2, Table 1.

Table 3
Properties of Superheated Steam and Compressed Water (Temperature and Pressure)[1]

Press., psia (sat. temp)		100	200	300	400	500	600	700	800	900	1000	1100	1200	1300	1400	1500
	v	0.0161	392.5	452.3	511.9	571.5	631.1	690.7								
1	H	68.00	1150.2	1195.7	1241.8	1288.6	1336.1	1384.5								
(101.74)	s	0.1295	2.0509	2.1152	2.1722	2.2237	2.2708	2.3144								
	v	0.0161	78.14	90.24	102.24	114.21	126.15	138.08	150.01	161.94	173.86	185.78	197.70	209.62	221.53	233.45
5	H	68.01	1148.6	1194.8	1241.3	1288.2	1335.9	1384.3	1433.6	1483.7	1534.7	1586.7	1639.6	1693.3	1748.0	1803.5
(162.24)	s	0.1295	1.8716	1.9369	1.9943	2.0460	2.0932	2.1369	2.1776	2.2159	2.2521	2.2866	2.3194	2.3509	2.3811	2.4101
	v	0.0161	38.84	44.98	51.03	57.04	63.03	69.00	74.98	80.94	86.91	92.87	98.84	104.80	110.76	116.72
10	H	68.02	1146.6	1193.7	1240.6	1287.8	1335.5	1384.0	1433.4	1483.5	1534.6	1586.6	1639.5	1693.3	1747.9	1803.4
(193.21)	s	0.1295	1.7928	1.8593	1.9173	1.9692	2.0166	2.0603	2.1011	2.1394	2.1757	2.2101	2.2430	2.2744	2.3046	2.3337
	v	0.0161	0.0166	29.899	33.963	37.985	41.986	45.978	49.964	53.946	57.926	61.905	65.882	69.858	73.833	77.807
15	H	68.04	168.09	1192.5	1239.9	1287.3	1335.2	1383.8	1433.2	1483.4	1534.5	1586.5	1639.4	1693.2	1747.8	1803.4
(213.03)	s	0.1295	0.2940	1.8134	1.8720	1.9242	1.9717	2.0155	2.0563	2.0946	2.1309	2.1653	2.1982	2.2297	2.2599	2.2890
	v	0.0161	0.0166	22.356	25.428	28.457	31.466	34.465	37.458	40.447	43.435	46.420	49.405	52.388	55.370	58.352
20	H	68.05	168.11	1191.4	1239.2	1286.9	1334.9	1383.5	1432.9	1483.2	1534.3	1586.3	1639.3	1693.1	1747.8	1803.3
(227.96)	s	0.1295	0.2940	1.7805	1.8397	1.8921	1.9397	1.9836	2.0244	2.0628	2.0991	2.1336	2.1665	2.1979	2.2282	2.2572
	v	0.0161	0.0166	11.036	12.624	14.165	15.685	17.195	18.699	20.199	21.697	23.194	24.689	26.183	27.676	29.168
40	H	68.10	168.15	1186.6	1236.4	1285.0	1333.6	1382.5	1432.1	1482.5	1533.7	1585.8	1638.8	1992.7	1747.5	1803.0
(267.25)	s	0.1295	0.2940	1.6992	1.7608	1.8143	1.8624	1.9065	1.9476	1.9860	2.0224	2.0569	2.0899	2.1224	2.1516	2.1807
	v	0.0161	0.0166	7.257	8.354	9.400	10.425	11.438	12.446	13.450	14.452	15.452	16.450	17.448	18.445	19.441
60	H	68.15	168.20	1181.6	1233.5	1283.2	1332.3	1381.5	1431.3	1481.8	1533.2	1585.3	1638.4	1692.4	1747.1	1802.8
(292.71)	s	0.1295	0.2939	1.6492	1.7134	1.7681	1.8168	1.8612	1.9024	1.9410	1.9774	2.0120	2.0450	2.0765	2.1068	2.1359
	v	0.0161	0.0166	0.0175	6.218	7.018	7.794	8.560	9.319	10.075	10.829	11.581	12.331	13.081	13.829	14.577
80	H	68.21	168.24	269.74	1230.5	1281.3	1330.9	1380.5	1430.5	1481.1	1532.6	1584.9	1638.0	1692.0	1746.8	1802.5
(312.04)	s	0.1295	0.2939	0.4371	1.6790	1.7349	1.7842	1.8289	1.8702	1.9089	1.9454	1.9800	2.0131	2.0446	2.0750	2.1041
	v	0.0161	0.0166	0.0175	4.935	5.588	6.216	6.833	7.443	8.050	8.655	9.258	9.860	10.460	11.060	11.659
100	H	68.26	168.29	269.77	1227.4	1279.3	1329.6	1379.5	1429.7	1480.4	1532.0	1584.4	1637.6	1691.6	1746.5	1802.2
(327.82)	s	0.1295	0.2939	0.4371	1.6516	1.7088	1.7586	1.8036	1.8451	1.8839	1.9205	1.9552	1.9883	2.0199	2.0502	2.0794
	v	0.0161	0.0166	0.0175	4.0786	4.6341	5.1637	5.6831	6.1928	6.7006	7.2060	7.7096	8.2119	8.7130	9.2134	9.7130
120	H	68.31	168.33	269.81	1224.1	1277.4	1328.1	1378.4	1428.8	1479.8	1531.4	1583.9	1637.1	1691.3	1746.2	1802.0
(341.27)	s	0.1295	0.2939	0.4371	1.6286	1.6872	1.7376	1.7829	1.8246	1.8635	1.9001	1.9349	1.9680	1.9996	2.0300	2.0592
	v	0.0161	0.0166	0.0175	3.4661	3.9526	4.4119	4.8585	5.2995	5.7364	6.1709	6.6036	7.0349	7.4652	7.8946	8.3233
140	H	68.37	168.38	269.85	1220.8	1275.3	1326.8	1377.4	1428.0	1479.1	1530.8	1583.4	1636.7	1690.9	1745.9	1801.7
(353.04)	s	0.1295	0.2939	0.4370	1.6085	1.6686	1.7196	1.7652	1.8071	1.8461	1.8828	1.9176	1.9508	1.9825	2.0129	2.0421
	v	0.0161	0.0166	0.0175	3.0060	3.4413	3.8480	4.2420	4.6295	5.0132	5.3945	5.7741	6.1522	6.5293	6.9055	7.2811
160	H	68.42	168.42	269.89	1217.4	1273.3	1325.4	1376.4	1427.2	1478.4	1530.3	1582.9	1636.3	1690.5	1745.6	1801.4
(363.55)	s	0.1294	0.2938	0.4370	1.5906	1.6522	1.7039	1.7499	1.7919	1.8310	1.8678	1.9027	1.9359	1.9676	1.9980	2.0273
	v	0.0161	0.0166	0.0174	2.6474	3.0433	3.4093	3.7621	4.1084	4.4505	4.7907	5.1289	5.4657	5.8014	6.1363	6.4704
180	H	68.47	168.47	269.92	1213.8	1271.2	1324.0	1375.3	1426.3	1477.7	1529.7	1582.4	1635.9	1690.2	1745.3	1801.2
(373.08)	s	0.1294	0.2938	0.4370	1.5743	1.6376	1.6900	1.7362	1.7784	1.8176	1.8545	1.8894	1.9227	1.9545	1.9849	2.0142
	v	0.0161	0.0166	0.0174	2.3598	2.7247	3.0583	3.3783	3.6915	4.0008	4.3077	4.6128	4.9165	5.2191	5.5209	5.8219
200	H	68.52	168.51	269.96	1210.1	1269.0	1322.6	1374.3	1425.5	1477.0	1529.1	1581.9	1635.4	1689.8	1745.0	1800.9
(381.80)	s	0.1294	0.2938	0.4369	1.5593	1.6242	1.6776	1.7239	1.7663	1.8057	1.8426	1.8776	1.9109	1.9427	1.9732	2.0025
	v	0.0161	0.0166	0.0174	0.0186	2.1504	2.4662	2.6872	2.9410	3.1909	3.4382	3.6837	3.9278	4.1709	4.4131	4.6546
250	H	68.66	168.63	270.05	375.10	1263.5	1319.0	1371.6	1423.4	1475.3	1527.6	1580.6	1634.4	1688.9	1744.2	1800.2
(400.97)	s	0.1294	0.2937	0.4368	0.5667	1.5951	1.6502	1.6976	1.7405	1.7801	1.8173	1.8524	1.8858	1.9177	1.9482	1.9776
	v	0.0161	0.0166	0.0174	0.0186	1.7665	2.0044	2.2263	2.4407	2.6509	2.8585	3.0643	3.2688	3.4721	3.6746	3.8764
300	H	68.79	168.74	270.14	375.15	1257.7	1315.2	1368.9	1421.3	1473.6	1526.2	1579.4	1633.3	1688.0	1743.4	1799.6
(417.35)	s	0.1294	0.2937	0.4307	0.5665	1.5703	1.6274	1.6758	1.7192	1.7591	1.7964	1.8317	1.8652	1.8972	1.9278	1.9572
	v	00161	0.0166	0.0174	0.0186	1.4913	1.7028	1.8970	2.0832	2.2652	2.4445	2.6219	2.7980	2.9730	3.1471	3.3205
350	H	68.92	168.85	270.24	375.21	1251.5	1311.4	1366.2	1419.2	1471.8	1524.7	1578.2	1632.3	1687.1	1742.6	1798.9
(431.73)	s	0.1293	0.2936	0.4367	0.5664	1.5483	1.6077	1.6571	1.7009	1.7411	1.7787	1.8141	1.8477	1.8798	1.9105	1.9400
	v	0.0161	0.0166	0.0174	0.0162	1.2841	1.4763	1.6499	1.8151	1.9759	2.1339	2.2901	2.4450	2.5987	2.7515	2.9037
400	H	69.05	168.97	270.33	375.27	1245.1	1307.4	1363.4	1417.0	1470.1	1523.3	1576.9	1631.2	1686.2	1741.9	1798.2
(444.60)	s	0.1293	0.2935	0.4366	0.5663	1.5282	1.5901	1.6406	1.6850	1.7255	1.7632	1.7988	1.8325	1.8647	1.8955	1.9250
	v	0.0161	0.0166	0.0174	0.0186	0.9919	1.1584	1.3037	1.4397	1.5708	1.6992	1.8256	1.9507	2.0746	2.1977	2.3200
500	H	69.32	169.19	270.51	375.38	1231.2	1299.1	1357.7	1412.7	1466.6	1520.3	1574.4	1629.1	1684.4	1740.3	1796.9
(467.01)	s	0.1292	0.2934	.04364	0.5660	1.4921	1.5595	1.6123	1.6578	1.6690	1.7371	1.7730	1.8069	1.8393	1.8702	1.8998

1. See Notes 1 and 2, Table 1.

Table 3
Properties of Superheated Steam and Compressed Water (Temperature and Pressure)[1]

Press., psia (sat. temp)		Temperature, F 100	200	300	400	500	600	700	800	900	1000	1100	1200	1300	1400	1500
600	v	0.0161	0.0166	0.0174	0.0186	0.7944	0.9456	1.0726	1.1892	1.3008	1.4093	1.5160	1.6211	1.7252	1.8284	1.9309
	H	69.58	169.42	270.70	375.49	1215.9	1290.3	1351.8	1408.3	1463.0	1517.4	1571.9	1627.0	1682.6	1738.8	1795.6
(486.20)	s	0.1292	0.2933	0.4362	0.5657	1.4590	1.5329	1.5844	1.6351	1.6769	1.7155	1.7517	1.7859	1.8184	1.8494	1.8792
700	v	0.0161	0.0166	0.0174	0.0186	0.0204	0.7928	0.9072	1.0102	1.1078	1.2023	1.2948	1.3858	1.4757	1.5647	1.6530
	H	69.84	169.65	270.89	375.61	487.93	1281.0	1345.6	1403.7	1459.4	1514.4	1569.4	1624.8	1680.7	1737.2	1794.3
(503.08)	s	0.1291	0.2932	0.4360	0.5655	0.6889	1.5090	1.5673	1.6154	1.6580	1.6970	1.7335	1.7679	1.8006	1.8318	1.8617
800	v	0.0161	0.0166	0.0174	0.0186	0.0204	0.6774	0.7828	0.8759	0.9631	1.0470	1.1289	1.2093	1.2885	1.3669	1.4446
	H	70.11	169.88	271.07	375.73	487.88	1271.1	1339.2	1399.1	1455.8	1511.4	1566.9	1622.7	1678.9	1735.0	1792.9
(518.21)	s	0.1290	0.2930	0.4358	0.5652	0.6885	1.4869	1.5484	1.5980	1.6413	1.6807	1.7175	1.7522	1.7851	1.8164	1.8464
900	v	0.0161	0.0166	0.0174	0.0186	0.0204	0.5869	0.6858	0.7713	0.8504	0.9262	0.9998	1.0720	1.1430	1.2131	1.2825
	H	70.37	170.10	271.26	375.84	487.83	1260.6	1332.7	1394.4	1452.2	1508.5	1564.4	1620.6	1677.1	1734.1	1791.6
(531.95)	s	0.1290	0.2929	0.4357	0.5649	0.6881	1.4659	1.5311	1.5822	1.6263	1.6662	1.7033	1.7382	1.7713	1.8028	1.8329
1000	v	0.0161	0.0166	0.0174	0.0186	0.0204	0.5137	0.6080	0.6875	0.7603	0.8295	0.8966	0.9622	1.0266	1.0901	1.1529
	H	70.63	170.33	271.44	375.96	487.79	1249.3	1325.9	1389.6	1448.5	1504.4	1561.9	1618.4	1675.3	1732.5	1790.3
(544.58)	s	0.1289	0.2928	0.4355	0.5647	0.6876	1.4457	1.5149	1.5677	1.6126	1.6530	1.6905	1.7256	1.7589	1.7905	1.8207
1100	v	0.0161	0.0166	0.0174	0.0185	0.0203	0.4531	0.5440	0.6188	0.6865	0.7505	0.8121	0.8723	0.9313	0.9894	1.0468
	H	70.90	170.56	271.63	376.08	487.75	1237.3	1318.8	1384.7	1444.7	1502.4	1559.4	1616.3	1673.5	1731.0	1789.0
(556.28)	s	0.1289	0.2927	0.4353	0.5644	0.6872	1.4259	1.4996	1.5542	1.6000	1.6410	1.6787	1.7141	1.7475	1.7793	1.8097
1200	v	0.0161	0.0166	0.0174	0.0185	0.0203	0.4016	0.4905	0.5615	0.6250	0.6845	0.7418	0.7974	0.8519	0.9055	0.9584
	H	71.16	170.78	271.82	376.20	487.72	1224.2	1311.5	1379.7	1440.9	1499.4	1556.9	1614.2	1671.6	1729.4	1787.6
(567.19)	s	0.1288	0.2926	0.4351	0.5642	0.6868	1.4061	1.4851	1.5415	1.5883	1.6298	1.6679	1.7035	1.7371	1.7691	1.7996
1400	v	0.0161	0.0166	0.0174	0.0185	0.0203	0.3176	0.4059	0.4712	0.5282	0.5809	0.6311	0.6798	0.7272	0.7737	0.8195
	H	71.68	171.24	272.19	376.44	487.65	1194.1	1296.1	1369.3	1433.2	1493.2	1551.8	1609.9	1668.0	1726.3	1785.0
(587.07)	s	0.1287	0.2923	0.4348	0.5636	0.6859	1.3652	1.4575	1.5182	1.5670	1.6096	1.6484	1.6845	1.7185	1.7508	1.7815
1600	v	0.0161	0.0166	0.0173	0.0185	0.0202	0.0236	0.3415	0.4032	0.4555	0.5031	0.5482	0.5915	0.6336	0.6748	0.7153
	H	72.21	171.69	272.57	376.69	487.60	616.77	1279.4	1358.5	1425.2	1486.9	1546.6	1605.6	1664.3	1723.2	1782.3
(604.87)	s	0.1286	0.2921	0.4344	0.5631	0.6851	0.8129	1.4312	1.4968	1.5478	1.5916	1.6312	1.6678	1.7022	1.7344	1.7657
1800	v	0.0160	0.0165	0.0173	0.0185	0.0202	0.0235	0.2906	0.3500	0.3988	0.4426	0.4836	0.5229	0.5609	0.5980	0.6343
	H	72.73	172.15	272.95	376.93	487.56	615.58	1261.1	1347.2	1417.1	1480.6	1541.1	1601.2	1660.7	1720.1	1779.7
(621.02)	s	0.1284	0.2918	0.4341	0.5626	0.6843	0.8109	1.4054	1.4768	1.5302	1.5753	1.6156	1.6528	1.6876	1.7204	1.7516
2000	v	0.0160	0.0165	0.0173	0.0184	0.0201	0.0233	0.2488	0.3072	0.3534	0.3942	0.4320	0.4680	0.5027	0.5365	0.5695
	H	73.26	172.60	273.32	377.19	487.53	614.48	1240.9	1353.4	1408.7	1474.1	1536.2	1596.9	1657.0	1717.0	1777.1
(635.80)	s	0.1283	0.2916	0.4337	0.5621	0.6834	0.8091	1.3794	1.4578	1.5138	1.5603	1.6014	1.6391	1.6743	1.7075	1.7389
2500	v	0.0160	0.0165	0.0173	0.0184	0.0200	0.0230	0.1681	0.2293	0.2712	0.3068	0.3390	0.3692	0.3980	0.4259	0.4529
	H	74.57	173.74	274.27	377.82	487.50	612.08	1176.7	1303.4	1386.7	1457.5	1522.9	1585.9	1647.8	1709.2	1770.4
(668.11)	s	0.1280	0.2910	0.4329	0.5609	0.6815	0.8048	1.3076	1.4129	1.4766	1.5269	1.5703	1.6094	1.6456	1.6796	1.7116
3000	v	0.0160	0.0165	0.0172	0.0183	0.0200	0.0228	0.0982	0.1759	0.2161	0.2484	0.2770	0.3033	0.3282	0.3522	0.3753
	H	75.88	174.88	275.22	378.47	487.52	610.08	1060.5	1267.0	1363.2	1440.2	1509.4	1574.8	1638.5	1701.4	1761.8
(695.33)	s	0.1277	0.2904	0.4320	0.5597	0.6796	0.8009	1.1966	1.3692	1.4429	1.4976	1.5434	1.5841	1.6214	1.6561	1.6888
3200	v	0.0160	0.0165	0.0172	0.0183	0.0199	0.0227	0.0335	0.1588	0.1987	0.2301	0.2576	0.2327	0.3065	0.3291	0.3510
	H	76.4	175.3	275.6	378.7	487.5	609.4	800.8	1250.9	1353.4	1433.1	1503.8	1570.3	1634.8	1698.3	1761.2
(705.08)	s	0.1276	0.2902	0.4317	0.5592	0.6788	0.7994	0.9708	1.3515	1.4300	1.4866	1.5335	1.5749	1.6126	1.6477	1.6806
3500	v	0.0160	0.0164	0.0172	0.0183	0.0199	0.0225	0.0307	0.1364	0.1764	0.2066	0.2326	0.2563	0.2784	0.2995	0.3198
	H	77.2	176.0	276.2	379.1	487.6	608.4	779.4	1224.6	1338.2	1422.2	1495.5	1563.3	1629.2	1693.6	1757.2
	s	0.1274	0.2899	0.4312	0.5585	0.6777	0.7973	0.9508	1.3242	1.4112	1.4709	1.5194	1.5618	1.6002	1.6358	1.6691
4000	v	0.0159	0.0164	0.0172	0.0182	0.0198	0.0223	0.0287	0.1052	0.1463	0.1752	0.1994	0.2210	0.2411	0.2601	0.2783
	H	78.5	177.2	277.1	379.8	487.7	606.9	763.0	1174.3	1311.6	1403.6	1481.3	1552.2	1619.8	1685.7	1750.6
	s	0.1271	0.2893	0.4304	0.5573	0.6760	0.7940	0.9343	1.2754	1.3807	1.4461	1.4976	1.5417	1.5812	1.6177	1.6516
5000	v	0.0159	0.0164	0.0171	0.0181	0.0196	0.0219	0.0268	0.0591	0.1038	0.1312	0.1529	0.1718	0.1890	0.2050	0.2203
	H	81.1	179.5	279.1	381.2	488.1	604.6	746.0	1042.9	1252.9	1364.6	1452.1	1529.1	1600.9	1670.0	1737.4
	s	0.1265	0.2881	0.4287	0.5550	0.6726	0.7880	0.9153	1.1593	1.3207	1.4001	1.4582	1.5061	1.5481	1.5863	1.6216
6000	v	0.0159	0.0163	0.0170	0.0180	0.0195	0.0216	0.0256	0.0397	0.0757	0.1020	0.1221	0.1391	0.1544	0.1684	0.1817
	H	83.7	181.7	281.0	382.7	488.6	602.9	736.1	945.1	1188.8	1323.6	1422.3	1505.9	1582.0	1654.2	1724.2
	s	0.1258	0.2870	0.4271	0.5528	0.6693	0.7826	0.9026	1.0176	1.2615	1.3574	1.4229	1.4748	1.5194	1.5593	1.5962
7000	v	0.0158	0.0163	0.0170	0.0180	0.0193	0.0213	0.0248	0.0334	0.0573	0.0816	0.1004	0.1160	0.1298	0.1424	0.1542
	H	86.2	184.4	283.0	384.2	489.3	601.7	729.3	901.8	1124.9	1281.7	1392.2	1482.6	1563.1	1638.6	1711.1
	s	0.1252	0.2859	0.4256	0.5507	0.6663	0.7777	0.8926	1.0350	1.2055	1.3171	1.3904	1.4466	1.4938	1.5355	1.5735

1. See Notes 1 and 2, Table 1.

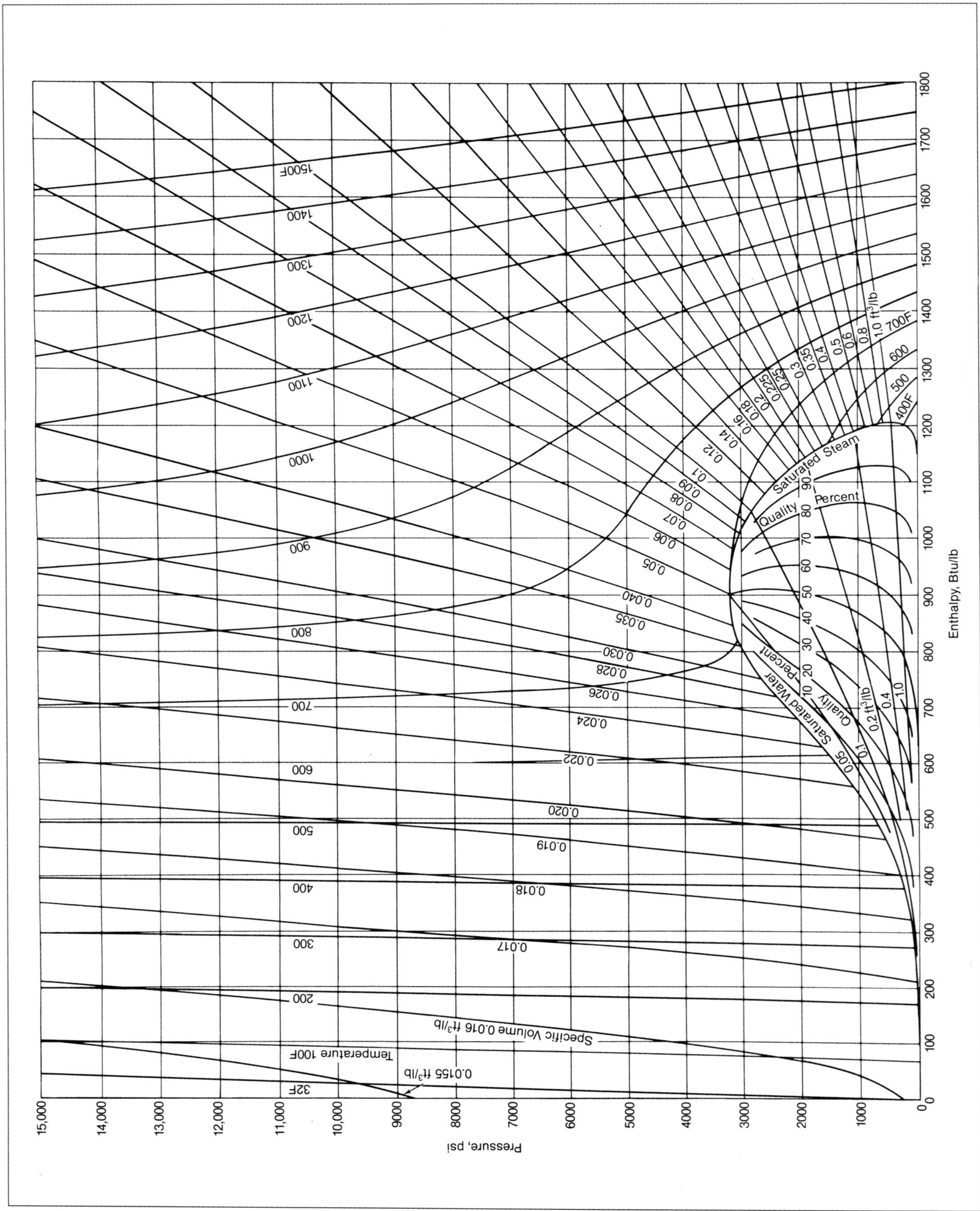

Fig. 1 Pressure-enthalpy chart for steam (English units).

where m_s is the mass of steam and m_w is the mass of water. The quality is frequently recorded as a percent steam by weight (% SBW) after multiplying by 100%. The mixture enthalpy (H) (see Note below), entropy (s) and specific volume (v) of a steam-water mixture can then be simply defined as:

$$H = H_f + x\left(H_g - H_f\right) \quad \textbf{(2a)}$$

$$s = s_f + x\left(s_g - s_f\right) \quad \textbf{(2b)}$$

$$v = v_f + x\left(v_g - v_f\right) \quad \textbf{(2c)}$$

where the subscripts f and g refer to properties at saturated liquid and vapor conditions, respectively. The difference in a property between saturated liquid and vapor conditions is frequently denoted by the subscript fg; for example, $H_{fg} = H_g - H_f$. With these definitions, if the pressure or temperature of a steam-water mixture is known along with one of the mixture properties, the quality can then be calculated. For example, if the mixture enthalpy is known, then:

$$x = \left(H - H_f\right) / H_{fg} \quad \textbf{(3)}$$

Engineering problems deal mainly with changes or differences in enthalpy and entropy. It is not necessary to establish an absolute zero for these properties, although this may be done for entropy. The Steam Tables indicate an arbitrary zero internal energy and entropy for the liquid state of water at the triple point corresponding to a temperature of 32.018F (0.01C) and a vapor pressure of 0.08865 psi (0.6112 kPa). The triple point is a unique condition where the three states of water (solid, liquid and vapor) coexist at equilibrium.

Properties of gases

In addition to steam, air is a common working fluid for some thermodynamic cycles. As with steam, well defined properties are important in cycle analysis. Air and many common gases used in power cycle applications can usually be treated as ideal gases. An ideal gas is defined as a substance that obeys the ideal gas law:

$$Pv = RT \quad \textbf{(4)}$$

where R is a constant which varies with gas species; P and T are the pressure and temperature, respectively. R is equal to the universal gas constant, **R** [1545 ft lb/lb-mole R (8.3143 kJ/kg mole K)], divided by the molecular weight of the gas. For dry air, R is equal to 53.34 lbf ft/lbm R (0.287 kJ/kg K). Values for other gases are summarized in Reference 3. The ideal gas law is commonly used in a first analysis of a process or cycle because it simplifies calculations. Final calculations often rely on tabulated gas properties for greater accuracy.

Tabulated gas properties are available from numerous sources. (See References 4 and 5 for examples.) Unfortunately, there is less agreement on gas properties than on those for steam. The United States (U.S.) boiler industry customarily uses 80F (27C) and 14.7 psia (101.35 kPa) as the zero enthalpy of air and combustion products. A more general reference is one atmosphere pressure, 14.696 psia (101.35 kPa), and 77F (25C). This is the standard reference point for heats of formation of compounds from elements in their standard states, latent heats of phase changes and free energy changes. Because of different engineering conventions, considerable care must be exercised when using tabulated properties. Selected properties for air and other gases are provided in Chapter 3, Table 3.

Conservation of mass and energy

Thermodynamic processes are governed by the laws of conservation of mass and conservation of energy except for the special case of nuclear reactions discussed in Chapter 47. These conservation laws basically state that the total mass and total energy (in any of its forms) can neither be created nor destroyed in a process. In an open flowing power system, where mass continually enters and exits a system such as Fig. 2, these laws can take the forms:

Conservation of mass

$$m_1 - m_2 = \Delta m \quad \textbf{(5)}$$

Conservation of energy

$$E_2 - E_1 + \Delta E = Q - W \quad \textbf{(6)}$$

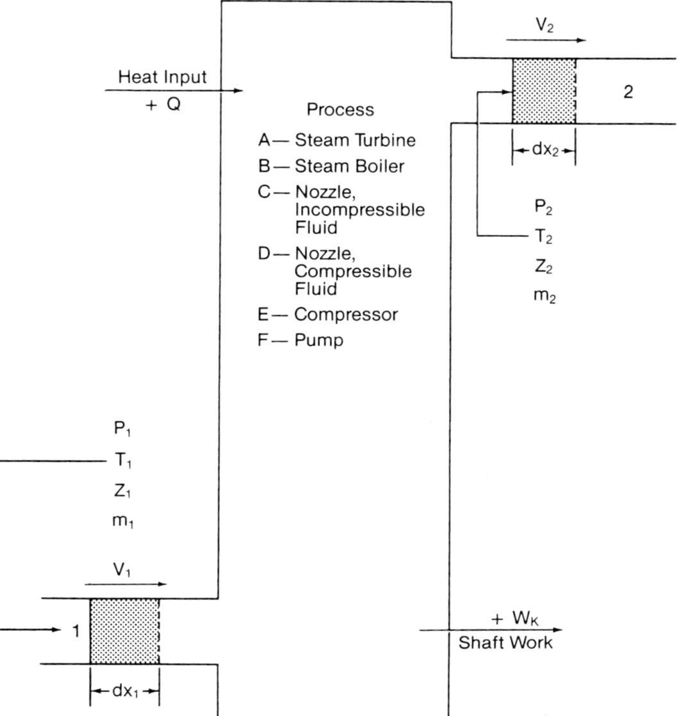

Fig. 2 Diagram illustrating thermodynamic processes.

Note: To avoid confusion with the symbol for heat transfer coefficient, enthalpy (Btu/lb or kJ/kg) is denoted by H in this chapter and the balance of *Steam* unless specially noted. Enthalpy is frequently denoted by h in thermodynamic texts.

where m is the mass flow, Δm is the change in internal system mass, E is the total energy flowing into or out of the processes, ΔE is the change in energy stored in the system with time, Q is the heat added to the system, W is the work removed, and the subscripts 1 and 2 refer to inlet and outlet conditions respectively. For steady-state conditions Δm and ΔE are zero.

The conservation of energy states that a balance exists between energy, work, and heat quantities entering and leaving the system. This balance of energy flow is also referred to as the first law of thermodynamics. The terms on the left side of Equation 6 represent stored energy entering or leaving the system as part of the mass flows and the accumulation of total stored energy within the system. The terms on the right side are the heat transferred to the system, Q, and work done by the system, W. The stored energy components, represented by the term E, consist of the internally stored energy and the kinetic and potential energy. In an open system, there is work required to move mass into the system and work done by the system to move mass out. In each case, the total work is equal to the product of the mass, the system pressure, and the specific volume. Separating this work from other work done by the system and including a breakdown of the stored energy, the energy conservation equation becomes:

$$m_2\left(u + Pv + \frac{V^2}{2g_c} + z\right)_2 - m_1\left(u + Pv + \frac{V^2}{2g_c} + z\right)_1 + \Delta E = Q - W_k \qquad (7)$$

where m is the mass, u is the internal stored energy, P is the system pressure, V is the fluid velocity, v is the specific volume, z is the elevation, and W_k is the sum of the work done by the system.

In this form, the work terms associated with mass moving into and out of the system (Pv) have been grouped with the stored energy crossing the system boundary. W_k represents all other work done by the system.

For many practical power applications, the energy equation can be further simplified for steady-state processes. Because the mass entering and leaving the system over any time interval is the same, dividing Equation 7 by the mass (m_2 or m_1 because they are equal) yields a simple balance between the change in stored energy due to inflow and outflow and the heat and work terms expressed on a unit mass basis. Heat and work expressed on a unit mass basis are denoted q and w, respectively. The unsteady term for system stored energy in Equation 7 is then set to zero. This yields the following form of the energy conservation equation:

$$\Delta u + \Delta(Pv) + \Delta\frac{V^2}{2g_c} + \Delta z\frac{g}{g_c} = q - w_k \qquad (8)$$

Each Δ term on the left in Equation 8 represents the difference in the fluid property or system characteristic between the system outlet and inlet. Δu is the difference in internally stored energy associated with molecular and atomic motions and forces. Internally stored energy, or simply internal energy, accounts for all forms of energy other than the kinetic and potential energies of the collective molecule masses. This is possible because no attempt is made to absolutely define u.

The term $\Delta(Pv)$ can be viewed as externally stored energy in that it reflects the work required to move a unit mass into and out of the system. The remaining terms of externally stored energy, $\Delta(V^2/2g_c)$ and Δz, depend on physical aspects of the system. $\Delta(V^2/2g_c)$ is the difference in total kinetic energy of the fluid between two reference points (system inlet and outlet). $\Delta zg/g_c$ represents the change in potential energy due to elevation, where g is the gravitational constant 32.17 ft/s^2 (9.8 m/s^2) and g_c is a proportionality constant for English units. The value of the constant is obtained from equivalence of force and mass times acceleration:

$$\text{Force} = \frac{\text{mass} \times \text{acceleration}}{g_c} \qquad (9)$$

In the English system, by definition, when 1 lb force (lbf) is exerted on a 1 lb mass (lbm), the mass accelerates at the rate of 32.17 ft/s^2. In the SI system, 1 N of force is exerted by 1 kg of mass accelerating at 1 m/s^2. Therefore, the values of g_c are:

$$g_c = 32.17 \text{ lbm ft / lbf s}^2 \qquad (10a)$$

$$g_c = 1 \text{ kg m/N s}^2 \qquad (10b)$$

Because of the numerical equivalency between g and g_c in the English system, the potential energy term in Equation 8 is frequently shown simply as Δz. When SI units are used, this term is often expressed simply as Δzg because the proportionality constant has a value of 1.

While many texts use the expression lbf to designate lb force and lbm to designate lb mass, this is not done in this text because it is believed that it is generally clear simply by using lb. As examples, the expression Btu/lb always means Btu/lb mass, and the expression ft lb/lb always means ft lb force/lb mass.

Application of the energy equation requires dimensional consistency of all terms and proper conversion constants are inserted as necessary. For example, the terms u and q, usually expressed in Btu/lb or J/kg, may be converted to ft lb/lb or N m/kg when multiplied by J, the mechanical equivalent of heat. This conversion constant, originally obtained by Joule's experiments between 1843 and 1878, is defined as:

$$J = 778.17 \text{ ft lbf / Btu} \qquad (11a)$$

$$J = 1 \text{ Nm/J} \qquad (11b)$$

Particular attention should be given to the sign convention applied to heat and work quantities. Originating with the steam engine analysis, heat quantities are defined as positive when entering the system and work (for example, shaft work) is positive when leaving the system.

Because u and Pv of Equation 8 are system properties, their sum is also a system property. Because these properties of state can not be changed independently of one another and because the combination ($u + Pv$) appears whenever mass enters or leaves the system, it is customary to consider the sum $(u + Pv)$ as a single property H, called enthalpy.

$$H = u + (Pv/J) \tag{12}$$

where Pv is divided by J to provide consistent units.

In steam applications, H is usually expressed in Btu/lb or J/kg. The examples in the following section illustrate the application of the steady-state open system energy Equation 7 and the usefulness of enthalpy in the energy balance of specific equipment.

Applications of the energy equation

Steam turbine

To apply the energy equation, each plant component is considered to be a system, as depicted in Fig. 2. In many cases, Δz, $\Delta(V^2/2g_c)$ and q from throttle (1) to exhaust (2) of the steam turbine are small compared to $(H_2 - H_1)$. This reduces Equation 8 to:

$$u_2 + (P_2 v_2 / J) - u_1 - (P_1 v_1 / J) = w_k / J \tag{13a}$$

or

$$H_2 - H_1 = w_k / J \tag{13b}$$

Equation 13 indicates that the work done by the steam turbine, w_k/J, is equal to the difference between the enthalpy of the steam entering and leaving. However, H_1 and H_2 are seldom both known and further description of the process is required for a solution of most problems.

Steam boiler

The boiler does no work, therefore $w_k = 0$. Because Δz and $\Delta(V^2/2g_c)$ from the feedwater inlet (1) to the steam outlet (2) are small compared to $(H_2 - H_1)$, the steady-state energy equation becomes:

$$q = H_2 - H_1 \tag{14}$$

Based on Equation 14 the heat added, q (positive), in the boiler per unit mass of flow in is equal to the difference between H_2 of the steam leaving and H_1 of the feedwater entering. Assuming that the pressure varies negligibly through the boiler and the drum pressure is known, Equation 14 can be solved knowing the temperature of the incoming feedwater.

Water flow through a nozzle

For water flowing through a nozzle, the change in specific volume is negligible. Commonly the change in elevation, Δz, the change in internal energy, Δu, the work done, w_k, and the heat added, q, are negligible and the energy equation reduces to:

$$(V_2^2 / 2g_c) - (V_1^2 / 2g_c) = (P_1 - P_2) v \tag{15}$$

The increase in kinetic energy of the water is given by Equation 15 for the pressure drop $(P_1 - P_2)$. If the approach velocity to the nozzle, V_1, is zero, Equation 15 becomes:

$$V_2 = \sqrt{2g_c (P_1 - P_2) v} \tag{16}$$

The quantity $(P_1 - P_2)\, v$ is often referred to as the static head.

Flow of a compressible fluid through a nozzle

In contrast to water flow, when steam, air or other compressible fluid flows through a nozzle, the changes in specific volume and internal energy are not negligible. In this case, assuming no change in elevation Δz, Equation 8 becomes:

$$(V_2^2 / 2g_c) - (V_1^2 / 2g_c) = (H_1 - H_2) J \tag{17}$$

If the approach velocity, V_1, is zero, this further simplifies to:

$$V_2 = \sqrt{2g_c J (H_1 - H_2)} \tag{18}$$

From this, it is evident that the velocity of a compressible fluid leaving a nozzle is a function of its entering and leaving enthalpies. Unfortunately, as with the steam turbine, H_1 and H_2 are seldom both known.

Compressor

If a compressible fluid moves through an adiabatic compressor ($q = 0$, a convenient approximation) and the change in elevation and velocity are small compared to $(H_2 - H_1)$, the energy equation reduces to:

$$-w_k / J = H_2 - H_1 \tag{19}$$

Note that w_k is negative because the compressor does work on the system. Therefore, the net effect of the compressor is expressed as an increase in fluid enthalpy from inlet to outlet.

Pump

The difference between a pump and a compressor is that the fluid is considered to be incompressible for the pumping process; this is a good approximation for water. For an incompressible fluid, the specific volume is the same at the inlet and outlet of the pump. If the fluid friction is negligible, then the internal energy changes, Δu, are set to zero and the energy equation can be expressed as:

$$-w_k = (P_2 - P_1) v \tag{20}$$

Because all real fluids are compressible, it is important to know what is implied by the term incompressible. The meaning here is that the isothermal compressibility, k_T, given by

$$k_T = -\frac{1}{v}\frac{\delta v}{\delta P} \tag{21}$$

is assumed to be arbitrarily small and approaching zero. Because neither v nor P is zero, δv must be zero and v must be a constant. Also, for isothermal conditions (by definition) there can be no change in internal energy, u, due to pressure changes only.

Entropy and its application to processes

The preceding examples illustrate applications of the energy balance in problems where a fluid is used for heat transfer and shaft work. They also demonstrate the usefulness of the enthalpy property. However, as was pointed out, H_1 and H_2 are seldom both known. Additional information is frequently provided by the first and second laws of thermodynamics and their consequences.

The first and second laws of thermodynamics

The first law of thermodynamics is based on the energy conservation expressed by Equation 6 and, by convention, relates the heat and work quantities of this equation to internally stored energy, u. Strictly speaking, Equation 6 is a complete form of the first law of thermodynamics. However, it is frequently useful to use the steady-state formulation provided in Equation 8 and further simplify this for the special case of 1) no change in potential energy due to gravity acting on the mass, and 2) no change in kinetic energy of the mass as a whole. In a closed system where only shaft work is permitted, these simplifying assumptions permit the energy Equation 8 for a unit mass to be reduced to:

$$\Delta u = q - \left(w_k / J\right) \tag{22a}$$

or in differential form

$$du = \delta q - \left(\delta w_k / J\right) \tag{22b}$$

The first law treats heat and work as being interchangeable, although some qualifications must apply. All forms of energy, including work, can be wholly converted to heat, but the converse is not generally true. Given a source of heat coupled with a heat-work cycle, such as heat released by high temperature combustion in a steam power plant, only a portion of this heat can be converted to work. The rest must be rejected to an energy sink, such as the atmosphere, at a lower temperature. This is essentially the *Kelvin* statement of the second law of thermodynamics. It can also be shown that it is equivalent to the *Clausius* statement wherein heat, in the absence of external assistance, can only flow from a hotter to a colder body.

Concept and definition of entropy

Heat flow is a function of temperature difference. If a quantity of heat is divided by its absolute temperature, the quotient can be considered a type of distribution property complementing the intensity factor of temperature. Such a property, proposed and named entropy by Clausius, is widely used in thermodynamics because of its close relationship to the second law.

Rather than attempt to define entropy (s) in an absolute sense, consider the significance of differences in this property given by:

$$S_2 - S_1 = \Delta S$$

$$= \int_1^2 \frac{\delta q_{rev}}{T} \times \text{total system mass} \tag{23}$$

where

ΔS = change in entropy, Btu/R (J/K)
q_{rev} = reversible heat flow between thermodynamic equilibrium states 1 and 2 of the system, Btu/lb (J/kg)
T = absolute temperature, R (K)

Entropy is an extensive property, i.e., a quantity of entropy, S, is associated with a finite quantity of mass, m. If the system is closed and the entire mass undergoes a change from state 1 to 2, an intensive property s is defined by S/m. The property s is also referred to as entropy, although it is actually specific entropy. If the system is open as in Fig. 2, the specific entropy is calculated by dividing by the appropriate mass.

Use of the symbol δ instead of the usual differential operator d is a reminder that q depends on the process and is not a property of the system (steam). δq represents only a small quantity, not a differential. Before Equation 23 can be integrated, q_{rev} must be expressed in terms of properties, and a reversible path between the prescribed initial and final equilibrium states of the system must be specified. For example, when heat flow is reversible and at constant pressure, $q_{rev} = c_p dT$. This may represent heat added reversibly to the system, as in a boiler, or the equivalent of internal heat flows due to friction or other irreversibilities. In these two cases, Δs is always positive.

The same qualifications for δ hold in the case of thermodynamic work. Small quantities of w similar in magnitude to differentials are expressed as δw.

Application of entropy to a reversible process

Reversible thermodynamic processes exist in theory only; however, they serve an important function of defining limiting cases for heat flow and work processes. The properties of a system undergoing a reversible process are constrained to be homogeneous because there are no variations among subregions of the system. Moreover, during interchanges of heat or work between a system and its surroundings, only corresponding potential gradients of infinitesimal magnitude may exist.

All actual processes are irreversible. To occur, they must be under the influence of a finite potential difference. A temperature difference supplies this drive and direction for heat flow. The work term, on the other hand, is more complicated, because there are as many different potentials (generalized forces) as there are forms of work. However, the main concern here is expansion work for which the potential is clearly a pressure difference.

Regardless of whether a process is to be considered reversible or irreversible, it must have specific beginning and ending points (limits) in order to be evaluated. To apply the first and second laws, the limits must be equilibrium states. Nonequilibrium thermodynamics is beyond the scope of this text. Because the limits of real processes are to be equilibrium states, any process can be approximated by a series of smaller reversible processes starting and ending at the same states as the real processes. In this way, only equilibrium conditions are considered and the substitute processes can be defined in terms of the system properties. The following lists the reversible processes for heat flow and work:

Reversible Heat Flow		Reversible Work	
Constant pressure,	$dP = 0$	Constant pressure,	$dP = 0$
Constant temperature,	$dT = 0$	Constant temperature,	$dT = 0$
Constant volume,	$dv = 0$	Constant entropy,	$ds = 0$
	$w = 0$		$q = 0$

The qualification of these processes is that each describes a path that has a continuous functional relationship on coordinate systems of thermodynamic properties.

A combined form of the first and second laws is obtained by substituting $\delta q_{rev} = Tds$ for δq in Equation 22b, yielding:

$$du = Tds - \delta w_k \tag{24}$$

Because only reversible processes are to be used, δw should also be selected with this restriction. Reversible work for the limited case of expansion work can be written:

$$\delta\left(w_{rev}\right) = Pdv \tag{25}$$

In this case, pressure is in complete equilibrium with external forces acting on the system and is related to v through an equation of state.

Substituting Equation 25 in 24, the combined expression for the first and second law becomes:

$$du = Tds - Pdv \tag{26}$$

Equation 26, however, only applies to a system in which the reversible work is entirely shaft work. To modify this expression for an open system in which flow work $d(Pv)$ is also present, the quantity $d(Pv)$ is added to the left side of Equation 25 and added as $(Pdv + vdP)$ on the right side. The result is:

$$du + d\left(Pv\right) = Tds - Pdv + Pdv + vdP \tag{27a}$$

or

$$dH = Tds + vdP \tag{27b}$$

The work term vdP in Equation 27 now represents reversible shaft work in an open system, expressed on a unit mass basis.

Because Tds in Equation 26 is equivalent to δq, its value becomes zero under adiabatic or zero heat transfer conditions ($\delta q = 0$). Because T can not be zero, it follows that $ds = 0$ and s is constant. Therefore, the maximum work from stored energy in an open system during a reversible adiabatic expansion is $\int vdP$ at constant entropy. The work done is equal to the decrease in enthalpy. Likewise for the closed system, the maximum expansion work is $-\int Pdv$ at constant entropy and is equal to the decrease in internal energy. These are important cases of an adiabatic isentropic expansion.

Irreversible processes

All real processes are irreversible due to factors such as friction, heat transfer through a finite temperature difference, and expansion through a process with a finite net force on the boundary. Real processes can be solved approximately, however, by substituting a series of reversible processes. An example of such a substitution is illustrated in Fig. 3, which represents the adiabatic expansion of steam in a turbine or any gas expanded from P_1 to P_2 to produce shaft work. T_1, P_1 and P_2 are known. The value of H_1 is fixed by T_1 and P_1 for a single-phase condition (vapor) at the inlet. H_1 may be found from the Steam Tables, a T-s diagram (Fig. 3) or, more conveniently, from an H-s (Mollier) diagram, shown in the chapter frontispiece. From the combined first and second laws, the maximum energy available for work in an adiabatic system is ($H_1 - H_3$), as shown in Fig. 3, where H_3 is found by the adiabatic *isentropic* expansion (expansion at constant entropy) from P_1 to P_2. A portion of this available energy, usually about 10 to 15%, represents work lost (w_L) due to friction and form loss, limiting ΔH for shaft work to ($H_1 - H_2$). The two reversible paths used to arrive at point b in Fig. 3 (path a to c at constant entropy, s, and path c to b at constant pressure) yield the following equation:

$$\left(H_1 - H_3\right) - \left(H_2 - H_3\right) = H_1 - H_2 \tag{28}$$

Point b, identified by solving for H_2, now fixes T_2; v_1 and v_2 are available from separate tabulated values of physical properties.

Note that $\Delta H_{2\text{-}3}$ can be found from:

$$\Delta H_{2-3} = \int_3^2 Tds \tag{29}$$

or, graphically, the area on the T-s diagram (Fig. 3) under the curve P_2 from points c to b. Areas bounded by reversible paths on the T-s diagram in general rep-

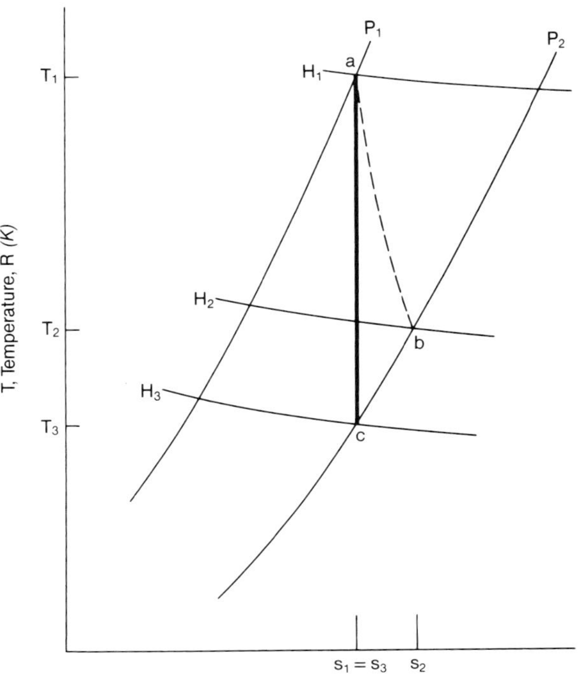

Fig. 3 Irreversible expansion, state a to state b.

resent q (heat flow per unit mass) between the system and its surroundings. However, the path a to b is irreversible and the area under the curve has no significance. The area under path c to b, although it has the form of a reversible quantity q, does not represent heat added to the system but rather its equivalent in internal heat flow. A similar situation applies to the relationship between work and areas under reversible paths in a pressure-volume equation of state diagram. Because of this important distinction between reversible and irreversible paths, care must be exercised in graphically interpreting these areas in cycle analysis.

Returning to Fig. 3 and the path a to b, w_L was considered to be a percentage of the enthalpy change along the path a to c. In general, the evaluation should be handled in several smaller steps (Fig. 4) for the following reason. Point b has a higher entropy than point c and, if expansion to a pressure lower than P_2 (Fig. 3) is possible, the energy available for this additional expansion is greater than that at point c. In other words, a portion of w_L (which has the same effect as heat added to the system) for the first expansion can be recovered in the next expansion or stage. This is the basis of the reheat factor used in analyzing expansions through a multistage turbine. Since the pressure curves are divergent on an H-s or T-s diagram, the sum of the individual ΔH_s values (isentropic ΔH) for individual increments of ΔP (or stages in an irreversible expansion) is greater than that of the reversible ΔH_s between the initial and final pressures (Fig. 4). Therefore the shaft work that can be achieved is greater than that calculated by a simple isentropic expansion between the two pressures.

Principle of entropy increase

Although entropy has been given a quantitative meaning in previous sections, there are qualitative aspects of this property which deserve special emphasis. An increase in entropy is a measure of that portion of process heat which is unavailable for conversion to work. For example, consider the constant pressure reversible addition of heat to a working fluid with the resulting increase in steam entropy. The minimum portion of this heat flow which is unavailable for shaft work is equal to the entropy increase multiplied by the absolute temperature of the sink to which a part of the heat must be rejected (in accordance with the second law). However, because a reversible addition of heat is not possible, incremental entropy increases also occur due to internal fluid heating as a result of temperature gradients and fluid friction.

Even though the net entropy change of any portion of a fluid moving through a cycle of processes is always zero because the cycle requires restoration of all properties to some designated starting point, the sum of all entropy increases has a special significance. These increases in entropy, less any decreases due to recycled heat within a regenerator, multiplied by the appropriate sink absolute temperature (R or K) are equal to the heat flow to the sink. In this case, the net entropy change of the system undergoing the cycle is zero, but there is an entropy increase of the surroundings. Any thermodynamic change that takes place,

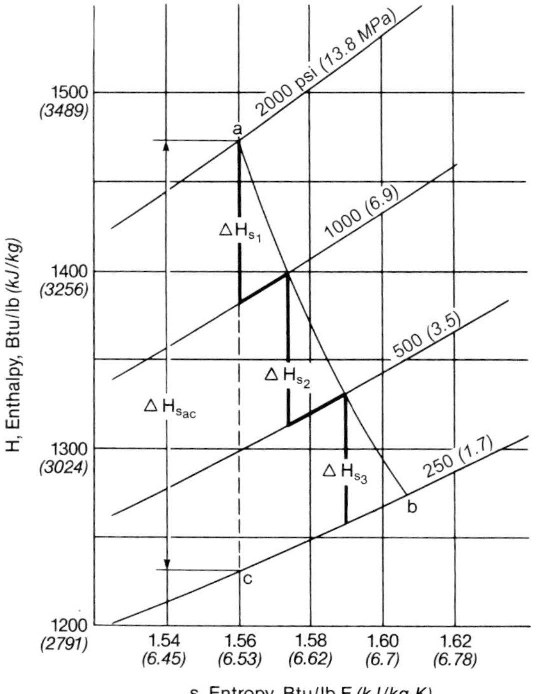

Fig. 4 Three-stage irreversible expansion – $\Delta H_{s1} + \Delta H_{s2} + \Delta H_{s3} > \Delta H_{sac}$.

whether it is a stand alone process or cycle of processes, results in a net entropy increase when both the system and its surroundings are considered.

Cycles

To this point, only thermodynamic processes have been discussed with minor references to the cycle. The next step is to couple processes so heat may be converted to work on a continuous basis. This is done by selectively arranging a series of thermodynamic processes in a cycle forming a closed curve on any system of thermodynamic coordinates. Because the main interest is steam, the following discussion emphasizes expansion or Pdv work. This relies on the limited differential expression for internal energy, Equation 26, and enthalpy, Equation 27. However, the subject of thermodynamics recognizes work as energy in transit under any potential other than differential temperature and electromagnetic radiation.

Carnot cycle

Sadi Carnot (1796 to 1832) introduced the concept of the cycle and reversible processes. The *Carnot cycle* is used to define heat engine performance as it constitutes a cycle in which all component processes are reversible. This cycle, on a temperature-entropy diagram, is shown in Fig. 5a for a gas and in Fig. 5b for a two-phase saturated fluid. Fig. 5c presents this cycle for a nonideal gas, such as superheated steam, on Mollier coordinates (entropy versus enthalpy).

Referring to Fig. 5, the Carnot cycle consists of the following processes:

1. Heat is added to the working medium at constant temperature ($dT = 0$) resulting in expansion work

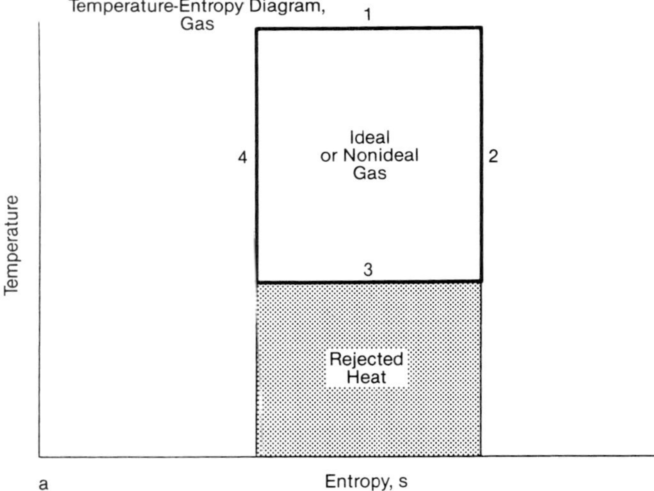

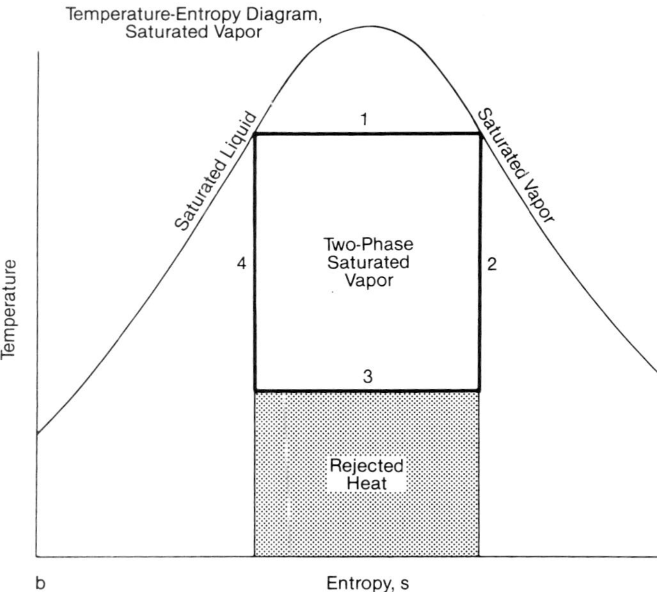

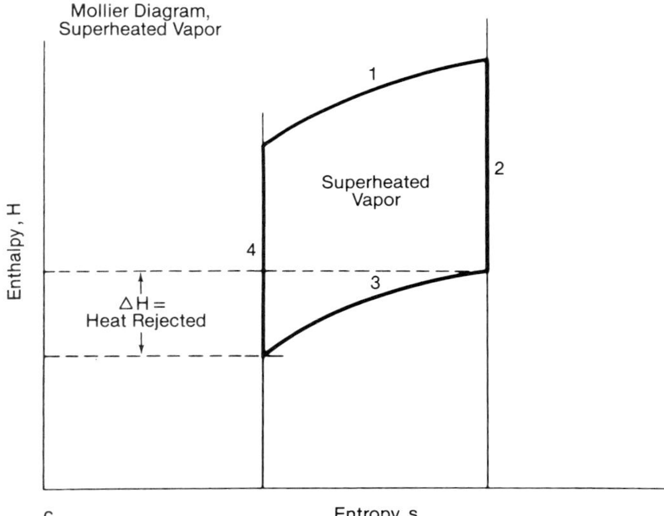

Fig. 5 Carnot cycles.

and changes in enthalpy. (For an ideal gas, changes in internal energy and pressure are zero and, therefore, changes in enthalpy are zero.)

2. Adiabatic isentropic expansion ($ds = 0$) occurs with expansion work and an equivalent decrease in enthalpy.
3. Heat is rejected to the surroundings at a constant temperature and is equivalent to the compression work and any changes in enthalpy.
4. Adiabatic isentropic compression occurs back to the starting temperature with compression work and an equivalent increase in enthalpy.

This cycle has no counterpart in practice. The only way to carry out the constant temperature processes in a one-phase system would be to approximate them through a series of isentropic expansions and constant pressure reheats for heat addition, and isentropic compressions with a series of intercoolers for heat rejections. Another serious disadvantage of a Carnot gas engine would be the small ratio of net work to gross work (net work referring to the difference between the expansion work and the compression work, and gross work being expansion work). Even a two-phase cycle, such as Fig. 5b, would be subject to the practical mechanical difficulties of wet compression and, to a lesser extent, wet expansion where a vapor-liquid mixture exists.

Nevertheless, the Carnot cycle illustrates the basic principles of thermodynamics and, because the processes are reversible, the Carnot cycle offers the maximum thermodynamic efficiency attainable between any given temperatures of heat source and sink. The efficiency of the cycle is defined as the ratio of the net work output to the total heat input. Various texts refer to this ratio as either thermodynamic efficiency, thermal efficiency, or simply efficiency. Using the *T-s* diagram for the Carnot cycle shown in Fig. 5a, the thermodynamic efficiency depends solely on the temperatures at which heat addition and rejection occur:

$$\eta = \frac{T_1 - T_2}{T_1} = 1 - \frac{T_2}{T_1} \tag{30}$$

where

η = thermodynamic efficiency of the conversion from heat into work
T_1 = absolute temperature of heat source, R (K)
T_2 = absolute temperature of heat sink, R (K)

The efficiency statement of Equation 30 can be extended to cover all reversible cycles where T_1 and T_2 are defined as mean temperatures found by dividing the heat added and rejected reversibly by Δs. For this reason, all reversible cycles have the same efficiencies when considered between the same mean temperature limits of heat source and heat sink.

Rankine cycle

Early thermodynamic developments were centered around the performance of the steam engine and, for comparison purposes, it was natural to select a reversible cycle which approximated the processes related to its operation. The *Rankine cycle* shown in Fig. 6, proposed independently by Rankine and Clausius, meets this objective. All steps are specified for the system only (working medium) and are carried out reversibly as the fluid cycles among liquid, two-phase and vapor states. Liquid is compressed isentropically from points a to b. From points b to c, heat is added reversibly in the compressed liquid, two-phase and finally superheat states. Isentropic expansion with shaft work output takes place from points c to d and unavailable heat is rejected to the atmospheric sink from points d to a.

The main feature of the Rankine cycle is that compression (pumping) is confined to the liquid phase, avoiding the high compression work and mechanical problems of a corresponding Carnot cycle with two-phase compression. This part of the cycle, from points a to b in Fig. 6, is greatly exaggerated, because the difference between the saturated liquid line and point b (where reversible heat addition begins) is too small to show in proper scale. For example, the temperature rise with isentropic compression of water from a saturation temperature of 212F (100C) and one atmosphere to 1000 psi (6.89 MPa) is less than 1F (0.6C).

If the Rankine cycle is closed in the sense that the fluid repeatedly executes the various processes, it is termed a *condensing cycle*. Although the closed, condensing Rankine cycle was developed to improve steam engine efficiency, a closed cycle is essential for any toxic or hazardous working fluid. Steam has the important advantage of being inherently safe. However, the close control of water chemistry required in high pressure, high temperature power cycles also favors using a minimum of makeup water. (Makeup is the water added to the steam cycle to replace leakage and other withdrawals.) Open steam cycles are still found in small units, some special processes, and heating load applications coupled with power. The condensate from process and heating loads is usually returned to the power cycle for economic reasons.

The higher efficiency of the condensing steam cycle is a result of the pressure-temperature relationship between water and its vapor state, steam. The lowest temperature at which an *open*, or *noncondensing*, steam cycle may reject heat is approximately 212F (100C), the saturation temperature corresponding to atmospheric pressure of 14.7 psi (101.35 kPa). The pressure of the condensing fluid can be set at or below atmospheric pressure in a closed cycle. This takes advantage of the much lower sink temperature available for heat rejection in natural bodies of water and the atmosphere. Therefore, the condensing temperature in the closed cycle can be 100F (38C) or lower.

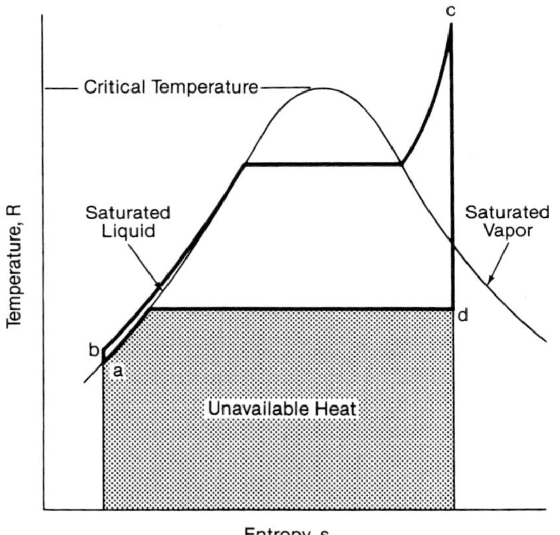

Fig. 6 Temperature-entropy diagram of the ideal Rankine cycle.

Fig. 7 illustrates the difference between an open and closed Rankine cycle. Both cycles are shown with nonideal expansion processes. Liquid compression takes place from points a to b and heat is added from points b to c. The work and heat quantities involved in each of these processes are the same for both cycles. Expansion and conversion of stored energy to work take place from points c to d´ for the open cycle and from c to d for the closed cycle. Because this process is shown for the irreversible case, there is internal fluid heating and an entropy increase. From points d´ to a, and d to a, heat is rejected in order to condense the steam. Because this last portion of the two cycles is shown as reversible, the shaded areas are proportional to the rejected heat. The larger amount of rejected heat for the open cycle is evident and is directly related to the lower amount of work that can be done by the expansion process.

Regenerative Rankine cycle

The reversible cycle efficiency given by Equation 30, where T_2 and T_1 are mean absolute temperatures for rejecting and adding heat respectively, indicates only three choices for improving ideal cycle efficiency: decreasing T_2, increasing T_1, or both. Little can be done to reduce T_2 in the Rankine cycle because of the limitations imposed by the temperatures of available rejected heat sinks in the general environment. Some T_2 reduction is possible by selecting variable condenser pressures for very large units with two or more exhaust hoods, because the lowest temperature in the condenser is set by the lowest temperature of the cooling water. On the other hand, there are many ways to increase T_1 even though the steam temperature may be limited by high temperature corrosion and allowable stress properties of the material.

One early improvement to the Rankine cycle was the adoption of *regenerative feedwater heating*. This is done by extracting steam from various stages in the turbine to heat the feedwater as it is pumped from the bottom of the condenser (hot well) to the boiler economizer.

Fig. 8 is a diagram of a widely used supercritical pressure steam cycle showing the arrangement of various components including the feedwater heaters. This cycle also contains one stage of steam reheat, which is another method of increasing the mean T_1. Regardless of whether the cycle is high temperature, high pressure or reheat, regeneration is used in all modern condensing steam power plants. It improves cycle efficiency and has other advantages, including lower volume flow in the final turbine stages and a convenient means of deaerating the feedwater. In the power plant heat balances shown in Fig. 8 and later in Fig. 10, several parameters require definition:

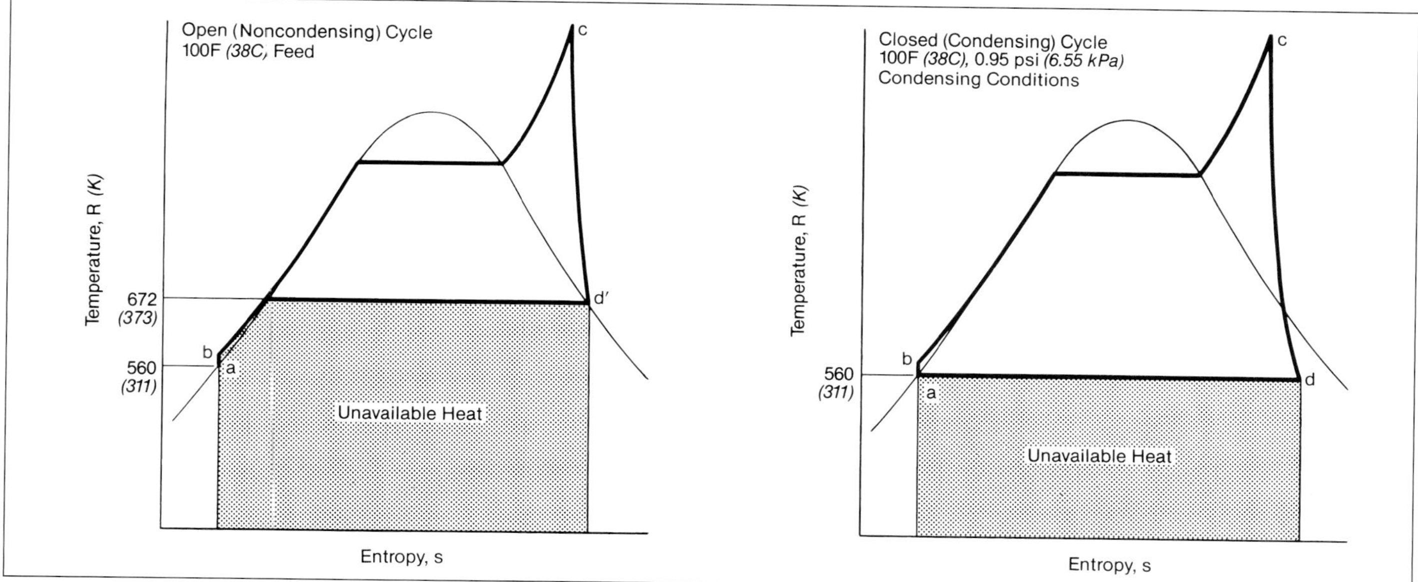

Fig. 7 Rankine cycles.

DC: In the feedwater heater blocks, this parameter is the drain cooler approach temperature or the difference between the shell-side condensate outlet (drain) temperature and the feedwater inlet temperature.

TD: In the feedwater heater blocks, this parameter is the terminal temperature difference or the difference between the shell-side steam inlet temperature and the feedwater outlet temperature.

P: In the feedwater heater blocks, this parameter is the nominal shell-side pressure.

The temperature-entropy diagram of Fig. 9 for the steam cycle of Fig. 8 illustrates the principle of regeneration in which the mean temperature level is increased for heat addition. Instead of heat input starting at the hot well temperature of 101.1F (38.4C), the water entering the boiler economizer has been raised to 502F (261C) by the feedwater heaters.

Fig. 9 also shows that the mean temperature level for heat addition is increased by reheating the steam after a portion of the expansion has taken place. Because maximum temperatures are limited by physical or economic reasons, reheating after partial expansion of the working fluid is also effective in raising the average T_1. The hypothetical case of an infinite number of reheat and expansion stages approaches a constant temperature heat addition of the Carnot cycle, at least in the superheat region. It would appear beneficial to set the highest temperature in the superheat reheat stage at the temperature limit of the working medium or its containment. However, merely increasing T_1 may not improve efficiency. If the entropy increase accompanying reheat causes the final expansion process to terminate in superheated vapor, the mean temperature for heat rejection, T_2, has also been increased unless the superheat can be extracted in a regenerative heater, adding heat to the boiler feedwater. Such a regenerative heater would have to operate at the expense of the very effective cycle. All of these factors, plus component design limitations, must be considered in a cycle analysis where the objective is to optimize the thermodynamic efficiency within the physical and economic constraints of the equipment. In addition, there are constraints imposed by the economics of fuel selection and the environmental impacts of fuel combustion that impact the design of the boiler/turbine regenerative cycle. Overall cycle characteristics, including efficiency, can also be illustrated by plotting the cycle on a Mollier chart. (See chapter frontispiece and Fig. 4.)

The procedure used in preparing Fig. 9 deserves special comment because it illustrates an important function of entropy. All processes on the diagram represent total entropies divided by the high pressure steam flow rate. Total entropies at any point of the cycle are the product of the mass flowing past that point in unit time and the entropy per pound (specific entropy) corresponding to the pressure, temperature, and state of the steam. Specific entropy values are provided by the Steam Tables such as those provided here in Tables 1, 2, and 3. If a point falls in the two-phase region, entropy is calculated in the same manner as enthalpy. That is, the value for evaporation is multiplied by the steam quality (fraction of uncondensed steam) and added to the entropy value of water at saturation conditions corresponding to the pressure at that point in the system.

Because there are different flow rates for the various cycle processes, small sections of individual *T-s* diagrams are superimposed in Fig. 9 on a base diagram that identifies saturated liquid and vapor parameters. However, the saturation parameters can only be compared to specific points on the *T-s* diagram. These points correspond to the parts of the cycle representing heat addition to high pressure steam and the expansion of this steam in a high pressure turbine. In these parts of the cycle, the specific entropy of the fluid and the value plotted in the diagram are the same. At each steam bleed point of the intermedi-

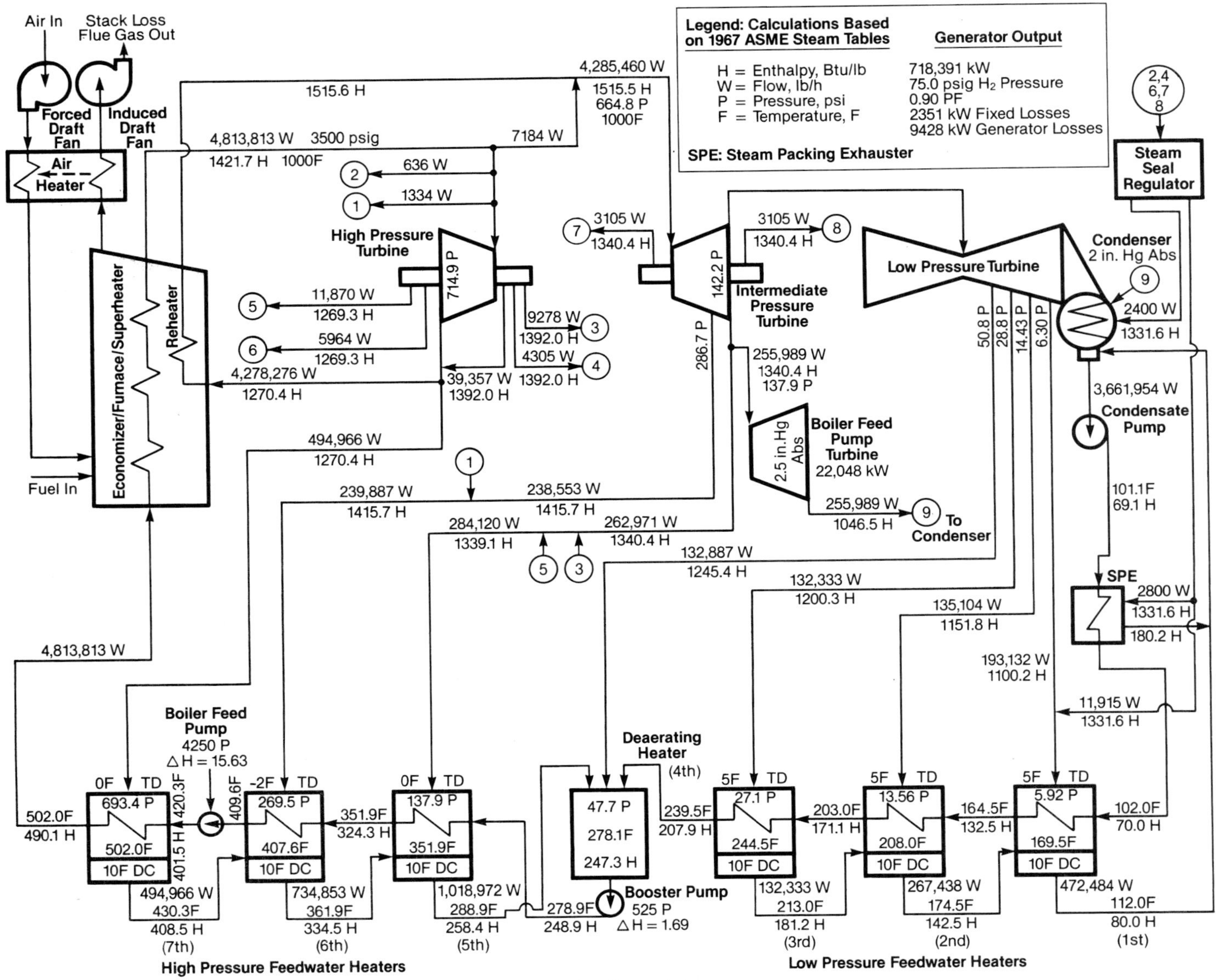

Fig. 8 Supercritical pressure, 3500 psig turbine cycle heat balance (English units).

ate and low pressure turbine, the expansion line should show a decrease in entropy due to reduced flow entering the next turbine stage. However, for convenience, the individual step backs in the expansion lines have been shifted to the right to show the reheated steam expansion as one continuous process.

Feedwater heating through the regenerators and compression by the pumps (represented by the zigzag lines in Fig. 9) result in a net entropy increase. However, two factors are involved in the net increase, an entropy increase from the heat added to the feedwater and a decrease resulting from condensing and cooling the bleed steam and drain flows from higher pressure heaters.

Consider an example in which the feedwater heater just before the deaerating heater increases the temperature of a 3,661,954 lb/h feedwater from 203.0 to 239.5F. From Table 1, this increases the enthalpy, H, of the feedwater from 171.2 to 208.0 and increases the entropy, s, from 0.2985 to 0.3526. The total entropy increase per lb of high pressure steam flowing at 4,813,813 lb/h is:

$$\frac{(s_2 - s_1)\,\dot{m}_{\text{feedwater}}}{\dot{m}_{\text{HP steam}}} = \frac{(0.3526 - 0.2985)\;3,661,954}{4,813,813} = 0.0412\ \text{Btu/lb F} \tag{31}$$

The feedwater temperature rises 36.5F and the total heat absorbed is:

$$(H_2 - H_1)\,\dot{m}_{\text{feedwater}} = (208.0 - 171.2)\;3,661,954 = 134,759,907\ \text{Btu/h} \tag{32}$$

On the heat source side of the balance, 132,333 lb/h of steam are bled from the low pressure turbine at 28.8 psig. This steam has an enthalpy of 1200.3 and an entropy of 1.7079. The steam is desuperheated and condensed according to the following equation:

$$H_2 = H_1 - \frac{\text{heat absorbed by feedwater}}{\dot{m}_{\text{LP steam}}}$$

$$= 1200.3 - \frac{134,759,907}{132,333} = 182.0 \text{ Btu/lb} \qquad (33)$$

Interpolating Table 1, the low pressure steam is cooled to 213.0F at H_f = 181.2 Btu/lb. The corresponding entropy of the heater drain is 0.3136 Btu/lb F. Therefore, the entropy decrease is:

$$\frac{(s_1 - s_2)\dot{m}_{\text{LP steam}}}{\dot{m}_{\text{HP steam}}}$$

$$= \frac{(1.7079 - 0.3136)\ 132,333}{4,813,813} \qquad (34)$$

$$= 0.0383 \text{ Btu/lb F}$$

This heater shows a net entropy increase of 0.0412 – 0.0383 = 0.0029 Btu/lb F.

Recall that an increase in entropy represents heat energy that is unavailable for conversion to work. Therefore, the net entropy increase through the feedwater heater is the loss of available energy that can be attributed to the pressure drop required for flow and temperature difference. These differences are necessary for heat transfer. The quantity of heat rendered unavailable for work is the product of the entropy increase and the absolute temperature of the sink receiving the rejected heat.

Available energy

From the previous feedwater heater example, there is a derived quantity, formed by the product of the corresponding entropy and the absolute temperature of the available heat sink, which has the nature of a property. The difference between H (enthalpy) and $T_o s$ is another derived quantity called available energy.

$$e = H - T_o s \qquad (35)$$

where

e = available energy, Btu/lb (kJ/kg)
H = enthalpy, Btu/lb (kJ/kg)
T_o = sink temperature, R (K)
s = entropy, Btu/lb R (kJ/kg K)

Available energy is not a property because it can not be completely defined by an equation of state; rather, it is dependent on the sink temperature. However, a combined statement of the first and second laws of thermodynamics indicates that the difference in the available energy between two points in a reversible process represents the maximum amount of work (on a unit mass basis) that can be extracted from the fluid due to the change of state variables H and s between the two points. Conceptually then, differences in the value of $T_o s$ represent energy that is unavailable for work.

The concept of available energy is useful in cycle analysis for optimizing the thermal performance of various components relative to overall cycle efficiency. In this way small, controllable changes in availability may be weighed against larger, fixed unavailable heat quantities which are inherent to the cycle. By comparing actual work to the maximum reversible work calculated from differences in available energy, the potential for improvement is obtained.

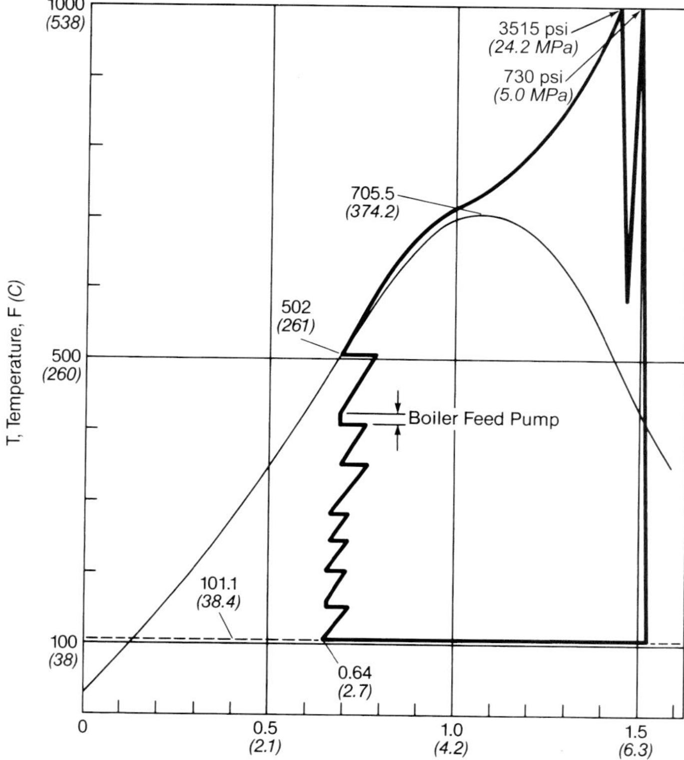

Fig. 9 Steam cycle for fossil fuel temperature-entropy diagram – single reheat, seven-stage regenerative feedwater heating – 3500 psig, 1000F/1000F steam.

Rankine cycle efficiency

As with the Carnot cycle efficiency, the Rankine cycle efficiency (η) is defined as the ratio of the net work ($W_{out} - W_{in}$) produced to the energy input (Q_{in}):

$$\eta = \frac{W_{out} - W_{in}}{Q_{in}} \qquad (36)$$

For the simple cycle shown in Fig. 7, the work terms and the energy input are defined as:

$$W_{out} = \eta_t\, \dot{m}_w (H_c - H_d) \qquad (37)$$

$$W_{in} = \dot{m}_w (H_b - H_a)/\eta_p$$
$$\cong \dot{m}_w v_a (P_b - P_a)/\eta_p \qquad (38)$$

$$Q_{in} = \dot{m}_w (H_c - H_b) \qquad (39)$$

where H_{a-d} are the enthalpies defined in Fig. 7, $\dot{m}_w$ is the water flow rate, P_{a-b} are the pressures at points a

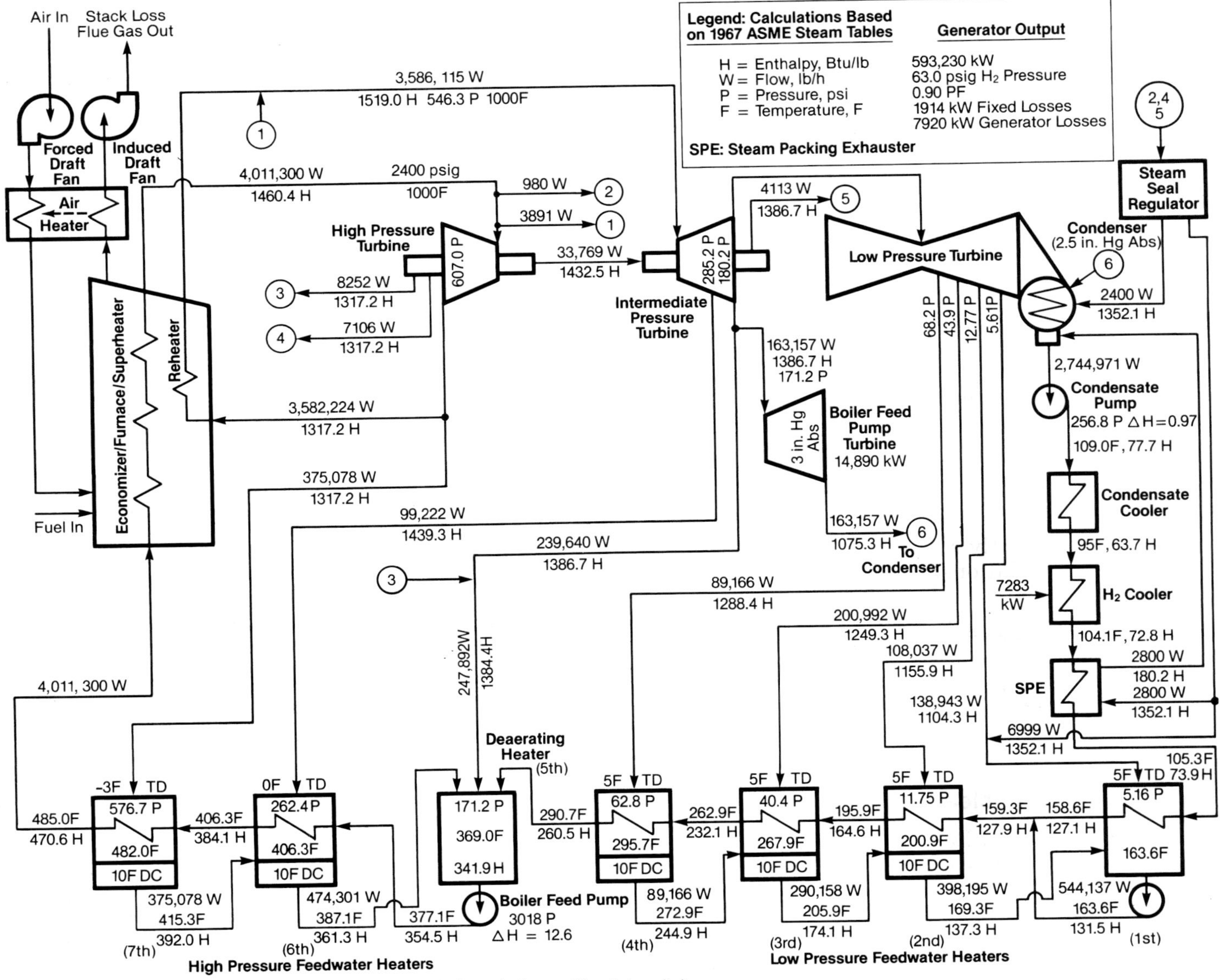

Fig. 10 Subcritical pressure, 2400 psig turbine cycle heat balance (English units).

and b, v_a is the water specific volume at point a, while η_t and η_p are the efficiencies of the turbine and boiler feed pump respectively.

Substituting Equations 37, 38 and 39 into Equation 36 and canceling the mass flow rate, $\dot{m}_w$, which is the same in all three cases provide the following overall thermodynamic efficiency (η_{th}):

$$\eta_{th} = \frac{\eta_t (H_c - H_d) - v_a (P_b - P_a)/\eta_p}{(H_c - H_b)} \tag{40}$$

In even a simple power producing facility using the Rankine cycle, several other factors must also be considered:

1. Not all of the chemical energy supplied to the boiler from the fuel is absorbed by the steam – typically 80 to 85% of the energy input is absorbed.
2. A variety of auxiliary equipment such as fans, sootblowers, environmental protection systems, water treatment equipment, and fuel handling systems, among others use part of the power produced.
3. Electrical generators and motors are not 100% efficient.

Incorporating these general factors into Equation 40 for a simple power cycle yields the net generating efficiency, η_{net}:

$$\eta_{net} = \frac{\eta_g \eta_t (H_c - H_d) - \left[v_1 (P_b - P_a)/\eta_p \eta_m \right] - w_{aux}}{(H_c - H_b)/\eta_b} \tag{41}$$

where w_{aux} is the auxiliary power usage, η_b is the boiler efficiency, while η_g and η_m are the electrical generator and motor efficiencies, both typically 0.98 to 0.99. The gross power efficiency can be evaluated from Equation 41 with w_{aux} set at zero.

The evaluation of efficiency in modern high pressure steam power systems is more complex. Provision in the evaluation must be made for steam reheat or double reheat, and turbine steam extraction for regenerative feedwater heating, among others. This evaluation is based upon a steam turbine heat balance or steam cycle diagram such as that shown in Fig. 8 for a 3500 psig (24.13 MPa) supercritical pressure fossil fuel-fired unit, or Fig. 10 for a 2400 psig (16.55 MPa) subcritical pressure unit. The subcritical pressure unit shown has a single reheat, six closed feedwater heaters and one open feedwater heater.

Rankine cycle heat rate

Heat rate is a term frequently used to define various power plant efficiencies. If the electrical generation used is the net output after subtracting all auxiliary electrical power needs, then Equation 42 defines the *net heat rate* using English units. If the auxiliary electrical usage is not deducted, Equation 42 defines the *gross heat rate*.

$$\text{Heat rate} = \frac{\text{Total fuel heat input (Btu/h)}}{\text{Electrical generation (kW)}} \tag{42}$$

Heat rate is directly related to plant efficiency, η, by the following relationships:

$$\text{Net heat rate} = \frac{3412.14 \text{ Btu/kWh}}{\eta_{net}} \tag{43a}$$

$$\text{Gross heat rate} = \frac{3412.14 \text{ Btu/kWh}}{\eta_{gross}} \tag{43b}$$

Steam cycle in a nuclear plant

Fig. 11 illustrates a Rankine cycle whose thermal energy source is a pressurized water nuclear steam system. High pressure cooling water is circulated from a pressurized water reactor to a steam generator. Therefore, heat produced by the fission of enriched uranium in the reactor core is transferred to feedwater supplied to the steam generator which, in turn, supplies steam for the turbine. The steam generators of a nuclear plant are shell and tube heat exchangers in which the high pressure reactor coolant flows inside the tubes and lower pressure feedwater is boiled outside of the tubes. For the pressurized water reactor system, the Rankine cycle for power generation takes place entirely in the nonradioactive water side (*secondary side*) that is boiling and circulating in the steam system; the reactor coolant system is simply the heat source for the power producing Rankine cycle.

The steam pressure at the outlet of the steam generator varies among plants due to design differences and ranges from 700 to 1000 psi (4.83 to 6.90 MPa). Nominally, a nuclear steam system by The Babcock & Wilcox Company (B&W) with a once-through steam generator provides slightly superheated steam at 570F (299C) and 925 psi (6.38 MPa). Steam flow from the once-through generator reaches the high pressure turbine at about 900 psi (6.21 MPa) and 566F (297C). More prevalent are nuclear steam systems that use a recirculating steam generator. In this design, feedwater is mixed with saturated water coming from the steam generator's separators before entering the tube bundle and boiling to generate steam. This boiling steam-water mixture reaches a quality of 25 to 33% at the end of the heat exchanger and enters the steam generator's internal separators. The separators return the liquid flow to mix with incoming feedwater and direct the saturated steam flow to the outlet of the steam generator. Inevitably, a small amount of moisture is formed by the time the steam flow reaches the high pressure turbine.

Even though the once-through steam generator is capable of providing superheated steam to the turbine, the pressure and temperature limitations of nuclear plant components must be observed. As a result, the expansion lines of the power cycle lie largely in the wet steam region. This is essentially a saturated or nearly saturated steam cycle. The expansion lines for the nuclear steam system shown in Fig. 11 (featuring a once-through steam generator) are plotted on an enthalpy-entropy or *H-s* diagram in Fig. 12.

The superheated steam is delivered to the turbine at a temperature only 34F (19C) above saturation. Although this superheat improves cycle efficiency, large quantities of condensed moisture still exist in the turbine. For example, if expansion from the initial conditions shown in Fig. 12 proceed down one step to the back pressure of 2.0 in. Hg [approximately 1.0 psi (6.9 kPa)] the moisture formed would exceed 20%. At best, steam turbines can accommodate about 15% moisture content. High moisture promotes erosion, especially in the turbine blades, and reduces expansion efficiency.

In addition to mechanical losses from momentum exchanges between slow moving condensate particles, high velocity steam and rotating turbine blades, there is also a thermodynamic loss resulting from the condensate in the turbine. The expansion of the steam is too rapid to permit equilibrium conditions to exist when condensation is occurring. Under this condition, the steam becomes subcooled, retaining a part of the available energy which would be released by condensation.

Fig. 11 indicates two methods of moisture removal used in this cycle and Fig. 12 shows the effect of this moisture removal on the cycle. After expansion in the high pressure turbine, the steam passes through a moisture separator, which is a low pressure drop separator external to the turbine. After passing through this separator, the steam is reheated in two stages, first by bleed steam and then by high pressure steam to 503F (262C), before entering the low pressure turbine. Here a second method of moisture removal, in which grooves on the back of the turbine blades drain the moisture from several stages of the low pressure turbine, is used. The separated moisture is carried off with the bleed steam.

Internal moisture separation reduces erosion and affords a thermodynamic advantage due to the divergence of the constant pressure lines with increasing enthalpy and entropy. This can be shown by the use of available energy, *e*, as follows. Consider the moisture removal stage at 10.8 psi in Fig. 12. After expansion to 10.8 psi, the steam moisture content is 8.9%. Internal separation reduces this to approximately 8.2%. Other properties are as follows:

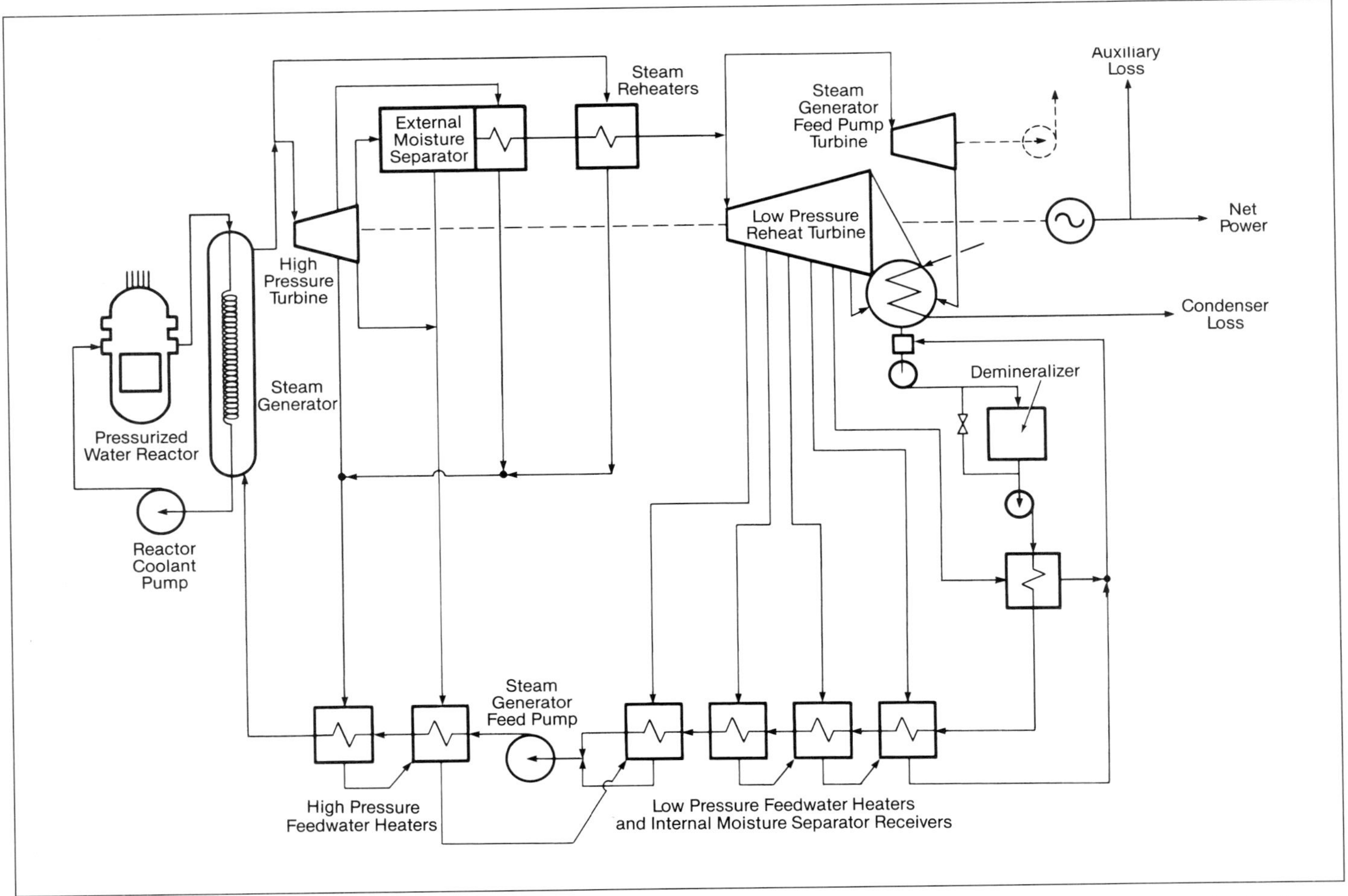

Fig. 11 Power cycle diagram, nuclear fuel: reheat by bleed and high pressure steam, moisture separation, and six-stage regenerative feedwater heating – 900 psi, 566F/503F (6.21 MPa, 297C/262C) steam.

	End of Expansion	After Moisture Extraction
P	10.8 psi	10.8 psi
H	1057.9 Btu/lb	1064.7 Btu/lb
s	1.6491 Btu/lb F	1.6595 Btu/lb F
T_o (at 2 in. Hg)	560.8 R	560.8 R
$T_o s$	924.8 Btu/lb	930.7 Btu/lb
$e = H - T_o s$	133.1 Btu/lb	134.0 Btu/lb

The increase in available energy, Δe, due to moisture extraction is 134.0 – 133.1 = 0.9 Btu/lb of steam.

The values of moisture and enthalpy listed are given for equilibrium conditions without considering the nonequilibrium effects that are likely to exist within the turbine. These effects can be empirically accounted for by the isentropic efficiency of the expansion line. An important point to observe from this example is the need to retain a sufficient number of significant digits in the calculations. Frequently, the evaluation of thermodynamic processes results in working with small differences between large numbers.

Supercritical steam cycles

As previously pointed out, cycle thermodynamic efficiency is improved by increasing the mean temperature of the heat addition process. This temperature can be increased when the feedwater pressure is increased because the boiler inlet pressure sets the saturation temperature in the Rankine cycle. If the pressure is increased above the critical point of 3200.1 psi (22.1 MPa), heat addition no longer results in the typical boiling process in which there is an interface between the steam and water. Rather, the fluid can be treated as a single phase as it passes through the process where properties change from those of a liquid to a gas without an interface. Additional heating superheats the steam and expansion in a first stage (high pressure) turbine can occur entirely in a superheated state. This is referred to as a supercritical steam cycle, originally given the name Benson Super Pressure Plant when first proposed in the 1920s. The first commercial unit featuring the supercritical cycle and two stages of reheat was placed in service in 1957.

The steam cycle of a typical supercritical plant is shown in Fig. 13. In this T-s diagram, point a represents the outlet of the condensate pump. Between points a and b, the condensate is heated in the low pressure feedwater heater using saturated liquid and/or steam extracted from the steam turbines. Point b corresponds to the high pressure feedwater pump inlet. The pump increases the pressure to 4200 psi (28.96 MPa), obtaining conditions of point c. Between points

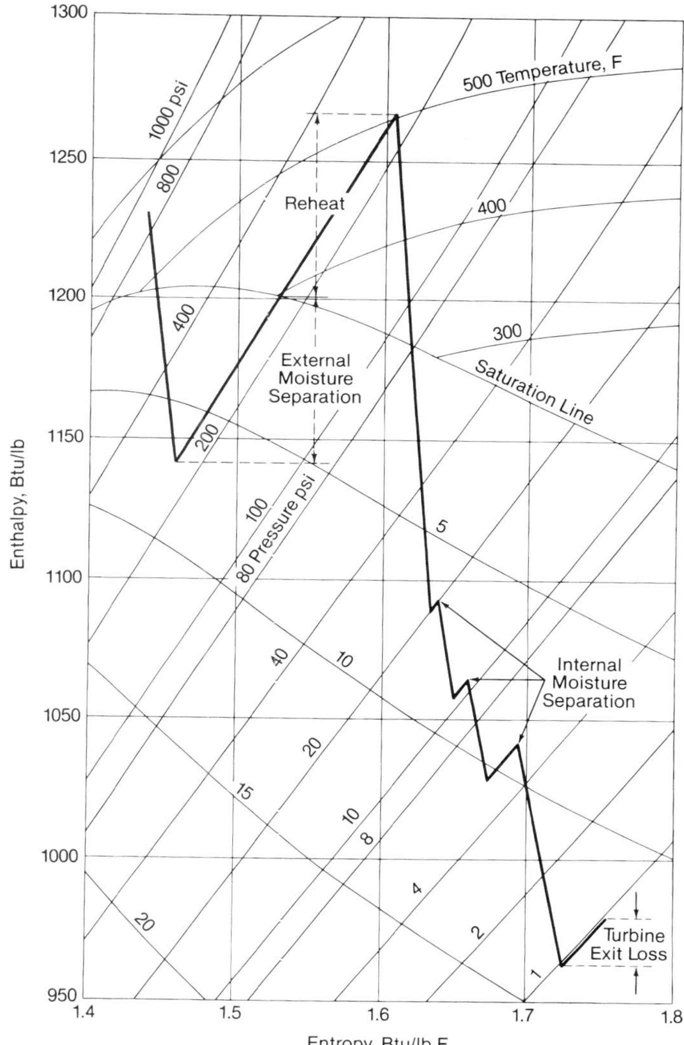

Fig. 12 Steam cycle for nuclear fuel on a Mollier chart: reheat by bleed and high pressure steam, moisture separation and six-stage regenerative feedwater heating – 900 psi, 566F/503F steam (English units).

c and d, additional feedwater heating is provided by steam extracted from the high and low pressure turbines. Point d corresponds to the supercritical boiler inlet. Due to the nature of the fluid, the supercritical boiler is a once-through design, having no need for separation equipment. The Universal Pressure, or UP®, boiler design used in the supercritical unit is described further in Chapter 26. For the supercritical cycle shown, the steam arrives at the high pressure turbine at 3500 psi (24.1 MPa) and 1050F (566C). Expansion in this turbine is complete at point f, which corresponds to a superheated condition. Steam exhausted from the high pressure turbine is then reheated in the boiler to approximately 1040F (560C), before entering the low pressure turbine at approximately 540 psi (3.7 MPa); this corresponds to point g on the *T-s* diagram. The low pressure turbine expands the steam to point h on the diagram. The cycle is completed by condensing the exhaust from the low pressure turbine to a slightly subcooled liquid, and a condensate pump delivers the liquid to the low pressure feedwater heater, which corresponds to point a in the *T-s* diagram.

The high pressure of the feedwater in the supercritical cycle requires a substantially higher power input to the feedwater pump than that required by the saturated Rankine cycle. In a typical Rankine cycle with a steam pressure of 2400 psi (16.55 MPa), the pump power input requires approximately 2% of the turbine output. This may increase to as much as 3% in the supercritical unit. However, this increase is justified by the improved thermodynamic efficiency of the cycle. In general, with equivalent plant parameters (fuel type, heat sink temperature, etc.), the supercritical steam cycle generates about 4% more net power output than the subcritical pressure regenerative Rankine steam cycle.

Process steam applications

In steam power plants generating only electric power, economically justifiable thermodynamic efficiencies range up to about 42% in fossil fuel plants (higher in combined cycle plants discussed later in this chapter) and 34% in nuclear plants. Therefore, typically more than half of the heat released from the fuel must be transferred to the environment.

Energy resources may be more efficiently used by operating multipurpose steam plants, where steam is exhausted or extracted from the cycle at a sufficient pressure for use in an industrial process or space heating application. With these arrangements, an overall thermal utilization of 65% or greater is possible. Combination power and process installations have been common for many years, but the demand for process steam is not sufficient to permit the use of these combined cycles in most central station electric power generating plants. However, in regions where waste disposal and renewable energy sources have become significant environmental issues, the use of cogeneration, biomass and waste-to-energy installations have been successfully tied together with district heating and

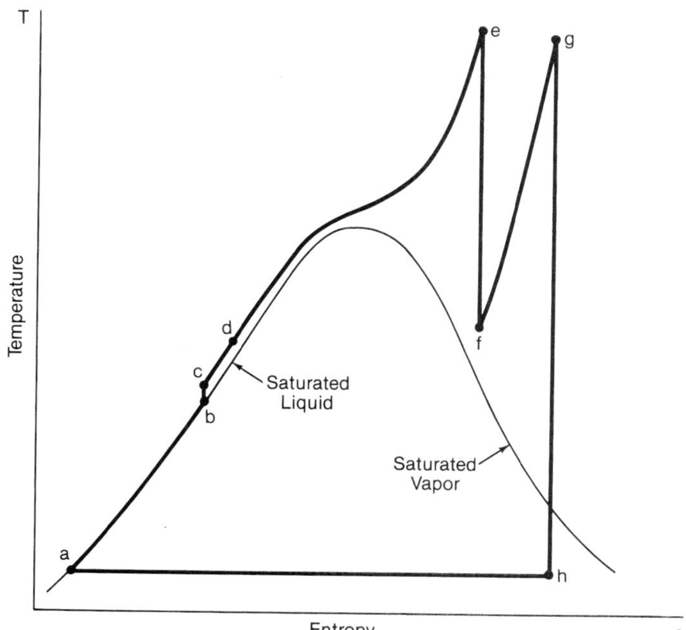

Fig. 13 Supercritical steam cycle with one reheat.

other process steam application projects. In recent years, the most successful of these have been in European municipalities.

Gas turbine cycle

In the thermodynamic cycles previously described, the working fluid has been steam used in a Rankine cycle. The Rankine cycle efficiency limit is dictated by the ratio of the current maximum and minimum cycle temperatures. The current maximum temperature of the steam Rankine cycle is approximately 1200F (649C), which is set primarily by material constraints at the elevated pressures of the steam cycles. One means of extending the efficiency limit is to replace the working fluid with air or gas. The gas turbine system in its simplest form consists of a compressor, combustor and turbine, as shown in Fig. 14. Because of its simplicity, low capital cost and short lead time, gas turbine systems are being used by some utilities to add capacity in smaller increments. Use of the gas turbine system in conjunction with the steam Rankine cycle is also an effective means of recovering some of the heat lost when combustion gases are released to the atmosphere at high temperatures.

In the simple gas turbine system shown in Fig. 14, air is compressed then mixed with fuel and burned in a combustor. The high temperature gaseous combustion products enter the turbine and produce work by expansion. A portion of the work produced by the turbine is used to drive the compressor and the remainder is available to produce power. The turbine exhaust gases are then vented to the atmosphere. To analyze the cycle, several simplifying assumptions are made. First, although the combustion process changes the composition of the working fluid, the fluid is treated as a gas of single composition throughout, and it is considered an ideal gas to obtain simple relationships between points in the system. Second, the combustion process is approximated as a simple heat transfer process in which the heat input to the working fluid is determined by the fuel heating values. A result of this approximation is that the mass flow rate through the system remains constant. The final approximation is to assume that each of the processes is internally reversible.

If the turbine expansion is complete with the exhaust gas at the same pressure as the compressor inlet air, the combination of processes can be viewed as a cycle. The simplifying assumptions above result in the idealized gas turbine cycle referred to as the air-standard *Brayton* cycle. Fig. 15 shows the cycle on *T-s* and *P-v* diagrams, which permit determining the state variables at the various cycle locations.

The idealized cycle assumes an isentropic process between points 1 and 2 (compression) and between 3 and 4 (expansion work). The temperature rise between points 2 and 3 is calculated by assuming the heat addition due to combustion is at a constant pressure. In the analysis, the pressure ratio between points 1 and 2 is given by the compressor design and is assumed to be known. To determine the temperature at point 2, a relationship between the initial and final states of an isentropic ideal gas process is obtained as follows.

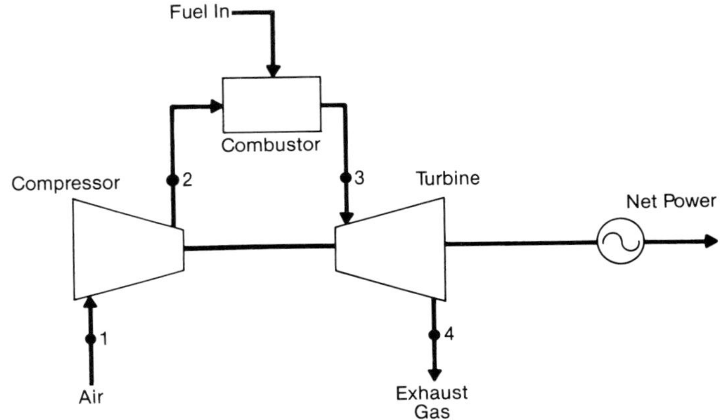

Fig. 14 Simple gas turbine system.

First, the general definitions of constant pressure and constant volume specific heats, respectively, are:

$$c_p = \left(\frac{\partial H}{\partial T}\right)_p \tag{44}$$

$$c_v = \left(\frac{\partial u}{\partial T}\right)_v \tag{45}$$

Strictly speaking, the specific heat values vary with temperature. In practice, however, they are assumed to be constant to facilitate the calculations. The two constants are related in that their difference equals the gas constant in the ideal gas law ($Pv = RT$):

$$c_p - c_v = R \tag{46}$$

The ratio of the constant pressure and constant volume specific heats is designated the specific heat ratio, k.

$$k = c_p / c_v \tag{47}$$

From these definitions, changes in enthalpy and internal energy for an ideal gas can be calculated from:

$$dH = c_p dT \tag{48}$$

$$du = c_v dT \tag{49}$$

Although expressed to relate differential changes in enthalpy and temperature, the concept of specific heat can be used to calculate finite enthalpy changes as long as the change in temperature is not excessive. When higher accuracy is required, tabulated enthalpy values should be used.

Recalling Equation 26, the combined expression of the first and second laws of thermodynamics, setting $ds = 0$ for the isentropic process, and inserting the change in internal energy given by Equation 49, the former equation may be written:

$$Tds = du + Pdv = c_v dT + Pdv = 0 \tag{50}$$

Substituting the ideal gas law (in differential form, $RdT = Pdv + vdP$) and using the specific heat ratio definition, Equation 50 becomes:

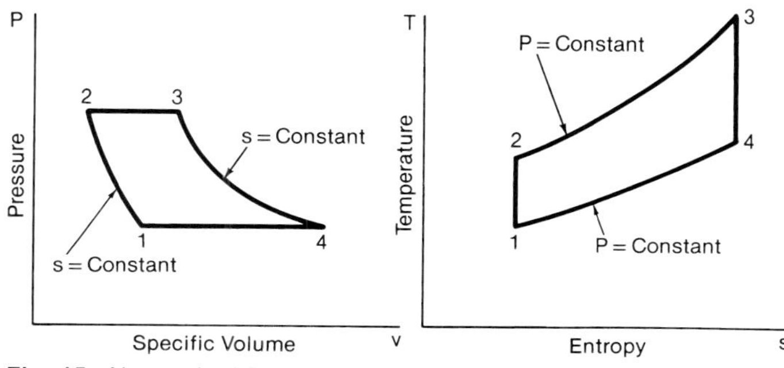

Fig. 15 Air-standard Brayton cycle.

$$\frac{dP}{P} + \frac{kdv}{v} = 0 \quad \textbf{(51)}$$

Integrating this yields:

$$Pv^k = \text{constant} \quad \textbf{(52)}$$

From Equation 52 and the ideal gas law, the following relationship between pressures and temperatures in an isentropic process is obtained, and the temperatures at points 2 and 4 are determined.

$$\frac{T_2}{T_1} = \frac{T_3}{T_4} = \left(\frac{P_2}{P_1}\right)^{(k-1)/k} \quad \textbf{(53)}$$

With this, the temperature and pressure (state variables) at all points in the cycle are determined. The turbine work output, w_t, required compressor work, w_c, and heat input to the process, q_b, are calculated as:

$$w_t = c_p(T_3 - T_4) \quad \textbf{(54)}$$

$$w_c = c_p(T_2 - T_1) \quad \textbf{(55)}$$

$$q_b = c_v(T_3 - T_2) \quad \textbf{(56)}$$

As in other cycle analyses described to this point, the cycle efficiency η is calculated as the net work produced divided by the total heat input to the cycle and is given by:

$$\eta = \frac{w_t - w_c}{q_b} \quad \textbf{(57)}$$

in which q_b is the heat input in the combustor (burner) per unit mass of gas (the working fluid) flowing through the system. For the ideal cycle, this can also be expressed in terms of gas temperatures by using Equation 48 to express the enthalpy change in the combustor and Equations 54 and 55 for the turbine and compressor work:

$$\eta = \left(1 - \frac{T_4 - T_1}{T_3 - T_2}\right)\frac{c_p}{c_v} \quad (58)$$

The actual gas turbine cycle differs from the ideal cycle due to inefficiencies in the compressor and turbine and pressure losses in the system. The effects of these irreversible aspects of the real gas turbine cycle are shown in the T-s diagram in Fig. 16. An isentropic compression would attain the point 2s, whereas the real compressor attains the pressure P_2, with an entropy corresponding to point 2 on the T-s diagram; likewise the turbine expansion attains point 4 rather than 4s. Constant-pressure lines on the diagram for pressures P_2 and P_3 illustrate the effect of pressure losses in the combustor and connecting piping, and the deviation of the process between points 4 and 1 from a constant-pressure process illustrates the effect of compressor inlet and turbine exhaust pressure losses on the cycle efficiency.

Points along the real cycle are determined by calculating the temperature T_{2s} from Equation 53 as:

$$T_{2s} = T_1\left(\frac{P_2}{P_1}\right)^{(k-1)/k} \quad \textbf{(59)}$$

Using the compressor efficiency provided by the manufacturer and solving for enthalpy or temperature at the compressor outlet yields the following:

$$\eta_c = \frac{T_{2s} - T_1}{T_2 - T_1} = \frac{H_{2s} - H_1}{H_2 - H_1} \quad \textbf{(60)}$$

Despite the significant advances in the mechanical efficiency of compressor and turbine designs (both on the order of 80% or greater), the overall cycle efficiency of a real gas turbine system is relatively low (30 to 35%) due to the high exhaust gas temperature and because a significant portion of the turbine output is used for compressor operation. The cycle efficiency may be increased by using a heat exchanger to preheat the air between the compressor and combustor. This heat is supplied by the turbine exhaust gas in a manner similar to that of the Rankine cycle regenerative heat exchangers. However, the higher efficiency is achieved in a system with a lower pressure ratio across the compressor and turbine, which in turn lowers the net work output for a given combustion system. The lower net output and extra hardware cost must be weighed in each case against the thermodynamic efficiency improvement.

One of the key benefits of the gas turbine cycle is its ability to operate at much higher temperatures than the Rankine steam cycle. Gas turbines typically operate with an inlet temperature of 1800 to 2200F (982 to 1204C) and some turbine designs with complex internal cooling systems have been operated as high as 2300F (1260C), raising the thermodynamic efficiency. With the ability to operate at elevated temperatures and to use combustion gases as a working fluid, some gas turbine systems are operated in conjunction with the steam Rankine cycle.

Combined cycles and cogeneration

As seen in the previous discussions of the Rankine and Brayton cycles, the gas turbine Brayton cycle efficiently uses high temperature gases from a combustion process but discharges its exhaust gas at a rela-

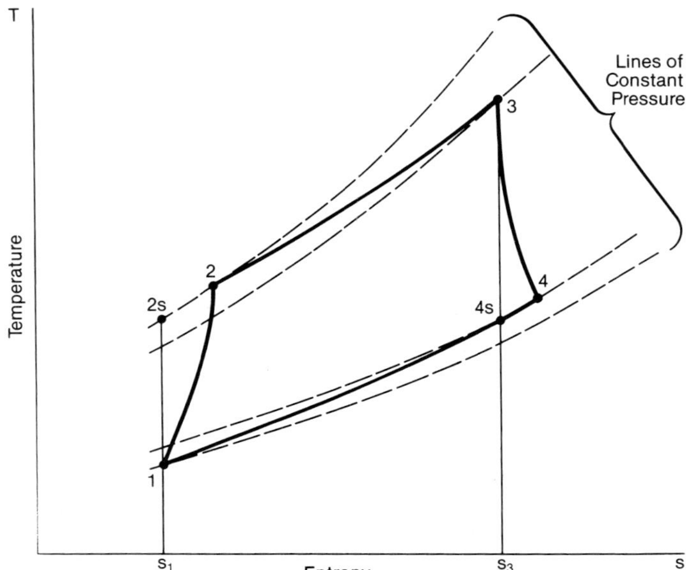

Fig. 16 *T-s* diagram of an actual gas turbine system.

tively high temperature; in the Brayton cycle, this constitutes wasted heat. On the other hand, the steam turbine Rankine cycle is unable to make full use of the highest temperatures. Combined cycles are designed to take advantage of the best features of these two cycles to improve the overall thermodynamic efficiency of the plant. Advanced combined cycles, in which the gas turbine exhaust is used as a heat source for a steam turbine cycle, can achieve overall thermal efficiencies in excess of 50%, generally representing a 15% improvement in the cycle efficiency compared to the gas turbine alone.

Waste heat boilers

In its simplest form, the combined cycle plant is a gas turbine (Brayton cycle) plant enhanced by passing the turbine exhaust through a steam generator, as shown in Fig. 17. The steam generator uses the hot turbine exhaust as a heat source for a steam turbine Rankine cycle. Electric power is generated from the mechanical work provided by the gas turbine and the steam turbine. In concept, the steam generator in the combined cycle is recovering the otherwise wasted heat from the gas turbine exhaust, and therefore it is referred to as a heat recovery steam generator or a waste heat boiler. (See Chapter 27.) More recent applications of the combined cycle have incorporated supplemental firing in the waste heat boiler to elevate the steam temperature and, therefore, to improve the steam cycle performance. Thermodynamic efficiency is defined as the work output of the two cycles divided by the total heat supplied (Q_{total}):

$$\eta = \left[\left(W_{out} - W_{in} \right)_{GT} + \left(W_{out} - W_{in} \right)_{ST} \right] / Q_{total} \qquad \textbf{(61)}$$

where the subscripts GT and ST refer to gas turbine and steam turbine, respectively.

Another approach to combining the gas and steam cycles, in which the steam generator serves as the combustion chamber for the gas turbine cycle, is shown in Fig. 18. In this arrangement, the principal heat source to the gas and steam cycle is the combustion process taking place in the steam generator. The gaseous combustion products are expanded in the gas turbine and the steam generated in the boiler tubes is expanded in the steam turbine. Although not shown in Fig. 18, the heat contained in the gas turbine exhaust may be recovered by using either a regenerative heat exchanger in the gas turbine cycle or a feedwater heater in the steam cycle. A pressurized fluidized-bed combustion combined cycle is a specific example of this approach to combining the gas and steam cycles. (See Chapter 17.)

Cogeneration

In the most general sense, cogeneration is the production of more than one useful form of energy (thermal, mechanical, electrical, etc.) simultaneously from a single fuel. In practice, cogeneration refers to generating electricity while principally performing an industrial function such as space heating, process heating or fuel gasification. Cogeneration systems are divided into two basic arrangements, topping and bottoming cycles.

A topping cycle is shown in Fig. 19. In this system the fuel is used for power generation in a steam boiler or gas turbine cycle combustor, and the waste heat from the power generation cycle supports an industrial process. The most common topping cycle is one in which a boiler generates steam at a higher pressure than that needed for the process or space conditioning application. The high pressure steam is then expanded in a turbine to a pressure that is appropriate for the application, generating electricity in the expansion process. Steam turbines, gas turbines and reciprocating engines are commonly used in topping cycles.

A bottoming cycle is most commonly associated with the recovery or waste heat boiler. In the bottoming cycle, fuel is not supplied directly to the power generating cycle. Rather, steam is generated from a waste

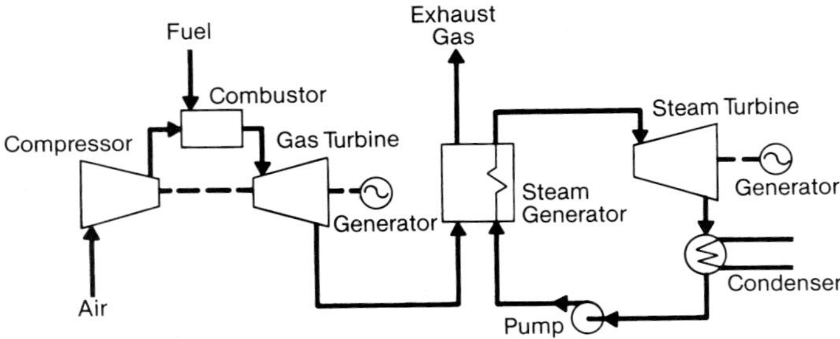

Fig. 17 Simple combined cycle plant.

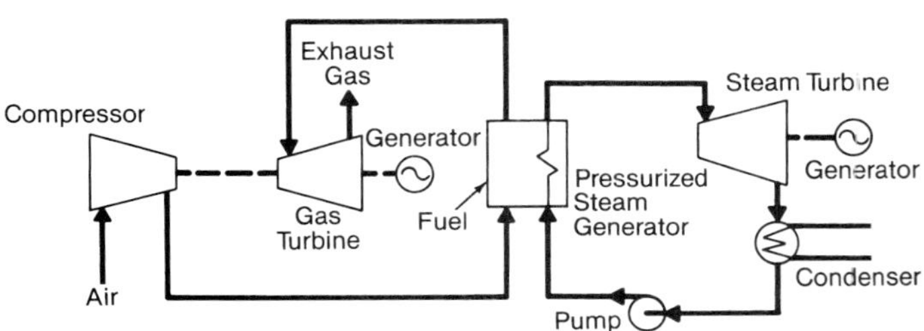

Fig. 18 Pressurized combustion combined cycle plant.

heat source and then expanded in a turbine to produce work or to generate electricity. Steam is frequently used in the bottoming cycle because of its ability to condense at low temperatures in the closed Rankine cycle. The bottoming cycle is shown in Fig. 20. The steam Rankine cycle used as a bottoming cycle has been illustrated previously in the descriptions of combined cycle plants.

Combustion processes

To this point, cycles have been compared based on the thermodynamic efficiency achieved, i.e., the net work produced divided by the total heat input to the cycle. To complete the evaluation of a combustion-based cycle, however, the performance must be expressed in terms of fuel consumption. In addition, the ability of the different machines to make full use of the combustion energy varies with temperatures reached in the combustion chamber and with dissociation of the combustion products.

The energy release during combustion is illustrated by considering the combustion of carbon (C) and oxygen (O_2) to form carbon dioxide (CO_2):

$$C + O_2 \rightarrow CO_2$$

If heat is removed from the combustion chamber and the reactants and products are maintained at 25C and 0.1 MPa (77F and 14.5 psi) during the process, the heat transfer from the combustion chamber would be 393,522 kJ per kmole of CO_2 formed. From the first law applied to the process, the heat transfer is equal to the difference in enthalpy between the reactants and products:

$$q - w = H_P - H_R \qquad \textbf{(62)}$$

The subscripts R and P refer to reactants and products, respectively. Assuming that no work is done in the combustion chamber and expressing the enthalpy of reactants and products on a per mole basis, this becomes:

$$Q = \sum n_P H_P - \sum n_R H_R \qquad \textbf{(63)}$$

The number of moles of each element or molecular species entering or leaving the chamber, n_R or n_P respectively, is obtained from the chemical reaction equation. By convention, the enthalpy of elements at 25C and 0.1 MPa (77F and 14.5 psi) are assigned the value of zero. Consequently, the enthalpy of CO_2 at these conditions is −393,522 kJ/kmole (the negative sign is due to the convention of denoting heat transferred from a control volume as negative). This is referred to as the enthalpy of formation and is designated by the symbol H_f^o. The enthalpy of CO_2 (and other molecular species) at other conditions is found by adding the change in enthalpy between the desired condition and the standard state to the enthalpy of formation. [Note that some tables may not use 25C and 0.1 MPa (77F and 14.5 psi) as the standard state when listing the enthalpy of formation.] Ideal gas behavior or tabulated properties are used to determine enthalpy changes from the standard state.

The stoichiometrically balanced chemical reaction equation provides the relative quantities of reactants and products entering and leaving the combustion chamber. The first law analysis is usually performed on a per mole or unit mass of fuel basis. The heat transfer from a combustion process is obtained from a first law analysis of the combustion process, given the pressure and temperature of the reactants and products. Unfortunately, even in the case of complete combustion as assumed in the previous example, the temperature of the combustion products must be determined by additional calculations discussed later in this section.

The combustion of fossil or carbon based fuel is commonly accompanied by the formation of steam or water (H_2O) as in the reaction:

$$CH_4 + 2O_2 \rightarrow CO_2 + 2H_2O$$

Again, the difference in enthalpy between the reactants and products is equal to the heat transfer from the combustion process. The heat transfer per unit mass of fuel (methane in this example) is referred to as the heating value of the fuel. If the H_2O is present as liquid in the products, the heat transferred is referred to as the *higher heating value* (HHV). The term *lower heating value* (LHV) is used when the H_2O is present as a vapor. The difference between these two values is frequently small (about 4% for most hydrocarbon fuels) but still significant. When the efficiency

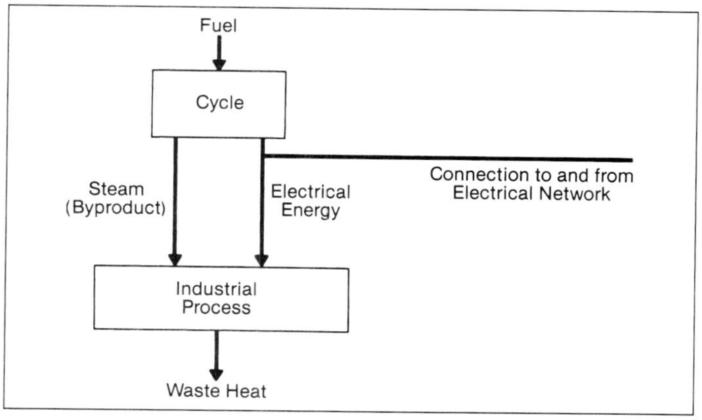

Fuel
Cycle
Steam (Byproduct)
Electrical Energy
Connection to and from Electrical Network
Industrial Process
Waste Heat

Fig. 19 Topping cycle.

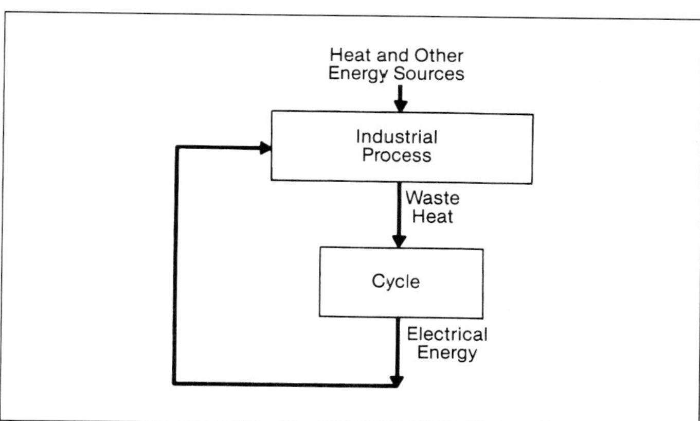

Fig. 20 Bottoming cycle.

of a cycle is expressed as a percentage of the fuel's heating value, it is important to know whether the HHV or LHV is used.

As noted, one of the difficulties in completing the first law analysis of the combustion process is determining the temperature of the products. In some applications an upper limit of the combustion temperature may be estimated. From the first law, this can occur if the combustion process takes place with no change in kinetic or potential energy, with no work and with no heat transfer (adiabatically). Under these assumptions, the first law indicates that the sum of the enthalpies of the reactants equals that of the products. The temperature of the products is then determined iteratively by successively assuming a product temperature and checking the equality of the reactant and product enthalpies. For a given fuel and reactants at the specified inlet temperature and pressure, this procedure determines the highest attainable combustion temperature, referred to as the *adiabatic combustion* or *flame temperature*.

Gibbs free energy

An important thermodynamic property derived from a combination of other properties (just as enthalpy was derived from u, P and v) is Gibbs free energy (g), which is also frequently referred to as *free energy*:

$$g = H - Ts \tag{64}$$

Free energy g is a thermodynamic potential similar to enthalpy and internal energy because in any thermodynamic process, reversible or irreversible, differences in this quantity depend only on initial and final states of the system.

The usefulness of free energy is particularly evident from the following expression of the combined first and second laws, expressed for a reversible process with negligible changes in kinetic and potential energy:

$$\begin{aligned} W_{rev} &= \sum m_1 (H_1 - T_o s_1) \\ &-\sum m_2 (H_2 - T_o s_2) \end{aligned} \tag{65}$$

When applied to a combustion process in which the reactants and products are in temperature equilibrium with the surroundings, this becomes:

$$W_{rev} = \sum n_R g_R - \sum n_P g_P \tag{66}$$

This equation indicates the maximum value of reversible work that can be obtained from the combustion of a given fuel. The reversible work is maximized when the reactants constitute a stoichiometrically balanced mixture (no excess air). The quantities n_R and n_P are obtained from the chemical reaction equation and g is expressed on a per mole basis in Equation 66.

From this, one might expect to express the efficiency of a cycle that extracts energy from a combustion process as a percentage of the Gibbs free energy decrease, rather than in terms of the heating value of the fuel. It is uncommon for this to be done, however, because the difference between the free energy decrease and the heating value of hydrocarbon fuels is small and because the use of fuel heating value is more widespread.

Free energy is more commonly used to determine the temperature reached in burning fuel, including the effects of dissociation. The problem of dissociation is illustrated by again considering the combustion of carbon and oxygen to form CO_2. If the temperature of the combustion process is high enough, the CO_2 dissociates to form CO and O_2 according to the reaction:

$$CO_2 \leftrightarrow CO + \frac{1}{2} O_2$$

As the dissociation reaction occurs from left to right (from all CO_2 to none), the sum of the reactant free energies and that of the products vary. Equilibrium of this reaction is reached when the sum of the free energies is a minimum. The equilibrium point (degree of dissociation) varies with the combustion temperature.

While the process of iteratively determining a minimum free energy point is suited for computer calculations, the equilibrium conditions of the dissociation reaction at an assumed temperature can also be determined using tabulated values of a constant relating the species involved in the reaction. This constant is known as the equilibrium constant K_{eq}, which for ideal gases is given by:

$$K_{eq} = \frac{(P_B)^b (P_C)^c}{(P_A)^a} \tag{67}$$

where P_A, P_B and P_C are the partial pressures, i.e., the products of total pressure and mole fractions in the mixture, of the reactants and products. The exponents represent the number of moles present for each species (A, B and C) in the stoichiometric balance equation as follows:

$$aA \leftrightarrow bB + cC \tag{68}$$

Equations 67 and 68 yield simultaneous equations for the mole fractions a, b and c. For nonideal gas reactions, the partial pressures are replaced by what are known as *fugacities* (the tendencies of a gas to expand or escape). Thermodynamic properties and relationships for the compounds and their elements encountered in the combustion process are available in the literature. One of the best sources for this information is the JANAF Thermochemical Tables, published by the U.S. Department of Commerce.[6] These tables include $\log_{10}$ values of the equilibrium constants for temperatures from 0 to 6000K.

To continue the carbon-oxygen combustion example, the overall chemical reaction, including dissociation, is now written as:

$$C + O_2 \rightarrow aCO_2 + bCO + cO_2$$

in which the coefficients a, b and c represent the mole fractions of the product components as determined by the solution to the dissociation reaction at the assumed combustion temperature. The overall reaction equation is now used to check the assumed temperature by adding the enthalpies of the combustion products at the assumed temperature, noting that the enthalpy per mole must be multiplied by the corresponding mole fraction a, b, or c for each product species. The combustion temperature is determined when the sum of product enthalpies minus that of the reactants equals the heat transfer to the surroundings of the combustion chamber. The convective and radiative heat transfer from the combustion products to the chamber and eventually to the working fluid of the cycle at the assumed combustion temperature are discussed in Chapter 4.

References

1. Parry, W.T., et al., *ASME International Steam Tables for Industrial Use*, Based on IAPWS-IF97, The American Society of Mechanical Engineers, New York, New York, January, 2000.

2. *ASME Steam Properties for Industrial Use*, Based on IAPWS-IF97, Professional Version 1.1, The American Society of Mechanical Engineers, New York, New York, 2003.

3. Weast, R.C., et al., *CRC Handbook of Chemistry and Physics*, 70th Ed., CRC Press, Inc., Boca Raton, Florida, 1989.

4. Keenan, J.H., Chao, J., and Kaye, J., *Gas Tables: Thermodynamic Properties of Air Products of Combustion and Component Gases Compressible Flow Functions*, Second Ed., John Wiley & Sons, New York, New York, June, 1983.

5. Vargaftik, N.B., *Tables on the Thermophysical Properties of Liquids and Gases: In Normal and Dissociated States*, Second Ed., John Wiley & Sons, New York, New York, November, 1975.

6. Chase, Jr., M.W., et al., *JANAF Thermochemical Tables*, Fourth Ed., American Chemical Society, American Institute of Physics, New York, New York, 1998.

Laser velocity measurements in a steam generator flow model.

Chapter 3

Fluid Dynamics

In the production and use of steam there are many fluid dynamics considerations. Fluid dynamics addresses steam and water flow through pipes, fittings, valves, tube bundles, nozzles, orifices, pumps and turbines, as well as entire circulating systems. It also considers air and gas flow through ducts, tube banks, fans, compressors and turbines plus convection flow of gases due to draft effect. The fluid may be a liquid or gas but, regardless of its state, the essential property of a fluid is that it yields under the slightest shear stress. This chapter is limited to the discussion of Newtonian liquids, gases and vapors where any shear stress is directly proportional to a velocity gradient normal to the shear force. The ratio of the shear stress to the velocity gradient is the property *viscosity* represented by the symbol μ.

Liquids and gases are recognized as states of matter. In the liquid state, a fluid is relatively incompressible, having a definite volume. It is also capable of forming a free surface interface between itself and its vapor or any other fluid with which it does not mix. On the other hand, a gas is highly compressible. It expands or diffuses indefinitely and is subject only to the limitations of gravitational forces or an enclosing vessel.

The term *vapor* generally implies a gas near saturation conditions where the liquid and the gas phase coexist at essentially the same temperature and pressure, during a process such as vaporization or boiling. In a similar sense the term *gas* denotes a highly superheated steam. Sometimes steam may be treated as an ideal gas and careful judgment is needed when doing so.

Fluid dynamics principles normally consider the fluid to be a continuous region of matter, a continuum, and a molecular model is not required except for rare instances. However, one property is noteworthy to consider due to the effect on steam generation fluid flow and due to intermolecular forces. *Surface tension*, σ, is a liquid property of the vapor-liquid interface and is the energy per unit area required to extend the interface. Surface tension is important in two-phase systems, such as a mixture flowing in a boiler tube, and relates to the shape and flow regime of the bubble interface and also to the heat transfer area of droplets. Vapor bubbles increase the resistance to fluid flow. The surface tension of water is dependent on temperature and its value goes to zero at the critical temperature (705.47 F, 374.15C). Supercritical water is considered single phase in fluid dynamic analysis due to zero surface tension.

The recommended correlation[1] for the surface tension of water and its vapor, σ, is:

$$\sigma = 235.8 \times 10^{-3}\, N/m \left[\frac{(T_c - T)}{T}\right]^{1.256} \left[1 - 0.625\left(\frac{T_c - T}{T}\right)\right] \quad (1)$$

where T_c = 647.15K and T is the fluid temperature in K.

Water in steam generators operating at supercritical pressure (above 3200.1 psia, 22.1 MPa) will behave as a single phase fluid converting from liquid to steam without creating bubbles. At the critical pressure and critical temperature, the density of water and steam are identical and there is no distinguishable interface at equilibrium conditions. Surface tension is also related to the latent heat of vaporization which also decreases to zero at the critical temperature.[2] This chapter discusses single phase fluid flow. Chapter 5 pertains to two-phase fluid flow that occurs in boiling tube circuits.

Fundamental relationships

Three fundamental laws of conservation apply to fluid dynamic systems: conservation of mass, momentum and energy. With the exception of nuclear reactions where minute quantities of mass are converted into energy, these laws must be satisfied in all flowing systems. Fundamental mathematical relationships for these principles are presented in several different forms that may be applied in particular fluid dynamic situations to provide an appropriate solution method. However, full analytical solutions are frequently too complex without the use of a computer. Simplified forms of the full equations can be derived by applying engineering judgment to drop negligible terms and consider only terms of significant magnitude for certain classes of problems. Fluid dynamics problems can be classified as compressible or incompressible, viscous or inviscid. Engineering practice is based upon applying various assumptions and empirical relationships in order to obtain a practical method of solution. A more complete discussion of the derivation of these conservation law relationships and vector notation representing three dimensional spaces may be found in References 3, 4, 5 and 6.

Conservation of mass

The *law of conservation of mass* simply states that the rate of change in mass stored in a system must equal the difference in the mass flowing into and out of the system. The *continuity equation* of mass for one dimensional single phase flow in a variable area channel or stream tube is:

$$A\frac{\partial \rho}{\partial t} + AV\frac{\partial \rho}{\partial x} + \rho V\frac{\partial A}{\partial x} + \rho A\frac{\partial V}{\partial x} = 0 \tag{2}$$

In its simplest form in x, y and z three dimensional Cartesian coordinates, conservation of mass for a small fixed control volume is:

$$\frac{\partial}{\partial x}\rho u + \frac{\partial}{\partial y}\rho v + \frac{\partial}{\partial z}\rho w = -\frac{\partial \rho}{\partial t} \tag{3}$$

where u, v and w are the fluid velocities in the x, y and z coordinate directions; t is time and ρ is the fluid density. An important form of this equation is derived by assuming steady-state ($\partial / \partial t = 0$) and incompressible (constant density) flow conditions:

$$\frac{\partial u}{\partial x} + \frac{\partial v}{\partial y} + \frac{\partial w}{\partial z} = 0 \tag{4}$$

Although no liquid is truly incompressible, the assumption of incompressibility simplifies problem solutions and is frequently acceptable for engineering practice considering water and oils.

Another relationship useful in large scale pipe flow systems involves the integration of Equation 3 around the flow path for constant density, steady-state conditions. For only one inlet (subscript 1) and one outlet (subscript 2):

$$\dot{m} = \rho_1 A_1 V_1 = \rho_2 A_2 V_2 \tag{5}$$

where ρ is the average density, V is the average velocity, A is the cross-sectional area, and $\dot{m}$ is the mass flow rate.

Conservation of momentum

The *law of conservation of momentum* is a representation of Newton's Second Law of Motion – the mass of a particle times its acceleration is equal to the sum of all of the forces acting on the particle. In a flowing system, the equivalent relationship for a fixed (control) volume becomes: the rate of change in momentum entering and leaving the control volume is equal to the sum of the forces acting on the control volume.

The conservation of momentum for one dimensional single phase flow in a variable area channel or stream tube is:

$$\frac{1}{g_c}\left[\frac{\partial G}{\partial t} + \frac{1}{A}\frac{\partial}{\partial x}\left(\frac{G^2 A}{\rho}\right)\right] + \frac{\tau\ P_f}{A} + \frac{g}{g_c}\rho\sin\theta + \frac{\partial P}{\partial x} = 0 \tag{6}$$

where

P = pressure, psia (MPa)
G = mass flux, $G = \rho V$, lb/h ft^2 (kg/s m^2)
A = flow area of channel ft^2 (m^2)
ρ = density lb/ft^3 (kg/m^3)
τ = wall shear stress, lb/ft^2 (N/m^2) (refer to Equation 26)
P_f = channel wetted perimeter, ft (m)
g = 32.17 ft /s^2 (9.8 m /s^2)
g_c = 32.17 lbm ft/lbf s^2 (1 kg m /N s^2)
θ = angle of channel inclination for x distance

This relationship is useful in calculating steam generator tube circuit pressure drop.

The conservation of momentum is a vector equation and is direction dependent, resulting in one equation for each coordinate direction (x, y and z for Cartesian coordinates), providing three momentum equations for each scaler velocity component, u, v and w.

The full mathematical representation of the momentum equation is complex and is of limited direct use in many engineering applications, except for numerical computational models. As an example, in the x coordinate direction, the full momentum equation becomes:

$$\begin{aligned} &\rho\left(\frac{\partial u}{\partial t} + u\frac{\partial u}{\partial x} + v\frac{\partial u}{\partial y} + w\frac{\partial u}{\partial z}\right) && \text{Term 1} \\ &= \rho f_x && \text{Term 2} \\ &- \frac{\partial P}{\partial x} && \text{Term 3} \\ &+ \frac{\partial}{\partial x}\left[\frac{2}{3}\mu\left(2\frac{\partial u}{\partial x} - \frac{\partial v}{\partial y} - \frac{\partial w}{\partial z}\right)\right] && \text{Term 4} \\ &+ \frac{\partial}{\partial y}\left[\mu\left(\frac{\partial v}{\partial x} + \frac{\partial u}{\partial y}\right)\right] \\ &+ \frac{\partial}{\partial z}\left[\mu\left(\frac{\partial w}{\partial x} + \frac{\partial u}{\partial z}\right)\right] \end{aligned} \tag{7}$$

where f_x is the body force in the x direction, P is the pressure, and μ is the viscosity. This equation and the corresponding equations in the y and z Cartesian coordinates represent the Navier-Stokes equations which are valid for all compressible Newtonian fluids with variable viscosity. *Term 1* is the rate of momentum change. *Term 2* accounts for body force effects such as gravity. *Term 3* accounts for the pressure gradient. The balance of the equation accounts for momentum change due to viscous transfer. *Term 1* is sometimes abbreviated as $\rho(Du/Dt)$ where Du/Dt is defined as the substantial derivative of u. For a function β (scaler or vector), D/Dt is the substantial derivative operator on function β defined as:

$$\frac{D\beta}{Dt} = \frac{\partial \beta}{\partial t} + u\frac{\partial \beta}{\partial x} + v\frac{\partial \beta}{\partial y} + w\frac{\partial \beta}{\partial z} = \frac{\partial \beta}{\partial t} + \mathbf{v} \cdot \nabla \beta \tag{8}$$

where the vector gradient or **grad** or **del** operator on function β is defined as:

$\nabla\beta$ or **grad** β or **del** $\beta = \mathbf{i}\ \partial\beta/\partial x + \mathbf{j}\ \partial\beta/\partial y + \mathbf{k}\ \partial\beta/\partial z$

For the special case of constant density and viscosity, this equation reduces to (for the x coordinate direction):

$$\frac{Du}{Dt} = f_x - \frac{1}{\rho}\frac{\partial P}{\partial x} + \frac{\mu}{\rho}\left(\frac{\partial^2 u}{\partial x^2} + \frac{\partial^2 u}{\partial y^2} + \frac{\partial^2 u}{\partial z^2}\right) \tag{9}$$

The y and z coordinate equations can be developed by substituting appropriate parameters for velocity u, pressure gradient $\partial P/\partial x$, and body force f_x. Where viscosity effects are negligible ($\mu = 0$), the Euler equation of momentum is produced (x direction only shown):

$$\frac{Du}{Dt} = f_x - \frac{1}{\rho}\frac{\partial P}{\partial x} \tag{10}$$

Energy equation (first law of thermodynamics)

The *law of conservation of energy* for nonreacting fluids states that the energy transferred into a system less the mechanical work done by the system must be equal to the rate of change in stored energy, plus the energy flowing out of the system with a fluid, minus the energy flowing into the system with a fluid. A single scaler equation results. The one dimensional single phase flow energy equation for a variable area channel or stream tube is:

$$\rho\frac{\partial H}{\partial t} + G\frac{\partial H}{\partial x} = q''\frac{P_H}{A} + q''' + \frac{1}{J}\frac{\partial P}{\partial \tau} \tag{11}$$

where

P = pressure, psia (MPa)
G = mass flux, lb/h ft^2 (kg/s m^2)
A = flow area of channel, ft^2 (m^2)
ρ = density, lb/ft^3 (kg/m^3)
τ = wall shear stress, lb/ft^2 (N/m^2)
P_H = channel heated area, ft^2 (m^2)
x = channel distance, ft (m) for x distance
H = enthalpy, Btu/lb (kJ/kg)
J = mechanical equivalent of heat = 778.17 ft lbf/Btu (1 N m/J)
q'' = heat flux at boundary, Btu/h ft^2 (W/m^2)
q''' = internal heat generation, Btu/h ft^3 (W/m)

A general form of the energy equation for a flowing system using an enthalpy based formulation and vector notation is:

$$\rho\frac{DH}{Dt} = q''' + \frac{DP}{Dt} + \nabla \cdot k\nabla T + \frac{\mu}{g_c}\Phi \tag{12}$$

Term 1 Term 2 Term 3 Term 4 Term 5

where ρ is the fluid density, H is the enthalpy per unit mass of a fluid, T is the fluid temperature, q''' is the internal heat generation, k is the thermal conductivity, and Φ is the dissipation function for irreversible work.[6] *Term 1* accounts for net energy convected into the system, *Term 2* accounts for internal heat generation, *Term 3* accounts for work done by the system, *Term 4* addresses heat conduction, and *Term 5* accounts for viscous dissipation.

As with the momentum equations, the full energy equation is too complex for most direct engineering applications except for use in numerical models. (See Chapter 6.) As a result, specialized forms are based upon various assumptions and engineering approximations. As discussed in Chapter 2, the most common form of the energy equation for a simple, inviscid (i.e., frictionless) steady-state flow system with flow in at location 1 and out at location 2 is:

$$JQ - W = J(u_2 - u_1) + (P_2 v_2 - P_1 v_1) + \frac{1}{2g_c}(V_2^2 - V_1^2) + (Z_2 - Z_1)\frac{g}{g_c} \tag{13a}$$

or

$$JQ - W = J(H_2 - H_1) + \frac{1}{2g_c}(V_2^2 - V_1^2) + (Z_2 - Z_1)\frac{g}{g_c} \tag{13b}$$

where

Q = heat added to the system, Btu lbm (J/kg) (See Note below)
W = work done by the system, ft-lbf/lbm (N m/kg)
J = mechanical equivalent of heat = 778.17 ft lbf/Btu (1 N m/J)
u = internal energy, Btu/lbm (J/kg)
P = pressure, lbf/ft^2 (N/m^2)
v = specific volume, ft^3/lbm (m^3/kg)
V = velocity, ft /s (m/s)
Z = elevation, ft (m)
H = enthalpy = $u + Pv/J$, Btu/lbm (J/kg)
g = 32.17 ft /s^2 (9.8 m /s^2)
g_c = 32.17 lbm ft/lbf s^2 (1 kg m /N s^2)

Note: Where required for clarity, the abbreviation lb is augmented by f (lbf) to indicate pound force and by m (lbm) to indicate pound mass. Otherwise lb is used with force or mass indicated by the context.

Energy equation applied to fluid flow (pressure loss without friction)

The conservation laws of mass and energy, when simplified for steady, frictionless (i.e., inviscid) flow of an incompressible fluid, result in the mechanical energy balance referred to as Bernoulli's equation:

$$P_1 v + Z_1 \frac{g}{g_c} + \frac{V_1^2}{2g_c} = P_2 v + Z_2 \frac{g}{g_c} + \frac{V_2^2}{2g_c} \quad \textbf{(14)}$$

The variables in Equation 14 are defined as follows with the subscripts referring to location 1 and location 2 in the system:

P = pressure, lbf/ft^2 (N/m^2)
v = specific volume of fluid, ft^3/lbm (m^3/kg)
Z = elevation, ft (m)
V = fluid velocity, ft/s (m/s)

Briefly, Equation 14 states that the total mechanical energy present in a flowing fluid is made up of pressure energy, gravity energy and velocity or kinetic energy; each is mutually convertible into the other forms. Furthermore, the total mechanical energy is constant along any stream-tube, provided there is no friction, heat transfer or shaft work between the points considered. This stream-tube may be an imaginary closed surface bounded by stream lines or it may be the wall of a flow channel, such as a pipe or duct, in which fluid flows without a free surface.

Applications of Equation 14 are found in flow measurements using the velocity head conversion resulting from flow channel area changes. Examples are the venturi, flow nozzle and various orifices. Also, pitot tube flow measurements depend on being able to compare the total head, $Pv + Z + (V^2/2\,g_c)$, to the static head, $Pv + Z$, at a specific point in the flow channel. Descriptions of metering instruments are found in Chapter 40. Bernoulli's equation, developed from strictly mechanical energy concepts some 50 years before any precise statement of thermodynamic laws, is a special case of the conservation of energy equation or first law of thermodynamics in Equations 13a and b.

Applications of Equation 13 to fluid flow are given in the examples on water and compressible fluid flow through a nozzle under the *Applications of the Energy Equation* section in Chapter 2. Equation 18, Chapter 2 is:

$$V_2 = \sqrt{2g_c J(H_1 - H_2)} = C\sqrt{H_1 - H_2} \quad \textbf{(15)}$$

where

V_2 = downstream velocity, ft/s (m/s)
g_c = 32.17 lbm ft/lbf s^2 = 1 kg m/Ns^2
J = 778.26 ft lbf/Btu = 1 Nm/J
H_1 = upstream enthalpy, Btu/lb (J/kg)
H_2 = downstream enthalpy, Btu/lb (J/kg)
C = $223.8\sqrt{\text{lbm/Btu}}$ × ft/s ($1.414\sqrt{\text{kg/J}}$ × m/s)

This equation relates fluid velocity to a change in enthalpy under *adiabatic* (no heat transfer), steady, *inviscid* (no friction) flow where no work, local irreversible flow pressure losses, or change in elevation occurs. The initial velocity is assumed to be zero and compressible flow is permitted. If the temperature (T) and pressure (P) of steam or water are known at points 1 and 2, Equation 15 provides the exit velocity using the enthalpy *(H)* values provided in Tables 1, 2 and 3 of Chapter 2. If the pressure and temperature at point 1 are known but only the pressure at point 2 is known, the outlet enthalpy (H_2) can be evaluated by assuming constant entropy expansion from points 1 to 2, i.e., $S_1 = S_2$.

Ideal gas relationships

There is another method that can be used to determine velocity changes in a frictionless adiabatic expansion. This method uses the ideal gas equation of state in combination with the pressure-volume relationship for constant entropy.

From the established gas laws, the relationship between pressure, volume and temperature of an ideal gas is expressed by:

$$Pv = RT \quad \textbf{(16a)}$$

or

$$Pv = \frac{\mathbf{R}}{M} T \quad \textbf{(16b)}$$

where

P = absolute pressure, lb/ft^2 (N/m^2)
v = specific volume, ft^3/lb of gas (m^3/kg)
M = molecular weight of the gas, lb/lb-mole (kg/kg-mole)
T = absolute temperature, R (K)
R = gas constant for specific gas, ft lbf/lbm R (N m/kg K)
MR = $\mathbf{R}$ = the universal gas constant
= 1545 ft lb/lb-mole R (8.3143 kJ/kg-mole K)

The relationship between pressure and specific volume along an expansion path at constant entropy, i.e., isentropic expansion, is given by:

$$Pv^k = \text{constant} \quad \textbf{(17)}$$

Because P_1 and v_1 in Equation 13 are known, the constant can be evaluated from $P_1 v_1{}^k$. The exponent k is constant and is evaluated for an ideal gas as:

$$k = c_p / c_v = \text{specific heat ratio} \quad \textbf{(18)}$$

where

c_p = specific heat at constant pressure, Btu/lb F (J/kg K)
c_v = specific heat at constant volume, Btu/lb F (J/kg K)
= $(u_1 - u_2)/(T_1 - T_2)$

For a steady, adiabatic flow with no work or change in elevation of an ideal gas, Equations 13, 16, 17 and 18 can be combined to provide the following relationship:

$$V_2^2 - V_1^2 = 2g_c \left(\frac{k}{k-1}\right) P_1 v_1 \left\{1 - \left(\frac{P_2}{P_1}\right)^{\frac{k-1}{k}}\right\} \quad \textbf{(19)}$$

When V_1 is set to zero and using English units Equation 19 becomes:

$$V_2 = 8.02\sqrt{\left(\frac{k}{k-1}\right)P_1 v_1 \left\{1-\left(\frac{P_2}{P_1}\right)^{\frac{k-1}{k}}\right\}}\text{, ft/s} \quad \textbf{(20)}$$

Equations 19 and 20 can be used for gases in pressure drop ranges where there is little change in k, provided values of k are known or can be calculated. Equation 20 is widely used in evaluating gas flow through orifices, nozzles and flow meters.

It is sufficiently accurate for most purposes to determine velocity differences caused by changes in flow area by treating a compressible fluid as incompressible. This assumption only applies when the difference in specific volumes at points 1 and 2 is small compared to the final specific volume. The accepted practice is to consider the fluid incompressible when:

$$(v_2 - v_1) / v_2 < 0.05 \quad \textbf{(21)}$$

Because Equation 14 represents the incompressible energy balance for frictionless adiabatic flow, it may be rearranged to solve for the velocity difference as follows:

$$V_2^2 - V_1^2 = 2g_c\left[\Delta(Pv) + \Delta Zg / g_c\right] \quad \textbf{22)}$$

where

$\Delta(Pv)$ = pressure head difference between locations 1 and 2 = $(P_1 - P_2)\,v$, ft (m)
ΔZ = head (elevation) difference between locations 1 and 2, ft (m)
V = velocity at locations 1 and 2, ft/s (m/s)

When the approach velocity is approximately zero, Equation 22 in English units becomes:

$$V_2 = \sqrt{2gh} = 8.02\sqrt{h}\text{, ft/s} \quad \textbf{(23)}$$

In this equation, h, in ft head of the flowing fluid, replaces $\Delta(Pv) + \Delta Z$. If the pressure difference is measured in psi, it must be converted to lb/ft^2 to obtain Pv in ft.

Pressure loss from fluid friction

So far, only pressure changes associated with the kinetic energy term, $V^2/2\,g_c$, and static pressure term, Z, have been discussed. These losses occur at constant flow where there are variations in flow channel cross-sectional area and where the inlet and outlet are at different elevations. Fluid friction and, in some cases heat transfer with the surroundings, also have important effects on pressure and velocity in a flowing fluid. The following discussion applies to fluids flowing in channels without a free surface.

When a fluid flows, molecular diffusion causes momentum interchanges between layers of the fluid that are moving at different velocities. These interchanges are not limited to individual molecules. In most flow situations there are also bulk fluid interchanges known as eddy diffusion. The net result of all inelastic momentum exchanges is exhibited in shear stresses between adjacent layers of the fluid. If the fluid is contained in a flow channel, these stresses are eventually transmitted to the walls of the channel. To counterbalance this wall shear stress, a pressure gradient proportional to the bulk kinetic energy, $V^2 / 2\,g_c$, is established in the fluid in the direction of the bulk flow. The force balance is:

$$\pi\frac{D^2}{4}(dP) = \tau_w \pi\, D(dx) \quad \textbf{(24)}$$

where

D = tube diameter or hydraulic diameter D_h ft (m)
D_h = 4 × (flow area)/(wetted perimeter) for circular or noncircular cross-sections, ft (m)
dx = distance in direction of flow, ft (m)
τ_w = shear stress at the tube wall, lb/ft^2 (N/m^2)

Solving Equation 24 for the pressure gradient (dP / dx):

$$\frac{dP}{dx} = \frac{4}{D}\tau_w \quad \textbf{(25)}$$

This pressure gradient along the length of the flow channel can be expressed in terms of a certain number of velocity heads, f, lost in a length of pipe equivalent to one tube diameter. The symbol f is called the friction factor, which has the following relationship to the shear stress at the tube wall:

$$\tau_w = \frac{f}{4}\frac{1}{v}\frac{V^2}{2g_c} \quad \textbf{(26)}$$

Equation 25 can be rewritten, substituting for τ_w from Equation 26 as follows:

$$\frac{dP}{dx} = \frac{4}{D}\left(\frac{f}{4}\frac{1}{v}\frac{V^2}{2g_c}\right) = \frac{f}{D}\frac{1}{v}\frac{V^2}{2g_c} \quad \textbf{(27)}$$

The general energy equation, Equation 13, expressed as a differential has the form:

$$du + \frac{VdV}{g_c} + d(Pv) = dQ - dW_k \quad \textbf{(28a)}$$

or

$$du + \frac{VdV}{g_c} + Pdv + vdP = dQ - dW_k \quad \textbf{(28b)}$$

Substituting Equation 26 of Chapter 2 ($du = Tds - Pdv$) in Equation 28 yields:

$$Tds + \frac{VdV}{g_c} + vdP = dQ - dW_k \quad \textbf{(29)}$$

The term Tds represents heat transferred to or from the surroundings, dQ, and any heat added internally to the fluid as the result of irreversible processes. These processes include fluid friction or any irreversible pressure losses resulting from fluid flow. (See Equation 29 and explanation, Chapter 2.) Therefore:

$$Tds = dQ + dQ_F \tag{30}$$

where dQ_F is the heat equivalent of fluid friction and any *local irrecoverable pressure losses* such as those from pipe fittings, bends, expansions or contractions.

Substituting Equation 30 into Equation 29, canceling dQ on both sides of the equation, setting dW_k equal to 0 (no shaft work), and rearranging Equation 29 results in:

$$dP = -\frac{VdV}{\upsilon g_c} - \frac{dQ_F}{\upsilon} \tag{31}$$

Three significant facts should be noted from Equation 31 and its derivation. First, the general energy equation does not accommodate pressure losses due to fluid friction or geometry changes. To accommodate these losses Equation 31 must be altered based on the first and second laws of thermodynamics (Chapter 2). Second, Equation 31 does not account for heat transfer except as it may change the specific volume, υ, along the length of the flow channel. Third, there is also a pressure loss as the result of a velocity change. This loss is independent of any flow area change but is dependent on specific volume changes. The pressure loss is due to acceleration which is always present in compressible fluids. It is generally negligible in incompressible flow without heat transfer because friction heating has little effect on fluid temperature and the accompanying specific volume change.

Equation 27 contains no acceleration term and applies only to friction and local pressure losses. Therefore, dQ_F/υ in Equation 31 is equivalent to dP of Equation 27, or:

$$\frac{dQ_F}{\upsilon} = f\frac{dx}{D}\frac{V^2}{\upsilon 2g_c} \tag{32}$$

Substitution of Equation 32 into Equation 31 yields:

$$dP = -\frac{VdV}{\upsilon g_c} - \frac{f}{D}\frac{V^2}{\upsilon 2g_c}dx \tag{33}$$

From Equation 5, the continuity equation permits definition of the mass flux, G, or mass velocity or mass flow rate per unit area [lb/h ft^2 (kg/m^2 s)] as:

$$\frac{V}{\upsilon} = G = \text{constant} \tag{34}$$

Substituting Equation 34 into Equation 33 for a flow channel of constant area:

$$dP = -2\frac{G^2}{2g_c}d\upsilon - f\frac{G^2}{2g_c}\frac{\upsilon}{D}dx \tag{35}$$

Integrating Equation 35 between points 1 and 2, located at x = 0 and x = L, respectively:

$$P_1 - P_2 = 2\frac{G^2}{2g_c}(\upsilon_2 - \upsilon_1) + f\frac{G^2}{2g_c}\frac{1}{D}\int_0^L \upsilon dx \tag{36}$$

The second term on the right side of Equation 36 may be integrated provided a functional relationship between υ and x can be established. For example, where the heat absorption rate over the length of the flow channel is constant, temperature T is approximately linear in x, or:

$$dx = \frac{L}{T_2 - T_1}dT \tag{37}$$

and

$$\int_0^L \upsilon dx = \frac{L}{T_2 - T_1}\int_1^2 \upsilon dT = L\upsilon_{av} \tag{38}$$

The term υ_{av} is an average specific volume with respect to temperature, T.

$$\upsilon_{av} = \phi(\upsilon_2 + \upsilon_1) = \phi\upsilon_1(\upsilon_R + 1) \tag{39}$$

where

$\upsilon_R = \upsilon_2/\upsilon_1$

ϕ = averaging factor

In most engineering evaluations, υ is almost linear in T and $\phi \approx 1/2$. Combining Equations 36 and 37, and rewriting $\upsilon_2 - \upsilon_1$ as $\upsilon_1(\upsilon_R - 1)$:

$$P_1 - P_2 = 2\frac{G^2}{2g_c}\upsilon_1(\upsilon_R - 1) + f\frac{L}{D}\frac{G^2}{2g_c}\upsilon_1\phi(\upsilon_R + 1) \tag{40}$$

Equation 40 is completely general. It is valid for compressible and incompressible flow in pipes of constant cross-section as long as the function $T = F(x)$ can be assigned. The only limitation is that dP/dx is negative at every point along the pipe. Equation 33 can be solved for dP/dx making use of Equation 34 and the fact that $P_1\upsilon_1$ can be considered equal to $P_2\upsilon_2$ for adiabatic flow over a short section of tube length. The result is:

$$\frac{dP}{dx} = \frac{Pf/2D}{1 - \frac{g_c P\upsilon}{V^2}} \tag{41}$$

At any point where $V^2 = g_c P\upsilon$, the flow becomes choked because the pressure gradient is positive for velocities greater than $(g_c P\upsilon)^{0.5}$. The flow is essentially choked by excessive stream expansion due to the drop in pressure. The minimum downstream pressure that is effective in producing flow in a channel is:

$$P_2 = V^2/\upsilon_2 g_c = \upsilon_2 G^2/g_c \tag{42}$$

Dividing both sides of Equation 40 by $G^2\upsilon_1/2g_c$, the pressure loss is expressed in terms of velocity heads. One velocity head equals:

$$\Delta P \text{ (one velocity head)} = \frac{V^2}{2g_c C\upsilon} = \frac{\rho V^2}{2g_c C} \tag{43}$$

where

ΔP = pressure drop equal to one velocity head, lb/in.^{2} (N/m^2)
V = velocity, ft/s (m/s)
υ = specific volume, ft^3/lb (m^3/kg)
g_c = 32.17 lbm ft/lbf s^2 = 1 kg m/N s^2
C = 144 in.^{2}/ft^2 (1 m^2/m^2)
ρ = density, lb/ft^3 (kg/m^3)

In either case, f represents the number of velocity heads (N_{vh}) lost in each diameter length of pipe.

The dimensionless parameter defined by the pressure loss divided by twice Equation 43 is referred to as the Euler number:

$$Eu = \Delta P / \left(\rho V^2 / g_c\right) \quad \textbf{(44)}$$

where ρ is the density, or $1/\upsilon$.

Two other examples of integrating Equation 35 have wide applications in fluid flow. First, adiabatic flow through a pipe is considered. Both H and D are constant and $P_1\upsilon_1^m = P_2\upsilon_2^m$ where m is the exponent for constant enthalpy. Values of m for steam range from 0.98 to 1.0. Therefore, the assumption $P\upsilon$ = constant = $P_1\upsilon_1$ is sufficiently accurate for pressure drop calculations. This process is sometimes called isothermal pressure drop because a constant temperature ideal gas expansion also requires a constant enthalpy. For $P\upsilon = P_1\upsilon_1$, the integration of Equation 35 reduces to:

$$P_1 - P_2 = 2\frac{G^2}{2g_c}\frac{2\upsilon_1\upsilon_2}{\upsilon_1+\upsilon_2}\ell n\left(\frac{\upsilon_2}{\upsilon_1}\right) + f\frac{L}{D}\frac{G^2}{2g_c}\frac{2\upsilon_1\upsilon_2}{\upsilon_1+\upsilon_2} \quad \textbf{(45)}$$

Neither P_2 nor υ_2 are known in most cases, therefore Equation 45 is solved by iteration. Also, the term $2\upsilon_1\upsilon_2/(\upsilon_1+\upsilon_2)$ can usually be replaced by the numerical average of the specific volumes – $\upsilon_{av} = {}^1/_2\,\upsilon_1(P_R+1)$ where $P_R = P_1/P_2 = \upsilon_2/\upsilon_1$. The maximum high side error at P_R = 1.10 is 0.22% and this increases to 1.3% at P_R = 1.25. It is common practice to use a numerical average for the specific volume in most fluid friction pressure drop calculations. However, where the lines are long, P_2 should be checked by Equation 42. Also, where heat transfer is taking place, P_2 is seldom constant along the flow channel and appropriate averaging factors should be used. Computation using small zone subdivisions along the length of the tube circuit is recommended to limit errors in widely varying property values.

The second important example considering flow under adiabatic conditions assumes an almost incompressible fluid, i.e., υ_1 is approximately equal to υ_2. (See Equation 21.) Substituting υ for υ_1 and υ_2 in Equation 45, the result is:

$$P_1 - P_2 = f\frac{L}{D}\frac{G^2}{2g_c}\upsilon \quad \textbf{(46)}$$

All terms in Equations 45 and 46 are expressed in consistent units. However, it is general practice and often more convenient to use mixed units. For example, a useful form of Equation 46 in English units is:

$$\Delta P = f\frac{L}{D_e}\upsilon\left(\frac{G}{10^5}\right)^2 \quad \textbf{(47)}$$

where

ΔP = fluid pressure drop, psi
f = friction factor from Fig. 1, dimensionless
L = length, ft
D_e = equivalent diameter of flow channel, in. (note units)
υ = specific volume of fluid, ft^3/lb
G = mass flux of fluid, lb/h ft^2

Friction factor

The friction factor (f) introduced in Equation 26, is defined as the dimensionless fluid friction loss in velocity heads per diameter length of pipe or equivalent diameter length of flow channel. Earlier correlators in this field, including Fanning, used a friction factor one fourth the magnitude indicated by Equation 26. This is because the shear stress at the wall is proportional to one fourth the velocity head. All references to f in this book combine the factor 4 in Equation 25 with f as has been done by Darcy, Blasius, Moody and others.

The friction factor is plotted in Fig. 1 as a function of the Reynolds number, a dimensionless group of variables defined as the ratio of inertial forces to viscous forces. The Reynolds number (Re) can be written:

$$\text{Re} = \frac{\rho V D_e}{\mu} \text{ or } \frac{V D_e}{\nu} \text{ or } \frac{G D_e}{\mu} \quad \textbf{(48)}$$

where

ρ = density of fluid, lbm/ft^3 (kg/m^3)
ν = kinematic viscosity = μ/ρ, ft^2/h (m^2/s)
μ = viscosity of fluid, lbm/ft h (kg/m s)
V = velocity of fluid, ft/h (m/s)
G = mass flux of fluid, lb/h ft^2 (kg/m^2 s)
D_e = equivalent diameter of flow channel, ft (m)

Fluid flow inside a closed channel occurs in a viscous or laminar manner at low velocity and in a turbulent manner at high velocities. Many experiments on fluid friction pressure drop, examined by dimensional analysis and the laws of similarity, have shown that the Reynolds number can be used to characterize a flow pattern. Examination of Fig. 1 shows that flow is laminar at Reynolds numbers less than 2000, generally turbulent at values exceeding 4000 and completely turbulent at higher values. Indeterminate conditions exist in the critical zone between Reynolds numbers of 2000 and 4000.

Fluid flow can be described by a system of simultaneous partial differential equations. (See earlier *Fundamental relationships* section.) However, due to the complexity of these equations, solutions are generally only available for the case of laminar flow, where the only momentum changes are on a molecular basis. For laminar flow, integration of the Navier-Stokes equation with velocity in the length direction only gives the following equation for friction factor:

$$f = 64/\text{Re} \quad \textbf{(49)}$$

The straight line in the laminar flow region of Fig. 1 is a plot of this equation.

It has been experimentally determined that the friction factor is best evaluated by using the Reynolds number to define the flow pattern. A factor ε/D_e is then introduced to define the relative roughness of the channel surface. The coefficient ε expresses the average height of roughness protrusions equivalent to the sand grain roughness established by Nikuradse.[6] The friction factor values in Fig. 1 and the ε/D_e values in Fig. 2 are taken from experimental data as correlated by Moody.[7]

Laminar flow

Laminar flow is characterized by the parallel flowing of individual streams like layers sliding over each other. There is no mixing between the streams except for molecular diffusion from one layer to the other. A small layer of fluid next to the boundary wall has zero velocity as a result of molecular adhesion forces. This establishes a velocity gradient normal to the main body of flow. Because the only interchanges of momentum in laminar flow are between the molecules of the fluid, the condition of the surface has no effect on the velocity gradient and therefore no effect on the friction factor. In commercial equipment, laminar flow is usually encountered only with more viscous liquids such as the heavier oils.

Turbulent flow

When turbulence exists, there are momentum interchanges between masses of fluid. These interchanges are induced through secondary velocities, irregular fluctuations or eddys, that are not parallel to the axis of the mean flow velocity. In this case, the condition of the boundary surface, roughness, does have an effect on the velocity gradient near the wall, which in turn affects the friction factor. Heat transfer is substantially greater with turbulent flow (Chapter 4) and, except for viscous liquids, it is common to induce turbulent flow with steam and water without

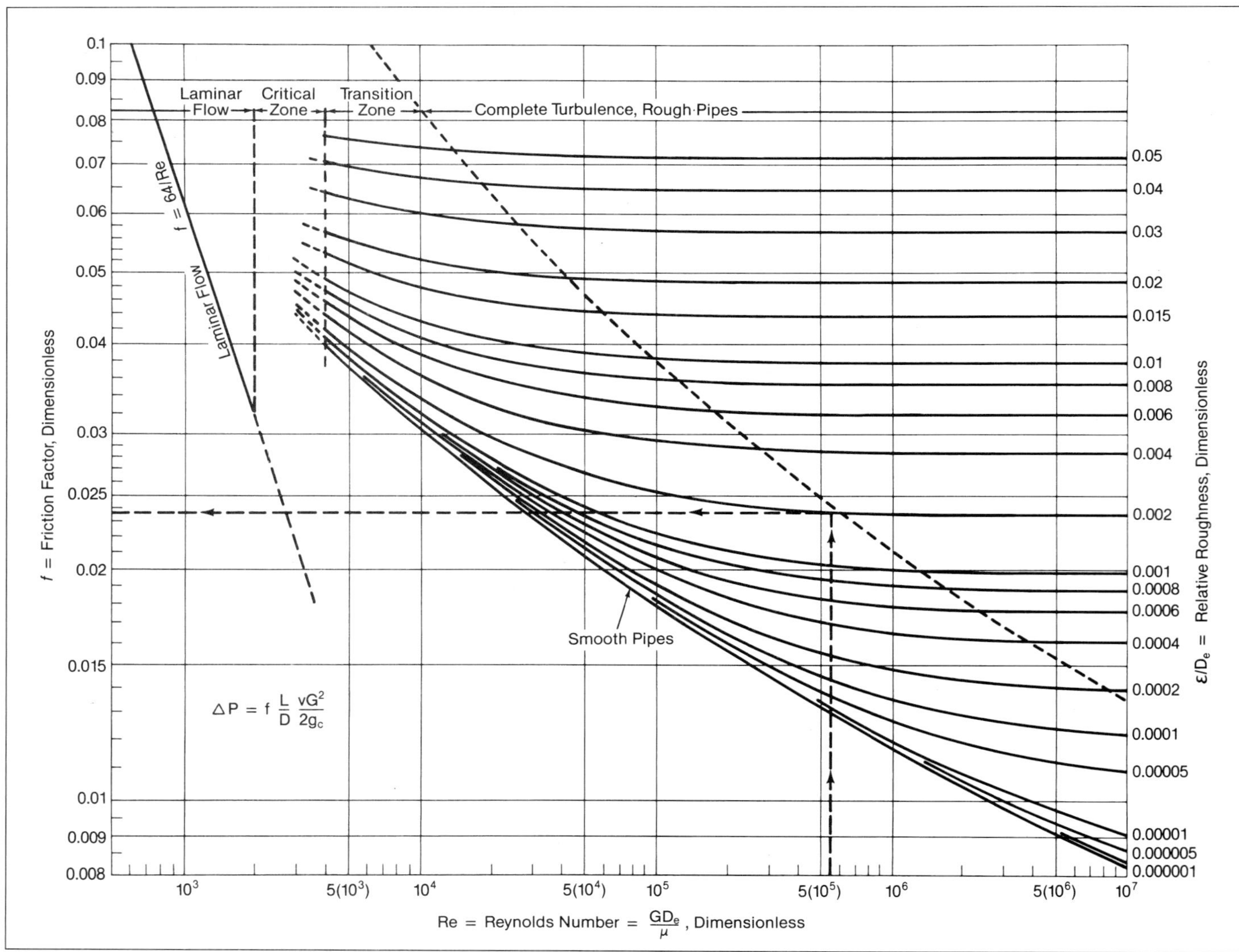

Fig. 1 Friction factor/Reynolds number relationship for determining pressure drop of fluids flowing through closed circuits (pipes and ducts).

excessive friction loss. Consequently, it is customary to design for Reynolds numbers above 4000 in steam generating units.

Turbulence fluctuations in the instantaneous velocity introduce additional terms to the momentum conservation equation called Reynolds stresses. These fluctuations influence the mean motion and increase the flow resistance in a manner producing an increase in the *apparent viscosity*. Analysis of turbulent flow must consider the impact of the fluctuating velocity component along with the mean flow velocity or resort to empirical methods that account for the additional momentum dissipation.[4,6,8]

Velocity ranges

Table 1 lists the velocity ranges generally encountered in the heat transfer equipment as well as in duct and piping systems of steam generating units. These values, plus the specific volumes from the ASME Steam Tables (see Chapter 2) and the densities listed in Tables 2 and 3 in this chapter, are used to establish mass velocities for calculating Reynolds numbers and fluid friction pressure drops. In addition, values of viscosity, also required in calculating the Reynolds number, are given in Figs. 3, 4 and 5 for selected liquids and gases. Table 4 lists the relationship between various units of viscosity.

Resistance to flow in valves and fittings

Pipelines and duct systems contain many valves and fittings. Unless the lines are used to transport fluids over long distances, as in the distribution of process steam at a factory or the cross country transmission of oil or gas, the straight runs of pipe or duct are relatively short. Water, steam, air and gas lines in a power plant have relatively short runs of straight pipe and many valves and fittings. Consequently, the flow resistance due to valves and fittings is a substantial part of the total resistance.

Methods for estimating the flow resistance in valves and fittings are less exact than those used in establishing the friction factor for straight pipes and ducts. In the latter, pressure drop is considered to be the result of the fluid shear stress at the boundary walls of the flow channel; this leads to relatively simple boundary value evaluations. On the other hand, pressure losses associated with valves, fittings and bends are mainly the result of impacts and inelastic exchanges

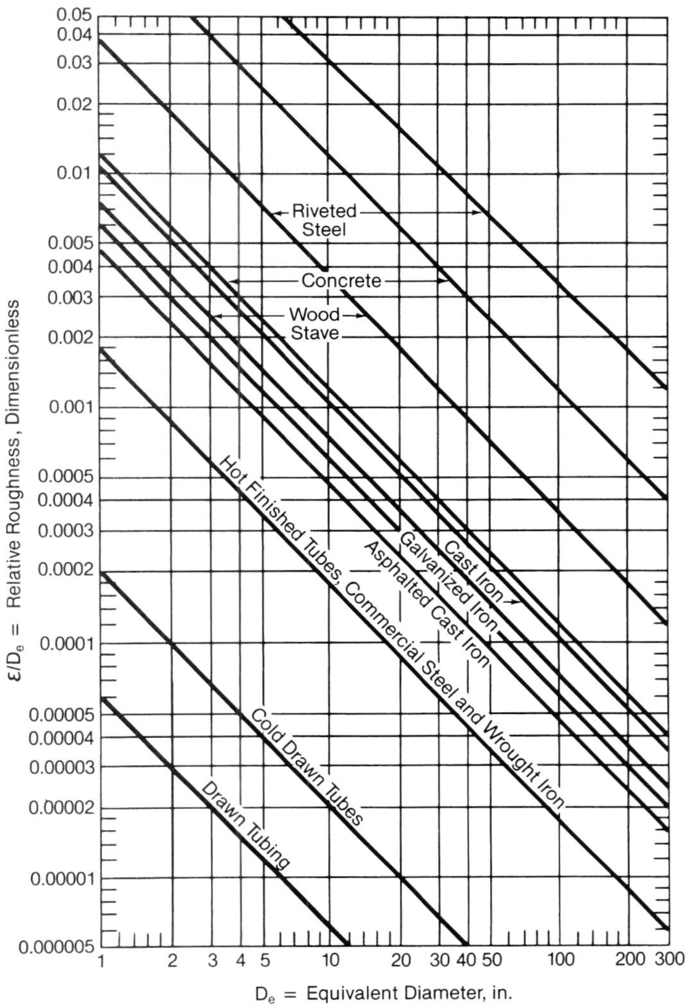

Fig. 2 Relative roughness of various conduit surfaces. (SI conversion: mm = 25.4 X in.)

Table 1
Velocities Common in Steam Generating Systems

Nature of Service	Velocity ft/min	Velocity m/s
Air:		
Air heater	1000 to 5000	5.1 to 25.4
Coal and air lines, pulverized coal	3000 to 4500	15.2 to 22.9
Compressed air lines	1500 to 2000	7.6 to 10.2
Forced draft air ducts	1500 to 3600	7.6 to 18.3
Forced draft air ducts, entrance to burners	1500 to 2000	7.6 to 10.2
Ventilating ducts	1000 to 3000	5.1 to 15.2
Crude oil lines [6 to 30 in. (152 to 762 mm)]	60 to 3600	0.3 to 18.3
Flue gas:		
Air heater	1000 to 5000	5.1 to 25.4
Boiler gas passes	3000 to 6000	15.2 to 30.5
Induced draft flues and breaching	2000 to 3500	10.2 to 17.8
Stacks and chimneys	2000 to 5000	10.2 to 25.4
Natural gas lines (large interstate)	1000 to 1500	5.1 to 7.6
Steam:		
Steam lines		
High pressure	8000 to 12,000	40.6 to 61.0
Low pressure	12,000 to 15,000	61.0 to 76.2
Vacuum	20,000 to 40,000	101.6 to 203.2
Superheater tubes	2000 to 5000	10.2 to 25.4
Water:		
Boiler circulation	70 to 700	0.4 to 3.6
Economizer tubes	150 to 300	0.8 to 1.5
Pressurized water reactors		
Fuel assembly channels	400 to 1300	2.0 to 6.6
Reactor coolant piping	2400 to 3600	12.2 to 18.3
Water lines, general	500 to 750	2.5 to 3.8

Table 2
Physical Properties of Liquids at 14.7 psi (0.101 MPa)

Liquid	Temperature F (C)	Density lb/ft^3 (kg/m^3)	Specific Heat Btu/lb F (kJ/kg C)
Water	70 (21)	62.4 (999.4)	1.000 (4.187)
	212 (100)	59.9 (959.3)	1.000 (4.187)
Automotive oil	70 (21)		
SAE 10		55 to 57 (881 to 913)	0.435 (1.821)
SAE 50		57 to 59 (913 to 945)	0.425 (1.779)
Mercury	70 (21)	846 (13,549)	0.033 (0.138)
Fuel oil, #6	70 (21)	60 to 65 (961 to 1041)	0.40 (1.67)
	180 (82)	60 to 65 (961 to 1041)	0.46 (1.93)
Kerosene	70 (21)	50 to 51 (801 to 817)	0.47 (1.97)

of momentum. These losses are frequently referred to as local losses or local nonrecoverable pressure losses. Even though momentum is conserved, kinetic energies are dissipated as heat. This means that pressure losses are influenced mainly by the geometries of valves, fittings and bends. As with turbulent friction factors, pressure losses are determined from empirical correlations of test data. These correlations may be based on equivalent pipe lengths, but are preferably defined by a multiple of velocity heads based on the connecting pipe or tube sizes. Equivalent pipe length calculations have the disadvantage of being dependent on the relative roughness (ε/D) used in the correlation. Because there are many geometries of valves and fittings, it is customary to rely on manufacturers for pressure drop coefficients.

It is also customary for manufacturers to supply valve flow coefficients (C_V) for 60F (16C) water. These are expressed as ratios of weight or volume flow in the fully open position to the square root of the pressure drop. These coefficients can be used to relate velocity head losses to a connecting pipe size by the following expression:

$$N_{\upsilon} = kD^4 / C_V^2 \tag{50}$$

Table 3
Physical Properties of Gases at 14.7 psi (0.101 MPa)**

Gas	Temperature F	Density, lb/ft^3	Instantaneous Specific Heat c_p Btu/lb F	c_v Btu/lb F	k, c_p/c_v
Air	70	0.0749	0.241	0.172	1.40
	200	0.0601	0.242	0.173	1.40
	500	0.0413	0.248	0.180	1.38
	1000	0.0272	0.265	0.197	1.34
CO_2	70	0.1148	0.202	0.155	1.30
	200	0.0922	0.216	0.170	1.27
	500	0.0634	0.247	0.202	1.22
	1000	0.0417	0.280	0.235	1.19
H_2	70	0.0052	3.440	2.440	1.41
	200	0.0042	3.480	2.490	1.40
	500	0.0029	3.500	2.515	1.39
	1000	0.0019	3.540	2.560	1.38
Flue gas*	70	0.0776	0.253	0.187	1.35
	200	0.0623	0.255	0.189	1.35
	500	0.0429	0.265	0.199	1.33
	1000	0.0282	0.283	0.217	1.30
CH_4	70	0.0416	0.530	0.406	1.30
	200	0.0334	0.575	0.451	1.27
	500	0.0230	0.720	0.596	1.21
	1000	0.0151	0.960	0.836	1.15

* From coal; 120% total air; flue gas molecular weight 30.
** SI conversions: T, C = 5/9 (F-32); ρ, kg/m^3 = 16.02 x lbm/ft^3; c_p, kJ/kg K = 4.187 x Btu/lbm F.

Table 4
Relationship Between Various Units of Viscosity

Part A: Dynamic (or Absolute) Viscosity, μ

Pa s $\frac{N\,s}{m^2} = \frac{kg}{m\,s}$	Centipoise $\frac{0.01\,g}{cm\,s}$	$\frac{lbm}{ft\,s}$	$\frac{lbm}{ft\,h}$	$\frac{lbf\,s}{ft^2}$
1.0	1000	672 x 10^{-3}	2420	20.9 x 10^{-3}
0.001	1.0	672 x 10^{-6}	2.42	20.9 x 10^{-6}
1.49	1488	1.0	3600	0.0311
413 x 10^{-6}	0.413	278 x 10^{-6}	1.0	8.6 x 10^{-6}
47.90	47,900	32.2	115,900	1.0

Part B: Kinematic Viscosity, ν = μ/ρ

$\frac{m^2}{s}$	Centistoke $\frac{0.01\,cm^2}{s}$	$\frac{ft^2}{s}$	$\frac{ft^2}{h}$
1.0	10^6	10.8	38,800
10^{-6}	1.0	10.8 x 10^{-6}	0.0389
92.9 x 10^{-3}	92,900	1.0	3600
25.8 x 10^{-6}	25.8	278 x 10^{-6}	1.0

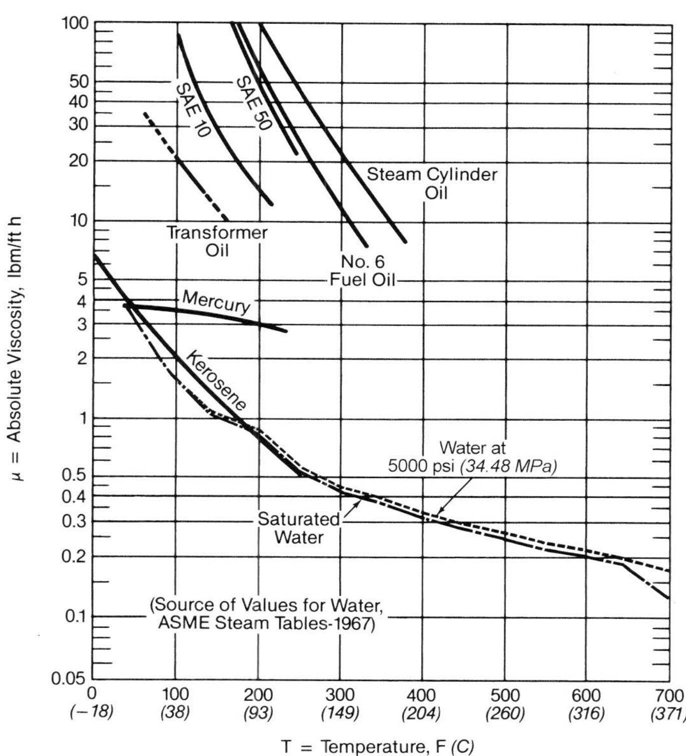

Fig. 3 Absolute viscosities of some common liquids (Pa s = 0.000413 X lbm/ft h).

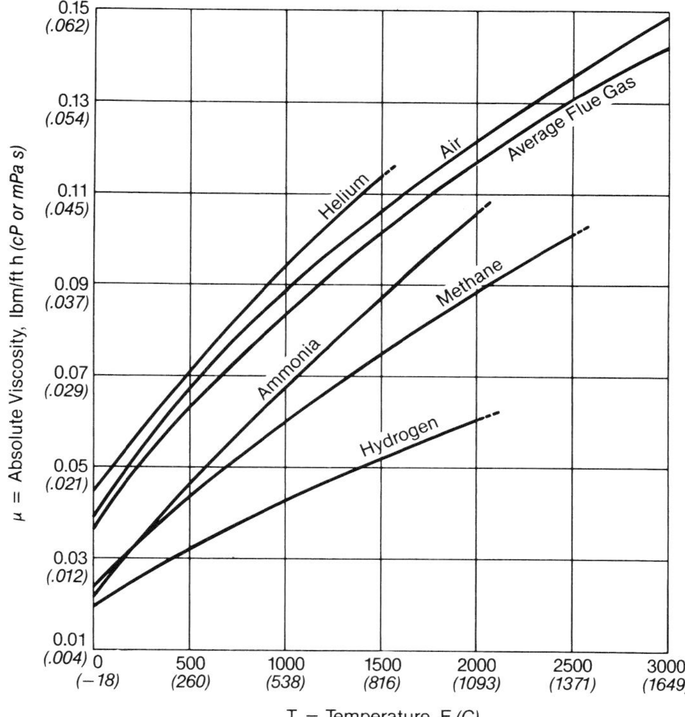

Fig. 4 Absolute viscosities of some common gases at atmospheric pressure.

where

N_υ = number of velocity heads, dimensionless
k = units conversion factor: for C_V based upon gal/min/$(\Delta\rho)^{1/2}$, $k = 891$
D = internal diameter of connecting pipe, in. (mm)
C_V = flow coefficient in units compatible with k and D: for $k = 891$, C_V = gal/min/$(\Delta\rho)^{1/2}$

C_V and corresponding values of N_υ for valves apply only to incompressible flow. However, they may be extrapolated for compressible condition using an average specific volume between P_1 and P_2 for ΔP values as high as 20% of P_1. This corresponds to a maximum pressure ratio of 1.25. The ΔP process for valves, bends and fittings is approximately isothermal and does not require the most stringent limits set by Equation 21.

When pressure drop can be expressed as an equivalent number of velocity heads, it can be calculated by the following formula in English units:

$$\Delta P = N_\upsilon \frac{\upsilon}{12}\left(\frac{G}{10^5}\right)^2 \qquad \textbf{(51)}$$

where

ΔP = pressure drop, lb/in.^{2}
N_υ = number of equivalent velocity heads, dimensionless
υ = specific volume, ft^3/lb
G = mass flux, lb/ft^2 h

Another convenient expression, in English units only, for pressure drop in air (or gas) flow evaluations is:

$$\Delta P = N_\upsilon \frac{30}{B} \frac{T + 460}{1.73 \times 10^5}\left(\frac{G}{10^3}\right)^2 \qquad \textbf{(52)}$$

where

ΔP = pressure drop, in. wg
B = barometric pressure, in. Hg
T = air (or gas) temperature, F

Equation 52 is based on air, which has a specific volume of 25.2 ft^3/lb at 1000R and a pressure equivalent to 30 in. Hg. This equation can be used for other gases by correcting for specific volume.

The range in pressure drop through an assortment of commercial fittings is given in Table 5. This resistance to flow is presented in equivalent velocity heads based on the internal diameter of the connecting pipe. As noted, pressure drop through fittings may also be expressed as the loss in equivalent lengths of straight pipe.

Contraction and enlargement irreversible pressure loss

The simplest sectional changes in a conduit are converging or diverging boundaries. Converging boundaries can stabilize flow during the change from pressure energy to kinetic energy, and local irrecoverable flow losses (inelastic momentum exchanges) can be practically eliminated with proper design. If the included angle of the converging boundaries is 30 deg (0.52 rad) or less and the terminal junctions are smooth and tangent, any losses in mechanical energy are largely due to fluid friction. It is necessary to consider this loss as 0.05 times the velocity head, based on the smaller downstream flow area.

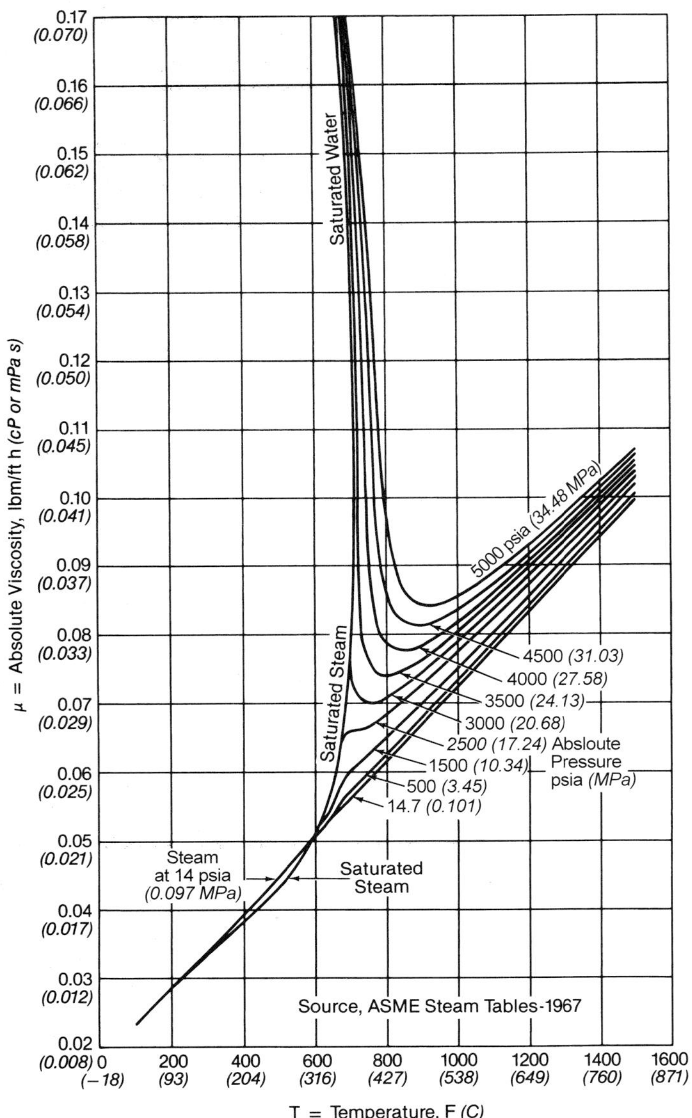

Fig. 5 Absolute viscosities of saturated and superheated steam.

Table 5
Resistance to Flow of Fluids Through Commercial Fittings*

Fitting	Loss in Velocity Heads
L-shaped, 90 deg (1.57 rad) standard sweep elbow	0.3 to 0.7
L-shaped, 90 deg (1.57 rad) long sweep elbow	0.2 to 0.5
T-shaped, flow through run	0.15 to 0.5
T-shaped, flow through 90 deg (1.57 rad) branch	0.6 to 1.6
Return bend, close	0.6 to 1.7
Gate valve, open	0.1 to 0.2
Check valve, open	2.0 to 10.0
Globe valve, open	5.0 to 16.0
Angle valve, 90 deg (1.57 rad) open	3.0 to 7.0
Boiler nonreturn valve, open	1.0 to 3.0

* See Fig. 9 for loss in velocity heads for flow of fluids through pipe bends.

When the elevation change ($Z_2 - Z_1$) is zero, the mechanical energy balance for converging boundaries becomes:

$$P_1 v + \frac{V_1^2}{2g_c} = P_2 v + \frac{V_2^2}{2g_c} + N_c \frac{V_2^2}{2g_c} \tag{53}$$

Subscripts 1 and 2 identify the upstream and downstream sections. N_c, the contraction loss factor, is the number of velocity heads lost by friction and local nonrecoverable pressure loss in contraction. Fig. 6 shows values of this factor.

When there is an enlargement of the conduit section in the direction of flow, the expansion of the flow stream is proportional to the kinetic energy of the flowing fluid and is subject to a pressure loss depending on the geometry. Just as in the case of the contraction loss, this is an irreversible energy conversion to heat resulting from inelastic momentum exchanges. Because it is customary to show these losses as coefficients of the higher kinetic energy term, the mechanical energy balance for enlargement loss is:

$$P_1 v + \frac{V_1^2}{2g_c} = P_2 v + \frac{V_2^2}{2g_c} + N_e \frac{V_1^2}{2g_c} \tag{54}$$

The case of sudden enlargement [angle of divergence $\beta = 180$ deg (π rad)] yields an energy loss of $(V_1 - V_2)^2/2g_c$. This can also be expressed as:

$$N_e = \left(1 - \frac{A_1}{A_2}\right)^2 \tag{55}$$

where A_1 and A_2 are the upstream and downstream cross-sectional flow areas, respectively and ($A_1 < A_2$). Even this solution, based on the conservation laws, depends on qualifying assumptions regarding static

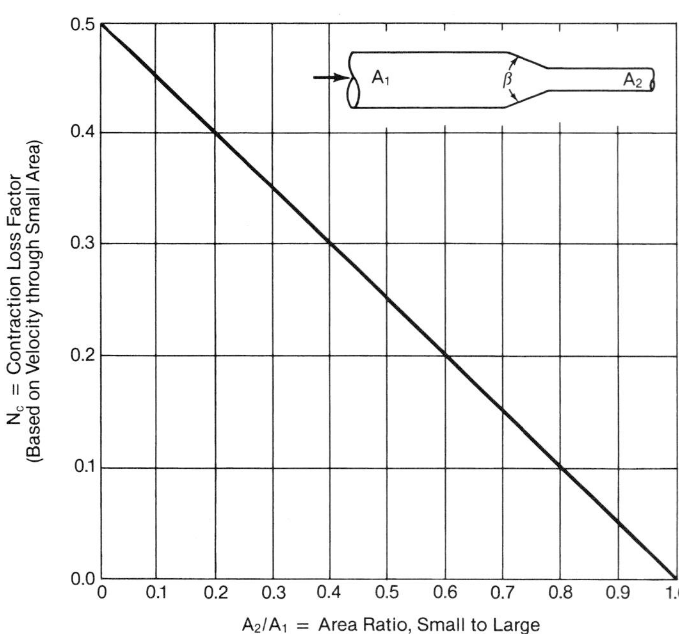

Fig. 6 Contraction loss factor for β>30 deg (N_c = 0.05 for $\beta \leq$30 deg).

pressures at the upstream and downstream faces of the enlargement.

Experimental values of the enlargement loss factor, based on different area ratios and angles of divergence, are given in Fig. 7. The differences in static pressures caused by sudden and gradual changes in section are shown graphically in Fig. 8. The pressure differences are shown in terms of the velocity head at the smaller area plotted against section area ratios.

Flow through bends

Bends in a pipeline or duct system produce pressure losses caused by both fluid friction and momentum exchanges which result from a change in flow direction. Because the axial length of the bend is normally included in the straight length friction loss of the pipeline or duct system, it is convenient to subtract a calculated equivalent straight length friction loss from experimentally determined bend pressure loss factors. These corrected data form the basis of the empirical bend loss factor, N_b.

The pressure losses for bends in round pipe in excess of straight pipe friction vary slightly with Reynolds numbers below 150,000. For Reynolds numbers above this value, they are reasonably constant and depend solely on the dimensionless ratio r/D, the ratio of the centerline radius of the bend to the internal diameter of the pipe. For commercial pipe, the effect of Reynolds number is negligible. The combined effect of radius ratio and bend angle, in terms of velocity heads, is shown in Fig. 9.

Flow through rectangular ducts

The loss of pressure caused by a direction change in a rectangular duct system is similar to that for cylindrical pipe. However, an additional factor, the shape

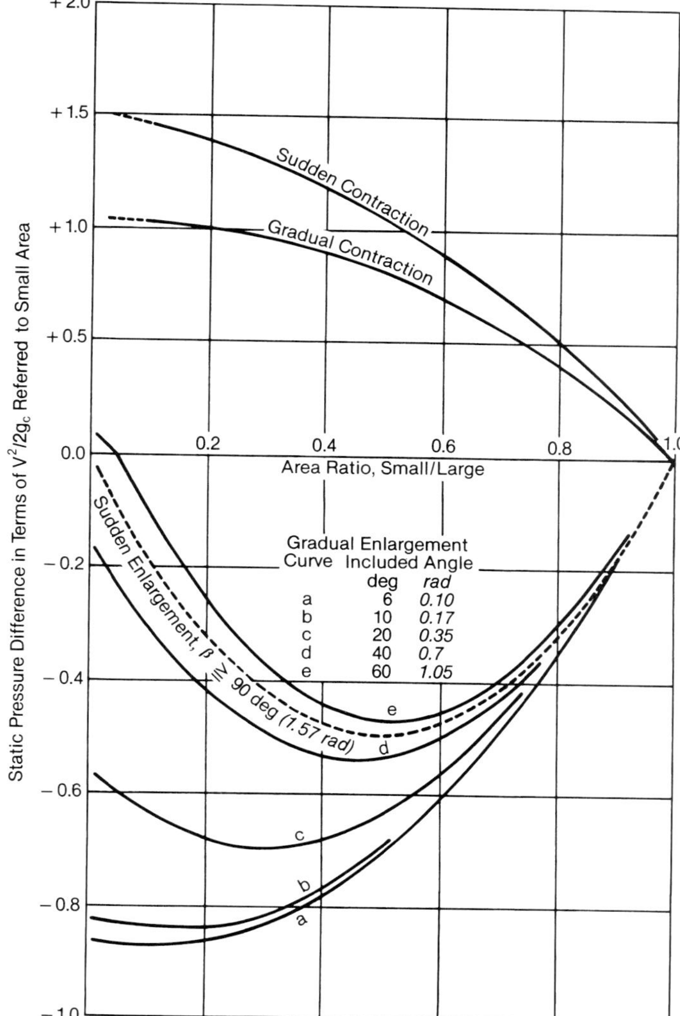

Fig. 8 Static pressure difference resulting from sudden and gradual changes in section.

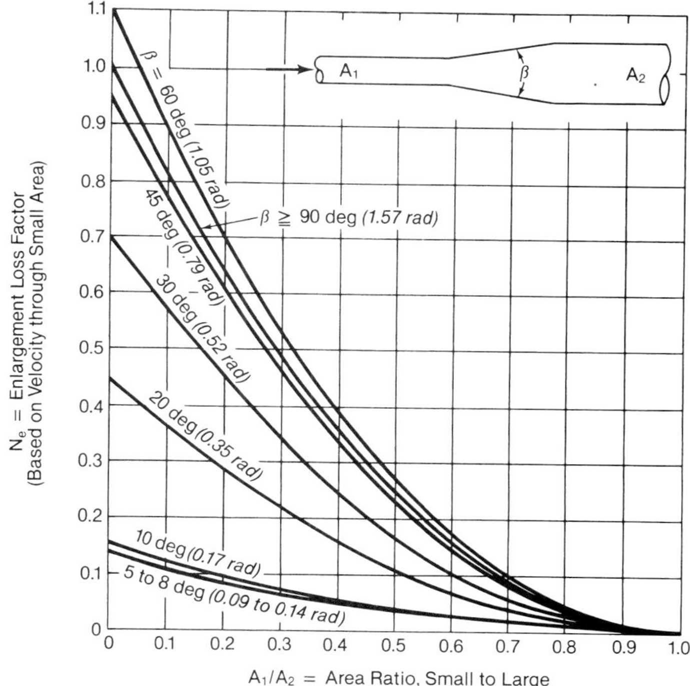

Fig. 7 Enlargement loss factor for various included angles.

of the duct in relation to the direction of bend, must be taken into account. This is called the aspect ratio, which is defined as the ratio of the width to the depth of the duct, i.e., the ratio b/d in Fig. 10. The bend loss for the same radius ratio decreases as the aspect ratio increases, because of the smaller proportionate influence of secondary flows on the stream. The combined effect of radius and aspect ratios on 90 deg (1.57 rad) duct bends is given in terms of velocity heads in Fig. 10.

The loss factors shown in Fig. 10 are average values of test results on ducts. For the given range of aspect ratios, the losses are relatively independent of the Reynolds number. Outside this range, the variation with Reynolds number is erratic. It is therefore recommended that N_b values for b/d = 0.5 be used for all aspect ratios less than b/d = 0.5, and values for b/d = 2.0 be used for ratios greater than b/d = 2.0. Losses for bends other than 90 deg (1.57 rad) are customarily considered to be proportional to the bend angle.

Turning vanes

The losses in a rectangular elbow duct can be reduced by rounding or beveling its corners and by in-

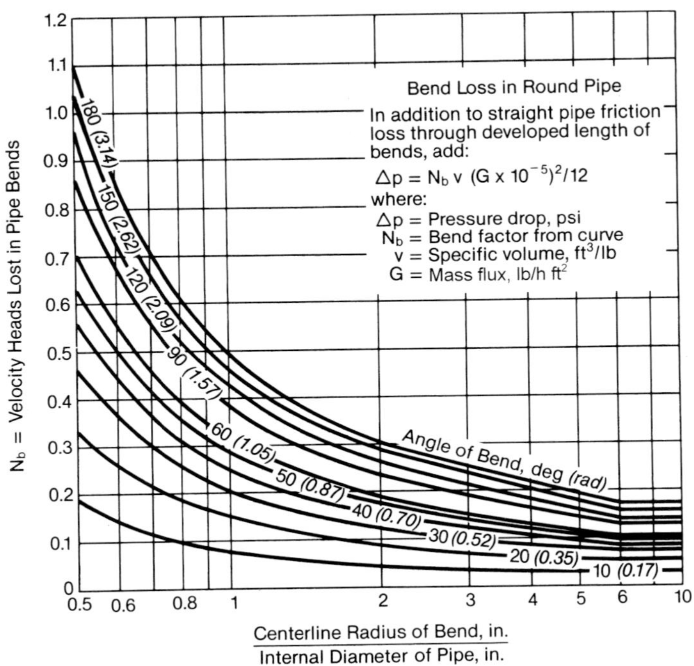

Fig. 9 Bend loss for round pipe, in terms of velocity heads.

stalling turning vanes. With rounding or beveling, the overall size of the duct can become large; however, with turning vanes, the compact form of the duct is preserved.

A number of turning vane shapes can be used in a duct. Fig. 11 shows four different arrangements. Segmented shaped vanes are shown in Fig. 11a, simple curved thin vanes are shown in Fig. 11b, and concentric splitter vanes are shown in Fig. 11c. In Fig. 11c, the vanes are concentric with the radius of the duct. Fig. 11d illustrates simple vanes used to minimize flow separation from a square edged duct.

The turning vanes of identical shape and dimension, Fig. 11b, are usually mounted within the bend of an elbow. They are generally installed along a line or section of the duct and are placed from the inner corner to the outside corner of the bend. Concentric turning vanes, Fig. 11c, typically installed within the bend of the turn, are located from one end of the turn to the other end.

The purpose of the turning vanes in an elbow or turn is to deflect the flow around the bend to the inner wall of the duct. When the turning vanes are appropriately designed, the flow distribution is improved by reducing flow separation from the walls and reducing the formation of eddy zones in the downstream section of the bend. The velocity distribution over the downstream cross-section of the turn is improved (see Fig. 12), and the pressure loss of the turn or elbow is decreased.

The main factor in decreasing the pressure losses and obtaining equalization of the velocity field is the elimination of an eddy zone at the inner wall of the turn. For a uniform incoming flow field, the largest effect of decreasing the pressure losses and establishing a uniform outlet flow field for a turn or elbow is achieved by locating the turning vanes closer to the inner curvature of the bend. (See Figs. 11d and 12c.)

For applications requiring a uniform velocity distribution directly after the turn, a full complement or normal arrangement of turning vanes (see Fig. 12b) is required. However, for many applications, it is sufficient to use a reduced number of vanes, as shown in Fig. 12c.

For nonuniform flow fields, the arrangement of turning vanes is more difficult to determine. Many times, numerical modeling (see Chapter 6) and flow testing of the duct system must be done to determine the proper vane locations.

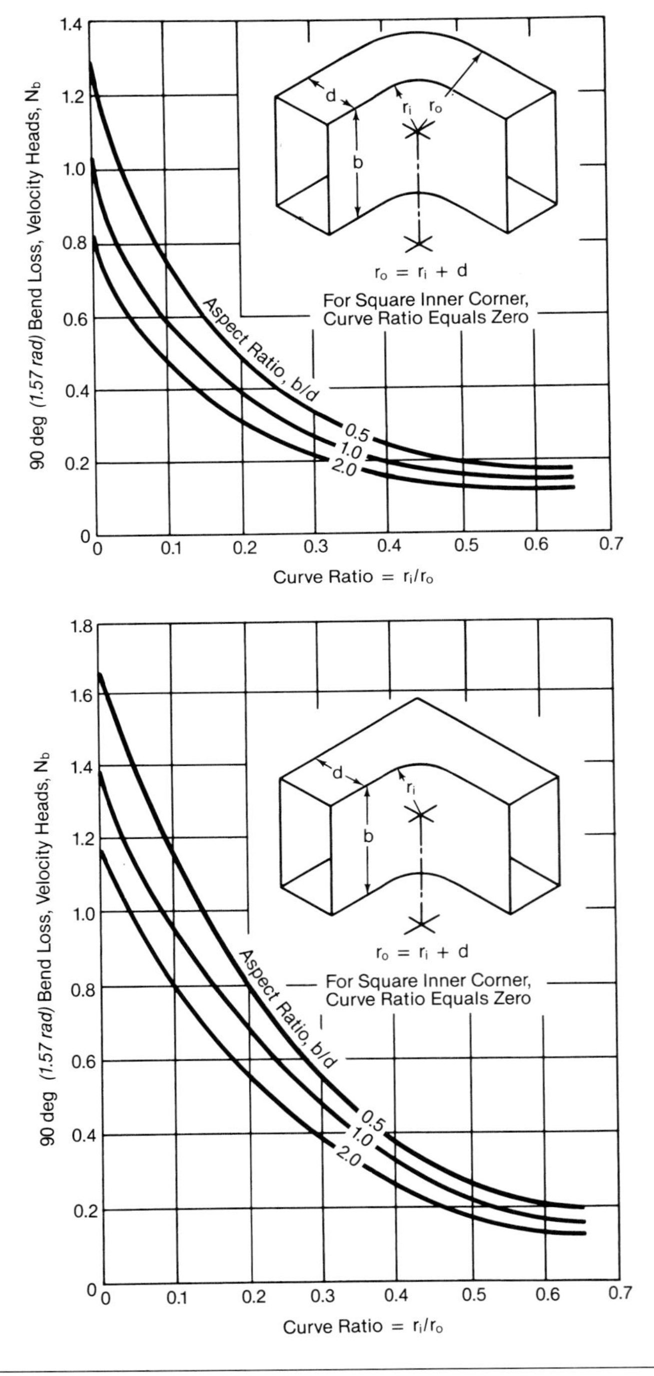

Fig. 10 Loss for 90 deg (1.57 rad) bends in rectangular ducts.

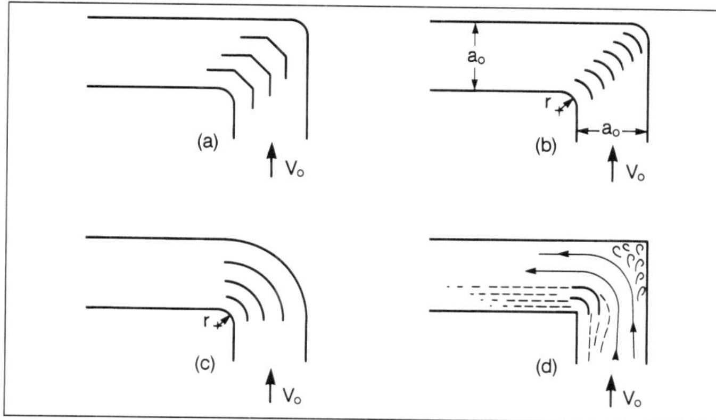

Fig. 11 Turning vanes in elbows and turns: a) segmented, b) thin concentric, c) concentric splitters, and d) slotted *(adapted from Idelchik, Reference 12)*.

Pressure loss

A convenient chart for calculating the pressure loss resulting from impact losses in duct systems conveying air (or flue gas) is shown in Fig. 13. When mass flux and temperature are known, a base velocity head in inches of water at sea level can be obtained.

Flow over tube banks

Bare tube The transverse flow of gases across tube banks is an example of flow over repeated major cross-sectional changes. When the tubes are staggered, sectional and directional changes affect the resistance. Experimental results and the analytical conclusions of extensive research by The Babcock & Wilcox Company (B&W) indicate that three principal variables other than mass flux affect this resistance. The primary variable is the number of major restrictions, i.e., the number of tube rows crossed, N. The second variable is the friction factor f which is related to the Reynolds number (based on tube diameter), the tube spacing diameter ratios, and the arrangement pattern (in-line or staggered). The third variable is the depth factor, F_d (Fig. 14), which is applicable to banks less than ten rows deep. The friction factors f for various in-line tube patterns are given in Fig. 15.

The product of the friction factor, the number of major restrictions (tube rows) and the depth factor is, in effect, the summation of velocity head losses through the tube bank.

$$N_\upsilon = f N F_d \tag{56}$$

The N_υ value established by Equation 56 may be used in Equations 51 or 52 to find the tube bank pressure loss. Some test correlations indicate f values higher than the isothermal case for cooling gas and lower for heating gas.

Finned tube In some convective boiler design applications, extended surface tube banks are used. Many types of extended surface exist, i.e., solid helical fin, serrated helical fin, longitudinal fin, square fin and different types of pin studs. For furnace applications, the cleanliness of the gas or heat transfer medium dictates whether an extended surface tube bank can be used and also defines the type of extended surface.

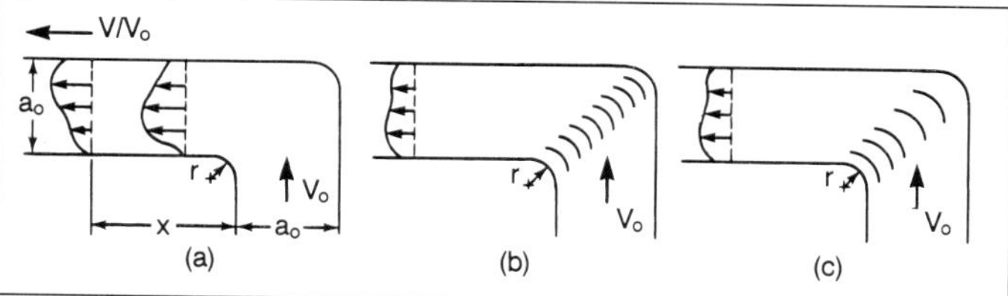

Fig. 12 Velocity profiles downstream of an elbow: a) without vanes, b) with typical vanes, and c) with optimum vanes *(adapted from Idelchik, Reference 12)*.

Several different tube bank calculation methods exist for extended surface, and many are directly related to the type of extended surface that is used. Various correlations for extended surface pressure loss can be found in References 9 through 15. In all cases, a larger pressure loss per row of bank exists with an extended surface tube compared to a bare tube. For in-line tube bundles, the finned tube resistance per row of tubes is approximately 1.5 times that of the bare tube row. However, due to the increased heat transfer absorption of the extended surface, a smaller number of tube rows is required. This results in an overall bank pressure loss that can be equivalent to a larger but equally absorptive bare tube bank.

Flow through stacks or chimneys

The flow of gases through stacks or chimneys is established by the natural draft effect of the stack and/or the mechanical draft produced by a fan. The resistance to this flow, or the loss in mechanical energy be-

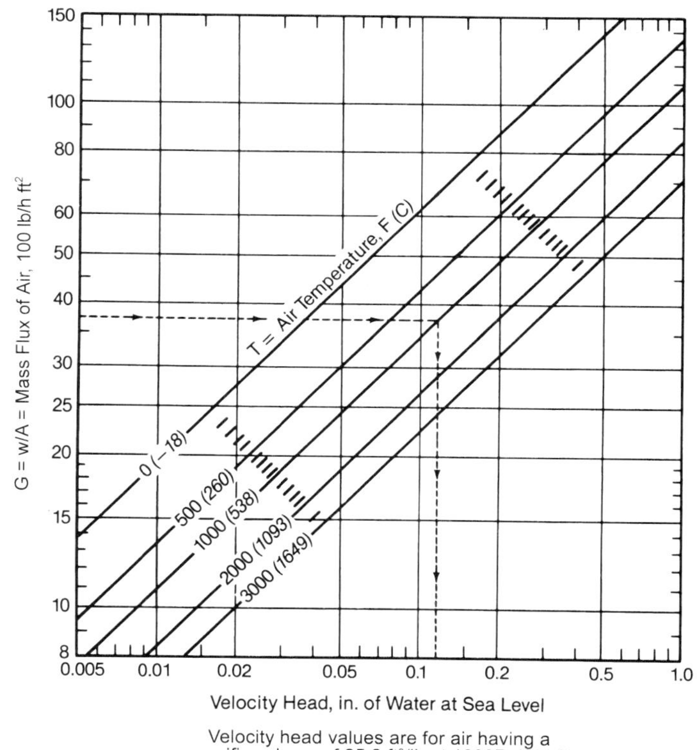

$$\frac{V^2}{2g_c} = \left(\frac{V^2}{2g_c}\text{ for Air}\right)\left(\frac{\text{Flue Gas Sp Vol}}{\text{Air Sp Vol}}\right)$$

Fig. 13 Mass flux/velocity head relationship for air.

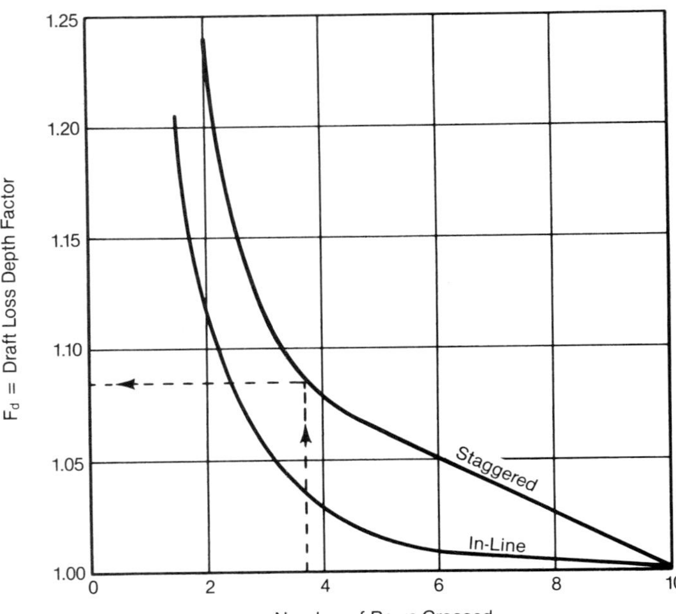

Fig. 14 Draft loss depth factor for number of tube rows crossed in convection banks.

tween the bottom and the top of the stack, is a result of the friction and stack exit losses. Application examples of these losses are given in Chapter 25.

Pressure loss in two-phase flow

Evaluation of two-phase steam-water flows is much more complex. As with single-phase flow, pressure loss occurs from wall friction, acceleration, and change in elevation. However, the relationships are more complicated. The evaluation of friction requires the assessment of the interaction of the steam and water phases. Acceleration is much more important because of the large changes in specific volume of the mixture as water is converted to steam. Finally, large changes in average mixture density at different locations significantly impact the static head. These factors are presented in detail in Chapter 5.

Entrainment by fluid flow

Collecting or transporting solid particles or a second fluid by the flow of a primary fluid at high velocity is known as entrainment. This is usually accomplished with jets using a small quantity of high pressure fluid to carry large quantities of another fluid or solid particles. The pressure energy of the high pressure fluid is converted into kinetic energy by nozzles, with a consequent reduction of pressure. The material to be transported is drawn in at the low pressure zone, where it meets and mixes with the high velocity jet. The jet is usually followed by a parallel throat section to equalize the velocity profile. The mixture then enters a diverging section where kinetic energy is partially reconverted into pressure energy. In this case, major fluid flow mechanical energy losses are an example of inelastic momentum exchanges occurring within the fluid streams.

The *injector* is a jet pump that uses condensing steam as the driving fluid to entrain low pressure water for delivery against a back pressure higher than the pressure of the steam supplied. The *ejector,* similar to the injector, is designed to entrain gases, liquids, or mixtures of solids and liquids for delivery against a pressure less than that of the primary fluid. In a water-jet *aspirator,* water is used to entrain air to obtain a partial vacuum. In the Bunsen type burner, a jet of gas entrains air for combustion. In several instances, entrainment may be detrimental to the operation of steam boilers. Particles of ash entrained by the products of combustion, when deposited on heating surfaces, reduce thermal conductance, erode fan blades, and add to pollution when discharged into the atmosphere. Moisture carrying solids, either in suspension or in solution, are entrained in the stream. The solids may be carried through to the turbine and deposited on the blades, decreasing turbine capacity and efficiency. In downcomers or supply tubes, steam bubbles are entrained in the water when the drag on the bubbles is greater than the buoyant force. This reduces the density in the pumping column of natural circulation boilers.

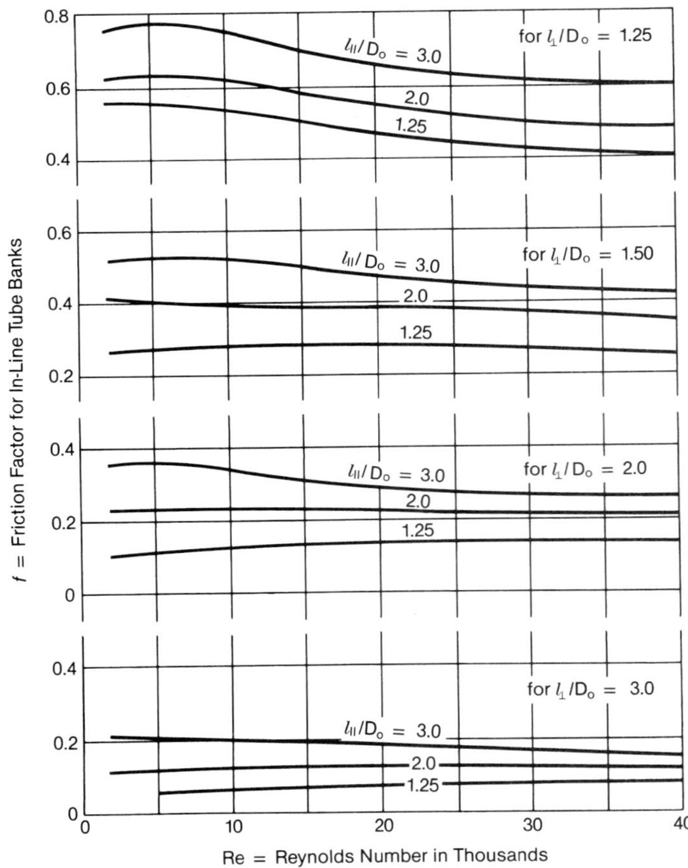

Fig. 15 Friction factor (f) as affected by Reynolds number for various in-line tube patterns; crossflow gas or air.

Boiler circulation

An adequate flow of water and steam-water mixture is necessary for steam generation and control of tube metal temperatures in all circuits of a steam generating unit. At supercritical pressures this flow is produced mechanically by pumps. At subcritical pressures, circulation is produced by the force of gravity or pumps, or a combination of the two. The elements of single-phase flow discussed in this chapter, two-phase flow discussed in Chapter 5, heat input rates, and selected limiting design criteria are combined to evaluate the circulation in fossil-fired steam generators. The evaluation procedures and key criteria are presented in Chapter 5.

References

1. Meyer, C.A., et al., *ASME Steam Tables*, Sixth Ed., American Society of Mechanical Engineers, New York, New York, 1993.
2. Tabor, D., *Gases, Liquids and Solids: and Other States of Matter,* First Ed., Penguin Books, Ltd., Harmondsworth, England, United Kingdom, 1969.
3. Lahey, Jr., R.T., and Moody, F.J., *The Thermal-Hydraulics of a Boiling Water Nuclear Reactor,* American Nuclear Society, Hinsdale, Ilinois, 1993.
4. Rohsenow, W., Hartnett, J., and Ganic, E., *Handbook of Heat Transfer Fundamentals,* McGraw-Hill Company, New York, 1985.
5. Burmeister, L.C., *Convective Heat Transfer,* Second Ed., Wiley-Interscience, New York, New York, 1993.
6. Schlichting, H.T. Gersten, K., and Krause, E., *Boundary-Layer Theory,* Eighth Ed., Springer-Verlag, New York, New York, 2000.
7. Moody, L.F., "Friction Factors for Pipe Flow," *Transactions of the American Society of Mechanical Engineers (ASME),* Vol. 66, 8, pp. 671-684, November, 1944.
8. Hinze, J.O., *Turbulence: An Introduction to Its Mechanism and Theory,* Second Ed., McGraw-Hill Company, New York, New York, 1975.
9. Briggs, D.E., and Young, E.H., "Convective heat transfer and pressure drop of air flowing across triangular pitch banks of finned tubes," *Chemical Engineering Progress Symposium Series (Heat Transfer),* AIChE, Vol. 41, No. 41, pp. l-10, Houston, Texas, 1963.
10. Grimison, E.D., "Correlation and utilization of new data on flow resistance and heat transfer for crossflow of gases over tube banks," *Transactions of ASME, Process Industries Division*, Vol. 59, pp. 583-594, New York, New York, 1937.
11. Gunter, A.Y., and Shan, W.A., "A general correlation of friction factors for various types of surfaces in crossflow," *Transactions of ASME,* Vol. 67, pp. 643-660, 1945.
12. Idelchik, I.E., *Handbook of Hydraulic Resistance,* Third Ed., Interpharm/CRC, New York, New York, November, 1993.
13. Jakob, M., Discussion appearing in *Transactions of ASME,* Vol. 60, pp. 384-386, 1938.
14. Kern, D.Q., *Process Heat Transfer,* p. 555, McGraw-Hill Company, New York, New York, December, 1950.
15. Wimpress, R.N., *Hydrocarbon Processing and Petroleum Refiner,* Vol. 42, No. 10, pp. 115-126, Gulf Publishing Company, Houston, Texas, 1963.

Wall-fired utility boiler furnace under construction.

Chapter 4

Heat Transfer

Heat transfer deals with the transmission of thermal energy and plays a central role in most energy conversion processes. Heat transfer is important in fossil fuel combustion, chemical reaction processes, electrical systems, nuclear fission and certain fluid systems. It also occurs during everyday activities including cooking, heating and refrigeration, as well as being an important consideration in choosing clothing for different climates.

Although the fundamentals of heat transfer are simple, practical applications are complex because real systems contain irregular geometries, combined modes of heat transfer and time dependent responses.

Fundamentals

Basic modes of heat transfer

There are three modes of heat transfer: conduction, convection and radiation. One or more of these modes controls the amount of heat transfer in all applications.

Conduction Temperature is a property that indicates the kinetic energy possessed by the molecules of a substance; the higher the temperature the greater the kinetic energy or molecular activity of the substance. Molecular conduction of heat is simply the transfer of energy due to a temperature difference between adjacent molecules in a solid, liquid or gas.

Conduction heat transfer is evaluated using Fourier's law:

$$q_c = -kA\frac{dT}{dx} \tag{1}$$

The flow of heat, q_c, is positive when the temperature gradient, dT/dx, is negative. This result, consistent with the second law of thermodynamics, indicates that heat flows in the direction of decreasing temperature. The heat flow, q_c, is in a direction normal (or perpendicular) to an area, A, and the gradient, dT/dx, is the change of temperature in the direction of heat flow. The thermal conductivity, k, a property of the material, quantifies its ability to conduct heat. A range of thermal conductivities is listed in Table 1. The discrete form of the conduction law is written:

$$q = \frac{kA}{L}(T_1 - T_2) \tag{2}$$

Fig. 1 illustrates positive heat flow described by this equation and shows the effect of variable thermal conductivity on the temperature distribution. The grouping kA/L is known as the thermal conductance, K_c; the inverse L/kA is known as the thermal resistance, R_c and $K_c = 1/R_c$.

A special case of conduction is the thermal contact resistance across a joint between solid materials. At the interface of two solid materials the surface to surface contact is imperfect from the gap that prevails due to surface roughness. In nuclear applications with fuel pellets and fuel cladding, surface contact resistance can have a major impact on heat transfer. If one dimensional steady heat flow is assumed, the heat transfer across a gap is defined by:

$$q = \frac{T_1 - T_2}{R_{ct}} \tag{3}$$

where the quantity R_{ct} is called the thermal contact resistance, $1/h_{ct}A$, and h_{ct} is called the contact coefficient. T_1 and T_2 are the average surface temperatures on each side of the gap. Tabulated values of the contact coefficient are presented in References 1 and 2.

Table 1
Thermal Conductivity, k, of Common Materials

Material	Btu/h ft F	W/m C
Gases at atmospheric pressure	0.004 to 0.70	0.007 to 1.2
Insulating materials	0.01 to 0.12	0.02 to 0.21
Nonmetallic liquids	0.05 to 0.40	0.09 to 0.70
Nonmetallic solids (brick, stone, concrete)	0.02 to 1.5	0.04 to 2.6
Liquid metals	5.0 to 45	8.6 to 78
Alloys	8.0 to 70	14 to 121
Pure metals	30 to 240	52 to 415

Nomenclature

A surface area, ft² (m²)
c_p specific heat at constant pressure, Btu/lb F (J/kg K)
C_f cleanliness factor, dimensionless
C_t thermal capacitance, Btu/ft³ F (J/m³ K)
C electrical capacitance, farad
D diameter, ft (m)
D_e equivalent diameter, ft (m)
E_b blackbody emissive power, Btu/h ft² (W/m²)
F radiation configuration factor, dimensionless
F heat exchanger arrangement factor, dimensionless
F_a crossflow arrangement factor, dimensionless
F_d tube bundle depth factor, dimensionless
F_{pp} fluid property factor, see text
F_T fluid temperature factor, dimensionless
$\mathcal{F}$ total radiation exchange factor, dimensionless
g acceleration of gravity, 32.17 ft/s² (9.8 m/s²)
G incident thermal radiation, Btu/h ft² (W/m²)
G mass flux or mass velocity, lb/h ft² (kg/m² s)
h heat transfer coefficient, Btu/h ft² F (W/m² K)
h_c crossflow heat transfer coef., Btu/h ft² F (W/m² K)
h_{ct} contact coefficient, Btu/h ft² F (W/m² K)
h_c' crossflow velocity and geometry factor, see text
h_l longitudinal heat transfer coef., Btu/h ft² F (W/m² K)
h_l' longitudinal flow velocity and geometry factor, see text
H enthalpy, Btu/lb (J/kg)
H_{fg} latent heat of vaporization, Btu/lb (J/kg)
I electrical current, amperes
J radiosity, Btu/h ft² (W/m²)
k thermal conductivity, Btu/h ft F (W/m K)
K thermal conductance, Btu/h F (W/K)
K_y mass transfer coefficient, lb/ft² s (kg/m² s)
L beam length, ft (m)
L, ℓ length or dimension, ft (m)
L_h fin height, ft (m)
L_t fin spacing, ft (m)
$\dot{m}$ mass flow rate, lb/h (kg/s)
p pressure or partial pressure, atm (Pa)
P temp. ratio for surface arrgt. factor, dimensionless
q heat flow rate, Btu/h (W)
q''' volumetric heat generation rate, Btu/h ft³ (W/m³)
q_{rel} heat release, Btu/h ft³ (W/m³)
r radius, ft (m)
R electrical resistance (ohms)
R temp. ratio for surface arrgt. factor, dimensionless
R thermal resistance, h F/Btu (K/W)
R radiative resistance, 1/ft² (1/m²)
R_f fouling factor, h ft² F/Btu (m² K/W)
S total exposed surface area for a finned surface, ft² (m²)
S_f fin surface area; sides plus peripheral area, ft² (m²)
S_H source term for internal heat generation, Btu/h ft³ (W/m³)
t time, s or h, see text (s)
T temperature, F or R (C or K)
T° temperature at initial time, F (C)
Δt time interval, s
ΔT temperature difference, F (C)
ΔT_{LMTD} log mean temperature difference, F (C)
u,v,w velocity in x, y, z coordinates respectively, ft/s (m/s)
U overall heat transfer coef., Btu/h ft² F (W/m² K)
V electrical voltage, volts
V velocity, ft/s (m/s)
V volume, ft³ (m³)
x dimension, ft (m)
x,y,z dimensions in Cartesian coordinate system, ft (m)
Δx change in length, ft (m)
Y Schmidt fin geometry factor, dimensionless
Y_g concentration in bulk fluid, lb/lb (kg/kg)
Y_i concentration at condensate interface, lb/lb (kg/kg)
Z Schmidt fin geometry factor, dimensionless
α absorptivity, or total absorptance, dimensionless
β volume coefficient of expansion, 1/R (1/K)
Γ effective diffusion coefficient, lb/ft s (kg/m s)
ε emissivity, or total emittance, dimensionless
η Schmidt fin efficiency, dimensionless
μ dynamic viscosity, lbm/ft s (kg/m s)
ρ density, lb/ft³ (kg/m³)
ρ reflectivity, dimensionless
σ Stefan-Boltzmann constant, 0.1713×10^{-8} Btu/h ft² R⁴ (5.669×10^{-8} W/m² K⁴)
τ transmissivity, dimensionless

Subscripts:

b bulk
c conduction
ct contact
cv convection
e node point east
eff effective
f film or fin
fd fully developed
g gas
i inside or *ith* parameter
j *jth* parameter
o outside
p node point under evaluation
r radiation
s surface
sg surface to gas
w node point west
w wall
δ liquid film surface (gas liquid interface)
∞ free stream conditions
$\perp$ perpendicular to flow
$\|$ parallel to flow

Dimensionless groups:

$$Gr = \frac{g\beta(T_s - T_\infty)\rho^2 L^3}{\mu^2}$$ Grashof number

$$\text{Nu} = \frac{hL}{k}$$ Nusselt number

$$\text{Pe} = \text{Re Pr}$$ Peclet number

$$\text{Pr} = \frac{c_p \mu}{k}$$ Prandtl number

$$\text{Ra} = \text{Gr Pr}$$ Rayleigh number

$$\text{Re} = \frac{\rho V L}{\mu} = \frac{GL}{\mu}$$ Reynolds number

$$\text{St} = \frac{\text{Nu}}{\text{Re Pr}} = \frac{h}{c_p G}$$ Stanton number

Examples include 300 Btu/h ft^2 F (1.7 kW/m^2 K) between two sections of ground 304 stainless steel in air and 25,000 Btu/h ft^2 F (142 kW/m^2 K) between two sections of ground copper in air. The factors are usually unknown for specific applications and estimates need to be made. There are two principal contributions across the gap – solid to solid conduction at the points of contact and thermal conduction through the entrapped gases in the void spaces.

Convection Convection heat transfer within a fluid (gas or liquid) occurs by a combination of molecular conduction and macroscopic fluid motion. Convection occurs adjacent to heated surfaces as a result of fluid motion past the surface as shown in Fig. 2.

Natural convection occurs when the fluid motion is due to buoyancy effects caused by local density differences. In the top portion of Fig. 2, the fluid motion is due to heat flow from the surface to the fluid; the fluid density decreases causing the lighter fluid to rise and be replaced by cooler fluid. Forced convection results when mechanical forces from devices such as fans give motion to the fluids. The rate of heat transfer by convection, q_{cv}, is defined:

$$q_{cv} = hA\left(T_s - T_f\right) \tag{4}$$

where h is the local heat transfer coefficient, A is the surface area, T_s is the surface temperature and T_f is the fluid temperature. Equation 4 is known as Newton's Law of Cooling and the term hA_s is the convection conductance, K_{cv}. The heat transfer coefficient, h, is also termed the unit conductance, because it is defined as the conductance per unit area. Average heat transfer coefficients over a surface are used in most engineering applications. This convective heat transfer coefficient is a function of the thermal and fluid dynamic properties and the surface geometry. Approximate ranges are shown in Table 2.

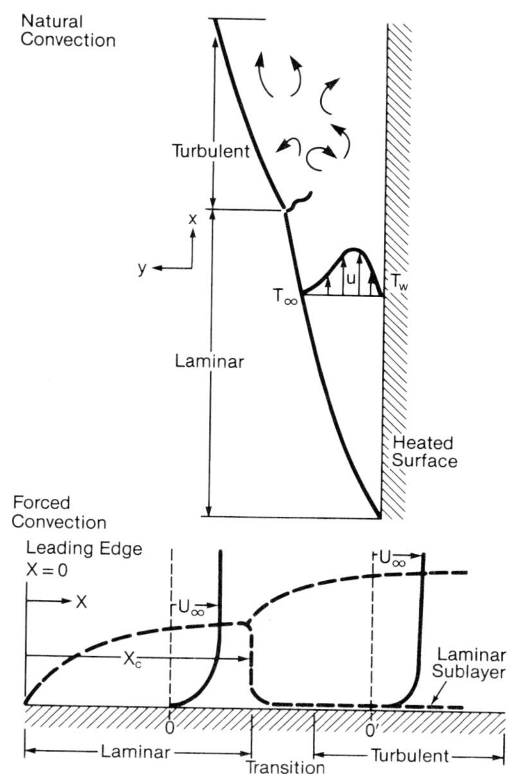

Fig. 2 Natural and forced convection. Above, boundary layer on a vertical flat plate. Below, velocity profiles for laminar and turbulent boundary layers in flow over a flat plate. (Vertical scale enlarged for clarity.)

Radiation Radiation is the transfer of energy between bodies by electromagnetic waves. This transfer, unlike conduction or convection, requires no intervening medium. The electromagnetic radiation, in the wavelength range of 0.1 to 100 micrometers, is produced solely by the temperature of a body. Energy at the body's surface is converted into electromagnetic waves that emanate from the surface and strike another body. Some of the thermal radiation is absorbed by the receiving body and reconverted into internal energy, while the remaining energy is reflected from or transmitted through the body. The fractions of radiation reflected, transmitted and absorbed by a surface are known respectively as the reflectivity, ρ, transmissivity, τ, and absorptivity α. The sum of these fractions equals one:

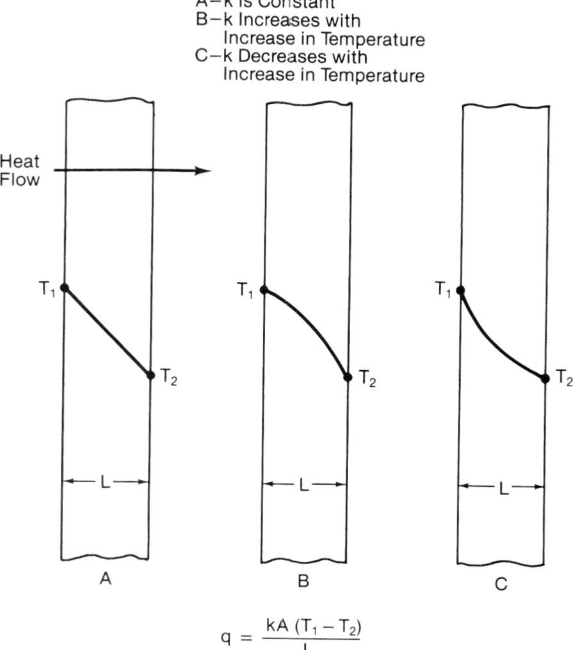

Fig. 1 Temperature-thickness relationships corresponding to different thermal conductivities, *k*.

Table 2
Typical Convective Heat Transfer Coefficients, h

Condition	Btu/h ft^2 F	W/m^2 C
Air, free convection	1 to 5	6 to 30
Air, forced convection	5 to 50	30 to 300
Steam, forced convection	300 to 800	1800 to 4800
Oil, forced convection	5 to 300	30 to 1800
Water, forced convection	50 to 2000	300 to 12,000
Water, boiling	500 to 20,000	3000 to 120,000

$$\rho + \tau + \alpha = 1 \quad (5)$$

All surfaces absorb radiation, and all surfaces whose temperatures are above absolute zero emit thermal radiation. Surfaces in boilers are typically opaque, which do not allow transmission of any radiation ($\tau = 0$).

Thermal radiation generally passes through gases such as dry air with no absorption taking place. These nonabsorbing, or nonparticipating, gases do not affect the radiative transfer. Other gases, like carbon dioxide, water vapor and carbon monoxide, to a lesser degree, affect radiative transfer and are known as participating gases. These gases, prevalent in the flue gases of a boiler, affect the heat transfer to surfaces and the distribution of energy absorbed in the boiler.

All bodies continuously emit radiant energy in amounts which are determined by the temperature and the nature of the surface. A perfect radiator, or *blackbody,* absorbs all of the incident thermal radiation, G, reaching its surface:

$$q_r^+ = A\,G \quad (6)$$

and emits radiant energy at the maximum theoretical limit according to the Stefan-Boltzmann law:

$$q_r^- = A\,\sigma\,T_s^4 \quad (7)$$

σ is the Stefan-Boltzmann constant 0.1713×10^{-8} Btu/h ft^2 R^4 (5.669×10^{-8} W/m^2 K^4), and T_s is the absolute temperature of the surface, R (K). The product $\sigma\,T_s^4$ is also known as the blackbody emissive power, E_b. The net radiative heat transfer of a blackbody is the difference between absorbed and emitted radiant energy:

$$q_r = q_r^+ - q_r^- = A\,(G - \sigma\,T_s^4) \quad (8)$$

The radiation from a blackbody extends over the whole range of wavelengths, although the bulk of it in boiler applications is concentrated in a band from 0.1 to 20 micrometers. The wavelength at which the maximum radiation intensity occurs is inversely proportional to the absolute temperature of the body; this is known as Wien's law.

A real radiator absorbs less than 100% of the energy incident on it and emits less than the maximum theoretical limit. The net heat transfer by radiation from a real surface can be expressed by:

$$q_r = A\,(\alpha\,G - \varepsilon\,\sigma\,T_s^4) \quad (9)$$

where ε is the total emissivity and α is the total absorptivity. If the emissivity and absorptivity are independent of wavelength, the surface is termed a nonselective radiator, or *gray* surface. According to Kirchoff's law, emissivity and absorptivity are always equal for a gray surface:

$$\varepsilon = \alpha \quad (10)$$

Therefore, α can be eliminated from Equation 9, and emissivity is all that is needed to describe the radiation properties of the surface. Table 3 shows some representative values of emissivity. If all surfaces are assumed to be gray, a simpler treatment of radiation is

Table 3
Representative Values of Emissivity

Surface	Emissivity
Polished metals	$0.01 < \varepsilon < 0.08$
Metals, as-received	$0.1 < \varepsilon < 0.2$
Metals oxidized	$0.25 < \varepsilon < 0.7$
Ceramic oxides	$0.4 < \varepsilon < 0.8$
Special paints	$0.9 < \varepsilon < 0.98+$

possible. For two surface enclosures, this treatment involves introducing a total exchange factor, $\mathcal{F}_{12}$, which depends on the configuration (geometry), the emissivities and the surface areas.[3]

If the emissivity depends on wavelength, the surface is termed a selective or non-gray radiator. According to Kirchhoff's law, spectral emissivity and spectral absorptivity are always equivalent, $\varepsilon_\lambda = \alpha_\lambda$, for non-gray surfaces. Total emissivity is the integrated average of ε_λ over the spectrum of emitted radiation, and total absorptivity is the integrated average of α_λ over the spectrum of incident radiation. The terms *emissivity* and *emittance* (and corresponding terms *absorptivity* and *absorptance*) are commonly interchanged in the literature. For convenience, the term *emittance* is used here for *total emissivity* and *absorptance* is used for *total absorptivity*, of non-gray surfaces.

For non-gray surfaces, the emittance can be expressed as a function of the surface temperature $\varepsilon\,(T_s)$, and absorptance as a function of the incident radiation or flame temperature, $\alpha\;(T_f)$. Based on Kirchhoff's law, plots of ε vs T_s may be interpreted as plots of α vs T_f if the physical state of the surface is unchanged. An analysis of non-gray conditions requires temperature dependent emittance and absorptance, or spectral property calculations which are more complicated. An example of non-gray radiators in a boiler are the ash deposits on waterwall heating surfaces.

The net radiation heat transfer between two blackbody surfaces which are separated by a vacuum or nonparticipating gas is written:

$$q_{12} = A_1\,F_{12}\,\sigma\left(T_1^4 - T_2^4\right) \quad (11)$$

A_1 is the surface area; F_{12} is the geometric shape factor and represents the fraction of radiant energy leaving surface 1 that directly strikes surface 2. As will be discussed later for radiation between two surfaces, F_{12} is the exchange factor for two surfaces based on the geometric arrangement only, and $\mathcal{F}_{12}$ is the exchange factor that includes the effects of emissivity for gray surfaces, and participating media between the surfaces. For blackbody surfaces ($\varepsilon_1 = \varepsilon_2 = 1$) and nonparticipating media, $\mathcal{F}_{12} = F_{12}$. T_1 and T_2 are the surface temperatures. Since the net energy at surface 1 must balance the net energy at surface 2, we can write:

$$q_{12} = -q_{21} \quad (12)$$

Using Equations 11 and 12, the following results:

$$A_1\,F_{12} = A_2\,F_{21} \quad (13)$$

This equation, known as the principle of reciprocity, guarantees conservation of the radiant heat transfer between two surfaces. The following rule applies to the surfaces of an enclosure:

$$\sum_j F_{ij} = 1 \tag{14}$$

stating that the total fraction of energy leaving surface i to all other (j) surfaces must equal 1. Many texts include the calculation of geometric shape factors, commonly named shape factors or configuration factors.[1,2] Radiation balances for participating and nonparticipating media are presented later in the chapter.

Governing equations

Energy balances The solution of a heat transfer problem requires defining the system which will be analyzed. This usually involves idealizing the actual system by defining a schematic *control volume* of the modeled system. A net energy balance on the control volume reflects the first law of thermodynamics and can be stated:

$$\text{energy in} - \text{energy out} = \text{stored energy} \tag{15}$$

For a steady flow of heat, the balance simplifies to:

$$\text{heat in} = \text{heat out} \tag{16}$$

The laws governing the flow of heat are used to obtain equations in terms of material temperature or fluid enthalpy.

Steady-state conduction The basic laws for each heat transfer mode and the energy balance provide the tools needed to write the governing equations for rectangular and cylindrical heat transfer systems. For example, for the plane wall shown in Fig. 1, the steady flow energy balance for a slice of thickness, dx, is:

$$q_1 - q_2 = q - \left(q + \frac{dq}{dx} dx \right) = 0 \tag{17}$$

After substituting Equation 1, this is rewritten as:

$$\frac{d}{dx}\left(kA \frac{dT}{dx} \right) = 0 \tag{18}$$

The conditions at the boundaries, $T = T_1$ at $x = 0$ and $T = T_2$ at $x = L$, provide closure. The general symbolic form of the equation in three dimensions can be represented, vector notation:

$$\nabla \cdot (k \nabla T) = 0 \tag{19}$$

or in x, y, z Cartesian coordinates:

$$\frac{\partial}{\partial x}\left(k \frac{\partial T}{\partial x} \right) + \frac{\partial}{\partial y}\left(k \frac{\partial T}{\partial y} \right) + \frac{\partial}{\partial z}\left(k \frac{\partial T}{\partial z} \right) = 0 \tag{20}$$

This assumes there is no net heat storage or heat generation in the wall.

Unsteady-state conduction So far only steady-state conduction, where temperatures vary from point to point but do not change with time, has been discussed. All unsteady-state conduction involves heat storage. For instance, in heating a furnace, enough heat must be supplied to bring the walls to the operating temperature and also to make up for the steady-state losses of normal operation. In large power boilers that run for long periods of time, heat storage in the walls and boiler metal is an insignificant fraction of the total heat input. In small boilers with refractory settings that are operated only part time, or in furnaces that are frequently heated and cooled in batch process work, heat stored in the walls during startup may be a considerable portion of the total heat input.

Unsteady-state conduction is important when equalizing boiler drum temperature during pressure raising and reducing periods. When boiler pressure is raised, the water temperature rises. The inner surface of the steam drum is heated by contact with the water below the water line and by the condensation of steam above the water line. The inside and outside drum temperatures are increased by unsteady-state conduction. During this transient heatup period, temperature differentials across the drum wall (or thermal gradients) will be larger than during steady-state operation. Larger thermal gradients result in higher thermal stresses as discussed in Chapter 8. The rate of temperature and pressure increase must therefore be controlled to maintain the thermal stresses within acceptable levels in order to protect the drum. During pressure reducing periods, the inside of the drum below the water line is cooled by boiler water while the top of the drum is cooled by radiation to the water, by the steam flow to the outlet connections, and by unsteady-state conduction through the drum walls.

Unsteady-state conduction occurs in heating or cooling processes where temperatures change with time. Examples include heating billets, quenching steel, operating regenerative heaters, raising boiler pressure, and heating and cooling steam turbines. By introducing time as an additional variable, conduction analyses become more complicated. For unsteady heat flow, the one dimensional thermal energy equation becomes:

$$\rho c_p \frac{\partial T}{\partial t} = \frac{\partial}{\partial x}\left(k \frac{\partial T}{\partial x} \right) \tag{21}$$

The left side of the equation represents the rate of energy storage. The two boundary temperatures at $x = 0$ and L, and the initial temperature, $T = T^o$, are sufficient to find a solution. Other boundary conditions involving radiation, convection, or specified heat flux at $x = 0$ or L can also be applied.[3]

A general form of the energy equation for multidimensional applications is:

$$\rho c_p \frac{\partial T}{\partial t} = \nabla \cdot (k \nabla T) \tag{22}$$

where $\nabla \cdot (k \nabla T)$ is defined in Equations 19 and 20. Conditions on the boundary as a function of time, and initial temperature of the system, are sufficient to find a solution.

Electrical analogy The basic laws of conduction, convection and radiation can frequently be rearranged into equations of the form:

$$q = \frac{T_1 - T_2}{R_t} \tag{23}$$

This equation can be compared to Ohm's law for electrical circuits ($I = V/R$). The heat transfer or *heat flow* from point 1 to 2 (q) is analogous with current (I), the temperature difference ($T_1 - T_2$) is analogous with voltage (V), and the thermal resistance (R_t) is analogous with electrical resistance (R). Thermal resistance is defined as the reciprocal of thermal conductance, K_t. Table 4 contains analog thermal resistances used in many applications.

For systems with unsteady-state conduction governed by Equation 21 or 22, an electrical analogy can be written:

$$q = C_t \frac{dT}{dt} \tag{24}$$

where C_t is the thermal capacitance, $\rho c_p V$. This equation can be compared with its electrical equivalent:

$$I = C\frac{dV}{dt} \tag{25}$$

where C is the electrical capacitance. Kirchhoff's law for electrical circuits provides the last analogy needed. In heat transfer notation this would be:

$$\sum q = q_{\text{stored}} \tag{26}$$

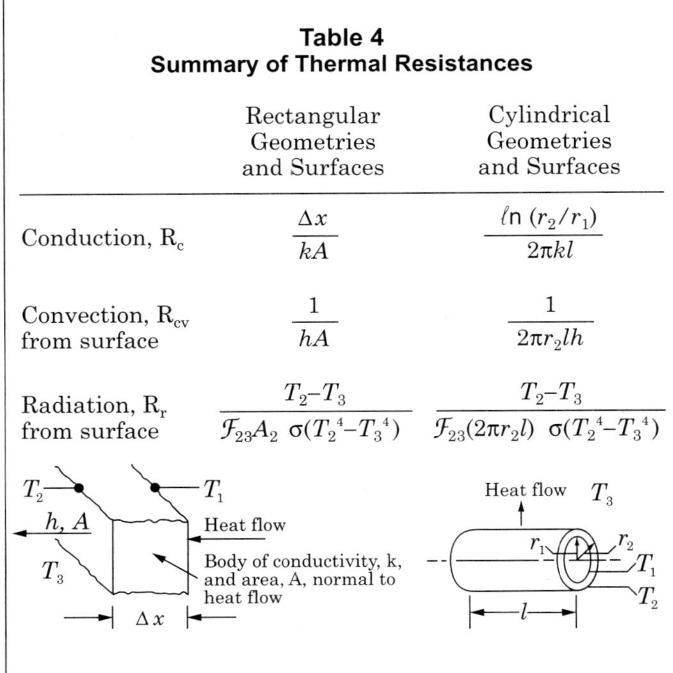

Table 4
Summary of Thermal Resistances

	Rectangular Geometries and Surfaces	Cylindrical Geometries and Surfaces
Conduction, R_c	$\frac{\Delta x}{kA}$	$\frac{\ln(r_2/r_1)}{2\pi kl}$
Convection, R_{cv} from surface	$\frac{1}{hA}$	$\frac{1}{2\pi r_2 lh}$
Radiation, R_r from surface	$\frac{T_2-T_3}{\mathcal{F}_{23}A_2\,\sigma(T_2^4-T_3^4)}$	$\frac{T_2-T_3}{\mathcal{F}_{23}(2\pi r_2 l)\,\sigma(T_2^4-T_3^4)}$

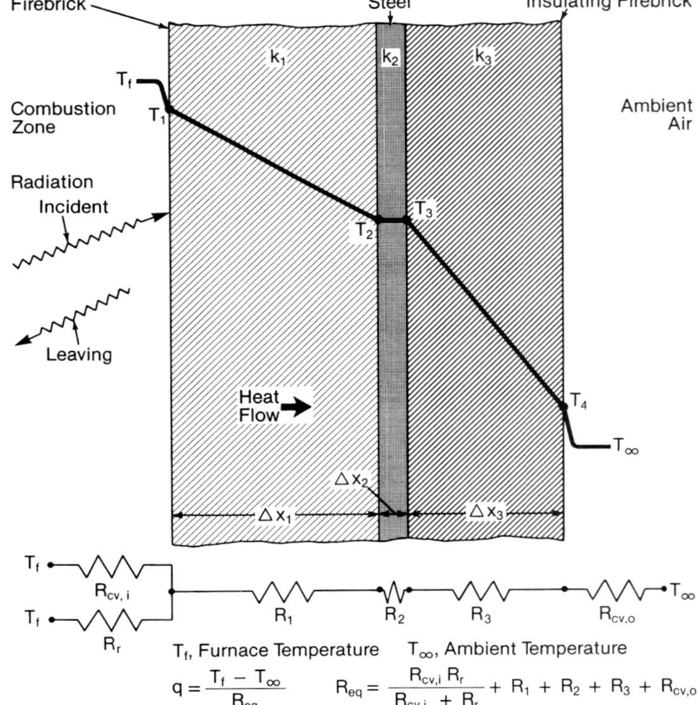

$$q = \frac{T_f - T_\infty}{R_{eq}} \qquad R_{eq} = \frac{R_{cv,i}\,R_r}{R_{cv,i} + R_r} + R_1 + R_2 + R_3 + R_{cv,o}$$

Fig. 3 Temperature distribution in composite wall with fluid films.

This is an expression of the first law of thermodynamics which states that all heat flows into a point equal the rate of energy storage.

Consider the composite system and the equivalent thermal circuit shown in Fig. 3. The concepts of resistance and conductance are particularly useful when more than one mode of heat transfer or more than one material or boundary is involved. When two modes of heat transfer, such as convection and radiation, occur simultaneously and independently, the combined conductance, K, is the sum of the individual conductances, K_{cv} and K_r. These individual conductances are essentially heat flows in parallel. When the heat flows are in series, the resistances, not the conductances, are additive. The total or equivalent thermal resistance can then be substituted into Equation 23 to calculate the total heat flow.

Flowing systems Boilers have complex distributions of flow, temperature and properties. In the basic example depicted in Fig. 4, there is steady flow into and out of the system, which is lumped into a single control volume. This leads to a balance of energy written as:

$$\dot{m}_1 H_1 + \dot{m}_2 H_2 + \dot{m}_3 H_3 = \dot{m}_4 H_4 \tag{27}$$

where $\dot{m}$ is the mass flow rate at each inlet or outlet and H is the fluid enthalpy.

As discussed in Chapter 3, the full energy equation sets the net energy entering a system (from mass flow into and out of the system) equal to the internal heat generation, plus the work done by the system, plus energy conducted into the system, plus a viscous dissipation term (see Chapter 3, Equation 12). Viscous dissipation and work done in the boiler system can both usually be neglected. For steady-state conditions,

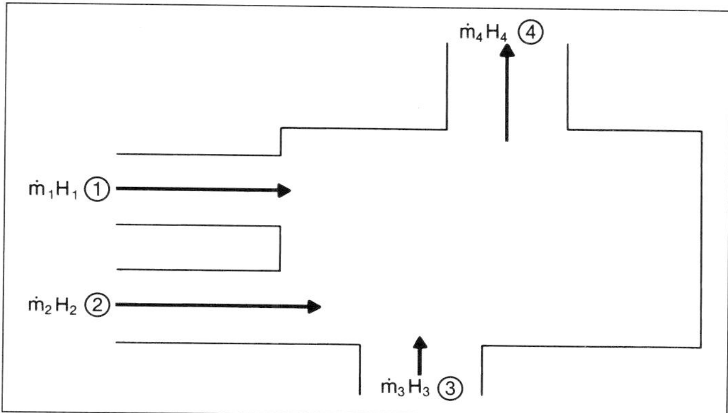

Fig. 4 Energy balance for a flowing system.

the energy equation in terms of fluid enthalpy and in vector notation can then be written as:

$$\underset{\text{Convection}}{\nabla \cdot (\rho \vec{u} H)} = \underset{\text{Conduction}}{\nabla \cdot (\Gamma \nabla H)} + \underset{\text{Internal heat generation}}{S_H} \quad \textbf{(28)}$$

The parameter Γ is an effective diffusion coefficient that includes molecular diffusion and turbulent diffusion. S_H is internal heat generation per unit volume. Rearranging Equation 28 and replacing enthalpy with temperature ($dH = c_p\, dT$) the energy equation becomes:

$$\rho c_p \vec{u} \cdot \nabla T = \nabla \cdot (k_{eff} \nabla T) + S_H \quad \textbf{(29)}$$

The parameter $k_{eff} = c_p \Gamma$ is effective thermal conductivity. Using x, y, z Cartesian coordinates, the equation is expressed:

$$\rho c_p \left(u \frac{\partial T}{\partial x} + v \frac{\partial T}{\partial y} + w \frac{\partial T}{\partial z} \right)$$
$$= \frac{\partial}{\partial x}\left(k_{eff} \frac{\partial T}{\partial x} \right) + \frac{\partial}{\partial y}\left(k_{eff} \frac{\partial T}{\partial y} \right) + \frac{\partial}{\partial z}\left(k_{eff} \frac{\partial T}{\partial z} \right) + S_H \quad \textbf{(30)}$$

In boiler applications, internal heat generation (S_H) includes radiation absorption and emission from participating gases, and heat release from combustion. The complete velocity field and thermal boundary conditions are necessary to find a solution to the fluid temperature field.

Equations 28 or 29 can be used with single phase flow or can be used with multiple phase flow (gas-solid or steam-water) by using mass averaged enthalpy or temperature. The development and application of the energy equation in rectangular, cylindrical and spherical coordinates are discussed in References 1 and 2.

Most boiler applications are too complex for an algebraic solution of the energy equation. However, the continuity and momentum equations discussed in Chapter 3 (Equations 3 to 10) are combined with the energy equation to form a fundamental part of the computational models discussed in Chapter 6. A numerical solution to this equation can then be readily achieved for many complex problems involving radiation and combustion.

Radiation balances for enclosures

Nonparticipating media Referring to Equation 11, the net radiation between two blackbody surfaces can be written:

$$q_{12} = A_1 F_{12} \sigma \left(T_1^4 - T_2^4 \right) \quad \textbf{(31)}$$

The term F_{12} is the geometric shape factor and is shown for two common geometries in Fig. 5. Use of the tabulated values for more complex problems is demonstrated in the example problems under *Applications.* Equation 31 has limited value in boilers, because most fireside surfaces are not blackbodies. This equation is better used to obtain estimates of radiation heat transfer, because it describes the maximum theoretical rate of energy transfer between two surfaces.

For the theory of radiation heat transfer in enclosures, see Reference 4. The energy striking a surface, called the *incident energy, G,* is the total energy striking a surface from all other surfaces in the enclosure. The energy leaving a surface, called the *radiosity, J,* is comprised of the energy emitted from the surface (E_b)

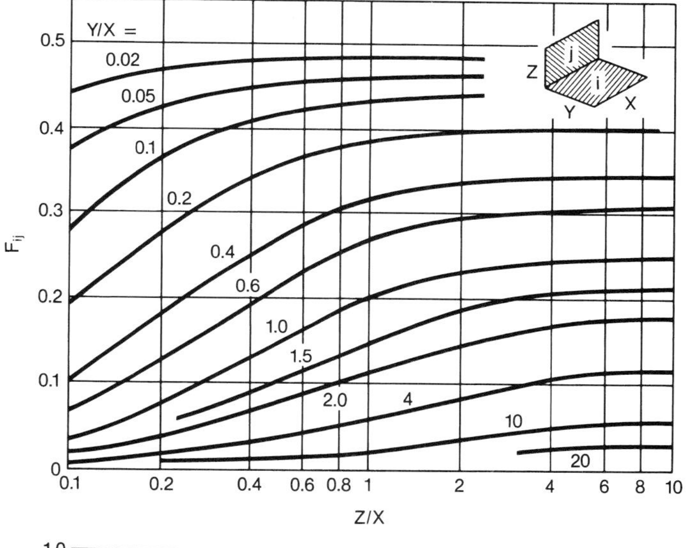

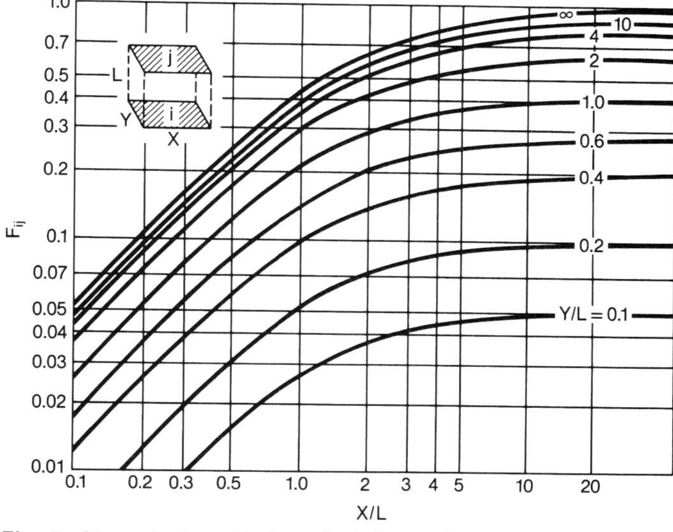

Fig. 5 Shape factors, F_{ij}, for calculating surface-to-surface radiation heat transfer.

and the reflected incident energy. These terms are related by:

$$J = \varepsilon E_b + \rho G \tag{32}$$

The net radiation heat transfer from a surface is found as follows:

$$q = A(J - G) \tag{33}$$

Combining Equations 32 and 33 leads to the electrical analogy listed first in Table 5. This circuit equivalent describes the potential difference between the surface at an emissive power, evaluated at T_s, and the surface radiosity. To evaluate the radiative heat transfer, the radiosity must first be determined. The net energy between surface i and surface j is the difference between the outgoing radiosities:

$$q_{ij} = A_i F_{ij} \left(J_i - J_j\right) \tag{34}$$

The electrical analogy of the net exchange between two surfaces is listed in Table 5. The sum of similar terms for all surfaces in the enclosure yields the circuit diagram in the table and the equation:

$$q_i = \sum_{j=1}^{N} A_i F_{ij} \left(J_i - J_j\right) \tag{35}$$

The rules for electrical circuits are useful in finding the net radiation heat transfer. Consider the electrical circuit for radiation heat transfer between two gray walls shown in Fig. 6.

The rules for series circuits can be used to determine the net heat transfer:

$$q_{12} = \frac{E_{b2} - E_{b1}}{R_1 + R_2 + R_3} \tag{36}$$

Table 5
Network Equivalents for Radiative Exchange in Enclosures with Gray Surfaces

Description	Circuit Equivalent	Resistance
Net exchange at surface	E_{bi} —R_i— J_i	$R_i = \dfrac{1-\varepsilon_i}{A_i \varepsilon_i}$
Net exchange between surfaces i and j	J_i —R_{ij}— J_j	$R_{ij} = \dfrac{1}{A_i F_{ij}}$
Net exchange between surface i and all other surfaces	J_i — J_1, J_2, J_k, J_j	$R_{ik} = \dfrac{1}{A_i F_{ik}}$

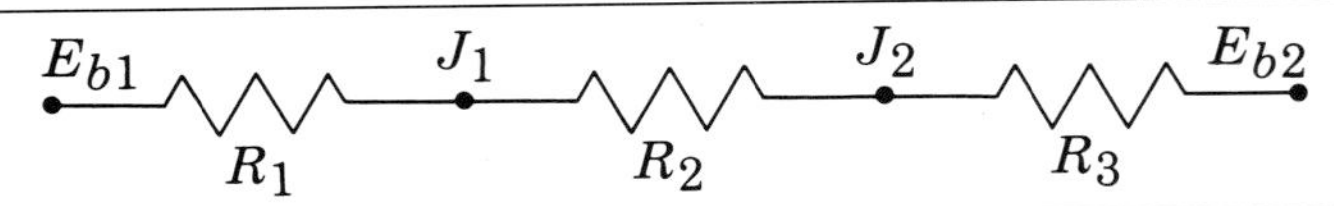

Fig. 6 Electric circuit analogy for thermal radiation.

where

$$R_1 = \frac{1-\varepsilon_1}{\varepsilon_1 A_1},\; R_2 = \frac{1}{A_1 F_{12}},\; R_3 = \frac{1-\varepsilon_2}{\varepsilon_2 A_2}$$

Total radiation exchange factors for common two-surface geometries encountered in boiler design are listed in Table 6.

Participating media On the fire side of the boiler, the mixture of gases and particles absorbs, emits, and scatters radiant energy. When a uniform temperature-bounding surface encloses an isothermal gas volume, the radiant heat transfer can be treated as one zone, with absorption and emission from the participating gas mixture. The incident radiation on the surfaces is made up of the emitted energy from the gas, $\varepsilon_g E_{bg}$, and incoming energy from the surrounding walls, $(1 - \alpha_g) J_s$. Therefore, the incident radiation is defined by:

$$G_s = \varepsilon_g E_{bg} + \left(1 - \alpha_g\right) J_s \tag{37}$$

Table 6
Common Gray Two-Surface Enclosures

Geometry	Relations	Exchange factor
Large (Infinite) Parallel Planes A_1, T_1, ε_1 A_2, T_2, ε_2	$A_1 = A_2 = A$ $F_{12} = 1$	$\mathcal{F}_{12} = \dfrac{1}{\dfrac{1}{\varepsilon_1} + \dfrac{1}{\varepsilon_2} - 1}$
Long (Infinite) Concentric Cylinders 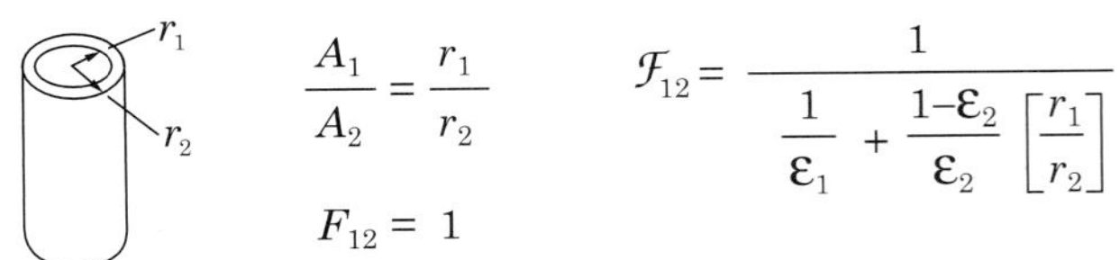 r_1, r_2	$\dfrac{A_1}{A_2} = \dfrac{r_1}{r_2}$ $F_{12} = 1$	$\mathcal{F}_{12} = \dfrac{1}{\dfrac{1}{\varepsilon_1} + \dfrac{1-\varepsilon_2}{\varepsilon_2}\left[\dfrac{r_1}{r_2}\right]}$
Concentric Spheres r_1, r_2	$\dfrac{A_1}{A_2} = \dfrac{r_1^2}{r_2^2}$ $F_{12} = 1$	$\mathcal{F}_{12} = \dfrac{1}{\dfrac{1}{\varepsilon_1} + \dfrac{1-\varepsilon_2}{\varepsilon_2}\left[\dfrac{r_1}{r_2}\right]^2}$
Small Convex Object in a Large Cavity A_1, T_1, ε_1 A_2, T_2, ε_2	$\dfrac{A_1}{A_2} \approx 0$ $F_{12} = 1$	$\mathcal{F}_{12} = \varepsilon_1$

Note: The net heat flow is calculated from $q_{12} = \sigma A_1 \mathcal{F}_{12}(T_1^4 - T_2^4)$

The energy leaving the surface is made up of direct emission, $\varepsilon_s E_{bs}$, and reflected incident energy, $(1-\varepsilon_s)G_s$. Therefore, the radiosity is:

$$J_s = \varepsilon_s E_{bs} + (1 - \varepsilon_s) G_s \quad \textbf{(38)}$$

The solution of Equations 37 and 38 yields values for the incoming and the outgoing heat fluxes (q / A_s). The net heat transfer between the surface and gas becomes:

$$q_{sg} = J_s - G_s = \frac{A_s \varepsilon_s (\varepsilon_g E_{bg} - \alpha_g E_{bs})}{1 - (1 - \alpha_g)(1 - \varepsilon_s)} \quad \textbf{(39)}$$

The calculation of absorptivity is described in the examples at the end of the chapter. When the surfaces are radiatively black, $\varepsilon_s = 1$ and Equation 39 becomes:

$$q_{sg} = A_s (\varepsilon_g E_{bg} - \alpha_g E_{bs}) \quad \textbf{(40)}$$

The procedures presented here provide the basis for engineering estimates which are described with examples at the end of the chapter. However, boiler enclosure wall and gas temperatures generally vary from wall to wall and even from point to point. A multi-zone, three-dimensional numerical analysis is then required, because simple expressions of surface heat transfer and participating media can not be solved analytically.

Numerical analysis of radiation heat transfer in multi-dimensional applications with absorption, emission and scattering is routinely applied with commercially available computational fluid dynamics (CFD) software. This involves the solution of an integral-differential equation for radiation intensity as a function of position, direction, and wavelength. The radiative transport equation (RTE) accounts for loss in intensity by absorption and scattering and gain in intensity by emission and scattering. Boundary conditions are applied for absorption, emission, and reflection at the surface. Integration of the equation over the blackbody spectrum simplifies the transport equation by eliminating the dependence on wavelength. The RTE for total (spectrally integrated) radiation intensity uses total radiative properties for gases, particles, and surfaces.

Heat transfer properties and correlations

Thermal conductivity, specific heat and density

Thermal conductivity, k, is a material property that is expressed in Btu/h ft F (W/m K) and is dependent on the chemical composition and physical characteristics of the substance. The relative order of magnitude of values for various substances is shown in Table 7. Thermal conductivities are generally highest for solids, lower for liquids and lower yet for gases. Insulating materials have the lowest conductivities of solid materials.

Thermal conductivities of pure metals generally decrease with an increase in temperature, while alloy conductivities may either increase or decrease. (See Fig. 7.) Conductivities of several steels and alloys are shown in Table 7. Thermal conductivities of various refractory materials are shown in Chapter 23, Fig. 10. For many heat transfer calculations it is sufficiently accurate to assume a constant thermal conductivity that corresponds to the average temperature of the material.

The effective thermal conductivity of ash deposits on water wall heating surfaces varies widely depending on temperature, composition, heating cycle and physical characteristics of the deposits. The lower limit is close to the thermal conductivity of air or lower (0.03 Btu/h ft F or 0.05 W/m K), and the upper limit does not exceed values for refractory materials (1.4 Btu/h ft F or 2.4 W/m K). The effective thermal conductivity of a friable particulate layer is near the lower limit and is fairly independent of temperature below 1650 to 2200F (899 to 1204C) at which sintering usually occurs. Above this temperature, particles fuse together and thermal contact between particles increases, resulting in a sharp increase in thermal conductivity. The highest conductivity is achieved with complete melting. The physical changes caused by fusion and melting are irreversible upon cooling, and thermal conductivity of fused deposits decreases with decreasing temperature.

Table 7
Properties of Various Substances at Room Temperature (see Note 1)

	ρ lb/ft³	c_p Btu/lb F	k Btu/h ft F
METALS			
Copper	559	0.09	223
Aluminum	169	0.21	132
Nickel	556	0.12	52
Iron	493	0.11	42
Carbon steel	487	0.11	25
Alloy steel 18Cr 8Ni	488	0.11	9.4
NONMETAL SOLIDS			
Limestone	105	~0.2	0.87
Pyrex® glass	170	~0.2	0.58
Brick K-28	27	~0.2	0.14
Plaster	140	~0.2	0.075
Kaowool	8	~0.2	0.016
GASES			
Hydrogen	0.006	3.3	0.099
Oxygen	0.09	0.22	0.014
Air	0.08	0.24	0.014
Nitrogen	0.08	0.25	0.014
Steam (see Note 2)	0.04	0.45	0.015
LIQUIDS			
Water	62.4	1.0	0.32
Sulfur dioxide (liquid)	89.8	0.33	0.12

Notes:
1. SI conversions: ρ, 1 lb/ft³ = 16.018 kg/m³; c_p, 1 Btu/lb F = 4.1869 kJ/kg K; k, 1 Btu/h ft F = 1.7307 W/m K.
2. Reference temperature equals 32F (0C) except for steam which is referenced at 212F (100C).

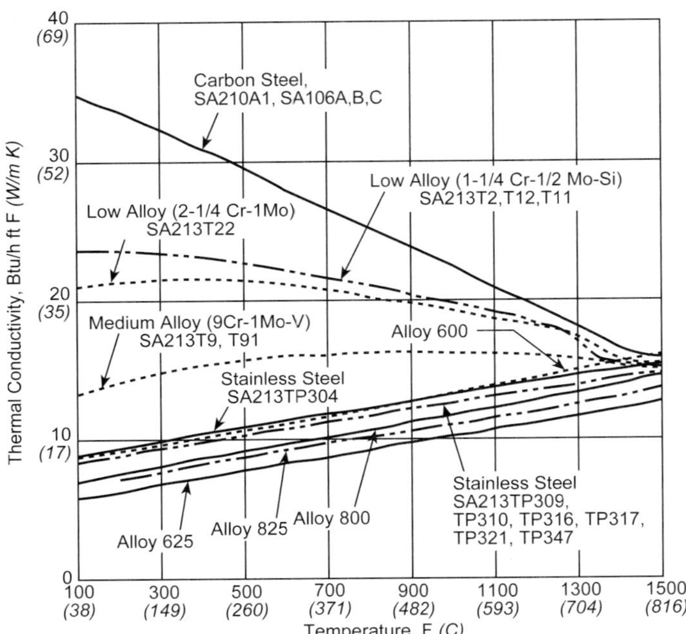

Fig. 7 Thermal conductivity, *k*, of some commonly used steels and alloys. (1 Btu/h ft F = 1.7307 W/m K)

Thermal conductance of ash deposits (k/x) is less sensitive to changing conditions than thermal conductivity. As the deposit grows in thickness (x), thermal conductivity (k) also increases due to fusion and slagging. The net effect is that unit thermal conductance may only vary by a factor of four, 25 to 100 Btu/h ft^2 F (142 to 568 W/m^2 C), while variations in thermal conductivity are an order of magnitude larger. The thermal effects of coal-ash deposits are further described by Wall et al.[5]

The thermal conductivity of water ranges from 0.33 Btu/h ft F (0.57 W/m K) at room temperature to 0.16 Btu/h ft F (0.28 W/m K) near the critical point. Water properties are relatively insensitive to pressure, particularly at pressures far from the critical point. Most other nonmetallic liquid thermal conductivities range from 0.05 to 0.15 Btu/h ft F (0.09 to 0.26 W/m K). In addition, thermal conductivities of most liquids decrease with temperature.

The thermal conductivities of gases increase with temperature and are independent of pressure at normal boiler conditions. These conductivities generally decrease with increasing molecular weight. The relatively high conductivity of hydrogen (a low molecular weight gas) makes it a good cooling medium for electric generators. The relatively low conductivity of argon (a high molecular weight gas) makes a good insulating medium for thermal pane windows.

When calculating the conductivity of nonhomogeneous materials, the designer must use an apparent thermal conductivity to account for the porous or layered construction materials. In boilers and furnaces with refractory walls, thermal conductivity may vary from site to site due to variations in structure, composition, density, or porosity when the materials were installed. The thermal conductivities of these materials are strongly dependent on their apparent bulk density (mass per unit volume). For higher temperature insulations, the apparent thermal conductivity of fibrous insulations and insulating firebrick decreases as bulk density increases, because the denser material attenuates the radiation. However, an inflection occurs at some point at which a further increase in density increases the thermal conductivity due to conduction in the solid material.

Theory shows that specific heats of solids and liquids are generally independent of pressure. Table 7 lists specific heats of various metals, alloys and nonhomogeneous materials at 68F (20C). These values may be used at other temperatures without significant error.

The temperature dependence of the specific heat for gases is more pronounced than for solids and liquids. In boiler applications, pressure dependence may generally be neglected. Tables 8a and 8b give specific heat data for air and other gases.

In the case of steam and water, property variations (specific heat and thermal conductivity) can be significant over the ranges of temperature and pressure found in boilers. It is therefore recommended that the properties as compiled in the American Society of Mechanical Engineers (ASME) Steam Tables[6] be used.

Radiation properties

Bodies that are good radiation absorbers are equally good emitters and Kirchhoff's law states that, for gray surfaces at thermal equilibrium, their emissivities are equal to their absorptivities. A *blackbody* is one which absorbs all incident radiant energy while reflecting or transmitting none of it. The absorptivity and emissivity of a blackbody are, by definition, each equal to one. This terminology does not necessarily mean that the body appears to be black. Snow, for instance, absorbs only a small portion of the incident visible light, but to the longer wavelengths (the bulk of thermal radiation), snow is almost a blackbody. At a temperature of 2000F (1093C) a blackbody glows brightly, because a non-negligible part of its radiation is in the visible range. Bodies are never completely black, but a hole through the wall of a large enclosure can be used to approximate blackbody conditions, because radiation entering the hole undergoes multiple reflections and absorptions. As a result, most of the radiation is retained in the enclosure, and surfaces are treated as gray.

Fortunately, a number of commercial surfaces, particularly at high temperatures, have emissivities of 0.80 to 0.95 and behave much like blackbodies. Typical average emissivity values are noted in Table 9. Although emissivity depends on the surface composition and roughness and wavelength of radiation, the wavelength dependence is often neglected in practical boiler calculations and surfaces are treated as gray.

Ash deposits The emittance and thermal properties of furnace ash deposits have a large effect on boiler heat transfer. The emittance depends on the temperature, chemical composition, structure and porosity of the particulate layer, and whether deposits are partially fused or molten. The same ash at different locations within the same boiler (or the same location in different boilers) may have significantly different values of surface emittance. Reported values in the

Table 8a
Properties of Selected Gases
at 14.696 psi (101.33 kPa) (see Note 1)

T F	ρ lb/ft³	c_p Btu/ lb F	k Btu/ h ft F	μ lbm/ ft h	Pr
Air					
0	0.0860	0.239	0.0133	0.0400	0.719
100	0.0709	0.240	0.0154	0.0463	0.721
300	0.0522	0.243	0.0193	0.0580	0.730
500	0.0413	0.247	0.0231	0.0680	0.728
1000	0.0272	0.262	0.0319	0.0889	0.730
1500	0.0202	0.276	0.0400	0.1080	0.745
2000	0.0161	0.286	0.0471	0.1242	0.754
2500	0.0134	0.292	0.0510	0.1328	0.760
3000	0.0115	0.297	0.0540	0.1390	0.765
Carbon Dioxide (CO_2)					
0	0.1311	0.184	0.0076	0.0317	0.767
100	0.1077	0.203	0.0100	0.0378	0.767
300	0.0793	0.226	0.0149	0.0493	0.748
500	0.0628	0.247	0.0198	0.0601	0.750
1000	0.0413	0.280	0.0318	0.0828	0.729
1500	0.0308	0.298	0.0420	0.1030	0.731
2000	0.0245	0.309	0.0500	0.1188	0.734
2500	0.0204	0.316	0.0555	0.1300	0.739
3000	0.0174	0.322	0.0610	0.1411	0.745
Water Vapor (H_2O)					
212	0.0372	0.451	0.0145	0.0313	0.974
300	0.0328	0.456	0.0171	0.0360	0.960
500	0.0258	0.470	0.0228	0.0455	0.938
1000	0.0169	0.510	0.0388	0.0691	0.908
1500	0.0127	0.555	0.0570	0.0889	0.866
2000	0.0100	0.600	0.0760	0.1091	0.861
2500	0.0083	0.640	0.0960	0.1289	0.859
3000	0.0071	0.670	0.1140	0.1440	0.846
Oxygen (O_2)					
0	0.0953	0.219	0.0131	0.0437	0.730
100	0.0783	0.220	0.0159	0.0511	0.707
300	0.0577	0.227	0.0204	0.0642	0.715
500	0.0457	0.235	0.0253	0.0759	0.705
1000	0.0300	0.253	0.0366	0.1001	0.691
1500	0.0224	0.264	0.0465	0.1195	0.677
2000	0.0178	0.269	0.0542	0.1414	0.701
2500	0.0148	0.275	0.0624	0.1594	0.703
3000	0.0127	0.281	0.0703	0.1764	0.703
Nitrogen (N_2)					
0	0.0835	0.248	0.0132	0.0380	0.713
100	0.0686	0.248	0.0154	0.0440	0.710
300	0.0505	0.250	0.0193	0.0547	0.710
500	0.0400	0.254	0.0232	0.0644	0.704
1000	0.0263	0.269	0.0330	0.0848	0.691
1500	0.0196	0.284	0.0423	0.1008	0.676
2000	0.0156	0.292	0.0489	0.1170	0.699
2500	0.0130	0.300	0.0565	0.1319	0.700
3000	0.0111	0.305	0.0636	0.1460	0.701

Note:
1. SI conversions: T(C) = [T(F) – 32]/1.8; ρ, 1 lb/ft³ = 16.018 kg/m³; c_p, 1 Btu/lb F = 4.1869 kJ/kg K; k, 1 Btu/h ft F = 1.7307 W/m K; μ, 1 lbm/ft h = 0.0004134 kg/m s.

Table 8b
Properties of Selected Gases
at 14.696 psi (101.33 kPa) (see Note 1)

T F	ρ lb/ft³	c_p Btu/ lb F	k Btu/ h ft F	μ lbm/ ft h	Pr
Flue gas – natural gas (see Note 2)					
300	0.0498	0.271	0.0194	0.0498	0.694
500	0.0394	0.278	0.0237	0.0593	0.694
1000	0.0259	0.298	0.0345	0.0803	0.694
1500	0.0193	0.317	0.0452	0.0989	0.693
2000	0.0154	0.331	0.0555	0.1160	0.692
2500	0.0128	0.342	0.0651	0.1313	0.691
3000	0.0109	0.351	0.0742	0.1456	0.689
Flue gas – fuel oil (see Note 3)					
300	0.0524	0.259	0.0192	0.0513	0.692
500	0.0415	0.266	0.0233	0.0608	0.694
1000	0.0273	0.287	0.0336	0.0817	0.696
1500	0.0203	0.304	0.0436	0.1001	0.697
2000	0.0162	0.316	0.0531	0.1169	0.697
2500	0.0134	0.326	0.0618	0.1318	0.696
3000	0.0115	0.334	0.0700	0.1459	0.695
Flue gas – coal (see Note 4)					
300	0.0537	0.254	0.0191	0.0519	0.691
500	0.0425	0.261	0.0232	0.0615	0.693
1000	0.0279	0.282	0.0333	0.0824	0.697
1500	0.0208	0.299	0.0430	0.1007	0.699
2000	0.0166	0.311	0.0521	0.1173	0.700
2500	0.0138	0.320	0.0605	0.1322	0.701
3000	0.0118	0.328	0.0684	0.1462	0.701

Notes:
1. SI conversions: T(C) = [T(F) – 32]/1.8; ρ, 1 lb/ft³ = 16.018 kg/m³; c_p, 1 Btu/lb F = 4.1869 kJ/kg K; k, 1 Btu/h ft F = 1.7307 W/m K; μ, 1 lbm/ft h = 0.0004134 kg/m s.
2. Flue gas composition by volume (natural gas, 15% excess air): 71.44% N_2, 2.44% O_2, 8.22% CO_2, 17.9% H_2O.
3. Flue gas composition by volume (fuel oil, 15% excess air): 74.15% N_2, 2.54% O_2, 12.53% CO_2, 0.06% SO_2, 10.72% H_2O.
4. Flue gas composition by volume (coal, 20% excess air): 74.86% N_2, 3.28% O_2, 13.97% CO_2, 0.08% SO_2, 7.81% H_2O.

literature claim emittances between 0.5 and 0.9 for most ash and slag deposits.

The effect of coal ash composition, structure, and temperature on deposit emittance[5,7] is shown in Fig. 8. A friable particulate material has low emittance because radiation is scattered (and reflected) from individual particles and does not penetrate beyond a thin layer (~1 mm) near the surface. Emittance of friable ash deposits decreases with increasing surface temperature, until sintering and fusion changes the structure of the deposit. A sharp increase in emittance is associated with ash fusion as particles grow together (pores close) and there are fewer internal surfaces to scatter radiation. Completely molten ash or slag is partially transparent to radiation, and emittance may depend upon substrate conditions. The emittance of completely fused deposits (molten or frozen slag) on oxidized carbon steel is about 0.9. Emittance increases

Table 9
Normal Emissivities, ε, for Various Surfaces[13] (see Note 1)

Material	Emissivity, ε	Temp., F	Description
Aluminum	0.09	212	Commercial sheet
Aluminum oxide	0.63 to 0.42	530 to 930	
Aluminum paint	0.27 to 0.67	212	Varying age and Al content
Brass	0.22	120 to 660	Dull plate
Copper	0.16 to 0.13	1970 to 2330	Molten
Copper	0.023	242	Polished
Cuprous oxide	0.66 to 0.54	1470 to 2012	
Iron	0.21	392	Polished, cast
Iron	0.55 to 0.60	1650 to 1900	Smooth sheet
Iron	0.24	68	Fresh emeried
Iron oxide	0.85 to 0.89	930 to 2190	
Steel	0.79	390 to 1110	Oxidized at 1100F
Steel	0.66	70	Rolled sheet
Steel	0.28	2910 to 3270	Molten
Steel (Cr-Ni)	0.44 to 0.36	420 to 914	18-8 rough, after heating
Steel (Cr-Ni)	0.90 to 0.97	420 to 980	25-20 oxidized in service
Brick, red	0.93	70	Rough
Brick, fireclay	0.75	1832	
Carbon, lamp-black	0.945	100 to 700	0.003 in. or thicker
Water	0.95 to 0.963	32 to 212	

Note:
1. SI conversion: T(C) = [T(F) − 32]/1.8; 1 in. = 25.4 mm.

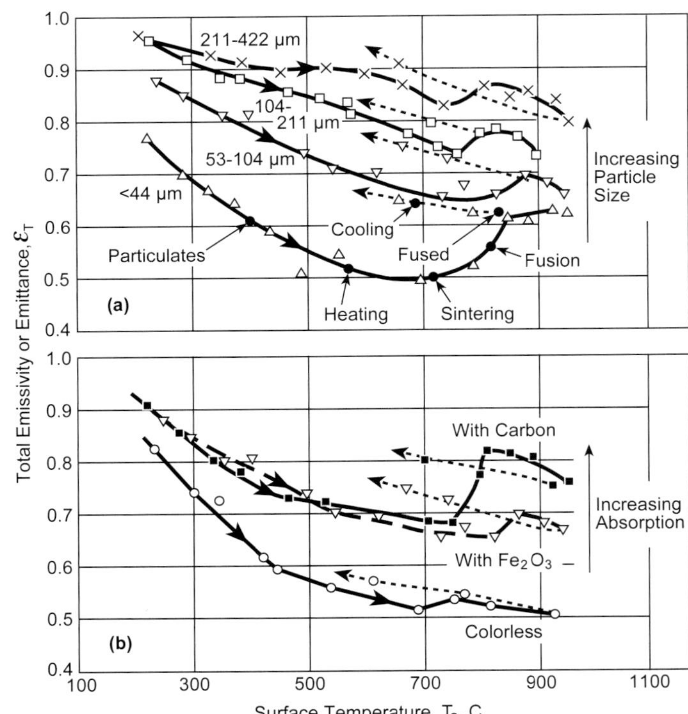

Fig. 8 Effect of coal ash composition, structure and temperature on deposit emittance.[5,7]

with increasing particle size of friable particulate deposits (Fig. 8a), because larger particles have less capacity to back-scatter incident radiation. Emittance increases with increasing iron oxide (Fe_2O_3) and unburned carbon content of the ash (Fig. 8b) because these components have a greater capacity to absorb radiation. Low emittance of some lignitic ash deposits, known as reflective ash, may be attributed to low Fe_2O_3 content, although this alone is not a reliable indicator of a reflective ash. Emittance is also indirectly dependent upon oxidizing and reducing environment of the flue gas, due to the effect on the melting characteristics and unburned carbon content in the ash. The thermal and radiative effects of coal-ash deposits are further described by Wall et al.[5]

Combustion gases Although many gases, such as oxygen and nitrogen, absorb or emit only insignificant amounts of radiation, others, such as water vapor, carbon dioxide, sulfur dioxide and carbon monoxide, substantially absorb and emit. Water vapor and carbon dioxide are important in boiler calculations because of their presence in the combustion products of hydrocarbon fuels. These gases are selective radiators. They emit and absorb radiation only in certain bands (wavelengths) of the spectrum that lie outside of the visible range and are consequently identified as nonluminous radiators. Whereas the radiation from a furnace wall is a surface phenomenon, a gas radiates and absorbs (within its absorption bands) at every point throughout the furnace. Furthermore, the emissivity of a gas changes with temperature, and the presence of one radiating gas may have characteristics that overlap with the radiating characteristics of another gas when they are mixed. The energy emitted by a radiating gaseous mixture depends on gas temperature, the partial pressures, p, of the constituents and a beam length, L, that depends on the shape and dimensions of the gas volume. An estimate of the mean beam length is L = 3.6 V/A for radiative transfer from the gas to the surface of the enclosure, where V is the enclosure volume and A is the enclosure surface area. The factor 3.6 is approximate, and values between 3.4 to 3.8 have been recommended depending on the actual geometry.[4]

Figs. 9 and 10 show the emissivity for water vapor and carbon dioxide.[8] The accuracy of these charts has gained greater acceptance than the more widely known charts of Hottel,[4] particularly at high temperatures and short path lengths. The effective emissivity of a water vapor-carbon dioxide mixture is calculated as follows:

$$\varepsilon = \varepsilon_{H_2O} + \varepsilon_{CO_2} - \Delta\varepsilon \qquad \textbf{(41)}$$

where $\Delta\varepsilon$ is a correction factor that accounts for the effect of overlapping spectral bands. This equation neglects pressure corrections and considers boilers operating at approximately 1 atm. The factors shown in Fig. 11 depend on temperature, the partial pressures, p, of the constituents and the beam length, L. The presence of carbon monoxide and sulfur dioxide can typically be neglected in combustion products, because CO and SO_2 are weakly participating and overlap with the infrared spectrum of H_2O and CO_2.

When using Figs. 9 to 11 to evaluate absorptivity, α, of a gas, Hottel[4] recommends modification of the pL product by a surface to gas temperature ratio. This is illustrated in Example 6 at the end of this chapter.

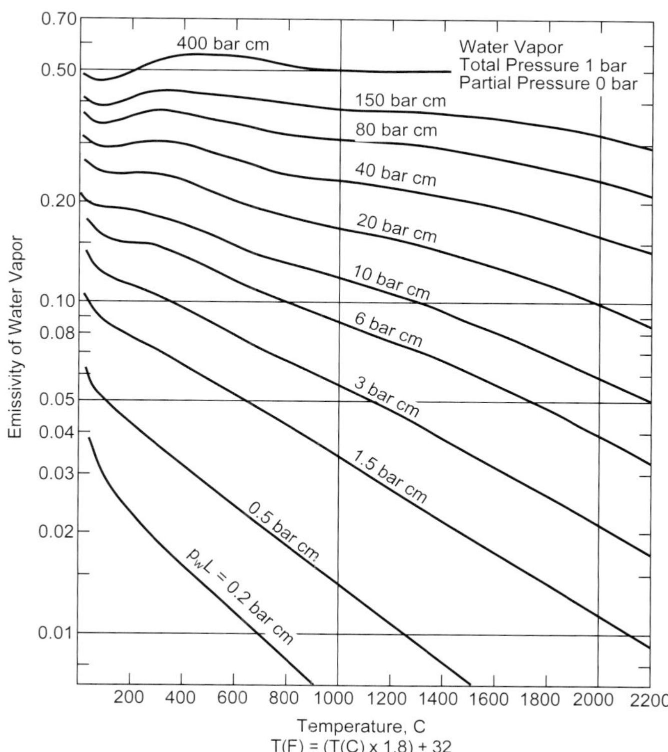

Fig. 9 Emissivity of water vapor at one atmosphere total pressure: p_wL= partial pressure in atmospheres x mean beam length in feet.[8] (1 bar-cm = 0.0324 ft-atm; T(F) = [T(C) x 1.8] + 32)

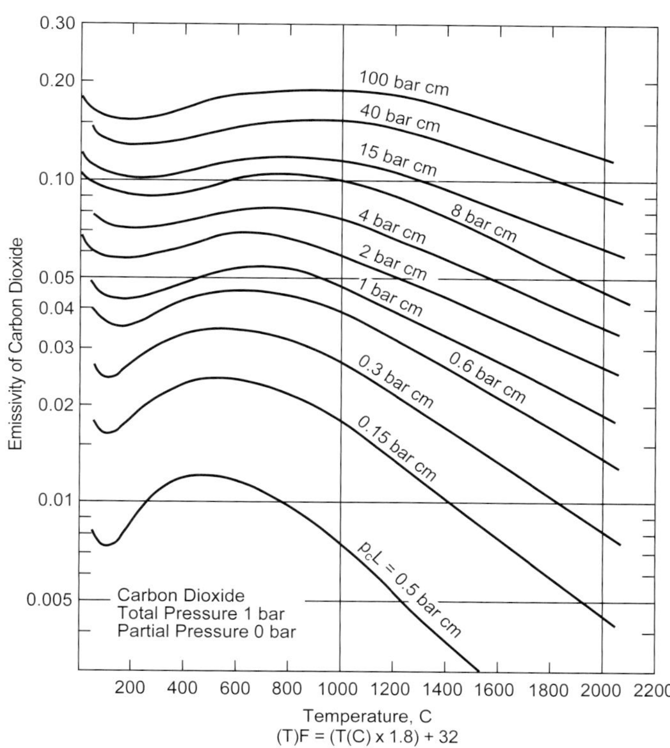

Fig.10 Emissivity of carbon dioxide at one atmosphere total pressure: p_cL= partial pressure in atmospheres x mean beam length in feet.[8] (1 bar-cm = 0.0324 ft-atm; T(F) = [T(C) x 1.8] + 32)

Radiation properties of gases can be calculated more accurately based on fundamental models for spectral gas radiation. The exponential wide band model predicts spectral absorption and emission properties of single and multi-component gases including H_2O, CO_2, CO, CH_4, NO, and SO_2 as a function of temperature and pressure. Diatomic gases N_2, O_2 and H_2 may contribute to the total gas volume and pressure of the mixture, but are considered transparent to infrared radiation. Radiation properties are conveniently expressed as emission and absorption coefficients that depend on local variations in gas composition, temperature, and pressure. This approach is suitable for numerical modeling of radiation with participating media, which requires frequent evaluation of gas properties at a large number of control volumes.

Entrained particles Combustion usually involves some form of particulate that is entrained in combustion gases. Particles are introduced as the fuel which undergo transformations of combustion and/or are formed by the processes of condensation and agglomeration of aerosol particles. Entrained particles have a significant role in radiation heat transfer because they absorb, emit, and scatter radiation. Scattering effectively extends the beam length of radiation in an enclosure, because the beam changes direction many times before it reaches a wall. Radiation from entrained particles depends on the particle shape, size distribution, chemical composition, concentration, temperature, and the wavelength of incident radiation.

Particulates in boilers are comprised of unreacted fuel (coal, oil, black liquor), char, ash, soot, and other aerosols. Soot is an example of an aerosol that con-

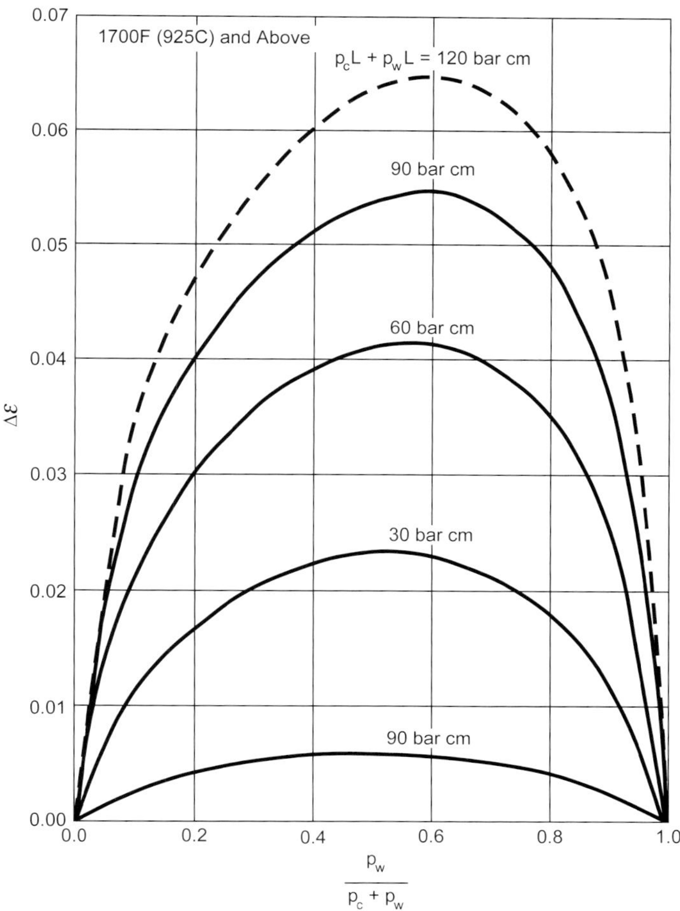

Fig. 11 Radiation heat transfer correction factor associated with mixtures of water vapor and carbon dioxide.[8] (1 bar-cm = 0.0324 ft-atm)

tributes to radiation from gas flames in boilers. Neglecting the effect of soot on radiation heat transfer in the flame could lead to significant errors in the calculated flame temperature, and radiation heat transfer to the furnace walls in the flame zone. Ash is an example of particulate that contributes to radiation in coal-fired boilers. Scattering by ash particles effectively redistributes radiation in the furnace, and smooths out variations in radiation heat flux, analogous to the way a cloud distributes solar radiation on the earth. The absorption and emission characteristics of flyash particles increase, and scattering decreases with the relative amount of iron oxide or residual carbon, which acts as a coloring agent in the ash.

Analytical methods such as Equation 39 that depend upon emissivity and absorptivity of the participating media are inaccurate when particles other than soot are involved, because the effects of scattering are neglected. Numerical methods which solve the general form of the radiative transport equation include the effects of scattering (see *Numerical methods*).

Mie Theory[10] is a general method for calculating the radiation properties of spherical particles as a function of particle composition, concentration, diameter and wavelength. Rigorous calculations by this method can only be performed with the aid of a computer and require that optical properties (complex refractive index as a function of wavelength) of the particle materials are known. The complex refractive index of lignite, bituminous, and anthracite coals, and corresponding properties of char and ash have been measured, as well as other materials that are typically encountered in combustion systems. Radiation properties of particles are conveniently expressed as total emission, absorption, and scattering efficiencies that depend on particle composition, diameter and temperature. Particle properties must be combined with gas properties in an analysis of radiation with participating media.

Working formulas for convection heat transfer

Heat transfer by convection between a fluid (gas or liquid) and a solid is expressed by Equation 4. This equation is a definition of the heat transfer coefficient but is inadequate in describing the details of the convective mechanisms. Only a comprehensive study of the flow and heat transfer would define the dependence of the heat transfer coefficient along the surface. In the literature, simple geometries have been modeled and predictions agree well with experimental data. However, for the more complex geometries encountered in boiler analysis, correlations are used that have been developed principally from experimental data.

Convective heat transfer near a surface takes place by a combination of conduction and mass transport. In the case of heat flowing from a heated surface to a cooler fluid, heat flows from the solid first by conduction into a fluid element, raising its internal energy. The heated element then moves to a cooler zone where heat flows from it by conduction to the cooler surrounding fluid.

Fluid motion can occur in two ways. If the fluid is set in motion due to density differences arising from temperature variations, free or natural convection occurs. If the motion is externally induced by a pump or fan, the process is referred to as forced convection.

Convective heat transfer can occur in laminar or turbulent flows. For laminar flow, the fluid moves in layers, or lamina, with each element following an orderly path. In turbulent flow, prevalent in boiler passages, the local motion of the fluid is chaotic and statistical treatment is used to establish average velocity and heat transfer values.

Experimental studies have confirmed that a flow field can be divided into two zones: a viscous zone adjacent to the surface and a nonviscous zone removed from the heat transfer surface. The viscous, heated zone is termed the boundary layer region. The hydrodynamic boundary layer is defined as the distance from the wall at which the local velocity reaches 99% of the velocity far from the wall.

At the entrance of a pipe or duct, the boundary layer begins to grow; this flow portion is called the developing region. Downstream, when the viscous region fills the pipe core or grows to a maximum, the flow is termed fully developed. Developing region heat transfer coefficients are larger than the fully developed values. In many applications it is sufficient to assume that the hydrodynamic and thermal boundary layers start to grow at the same location, although this is not always the case.

Flow over a body (around a circular cylinder) is termed external flow, while flow inside a confined region, like a pipe or duct, is termed internal flow.

Natural or free convection

A fluid at rest, exposed to a heated surface, will be at a higher temperature and lower density than the surrounding fluid. The differences in density, because of this difference in temperature, cause the lighter, warmer fluid elements to circulate and carry the heat elsewhere. The complex relationships governing this type of convective heat transfer are covered extensively in other texts.[1] Experimental studies have confirmed that the main dimensionless parameters governing free convection are the Grashof and Prandtl numbers:

$$\text{Gr} = \frac{g\beta\left(T_s - T_\infty\right)\rho^2 L^3}{\mu^2} \tag{42}$$

$$\text{Pr} = \frac{c_p \mu}{k} \tag{43}$$

The Grashof number is a ratio of the buoyant to viscous forces. The Prandtl number is the ratio of the diffusion of momentum and heat in the fluid. The product, Gr Pr, is also called the Rayleigh number, Ra.

In boiler system designs, air and flue gases are the important free convection heat transfer media. For these designs, the equation for the convective heat transfer coefficient h is:

$$h = C\left(T_s - T_\infty\right)^{1/3} \tag{44}$$

This correlation is applicable when the Rayleigh number, Ra, is greater than 10^9, which is generally recognized as the transition between laminar and turbulent flow. Values of the constant C in the equation are listed below:

Geometry	$\frac{Btu}{h\ ft^2\ F^{4/3}}$	$\frac{W}{m^2\ K^{4/3}}$
Horizontal plate facing upward	0.22	1.52
Vertical plates or pipes more than 1 ft (0.3 m) high	0.19	1.31
Horizontal pipes	0.18	1.24

The correlation generally produces convective heat transfer coefficients in the range of 1 to 5 Btu/h ft^2 F (5.68 to 28.39 W/m^2 K).

Forced convection

Dimensionless numbers Forced convection implies the use of a fan, pump or natural draft stack to induce fluid motion. Studies of many heat transfer systems and numerical simulation of some simple geometries confirm that fluid flow and heat transfer data may be correlated by dimensionless numbers. Using these principles, scale models enable designers to predict field performance. For simple geometries, a minimum of dimensionless numbers is needed for modeling. More complex scaling requires more dimensionless groups to predict unit performance.

The Reynolds number is used to correlate flow and heat transfer in closed conduits. It is defined as:

$$Re = \frac{\rho V L}{\mu} = \frac{GL}{\mu} \quad (45)$$

where L is a characteristic length of the conduit or an obstacle in the flow field. This dimensionless group represents the ratio of inertial to viscous forces.

The Reynolds number is only valid for a continuous fluid filling the conduit. The use of this parameter generally assumes that gravitational and intermolecular forces are negligible compared to inertial and viscous forces.

The characteristic length, termed equivalent hydraulic diameter, is different for circular and noncircular conduits. For circular conduits, the inside diameter (ID) is used. For noncircular ducts, the equivalent diameter becomes:

$$D_e = 4 \times \frac{\text{Flow cross-sectional area}}{\text{Wetted perimeter}} \quad (46)$$

This approach, used to compare dynamically similar fluids in geometrically similar conduits of different size, yields equal Reynolds numbers for the flows considered.

At low velocities, the viscous forces are strong and laminar flow predominates, while at higher velocities, the inertial forces dominate and there is turbulent flow. In closed conduits, such as pipes and ducts, the transition to turbulent flow occurs near Re = 2000. The generally accepted range for transition to turbulent flow under common tube flow conditions is 2000 < Re < 4000.

For fluid flow over a flat external surface, the characteristic length for the Reynolds number is the surface length in the direction of the flow, x. Transition to turbulence is generally considered for Re ≥ 10^5. In the case of flow over a tube, the outside diameter (OD), D, is the characteristic length. In tube bundles with crossflow, transition generally occurs at Re > 100.

Experimental studies have confirmed that the convective heat transfer coefficient can be functionally characterized by the following dimensionless groups:

$$Nu = f(Re, Pr) \quad (47)$$

where Nu is the Nusselt number, Re is the Reynolds number and Pr is the Prandtl number.

The Nusselt number, a ratio of the wall temperature gradient to reference gradients, is defined as follows:

$$Nu = \frac{hL}{k} \quad (48)$$

The previously discussed Prandtl number, representing a ratio of the diffusion of momentum and heat in the fluid, is also the ratio of the relative thickness of viscous and thermal boundary layers. For air and flue gases, Pr < 1.0 and the thermal boundary layer is thicker than the viscous boundary layer.

In the literature, correlations are also presented using other dimensionless groups; the Peclet and Stanton numbers are the most common. The Peclet number is defined as follows:

$$Pe = Re\ Pr \quad (49)$$

The Stanton number is defined in terms of the Nusselt, Reynolds and Prandtl numbers:

$$St = \frac{Nu}{Re\ Pr} \quad (50)$$

Laminar flow inside tubes For heating or cooling of viscous fluids in horizontal or vertical tubes with constant surface temperature and laminar flow conditions (Re < 2300), the heat transfer coefficient, or film conductance, can be determined by the following equation:[11]

$$Nu = 1.86 \left(Re\ Pr \frac{D}{L} \right)^{1/3} \left(\frac{\mu_b}{\mu_w} \right)^{0.14} \quad (51)$$

or

$$h = 1.86 \frac{k_b}{D} \left(\frac{GD}{\mu} \frac{c_p \mu}{k} \frac{D}{L} \right)_b^{1/3} \left(\frac{\mu_b}{\mu_w} \right)^{0.14} \quad (52)$$

where the parameter $G = \rho V$ is defined as the mass flux or mass flow rate per unit area and tube diameter, D, is the characteristic length used in the evaluation of the Reynolds number. The ratio of viscosities (μ_b/μ_w) is a correction factor that accounts for temperature dependent fluid properties. Properties in Equations 51 and 52 are evaluated at an average bulk fluid temperature, except μ_w which is evaluated at the wall temperature.

For low viscosity fluids, such as water and gases, a more complex equation is required to account for the effects of natural convection at the heat transfer surface. This refinement is of little interest in industrial practice because water and gases in laminar flow are rarely encountered.

Turbulent flow Studies of turbulent flow indicate several well defined regions as shown in Fig. 12. Next to the heat transfer surface is a very thin laminar flow region, less than 0.2% of the characteristic length, where the heat flow to or from the surface is by molecular conduction. The next zone, known as the buffer layer, is less than 1% of the characteristic length and is a mixture of laminar and turbulent flow. Here the heat is transferred by a combination of convection and conduction. In the turbulent core, which comprises roughly 98% of the cross-section, heat is transferred mainly by convection.

In turbulent flow, the local but chaotic motion of the fluid causes axial and radial motion of fluid elements. This combination of motions sets up eddies, or local swirling motions, augmenting the heat transfer from the core to the laminar sublayer. The laminar flow in the sublayer and the laminar component in the buffer layer act as a barrier, or film, to the heat transfer process. Increasing the fluid velocity has been found to decrease this film thickness, reducing the resistance to heat transfer.

Turbulent flow in tubes The distance required to obtain hydrodynamically and thermally fully developed turbulent flow is shorter than that for laminar flow. The flow length needed to achieve hydrodynamically fully developed conditions is variable and depends upon the specific Reynolds number (operating conditions) and surface geometry. It typically varies from 6 to 20 diameters (x/D). Fully developed thermal flow for gases and air, important in boiler analysis, occurs at similar x/D ratios. However, for liquids, the ratio is somewhat higher and increases with the Prandtl number.

Extensive research data using low viscosity gases and liquids have been correlated. The following equation[12] is recommended for fully developed flow with small to moderate temperature differences:

$$\mathrm{Nu}_{fd} = 0.023\,\mathrm{Re}^{0.8}\,\mathrm{Pr}^{n} \tag{53}$$

with n = 0.4 for heating of the fluid and n = 0.3 for cooling of the fluid, and properties evaluated at the bulk temperature. Equation 53 applies to gases and liquids in the range $0.7 < \mathrm{Pr} < 160$, which covers all fluids in boiler analysis. If the conditions are not fully developed, the correlation is corrected as shown below:[13]

$$\mathrm{Nu} = \mathrm{Nu}_{fd}\left[1 + (D/x)^{0.7}\right] \tag{54}$$

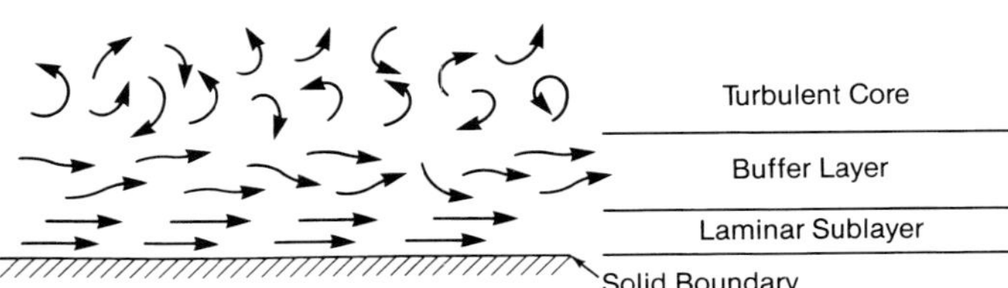

Fig. 12 Structure of turbulent flow field near a solid boundary.

with the stipulation that $2 \le x/D \le 20$. These correlations should only be used for small to moderate temperature differences.

A correlation by Seider and Tate[11] is widely used for heating or cooling of a fluid and larger temperature differences. All of the properties are evaluated at the bulk temperature, except μ_w which is evaluated at the wall temperature:

$$\mathrm{Nu}_{fd} = 0.027\,\mathrm{Re}^{0.8}\,\mathrm{Pr}^{1/3}\left(\frac{\mu_b}{\mu_w}\right)^{0.14} \tag{55}$$

The foregoing correlations may be applied for both constant surface temperature and heat flux conditions to a good approximation.

For boiler applications involving turbulent flow in tubes, Equation 53 is rewritten with the temperature ratio added to convert the properties from a bulk to film temperature basis:

$$\mathrm{Nu}_{fd} = 0.023\,\mathrm{Re}_f^{0.8}\,\mathrm{Pr}_f^{0.4}\left(\frac{T_b}{T_f}\right)^{0.8} \tag{56}$$

All properties are evaluated at the film temperature (T_f), which is defined as the arithmetic mean temperature between the wall temperature (T_w) and the bulk fluid temperature (T_b): $T_f = (T_w + T_b)/2$ with all temperatures in absolute units (R or K). Equation 56 is rewritten using parametric groupings:

$$h_l = \left[0.023\,\frac{G^{0.8}}{D_e^{0.2}}\right]\left[\frac{c_p^{0.4}\,k^{0.6}}{\mu^{0.4}}\right]_f\left[\frac{T_b}{T_f}\right]^{0.8} \tag{57}$$

which can be expressed in the form:

$$h_l = h_l'\,F_{pp}\,F_T \tag{58}$$

Figs. 13 to 17 display the various factors that make up the right side of Equation 58. Unlike non-dimensional parameters (Nu, Re, Pr), these terms do not have any physical significance and are dependent upon the choice of engineering units. The physical properties factor, F_{pp}, combines all of the properties of the fluid into one term, and is evaluated at the gas film temperature for a particular fluid (gas, air or steam). Note that if F_{pp} for steam can not be obtained from Fig. 16, it can be calculated with values of c_p, k and μ evaluated at the film temperature from the ASME Steam Tables.[6]

Turbulent cross flow around tubes The most important boiler application of convection is heat transfer from the combustion gases to the tubular surfaces in the convection passes. Perhaps the most complete and authoritative research on heat transfer of tubes in crossflow was completed in an extensive program conducted by The Babcock & Wilcox Company (B&W).[14] The following correlation was adapted from this study for different fluids:

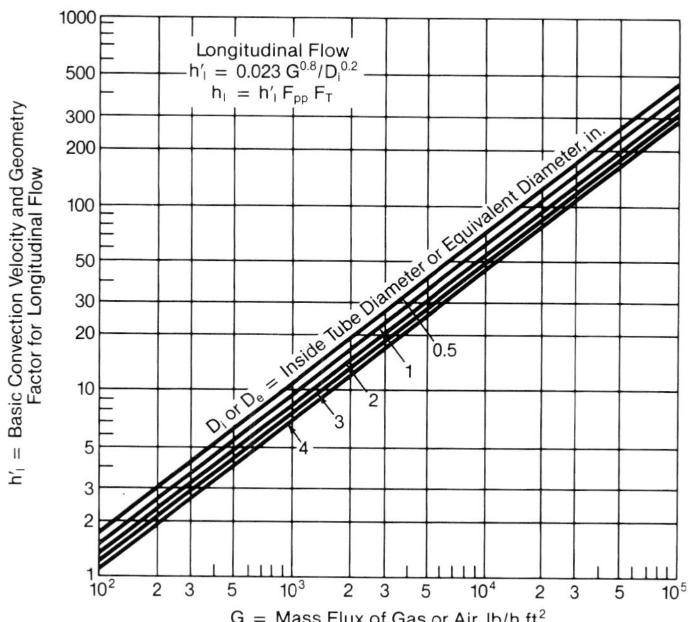

Fig. 13 Basic convection velocity and geometry factor, h'_l, for air, gas or steam; turbulent flow inside tubes or longitudinal flow over tubes (English units only).

$$\text{Nu} = 0.321\,\text{Re}_f^{0.61}\,\text{Pr}_f^{0.33}\,F_a\,F_d \qquad \textbf{(59)}$$

The last terms are an arrangement factor, F_a, and a depth factor F_d, that correct the results from the base configuration ($\ell_{\parallel}/D_O = 2.0$, $\ell_{\perp}/D_O = 1.75$, number of rows ≥ 10) which by definition $F_a = F_d = 1$. The equation applies to heating and cooling of fluids for clean tubes in crossflow. Equation 59 is rewritten using parametric groupings shown below:

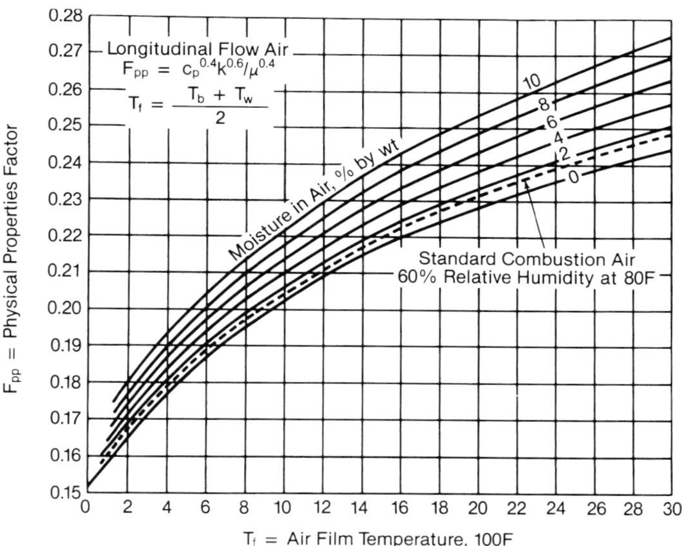

Fig. 15 Effect of film temperature, T_f, and moisture on the physical properties factor, F_{pp}, for air; turbulent flow inside tubes or longitudinal flow over tubes (English units only).

$$h_c = \left[\frac{0.321\,G^{0.61}}{D^{0.39}}\right]\left[\frac{c_p^{0.33}\,k^{0.67}}{\mu^{0.28}}\right]_f F_a\,F_d \qquad \textbf{(60)}$$

which can be expressed in the form:

$$h_c = h'_c\;F_{pp}\;F_a\;F_d \qquad \textbf{(61)}$$

Figs. 18 to 23 display the various factors that make up the right side of Equation 61. Unlike non-dimensional parameters (Nu, Re, Pr), these terms do not

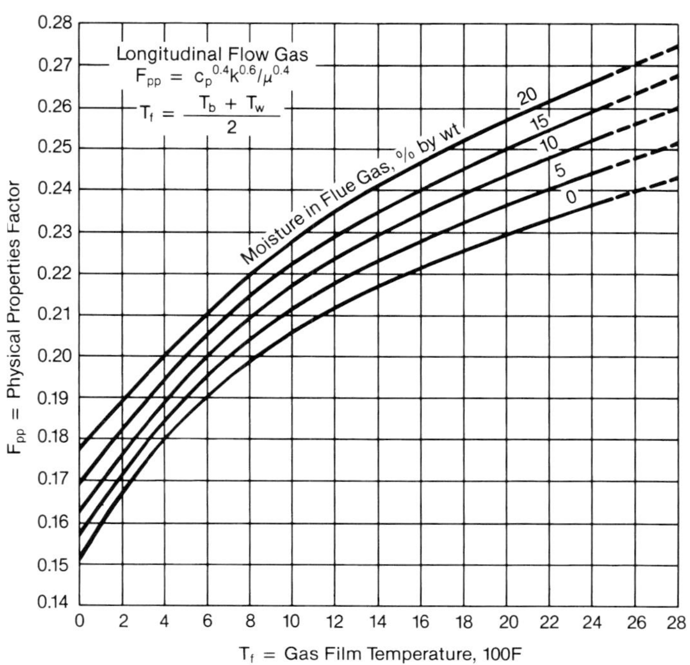

Fig. 14 Effect of film temperature, T_f, and moisture on the physical properties factor, F_{pp}, for gas; turbulent flow inside tubes or longitudinal flow over tubes (English units only).

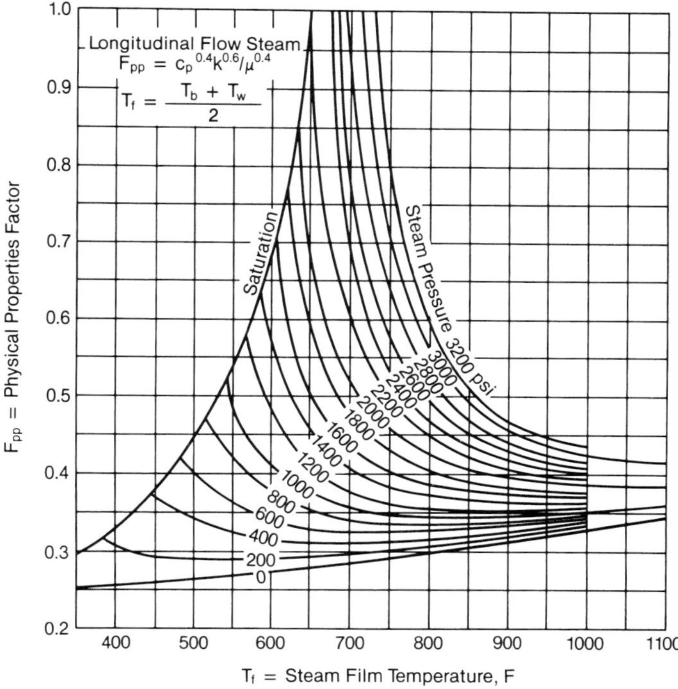

Fig. 16 Effect of film temperature, T_f, and pressure on the physical properties factor, F_{pp}, for steam; turbulent flow inside tubes or longitudinal flow over tubes (English units only).

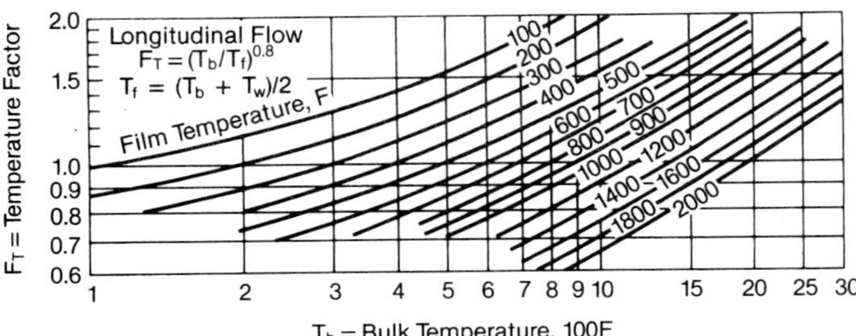

Fig. 17 Temperature factor, F_T, for converting mass velocity from bulk to film basis for air, gas or steam; turbulent flow inside tubes or longitudinal flow over tubes.

have any physical significance and are dependent upon the choice of engineering units. The physical properties factor, F_{pp}, similar to the one previously defined, is evaluated at the gas film temperature for a particular fluid (gas or air). The mass flux or mass flow per unit area, G, and the Reynolds numbers used in Equations 59 and 60 and Figs. 18, 21 and 22 are calculated based on flow conditions at the minimum free area (maximum velocity) between tubes.

The arrangement factor, F_a, depends on the geometric configuration of tubes, the ratio of tube spacing to diameter, Reynolds number, and the presence of ash in the flue gas. Values of F_a for clean tube conditions with air or flue gas without ash are given in Fig. 21. Values of F_a for commercially clean tube conditions with ash-laden flue gas are given in Fig. 22.

The depth factor, F_d, accounts for entrance effects for banks of tubes which are less than ten rows deep in the direction of gas flow. For undisturbed flow [flow that is straight and uninterrupted for at least 4 ft (1.2 m) before entering a tube bank] approaching a bank of less than ten rows, the film conductance must include the correction factor, F_d, shown in Fig. 23. F_d is unity when the tube bank is preceded by a bend, screen, damper or another tube bank in close proximity.

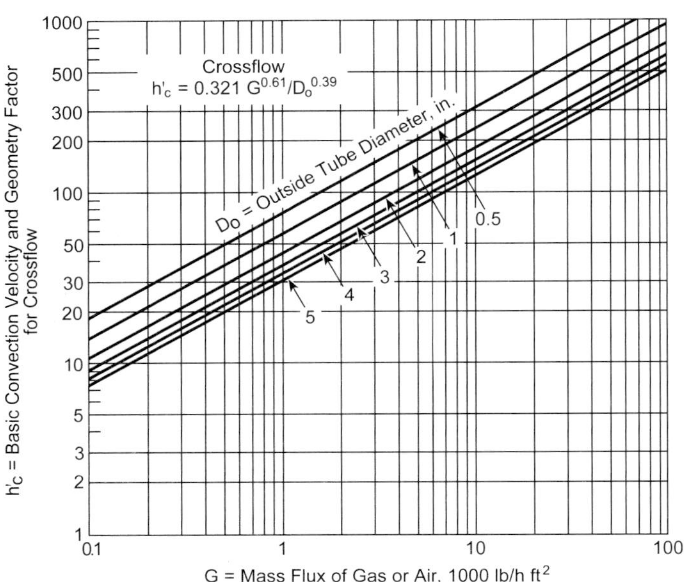

Fig. 18 Basic crossflow convection velocity and geometry factor, h'_c, for gas or air (English units only).

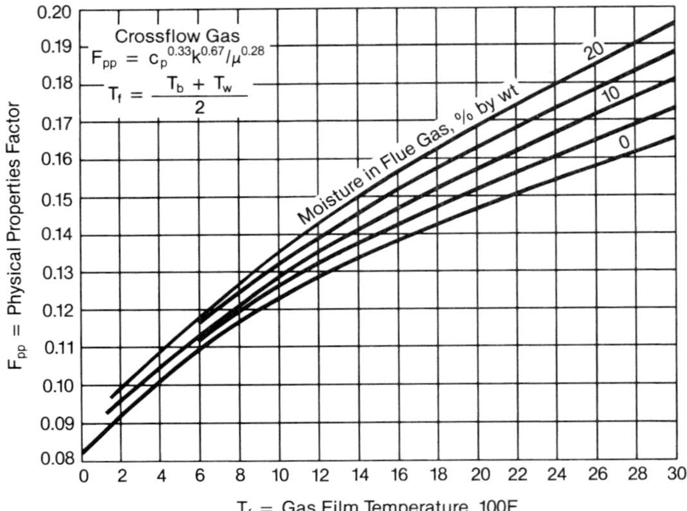

Fig. 19 Effect of film temperature, T_f, and moisture on the physical properties factor, F_{pp}, for gas in turbulent crossflow over tubes (English units only).

Turbulent longitudinal flow around tubes Correlations that were developed based on turbulent flow in tubes (Equations 56 and 57, and Figs. 13 to 17) can also be applied for external flow parallel to tubes. In this case, the equivalent diameter D_e (defined by Equation 46) is used in the evaluation of Reynolds number. For flow parallel to a bank of circular tubes arranged on rectangular spacing, the equivalent diameter becomes:

$$D_e = \frac{4\ell_1 \ell_2}{\pi D_o} - D_o \qquad \textbf{(62)}$$

where D_o is the tube outside diameter and ℓ_1 and ℓ_2 are the centerline spacing between tubes. The mass flux or mass flow per unit area, G, in Equations 56 and 57, and Fig. 13 is calculated based on the free area between tubes.

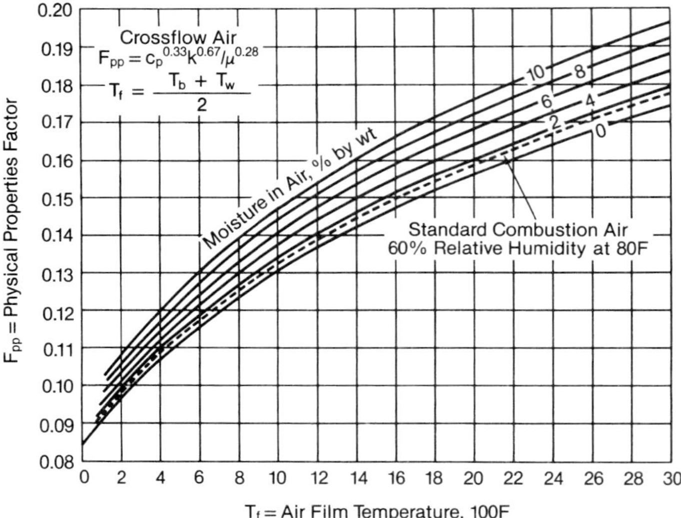

Fig. 20 Effect of film temperature, T_f, and moisture on the physical properties factor, F_{pp}, for air in crossflow over tubes (English units only).

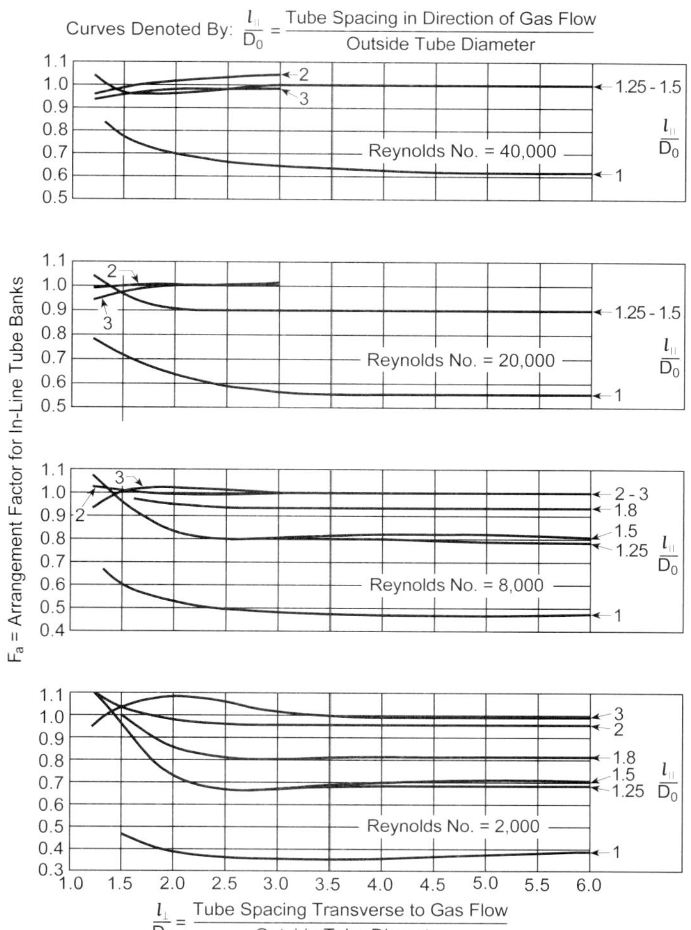

Fig. 21 Arrangement factor, F_a, as affected by Reynolds number for various in-line tube patterns, clean tube conditions for crossflow of air or natural gas combustion products.

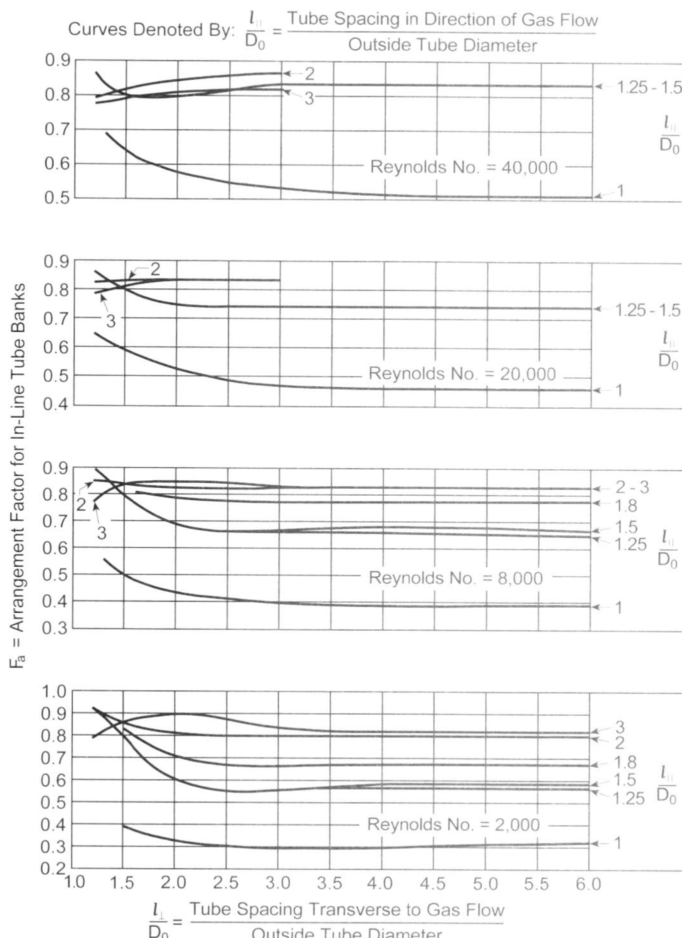

Fig. 22 Arrangement factor, F_a, as affected by Reynolds number for various in-line tube patterns, commercially clean tube conditions for crossflow of ash-laden gases.

General heat transfer topics

Heat exchangers

Boiler systems contain many heat exchangers. In these devices, the fluid temperature changes as the fluids pass through the equipment. With an energy balance specified between two locations, 1 and 2:

$$q = \dot{m} c_p (T_2 - T_1) \tag{63}$$

the change in fluid temperature can be calculated:

$$T_2 = T_1 + \left(q / \dot{m} c_p\right) \tag{64}$$

It is therefore appropriate to define a mean effective temperature difference governing the heat flow. This difference is determined by performing an energy balance on the energy lost by the hot fluid and that energy gained by the cold fluid. An equation of the form:

$$q = UA\,F\,\Delta T_{LMTD} \tag{65}$$

is obtained where the parameters U, A and F define the overall heat transfer coefficient, surface area, and arrangement correction factor, respectively. The term ΔT_{LMTD}, known as the log mean temperature difference, is defined as:

$$\Delta T_{LMTD} = \frac{\Delta T_1 - \Delta T_2}{\ell n\left(\Delta T_1 / \Delta T_2\right)} \tag{66}$$

ΔT_1 is the initial temperature difference between the hot and cold fluids (or gases), while ΔT_2 defines the final temperature difference between these media. The pa-

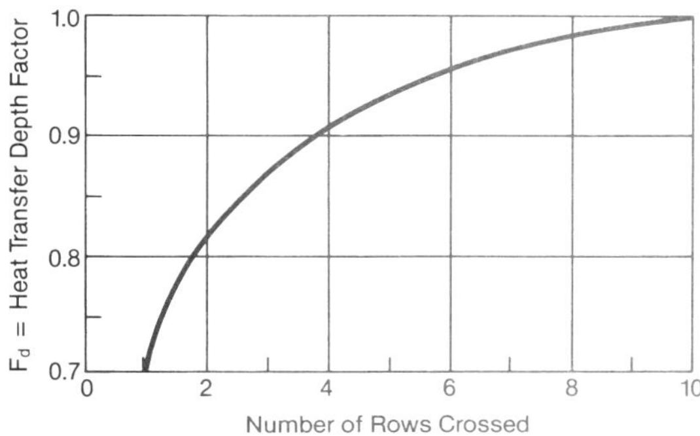

Fig. 23 Heat transfer depth factor for number of tube rows crossed in convection banks. (F_d = 1.0 if tube bank is immediately preceded by a bend, screen or damper.)

rameter U in Equation 65 defines the overall heat transfer coefficient for clean surfaces and represents the unit thermal resistance between the hot and cold fluids:

$$\frac{1}{UA_{clean}} = \frac{1}{h_i A_i} + R_w + \frac{1}{h_o A_o} \tag{67}$$

For surfaces that are fouled, the equation is written:

$$\frac{1}{UA} = \frac{R_{f,i}}{A_i} + \frac{1}{UA_{clean}} + \frac{R_{f,o}}{A_o} \tag{68}$$

where $R_{f,i}$ is the reciprocal effective heat transfer coefficient of the fouling on the inside surface, $(1 / UA)$ is the thermal resistance and $R_{f,o}$ is the reciprocal heat transfer coefficient of the fouling on the outside surface. Estimates of overall heat transfer coefficients and fouling factors are listed in Tables 10 and 11. Actual fouling factors are site specific and depend on water chemistry and other deposition rate factors. Overall heat transfer coefficients can be predicted using: 1) the fluid conditions on each side of the heat transfer surface with either Equation 56 or 59, 2) the known materials of the heat transfer surface, and 3) the fouling factors listed in Table 11. Often the heat exchanger tube wall resistance (R_w) is small compared to the surface resistances and can be neglected, leading to the following equation for a clean surface:

$$U = \frac{h_i h_o}{h_i + (h_o D_o / D_i)} \tag{69}$$

This equation assumes that area, A in product UA, is based on the outside diameter of the tube, D_o.

The difficulty in quantifying fouling factors for gas-, oil- and coal-fired units has led to use of a *cleanliness factor*. This factor provides a practical way to provide extra surface to account for the reduction in heat transfer due to fouling. In gas-fired units, experience indicates that gas-side heat transfer coefficients are higher as a result of the cleanliness of the surface. In oil- and coal-fired units that are kept free of slag and deposits, a lower value is used. For units with difficult to remove deposits, values are reduced further.

Table 10
Approximate Values of Overall Heat Transfer Coefficients

Physical Situation	Btu/h ft^2 F	W/m^2 K
Plate glass window	1.10	6.20
Double plate glass window	0.40	2.30
Steam condenser	200 to 1000	1100 to 5700
Feedwater heater	200 to 1500	1100 to 8500
Water-to-water heat exchanger	150 to 300	850 to 1700
Finned tube heat exchanger, water in tubes, air across tubes	5 to 10	30 to 55
Water-to-oil heat exchanger	20 to 60	110 to 340
Steam-to-gas	5 to 50	30 to 300
Water-to-gas	10 to 20	55 to 110

Table 11
Selected Fouling Factors

Type of Fluid	h ft^2 F/Btu	m^2 K/W
Sea water above 125F (50C)	0.001	0.0002
Treated boiler feedwater above 125F (50C)	0.001	0.0002
Fuel oil	0.005	0.0010
Alcohol vapors	0.0005	0.0001
Steam, non-oil bearing	0.0005	0.0001
Industrial air	0.002	0.0004

There are three general heat transfer arrangements: parallel flow, counterflow and crossflow, as shown in Fig. 24. In parallel flow, both fluids enter at the same relative location with respect to the heat transfer surface and flow in parallel paths over the heating surface. In counterflow, the two fluids enter at opposite ends of the heat transfer surface and flow in opposite directions over the surface. This is the most efficient heat exchanger although it can also lead to the highest tube wall metal temperatures. In crossflow, the paths of the two fluids are, in general, perpendicular to one another.

Fig. 24 shows the flow arrangements and presents Equation 66 written specifically for each case. The arrangement correction factor, F, is 1.0 for parallel and counterflow cases. For crossflow and multi-pass arrangements, the correction factors are shown in Figs. 25 and 26.

Extended surface heat transfer

The heat absorption area in boilers can be increased using longitudinally and circumferentially finned tubes. Finned, or extended, tube surfaces are used on the flue gas side. In regions prone to fouling, the fins must be spaced to permit cleaning. Experimental data on actual finned or extended surfaces are preferred for design purposes; the data should be collected at conditions similar to those expected to be encountered. However, in place of these data, the method by Schmidt[15] generally describes the heat transfer across finned tubes. It is based on heat transfer to the underlying bare tube configuration, and it treats the tube as if it has zero fin height. Schmidt's correlation for the gas-side conductance to tubes with helical, rectangular, circular, or square fins is as follows:

$$h_f = h_c Z \left\{ 1 - (1 - \eta_f) \left(\frac{S_f}{S} \right) \right\} \tag{70}$$

where h_c is the heat transfer coefficient of the bare tubes in crossflow defined by Equations 59 and 60, and Z is the geometry factor defined as:

$$Z = 1 - 0.18 \left(\frac{L_h}{L_t} \right)^{0.63} \tag{71}$$

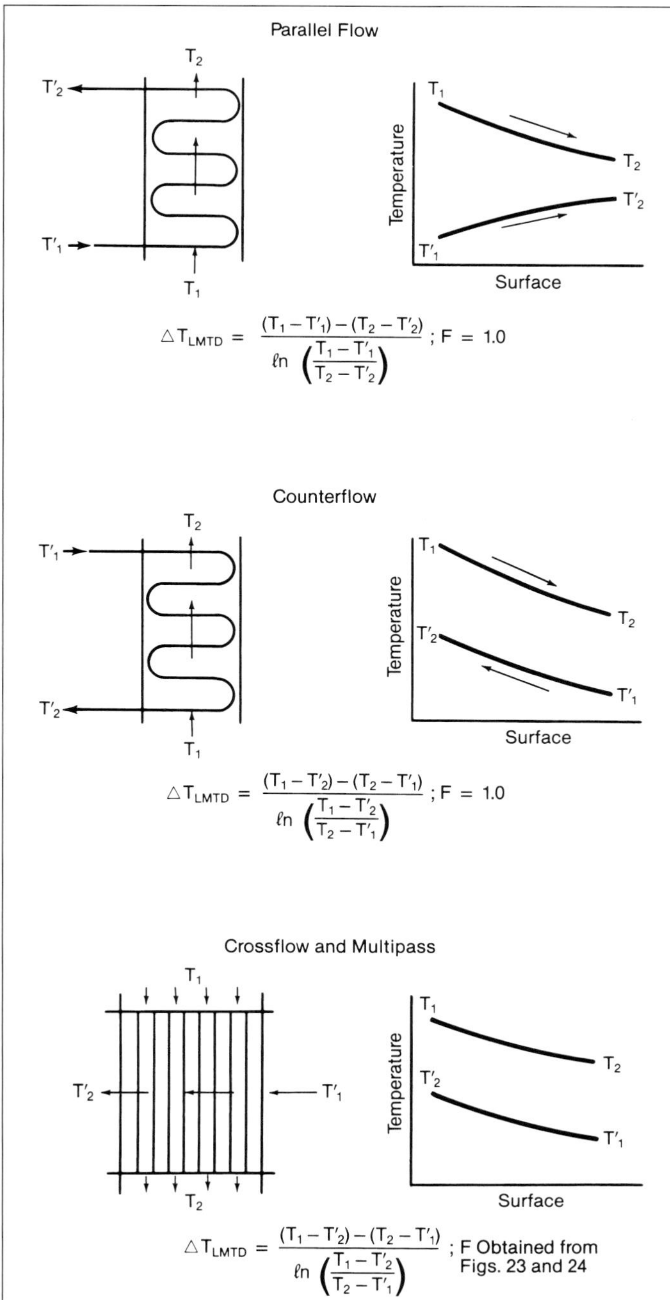

Fig. 24 Mean effective temperature difference.

S_f represents the fin surface area including both sides and the peripheral area, while S represents the exposed bare tube surface between the fins plus the fin surface, S_f. The ratio L_h/L_t is the fin height divided by the clear spacing between fins. Fin efficiency, η_f, is shown in Fig. 27 as a function of the parameter X, defined as:

$$X = L_h \sqrt{2Zh_c / \left(k_f L_t\right)} \tag{72}$$

for helical fins, and

$$X = r\,Y \sqrt{2Zh_c / \left(k_f L_t\right)} \tag{73}$$

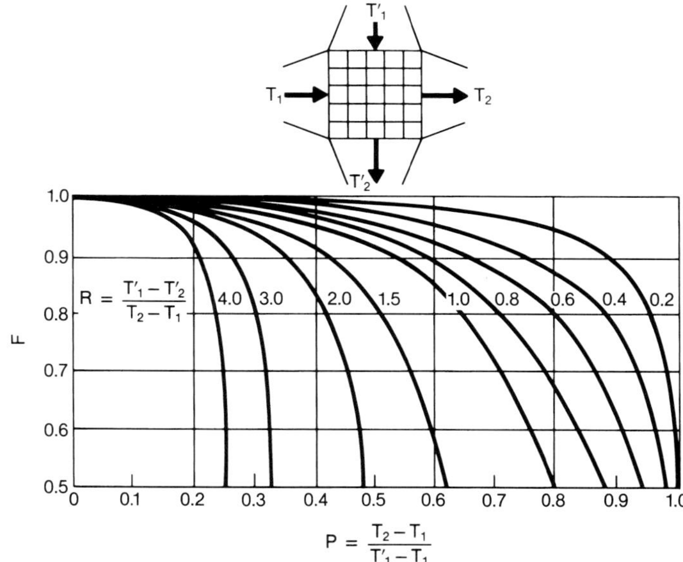

Fig. 25 Arrangement correction factors for a single-pass, crossflow heat exchanger with both fluids unmixed.

for rectangular, square or circular fins. The parameter Y is defined in Fig. 28.

The overall conductance can be written:

$$\frac{1}{UA} = \frac{1}{C_f A_o h_{f,o}} + R_w + \frac{1}{A_i h_{c,i}} \tag{74}$$

The parameter C_f is the surface cleanliness factor.

NTU method

There are design situations for which the performance of the heat exchanger is known, but the fluid temperatures are not. This occurs when selecting a unit for which operating flow rates are different than those

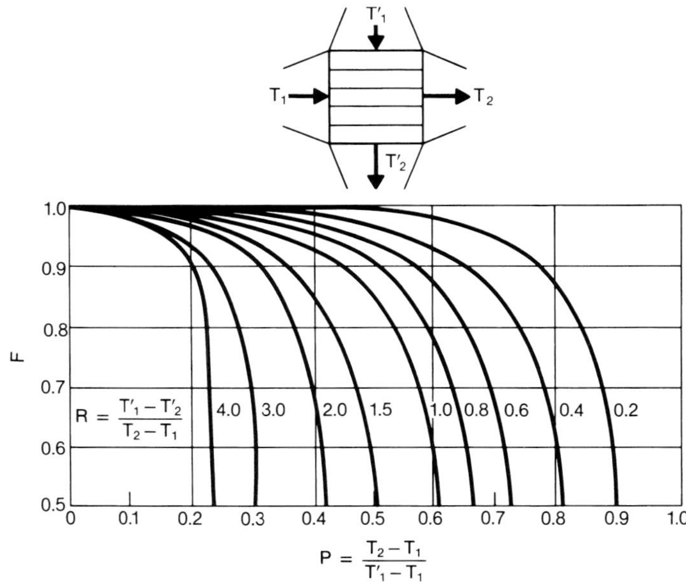

Fig. 26 Arrangement correction factors for a single-pass, crossflow heat exchanger with one fluid mixed and the other unmixed (typical tubular air heater application).

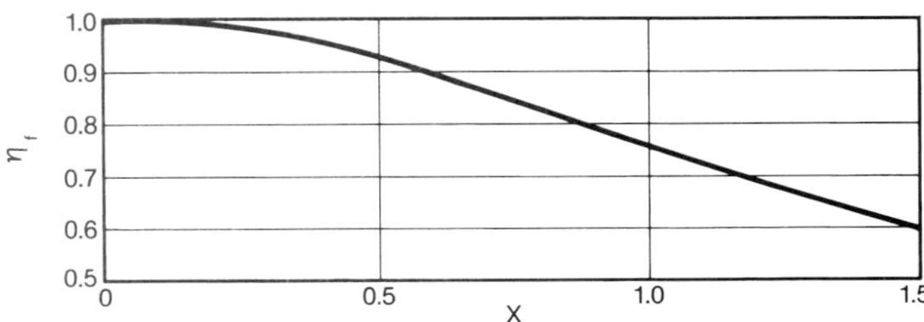

Fig. 27 Fin efficiency as a function of parameter *X*.

previously tested. The outlet temperatures can only be found by trial and error using the methods previously presented. These applications are best handled by the net transfer unit (NTU) method that uses the heat exchanger effectiveness (see Reference 16).

Heat transfer in porous materials

Porosity is an important factor in evaluating the effectiveness of insulation materials. In boiler applications, porous materials are backed up by solid walls or casings, so that there is minimal flow through the pores.

Heat flow in porous insulating materials occurs by conduction through the material and by a combination of conduction and radiation through the gas-filled voids. In most refractory materials, the Grashof-Prandtl (Raleigh) number is small enough that negligible convection exists although this is not the case in low density insulations [< 2 lb/ft^3 (32 kg/m^3)]. The relative magnitudes of the heat transfer mechanisms depend, however, on various factors including porosity of the material, gas density and composition filling the voids, temperature gradient across the material, and absolute temperature of the material.

Analytical evaluation of the separate mechanisms is complex, but recent experimental studies at B&W have shown that the effective conductivity can be approximated by:

$$k_{eff} = a + bT + cT^3 \qquad \textbf{(75)}$$

Experimental data can be correlated through this form, where a, b and c are correlation coefficients. The heat flow is calculated using Equation 1; k is replaced by k_{eff} and T is the local temperature in the insulation.

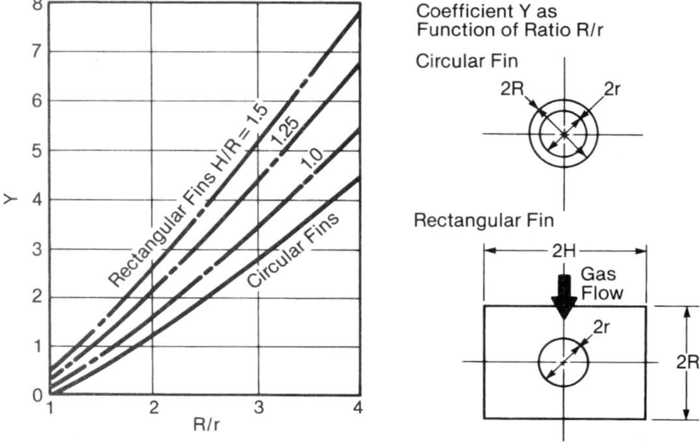

Fig. 28 Coefficient *Y* as a function of ratio *R/r* for fin efficiency.

In high temperature applications, heat transfer across the voids occurs mainly by radiation and the third term of Equation 75 dominates. In low temperature applications, heat flow by conduction dominates and the first two terms of Equation 75 are controlling.

Film condensation

When a pure saturated vapor strikes a surface of lower temperature, the vapor condenses and a liquid film is formed on the surface. If the film flows along the surface because of gravity alone and a condition of laminar flow exists throughout the film thickness, then heat transfer through this film is by conduction only. As a result, the thickness of the condensate film has a direct effect on the quantity of heat transferred. The film thickness, in turn, depends on the flow rate of the condensate. On a vertical surface, because of drainage, the thickness of the film at the bottom will be greater than at the top. Film thickness increases as a plate surface is inclined from the vertical position.

As the film temperature increases, its thickness decreases primarily due to increased drainage velocity. In addition, the film thickness decreases with increasing vapor velocity in the direction of drainage.

Mass diffusion and transfer

Heat transfer can also occur by diffusion and mass transfer. When a mixture of a condensable vapor and a noncondensable gas is in contact with a surface that is below the dew point of the mixture, some condensation occurs and a film of liquid is formed on the surface. An example of this phenomenon is the condensation of water vapor on the outside of a metal container. As vapor from the main body of the mixture diffuses through the vapor-lean layer, it is condensed on the cold surface as shown in Fig. 29. The rate of condensation is therefore governed by the laws of gas diffusion. The heat transfer is controlled by the laws of conduction and convection.

The heat transferred across the liquid layer must equal the heat transferred across the gas film plus the latent heat given up at the gas-liquid interface due to condensation of the mass transferred across the gas film. An equation relating the mass transfer is:

$$h_\delta \left(T_i - T_\delta\right) = h_g \left(T_g - T_i\right) + K_y H_{fg} \left(Y_g - Y_i\right) \qquad \textbf{(76)}$$

where T and Y define the temperatures and concentrations respectively identified in Fig. 29, h_δ is the heat transfer coefficient across the liquid film, h_g is the heat transfer coefficient across the gas film, and K_y is the mass transfer coefficient. H_{fg} is the latent heat of vaporization.

Heat transfer due to mass transfer is important in designing cooling towers and humidifiers, where mixtures of vapors and noncondensable gases are encountered.

Evaporation or boiling

The phenomenon of boiling is discussed in Chapters 1 and 5, where the heat transfer advantages of nucleate boiling are noted. Natural-circulation fossil fuel boilers are designed to operate in the boiling range. In this range, the heat transfer coefficient var-

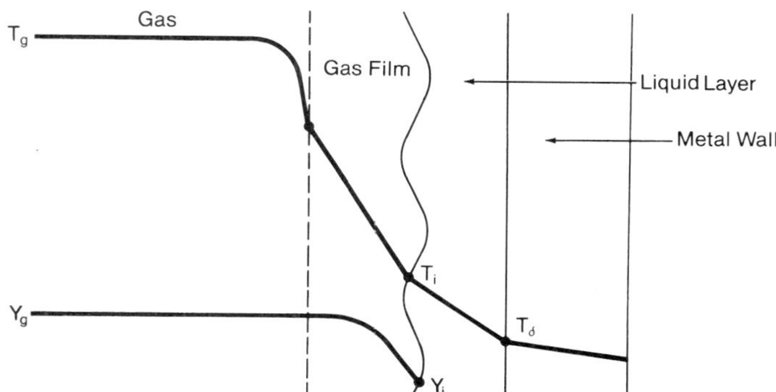

Fig. 29 Simultaneous heat and mass transfer in the dehumidification of air.

ies from 5000 to 20,000 Btu/h ft^2 F (28,392 to 113,568 W/m^2 K). This is not a limiting factor in the design of fossil fuel boilers provided scale and other deposits are prevented by proper water treatment, and provided the design avoids critical heat flux (CHF) phenomena. (See Chapter 5.)

In subcritical pressure once-through boilers, water is completely evaporated (dried out) in the furnace wall tubes which are continuous with the superheater tubes. These units must be designed for subcooled nucleate boiling, nucleate boiling, and film boiling, depending on fluid conditions and expected maximum heat absorption rates.

Fluidized-bed heat transfer

The heat transfer in gas-fluidized particle beds used in some combustion systems is complex, involving particle-to-surface contact, general convection and particle-to-surface thermal radiation. Correlations for heat transfer to tube bundles immersed in fluidized beds are summarized in Chapter 17.

Numerical modeling

Advances in computers have enabled B&W to mathematically model complex heat transfer systems. These models provide a tool for analyzing thermal systems inexpensively and rapidly. Although empirical methods and extensive equipment testing continue to provide information to designers, numerical simulation of boiler components, e.g., membrane walls, will become increasingly important as computer technology evolves. (See Chapter 6.)

Conduction

The energy equation for steady-state heat flow was previously defined as Equation 19 and is more generally written as Equation 20. Solutions of these equations for practical geometries are difficult to obtain except in idealized situations. Numerical methods permit the consideration of additional complex effects including irregular geometries, variable properties, and complex boundary conditions. Conduction heat flow through boiler membrane walls, refractory linings with several materials, and steam drum walls are several applications for these methods. The approach is to divide the heat transfer system into subvolumes called control volumes (Fig. 30). (See References 16 and 17.) The governing equation is integrated, or averaged, over the subvolume, leading to an expression of the form:

$$\frac{T_e - T_p}{R_{pe}} + \frac{T_w - T_p}{R_{pw}} + q_p''' V_p = c_p \frac{T_p - T_p^o}{\Delta t} \tag{77}$$

where the subscripts denote the neighbor locations as points on a compass. If the steady-state solution is desired, the right hand side of the equation, $c_p\left(T_p - T_p^o\right) / \Delta t$, which accounts for changes in stored energy is set to zero. A solution is then obtained numerically. Equation 77 is a discrete form of the continuous differential equation. The modeled geometry is subdivided and equations of this form are determined for each interior volume. The electrical analogy of the equation is apparent. First, each term is an expression of heat flow into a point using Fourier's law by Equation 1 and, second, Kirchhoff's law for electrical circuits, Equation 26, is used to determine the net flow of heat into any point. The application of the electrical analogy is straightforward for any interior volume once the subvolumes are defined. At the boundaries, temperature or heat flow is defined.

For unsteady-state problems, a sequence of solutions is obtained for the time interval Δt, with T_p^o being the node temperature at the beginning of the interval and T_p being the temperature at the end of the interval. References 16 and 17 explore these models in depth. Various computer codes are commercially available to perform the analysis and display the results.

Radiation

Numerical methods provide accurate estimates of radiative transfer in the absorbing and scattering media that is ubiquitous in the combustion and post-

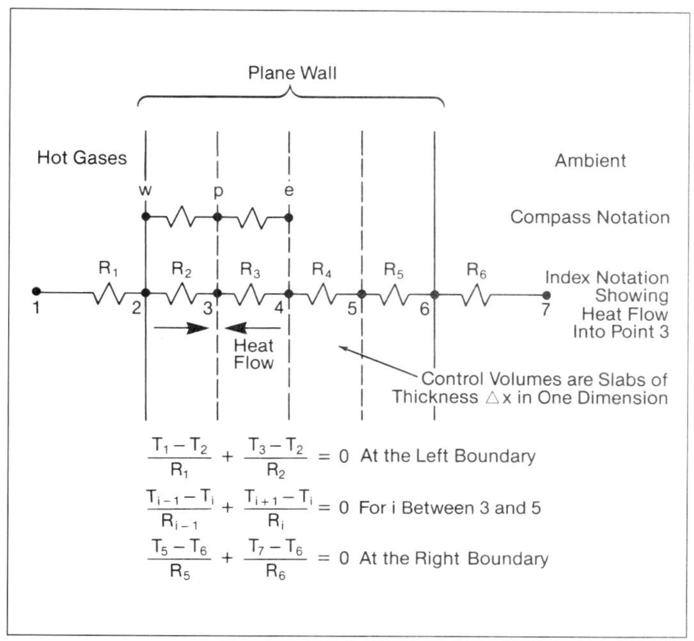

Fig. 30 Control volume layout for a plane wall with notation for heat flow to node 3 and steady-state solution.

combustion zones of a boiler. Many numerical methods have the advantage of incorporating complex geometrical description of the enclosure walls. These methods solve the radiation transport equation (RTE) in two and three dimensions and include the effects of absorption, emission, and scattering media, for gray or non-gray enclosure walls. Numerical models start by dividing the volume and surfaces of an enclosure into multiple control volumes and surface elements (as described in Chapter 6). Radiation properties of gases and particles are evaluated at each control volume, and may vary with local composition and temperature of the combustion mixture. Emissivity and temperature of the walls may also vary with local conditions, for each surface element of the enclosure. The solution of the RTE is generally carried out for the entire radiative spectrum for greatest numerical efficiency. However, if increased accuracy is required, the spectrum can be divided into discrete bands corresponding with the bands of gaseous radiation and the RTE is solved for each band separately.

Radiative heat transfer in furnaces can be calculated by one of several numerical methods, which are described in Reference 10. The simplest of these is the zonal method,[4] an extension of the network exchange method described previously in this chapter. The enclosure is divided into a finite number of isothermal volume and surface area zones. Exchange factors for all combinations of volume to volume, volume to surface, and surface to surface exchange are precalculated. This analysis leads to a set of simultaneous equations for unknown radiant heat fluxes, which are solved numerically.

The discrete ordinates method is perhaps the most robust approach for numerical analysis of radiation heat transfer in boilers. The angular dependence of radiation is first expressed in spherical coordinates, and is divided into a finite number of discrete directions for solving the RTE. The equations are transformed into a set of simultaneous partial differential equations, one for each direction, that are solved numerically. The accuracy of the method increases with the number of directions that are used in the approximation (typically 12, 24, or 48). Discrete ordinates was developed and optimized for thermal radiation in multi-dimensional geometries by the pioneering work at B&W.[18] Since then, it has gained in popularity, and is now used in many commercial computational fluid dynamics (CFD) codes.

The numerical solution for radiation leads to the distribution of radiant intensity or radiant heat flux, for a given temperature field. This solution is coupled to the energy equation and temperature of the gas-particle mixture. The energy equation (Equation 28 or 29) can be solved numerically for gas-particle temperature field using the methods described in Chapter 6. Radiation absorption and emission is represented by the internal heat generation term, S_H. Wall temperature is determined from an energy balance for convection and radiation heat transfer to the surface, and heat conduction through the wall. Several iterations between radiation, gas-particle energy, and wall temperature will ultimately yield a converged solution in which an overall energy balance is achieved.

Design considerations

Furnaces

Fossil-fuel fired boiler designers need to evaluate furnace wall temperature and heat flux, flue gas temperature, and furnace exit gas temperature. These parameters are required to determine materials and their limits, and to size heat transfer surface.

An analytical solution for heat transfer in a steam generating furnace is extremely complex. It is not possible to calculate furnace outlet temperatures by theoretical methods alone. Nevertheless, this temperature must be correctly predicted because it determines the design of the superheater and other system components.

In a boiler, all of the principal heat transfer mechanisms take place simultaneously. These mechanisms are intersolid radiation between suspended solid particles, tubes, and refractory materials; nonluminous gas radiation from the products of combustion; convection from the gases to the furnace walls; and conduction through ash deposits on tubes.

Fuel variation is significant. Pulverized coal, gas, oil or waste-fuel firing may be used. In addition, different types of the same fuel also cause variations. Coal, for example, may be high volatile or low volatile, and may have high or low ash and moisture contents. The ash fusion temperature may also be high or low, and may vary considerably with the oxidizing properties of the furnace atmosphere.

Furnace geometry is complex. Variations occur in the burner locations and spacing, in the fuel bed size, in the ash deposition, in the type of cooling surface, in the furnace wall tube spacing, and in the arch and hopper arrangements. Flame shape and length also affect the distribution of radiation and heat absorption in the furnace.

High intensity, high mixing burners produce bushy flames and promote large high temperature zones in the lower furnace. Lower intensity, controlled mixing burners frequently have longer flames that delay combustion while controlling pollutant formation.

Surface characteristics vary. The enclosing furnace walls may include any combination of fuel arrangements, refractory material, studded tubes, spaced tubes backed by refractory, close-spaced tubes, membrane construction or tube banks. Emissivities of these surfaces are different. The water-cooled surface may be covered with fluid slag or dry ash in any thickness, or it may be clean.

Temperature varies throughout the furnace. Fuel and air enter at relatively low temperatures, reach high temperatures during combustion, and cool again as the products of combustion lose heat to the furnace enclosure. All temperatures change with load, excess air, burner adjustment and other operating conditions.

Accurate estimates of furnace exit gas temperature are important. For example, high estimates may lead to over-estimating the heat transfer surface, while low estimates may cause operational problems. These are discussed in Chapter 19.

Empirical methods Considering the fuel type, firing rate and furnace configuration, empirical methods as illustrated in Fig. 31 have long been used to predict local absorption rates in the furnace. These methods, although largely empirical, contain engineering models which are based on fundamentals. Data and operating experience are used to tune the models employed in the design envelope. Fig. 31 shows typical vertical and horizontal heat flux distributions for furnace walls.

Deviations in the heat flux distribution are caused by unbalanced firing, variations in tube surface condition, differences in slagging, load changes, sootblower operation and other variations in unit operation. A typical upset heat flux distribution is shown in Fig. 31. These upset factors are typically a function of vertical/horizontal location, firing method and fuel, and furnace configuration. They are derived from operating experience.

The heat flux applied to the tubes in the furnace wall is also nonuniform in the circumferential direction. As shown in Fig. 32, the membrane wall is exposed to the furnace on one side while the opposite side is typically insulated to minimize heat loss. The resulting heat flux distribution depends upon the tube outside diameter, wall thickness, and spacing, as well as the web thickness and materials. The fluid temperature and inside heat transfer coefficient have secondary effects. This distribution can be evaluated using commercially available computer codes.

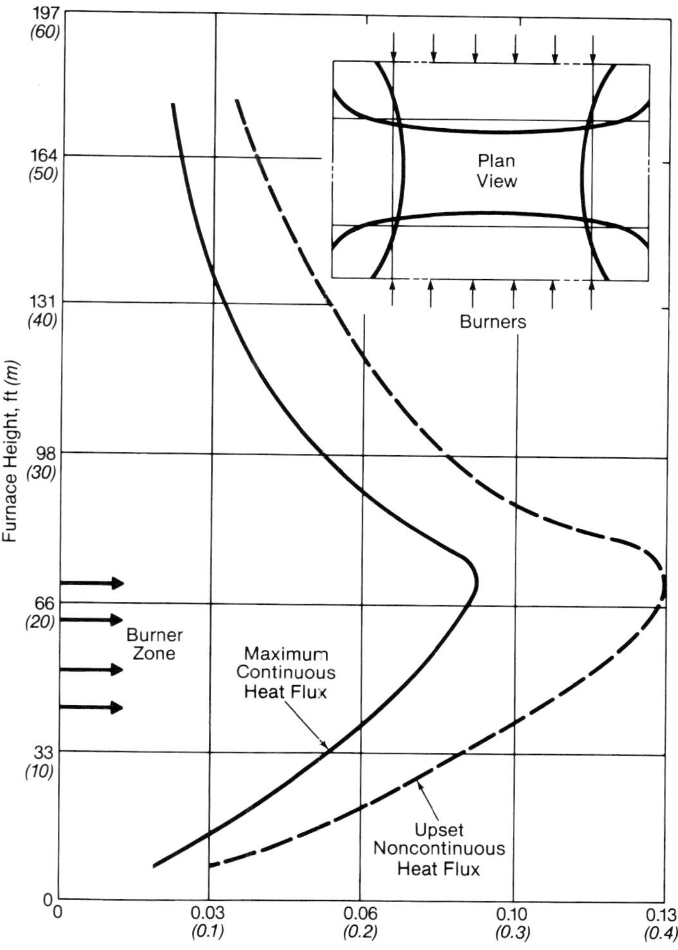

Fig. 31 Typical vertical and horizontal heat flux distributions for furnace walls.

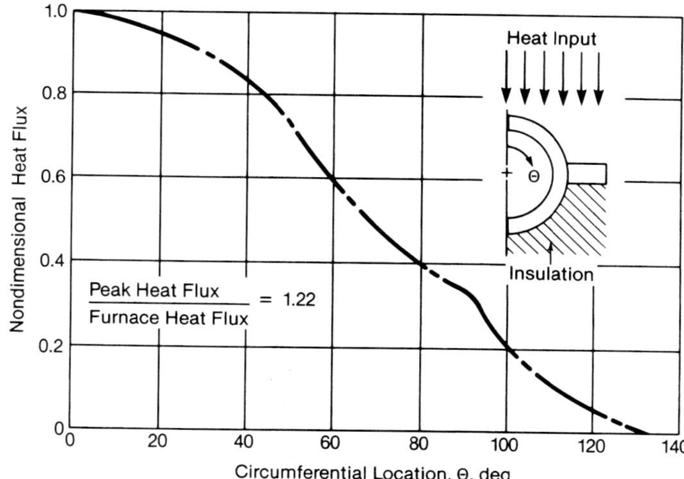

Fig. 32 Typical circumferential heat flux distribution for a furnace membrane wall panel tube.

To correlate data and calculations for different furnaces, methods for comparing the relative effectiveness of different furnace wall surfaces are needed. The effectiveness and spacing of tubes compared to a completely water-cooled surface are shown in Fig. 33. A wall of flat-studded tubes is considered completely water-cooled. The effectiveness of expected ash covering, compared with completely water-cooled surfaces, can also be estimated. The entire furnace envelope can then be evaluated in terms of equivalent cold surface.

The heat energy supplied by the fuel and by the preheated combustion air, corrected for unburned combustible loss, radiation loss, and moisture from the fuel, may be combined into a single variable, known as *heat available*. The heat available divided by the equivalent flat projected furnace enclosure plus furnace

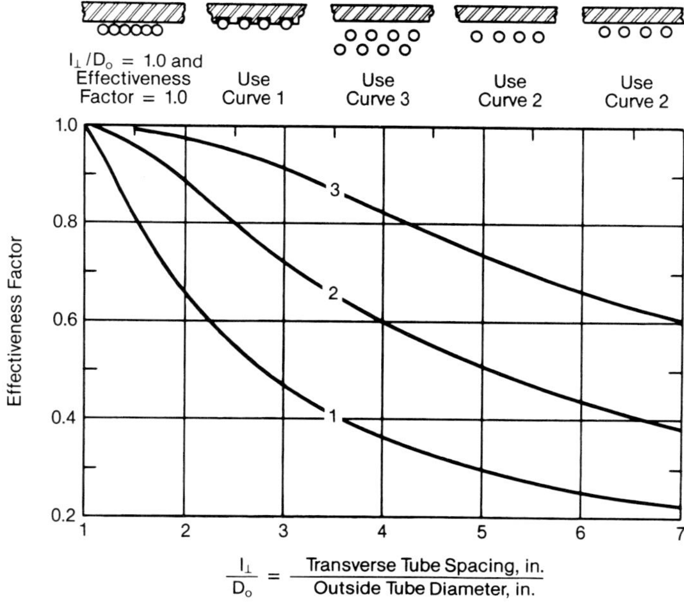

Fig. 33 Furnace wall area effectiveness factor (1.0 for completely water-cooled surface). A reduced area (equivalent cold surface) is determined from these curves for walls not completely water cooled. (Adapted from Hottel[4].)

platen area is called the *furnace heat release rate*. The heat input from fuel divided by the furnace volume is called the *furnace liberation rate*. The *furnace exit plane* defines the boundary of the furnace volume and flat projected furnace enclosure area. The furnace exit plane area and back spacing between the furnace platen tubes are included in the flat projected area calculation. For furnace platens and membrane wall furnace enclosure, the effectiveness factor for all examples given in this and other chapters is equal to 1.00.

Furnace exit gas temperature (FEGT) is primarily a function of heat release rate rather than liberation rate. The *furnace exit* is commercially defined as being located at the face of the first tube bank having a tube spacing of less than 15 in. (38.1 cm) side centers because, as can be inferred from Fig. 39, convection conductance typically becomes the predominant heat transfer mode at this side spacing. The furnace exit plane, generally used for the accurate calculation of overall heat transfer, is normally set at the face of the first tube bank having a tube spacing of 36 in. (91.4 cm) side centers or less in order to include the convection conductance in the heat transfer calculations. At tube side centers of 36 in. (91.4 cm) or less, the convection conductance is too significant to ignore as a portion of total heat transfer. The approximate relation of FEGT to heat release rate at the furnace exit plane for a typical pulverized bituminous coal is given in Fig. 34.

Furnace exit gas temperatures and related heat absorption rates, as functions of furnace heat release rate for most pulverized coal-fired furnaces, lie within the shaded bands shown in Figs. 35 and 36. The limits indicated serve only as a general guide and may vary due to combustion system type, burner and air port placement, stoichiometry, fuel characteristics and cleaning cycle. The bands for dry ash and for slag-tap furnaces overlap between 100,000 and 150,000 Btu/h ft^2 (315,460 to 473,190 W/m^2), but different types of coal are involved. To be suitable for a slag-tap furnace, a bituminous coal should have an ash viscosity of 250 poises at 2450F (1343C) or lower. In the overlapping range, dry ash and slag-tap both have about the same heat absorption rate, or dirtiness factor, as shown in Fig. 36. Both bands are rather broad, but they cover a wide range of ash characteristics and a considerable diversity in waterwall construction and dirtiness.

The heat leaving the furnace is calculated from the exiting gas flow rate (the gas enthalpy values evaluated at the furnace exit gas temperature) plus the net radiative transfer at the furnace exit. The heat absorbed in the furnace is the difference between the heat available from the fuel, including the preheated combustion air, and heat leaving the furnace.

Numerical methods Empirical design methods are gradually being supplemented with numerical methods, as the level of detail increases and confidence is improved. Radiation heat transfer in furnace enclosures can now be solved on computers, in combination with turbulent flow, energy, and combustion. Radiation properties of gases, particles, and fuel specific properties of ash deposits can be included in the analysis with more advanced engineering models and correlations. The effects of spectral radiation from gases and particles can also be included to improve accuracy of the analysis. Detailed results include the three-dimensional distribution of radiation intensity, gas temperature, and heat flux on the furnace walls. Numeri-

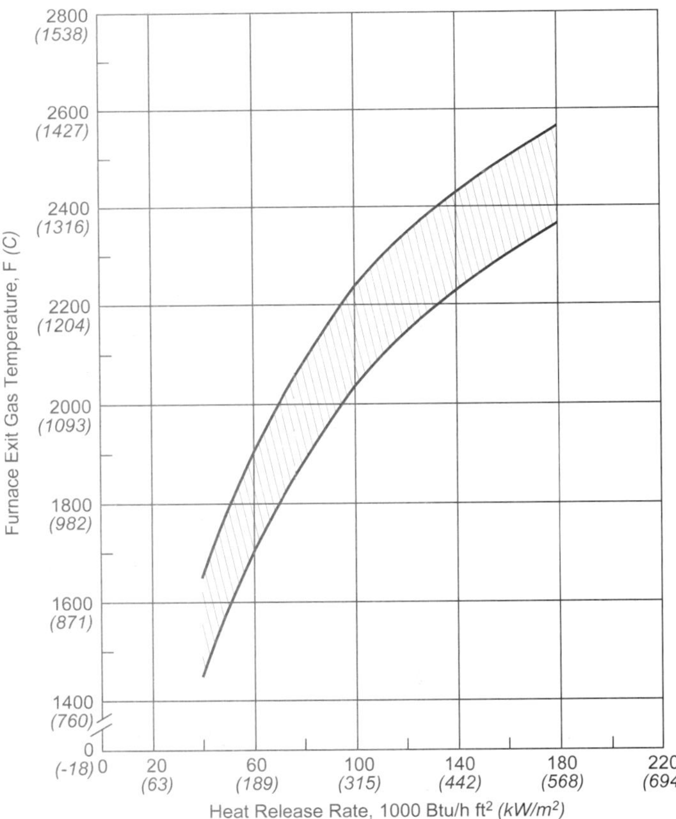

Fig. 34 Approximate relationship of furnace exit gas temperatures to heat release rate for a typical pulverized bituminous coal.

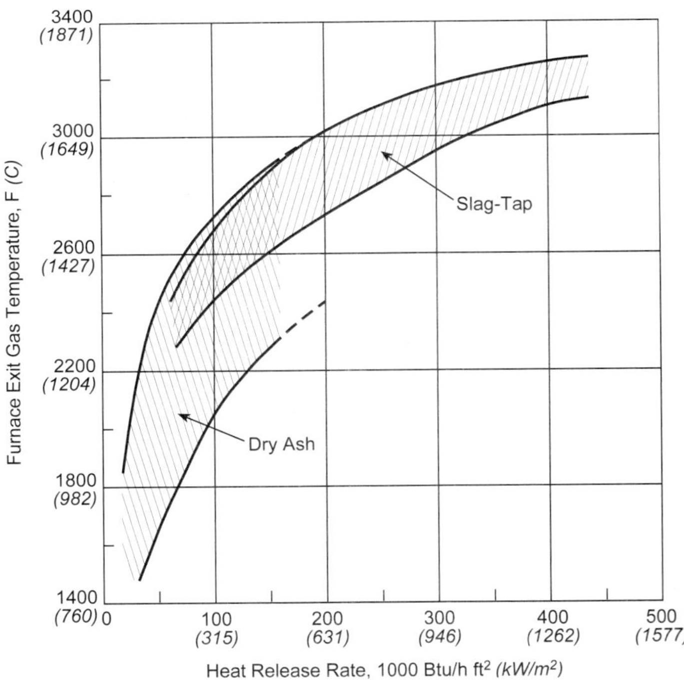

Fig. 35 General range of furnace exit gas temperature for dry ash and slag-tap pulverized coal-fired furnaces.

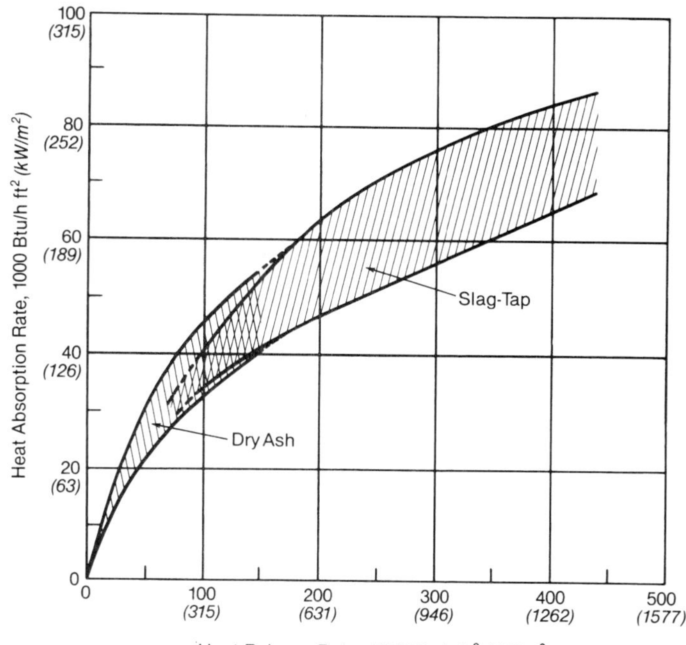

Fig. 36 General range of furnace heat absorption rates for dry ash and slag-tap pulverized coal-fired furnaces.

cal methods have the potential for more accurate prediction of heat flux distribution on furnace walls and convective surfaces. However, further validation of results and improvements in computational efficiency are needed to make numerical methods more practical for routine engineering applications.

A numerical model was created for the furnace of a 560 MW supercritical steam pressure boiler firing high volatile eastern United States bituminous coal. A schematic of the furnace is shown in Fig. 37. The sloping furnace walls of the ash hopper, the furnace nose, and the horizontal section of the convection pass were included in the model. Inlet fuel, inlet air and exit streams were properly located around the boundaries. An example of the predicted heat flux distribution is shown in Fig. 38. The predicted furnace exit gas temperature for this case was 2242F (1228C), while the observed average value was 2276F (1247C). Relative magnitudes of convective and radiative heat transfer at various locations are shown in Fig. 39 for a 650 MW boiler. The furnace area is dominated by radiation while the back-end heat transfer surfaces in the direction of flow are increasingly dominated by convection.

Convection banks

Tube spacing and arrangement In addition to heat absorption and resistance to gas flow, other important factors must be considered in establishing the optimum tube spacing and arrangement for a convection surface. These are slagging or fouling of surfaces, accessibility for cleaning, and space occupied. A large longitudinal spacing relative to the transverse spacing is usually undesirable because it increases the space requirement without improving performance. These are discussed further in Chapter 21.

Tube diameter For turbulent flow, the heat transfer coefficient is inversely proportional to a power of the tube diameter. In Equations 57 and 60 the exponent for longitudinal flow is 0.20; for cross flow it is 0.39. These equations indicate that the tube diameter should be minimized for the most effective heat transfer. However, this optimum tube diameter may require

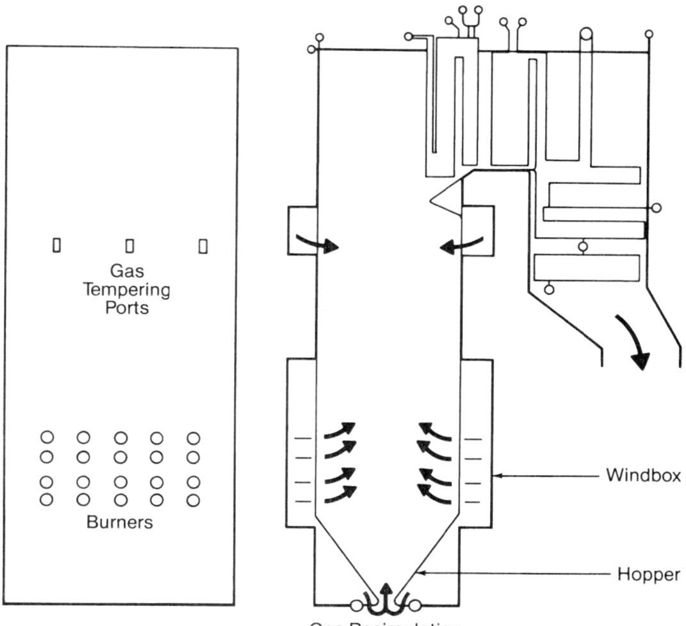

Fig. 37 560 MW utility boiler schematic used for numerical model (see Fig. 38).

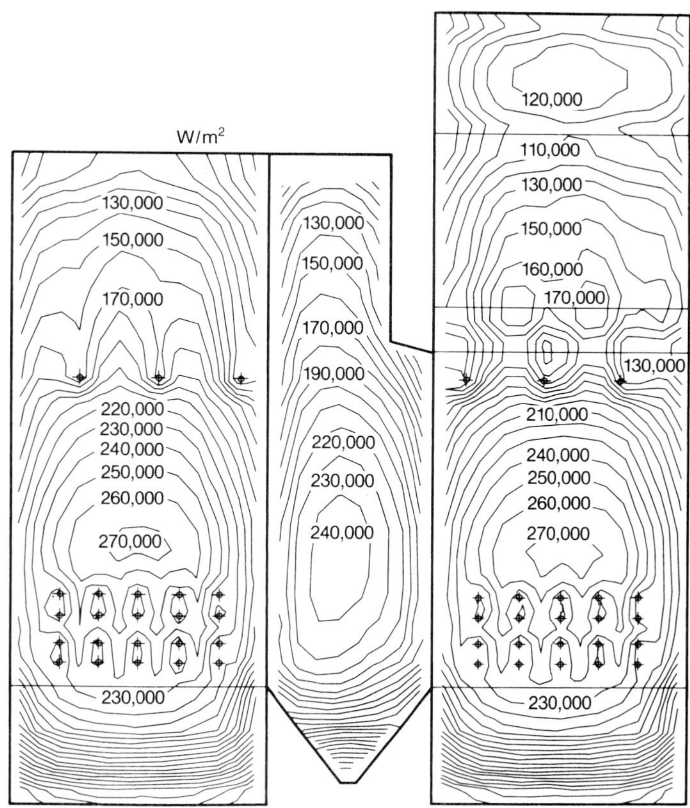

Fig. 38 Numerical model – predicted furnace wall flat projected heat flux distribution. (1 W/m² = 0.317 Btu/h ft²)

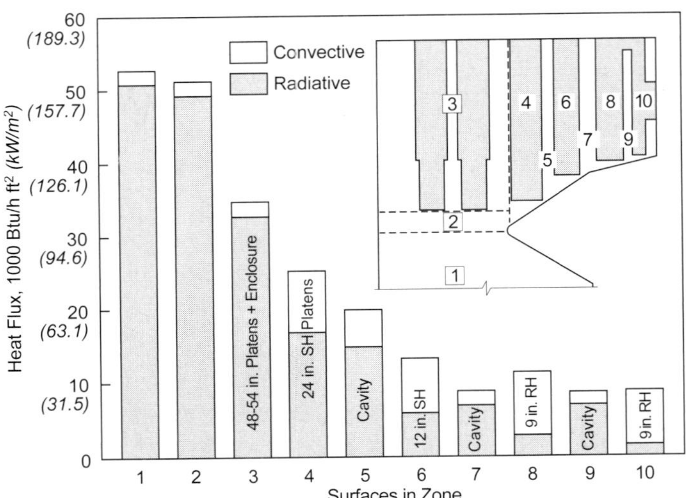

Fig. 39 Comparison of radiative and convective heat transfer contributions to absorption in various locations within a large utility boiler (SH = superheater; RH = reheater; 1 in. = 2.54 cm).

an arrangement that is expensive to fabricate, difficult to install, or costly to maintain. A compromise between heat transfer effectiveness and manufacturing, erection, and service limitations is therefore necessary in selecting tube diameter.

Penetration of radiation A convection bank of tubes bordering a furnace or a cavity acts as a blackbody radiant heat absorber. Some of the impinging heat, however, radiates through the spaces between the tubes of the first row and may penetrate as far as the fourth row. The quantity of heat penetration can be established by geometric or analytical methods. The effect of this penetration is especially important in establishing tube temperatures for superheaters located close to a furnace or high temperature cavity. Consider 2.0 in. (50.8 mm) OD tubes placed in an array of tubes on a 6.0 in. (152.4 mm) pitch. Fig. 33, curve 1 can be used to estimate the remaining radiation. For a given radiant heat flux, 45% is absorbed in the first tube row, and 55% passes to the second row. 45% of this reduced amount is again absorbed in the second tube row. After the fourth row, less than 10% of the initial radiation remains.

Effect of lanes Lanes in tube banks, formed by the omission of rows of tubes, may decrease the heat absorption considerably. These passages act as bypasses for flowing hot gases and radiation losses. Although the overall efficiency decreases, the high mass flow through the lanes increases the absorption rate of the adjacent tubes. Critical tube temperatures in superheaters or steaming conditions in economizers may develop. Whenever possible, lanes should be avoided within tube banks and between tube banks and walls; however, this is not always possible. A calculation accounting for the lanes is necessary in such cases.

Heat transfer to water

Water heat transfer coefficient The heat transfer coefficient for water in economizers is so much higher than the gas-side heat transfer coefficient that it can be neglected in determining economizer surface.

Boiling water heat transfer coefficient The combined gas-side heat transfer coefficient (convection plus intertube radiation) seldom exceeds 30 Btu/h ft² F (170 W/m² K) in boiler design practice. The heat transfer coefficient for boiling water [10,000 Btu/h ft² F (56,784 W/m² K)] is so much larger that it is generally neglected in calculating the resistance to heat flow, although Equation 4 in Chapter 5 can be used to calculate this value.

Effect of scale Water-side and steam-side scale deposits provide high resistance to heat flow. As scale thickness increases, additional heat is required to maintain a given temperature inside a furnace tube. This leads to high metal temperatures and can cause tube failure. Deposition of scale and other contaminants is prevented by good feedwater treatment and proper operating practices.

Heat transfer to steam In superheaters, the steam-side convection constitutes a significant resistance to heat flow. Although this resistance is much lower than the gas-side resistance, it can not be neglected in computing the overall heat flow resistance or the heat transfer rate. It is particularly significant in calculating superheater tube temperatures, because the mean tube wall temperature is equal to the steam temperature plus the temperature drop through the steam film plus half of the metal temperature drop.

The steam-side heat transfer coefficient is calculated from Equation 58 using information from Figs. 13, 16 and 17. If the steam heat transfer coefficient is designated as h, the film temperature drop, ΔT_f, is $q/(hA)$, using the outside surface area of the tube as the base in each expression.

It is imperative to prevent scale deposits in superheater tubes. Because of its high resistance to heat flow and due to the elevated temperatures, even a thin layer of scale may be sufficient to overheat and fail a tube.

Cavities

Cavities are necessary between tube banks of steam generating units for access, for sootblowers, and for possible surface addition. Hot flue gas radiates heat to the boundary surfaces while passing through the cavity. The factors involved in calculating heat transfer in cavities are as follows.

Temperature level Radiation from nonluminous gases to boundary surfaces and radiation to the gas by the surroundings increase approximately by the fourth power of their respective absolute temperatures. Remembering that $E_b = \sigma T^4$, Equations 39 and 40 illustrate this relationship.

Gas composition Carbon dioxide and water vapor are the normal constituents of flue gases which emit nonluminous radiation in steam generating units. The concentrations of these constituents depend on the fuel burned and the amount of excess air.

Particles in the gas The particles carried by flue gases receive heat from the gas by radiation, convection, and conduction, and emit heat by radiation to the furnace enclosure.

Size of cavity The heat transfer rate increases with cavity size. Thick layers of gas radiate more vigorously than thin layers. The shape of the cavity can also complicate heat transfer calculations.

Receiving surface A refractory surface forming part of a cavity boundary reaches a high temperature by convection and radiation from the flue gas. It also reradiates heat to the gas and to the other walls of the enclosure. Reradiation from a clean, heat-absorbing surface is small unless the receiving surface temperature is high, as is the case with superheaters and reheaters. Ash or slag deposits on the tube reduce heat absorption and increase reradiation.

In boiler design, there are two significant effects of cavity radiation: 1) the temperature of flue gas drops, from several degrees up to 40F (22C), in passing across a cavity, and 2) gas radiation increases the heat absorption rates for the tubes forming the cavity boundaries. The second effect influences superheater tube temperatures and the selection of alloys.

Insulation

The calculation of heat transfer through insulation follows the principles outlined for conduction through a composite wall. Refer to Chapter 23 for more information regarding insulating materials.

Hot face temperature In a furnace with tube-to-tube walls, the hot face temperature of the insulation is the saturation temperature of the water in the tubes. If the inner face of the furnace wall is refractory, the hot face insulation temperature must be calculated using radiation and convection heat transfer principles on the gas side of the furnace wall, or estimated using empirical data.

Heat loss and cold face temperature The heat loss to the surroundings and the cold face temperature decrease as the insulation thickness increases. However, once an acceptable layer of insulation is applied, additional amounts are not cost effective. Standard commercial insulation thicknesses should be used in the composite wall.

The detailed calculation of overall heat loss by radiation and convection from the surfaces of a steam generating unit (usually called radiation loss) is tedious and time consuming. A simple approximate method is provided by the chart prepared from the American Boiler Manufacturers Association (ABMA) original. (See Chapter 23, Fig. 12.)

Ambient air conditions Low ambient air temperature and high air velocities reduce the cold face temperature. However, they have only a small effect on total heat loss, because surface film resistance is a minor part of the total insulation resistance. Combined heat loss rates (radiation plus convection) are given in Chapter 23, Fig. 11, for various temperature differences and air velocities. The effect of surface film resistance on casing temperature and on heat loss through casings is shown in Chapter 23, Fig. 15.

Temperature limits and conductivities Refractory or insulating material suitable for high temperature applications is usually more expensive and less effective than low temperature materials. It is therefore customary to use several layers of insulation. The lower cost, more effective insulation is used in the cool zones; the higher cost materials are used only where demanded by high operating temperatures. Thermal conductivities for refractory and insulating materials, and temperatures for which they are suitable, are shown in Chapter 23, Fig. 10.

Applications

Example 1 – Conduction through a plane wall

If a flat plate is heated on one side and cooled on the other, the heat flow rate in the wall, shown in Fig. 1, is given by Equation 2. The rate of heat flow through a 0.25 in. thick steel plate with 1 ft² surface area and ΔT = 25F may be evaluated with Equation 2 as follows:

$$q = kA\frac{\Delta T}{L} = \frac{30 \times 1 \times 25}{0.25/12} = 36{,}000 \text{ Btu/h} \qquad (78)$$

where the thermal conductivity, k, for steel is 30 Btu/h ft F.

Example 2 – Heat flow in a composite wall with convection

The heat flow through a steel wall which is insulated on both sides is shown in Fig. 3. This example demonstrates the procedure for combining thermal resistances. In addition to the thermal resistance of the firebrick, steel, and insulation, the heat flow is impeded by the surface resistances. Consider a 600 ft² surface with gases at 1080F or 1540R on the inside exposed to an ambient temperature of 80F on the outside. The thermal conductivities of the firebrick, steel flue and insulation are assumed to be $k_1 = 0.09$, $k_2 = 25$, and $k_3 = 0.042$ Btu/h ft F, respectively. These assumptions are verified later. The layer thicknesses are $\Delta x_1 = 4$ in., $\Delta x_2 = 0.25$ in., and $\Delta x_3 = 3$ in. The heat transfer coefficients for convection are $h_{cv,i} = 5.0$ Btu/h ft² F on the inside surface $h_{cv,o} = 2.0$ Btu/h ft² F on the outside surface. Where the temperature difference between the radiating gas, T_g, and a surface, T_s, is small, the radiation heat transfer coefficient can be estimated by:

$$h_r \cong 4.0\,\sigma\,\varepsilon\,F\left[\left(T_g + T_s\right)/2\right]^3 \approx 4\,\sigma\,\varepsilon\,F\,T_g^3 \qquad (79a)$$

where T_g and T_s are the absolute temperatures, R (K). In this example, the surface emissivity is assumed close to 1.0 and $F = 1.0$ resulting in:

$$\begin{aligned} h_r &= 4.0\left(0.1713 \times 10^{-8}\right)\left(1080 + 460\right)^3 \\ &= 25 \text{ Btu/h ft}^2\text{ F} \end{aligned} \qquad (79b)$$

Using the R_{eq} shown in Fig. 3 and values of R evaluated using Table 4:

$$\begin{aligned} R_{eq}A &= \frac{\left(\frac{1}{5}\right)\left(\frac{1}{25}\right)}{\frac{1}{5}+\frac{1}{25}} + \frac{\frac{4}{12}}{0.09} + \frac{\frac{0.25}{12}}{25} + \frac{\frac{3}{12}}{0.042} + \frac{1}{2} \\ &= 0.033 + 3.70 + 0.000833 + 5.95 + 0.5 \qquad (80) \\ &= 10.18\,\frac{\text{h ft}^2\text{ F}}{\text{Btu}} \end{aligned}$$

It is clear that the firebrick and insulation control the overall resistance; the steel resistance can be neglected. If the successive material layers do not make good thermal contact with each other, there will be interface resistances due to the air space or film. These resistances may be neglected in composite walls of insulating materials. However, they must be included in calculations if the layer resistances are small compared to the interface resistances. An example of this is heat transfer through a boiler tube with internal oxide deposits.

The heat flow can be computed using Equation 23:

$$q = \frac{T_f - T_\infty}{R_{eq}} = A\frac{T_f - T_\infty}{R_{eq}\,A}$$
$$= 600\left(\frac{1080-80}{10.18}\right) = 58{,}939\,\frac{\text{Btu}}{\text{h}} \quad \textbf{(81)}$$

To determine if the correct thermal conductivities were assumed and if the temperature levels are within allowable operating limits of the material, it is necessary to calculate the temperatures at the material interfaces. Solving Equation 81 for temperature and substituting individual resistances and local temperatures:

$$T_1 = T_f - \frac{q}{A}(R\,A)_f = 1080 - \frac{(59{,}939)(0.033)}{600} = 1077\text{F} \quad \textbf{(82)}$$

$$T_2 = T_1 - \frac{q}{A}(R\,A)_1 = 1077 - \frac{(58{,}939)(3.70)}{600} = 713\text{F} \quad \textbf{(83)}$$

$$T_3 = T_2 - \frac{q}{A}(R\,A)_2 = 713 - \frac{(58{,}939)(0.000833)}{600} = 713\text{F} \quad \textbf{(84)}$$

$$T_4 = T_3 - \frac{q}{A}(R\,A)_3 = 713 - \frac{(58{,}939)(5.95)}{600} = 129\text{F} \quad \textbf{(85)}$$

$$T_\infty = T_4 - \frac{q}{A}(R\,A)_\infty = 129 - \frac{(58{,}939)(0.5)}{600} = 80\text{F} \quad \textbf{(86)}$$

The small temperature difference between the radiating gas and the surface ($T_f - T_1 = 3\text{F}$) verifies the assumption for using Equation 79a. Had the temperature difference been larger, then the full equation for the radiation resistance in Table 4 may have been required.

The negligible resistance of the steel flue is reflected in the temperature drop $T_2 - T_3 = 0$. If the calculated interface temperatures indicate the conductivity was chosen improperly, new conductivities are defined using the mean temperature of each material. For example, a new firebrick conductivity is determined using $0.5\,(T_1 + T_2)$.

Example 3 – Heat flow in an insulated pipe

Heat flow in cylindrical geometries is important in evaluating boiler heat transfer. Refer to the example steam line shown in Fig. 40. The resistances in Table 4 for cylindrical geometries must be used. The thermal analogy for the pipe in Fig. 40 can be written:

$$R_{eq} = \frac{1}{h_i\,(2\pi\,r_1\,l)} + \frac{ln(r_2/r_1)}{2\pi\,k_2\,l} + \frac{ln(r_3/r_2)}{2\pi\,k_3\,l} + \frac{1}{h_0\,(2\pi\,r_3\,l)} \quad \textbf{(87)}$$

A 3 in. Schedule 40 steel pipe (k = 25 Btu/h ft F) is covered with 0.75 in. insulation of k = 0.10 Btu/h ft F. This pipe has a 3.07 in. ID and a 3.50 in. OD. The pipe transports fluid at 300F and is exposed to an ambient temperature of 80F. With an inside heat transfer coefficient of 50 Btu/h ft^2 F and an outside heat transfer coefficient of 4 Btu/h ft^2 F, the thermal resistance and heat flow per unit length are:

$$R_{eq}\,l = \frac{1}{50(2\pi)\left(\frac{3.07/2}{12}\right)} + \frac{ln\left(\frac{3.50}{3.07}\right)}{2\pi\,(25)} + \frac{ln\left(\frac{5.0}{3.5}\right)}{2\pi\,(0.1)} + \frac{1}{4(2\pi)\left(\frac{5.0/2}{12}\right)} \quad \textbf{(88)}$$

$$R_{eq}\,l = 0.0249 + 0.000834 + 0.568 + 0.191$$
$$= 0.785\,\frac{\text{h ft F}}{\text{Btu}} \quad \textbf{(89)}$$

The overall resistance is dominated by the insulation resistance and that of the outer film boundary layer. The resistance of the metal pipe is negligible.

$$\frac{q}{l} = \frac{300-80}{0.785} = 280 \text{ Btu/h ft} \quad \textbf{(90)}$$

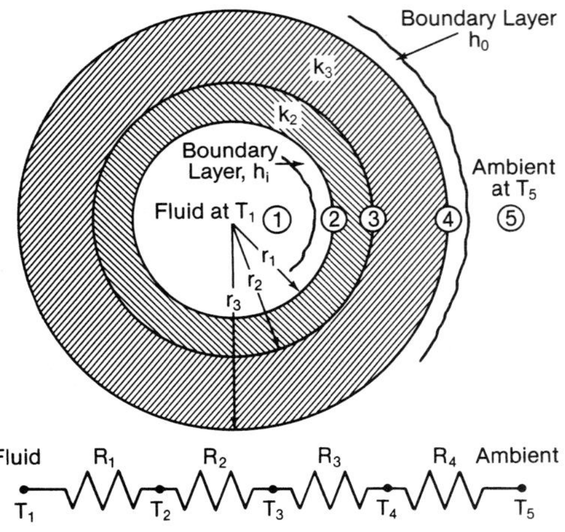

Fig. 40 Example of heat flow in an insulated pipe.

Example 4 – Heat flow between a small object and a large cavity

Consider an unshielded thermocouple probe with an emissivity of 0.8 inserted in a duct at 240F carrying combustion air. If the thermocouple indicates a temperature of 540F and the surface heat transfer coefficient, h, between the thermocouple and gas is 20 Btu/h ft^2 F, the true gas temperature can be estimated. The thermocouple temperature must be below the gas temperature because heat is lost to the walls. Under steady-state conditions, an energy balance equates the radiant heat loss from the thermocouple to the wall and the rate of heat flow from the gas to the thermocouple.

Using Table 6, the heat flow between the thermocouple and the cavity becomes:

$$\frac{q}{A} = 0.8\left(0.1713\times10^{-8}\right) \times\left[\left(540+460\right)^4 - \left(240+460\right)^4\right] = 1041.4 \text{ Btu/h ft}^2 \quad \textbf{(91)}$$

The true gas temperatures becomes:

$$T_g = \frac{q/A}{h} + T_t = \frac{1041.4}{20} + 540 = 592\text{F} \quad \textbf{(92)}$$

Similar analyses can be performed for thermocouples shielded with reflective foils in high temperature environments. This practice prevents thermocouple heat losses and incorrect temperature readings. The heat flow from a shielded thermocouple is calculated as follows:

$$q_{\text{shielded}} = \frac{1}{(N+1)} q_{\text{no shield}} \quad \textbf{(93)}$$

where N is the number of concentric layers of material.

Example 5 – Heat flow between two surfaces

An estimate of the maximum radiant heat transfer between two surfaces can be determined using Equation 11. This approximation is valid when the walls are considered black and any intervening absorbing gases are neglected. If two 5 × 10 ft black rectangles, directly opposed, are spaced 10 ft apart with temperatures of 940 and 1040F, the energy exchange is estimated as follows.

The energy from surface 1 directly striking surface 2 is defined by the shape factor F_{12}. Referring to Fig. 5, this factor F_{12} is 0.125, indicating that 87.5% of the energy leaving surface 1 strikes a surface other than surface 2. The net heat flow is:

$$\begin{aligned} q_{12} &= A_1 F_{12}\sigma\left(T_1^4 - T_2^4\right) \\ &= 50(0.125)\left(0.1713\times10^{-8}\right) \\ &\quad \times\left[(1040+460)^4 - (940+460)^4\right] \quad \textbf{(94)} \\ q_{12} &= 13{,}071 \text{ Btu/h} \end{aligned}$$

Intervening gases and/or gray walls further reduce the net heat flow.

Example 6 – Radiation from a hot gas to furnace walls

Consider a furnace with a volume of 160,000 ft^3 and a heat transfer surface area of 19,860 ft^2. The gas traversing the furnace (T_g) is at 2540F (1393C) and the furnace walls (T_s) are at 1040F (560C). The radiant heat transfer rate can be estimated using Equation 40, assuming the walls are radiatively black ($\varepsilon_s = 1$). If the products of combustion at one atmosphere consist of 10% carbon dioxide, 5% water vapor, and 85% nitrogen, Figs. 9 to 11 can be used to estimate the gas emissivity and absorptivity. The beam length is L = 3.6 V/A = 29.0 ft. Then for H_2O, $p_w L$ = (29.0) (0.05) = 1.45 ft-atm (45 bar-cm) and from Fig. 9 at 1393C the emissivity is found to be 0.22. For CO_2, $p_c L$ = (29.0) (0.10) = 2.90 ft-atm (89 bar-cm) and from Fig. 10, at 1393C, the emissivity is found to be 0.16. The correction $\Delta\varepsilon$ is determined from Fig. 11. The total gas emissivity is then found from Equation 41:

$$\varepsilon_g = \varepsilon_{H_2O} + \varepsilon_{CO_2} - \Delta\varepsilon_g \quad \textbf{(95)}$$

$$\varepsilon_g = 0.22 + 0.16 - 0.06 = 0.32 \quad \textbf{(96)}$$

Hottel[4] suggests calculating the absorptivity of the gas using modified pressure length parameters:

$$\begin{aligned} F_w &= p_w L \frac{T_s}{T_g} = (0.05)(29)\frac{1040+460}{2540+460} \\ &= 0.73 \text{ ft-atm (or 22 bar-cm)} \end{aligned} \quad \textbf{(97)}$$

$$\begin{aligned} F_c &= p_c L \frac{T_s}{T_g} = (0.10)(29)\frac{1040+460}{2540+460} \\ &= 1.45 \text{ ft-atm (or 45 bar-cm)} \end{aligned} \quad \textbf{(98)}$$

$$\begin{aligned} \alpha_{H_2O} &= \varepsilon_{H_2O}\left(F_w, T_s\right) \times \left(\frac{T_g}{T_s}\right)^{0.45} \\ &= 0.21\left(\frac{2540+460}{1040+460}\right)^{0.45} = 0.29 \end{aligned} \quad \textbf{(99)}$$

$$\begin{aligned} \alpha_{CO_2} &= \varepsilon_{CO_2}\left(F_c, T_s\right) \times \left(\frac{T_g}{T_s}\right)^{0.65} \\ &= 0.15\left(\frac{2540+460}{1040+460}\right)^{0.65} = 0.24 \end{aligned} \quad \textbf{(100)}$$

$$\Delta\alpha_g = \Delta\varepsilon\left(T_s\right) = 0.04 \quad \textbf{(101)}$$

$$\alpha_g = \alpha_{H_2O} + \alpha_{CO_2} - \Delta\alpha_g = 0.49 \quad \textbf{(102)}$$

The net rate of heat flow calculated from Equation 40:

$$\begin{aligned} q_{sg} &= A\left(\varepsilon_g E_{bg} - \alpha_g E_{bs}\right) \\ &= 19{,}860\left(0.32\ \sigma\ 3000^4 - 0.49\ \sigma\ 1500^4\right) \\ &= 797\times10^6 \text{ Btu/h} \end{aligned} \quad \textbf{(103)}$$

In estimating boiler heat transfer, the beam lengths are large, effecting large pL values. Proprietary data are used to estimate the heat transfer for these values, and extrapolation of the curves in Figs. 9 to 11 is not recommended.

Example 7 – Radiation in a cavity

Radiation in a cavity containing absorbing gases can be analyzed with the concepts previously presented. These concepts are useful in analyzing surface to surface heat transfer. Examples include boiler wall to boiler wall, platen to platen, and boiler wall to boiler enclosure heat exchanges. Table 5 (and Reference 4) contains the thermal resistances used in constructing the thermal circuit in Fig. 41. Note the resistance between surface 1 and 2 decreases as the transmission in the gas, τ_{12}, increases to a transparent condition $\tau_{12} = 1$. At τ_{12} near zero, the gases are opaque, and the resistance is very large. As the gas emissivity decreases, the thermal circuit reduces to Fig. 6 and Equation 36. The solution of the circuit in Fig. 41 is found using Kirchhoff's rule for nodes J_1 and J_2. The equations are solved simultaneously for J_1 and J_2:

$$\frac{E_{b1} - J_1}{\dfrac{1-\varepsilon_1}{\varepsilon_1 A_1}} + \frac{J_2 - J_1}{\dfrac{1}{A_1 F_{12} \tau_{12}}} + \frac{E_{bg} - J_1}{\dfrac{1}{A_1 \varepsilon_{g1}}} = 0 \quad \textbf{(104)}$$

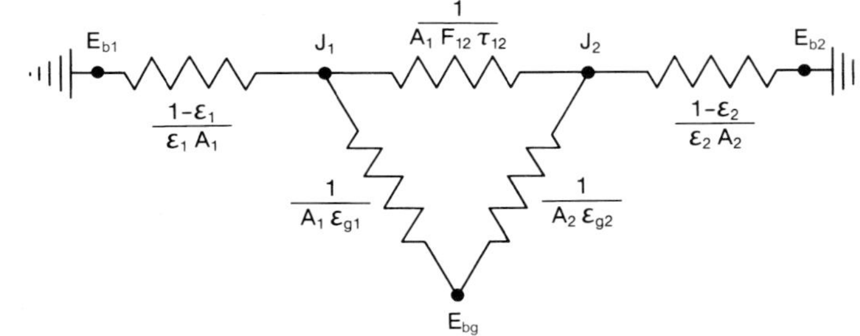

Fig. 41 Example of radiation in a cavity.

$$\frac{E_{b2} - J_2}{\dfrac{1-\varepsilon_2}{\varepsilon_2 A_2}} + \frac{J_1 - J_2}{\dfrac{1}{A_1 F_{12} \tau_{12}}} + \frac{E_{bg} - J_{b2}}{\dfrac{1}{A_1 \varepsilon_{g2}}} = 0 \quad \textbf{(105)}$$

The net heat flow between the surfaces is:

$$q_{12} = A_1 F_{12} \tau_{12} \left(J_1 - J_2\right) \quad \textbf{(106)}$$

Hottel[4] demonstrates the procedures for finding the beam length to determine F_{12}, ε_{g1}, and ε_{g2}.

References

1. Roshenow, W.M., Hartnett, J.P., and Ganic, E.N., *Handbook of Heat Transfer Fundamentals,* Second Ed., McGraw-Hill, Inc., New York, New York, 1985.

2. Roshenow, W.M., Hartnett, J.P., and Ganic, E.N., *Handbook of Heat Transfer Applications,* Second Ed., McGraw-Hill, Inc., New York, New York, 1985.

3. Kreith, F., and Bohn, M.S., *Principles of Heat Transfer,* Fourth Ed., Harper and Row, New York, New York, 1986.

4. Hottel, H.C., and Sarofim, A.F., *Radiative Transfer,* McGraw-Hill, Inc., New York, New York, 1967.

5. Wall, T.F., Bhattacharya, S.P., Zhang, D.K., et al., "The Properties and Thermal Effects of Ash Deposits in Coal-Fired Furnaces," *Progress in Energy and Combustion Science*, Vol. 19, pp. 487-504, 1993.

6. Meyer, C.A., et al., *ASME Steam Tables: Thermodynamic and Transport Properties of Steam,* Sixth Ed., American Society of Mechanical Engineers, New York, New York, 1993.

7. Boow, J., and Goard, P.R.C., "Fireside Deposits and Their Effect on Heat Transfer in a Pulverized Fuel-Fired Boiler: Part III. The Influence of the Physical Characteristics of the Deposit on its Radiant Emittance and Effective Thermal Conductance," *Journal of the Institute of Fuel*, pp. 412-419, Vol. 42, No. 346, 1969.

8. Leckner, B., "Spectral and Total Emissivity of Water Vapor and Carbon Dioxide," *Combustion and Flame*, Vol. 19, pp. 33-48, 1972.

9. Edwards, D.K., "Molecular Gas Band Radiation," *Advances in Heat Transfer*, Vol. 12, Academic Press, New York, New York, pp. 115-193, 1964.

10. Modest, M.F., *Radiative Heat Transfer*, McGraw-Hill, Inc., New York, New York, 1993.

11. Sieder, E.N., and Tate, G.E., "Heat Transfer and Pressure Drop of Liquids in Tubes," *Industrial & Engineering Chemistry Research (I&EC)*, Vol. 28, p. 1429, 1936.

12. Dittus, F.W., and Boelter, L.M.K., *University of California Publications on Engineering*, Vol. 2, p. 443, Berkeley, California, 1930.

13. McAdams, W., *Heat Transmission,* Third Ed., McGraw-Hill, Inc., New York, New York, 1954.

14. Grimison, E.D., "Correlation and Utilization of New Data on Flow Resistance and Heat Transfer for Crossflow of Gases over Tube Banks," Transactions of the *American Society of Mechanical Engineers*, Vol. 59, pp. 583-594, 1937.

15. Schmidt, T.F., "Wärme leistung von berippten Flächen," *Mitt. des Kältetechn. Institut der T.H. Karlshruhe,* Vol. 4, 1949.

16. Incropera, F., and DeWitt, D.P., *Fundamentals of Heat and Mass Transfer*, Third Ed., John Wiley & Sons, New York, New York, 1990.

17. Patankar, S., *Numerical Heat Transfer and Fluid Flow,* McGraw-Hill, Inc., New York, New York, 1980.

18. Fiveland, W. A., "Discrete-Ordinates Solutions of the Radiative Transport Equation for Rectangular Enclosures," Transactions of the *American Society of Mechanical Engineers Journal of Heat Transfer,* Vol. 106, pp. 699-706, 1984.

Two-phase flow void fraction measurements.

Chapter 5

Boiling Heat Transfer, Two-Phase Flow and Circulation

A case of heat transfer and flow of particular interest in steam generation is the process of boiling and steam-water flow. The boiling or evaporation of water is a familiar phenomenon. In general terms, boiling is the heat transfer process where heat addition to a liquid no longer raises its temperature under constant pressure conditions; the heat is absorbed as the liquid becomes a gas. The heat transfer rates are high, making this an ideal cooling method for surfaces exposed to the high heat input rates found in fossil fuel boilers, concentrated solar energy collectors and the nuclear reactor fuel bundles. However, the boiling phenomenon poses special challenges such as: 1) the sudden breakdown of the boiling behavior at very high heat input rates, 2) the potential flow rate fluctuations which may occur in steam-water flows, and 3) the efficient separation of steam from water. An additional feature of boiling and two-phase flow is the creation of significant density differences between heated and unheated tubes. These density differences result in water flowing to the heated tubes in a well designed boiler natural circulation loop.

Most fossil fuel steam generators and all commercial nuclear steam supply systems operate in the pressure range where boiling is a key element of the heat transfer process. Therefore, a comprehensive understanding of boiling and its various related phenomena is essential in the design of these units. Even at operating conditions above the critical pressure, where water no longer boils but experiences a continuous transition from a liquid-like to a gas-like fluid, boiling type behavior and special heat transfer characteristics occur.

Boiling process and fundamentals

Boiling point and thermophysical properties

The boiling point, or saturation temperature, of a liquid can be defined as the temperature at which its vapor pressure is equal to the total local pressure. The saturation temperature for water at atmospheric pressure is 212F (100C). This is the point at which net vapor generation occurs and free steam bubbles are formed from a liquid undergoing continuous heating. As discussed in Chapter 2, this saturation temperature (T_{sat}) is a unique function of pressure. The American Society of Mechanical Engineers (ASME) and the International Association for the Properties of Steam (IAPS) have compiled extensive correlations of thermophysical characteristics of water. These characteristics include the enthalpy (or heat content) of water, the enthalpy of evaporation (also referred to as the latent heat of vaporization), and the enthalpy of steam. As the pressure is increased to the critical pressure [3200 psi (22.1 MPa)], the latent heat of vaporization declines to zero and the bubble formation associated with boiling no longer occurs. Instead, a smooth transition from liquid to gaseous behavior occurs with a continuous increase in temperature as energy is applied.

Two other definitions are also helpful in discussing boiling heat transfer:

1. *Subcooling* For water below the local saturation temperature, this is the difference between the saturation temperature and the local water temperature ($T_{sat} - T$).
2. *Quality* This is the flowing mass fraction of steam (frequently stated as percent steam by weight or %SBW after multiplying by 100%):

$$x = \frac{\dot{m}_{steam}}{\dot{m}_{water} + \dot{m}_{steam}} \quad (1)$$

where

$\dot{m}_{steam}$ = steam flow rate, lb/h (kg/s)
$\dot{m}_{water}$ = water flow rate, lb/h (kg/s)

Thermodynamically, this can also be defined as:

$$x = \frac{H - H_f}{H_{fg}} \quad or \quad \frac{H - H_f}{H_g - H_f} \quad (2)$$

where

H = local average fluid enthalpy, Btu/lb (J/kg)
H_f = enthalpy of water at saturation, Btu/lb (J/kg)
H_g = enthalpy of steam at saturation, Btu/lb (J/kg)
H_{fg} = latent heat of vaporization, Btu/lb (J/kg)

When boiling is occurring at saturated, thermal equilibrium conditions, Equation 2 provides the fractional steam flow rate by mass. For subcooled conditions where $H < H_f$, quality (x) can be negative and is an indication of liquid subcooling. For conditions where $H > H_g$, this value can be greater than 100% and represents the amount of average superheat of the steam.

Boiling curve

Fig. 1 illustrates a boiling curve which summarizes the results of many investigators. This curve provides the results of a heated wire in a pool, although the characteristics are similar for most situations. The heat transfer rate per unit area, or *heat flux*, is plotted versus the temperature differential between the metal surface and the bulk fluid. From points A to B, convection heat transfer cools the wire and boiling on the surface is suppressed. Moving beyond point B, which is also referred to as the *incipient boiling point*, the temperature of the fluid immediately adjacent to the heated surface slightly exceeds the local saturation temperature of the fluid while the bulk fluid remains subcooled. Bubbles, initially very small, begin to form adjacent to the wire. The bubbles then periodically collapse as they come into contact with the cooler bulk fluid. This phenomenon, referred to as *subcooled boiling*, occurs between points B and S on the curve. The heat transfer rate is quite high, but no net steam generation occurs. From points S to C, the temperature of the bulk fluid has reached the local saturation temperature. Bubbles are no longer confined to the area immediately adjacent to the surface, but move into the bulk fluid. This region is usually referred to as the *nucleate boiling* region, and as with subcooled boiling, the heat transfer rates are quite high and the metal surface is only slightly above the saturation temperature.

As point C is approached, increasingly large surface evaporation rates occur. Eventually, the vapor generation rate becomes so large that it restricts the liquid return flow to the surface. The surface eventually becomes covered (blanketed) with an insulating layer of steam and the ability of the surface to transfer heat drops. This transition is referred to as the *critical heat flux (CHF)*, *departure from nucleate boiling (DNB)*, *burnout*, *dryout*, *peak heat flux*, or *boiling crisis*. The temperature response of the surface under this condition depends upon how the surface is being heated. In fossil fuel boiler furnaces and nuclear reactor cores, the heat input is effectively independent of surface temperature. Therefore, a reduction in the heat transfer rate results in a corresponding increase in surface temperature from point D to D′ in Fig. 1. In some cases, the elevated surface temperature is so high that the metal surface may melt. If, on the other hand, the heat input or heat transfer rate is dependent upon the surface temperature, typical of a nuclear steam generator, the average local temperature of the surface increases as the local heat transfer rate declines. This region, illustrated in Fig. 1 from points D to E, is typically referred to as *unstable film boiling* or *transition boiling*. Because a large surface temperature increase does not occur, the main consequences are a decline in heat transfer performance per unit surface area and less overall energy transfer. The actual local phenomenon in this region is quite complex and unstable as discrete areas of surface fluctuate between a wetted boiling condition and a steam blanketed, or dry patch, condition. From position E through D′ to F, the surface is effectively blanketed by an insulating layer of steam or vapor. Energy is transferred from the solid surface through this layer by radiation, conduction and microconvection to the liquid-vapor interface. From this interface, evaporation occurs and bubbles depart. This heat transfer region is frequently referred to as stable *film boiling*.

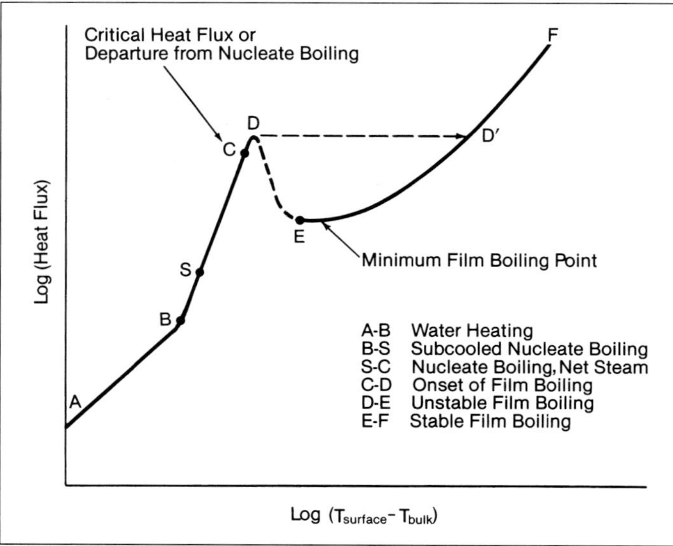

Fig. 1 Boiling curve – heat flux versus applied temperature difference.

In designing steam generating systems, care must be exercised to control which of these phenomena occur. In high heat input locations, such as the furnace area of fossil fuel boilers or nuclear reactor cores, it is important to maintain nucleate or subcooled boiling to adequately cool the surface and prevent material failures. However, in low heat flux areas or in areas where the heat transfer rate is controlled by the boiling side heat transfer coefficient, stable or unstable film boiling may be acceptable. In these areas, the resultant heat transfer rate must be evaluated, any temperature limitations maintained and only allowable temperature fluctuations accepted.

Flow boiling

Flow or *forced convective boiling*, which is found in virtually all steam generating systems, is a more complex phenomenon involving the intimate interaction of two-phase fluid flow, gravity, material phenomena and boiling heat transfer mechanisms. Fig. 2 is a classic picture of boiling water in a long, uniformly heated, circular tube. The water enters the tube as a subcooled liquid and convection heat transfer cools the tube. The point of incipient boiling is reached (point 1 in Fig. 2). This results in the beginning of subcooled boiling and bubbly flow. The fluid temperature continues to rise until the entire bulk fluid reaches the saturation temperature and nucleate boiling occurs, point 2. At this location, flow boiling departs somewhat from the simple pool boiling model previously discussed. The steam-water mixture progresses through a series of flow structures or patterns: bubbly, intermediate and annular. This is a result of the complex interaction of surface tension forces, interfacial phenomena, pressure drop, steam-water densities and momentum effects coupled with the surface boiling behavior. While boiling heat transfer continues throughout, a point is reached in the annular flow regime where the liquid film on the wall becomes so thin that nucleation in the film is suppressed, point 3. Heat transfer then occurs through conduction and convection across the thin annular film with surface evaporation at the steam-water interface. This heat transfer mechanism, called *convective boiling*, also results in high heat transfer rates. It should also be noted that not all of the liquid is on the tube wall. A portion is entrained in the steam core as dispersed droplets.

Eventually, an axial location, point 4, is reached where the tube surface is no longer wetted and CHF or dryout occurs. This is typically associated with a temperature rise. The exact tube location and magnitude of this temperature, however, depend upon a variety of parameters, such as the heat flux, mass flux, geometry and steam quality. Fig. 3 illustrates the effect of heat input rate, or heat flux, on CHF location and the associated temperature increase. From points 4 to 5 in Fig. 2, post-CHF heat transfer, which is quite complex, occurs. Beyond point 5, all of the liquid is evaporated and simple convection to steam occurs.

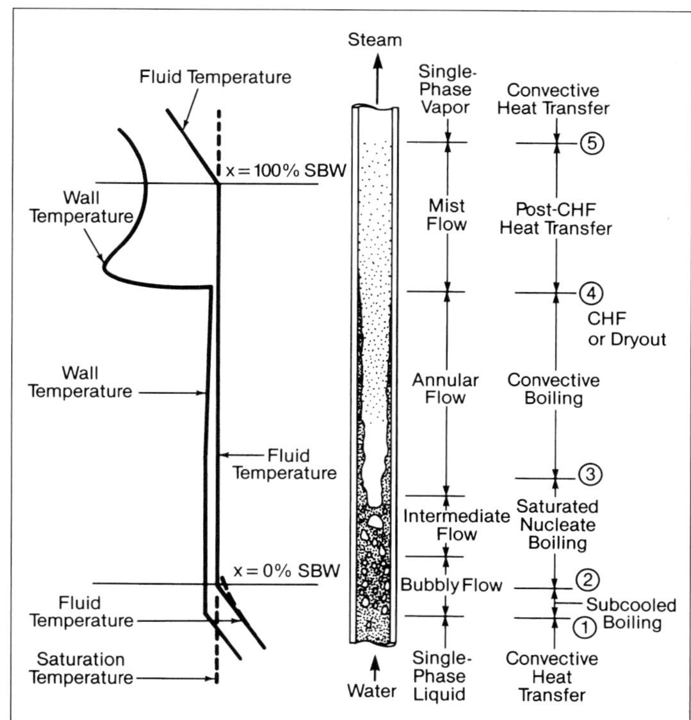

Fig. 2 Simplified flow boiling in a vertical tube (adapted from Collier[1]).

Boiling heat transfer evaluation

Engineering design of steam generators requires the evaluation of water and steam heat transfer rates under boiling and nonboiling conditions. In addition, the identification of the location of critical heat flux (CHF) is important where a dramatic reduction in the heat transfer rate could lead to: 1) excessive metal temperatures potentially resulting in tube failures, 2) an unacceptable loss of thermal performance, or 3) unacceptable temperature fluctuations leading to thermal fa-

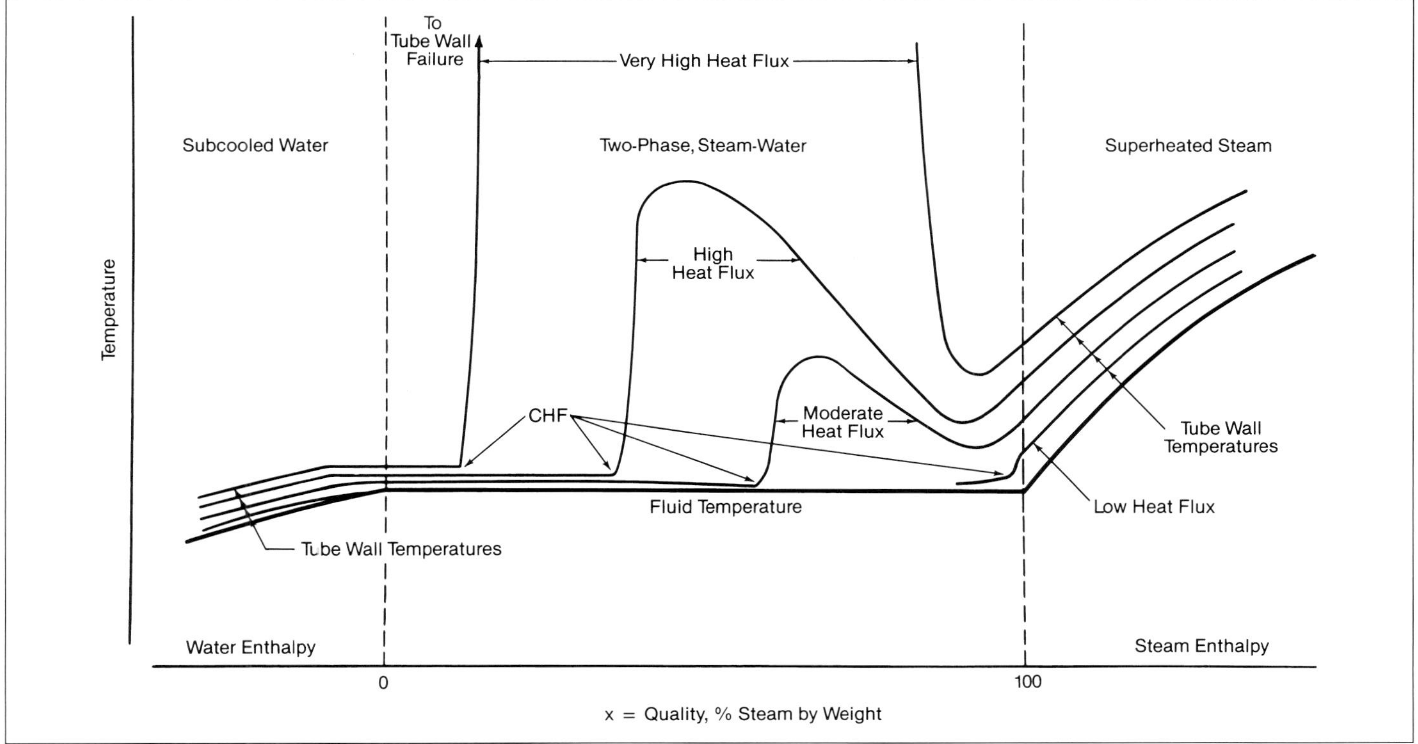

Fig. 3 Tube wall temperatures under different heat input conditions.

tigue failures. Data must also be available to predict the rate of heat transfer downstream of the dryout point. CHF phenomena are less important than the heat transfer rates for performance evaluation, but are more important in defining acceptable operating conditions. As discussed in Chapter 4, the heat transfer rate per unit area or heat flux is equal to the product of temperature difference and a heat transfer coefficient.

Heat transfer coefficients

Heat transfer correlations are application (surface and geometry) specific and The Babcock & Wilcox Company (B&W) has developed extensive data for its applications through experimental testing and field experience. These detailed correlations remain proprietary to B&W. However, the following generally available correlations are provided here as representative of the heat transfer relationships.

Single-phase convection Several correlations for forced convection heat transfer are presented in Chapter 4. Forced convection is assumed to occur as long as the calculated forced convection heat flux is greater than the calculated boiling heat flux (point 1 in Fig. 2):

$$q''_{\text{Forced Convection}} > q''_{\text{Boiling}} \tag{3}$$

While not critical in most steam generator applications, correlations are available which explicitly define this onset of subcooled boiling and more accurately define the transition region.[1]

Subcooled boiling In areas where subcooled boiling occurs, several correlations are available to characterize the heat transfer process. Typical of these is the Jens and Lottes[2] correlation for water. For inputs with English units:

$$\Delta T_{sat} = 60\left(q''/10^6\right)^{1/4} e^{-P/900} \tag{4a}$$

and for inputs with SI units:

$$\Delta T_{sat} = 25\left(q''\right)^{1/4} e^{-P/6.2} \tag{4b}$$

where

ΔT_{sat} = $T_w - T_{sat}$, F (C)
T_w = wall temperature, F (C)
T_{sat} = saturated water temperature, F (C)
q'' = heat flux, Btu/h ft^2 (MW$_t$/m^2)
P = pressure, psi (MPa)

Another relationship frequently used is that developed by Thom.[3]

Nucleate and convective boiling Heat transfer in the saturated boiling region occurs by a complex combination of bubble nucleation at the tube surface (nucleate boiling) and direct evaporation at the steam-water interface in annular flow (convective boiling). At low steam qualities, nucleate boiling dominates while at higher qualities convective boiling dominates. While separate correlations are available for each range, the most useful relationships cover the entire saturated boiling regime. They typically involve the summation of appropriately weighted nucleate and convective boiling components as exemplified by the correlation developed by J.C. Chen and his colleagues.[4] While such correlations are frequently recommended for use in saturated boiling systems, their additional precision is not usually required in many boiler or reactor applications. For general evaluation purposes, the subcooled boiling relationship provided in Equation 4 is usually sufficient.

Post-CHF heat transfer As shown in Fig. 3, substantial increases in tube wall metal temperatures are possible if boiling is interrupted by the CHF phenomenon. The maximum temperature rise is of particular importance in establishing whether tube wall overheating may occur. In addition, the reliable estimation of the heat transfer rate may be important for an accurate assessment of thermal performance. Once the metal surface is no longer wetted and water droplets are carried along in the steam flow, the heat transfer process becomes more complex and includes: 1) convective heat transfer to the steam which becomes superheated, 2) heat transfer to droplets impinging on the surface from the core of the flow, 3) radiation directly from the surface to the droplets in the core flow, and 4) heat transfer from the steam to the droplets. This process results in a nonequilibrium flow featuring superheated steam mixed with water droplets. Current correlations do not provide a good estimate of the heat transfer in this region, but computer models show promise. Accurate prediction requires the use of experimental data for similar flow conditions.

Reflooding A key concept in evaluating emergency core coolant systems for nuclear power applications is *reflooding*. In a loss of coolant event, the reactor core can pass through critical heat flux conditions and can become completely dry. Reflooding is the term for the complex thermal-hydraulic phenomena involved in rewetting the fuel bundle surfaces as flow is returned to the reactor core. The fuel elements may be at very elevated temperatures so that the post-CHF, or steam blanketed, condition may continue even in the presence of returned water flow. Eventually, the surface temperature drops enough to permit a rewetting front to wash over the fuel element surface. Analysis includes transient conduction of the fuel elements and the interaction with the steam-water heat transfer processes.

Critical heat flux phenomena

Critical heat flux is one of the most important parameters in steam generator design. CHF denotes the set of operating conditions (mass flux, pressure, heat flux and steam quality) covering the transition from the relatively high heat transfer rates associated with nucleate or forced convective boiling to the lower rates resulting from transition or film boiling (Figs. 1 and 2). These operating conditions have been found to be geometry specific. CHF encompasses the phenomena of departure from nucleate boiling (DNB), burnout, dryout and boiling crisis. One objective in recirculating boiler and nuclear reactor designs is to avoid CHF conditions. In once-through steam generators, the objective is to design to accommodate the temperature increase at the CHF locations. In this process, the heat flux profile, flow passage geometry, operating pressure and inlet enthalpy are usually fixed, leaving mass flux, local quality, diameter and some surface effects as the more easily adjusted variables.

Factors affecting CHF Critical heat flux phenomena under flowing conditions found in fossil fuel and nuclear steam generators are affected by a variety of parameters.[5] The primary parameters are the operating conditions and the design geometries. The operating conditions affecting CHF are pressure, mass flux and steam quality. Numerous design geometry factors include flow passage dimensions and shape, flow path obstructions, heat flux profile, inclination and wall surface configuration. Several of these effects are illustrated in Figs. 3 through 7.

Fig. 3 illustrates the effect of increasing the heat input on the location of the temperature excursion in a uniformly heated vertical tube cooled by upward flowing water. At low heat fluxes, the water flow can be almost completely evaporated to steam before any temperature rise is observed. At moderate and high heat fluxes, the CHF location moves progressively towards the tube inlet and the maximum temperature excursion increases. At very high heat fluxes, CHF occurs at a low steam quality and the metal temperature excursion can be high enough to melt the tube. At extremely high heat input rates, CHF can occur in subcooled water. Avoiding this type of CHF is an important design criterion for pressurized water nuclear reactors.

Many large fossil fuel boilers are designed to operate between 2000 and 3000 psi (13.8 and 20.7 MPa). In this range, pressure has a very important effect, shown in Fig. 4, with the steam quality limit for CHF falling rapidly near the critical pressure; i.e., at constant heat flux, CHF occurs at lower steam qualities as pressure rises.

Many CHF correlations have been proposed and are satisfactory within certain limits of pressure, mass velocity and heat flux. Fig. 5 is an example of a correlation which is useful in the design of fossil fuel natural circulation boilers. This correlation defines safe and unsafe regimes for two heat flux levels at a given pressure in terms of steam quality and mass velocity. Additional factors must be introduced when tubes are used in membrane or tangent wall construction, are inclined from the vertical, or have different inside diameter or surface configuration. The inclination of the flow passage can have a particularly dramatic effect on the CHF conditions as illustrated in Fig. 6.[6]

Ribbed tubes Since the 1930s, B&W has investigated a large number of devices, including internal twisters, springs and grooved, ribbed and corrugated tubes to delay the onset of CHF. The most satisfactory overall performance was obtained with tubes having helical ribs on the inside surface.

Two general types of rib configurations have been developed:

1. single-lead ribbed (SLR) tubes (Fig. 8a) for small internal diameters used in once-through subcritical pressure boilers, and
2. multi-lead ribbed (MLR) tubes (Fig. 8b) for larger internal diameters used in natural circulation boilers.

Both of these ribbed tubes have shown a remarkable ability to delay the breakdown of boiling. Fig. 7

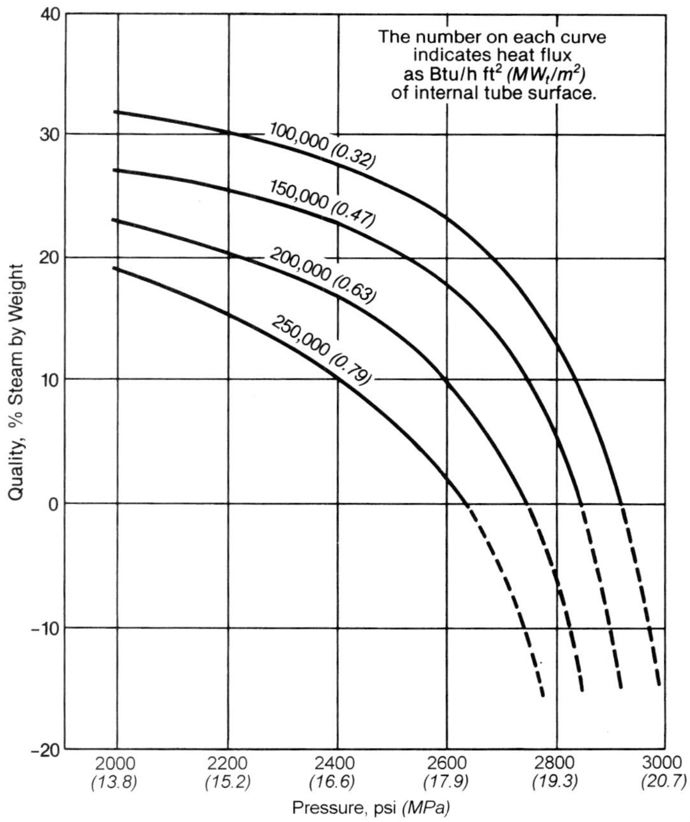

Fig. 4 Steam quality limit for CHF as a function of pressure.

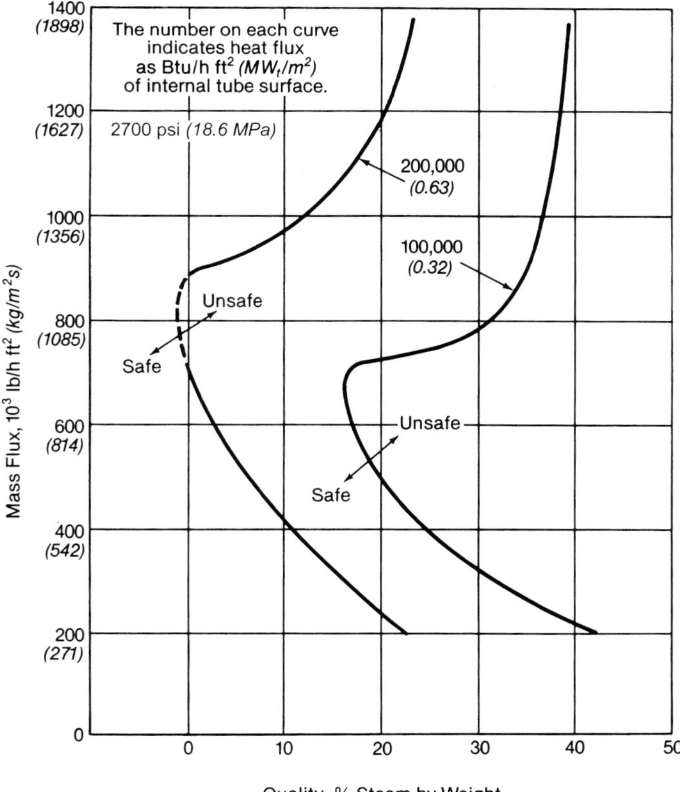

Fig. 5 Steam quality limit for CHF as a function of mass flux.

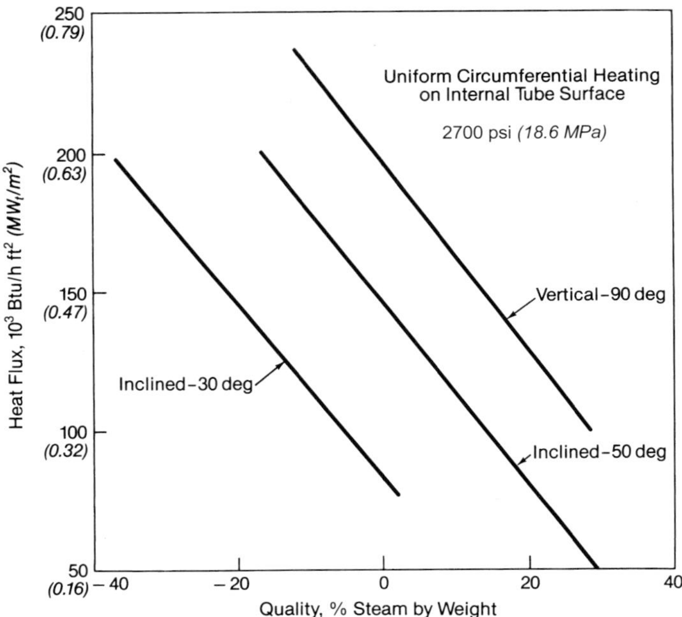

Fig. 6 Effect of inclination on CHF at 700,000 lb/h ft² (950 kg/m² s).[6]

compares the effectiveness of a ribbed tube to that of a smooth tube in a membrane wall configuration. This plot is different from Fig. 5 in that heat flux is given as an average over the flat projected surface. This is more meaningful in discussing membrane wall heat absorption.

The ribbed bore tubes provide a balance of improved CHF performance at an acceptable increase in pressure drop without other detrimental effects. The ribs generate a swirl flow resulting in a centrifugal action which forces the water to the tube wall and retards entrainment of the liquid. The steam blanketing and film dryout are therefore prevented until substantially higher steam qualities or heat fluxes are reached.

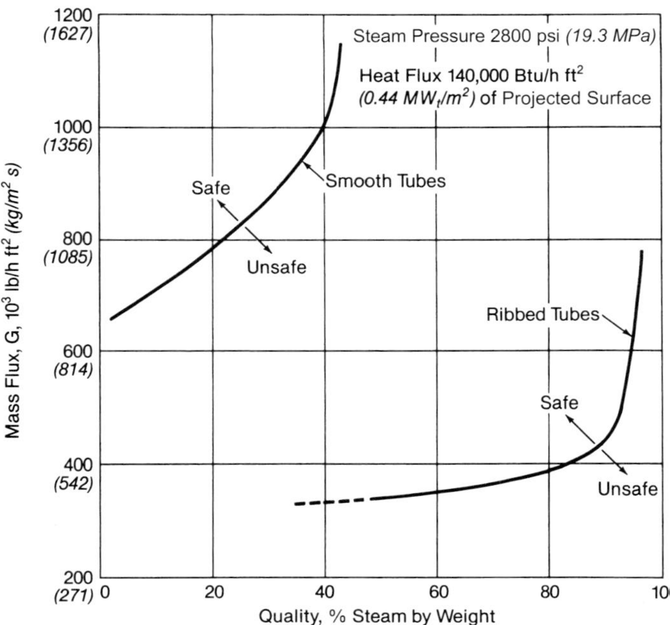

Fig. 7 Steam quality limit for CHF in smooth and ribbed bore tubes.

Fig. 8a Single-lead ribbed tube.

Fig. 8b Multi-lead ribbed tube.

Because the ribbed bore tube is more expensive than a smooth bore tube, its use involves an economic balance of several design factors. In most instances, there is less incentive to use ribbed tubes below 2200 psi (15.2 MPa).

Evaluation CHF is a complex combination of thermal-hydraulic phenomena for which a comprehensive theoretical basis is not yet available. As a result, experimental data are likely to continue to be the basis for CHF evaluations. Many data and correlations define CHF well over *limited* ranges of conditions and geometries. However, progress is being made in developing more general evaluation procedures for at least the most studied case – a uniformly heated smooth bore tube with upward flowing water.

To address this complex but critical phenomenon in the design of reliable steam generating equipment, B&W has developed an extensive proprietary database and associated correlations. A graphical example is shown in Fig. 5 for a fossil fuel boiler tube. A B&W correlation[7] for nuclear reactor fuel rod bundle subchannel analysis is shown in Table 1.

CHF criteria A number of criteria are used to assess the CHF margins in a particular tube or tube bundle geometry.[8] These include the CHF ratio, flow ratio and quality margin, defined as follows:

1. $\text{CHF ratio} = \text{minimum value of } \dfrac{\text{CHF heat flux}}{\text{upset heat flux}}$

2. $\text{flow ratio} = \text{minimum value of } \dfrac{\text{min. design mass flux}}{\text{mass flux at CHF}}$

3. quality margin = CHF quality – max. design quality

The CHF ratios for a sample fossil fuel boiler are illustrated in Fig. 9 for a smooth bore tube (q''_B / q''_A) and a ribbed bore tube (q''_C / q''_A). The graph indicates the relative increase in local heat input which can be tolerated before the onset of CHF conditions. A similar relationship for a nuclear reactor fuel rod application is shown in Fig. 10.

Supercritical heat transfer

Unlike subcritical pressure conditions, fluids at supercritical pressures experience a continuous transition from water-like to steam-like characteristics. As a result, CHF conditions and boiling behavior would not be expected. However, at supercritical pressures, especially in the range of $1 < P/P_c < 1.15$ where P_c is the critical pressure, two types of boiling-like behavior have been observed: pseudo-boiling and pseudo-film boiling. Pseudo-boiling is an increase in heat transfer coefficient not accounted for by traditional convection relationships. In pseudo-film boiling, a dramatic reduction in the heat transfer coefficient is observed at high heat fluxes. This is similar to the critical heat flux condition at subcritical pressures.

These behaviors have been attributed to the sharp changes in fluid properties as the transition from water-like to steam-like behavior occurs.

Fluid properties In the supercritical region, the thermophysical properties important to the heat transfer process, i.e., conductivity, viscosity, density and specific heat, experience radical changes as a certain pressure-dependent temperature is approached and exceeded. This is illustrated in Fig. 11. The transition temperature, referred to as the pseudo-critical temperature, is defined as the temperature where the specific heat, c_p, reaches its maximum. As the operating pressure is increased, the pseudo-critical temperature increases and the dramatic change in the thermophysical properties declines as this temperature is approached and exceeded.

Heat transfer rates Because of the significant changes in thermophysical properties (especially in specific heat) near the pseudo-critical temperature, a modified approach to evaluating convective heat transfer is needed. A number of correlations have been developed and a representative relationship for smooth bore tubes is:[9]

$$\frac{hD_i}{k_w} = 0.00459\left[\frac{D_i G}{\mu_w}\right]^{0.923} \times \left[\left(\frac{H_w - H_b}{T_w - T_b}\right)\left(\frac{\mu_w}{k_w}\right)\right]^{0.613}\left[\frac{\upsilon_b}{\upsilon_w}\right]^{0.231} \quad (5)$$

Table 1
B&W2 Reactor Rod Bundle Critical Heat Flux (CHF) Correlation[7]

$$q''_{CHF} = \frac{(a - bD_i)\left[A_1 (A_2 G)^{A_3+A_4(P-2000)} - A_9 G x_{CHF} H_{fg}\right]}{A_5 (A_6 G)^{A_7+A_8(P-2000)}}$$

where

a = 1.15509	A = area, in.^{2}
b = 0.40703	D_i = equivalent diameter = $4A/Per$
A_1 = 0.37020 x 10^8	G = mass flux, lb/h ft^2
A_2 = 0.59137 x 10^{-6}	H_{fg} = latent heat of vaporization, Btu/lb
A_3 = 0.83040	
A_4 = 0.68479 x 10^{-3}	P = pressure, psi
A_5 = 12.710	Per = wetted perimeter, in.
A_6 = 0.30545 x 10^{-5}	x_{CHF} = steam quality at CHF conditions, fraction steam by weight
A_7 = 0.71186	
A_8 = 0.20729 x 10^{-3}	q''_{CHF} = heat flux at CHF conditions, Btu/h ft^2
A_9 = 0.15208	

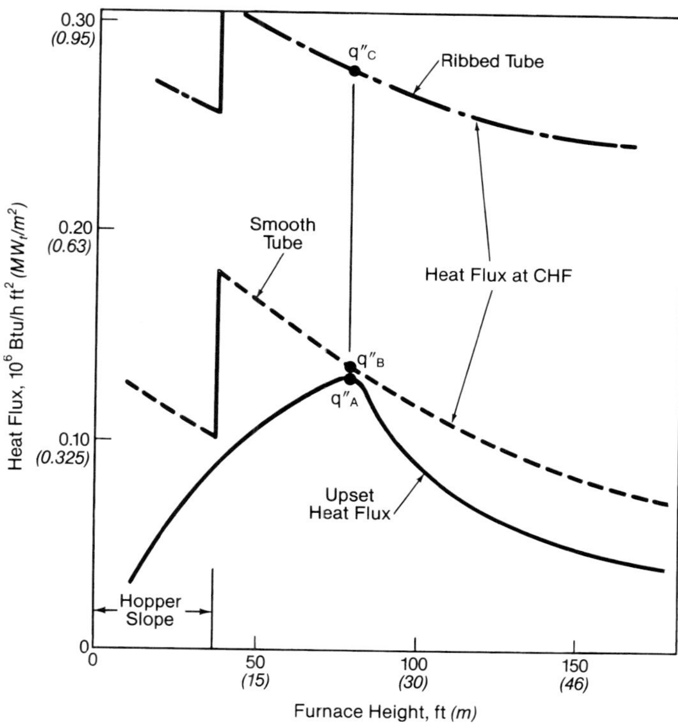

Fig. 9 Fossil boiler CHF ratio = minimum value of critical heat flux divided by upset heat flux.

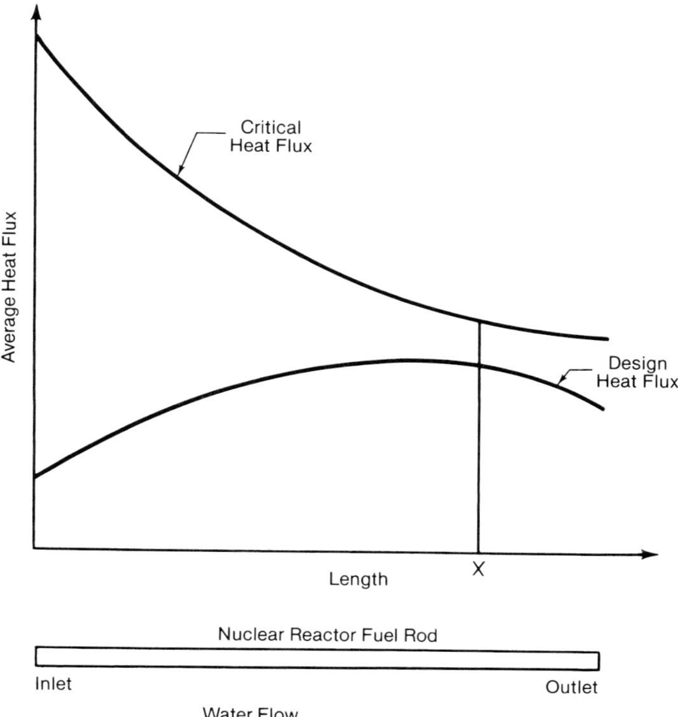

Fig. 10 Nuclear reactor CHF ratio = minimum value of critical heat flux divided by design heat flux.

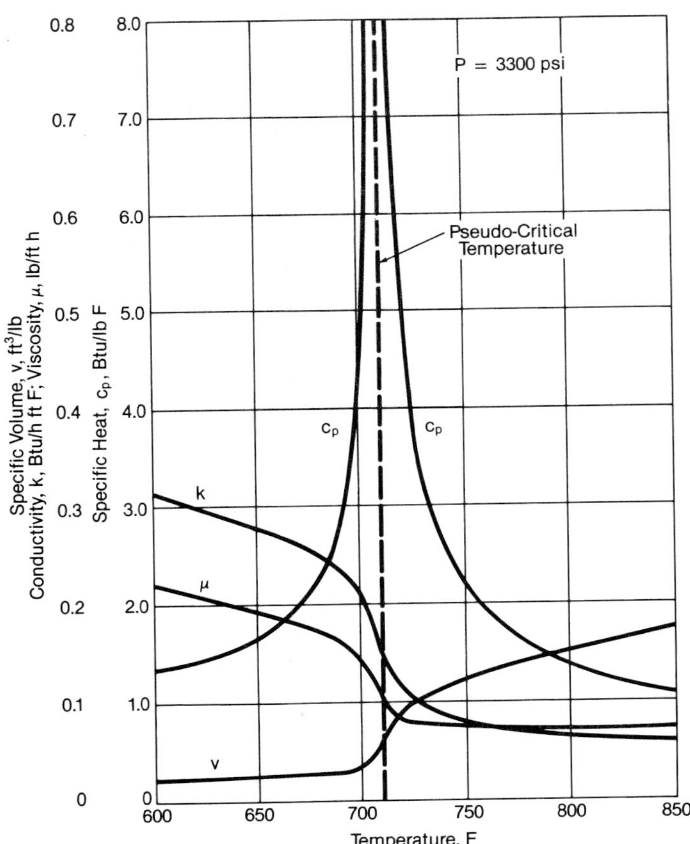

Fig. 11 Thermophysical properties of water (English units).

where

h = heat transfer coefficient, Btu/h ft² F (W/m² K)
k = thermal conductivity, Btu/h ft F (W/m K)
D_i = inside tube diameter, ft (m)
G = mass flux, lb/h ft² (kg/m² s)
μ = viscosity, lb/ft h (kg/m s)
H = enthalpy, Btu/lb (J/kg)
T = temperature, F (C)
v = specific volume, ft³/lb (m³/kg)

The subscripts b and w refer to properties evaluated at the bulk fluid and wall temperatures respectively.

This correlation has demonstrated reasonable agreement with experimental data from tubes of 0.37 to 1.5 in. (9.4 to 38.1 mm) inside diameter and at low heat fluxes.

Pseudo-boiling For low heat fluxes and bulk fluid temperatures approaching the pseudo-critical temperature, an improvement in the heat transfer rate takes place. The enhanced heat transfer rate observed is sometimes referred to as pseudo-boiling. It has been attributed to the increased turbulence resulting from the interaction of the water-like and steam-like fluids near the tube wall.

Pseudo-film boiling Potentially damaging temperature excursions associated with a sharp reduction in heat transfer can be observed at high heat fluxes. This temperature behavior is similar to the CHF phenomenon observed at subcritical conditions and is referred to as pseudo-film boiling. This phenomenon has been attributed to a limited ability of the available turbulence to move the higher temperature steam-like fluid away from the tube wall into the colder, higher density (water-like) fluid in the bulk stream. A phenomenon similar to steam blanketing occurs and the wall temperature increases in response to the relatively constant applied heat flux.

Single-lead ribbed (SLR) bore tubes are very effective in suppressing the temperature peaks encountered in smooth bore tubes.[10]

Two-phase flow

Flow patterns

As illustrated in Fig. 2, two-phase steam-water flow may occur in many regimes or structures. The transition from one structure to another is continuous rather than abrupt, especially under heated conditions, and is strongly influenced by gravity, i.e., flow orientation. Because of the qualitative nature of flow pattern identification, there are probably as many flow pattern descriptions as there are observers. However, for vertical, heated, upward, co-current steam-water flow in a tube, four general flow patterns are generally recognized (see Fig. 12):

1. *Bubbly flow* Relatively discrete steam bubbles are dispersed in a continuous liquid water phase. Bubble size, shape and distribution are dependent upon the flow rate, local enthalpy, heat input rate and pressure.
2. *Intermediate flow* This is a range of patterns between bubbly and annular flows; the patterns are also referred to as slug or churn flow. They range from: a) large bubbles, approaching the tube size in diameter, separated from the tube wall by thin annular films and separated from each other by slugs of liquid which may also contain smaller bubbles, to b) chaotic mixtures of large nonsymmetric bubbles and small bubbles.
3. *Annular flow* A liquid layer is formed on the tube wall with a continuous steam core; most of the liq-

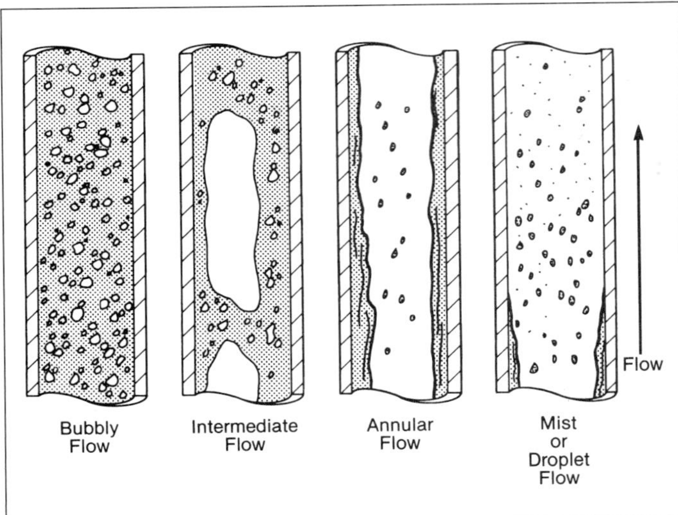

Fig. 12 Flow pattern – upward, co-current steam-water flow in a heated vertical tube.

uid is flowing in the annular film. At lower steam qualities, the liquid film may have larger amplitude waves adding to the liquid droplet entrainment and transport in the continuous steam core. At high qualities, the annular film becomes very thin, bubble generation is suppressed and the large amplitude waves disappear.

4. *Mist flow* A continuous steam core transports entrained water droplets which slowly evaporate until a single-phase steam flow occurs. This is also referred to as droplet or dispersed flow.

In the case of inclined and horizontal co-current steam-water flow in heated tubes, the flow patterns are further complicated by stratification effects. At high flow rates, the flow patterns approach those of vertical tubes. At lower rates, additional distinct flow patterns (wavy, stratified and modified plug) emerge as gravity stratifies the flow with steam concentrated in the upper portion of the tube. This can be a problem where inclined tubes are heated from the top. CHF or dryout conditions occur at much lower steam qualities and lower heat input rates in such inclined or horizontal tubes.

Additional complexity in patterns is observed when two-phase flow occurs in parallel or crossflow tube bundles. The tubes, baffles, support plates and mixing devices further disrupt the flow pattern formation.

Flow maps The transitions from one flow regime to another are quite complex, with each transition representing a combination of factors. However, two dimensional flow maps provide at least a general indication of which flow pattern is likely under given operating conditions. The maps generally are functions of superficial gas and liquid velocities. An example for vertical, upward, steam-water co-current flow is provided in Fig. 13.[11] The axes in this figure represent the superficial momentum fluxes of the steam (y-axis) and water (x-axis). A sample flow line is shown beginning at nearly saturated water conditions and ending with saturated steam conditions. The tube experiences bubbly flow only near its inlet. This is followed by a brief change to intermediate flow before annular flow dominates the heated length.

Other flow maps are available for arrangements such as downflow tubes, inclined tubes and bundles. Flow maps, however, are only approximations providing guidance in determining the relevant flow structure for a given situation.

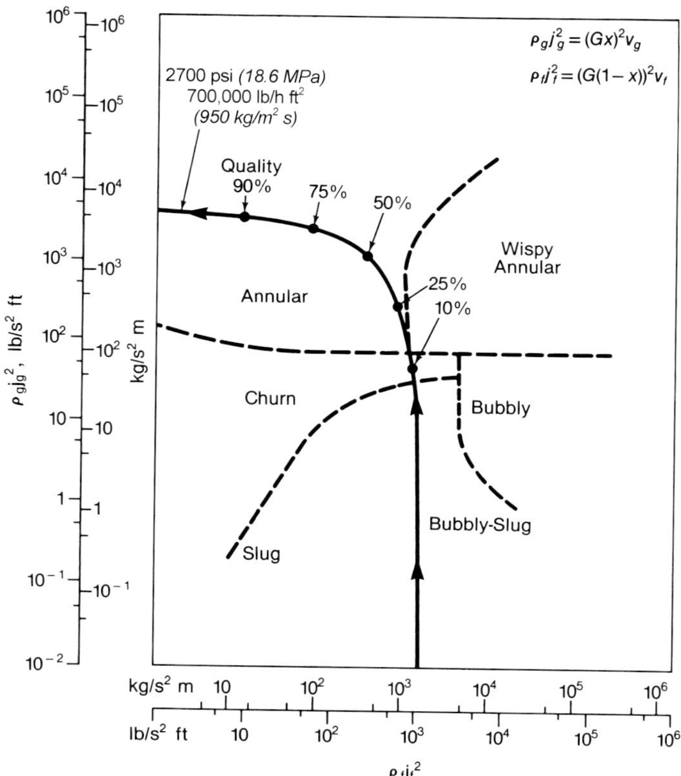

Fig. 13 Flow pattern map for vertical upward flow of water.[11]

Pressure loss

The local pressure loss, ΔP [lb/ft² (Pa)] or gradient $\delta P/\delta l$ [lb/ft²/ft (Pa/m)] in a two-phase steam-water system may be represented by:

$$\Delta P = \Delta P_f + \Delta P_a + \Delta P_g + \Delta P_l \tag{6a}$$

or

$$-\frac{\delta P}{\delta l} = -\left(\frac{\delta P}{\delta l}\right)_f - \left(\frac{\delta P}{\delta l}\right)_a - \left(\frac{\delta P}{\delta l}\right)_g + \Delta P_l \tag{6b}$$

The ΔP_f and $-(\delta P/\delta l)_f$ terms account for local wall friction losses. The ΔP_a and $-(\delta P/\delta l)_a$ terms address the momentum or acceleration loss incurred as the volume increases due to evaporation. The hydraulic or static head loss is accounted for by ΔP_g and $-(\delta P/\delta l)_g$. Finally, all of the local losses due to fittings, contractions, expansions, bends, or orifices are included in ΔP_l. The evaluation of these parameters is usually made using one of two models: homogeneous flow or separated flow.

A parameter of particular importance when evaluating the pressure loss in steam-water flows is void fraction. The void fraction can be defined by time-averaged flow area ratios or local-volume ratios of steam to the total flow. The area-based void fraction, α, can be defined as the ratio of the time-averaged steam flow cross-sectional area (A_{steam}) to the total flow area ($A_{steam} + A_{water}$):

$$\alpha = \frac{A_{steam}}{A_{steam} + A_{water}} \tag{7}$$

Using the simple continuity equation, the relationship between quality, x, and *void fraction* is:

$$\alpha = \frac{x}{x + (1-x)\dfrac{\rho_g}{\rho_f} S} \tag{8}$$

where

S = ratio of the average cross-sectional velocities of steam and water (referred to as *slip*)
ρ_g = saturated steam density, lb/ft³ (kg/m³)
ρ_f = saturated water density, lb/ft³ (kg/m³)

If the steam and water are moving at the same velocity, $S = 1$ (no slip). Obviously, the relationship between void fraction and quality is also a strong function of system pressure. This relationship is illustrated in Fig. 14. The difference between the homogeneous and separated flow models is illustrated by the shaded band. The upper bound is established by the homogeneous model and the lower bound by the separated flow model.

Homogeneous model The homogeneous model is the simpler approach and is based upon the premise that the two-phase flow behavior can be directly modeled after single-phase behavior (see Chapter 3) if appropriate average properties are determined. The temperature and velocities of steam and water are assumed equal. The mixed weight averaged specific volume (υ) or the inverse of the homogeneous density ($1/\rho_{hom}$) is used:

$$\upsilon = \upsilon_f (1-x) + \upsilon_g x \tag{9a}$$

or

$$\frac{1}{\rho_{hom}} = \frac{(1-x)}{\rho_f} + \frac{x}{\rho_g} \tag{9b}$$

where

υ_f = saturated water specific volume, ft³/lb (m³/kg)
υ_g = saturated steam specific volume, ft³/lb (m³/kg)
ρ_f = saturated water density, lb/ft³ (kg/m³)
ρ_g = saturated steam density, lb/ft³ (kg/m³)
x = steam quality

This model provides reasonable results when high or low steam qualities exist, when high flow rates are present, or at higher pressures. In these cases, the flow is reasonably well mixed.

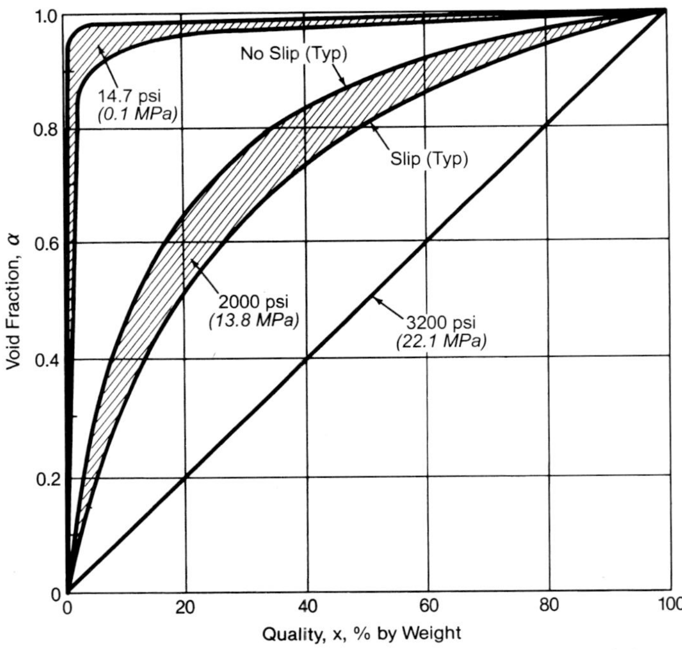

Fig. 14 Void fraction – quality relationship (homogeneous model, upper bound; separated flow model, lower bound).

The friction pressure drop (ΔP_f) can be evaluated by the equations provided in Chapter 3 using the mixture thermophysical properties. The pressure difference due to elevation (ΔP_g) can be evaluated as:

$$\Delta P_g = \pm \rho_{hom} \left(\frac{g}{g_c}\right) L \sin\theta \tag{10}$$

where

g = acceleration of gravity, ft/s² (m/s²)
g_c = 32.17 lbm ft/lbf s² (1 kg m/N s²)
L = length, ft (m)
θ = angle from the horizontal

The constant g_c is discussed in Chapter 2. A pressure gain occurs in downflow and a pressure loss occurs in upflow. The acceleration loss can be evaluated by:

$$\Delta P_a = \frac{G^2}{g_c}\left(\frac{1}{\rho_{out}} - \frac{1}{\rho_{in}}\right) \tag{11}$$

where

G = mass flux, lb/s ft² (kg/m²s)
ρ_{out} = outlet homogeneous density, lb/ft³ (kg/m³)
ρ_{in} = inlet homogeneous density, lb/ft³ (kg/m³)

Separated flow model In the steady-state separated flow model, the steam and water are treated as separate streams under the same pressure gradient but different velocities and differing properties. When the actual flow velocities of steam and water are equal, the simplest separated flow models approach the homogeneous case. Using one of several separated flow models[1] with unequal velocities, the pressure drop components (in differential form) are:

$$-\left(\frac{\delta P}{\delta l}\right)_f = -\left(\frac{\delta P}{\delta l}\right)_{LO} \phi^2_{LO} \quad \text{(friction)} \tag{12}$$

$$-\left(\frac{\delta P}{\delta l}\right)_{LO} = \frac{f}{D_i}\frac{G^2\upsilon_f}{2g_c} \quad \text{(single-phase friction)} \tag{13}$$

$$-\left(\frac{\delta P}{\delta l}\right)_a = \frac{G^2}{g_c}\frac{\delta}{\delta l}\left(\frac{x^2\upsilon_g}{\alpha} + \frac{(1.0-x)^2\upsilon_f}{(1.0-\alpha)}\right) \quad \text{(acceleration)} \tag{14}$$

$$-\left(\frac{\delta P}{\delta l}\right)_g = \frac{g}{g_c}\sin\theta\left(\frac{\alpha}{\upsilon_g} + \frac{(1.0-\alpha)}{\upsilon_f}\right) \quad \text{(static head)} \tag{15}$$

$$\Delta P_l = \Phi K \frac{G^2\upsilon_f}{2g_c} \quad \text{(local losses)} \tag{16}$$

where

Φ and ϕ^2_{LO} = appropriate two-phase multipliers
G = mass flux, lb/s ft² (kg/m² s)
f = fanning friction factor (see Chapter 3)
D_i = tube inside diameter, ft (m)
g = acceleration of gravity, ft/s² (m/s²)
g_c = 32.17 lbm ft/lbf s² (1 kg m/N s²)

υ_f = liquid specific volume, ft³/lb (m³/kg)
υ_g = vapor specific volume, ft³/lb (m³/kg)
x = steam quality
α = void fraction
θ = angle from the horizontal
K = loss coefficient

While ΔP_l usually represents just the irreversible pressure loss in single-phase flows, the complexity of two-phase flows results in the loss of ΔP_l typically representing the reversible and irreversible losses for fittings.

To evaluate the individual pressure losses from Equations 12 through 16 and Equation 6b, it is necessary to calculate ϕ^2_{LO}, α and Φ. Unfortunately, these factors are not well defined.

Specific correlations and evaluations can only be used where experimental data under similar conditions provide confidence in the prediction. Proprietary correlations used by B&W are based upon experimental data and practical experience.

For straight vertical tubes, generally available representative relationships include:

1. *Acceleration loss* The void fraction can frequently be evaluated with the homogeneous model ($S = 1$ in Equation 8).
2. *Friction loss and void fraction* Typical two-phase multiplier, ϕ^2_{LO}, and void fraction, α, relationships are presented by Thom,[12] Martinelli-Nelson,[13] Zuber-Findlay[14] and Chexal-Lellouche.[15] For illustration purposes the correlations of Thom are presented in Figs. 15 and 16. These curves can be approximated by:

$$\phi^2_{LO} = \left\{ \begin{aligned} &\left[0.97303(1-x) + x\left(\frac{\upsilon_g}{\upsilon_f}\right)\right]^{0.5} \\ &\times \left[0.97303(1-x)+x\right]^{0.5} + 0.027(1-x) \end{aligned} \right\}^{2.0} \quad (17)$$

and

$$\alpha = \frac{\gamma x}{1 + x(\gamma - 1)} \quad (18)$$

where

γ = $(\upsilon_g/\upsilon_f)^n$
n = $(0.8294 - 1.1672/P)$
P = pressure, psi
υ_g = saturated steam specific volume, ft³/lb
υ_f = saturated liquid specific volume, ft³/lb
x = steam quality

Instabilities

Instability in two-phase flow refers to the set of operating conditions under which sudden changes in flow direction, reduction in flow rate and oscillating flow rates can occur in a single flow passage. Often in manifolded multi-channel systems, the overall mass flow rate can remain constant while oscillating flows in individual channels still may occur. Such unstable conditions in steam generating systems can result in:

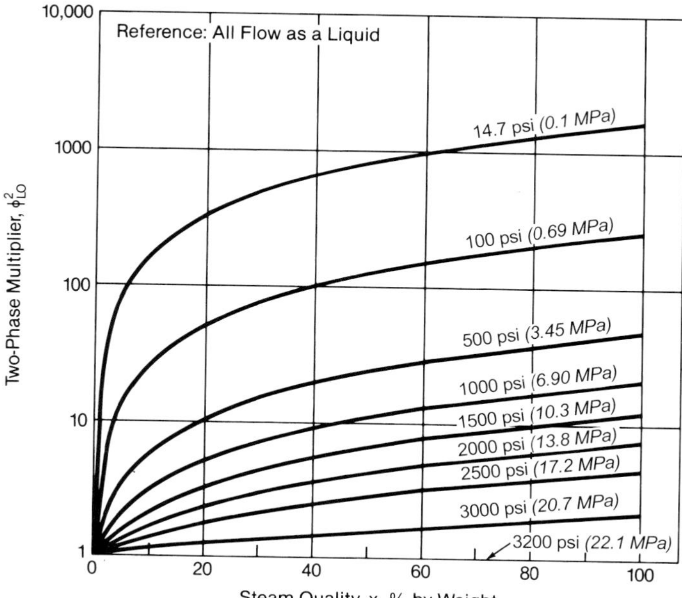

Fig. 15 Thom two-phase friction multiplier.[12]

1. unit control problems, including unacceptable variations in steam drum water level,
2. CHF/DNB/dryout,
3. tube metal temperature oscillation and thermal fatigue failure, and
4. accelerated corrosion attack.

Two of the most important types of instabilities in steam generator design are excursive instability, including Ledinegg and flow reversal, and density wave/pressure drop oscillations. The first is a static instability evaluated using steady-state equations while the last is dynamic in nature requiring the inclusion of time dependent factors.

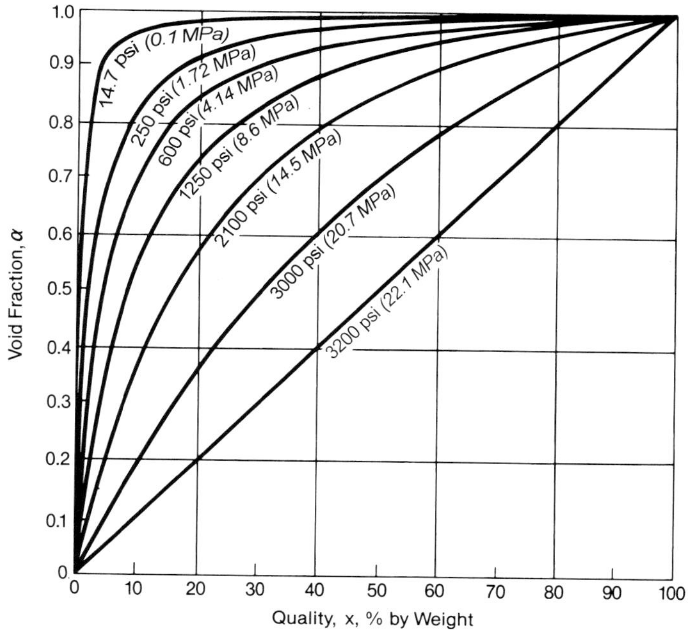

Fig. 16 Thom void fraction correlation (>3% SBW).[12]

Excursive and flow reversal instability evaluation The excursive instability is characterized by conditions where small perturbations in operating parameters result in a large flow rate change to a separate steady-state level. This can occur in both single channel and multi-channel manifolded systems. Excursive instabilities can be predicted by using the Ledinegg criteria.[16] Instability may occur if the slope of the pressure drop versus flow characteristic curve (internal) for the tube becomes less than the slope of the supply (or applied) curve at any intersection point:

$$\left(\frac{\delta \Delta P}{\delta G}\right)_{\text{internal}} \leq \left(\frac{\delta \Delta P}{\delta G}\right)_{\text{applied}} \tag{19}$$

The stable and unstable situations are illustrated in Fig. 17. As shown in the figure for unstable conditions, if the mass flow rate drops below point B then the flow rate continues to fall dramatically because the applied pumping head is less than that needed to move the fluid. For slightly higher mass flow rates (higher than point B), a dramatic positive flow excursion occurs because the pumping head exceeds the flow system requirement.

In most systems, the first term in Equation 19 is generally positive and the second is negative. Therefore, Equation 19 predicts stability. However, in two-phase systems, thermal-hydraulic conditions may combine to produce a local area where $(\delta\Delta P/\delta G)_{\text{internal}}$ is negative and the potential for satisfying Equation 19 and observing an instability exists. A heated tube flow characteristic showing a potential region of instability is illustrated in Fig. 18 where multiple flow rates can occur for a single applied pressure curve. Operating at point B is unstable with small disturbances resulting in a shift to point A or point C. More intense disturbances could result in flow shifts between A and C.

For the relatively small subcooling found at the entrance to tube panels in recirculating drum boilers and due to the relatively low exit steam qualities, negative slope regions in the pressure drop versus flow curves are typically not observed for positive flow cases. However, for once-through fossil fuel boilers and nuclear steam generators with high subcooling at the inlet and evaporation to dryness, negative slope regions in the upflow portion of the pressure drop characteristic may occur. Steps can be taken to avoid operation in any region where the circuit internal $\delta\Delta P/\delta G \leq 0$. General effects of operating and design parameters on the pressure drop versus mass flow curves include:

Parameter Increased	Effect on ΔP	Comment
heat input	decrease	more stable
inlet ΔP	increase	more stable
pressure	increase	more stable

In situations where static instability may occur, the inlet pressure drop can be increased by adding an orifice or flow restriction to modify the overall flow characteristic as shown in Fig. 18.

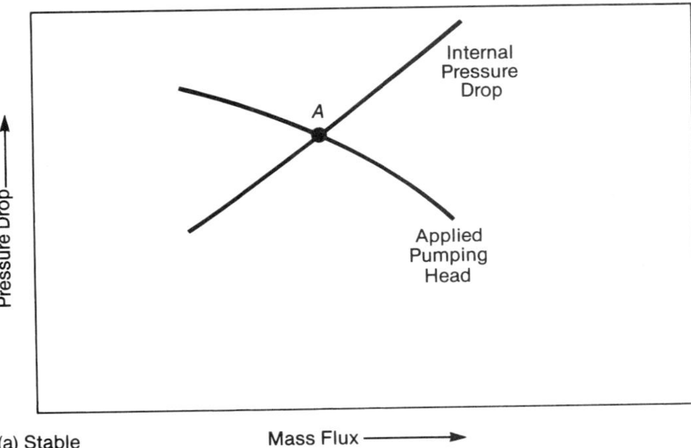

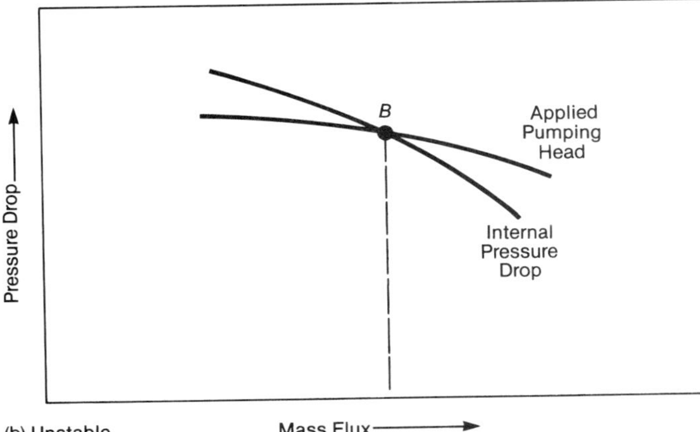

Fig. 17 Stable and unstable flow-pressure drop characteristics.

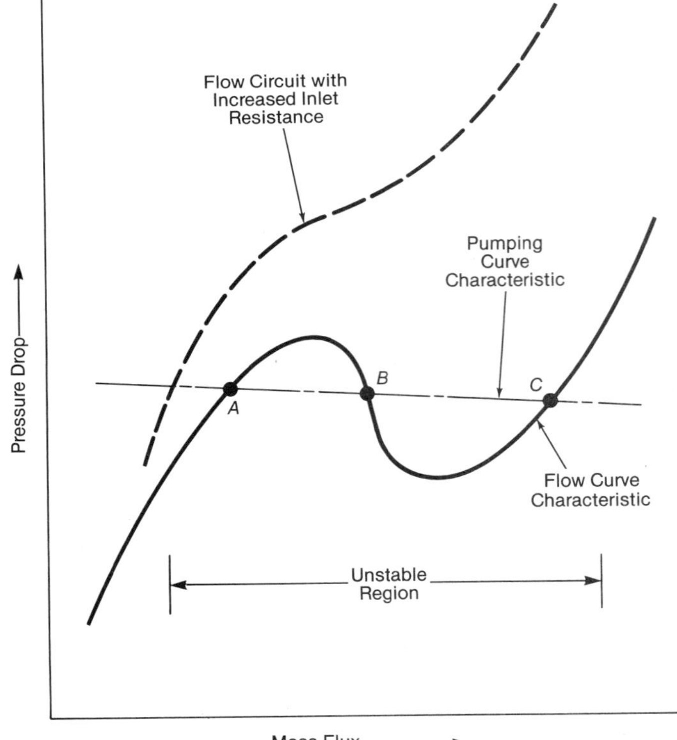

Fig. 18 Pressure drop characteristics showing unstable region.

Density wave/pressure drop instability Density wave instabilities involve kinematic wave propagation phenomena. Regenerative feedback between flow rate, vapor generation rate and pressure drop produce self sustaining alternating waves of higher and lower density mixture that travel through the tube. This dynamic instability can occur in single tubes that contain two-phase flows. In addition, when multiple tubes are connected by inlet and outlet headers, a more complex coupled channel instability, which is driven by density wave oscillations, may occur. Vertical heat flux distribution is a particularly sensitive parameter in dynamic instability evaluation.

Density wave oscillations can be predicted by the application of feedback control theory. A number of computer codes have been developed to provide these predictions. In addition, instability criteria, which use a series of dimensionless parameters to reduce the complexity of the evaluation, have been developed.

Effects of operating and design parameters on the density wave instability include:

Parameter Increased	Change in stability
mass flux	improved
heat flux	reduced
pressure	improved
inlet ΔP	improved
inlet subcooling	improved (large subcooling) reduced (small subcooling)

Steam-water separation

Subcritical pressure recirculating boilers and steam generators are equipped with large cylindrical vessels called *steam drums*. Their primary objective is to permit separation of the saturated steam from the steam-water mixture leaving the boiling heat transfer surfaces. The steam-free water is recirculated with the feedwater to the heat absorbing surfaces for further steam generation. The saturated steam is discharged through a number of outlet nozzles for direct use or further heating. The steam drum also serves to:

1. mix the feedwater with the saturated water remaining after steam separation,
2. mix the corrosion control and water treatment chemicals (if used),
3. purify the steam to remove contaminants and residual moisture,
4. remove part of the water (blowdown) to control the boiler water chemistry (solids content), and
5. provide limited water storage to accommodate rapid changes in boiler load.

However, the primary function of the steam drum is to permit the effective separation of steam and water. This may be accomplished by providing a large steam-water surface for natural gravity-driven separation or by having sufficient space for mechanical separation equipment.

High efficiency separation is critical in most boiler applications in order to:

1. prevent water droplet *carryover* into the superheater where thermal damage may occur,
2. minimize steam *carryunder* in the water leaving the drum where residual steam can reduce the effective hydraulic pumping head, and
3. prevent the carryover of solids dissolved in the steam-entrained water droplets into the superheater and turbine where damaging deposits may form.

The last item is of particular importance. Boiler water may contain contaminants, principally in solution. These arise from impurities in the makeup water, treatment chemicals and condensate system leaks, as well as from the reaction of the water and contaminants with the boiler and preboiler equipment materials. Even low levels of these solids in the steam (less than 0.6 ppm) can damage the superheater and turbine. Because the solubility of these solids is typically several orders of magnitude less in steam than in water (see Chapter 42), small amounts of water droplet carryover (greater than 0.25% by weight) may result in dramatically increased solids carryover and unacceptable deposition in the superheater and turbine. The deposits have caused turbine damage as well as superheater tube temperature increases, distortion and burnout.

A cross-section of a horizontal steam drum found on a modern high capacity fossil fuel boiler is shown in Fig. 19. This illustrates the general arrangement of the baffle plates, primary cyclone separators, secondary separator elements (scrubbers), water discharger (downcomer) and feedwater inlets. The blowdown (water removal) connections are not shown. The steam-water separation typically takes place in two stages. The primary separation removes nearly all the steam from the water so that very little steam is recirculated from the bottom of the drum through the outlet connection (downcomer) towards the heated tubes. The steam leaving the primary separators in high pressure boilers still typically contains too much liquid in the form of contaminant-containing droplets for satisfactory superheater and turbine performance. Therefore, the steam is passed through a secondary set of separators, or scrubber elements (usually closely spaced, corrugated parallel plates) for final water droplet removal. The steam is then exhausted through several connections. As this figure indicates, successful steam-water separation involves the integrated operation of primary separators, secondary scrubbers and general drum arrangement.

Factors affecting steam separation

Effective steam separation from the steam-water mixture relies on certain design and operating factors. The design factors include:

1. pressure,
2. drum length and diameter,
3. rate of steam generation,
4. average inlet steam quality,
5. type and arrangement of mechanical separators,
6. feedwater supply and steam discharge equipment arrangement, and
7. arrangement of downcomer and riser connections to the steam drum.

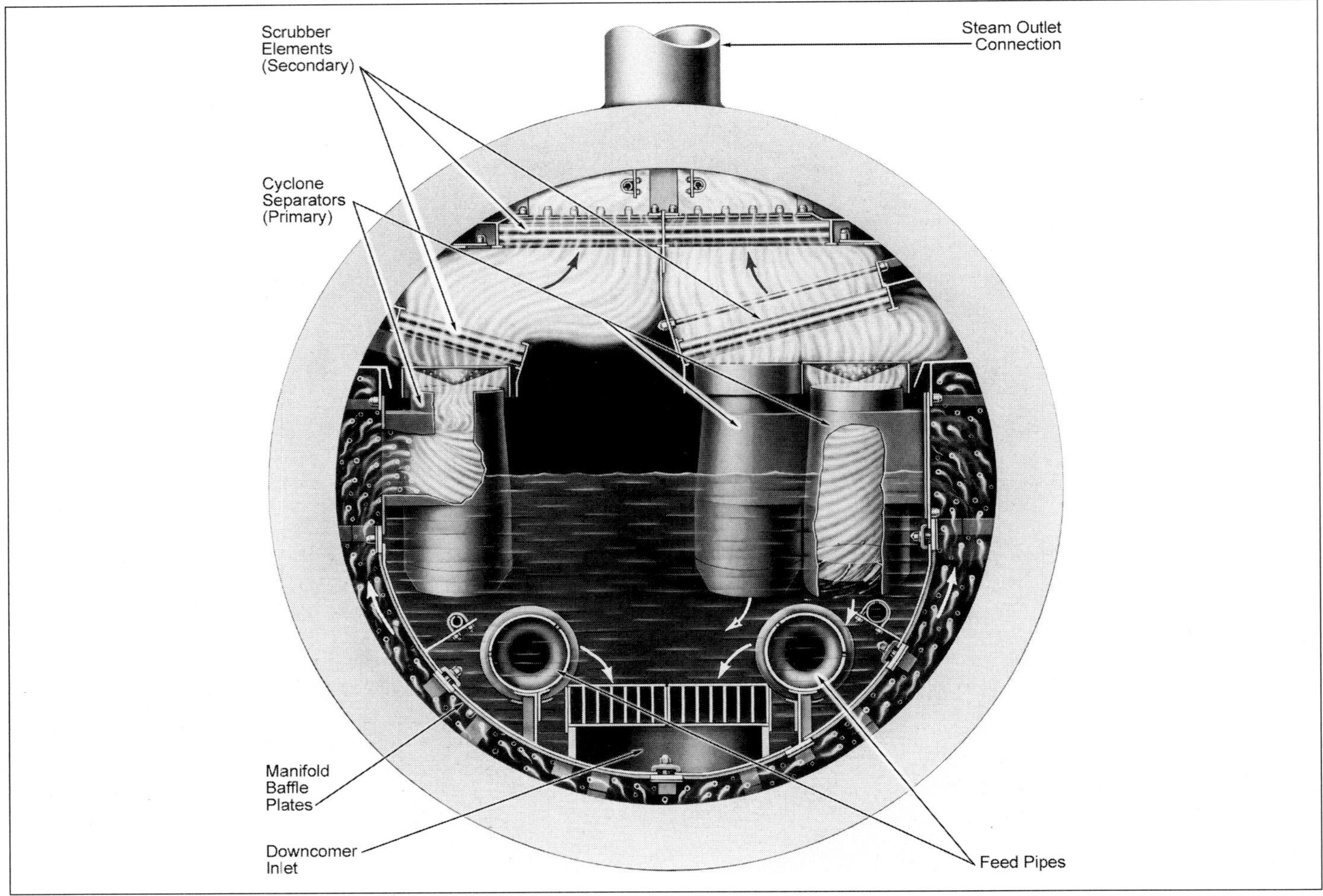

Fig. 19 Steam drum with three rows of primary cyclone separators.

The operating factors include:

1. pressure,
2. boiler load (steam flow),
3. type of steam load,
4. chemical analysis of boiler water, and
5. water level.

Primary separation equipment generally takes one of three forms:

1. natural gravity-driven separation,
2. baffle-assisted separation, and
3. high capacity mechanical separation.

Natural gravity-driven separation

While simple in concept, natural steam-water separation is quite complex. It is strongly dependent upon inlet velocities and inlet locations, average inlet steam quality, water and steam outlet locations, and disengagement of liquid and steam above the nominal water surface. Some of these effects are illustrated in Figs. 20 and 21.

For a low rate of steam generation, up to about 3 ft/s (0.9 m/s) velocity of steam leaving the water surface, there is sufficient time for the steam bubbles to separate from the mixture by gravity without being drawn into the discharge connections and without carrying entrained water droplets into the steam outlet (Fig. 20a). However, for the same arrangement at a higher rate of steam generation (Fig. 20b), there is insufficient time to attain either of these desirable results. Moreover, the dense upward traffic of steam bubbles in the mixture may also cause a false water level indication, as shown.

The effect of the riser or inlet connection locations in relation to the water level is illustrated in diagrams a and b of Fig. 21. Neither arrangement is likely to yield desirable results in a drum where gravity alone is used for separation.

From an economic standpoint, the diameter of a single drum may become prohibitive. To overcome this limitation, several smaller steam drums may be used, as shown in Fig. 22a, although this is no longer common. However, in most boiler applications, natural gravity-driven separation alone is generally uneconomical, leading to the need for separation assistance.

Baffle-assisted primary separation

Simple screens and baffle arrangements may be used to greatly improve the steam-water separation process. Three relatively common baffle arrangements are illustrated in Fig. 22. In each case, the baffles provide: 1) changes in direction, 2) more even distri-

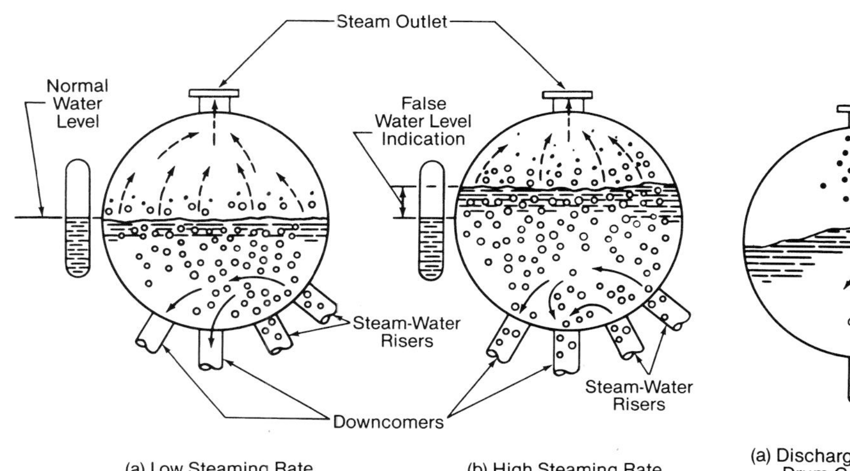

Fig. 20 Effect of rate of steam generation on steam separation in a boiler drum without separation devices.

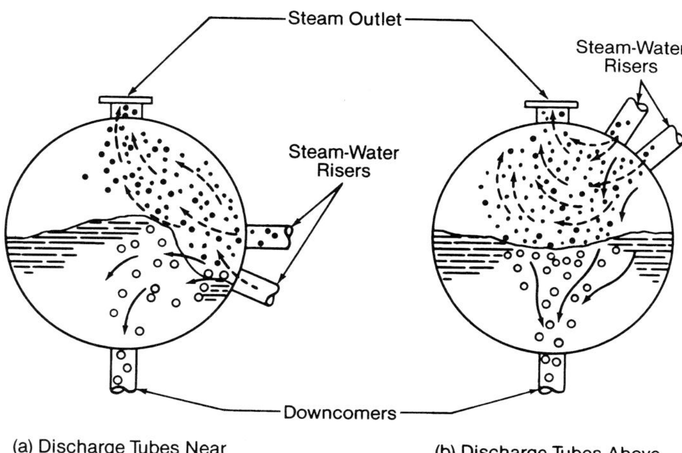

Fig. 21 Effect of location of discharge from risers on steam separation in a boiler drum without separation devices.

bution of the steam-water mixture, 3) added flow resistance, and 4) the maximum steam flow travel length to enhance the gravity-driven separation process. Various combinations of perforated plates have also been used. The performance of these devices must be determined by experimental evaluations and they are typically limited to smaller, low capacity boilers.

Mechanical primary separators

Centrifugal force or radial acceleration is used almost universally for modern steam-water separators. Three types of separators are shown in Fig. 23: the conical cyclone, the curved arm and the horizontal cyclone. The B&W vertical cyclone steam separator is shown in more detail in Fig. 24. Vertical cyclones are arranged internally in rows along the length of the drum and the steam-water mixture is admitted tangentially as shown in Fig. 19. The water forms a layer against the cylinder walls and the steam moves to the core of the cylinder then upward. The water flows downward in the cylinder and is discharged through an annulus at the bottom, below the drum water level. With the water returning from drum storage to the downcomers virtually free of steam bubbles, the maximum net pumping head is available for producing flow in the circuits. The steam moving upward from the cylinder passes through a small primary corrugated scrubber at the top of the cyclone (see Fig. 24) for additional separation. Under many operating conditions, no further separation is required.

When wide load fluctuations and water analysis variations are expected, large corrugated secondary scrubbers may be installed at the top of the drum (see Fig. 19) to provide very high steam separation. These scrubbers are also termed secondary separators. They provide a large surface which intercepts water droplets as the steam flows sinuously between closely fitted plates. Steam velocity through the corrugated plate assembly is very low, so that water re-entrainment is avoided. The collected water is drained from the bottom of the assembly to the water below.

One to four rows of cyclone separators are installed in boiler drums, with ample room for access. For smaller boilers at lower pressures [100 psig (0.7 MPa gauge)], the separation rate of clean steam by single and double rows of cyclone separators is approximately

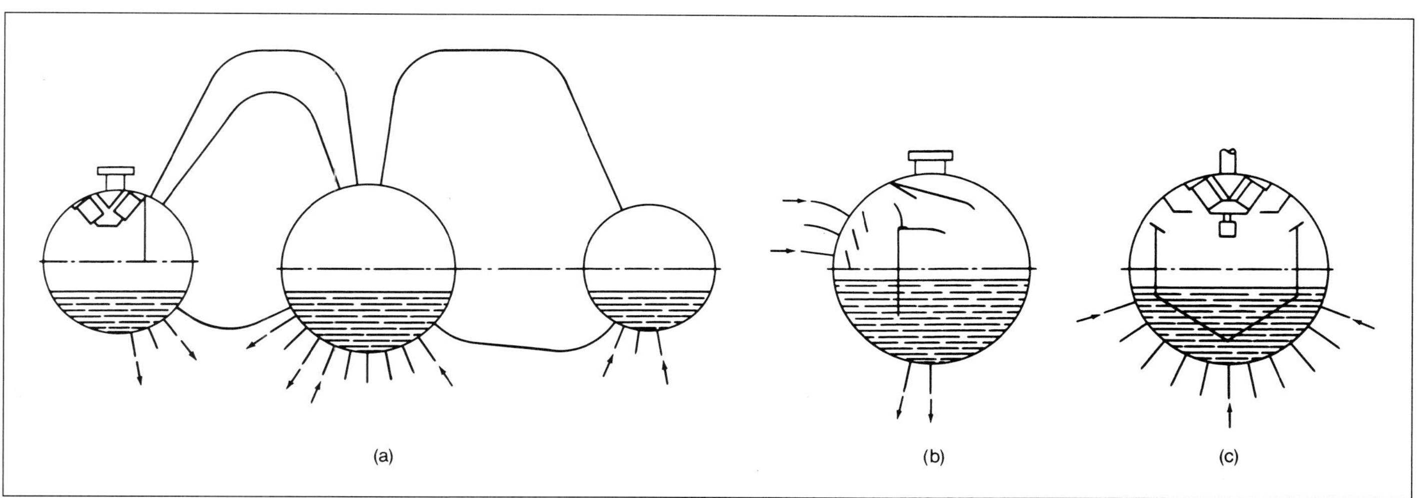

Fig. 22 Simple types of primary steam separators in boiler drums: a) deflector baffle, b) alternate deflector baffle, and c) compartment baffle.

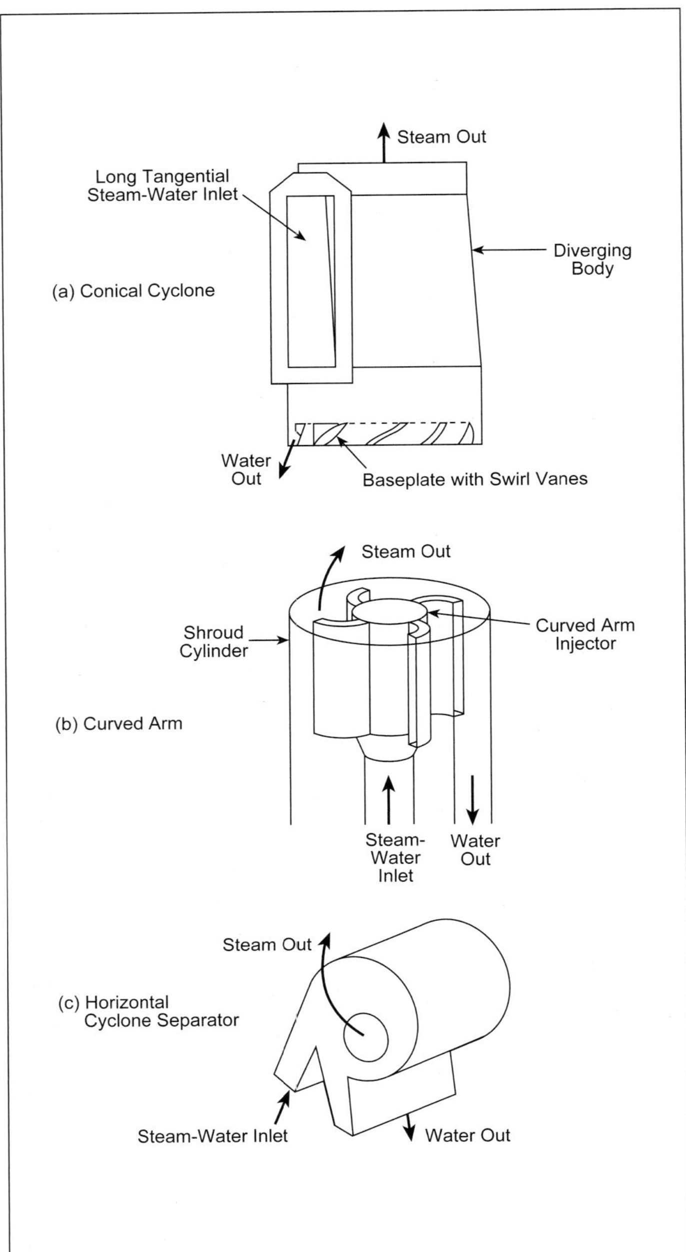

Fig. 23 Typical primary steam-water separators.

4000 and 6000 lb, respectively, per hour per foot of drum length (1.7 and 2.5 kg/s m). At pressures near 1050 psig (7.24 MPa gauge), these values increase to 9000 and 15,000 lb/h ft (3.7 and 6.2 kg/s m), respectively. For large utility boilers operating at 2800 psig (19.3 MPa gauge), separation can be as high as 67,000 lb/h ft (28 kg/s m) of steam with four rows of cyclone separators.

This combination of cyclone separators and scrubbers provides a steam purity of less than 1.0 ppm solids content under a wide variety of operating conditions. This purity is generally adequate in commercial practice. However, further refinement in steam purification is required where it is necessary to remove boiler water salts, such as silica, which are entrained in the steam by a vaporization or solution mechanism. Washing the steam with condensate or feedwater of acceptable purity may be used for this purpose.

Specialized vertical steam-water separators can be used in once-through fossil fueled boiler systems which are designed for part-load recirculation of water during startup and low-load operation. These are basically vertical cylindrical pressure vessels (see Fig. 25) where the steam-water mixture enters through multiple tangential inlets in the vertical vessel wall. The resulting centrifugal acceleration creates a cyclone action similar to that in the primary cyclone separators (Fig. 24) which separates the water from the steam. Water is returned to the boiler circuitry for further heating and steam generation while the steam is sent to the superheating circuits.

Mechanical separator performance

The overall performance of mechanical separators is defined by: 1) the maximum steam flow rate at a specified average inlet quality per cyclone which meets droplet carryover limits, and 2) the predicted pressure loss. In addition, the maximum expected steam carryunder (% steam by weight) should also be known. These parameters are influenced by total flow rate, pressure, separator length, aperture sizes, drum water level, inlet steam quality, interior separator finish and overall drum arrangement. Performance characteristics are highly hardware-specific. The general trends are listed in Table 2.

Steam separator evaluation To date, theoretical analyses alone do not satisfactorily predict separation performance. Therefore, extensive experimental investigations are performed to characterize individual steam-water primary separator designs.

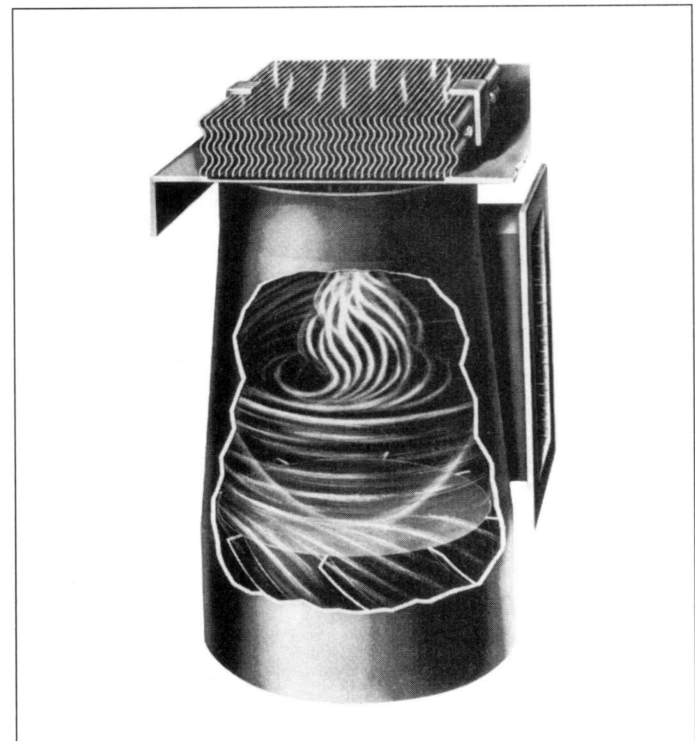

Fig. 24 Vertical cyclone separator.

Fig. 25 Vertical steam-water separator in a spiral wound universal pressure (SWUP™) boiler startup system.

Pressure drop of two-phase flow through a separator is extremely complex. An approximation involves using the homogeneous model two-phase multiplier, Φ, and a dimensionless loss coefficient, K_{ss}, as follows:

$$\Delta P_{separator} = K_{ss}\,\Phi\,\frac{G^2 \upsilon_f}{2g_c} \qquad \textbf{(20)}$$

where

$$\Phi = 1.0 + \left(\frac{\upsilon_g - \upsilon_f}{\upsilon_f}\right) x$$

The variable K_{ss} is a unique function of pressure for each steam separator design. The other variables are defined after Equation 16.

The maximum steam flow per primary separator defines the minimum number of standard units required, while the ΔP is used in the circulation calculations. Given the unique design of each separator, B&W has acquired extensive experimental performance data under full-scale, full-flow and full-pressure conditions for its equipment.

Table 2
Mechanical Separator Performance Trends

Moisture carryover with steam
1. increases gradually with steam flow rate until a breakaway point is reached where a sudden rise in carryover occurs,
2. increases with water level until flooding occurs, and
3. increases with steam quality.

Carryunder of steam with water
1. declines with increasing water level, and
2. declines with decreasing inlet steam quality.

Pressure drop ($P_{in} - P_{drum}$)
1. increases with mass flow and steam quality.

Steam drum capacity

Given the flow capabilities of standardized steam-water separation equipment, the boiler drum is sized to accommodate the number of separators necessary for the largest expected boiler load (maximum steam flow rate) and to accommodate the changes in water level that occur during the expected load changes. The drum diameter, in incremental steps, and length are adjusted to meet the space requirements at a minimum cost.

An evaluation limit in steam drum design is the maximum steam carryunder into the downcomer. Carryunder, or transport of steam into the downcomers, is not desirable because it reduces the available thermal pumping force by reducing the density at the top of the downcomer. Carryunder performance is a function of physical arrangement, operating pressure, feedwater enthalpy, free-water surface area, drum water level and separator efficiency. Empirical correction factors for specific designs are developed and used in the circulation calculations to account for the steam entering the downcomers. The steam is eventually completely condensed after it travels a short distance into the downcomer. However, the average density in the top portion of the downcomer is still lower than thermal equilibrium would indicate.

A rapid increase in steam demand is usually accompanied by a temporary drop in pressure until the firing rate can be sufficiently increased. During this interval, the volume of steam throughout the boiler is increased and the resulting swell raises the water level in the drum. The rise depends on the rate and magnitude of the load change and the rate at which the heat and feed inputs can be changed to meet the load demand. Steam drums are designed to provide the necessary volume, in combination with the controls and firing equipment, to prevent excessive water rise into the steam separators. This, in turn, prevents water carryover with the steam.

Circulation

The purpose of the steam-water flow circuitry is to provide the desired steam output at the specified temperature and pressure. The circuitry flow also ensures effective cooling of the tube walls under expected operating conditions, provided the unit is properly operated and maintained. A number of methods have been developed. Four of the most common systems are illustrated in Fig. 26. These systems are typically classified as either *recirculating* or *once-through*.

In recirculating systems, water is only partially evaporated into steam in the boiler tubes. The residual water plus the makeup water supply are then recirculated to the boiler tube inlet for further heating and steam generation. A steam drum provides the space required for effective steam-water separation. Once-through systems provide for continuous evaporation of slightly subcooled water to 100% steam without steam-water separation. Steam drums are not required. These designs use forced circulation for the necessary water and steam-water flow. In some cases, a combination of these approaches is used. At low loads, recirculation maintains adequate tube wall cool-

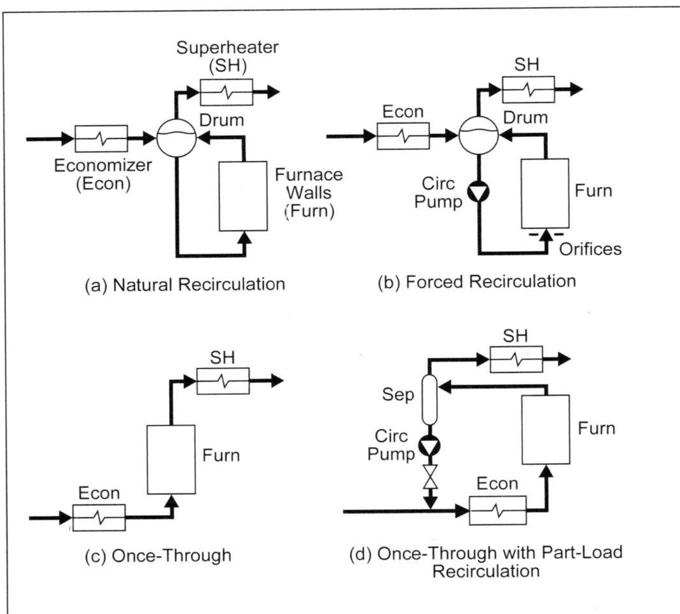

Fig. 26 Common fossil fuel boiler circulation systems.

ing while at high loads, high pressure once-through operation enhances cycle efficiency.

Natural circulation

In *natural circulation*, gravity acting on the density difference between the subcooled water in the downcomer and the steam-water mixture in the tube circuits produces the driving force or pumping head to drive the flow. As shown in Fig. 27, a simplified boiler circuit consists of an unheated leg or downcomer and heated boiler tubes. The water in the downcomer is subcooled through the mixing of the low temperature feedwater from the economizer with the saturation-temperature water discharged from the steam-water separators. Steam-water, two-phase flow is created in the boiler tubes as a result of the heat input. Because the steam-water mixture has a lower average density than the single-phase downcomer flow, a pressure differential or pumping pressure is created by the action of gravity and the water flows around the circuit. The flow increases or decreases until the pressure losses in all boiler circuits are balanced by the available pumping pressure. For steady-state, incompressible flow conditions, this balance takes the form:

$$\left(Z\bar{\rho}_d - \int_0^Z \rho(z)dz\right)\left(\frac{g}{g_c}\right) = \left(\Delta P_{\text{friction}} + \Delta P_{\text{acceleration}} + \Delta P_{\text{local}}\right) \quad \textbf{(21)}$$

where

Z = total vertical elevation, ft (m)
z = incremental vertical elevations, ft (m)
$\rho(z)$ = heated tube local fluid density, lb/ft^3 (kg/m^3)
$\bar{\rho}_d$ = average downcomer fluid density, lb/ft^3 (kg/m^3)
g = acceleration of gravity, ft/s^2 (m/s^2)

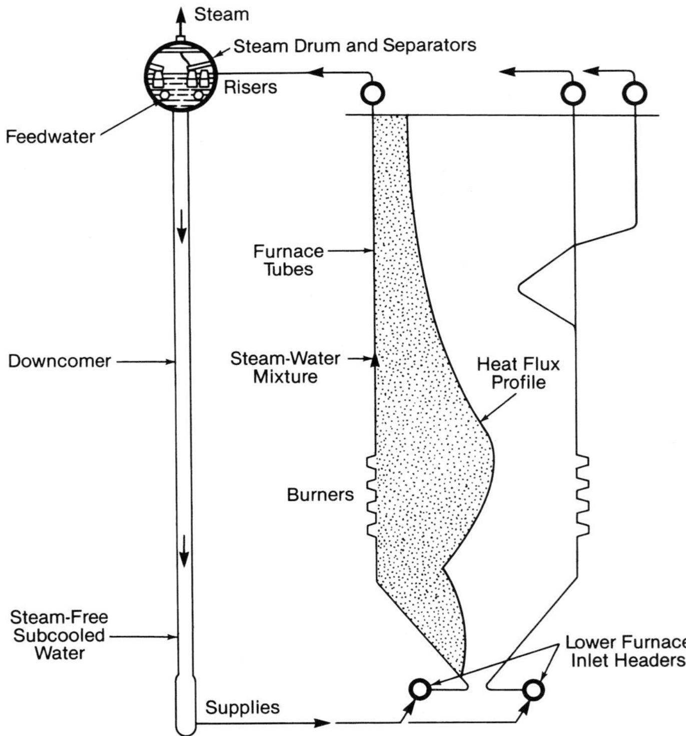

Fig. 27 Simple furnace circulation diagram.

g_c = 32.17 lbm ft/lbf s^2 (1 kg m/N s^2)
ΔP = circuitry pressure loss due to friction, fluid acceleration and local losses, lb/ft^2 (Pa)

As the heat input increases, circulation rate increases until a maximum flow rate is reached (Fig. 28). If higher heat inputs occur, they will result in larger pressure losses in the heated tubes without corresponding increases in pressure differential. As a result, the flow rate declines.

Natural circulation boilers are designed to operate in the region where increased heat input results in an increase in flow for all specified operating conditions. In this mode, a natural circulation system tends to be self compensating for numerous variations in heat absorption. These can include sudden changes in load,

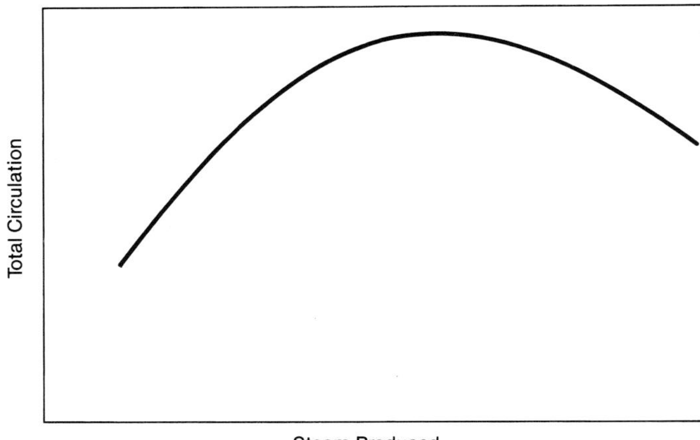

Fig. 28 Typical relationship between circulation at a given pressure and steam production (arbitrary scale).

changes in heating surface cleanliness and changes in burner operation.

Natural circulation is most effective where there is a considerable difference in density between steam and water phases. As shown in Fig. 29, the potential for natural circulation flow remains very high even at pressures of 3100 psi (21.4 MPa).

Forced circulation

In recirculating or once-through *forced circulation* systems, mechanical pumps provide the driving head to overcome the pressure losses in the flow circuitry. Unlike natural circulation, forced circulation does not enjoy an inherent flow-compensating effect when heat input changes, i.e., flow does not increase significantly with increasing heat input. This is because a large portion of the total flow resistance in the boiler tubes arises from the flow distribution devices (usually orifices) used to balance flow at the circuit inlets. The large resistance of the flow distributors prevents significant increases in flow when heat absorption is increased.

Forced circulation is, however, used where the boilers are designed to operate near or above the critical pressure [3200 psi (22.1 MPa)]. There are instances in the process and waste heat fields and in some specialized boiler designs where the use of circulating pumps and forced circulation can be economically attractive. At pressures above 3100 psi (21.4 MPa) a natural circulation system becomes increasingly large and costly and a pump can be more economical. In addition, the forced circulation principle can work effectively in both the supercritical and subcritical pressure ranges.

In forced recirculation there is a net thermal loss because of the separate circulating pump. While practically all the energy required to drive the pumps reappears in the water as added enthalpy, this energy originally came from the fuel at a conversion to useful energy factor of less than 1.0. If an electric motor drive is used, the net energy lost is about twice the energy supplied to the pump motor for typical fossil fuel systems.

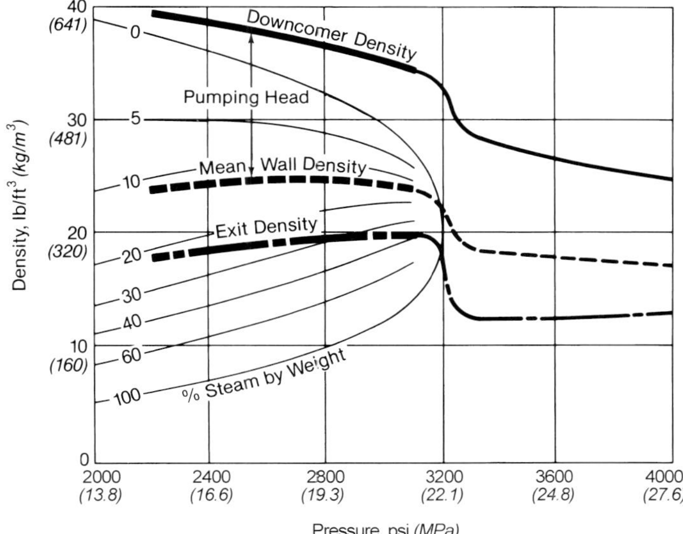

Fig. 29 Effect of pressure on pumping head.

Circulation design and evaluation

The furnace wall enclosure circuits are very important areas in a boiler. High constant heat flux conditions make uninterrupted cooling of furnace tubes essential. Inadequate cooling can result in rapid overheating, cycling thermal stress failure, or material failures from differential tube expansion. Sufficient conservatism must be engineered into the system to provide adequate cooling even during transient upset conditions. Simultaneously, the rated steam flow conditions must be maintained at the drum outlet. Any of the circulation methods discussed may be used to cool the furnace waterwall tubes. In evaluating the circulation method selected for a particular situation, the following general procedure can be used:

1. The furnace geometry is set by the fuel and combustion system selected. (See Chapters 11, 14, 19 and 21.)
2. Standardized components (furnace walls, headers, drums, etc.) are selected to enclose the furnace arrangement as needed. (See Chapters 19 and 21.)
3. The local heat absorption is evaluated based upon the furnace geometry, fuel and firing method. Local upset factors are evaluated based upon past field experience. (See Chapter 4.)
4. Circulation calculations are performed using the pressure drop relationships.
5. The calculated circulation results (velocities, steam qualities, etc.) are compared to the design criteria.
6. The flow circuitry is modified and the circulation re-evaluated until all of the design criteria are met.

Some of the design criteria include:

1. *Critical heat flux limits* For recirculating systems, CHF conditions are generally avoided. For once-through systems, the temperature excursions at CHF are accommodated as part of the design.
2. *Stability limits* These limits generally indicate acceptable pressure drop versus mass flow relationships to ensure positive flow in all circuits and to avoid oscillating flow behavior.
3. *Steam separator and steam drum limits* These indicate maximum steam and water flow rates to individual steam-water separators and maximum water flow to the drum downcomer locations to ensure that steam carryunder and water carryover will not be problems.
4. *Minimum velocity limits* Minimum circuit saturated velocities assure that solids deposition, potentially detrimental chemistry interactions, and selected operating problems are minimized.
5. *Sensitivity* The system flow characteristic is checked to ensure that flow increases with heat input for all expected operating conditions.

Circulation is analyzed by dividing the boiler into individual simple circuits – groups of tubes or circuits with common end points and similar geometry and heat absorption characteristics. The balanced flow condition is the simultaneous solution of the flow characteristics of all boiler circuits.

At the heart of a B&W circulation evaluation is a circulation computer program that incorporates tech-

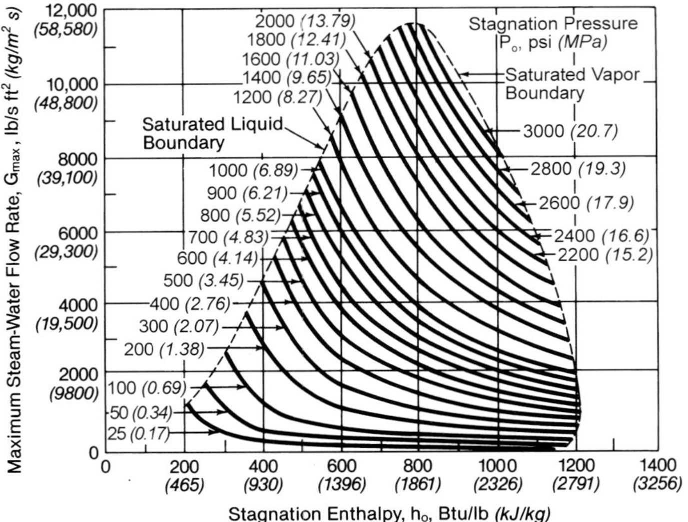

Fig. 30 Moody critical flow model for maximum steam-water flow rate.[17]

niques for calculating the single- and two-phase heat transfer and flow parameters discussed above and in Chapters 3 and 4. With this program, a circulation model of the entire boiler is developed. Input into the program is a geometric description of each boiler circuit including descriptions of downcomers, supplies, risers, orifices, bends and swages, as well as individual tubes. Each of the circuits within the boiler is subjected to the local variation in heat transfer through inputs based upon the furnace heat flux distribution. (See Chapter 4.) Given the geometry description and heat absorption profile, the computer program determines the balanced steam-water flow to each circuit by solving the energy, mass and momentum equations for the model. The results of the program provide the detailed information on fluid properties, pressure drop and flow rates for each circuit so that they can be compared to the design criteria. Adjustments frequently made to improve the individual circuit circulation rates can include: changing the number of riser and supply connections, changing the number or type of steam separators in the drum, adding orifices to the inlets to individual tubes, changing the drum internal baffling, changing the operating pressure (if possible) and lowering the feedwater temperature entering the drum. Once the steam-water circuitry is finalized, the detailed mechanical design proceeds.

Critical flow

A two-phase flow parameter of particular importance in nuclear reactor safety analysis and in the operation of valves in many two-phase flow systems is the *critical flow rate*. This is the maximum possible flow rate through an opening when the flow becomes choked and further changes in upstream pressure no longer affect the rate. For single-phase flows, the critical flow rate is set by the sonic velocity. The analysis is based upon the assumption that the flow is one dimensional, homogeneous, at equilibrium and isentropic. These assumptions result in the following relationships:

$$\text{Sonic velocity} = C = \sqrt{\left(\frac{dP}{d\rho}\right)_s g_c} \tag{22}$$

$$\text{Critical flow} = G_{max} = \rho\sqrt{\left(\frac{dP}{d\rho}\right) g_c} \tag{23}$$

where

C = velocity, ft/s (m/s)
P = pressure, lb/ft² (Pa)
ρ = fluid density, lb/ft³ (kg/m³)
g_c = 32.17 lbm ft/lbf s² (1 kg m/N s²)
G_{max} = mass flux, lb/s ft² (kg/m² s)

However, when saturated water or a two-phase steam-water mixture is present, these simplifying assumptions are no longer valid. The flow is heterogeneous and nonisentropic with strong interfacial transport and highly unstable conditions.

Moody's analysis[17] of steam-water critical flow is perhaps the most frequently used. It is based upon an annular flow model with uniform axial velocities of each phase and equilibrium between the two phases. A key element of the analysis involves maximizing the flow rate with respect to the slip ratio and the pressure. The results are presented in Fig. 30. The critical steam-water flow rate is presented as a function of the stagnation condition. Compared to experimental observations, this correlation slightly overpredicts the maximum discharge at low qualities ($x < 0.1$) and predicts reasonably accurately at moderate qualities ($0.2 < x < 0.6$), but tends to underpredict at higher qualities ($x > 0.6$).

References

1. Collier, J.G., and Thome, J.R., *Convective Boiling & Condensation,* Third Ed., Oxford University Press, Oxford, United Kingdom, 1994.

2. Jens, W.H., and Lottes, P.A., "Analysis of heat transfer, burnout, pressure drop, and density data for high pressure water," Argonne National Laboratory Report ANL-4627, May, 1951.

3. Thom, J.R.S., *et al.,* "Boiling in subcooled water during flow up heated tubes or annuli," *Proceedings of Institute of Mechanical Engineers,* Vol. 180, pp. 226-246, 1966.

4. Chen, J.C., "Correlation for boiling heat transfer to saturated liquids in convective flow," *Industrial & Engineering Chemistry Process & Design Development,* Vol. 5, pp. 322-329, 1966.

5. Kitto, J.B., and Albrecht, M.J., "Elements of two-phase flow in fossil boilers," *Two-Phase Flow Heat Exchangers,* Kakaç, S., Bergles, A.E. and Fernandes, E.O., Eds., Kluwer Academic Publishers, Dordrecht, The Netherlands, pp. 495-552, 1988.

6. Watson, G.B., Lee, R.A., and Wiener, M., "Critical heat flux in inclined and vertical smooth and ribbed tubes," *Proceedings of The Fifth International Heat Transfer Conference,* Vol. 4, Japan Society of Mechanical Engineers, Tokyo, Japan, pp. 275-279, 1974.

7. Gellerstedt, J.S., *et al.,* "Correlation of critical heat flux in a bundle cooled by pressurized water," *Two-Phase Flow and Heat Transfer in Rod Bundles,* Schock, V.E., Ed., American Society of Mechanical Engineers (ASME), New York, New York, pp. 63-71, 1969.

8. Wiener, M., "The latest developments in natural circulation boiler design," *Proceedings of The American Power Conference,* Vol. 39, pp. 336-348, 1977.

9. Swenson, H.S., Carver, J.R., and Kakarala, C.R., "Heat transfer to supercritical water in smooth-bore tubes," *Journal of Heat Transfer,* Vol. 87, pp. 477-484, 1965.

10. Ackerman, J.W., "Pseudoboiling heat transfer to supercritical pressure water in smooth and ribbed tubes," *Journal of Heat Transfer,* Vol. 92, pp. 490-498, 1970.

11. Hewitt, G.F., and Roberts, D.W., "Studies of two-phase flow patterns by simultaneous x-ray and flash photography," Atomic Energy Research Establishment Report M2159, HMSO, London, England, United Kingdom, 1969.

12. Thom, J.R.S., "Prediction of pressure drop during forced circulation boiling of water," *International Journal of Heat and Mass Transfer,* Vol. 7, pp. 709-724, 1964.

13. Martinelli, R.C., and Nelson, D.B., "Prediction of pressure drop during forced-circulation boiling of water," Transactions of the American Society of Mechanical Engineers (ASME), pp. 695-702, 1948.

14. Zuber, N., and Findlay, J.A., "Average volumetric concentration in two-phase flow systems," *Journal of Heat Transfer,* Vol. 87, pp. 453-468, 1965.

15. Chexal, B.J., Horowitz, J., and Lellouche, G.S., "An assessment of eight void fraction models for vertical flows," Electric Power Research Institute Report NSAC-107, December, 1986.

16. Ledinegg, M., "Instability of flow during natural and forced circulation," *Die Wärme,* Vol. 61, No. 48, pp. 891-898, 1938 (AEC-tr-1861, 1954).

17. Moody, F.J., "Maximum flow rate of a single component, two-phase mixture," *Journal of Heat Transfer,* Vol. 87, pp. 134-142, 1965.

Bibliography

Bergles, A.E., *et al., Two-Phase Flow and Heat Transfer in the Power and Process Industries*, Hemisphere, Washington, D.C., August, 1981.

Butterworth, D., and Hewitt, G.F., Eds., *Two-Phase Flow and Heat Transfer*, Oxford University Press, Oxford, England, United Kingdom, 1977.

Chen, J.C., Ed., *Flow Boiling*, Taylor and Francis Group, New York, New York, 1996.

Hsu, Y-Y, and Graham, R.W., *Transport Processes in Boiling and Two-Phase Systems*, Hemisphere, Washington, D.C., 1976.

Kakaç, S., *Boilers, Evaporators and Condensers*, John Wiley & Sons, New York, New York, 1991.

Kitto, J.B., "Steam Generators," *Standard Handbook of Powerplant Engineering*, Second Ed., Elliot, T.C., Chen, K., and Swanekamp, R.C., McGraw-Hill, New York, New York, 1998.

Lahey, R.T., and Moody, F.J., *Thermal-Hydraulics of a Boiling Water Nuclear Reactor*, Second Ed., American Nuclear Society (ANS), Hinsdale, Illinois, 1993.

Lokshin, V.A., Peterson, D.F., and Schwarz, A.L., *Standard Methods of Hydraulic Design for Power Boilers*, Hemisphere Publishing, New York, New York, 1988.

Tong, L.S., *Boiling Heat Transfer and Two-Phase Flow*, John Wiley & Sons, New York, New York, 1965.

Wallis, G.B., *One-Dimension Two-Phase Flow*, McGraw-Hill, New York, New York, 1969.

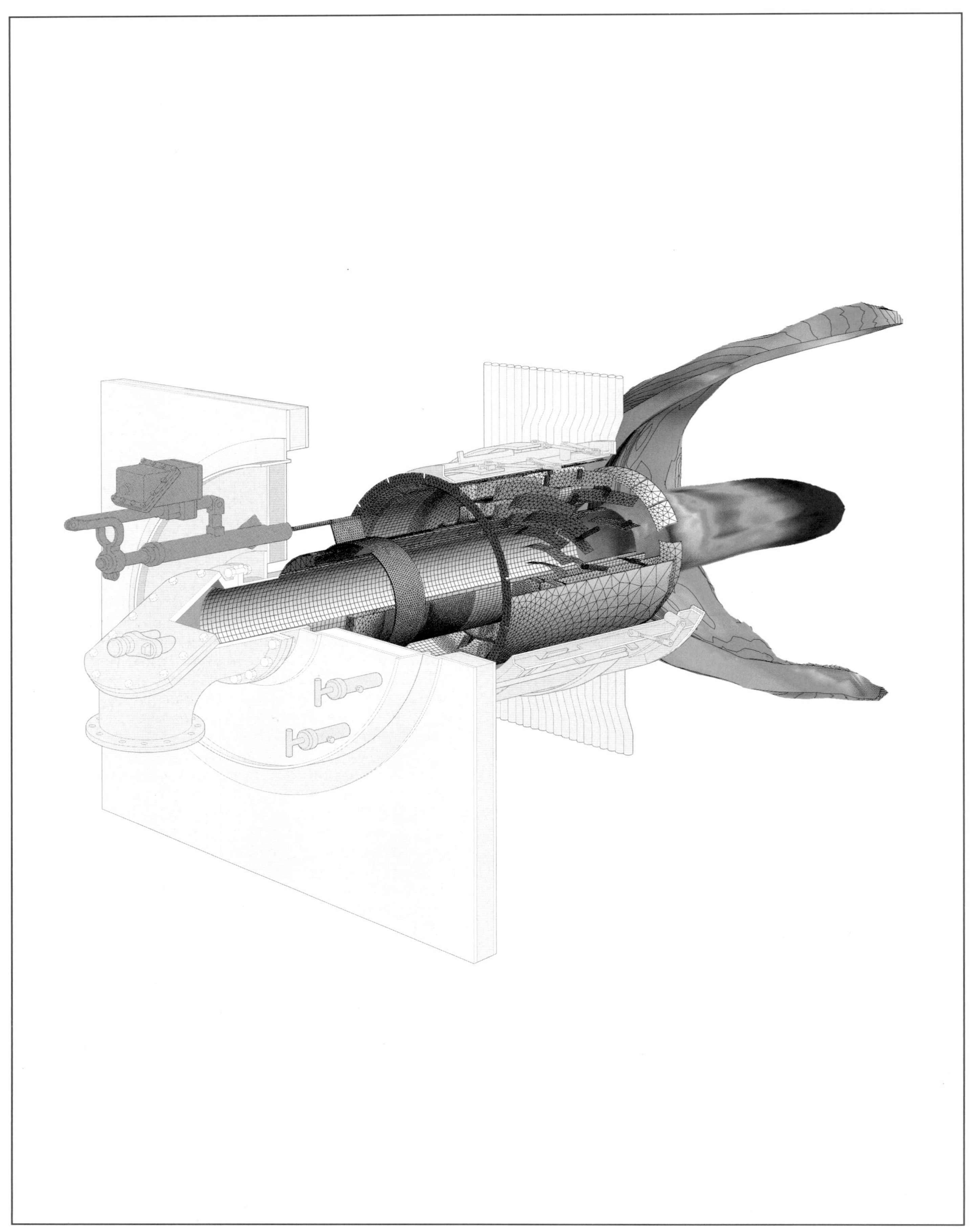

Advanced computational numerical modeling of a pulverized coal burner.

Chapter 6

Numerical Modeling for Fluid Flow, Heat Transfer, and Combustion

Numerical modeling – an overview

Continuous and steady advances in computer technology have changed the way engineering design and analyses are performed. These advances allow engineers to deal with larger-scale problems and more complex systems, or to look in more detail at a specific process. Indeed, through the use of advanced computer technology to perform engineering analysis, numerical modeling has emerged as an important field in engineering. While this chapter focuses on fluid flow and heat transfer, Chapter 8 provides a brief discussion of numerical modeling for structural analysis.

In general, the term *numerical method* describes solving a mathematical description of a physical process using a numerical rather than an analytical approach. This may be done for a number of reasons, including the following:

1. An analytical means of solving the equations that describe the system may not exist.
2. Even though an analytical method is available, it may be necessary to repeat the calculation many times, and a numerical method can be used to accelerate the overall process.

A small-scale replica of an apparatus is considered a *physical model* because it describes the full-size apparatus on a smaller scale. This model can incorporate varying levels of detail depending on need and circumstances. A mathematical description of a physical system (referred to as a *mathematical model*) can also incorporate varying levels of detail. Similar to a physical model, the amount of detail is often determined by the accuracy required and the resources available to use the model. This creates a need to strike a balance between accuracy, complexity and efficiency.

There are two basic approaches to mathematical modeling.

1. *Model the behavior of a system*. Network flow models and heat exchanger heat transfer correlations are examples of a system model.
2. *Model the fundamental physics of a system to determine the behavior*. Computational fluid dynamics (CFD) and chemical reaction models fall into this category.

The term *numerical modeling* usually refers to the use of numerical methods on high-powered computers to solve a complex system of mathematical models based on the fundamental physics of the system. In this respect, it describes the second approach identified above.

As an example, consider analysis of hot air moving through a length of duct composed of several different components all in a cold environment.

The first type of analysis would involve a network model. This model would describe the pressure drop and heat loss along the duct based on the length, shape, number of turns, etc. This model is based on extensive flow measurements taken on the individual components (i.e., straight sections, turns, reductions, etc.) that make up the duct. A set of empirical and fundamental correlations is used to analyze the flow rate through the duct. The computation can be set up quickly and with minimal effort. Results and multiple variations can be rapidly obtained. While results are reasonably accurate, they are limited to the components for which a flow correlation already exists. A unique component design that has not been described by a correlation may not be accurately evaluated with this type of model.

The second type of analysis would involve a CFD model of the same duct. The detailed behavior of the flow through the entire duct is modeled. From this information, pressure drop and heat loss along the length of the duct may be determined. However, unlike the first analysis, this type of model provides additional details. For example, the first model does not consider how the flow through a bend differs if it is followed by another bend or a straight section; the first model may result in the same pressure drop regardless of how the components are arranged. The second analysis would account for these differences. In addition, variation in heat loss from one side of the duct to the other can be determined. Most importantly, this model is not restricted to duct components where extensive experimental data is available. New concepts can easily be evaluated.

These two approaches have both benefits and limitations. The appropriate use of each is determined by the information needed and the information available. While both approaches are important engineering tools, the remaining discussion here will focus on the second, specifically on CFD and combustion modeling, and how they relate to furnaces, boilers and accessory equipment.

Benefits

There are numerous benefits to using a sophisticated tool such as a numerical model for engineering analysis. These tools can often provide information that can only be obtained through expensive experiments or may not be available any other way. Numerical modeling may often be used to obtain needed information quickly and at a reduced cost.

While it is important to understand the advantages of using numerical modeling, it is equally important to understand that it is only one means of obtaining the required information. Engineering has long relied on *theory* and *experiments* for design and analysis. *Numerical modeling* adds a third approach. Each approach offers different insights with different benefits.

Increased understanding

The primary purpose of using numerical modeling is to increase understanding of a physical process. As such, it is often used in addition to or in conjunction with other available tools. Consider the duct example described above. It is possible to use a network model on a large number of duct designs to narrow the possibilities to a few candidate designs. A full CFD model could then be used to analyze each of the candidate designs to gain a better understanding of the strengths and weaknesses of each design.

Exploration of unfamiliar conditions

As previously described, it is possible that a component of the duct can not be accurately described within the context of a flow network model. A conservative approximation can be used but may result in an overly conservative solution. A CFD model of the new component can provide the missing information, or a CFD model of the entire system can be performed. The model allows the exploration and analysis of new equipment and systems.

Design validation/examination of interactions

Traditional methods of analysis and design are often focused on individual system components such as the burners, air system, or heat transfer surfaces in a furnace. A full accounting of the complex interaction between the components is often not given. Numerical modeling provides a vehicle to evaluate the interactions and validate the system design.

Troubleshooting

Engineering analysis often investigates the behavior of existing systems. This is particularly true when the behavior does not agree with the expectations. Numerical modeling can play a vital role in determining the nature of the problem and suggesting solutions.

Flexibility

A distinct feature of numerical modeling is that it is a flexible method of analysis. Modeling can be used to look at any number of different geometries or operating conditions. In addition, the level of detail used in the model can vary from use to use. A high level of detail may be required to model flow near a fuel inlet to a burner, but the same level of detail may not be necessary for flow in a duct. The complexity is often dictated by the problem.

Historical perspectives

In many ways, the history of numerical modeling in the context of CFD has followed the development of computational capabilities. Early efforts in CFD started in the 1960s, when computers first became commercially available, and when many of the concepts and ideas that form the basis of current techniques in CFD were first developed. One example is the way much of the turbulent flow is modeled today. Early efforts were often limited to simple two-dimensional laminar flows. The resolution of the geometry was also very limited.

It was not until the 1970s that CFD saw substantial successes. It was during this time that CFD began to be used for general engineering problems. Progress included turbulence modeling, two-dimensional reacting flows and three-dimensional flows.

As further advances were made in computational technology, more sophisticated and detailed numerical models, as well as increased resolution, became possible. This increased the acceptance of CFD as a useful engineering tool and gave it a much wider application base. Soon, large comprehensive combustion CFD models were developed. These were fully three-dimensional turbulent-reacting flow models. Sub-models of detailed physics for specific applications were included, such as pulverized coal combustion and radiation heat transfer models. Improvements continue to be made today that promise to increase the utility of combustion CFD modeling.

Modeling process

In its simplest terms, a numerical model is provided with input data that is used to fix specific operating parameters and return results. Without further understanding, this simplistic view of modeling can lead to unsatisfactory results. More appropriately, a multi-step process is used:

1. Obtain a complete situational description including physical geometry, process flow, physical property data and the level of detail needed. It is important to obtain detailed information because seemingly small differences can have a significant effect on numerical solutions.
2. Define the modeling assumptions appropriate for the specific flow system and computer model selected while making appropriate tradeoffs; cost and time are balanced against level of detail and information required.
3. Prepare the input data by converting the general technical information obtained in step 1 into the detailed inputs required by the computational model selected. Much of this is accomplished with the use of various computer programs such as computer assisted drafting (CAD) software and mesh generation software. Verification of the input data is an important part of this process.

4. Run the numerical computational model until an acceptable solution is obtained.
5. Analyze the results to verify the initial model assumptions, to check the results against known trends, to benchmark the output with known field data, and to present the results in a usable form.

In application, the computer programs or software used to perform the modeling function are broken down into three general groups that work together to complete the analysis:

1. Pre-processing: generation of the calculational mesh or grid representing all boundary conditions (part of step 3 above and discussed later under *Mesh generation*),
2. Solution: execution of the numerical model to derive an acceptable solution (step 4 above), and
3. Post-processing: generation of typically graphical or tabular key results from the numerical model to permit interpretation and evaluation of the results (part of step 5 above).

Limitations

Despite recent advances in technology, increased understanding of physics, and improvements in describing input conditions, limitations remain in applying numerical modeling to engineering problems. Numerical modeling can only be applied where there is an adequate understanding of the physics involved. In situations where there is not an appropriate mathematical description of the physics, numerical modeling is not possible. Even when a description exists, it may be too complex to be readily used in a model and a simplified approach is required. In this case, results will reflect the simplifying assumptions of the model.

Computer technology continues to limit the level of detail that can be modeled with numerical methods. Our understanding of the physics of systems that are routinely modeled with CFD far exceeds the computational resources (size and speed) that are available to model them. A considerable amount of effort is expended on developing simplified descriptions of the physics to make the problem manageable with current computer technology.

The precision and accuracy of the input data also represents a significant limitation to numerical modeling. Sources where this may be significant include the level at which the geometry is described and represented, the accuracy of imposing an inlet condition, and the assumptions made in specifying other boundary conditions and modeling parameters.

Despite these limitations, numerical modeling can be used in conjunction with other engineering analyses. When applied appropriately, numerical modeling can provide invaluable information.

Uses

Many applications for CFD and combustion modeling exist within the design and evaluation of steam generators (or boilers) and related equipment. Numerical models of the flue gas and steam-water flows are used to predict boiler behavior, evaluate design modifications, or investigate localized phenomena. Examples of flue gas applications include predicting temperature distributions within a furnace, evaluating fluid mixing due to the retrofit of systems to control nitrogen oxides (NO_x) emissions, and improving air heater flow distributions to increase heat absorption. Water-side applications include determining flow rates for boiler furnace circulation systems and evaluating system stability, among others. Many of the uses are summarized in Table 1.

Theory

The foundation of numerical modeling is the development of a mathematical description of the physical system to be modeled. Whether this is as simple as heat transfer through a wall or as complex as a pul-

Table 1
Sample Numerical Model Applications

Application	Purpose
Windboxes	Evaluate flow field within windbox, determine expected air distribution to combustion equipment, and determine pressure losses throughout system
Burners	Accurately determine boundary conditions for furnace models, evaluate flame and burner flow characteristics
Overfire air ports	Accurately determine boundary conditions for furnace models, determine flow characteristics and pressure losses through port
Pulverized coal-fired boilers	Examine combustion characteristics throughout the entire furnace; evaluate fuel/air mixing, furnace performance, heat transfer, emissions and flow characteristics
Recovery boilers	Examine combustion characteristics throughout the entire furnace; evaluate fuel/air mixing, furnace performance, heat transfer, emissions, carryover and flow characteristics within the furnace
Waste-to-energy boilers	Examine combustion characteristics of the entire furnace; evaluate fuel/air mixing, furnace performance, heat transfer, emissions and flow characteristics
Selective catalytic reduction systems	Determine inlet flow and temperature distributions; evaluate flow correction devices to meet specified velocity and temperature criteria
Wet scrubbers	Determine flow and pressure drop conditions, evaluate scrubber emission removal performance

verized coal flame, the first step is to adequately define the mathematical description.

The description is derived from first principles and physical laws and is primarily based on a set of conservation relationships that result in a series of ordinary and partial differential equations (ODE and PDE). The PDEs describe such things as the conservation of mass, momentum, energy, and others. In addition, fundamental relationships are used to complete the description of the system. The complete description is made up of these PDEs and algebraic relationships.

Combustion modeling results in a particularly complex mathematical description of the overall process. Each physical process involved in a combustion system is described individually; however, they interact with other physical processes. This interaction creates a coupling between all the descriptions of the individual processes.

To demonstrate this coupling, consider a simple diffusion flame. Fluid dynamics describe the process of mixing two streams of reactants. The resulting reaction alters the constituents of the fluid, and heat release from the reaction increases the local temperature. The change in temperature and chemical composition has a strong effect on local density. This change in density, in turn, has a strong effect on the fluid flow.

The system of processes, equations and interrelationships in a coal-fired boiler is far more complex, as shown in Fig. 1. Five fundamental processes must be addressed while providing for all key interactions:

1. *Fluid transport*: fluid motion, component mass and energy transport in a turbulent mixing environment.
2. *Particle transport*: particle (in this case coal) or *discrete phase* motion in a fluid.
3. *Homogeneous chemical reactions*: gaseous species combustion.
4. *Heterogeneous chemical reactions*: particle combustion.
5. *Radiative heat transfer*: radiative heat transfer in a particle-laden participating media.

The second step to modeling the system is to use an appropriate technique to solve the set of equations that has been chosen to describe the physical system. It is not possible to analytically solve the partial differential equations typically encountered in modeling combustion systems. Thus, the differential equations are *discretized* to obtain a set of non-linear algebraic equations that can be solved with known numerical techniques. The last step in the process is to obtain the final solution.

Following is a more detailed description of each of these processes.

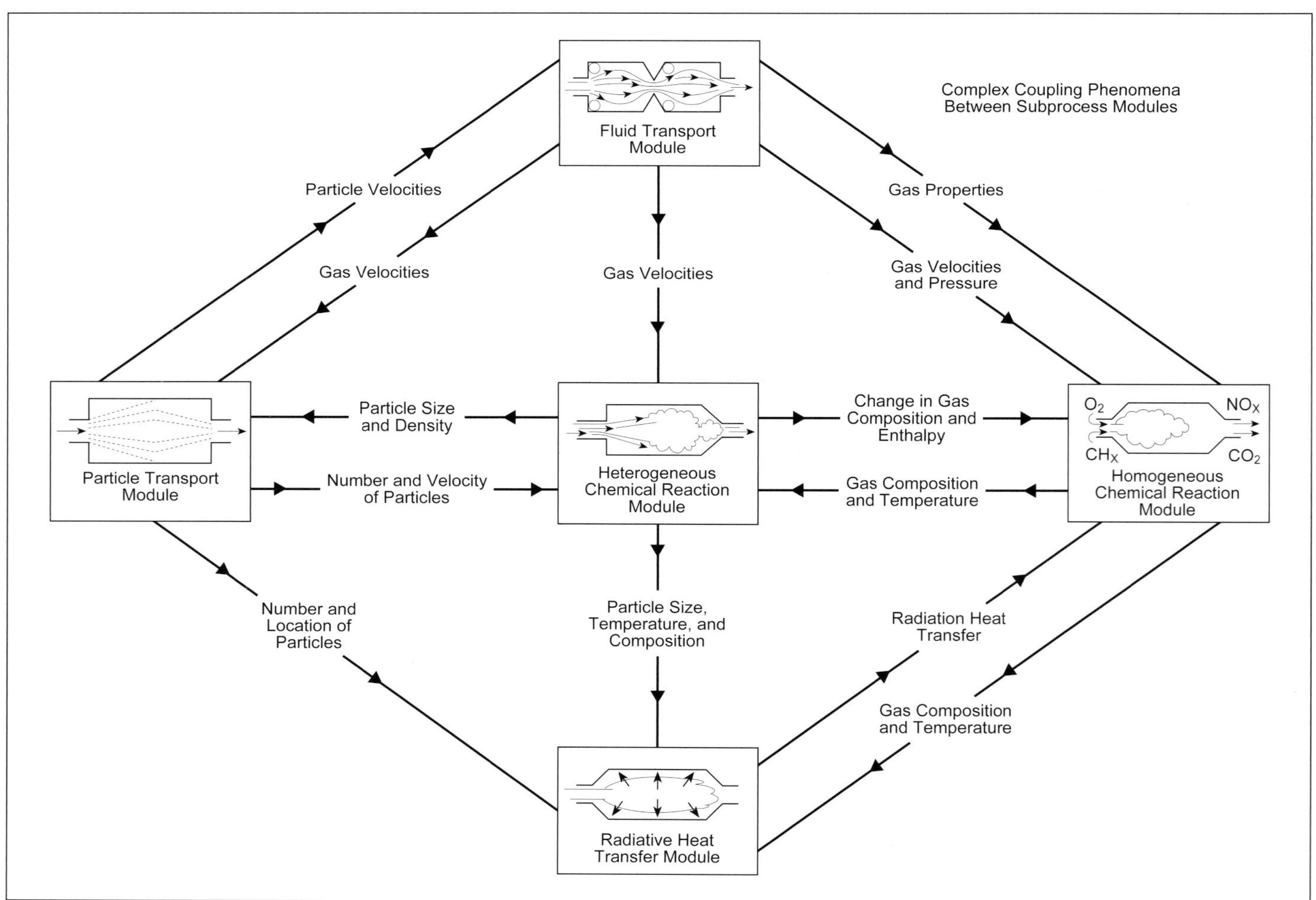

Fig. 1 Model for the evaluation of pulverized coal-fired combustion based upon five fundamental processes.

Fundamental equations

Combustion systems involve a complex interaction of many different physical processes. This includes fluid flow, heat transfer, chemical reactions, and potentially fluid-solid interactions. Some of the fundamental equations that describe these processes are introduced in Chapters 3 and 4. Each of these processes is briefly described below in the context of numerical modeling.

Representation of turbulence

Large-scale combustion systems are typically characterized by turbulent, reacting flow conditions. The effect of turbulent flow (turbulence) on combustion processes is significant and must be considered to account for this effect. As yet, it is not practical to model the full detail of the temporal and spatial fluctuations that are associated with turbulence. As computing resources become more powerful and our ability to handle the enormous amount of information that will be generated increases, it may one day be possible to model the details of turbulent flow on industrial combustion systems. Until that day, a simplified model representation of turbulence must be used.

Often, dealing with turbulence involves time-averaging the fundamental equations to eliminate the turbulent fluctuations and utilizing a separate turbulence model to account for the influence of turbulent fluctuations on the flow. The fundamental equations can then be solved for the mean quantities. Alternatively, large scale turbulent fluctuations can be directly solved while utilizing a turbulence model for the small scale fluctuations. This technique, called large eddy simulation (LES), is an important advancement in turbulence modeling but requires large computational resources compared to time-averaging.

Time-averaging is typically done either with Reynolds averaging, the conventional time-averaging, or with Favre averaging, a density-weighted averaging. The latter is better suited to handle the large density variations experienced in combustion applications. Averaging of the conservation equations is accomplished by first assuming that instantaneous quantities are represented by mean and fluctuating portions as shown in Equation 1. By allowing ϕ to represent the dependant variable, this can be expressed as:

$$\phi = \bar{\phi} + \phi' \tag{1}$$

where ϕ is the instantaneous value, $\bar{\phi}$ is the mean portion and ϕ' is the fluctuating portion. Density-weighted averaging offers advantages over conventional time-averaging for combustion-related flows since it simplifies the treatment of large density changes. The density-weighted mean value, $\tilde{\phi}$, is defined as:

$$\tilde{\phi} = \frac{\overline{\rho\phi}}{\bar{\rho}} \tag{2}$$

where $\overline{\rho\phi}$ is the time-averaged product of the instantaneous density (ρ) and instantaneous value (ϕ) and $\bar{\rho}$ is the time-averaged density. The instantaneous value may then be written as the sum of the density-weighted average and the fluctuating value ϕ'':

$$\phi = \tilde{\phi} + \phi'' \tag{3}$$

Equation 3 can be substituted into the transport equation and then time-averaged to derive equations in terms of the mean quantities. While it is not important to detail the process here, it is important to note that the results produce additional terms in the resulting equations. These extra terms are known as Reynolds stresses in the equations of motion and turbulent fluxes in the other conservation equations. Turbulence models are generally required to model these extra terms, closing the system of equations.

Fluid flow and heat transfer

Gas-phase transport in combustion systems is governed by PDEs that describe the conservation of mass, momentum, component mass and energy. The conservation of mass or continuity equation is discussed in Chapter 3. The conservation of momentum is represented by the Navier-Stokes equations that are also briefly discussed in Chapter 3. The Cartesian form of Navier-Stokes equations, as well as the continuity equation, can be found in the first four equations in Table 2. In these four equations, ρ is the density, u, v, and w are the velocity components, and x, y, and z are the coordinate directions, μ is the dynamic viscosity, P is the pressure, and g is the body force due to gravity.

The remaining conservation equations used to describe the gas-phase transport are the energy and component mass equations, expressed in Table 2 in terms of specific enthalpy and component mass fraction. The energy source terms are on a volumetric basis and represent the contribution from radiative heat transfer, $-\nabla \cdot q_r$, energy exchange with the discrete phase particles, S_H^{part}, and viscous dissipation, S_H. The component mass source terms include the mean production rate due to gas-phase reactions, R_i, and the net species production rate from heterogeneous reactions, S_i^{part}.

Turbulence model

As previously mentioned, the process of time-averaging the conservation equations introduces extra terms into the equations. Numerous turbulence models have been developed over the years to determine the values of these extra terms. One of the most common and widely accepted approaches, known as the Boussinesq hypothesis, is to assume that the Reynolds stresses are analogous to viscous dissipation stresses. This approach introduces the turbulent viscosity μ and a turbulent transport coefficient σ into each equation.

Most of the turbulence models currently used for fluid flow and combustion are focused on determining μ. In the k-epsilon model (one of the most widely used and accepted), the turbulent viscosity is given as:

$$\mu_t = \frac{\bar{\rho} C_\mu k^2}{\varepsilon} \tag{4}$$

Table 2
Summary of Fundamental Differential Equations

General form of the transport equation:

$$\frac{\partial}{\partial t}(\bar{\rho}\phi) + \frac{\partial}{\partial x}(\bar{\rho}\tilde{u}\phi) + \frac{\partial}{\partial y}(\bar{\rho}\tilde{v}\phi) + \frac{\partial}{\partial z}(\bar{\rho}\tilde{w}\phi) = \frac{\partial}{\partial x}\left(\Gamma_\phi \frac{\partial \phi}{\partial x}\right) + \frac{\partial}{\partial y}\left(\Gamma_\phi \frac{\partial \phi}{\partial y}\right) + \frac{\partial}{\partial z}\left(\Gamma_\phi \frac{\partial \phi}{\partial z}\right) + S_\phi$$

Equation	Physical Parameter ϕ	Transport Coefficient Γ_ϕ	Source Term S_ϕ
Continuity	1	0	S_m^{part}
X-Momentum	$\tilde{u}$	μ_e	$-\frac{\partial P}{\partial x} + \frac{\partial}{\partial x}\left(\mu_e \frac{\partial \tilde{u}}{\partial x}\right) + \frac{\partial}{\partial y}\left(\mu_e \frac{\partial \tilde{v}}{\partial x}\right) + \frac{\partial}{\partial z}\left(\mu_e \frac{\partial \tilde{w}}{\partial x}\right) + \bar{\rho} g_x + S_u^{part}$
Y-Momentum	$\tilde{v}$	μ_e	$-\frac{\partial P}{\partial y} + \frac{\partial}{\partial x}\left(\mu_e \frac{\partial \tilde{u}}{\partial y}\right) + \frac{\partial}{\partial y}\left(\mu_e \frac{\partial \tilde{v}}{\partial y}\right) + \frac{\partial}{\partial z}\left(\mu_e \frac{\partial \tilde{w}}{\partial y}\right) + \bar{\rho} g_y + S_v^{part}$
Z-Momentum	$\tilde{w}$	μ_e	$-\frac{\partial P}{\partial z} + \frac{\partial}{\partial x}\left(\mu_e \frac{\partial \tilde{u}}{\partial z}\right) + \frac{\partial}{\partial y}\left(\mu_e \frac{\partial \tilde{v}}{\partial z}\right) + \frac{\partial}{\partial z}\left(\mu_e \frac{\partial \tilde{w}}{\partial z}\right) + \bar{\rho} g_z + S_w^{part}$
Enthalpy	$\tilde{H}$	$\frac{\mu_e}{\sigma_H}$	$-\nabla \cdot q_r + S_H + S_H^{part}$
Turbulent Energy	k	$\frac{\mu_e}{\sigma_k}$	$G - \bar{\rho}\varepsilon$
Dissipation Rate	ε	$\frac{\mu_e}{\sigma_\varepsilon}$	$\frac{\varepsilon}{k}(c_1 G - c_2 \bar{\rho}\varepsilon)$
Species	$\tilde{Y}_i$	$\frac{\mu_e}{\sigma_i}$	$\bar{\rho} R_i + S_i^{part}$

Other terms appearing in general form:

$$G = \mu_e \left\{ 2\left[\left(\frac{\partial \tilde{u}}{\partial x}\right)^2 + \left(\frac{\partial \tilde{v}}{\partial y}\right)^2 + \left(\frac{\partial \tilde{w}}{\partial z}\right)^2\right] + \left(\frac{\partial \tilde{u}}{\partial y} + \frac{\partial \tilde{v}}{\partial x}\right)^2 + \left(\frac{\partial \tilde{u}}{\partial z} + \frac{\partial \tilde{w}}{\partial x}\right)^2 + \left(\frac{\partial \tilde{v}}{\partial z} + \frac{\partial \tilde{w}}{\partial y}\right)^2 \right\}$$

$$\mu_e = \mu + \mu_t$$

Nomenclature

S^{part}	=	source term accounting for exchange between discrete phase particles and gas phase
u, v, w	=	velocity components
H	=	enthalpy
k	=	turbulent kinetic energy
ε	=	turbulent kinetic energy dissipation
μ_e	=	effective viscosity
μ_t	=	turbulent viscosity
g	=	gravitational vector (x, y, z)
ρ	=	density
c_1, c_2	=	model constants
R_i	=	reaction rate
Y_i	=	species mass fraction

Subscripts/Superscripts

e	=	effective
t	=	turbulent
x, y, z	=	directional component
i	=	ith chemical specie
~	=	Favre (density weighted) average
–	=	time-average
part	=	discrete phase particle component

where C_μ is a model parameter, k is the turbulent kinetic energy, and ε is the turbulent kinetic energy dissipation. The turbulent kinetic energy and the dissipation are determined by solving an additional partial differential equation for each quantity as given in Table 2.

Discrete phase transport

Many combustion applications, including pulverized coal, oil, black liquor and even wood involve small solid or liquid particles moving through the combustion gases. The combustion gases are described by assuming that they represent a continuum, whereas a description of the solid and liquid fuel involves discrete particles. Describing the motion of this discrete phase presents unique modeling challenges. There are two basic reference frames that can be used to model the transport of the discrete phase particles, Eulerian and Lagrangian.

The Eulerian reference frame describes a control volume centered at a fixed point in space. Conservation equations similar to the ones used for gas transport are used to describe the transport mass and energy of particles passing through this control volume. The interaction of the particle phase and the gas phase is accomplished through source terms in the respective transport equations.

The Lagrangian reference frame considers a control volume centered on a single particle. This approach tracks the particle on its trajectory as it travels through space and interacts with the surrounding gases. The motion of a particle can be described by:

$$m_{part} \frac{d\vec{u}_{part}}{dt} = \vec{F}_D + \vec{F}_g \qquad \textbf{(5)}$$

where m_{part} represents the mass of the particle, $\vec{u}_{part}$ is the particle vector velocity, t is time, and $\vec{F}_D$ and $\vec{F}_g$ represent drag and gravitational forces. Aerodynamic drag is a function of the relative differences between particle and gas velocities, Reynolds number and turbulent fluctuations in the gas. Consideration is also given for mass loss from the particle due to combustion.[1,2,3]

Turbulence has the effect of dispersing or diffusing the particles. This dispersion effect has been identified with the ratio of the particle diameter to turbulence integral scale. For large particle sizes, particle migration will be negligible, while at small sizes particles will follow the motion of the gas phase. This effect can be modeled using the Lagrangian stochastic deterministic (LSD) model.[4] The LSD model computes an instantaneous gas velocity which is the sum of the mean gas velocity and a fluctuating component. The instantaneous gas velocity is used in computing the right-hand side of Equation 5.

From the particle velocity the particle position, x_{part}, is expressed as:

$$\frac{d\vec{x}_{part}}{dt} = \vec{u}_{part} \qquad \textbf{(6)}$$

This equation, along with appropriate initial conditions, describes the particle trajectory within the computational domain.

Combustion

Homogeneous chemical reactions Homogeneous or gas-phase combustion involves the transport and chemical reaction of various gas species. During this process, heat is released and combustion product species are formed. As mentioned, a transport equation for each of the chemical species involved is solved. The main objective of a gas-phase combustion model is to determine the mean production rate, R_i for turbulent combustion.

Various methods can be used to determine the production rate. One common method known as the Eddy Dissipation Combustion Model (EDM) was developed by Magnussen and Hjertager[5] and is based on the eddy break-up model.[6] This model assumes that the rate of combustion is controlled by the rate of mixing of the reactants on a molecular scale. The reaction rate is given by:

$$R_i = \frac{W_i \dot{\omega}_i}{\bar{\rho}} =$$

$$\underbrace{W_i \sum_{j=1}^{N_{rc}} (\upsilon''_{ij} - \upsilon'_{ij})}_{\text{Term 1}} \; \underbrace{C_A \frac{\varepsilon}{\bar{\rho} k}}_{\text{Term 2}} \; \underbrace{\min \left(\frac{Y_k / W_k}{\upsilon'_{kj}} : k \in RCT_j \right)}_{\text{Term 3}} \qquad \textbf{(7)}$$

where W_i is the component molecular weight, υ'_{ij} and υ''_{ij} are the reactant and product stoichiometric coefficients for the ith species and the jth reaction, ε is the turbulent dissipation, k is the turbulent kinetic energy, C_A is the model dependent mixing constant and RCT_j denotes the set of species that are reactants for the jth reaction. Term 1 represents the stoichiometric coefficients in the particular reaction, Term 2 represents the molecular mixing rate, and Term 3 limits the reaction to the availability of individual reactants.

Magnussen[7] later proposed the eddy dissipation concept (EDC) to overcome some limitations of other models. Specifically, the EDC model is applicable to non-premixed and premixed combustion and can be used with simplified or detailed chemistry to describe the reaction process. A detailed description of the EDC model can be found in Magnussen,[7] Lilleheie *et al.*,[8] Magnussen[9] and Lilleheie *et al.*[10]

Magnussen's premise is that chemical reactions occur in the fine structures of turbulence where the turbulent energy is being dissipated. Within these structures, molecular mixing occurs and the reactions can be treated at the molecular level. The EDC model is based on the concept of a reactor defined by a reaction zone in these fine turbulence structures. The length and time scales from the turbulence model are used to characterize these fine turbulence structures. The reaction rates within these fine structures can be defined with the specification of an appropriate chemical kinetics mechanism. These reaction rates are then related to the average reaction rates in the bulk fluid and then applied to the time-averaged transport equations.

While some of the simpler models mentioned above have been utilized extensively, the EDC provides a means of more accurately treating the complexities of coal combustion and modern combustion systems. This is particularly important as the sophistication of the heterogeneous combustion models improves.

Heterogeneous chemical reactions Simulation of coal combustion must account for a complex set of physical processes including drying, devolatilization, and char oxidation. When a coal particle enters the combustion zone, the rapid heatup causes moisture to evaporate.

$$\text{Coal} \rightarrow \text{Dry Coal} + \text{Water Vapor} \tag{8}$$

Evaporation is followed by devolatilization to produce volatiles and char.

$$\text{Dry Coal} \rightarrow \text{Gaseous Fuel} + \text{Char} \tag{9}$$

The volatiles consist of light gases (primarily hydrogen, carbon monoxide, carbon dioxide, and methane), tars and other residues. The devolatilization rate can not be adequately represented with a single first-order kinetic expression. Ubhayaker *et al.*[11] suggested a two first-order kinetic rate expression:

$$\begin{aligned} &\text{Dry Coal} \xrightarrow{K_i^d} \alpha_i \text{ Gaseous Fuel}_i \\ &\quad + (1 - \alpha_i) \text{ Char}_i \qquad i = 1,2 \end{aligned} \tag{10}$$

where K_i^d is the kinetic rate of reaction and α_i is the volatiles' mass fraction. The kinetic rates are first-order in the mass of coal remaining and are expressed in an Arrhenius form. The total devolatilization rate becomes:

$$K^d = \sum_i \alpha_i K_i^d \tag{11}$$

A more advanced model known as the Chemical Percolation Devolatilization (CPD)[12,13,14] has been developed and is described elsewhere. Unlike the empirical formulation of Ubhayaker *et al.*,[11] the CPD model is based on characteristics of the chemical structure of the parent coal.

Following devolatilization the remaining particle consists of char residue and inert ash. Char is assumed to react heterogeneously with the oxidizer:

$$\text{Char} + \text{Oxidant} \rightarrow \text{Gaseous Products} + \text{Ash} \tag{12}$$

A basic approach to char oxidation was described by Field.[15] The effective char oxidation rate is a function of the kinetic rate of the chemical reaction and the diffusion rate of the oxidizer to the particle.[15,16]

$$\begin{aligned} &\text{Char} + \text{Oxidant}_i \xrightarrow{K_i^{ch}} \text{Gaseous Products}_i \\ &\quad + \text{Ash} \qquad i = 1,2 \end{aligned} \tag{13}$$

where K_i^{ch} is the effective char oxidation rate. The total char oxidation rate is expressed as:

$$K^{ch} = \sum_i K_i^{ch} \tag{14}$$

The Carbon Burnout Kinetic (CBK) model has been developed by Hurt *et al.*[17] specifically to model the details of carbon burnout. The model has a quantitative description of thermal annealing, statistical kinetics, statistical densities, and ash inhibition in the late stages of combustion.

Radiative heat transfer

Radiative heat transfer in combustion systems is an important mode of heat transfer and is described by the radiative transfer equation (RTE):

$$\begin{aligned} (\bar{\Omega} \bullet \nabla) I_\lambda (\bar{r}, \bar{\Omega}) &= -(\kappa_\lambda + \sigma_\lambda) I_\lambda (\bar{r}, \bar{\Omega}) \\ &+ \kappa_\lambda I_{b\lambda}(\bar{r}) + \frac{\sigma}{4\pi} \int_\Omega \Phi(\bar{\Omega}' \rightarrow \bar{\Omega}) I_\lambda (\bar{r}, \bar{\Omega}') \, d\bar{\Omega}' \end{aligned} \tag{15}$$

where κ_λ is the spectral absorption coefficient, σ_λ is the scattering coefficient, and $I_{b\lambda}$ is the black body radiant intensity.

This equation describes the change in radiant intensity, $I_\lambda(\bar{r}, \bar{\Omega})$, at location $\bar{r}$ in direction $\bar{\Omega}$. The three terms on the right-hand side represent the decrease in intensity due to absorption and out-scattering, the increase in intensity due to emission, and the increase in intensity due to in-scattering.

Radiative heat transfer information is obtained by solving the RTE (Equation 15) which is coupled with the thermal energy equation by the divergence of the radiant flux vector $-\nabla \bullet \Im$. The divergence can be obtained from:

$$\nabla \bullet \Im = 4 \int_0^\infty \kappa_\lambda E_{b\lambda}(T) d\lambda - \int_0^\infty \kappa_\lambda \left[\int_0^{4\pi} I_\lambda (\bar{\Omega}) d\Omega \right] d\lambda \tag{16}$$

The two terms on the right-hand side account for emission and absorption, respectively.

Discretization of equations

In the preceding sections, a mathematical description of combustion modeling, consisting of a fundamental set of algebraic relations and differential equations of various forms, has been described. This includes fluid transport, particle transport, combustion and radiative heat transfer. Because this system of equations is too complex to solve with analytic methods, a numerical method must be employed. The methods of discretizing the fluid transport and radiative heat transfer are of particular interest and are presented here.

Finite volume approach

It should be recognized that many of the partial differential equations are of a single general form as provided in Table 2 and can be expressed as:

$$\begin{aligned} &\frac{\partial}{\partial t}(\bar{\rho}\phi) + \frac{\partial}{\partial x}(\bar{\rho}\tilde{u}\phi) + \frac{\partial}{\partial y}(\bar{\rho}\tilde{v}\phi) + \frac{\partial}{\partial z}(\bar{\rho}\tilde{w}\phi) \\ &= \frac{\partial}{\partial x}\left(\Gamma_\phi \frac{\partial \phi}{\partial x}\right) + \frac{\partial}{\partial y}\left(\Gamma_\phi \frac{\partial \phi}{\partial y}\right) + \frac{\partial}{\partial z}\left(\Gamma_\phi \frac{\partial \phi}{\partial z}\right) + S_\phi \end{aligned} \tag{17}$$

Since many of the equations share this form, a single method can be used to solve all of the associated equations. Most of these methods involve dividing the physical domain into small sub-domains and obtaining a solution only at discrete locations, or grid points, throughout the domain. The well-known *finite difference method* is one such method. Another very powerful method, that is particularly suited for use in combustion modeling, is the *finite volume approach.*

The basic idea of the finite volume approach is very straightforward and is detailed in Patankar.[18] The entire domain is divided into non-overlapping control volumes with a grid point at the center of each. The differential equation in the form of Equation 17 is integrated over the entire control volume and after some rearrangement becomes:

$$\iiint_V \frac{\partial}{\partial t}(\rho\phi)\,dV + \oiint_S (\rho\tilde{u}\phi, \rho\tilde{v}\phi, \rho\tilde{w}\phi)\cdot\hat{n}\,dS + \oiint_S \left(-\Gamma_\phi \frac{\partial\phi}{\partial x}, -\Gamma_\phi \frac{\partial\phi}{\partial y}, -\Gamma_\phi \frac{\partial\phi}{\partial z}\right)\cdot\hat{n}\,dS = \iiint_V S_\phi\,dV \quad (18)$$

Carrying out the integrations, the resulting equation is:

$$\frac{\partial}{\partial t}(\rho\phi)\,\Delta V + \sum_f \left(C_f\phi_f - D_f(\phi)\right) = S_\phi\,\Delta V \quad (19)$$

where ΔV is the volume of the control volume, C_f is the mass flow rate out of the control volume, D_f is the diffusive flux into the control volume, and the summation is made over all the control volume faces, f. The temporal derivative in the first term of Equation 19 can be expressed using a first-order backward difference scheme:

$$\frac{\partial}{\partial t}(\rho\phi) = \rho^t\left(\frac{\phi^{t+\Delta t} - \phi^t}{\Delta t}\right) \quad (20)$$

The mass flow rate C_f is determined from the solution of the mass and momentum equations while the diffusive flux D_f is based on the effective diffusivity and the gradient at the control volume face. Combining Equations 19 and 20 with the definitions of C_f and D_f and an interpolated value for ϕ_f results in an algebraic expression in terms of the dependant variable ϕ_i at grid point i and the neighboring grid points. This is expressed as:

$$a_i\phi_i = \sum_n a_n\phi_n + b_i \quad (21)$$

where a_i and a_n are coefficients for the control volume and its neighbors respectively and b_i represents the remaining terms. The number of neighboring values that appear in Equation 21 is a function of the mesh, the method used to interpolate the dependant variable to the control volume face, and the method used to determine gradients at the control volume face.

Following this procedure for each grid point in the entire domain produces a coupled set of algebraic equations. This set of equations can be solved with an appropriate method from linear algebra. Many different techniques are possible and can be found in a reference on numerical methods.

There are two advantages to the finite volume approach. First, the dependant variable in the resultant discretized equation is a quantity of fundamental interest such as enthalpy, velocity or species mass fraction, and the physical significance of the individual terms is maintained. Second, this approach expresses the conservation principle for the dependant variable over a finite control volume in the same way the conservation equation expresses it for an infinitesimal control volume. By so doing, conservation is maintained over any collection of control volumes and is enforced over the entire domain.

Discrete ordinates method

Several radiative heat transfer models have been developed and many are described by Brewster[19] and Modest.[20] A recent review of radiative heat transfer models[21] states that the discrete ordinates method coupled with an appropriate spectral model provide the necessary detail to accurately model radiative heat transfer in combustion systems. This is one of the most common methods currently used to model radiative heat transfer.

The discrete ordinates method (DOM)[22,23] solves the radiative transport equation for a number of ordinate directions. The integrals over direction are replaced by a quadrature and a spectral model is used to determine radiative properties of κ and σ. This results in a set of partial differential equations given by:

$$\mu_m \frac{\partial I_m}{\partial x} + \eta_m \frac{\partial I_m}{\partial y} + \xi_m \frac{\partial I_m}{\partial z} = -(\kappa+\sigma)\,I_m + \kappa I_b + \frac{\sigma}{4\pi}S_m \quad (22)$$

where μ_m, η_m, ξ_m are the direction cosines of the chosen intensity I_m and S_m is the angular integral. This set of equations is solved by a method outlined by Fiveland[22] to find the radiative intensities throughout the combustion space. The source term for the energy equation can be found by summing over all directions:

$$\nabla\cdot q_r = 4\kappa\tilde{\sigma}T^4 - \kappa\sum_{m'} w_{m'}\,I_{m'} \quad (23)$$

Mesh generation

Once discretized, the transport equations must be solved at individual points throughout the domain. This requires that the individual points be specified and the relationship between other points be identified. Displaying the points along with the connections between them creates a pattern that looks something like a woven mesh. The process of creating the mesh is therefore known as *mesh generation.*

Mesh generation is an important and often challenging step in the overall modeling effort. The first criterion in mesh generation is to accurately represent the geometry being modeled. Secondly, adequate detail must be placed throughout the domain to obtain an accurate solution. Other criteria include mesh quality and total mesh size. A discussion on these criteria can be found elsewhere.[24]

Cell types

The basic unit in a mesh is the control volume or cell. The cells are arranged such that they cover the entire domain without overlapping. Common cell types are shown in Fig. 2. The mesh may be made of a single cell type (homogeneous mesh) or possibly a combination of different types (hybrid mesh).

Structured mesh

Structured meshes consist of cells placed in a regular arrangement such that adjacent cells can be identified simply by their order in a list. Fig. 3 shows how the neighboring cells are identifiable simply by incrementing an index that is typically aligned with the coordinate directions. This greatly simplifies the task of retrieving information from neighboring cells. For simple geometries, a structured mesh is both simple to generate and efficient when solving the problem. However, complex geometries highlight particular challenges with this approach. This is illustrated in Fig. 4. Two common techniques of dealing with irregularities in geometry are 1) a cartesian stair-stepped mesh and 2) body-fitted mesh. The stair-stepped mesh can place cells in areas outside of the domain and approximates boundaries by stair-stepping the mesh (Fig. 4a). The cells outside of the domain are maintained as part of the mesh structure but are unused during the computation. The body-fitted mesh follows geometric features and the cell shape changes to accommodate these physical features (Fig. 4b).

Slight variations of the simple structured mesh can be used. Block structured meshes provide more geometric flexibility since the entire mesh is a composite of smaller structured meshes.

Unstructured mesh

An unstructured mesh provides the maximum flexibility for complex geometries. While the ease of obtaining neighboring cell information has been lost, the ability to place cells anywhere in the computational domain increases the ease by which the geometry can be accurately represented. Fig. 4 compares an unstructured mesh with two structured mesh approaches. With an unstructured mesh approach there is greater control over the level of detail in the mesh for different parts of the domain.

Embedding and adaption

More detail can be obtained in portions of the domain through increasing the mesh resolution by embedding more cells. This is accomplished by splitting an existing cell in some fashion to create additional cells. By splitting a cell in each of the cell's parametric coordinate directions, a single hexahedron would become eight. This greatly increases the resolution in this region.

Often it is not possible to know where resolution is needed a priori. The process of adaption can be used to increase the resolution based on the actual solution. For example, high discretization error may often be related to high solution gradients across cells. In this case the velocity gradient from the solution can be used to discover where more cells are needed. This is an especially powerful tool when computational resources are limited.

Example applications

Wet scrubbers

Situation The two-phase flow in a wet flue gas desulfurization (WFGD) scrubber tower is a complex process involving spray atomization, liquid entrainment, droplet disengagement and phase separation. The physical arrangement of a basic WFGD scrubber module is shown in Chapter 35, Fig. 2. With a tray, there is a bubbly froth due to countercurrent flow of liquid and gas with holdup of liquid on the tray. The various two-phase flow regimes complicate the calculation of pressure drop and gas velocity distribution in a wet scrubber. Prediction of two-phase flow is essential since liquid residence time and total interfacial liquid/gas area are important factors in determining the amount of SO_2 absorption. Therefore, The Babcock & Wilcox Company (B&W) has implemented a multi-dimensional two-phase flow model for wet scrubbers based on CFD analysis.

A multi-dimensional hydraulic model solves separate equations for mass and momentum for both the liquid and gas phases. An interfacial drag law calculates the resistance of liquid to the gas flow and vice versa. These interfacial drag laws depend primarily on droplet diameter. However, alternate drag equations can be implemented in the multi-dimensional model for the various two-phase flow regimes. By using the fundamental relations for interfacial drag, the model can calculate separate three-dimensional velocity fields for the liquid and gas phases. Both liquid and gas momentum equations share a common static pressure field.

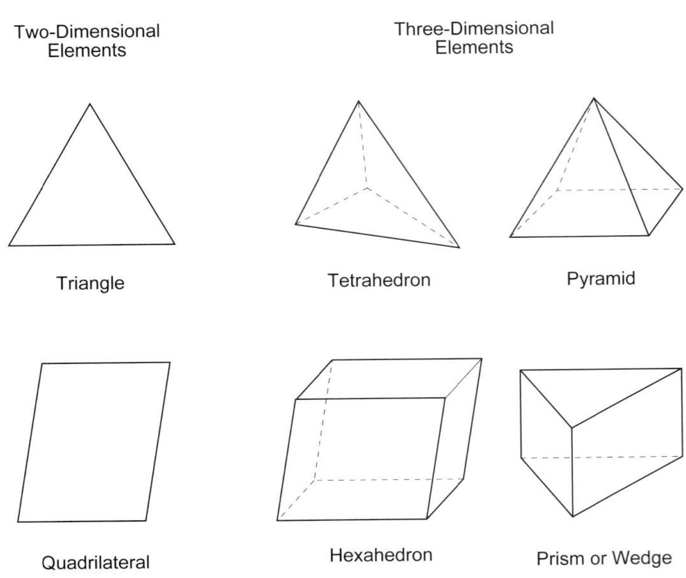

Fig. 2 Cell or control volume types used in numerical modeling grids or meshes.

The numerical model is used as an analysis tool to compare and contrast various design configurations for wet scrubbers. The design features considered by the model include:

1. an overall cylindrical geometry,
2. a tray model with baffles and porous plates,
3. nozzles located at various elevations,
4. conventional or interspatial headers,
5. separate air and water outlets,
6. multiple mist eliminator heights and elevations,
7. conical or straight inlets with or without inlet awnings, and
8. cylindrical or conical outlet ducts.

These design features are adjustable, thereby permitting a wide range of scrubber configurations for utility boiler applications. The boundary conditions at the gas inlet and spray nozzles can be adjusted to cover all scrubber gas velocities and liquid mass fluxes.

Analysis The model was initially validated against hydraulic data from a one-eighth scale laboratory wet scrubber. By comparing model predictions to scale model pressure drop data, confidence was built in the two-phase flow modeling capability. Once validated, the model was tested for full-scale application by comparing results to company design standards. Although the model compared favorably to data and standards, absolute prediction of wet scrubber performance is not the primary purpose. Instead, comparative studies are done to predict relative performance of various design options. The numerical model excels at looking at new design configurations that fall outside of existing design standards.

Results As discussed in Chapter 35, *Sulfur Dioxide (SO_2) Control*, the flue gas enters the side of the scrubber tower and turns upward to flow through the tower while the reagent slurry flows countercurrent downward, removing SO_2. A uniform gas velocity profile across the tower diameter maximizes removal efficiency as the reagent slurry and flue gas flow are uniformly mixed. Fig. 5 shows the CFD modeling results as vertical velocity profiles at several plains through the tower, and illustrates the impact of the B&W tray design in producing uniform flue gas flow

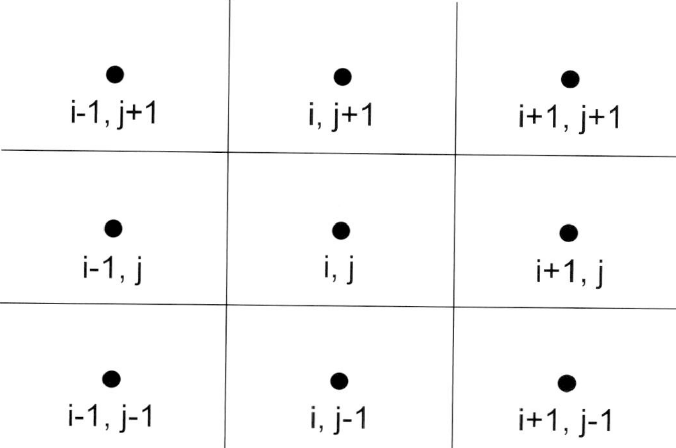

Fig. 3 Unit cell identification in a rectangular arrangement.

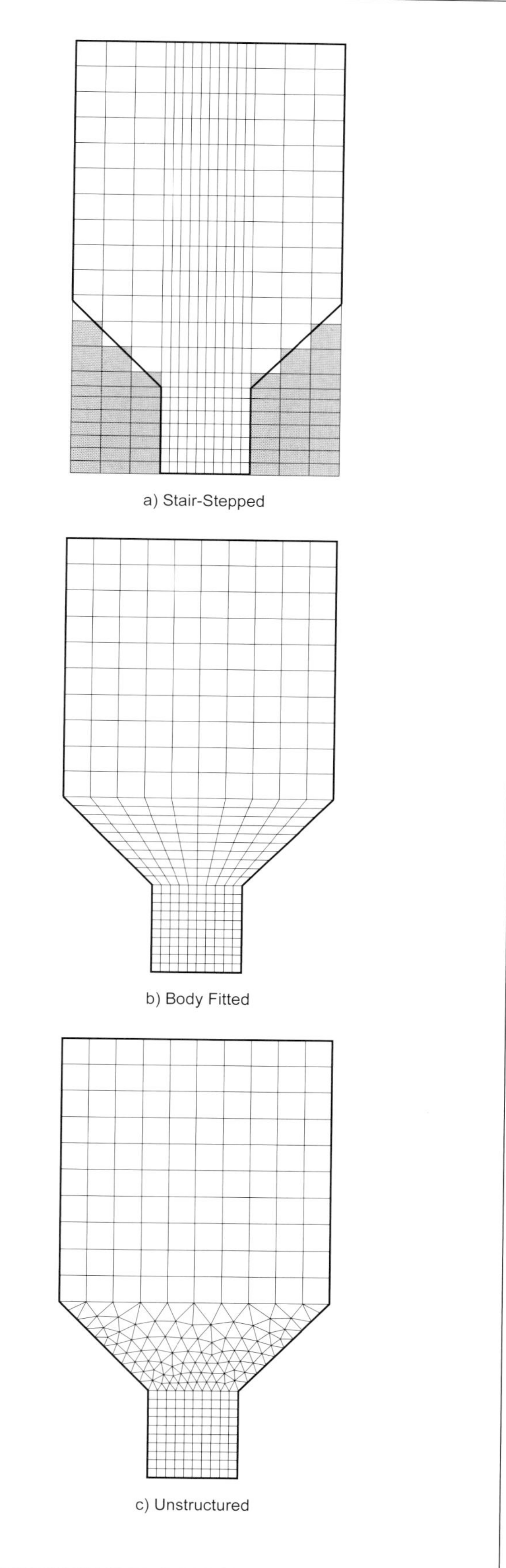

Fig. 4 How mesh or grid structure approximates geometric features.

through the unit. Fig. 5a shows the case without the tray. The lowest profile shows how non-uniform flow develops as the high velocity flue gas is introduced into the tower, is decelerated, and makes a sharp right-angle turn to flow up the tower. In the absence of a tray, the high velocity (red) and low velocity (blue) regions persist as the flue gas moves through the middle of the tower (middle velocity profile) entering the first level of spray headers. Some of the non-uniformity persists even up to the mist eliminators. With the addition of the tray (Fig. 5b), the large high and low velocity regions are effectively eliminated. The resulting more-uniform velocity profile and the gas/reagent mixing on top of the tray permit higher levels of SO_2 control at reduced slurry recirculation rates.

This model has also been used to explore design changes to meet site-specific new and retrofit requirements.[25] These have included alternate flue gas exit geometries, flue gas inlet conditions, tower diameter transitions, header locations, slurry recirculation rates or other factors while still achieving the desired performance. It has also been used to investigate internal design alternatives to boost performance and reduce pressure drop.

Popcorn ash

Situation Popcorn, or large particle, ash forms under certain conditions from the combustion of coal and is light, porous, irregularly shaped, and often forms in the upper boiler furnace or on the convective heat transfer surface. This ash can plug the top catalyst layer in selective catalyst reduction (SCR) NO_x control systems, increasing pressure drop and decreasing catalyst performance. Modifications to both the economizer outlet hoppers and the ash removal systems can increase ash capture to address this situation.

Accurately predicting how the popcorn ash behaves within the economizer gas outlet requires detailed knowledge of the aerodynamic properties of the ash particles and sophisticated modeling techniques. Key ash properties include the particle density, drag coefficient, coefficients of restitution, and its coefficient of friction with a steel plate. CFD models involve solving the gas flow solution, then calculating the particle trajectories using B&W's proprietary CFD software.

Analysis Most CFD programs that handle particle-to-wall interactions are not adequate to accurately predict the complex behavior seen in the popcorn ash physical experiments. These deficiencies have been remedied by adding capabilities to B&W's proprietary CFD software. First, the coefficient of restitution is separated into its normal and tangential components. Next, a particle-to-wall friction model is used for particles sliding along the wall and experiencing a friction force proportional to the coefficient of friction measured in the physical tests. Also, the ability to set

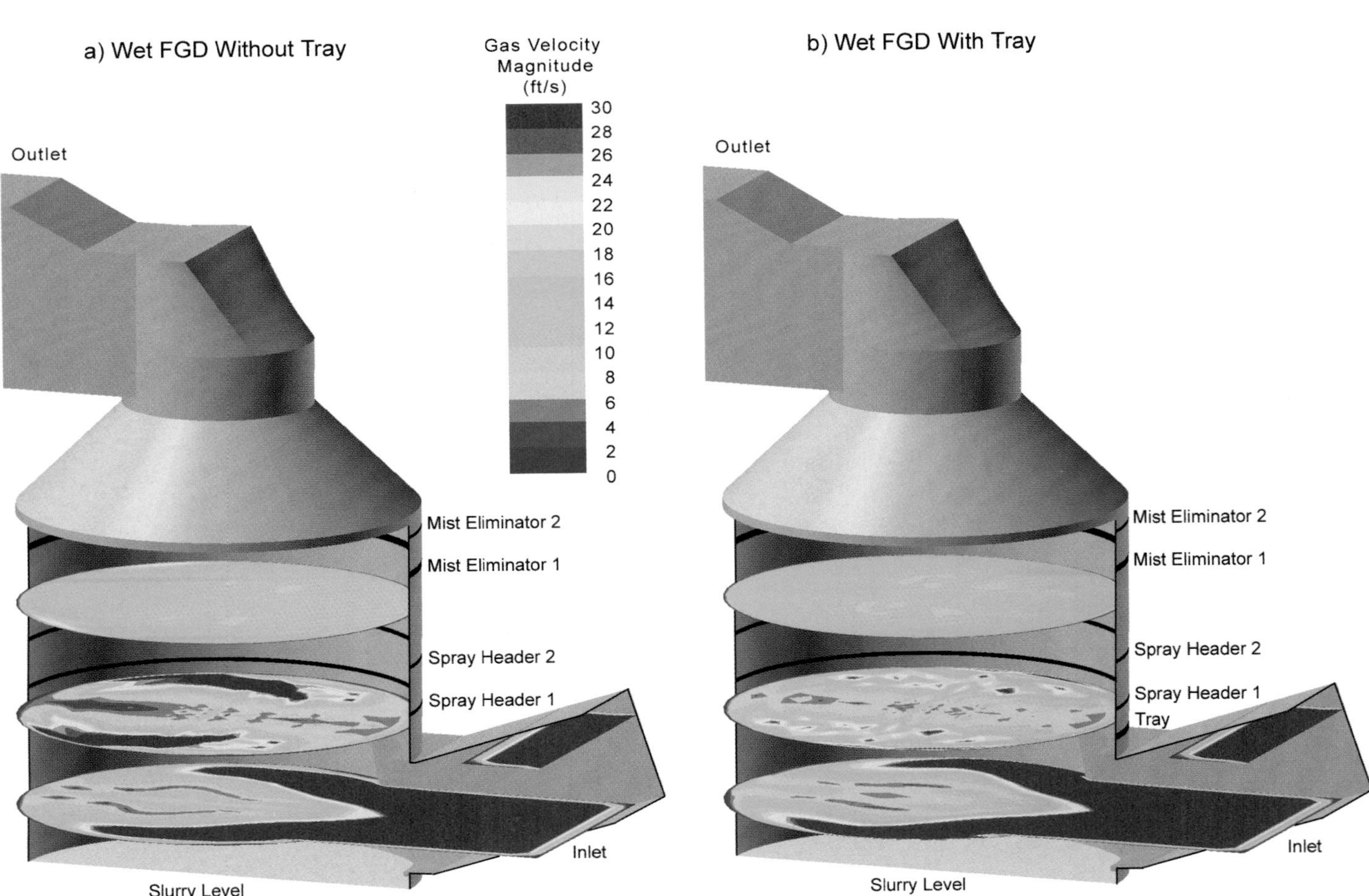

Fig. 5 Effect of B&W's tray design on gas velocities through a wet flue gas desulfurization system – numerical model results on a 650 MW absorber.

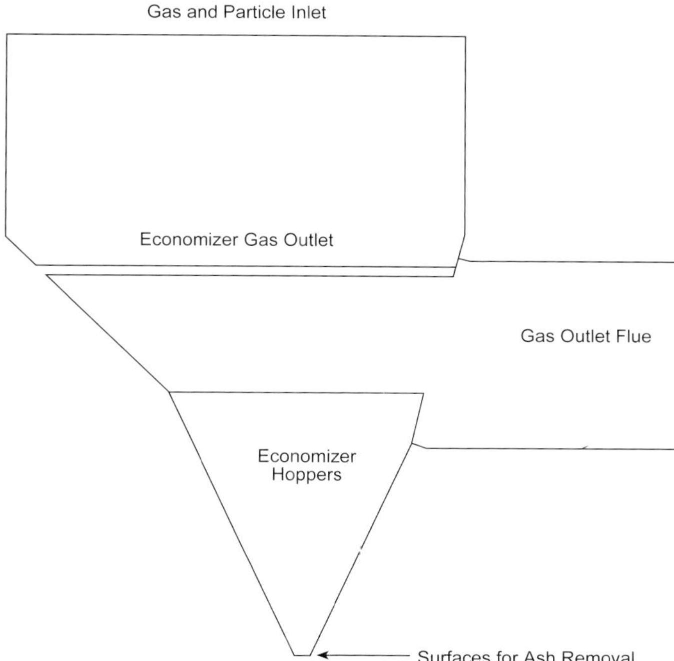

Fig. 6 Profile of popcorn ash evaluation numerical model.

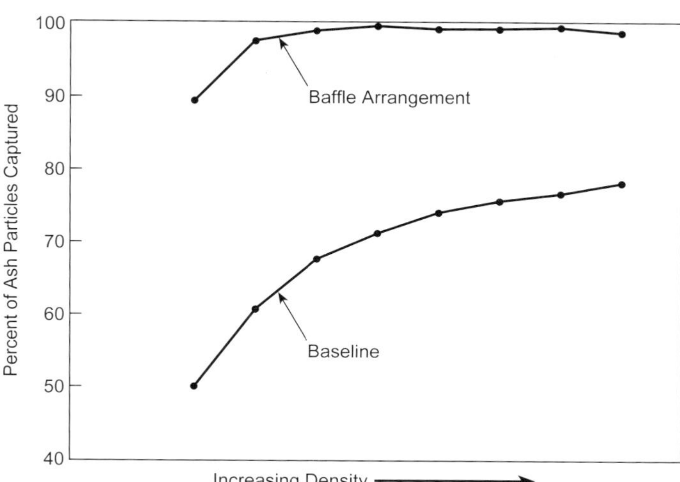

Fig. 8 Comparison of sensitivity to particle density between base case and baffled numerical models.

up user-defined planes through which flue gas could flow, but off which particles would reflect, has been developed to accurately model particle interaction with wire mesh screens.

Results The numerical models used in popcorn ash analysis normally extend from just above the bottom of the economizer (providing a reasonably uniform inlet flow distribution) to just beyond the opening of the economizer gas outlet flue (see Fig. 6). For this sample geometry, the baseline particle trajectories from the numerical model are shown in Fig. 7. Over the ash particle size range typical of this application, 20 to 50 % of the particles (Fig. 8) pass through the economizer hopper into the downstream equipment depending on particle size, potentially causing the plugging problems in the SCR or air heater. Several solutions were evaluated for this sample geometry including a design that relies on the aerodynamic separation of the particles from the flue gas and another design that involves physical barriers to the particles using a wire mesh screen. The aerodynamic solution was selected and a baffle was designed and installed. The general baffle location and the particle trajectories from the numerical model are shown in Fig. 9. The fully three-dimensional model predicted a dramatic improvement in the particle collection efficiency with more than 90% of particles collected for the range of particle sizes evaluated and virtually 100% above a certain cut size

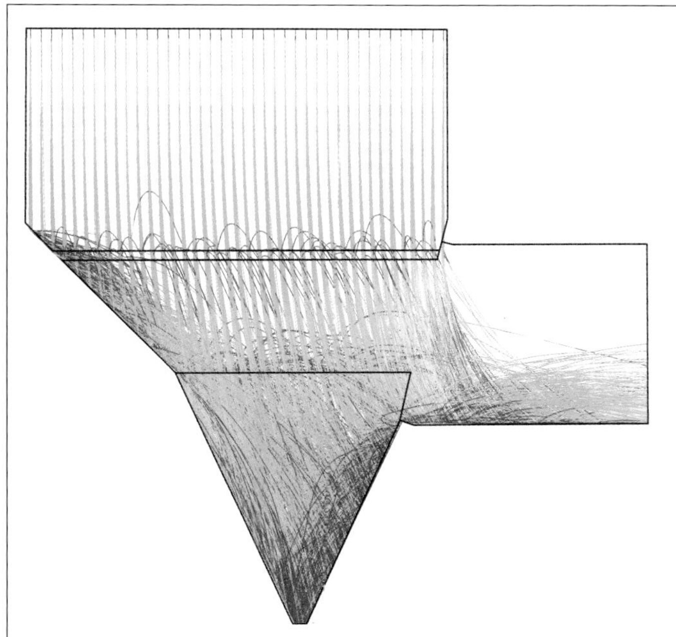

Fig. 7 Base case particle trajectories from popcorn ash evaluation.

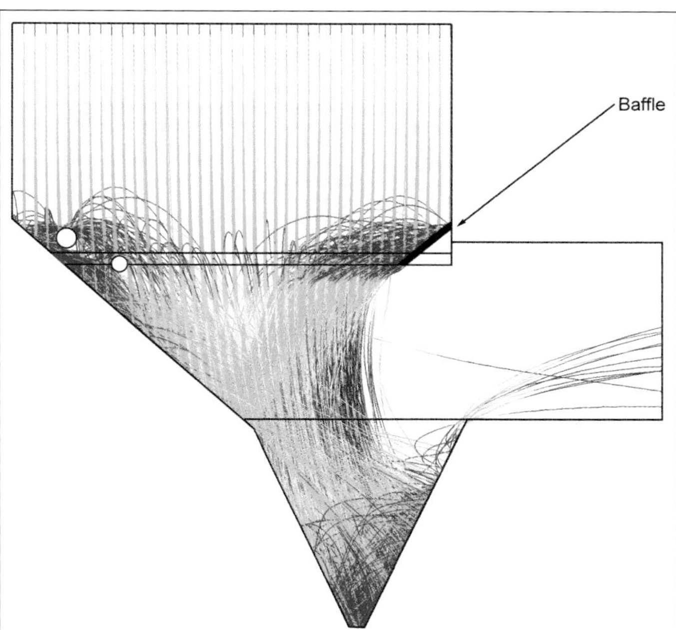

Fig. 9 Particle trajectories from popcorn ash evaluation numerical model with baffle.

(see Fig. 8). Using numerical models also permitted an optimization of the baffle position to achieve the greatest capture rate while minimizing the pressure drop. In other cases, where an aerodynamic solution is not obtainable, barriers made from wire mesh screens have been recommended. In these cases, the screen openings would be smaller than the openings in the catalyst.[26]

Kraft recovery boilers

Situation A kraft process recovery boiler, as its name implies, recovers energy and chemicals from black liquor, a byproduct of the papermaking process (see Chapter 28). Air and liquor delivery systems control several complex and interacting combustion processes (black liquor spray, deposition and burning on furnace walls, char bed burning, smelt flow) that affect boiler performance (capacity, reliability, emissions, chemical recovery, and energy efficiency). Good air jet penetration and effective mixing of secondary and tertiary air are desirable for complete combustion and reduced emissions of carbon monoxide (CO) and hydrogen sulfide. Distribution of air to three or more air injection levels produces fuel-rich conditions in the lower furnace that are desirable for smelt reduction and reduced emissions of NO_x. Flow and temperature uniformity in the furnace minimize carryover of inorganic salts, provide an even heat load, and minimize deposition on convection surfaces at the furnace exit. Uniform distribution of liquor spray ensures adequate drying of liquor spray, minimum carryover, and stable char bed combustion.

Analysis Detailed combustion models for black liquor have been developed[27,28] and are used in conjunction with CFD modeling. Black liquor combustion is simulated for individual droplets as they heat up and burn in suspension. Stages of combustion along a single trajectory include drying, devolatilization, char burning, smelt oxidation, and molten salt formation. The trajectories of thousands of particles determine the distribution of liquor spray in the furnace as shown in Fig. 10 for a range of droplet sizes. Combustion processes on the walls and char bed are also simulated with particle deposition, char burning, smelt flow and char accumulation. These capabilities are useful for evaluating the effect of air and liquor delivery systems on combustion processes in the furnace and for predicting the quantity and composition of particulate that leaves the furnace.

Results Fig. 11 shows gas velocity vectors at selected planes that cross-sect the furnace. The char bed shape is approximated so its impact on flow in the lower furnace can be evaluated with the model. Jets of air penetrate across the furnace to produce uniform upward flow and effective mixing with combustion gases. Three-dimensional computer-generated images can be examined interactively to help visualize air jet penetration and the interaction of jets from neighboring air ports. Gas temperature distribution predictions, shown in Fig. 12, are used to analyze heat transfer in the furnace and convection pass. Other information such as char bed surface temperature and burning rates, gas species concentrations (i.e., O_2, CO, NO_x), and wall heat flux distribution are also generated. Results are used by boiler designers and operators to evaluate air system designs, liquor spraying systems, liquor firing capacity, char bed combustion instabilities, convection pass fouling, furnace wall corrosion, and CO and NO_x emissions. The results shown were created by B&W's proprietary CFD software.

Wall-fired pulverized-coal boiler furnaces

Situation Within a staged, wall-fired furnace, the mixing between the upward-flowing partially-reacted fuel and the jets from the overfire air (OFA) ports is a complex, three-dimensional process. This mixing process can have a significant impact on the distribution and magnitude of CO emissions. While proprietary technology standards can initially be used to set effective OFA port arrangements for a staged combustion system, numerical modeling is often used to confirm this design and suggest alternatives to improve performance. Modeling is especially useful when there are physical obstructions that prevent OFA port placement in the optimal locations. In these circumstances, compromises must be made and determining the best available port layout may not be obvious.

Analysis In this example, a numerical model has been used to predict the steady-state flow, heat transfer, and combustion processes within a wall-fired pulverized-coal boiler being upgraded with low NO_x burners and OFA ports. As part of the design process, many

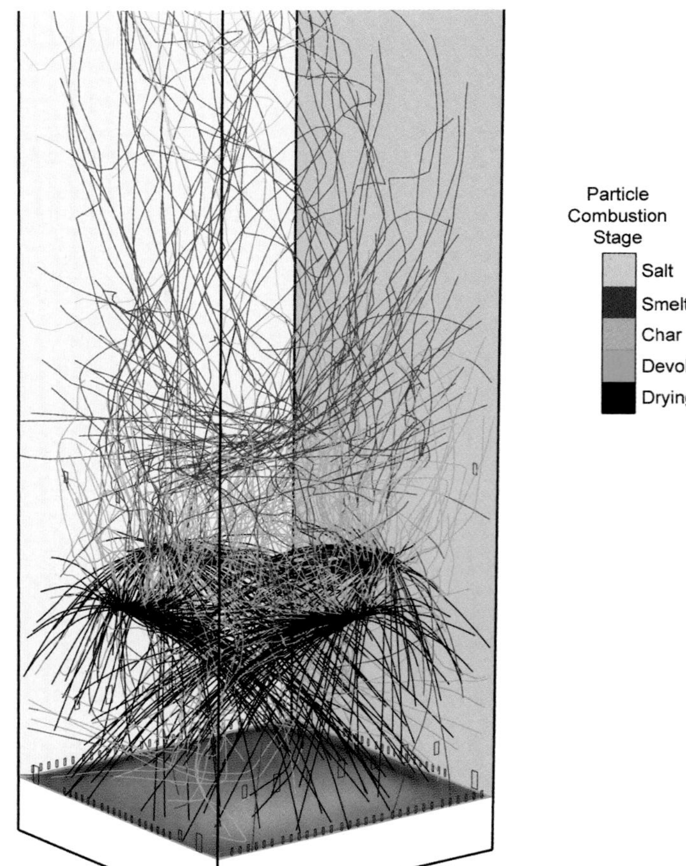

Fig. 10 Liquor spray distribution in the lower furnace of a recovery boiler.

configurations (number and location) of OFA ports were modeled, and the results were compared to determine the best port configuration. Boiler geometry, including a portion of the convection tube banks, was approximated using a collection of control volumes, also called a *computational grid* or *mesh,* for one of the configurations considered (see Fig. 13). Local refinement of the mesh was used as needed to better resolve the solution, such as within the OFA region. The coal analysis and boiler operating conditions including burner and OFA port settings were used to set inlet and boundary conditions for the model.

Results The model produces tabular (integrated species concentrations, gas temperatures, gas flow rates, emissions) and graphical (color contour plots of gas speed, gas temperature, or species; coal particle trajectories; gas streamlines) output that are used to evaluate each configuration. As an example, Fig. 14 compares contours of CO concentration throughout the boiler for two different OFA arrangements for a 775 MW wall-fired pulverized coal boiler. Arrangement 1 has the OFA ports directly above the burner openings and directly across from the ports on the opposing wall, while arrangement 2 uses horizontally offset ports which provide better mixing and cross-sectional coverage. As shown in the figure, OFA arrangement 2 results in lower CO concentrations in the upper furnace than the OFA arrangement 1 (15% lower at the arch, and 23% lower at the furnace exit). The results for this example were created by the B&W-developed computer software.

The numerical model described above also provides

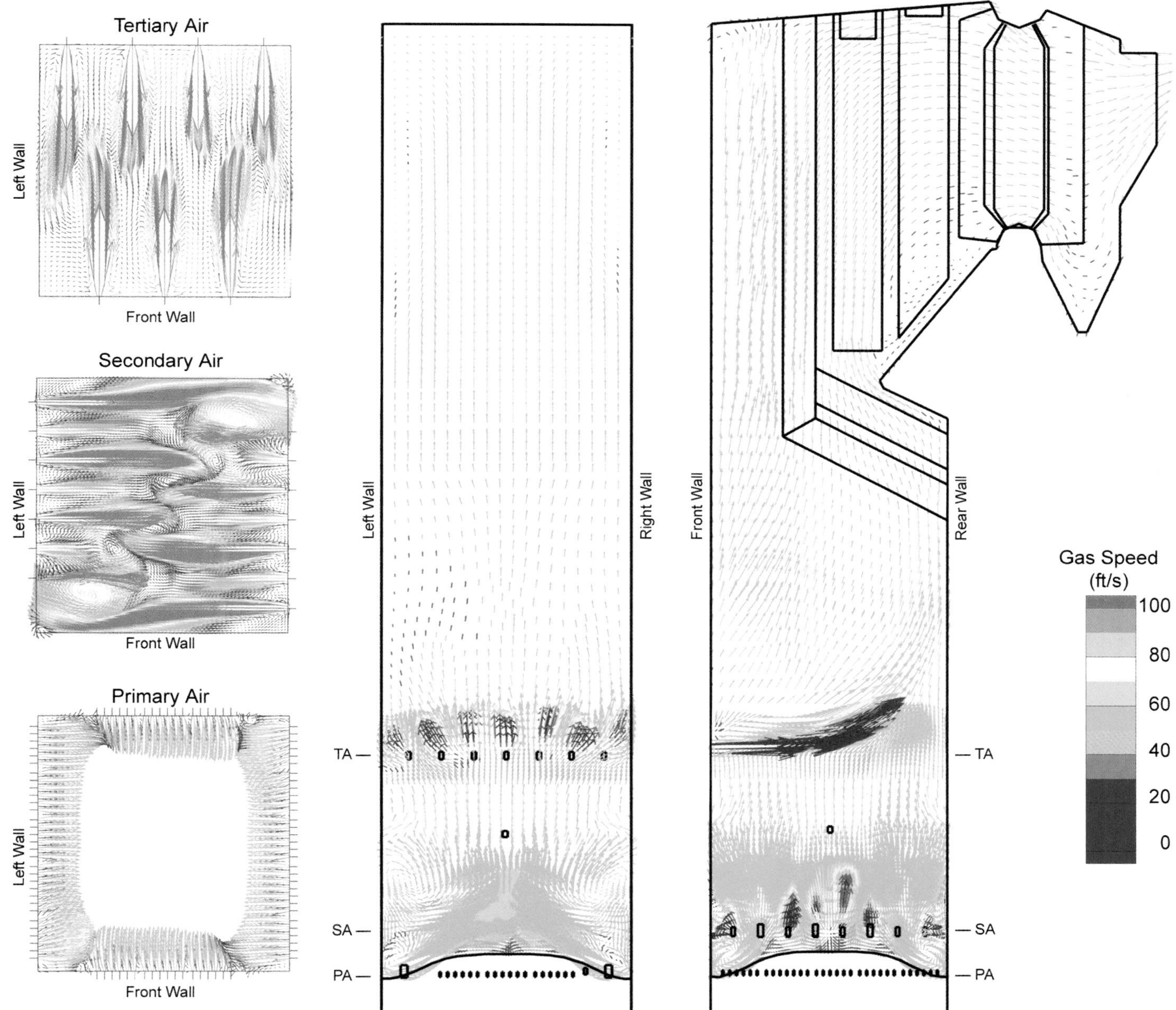

Fig. 11 Velocity vectors at selected planes that cross-sect a recovery furnace – horizontal planes at primary, secondary and tertiary levels (left); vertical planes at center of furnace (center and right).

a wealth of other information for the boiler designer. Fig. 15 provides a flue gas temperature profile through the center of the furnace and a horizontal section profile across the furnace exit gas pass. As noted in Chapter 19, the average or integrated furnace exit gas temperature (FEGT) is a critical design parameter in boiler sizing for performance while mitigating slagging and fouling. Flow areas with excessively high local temperatures identified by such numerical models may be more prone to slagging in the furnace or fouling in the convection pass. Additional parameters of interest provided by the numerical models include, but are not limited to, local velocity profiles for performance enhancement and erosion evaluation, furnace heat flux profiles for steam-water circulation evaluation,[29] variation in local chemical constituents such as oxygen for studying combustion optimization, and many others. Numerical boiler furnace models continue to evolve and more closely simulate field conditions. While current models as of this publication are not sufficient alone for final boiler design, they offer an additional tool to: 1) aid in design optimization, 2) address non-standard conditions, 3) evaluate the relative impact of fuel changes, 4) highlight areas for design improvement, 5) help investigate the root causes of unusual field observations, and 6) screen potential approaches to address design issues. Numerical modeling will become an increasingly important tool in boiler engineering.

Windbox

Situation The problems encountered in a windbox analysis deal with air flow imbalance and/or excess system pressure loss. Difficulties in tuning burner combustion performance can be frequently attributed to the flow distribution within the windbox. Therefore, creating a uniform flow distribution to each burner is highly desirable to obtain optimum emissions performance. The flow imbalance problem can be between the front and rear walls of a furnace, compartments in a

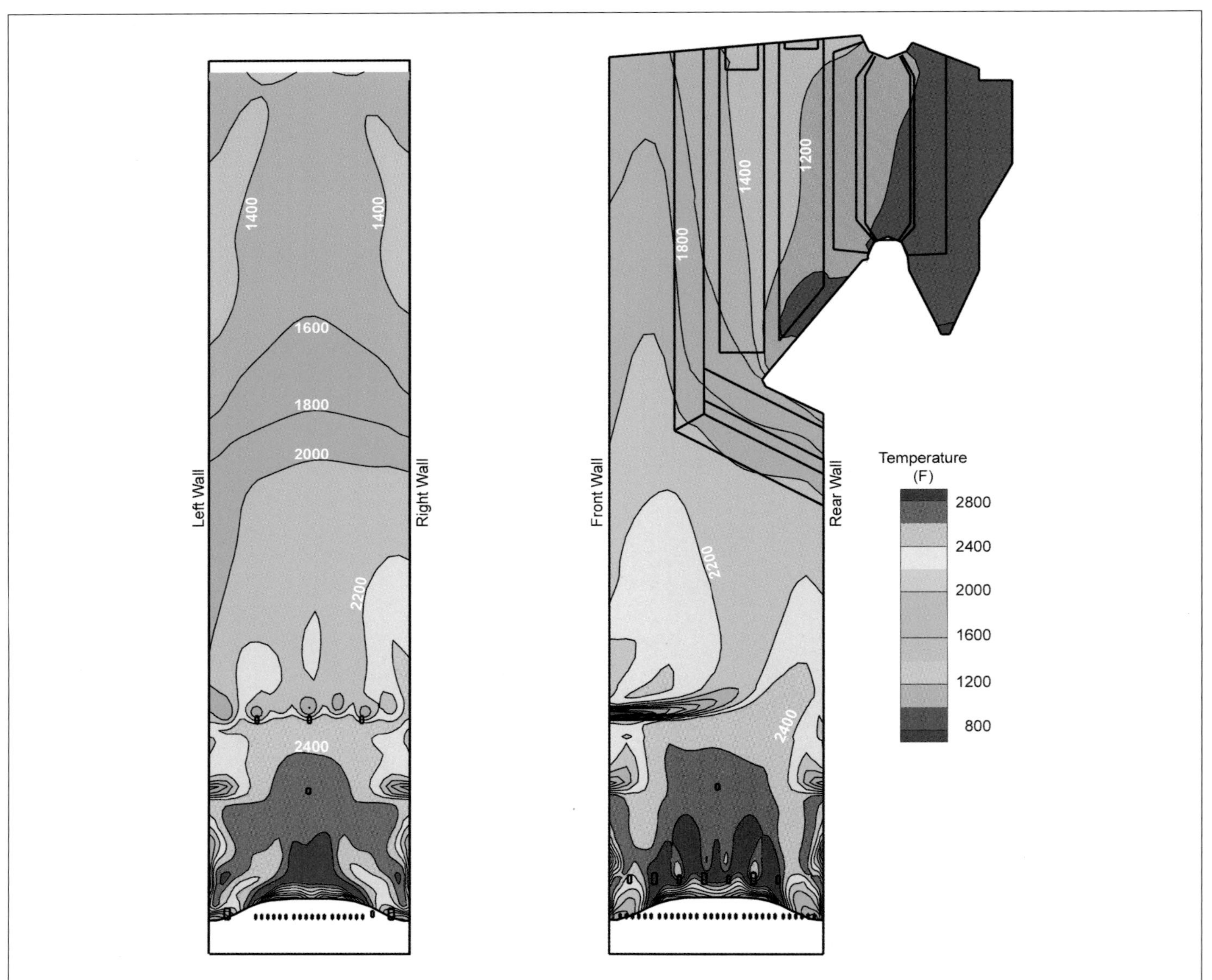

Fig. 12 Gas temperature contours at vertical planes at the center of a recovery boiler furnace.

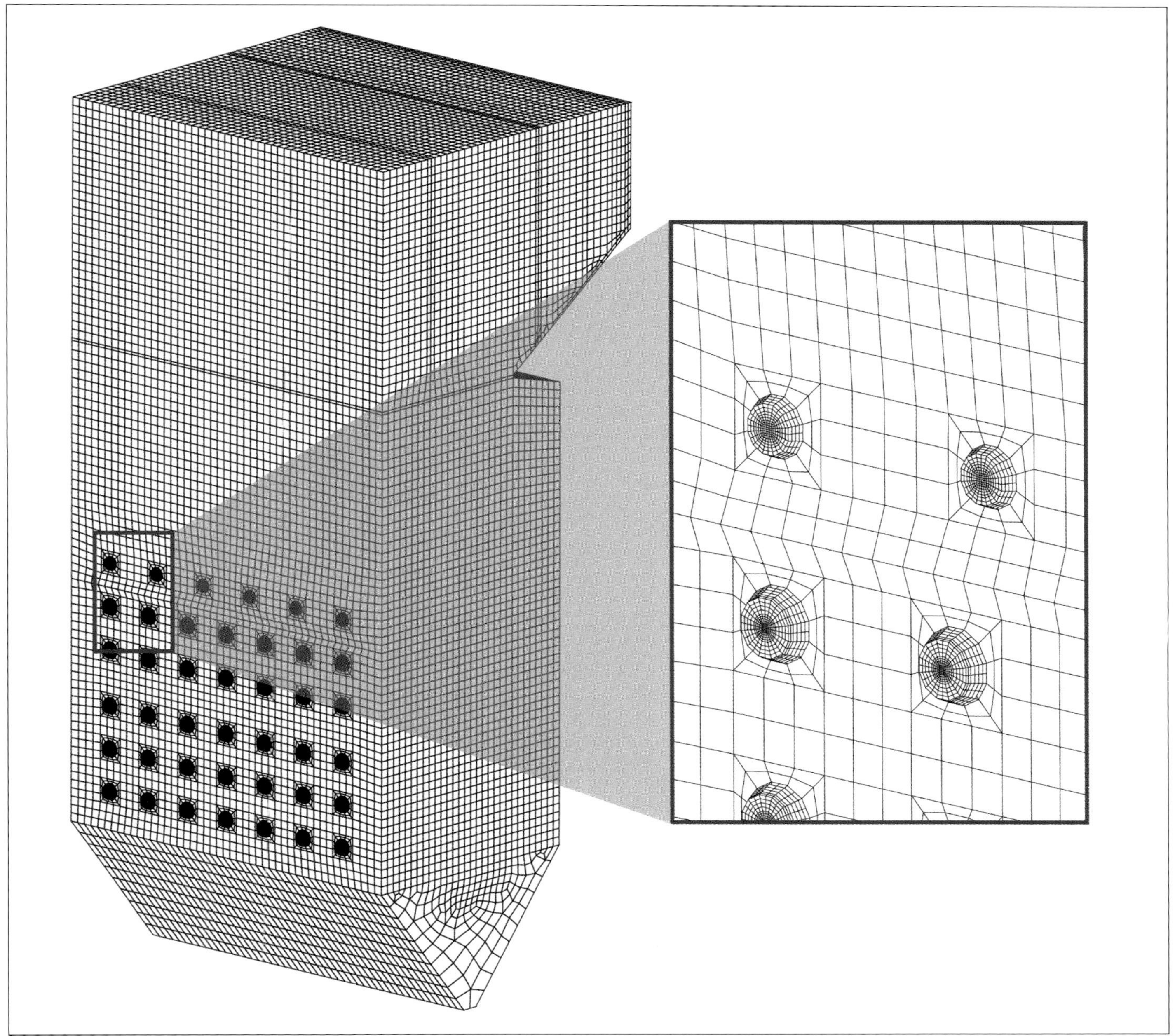

Fig. 13 Computational mesh on wall-fired boiler surface – full mesh (left) and enlarged view of upper burners and OFA ports (right).

windbox, or individual burners and/or ports. Any imbalance or maldistribution causes non-uniform air introduction into the furnace. This imbalance can lead to poor furnace combustion and potentially higher gas emissions. The system can be modeled to reduce air flow imbalance and reduce system pressure loss, which allows more flexibility in combustion tuning of a single burner.

Analysis A computer model that describes the details of the windbox (walls, bends, etc.) must first be built (see Fig. 16). This requires both flow and geometric design information. Care must be taken to ensure accurate representation of the entire air flow path including any significantly-sized internal obstructions. The inlet of the model is usually the outlet of the air heater. This is done for two reasons. An accurate and simple air flow distribution is usually known at this location, and it is far enough upstream to capture all the resulting flow disturbances. The burners and ports must also be modeled accurately to ensure precise flow results. Boundary conditions are the final and very important step, to be placed accurately in the model to exactly represent the windbox/duct flow conditions.

Results Once the model has been built, it is checked to make sure grid characteristics are acceptable. This step ensures that there is enough grid resolution to accurately represent the flow conditions in any area (i.e., turns, ducts, plenums) and around any objects (i.e., turning vanes, perforated plates, air foils). The model is then run using CFD software. These calculations yield an accurate representation of the air flowing in the space inside the ducts and windbox.

Fig. 17 shows the plan view of the secondary air

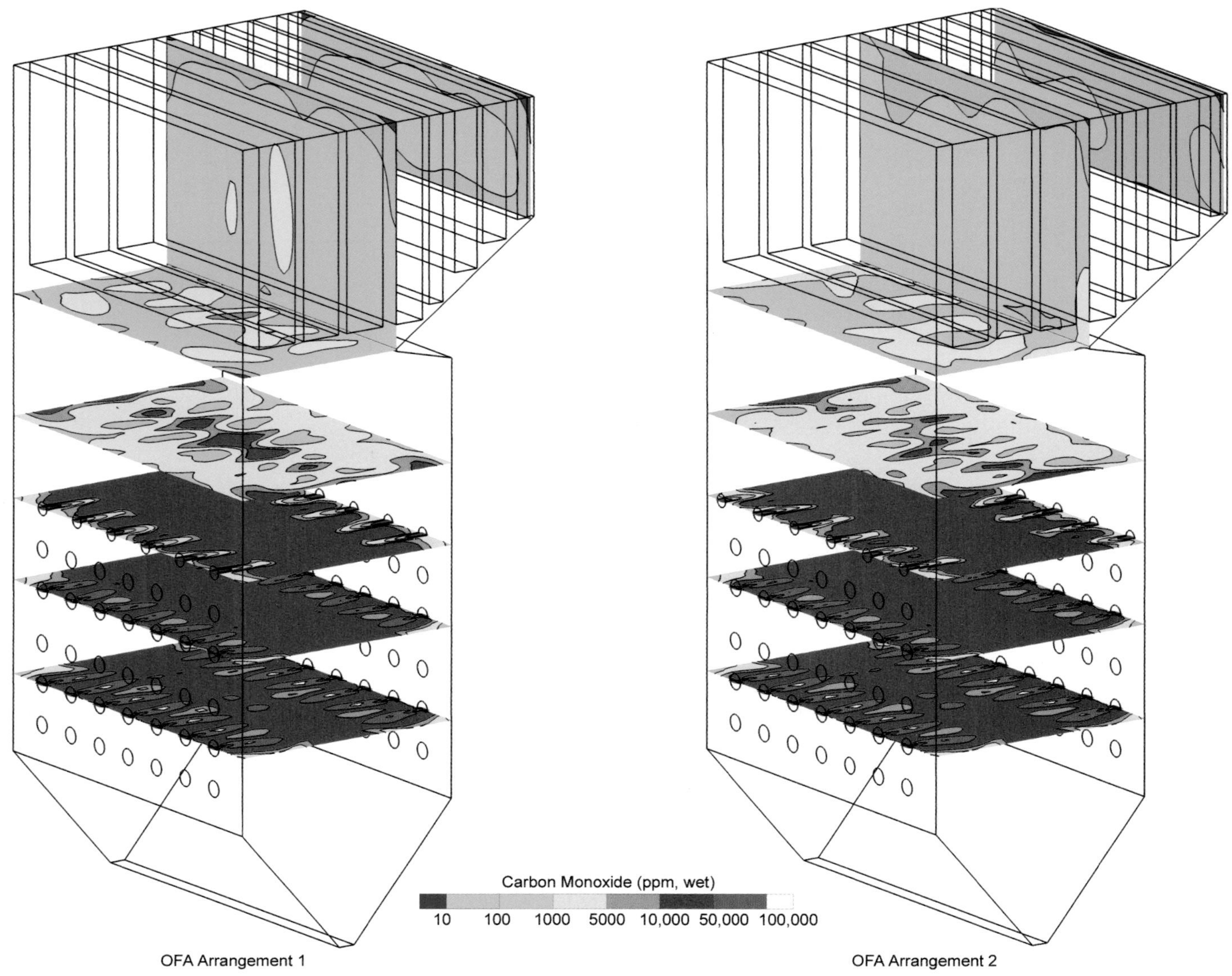

Fig. 14 Carbon monoxide concentration contours at various elevations – comparison between two OFA arrangements.

ducts and the windbox with the velocity-vector flow field at the middle of the duct system. Each arrow provides the direction and the magnitude (arrow length) of the local air flow. Fig. 18 is Section A-A through the windbox plan view of Fig. 17, looking into the furnace. Fig. 18a shows the original design which included a simple windbox with a large horizontal perforated plate intended to provide uniform flow to the bottom three burner rows. The numerical model results indicated a very high velocity zone (large red arrows) in the upper windbox which forced much of the air to bypass the upper burner row and over-supply the bottom two rows. The 30% flow variation between highest and lowest flow burners was too high and could lead to poor emissions performance and incomplete combustion. Several numerical modeling iterations using CFD computer software suggested the optimized solution shown in Fig. 18b. Eliminating the original large horizontal perforated plate plus adding two turning vanes, a vertical solid plate in the top of the windbox, and ten short vertical perforated plates dramatically improved the burner-to-burner flow distribution to within normal design tolerances.

The numerical model permitted testing of ten alternatives prior to selecting the low-cost solution which would also achieve the desired performance results.

SCR systems with economizer bypass

Situation A selective catalyst reduction (SCR) system with an economizer bypass is designed to reduce NO_x emissions by a chemical reaction between NO_x and added ammonia in the presence of a catalyst. (See Chapter 34.) To optimize the chemical reaction at low and intermediate loads, an economizer bypass is needed to increase the temperature of the economizer exit flue gas. The ammonia injection grid (AIG) distributes ammonia uniformly into the exit gas for the correct molar ratio of ammonia to NO_x. Finally, the catalyst is used to aid in the chemical reaction.

CFD modeling of the SCR system includes full-scale representation, multiple temperature gas paths, heat absorption modeling capability, multi-point testing

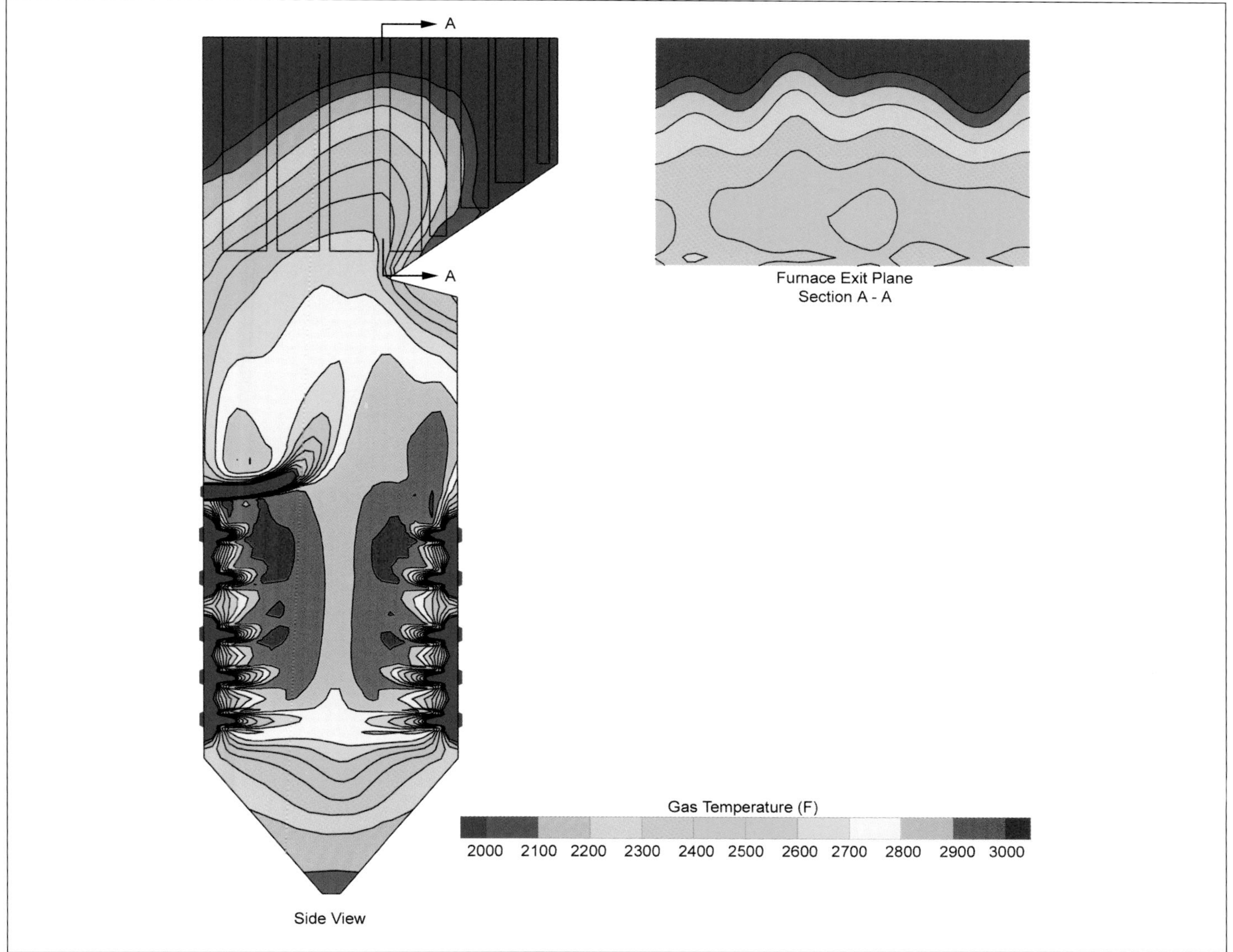

Fig. 15 Numerical modeling results of furnace temperature profiles for a typical 775 MW bituminous coal-fired boiler.

over an entire grid plane or discrete point testing, chemical species tracking, and rapid flow device testing for proper mixing, flow distribution, and minimal pressure drop.

Analysis The numerical model is constructed and tuned for actual conditions of an unmodified system according to existing economizer exit flue gas conditions. Drawings of future construction and design of the flue work, AIG, and information supplied by the catalyst vendor are used to establish a base operating condition. The chemical species are tracked for accurate mixing of ammonia and NO_x reagents. The data collected at specific planes in the grid are evaluated against established criteria for efficient NO_x removal such as velocity distribution, ammonia-to-NO_x ratio, and average temperature entering the catalyst. Internal corrective devices such as turning vanes, flow distribution diverters, static mixers and porous plates, are used to precondition the flue gas to meet the criteria for NO_x reduction. Grid refinement may be necessary to accurately predict the physical characteristics of internal objects, flue bends and flow distribution devices.

Results One such design involves a unit operating at three loads with an economizer bypass taken off the reheat side of the back wall convection pass to achieve adequate remix temperatures for the chemical reaction. For bypass operation, three gas paths are considered in the design process: superheat, reheat, and economizer bypass. Because of physical constraints and potential changes in the economizer outlet temperature with reheater or superheater bypass arrangements, a bypass around the economizer surface was selected. Fig. 19 shows the velocity flow field for the numerical evaluation from the superheater through the exit of the SCR. Fig. 20 shows the detailed velocity field and physical geometry at the bypass location. A key issue was the complete mixing of the high temperature bypass flow with the main flue gas flow exiting the economizer in order to provide an acceptably uniform flue gas temperature entering the SCR catalyst. To achieve the desired mixing, a series of turning vanes and mixing devices for the economizer

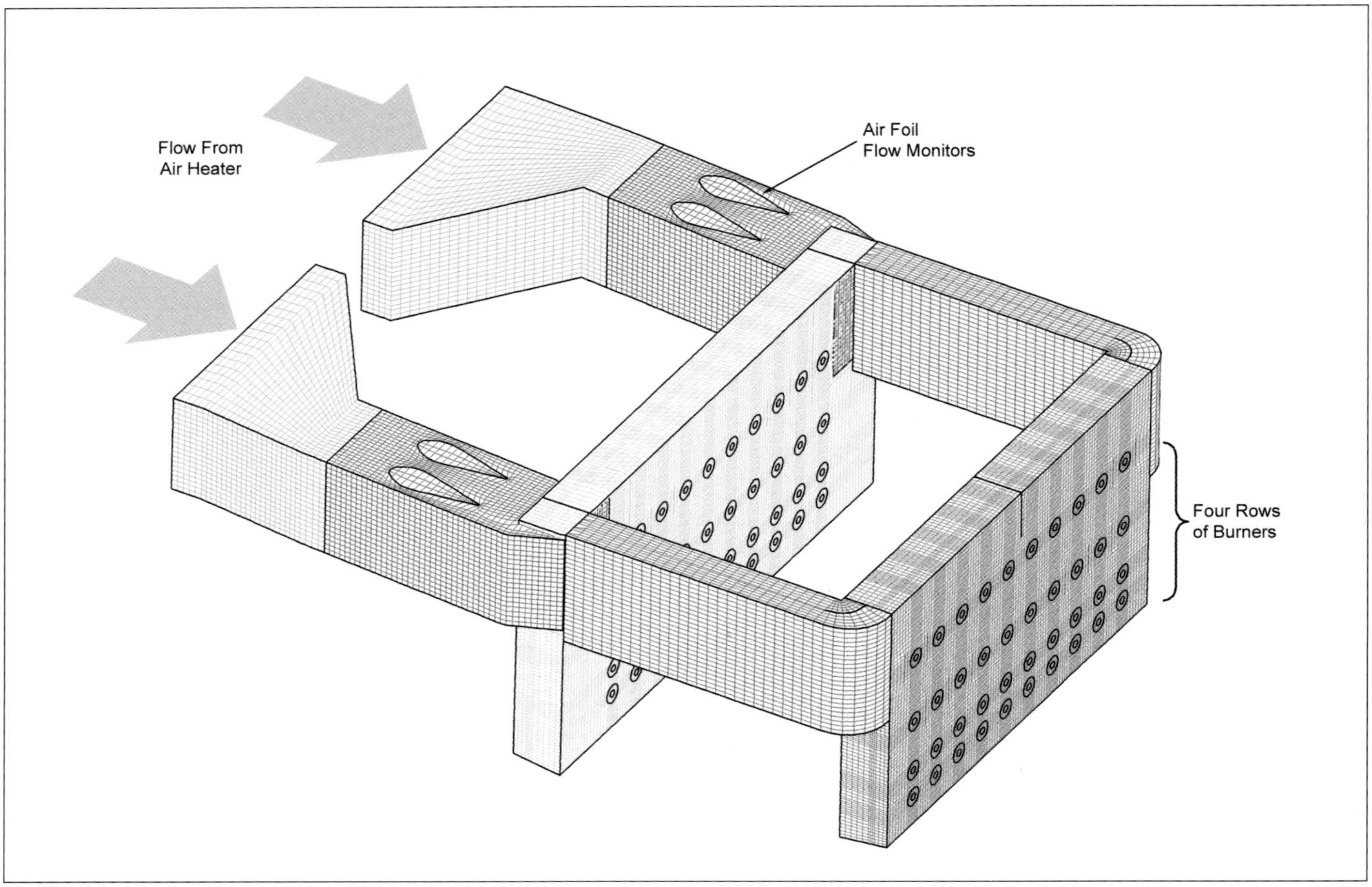

Fig. 16 Numerical model of 1100 MW coal-fired boiler windbox and secondary air system.

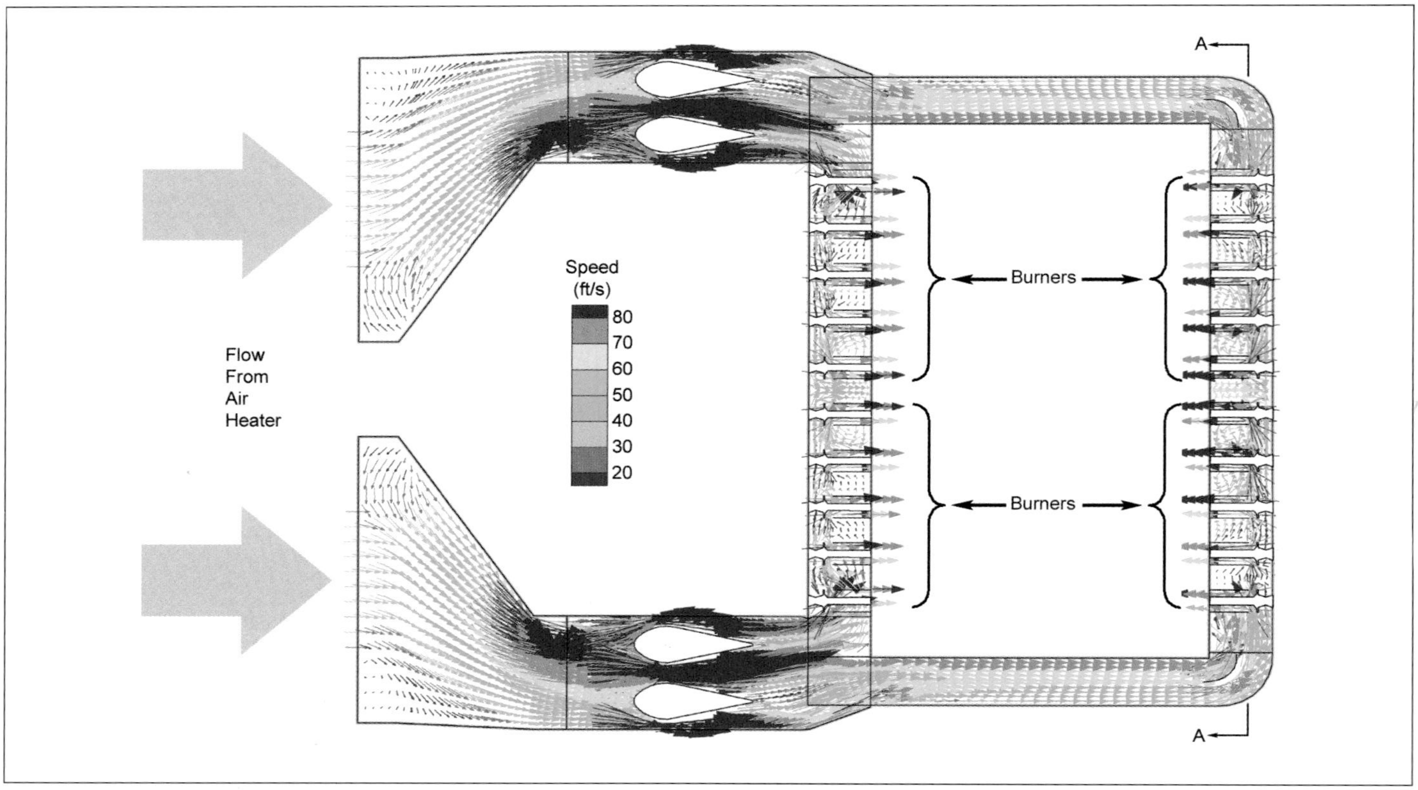

Fig. 17 Plan view of secondary air system flow model results – velocity vectors.

outlet hopper were developed with the aid of the numerical model to provide the high velocity bypass jet that would adequately penetrate the main economizer outlet flue gas flow. Success in the design iteration process was achieved when the velocity profile entering the AIG and the temperature profile entering the SCR achieved the specified uniformity.

Waste-to-energy systems

Situation Effective combustion of municipal solid waste (MSW) and biomass fuels has become more challenging over time as emissions regulations have been tightened and the variation of fuel characteristics has increased. As part of the process to meet these more demanding requirements, numerical modeling has become a routine engineering tool used in the installation of emissions control systems such as selective non-catalytic reduction (SNCR) NO_x control systems (see Chapter 34), refinement of the design and operation of the stoker/grate combustion system with auxiliary burners (Chapter 16) and overall design of the boiler (Chapters 29 and 30). Control of the flue gas in the furnace in terms of chemical species, particles, temperature and flow is important where a *good* furnace design results in more uniform velocity profiles. High velocity regions can cause: 1) increased wall deterioration from the hot corrosive flue gas with premature component replacement, 2) sub-optimal emissions control without adequate residence time at temperature, and 3) incomplete burnout of the fuel.

Analysis A numerical evaluation of the furnace was conducted as part of the design of a 132 ton per day (120 t_m/d) mass burn MSW stoker-fired system. Fig. 21 shows the sectional side view of a European waste-to-energy plant design supplied by B&W. A complete flow field evaluation of the furnace design using numerical modeling was conducted to determine the physical furnace modifications necessary to minimize high velocity areas.

Results Fig. 22 shows the numerically evaluated velocity vector flow field before (a) and after (b) the design changes. The flow field is represented by arrows that show the local velocity direction and magnitude (arrow length). In Fig. 22a, a high velocity jet region impinges on the top of the grate, and high velocity regions exist along the first (up) and second (down) pass furnace walls. The addition of *noses* at the bottom of the first pass and the top of the second pass walls as shown in Fig. 22b significantly reduce the velocities throughout the furnace and reduce the peak velocity regions near the grate and along the furnace walls. The more moderate velocity in the first pass results in less particle impingement and longer overall residence time. The maximum velocity in the second pass is reduced from 13 m/s to 9 m/s (42.6 to 29.5 ft/s), which reduced the thermal load on the back wall of the second pass.

Advanced burner development

Situation Advanced burner and combustion system development are increasingly relying on the use of numerical modeling as an integral tool in the quest for new hardware and concepts to improve the NO_x reduction performance of coal-fired burners. While traditional experimental methods of burner development have been able to dramatically reduce NO_x emission levels from bituminous pulverized coal burners below 0.4 lb/10^6 Btu (492 mg/Nm^3), increasingly more stringent emission reduction regulations are pushing specified combustion emission limits to well below 0.15 lb/10^6 Btu (184 mg/Nm^3). To develop such equipment, it is becoming even more necessary to understand not only what is happening at the macro-level (which can be observed and tested) but also with small-scale interactions deep within the flame and initial ignition zone. Numerical modeling studies of detailed burner designs offer a valuable tool by combining fundamental knowledge of combustion with complex fluid and thermal dynamics to better understand how to further reduce NO_x emissions and improve combustion efficiency. When combined with small-scale and large-scale tests with advanced test instrumentation, nu-

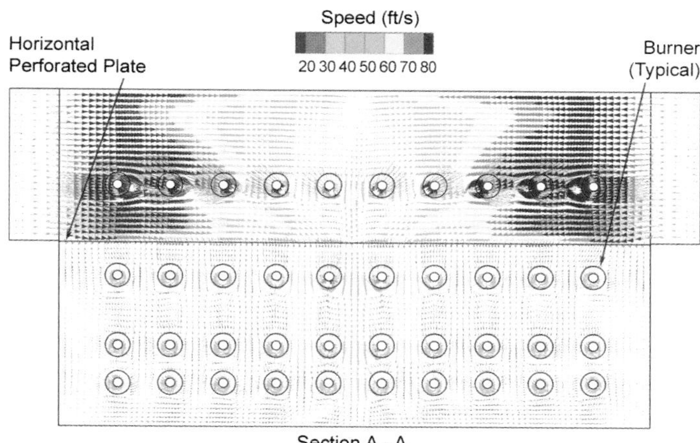

Fig. 18a Before – high velocity zone in upper windbox (red arrows) under-supplies top row of burners and over-supplies other rows.

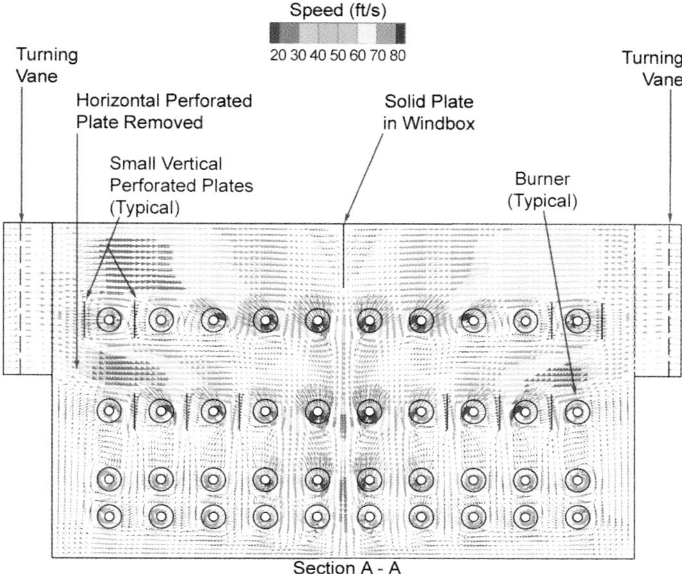

Fig. 18b After – removal of the large horizontal perforated plate plus the addition of two turning vanes and 10 small vertical perforated plates provides more uniform flow to the burners.

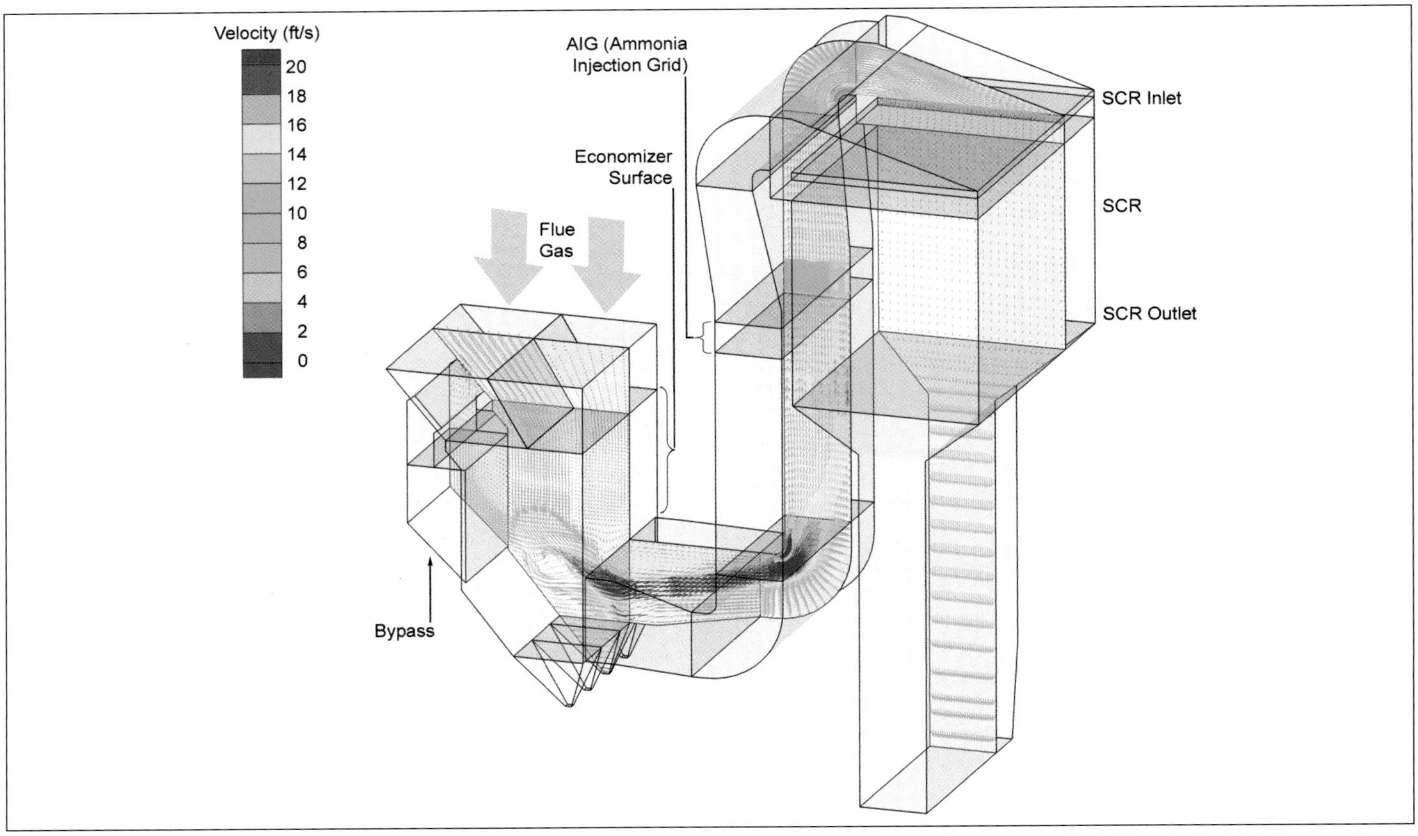

Fig. 19 Velocity field numerical model output – SCR system from the boiler convection pass to SCR outlet. The high velocity, high temperature bypass flow is visible as high jet penetration is needed to achieve good thermal mixing by the SCR inlet. See also Fig. 20.

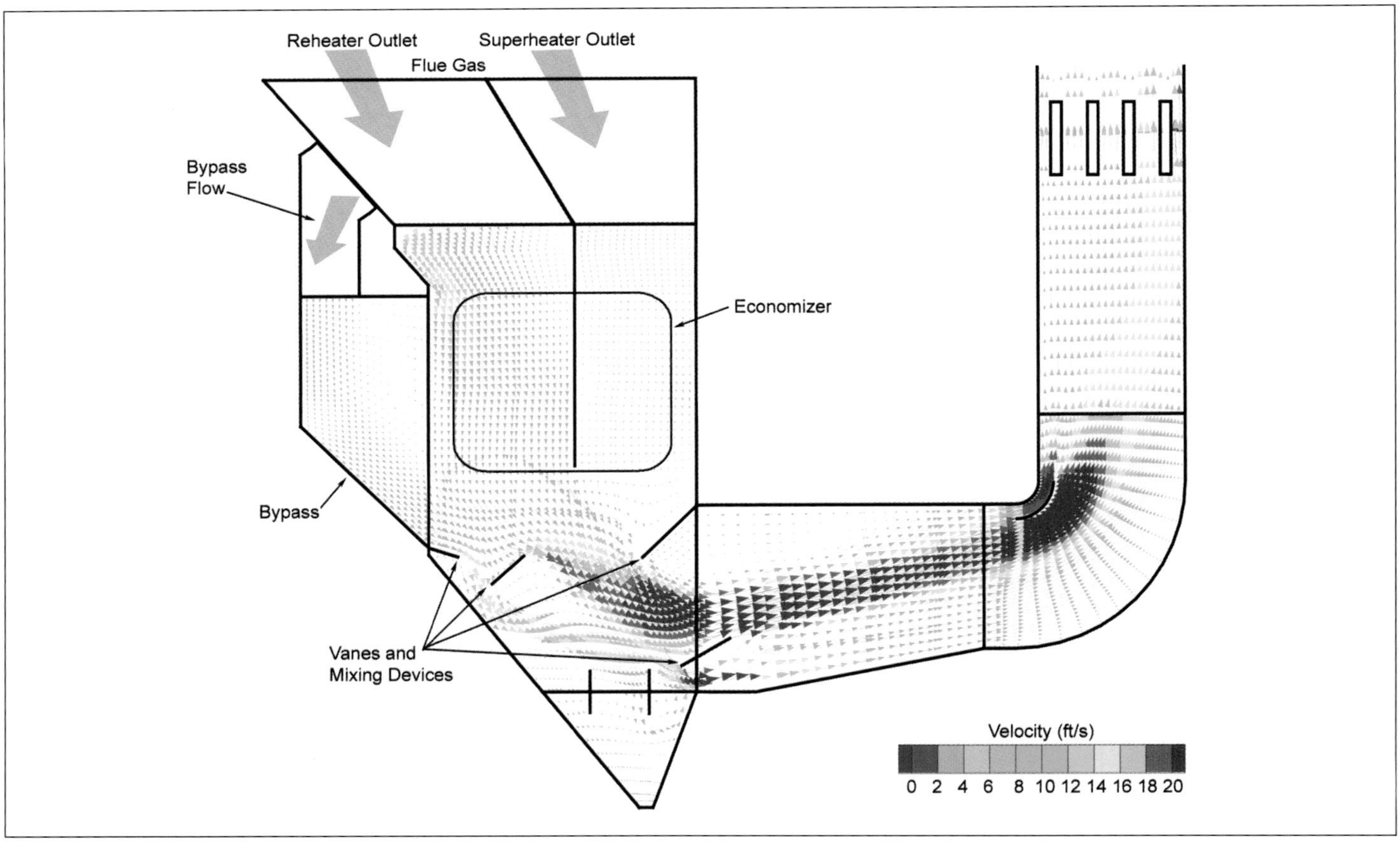

Fig. 20 SCR detail at the bypass flue location – turning vanes and mixing devices provide adequate bypass flow penetration for optimal mixing.

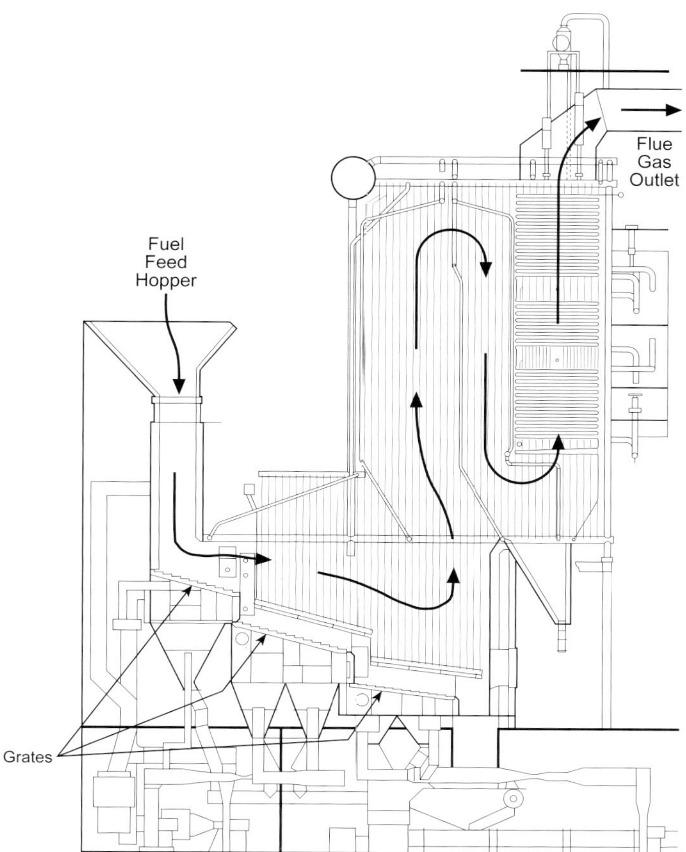

Fig. 21 Sectional view of a 132 t/d (120 t_m/d) mass burn municipal solid waste (MSW) boiler for European application.

merical modeling is helping identify techniques to burn fuels more cleanly.

Analysis Numerical models, with enhanced resolution (see Fig. 23), have been developed that accurately represent the critical details of physical burners. Experimental studies provide inlet flow boundary conditions that can offer the starting point for the analysis. These are combined with the results from fundamental studies of fuel devolatilization, burning of gaseous species, gaseous diffusion, combustion of solid material and other factors to develop numerical models that begin to simulate the complex combustion process in commercial coal-fired burners. Physical testing, validation and adjustments to the model can produce numerical tools that can be used for advanced burner development.

Results Fig. 24 shows detailed gas velocity fields for an advanced burner design. In this case, analysis of the numerical model predictions helped identify the value of an additional burner air supply zone to induce recirculation of nitrogen oxide (NO) formed in the outer oxygen-rich portions of the flame into the fuel-rich internal recirculation zone where NO is reduced. See Chapter 14 for further discussion of coal-fired burners and combustion systems.

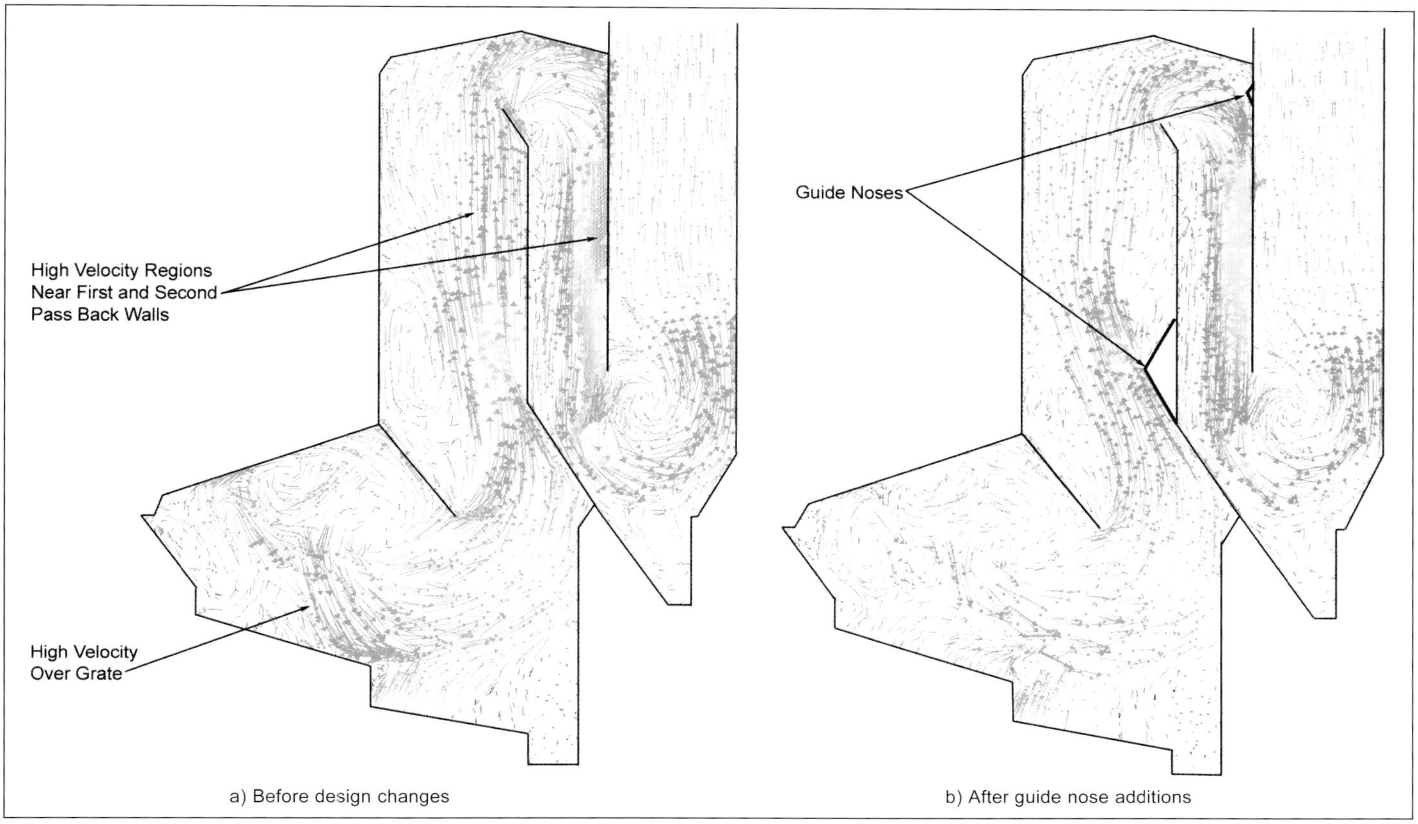

a) Before design changes

b) After guide nose additions

Fig. 22 MSW boiler from Fig. 21 showing flow field before and after the addition of guide noses in the furnace wall.

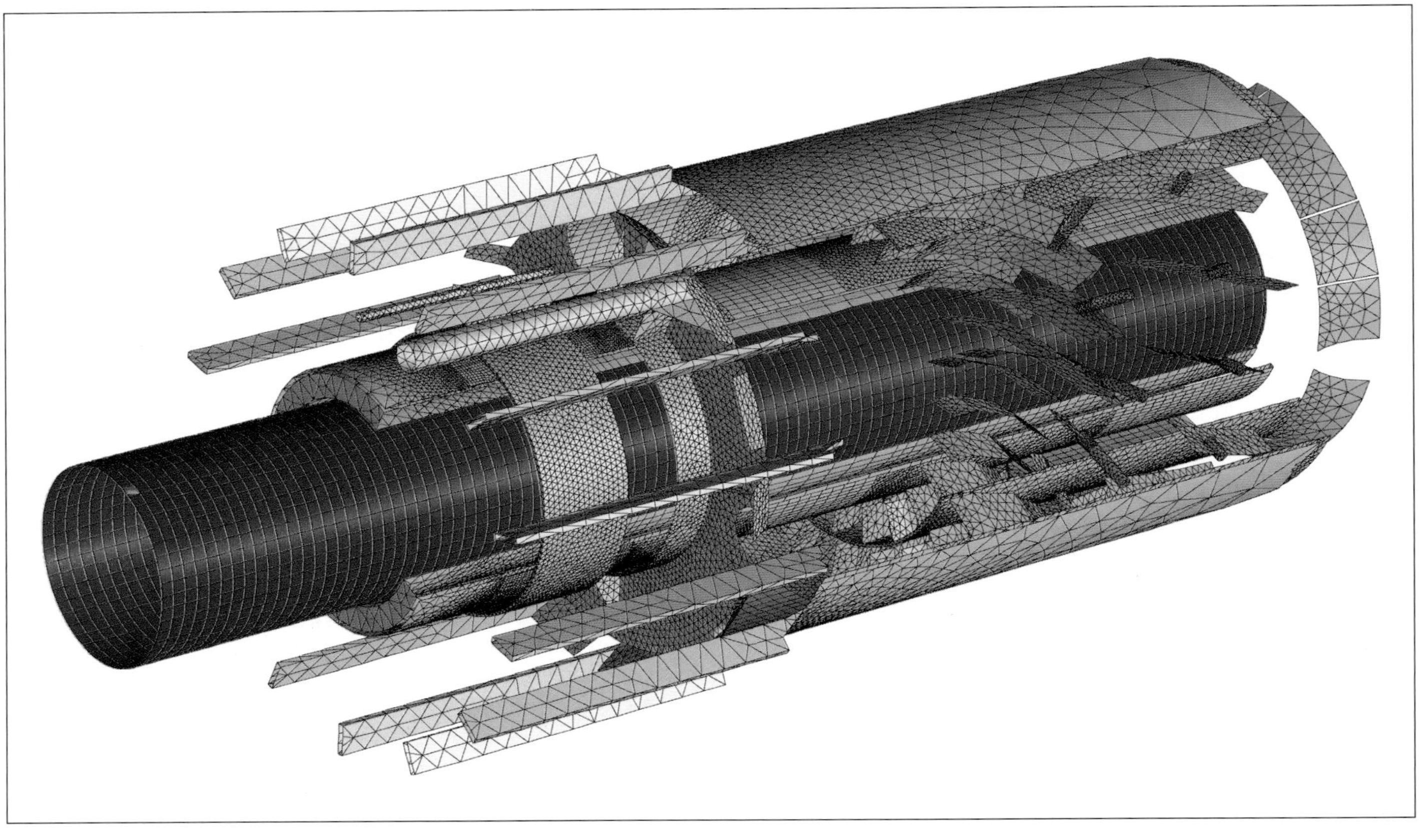

Fig. 23 Detailed numerical model evaluation grid for an advanced coal burner.

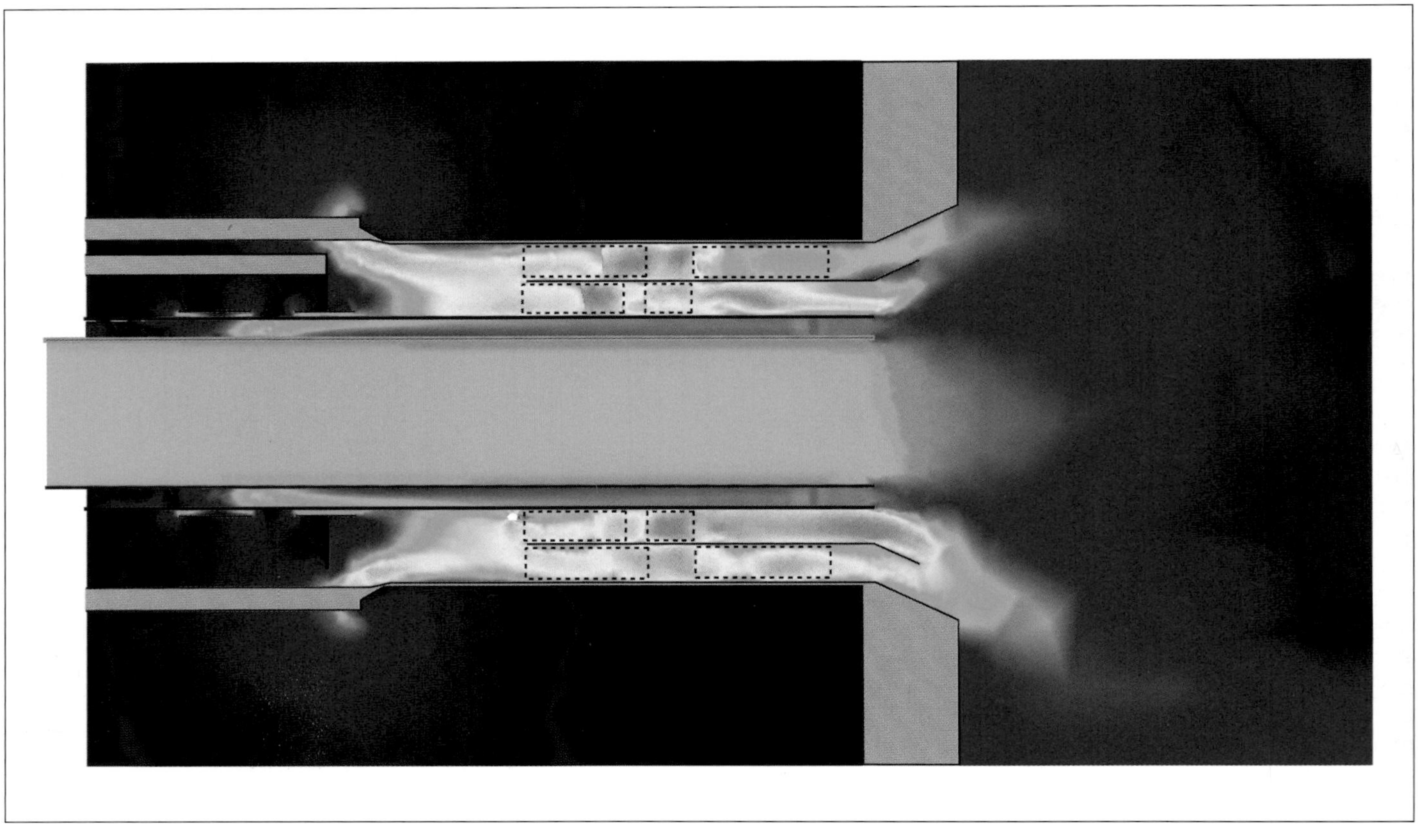

Fig. 24 Gas velocity model for the coal burner shown in Fig. 23.

References

1. Wallis, G.B., *One-Dimensional Two-Phase Fundamental*, McGraw-Hill Company, New York, New York, 1969.

2. Crowe, C.T., Smoot, L.D., Pratt, D.T., Eds., "Gas Particle Flow," *Pulverized Coal Combustion and Gasification*, Plenum Press, New York, New York, 1979.

3. Bailey, G.H., Slater, I.W., Eisenblam, P., "Dynamics Equations and Solutions for Particles Undergoing Mass Transfer," *British Chemical Engineering*, Vol. 15, p. 912, 1970.

4. Milojevic, D., "Lagrangian Stochastic-Deterministic (LSD) Predictions of Particle Dispersion in Turbulence," *Journal of Particles and Particle Systems Characterization*, Vol. 7, pp. 181-190, 1990.

5. Magnussen, B. F., and Hjertager, B. H., "On mathematical modeling of turbulent combustion with emphasis on soot formation and combustion," *Proceedings of the 16th International Symposium on Combustion, 719-729*, The Combustion Institute, Pittsburgh, Pennsylvania, 1976.

6. Spalding, D. B., "Mixing and Chemical Reaction in Steady Confined Turbulent Flames," *Proceedings of the 13th International Symposium on Combustion*. The Combustion Institute, Pittsburgh, Pennsylvania, 1971.

7. Magnussen, B. F., "On the structure of turbulence and the generalized Eddy Dissipation Concept for turbulent reactive flows," *Proceedings of the 19th American Institute of Aeronautics and Astronautics Aerospace Science Meeting*, St. Louis, Missouri, 1981.

8. Lilleheie, N. I., Ertesvåg, I., Bjosrge, T., et al., "Modeling and Chemical Reactions," SINTEF Report STF1s-A89024, 1989.

9. Magnussen, B. F., "The Eddy Dissipation Concept," XI Task Leaders Meeting: Energy Conservation in Combustion, IEA, 1989.

10. Lilleheie, N. I., Byggstøyl, B., Magnussen, B. F., et al., "Modeling Natural Gas Turbulent Jet Diffusion Flames with Full and Reduced Chemistry," *Proceedings from the 1992 International Gas Research Conference*, Orlando, Florida, November 2-5, 1992.

11. Ubhayakar, S.K., Stickler, D.B., Von Rosenburg, C.W., et al., "Rapid Devolatilization of Pulverized Coal in Hot Combustion Gases," *16th International Symposium on Combustion*, The Combustion Institute, Pittsburgh, Pennsylvania, 1975.

12. Grant, D. M., Pugmire, R. J., Fletcher, T. H., et al., "A Chemical Model of Coal Devolatilization Using Percolation Lattice Statistics," *Energy and Fuels*, Vol. 3, p. 175, 1989.

13. Fletcher, T. H., Kerstein, A. R., Pugmire, R. J., et al., "A Chemical Percolation Model for Devolatilization: Milestone Report," Sandia report SAND92-8207, available National Technical Information Service, May, 1992.

14. Perry, S., "A Global Free-Radical Mechanism for Nitrogen Release During Devolatilization Based on Coal Chemical Structure," Ph.D. dissertation for the Department of Chemical Engineering, Brigham Young University, Provo, Utah, United States, 1999.

15. Field, M.A., Grill, D.W., Morgan, B.B., et al., *Combustion of Pulverized Coal*, The British Coal Utilization Research Association, Leatherhead, Surrey, England, United Kingdom, 1967.

16. Fiveland, W.A., Jamaluddin, A.S., "An Efficient Method for Predicting Unburned Carbon in Boilers," *Combustion Science and Technology*, Vol. 81, pp. 147-167, 1992.

17. Hurt, R., Sun, J.K., Lunden, L., "A Kinetic Model of Carbon Burnout in Pulverized Coal Combustion," *Combustion and Flame*, Vol. 113, pp. 181-197, 1998.

18. Patakanar, S., *Numerical Heat Transfer and Fluid Flow*, Hemisphere Publishing Corporation, New York, New York, 1980.

19. Brewster, M. Q., *Thermal Radiative Transfer and Properties*, John Wiley & Sons, Inc. New York, New York, 1992.

20. Modest, M. F., *Radiative Heat Transfer*, McGraw-Hill, Inc., New York, New York, 1993.

21. Viskanta, R., "Overview of computational radiation transfer methods for combustion systems," *Proceedings of the Third International Conference on Computational Heat and Mass Transfer*, Banff, Alberta, Canada, 2003.

22. Fiveland, W. A., "Discrete-ordinates solutions of the radiative transport equations for rectangular enclosures," *Transactions of American Society of Mechanical Engineers Journal of Heat Transfer*, 106, pp. 699-706, 1984.

23. Jessee, J.P. and Fiveland, W.A., "Bounded, High-Resolution Differencing Schemes Applied to the Discrete Ordinates Method," *Journal of Thermophysics and Heat Transfer*, Vol.11, No. 4. October-December, 1997.

24. Thompson, J.F., Soni, B., Weatherhill, N., Eds., *Handbook of Grid Generation*, CRC Press, New York, New York, 1999.

25. Dudek, S.A., Rodgers, J.A. and Gohara, W.F., "Computational Fluid Dynamics (CFD) Model for Predicting Two-Phase Flow in a Flue-Gas-Desulfurization Wet Scrubber," EPRI-DOE-EPA Combined Utility Air Pollution Control Symposium, Atlanta, Georgia, United States, August 16-20, 1999 (BR-1688).

26. Ryan, A. and St. John B., "SCR System Design Considerations for 'Popcorn' Ash," EPRI-DOE-EPA-AWMA Combined Power Plant Air Pollutant Control Mega Symposium, Washington, D.C., May 19-22, 2003 (BR-1741).

27. Verrill, C.L., Wessel, R.A., "Detailed Black Liquor Drop Combustion Model for Predicting Fume in Kraft Recovery Boilers," *TAPPI Journal*, 81(9):139, 1998.

28. Wessel, R.A., Parker, K.L., Verrill, C.L., "Three-Dimensional Kraft Recovery Furnace Model: Implementation and Results of Improved Black Liquor Combustion Models," *TAPPI Journal*, 80(10):207, 1997.

29. Albrecht, M.J., "Enhancing the Circulation Analysis of a Recovery Boiler through the Incorporation of 3-D Furnace Heat Transfer Results from COMO™," TAPPI Fall Technical Conference, San Diego, California, September 8-12, 2002.

Application of protective coating to boiler tubes.

Chapter 7

Metallurgy, Materials and Mechanical Properties

Boilers, pressure vessels and their associated components are primarily made of metals. Most of these are various types of steels. Less common, but still important, are cast irons and nickel base alloys. Finally, ceramics and refractories, coatings, and engineered combinations are used in special applications.

Metallurgy

Crystal structure

The smallest unit of a metal is its atom. In solid structures, the atoms of metals follow an orderly arrangement, called a *lattice*. An example of a *simple point lattice* is shown in Fig. 1a and the unit cell is emphasized. The lengths of the unit cell axes are defined by a, b and c, and the angles between them are defined by α, β and γ in Fig. 1b. The steels used in boilers and pressure vessels are mainly limited to two different lattice types: *body-centered cubic* (BCC) and *face-centered cubic* (FCC). (See Fig. 2.) Where changes in the structure or interruptions occur within a crystal, these are referred to as *defects*. *Crystal* (or *grain*) *boundaries* are a type of crystal defect. A few useful structures are composed of a single crystal in which all the unit cells have the same relationship to one another and have few defects. Some high performance jet engine turbine blades have been made of single crystals. These structures are difficult to make, but are worthwhile; their strength is very high as it is determined by close interactions of the atomic bonds in their optimum arrangement. The behavior of all other metallic structures, which make up the huge majority of engineering metallic materials, is determined by the nature and extent of the defects in their structures. Structures are made of imperfect assemblies of imperfect crystals, and their strengths are orders of magnitude lower than the theoretical strengths of perfect single crystals.

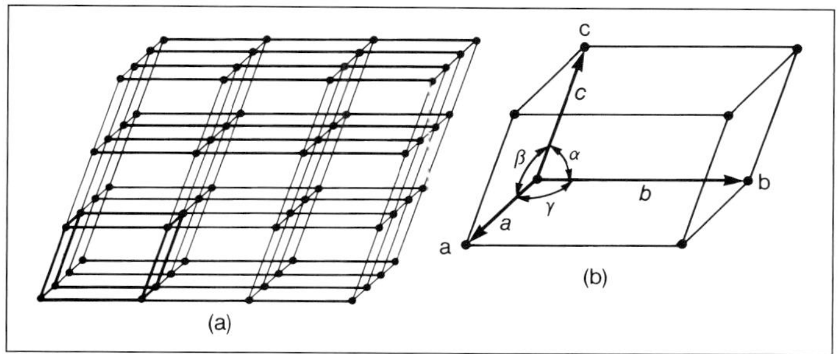

Fig. 1 Simple point lattice and unit cell (*courtesy of Addison-Wesley*).[1]

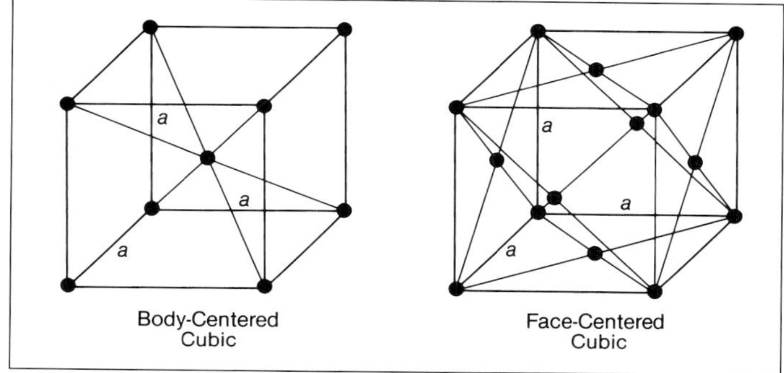

Fig. 2 Two Bravais lattices.[1]

Defects in crystals

Perfect crystals do not exist in nature. The imperfections found in metal crystals and their interactions control their material properties.

Point defects Point defects include missing atoms (vacancies), atoms of a different element occurring on crystal lattice points (substitutionals), and atoms of a different element occurring in the spaces between crystal lattice points (interstitials). Thermally created vacancies are always present, because they reduce the free energy of the crystal structure by raising its entropy. There is an equilibrium number of vacancies present; this number varies with the temperature of the crystal. The presence of such vacancies permits *diffusion* (the transport of one species of metal atom through the lattice of another) and helps facilitate some forms of time dependent deformation, such as *creep*.

Vacancies can also be created by irradiation damage and plastic deformations, and the thermodynamically controlled processes of diffusion and creep can also be affected by these other processes.

When atoms of two metals are mixed in the molten state and then cooled to solidification, the atoms of one metal may take positions in the lattice of the other, forming a substitutional *alloy*. Because the atoms may be different sizes and because the bond strength between unlike atoms is different from that of like atoms, the properties of the alloy can be quite different from those of either pure metal.

Atoms of carbon, oxygen, nitrogen and boron are much smaller than metal atoms and they can fit in the spaces, or interstices, between the metal atoms in the lattice structure. The diffusion of an interstitial in a metal lattice is also affected by temperature, and is much more rapid at higher temperatures. Interstitial elements are often only partly soluble in metal lattices. Certain atoms such as carbon in iron are nearly insoluble, so their presence in a lattice produces major effects.

Several crystal defects are illustrated in Fig. 3. This is a two-dimensional schematic of a cubic iron lattice containing point defects (vacancies, substitutional foreign atoms, interstitial atoms), linear and planar defects (dislocations, sub-boundaries, grain boundaries), and volume defects (voids, and inclusions or precipitates of a totally different structure).[2] *Dislocations* are linear defects formed by a deformation process called *slip*, the sliding of two close-packed crystal structure planes over one another.

Grain boundaries Grain boundaries are more complex interfaces between crystals (grains) of significantly different orientations in a metal. They are arrays of dislocations between misoriented crystals. Because the atomic bonds at grain boundaries and at other planar crystal defects are different from those in the body of the more perfect crystal, they react differently to heat and chemical reagents. This difference appears as grain boundaries on polished and etched metal surfaces under a microscope. Grain size can have positive or negative effects on metal properties. At lower temperatures, a steel with very small grains (fine grain size) may be stronger than the same steel with fewer large grains (coarse grain size) because the grain boundaries act as barriers to deformation due to slip. At higher temperatures, where thermally activated deformation such as creep can occur, a fine grain structure material may be weaker because the irregular structure at the grain boundaries promotes local creep due to a mechanism known as grain boundary sliding.

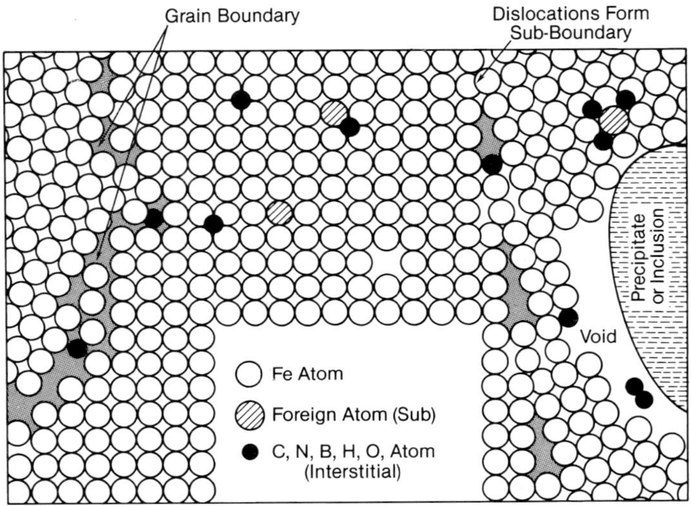

Fig. 3 Some important defects and defect complexes in metals (*courtesy of Wiley*).[3]

Volume defects Volume defects can be voids formed by coalescence of vacancies or separation of grain boundaries. More common volume defects are inclusions of oxides, sulfides and other compounds, or other phases that form during solidification from the molten state.

Physical metallurgy of steel

Phases A phase is a homogeneous body of matter existing in a prescribed physical form. Metallurgists use a graph, called a *phase diagram*, to plot the stable phases at temperature versus composition of any metal composed of two or more elements.

When more than one element is involved, even for binary alloys, a variety of phases can result. One type is the binary isomorphous system, typified by only a few combinations: copper-nickel (Cu-Ni), gold-silver (Au-Ag), gold-platinum (Au-Pt) and antimony-bismuth (Sb-Bi). The phase diagram for one of these simple systems illustrates two characteristics of all solid solutions: 1) a range of composition can coexist in liquid/solid solutions, and 2) the change of phase (in these systems, from liquid to solid) takes place over a range of temperatures (unlike water and pure metals which freeze and change structure at a single temperature). Fig. 4 is a portion of the phase diagram for Cu-Ni, which shows what species precipitate out of solution when the liquid is slowly cooled.[4] (In the remainder of this chapter, chemical symbols are often used to represent the elements. See *Periodic Table*, Appendix 1.)

Alloy systems in which both species are infinitely soluble in each other are rare. More often the species are only partly soluble and mixtures of phases precipitate on cooling. Also common is the situation in which the species attract each other in a particular ratio and form a chemical compound. These intermetallic compounds may still have a range of compositions, but it is much narrower than that for solid solutions. Two systems that form such intermetallic compounds are chromium-iron (Cr-Fe) and iron-carbon (Fe-C).

Iron-carbon phase diagram Steel is an iron base alloy containing manganese (Mn), carbon and other alloying elements. Virtually all metals used in boilers and pressure vessels are steels. Mn, usually present at about 1% in carbon steels, is a substitutional solid solution element. Because its atomic size and electronic structure are similar to those of Fe, it has little effect on the Fe lattice or phase diagram in these low concentrations. Carbon, on the other hand, has significant effects; by varying the carbon content and heat treatment of Fe, an enormous range of mechanical properties can be obtained. These effects can best be understood using the Fe-C equilibrium phase diagram, shown in Fig. 5. This shows that the maximum solu-

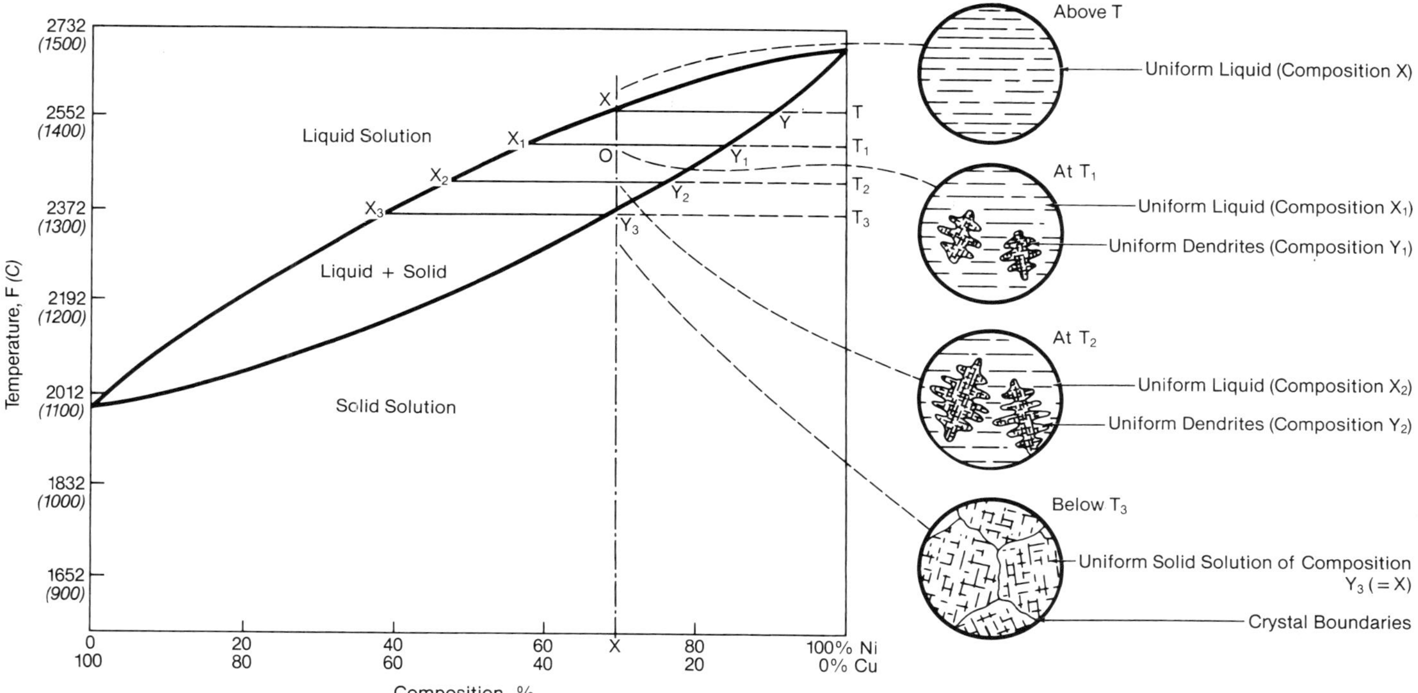

Fig. 4 The copper-nickel equilibrium diagram (*courtesy of Hodder and Staughton*).[4]

bility of carbon in α (BCC) iron is only about 0.025%, while its solubility in γ (FCC) iron is slightly above 2.0%. Alloys of Fe-C up to 2% C are malleable and are considered steels. Iron alloys containing more than 2% C are decidedly inferior to steels in malleability, strength, toughness and ductility. They are usually used in cast form and are called *cast irons*.

Carbon atoms are substantially smaller than iron atoms, and in BCC iron, they fit at the midpoints of the cube edges and face centers. This structure is called *ferrite*. In FCC iron, the carbon atoms fit at the midpoints of the cube edges at the cube center. This structure is called *austenite*. In both structures, the interstitial spaces are smaller than the carbon atom, leading to local distortion of the lattice and resulting in limited carbon solubility in iron. The interstices are larger in austenite than in ferrite, partly accounting for the higher solubility of carbon in austenite. If austenite containing more than 0.025% C cools slowly and transforms to ferrite, the carbon in excess of 0.025% precipitates from the solid solution. However, it is not precipitated as pure carbon (graphite) but as the intermetallic compound Fe_3C, *cementite*. As with most metallic carbides, this is a hard substance. Therefore, the hardness of steel generally increases with carbon content even without heat treatment.

Critical transformation temperatures The melting point of iron is reduced by the addition of carbon up to about 4.3% C. At the higher temperatures, solid and liquid coexist. The BCC δ iron range is restricted and finally eliminated as a single phase when the carbon content reaches about 0.1%. Some δ iron remains up to about 0.5% C but is in combination with other phases. Below the δ iron region, austenite exists and absorbs carbon up to the composition limits (Fig. 5), the limiting solid-solution solubility. The temperature at which only austenite exists decreases as the carbon increases (line G-S) to the eutectoid point: 0.80% C at 1333F (723C). Then, the temperature increases along line S-E with the carbon content because the

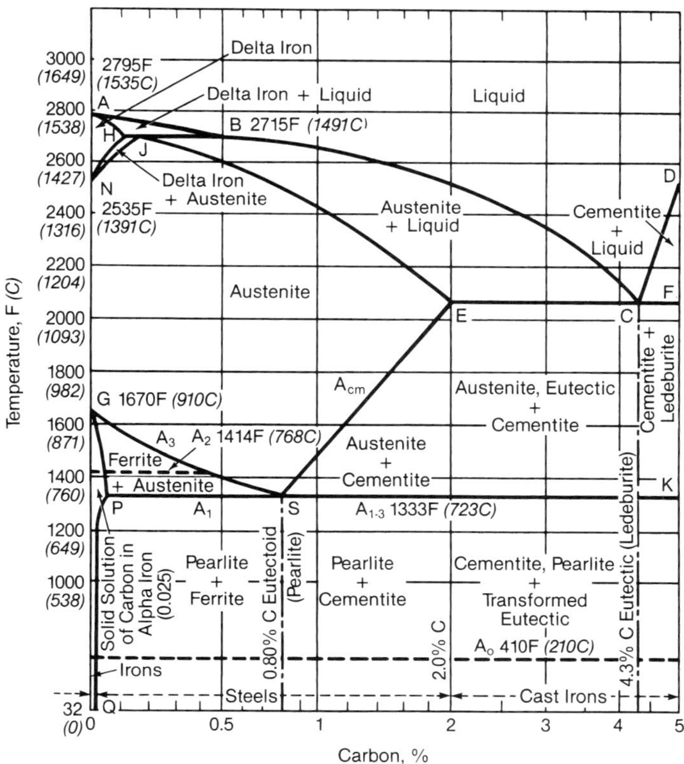

Fig. 5 Carbon-iron equilibrium diagram showing phase solubility limits.

austenite is unable to absorb additional carbon, except at higher temperatures.

Any transformation in which a single solid phase decomposes into two new phases on cooling, and in which the reverse reaction takes place on heating, is called a *eutectoid* reaction. At the eutectoid composition of 0.80% C, only austenite exists above 1333F (723C) and only ferrite and Fe_3C carbide exist below that temperature. This is the *lower critical transformation temperature*, A_1. At lower carbon contents, in the *hypoeutectoid* region, as austenite cools and reaches A_3, the *upper critical transformation temperature*, ferrite precipitates first. As the temperature is further reduced to 1333F (723C) at A_1, the remaining austenite is transformed to ferrite and carbide. In the *hypereutectoid* region, above 0.80% C, cementite precipitates first when austenite cools to the thermal arrest line (A_{cm}). Again, the remaining austenite transforms to ferrite and carbide when it cools to 1333F (723C). For a given steel composition, A_3, A_1 and A_{cm} represent the *critical transformation temperatures*, or critical points. A_2 is the Curie point, the temperature at which iron loses its ferromagnetism.

At the A_1 temperature, on cooling, all the remaining austenite must transform to ferrite and carbide. Because there is not time for the carbon to go very far as it is rejected from the forming ferrite matrix, the resulting structure is one of alternating thin plates, or lamellae, of ferrite and carbide. This lamellar structure is typical of all eutectoid decomposition reactions. In steel, this structure is called *pearlite*, which always has the eutectoid composition of 0.8% C.

When pearlite is held at a moderately high temperature, such as 950F (510C), for a long time (years), the metastable cementite eventually decomposes to ferrite and graphite. First, the Fe_3C lamellae agglomerate into spheres. The resulting structure is considered *spheroidized*. Later, the iron atoms are rejected from the spheres, leaving a *graphitized* structure. Graphitized structures are shown in Fig. 6.

Isothermal transformation diagrams The transformation lines on the equilibrium diagram, Fig. 5, are subject to displacement when the austenite is rapidly cooled or when the pearlite and ferrite, or pearlite and cementite, are rapidly heated. This has led to the refinement of A_1 and A_3 into A_{c1} and A_{c3} on heating (c, from the French *chauffage*, heating) and into A_{r1} and A_{r3} for the displacement on cooling (r, from *refroidissement*, cooling). Because these are descriptions of dynamic effects, they distort the meaning of an equilibrium diagram which represents prevailing conditions given an infinite time for reactions to occur. Because fabrication processes involve times ranging from seconds (laser welding) to several days (heat treatment of large vessels), the effect of time is important. Isothermal transformation experiments are used to determine phase transformation times when the steel is cooled very rapidly to a particular temperature. The data are plotted on time-temperature-transformation (TTT) diagrams.

The isothermal transformation diagram in Fig. 7, for a hypoeutectoid steel, shows the time required for transformation from austenite to other constituents at the various temperature levels. The steel is heated to about 1600F (871C) and it becomes completely austenitic. It is then quickly transferred to and held in a furnace or bath at 700F (371C). Fig. 5 shows that ferrite and carbides should eventually exist at this temperature and Fig. 7 indicates how long this reaction takes. By projecting the time intervals during the transformation, as indicated in the lower portion of Fig. 7, to the top portion of the diagram, the austenite is predicted to exist for about three seconds before transformation. Then, at about 100 s, the transformation is 50% complete. At 700 s, the austenite is entirely replaced by an agglomerate of fine carbides and ferrite.

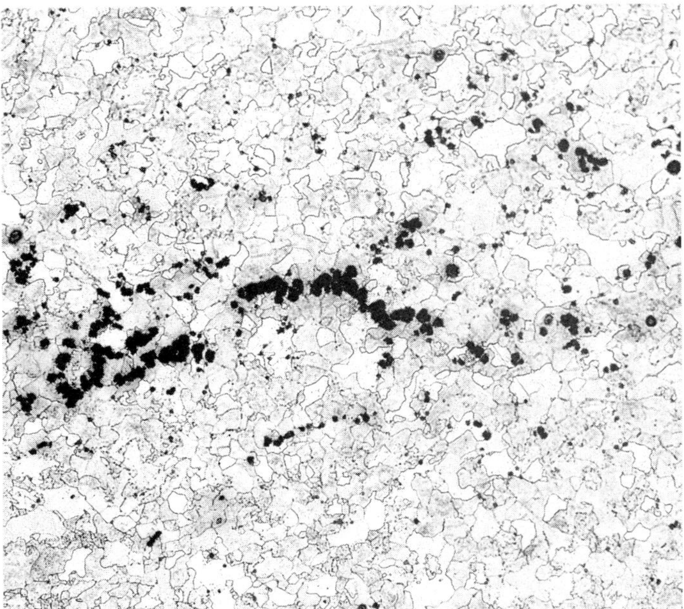

Fig. 6 Chain graphitization (black areas) in carbon-molybdenum steel, 200 X magnification.

For this particular steel, at temperatures below about 600F (316C) austenite transforms to *martensite*, the hardest constituent of heat treated steels. The temperature at which martensite starts to form is denoted M_s. It decreases with increasing carbon content of the austenite. The nose of the left curve in Fig. 7, at about 900F (482C), is of prime significance because the transformation at this temperature is very rapid. Also, if this steel is to be quenched to form martensite (for maximum hardness), it must pass through about 900F (482C) very rapidly to prevent some of the austenite from transforming to pearlite (F + C), which is much softer.

Martensite is therefore a supercooled metastable structure that has the same composition as the austenite from which it forms. It is a solution of carbon in iron, having a *body-centered tetragonal* (BCT) crystal structure. (See Reference 1.) Because martensite forms with no change of composition, diffusion is not required for the transformation to occur. It is for this reason that martensite can form at such low temperatures. Its hardness is due to the high, supersaturated carbon content, to the great lattice distortion caused by trapping excess carbon, and to the volume change of the transformation. The specific volume of martensite is greater than that of the austenite.

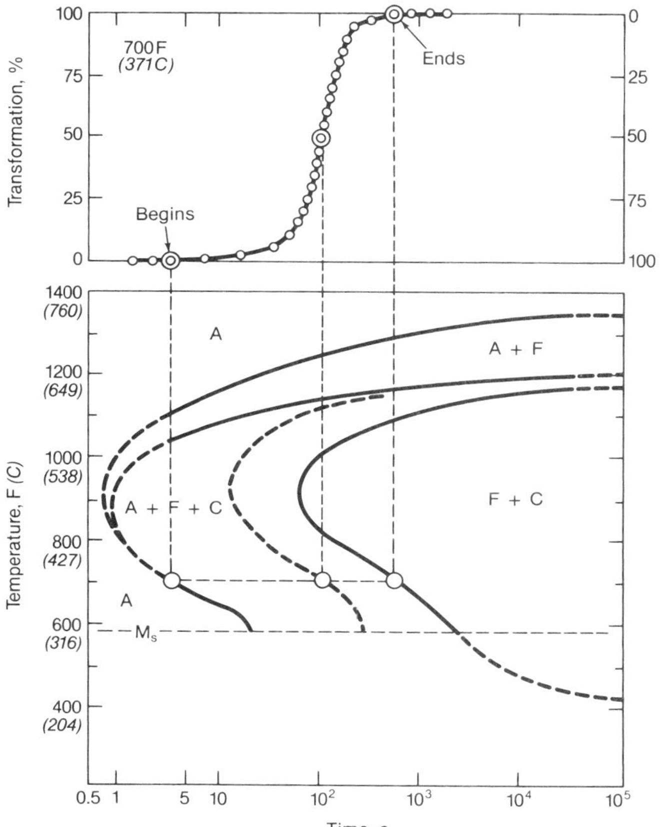

Fig. 7 Typical isothermal transformation diagram. Time required in a specific steel at 700F (371C) taken as an example.

The formation of martensite does not occur by nucleation and growth. It can not be suppressed by quenching and it is athermal. Austenite begins to form martensite at a temperature M_s. As the temperature is lowered, the relative amount of martensite in the structure increases. Eventually, a temperature (M_f) is reached where the transformation to martensite is complete. At any intermediate temperature, the amount of martensite characteristic of that temperature forms instantly and holding at that temperature results in no further transformation. The M_s and M_f temperatures, therefore, are shown on the isothermal transformation diagram (Fig. 8) as horizontal lines. Under the microscope, martensite has the appearance of acicular needles. Each needle is a martensite crystal.

Bainite is produced when the eutectoid (0.8% C) transformation takes place at a lower temperature (but above the M_s temperature for the alloy). The temperature regions of the TTT curve in which pearlite, bainite and martensite form are shown in Fig. 8. In the pearlite transformation, the cementite and the ferrite form in a fine lamellar pattern of alternating layers of ferrite and cementite.

Effects of alloying elements on the Fe-Fe$_3$C phase diagram Adding one or more elements to the Fe-C alloy can have significant effects on the relative size of the phase fields in the Fe-Fe$_3$C phase diagram. The elements Ni, Mn, Cu and cobalt (Co) are called *austenite formers* because their addition to the Fe-C alloy system raises the temperature at which austenite transforms to δ ferrite and lowers A_3 in Fig. 5. Adding a sufficient amount of these elements increases the size of the austenite field and the FCC structure may become stable at room temperature. Because most of these elements do not form carbides, the carbon stays in solution in the austenite. Many useful material properties result, including high stability, strength and ductility, even at high temperatures. The elements Cr, molybdenum (Mo), tungsten (W), vanadium (V), aluminum (Al) and silicon (Si) have the opposite effect and are considered *ferrite formers*. They raise the A_3 temperature and some of them form very stable carbides, promoting the stability of BCC ferrite, even at very high temperatures.

Specific effect of alloying elements

Steel alloys are the chief structural materials of modern engineering because their wide range in properties suits so many applications. These properties are affected directly not only by the characteristics and the amounts of the elements which, either alone or in combination, enter into the composition of the steel, but also by their reaction as constituents under various conditions of temperature and time during fabrication and use. For example, Cr increases resistance to corrosion and scaling, Mo increases creep strength at elevated temperatures, and Ni (in adequate amounts) renders the steel austenitic. The specific effects of the most important elements found in steel are as follows.

Carbon (C) is the most important alloying element in steel. In general, an increase in carbon content produces higher ultimate strength and hardness but lowers the ductility and toughness of steel alloys. The curves in Fig. 9 indicate the general effect of carbon on the mechanical properties of hot rolled carbon steel. Carbon also increases air hardening tendencies and

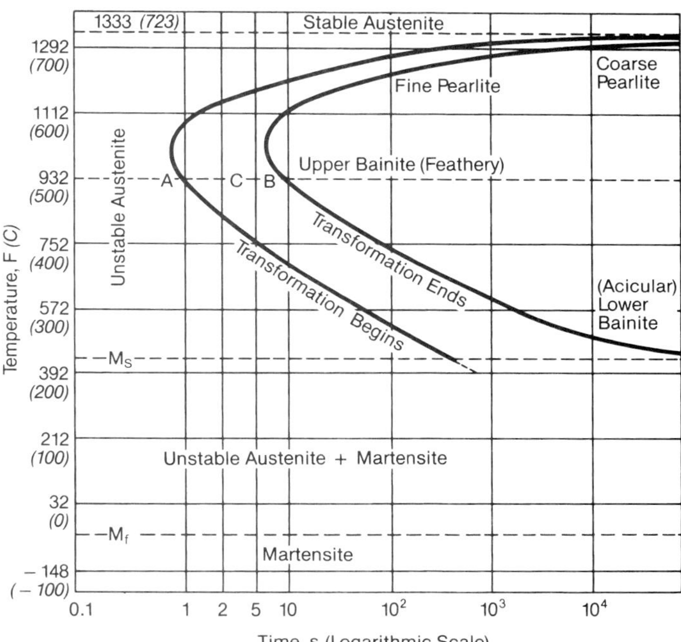

Fig. 8 Time-temperature-transformation curves for a 0.8% plain carbon steel.[4]

weld hardness, especially in the presence of Cr. In low alloy steel for high temperature applications, the carbon content is usually restricted to a maximum of about 0.15% to ensure optimum ductility for welding, expanding and bending operations, but it should be no lower than 0.07% for optimum creep strength. To minimize intergranular corrosion caused by carbide precipitation, the carbon content of austenitic stainless steel alloys is limited to 0.10%. This maximum may be reduced to 0.03% in extremely low carbon grades used in certain corrosion resistant applications. However, at least 0.04% C is required for acceptable creep strength. In plain, normalized carbon steels, the creep resistance at temperatures below 825F (441C) increases with carbon content up to 0.4% C; at higher temperatures, there is little variation of creep properties with carbon content. An increase in carbon content also lessens the thermal and electrical conductivities of steel and increases its hardness on quenching.

Manganese (Mn) is infinitely soluble in austenite and up to about 10% soluble in ferrite. It combines with residual sulfur while the steel is molten to form manganese sulfides, which have a much higher melting point than iron sulfides. Without the Mn, iron sulfides, which melt at about 1800F (982C), would form. This would lead to *hot-shortness*, a brittle-failure mechanism, during hot forming operations. The Mn therefore produces the malleability that differentiates steel from cast iron.

Mn is a good solid solution strengthener, better than Ni and about as good as Cr. In alloy steels, manganese decreases the critical cooling rate to cause martensitic structure and thus contributes to deep hardening. It can also be used in austenitic stainless steels to replace Ni as the austenite stabilizer at lower cost.[5]

Molybdenum (Mo), when added to steel, increases its strength, elastic limit, resistance to wear, impact qualities and hardenability. Mo contributes to high temperature strength and permits heating steel to a red hot condition without loss of hardness. It also increases the resistance to softening on tempering and restrains grain growth. Mo makes chromium steels less susceptible to temper embrittlement and it is the most effective single additive that increases high temperature creep strength.

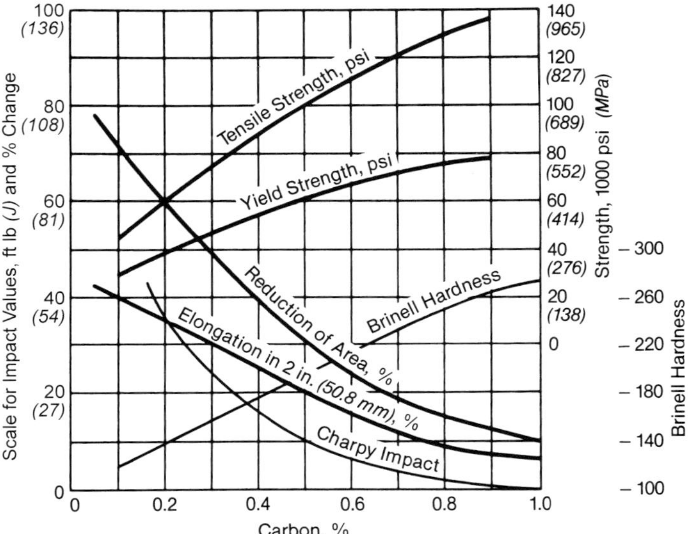

Fig. 9 General effect of carbon on the mechanical properties of hot-rolled carbon steel.

An important use of Mo is for corrosion resistance improvement in austenitic stainless steels. It enhances the inherent corrosion resistance of these steels in reducing chemical media and it increases their passivity under mildly oxidizing conditions. Under certain conditions, molybdenum reduces the susceptibility of stainless steels to pitting.

Chromium (Cr) is the essential constituent of stainless steel. While other elements are stronger oxide formers, Cr is the only one that is highly soluble in iron (about 20% in austenite and infinite in ferrite) and forms a stable, tightly adherent oxide. It is virtually irreplaceable in resisting oxidation at elevated temperatures.

Cr raises the yield and ultimate strength, hardness, and toughness of ferritic steel at room temperature. It also contributes to high temperature strength. The optimum chromium content for creep strength in annealed low alloy steels is about 2.25%.

A steady improvement in resistance to atmospheric corrosion and to attack by many reagents is also noted when the chromium content is increased. A steel with 12% or more Cr is considered stainless, i.e., the Cr_2O_3 film is sufficient to prevent surface rust (hydrated iron oxide) formation. The chemical properties of the steel, however, are affected by the carbon content. Higher chromium and lower carbon levels generally promote increased corrosion resistance.

Adding more than 1% of chromium may cause appreciable air hardening in the steel. Up to about 13.5% Cr, air hardening is a direct function of chromium and carbon content. Low carbon alloy steels containing more than 12% Cr can become nonhardening, but the impact strength is reduced and the ductility is poor. Cr lessens thermal and electrical conductivities. The addition of sufficient Cr prevents *graphitization* during long-term high temperature service of ferritic steels.

Nickel (Ni) increases toughness when added to steel, particularly in amounts over 1%. Improved resistance to corrosion by some media is attained with Ni contents over 5%. Ni dissolves in the iron matrix in all proportions and, therefore, raises the ultimate strength without impairing the ductility of the steel. Ni is particularly effective in improving impact properties, especially at low temperature.

The most important use of nickel as an alloying element in steel is its combination with chromium in amounts of 8% Ni or more. Ni is such a strong austenite former that the high chromium Fe-Ni-C alloys are austenitic at room temperature. The various combinations of chromium and nickel in iron produce alloy properties that can not be obtained with equivalent amounts of a single element. Common combinations are 18% Cr - 8% Ni, 25% Cr - 12% Ni, 25% Cr - 20% Ni, and 20% Cr - 30% Ni. These steels are resistant to atmospheric corrosion and to oxidation at high temperatures. In addition, they offer greatly enhanced creep strength.

Ni is only slightly beneficial to creep properties of

low alloy ferritic steels. It reduces the coefficient of thermal expansion and diminishes the electrical and thermal conductivities. It is attacked by sulfur compounds at elevated temperatures.

Cobalt (Co) suppresses hardenability in steels. However, when added to austenite, it is a strong solution strengthener and a carbide former. It also significantly improves creep strength. Binary Fe-Co alloys have the highest magnetic saturation induction of any known materials. Therefore, such alloys are often used in permanent magnets.

Tungsten (W) acts similarly to molybdenum. It is a very strong carbide former and solid solution strengthener. It forms hard, abrasion resisting carbides in tool steels, develops high temperature hardness in quenched and tempered steels, and contributes to creep strength in some high temperature alloys.[5]

Vanadium (V) is a degasifying and deoxidizing agent, but it is seldom used in that capacity because of high cost. It is applied chiefly as an alloying element in steel to increase strength, toughness and hardness. It is essentially a carbide forming element which stabilizes the structure, especially at high temperatures. Vanadium minimizes grain growth tendencies, thereby permitting much higher heat treating temperatures. It also intensifies the properties of other elements in alloy steels. Small additions of vanadium (0.1 to 0.5%), accompanied by proper heat treatment, give steels containing 0.5 to 1.0% molybdenum pronounced improvement in high temperature creep properties.

Titanium (Ti) and *columbium* (Cb) (also known as *niobium*) are the most potent carbide forming elements. Ti is also a good deoxidizer and denitrider. These elements are most effective in the Cr-Ni austenitic alloys, where they react more readily with carbon than does Cr. This allows the Cr to remain in solid solution and in the concentrations necessary to maintain corrosion resistance. Ti and Cb [or Cb plus tantalum (Ta)] are used to reduce air hardening tendencies and to increase oxidation resistance in steel containing up to 14% Cr. These elements have a beneficial effect on the long term, high temperature properties of Cr-Ni stainless steels because of the stability of their carbides, nitrides and carbonitrides. Cb and Ti have also been used in some of the super alloys to improve high temperature properties. Ti forms an intermetallic compound with Ni in these alloys, Ni_3Ti, called gamma prime (γ'), which is a potent strengthening phase.

Copper (Cu), when added to steel in small amounts, improves its resistance to atmospheric corrosion and lowers the attack rate in reducing acids. Cu, like Ni, is not resistant to sulfur compounds at elevated temperatures. Consequently, it is not ordinarily used in low alloy steels intended for high temperature service where sulfur is a major component of the environment, as in combustion gases. Cu is added (up to 1%) in low alloy constructional steels to improve yield strength and resistance to atmospheric corrosion. Its presence in some of the high alloy steels increases corrosion resistance to sulfuric acid.

Boron (B) is usually added to steel to improve hardenability, that is, to increase the depth of hardening during quenching of alloy steels. When combined with Mo, it is a strong bainite stabilizer. Small amounts of boron in the presence of Mo suppress the formation of martensite, leading to the complete transformation to bainite before the M_s temperature is reached. This substantially improves the strength and stability of Cr-Mo pressure vessel steels. The B-10 isotope of boron has a very high neutron-capture cross-section, so it is added to steels used for containment and storage vessels of nuclear fuels and waste products.

Nitrogen (N) has two primary functions as an alloying agent in steels. In carbon and low alloy steels, it is used in case hardening, in which nascent nitrogen is diffused into the steel surface. Nitrogen and carbon are interstitial solid solution strengtheners. In the presence of Al or Ti, additional strengthening results by precipitate formations of the respective nitrides or carbonitrides. In austenitic stainless steels, nitrogen provides the same interstitial strengthening as carbon, but does not deplete the austenite of chromium as does carbon by the formation of carbides. The strength of nitrogen-containing stainless steels is therefore equivalent to that of the carbon-containing stainlesses. This strength is achieved without the susceptibility to corrosive attack that results from local carbide formation at grain boundaries of these steels.

Oxygen (O) is not normally considered to be an alloying element. It is present in steel as a residual of the steel making process.[6,7] However, a few oxides are so hard and stable, notably those of Al, Ti and thorium (Th), that they are potent strengtheners when dispersed as fine particles throughout an alloy. This can be accomplished by internal oxidation in an oxygen-containing atmosphere or by powder metallurgical techniques.

Aluminum (Al) is an important minor constituent of low alloy steels. It is an efficient deoxidizer and grain refiner, and is widely used in producing killed steel. When added to steel in appreciable quantities, Al forms tightly adhering refractory oxide scales and therefore, increases resistance to scaling. It is difficult, however, to add appreciable amounts of this element without producing undesirable effects. In the amounts customarily added (0.015 to 0.080%), Al does not increase resistance to ordinary forms of corrosion. Because of their affinity for oxygen, high-aluminum steels generally contain numerous alumina inclusions which can promote pitting corrosion. The refined grain size does improve room temperature toughness and ductility of carbon steels.

An excessive quantity of aluminum has a detrimental effect on creep properties, particularly in plain carbon steel. This is attributable to its grain refining effect and to its acceleration of graphitization of the carbide phase.

Silicon (Si) greatly contributes to steel quality because of its deoxidizing and degasifying properties. When added in amounts up to 2.5%, the ultimate strength of steel is increased without loss in ductility. Si in excess of 2.5% causes brittleness and amounts higher than 5% make the steel nonmalleable.

Resistance to oxidation of steel is increased by adding silicon. Si increases the electrical conductivity of steel and decreases hysteresis losses. Si steels are, therefore, widely used in electrical apparatus.

Killing agents, such as Si and Al, are added to steel for deoxidation; the latter is used for grain size control. Calcium and rare earth metals, when added to the melt, have the same effects. Additionally, these elements form complex oxides or oxysulfides and can significantly improve formability by controlling the sulfide shape.

Phosphorus (P) is a surprisingly effective hardener when dissolved in quantities of up to 0.20%.[5] However, a high phosphorus content can notably decrease the resistance of carbon steel to brittle fracture and reduce ductility when the metal is cold worked. This embrittling effect is referred to as *cold-shortness*. The detrimental effect of phosphorus increases with carbon content.

Phosphorus is effective in improving the machinability of free-cutting steels. This is related to its embrittling effect, which permits easier chip formation on machining. In alloy steels intended for boiler applications, the permissible phosphorus content is less than that for machining steels and its presence is objectionable for welding. Phosphorus is used as an alloying element (up to 0.15%) in proprietary low alloy, high strength steels, where increased yield strength and atmospheric corrosion resistance are primary requirements. In the presence of certain acids, however, a high phosphorus content may increase the corrosion rate.

Sulfur (S) is generally undesirable in steel and many processes have been developed to minimize its presence. However, sulfur is sometimes added to steel to improve its machinability, as are phosphorus and other free-machining additives: calcium, lead, bismuth, selenium and tellurium. Several of these elements are virtually insoluble in steel and have low melting points, or they form low melting temperature compounds. These compounds can lead to cracking due to liquid metal embrittlement or hot-shortness at even moderately elevated temperatures. Hot-shortness occurs when liquid iron sulfide forms at grain boundaries during hot-working and heat treatment of steels.[6] Because the fastener industry favors free machining steels due to their beneficial production effects, boiler and pressure vessel manufacturers must exercise care in applying threaded fasteners containing these elements at high temperatures.

Heat treating practices

Steel can be altered by modifying its microstructure through heat treatment. Various heat treatments may be used to meet hardness or ductility requirements, improve machinability, refine grain structure, remove internal stresses, or obtain high strength levels or impact properties. The more common heat treatments, annealing, normalizing, spheroidizing, hardening (quenching) and tempering, are briefly described.

Annealing is a general term applied to several distinctly different methods of heat treatment. These are full, solution, stabilization, intercritical, isothermal, and process annealing.

Full annealing is done by heating a ferritic steel above the upper critical transformation temperature (A_3 in Fig. 5), holding it there long enough to fully transform the steel to austenite, and then cooling it at a controlled rate in the furnace to below 600F (316C). A full anneal refines grain structure and provides a relatively soft, ductile material that is free of internal stresses.

Solution annealing is done by heating an austenitic stainless steel to a temperature that puts most of the carbides into solution. The steel is held at this temperature long enough to achieve grain growth. It is then quenched in water or another liquid for fast cooling, which prevents most of the carbides from reprecipitating. This process achieves optimum creep strength and corrosion resistance. For many boiler applications, austenitic stainless steels require the high creep strength of a coarse grain structure but do not require aqueous corrosion resistance, because they are only exposed to dry steam and flue gases. *Solution treatment*, used to achieve grain growth, is required for these applications, but the quenching step is not required.

Stabilization annealing is performed on austenitic stainless steels used in severe aqueous corrosion environments. The steel is first solution annealed, then reheated to about 1600F (871C) and held there. Initially, chromium carbides precipitate at the grain boundaries in the steel. Because these are mostly of the complex $M_{23}C_6$ type, which are very high in Cr, the austenite adjacent to the grain boundaries is depleted of chromium. This would normally leave the steel susceptible to corrosive attack, but holding it at 1600F (871C) permits the Cr remaining in the austenite solution to redistribute within the grains, restoring corrosion resistance, even adjacent to the grain boundaries.

Intercritical annealing and *isothermal annealing* are similar. They involve heating a hypoeutectoid ferritic steel above the lower critical transformation temperature (A_1 in Fig. 5) but below the upper critical temperature, A_3. This dissolves all the iron carbides but does not transform all the ferrite to austenite. Cooling slowly from this temperature through A_1 produces a structure of ferrite and pearlite that is free of internal stresses. In intercritical annealing, the steel continues to cool slowly in the furnace, similarly to full annealing. In isothermal annealing, cooling is stopped just below A_1, assuring complete transformation to ferrite and pearlite, and eliminating the potential for bainite formation.

Process annealing, sometimes called subcritical annealing or stress relieving, is performed at temperatures just below the lower critical temperature A_1, usually between 950 and 1300F (510 and 704C). Process annealing neither refines grains nor redissolves cementite, but it improves the ductility and decreases residual stresses in work hardened steel.

Normalizing is a variation of full annealing. Once it has been heated above the upper critical temperature, normalized steel is cooled in air rather than in a controlled furnace atmosphere. Normalizing is sometimes used as a homogenization procedure; it assures that any prior fabrication or heat treatment history of the material is eliminated. Normalizing relieves the internal stresses caused by previous working and, while it produces sufficient softness and ductility for many purposes, it leaves the steel harder and with higher tensile strength than full annealing. To remove cooling stresses, normalizing is often followed by tempering.

Spheroidizing is a type of subcritical annealing used to soften the steel and to improve its machinability. Heating fine pearlite just below the lower critical temperature of the steel, followed by very slow cooling, causes spheroidization.

Hardening (quenching) occurs when steels of the higher carbon grades are heated to produce austenite and then cooled rapidly (quenched) in a liquid such as water or oil. Upon hardening, the austenite transforms into martensite. Martensite is formed at temperatures below about 400F (204C), depending on the carbon content and the type and amount of alloying elements in the steel. It is the hardest form of heat treated steels and has high strength and abrasion resistance.

Tempering is applied after normalizing or quenching some air hardening steels. These preliminary treatments impart a degree of hardness to the steel but also make it brittle. The object of tempering, a secondary treatment, is to remove some of that brittleness by allowing certain transformations to proceed in the hardened steel. It involves heating to a predetermined point below the lower critical temperature, A_1, and is followed by any desired rate of cooling. Some hardness is lost by tempering, but toughness is increased, and stresses induced by quenching are reduced or eliminated. Higher tempering temperatures promote softer and tougher steels. Some steels may become embrittled on slow cooling from certain tempering temperatures. These steels are said to be temper brittle. To overcome this difficulty, they are quenched from the tempering temperature.

Post fabrication heat treatments are often applied to restore more stable, stress free conditions. These include post weld and post forming heat treatments and solution treatment.

Fabrication processes

Any mechanical work applied to the metal below its recrystallization temperature is *cold* work. Mechanical work performed above the recrystallization temperature is *hot* work and the simultaneous annealing that occurs at that temperature retards work-hardening. The recrystallization temperature is dependent on the amount of deformation. If a material is formed at a temperature significantly above room temperature, but below its recrystallization temperature, the process is sometimes referred to as *warm working*.

The temperature at which steel is mechanically worked has a profound effect on its properties. Cold work increases the hardness, tensile strength and yield strength of steel, but its indices of ductility – elongation and reduction of area – are decreased. The extent of the work-hardening, with progressive elongation of the grains in the direction of working, depends on the amount of cold work and on the material. If the work-hardening caused by the necessary shaping operation becomes excessive, the ductility may be exhausted and further work can cause fracture.

Hot working variations include forging, rolling, pressing, extruding, piercing, upsetting and bending. Most of these are largely compressive operations, in which the metal is squeezed into a desired shape. They introduce some degree of orientation to the internal structure. Even if the metal experiences phase transformations or other recrystallization processes, some degree of orientation is maintained in the pattern retained by the oxides, sulfides, and other inclusions that do not dissolve during hot working or heat treatment. Depending on the application, the resultant orientation may have no effect, be useful, or be harmful. Rolled plates, for instance, often have inferior properties in the through-thickness direction due to retention of mid-plane segregated inclusions and to the predominant grain orientation in the longitudinal and transverse directions. This can result in a failure mode known as lamellar tearing if not addressed.

Hot rolling of carbon steel and low alloy steel into drum or pressure vessel sections is often done at temperatures above A_3. Temperatures and times of heating before forming need to be controlled to ensure that the resulting product retains the desired fine grain size and consequent good toughness, and to ensure that excessive plate surface oxidation does not occur.

Cold working operations used in manufacturing boiler components are rolling, forging, bending and swaging. Detailed information about these processes and their effects on materials can be found elsewhere. (See References 6 and 7.)

Cold rolling of plate to make shells for drums is limited only by the capacity and diameter of available rolling equipment and the inherent ductility of the steel. This process is most often applied to carbon steel, and any post forming heat treatment performed is usually combined with post weld heat treatment of the completed drum. In some low pressure applications, tube-to-header or tube-to-drum connections may be made by roll expanding the tube into an internally grooved socket in the shell. The strength of the connection depends on the mechanical interference between the roll expanded tube, which generally deforms plastically, and the hole in the shell, which mostly deforms elastically.

Cold forging of boiler components is usually limited to final size forming of shells. Threaded fasteners used in boilers may have been cold headed or may have had their threads cold rolled. Effects of such forming operations are normally mitigated by heat treatments required by the specification, but occasionally this heat treatment does not eliminate microstructural differences between the cold formed portion and the remainder of the part. This is particularly true of austenitic stainless steel or nickel alloy bolts, which do not transform during heat treatment. These bolts may be susceptible to cracking at the interface between the cold formed head and the shank in certain aqueous environments.

Cold bending is performed on many configurations of tubes and pipes for boilers. Boiler designers consider the effects of this process on the geometry and properties of the finished product.

Austenitic stainless steels and nickel alloys used in high pressure boilers are often exposed to temperatures at which the strain energy of the cold bending is sufficient to cause polygonization and recrystallization to a fine grain size during service. The service temperature is insufficient to produce grain growth and the fine grain size material has lower high temperature (creep) strength. To prevent this from happening, cold bends in these alloys are given a high temperature (solution) heat treatment to stabilize the coarse grain structure.

Most carbon and low alloy ferritic steel tube and pipe alloys may be used in the cold bent condition, unless the amount of cold strain imparted is very high. If strain in excess of about 30% is developed in bending, the resulting structure and low residual ductility can render the bends susceptible to strain aging and breakage during subsequent handling and service.

Cold working of carbon steels has also been shown to render them susceptible to creep crack growth of minor surface flaws and imperfections when these steels are operating at temperatures where the cracking mechanism is operative.

Certain Cr-Mo high strength ferritic steels can also experience significant degradation of creep strength if they are cold worked to levels near and above about 20% strain. In all of these cases, post fabrication heat treatments must be applied to recover acceptable properties. This most often requires re-annealing, normalization, or normalizing and tempering as is appropriate for the given alloy. In some cases, simple subcritical tempering or stress relief heat treatment may be sufficient.

Welding

Joining of boiler pressure parts and of nonpressure parts to pressure parts is almost always accomplished by welding. This is particularly true of high temperature, high pressure boilers, whose service conditions are too severe for most mechanical joints (bolted flanges with gaskets) and brazed joints.

Welding is the joining of two or more pieces of metal by applying heat or pressure, or both, with or without the addition of filler metal, to produce a localized union through fusion across the interface.[8] There are many welding processes, but the most widely used for joining pressure parts is fusion welding with the addition of filler metal, using little or no pressure. Fig. 10 indicates the variety of processes.

Weld morphology Because of the heat distribution characteristics of the welding process, the weld joint is usually a chemically and mechanically heterogeneous composite consisting of up to six metallurgically distinct regions: a composite zone, the unmixed zone, the weld interface, the partially melted zone, the heat-affected zone (HAZ) and the unaffected base metal. These zones are shown in Fig. 11. The *composite zone* is the completely melted mixture of filler metal and melted base metal. The narrow region surrounding the composite zone is the *unmixed zone*, which is a boundary layer of melted base metal that solidifies before mixing in the composite zone. This layer is at the edges of the weld pool, with a composition essentially identical to the base metal. The composite zone and the unmixed zone together make up what is commonly referred to as the fusion zone. The third region is the *weld interface*, or the boundary between the unmelted base metal on one side and the solidified weld metal on the other. The *partially melted zone* occurs in the base metal immediately adjacent to the weld interface, where some localized melting of lower melting temperature constituents, inclusions or impurities may have occurred. Liquation, for instance, of manganese sulfide inclusions can result in hot cracking or microfissuring. The *heat-affected zone* is that portion of the base material in the weld joint that has been subjected to peak temperatures high enough to produce solid state microstructural changes, but not high

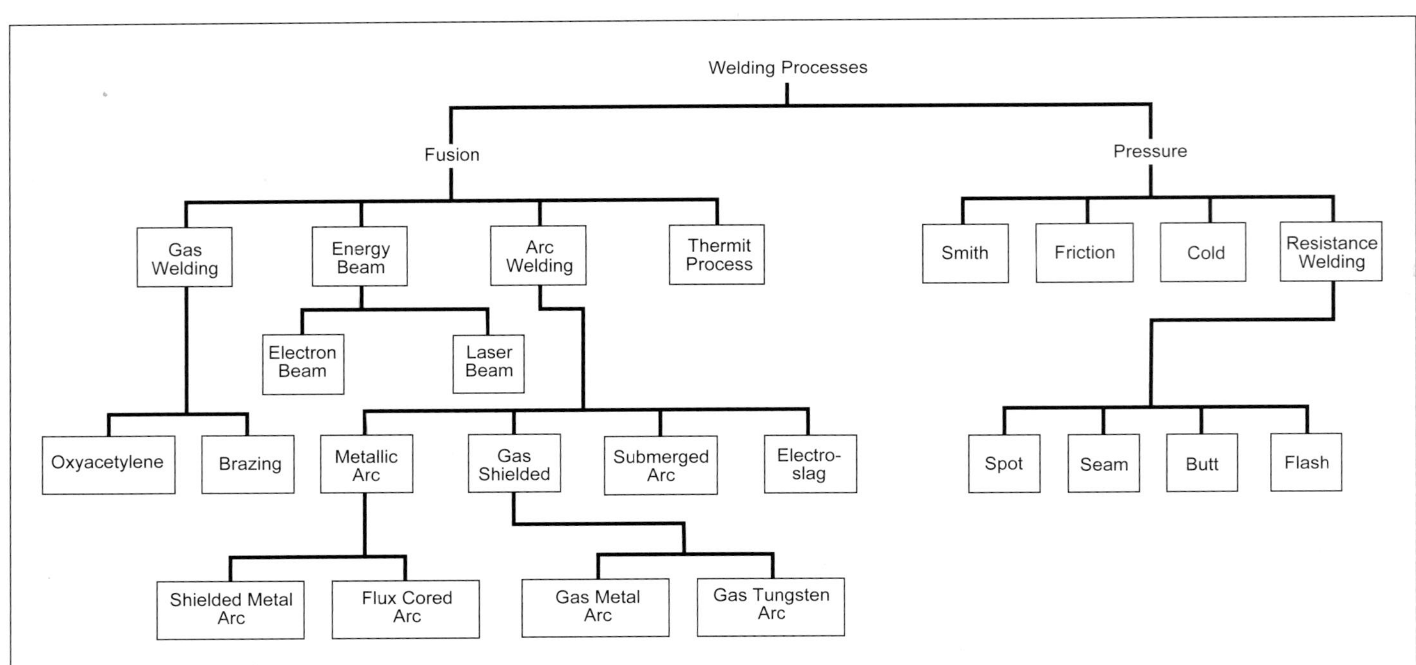

Fig. 10 Classification of welding processes.[4]

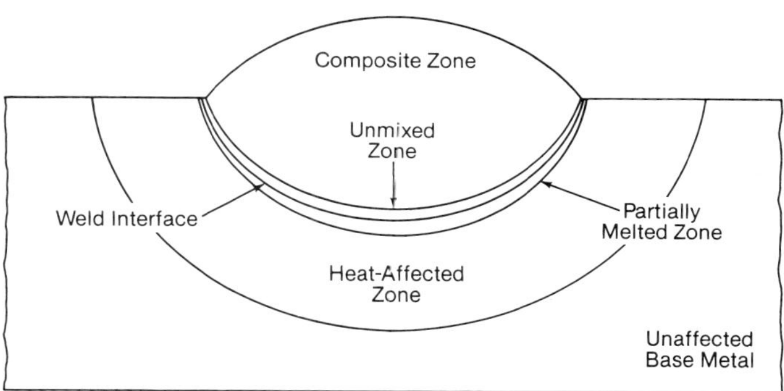

Fig. 11 Metallurgical zones developed in a typical weld (*courtesy of ASM*).[9]

enough to cause melting. Finally, the last part of the work piece that has not undergone a metallurgical change is the *unaffected base metal.*

Weld quality issues

The general subject of welding and weldability is a very broad subject, addressed by a large number of excellent references.[10,10a] The following addresses only a few issues that are particularly unique and important to boilers and pressure vessels.

Ferrite content Austenitic stainless steel weld metals are susceptible to hot cracking or microfissuring as they cool from the solidus to about 1800F (982C). The microfissuring can be minimized by providing a small percentage of ferrite in the as-deposited welds.

Graphitization The shrinkage of the weld on freezing results in plastic deformation and high residual stresses in the weld joint. In carbon and carbon-molybdenum steels containing no stronger carbide-forming elements, the areas of localized strain adjacent to the heat-affected weld zones provide sites where the volume increase of the cementite to graphite decomposition can be more readily accommodated. At about 900F (482C), graphite nodules can precipitate on these planes of deformation. When samples of such materials are viewed in cross-section, the nodules appear to be arranged in rows or chains, and this condition has been termed *chain graphitization* (Fig. 6). The interfacial bond between the graphite and the ferrite matrix in such weldments is very low, much lower than that between ferrite and pearlite or ferrite and cementite. In the early 1950s, several failures of carbon-molybdenum main steam piping weldments occurred due to this phenomenon. The ruptures occurred with little warning because they were not preceded by swelling of the joints and, as a result, significant damage resulted. In consequence, the use of carbon-molybdenum main steam piping has been essentially eliminated and the maximum use temperature of carbon steel piping has been significantly restricted.

Post weld heat treatment When cooling is complete, the welded joint contains residual stresses comparable to the yield strength of the base metal at its final temperature. The thermal relief of residual stresses by post weld heat treatment (PWHT) is accomplished by heating the welded structure to a temperature high enough to reduce the yield strength of the steel to a fraction of its magnitude at ambient temperature. Because the steel can no longer sustain the residual stress level, it undergoes plastic deformation until the stresses are reduced to the at-temperature yield strength. Fig. 12 shows the effect of stress relief on several steels. The temperature reached during the treatment has a far greater effect in relieving stresses than the length of time the weldment is held at temperature. The closer the temperature is to the critical or recrystallization temperature, the more effective it is in removing residual stresses, provided the proper heating and cooling cycles are used.[10]

Lamellar tearing Weld defect causes and inspection procedures are covered more extensively in References 9 and 11. However, one metallurgical effect of residual stresses should be mentioned in the context of boilers and pressure vessels: *lamellar tearing.* Lamellar tearing may result when an attachment is welded to a plate in the T-shaped orientation shown in Fig. 13, particularly if the plate contains shrinkage voids, inclusions, or other internal segregation parallel to the plate surface. In such an instance, the residual shrinkage stresses may be sufficient to open a tear or tears parallel to the plate surface to which the T-portion is welded.

Joining dissimilar metals It may be necessary to join austenitic and ferritic steels. Weld failures have occurred in these welds since the introduction of austenitic stainless steel superheater tubing materials. Nickel base filler metals have long been used to mitigate these problems, but these do not offer a permanent solution. Additional system stresses from component location, system expansion and bending can increase the potential for such failures.

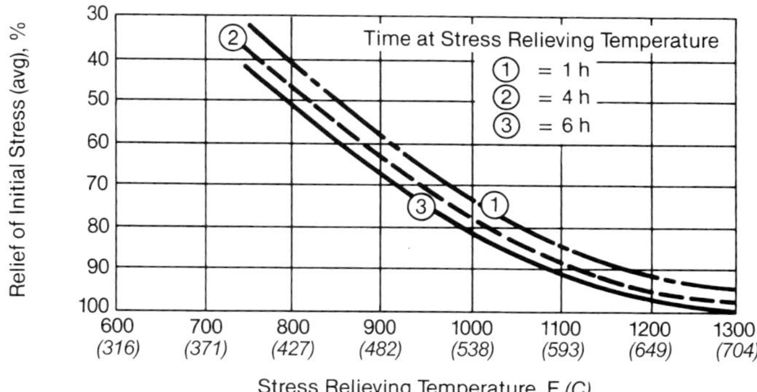

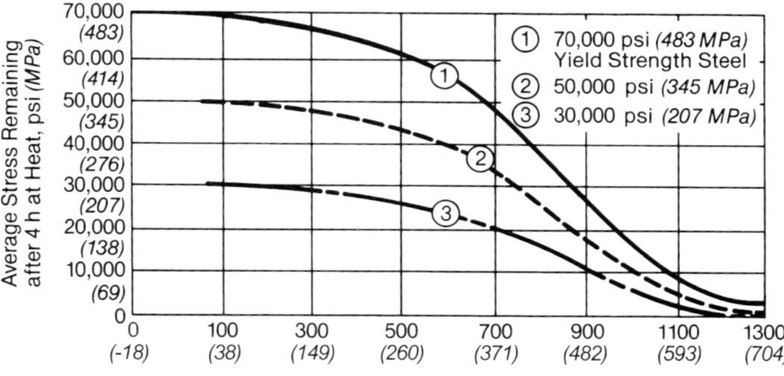

Fig. 12 Effect of temperature and time on stress relief in carbon steel (upper graph) and steels with varying as-welded strengths (*courtesy of AWS*).[11]

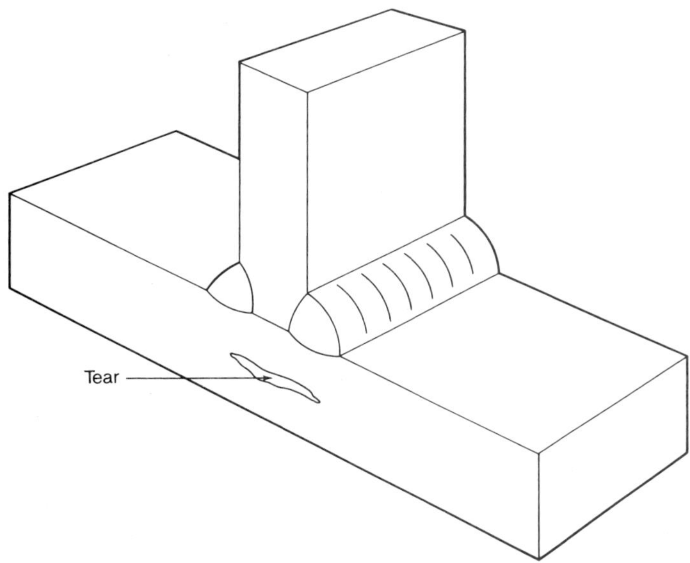

Fig. 13 Lamellar tearing.

Research is continuing toward the development of filler metals less likely to permit failures but none has become commercially available. The best alternative is to avoid dissimilar metal welds by using higher strength ferritic alloy materials, such as modified 9Cr-1Mo-V tubing and piping, when design conditions permit.

Materials

Almost all of the materials used in constructing boilers and pressure vessels are steels and the vast majority of components are made of *carbon steels*. Carbon steels are used for most types of pressure and nonpressure parts: drums, headers, piping, tubes, structural steel, flues and ducts, and lagging.

Carbon steels may be defined by the amount of carbon retained in the steel or by the steelmaking practice. These steels are commonly divided into four classes by carbon content: *low carbon*, 0.15% C maximum; *medium-low carbon*, between 0.15 and 0.23% C; *medium-high carbon*, between 0.23 and 0.44% C; and *high carbon*, more than 0.44% C. However, from a design viewpoint, high carbon steels are those over 0.35%, because these can not be used as welded pressure parts. Low carbon steels see extensive use as pressure parts, particularly in low pressure applications where strength is not a significant design issue. For most structural applications and the majority of pressure parts, medium carbon steels, with carbon contents between 0.20 and 0.35%, predominate.

Carbon steels are also referred to as killed, semikilled, rimmed and capped, depending on how the carbon-oxygen reaction of the steel refining process was treated. During the steelmaking process, oxygen, introduced to refine the steel, combines with carbon to form carbon monoxide or carbon dioxide, and also exists as excess oxygen. If the oxygen introduced is not removed or combined prior to or during casting by the addition of Si, Al, or some other deoxidizing agent, the gaseous products are trapped during solidification of the metal in the mold. The amount of gas evolved during solidification determines the type of steel and the amount of carbon left in the steel. If no gas is evolved and the liquid lies quietly in the mold, it is known as killed steel. With increasing degrees of gas evolution, the products are known as semi-killed and rimmed steels. Virtually all steels used in boilers today are fully killed.

Microalloyed steels are carbon steels to which small amounts (typically less than 1%) of alloying elements have been added to achieve higher strength. Common additions are vanadium and boron. Such steels are seldom used in pressure part applications, but they are gaining acceptance as structural steels.

Residual elements are present in steels in small amounts and are elements other than those deliberately added as alloying or killing agents during the steelmaking process. Their source is the scrap or pig iron used in the furnace charge. Cu, Ni, Cr, V and B are typical examples of residuals often found in carbon steels. S and P, also considered to be residual elements, usually are reported in chemical analyses of steels, and their concentrations are limited by specification because they degrade ductility. The residual elements S, P, Sb and tin (Sn) are also important contributors to temper embrittlement in steels.

Historically, residual elements other than S and P were neither limited nor reported. This practice is changing, however, and several residuals have established limits.

Low and medium alloy steels are the next most important category of steels used in boilers. These are characterized by Cr contents less than 11.5% and lesser amounts of other elements. The most common alloy combinations in this group encountered in boilers are: C-1/2Mo, 1/2Cr-1/2Mo, 1Cr-1/2Mo, 1-1/4Cr-1/2Mo-Si, 2-1/4Cr-1Mo, and 9Cr-1Mo-V. Other less common alloys in this group are 3Cr-1Mo, 5Cr-1/2Mo and 9Cr-1Mo.

Because of the exceptional strength-enhancing capability of Mo in carbon steel, it is not surprising that C-1/2Mo steel has many applications for pressure parts, particularly in the temperature range of about 700 to 975F (371 to 524C). C-Mo steels, however, are particularly prone to graphitization at temperatures above about 875F (468C). Inside the boiler, where graphitization failures do not present a safety hazard, C-Mo tubing has many uses up to 975F (524C), its oxidation limit. Because Al content promotes graphitization, C-Mo steel is usually Si-killed and it has a coarse grain structure as a consequence. Therefore, C-Mo components are somewhat prone to brittle failures at low temperatures. This is not a problem in service, because the design application range of this alloy is at high temperature.

The oxidation resistance of low alloy steels increases with Cr content. The first common alloy in the Cr-Mo family is 1/2Cr-1/2Mo. This steel was developed in response to the graphitization failures of C-Mo piping. It was found that the addition of about 0.25% Cr was sufficient to make the alloy immune to graphitization. Furthermore, 1/2Cr-1/2Mo has essentially the same strength as C-Mo and has therefore displaced it in many applications. Because the application of 1/2Cr-1/2Mo is virtually unique to the boiler industry, it is less readily available in certain sizes and product forms.

The next alloys in this series are the nearly identical 1Cr-1/2Mo and 1-1/4Cr-1/2Mo-Si; the Si-containing version is slightly more oxidation resistant. However, extensive analyses of the databases indicate that the 1Cr-1/2Mo version is stronger over the temperature range of 800 to 1050F (427 to 566C). As a result, this alloy is rapidly displacing 1-1/4Cr-1/2Mo-Si in most applications in this temperature regime.

Absent the addition of other alloying elements, the 2-1/4Cr-1Mo composition is the optimum alloy for high temperature strength. Where the need for strength at temperatures between 975 and 1115F (524 and 602C) is the dominant design requirement, 2-1/4Cr-1Mo is the industry workhorse alloy. The 3Cr through 9Cr alloys are less strong, but they have application where improved oxidation resistance is desired and lower strength can be tolerated. The increasing air-hardenability of these alloys with increasing Cr content makes fabrication processes more complex and their use is somewhat more costly as a result.

Mn-Mo and Mn-Mo-Ni alloys have limited use in fossil-fueled boilers. Their slightly higher strength compared to carbon steels promotes their application in very large components, where the strength to weight ratio is an important consideration. Their generally superior toughness has made them a popular choice for nuclear pressure vessels.

The *heat treatable low alloy steels*, typified by the AISI-SAE 4340 grade (nominally 0.40C-0.80Cr-1.8Ni-0.25Mo), are used for relatively low temperature structural applications in boilers and nuclear pressure vessels. The instability of their microstructures and therefore, of their strength, with long exposure at elevated temperatures, has eliminated them from consideration for boiler pressure parts.

Higher Cr-Mo alloys Because of their tendency toward embrittlement, the martensitic 9Cr-1Mo and 12Cr-1Mo steels have not been widely used for pressure vessel and piping applications in North America prior to the 1980s. However, in the early 1970s, the United States (U.S.) Department of Energy (DOE) sponsored research to develop a 9Cr-1Mo steel with improved strength, toughness and weldability[12] for tubing in steam generators for liquid metal fast breeder nuclear reactors.[13] The alloy is 9Cr-1Mo-V, commonly called Grade 91. It has exceptional strength, toughness and stability at temperatures up to 1150F (621C). Because it is nearly twice as strong as 2-1/4Cr-1Mo at 1000F (538C), it is displacing that alloy in high pressure header applications. The resultant thinner vessels have significantly reduced thermal stresses and associated creep-fatigue failures compared to 2-1/4Cr-1Mo and 1-1/4Cr-1/2Mo-Si headers.[14,15] Because Grade 91 is stronger than austenitic stainless steel up to about 1125F (607C), it is also displacing that alloy class in high pressure tubing applications. It has the added advantage of being a ferritic alloy, eliminating the need for many dissimilar metal welds between pressure parts. The Grade 91 gains its strength, toughness, and stability from its alloy additions and features the fully bainitic microstructure resulting from careful normalizing and tempering.

While not popular in North America because of the care necessary in handling very air-hardenable alloys during fabrication, the 12Cr-Mo and 12Cr-Mo-V alloys have had wide use in the European boiler industry. In addition, the experience being gained with Grade 91 may eventually enhance the acceptance of the 12Cr group.

Austenitic stainless steel Every attempt is made to minimize the use of stainless steels in boilers because of their high cost, but the combination of strength and corrosion resistance they provide makes them the favored choices in certain applications. They are virtually the only choices for service above 1150F (621C). At lower temperatures, down to about 1050F (566C), they often displace the Cr-Mo ferritic steels, where the lower pressure drop afforded by the thinner stainless steel component wall is important.

The common alloys of stainless steels used in boiler pressure parts are 18Cr-8Ni, 18Cr-8Ni-Ti, 18Cr-8Ni-Cb, 16Cr-12Ni-2Mo, 25Cr-12Ni, 25Cr-20Ni, and 20Cr-30Ni. The last alloy in this group is technically a nonferrous alloy, because it has less than 50% Fe when its other minor alloying constituents are considered. However, because it is so similar to the other austenitic stainless steels, it may be considered one of them. These alloys are commonly designated the 300 series: 304, 321, 347, 316, 309 and 310 stainless steels, respectively. The 20Cr-30Ni alloy is commonly known as Alloy 800. Because the strength of these materials at high temperature is dependent on a moderate carbon content and usually on a coarse grain size, materials with those qualities are often specified for high temperature service. They carry the added designation of the letter H, e.g., 304H or 800H.

Of these alloys, 304H is the most commonly used. It provides an excellent balance of strength and oxidation resistance at the lowest cost of any alloy in this group. However, if severe aqueous corrosive conditions may exist either before or during service, especially in solutions containing halogens, other stainless steel alloys should be substituted. The 304H alloy is susceptible to sensitization at grain boundaries and, thus, may suffer stress corrosion cracking or intergranular attack in that environment. In these cases, a stabilized stainless steel composition such as 347H should be used. Further details on how the American Society of Mechanical Engineers (ASME) Code establishes allowable design stresses for materials can be found in the Appendix to ASME Section II, Part D.

All of the 300 series alloys are susceptible to sigma phase formation after long exposure at temperatures of 1050 to 1700F (566 to 927C). Those with some initial ferrite, such as 309, can form the sigma phase earlier, but all eventually do so. This phase formation decreases toughness and ductility but has no effect on strength or corrosion resistance. It has been a problem in heavy-section piping components made of 316 stainless steels, but it is not a design consideration for smaller (tubing) components.

The 321 type is not as strong as the others in this series. While it is a stabilized grade and has important low temperature applications, the stability of the titanium carbide makes it extremely difficult to heat treat type 321 in one thermal treatment and obtain a resulting structure that is both coarse grained, for high temperature creep strength, and has stabilized carbides for sensitization resistance. It is possible to apply a lower temperature stabilizing heat treatment, at about 1300F (704C), following the solution treatment to achieve a stabilized condition and good creep strength. The stability of the columbium (niobium) carbides in type 347 is better, and this grade can be heat treated to obtain creep strength and sensitization resistance. This 18Cr-8Ni-Cb alloy is widely used at high temperatures because of its superior creep strength.

The Mo content of the 316 type increases its pitting resistance at lower temperatures. While this alloy has good creep strength, it is not often used because of its higher cost.

These alloys are susceptible to stress corrosion cracking in certain aqueous environments. The 300 series alloys are particularly sensitive to the presence of halide ions. As a result, their use in water-wetted service is usually prohibited. The stress corrosion cracking experience with Alloy 800 has been mixed and, while this grade is permitted in water-wetted service, it is not common practice.

Types 309, 25Cr-12Ni, and 310, 25Cr-20Ni, have virtually identical strengths and corrosion resistance. They are not as strong as 304 or 347 but are more oxidation resistant. The high Ni alloys, like Alloy 800, are somewhat more affected by sulfidation attack. They have been used as nonpressure fluidized-bed boiler components designed to remove particulate from hot gas streams.

Most of these alloys are available in a multiplicity of minor variations: H grades, with 0.04 to 0.10% C and a required high temperature anneal and coarse grain size for creep strength; L grades, with 0.035% maximum C for sensitization resistance; N grades, with 0.010% minimum N added for strength; LN grades, with 0.035% maximum C and 0.010% minimum N for sensitization resistance and strength; and straight (no suffix) grades, with 0.08% maximum C.

Ferritic stainless steels contain at least 10% Cr and have a ferrite-plus-carbide structure. *Martensitic stainless steels* are ferritic in the annealed condition but are martensitic after rapid cooling from above the critical temperature. They usually contain less than 14% Cr.[16] *Precipitation hardened stainless steels* are more highly alloyed and are strengthened by precipitation of a finely dispersed phase from a supersaturated solution on cooling. None of these steels are used for pressure parts or load carrying components in boilers because, at the high temperatures at which their oxidation resistance is useful, they are subject to a variety of embrittling, phase precipitation reactions, including 885F (474C) embrittlement and sigma phase formation. They are used as studs for holding refractories and heat absorbing projections and as thermal shields. These alloys are also difficult to weld without cracking.

Duplex alloys, with mixed austenitic-ferritic structures, have been developed. They are useful in corrosive lower temperature applications such as those found in wet desulfurization equipment used as boiler flue gas scrubbers.

Bimetallic materials Weld cladding of one alloy with another has been available for many years. A more recent development has been the proliferation of *bimetallic* components, such as tubes and plate containing a load carrying alloy for their major constituent covered with a layer of a corrosion resistant alloy. The first bimetallic tubes to see wide use in boilers were made from Alloy 800H clad with a 50Cr-50Ni alloy (Alloy 671) for coal ash corrosion resistance. The combination in widest use today is carbon steel clad with 304L, used in pulp and paper process recovery (PR) boilers. One of the latest to be developed is carbon steel or 1/2Cr-1/2Mo clad with Alloy 825 (42Ni-21.5Cr-5Mo-2.3Cu) used in PR and refuse-fired boilers.[17] Other combinations that have been used are 1/2Cr-1/2Mo and 2-1/4Cr-1Mo clad with 309.

Cast irons Cast irons and cast steels (containing more than 2% or less than 2% C, respectively) have long had wide acceptance as wear resistant and structural components in boilers. Cast steels are also used for boiler pressure parts. The three types of cast iron used in boilers are white, gray and ductile iron.

White iron White cast iron is so known because of the silvery luster of its fracture surface. In this alloy, the carbon is present in combined form as the iron carbide cementite (Fe_3C). This carbide is chiefly responsible for the hardness, brittleness and poor machinability of white cast iron. Chilled iron differs from white cast iron only in its method of manufacture and it behaves similarly. This type of iron is cast against metal blocks, or chills, that cause rapid cooling at the adjacent areas, promoting the formation of cementite. Consequently, a white or mottled structure, which is characterized by high resistance to wear and abrasion, is obtained. Elverite® alloys, a series of white iron, Ni-enriched cast materials developed by The Babcock & Wilcox Company (B&W) for use in pulverizers and other wear resistant parts, have long been noted for their uniformity and high quality.

VAM® 20, a more recent development, is a 20% Cr white iron with a carbide-in-martensite matrix, very high hardness and good toughness (compared to other white irons). The hardness and wear resistance of VAM 20 are superior to those of the Elverites and similar alloys. It is always used in the heat treated condition, which accounts for its good toughness and uniformity. VAM 20 is used in grinding elements of coal pulverizers.

Malleable cast iron is white cast iron that has been heat treated to change its combined carbon (cementite) into free, or temper carbon (nodules of graphite). The iron becomes malleable because, in this condition, the carbon no longer forms planes of weakness.

Gray iron Gray cast iron is by far the most widely used cast metal. In this alloy, the carbon is predominantly in the free state in the form of graphite flakes, which form a multitude of notches and discontinuities in the iron matrix. The fracture appearance of this iron is gray because the graphite flakes are exposed. Gray iron's strength depends on the size of the graphite crystals and the amount of cementite formed with the graphite. The strength of the iron increases as the graphite crystal size decreases and the amount of cementite increases. Gray cast iron is easily machinable because the graphite acts as a lubricant. It also provides discontinuities that break the chips as they are formed. Modern gray iron having a wide range of tensile strength, from 20,000 to 90,000 psi (138 to 621 MPa), can be made by suitable alloying with Ni, Cr, Mo, V and Cu.

Ductile iron Another member of the cast iron family is *ductile cast iron*. It is a high carbon, Mg-treated ferrous product containing graphite in the form of spheroids or impacted particles. Ductile cast iron is similar to gray cast iron in melting point, fluidity and machinability, but it possesses superior mechanical properties. This alloy is especially suited for pressure castings. By special procedures (casting against a chill), it is possible to obtain a carbide-containing abrasion resistant surface with an interior of good ductility.

Cast iron was used extensively in early steam boilers for tubes and headers. This material is no longer used in the pressure parts of modern power boilers but is used in related equipment such as stoker parts and the grinding elements of coal pulverizers.

Cast alloys Cast steels and non-ferrous alloys are used for many support and alignment applications in boilers, and for some pressure parts having complex shapes. The alloys range from carbon steel and 2-1/4Cr-1Mo to 25Cr-12Ni and 50Cr-50Ni.

Ceramics and refractory materials Ceramics and refractory materials are used for their insulating and erosion resisting properties. Brick furnace walls have mostly been replaced by steel membrane panels. (See Chapter 23.) However, in many applications, these walls may still have a rammed, troweled or cast refractory protection applied. Refractory linings are still important features of some furnaces, particularly those exposed to molten slag. In Cyclone furnaces (see Chapter 15) and other wet-bottom boilers, gunned and troweled alumina and silicon carbide refractory products are generally used. Chromium-containing refractories are no longer in general use since being classified as a hazardous material.

Cera-VAM® is a high density alumina ceramic used as an erosion liner in coal-air pipeline elbows, coal pulverizer internals, and pulverizer swing valves to reduce erosion and the associated maintenance costs. (See Chapter 13.) Structural ceramics have also been introduced as hot gas filters. These filters remove particulates from the flue gas of fluidized-bed boilers before the gas enters the high temperature gas turbine of combined cycle plants. (See Chapter 17.)

Coatings Many types of coatings are applied to boiler metal parts. In addition to the cast, gunned and troweled types mentioned above, thinner carbide-containing, metallic matrix coatings are sprayed onto surfaces in boilers exposed to high velocity particulate erosion. Metallic coatings are sprayed on boiler parts exposed to erosion and corrosion wastage by the *flame spraying, twin-wire electric arc, plasma* and *high velocity oxy-fuel* processes. These are shop- and field-applied maintenance processes that protect and repair components that experience wastage. Proper surface preparation and process control must be exercised to ensure that these coatings adhere, have the proper density, and achieve the recommended thickness on all surfaces.

Chromizing In the mid 1970s, B&W pioneered the use of chemical vapor deposition (CVD) coatings for boiler components. *Chromizing*, a process previously applied to aircraft jet engine components, is applied to large surfaces on the interior of tubing and piping. The purpose of this process is to develop a high Cr-containing surface that is resistant to oxidation and subsequent exfoliation. High temperature steam carrying pressure parts suffer from oxidation on their internal surfaces. When the oxide layer becomes thick enough, it spalls off the surface and the particles are carried to the steam turbine, where the resulting erosion damage causes loss of efficiency and creates a risk of mechanical damage. Perfect coverage of tube inside diameter (ID) surfaces is not necessary to reduce this condition. If 95% of the susceptible tube surface is chromized, a twenty-fold reduction in exfoliate particles will result.

In CVD processes, such as chromizing, the surfaces to be coated are usually covered with or embedded in a mixture containing powdered metal of the coating element, e.g., Cr, a halide salt, and a refractory powder, often alumina. When the parts and the mixture are heated to a sufficiently high temperature, the salt decomposes and the metal powder reacts with the halide ion to form a gas, e.g., $CrCl_2$ or $CrBr_2$. At the surface of the part being coated, an exchange reaction takes place. An Fe atom replaces the Cr in the gas and the Cr atom is deposited on the surface. The process is conducted at sufficient time and temperature to permit the Cr to diffuse into the base material. At the chromizing temperature, 2-1/4Cr-1Mo, for example, is fully austenitic. However, as Cr atoms are deposited on the surface, the Cr increases the stability of the ferrite phase. As a result, the diffusion front advances into the matrix concurrently with the phase transformation front. This results in a diffusion zone with a nearly constant Cr content. (See Fig. 14.) Typical depths of this zone range from 0.002 to 0.025 in. (0.051 to 0.64 mm). The diffusion layer on a 2-1/4Cr-1Mo substrate has a Cr content range of 30 to 13%.

Chromizing, though first developed to reduce solid particle erosion of turbines, is now being applied to external surfaces of boiler pressure parts to reduce or prevent corrosion and corrosion-fatigue damage. In these applications, near perfect continuity and integrity of the coating is required. A thicker coating is necessary to resist the more hostile external environments. Improvements in chromized coating composition and processing have been achieved. Co-diffusion of Cr and Si, or Cr and Al, is now possible, improving the corrosion resistance of the coating. Process improvements have allowed for shorter times at diffusing temperature, thus resulting in less undercoating decarburization and better material properties.

Aluminizing Aluminizing, a similar CVD process, has been used for many years to protect components in petrochemical process pressure vessels. However, alumina, as silica, is soluble in high temperature, high pressure steam and it can be carried to the turbine, where pressure and temperature drops cause it to precipitate on the turbine components; this is undesirable. Aluminizing, either with diffusion processes

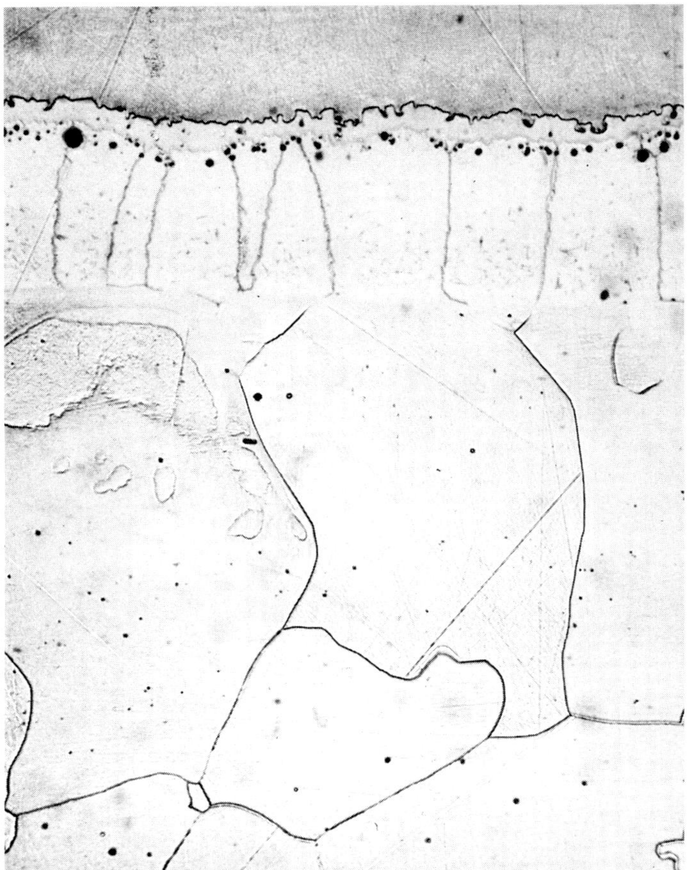

Fig. 14 Chromized 2-1/4Cr-1Mo at 400 X magnification.

or spray metallizing, has also seen some use as an external, fireside surface protective coating. The diffusion coating over iron-based alloys does create a brittle iron aluminide phase that can lead to premature loss of the coating, but it has had long term success in a number of petrochemical applications, especially sulfuric acid service, and where carburization (metal dusting) must be avoided.

Fused coatings Tungsten carbide/chromium carbide fused metallic coatings are also used for erosion protection of tube membrane panels, for example, in basic oxygen furnace steelmaking furnace hoods. Fused coatings differ from sprayed coatings in possessing higher density and achieving better bond strength due to the brazing-type action of the application process, which includes a high temperature heat treatment following the coating application using a conventional metallizing process.

Galvanizing More mundane coatings, such as galvanizing, painting and organic rust prevention coatings, are also used on boiler components.

Galvanizing, a zinc coating usually applied by dipping in molten metal or by electroplating, is usually used on structural components external to the boiler, when erection is near a seashore or a petrochemical complex.[18] Galvanized components must be kept out of high temperature areas to avoid structural damage due to zinc grain boundary embrittlement, generally believed to occur at temperatures above 450F (232C).

Mechanical properties

Low temperature properties

Steels of different properties are used in boilers, each selected for one or more specific purposes. Each steel must have properties for both manufacturing and satisfactory service life. Each particular type, or grade, of steel must be consistent in its properties, and tests are normally run on each lot to demonstrate that the desired properties have been achieved.

Specifications standardizing all the conditions relating to test specimens, methods and test frequency have been formulated by the American Society for Testing and Materials (ASTM) and other authorities.

Tensile test

In the tensile test, a gradually applied unidirectional pull determines the maximum load that a material can sustain before breaking. The relationship between the stress (load per unit area) and the corresponding strain (change of length as a percent of the original length) in the test piece is illustrated in the stress-strain diagrams of Figs. 15 and 16. The metal begins to stretch as soon as the load is applied and, for some range of increasing load, the strain is proportional to the stress. This is the elastic region of the stress-strain curve, in which the material very closely follows Hooke's Law: strain, ε, is proportional to stress, σ. The proportionality constant may be considered as a spring constant and is called Young's modulus, E. Young's modulus is a true material property, characteristic of each alloy. Young's modulus for steel is approximately 30×10^6 psi (206.8×10^6 kPa) at room temperature.

If the stress is released at any point in this region, the test specimen will return to very nearly its initial dimensions. However, if the stress is increased beyond a certain point, the metal will no longer behave elastically; it will have a permanent (plastic) elongation, and the linear relationship between stress and strain ceases. This value is known as the proportional limit of the material and, in this discussion, may be considered practically the same as the elastic limit, which may be defined as the maximum stress that can be developed just before permanent elongation occurs.

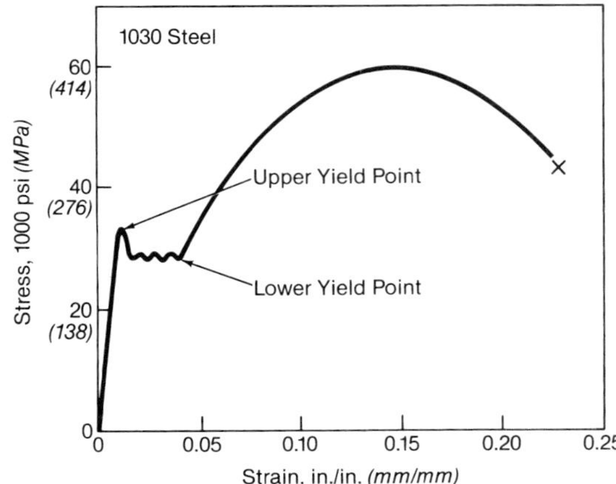

Fig. 15 Engineering stress-strain curve for 1030 carbon steel (*courtesy of Wiley*).[19]

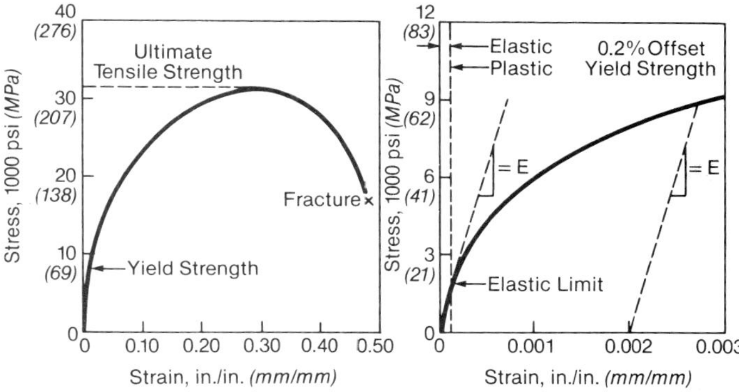

Fig. 16 Engineering stress-strain diagram for polycrystalline copper. Left, complete diagram. Right, elastic region and initial plastic region showing 0.2% offset yield strength.[19]

When a material has a well defined point at which it continues to elongate without further increase in load, this point is called the yield point. Many steels do not have a yield point and even in those that do, neither it nor the proportional or elastic limits can be determined with accuracy. By convention, therefore, engineers have adopted an arbitrary but readily measurable concept: the *yield strength* of a metal. This is defined as the stress at which the strain reaches 0.2% of the gauge length of the test specimen. This is illustrated in Fig. 16. (Other values, 0.1% or 0.5% are occasionally used, but 0.2% is most common.)

If the loading is continued after yielding begins, a test specimen of a ductile material with homogeneous composition and uniform cross-section will be elongated uniformly over its length, with a corresponding reduction in area. Eventually, a constriction or necking may occur. In some materials, localized necking may not occur, but the cross-section may reduce more or less uniformly along the full gauge length to the instant of rupture. In all ductile materials, however, an appreciable increase in elongation occurs in the reduced area of the specimen. The more ductile the steel, the greater is the elongation before rupture. The maximum applied load required to pull the specimen apart, divided by the area of the original cross-section, is known as the *ultimate tensile strength*. Brittle materials do not exhibit yielding or plastic deformation, and their yield point and ultimate tensile strength are nearly coincident.

The ductility of the metal is determined by measuring the increase in length (total elongation) and the final area at the plane of rupture after the specimen has broken, and is expressed as percent elongation or percent reduction of area.

Hardness test

Hardness may be defined as resistance to indentation under static or dynamic loads and also as resistance to scratching, abrasion, cutting or drilling. To the metallurgist, hardness is important as an indicator of the effect of heat treatment, fabrication processes, or service exposure. Hardness values are roughly indicative of the ultimate tensile strength of steels. Hardness tests are also used as easy acceptance tests and to explore local variations in properties.

Hardness is usually determined by using specially designed and standardized machines: Rockwell, Brinell, Vickers (diamond pyramid), or Tukon. These all measure resistance to indentation under static loads. The pressure is applied using a fixed load and for a specified time, and the indentation is measured either with a microscope or automatically. It is expressed as a hardness number, by reference to tables. Hardness can also be determined by a scleroscope test, in which the loss in kinetic energy of a falling metal weight, absorbed by indentation upon impact of the metal being tested, is indicated by the height of the rebound.

Toughness tests

Toughness is a property that represents the ability of a material to absorb local stresses by plastic deformation and thereby redistribute the stresses over a larger volume of material, before the material fails locally. It is therefore dependent on the rate of application of the load and the degree of concentration of the local stresses. In most steels, it is also temperature dependent, increasing with increasing temperature (although not linearly). Toughness tests are of two types, relative and absolute.

Notched bar impact tests are an example of the relative type. The most common is the Charpy test, in which a simple horizontal beam, supported at both ends, is struck in the center, opposite a V-shaped notch, by a single blow of a swinging pendulum. A Charpy specimen is illustrated in Fig. 17a. The energy absorbed by the breaking specimen can be read directly on a calibrated scale and is expressed in ft lb units. The specimen is also examined to determine how much it has spread laterally and how much of its fracture surface deformed in shear versus cleavage. The toughness is expressed in units of absorbed energy (ft lb or J), mils (thousandths of an inch or mm) lateral expansion and percent shear. The values are characteristic of not only the material and temperature, but also of the specimen size. Therefore, comparison between materials and tests have meaning only when specimen geometries and other test conditions are identical. Specimens are inexpensive and the test is easy to do. Often, vessel designers are interested in the variation of toughness with temperature. Fig. 18 illustrates the variation in toughness with temperature of 22 heats of a fine grained carbon steel, SA-299, as determined by Charpy testing. This material displays a gradual transition from higher to lower toughness.

Another toughness test, and one that provides a more sharply defined transition, is the drop-weight test. The specimen for this test is shown in Fig. 17b. A known weight is dropped from a fixed height and impacts the specimen. This is a pass or fail test and is performed on a series of specimens at varying temperatures, selected to bracket the break versus no-break temperature within 10F (6C). If the impact causes a crack to propagate to either edge of the specimen from the crack-starter notch in the brittle weld bead deposited on the face of the specimen, the specimen is considered to have broken at that temperature. The lowest temperature at which a specimen fails determines the nil-ductility transition temperature

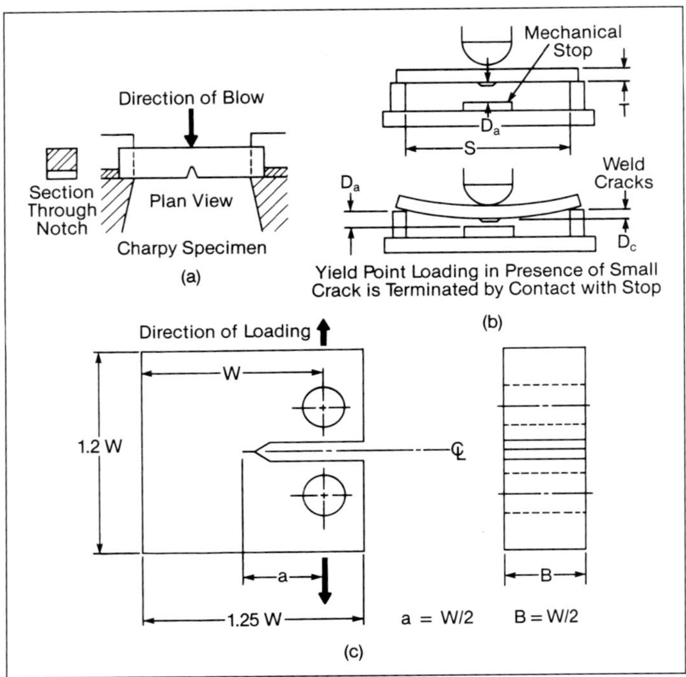

Fig. 17 (a) Charpy specimen, (b) nil ductility transition temperature (drop weight) test specimen, (c) compact tension specimen (*courtesy of Prentice-Hall*).[20]

(NDTT). Fig. 19 shows a histogram of NDTTs from 20 heats of fine grained SA-299.

Fracture toughness tests measure true characteristics of a given metal. They are more complex and specimens are more costly. However, they produce values that can be used in analytical stress calculations to determine critical flaw sizes above which flaws or cracks may propagate with little or no increase in load. A typical fracture toughness specimen is shown in Fig. 17c. Variations of fracture toughness tests involve testing under cyclic rather than monotonically increasing load (fatigue crack growth testing) and testing in various environments to determine crack growth rates as a function of concurrent corrosion processes. The same specimen is used to determine fatigue crack growth behavior. Fig. 20 illustrates the difference in crack growth rate in air and in a salt solution for 4340 steel tempered to two strength levels.

Formability tests

Several different types of deformation tests are used to determine the potential behavior of a material in fabrication. These include bending, flattening, flaring and cupping tests. They furnish visual evidence of the capability of the material to withstand various forming operations. They are only a rough guide and are no substitute for full scale testing on production machinery.

High temperature properties

Tensile or yield strength data determined at ambient temperatures can not be used as a guide to the mechanical properties of metals at higher temperatures. Even though such tests are made at the higher temperatures, the data are inadequate for designing equipment for long term service at these temperatures. This is true because, at elevated temperatures, continued application of load produces a very slow continuous deformation, which can be significant and measurable over a period of time and may eventually lead to fracture, depending on the stress and temperatures involved. This slow deformation (creep) occurs for temperatures exceeding about 700F (371C) for ferritic steels and about 1000F (538C) for austenitic steels.

Tensile strength

Although the design of high temperature equipment generally requires use of creep and creep-rupture test data, the short time tensile test does indicate the strength properties of metals up to the creep range of the material. This test also provides information on ductility characteristics helpful in fabrication.

The ultimate strength of plain carbon steel and a number of alloy steels, as determined by short time tensile tests over a temperature range of 100F (38C) to 1300 to 1500F (704 to 816C), is shown in Fig. 21. In general, the results of these tests indicate that strength decreases with increase in temperature, although there is a region for the austenitic alloys between 400 and 900F (204 and 482C) where strength is fairly constant. An exception to the general rule is

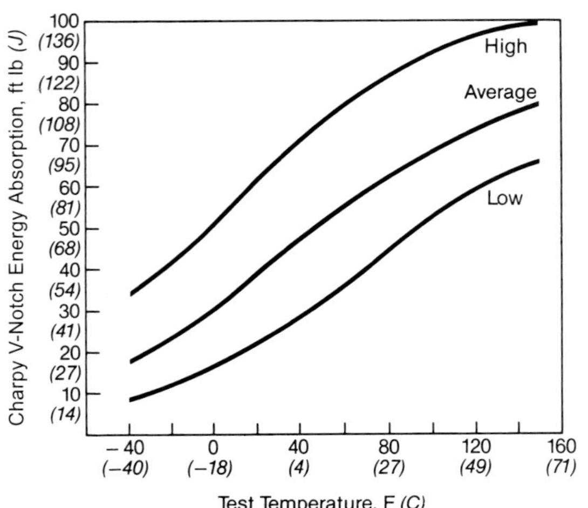

Fig. 18 Charpy V-notch impact energy versus test temperature for fine grained SA-299 plate material.

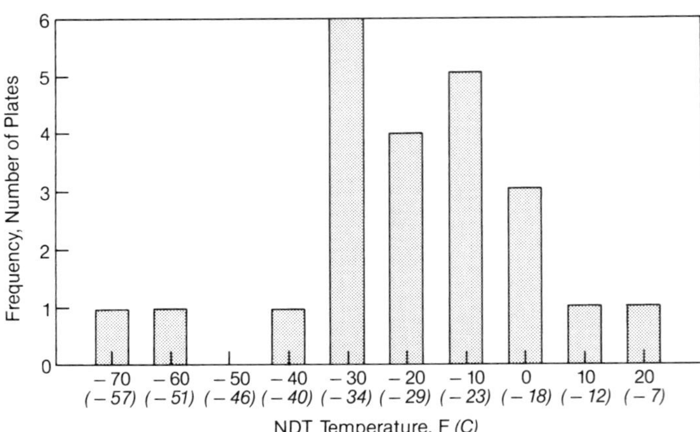

Fig. 19 Drop weight nil-ductility transition temperature (NDTT) frequency distribution for 20 heats of fine grained SA-299 plate material.

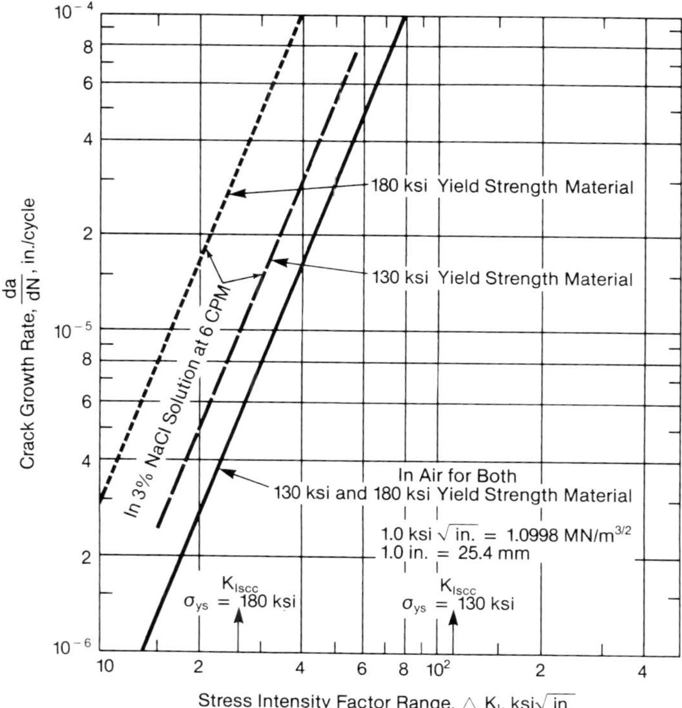

Fig. 20 Corrosion fatigue crack growth rates for 4340 steel.[20]

the increase in strength over that at room temperature of carbon and many low alloy steels (with corresponding decrease in ductility) over the temperature range of 100 to 600F (38 to 316C). As the temperature is increased beyond 600 to 750F (316 to 399C), the strength of the carbon and most of the low alloy steels falls off from that at room temperature with a corresponding increase in ductility.

Creep and creep-rupture test

It has long been known that certain nonmetallic materials, such as glass, undergo slow and continuous deformation with time when subjected to stress. The concept of creep in metallic materials, however, did not attract serious attention until the early 1920s. Results of several investigations at that time demonstrated that rupture of a metallic material could occur when it is subjected to a stress at elevated temperatures for a sufficiently long time, even though the load applied is considerably lower than that necessary to cause rupture in the short time tensile test at the same temperature.

The earliest investigations of creep in the U.S. were sponsored by B&W in 1926. Many steels now used successfully in power generating units and in the petroleum refining and chemical industries were tested and proved in the course of these investigations, using the best equipment available at the time.

The creep-rupture test is used to determine both the rate of deformation and the time to rupture at a given temperature. The test piece, maintained at constant temperature, is subjected to a fixed static tensile load. The deformation of the test sample is measured during the test and the time to rupture is determined. The duration of the test may range from 1000 to 10,000 h, or even longer. A diagrammatic plot of the observed length of the specimen against elapsed time is often of the form illustrated in Fig. 22.

The curve representing classical creep is divided into three stages. It begins after the initial extension (0-A), which is simply the measure of deformation of the specimen caused by the loading. The magnitude of this initial extension depends on test conditions, varying with load and temperature and normally increasing with increases in temperature and load. The first stage of creep (A-B), referred to as primary creep, is characterized by a decreasing rate of deformation during the period. The second stage (B-C), referred to as secondary creep, is usually characterized by extremely small variations in rate of deformation; this period is essentially one of constant rate of creep. The third stage (C-D), referred to as tertiary creep, is characterized by an accelerating rate of deformation leading to fracture. Some alloys, however, display a very limited (or no) secondary creep and spend most of their test life in tertiary creep.

To simplify the practical application of creep data it is customary to establish two values of stress (for a material at a temperature) that will produce two corresponding rates of creep (elongation): 1.0% per 10,000 h and 100,000 h, respectively.

For any specified temperature, several creep-rupture tests must be run under different loads. The creep rate during the period of secondary creep is determined from these curves and is plotted against the stress. When these data are plotted on logarithmic scales, the points for each specimen often lie on a line

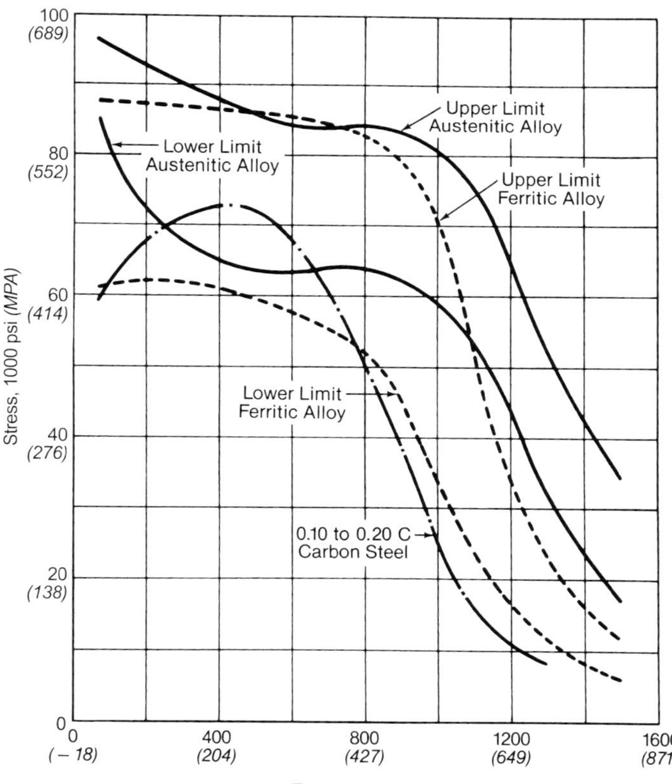

Fig. 21 Tensile strength of various steels at temperatures to 1500F (816C).

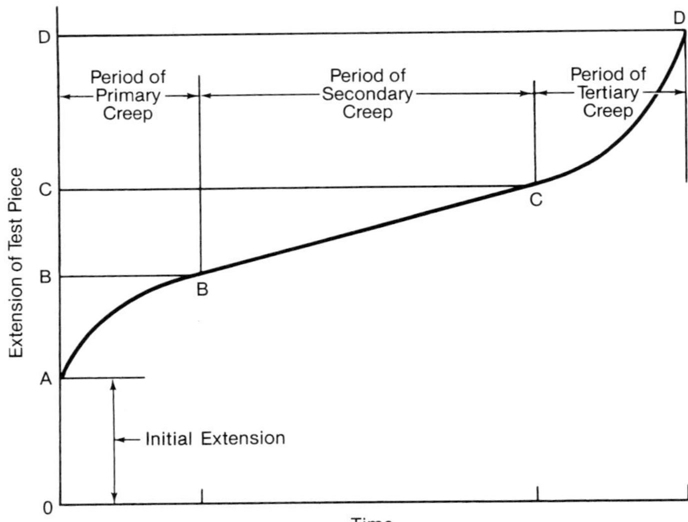

Fig. 22 Classic (diagrammatic) creep test at constant load and temperature.

with a slight curvature. The minimum creep rate for any stress level can be obtained from this graph, and the curve can also be extrapolated to obtain creep rates for stresses beyond those for which data are obtained. Fig. 23 presents such creep rate curves for 2-1/4Cr-1Mo steel at 1000, 1100 and 1200F (538, 593 and 649C). The shape of the creep curve depends on the chemical composition and microstructure of the metal as well as the applied load and test temperature.

Creep-rupture strength is the stress (initial load divided by initial area) at which rupture occurs in some specified time, in an air atmosphere, in the temperature range in which creep takes place. The time for rupture at any temperature is a function of the applied load. A logarithmic-scale plot of stress versus time for fracture of specimens generally takes the form of the curves shown for 2-1/4Cr-1Mo steel in Fig. 24.

In general, rapid rates of elongation indicate a transgranular (ductile) fracture and slow rates of elongation indicate an intergranular (brittle) fracture. As a rule, surface oxidation is present when the fracture is transgranular, while visible intercrystalline oxidation may or may not be present when the fracture is intergranular. Because of the discontinuities produced by the presence of intercrystalline oxides, the time to rupture at a given temperature-load relationship may be appreciably reduced. In Fig. 24, the slope of the data at 1200F (649C) is steeper than those for lower temperatures. This is to be expected, because 1200F (649C) is above the usual temperature limit for maximum resistance to oxidation of 2-1/4Cr-1Mo. Therefore, excessive scaling occurs in the long time rupture tests conducted at 1200F (649C).

A complete creep-rupture test program for a given steel actually consists of a series of tests at constant temperature with each specimen loaded at a different level. Because tests are not normally conducted for more than 10,000 h, the values for rupture times longer than this are determined by extrapolation. The ASME Boiler and Pressure Vessel Code Committee uses several methods of extrapolation, depending on the behavior of the particular alloy for which design values are being established and on the extent and quality of the database that is available. Several informative discussions on these methods may be found in ASME publications. (See Appendix 2.)

Material applications in boilers

ASME specifications and allowable stresses

The ASME Boiler and Pressure Vessel Code Subcommittee on Materials is responsible for identifying and approving material specifications for those metals deemed suitable for boiler and pressure vessel construction and for developing the allowable design values for metals as a function of temperature. Most industrial and all utility boilers are designed to Section I of the Code, *Power Boilers*, which lists those material specifications approved for boiler construction. The specifications themselves are listed in Section II, Part A, *Ferrous Materials*, and Section II, Part B, *Non-Ferrous Materials*. The design values are listed in Section II, Part D, *Properties*. (Section II, Part C, *Specifications for Welding Rods, Electrodes, and Filler Metals*, contains approved welding materials.)

For many years, there were relatively few changes in either specifications or design values. Over the last five or ten years, however, many new alloys have been introduced and much new data have become available. The restructuring of the North American steel industry and the globalization of sources of supply and markets are partly responsible for this rapid rate of change. As a result, detailed tables of design values in a text such as this become obsolete much more rapidly than was once the case. A few examples of current allowable stresses are presented below for illustrative purposes. However, the reader is encouraged to consult Section II for an exposure to material specifications and the latest design values.

The maximum allowable working stresses for materials in power boilers, set by the American Society of Mechanical Engineers (ASME), are based on both time-independent and time-dependent properties.

The ASME Boiler and Pressure Vessel Code, Section I, *Power Boilers*, has established the maximum

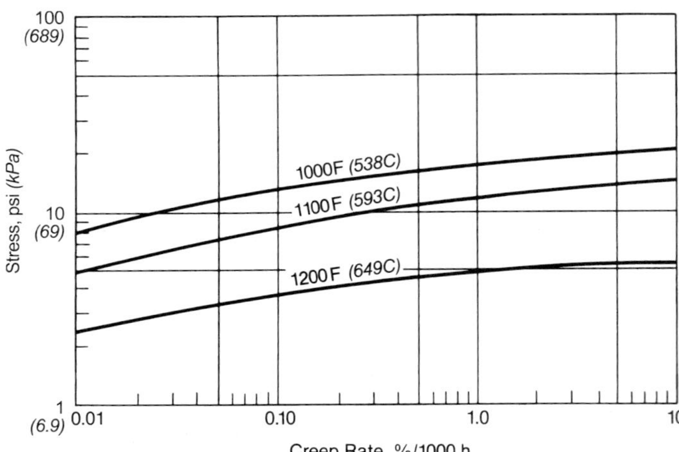

Fig. 23 Creep rate curves for 2-1/4Cr-1Mo steel.

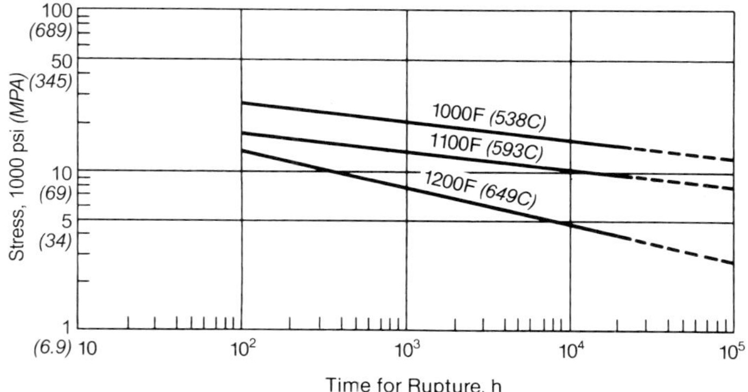

Fig. 24 Typical creep rupture curves for 2-1/4Cr-1Mo steel.

allowable design stress values for pressure parts to be no higher than the lowest of:

1. 1 / 3.5 × the minimum specified ultimate tensile strength,
2. 1.1 / 3.5 × the tensile strength at temperature,
3. 67% of the specified minimum yield strength at room temperature,
4. 67% of the yield strength at temperature, for ferritic steels; or 90% of the yield strength at temperature of austenitic steels and nickel base alloys,
5. a conservative average of the stress to give a creep rate of 0.01% in 1000 hours (1% in 100,000 hours), or
6. 67% of the average or 80% of the minimum stress to produce rupture in 100,000 hours.

Furthermore, the allowable stress at a higher temperature can not exceed that at a lower temperature, so no advantage is taken of strain aging behavior. The allowable stress is therefore the lower bound envelope of all these criteria. The tensile and yield strengths at temperature have a particular meaning in Code usage.

For austenitic materials that possess allowable design properties above 1500F (816C), the ASME Code has recently adopted an additional criterion applied to creep rupture data of the specific alloy. The criterion involves a statistical evaluation of the data and ensures that a consistent safety factor exists at the very high temperatures.

Pressure part applications

The metal product forms used in boiler pressure parts are tubes and pipe (often used interchangeably), plate, forgings and castings. Tubular products compose the greatest part of the weight. The matrix in Table 1 shows the common pressure part material specifications used in fossil fuel fired boilers today, their minimum specified properties, recommended maximum use temperatures, and their applications. This list is not meant to be all inclusive, as there are many other specifications permitted by Section I and several of them see occasional use. Neither is it meant to be exclusive, as several of these specifications are used occasionally for components not checked in Table 1. Finally, the recommended maximum use temperatures represent one or more of a variety of limits. The temperature listed may be the highest for which stresses are listed in Section I, the oxidation limit for long-term service, a temperature at which graphitization may be expected, or current commercial practice, whichever is least.

Boiler, furnace waterwall, convection pass enclosures and economizers

Boiler, furnace waterwall, and convection pass enclosure surfaces are generally made of carbon steel, C-Mo, and 1/2Cr-1/2Mo seamless or electric-resistance-welded (ERW) tubes. Lower carbon grades and 1/2Cr-1/2Mo alloy are used in high heat input regions to avoid the risk of graphitization in this region where tube metal temperatures may be subject to more fluctuation and uncertainty. Higher carbon grades and C-Mo are used in furnace floors, upper furnace walls, convection pass enclosures, and economizers.

The boiler industry is seeing increasing use of higher Cr-Mo grades in these applications, especially as temperatures and pressures increase with the latest designs. 1Cr-1/2Mo, 1-1/4 Cr-1/2Mo-Si, and even 2-1/4 Cr-1Mo will see increased use. Even higher alloys present formidable manufacturing challenges that will likely restrict these applications to alloys possessing less than 3% Cr.

Superheaters and reheaters

The highest metal temperatures of pressure parts in the steam generating unit occur in the superheater and reheater. Consequently, these tubes are made of material having superior high temperature properties and resistance to oxidation. Carbon steel is a suitable and economical material to about 850 to 950F (454 to 510C) metal temperature, depending on pressure. Above this range, alloy and stainless steels are required because of the low oxidation resistance and the low allowable stresses of carbon steel. Usually two or more alloys are used in the construction of the superheater. The lower alloys, such as carbon and C-Mo steels, are used toward the inlet section, while the low and intermediate alloy Cr-Mo steels are used toward the outlet, where the steam and metal temperatures are higher. (See Chapter 19.)

Stainless steel tubes have been required in the hottest sections of the superheater. However, stainless steels are being replaced in many applications by 9Cr-1Mo-V. This high strength ferritic steel was developed initially by The Oak Ridge National Laboratory for fast breeder nuclear reactor components. However, it has found many applications in fossil fuel-fired boilers because of its high strength and excellent toughness. Because it is ferritic, its use in place of stainless steel eliminates dissimilar metal weld failures.

New alloys, both ferritic and austenitic, are constantly being developed and are appearing in boilers around the world. The most promising ferritic alloys use alloying additions of elements such as V, Cb (Nb), W and N which, when combined with controlled normalizing and tempering heat treatment, result in materials possessing creep strength far superior to the traditional Cr-Mo alloy grades in the 2.25 to 12% Cr range. Newer austenitic alloys use modified alloying with elements such as Cu, Cb (Nb) and N, sometimes combined with special thermal mechanical processing,

Table 1
Boiler Materials and Typical Applications (English Units)

Specification	Nominal Composition	Product Form	Min Tensile, ksi	Min Yield, ksi	High Heat Input Furn Walls	Other Furn Walls and Enclosures	SH RH Econ	Unheated Conn Pipe <10.75 in. OD	Headers and Pipe >10.75 in. OD	Drums	Recomm Max Use Temp, F	Notes
SA-178A	C-Steel	ERW tube	(47.0)	(26.0)	X	X	X				950	1,2
SA-192	C-Steel	Seamless tube	(47.0)	(26.0)	X	X	X	X			950	1
SA-178C	C-Steel	ERW tube	60.0	37.0		X	X				950	2
SA-210A1	C-Steel	Seamless tube	60.0	37.0	X	X	X	X			950	
SA-106B	C-Steel	Seamless pipe	60.0	35.0				X	X		950	3
SA-178D	C-Steel	ERW tube	70.0	40.0	X	X	X				950	2
SA-210C	C-Steel	Seamless tube	70.0	40.0		X	X	X			950	
SA-106C	C-Steel	Seamless pipe	70.0	40.0				X	X		950	3
SA-216WCB	C-Steel	Casting	70.0	36.0		X	X	X	X		950	
SA-105	C-Steel	Forging	70.0	36.0		X	X	X	X		950	3
SA-181-70	C-Steel	Forging	70.0	36.0		X	X	X	X		950	3
SA-266Cl2	C-Steel	Forging	70.0	36.0					X		800	
SA-516-70	C-Steel	Plate	70.0	38.0					X	X	800	
SA-266Cl3	C-Steel	Forging	75.0	37.5					X		800	
SA-299	C-Steel	Plate	75.0	40.0						X	800	
SA-250T1a	C-Mo	ERW tube	60.0	32.0		X	X				975	4,5
SA-209T1a	C-Mo	Seamless tube	60.0	32.0		X	X	X			975	4
SA-250T2	1/2Cr-1/2Mo	ERW tube	60.0	30.0		X	X		X		1025	6
SA-213T2	1/2Cr-1/2Mo	Seamless tube	60.0	30.0	X	X	X				1025	6
SA-250T12	1Cr-1/2Mo	ERW tube	60.0	32.0		X	X				1050	5
SA-213T12	1Cr-1/2Mo	Seamless tube	60.0	32.0	X	X	X				1050	
SA-335P12	1/2Cr-1/2Mo	Seamless pipe	60.0	32.0					X		1050	
SA-250T11	1-1/4Cr-1/2Mo-Si	ERW tube	60.0	30.0		X	X				1050	5
SA-213T11	1-1/4Cr-1/2Mo-Si	Seamless tube	60.0	30.0	X	X	X				1050	
SA-335P11	1-1/4Cr-1/2Mo-Si	Seamless pipe	60.0	30.0				X	X		1050	
SA-217WC6	1-1/4Cr-1/2Mo	Casting	70.0	40.0		X	X	X	X		1100	
SA-250T22	2-1/4Cr-1Mo	ERW tube	60.0	30.0			X				1115	5
SA-213T22	2-1/4Cr-1Mo	Seamless tube	60.0	30.0			X				1115	
SA-213T23	2-1/4Cr-W-V	Seamless tube	74.0	58.0	X		X				1115	
SA-335P22	2-1/4Cr-1Mo	Seamless pipe	60.0	30.0				X	X		1100	
SA-217WC9	2-1/4Cr-1Mo	Casting	70.0	40.0			X	X	X		1115	
SA-182F22Cl1	2-1/4Cr-1Mo	Forging	60.0	30.0			X		X		1115	
SA-336F22Cl1	2-1/4Cr-1Mo	Forging	60.0	30.0					X		1100	
SA-213T91	9Cr-1Mo-V	Seamless tube	85.0	60.0			X				1150	
SA-335P91	9Cr-1Mo-V	Seamless pipe	85.0	60.0				X	X		1150	
SA-217C12A	9Cr-1Mo-V	Casting	85.0	60.0			X		X		1200	
SA-182F91	9Cr-1Mo-V	Forging	85.0	60.0			X				1150	
SA-336F91	9Cr-1Mo-V	Forging	85.0	60.0					X		1150	
SA-213T92	9Cr-2W	Seamless tube	90.0	64.0			X		X		1200	
SA-213TP304H	18Cr-8Ni	Seamless tube	75.0	30.0			X				1400	
SA-213TP347H	18Cr-10Ni-Cb	Seamless tube	75.0	30.0			X				1400	
SA-213TP310H	25Cr-20Ni	Seamless tube	75.0	30.0			X				1500	
SB-407-800H	Ni-Cr-Fe	Seamless tube	65.0	25.0			X				1500	
SB-423-825	Ni-Fe-Cr-Mo-Cu	Seamless tube	85.0	35.0			X				1000	

Notes:
1. Values in parentheses are not required minimums, but are expected minimums.
2. Requires special inspection if used at 100% efficiency above 850F.
3. Limited to 800F maximum for piping 10.75 in. OD and larger and outside the boiler setting.
4. Limited to 875F maximum for applications outside the boiler setting.
5. Requires special inspection if used at 100% efficiency.
6. Maximum OD temperature is 1025F. Maximum mean metal temperature for Code calculations is l000F.

to enhance creep strength. A new family of Ni and Ni-Cr-Co alloys is also available, specifically for advanced supercritical and advanced supercritical boiler designs.

Fuel ash corrosion considerations might dictate the use of higher alloys at lower temperatures. This is common in process recovery and refuse-fired boilers with very corrosive flue gas and ash. For example, SB-407-825 (42Ni-21.5Cr-3Mo-2.25Cu-0.9Ti-bal Fe) is used in the highly corrosive regions of refuse boiler superheaters, even at temperatures below 1000F (538C). In extreme cases, bimetallic tubes, with a core of a Code material for pressure retention and a cladding of a corrosion resistant alloy, are used for both furnace wall and superheater applications. Some common combinations are SA-210A1/304L, SA-210A1/Alloy 825, and SB-407-800H/50Cr-50Ni.

Selection factors

Many factors influence material selection in a superheater. These include cost as well as performance factors (heat transfer surface area required, final steam temperature, total mass flow through the tubes, and flow balancing among circuits); mechanical factors (internal pressure, design temperature, support systems and relative thermal expansion stresses); environmental factors (resistance to steam oxidation and out of service pitting corrosion on the ID, and oxidation, fuel ash corrosion, and erosion on the outside diameter/OD); and manufacturing process and equipment limitations and considerations, such as weldability.

Cost

Material cost is usually the single largest factor affecting material selection when more than one material candidate exists for a given set of boiler application conditions. Raw material cost is established by taking into account each material's allowable design stress, at design temperature and pressure, and required mass flow requirements for the water or steam. Pressure part sizes are established, average weight determined, and material cost is estimated when knowing the raw material cost offered by the selected raw material supplier. Other factors are also considered such as required corrosion allowance, if any, and unique manufacturing costs and risks. Once this evaluation is accomplished and comparison of materials is completed, it is quite common to find that stronger, higher alloys become economically attractive over lower strength, less costly steels at operating conditions usually acceptable and appropriate for the lower alloy steel.

Headers and piping

Specifications for most of the commonly used pipe materials are listed in Table 1. As these components are usually not in the gas stream and are unheated, the major design factor, other than strength at temperature, is steam oxidation resistance. Carbon steels are not used above 800F (427C) outside the boiler setting, and C-Mo is limited to applications of small sizes [less than 10.75 in. (273 mm) OD] and below 875F (468C) to avoid graphitization.

9Cr-1Mo-V has replaced 2-1/4Cr-1Mo for many superheater outlet header applications (see Fig. 25). This material is not operating in the creep range even at the 1000 to 1050F (538 to 566C) design temperatures of most such components. This factor and its very high strength allow thinner components which are much less susceptible to the creep-fatigue failures observed in older 1-1/4Cr-1/2Mo-Si and 2-1/4Cr-1Mo headers. The use of forged outlet nozzle *tee* sections in place of welded nozzles has also reduced the potential for failure of these large piping connections.

Drums

Carbon steel plate is the primary material used in drums. SA-299, a 75,000 psi (517.1 MPa) tensile strength material, ordered to fine grain melting practice for improved toughness, is used for heavy section

Fig. 25 The 9Cr-1Mo-V superheater outlet header features high strength and thin material less susceptible to creep-fatigue failure.

drums, those more than about 4 in. (101.6 mm) in thickness. SA-516 Gr 70, a fine grained 70,000 psi (482.7 MPa) tensile strength steel, is used for applications below this thickness, down to 1.5 in. (38.1 mm) thick shells. SA-515 Gr 70, a coarse grain melting practice steel, is used for thinner shells. Steel grades of 80,000 psi (552 MPa) and higher are available. However, only in rare cases, where crane lifting capacity or long distance shipping costs are important considerations, are higher strength steels used due to increased manufacturing difficulties with the higher strength steels.

Heat resistant alloys for nonpressure parts

High alloy heat resistant materials must be used for certain boiler parts that are exposed to high temperature and can not be water or steam cooled. These parts are made from alloys of the oxidation resistant, relatively high strength Cr-Ni-Fe type, many of them cast to shape as baffles, supports and hanger fittings. Oil burner impellers, sootblower clamps and hangers are also made of such heat resisting alloy steels.

Deterioration of these parts may occur through conversion of the surface layers to oxides, sulfides and sulfates. Experience indicates that 25Cr-12Ni and 25Cr-20Ni steels give reasonably good service life, depending on the location of the part in the flue gas stream and on the characteristics of the fuel. Temperatures to which these metal parts are exposed may range from 1000 to 2800F (538 to 1538C). Welding of such austenitic castings to ferritic alloy tubes presents a dissimilar metal weld that is susceptible to failure. Normal practice is to use a nickel-base filler metal that better matches the thermal expansion properties of the ferritic tube than would an austenitic stainless weld composition. If possible, a ferritic alloy weld rod should be used to further reduce stresses against the ferritic pressure part, provided there is adequate weldability and operating conditions at the weld location are acceptable. Special patented nonwelded constructions are also used where such combinations are required.

Life may be shortened if these steels are exposed to flue gases from fuel oil containing vanadium compounds. Sulfur compounds formed from combustion of high sulfur fuels are also detrimental and act to reduce life. These may react in the presence of V and cause greatly accelerated rates of attack, especially when the temperature of the metal part exceeds 1200F (649C). Combinations of Na, S and V compounds are reported to melt at as low as 1050F (566C). Such deposits are extremely corrosive when molten because of their slagging action. In these circumstances, 50Cr-50Ni or 60Cr-40Ni castings are used to resist corrosion.

References

1. Cullity, B.D., *Elements of X-Ray Diffraction,* Second Ed., Addison-Wesley Publishing Company, Inc., Reading, Massachusetts, 1978.

2. Darken, L.S., *The Physical Chemistry of Metallic Solutions and Intermetallic Compounds,* Her Majesty's Stationery Office, London, England, United Kingdom, 1958.

3. Swalin, R.A., *Thermodynamics of Solids,* Wiley & Sons, New York, New York, 1972.

4. Higgins, R.A., *Properties of Engineering Materials,* Hodder and Staughton, London, England, United Kingdom, 1979.

5. Bain, E.C., and Paxton, H.W., *Alloying Elements in Steel,* Second Ed., American Society for Metals, Metals Park, Ohio, 1966.

6. McGannon, H.E., Ed., *The Making, Shaping and Treating of Steel,* Ninth Ed., United States Steel, Pittsburgh, Pennsylvania, 1970.

7. Lankford, W.T., Jr., et al., *The Making, Shaping and Treating of Steel,* Tenth Ed., Association of Iron and Steel Engineers, Pittsburgh, Pennsylvania, 1985.

8. Long, C.J., and DeLong, W.T., "The ferrite content of austenitic stainless steel weld metal," *Welding Journal,* Research Supplement, pp. 281S-297S, Vol. 52 (7), 1973.

9. Boyer, H.E., and Gall, T.L., Eds., *Metals Handbook: Desk Edition,* American Society for Metals, Metals Park, Ohio, 1985.

10. Connor, L., Ed., *Welding Handbook*, Eighth Ed., American Welding Society, Vol. 1, Miami, Florida, 1987.

10a. Weisman, C., Ed., *Welding Handbook,* Seventh Ed., American Welding Society, p. 272, Vol. 1, Miami, Florida, 1981.

11. Weisman, C., Ed., *Welding Handbook,* Seventh Ed., American Welding Society, p. 229, Vol. 1, Miami, Florida, 1981.

12. Sikka, V.K., et al., "Modified 9Cr-1Mo steel: an improved alloy for steam generator application," *Ferritic Steels for High Temperature Applications,* Proceedings of the ASM International Conference on Production, Fabrication, Properties and Application of Ferritic Steels for High Temperature Service, pp. 65-84, Warren, Pennsylvania, October 6-8, 1981, Khare, A.K., Ed., American Society for Metals, Metals Park, Ohio, 1963.

13. Swindeman, R.W., and Gold, M., "Developments in ferrous alloy technology for high temperature service," Widera, G.E.O., Ed., Transactions of the *ASME: J. Pressure Vessel Technology,* p. 135, American Society of Mechanical Engineers (ASME), New York, New York, May, 1991.

14. Rudd, A.H., and Tanzosh, J.M., "Developments applicable to improved coal-fired power plants," presented at the First EPRI International Conference on Improved Coal-Fired Power Plants, Palo Alto, California, November 19-21, 1986.

15. Viswanathan, R., et al., "Ligament cracking and the use of modified 9Cr-1Mo alloy steel (P91) for boiler headers," presented at the 1990 American Society of Mechanical Engineers (ASME) Pressure Vessels and Piping Conference, Nashville, Tennessee, June 17-21, 1990, Prager, M., and Cantzlereds, C., *New Alloys for Pressure Vessels and Piping,* pp. 97-104, ASME, New York, New York, 1990.

16. Benjamin, D., et al., "Properties and selection: stainless steels, tool materials and special purpose metals," *Metals Handbook,* Ninth Ed., Vol. 3, American Society for Metals, Metals Park, Ohio, p. 17, 1980.

17. Barna, J.L., et al., "Furnace wall corrosion in refuse-fired boilers," presented to the ASME 12th Biennial National Waste Processing Conference, Denver, Colorado, June 1-4, 1986.

18. Morro, H., III, "Zinc," *Metals Handbook, Desk Edition,* pp. 11-1 to 11-3, Boyer, H.E., and Gall, T.L., Eds., American Society for Metals, Metals Park, Ohio.

19. Hayden, H.W., et al., *The Structure and Properties of Materials,* Vol. III, Mechanical Behavior, Wiley & Sons, New York, New York, 1965.

20. Barsom, J.M., and Rolfe, S.T., *Fracture and Fatigue Control in Structures*, Second Ed., Prentice-Hall, Englewood Cliffs, New Jersey, 1987.

A large steam drum is being lifted within power plant structural steel.

Chapter 8

Structural Analysis and Design

Equipment used in the power, chemical, petroleum and cryogenic fields often includes large steel vessels. These vessels may require tons of structural steel for their support. Steam generating and emissions control equipment, for example, may be comprised of pressure parts ranging from small diameter tubing to vessels weighing more than 1000 t (907 t_m). A large fossil fuel boiler may extend 300 ft (91.4 m) above the ground, requiring a steel support structure comparable to a 30 story building. To assure reliability, a thorough design analysis of pressure parts and their supporting structural components is required.

Pressure vessel design and analysis

Steam generating units require pressure vessel components that operate at internal pressures of up to 4000 psi (27.6 MPa) and at steam temperatures up to 1100F (566C). Even higher temperature and pressure conditions are possible in advanced system designs. Maximum reliability can be assured only with a thorough stress analysis of the components. Therefore, considerable attention is given to the design and stress analysis of steam drums, superheater headers, heat exchangers, pressurizers and nuclear reactors. In designing these vessels, the basic approach is to account for all unknown factors such as local yielding and stress redistribution, variability in material properties, inexact knowledge of loadings, and inexact stress evaluations by using allowable working stresses that include appropriate factors of safety.

The analysis and design of complex pressure vessels and components such as the reactor closure head, shown in Fig. 1, and the fossil boiler steam drum, shown in Fig. 2, requires sophisticated principles and methods. Mathematical equations based on the theory of elasticity are applied to regions of discontinuities, nozzle openings and supports. Advanced computerized structural mechanics methods, such as the finite element method, are used to determine complex vessel stresses.

In the United States (U.S.), pressure vessel construction codes adopted by state, federal and municipal authorities establish safety requirements for vessel construction. The most widely used code is the American Society of Mechanical Engineers (ASME) *Boiler and Pressure Vessel Code*. Key sections include Sections I, *Rules for Construction of Power Boilers*; III, *Rules for Construction of Nuclear Power Plant Components*; and VIII, *Rules for Construction of Pressure Vessels*. A further introduction to the ASME Code is presented in Appendix 2.

Stress significance

Stress is defined as the internal force between two adjacent elements of a body, divided by the area over which it is applied. The main significance of a stress is its magnitude; however, the nature of the applied

Fig. 1 Head of nuclear reactor vessel.

Fig. 2 Fossil fuel boiler steam drum.

load and the resulting stress distribution are also important. The designer must consider whether the loading is mechanical or thermal, whether it is steady-state or transient, and whether the stress pattern is uniform.

Stress distribution depends on the material properties. For example, yielding or strain readjustment can cause redistribution of stresses.

Steady-state conditions An excessive steady-state stress due to applied pressure results in vessel material distortion, progresses to leakage at fittings and ultimately causes failure in a ductile vessel. To prevent this type of failure a safety factor is applied to the material properties. The two predominant properties considered are yield strength, which establishes the pressure at which permanent distortion occurs, and tensile strength, which determines the vessel bursting pressure. ASME Codes establish pressure vessel design safety factors based on the sophistication of quality assurance, manufacturing control, and design analysis techniques.

Transient conditions When the applied stresses are repetitive, such as those occurring during testing and transient operation, they may limit the fatigue life of the vessel. The designer must consider transient conditions causing fatigue stresses in addition to those caused by steady-state forces.

Although vessels must have nozzles, supports and flanges in order to be useful, these features often embody abrupt changes in cross-section. These changes can introduce irregularities in the overall stress pattern called local or peak stresses. Other construction details can also promote stress concentrations which, in turn, affect the vessel's fatigue life.

Strength theories

Several material strength theories are used to determine when failure will occur under the action of multi-axial stresses on the basis of data obtained from uni-axial tension or compression tests. The three most commonly applied theories which are used to establish elastic design stress limits are the maximum (principal) stress theory, the maximum shear stress theory, and the distortion energy theory.

Maximum stress theory The maximum stress theory considers failure to occur when one of the three principal stresses (σ) reaches the material yield point ($\sigma_{y.p.}$) in tension:

$$\sigma = \sigma_{y.p.} \tag{1}$$

This theory is the simplest to apply and, with an adequate safety factor, it results in safe, reliable pressure vessel designs. This is the theory of strength used in the ASME Code, Section I, Section VIII Division 1, and Section III Division 1 (design by formula Subsections NC-3300, ND-3300 and NE-3300).

Maximum shear stress theory The maximum shear stress theory, also known as the Tresca theory,[1] considers failure to occur when the maximum shear stress reaches the maximum shear stress at the yield strength of the material in tension. Noting that the maximum shear stress (τ) is equal to half the difference of the maximum and minimum principal stresses, and that the maximum shear stress in a tension test specimen is half the axial principal stress, the condition for yielding becomes:

$$\begin{aligned} \tau &= \frac{\sigma_{max} - \sigma_{min}}{2} = \frac{\sigma_{y.p.}}{2} \\ 2\tau &= \sigma_{max} - \sigma_{min} = \sigma_{y.p.} \end{aligned} \tag{2}$$

The value 2τ is called the shear stress intensity. The maximum shear stress theory predicts ductile material yielding more accurately than the maximum stress theory. This is the theory of strength used in the ASME Code, Section VIII Division 2, and Section III Division 1, Subsection NB, and design by analysis, Subsections NC-3200 and NE-3200.

Distortion energy theory The distortion energy theory (also known as the Mises criterion[1]) considers yielding to occur when the distortion energy at a point in a stressed element is equal to the distortion energy in a uni-axial test specimen at the point it begins to yield. While the distortion energy theory is the most accurate for ductile materials, it is cumbersome to use and is not routinely applied in pressure vessel design codes.

Design criteria[1]

To determine the allowable stresses in a pressure vessel, one must consider the nature of the loading and the vessel response to the loading. Stress interpretation determines the required stress analyses and the allowable stress magnitudes. Current design codes establish the criteria for safe design and operation of pressure vessels.

Stress classifications Stresses in pressure vessels have three major classifications: primary, secondary and peak.

Primary stresses (P) are caused by loadings which are necessary to satisfy the laws of equilibrium with applied pressure and other loads. These stresses are further divided into general primary membrane (P_m), local primary membrane (P_L) and primary bending (P_b) stresses. A primary stress is not self-limiting, i.e., if the material yields or is deformed, the stress is not reduced. A good example of this type of stress is that produced by internal pressure such as in a steam drum. When it exceeds the vessel material yield strength, permanent distortion appears and failure may occur.

Secondary stresses (Q), due to mechanical loads or differential thermal expansion, are developed by the constraint of adjacent material or adjacent components. They are self-limiting and are usually confined to local areas of the vessel. Local yielding or minor distortion can reduce secondary stresses. Although they do not affect the static bursting strength of a vessel, secondary stresses must be considered in establishing its fatigue life.

Peak stresses (F) are concentrated in highly localized areas at abrupt geometry changes. Although no appreciable vessel deformations are associated with them, peak stresses are particularly important in evaluating the fatigue life of a vessel.

Code design/analysis requirements Allowable stress limits and design analysis requirements vary with pressure vessel design codes.

According to ASME Code, Section I, the minimum vessel wall thickness is determined by evaluating the general primary membrane stress. This stress, limited to the allowable material tension stress S, is calculated at the vessel design temperature. The Section I regulations have been established to ensure that secondary and peak stresses are minimized; a detailed analysis of these stresses is normally not required.

The design criteria of ASME Code, Sections VIII Division 1, and Section III Division 1 (design by formula Subsections NC-3300, ND-3300 and NE-3300), are similar to those of Section I. However, they require cylindrical shell thickness calculations in the circumferential and longitudinal directions. The minimum required pressure vessel wall thickness is set by the maximum stress in either direction. Section III Division 1 and Subsections NC-3300 and ND-3300 permit the combination of primary membrane and primary bending stresses to be up to 1.5 S at design temperature. Section VIII Division 1 permits the combination of primary membrane and primary bending stresses to be 1.5 S at temperatures where tensile or yield strength sets the allowable stress S, and a value smaller than 1.5 S at temperatures where creep governs the allowable stress.

ASME Code, Section VIII Division 2 provides formulas and rules for common configurations of shells and formed heads. It also requires detailed stress analysis of complex geometries with unusual or cyclic loading conditions. The calculated stress intensities are assigned to specific categories. The allowable stress intensity of each category is based on a multiplier of the Code allowable stress intensity value. The Code allowable stress intensity, S_m, is based on the material yield strength, S_y, or tensile strength, S_u. (See Table 1.)

The factor k varies with the type of loading:

k	Loading
1.0	sustained
1.2	sustained and transient
1.25	hydrostatic test
1.5	pneumatic test

Table 1
Code-Allowable Stress Intensity

Stress Intensity Category	Allowable Value	Basis for Allowable Value at k = 1.0 (Lesser Value)
General primary membrane (P_m)	kS_m	2/3 S_y or 1/3 S_u
Local primary membrane (P_L)	3/2 kS_m	S_y or 1/2 S_u
Primary membrane plus primary bending ($P_m + P_b$)	3/2 kS_m	S_y or 1/2 S_u
Range of primary plus secondary ($P_m' + P_b + Q$)	3 S_m	2 S_y or S_u

The design criteria for ASME Code, Section III Division 1, Subsection NB and design by analysis Subsections NC-3200 and NE-3200 are similar to those for Section VIII Division 2 except there is less use of design formulas, curves, and tables, and greater use of design by analysis in Section III. The categories of stresses and stress intensity limits are the same in both sections.

Stress analysis methods

Stress analysis of pressure vessels can be performed by analytical or experimental methods. An analytical method, involving a rigorous mathematical solution based on the theory of elasticity and plasticity, is the most direct and inexpensive approach when the problem is adaptable to such a solution. When the problem is too complex for this method, approximate analytical structural mechanics methods, such as finite element analysis, are applied. If the problem is beyond analytical solutions, experimental methods must be used.

Mathematical formulas[2] Pressure vessels are commonly spheres, cylinders, ellipsoids, tori or composites of these. When the wall thickness is small compared to other dimensions, vessels are referred to as membrane shells. Stresses acting over the thickness of the vessel wall and tangential to its surface can be represented by mathematical formulas for the common shell forms.

Pressure stresses are classified as primary membrane stresses since they remain as long as the pressure is applied to the vessel. The basic equation for the longitudinal stress σ_1 and hoop stress σ_2 in a vessel of thickness h, longitudinal radius r_1, and circumferential radius r_2, which is subject to a pressure P, shown in Fig. 3 is:

$$\frac{\sigma_1}{r_1} + \frac{\sigma_2}{r_2} = \frac{P}{h} \quad (3)$$

From this equation, and by equating the total pressure load with the longitudinal forces acting on a transverse section of the vessel, the stresses in the commonly used shells of revolution can be found.

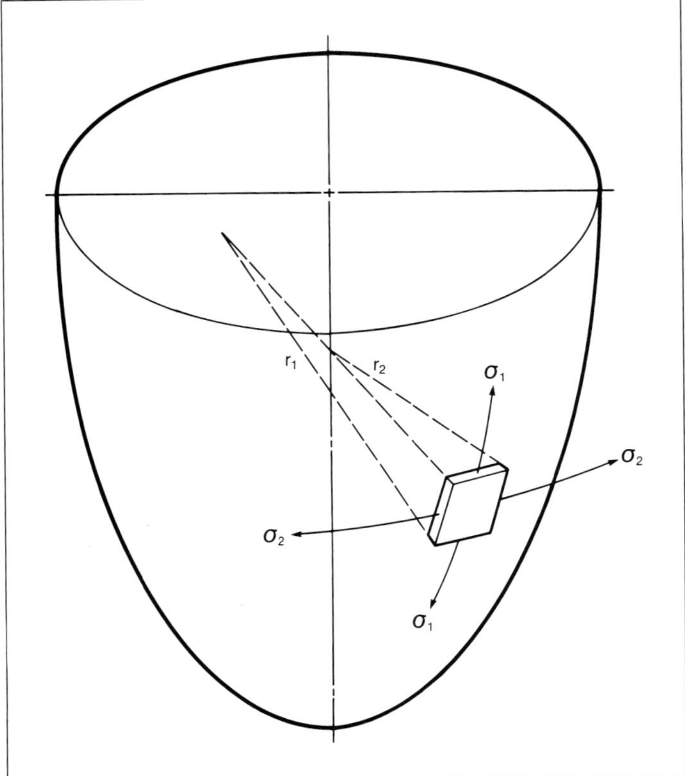

Fig. 3 Membrane stress in vessels (*courtesy Van Nostrand Reinhold*).[2]

1. Cylindrical vessel – in this case, $r_1 = \infty$, $r_2 = r$, and

$$\sigma_1 = \frac{Pr}{2h} \tag{4}$$

$$\sigma_2 = \frac{Pr}{h} \tag{5}$$

2. Spherical vessel – in this case, $r_1 = r_2 = r$, and

$$\sigma_1 = \frac{Pr}{2h} \tag{6}$$

$$\sigma_2 = \frac{Pr}{2h} \tag{7}$$

3. Conical vessel – in this case, $r_1 = \infty$, $r_2 = r/\cos\alpha$ where α is half the cone apex angle, and

$$\sigma_1 = \frac{Pr}{2h\cos\alpha} \tag{8}$$

$$\sigma_2 = \frac{Pr}{h\cos\alpha} \tag{9}$$

4. Ellipsoidal vessel – in this case (Fig. 4), the instantaneous radius of curvature varies with each position on the ellipsoid, whose major axis is a and minor axis is b, and the stresses are given by:

$$\sigma_1 = \frac{Pr_2}{2h} \tag{10}$$

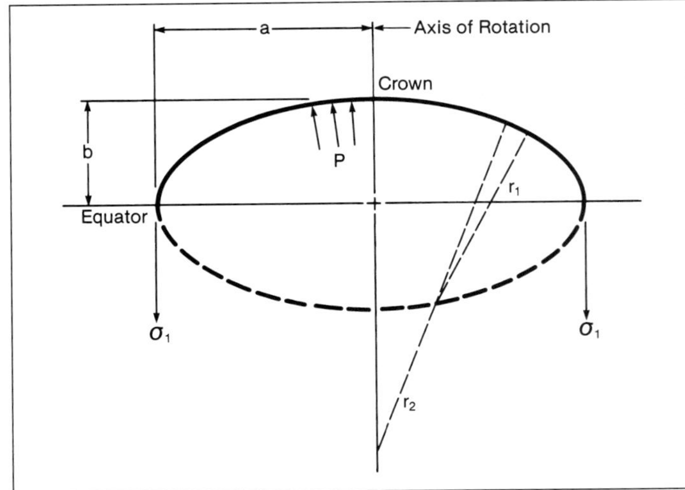

Fig. 4 Stress in an ellipsoid.[2]

$$\sigma_2 = \frac{P}{h}\left(r_2 - \frac{r_2^2}{2r_1}\right) \tag{11}$$

At the equator, the longitudinal stress is the same as the longitudinal stress in a cylinder, namely:

$$\sigma_1 = \frac{Pa}{2h} \tag{12}$$

and the hoop stress is:

$$\sigma_2 = \frac{Pa}{h}\left(1 - \frac{a^2}{2b^2}\right) \tag{13}$$

When the ratio of major to minor axis is 2:1, the hoop stress is the same as that in a cylinder of the same mating diameter, but the stress is compressive rather than tensile. The hoop stress rises rapidly when the ratio of major to minor axis exceeds 2:1 and, because this stress is compressive, buckling instability becomes a major concern. For this reason, ratios greater than 2:1 are seldom used.

5. Torus – in this case (Fig. 5), R_o is the radius of the bend centerline, θ is the angular hoop location from this centerline and:

$$\sigma_1 = \frac{Pr}{2h} \tag{14}$$

$$\sigma_2 = \frac{Pr}{2h}\left(\frac{2R_o + r\sin\theta}{R_o + r\sin\theta}\right) \tag{15}$$

The longitudinal stress remains uniform around the circumference and is the same as that for a straight cylinder. The hoop stress, however, varies for different points in the torus cross-section. At the bend centerline, it is the same as that in a straight cylinder. At the outside of the bend, it is less than this and is at its minimum. At the inside of the bend, or crotch, the value is at its maximum. Hoop stresses are dependent on the sharpness of the bend and are inversely proportional to bend radii. In pipe bending operations, the material thins at the outside and becomes thicker at the crotch of the bend. This is an offsetting factor for the higher hoop stresses that form with smaller bend radii.

Thermal stresses result when a member is restrained as it attempts to expand or contract due to a temperature change, ΔT. They are classified as secondary stresses because they are self-limiting. If the material is restricted in only one direction, the stress developed is:

$$\sigma = \pm E\alpha\Delta T \qquad \textbf{(16)}$$

where E is the modulus of elasticity and α is the coefficient of thermal expansion. If the member is restricted from expanding or contracting in two direc-

ASME Code calculations

In most U.S. states and Canadian provinces laws have been established requiring that boilers and pressure vessels comply with the rules for the design and construction of boilers and pressure vessels in the ASME Code. The complexity of these rules and the amount of analysis required are inversely related to the factors of safety which are applied to the material properties used to establish the allowable stresses. That is, when the stress analysis is simplified, the factor of safety is larger. When the stress analysis is more complex, the factory of safety is smaller. Thus, overall safety is maintained even though the factor of safety is smaller. For conditions when material tensile strength establishes the allowable stress, ASME Code, Section IV, *Rules for Construction of Heating Boilers,* requires only a simple thickness calculation with a safety factor on tensile strength of 5. ASME Code, Section I, *Rules for Construction of Power Boilers* and Section VIII, Division 1, *Rules for Construction of Pressure Vessels,* require a more complex analysis with additional items to be considered. However, the factor of safety on tensile strength is reduced to 3.5. Section III, *Rules for Construction of Nuclear Components* and Section VIII, Division 2, *Rules for Construction of Pressure Vessels* require extensive analyses which are required to be certified by a registered professional engineer. In return, the factor of safety on tensile strength is reduced even further to 3.0.

When the wall thickness is small compared to the diameter, membrane formulas (Equations 4 and 5) may be used with adequate accuracy. However, when the wall thickness is large relative to the vessel diameter, usually to accommodate higher internal design pressure, the membrane formulas are modified for ASME Code applications.

Basically the minimum wall thickness of a cylindrical shell is initially set by solving the circumferential or hoop stress equation assuming there are no additional loadings other than internal pressure. Other loadings may then be considered to determine if the initial minimum required wall thickness has to be increased to keep calculated stresses below allowable stress values.

As an example, consider a Section VIII, Division 1, pressure vessel with no unreinforced openings and no additional loadings other than an internal design pressure of 1200 psi at 500F. The inside diameter is 10 in. and the material is SA-516, Grade 70 carbon steel. There is no corrosion allowance required by this application and the butt weld joints are 100% radiographed. What is the minimum required wall thickness needed? The equation for setting the minimum required wall thickness in Section VIII, Division 1, of the Code (paragraph UG-27(c)(l), 2001 Edition) is:

$$t = \frac{PR}{SE - 0.6P}$$

where

t = minimum required wall thickness, in.
P = internal design pressure, psi
R = inside radius, in.
S = allowable stress at design temperature, psi (Section II, Part D) = 20,000 psi
E = lower of weld joint efficiency or ligament efficiency (fully radiographed with manual penetrations) = 1.0

For the pressure vessel described above:

P = 1200 psi
R = 5 in.
S = 20,000 psi
E = 1.0

$$t = \frac{(1200)(5)}{(20{,}000)(1.0) - (0.6)(1200)} = 0.311 \text{ in.}$$

Using commercial sizes, this plate thickness probably would be ordered at 0.375 in.

If Equation 5 for simple hoop stress (see Figure below) is used alone to calculate the plate thickness using the specified minimum tensile strength of SA-516, Grade 70 of 70,000 psi, the thickness h would be evaluated to be:

$$h = \frac{1200\,(5)}{70{,}000} = 0.0857$$

Therefore, the factor of safety (FS) based on tensile strength is:

$$\text{FS} = \frac{0.311}{0.0857} = 3.6$$

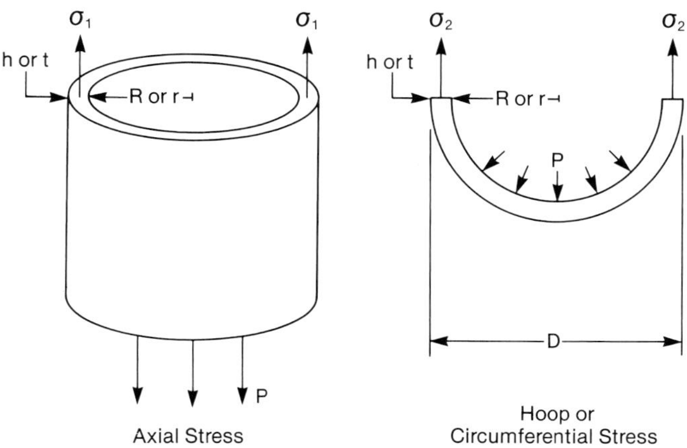

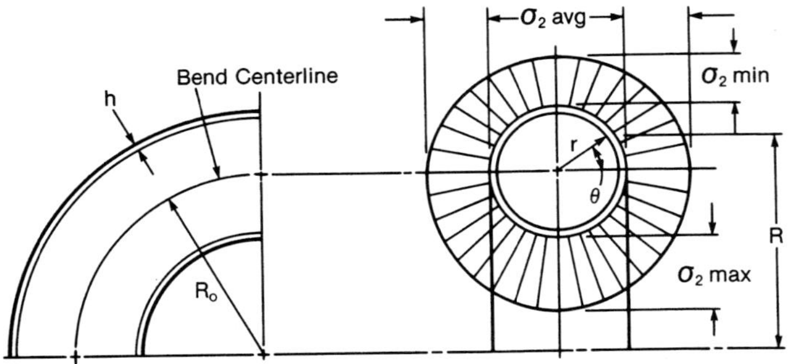

Fig. 5 Hoop stress variation in a bend.[2]

tions, as is the case in pressure vessels, the resulting stress is:

$$\sigma = \pm \frac{E\alpha\,\Delta T}{1-\mu} \tag{17}$$

where μ is Poisson's ratio.

These thermal stress equations consider full restraint, and therefore are the maximum that can be created. When the temperature varies within a member, the natural growth of one fiber is influenced by the differential growth of adjacent fibers. As a result, fibers at high temperatures are compressed and those at lower temperatures are stretched. The general equations for radial (σ_r), tangential (σ_t), and axial (σ_z) thermal stresses in a cylindrical vessel subject to a radial thermal gradient are:

$$\sigma_r = \frac{\alpha E}{(1-\mu)r^2}\left[\frac{r^2-a^2}{b^2-a^2}\int_a^b Trdr - \int_a^r Trdr\right] \tag{18}$$

$$\sigma_t = \frac{\alpha E}{(1-\mu)r^2}\left[\frac{r^2+a^2}{b^2-a^2}\int_a^b Trdr + \int_a^r Trdr - Tr^2\right] \tag{19}$$

$$\sigma_z = \frac{\alpha E}{(1-\mu)}\left[\frac{2}{b^2-a^2}\int_a^b Trdr - T\right] \tag{20}$$

where

E = modulus of elasticity
μ = Poisson's ratio
r = radius at any location
a = inside radius
b = outside radius
T = temperature

For a cylindrical vessel in which heat is flowing radially through the walls under steady-state conditions, the maximum thermal stresses are:

$$\sigma_{ta}(inside) = \frac{\alpha E T_a}{2(1-\mu)\ell n\left(\frac{b}{a}\right)}\left[1-\frac{2b^2}{b^2-a^2}\ell n\left(\frac{b}{a}\right)\right] \tag{21}$$

$$\sigma_{tb}(outside) = \frac{\alpha E T_a}{2(1-\mu)\ell n\left(\frac{b}{a}\right)}\left[1-\frac{2a^2}{b^2-a^2}\ell n\left(\frac{b}{a}\right)\right] \tag{22}$$

For relatively thin tubes and $T_a > T_b$, this can be simplified to:

$$\sigma_{ta} = \frac{-\alpha E\,\Delta T}{2(1-\mu)} \tag{23}$$

$$\sigma_{tb} = \frac{\alpha E\,\Delta T}{2(1-\mu)} \tag{24}$$

To summarize, the maximum thermal stress for a thin cylinder with a logarithmic wall temperature gradient is one half the thermal stress of an element restrained in two directions and subjected to a temperature change ΔT (Equation 17). For a radial thermal gradient of different shape, the thermal stress can be represented by:

$$\sigma = K\frac{E\alpha\,\Delta T}{1-\mu} \tag{25}$$

where K ranges between 0.5 and 1.0.

Alternating stresses resulting from cyclic pressure vessel operation may lead to fatigue cracks at high stress concentrations. Fatigue life is evaluated by comparing the alternating stress amplitude with design fatigue curves (allowable stress versus number of cycles or σ-N curves) experimentally established for the material at temperature. A typical σ-N design curve for carbon steel is shown in Fig. 6 and can be expressed by the equation:

$$\sigma_a = \frac{E}{4\sqrt{N}}\ell n\left(\frac{100}{100-d_a}\right) + .01(TS)d_a \tag{26}$$

where

σ_a = allowable alternating stress amplitude
E = modulus of elasticity at temperature
N = number of cycles
d_a = percent reduction in area
TS = tensile strength at temperature

The two controlling parameters are tensile strength and reduction in area. Tensile strength is controlling in the high cycle fatigue region, while reduction in area is controlling in low cycle fatigue. The usual di-

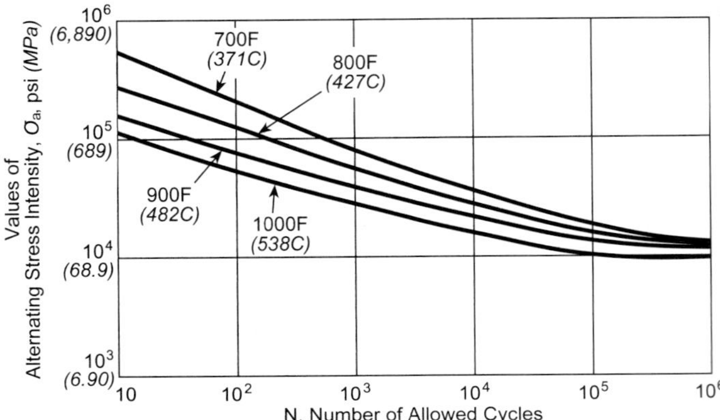

Fig. 6 Design fatigue curve.

vision between low and high cycle fatigue is 10^5 cycles. Pressure vessels often fall into the low cycle fatigue category, thereby demonstrating the importance of the material's ability to deform in the plastic range without fracturing. Lower strength materials, with their greater ductility, have better low cycle fatigue resistance than do higher strength materials.

Practical operating service conditions subject many vessels to the random occurrence of a number of stress cycles at different magnitudes. One method of appraising the damage from repetitive stresses to a vessel is the criterion that the cumulative damage from fatigue will occur when the summation of the increments of damage at the various stress levels exceeds unity. That is:

$$\sum \frac{n}{N} = 1 \tag{27}$$

where n = number of cycles at stress σ, and N = number of cycles to failure at the same stress σ. The ratio n/N is called the cycle damage ratio since it represents the fraction of the total life which is expended by the cycles that occur at a particular stress value. The value N is determined from σ-N curves for the material. If the sum of these cycle ratios is less than unity, the vessel is considered safe. This is particularly important in designing an economic and safe structure which experiences only a relatively few cycles at a high stress level and the major number at a relatively low stress level.

Discontinuity analysis method At geometrical discontinuities in axisymmetric structures, such as the intersection of a hemispherical shell element and a cylindrical shell element (Fig. 7a), the magnitude and characteristic of the stress are considerably different than those in elements remote from the discontinuity. A linear elastic analysis method is used to evaluate these local stresses.

Discontinuity stresses that occur in pressure vessels, particularly axisymmetric vessels, are determined by a discontinuity analysis method. A discontinuity stress results from displacement and rotation incompatibilities at the intersection of two elements. The forces and moments at the intersection (Fig. 7c) are redundant and self-limiting. They develop solely to ensure compatibility at the intersection. As a consequence, a discontinuity stress can not cause failure in ductile materials in one load application even if the maximum stress exceeds the material yield strength. Such stresses must be considered in cyclic load applications or in special cases where materials can not safely redistribute stresses. The ASME Code refers to discontinuity stresses as secondary stresses. The application to the shell of revolution shown in Fig. 7 outlines the major steps involved in the method used to determine discontinuity stresses.

Under internal pressure, a sphere radially expands approximately one half that of a cylindrical shell (Fig. 7b). The difference in free body displacement results in redundant loadings at the intersection if Elements (1) and (2) are joined (Fig. 7c). The final displacement and rotation of the cylindrical shell are equal to the

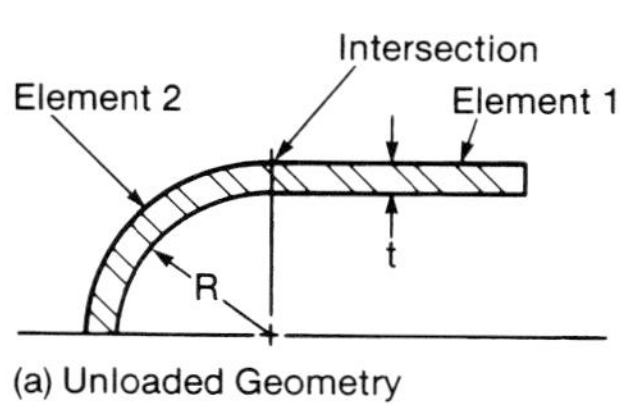

(a) Unloaded Geometry

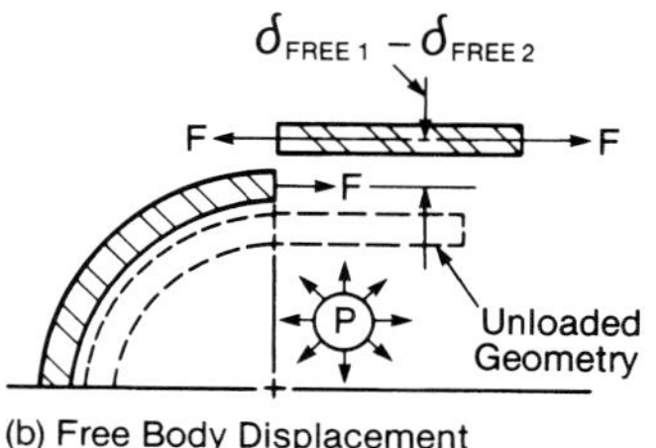

(b) Free Body Displacement

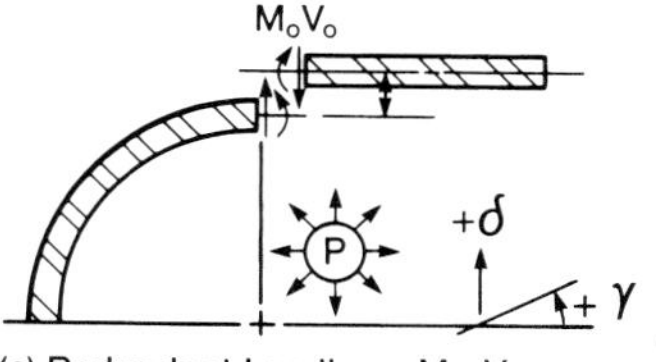

(c) Redundant Loadings, M_o, V_o

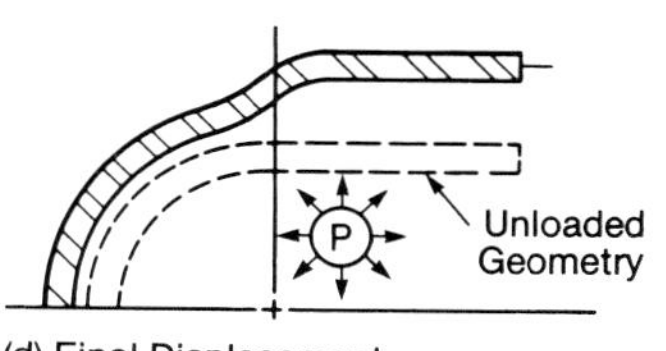

(d) Final Displacement

Fig. 7 Discontinuity analysis.

free body displacement plus the displacements due to the redundant shear force V_o and redundant bending moment M_o (Fig. 7d).

The direction of the redundant loading is unknown and must be assumed. A consistent sign convention must be followed. In addition, the direction of loading on the two elements must be set up consistently because Element (1) reacts Element (2) loading and vice versa. If M_o or V_o as calculated is negative, the correct direction is opposite to that assumed.

In equation form then, for Element (1):

$$\delta_{\text{FINAL}\,1} = \delta_{\text{FREE}\,1} - \beta_{\delta V1} V_o + \beta_{\delta M1} M_o \tag{28}$$

$$\gamma_{\text{FINAL}\,1} = \gamma_{\text{FREE}\,1} + \beta_{\gamma V1} V_o - \beta_{\gamma M1} M_o \tag{29}$$

Similarly for Element (2):

$$\delta_{\text{FINAL}\,2} = \delta_{\text{FREE}\,2} + \beta_{\delta V2} V_o + \beta_{\delta M2} M_o \tag{30}$$

$$\gamma_{\text{FINAL}\,2} = \gamma_{\text{FREE}\,2} + \beta_{\gamma V2} V_o + \beta_{\gamma M2} M_o \tag{31}$$

where

$$\delta_{\text{FREE}\,1} = \frac{\text{PR}^2}{Et}\left(1 - \frac{\mu}{2}\right) \tag{32}$$

$$\delta_{\text{FREE}\,2} = \frac{\text{PR}^2}{2Et}(1 - \mu)$$
$$\gamma_{\text{FREE}\,1} = \gamma_{\text{FREE}\,2} = 0 \text{ in this case} \tag{33}$$

The constants β are the deflections or rotations due to loading per unit of perimeter, and are referred to as influence coefficients. These constants can be determined for a variety of geometries, including rings and thin shells of revolution, using standard handbook solutions. For example:

$\beta_{\delta V1}$ = radial displacement of Element (1) due to unit shear load

$\beta_{\delta M1}$ = radial displacement of Element (1) due to unit moment load
$\beta_{\gamma V1}$ = rotation of Element (1) due to unit shear load
$\beta_{\gamma M1}$ = rotation of Element (1) due to unit moment load

Because $\delta_{FINAL\,1} = \delta_{FINAL\,2}$ and $\gamma_{FINAL\,1} = \gamma_{FINAL\,2}$ from compatibility requirements, Equations 28 through 31 can be reduced to two equations for two unknowns, V_o and M_o, which are solved simultaneously. Note that the number of equations reduces to the number of redundant loadings and that the force F can be determined by static equilibrium requirements.

Once V_o and M_o have been calculated, handbook solutions can be applied to determine the resulting membrane and bending stresses. The discontinuity stress must then be added to the free body stress to obtain the total stress at the intersection.

Although the example demonstrates internal pressure loading, the same method applies to determining thermally induced discontinuity stress. For more complicated geometries involving four or more unknown redundant loadings, commercially available computer programs should be considered for solution.

Finite element analysis When the geometry of a component or vessel is too complex for classical formulas or closed form solutions, finite element analysis (FEA) can often provide the required results. FEA is a powerful numerical technique that can evaluate structural deformations and stresses, heat flows and temperatures, and dynamic responses of a structure. Because FEA is usually more economical than experimental stress analysis, scale modeling, or other numerical methods, it has become the dominant sophisticated stress analysis method.

During product development, FEA is used to predict performance of a new product or concept before building an expensive prototype. For example, a design idea to protect the inside of a burner could be analyzed to find out if it will have adequate cooling and fatigue life. FEA is also used to investigate field problems.

To apply FEA, the structure is modeled as an assembly of discrete building blocks called elements. The elements can be linear (one dimensional truss or beam), plane (representing two dimensional behavior), or solid (three dimensional bricks). Elements are connected at their boundaries by nodes as illustrated in Fig. 8.

Except for analyses using truss or beam elements, the accuracy of FEA is dependent on the mesh density. This refers to the number of nodes per modeled volume. As mesh density increases, the result accuracy also increases. Alternatively, in *p-method* analysis, the mesh density remains constant while increased accuracy is attained through mathematical changes to the solution process.

A computer solution is essential because of the numerous calculations involved. A medium sized FEA may require the simultaneous solution of thousands of equations, but taking merely seconds of computer time. FEA is one of the most demanding computer applications.

FEA theory is illustrated by considering a simple structural analysis with applied loads and specified node displacements. The mathematical theory is essentially as follows.

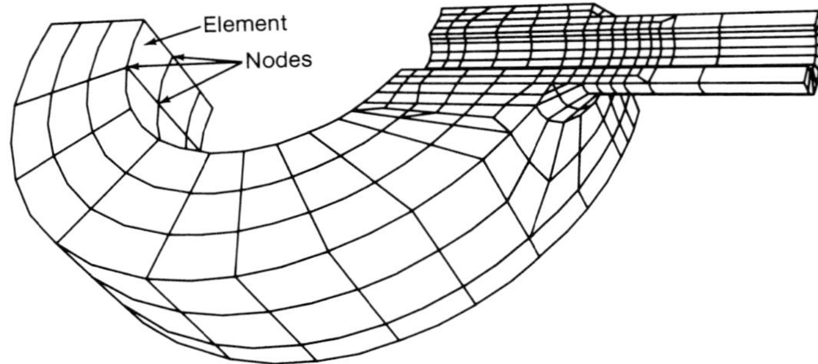

Fig. 8 Finite element model composed of brick elements.

For each element, a stiffness matrix satisfying the following relationship is found:

$$[k]\{d\} = \{r\} \tag{34}$$

where

$[k]$ = an element stiffness matrix. It is square and defines the element stiffness in each direction (degree of freedom)
$\{d\}$ = a column of nodal displacements for one element
$\{r\}$ = a column of nodal loads for one element

The determination of $[k]$ can be very complex and its theory is not outlined here. Modeling the whole structure requires that:

$$[K]\{D\} = \{R\} \tag{35}$$

where

$[K]$ = structure stiffness matrix; each member of $[K]$ is an assembly of the individual stiffness contributions surrounding a given node
$\{D\}$ = column of nodal displacements for the structure
$\{R\}$ = column of nodal loads on the structure

In general, neither $\{D\}$ nor $\{R\}$ is completely known. Therefore, Equation 35 must be partitioned (rearranged) to separate known and unknown quantities. Equation 35 then becomes:

$$\left[\begin{array}{c|c} K_{11} & K_{12} \\ K_{21} & K_{22} \end{array}\right]\begin{Bmatrix} D_s \\ D_o \end{Bmatrix} = \begin{Bmatrix} R_o \\ R_s \end{Bmatrix} \tag{36}$$

where

D_s = unknown displacements
D_o = known displacements
R_s = unknown loads
R_o = known loads

Equation 36 represents the two following equations:

$$[K_{11}]\{D_s\} + [K_{12}]\{D_o\} = \{R_o\} \tag{37}$$

$$[K_{21}]\{D_s\} + [K_{22}]\{D_o\} = \{R_s\} \tag{38}$$

Equation 37 can be solved for D_s and Equation 38 can then be solved for R_s.

Using the calculated displacements $\{D\}$, $\{d\}$ can be

found for each element and the stress can be calculated by:

$$\{\sigma\} = [E][B]\{d\} \quad \textbf{(39)}$$

where

$\{\sigma\}$ are element stresses
$[E]$ and $[B]$ relate stresses to strains and strains to displacements respectively

FEA theory may also be used to determine temperatures throughout complex geometric components. (See also Chapter 4.) Considering conduction alone, the governing relationship for thermal analysis is:

$$[C]\{\dot{T}\} + [K]\{T\} = \{Q\} \quad \textbf{(40)}$$

where

$[C]$ = system heat capacity matrix
$\{\dot{T}\}$ = column of rate of change of nodal temperatures
$[K]$ = system thermal conductivity matrix
$\{T\}$ = column of nodal temperatures
$\{Q\}$ = column of nodal rates of heat transfer

In many respects, the solution for thermal analysis is similar to that of the structural analysis. One important difference, however, is that the thermal solution is iterative and nonlinear. Three aspects of a thermal analysis require an iterative solution.

First, thermal material properties are temperature dependent. Because they are primary unknowns, temperature assumptions must be made to establish the initial material properties. Each node is first given an assumed temperature. The first thermal distribution is then obtained, and the calculated temperatures are used in a second iteration. Convergence is attained when the calculated temperature distributions from two successive iterations are nearly the same.

Second, when convective heat transfer is accounted for, heat transfer at a fluid boundary is dependent on the material surface temperature. Again, because temperatures are the primary unknowns, the solution must be iterative.

Third, in a transient analysis, the input parameters, including boundary conditions, may change with time, and the analysis must be broken into discrete steps. Within each time step, the input parameters are held constant. For this reason, transient thermal analysis is sometimes termed quasi-static.

FEA applied to dynamic problems is based upon the differential equation of motion:

$$[M]\{\ddot{D}\} + [C]\{\dot{D}\} + [K]\{D\} = \{R\} \quad \textbf{(41)}$$

where

$[M]$ = structure lumped mass matrix
$[C]$ = structure damping matrix
$[K]$ = structure stiffness matrix
$\{R\}$ = column of nodal forcing functions

$\{D\}$, $\{\dot{D}\}$ and $\{\ddot{D}\}$ are columns of nodal displacements, velocities, and accelerations, respectively.

Variations on Equation 41 can be used to solve for the natural frequencies, mode shapes, and responses due to a forcing function (periodic or nonperiodic), or to do a dynamic seismic analysis.

Limitations of FEA involve computer and human resources. The user must have substantial experience and, among other abilities, he must be skilled in selecting element types and in geometry modeling.

In FEA, result accuracy increases with the number of nodes and elements. However, computation time also increases and handling the mass of data can be cumbersome.

In most finite element analyses, large scale yielding (plastic strain) and deformations (including buckling instability), and creep are not accounted for; the material is considered to be linear elastic. In a linear structural analysis, the response (stress, strain, etc.) is proportional to the load. For example, if the applied load is doubled, the stress response would also double. For nonlinear analysis, FEA can also be beneficial. Recent advancements in computer hardware and software have enabled increased use of nonlinear analysis techniques.

Although most FEA software has well developed three dimensional capabilities, some pressure vessel analyses are imprecise due to a lack of acceptance criteria.

Computer software consists of commercially available and proprietary FEA programs. This software can be categorized into three groups: 1) preprocessors, 2) finite element solvers, and 3) postprocessors.

A preprocessor builds a model geometry and applies boundary conditions, then verifies and optimizes the model. The output of a finite element solver consists of displacements, stresses, temperatures, or dynamic response data.

Postprocessors manipulate the output from the finite element solver for comparison to acceptance criteria or to make contour map plots.

Application of FEA Because classical formulas and shell analysis solutions are limited to simple shapes, FEA fills a technical void and is applied in response to ASME Code requirements. A large portion of The Babcock & Wilcox Company's (B&W) FEA supports pressure vessel design. Stresses can be calculated near nozzles and other abrupt geometry changes. In addition, temperature changes and the resulting thermal stresses can be predicted using FEA.

The raw output from a finite element solver can not be directly applied to the ASME Code criteria. The stresses or strains must first be classified as membrane, bending, or peak (Fig. 9). B&W pioneered the classification of finite element stresses and these procedures are now used throughout the industry.

Piping flexibility, for example, is an ideal FEA application. In addition, structural steel designers rely on FEA to analyze complex frame systems that support steam generation and emissions control equipment.

Finite element analysis is often used for preliminary review of new product designs. For example, Fig. 10 shows the deflected shape of two economizer fin configurations modeled using FEA.

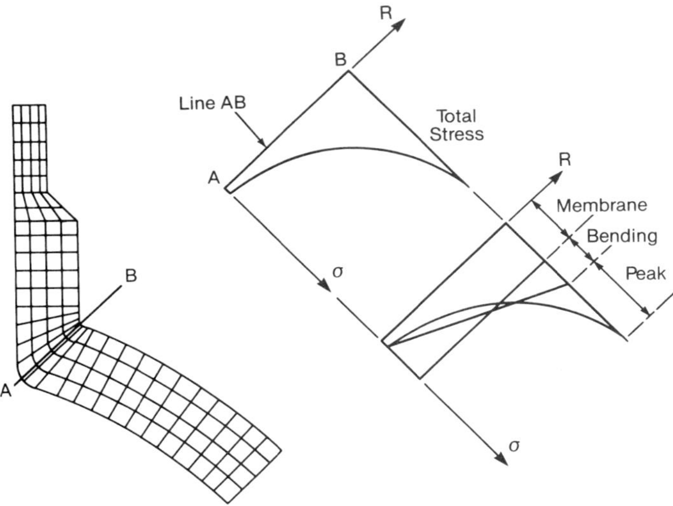

Fig. 9 Classification of finite element stress results on a vessel cross-section for comparison to code criteria.[3]

Fracture mechanics methods

Fracture mechanics provides analysis methods to account for the presence of flaws such as voids or cracks. This is in contrast to the stress analysis methods discussed above in which the structure was considered to be free of those kinds of defects. Flaws may be found by nondestructive examination (NDE) or they may be hypothesized prior to fabrication. Fracture mechanics is particularly useful to design or evaluate components fabricated using materials that are more sensitive to flaws. Additionally, it is well suited to the prediction of the remaining life of components under cyclic fatigue and high temperature creep conditions.

During component design, the flaw size is hypothesized. Allowable design stresses can be determined knowing the lower bound material toughness from accepted design procedures in conjunction with a factor of safety.

Fracture mechanics can be used to evaluate the integrity of a flawed existing structure. The defect, usually found by NDE, is idealized according to accepted ASME practices. An analysis uses design or calculated stresses based on real or hypothesized loads, and material properties are found from testing a specimen of similar material. Determining allowable flaw sizes strongly relies on accurate material properties and the best estimates of structural stresses. Appropriate safety factors are then added to the calculations.

During inspection of power plant components, minor cracks or flaws may be discovered. However, the flaws may propagate by creep or fatigue and become significant. The remaining life of components can not be accurately predicted from stress/cycles to failure (σ/N) curves alone. These predictions become possible using fracture mechanics.

Linear elastic fracture mechanics The basic concept of linear elastic fracture mechanics (LEFM) was originally developed to quantitatively evaluate sudden structural failure. LEFM, based on an analysis of the stresses near a sharp crack, assumes elastic behavior throughout the structure. The stress distribution near the crack tip depends on a single quantity, termed the stress intensity factor, K_I. LEFM assumes that unstable propagation of existing flaws occurs when the stress intensity factor becomes critical; this critical value is the fracture toughness of the material K_{IC}.

The theory of linear elastic fracture mechanics, LEFM, is based on the assumption that, at fracture, stress σ and defect size a are related to the fracture toughness K_{IC}, as follows:

$$K_I = C\sigma\sqrt{\pi a} \qquad \textbf{(42a)}$$

and

$$K_I \geq K_{IC} \text{ at failure} \qquad \textbf{(42b)}$$

The critical material property, K_{IC}, is compared to the stress intensity factor of the cracked structure, K_I, to identify failure potential. K_I should not be confused with the *stress intensity* used elsewhere in ASME design codes for analysis of unflawed structures. The term C, accounting for the geometry of the crack and structure, is a function of the crack size and relevant structure dimensions such as width or thickness.

C is exactly 1.0 for an infinitely wide center cracked panel with a through-wall crack length 2a, loaded in tension by a uniform remote stress σ. The factor C varies for other crack geometries illustrated in Fig. 11. Defects in a structure due to manufacture, in-service environment, or in-service cyclic fatigue are usually assumed to be flat, sharp, planar discontinuities where the planar area is normal to the applied stress.

ASME Code procedures for fracture mechanics design/analysis are presently given in Sections III and XI which are used for component thicknesses of at least 4 in. (102 mm) for ferritic materials with yield strengths less than 50,000 psi (344.7 MPa) and for simple geometries and stress distributions. The basic concepts of the Code may be extended to other ferritic materials (including clad ferritic materials) and more complex geometries; however, it does not apply to austenitic or high nickel alloys. These procedures provide methods for designing against brittle fractures in structures and for evaluating the significance of flaws found during in-service inspections.

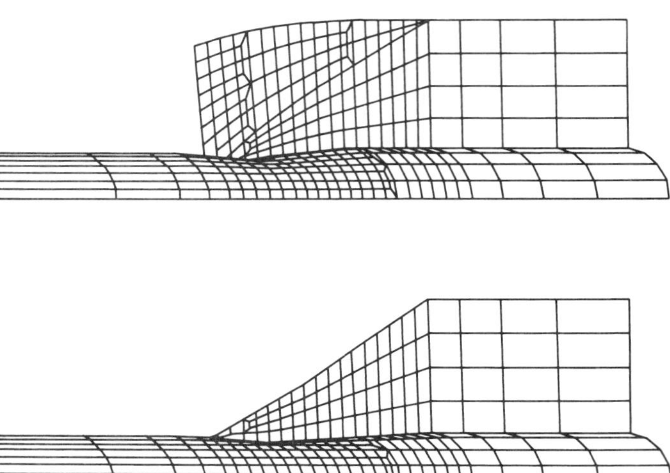

Fig. 10 Economizer tube and fin (quarter symmetry model) deformed shape plots before (upper) and after (lower) design modifications.

The ASME Code, Section III uses the principles of linear elastic fracture mechanics to determine allowable loadings in ferritic pressure vessels with an assumed defect. The stress intensity factors (K_I) are calculated separately for membrane, bending, and thermal gradient stresses. They are further subdivided into primary and secondary stresses before summing and comparison to the allowable toughness, K_{IR}. K_{IR} is the reference critical stress intensity factor (toughness). It accounts for temperature and irradiation embrittlement effects on toughness. A safety factor of 2 is applied to the primary stress components and a factor of 1 is applied to the secondary components.

To determine an operating pressure that is below the brittle fracture point, the following approach is used:

1. A maximum flaw size is assumed. This is a semi-elliptical surface flaw one fourth the pressure vessel wall thickness in depth and 1.5 times the thickness in length.
2. Knowing the specific material's nil ductility temperature, and the design temperature K_{IR} can be found from the Code.
3. The stress intensity factor is determined based on the membrane and bending stresses, and the appropriate correction factors. Additional determinants include the wall thickness and normal stress to yield strength ratio of the material.
4. The calculated stress intensity is compared to K_{IR}.

The ASME Code, Section XI provides a procedure to evaluate flaw indications found during in-service inspection of nuclear reactor coolant systems. If an indication is smaller than certain limits set by Section XI, it is considered acceptable without further analysis. If the indication is larger than these limits, Section XI provides information that enables the following procedure for further evaluation:

1. Determine the size, location and orientation of the flaw by NDE.
2. Determine the applied stresses at the flaw location (calculated without the flaw present) for all normal (including upset), emergency and faulted conditions.
3. Calculate the stress intensity factors for each of the loading conditions.
4. Determine the necessary material properties, including the effects of irradiation. A reference temperature shift procedure is used to normalize the lower bound toughness versus temperature curves. These curves are based on crack arrest and static initiation values from fracture toughness tests. The temperature shift procedure accounts for heat to heat variation in material toughness properties.
5. Using the procedures above, as well as a procedure for calculating cumulative fatigue crack growth, three critical flaw parameters are determined:
   a_f = maximum size to which the detected flaw can grow during the remaining service of the component
   a_{crit} = maximum critical size of the detected flaw under normal conditions
   a_{init} = maximum critical size for nonarresting growth initiation of the observed flaw under emergency and faulted conditions

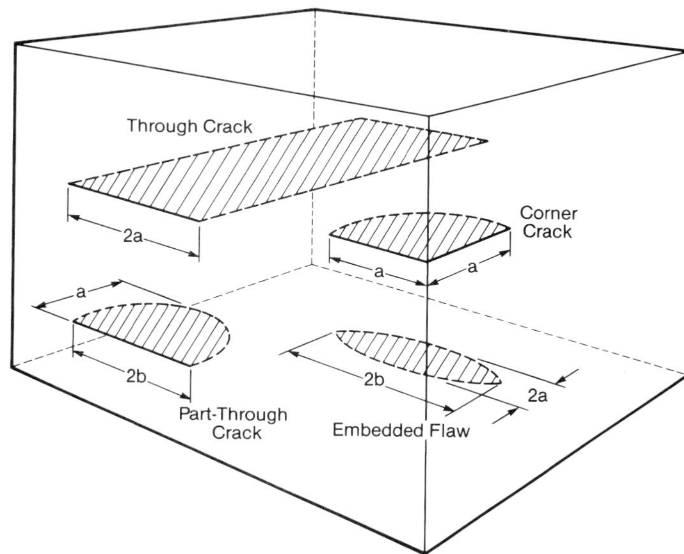

Fig. 11 Types of cracks.

6. Using these critical flaw parameters, determine if the detected flaw meets the following conditions for continued operation:

$$a_f < 0.1\, a_{crit}$$
$$a_f < 0.5\, a_{init} \quad \textbf{(43)}$$

Elastic-plastic fracture mechanics (EPFM) LEFM provides a one parameter failure criterion in terms of the crack tip stress intensity factor (K_I), but is limited to analyses where the plastic region surrounding the crack tip is small compared to the overall component dimensions. As the material becomes more ductile and the structural response becomes nonlinear, the LEFM approach loses its accuracy and eventually becomes invalid.

A direct extension of LEFM to EPFM is possible by using a parameter to characterize the crack tip region that is not dependent on the crack tip stress. This parameter, the path independent J-integral, can characterize LEFM, EPFM, and fully plastic fracture mechanics. It is capable of characterizing crack initiation, growth, and instability. The J-integral is a measure of the potential energy rate of change for nonlinear elastic structures containing defects.

The J-integral can be calculated from stresses around a crack tip using nonlinear finite element analysis. An alternate approach is to use previously calculated deformation plasticity solutions in terms of the J-integral from the Electric Power Research Institute (EPRI) *Elastic-Plastic Fracture Analysis Handbook*.[4]

The onset of crack growth is predicted when:

$$J_I \geq J_{IC} \quad \textbf{(44)}$$

The material property J_{IC} is obtained using American Society for Testing and Materials (ASTM) test E813-89, and J_I is the calculated structural response.

Stable crack growth occurs when:

$$J_I(a,P) = J_R(\Delta a)$$

and

$$a = a_o + \Delta a \tag{45}$$

where

a = current crack size
P = applied remote load
$J_R(\Delta a)$ = material crack growth resistance (ASTM test standard E1152-87)
Δa = change in crack size
a_o = initial crack size

For crack instability, an additional criterion is:

$$\partial J / \partial a \geq \partial J_R / \partial a \tag{46}$$

Failure assessment diagrams Failure assessment diagrams are tools for the determination of safety margins, prediction of failure or plastic instability and leak-before-break analysis of flawed structures. These diagrams recognize both brittle fracture and net section collapse mechanisms. The failure diagram (see Fig. 12) is a safety/failure plane defined by the stress intensity factor/toughness ratio (K_r) as the ordinate and the applied stress/net section plastic collapse stress ratio (S_r) as the abscissa. For a fixed applied stress and defect size, the coordinates K_r, S_r are readily calculable. If the assessment point denoted by these coordinates lies inside the failure assessment curve, no crack growth can occur. If the assessment point lies outside the curve, unstable crack growth is predicted. The distance of the assessment point from the failure assessment curve is a measure of failure potential of the flawed structure.

In a leak-before-break analysis, a through-wall crack is postulated. If the resulting assessment point lies inside the failure assessment curve, the crack will leak before an unstable crack growth occurs.

The deformation plasticity failure assessment diagram (DPFAD)[5] is a specific variation of a failure assessment diagram. DPFAD follows the British PD 6493 R-6[6] format, and incorporates EPFM deformation plasticity J-integral solutions. The DPFAD curve is determined by normalizing the deformation plasticity J-integral response of the flawed structure by its elastic response. The square root of this ratio is denoted by K_r. The S_r coordinate is the ratio of the applied stress to the net section plastic collapse stress. Various computer programs are available which automate this process for application purposes.

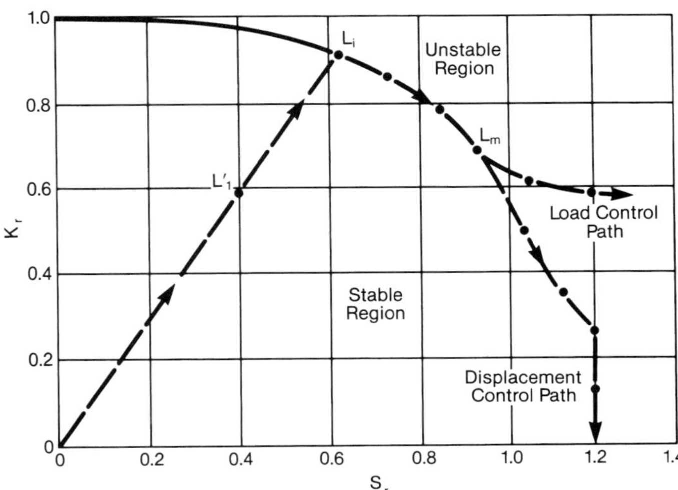

Fig. 12 Deformation plasticity failure assessment diagram in terms of stable crack growth.

Subcritical crack growth Subcritical crack growth refers to crack propagation due to cyclic fatigue, stress corrosion cracking, creep crack growth or a combination of the three. Stress corrosion cracking and creep crack growth are time based while fatigue crack growth is based on the number of stress cycles.

Fatigue crack growth Metal fatigue, although studied for more than 100 years, continues to plague structures subjected to cyclic stresses. The traditional approach to prevent fatigue failures is to base the allowable fatigue stresses on test results of carefully made laboratory specimens or representative structural components. These results are usually presented in cyclic stress versus cycles to failure, or σ/N, curves.

The significant events of metal fatigue are crack initiation and subsequent growth until the net section yields or until the stress intensity factor of the structure exceeds the material resistance to fracture. Traditional analysis assumes that a structure is initially crack free. However, a structure can have cracks that originate during fabrication or during operation. Therefore, fatigue crack growth calculations are required to predict the service life of a structure.

Fatigue crack growth calculations can 1) determine the service life of a flawed structure that (during its lifetime) undergoes significant in-service cyclic loading, or 2) determine the initial flaw size that can be tolerated prior to or during a specified operating period of the structure.

The most useful way of presenting fatigue crack growth rates is to consider them as a function of the stress intensity difference, ΔK, which is the difference between the maximum and minimum stress intensity factors.

To calculate fatigue crack growth, an experimentally determined curve such as Fig. 13 is used. The vertical axis, da/dN, is the crack growth per cycle. ASME Code, Section XI contains similar growth rate curves for pressure vessel steels.

Creep crack growth Predicting the remaining life of fossil power plant components from creep rupture data alone is not reliable. Cracks can develop at critical locations and these cracks can then propagate by creep crack growth.

At temperatures above 800F (427C), creep crack growth can cause structural components to fail. Operating temperatures for certain fossil power plant components range from 900 to 1100F (482 to 593C). At these temperatures, creep deformation and crack growth become dependent on strain rate and time exposure. Macroscopic crack growth in a creeping material occurs by nucleation and joining of microcavities in the highly strained region ahead of the crack tip. In time dependent fracture mechanics (TDFM), the energy release rate (power) parameter C_t correlates[7] creep crack growth through the relationship:

$$da / dt = bC_t^q \tag{47}$$

By using the energy rate definition, C_t can be determined experimentally from test specimens. The constants b and q are determined by a curve fit technique. Under steady-state creep where the crack tip stresses no longer change with time, the crack growth can be characterized solely by the path independent energy rate line integral C^*, analogous to the J-integral.

C^* and C_t can both be interpreted as the difference in energy rates (power) between two bodies with incrementally differing crack lengths. Furthermore, C^* characterizes the strength of the crack tip stress singularity in the same manner as the J-integral characterizes the elastic-plastic stress singularity.

The fully plastic deformation solutions from the EPRI *Elastic-Plastic Fracture Handbook* can then be used to estimate the creep crack tip steady-state parameter, C^*.

Significant data support C_t as a parameter for correlating creep crack growth behavior represented by Equation 47. An approximate expression[8] for C_t is as follows:

$$C_t = C^* \left[\left(t_T / t \right)^{\frac{n-3}{n-1}} + 1 \right] \tag{48}$$

where t_T is the transition time given by:

$$t_T = \frac{\left(1-\mu^2\right) K_I^2}{(n+1) EC^*} \tag{49}$$

and μ is Poisson's ratio, and n is the secondary creep rate exponent.

For continuous operation, Equation 48 is integrated over the time covering crack growth from the initial flaw size to the final flaw size. The limiting final flaw size is chosen based on fracture toughness or instability considerations, possibly governed by cold startup conditions. For this calculation, fracture toughness data such as K_{IC}, J_{IC} or the J_R curve would be used in a failure assessment diagram approach to determine the limiting final flaw size.

Construction features

All pressure vessels require construction features such as fluid inlets and outlets, access openings, and structural attachments at support locations. These shell areas must have adequate reinforcement and gradual geometric transitions which limit local stresses to acceptable levels.

Openings Openings are the most prevalent construction features on a vessel. They can become areas of weakness and may lead to unacceptable local distortion, known as bell mouthing, when the vessel is pressurized. Such distortions are associated with high local membrane stresses around the opening. Analytical studies have shown that these high stresses are confined to a distance of approximately one hole diameter, d, along the shell from the axis of the opening and are limited to a distance of $0.37\ (dt_{nozzle})^{1/2}$ normal to the shell.

Reinforcement to reduce the membrane stress near an opening can be provided by increasing the vessel wall thickness. An alternate, more economical stress reduction method is to thicken the vessel locally around the nozzle axis of symmetry. The reinforcing material must be within the area of high local stress to be effective.

The ASME Code provides guidelines for reinforcing openings. The reinforcement must meet requirements for the amount and distribution of the added material. A relatively small opening [approximately $d<0.2\ (Rt_s)^{1/2}$ where R is mean radius of shell and t_s is thickness of shell] remote from other locally stressed areas does not require reinforcement.

Larger openings are normally reinforced as illustrated in Figs. 14a and 14b. It is important to avoid excessive reinforcement that may result in high secondary stresses. Fig. 14c shows an opening with over reinforcement and Fig. 14a shows one with well proportioned reinforcement. Fig. 14b also shows a balanced design that minimizes secondary stresses at the nozzle/shell juncture. Designs a and b, combined with generous radii r, are most suitable for cyclic load applications.

The *ligament efficiency* method is also used to compensate for metal removed at shell openings. This method considers the load carrying ability of an area between two points in relation to the load carrying ability of the remaining ligament when the two points become the centers of two openings. The ASME Code guidelines used in this method only apply to cylindri-

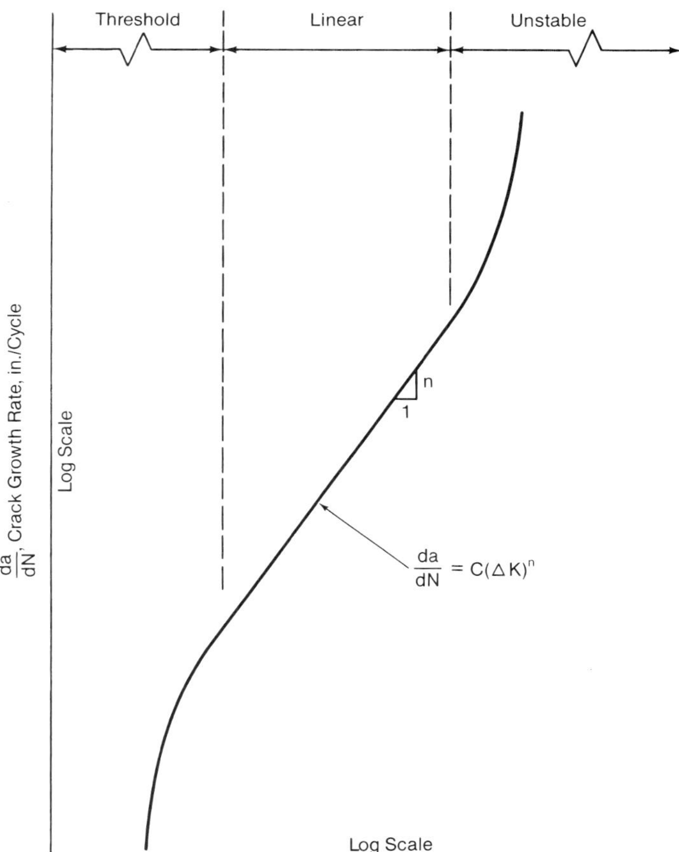

Fig. 13 Relationship between da/dN and ΔK as plotted on logarithmic coordinates.

cal pressure vessels where the circumferential stress is twice the longitudinal stress. In determining the thickness of such vessels, the allowable stress in the thickness calculation is multiplied by the ligament efficiency.

Nozzle and attachment loadings When external loadings are applied to nozzles or attachment components, local stresses are generated in the shell. Several types of loading may be applied, such as sustained, transient and thermal expansion flexibility loadings. The local membrane stresses produced by such loadings must be limited to avoid unacceptable distortion due to a single load application. The combination of local membrane and bending stresses must also be limited to avoid incremental distortion under cyclic loading.

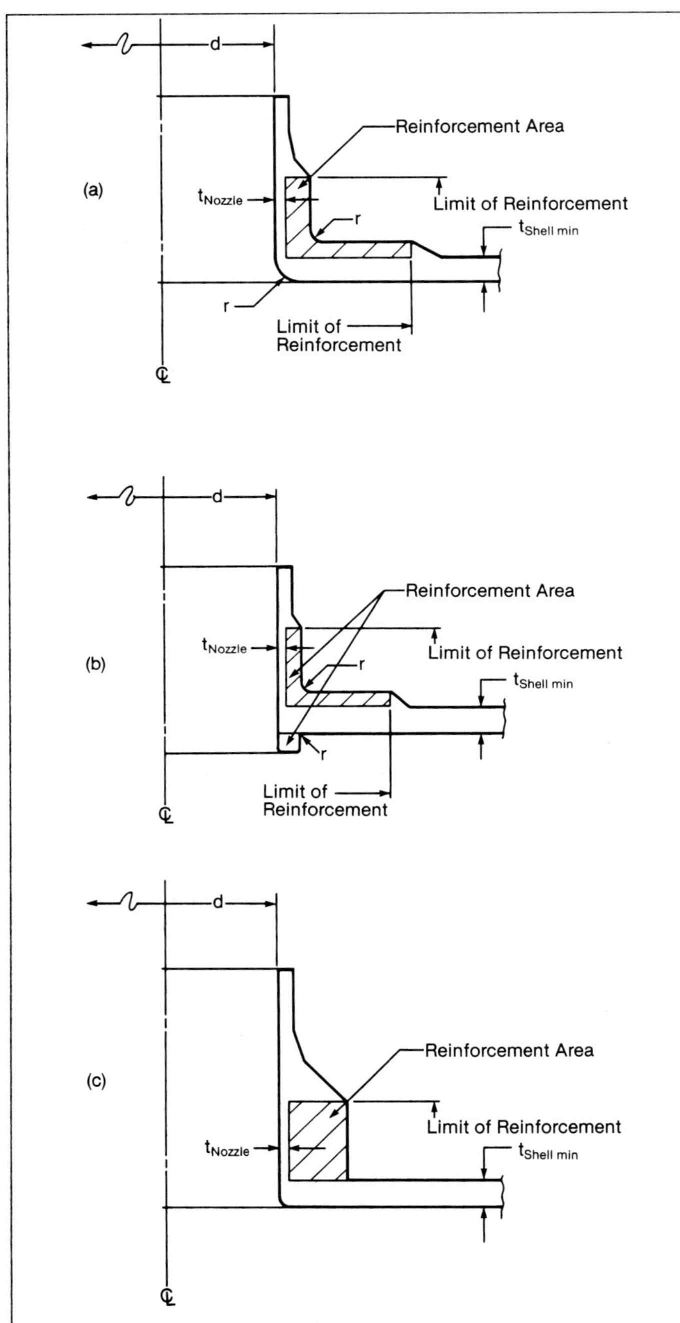

Fig. 14 Nozzle opening reinforcements.

Finally, to prevent cyclic load fatigue failures, the nozzle or attachment should include gradual transitions which minimize stress concentrations.

Pressure vessels may require local thickening at nozzles and attachments to avoid yielding or incremental distortion due to the combined effects of external loading, internal pressure, and thermal loading. Simple procedures to determine such reinforcement are not available, however FEA methods can be used. The Welding Research Council (WRC) Bulletin No. 107 also provides a procedure for determining local stresses adjacent to nozzles and rectangular attachments on cylindrical and spherical shells.

The external loadings considered by the WRC are longitudinal moment, transverse moment, torsional moment, and axial force. Stresses at various inside and outside shell surfaces are obtained by combining the stresses from the various applied loads. These external load stresses are then combined with internal pressure stresses and compared with allowable stress limits.

Use of the WRC procedure is restricted by limitations on shell and attachment parameters; however, experimental and theoretical work continues in this area.

Structural support components

Pressure vessels are normally supported by saddles, cylindrical support skirts, hanger lugs and brackets, ring girders, or integral support legs. A vessel has concentrated loads imposed on its shell where these supports are located. Therefore, it is important that the support arrangements minimize local stresses in the vessel. In addition, the components must provide support for the specified loading conditions and withstand corresponding temperature requirements.

Design criteria

Structural elements that provide support, stiffening, and/or stabilization of pressure vessels or components may be directly attached by welding or bolting. They can also be indirectly attached by clips, pins, or clamps, or may be completely unattached thereby transferring load through surface bearing and friction.

Loading conditions In general, loads applied to structural components are categorized as dead, live, or transient loads. Dead loads are due to the force of gravity on the equipment and supports. Live loads vary in magnitude and are applied to produce the maximum design conditions. Transient loads are time dependent and are expected to occur randomly for the life of the structural components. Specific loadings that are considered in designing a pressure component support include:

1. weight of the component and its contents during operating and test conditions, including loads due to static and dynamic head and fluid flow,
2. weight of the support components,
3. superimposed static and thermal loads induced by the supported components,
4. environmental loads such as wind and snow,
5. dynamic loads including those caused by earthquake, vibration, or rapid pressure change,
6. loads from piping thermal expansion,

7. loads from expansion or contraction due to pressure, and
8. loads due to anchor settlement.

Code design/analysis requirements Code requirements for designing pressure part structural supports vary. The ASME Code, Section I, only covers pressure part attaching lugs, hangers or brackets. These must be properly fitted and must be made of weldable and comparable quality material. Only the weld attaching the structural member to the pressure part is considered within the scope of Section I. Prudent design of all other support hardware is the manufacturer's responsibility.

The ASME Code, Section VIII, Division 1, does not contain design requirements for vessel supports; however, suggested rules of good practice are presented. These rules primarily address support details which prevent excessive local shell stresses at the attachments. For example, horizontal pressure vessel support saddles are recommended to support at least one third of the shell circumference. Rules for the saddle design are not covered. However, the Code refers the designer to the *Manual of Steel Construction*, published by the American Institute of Steel Construction (AISC). This reference details the allowable stress design (ASD) method for structural steel building designs. When adjustments are made for elevated temperatures, this specification can be used for designing pressure vessel support components. Similarly, Section VIII, Division 2, does not contain design methods for vessel support components. However, materials for structural attachments welded to pressure components and details of permissible attachment welds are covered.

Section III of the ASME Code contains rules for the material, design, fabrication, examination, and installation of certain pressure component and piping supports. The supports are placed within three categories:

1. plate and shell type supports, such as vessel skirts and saddles, which are fabricated from plate and shell elements,
2. linear supports which include axially loaded struts, beams and columns, subjected to bending, and trusses, frames, rings, arches and cables, and
3. standard supports (catalog items) such as constant and variable type spring hangers, shock arresters, sway braces, vibration dampers, clevises, etc.

The design procedures for each of these support types are:

1. design by analysis including methods based on maximum shear stress and maximum stress theories,
2. experimental stress analysis, and
3. load rating by testing full size prototypes.

The analysis required for each type of support depends on the class of the pressure component being supported.

Typical support design considerations

Design by analysis involves determining the stresses in the structural components and their connections by accepted analysis methods. Unless specified in an applicable code, choosing the analysis method is the designer's prerogative. Linear elastic analysis (covered in depth here), using the maximum stress or maximum shear stress theory, is commonly applied to plate, shell type and linear type supports. As an alternate, the method of limit (plastic) analysis can be used for framed linear structures when appropriate load adjustment factors are applied.

Plate and shell type supports Cylindrical shell skirts are commonly used to support vertical pressure vessels. They are attached to the vessel with a minimum offset in order to reduce local bending stresses at the vessel skirt junction. This construction also permits radial pressure and thermal growth of the supported vessel through bending of the skirt. The length of the support is chosen to permit this bending to occur safely. See Fig. 15 for typical shell type support skirt details.

In designing the skirt, the magnitudes of the loads that must be supported are determined. These normally include the vessel weight, the contents of the vessel, the imposed loads of any equipment supported from the vessel, and loads from piping or other attachments. Next a skirt height is set and the forces and moments at the skirt base, due to the loads applied, are determined. Treating the cylindrical shell as a beam, the axial stress in the skirt is then determined from:

$$\sigma = \frac{-P_v}{A} \pm \frac{Mc}{I} \tag{50}$$

where

σ = axial stress in skirt
P_v = total vertical design load
A = cross-sectional area
M = moment at base due to design loads
c = radial distance from centerline of skirt
I = moment of inertia

For thin shells ($R/t > 10$), the equation for the axial stress becomes:

$$\sigma = \frac{-P_v}{2\pi Rt} \pm \frac{M}{\pi R^2 t} \tag{51}$$

where

R = mean radius of skirt
t = thickness of skirt

Because the compressive stress is larger than the tensile stress, it usually controls the skirt design. Using the maximum stress theory for this example, the skirt thickness is obtained by:

$$t = \frac{P_v}{2\pi RF_A} + \frac{M}{\pi R^2 F_A} \tag{52}$$

where

F_A = allowable axial compressive stress

The designer must also consider stresses caused by transient loadings such as wind or earthquakes. Finally, skirt connections at the vessel and support base must be checked for local primary and secondary

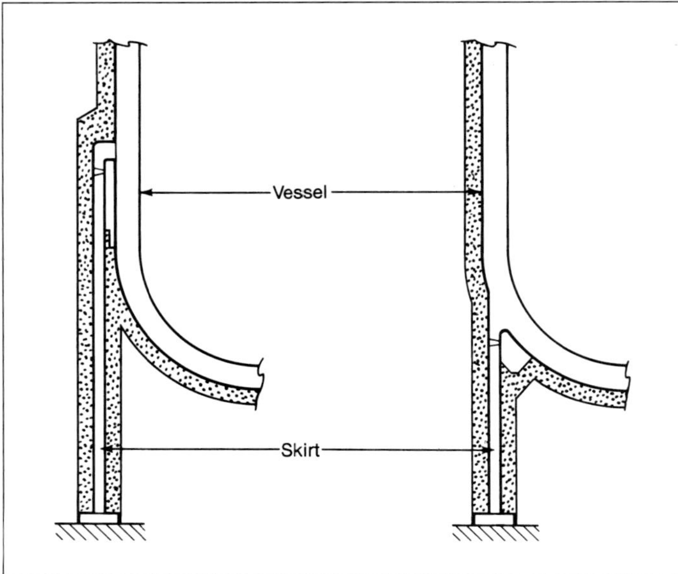

Fig. 15 Support skirt details.[2]

bending stresses. The consideration of overall stress levels provides the most accurate design.

Local thermal bending stresses often occur because of a temperature difference between the skirt and support base. The magnitudes of these bending stresses are dependent upon the severity of this axial thermal gradient; steeper gradients promote higher stresses. To minimize these stresses, the thermal gradient at the junction can be reduced by full penetration welds at the skirt to shell junction, which permit maximum conduction heat flow through the metal at that point, and by selective use of insulation in the crotch region to permit heat flow by convection and radiation. Depending on the complexity of the attachment detail, the discontinuity stress analysis or the linear elastic finite element method is used to solve for the thermal bending stresses.

Linear type supports Utility fossil fuel-fired steam generators contain many linear components that support and reinforce the boiler pressure parts. For example, the furnace enclosure walls, which are constructed of welded membraned tube panels, must be reinforced by external structural members (buckstays) to resist furnace gas pressure as well as wind and seismic forces. (See Chapter 23.) Similarly, chambers, such as the burner equipment enclosure (windbox), require internal systems to support the enclosure and its contents as well as to reinforce the furnace walls. The design of these structural systems is based on linear elastic methods using maximum stress theory allowable limits.

The buckstay system is typically comprised of horizontally oriented beams or trusses which are attached to the outside of the furnace membraned vertical tube walls. As shown in Fig. 16, the buckstay ends are connected to tie bars that link them to opposing wall buckstays thereby forming a self-equilibrating structural system. The furnace enclosure walls are continuously welded at the corners creating a water-cooled, orthotropic plate, rectangular pressure vessel. The strength of the walls in the horizontal direction is considerably less than in the vertical direction, therefore the buckstay system members are horizontally oriented.

The buckstay spacing is based on the ability of the enclosure walls to resist the following loads:

1. internal tube design pressure (P),
2. axial dead loads (DL),
3. sustained furnace gas pressure (PL_s),
4. transient furnace gas pressure (PL_T),
5. wind loads (WL), and
6. seismic loads (EQ).

The buckstay elevations are initially established based on wall stress checks and on the location of necessary equipment such as sootblowers, burners, access doors, and observation ports. These established buckstay elevations are considered as horizontal supports for the continuous vertical tube wall. The wall is then analyzed for the following load combinations using a linear elastic analysis method:

1. $DL + PL_s + P$,
2. $DL + PL_s + WL + P$,
3. $DL + PL_s + EQ + P$, and
4. $DL + PL_T + P$.

Buckstay spacings are varied to assure that the wall stresses are within allowable design limits. Additionally, their locations are designed to make full use of the structural capability of the membraned walls.

The buckstay system members, their end connections, and the wall attachments are designed for the maximum loads obtained from the wall analysis. They are designed as pinned end bending members according to the latest AISC ASD specification. This specification is modified for use at elevated temperatures and uses safety factors consistent with ASME Code, Sections I and VIII. The most important design considerations for the buckstay system include:

1. stabilization of the outboard beam flanges or truss chords to prevent lateral buckling when subjected to compression stress,
2. the development of buckstay to tie bar end connections and buckstay to wall attachments that provide load transfer but allow differential expansion between connected elements, and
3. providing adequate buckstay spacing and stiffness to prevent resonance due to low frequency combustion gas pressure pulsations common in fossil fuel-fired boilers.

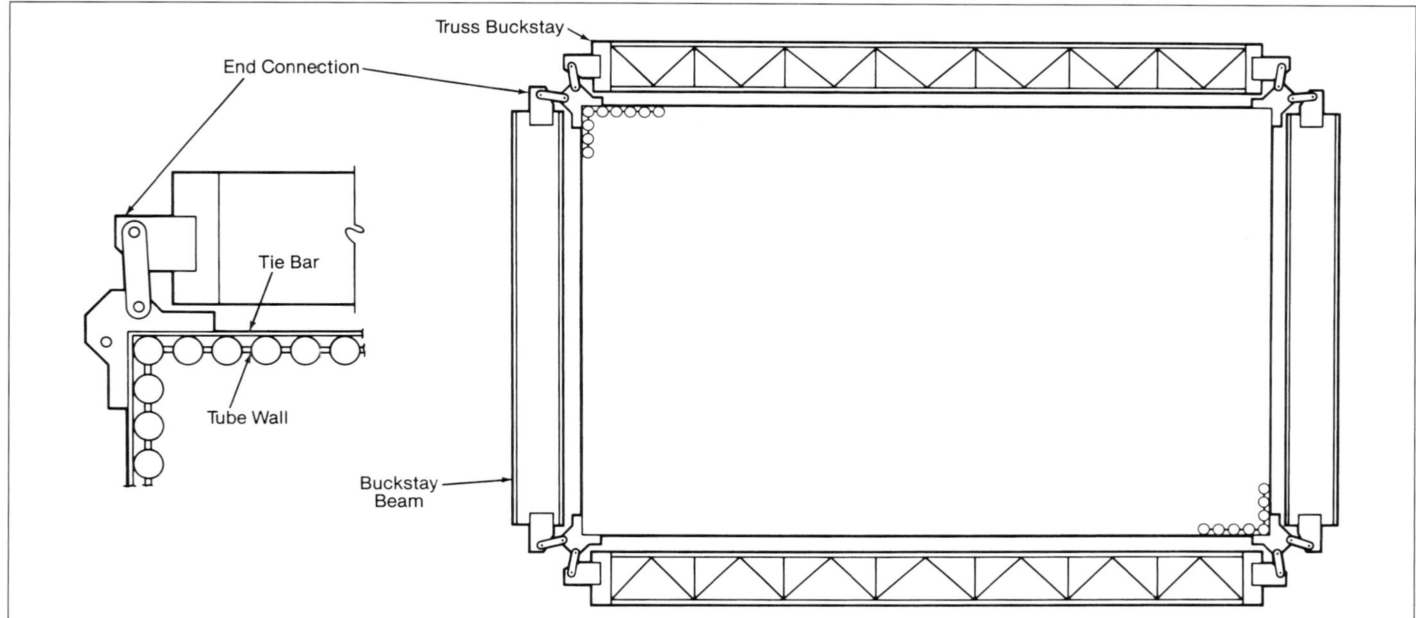

Fig. 16 Typical buckstay elevation, plan view.

References

1. Farr, J.R., and Jawad, M.H., *Structural Analysis and Design of Process Equipment*, Second Ed., John Wiley and Sons, Inc., New York, New York, January, 1989.

2. Harvey, J.F., *Theory and Design of Pressure Vessels*, Van Nostrand Reinhold Company, New York, New York, 1985.

3. Kroenke, W.C., "Classification of Finite Element Stresses According to ASME Section III Stress Categories," *Pressure Vessels and Piping, Analysis and Computers*, American Society of Mechanical Engineers (ASME), June, 1974.

4. Kumar, V., et al., "An Engineering Approach for Elastic-Plastic Fracture Analysis," Report EPRI NP-1931, Electric Power Research Institute (EPRI), Palo Alto, California, July, 1981.

5. Bloom, J.M., "Deformation Plasticity Failure Assessment Diagram," *Elastic Plastic Fracture Mechanics Technology,* ASTM STGP 896, American Society for Testing and Materials, Philadelphia, Pennsylvania, 1985.

6. "Guidance on Methods for Assessing the Acceptability of Flaws in Fusion Welded Structure," PD 6493:1991 Welding Standards Committee, London, England, United Kingdom, August 30, 1991.

7. Saxena, A., "Creep Crack Growth Under Non-Steady-State Conditions," *Fracture Mechanics,* Vol. 17, ASTM STP 905, Philadelphia, Pennsylvania, 1986.

8. Bassani, J.L., Hawk, D.E., and Saxena, A., "Evaluation of the C_t Parameter for Characterizing Creep Crack Growth Rate in the Transient Region," Third International Symposium on Nonlinear Fracture Mechanics, ASTM STP 995, Philadelphia, Pennsylvania, 1989.

Bibliography

Manual of Steel Construction (M016): Includes Code of Standard Practice, Simple Shears, and Specification for Structural Joints Using ASTM A325 or A490 Bolts, Ninth Ed., American Institute of Steel Construction, July 1, 1989.

Cook, R.D. et al., *Concepts and Applications of Finite Element Analysis*, Fourth Ed., Wiley Publishers, New York, New York, October, 2001.

Harvey, J.F., *Theory and Design of Pressure Vessels*, Second Ed., Chapman and Hall, London, England, United Kingdom, December, 1991.

Mershon, J.L., et al., "Local Stresses in Cylindrical Shells Due to External Loadings on Nozzles," Welding Research Council (WRC) Bulletin No. 297, Supplement to WRC Bulletin No. 107 (Revision 1), August, 1984, revised September, 1987.

Thornton, W.A., *Manual of Steel Construction: Load and Resistance Factor Design (Manual of Steel Construction)*, Third Ed., American Institute of Steel Construction (AISC), November 1, 2001.

Wichman, K.R., Hopper, A.G., and Mershon, J.L., "Local Stresses in Spherical and Cylindrical Shells Due to External Loadings," Welding Research Council (WRC), Bulletin No. 107, August, 1965, revised March, 1979, updated October, 2002.

Section II
Steam Generation from Chemical Energy

This section containing 17 chapters applies the fundamentals of steam generation to the design of boilers, superheaters, economizers and air heaters for steam generation from chemical or fossil fuels (coal, oil and natural gas). As discussed in Chapter 1, the fuel and method of combustion have a dramatic impact on the size and configuration of the steam producing system. Therefore, Chapters 9 and 10 begin the section by exploring the variety and characteristics of chemical and fossil fuels, and summarize the combustion calculations that are the basis for system design.

The variety of combustion systems available to handle these fuels and the supporting fuel handling and preparation equipment are then described in Chapters 11 through 18. These range from the venerable stoker in its newest configurations to circular burners used for pulverized coal, oil and gas, to fluidized-bed combustion and coal gasification. A key element in all of these systems is the control of atmospheric emissions, in particular oxides of nitrogen (NO_x) which are byproducts of the combustion process. Combustion NO_x control is discussed as an integral part of each system. It is also discussed in Section IV, Chapter 34.

Based upon these combustion systems, Chapters 19 through 22 address the design and performance evaluation of the major steam generator heat transfer components: boiler, superheater, reheater, economizer and air heater. These are configured around the combustion system selected with special attention to properly handling the high temperature, often particle-laden flue gas. The fundamentals of heat transfer, fluid dynamics, materials science and structural analysis are combined to provide the tradeoffs necessary for an economical steam generating system design. The boiler setting and auxiliary equipment, such as sootblowers, ash handling systems and fans, which are key elements in completing the overall steam system, conclude this section in Chapters 23 through 25.

Coal remains the dominant fuel source for electric power generation worldwide.

Chapter 9

Sources of Chemical Energy

World energy consumption continues to grow with the primary resources being the fossil fuels. Between 1991 and 2000, world production of primary energy increased at an annual rate of 1.4%. Production of primary energy increased from 351×10^{15} Btu (370×10^{18} J) in 1991 to 397×10^{15} Btu (419×10^{18} J) in 2000. The trend in energy production by source from 1970 to 2000 is shown in Fig. 1. World energy production and fossil fuel reserves by region are shown in Figs. 2 and 3.

The United States (U.S.), former Soviet Union (FSU) and China were the leading producers and consumers of world energy in 2000. They produced 38% and consumed 41% of the world's energy. Energy use in the developing world is expected to continue to increase with demand in developing Asia and Central and South America more than doubling between 1999 and 2020. Projected world energy consumption through the year 2025 is shown in Fig. 4.

Annual energy production in the U.S. rose to 71.6 $\times 10^{15}$ Btu (75.5×10^{18} J) in 2000, which is about 18% of world production. Approximately 81% of this energy is in the form of fossil fuels. U.S. energy production by source is given in Fig. 5.

The relative U.S. production of coal compared to other fossil fuels has increased since 1976, when 26% was coal, 29% was crude oil and 33% was natural gas. In 1999, coal production accounted for 32%, crude oil was 17% and natural gas was 28%. Coal production for 1999 and 2000 represented the first time in forty years that production declined for two consecutive years. On an annual basis, the average utility price per ton of coal delivered to utilities dropped by 1.8% in 2000, continuing a downward trend started in 1978.

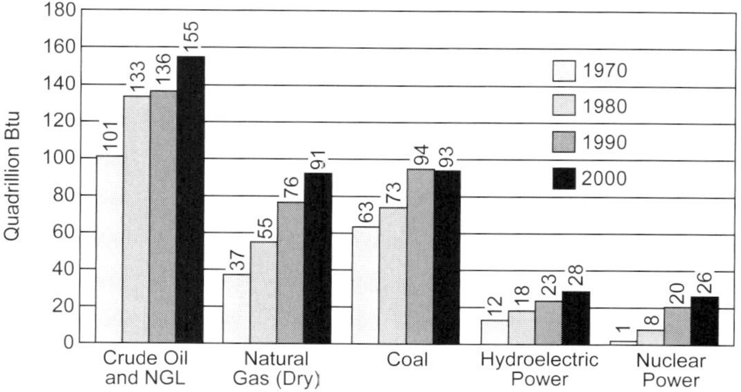

Fig. 1 Trends in world energy production by source (NGL = natural gas liquids).

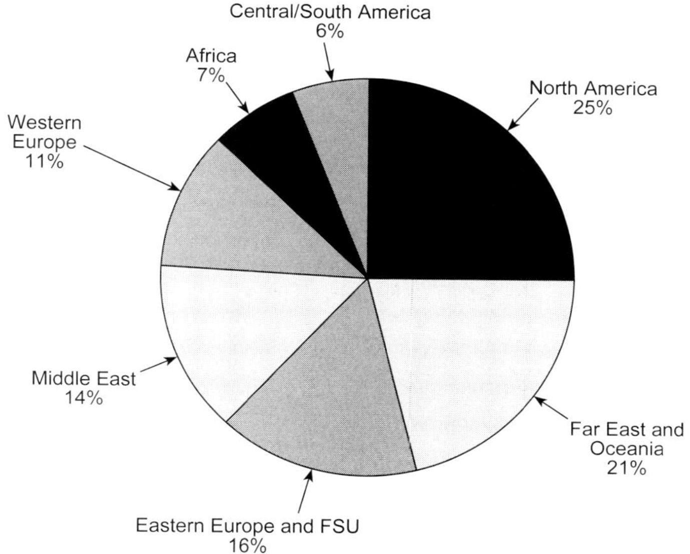

Fig. 2 World primary energy production by region, 2001.

Overall energy consumption in the U.S. was approximately 99×10^{15} Btu (104×10^{18} J) in 2000. About 28% of this energy was consumed by electric utilities in the form of fossil fuels.

Overall U.S. fossil fuel consumption continues to increase and grew to 84×10^{15} Btu (88.6×10^{18} J) in 2000. In spite of the decline in the cost of crude oil in the 1980s, it continues to be the most dominant and costly fuel in the fossil fuel mix. The trends in coal, oil and natural gas prices are given in Fig. 6.

World availability of coal

Coal is the second leading source of fuel, supplying 23% of the world's primary energy in 2000. It is also the most used fossil fuel for utility and industrial power gen-

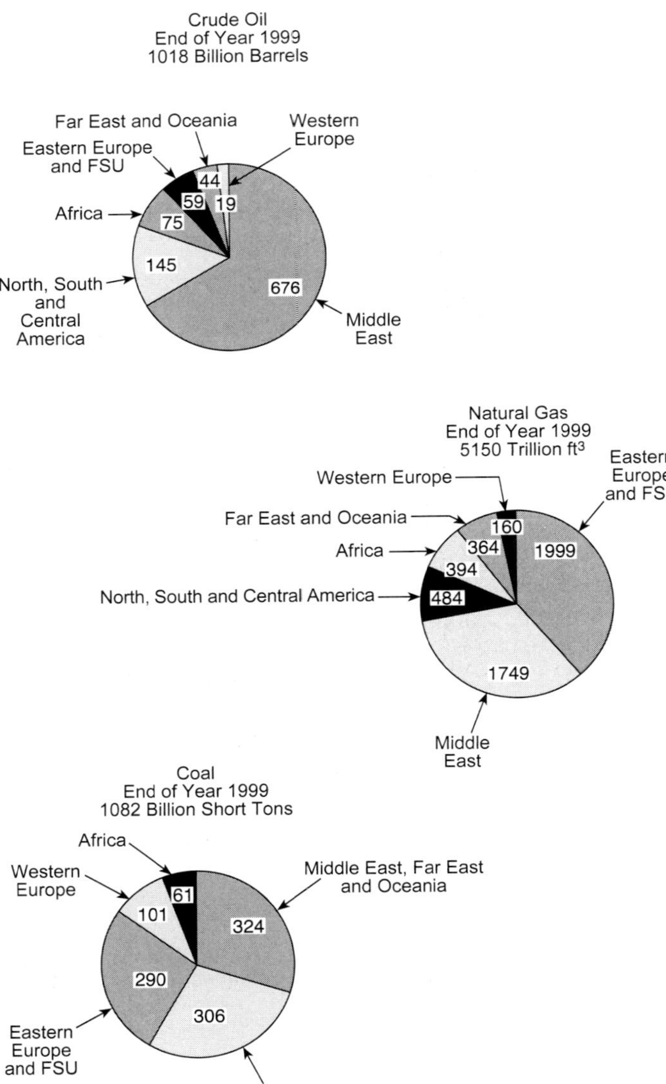

Fig. 3 Fossil fuel reserves by world region.

eration. Major reserves by coal type and location are lignite in the U.S. and the former Soviet Union (FSU); subbituminous in China, the FSU, Australia and Germany; and bituminous in China, the U.S. and the FSU.

Reserves of coal by regions of the world are given in Fig. 3. Of those regions, China consumed the most (25%) in 2000, followed by the U.S. (21%) and the FSU (9%). Because of its worldwide availability and low price, the demand for coal has grown and world coal trade has expanded by about 40% since 1980. The largest coal exporters are Australia, China, Indonesia, South Africa, the U.S., Canada, the FSU and Poland.[1]

U.S. availability of coal

The coal reserves of the U.S. constitute a vast energy resource, accounting for about 25% of the world's total recoverable coal.[2] According to the Energy Information Administration (EIA), the national estimate of the Demonstrated Reserve Base coal resources remaining as of 2002, is 498 billion short tons. Reserves that are likely to be mined are estimated at 275×10^9 t (249×10^9 t_m).[3] The U.S. produced 1.074×10^9 t (0.974×10^9 t_m) of coal in 2000. Fig. 7 summarizes U.S. production from 1978 to 2000. U.S. coal consumption has steadily increased from 0.7 billion short tons in 1980 to 1.05 billion in 1999. The states with the largest coal reserves in the ground as of January, 2000, are shown in Table 1.[4] States with large reserves, such as Montana and Illinois, do not necessarily rank as high in production as Wyoming, Kentucky or West Virginia.

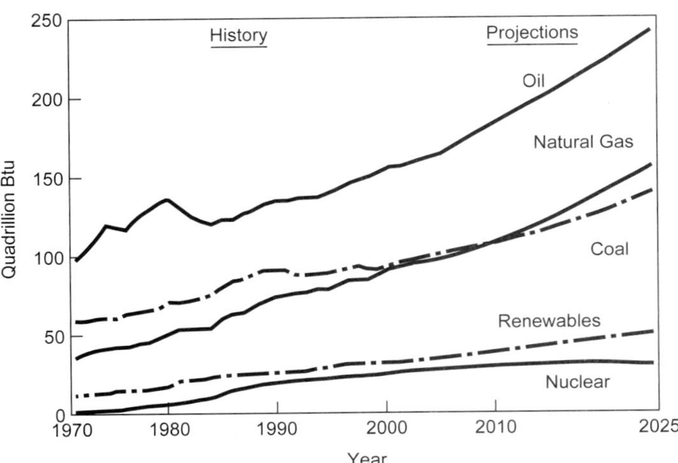

Fig. 4 World primary energy consumption by fuel.[1]

Because of the resulting sulfur dioxide (SO_2) emissions, coal sulfur levels are important production criteria and have been a factor in the growth of production from the western region, particularly the Powder River Basin. Table 2 shows the distribution of coal reserves by state at various sulfur levels.

Coal fields in the U.S. are shown in Fig. 8. The two largest producing regions are the western region consisting of Arizona, Colorado, Montana, New Mexico, North Dakota, Utah, Washington, and Wyoming and the Appalachian region including Pennsylvania, West Virginia, Ohio, western Maryland, eastern Kentucky, Virginia, Tennessee and Alabama. In 2000, these regions produced 510.7×10^6 t (463×10^6 t_m) and 419×10^6 t (380×10^6 t_m), respectively. Two-thirds of the re-

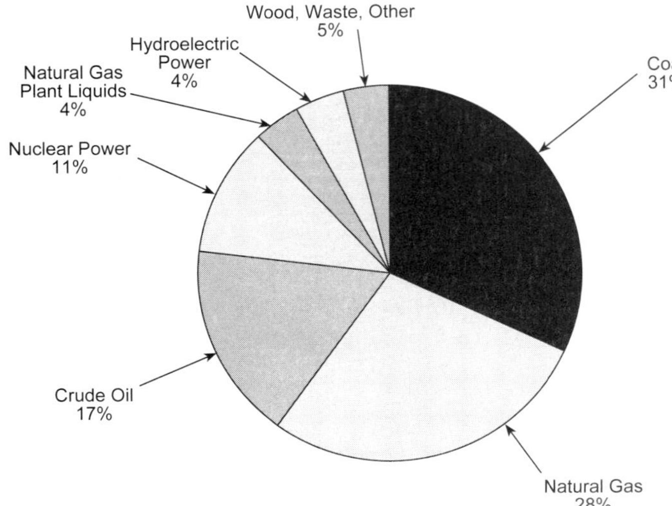

Fig. 5 U.S. energy production by source, 2002.

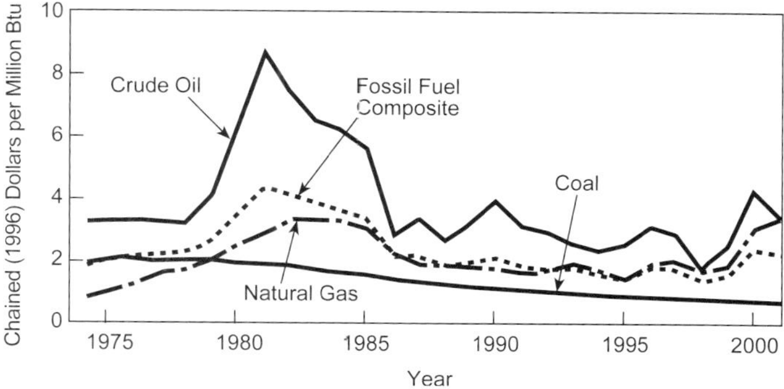

Fig. 6 Trends in U.S. fossil fuel prices

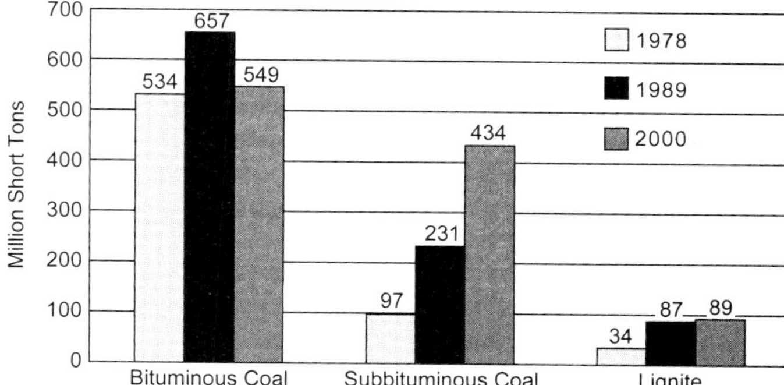

Fig. 7 U.S. coal production trends.

serves lie in the Great Plains, the Rocky Mountains and the western states. These coals are mostly subbituminous and lignitic, which have low sulfur content. Therefore, these fields have been rapidly developed to meet the increasing demands of electric utilities. The low sulfur coal permits more economical conformance to the Federal Clean Air Act, its Amendments, and acid rain legislation. (See Chapter 32.)

U.S. electric utilities used coal to generate 51% of the net electrical power in 2000, and remain the largest coal consumers. Continuing the downward trend since 1982, the average delivered cost of coal decreased 27% in current dollars per million Btu.

Environmental concerns about SO_2, nitrogen oxides (NO_x), carbon dioxide (CO_2) and mercury (Hg) emissions could limit the growth of coal consumption. However, the U.S., as well as Japan and several European countries, is researching clean coal technologies to reduce these emissions while boosting power production efficiency. These technologies are rapidly approaching commercialization in the U.S. They are expected to be integrated into current and future power plants.

Table 1
U.S. Energy Information Administration States with Largest Demonstrated Coal Reserves (x 10^9 t)*

State	Total Reserves		Underground Reserves		Surface Reserves		% Total U.S.
	t	(t_m)	t	(t_m)	t	(t_m)	
Montana	120	109	71	64	49	44	23.9
Illinois	105	95	88	80	17	15	20.9
Wyoming	67	61	43	39	24	22	13.3
West Virginia	35	32	30	27	4	3.6	7.0
Kentucky	31	28	18	16	14	13	6.2
Pennsylvania	28	25	24	22	4	3.6	5.5
Ohio	24	22	18	16	6	5	4.8
Colorado	17	15	12	11	5	4.5	3.4
Texas	13	12	0	0	13	11.8	2.6
New Mexico	12	11	6	5	6	5	2.4
Indiana	10	9	9	8	1	0.9	2.0
All others	41	37	20	18	21	19	8.2
Total U.S.	503	456	339	306	164	147	100.0

* Figures are rounded and include anthracite.

How coal is formed

Coal is formed from plants by chemical and geological processes that occur over millions of years. Layers of plant debris are deposited in wet or swampy regions under conditions that prevent exposure to air and complete decay as the debris accumulates. Bacterial action, pressure and temperature act on the organic matter over time to form coal. The geochemical process that transforms plant debris to coal is called *coalification*. The first product of this process, peat, often contains partially decomposed stems, twigs, and bark

Table 2
Sulfur Content and Demonstrated Total Underground and Surface Coal Reserve Base of the U.S. (Million tons)

State	Sulfur Range, % <1.0	1.1 to 3.0	>3.0	Unknown	Total*
Alabama	624.7	1,099.9	16.4	1,239.4	2,981.8
Alaska	11,458.4	184.2	0.0	0.0	11,645.4
Arizona	173.3	176.7	0.0	0.0	350.0
Arkansas	81.2	463.1	46.3	74.3	665.7
Colorado	7,475.5	786.2	47.3	6,547.3	14,869.2
Georgia	0.3	0.0	0.0	0.2	0.5
Illinois	1,095.1	7,341.4	42,968.9	14,256.2	65,664.8
Indiana	548.8	3,305.8	5,262.4	1,504.1	10,622.6
Iowa	1.5	226.7	2,105.9	549.2	2,884.9
Kansas	0.0	309.2	695.6	383.2	1,388.1
Kentucky-East	6,558.4	3,321.8	299.5	2,729.3	12,916.7
Kentucky-West	0.2	564.4	9,243.9	2,815.9	12,623.9
Maryland	135.1	690.5	187.4	34.6	1,048.2
Michigan	4.6	85.4	20.9	7.0	118.2
Missouri	0.0	182.0	5,226.0	4,080.5	9,487.3
Montana	101,646.6	4,115.0	502.6	2,116.7	108,396.2
New Mexico	3,575.3	793.4	0.9	27.5	4,394.8
North Carolina	0.0	0.0	0.0	31.7	31.7
North Dakota	5,389.0	10,325.4	268.7	15.0	16,003.0
Ohio	134.4	6,440.9	12,534.3	1,872.0	21,077.2
Oklahoma	275.0	326.6	241.4	450.5	1,294.2
Oregon	1.5	0.3	0.0	0.0	1.8
Pennsylvania	7,318.3	16,913.6	3,799.6	2,954.2	31,000.6
South Dakota	103.1	287.9	35.9	1.0	428.0
Tennessee	204.8	533.2	156.6	88.0	986.7
Texas	659.8	1,884.6	284.1	444.0	3,271.9
Utah	1,968.5	1,546.7	49.4	478.3	4,042.5
Virginia	2,140.1	1,163.5	14.1	330.0	3,649.9
Washington	603.5	1,265.5	39.0	45.1	1,954.0
West Virginia	14,092.1	14,006.2	6,823.3	4,652.5	39,589.8
Wyoming	33,912.3	14,657.4	1,701.1	3,060.3	53,336.1
Total*	200,181.4	92,997.5	92,571.5	50,788.0	436,725.7

*Data may not add to totals shown due to independent rounding. Source, Bureau of Mines Bulletin, Coal—Bituminous and Lignite, 1974.

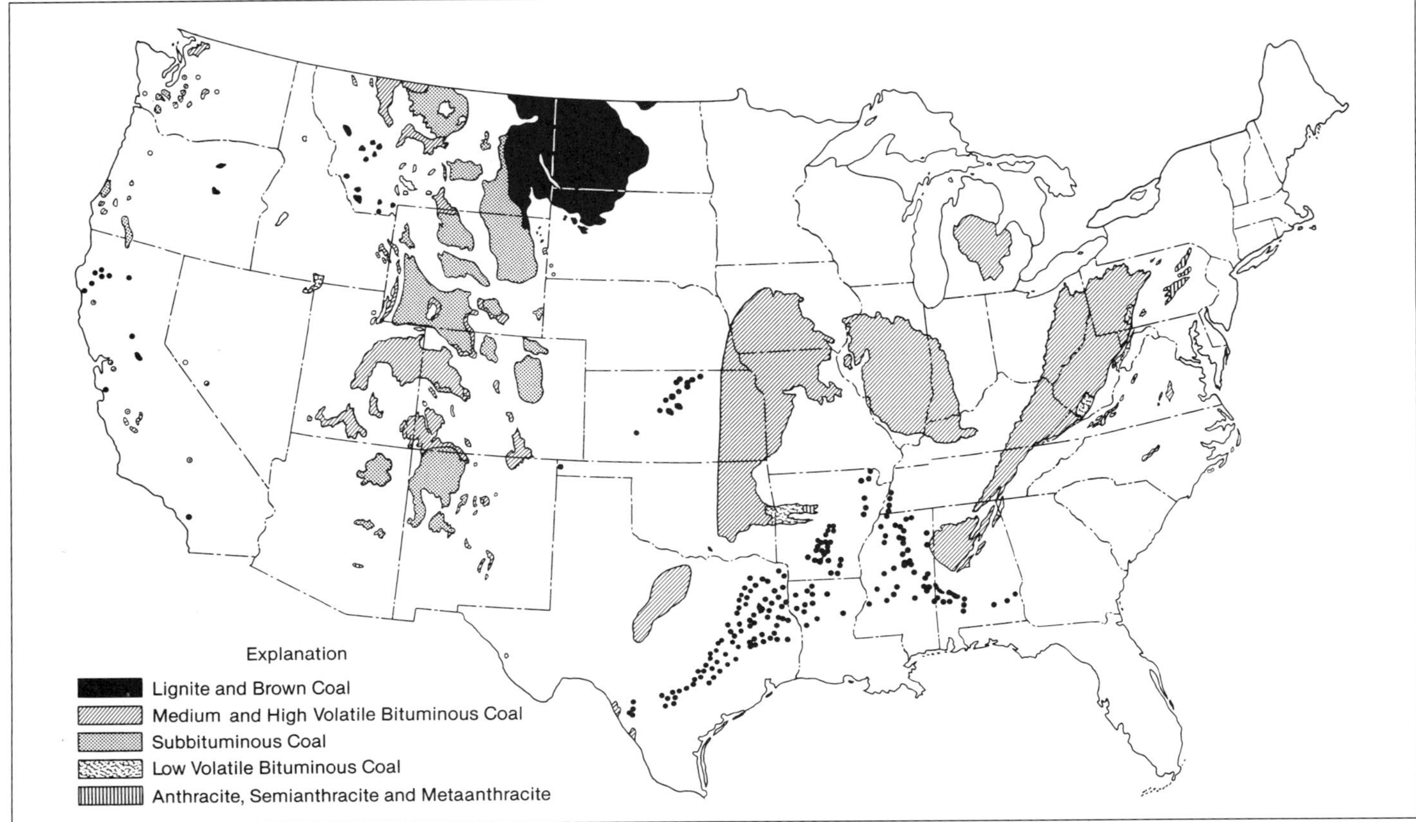

Fig. 8 U.S. coal reserves.

and is not classified as coal. However, peat is progressively transformed to lignite that eventually can become anthracite, given the proper progression of geological changes.

Various physical and chemical processes occur during coalification. The heat and pressure to which the organic material is exposed cause chemical and structural changes. These changes include an increase in carbon content; loss of water, oxygen and hydrogen; and resistance to solvents. The coalification process is shown schematically in Fig. 9.

Coal is very heterogeneous and can vary in chemical composition by location. In addition to the major organic ingredients (carbon, hydrogen and oxygen), coal also contains impurities. The impurities that are of major concern are ash and sulfur. The ash results from mineral or inorganic material introduced during coalification. Ash sources include inorganic substances, such as silica, that are part of the chemical structure of the plants. Dissolved inorganic ions and mineral grains found in swampy water are also captured by the organic matter during early coalification. Mud, shale and pyrite are deposited in pores and cracks of the coal seams.

Sulfur occurs in coal in three forms: 1) organic sulfur, which is part of the coal's molecular structure, 2) pyritic sulfur, which occurs as the mineral pyrite, and 3) sulfate sulfur, primarily from iron sulfate. The principal sulfur source is sulfate ion, found in water. Fresh water has a low sulfate concentration while salt water has a high sulfate content. Therefore, bituminous coal, deposited in the interior of the U.S. when seas covered this region, are high in sulfur. Some Iowa coals contain as much as 8% sulfur.

Although coal is a complex, heterogeneous mixture and not a polymer or biological molecule, it is sometimes useful for chemists to draw an idealized structural formula. These formulas can serve as models that illustrate coal reactions. This can aid the further development of coal processes such as gasification, combustion and liquefaction.

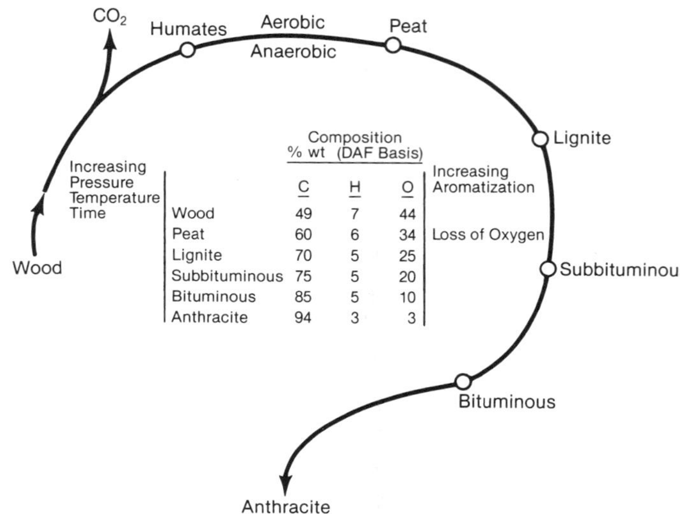

Fig. 9 The coalification process (DAF = dry ash-free).

Classifying coal

A coal classification system is needed because coal is a heterogeneous substance with a wide range of composition and properties. Coals are typically classified by rank. This indicates the progressive alteration in the coalification process from lignite to subbituminous, bituminous and anthracite coals. The rank indicates a coal's geological history and broad characteristics.

ASTM classification by rank

The system used in the U.S. for classifying coal by rank was established by the American Society for Testing and Materials (ASTM).[5] ASTM classification is a system that uses the volatile matter (VM) and fixed carbon (FC) results from the proximate analysis and the heating value of the coal as ranking criteria. This system aids in identifying commercial uses of coals and provides basic information regarding combustion characteristics.

The classification system is given in Table 3 and described in section D 388 of the ASTM standards. Proximate analysis is based on the laboratory procedure described in ASTM D 3172. In this procedure, moisture content, ash remaining after complete burning, amount of gases released when heated to a prescribed temperature, and fixed carbon remaining after volatilization are determined.

Table 4 gives a typical as-received proximate analysis of a West Virginia coal. An as-received analysis includes the total moisture content of the coal as it is received at the power plant.

For older or higher rank coals, FC and VM are used as the classifying criteria. These criteria are determined on a dry, mineral-matter-free basis using formulas developed by S.W. Parr in 1906 (shown in Equations 1 through 6).[6] The younger or low rank coals are classified by Btu content on a moist, mineral-matter-free basis. Agglomerating or weathering indices, as described in ASTM D 388, are used to differentiate adjacent groups.

Parr Formulas

Dry, mineral-free FC =

$$\frac{FC - 0.15\,\mathrm{S}}{100 - \left(M + 1.08A + 0.55\,\mathrm{S}\right)} \times 100,\ \% \tag{1}$$

Table 3
Classification of Coals by Rank[a] (ASTM D 388)

Class	Group	Fixed Carbon Limits, % (Dry, Mineral-Matter-Free Basis)		Volatile Matter Limits, % (Dry, Mineral-Matter-Free Basis)		Calorific Value Limits, Btu/lb (Moist,[b] Mineral-Matter-Free Basis)		Agglomerating Character
		Equal or Greater Than	Less Than	Greater Than	Equal or Less Than	Equal or Greater Than	Less Than	
I. Anthracitic	1. Meta-anthracite	98	–	–	2	–	–	Nonagglomerating
	2. Anthracite	92	98	2	8	–	–	Nonagglomerating
	3. Semianthracite[c]	86	92	8	14	–	–	Nonagglomerating
II. Bituminous	1. Low volatile bituminous coal	78	86	14	22	–	–	Commonly agglomerating[e]
	2. Medium volatile bituminous coal	69	78	22	31	–	–	Commonly agglomerating[e]
	3. High volatile A bituminous coal	–	69	31	–	14,000[d]	–	Commonly agglomerating[e]
	4. High volatile B bituminous coal	–	–	–	–	13,000[d]	14,000	Commonly agglomerating[e]
	5. High volatile C bituminous coal	–	–	–	–	11,500	13,000	Commonly agglomerating[e]
						10,500[e]	11,500	Agglomerating
III. Subbituminous	1. Subbituminous A coal	–	–	–	–	10,500	11,500	Nonagglomerating
	2. Subbituminous B coal	–	–	–	–	9,500	10,500	Nonagglomerating
	3. Subbituminous C coal	–	–	–	–	8,300	9,500	Nonagglomerating
IV. Lignitic	1. Lignite A	–	–	–	–	6,300	8,300	Nonagglomerating
	2. Lignite B	–	–	–	–	–	6,300	Nonagglomerating

[a] This classification does not include a few coals, principally nonbanded varieties, which have unusual physical and chemical properties and which come within the limits of fixed carbon or calorific value of the high volatile bituminous and subbituminous ranks. All of these coals either contain less than 48% dry, mineral-matter-free fixed carbon or have more than 15,500 moist, mineral-matter-free Btu/lb.

[b] Moist refers to coal containing its natural inherent moisture but not including visible water on the surface of the coal.

[c] If agglomerating, classify in low volatile group of the bituminous class.

[d] Coals having 69% or more fixed carbon on the dry, mineral-matter-free basis shall be classified according to fixed carbon, regardless of calorific value.

[e] It is recognized that there may be nonagglomerating varieties in these groups of the bituminous class, and there are notable exceptions in high volatile C bituminous group.

Table 4
Coal Analyses on As-Received Basis
(Pittsburgh Seam Coal, West Virginia)

Proximate Analysis Component	% by wt	Ultimate Analysis Component	% by wt
Moisture	2.5	Moisture	2.5
Volatile matter	37.6	Carbon	75.0
Fixed carbon	52.9	Hydrogen	5.0
Ash	7.0	Sulfur	2.3
Total	100.0	Nitrogen	1.5
		Oxygen	6.7
Heating value,		Ash	7.0
Btu/lb	13,000	Total	100.0
(kJ/kg)	(30,238)		

$$\text{Dry, mineral-free } VM = 100 - \text{Dry, mineral-free } FC, \% \quad \textbf{(2)}$$

$$\text{Moist, mineral-free Btu} = \frac{\text{Btu} - 50\,\text{S}}{100 - (1.08A + 0.55\,\text{S})} \times 100, \text{ per lb} \quad \textbf{(3)}$$

Approximation Formulas

$$\text{Dry, mineral-free } FC = \frac{FC}{100 - (M + 1.1A + 0.1\,\text{S})} \times 100, \% \quad \textbf{(4)}$$

$$\text{Dry, mineral-free } VM = 100 - \text{Dry, mineral-free } FC, \% \quad \textbf{(5)}$$

$$\text{Moist, mineral-free Btu} = \frac{\text{Btu}}{100 - (1.1A + 0.1\,\text{S})} \times 100, \text{ per lb} \quad \textbf{(6)}$$

where

Btu = heating value per lb (kJ/kg = 2.326 × Btu/lb)
FC = fixed carbon, %
VM = volatile matter, %
M = bed moisture,
A = ash, %
S = sulfur, %

all for coal on a moist basis.

Table 5 lists 16 selected U.S. coals, arranged in order of ASTM classification. The following descriptions briefly summarize the characteristics of each coal rank.

Peat Peat, the first product in the formation of coal, is a heterogeneous material consisting of partially decomposed plant and mineral matter. Its color ranges from yellow to brownish-black, depending on its geologic age. Peat has a moisture content up to 70% and a heating value as low as 3000 Btu/lb (6978 kJ/kg).

Lignite Lignite is the lowest rank coal. Lignites are relatively soft and brown to black in color with heating values of less than 8300 Btu/lb (19,306 kJ/kg). The deposits are geologically young and can contain recognizable remains of plant debris. The moisture content of lignites is as high as 30% but the volatile content is also high; consequently, they ignite easily. Lignite coal dries when exposed to air and spontaneous combustion during storage is a concern. Long distance shipment of these coals is usually not economical because of their high moisture and low Btu content. The largest lignite deposit in the world spreads over the regions of North and South Dakota, Wyoming, and Montana in the U.S. and parts of Saskatchewan and Manitoba in Canada.

Subbituminous Subbituminous coals are black, having little of the plant-like texture and none of the brown color associated with the lower rank lignite coal. Subbituminous coals are non-coking (undergo little swelling upon heating) and have a relatively high moisture content which averages from 15 to 30%. They also display a tendency toward spontaneous combustion when drying.

Although they are high in VM content and ignite easily, subbituminous coals generally have less ash and are cleaner burning than lignite coals. Subbituminous coals in the U.S. in general have a very low sulfur content, often less than 1%. Because they have reasonably high heating values [8300 to 11,500 Btu/lb (19,306 to 26,749 kJ/kg)] and low sulfur content, switching to subbituminous coal has become an attractive option for many power plants to limit SO_2 emissions.

Bituminous Bituminous coal is the rank most commonly burned in electric utility boilers. In general, it appears black with banded layers of glossy and dull black. Typical bituminous coals have heating values of 10,500 to 14,000 Btu/lb (24,423 to 32,564 kJ/kg) and a fixed carbon content of 69 to 86%. The heating value is higher, but moisture and volatile content are lower than the subbituminous and lignite coals. Bituminous coals rarely experience spontaneous combustion in storage. Furthermore, the high heating value and fairly high volatile content enable bituminous coals to burn easily when pulverized to a fine powder. Some types of bituminous coal, when heated in the absence of air, soften and release volatiles to form the porous, hard, black product known as *coke*. Coke is used as fuel in blast furnaces to make iron.

Anthracite Anthracite, the highest rank of coal, is shiny black, hard and brittle, with little appearance of layers. It has the highest content of fixed carbon, 86 to 98%. However, its low volatile content makes it a slow burning fuel. Most anthracites have a very low moisture content of about 3%; heating values of 15,000 Btu/lb (34,890 kJ/kg) are slightly lower than the best quality bituminous coals. Anthracite is low in sulfur and volatiles and burns with a hot, clean flame. These qualities make it a premium fuel used mostly for domestic heating.

Other classification systems

There are other classifications of coal that are currently in limited use in Europe. These are the International Classification of Hard Coals by Type and the

Table 5
Sixteen Selected U.S. Coals Arranged in Order of ASTM Classification

No.	Coal Rank Class	Coal Rank Group	State	County	M	VM	FC	A	S	Btu	Rank FC	Rank Btu
					Coal Analysis, Bed Moisture Basis							
1	I	1	Pa.	Schuylkill	4.5	1.7	84.1	9.7	0.77	12,745	99.2	14,280
2	I	2	Pa.	Lackawanna	2.5	6.2	79.4	11.9	0.60	12,925	94.1	14,880
3	I	3	Va.	Montgomery	2.0	10.6	67.2	20.2	0.62	11,925	88.7	15,340
4	II	1	W.Va.	McDowell	1.0	16.6	77.3	5.1	0.74	14,715	82.8	15,600
5	II	1	Pa.	Cambria	1.3	17.5	70.9	10.3	1.68	13,800	81.3	15,595
6	II	2	Pa.	Somerset	1.5	20.8	67.5	10.2	1.68	13,720	77.5	15,485
7	II	2	Pa.	Indiana	1.5	23.4	64.9	10.2	2.20	13,800	74.5	15,580
8	II	3	Pa.	Westmoreland	1.5	30.7	56.6	11.2	1.82	13,325	65.8	15,230
9	II	3	Ky.	Pike	2.5	36.7	57.5	3.3	0.70	14,480	61.3	15,040
10	II	3	Ohio	Belmont	3.6	40.0	47.3	9.1	4.00	12,850	55.4	14,380
11	II	4	Ill.	Williamson	5.8	36.2	46.3	11.7	2.70	11,910	57.3	13,710
12	II	4	Utah	Emery	5.2	38.2	50.2	6.4	0.90	12,600	57.3	13,560
13	II	5	Ill.	Vermilion	12.2	38.8	40.0	9.0	3.20	11,340	51.8	12,630
14	III	2	Wyo.	Sheridan	25.0	30.5	40.8	3.7	0.30	9,345	57.5	9,745
15	III	3	Wyo.	Campbell	31.0	31.4	32.8	4.8	0.55	8,320	51.5	8,790
16	IV	1	N.D.	Mercer	37.0	26.6	32.2	4.2	0.40	7,255	55.2	7,610

Notes: For definition of Rank Classification according to ASTM requirements, see Table 3.

Data on Coal (Bed Moisture Basis)

M = equilibrium moisture, %; VM = volatile matter, %;
FC = fixed carbon, %; A = ash, %; S = sulfur, %;
Btu = Btu/lb, higher heating value.

Rank FC = dry, mineral-matter-free fixed carbon, %;
Rank Btu = moist, mineral-matter-free Btu/lb.
Calculations by Parr formulas.

International Classification of Brown Coals. These systems were developed by the Coal Committee of the Economic Commission for Europe in 1949.

Coal characterization

As previously described, the criteria for ranking coal are based on its proximate analysis. In addition to providing classifications, coal analysis provides other useful information. This includes assistance in selecting coal for steam generation, evaluation of existing handling and combustion equipment, and input for design. The analyses consist of standard ASTM procedures and special tests developed by The Babcock & Wilcox Company (B&W). The following briefly summarizes some of these tests.

Standard ASTM analyses[5,7]

Bases for analyses Because of the variability of moisture and ash content in coals, the composition determined by proximate analysis can be reported on several bases. The most common include as-received, moisture-free or dry, and mineral-matter-free. The as-received analysis reports the percentage by weight of each constituent in the coal as it is received at the laboratory. As-received samples contain varying levels of moisture. For analysis on a dry basis, the moisture of the sample is determined and then used to correct each constituent to a common dry level. As previously mentioned, the ash in coal as determined by proximate analysis is different than the mineral matter in coal. This can cause problems when ranking coals by the ASTM method. Formulas used to correct for the mineral matter and to determine volatile matter, fixed carbon and heating value on a mineral-matter-free basis are provided in Equations 1 to 6 above.

Moisture determination Coal received at an electric power plant contains varying amounts of moisture in several forms. There is *inherent* and *surface* moisture in coal. Inherent moisture is that which is a naturally combined part of the coal deposit. It is held tightly within the coal structure and can not be removed easily when the coal is dried in air. The surface moisture is not part of the coal deposit and has been added externally. Surface moisture is more easily removed from coal when exposed to air. It is not possible to distinguish, by analysis, inherent and surface moisture.

There are many other moistures that arise when characterizing coal including equilibrium, free and air dry moisture. Their definitions and use depend on the application. Equilibrium moisture is sometimes used as an estimate of bed moisture. The ASTM standard terminology of coal and coke, D 121, defines the total coal moisture as the loss in weight of a sample under controlled conditions of temperature, time and air flow. Using ASTM D 3302, the total moisture is calculated from the moisture lost or gained in air drying and the residual moisture. The residual moisture is determined by oven drying the air dried sample. Because subsequent ASTM analyses (such as proximate and ultimate) are performed on an air dried sample, the residual moisture value is required to convert these results to a dry basis. In addition, the moisture lost on air drying provides an indication of the drying required in the handling and pulverization portions of the boiler coal feed system.

Proximate analysis Proximate analysis, ASTM D 3172, includes the determination of volatile matter, fixed carbon and ash. Volatile matter and fixed carbon, exclusive of the ash, are two indicators of coal rank. The amount of volatile matter in a coal indicates ease of ignition and whether supplemental flame stabilizing fuel is required. The ash content indicates the load under which the ash collection system must operate. It also permits assessing related shipping and handling costs.

Ultimate analysis Ultimate analysis, described in ASTM D 3176, includes measurements of carbon, hydrogen, nitrogen and sulfur content, and the calculation of oxygen content. Used with the heating value of the coal, combustion calculations can be performed to determine coal feed rates, combustion air requirements, heat release rates, boiler performance, and sulfur emissions from the power plant. (See Table 4.)

Heating value The gross calorific value of coal, determined using an adiabatic bomb calorimeter as described in ASTM D 2015, is expressed in Btu/lb (kJ/kg) on various bases (dry, moisture and ash free, etc.).

This value determines the maximum theoretical fuel energy available for the production of steam. Consequently, it is used to determine the quantity of fuel which must be handled, pulverized and fired.

Gross (higher) heating value (HHV) is defined as the heat released from combustion of a unit fuel quantity (mass), with the products in the form of ash, gaseous CO_2, SO_2, nitrogen and liquid water, exclusive of any water added as vapor. The net (lower) heating value (LHV) is calculated from the HHV. It is the heat produced by a unit quantity of fuel when all water in the products remains as vapor. This LHV calculation (ASTM Standard D 407) is made by deducting 1030 Btu/lb (2396 kJ/kg) of water derived from the fuel, including the water originally present as moisture and that formed by combustion. In the U.S., the gross calorific value is commonly used in heat balance calculations, while in Europe the net value is generally used.

Grindability The Hardgrove Grindability Test, developed by B&W, is an empirical measure of the relative ease with which coal can be pulverized. The ASTM D 409 method has been used for the past 30 years to evaluate the grindability of coals. The method involves grinding 50 g of air-dried 16 × 30 mesh (1.18 mm × 600 μm) test coal in a small ball-and-race mill. The mill is operated for 60 revolutions and the quantity of material that passes a 200 mesh (75 micron) screen is measured. From a calibration curve relating –200 mesh (–75 micron) material to the grindability of standard samples supplied by the U.S. Department of Energy, the Hardgrove Grindability Index (HGI) is determined for the test coal. Pulverizer manufacturers have developed correlations relating HGI to pulverizer capacity at desired levels of fineness.

Sulfur forms The sulfur forms test, ASTM D 2492, measures the amounts of sulfate sulfur, pyritic sulfur and organically bound sulfur in a coal. This is accomplished by measuring the total sulfur, sulfate, and pyritic sulfur contents and obtaining the organic sulfur by difference. The quantity of pyritic sulfur is an indicator of potential coal abrasiveness.

Free swelling index The free swelling index can be used to indicate caking characteristics. The index is determined by ASTM D 720 which consists of heating a one gram coal sample for a specified time and temperature. The shape of the sample or button formed by the swelling coal is then compared to a set of standard buttons. Larger formed buttons indicate higher free swelling indices. Oxidized coals tend to have lower indices. The free swelling index can be used as a relative measurement of a coal's caking properties and extent of oxidation.

Ash fusion temperatures Coal ash fusion temperatures are determined from cones of ash prepared and heated in accordance with ASTM method D 1857. The temperatures at which the cones deform to specific shapes are determined in oxidizing and reducing atmospheres. Fusion temperatures provide ash melting characteristics and are used for classifying the slagging potentials of the lignitic-type ashes.

Ash composition Elemental ash analysis is conducted using a coal ash sample produced by the ASTM D 3174 procedure. The elements present in the ash are determined and reported as oxides. Silicon dioxide (SiO_2), aluminum oxide (Al_2O_3), titanium dioxide (TiO_2), ferric hydroxide (Fe_2O_3), calcium oxide (CaO), magnesium oxide (MgO), sodium oxide (Na_2O) and potassium oxide (K_2O) are measured using atomic absorption per ASTM D 3682. The results of the ash analyses permit calculations of fouling and slagging indices and slag viscosity versus temperature relationships. The nature, composition and properties of coal ash and their effects on boiler performance are described further in Chapter 21.

Special B&W tests[7]

Burning profiles The burning profile technique was originated by B&W for predicting the relative combustion characteristics of fuels. The technique and application of results were described by Wagoner and Duzy,[8] and are routinely applied to liquid and solid fuels. The test uses derivative thermogravimetry in which a sample of fuel is oxidized under controlled conditions. A 300 mg sample of solid fuel with a particle size less than 60 mesh (250 microns) is heated at 27F/min (15C/min) in a stream of air. Weight change is measured continuously and the burning profile is the resulting plot of rate of weight loss versus furnace temperature.

Coals with similar burning profiles would be expected to behave similarly in large furnaces. By comparing the burning profile of an unknown coal with that of a known sample, furnace design, residence time, excess air and burner settings can be predicted. In comparing profiles, key information is provided by the start and completion temperatures of oxidation. The area under the temperature curve is proportional to the amount of combustible material in the sample; the height of the curve is a measure of the combustion intensity. Burning profiles are particularly useful for preliminary evaluations of new boiler fuels such as chars, coal-derived fuels and processed refuse. Fig. 10 shows burning profiles of coals of various ranks.

Abrasiveness index The abrasiveness of coal affects pulverizer grinding element life, and quartz particles in the coal can significantly contribute to its abrasiveness. A procedure for determining a coal's quartz count has been developed at B&W. This procedure consists of burning the coal, collecting and washing the ash to remove acid soluble constituents, and screening to separate size fractions. In each size fraction, 1000 particles are counted and the number of quartz particles is determined by a microscopic technique. From these data, the relative quartz value, an indicator of the coal's relative abrasiveness, is calculated.

Another abrasion index is determined using the Yancey-Geer Price apparatus. In this test, a sample of coal, sized 0.25 in. × 0 (6.35 mm × 0), is placed in contact with four metal test samples or *coupons* attached to a rotating shaft. The shaft is rotated at 1440 rpm (150.8 rad/s) for a total of 12,000 revolutions (75,400 rad). The weight loss of the metal coupons is then determined, from which a relative abrasion index is calculated. Indices from the test coals can be compared to those for other fuels. B&W has used the Yancey-Geer Price Index to determine wear in full scale pulverizers. The quartz count procedure and the Yancey-Geer Price procedures can provide some relative information and insight when comparing the abrasiveness of different coals; however, they have limited value in predicting actual field wear rates. (See Chapter 13.)

Erosiveness index Erosion occurs in boilers due to the impact of pulverized particles on burner lines and other components between the pulverizers and burners. The erosiveness test, developed by B&W, subjects a steel coupon to a stream of pulverized coal under controlled conditions. The measured weight loss of the coupon indicates the erosiveness of the coal.

Slag viscosity The viscosity of a coal ash slag is measured at various temperatures under oxidizing and reducing conditions using a high temperature rotational bob viscometer. This viscometer and its application are described in more detail in Chapter 21. The data obtained from slag viscosity measurements are used to predict a coal's slagging behavior in pulverized coal-fired boiler applications. The results also indicate the suitability of a coal for use in B&W's slagging and Cyclone furnaces.

Properties of selected coals

Table 6 gives basic fuel characteristics of typical U.S. coals. The coals are identified by state and rank, and the analytical data include proximate and ultimate analyses and HHVs. Table 7 provides similar fuel properties of coals mined outside the U.S. The source of this information, B&W's Fuels Catalogue, contains more than 10,000 fuel analyses performed and compiled since the 1950s.

Fuels derived from coal

Because of abundant supplies and low prices, the demand for coal as the prime or substitute fuel for utility boilers will most likely continue to increase. In addition, the future use of coal-derived fuels, such as coal refined liquids and gases, coal slurries, and chars, as inexpensive substitutes for oil and natural gas, is also possible. Therefore, methods to obtain clean and efficiently burning fuels derived from coal are continually being investigated. A few of these fuels that apply to steam generation are discussed below.

Coke

When coal is heated in the absence of air or with a large deficiency of air, the lighter constituents are volatilized and the heavier hydrocarbons crack, liberating gases and tars and leaving a residue of carbon. Some of the volatilized portions crack on contact with the hot carbon, leaving an additional quantity of carbon. The carbonaceous residue containing the ash and some of the original coal sulfur is called *coke*. The amount of sulfur and ash in the coke mainly depends on the coal from which it is produced and the coking process used. The principal uses for coke are the production of pig iron in blast furnaces and the charging of iron foundry cupolas. Because it is smokeless when burned, considerable quantities have been used for space heating.

Undersized coke, called *coke breeze*, usually passing a 0.625 in. (15.875 mm) screen, is unsuitable for charging blast furnaces and is often used for steam generation. A typical analysis of coke breeze appears in Table 8. Approximately 4.5% of the coal supplied to slot-type coke ovens is recovered as coke breeze. A portion of the coal tars produced as byproducts of the various coking processes may be burned in equipment similar to that used for heavy petroleum oil.

Gaseous fuels from coal

A number of gaseous fuels are derived from coal as process byproducts or from gasification processes. (See

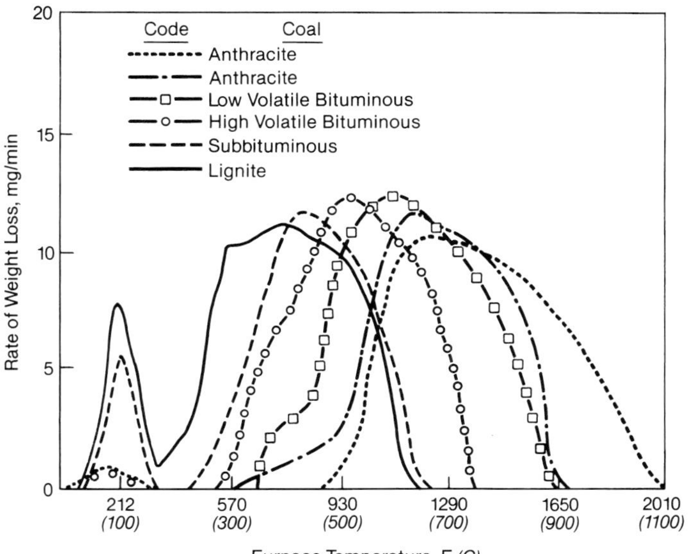

Fig. 10 Coal burning profiles.

Table 6
Properties of U.S. Coals

	Anthracite	Pittsburgh #8 HV Bituminous	Illinois #6 HV Bituminous	Upper Freeport MV Bituminous	Spring Creek Subbitu-minous	Decker Subbitu-minous	Lignite	Lignite (S.Hallsville)	Lignite (Bryan)	Lignite (San Miguel)
State	—	Ohio or Pa.	Illinois	Pennsylvania	Wyoming	Montana	North Dakota	Texas	Texas	Texas
Proximate:										
Moisture	7.7	5.2	17.6	2.2	24.1	23.4	33.3	37.7	34.1	14.2
Volatile matter, dry	6.4	40.2	44.2	28.1	43.1	40.8	43.6	45.2	31.5	21.2
Fixed carbon, dry	83.1	50.7	45.0	58.5	51.2	54.0	45.3	44.4	18.1	10.0
Ash, dry	10.5	9.1	10.8	13.4	5.7	5.2	11.1	10.4	50.4	68.8
Heating value, Btu/lb:										
As-received	11,890	12,540	10,300	12,970	9,190	9,540	7,090	7,080	3,930	2,740
Dry	12,880	13,230	12,500	13,260	12,110	12,450	10,630	11,360	5,960	3,200
MAF	14,390	14,550	14,010	15,320	12,840	13,130	11,960	12,680	12,020	10,260
Ultimate:										
Carbon	83.7	74.0	69.0	74.9	70.3	72.0	63.3	66.3	33.8	18.4
Hydrogen	1.9	5.1	4.9	4.7	5.0	5.0	4.5	4.9	3.3	2.3
Nitrogen	0.9	1.6	1.0	1.27	0.96	0.95	1.0	1.0	0.4	0.29
Sulfur	0.7	2.3	4.3	0.76	0.35	0.44	1.1	1.2	1.0	1.2
Ash	10.5	9.1	10.8	13.4	5.7	5.2	11.1	10.4	50.4	68.8
Oxygen	2.3	7.9	10.0	4.97	17.69	16.41	19.0	16.2	11.1	9.01

Ash fusion temps, F	Anthracite		Pittsburgh #8		Illinois #6		Upper Freeport		Spring Creek		Decker		Lignite		Lignite (S.Hallsville)		Lignite (Bryan)		Lignite (San Miguel)	
Reducing/Oxidizing:	Red	Oxid	Red	Oxid	Red	Oxid	Red	Oxid	Red	Oxid	Red	Oxid	Red	Oxid	Red	Oxid	Red	Oxid	Red	Oxid
ID	—	—	2220	2560	1930	2140	2750+	2750+	2100	2180	2120	2420	2030	2160	2000	2210	2370	2470	2730	2750+
ST Sp.	—	—	2440	2640	2040	2330	2750+	2750+	2160	2300	2250	2470	2130	2190	2060	2250	2580	2670	2750+	2750+
ST Hsp.	—	—	2470	2650	2080	2400	2750+	2750+	2170	2320	2270	2490	2170	2220	2090	2280	2690	2760	2750+	2750+
FT 0.0625 in.	—	—	2570	2670	2420	2600	2750+	2750+	2190	2360	2310	2510	2210	2280	2220	2350	2900+	2900+	2750+	2750+
FT Flat	—	—	2750+	2750+	2490	2700	2750+	2750+	2370	2700	2380	2750+	2300	2300	2330	2400	2900+	2900+	2750+	2750+

Ash analysis:	Anthracite	Pittsburgh #8	Illinois #6	Upper Freeport	Spring Creek	Decker	Lignite	Lignite (S.Hallsville)	Lignite (Bryan)	Lignite (San Miguel)
SiO_2	51.0	50.58	41.68	59.60	32.61	23.77	29.80	23.32	62.4	66.85
Al_2O_3	34.0	24.62	20.0	27.42	13.38	15.79	10.0	13.0	21.5	23.62
Fe_2O_3	3.5	17.16	19.0	4.67	7.53	6.41	9.0	22.0	3.0	1.18
TiO_2	2.4	1.10	0.8	1.34	1.57	1.08	0.4	0.8	0.5	1.46
CaO	0.6	1.13	8.0	0.62	15.12	21.85	19.0	22.0	3.0	1.76
MgO	0.3	0.62	0.8	0.75	4.26	3.11	5.0	5.0	1.2	0.42
Na_2O	0.74	0.39	1.62	0.42	7.41	6.20	5.80	1.05	0.59	1.67
K_2O	2.65	1.99	1.63	2.47	0.87	0.57	0.49	0.27	0.92	1.57
P_2O_5	-	0.39	-	0.42	0.44	0.99	-	-	-	-
SO_3	1.38	1.11	4.41	0.99	14.56	18.85	20.85	9.08	3.50	1.32

Note: HV = high volatile; MV = medium volatile; ID = initial deformation temp; ST = softening temp; FT = fluid temp; Sp. = spherical; Hsp. = hemispherical.

Chapter 18.) Table 9 lists selected analyses of these gases. They have currently been largely supplanted by natural gas and oil. However, improvements in coal gasification and wider use of coal in the chemical and liquid fuel industries could reverse this trend.

Coke oven gas A considerable portion of coal is converted to gases in the production of coke. Valuable products recovered from these gaseous portions include ammonium sulfate, oils and tars. The non-condensable portion is called *coke oven gas*. Constituents depend on the nature of the coal and the coking process used (Table 9).

Part of the sulfur from coal may be present in coke oven gas as hydrogen sulfide and carbon disulfide. These may be removed by scrubbing. Coke oven gas often contains other impurities that deposit in pipelines and burners. The gas burns readily because of its high free hydrogen content and presents minimal problems when used as steam generation fuel.

Blast furnace gas The gas discharged from steel mill blast furnaces is used at the mills in furnaces, in gas engines and for steam generation. Blast furnace gas has variable quality but generally has a high carbon monoxide (CO) content and low heating value (Table 9). This gas may be burned for steam generation. However, blast furnace gas deposits adhere firmly and provisions must be made for cleaning boiler heating surfaces.

Water gas The gas produced by passing steam through a bed of hot coke is known as *water gas*. Carbon in the coke combines with the steam to form H_2 and CO. This is an endothermic reaction that cools the coke bed. Water gas is often enriched with oil by passing the gas through a checkerwork of hot bricks sprayed with oil. The oil, in turn, is cracked to a gas by the heat. Refinery gas is also used for enrichment. It may be mixed with the steam and passed through the coke bed or may be mixed directly with the water gas. Such enriched gas is called *carbureted water gas*

Table 7
Properties of Selected International Coals

Source	Australia	China	France	S. Africa	Indonesia	Korea	Spain
Ultimate:							
Carbon	56.60	62.67	74.60	69.70	56.53	68.46	37.02
Hydrogen	3.50	3.86	4.86	4.50	4.13	0.90	2.75
Nitrogen	1.22	0.83	1.39	1.60	0.88	0.20	0.88
Sulfur	0.35	0.46	0.79	0.70	0.21	2.09	7.46
Ash	24.00	4.71	8.13	10.10	1.77	23.48	38.69
Oxygen	7.43	10.34	9.42	9.10	12.58	4.38	11.39
Proximate:							
Moisture	6.90	17.13	0.80	4.30	23.90	0.50	1.80
Volatile matter, dry	24.80	30.92	36.11	35.30	45.57	7.46	45.27
Fixed carbon, dry	44.30	47.24	54.96	50.30	28.76	68.56	14.24
Ash, dry	24.00	4.71	8.13	10.10	1.77	23.48	38.69
Higher heating value, Btu/lb	9660	10,740	13,144	12,170	9,840	9,443	6,098
Ash analysis:							
SiO_2	57.90	22.70	44.60	44.00	71.37	55.00	14.50
Al_2O_3	32.80	9.00	29.90	32.70	13.32	17.00	8.20
Fe_2O_3	6.20	15.68	13.10	4.60	7.00	12.50	2.70
TiO_2	1.00	0.43	0.60	1.20	0.57	1.40	0.30
CaO	0.60	28.88	–	5.70	2.88	0.10	45.00
MgO	0.80	2.00	3.50	1.30	0.53	0.10	1.20
Na_2O	0.10	0.70	3.10	0.10	0.34	0.10	0.10
K_2O	0.50	0.46	–	0.30	0.25	3.10	0.40
P_2O_5	–	0.09	–	2.20	0.16	–	–
SO_3	0.80	20.23	2.80	4.60	3.90	–	–

Ash fusion temps, F	Australia		China		France		S. Africa		Indonesia		Korea		Spain	
Reducing/Oxidizing:	Red	Oxid	Red	Oxid	Red	Oxid	Red	Oxid	Red	Oxid	Red	Oxid	Red	Oxid
ID	2740	2750+	2200	2220	2190	2300	2620	2670	2140	2410	2350	2600	2530	2520
ST Sp.	2750+	2750+	2240	2270	2310	2500	2750	2750+	2400	2490	2630	2730	2700	2670
ST Hsp.	2750+	2750+	2250	2280	–	–	2750+	2750+	2450	2540	–	–	–	–
FT 0.0625 in.	2750+	2750+	2280	2290	2670	2820	2750+	2750+	2630	2680	2900	2900	2730	2740
FT Flat	2750+	2750+	2340	2320	–	–	2750+	2750+	2750	2750+	–	–	–	–

(Table 9). In many areas, carbureted water gas has been replaced by natural gas.

Producer gas When coal or coke is burned with a deficiency of air and a controlled amount of moisture (steam), a product known as *producer gas* is obtained. This gas, after removal of entrained ash and sulfur compounds, is used near its source because of its low heating value (Table 9).

Byproduct gas from gasification

Coal gasification processes are a source of synthetic natural gas. There are many processes under development. The effluent gas from steam-oxygen coal gasification consists principally of H_2, CO, CH_4, CO_2 and unreacted steam. The gas will also be diluted with N_2 if air is used as the oxygen source. Although the competing chemical reactions that coal undergoes during gasification are complex, they usually include the reaction of steam and carbon to produce H_2 and CO. Some CH_4 is produced by the reaction of carbon with H_2 and by thermal cracking of the heavy hydrocarbons in the coal. CO_2 and heat needed for the process are produced by reaction of carbon with O_2. Final gas composition is modified by reaction between CO and steam to produce H_2 and CO_2.

The products of coal gasification are often classified as low, intermediate and high Btu gases. Low Btu gas has a heating value of 100 to 200 Btu/SCF (3.9 to 7.9 MJ/Nm^3) and is produced by gasification with air rather than oxygen. Typically, the gas is used as a boiler fuel at the gasification plant site or as feed to a turbine in combined cycles. Intermediate Btu gas has a heating value of 300 to 450 Btu/SCF (11.8 to 17.7 MJ/Nm^3) and is produced by gasification with oxygen or by a process that produces a nitrogen-free product. The applications of intermediate Btu gas are similar to low Btu gas. High Btu gas has a heating value greater than 900 Btu/SCF (35.4 MJ/Nm^3) and is used as a fuel

Table 8
Analyses — Bagasse and Coke Breeze

Analyses (as-fired), % by wt	Bagasse	Coke Breeze
Proximate		
Moisture	52.0	7.3
Volatile matter	40.2	2.3
Fixed carbon	6.1	79.4
Ash	1.7	11.0
Ultimate		
Hydrogen, H_2	2.8	0.3
Carbon, C	23.4	80.0
Sulfur, S	trace	0.6
Nitrogen, N_2	0.1	0.3
Oxygen, O_2	20.0	0.5
Moisture, H_2O	52.0	7.3
Ash	1.7	11.0
Heating value, Btu/lb	4000	11,670
(kJ/kg)	(9304)	(27,144)

in place of natural gas. High Btu gas is produced by the same gasification process as intermediate Btu gas and then upgraded by methanation. (See also Chapter 18.)

Fuel oil

One of the most widely accepted theories explaining the origin of oil is the organic theory. Over millions of years, rivers carried mud and sand that deposited and ultimately became sedimentary rock formations. Along with this inorganic material, tiny marine organisms were buried with the silt. Over time, in an airless and high pressure environment, the organic material containing carbon and hydrogen was converted to the hydrocarbon molecules of petroleum (oil). Because of the porosity of sedimentary rock formations, the oil flowed and collected in *traps*, or locations where crude oil is concentrated. This phenomenon greatly assists the economic recovery of crude oil.

Table 9
Selected Analyses of Gaseous Fuels Derived from Coal

	Coke Oven Gas	Blast Furnace Gas	Carbureted Water Gas	Producer Gas
Analysis No.	1	2	3	4
Analyses, % by volume				
Hydrogen, H_2	47.9	2.4	34.0	14.0
Methane, CH_4	33.9	0.1	15.5	3.0
Ethylene, C_2H_4	5.2	—	4.7	—
Carbon monoxide, CO	6.1	23.3	32.0	27.0
Carbon dioxide, CO_2	2.6	14.4	4.3	4.5
Nitrogen, N_2	3.7	56.4	6.5	50.9
Oxygen, O_2	0.6	—	0.7	0.6
Benzene, C_6H_6	—	—	2.3	—
Water, H_2O	—	3.4	—	—
Specific gravity (relative to air)	0.413	1.015	0.666	0.857
HHV — Btu/ft^3 (kJ/m^3)				
at 60F (16C) and 30 in. Hg (102 kPa)	590 (21,983)	— —	534 (19,896)	163 (6,073)
at 80F (27C) and 30 in. Hg (102 kPa)	— (3,122)	83.8 (3,122)	—	—

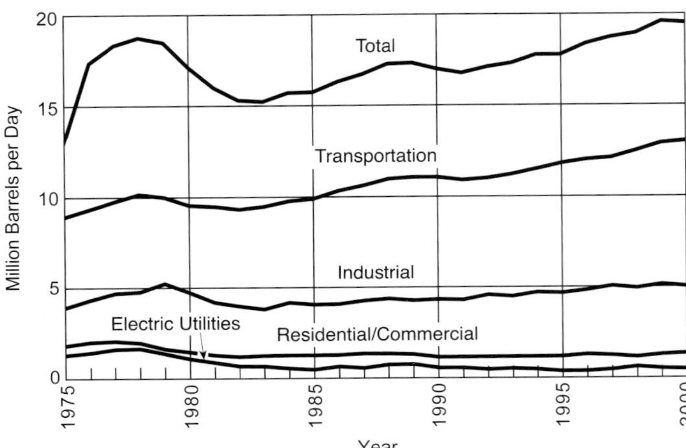

Fig. 11 U.S. petroleum end users.

Fuel oil consumption for steam generation accounts for a minor share of U.S. domestic petroleum fuel usage. Industrial users, excluding transportation, account for about 25% of all petroleum use; electric utilities consume about 2% of the total.[2] The end users of petroleum products for the years 1975 to 2000 are shown in Fig. 11. Crude oil reserves and world petroleum consumption are shown in Figs. 12 and 13.

Compared to coal, fuel oils are relatively easy to handle and burn. There is less bulk ash to dispose of and the ash discharged is correspondingly small. In most oil burners, the fuel is atomized and mixed with combustion air. In the atomized state, the characteristics of oil approach those of natural gas. (See Chapter 11.)

Because of its relatively low cost, No. 6 fuel oil is the most widely used for steam generation. It can be considered a byproduct of the refining process. Its ash content ranges from 0.01 to 0.5% which is very low compared to coal. However, despite this low ash content, compounds of vanadium, sodium and sulfur in the ash can pose operating problems. (See Chapter 21.)

Fuel oil characterization

Fuel oils include virtually all petroleum products that are less volatile than gasoline. They range from light oils, suitable for use in internal combustion or turbine engines, to heavy oils requiring heating. The

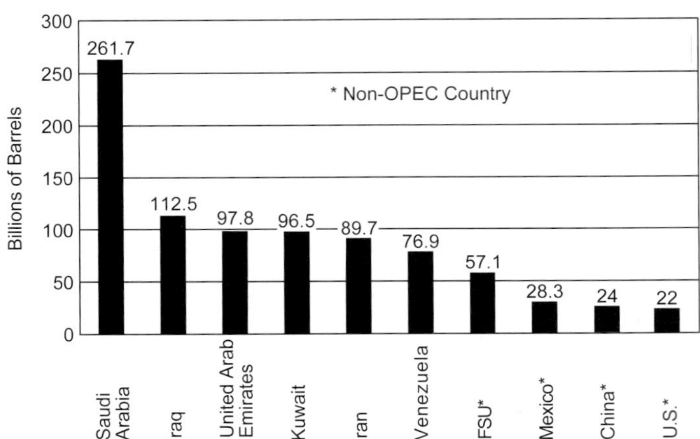

Fig. 12 Major world crude oil reserves, 2000 (OPEC = Organization of Petroleum Exporting Countries).

heavier fuels are primarily suited for steam generation boilers. The ASTM specifications for fuel oil properties are given in Table 10.

Fuel oils can be divided into two classes: distillate and residual. Distillate fuels are those that are vaporized in a petroleum refining operation. They are typically clean, essentially free of sediment and ash, and relatively low in viscosity. These fuels fall into the No. 1 or No. 2 category in ASTM D 396. Although No. 2 oil is sometimes used as a premium steam generation fuel, it best lends itself to applications where cleanliness and ease of handling outweigh its cost. Examples include home heating and industrial applications where low ash and/or sulfur are important. Steam generating applications are primarily limited to use as a startup or support fuel.

The residual fuel oils are those that are not vaporized by heating. They contain virtually all the inorganic constituents present in the crude oil. Frequently, residual oils are black, high viscosity fluids that require heating for proper handling and combustion.

Fuel oils in grades No. 4 and 5 are less viscous and therefore more easily handled and burned than is No. 6 oil. Depending on the crude oil used, a fuel meeting the No. 4 specification may be a blend of residual oil and lighter distillate fractions. This oil does not usually require heating for pumping and handling.

No. 5 oils may require heating, depending on the firing equipment and the ambient temperature. No. 6 oils usually require heating for handling and burning. (See Chapter 11 for oil storage, handling and use requirements.)

Table 10
ASTM Standard Specifications for Fuel Oils[a]

No. 1 A distillate oil intended for vaporizing pot-type burners and other burners requiring this grade of fuel

No. 2 A distillate oil for general purpose domestic heating for use in burners not requiring No. 1 fuel oil

No. 4 Preheating not usually required for handling or burning

No. 5 (Light) Preheating may be required depending on climate and equipment

No. 5 (Heavy) Preheating may be required for burning and, in cold climates, may be required for handling

No. 6 Preheating required for burning and handling

Grade of Fuel Oil[b]	Flash Point, F *(C)*	Pour Point, F *(C)*	Water and Sediment, % by vol	Carbon Residue on 10% Bottoms, %	Ash % by wt	Distillation Temperatures, F *(C)* 10% Point	90% Point		Saybolt Viscosity, s Universal at 100F *(38C)*		Furol at 122F *(50C)*		Kinematic Viscosity, centistokes At 100F *(38C)*		At 122F *(50C)*		Gravity, deg API	Copper Strip Corrosion
	Min	Max	Max	Max	Max	Max	Min	Max	Min	Max	Min	Max	Min	Max	Min	Max	Min	Max
No. 1	100 or legal *(38)*	0	trace	0.15	—	420 *(216)*	—	550 *(288)*	—	—	—	—	1.4	2.2	—	—	35	No. 3
No. 2	100 or legal *(38)*	20[c] *(-7)*	0.10	0.35	—	d	540[e] *(282)*	640 *(338)*	*(32.6)*[f]	*(37.93)*	—	—	2.0[e]	3.6	—	—	30	—
No. 4	130 or legal *(55)*	20 *(-7)*	0.50	—	0.10	—	—	—	45	125	—	—	(5.8)	(26.4)	—	—	—	—
No. 5 (Light)	130 or legal *(55)*	—	1.00	—	0.10	—	—	—	150	300	—	—	(32)	(65)	—	—	—	—
No. 5 (Heavy)	130 or legal *(55)*	—	1.00	—	0.10	—	—	—	350	750	(23)	(40)	(75)	(162)	(42)	(81)	—	—
No. 6	150 *(65)*	—	2.00[g]	—	—	—	—	—	(900)	(9000)	45	300	—	—	(92)	(638)	—	—

Notes:

a. Recognizing the necessity for low sulfur fuel oils used in connection with heat treatment, nonferrous metal, glass, and ceramic furnaces and other special uses, a sulfur requirement may be specified in accordance with the following table:

Grade of Fuel Oil	Sulfur, Max, %
No. 1	0.5
No. 2	0.7
No. 4	no limit
No. 5	no limit
No. 6	no limit

Other sulfur limits may be specified only by mutual agreement between the purchaser and the seller.

b. It is the intent of these classifications that failure to meet any requirement of a given grade does not automatically place an oil in the next lower grade unless, in fact, it meets all requirements of the lower grade.

c. Lower or higher pour points may be specified whenever required by conditions of storage or use.

d. The 10% distillation temperature point may be specified at 440F *(226C)* maximum for use in other than atomizing burners.

e. When pour point less than 0F is specified, the minimum viscosity shall be 1.8 cs (32.0 s, Saybolt Universal) and the minimum 90% point shall be waived.

f. Viscosity values in parentheses are for information only and not necessarily limiting.

g. The amount of water by distillation plus the sediment by extraction shall not exceed 2.00%. The amount of sediment by extraction shall not exceed 0.50%. A deduction in quantity shall be made for all water and sediment in excess of 1.0%.

Source, ASTM D 396.

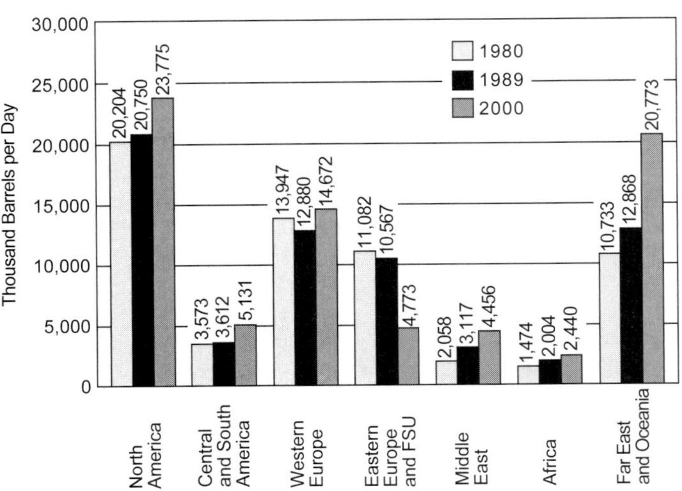

Fig. 13 Major petroleum consumption.

Fuel analyses

A typical analysis of a fuel oil or waste liquid contains the following information:

1. ultimate analysis
2. API gravity
3. heating value
4. viscosity
5. pour point
6. flash point
7. water and sediment

Ultimate analysis The ultimate analysis for an oil is similar to that for a coal. The results indicate the quantities of sulfur, hydrogen, carbon, nitrogen, oxygen and ash. Ultimate analyses for various fuel oils are given in Table 11.

The sulfur content of the oil is an indicator of its corrosiveness and is oxidized to sulfur oxides during combustion. These oxides can react with water vapor or ash constituents to form corrosive acids, salts, or boiler fouling potassium sulfate. When molten, these ash deposits are corrosive. Furthermore, vanadium can combine with the sulfur oxides to form a corrosive product. (See Chapter 21.)

API gravity The petroleum industry uses the API gravity scale to determine the relative density of oil. The scale was devised jointly by the American Petroleum Institute (API) and the National Bureau of Standards. The relationship between the *API gravity* and the *specific gravity* is given by the following formula:

$$\text{Deg API Gravity} = \frac{141.5}{\text{Specific gravity at } 60/60\text{F}} - 131.5$$

Given this relationship, heavier liquid fuels are denoted by lower API gravity values.

Heating value The heating value of a liquid fuel indicates the heat released by the complete combustion of one unit of fuel [lb (kg)]. As for coal, there are two calculated heating values, higher (HHV) and lower (LHV). In computing the HHV, it is assumed that any water vapor formed by burning the hydrogen constituent is condensed and cooled to its initial temperature. Therefore, the heat of vaporization of the water formed is included in the HHV. For the LHV, it is assumed that none of the water vapor condenses. Both heating values are determined by using an oxygen bomb calorimeter.

Viscosity The viscosity of a liquid is the measure of its internal resistance to flow. Although there are numerous viscosity scales, those most commonly used in the U.S. are:

1. Saybolt Universal Seconds (SUS),
2. Saybolt Furol Seconds (SFS),
3. absolute viscosity (centipoise), and
4. kinematic viscosity (centistokes).

The *kinematic viscosity* of oil is related to the *absolute viscosity* by the following formula:

$$\text{Kinematic viscosity (centistokes)} = \frac{\text{Absolute viscosity (centipoise)}}{\text{Specific gravity}}$$

Pour point The pour point is the lowest temperature at which a liquid fuel flows under standardized conditions.

Flash point The flash point is the temperature to which a liquid must be heated to produce vapors that flash but do not burn continuously when ignited. There are two instruments used to determine the flash point: the Pensky-Martens or closed cup flash tester, and the Cleveland or open cup tester. The closed cup tester indicates a lower flash point because it retains light vapors which are lost by the open cup unit.

Water and sediment The water and sediment level, also called bottom sediment and water (BSW), is a measure of the contaminants in a liquid fuel. The sediment normally consists of calcium, sodium, magnesium and iron compounds. For heavy fuels, the sediment may also contain carbon.

The basic analyses described are important in designing oil-fired boilers. The HHV determines the quantity of fuel required to reach a given heat input. The ultimate analysis determines the theoretical air required for complete combustion and therefore indicates the size of the burner throat. Also available from the ultimate analysis is the carbon/hydrogen ratio, which shows the ease with which a fuel burns. This ratio also indicates the expected level of carbon particulate emissions. A carbon/hydrogen ratio in excess of 7.5 is usually indicative of troublesome burning.

Considering the percentages of nitrogen and sulfur in conjunction with the HHV, an estimate of NO_x and SO_2 emissions can be made. The ash percentage has a similar bearing on particulate emissions. The ash constituent analysis and ash content indicate fouling and corrosion tendencies.

Additional information, which is often required when designing a boiler, includes:

1. carbon residue,
2. asphaltenes,
3. elemental ash analysis,
4. burning profile, and
5. distillation curve.

Table 11
Analyses of Fuel Oils

Grade of Fuel Oil	No. 1	No. 2	No. 4	No. 5	No. 6
% by weight:					
Sulfur	0.01 to 0.5	0.05 to 1.0	0.2 to 2.0	0.5 to 3.0	0.7 to 3.5
Hydrogen	13.3 to 14.1	11.8 to 13.9	(10.6 to 13.0)*	(10.5 to 12.0)*	(9.5 to 12.0)*
Carbon	85.9 to 86.7	86.1 to 88.2	(86.5 to 89.2)*	(86.5 to 89.2)*	(86.5 to 90.2)*
Nitrogen	nil to 0.1	nil to 0.1	—	—	—
Oxygen	—	—	—	—	—
Ash	—	—	0 to 0.1	0 to 0.1	0.01 to 0.5
Gravity:					
Deg API	40 to 44	28 to 40	15 to 30	14 to 22	7 to 22
Specific	0.825 to 0.806	0.887 to 0.825	0.966 to 0.876	0.972 to 0.922	1.022 to 0.922
lb/gal	6.87 to 6.71	7.39 to 6.87	8.04 to 7.30	8.10 to 7.68	8.51 to 7.68
Pour point, F	0 to −50	0 to −40	−10 to +50	−10 to +80	+15 to +85
Viscosity:					
Centistokes at 100F	1.4 to 2.2	1.9 to 3.0	10.5 to 65	65 to 200	260 to 750
SUS at 100F	—	32 to 38	60 to 300	—	—
SFS at 122F	—	—	—	20 to 40	45 to 300
Water and sediment, % by vol	—	0 to 0.1	tr to 1.0	0.05 to 1.0	0.05 to 2.0
Heating value, Btu/lb gross (calculated)	19,670 to 19,860	19,170 to 19,750	18,280 to 19,400	18,100 to 19,020	17,410 to 18,990

*Estimated

Properties of fuel oils

Analytical results for various fuel oil properties are given in Table 11.

Fuel oil heating values are closely related to their specific gravities. The relationships between the *HHV* of various fuel oils and their API gravities are shown in Fig. 14.

A more accurate estimate of the heating value for an oil is obtained by correcting the *HHV* from Fig. 14 as follows:

$$\text{Apparent heating value} = \frac{HHV\left[100 - (A + M + S)\right]}{100} + 40.5\,S \quad \textbf{(7)}$$

where

A = % weight of ash
M = % weight of water
S = % weight of sulfur

The volume percentages of water and sediment can be used without appreciable error in place of their weight percentages.

Fuel oils are generally sold on a volume basis using 60F (16C) as the base temperature. Correction factors are given in Fig. 15 for converting volumes at

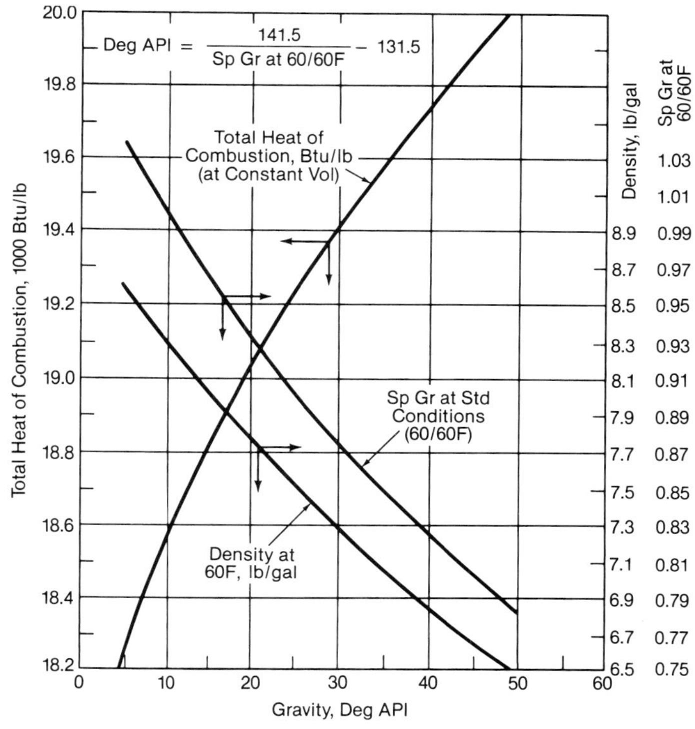

Fig. 14 Relationship between HHV of various fuel oils and their API gravities.

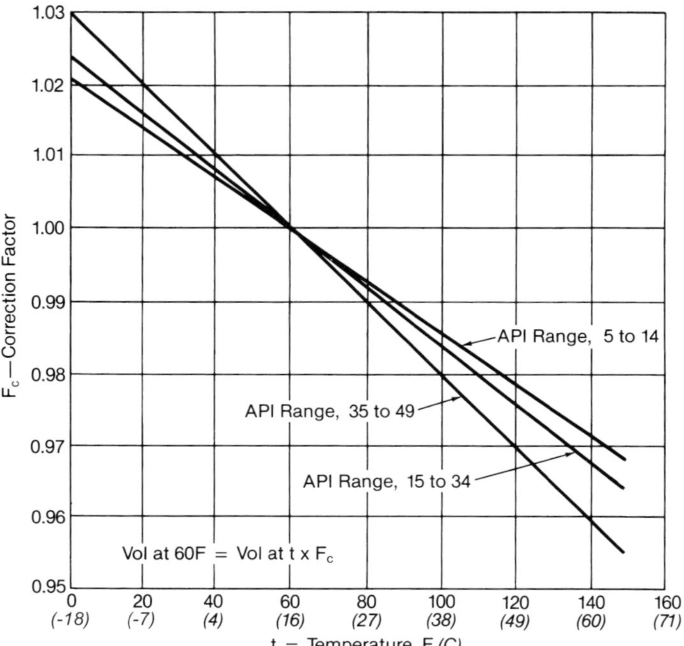

Fig. 15 Oil volume-temperature correction factors.

other temperatures to this standard base. This correction is also dependent on the API gravity range, as illustrated by the three lines of Fig. 15.

Handling and burning equipment are usually designed for a maximum oil viscosity. If the viscosities of heavy oils are known at two temperatures, their viscosities at other temperatures can be closely predicted by a linear interpolation between these two values on the standard ASTM chart (Fig. 16). Viscosity-temperature variations for certain light oils can also be found using the ASTM chart. In this case, however, the designer only needs to know the viscosity at one temperature. For example, the viscosity of a light oil at a given temperature within the No. 2 fuel oil range can be found by drawing a line parallel to the No. 2 boundary lines through the point of known temperature.

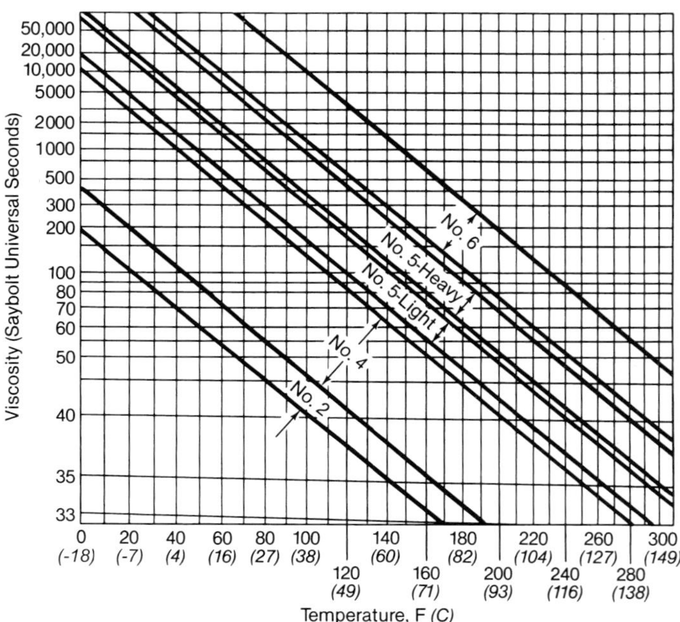

Note: On the Y axis, find the SUS viscosity at 100F (standard test temperature) for the given oil; move horizontally to the vertical line for 100F. From this intersection, move parallel to the diagonal lines to the viscosity required for atomization; the temperature necessary to achieve this viscosity can be read on the X axis. The chart, based on U.S. Commercial Standard 12-48, has been developed from data for many fuels and should be sufficiently accurate for most applications.

Fig. 16 Approximate viscosity of fuel oil at various temperatures (*courtesy of ASTM*).

Natural gas

Past consumption and availability

Natural gas is found in porous rock in the earth's crust. World natural gas production for 1999 is shown in Fig. 17.

Electric power generation is the fastest growing segment of U.S. natural gas consumption. By 2000, electric generators had overtaken the residential segment as the second largest user of natural gas with a 22% share of U.S. consumption (Table 12). Environmental regulations, higher efficiency gas turbines, and a large base of simple and combined cycle gas turbine plants installed in the late 1990s and early 2000s drove the annual usage of natural gas in the U.S., for electric power generation, from 3.8 trillion cubic feet in 1996 to 5.5 trillion cubic feet in 2002. The Department of Energy (DOE) expects that the volatile price of natural gas will hold growth to about 1.8% per year to 2025.

Natural gas characteristics

Natural gas can be found with petroleum reserves or in separate reservoirs. Methane is the principal component of natural gas; smaller components include ethane, propane and butane. Other hydrocarbons, such as pentane through decane, can also be found in natural gas. Furthermore, other gases such as CO_2, nitrogen, helium and hydrogen sulfide (H_2S) may be present.

Gas containing mostly methane is referred to as *lean* gas. *Wet* gas contains appreciable amounts of the higher hydrocarbons (5 to 10% C). Gas containing H_2S is *sour* gas; conversely, *sweet* gas contains little or no H_2S.

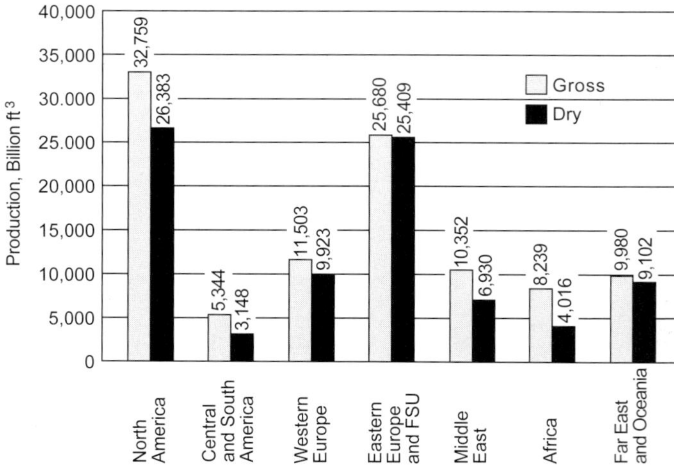

Fig. 17 World natural gas production, 1999.

Table 12
U.S. Natural Gas Consumption (Trillion ft³)

Year	Residential	Commercial	Industrial	Elec. Power	Transportation	Total
1989	4.78	2.72	7.89	3.11	0.63	19.12
1990	4.39	2.62	8.26	3.25	0.66	19.17
1991	4.56	2.73	8.36	3.32	0.60	19.56
1992	4.69	2.80	8.70	3.45	0.59	20.23
1993	4.96	2.86	8.87	3.47	0.63	20.79
1994	4.85	2.90	8.91	3.90	0.69	21.25
1995	4.85	3.03	9.38	4.24	0.71	22.21
1996	5.24	3.16	9.69	3.81	0.72	22.61
1997	4.98	3.22	9.71	4.07	0.76	22.74
1998	4.52	3.00	9.49	4.57	0.65	22.25
1999	4.73	3.05	9.16	4.82	0.66	22.41
2000	5.00	3.22	9.29	5.21	0.66	23.37
2001	4.78	3.04	8.45	5.34	0.64	22.25
2002	4.91	3.11	8.23	5.55	0.65	22.46

Note: Total may not equal sum of components due to independent rounding. Source: Energy Information Administration, *Annual Energy Review*, 2003.

Of all chemical fuels, natural gas is considered to be the most desirable for steam generation. It is piped directly to the consumer, eliminating the need for storage. It is substantially free of ash and mixes easily with air, providing complete combustion without smoke. Although the total hydrogen content of natural gas is high, its free hydrogen content is low. Because of this, natural gas burns less easily than some manufactured gases with high free hydrogen content.

The high hydrogen content of natural gas compared to that of oil or coal results in more water vapor being produced in the combustion gases. This results in a correspondingly lower efficiency of the steam generating equipment. (See Chapter 10.) This can readily be taken into account when designing the equipment.

Properties of natural gas

Analyses of natural gas from several U.S. fields are given in Table 13.

Table 13
Selected Samples of Natural Gas from U.S. Fields

Sample No.	1	2	3	4	5
Source:	Pa.	S.C.	Ohio	La.	Ok.
Analyses:					
Constituents, % by vol					
H_2, Hydrogen	—	—	1.82	—	—
CH_4, Methane	83.40	84.00	93.33	90.00	84.10
C_2H_4, Ethylene	—	—	0.25	—	—
C_2H_6, Ethane	15.80	14.80	—	5.00	6.70
CO, Carbon monoxide	—	—	0.45	—	—
CO_2, Carbon dioxide	—	0.70	0.22	—	0.80
N_2, Nitrogen	0.80	0.50	3.40	5.00	8.40
O_2, Oxygen	—	—	0.35	—	—
H_2S, Hydrogen sulfide	—	—	0.18	—	—
Ultimate, % by wt					
S, Sulfur	—	—	0.34	—	—
H, Hydrogen	23.53	23.30	23.20	22.68	20.85
C, Carbon	75.25	74.72	69.12	69.26	64.84
N, Nitrogen	1.22	0.76	5.76	8.06	12.90
O, Oxygen	—	1.22	1.58	—	1.41
Specific gravity (rel to air)	0.636	0.636	0.567	0.600	0.630
HHV					
Btu/ft³ at 60F and 30 in. Hg	1,129	1,116	964	1,022	974
(kJ/m³ at 16C and 102 kPa)	(42,065)	(41,581)	(35,918)	(38,079)	(36,290)
Btu/lb(kJ/kg) of fuel	23,170 (53,893)	22,904 (53,275)	22,077 (51,351)	21,824 (50,763)	20,160 (46,892)

Other fuels

While coal, oil and gas are the dominant fuel sources, other carbonaceous fuels being used for boiler applications include petroleum byproducts and heavy hydrocarbon emulsions; wood, its byproducts and wastes from wood processing industries; certain types of vegetation, particularly bagasse; and municipal solid waste.

Coke from petroleum

The heavy residuals from petroleum cracking processes are presently used to produce a higher yield of lighter hydrocarbons and a solid residue suitable for fuel. Characteristics of these residues vary widely and depend on the process used. Solid fuels from oil include delayed coke, fluid coke and petroleum pitch. Some selected analyses are given in Table 14.

The delayed coking process uses residual oil that is heated and pumped to a reactor. Coke is deposited in the reactor as a solid mass and is subsequently stripped, mechanically or hydraulically, in the form of lumps and granular material. Some cokes are easy to pulverize and burn while others are difficult.

Fluid coke is produced by spraying hot residual feed onto externally heated seed coke in a fluidized bed. The fluid coke is removed as small particles, which are built up in layers. This coke can be pulverized and burned, or it can be burned in a Cyclone furnace or in a fluidized bed. All three types of firing require supplemental fuel to aid ignition.

The petroleum pitch process is an alternate to the coking process and yields fuels of various characteristics. Melting points vary considerably, and the physi-

Table 14
Selected Analyses of Solid Fuels Derived from Oil

Analyses (dry basis) % by wt	Delayed Coke		Fluid Coke	
Proximate:				
VM	10.8	9.1	6.0	6.7
FC	88.5	90.8	93.7	93.2
Ash	0.7	0.1	0.3	0.1
Ultimate:				
Sulfur	9.9	1.5	4.7	5.7
Heating value,				
Btu/lb	14,700	15,700	14,160	14,290
(kJ/kg)	(34,192)	(36,518)	(32,936)	(33,239)

cal properties vary from soft and gummy to hard and friable. The low melting point pitches may be heated and burned like heavy oil, while those with higher melting points may be pulverized or crushed and burned.

Oil emulsions

With the discovery of large reserves of heavy hydrocarbon and bitumen in Venezuela, considerable effort has been devoted to developing these sources as commercial fuels. This has led to the formulation of bitumen oil emulsions. Generically, these emulsions are liquid fuels composed of micron-size oil droplets dispersed in water. Droplet coalescence is prevented by adding a small amount of a proprietary chemical. The fuel is characterized by relatively high levels of sulfur, asphaltenes and metals. The heating value, ash content and viscosity of the emulsions are similar to residual fuel oil as are their handling and combustion performance characteristics. The emulsions can contain vanadium which forms corrosive compounds during combustion. Vanadium can also catalyze the oxidation of SO_2 to SO_3 and require the use of specific emission controls to avoid stack plumes of sulfuric acid aerosol.

Orimulsion® is the trade name of a proprietary bitumen emulsion produced by Bitor, a part of Petroleos de Venezuela SA (PDVSA), the Venezuelan national oil company. It is prepared from approximately 30% water and 70% bitumen from the Orinoco basin. Combined with other performance enhancing chemicals, a stable emulsified fuel for application in boilers and other combustion equipment is produced. A typical Orimulsion composition compared to fuel oil is shown in Table 15.

As of 2000, Orimulsion was being utilized worldwide at the rate of 6.2 million tons per year at boiler installations in Denmark, Japan, Italy and Canada. The fuel can offer cost advantage over No. 6 fuel oil. Environmental control equipment is required to address sulfur oxides, nitrogen oxides, and particulate emissions.

Wood

Selected analyses and heating values of wood and wood ash are given in Table 16. Wood is composed primarily of carbohydrates. Consequently, it has a relatively low heating value compared with bituminous coal and oil.

Wood bark may pick up impurities during transportation. It is common practice to drag the rough logs to central loading points and sand is often picked up. Where the logs are immersed in salt water, the bark can absorb the salt. Combustion temperatures from burning dry bark may be high enough for these impurities to cause fluxing of refractory furnace walls and fouling of boiler heating surfaces, unless sufficient furnace cooling surface is provided. Sand passing through the boiler banks can cause erosion of the tubes, particularly if the flue gas sand loading is increased by returning collected material to the furnace. Such collectors may be required with some bark burning equipment to reduce the stack discharge of incompletely burned bark.

Wood or bark with a moisture content of 50% or less burns quite well; however, as the moisture increases above this amount, combustion becomes more difficult. With a moisture content above 65%, a large part of the heat is required to evaporate the inherent moisture and little remains for steam generation. Burning this wet bark becomes a means of disposal rather than a source of energy.

Table 15
Composition of Orimulsion® 400

	Orimulsion® 400	No. 6 Fuel Oil
Carbon (%)	60.20	85.71
Hydrogen (%)	7.20	10.14
Sulfur (%)	2.85	2.63
Oxygen (%)	0.18	0.92
Nitrogen (%)	0.50	0.51
Water (%)	29.00	0
Ash (%)	0.07	0.09
HHV (Btu/lb)	12,984	18,192

Table 16
Analyses of Wood and Wood Ash

Wood analyses (dry basis), % by wt	Pine Bark	Oak Bark	Spruce Bark*	Redwood Bark*
Proximate analysis, %				
Volatile matter	72.9	76.0	69.6	72.6
Fixed carbon	24.2	18.7	26.6	27.0
Ash	2.9	5.3	3.8	0.4
Ultimate analysis, %				
Hydrogen	5.6	5.4	5.7	5.1
Carbon	53.4	49.7	51.8	51.9
Sulfur	0.1	0.1	0.1	0.1
Nitrogen	0.1	0.2	0.2	0.1
Oxygen	37.9	39.3	38.4	42.4
Ash	2.9	5.3	3.8	0.4
Heating value, Btu/lb	9,030	8,370	8,740	8,350
(kJ/kg)	(21,004)	(19,469)	(20,329)	(19,422)
Ash analysis, % by wt				
SiO_2	39.0	11.1	32.0	14.3
Fe_2O_3	3.0	3.3	6.4	3.5
TiO_2	0.2	0.1	0.8	0.3
Al_2O_3	14.0	0.1	11.0	4.0
Mn_3O_4	Trace	Trace	1.5	0.1
CaO	25.5	64.5	25.3	6.0
MgO	6.5	1.2	4.1	6.6
Na_2O	1.3	8.9	8.0	18.0
K_2O	6.0	0.2	2.4	10.6
SO_3	0.3	2.0	2.1	7.4
Cl	Trace	Trace	Trace	18.4
Ash fusibility temp, F				
Reducing				
Initial deformation	2180	2690		
Softening	2240	2720		
Fluid	2310	2740		
Oxidizing				
Initial deformation	2210	2680		
Softening	2280	2730		
Fluid	2350	2750		

* Salt water stored.

Hogged wood and bark are very bulky and require relatively large handling and storage equipment. Uninterrupted flow from bunkers or bins through chutes is difficult to maintain. (Also see Chapter 30.)

Wood wastes There are several industries using wood as a raw material where combustible byproducts or wastes are available as fuels. The most important of these are the pulp and turpentine industries. The nature and methods of utilization of the combustible byproducts from the pulp industry are discussed in Chapter 28.

The residue remaining after the steam distillation of coniferous woods for the production of turpentine is usable as a fuel. Some of the more easily burned constituents are removed in the distillation process; as a result, the residue is somewhat more difficult to burn. Other than this, fuel properties are much the same as those of the raw wood and the problems involved in utilization are similar.

Bagasse

Mills grinding sugar cane commonly use bagasse for steam production. Bagasse is the dry pulp remaining after the juice has been extracted from sugar cane. The mills normally operate 24 hours per day during the grinding season. The supply of bagasse will easily meet the plant steam demands in mills where the sugar is not refined. Consequently, where there is no other market for the bagasse, no particular effort is made to burn it efficiently, and burning equipment is provided that will burn the bagasse as-received from the grinders. In refining plants, supplemental fuels are required to provide the increased steam demands. Greater efforts to obtain higher efficiency are justified in these plants. A selected analysis of bagasse is given in Table 8.

Other vegetation wastes

Food and related industries produce numerous vegetable wastes that are usable as fuels. They include such materials as grain hulls, the residue from the production of furfural from corn cobs and grain hulls, coffee grounds from the production of instant coffee, and tobacco stems. Fuels of this type are available in such small quantities that they are relatively insignificant in total energy production.

Table 17
Analyses of MSW and RDF Compared to Bituminous Coal

Constituent	Analyses, % by wt MSW	RDF	Bituminous Coal
Carbon	27.9	36.1	72.8
Hydrogen	3.7	5.1	4.8
Oxygen	20.7	31.6	6.2
Nitrogen	0.2	0.8	1.5
Sulfur	0.1	0.1	2.2
Chlorine	0.1	0.1	0
Water	31.3	20.2	3.5
Ash	16.0	6.0	9.0
HHV (wet), Btu/lb	5,100	6,200	13,000
(kJ/kg)	(11,863)	(14,421)	(30,238)

Municipal solid waste

Municipal solid waste (MSW), or refuse, is an energy source in the U.S., Europe and Japan. MSW is the combined residential and commercial waste generated in a given municipality. It is burned as-received, called mass burning, or processed using size reduction and material recovery techniques to produce refuse-derived fuel (RDF). Much MSW continues to be landfilled, since siting and acceptance of waste-to-energy boilers have been greatly limited by the public's concern over environmental issues.

Table 17 shows a typical analysis of raw refuse and RDF compared to bituminous coal. The relatively low calorific value and high heterogeneous nature of MSW provide a challenge to the combustion system design engineer. The design of MSW handling and combustion systems is discussed in Chapter 29.

References

1. International Energy Outlook 2003, Report DOE/EIA-0484 (2003), United States (U.S.) Energy Information Administration, Washington, D.C., May, 2003.

2. Annual Energy Review 2001, Report DOE/EIA-0384 (2001), U.S. Energy Information Administration, Washington, D.C., November, 2002.

3. *2001 Survey of Energy Resources,* World Energy Congress, London, England, 2001.

4. Coal Industry Annual 2000, Report DOE/EIA-0584 (2000), U.S. Energy Information Administration, Washington, D.C., 2001.

5. "Gaseous Fuels; Coal and Coke," Vol. 05.05, *Annual Book of ASTM Standards,* American Society for Testing and Materials, West Conshohocken, Pennsylvania, 1999.

6. Parr, S.W., "The Classification of Coal," Bulletin No. 180, Engineering Experiment Station, University of Illinois, Chicago, Illinois, 1928.

7. Vecci, S.J., Wagoner, C.L., and Olson, G.B., "Fuel and Ash Characterization and Its Effect on the Design of Industrial Boilers," *Proceedings of the American Power Conference,* Vol. 40, pp. 850-864, 1978.

8. Wagoner, C.L., and Duzy, A.F., "Burning Profiles for Solid Fuels," Technical Paper 67-WA-FU-4, American Society of Mechanical Engineers, New York, New York, 1967.

450 MW midwest power station firing pulverized subbituminous coal.

Chapter 10

Principles of Combustion

A boiler requires a source of heat at a sufficient temperature to produce steam. Fossil fuel is generally burned directly in the boiler furnace to provide this heat although waste energy from another process may also be used.

Combustion is defined as the rapid chemical combination of oxygen with the combustible elements of a fuel. There are just three combustible elements of significance in most fossil fuels: carbon, hydrogen and sulfur. Sulfur, usually of minor significance as a heat source, can be a major contributor to corrosion and pollution problems. (See Chapters 21 and 32.)

The objective of good combustion is to release all of the energy in the fuel while minimizing losses from combustion imperfections and excess air. System requirement objectives include minimizing nitrogen oxides (NO_x), carbon monoxide (CO), volatile organic compounds (VOC) and, for more difficult to burn fuels, minimizing unburned carbon (UBC) and furnace corrosion. The combination of the combustible fuel elements and compounds in the fuel with all the oxygen requires *temperatures* high enough to ignite the constituents, mixing or *turbulence* to provide intimate oxygen-fuel contact, and sufficient *time* to complete the process, sometimes referred to as the *three Ts of combustion*.

Table 1 lists the chemical elements and compounds found in fuels generally used in commercial steam generation.

Concept of the mole

The mass of a substance in pounds equal to its molecular weight is called a pound-mole (lb-mole) of the substance. The molecular weight is the sum of the atomic masses of a substance's constituent atoms. For example, pure elemental carbon (C) has an atomic mass and molecular weight of 12 and therefore a lb-mole is equal to 12. In the case of carbon dioxide (CO_2), carbon still has an atomic mass of 12 and oxygen has an atomic mass of 16 giving CO_2 a molecular weight and a lb-mole equal to $(1 \times 12) + (2 \times 16)$ or 44. In SI, a similar system is based upon the molecular weight in kilograms expressed as kg-mole or kmole. In the United States (U.S.) power industry it is common practice to replace lb-mole with mole.

In the case of a gas, the volume occupied by one mole is called the molar volume. The volume of one mole of an ideal gas (a good approximation in most combustion calculations) is a constant regardless of its composition for a given temperature and pressure. Therefore, one lb-mole or mole of oxygen (O_2) at 32 lb and one mole of CO_2 at 44 lb will occupy the same volume equal to 394 ft^3 at 80F and 14.7 psi. The volume occupied by one mole of a gas can be corrected to other pressures and temperatures by the ideal gas law.

Because substances combine on a molar basis during combustion but are usually measured in units of mass (pounds), the lb-mole and molar volume are important tools in combustion calculations.

> For clarity, this chapter is provided in English units only. Appendix 1 provides a comprehensive list of conversion factors. Selected factors of particular interest here include: Btu/lb × 2.326 = kJ/kg; 5/9 (F-32) = C; lb × 0.4536 = kg. Selected SI constants include: universal gas constant = 8.3145 kJ/kmole K; one kmole at 0C and 1.01 bar = 22.4 m^3.

Fundamental laws

Combustion calculations are based on several fundamental physical laws.

Conservation of matter

This law states that matter can not be destroyed or created. There must be a mass balance between the sum of the components entering a process and the sum of those leaving: X pounds of fuel combined with Y pounds of air always results in X + Y pounds of products (see Note below).

Conservation of energy

This law states that energy can not be destroyed or created. The sum of the energies (potential, kinetic, thermal, chemical and electrical) entering a process must equal the sum of those leaving, although the proportions of each may change. In combustion, chemical energy is converted into thermal energy (see Note below).

Note: While the laws of conservation of matter and energy are not rigorous from a nuclear physics standpoint (see Chapter 47), they are quite adequate for engineering combustion calculations. When a pound of a typical coal is burned releasing 13,500 Btu, the equivalent quantity of mass converted to energy amounts to only 3.5×10^{-10} lb.

Table 1 – Combustion Constants – Reference 1

No.	Substance	Formula	Molecular Weight[a]	Density,[b] lb per ft³	Specific Volume[b] ft³ per lb	Specific Gravity[b] (air=1)	Heat of Combustion[c] Btu per ft³ Gross[d]	Btu per ft³ Net[e]	Btu per lb Gross[d]	Btu per lb Net[e]	ft³ per ft³ of Combustible, Required for Combustion O_2	N_{2a}	Air	Flue Products CO_2	H_2O	N_{2a}	lb per lb of Combustible, Required for Combustion O_2	N_{2a}	Air	Flue Gas Products CO_2	H_2O	N_{2a}	Theor air lb/ 10,000 Btu
1	Carbon	C	12.0110	—	—	—	—	—	14,093	14,093	1.0	3.773	4.773	1.0	—	3.773	2.664	8.846	11.510	3.664	—	8.846	8.167
2	Hydrogen	H_2	2.0159	0.0053	188.245	0.0696	324.2	273.9	61,029	51,558	0.5	1.887	2.387	—	1.0	1.887	7.936	26.353	34.290	—	8.937	26.353	5.619
3	Oxygen	O_2	31.9988	0.0844	11.850	1.1053	—	—	—	—	—	—	—	—	—	—	—	—	—	—	—	—	—
4	Nitrogen	N_2	28.0134	0.0738	13.543	0.9671	—	—	—	—	—	—	—	—	—	—	—	—	—	—	—	—	—
4	Nitrogen (atm.)[f]	N_{2a}	28.1580	0.0742	13.474	0.9720	—	—	—	—	—	—	—	—	—	—	—	—	—	—	—	—	—
5	Carbon Monoxide	CO	28.0104	0.0738	13.542	0.9672	320.6	320.6	4342	4342	0.5	1.887	2.387	1.0	—	1.887	0.571	1.897	2.468	1.571	—	1.897	5.684
6	Carbon Dioxide	CO_2	44.0098	0.1166	8.574	1.5277	—	—	—	—	—	—	—	—	—	—	—	—	—	—	—	—	—
	Parafin series C_nH_{2n+2}																						
7	Methane	CH_4	16.0428	0.0424	23.608	0.5548	1012	911	23,891	21,511	2.0	7.547	9.547	1.0	2.0	7.547	3.989	13.246	17.235	2.743	2.246	13.246	7.214
8	Ethane	C_2H_6	30.0697	0.0799	12.514	1.0466	1785	1634	22,334	20,429	3.5	13.206	16.706	2.0	3.0	13.206	3.724	12.367	16.092	2.927	1.797	12.367	7.205
9	Propane	C_3H_8	44.0966	0.1183	8.456	1.5489	2561	2359	21,653	19,921	5.0	18.866	23.866	3.0	4.0	18.866	3.628	12.047	15.676	2.994	1.634	12.047	7.239
10	n-Butane	C_4H_{10}	58.1235	0.1585	6.310	2.0758	3376	3124	21,299	19,657	6.5	24.526	31.026	4.0	5.0	24.526	3.578	11.882	15.460	3.029	1.550	11.882	7.259
11	Isobutane	C_4H_{10}	58.1235	0.1580	6.328	2.0699	3355	3104	21,231	19,589	6.5	24.526	31.026	4.0	5.0	24.526	3.578	11.882	15.460	3.029	1.550	11.882	7.282
12	n-Pentane	C_5H_{12}	72.1504	0.2019	4.952	2.6450	4258	3956	21,085	19,498	8.0	30.186	38.186	5.0	6.0	30.186	3.548	11.781	15.329	3.050	1.498	11.781	7.270
13	Isopentane	C_5H_{12}	72.1504	0.2001	4.999	2.6202	4210	3908	21,043	19,455	8.0	30.186	38.186	5.0	6.0	30.186	3.548	11.781	15.329	3.050	1.498	11.781	7.284
14	Neopentane	C_5H_{12}	72.1504	0.1984[g]	5.040[g]	2.5989[g]	4159[g]	3857	20,958[g]	19,370	8.0	30.186	38.186	5.0	6.0	30.186	3.548	11.781	15.329	3.050	1.498	11.781	7.314
15	n-Hexane	C_6H_{14}	86.1773	0.2508	3.987	3.2849	5252	4900	20,943	19,392	9.5	35.846	45.346	6.0	7.0	35.846	3.527	11.713	15.240	3.064	1.463	11.713	7.277
	Olefin series C_nH_{2n}																						
16	Ethylene	C_2H_4	28.0538	0.0744	13.447	0.9740	1609	1509	21,643	20,282	3.0	11.320	14.320	2.0	2.0	11.320	3.422	11.362	14.784	3.138	1.284	11.362	6.831
17	Propylene	C_3H_6	42.0807	0.1127	8.874	1.4760	2371	2220	21,039	19,678	4.5	16.980	21.480	3.0	3.0	16.980	3.422	11.362	14.784	3.138	1.284	11.362	7.027
18	n-Butene (Butylene)	C_4H_8	56.1076	0.1524[g]	6.560[g]	1.9966[g]	3175[g]	2974	20,831[g]	19,470	6.0	22.640	28.640	4.0	4.0	22.640	3.422	11.362	14.784	3.138	1.284	11.362	7.097
19	Isobutene	C_4H_8	56.1076	0.1524[g]	6.561[g]	1.9964[g]	3156[g]	2955	20,704[g]	19,343	6.0	22.640	28.640	4.0	4.0	22.640	3.422	11.362	14.784	3.138	1.284	11.362	7.141
20	n-Pentene	C_5H_{10}	70.1345	0.1947[h]	5.135[h]	2.5508[h]	4032[g]	3781	20,704[g]	19,343	7.5	28.300	35.800	5.0	5.0	28.300	3.422	11.362	14.784	3.138	1.284	11.362	7.140
	Aromatic series C_nH_{2n-6}																						
21	Benzene	C_6H_6	78.1137	0.2213	4.518	2.8989	4024	3873	18,179	17,446	7.5	28.300	35.800	6.0	3.0	28.300	3.072	10.201	13.274	3.380	0.692	10.201	7.302
22	Toluene	C_7H_8	92.1406	0.2750[h]	3.637[h]	3.6016[h]	5068[g]	4867	18,430[g]	17,602	9.0	33.959	42.959	7.0	4.0	33.959	3.125	10.378	13.504	3.343	0.782	10.378	7.327
23	Xylene	C_8H_{10}	106.1675	0.3480[h]	2.874[h]	4.5576[h]	6480[g]	6228	18,622[g]	17,723	10.5	39.619	50.119	8.0	5.0	39.619	3.164	10.508	13.673	3.316	0.848	10.508	7.342
	Miscellaneous																						
24	Acetylene	C_2H_2	26.0379	0.0691	14.480	0.9046	1484	1433	21,482	20,749	2.5	9.433	11.933	2.0	1.0	9.433	3.072	10.201	13.274	3.380	0.692	10.201	6.179
25	Naphthalene	$C_{10}H_8$	128.1736	0.3384[h]	2.955[h]	4.4323[h]	5866	5665	17,335	16,739	12.0	45.279	57.279	10.0	4.0	45.279	2.995	9.947	12.943	3.434	0.562	9.947	7.467
26	Methyl alcohol	CH_3OH	32.0422	0.0846[h]	11.820[h]	1.1081[h]	868[g]	768	10,265[g]	9073	1.5	5.660	7.160	1.0	2.0	5.660	1.498	4.974	6.472	1.373	1.124	4.974	6.305
27	Ethyl alcohol	C_2H_5OH	46.0691	0.1216[h]	8.224[h]	1.5927[h]	1602[g]	1451	13,172[g]	11,929	3.0	11.320	14.320	2.0	3.0	11.320	2.084	6.919	9.003	1.911	1.173	6.919	6.835
28	Ammonia	NH_3	17.0306	0.0454[g]	22.008[g]	0.5951[g]	440[g]	364	9680[g]	7998	0.75	2.830	3.580	—	1.5	3.330	1.409	4.679	6.088	—	1.587	5.502	6.290
29	Sulfur	S	32.0660	—	—	—	—	—	3980	3980	1.0	3.773	4.773	SO_2 1.0	—	3.773	1.000	3.320	4.310	SO_2 1.998	—	3.320	10.829
30	Hydrogen sulfide	H_2S	34.0819	0.0907	11.030	1.1875	643	593	7094	6534	1.5	5.660	7.160	SO_2 1.0	1.0	5.660	1.410	4.682	6.093	SO_2 1.880	0.529	4.682	8.576
31	Sulfur dioxide	SO_2	64.0648	0.1722[g]	5.806[g]	2.2558[g]	—	—	—	—	—	—	—	—	—	—	—	—	—	—	—	—	—
32	Water vapor	H_2O	18.0153	0.0503	19.863	0.6594	50.312	0.0	1059.8	0.0	—	—	—	—	—	—	—	—	—	—	—	—	—
33	Air[f]	—	28.9625	0.0763	13.098	1.0000	—	—	—	—	—	—	—	—	—	—	—	—	—	—	—	—	—

All gas volumes corrected to 60F and 14.696 psi dry.

a 1987 Atomic Weights: C=12.011, H=1.00794, O=15.9994, N=14.0067, S=32.066.

b Densities calculated from ideal values and compressibility factor given in ASTM D 3588-98. Some of the materials can not exist as gases at 60F and 14.696 psi, in which case the values are theoretical ones. Under the actual concentrations in which these materials are present, their partial pressure is low enough to keep them as gases.

c For gases saturated with water at 60F and 14.696 psi, 1.74% of the Btu value must be deducted. Reference 2.

d Reference 2, ASTM 3588-98.

e Correction from gross to net heating value determined by deducting the HV shown for water vapor times the moles of H_2.

f Reference 3, Jones, F.E.

g Gas Processors Suppliers Association (GPSA) Data Book, Fig 23-2, Physical Constants, 1987.

h Either the density or the compressibility factor has been assumed.

Ideal gas law

This law states that the volume of an ideal gas is directly proportional to its absolute temperature and inversely proportional to its absolute pressure.

The proportionality constant is the same for one mole of any ideal gas, so this law may be expressed as:

$$v_M = \frac{\mathbf{R}T}{P} \qquad (1)$$

where

v_M = volume, ft^3/mole
$\mathbf{R}$ = universal gas constant, 1545 ft lb/mole R
T = absolute temperature, R = F + 460
P = absolute pressure, lb/ft^2

Most gases involved in combustion calculations can be approximated as ideal gases.

Law of combining weights

This law states that all substances combine in accordance with simple, definite weight relationships. These relationships are exactly proportional to the molecular weights of the constituents. For example, carbon (molecular weight = 12) combines with oxygen (molecular weight of O_2 = 32) to form carbon dioxide (molecular weight = 44) so that 12 lb of C and 32 lb of O_2 unite to form 44 lb of CO_2. (See *Application of fundamental laws* below.)

Avogadro's law

Avogadro determined that equal volumes of different gases at the same pressure and temperature contain the same number of molecules. From the concept of the mole, a pound mole of any substance contains a mass equal to the molecular weight of the substance. Therefore, the ratio of mole weight to molecular weight is a constant and a mole of any chemically pure substance contains the same number of molecules. Because a mole of any ideal gas occupies the same volume at a given pressure and temperature (ideal gas law), equal volumes of different gases at the same pressure and temperature contain the same number of molecules.

Dalton's law

This law states that the total pressure of a mixture of gases is the sum of the partial pressures which would be exerted by each of the constituents if each gas were to occupy alone the same volume as the mixture. Consider equal volumes V of three gases (a, b and c), all at the same temperature T but at different pressures (P_a, P_b and P_c). When all three gases are placed in the space of the same volume V, then the resulting pressure P is equal to $P_a + P_b + P_c$. Each gas in a mixture fills the entire volume and exerts a pressure independent of the other gases.

Amagat's law

Amagat determined that the total volume occupied by a mixture of gases is equal to the sum of the volumes which would be occupied by each of the constituents when at the same pressure and temperature as the mixture. This law is related to Dalton's law, but it considers the additive effects of volume instead of pressure. If all three gases are at pressure P and temperature T but at volumes V_a, V_b and V_c, then, when combined so that T and P are unchanged, the volume of the mixture V equals $V_a + V_b + V_c$.

Application of fundamental laws

Table 2 summarizes the molecular and weight relationships between fuel and oxygen for constituents commonly involved in combustion. The heat of combustion for each constituent is also tabulated. Most of the weight and volume relationships in combustion calculations can be determined by using the information presented in Table 2 and the seven fundamental laws.

The combustion process for C and H_2 can be expressed as follows:

C	+	O_2	=	CO_2
1 molecule	+	1 molecule	→	1 molecule
1 mole	+	1 mole	=	1 mole
(See Note below)	+	1 ft^3	→	1 ft^3
12 lb	+	32 lb	=	44 lb

$2H_2$	+	O_2	=	$2H_2O$
2 molecules	+	1 molecule	→	2 molecules
2 moles	+	1 mole	=	2 moles
2 ft^3	+	1 ft^3	→	2 ft^3
4 lb	+	32 lb	=	36 lb

Note: When 1 ft^3 of oxygen (O_2) combines with carbon (C), it forms 1 ft^3 of carbon dioxide (CO_2). If carbon were an ideal gas instead of a solid, 1 ft^3 of carbon would be required.

It is important to note that there is a mass or weight balance according to the law of combining weights but there is not necessarily a molecular or volume balance.

Molar evaluation of combustion

Gaseous fuel

Molar calculations have a simple and direct application to gaseous fuels, where the analyses are usually reported on a percent by volume basis. Consider the following fuel analysis:

Fuel Gas Analysis, % by Volume	
CH_4	85.3
C_2H_6	12.6
CO_2	0.1
N_2	1.7
O_2	0.3
Total	100.0

The mole fraction of a component in a mixture is the number of moles of that component divided by the total number of moles of all components in the mixture. Because a mole of every ideal gas occupies the same volume, by Avogadro's Law, the mole fraction of a component in a mixture of ideal gases equals the volume fraction of that component.

Table 2
Common Chemical Reactions of Combustion

Combustible	Reaction	Moles	Mass or weight, lb	Heat of Combustion (High) Btu/lb of Fuel
Carbon (to CO)	$2C + O_2 = 2CO$	2 + 1 = 2	24 + 32 = 56	3,967
Carbon (to CO_2)	$C + O_2 = CO_2$	1 + 1 = 1	12 + 32 = 44	14,093
Carbon monoxide	$2CO + O_2 = 2CO_2$	2 + 1 = 2	56 + 32 = 88	4,342
Hydrogen	$2H_2 + O_2 = 2H_2O$	2 + 1 = 2	4 + 32 = 36	61,029
Sulfur (to SO_2)	$S + O_2 = SO_2$	1 + 1 = 1	32 + 32 = 64	3,980
Methane	$CH_4 + 2O_2 = CO_2 + 2H_2O$	1 + 2 = 1 + 2	16 + 64 = 80	23,891
Acetylene	$2C_2H_2 + 5O_2 = 4CO_2 + 2H_2O$	2 + 5 = 4 + 2	52 + 160 = 212	21,482
Ethylene	$C_2H_4 + 3O_2 = 2CO_2 + 2H_2O$	1 + 3 = 2 + 2	28 + 96 = 124	21,643
Ethane	$2C_2H_6 + 7O_2 = 4CO_2 + 6H_2O$	2 + 7 = 4 + 6	60 + 224 = 284	22,334
Hydrogen sulfide	$2H_2S + 3O_2 = 2SO_2 + 2H_2O$	2 + 3 = 2 + 2	68 + 96 = 164	7,094

$$\frac{\text{Moles of component}}{\text{Total moles}} = \frac{\text{Volume of component}}{\text{Volume of mixture}} \tag{2}$$

This is a valuable concept because the volumetric analysis of a gaseous mixture automatically gives the mole fractions of the components.

Accordingly, the previous fuel analysis may be expressed as 85.3 moles of CH_4 per 100 moles of fuel, 12.6 moles of C_2H_6 per 100 moles of fuel, etc.

The elemental breakdown of each constituent may also be expressed in moles per 100 moles of fuel as follows:

C in CH_4	=	85.3 × 1	=	85.3 moles
C in C_2H_6	=	12.6 × 2	=	25.2 moles
C in CO_2	=	0.1 × 1	=	0.1 moles
Total C per 100 moles fuel			=	110.6 moles
H_2 in CH_4	=	85.3 × 2	=	170.6 moles
H_2 in C_2H_6	=	12.6 × 3	=	37.8 moles
Total H_2 per 100 moles fuel			=	208.4 moles
O_2 in CO_2	=	0.1 × 1	=	0.1 moles
O_2 as O_2	=	0.3 × 1	=	0.3 moles
Total O_2 per 100 moles fuel			=	0.4 moles
Total N_2 per 100 moles fuel			=	1.7 moles

The oxygen/air requirements and products of combustion can now be calculated for each constituent on an elemental basis. These requirements can also be calculated directly using Table 1. Converting the gaseous constituents to an elemental basis has two advantages. It provides a better understanding of the combustion process and it provides a means for determining an elemental fuel analysis on a mass basis. This is boiler industry standard practice and is convenient for determining a composite fuel analysis when gaseous and solid/liquid fuels are fired in combination.

The following tabulation demonstrates the conversion of the gaseous fuel constituents on a moles/100 moles gas basis to a lb/100 lb gas (percent mass) basis.

Constituent	Moles/ 100 Moles	Mol Wt lb/ Mole	lb/ 100 Moles	lb/ 100 lb
C	110.6 ×	12.011	= 1328.4	/1808.9 × 100 = 73.5
H_2	208.4 ×	2.016	= 420.1	/1808.9 × 100 = 23.2
O_2	0.4 ×	31.999	= 12.8	/1808.9 × 100 = 0.7
N_2	1.7 ×	28.013	= 47.6	/1808.9 × 100 = 2.6
Total			1808.9	100.0

Solid/liquid fuel

The ultimate analysis of solid and liquid fuels is determined on a percent mass basis. The mass analysis is converted to a molar basis by dividing the mass fraction of each elemental constituent by its molecular weight.

$$\frac{\dfrac{\text{lb Constituent}}{\text{100 lb Fuel}}}{\dfrac{\text{lb Constituent}}{\text{Mole constituent}}} = \frac{\text{Mole constituent}}{\text{100 lb Fuel}} \tag{3}$$

The calculation is illustrated in Table 3.

The products of combustion and moles of oxygen required for each combustible constituent are shown. Note that when a fuel contains oxygen, the amount of theoretical O_2/air required for combustion is reduced (as designated by the brackets).

Composition of air

So far, combustion has been considered only as a process involving fuel and oxygen. For normal combustion and steam generator applications, the source of oxygen is air. Atmospheric air is composed of oxygen, nitrogen and other minor gases. The calculations and derivation of constants which follow in this text are based upon a U.S. standard atmosphere[3] composed of 0.20946 O_2, 0.78102 N_2, 0.00916 argon (Ar) and 0.00033 CO_2 moles per mole of dry air, which has an

Table 3
Calculation of Combustion Products and Theoretical Oxygen Requirements – Molar Basis

Fuel constituent (1)	% by wt (2)	Molecular weight (3)	Moles/100 lb fuel (2 ÷ 3) (4)	Combustion product (5)	Moles theoretical O_2 required (6)
C	72.0	12.011	= 5.995	CO_2	5.995
H_2	4.4	2.016	= 2.183	H_2O	1.091*
S	1.6	32.066	= 0.050	SO_2	0.050
O_2	3.6	31.999	= 0.113		(0.113)
N_2	1.4	28.013	= 0.050	N_2	0.000
H_2O	8.0	18.015	= 0.444	H_2O	0.000
Ash	9.0				
Total	100.0		8.835		7.023

* Column 6 is based upon moles of oxygen as O_2 needed for combustion. Therefore, the moles of H_2O need to be divided by 2 to obtain equivalent moles of O_2.

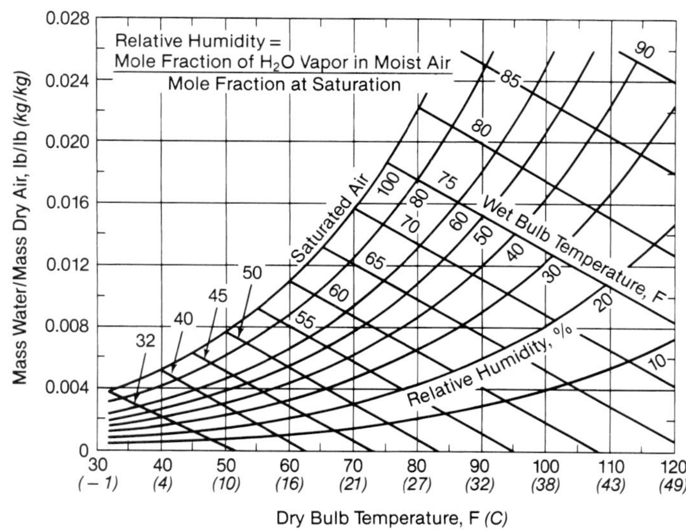

Fig. 1 Psychrometric chart – water content of air for various wet and dry bulb temperatures.

average molecular weight of 28.9625. To simplify the calculations, N_2 includes argon and other trace elements; it is referred to as atmospheric nitrogen (N_{2a}) having an equivalent molecular weight of 28.158. (See Table 4.)

Air normally contains some moisture. As standard practice, the American Boiler Manufacturers Association (ABMA) considers moisture content to be 0.013 lb water/lb dry air, which corresponds to approximately 60% relative humidity at 80F. For combustion calculations on a molar basis, multiply the mass basis moisture by 1.608 (molecular weight of air divided by molecular weight of water). Therefore, 0.013 lb water/lb dry air becomes 0.0209 moles water/mole dry air.

The moisture content in air is normally determined from wet and dry bulb temperatures or from relative humidity using a psychrometric chart, as shown in Fig. 1. Air moisture may also be calculated from:

$$MFWA = 0.622 \times \frac{P_v}{(P_b - P_v)} \quad (4)$$

where

$MFWA$ = moisture content in air, lb/lb dry air
P_b = barometric pressure, psi
P_v = partial pressure of water vapor in air, psi = 0.01 (RH) (P_{vd}), psi
P_{vd} = saturation pressure of water vapor at dry bulb temperature, psi
RH = relative humidity, %

P_v may also be calculated from Carrier's equation:

Table 4
Air Composition

	Composition of Dry Air % by vol	% by wt
Oxygen, O_2	20.95	23.14
Atmospheric nitrogen, N_{2a}	79.05	76.86

$$P_v = P_{vw} - \frac{(P_b - P_{vw})(T_d - T_w)}{2830 - (1.44T_w)} \quad (5)$$

where

T_d = dry bulb temperature, F
T_w = wet bulb temperature, F
P_{vw} = saturation pressure of water vapor at wet bulb temperature, psi

The following constants, with values from Table 4, are frequently used in combustion calculations:

moles air/mole O_2 or ft³ air/ft³ O_2 $= \frac{100}{20.95} = 4.77$

moles N_{2a} /mole O_2 $= \frac{79.05}{20.95} = 3.77$

lb air (dry)/lb O_2 $= \frac{100}{23.14} = 4.32$

lb N_{2a} /lb O_2 $= \frac{76.86}{23.14} = 3.32$

The calculations in Table 2 can be converted to combustion with air rather than oxygen by adding 3.77 moles of N_{2a}/mole of O_2 to the left and right side of each equation. For example, the combustion of carbon monoxide (CO) in air becomes:

$$2\,CO + O_2 + 3.77\,N_{2a} = 2CO_2 + 3.77\,N_{2a}$$

or for methane, CH_4:

$$CH_4 + 2O_2 + 2(3.77)\,N_{2a} = CO_2 + 2H_2O + 7.54\,N_{2a}$$

Theoretical air requirement

Theoretical air is the minimum air required for complete combustion of the fuel, i.e., the oxidation of carbon to CO_2, hydrogen to water vapor (H_2O) and sulfur to sulfur dioxide (SO_2). In the combustion process, small amounts of sulfur trioxide (SO_3), nitrogen oxides (NO_x), unburned hydrocarbons and other minor species may be formed. While these may be of concern as pollutants, their impact is negligible with regard to the quantity of air and combustion products and, therefore, they are not normally considered in these calculations.

In practice, it is necessary to use more than the theoretical amount of air to assure complete combustion of the fuel. For the example shown in Table 3, consider completing the combustion calculations on a molar basis using 20% excess air. These calculations are summarized in Table 5.

Now consider the portion of the combustion products attributable to the air. The oxygen in the theoretical air is already accounted for in the products of combustion: CO_2, H_2O (from the combustion of hydrogen) and SO_2. That leaves N_{2a} in the theoretical air, N_{2a} in the excess air, O_2 in the excess air and H_2O in air (as calculated in Table 5) as the products in the combustion gas attributable to the wet combustion air. These constituents are in addition to the combustion products from fuel shown in Table 3.

Table 5
Calculation of Wet Air Requirements for Combustion – Molar Basis

Line No.	Description	Source	Quantity (mole/ 100 lb fuel)
1	Theoretical combustion O_2	From Table 3	7.023
2	Molar fraction O_2 in dry air	O_2 Vol fraction from Table 4	0.2095
3	Theoretical dry combustion air	Line 1/Line 2	33.523
4	Excess air at 20%	Line 3 x 0.20	6.705
5	Total dry combustion air	Line 3 + Line 4	40.228
6	Molar fraction of H_2O in dry air	*	0.0209
7	H_2O in total dry air	Line 5 x Line 6	0.841
8	Molar fraction of N_{2a} in dry air	N_{2a} Vol fraction from Table 4	0.7905
9	N_{2a} in theoretical dry air	Line 3 x Line 8	26.500
10	N_{2a} in dry excess air	Line 4 x Line 8	5.300
11	O_2 in dry excess air	Line 2 x Line 4	1.405

* Standard combustion air: 80F, 60% relative humidity; 0.013 lb H_2O/lb dry air; 0.0209 moles H_2O/mole dry air.

Products of combustion – mass/mass fuel basis

Table 6 shows a tabulation of the flue gas products and combustion air on a molar (or volumetric) basis and the conversion to a mass basis (wet and dry). The products of combustion calculated on a molar basis in Tables 3 and 5 are itemized in column A. The moisture (H_2O) sources are separated from the dry products for convenience of calculating the flue gas composition on a wet and dry basis.

The water products shown in column A are from the combustion of hydrogen in the fuel, from moisture in the fuel and from moisture in the air. The N_{2a} is the sum of nitrogen in the theoretical air plus the nitrogen in the excess air. The N_{2a} is tabulated separately from the elemental nitrogen in the fuel to differentiate the molecular weight of the two. In practice, the nitrogen in the fuel is normally small with respect to the N_{2a} and can be included with the nitrogen in air. For manufactured gases that are formed when combustible products oxidize with air (blast furnace gas, for example), the nitrogen in the fuel is predominately atmospheric nitrogen.

Flue gas products are normally measured on a volumetric basis. If the sample includes water products, it is measured on a *wet basis*, typical of in situ analyzers. Conversely, if water products are excluded, measurements are done on a *dry basis*, which is typical of extractive gas sample systems. (See *Flue gas analysis*.) Note that the flue gas products are summed on a dry and wet basis to facilitate calculation of the flue gas constituents on a dry and wet percent by volume basis in columns B and C. The molecular weight of each constituent is given in column D. Finally, the mass of each constituent on a lb/100 lb fuel basis is the product of the moles/100 lb fuel and the molecular weight.

The calculation of the mass of air on a lb/100 lb fuel basis, shown at the bottom of Table 6, follows the same principles as the flue gas calculations.

For most engineering calculations, it is common U.S. practice to work with air and flue gas (combustion products) on a mass basis. It is usually more convenient to calculate these products on a mass basis directly as discussed later. The mole method described above is the fundamental basis for understanding and calculating the chemical reactions. It is also the basis for deriving certain equations that are presented later. For those who prefer using the mole method, Table 7 presents this method in a convenient calculation format.

Alternate units – Btu method

It is customary within the U.S. boiler industry to use units of mass rather than moles for expressing the quantity of air and flue gas. This is especially true for heat transfer calculations, where the quantity of the working fluid (usually steam or water) is expressed on a mass basis and the enthalpy of the hot and cold fluids is traditionally expressed on a Btu/lb basis. Therefore, if the combustion calculations are performed on the mole basis, it is customary to convert the results to lb/100 lb fuel.

Table 6
Calculation of Flue Gas and Air Quantities – Mass Basis

	Constituent	A (From Tables 3 and 5) Moles/100 lb Fuel	B (A/A6) % Vol dry	C (A/A11) % Vol wet	D (From Table 1) Molecular weight	E (A x D) lb/100 lb Fuel
		Flue Gas or Combustion Product				
1	CO_2	5.995	15.25	14.02	44.010	263.8
2	SO_2	0.050	0.13	0.115	64.065	3.2
3	N_2 (fuel)	0.050	0.13	0.115	28.013	1.4
4	N_{2a} (air)	31.800 (26.500 + 5.300)	80.92	74.35	28.158	895.4
5	O_2	1.405	3.57	3.29	31.999	45.0
6	Total, dry combustion products	39.300	100.00			1208.8
7	H_2O combustion	2.183				
8	H_2O fuel	+0.444				
9	H_2O air	+0.841				
10	Total H_2O	3.468		8.11	18.015	62.5
11	Total, wet combustion products	42.768		100.00		1271.3
Air						
12	Dry air	40.228			28.963	1165.1
13	H_2O	0.841			18.015	15.2
14	Total wet air					1180.3

Items that are expressed on a unit of fuel basis (mole/100 lb fuel, mass/mass fuel, etc.) can be normalized by using an input from fuel basis. For example, knowing that a coal has 10% ash only partly defines the fuel. For a 10,000 Btu/lb fuel, there are 10 lb ash per million Btu input, but for a 5000 Btu/lb fuel there would be 20 lb ash per million Btu input. Considering that fuel input for a given boiler load does not vary significantly with heating value, a boiler firing the lower heating value fuel would encounter approximately twice the amount of ash.

The mass per unit input concept is valuable when determining the impact of different fuels on combustion calculations. This method is particularly helpful in theoretical air calculations. Referring to Table 8, in the first column, theoretical air has been tabulated for various fuels on a mass per mass of fuel basis. The resulting values have little significance when comparing the various fuels. However, when the theoretical air is converted to a mass per unit heat input from fuel basis, the theoretical air varies little between fuels. Also refer to the discussion on the Btu Method in the *Combustion calculations* section. The common units are lb/10,000 Btu, abbreviated as lb/10KB. The fuel labeled MSW/RDF refers to municipal solid waste and refuse derived fuel. Note that the theoretical air is in the same range as that for fossil fuels on a heat input basis. Carbon and hydrogen, the principal combustible fuel elements, are shown for reference. Note that the coals listed in the table are limited to those with a volatile matter (moisture and ash free) greater than 30%. As volatile matter decreases, the carbon content increases and requires more excess air. To check the expected theoretical air for low volatile coals, refer to Fig. 2. The theoretical air of all coals should fall within plus or minus 0.2 lb/10,000 Btu of this curve. Table 9 provides the fuel analysis and theoretical air requirements for a typical fuel oil and natural gas.

Heat of combustion

In a boiler furnace (where no mechanical work is done), the heat energy evolved from combining combustible elements with oxygen depends on the ultimate products of combustion; it does not rely on any intermediate combinations that may occur.

For example, one pound of carbon reacts with oxygen to produce about 14,093 Btu of heat (refer to Table 2). The reaction may occur in one step to form CO_2 or, under certain conditions, it may take two steps. In this process, CO is first formed, producing only 3967 Btu per pound of carbon. In the second step, the CO joins with additional oxygen to form CO_2, releasing 10,126 Btu per pound of carbon (4342 Btu per pound of CO). The total heat produced is again 14,093 Btu per pound of carbon.

Measurement of heat of combustion

In boiler practice, a fuel's *heat of combustion* is the amount of energy, expressed in Btu, generated by the complete combustion, or oxidation, of a unit weight of fuel. *Calorific value, fuel Btu value* and *heating value* are terms also used.

The amount of heat generated by complete combustion is a constant for a given combination of combustible elements and compounds. It is not affected by the manner in which the combustion takes place, provided it is complete.

A fuel's heat of combustion is usually determined by direct calorimeter measurement of the heat evolved. Combustion products within a calorimeter are cooled to the initial temperature, and the heat absorbed by the cooling medium is measured to determine the higher, or gross, heat of combustion (typically referred to as the higher heating value, or HHV).

For all solid and most liquid fuels, the bomb type calorimeter is the industry standard measurement device. In these units, combustible substances are burned in a constant volume of oxygen. When they

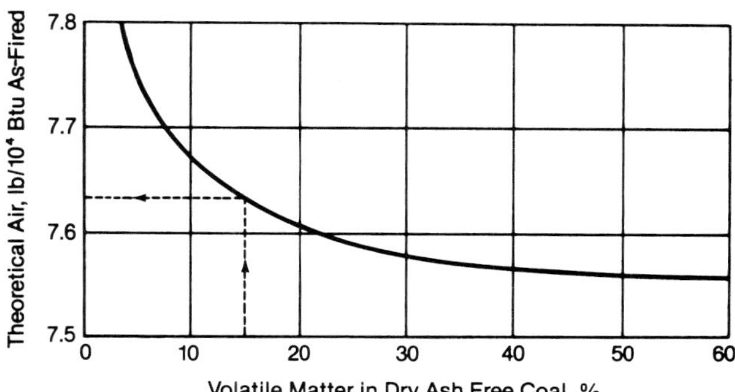

Fig. 2 Theoretical air in lb/10,000 Btu heating value of coal with a range of volatile matter.

Table 7
Combustion Calculations – Molar Basis

	INPUTS (see also lightly shaded blocks)			FUEL – *Bituminous coal, Virginia*	
1	Excess air: at burner/at boiler/econ, %	*20/20*	4	Fuel input, 1,000,000 Btu/h	*330.0*
2	Moisture in air, lb/lb dry air	*0.013*	5	Unburned carbon loss, % efficiency	*0.40*
3	Fuel heating value, Btu/lb	*14,100*	6	Unburned carbon (UBC), [5] x [3] / 14,500	*0.39*

COMBUSTION PRODUCTS CALCULATIONS

7	Ultimate Analysis, % mass			8 Molecular Weight lb/mole	9 Moles /100 lb Fuel [7] / [8]	10 Moles O_2 /Mole Fuel Constituent	11 Moles Theo. O_2/100 lb Fuel [9] x [10]	12 Combustion Product
	Fuel Constituent	As-Fired	Carbon Burned(CB)					
A	C	*80.31*	*80.31*					
B	UBC [6]		*0.39*					
C	CB [A] – [B]		*79.92*	12.011	*6.654*	1.0	*6.654*	CO_2
D	S	*1.54*		32.066	*0.048*	1.0	*0.048*	SO_2
E	H_2	*4.47*		2.016	*2.217*	0.5	*1.109*	H_2O
F	H_2O	*2.90*		18.015	*0.161*			H_2O
G	N_2	*1.38*		28.013	*0.049*			N_2 (fuel)
H	O_2	*2.85*		31.999	*0.089*	–1.0	*–0.089*	
I	Ash	*6.55*						
K	Total	*100.00*			*9.218*		*7.722*	

	AIR CONSTITUENTS, Moles/100 lb Fuel		At Burner	At Blr/Econ
13	O_2 – excess	[11K] x [1] / 100	*1.544*	*1.544*
14	O_2 – total	[13] + [11K]	*9.266*	*9.266*
15	N_{2a} – air	[14] x 3.77	*34.933*	*34.933*
16	Air (dry)	[14] + [15]	*44.199*	*44.199*
17	H_2O – air	[16] x [2] x 1.608	*0.924*	*0.924*
18	Air (wet)	[16] + [17]	*45.123*	*45.123*

	FLUE GAS CONSTITUENTS		19 Moles /100 lb Fuel	20 Vol % Dry 100 x [19] / [19G]	21 Vol % Wet 100 x [19] / [19H]	22 Molecular Weight lb/mole	23 Flue Gas lb/100 lb Fuel [19] x [22]
A	CO_2	[9C]	*6.654*	*15.39*	*14.30*	44.010	*292.8*
B	SO_2	[9D]	*0.048*	*0.11*	*0.10*	64.065	*3.1*
C	O_2	[13]	*1.544*	*3.57*	*3.32*	31.999	*49.4*
D	N_2 (fuel)	[9G]	*0.049*	*0.11*	*0.11*	28.013	*1.4*
E	N_{2a} (air)	[15]	*34.933*	*80.82*	*75.07*	28.158	*983.6*
F	H_2O	[9E] + [9F] + [17]	*3.302*		*7.10*	18.015	*59.5*
G	Total dry	Sum [A] through [E]	*43.228*	*100.00*			*1330.3*
H	Total wet	Sum [A] through [F]	*46.530*		*100.00*		*1389.8*

	KEY PERFORMANCE PARAMETERS		At Burner	At Blr/Econ
24	Molecular weight wet flue gas, lb/mole	[23H] / [19H]		*29.869*
25	H_2O in wet gas, % by wt	100 x [23F] / [23H]		*4.28*
26	Dry gas weight, lb/10,000 Btu	100 x [23G] / [3]		*9.435*
27	Wet gas weight, lb/10,000 Btu	100 x [23H] / [3]		*9.857*
28	Wet gas weight, 1000 lb/h	[27] x [4] / 10		*325.3*
29	Air flow (wet), lb/100 lb fuel	[16] x 28.966 + [17] x 18.015	*1296.9*	
30	Air flow (wet), lb/10,000 Btu	100 x [29] / [3]	*9.198*	
31	Air flow (wet), 1000 lb/h	[30] x [4] / 10	*303.5*	

Table 8
Theoretical Air Required for Various Fuels

Fuel	Theoretical Air, lb/lb Fuel	HHV Btu/lb	Theoretical Air Typical lb/10^4 Btu	Theoretical Air Range lb/10^4 Btu
Bituminous coal (VM* >30%)	9.07	12,000	7.56	7.35 to 7.75
Subbituminous coal (VM* >30%)	6.05	8,000	7.56	7.35 to 7.75
Oil	13.69	18,400	7.46	7.35 to 7.55
Natural Gas	15.74	21,800	7.22	7.15 to 7.35
Wood	3.94	5,831	6.75	6.60 to 6.90
MSW* and RDF*	4.13	5,500	7.50	7.20 to 7.80
Carbon	11.51	14,093	8.16	—
Hydrogen	34.29	61,029	5.62	—

* VM = volatile matter, moisture and ash free basis
MSW = municipal solid waste
RDF = refuse-derived fuel

are properly operated, combustion is complete and all of the heat generated is absorbed and measured. Heat from external sources can be excluded or proper corrections can be applied.

For gaseous fuels of 900 to 1200 Btu/ft^3, continuous or constant flow type calorimeters are industry standards. The principle of operation is the same as for the bomb calorimeter; however, the heat content is determined at constant pressure rather than at constant volume.

For most fuels, the difference between the constant pressure and constant volume heating values is small and is usually neglected. However, because fuel is burned under essentially constant pressure conditions, the constant pressure heating value is the technically correct value.

For solid or liquid fuels, to convert the constant volume higher heating value (HHV_{CV}) measured in the bomb calorimeter to constant pressure (HHV_{CP}), an adjustment for the volume change is required. During the constant pressure combustion process:

1. Every mole of carbon combines with one mole of oxygen to form one mole of carbon dioxide. Therefore, there is no volume change.
2. Every mole of sulfur combines with one mole of oxygen to form one mole of sulfur dioxide. Therefore, there is no volume change.
3. Every mole of hydrogen combines with 1/2 mole of oxygen to form one mole of water vapor. Therefore, there is a net increase of 1/2 mole of gas produced. When the water vapor is condensed to a liquid in the bomb calorimeter, there is a net decrease of 1/2 mole of gas for each mole of hydrogen.
4. For every mole of oxygen in the fuel, there is one mole of oxygen gas produced. Therefore, there is a net increase of one mole of gas produced for each mole of oxygen in the fuel.
5. The solid or liquid nitrogen in the fuel is released as a gas. Therefore, there is a net increase of one mole of gas produced for every mole of nitrogen.

Using the ideal gas law, the energy change due to the volume change is as follows:

$$\Delta HHV = \sum_{n=1}^{k} N_k \left(\frac{\mathbf{R} \times T}{J} \right), \text{ Btu/lb} \tag{6}$$

where

N_k = number of moles of constituent k
$\mathbf{R}$ = universal gas constant, 1545 ft-lb/mole-R
T = absolute reference temperature for the bomb calorimeter, 537R
J = mechanical equivalent of heat, 778.2 ft lbf/Btu

Substituting:

$$\Delta HHV = \left(\frac{-O_2}{31.9988} + \frac{-N_2}{28.0134} + \frac{0.5\,H_2}{2.0159} \right) \times \frac{\mathbf{R}\,T}{778.2} \tag{7}$$

where

O_2 = mass fraction of oxygen in fuel
N_2 = mass fraction of nitrogen in fuel
H_2 = mass fraction of hydrogen in fuel

The corrections for nitrogen and oxygen in typical solid and liquid fuels are small, generally less than 1 and 2 Btu/lbm respectively, and are generally considered negligible. The fuel heating value correction from constant volume to constant pressure then becomes:

$$HHV_{CP} = HHV_{CV} + 264.4\ H_2, \text{ Btu/lb} \tag{8}$$

Gas chromatography is also commonly used to determine the composition of gaseous fuels. When the composition of a gas mixture is known, its heat of combustion may be determined as follows:

Table 9
Fuel Analysis and Theoretical Air for Typical Oil and Gas Fuels

Heavy Fuel Oil, % by wt		Natural Gas, % by vol	
S	1.16	CH_4	85.3
H_2	10.33	C_2H_6	12.6
C	87.87	CO_2	0.1
N_2	0.14	N_2	1.7
O_2	0.50	O_2	0.3
		Sp Gr	0.626
		Btu/ft^3, as-fired	1090
Btu/lb, as-fired	18,400	Btu/lb, as-fired	22,379

Theoretical Air, Fuel and Moisture

Theoretical air, lb/10,000 Btu	7.437	Theoretical air, lb/10,000 Btu	7.206
Fuel, lb/ 10,000 Btu	0.543	Fuel, lb/ 10,000 Btu	0.440
Moisture, lb/ 10,000 Btu	0.502	Moisture, lb/ 10,000 Btu	0.912

$$hc_{mix} = v_a hc_a + v_b hc_b + \ldots + v_x hc_x \quad (9)$$

where

hc_{mix} = heat of combustion of the mixture
v_x = volume fraction of each component
hc_x = heat of combustion of each component

For an accurate heating value of solid and liquid fuels, a laboratory heating value analysis is required. Numerous empirical methods have been published for estimating the heating value of coal based on the proximate or ultimate analyses. (See Chapter 9.) One of the most frequently used correlations is Dulong's formula which gives reasonably accurate results for bituminous coals (within 2 to 3%). It is often used as a routine check of calorimeter-determined values.

$$HHV = 14{,}544\,C + 62{,}028\left[H_2 - (O_2/8)\right] + 4050\,S \quad (10)$$

where

HHV = higher heating value, Btu/lb
C = mass fraction carbon
H_2 = mass fraction hydrogen
O_2 = mass fraction oxygen
S = mass fraction sulfur

A far superior method for checking whether the heating value is reasonable in relation to the ultimate analysis is to determine the theoretical air on a mass per Btu basis. (See *Alternate units – Btu method*.) Table 8 indicates the range of theoretical air values. The equation for theoretical air can be rearranged to calculate the higher heating value, *HHV*, where the median range for theoretical air for the fuel from Table 8, *MQTHA*, is used:

$$HHV = 100 \times \frac{11.51\,C + 34.29\,H_2 + 4.31\,S - 4.32\,O_2}{MQTHA} \quad (11)$$

where

HHV = higher heating value, Btu/lb
C = mass percent carbon, %
H_2 = mass percent hydrogen, %
S = mass percent sulfur, %
O_2 = mass percent oxygen, %
$MQTHA$ = theoretical air, lb/10,000 Btu

Higher and lower heating values

Water vapor is a product of combustion for all fuels that contain hydrogen. The heat content of a fuel depends on whether this vapor remains in the vapor state or is condensed to liquid. In the bomb calorimeter, the products of combustion are cooled to the initial temperature and all of the water vapor formed during combustion is condensed to liquid. This gives the *HHV* or gross calorific value (defined earlier) of the fuel, and the heat of vaporization of water is included in the reported value. For the lower heating value (*LHV*) or net calorific value (net heat of combustion at constant pressure), all products of combustion including water are assumed to remain in the gaseous state, and the water heat of vaporization is not available.

While the high, or gross, heat of combustion can be accurately determined by established American Society for Testing and Materials (ASTM) procedures, direct determination of the lower heating value is difficult. There is no international standard for calculation of LHV from the measured HHV. The constants used for heats of combustion, and the temperature used to calculate the latent heat of vaporization (H_{FG}), may vary slightly between references. It is important that the temperature used for the calculation of H_{FG} be consistent with the basis for the boiler efficiency calculations, otherwise there can be errors in calculated fuel flow for a given boiler output. ASME Performance Test Code PTC 4 specifies a reference temperature of 77F (25C). The value given for H_{FG} at 77F in the ASME International Steam Tables for Industrial Use, based on IAWPS-IF97 is 1049.7 Btu/lb. Calculation of LHV at constant pressure from HHV at constant pressure is then as follows:

$$LHV_{CP} = HHV_{CP} - 1049.7\left(H_2 \times 8.937 + M\right),\ \text{Btu/lb} \quad (12)$$

where

H_2 = mass fraction of hydrogen in fuel
M = mass fraction of water in fuel

In some references, the calculation of LHV includes a correction for the difference between constant volume and constant pressure combustion. Combining Equations 8 and 12, the calculation of LHV at constant pressure from HHV at constant volume is as follows:

$$LHV_{CP} = HHV_{CV} - 1049.7\left(H_2 \times 8.937 + M\right) + 264.4H_2,\ \text{Btu/lb} \quad (13)$$

Ignition temperatures

Ignition temperatures of combustible substances vary greatly, as indicated in Table 10. This table lists minimum temperatures and temperature ranges in air for fuels and for the combustible constituents of fuels commonly used in the commercial generation of heat. Many factors influence ignition temperature, so any tabulation can be used only as a guide. Pressure, velocity, enclosure configuration, catalytic materials, air/fuel mixture uniformity and ignition source are examples of the variables. Ignition temperature usually decreases with rising pressure and increases with increasing air moisture content.

The ignition temperatures of coal gases vary considerably and are appreciably higher than those of the fixed carbon in the coal. However, the ignition temperature of coal may be considered as the ignition temperature of its fixed carbon content, because the gaseous constituents are usually distilled off, but not ignited, before this temperature is attained.

Table 10
Ignition Temperatures of Fuels in Air (Approximate Values or Ranges at Atmospheric Pressure)

Combustible	Formula	Temperature, F
Sulfur	S	470
Charcoal	C	650
Fixed carbon (bituminous coal)	C	765
Fixed carbon (semi-anthracite)	C	870
Fixed carbon (anthracite)	C	840 to 1115
Acetylene	C_2H_2	580 to 825
Ethane	C_2H_6	880 to 1165
Ethylene	C_2H_4	900 to 1020
Hydrogen	H_2	1065 to 1095
Methane	CH_4	1170 to 1380
Carbon monoxide	CO	1130 to 1215
Kerosene	—	490 to 560
Gasoline	—	500 to 800

Adiabatic flame temperature

The *adiabatic flame temperature* is the maximum theoretical temperature that can be reached by the products of combustion of a specific fuel and air (or oxygen) combination, assuming no loss of heat to the surroundings and no dissociation. The fuel's heat of combustion is the major factor in the flame temperature, but increasing the temperature of the air or the fuel also raises the flame temperature. This adiabatic temperature is a maximum with zero excess air (only enough air chemically required to combine with the fuel). Excess air is not involved in the combustion process; it only acts as a dilutant and reduces the average temperature of the products of combustion.

The adiabatic temperature is determined from the adiabatic enthalpy of the flue gas:

$$H_g = \frac{HHV - \text{Latent heat } H_2O + \text{Sensible heat in air}}{\text{Wet gas weight}} \quad (14)$$

where

H_g = adiabatic enthalpy, Btu/lb

Knowing the moisture content and enthalpy of the products of combustion, the theoretical flame or gas temperature can be obtained from Fig. 3 (see pages 12 and 13).

The adiabatic temperature is a fictitiously high value that can not exist. Actual flame temperatures are lower for two main reasons:

1. Combustion is not instantaneous. Some heat is lost to the surroundings as combustion takes place. Faster combustion reduces heat loss. However, if combustion is slow enough, the gases may be cooled sufficiently and incomplete combustion may occur, i.e., some of the fuel may remain unburned.
2. At temperatures above 3000F, some of the CO_2 and H_2O in the flue gases dissociates, absorbing heat in the process. At 3500F, about 10% of the CO_2 in a typical flue gas dissociates to CO and O_2. Heat absorption occurs at 4342 Btu/lb of CO formed, and about 3% of the H_2O dissociates to H_2 and O_2, with a heat absorption of 61,029 Btu/lb of H_2 formed. As the gas cools, the dissociated CO and H_2 recombine with the O_2 and liberate the heat absorbed in dissociation, so the heat is not lost. However, the overall effect is to lower the maximum actual flame temperature.

The term *heat available* (Btu/h) is used throughout this text to define the heat available to the furnace. This term is analogous to the energy term (numerator) in the adiabatic sensible heat equation above except that one half of the radiation heat loss and the manufacturer's margin portion are not considered available to the furnace.

Practical combustion application issues

In addition to the theoretical combustion evaluation methodologies addressed above, several application issues are very important in accurate combustion calculations of actual applications. These include the impact of the injection of SO_2 sorbents and other chemicals into the combustion process, solid ash or residue, unburned carbon and excess air.

Sorbents and other chemical additives

In some combustion systems, chemical compounds are added to the gas side of the steam generator to reduce emissions. For example, limestone is used universally in fluidized-bed steam generators to reduce SO_2 emissions. (See Chapter 17.)

Limestone impacts the combustion and efficiency calculations by: 1) altering the mass of flue gas by reducing SO_2 and increasing CO_2 levels, 2) increasing the mass of solid waste material (ash residue), 3) increasing the air required in forming SO_3 to produce calcium sulfate, $CaSO_4$, 4) absorbing energy (heat) from the fuel to calcine the calcium and magnesium carbonates, and 5) adding energy to the system in the sulfation reaction ($SO_2 + \frac{1}{2} O_2 + CaO \rightarrow CaSO_4$). The impact of sorbent/limestone is shown as a correction to the normal combustion calculations presented later.

The limestone constituents that are required in the combustion and efficiency calculations are:

Reactive constituents:
- Calcium carbonate ($CaCO_3$)
- Magnesium carbonate ($MgCO_3$)

Water
Inerts

Some processes may use sorbents derived from limestone. These sorbents contain reactive constituents such as calcium hydroxide [$Ca(OH)_2$] and magnesium hydroxide [$Mg(OH)_2$].

For design purposes, the amount of sorbent is determined from the design calcium to sulfur molar ratio, *MOFCAS*. The sorbent to fuel ratio, *MFSBF*, is a

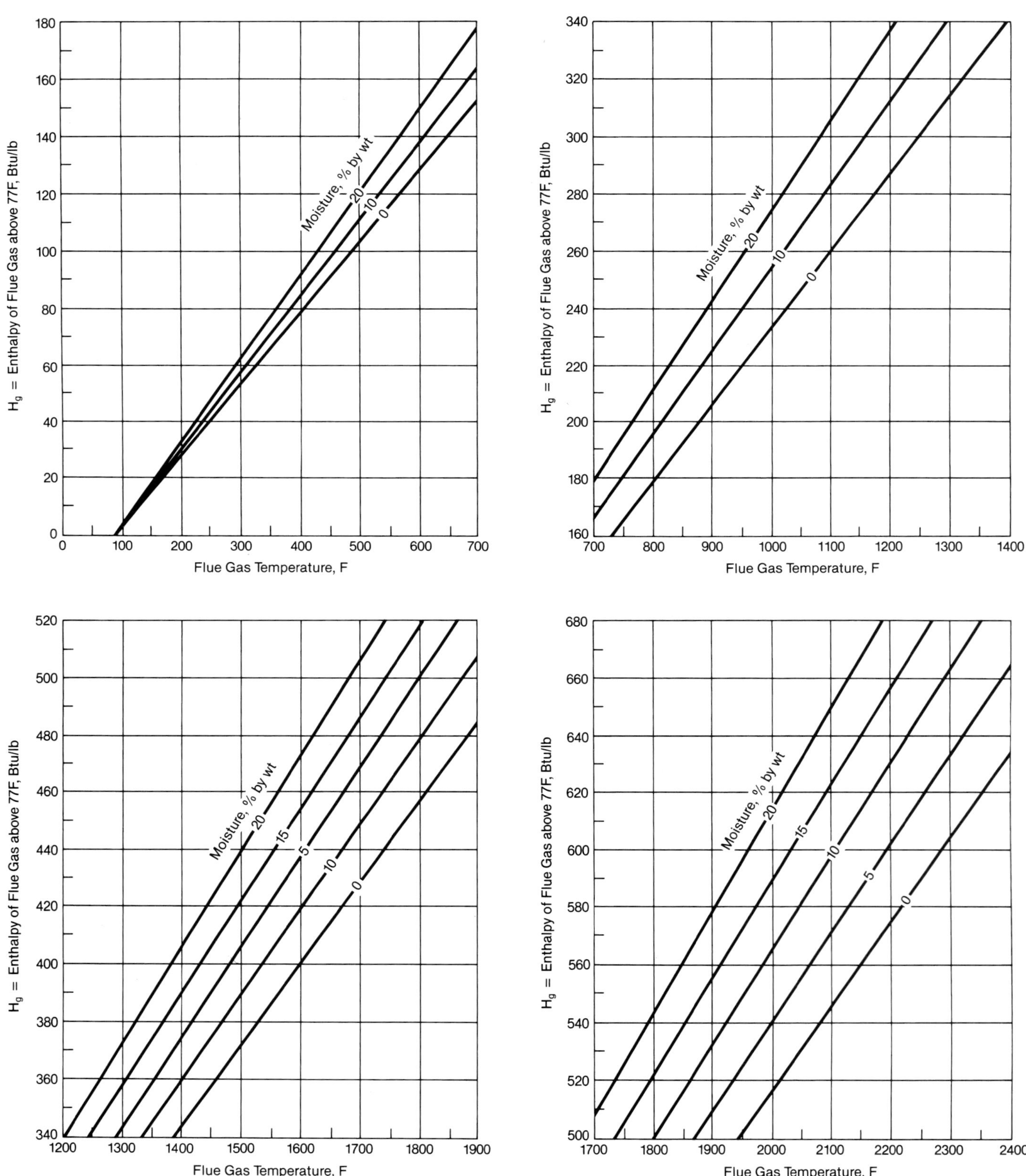

Fig. 3 Enthalpy of flue gas above 77F at 30 in. Hg.

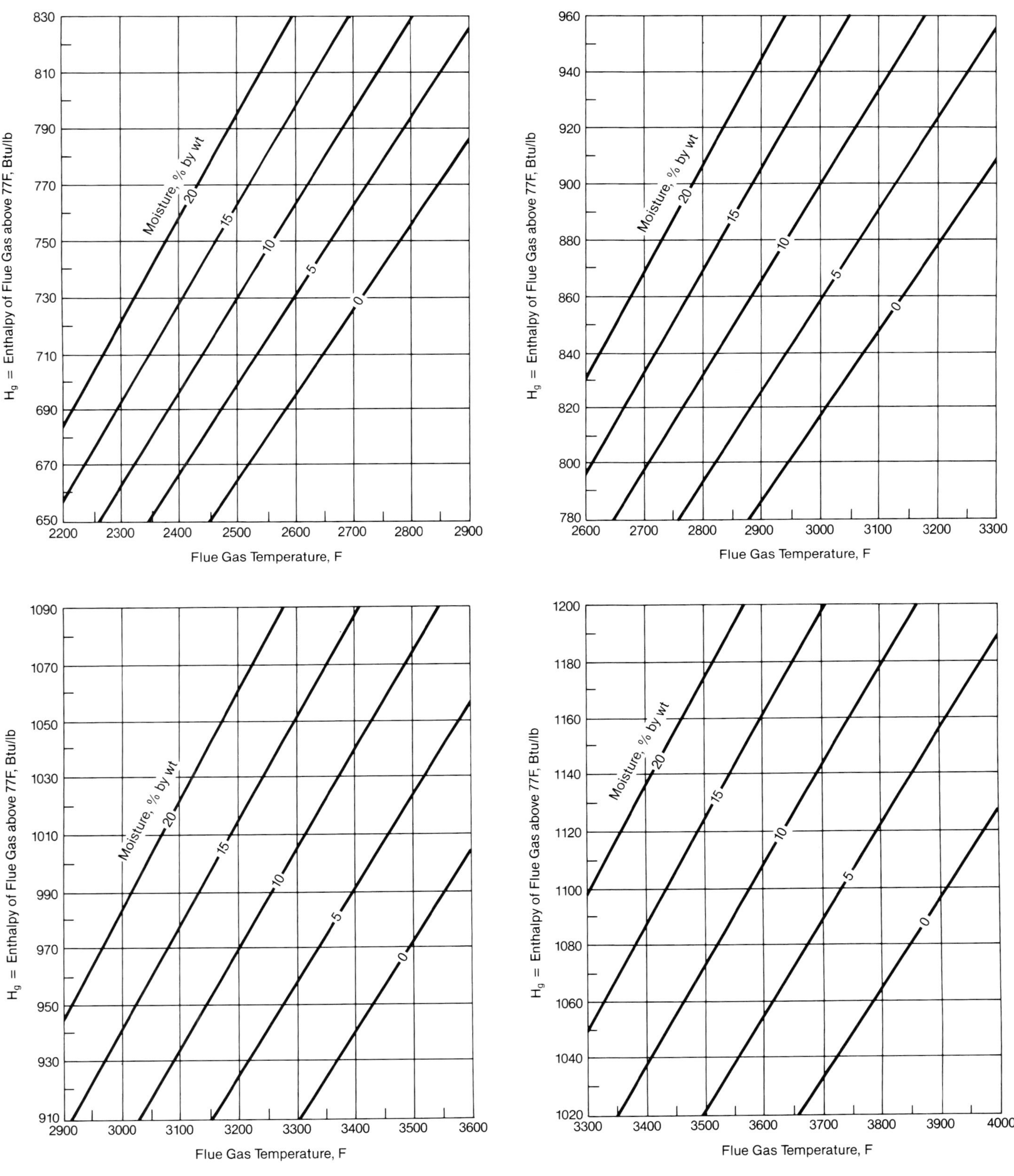

830
810
790
770
750
730
710
690
670
650
H_g = Enthalpy of Flue Gas above 77F, Btu/lb
Moisture, % by wt
20
15
10
5
0
2200
2300
2400
2500
2600
2700
2800
2900
Flue Gas Temperature, F
960
940
920
900
880
860
840
820
800
780
H_g = Enthalpy of Flue Gas above 77F, Btu/lb
Moisture, % by wt
20
15
10
5
0
2600
2700
2800
2900
3000
3100
3200
3300
Flue Gas Temperature, F
1090
1070
1050
1030
1010
990
970
950
930
910
H_g = Enthalpy of Flue Gas above 77F, Btu/lb
Moisture, % by wt
20
15
10
5
0
2900
3000
3100
3200
3300
3400
3500
3600
Flue Gas Temperature, F
1200
1180
1160
1140
1120
1100
1080
1060
1040
1020
H_g = Enthalpy of Flue Gas above 77F, Btu/lb
Moisture, % by wt
20
15
10
5
0
3300
3400
3500
3600
3700
3800
3900
4000
Flue Gas Temperature, F

convenient equation that converts sorbent products to a mass of fuel or input from fuel basis.

$$MFSBF = \frac{MOFCAS \times S}{MOPCA \times 32.066} \quad \textbf{(15)}$$

and

$$MOPCA = \left(\frac{CaCO_3}{100.089} + \frac{Ca(OH)_2}{74.096} \right) \quad \textbf{(16)}$$

where

$MFSBF$ = mass ratio of sorbent to fuel, lb/lb
$MOFCAS$ = calcium to sulfur molar ratio
S = mass percent sulfur in fuel, %
$MOPCA$ = calcium in sorbent molar basis, moles/100 lb sorbent
$CaCO_3$ = mass percent calcium carbonate in sorbent, %
$Ca(OH)_2$ = mass percent calcium hydroxide in sorbent, %

When calcium carbonate and magnesium carbonate are heated, they release CO_2, which adds to the flue gas products. This is referred to as *calcination*. Magnesium carbonate calcines readily; however, at the operating temperatures typical of atmospheric pressure fluidized beds, not all of the calcium carbonate is calcined. For design purposes, 90% calcination is appropriate for atmospheric fluidized-bed combustion. On an operating unit, the mass fraction of calcination can be determined by measuring the CO_2 in the ash residue and by assuming it exists as $CaCO_3$. The quantity of CO_2 added to the flue gas may be calculated from:

$$MQGSB = 44.01 \times MOGSB \times \frac{100}{HHV} \quad \textbf{(17)}$$

and

$$MOGSB = MFSBF\left(\frac{MFCL \times CaCO_3}{100.089} + \frac{MgCO_3}{58.320} \right) \quad \textbf{(18)}$$

where

$MQGSB$ = incremental CO_2 from sorbent, lb/10,000 Btu
$MOGSB$ = moles CO_2 from sorbent, moles/100 lb sorbent
HHV = higher heating value, Btu/lb fuel
$MFCL$ = fraction of available $CaCO_3$ calcined, lb/lb
$CaCO_3$ = mass percent calcium carbonate in sorbent, %
$MgCO_3$ = mass percent magnesium carbonate in sorbent, %

The water added to the flue gas, *MQWSB*, includes the free water and water evaporated due to dehydration of calcium and magnesium hydroxide products.

$$MQWSB = 18.015 \times MOWSB \times \frac{100}{HHV} \quad \textbf{(19)}$$

and

$$MOWSB = H_2O + \left(\frac{Ca(OH)_2}{74.096} + \frac{Mg(OH)_2}{84.321} \right) \quad \textbf{(20)}$$

where

$MQWSB$ = water added to flue gas from sorbent, lb/10,000 Btu
$MOWSB$ = moles of water from sorbent, moles/100 lb sorbent
HHV = higher heating value, Btu/lb fuel
H_2O = free water from sorbent, moles/100 lb sorbent
$Ca(OH)_2$ = mass percent calcium hydroxide in sorbent, %
$Mg(OH)_2$ = mass percent magnesium hydroxide in sorbent, %

Spent sorbent refers to the solid products remaining due to the use of limestone. Spent sorbent is the sum of the inerts in the limestone, the mass of the reactive constituents after calcination ($CaCO_3$, CaO and MgO), and the SO_3 formed in the sulfation reaction.

$$MQSSB = MFSSB \times \frac{10{,}000}{HHV} \quad \textbf{(21)}$$

and

$$MFSSB = MFSBF - (0.4401 \times MOGSB) - (0.18015 \times MOWSB) + (250 \times S \times MFSC) \quad \textbf{(22)}$$

where

$MQSSB$ = solids added to flue gas, lb/10,000 Btu
$MFSSB$ = solids added to flue gas, lb/lb fuel
HHV = higher heating value, Btu/lb fuel
$MFSBF$ = mass ratio of sorbent to fuel, lb/lb
$MOGSB$ = moles CO_2 from sorbent, moles/100 lb sorbent
$MOWSB$ = moles H_2O from sorbent, moles/100 lb sorbent
S = mass percent sulfur in fuel, %
$MFSC$ = mass fraction of sulfur in fuel captured, lb/lb

The combustion and efficiency values related to limestone (sorbent) are calculated separately in this text; they are treated as a supplement to the basic calculations. (See *Combustion and efficiency calculations*.)

Residue versus refuse

The term *residue* is used within this text to refer to the solid waste products that leave the steam generator envelope. This replaces the term *refuse* which is now used to refer to municipal solid waste fuels and their derivatives.

Unburned carbon

In commercial solid fuel applications, it is not always practical to completely burn the fuel. Some of the fuel may appear as unburned carbon in the residue or CO in the flue gas, although the hydrogen in the fuel is usually completely consumed. The capital and operating (energy) costs incurred to burn this residual fuel are usually far greater than the energy lost. In addition, the evolution of combustion equipment to reduce NO_x emissions has resulted in some tradeoffs with increases in unburned carbon and CO.

Unburned carbon impacts the combustion calculations and represents an efficiency loss. Therefore, unburned carbon must be measured when present. The preferred procedure is to determine the quantity of combustible carbon in the boiler flyash and bottom ash in accordance with ASTM D-6316, *Determination of Total, Combustible, and Carbonate Carbon in Solid Residues from Coal and Coke.*

The unburned carbon determined by ASTM D-6316 is on the basis of percent carbon in the flyash/bottom ash (lb carbon/100 lb residue). The combustion calculations require unburned carbon on a lb/100 lb fuel basis (percent unburned carbon from fuel) calculated from the following equations:

$$UBC = MPCR \times MFR \qquad \textbf{(23)}$$

and

$$MFR = \frac{AF + (100 \times MFSSB)}{(100 - MPCR)} \qquad \textbf{(24)}$$

where

UBC = unburned carbon, lb/100 lb fuel
$MPCR$ = unburned carbon in residue (measured or reference), mass %
MFR = mass fraction residue, fuel basis, lb/lb fuel
AF = mass percent ash from fuel, %
$MFSSB$ = mass fraction of spent sorbent, lb/lb fuel

The quantity of ash in solid fuels is variable, and therefore it is sometimes desirable to correct the measured percent unburned carbon in residue to a reference (or baseline) fuel ash content in order to evaluate combustion system performance. For given boiler operating conditions, the heat loss due to unburned carbon (UBCL) is assumed to remain constant for typical variations in fuel ash (and spent sorbent if applicable). The unburned carbon as it would appear in a residue produced by the *reference* fuel (and sorbent flow) may be calculated using the following equation:

$$MPCR_{REF} = \frac{100}{\left[\dfrac{14{,}500 \times (AF_{REF} + 100 \times MFSSB_{REF})}{UBCL_{MEAS} \times HHV_{REF}}\right] + 1} \qquad \textbf{(25)}$$

where

$MPCR_{REF}$ = unburned carbon in residue corrected to reference fuel ash (and spent sorbent), %
AF_{REF} = mass percent ash from reference fuel, %
$MFSSB_{REF}$ = reference mass fraction of spent sorbent, lb/lb fuel
$UBCL_{MEAS}$ = measured heat loss due to unburned carbon, %
HHV_{REF} = higher heating value of reference fuel, Btu/lb

Equation 25 assumes that the unburned carbon loss, UBCL, is constant between the test and reference conditions.

Excess air

For commercial applications, more than theoretical air is needed to assure complete combustion. This *excess air* is needed because the air and fuel mixing is not perfect. Because the excess air that is not used for combustion leaves the unit at stack temperature, the amount of excess air should be minimized. The energy required to heat this air from ambient to stack temperature usually serves no purpose and is lost energy. Typical values of excess air required at the burning equipment are shown in Table 11 for various fuels and methods of firing. When substoichiometric firing is used in the combustion zone, i.e., less than the theoretical air is used, the values shown would apply to the furnace zone where the final air is admitted to

Table 11
Typical Excess Air Requirements at Fuel Burning Equipment

Fuel	Type of Furnace or Burners	Excess Air % by wt
Pulverized coal	Completely water-cooled furnace — wet or dry ash removal	15 to 20
	Partially water-cooled furnace	15 to 40
Crushed coal	Cyclone furnace — pressure or suction	13 to 20
	Fluidized-bed combustion	15 to 20
Coal	Spreader stoker	25 to 35
	Water-cooled vibrating grate stoker	25 to 35
	Chain grate and traveling grate	25 to 35
	Underfeed stoker	25 to 40
Fuel oil	Register type burners	3 to 15
Natural, coke oven and refinery gas	Register type burners	3 to 15
Blast furnace gas	Register type burners	15 to 30
Wood/bark	Traveling grate, water-cooled vibrating grate	20 to 25
	Fluidized-bed combustion	5 to 15
Refuse-derived fuels (RDF)	Completely water-cooled furnace — traveling grate	40 to 60
Municipal solid waste (MSW)	Water-cooled/refractory covered furnace reciprocating grate	80 to 100
	Rotary kiln	60 to 100
Bagasse	All furnaces	25 to 35
Black liquor	Recovery furnaces for Kraft and soda pulping processes	15 to 20

complete combustion. The amount of excess air at the exit of the pressure parts (where it is usually monitored) must be greater than the air required at the burning equipment to account for setting infiltration on balanced draft units (or seal air on pressure-fired units). On modern units with membrane wall construction, this is usually only 1 or 2% excess air at full load. On older units, however, setting infiltration can be significant, and operating with low air at the steam generator exit can result in insufficient air at the burners. This can cause poor combustion performance.

For units with air heaters, excess air must be measured at the air heater gas inlet to determine efficiency. When equipment such as selective catalytic reduction (SCR) systems or dust collection equipment is located between the exit of the pressure parts and air heater gas inlet, additional air infiltration may occur, including SCR dilution air for ammonia transport. A typical value for SCR dilution air is 0.8% excess air.

Combustion and efficiency calculations

The combustion calculations are the starting point for all design and performance calculations for boilers and their related component parts. They establish the quantities of the constituents involved in the combustion process chemistry (air, flue gas, residue and sorbent), the efficiency of the combustion process and the quantity of heat released.[4]

The units used for the combustion and efficiency calculations are lb/10,000 Btu. The acronym MQxx also refers here to constituents on a mass per 10,000 Btu basis. For gaseous fuels, the volumetric analysis is converted to an elemental mass basis, as described in *Molar evaluation of combustion*.

Combustion air – theoretical air

The combustion air is the total air required for the burning equipment; it is the theoretical air plus the excess air. Theoretical air is the minimum air required for complete conversion of the carbon, hydrogen and sulfur in the fuel to standard products of combustion. For some fuels and/or combustion processes, all of the carbon is not converted. In addition, when limestone or other additives are used, some of the sulfur is not converted to sulfur dioxide. However, additional air is required for the conversion of sulfur dioxide to sulfur trioxide in the sulfation reaction ($CaO + SO_2 + {}^1/_2 O_2 \rightarrow CaSO_4$). Because the actual air required is the desired calculation result, the theoretical air is corrected for unburned carbon and sulfation reactions.

$$MQTHAC = THAC \times \frac{100}{HHV} \qquad (26)$$

and

$$THAC = 11.51 \times CB + 34.29 \times H_2 + \left[4.31 \times S \times (1 + 0.5 \times MFSC)\right] - 4.32 \times O_2 \qquad (27)$$

where

$MQTHAC$ = theoretical air, corrected, lb/10,000 Btu
$THAC$ = theoretical air, corrected, lb/100 lb fuel
HHV = higher heating value, Btu/lb
CB = mass percent carbon burned = percent carbon in fuel – UBC, %
H_2 = mass percent hydrogen in fuel, %
S = mass percent sulfur in fuel, %
$MFSC$ = mass fraction sulfur captured by furnace sorbent, lb/lb sulfur
O_2 = mass percent oxygen in fuel, %
UBC = unburned carbon percent from fuel, %

For test purposes, the unburned carbon is measured. For design calculations, the unburned carbon may be calculated from the estimated unburned carbon loss, *UBCL*:

$$UBC = UBCL \times \frac{HHV}{14{,}500} \qquad (28)$$

MFSC is the sulfur capture/retention ratio or mass of sulfur captured per mass sulfur available from the fuel. It is zero unless a sorbent, e.g., limestone, is used in the furnace to reduce SO_2 emissions. See *Flue gas analysis* to determine *MFSC* for test conditions.

The mass of dry air, *MQDA*, water in air, *MQWA*, and wet air, *MQA*, are calculated from the following equations:

$$MQDA = MQTHAC \times \left(1 + \frac{PXA}{100}\right) \qquad (29)$$

$$MQWA = MA \times MQDA \qquad (30)$$

$$MQA = MQDA + MQWA = MQDA \times (1 + MA) \qquad (31)$$

where

$MQDA$ = mass dry air, lb/10,000 Btu
$MQTHAC$ = theoretical air, lb/10,000 Btu
PXA = percent excess air, %
$MQWA$ = mass of moisture in air, lb/10,000 Btu
MA = moisture in air, lb/lb dry air
MQA = mass of wet air, lb/10,000 Btu

Flue gas

The total gaseous products of combustion are referred to as *wet flue gas*. Solid products or residue are excluded. The wet flue gas flow rate is used for heat transfer calculations and design of auxiliary equipment. The total gaseous products excluding moisture are referred to as *dry flue gas*; this parameter is used in the efficiency calculations and determination of flue gas enthalpy.

The wet flue gas is the sum of the wet gas from fuel (fuel less ash, unburned carbon and sulfur captured), combustion air, moisture in the combustion air, additional moisture such as atomizing steam and, if sorbent is used, carbon dioxide and moisture from sorbent. Dry flue gas is determined by subtracting the summation of the moisture terms from the wet flue gas.

Wet gas from fuel is the mass of fuel less the ash in the fuel, less the percent unburned carbon and, when sorbent is used to reduce SO_2 emissions, less the sulfur captured:

$$MQGF = \left[100 - AF - UBC - \left(MFSC \times S\right)\right] \times \frac{100}{HHV} \quad \textbf{(32)}$$

where

$MQGF$ = wet gas from fuel, lb/10,000 Btu
AF = mass percent ash in fuel, %
UBC = unburned carbon as mass percent in fuel, %
$MFSC$ = mass fraction of sulfur captured, lb/lb sulfur
S = mass percent sulfur in fuel, %
HHV = higher heating value, Btu/lb

Water from fuel is the sum of the water in the fuel, H_2O and the water produced from the combustion of hydrogen in the fuel, H_2:

$$MQWFF = \left[\left(8.937 \times H_2\right) + H_2O\right] \times \frac{100}{HHV} \quad \textbf{(33)}$$

where

$MQWFF$ = water from fuel, lb/10,000 Btu
H_2 = mass percent hydrogen in fuel, %
H_2O = mass percent moisture in fuel, %

Refer to *Sorbents and other chemical additives* for calculating gas from sorbent (CO_2), $MQGSB$, and water from sorbent, $MQWSB$. The total wet gas weight, MQG, is then the sum of the dry air, water in air, wet gas from fuel and, when applicable, additional water, gas from sorbent (CO_2), and water from sorbent:

$$MQG = MQDA + MQWA + MQGF + MQWAD + MQGSB + MQWSB \quad \textbf{(34)}$$

where

MQG = total wet gas weight, lb/10,000 Btu
$MQDA$ = mass dry air, lb/10,000 Btu
$MQWA$ = mass of moisture in air, lb/10,000 Btu
$MQGF$ = wet gas from the fuel, lb/10,000 Btu
$MQWAD$ = additional water such as atomizing steam, lb/10,000 Btu
$MQGSB$ = gas from the sorbent, lb/10,000 Btu
$MQWSB$ = water from the sorbent, lb/10,000 Btu

The total moisture in the flue gas, $MQWG$, is the sum of the water from fuel, water in air and, if applicable, additional water and water from sorbent.

$$MQWG = MQWFF + MQWA + MQWAD + MQWSB \quad \textbf{(35)}$$

Dry flue gas, $MQDG$ in lb/10,000 Btu, is the difference between the wet flue gas and moisture in the flue gas:

$$MQDG = MQG - MQWG \quad \textbf{(36)}$$

The percent moisture in flue gas is a parameter required to determine flue gas energy heat content or enthalpy (see *Enthalpy of air and gas*) and is calculated as follows:

$$MPWG = 100 \times \frac{MQWG}{MQG}, \% \quad \textbf{(37)}$$

For most fuels, the mass of solids, or residue, in the flue gas is insignificant and can be ignored. Even when the quantity is significant, solids do not materially impact the volume flow rate of flue gas. However, solids add to the heat content, or enthalpy, of flue gas and should be accounted for when the ash content of the fuel is greater than 0.15 lb/10,000 Btu or when sorbent is used.

The mass of residue from fuel, $MQRF$ in lb/10,000 Btu, is calculated from the following equation:

$$MQRF = \left(AF + UBC\right) \times \frac{100}{HHV} \quad \textbf{(38)}$$

where

$MQRF$ = residue from fuel, lb/10,000 Btu
AF = mass percent ash in fuel, %
UBC = unburned carbon as mass percent in fuel, %

The mass percent of solids or residue in the flue gas is then:

$$MPRG = 100 \times \frac{MQRF + MQSSB}{MQG} \quad \textbf{(39)}$$

where

$MPRG$ = mass percent solids or residue in flue gas, %
$MQSSB$ = spent sorbent, lb/10,000 Btu
MQG = mass of gaseous combustion products excluding solids, lb/10,000 Btu

Efficiency

Efficiency is the ratio of energy output to energy input and is usually expressed as a percentage. The output term for a steam generator is the energy absorbed by the working fluid that is not recovered within the steam generator envelope. It includes the energy added to the feedwater and desuperheating water to produce saturated/superheated steam, reheat steam, auxiliary steam and blowdown. It does not include the energy supplied to preheat the entering air such as air preheater coil steam supplied by the steam generator. The energy input term is the maximum energy available when the fuel is completely burned, i.e., the mass flow rate of fuel, MRF, multiplied by the higher heating value of the fuel. This is conventionally expressed as:

$$\eta_f = 100 \times \frac{\text{Output}}{\text{Input fuel}} = 100 \times \frac{\text{Output}}{MRF \times HHV}, \% \quad \textbf{(40)}$$

and is commonly referred to as steam generator fuel efficiency. In the U.S., it is customary to express steam generator efficiency on a higher heating value basis. Steam generator efficiency may also be expressed on a lower heating value basis (common in Europe). For the same mass flow rate of fuel, the LHV efficiency may be 3 to 10 percentage points higher than the HHV efficiency, depending upon the amount of H_2 and H_2O in the fuel. When comparing steam generator efficiency and/or plant heat rate, they must be on the same basis, i.e., HHV or LHV.

Efficiency may be determined by measuring the mass flow rate of fuel and steam generator output, which is referred to as the input-output method, or by the energy balance method. The energy balance method is generally the preferred method. It is usually more accurate than the input-output method and is discussed below.

According to the law of conservation of energy, for steady-state conditions, the energy balance on the steam generator envelope can be expressed as:[5]

$$QRF = QRO + QHB, \text{Btu/h} \tag{41}$$

where QRF is the input from fuel, Btu/h, QRO is the steam generator output, Btu/h, and QHB is the energy required by heat balance for closure, Btu/h. The heat balance energy associated with the streams entering the steam generator envelope and the energy added from auxiliary equipment power are commonly referred to as heat credits, QRB (Btu/h). The heat balance energy associated with streams leaving the steam generator and the heat lost to the environment are commonly referred to as heat losses, QRL (Btu/h). This steam generator energy balance may be written as:

$$\begin{aligned} QRF &= QRO + QHB \\ &= QRO + QRL - QRB, \text{Btu/h} \end{aligned} \tag{42}$$

and the efficiency may be expressed as:

$$\eta_f = 100 \times \frac{QRO}{QRO + QRL - QRB}, \% \tag{43}$$

When losses and credits are expressed as a function of percent input from fuel, QPL and QPB, the efficiency may be calculated from:

$$\eta_f = 100 - QPL + QPB, \% \tag{44}$$

Most losses and credits are conveniently calculated on a percent input from fuel basis. However, some losses are more conveniently calculated on a Btu/h basis. The following expression for efficiency allows the use of mixed units; some of the losses/credits are calculated on a percent basis and some on a Btu/h basis.

$$\begin{aligned} \eta_f &= (100 - QPL + QPB) \\ &\times \left(\frac{QRO}{QRO + QRL - QRB} \right), \% \end{aligned} \tag{45}$$

For a more detailed understanding of losses and credits, refer to the American Society of Mechanical Engineers (ASME) Performance Test Code, PTC 4, for steam generators.

The general form for calculating losses (QPL_k) using the mass per unit of heat input basis to express the percent heat loss for individual constituents is:

$$\begin{aligned} QPL_k &= \frac{MQ_k \times MCP_k \times (TO_k - TR)}{100} \\ &= \frac{MQ_k \times (HO_k - HR_k)}{100}, \% \end{aligned} \tag{46}$$

where

MQ_k = mass of constituent k, lb/10,000 Btu
MCP_k = mean specific heat between TO_k and TR, Btu/lb F
TO_k = outlet temperature, F
TR = reference temperature, F
HO_k = outlet enthalpy, Btu/lb
HR_k = reference enthalpy, Btu/lb

For units with gas to air heat exchangers, there is usually some air leakage from the air inlet to the gas outlet. This leakage lowers the gas temperature leaving the air heater (measured gas temperature) without performing any useful work. It is recommended that the calculated gas temperature leaving the air heater without leakage be used for TO_k above, in accordance with PTC 4 (see Chapter 20 for calculation). For this case, the dry gas weight is based on the excess air entering the air heater. Other codes, including the older PTC 4.1, may use the measured gas temperature leaving the air heater, in which case, the dry gas weight must be based on the excess air leaving the air heater.

The reference temperature for PTC 4 is 77F (25C) and the calculation of both losses and credits are required to determine efficiency. The energy credit will be negative for any stream entering the steam generator envelope at a temperature lower than the reference temperature. The most significant credit is generally the energy in the entering air. The entering air temperature (air temperature entering the boundary) is the air temperature leaving the forced draft fans or leaving the air pre-heater coils (entering an air to gas heat exchanger) if the source of energy (steam) is external to the steam generator. When air pre-heater coils are used and the energy is supplied by steam from the steam generator, the entering air temperature is the air temperature entering the pre-heater coils. The air temperature entering the fan(s) is usually taken as the design ambient condition, but may be some other specified condition such as when the fan inlets are supplied by air from within the building. The fan compression energy (typically 1/2 degree F per 1 in. wg fan pressure rise) may be considered to establish the fan discharge temperature. Some test codes, including the older PTC 4.1, may use some other arbitrary reference temperature or the entering air temperature as the reference temperature. An advantage of using the entering air temperature as the reference temperature is that it eliminates the need to calculate credits for entering air and moisture in air.

The general form for calculating credits (QPB_k) using the mass per unit of input basis to express the quantity of individual constituents is:

$$QPB_k = \frac{MQ_k \times MCP_k \times (TI_k - TR)}{100} = \frac{MQ_k \times (HI_k - HR_k)}{100}, \% \quad \textbf{(47)}$$

where

TI_k = inlet temperature, F
HI_k = inlet enthalpy, Btu/lb

and other terms were defined in Equation 46.

The terms used to calculate losses and credits that are a function of fuel input have been discussed previously. The other losses and credits are described below.

Surface radiation and convection loss

This is the heat lost to the atmosphere from the boiler envelope between the first and the last heat trap (commonly between the steam generator air inlet and the boiler exit or air heater exit). Surfaces include the boiler casing, flues and ducts, piping and other surfaces above ambient temperature as a result of the energy entering the unit. It is a function of the average velocity and the difference between the average surface temperature and average ambient temperature. The U.S. industry and PTC 4 standard for calculating this heat loss use a temperature differential of 50F (for insulated surfaces) and a surface velocity of 100 ft/min. For PTC 4, the heat loss is based on the actual flat projected area of the unit and standard ASME Performance Test Code heat transfer coefficients. For convenience, the American Boiler Manufacturers Association (ABMA) standard radiation loss chart, shown in Chapter 23, may be used for an approximation. The ABMA curve expresses the radiation loss on a percent of gross heat input basis as a function of steam generator output (percent gross heat input may be interpreted as heat input from fuel for most applications). This curve is the basis for the surface radiation and convection loss prior to the release of PTC 4 and is approximately the same as PTC 4 for oil- and gas-fired units. For coal-fired units, due to the requirement for a larger furnace and convection surface area due to the requirement for lower gas velocities, the PTC 4 radiation loss is typically on the order of 2 to 2.5 times greater than the ABMA curve.

Unburned carbon loss

For design of a unit, this is normally estimated based on historical data and/or combustion models. For an efficiency test, this item is calculated from measured unburned carbon in the residue. (See *Unburned carbon.*)

Other losses and manufacturers' margins

When testing a unit, it is usually only economically practical to measure the major losses and credits. The other minor losses (and credits) are estimated or based on historical data. Accordingly, when designing a unit, the individual losses and credits to be tested are itemized separately and the estimated losses (and credits) are grouped together and referred to as *Other* losses and credits (also referred to as *Unaccounted for* or *Unmeasured* losses). The most typical Other losses are CO (0.05% loss for 145 ppm or 0.12 lb/10^6 Btu), NO_x (0.01% for 50 ppm or 0.07 lb/10^6 Btu), radiation to the furnace ash pit (0.03% loss for a typical radiation rate of 10,000 Btu/ft^2 h), pulverizer rejects (0.02% loss for a reject rate of 0.25% of fuel flow at a higher heating value of 1000 Btu/lb and 170F/77C mill outlet temperature), and unburned hydrocarbons/VOCs (normally negligible and assumed to be zero). In addition, the manufacturer normally adds a margin, or safety factor, to the losses to account for unexpected performance deviations and test measurement uncertainty. Typical design values for these margins are 0 to 0.5% of heat input for gas, oil and coals with good combustion characteristics and slagging/fouling properties to 0 to 1.5% of heat input or higher for fuels with poor combustion characteristics and poor slagging/fouling characteristics. In the evaluation of actual unit efficiency, the minor or Other losses that are not measured should be estimated and agreed to.

Enthalpy

Enthalpy of air and gas

Enthalpy, *H*, in Btu/lb is an indication of the relative energy level of a material at a specific temperature and pressure. It is used in thermal efficiency, heat loss, heat balance and heat transfer calculations (see Chapter 2). Extensive tabulated and graphical data are available such as the ASME Steam Tables summarized in Chapter 2. Except for steam and water at high pressure, the pressure effect on enthalpy is negligible for engineering purposes.

Enthalpies of most gases used in combustion calculations can be curve-fitted by the simple second order equation:

$$H = aT^2 + bT + c \quad \textbf{(48)}$$

where

H = enthalpy in Btu/lb
T = temperature in degrees, F

To determine the enthalpy of most gases used in combustion calculations at a temperature, *T*, Equation 48 can be used with the coefficients summarized in Table 12. Reference 6 is the source for the properties and the curve fits are in accordance with Reference 7. The curve fits are within plus or minus 0.2 Btu/lb for enthalpies less than 40 Btu/lb and within plus or minus 0.5% for larger values. If the enthalpy of a fluid is known, the temperature in degrees F can be evaluated from the quadratic equation:

$$T = \frac{-b + \sqrt{b^2 - 4a(c - H)}}{2a} \quad \textbf{(49)}$$

For mixtures of gases, such as dry air and water vapor or flue gas and water vapor, Equation 48 coefficient, *a*, *b* and *c* can be determined by a simple mass average:

$$n_{mix} = \sum x_i \, n_i \qquad \textbf{(50)}$$

where

n_{mix} = equivalent coefficient a, b or c of the mixture
x_i = mass fraction of constituent i
n_i = coefficient a, b or c for constituent i

For convenience, Table 12 lists coefficients for a number of gas mixtures including standard wet air with 0.013 lb H_2O per lb dry air. In addition, Figs. 3 and 4 provide graphical representations of flue gas and standard air enthalpy.

Another method of evaluating the change in specific enthalpy of a substance between conditions 1 and 2 is to consider the specific heat and temperature difference:

$$H_2 - H_1 = c_p \left(T_2 - T_1\right) \qquad \textbf{(51)}$$

where

H = enthalpy, Btu/lb
c_p = specific heat at constant pressure, Btu/lb F
T = temperature, F

Enthalpy of solids and fuels

Enthalpy of coal, limestone and oil can be evaluated from the following relationships:

Coal:[8]

$$H = \left[\left(1 - W_F\right)\left(0.217 + 0.00248\,VM\right) + W_F\right]\left(T - 77\right) \qquad \textbf{(52)}$$

Limestone:

$$H = \left[\left(1 - W_F\right) H_{LS} + W_F\right]\left(T - 77\right) \qquad \textbf{(53)}$$

and

$$H_{LS} = \left(0.179\,T\right) + \left(0.1128 \times 10^{-3}\,T^2\right) - 14.45 \qquad \textbf{(54)}$$

Oil:[9]

$$H = C_1 + C_2\left(API\right) + C_3 T + C_4\left(API\right) T + \left[C_5 + C_6\left(API\right)\right] T^2 \qquad \textbf{(55)}$$

and

$$API = \left(141.5 - 131.5\,SPGR\right) / \,SPGR \qquad \textbf{(56)}$$

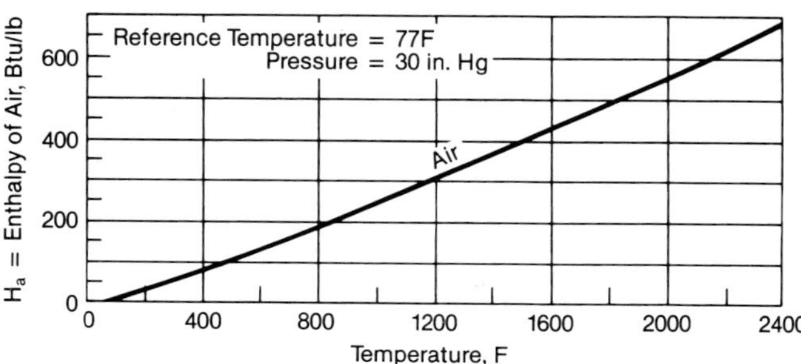

Fig. 4 Enthalpy of air assuming 0.987 mass fraction dry air plus 0.013 mass fraction of water vapor.

Table 12
Enthalpy Coefficients for Equation 48

Coefficient	a	b	c
Dry air (a)			
0 to 500	8.299003E-06	0.2383802	−18.43552
500 to 1500	1.474577E-05	0.2332470	−17.48061
1500 to 2500	8.137865E-06	0.2526050	−31.64983
2500 to 4000	4.164187E-06	0.2726073	−56.82009
Wet air (b)			
0 to 500	8.577272E-06	0.2409682	−18.63678
500 to 1500	1.514376E-05	0.2357032	−17.64590
1500 to 2500	8.539973E-06	0.2551066	−31.89248
2500 to 4000	4.420080E-06	0.2758523	−58.00740
Water vapor			
0 to 500	2.998261E-05	0.4400434	−34.11883
500 to 1500	4.575975E-05	0.4246434	−30.36311
1500 to 2500	3.947132E-05	0.4475365	−50.55380
2500 to 4000	2.413208E-05	0.5252888	−149.06430
Dry flue gas (c)			
0 to 500	1.682949E-05	0.2327271	−18.03014
500 to 1500	1.725460E-05	0.2336275	−18.58662
1500 to 2500	8.957486E-06	0.2578250	−36.21436
2500 to 4000	4.123110E-06	0.2821454	−66.80051
Dry turbine exhaust gas (d)			
0 to 500	1.157682E-05	0.2369243	−18.35542
500 to 1500	1.553788E-05	0.2343280	−18.04780
1500 to 2500	8.510000E-06	0.2550950	−33.38583
2500 to 4000	4.168439E-06	0.2768102	−60.53935
Ash/SiO_2			
0 to 500	7.735829E-05	0.1702036	−13.36106
500 to 1500	2.408712E-05	0.2358873	−32.88512
1500 to 2500	1.394202E-05	0.2324186	−4.85559
2500 to 4000	1.084199E-05	0.2460190	−19.48141
N_{2a} – Atmospheric nitrogen (e)			
0 to 500	5.484935E-06	0.2450592	−18.93320
500 to 1500	1.496168E-05	0.2362762	−16.91089
1500 to 2500	8.654128E-06	0.2552508	−31.18079
2500 to 4000	3.953408E-06	0.2789019	−60.92904
O_2 – Oxygen			
0 to 500	1.764672E-05	0.2162331	−16.78533
500 to 1500	1.403084E-05	0.2232213	−19.37546
1500 to 2500	6.424422E-06	0.2438557	−33.21262
2500 to 4000	4.864890E-06	0.2517422	−43.18179
CO_2 – Carbon dioxide			
0 to 500	5.544506E-05	0.1943114	−15.23170
500 to 1500	2.560224E-05	0.2270060	−24.11829
1500 to 2500	1.045045E-05	0.2695022	−53.77107
2500 to 4000	4.595554E-06	0.2989397	−90.77172
SO_2 – Sulfur dioxide			
0 to 500	3.420275E-05	0.1439724	−11.25959
500 to 1500	1.366242E-05	0.1672132	−17.74491
1500 to 2500	4.470094E-06	0.1923931	−34.83202
2500 to 4000	2.012353E-06	0.2047152	−50.27639
CO – Carbon monoxide			
0 to 500	5.544506E-05	0.1943114	−15.23170
500 to 1500	2.559673E-05	0.2269866	−24.10722
1500 to 2500	1.044809E-05	0.2695040	−53.79888
2500 to 4000	4.630355E-06	0.2987122	−90.45853

Notes:
(a) Dry air composed of 20.946% O_2, 78.105% N_2, 0.916% Ar and 0.033% CO_2 by volume.
(b) Wet air contains 0.013 lb H_2O/lb dry air.
(c) Dry gas composed of 3.5% O_2, 15.3% CO_2, 0.1% SO_2 and 81.1% N_{2a} by volume.
(d) Dry turbine exhaust gas (TEG) composed of 11.48% O_2, 5.27% CO_2 and 83.25% N_{2a} by volume (natural gas with 110% excess air). See PTC 4.4 for a rigorous determination of TEG enthalpy.
(e) N_{2a} composed of the atomic nitrogen, Ar and CO_2 in standard air.

Source: JANAF Thermochemical Tables, 2nd Ed., NSRDS-NBS 37, 1971. Curve fits developed from NASA SP-273, 1971 correlations.

where

H = enthalpy of coal, limestone or oil at T, Btu/lb
W_F = mass fraction free moisture in coal or limestone, lb/lb
VM = volatile matter on a moisture and ash free basis, %
T = temperature, F
H_{LS} = enthalpy of dry limestone, Btu/lb
API = degrees API
$SPGR$ = specific gravity, dimensionless
= density in lb/ft^3 divided by 62.4 at 60F
C_1 = −30.016
C_2 = −0.11426
C_3 = 0.373
C_4 = 0.143×10^{-2}
C_5 = 0.2184×10^{-3}
C_6 = 7.0×10^{-7}

Measurement of excess air

One of the most critical operating parameters for attaining good combustion is excess air. Too little air can be a source of excessive unburned combustibles and can be a safety hazard. Too much excess air increases stack gas losses.

Flue gas analysis

The major constituents in flue gas are CO_2, O_2, N_2 and H_2O. Excess air is determined by measuring the O_2 and CO_2 contents of the flue gas. Before proceeding with measuring techniques, consider the form of the sample. A flue gas sample may be obtained on a wet or dry basis. When a sample is extracted from the gas stream, the water vapor normally condenses and the sample is considered to be on a dry basis. The sample is usually drawn through water near ambient temperature to ensure that it is dry. The major constituents of a dry sample do not include the water vapor in the flue gas. When the gas is measured with an in situ analyzer or when precautions are taken to keep the moisture in the sample from condensing, the sample is on a wet basis.

The amount of O_2 in the flue gas is significant in defining the status of the combustion process. Its presence always means that more oxygen (excess air) is being introduced than is being used. Assuming complete combustion, low values of O_2 reflect moderate excess air and normal heat losses to the stack, while higher values of O_2 mean needlessly higher stack losses.

The quantity of excess O_2 is very significant since it is a nearly exact indication of excess air. Fig. 5 is a dry flue gas volumetric combustion chart that is universally used in field testing; it relates O_2, CO_2 and N_{2a} (by difference). For complete uniform combustion of a specific fuel, all points should lie along a straight line drawn through the pivot point. This line is referred to as the combustion line. The combustion line should be determined by calculating the CO_2 content at zero O_2 for the test fuel (see Table 15). Lines indicating constant excess air have been superimposed on the volumetric combustion chart. Note that excess air is essentially constant for a given O_2 level over a wide range of fuels. The O_2 is an equally constant indication of excess air when the gas is sampled on a wet or in situ basis because the calculated excess air result is insensitive to variations in moisture for specific types/sources of fuel.

The current industry standard for boiler operation is continuous monitoring of O_2 in the flue gas with in situ analyzers that measure oxygen on a wet basis.

For testing, the preferred instrument is an electronic oxygen analyzer. The Orsat unit, which measures ($CO_2 + SO_2$) and O_2 on a dry volumetric basis, remains a trusted standard for verifying the performance of electronic equipment. The Orsat uses chemicals to absorb the ($CO_2 + SO_2$) and O_2, and the amount of each is determined by the reduction in volume from the original flue gas sample. When an Orsat is used, the dry flue gas volumetric combustion chart should be used to plot the results. Valid results for any test with a consistent fuel should fall on a single combustion line (plus or minus 0.2 points of O_2/CO_2 is a reasonable tolerance). The Orsat has several disadvantages. It lacks the accuracy of more refined devices, an experienced operator is required, there are a limited number of readings available in a test, and the results do not lend themselves to electronic recording. Electronic CO_2 analyzers may be used in addition to oxygen analyzers to relate the O_2/CO_2 results to the fuel line on the volumetric combustion chart. When CO_2 is measured, by Orsat or a separate electronic analyzer, it is best to calculate excess air based on the O_2 result due to the insensitivity of excess air versus O_2 results in the fuel analysis.

Depending upon whether O_2 is measured or excess air known, the corresponding excess air, O_2, CO_2 and SO_2 can be calculated using procedures provided in ASME Performance Test Code 4, *Steam Generators.*[5] The calculations are summarized in Table 15 at the end of this chapter in the *Combustion calculations – examples* section.

Flue gas sampling

To ensure a representative average gas sample, samples from a number of equal area points should be taken. Reference the U.S. Environmental Protection Agency (EPA) Method 1 standards and ASME Performance Test Code PTC 19.10. For normal performance testing, equal areas of approximately 9 ft^2 (0.8 m^2) up to 24 points per flue are adequate.

For continuous monitoring, the number of sampling points is an economic consideration. Strategies for locating permanent monitoring probes should include point by point testing with different burner combinations. As a guideline, four probes per flue located at quarter points have been used successfully on large pulverized coal-fired installations.

Testing heterogeneous fuels

When evaluating the performance of a steam generator firing a heterogeneous fuel such as municipal solid waste (MSW) (see Chapter 29), it is generally not possible to obtain a representative fuel sample. Waste fuel composition may vary widely between samples and is usually not repeatable.

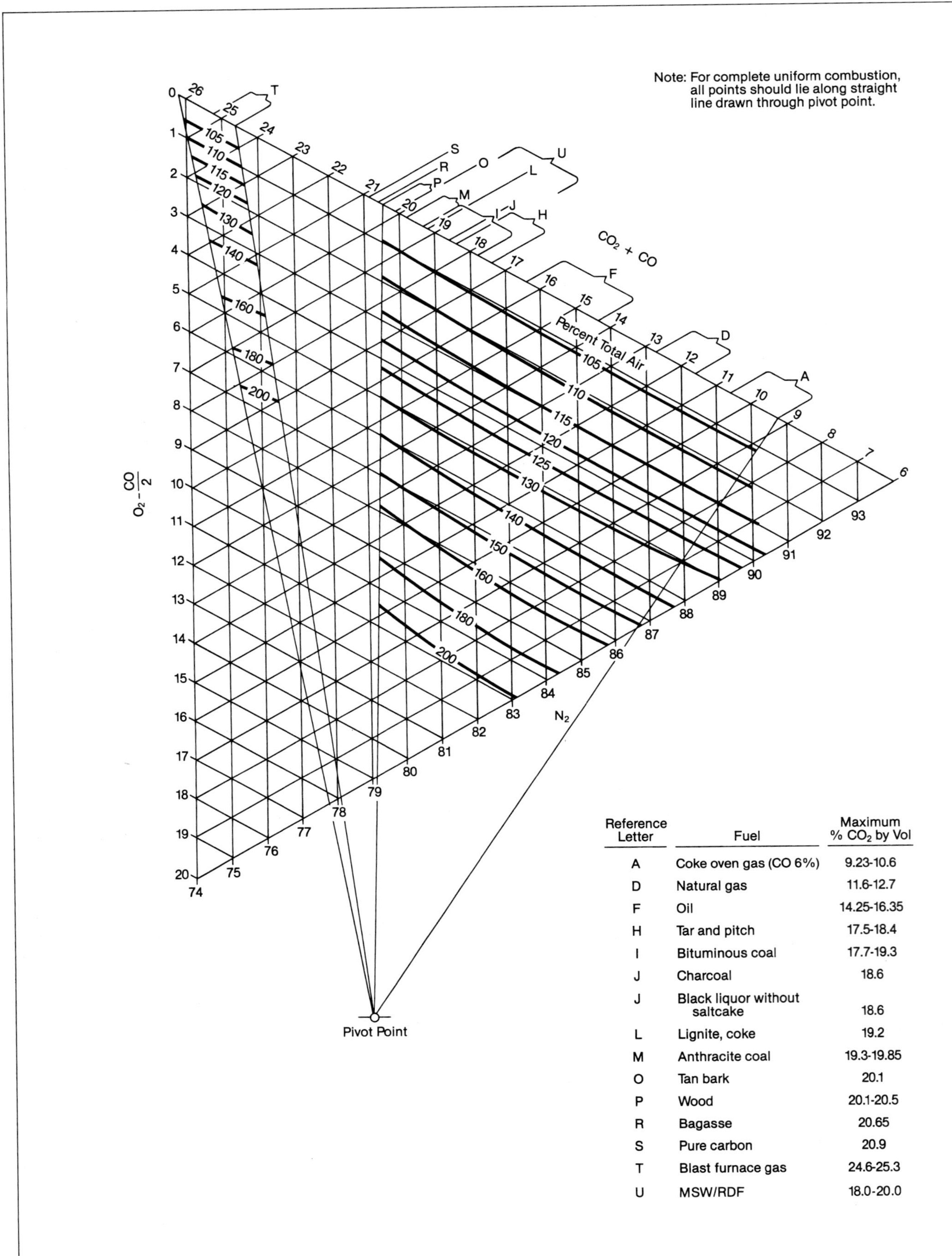

Reference Letter	Fuel	Maximum % CO_2 by Vol
A	Coke oven gas (CO 6%)	9.23-10.6
D	Natural gas	11.6-12.7
F	Oil	14.25-16.35
H	Tar and pitch	17.5-18.4
I	Bituminous coal	17.7-19.3
J	Charcoal	18.6
J	Black liquor without saltcake	18.6
L	Lignite, coke	19.2
M	Anthracite coal	19.3-19.85
O	Tan bark	20.1
P	Wood	20.1-20.5
R	Bagasse	20.65
S	Pure carbon	20.9
T	Blast furnace gas	24.6-25.3
U	MSW/RDF	18.0-20.0

Fig. 5 Dry flue gas volumetric combustion chart.

For boiler design, an ultimate analysis for an average fuel and a range of the most significant components, such as moisture and ash, are used. Therefore, the design calculations are the same as those for homogeneous fuels.

When firing a heterogeneous fuel, the current industry practice used to evaluate average fuel properties and determine boiler efficiency is to test using the boiler as a calorimeter (BAC). The BAC method features the same principles for determining efficiency as those used when the fuel analysis is known. The significant difference is that the mass/volume flow rate of flue gas and moisture in the flue gas are measured directly rather than being calculated based upon the measured fuel analysis and O_2 in the flue gas.

The additional measurements that are required for the BAC test method versus conventional test methods are flue gas flow, moisture in flue gas, O_2 and CO_2 in the flue gas, and residue mass flow rates from the major extraction points.

BAC calculation method

This section describes how to calculate excess air, dry gas weight and water from fuel (water evaporated). The results are on a mass per unit of time basis and losses and credits, therefore, are calculated as Btu/h. Refer to the basic efficiency equations for application. (See Equations 42 and 43.)

The wet gas weight and water in the wet gas are measured. The dry gas weight is then calculated as the difference of the two.

The composition of flue gas is determined by measuring O_2 and CO_2. N_{2a} is determined by difference from 100%. The nitrogen in flue gas is considered to be atmospheric with a molecular weight of 28.158 lb/mole. Because waste fuel combustors operate at high levels of excess air and the nitrogen in the fuel is small, this nitrogen can be ignored.

The moisture in the flue gas may be from vapor or liquid sources. Vapor sources include moisture in the air and atomizing steam. Water sources are moisture in the fuel, moisture formed by combustion of H_2, water from ash quenching systems, and fuel pit water spray. The moisture in air and that from other vaporous sources must be measured, so the sensible heat efficiency loss may be differentiated from the water evaporated loss. The water evaporated is the total moisture in the flue gas less the vaporous sources. The water evaporation loss is calculated in the same manner as the water from fuel loss and is analogous to the total water from fuel loss if miscellaneous water sources are accounted for.

The total dry air flow at the point of flue gas measurement is calculated from the nitrogen in the flue gas. Excess air is determined from the measured O_2 and theoretical air is calculated by difference from the total air flow. The percent excess air is calculated from the excess air and theoretical air weight flow rates.

Combustion calculations – examples

The detailed steps in the solution of combustion problems are best illustrated by examples. The examples in this section are presented through calculation forms which are a convenient method for organizing the calculations in a logical sequence. The input required to complete the forms is located at the top of the form. An elemental fuel analysis on a mass basis is used for all of the examples. For gaseous fuels, the analysis on a volume basis must be converted to an elemental mass basis as described in *Molar evaluation of combustion*. The calculations required are shown as a combination of item numbers (enclosed in brackets) and constants.

Mole method

The mole method is the fundamental basis for all combustion calculations. It is the source for the constants used in other more simplified methods. The only constants the user needs are the molecular weights of the fuel and air constituents. The reader should understand the mole method before proceeding with the Btu method.

Table 7 is an example of the combustion calculations for a bituminous coal on a molar basis. Items 1 through 6 are the required input. If the unburned carbon is known (Item 6), the unburned carbon loss (Item 5) is calculated. Provision is made for entering the excess air to the burners and excess air leaving the boiler if the user desires to account for setting infiltration (Item 1). For this example, the excess air to the burners is assumed to be the same as that leaving the boiler. An intermediate step in the calculations on a molar basis is the volumetric flue gas analysis (Items 20 and 21). Air and gas mass flow rates are shown on a lb/10,000 Btu basis as well as a 1000 lb/h basis.

Btu method

Once the reader understands the principles of the combustion calculations on a mole basis, the Btu method is the preferred method for general combustion calculations. The calculations provided in Table 13A are more comprehensive than the simple calculation of air and gas weights shown in Table 7. Provision is made for handling the impact of sorbent on the combustion calculations, the calculation of efficiency and finally heat available to the furnace.

The inputs to Table 13A are similar to those used in Table 7. The same fuel analysis and excess air are used and the calculated input from fuel is very nearly the same. These inputs are also the same as those used in the example performance problem in Chapter 22. Items 1 through 19 are the inputs and initial calculations required for the combustion calculations. For the efficiency calculations, Items 44 through 46 must be provided. If sorbent is used, Table 14, *Combustion Calculations – Sorbent*, must be completed first. (See Items 11 through 14 and 46). Because the entering air temperature and fuel temperature are the same as the reference temperature selected (80F), the efficiency credits are zero. The total fuel heat is calculated from the efficiency, Item 53, and steam generator output, Item 10. Flue gas and air flow rates are calculated from the fuel input and the results of the combustion gas calculations.

Table 13A shows the calculation results for a typical eastern U.S. coal. A similar set of calculations can be made for a typical western subbituminous coal which has an increased moisture content (30% by weight) and reduced LHV (8360 Btu/lb).

For the same boiler rating and other boundary conditions, the results can be compared on a lb per 10,000 Btu basis:

	Eastern Bit.	Western Subbit.	No. 6 Oil	Natural Gas
Theoretical air	7.572	7.542	7.437	7.220
Dry air	9.086	9.050	8.924	8.664
Dry gas weight	9.442	9.463	8.965	8.194
Wet gas weight	9.864	10.303	9.583	9.236
H_2O in gas	0.422	0.840	0.618	1.042
Efficiency, %	86.91	82.10	85.30	80.79

The theoretical air, dry air and resulting dry gas weight are approximately the same for each coal. The wet gas weight and H_2O in gas are higher for the subbituminous coal due to the higher moisture content. Referring to the efficiency calculations and losses, the efficiency is lower for the subbituminous coal essentially due to the higher moisture content, not the lower heating value. However, if the actual mass flow rates for subbituminous coal versus an eastern bituminous coal are compared, it will be found that a higher air weight is required primarily due to the lower efficiency, while a higher gas weight is required due to the higher moisture in the fuel and the lower efficiency.

Table 13B is the same example as shown in Table 13A except that it is assumed that a limestone sorbent is used in a fluidized bed at a calcium to sulfur molar ratio of 2.5. A sulfur capture of 90% is expected. A higher unburned carbon loss is used, typical of this combustion process. It is necessary to complete the calculations shown in Table 14 to develop input for this Table. The net losses due to sorbent, Item 46 in Table 13B, are not overly significant. Therefore, the difference in efficiency from the example in Table 13A is primarily due to the difference in the assumed unburned carbon loss.

When testing a boiler, the excess air required for the combustion calculations is determined from measured O_2 in the flue gas. Table 15A, *Excess Air Calculations from Measured* O_2, demonstrates the calculation of excess air from O_2 on a wet basis. The fuel analysis and unburned carbon are the same as in Tables 7 and 13A. These tables can also be used to determine the volumetric composition of wet or dry flue gas when excess air is known (Items 25 through 32). These values can be compared to the flue gas composition calculated on a molar basis, Table 7.

Table 15B is an example of calculating excess air from O_2 when a sorbent is used. All of the sulfur in the fuel will not be converted to sulfur dioxide. Therefore, the sulfur capture must first be determined from Table 16, *Sulfur Capture Based on Gas Analysis*. The example presented in Tables 13B and 14 is used as the basis for this example. The flue gas composition in Tables 13A and 13B can be compared to assess the impact of adding the sorbent.

When units firing municipal solid waste or refuse-derived fuels are tested, it is not practical to determine the ultimate analysis of the fuel. Table 17, *Combustion Calculations – Measured Gas Weight*, shows the combustion calculations for determining dry gas weight, water evaporated and excess air using measured gas weight.

References

1. American Gas Association, Segeler, C.G., Ed., *Gas Engineers Handbook,* Industrial Press, Inc., New York, New York, 1965.

2. "Standard Practice for Calculating Heat Value, Compressibility Factor, and Relative Density of Gaseous Fuel," ASTM 3588-98, Annual Book of ASTM Standards, Vol. 05.06, September, 2003.

3. Jones, F.E., "The Air Density Equation and the Transfer of Mass Unit," *Journal of Research of the National Bureau of Standards,* Vol. 83, No. 5, September-October, 1978.

4. Gerhart, P.M., Heil, T.C., and Phillips, J.T., "Steam Generator Performance Calculation Strategies for ASME PTC 4," Technical Paper 91-JPGC-PTC-1, American Society of Mechanical Engineers, New York, New York, October, 1991.

5. Entwistle, J., Heil, T.C., and Hoffman, G.E., "Steam Generation Efficiency Revisited," Technical Paper 88-JPGCLPTC-3, American Society of Mechanical Engineers, New York, New York, September, 1988.

6. *JANAF Thermochemical Tables,* Second Ed., Publication NSRDS-NBS 37, United States National Bureau of Standards (now National Institute of Standards and Technology), Washington, D.C., 1971.

7. Taken from *NASA Publication SP-273,* Chemical Equilibrium Code, 1971.

8. Elliot, M.A., Ed., *Chemistry of Coal Utilization,* Second Supplemental Volume, Wiley, New York, New York, 1981.

9. Dunstan, A.E., *The Science of Petroleum,* Oxford University Press, Oxford, United Kingdom, 1938.

Bibliography

ASME Steam Properties for Industrial Use, Based on IAPWS-IF97, Professional Version 1.1, The American Society of Mechanical Engineers, New York, New York, 2003.

International Boiler & Pressure Vessel Code, "ASME Performance Test Code PTC4," The American Society of Mechanical Engineers, New York, New York, 2004.

Parry, W.T., et al., *ASME International Steam Tables for Industrial Use*, Based on IAPWS-IF97, The American Society of Mechanical Engineers, New York, New York, January, 2000.

See the following pages for Tables 13 through 17.

Table 13A
Combustion Calculations (Efficiency per PTC 4.1) – Btu Method

	INPUT CONDITIONS – BY TEST OR SPECIFICATION		
1	Excess air: at burners; leaving boiler/econ/entering AH, % by wt.		20/20
2	Entering air temperature, F		80
3	Reference temperature, F (tRA = 77 for PTC 4)		80
4	Fuel temperature, F		80
5	Air temperature leaving air heater, F		350
6	Flue gas temperature leaving (excluding leakage), F		390
7	Moisture in air, lb/lb dry air		0.013
8	Additional moisture, lb/100 lb fuel		0
9	Residue leaving boiler/econ/entering AH, % Total		85
10	Output, 1,000,000 Btu/h		285.5
	Corrections for sorbent (if used)		
11	Sulfur capture, lbm/lbm sulfur	Table 16, Item [24]	0
12	CO_2 from sorbent, lb/10,000 Btu	Table 14, Item [19]	0
13	H_2O from sorbent, lb/10,000 Btu	Table 14, Item [20]	0
14	Spent sorbent, lb/10,000 Btu	Table 14, Item [24]	0

FUEL – *Bituminous coal, Virginia: no sorbent*

	15 Ultimate Analysis		16 Theo Air, lb/100 lb fuel		17 H_2O, lb/100 lb fuel	
	Constituent	% by weight	K1	[15] x K1	K2	[15] x K2
A	C	80.31	11.51	924.4		
B	S	1.54	4.31	6.6		
C	H_2	4.47	34.29	153.3	8.94	39.96
D	H_2O	2.90			1.00	2.90
E	N_2	1.38				
F	O_2	2.85	–4.32	–12.3		
G	Ash	6.55				
H	Total	100.00	Air	1072.0	H_2O	42.86

18	Higher heating value (HHV), Btu/lb		14,100
19	Unburned carbon loss, % fuel input		0.40
20	Theoretical air, lb/10,000 Btu	[16H] x 100 / [18]	7.603
21	Unburned carbon, % of fuel	[19] x [18] / 14,500	0.39

COMBUSTION GAS CALCULATIONS, Quantity per 10,000 Btu Fuel Input

22	Theoretical air (corrected), lb/10,000 Btu	[20] – [21] x 1151 / [18] + [11] x [15B] x 216 / [18]						7.571
23	Residue from fuel, lb/10,000 Btu	([15G] + [21]) x 100 / [18]					0.049	
24	Total residue, lb/10,000 Btu	[23] + [14]						0.049
			A At Burners	B Infiltration	C Leaving Furnace		D Leaving Blr/Econ/Entering AH	
25	Excess air, % weight		20.0	0.0		20.0		20.0
26	Dry air, lb/10,000 Btu	(1 + [25] / 100) x [22]				9.085		9.085
27	H_2O from air, lb/10,000 Btu	[26] x [7]			0.118	0.118	0.118	0.118
28	Additional moisture, lb/10,000 Btu	[8] x 100 / [18]			0.000	0.000	0.000	0.000
29	H_2O from fuel, lb/10,000 Btu	[17H] x 100 / [18]			0.304		0.304	
30	Wet gas from fuel, lb/10,000 Btu	(100 – [15G] – [21] – [11] x [15B]) x 100 / [18]				0.660		0.660
31	CO_2 from sorbent, lb/10,000 Btu	[12]				0.000		0.000
32	H_2O from sorbent, lb/10,000 Btu	[13]			0.000	0.000	0.000	0.000
33	Total wet gas, lb/10,000 Btu	Summation [26] through [32]				9.863		9.863
34	Water in wet gas, lb/10,000 Btu	Summation [27] + [28] + [29] + [32]			0.422	0.422	0.422	0.422
35	Dry gas, lb/10,000 Btu	[33] – [34]				9.441		9.441
36	H_2O in gas, % by weight	100 x [34] / [33]				4.28		4.28
37	Residue, % by weight (zero if < 0.15 lbm/10KB)	[9] x [24] / [33]				0.00		0.00

EFFICIENCY CALCULATIONS, % Input from Fuel

	Losses				
38	Dry gas, %		[35D] x (HFg[6] – HFg[3]) / 100	9.441 x (75.3 – 0.7) / 100	7.04
39	Water from fuel, as fired %	Enthalpy of steam at 1 psia, T = [6]	H_1 = (3.958E – 5 x T + 0.4329) x T + 1062.2	1237.1	
40		Enthalpy of water at T = [3]	H_2 = [3] – 32	48.0	
41			[29] x ([39] – [40]) / 100	0.304 x 1189.1 / 100	3.61
42	Moisture in air, %		[27D] x (HWv[6] – HWv[3]) / 100	0.118 x (142.0 – 1.3) / 100	0.17
43	Unburned carbon, %		[19] or [21] x 14,500 / [18]	0.39 x 14,500 / 14,100	0.40
44	Surface radiation and convection		See surface radiation and convection loss		0.40
45	Other, % (include manufacturers margin if applicable)				1.50
46	Sensible heat of residue, % (PTC 4)		[24] x (100 – [9]) x 516 + [9] x HRs[6] / 10,000	HRs[6] = 0 (or Table 14, Item [40])	0.00
47	Sorbent net losses, % if sorbent used		From Table 14, Items ([30] – [31] + [37])		0.00
48	Summation of losses, %		Summation [38] through [46]		13.12
	Credits				
49	Entering dry air, %		[26D] x (HDA[2] – HDA[3]) / 100	9.085 x (0.7 – 0.7) / 100	0.00
50	Moisture in entering air, %		[27D] x (HWv[2] – HWv[3]) / 100	0.118 x (1.3 – 1.3) / 100	0.00
51	Sensible heat in fuel, %		100 x (HF[4] – HF[3]) / [18]	100 x (1.0 – 1.0) / 14,100	0.00
52	Other, %				0.00
53	Summation of credits, %		Summation [48] through [51]		0.00
54	**Efficiency, %**		100 – [48] + [53]		86.88

KEY PERFORMANCE PARAMETERS

			Leaving Furnace	Leaving Blr/Econ/Entering AH
55	Input from fuel, 1,000,000 Btu/h	100 x [10] / [54]		328.6
56	Fuel rate, 1000 lb/h	1000 x [55] / [18]		23.3
57	Wet gas weight, 1000 lb/h	[55] x [33] / 10	324.1	324.1
58	Air to burners (wet), lb/10,000 Btu	(1 + [7]) x (1 + [25A] / 100) x [22]	9.203	
59	Air to burners (wet), 1000 lb/h	[55] x [58] / 10	302.4	
60	Heat available, 1,000,000 Btu/h	[55] x {([18] – 10.30 x [17H]) / [18] – 0.005		
	Ha = 66.0 Btu/lb	x ([44] + [45]) + Ha[5] x [58] / 10,000}	335.2	
61	Heat available/lb wet gas, Btu/lb	1000 x [60] / [57]	1034.2	
62	Adiabatic flame temperature, F	From Fig. 3 at H = [61], % H_2O = [36C]	3560	

Table 13B
Combustion Calculations (Efficiency per PTC 4) – Btu Method (with Sorbent)

INPUT CONDITIONS – BY TEST OR SPECIFICATION

No.	Item		Value
1	Excess air: at burners; leaving boiler/econ/entering AH, % by wt.		18/20
2	Entering air temperature, F		80
3	Reference temperature, F (tRA = 77 for PTC 4)		77
4	Fuel temperature, F		80
5	Air temperature leaving air heater, F		350
6	Flue gas temperature leaving (excluding leakage), F		390
7	Moisture in air, lb/lb dry air		0.013
8	Additional moisture, lb/100 lb fuel		0
9	Residue leaving boiler/econ/entering AH, % Total		90
10	Output, 1,000,000 Btu/h		285.5
	Corrections for sorbent (if used)		
11	Sulfur capture, lbm/lbm sulfur	Table 16, Item [24]	0.9000
12	CO_2 from sorbent, lb/10,000 Btu	Table 14, Item [19]	0.0362
13	H_2O from sorbent, lb/10,000 Btu	Table 14, Item [20]	0.0015
14	Spent sorbent, lb/10,000 Btu	Table 14, Item [24]	0.0819

FUEL – *Bituminous coal, Virginia: with sorbent*

	15 Ultimate Analysis		16 Theo Air, lb/100 lb fuel		17 H_2O, lb/100 lb fuel	
	Constituent	% by weight	K1	[15] x K1	K2	[15] x K2
A	C	80.31	11.51	924.4		
B	S	1.54	4.31	6.6		
C	H_2	4.47	34.29	153.3	8.94	39.96
D	H_2O	2.90			1.00	2.90
E	N_2	1.38				
F	O_2	2.85	–4.32	–12.3		
G	Ash	6.55				
H	Total	100.00	Air	1072.0	H_2O	42.86

No.	Item	Formula	Value
18	Higher heating value (HHV), Btu/lb		14,100
19	Unburned carbon loss, % fuel input		2.50
20	Theoretical air, lb/10,000 Btu	[16H] x 100 / [18]	7.603
21	Unburned carbon, % of fuel	[19] x [18] / 14,500	2.43

COMBUSTION GAS CALCULATIONS, Quantity per 10,000 Btu Fuel Input

No.	Item	Formula	Value
22	Theoretical air (corrected), lb/10,000 Btu	[20] – [21] x 1151 / [18] + [11] x [15B] x 216 / [18]	7.426
23	Residue from fuel, lb/10,000 Btu	([15G] + [21]) x 100 / [18]	0.064
24	Total residue, lb/10,000 Btu	[23] + [14]	0.146

No.	Item	Formula	A At Burners	B Infiltration	C Leaving Furnace		D Leaving Blr/Econ/Entering AH	
25	Excess air, % weight		18.0	1.0		19.0		20.0
26	Dry air, lb/10,000 Btu	(1 + [25] / 100) x [22]				8.837		8.911
27	H_2O from air, lb/10,000 Btu	[26] x [7]			0.115	0.115	0.116	0.116
28	Additional moisture, lb/10,000 Btu	[8] x 100 / [18]			0.000	0.000	0.000	0.000
29	H_2O from fuel, lb/10,000 Btu	[17H] x 100 / [18]			0.304		0.304	
30	Wet gas from fuel, lb/10,000 Btu	(100 – [15G] – [21] – [11] x [15B]) x 100 / [18]				0.636		0.636
31	CO_2 from sorbent, lb/10,000 Btu	[12]				0.036		0.036
32	H_2O from sorbent, lb/10,000 Btu	[13]			0.002	0.002	0.002	0.002
33	Total wet gas, lb/10,000 Btu	Summation [26] through [32]				9.626		9.701
34	Water in wet gas, lb/10,000 Btu	Summation [27] + [28] + [29] + [32]			0.421	0.421	0.422	0.422
35	Dry gas, lb/10,000 Btu	[33] – [34]				9.205		9.279
36	H_2O in gas, % by weight	100 x [34] / [33]				4.37		4.35
37	Residue, % by weight (zero if < 0.15 lbm/10KB)	[9] x [24] / [33]				1.37		1.35

EFFICIENCY CALCULATIONS, % Input from Fuel

No.	Item	Formula	Calculation	Value
	Losses			
38	Dry gas, %	[35D] x (HFg[6] – HFg[3]) / 100	9.279 x (75.3 – 0.0) / 100	6.99
39	Water from fuel, as fired %: Enthalpy of steam at psia, T = [6]	H_1 = (3.958E – 5 x T + 0.4329) x T + 1062.2	1237.1	
40	Enthalpy of water at T = [3]	H_2 = [3] – 32	45.0	
41		[29] x ([39] – [40]) / 100	0.304 x 1192.1 / 100	3.62
42	Moisture in air, %	[27D] x (HWv[6] – HWv[3]) / 100	0.116 x (142.0 – 0.0) / 100	0.16
43	Unburned carbon, %	[19] or [21] x 14,500 / [18]	2.43 x 14,500 / 14,100	2.50
44	Surface radiation and convection	See surface radiation and convection loss		0.40
45	Other, % (include manufacturers margin if applicable)			1.50
46	Sensible heat of residue, % (PTC 4)	[24] x (100 – [9]) x 516 + [9] x HRs[6] / 10,000	HRs[6] = 65.1 (or Table 14, Item [40])	0.15
47	Sorbent net losses, % if sorbent used	From Table 14, Items ([30] – [31] + [37])		–0.03
48	Summation of losses, %	Summation [38] through [46]		15.45
	Credits			
49	Entering dry air, %	[26D] x (HDA[2] – HDA[3]) / 100	8.911 x (0.7 – 0.0) / 100	0.06
50	Moisture in entering air, %	[27D] x (HWv[2] – HWv[3]) / 100	0.116 x (1.3 – 0.0) / 100	0.00
51	Sensible heat in fuel, %	100 x (HF[4] – HF[3]) / [18]	100 x (1.0 – 0.0) / 14,100	0.01
52	Other, %			0.00
53	Summation of credits, %	Summation [49] through [51]		0.07
54	**Efficiency, %**	100 – [48] + [53]		84.78

KEY PERFORMANCE PARAMETERS

No.	Item	Formula	Leaving Furnace	Leaving Blr/Econ/Entering AH
55	Input from fuel, 1,000,000 Btu/h	100 x [10] / [54]		336.8
56	Fuel rate, 1000 lb/h	1000 x [55] / [18]		23.9
57	Wet gas weight, 1000 lb/h	[55] x [33] / 10	324.2	326.7
58	Air to burners (wet), lb/10,000 Btu	(1 + [7]) x (1 + [25A] / 100) x [22]	8.877	
59	Air to burners (wet), 1000 lb/h	[55] x [58] / 10	299.0	
60	Heat available, 1,000,000 Btu/h	[55] x {([18] – 10.30 x [17H]) / [18] – 0.005		
	Ha = 66.0 Btu/lb	x ([44] + [45]) + Ha[5] x [58] / 10,000}	342.8	
61	Heat available/lb wet gas, Btu/lb	1000 x [60] / [57]	1057.4	
62	Adiabatic flame temperature, F	From Fig. 3 at H = [61], % H_2O = [36C]	3627	

Table 14
Combustion Calculations – Sorbent

	INPUTS (see also lightly shaded blocks)			FUEL – *Bituminous coal, Virginia*	
1	Sulfur in fuel, % by weight	*1.54*	6	Sulfur capture, lb/lb sulfur	*0.90*
2	Ash in fuel, % by weight	*6.55*	7	Reference temperature, F	*80.0*
3	HHV of fuel, Btu/lb	*14,100*	8	Exit gas temperature (excluding leakage), F	*390.0*
4	Unburned carbon loss, % fuel input	*2.5*	9	Sorbent temperature, F	*80.0*
5	Calcium to sulfur molar ratio	*2.5*			

SORBENT PRODUCTS

		[10] Chemical Analysis % Mass	[11] Molecular Weight lb/mole	[12] Ca mole/100 lb sorb [10] / [11]	[13] Calcination Fraction	[14] Molecular Weight lb/mole	[15] CO_2 lb/100 lb sorb [10]x[13]x[14]/[11]	[16] H_2O lb/100 lb sorb [10]x[13]x[14]/[11]
A	$CaCO_3$	*89.80*	100.089	*0.897*	*0.90*	44.010	*35.529*	
B	$MgCO_3$	*5.00*	84.321		1.00	44.010	*2.610*	
C	$Ca(OH)_2$	*0.00*	74.096	*0.000*	1.00	18.015		*0.000*
D	$Mg(OH)_2$	*0.00*	58.328		1.00	18.015		*0.000*
E	H_2O	*1.60*	18.015		1.00	18.015		*1.600*
F	Inert	*3.60*						
G	Total Ca, mole/100 lb sorbent			*0.897*		Total	*38.139*	*1.600*

SORBENT/GAS CALCULATIONS, lb/10,000 Btu Except as Noted

17	Sorbent, lb/lb fuel	[1] x [5] / [12G] / 32.066	*0.1339*
18	Sorbent, lb/10,000 Btu	10,000 x [17] / [3]	*0.0950*
19	CO_2 from sorbent, lb/10,000 Btu	[15G] x [18] / 100	*0.0362*
20	H_2O from sorbent, lb/10,000 Btu	[16G] x [18] / 100	*0.0015*
21	Additional theoretical air, lb/10,000 Btu	216 x [1] x [6] / [3]	*0.0212*
22	SO_2 reduction, lb/10,000 Btu	200 x [1] x [6] / [3]	*0.0197*
23	SO_3 formed, lb/10,000 Btu	0.2314 x [21] + [22]	*0.0246*
24	Spent sorbent, lb/10,000 Btu	[18] – [19] – [20] + [23]	*0.0819*
25	Unburned carbon, lb/10,000 Btu	[4] x 100 / 14,500	*0.0172*
26	Residue from fuel, lb/10,000 Btu	[2] x 100 / [3] + [25]	*0.0637*
27	Total residue, lb/10,000 Btu	[24] + [26]	*0.1456*

LOSSES DUE TO SORBENT, % Input from Fuel

28	H_2O from sorbent, %	H of steam at 1 psi, T = [8]	H_1 = (3.958E – 5 x [8] + 0.4329) x [8] + 1062.2	*1237.1*	
29		H of water	H_2 = [9] – 32	*48.0*	
30		0.01 x [20] x ([28] – [29])			*0.018*
31	Sensible heat sorbent (dry), %	[18] x (1.0 – [10E] / 100) x (H at T = [9] – H at T = [7]) / 100			
		H of limestone (dry) = (0.1128E – 3 x T + 0.179) x T – 14.45			*0.000*

Calcination/Dehydration, %

32	$CaCO_3$, %	[10A] x [13A] x [18] x 766 / 10,000	*0.588*	
33	$MgCO_3$, %	[10B] x 1.0 x [18] x 652 / 10,000	*0.031*	
34	$Ca(OH)_2$, %	[10C] x 1.0 x [18] x 636 / 10,000	*0.000*	
35	$Mg(OH)_2$, %	[10D] x 1.0 x [18] x 625 / 10,000	*0.000*	
36	Heat gain due to sulfation, %	[6] x [1] x 6733 / [3]	*0.662*	
37	Total of losses due to chemical reactions, %	[32] + [33] + [34] + [35] – [36]		*–0.043*

Sensible Heat of Residue Loss, %

	Location	[38] Temp Residue, F	[39] Mass Flow Rate, % Total	x	[27] lb/10,000 Btu	x	(H at T = [38] Btu/lb	–	H at T = [7] Btu/lb	) / 10,000	=	Loss %
A	Bed drain	*1500*	*10*	x	*0.1456*	x	(376.3	–	0.5	) / 10,000	=	*0.055*
B	Economizer	*600*	*10*	x	*0.1456*	x	(116.2	–	0.5	) / 10,000	=	*0.017*
C	Flyash	*390*	*80*	x	*0.1456*	x	(64.3	–	0.5	) / 10,000	=	*0.074*
	H Residue = ((–2.843E – 8 x T + 1.09E – 4) x T + 0.16) x T – 12.95								[40] Total			*0.146*
41	Summation losses due to sorbent, %						[30] – [31] + [37] + [40]					*0.121*

Table 15A
Excess Air Calculations from Measured O_2
Bituminous coal, Virginia – O_2 on wet basis

	INPUTS (see also lightly shaded blocks)			SORBENT DATA (if applicable)	
1	Moisture in air, lb/lb dry air	*0.013*	6	CO_2 from sorbent, moles/100 lb fuel, Table 16 [17]	*0*
2	Additional moisture, lb/100 lb fuel	*0.00*	7	H_2O from sorbent, moles/100 lb fuel, Table 16 [16]	*0*
3	HHV fuel, Btu/lb	*14,100*	8	Sulfur capture, lb/lb sulfur fuel, Table 16 [24]	*0*
4	Unburned carbon loss, % fuel input	*0.40*			
5	Unburned carbon (UBC), [3] x [4] / 14,500	*0.39*			

COMBUSTION PRODUCTS

9	Ultimate Analysis, % Mass			10 Theoretical Air lb/100 lb Fuel		11 Dry Products from Fuel mole/100lb Fuel		12 Wet Products from Fuel mole/100 lb Fuel	
	Fuel Constituent	As-Fired	Carbon Burned (CB)	K1	[9] x K1	K2	[9] / K2	K3	[9] / K3
A	C	*80.31*	*80.31*						
B	UBC [5]		*0.39*						
C	CB [A] – [B]		*79.92*	11.51	*919.9*	12.011	*6.654*		
D	S	*1.54*		4.31	*6.6*	32.066	*0.048*		
E	H_2	*4.47*		34.29	*153.3*			2.016	*2.217*
F	H_2O	*2.90*						18.015	*0.161*
G	N_2	*1.38*				28.013	*0.049*		
H	O_2	*2.85*		–4.32	*–12.3*				
I	Ash	*6.55*							
K	Total	*100.00*			*1067.5*		*6.751*		*2.378*

13	Dry products of combustion, mole/100 lb fuel	[11K] – [11D] x [8] + [6]	*6.751*
14	Wet products of combustion, mole/100 lb fuel	[12K] + [13] + [7]	*9.129*
15	Theoretical air (corrected), mole/100 lb fuel	([10K] + [8] x [9D] x 2.16) / 28.963	*36.857*

EXCESS AIR WHEN O_2 KNOWN

16	O_2, % volume (input)					*3.315*
17	O_2 measurement basis	0 = Dry 1 = Wet	*1*	Dry	Wet	
18	Moisture in air, mole/mole dry air			0.0	[1] x 1.608	*0.021*
19	Dry/wet products of combustion, mole/100 lb fuel			[13]	[14]	*9.129*
20	Additional moisture, mole/100 lb fuel			0.0	[2] / 18.016	*0.000*
21	Intermediate calculation, step 1	[15] x (0.7905 + [18])				*29.909*
22	Intermediate calculation, step 2	[19] + [20] + [21]				*39.038*
23	Intermediate calculation, step 3	20.95 – [16] x (1 + [18])				*17.565*
24	Excess air, % by weight	100 x [16] x [22] / [15] / [23]				*20.0*

O_2, CO_2, SO_2 WHEN EXCESS AIR KNOWN

25	Excess air, % by weight				*20.0*
26	Dry gas, mole/100 lb fuel	[13] + [15] x (0.7905 + [25] / 100)			*43.258*
27	Wet gas, mole/100 lb fuel	[14] + [15] x (0.7905 + [18] + (1 + [18]) x [25] / 100) + [20]			*46.565*
			Dry	Wet	
28	O_2, % by volume	[25] x [15] x 0.2095 / ([26] or [27])	[26]	[27]	*3.32*
29	CO_2, % by volume	100 x ([11C] + [6]) / ([26] or [27])	[26]	[27]	*14.29*
30	SO_2, % by volume	100 x (1 – [8]) x [11D] / ([26] or [27])	[26]	[27]	*0.1031*
31	H_2O, % by volume	H_2O = 0.0 if dry or 100 x ([27] – [26]) / [27]	NA	[27]	*7.10*
32	N_2 (fuel), % by volume	100 x [11G] / ([26] or [27])	[26]	[27]	*0.11*
33	N_{2a} (air), % by volume	100 – [28] – [29] – [30] – [31] – [32]			75.08
34	MW wet flue gas, lbm/mole	0.32 x [28] + 0.4401 x [29] + 0.64064 x [30] + 0.18015 x [31] + 0.28013 x [32] + 0.28158 x [55]			*29.868*
35	Density flue gas, lbm/ft^3 at 60F and 29.92 in. Hg	0.0026356 x [34]		Wet basis	*0.07872*

Table 15B
Excess Air Calculations from Measured O_2
Bituminous coal, Virginia: with sorbent – O_2 on wet basis

	INPUTS (see also lightly shaded blocks)			SORBENT DATA (if applicable)	
1	Moisture in air, lb/lb dry air	*0.013*	6	CO_2 from sorbent, moles/100 lb fuel, Table 16 [17]	*0.116*
2	Additional moisture, lb/100 lb fuel	*0.00*	7	H_2O from sorbent, moles/100 lb fuel, Table 16 [16]	*0.012*
3	HHV fuel, Btu/lb	*14,100*	8	Sulfur capture, lb/lb sulfur fuel, Table 16 [24]	*0.90*
4	Unburned carbon loss, % fuel input	*2.50*			
5	Unburned carbon (UBC), [3] x [4] / 14,500	*2.43*			

COMBUSTION PRODUCTS

9	Ultimate Analysis, % Mass			10 Theoretical Air lb/100 lb Fuel		11 Dry Products from Fuel mole/100lb Fuel		12 Wet Products from Fuel mole/100 lb Fuel	
	Fuel Constituent	As-Fired	Carbon Burned (CB)	K1	[9] x K1	K2	[9] / K2	K3	[9] / K3
A	C	*80.31*	*80.31*						
B	UBC [5]		*2.43*						
C	CB [A] – [B]		*77.88*	11.51	*896.4*	12.011	*6.484*		
D	S	*1.54*		4.31	*6.6*	32.066	*0.048*		
E	H_2	*4.47*		34.29	*153.3*			2.016	*2.217*
F	H_2O	*2.90*						18.015	*0.161*
G	N_2	*1.38*				28.013	*0.049*		
H	O_2	*2.85*		–4.32	*–12.3*				
I	Ash	*6.55*							
K	Total	*100.00*			*1044.0*		*6.581*		*2.378*

13	Dry products of combustion, mole/100 lb fuel	[11K] – [11D] x [8] + [6]	*6.654*
14	Wet products of combustion, mole/100 lb fuel	[12K] + [13] + [7]	*9.044*
15	Theoretical air (corrected), mole/100 lb fuel	([10K] + [8] x [9D] x 2.16) / 28.963	*36.149*

EXCESS AIR WHEN O_2 KNOWN

16	O_2, % volume (input)				*3.315*
17	O_2 measurement basis	0 = Dry 1 = Wet *1*	Dry	Wet	
18	Moisture in air, mole/mole dry air		0.0	[1] x 1.608	*0.021*
19	Dry/wet products of combustion, mole/100 lb fuel		[13]	[14]	*9.044*
20	Additional moisture, mole/100 lb fuel		0.0	[2] / 18.016	*0.000*
21	Intermediate calculation, step 1	[15] x (0.7905 + [18])			*29.335*
22	Intermediate calculation, step 2	[19] + [20] + [21]			*38.379*
23	Intermediate calculation, step 3	20.95 – [16] x (1 + [18])			*17.565*
24	Excess air, % by weight	100 x [16] x [22] / [15] / [23]			*20.0*

O_2, CO_2, SO_2 WHEN EXCESS AIR KNOWN

25	Excess air, % by weight				*20.0*
26	Dry gas, mole/100 lb fuel	[13] + [15] x (0.7905 + [25] / 100)			*42.460*
27	Wet gas, mole/100 lb fuel	[14] + [15] x (0.7905 + [18] + (1 + [18]) x [25] / 100) + [20]			*45.761*
			Dry	Wet	
28	O_2, % by volume	[25] x [15] x 0.2095 / ([26] or [27])	[26]	[27]	3.31
29	CO_2, % by volume	100 x ([11C] + [6]) / ([26] or [27])	[26]	[27]	*14.42*
30	SO_2, % by volume	100 x (1 – [8]) x [11D] / ([26] or [27])	[26]	[27]	*0.0105*
31	H_2O, % by volume	H_2O = 0.0 if dry or 100 x ([27] – [26]) / [27]	NA	[27]	*7.21*
32	N_2 (fuel), % by volume	100 x [11G] / ([26] or [27])	[26]	[27]	*0.11*
33	N_{2a} (air), % by volume	100 – [28] – [29] – [30] – [31] – [32]			74.94
34	MW wet flue gas, lbm/mole	0.32 x [28] + 0.4401 x [29] + 0.64064 x [30] + 0.18015 x [31] + 0.28013 x [32] + 0.28158 x [55]			*29.843*
35	Density flue gas, lbm/ft^3 at 60F and 29.92 in. Hg	0.0026356 x [34]		Wet basis	*0.07866*

Table 16
Sulfur Capture Based on Gas Analysis
Bituminous coal, Virginia

	INPUTS							
1	SO_2, ppm	105	/ 10,000 = %	0.0105	2	O_2 Flue gas at location SO_2 measured, %		3.31
	Data from Table 15, Excess Air Calculations from Measured O_2							
3	Moisture in air, lb/lb dry air	[1]		0.013	7	Theoretical air, lb/100 lb fuel	[10K]	1044.0
4	Additional moisture, lb/100 lb fuel	[2]		0	8	Dry products of fuel, mole/100 lb fuel	[11K]	6.581
5	Sulfur in fuel, % by weight	[9D]		1.54	9	Wet products of fuel, mole/100 lb fuel	[12K]	2.378
6	HHV fuel, Btu/lb fuel	[3]		14,100				
	Data from Table 14, Combustion Calculations - Sorbent							
10	CO_2 from sorbent, lb/100 lb sorbent	[15G]		38.139	12	Sorbent, lb sorbent/lb fuel	[17]	0.134
11	H_2O from sorbent, lb/100 lb sorbent	[16G]		1.600				

	CALCULATIONS, Moles/100 lb Fuel Except As Noted				
	SO_2 / O_2 Measurement basis	0 = Dry 1 = Wet 1	Dry	Wet	
13	Moisture in air, mole/mole dry air		0.0	[3] x 1.608	0.0209
14	Additional moisture		0.0	[4] / 18.015	0.000
15	Products of combustion from fuel		[8]	[8] + [9]	8.959
16	H_2O from sorbent	[11] x [12] / 18.015	0.0	Calculate	0.012
17	CO_2 from sorbent	[10] x [12] / 44.01			0.116
18	Intermediate calculation, step 1	(0.7905 + [13]) x [7] / 28.963			29.245
19	Intermediate calculation, step 2	Summation [14] through [18]			38.332
20	Intermediate calculation, step 3	1.0 – (1.0 + [13]) x [2] / 20.95			0.8387
21	Intermediate calculation, step 4	(0.7905 + [13]) x 2.387 – 1.0			0.9368
22	Intermediate calculation, step 5	[1] x [19] x 32.066 / [5] / [20]			9.992
23	Intermediate calculation, step 6	[21] x [1] / [20]			0.0117
24	Sulfur capture, lb/lb sulfur	(100 – [22]) / (100 + [23])			0.90
25	SO_2 released, lb/1,000,000 Btu	20,000 x (1.0 – [24]) x [5] / [6]			0.22

Table 17
Combustion Calculations – Measured Gas Weight

	INPUTS			
		A Wet Analysis (not required)		B Dry Analysis
1	O_2, % volume	9.28	Measured dry or 100 / (100 – [3A]) x [1A]	10.55
2	CO_2, % volume	8.56	Measured dry or 100 / (100 – [3A]) x [2A]	9.73
3	H_2O, % volume	12.00		
4	Mass flow wet gas, 1000 lb/h			539.2
5	Moisture in wet gas, lb/lb wet gas			0.0754
6	Moisture in air, lb/lb dry air			0.0130
7	Additional moisture (sources other than fuel and air), 1000 lb/h			0
	CALCULATIONS			
8	Water in wet gas, 1000 lb/h	[4] x [5]		40.7
9	Dry gas weight, 1000 lb/h	[4] – [8]		498.5
10	N_{2a} in dry gas, % dry volume	100 – [1B] – [2B]		79.72
11	Molecular weight of dry gas, lb/mole	0.32 x [1B] + 0.4401 x [2B] + 0.28158 x [10]		30.11
12	Dry gas, 1000 moles/h	[9] / [11]		16.56
13	Dry air weight, 1000 lb/h	0.28161 x [10] x [12] / 0.7685		483.8
14	Water in dry air, 1000 lb/h	[13] x [6]		6.3
15	Water evaporated, 1000 lb/h	[8] – [7] – [14]		34.4
16	Excess air, 1000 lb/h	[1B] x [9] x 0.32 / 0.2314 / [11]		241.5
17	Theoretical air, 1000 lb/h	[13] – [16]		242.3
18	Excess air, % by weight	100 x [16] / [17]		99.7

Fig. 1 Deepwater oil and gas fields require new floating technology facilities like this Spar being installed in the Gulf of Mexico.

Chapter 11

Oil and Gas Utilization

Before the industrial revolution, distilled petroleum products were used primarily as a source of illumination. Today, petroleum finds its primary importance as an energy source and greatly influences the world's economy. The following discusses the use of petroleum products and natural gas as energy sources for steam generation.

Fuel oil

Preparation

Petroleum or crude oil is the source of various fuel oils used for steam generation (Fig. 1, facing page). Most petroleum is refined to some extent before use although small amounts are burned without processing. Originally, refining petroleum was simply the process of separating the lighter compounds, higher in hydrogen, from the heavier compounds by fractional distillation. This yielded impure forms of kerosene, gasoline, lubricating oils and fuel oils. Through the development of refining techniques, such as thermal cracking and reforming, catalytic reforming, polymerization, isomerization and hydrogenation, petroleum is now regarded as a raw material source of hydrogen and carbon elements that can be combined as required to meet a variety of needs.

In addition to hydrocarbons, crude oil contains compounds of sulfur, oxygen and nitrogen and traces of vanadium, nickel, arsenic and chlorine. Processes are used during petroleum refinement to remove impurities, particularly compounds of sulfur. Purification processes for petroleum products include sulfuric acid treatment, sweetening, mercaptan extraction, clay treatment, hydrogen treatment and the use of molecular sieves.

The refining of crude oil yields a number of products having many different applications. Those used as fuel include gasoline, distillate fuel, residual fuel oil, jet fuels, still gas, liquefied gases, kerosene and petroleum coke. Products for other applications include lubricants and waxes, asphalt, road oil, and petrochemical feedstock.

Fuel oils for steam generation consist primarily of residues from the distillation of crude oil. As refinery methods improve, the quality of residual oil available for utility and industrial steam generation is deteriorating. High sulfur fuels containing heavy components create challenges during combustion that range from high particulate and sulfur oxide emissions to higher maintenance costs due to the corrosive constituents in the flue gas.

Transportation, storage and handling

The high heating value per unit of volume of oil, its varied applications, and its liquid form have fostered a worldwide system of distribution. The use of supertankers for the transportation of crude oil has significantly reduced transportation costs and has allowed refineries to be located near centers of consumption rather than adjacent to the oil fields. Large supertankers, up to 250,000 t (227,000 t_m), are capable of transporting nearly 2,000,000 bbl (318,000 m^3) of crude oil at a time to deepwater ports.

Tanker and barge shipments on coastal and inland waterways are by far the cheapest method of transporting the various grades of oil. With the depletion of oil fields in the eastern United States (U.S.), crude oil trunk lines were developed in the early 1900s to transport oil from points west of the Mississippi River to the east coast refineries. Today, more than 170,000 mi (274,000 km) of pipeline, including small feeder lines, are used for the transportation of oil within the U.S. Much smaller quantities of oil are shipped overland by rail and truck because of the higher cost of haulage.

Fuel oil systems require either underground or surface storage tanks. Oil is usually stored in cylindrical shaped steel tanks to eliminate evaporation loss. Loss in storage of the relatively nonvolatile heavy fuel oils is negligible. Lighter products, such as gasoline, may volatilize sufficiently in warm weather to cause appreciable loss. In this instance, storage tanks with floating roofs are used to eliminate the air space above the fuel where vapors can accumulate. The National Fire Protection Association (NFPA) has prepared a standard set of codes for the storage and handling of oils (NFPA 30 and 31). These codes serve as the basis for many local ordinances and are required for the safe transportation and handling of fuel products.

Extensive piping and valving and suitable pumping and heating equipment are necessary for the transportation and handling of fuel oil. Storage tanks, piping and heaters for heavy oils must be cleaned periodically because of fouling or sludge accumulation.

Fuel properties

Safe and efficient transportation, handling and combustion of fuel oil requires a knowledge of fuel characteristics. Principal physical properties of fuel oils, important to boiler applications, are summarized below (see Chapter 9 for typical fuel oil physical property values):

Viscosity The viscosity of an oil is the measure of its resistance to internal movement, or flow. Viscosity is important because of its effect on the rate at which oil flows through pipelines and on the degree of atomization obtained by oil firing equipment.

Ultimate analysis An ultimate analysis is used to determine theoretical air requirements for combustion of the fuel and also to identify potential environmental emission characteristics.

Heating value The heating value of a liquid fuel is the energy produced by the complete combustion of one unit of fuel [Btu/lb (J/kg)]. Heating value can be reported either as the gross or higher heating value (HHV) or the net or lower heating value (LHV). To determine HHV, it is assumed that any water vapor formed during combustion is condensed and cooled to the initial temperature (i.e., all of the chemical energy is available). The heat of vaporization of the water formed is included in the HHV. For LHV, it is assumed that the water vapor does not condense and is not available. The heating value determines the quantity of fuel necessary to achieve a specified heat input.

Specific gravity Specific gravity (sp gr) is the ratio of the density of oil to the density of water. It is important because fuel is purchased by volume, in gallons (l) or barrels (m^3). The most widely used fuel oil gravity scale is degree API devised by the American Petroleum Institute; its use is recommended by the U.S. Bureau of Standards and the U.S. Bureau of Mines. The scale is based on the following formula:

$$\text{degrees API} = \frac{141.5}{\text{sp gr at 60/60F (16/16C)}} - 131.5$$

[sp gr at 60/60F (16/16C) means when both oil and water are at 60F (16C)]

Flash and *fire point* Flash point is the lowest temperature at which a volatile oil will give off explosive or ignitable vapors. It is important in determining oil handling and storage requirements. The fire point is the temperature to which a liquid must be heated to produce vapors sufficient for continuous burning when ignited by an external flame.

Pour point The pour point is the temperature at which a liquid fuel will first flow under standardized conditions.

Distillation Distillation determines the quantity and number of fractions which make up the liquid fuel.

Water and sediment Water and sediment are a measure of the contaminates in a liquid fuel. The sediment normally consists of calcium, sodium, magnesium, and iron compounds. Impurities in the fuel provide an indication of the potential for plugging of fuel handling and combustion equipment.

Carbon residue Residue that remains after a liquid fuel is heated in the absence of air is termed carbon residue. The tests commonly used to determine carbon residue are the Conradson Carbon Test and the Ramsbottom Carbon Test. Carbon residue gives an indication of the coking tendency of a particular fuel (i.e. the tendency of oil, when heated, to form solid compounds).

Asphaltene content Asphaltenes are long chain, high molecular weight hydrocarbon compounds. The asphaltene content of a petroleum product is the percentage by weight of wax free material insoluble in n-heptane but soluble in hot benzene. Their structure requires high temperatures and high atomization energy for the fuel to burn completely. Higher asphaltene content indicates a higher potential to produce particulate emissions.

Burning profile Burning profile is a plot of the rate at which a sample of fuel burns under standard conditions as temperature is increased at a fixed rate. The burning profile is a characteristic fingerprint of the fuel oxidized under standard conditions and is not intended to provide absolute kinetic and thermodynamic data. It helps evaluate combustion characteristics of various fuels on a relative basis to determine excess air and residence time necessary for complete combustion.

Natural gas

Preparation

Natural gas, found in crude oil reservoirs, either dissolved in the oil or as a gas cap above the oil, is called associated gas. Natural gas is also found in reservoirs that contain no oil and is termed non-associated gas.

Natural gas, directly from the well, must be treated to produce commercially marketable fuels. Initially, natural gas undergoes a process to remove condensate which is distilled to produce butane, propane and stabilized gasoline. Propane and butane are widely used as bottle gas. They are distributed and stored liquefied under pressure. When the pressure is released, the liquid boils, producing a gaseous fuel.

Natural gas may contain enough sand or gaseous sulfur compounds to be troublesome. The sand is usually removed at the source. Natural gas containing excessive amounts of hydrogen sulfide, commonly known as sour gas, can be treated by a process known as sweetening. Sweetening removes hydrogen sulfide as well as carbon dioxide. Additional treatments include the removal of mercaptan by soda fixation and the extraction of long chain hydrocarbons.

Where natural gas is used to replace or supplement manufactured gas, it is sometimes reformed to bring its heating value in line with the manufactured gas. Natural gas may also be mixed directly with manufactured gas to increase the heating value of the final product.

Transportation, storage and handling

Pipelines are an economical means of transporting natural gas in its gaseous form. The rapid increase in consumption of natural gas in areas far from the source has resulted in an extensive system of long distance pipelines. Natural gas can also be transported by tanker when liquefied under pressure producing liquefied natural gas (LNG).

The distribution of natural gas is subject to some practical limitations because of the energy required for transportation. High pressures, in the order of 1000 psig (6895 kPa), are necessary for economic pipeline transportation over long distances. Compression stations are needed at specified intervals to boost the pressure due to losses in the line.

In general, it is not practical to vary the supply of natural gas to accommodate the hourly or daily fluctuations in consumer demand. For economic reasons, long distance pipelines operate with a high load factor. The rate of withdrawal from the wells may often be limited for conservation reasons, and the cost of the pipeline to provide the peak rate would be prohibitive. Therefore, to meet fluctuations in demand, it is usually necessary to provide local storage or to supplement the supply with manufactured gas for brief periods.

Above ground methods of storage include: 1) large water seal tanks, 2) in-pipe holders laid parallel to commercial gas lines, and 3) using the trunk transmission line as a reservoir by building up the line pressure. In consumer areas where depleted or partially depleted gas and oil wells are available, underground storage of gas pumped back into these wells provides, at minimum cost, the large storage volume required to meet seasonal variations in demand. In liquid form, natural gas can be stored in insulated steel tanks or absorbed in a granular substance, released by passing warm gas over the grains.

Fuel properties

Natural gas is comprised primarily of methane and ethane. Physical properties of practical importance to boiler applications include constituents by volume percent, heating value, specific gravity, sulfur content and flammability (see Chapter 9 for typical natural gas physical property values).

Other liquid and gaseous fuels

Numerous combustion system applications utilize liquid or gaseous fuels other than conventional fuel oils or natural gas. These fuels include Orimulsion®, blast furnace gas, coke oven gas, refinery gas, regenerator offgas, landfill gas, and other byproduct gases.

The large heavy hydrocarbon and bitumen reserves available in Venezuela have led to a bitumen oil emulsion fuel that has gained acceptability. Orimulsion is the trade name for a commercially established fossil fuel oil emulsion. It consists of natural bitumen dispersed in water, in approximately a 70/30 proportion split. The resulting emulsion is stabilized by a surfactant package. Orimulsion can be transported over land or water and stored for extended periods while maintaining a consistent quality. Although it can be handled using most of the equipment and systems originally designed for heavy fuel oil, Orimulsion requires some special handling and combustion considerations because of its emulsified state. In addition, although the fuel exhibits very good combustion characteristics, it contains relatively high levels of sulfur, nitrogen, ash, asphaltenes, vanadium and other metals. Thus, careful design and cleanup considerations are important when firing a fuel with high levels of these constituents.

Steel mill blast furnaces generate a byproduct gas containing about 25% carbon monoxide by volume. This fuel can be burned to produce steam for mill heating and power applications. Many mills also have their own coke producing plant, another source of byproduct fuel. Coke oven gas is an excellent fuel that burns readily because of its high free hydrogen content. With these gases, available supply pressures and the volumetric heating value of fuel may be different from that of natural gas. Therefore, gas components must be designed to accommodate the particular characteristics of the gas to be burned.

In the petroleum industry, refinery gas and regenerator offgas are frequently used as energy sources for boilers. Refinery gas is a mixture of gaseous hydrocarbon streams from various refinery processes. Depending on economic and technical considerations within the refinery, the compositions of these individual streams vary with process modifications and thus, the resultant refinery gas can change over time. Combustion equipment and controls for refinery gas must be suitably designed for this variability. Regenerator offgas, or CO (carbon monoxide) gas, is a high-temperature gas produced in catalytic cracking units. CO boilers have been developed to reclaim the thermal energy present in this gas (see Chapter 27).

Landfill gas is a combustible gas recovered by a gas collection system at a landfill. Its primary constituents are methane and carbon dioxide. Landfill gas processing systems filter suspended particulates and condensate from the gas stream. Additional processing may be done to further purify the gas, but trace contaminants that typically remain in the gas require special attention when designing fuel handling systems to minimize corrosion concerns.

Oil and gas combustion – system design

The burner is the principal equipment component for the combustion of oil and natural gas (Fig. 2). In utility and industrial steam generating units (both wall and corner-fired designs), the burner admits fuel and air to the furnace in a manner that ensures safe and efficient combustion while realizing the full capability of the boiler. Burner design determines mixing characteristics of the fuel and air, fuel particle size and distribution, and size and shape of the flame envelope.

The means of transporting, measuring and regulating fuel and air to the furnace, together with the burners, igniters and flame safety equipment, comprises the overall combustion system. The following factors must be considered when designing the combustion system and when establishing overall performance requirements:

1. the rate of feed of the fuel and air to comply with load demand on the boiler over a predetermined operating range,
2. the types of fuel to be fired including elemental constituents and characteristic properties of each fuel,

Fig. 2 Typical oil and gas utility boiler burner front.

3. the efficiency of the combustion process to minimize unburned combustibles and excess air requirements,
4. imposed limitations on emissions,
5. physical size and complexity of the furnace and burners to establish the most efficient and economic design,
6. hardware design and material properties of the combustion equipment to ensure reliable uninterrupted service for long firing periods, and
7. safety standards and procedures for control of the burners and boiler, including starting, stopping, load changes and variations in fuel.

The combustion system must be designed for optimum flexibility of operation, including the potential for variations in fuel type, fuel firing rate and combinations of burners in and out of service. Control must be simple and direct to ensure rapid response to varying load demands.

Combustion air is typically conveyed to the burners by forced draft fans. To improve both thermal and combustion efficiency and further ensure burner stability, combustion air is normally preheated to a temperature of 400 to 600F (204 to 316C) by air preheaters located downstream of the fans. The fans must be capable of delivering adequate quantities of air for complete combustion at a pressure sufficient to overcome losses across the air preheaters, burners, control dampers, and intervening duct work. The total combustion air is that required to theoretically burn all the fuel plus excess air necessary for complete combustion. (See Chapter 10.)

The fuel delivery system must be able to regulate fuel pressure and flow to the burners and must be safeguarded in accordance with applicable fire protection codes. Proper distribution of fuel to the burners, in multiple burner applications, is critical to safe and efficient operation of the combustion system. Piping and valves must be designed for allowable velocity limits, absolute pressure requirements, and pressure losses.

Performance requirements

Excess air

Excess air is the air supplied for combustion and cooling of idle burners in excess of that theoretically required for complete oxidation of the fuel. Excess air is generally required to compensate for imperfections in the air delivery system that results in maldistribution of combustion air to the burners. Excess air also helps compensate for imperfect mixing of the air and fuel in the furnace. At full load, with all burners in service, excess air required for gas and oil firing, expressed as a percent of theoretical air, is typically in the range of 5 to 10%, depending upon fuel type and the requirements of the combustion system. Operation at excess air levels below these values is possible if combustion efficiency does not deteriorate. Combustion efficiency is measured in terms of carbon monoxide, unburned combustibles in the ash, soot, particulate matter and stack opacity. Through careful design of the burners and the air delivery system, excess air can be held to a minimum, thereby minimizing sensible heat loss to the stack.

Operation at partial load requires additional excess air. When operating with all burners in service at reduced load, lower air velocity at the burners results in reduced mixing efficiency of the fuel and air. Increasing the excess air improves combustion turbulence and maintains overall combustion efficiency. Additional excess air and improved burner mixing also compensate for lower furnace temperature during partial load operation. In some instances, boiler performance dictates the use of higher than normal excess air at reduced loads to maintain steam temperature or to minimize cold end corrosion.

Additional excess air is also necessary when operating with burners out of service. Sufficient cooling air must be provided to idle burners to prevent overheat damage. Permanent thermocouples installed on selected burners measure metal temperatures and establish the minimum excess air necessary to maintain burner temperatures below the maximum use limits of the steel. Excess air for burner cooling varies with the percentage of burners out of service.

Stability and turndown

Proper burner and combustion system design will permit stable operation of the burners over a wide operating range. A stable burner, best determined through visual observation, is one where the flame front remains relatively stationary and the root of the flame is securely anchored near the burner fuel element. To ensure stable combustion, the burner must be designed to prevent blowoff or flashback of the flame for varying rates of fuel and air flow.

It is often desirable to operate over a wide boiler load range without taking burners out of service. This reduces partial load excess air requirements to cool idle burners. The burners must therefore be capable of operating in a turned down condition. Burner *turndown* is defined as the ratio of full load fuel input to partial load input while still maintaining stable combustion. Limitations in burner turndown are generally dictated by fuel characteristics, fuel and air velocity, full load to partial load fuel pressures, and adequacy of the flame safety system. Automated and reliable flame safety supervision, with proper safeguards, must be available to achieve high burner turndown ratios.

With gas firing, a turndown ratio of 10:1 is not uncommon. Natural gas is easily burned and relatively easy to control. Residual oil, on the other hand, is more difficult to burn. Combustion characteristics are highly sensitive to particle size distribution, excess air and burner turbulence. A typical turndown ratio for oil is in the order of 6:1, depending upon fuel characteristics, flexibility of the delivery system and atomization technique.

Burner pulsation

Burner pulsation is a phenomenon frequently associated with natural gas firing and, to a lesser degree, with oil firing. Pulsation is thought to occur when fuel rich pockets of gas suddenly and repeatedly ignite within the flame envelope. The resultant pulsating burner flame is often accompanied by a noise referred to as *combustion rumble*. Combustion rumble may transmit frequencies that coincide with the natural frequency of the furnace enclosure resulting in apparent boiler vibration. In some instances, these vibrations may become alarmingly violent.

Boiler vibration on large furnaces can sometimes be attributed to a single burner. Minor air flow adjustment to a given burner, or removing select burners from service, may suddenly start or stop pulsations. Pulsation problems can be corrected through changes to burner hardware that affect mixing patterns of the fuel and air. Changes to the burner throat profile to correct anomalies in burner aerodynamics or changes to the fuel element discharge ports have successfully eliminated pulsation.

Historical operating data has enabled the development of empirical curves that are useful in designing burners to avoid pulsation. These curves relate the potential for burner pulsation to the ratio of burner fuel to air velocity. Together with careful consideration of furnace geometry, burner firing patterns and burner aerodynamics, problems with burner pulsation are becoming less common.

Combustion efficiency

Many factors influence combustion efficiency including excess air, burner mixing, fuel properties, furnace thermal environment, residence time, and particle size and distribution. Complete combustion occurs when all combustible elements and compounds of the fuel are entirely oxidized. In utility and industrial boilers, the goal is to achieve the highest degree of combustion efficiency with the lowest possible excess air. Thermal efficiency decreases with increasing quantities of excess air. Combustion performance is then measured in terms of the boiler efficiency loss due to incomplete combustion together with the efficiency loss due to sensible heat in the stack gases.

From the standpoint of optimum combustion efficiency, the following factors are critical to proper design:

1. careful distribution and control of fuel and air to the burners,
2. burner and fuel element design that provides thorough mixing of fuel and air and promotes rapid, turbulent combustion, and
3. proper burner arrangement and furnace geometry to provide sufficient residence time to complete chemical reactions in a thermal environment conducive to stable and self-sustained combustion.

In most cases, boiler efficiency loss due to unburned carbon loss (UCL) when firing oil and natural gas is virtually negligible. However, depending on fuel oil properties and the condition of the combustion system, the percent UCL can be in the order of 0.10% while firing oil. Combustion efficiency with these fuels is usually measured in terms of carbon monoxide (CO) emissions, particulate emissions and stack opacity. Generally, CO levels less than 200 ppm (corrected to 3% O_2) are considered satisfactory.

Emission control techniques

Ever increasing concern over atmospheric pollutants is changing the focus of wall and corner-fired boiler and combustion system designs. The combustion of fossil fuels produces emissions that have been attributed to the formation of acid rain, smog, changes to the ozone layer, and the so-called *greenhouse* effect. To mitigate these problems, federal and local regulations are currently in place that limit oxides of nitrogen, oxides of sulfur, particulate matter and stack opacity. While emission limits vary depending upon state and local regulations, the trend is toward more stringent control. (See also Chapter 32.)

Many combustion control techniques have emerged to reduce fossil fuel emissions. These techniques generally focus on the reduction of nitrogen oxides (NO_x), as changes to the combustion process can greatly influence NO_x formation and destruction.

Oxides of nitrogen

Nitrogen oxides in the form of NO and NO_2 are formed during combustion by two primary mechanisms: thermal NO_x and fuel NO_x. A secondary mechanism called prompt NO_x can also contribute to overall NO_x formation.

Thermal NO_x results from the dissociation and oxidation of nitrogen in the combustion air. The rate and degree of thermal NO_x formation is dependent upon oxygen availability during the combustion process and is exponentially dependent upon combustion temperature. Thermal NO_x reactions occur rapidly at combustion temperatures in excess of 2800F (1538C). Thermal NO_x is the primary source of NO_x formation from natural gas and distillate oils because these fuels are generally low in or devoid of fuel-bound nitrogen.

Fuel NO_x, on the other hand, results from oxidation of nitrogen organically bound in the fuel and is the primary source of NO_x formation from heavy fuel oil. Fuel bound nitrogen in the form of volatile compounds is intimately tied to the fuel hydrocarbon chains. For this reason, the formation of fuel NO_x is linked to both fuel nitrogen content and fuel volatility. Inhibiting oxygen availability during the early stages of combustion, where the fuel devolatilizes, is the most effective means of controlling fuel NO_x formation.

Prompt NO_x is formed during the early, low temperature stages of combustion. Hydrocarbon fragments may react with atmospheric nitrogen under fuel-rich conditions to yield fixed nitrogen species. These, in turn, can be oxidized to NO in the lean zone of the flame. In most flames, especially those from nitrogen-containing fuels, the prompt mechanism is responsible for only a small fraction of the total NO_x.

Numerous combustion process NO_x control techniques are commonly used. These vary in effectiveness and cost. In all cases, control methods are mainly aimed at reducing either thermal NO_x, fuel NO_x, or a combination of both. A range of typical anticipated NO_x emission levels relative to various NO_x control mechanisms is shown in Fig. 3.

Low excess air Low excess air (LEA) effectively reduces NO_x emissions with little, if any, capital expenditure. LEA is a desirable method of increasing thermal efficiency and has the added benefit of inhibiting thermal NO_x. If burner stability and combustion efficiency are maintained at acceptable levels, lowering the excess air may reduce NO_x by as much as 5 to 15% from an uncontrolled baseline. The success of this method depends largely upon fuel properties and the ability to carefully control fuel and air distribution to the burners. Operation may require more sophisticated methods of measuring and regulating fuel and air flow to the burners and modifications to the air delivery system to ensure equal distribution of combustion air to all burners.

Burners out of service Essentially a simple form of two-stage combustion, burners out of service (BOOS) is a simple and direct method of reducing NO_x emissions. When removing burners from service in multiple burner applications, active burner inputs are typically increased to maintain load. Without changing total air flow, increased fuel input to the active burners results in a fuel rich mixture, effectively limiting oxygen availability and thereby limiting both fuel and thermal NO_x formation. Air control registers on the out of service burners remain open, essentially serving as staging ports.

While a fairly significant NO_x reduction is possible with this method, lower NO_x is frequently accompanied by higher levels of CO in the flue gas and boiler back-end oxygen (O_2) imbalances. With oil firing, an increase in particulate emissions and increased stack opacity are likely. Through trial and error, some patterns of burners out of service may prove more successful than others. A limiting factor is the ability of existing burners to handle the increased input necessary to maintain full load operation. Short of derating the unit, changes to fuel element sizes may be required.

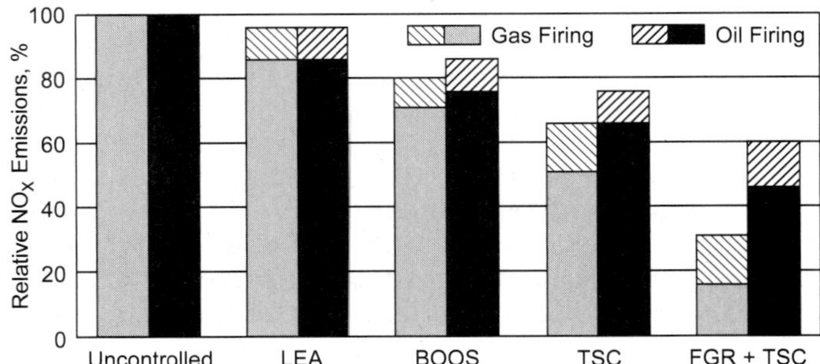

Fig. 3 Approximate NO_x emission reductions for oil and gas burners using various control techniques. (LEA = low excess air; BOOS = burner out of service; TSC = two-stage combustion; FGR = flue gas recirculation.)

Two-stage combustion Two-stage combustion is a relatively long standing and accepted method of achieving significant NO_x reduction. Combustion air is directed to the burner zone in quantities less than that required to theoretically burn the fuel, with the remainder of the air introduced through overfire air ports. By diverting combustion air away from the burners, oxygen concentration in the lower furnace is reduced, thereby limiting the oxidation of chemically bound nitrogen in the fuel. By introducing the total combustion air over a larger portion of the furnace, peak flame temperatures are also lowered.

Appropriate design of a two-stage combustion system can reduce NO_x emissions by as much as 50% and simultaneously maintain acceptable combustion performance. The following factors must be considered in the overall design of the system.

1. *Burner zone stoichiometry* The fraction of theoretical air directed to the burners is predetermined to allow proper sizing of the burners and overfire air ports. Normally a burner zone stoichiometry in the range of 0.85 to 0.90 will result in desired levels of NO_x reduction without notable adverse effects on combustion stability and turndown.
2. *Overfire air port design* Overfire air ports must be designed for thorough mixing of air and combustion gases in the second stage of combustion. Ports must have the flexibility to regulate flow and air penetration to promote mixing both near the furnace walls and toward the center of the furnace. Mixing efficiency must be maintained over the anticipated boiler load range and the range in burner zone stoichiometries.
3. *Burner design* Burners must be able to operate at lower air flow rates and velocities without detriment to combustion stability. In a two-stage combustion system, burner zone stoichiometry is typically increased with decreasing load to ensure that burner air velocities are maintained above minimum limits. This further ensures positive windbox-to-furnace differential pressures at reduced loads.
4. *Overfire air port location* Sufficient residence time from the burner zone to the overfire air ports and from the ports to the furnace exit is critical to proper system design. Overfire air ports must be located to optimize NO_x reduction and combustion efficiency and to limit change to furnace exit gas temperatures.
5. *Furnace geometry* Furnace geometry influences burner arrangement and flame patterns, residence time and thermal environment during the first and second stages of combustion. Liberal furnace sizing is generally favorable for lower NO_x as combustion temperatures are lower and residence times are increased.

6. *Air flow control* Ideally, overfire air ports are housed in a dedicated windbox compartment. In this manner, air to the NO_x ports can be metered and controlled separately from air to the burners. This permits operation at desired stoichiometric levels in the lower furnace and allows for compensation to the flow split as a result of air flow adjustments to individual burners or NO_x ports.

Additional flexibility in controlling burner fuel and air flow characteristics is required to optimize combustion under a two-stage system. Improved burner designs have addressed these needs.

In the reducing gas of the lower furnace, sulfur in the fuel forms hydrogen sulfide (H_2S) rather than sulfur dioxide (SO_2) and sulfur trioxide (SO_3). The corrosiveness of reducing gas and the potential for increased corrosion of lower furnace wall tubes is highly dependent upon H_2S concentration. Two-stage combustion is therefore not normally recommended when firing high sulfur residual fuel oils except when extra furnace wall protection measures are included.

Flue gas recirculation Flue gas recirculation (FGR) to the burners is instrumental in reducing NO_x emissions when the contribution of fuel nitrogen to total NO_x formation is small. For this reason, the use of gas recirculation is generally limited to the combustion of natural gas and fuel oils. By introducing flue gas from the economizer outlet into the combustion air stream, burner peak flame temperatures are lowered and NO_x emissions are significantly reduced. (See Fig. 4.)

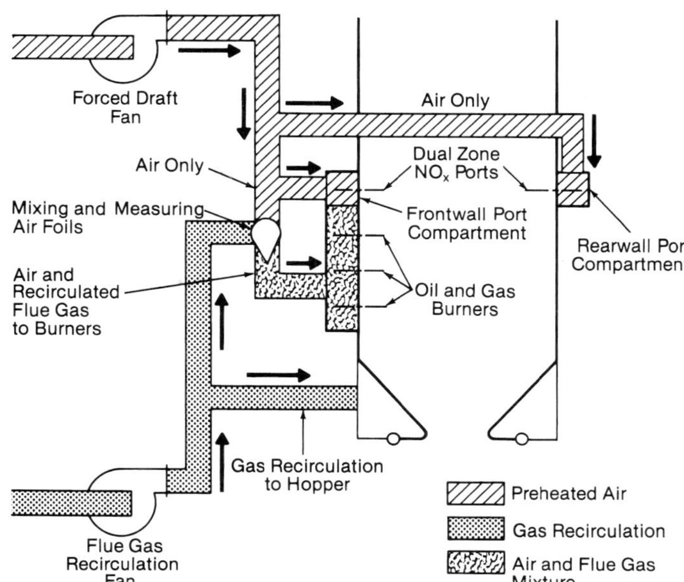

Fig. 4 Flue gas recirculation low NO_x system for oil and gas firing.

Air foils are commonly used to mix recirculated flue gas with the combustion air. Flue gas is introduced in the sides of the secondary air measuring foils and exits through slots downstream of the air measurement taps. This method ensures thorough mixing of flue gas and combustion air before reaching the burners and does not affect the air flow metering capability of the foils.

In general, increasing the rate of flue gas recirculation to the burners results in an increasingly significant NO_x reduction. Target NO_x emission levels and limitations on equipment size and boiler components dictate the practical limit of recirculated flue gas for NO_x control. Other limiting factors include burner stability and oxygen concentration of the combustion air. Typically, oxygen content must be maintained at or above 17% on a dry basis for safe and reliable operation of the combustion equipment.

The expense of a flue gas recirculation system can be significant. Gas recirculation (GR) fans may be required for the desired flow quantities at static pressures capable of overcoming losses through the flues, ducts, mixing devices and the burners themselves. Additional controls and instruments are also necessary to regulate GR flow to the windbox at desired levels over the load range. In retrofit applications, significant cost is associated with routing of flues and ducts to permit mixing of the flue gas with combustion air. Also, the accompanying increase in furnace gas weight at full load operation may require modifications to convection pass surfaces or dictate changes to standard operating procedures.

From an operational standpoint, the introduction of flue gas recirculation as a retrofit NO_x control technique must, in virtually all cases, be accompanied by the installation of overfire air ports. Oil and gas burners, initially designed without future consideration to FGR, are not properly sized to accommodate the increase in burner mass flow as a result of recirculated flue gas. The quantity of flue gas necessary to significantly reduce NO_x emissions will, in all likelihood, result in burner throat velocities that exceed standard design practices. This, in turn, may cause burner instability, prohibitive burner differentials and in the case of gas firing, undesirable pulsation. Therefore, the installation of overfire air ports in conjunction with FGR serves two useful purposes, 1) lower NO_x emissions through two-stage combustion, and 2) a decrease in mass flow of air to the burners to accommodate the increased burden of recirculated flue gas.

When employing flue gas recirculation in combination with overfire air, it is desirable to house the overfire air ports in a dedicated windbox compartment separate from the burners. In this manner, it is possible to introduce recirculated flue gas to the burners only. This permits more efficient use of the GR fans and overall system design as only that portion of flue gas introduced through the burners is considered effective in controlling NO_x emissions.

An inexpensive means of recirculating lesser amounts of flue gas is *induced* FGR, or IFGR. Here, flue gas is introduced through the forced draft fan(s) and is restricted by the fans' capacity for flue gas. The effectiveness of IFGR is, as a result, limited.

Reburning Reburning is an in-furnace NO_x control technique that divides the furnace into three distinct zones (main, reburn, and burnout). By effectively staging both fuel and combustion air, NO_x emission reductions of 50 to 75% from baseline levels can be achieved. Heat input is spread over a larger portion of the furnace, with combustion air carefully regulated to the various zones to achieve optimum NO_x reduction (Fig. 5).

In reburning, the lower furnace or main burner zone provides the major portion of the total heat input to the furnace. Depending on the percent NO_x reduction and the specific combustion system requirements, the main zone burners can be designed to operate at less than theoretical air to normal excess air levels. Combustion gases from the main burner zone then pass through a second combustion zone termed the reburning zone. Here, burners provide the remaining heat input to the furnace to achieve full load operation but at a significantly lower stoichiometry. By injecting reburn fuel above the main burner zone, a NO_x reducing region is produced in the furnace where hydrocarbon radicals from the partially oxidized reburn fuel strip oxygen from the NO molecules, forming nitrogen compounds and eventually molecular nitrogen (N_2). Overfire air ports are installed above the reburning zone where the remainder of air is introduced to complete combustion in an environment both chemically and thermally non-conducive to NO_x formation.

Application of this technology must consider a number of variables. System parameters requiring definition include: fuel split between the main combustion zone and the reburn zone, stoichiometry to the main and reburn burners, overall stoichiometry in the reburn and burnout zones of the furnace, residence time in the reburn zone, and residence time required above the overfire air ports to complete combustion. An optimum range of values has been defined for each of these parameters through laboratory tests and field application and is largely dependent upon the type of fuel being fired. For example, fuels with high sulfur contents (Orimulsion or some heavy fuel oils) are not as suitable in applications where operating the main combustion zone under low sub-stoichiometric conditions is required to reduce NO_x levels due to corrosion concerns. For these fuels, reburning technology can be effectively used by operating the main zone at higher stoichiometries, thus minimizing corrosion concerns while still achieving good NO_x reduction results.

Although implementation of the reburning technology adds complexity to operation and maintenance of the overall combustion system, it also provides considerable emission performance optimization flexibility. In addition, higher initial costs for a reburn system as compared to other combustion techniques need to be factored into the evaluation process. From an economic standpoint, the potential benefits and technical merit of the reburning process must be commensurate with long term goals for NO_x abatement.

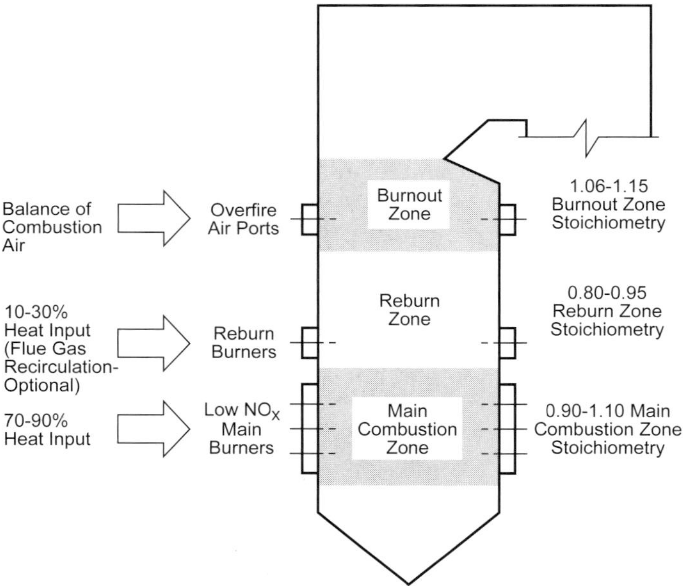

Fig. 5 Boiler side view showing reburn principle and combustion zones for a utility boiler.

Oxides of sulfur

The sulfur content of fuel oils can range anywhere from a fraction of a percent for lighter oils to 3.5% for some residual oils. During the combustion process, sulfur contained in the fuel is converted to either sulfur dioxide, SO_2, or sulfur trioxide, SO_3 (SO_x emissions). The control of SO_x emissions is a key environmental concern and sulfur compounds in the flue gas can also cause corrosion problems in the boiler and downstream equipment.

SO_3 will form sulfuric acid when cooled in the presence of water vapor. In addition to corrosion problems, it can produce emissions of acid smut and visible plume opacity from the stack. Emissions of SO_3 are best controlled during combustion through low excess air operation and can also be reduced by use of magnesium based fuel additives.

Techniques to control sulfur oxides during the combustion process have been investigated in laboratory and pilot scale tests with varying degrees of success. At present, however, the most effective and commercially accepted method, short of firing low sulfur fuels, is to install flue gas cleanup equipment. (See Chapter 35.)

Particulate matter

Particulate matter in the form of soot or coke is a byproduct of the combustion process resulting from carryover of inert mineral matter in the fuel and from incomplete combustion. Primarily a concern on oil-fired units, particulate matter becomes apparent when fuel oil droplets undergo a form of fractional distillation during combustion, leaving relatively large carbonaceous particles known as cenospheres.

Cenospheres are porous, hollow particles of carbon that are virtually unaffected by further combustion in a conventional furnace environment. Cenospheres can also absorb sulfur oxides in the gaseous phase and thereby further contribute to the formation of ash particles contaminated with acid.

Particulate matter from the combustion of fuel oil is, in large part, a function of the fuel properties. Ash content of the fuel oil plays a significant role in forming submicron particulate emissions. These ultrafine particulate emissions are potentially more dangerous to the environment than larger particles as they tend to stay suspended in the atmosphere. The fuel oil property most closely linked to forming cenospheres during the combustion process is asphaltene concentration. Asphaltenes are high molecular weight hydrocarbons that do not vaporize when heated. The combustion of fuel oils high in asphaltene content produces a greater quantity of large and intermediate size particulate emissions. Carbon residue, commonly determined by the Conradson Carbon Test, is also a means of evaluating the tendency to form particulate matter during combustion.

Control of particulate emissions is best achieved through proper atomization of the fuel oil and careful design of the burners and combustion control system to ensure thorough and complete mixing of fuel and air. Liberal residence time in a high temperature environment is favorable for complete combustion of fuel and low particulate emissions, although not conducive to low NO_x emissions. Oil emulsification also has a favorable impact on reducing particulate emissions. Oil emulsified with water further breaks up individual oil droplets when the water vaporizes during combustion. In addition, many fuel additives are now available to promote carbon burnout. These additives are primarily transition metals such as iron, manganese, cobalt and nickel that act as catalysts for the further oxidation of carbon particles.

Opacity

A visible plume emanating from the stack is undesirable from both a regulatory and public relations standpoint. Stack opacity is controlled in much the same manner as particulate emissions. Careful selection of fuels and complete combustion are keys to minimizing plume visibility.

Dark plumes are generally the result of incomplete combustion and can be controlled by careful attention to the combustion process and through transition metal-based additives. White plumes are frequently the result of sulfuric acid in the flue gas and can be controlled by low excess air operation or alkaline-based fuel additives that neutralize the acid.

The simultaneous control of all criteria pollutants poses a significant challenge to the burner and boiler designer. Techniques, or operating conditions, effective in reducing one form of atmospheric contaminant are frequently detrimental to controlling others or to unit performance. For this reason, hardware design and combustion controls are becoming increasingly complex. Modern boilers must have the added flexibility necessary to optimize thermal efficiency and combustion performance in conjunction with sound environmental practices.

Burner selection and design

As environmental concerns continue to dictate boiler and combustion system design, higher standards of performance are imposed on the fuel burning equipment. Control techniques to reduce NO_x emissions can conflict with proven methods of good combustion performance, e.g., time, temperature and turbulence.

Potential increases in carbon monoxide emissions, particulate emissions and stack opacity as a result of low NO_x operation suggest that the burners must be capable of continued reliable mechanical operation and of providing the flexibility necessary to optimize combustion under a variety of operating conditions.

Circular burner

Shown in Fig. 6, the circular type burner has long been the standard design for wall-fired oil and gas applications. The tangentially disposed doors of the circular burner air register provide the turbulence necessary to mix the fuel and air and produce short, compact flames. This burner typically operates with high secondary air velocities providing rapid, turbulent combustion for high combustion efficiency. Fuel is introduced to the burner in a fairly dense mixture in the center. The direction and velocity of the air and dispersion of the fuel result in complete and thorough mixing of fuel and air.

XCL-S® type burner

The XCL-S® type oil and gas combination burner shown in Fig. 7 was originally developed to replace the circular burner. It not only met the demand for added flexibility and improved control of combustion air flow, but also incorporated the necessary design features to result in producing low NO_x emission levels. The XCL-S type burner incorporates several design features not available with the circular register. This burner has two air zones: the inner or core zone and the outer secondary air zone. When firing natural gas or oil, combustion air is introduced to the core zone through slots located around the periphery of the inner sleeve. An oil impeller or swirler is unnecessary with this burner; control of the flow entering the core zone ensures stable ignition. The core zone houses the main gas fuel element(s) and the main fuel oil atomizer.

The majority of combustion air enters the burner through the outer air zone. Axially disposed spin vanes are located in the outer sleeve to impart swirl to the combustion air and a sliding disk separately controls total air to the burner independent of swirl. An air zone swirler (AZS) device can also be incorporated into this zone to help reduce excess air and associated emission levels. The burner is equipped with an air measuring device upstream of the spin vanes. This device provides a relative indication of air flow to each burner and, on multiple burner applications, permits balancing of air flow from burner to burner. The outer zone also houses the igniter and flame detection equipment.

Because it can measure air flow to individual burners and regulate total air flow independent of swirl, the XCL-S type burner provides the added flexibility

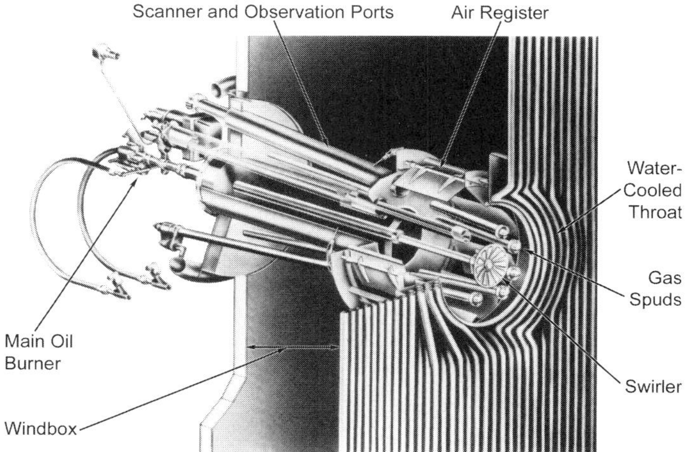

Fig. 6 Circular register burner with water-cooled throat for oil and gas firing.

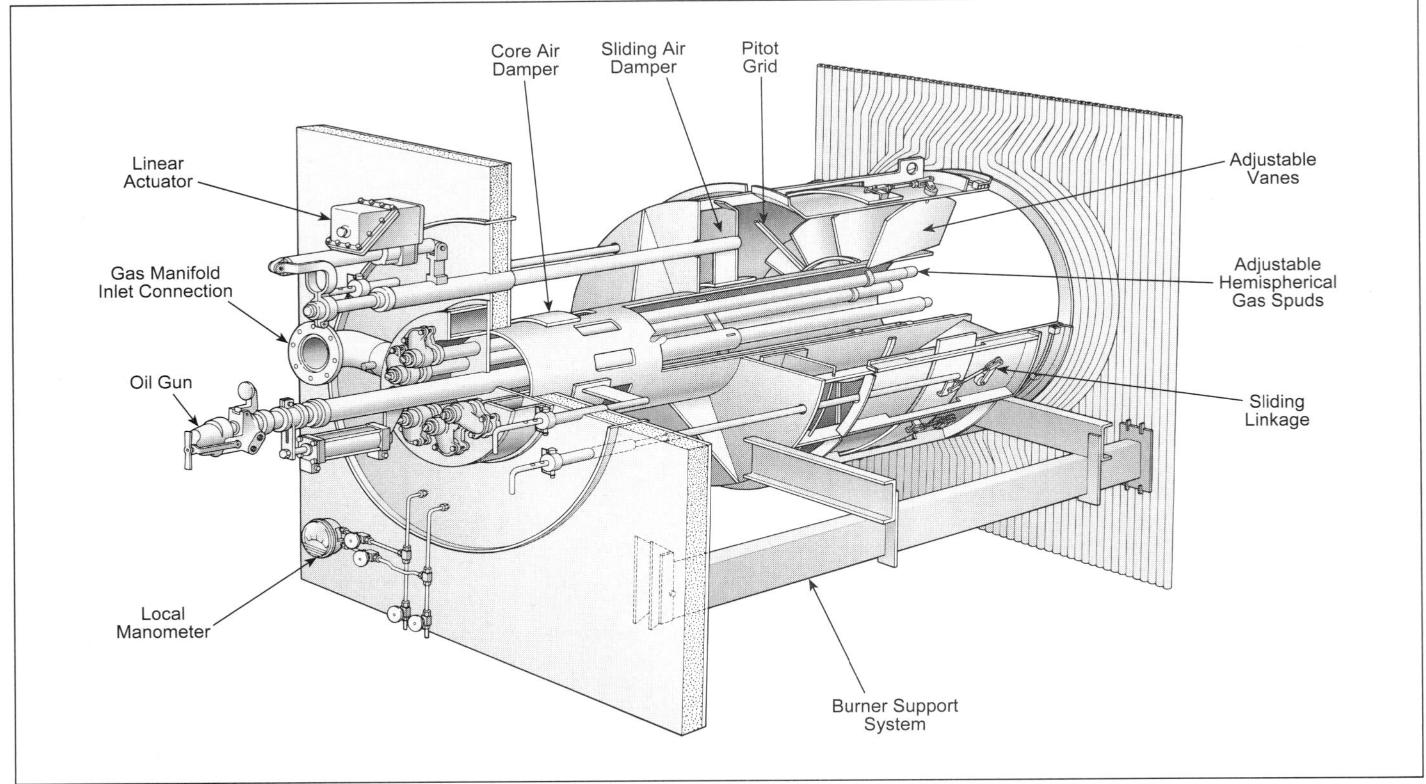

Fig. 7 XCL-S® low NO_x burner for oil and gas.

needed when employing combustion control techniques aimed at reducing NO_x emissions. This burner is ideally suited as an upgrade to existing circular (or other wall-fired type) burners as burner throat pressure part modifications can typically be avoided.

This advanced low NO_x burner achieves superior emission performance in burner only applications, as well as situations using overfire air and/or flue gas recirculation. In applications using overfire air ports, The Babcock & Wilcox Company's (B&W's) DualAir Zone port provides maximum operating flexibility to optimize overall performance. See Chapter 14 for more details.

Corner-firing system

The general theory and configuration for corner firing of oil and/or natural gas is the same as that for pulverized coal firing described in Chapter 14.

Fuel and combustion air are introduced and distributed in each corner of the furnace through windbox assemblies similar to that shown in Fig. 8. The general arrangement of the windbox for natural gas or oil firing is essentially the same as that for coal firing in that the windbox is divided along its height into alternating compartments of fuel and air. For oil firing, a retractable oil gun is inserted into the center air nozzle of each fuel compartment. For natural gas firing, a stationary manifold gas nozzle attached to the inside of the windbox fuel compartment distributes fuel into the bellmouth of the air nozzle.

For conventional firing applications, approximately half of the total combustion air is introduced through the fuel compartments. This is known as fuel air and is used to support combustion and stabilize flames. The balance of the air, known as auxiliary air, is used to complete combustion and is introduced through the compartments located above and below each fuel elevation. All of the combustion air enters through the corner windbox assemblies and is distributed in the windbox by a system of dampers at the inlet of each compartment. The fuel air dampers are controlled based on fuel flow rate whereas the auxiliary air dampers are controlled based on windbox to furnace differential pressure.

The most cost effective method of reducing NO_x in oil and/or natural gas corner-fired boilers is by addition of overfire air which uses the air staging or two stage combustion principle commonly used with pulverized coal firing. Application of overfire air to a corner fired system involves diverting a portion of the auxiliary air to an overfire air zone located above the combustion zone (admitted at the top of the windbox through two or more compartments and/or through separate ports located several feet above the top fuel elevation). The overfire air port compartments are typically equipped with individual dampers to control air flow, tilting mechanisms to allow for variation in separation distance for improved NO_x performance, and air flow measuring devices (e.g., pitot tubes) to facilitate air flow control. The design and arrangement of the B&W overfire air system and corresponding windbox modifications for oil and/or natural gas firing are similar to that described in Chapter 14 for pulverized coal firing.

Also, flue gas recirculation (FGR) has been used successfully and has provided significant NO_x reductions in units firing natural gas and/or heavy fuel oil. With FGR, a portion of the flue gas (typically 10 to 15% by weight) is taken from the boiler exhaust and is mixed with the combustion air. However, the use of FGR requires careful analysis as its effects on boiler operation and maintenance costs can be significant.

Fuel oil equipment

Oil for combustion must be atomized into the furnace as a fine mist and dispersed into the combustion air stream. Proper atomization is the key to efficient combustion and reduced particulate emissions. Atomization quality is measured in terms of droplet size and droplet size distribution. High quality atomization occurs when oil droplets are small, producing high surface to volume ratios and thereby exposing more surface to the combustion air. A convenient means of expressing and comparing atomization quality produced by various atomizer designs is the Sauter mean diameter (D_{sm}). This is the ratio of the mean volume of the oil droplets over the mean surface area, expressed in microns. The lower the Sauter mean diameter, the better the atomization.

For proper atomization, oil grades heavier than No. 2 must be heated by means of steam or electric heaters to reduce their viscosity to between 100 and 150 SUS (Saybolt Universal Seconds). In heating fuel oils, caution must be observed to ensure that temperatures are not raised to the point where vapor lock may occur. Vapor lock results when volatile fractions of the fuel separate in the fuel supply system causing flow interruptions and subsequent loss of ignition. Fuel supplied to the atomizers must also be free of acid, grit and fibrous or other foreign matter likely to clog or damage hardware components.

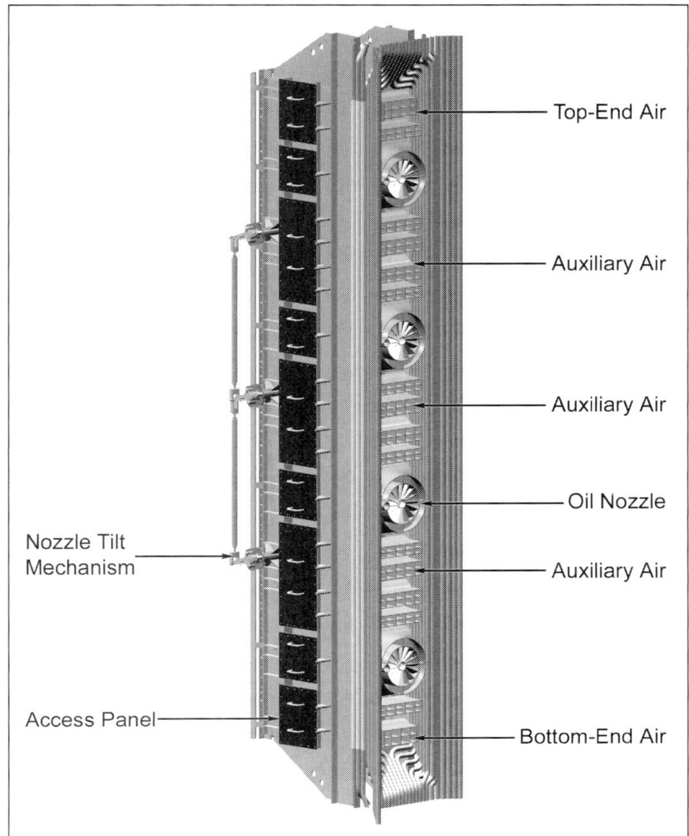

Fig. 8 Windbox assembly for corner-firing system; compartment dampers on opposite side not shown (*courtesy of R-V Industries, Inc.*).

Fuel oil is atomized by either mechanical or dual fluid atomizers that use steam or air as the medium. The choice of whether to use mechanical or steam atomization is determined by the boiler design and operating requirements. In general, steam assisted atomizers produce a higher quality spray and are more appropriate for low NO_x applications or where particulate emissions and stack opacity are of primary concern. At some installations, however, the heat balance is such that steam can not be used economically for oil atomization. Another factor to consider is the necessity to conserve boiler feedwater. In these instances, mechanical atomization may be more appropriate.

B&W's atomization test facility has been used to quantitatively and qualitatively characterize numerous atomizer designs through state-of-the-art laser diagnostic techniques. B&W's family of atomizers range from mechanical Return Flow to various dual fluid (Racer®, Y-Jet, T-Jet, and I-Jet) atomizer types. A summary of their operating characteristics is provided in Fig. 9.

Mechanical atomizers With mechanical atomizers, the pressure of the fuel itself provides the energy necessary for atomization. These atomizers require relatively high pressure oil for proper performance. Three conventional types are in common use: the Uniflow, the Return Flow and the Steam Mechanical atomizer.

The Uniflow atomizer is used in small and medium sized stationary power plants, as well as in naval and merchant marine boilers. This atomizer is simple to operate and made with as few parts as possible. Fuel is introduced into ports that discharge tangentially into a whirl chamber. The fuel is spun out of the whirl chamber and passes through an orifice into the combustion chamber as a well atomized conical spray. Required oil pressure at the atomizer is above 300 psig (2.07 MPa gauge) for heat inputs of 70 to 80 × 10^6 Btu/h (20.5 to 23.4 MW_t).

The Return Flow atomizer (Fig. 10) is used on either stationary or marine boilers that require wide capacity ranges. This atomizer is designed to minimize, or eliminate entirely, the need for changing sprayer plates or the number of burners in service during normal operation. A wide range of operation is possible by maintaining a high flow through the sprayer plate slots even at reduced firing rates. At reduced firing rates, because the quantity of oil supplied is greater than the required firing rate, the excess oil is returned to a low pressure point in the piping system. The required oil pressure at the atomizer must be either 600 or 1000 psig (4.14 or 6.90 MPa gauge) depending upon fuel, capacity and load range requirements. Maximum input is in the order of 200 × 10^6 Btu/h (58.6 MW_t).

The Steam Mechanical atomizer combines the features of steam and mechanical atomization permitting operation over wide ranges, including low loads with cold furnaces. At high loads it can be operated as a mechanical atomizer. At reduced loads mechanical

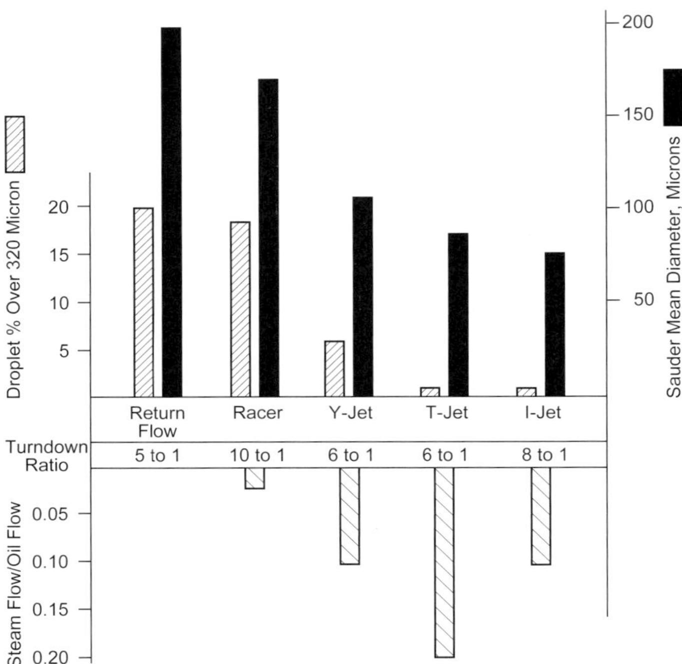

Fig. 9 Operating characteristics of B&W atomizers.

atomization is augmented by the use of steam. Required oil pressure is 200 to 300 psig (1.38 to 2.07 MPa gauge) depending upon capacity requirements. Steam pressure at the atomizer must be 10 to 15 psi (0.07 to 0.10 MPa) above oil pressure not to exceed 125 psig (0.86 MPa gauge). Maximum capacity of this atomizer is in the order of 80 to 90 × 10^6 Btu/h (23.4 to 26.4 MW_t).

Residual fuel oils contain heavy residues that may condense on cold surfaces creating potentially hazardous conditions. Improving atomization quality can help avoid this situation. A certain number of steam or air atomizers could be provided for satisfactory operation when boiling and drying out new units and for continued operation at very low capacities.

Steam atomizers Due to better operating and safety characteristics, steam assisted atomization is preferred. Steam atomization generally produces a finer spray. The steam-fuel emulsion produced in a dual fluid atomizer reduces oil droplet size when released into the furnace through rapid expansion of the steam. As a minimum, dry saturated steam at the specified pressure must be used for fuel oil atomization to avoid the potential for burner pulsation. Typically, 20 to 40F (11.1 to 22.2C) superheat is recommended to avoid such problems. If necessary, moisture free compressed air can be substituted.

Several steam atomizer designs are available in sizes up to 300 × 10^6 Btu/h (87.9 MW_t) or about 16,500 lb/h (2.08 kg/s) of fuel. Required oil pressure is much lower than for mechanical atomizers. Steam and oil pressure requirements are dependent upon the specific atomizer design. Maximum oil pressure can be as much as 300 psig (2.07 MPa gauge) with steam pressures as high as about 200 psig (1.38 MPa gauge). The four most common steam atomizer designs are the Y-jet, Racer, T-jet and I-jet. (See Fig. 11.) Each design is characterized by different operating ranges, steam consumption rates, and atomization quality based on resultant oil droplet particle size distributions. Fig. 9 summarized these various characteristics for each of the atomizers.

The Y-jet is designed for a wide firing range without changing the number of burners in service or the size of the sprayer plate. It can be used on any type of boiler with either steam or air atomization. Fuel oil and atomizing medium flow through separate channels to the atomizer sprayer plate assembly where they mix immediately before discharging into the furnace. Required oil pressure at the atomizer for maximum capacity ranges from 65 to 90 psig (0.45 to 0.62 MPa gauge). The Y-jet is a constant differential atomizer design requiring steam pressure to be maintained at 40 psi (0.28 MPa) over the oil pressure throughout the normal operating range. Steam consumption with this design is in the order of 0.1 lb steam/lb oil.

The Racer atomizer is a refinement of the Y-jet. It was developed primarily for use where high burner turndown and low atomizing steam consumption are required. The Racer name refers to the class of merchant ship where this atomizer was first used. Required design oil pressure for maximum capacity is 300 psig (2.07 MPa gauge). Steam pressure is held constant throughout the load range at 150 psig (1.03 MPa gauge). Atomizing steam consumption at full load is approximately 0.02 lb steam/lb oil.

The T-jet and I-jet atomizers also allow a wide range of operation without the need for excessive oil pressure. These atomizer designs are unique in that the steam and oil are mixed in a chamber prior to discharging through the sprayer cap (Fig. 11). Rated capacity oil pressure for the T-jet may range from 90 to 110 psig (0.62 to 0.76 MPa gauge) depending upon fuel and capacity requirement. The I-jet atomizer rated capacity oil pressure ranges from 70 to about 170 psig (0.48 to 1.17 MPa gauge). Of the constant differential variety, the I-jet and T-jet require a steam pressure of approximately 20 to 40 psi (0.14 to 0.28 MPa) above oil pressure. Steam pressures may be adjusted to obtain optimum combustion performance. Steam consumption rates vary with these atomizer designs depending upon actual operating pressures. The standard design steam consumption rates at full atomizer

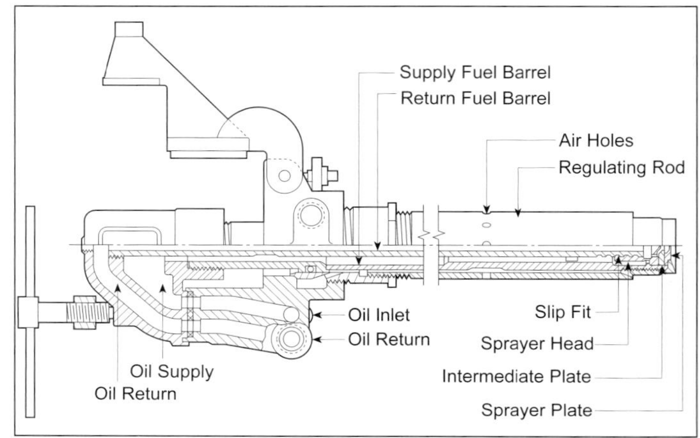

Fig. 10 Mechanical Return Flow oil atomizer assembly.

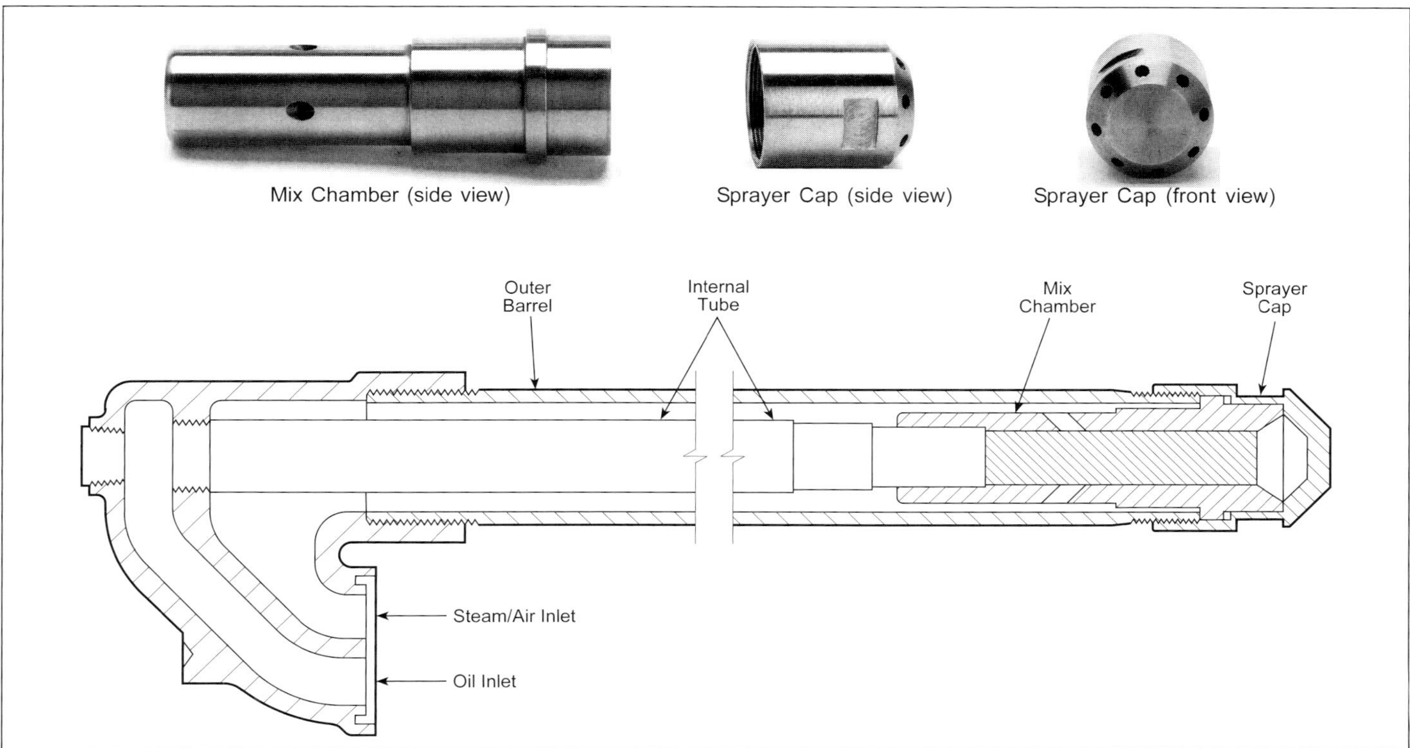

Fig. 11 I-jet atomizer assembly.

capacity for the T-jet and I-jet are 0.20 and 0.10 lb steam/lb oil, respectively. In general, higher steam consumption rates result in improved atomization quality.

A disadvantage of the steam atomizer is its consumption of steam. For a large unit, this can amount to a sizable quantity of steam and consequent heat loss to the stack. When the boiler supplies a substantial amount of steam for a process where condensate recovery is small, the additional makeup for the steam atomizer is inconsequential. However, in a large utility boiler where turbine losses are low and there is little makeup, the use of atomizing steam can have a significant effect.

Natural gas equipment

Three common types of gas element assemblies available for the combustion of commercial high Btu natural gas are the variable mix multi-spud, hemispherical multi-spud and the radial spud. The variable mix multi-spud gas element was developed for use with the circular type burner. The goal was to improve ignition stability when flue gas recirculation or two-stage combustion are implemented for NO_x control. This design uses a manifold outside the furnace, with a number of individual gas elements, or spuds, projecting through the windbox into the throat. The following is a list of the distinctive features of this gas element:

1. Individual gas spuds are removable with the boiler in service to enable cleaning or redrilling of the gas nozzles as required.
2. Individual spuds can be rotated to orient the discharge holes for optimum firing conditions with the burner in service.
3. The location of the spud tip, with respect to the burner throat, can be varied to a limited degree for optimum firing conditions.
4. Individual gas spud flame retainers provide added stability and are more conducive to lower NO_x emissions.

This gas element design adds needed flexibility for optimizing combustion under a variety of operating conditions. With proper selection of control equipment, the operator can change from one fuel to another without a drop in load or boiler pressure. Simultaneous firing of natural gas and oil in the same burner is acceptable on burners equipped with the variable-mix multi-spud arrangement.

The hemispherical gas element (Fig. 12) was developed for use with the low NO_x XCL-S and FM-XCL (package boiler application version of the XCL-S) type burners. With the exception of the flame retainers, the hemispherical gas element offers the same features as the variable mix element design. This gas element, however, can achieve lower NO_x emissions by virtue of the tip profile, drilling pattern and location within the burner.

The radial spud design consists of a manifold located inside the windbox with individual gas elements projecting into the burner throat. Unlike the variable mix and hemispherical design, the radial spud is not conducive to low NO_x emissions. Use of the radial spud is reserved for applications where coal and gas are fired in the same burner and where stringent NO_x emission limits are not an issue.

For low NO_x gas firing capability in a burner that must also be capable of firing pulverized coal, a single, high-capacity gas element (HCGE) is used (Fig. 13).

The HCGE's drilling pattern and its location on the axial centerline of the burner inside the coal nozzle promote lower NO_x emissions as a result of slower mixing of fuel and air. The gas flame associated with the HCGE tends to be somewhat longer than gas flames associated with burners having multiple spud configurations. Consequently, burner capacity and furnace size must be appropriately matched. If adequate furnace depth is available, the HCGE can also be used with the low NO_x XCL-S and FM-XCL gas or gas/oil burners to achieve lower NO_x emissions than can be achieved with hemispherical multi-spuds.

The maximum practical limit for gas input per burner is in the order of 200×10^6 Btu/h (58.6 MW_t). Frequently, physical arrangement of the gas hardware is a limiting factor in fuel input capability. With all gas element designs, the spud tip drilling is determined by adhering to empirical curves aimed at eliminating the potential for burner pulsation. Allowable gas discharge velocity criteria are established and the resulting required manifold pressure is determined. Gas pressure at the burner at rated capacity is generally in the range of 8 to 12 psig (55 to 82 kPa) for multiple spud configurations and as high as 20 psig (137 kPa) for the HCGE.

In many respects, natural gas is an ideal fuel since it requires no preparation for rapid and intimate mixing with the combustion air. However, this characteristic of easy ignition under most operating conditions has, in some cases, led to operator carelessness and damaging explosions.

To ensure safe operation, gas flames must remain anchored to the gas element discharge ports throughout the full range of allowable gas pressures and air flow conditions. Ideally, stable ignition should be possible at minimum load with full load air flow through the burner and at full load with as much as 25% excess air. With this latitude in air flow it is not likely that ignition can be lost, even momentarily, during upset conditions.

Byproduct gases

Many industrial applications utilize blast furnace gas, coke oven gas, refinery gas, regenerator offgas, landfill gas, or other byproduct gases to produce steam. (See Chapter 27.)

Burners are designed specifically to fire these byproduct fuels. For example, blast furnace gas is a low heating value fuel that typically is available at relatively low supply pressure. Accordingly, a blast furnace gas burner utilizes a nozzle designed for low gas velocity to promote flame stability and to accommodate a high gas flow rate with minimal pressure loss. A scroll located at the nozzle inlet imparts swirl to the incoming gas stream thereby improving subsequent mixing of fuel and air. Cleanout doors in the scroll permit removal of accumulated deposits common to blast furnace gas. Similarly, the fuel element for any other byproduct gas must be designed for its particular characteristics, ignition stability, and appropriate load range.

Some byproduct gases, such as CO gas, are not of sufficient fuel quality to sustain their own flames. In CO boilers, supplementary gas and/or oil burners are used to raise the temperature of the CO gas to the ignition point and to assure complete burning of the combustibles in the CO gas stream. Supplementary fuel burners and CO gas ports are positioned to ensure thorough mixing and to promote rapid, complete combustion.

Fig. 12 Furnace view of hemispherical gas spuds (oil atomizer not shown).

Cold start and low load operation

Cold boiler startup requires firing at low heat input for long periods of time to avoid expansion difficulties and possible overheat of superheaters or reheaters. Low pressure [200 psig (1.38 MPa gauge)] boilers without superheaters may need only about an hour of low load operation. Larger high pressure units, however, may need four to six hours for startup. During startup, combustion efficiency is typically poor, especially with residual oils, due largely to low furnace and combustion air temperatures.

Low load operation may also result in poor distribution of air to the burners in a common windbox. Low air flow and accompanying low windbox to furnace pressure differentials may result in significant stratification of air in the windbox and adversely affect combustion performance. Stack effect between the bottom and top rows of burners may further alter the quantity of air reaching individual burners. At extremely low loads, firing oil or gas, the fuel piping itself can also introduce poor distribution and accordingly affect burner operation. For a given supply pressure, some burners may have adequate fuel while others do not receive enough fuel for continuous operation.

The low ignition temperature and clean burning characteristics of natural gas make it an ideal fuel for startup and low load operation. However, extreme care must be taken to avoid a momentary interruption in the flame. Large quantities of water are generated by the combustion of gaseous fuels with a high percentage of hydrogen. For units equipped with regenerative type air heaters, the air heater should not be placed in service until the flue gas temperature to the air heater has reached 400F (204C). This prevents condensation of water from the flue gas on the air heater surface and subsequent transport of this wa-

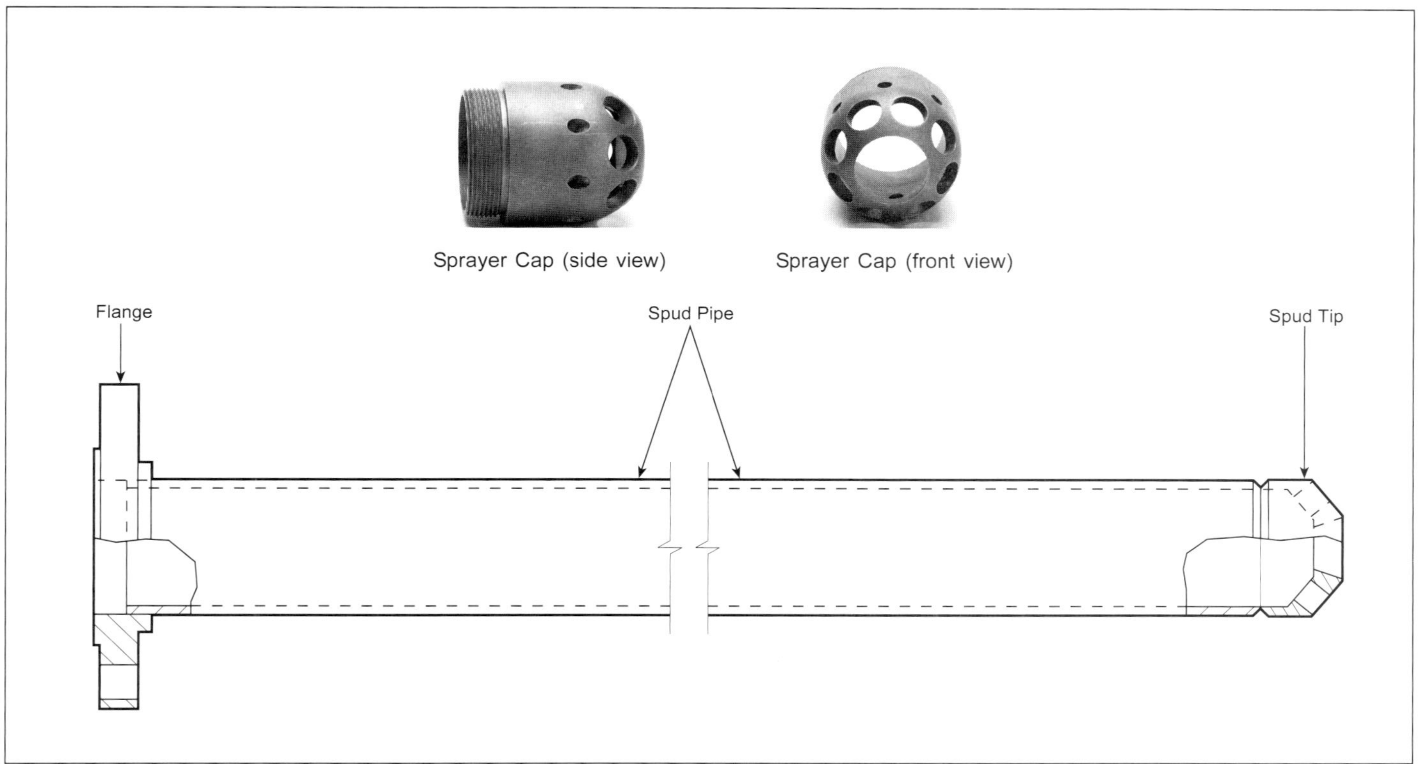

Fig. 13 High-capacity gas element (HCGE).

ter to the burners by the combustion air. Water in the combustion air may cause loss of ignition.

Fuel oil is potentially more hazardous than gas for startup and low load operation. Although ignition stability is generally not a problem, the low temperature furnace environment may result in the accumulation of soot and carbon particles on air heater and economizer surfaces and may also create a visible plume. If deposits accumulate over an extended period of time they pose a fire hazard. To reduce the incidence of fires, some operators use sootblowers that blow steam continuously on the surfaces of regenerative type air heaters as they rotate in the flue gas stream.

When oil must be used for startup, the following firing methods are listed in order of preference:

1. Use of dual fluid atomizers and light fuel oil, No. 1 or 2, with steam or compressed air as the atomizing medium. This will provide a clean stack and limit the deposit of carbonaceous residue on boiler back-end surfaces.
2. Use of dual fluid atomizers with steam atomization and fuel oil no heavier than No. 6, heated as required for reduced viscosity and proper atomization.
3. Use of dual fluid atomizers with compressed air as the atomizing medium and fuel oil no heavier than No. 6, heated as required.

Startup with oil using mechanical atomizers is not recommended due to the inherent lower atomization quality. Stationary boilers equipped with mechanical atomizers for oil firing at loads between 20 and 100% are generally started up with natural gas or a burner level of dual-fluid startup atomizers.

Igniters

Lighting a burner requires an independent source of ignition. The igniters used to provide the necessary ignition energy range from hand held torches inserted temporarily, to direct ignition electrical igniters, to automatically spark ignited oil or gas fired igniters.

Today, the boiler codes require permanently installed igniters in each burner. These are most often oil or gas fired designs that are ignited by means of high voltage or high energy spark ignition systems. The capacity of modern igniters can reach 10% or more of the main burner. This permits their use in stabilizing the combustion of difficult to burn fuels or in providing support ignition for abnormal main fuel conditions. Their high input capacity may also be used to warm the furnace prior to starting the main burner system or to synchronize the turbine. Although the igniter capacity is tailored to meet the specific requirements of the burner and flame safety system, a maximum input of approximately 25×10^6 Btu/h (7.3 MW_t) is typical.

The National Fire Protection Association (NFPA) categorizes igniters according to their input capacity relative to the burner input. Class I igniters generally provide 10% or more of the main burner heat release and may be operated continuously for support ignition when required under all boiler conditions. Specifically, the Class I igniter location and capacity will provide sufficient ignition energy to raise any credible combination of burner inputs of both fuel and air above the minimum ignition temperature. Class II igniters generally provide 4 to 10% of the burner heat release and may be operated continuously for ignition support under well defined boiler conditions. Class III igniters generally provide a heat release less than 4% of the burner and may not be operated continuously to provide support ignition for the burner. Most modern utility boilers employ Class I igniters to warm the boiler, provide support for difficult fuels and/or poor fuel conditions, and to minimize the complexity of the flame detection system.

Gas provides the most reliable and cleanest burning fuel for high capacity igniters. Gas igniters may be operated with either natural gas or liquefied petroleum gas (LPG). When operated with LPG, it is critical that the LPG supply system provide a reliable, vapor-only flow at the required rate. Oil fired igniters may be used with both distillate and residual fuel oils. Light oils are favored as they avoid complexities of oil heating/recirculation, which can impair reliability. Distillate oil can be used with either air or steam atomization with air being preferred. Residual oil igniters require superheated steam for optimum performance.

Gas and oil igniters are available in both fixed and retractable designs. Retractable igniters are inserted to their firing position for ignition of the main burner and then retracted back into the burner when they are no longer needed. Retracting the igniter sufficiently reduces radiant heat exposure such that cooling air is not required. Fig. 14 is an example of a typical retractable igniter (B&W CFS gas igniter is shown and an oil fired version is also available).

Fixed igniters are permanently positioned in the burner throat at the proper location for ignition of the burner. Fig. 15 is an example of a stationary igniter (FPS oil igniter is shown and a gas fired version is also available). A dedicated supply of air is provided to the FPS igniter to improve combustion performance and to cool the igniter when it is out of service.

In addition to these fueled igniters, there is a special Class III igniter which does not burn fuel. These ignite the burner directly with electrical energy. They are usually high energy, capacitive discharge, spark ignition systems similar to those that ignite oil and gas fired igniters, but usually they operate with a higher energy level. Direct electric ignition can be used on oil burners which burn distillate and residual fuel oils having specific characteristics. Direct ignition of gas burners is recommended only for burners with a single gas element. Reliable ignition of gas burners having multiple gas elements spread around the burner can often be difficult, resulting in potentially hazardous conditions, and is not normally recommended.

Direct electric ignition spark systems generally provide an ignition energy level of at least 8 joules, with 10 to 12 joules used in most applications. The small spark produced by the direct ignition spark rod must be carefully positioned in the fuel stream to ensure rapid ignition of the fuel upon opening of the burner fuel valve. Direct electric ignition spark systems must always be retractable.

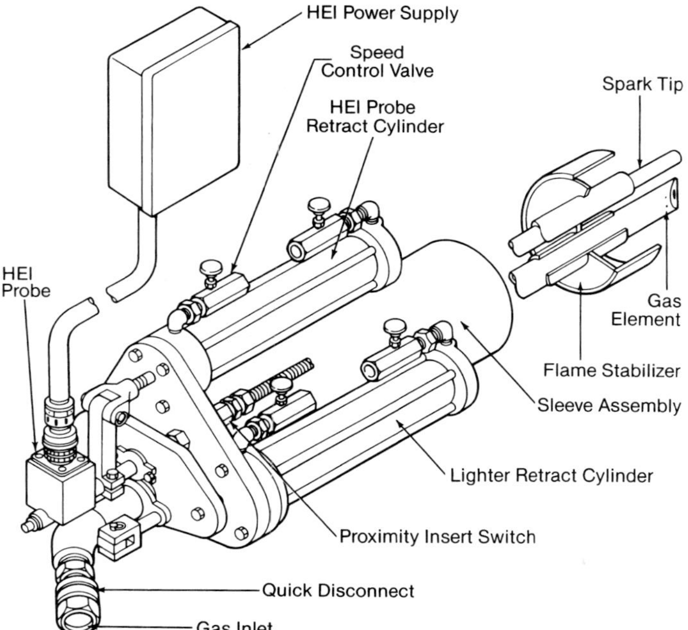

Fig. 14 CFS retractable gas igniter with high energy spark probe.

Safety precautions

NFPA 85 is a standard set of operating codes (procedures, interlocks, and trips) for the safe and reliable operation of gas and oil combustion processes. Recommendations of this standard along with governing local codes and ordinances are followed in the design of gas and oil burner systems, burner interlock and trip systems, and burner sequence control systems.

Five rules of prime importance in safe and reliable operation of gas and oil fired combustion systems, whether employing a manual or automatic control system, are described below.

1. Never allow oil or gas to accumulate anywhere, other than in a tank or lines that form part of the fuel delivery system. The slightest odor of gas must be cause for alarm. Steps should be taken immediately to ventilate the area thoroughly and locate the source of the leak.
2. A minimum purge rate air flow not less than 25% of full load volumetric air flow must be maintained during all stages of boiler operation. This includes pre-purging the setting and lighting of the igniters and burners until the firing rate air requirement exceeds the purge rate air flow.
3. A spark producing device or lighted torch must be in operation before introducing any fuel into the furnace. The ignition source must be properly placed with respect to the burner and must continuously provide a flame or spark of adequate size until a stable main flame is established.
4. A positive air flow through the burners into the furnace and up the stack must be maintained at all times.
5. Adequate fuel pressure for proper burner operation must be maintained at all times. In the case of oil firing, fuel pressure and temperature must be maintained for proper atomization. In dual fluid atomizer applications, adequate steam or air pressure must be available at the atomizer.

Equipment requirements, sequence of operation, interlock systems and alarm systems are equally im-

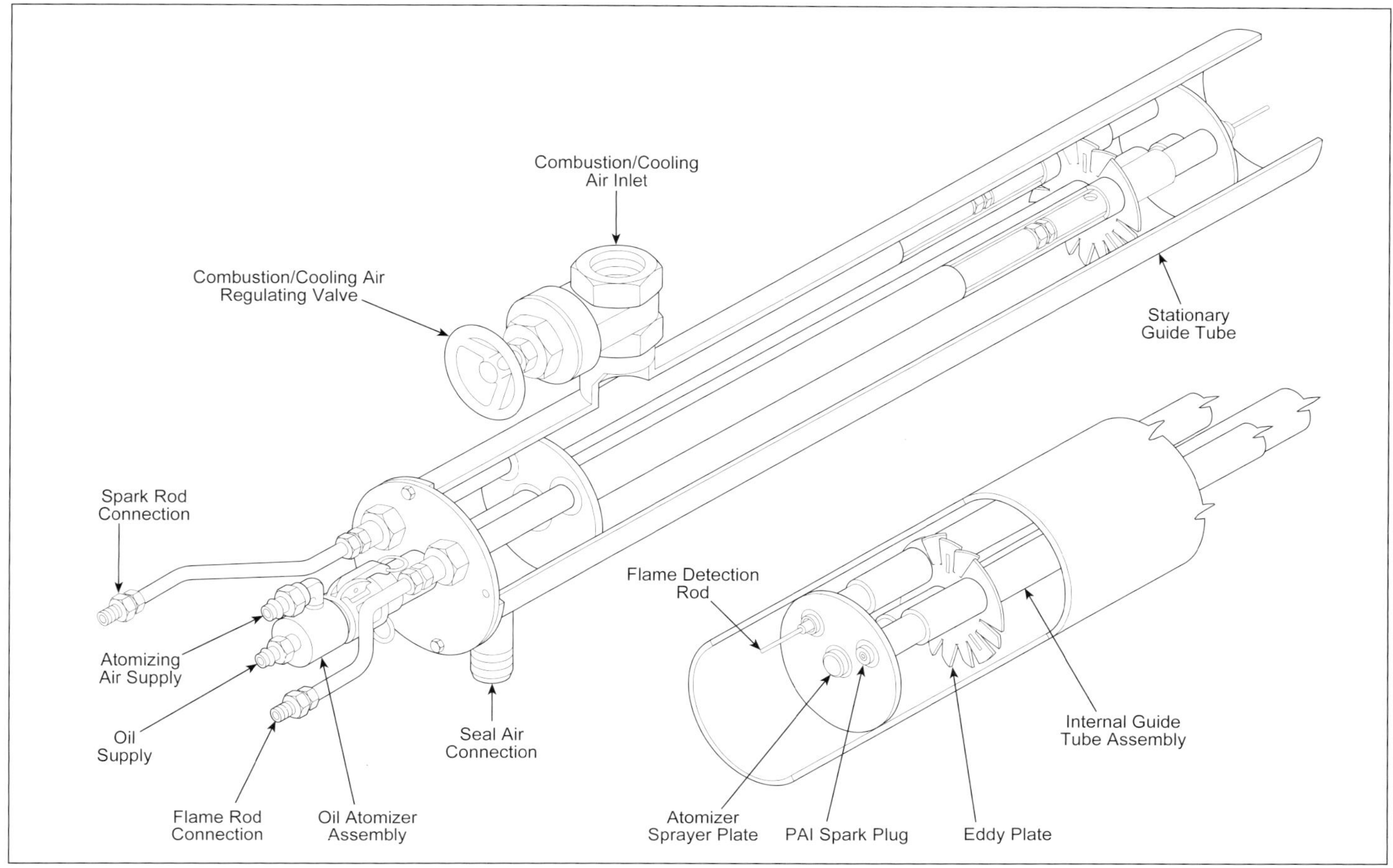

Fig. 15 FPS oil igniter.

portant, if not more so, when implementing combustion control techniques to reduce NO_x emissions.

In observance of the recommended rules of operation, an automatic control system should include the following:

1. purge interlocks requiring a specified minimum air flow for a specific time period to purge the setting before the fuel trip valve can be opened,
2. flame detectors on each burner, connected to an alarm and interlocked to shut off the burner fuel valve upon loss of flame,
3. closed position limit switches for burner shutoff valves, requiring that individual burner shutoff valves be closed to permit opening the fuel trip valve,
4. shutoff of fuel on failure of the forced or induced draft fan,
5. shutoff of fuel in the event of low fuel pressure (and low steam or air pressure to oil atomizers),
6. shutoff of fuel in the event of low oil temperature, and
7. shutoff of fuel in gas fired units in the event of excessive fuel gas pressure.

Any of the commonly used fuels can be safely burned when using the proper equipment and operating skill. Hazards are introduced when, through carelessness or mis-operation of equipment, the fuel is no longer burned in a safe manner. While a malfunction should be corrected promptly, panic must be avoided. Investigation of explosions of boiler furnaces equipped with good recording apparatus reveal that conditions leading to the explosion had, in most cases, existed for a considerable period of time. Sufficient time was available for someone to have taken unhurried corrective action before the accident. The notion that malfunctions of safety equipment cause frequent, unnecessary unit trips is not consistent with the facts.

FPS is a trademark of Fossil Power Systems, Inc.

Orimulsion is a trademark of Bitumenes Orinoco, S.A.

Coal storage and handling facilities at a utility power plant.

Chapter 12

Solid Fuel Processing and Handling

Coal remains the dominant worldwide source of energy for steam generation. However, additional solid fuels such as wood by-products and municipal wastes are also in use. The large scale continuous supply of such solid fuels for cost-effective and reliable steam power generation requires the effective integration of recovery (e.g., mining), preparation, transportation and storage technologies. The relationships between these are illustrated in Fig. 1. While each fuel offers unique challenges, a discussion of the processing and handling of coal helps identify many of the common issues and considerations for all solid fuels. Selected additional topics about the special aspects of some other solid fuels are covered in Chapters 28 through 30.

Mining is the first step in producing coal. Raw coal can be treated to remove impurities and to provide a more uniform feed to the boiler. The resulting reduction in ash and sulfur can significantly improve overall boiler performance and reduce pollutant emissions. The transportation of coal to the plant may represent a major portion of the plant's total fuel cost, although mine-mouth generating stations can minimize these transportation costs. Storage and handling of large coal quantities at the plant site require careful planning to avoid service disruptions.

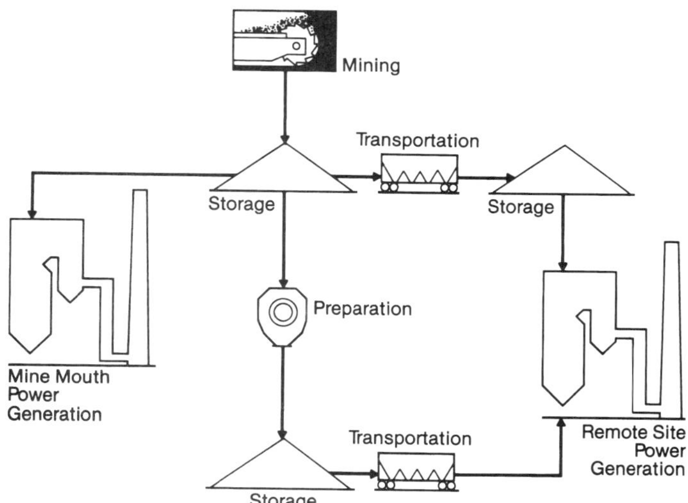

Fig. 1 Fuel supply chains for coal-fired power generation.

The changing nature of electric utility industry regulation, and the increasingly stringent emissions control requirements have dramatically changed the way coal is used to generate electric power in the United States (U.S.). Where once a new boiler was dedicated to one or two sources of coal over its lifetime, today active fuel management with multiple sources of coal and transport options is a core part of power plant competition and profitability. The plant objective is to meet the emission control requirements at the lowest overall power production cost.

Coal source flexibility has become necessary for economic survival. Plant owners have had to meet these new challenges by employing multiple coal sources, some very remote from the plant. This typically has required the installation of new and more flexible coal handling equipment in the plant storage facilities.

Coal mining

As discussed in Chapter 9, electric utility coal consumption for power generation dominates the market for coal produced in the U.S. and in the rest of the world. Coal production in the U.S. reached approximately 1.1 billion short tons per year at the turn of the 21st century. At the same time, electric utilities accounted for 92% of the coal consumed. Overall worldwide production and consumption of coal has been steadily increasing since the late 1980s, with the major producers being China, the U.S., India and Australia. Together these countries produced more than 65% of the total amount of coal mined.[1-3]

U.S. coal production is split between surface mining, which produces 66% of the total and underground mining, which produces 34% of the total.[2] Because of the geologic locations of coal deposits or *seams*, and cost considerations, surface mining dominates coal production west of the Mississippi River and underground mining dominates in the East. In 2003, Wyoming, West Virginia and Kentucky led the nation in coal production with a 59% combined total. The emergence of large, high volume surface mines producing low sulfur coal in the west has resulted in a continuing shift of coal production from the eastern U.S. to the west.[1]

Surface mining

Coal may be recovered from relatively shallow seams by removing the overlying earth, or *overburden,* to expose the coal seam. Typically, topsoil is first removed and stored for later use in reclamation of the site. The remaining overburden is drilled and blasted to loosen the rock for removal with a dragline or excavating shovel. The overburden is then methodically stripped away and stored for restoration of the land to the original contour following removal of the coal. A dragline is commonly used to expose the coal seam (Fig. 2). The coal may then be removed using a bulldozer and front-end loader or a shovel. A mobile crusher and screen may be set up in the mine for initial sizing of the raw coal. The coal is then loaded into trucks for delivery to a cleaning facility or steam generating plant.

The *stripping ratio* is defined as the unit amount of overburden that must be removed to access a unit amount of coal. In general, surface mines in the western U.S. have lower stripping ratios than mines in the east.

Strict environmental regulations limit the amount of land surface area that may be exposed at any one time, control water runoff, and establish land reclamation procedures.

Underground mining

Many coal reserves are accessible only by underground mining. (See Chapter 9.) Continuous mining, the most prevalent form of underground mining, makes use of machines like those in Fig. 3. It accounts for approximately 55% of the total underground production of coal.

Room and pillar mining Most underground U.S. coal is produced using a technique known as room and pillar mining. A series of headings or parallel entries are cut into the coal seam to provide passages for the mining machinery, uncontaminated ventilation air, and conveying equipment. These headings are typically 18 to 20 ft (5.5 to 6.1 m) wide and may be several miles (km) long. Rooms are driven off of the main headings to the property limits of the mine. Typically, only about 50% of the coal is removed when the rooms are first mined. The remaining coal is left in place to support the roof. Roof bolts are installed where the coal has been removed for additional roof support. At the completion of the development cycle, the remaining coal may be removed by retreat mining, and the unsupported roof is allowed to collapse.

Fig. 2 Large coal dragline in operation (*courtesy of National Coal Association*).

Fig. 3 Continuous mining machine in operation (*courtesy of National Coal Association*).

Longwall mining In longwall mining, shearers or plows are pulled back and forth across a panel of coal to break it loose from the seam. The coal falls onto a flight conveyor and is transported to the main haulage line. Because essentially all of the coal is removed, artificial roof supports, known as shields or chocks, are used to cover the plow and conveyor. The plow, conveyor and roof supports are advanced using hydraulic jacks as the coal is removed from the mining face. The unsupported roof is allowed to cave in behind the supports.

Although more underground mines use the room and pillar method of mining, longwall mines generally produce more coal because they are larger operations overall. Since the early 1980s, there has been a significant increase in productivity using longwall mining. Today, more than one-half of the longwall mines in the U.S. are located in Appalachia, but this type of mining is being used more widely in western coal areas as well.

A key reason for the increase in productivity involves increased panel size of longwall mining sections. The average panel face width is 759 ft (231 m), and the average panel length is 6853 ft (2089 m). Some of these operating units reaching 10,000 ft (3048 m).[4]

Longwall mining can be a very high capacity production system provided the geologic conditions are suitable and the longwall has been carefully integrated with the existing panel development scheme and coal haulage infrastructure. Production capacities exceeding 6000 t of raw coal per shift have been reported. The annual production capacity for a longwall is a function of the seam height, panel width and length, and the time required to move the equipment between operating panels.

Raw coal size reduction and classification

Sizing requirements

Size reduction operations at steam power plants are usually confined to crushing and pulverizing, although it is sometimes more economical to purchase pre-crushed coal for smaller plants, especially stoker-fired units. Screening at the plant is generally not required, except to remove large impurities and trash using simple grids (or grizzlies) and rotary breakers. Techniques for determining coal particle size distribution are discussed in Chapter 9.

Coal particle size degradation occurs in transportation and handling and must be considered when establishing supply size specifications. This may be critical where the maximum quantity of coal fines is set by firing equipment limitations.

For stoker-fired installations, it is customary to specify purchased coal sized to suit the stoker (see Chapter 16), so that no additional sizing is required at the plant [typically 1.5 in. × 0 (38.1 mm × 0)].

For pulverized coal-fired boilers, a maximum delivered top size is usually specified with no limitation on the percentage of fines, so that the delivered coal is suitable for crushing and pulverizing in the available equipment. The coal is crushed to reduce particle size and then ground to a very fine size in the pulverizer. (See Chapter 13.)

Crushers alone may be used to provide the relatively coarser sizes required for Cyclone furnaces. (See Chapter 15.) A properly sized crusher efficiently reduces particle size while producing minimal fines.

Size reduction equipment selection

Size reduction equipment is generally characterized by the maximum acceptable feed size and the desired product top size. The reduction ratio is defined as follows:

$$\frac{p_{\text{feed}}}{p_{\text{product}}} \qquad \textbf{(1)}$$

where

p_{feed} = feed particle size in which 80% of the particles pass a given screen size or mesh

p_{product} = product particle size in which 80% of the particles pass a given screen size or mesh

Once-through crushing devices, that discharge the fines without significant re-crushing, are used to minimize the production of fines. Rotary breakers and roll crushers are commonly used to reduce the coal top size without producing a significant amount of fines.

The *rotary breaker*, illustrated in Fig. 4, reduces the coal to a predetermined maximum size and rejects larger refuse, mine timbers, trash, and some tramp metal. This equipment also breaks apart frozen coal prior to further coal processing. It consists of a large cylinder of steel screen plates that rotates at approximately 20 rpm. The size of the screen openings determines the top size of the coal. The coal feed at one end of the cylinder is picked up by lifting shelves and is carried up until the angle of the shelf permits the coal to drop onto the screen plate. The coal shatters and is discharged through the screen openings. The harder rock does not break as readily; it travels along the screen and is rejected at the discharge end. Wood, large rocks, and other trash that do not pass through the screen are separated from the coal. Rotary breakers may be installed at the mine, preparation plant, or steam generating plant.

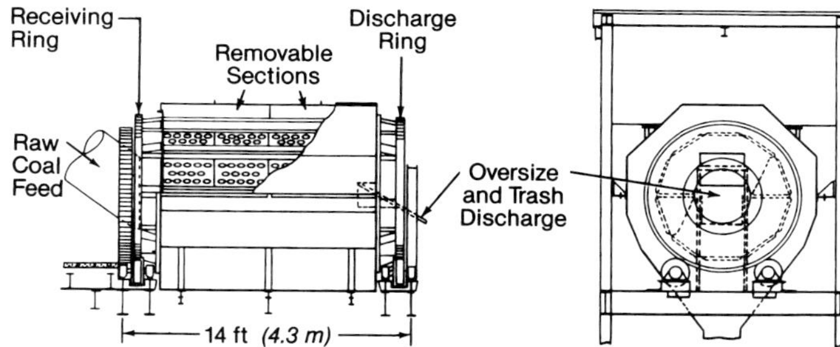

Fig. 4 Rotary breaker for use at the mine and power plant.

The elements of a *single-roll crusher* are illustrated in Fig. 5. This crusher consists of a single, toothed roll that forces the coal against a plate to produce the crushing action. The maximum product particle size is determined by the gap between the roll and the plate. To prevent jamming by large impurities such as tramp metal, the roll is permitted to rise or the plate can swing away, allowing the impurities to pass through. This is an old type of crusher commonly used for reducing run-of-mine bituminous coal to a maximum product of 1.25 to 6 in. (31.8 to 152 mm). The abrasive action between the coal and the plate produces some fines; however, they are discharged with minimal re-breakage.

In a *double-roll crusher,* the coal is forced between two counter-rotating toothed rolls (Fig. 6). The mating faces of both rolls move in a downward direction, pulling the coal through the crusher. The size of the roll teeth and the spacing between rolls determine the

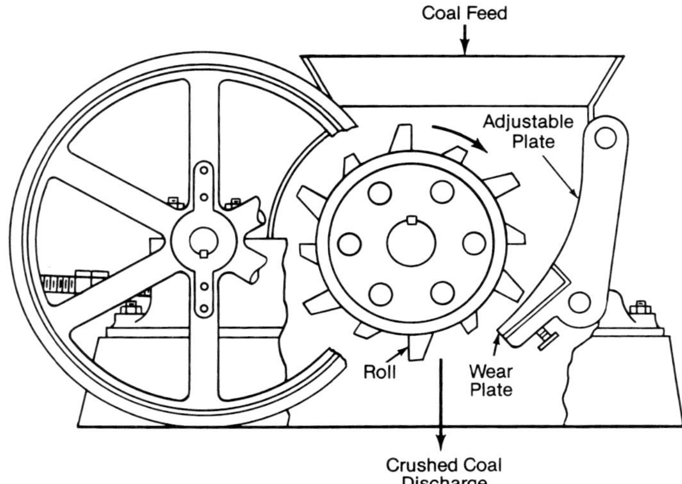

Fig. 5 Single-roll crusher – diagrammatic section.

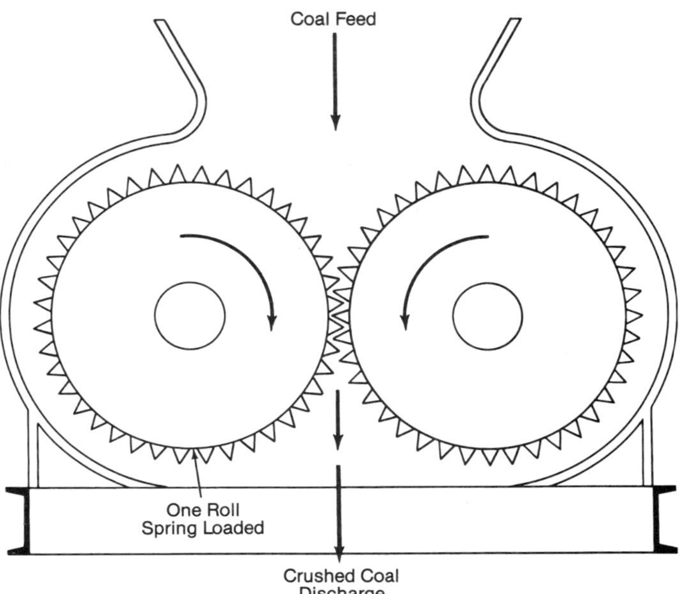

Fig. 6 Double-roll crusher – diagrammatic section.

product top size. One of the rolls may be spring loaded to provide a means for passing large, hard impurities. Double-roll crushers are used for reducing run-of-mine coal to smaller sizes at preparation and steam generating plants.

In *retention crushers* such as *hammer mills* and *ring crushers*, coal is retained in the breakage zone until it is sufficiently fine to pass through a screen to the discharge. The re-breakage action produces considerable fines, consequently these mills are not used in applications where fines are objectionable. They are often used to reduce run-of-mine coal to an acceptable size for feed to a stoker or pulverizer, e.g., 0.75 in. × 0 (19 mm × 0).

A *hammer mill* is depicted in Fig. 7. In this mill, the coal is broken by impact with the hammers mounted on a central shaft and permitted to swing freely as the shaft is rotated. The coal is fed at the top of the mill and is forced down and outward to the grate bars as it is struck by the hammers. The spacing of the bars determines the maximum size of the finished product. The coal remains in the mill, and breakage continues until the particles are fine enough to pass through the grate. A trap is usually provided for collection and removal of tramp metal.

Screen selection

Screening is usually performed at the mine or preparation plant to remove unwanted material and size coal for various uses prior to shipment to the steam generating plant. The run-of-mine coal is usually passed over a grid of steel bars, or grizzly, to remove mine timbers, rocks and other trash. The raw coal may be separated into various size fractions to meet contract specifications or for further processing in a preparation plant by passing the coal through various screens. Common screen types include gravity bar screens, trommels or revolving screens, shaker screens, and vibrating screens.

A *gravity bar screen* consists of a number of sloped parallel bars. The gaps between bars, the slope, and the length of the bars determine the separating size. The bars have tapered cross-sections; the gaps between the bars are smaller on the top side than on the bottom. This design reduces plugging.

A *revolving screen* consists of a slowly rotating cylinder with a slight downward slope parallel to the axis of coal flow. The cylinder is comprised of a perforated plate or a wire cloth, and the size of the openings determines the separating size. Because of the repeated tumbling as the coal travels along the cylinder, considerable breakage can occur. For this reason, revolving screens are not used for sizes larger than about 3 in. (76.2 mm). Because only a small portion of the screen surface is covered with coal, the capacity per area of screen surface is low.

A *shaker screen* consists of a woven wire mesh mounted in a rectangular frame that is oscillated back and forth. This screen may be horizontal or sloped slightly downward from the feed end to the discharge end. If the screen is horizontal, it is given a differential motion to help move the coal along its surface.

Bituminous coal generally fractures into roughly cubicle shapes, while commonly associated slate and shale impurities fracture to form relatively thin slabs. This shape difference enables the impurities to be separated with a *slotted shaker*.

Vibrating screens are similar to shaker screens except that an electric vibrator is used to apply a high frequency, low magnitude vibration to the screen. The screen surface is sloped downward from the feed to the discharge end. The vibration helps to keep the mesh openings clear of wedged particles and helps to stratify the coal so that fine particles come in contact with the screen surface. For screening fine, wet coal, water sprays are used to wash fine particles through the coal bed and the screen surface. Vibrating screens are the most widely used types for sizing and preliminary dewatering.

It is common practice to separate the fines from the coarse coal to improve the efficiency of subsequent cleaning and dewatering processes. The fines may be discarded, cleaned separately, or bypassed around the cleaning process and then blended back into the coarse clean coal product.

Coal cleaning and preparation

The demand for coal cleaning has increased in response to environmental regulations restricting sulfur dioxide (SO_2) emissions from coal-fired boilers. The demand is also due to a gradual reduction in run-of-mine coal quality as higher quality seams are depleted and continuous mining machines are used to increase production. Approximately 70% of coal mined for electric utility use is cleaned in some way.

Coal cleaning and preparation cover a broad range of intensity, from a combination of initial size reduction, screening to remove foreign material, and sizing discussed previously, to more extensive processing to remove additional ash, sulfur and moisture more intimately associated with the coal.

The potential benefits of coal cleaning must be bal-

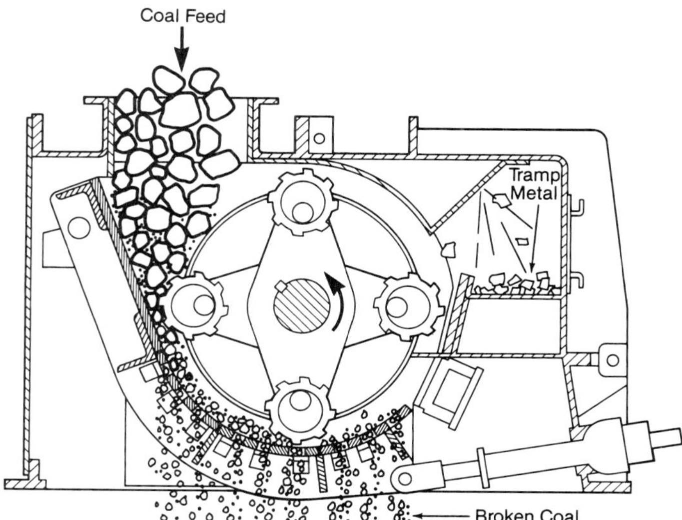

Fig. 7 Ring hammer mill crusher – diagrammatic section (*courtesy of Pennsylvania Crusher Corporation*).

anced against the associated costs. The major costs to consider, in addition to the cleaning plant capital and operating costs, include the value of the coal lost to the refuse product through process-related inefficiencies and the cost of disposing of the refuse product. Generally, the quantity of coal lost increases with the degree of desired ash and sulfur reduction. An economic optimum level of ash and sulfur reduction can be established by balancing shipping and post-combustion cleanup costs against pre-combustion coal cleaning costs.

Coal characterization

Coal is a heterogeneous mixture of organic and inorganic materials as described in detail in Chapter 9. Coal properties vary widely between seams and within a given seam at different elevations and locations. The impurities associated with coal can generally be classified as inherent or extraneous. *Inherent impurities* such as organic sulfur can not be separated from the coal by physical processes. *Extraneous impurities* can be partly segregated from the coal and removed by physical coal cleaning processes. The extent to which these impurities can be economically removed is determined by the degree of material dissemination throughout the coal matrix, the degree of liberation possible at the selected processing particle size distribution, and physical limitations of the processing equipment.

Mineral matter associated with the raw coal forms ash when the coal is burned. Ash-forming mineral matter may also be classified as inherent or extraneous. Inherent mineral matter consists of chemical elements from plant material organically combined with coal during its formation. This mineral matter generally accounts for less than 2% of the total ash. Extraneous mineral matter consists of material introduced into the deposit during or after the coalification process, or is extracted with the coal in the mining process.

Sulfur is always present in coal and forms SO_2 when the coal is burned. If the sulfur is not removed before combustion, the SO_2 that forms is exhausted through the stack or removed by post-combustion flue gas treatment, discussed in Chapters 32 and 35. Sulfur is generally present in coal in three forms: pyritic, organic or sulfate.

Pyritic sulfur refers to sulfur combined with iron in the minerals pyrite (FeS_2) or marcasite. Pyrite may be present as lenses, bands, balls or as finely disseminated particles. *Organic sulfur* is chemically combined with molecules in the coal structure. *Sulfate sulfur* is present as calcium or iron sulfates. The sulfate sulfur content of coal is generally less than 0.1%.

The total sulfur in U.S. coals can vary from a few tenths of a percent to more than 8% by weight. The pyritic portion may vary from 10 to 80% of the total sulfur and is usually less than 2% of the coal by weight (Table 1).

The larger pyritic sulfur particles can generally be removed by physical cleaning, but finely disseminated pyritic sulfur and organic sulfur can not. Advanced physical and chemical cleaning technologies are under development to remove these sulfur forms.

Moisture can also be considered an impurity because it reduces the heating value of raw coal. Inherent moisture varies with coal rank, increasing from 1 to 2% in anthracite to 45% or more in lignite. Surface moisture can generally be removed by mechanical or thermal dewatering. This drying requires an energy expense at the cleaning plant or the steam generating plant (in pre-drying or during combustion). Drying before shipment reduces transportation costs on a per-Btu basis. When pre-drying is used, atmospheric oxidation will tend to be increased for low rank coals because of the exposure of additional oxidation sites in the particles.

The distribution of ash and sulfur in a coal sample can be characterized by performing a *washability analysis*. This analysis consists of separating the raw coal into relatively narrow size fractions and then dividing each fraction into several specific gravity fractions. The coal in each size/specific gravity fraction is

Table 1
Distribution of Sulfur Forms in Various Coals (%)

Mine Location County, State	Coal Seam	Total Sulfur	Pyritic Sulfur	Organic Sulfur
Henry, MO	Bevier	8.20	6.39	1.22
Henry, MO	Tebo	5.40	3.61	1.80
Muhlenburg, KY	Kentucky #11	5.20	3.20	2.00
Coshocton, OH	Ohio #6	4.69	2.63	2.06
Clay, IN	Indiana #3	3.92	2.13	1.79
Clearfield, PA	Upper Freeport	3.56	2.82	0.74
Franklin, IL	Illinois #6	2.52	1.50	1.02
Meigs, OH	Ohio #8A	2.51	1.61	0.86
Boone, WV	Eagle	2.48	1.47	1.01
Walker, AL	Pratt	1.62	0.81	0.81
Washington, PA	Pittsburgh	1.13	0.35	0.78
Mercer, ND	Lignite	1.00	0.38	0.62
McDowell, WV	Pocahontas #3	0.55	0.08	0.46
Campbell, WY	Gillette	0.46	0.14	0.32
Pike, KY	Freeburn	0.46	0.13	0.33
Kittitas, WA	Big Dirty	0.40	0.09	0.31

then analyzed for ash, sulfur and heating value content. The hardness and distribution of the impurities relative to the coal determine if the impurities are concentrated in the larger or smaller size fractions. Relatively soft impurities are generally found in the finer size fractions. In general, the lowest specific gravity fractions have the lowest ash content, as indicated in Table 2.

The information generated by these *float/sink* characterization tests can be used to predict the degree of ash and sulfur reduction possible using various specific gravity based cleaning technologies discussed below. In general, the more material that is present near the desired specific gravity of separation, the more difficult it is to make an efficient separation.

Coal cleaning and preparation operations

The initial steps in the coal cleaning process include removal of trash, crushing the run-of-mine coal, and screening for size segregation. These preliminary operations and associated hardware were discussed previously. The following operations are then used to produce and dewater a reduced ash and sulfur product. Fig. 8 provides a general layout of coal cleaning unit operations.

Gravity concentration Concentration by specific gravity and the subsequent separation into multiple products is the most common means of mechanical coal cleaning. Concentration is achieved because heavier particles settle farther and faster than lighter particles of the same size in a fluid medium. Coal and impurities may be segregated by their inherent differences in specific gravity, as indicated in Table 3.

The fluid separating medium may consist of a suspension of the raw coal in water or air, a mixture of sand and water, a slurry of finely ground magnetite (iron oxide, FE_3O_4), or an organic liquid with an intermediate specific gravity. Aqueous slurries of raw coal and magnetite are currently the most common separating media.

If the effective separating specific gravity of the media is 1.5, particles with a lower specific gravity are concentrated in the clean coal product and heavier particles are in the reject or refuse product. Several factors prevent ideal separation in practice.

Gravity separation processes concentrate particles by mass. The mass of a particle is determined by its specific gravity and particle size. Raw coal consists of particles representing a continuous distribution of specific gravities and sizes. It is quite possible for a larger, less dense particle to behave similarly to a smaller particle with a higher specific gravity. For example, a relatively smaller pyrite particle may settle at a similar rate as a larger coal particle. The existence of *equal settling* particles can lead to separating process inefficiency. Fine pyrite in the clean coal product and coarse coal in the refuse are commonly referred to as *misplaced material*. The amount of misplaced material is determined by the quantity and distribution of the raw coal impurities, the specific gravity of separation, and the physical separation efficiency of the segregated material.

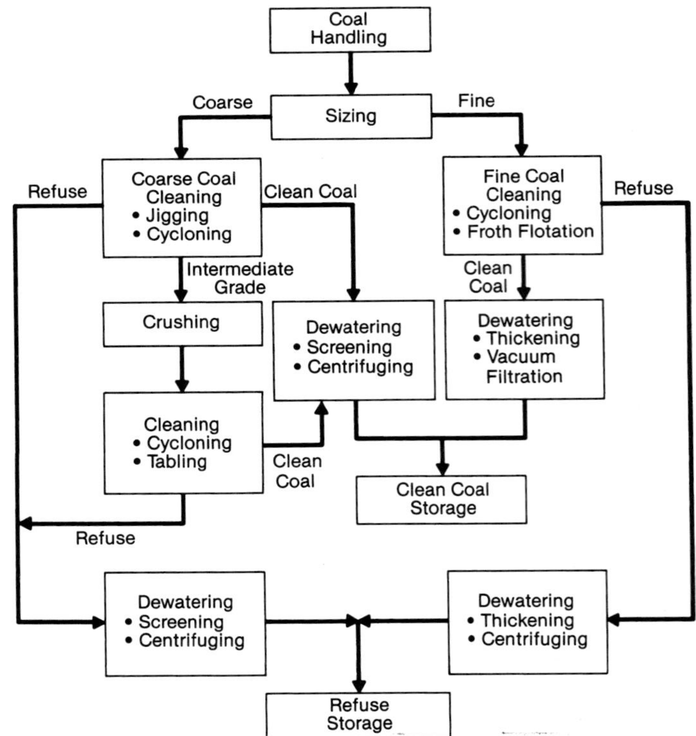

Fig. 8 General layout of coal cleaning operations.

Table 2
Typical Ash Contents of Various Bituminous Coal Specific Gravity Fractions

Specific Gravity Fraction	Ash Content % by wt
1.3 to 1.4	1 to 5
1.4 to 1.5	5 to 10
1.5 to 1.6	10 to 35
1.6 to 1.8	35 to 60
1.8 to 1.9	60 to 75
Above 1.9	75 to 90

A significant amount of material with a specific gravity close to the desired specific gravity of separation results in a more inefficient separation. If the amount of *near gravity material* exceeds approximately 15 to 20% of the total raw coal, efficient gravity separation is difficult.

The most common wet gravity concentration techniques include jigging, tabling, and dense media separation processes. Each technique offers technical and economic advantages.

Jigging In a coal jig, a pulsating current of water is pushed upward in a regular, periodic cycle through a bed of raw coal supported on a screen plate. This upward or pulsion stroke of the cycle causes the bed to expand into a suspension of individual coal and refuse particles. The particles are free to move and generally separate by specific gravity and size, with the lighter and smaller pieces of coal moving to the upper region of the expanded bed. In the downward or suction stroke of the cycle, the bed collapses, and the separation is enhanced as the larger and heavier pieces of rock settle faster than the coal. The pulsion/suction

Table 3
Typical Specific Gravities of Coal and Related Impurities

Material	Specific Gravity
Bituminous coal	1.10 to 1.35
Bone coal	1.35 to 1.70
Carbonaceous shale	1.60 to 2.20
Shale	2.00 to 2.60
Clay	1.80 to 2.20
Pyrite	4.80 to 5.20

cycle is repeated continuously. The separated layers are split at the discharge end of the jig to form a clean coal and a refuse product. The bed depth where the cut is made determines the effective specific gravity of separation.

The upward water pulsation can be induced by using a diaphragm or by the controlled release of compressed air in an adjacent compartment. Operation of a Baum jig is illustrated in Fig. 9. This type of jig may be used to process a wide feed size range. Typically, the specific gravity of separation ranges from 1.4 to 1.8. The separation efficiency may be enhanced by pre-screening the feed to remove the fines for separate processing.

Tabling A concentrating, pitched table is mounted so that it may be oscillated at a variable frequency and amplitude. A slurry of coal and water is continuously fed to the top of the table and is washed across it by the oncoming feed. Diagonal bars, or *riffles*, are spaced perpendicular to the flow of particles. The coal-water mixture and oscillating motion of the table create a *hindered settling* environment where the lower gravity particles rise to the surface. Higher specific gravity particles are caught behind the riffles and transported to the edge of the table, away from the clean coal discharge.

Tables are generally used to treat 0.375 in. × 0 (9.53 mm × 0) coal. Three or four tables may be stacked vertically to increase throughput while minimizing plant floor space requirements.

Dense media separation In dense or heavy media separation processes, the raw coal is immersed in a fluid with a specific gravity between that of the coal and the refuse. The specific gravity differences cause

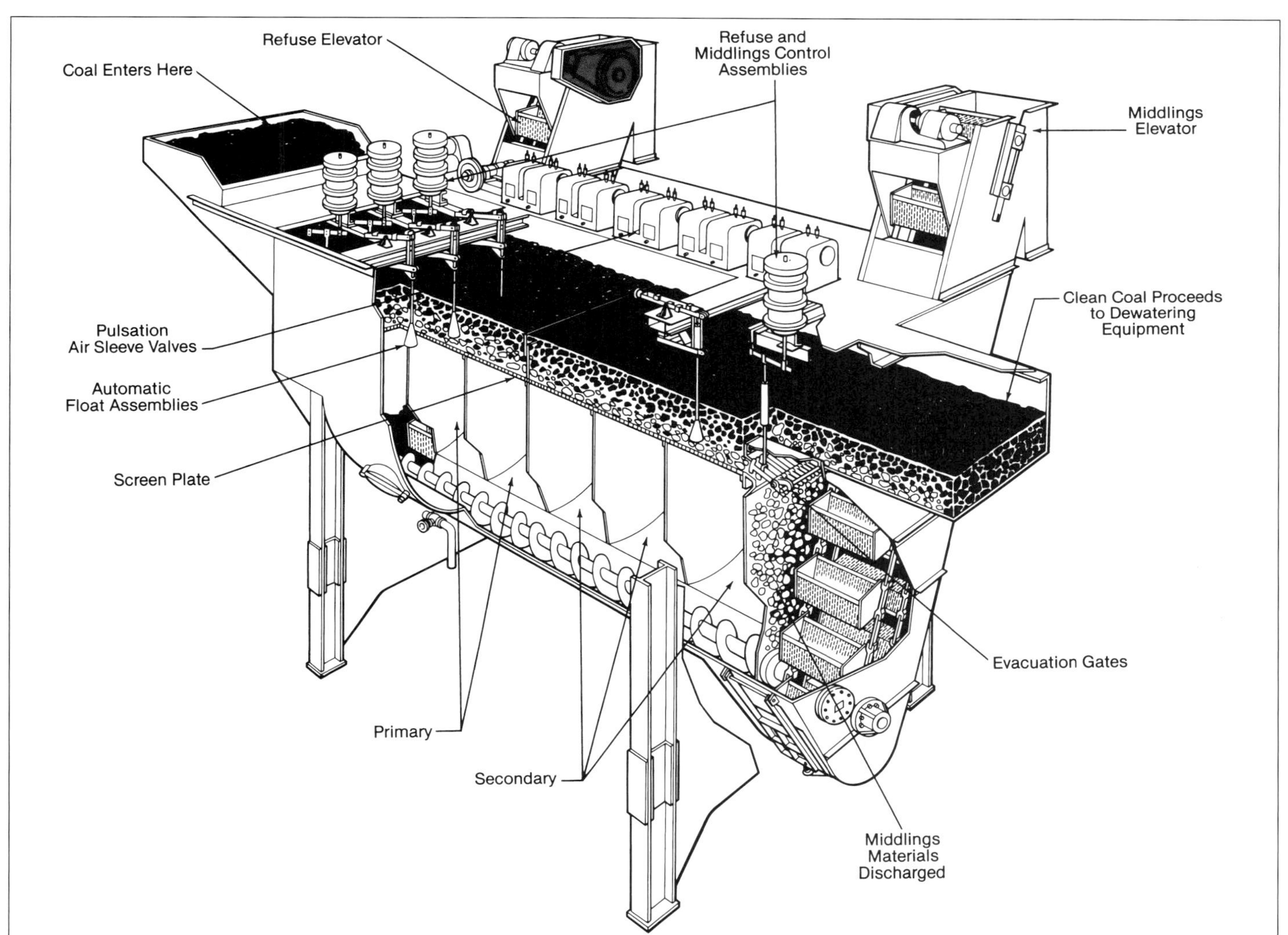

Fig. 9 Typical Baum jig for coal preparation.

the coal and refuse to migrate to opposite regions in the separation vessel. In coal preparation, the heavy media fluid is usually an aqueous suspension of fine magnetite in water.

Flotation Coal and refuse separation by *froth flotation* is accomplished by exploiting differences in coal and mineral matter surface properties rather than specific gravities. Air bubbles are passed through a suspension of coal and mineral matter in water. This suspension is agitated to prevent particles from settling out. Air bubbles preferentially attach to the coal surfaces that are generally more hydrophobic, or difficult to wet. The coal then rises to the surface where it is concentrated in a froth on top of the water. The mineral matter remains dispersed (Fig. 10). Chemical reagents, referred to as collectors and frothers, are added to enhance the selective attachment of the air bubbles to the coal and to permit a stable froth to form.

Flotation is generally used for cleaning coal finer than 48 mesh (300 microns). The efficiency of the process can be enhanced by carefully selecting the type and quantity of reagents, fine grinding to generate discrete coal and refuse particles, and generating fine air bubbles.

Dry processing Dry coal preparation processes account for a small percentage of the total coal cleaned in the U.S. In general, pneumatic processing is only applied to coal less than 0.5 in. (12.7 mm) in size with low surface moisture.

Dewatering Dewatering is a key step in the preparation of coal. Reducing the fuel's moisture content increases its heating value per unit weight. Because coal shipping charges are based on tonnage shipped, a reduction in moisture content results in lower shipping costs per unit heating value.

Coarse coal, greater than 0.375 in. (9.53 mm) particle size, can be sufficiently dewatered using vibrating screens. Intermediate size coal, 0.375 in. (9.53 mm) by approximately 28 mesh (600 microns), is normally dewatered on vibrating screens followed by centrifuges.

Fine coal dewatering often involves the use of a thickener to increase the solids content of the feed to a vacuum drum, vacuum disc filter, or high gravity centrifuge. The filter cake may be mixed with the coarser size fractions to produce a composite product satisfying the specifications. Fine coal dewatering also serves to clarify the water for reuse in the coal preparation plant. Fines must be separated from the recycled water to maximize the efficiency of the separation processes.

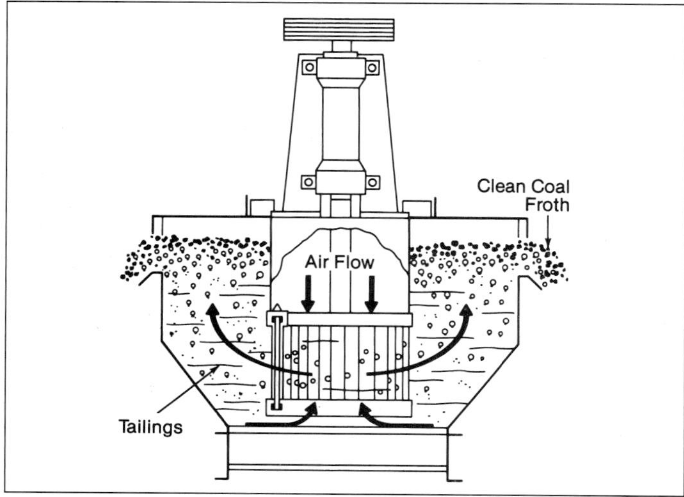

Fig. 10 Flotation cell.

Thermal dewatering may be necessary to meet product moisture specifications when the raw coal is cleaned at a fine size to maximize ash and sulfur rejection. The various types of thermal dryers include rotary, cascade, reciprocating screen, suspension and fluidized-bed dryers. Cyclones or bag filters are used to prevent fine dust emissions from the dryer. The collected fine coal may be recycled to support dryer operation. Thermal drying represents an economic tradeoff of reduced product moisture content versus heat required to fire the dryer.

Impact on steam generator system operations

The principal benefit of coal cleaning is the reduction in ash and sulfur content. Reduced ash content results in lower shipping costs and reduced storage and handling requirements at the plant on a cost per unit heating value basis. Boiler heat transfer effectiveness may increase as a result of reduced ash deposition on tube surfaces. A reduction in sulfur content leads directly to reduced SO_2 emissions. Lower sulfur feed coal may preclude the need for or reduce the performance requirements of post-combustion SO_2 emission control systems. A reduction in sulfur content may also reduce spontaneous combustion during storage, and corrosion in coal handling and storage equipment. Reduced ash content can result in reduced maintenance through removal of abrasive pyrite and quartz from the coal. Reduction of clay in the coal can improve handling and bunker or silo storage characteristics, but this may be offset by the effects of higher fines content and higher surface moisture on cleaned coal.

Coal transportation

The means of transportation and the shipping distance significantly influence the total fuel cost, reliability of supply, and fuel uniformity at the power plant. In some cases where western U.S. coal is shipped over an extended distance, freight costs may represent 75 to 80% of the total delivered fuel cost. At the other extreme, transportation costs may be negligible for mine-mouth generating stations. In transit, the coal's handling characteristics may be changed by freezing, increased moisture content, or size degradation. When open rail car, truck or barge transport is used, the moisture content of the delivered coal depends on the initial moisture level, the weather conditions in transit, and the particle size distribution. Size degradation during shipping is dependent on the coal friability (ease of crumbling) and the techniques and number of transfers. As previously stated, for pulverized coal applications, size degradation is generally not a concern.

Coal is primarily shipped by rail, barge, truck and conveyor, but can also be transported by pipelines and tramway systems. The volume and distribution of coal transported by these various means is summarized in Table 4.[5] Combinations of these methods are often used to obtain the lowest delivery cost. Available transportation infrastructure, haulage distance, required flexibility, capital cost and operating cost are important factors in selection of a system for delivering coal to the power plant.

In general, barge transport represents the lowest unit cost per ton per mile followed by rail, truck and conveyor in terms of increasing cost. Combinations of these four transportation systems may be used to move coal to loading docks for overseas shipment.

Transportation systems are generally designed to minimize intermediate storage of coal to control inventory costs, reduce insurance costs, and minimize the effects of changes that can reduce the commercial value of coal. Potentially harmful changes include a reduction of heating value, particle size degradation, and loss due to self-ignition or wind and water erosion.

Rail

Railroads delivered approximately 70% of the coal transported to electric generating power plants in the U.S. in 2002, and nearly 95% of this total was shipped in unit trains. Unit trains travel from the loading facility to the customer without stopping and normally consist of 50 or more cars with a total of 10,000 t of coal or more. Bottom dump and rotary dump rail cars [100 t (91 t_m) capacity] are typically used. These high capacity rail cars are generally not uncoupled from the time they are loaded at the mine until they arrive at the plant. In 1999, coal accounted for 44% of the rail industry's total freight tonnage and 22% of the revenues for Class I railroads.[6]

Rail transport provides for the movement of large quantities of coal over distances ranging from 10 to 1500 mi (16 to 2414 km).[7] Dedicated service between one mine and the steam generating plant simplifies management of coal deliveries. Improvements in coal-carrying capacity in rail cars and more powerful locomotives have resulted in higher efficiency in rail transport. In comparison to rail cars in the 1930s, today's coal car carries double the capacity.

The advantages of rail transport are offset slightly by the restricted rail access. Generally only one rail line is available to transport coal from a mine or to a specific steam generating plant. The installation of dedicated rail lines must be included as part of the cost of the coal handling and storage system. Rail spurs to a specific mine location are useful only for the life of the mining activity. Transit time is typically on the order of 4 to 20 days.[7] The rail car unloading system and intermediate storage facilities must be designed to quickly process the cars to avoid demurrage (delay) charges at the plant.

Railroads will continue to play a significant part in coal transportation as long as coal is used to generate electricity and remains the greatest and single-most important commodity.

Table 4
Distribution of Coal Transportation Methods for Electricity Generation (Thousand Short Tons) in the U.S. (2002)[5]

Method	10^6 t/yr	10^6 t_m/yr	% of Total
Rail	615	558	70
Barge	97	88	11
Truck	81	74	9
Conveyor/pipeline/ tramway	79	72	9
Other water shipping methods	11	10	1
U.S. totals	883	802	100

Barge

Barge transport of coal is the most cost-effective alternative to rail or truck. Approximately 11% of all coal shipped to electric generating plants in the U.S. is delivered by barge. According to the U.S. Army Corps of Engineers (COE), coal is the largest single barge commodity. Coal traffic accounted for 176 million tons of the total annual barge tonnage in 1990 (623 million tons).[8]

The cargo capacity difference between the three major transportation methods for carrying coal is shown below:

1 Barge = 15 Rail Cars = 58 Trucks

A key to barge efficiency is the ability of barges to carry cargo many times their own weight. Additionally, less energy is expended to move cargo by barge than by rail car or truck. There are several styles of barges used for transporting various types of cargo, but standard open-topped jumbo hopper barges are commonly applied to coal transport.[9] (See Fig. 11.)

Export coal is shipped on the Great Lakes by large bulk carrier ships called colliers, but barges transport coal within the U.S. by using inland and intracoastal waterways. The major waterways for coal traffic in the U.S. are the Ohio, Mississippi and Black Warrior-Tombigbee Rivers. The quantity of coal shipped in a

Fig. 11 Typical barge transport for large quantities of coal.

single tow or string of barges is determined by the lock requirements of the river system being navigated. For example, on the Ohio River system, a tow of three barges wide by five barges long is commonly used because of the River's lock requirements.[10] However, on the relatively unobstructed lower Mississippi, tows of 30 barges are not uncommon.

There is a significant degree of competition and transport prices are generally stable. Some cost differences between upstream and downstream travel are common.

Barge transportation of coal to steam generating plants is constrained by the location and characteristics of the available river systems. Close proximity to waterways for direct loading and unloading is needed for efficient barge transportation. Barge delivery must be supported by truck, rail, or belt conveyor transloading at the mine location or the steam generating plant. The natural river network is not always the most direct route and may result in increased delivery time. The channel width and seasonal variability in water level are natural limitations for barge traffic. River lock sizes and condition of repair may restrict the maximum permitted tow size. Delays due to deteriorating locks and congestion may be significant on some river systems. In some areas, lock repair costs are recovered through a surcharge on tonnage shipped through the lock.

Barges are not self-unloading. Barge unloading can consist of a simple clam-shell crane discharging onto a take-away conveyor. For large capacities, high-rate automated bucket-elevator unloaders are often used. Capacities of these systems range from several hundred to several thousand tons per hour. System capacity must be considered to limit demurrage time for barge delivery and unloading at large power stations. The capital investment required for the unloading facilities may restrict barge deliveries to plants using more than 50,000 t/yr.[7]

Truck

For power plants located near mines, trucks loaded at the mine deliver coal directly to the power plant storage site. In 2002, trucking accounted for 9% of the total tonnage of coal delivered to electric generating plants.[5] The truck deliveries may unload directly onto the coal storage pile or into hoppers feeding the automated conveying system for distribution to the storage pile or silos.

Trucking also plays an important role in both rail and barge transport. Many times coal transported by rail or barge is first trucked to the loading dock or involves truck transfer at the mine. Highway trucks typically carry 15 to 30 t (14 to 27 t_m) of coal over distances up to 70 mi (113 km). Off-road vehicles can handle 100 to 200 t (91 to 181 t_m) over a range of 5 to 20 mi (8 to 32 km) at mine-mouth generating stations.

Trucking is the most flexible mode of coal transportation. It is relatively easy to adjust to changes in demand to meet the generating plant's variable supply requirements. The short haulage distances, and therefore short delivery times, can be used to minimize storage requirements at the generating plant. Trucks are simple to unload and a minimum of on-site handling and distribution is needed. Use of the existing highway infrastructure provides for flexible delivery routes and reduces travel restrictions associated with rail and river transport. Trucks are very efficient for short haulage distances and for smaller generating plants. Trucking is the least capital intensive mode of transporting coal and a high degree of competition exists.

Truck transportation is characterized by a high operating cost per ton mile relative to barge or rail transport. Practical haulage distances are usually limited to 50 mi (80 km). State and local transportation regulations often limit loads to 25 t (23 t_m) or less. A large generating plant would require a significant amount of truck traffic and congestion at the delivery site may be severe. Truck deliveries require the highest degree of monitoring at the plant. Frequently, every truck must be weighed.

Continuous transport

Coal may be transported from the mine to the generating plant by continuous belt conveyors, slurry pipelines or tramways. In 2002, continuous transport systems accounted for approximately 9% of the total coal deliveries.[5] Belt conveyors are normally limited to lengths of 5 to 15 mi (8 to 24 km). The coal delivery rate is a function of the belt width, operating speed, and the number of transfer points. Only one major coal slurry pipeline is in operation. The 273 mi (439 km) long Black Mesa Pipeline runs from a mine in Arizona to a generating plant in Nevada. The coal transport rate is determined by the pipe diameter, slurry velocity and solids loading.[11]

Continuous systems can move large amounts of coal cost-effectively over short distances. Often, continuous systems can be used where the terrain limits the use of other modes of transport. Social and environmental impacts are minimal.

The application of continuous transportation systems is limited by the proximity to the generating plant, a low degree of operating flexibility due to the fixed carrying capacity, the inflexibility of the loading and discharge locations, high capital cost, and a relatively high energy consumption per ton mile of coal delivered. Pipeline builders must overcome significant opposition in obtaining rights of way and water resource allocation. The added costs associated with dewatering the coal at the generating plant must also be considered.

Coal handling and storage at the power plant

Bulk storage of coal at the power plant is necessary to provide an assured continuous supply of fuel. The tonnage of coal stored at the site is generally proportional to the size of the boiler. A 100 MW plant burns approximately 950 t/d (862 t_m/d), while a 1300 MW plant requires approximately 12,000 t/d (10,887 t_m/d). For most power plants, a 30 to 90 day supply is stored at the plant to assure adequate supply through any delivery interruptions. Public utilities in the U.S. are required by law to maintain minimum supplies. However, stored coal represents substantial working capital and requires land that may be otherwise productive. Economic considerations are a key factor in determining when to purchase coal and how much coal to store at the plant. Additional considerations, such as the changes to coal characteristics due to weathering, restrict the maximum amount of coal stored on site.

For smaller, industrial boiler applications, bin or silo storage may be preferred over stockpile storage. The advantages of bin storage include shelter from the weather and ease of reclamation. Prefabricated bins with capacities holding several hundred tons are commercially available as are large, field-erected concrete silos of several thousand tons capacity. In regions with severe winter weather, even large power stations use silo storage to facilitate easier reclamation.

The complexity of the coal storage and handling operations increases in proportion to the size of the steam generating plant. Efficient techniques have been developed for large and small plants. The components of a sophisticated coal storage and handling system for a large, 1000 MW electric generating plant are illustrated in Fig. 12. Coal is delivered in self-unloading rail cars and is transferred to a large stock-

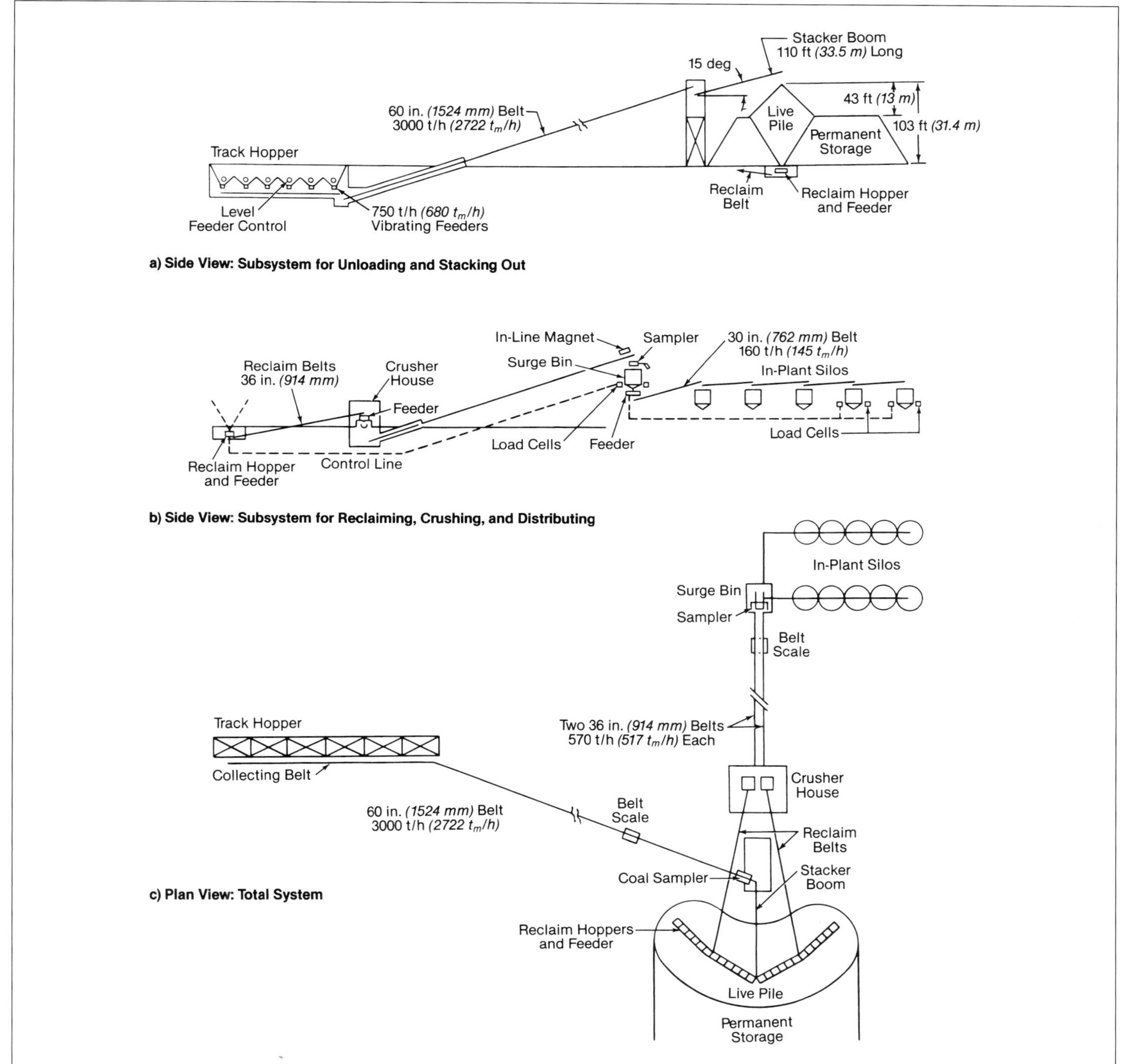

Fig. 12 Typical coal handling system and subsystems for a 1000 MW coal-fired power generation plant.

pile. An automatic reclaim system recovers coal from the stockpile for crushing and distribution to in-plant storage silos. The system is automated and a two person crew can handle 7000 t/d (6350 t_m/d) of coal. All the equipment from the reclaim feeders to the in-plant silos to the boiler is controlled by the central control room operator.

The storage and handling operations of a utility boiler are depicted in the chapter frontispiece.

Raw coal handling

An extensive array of equipment is available for unloading coal at the plant site and distributing it to stockpile and bin storage locations. Equipment selection is generally based on the method of coal delivery to the plant, the boiler type, and the required coal capacity. For small plants, portable conveyors may be used to unload rail cars, to reclaim coal from yard storage piles, and to fill bunkers. Larger plants require dedicated handling facilities to meet the demand for a continuous fuel supply. However, even relatively small plants may benefit from the improved plant appearance, cleanliness and reduced coal handling labor requirements associated with mechanical handling systems.

The coal handling system components are determined by the design and requirements of the boiler. If coal particle size is not specified to the coal supplier, then a crusher is normally integrated into the system to generate a uniform top size coal feed to the pulverizer or feeder equipment of circulating fluidized bed (CFB) or stoker-fired boilers. The system normally includes a magnetic separator to remove misplaced mining tools, roof support bolts, and other metallic debris that could damage the pulverizer or feeder equipment. Crushing and tramp metal removal needs are generally less critical for stoker-fired boilers than for pulverized coal and CFB units.

The coal handling system capacity is determined by the boiler's rate of coal use, the frequency of coal deliveries to the plant, and the time allowed for unloading. In most large plants, only four to six hours per day are dedicated to unloading of coal deliveries.

Rail car unloading

Rail cars can either be of the bottom discharge or rotary dump type. In automatic rotary car dumping systems, the rail cars are hydraulically or mechanically clamped in a cradle, and the cradle is rotated about a rotating track section so the coal falls into a hopper below the tracks. The rail cars have special swivel couplings to allow the dumping to be completed without uncoupling the cars. The rotary dump system advantages of short cycle time and high capacity are offset by a relatively high capital cost.

With bottom dump cars, unloading is relatively simple when the coal is dry and free flowing. However, high surface moisture can cause the coal to adhere within the car and, in cold weather, freeze into a solid mass. In hot, dry weather, high winds can create severe dust clouds at the unloading station unless special precautions are taken. The coal supplier frequently sprays the coal with oil or an anti-freezing chemical such as ethylene glycol as the car is loaded to settle the fines and to improve handling in freezing weather. The treatment does not appreciably affect combustion or cause problems in the pulverizers. There is also some evidence that the treatment may reduce adherence and hangups in bunkers and chutes.

A simple bottom dump rail car unloading system that includes a crusher and magnetic separator is illustrated in Fig. 13. A screw conveyor is used to distribute coal along the length of the bunker. The capacity of the bucket elevator generally limits this system to relatively small plants.

A rotary dump rail car unloading and handling sys-

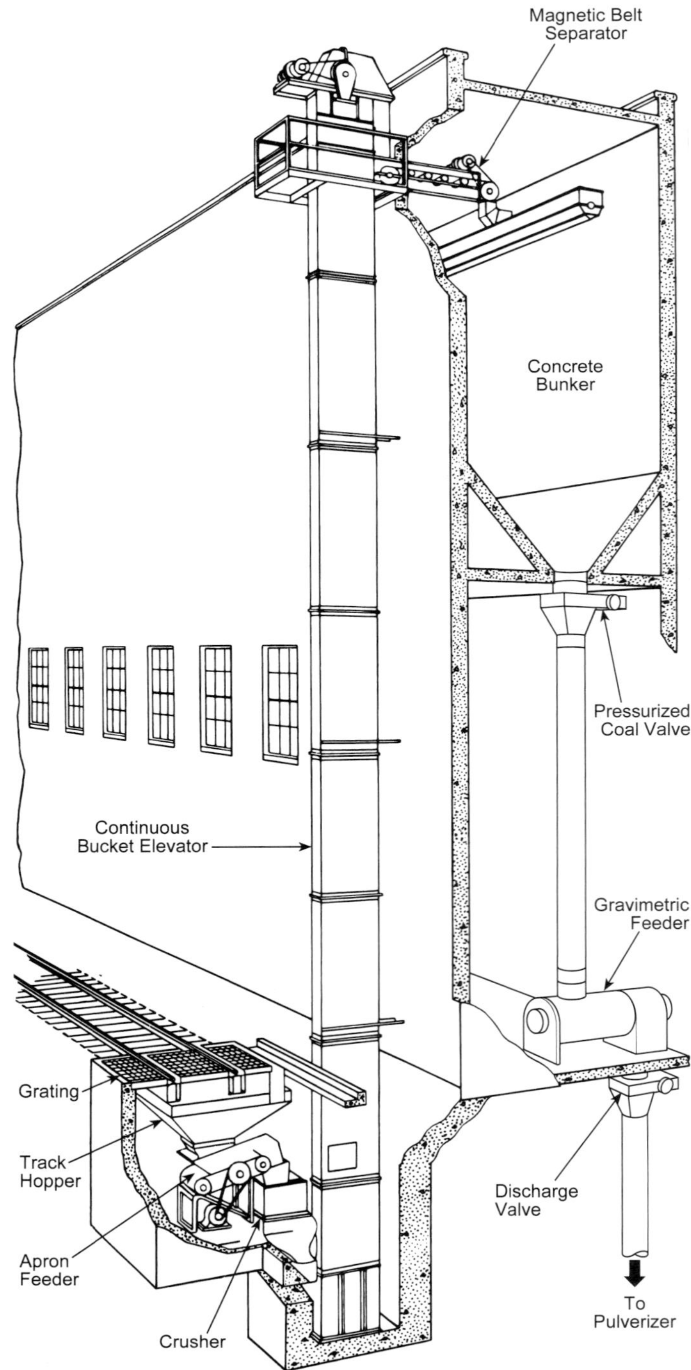

Fig. 13 Rail car dumping system for a small power plant.

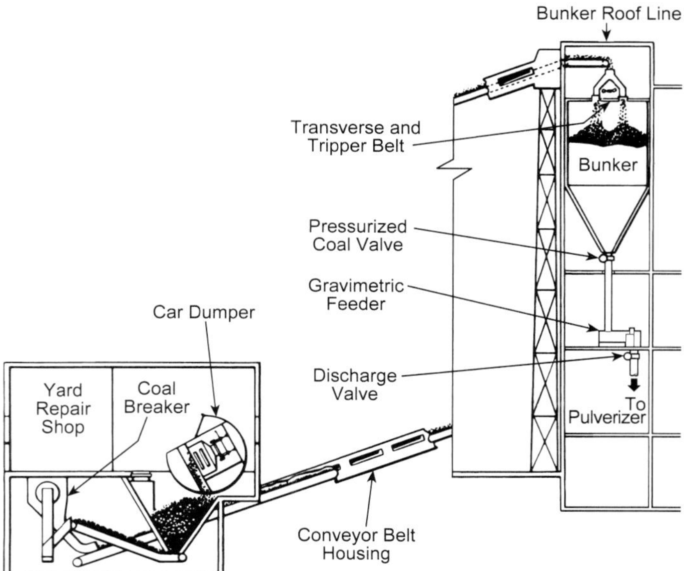

Fig. 14 Rail car unloading and coal handling system for a large power plant.

tem for a larger plant is shown in Fig. 14. Coal from the car dump hopper is fed to a rotary breaker, where the coal breaks into smaller pieces as it is tumbled and passes through a screen shell. The broken coal can then be conveyed directly to the storage bunkers, or to the stockpile.

Barge unloading

The simplest barge unloader consists of a clamshell bucket mounted on a fixed tower. The barge is positioned under the bucket and is moved as necessary to allow emptying. With this type of unloader, the effective grab capacity of the buckets is only 40 to 50% of the nominal bucket capacity. A shore mounted bucket wheel or elevator unloader can increase the efficiency and capacity of this unloading operation. Modern ocean-going vessels are often equipped with a bucket wheel for self-unloading.

Truck unloading

Trucks may dump coal through a grid into a storage hopper. This grid separates large pieces of wood and other trash from the coal. At some plants, the trucks are directed to a temporary storage area, where coal from various mines can be blended prior to crushing or feeding to the boiler.

An effective truck delivery and coal handling system for a small to medium size (30 to 300 t/d) stoker coal-fired boiler is illustrated in Fig. 15. This system provides site preparation of the coal along with tramp metal removal.

Stockpile storage

Careful consideration should be given to storage pile location. The site must be conveniently accessible by barge, rail or truck. Frequently, provisions must be made for more than one method of coal delivery. The site should be free of underground power lines. Other underground utilities that would not be accessible after the storage pile is constructed must also be avoided. A thorough evaluation and environmental survey of the proposed site topography should include analysis of the soil characteristics, bedrock structure, local drainage patterns, and flood potential. Climatic data, such as precipitation records and prevailing wind patterns, should also be evaluated. Protection from tidal action or salt water spray may be needed in coastal areas. The potential effects of water runoff and dust emissions from the pile must be considered. Site preparation includes removing foreign material, grading for drainage, compacting the soil, and providing for collection of site drainage.

The shape of a stockpile is generally dependent on the type of equipment used for pile construction and for reclaiming coal from the pile. Conical piles are generally associated with a fixed stacker while a radial stacker generates a kidney shaped pile. A rail mounted traveling stacker can be used to form a rectangular pile. Regardless of the shape of the pile, the sides should have a shallow slope.

Bituminous coal, subbituminous coal and lignite should be stockpiled in multiple horizontal layers. To reduce the potential for spontaneous combustion, coal piles are frequently compacted to minimize air channels. These channels can function as chimneys that promote increased air flow through the pile as the coal heats. For bituminous coal, an initial layer, 1 to 2 ft (0.3 to 0.6 m) thick, is spread and thoroughly packed to eliminate air spaces. A thinner layer is required for subbituminous coal and lignite to assure good compaction. Care should be taken to avoid coal pile size segregation by blending coal during pile preparation.

For long-term storage (see Fig. 16), the top of the pile may be slightly crowned to permit even rain runoff. All exposed sides and the top may be covered with a 1 ft (0.3 m) thick compacted layer of fines and then capped with a 1 ft (0.3 m) layer of screened lump coal. It is not practical to seal subbituminous and lignite piles with coarse coal because the coarse coal would

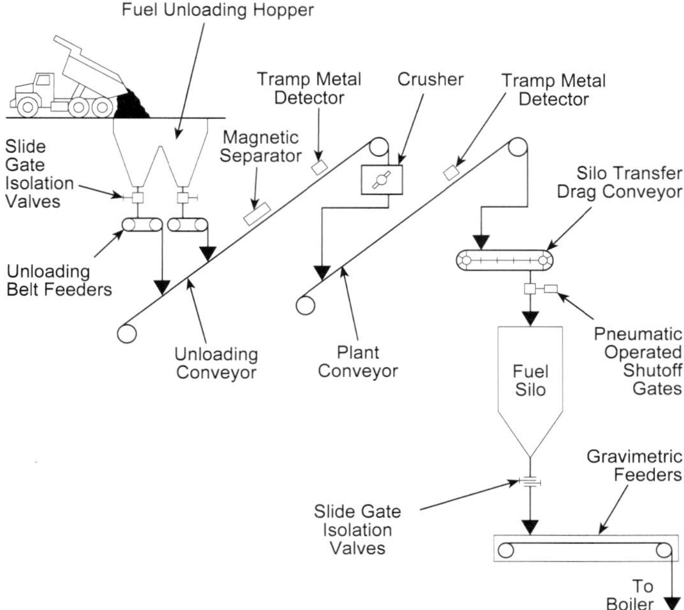

Fig. 15 Typical coal handling system for truck delivery.

Fig. 16 Long-term coal storage – typical example of thorough packing with minimal size segregation. This pile contains about 200,000 t of coal.

weather and break apart to a smaller size in a short period of time. At smaller industrial plants, where heavy equipment for compaction can not be justified, a light coating of diesel oil can help to seal off the outer surface of the pile.

The quantity of coal stored in a stockpile can be estimated using geometry and some assumptions about the characteristics of coal. The volume of the pile can be estimated based on its shape. Approximate values for the material's bulk density and its angle of repose are required to complete tonnage calculations. Both of these parameters are particle size dependent. For typical utility storage pile applications, a loose coal bulk density of 50 lb/ft^3 (801 kg/m^3) and a 40 deg angle of repose may be used. For well compacted piles, a bulk density of 65 to 72 lb/ft^3 (1041 to 1153 kg/m^3) is more appropriate.

Storage pile inspection and maintenance Visual inspections for hot spots should be made daily. In wet weather, a hot area can be identified by the lighter color of the surface coal dried by escaping heat. On cold or humid days, streams of water vapor and the odor of burning coal are signs of heating or air flowing through the pile. Hot spots may also be located by probing the pile with a metal rod. If the portion of the rod in contact with the coal is too hot to be held as it is withdrawn, the coal temperature is dangerously high.

It is also important to rotate areas of the storage pile from long to short term storage on a planned schedule. This minimizes harmful degradation of the coal.

Bulk storage reclaim and transfer The specific system for reclaiming the stored coal from stockpiles depends upon plant size and economic evaluation of operating and capital costs. For higher capacity systems of around 200 t/h (181 t_m) and greater, automated under-pile or over-pile reclaim systems are used. Under-pile reclaim systems can consist of under-pile hoppers with vibrating or belt feeders discharging onto a take-away belt conveyor. The belt conveyor travels at an incline through an underground tunnel until exiting the ground surface and is routed to the desired elevation. Over-pile reclaimers can be a variety of traveling bucket wheels or flight reclaimers that discharge onto an above ground take-away belt conveyor. For lower capacity systems, reclaim can be by bulldozer or front loader into a hopper feeding a take-away conveyor.

A variety of mechanical conveyors are used for transfer conveying. These include belt conveyors, drag (or flight) conveyors, bucket elevators, and screw conveyors. Belt conveyors are generally limited to an incline angle of 15 to 16 deg from horizontal to avoid roll-back of coal lumps on the traveling belt. Advancements in belt conveyor technology have yielded designs capable of greater incline angles and lateral routing, however at greater capital cost. Bucket and drag conveyors provide greater elevation capabilities over short horizontal distances, but mechanical conveyors inherently have greater wear and associated maintenance costs than belt conveyors. Screw conveyors are utilized primarily for horizontal, short-length conveying in low capacity applications. Wear and maintenance costs are also higher with screw conveyors.

Silo storage

Silos are often used where severe winter conditions can justify the added cost. Power plants with limited land space for large storage piles can also justify the expense. Silos are typically large field-constructed concrete structures of the slip-form or jump-form design. Large silos at central power stations can store 10,000 to 20,000 t (9710 to 18,144 t_m) and can be grouped together for necessary storage capacity. Multiple silos also provide the ability to empty one and perform maintenance while supporting operations. The silo hopper design must ensure adequate coal flow to avoid hide-out and coal hang-ups in the silo.

Bunker storage

Coal bunkers provide intermediate, short term storage, ahead of the pulverizers or other boiler coal feed equipment. Bunkers (also known as *day-bins*) are normally sized for 8 to 24 hours of fuel supply at maximum boiler operation. Normally, there is one bunker for each pulverizer and boiler coal feed point, but it is not uncommon to see one bunker with two or more outlets, feeding multiple points.

Bunker design

Once constructed, the coal bunkers are an integral part of the boiler house structure. Any redesign or modification to the bunkers can be expensive and difficult, if not impossible. The bunkers also must be designed for reliable flow of the stored fuel. (See Fig. 17.)

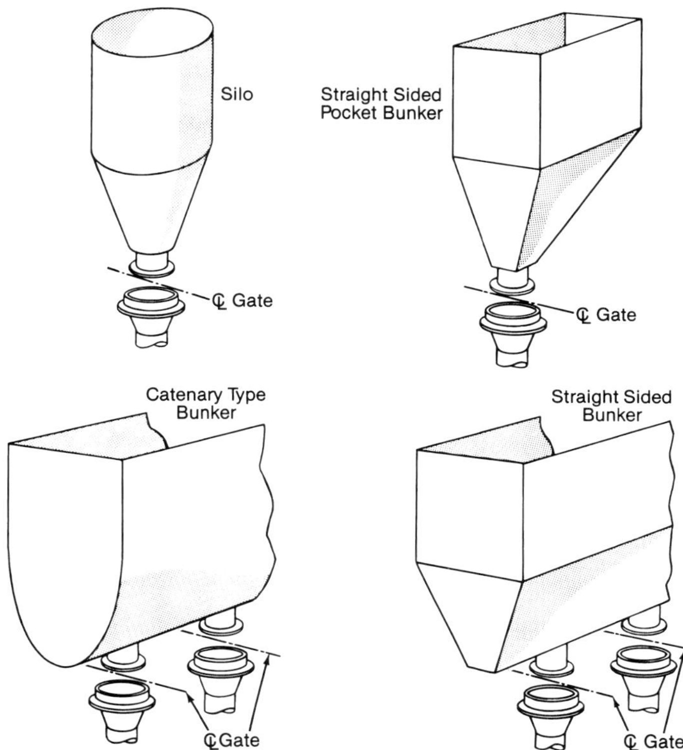

Fig. 17 Four commonly used shapes in coal bunker design.

The flowability characteristics of coal are important parameters in the design of the hoppers and chutes in the coal handling system and boiler feed systems. A *shear cell* test device measures the resistance of coal to slide in shear action against itself, and measures the resistance of the material to slide on a surface such are steel, stainless steel or other common materials. This testing can be performed after the coal has consolidated for several days, to simulate storage of the coal in a silo or bunker. From this data, the optimum combination of hopper slope angle and minimum outlet opening can be determined to assure reliable flow. This data can also be used to determine critical chute angles to assure continuous sliding flow through the chute, regardless of the cohesive nature of the coal and tendency to adhere to chute walls. Other factors that are utilized in optimizing the hoppers or chutes are the flowability of the coal on varying surfaces. This information is used in determining the most economic material to use for fabrication of the hoppers or chutes, or the liners applied. Materials such as carbon steel, polished and unpolished stainless steel, abrasion resistant steels, abrasion resistant plastic liner materials, and ceramic are just some of the materials routinely tested.

In bunkers or silos without proper design, reduced flow or even pluggage can occur. This is the result of the coal adhering to the hopper walls, especially at the transition from the straight walls to the angled hopper. This build-up can form a channel of limited cross-section through which the material flows. As the material continues to build-up, this flow channel of small cross section can bridge over and total pluggage occurs. Flow aids such as air blasters and bin vibrators can reduce this pluggage, but many times have limited success. Even with continuous flow, some residual material will remain in an emptied bunker due to the adherence in the straight wall to sloped hopper zone. This residual material can result in bunker fires due to spontaneous combustion. This is especially problematic with higher volatile coals, such as Powder River Basin coals.

With proper flow testing and design, complete emptying of the bunker or silo is assured each time the level is drawn down. The bunker or silo will be designed for *funnel flow* or *mass flow*. In funnel flow there is a stagnant zone (where the material does not flow) where the bunker vertical wall transitions to the sloped hopper wall. However, when the level in the vessel is drawn below this region and the pressure is reduced on this material, it releases from the hopper wall and collapses on the material level in the hopper. In a mass flow design, the material draws down uniformly throughout the vessel, in a pattern down the vertical walls, then transcends down the sloped hopper walls. Mass flow requires much steeper hopper slopes. The cohesive nature of the coal and ability to adhere to the vessel walls determine the design. A third design, called expanded flow, combines these two. The hopper is in two sections, the upper section funnel flow and the lower section mass flow.

Downspout design Flow testing data can also be used to optimize downspout design. Downspouts should be circular, short and as steep as possible, preferably vertical. Reductions in cross-sectional area and sudden changes in direction should be avoided.

In pressurized pulverizer feed applications, or coal feed to the pressurized location of a CFB boiler, special consideration must be given to downspout design because the coal inside the downspout also serves as the seal against the pressurized location and air or gas backflow to the bunker. A minimum required height (*seal height*) and a sealed downspout/feeder system are required. A typical bunker to pulverizer system is shown in Fig. 18. Vertical, constant-diameter downspouts connect the bunker to the feeder and the feeder to the pulverizer. Appropriate couplings and valves complete the system.

Feeder design

Feeders are used to control coal flow from the storage bunker at a uniform rate. Feeder selection should be based on an analysis of the material properties (maximum particle size, particle size distribution, bulk density, moisture content and abrasiveness), the desired flow rate, and the degree of flow control required. A variety of feeder designs have been used for coal-fired applications with increasing sophistication as more accurate control of coal flow has become necessary. Feeders for modern pulverized coal and CFB boiler applications can generally be classified as *volumetric* or *gravimetric*.

Volumetric feeders, as the name implies, are designed to provide a controlled volume rate of coal to the pulverizer. Typical examples include drag, table, pocket, apron and belt feeders. Belt feeders, perhaps the most accurate type, have a *level bar* to maintain

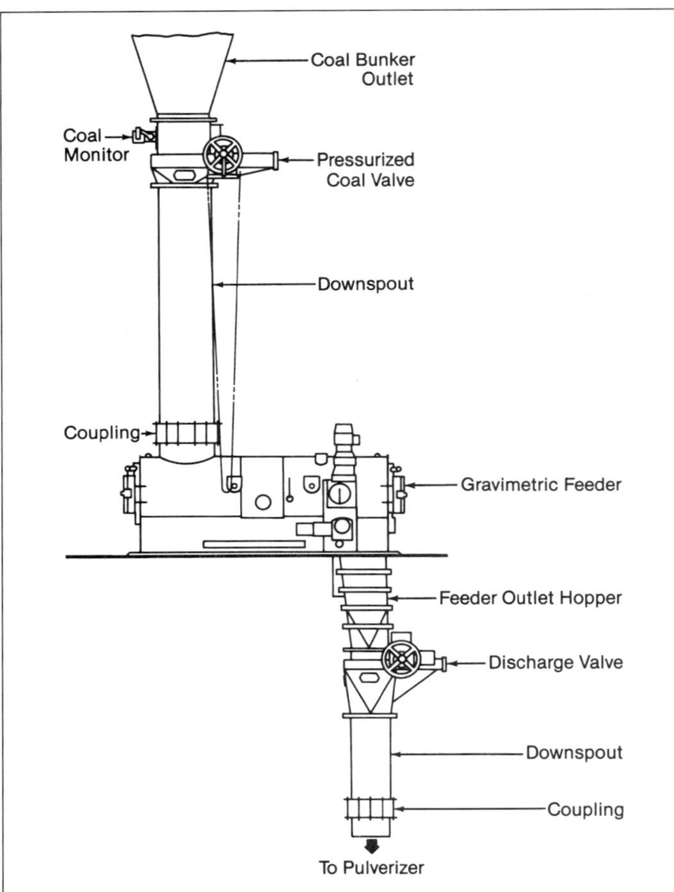

Fig. 18 Arrangement of bunker discharge to pulverizer showing typical feed system.

the flow of coal at a constant height and width while the belt speed sets the velocity of the coal through the opening. As with all volumetric designs, however, the belt feeder does not compensate for changes in coal bulk density. This results in variations in the energy input to the pulverizer and ultimately to the burners, for a pulverized coal unit, or to the bed of a CFB unit.

Gravimetric feeders (see Fig. 19) compensate for variations in bulk density due to moisture, coal size and other factors. They provide a more precise weight flow rate of coal to the pulverizer or CFB boiler bed feed point and, therefore, more accurate heat input to the burners and boiler. Even variations in coal moisture have a larger relative impact on coal bulk density than on heating value. Therefore, modern gravimetric feeders offer an accurate, commercially accepted technology to control fuel and heat input to the burners and boiler. This can be a very significant issue where more accurate control of fuel/air ratios is needed to: 1) minimize the formation of nitrogen oxides (NO_x), 2) control furnace slagging, and 3) maximize boiler thermal efficiency by reducing excess air levels. (See Chapter 14.)

In the most common gravimetric feeder system, coal is carried on a belt over a *load cell* that monitors the coal weight on the belt. The feedback signal is used to maintain the weight flow by either: 1) adjusting the height of a leveling bar to control the cross-sectional coal flow area while the belt speed remains constant, or 2) fine tuning the belt speed while the cross-sectional area remains constant. The overall set point flow rate is adjusted by varying the base belt speed.

Coal blending

When coals from two or more sources fuel a single boiler, effective coal blending or mixing is required to provide a uniform feed to the boiler. The use of multiple coals can be driven by economics, coal sulfur content to meet emission requirements, and/or the effects of different coals on boiler operation. The goal of effective blending is to provide a coal supply with reasonably uniform properties that meet the blend specification typically including sulfur content, heating value, moisture content, and grindability.

Coal blending may occur at a remote location, such as the mine or coal washing/preparation facility, or at the steam generating plant. Off-site blending eliminates the need for separate coal storage and additional fuel blending facilities. Steam plant on-site blending may be accomplished through a variety of techniques. It may be sufficient to provide separate stockpiles for each coal source and use front-end loaders to transfer the appropriate quantities to a common pile or hopper for blending prior to crushing. Coal may also be reclaimed from the various stockpiles using under-pile or over-pile reclaim systems transferring to a common conveyor belt. Coal from the various sources may be stored in separate bins with a feeder from each bin used to meter the desired quantities onto a common transfer belt. On-site blending provides more flexibility in coal sourcing and in adjusting to actual on-site coal variations.

Fig. 19 Typical gravimetric feeder (*courtesy of Stock Equipment Company, Inc.*).

Particular care must be maintained to ensure proper and complete blending. Significant variations in the blended coal can have a major impact on operation of the pulverizers, burners and sootblowers of a pulverized coal unit, and on bed temperature control of a CFB boiler. If uniform blending does not occur, pulverizer performance can deteriorate (see Chapter 13), the boiler may experience excessive slagging and fouling, and electrostatic precipitator particulate collection efficiency may decline, among other potential problems.

Resolution of common coal handling problems

Dust suppression

Water, oil and calcium chloride ($CaCl_2$) are common agents used to suppress fugitive dust emissions on open coal stock piles and during loading into open ground hoppers. A water or oil mist may be sprayed into the discharge area of rail cars as well as open discharges of conveyors onto storage piles. The water or oil spray reduces dust emissions by causing the dust to adhere to larger pieces of coal and by forming agglomerates that are less likely to become airborne. Use of $CaCl_2$ should be limited because of its potentially harmful side effects on boiler operation. (See Chapter 21.)

Dust collection at the transfer points from conveyor to conveyor in the coal transport system is performed by a vacuum collection system. These systems consist of a blower that draws the airborne dust from the discharge chutes, transfer chutes, and loading chutes of a transfer point. The enclosed chute arrangement is essentially maintained at a negative pressure. The dust laden air flow is routed through a bag filter to collect the dust, with the air discharging to the atmosphere. The collected dust is discharged from the bag filter hopper through a rotary seal to the continuing coal conveyer, silo or bunker. For coals with high volatility, the dust is collected with water and sluiced to waste or reclaim to avoid fire or explosion. These dust collection systems can also evacuate the methane that is generated in bunkers and silos.

Oxidation

Coal constituents begin to oxidize when exposed to air. This oxidation may be considered as a very slow, low temperature combustion process, because the end products, carbon dioxide (CO_2), carbon monoxide, water and heat, are the same as those from furnace coal combustion. Furnace combustion of coal may be viewed as a very rapid oxidation process. Although there is evidence that bacterial action causes coal heating, the heating primarily occurs through a chemical reaction process. If spontaneous combustion is to be avoided, heat from the oxidation should be minimized by retarding oxidation or removing the generated heat.

Coal oxidation is primarily a surface action. Finer coal particles have more surface area for a given volume and, therefore, oxidize more rapidly. Freshly crushed coal also has a high oxidation rate. Coal's oxygen absorption rate at constant temperature decreases with time. Once a safe storage pile has been established through compaction to minimize entrapped air, the rate of oxidation has been slowed considerably. Coal should be kept in dead storage undisturbed until it is to be used.

The rate of oxidation also increases with moisture content. High-moisture western coals are particularly susceptible to self-heating.

Frozen coal

The difficulties associated with handling frozen coal may be avoided by thermally or mechanically drying the fines following coal preparation. Spraying the coal with an oil or anti-freezing solution mist is one method, but this can have a negative effect on combustion and emissions performance.

Permanent installations for thawing frozen coal in rail cars include steam-heated thawing sheds, oil-fired thawing pits, and radiant electric thawing systems. Steam-heated systems are reliable and efficient, but are relatively expensive. Oil-fired systems that prevent direct flame impingement on the cars provide reliable operation and rapid thawing. Electric thawing systems are used at many plants that handle unit train coal shipments.

The coal handling system must also be protected. The chutes and bunker/silo hoppers can be fitted with heating coils to avoid hang-up due to coal freezing on the metal surfaces. Heating of the buildings enclosing these transfer chutes or hoppers should also be considered.

Frost and ice can form on belt conveyor surfaces. As coal is loaded onto the conveyor, it can skid on the ice and not convey. Systems using ethylene glycol or salt applications can help alleviate this problem.

Coal pile fires

A primary concern in coal storage is the potential for spontaneous combustion in the pile as a result of self-heating properties.

A coal pile fire may be handled in several ways depending on its size or severity. The hot region should be isolated from the remainder of the pile. This may be accomplished by trenching and sealing the sides and top of the hot area with an air tight coating of road tar or asphalt. Caution should be used in working the hot area with heavy equipment as subsurface coal combustion can affect the stability and load bearing characteristics of the pile. Water should not be used unless it is necessary to control flames. Pouring water on a smoldering pile induces more pronounced channeling and promotes greater air flow through the pile.

Bunker flow problems

Bunkers may be equipped with ports located near the outlet. These ports permit the use of air lances for restoring flow. Air lances may also be effectively used from the top of the bunker. Small boring machines can be mounted above the bunker to loosen coal jams at the outlet. Service companies can be contracted to remove flow obstructions using boring tools. Air blasters or air cannons have been successfully used to promote flow of coal in bunker hoppers. If there is any possibility of fire, these devices must be charged with nitrogen or carbon dioxide to prevent triggering a dust explosion. Bin vibrators can be used to promote flow similar to air blasters and cannons. Air blasters and cannons along with bin vibrators must be carefully evaluated prior to installation as these devices can further consolidate coal accumulated in the vessel.

Flow testing can be performed to determine if the current hopper design can be economically modified for reliable flow of a poorer quality coal. The application of different hopper liner materials can provide a simple solution to coal flow problems.

Bunker fires

A fire in a coal bunker is a serious danger to personnel and equipment and must be dealt with promptly. The coal feed to the bunker should be stopped. An attempt should be made to smother the fire while quickly discharging the coal. Continuity and uniformity of the hot or burning coal discharge from the bunker is especially important; interruption of coal flow aggravates the danger. The bunker should be emptied completely; no fresh coal should be added until the bunker has cooled and the cause of the fire determined.

The fire may be smothered using steam or carbon dioxide. CO_2 settles through the coal and displaces oxygen from the fire zone because it is heavier than air. Permanent piping connections to the bottom of the bunker may be made to supply CO_2 on demand. The CO_2 should fill the bunker, displace the air and smother the fire.

It is highly desirable to completely extinguish the fire before emptying the bunker. This is rarely possible because of boiler load demands and the difficulty of eliminating air flow to the fire. However, the use of steam or CO_2 to smother the fire can minimize the danger.

Bunker flow problems that result in dead zones may contribute to fires. Thermocouples installed in the bunker can monitor the temperature of the stored coal. The coal feed to the bunkers may also be monitored to prevent loading the bunker with hot coal.

Additional remarks on dealing with bunker fires in pulverized coal plants are provided in Chapter 13.

Environmental concerns

Water percolating through the coal storage pile can become a source of acidic drainage that may contaminate local streams. Runoff water must be isolated by directing the drainage to a holding pond where the pH may be adjusted. (See also Chapter 32.)

Airborne fugitive dust from stockpiles and adjacent haul roads creates a public nuisance, has potentially harmful effects on surrounding vegetation, and may violate regulated dust emission standards. To satisfy environmental regulations, plant haul roads are normally watered frequently during hot, dry, dust seasons.

Alternate solid fuel handling

Economic and environmental concerns have led to increasing steam generation from solid fuels derived from residential, commercial and industrial by-products and wastes. Key among these are municipal solid waste (MSW), wood and biomass as discussed in Chapters 29 and 30. The properties of these solid fuels require storage, handling and separation considerations different from those applied to coal.

MSW can either be burned with little pre-combustion processing (mass burn) or as a refuse-derived fuel (RDF). As-received refuse for mass-burn units is delivered to the tipping area and stored in an open concrete storage pit. The refuse pit is usually enclosed and kept under a slightly negative pressure to control odors and dust emissions. The tipping bay is designed to facilitate traffic flow based upon the frequency of deliveries and the size of the delivery trucks or trailers. The pit is usually equipped with a water spray system to suppress fires that may arise in part due to heat generated from decomposition of the refuse. An overhead crane is used to mix the raw MSW in the storage pit, to remove bulky items and to transfer material to the boiler feed charging hoppers. Large objects and potentially explosive containers are located and removed prior to combustion. (See Chapter 29.) A full capacity spare crane is recommended. Storage capacity is typically three to five days to accommodate weekends, holidays and other periods when refuse delivery may not be available. Longer term storage of refuse is not normally recommended.

MSW may be processed to yield a higher and more uniform Btu, lower ash RDF. The degree of processing required is determined by economics and by the fuel properties necessary for efficient boiler operation. MSW is usually delivered to an enclosed receiving floor. Front-end loaders can be used to spread the refuse, remove oversized and potentially dangerous items and feed the MSW to the RDF processing system as needed. RDF processing includes an integrated system of conveying, size reduction, separation, ferrous and non-ferrous metal recovery, sizing and other equipment discussed in depth in Chapter 29. MSW may be processed into RDF at the power plant site or at a remote location. The selection is based upon a number of economic factors, but operation of the RDF processing system at the boiler site typically will enhance availability to support uninterrupted steam generation.

Wood, forestry and agricultural wastes generally consist of bark, sawdust, saw mill shavings, lumber rejects, raw tree trunks and prunings, and straw. Material is generally shipped by truck to the steam generator site near the source. Material can be dumped directly on the storage pile or an unloading facility can be used. The unloading and handling equipment must be designed to handle very abrasive material under extremely dusty conditions. Wood products can be stored in large outdoor piles or inside bins or silos. This fuel is not typically stored in piles for more than six months or in silos or bins for more than three to five days. Wood is typically screened to remove oversized material for further size reduction. Oversized material is reduced by a shredding machine, or hog, and either returned to storage or sent directly to the combustor. Mechanical belt conveyors are the most popular method of transporting the fuel on site, although pneumatic systems can be effective with a finely ground, clean fuel such as sawdust. Tramp iron is usually removed by a magnetic separator. This material is fibrous and very stringy, so careful evaluation must go into the design and selection of the fuel handling and storage system. While most modern wood-fired boilers can burn materials with a moisture content of up to 65% as-received (see Chapter 30), predrying may be required. Mechanical hydraulic presses or hot gas drying, or both, are used.

Economics

The selection of the fuel source, degree of cleaning, and transportation system are closely tied to providing the lowest plant fuel cost. The selections must not be made in isolation, but in concert with evaluating the impact of the specific fuel on the boiler and bypass system operation, and the environment. For example, use of a new less expensive fuel may result in significant deterioration in boiler performance and availability due to more severe slagging and fouling tendencies of the flyash. (See Chapter 21.) The relative contributions of coal cleaning, transportation and base fuel price vary widely.

References

1. Freme, F., "U.S. Coal Supply and Demand: 2003 Review," United States Energy Information Agency, Department of Energy, Washington, D.C., 2004.

2. Bonskowski, R., "The U.S. Coal Industry in the 1990s: Low Prices and Record Production," United States Department of Energy, Washington, D.C., September, 1999.

3. International Energy Annual 2001, United States Department of Energy, Energy Information Agency, Washington D.C., March, 2003.

4. "Report on Longwall Mining," United States Department of Energy, Energy Information Agency, Washington, D.C., 1994.

5. "Coal Distribution Report," Form E1A-6A, United States Department of Energy, Energy Information Agency, Washington, D.C., 2002.

6. "The Rail Transportation of Coal," Vol. No. 3, Policy and Economics Department of the Association of the American Railroads, January, 2001.

7. Wilbur, L.C., Ed., *Handbook of Energy Systems Engineering: Production and Utilization,* See chapter entitled, "Coal Transportation," by R.D. Bessett, John Wiley & Sons, New York, New York, October, 1985.

8. "Domestic Shipping Inland Waterways," United States Maritime Administration (MARAD), Washington, D.C., 2003.

9. "Advantages of Inland Barge Transportation," Coosa-Alabama River Improvement Association, Inc. (CARIA) and the U.S. Department of Transportation Maritime Administration, www.caria.org/waterway-facts.

10. Mahr, D., "Coal Transportation and Handling," *Power Engineering,* pp. 38-43, November, 1985.

11. Edgar, T.F., *Coal Processing and Pollution Control*, Gulf Publishing, Houston, Texas, 1983.

Bibliography

Elliott, T.C., Chen, K., Swanekamp, R.C., *Standard Handbook of Power Plant Engineering,* Second Ed., pp. 1.20-1.44, McGraw-Hill Company, New York, New York, 1998.

Energy Information Administration (EIA), United States (U.S.) Department of Energy (DOE), www.eia.doe.gov, 2004.

Given, I.A., Ed., *Society of Mining Engineers Mining Engineering Handbook,* The American Institute of Mining, Metallurgical and Petroleum Engineers, New York, New York, 1973.

Heidrich, K., "Mine-Mouth Power Plants: Convenient Coal Not Always a Simple Solution," *Coal Age,* June 1, 2003.

Leonard, J.W., Ed., *Coal Preparation,* The American Institute of Mining, Metallurgical and Petroleum Engineers, New York, New York, 1979.

Lotz, C.W., *Notes on the Cleaning of Bituminous Coal,* West Virginia University, Charleston, West Virginia, 1960.

Pfleider, E.P., Ed., *Surface Mining,* The American Institute of Mining, Metallurgical, and Petroleum Engineers, New York, New York, 1968.

Coal pulverizers at a modern utility power station.

Chapter 13

Coal Pulverization

The development and growth of coal pulverization closely parallels the development of pulverized coal-firing technology. Early systems used ball-and-tube mills to grind coal and holding bins to temporarily store the coal before firing. Evolution of the technology to eliminate the bins and direct fire the coal pneumatically transported from the pulverizers required more responsive and reliable grinding equipment. Vertical air-swept pulverizers met this need.

The first vertical air-swept pulverizer by The Babcock & Wilcox Company (B&W) was designated the type E pulverizer, introduced in 1937. Today, B&W offers a broad line of proven B&W Roll Wheel™ pulverizers (see Fig. 1) to meet utility needs, and ball-and-race mills designated EL pulverizers (see Fig. 5) to meet lower load utility and industrial requirements. Both use rolling elements on rotating tables to finely grind the coal which is swept from the mill by air for pneumatic transport directly to the burners.

Reliable coal pulverizer performance is essential for sustained full load operation of modern pulverized coal-fired electric generating stations. Also, an effective pulverizer must be capable of handling a wide variety of coals and accommodating load swings. The B&W Roll Wheel pulverizer, through conservative design, reserve capacity and long grinding element life, has set the standard for high availability, reliability and low maintenance, contributing to stable boiler performance.

A key difference between much of the boiler system and the pulverizer is that the pulverizer is sized and operated as a mass flow machine while the boiler is a thermal driven machine. Therefore, the heating value of the fuel plays a key role in integrating these two components.

Vertical air-swept pulverizers

Principles of operation

The elements of a rolling action grinding mechanism are shown in Fig. 2. The roller passes over a layer of granular material, compressing it against a moving table. The movement of the roller causes motion between particles, while the roller pressure creates compressive loads between particles. Motion under applied pressure within the particle layer causes attrition (particle breakup by friction) which is the dominant size reduction mechanism. The compressed granular layer has a cushioning influence which reduces grinding effectiveness but also reduces the rate of roller wear dramatically. When working surfaces in a grinding zone are close together, near the dimensions of single product particles, wear is increased by three body contact (roller, particle and table). Wear rates can be as much as much as 100 times those found

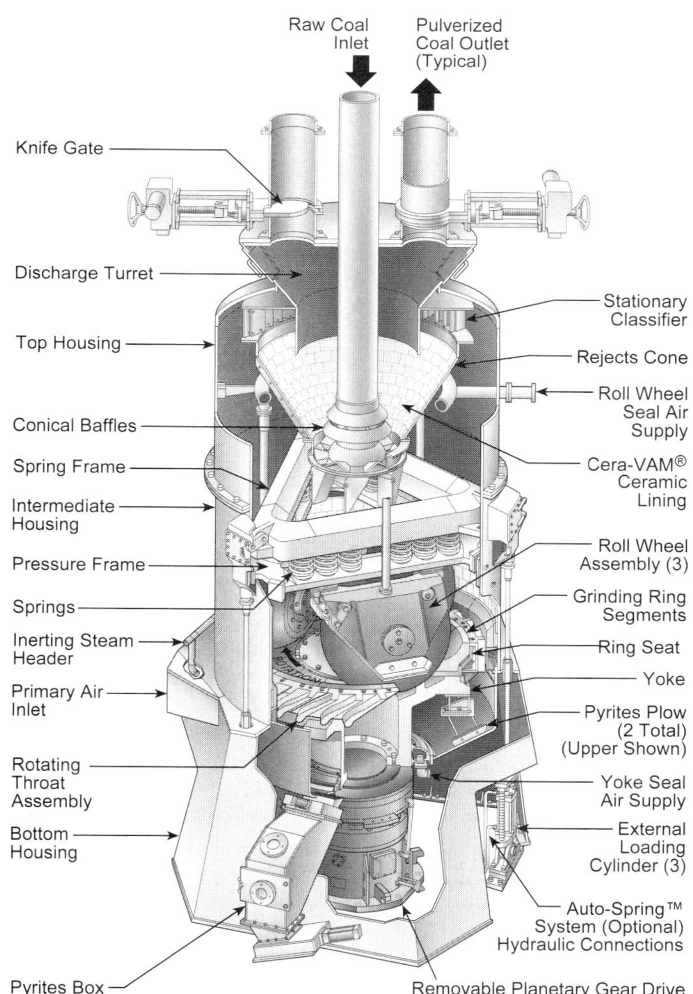

Fig. 1 B&W Roll Wheel™ pulverizer.

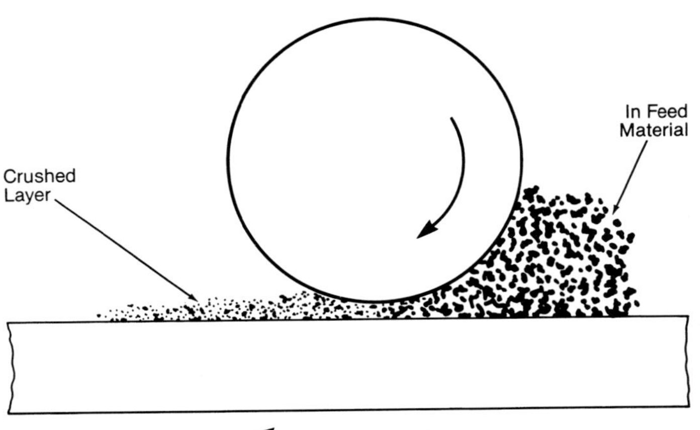

Fig. 2 Roller mill grinding mechanism.

in normal pulverizer field experience. Wear from the three body contact has also been observed in operating mills when significant amounts of quartz bearing rock are present in sizes equal to or greater than the grinding layer thickness.

As grinding proceeds, fine particles are removed from the process to prevent excessive grinding, power consumption and wear. Fig. 3 presents a simplified B&W Roll Wheel pulverizer, showing the essential elements of a vertical air-swept design. A table is turned from below and elements, in this case called *roll wheels*, rotate against the table. Raw coal is fed into the mill from above and passes between the rollers and the rotating table. Each passage of the particles under the rollers reduces the size of the coal. The combined effects of centrifugal force and displacement of the coal layer by the rollers spills partly ground coal off the outside edge of the table. An upward flow of air fluidizes and entrains this coal.

The point where air is introduced is often called the throat, air port ring, or nozzle ring. Rising air flow, mixed with the coal particles, creates a fluidized particle bed just above the throat. The air velocity is low enough so that it entrains only the smaller particles and percolates with them through the bed. This selective entrainment is the initial stage of size separation or classification. The preheated air stream also dries the coal to facilitate pulverization and enhance the combustion process.

Vertical pulverizers are effective drying devices. Coals with moisture content up to 40% have been successfully handled in vertical mills. Higher moisture coals could be pulverized, but the very high primary air temperature needed would require special structural materials and would increase the chance of pulverizer fires. A practical moisture limit is 40%, by weight, requiring air temperatures up to 750F (399C).

As the air-solids mixture flows upward, the flow area increases and velocity decreases, allowing gravity to return the larger particles to the grinding zone for the second stage of size separation. The final stage of size separation is provided by the classifier located at the top of the pulverizer. This device is a centrifugal separator. The coal-air mixture flows through openings angled to impart spin and induce centrifugal force. The larger particles impact the perimeter, come out of suspension and fall back into the grinding zone. The finer particles remain suspended in the air mixture and exit to the fuel conduits.

Pulverizer control

There are two input streams into an air-swept pulverizer, air and coal. Both must be controlled for satisfactory operation. Many older methods of coal flow control are still used successfully. Either volumetric or gravimetric belt feeders are the currently preferred method of coal flow regulation. Accurate measurement of coal flow has allowed parallel control of air and coal (Fig. 4).

Pulverizer design requirements

Many different pulverizer designs have been applied to coal firing. Successful designs have met certain fundamental goals and requirements:

1. optimum fineness for design coals over the entire pulverizer operating range,
2. rapid response to load changes,
3. stable and safe operation over the entire load range,
4. continuous service over long operating periods,
5. acceptable maintenance requirements, particularly grinding elements, over the pulverizer life,
6. ability to handle variations in coal properties,
7. ease of maintenance (minimum number of moving parts and adequate access), and
8. minimum building volume.

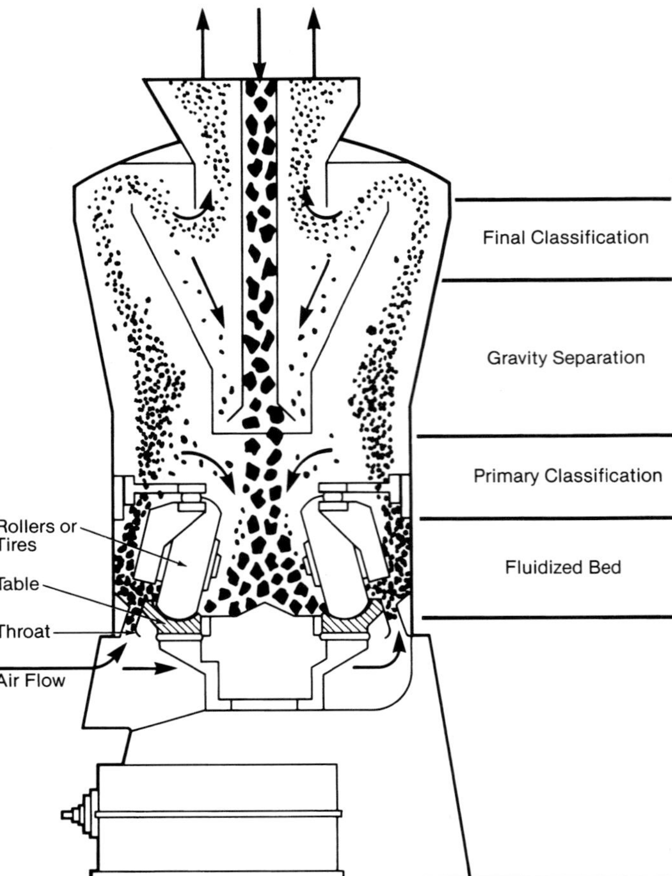

Fig. 3 Pulverizer coal recirculation.

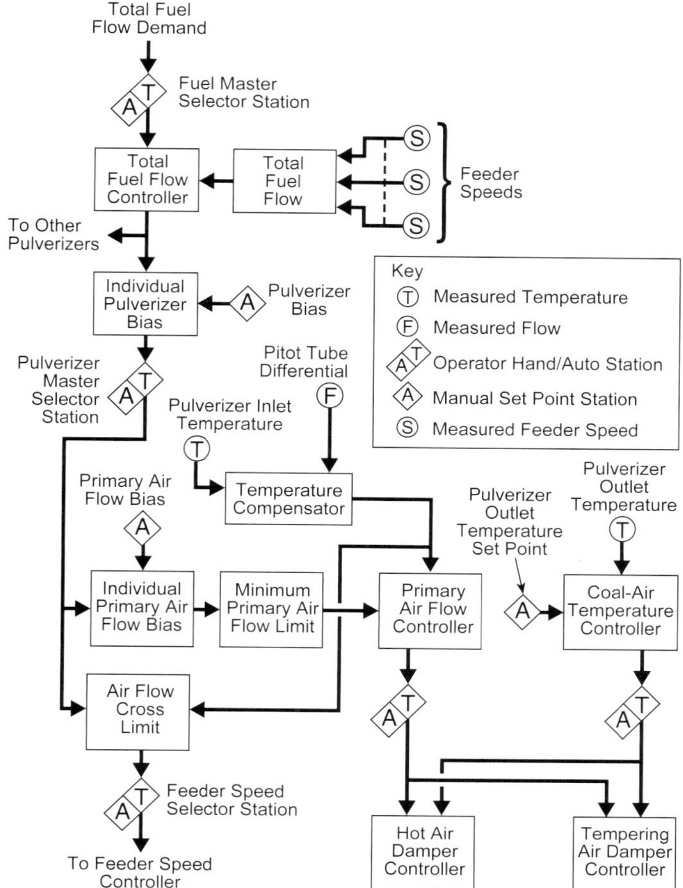

Fig. 4 Schematic of parallel cross limited pulverizer control system.

Pulverizer designs

Mill development efforts by B&W were, for many years, based on variations of ball-and-ring (ball-and-race) grinding elements. In the 1930s, B&W introduced the type E pulverizer. During the early 1950s, a redesign was installed and tested at a textile plant in Lancaster, South Carolina. This design, which uses two vertical axis horizontal rings, was designated E (Lancaster) or EL (Fig. 5). The lower ring rotates while the upper ring is stationary and is spring loaded to create grinding pressure. A set of balls is placed between the rings. The force from the upper ring pushes the balls against the coal layer on the lower ring.

Type E and EL mills have size designations which indicate approximately the centerline diameter of the grinding track. Type E mills have been built as large as the E-70 with a capacity of 12.5 t/h (11.4 t_m/h). Type EL mills have been built as large as the EL-76 with a nominal capacity of 20 t/h (18 t_m/h), and with modern enhancements, a maximum capacity of 23 t/h (21 t_m/h) is possible. Approximately 1600 E and EL mills have been installed, with more than 1000 still in service. The major wear parts of E and EL pulverizers are the two rings and the balls which are made of abrasion resistant alloys and are easily replaced.

B&W introduced the roll wheel style coal pulverizer to the North American market in the 1970s and installed more than 1100 units during the next three decades. B&W has continuously advanced the design, developing a unique product with enhanced performance and reduced operating costs. Having completed the license agreement, B&W retains all of the original technology as well as all of the design advancements and has changed the name to the B&W Roll Wheel™ pulverizer. A current configuration is shown in Fig. 1. B&W continues to fully support the full fleet of units with state of the art replacement parts and service.

The B&W Roll Wheel pulverizer is an air-swept, roller type vertical pulverizer; however, it differs in significant ways from other roller mills. The rollers (roll wheels) of the B&W mill are supported and loaded by a unique system which loads the three rollers simultaneously. The common loading system allows independent radial movement of each roller. This, in turn, allows continuous realignment with the grinding track as rollers wear. The rollers can also accommodate large foreign objects such as tramp iron or rocks which inadvertently enter the grinding zone. The B&W mill can maintain design performance with as much as 40% of the roll wheel tires' weight lost due to wear.

Roll wheel mills operate at speeds which produce centrifugal force at the grinding track centerline of about 0.8 times the force of gravity. This very low speed contributes to low vibration levels and the ability to handle large foreign objects.

This mill also introduced a gear drive module to coal pulverizer design, where the self-contained gear box is separate from the pulverizer housing structure. This feature is now popular with most mill designers in the United States (U.S.). When the gear drive module needs repair or preventive overhaul, the pulverizer is only disabled for the brief time it takes to remove the drive and install a spare unit. This allows thorough and economic gear drive maintenance, independent of production demands.

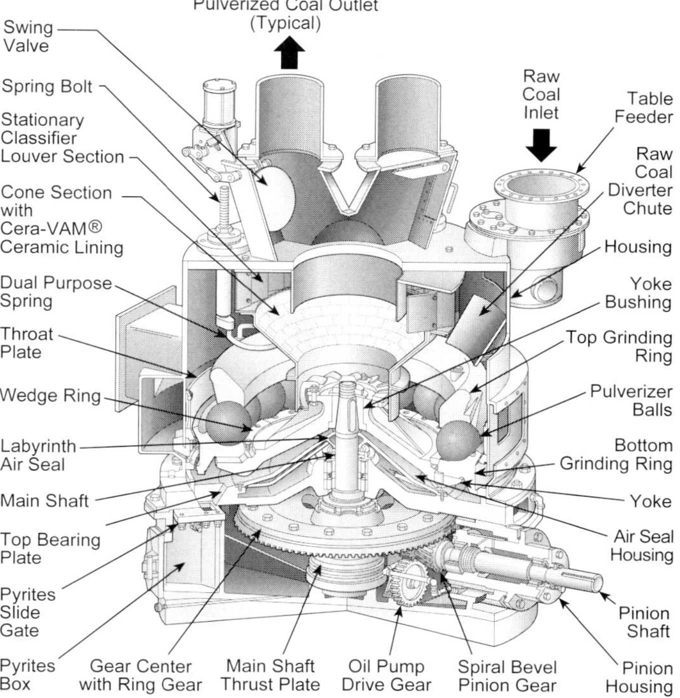

Fig. 5 B&W EL ball-and-race pulverizer.

The major replaceable wear parts of B&W pulverizers include the tires and ring segments plus minor items such as wear guide plates, ceramic linings, and parts for the throat or air port ring. Wear life of all parts in contact with the coal is dependent upon the coal abrasiveness, and ranges from less than 10,000 to more than 100,000 hours.

The B&W mill size designations are also based upon the grinding track centerline diameter, in inches. Units have been furnished ranging from the B&W-56™ to the B&W-118™ designs with capacities ranging from 15 to 115 t/h (13 to 105 tm/h) respectively.

Table 1 lists important features and characteristics of B&W Roll Wheel and EL mills for comparison.

Horizontal air-swept pulverizers

High speed pulverizers

One high speed design which has been used in the U.S. serves the same applications as the vertical types. This machine operates at about 600 rpm and grinds by both impact and attrition. Coal and heated air enter at the impact section for initial size reduction then pass between moving and stationary parts for final size reduction (Fig. 6). At the final stage, an exhauster fan draws the air and coal through the mill for drying and transport. There is no classifier and the design relies upon the action of the rotor and stator to achieve size control in a single pass of the coal. The coal passes through quickly so there is little storage in the mill. This pulverizer is limited to coals with moisture content of 20% or less because of the short residence time for drying.

Beater mills are used outside North America (Fig. 7). These are similar in some respects to the attrition mills described above. These mills are used for grinding and drying the fossil fuels referred to as brown coals. These are very low grade lignite type coals with high ash and moisture contents and low heating values. Moisture content of brown coal often exceeds 50% and combined ash and moisture contents may exceed 60%.

Beater mills grind by impact and attrition with very rapid drying. They achieve drying despite short residence time by replacing primary air with extremely high temperature gas extracted from the upper furnace at temperatures about 1900F (1038C). Flue gas from a cooler location such as the air heater outlet may be mixed with the hot gas for mill outlet temperature control. Coal is mixed with the gas stream ahead of the mill for initial drying to reduce the inlet temperature. The final stage is the fan section which maintains a negative pressure in the mill at all times. Beater mills operate at variable speed to control grinding performance at varying coal feed rates. One and two shaft machines are used, trading simplicity in the former for more flexibility in the latter. Because of the low coal quality, beater mills must handle huge amounts of coal and are made in very large sizes. There are very few coal deposits in North America which would require this grinding/drying technology. However, significant deposits in Germany, Eastern Europe, Turkey

Table 1
Characteristics of B&W EL and B&W Roll Wheel™ Pulverizers

	Type EL Ball-and-Race	Type B&W Roll-and-Race
Size range	EL-17 to EL-76	B&W-56 to B&W-118
Capacity, t/h (t_m/h)	1.5 to 23 (1.4 to 21)	17 to 115 (15 to 104)
Motor size, hp (kW)	25 to 350 (18 to 260)	200 to 1500 (149 to 1120)
Speed	Medium	Slow
Table, rpm	231 to 90	32 to 21
Operates under	Pressure	Pressure
Classifier	Internal, centrifugal	Internal, centrifugal
Classification adjustment	Internal	Internal (standard) External (special)
Optional dynamic classifier	Internal, dual stage, variable speed	Internal, dual stage, variable speed
Drying limit	40% H_2O or 700F (371C) maximum Primary air temperature	40% H_2O or 750F (399C) maximum Primary air temperature
Moisture load correction	None up to temperature limit	Load correction above 4% surface moisture
Maximum exit temperature limit	250F (121C)	210F (99C)
Effect of wear on performance	None if fill-in balls added	Power increase up to 15% at fully worn condition
Air-coal control system	Mill level with table feeders, parallel control with belt feeders	Parallel coal and air flow control
Air/coal weight ratio	1.75:1 at full load	1.75:1 at full load
Internal inventory	Medium, 2 to 3 min of output	High, 5 to 6 min of output
Load response	> 10%/min	> 10%/min
Specific power, kWh/t (kWh/t_m)	Low, 14 (15) including primary air fan	Low, 14 (15) including primary air fan
Noise level	Above 90 dBA	Above 90 dBA, 85 dBA attenuated
Vibration	Moderate	Low

Note: Capacities are those for bituminous coal with a 50 HGI and 70% fineness passing through 200 mesh (74 microns). Specific power is that at full load.

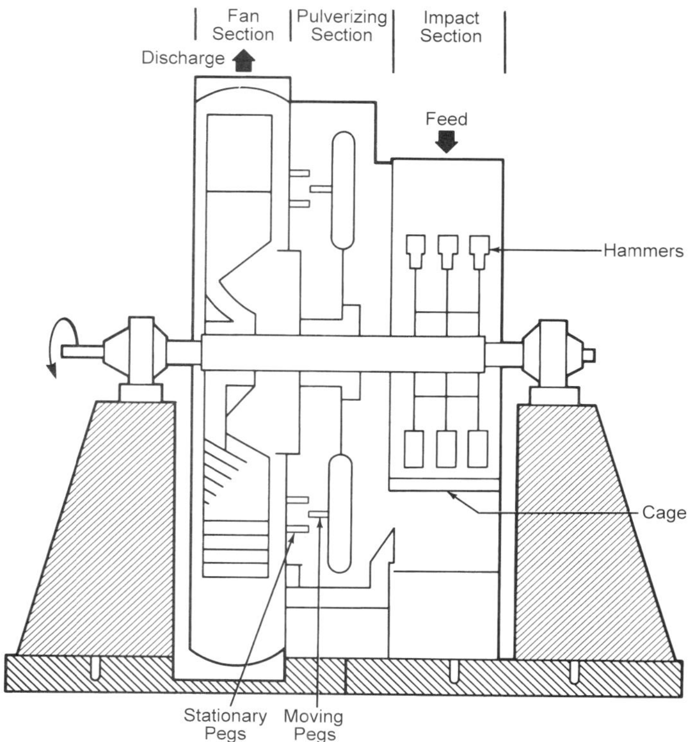

Fig. 6 Horizontal high-speed pulverizer.

and Australia make brown coal a very important fuel for power generation throughout these regions.

Low speed pulverizers

The oldest pulverizer design still in frequent use is the ball-and-tube mill. This is a horizontal cylinder, partly filled with small diameter balls (Fig. 8). The cylinder is lined with wear resistant material contoured to enhance the action of the tumbling balls. The balls fill 25 to 30% of the cylinder volume. The rotational speed is 80% of that at which centrifugal force would overcome gravity and cause the balls to cling to the shell wall. Grinding is caused by the tumbling action which traps coal particles between balls as they impact.

Ball-and-tube mills may be either single or double ended. In the former, air and coal enter through one end and exit the opposite. Double ended mills are fed coal and air at each end and ground-dried coal is extracted from each end. In both types, classifiers are external to the mill and oversize material is injected back to the mill with the raw feed. Ball-and-tube mills do not develop the fluidized bed which is characteristic of vertical mills and the poor mixing of air and coal limits the drying capability. When coals with moisture above 20% must be ground in ball-and-tube mills, auxiliary equipment, usually crusher dryers, must be used.

Ball-and-tube mills have largely been supplanted by vertical air-swept pulverizers for new boilers. They typically require larger building volume and higher specific power consumption than the vertical air-swept pulverizers. They are also more difficult to control and have higher metal wear rates. They are, however, well suited for grinding extremely abrasive, low moisture, and difficult materials such as petroleum coke. Their long coal residence time makes them effective for fine grinding.

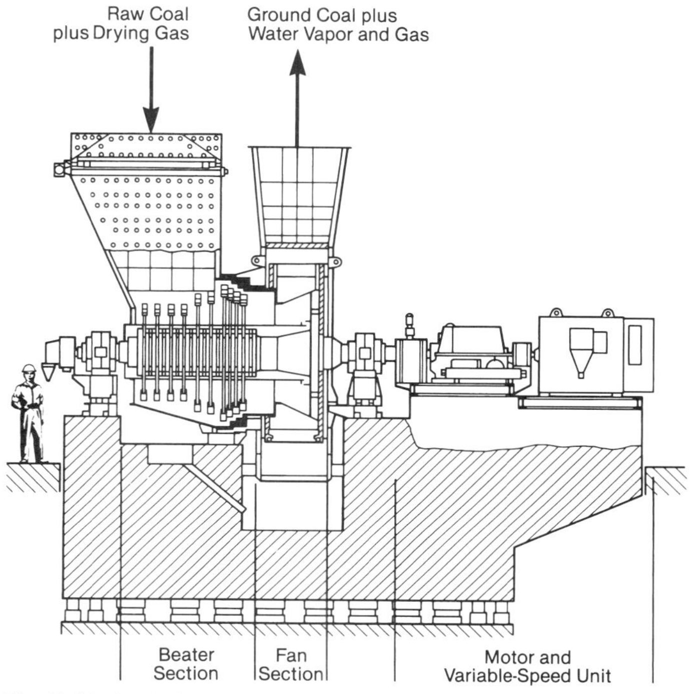

Fig. 7 Single-shaft beater mill.

Application engineering

The arrangement of coal-fired system components must be determined according to economic factors as well as the attributes of the coal. The performance in terms of product fineness, mill outlet temperature, and

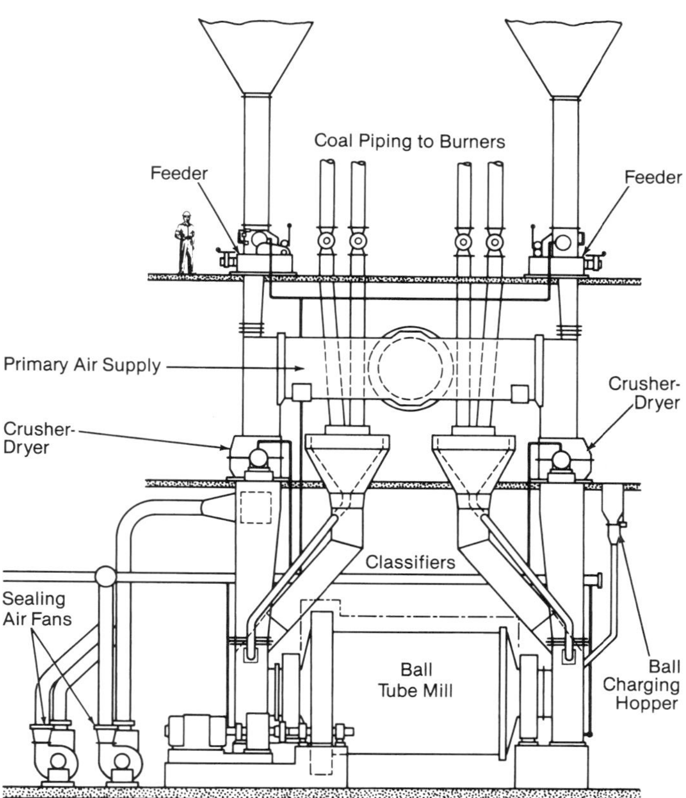

Fig. 8 Typical pressurized ball-and-tube coal pulverizing system.

air-coal ratio must all be determined as part of overall combustion system design.

Pulverizer systems

Pulverizers are part of larger systems, normally classified as either direct-fired or storage. In direct firing, coal leaving each mill goes directly to the combustion process. The air, evaporated moisture, and the thermal energy which entered the mill, along with the ground coal, all become part of the combustion process. Storage systems separate the ground coal from the air or gas, evaporated moisture and the thermal energy prior to the combustion process. Stored ground coal is then injected with new transport air or gas to the combustion process. Bin storage systems are seldom used in steam generation today, but are still used with special technologies such as coal gasification and blast furnace coal injection. More than 99% of the B&W Roll Wheel pulverizers in service in the U.S. are used in direct-fired systems.

The essential elements of a direct-fired system are:

1. a raw coal feeder that regulates the coal flow from a silo or bunker to the pulverizer,
2. a heat source that preheats the primary air for coal drying,
3. a pulverizer (primary air) fan that is typically located ahead of the mill (pressurized mill) as a blower, or after the mill (suction mill) as an exhauster,
4. a pulverizer, configured as either a pressurized or suction unit,
5. piping that directs the coal and primary air from the pulverizer to the burners,
6. burners which mix the coal and the balance of combustion air, and
7. controls and regulating devices.

These components can be arranged in several ways based on project economics. With pressurized pulverizers, the choice must be made between hot primary air fans with a dedicated fan for each mill, or cold fans located ahead of a dedicated air heater and a hot air supply system with branches to the individual mills. Hot fan systems have a lower capital cost because a dedicated primary air heater is not required. Cold fan systems have lower operating costs which, on larger systems, may offset the higher initial cost. Figs. 9 and 10 show these systems.

The terminology for air-swept pulverizers refers to the air introduced for drying and transport as primary air. Control of primary air is of vital importance to proper pulverizer system operation. Primary air must be controlled for flow rate and pulverizer outlet temperature. This control can be achieved by three interrelated dampers. Hot and cold air dampers regulate air temperature to the mill and these dampers are often linked so that as one opens, the other closes. The third damper independently controls air volume. On cold fan systems, modern controls allow flow and temperature control with just the hot and cold air dampers, eliminating the need for the independent flow control damper, thus reducing cost with no loss of controllability.

Because direct-fired pulverizers are closely linked to the firing system, engineering must coordinate the design performance of the mills and burners. A set of curves can be used to relate important operating characteristics of volume flow, velocities at critical locations, and system pressure losses through the load range of the boiler. The curves consider the number of mills in service and the output range of the individual mills. An example set of curves is shown in Fig. 11. Study of these curves provides information on many aspects

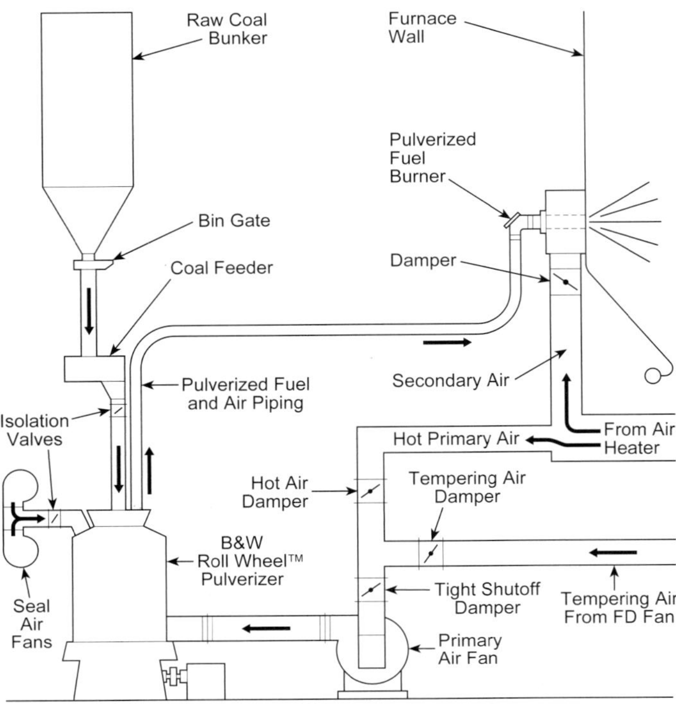

Fig. 9 Direct-fired, hot fan system for pulverized coal.

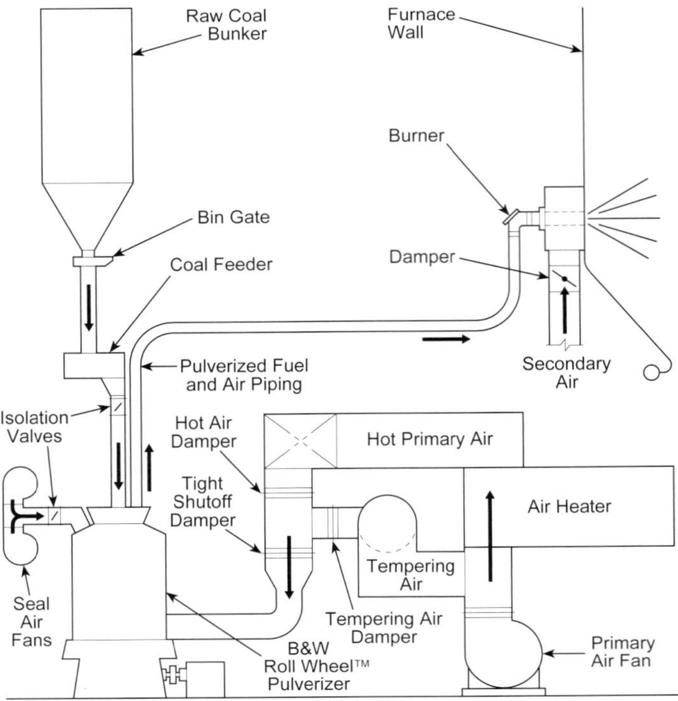

Fig. 10 Direct-fired, cold fan, fuel-air system for pulverized coal.

of pulverizer and system operation. The lower curve, labeled A, shows boiler steam flow versus coal output per pulverizer. The individual lines show the number of mills in operation. In this example, a full boiler load of 3.6×10^6 lb/h (454 kg/s) steam flow can be reached with five mills in service at about 89,000 lb/h (11.2 kg/s) each. The maximum load also can be reached with four mills at about 111,000 lb/h (14.0 kg/s) each. The maximum steam flow with three mills is just over 3.0 $\times 10^6$ lb/h (378 kg/s). The minimum load line shown is determined by Curve C. The indicated 36,000 lb/h (4.54 kg/s) is a very good turndown ratio, but the practical minimum may be higher based on ignition stability or the onset of mechanical vibration.

Curve B shows primary air flow at mill exit conditions. Maximum flow is 55,800 ft^3/min (26.34 m^3/s) for a B&W-89™N mill, corresponding to 124,000 lb/h (15.6 kg/s) coal flow. The minimum equipment design flow is approximately 65% of this or 36,300 ft^3/min (17.13 m^3/s). However, the exact minimum flow may need adjustment to permit stable burner operation. This is illustrated by 39,000 ft^3/min (18.41 m^3/s) minimum shown in Fig. 11, curve B, and the 3000 ft/min (15.2 m/s) minimum in curve D.

Curve C is the air/fuel ratio expressed in ft^3 of air per lb of coal. This ratio is critical to stable ignition at low loads and is influenced by coal rank and fineness. For this example, the maximum air to fuel ratio is 65 ft^3 air per pound fuel, typical of bituminous coal. Dividing the minimum air flow by this ratio establishes the minimum pulverizer load.

Curve D shows primary air velocity in the burner pipes versus pulverizer coal flow. This curve is plotted by dividing air volume at mill exit conditions by the total flow area of the pipes connecting the mill to the burners. The minimum velocity allowed is 3000 ft/min (15.2 m/s), at which the pulverized coal can be kept entrained in the primary air stream. The mini-

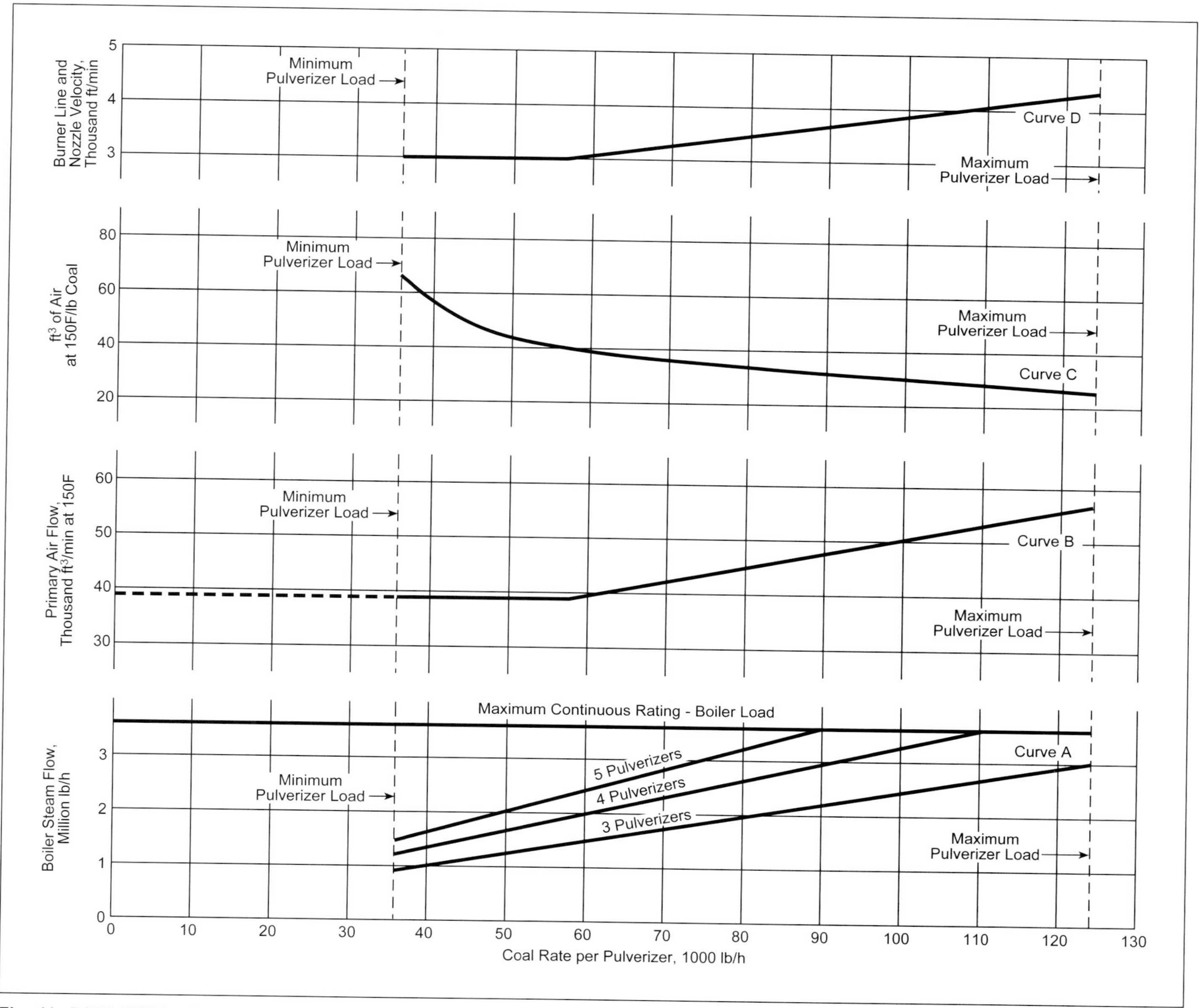

Fig. 11 B&W-89™N pulverizer-burner coordination curves.

mum, or saltation, velocity is particularly important in long horizontal spans of pipe. This velocity limit is influenced by air density and viscosity, particle loading and particle size. For the relatively narrow range of density, viscosity and particle size distribution in burner pipes, the minimum velocity limit is typically adjusted only for particle loading (see Fig. 12).

An important pulverizer performance requirement is particle size leaving the mill. There is a mixture of particle sizes with any pulverizer because of the statistical distribution of particle sizes produced by the pulverizing process. The most frequently referenced size fraction for coal combustion is that portion smaller than a 200 U.S. standard sieve opening (mesh), or 74 microns. The portion passing this sieve size has been used as a gauge of pulverizer capacity as well as for assuring good combustion and good carbon burnout. The required particle size (fineness) is determined by the combination of coal combustion characteristics and the combustion chamber.

For bituminous coals burned in water-cooled enclosures, a fineness of 70% passing through 200 mesh (74 microns) or better is traditionally required. The nominal capacity rating of pulverizers in the U.S. is also based on 70% fineness. This value is suitable for good combustion efficiency with conventional combustion systems, but may not be sufficient for low nitrogen oxides (NO_x) systems which delay combustion. Coarse particles, those larger than 100 mesh (150 microns), are primary contributors to unburned carbon loss. With traditional stationary classifiers it may be necessary to maintain fineness above 70% passing 200 mesh for acceptable unburned carbon loss. With modern rotating classifiers, the slope of the particle distribution curve is changed such that very low coarse particle delivery and acceptable unburned carbon losses can be achieved with fineness at or below 70%.

Outlet temperature

Coal with low volatile content may require higher air-coal temperatures to assure stable combustion, especially at lower burner inputs or low furnace loads. The typical pulverizer exit temperature is 150F (66C). Higher temperatures, up to 210F (99C) for coal, may be used if needed. Higher temperatures require care in the selection of internal lubricants and soft seal materials. In addition, high outlet temperatures require high inlet temperatures which increase the risk of pulverizer fires. Some pulverizer applications, notably bin storage systems, will require high outlet temperatures to assure complete drying and prevent handling problems with fine coal. Usually, 180F (82C) is adequate to assure drying of coal having raw coal moisture up to 10%. For direct firing, the outlet temperature requirements are determined by volatile content and the need for stable combustion. Table 2 lists mill outlet temperatures for various coal types.

Effects of coal properties

Grindability When determining pulverizer size, the most important physical coal characteristic to consider is *grindability*. This characteristic is indicative of the ease with which coal can be ground. Higher grindabilities indicate coals which are easier to pulverize. The test procedure to determine the commonly used Hardgrove Grindability Index (HGI) is described in The American Society for Testing and Materials (ASTM) D 409. The operative principle with this procedure is the application of a fixed, predetermined amount of grinding effort or work on a prepared and sized sample. The apparatus shown in Fig. 13 is used to determine grindability. By rotating exactly 60 revolutions, this miniature pulverizer does a fixed amount of work on each sample. The amount of new, fine material produced is a measure of the ease of grinding. The neutral value of grindability used by B&W is HGI = 50, i.e., at HGI = 50, the capacity correction factor is 1.0. The index is open ended with no upper limit on the HGI scale.

Grindability is not strictly a matter of hardness. Some materials, fibrous in nature, are not hard but are very difficult to grind. Sticky or plastic materials can also defy grinding.

The procedure described in ASTM D 409 was originally developed as a purely empirical method but provides a valid capacity calculation for vertical pulverizers when grinding bituminous coal. When lower rank coals are tested in the Hardgrove apparatus, the correlation between laboratory tests and field operating equipment is often quite poor. There is a strong but unpredictable effect of sample moisture on test re-

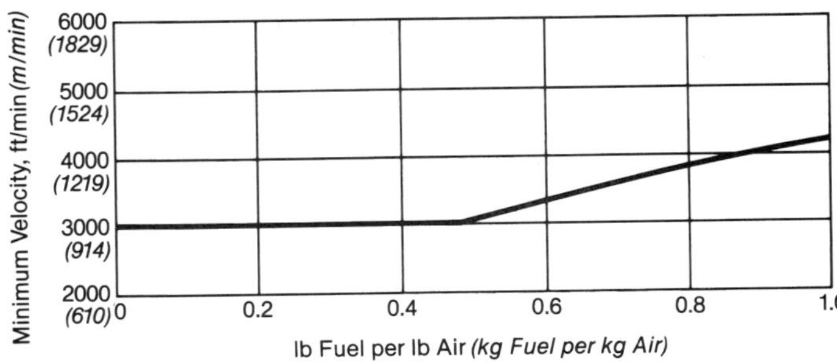

Fig. 12 Correction curve for minimum velocity.

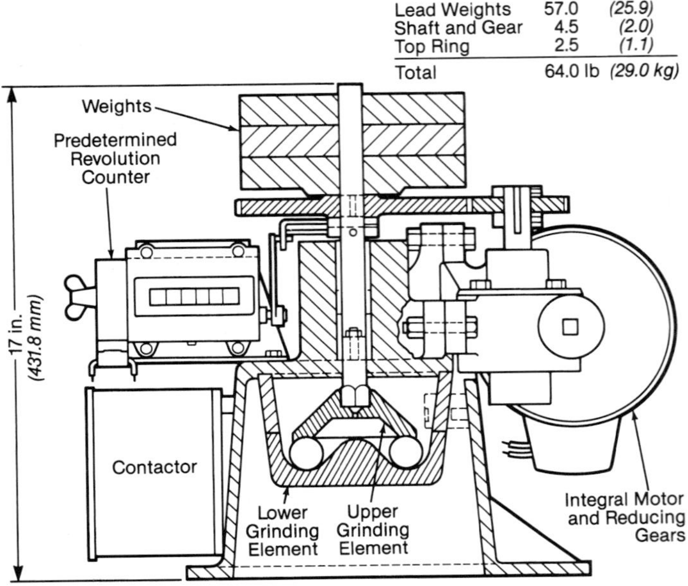

Fig. 13 Hardgrove grindability testing machine.

Table 2
Typical Pulverizer Outlet Temperature

Fuel Type	Volatile Content, %*	Exit Temperature F (C)**
Lignite	All values	120 to 140 (49 to 60)
Subbituminous	All values	130 to 150 (54 to 66)
High volatile bituminous	Above 31	140 to 175 (60 to 79)
Medium and low volatile bituminous	14 to 31	160 to 200 (71 to 93)
Anthracite and coal waste	0 to 14	200 to 210 (93 to 99)
Petroleum coke	0 to 8	200 to 250 (93 to 121)

* Volatile content is on a dry, mineral-matter-free basis.
** The capacity of pulverizer is adversely affected for exit temperatures below 125F (52C) when grinding high moisture lignites.

sults, illustrated in Fig. 14. For high ash content coals, the possibility of mineral accumulation in the pulverizer, which hinders the grinding process, is not accounted for in the HGI procedure. A more accurate index of grindability for low rank coals is determined in a laboratory sized roll wheel pulverizer. This equipment has all the features of a field installation and can simulate field conditions. The capacity of the mill is about 800 lb/h (0.10 kg/s) and a coal capacity test requires a 3000 lb (1361 kg) sample. Fig. 15 is an arrangement diagram of this test apparatus. Results are interpreted as apparent grindability and the values are valid for capacity calculations.

Wear properties Wear in coal pulverizers results from the combined effects of abrasion and erosion. The mechanism which dominates in the wear of a particular machine component depends upon the designed function of the component and on the properties of the coal being ground. Abrasiveness and erosiveness are the important properties to be considered when evaluating the influence of a candidate coal on expected maintenance cost. Unfortunately, these two important properties are inherently difficult to measure, especially with a material as variable as coal.

Abrasiveness Abrasion is the removal of material by friction due to relative motion of two surfaces in contact. A coal's ability to remove metal is primarily related to the quantity and size distribution of quartz and pyrite found in the as-received coal, especially for particles larger than 100 mesh (150 microns). While standard and accepted procedures are available to determine the quantity of these two minerals in a sample (ASTM D 2492 and ASTM D 2799 for pyrite and quartz respectively), procedures to establish quartz and pyrite particle sizes are not widely accepted.

Unfortunately, direct small-scale laboratory abrasiveness testing is not satisfactory for correlation to field wear conditions because of operating differences between small laboratory equipment and power plant pulverizers. However, satisfactory abrasiveness measurements have been made using the laboratory sized roll wheel pulverizer with 4000 lb (1814 kg) samples. Although expensive, wear life testing with this procedure can typically predict field performance with an error of 10% or less. While other abrasiveness tests such as Yancy-Geer Price apparatus are used to provide some insight into relative wear rates, they are of very limited value in predicting actual field wear rates.

Erosiveness Erosion is not the same as abrasion. It can be defined as the progressive removal of material from a target on which a fluid borne stream of solids impinges. For a given combination of particles and target material, wear rate increases with increasing ve-

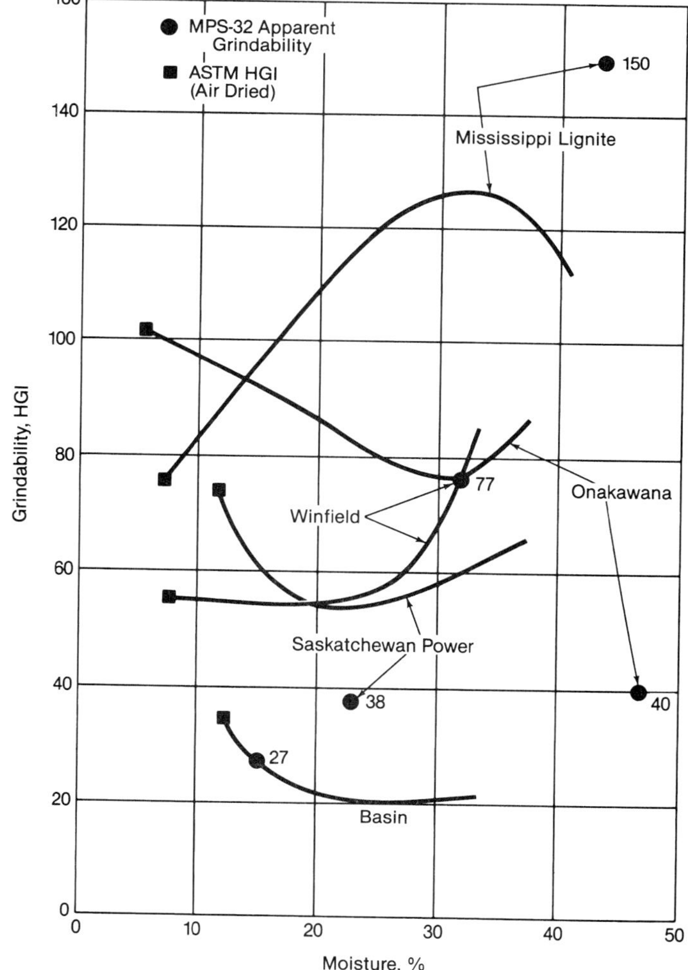

Fig. 14 Hardgrove Grindability Index versus fuel moisture.

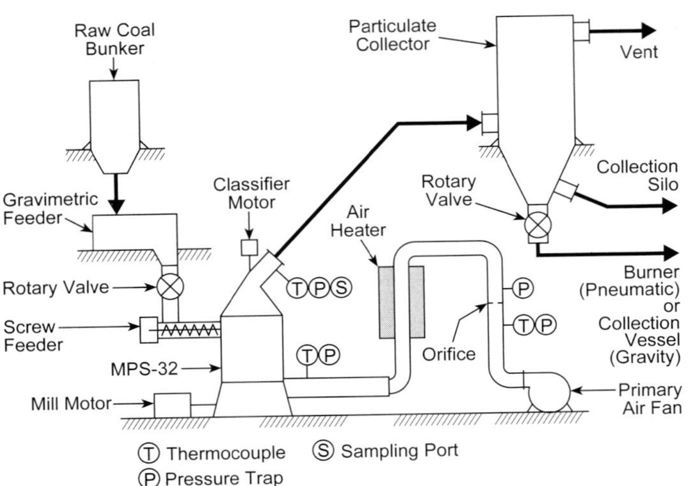

Fig. 15 B&W's test apparatus for low rank coals.

locity, particle size, and the solids content of the stream. For ductile target materials, the wear rate increases for an impingement angle from parallel up to approximately 45 deg, at which point wear begins to decline.[1] For brittle materials, the rate of erosion increases up to the angle of approximately 70 deg, where rebounding material interferes with the impinging stream and measurement becomes erratic.

Moisture Moisture content in the fuel is a key design parameter. Coal moisture is highly variable and depends more on coal type than on the amount of water introduced after mining. Coals used in the U.S. range from inherent moisture levels of 2% for Appalachian bituminous to near 40% for lignites. Brown coals may range up to 70%. Moisture in coal is determined according to the procedures in ASTM D 3302. For pulverizer capacity corrections and primary air system design, the total moisture must be separated into air dry moisture (air dry loss) and residual moisture (moisture remaining after air drying). For capacity correction, since the air dry moisture is evaporated from the coal sample before performing the HGI grindability test, an additional correction factor for air dry moisture must be included. For the primary air system design, residual moisture requires just enough heat to raise it from raw coal temperature to mill outlet temperature. The air dry moisture will be evaporated in the pulverizer, so it requires much more heat.

Pulverizer size selection

The ultimate task in pulverizer application is the selection of the size and number of mills for the proposed project. The total boiler heat input and coal flow requirements are established from the combustion calculations (see Chapter 10) and the specified boiler steam flow requirements. The coal flow rate is then divided by the capacity correction factor to establish the equivalent required pulverizer capacity. The correction factor not only includes the fineness and grindability factors shown in Fig. 16, but also the appropriate fuel moisture correction.

The number of pulverizers is then evaluated by dividing the equivalent required capacity by the unit pulverizer rated capacity based upon 70% passing through a 200 mesh (74 microns) screen and a grindability of 50. The unit pulverizer rated capacity is based upon pulverizer size, type and manufacturer. The tradeoff between fewer larger mills and more smaller mills is based upon balancing total capital cost with the proposed operating requirements such as boiler turndown. Frequently an extra pulverizer is specified to permit the boiler to operate at full load while one mill is out of service for maintenance.

Pulverizer selection and sizing are complicated by the frequent need to consider various coal sources and emission requirements. If the coal that determines the pulverizer size is greatly different from the intended primary use coal, the result may be oversized mills and limits on turndown on the primary coal. In such a case, it may be necessary to consider eliminating this coal, or accepting a load reduction when burning it.

Performance testing

Acceptance testing will usually be guided by the procedures in The American Society of Mechanical Engineers (ASME) Performance Test Code, PTC 4.2 which references other applicable codes and standards. Testing may also be done as part of an overall boiler test, or simply to learn whether maintenance or adjustments are needed.

Fineness testing

The method of fineness sampling is well described in PTC 4.2 and ASTM D 197. While this procedure has inherent weaknesses, it is the basis for measuring one of the fundamental performance parameters to verify pulverizer capacity. Periodically, alternative sampling procedures or new hardware are proposed and they may be more accurate or easier to use than the ASTM method. However, the ASTM probe is the accepted standard on which capacity correction factors for fineness are based. Any potential improvement in absolute accuracy does not justify the abandonment of decades of data based on the ASTM method, nor the cost to establish a new database.

When tests are run to determine the condition of the pulverizer system or to evaluate a proposed new coal, comparative values between successive tests may be adequate. In such cases, alternatives, such as the ISO probe, may serve well if they provide acceptable repeatability and are not sensitive to variables such as velocity or temperature. Panel board instruments, if they can be read with sufficient precision, will serve as well as calibrated instruments for periodic testing.

Evaluating test results

The evaluation approach depends on the purpose of the tests. If the parties have agreed upon test protocols as prescribed by PTC 4.2, results evaluation is simply a calculation process. When results indicate that the performance requirements have not been met but that the shortfall is slight, reference to the ASTM standards, D 409 and D 197, may show that the shortfall could lie within the limits of laboratory analysis repeatability. The protocols should provide for resolution of small deviations in the test results.

Most testing is to evaluate the equipment or its systems. It is a mistake to try to read excessive precision into test results. It is more important to track relative performance to indicate the need for corrective action or the frequency of retesting. It is useful to develop checklists or troubleshooting guides to note what each abnormality indicates. For example, acceptable fineness but excessive pressure differential probably indicates a need to adjust the pulverizer load springs. Acceptable fine fraction fineness but poor coarse fraction fineness indicates erosion damage and internal short circuiting through the classifier. Also, it is common to discover that recalibration of primary measuring devices is needed on a regular schedule.

When equipment tests are first undertaken, it is important to collect all data which could possibly prove useful. As test experience grows, extraneous data can be removed. It is usually only a short time before standard data sheets evolve and testing becomes a routine task for an individual plant.

Operations

The main costs of coal pulverizer operation are capital, power consumption, and maintenance. The capital cost is usually an annual levelized charge including capital, taxes, and several other factors. Power and maintenance are often expressed as costs per ton of coal ground, and are influenced by plant operating practices.

Power consumption

When comparing pulverizer designs, it is common to combine primary air fan and pulverizer power consumption. The power consumption for B&W Roll Wheel or EL pulverizers and their respective fans is about 14 kWh/t (15 kWh/t_m) with coal at 50 HGI and 70% fineness, operating at rated output and with new grinding parts. These conditions seldom exist for any length of time during operation. Most modern installations include extra capacity for future coal variation. For most mills, there will be either a decline in capacity or an increase in power consumed as wear progresses. Therefore, because of the combined effect of pulverizer oversizing and wear, a realistic power value might be 20 kWh/t (22 kWh/t_m) for a well sized mill on a base loaded boiler or 22 kWh/t (24 kWh/t_m) on a load following unit.

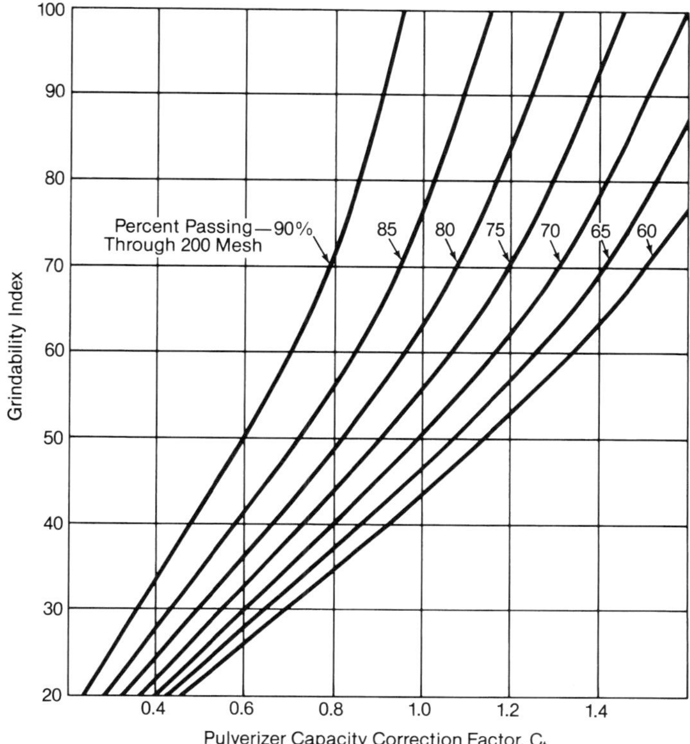

Fig. 16 Capacity correction factor.

There are various ways of calculating the cost of auxiliary power. The lowest cost which can be calculated is the cost of coal used to generate the power. This is a heat rate cost and, for a hypothetical boiler using coal with a 13,000 Btu/lb (30,238 kJ/kg) heating value and a thermal heat rate of 10,000 Btu/kWh (34.1% efficiency), 22 kWh/t (24 kWh/tm) is about 0.85% of the fuel energy input to the boiler. As heating value declines, more power is spent grinding and moving material which does not produce thermal input, so the 22 kWh/t (24 kWh/t_m) becomes a higher percentage of thermal input.

Power can be expected to increase as the grinding parts wear. The depth of the grinding track may play a part in the power increase as well. There is little information to show whether power increases uniformly with time or if it takes place mostly at the end of wear life.

Maintenance

Costs for material and labor depend on wear life which varies with the abrasiveness of the coal. Wear life on B&W Roll Wheel mills grinding U.S. coals ranges from less than 10,000 to more than 100,000 hours, and normally between 25,000 and 60,000 hours. Therefore, average annual maintenance costs can vary widely. Costs will also vary by mill size, where part costs are proportionately higher for smaller mills. For a given set of assumptions, the B&W-67™ mill maintenance will cost about 60% of the power costs cited above at a wear life of 25,000 hours and as little as 20% at 60,000 hours. Respective values for B&W-89 are 40% and 10%. For a very rough estimate, maintenance costs will be about one half the power costs for small mills and about one third for larger mills.

Maintenance and power costs are also influenced by operating practices. For example, experienced operators may notice that an increase in pulverizer pressure differential or pulverizer motor power indicates a need for spring readjustment. There may be cost effective decisions between maintenance and operations regarding rebuild frequency to minimize the costly power increases near the end of wear life.

Ceramics The use of ceramic tiles for erosion resistance has reduced repair costs. Erosion resistance of ceramic is at least eight times greater than an equivalent thickness of carbon steel. Ceramic lined panels are used in classifier cones, in the housing above the throat, and in the discharge turret in B&W Roll Wheel mills. Ceramic lining is used extensively in burner pipe elbows and for wear shields on roller brackets. When ceramic tiles are used, they must be carefully fitted to avoid undercutting and loss of the lining. Ceramic tiles are not suitable for all forms of erosion protection. High angles of impingement are destructive to brittle materials.

Record keeping Records are useful for pulverizer maintenance. Critical dimensions as recommended by the manufacturer should be measured and recorded during internal inspections. Suspicious wear data should be recorded and given special attention at the next inspection. When lubricants are drained, the amount should be noted and recorded and a sample obtained for analysis. Many owners are now including periodic gearbox vibration measurements in their data collection and record keeping. Such data may provide early warning of bearing or gear failure.

Fires and explosions

It must always be remembered that a coal pulverizer grinds fuel to a form suitable for good combustion. Combustion can and does occur at unplanned locations with costly results. Pulverizer fires are serious and should be treated with preplanned emergency procedures. The surprising speed with which a fire can develop requires prompt action.

Pulverizer fires can develop in the high temperature areas of mills or in the coal rich low temperature areas. Usually, there should be no coal in the areas upstream of the throat – the high temperature areas. Excessive spillage, because of insufficient air flow or throat wear, can cause coal accumulation in the plenum or windbox of the pulverizer. If this coal is not rejected promptly to the pyrite removal system, it can be ignited by the high temperature primary air. Coal will also enter this area when a mill is tripped or shut down quickly. Coal in the windbox may burn out without damage and, indeed, this form of fire may go undiscovered. Fires in the cooler, fuel rich zones may develop more slowly, but these fires are fed by enormous amounts of coal and can quickly destroy the internal pulverizer parts.

The most common fire indicator is mill outlet temperature. Temperature indicators are slow in response because they are sheltered by erosion protection, but they are inherently reliable and their indications should be taken seriously.

Control room operators can gain useful information about potential hazards by following consistent operating practices, especially on multiple mill installations. If, for example, all mills are at the same coal feed rate, and if primary air control is in the automatic mode, all mills should have the same air/coal ratio. Under these conditions, a significant temperature difference from inlet to outlet in one mill may indicate a fire and should be investigated immediately. Of course, there are other possible causes of inlet temperature differences such as faulty feeder calibration, faulty air flow calibration, or coal moisture variation between the respective mills. The significant point is that there is important information to be drawn from careful observation of the vital signs of pulverizer system performance. Use of all vital signs as leading indicators of potential hazards is important to safe operations.

Severe fire damage has also been caused by feeding burning coal from silos or bunkers. All pulverizer manufacturers prohibit this practice. Silo fires must be dealt with separately by emptying the fire through special diverter chutes or by extinguishing agents applied to the coal surface.

In mill trips, the sudden loss of air flow allows the collapse of the fluid bed above the throat. The coal then falls into the windbox and contacts the hot metal surfaces. Low rank coals can begin smoldering within a few minutes at temperatures of 450F (232C) or more, and, upon agitation, dust clouds can be created leading to explosions. The agitation may be from mechanical action of the pyrite plows upon restart, or from the start of primary air flow before restart. The danger of explosion is minimized by removing the coal under rigidly controlled conditions. It is first necessary to assure personnel safety by evacuating all workers from places which could be affected by an explosion if the enclosures of the mill, ducts or burner pipes are breached. The mill is isolated from all air flow and an inert internal atmosphere is created by injecting inert gas or vapor. Brief operation of the mill, while isolated, rejects the coal to the pyrite removal system. The inert atmosphere prevents any smoldering coal from igniting the dust cloud raised internally by mill operation.

When a mill explosion occurs it is nearly always during changes in operating status, such as during startup and shutdown.[2] During startup, the air-coal mixture in the mill passes from extremely fuel lean to fuel rich. There is a transitional mixture which is ideal for a dust explosion if a sufficiently strong ignition source exists. One recurring source of ignition is smoldering fires which develop in the residual coal left in the hot zone of a mill following a trip. The coal removal procedure just described was devised to avoid or safely remove this ignition source.

Undiscovered fires can cause dust explosions during otherwise normal shutdowns. These are not common, but experience has led many operators to actuate audible and visible alarms before routine starts and stops. This is a recommended safety precaution for all coal pulverizer operators.

Controls and interlocks

The most widely followed standard for design and application of coal pulverizer systems is the National Fire Protection Association Standard, currently NFPA-85. This sets the minimum standards listed below for safety interlocks to be used with coal pulverizers:

1. failure of primary air flow trips the pulverizer system,
2. failure of the pulverizer trips the coal feeder and primary air flow,
3. closing of all burner pipe valves trips the pulverizer, the feeder, and the primary air flow, and
4. primary air flow below the manufacturer's minimum trips the pulverizer.

Other interlocks not mandated by NFPA but required by B&W are:

1. loss of the lube oil pump, or low oil pressure, trips the pulverizer (B&W Roll Wheel mill only), and
2. flame safety systems trip the pulverizer if minimum flame detector indications are not met.

Additional indications of malfunctions which may require manually tripping a mill include:

1. excessively high or rapidly rising mill exit temperature probably indicates a fire,
2. inability to maintain an exit temperature of at least 125F (52C) may lead to loss of grinding capacity,
3. loss of seal air pressure or flow will eventually lead to bearing failure, and
4. high lube oil temperature indicates a loss of cooling water flow which may lead to gear drive damage.

Items 2, 3 and 4 above are not safety related. They are examples of the designers' choices to avoid unnecessary mill trips and the resultant special safety procedures required to restart tripped mills as well as to protect the equipment from damage.

Special safety systems

The procedure for removing coal after a pulverizer trip requires a separate control system. This system must bypass the control logic which prevents operation of the pulverizer drive motor without adequate primary air flow. To prevent misuse of this bypass feature, the coal removal logic (referred to as inert and clear logic) is enabled by actuation of the mill trip logic. A typical schematic logic diagram is shown in Fig. 17.

Another special system has been developed to deal with a phenomenon which has been observed in ducts after trips. A fine layer of dust is often seen in the clean air portion of the primary air system. This is thought to be due to a turbulent dust cloud which persists after collapse of the fluidized bed of coal above the throat. An array of water fog nozzles under the throat will intercept this cloud if actuated with the mill trip signal. Hardware for tangential wash nozzles to enhance the coal removal action of the pyrite plows during the inert and clear cycle can be installed adjacent to the fogging nozzles. The wash nozzles are actuated manually at the appropriate time in the cycle.

Advanced designs

Rotating classifiers

As noted under *Application engineering*, product fineness is under review after being generally standardized at 70% passing through 200 mesh (74 microns) for direct firing of bituminous coal. Advanced low NO_x combustion systems benefit incrementally from higher 200 mesh fineness, but very high 100 mesh fineness can sharply reduce unburned carbon losses. Modern independently driven rotating classifiers can provide up to 95% passing 200 mesh fineness but at a significant reduction in coal flow capacity. Alternately, they can provide excellent 100 mesh fineness at 70% or less though 200 mesh, with little or no capacity reduction. While main shaft driven rotating classifiers were part of the earliest E mill de-

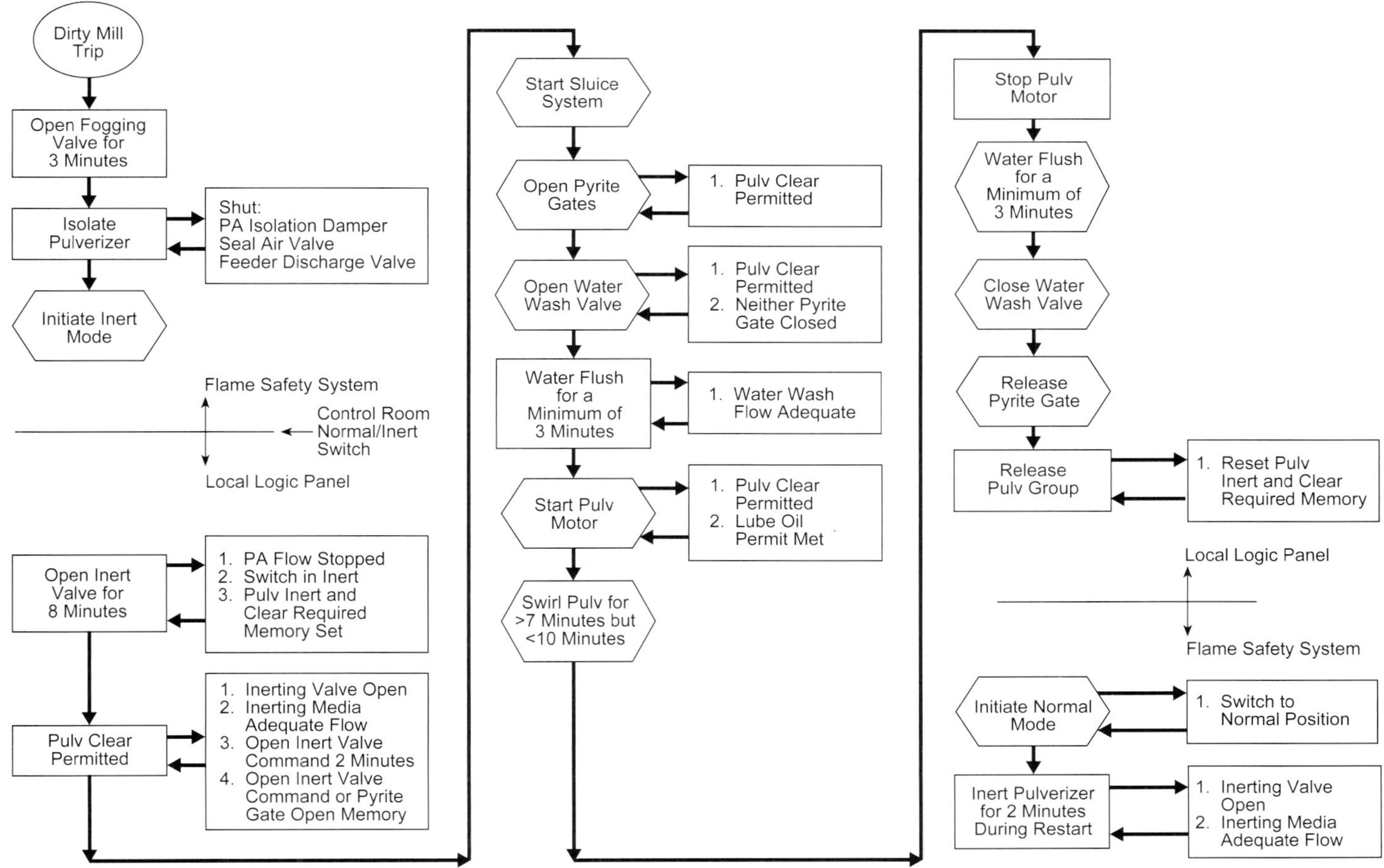

Fig. 17 Typical safety system logic diagram.

sign, independently driven variable speed classifiers were developed in the 1970s, and the two stage design shown in the B&W Roll Wheel mill in Fig. 18 was developed in the 1990s. A similar dynamically staged variable speed (DSVS®) classifier for EL pulverizers, is shown in Fig. 19. The improved particle distribution achieved with the DSVS classifier is shown in Fig. 20.

Rotating throats

Rotating throat rings were introduced for B&W Roll Wheel mills in the 1980s, initially to reduce erosive wear. The rotating throat designs have been further enhanced to provide better air distribution and lower air resistance, as well as better reliability and easier installation.

Variable spring loading

Also for the B&W Roll Wheel mills, a variable online spring load adjusting system is available. With the traditional spring loading system, the applied grinding force is a compromise between efficient high load operation, and smooth operation at low loads, including start and stop. With the Auto Spring™ system, the spring load can be increased for optimum high load performance, then reduced for stable low load operation as well as grind out and startup. The typical spring load control will vary pressure with coal flow, but it can also vary loading to accommodate wide variations in coal grindability, as shown in Fig. 21.

Fine grinding

Vertical roller mills have a practical limit of about 95% passing through 200 mesh (74 microns) for product fineness. As the product becomes finer, the recirculated material in the grinding chamber also becomes finer and the compressed layer becomes more fluid and less stable. For very fine grinding other machines are needed. One of these is the ball-and-tube mill, offering the long residence time needed for an extremely fine product. In addition, ball-and-tube mills are built in the large sizes needed for direct firing.

The high capacity of ball-and-tube mills may not be needed to meet fine grinding applications. One potential application of fine grinding is to provide fuel for coal-fired lighters to reduce oil consumption for startup and low load flame stabilization. Another application is firing very fine coal in smaller furnaces, originally intended for oil or natural gas and not suit-

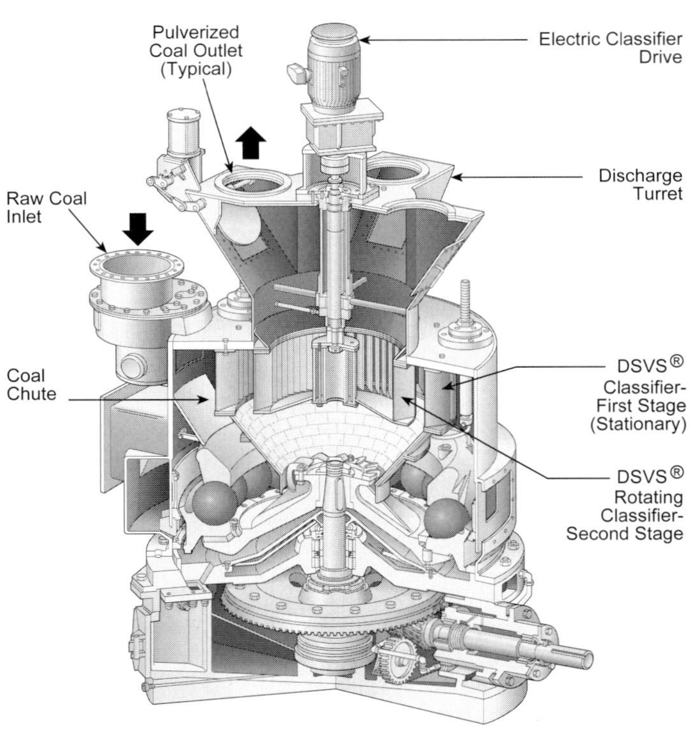

Fig. 19 DSVS® rotating classifier for EL mill.

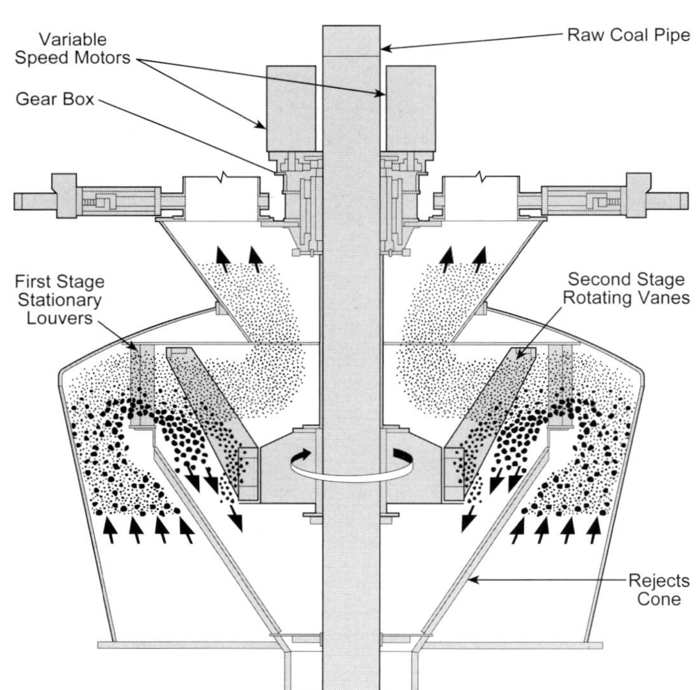

Fig. 18 DSVS® rotating classifier for B&W Roll Wheel™ pulverizer.

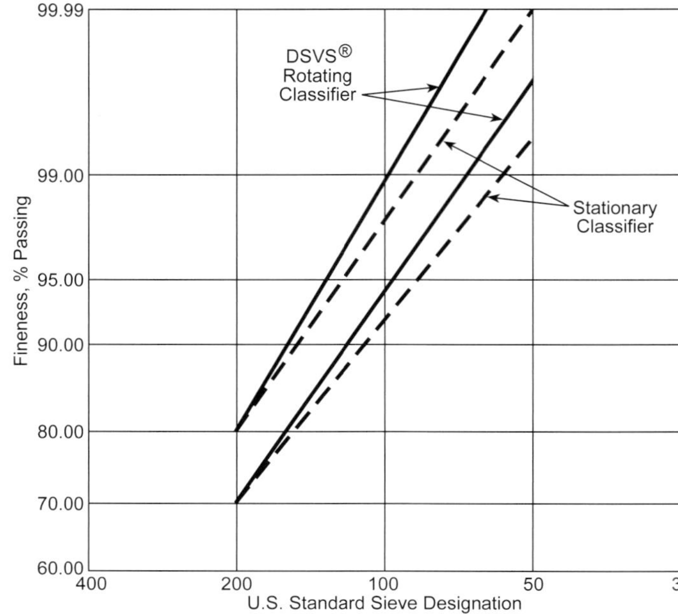

Fig. 20 Particle size distribution.

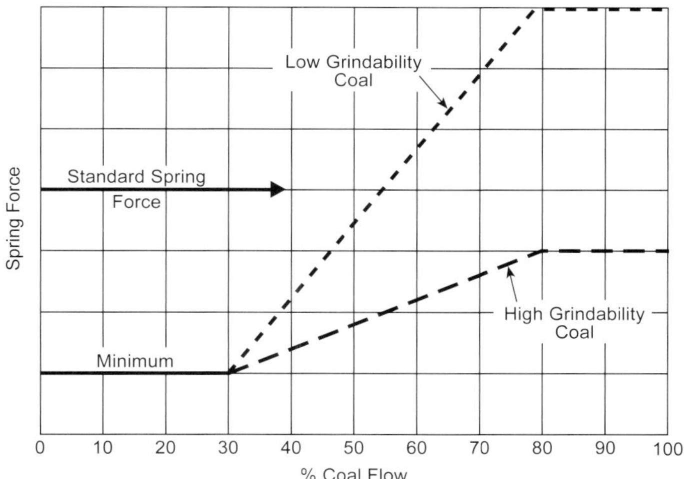

Fig. 21 B&W Roll Wheel™ pulverizer variable spring force versus coal flow.

able for traditional coal firing. Smaller capacity mills could be used for this application.

Mills for fine grinding have been used to grind pigments, food and pharmaceuticals. These are small mills, generally 5 t/h (4.5 t_m/h) or less. Grinding principles include fluid energy, stirred ball, and high speed attrition. Fluid energy mills cause solids bearing fluid streams to impinge; particle impacts cause the size reduction. Stirred ball mills, usually wet grinders, use a charge of small balls in a cylindrical vessel which are agitated by a rotor, grinding the solids flowing through the charge.

High speed attrition mills grind by the combination of first stage impact and final stage attrition. It is these mills which have begun to attract some attention for preparation of oil or natural gas replacement fuel. The mill's rotor is a series of disks, each with impact bars, and each forming the driver for a grinding stage. The grinding stages are separated by diaphragms and coal must move inward against centrifugal force to pass from one stage to the next. Coal captured in each stage is said to provide an effective wearing surface to protect the mill shell from rapid wear. These mills are made in capacities up to about 5 t/h (4.5 t_m/h) and may be once-through or be equipped with external centrifugal classifiers for size control. Manufacturers of these mills refer to the product as micronized coal. When the term was adopted in the early 1980s, micronized coal referred to coal which was 100% finer than a 325 mesh (44 microns). The output of vertical high speed attrition mills is about 98% passing through 200 mesh (74 microns), 86% passing through 325 mesh (44 microns). However, this coal is reported to be fine enough to be easily ignited without the need for preheated primary or combustion air. These mills are high in power consumption and motor power is said to be sufficient to provide the necessary energy for drying most coals. The economic attractiveness of finely ground coal as a stabilizing and startup fuel depends on oil and natural gas prices.

References

1. Johnson, T.D., et al., "Pulverized Fuel System Erosion," presented at the Electric Power Research Institute (EPRI) Conference on Coal Pulverizers, pp. 18-22, Central Electricity Generating Board, England, 1985.

2. Zalosh, R.G., "Review of coal pulverizer fire and explosion incidents," ASTM STP 958, Cashdollar, K.L., and Hertzberg, M., eds., American Society for Testing and Materials, Philadelphia, Pennsylvania, p. 194, 1987.

Bibliography

"Coal: Sampling of Pulverized Coal Conveyed by Gases in Direct Fired Coal Systems," pp. 18-22, International Standards Organization, 1991.

Donais, R.T., Tyler, A.L., and Bakker, W.T., "The effect of quartz and pyrite on abrasive wear in coal pulverizers," *Proceedings of the International Conference on Advances in Material for Fossil Power Plants,* American Society for Metals, Metals Park, Ohio, p. 643, 1987.

Piepho, R.R., and Dougan, D.R., "Grindability measurements on low rank fuels," *Proceedings, Coal Technology '81,* Vol. 3, pp. 111-121, 1981.

Wiley, A.C., *et al.*, "Micronized coal for boiler upgrade/retrofit," *Proceedings of POWER-GEN International '90*, PennWell, Houston, Texas, pp. 1155-1170, 1990.

Advanced low NO_x B&W DRB-4Z® burners ready for installation.

Chapter 14

Burners and Combustion Systems for Pulverized Coal

Advanced low NO_x (nitrogen oxides) pulverized coal (PC) combustion systems, widely in use today in utility and industrial boilers, provide dramatic reductions in NO_x emissions in a safe, efficient manner. These systems have been retrofitted to many existing units and are reducing NO_x emissions to levels which in some cases rival the most modern units. The challenges are considerable, given that the older units were not built with any thought of adding low NO_x systems in the future. Low NO_x combustion systems can reduce NO_x emissions by up to 80% from uncontrolled levels, with minimal impact otherwise on boiler operation; and they do so while regularly exceeding 99% efficiency in fuel utilization. These advanced low NO_x systems, depicted in Fig. 1, start with fuel preparation that consistently provides the necessary coal fineness while providing uniform fuel flow to the multiple burners. Low NO_x burners form the centerpiece of the system, and are designed and arranged to safely initiate combustion and control the process to minimize NO_x. An overfire air (OFA) system supplies the remaining air to complete combustion while minimizing emissions of NO_x and unburned combustibles. Distributed control systems (DCS) manage all aspects of fuel preparation, air flow measurement and distribution, flame safety, and also monitor emissions. Cutting edge diagnostic

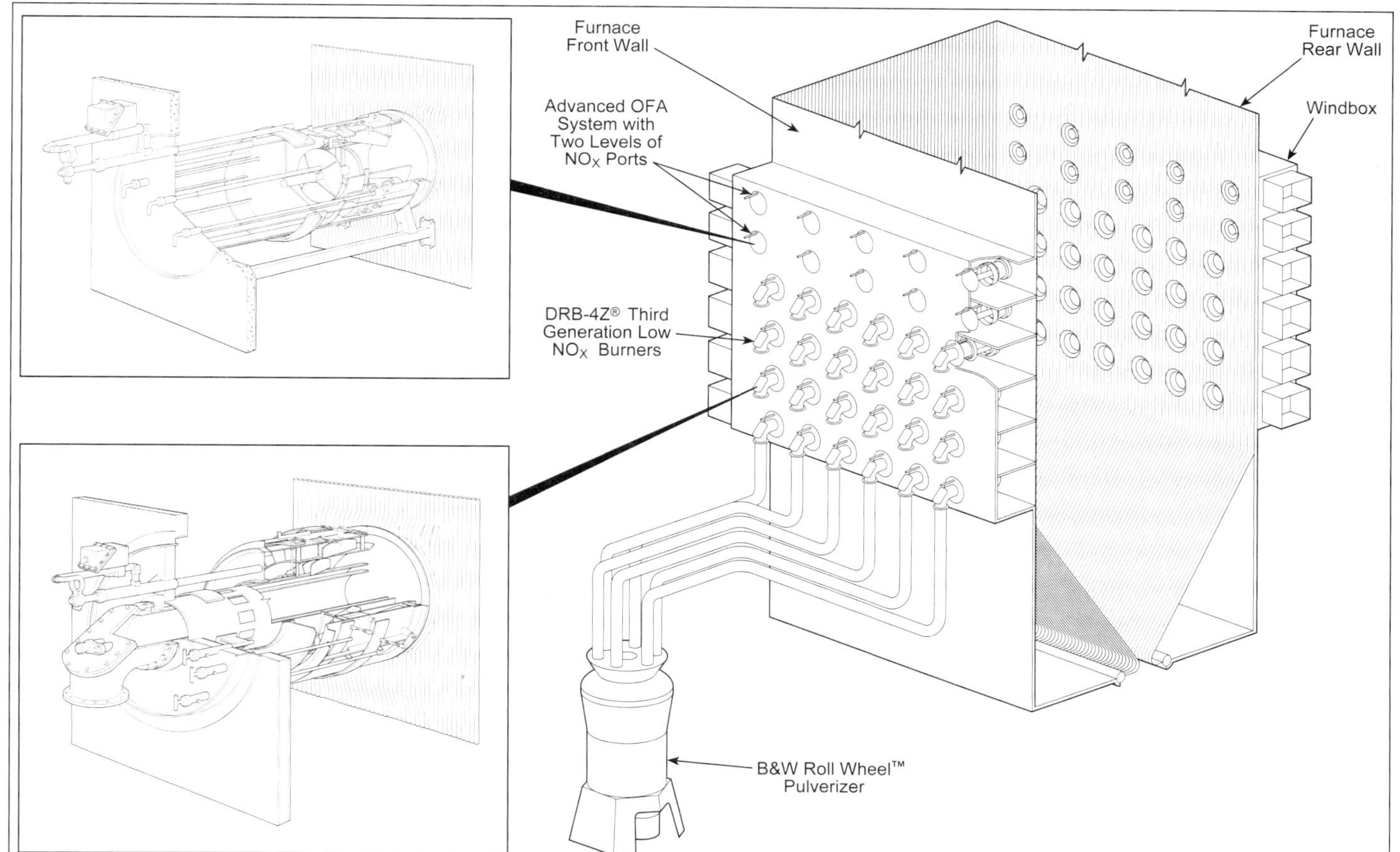

Fig. 1 Advanced low NO_x pulverized coal-fired combustion system.

and control techniques, using neural networks and chaos theory, assist operators in maintaining performance at peak levels. Like automobiles, the fundamentals go back more than 100 years, but the modern product is dramatically more reliable, more efficient, and lower in emissions.

The greatest use of coal worldwide is for the energy source in electric power generation. Coal remains the dominant fuel for electrical generation in the United States (U.S.) and in many other countries as well. This is attributable to both the abundance and relatively low cost of this fossil fuel. It can be burned in a number of ways depending upon the characteristics of the coal and the particular boiler application. Cyclone, stoker, and fluidized-bed firing methods are used for a variety of applications as discussed in Chapters 15 through 17. However, pulverized coal (PC) firing – burning coal as a fine powder suspension in an open furnace – is the most prevalent method in use today. PC firing has made possible the large, efficient utility boilers used as the foundation for power generation in many utilities worldwide.

Pulverized coal firing differs from the other coal combustion technologies primarily through the much smaller particle size used and the resulting high combustion rates. The combustion rate of coal as a solid fuel is, to a large extent, controlled by the total particle surface area. By pulverizing coal to a nominal 50 micron diameter or smaller (see Chapter 13), the coal can be completely burned in approximately one to two seconds. This approaches the rate for oil and gas. In contrast, the other technologies discussed in subsequent chapters use crushed coal of a larger size and require substantially longer combustion zone residence times (up to 60 seconds or longer).

Pulverized coal was first used in the 1800s as a cost effective fuel for cement kilns.[1] The ash content enhanced the properties of the cement and the low cost resulted in the rapid displacement of oil and gas as a fuel. However, early pulverizing equipment was not very reliable and, as a result, an indirect bin system was developed to temporarily store the pulverized coal prior to combustion, providing a buffer between pulverization and combustion steps. Use in the steel industry followed closely. In this application, the importance of coal drying, particle size control and uniform coal feed were recognized.

Success for boiler PC firing applications required modification of the boiler furnace geometry to effectively use the PC technology. In the early 1900s, fireboxes and crowded tube banks had to be expanded to accommodate PC firing. By the late 1920s, the first water-cooled furnaces were used with pulverized coal-fired systems. Burner design progressed, providing improved flame stability, better mixing, and higher combustion efficiency. Coupled with improvements in pulverizer design and reliability, improved PC burners permitted the use of direct firing with coal transported directly from the pulverizer to the burner. (See Fig. 2.)

The exponential increase in the U.S. electric power generation and the increase in boiler size from 100 MW to 1300 MW could not have been achieved economically without the development and refinement of pulverized coal firing. Today, nearly all types of coal from anthracite to lignite can be burned through pulverized firing. Combustion efficiencies of most coals approach those of oil and gas. Current research is focusing on further reduction in emissions of nitrogen oxides (NO_x) for environmental protection without sacrificing boiler performance or availability. (See also Chapter 34.)

Fig. 2 Early pulverized coal-fired radiant boiler.

Combustion

The manner in which a coal particle burns depends on how it was pulverized, its inherent characteristics, and conditions in the furnace. Pulverization obviously produces fine coal particles, but is also accompanied by introduction of heated primary air (PA) to the coal to facilitate grinding and drying, and for transport of the coal from the pulverizer. The quantity and temperature of PA varies significantly with the type of pulverizer, grinding rate, and coal properties. Hot PA causes some of the moisture in the coal to evaporate and raises the temperature of the coal from ambient to nominally 150F (66C). In a direct fired system, the pulverized coal is immediately transported to the burners by the moisture laden PA. Secondary air is introduced through a burner to the PA/PC mixture, in a controlled manner to induce air-fuel mixing in the furnace. As a coal particle enters the furnace (see Fig. 3), its surface temperature increases due to radiative and convective heat transfer from furnace gases and other burning particles. As particle temperature in-

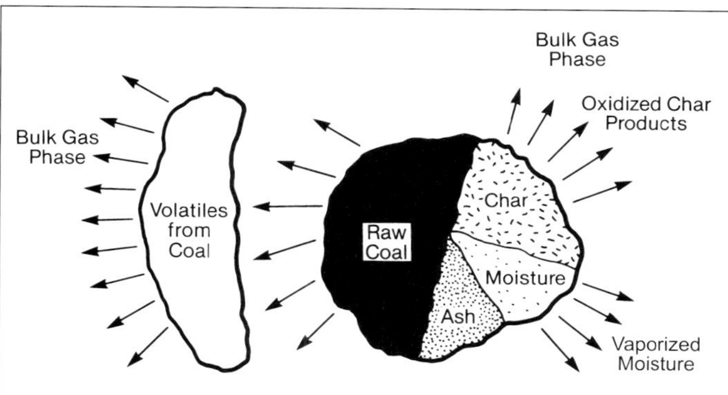

Fig. 3 Coal particle combustion.

creases, the remaining moisture is vaporized and volatile matter is released. This volatile matter, which ignites and burns almost immediately, further raises the temperature of the char particle, which is primarily composed of carbon and mineral matter. The char particle is then consumed at high temperature, leaving the ash content and a small amount of unburned carbon. The volatile matter, fixed carbon (char precursor), moisture and ash content of the fuel are identified on a percentage basis as part of the proximate analysis discussed in Chapter 9.

Volatile matter content

Volatile matter is critical for maintaining flame stability and accelerating char burnout. Coals with minimal volatile matter, such as anthracites and low volatile bituminous, are more difficult to ignite and require specially designed combustion systems. The amount of volatile matter evolved from a coal particle depends on coal composition, the temperature to which it is exposed, and the time of this exposure. The American Society for Testing and Materials (ASTM) Method D 3175 stipulates a temperature of 950 ±20C for seven minutes for volatile matter content determination.[2] Raising the temperature would increase volatile yield with other factors held constant. The heat contribution from volatile matter, as a percentage of total coal heating value, can be less than 10% for anthracites, reaches 40% for high volatile bituminous and subbituminous coals, and can exceed 50% for lignite. Coals with higher volatile matter content also benefit from more effective NO_x control by combustion methods, as discussed later. Ignition is influenced by the quality and the quantity of volatile matter. Volatile matter from bituminous and higher rank coals is rich in hydrocarbons and high in heating value. Volatile matter from lower rank coals includes larger quantities of carbon monoxide and moisture (from thermal decomposition) and consequently has a lower heating value. Volatile matter from higher rank coals can provide twice the heating value per unit weight as that from low grade coals.

Char particles

The rate of char particle combustion depends on several factors including composition, particle size, porosity, thermal environment, and oxygen partial pressure. Char reactions often begin as the coal particle is heated and devolatilizes, but they continue long after devolatilization is complete. Devolatilization is mostly completed after 0.1 seconds, but char-based reactions continue for one to two seconds. The char particle retains a fraction of the hydrocarbons. Char reactivity lessens with increasing coal rank due to accompanying changes in composition and structure of the coal. Younger coals benefit from higher inherent reactivity and from a structure which more closely resembles fibrous plant matter than rock. Many coals go through plastic deformation and swell by 10 to 15% when heated. These changes can significantly impact the porosity and reactivity of the coal particle. Char reactivity drops during the combustion process, which further extends the time for char burnout.

Char oxidation requires oxygen to reach the carbon in the particle and the carbon surface area is primarily within the particle interior structure. Char combustion generally begins at relatively low particle temperatures. Reaction rates are primarily dependent upon local temperature as well as oxygen diffusion and char reactivity. Small char particles, with 10 to 20 micron diameters, benefit from high surface to mass ratios and heat up rapidly, while coarse particles heat more slowly. Rapid heat transfer to and combustion of smaller particles lead to higher particle temperatures. Reaction rates increase exponentially with temperature, and oxygen (O_2) for diffusion into the particle becomes the controlling parameter. Particle diameter and density change in the process. At higher particle temperatures, char reactions are so fast that oxygen is consumed before it can penetrate the particle surface. The particle shrinks as the outer portions are consumed, and transport of oxygen from the surroundings to the particle is the factor governing combustion rate.

For larger particles, the solid mass is reduced as carbon monoxide (CO) and carbon dioxide (CO_2) form, but particle volume is maintained. Coarse particles, more than 100 micron diameter, burn out slowly as a result of their lower surface to mass ratios. Longer burnout times cause these larger char particles to continue reacting downstream where the flame temperature and oxygen concentration have moderated. A portion of the carbon may not be oxidized depending on the residence time and combustion environment. This unburned carbon amounts to an efficiency loss for the boiler.

Effect of moisture content

The moisture content of the coal also influences combustion behavior. Direct pulverized coal-fired systems convey the moisture evaporated during pulverization to the burners. This moisture plus that remaining in the coal particles present a burden to coal ignition. The water must be vaporized and superheated during the early stages of the combustion process. Further energy is absorbed at elevated temperatures as the water molecules dissociate.

Moisture content increases as rank decreases as discussed in Chapter 9. 15% moisture is common in high volatile bituminous coals, 30% is seen in subbituminous, and more than 40% is common in some lignites. Moisture contents in excess of 40% exceed the ignition capability of conventional PC-fired systems. Alternate systems are then required to boost drying during fuel preparation and/or divert a portion of the evaporated moisture from the burners. Char burnout is impaired by moisture which depresses the flame temperature. This is largely compensated for by the generally higher inherent reactivities and porosities of the higher moisture coals.

Effect of mineral matter content

The mineral matter, or resulting ash, of the coal is inert and dilutes the coal's heating value. Consequently, more fuel by weight is required as ash content increases in order to reach the furnace net heat input. Mineral matter content averages about 10% for U.S. coals but can exceed 30% in some U.S. and a number of international coals.

The ash absorbs heat and interferes with radiative heat transfer to coal particles, inhibiting the combustion process noticeably with high ash coals. The forms of mineral matter determine the slagging tendencies of the ash. These impact combustion by influencing the furnace size and burner arrangement and the resulting flame temperatures and residence time. The additional effects of ash on overall boiler design are discussed in Chapter 21.

Pulverized coal combustion system

Effect of coal type

A variety of laboratory tests and correlations (Table 1) have been developed to predict combustion performance of coal and other PC-fired hydrocarbons such as coke. The tests provide basic information on coal properties and combustion behavior using small samples of coal. Coal heating value, proximate and ultimate analyses are fundamental requirements and essential for performance calculations and emission predictions. Other tests are performed to better characterize combustion behavior, particularly for unfamiliar coals. Correlations predict key aspects of combustion such as ignitability and char reactivity. The correlations are often functions of proximate and ultimate analyses as these are nearly universally available. They relate experience to coal properties and assist designers in formulating an appropriate combustion system for the coals in question.

Conventional PC Pulverized coal firing is adaptable to most types of coal. Its versatility has made it the most widely used method of coal firing for power generation worldwide. Most PC-fired systems have the burners located in the lower portion of the furnace. Wall-fired systems usually have the burners positioned on either one wall or on two walls in an opposed arrangement as shown in Fig. 4. The other common arrangement positions the burners in the corners of the lower furnace, and is referred to as corner or tangentially-fired. Primary air typically at 130 to 200F (54 to 93C) conveys pulverized coal directly to the burners at a rate set by the combustion controls based on steam generation requirements. Secondary air is supplied by the forced draft fans and is typically preheated to about 600F (316C). All or most of the secondary air is supplied to the windboxes enclosing the burners. A portion of the secondary air may be diverted from the burners to overfire air (OFA) ports (discussed later) in order to control the formation of NO_x. The secondary air supplied to the burners is mixed with the pulverized coal in the throat of the burner. This permits the coal to ignite and burn.

The combustion process continues as the gases and unburned fuel move away from the burner and up

Table 1
Key Tests and Correlations to Evaluate Coal

Test	Information
Proximate analysis	Combustibles: volatile matter, char Inerts: moisture, ash
Ultimate analysis	Major elements: carbon, hydrogen, sulfur, nitrogen, oxygen
Drop tube furnace tests	Ignition and char burnout behavior
Thermogravimetric analysis	Ignition, evaporation and char burnout behavior
Free swelling index	Char reactivity
Petrographic examination	Char reactivity

Correlation	Information
Dulong's heating value	Estimated heating value of coal
Ignition factor (B&W)	Ignitability of coal
(Volatile matter x coal oxygen) dry, ash-free	Char reactivity prediction

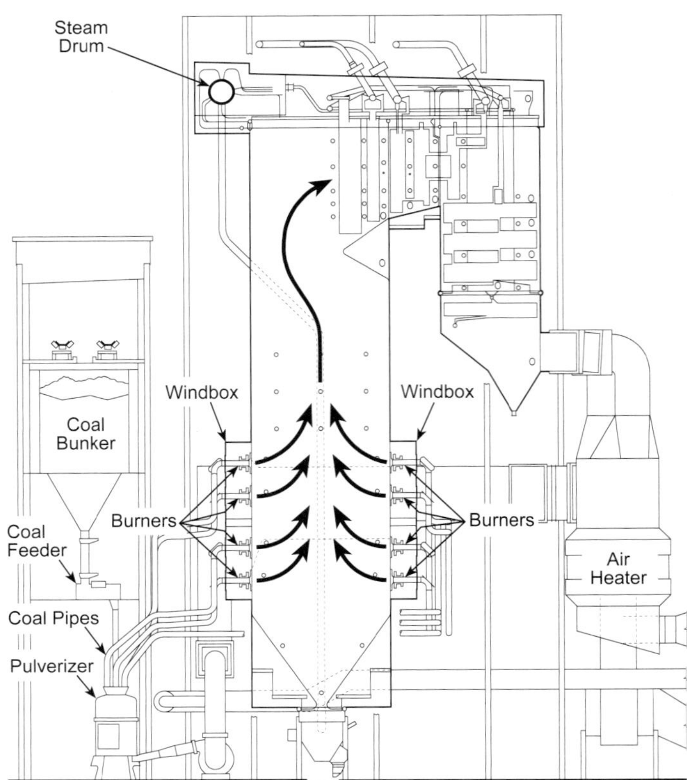

Fig. 4 Conventional pulverized coal-fired boiler.

the furnace shaft. Final burnout of the char depends on the coal properties, particle fineness, excess air, air-fuel mixing, and thermal environment. The products of combustion proceed out of the furnace after being cooled sufficiently and enter the convection pass. Properly controlled furnace exit gas temperature is critical for achieving the required steam conditions and efficiency, and to limit slagging and fouling on downstream tube surfaces.

Low volatile coals While the majority of bituminous, subbituminous, and lignite PC-fired units are arranged in the manner described above, alternate designs are required to accommodate other coals. Coals with low volatile matter content, notably anthracites, are difficult to ignite. Their volatile matter provides insufficient ignition energy for stable combustion with a conventionally configured system. In addition, their advanced coalification (see Chapter 9) has reduced char reactivity, resulting in elevated char ignition and burnout temperatures. These factors require hotter furnaces to sustain combustion. A downshot firing system (see Fig. 5) accomplishes this in combination with an enlarged, refractory-lined lower furnace. The downshot arrangement causes the flame to travel down and then turn upward to leave the lower furnace. The char reactions, which build in intensity some distance down from the burners, return their heat near the burners as the gases turn upward to exit the lower furnace. This heat supplies energy to ignite the fuel introduced at the burners. The refractory lining in the furnace inhibits heat transfer and elevates gas temperatures to sustain combustion. The enlarged lower furnace also provides increased residence time to accommodate the slower burning char. In addition, very high coal fineness is used to accelerate char reactions and reduce unburned carbon loss. Indirect firing systems can provide hotter, fuel-rich mixtures to the burners to further improve ignition performance with anthracites. Economics and NO_x emission regulations have limited downshot fired units in the U.S., although they are applied internationally where low volatile coals are a major fuel source.

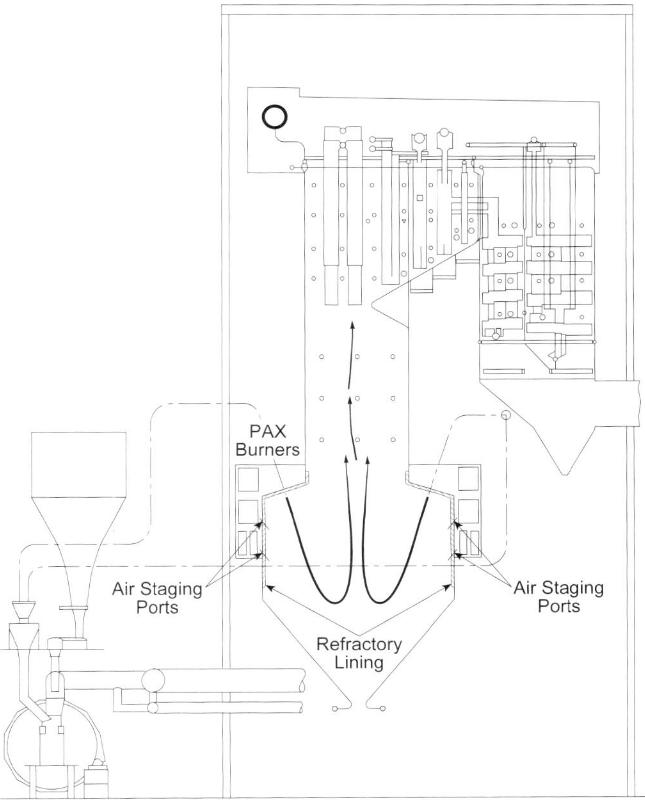

Fig. 5 Downshot-fired unit with refractory-lined furnace for low volatile coals and primary air exchange (PAX) burners.

High moisture coals The other main category of nonconventional PC-fired units is that used for lignites with very high moisture content. Moisture content in these coals typically ranges from 50 to 70% by weight, which is well beyond the level that air-swept pulverizers can adequately dry. (See Chapter 13.) Instead, hot gases, removed from the upper portion of the furnace, are used in combination with beater mills to dry and prepare the fuel. These hot gases, near 1832F (1000C), are more suitable for coal drying than preheated air, with a practical upper limit of 752F (400C). Furthermore, the low oxygen content of the flue gas provides a relatively inert atmosphere in the grinding equipment, lessening the threat of fire or explosion as the hot gas mixes with these reactive coals. The fuel is conveyed from the mills to the burners located on the walls, and combustion proceeds as secondary air mixes with the fuel in the furnace. Units of this design are prevalent in parts of Europe and Australia where high moisture brown coals are used.

Combustion system and boiler integration

The most fundamental factors that determine the boiler design are the steam production requirements and the coal to be fired. The thermal cycle defines the heat absorption requirements of the boiler which supplies the rated steam flow at design temperature and pressure. Gas-side parameters, based on the design coal, are used to estimate boiler efficiency. Heat input requirements from the coal can then be determined, setting the coal firing rate at maximum load. The number and size of pulverizers are then selected. Frequently, the pulverizer selection is based on meeting maximum requirements with one mill out of operation. This permits maintenance on one pulverizer without limiting boiler load.

The size and configuration of the furnace are designed to accommodate the combustion characteristics and the slagging tendencies of the coal ash as discussed in Chapter 21. NO_x emission control factors are also incorporated into the layouts of modern combustion systems. Furnace volume must provide sufficient residence time to complete combustion. Heat input per plan area is a key parameter to gauge thermal loading, and has to be moderated with high or severe slagging coals. Peak flame temperatures can exceed 3000F (1649C) in the burner zone, well in excess of ash fusion temperatures. Molten ash accumulations are controlled by providing a furnace of appropriate size, and by inclusion of sootblowers. Furnace width strongly influences the burner arrangement, furnace depth is critical to gas-side mixing and OFA system performance, and height is elemental for setting vertical burner spacing and for residence time. Increasing the height of the burner zone reduces thermal loading and tends to reduce NO_x formation. However, taller burner zones tend to reduce upper furnace residence time and pose problems for completing combustion for a given furnace volume. The best overall furnace design takes into consideration many cross-related factors and is strongly influenced by experience.

The number of burners and the heat input per burner are selected to minimize flame impingement on the furnace walls. Multiple burners at moderate inputs, in an opposed-fired arrangement, are favored to provide uniform heat distribution across the furnace and to inhibit localized slagging. The burner design is based on fuel parameters and NO_x emission control requirements in order to provide complete combustion and minimum pollutant emissions. An OFA system may be included to further reduce NO_x, depending on emission limits and coal sulfur content. High sulfur coals pose corrosion risks to common furnace tube materials under fuel-rich conditions. OFA can still be safely employed in such cases by use of alloy weld overlay or similar measures to protect tubes in the burner zone.

Theoretical and excess air quantities are determined for the required coal input rate from combustion air calculations as discussed in Chapter 10. Fans, ductwork, air heaters, windboxes, and burners are sized to satisfy flow, pressure loss, air preheat and velocity requirements for the unit. OFA ports are designed to satisfy jet penetration and dispersion criteria with acceptable pressure drop.

Burner integration

Burners are the central element of effective combustion system designs which include fuel preparation, air-fuel distribution, furnace design and combustion control. Burners can be categorized as pre-mix and throat-mix types. Pre-mix burners combine secondary air and fuel upstream of the combustion chamber. Throat-mix burners, including essentially all PC-fired burners, introduce secondary air to the fuel in the burner throat at the entrance to the furnace (Fig. 6). The secondary air is supplied in a manner to induce air-fuel mixing and thereby sustain ignition and produce a stable flame. The mixing rate of air and fuel directly affects flame stability, flame shape, and emissions.

Fuel preparation influences Pulverizers are designed to grind a specific quantity of coal, with given grinding characteristics, to a prescribed level of fineness (see Chapter 13). The pulverizer type and operating conditions determine the quantity and fineness of pulverized coal supplied to the burner, as well as the required quantity of primary air. The number of burner lines per pulverizer varies with the furnace size and design philosophy, with four to eight as the most common range for medium and large utility boilers. In practice this results in burner maximum coal rates typically in the range of 10,000 to 30,000 lb/h (1.26 to 3.79 kg/s). Given the normal range of coal heating values, this translates into maximum burner heat inputs of 100 to 300 million Btu/hr (29 to 88 MW_t). In contrast, small utility and industrial boilers have PC-fired burners with maximum inputs commonly in the 50 to 100 million Btu/hr (15 to 29 MW_t) range, and may have only 2 or 3 burners per pulverizer. Pulverizers can typically operate at grinding rates down to 25 to 40% of maximum capacity, depending on the design, coal variability, and control system. Burners are expected to operate with stable flames over the normal turndown range of the associated pulverizer. This can usually be accomplished with good coal, but can be problematic with difficult to ignite coals.

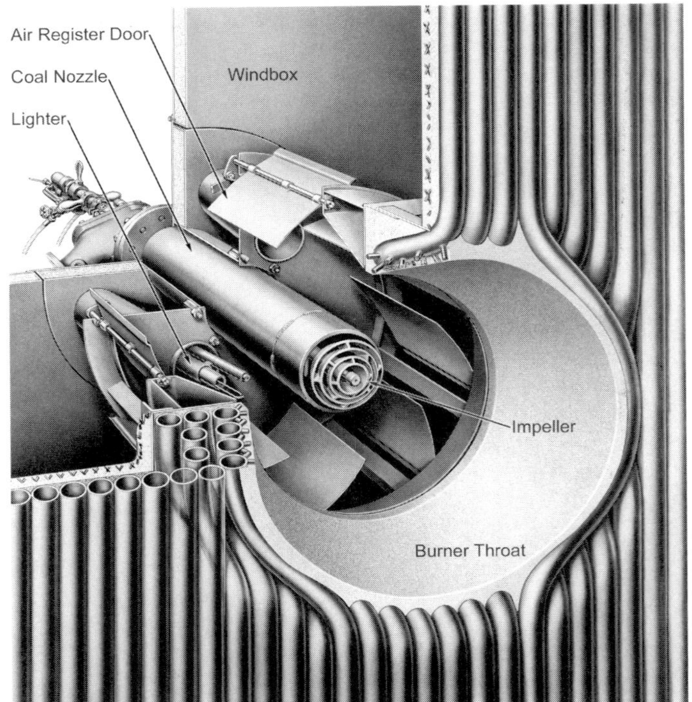

Fig. 6 Circular register pulverized coal burner with water-cooled throat.

Primary air is supplied to the pulverizer to facilitate coal drying, for circulation within the mill, for classification of coal particles, and for transport of the coal through the coal pipes to the burners. Primary air temperature to the mill depends on the air preheat capability, moisture content in the coal, primary air to coal ratio in the mill, and the required mill outlet temperature. Mill inlet temperatures may drop below 300F (149C) for dry coals, and reach 700F (371C) with lower rank coals. Resulting mill outlet temperatures, approximately the fuel temperature supplied to the burners, range from 130F (54C) with low rank coals to 200F (93C) with high rank coals. Primary air leaving the mill is saturated with moisture with low rank coals, posing an ignition burden. Pulverizers typically require 40 to 70% of their full load primary air requirements at their minimum output level. In addition, the PA/PC mixture traveling to the burners must be transported at a minimum of 3,000 ft/min (15 m/s). This velocity serves to prevent the coal particles from dropping out of suspension in horizontal runs of coal pipe. Minimum primary air flow is the greater of the minimum PA flow required for the pulverizer or the minimum required to satisfy coal pipe velocity limits.

Stoichiometry refers to the ratio of air supplied to

the calculated theoretical air requirement. The primary stoichiometry, from the primary air, can vary by a factor of three depending on the coal and pulverizer conditions. Primary stoichiometry directly influences combustion conditions in the root of the burner flame, with strong impacts on NO_x formation as discussed later. Further, for a given burner stoichiometry, increasing PA reduces the available secondary air (SA). The SA/PA ratio is also seen to vary greatly with coal and pulverizer conditions. See Table 2. Higher SA/PA ratios provide greater aerodynamic control of air-fuel mixing and improved burner performance. Low SA/PA ratios can result in flame instability. In summary, fuel preparation issues dramatically vary the condition of the fuel stream supplied to the burner, and greatly influence the burner performance.

The quantity of coal supplied to the pulverizer is regulated by a coal feeder (see Chapter 12). Systems lacking accurate PA or coal measurement and control, as is commonly the case with ball tube mills, restrain optimization of combustion equipment and cause operational inconsistencies.

Primary air and pulverized coal exit a pulverizer and travel through coal pipes to the associated burners. The distribution of the PA/PC mixture among coal pipes is dependent on air and solids flow patterns at the pulverizer exit, and by flow resistance in the individual coal pipes. Restrictors or orifices in the coal piping from the pulverizer to its burners serve to more evenly distribute coal among several burners. As burner nozzle velocity increases, the ignition point gradually moves farther from the burner. Excessive burner nozzle velocity can result in blowoff of the flame, a potentially hazardous condition where coal ignition and flame stability are lost. High burner nozzle velocities also result in accelerated erosion of burner hardware.

Secondary air supply Forced draft (FD) fan(s) supply the secondary air (SA) for combustion. Total air to the boiler is basically the sum of primary air and secondary air. The FD fans must supply secondary air in sufficient quantity to achieve the intended excess air at the boiler outlet. Modern PC-fired boilers typically operate with 15 to 20% excess air at full load, and with higher excess air as load is reduced. Excess air, that is air beyond theoretical requirements, is necessary to assure complete combustion in practical systems given imperfections in air-fuel balance and unpredictable variations in coal quality. The FD fans also supply the motive force to pump secondary air through the various downstream equipment. In pressurized systems, the FD fans provide the pressure to overcome system resistance from the fans, through the boiler and all backend equipment. In balanced draft systems, the FD fans supply sufficient pressure for the SA to enter the furnace, then are augmented by Induced Draft (ID) fans which draw gases through the boiler and backend equipment.

Dampers and variable speed motor drives at the FD fans are used to regulate air flow. Steam coil or glycol heaters boost air temperature from the FD fans. Tubular or regenerative air heaters (Chapter 20) transfer heat from flue gas leaving the boiler and raise air temperature to about 600F (316C). In some systems, a portion of this preheated air is diverted to hot PA fans to boost pressure for use in the pulverizers. Other systems use cold PA fans and separate or tri-sector air heaters to supply PA for the pulverizers. The secondary air flow is measured downstream of the air heaters, then ducted to the windbox(es) which enclose the burners. The windbox may be partitioned or function as an open plenum, as discussed later. An overfire air system, if used, will extract a portion of the secondary air for the OFA ports.

Distribution of SA among burners in a windbox occurs by several means and to varying degrees of uniformity. The physical size of the windbox and the burner arrangement result in variations in SA pressure and velocity within the windbox which lead to SA maldistribution. The air flow resistance of the burners acts to improve distribution among burners to some extent. The burners may be equipped with adjustable dampers or registers which provide a means to further balance SA distribution. Modern burners are equipped with air flow measurement equipment to facilitate adjustments for uniform SA distribution.

Table 2
Pulverizer Impact on Burners (Unstaged)

	Pulverizer Type				
	Vertical Spindle			Attrita	Ball Tube
Pulverizer capacity, %	100	100	100	100	100
PA/PC (pulverizer), lb/lb	1.8	1.8	1.8	1.5	1.2
Coal heating value, Btu/lb	12,000	8000	6000	12,000	12,000
Coal theoretical air, lb/10^4 Btu	7.5	7.5	7.5	7.5	7.5
Theoretical lb air/lb coal, lb/lb	9.0	6.0	4.5	9.0	9.0
Stoichiometry to burner nozzle	0.20	0.30	0.40	0.17	0.13
Total burner stoichiometry	1.15	1.15	1.15	1.15	1.15
SA/PA ratio	4.8	2.8	1.9	5.8	7.6

Burner functions

Flame stability A burner introduces the primary air and pulverized coal to the secondary air in a manner that establishes a stable flame. This involves producing a flame front close to the burner over a range of operating conditions. Igniters are used to sustain combustion when flames would otherwise be unstable. Igniters are always required to initiate combustion as coal is first introduced to the burner, and during a normal burner shutdown. The burner normally sustains a stable flame by using heat from coal combustion to ignite the incoming pulverized coal. A flame safety system, included with modern burners, electronically scans the flame to verify stability and triggers corrective action if the flame becomes unstable.

Air-fuel mixing The burner initiates the mixing process of secondary air with the PA/PC mixture. However, overall air-fuel mixing is a combination of several factors. An air staging OFA system, if so equipped, supplies the remaining secondary air to the process. Flue gas recirculation (FGR) to the furnace hopper or tempering ports, if so equipped for steam temperature control, adds additional flow and mixing dynamics to the furnace. Furnace mixing results from the expansion of high-temperature combustion products and flow energies of PA, SA, OFA and FGR systems; and from resulting flow conditions within the specific size and structure of the furnace enclosure.

The amount of air-fuel mixing produced by the burner is critical to combustion performance, and varies considerably with burner type. The simplest burners inject the fuel and air in parallel or concentric streams without the benefit of other burner-induced mixing. Entrainment of adjacent flow streams occurs as the jets develop and due to jet expansion from combustion. Corner-fired furnaces operate in this manner, as do some roof-fired designs. Combustion performance greatly depends on mixing subsequently generated in the furnace. Mixing effectiveness decreases with boiler load as air-fuel flow rates drop. This can result in poor combustion performance at reduced boiler loads.

Wall-fired burners may incorporate a variety of air-fuel mixing techniques. Burner mixing can be induced by using the primary air/pulverized coal (PA/PC) stream, by using the secondary air, or by a combination of the two. Looking first at the PA/PC side, the most frequently used burner mixing devices are deflectors, bluff bodies, and swirl generators. Deflectors are frequently installed near the exit of the burner nozzle to cause the PA/PC stream to disperse into the secondary air. These deflectors, or impellers, also reduce axial momentum of the fuel jet, reducing flame length. Impellers may also induce radially-pitched swirl of the fuel jet to further accelerate mixing. Bluff bodies are sometimes used in or adjacent to the burner nozzle exit. Flow locally accelerates around the upstream side of the bluff body and recirculates on the downstream side. The recirculation promotes mixing. The bluff body can also be used to increase residence time for a portion of the fuel near the burner, thereby improving flame stability. The PA/PC may also be divided and injected as multiple streams in order to increase the surface area of the fuel jets and increase the rate of combustion.

Secondary air swirl is the most common way to induce air-fuel mixing in circular throat burners. Swirl generators are used upstream of the burner throat to impart rotating motion to the secondary air. This air leaves the burner throat with tangential, radial and axial velocity components. Radial and axial pressure gradients form in the flow field downstream of the throat, with the lowest pressure being near the center of the throat.[3] As swirl is increased, the pressure gradients increase. This causes the flow to reverse and travel along the axis of the flame toward the low pressure zone. A recirculating flow pattern is then generated near the burner. Deflectors and bluff bodies are also used to induce localized mixing from the secondary air. Modern low NO_x burners frequently subdivide the secondary air into two or more streams. Swirl may be imposed at different rates for these SA streams, to satisfy flame stabilizing functions while limiting the overall air mixing rate.

Burner mixing intensity is directly related to burner throat velocity. This is a key design parameter and varies considerably from burner type to burner type. Higher throat velocities promote furnace mixing at the expense of burner pressure drop. Higher mixing rates tend to reduce unburned combustibles at the expense of higher NO_x emissions. Throat velocity decreases with decreasing burner firing rate. Optimal air-fuel mixing is achieved at reduced boiler loads by operating with fewer burners in service and by operating those burners at higher firing rates.

Emission control Modern PC-fired burners play a major role in controlling emissions of NO_x (NO and NO_2), in addition to carbon monoxide (CO) and unburned carbon. Unburned combustibles have traditionally been a criterion for efficient burner operation. Combustion efficiency depends on numerous other factors, including coal reactivity and fineness, burner design, OFA design, furnace design, excess air, and operating conditions. Unburned combustibles are minimized by rapid thorough mixing of air and fuel in the furnace. NO_x emissions from PC are minimized in the opposite manner, by gradually combining air and fuel. Many techniques which serve to lower NO_x tend to raise unburned combustibles and vice versa. The practical application of these techniques to combustion system design in order to effectively control emissions is considered in detail below.

Performance requirements

Pulverized coal-fired equipment should meet the following performance conditions:

1. The coal and air feed rates must comply with the load demand over a predetermined operating range. The burners have to operate in a reliable manner with stable flames. For modern applications with high volatile bituminous or subbituminous coal, flames should be stable without the use of igniters from about 30 to 100% boiler load. The minimum load depends on the coal, burner design, and pulverizer load; operation below this load is performed in combination with igniters in service.
2. Boiler emissions of NO_x, CO, and unburned carbon comply with design expectations for the particular application. The actual limits vary considerably due to regulations, the coal and combustion conditions, the use of downstream equipment to further reduce emissions, and ash disposal requirements.
3. The burner should not require continual adjustment to maintain performance. This fundamentally requires consistency in the fuel preparation system and SA systems supplying fuel and air to the burners. The unit should be designed to avoid the formation of localized slag deposits that may interfere with burner performance or damage the boiler.
4. Only minor maintenance should be necessary during scheduled outages. To avoid high temperature damage, alloy steel should be used for burner parts exposed to furnace radiative heat transfer. The burner structure needs to accommodate differen-

tial expansion and maintain integrity at high temperature. Burner parts subject to PC erosion should be protected with wear resistant ceramic materials.

5. Safety must be paramount under all operating conditions. Automated flame safety and combustion control systems are recommended, and often required.

Conventional PC burners

Prior to NO_x emission regulations introduced in 1971 in the U.S., the primary focus of combustion system development was to permit the design of compact, cost effective boilers. As a result, the burner systems developed focused on maximizing heat input per unit volume to enable smaller furnace volumes using rapid mixing burners which generated very high flame temperatures. An unintended side effect was the production of high levels of NO_x. Burners used on such boilers include the conventional circular burner, the cell burner, and the S-type burner. Most of these have been replaced with low NO_x burners in the U.S., but an examination of their designs is useful as it forms a foundation for the more advanced designs which followed.

Conventional circular burner

The circular burner (Fig. 6) was one of the earliest forms of swirl-stabilized PC-fired burners. Due to its success, this burner was used for more than six decades firing a variety of coals in many boiler sizes. The burner is composed of a central nozzle to which PA/PC is supplied. The nozzle is equipped with an impeller at the tip to rapidly disperse the coal into the secondary air. Secondary air is admitted to the burner through a register. The register consists of interlinked doors arranged in a circular pattern between two plates. The doors are closed to cooling position when the burner is out of service, are partially open for lightoff and are more fully open for normal operation.

The registers influence SA swirl as well as quantity. Opening the register doors allows more air into the burner, but reduces swirl, and vice versa. Moderate swirl produces a well shaped flame plume. Excessive swirl can cause the flame to flatten against the firing wall and cause localized slagging. Inadequate swirl results in flame instability. Register adjustments to modify flame shape can result in improper SA supply to the burner in multi-burner applications. High combustion efficiency depends on uniform air-fuel distribution.

The circular burner can fire natural gas, oil or pulverized coal. Simultaneous firing of more than one fuel is not recommended due to the potential for over-firing the burners and the resulting combustion problems. Variations of the circular burner are in operation at inputs from 50 to 300 × 10^6 Btu/h (15 to 88 MW_t).

Cell burner

The cell burner combines two or three circular burners into a vertically stacked assembly that operates as a single unit (Fig. 7). In the 1960s and 1970s, the cell burner was applied to numerous utility boilers which had compact burner zones. While highly efficient, the cell burners produced high levels of NO_x emissions and tended to be mechanically unreliable.

S-type burner

The S-type burner was developed in the early 1980s as a functional and mechanical upgrade for the circular burner. The S-type burner separates the functional attributes of the circular burner for improved SA control in a mechanically superior configuration. (See Fig. 8.) The burner nozzle is generally the same as that in the circular burner. However, secondary air flow and swirl are separately controlled. Secondary air quantity is controlled by a sliding disk as it moves closer to or farther from the burner barrel. Secondary air swirl is provided by adjustable spin vanes positioned in the burner barrel. An air-measuring pitot tube grid is installed in the barrel ahead of the spin vanes. This provides a local indication of relative secondary air flow to facilitate sliding disk adjustments to balance SA among burners. Swirl control for flame shaping is controlled separately by spin vanes. The S-type burner provides higher combustion efficiency and mechanical reliability than the circular unit and requires no pressure part replacement.

Low NO_x combustion systems

NO_x formation

NO_x is an unintended byproduct from the combustion of fossil fuels and its emissions are regulated in the U.S. and many other parts of the world. While a number of options are available to control and reduce NO_x emissions from boilers, as discussed in Chapter 34, the most cost effective means usually involves the use of low NO_x combustion technology, either alone or in combination with other techniques. The effectiveness of the pulverized coal NO_x control technology depends primarily upon the fuel characteristics and the combustion system design. For pulverized coal wall-fired units, NO_x emissions from older conventional combustion systems discussed above typically range from 0.8 to

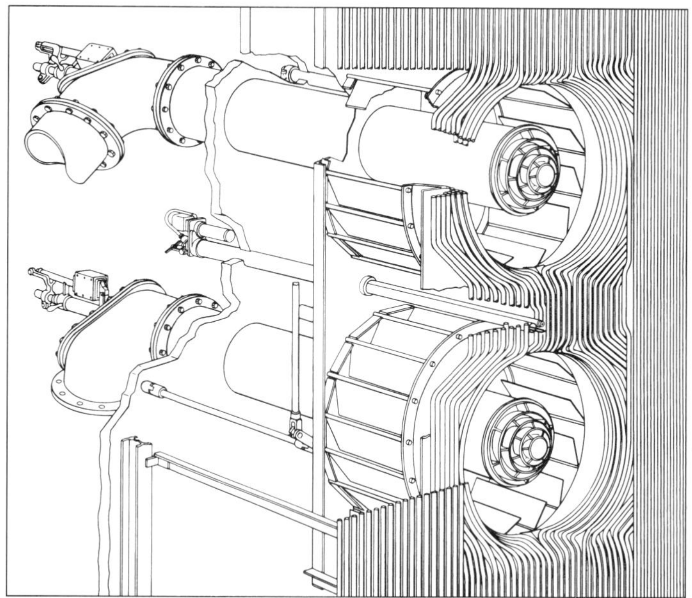

Fig. 7 Conventional two-nozzle cell burner.

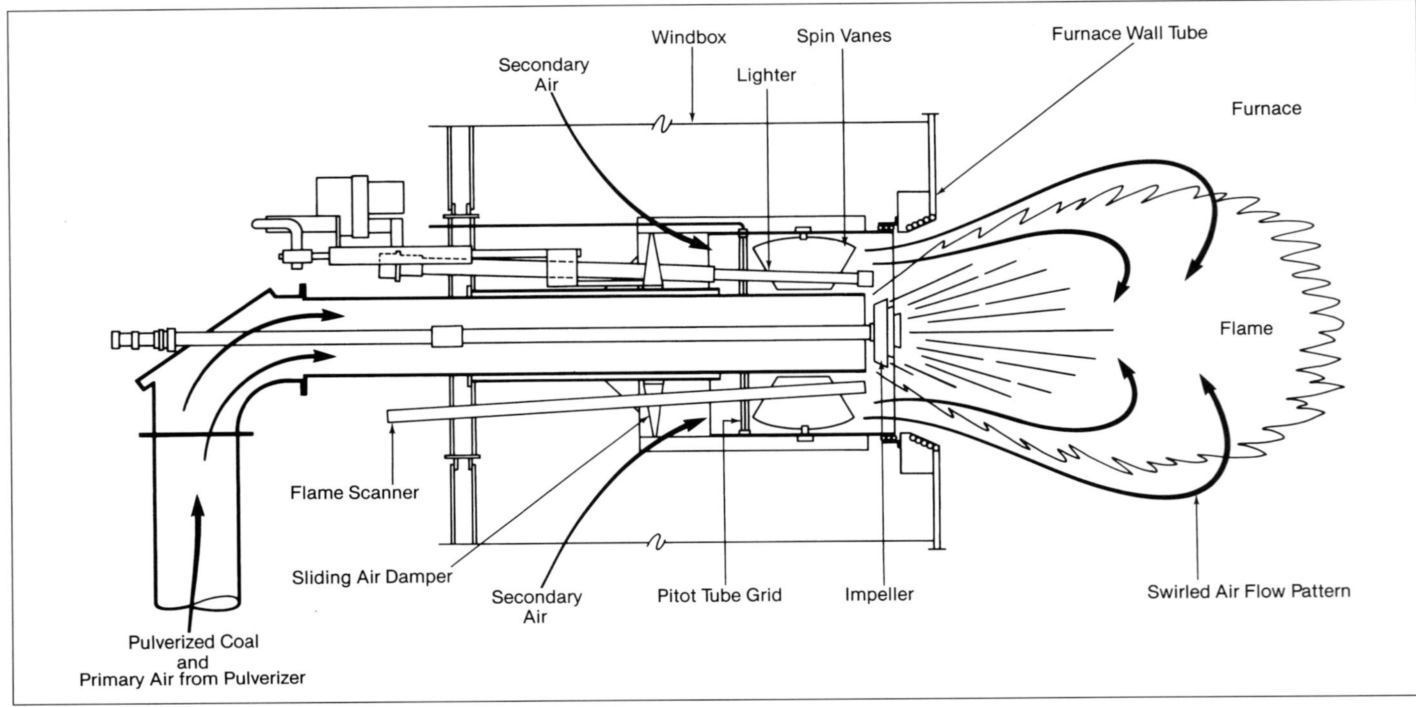

Fig. 8 S-type burner and components.

1.6 lb/10^6 Btu (984 to 1968 mg/Nm^3: see Note below). Low NO_x PC combustion systems are capable of reducing NO_x down to 0.15 to 0.5 lb/10^6 Btu (185 to 615 mg/Nm^3).

NO_x control by combustion

Approximately 75% of the NO_x formed during conventional PC firing is fuel NO_x; the remainder is primarily thermal NO_x. Consequently, the most effective combustion countermeasures are those limiting fuel NO_x formation. Fuel NO_x is formed by oxidation of fuel-bound nitrogen during devolatilization and char burnout. Coal typically contains 0.5 to 2.0% nitrogen bound in its organic matter. High oxygen availability and high flame temperatures during devolatilization encourage the conversion of volatile-released nitrogen to NO_x. Nitrogen retained in the char has a lower conversion efficiency to NO_x, primarily due to lower oxygen availability during char burnout. Reactive coals with high volatile matter and low fixed carbon/volatile matter (FC/VM) ratios tend to be the most amenable to NO_x control by combustion modification. Coals with higher FC/VM and nitrogen content tend to produce higher NO_x emissions.

The most effective means of reducing fuel-based NO_x formation is to reduce oxygen (air) availability during the critical step of devolatilization. Additional air must then be added later in the process to complete char reactions and maintain high combustion efficiency.

Oxygen availability can be reduced during devolatilization in two ways. One method is to design the burner(s) to supply all of the combustion air, but in a manner which limits the rate of air introduction to the flame. Only a fraction of the air is permitted to mix with the coal during devolatilization. The remaining air is then mixed downstream in the flame to complete combustion. Burners employing this technique (low NO_x burners) can reduce NO_x emissions by 30 to 60% relative to uncontrolled levels. However, overall air-fuel mixing is reduced to some extent and the flame envelope is larger compared to rapid mixing conventional burners. Reducing the rate of air-fuel mixing also tends to increase the quantity of unburned combustibles from the process. This strong tendency can be counteracted by careful control of burner aerodynamics and overall combustion system design. In some cases, improvements in the fuel preparation/delivery system and increased coal fineness are required to limit unburned combustibles to acceptable levels.

A second method of reducing oxygen availability during the early phase of combustion is to remove a portion of the combustion air from the burners and introduce it elsewhere in the furnace. This method is referred to as air staging. Air staging systems are frequently used in combination with low NO_x burners to multiply the NO_x reduction effect. PC combustion systems typically operate with 15 to 20% excess air at maximum firing rate. (See Chapter 10.) In some staging situations, reducing the air flow by 10% at the burners is sufficient for a given level of NO_x emissions control. This permits the burners to continue to operate in an excess air condition. Reducing the air flow to less than stoichiometric (theoretically required for complete combustion air flow) conditions can further reduce NO_x emissions. This deeper staging minimizes NO_x formation, but inevitably results in more unburned or partially burned fuel leaving the burner zone. The remaining theoretical air and excess air must then be

Note: The International Energy Agency conversion has been adopted for NO_x emission rates: 1230 mg/Nm^3 equals 1 lb/10^6 Btu for dry flue gas, 6% O_2, 350 Nm^3/GJ for coal.

supplied through air staging ports (NO_x ports or OFA ports). This can amount to 25% or more of the total air supplied to the boiler. The introduction of large quantities of staging air to the unburned fuel subsequently causes NO_x reformation to occur. The rate of introduction and mixing of OFA in the furnace are critical to minimizing NO_x reformation in deeply staged systems.

The application of an OFA system can increase unburned combustibles, slagging in the combustion zone and furnace, and the risks of furnace tube corrosion. These risks increase with coals having high sulfur content and/or high slagging characteristics, and with less reactive coals. While technologies exist to address these risks, the use of OFA may be limited or avoided with such coals for economic reasons in favor of alternate NO_x reduction methods.

Another method of reducing NO_x emissions is referred to as reburning or fuel staging. This control technique destroys NO_x after it has formed. Fuel staging involves introducing the fuel into the furnace in steps. Typically, the bulk of the fuel is burned in the furnace at near stoichiometric conditions. The balance of the fuel, with a limited amount of air, is then injected to create a reducing zone part way through the combustion process. The reducing conditions form hydrocarbon radicals which strip the oxygen from previously formed NO_x, thereby reducing overall NO_x emissions. The balance of the air necessary to complete the combustion is then added. Some fuel-staged systems involve separate burners or fuel injectors followed by NO_x ports which supply the remaining combustion air. The reburn fuel can be coal, but this calls for special measures to limit unburned combustibles from the process. Lighter fossil fuels, especially natural gas, generate more hydrocarbon radicals and can improve emissions reduction, but their cost or availability can deter usage. Fuel staged systems have generally not proven to be as cost-effective as other NO_x reduction techniques with PC-firing. However, some advanced low NO_x burners embody fuel staging while supplying all the fuel to the burners. A portion of the fuel is introduced in a manner to generate hydrocarbon radicals; these are mixed into the flame later to reduce previously formed NO_x. These burners can achieve NO_x reductions of 50 to 70% from baseline uncontrolled levels.

Dual register burner

The Babcock & Wilcox Company (B&W) began experimenting with techniques to reduce NO_x from PC-fired burners in the 1950s. A slot burner, which incorporated controlled air-fuel mixing, was first developed.[4] The air introduced through the burner throat mixed rapidly in the flame while air diverted to the external slots mixed gradually. While effective, the design with slots external to the burner throat was difficult to apply to furnaces. In 1972, B&W developed the dual register burner (DRB) for firing pulverized coal. The DRB (Fig. 9) provided two air zones, each controlled by a separate register, around an axially positioned coal nozzle. A portion of the secondary air was admitted through the inner air zone and highly swirled to aid ignition and flame stability. The remainder passed through the outer air zone with moderate swirl and was gradually mixed downstream in the flame. The burner nozzle was equipped with a venturi to disperse the primary air-coal mixture. The mixture was then injected into the throat without deflection. The results were a stable flame with a fuel-rich core and gradually completed mixing downstream. The first unit retrofit was completed in 1973 and achieved 50% NO_x reduction relative to prior circular burners.

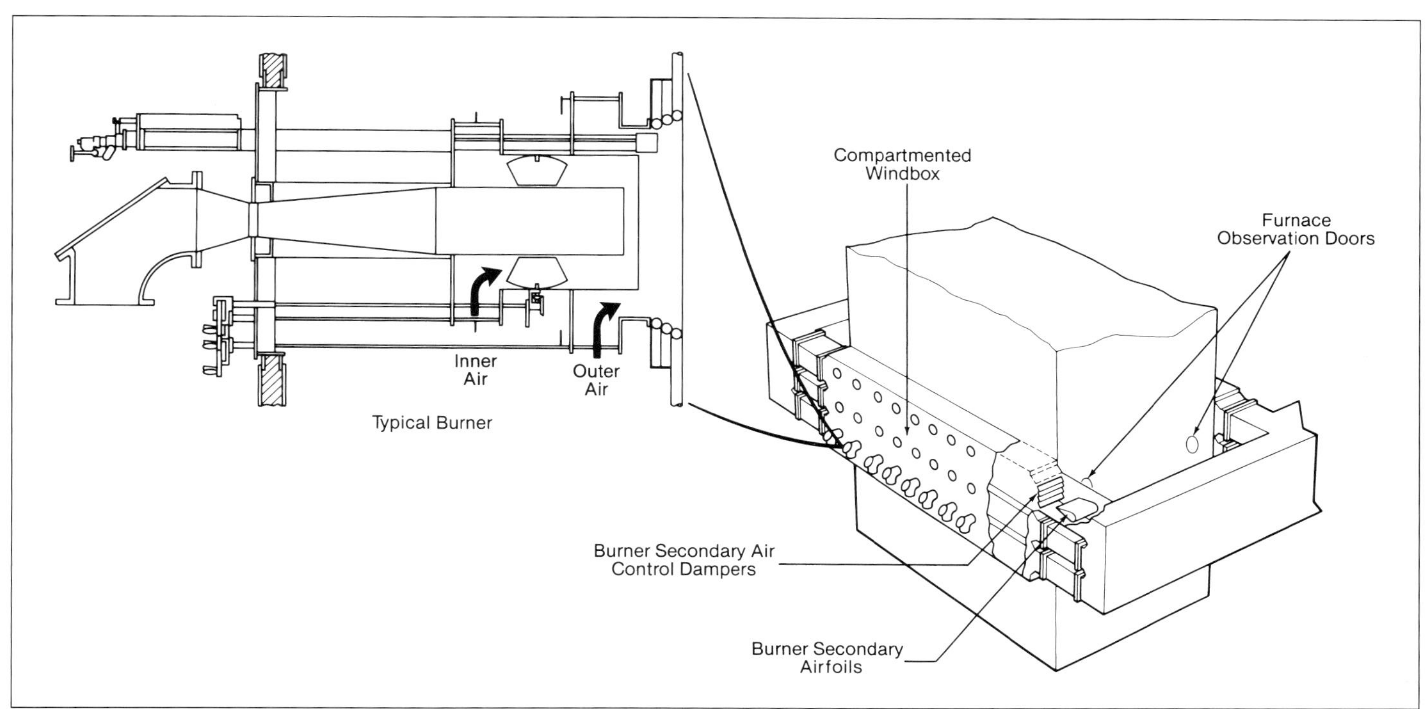

Fig. 9 Dual register burner (DRB) and compartmented windbox.

This retrofit required no modifications to boiler pressure parts, fans, or pulverizers, and NO_x ports were not used. The U.S. government granted a patent to B&W for the DRB in January 1974. Numerous variations have been used by B&W since its introduction.

The DRB was generally used in combination with enlarged furnace combustion zones to reduce thermal NO_x and with a compartmented windbox to better control air distribution (Fig. 9). This design has the burners coupled with each pulverizer situated in an individual windbox compartment. Secondary air was metered and controlled separately for each compartment. This approach corrected some of the SA flow imbalances among burners that were prevalent in open windbox designs.

The DRB typically reduced NO_x 40 to 60% from uncontrolled levels. This efficiency enabled its use on new boilers without NO_x ports or other air diversion systems. Utility boiler NO_x emissions with the DRB ranged from 0.27 to 0.70 lb/10^6 Btu (332 to 861 mg/Nm^3).

DRB-XCL® burner

The Dual Register XCL burner is a second generation low NO_x burner which incorporates fuel staging technology along with the air staging technology found in the original DRB. The DRB-XCL® is physically arranged as shown in Fig. 10, and draws heavily on the proven mechanical design of the S-type burner. Air flow to the burner is regulated by a sliding disk. An impact/suction pitot tube grid is located in the burner barrel to provide a local indication of secondary air flow. With this information, the secondary air can be more uniformly distributed among all burners by adjusting the sliding disks or other means. Balanced air and fuel distribution among all burners is critical to combustion efficiency, particularly with low NO_x systems. Downstream of the pitot tube grid are the inner and outer air zones. Adjustable vanes in the inner zone impart sufficient swirl to stabilize ignition at the burner nozzle tip. The outer zone, the main SA path, is equipped with two stages of vanes. The upstream set is fixed and improves peripheral air distribution within the burner. The downstream set is adjustable and provides proper mixing of this secondary air into the flame. The burner nozzle is equipped with hardware to accomplish the necessary fuel mixing. This may include a conical diffuser and flame stabilizing ring or an impeller. These combine to improve flame stability while controlling NO_x and flame length. The DRB-XCL burner reduces NO_x 50 to 70%[5] from uncontrolled levels without using NO_x ports. This added level of NO_x reduction is possible by the fuel staging effects generated in the near-flame field.

A plug-in version of the DRB-XCL enables its use in many retrofit applications without requiring pressure part modifications, and has demonstrated reduced unburned combustibles.[6] The burner can be equipped for firing natural gas and/or fuel oil when multi-fuel capability is desired.

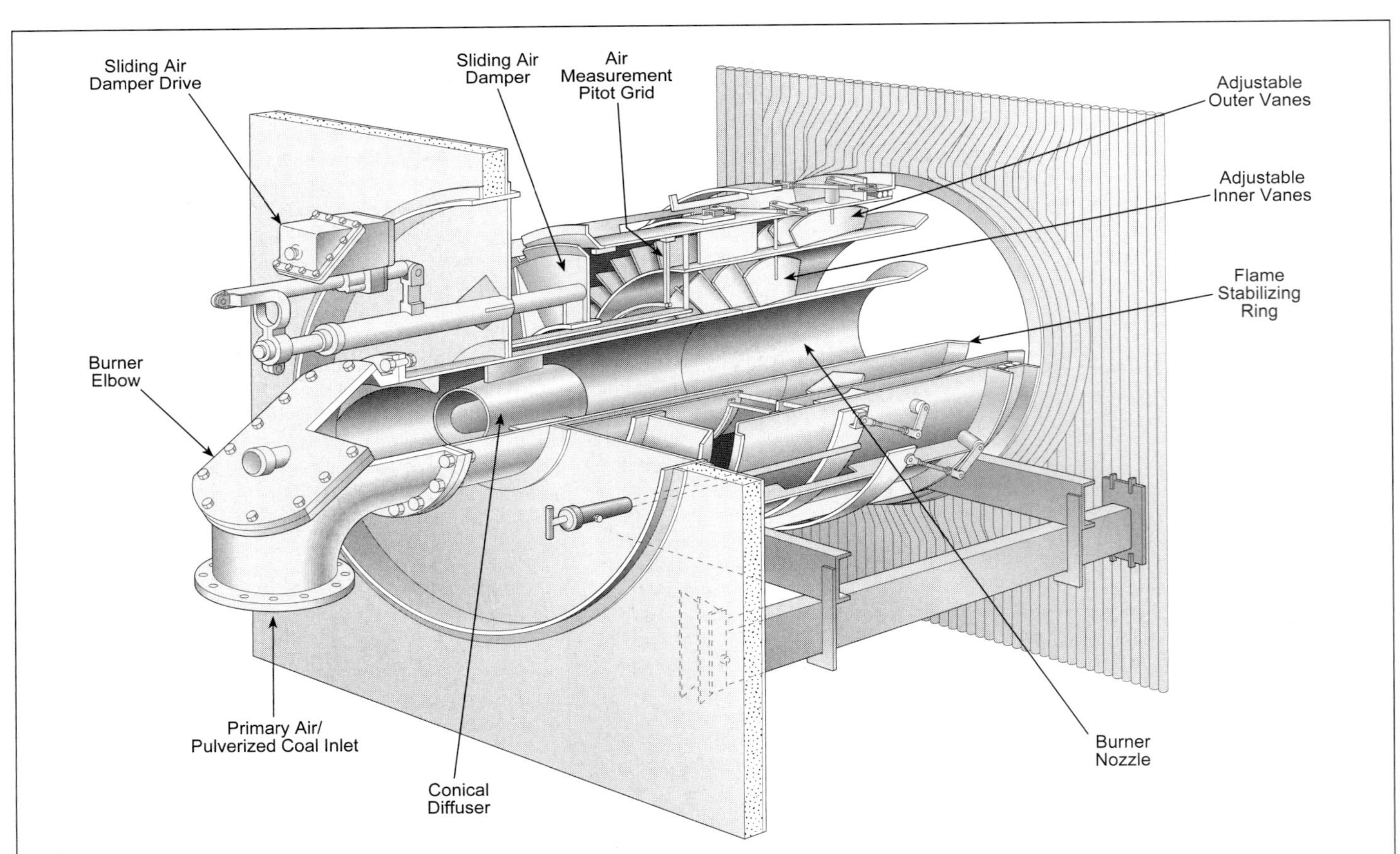

Fig. 10 DRB-XCL® low NO_x burner for pulverized coal firing.

DRB-4Z® burner

B&W's third generation low NO_x burner, the DRB-4Z® burner, incorporates even more advanced emission control technology in a well proven mechanical configuration. The patented DRB-4Z design combines primary zone stoichiometry control with fuel staging and air staging technology to achieve very low NO_x emissions along with improved combustion efficiency. Conceptual development, using Computational Fluid Dynamics (computer modeling), indicated the benefits of an additional air zone separating the coal nozzle from the principal air zones of the burner.[7] These concepts were proven through extensive testing[8] during the development of the burner in B&W's Clean Environment Development Facility.[9]

This additional air zone, the *transition zone*, acts as a buffer between the high temperature, fuel-rich flame core and the secondary air streams (see Fig. 11). The flow field produced by the transition zone draws gases from the outer portions of the flame inward toward the flame core. NO_x formed in the oxygen-rich outer flame region is reduced back to other nitrogenous species in the process. A further advantage of the transition zone is the ability to alter primary zone stoichiometry.

Table 2 indicates the variations in primary stoichiometry which are inherent to different coals and pulverizer designs. Inadequate primary zone stoichiometry inhibits combustion during the critical stage of devolatilization. Air supplied by the transition zone can compensate for this without flooding the flame core with secondary air from the main air zones. SA swirled through the inner air zone serves to anchor and stabilize the flame. SA admitted through the outer air zone is staged to gradually mix into the flame. NO_x reductions of up to 80% from uncontrolled levels are possible with the DRB-4Z. The burner is well-suited for use with an OFA system when the lowest NO_x emissions are required.[10,11]

The physical arrangement of the DRB-4Z is depicted in Fig. 12. The mechanical construction was adapted from the proven integrity of the DRB-XCL, while incorporating the advanced aerodynamic features. The coal nozzle strongly resembles the predecessors and is interchangeable in some situations. The transition zone is supplied with SA by means of slots and a manually adjustable sliding sleeve. A sliding air damper is used to regulate SA to the main air zones. A linear actuator positions this damper in open windbox applications in order to automatically control the SA flow to the burner. A pitot tube grid is installed downstream of the damper to provide a local indication of SA flow. Manually adjustable vanes are located in the inner and outer air zones to enable optimization of swirl settings during burner commissioning. Fixed vanes are supplied to improve air flow distribution and reduce pressure drop through these air zones. The DRB-4Z plugs into wall openings from most existing burners to facilitate its use as a replacement upgrade. Improved aerodynamics provide better overall air-fuel mixing which results in significantly lower unburned combustibles than earlier low NO_x burner designs.

The shop-assembled construction of the DRB-4Z is illustrated in Fig. 13. The one-piece assembly simplifies and quickens installation. Front end components are fabricated from heavy, high quality stainless steel and mechanically stiffened in a manner which accommodates differential expansion at elevated temperature while maintaining product integrity. Permanent thermocouples are supplied to assure components are

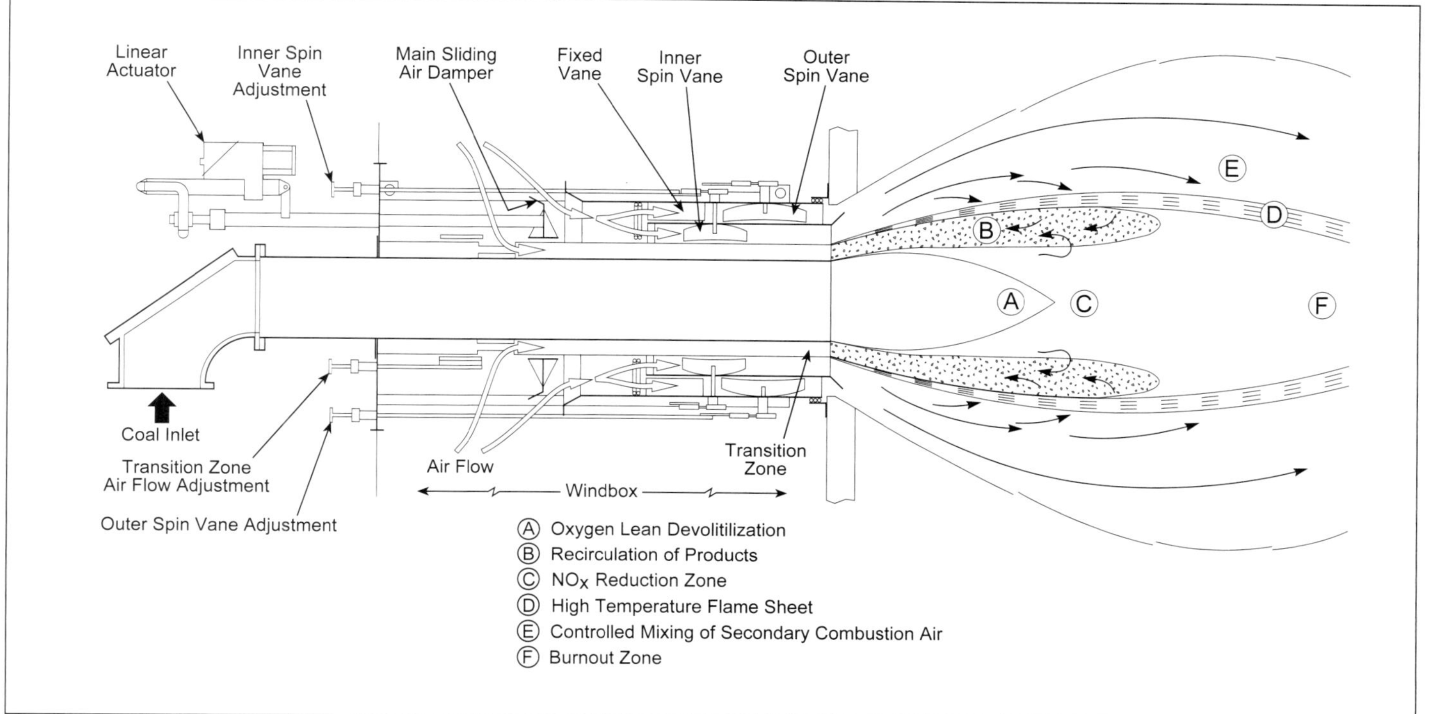

Fig. 11 DRB-4Z® low NO_x combustion zones.

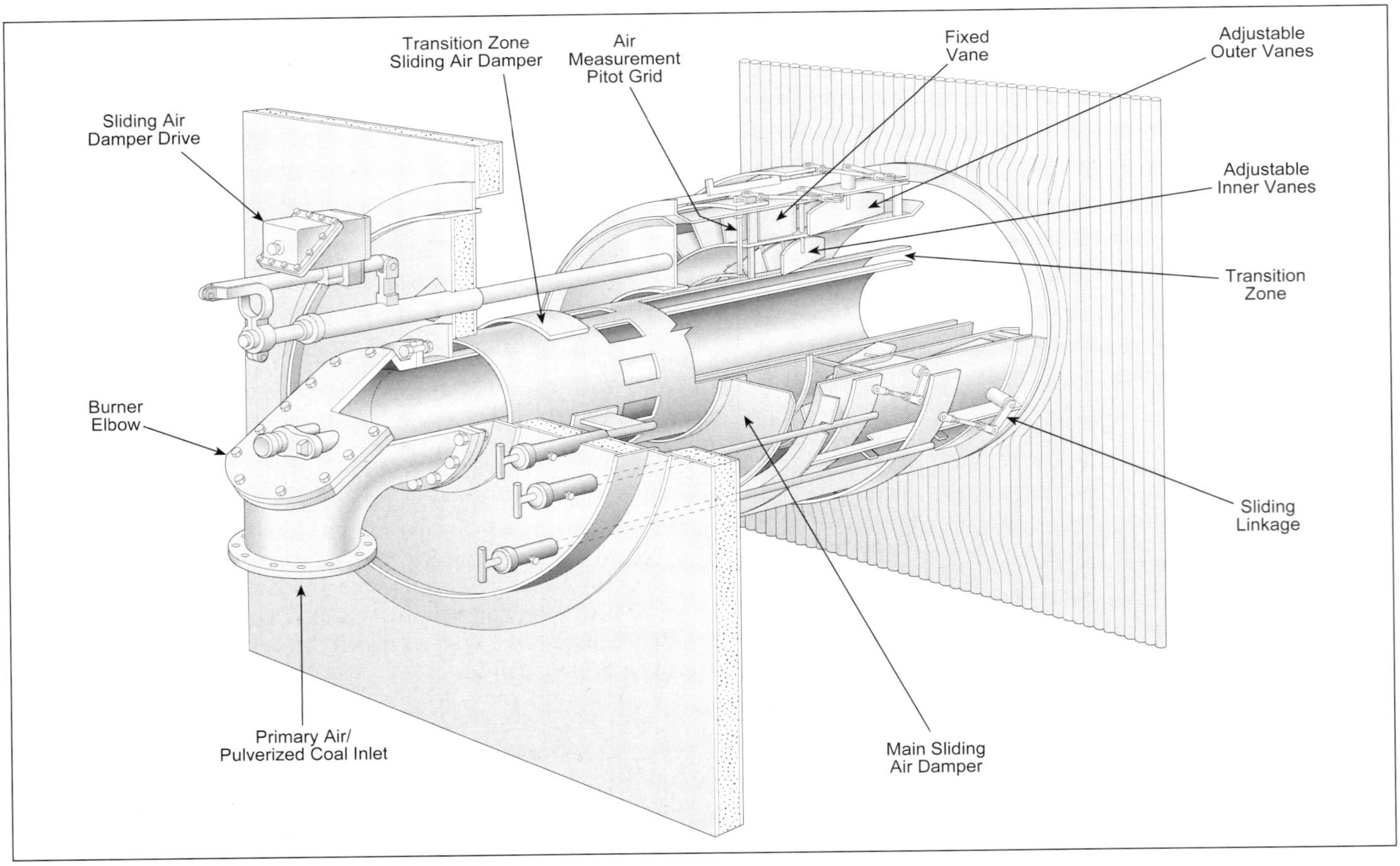

Fig. 12 DRB-4Z® low NO_x burner for pulverized coal applications.

kept within safe operating temperatures. The burner is also capable of low NO_x firing natural gas, or can be equipped to fire fuel oil as a primary fuel.

Low NO_x cell burner

The cell burners discussed earlier produced high NO_x emissions, but their unique arrangement (Fig. 7) made it difficult to fit low NO_x burners in the existing wall openings. The Low NO_x Cell burner (LNCB) was developed to replace two nozzle cell burners in a plug-in design. In essence, all the coal is supplied to one burner throat with a portion of the secondary air. The balance of the secondary air is supplied to the other burner throat, which serves as an integral NO_x port for each burner location. NO_x levels can be reduced by 50% with the LNCB while maintaining high combustion efficiency. Advanced second and third generation low NO_x burners are designed for plug-in use in cell burners and now provide lower NO_x alternatives to the LNCB.

Air staged systems

Air staging involves removing a portion of the air from the burners to reduce oxygen availability early in the combustion process, and reintroducing it later in the combustion process. Often the physical arrangement dictates replacing the staged air through ports located above the combustion zone, hence the name overfire air (OFA) is commonly applied to such systems. The layout of a combustion system and furnace, however, may necessitate supplying staged air at the same elevation or below the burner zone, such that OFA is something of a misnomer. OFA is used here in its generic sense to refer to any air staged system. The ports through which the OFA is injected are called OFA ports or NO_x ports.

B&W pioneered the development and application of NO_x ports and air staging as a NO_x emission control technique. This work, which began in the 1950s, was initially directed at NO_x reduction for oil- and gas-fired boilers in California. R.M. Hardgrove of B&W filed a patent for a *Method for Burning Fuel* on June

Fig. 13 Fully assembled DRB-4Z® burner ready for shipment.

18, 1959, disclosing key aspects of air-staging influences on NO_x emissions. The patent was granted in 1962, and NO_x ports have been used ever since by B&W. Today, the NO_x reduction effectiveness of OFA systems is attested to by their widespread use in coal, oil and gas units, including wall-fired, corner-fired and Cyclone designs.

The degree of NO_x reduction achieved by air staging depends on numerous factors, including fuel properties, burner design and associated NO_x emissions, furnace design, operating conditions, and the OFA system design. Key design issues are the quantity of OFA and the arrangement and NO_x port design. NO_x reduction is a function of the percent combustion air removed from the burners for use as OFA. Consider a typical PC-fired situation for a unit operating at 20% excess air at full load, that is, with 120% theoretical air (Fig. 14).

1. A low NO_x burner (LNB) would produce a NO_x emission level noted as a_1 for this example.
2. Adding a moderate amount of air staging, in this case 15 percentage points of air, would still leave the combustion zone under oxidizing conditions (above 100% theoretical air). Burner zone stoichiometry (BZS) is 1.05 in this case. NO_x emissions are reduced to level b_1. Removing this 15 percentage points of secondary air from an existing combustion system would reduce burner mixing effectiveness to some degree, impairing flame stability, and may add to slagging tendencies in the burner zone.
3. Increasing the air staging to 30 percentage points drops the combustion zone to reducing conditions (less than 100% theoretical air). BZS is now 0.90, and the greater air staging reduced NO_x emissions to point c_1. Notice that some NO_x reformation occurred from the introduction of the OFA such that the exit NO_x c_2 was somewhat higher than c_1. Removing 30 percentage points of SA from an existing combustion system requires new or revised burners to produce sufficient mixing with lesser SA. A BZS of 0.90 means only 90% of theoretical air is supplied to the burners, so at least 10% of the fuel remains unburned leaving the burner zone. This raises risk of increased unburned combustibles even after addition of OFA. Reducing conditions in the burner zone also raise risks of furnace tube corrosion with high sulfur coals, and can be problematic for slagging.
4. Increasing air staging to 40 percentage points drops BZS to 0.80, meaning at least 20 percent of the fuel is unburned leaving the combustion zone. Actual constraints inherent to the process cause considerably more than 20% of the fuel to leave the burner zone unburned. NO_x leaving the burner zone dropped to point d_1. In this example, the rapid addition of 40 percentage points of OFA to the significant amount of unburned fuel resulted in major NO_x reformation. Consequently the exit NO_x d_2 actually exceeded the lesser staged case c_2. However, NO_x reformation could be lessened by gradually reintroducing the OFA, and limiting O_2 concentrations and combustion rates in the latter stages of combustion. This results in the lowest level of exit NO_x noted as d_3.

OFA with wall-fired systems The description above illustrates the advantages of air staging for NO_x reduction, but also reveals potential problems which may result. Application of OFA to an existing wall-fired system reduces burner-induced mixing, which tends to impair flame stability and slows the combustion process. This occurs due to lower burner throat velocities and lower secondary air to primary air ratio. Resulting reductions in burner SA momentum and mixing energy can best be compensated for by raising burner throat velocity. Deep staged OFA systems therefore necessitate the use of new, smaller burners for good combustion performance. Greater amounts of swirl, for at least a portion of the SA, further assist in flame stabilization. The burner arrangement and/or swirl orientation may need to be altered to present a more uniform flow field to the OFA system. Advanced low NO_x burners with improved SA control measures can compensate for these issues with air staged systems, and have demonstrated equal or less furnace slagging tendencies compared to some unstaged conventional designs.

After properly dealing with the wall-fired burners, the design of the air staging system can proceed. The major functional issue concerns proper distribution of the OFA within the furnace, in a manner which minimizes NO_x reformation, CO, and unburned char. Minimizing unburned combustibles calls for rapid mixing of OFA, while minimizing NO_x calls for gradual mixing. The OFA port jets have to travel cross flow to the furnace gases and deliver air in time for combustion reactions to be completed in the furnace. This favors fewer ports with high mass flow and velocity. But OFA must also be adequately dispersed to cover the entire cross-section of the furnace. This tends to favor a large number of smaller ports. Ports located closer to the burners provide more time for mixing and lessen unburned combustibles, while locating ports further away tends to maximize NO_x reduction. Effective implementation of OFA systems requires the combination of experience, computer modeling, and on-line adjustability. Computer modeling (see Chapter 6) pro-

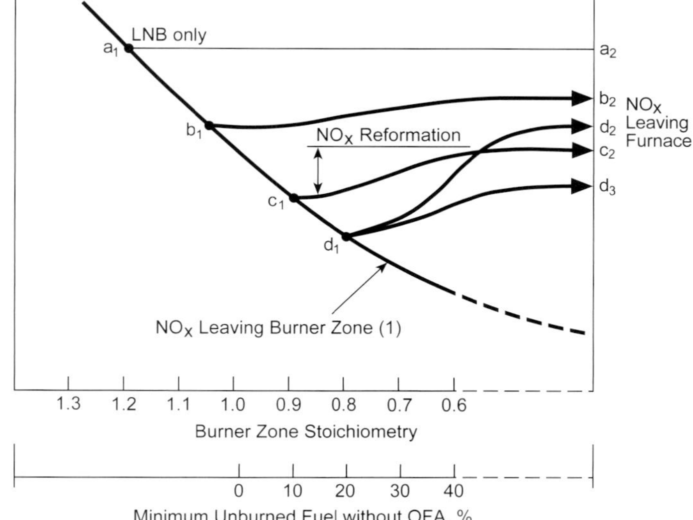

Fig. 14 The effects of OFA design on NO_x emissions.

vides a means to evaluate the mixing effectiveness of alternate OFA system designs, with regard to port size, quantity, placement, and design.[12] High grid resolutions near the burners and OFA ports enable more accurate assessment of reactions and flow fields, but require larger model size and computational resources. Validation of model predictions to field experience is essential. Ultimately, supply of equipment with on-line adjustability provides the means to tune the OFA system to the actual conditions and minimize emissions in the field.

The Dual Zone NO_x port (Fig. 15) provides the means to vary on-line the jet penetration versus dispersion characteristics of OFA supplied through this port. A sliding sleeve damper with linear actuator provides the means to regulate OFA to each port. This enables varying OFA among the various ports to improve distribution, and to vary OFA versus load for NO_x control. A relative indication of air flow is provided locally at each port by a pitot grid installed inside the port. The inner zone of the Dual Zone port is comprised of a core pipe with a manually adjustable sliding disk to adjust air flow. The core pipe produces a high velocity air jet for penetration across the furnace. The outer zone of the port is equipped with spin vanes. Increasing the swirl of this air causes more localized mixing by entraining furnace gases in the recirculation. The vanes can be manually adjusted to vary swirl over a wide range. The combination of these adjustments has a strong effect on mixing rates and emissions. Dual Zone ports, or other types, can be enclosed in windboxes in a manner similar to the burners. In some situations, it is more economical to supply individual air ducts to each port. Performance and repeatability benefit from the inclusion of equipment to measure and regulate air flow to the OFA system distinctly from flow to the burners.

Maximum NO_x reduction with air staging requires operation at low burner zone stoichiometries, e.g. 0.85 or less. NO_x reformation becomes a problem if all the associated OFA is rapidly reintroduced to the furnace. To avoid this, it is beneficial to employ two levels of OFA. The level closest to the burner zone supplies a lesser portion of the OFA to bring the BZS close to 1.00. This is followed by the second level of OFA which supplies the balance of secondary air. This arrangement reduces peak O_2 levels and moderates combustion rates in the upper furnace to control NO_x, while providing the mixing necessary to limit unburned combustibles.

Most OFA systems draw hot SA from the ducts or plenums which supply air to the burners. Use of preheated air supplied from the FD fans through the air heaters helps maintain boiler efficiency, but limits the available static pressure for the OFA system. The location, quantity, size and design of the OFA ports have to be compatible with static pressure constraints. Booster fans can be used to raise static pressure, but can become massive to accommodate the quantities and temperature of OFA involved with many utility boiler applications. Complications with locating such fans, along with their expense and operating costs, detract from their use. Most often the FD fans are suitable as motive sources for proper system design.

OFA with corner-fired systems Overfire air has proven to be very effective in corner-fired boilers and is the dominant method of NO_x control with modern systems.[13] It is useful to first consider the functionality of the corner-fired combustion system before considering the application of overfire air with such

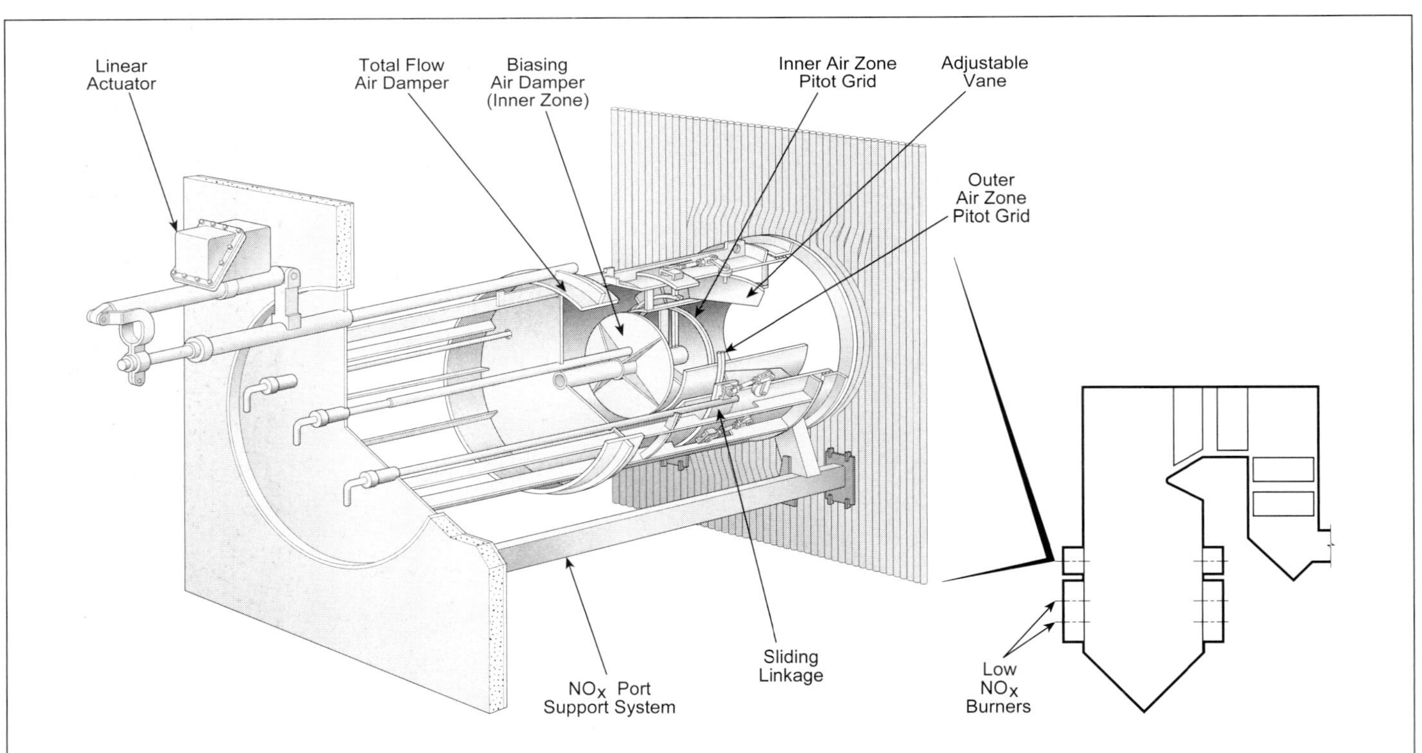

Fig. 15 Dual Zone NO_x port.

equipment. In a corner-fired system, coal and air are introduced into the furnace at distinct elevations through windbox assemblies located in the corners (for four corner furnaces) and at protrusions midway along the front and rear walls (for eight corner furnaces). The windbox nozzles direct the coal and air streams at slight angles off of the diagonals and tangent to an imaginary firing circle in the center of the furnace as shown in Fig. 16. The combination of the firing angles and momentum of the fuel and air streams create a rotating or cyclonic fireball that fills the plan area of the furnace. In this firing system, the furnace acts like a large burner where each corner provides the ignition energy to the corner downstream and the swirling action of the fireball stabilizes the flames. For steam temperature control at lower loads, the fuel and air nozzles are designed to tilt up or down in unison to raise or lower the fireball and thus increase or decrease furnace exit gas temperatures.

Each windbox is divided along its height into alternating compartments of air and coal as shown in Fig. 17. Each pulverizer serves one complete coal elevation (i.e., all four or eight corners). The windboxes distribute secondary combustion air as fuel air and auxiliary air. Fuel air is the portion of the secondary air admitted to the furnace through the coal compartments and air annulus around each coal nozzle tip and is used to support combustion and stabilize flames. Auxiliary air is the balance of the secondary air which is admitted into the furnace through air nozzles located in the compartments above and below each coal elevation to complete the combustion. Primary air is the balance of air entering through the coal pipes and nozzles and serves in the same manner as in wall-fired burner arrangements. Secondary air is distributed in the windbox by a system of dampers located at the inlet of each compartment as shown. Fuel air dampers are controlled based on pulverizer feed rate whereas auxiliary air dampers are controlled based on windbox to furnace differential pressure. In conventional applications, all of the secondary air enters through the corner windbox assemblies. However, these corner-fired combustion systems embody air staging technology by virtue of the layered introduction of fuel and air. This lowers NO_x by slowing the mixing of air and fuel as they proceed into the flame vortex.

Application of overfire air to a corner-fired system involves diverting a portion of the auxiliary air to an overfire air zone located above the combustion zone. Depending on the required NO_x reduction, OFA can be introduced at the top of the windbox through two or more air compartments and/or through separate ports located some distance above the top coal elevation. Increasing the separation between the combustion zone and the OFA ports reduces NO_x emissions most effectively. Therefore, for a given quantity of OFA, separated OFA is generally twice as effective as windbox OFA for reducing NO_x. Also, since separated OFA ports require new openings in the upper furnace, they can be made larger than windbox overfire air for greater NO_x reduction, as windbox compartment sizes are limited by the existing windbox width and compartment heights.

A proven approach for retrofitting a separated OFA low NO_x system involves the addition of one level of four to ten ports depending on furnace type. For smaller four-corner furnaces, the ports are typically located on furnace corners since adequate coverage

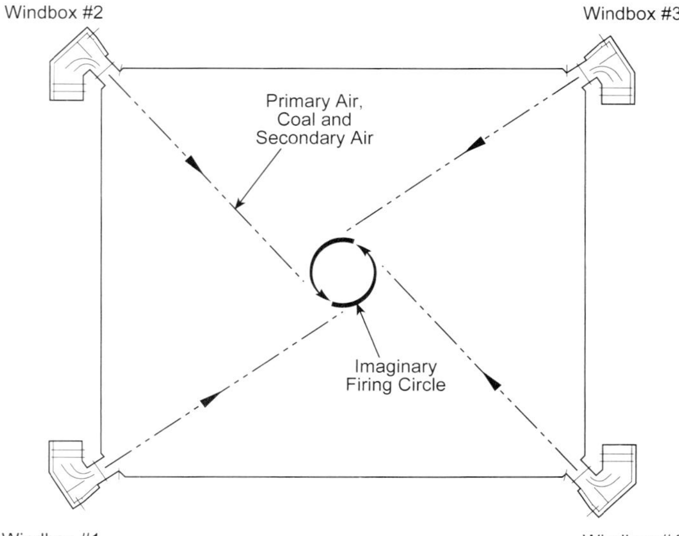

Fig. 16 Plan view of corner-firing arrangement.

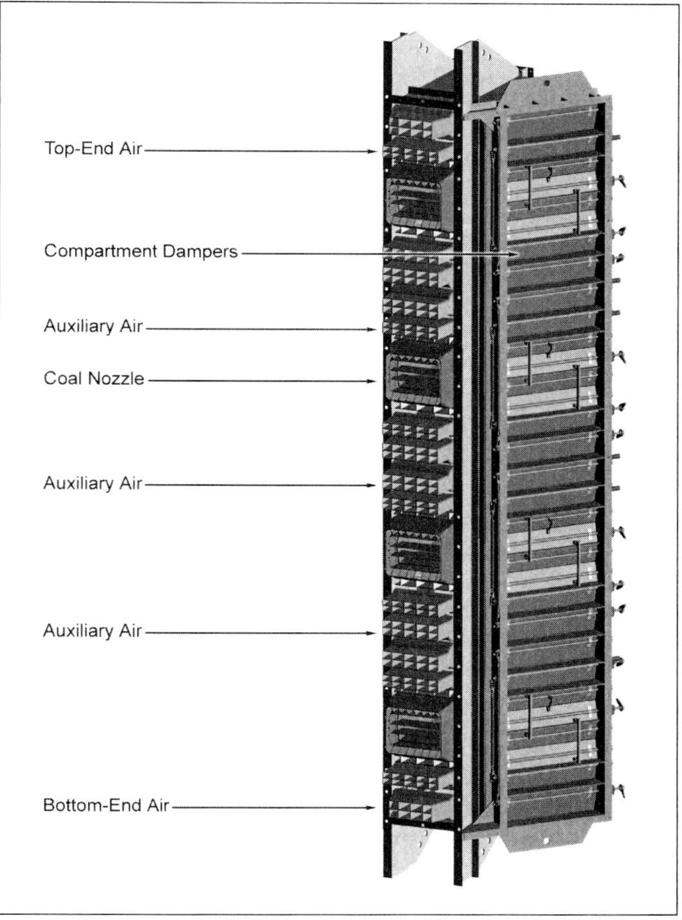

Fig. 17 Corner-fired windbox assembly for coal firing (*courtesy of R-V Industries, Inc.*).

can be achieved as these furnaces have smaller, more square cross-sections than the larger eight-corner units. Also, ductwork for four-corner units is simpler when the ports are located on the corners. For larger eight-corner furnaces with center division walls, the ports may be located on the front and rear walls. They are arranged in an interlaced fashion from front to rear and offset in the direction of fireball rotation. This effectively reduces NO_x while providing more complete coverage of the furnace plan area and corners for better carbon burnout and CO control. In both furnace types, the ports are located several feet above the highest coal elevation to allow separation distance for effective NO_x control but low enough to allow ample furnace residence time for good carbon burnout after overfire air injection. The location and sizing of ports are also set to avoid or minimize buckstay modifications when possible. If necessary, computer modeling serves to optimize port arrangement and performance.

The typical B&W separated OFA port for corner firing shown in Fig. 18 uses an all welded (no refractory), deep throat, perpendicular opening to cool the air nozzles from the intense radiant heat in the OFA zone. The port normally has two air compartments with integral turning vanes and independent flow control dampers to maximize velocity and penetration over the load range. The air nozzles can be manually tilted and/or yawed on line for enhanced emissions control. A small pitot tube grid is located near the inlet of each nozzle for relative air flow indication and balancing.

Addition of an OFA system reduces secondary air flow to the main windbox. To compensate for this, the existing auxiliary air compartment and damper sizes are reduced using blocking plates and the auxiliary air nozzles are replaced with smaller ones. These modifications are required to maintain existing injection velocity, pressure drop and thus, damper controllability over the load range. Adding only OFA has proven to be the most cost effective and reliable combustion system modification for reducing NO_x emissions in corner-fired boilers.

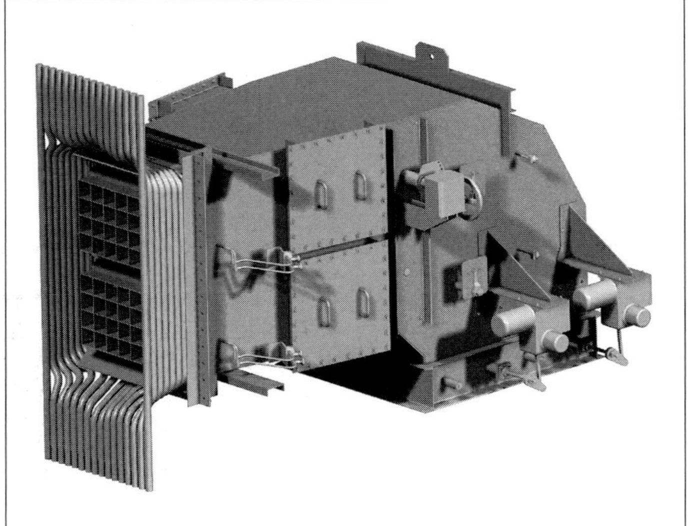

Fig. 18 Corner-fired overfire air port (*courtesy of R-V Industries, Inc.*).

NO_x reduction for units with only windbox overfire air typically ranges from 20 to 30%. For additional NO_x reduction, separated OFA can be added to units with or without windbox overfire air. NO_x reduction with these arrangements typically ranges from 40 to 60%, depending on initial NO_x levels, boiler size, furnace type, heat release rate, separation distance and coal reactivity.

Typical of other air staged systems, NO_x reduction on corner-fired units may be accompanied by some increase in CO emissions and unburned carbon. However, with careful system design and field tuning, existing CO emissions can normally be maintained with reasonable unburned carbon (UBC) results.

Burners for difficult coals

A few coals have proven to be considerably more difficult to burn than others, including unreactive coals and reactive coals which would burn readily except for an overabundance of moisture and/or ash. B&W has developed burners specifically for use with these problem coals. The selection of burner type is based on an empirical ignition factor derived from full scale experience with difficult coals. This factor relates variations in fuel ratio (FC/VM), moisture content, ash content, and coal heating value to ignition behavior. Burner choices for these applications include conventional and low NO_x burners discussed above, the enhanced ignition (EI) burner, and the primary air exchange (PAX) burner.

Enhanced ignition burner

The enhanced ignition (EI) burner resembles the DRB-XCL burner. The nozzle is equipped with a conical diffuser which concentrates the coal particles into a fuel-rich ring near the walls of the fuel nozzle. However, the EI burner uses a larger coal nozzle tip to reduce the primary air and coal mixture velocity as it enters the furnace. In addition, the dimensions of the inner and outer air zones are adjusted relative to the DRB-XCL burner to produce greater recirculation of hot gases along the fuel jet. These hot gases accelerate ignition of the dense, low velocity fuel jet and significantly improve burner flame stability. The EI burner is well suited for firing high moisture (>35%) lignites and low volatile bituminous coal.

Primary air exchange burner

The primary air exchange (PAX) burner is designed to fire anthracite and similar coals. Some anthracites can be wall-fired conventionally with the PAX burner, while those lowest in volatile matter must be fired in a downshot arrangement (Fig. 5). The design of the air zones for the PAX burner is similar to that of the EI burner. However, in the case of downshot firing, air staging ports direct air away from the burner to better sequence air supply with the slower burning nature of these coals, thus elevating flame temperature.

The anthracite fineness is increased to 90% (or even higher) through 200 mesh to help accelerate ignition and to reduce unburned carbon. In addition, mill exit temperature is increased to the maximum for the pulverizer, often limited to 200F (93C) for mechanical reasons. Tests have shown the benefit of higher temperatures on ignition performance. The PAX burner raises the primary air and pulverized coal (PA/PC) temperature by a patented design which separates half of the primary air from the PA/PC mixture and vents it into the furnace. This primary air is replaced by secondary air near 700F (371C), significantly raising fuel temperature as it enters the furnace. In some situations, indirect firing systems are used for firing anthracite. Pulverized coal from the storage bin can be transferred to the burners with hot secondary air to provide a hot, dense fuel mixture. In either case, the PAX nozzle tip is sized for low velocity to further assist ignition. Even with these techniques, good combustion is dependent upon high furnace temperatures. Low load operation still requires igniters for flame stabilization.

Unburned combustibles

Unburned carbon (UBC) in solid form, as distinguished from partially oxidized carbon in gaseous form (CO), results to varying degrees from all pulverized coal systems. Unburned combustibles represent an efficiency loss to the process which can vary from 0.05% for lignite, to 0.5% for bituminous coal, and to 5.0% or higher for anthracite. These order of magnitude variations are largely a consequence of inherent differences in coal reactivity. Variations around these values can occur due to gradients in coal reactivity, variations in coal fineness, and variations in how oxygen is made available to the coal during the combustion process.

Fineness

Given a coal of lesser reactivity, one of the most effective techniques to lower UBC is to reduce the quantity of coarse coal particles from the pulverizers. Pulverizer power is dependent on the mean particle size, which is heavily influenced by the fine particles. Therefore, the most efficient means of reducing UBC is to selectively reduce the coarse particles without over-grinding the fine particles. Advanced pulverizer classifiers, especially dynamic versions, can essentially eliminate the large 50 mesh particles and sharply reduce the 100 mesh fraction, with little change to the dominant smaller 200 mesh size. Such fineness improvements have accomplished significant reductions in UBC with bituminous and higher rank coals.

Oxygen availability

The influence of oxygen availability on carbon burnout depends on each coal particle's history regarding O_2 concentration, and the corresponding temperature and time exposures. Global factors include the excess air level for the process, rate and uniformity of air-fuel mixing, and residence time. Increasing excess air increases overall oxygen availability and reduces UBC, but with some offset in boiler efficiency due to increased dry gas loss. Increasing excess air also raises NO_x, placing further constraint on this countermeasure. Air infiltration to a boiler tends to indirectly increase unburned combustibles. Balanced draft units actually operate under a small vacuum which draws air into the boiler through cracks, tears, and miscellaneous openings in the structure. This air can enter the process once gases have cooled and it does not mix well, such that it provides little or no benefit to combustion. However it is measured as excess air by instruments located at the boiler outlet, which in turn cause reduction in air to the combustion equipment to maintain the target excess O_2. Inadequate excess air to the combustion equipment then causes UBC and CO to increase. Thus, minimizing air infiltration is important for controlling unburned combustibles.

Air/fuel imbalances

Air/fuel ratio and mixing rates are a consequence of many factors and vary considerably even within a given furnace. Fuel imbalance can occur because of variations in raw coal feed rate to the multiple pulverizers.

Another aspect of fuel imbalance concerns non-uniformities in coal distribution among the multiple burners on a furnace. Techniques for measuring burner line coal distribution have expanded, but retain a level of uncertainty and can be time consuming and expensive. This complicates implementing and verifying the effectiveness of corrective measures. Burner line restrictors are a proven method of improving coal pipe fuel distribution, but new techniques are under development to correct fuel distribution exiting the pulverizers. Air imbalances among burners can act to magnify air/fuel imbalances. Many modern burners are equipped with air measuring and adjustment hardware to correct this. Air-fuel mixing rates are directly influenced by interactions induced by the burners and OFA (if so equipped), and consequential flow patterns in the furnace. Adjustments which increase mixing rates will tend to reduce UBC while increasing NO_x.

Residence time

Furnace residence times combined with oxygen availability, thermal environment, and coal reactivity ultimately control unburned combustibles. Full load furnace residence times can vary from 1 to 3 seconds depending on burner location, furnace geometry, and operating conditions. Adding an OFA system to an existing furnace further reduces the residence time for final carbon burnout. Increased air-fuel mixing rates can help compensate for marginal residence time and reduce UBC and CO, but with some increase in NO_x.

Gaseous losses

CO emissions result from insufficient oxygen and/or temperature to complete combustion. Localized or global air deficiency and flame quenching are common causes for CO emissions in practical systems. Air/fuel balance is difficult to maintain over all combustion conditions. Combustion tuning, to some degree, involves compensating for errors. Fuel excesses from one burner are offset by air excesses on another. Changes in firing conditions, inevitable with different combinations of pulverizers/burners in service, cause shifts in air-fuel distribution with resultant increases in CO.

Advanced techniques

New diagnostic tools and control systems are providing improved means to reduce unburned combustibles while maintaining low NO_x emissions. Flame diagnostic equipment such as the Flame Doctor® system use non-linear analysis (chaos theory) to provide a means to better tune individual burners. Diagnostic instrumentation in the upper furnace or boiler outlet can more accurately assess excess air and emission levels, and identify localized areas with high emissions. Automated flow control devices on the burners and OFA ports enable adjustments with conventional or neural network control systems to dynamically minimize emissions.

Auxiliary equipment

Oil/gas firing equipment

In some cases, PC-fired furnaces are required to burn fuel oil or natural gas up to full load firing rates. These fuels can be used when the PC system is not available for use early in the life of the unit, due to an interruption in coal supply, or as a NO_x control strategy. Some operators avoid installing a spare pulverizer by firing oil or gas when a mill is out of service.

To fire oil, an atomizer is installed axially in the burner nozzle. Erosion protection for the atomizer is recommended. A source of air is needed when firing oil or gas to purge the nozzle and improve combustion. This air system must be sealed off prior to PC firing. Steam-assisted atomizers are recommended for best performance.

Gas elements of several designs can be used in PC-fired burners. These include designs with manifolds which supply multiple elements in the air zone, or single element designs mounted axially in the coal nozzle. Gas manifolds can be located inside or outside of the windbox. Installation in the windbox clears the burner front of considerable hardware but prevents on-line adjustment or repair of gas elements. External manifolds are more complex, but enable rotational adjustment or replacement of the gas injection spuds. On-line adjustability is very useful for optimizing emissions and to counteract any tendency toward burner rumble. A single axial high capacity gas element (HCGE) in the coal nozzle offers the most effective NO_x control in use with advanced low NO_x PC-fired burners. The HCGE (see Chapter 11) has to be inserted for operation and has to be retracted and supplied with seal air when out of service. Regardless of gas element type, the issue of compatibility with the combustion system has to be considered. Retrofit of gas or oil elements increases the forced draft fan load in many cases. Primary air is eliminated which results in higher quantities of secondary air compared to PC firing. As a result, fan flow and static margins need to be considered.

See Chapter 11 for more information on oil and gas firing equipment.

Igniters

An igniter is required to initiate combustion as pulverized coal is first introduced to the burner, as the burners are being normally shut down, and as otherwise required for flame stability. Additionally, the igniters may be used to warm the furnace and combustion air prior to starting the first pulverizer. In some cases they are used to synchronize the turbine prior to firing coal. Igniters typically use a High Energy Ignition spark system to ignite the fuel, which is usually natural gas or No. 2 fuel oil. Igniters on utility boilers and most large industrial boilers are operated with an input capacity of approximately 10% or more of the main PC-fired burner. Such igniters are categorized by NFPA (National Fire Protection Association) as Class I. The use of Class I igniters reduces the complexity of the flame detection system and permits operation of the igniters as desired under any boiler conditions. Modern boilers have automated controls to start, operate, purge, and shut down the igniters from the control room. (See also Chapter 11.)

Flame safety system

Modern PC-fired boilers are equipped with a flame safety system (FSS). The FSS uses one or more flame scanners at each burner to continuously monitor flame conditions electronically. The flame scanners are used to evaluate several characteristics of the igniter and/or main flames. The lack of satisfactory flame signals causes the FSS to automatically initiate actions to prevent hazardous operation of the burner and boiler. The specific actions taken by the FSS can vary with the boiler design, boiler load, fuel in service, and the status of the igniters in a burner group. The steps taken, and their sequence, are designed to prevent unburned fuel from entering the furnace and avoid large swings in the total furnace fuel-to-air ratio, thereby significantly reducing the risk of an explosion.

Safety and operation

Uncontrolled ignition of pulverized coal may result in an explosion which can severely harm personnel and destroy equipment. Explosions result from ignition of an accumulated combustible mixture within the furnace or associated boiler passes, ducts, and fans. The magnitude and intensity of the explosion depend on the quantity of accumulated combustibles and the air-fuel mixture at ignition. Explosions result from improper operation, design, or malfunction of the combustion system or controls. The National Fire Protection Association (NFPA) publishes codes to address these concerns. These have been consolidated into the NFPA 85 Boiler and Combustion Systems Hazards Code, which is periodically updated.[14] This serves as an excellent source of information concerning design and equipment issues, operation and control of PC-fired equipment.

Another key source of information is the operating instructions. Operators of PC-fired equipment should expect the instructions to provide specific information concerning the purpose, design, calibration and adjustment, startup, operation and shutdown of the various equipment. The instructions should also provide maintenance and troubleshooting information.

Other PC applications

Pulverized coal in metal and cement industries

The application of pulverized coal firing to copper- and nickel-ore smelting and refining has been standard practice for many years. With the use of pulverized coal, high purity metal can be obtained because the furnace atmosphere and temperature can be easily controlled. Pulverized coal may be favored over other fuels for several reasons: it may be less expensive, it offers a high rate of smelting and refining, and it readily oxidizes sulfur.

Copper reverberatory furnaces for smelting and refining are fitted with waste heat boilers for steam generation. These boilers supply a substantial portion of power for auxiliary equipment and also provide the means for cooling the gases leaving the furnace.

In producing cement, the fuel cost is a major expense item. Except in locations where the cost favors oil or gas, pulverized coal is widely used in the industry. Direct firing, with a single pulverizer delivering coal to a single burner, is common practice in the majority of U.S. cement plants. The waste heat air from the clinker cooler, taken from the top of the kiln hood through a dust collector, usually serves as the preheated air for drying the coal in the pulverizer.

Pulverized coal is being applied effectively in the steel industry where it is injected into the blast furnace tuyères. The pulverized coal replaces a similar weight of higher priced coke. Theoretically, 40% of the coke could be replaced by pulverized coal.

References

1. Lowry, H.H., Ed., *Chemistry of Coal Utilization*, Horizon Publishers & Distributors, Inc., New York, New York, pp. 1522-1567, January, 1945. See Chapter 34, "The Combustion of Pulverized Coal," by A. A. Orning.

2. "Standard test method for volatile matter in the analysis sample of coal and coke," D3175-89a, *Annual Book of ASTM Standards*, Vol. 5.05, American Society for Testing and Materials, Philadelphia, Pennsylvania, pp. 329-331, 1991.

3. Beer, J.M., and Chigier, N.A., *Combustion Aerodynamics*, Chapter 5, Krieger Publishing, Malabar, Florida, 1983.

4. Brackett, C.E., and Barsin, J.A., "The dual register pulverized coal burner," presented at The Electric Power Research Institute NO_x Control Technology Seminar, San Francisco, California, February 1976.

5. LaRue, A.D., "The XCL Burner – Latest Developments and Operating Experience," presented to Joint Symposium on Stationary NO_x Control, San Francisco, California, March 1989.

6. LaRue, A.D. and Nikitenko, G., "Lower NO_x/Higher Efficiency Combustion Systems," presented to the EPA/DOE/EPRI MegaSymposium, Chicago, Illinois, August 2001.

7. Sivy, J.L., Kaufman, K.C., and McDonald, D.K., "Development of a Combustion System for B&W's Advanced Coal-Fired Low-Emission Boiler System," presented at the 22nd International Technical Conference on Coal Utilization and Fuel Systems, Clearwater, Florida, March 1997.

8. Sivy, J.L., Sarv, H., and Koslosky, J.V., "NO_x Subsystem Evaluation of B&W's Advanced Coal-Fired Low-Emission Boiler System at 100 MBtu/hr," *Proceedings of the EPRI-DOE-EPA Combined Utility Air Pollutant Control Symposium*, Washington, D.C., August 1997.

9. Flynn, T.J., LaRue, A.D., Nolan, P.S., "Introduction to Babcock & Wilcox's 100 MBtu/h Clean Environment Development Facility," Presented to American Power Conference, Chicago, Illinois, April 1994.

10. LaRue, A.D., Bryk, S.A., Kleisley, R.J., Hoh, R.H., Blinka, H.S., "First DRB-4Z Burner Retrofit Achieves Ultra-Low NO_x Emission Goal," Presented to POWER-GEN International, Orlando, Florida, November 2000.

11. LaRue, A.D., Costanzo, M.A., Fulmer, J.A., Wohlwend, K.J., "Update of B&W's Low NO_x Burner Experience," Presented to ASME International Joint Power Generation Conference, Miami, Florida, July 2000.

12. Latham, C.E., LaRue, A.D., "Designing Air Staging Systems with Mathematical Modeling," Presented to AFRC/JFRC, Maui, Hawaii, October 1994.

13. Kokkinos, A., Wasyluk, D., Boris, M., "Retrofit Low NO_x Experience for Tangentially-Fired Boilers – 2002 Update," Presented to ICAC, Houston, Texas, February 2002.

14. NFPA 85–Boiler and Combustion Systems Hazards Code 2001 Edition, National Fire Protection Association, Quincy, Massachusetts, 2001.

Flame Doctor is a trademark of the Electric Power Research Institute, Inc.

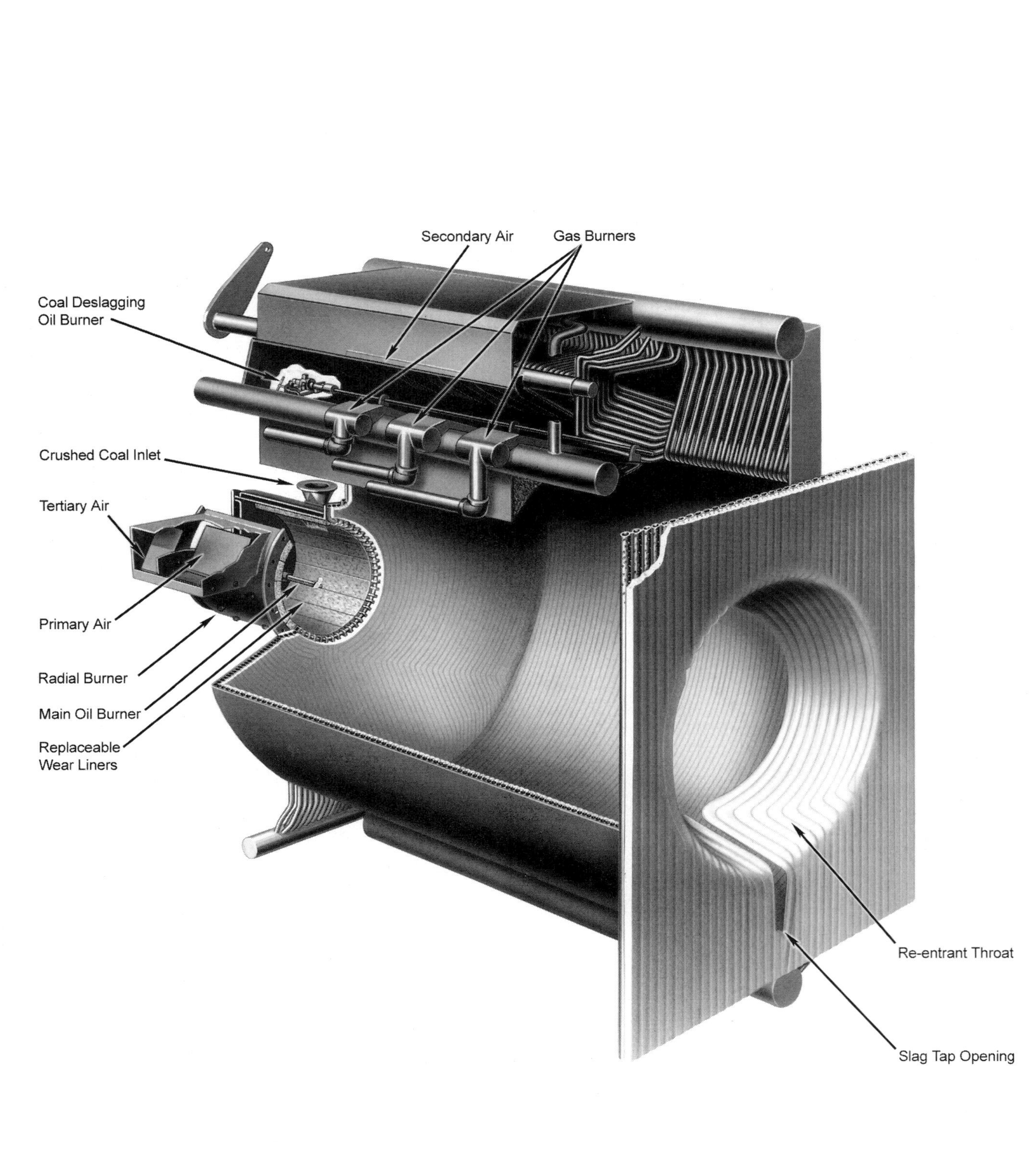

B&W Cyclone™ furnace.

Chapter 15

Cyclone™ Furnaces

The Babcock & Wilcox Company (B&W) developed the Cyclone™ furnace concept in the 1940s to burn coal grades that are not well suited for pulverized coal (PC) combustion. The ash from these coals has a low melting (fusion) temperature and would enter the superheaters of PC units in a molten state, creating severe slagging.

The Cyclone furnace was originally designed to take advantage of:

1. lower fuel preparation capital and operating costs (crushers only),
2. a smaller furnace, and
3. less flyash and convection pass fouling (15 to 30% of the fuel ash enters the convection pass instead of 80% for PC firing).

Cyclone furnace arrangement and basic operation

As shown in Fig. 1 and the facing page illustration, the Cyclone furnace consists of a horizontal cylindrical barrel 6 to 10 ft (1.8 to 3.0 m) in diameter, attached to the side of the boiler furnace. The Cyclone barrel is of water-cooled tangent tube construction. Inside the Cyclone barrel, short pin studs are welded to the outside surface of the tubes in a very dense pattern. A refractory lining material is installed in between the studs. This insulation maintains the Cyclone at a high enough temperature to permit adequate slag tapping from the bottom of the unit and significantly reduces the potential for corrosion.

Crushed coal and some air (primary and tertiary) enter the Cyclone through specially designed burners on the front of the Cyclone. In the main Cyclone barrel, a swirling motion is created by the tangential addition of the secondary air in the upper Cyclone barrel wall. A unique combustion pattern and circulating gas flow structure result (discussed below). The products of combustion eventually leave the Cyclone furnace through the re-entrant throat. A molten slag layer develops and coats the inside surface of the Cyclone barrel. The slag drains to the bottom of the Cyclone and is discharged through the slag tap.

Principles of operation

To understand the Cyclone concept, the basics of solid fuel combustion must first be considered, particularly as they relate to PC firing. During coal combustion in a boiler furnace, the volatiles burn without difficulty; however, combustion of the fuel carbon char particles requires special measures that ensure a continuing supply of oxygen to unburned carbon particles. A thorough mixing of coal particles and air must occur with sufficient turbulence to remove combustion products and to provide fresh air at the particle surface.

PC firing achieves these requirements by reducing the coal to a very fine powder, nominally 70% of which passes through a 200 mesh (74 micron) screen. This powder then mixes with the turbulent combustion air. After this initial phase, the small particles are carried in the air stream with much less mixing. Without the

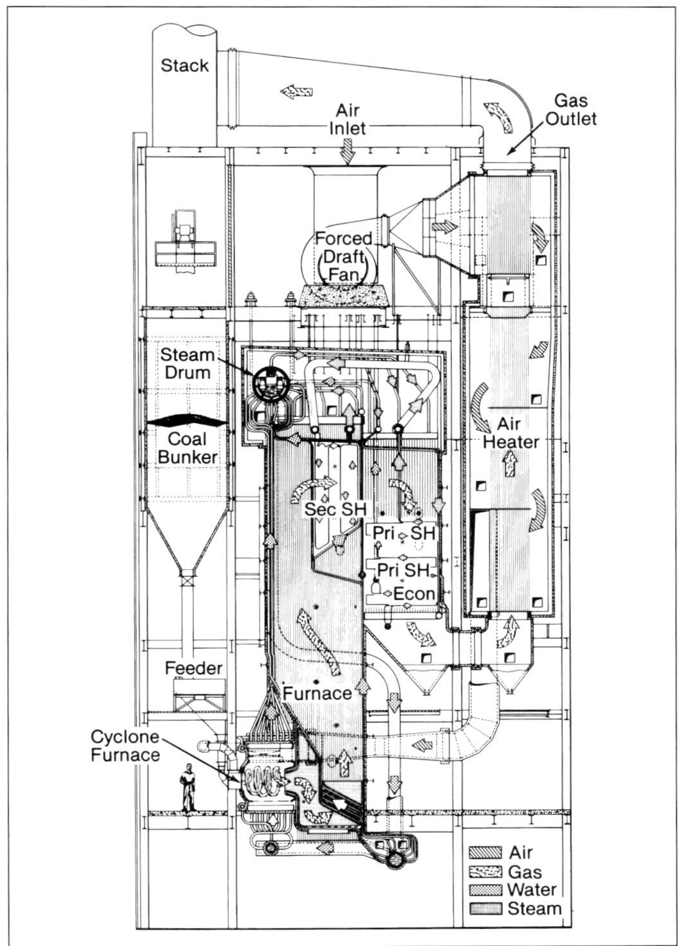

Fig. 1 Cyclone™ furnace boiler showing air, flue gas, water and steam flows.

continued rapid mixing action, the coal particulate combustion must be completed through diffusion of the oxygen and combustion products around the particle. The relatively large PC furnace provides sufficient residence time for oxygen to penetrate the combustion product's blanket around the particles as well as cooling of the ash to minimize convection pass fouling.

The Cyclone furnace, on the other hand, fires relatively large crushed coal particles of which approximately 95% pass through a 4 mesh screen (nominal 0.187 in. or 4.75 mm) and is dependent upon the fuel type being fired. Fuel of this size is too large to burn completely in suspension and would pass through the PC-fired boiler without burning all of the carbon. Therefore, the large particles must be retained in place with the air passing over the particle (air scrubbing) for complete combustion to occur. The Cyclone furnace accomplishes this by forming a molten sticky slag layer which captures and holds the heavier particles. While the large particles are trapped in the slag layer, the volatiles and fine coal particles (recommended at about 10+% passing 200 mesh) burn in suspension providing the intense radiant heat required for slag layer combustion. Ideally, all of the large coal particles become trapped in the molten slag where they complete carbon burnout, leaving behind ash to replenish the slag layer.

With most of the combustion occurring in the confines of the Cyclone furnace, the main boiler furnace can be relatively small compared to a pulverized coal furnace design.

Cyclone combustion

In the Cyclone, fuel is fired at high heat inputs and under very turbulent operating conditions to maximize combustion efficiency. Injecting the main combustion air tangentially at high velocities creates a swirling motion which throws the large coal particles against the Cyclone inside surface, where they are trapped in the slag layer and burn to completion. The hot gases then exit through the Cyclone core and depart through the re-entrant throat into the main boiler furnace.

Air distribution

The cyclonic gas flow does not follow a simple corkscrew path from entry to exit. (See Fig. 2.) The main combustion air (*secondary air*) enters the Cyclone furnace tangentially at high velocity and passes along the Cyclone periphery, also known as the *recirculation zone*. In this zone, the air-gas flow along the chamber is away from the exit opening and toward the burner, pulled along by the vortex vacuum created near the burner end of the Cyclone. Closer to the center of the Cyclone, the general gas flow is toward the exit (re-entrant throat). Eventually, the hot combustion gases flow from the recirculation zone to the Cyclone core, known as the *vortex*. Once in the vortex, the gases and particulate are immediately pulled out of the Cyclone and into the main furnace. The fuel residence time in the recirculation zone is affected directly by the velocity of the secondary air and inversely by the size of the re-entrant throat opening.

The balance of the combustion air is admitted as primary and sometimes tertiary air as discussed below. The air injection locations are identified in Fig. 2 and later in Fig. 5.

The *primary* air enters the burner tangentially in the same rotational direction as the secondary air, carrying the coal into the Cyclone. This primary air controls the coal distribution within the main Cyclone chamber. For optimized operation, the primary air is typically minimized to avoid throwing the raw coal too deep or near the vortex, but must not be set so low as to create a fuel buildup near or in the burner. Adequate velocity must be maintained to allow for proper coal distribution into the cyclone.

The *tertiary* air enters the center of the burner along the Cyclone axis, directly into the Cyclone vortex. It helps keep the cyclone burner door cool and controls the vortex vacuum which consequently determines the position of the main combustion zone, the primary source of radiant heat. As a negative effect, too high an amount of tertiary air reduces the vortex vacuum at the burner, allowing the main combustion zone to move deeper into the Cyclone toward the re-entrant throat and main boiler furnace.

Heat rates

Within the Cyclone, the fuel burns at a heat release rate of 450,000 to 800,000 Btu/h ft^3 (4.66 to 8.28 MW_t /m^3), developing gas temperatures of more than 3000F (1649C). Heat absorption rates by the water-cooled walls are relatively low as the unit has a relatively small surface protected by a refractory coating – 40,000 to 80,000 Btu/h ft^3 (414 to 828 kW/m^3). The high heat release and low heat absorption rates combine to ensure the very high temperatures needed to complete the combustion and maintain the slag layer in a molten state. The high temperature and high heat release rate characteristics of Cyclone firing also provide the conditions to produce very high uncontrolled nitrogen oxides (NO_x) emission levels. The Cyclone furnace traps and burns only as much coal as it can handle. The excess passes into the main furnace and boiler back-end as unburned carbon carryover. During overfiring, long term corrosion and/or erosion with associated tube deterioration could occur in the Cyclone and/or lower furnace regions. In addition, overfiring may cause problems in the main furnace,

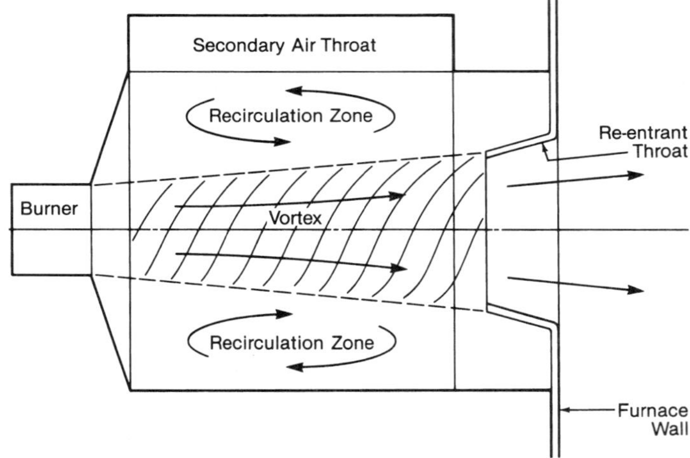

Fig. 2 Cyclone furnace gas recirculation pattern.

convection pass, economizer, and other downstream equipment components.

Slag layer

The intense radiant heat and high temperatures melt the ash into a liquid slag coating which covers the entire Cyclone interior surface except for the area immediately in front of the secondary air opening. The unit's refractory lining further assists this molten condition by limiting heat absorption to the water-cooled walls. The slag coating is kept hot and fluid by combustion and flows constantly from the Cyclone into the main furnace, where it drains through a floor tap opening into a water-filled tank. The slag layer must be constantly replenished by the ash from incoming coal. For this reason, bituminous coals fired in Cyclones need a minimum of 6% ash (dry basis); subbituminous coals must contain at least 4% ash (dry basis).

Optimal combustion conditions are generally achieved when excess air is maintained at about 10 to 13% at the Cyclone. Cyclone operating temperatures tend to decline at higher excess air levels. In addition, the adiabatic flame temperatures within the Cyclone will actually be higher at near theoretical air operating conditions.

A primary indicator of Cyclone combustion temperature is the slag temperature at the slag tap. Slag temperature is a function of radiant heat input to the Cyclone. The degree of radiant heat input indicates the portion of the combustion occurring within the Cyclone. Relative slag temperatures can be monitored with an optical or ultraviolet pyrometer from the furnace side ports or by sighting through an inactive Cyclone on opposed wall-fired units. Although the traditional method of aiming a pyrometer through the Cyclone front observation port has not proven to be as effective, new techniques are being investigated using existing scanners and/or other equipment viewing from the Cyclone front to help evaluate performance.

Under ideal combustion conditions, the Cyclone can capture approximately 70 to 75% of the original fuel ash as slag and drain it to the furnace for disposal in the slag tank. (See Fig. 3.) Smaller boilers can retain more of the ash by capturing additional quantities on the walls and screen tubes. Other details of the Cyclone furnace and boiler arrangements are discussed under *Design features*.

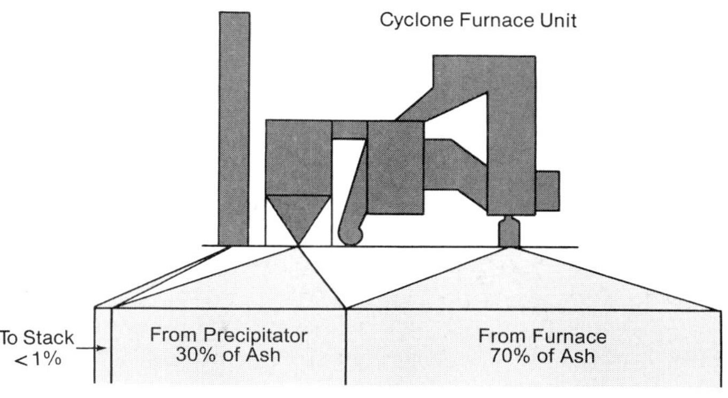

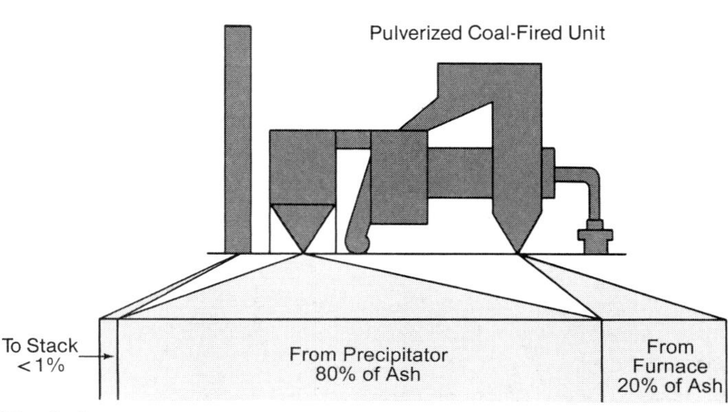

Fig. 3 Comparison of ash distribution from a large Cyclone furnace unit and typical pulverized coal unit.

Suitable fuels

The Cyclone furnace can handle a wide range of coals from low volatile bituminous to lignite, depending on the fuel preparation and delivery system. Cyclones have also successfully co-fired solid waste fuels such as wood chips, sawdust, bark, coal chars, refuse-derived fuel (RDF), petroleum coke, paper and sewage sludge, and tire-derived fuel (TDF). Fuel oils or gases (natural gas, coke oven gas, and others) can be burned in Cyclone furnaces as the primary, contingency or startup fuel.

On a continuous basis, the Cyclone must fire either a solid fuel or a liquid/gaseous fuel. Long term firing of oil or gas should only be pursued in a bare tube Cyclone because these fuels can destroy the refractory lining. The resulting loose slag and/or refractory debris would severely erode the tubing, leading to eventual pressure part failure. Cyclones switching from coal to oil or gas for an extended period should have all refractory within the cyclone removed, the slag tap direction reversed and all air ducts cleaned. The pin studding can remain, although a study should be performed to evaluate the change in magnitude on heat absorption profiles. When switching back to coal, the refractory must be reinstalled and the slag tap must be set in the correct direction.

Fuel criteria

Coals and co-fired fuels (solid or otherwise) must be evaluated against several criteria. Volatiles must be higher than 15% on a dry basis to ensure stable combustion within the Cyclone. The ash content must be at least 6% for bituminous coals or 4% for subbituminous coals, but should not exceed about 25% on a dry basis. Higher ash content fuels require special considerations. With low ash coals, the proper slag coating can not be developed, nor can it be maintained. Thin, low viscosity slag coatings do not protect the refractory. To compensate for this condition, the slag can be thickened by increasing excess air and reducing the Cyclone combustion temperature if other operating conditions permit.

The moisture content for standard bituminous coal firing Cyclones should not exceed approximately 20%. Higher moisture content requires better crushing and higher secondary air temperatures to dry the fuel for proper combustion. For subbituminous coals with moisture contents typically in the 25 to 30% region, other Cyclone modifications including reduced sized, tight gap re-entrant throats and segmented secondary air velocity dampers are required to maintain good Cyclone operating conditions. Improved coal fineness operation is also imperative to achieve good Cyclone combustion when firing subbituminous coals. Particularly high moisture fuels may require the pre-dry system used on lignite-fired Cyclone-equipped boilers.

Operating experience demonstrates that coal ash has a tendency to concentrate iron, initiate slag tapping problems, and/or result in a corrosive iron sulfide attack on the refractory and Cyclone furnace. This can be evaluated by comparing the total amount of sulfur to the iron/calcium and iron/magnesium ratios. (See Fig. 4.) Frozen iron puddles found on the Cyclone or furnace floor during outages provide an indication of the magnitude of the potential problems.

Slag viscosity factor – T_{250} value The most important evaluation consideration for Cyclone coals is the slag viscosity characteristic or T_{250} value. This ultimately separates coals into categories suitable and unsuitable for Cyclone operation. The T_{250} value denotes the temperature at which the coal slag has a viscosity of 250 poise. At this viscosity the slag flows on a horizontal surface. A slag with a higher viscosity would fail to flow steadily from the Cyclone and main furnace and would be too stiff to trap the unburned coal particles for proper combustion. The currently recommended maximum T_{250} value for all Cyclone bituminous coals is 2450F (1343C). With the advent of increasing low sulfur subbituminous coal use in the 1990s, B&W established a T_{250} limit of 2300F (1260C) for subbituminous coals fired in standard Cyclones. This value compensates for the effect of increased coal moisture on Cyclone combustion temperatures.

The preferred method of establishing the T_{250} value is by experimental measurements on actual fuel ash samples. However, due to the large database accumulated from past testing, the T_{250} value can also be estimated based upon calculations using the coal mineral ash analysis. (See Chapter 10.)

For some coals and/or other solid fuels, the T_{250} value of the slag can be lowered by altering its ash base to acid (B/A) constituent ratio (discussed in Chapter 21) through the addition of a fluxing agent (e.g. limestone, dolomite, etc.). Although B&W does not generally recommend this approach, sometimes it is the only alternative to improve operation. Fluxing system design considerations must incorporate the capability to accurately, uniformly, and continuously distribute the flux material with the fuel stream.

If a Cyclone unit must burn high fusion temperature fuel, the best approach is to adjust the T_{250} value by blending in a different coal or solid fuel; however, this usually represents higher fuel handling costs. Both fuel blending and fluxing create concerns with maintaining acceptable consistency.

Co-fired fuels must be evaluated on an individual basis with emphasis placed on the resulting ash content as a percentage of the total heat input.

The general characteristics of coals suitable for Cyclone furnace firing are summarized in Table 1.

Cyclone coal burner types

In the United States (U.S.), Cyclone furnaces have been equipped with three primary coal burner types – scroll, vortex and radial (two sub-types), as shown in Fig. 5. All types inject coal from the front end of the Cyclone and impart a swirl to the crushed coal in the same rotation as the secondary (main) combustion air. To protect the burner coal inlet from excessive erosion, all utilize a protective wear liner known as the *wear blocks*. The material used for these blocks is normally comprised of metal, ceramic, or a combination of the two. Finally, the burners are normally kept cool with the addition of a water-cooled jacket on the burner door and the burner scroll. The main purpose of the cooling jacket is personnel protection along with reducing the operating temperatures of the burner itself. Since untreated water is normally used for the cooling media, corrosion and pluggage can be encountered with time. Although not normally recommended, this water cooling capability has been removed by some

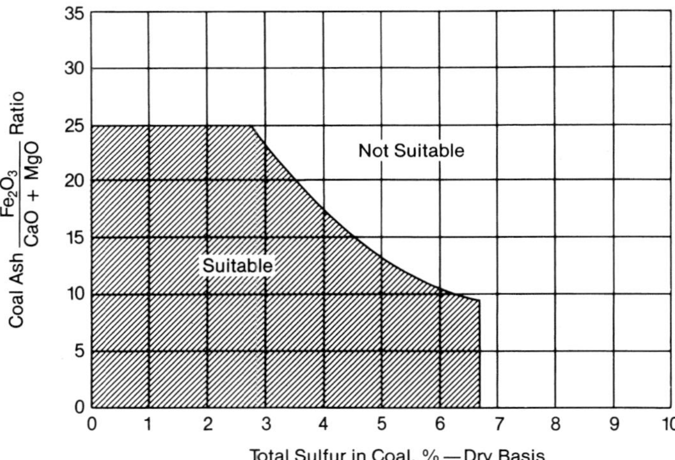

Fig. 4 Coal suitability for Cyclone furnaces based on tendency to form iron and iron sulfide.

Table 1
Summary of Cyclone Furnace Coal Suitability

Ash	Bituminous — minimum	> 6% dry basis
	Subbituminous — minimum	> 4% dry basis
	Maximum	< 25% dry basis
Moisture	Bin firing system (Bit.)	< 20% as-fired
	Direct firing system (Subb.)	< 30% as-fired
	Pre-dry firing system	< 42% as-fired
Volatile		> 15% dry basis
T_{250}	Bituminous	< 2450F (1343C)
	Subbituminous	< 2300F (1260C)
Coal type		Bituminous Subbituminous Lignite
Coal ash iron/sulfur tendencies		See Fig. 4

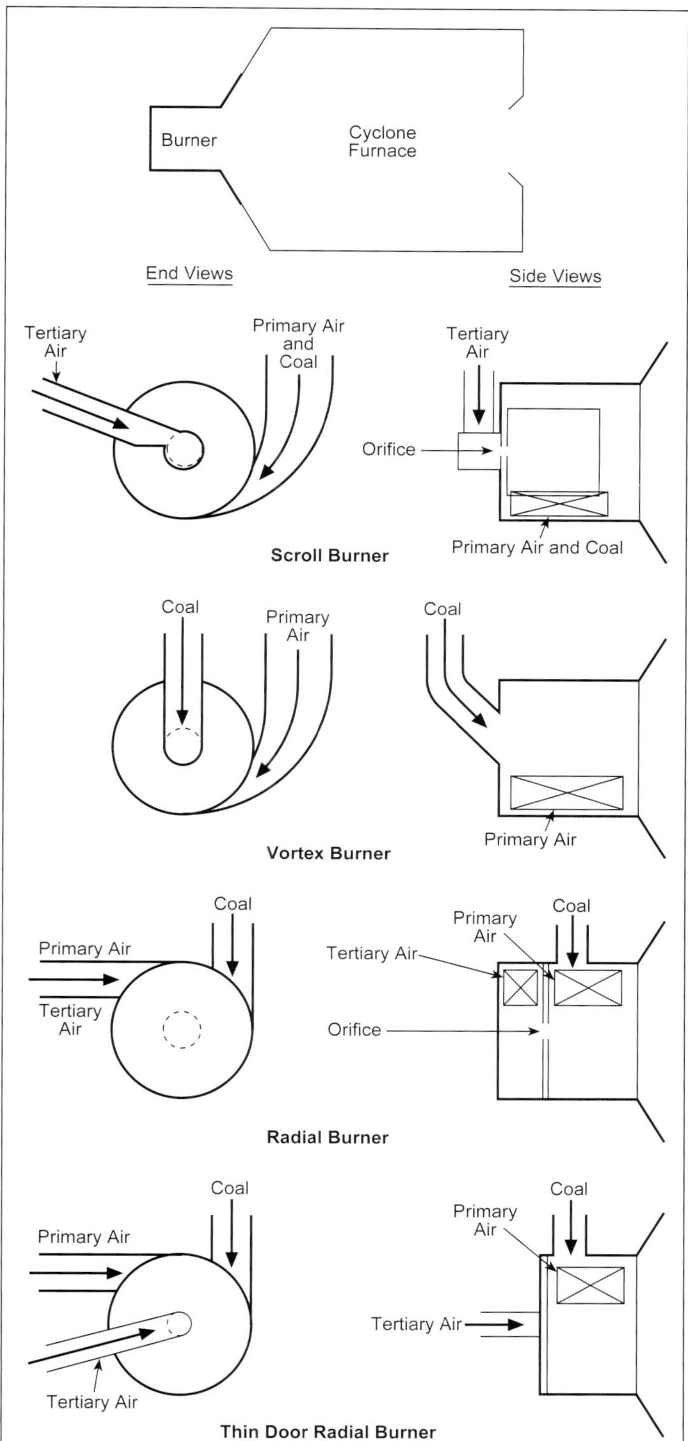

Fig. 5 Primary Cyclone coal burner types.

operators. If this is done, the burner material should be upgraded and insulation should be applied over the burner to provide personnel protection.

The *scroll* burner, which was used in the first Cyclones, combines the primary air and coal at the feeder outlet and injects the mixture into the Cyclone in a cone shaped distribution pattern. Tertiary air, admitted at the center of the burner, controls the position of the main flame within the Cyclone. However, this injection pattern resulted in a high concentration of recirculating coal particles around the burner outlet, leading to heavy burner wear-block erosion. The premixing of primary air and coal required the addition of a rotary seal to protect the feeder from backflow of hot combustion air. Later scroll burner applications eliminated the rotary valve by using a higher head of coal at the feeder inlet to prevent hot air backflow into the feeder and coal bunker.

The *vortex* burner eliminated the rotary seal by feeding coal and primary air to the burner separately. The primary air entered the Cyclone burner tangentially as with the scroll design, but the coal is gravity fed (along with feeder seal air) into the backside of the burner. Although this configuration eliminated the tertiary air, adjusting feeder seal air has the same effect on the Cyclone vortex zone. The vortex burner produced a similar cone shaped injection pattern as the scroll model. Advanced ceramic wear blocks and a new burner door/coal inlet design have significantly reduced erosion problems encountered with earlier designs.

The development of the *radial* burner eliminated the severe wear-block erosion. Like the vortex burner, the radial version does not combine coal and primary air until both enter the burner chamber. However, on a radial burner the coal is introduced tangentially in the same rotation as the primary air. The coal particles form a long rope (concentrated stream of coal) as they are swept across the burner wear blocks and enter the Cyclone. This approach greatly reduces the concentration of coal recirculating around the burner, effectively reducing wear-block erosion. Similar to the scroll burner, tertiary air is admitted using the same axial entry location.

A newer *thin door* radial burner version with the option to eliminate tertiary air (and/or redirect how it is introduced) has been offered since the late 1990s. The benefits of this design are the potential to minimize and/or eliminate tertiary air without adversely affecting performance, an improved range of view within the cyclone, and an overall smaller, easier to open door.

The radial burner remains the modern standard for coal-fired Cyclones. The scroll burner is utilized on modern lignite-fired Cyclone units because the predry system requires mixing of the primary air and coal during coal preparation, prior to the burner inlet. (See later discussion.) As the scroll, vortex and radial type burners have slightly different coal inlet locations, switching from one burner to another requires a change in the coal handling system.

Design features

Boiler furnace

Commercial Cyclone furnaces have been built in sizes ranging from 6 to 10 ft (1.8 to 3 m) in diameter, with standard maximum heat inputs ranging from 150 to 425 × 10^6 Btu/h (44 to 124.6 MW_t), respectively. Higher Cyclone heat inputs under peaking conditions are feasible with the existing equipment, but at the cost of various operational concerns (such as higher unburned carbon in the flyash, higher particulate loadings, increased maintenance, etc.). New larger diameter, higher capacity Cyclone designs are feasible and could be incorporated if the additional capital costs associated with such an overall retrofit can be justified versus the associated increased boiler/turbine output.

As illustrated in Fig. 6, the Cyclone furnace has been used on three general boiler arrangements – single wall firing with screen, open furnace single wall firing, and opposed wall firing. Full boiler drawings are provided in Figs. 1, 7 and 8, respectively. Boiler and Cyclone furnace arrangements have ranged from one 6 ft (1.8 m) Cyclone on a single wall unit to twenty-three 10 ft (3 m) Cyclones on an 1150 MW opposed wall unit. In all cases, the main furnace is relatively small to maintain furnace temperature over the furnace floor slag taps and to promote slag flow on the furnace walls. The lower furnace chamber also contains a protective refractory lining held in place by thousands of pin studs which are welded to the tubes. Generally, the Cyclone boiler can be operated continuously down to about 50% of total capacity. Below this point the slag freezes on the furnace floor and plugs the floor taps. Many units have been equipped with larger floor slag taps to improve slag removal.

Coal preparation

The Cyclone equipped boilers feature three fuel delivery systems – bin, direct-fired, and direct-fired with pre-drying configurations. (See Fig. 9.) The bin system, the simplest, least expensive and most common, uses a pair of large crushers in a central location to prepare the coal for overhead storage bunkers (bins). Because the Cyclone- required crushed coal has a relatively large particle size, the storage hazards as compared with pulverized coal systems are lessened, but not eliminated. Adequate venting to remove freshly released combustible gases from the crushed coal systems must still be maintained along with all associated dust control systems. Safety requirements are imperative for low rank, high volatile coals. With the bin system, a short crusher outage does not interrupt boiler operations. An existing bin system can be upgraded to fire high moisture coals by converting the coal piping into a simple pre-dry system (see Fig. 9).

The direct-fired system uses a smaller separate crusher, sometimes called a coal conditioner, between the coal feeder and burner on each individual Cyclone furnace. These crushers are swept by hot air which removes moisture from the freshly crushed coal. This produces an advantage by improving crusher performance and fuel ignition when firing coals with a moisture content up to 30%. The direct-fired system has also been used where the existing plant layout could not accommodate the bin system.

The direct-fired with pre-dry system represents the Cyclone fuel preparation system for firing lignite and other high moisture coals. In addition to the hot air-swept individual crushers, this system can also include crusher classifiers and mechanical cyclone moisture separators. The classifiers increase the coal fineness which results in more moisture extraction and better

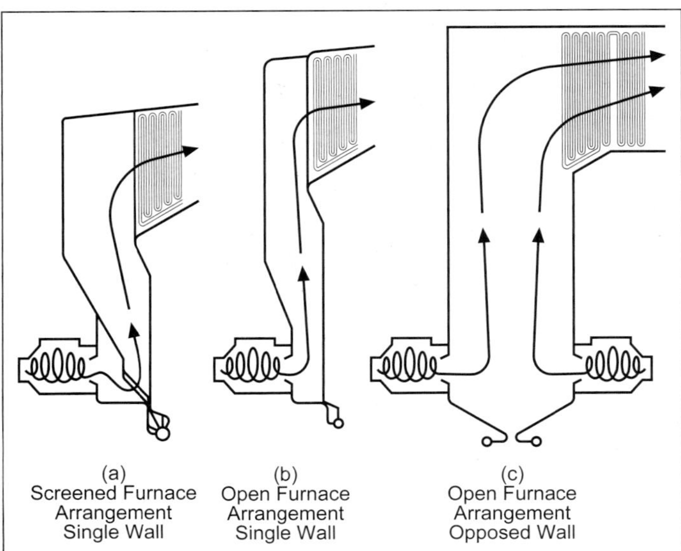

Fig. 6 Selected firing arrangements for Cyclone furnaces.

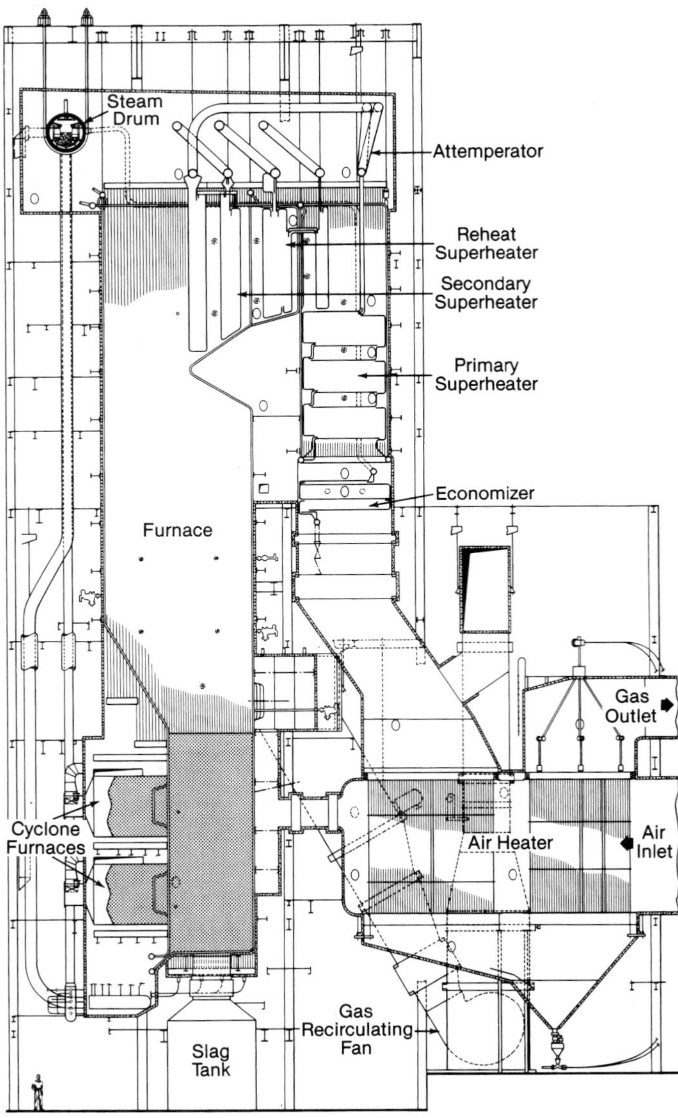

Fig. 7 Radiant boiler with Cyclone furnaces (one wall) and bin system for coal preparation and feeding.

combustion. The mechanical cyclone separators remove the moisture and fines from the coal-air mixture and vent the mixture directly to the boiler furnace through the gas recirculation plenum area. The exclusion of moisture from the main coal increases Cyclone combustion temperatures and subsequently enhances slag tapping. With moisture separators, this system, designed for lignites with moisture higher than 36%, handles poor Cyclone quality, low sulfur coals as easily as standard bin system Cyclones fire high grade bituminous coal.

The preferred coal size distribution for various coal grades is shown in Fig. 10. With high moisture, high ash fusion or other difficult-firing coals, every attempt should be made to produce the highest percentage of coal fines.

Coal feeders

A number of different feeders have been used with Cyclones since their inception. Early units had small table type feeders, followed by drag types, then volumetric belt types and now gravimetric belt types, some of which have been upgraded with microprocessor controls. With the introduction of each model, the control and continuity of the coal flow improved.

The preferred feeder type today is the gravimetric (weighing) type with a microprocessor control upgrade.

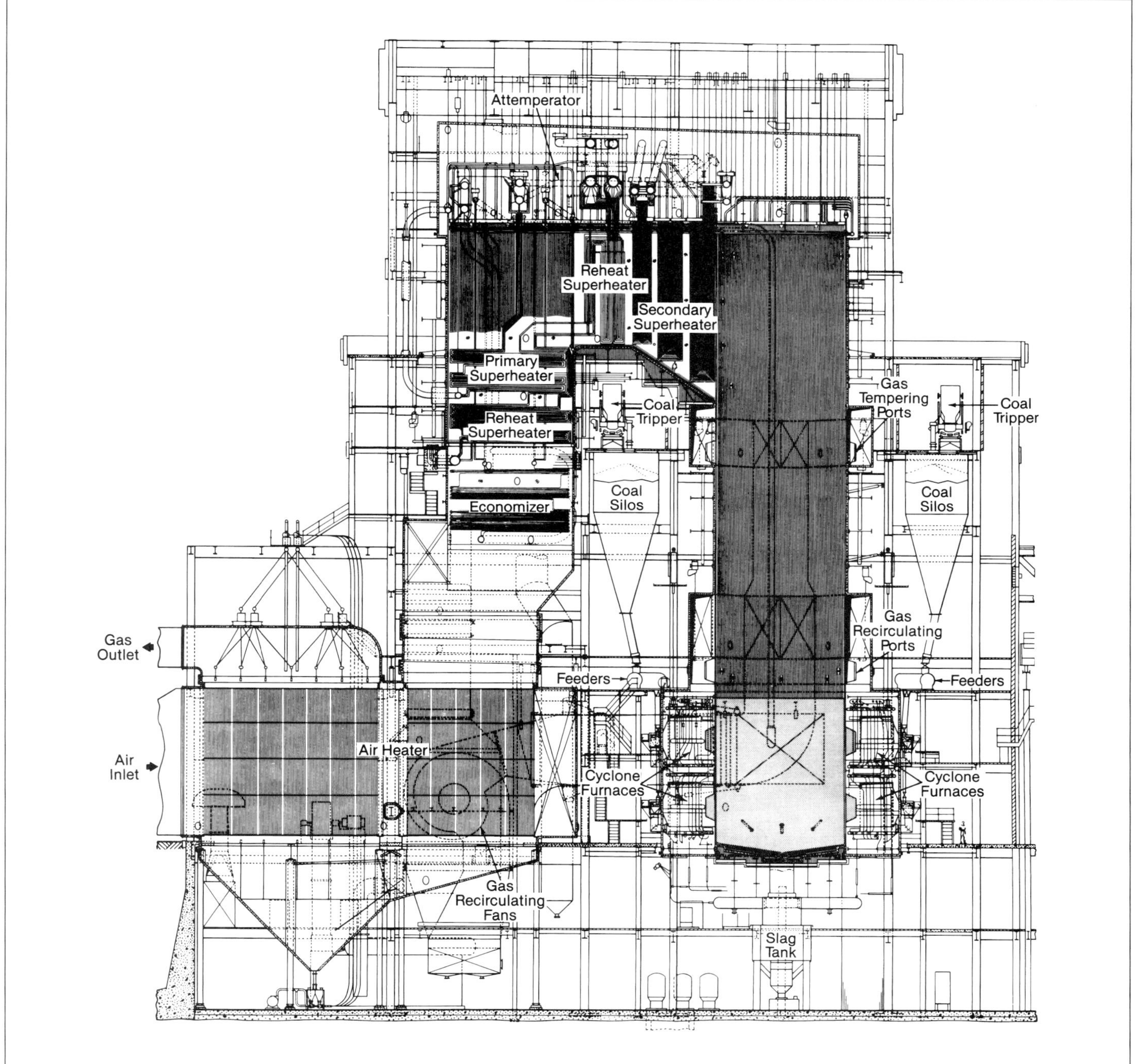

Fig. 8 Universal Pressure (UP®) boiler with opposed-wall Cyclone furnaces and bin system for coal preparation and feeding.

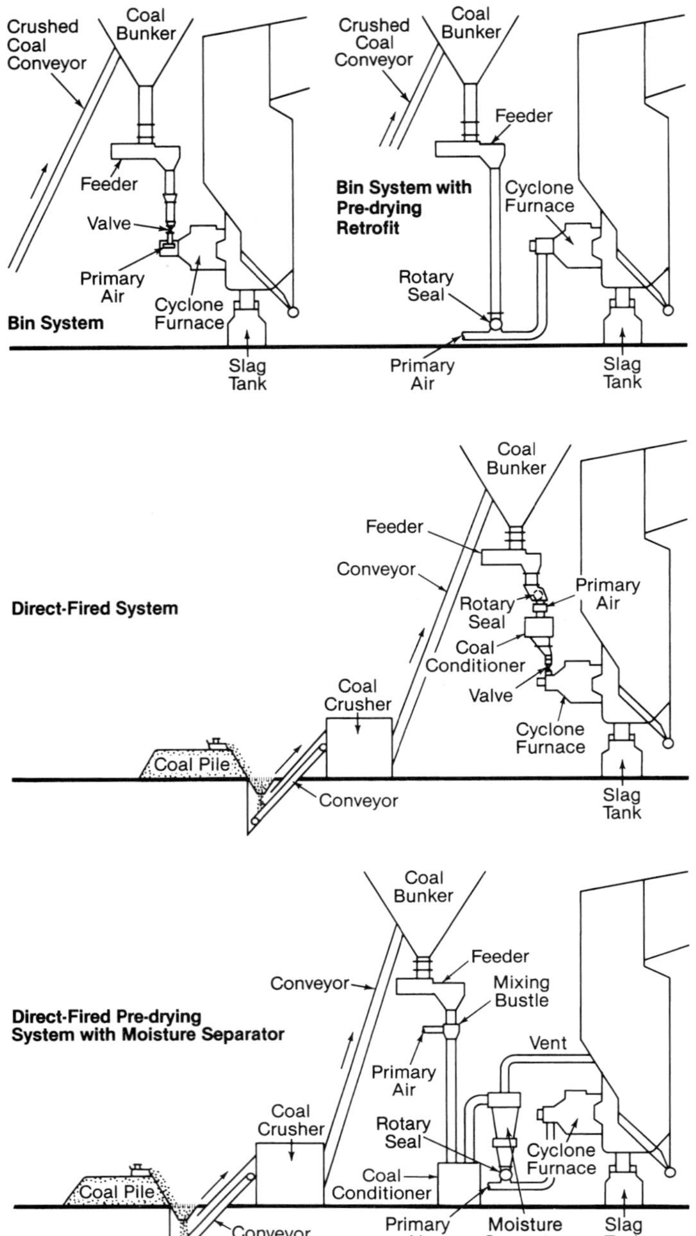

Fig. 9 Bin, direct-fired, and pre-drying bypass systems for coal preparation and feeding to a Cyclone furnace.

Accurate fuel-air control is needed to optimize Cyclone performance, requiring a modern gravimetric feeder.

Slag handling equipment

Cyclone boilers feature a batch removal system for disposing of slag after it is discharged from the main furnace. (See Fig. 11.) The molten slag continuously flows through the furnace floor tap and into a tank beneath the floor where it is quenched and solidified in a water bath. At intervals, the accumulated slag is broken down by a clinker grinder and removed in batches. However, combustible gases can build up at the top of the tank. Similarly, a plugged or partially closed furnace floor tap can result in carbon monoxide accumulation. Both potentially hazardous situations could be generally reduced by installing and properly maintaining a flue gas vent line on top of the slag tank. The main purpose of the slag tank vent line is to maintain a positive flow of hot furnace gases through the floor tap, which reduces the potential of the tap from plugging, especially at low boiler loads. The vent typically either runs from the top of the tank to the boiler economizer or to an aspirator which utilizes windbox air to vent the flow into the furnace.

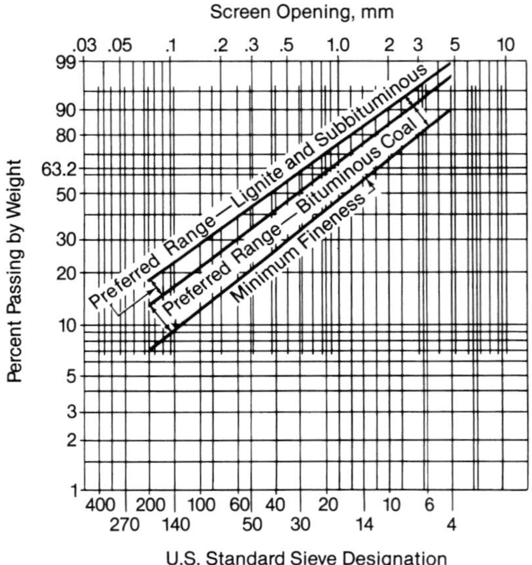

Fig. 10 Crushed coal sizing requirements for Cyclone furnaces.

Oil and gas burners

Although coal is the primary Cyclone fuel, oil and gas have been used as startup, auxiliary and main fuels using the combustion system shown in Fig. 12.

Originally, the oil burner could either be a simple transverse pipe across the secondary air inlet or a single oil gun through the coal burner door. The transverse pipe burner, or *roof burner*, has a single row of radial holes directed into the Cyclone tangent. This simple design has proven to be an excellent combustor. Many Cyclones are equipped with this burner and are typically capable of obtaining full Cyclone rated capacity. However, injecting the oil directly into the Cyclone furnace recirculation zone resulted in an extremely high heat release rate and Cyclone tubes would experience long term thermal damage. The oil roof burner continues to be used as an emergency Cyclone furnace deslagger for units firing coals with high T_{250} values. It is also used to deslag a Cyclone prior to a boiler outage. When coal flow to the Cyclone is interrupted, a heat input of up to about one half of the Cyclone rating is recommended for continuous load carrying purposes.

The alternate design uses an oil burner located at the Cyclone centerline firing directly into the Cyclone furnace vortex.

If continuous, long-term oil-only firing at full Cyclone rated capacity is required, splitting the heat input 50/50 between the roof burner and the center fired burner should be considered.

The Cyclone gas burners (capable of firing natural

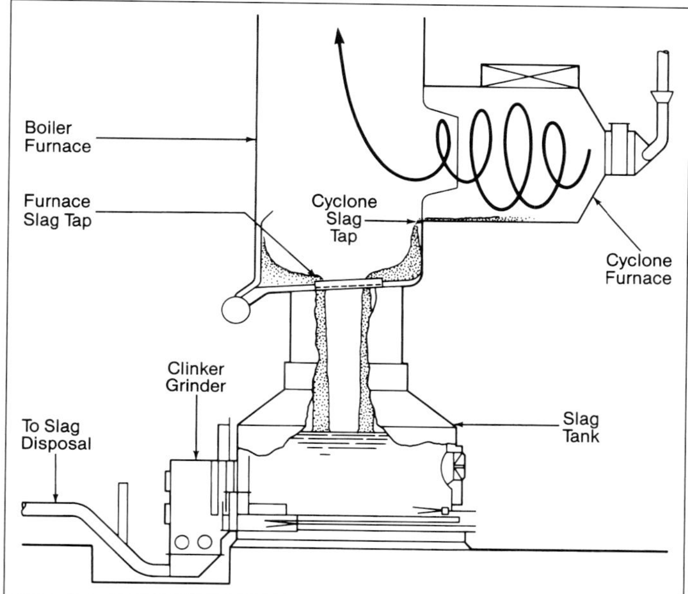

Fig. 11 Batch removal slag handling system for Cyclone furnace boiler.

gas, coke oven gas, or other gaseous fuels) are located in the secondary air throat, injecting gas normally through three flat nozzles. Although this also injects fuel directly into the recirculation zone, the lower radiant heating of natural gas produces a lower heat absorption rate within the Cyclone furnace. Therefore, long term thermal damage has not been observed with a gas burner. Natural gas assists in deslagging plugged Cyclones, but its low radiant heat transfer rate to the Cyclone does not heat slag deposits as well as oil firing.

Cyclone units firing coal may switch to oil or gas for a short period without serious damage. However, for long term gas and oil operation, the Cyclone should be stripped of all slag and refractory to prevent erosive scrubbing of the tubes as previously discussed.

Waste burning

Due to their high heat release rate, the Cyclone furnaces have always been suited for firing a number of waste fuels with coal. A nominal blend ratio tested has been in the range of 5 to 20% heat input from the co-fired waste, with a less than 10% level typically being considered acceptable. However, waste or refuse fuel should first be thoroughly tested on the prospective Cyclone equipped boiler to determine these limits.

Refuse-derived fuel (RDF) RDF consists of the lighter, more easily burned materials from municipal waste. The ideal point of injection is the secondary air throat to maximize fuel particle time in the Cyclone. A scaled-up insulated version of the gas burner nozzle provides the best injector. However, the resistance welded flat studs found in original Cyclones may not withstand the associated particulate erosion, and increased furnace slagging are potential negative consequences.

Tire-derived fuel (TDF) Along with stokers and wet slagging PC units, Cyclones have been viewed as one of the best proposed methods for disposing of scrap automobile tires. Tire rubber provides an excellent low sulfur fuel, but the material's elasticity prevents it from being crushed or pulverized like coal. It generally appears that a 0.5 to 1.0 in. (13 to 25 mm) sizing is close to optimum. Smaller grinds can be provided by the waste rubber industry, but the cost greatly exceeds that of coal. The TDF should fire easily when mixed with the coal. Although the steel content from bead and radial ply tires does not affect the iron sulfide factor noticeably, the bead wiring may be difficult to melt.

Petroleum coke A number of Cyclone operators have successfully blended petroleum coke with their coal. However, the lack of volatiles in petroleum coke can delay fuel burnout, increasing furnace exit gas temperature and resulting in superheater slagging.

Wood chips Wood chips, sawdust, bark and other similar solid waste products have long been co-fired in Cyclones. Moisture content and sizing of the prospective waste fuel helps determine the optimum blend ratios.

Various sludges Paper mill and sewage sludge materials have been successfully co-fired with coal and/or natural gas in new Cyclone-fired boilers. The Cyclones are designed to utilize a secant firing technique for the sludge material while the coal and/or natural gas are introduced according to standard arrangements. Secant firing introduces the fuel through dedicated openings in the Cyclone that are located directly beneath the secondary air inlet. This approach maximizes the retention time and temperature within the Cyclone itself to allow for successful operating characteristics at higher than typical waste material heat inputs (greater than 50% heat input on sludge). As with any Cyclone-fired application, fuel acceptability criteria must be closely followed to optimize operation (fuel T_{250} value, fuel fineness, minimal fuel moisture, etc.).

Combustion controls

The fuel/air ratio at each Cyclone remains the critical item in the combustion control system. When the Cyclone total air flow becomes too low or too high, operating problems can develop. The balancing of the fuel/air ratio is relatively easy to accomplish and monitor on smaller boilers with only one to five Cyclones. On these units, individual ducts supply the total combustion air flow to each Cyclone furnace. On larger

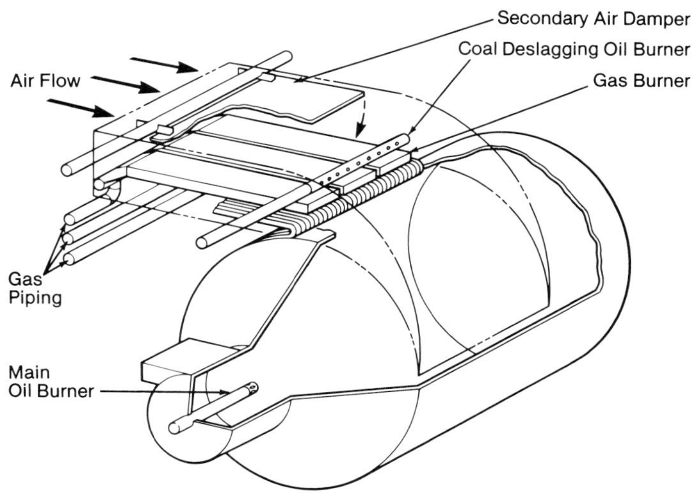

Fig. 12 Arrangement of gas and oil burners in Cyclone furnace.

units, multiple Cyclones are housed within common windboxes for the supply of secondary air. This arrangement inherently makes it more difficult to accurately monitor air to individual Cyclone furnaces. As a result, larger Cyclone equipped boilers have typically been operated at higher excess air levels than smaller units. Modern measurement and control techniques are improving the accuracy of air flow measurement. Although new devices are being designed with more repeatable measurements, it is still advisable to field-verify the equipment by performing air calibrations after installation. When retrofitted with accurate gravimetric coal feeders for each Cyclone, such devices permit tighter individual Cyclone furnace control.

With the advent of the Cyclone air staging technology to help reduce NO_x emission levels, the importance of maximizing the accuracy of the fuel/air flow measurements and controllability has increased. Improving this accuracy not only enhances the NO_x reduction capability of the system, but also maintains the required characteristics for acceptable Cyclone operation.

Operation

Igniters

Cyclones can be equipped with either No. 2 oil or natural gas igniters, with oil igniters being the more common option. On coal-fired Cyclones, both igniters are located at or near the Cyclone front in the secondary air throat outlet. The oil igniter is installed external to the Cyclone and is a retractable design to prevent overheating of the igniter and related problems with the oil atomizer. The gas igniter is a fixed design and installed inside the secondary air duct immediately ahead of the exit into the Cyclone. Original Cyclones were equipped with either a 10×10^6 Btu/h (3 MW_t) or 17×10^6 Btu/h (5 MW_t) oil igniter that employed mechanical atomization, or a 17×10^6 Btu/h (5 MW_t) stationary gas igniter. Both fuel igniter designs made use of high voltage ignition.

Demand for increased oil igniter heat release and decreased opacity, especially during cold start operations, led to the development of a larger, air atomized oil igniter with high energy spark ignition system. This new igniter, the CFS model, is capable of an input rating of 25×10^6 Btu/h (7.3 MW_t) under specific conditions. These higher input oil igniters require a slightly larger tube opening than was supplied on original Cyclones and would require some tube modifications if installed on these older units.

Minor design changes and field testing of the standard stationary gas igniter led to a revised igniter capacity on natural gas to approximately 30×10^6 Btu/h (8.8 MW_t), thereby providing similar capacity on either gas- or oil-fired Cyclone igniters.

Cyclones firing only oil and natural gas as the main fuels have additional igniter options with respect to location. The retractable CFS oil and/or gas igniter can be located axially down the center of the retired coal burner or through the modified burner door.

For further information on oil and gas igniters, refer to the igniter section in Chapter 11.

Sectional secondary air control dampers

Cyclone furnaces can be equipped with sectional dampers (2 or 3 blade sections) across the width of the secondary air inlet. This arrangement provides an additional level of control for the Cyclone furnace combustion. Biasing these dampers permits adjustment to the Cyclone combustion pattern and increases fuel retention time. The effectiveness of this additional control technique is greater when difficult to burn fuels are being fired (e.g., high moisture/low sulfur coal, higher ash fusion fuels, etc.).

Low load operations

Typically, Cyclone furnaces firing a good Cyclone bituminous coal can not operate below half load without the slag freezing. In addition, the main boiler furnace floor typically stops tapping slag below half load. When firing subbituminous type coals, this minimum low load level is higher due to the more difficult nature of firing that fuel in Cyclone boilers. In either case, the slag taps tend to plug solid. An individual Cyclone can continue to operate and should eventually start retapping. However, a boiler with a solidly plugged floor tap must be shut down for manual slag removal.

Power requirements

Cyclone furnace boilers have significantly different power requirements than do pulverized coal units. Fuel preparation power consumption for Cyclone boilers is very low because the coal is only crushed, not pulverized, and the primary air fan is not used. However, the forced draft fan power usage is substantially higher due to the relatively high Cyclone windbox to furnace pressure drop [typically 25 to 45 in. wg (6.2 to 11.2 kPa)]. This is illustrated in Fig. 13. The difference between Cyclone and PC firing is dependent upon the fuel type and heating value. PC firing has an advantage for high heat content and high grindability bituminous coal firing applications. Cyclone firing has an advantage for low heat content and lower rank fuels such as subbituminous, lignite and brown coals which are also harder to pulverize. In the case of the high heat content bituminous coals, the lower operating costs of PC firing must be balanced against the lower capital costs of Cyclone-based firing systems.

Maintenance

Corrosion and erosion within the Cyclone are the two most critical maintenance items. The Cyclone's wet slagging environment produces a potentially corrosive iron sulfide attack on the pressure part tubing.

Erosion is also a problem in an area opposite the secondary air throat where a protective slag coating can not form. The coal particles wear away the edges of the protective flat studding and can cut a channel between the studs and potentially damage the tubes.

Tubing – pin studs and refractory

In areas coated by molten slag, the tubes are protected by a refractory layer held in place by pin studs (Fig. 14). In addition to retaining the refractory, the

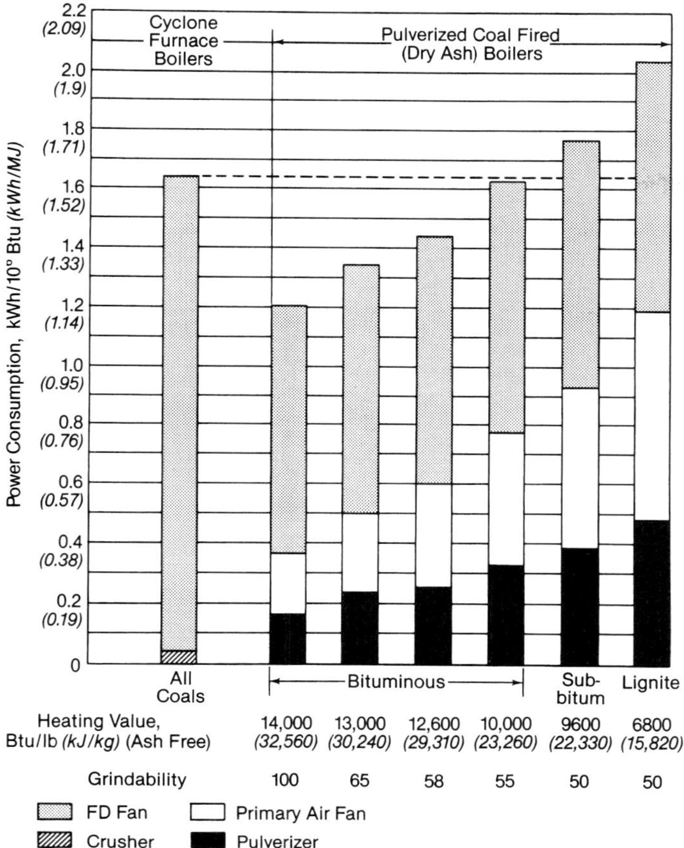

Fig. 13 Auxiliary power requirements for typical high-capacity, pressure-fired Cyclone furnace and pulverized coal units.

pin studs cool the refractory surface in contact with the corrosive slag and retard the corrosive chemical action. The pin studs protect the refractory and the refractory in turn protects the pin studs. Over time, experience has demonstrated that tighter, or denser, pin stud spacing improves refractory performance. The denser pin stud patterns have provided very good performance in their ability to protect refractory and resist corrosion (Figs. 15a and b).

Tubing – flat studs

An advanced design has been developed and installed by B&W to reduce maintenance (Fig. 16). The newer flat staggered stud design, using a hand applied fillet weld, offers the following advantages:

1. more precise stud manufacturing and closer spacing,
2. minimum potential for channeling and accelerated wear between studs,
3. excellent heat transfer which reduces metal temperature and erosion rates, and
4. thicker stud sizes to extend life.

Metallization

To enhance tube life, the plasma arc flame spraying of alloy metals onto the tube surface has been used. Despite a number of experiments with different metallization powders, the results generally remain inconclusive. Some applications using expensive coatings resulted in a nickel sulfide attack which consumed the coating in the same manner as iron sulfide corrodes studs and tube surfaces. Furthermore, the pin studding does not lend itself to consistent spray applications. Some coatings prohibit installing replacement pin studs unless the coating and any interacting tube surface have been removed.

Refractory

Field experience has demonstrated that corrosive slag in any form should be kept away from the tubes by a refractory coating. Experience on operating units has proven that the most durable refractories are ram-type high density formulations. The specific refractory selection may be contingent upon the specific plant fuel.

Key overall issues to achieve the best potential chance for increased refractory life include:

1. Ensure proper maintenance/application of studs (along with maximizing stud density).
2. Choose the proper refractory for the application (proven positive experience).
3. Use refractory that has not exceeded its shelf life.
4. Follow proper refractory installation and curing procedures.
5. Follow good Cyclone startup/operation procedures.

Any one of these items done incorrectly can cause early loss of refractory.

Coal crusher

Cyclone coal crushers have generally remained unchanged over the years. With the increased use of difficult to fire subbituminous low sulfur coals and the need to crush the fuel as fine as possible, a change in maintenance practice is recommended. Cages should be adjusted and mills should be reversed more frequently than with standard Cyclone furnace coals. In addition, the hammers should be discarded at their half life to maintain adequate striking mass. To improve fineness to a greater degree, fine grind cages can be installed. This normally reduces the original capacity of the crushers and requires a motor upgrade and/or additional installed crusher capacity to restore system capacity.

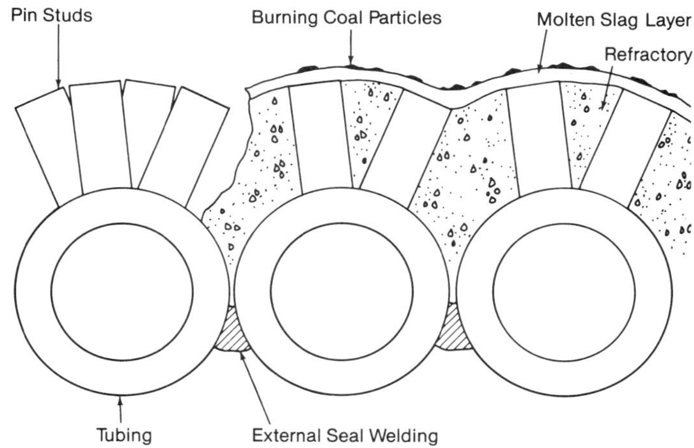

Fig. 14 Cyclone furnace stud and refractory section.

Fig. 15a High density Cyclone pin studding.

Fig. 15b Modern *super dense* pin stud pattern after five years without maintenance.

Additional advancements

Additional retrofit design features continue to be developed to improve the performance and reduce maintenance costs of Cyclone boilers. These features, based on site-specific needs, include:

1. contoured flat studs designed for the tubes that make up the Cyclone slag tap,
2. improved seal welding design,
3. tighter gapped and extra reduced size re-entrant throats,
4. continued wear block development,
5. stainless steel cooling water jackets for burners, and
6. improved radial and vortex burner design elements.

Air pollution control

Flyash

To meet modern particulate emission standards, a Cyclone equipped boiler precipitator must be about the same size as that for a pulverized coal unit. Fabric filters have also been applied. However, the production of more slag and less flyash can offer a significant disposal benefit. Furnace slag is much easier to dispose of than flyash in most cases. The constituent minerals are tightly bound in Cyclone slag, minimizing any leachate problems if landfilled. Cyclone slag physical properties have led to its reuse as road bed fill, shingles in the building industry, and sandblasting material (among other applications).

Fig. 16 State-of-the-art flat studs after five years of service.

Sulfur dioxide (SO_2) reduction

As with other combustion systems, SO_2 emissions from Cyclone units are a function of the sulfur content in the fuel. One frequently used method for reducing SO_2 emissions is to switch to lower sulfur coal. Changing fuels typically requires a complete review of the impacts on boiler operation (e.g. slagging, fouling, sootblower capability, main and reheat steam temperatures, economizer outlet gas temperature, flyash collection, etc.). In the case of Cyclone furnace boilers, additional care is required in such a change because Cyclone furnaces are more sensitive to the ash composition, ash quantity, heating value and especially moisture in the fuel. If high moisture subbituminous or lignite fuels are to be substituted for low moisture bituminous coal, installing the direct-fired pre-dry system could be considered to optimize performance. This system has been successfully used for lignite coal-fired Cyclone furnaces.

Provided that the fuel selected meets the T_{250} value discussed earlier, a standard Cyclone equipped boiler can potentially be modified to handle low sulfur, high moisture subbituminous or lignite coals through a series of hardware and operational changes. Key changes focus on improving and maintaining overall coal fineness, maximizing Cyclone furnace temperatures as high as possible, continuing an aggressive maintenance program, and making provisions for rapid deslagging. The actions required are unit specific and are usually established following an engineering study and test fuel burn.

Coal washing systems may be installed to remove pyritic sulfur. Although washing raises the T_{250} value and makes the coal more difficult to burn, SO_2 emissions will be reduced.

Nitrogen oxides (NO_x) reduction

Historically, Cyclone equipped boilers have produced relatively high uncontrolled levels of NO_x emissions, ranging from 0.70 to 2.6 lb of NO_x as NO_2 per million Btu (approximately 1050 to 3900 mg/Nm^3 corrected to 3% O_2). Based on ever increasing needs to reduce NO_x emission levels, various commercially available Cyclone equipped boiler NO_x reduction techniques are available and include, but are not limited to:

1. fuel switching from a bituminous to a subbituminous type coal,
2. air staging,
3. reburning,
4. selective catalytic reduction (SCR) technology, and
5. selective non-catalytic reduction (SNCR) technology.

Switching from a standard Cyclone bituminous coal to a subbituminous coal has shown that a nominal 10 to 25% NO_x reduction can be observed. These reduction levels are due to two factors:

1. reduced thermal NO_x component with the high moisture content of the fuel (lower Cyclone peak operating temperatures), and
2. lower fuel NO_x component due to the inherent constituents in the coal (lower fixed carbon to volatile ratio, lower nitrogen component, etc.).

The most widely used Cyclone boiler NO_x control technique is air staging. The basic theory of air staging is to reduce the fuel NO_x component within the burner zone by reducing oxygen availability. Additionally, numerical modeling activities have shown that due to the unique combustion characteristics of Cyclone equipped boilers, some reburning chemical reactions also help to reduce the overall NO_x emissions during air staging operation. This reburning phenomenon helps to explain the high percentage NO_x reduction that has been identified on numerous staged-air applications on Cyclone-fired units (up to 80% reduction).

Cyclone air staging employs multiple combustion zones within the furnace region (see Fig. 17), defined as the main combustion zone (Cyclone region) and the burnout zone (OFA ports to the furnace exit). The main combustion zone is designed to operate at nominal substoichiometric conditions. Operating at higher Cyclone stoichiometries is feasible, but at a cost of minimizing the overall NO_x reduction capabilities; actual operating Cyclone stoichiometries would be optimized during startup to provide the required NO_x reduction while maintaining optimum boiler operation. The balance of the required combustion air is introduced through overfire air (OFA) ports in the burnout zone. A satisfactory residence time within the burnout zone is required for complete combustion.

Experience indicates good NO_x reduction without major negative operational impacts while air staging on Cyclone units. However, there are potential negative issues that must be addressed. Reducing atmospheres accelerate corrosion, which can lead to significant maintenance problems. Corrosion and iron formation concerns are possible while operating under reducing/oxidizing conditions. Utilizing lower sulfur/iron fuels, such as lignite or low sulfur subbituminous coals, will help reduce, but not eliminate, corrosion and iron formation concerns. Additional operational issues may include higher unburned carbon levels, increased boiler flyash, increased boiler slagging/fouling, steam temperature reductions, and higher opacity levels.

Low NO_x Cyclone reburn technology has been specifically developed to reduce NO_x emissions levels from Cyclone equipped boilers (up to 70% reduction) while permitting successful Cyclone furnace operation. In this system, the Cyclone furnace can be operated under fully oxidizing conditions, but at a reduced load – typically 70 to 85% of full load air and fuel flow. The balance of the fuel is injected directly into the main boiler furnace with minimum air for fuel transport to create a reburning zone. In this zone, the reburn fuel creates an oxygen deficient or reducing zone where the NO_x created in the Cyclone furnace is reduced or decomposed into molecular nitrogen through a series of complex interactions with free hydrocarbon radicals. Overfire air ports located above the reburn zone permit injection of the balance of the air to produce a final stoichiometry of 1.15 to 1.20 and complete the fuel combustion. The use of coal, oil, or gas as the reburn fuel depends upon an economic evaluation balancing the higher capital cost of a coal based system against the fuel cost differential for other fuels.

Post-combustion technologies have also been successfully applied to Cyclone boilers and offer varying degrees of NO_x control capability (see Chapter 34).

Applications

The Cyclone boilers gained wide acceptance due to their ability to burn a substantial reserve of coals deemed unsuitable for pulverized coal firing. Also, the unique inherent combustion characteristics of Cyclone furnaces have proven to be an effective means to safely fire a variety of waste materials. New Cyclone boiler applications co-firing sludge materials have been and continue to be developed.

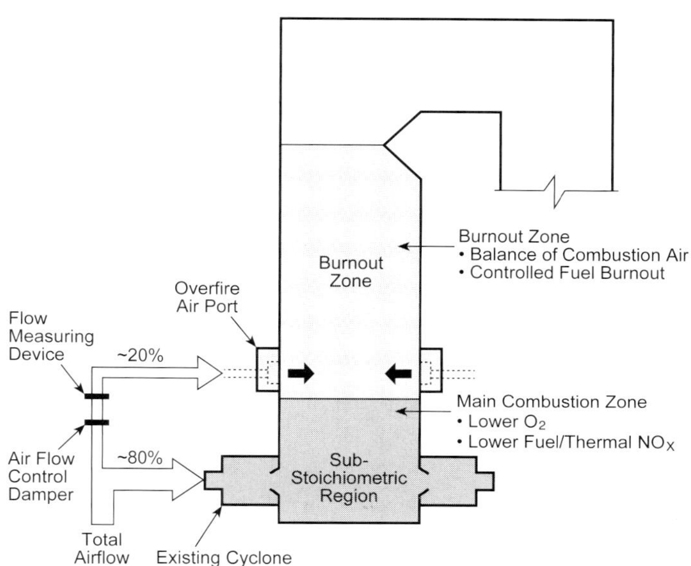

Fig. 17 Cyclone-equipped boiler air staging NO_x control system.

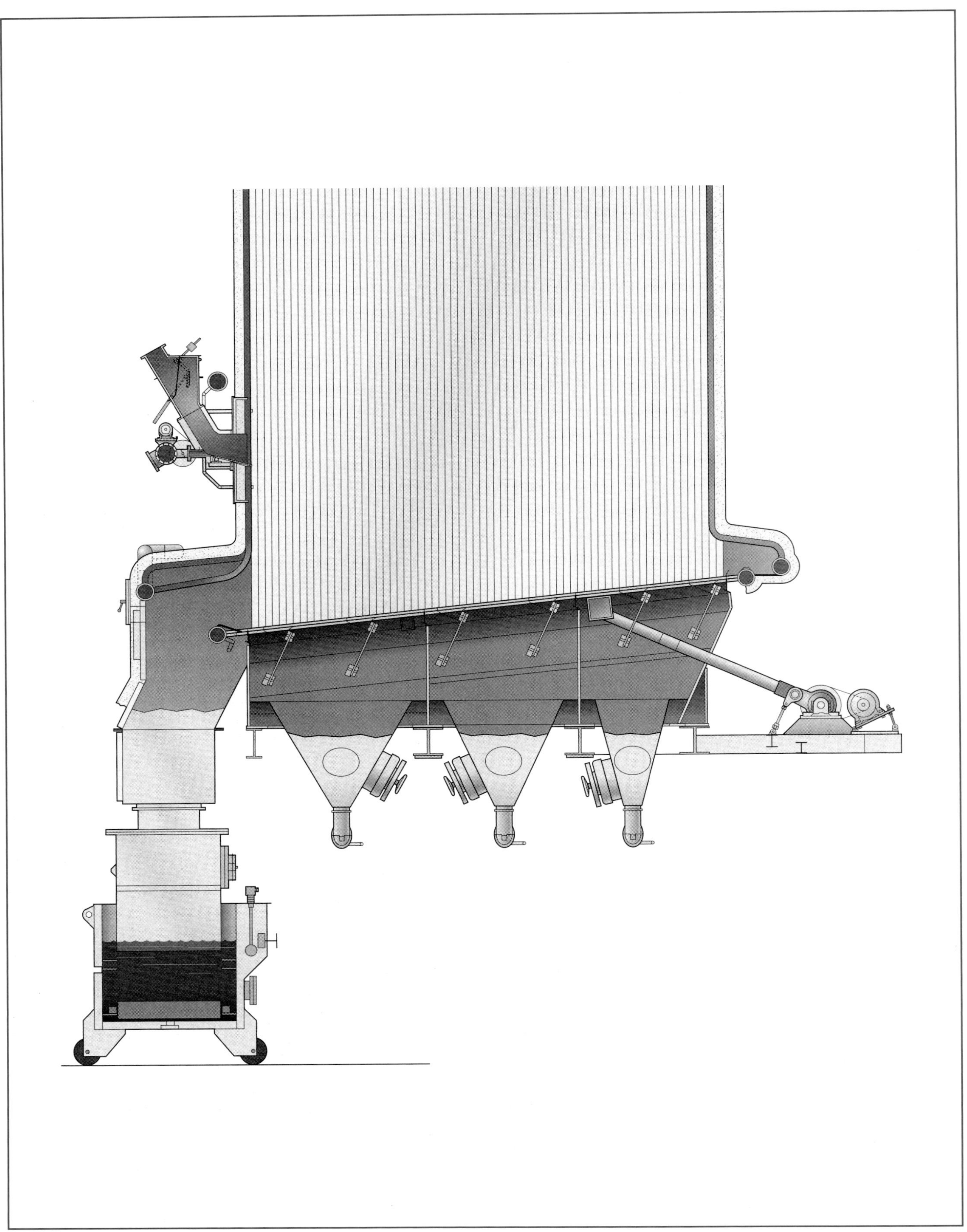

Vibrating grate stoker.

Chapter 16

Stokers

Many technologies have evolved to convert a wide variety of fuels to alternate forms of energy. One important technology is the mechanical stoker. All stokers are designed to feed fuel onto a grate where it burns with undergrate air (UGA) passing up through it, and overfire air (OFA) passing over it. The grate is located within the lower furnace and is designed to remove the ash residue after combustion. Evolving from the hand-fired boiler era, mechanical stoker/grate system designs are available to burn a wide range of fuels for industrial, small utility and cogeneration applications. In addition to burning all forms of coal, other fuels burned include sludge, wood waste and biomass (e.g., wood, bark, straw, bagasse, rice hulls, peach pits, almond shells, orchard prunings, coffee grounds), as well as residential, agricultural, industrial and commercial refuse. Modern mechanical stoker firing systems are composed of:

1. a *stoker* or fuel feeding system,
2. a stationary or moving *grate* assembly to support the burning mass of fuel and admit undergrate air to the fuel,
3. an OFA system to complete combustion and limit atmospheric pollutant emissions, and
4. an ash or residual discharge system.

These components are then integrated into the overall furnace design to optimize combustion and heat recovery while minimizing unburned fuel, atmospheric emissions and cost.

Despite the functional differences stated above, it is not uncommon to hear or see the term stoker referring to not only the feeding device but also the entire system, including the grate and air system. The feeding device can also be referred to as a *feeder*, *distributor* or *spout*, as well as a stoker.

A successful installation requires selecting the correct type and size of stoker for the fuel being used and for the load conditions and capacity being served. There are two general types of systems: *underfeed* and *overfeed*. Underfeed stokers supply both the fuel and air from under the grate while overfeed stokers supply fuel from above the grate and air from below the grate. Overfeed stokers are further divided into two types: *mass feed* and *spreader*. In the mass feed stoker, fuel is continuously fed to one end of the grate surface and travels horizontally across the grate as it burns. The residual ash is discharged from the opposite end. Combustion air is introduced from below the grate and moves up through the burning bed of fuel. In the spreader stoker, combustion air is again introduced primarily from below the grate but the fuel is thrown or spread uniformly across the grate area. The finer fraction of the fuel burns in suspension as it is lifted by the upward moving flue gas flow. The remaining heavier fraction of the fuel lands and burns on the grate surface with any residual ash removed from the discharge end of the grate.

There is little demand in today's market for the underfeed and small mass overfeed coal-fired units because of cost and environmental considerations. This market has been replaced with shop assembled oil- and gas-fired units and to some extent, by overfeed spreader stoker systems.

The stoker/grate systems are provided in many mechanical configurations depending upon the manufacturer. Table 1 summarizes several variations of basic stoker designs by type, fuel, heat release rate and approximate largest capacity available. Grate heat release rate is the fuel input divided by the *active* or *effective* area of the grate upon which fuel burning is intended to occur.

For a given boiler steam capacity, the typical fuel burning rates in Table 1 generally determine the plan area of the grate and furnace in which it is installed. Practical considerations limit stoker size and, consequently, the maximum steam generation rates. For coal firing, this maximum steam generation rate is about 390,000 lb/h (49.1 kg/s); for wood or biomass firing it is about 900,000 lb/h (113.4 kg/s).

Almost any coal can be burned on some type of stoker. Many other solid fuels such as refuse and biomass can be burned alone on a grate or in combination with another fuel such as pulverized coal, oil or natural gas.

The spreader stoker, in combination with the various grate types, is most commonly used for a steaming capacity range from 75,000 to 900,000 lb/h (9.5 to 113.4 kg/s). It responds rapidly to changes in steam

Table 1
Stoker/Grate System Overview

Stoker Type	Grate Type	Fuel	Typical Release Rate* 1000 Btu/h ft² (MW$_t$/m²)		Steam Capacity 1000 lb/h (kg/s)	
Underfeed:						
Single retort	—	Coal	425	(1.34)	25	(3.15)
Double retort	—	Coal	425	(1.34)	30	(3.78)
Multiple retort	—	Coal	600	(1.89)	500	(63.0)
Overfeed:						
Mass	Vibrating: water-cooled	Coal	400	(1.26)	125	(15.8)
	Vibrating: water-cooled	Straw	500	(1.58)	320	(40.3)
	Traveling chain	Coal	500	(1.58)	310	(39.1)
	Reciprocating	MSW**	300	(0.95)	350	(44.1)
Spreader	Vibrating: air-cooled	Coal	650	(2.05)	150	(18.9)
		Wood	1100	(3.47)	900	(113.4)
	water-cooled	Wood	1100	(3.47)	900	(113.4)
	Traveling	Coal	750	(2.37)	390	(49.1)
		Wood	1100	(3.47)	550	(69.3)
		RDF***	750	(2.37)	400	(50.4)

* Specific fuel characteristics and firing arrangements may allow higher heat release rates.
** Municipal solid waste
*** Refuse-derived fuel

demand, has good turndown capability and can use a wide variety of fuels. It is not, however, suitable for low volatile fuels such as anthracite and petroleum coke because of carbon burnout problems.

Underfeed stokers

There are two general types of underfeed stokers: the horizontal feed side ash discharge type shown in Fig. 1 and the gravity fed rear ash discharge type shown in Fig. 2.

In the side ash discharge type, coal is fed from a hopper to a central trough, called a *retort*, by a screw or ram pusher. Air is admitted through the tuyères or air nozzles as shown. In the larger units, a ram assisted by pusher blocks or a sliding retort bottom (fuel distributors) moves the fuel upward and into the retort. As the coal moves upward and over the retort edges and spreads out over the active grate area, it is exposed to air and radiant heat. Drying occurs and distillation of volatiles begins. As the coal moves to the sides and/or rear, the distillation is completed, leaving coke which is burned out near the edges or end of the grate. High pressure OFA is used to produce high turbulence and reduce smoke.

Burning coal in this fashion increases the probability of *clinkering* (producing large agglomerates of ash slag) or *matting* (layers of ash slag). To reduce this tendency, alternate fixed and moving grate sections are applied to the underfeed stoker design to agitate the fuel. Coal characteristics are critical to underfeed stoker performance. Table 2 outlines coal specifications for stationary and moving grates, although underfeed stokers have burned coals outside of these guidelines. To burn these coals, some deviation from the normal maximum grate release rate may be required. A reduction in the percentage of fines helps to keep the feed bed porous and extends the range of coals with a higher coking index.

With suitable coal, single and double retort units are generally limited to 25,000 to 30,000 lb/h (3.2 to 3.8 kg/s) steam flow. Typical grate release rates are 425,000 Btu/h ft² (1.34 MW$_t$/m²) with water-cooled walls and 300,000 Btu/h ft² (0.95 MW$_t$/m²) with refractory walls. Capacities up to 500,000 lb/h (63 kg/s) steam are possible with multiple retort rear ash discharge units similar to that shown in Fig. 2. Grate release rates up to 600,000 Btu/h ft² (1.89 MW$_t$/m²) are practical with a 20 to 25 deg (0.35 to 0.44 rad) grate inclination from the horizontal.

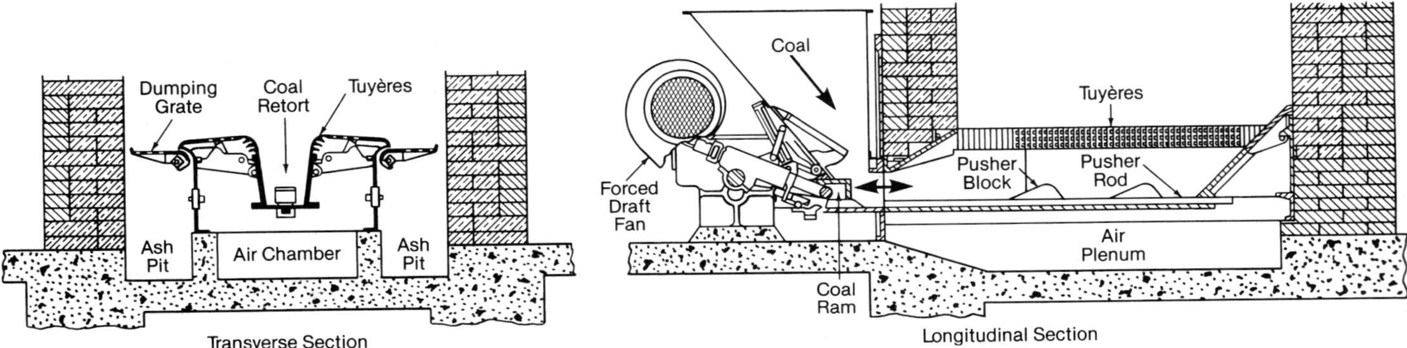

Fig. 1 Single-retort underfeed stoker with horizontal feed, side ash discharge.

Fig. 2 Multiple-retort, gravity-fed underfeed stoker with rear ash discharge.

Mass feed stokers

Two types of mass feed stokers are used for coal firing: the water-cooled vibrating grate and the moving (chain and traveling) grate stokers. Another mass feed stoker system used for municipal refuse is discussed in Chapter 29.

Mass feed stokers are characterized by the gravity feed of fuel onto a stoker via an adjustable gate that controls fuel bed height. The method of firing involves a fuel bed that moves along a grate with air being admitted under the grate perpendicular to the fuel flow. As it enters the furnace, the layer of coal is heated by furnace radiation to drive off volatiles and to promote ignition. The coal continues to burn as it is conveyed along the depth of the furnace. The fuel bed decreases in thickness until all the fuel has burned and cool ash discharges into a pit. With this method of fuel entry, the undergrate air must be sectionalized along the length of the grate because the quantities of air required for ignition, burning and burnout are different and must be regulated. This method of fuel entry and combustion inherently produces low ash carryover. However, it is more sensitive to variations in fuel characteristics that affect ignition without a larger ignition arch (discussed below). Mass feed stokers require nonsegregating coal feed hoppers. Without them, the fines migrate to the side walls and severe clinkering can occur.

A water-cooled vibrating grate stoker is shown in Fig. 3. The vibrating grate consists of tuyère grate surface mounted on and in intimate contact with a grid of water tubes. The grate is connected to the boiler or feedwater circulation system for cooling. The entire structure is supported by a number of flexing plates that allow the grid and its grate to move freely in a vibrating action that conveys the coal from the feed hopper to the ash discharge. Vibration of the grates is intermittent and is adjustable to convey fuel along its length and control ash bed thickness and discharge as needed. A rear arch extends over approximately the rear third of the grate as shown in Fig. 3. It assists burnout and directs the higher excess air gases forward to mix with the rich volatile gases from the ignition zone. A short front arch is adequate for most coals. If ignition is inadequate due to low volatile fuel, refractory can be added to the short arch to increase radiation and assist ignition. This is referred to as an ignition arch.

High pressure air, up to 30 in. wg (7.5 kPa), is injected through the front arch to promote turbulence and combustion. Water cooling of the grates makes this stoker more flexible with gaseous and liquid auxiliary fuels, because a shift to either does not require special grate protection other than the normal bed of ash left from coal firing. Burning rates of these stokers vary with different fuels; in general, the grate heat release rate should not exceed 400,000 Btu/h ft^2 (1.26 MW_t /m^2). Due to the limited number of moving parts, this stoker/grate system typically requires little maintenance.

Chain and traveling grate stokers, shown in Fig. 4, are similar to each other. Both form an endless belt arrangement that passes over drive and idler sprockets or return bends. They both convey coal from the hopper through the furnace to the ash discharge and return under the grate. In the chain grate design, the chain continues around to the return side. The traveling grate unit differs in that it uses a grate bar to provide better control of siftings (fine ash falling through the grate) when firing anthracite. The moving chain stoker requires more maintenance than the water-cooled vibrating grate.

Chain and traveling grate stokers can burn a wide range of solid fuels including peat, lignite, subbituminous, bituminous and anthracite coals and coke breeze. Typical coal characteristic ranges are provided in Table 2. Generally, these stokers use furnace arches (front and/or rear, not shown in Fig. 4) to improve combustion by reradiating heat to the fuel bed. When burning low volatile anthracite or coke breeze, rear arches direct the incandescent fuel particles and combustion gases toward the front of the stoker, where they assist ignition of the incoming fuel.

Burning rates on chain and traveling grate stokers vary with different fuels. The lower ash (8 to 12%) and lower moisture (10%) fuels permit rates to 500,000 Btu/h ft^2 (1.58 MW_t/m^2); higher moisture (20%) and higher ash (20%) fuels would limit the rate to 425,000 Btu/h ft^2 (1.34 MW_t/m^2). For low volatile anthracite, the rate should not exceed 350,000 Btu/h ft^2 (1.10 MW_t /m^2).

Spreader stokers

In the spreader stoker, the fuel is uniformly thrown into the furnace across the grate area. Fines ignite and burn in suspension; the coarser fuel particles fall to

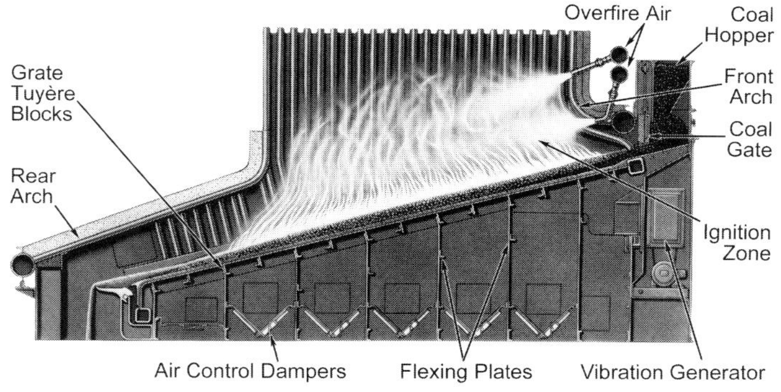

Fig. 3 Water-cooled vibrating grate stoker.

Table 2
Typical Coal Characteristics for Different Stoker Types

	Undergrate Feed Systems		Mass Feed Systems		Spreader Systems
	Stationary Grate	Moving Grate	Water-Cooled Vibragrate	Chain/ Traveling Grate	All
Moisture	0 to 10%	0 to 10%	0 to 10%	0 to 20%	25% max.***
Volatile matter	30 to 40%	30 to 40%	30 to 40%	30 to 40%	18% min.
Fixed carbon	40 to 50%	40 to 50%	40 to 50%	Remainder	65% max.
Ash	5 to 10%	5 to 10%	5 to 10%	6 to 20%	15% max.
Btu/lb (kJ/kg), as fired	12,500 (29,075) min.	12,500 (29,075) min.	12,500 (29,075) min.	10,500 (24,423) min.	—
Free swelling index	5 max.	7 max.	—	5 max.	—
Ash softening temperature*	2500F (1371C)**	2500F (1371C)**	2300F (1260C)	2100F (1149C)	2000F (1093C) min.
Coal size, in. (mm)	1 (25.4) top size x 0.25 (6.4)	Equal portions: -0.25, 0.25 to 0.5, 0.5 to 1 (-6.4, 6.4 to 12.7, 12.7 to 25.4)	1 to 0.75 (25.4 to 19.1) x 0	1 to 0 (25.4 to 0) top size	1.25 (31.8) max. top size; 0.75 (19.1) min. top size
Max. through 0.25 (6.4) screen	20%	—	40%	60%	40%
Iron oxide, % in ash	—	—	20% max.	20% max.	—

* The ash softening temperature here is the temperature at which the height of a molten globule is equal to half its width under reducing atmosphere conditions.
** Below 2500F (1371C) the moving grate is derated linearly to 70% of its rated capacity at 2300F (1260C) ash fusion temperature. Stationary grates are derated linearly to 70% at 2100F (1149C) ash fusion temperature and use steam for tempering below about 2400F (1316C) fusion temperature.
*** Higher moisture may require preheated combustion air.

the grate and combust on a thin, fast burning bed. Because the fuel is evenly distributed across the active grate area, the air is uniformly distributed under and through the grate. The undergrate air plenums may or may not be compartmented depending on the grate type and application. A portion of the total combustion air is admitted through ports above the grate as overfire air. The modern spreader stoker is the most versatile and the most commonly used stoker system.

Spreader coal firing

Fig. 5 shows a modern boiler equipped with a traveling grate spreader stoker and designed to fire an eastern bituminous coal. A straight, water-cooled membrane furnace wall construction minimizes refractory. For the typical stoker coal-fired application, ignition or combustion arches are not used. The installation consists of:

1. state-of-the-art feeder-distributor units that distribute fuel uniformly over the grate,
2. specifically designed air metering grates,
3. dust collection and reinjection equipment,
4. a combustion air system including forced draft fans for undergrate and overfire air, and
5. combustion controls to coordinate fuel and air supply with steam demand.

Feeders for spreader stokers

Spreader feeders have the capability to uniformly feed coal into a device that can propel it along the depth of a grate in an evenly distributed pattern. Many designs have been used successfully over the

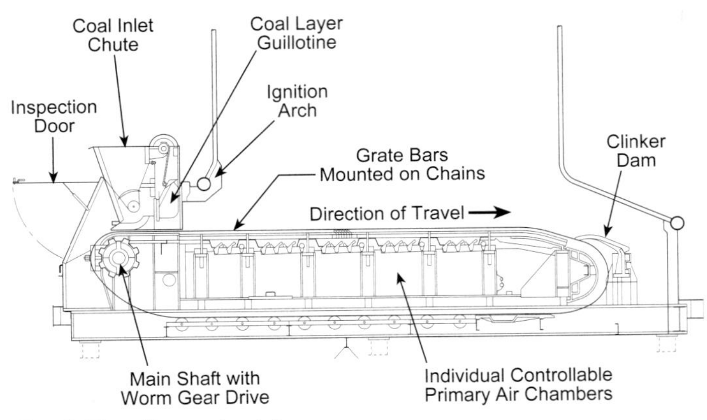

Fig. 4 Traveling grate stoker.

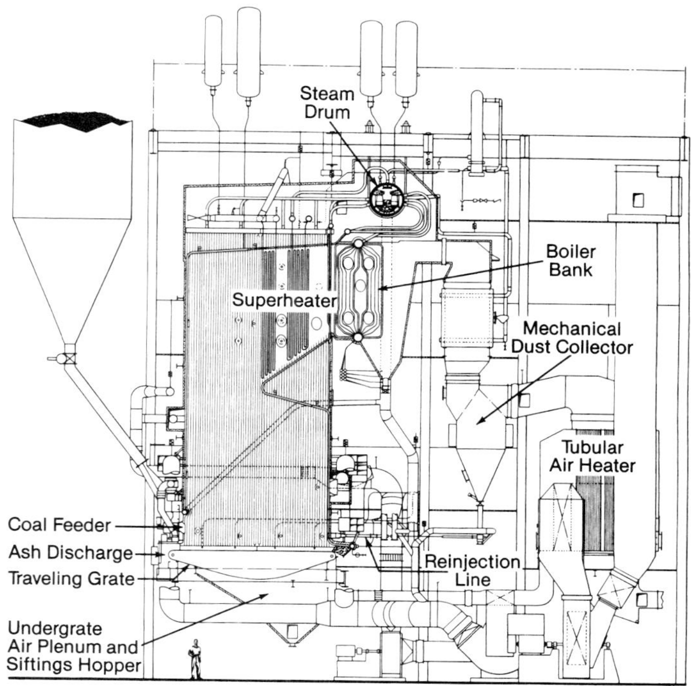

Fig. 5 Typical spreader stoker coal-fired boiler; 290,000 lb/h (36.5 kg/s) steam.

years. The coal feed mechanisms include gravity, reciprocating plates and metering chain conveyors. The mechanisms that propel the coal into the furnace include steam and air injection as well as underthrow and overthrow rotors. Steam or air assist can be used with the rotor systems. Fig. 6 shows a feeder-distributor with overthrow rotor. The metering chain moves coal from a small hopper to an overthrow rotor. The rotor is equipped with blades shaped to uniformly distribute coal over the grate area. Although spreader designs vary, the overthrow design has been the most common.

With increasingly stringent emissions regulations, stoker manufacturers have been studying design variations that reduce the formation of nitrogen oxides (NO_x) and carbon monoxide (CO). Fig. 7 depicts a state-of-the-art mechanical-pneumatic feeder-distributor with underthrow rotor. In this device, the coal is fed from a hopper by a metering chain into an underthrow rotor with air assist. This has improved the feed and distribution capabilities, and 10 to 15% reductions in NO_x have been recorded compared to the overthrow design.

Grates for spreader stokers

As with mass stokers, there has been a wide range of grates used in spreader stoker coal firing. Stationary and dumping-type units are no longer installed as new equipment. Traveling and/or vibrating air-cooled grates are the most common. The traveling grate shown in Fig. 5 is a high resistance air metering grate. The resistance air metering concept eliminates the need for undergrate air plenum compartmentation for good air distribution and control. Moving adjustable air seals are provided at the front and rear. The grates are bottom-supported and therefore require an expansion joint at the interface to the water-cooled furnace.

The grate is a continuous moving chain. It consists of a series of chains to which are attached the grate bars that contain the air metering holes. The grate bars are contoured to interface with adjacent bars, minimizing air leakage between bars. The chains and grate bars are supported by, and slide on, grate rails. The sliding interface between the grate bar and rail also serves as a seal to prevent excessive air from bypassing the air metering holes in the grate. The grate travels from rear to front, or towards the fuel feed end. This permits the optimum fuel distribution pattern and the maximum residence time for burnout of larger fuel particles. The ash is continuously conveyed and spills off the end of the grate into a hopper.

Although the traveling grate is a durable and proven design, it has many moving parts and is subject to wear. To minimize wear, the speed of the grate should generally be kept below 40 ft/h (12.2 m/h). This may limit its use in some high ash fuel applications or it may require limiting the grate plan area release rate and input per unit width of stoker.

The grate is typically driven through the front sprocket and shaft to keep the top of the chain in tension; mechanical and hydraulic drive systems are available. Some systems use a ratchet concept, which results in loading and unloading the chain at each stroke. This leads to higher grate component wear than does a continuous and uniformly torqued drive system.

Other types of traveling grates are similar to that shown in Fig. 4. Many of these consist of a chain with several links to form an endless belt. Air admission is through the gaps in the links. These types of traveling grates typically have compartmented undergrate air plenums to control air distribution.

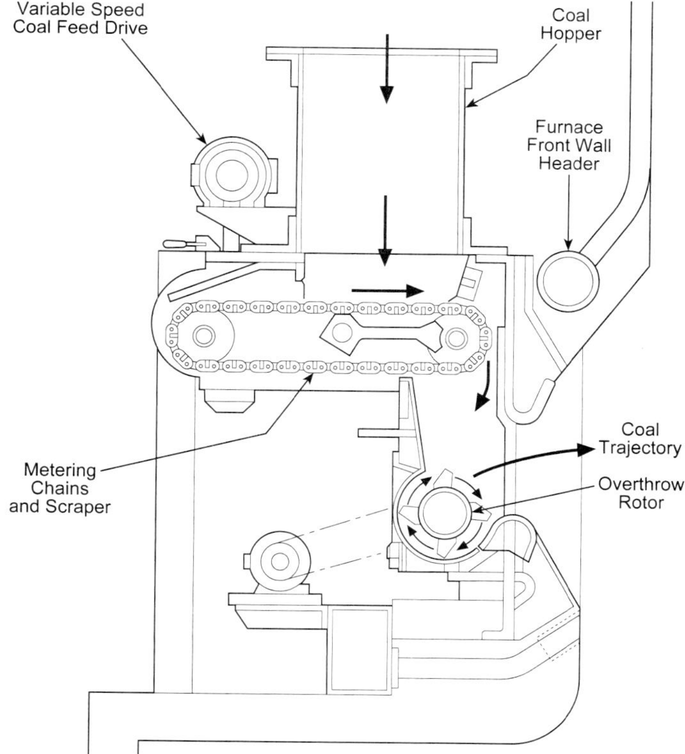

Fig. 6 Feeder-distributor with overthrow rotor.

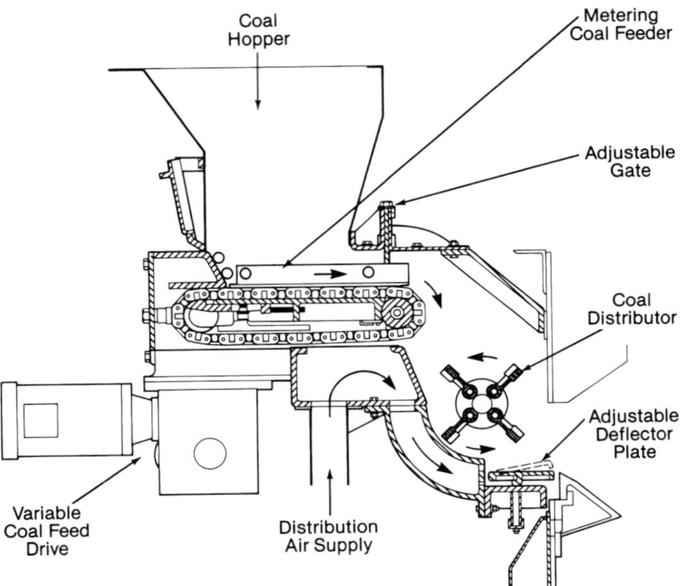

Fig. 7 Chain-type coal feeder with underthrow rotor (*courtesy of Detroit Stoker Company*).

Air-cooled vibrating grates, similar to those shown in Fig. 3, are also used in spreader stoker applications. Except for a smaller size limitation, the application to spreader coal firing is similar to that of the traveling grate.

Carbon reinjection systems

The high degree of suspension burning results in greater carryover of partially combusted fuel particles. To achieve the highest boiler efficiency, these particles are captured in a dust collector, as well as from generating bank, economizer and air heater hoppers, and returned to the furnace for complete combustion. The carbon reinjection system on the unit in Fig. 5 is pneumatic. The carbon particles and some ash are captured by a mechanical dust collector and routed to a pickup box, where air is injected to convey it back to the furnace. Multiple injection ports are located across the width of the unit to uniformly mix the unburned carbon into the combustion zone for enhanced burnout. Reinjection of significant ash is undesirable because it can contribute to boiler surface erosion and grate clinkering. Because a mechanical dust collector is more effective in collecting larger particles, and because the majority of the unburned carbon is the larger size fraction of the total carryover, the bias towards collecting the unburned particles can be effected by limiting the design efficiency of the mechanical collector. Carbon reinjection improves coal-fired boiler efficiency by 2 to 4%.

Combustion air system

In spreader stoker-fired units, 25% excess air is typically provided for stoker combustion at the design full load input. This air is split between the undergrate air, OFA and fuel distribution air. Due to the high degree of suspension burning, air is injected over the fuel bed for mixing to assist fuel burnout and to minimize smoking. This dictates that 15 to 20% of the total air be used as OFA. This air is injected at pressures of 15 to 30 in. wg (3.7 to 7.5 kPa) through a series of small nozzles arranged along the furnace front and rear walls. (See Fig. 8.)

Spreader stoker firing, with the air split between undergrate and overfire, is a form of staged combustion and is effective in reducing NO_x. By deeper staging, further reductions are possible. To take advantage of this characteristic, the modern unit shown in Fig. 5 is designed for a total excess air of 25%. The total air flow is split 65% undergrate and 35% overfire, which includes any air to the coal feeders. An additional level of overfire air nozzles is installed in the furnace front and rear walls above the coal feeder. (See Fig. 8.) This permits deeper staging and delays adding the remainder of combustion air until the hot fuel bed gases have radiated some heat to the waterwalls and are at a lower temperature. The OFA is admitted at a maximum static pressure of 30 in. wg (7.5 kPa) through nozzles designed for high penetration and mixing.

Spreader stoker coal characteristics

As noted earlier, the spreader method of feeding and combusting coal is versatile. It can operate satisfactorily on the full range of coals from lignite to bituminous. Fuels having less than 18% volatile matter are not generally suitable. Table 2 summarizes the range of coal properties for spreader stoker application.

Bituminous coals readily burn on a traveling grate without the need for preheated air. However, an air heater may be required in the unit design for improved efficiency. In these instances, the design air temperature should be limited to below 350F (177C). The use of preheated air may limit the selection of fuels to the lower iron, high fusion coals to prevent undesirable slagging and agglomerating on the grate. The use of preheated air at 350 to 400F (177 to 204C) is necessary for the higher moisture subbituminous coals and lignites.

Higher ash coals can also be satisfactorily burned. However, to keep grate speeds reasonable, it may be necessary to lower the grate heat release rate and/or reduce the input per unit frontal width.

Coal size segregation can present a problem on any stoker, but the spreader is more tolerant because the

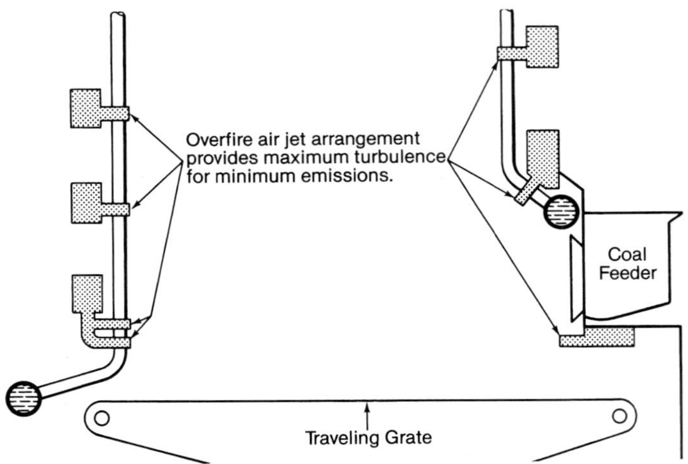

Fig. 8 Stoker overfire air system.

feeder rates can be adjusted; much of the fines burn in suspension and not on the bed. Proper selection of the coal feed equipment can significantly reduce the tendency for coal size segregation.

Selection of a spreader stoker

Spreader stoker-fired traveling grates for coal are typically designed with release rates summarized in Table 1. Higher heat release rates may be allowed for specific fuel characteristics and firing arrangements. The air-cooled vibrating grate is limited to 650,000 Btu/h ft^2 (2.05 MW$_t$/m^2). The length of the grate is limited to that over which the coal can be distributed. For vibrating grates, the length limit is 18 ft (5.5 m); for the traveling grate, it is 23 ft (7.0 m). The width of the grate is the main variable in providing sufficient total grate area. Sufficient width is also required to install enough feeders and to keep the heat input per foot of width below about 13.5 × 10^6 Btu/h ft (13.0 MW$_t$/m). A grate heat release rate of 750,000 Btu/h ft^2 (2.37 MW/m^2) or less also applies with full reinjection from a mechanical dust collector. Higher inputs tend to increase slagging potential and may cause excessive fuel entrainment and carryover.

Traveling grates are available up to 19 ft (5.8 m) wide with a single driveshaft. For higher capacities, two opposite hand grates provide the desired width. With a practical grate width limit of 38 ft (11.6 m) and a length limit of about 23 ft (7.0 m), the largest practical grate area is about 880 ft^2 (81.8 m^2).

Ash removal

When properly sized and operated, the ash discharging from a spreader stoker-fired unit is relatively cool and clinker free. Ash discharges into a hopper and may be removed by a conventional ash transport system, generally without the need for clinker grinders.

Spreader stoker firing of bark, wood and other biomass fuels

Several modern boiler designs are available to meet customer and project-specific requirements to fire bark, wood and other biomass fuels using a spreader stoker design. Fig. 9 shows a top supported Stirling® power boiler with a vibrating grate stoker. Fig. 10 shows a bottom-supported power boiler also with a vibrating grate. Both include state-of-the-art air-swept spouts in widths and numbers required to uniformly distribute the fuel over the grates. These units feature forced draft fans for undergrate and overfire air, mechanical dust collectors, and combustion controls to coordinate fuel and air supply with steam demand.

Several furnace configurations are used to meet project requirements and are available in either top or bottom-supported configurations. Fig. 9 shows the furnace with two arches separating the furnace into lower and upper zones (CCZ or Controlled Combustion Zone furnace) while Fig. 10 shows a straight wall furnace without arches, but including an advanced overfire air injection system. Selection of furnace type depends upon project requirements, fuel characteristics, and detailed numerical combustion modeling analyses.

Bark distributor feeders

Fig. 11 shows a modern air-swept spout. Bark is fed from a metering bin through a chute into the inlet of the spout. High pressure 20 in. wg (5.0 kPa) distribution air is introduced through an annulus across the width of the air-swept spout. In combination with the momentum of the falling bark, the air propels the bark into and across the grate depth. The distribution air is pulsed by rotating dampers or by a preprogrammed distributor to spread the bark uniformly fore-and-aft across the grate depth. A portion of the spout air is injected through small nozzles located at the furnace-side corners of the spout. This air assists the burning of fines and is adjustable to suit the fuel conditions.

The discharge of the air-swept spout is shaped to produce the optimum trajectory of the fuel particles. Some vendors provide an adjustable plate to control the discharge and subsequent distribution; some also supply mechanical bark feeders having features similar to those depicted earlier for stoker coal firing. Sufficient spouts are installed across the width of the unit to feed the necessary quantity of bark and to control the side-to-side distribution and grate coverage.

Of equal importance is the system that feeds the fuel to the air-swept spouts. Bark is conveyed across a series of smaller metering bins in a quantity greater than that needed for combustion; the excess is returned to storage. The individual metering bins are equipped with screw feeders that convey the bark in accordance with fuel demand. They are capable of

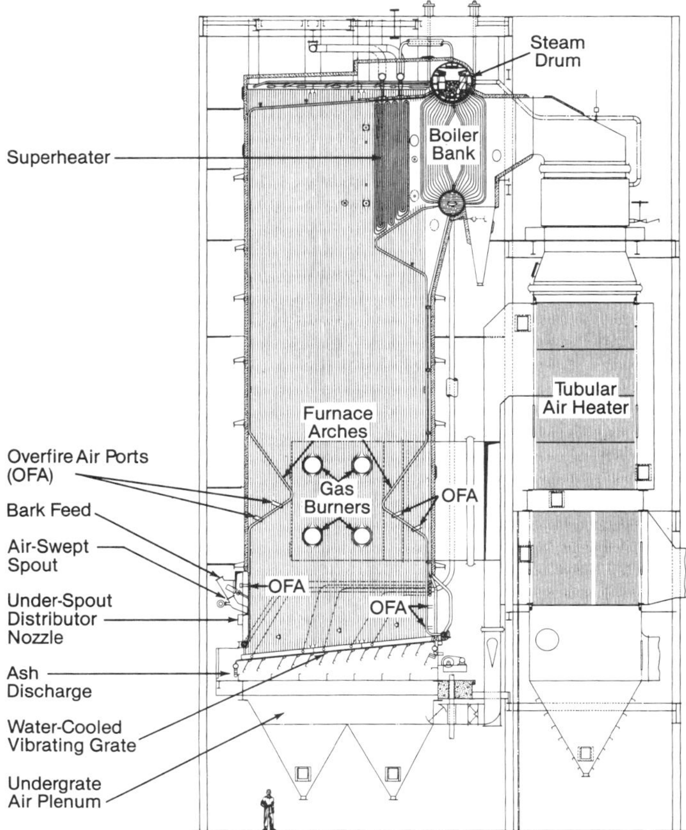

Fig. 9 Typical wood-fired stoker boiler with Controlled Combustion Zone (CCZ) furnace.

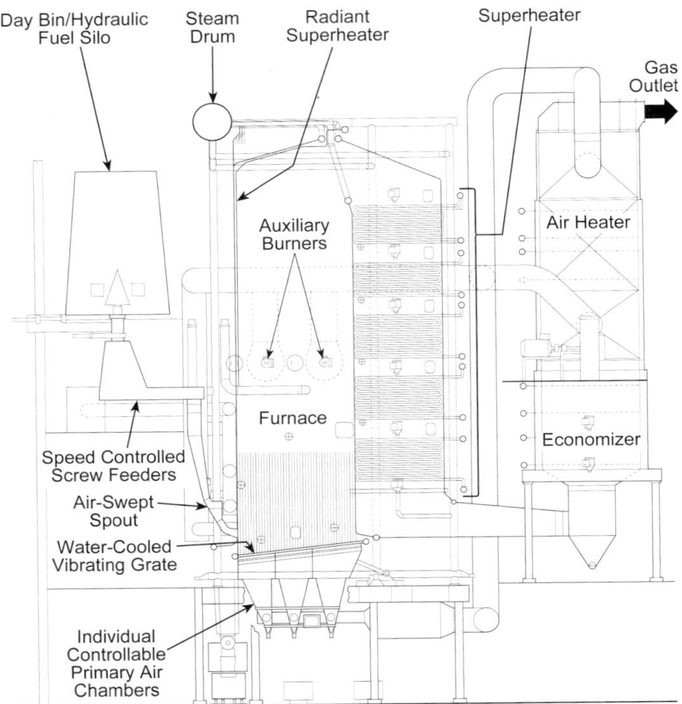

Fig. 10 Bottom-supported boiler with vibrating grate stoker – biomass firing.

speed biasing across the furnace width. This system minimizes fuel shortages at the boiler front, uses smaller bins, has a reduced tendency to have the fuel hang up or bridge in the bins and hoppers, and provides consistent fuel size distribution to the bins and across the width of the unit. Other types of feed systems, such as large live-bottom bins, are successful when applied with the proper fuel characteristics. Some bark and other biomass fuels are stringy and can agglomerate and segregate. These characteristics must be considered when designing feed systems for the air-swept spout.

Grates for firing bark and other biomass fuels

A traveling grate for bark firing is very similar to that used for spreader stoker coal firing. The bark is fed over the fuel bed and distributed uniformly across the grate area. This design is essentially an air-cooled grate; therefore, it is important to retain a layer of ash on the grate to shield the grate bars from furnace radiation. When firing low ash barks, it is common to operate the grates intermittently so that an inventory of ash can build and be retained. In addition, high alloy grate bars having higher resistance to thermal degradation can be used.

Bark is a high volatile fuel that, in combination with the fines, has a high degree of suspension burning. High volatile and low ash characteristics permit sizing traveling grates with heat release rates to 1,100,000 Btu/h ft^2 (3.47 MW$_t$/m^2). The width limitation of dual traveling grate stokers is also about 38 ft (11.6 m) and the depth is mechanically limited to about the equivalent furnace depth of 23 ft (7.0 m). At these size limits, steam flow capabilities are limited to about 550,000 lb/h (69.3 kg/s).

The ash in bark, predominately silica, is very abrasive. This, in combination with high temperature grate bar exposure, causes high maintenance. As a result, some vendors have reintroduced the water- or air-cooled vibrating grate stoker. A water-cooled version is shown in Fig. 12. The combustion concept is no different than that for traveling grates. A complement of grate bars with air metering holes is attached to and supported by a water-cooled tubular grid connected to the boiler or feedwater circuitry. A heat conducting cement may be used to enhance grate cooling. The water-cooled grid is then supported by a complement of flexible straps. To convey ash, a back-and-forth motion is imparted to the grid which, supported on the flexible straps, imparts a looping motion to the grate. This motion, at predetermined intervals, is sufficient to convey and discharge the ash to the hopper. A variant of the water-cooled grate is the air-cooled type. In the air-cooled grate, the water tube grid is replaced by a mechanical grid to support the grate bars; components are then cooled by the flow of undergrate air. The advantages of the water and air-cooled grates are their simplicity, minimal moving parts, and lower maintenance. Grate sections generally range in width from 4 to 11 ft (1.2 to 3.35 m) to meet project functional and economic requirements. Because of their

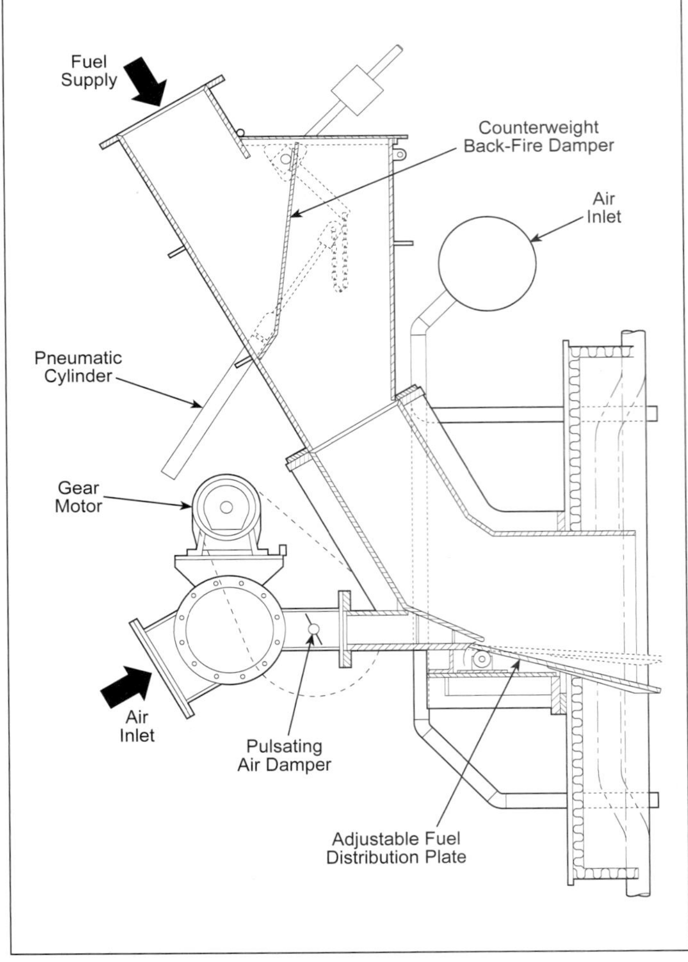

Fig. 11 Air-swept spout.

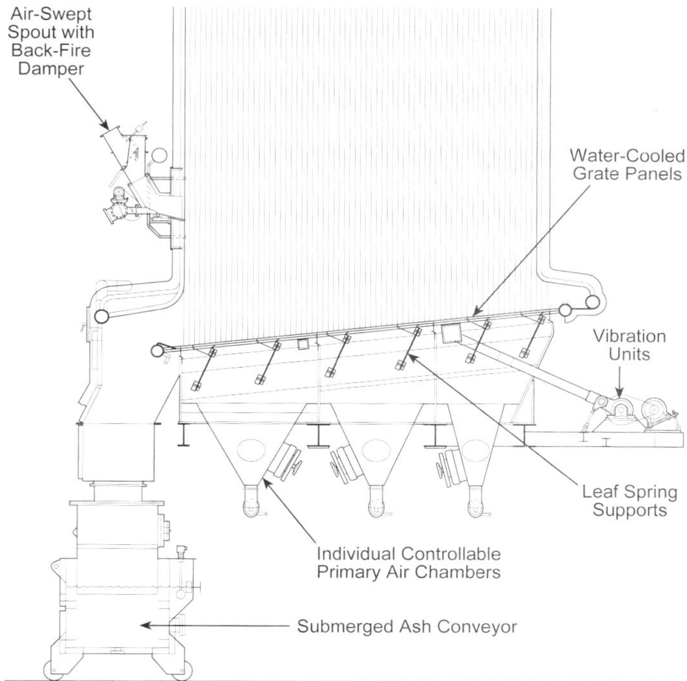

Fig. 12 Water-cooled vibrating grate stoker.

independent modular construction, vibrating grate modules can be installed side by side to the mechanical limits of the boiler. Furthermore, grate depths are generally permitted to the limit of fuel throw and distribution of 26 ft (7.9 m). The largest vibrating grate currently in operation is almost 45 ft (13.7 m) wide. When sized to grate release rate limits of 1,100,000 Btu/h ft^2 (3.47 MW_t/m^2), steam flow capacities of 800,000 to 900,000 lb/h (100.8 to 113.4 kg/s) are available on wood, bark and other biomass fuels.

Combustion air system

The excess air for bark, wood and most biomass fuels ranges from 15 to 50%, depending on stoker type and firing conditions. Because biomass fuels are typically highly volatile on a dry basis, heterogeneous in size and tend to burn more in suspension, combustion air systems are designed to provide more overfire air than that used for coal. Modern designs permit undergrate and overfire quantities of 40 to 60% of total stoker combustion air respectively, excluding any combustion air used for auxiliary burners. Overfire air system design and arrangement in combination with furnace geometry play an important role in completely burning the combustibles liberated from the grate. Fig. 9 shows a furnace design that promotes recirculation and mixing at and below the lower furnace bustle (arches). Multiple elevations of large overfire air nozzles allow high energy jets of air to penetrate and mix, enhancing combustion. The system is designed to permit flexibility in distributing the air within the overfire air system as fuel characteristics vary.

Numerical modeling techniques (see Chapter 6) have advanced OFA system design for biomass fuels. Today, fewer levels of OFA with very large ports are arranged to more effectively create turbulence for mixing gaseous combustibles with air without the need for a lower furnace bustle. Fewer ports mean lower initial cost and simplified operation. Port elevation(s) can be located to optimize CO and NO_x performance. The ports also serve as convenient injection sites for non-combustible/malodorous waste gases. Fig. 13 shows a horizontal rotary overfire air system designed to create a double rotating circulation zone within the furnace. Fig. 14 depicts a B&W PrecisionJet™ air system with ports arranged in an interlaced pattern. Both systems typically require air pressures of less than 20 in. wg (5.0 kPa) and result in both exceptional combustion performance and fairly uniform gas distribution leaving the combustion section. Because the highly turbulent air zone is located well above the fuel feed and grate area, less ash and char are carried out of the furnace.

Combustion carryover and flyash reinjection

Fig. 15 shows the recommended sizing distribution for wood or bark. The degree of grate or suspension burning varies with fuel size. A portion of the fuel that ignites in suspension may be too large to complete burning; it therefore goes out with the flue gas as unburned carbon. Many factors such as fuel sizing, grate release rates, moisture content and reinjection affect the unburned carbon loss. In addition, furnaces sized to a maximum liberation rate of about 18,000 Btu/h ft^3 (186 kW_t/m^3) can control unburned carbon losses to 1 to 3%. Further reductions can be attained with flyash reinjection in certain biomass-fired units. Because of high maintenance costs due to the high silica content and abrasiveness of wood and bark flyash, reinjection systems are not frequently used with these fuels unless sand classifiers are also installed. As one example, the unit shown in Fig. 9 does not feature a flyash reinjection system.

In cases where the unburned carbon carryover must be further controlled, mechanical particulate collectors can be installed downstream of the generating bank. These collectors are sized to take advantage of the fact that the carbon particles are typically larger and heavier than the inerts in the flyash. This permits a carbon-rich stream to be reinjected into the furnace for further combustion. This mechanical collection and reinjection reduces the ultimate carbon loading in the downstream equipment and the possibility of fires.

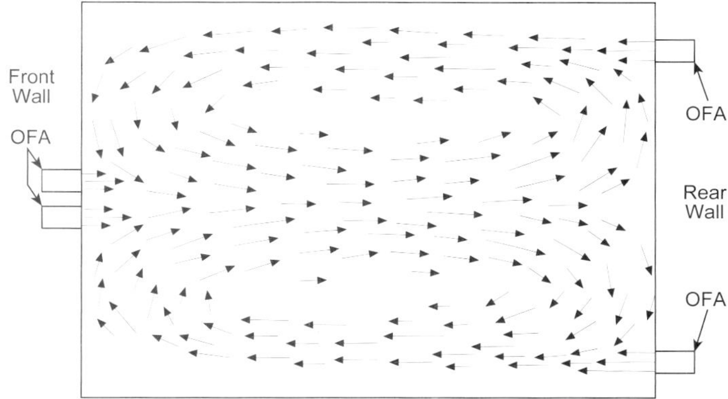

Fig. 13 Furnace plan view showing horizontal rotary overfire air system.

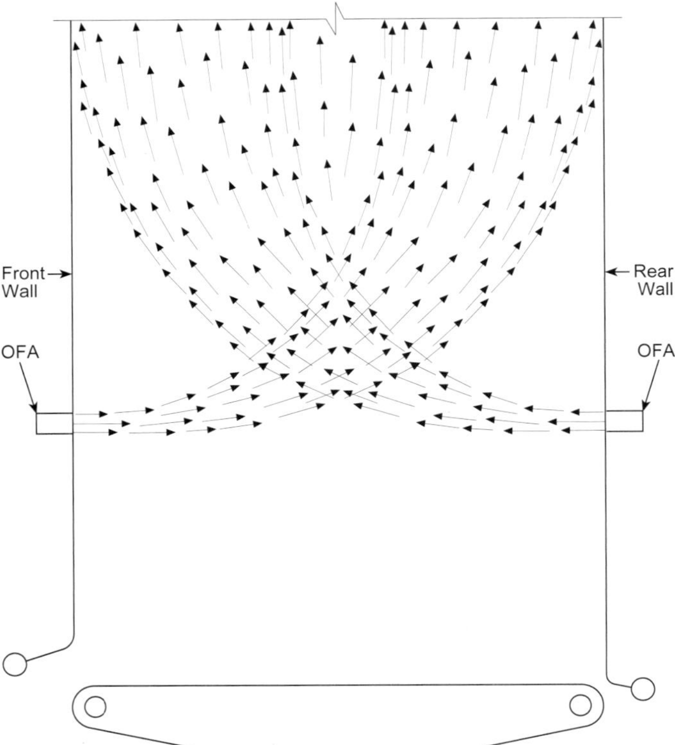

Fig. 14 Furnace sectional side view showing single level air system with PrecisionJet™ ports.

Grate sizing

For wood and bark firing with moisture contents at or below 50%, it is common practice to select a grate area that results in a design heat release rate of 1,100,000 Btu/h ft^2 (3.47 MW$_t$/m^2). At 35% moisture, the release rate may approach 1,250,000 Btu/h ft^2 (3.94 MW$_t$/m^2). At moisture levels higher than 55%, the fuel becomes more difficult to ignite and burn. Traveling grate stokers are mechanically limited to about 23 ft (7.0 m) of equivalent furnace depth and about 38 ft (11.6 m) in width. Grates are generally longer than they are wide because this design is the least expensive. For a water-cooled grate, the depth is limited by the feeder's ability to distribute the bark over the grate depth. This distance is about 26 ft (7.9 m). Because of its modular construction, the maximum width of a water-cooled grate assembly is limited only by mechanical limits of the boiler.

Combustion air temperature

When firing wood, bark and other biomass fuels, it is necessary to consider the fuel moisture content to determine the need for preheated combustion air. Undergrate air temperatures in excess of that needed for the moisture content of the as-fired fuel contribute to NO_x and particulate emissions. Biomass fuels with less than 35% moisture content do not require preheated air. With seasonal variation in fuel moisture and the growing interest in co-firing wood and bark with sludges, the air heating system should be sized and selected for the maximum expected moisture. The temperature limit for the air-cooled traveling grate with ductile iron bars is generally 550F (288C). If higher temperatures are needed, the water-cooled vibrating grate can withstand air temperatures to 650F (343C). In some cases, a steam or water coil air heater can provide variable undergrate air temperature when needed.

The decision of whether to use heated or unheated air for the OFA system must consider the overall project goals and economics. Hot OFA is preferred for new systems when maximum heat recovery is of singular importance. Hot air achieves greater penetration than cold air at the same mass flow rate, tends to burn out CO and hydrocarbons better, and reduces the chances for acid dewpoint corrosion of non-condensable gases in port boxes. Cold OFA systems are lower cost to retrofit because they do not require insulated ducts for personnel protection. Because cold OFA requires smaller ports, it is preferred where vertical clearance is tight.

There are many other biomass fuels such as rice hulls, bagasse, peach pits, coffee grounds and demolition debris, that may be fired on spreader-type stokers. Most of the criteria for sizing and using a spreader stoker are similar to those for wood and bark. Loose straw, however, should be mass fed. Use of a spreader stoker can cause uncontrolled carryover of burning lightweight particles into the convection section.

Ash removal

The ash content in wood and bark is only a few percent by weight and predominantly composed of silica. Because it is very abrasive, slow grate speeds and abrasion resistant materials should be used. Bottom ash typically drops into a refractory lined hopper from which it is intermittently removed. Submerged chain conveyors are also suitable. Due to the carbon content and the tendency for fires, the flyash from the hopper and dust collecting equipment is removed continuously. Wet sluice or inert gas (flue gas) systems are generally used to convey the flyash to the storage silos. Wet systems are preferred due to fire and dusting resistance.

Emissions

The installation of any new steam generator or a major rebuild or upgrade to an existing one requires environmental permitting. Therefore, the ability to predict and control the various emissions is important. Add-on pollution control equipment is generally required to meet the increasingly stringent regulated levels. However, it is also important to control or minimize the source emissions where possible to reduce the cost of this add-on equipment.

Table 3 lists typical uncontrolled emission values for spreader stoker firing of various coals and wood/bark. These values will vary with fuel composition and equipment selection.

NO_x is formed from the oxidation of the nitrogen compounds in the combustion air and in the fuel. With stoker firing it is believed that most of the NO_x is de-

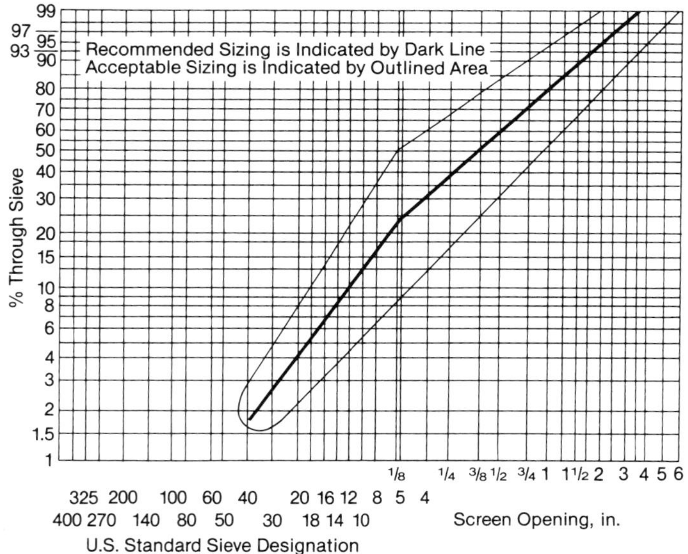

Fig. 15 Recommended sizing for wood or bark spreader stoker firing (*courtesy of Detroit Stoker Company*).

rived from fuel-bound nitrogen (fuel NO_x); the contribution due to oxidation of the nitrogen in the air (thermal NO_x) is small due to relatively low furnace temperatures. NO_x emissions can be effectively reduced by staging combustion, inherent in spreader stoker firing, and by controlling excess air levels. For both coal and wood/bark stoker firing, the excess air level in low NO_x stoker systems is about 25%. To reduce NO_x to the lower end of the range shown in Table 3, designers are now using deeper staging, i.e., lower undergrate and higher overfire air flows, as well as flue gas recirculation. For spreader firing, feeders are also designed to improve fuel distribution and combustion on the grates. Other factors that reduce NO_x formation include controlling the quantity of fines in the fuel and using ambient temperature combustion air.

Many waste fuels such as straw and other nonwood fibers have high fuel nitrogen content. Commercial wastes such as demolition debris are dry and will combust at higher temperatures, producing higher NO_x. It is therefore important that a thorough investigation of the fuel be made to assess its potential effect on emissions.

For most stoker-fired units burning coal or biomass that contains sulfur, sulfur dioxide (SO_2) will be present in the flue gas. For prediction purposes and for sizing of SO_2 control equipment, it is assumed that all the fuel sulfur becomes SO_2.

Carbon monoxide (CO) and volatile organic compound emissions are generally a function of the quality of the combustion process and the quantity and control of fines and excess air. CO will tend to increase as NO_x is reduced.

Table 3
Typical Post Combustion Emissions for Spreader Stoker Firing

Fuel	NO_x (as NO_2) lb/10^6 Btu	CO lb/10^6 Btu	Unburned Carbon Loss (% of Heat Input) With Reinjection	Without Reinjection
Bituminous	0.35 to 0.5	0.05 to 0.30	0.5 to 2.0	3 to 6
Subbituminous	0.3 to 0.5	0.05 to 0.30	0.5 to 1.5	3 to 5
Lignite	0.3 to 0.5	0.10 to 0.30	0.5 to 1.5	3 to 5
Wood/bark	0.1 to 0.35	0.05 to 0.50	0.2 to 2.0	2 to 5

Approximate conversion: 1 lb/10^6 Btu = 1230 mg/Nm^3 (dry flue gas 6% excess O_2, 350 Nm^3/GJ).
Includes applicable overfire air systems.

Circulating fluidized-bed power plant by B&W.

Chapter 17

Fluidized-Bed Combustion

In the 1970s, fluidized-bed combustion technology was first applied to large-scale utility boiler units to explore new ways of burning solid fuels, especially high-sulfur coal, in an environmentally acceptable and efficient manner. In concept, fluidized beds burn fuel in an air-suspended mass (or *bed*) of particles. By controlling bed temperature and using reagents such as limestone as bed material, emissions of nitrogen oxides (NO_x) and sulfur dioxide (SO_2) can be controlled. Additional benefits of fluidized-bed combustion include wide fuel flexibility and the ability to combust fuels such as biomass or waste fuels, which are difficult to burn in conventional systems because of their low heating value, low volatile matter, high moisture content or other challenging characteristics. In coal-fired systems, the fuel is burned in an air-suspended bed of limestone and inert ash particles where SO_2 is absorbed by the limestone, and NO_x formation is limited by lower operating temperatures and staged combustion, when used. This technology is now used in a variety of industrial and utility boiler applications.

Today, bubbling fluidized-bed (BFB) boilers, with a bed of fluidized particles that remain in the lower furnace, are used primarily in specialty fuel applications such as coal wastes and biomass fuels. Circulating fluidized-bed (CFB) boilers, with solids circulating through the entire furnace volume, address most larger steam generator applications and a broader range of fuels.

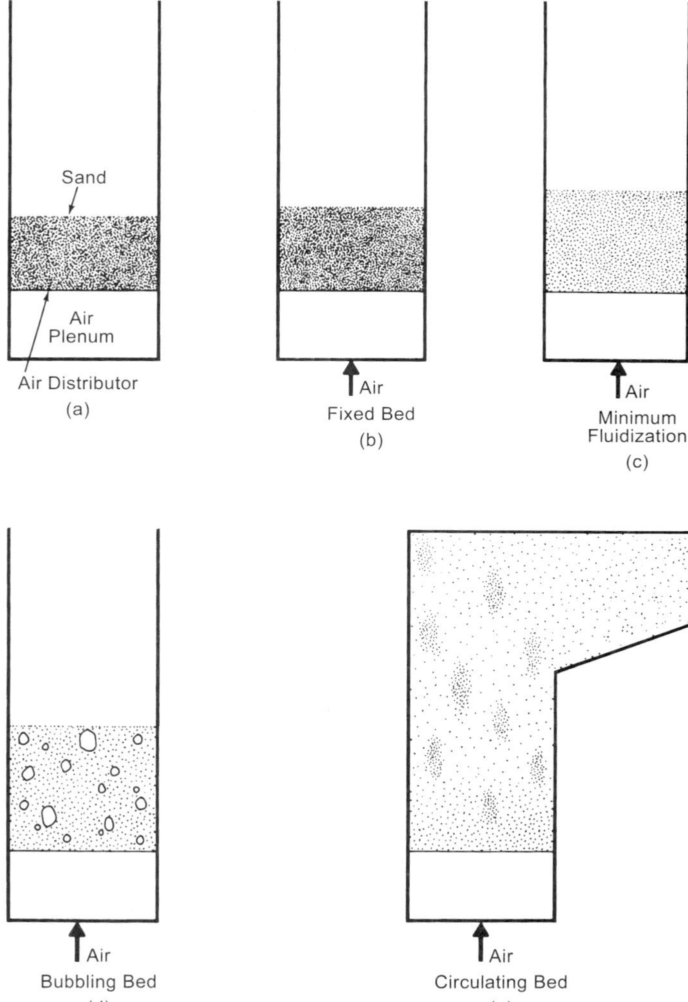

Fig. 1 Typical fluidized-bed conditions.

The fluidized-bed process

The fluidizing process induces an upward flow of a gas through a stacked height of solid particles. At high enough gas velocities, the gas/solids mass exhibits liquid-like properties, thus the term *fluidized* bed.

The following example helps illustrate the process. Fig. 1a shows a container with an air supply plenum at the bottom, an air distributor that promotes even air flow through the bed, and a chamber filled with sand or other granular material.

If a small quantity of air flows through the air distributor into the sand, it will pass through the voids of an immobile mass of sand. For low velocities, the air does not exert much force on the sand particles and they remain in place. This condition is called a *fixed* bed and is shown in Fig. 1b.

By increasing the air flow rate/velocity, the air exerts greater forces on the sand and reduces the contact forces between the sand particles caused by gravity. By increasing the air flow further, the drag forces on the particles will eventually counterbalance the gravitational forces, and the sand particles become suspended in the upward stream. The point where the bed starts to behave as a fluid is called the *minimum fluidization* condition. The increase in bed volume is insignificant when compared with the non-fluidized case (Fig. 1c).

As the air flow increases further, the bed becomes less uniform, bubbles of air start to form, and the bed becomes violent. This is called a *bubbling* fluidized bed (BFB), shown in Fig. 1d. The volume occupied by the air/solids mixture increases substantially. There is an obvious bed level and a distinct transition between the bed and the space above.

By increasing the air flow further, the bubbles become larger and begin to coalesce, forming large voids in the bed. The solids are present as interconnected groups of high solids concentrations. This condition is called a *turbulent* fluidized bed.

A further increase in air flow causes the particles to blow out of the bed and the container. If the solids are caught, separated from the air, and returned to the bed, they will circulate around a loop, defined as a *circulating* fluidized bed (Fig. 1e). Unlike the bubbling bed, the CFB has no distinct transition between the dense bed in the bottom of the container and the dilute zone above. The solids concentration gradually decreases between these two zones.

The pressure differential between the top and the bottom of the container changes with air flow, as shown in Fig. 2. At low air flow, the pressure differential increases with flow through the static bed until reaching the *minimum fluidization velocity*. At this point, the sand is supported by the air, and the pressure differential is determined only by the mass of bed material. The pressure differential is independent of further increases in air flow until the air velocity becomes high enough to convey material out of the container. Then the pressure differential decreases as mass is lost from the system, which is represented by the entrained flow portion of the curve in Fig. 2.

An important parameter for evaluating hydrodynamic and heat transfer performance of particle mixtures is the Sauter (also called volume-surface or harmonic) mean diameter (SMD):

$$SMD = \frac{1}{\frac{X_1}{D_1} + \frac{X_2}{D_2} + \ldots + \frac{X_N}{D_N}} \quad \textbf{(1)}$$

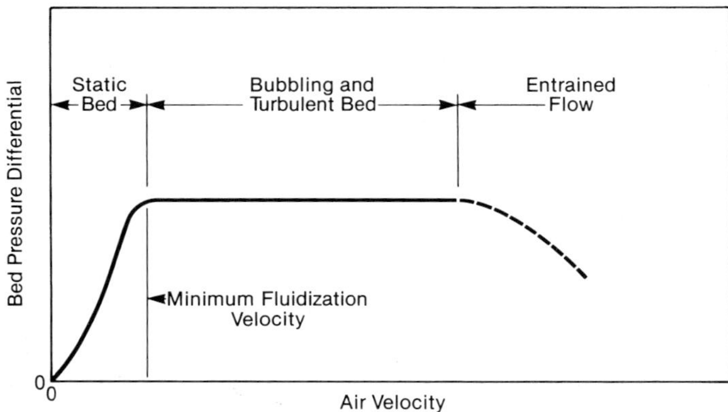

Fig. 2 Effect of velocity on bed pressure drop.

where

X_1 to X_N = weight fraction of first to last size cut
D_1 to D_N = average diameter of first to last size cut, microns

Of these fluidization conditions described above, only bubbling and circulating beds are currently used by the power industry to generate steam.

Bubbling fluidized-bed (BFB) boilers

Fig. 3 shows the main features of a bubbling fluidized-bed boiler while Fig. 4 shows a typical furnace bulk density profile curve. The sharp drop in density indicates the top of the bed.

The bottom of the furnace in a BFB boiler consists of a horizontal air distributor with bubble caps. This provides the fluidizing air to the bed material in the lower furnace. As discussed later, The Babcock & Wilcox Company (B&W) offers two air distributor systems depending upon the fuel and application. The bubble caps are closely spaced so that air flow is distributed uniformly over the furnace plan area. The lower furnace is filled with 2 ft (0.6 m) of sand or other non-combustible material such as crushed limestone or bed material from prior operation. Air flow is forced upward through the bed of material, and the bed expands to a depth of about 3 ft (0.9 m) taking on most of the characteristics of a fluid. The air flow through the bed is very uniform due to a high number of air distributors (bubble caps) and bed pressure drop. The

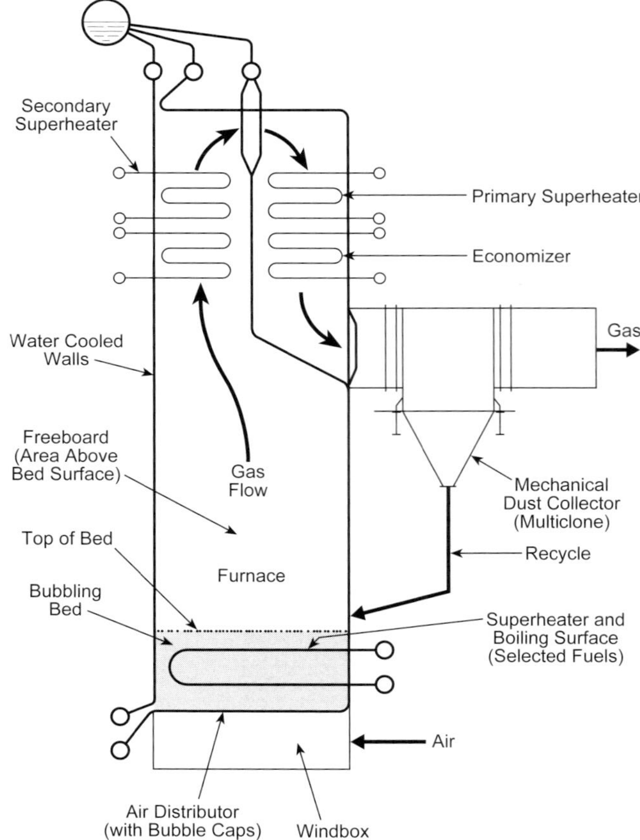

Fig. 3 Typical bubbling fluidized-bed boiler schematic.

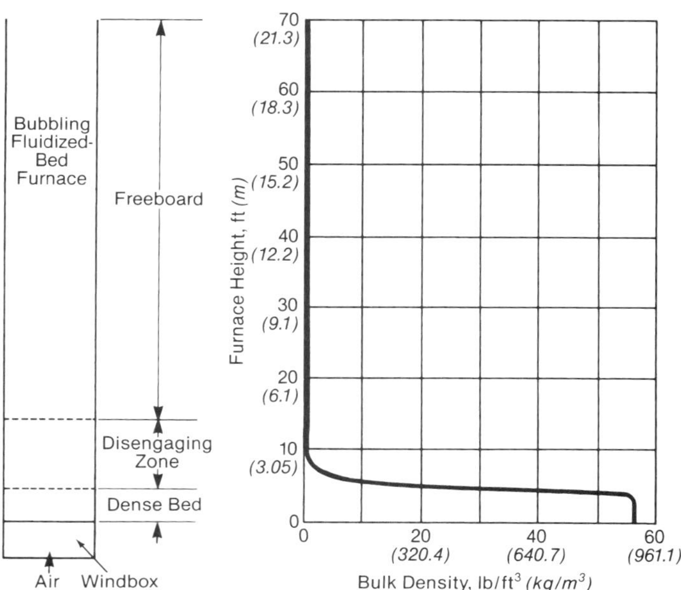

Fig. 4 Typical atmospheric pressure bubbling-bed furnace density profile.

typical slumped (non-fluidized) density of a sand bed is 90 lb/ft^3 (1442 kg/m^3). The voidage or volume between the particles in the fluidized bed is 65% and the bulk density is 58 lb/ft^3 (929 kg/m^3). The gas flow rate through the bed is defined as the *superficial bed velocity*, which is calculated by dividing the volumetric gas flow rate at the bed temperature by the plan area of the bed without the solids. A typical nominal design superficial velocity of 8 ft/sec (2.4 m/sec) is enough to fluidize BFB bed material with a particle size distribution between 500 and 1400 microns. The boiler enclosure is made of water-cooled membrane panels.

B&W offers two air distributor systems for its BFB boilers: *open bottom* and *flat floor* systems. The open bottom system shown in Fig. 5 is characterized by the fluidizing air bubble caps and pipes mounted on widely spaced distribution ducts (see also Fig. 6) located in the bottom of the BFB furnace. Stationary bed material fills the hoppers and furnace bottom up to the level of the bubble caps, above which the bed material is fluidized by the air flow. The open spacing is effective in removing larger rocks and debris from the active bed area as bed material moves down through hoppers. This design is particularly attractive in biomass and waste fuel applications, which contain non-combustible debris. In the flat floor system shown in Fig. 7, the floor of the furnace is formed by horizontal water-cooled membrane panels with bubble caps. Air passes from a windbox below the water-cooled panel through the bubble caps to enter and fluidize the bed material. Separate bed drains are provided. The membrane panel floor must form an airtight seal with the furnace walls, must support the weight of a slumped bed, and must resist the uplift generated from the air pressure drop during operation. This design is attractive for firing coal where there is much less large debris present.

Coal-fired bubbling-bed boilers normally incorporate a recycle system that separates the solids leaving the economizer from the gas and recycles them to the bed. This maximizes combustion efficiency and sulfur capture. Normally, the amount of solids recycled is limited to about 25% of the combustion gas weight. For highly reactive fuels such as biomass, this recycle system is usually omitted.

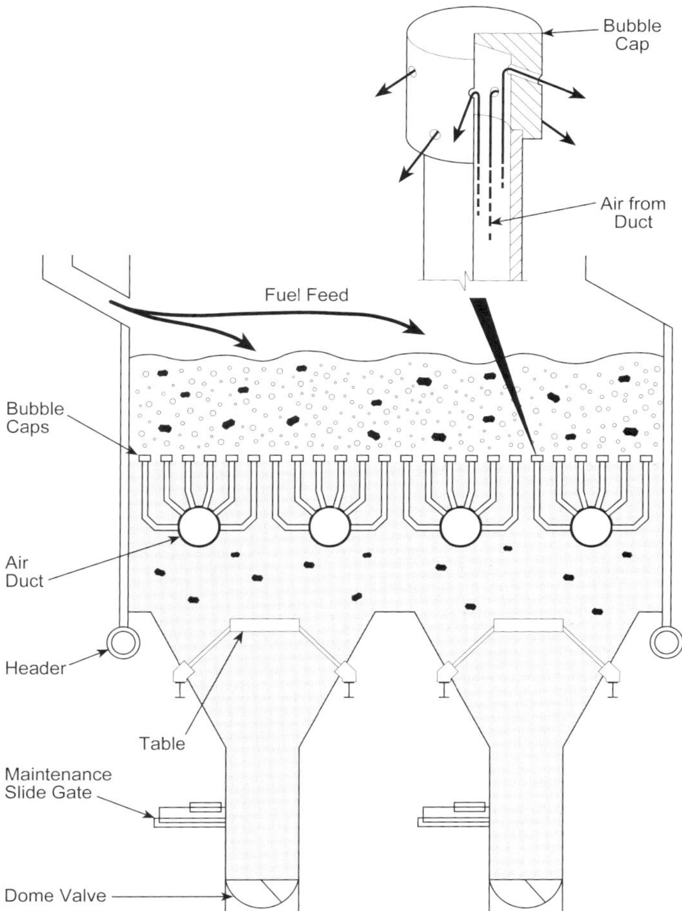

Fig. 5 Open bottom air distributor system with bubble caps.

The typical operating temperature range of a bubbling bed is 1350 to 1650F (732 to 899C), depending on the fuel moisture, ash content, and alkali content

Fig. 6 Air distributor ducts and bubble cap system.

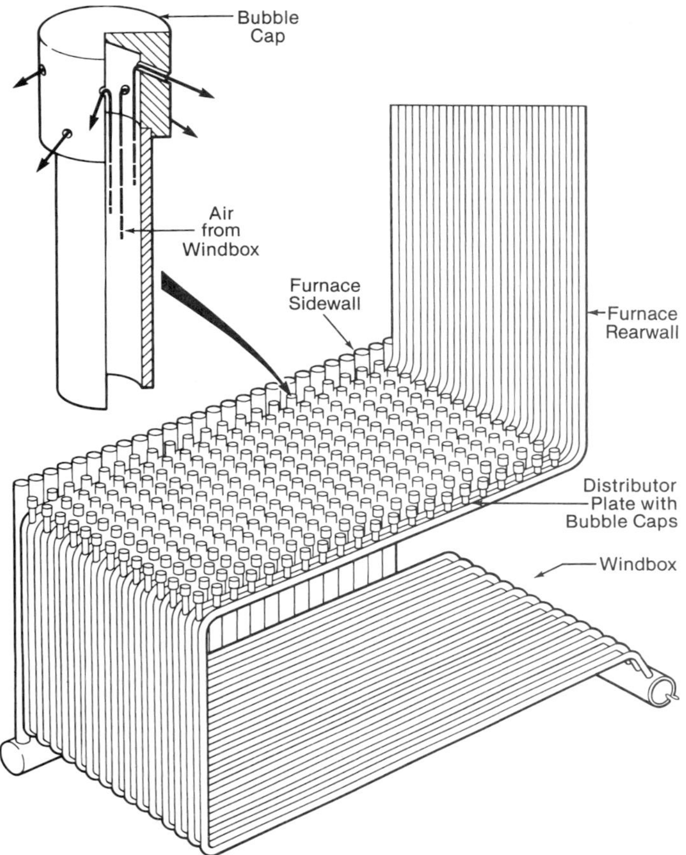

Fig. 7 Furnace distributor plate and bubble caps for a flat floor system.

in the ash. Even at these low combustion temperatures, the high convective and radiative heat transfer from bulk bed material to the fuel particles provides sufficient ignition energy to evaporate moisture, heat the ash, and still combust the remaining fuel without significantly changing the instantaneous bed temperature. This is why the bubbling bed can burn low-grade fuels, which burn at low combustion temperatures due to their high moisture and ash contents.

Heat transfer surface may be placed within the bed depending upon the fuel being burned. For biomass and other low heating value fuels, no in-bed surface is usually required because other methods of bed temperature control can be used. For coal firing with its high heat content and lower relative volatility, the heat transfer surface, in the form of a tube bundle, is placed in the bed to achieve the desired heat balance and bed operating temperature. The bed temperature is uniform, plus or minus 25F (14C), as a result of the vigorous mixing of gas and solids.

BFB combustion systems are attractive retrofits to older boiler designs where a change in fuels or wider fuel flexibility is required. Fig. 8 shows a sectional side view of a process recovery (PR) boiler that was retrofitted with a B&W open bottom BFB combustion system. The black liquor firing capacity was no longer needed and the owner wanted a power boiler capable of firing wood waste, primary clarifier sludge, and tire-derived fuel. Fig. 9 shows an isometric of the BFB combustion system in the bottom of the boiler. The air distributor, fluidized bed of material, and overfire air systems are clearly identified.

Fig. 10 shows a sectional side view of a small BFB boiler firing wood, wood waste and byproduct sludge. This Towerpak® boiler design is a version of the Stirling® power boiler (SPB, see Chapter 27) with two drums and bottom support for ease of installation. This unit is capable of supplying 120,000 lb/h of steam flow (15.12 kg/s).

All fossil fuel boilers require closely coordinated control of fuel, air, water and steam parameters. Fluidized-bed boilers require additional control. Selected key parameters for fluidized-bed boiler design include: bed temperature control (fuel dependent), bed inventory control, bed density (for heat transfer control), and emissions.

Bed temperature control and fuel feed

The most fundamental control function of a BFB combustion process is bed (combustion) temperature control. The control method depends upon the fuel

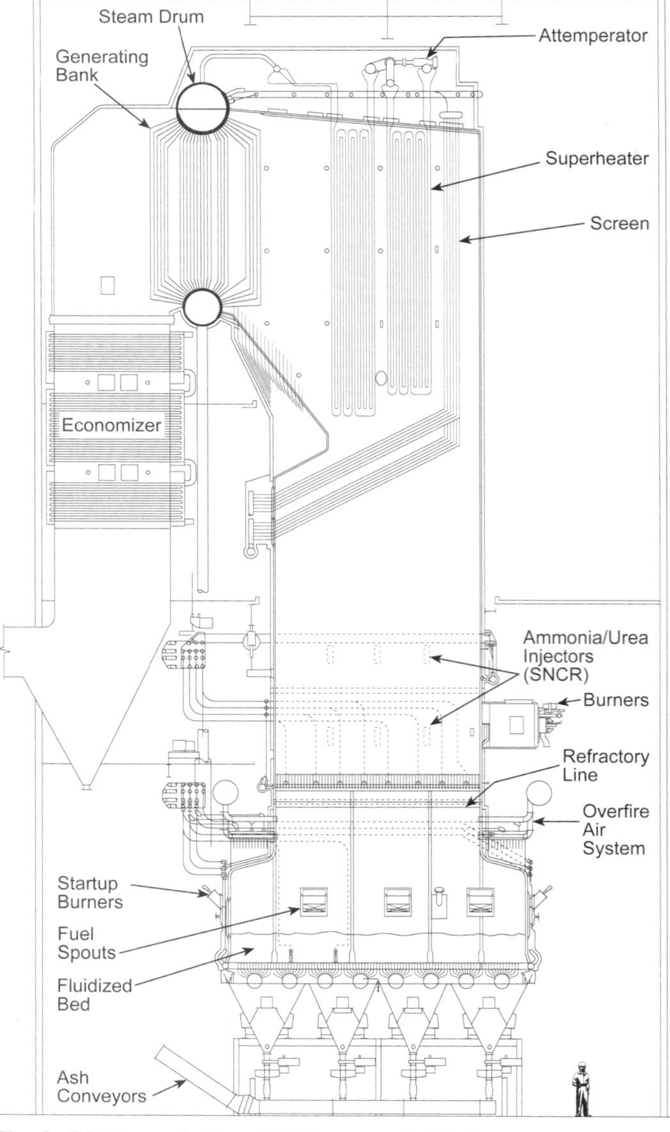

Fig. 8 B&W open bottom BFB furnace retrofit of a process recovery boiler.

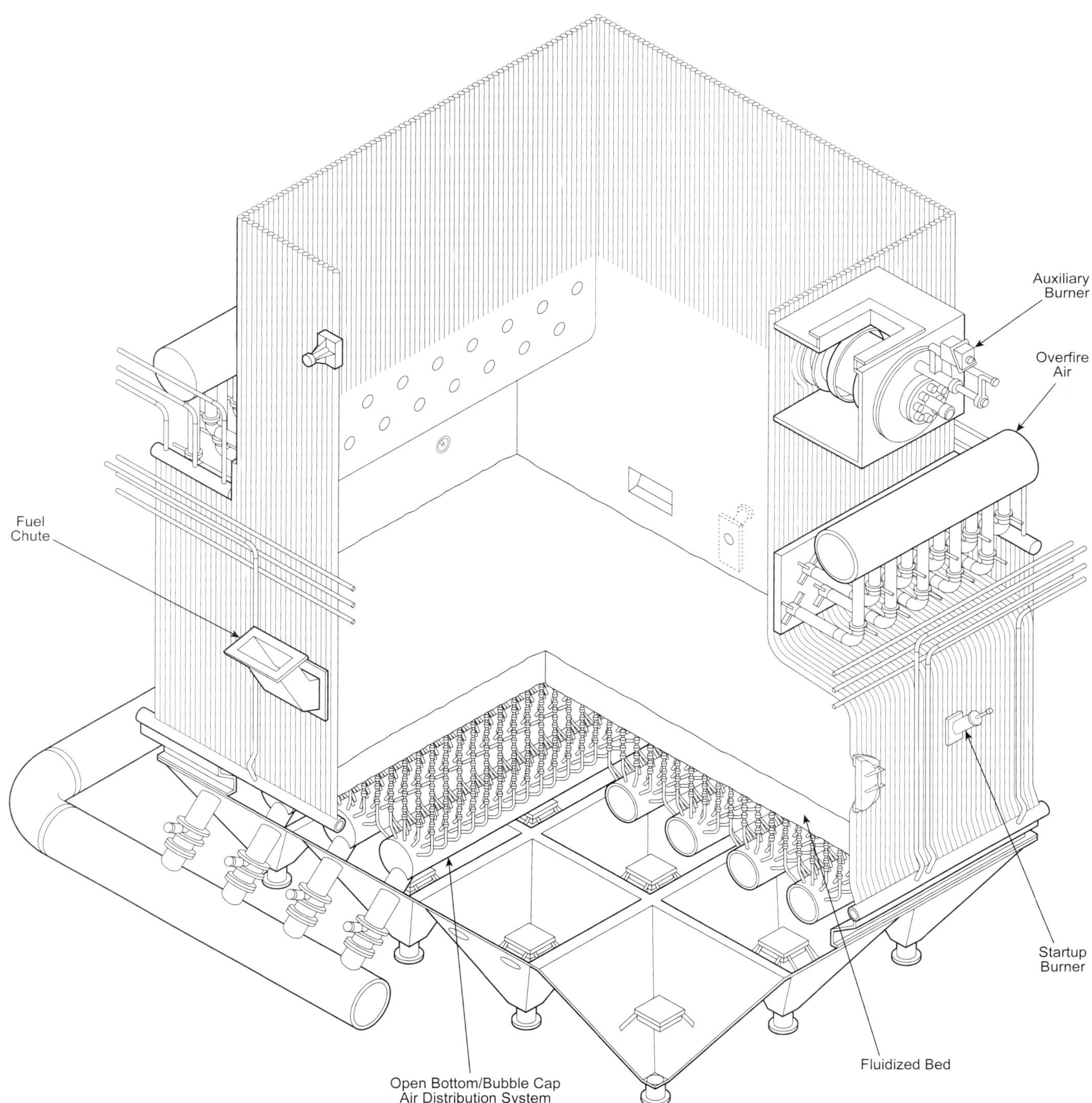

Fig. 9 Isometric of BFB furnace bottom.

being fired. Bed temperature is controlled to limit emissions of NO_x and SO_2, and to limit bed material agglomeration. Agglomeration is caused by sodium and potassium combining with alumina and silica to form low melting point eutectics that can coat the bed particles. If the alkali concentration in the coating is too high, the coating can start to melt and cause particles to stick together. As a result, the larger particles do not fluidize and, if the process continues, the bed can solidify and the combustion process can stop. The agglomeration process is quite temperature-sensitive and some fuels have strict requirements for the maximum peak bed temperatures. Also, some fuels have such high alkali contents that the fuel can not be fired in a fluidized bed combustion process. Such fuels are typically agro-based where the plants are fertilized or the ground is rich in alkali. Some process waste can also contain high alkali concentrations.

Biomass firing Biomass firing begins with an inert bed of solid particles in the bottom of the furnace preheated to 1500F (816C) using oil or gas startup burners. Air flows into this bed at 68 to 392F (20 to 200C) and fuel at 68F (20C). The bed material must heat the incoming air and fuel to the bed temperature and the

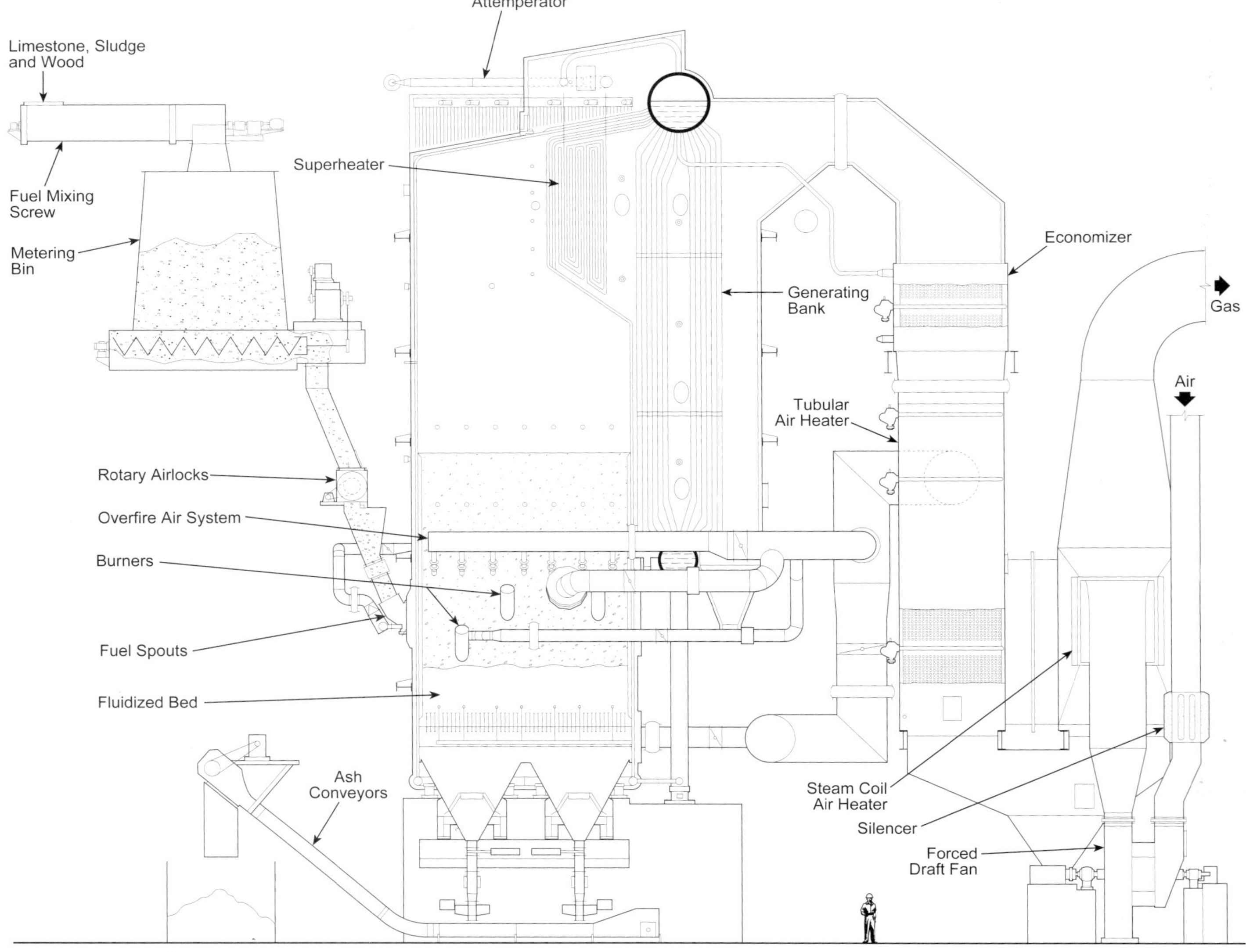

Fig. 10 Towerpak® bottom supported BFB.

fuel must release the same amount of heat back into the bed to maintain a constant temperature. Too much heat release increases bed temperature, while too little heat release reduces bed temperature. Three methods of bed temperature control are used in biomass firing: primary/overfire air split adjustment, flue gas recirculation (FGR), and fuel feeder selection and operation.

Primary/overfire air split adjustment Biomass bubbling beds operate substoichiometrically with less than theoretical combustion air. In the bed, all available oxygen is completely used. Any additional oxygen would oxidize more fuel and increase the in-bed heat release, while lower oxygen levels would have the reverse effect. Because of this repeatable relationship, an increase in air flow from the windbox through the bed will raise bed temperature while a decrease in air flow will lower bed temperature at steady-state conditions. Overfire air flow increases or decreases in the opposite direction to maintain constant total combustion air flow to the boiler.

Shifting the air flow between the bed and overfire air system shifts the combustion or heat release between the two. At steady-state conditions, the flue gas temperature above the overfire air nozzles can be varied by 360F (200C) just by shifting air between the bed and the overfire air. The bed temperature is rather slow to respond to changes in the bed air flow. The cycle of increasing the bed temperature 27F (15C) above the set point and returning back to the set point can take 20 minutes or more.

Flue gas recirculation (FGR) When changing the air flow through the bed is not sufficient or fast enough for acceptable bed temperature control, flue gas can be recirculated back from the induced draft (ID) fan outlet into the bed. This *flue gas recirculation* is another mass flow stream entering the bed at a temperature below bed operating temperature. Energy transferred from the bed material to the incoming flue gas to heat it to the new bed equilibrium temperature is greater than the heat released by incremental combustion from the low level of FGR oxygen. As a result, the equilibrium bed temperature declines. The bed temperature response rate for this control method is much more rapid and, at steady-state conditions, FGR can control the bed temperature to within 20F (11C) of the set point bed temperature.

Fuel feeder selection and operation Two types of feeders are commonly used for biomass fuels in BFB combustion: 1) chute-type feeders that deposit fuel in small areas of the bed, and 2) air distribution feeders that distribute fuel over a wider area. There is a relationship between in-bed heat release and distribution of fuel over the bed plan area. This is because the high volatile matter content and low-density char escape from the bed before the inherent churning motion of the bed material can distribute the fuel within the bed. If the fuel is not distributed evenly, some air simply passes through the bed without contributing to the in-bed combustion process. To obtain the highest amount of in-bed heat release, the fuel must be distributed as widely and uniformly as possible to match the uniform air flow. The lowest in-bed heat release results from dumping fuel to one spot. Therefore, the fuel feeder should be able to function as either a chute or as an air distribution feeder. When the fuel is dry, the feeder air flow can be reduced and the fuel falls into a small portion of the bed. When the fuel is wet, the feeder air flow can be increased and the fuel distributed over the plan area. The in-bed heat release change rate from low- to high-feeder air flows can be as much as 20% of the fuel's heating value.

Coal firing Bed stoichiometry adjustment is not effective for coal-fired BFB combustor temperature control because carbon builds up in the bed when there is less than theoretical combustion air. When the air flow is increased for load, the bed temperature increases rapidly and uncontrollably. In an extreme case, the pressure parts could be damaged, and more commonly, the bed could agglomerate. Therefore, BFBs firing coal are limited to no less than stoichiometric conditions in the bed.

The coal can be blown into the bed from below using pressurized feed pipes, or can be fed over the bed using common rotary flipper-type feeders. The underbed feed system is more expensive and the coal must be dry and less than 0.25 in. (6.35 mm). The feed pipes are prone to erosion and plugging, and the nozzles at the ends of the pipes erode. The in-bed heat release is higher and the carbon loss is lower using the underbed feed system, but the disadvantages typically cause the over-bed system to be preferred.

With the overfeed system, the in-bed heat release is 75 to 85% of the coal's heating value. For the coal falling on the bed, all of the fixed carbon and 70% of the volatile matter combusts while in the bed. The fines burn in suspension and reduce the overall in-bed heat release. With this high in-bed heat release, the adiabatic bed temperature would be significantly higher than the desired 1500 to 1600F (815 to 871C) range. To lower the bed temperature to the desired range, tube bundles are submerged in the bed. This in-bed surface can be either steam-generating surface or superheater surface. Water circulation in the steam-generating surface can be either natural or forced.

The fluidizing air is provided from a windbox mounted below the bed. The windbox is typically compartmentalized, with individual air flow control to each compartment. Bed temperature varies with the firing rate. Therefore, as the boiler load is reduced, the firing rate is reduced, and the bed temperature declines. When the bed temperature reaches a certain minimum, an outer portion of the bed is shut down by shutting off the air flow and fuel flow to that portion of the bed. The air flow and fuel flow to the remaining active bed area is increased which raises the bed temperature in the operating beds. Boiler load is increased by re-fluidizing out-of-service (slumped) compartments and firing to larger portions of the bed.

Bed inventory control

The air/flue gas pressure drop through the bed is directly related to the bed material mass when the bed is fluidized. Pressure drop through the bed is approximately 0.75 to 1.0 in. wg (0.19 to 0.25 kPa) per inch (2.54 cm) of bed height. For biomass-fired systems, the bed inventory is set to a pressure drop of 30 to 36 in. wg (7.5 to 9 kPa). This is sufficient mass to prevent the bed temperature from significantly changing in a short period following a sudden large change of fuel properties, such as moisture content. For a coal-fired BFB, the bed height must ensure that the in-bed surface is submerged. If the bed height is below the top of the in-bed tube bundle, heat transfer decreases significantly and bed temperature increases. Pressure taps in the bed enclosure walls provide the measurement of the bed pressure drop over a known height.

Bed inventory can change over time. The ash and tramp material that enter with the fuel can increase the overall inventory when these materials are too large to be blown out of the bed. The bed material can break down due to mechanical or thermal attrition and leave the bed with the flue gas. Oversized particles, such as rocks and metal, are removed by the bed drain system and attrited material is replaced with new (makeup) material when necessary.

Bed density and heat transfer (in-bed)

The following equations are for the bed pressure loss and the overall in-bed heat transfer coefficient used to size in-bed surface.

Equation 2 is used to calculate the flue gas bed pressure loss across a bubbling bed.

$$\Delta P = (C)(1-e)(\rho_s - \rho_g)(L) \qquad \textbf{(2)}$$

where

ΔP = pressure loss
C = units conversion constant
e = bed void fraction
ρ_s = particle density
ρ_g = gas density at bed conditions
L = bed height

The void fraction, *e*, is primarily a function of particle size, particle density, bed gas velocity and gas viscosity. Various methods are used to predict bed voidage or void fraction, including those proposed by Leva,[1] Babu,[2] and Staub and Canada.[3]

The equation for the overall heat transfer coefficient for any tube bundle immersed in the bed is:

$$U_o = \frac{1}{\frac{1}{h_c + h_r} + R_m + R_{ft}} \quad \textbf{(3)}$$

where

U_o = overall heat transfer coefficient, Btu/h ft^2 F (W/m^2 K)
h_c = convection heat transfer coefficient for the tube bank, Btu/h ft^2 F (W/m^2 K)
h_r = radiation heat transfer coefficient for the tube bank and walls, Btu/h ft^2 F (W/m^2 K)
R_m = metal wall resistance, h ft^2 F/Btu (m^2 K/W)
R_{ft} = tube fluid film resistance, h ft^2 F/Btu (m^2 K/W)

The convection heat transfer coefficient h_c is given by Equation 4. Two equations are used to evaluate the convective heat transfer coefficient (h_{st}) for an isolated or single tube depending upon average bed particle size. Equation 5 is a modified Vreedenberg[4] form and applies primarily to beds with particles less than an 800-micron average. Equation 6, a Glicksman-Decker[5] type, applies well when the average particle size in the bed exceeds 800 microns.

$$h_c = (h_{st})(FAB) \quad \textbf{(4)}$$

$$h_{st} = 900\,(1-e)\left(\frac{k}{d_t}\right)\left[\left(\frac{G d_t \rho_s}{\rho_g \mu}\right)\left(\frac{\mu^2}{D_p^3 \rho_s^2 g}\right)\right]^{0.326}(\mathrm{Pr})^{0.3} \quad \textbf{(5)}$$

for $D_p < 800$ microns

$$h_{st} = \frac{k(1-e)}{D_p}\left[C_1 + (C_2)\left(\frac{3600\, D_p \rho_g C_p V}{k}\right)\right] \quad \textbf{(6)}$$

for $D_p > 800$ microns

where

h_{st} = convective heat transfer coefficient for a single tube, Btu/h ft^2 F
e = bed voidage, dimensionless
k = gas thermal conductivity, Btu/h ft F
d_t = tube outside diameter, ft
G = mass velocity or flux of the gas, lb/s ft^2
ρ_s = particle density, lb/ft^3
μ = gas viscosity, lb/ft s
ρ_g = gas density, lb/ft^3
D_p = average particle diameter, ft
g = acceleration constant, 32.2 ft/s^2
Pr = Prandtl number, dimensionless
C_1 = experimental constant, dimensionless
C_2 = experimental constant, dimensionless
C_p = gas specific heat, Btu/lb F
V = nominal bed gas velocity, ft/s

To convert the single-tube heat transfer coefficients to those suitable for tube banks, the following equation is applied:

$$FAB = \left[1 - \left(\frac{D_o}{S_n}\right)\left(\frac{2D_o + S_p}{D_o + S_p}\right)\right]^{0.25} \quad \textbf{(7)}$$

where

FAB = bank arrangement factor (staggered arrangement only), dimensionless
D_o = tube outside diameter
S_n = tube spacing normal to flow
S_p = tube spacing parallel to flow

Other variables are as defined previously. The equation for *FAB* is as derived by Gel'perin.[6]

For the radiation heat transfer component, h_r, the following equation can be used:

$$h_r = (\sigma)(\varepsilon)\left[(T_b)^4 - (T_w)^4\right] / (T_b - T_w) \quad \textbf{(8)}$$

where

σ = 0.1713×10^{-8} Btu/h ft^2 R^4
ε = average overall emissivity, dimensionless
T_b = absolute bed gas temperature, R
T_w = absolute wall temperature, R

The average overall emissivity in bubbling beds will be about 0.8 depending on wall emissivity and particle size. Typically, the overall heat transfer coefficient (Equation 3) for an in-bed tube bundle is between 40 and 60 Btu/h ft^2 F (227 and 341 W/m^2 K).

BFB emissions

Biomass firing can require the control of carbon monoxide (CO), volatile organic compounds (VOCs), NO_x, SO_2 and hydrochloric acid (HCl). The CO and VOCs are controlled by good fluidization, uniform fuel distribution and high-velocity overfire air nozzles, and are easily controlled below 100 ppm and 10 ppm respectively at 7% O_2 dry. NO_x is controlled by bed temperature control with a two-stage overfire air system and/or a selective non-catalytic reduction (SNCR) system (see Chapter 34). The staged overfire air system provides approximately 15 to 25% reduction, and the SNCR system provides approximately 55 to 60% reduction. SO_2 and HCl are controlled by adding limestone to the bed. SO_2 can be reduced 80% with a significant Ca/S molar ratio because of the very low sulfur content in most biomass fuels. However, the air flow through the bed must be high enough to complete the sulfation process. HCl is reduced by the excess lime leaving the bed and by using a fabric filter baghouse for particulate collection, providing the proper temperature and solids/gas contact. See the *CFB emissions* section for the limestone and sulfur reactions.

Coal firing requires SO_2 emission control. SO_2 reduction can be as high as 90% with a high-solids recycle rate from a multi-cyclone dust collector (MDC). NO_x is only moderately low from a coal-fired BFB boiler. The bed and furnace temperatures are low, which reduces thermal NO_x formation. However, air flow through the bed is at or above theoretical combustion air. Therefore, the NO_x reduction possible by staging the combustion air is not available for coal-fired BFB boilers. Typical NO_x values are 0.4 to 0.5 lb/10^6 Btu (0.47 to 0.59 g/Nm^3 at 7% O_2 dry). Additional information on NO_x control is provided in the CFB discussion.

Circulating fluidized-bed (CFB) boilers

Fig. 11 shows the main features of a circulating fluidized-bed boiler and Fig. 12 shows the furnace density profile. 50 to 70% percent of the total combustion air enters the furnace through the windbox and air distributor with the balance of the combustion air injected through overfire air (OFA) ports. A typical flue gas superficial velocity at full load, thereby converting the process to a circulating bed above the OFA ports, is 16 to 17 ft/s (4.9 to 5.2 m/s). While single particles reaching the furnace exit could be up to 2000 microns in size, typical average particle size (SMD) is 100 to 200 microns in the upper furnace and 300 to 400 microns in the dense bed.

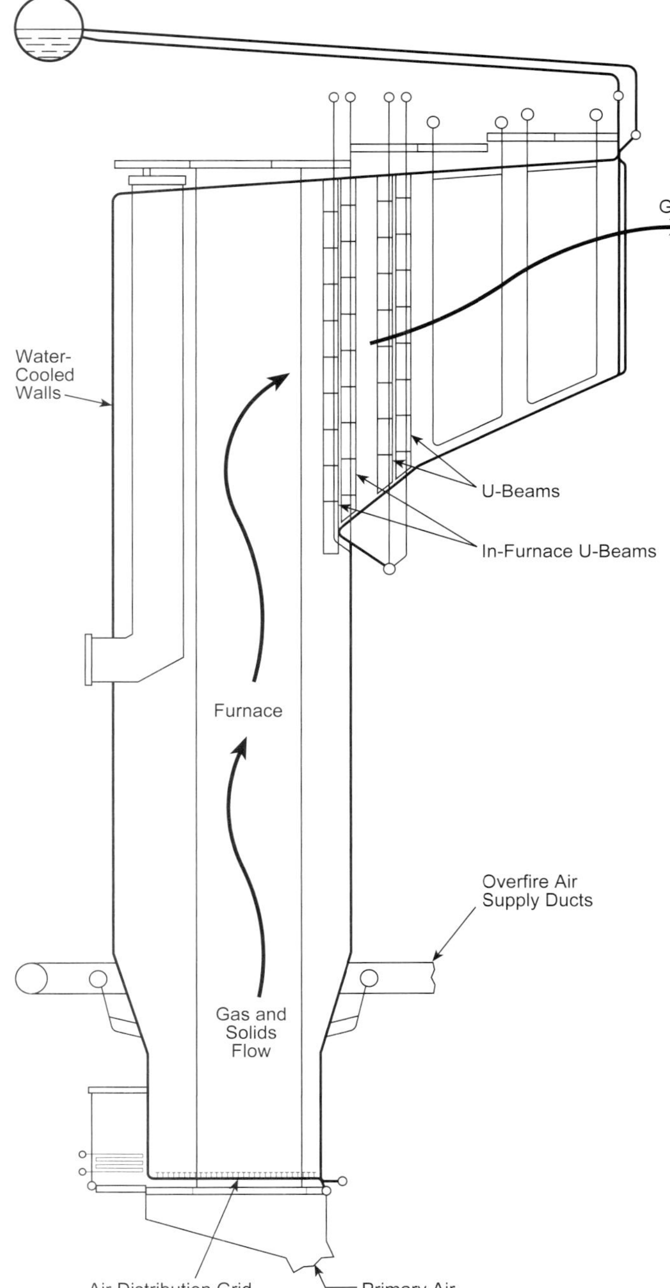

Fig. 11 Typical circulating-bed boiler schematic.

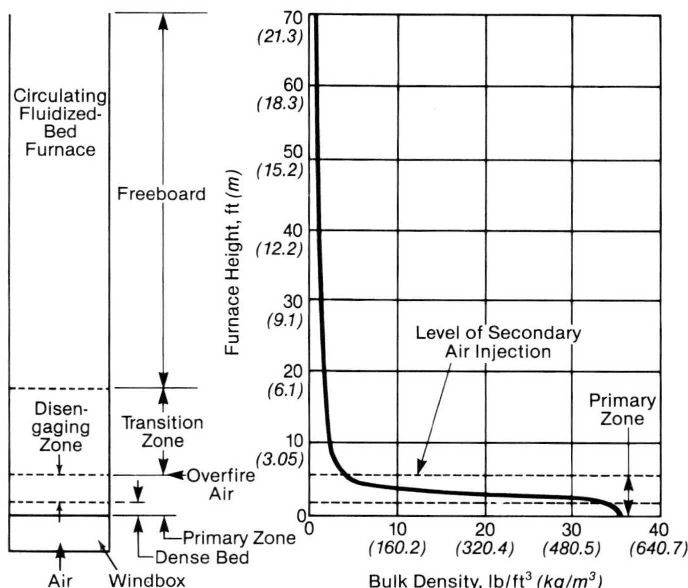

Fig. 12 Typical atmospheric pressure circulating-bed furnace density profile.

The upward flow of solids decreases with increased furnace height as heavier particles recirculate back down the furnace, resulting in declining local density with increasing furnace height. In the B&W internal recirculation (IR-CFB) design, U-beam collectors (see Fig. 13) located in the top of the furnace enclosure collect most (up to 97%) of the solids that remain in the flue gas and return them to the furnace (recirculation) while passing the flue gas to the convection pass heat transfer surfaces. Most of the remaining solids entrained in the flue gas are then collected in multi-cyclone dust collectors (MDC) located in the backpass, providing up to 99.7% overall particle recycle to the furnace. The furnace enclosure is water-cooled membrane panels.

Unlike coal-fired BFB boilers, the dense bed does not contain any in-bed tube bundle heating surface. The furnace enclosure and in-furnace heating surfaces (water-cooled panels or water/steam wing walls) provide the required heat removal surface. This is possible because of the large quantity of solids that are recycled internally within the furnace. Because the mass flow rate of recycled solids is many times the mass flow rate of the incoming fuel, limestone, air and resultant combustion gas, the bed solids temperature remains relatively uniform throughout the furnace height. Also, the heat transferred to the furnace walls is adequate to provide the heat absorption required to maintain the target bed temperature of 1500 to 1600F (816 to 871C).

B&W IR-CFB boilers use a single chute-type feeder installed above the air distributor. The upward action of the bed inventory and the intense mixing of the incoming fuel, with the large mass of active inventory, adequately distributes the fuel as compared to the BFB.

The B&W coal-fired CFB uses the flat floor air distributor system shown in Fig. 7, with the horizontal water-cooled membrane panel floor and bubble caps discussed previously. Separate bed drains are provided.

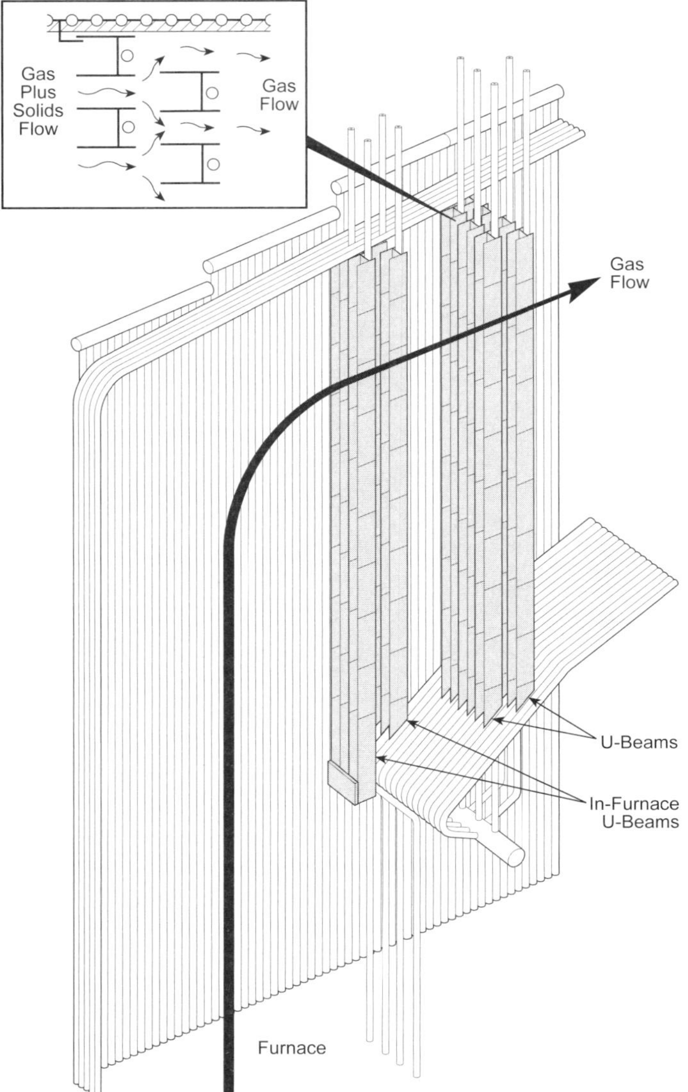

Fig. 13 CFB primary particle collection system.

Key operating requirements for a CFB are controlling furnace temperature and maintaining the vertical temperature distribution within a relatively narrow temperature window for SO_2 absorption (a temperature-dependent process). To accomplish this, the solids inventory is distributed throughout the furnace with a large amount of solids recirculation. A major portion of the heat transfer in a CFB is controlled by the solids inventory in the furnace. Therefore, by changing furnace inventory, the heat transfer from the gas and solids to the furnace walls can be varied and the furnace temperature can be controlled.

Fig. 14 shows the side view of a B&W 150 MW IR-CFB utility unit designed to burn bituminous coal, providing main and reheat steam to the turbine island. The unit contains full-height, water-cooled panels and steam-cooled wing walls in the upper furnace. Boiler design is based on a completely water-cooled setting. This feature provides a gas-tight enclosure suitable for operating with a positive pressure in the furnace. The B&W IR-CFB has no high-temperature refractory-lined flues, and thus has reduced furnace refractory maintenance. This construction is possible due to the U-beam particle collectors integrated into the boiler enclosure.

Fuel and sorbent are fed to the bed through the lower furnace walls. Ash and spent sorbent are removed through drain pipes in the floor or lower walls. Solids collected by the U-beams are returned directly to the furnace and solids collected in the multi-cyclones are returned to the lower furnace through the rear wall.

Primary air enters the furnace through the air distributor and secondary air is injected above the air distributor.

The lower furnace above the air distributor is covered by a thin layer of highly conductive refractory held to the water wall tubes by pin studs. Refractory in the lower furnace protects the tubes from corrosion and erosion. The remaining portion of the furnace enclosure consists of bare tubes.

Selected key issues for CFB design include: furnace temperature and heat transfer control, solids separation, solids inventory control, and emissions.

Furnace temperature and heat transfer control

A CFB furnace contains a substantial solids inventory that is distributed over the entire furnace volume. Fig. 15 shows the solids flow streams. Although the solids concentration or solids bulk density in the upper portion of the furnace is much less than that in the furnace bottom, the solids still represent a significant mass fraction of the gas/solids mixture at any given furnace location. For example, in a pulverized-coal (PC) furnace, the combustion gas carries a portion of the fuel ash as it flows through. In general, this PC fuel ash represents less than 10 lb (4.54 kg) of solids per 1000 lb (454 kg) of gas. Also, the PC heat transfer from the gas to the furnace enclosure walls is predominately by radiation. In a CFB furnace, the amount of solids in the gas leaving the furnace may

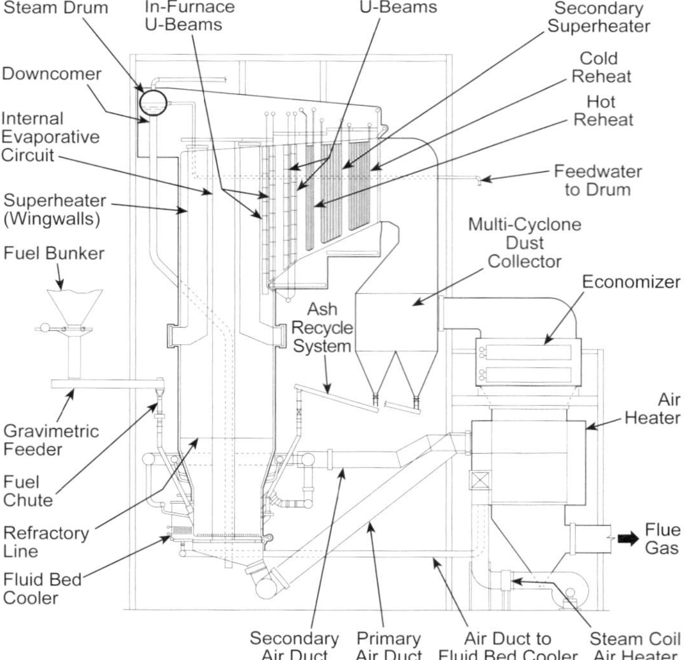

Fig. 14 150 MW utility reheat internal recirculation circulating fluidized-bed (IR-CFB) boiler.

exceed 5000 lb (2268 kg) of solids per 1000 lb (454 kg) of gas. As a result, the heat transfer to the walls of a circulating bed includes significantly higher solids/gas convection and less solids/gas radiation.

Furnace temperature is primarily controlled by furnace heat transfer to the pressure parts. Most of the furnace heat absorption is accomplished by the surface located in the zone above the overfire air level. The main factor controlling furnace heat transfer rate is local solids bulk density; the greater the bulk density, the higher the heat transfer rate (see Fig. 16) and the lower the furnace exit gas temperature. Most of the solids leaving the furnace are captured by the U-beams and returned to the upper portion of the furnace by gravity.

A secondary solids collection device, or multi-cyclone, is located downstream in the convection pass. The solids from the multi-cyclone hoppers are returned to the furnace, and this recycle stream controls the solids bulk density in the upper furnace. As more multi-cyclone solids are returned to the furnace, the upper furnace bulk density increases and the furnace exit gas temperature is reduced. At steady-state boiler operation, a balance is maintained between the solids flow to the multi-cyclone hoppers and the solids flow to the furnace, with any excess solids purged to the plant ash disposal system.

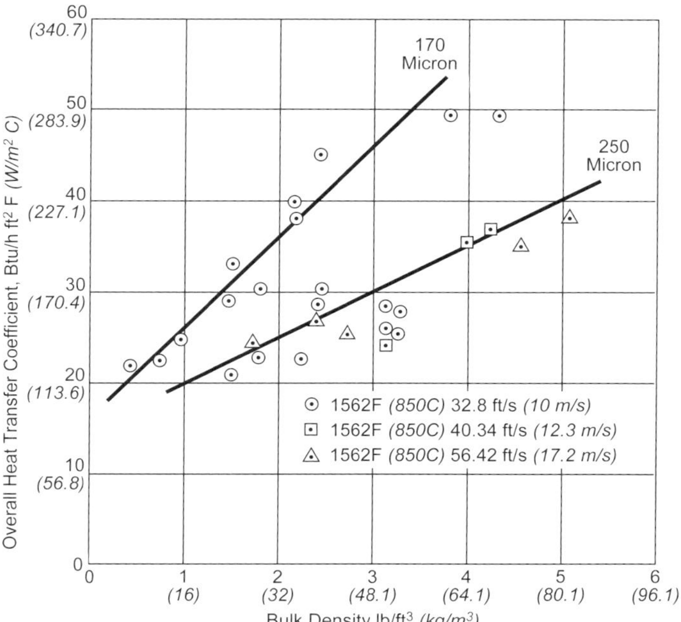

Fig. 16 Heat transfer coefficient versus density in a circulating fluidized bed for 170 and 250 micron mean diameter sand.

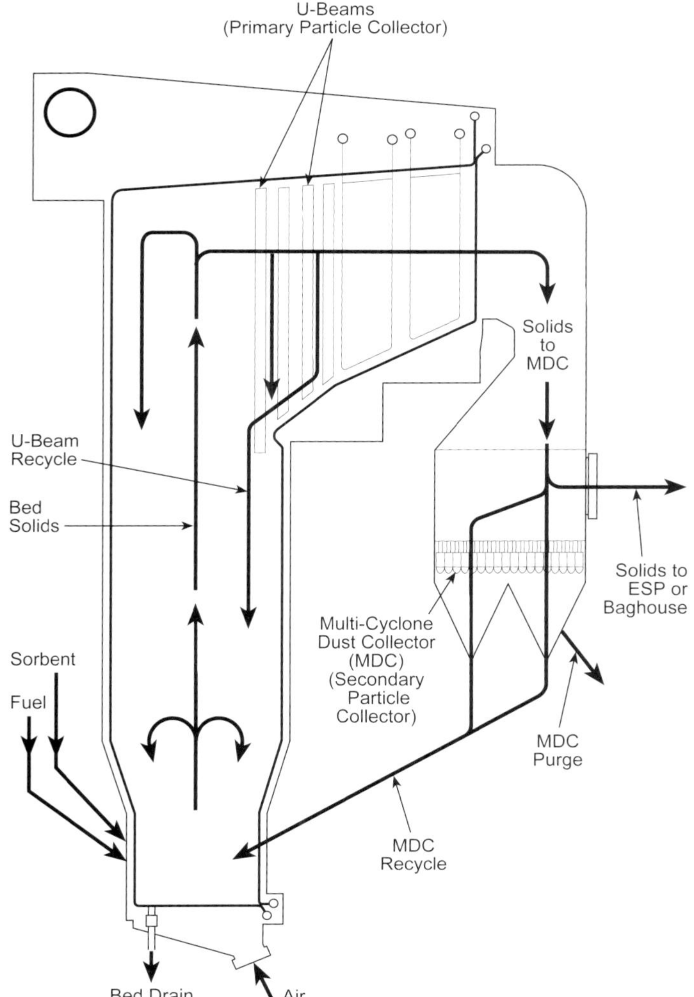

Fig. 15 CFB boiler solid material flows.

The solids recycle flow to the furnace can be limited by not having enough solids caught by the multi-cyclone. If, after reaching the maximum recycle rate, the upper furnace temperature is still high, increasing the primary air flow entrains more solids from the lower bed to the upper furnace. This increases the upper furnace bulk density and heat transfer rate. Changes in bed air flow can have an additional impact on the temperature in the lower furnace, which is fuel dependent. For some fuels, the increase in bed air flow causes an increase in the lower furnace temperature due to higher local fuel burnout. For others, the diluting effect of greater bed air flow prevails, causing the local temperature to decline.

Solids separation

A unique feature of B&W's CFB boiler design is the two-stage solids separation system shown in Fig. 15. The primary solids separator is an array of U-shaped beams (U-beams) located in the high temperature region at the furnace exit, with the secondary stage located downstream of the convection surfaces in a low gas temperature region, which varies from 400 to 900F (204C to 482C) depending upon unit design. As the gas flow passes the U-beams, the momentum of the particles causes a large fraction to be collected within the U-beam channel while the gas and remaining fine particulate fraction flows around the beams. As shown in Fig 13, the first two rows of U-beams are installed in the furnace (in-furnace) and the collected material is returned directly to the furnace, falling down along the rear wall. The second group of two to four U-beam rows after the furnace exit plane (external U-beams) collects additional material that falls down and also returns to the furnace.

The range of overall IR-CFB solids collection efficiency by particle size is shown in Fig. 17. The U-beams and MDC effectively collect and recycle all particles greater than 80 microns. The overall efficiency of the in-furnace and external U-beams is 97% or greater.

The fine particles fraction that passes the U-beams is collected in the secondary stage of solids separation, the mechanical dust collection multi-cyclones. These collect 90 to 95% of the remaining particles, for an overall collection efficiency of up to 99.7%.

Solids inventory control

The measure of furnace solids inventory is by air/flue gas pressure loss. The total furnace pressure loss or solids inventory is important. The typical overall furnace pressure loss is 42 in. wg (10.5 kPa). The upper furnace (above the air ports) pressure loss should be at least 17 in. wg (4.2 kPa), which results in adequate inventory in the upper furnace to maintain a constant furnace temperature profile.

Flue gas pressure loss in a CFB furnace conforms to the basic equation:

$$\Delta P = (C)\,(\rho_b)\,(L) \qquad \textbf{(9)}$$

where

ΔP = pressure loss
C = units conversion constant
ρ_b = average bulk density in the furnace segment associated with L
L = height of the furnace segment of interest

To use Equation 9, a density profile as shown in Fig. 12 is developed. This curve is a function of many variables and has been derived from empirical data. The important variables are:

D_p = average particle size above the dense bed zone
D_{DB} = average particle size in the dense bed zone
V = nominal gas velocity
T = nominal furnace temperature
W_S = external solids flux, lb/h ft^2 (kg/s m^2)
ρ_s = particle density
$\o$ = particle shape factor
D_e = furnace equivalent diameter

For a fluidized bed to operate properly, there must be a continuous and sufficient supply of particles of the proper size distribution. If the particles are too coarse, the lower furnace bed will be too deep and defluidize, or slump. If the particles are too fine, they will blow out of the furnace making it impossible to maintain an adequate overall solids inventory. There is a range of bed particle size distribution that is needed to maintain a stable fluidized-bed process in the lower furnace and an adequate inventory in the upper furnace to control the furnace exit gas temperature. The supply and retention of these particles must be controlled to provide the required inventory.

The solids inventory is maintained by adding makeup material, typically limestone or sand of the proper size distribution or bed material that had previously been drained and saved. When firing coal, limestone will normally be used for both the inventory control or makeup material. However, there needs to be enough sulfur to continuously generate material that will stay in the system. Calcium oxide is quite soft and does not hold up well in the fluidized bed environment. Once sulfated, the calcium sulfate is hard enough to maintain a solids inventory.

The amount of makeup material needed depends on whether, or to what degree, the fuel ash is contributing to the inventory and how well the inventory material maintains its integrity. For example, coal may contain very fine ash as well as large diameter particles or small rocks. A portion of the larger particles will remain in the lower portion of the furnace and provide stable fluidization in the dense bed portion. The small particles can be so small that the U-beams and multi-cyclone dust collector do not capture and return them back to the process. The intermediate size particles blow out of the dense bed, are large enough to be captured by the U-beams or dust collector, and return to the furnace for solids inventory. If the coal ash and reacted limestone particles continuously provide the proper amount of large and intermediate size particles, purchased makeup material for inventory control will not be required.

CFB emissions

Sulfur dioxide emissions control When sulfur-bearing fuels burn, most of the sulfur is oxidized to SO_2. Limestone can be added as a sorbent for sulfur capture. When limestone is added to the bed, it undergoes a transformation called calcination and then reacts with the SO_2 in the flue gas to form calcium sul-

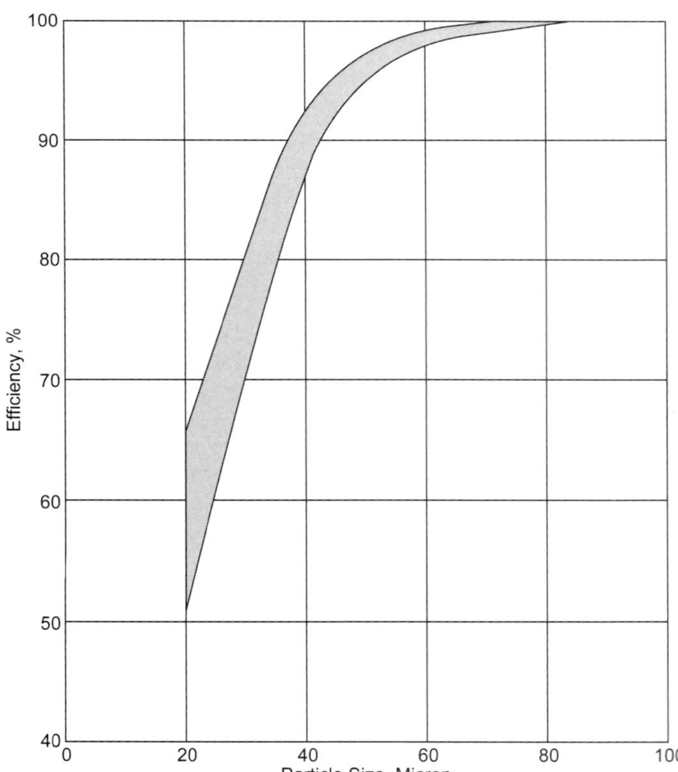

Fig. 17 Overall grade efficiency range of B&W IR-CFB solids collection system.

fate ($CaSO_4$) through *sulfation*. The calcining reaction is endothermic and is described by:

$$CaCO_3(s) + 766 \text{ Btu/lb} (\text{of } CaCO_3) \rightarrow CaO(s) + CO_2(g) \quad \textbf{(10)}$$

Once formed, solid CaO (lime) reacts with gaseous SO_2 and oxygen exothermically to form $CaSO_4$ according to the following reaction:

$$SO_2(g) + \tfrac{1}{2} O_2(g) + CaO(s) \rightarrow CaSO_4(s) + 6733 \text{ Btu/lb} (\text{of } S) \quad \textbf{(11)}$$

SO_2 reductions of 90 to 95% are typically achieved in a circulating bed with calcium to sulfur (Ca/S) mole ratios of 1:1.8 to 1:2.5, depending on the sulfur content of the fuel and the reactivity of the limestone. The lower the sulfur concentration in the fuel, the greater the calcium-to-sulfur mole ratio must be for a given removal in the furnace. Systems external to the CFB process can be applied to further enhance SO_2 capture and limestone utilization.

Certain temperature limitations affect the sulfur capture process. If the furnace temperature is below 1470F (799C), calcination of $CaCO_3$ is not complete and results in less CaO yield from a given amount of limestone, thereby increasing limestone consumption. If the furnace temperature is above 1650F (899C), $CaSO_4$ dissociates to CaO and SO_2 which again reduces sulfur capture efficiency. Therefore, it is important to maintain the furnace within a certain temperature window to reach minimum limestone consumption and not exceed a specified level of SO_2 emissions. This is a primary reason the IR-CFB combustion process controls furnace temperature over the entire furnace height.

Limestone properties such as reactivity, attrition, and size distribution have a major effect on limestone consumption. These parameters are carefully considered when providing recommendations for limestone selection and preparation. The IR-CFB process also allows for post-furnace SO_2 capture through various flyash recirculation schemes.

Nitrogen oxides emissions control The nitrogen oxides present in the flue gas come from two sources: the oxidation of nitrogen compounds in the fuel (fuel NO_x) and the reaction between nitrogen and oxygen in the combustion air (thermal NO_x). See Chapter 34. By maintaining the furnace temperature below 1650F (899C), thermal NO_x formation is low in CFB boilers.

Fuel NO_x can be reduced through combustion controls. The following operating parameters have a significant effect on emissions:

1. lowering primary air,
2. lowering overall excess air, and
3. minimizing limestone input or Ca/S molar ratio.

When burning fuels containing nitrogen, volatile nitrogen is released in a very unstable monatomic state. The monatomic nitrogen atom will either combine with another nitrogen atom to form N_2, or will react with oxygen to form NO_x. Part of the NO_x can be reduced back to N_2 when NO_x is in the presence of char and/or CO in a reducing atmosphere. Reducing the primary air to substoichiometric conditions increases the concentration of carbon and CO in the lower furnace and limits the availability of oxygen in the presence of monatomic nitrogen. This effect is more pronounced for high-volatile fuels.

Limestone is a catalyst for the nitrogen and oxygen reaction, which increases the fuel NO_x emissions. Controlling to a low Ca/S ratio minimizes this reaction.

Higher overall excess air means higher oxygen concentration in furnace gases and increased NO_x emissions. Limiting the overall excess air below 3% O_2 minimizes NO_x formation while providing low carbon loss due to high solids recycle to the furnace.

The combination of low temperatures and staged combustion allows fluidized-bed boilers to operate with low NO_x emissions. Typical uncontrolled values are within 0.1 to 0.15 lb/10^6 Btu (126 to 188 mg/Nm^3 at 6% O_2 dry) for coal-fired CFB boilers. Further NO_x can be controlled to lower values through the use of a selective noncatalytic reduction (SNCR) system consisting of ammonia injection near the U-beam elevation (see Chapter 34).

Carbon monoxide and hydrocarbons When designing a boiler, it is necessary to maximize combustion efficiency by minimizing unburned carbon and the quantity of CO and hydrocarbons in the flue gas. This is accomplished by choosing the proper number of fuel feed points, by proper design of the overfire air system, and by providing sufficient furnace residence time for combustion. Typical flue gas concentrations are less than 0.15 lb/10^6 Btu (188 mg/Nm^3) for CO and 0.005 lb/10^6 Btu (6.3 mg/ Nm^3) for hydrocarbons, both at 6% O_2 dry, in a CFB boiler burning coal.

Pressurized fluidized-bed combustion (PFBC)

BFB and CFB boilers discussed above operate effectively at atmospheric pressure and are generally referred to as atmospheric fluidized-bed combustors (AFBC). Pressurized fluidized-bed combustors (PFBC) are an outgrowth of this technology. The underlying concept is to create a combined cycle plant where a coal-fired system with a steam turbine is combined with a gas turbine to increase overall cycle efficiency. A simplified schematic and process explanation is shown in Fig. 18. In its simplest form, a compressor pressurizes the combustion air to 12 to 20 atmospheres (174 to 232 psi or 1.2 to 2.0 MPa). This air is then fed to a pressure vessel that fully encloses a fluidized-bed combustor where coal is burned in a bed of limestone (removing the SO_2) to produce steam plus a pressurized gas stream. The steam is used in a conventional steam turbine cycle to generate part of the electric power, and the pressurized gas is cleaned sufficiently of particulate and other key contaminants and used in a gas turbine to generate the balance of the electric power and compressed air for the combustion process.

Three basic cycle configurations have been proposed: 1) turbocharged cycle, 2) combined cycle, and 3) advanced combined cycle. The major differences involve how the gas from the pressurized fluidized-bed combustor is treated prior to use in the gas tur-

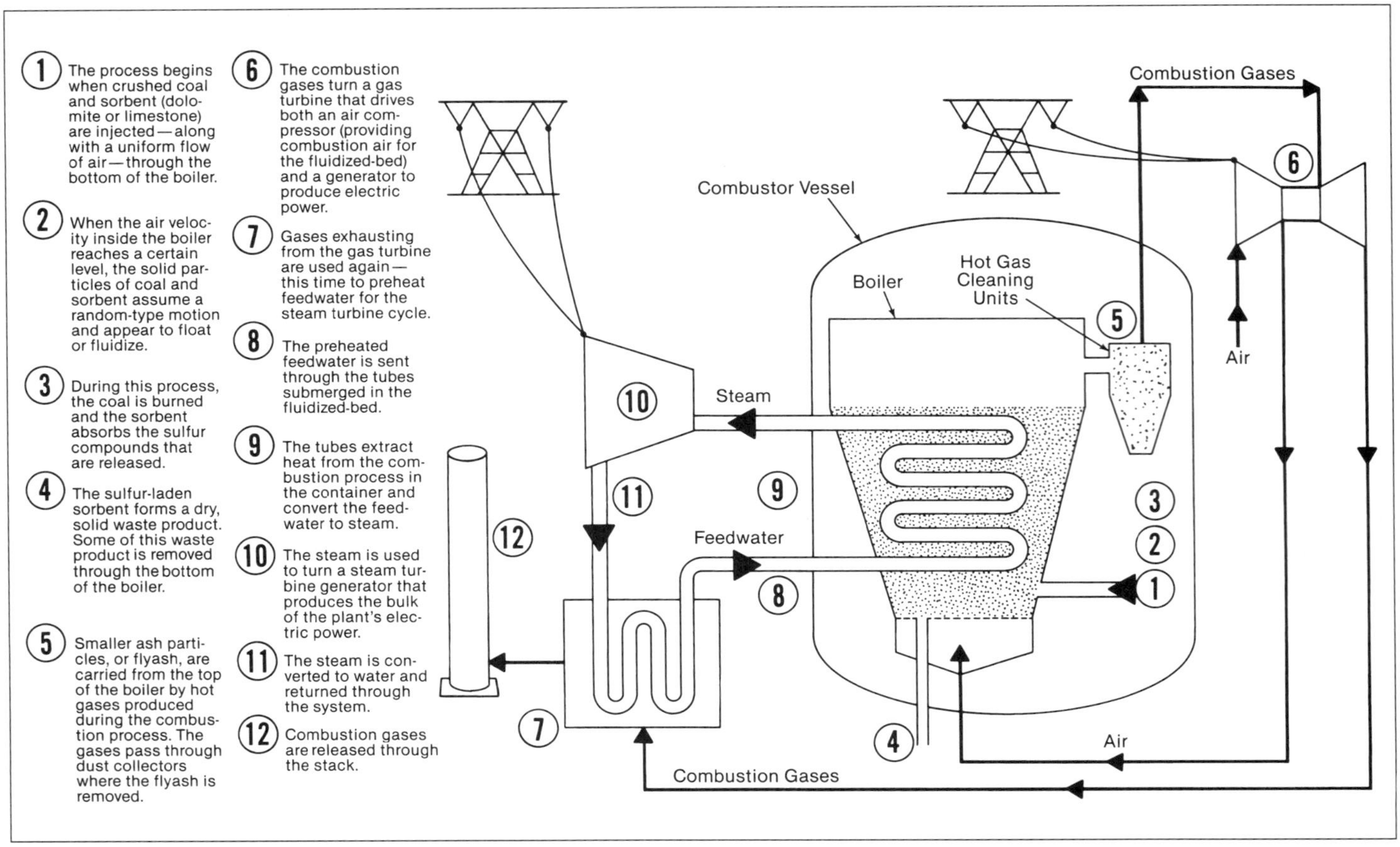

1. The process begins when crushed coal and sorbent (dolomite or limestone) are injected—along with a uniform flow of air—through the bottom of the boiler.
2. When the air velocity inside the boiler reaches a certain level, the solid particles of coal and sorbent assume a random-type motion and appear to float or fluidize.
3. During this process, the coal is burned and the sorbent absorbs the sulfur compounds that are released.
4. The sulfur-laden sorbent forms a dry, solid waste product. Some of this waste product is removed through the bottom of the boiler.
5. Smaller ash particles, or flyash, are carried from the top of the boiler by hot gases produced during the combustion process. The gases pass through dust collectors where the flyash is removed.
6. The combustion gases turn a gas turbine that drives both an air compressor (providing combustion air for the fluidized-bed) and a generator to produce electric power.
7. Gases exhausting from the gas turbine are used again—this time to preheat feedwater for the steam turbine cycle.
8. The preheated feedwater is sent through the tubes submerged in the fluidized-bed.
9. The tubes extract heat from the combustion process in the container and convert the feedwater to steam.
10. The steam is used to turn a steam turbine generator that produces the bulk of the plant's electric power.
11. The steam is converted to water and returned through the system.
12. Combustion gases are released through the stack.

Combustion Gases
Combustor Vessel
Boiler
Hot Gas Cleaning Units
Steam
Feedwater
Air
Combustion Gases
Air

Fig. 18 Simplified combined cycle PFBC diagram and process overview.

bine. In the turbocharged cycle, the gas is cooled to between 800 and 1000F (427 and 538C) so that current state-of-the-art hot gas cleanup technology can be used to remove the particulate and other contaminants. This increases the steam production capacity but reduces the power from the gas turbine. This cycle incorporates more conventional technology at the expense of cycle efficiency. In the full combined cycle system, advanced hot gas cleanup technology is used to clean the gas from the fluidized bed at approximately 1580F (860C) so that hotter and higher pressure gas can be sent to the gas turbine. In this case, 80% of the power is produced in the steam turbine and the balance is produced in the gas turbine. Finally, in the advanced combined cycle, a partial coal gasification process is added. The char from gasification is burned in the fluidized-bed combustor. The clean gas leaving the PFBC boiler, which still contains oxygen, can be mixed with the fuel gas (from the partial gasifier) and then burned in a gas turbine combustor at temperatures of 1800 to 2500F (982 to 1371C) for even higher efficiencies.

The bubbling fluidized bed operating inside the pressure vessel behaves basically the same as the atmospheric fluidized bed although the higher pressure (and higher gas density) combustion affects three key areas: fluidization, heat transfer, and combustion. The overall effect is to provide a very compact and cleaner combustion system. A more complete summary of this technology is provided in the 40th edition of this text.[7]

B&W designed, fabricated and installed a 70 MW PFBC system as a demonstration to repower one unit at the American Electric Power, Ohio Power Company Tidd plant near Brilliant, Ohio, in the United States (Fig. 19). The design was based on the combined cycle configuration utilizing the bubbling fluidized-bed combustion process. This PFBC combined cycle system consisted of a gas turbine, boiler, and associated systems including hot-gas cleaning, load control, fuel

Fig. 19 Pressure vessel arrival at Tidd site.

preparation and feeding, sorbent preparation and feeding, and ash removal. Many of these systems were either partially or totally contained in the pressure vessel. This system operated on an experimental basis in the 1990s to demonstrate the feasibility of the process and to define and overcome the challenges in a new first-of-a-kind engineered power system.

PFBC technology offers the prospect of improving cycle efficiency and reducing emissions while reducing power plant size and initial plant cost. In concept, PFBC boiler modules could be barge-shipped to sites effectively ready for installation or in a few large pieces to minimize erection expense. This technology awaits further development and evaluation.

References

1. Leva, M., *Canadian Journal of Chemical Engineering,* Vol. 35, pp. 71-76, August, 1957.
2. Babu, S.P., Shah, B., and Talwalkar, A., "Fluidization Engineering," Second Ed., AIChE Symposium Series No. 176, Vol. 74, pp. 176-186, American Institute of Chemical Engineers, and Kunii, D., Levenspiel, O., Butterworth-Heinemann Series in Chemical Engineering, 1991.
3. Staub, F.W., and Canada, G.S., *Fluidization,* Cambridge University Press, London, United Kingdom, pp. 339-344, 1978.
4. Vreedenberg, H.A., *Chemical Engineering Science,* Vol. 9, pp. 52-60, 1958.
5. Glicksman, L.R., and Decker, N.A., *Proceedings of the Sixth International Fluidized-Bed Combustion Conference*, Atlanta, Georgia, pp. 1152-1158, 1980.
6. Gel'perin, N.I., Ainshtein, V.G., and Korotyanskaya, L.A., *International Chemical Engineering,* Vol. 9, No. 1, pp. 137-142, January, 1969.
7. Stultz, S.C., and Kitto, J.B., Eds., *Steam/its generation and use*, 40th Ed., The Babcock & Wilcox Company, Barberton, Ohio, 1992.

Portion of an entrained-flow gasifier under fabrication at B&W.

Chapter 18

Coal Gasification

What is gasification

Coal gasification is a process that converts coal from a solid to a gaseous fuel through partial oxidation. Once the fuel is in the gaseous state, undesirable substances, such as sulfur compounds and coal ash, may be removed from the gas by various techniques to produce a clean, transportable gaseous energy source.

When coal is burned, its potential chemical energy is released in the form of heat. Oxygen from the air combines with the carbon and hydrogen in the coal to produce gaseous carbon dioxide (CO_2) and water plus thermal energy. Under normal conditions, when there is sufficient oxygen available, nearly all of the chemical energy in the coal is converted into heat and the process is called *combustion*.

However, if the available oxygen is reduced, less energy is released from the coal and new gaseous reaction products appear from the incomplete combustion. These byproducts, including hydrogen (H_2), carbon monoxide (CO) and methane (CH_4), contain unreleased or potential chemical energy which can be transported and used to provide energy to other processes such as combustion to power a gas turbine, chemical synthesis for use in fuel cells, or as a substitute for natural gas. If the objective is to maximize the chemical energy in the gas byproducts, it would appear logical to continue decreasing the available oxygen. However, a point is reached where an increasing percentage of the carbon in the coal remains unreacted, and less is converted to gas, rendering the process inefficient. When the oxygen supply is controlled such that both heat and a new gaseous fuel are produced as the coal is consumed, the process is called *gasification*.

Coal gasification has been practiced for more than 200 years and until the 1940s was a major industry. The most significant advances leading to current gasification technologies occurred during World War II in Germany, where coal was used to derive liquid fuels for military use. Following this, the increased availability of natural gas and petroleum products in the industrialized nations resulted in a decline of gasification until the oil embargo of 1973 and concerns about natural gas supply. Major efforts to develop gasification systems to produce synthetic natural gas (SNG) and liquid fuels were made in the United States (U.S.) and Europe during the 1970s and 1980s, but declines in gasoline and natural gas prices curtailed many of these efforts as economic justification declined. Recently, a desire for high-efficiency electric power generation to reduce emissions, availability of high-performance gas turbines, and the introduction of fuel cells have resulted in a resurgence of interest in gasification systems.

By recovering waste heat from a gasifier and gas turbine exhaust for steam production, a gas turbine (Brayton) cycle and steam turbine (Rankine) cycle can be efficiently combined into an *integrated gasification combined cycle* (IGCC). Although technically viable, this approach to power generation from coal has been demonstrated but not applied commercially because conventional combustion systems are less costly and less complicated. However, further development continues as increasingly stringent environmental regulations and interest in production of hydrogen for fuel cells and other co-products have renewed interest in IGCC.

Basic gasification reactions

Combustible gases such as carbon monoxide and hydrogen are the most common products of incomplete combustion. With partial combustion, only a fraction of the carbon in the coal is completely oxidized to CO_2. The heat released provides most of the energy necessary to break chemical bonds in the coal and raise the products to reaction temperature.

Although the chemistry of coal gasification is complex, the following are the major reactions involved:

Exothermic reactions – releasing heat

Carbon combustion: $$C + O_2 = CO_2 \quad (1)$$

$$C + 1/2\ O_2 = CO \quad (2)$$

Water-gas shift: $$CO + H_2O = CO_2 + H_2 \quad (3)$$

Methanation: $$CO + 3H_2 = CH_4 + H_2O \quad (4)$$

Direct Methanation: $$C + 2H_2 = CH_4 \quad (5)$$

Endothermic reactions – absorbing heat

Boudouard reaction: $$C + CO_2 = 2CO \quad (6)$$

Steam-carbon reaction: $$C + H_2O = H_2 + CO \quad (7)$$

Hydrogen liberation: $$2H\ \text{(in coal)} = H_2\ \text{(gas)} \quad (8)$$

Many fuel and operating variables, such as fuel composition, operating pressure, and steam and oxygen inputs influence the final gas composition. Gasifiers are designed for pressures ranging from atmospheric to 600 psi (4140 kPa) and higher. The methanation reactions are important in lower temperature systems and are favored by high pressures as shown in Fig. 1.[1] The other reactions are more prominent in high temperature, lower pressure systems. The combination of these reactions is autothermic, meaning sufficient heat is initially released by the carbon combustion reactions to provide energy for the endothermic reactions.

Sulfur in coal is converted primarily to hydrogen sulfide (H_2S) and a small amount of carbonyl sulfide (COS). High temperatures and low pressures favor coal nitrogen conversion to N_2, while the opposite conditions favor some ammonia (NH_3) formation and small amounts of hydrogen cyanide (HCN). In lower temperature processes [< 1200F (<649C)], tars, oils and phenols are not destroyed and therefore exit with the raw gas. Many other minor impurities are produced depending on their presence in the raw fuel. These include chlorides, vaporous alkali species and heavy metals. Heavy hydrocarbons are also produced in lower temperature gasifier processes.

The gasification reactions occur in two main stages: devolatilization and char gasification. Devolatilization is a transition stage in which coal becomes char, or fixed carbon, as the temperature rises. Weaker chemical bonds are broken and tars, oils, phenols and hydrocarbon gases are formed. The char that remains after devolatilization is gasified by reaction with oxygen, steam, carbon dioxide and hydrogen.

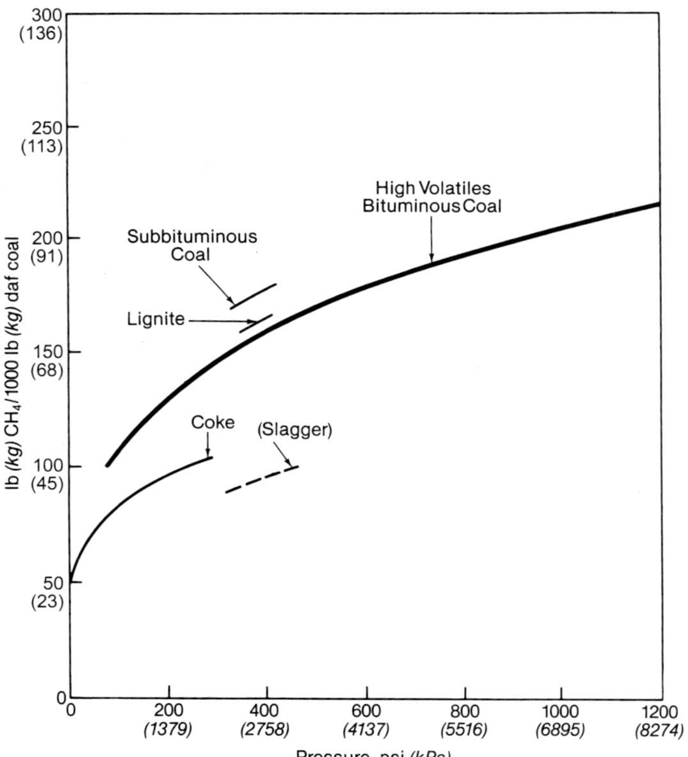

Fig. 1 Methane formation with pressure.[1]

The type of gasification process has a strong bearing on the devolatilization products. In fixed-bed or moving-bed gasifiers (see below), these products exit the gasifier with the product gas due to the low temperatures and lack of oxygen. In fluidized-bed and entrained-flow processes, uniform high temperature causes hydrocarbon cracking, or breaking down the more complex molecules to simpler ones. Also, oxygen is available to react with the devolatilization products, producing H_2, CO and CO_2. These reactions are most complete in an entrained-flow process.

The basic objective of gasification is to convert coal into a combustible gas containing the maximum remaining heating value. If the gas is used for chemical synthesis rather than fuel for combustion, its composition must be adjusted for the stoichiometry of the synthesized product. Hydrogen for synthesis gas (syngas) production must be provided by the steam-carbon reaction or by the water-gas shift. The influence of the end product may affect the choice of gasification process. The water-gas shift reaction can be promoted by supplying extra steam in the fuel bed or in a separate catalytic conversion stage downstream. This produces increased hydrogen. Where hydrogen is the only product required, CO_2, which is an inevitable product of splitting water by reaction with carbon, must be removed. Limestone may also be injected along with the coal and/or steam to capture sulfur.

In the case of synthetic natural gas (SNG) production with low temperature, high pressure moving-bed processes, much of the methane can be produced in the fuel bed by carbon hydrogenation (Reaction 5). However, all of the methane can not be produced by direct hydrogenation. Some must be produced indirectly in a separate catalytic stage, e.g., by Reaction 4.

Oxygen for the gasification process can be supplied either in pure form (oxygen-blown) or in air (air-blown). There are advantages and disadvantages to both methods. Oxygen-blown gasification results in smaller components, higher cold gas efficiency and purer gas composition, but requires additional oxygen separation and safety systems. The cold gas efficiency is defined as the ratio of the heating value of sulfur-free gas to coal heat content expressed as:

$$\frac{\text{HHV of sulfur-free gas at 60F}}{\text{HHV of feed coal}} \times 100 \qquad \textbf{(9)}$$

where

HHV = higher heating value

Raw gas from an oxygen-blown unit consists primarily of CO, H_2, CO_2, CH_4 and H_2O, while N_2 is a major constituent when air is used as the oxidant.

Coal gas has traditionally been classified by heating value, expressed in Btu per dry standard cubic foot [Btu/DSCF (MJ/Nm^3)].[1] Air-blown gasification produces a low Btu gas due to nitrogen dilution. This gas ranges from 90 to 200 Btu/DSCF (3.5 to 7.9 MJ/Nm^3). Oxygen-blown units produce a medium Btu gas at the lower end of the 270 to 600 Btu/DSCF (10.6 to 23.6 MJ/Nm^3) range. Post-gasifier shift and methanation reactions are required to produce heating values at the higher end of the medium Btu range and ultimately to produce high Btu gas of approximately 1000 Btu/DSCF (39.3 MJ/Nm^3). High Btu gas is pipeline quality and is therefore referred to as synthetic natural gas (SNG).

Raw gas composition as a function of gasifier type and oxidant is listed in Table 1. Heating values vary widely between product gases from air- and oxygen-blown moving-bed gasifiers, which can influence the combustion characteristics of the gas. Raw gas from a moving-bed gasifier contains significant quantities of ammonia and heavy condensable organics such as tars, oils and phenols. It also has a higher heating value than gas from entrained-flow processes due to the increased methane content. The two advanced entrained-flow gasifiers produce negligible quantities of organics due to their high temperature operation. The slurry-fed process produces a product gas lower in CO and higher in H_2, CO_2 and H_2O compared to the dry feed process. The data presented in Table 1 also contrast different gasification conditions.

Gasifier types and applications

Gasifier types (see Table 2) are described in terms of two parameters: 1) the direction of flow of the incoming oxidant relative to the outlet gas flow (up-flow, down-flow and cross-flow), and 2) the relative velocity of the solids (fixed or moving, fluidized, entrained, and transport).

Up-flow (countercurrent)

The oldest and simplest type of gasifier is the up-flow (countercurrent) gasifier as shown in Fig. 2a and 2d. The oxidant enters at the bottom of the gasifier vessel and the gas leaves at the top (up-flow), while the fuel flows in the opposite direction (countercurrent) of the gas. The combustion reactions occur near the grate at the bottom, followed by reduction reactions somewhat higher up in the gasifier. In the upper part of the gasifier, heating and pyrolysis of the feedstock occur as a result of heat transfer by forced convection and radiation from the lower zones. The tars and volatiles

Table 1
Raw Gas Composition for Selected Gasifier Types

Gasifier Type	Moving Bed, Dry[3]	Moving Bed, Slagging[2]	Fluidized Bed[2]	Entrained Flow (Slurry)[2]	Entrained Flow (Dry Feed)[2]
Oxidant:	Air	Oxygen	Oxygen	Oxygen	Oxygen
Fuel type:	Subbituminous	Bituminous	Lignite Note 3	Bituminous	Bituminous Note 3
Fuel analysis, % by wt					
C	41.1	61.2	56.9	61.2	66.1
H	4.6	4.7	3.8	4.7	5.0
N	0.8	1.1	0.8	1.1	1.2
O	20.5	8.8	15.9	8.8	9.5
S	0.6	3.4	1.0	3.4	3.7
Ash	16.1	8.8	9.6	8.8	9.5
Moisture	16.3	12.0	12.0	12.0	5.0
HHV, Btu/lb	11,258	11,235	9,914	11,235	12,128
(kJ/kg)	(26,189)	(26,136)	(23,063)	(26,136)	(28,213)
Fuel feed method	Dry	Dry	Dry	Slurry (66.5 wt % solids)	Dry
Operating pressure, psi	295	465	145	615	365
(kPa)	(2034)	(3206)	(1000)	(4240)	(2516)
Raw gas composition, % by vol					
CO	17.4	46.0	48.2	41.0	60.3
H_2	23.3	26.4	30.6	29.8	30.0
CO_2	14.8	2.9	8.2	10.2	1.6
H_2O	-- Note 1	16.3	9.1	17.1	2.0
N_2	38.5	2.8	0.7 Note 4	0.8 Note 4	4.7 Note 4
CH_4 + CnHm	5.8	4.2	2.8	0.3	--
H_2S + COS	0.2	1.1	0.4	1.1	1.3
NH_3+ HCN	-- Note 2	0.3	-- Note 2	0.2	0.1
HHV, Btu/dscf	196	333	309	278	297
(MJ/Nm^3)	(7.7)	(13.1)	(12.2)	(10.9)	(11.7)

Notes:
1. Dry analysis
2. Not reported
3. Dried fuel
4. Includes argon

Table 2
Gasifier Comparison

Parameter	Fixed (Moving) Bed	Fluidized Bed	Entrained Flow	Transport
Gas flow	Up-flow (countercurrent)	Up-flow (countercurrent)	Up-flow or down-flow (counter or co-current)	Up-flow (countercurrent)
Fuel size	Coarse < 2 in. (5 cm)	< 0.25 in. (6 mm)	Pulverized	
Coal rank	Low	Low	Any	Any
Oxygen required	Low	Moderate	Large	Moderate
Steam required	High	Moderate	Low	Moderate
Gas velocity	Low	Moderate	Medium	High
Typical temperature	800 to 1200F (427 to 649C)	1700 to 1900F (927 to 1038C)	2300F (1260C)	1600F (871C)
Types	Dry ash or slagging	Dry ash or agglomerating	Slagging	Dry ash
Key technical challenges	Utilization of fines and hydrocarbon liquids	Carbon conversion	Raw gas cooling	Coal feeding Raw gas cooling
Key developers	Lurgi, British Gas/Lurgi	Winkler, Kellogg Rust Westinghouse	Up-flow: Dow, Uhde, B&W Down-flow: Texaco, Shell	Kellogg Brown & Root (formerly MW Kellogg)

produced during this process remain in the gas stream while ashes are removed from the bottom.

The advantages of up-flow gasifiers are: 1) simplicity, 2) fuel flexibility, 3) high cold gas efficiency, and 4) high char burnout and internal heat exchange leading to low gas exit temperatures. Major drawbacks are: 1) potential for *channeling* that can lead to oxygen breakthrough and dangerous, explosive conditions, and 2) difficulty in disposing of tar-containing condensates resulting from the gas cleaning (except when the gas is used for direct heat applications where the tars are burned).

Down-flow (co-current)

In down-flow (co-current) gasifiers, the oxidant and fuel are introduced at the top of the gasifier. The product gas and ash or slag are removed at the bottom of the vessel. This solves the problem of tar entrainment in the gas stream. The down-flow configuration is shown in Fig. 2b. Depending upon where the oxidant is introduced (below the coal feed or with the coal feed as shown), there may be some or no (plug-flow) mixing of the gas with the particles.

As the gas flows downward, the acid and tarry distillation products from the fuel pass through the bed of burning char where they are converted to hydrogen, carbon dioxide, carbon monoxide and methane gases. The efficiency of this conversion depends on the temperature and residence time.

The advantage of down-flow gasifiers is the potential to produce tar-free gas. However, it is very difficult to achieve this goal over the entire operating range and turndown ratios of 3:1 are common; 5-6:1 is considered excellent. There are also some environmental advantages because the down-flow gasifiers produce a lower level of organic compounds in the condensate. The disadvantages are: 1) reduced fuel flexibility due to co-current flow, 2) greater susceptibility to slagging with high ash fuels, 3) lower cold gas efficiency due to poorer internal heat exchange resulting in a lower heating value of the gas, and 4) the need to maintain uniform high temperatures over the gasifier cross-sectional area.

Cross-flow

Cross-flow gasifiers, shown in Fig. 2c, are an adaptation for gasification of high fixed carbon, low volatile coals or char. Char gasification produces very high temperatures [2730F (1499C) and higher] in the oxidation zone, which can be a problem for materials of construction. In cross-flow gasifiers, insulation against these high temperatures is provided by the fuel (char) itself. Due to cross-flow gasifiers' inability to convert tars, only low volatile fuels can be used, making this type of gasifier unsuitable for power generation applications.

Fixed (Moving) bed

In a fixed-bed gasifier (Fig. 2a), also referred to as moving-bed, a column or bed of crushed coal is supported by a grate while the oxidant flows upward through the coal column at a velocity relative to the particle size and insufficient to lift or fluidize the particles. The gasification process involves a series of countercurrent reactions. At the top, the coal is heated and dried while cooling the product gas. The coal is further heated and devolatilized as it descends through the carbonization zone. Below this area, the devolatilized coal is gasified by reaction with steam and CO_2. The highest temperatures are reached in the combustion zone near the bottom of the gasifier. The char-steam reaction together with the presence of excess steam keeps the temperature in the combustion zone below the ash slagging temperature. Both the Lurgi (dry ash) and the British Gas/Lurgi (slagging) designs are fixed-bed gasifiers.

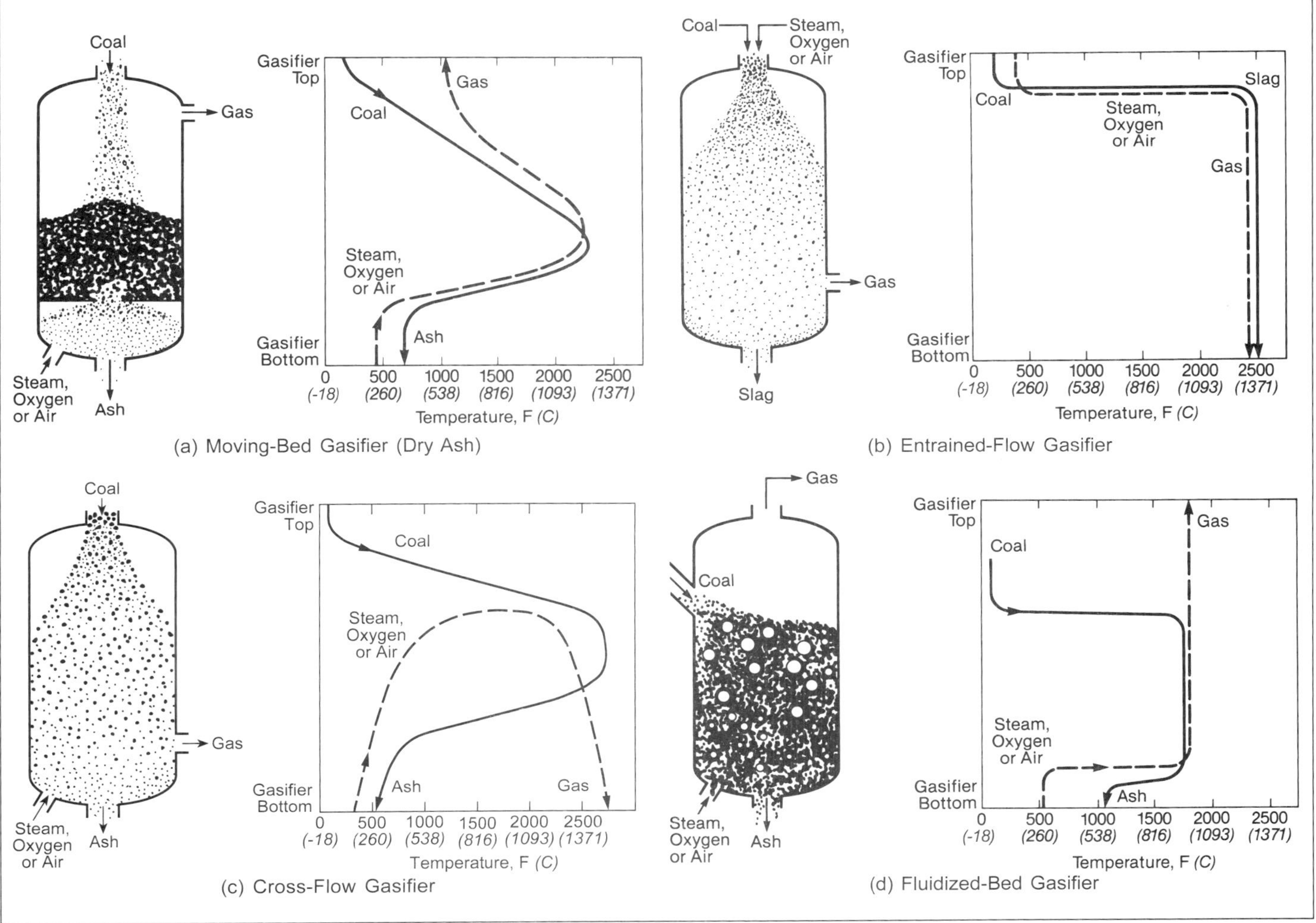

Fig. 2 Typical reactors for coal gasification (*courtesy of Electric Power Research Institute*).[2,3]

The distinguishing characteristics of a fixed-bed gasifier are:

1. produces hydrocarbon liquids such as tars and oils,
2. has limited ability to handle fine particles,
3. requires special steps to handle caking coals,
4. produces relatively high gas methane content, and
5. has low oxidant requirements.

The main differences among these gasifiers are the ash conditions (dry or slagging) and design provisions for handling fines, caking coals and hydrocarbon liquids. The first full scale fixed (moving)-bed coal gasification plant was constructed in 1936. Since then, 164 Lurgi gasifiers have been built by a number of companies. All of these gasifiers except one have been oxygen-blown units. The SASOL plant in South Africa uses 97 units to produce motor fuel from coal. In addition, the Dakota Gasification Company (Great Plains Synfuels Plant) at Beulah, North Dakota, has 14 gasifiers.

The configuration of the moving-bed dry ash gasifier is shown in Fig. 3. It is a high pressure cylindrical unit, operating at 350 to 450 psig (2410 to 3100 kPa). The main gasifier shell is surrounded by a cooling water jacket. Sized coal enters the top through a lock hopper and moves down the bed under the control of a rotating grate. The temperature in the combustion zone near the bottom is about 2000F (1093C), whereas the gas leaving the drying and devolatilization zone near the top is approximately 1000F (538C). This moving-bed process is capacity limited. Almost all of these units have a diameter of 13.1 ft (4 m) and a nominal dry gas capacity rating of 33,333 SCFM (14.9 Nm^3/s). This is equivalent to about 650 t/d (590 t_m/d) of MAF (moisture- and ash-free) coal. Lurgi and SASOL are operating a 16.4 ft (5 m) diameter Mark V gasifier having a capacity of 1000 t/d (907 t_m/d) and generating about 50,000 SCFM (22.35 Nm^3/s).

Fluidized bed

In a fluidized-bed gasifier (Fig. 2d), the oxidant flows upward through the feed coal particles at a sufficient velocity relative to the particle size that, although a discrete bed level exists, the particles are suspended or fluidized by the gas flow, producing good mixing and constant motion. The fluidized bed is maintained below the ash fusion temperature to avoid clinkers which could cause defluidization. (A clinker is a large solid mass of coal ash agglomerated by ash slagging.) Char particles entrained with the hot, raw gas are recovered and recycled to the gasifier. Fluid-

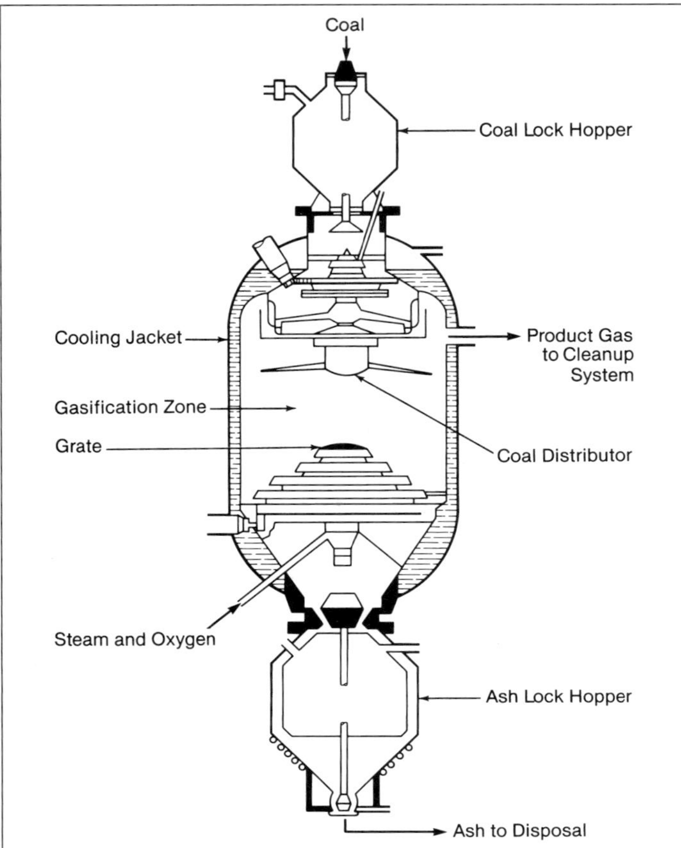

Fig. 3 Moving-bed dry ash gasifier (*courtesy of Dravo Corporation*).

ized-bed gasifiers, such as KRW's (Kellogg Rust Westinghouse) design, are always up-flow.

The distinguishing characteristics of a fluidized-bed gasifier are:

1. has large char recycle,
2. requires special steps to obtain high carbon conversion of high rank coals,
3. requires special steps to handle caking coals,
4. features uniform and moderate temperature, and
5. requires moderate oxygen and steam inputs.

The main differences among fluidized-bed gasifiers are the ash conditions (dry or agglomerated) and design provisions for char recycle. An agglomerated fluidized bed contains a hot zone where the ash particles are caused to cluster into small pellets prior to removal. Agglomerated ash operation enables the gasification of high rank coals. Dry ash fluidized-bed gasifiers operate most efficiently on low rank coals.

The fluidized-bed gasifier was the first commercial coal gasification unit and was the initial application of fluidized-bed technology. The first of these gasifiers was put into operation in Leuna, Germany in 1926. Since then, about 70 Winkler gasifiers have been constructed worldwide. However, these units were superseded by entrained-flow gasifiers and by pressurized moving-bed units. Only three of these gasifiers at two plants remain in operation today.[4] Low capacity and high operating costs have limited further use of the conventional Winkler gasifier.

Entrained flow

In an entrained-flow gasifier (Fig. 2b) fine coal particles are fed with the oxidant in such a way that they are suspended, or almost sprayed, co-currently into the process and mixed similar to a pulverized coal burner. The fine pulverized coal particles react with steam and oxidant with very short residence time. They can be up-flow (usually slurry fed) such as the ConocoPhillips/E-Gas [formerly Destec (DOW)], Uhde (Prenflo) and The Babcock & Wilcox Company (B&W) processes or down-flow such as the General Electric (GE)/Texaco and Royal Dutch Shell processes.

An entrained-flow gasifier has the following characteristics:

1. can gasify all coals regardless of rank, caking characteristics or amount of fines,
2. requires substantial heat recovery due to the large amount of sensible heat in the raw gas,
3. is a high temperature slagging operation,
4. has a large oxidant requirement, and
5. requires special steps to avoid molten slag carryover to downstream heat recovery surfaces.

The main differences among entrained-flow gasifiers are the coal feed system (slurry or dense phase) and the design configurations for raw gas cooling and sensible heat recovery. This unit is also called a *suspension* gasifier because it consists of a two-phase system of finely divided solids dispersed in a gas. No distinct bed level exists.

The entrained-flow technology began as an atmospheric pressure process for producing synthesis or fuel gas from solid or liquid carbonaceous fuels. The original bench-scale development of this gasifier type was done by Dr. Friedrich Totzek in the late 1930s while working for Heinrich Koppers GmbH of Essen, Germany. The process was jointly piloted in the U.S. by H. Koppers GmbH; Koppers Company, Inc. of Pittsburgh; and the U.S. Bureau of Mines. The first commercial entrained-flow gasification plant was built in France in 1949. Since then, 39 units have been built in various countries with most still in operation. The primary application is hydrogen production for ammonia synthesis.

Fig. 4 shows the entrained-flow atmospheric pressure gasifier. The coal is pulverized and fed by screw conveyors through opposing burners into a horizontal, elliptically-shaped gasifier. The fuel is oxidized, producing a flame zone temperature of about 3500F (1927C). Heat losses and endothermic reactions reduce the gas temperature to about 2700F (1482C). The hot product gas is further cooled by direct water quenching to about 1700F (927C) to solidify entrained liquid slag particles before they enter the heat recovery (waste heat) boiler.

The cold gas efficiency of the original Koppers-Totzek oxygen-blown gasifier was 67%. Although this appears to be low, an additional portion of the coal's energy was converted into heat which was recovered by the cooling jacket and heat recovery boiler. Combining these energy sources yielded an overall efficiency of 85%. More recent improvements in modern entrained-flow gasification systems can increase the

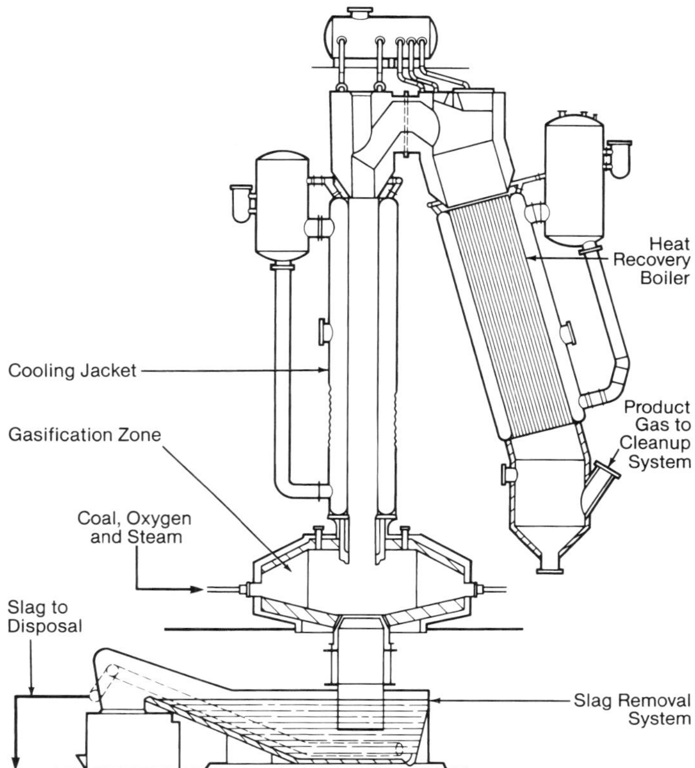

Fig. 4 Entrained-flow gasifier (*courtesy of Dravo Corporation*).

total efficiency into the high 90s. Most gasifiers use a design featuring two opposing burners. These have a total capacity of 10,583 SCFM (4.73 Nm^3/s), which is equivalent to about 230 t/d (209 t_m/d) of MAF coal. Two plants with four headed designs have about twice this capacity each. Current demonstration plants are exceeding 2500 t/d (2268 t_m/d) coal.

Transport

A transport gasifier is a circulating fluidized-bed gasifier in which the oxidant flows upward through the feed coal particles at a velocity high enough that no discrete bed level exists. The particles are fully suspended or highly fluidized by the gas flow, producing excellent mixing. As with the fluidized bed, the process is maintained below the ash fusion temperature to avoid clinkers which could cause defluidization. Due to the greater degree of mixing and the circulating loop inherent in the design, and particle size, char burnout is excellent. This type of gasifier, initially developed by MW Kellogg (now Kellogg Brown & Root), is now being tested at the U.S. Department of Energy's (DOE) Power Systems Development Facility (PSDF) near Wilsonville, Alabama. (See Fig. 5.)

The distinguishing characteristics of a transport gasifier are:

1. high carbon conversion (> 95%),
2. cold gas efficiency of 70-75%,
3. handles a broad range of coals,
4. uniform and moderate temperature [1600F (871C)], and
5. requires moderate oxygen and steam inputs.

Fig. 5 U.S. Department of Energy's (DOE) Power Systems Development Facility.

Major gasifier process developers

GE/Texaco

The GE/Texaco entrained-flow coal gasification process is shown in Fig. 6. The downflow gasifier is fed with a coal-water slurry of 60 to 65% solids by weight, and oxygen. It operates at up to 900 psi (6205 kPa) and is refractory lined. The raw gas leaves the unit at 2300 to 2700F (1260 to 1482C) and is separated from the slag. The syngas is cooled by a radiant boiler, followed by a convection boiler, which generates 1600 psi (11,031 kPa) saturated steam. These boilers are called *syngas coolers*.

There are several options for gas cooling. One is to use the radiant and convection coolers (as shown in Fig. 6) which results in maximum efficiency. A second

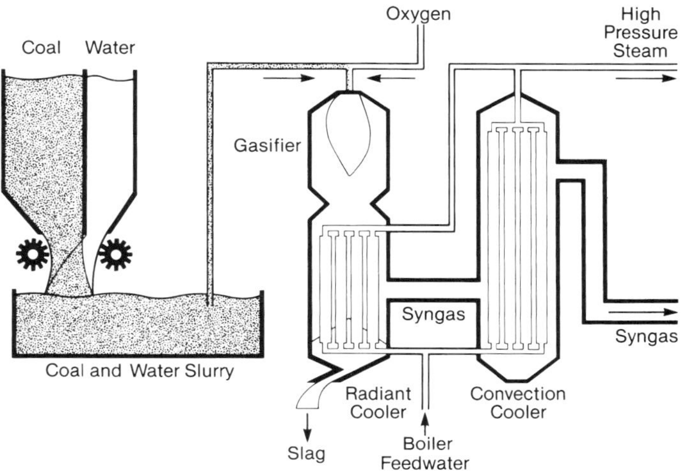

Fig. 6 Typical entrained-flow process (*courtesy of Texaco Development Corporation*).

involves replacing the radiant cooler with a direct water syngas quencher and eliminating the convection cooler, which minimizes cost. A third option is to use the radiant cooler only, which provides partial recovery of the syngas heat and is intermediate in efficiency and cost.

The cold gas efficiency for the process shown in Fig. 6 is 77%. If the energy of the steam produced is added to that of the fuel gas, the unit efficiency increases to 95%. Overall carbon conversions of 96.9% and 97.8%, respectively, are obtained with Illinois No. 6 and Pittsburgh No. 8 coals.[2,5] Carbon conversion is defined as the percent of carbon in the coal converted to gases or liquid (tar) products.

Royal Dutch Shell

The Shell gasification process is shown in Fig. 7. The coal is pulverized, dried, and fed to lock hoppers for pressurization. An operating pressure of 350 psig (2410 kPa) is lower than that of the GE/Texaco unit. The membraned walls are water cooled. Burners are opposed and in a reactor configuration similar to that of the Koppers-Totzek design. The raw gas leaves the unit at 2500 to 3000F (1371 to 1649C) and most of the coal ash exits through the slag tap in molten form. The syngas contains a small quantity of unburned carbon and a significant fraction of molten ash. To keep the ash particles from sticking together, the hot exiting gas is quenched with cold recycle gas. Further cooling takes place in the syngas cooler, consisting of radiant and convective sections. Some steam superheating is achieved with the hot, raw syngas.

The Shell gasifier cold gas efficiency is at least 80%. The combined chemical and thermal energy recovery is at least 97%. Feed coal drying reduces the net efficiency for low rank coals. The efficiency is also reduced when direct coal combustion is not used for drying. The effect of coal drying could lower the efficiency to approximately 94%. Finally, the net efficiency is about 84% when the energy required for oxygen production is also included.[2,6] Better than 99% carbon burnout has been obtained for most coals.[7]

ConocoPhillips/E-Gas [Destec (Dow)]

The ConocoPhillips/E-Gas gasifier is a two-stage, slurry feed, entrained-flow, slagging gasifier unit. Coals are slurried with water to produce a solids loading of 50 to 55% by weight. About 75% of the slurry is gasified with oxygen in the first stage. The hot gas leaving this stage at 2400 to 2600F (1316 to 1427C) is used to gasify the remaining 25% of the coal in the second stage. Both stages are refractory lined and uncooled. The first stage reactor is similar to the Koppers-Totzek unit with two opposing burners and slag removal. This arrangement assures high carbon conversion and optimal slag removal.

The direct injection of coal slurry at the second stage entrance quenches the hot gas and gasifies the additional coal, providing an exit gas temperature of about 1900F (1038C). This eliminates the need for a radiant boiler and requires reduced heat recovery downstream. There is no requirement for quench recycle gas or the compression energy associated with recycling large volumes of cooled gas. The ConocoPhillips/E-Gas unit includes a fire tube boiler followed by a steam superheater and economizer to recover heat from the raw product gas.[8]

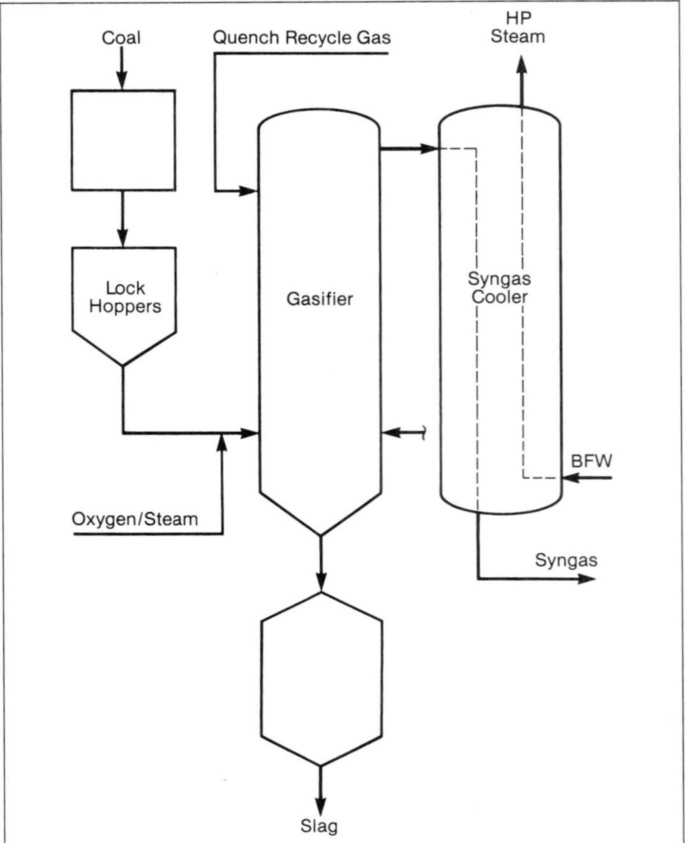

Fig. 7 Typical process with syngas cooler (*courtesy of Synfuels Business Development, a division of Shell Oil Company*).

British Gas/Lurgi

Lurgi has worked with the British Gas Corporation (BGC) in developing the BGC/Lurgi process. This gasifier, shown in Fig. 8, is very similar to the conventional dry ash Lurgi process. The key difference is that the BGC/Lurgi process slags the coal ash. This process partly alleviates the coal acceptability problems associated with the dry ash Lurgi process by permitting the use of higher rank, low ash coals. Tars and oils are re-injected, providing a cold gas efficiency of 88%. The value rises to about 90% when considering additional hydrocarbon liquids.[2] A substantial advantage of slagging the coal is that the steam requirement is only about 15% of that for the conventional Lurgi process when gasifying bituminous coal.

Winkler

The High Temperature Winkler (HTW) process is an extension of the old Winkler fluidized-bed technology. The HTW process is being developed by Rheinbraun (Rheinische Braunkohlenwerke AG) in Cologne, Germany in cooperation with its engineering partner, Uhde GmbH, Dortmund. The dry ash gasification of reactive German brown coal along with peat and wood is being researched. Recycling of fines entrained with the raw gas results in better carbon conversion. Furthermore, by operating at an increased pressure of 130 psig (896 kPa), the synthesis gas production rate for a given gasifier diameter more than doubles. A 580 t/d (526 t_m/d) HTW plant has been operating since 1986 at Huerth (Berrenrath brown coal plant), Germany, producing 22,917 SCFM (10.24 Nm^3/s) of synthesis gas from brown coal.

The HTW unit relies on the use of non-caking coals to achieve high carbon conversion. Approximately 95% conversion can be reached with very reactive German brown coals. The bed operates at temperatures of 1400 to 1500F (760 to 816C). Cold gas efficiency is about 82% for the oxygen-blown HTW; the efficiency increases to 85% if the net steam generation (deducting energy for coal drying and steam fed to gasifier) is added.[2] There is a high energy requirement for drying the high moisture, low rank coal.

Babcock & Wilcox gasifier development

As the firing and operation of an entrained-flow gasifier are somewhat similar to that of a pulverized coal-fired boiler[9,10] all of B&W's gasification processes are entrained-flow gasification systems featuring pulverized coal feed. As a result, there are inherent advantages in using pulverized coal as a fuel for entrained-flow gasifiers. A pressurized 850 to 1000 t/d (771 to 907 t_m/d) gasifier design with a waste heat recovery boiler is shown in Fig. 9.

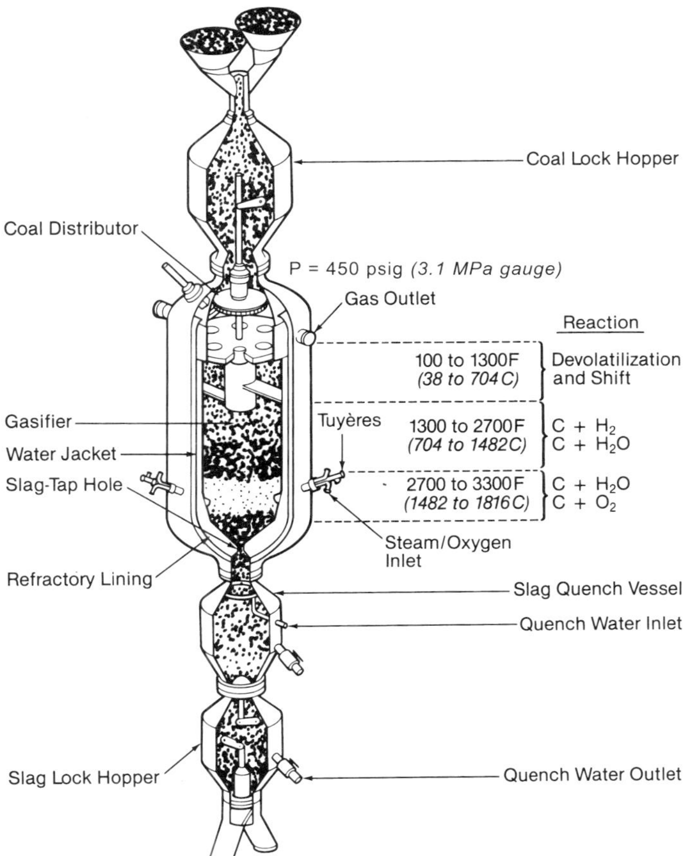

Fig. 8 Slagging gasifier (*courtesy of British Gas PLC*).

The gas-tight membrane wall enclosure provides an annular space which separates the gasification reactor and the pressure vessel. Therefore, the pressure vessel, at a relatively low temperature, is not in contact with the corrosive raw gas. In the lower portion of the unit (the gasification zone), the tubes are stud-

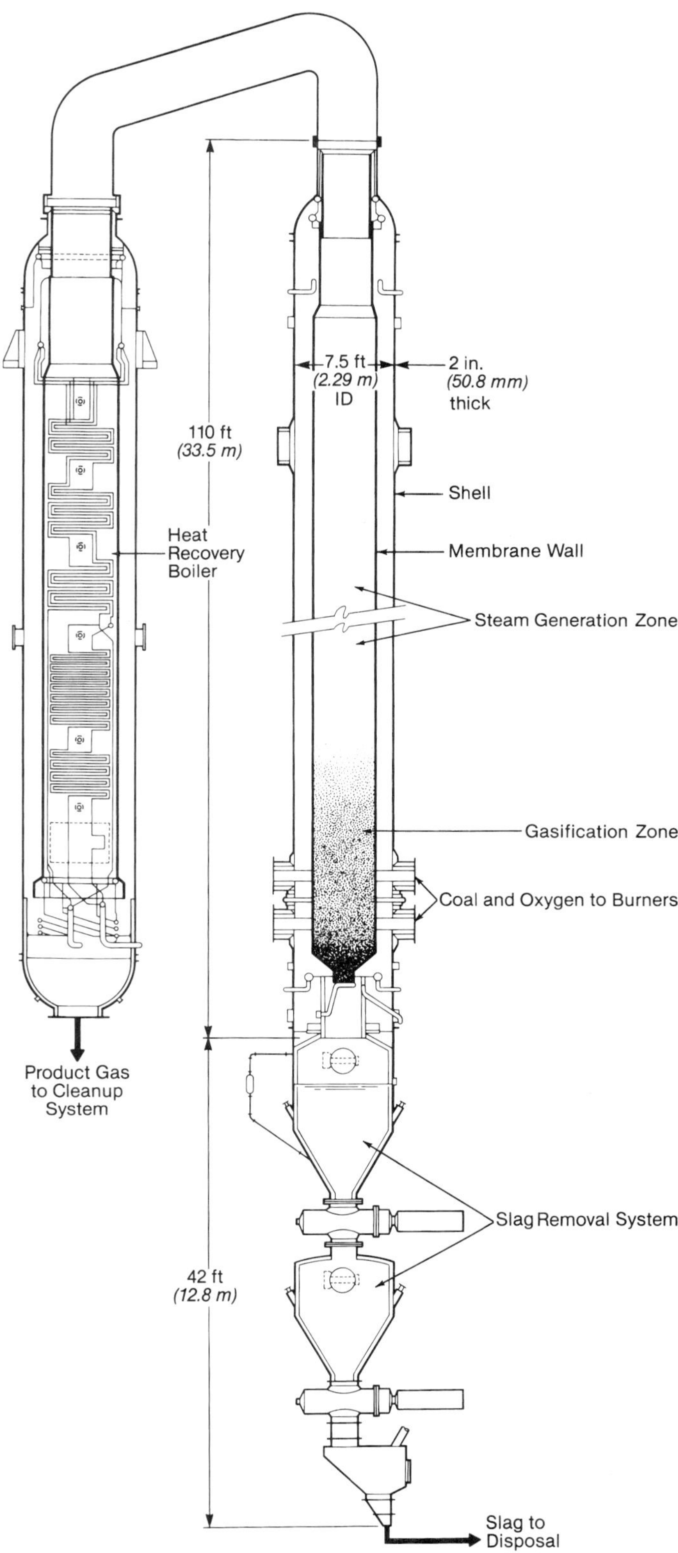

Fig. 9 Typical 1000 t/d entrained-flow design.

ded and covered with refractory to maintain the high temperatures necessary to keep the slag fluid. In the upper portion of the gasifier (the freeboard zone), the tubes are exposed to provide maximum cooling. Use in Cyclone wet bottom boilers (see Chapter 15) has proven the durability of this construction. This gasifier, featuring a 2 in. (51 mm) thick vessel wall, is designed for relatively low pressure operation [50 psig (345 kPa)]. Temperature in the combustion zone is about 3400F (1871C) and the gas leaving the unit is at 1800F (982C).

The advantages of the dry feed, entrained-flow gasifier include:

1. Insensitivity to coal characteristics – Unlike moving-bed or fluidized-bed gasifiers, it can accept caking coals, fines and any rank coal. With dense-phase coal feed, as contrasted to slurried coal, it can accommodate the wide range of coal moisture contents.
2. Larger unit size potential due to ease of handling and mixing of the coal – Basically, as in pulverized coal-fired boilers, unit size can be increased by adding burners. The higher throughput velocity allows a high quantity of coal per hour to be treated in a given size gasifier.
3. Rapid control response – This is due to its low solids inventory compared to other gasifiers.
4. Easy disposal of the dense granular slag – Similar material from slag-tap boilers has been used as road fill. It does not create a dust or water pollution problem.
5. High yield of synthesis gas ($CO + H_2$) – Conversely, the yield of CO_2 and H_2O is low. For example, the dry-bottom, moving-bed product gas can contain 60% (by volume) H_2O and 10% CO_2 as contrasted to about 2% each of these components in a dry feed entrained gasifier.
6. No hydrocarbon liquids generated – Tars, phenols and oils are absent. Special handling problems associated with production of such chemicals and special treatment requirements for scrubber water are avoided. Re-injection of unreacted char is considerably simpler than such tar-like streams.
7. High reliability (no moving parts in furnace) – The requirement for mechanical devices, such as a rotating grate, is eliminated because the pulverized coal is dispersed through the gasifier and rapidly heated through the plastic temperature range without coming into contact with adjacent coal particles.
8. Reduced maintenance due to the cooled membrane wall construction – The durability of the studded tube, refractory-coated gasifier cooling enclosure has been proven in Cyclone furnace and slag-tap boilers. This contrasts to the higher maintenance requirements of non-cooled, refractory-lined gasifiers.

The disadvantages are:

1. May not be economical for very small sizes – With its high gasification capacity for a given size vessel, fewer gasifiers are needed to supply the required gas. Scaledown would require a detailed evaluation of the scaling factors for the pressure vessel, up front coal preparation plant (pulverizers), and downstream heat recovery equipment.
2. Generates negligible methane (if methane is desired) – Methane production is generally favored in a lower temperature, high pressure, moving-bed type gasifier.
3. Control of coal feed rate is more complex – Maintenance of uniform heat release and slagging conditions requires precise control of coal flows. Dense-phase pulverized coal feed systems require inerting for safety. Coal feed and char recycle are simplified with coal slurries and higher pressures are possible, but there are associated penalties in carbon burnout and flexibility of coal feed sourcing.
4. Turndown may be limited with some coals – Turndown to approximately 25% of capacity should generally be possible, but certain high ash-fusion temperature coals may constrain proper slagging at low loads.
5. Relatively high oxygen and waste heat recovery duty – These characteristics are associated with the high temperature of the entrained-flow process. Heat recovery costs exceed those of the fluidized or moving bed. Because the gasifier walls and heat recovery boiler absorb about 85% of the energy released as sensible heat (25% of the energy in the coal), it has been estimated that the total recovered and chemical energy efficiency of the gasifier can exceed 96% for most coals.

B&W gasifier development dates back to 1951. Several pilot plants were built in-house and a full scale commercial gasifier was built to produce synthesis gas.

The first gasifier designed and constructed by B&W (Fig. 10) was put into operation in 1951 at the U.S. Bureau of Mines station at Morgantown, West Virginia (currently DOE's National Energy and Technology Laboratory). Operating at atmospheric pressure, this unit was oxygen-blown and was capable of gasifying 500 lb/h (0.063 kg/s) of coal, producing about 330,000 ft^3/d (147.3 Nm^3/h) of synthesis gas. It was refractory lined and was comprised of a primary and secondary reaction zone. The primary zone produced temperatures exceeding 3000F (1649C) and the secondary zone operated at about 2200F (1204C). The unit was operated successfully for more than 1200 hours.

B&W also assisted the U.S. Bureau of Mines in its early pressurized gasification studies. One unit designed by B&W featured a down-fired axial coal feed [500 lb/h (0.063 kg/s)] with tangential steam and oxygen feeds. This unit was designed for operation at 450 psig (3100 kPa). It was operated during the 1950s and 1960s and was occasionally used in the 1980s.

A reduced-scale gasifier was constructed at Belle, West Virginia in 1951.[11] This unit was rated at 3000 lb/h (0.38 kg/s) of coal and produced 1389 SCFM (0.62 Nm^3/s) of synthesis gas. When installed, it was the largest pulverized coal gasifier in the U.S. The unit operated for more than 5000 hours and established the design basis for a full scale gasifier at the same plant.

In 1955, a full scale gasifier began operation at Belle, West Virginia. The unit was 15 ft (4.6 m) in diameter and 88 ft (26.8 m) tall. It was designed to gasify 17 t/h (15.4 t_m/h) of coal, producing 17,360 SCFM (7.75 Nm^3/s) of carbon monoxide and hydrogen. The gasification zone was refractory lined and the floor

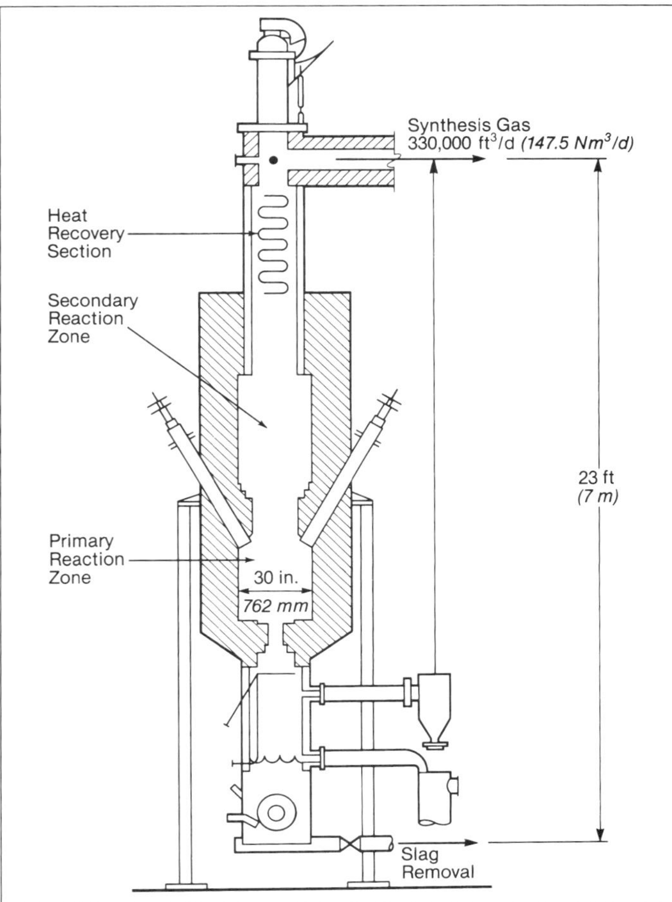

Fig. 10 Early B&W oxygen-blown gasifier.

and walls were water cooled. The gasifier operated for more than a year, then shut down when low cost natural gas became available.

In the early 1960s, another pilot unit was constructed by B&W at its research facility in Alliance, Ohio (Fig. 11). This air-blown unit was used in a cooperative program with General Electric to study combined gas turbine/steam turbine cycles. It consisted of a 3 t/h (2.7 t_m/h) gasifier with char recycle, mechanical gas cleanup equipment, and a gas-cooled cyclone combustor. This facility was operated for three years beginning in 1960, and horizontal and vertical slag-tap, vortex-fired gasifiers were studied.

B&W also built a 5 t/h (4.5 t_m/h), 1500 psig (10,340 kPa) gasifier for the Bi-Gas pilot plant at Homer City, Pennsylvania (Figs. 12 and 13). This plant, which opened in 1976, demonstrated the production of synthetic natural gas from coal. The project was sponsored by the Office of Coal Research, the U.S. Department of Interior, and the American Gas Association.

Development issues

Although there are other issues such as removal of acids from the gas, as well as solid and liquid effluents, there are three primary areas where further development is needed for integrated gasification combined cycle (IGCC) to achieve its potential. Due to materials issues, current systems must significantly cool the gas leaving the gasifier to remove sulfur and particulates prior to entering the gas turbine (resulting in an efficiency penalty). In addition, these coolers have experienced failures resulting in plant downtime. Ideally, the sulfur and particulates would be removed with minimal cooling of the gas.

Gas cooling

Syngas cooler material failures have accounted for a substantial percentage of downtime at the IGCC demonstration plants. Material selection in syngas coolers requires special attention.[12,13]

There are major differences between syngas coolers and coal-fired boilers:

1. The highly reducing and corrosive raw syngas presents a different environment for material qualification. Protective oxide scales which usually form on low alloy boiler steels do not readily form in syngas coolers, where H_2S and HCl are present.
2. Because of the high operating pressure, generally 300 to 600 psi (2070 to 4140 kPa), the syngas dew point is generally much higher than that in atmospheric pressure boilers. This is especially true when the steam content of the syngas is high, as with a coal-water slurry feed system. Precautions are required to prevent operation below the acid dew point and to avoid aqueous corrosion during downtime.
3. Pressurized operation, requiring a vessel enclosure for the exchangers, constrains the design layout. This requires special attention to assure accessibility for inspection and repair.

Low alloy steels are not suitable for syngas coolers because of excessive corrosion rates caused primarily by sulfidation, and also by chlorination. For low alloy steels, the sulfidation rate is dependent on the H_2S partial pressure and the rate is higher with high sulfur coals. Iron sulfide (FeS) and mixed sulfide-oxide

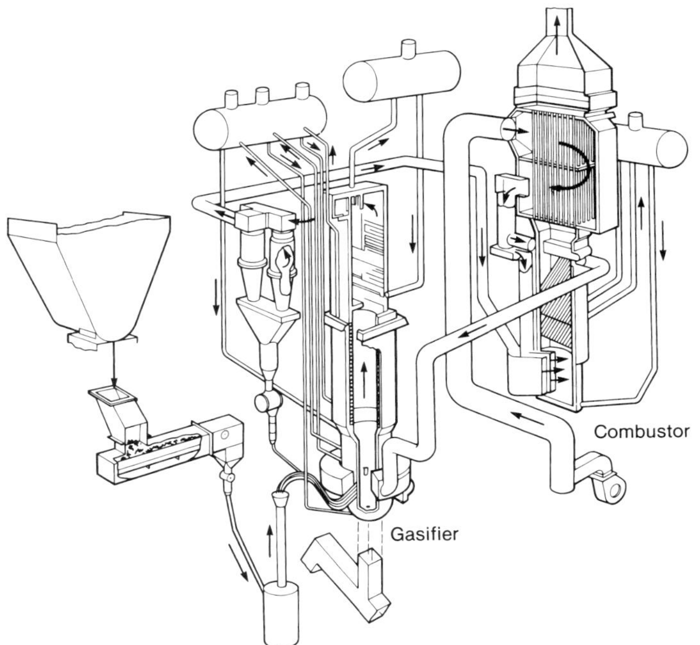

Fig. 11 Pilot plant for gasification research.

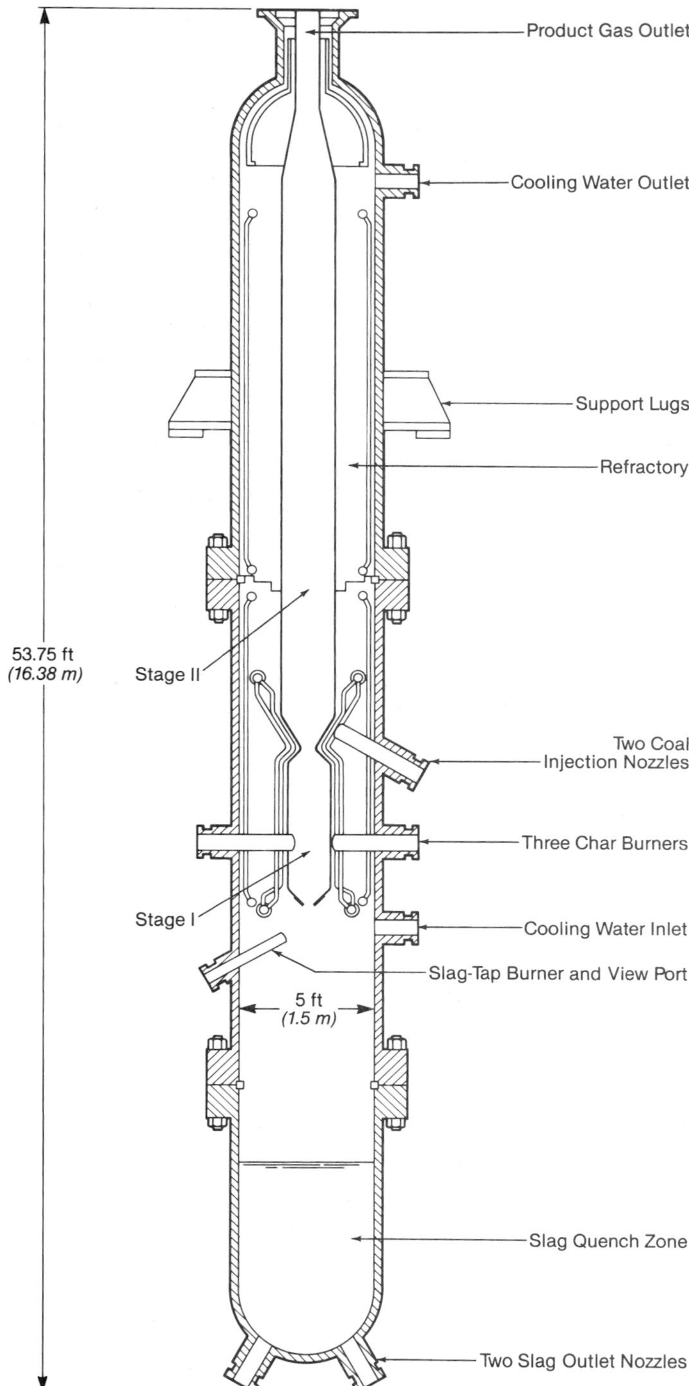

Fig. 12 B&W bi-gas gasifier.

scales form; these are generally less protective than oxide scales. Because low alloy steels experience excessive corrosion rates, stainless steels are used. Corrosion rates must be less than about 4 mils/yr (0.1 mm/yr) to obtain a service life of 25 years.

The effects of syngas chlorides are not clearly defined. Sulfidation rates in syngas with 300 to 500 ppm HCl become variable and are generally higher, presumably because of $FeCl_2$ formation and spalling during thermal cycling. Chlorides greatly accelerate aqueous corrosion during downtime if condensation occurs. The presence of HCl in syngas increases its dew point 54 to 90F (30 to 50C).

Preliminary material selection guidelines aimed at a 25 year service life have been formulated by the Electric Power Research Institute (EPRI) based on currently available laboratory and field data.[12,13] Actual service life will depend on the severity of the conditions to which the individual material components are exposed, which in turn are dependent on the specific design, fuel used, and operating conditions.

Particulate removal

Particulate removal at as high a temperature as practical is desirable to avoid significant heat losses and resulting loss of efficiency. Rigid, porous ceramic barrier filters have been the focus of development for hot particulate removal at temperatures around 1800F (982C). The ceramic material must be porous enough to allow gas penetration while capturing fine particulate, permitting solids removal and pressure drop recovery. It must also be resistant to chemical attack from the high temperature reducing environment and have sufficient ductility to resist the thermal and mechanical shocks imposed during operation. A variety of filter arrangements have been developed with these objectives including candle, granular bed, cross-flow and tubular designs (see Fig. 14). Although much testing has been done and applications have been attempted at a few of the IGCC demonstration facilities, ceramic filtration remains incapable of commercial reliability and all of the gasification plants to date cool the product gas to 1000F (538C) or lower to utilize lower temperature sintered metal filters to remove particulates.

Candle filters consist of thousands of ceramic or sintered metal *candles*, which are hollow tubes closed at one end, with the other end suspended from a tube sheet within a pressure vessel. The dirty gas enters the pressure vessel and flows around the outside of the candles and through the filter wall to the inside, leaving the entrained dust on the outside of the candle. The clean gas flows up through a common collector and out the top of the vessel. When the pressure drop through the candles reaches a prescribed value, pressurized gas is *pulsed* in the reverse direction to the normal gas flow (back-pulse), loosening the filter cake from the outside of the candles and allowing it to fall into an ash hopper. It is then removed through lock hoppers. High temperature ceramic candle filters have received a great amount of testing. The greatest hurdles to reliability have been cracking due to thermal shock from back pulsing, from thermal upsets caused by carbon carryover, or from ash that has been allowed to accumulate until it reached the candles. Most ceramic filter medium is monolithic and very brittle. In the late 1980s, B&W developed a continuous fiber ceramic composite (CFCC) material to provide more ductility than monolithic ceramic tubes. CFCC is woven from ceramic fibers and offers predictable permeability with considerable flexibility.

In granular bed filters, dirty gas is forced to pass through a moving bed of solids that traps the particulate. The bed material enters at the top and is removed

Fig. 13 Bi-gas gasifier prepared for shipment.

at the bottom. The ash is then cleaned and the filtration medium is returned to the process.

Cross-flow filters are composed of an array of ceramic structures much like a honeycomb with passages perpendicular to each other separated by ceramic walls. The dirty gas flows into the filter through one set of passages, permeates the ceramic wall to the perpendicular path, and flows out. Cross-flow filters have been arranged in several ways, usually within a pressure vessel utilizing a back-pulse cleaning system as previously described.

Tubular filters are made up of hundreds of ceramic tubes, open at both ends and sealed at both ends to a tube sheet. Dirty gas enters the vessel surrounding the outside of the tubes, flows through the ceramic tube wall and leaves the entrained dust on the surface of the tube. The clean gas passes up the center of the tube and out the top of the vessel. As with candle filters, the tubes are cleaned by back pulsing.

Until there is a breakthrough in hot gas filtration, gasification processes will be forced to cool the gas and suffer an efficiency penalty.

Sulfur removal

The U.S. DOE has funded research on hot gas desulfurization since the 1970s, testing mixed metal oxide sorbents such as zinc ferrite ($ZnFe_2O_4$), titanate ($ZnTiO_3$), copper oxide (CuO) and iron oxide (Fe_2O_3) as candidates for regenerable systems. At small scale, most have shown high removals of gas phase sulfur species and can still be regenerable.[14]

The general process consists of metal oxide pellets housed in a reactor vessel through which the sulfur-laden acid gas passes. Some systems are fixed beds in parallel, with hot valves diverting gas between beds as they absorb and regenerate. Other systems use a moving-bed or a fluidized concept, as shown in Fig. 15. The sorbent is regenerated by passing a controlled flow of air through the pellet bed; the air reacts exothermically to produce a concentrated SO_2 off gas which must be further treated. Depending on the metal oxide used, the temperature must be carefully controlled between 1200F (649C) and 1400F (760C) to prevent potentially hazardous zinc vaporization and sintering.[15] Sorbent life is the critical operating consideration due to its high cost.

Zinc ferrite has been tested in bulk removal and polishing modes. Bulk removal uses the sorbent to perform the total sulfur removal duty. In the polishing mode, the sorbent is a secondary means of sulfur removal; primary removal is achieved within the gasifier. Fluidized-bed gasifiers using a throwaway sorbent such as limestone or dolomite can capture some sulfur. Entrained-flow units have been fed with iron oxide or dolomite to capture sulfur in the slag.[16]

Many development issues must be resolved before hot gas cleanup systems can be offered commercially. In addition to particulate and sulfur removal issues,

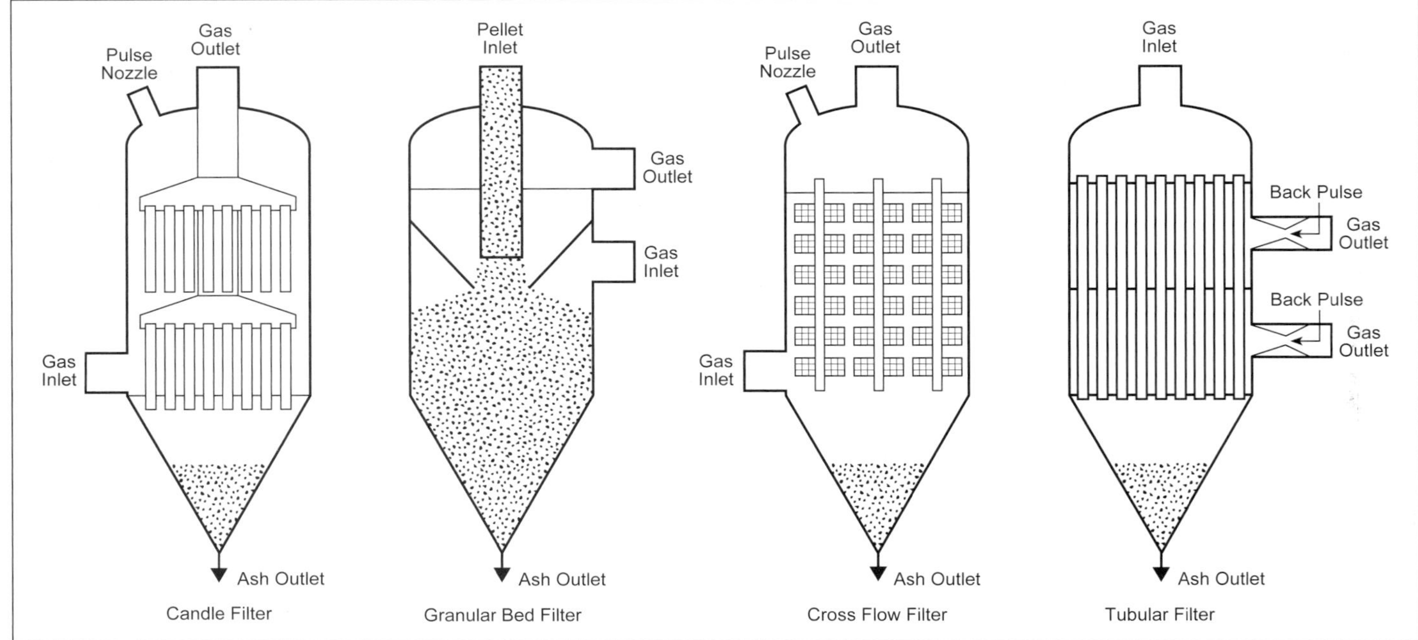

Fig. 14 Ceramic filter arrangements.

the effects of raw gas chlorides, alkali species and other trace volatile elements on the performance of downstream equipment must be addressed. Ammonia in the gas will not be condensed and may result in high NO_x emissions from a gas turbine combustor. The challenge is to achieve the excellent environmental performance of conventional processes while being reliable, efficient and economical.

IGCC for power generation

Production of power from gasifying coal and combusting the fuel gas in a combined cycle power plant requires a high degree of component integration, as shown in Fig. 16. In simple terms, the hot, combusted fuel gas is expanded in a gas turbine to drive the air compressor and generator, with some of the compressed air used to gasify the coal. The byproducts, such as char, are often burned in secondary processes to produce additional steam and/or power. The gas turbine exhaust, hotter than 1000F (538C) with modern machines, passes through a heat recovery steam generator (HRSG) to produce superheated steam to drive a steam turbine-generator (see Chapter 27) comprising a combined cycle. The gasification process releases significant thermal energy which must also be recovered into the steam cycle to achieve high overall plant efficiency.

Demonstrations of coal gasification processes in the 1980s and early 1990s include plants at Plaquemine, Louisiana (155 MW Destec or Dow process), Cool Water at Daggett, California (92 MW Texaco process), Puertollano, Spain (288 MW ELCOGAS or Prenflo process), and the first fully integrated IGCC plant at Buggenum in the Netherlands (250 MW Shell process). Significant investment was made to commercialize IGCC plants for power generation in the DOE's Clean Coal program from in the mid-1980s to 1990s including plants at Wabash River near Terra Haute, Indiana (262 MW Destec/Dow process), the Polk Power Station near Tampa, Florida (250 MW Texaco process), the Piñon Pine project at Sierra Pacific Power's Tracy station near Reno, Nevada (99 MW KRW process), and the Kentucky Pioneer Energy project near Trapp, Kentucky (400 MW British Gas/Lurgi process).

The design of an IGCC system is quite complex, with many factors to consider to achieve the right balance of capital cost, plant efficiency, operability, and environmental requirements for a given application. For example, the choice of gasifier type affects the amount of fuel gas heat recovery duty.

An oxygen-blown, entrained-flow gasifier with its high operating temperature requires more raw gas cooling in efficiently designed cycles. These coolers

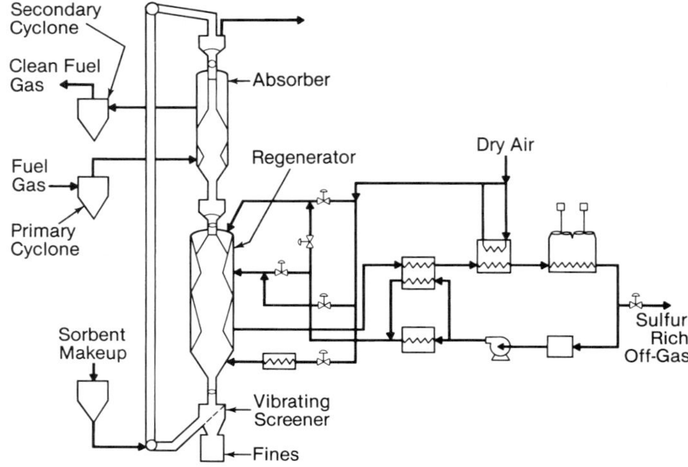

Fig. 15 Moving-bed zinc ferrite process schematic (*courtesy of GE Environmental Services, Inc.*).

must perform in a harsh gas environment and are a critical component in plant reliability. Integrating this large quantity of lower level steam energy (limited by metal temperatures) into the steam cycle tends to complicate plant control and operation. In contrast, cooling the raw gas by quenching eliminates the capital cost and complexity of heat recovery, but at a substantial efficiency penalty (as much as 10% less efficient than with total heat recovery).

Oxygen plant integration is another design issue. Fig. 16 shows the turbine-driven air compressor directly feeding the oxygen plant, which produces oxygen more efficiently by reducing external compression requirements. However, this imposes more demand on turbine controls due to the large system volume between the compressor and turbine.

Other technical factors, such as fuel characteristics, influence process design. A high inherent moisture fuel such as lignite may not be a good candidate for a coal-water slurry fed gasifier due to reduced efficiency. Gas turbines have varied abilities concerning fuel gas minimum heating value, emissions performance, and compressor range for handling the larger mismatch between air flow and gas flow when firing relatively high volume, low Btu coal gas. Ambient temperature effects and utility load demand characteristics add to the list of design considerations.

Plant size is determined by the available size of the gas turbine. A single gas turbine/steam turbine installation can produce more than 250 MW, with about 60% of the power from the gas turbine. Larger plants require multiple gas turbines, with economies of scale realized in maintaining a single steam turbine and in larger balance-of-plant systems such as fuel handling, electrical and controls, and water treatment. Phased construction of an IGCC plant by incrementally installing a natural gas-fired turbine (simple cycle), then steam turbine and HRSG (combined cycle) and finally a gasification system (IGCC) has been proposed to provide owner flexibility in meeting changing power demands from peaking to base load and accommodating a relative rise in fuel costs of natural gas versus coal. However, this apparently simple concept has proven impractical due to the significant modifications to the gas turbine and steam side heat balance necessary to switch from natural gas to coal gas.

The myriad of technologies available and the external design factors combine to make IGCC plant design a complex, site- and fuel-specific process. Concurrently, this flexibility offers opportunity to produce a highly optimized system for each application.

Future IGCC development

The major force behind development and implementation of IGCC is the demand for a cleaner environment. Conventional pulverized coal (PC) plants continue to add environmental controls to meet ever-tightening emissions requirements (see Chapters 32 through 36). An IGCC plant has the potential to achieve very low emissions. Oxygen-blown gasifiers with a cold gas cleanup system are particularly well suited to deliver low emissions of SO_2, NO_x, solid wastes, and air toxins such as mercury.

Sulfur removal efficiencies in excess of 99% are possible with an IGCC plant equipped with a H_2S removal process, a Claus (elemental sulfur recovery) plant, and a tail-gas cleanup system. Wet scrubbers in a PC plant can be practically designed with equivalent performance through use of more enhanced sorbents and/or increased sorbent-to-gas interaction.

With fuel nitrogen compounds removed in the cold gas cleanup system, NO_x emissions from an IGCC system are strictly determined by gas turbine combustor performance. NO_x values less than 0.05 lb/10^6 Btu (21.5 g/GJ) are feasible with state-of-the-art combustors, depending on fuel gas fired heating value (moisture and/or nitrogen can be added) and turbine inlet temperature, both of which affect peak flame temperature and thermal NO_x generation. By comparison, PC plants can also reach such values through low NO_x combustion techniques (see Chapter 14) in combination with post-combustion NO_x controls (see Chapter 34).

Solid byproducts from an IGCC plant with an entrained-flow gasifier consist primarily of non-reactive quenched slag and elemental sulfur. If the sulfur is sold as a byproduct, the disposal stream is reduced to the fuel ash. Depending on fuel sulfur content, a PC plant may produce up to twice as much solid waste due to sulfur dioxide (SO_2) byproducts. (See Chapters 1 and 32.) However, this may be partially offset by the production of saleable gypsum in conventional scrubber systems or other byproducts from regenerable scrubber systems (such as fertilizer from sodium-based systems).

The environmental impacts of heavy metals in fuel

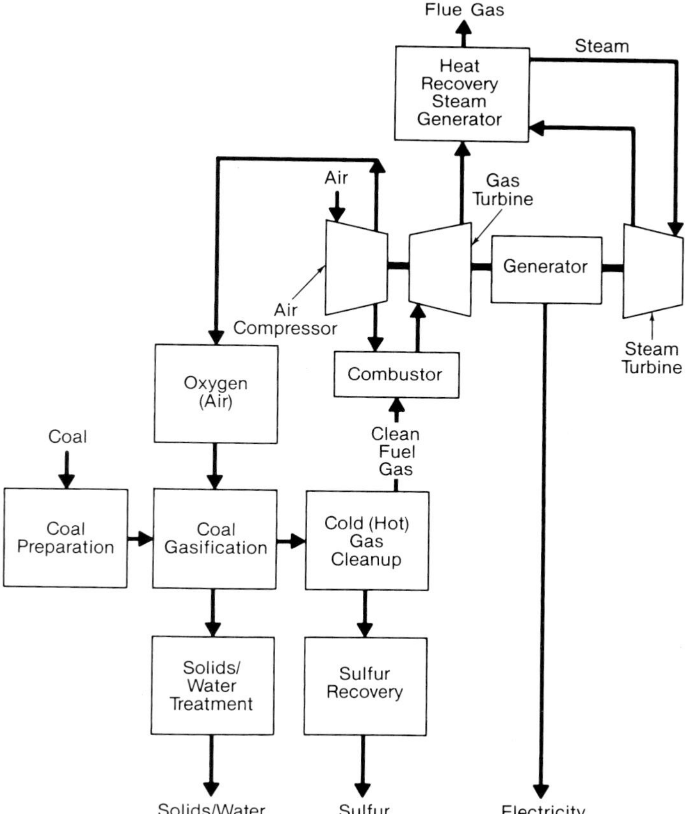

Fig. 16 Integrated coal gasification combined cycle schematic.

ash, emissions of air toxins, especially mercury, and extremely fine particulate matter are now being scrutinized. Techniques for mercury removal in PC plants are nearing commercial application and use of wet electrostatic precipitation can reduce sulfuric acid and fine particulate emissions to very low amounts, similar to what is possible in IGCC systems with high efficiency filters.

Because a gas turbine is employed in IGCC, the net plant efficiency is usually expressed in terms of the coal's lower heating value (LHV). It should be noted that due to the difference in the way the heat of vaporization of water is considered in the heat rate calculation, there is a significant difference (2-3%) between efficiencies on a lower versus higher heating value basis (HHV being the lesser). Net plant efficiencies as high as 45% LHV (42 to 43% HHV) have been reported, but most plants have actually achieved efficiencies in the range of 38 to 43% depending on capital cost tradeoffs, degree of plant integration, fuel type, and other specifics. Higher efficiencies are possible as gas turbines with higher firing temperatures are developed. Current projections of efficiency using advanced high temperature gas turbines not yet commercially available and supercritical heat recovery steam cycles are in the mid to upper 50% LHV range (high 40% range with HHV). Operational success of IGCC plants has varied widely, primarily depending on the degree of integration and the method of gas cleaning (hot or cold).

With heightened awareness worldwide of CO_2 emissions, efficiency is becoming as much an environmental issue as an economic factor related to fuel costs. A large state-of-the-art supercritical Rankine cycle steam plant operating at steam temperatures of 1100F (593C) (single reheat) with environmental controls can achieve an efficiency of 40% (HHV) or higher (depending on cooling water temperature and condenser backpressure and turbine cycle configuration), while modern subcritical steam cycles have achieved efficiencies of 37% (HHV).

Modern supercritical pulverized coal fired plants with advanced steam conditions provide very attractive economics, heat rate (efficiency) similar to current state-of-the art IGCC plants (which determines CO_2 emissions), and comparable emissions of SO_x, NO_x, mercury and particulates. Considering the state-of-the-art and the fact that development of gasification systems has been driven by chemical production rather than the utility power generation industry, coal gasification is likely to develop as a co-production plant. The DOE's Vision 21 program promotes this approach of using gasification systems to produce high value chemicals with electricity as a byproduct rather than a workhorse technology for power generation. Ultimately, the cost of producing power with an IGCC system, advanced Rankine cycle steam plants and other competing systems, such as pressurized fluidized-bed power systems, will determine the extent to which each system will be used in the future.

References

1. Meyers, R.A., Ed., *Handbook of Synfuels Technology, Part 3*, McGraw-Hill, New York, New York, 1984.

2. "Coal Gasification Systems: A Guide to Status, Applications, and Economics," Electric Power Research Institute (EPRI) Report AP-3109, June 1983.

3. Elliot, M.A., Ed., *Chemistry of Coal Utilization*, Second Supplementary Volume, Chapters 23, 24, 25, Wiley, New York, New York, 1981.

4. Huebler, J., and Janka, J.C., "Fuels, Synthetic," *Encyclopedia of Chemical Technology*, Third Ed., Volume 11, Wiley-Interscience, New York, New York, 1983.

5. "Cool Water Coal Gasification Program: Fifth Progress Report," EPRI Report AP-5931, Palo Alto, California, October 1988.

6. "Dry Feed Entrained-Flow Coal Gasification Processes: Technology Status," *The SFA Quarterly Report*, SFA Pacific, Inc., Mountain View, California, December 1988.

7. Krewinghaus, A.B., and Richards, P.C., "Coal Flexibility of the Shell Coal Gasification Process," Pittsburgh Coal Conference, Pittsburgh, Pennsylvania, September 25-29, 1989.

8. "Dow Coal Gasification Project and Technology Analysis," *The SFA Quarterly Report*, SFA Pacific, Inc., Mountain View, California, April 1987.

9. Probert, P.B., "Industrial Fuel Gas from Coal," American Ceramic Society, Washington, D.C., May 5-8, 1975.

10. James, D.E., and Peterson, M.W., "Suspension Type Gasifiers," American Institute of Chemical Engineers (AIChE), Los Angeles, California, November 16-20, 1975.

11. Grossman, P.R., and Curtis, R.W., "Pulverized-Coal-Fired Gasifier for Production of Carbon Monoxide and Hydrogen," Transactions of the American Society of Mechanical Engineers (ASME), Paper No. 53-A-49, July 1953.

12. Bakker, W.T., "Materials for Coal Gasification, an EPRI Perspective," EPRI Conference on Gasification Power Plants, Palo Alto, California, October 18, 1990.

13. Bakker, W.T., "Materials for Syngas Coolers in Integrated Gasification-Combined Cycle (IGCC) Power Plants," EPRI Technical Brief RP2048, 1988.

14. Bajura, R.A., and Bechtel, T.F., "Update on DOE's IGCC Programs," Ninth Annual EPRI Conference on Coal Gasification Power Plants, Palo Alto, California, October 17-19, 1990.

15. Notestein, J.E., "Update on Department of Energy Hot Gas Cleanup Programs," Eighth Annual EPRI Conference on Coal Gasification, Palo Alto, California, October 19-20, 1988.

16. Robin, A.M., "Integration and Testing of Hot Desulfurization and Entrained-Flow Gasification for Power Generation Systems," Tenth Annual Gasification and Gas Stream Cleanup Systems Contractors Review Meeting, Morgantown, West Virginia, August 28-30, 1990.

Bibliography

Coal: Energy for the Future, National Research Council, Committee on the Strategic Assessment of the U.S. Department of Energy, National Academies Press, August 1, 1995.

Handbook of Gasifiers and Gas Treatment Systems, USERDA Report FE-1772-11, Dravo Corporation, February, 1976.

Kirk, R.E., and Othmer, D.F., *Kirk-Othmer Encyclopedia of Chemical Technology*, Fifth Ed., John Wiley & Sons, New York, New York, 2003.

May, M.P., "Suspension Gasifier Offers Economics of Scale," Iron and Steel Engineer, pp. 46-52, November, 1977.

"Piñon Pine IGCC Power Project, A DOE Assessment," DOE/NETL-2003/1183, United States (U.S.) Department of Energy (DOE), Morgantown, West Virginia, December, 2002. See also www.netl.doe.gov.

"Wabash River Coal Gasification Repowering Project: A DOE Assessment," DOE/NETL-2002-1164, U.S. DOE, Morgantown, West Virginia, January, 2002. See also www.netl.doe.gov.

Superheater installation in a coal-fired utility boiler.

Chapter 19

Boilers, Superheaters and Reheaters

In a modern steam generator, various components are arranged to efficiently absorb heat from the products of combustion and provide steam at the rated temperature, pressure, and capacity. These components include the boiler surface, superheater, reheater, economizer, and air heater. They are supplemented by systems for steam-water separation (see Chapter 5) and the control of steam outlet temperature. The steam generator can be divided into two general sections, the furnace and the convection pass. The furnace provides a large open volume with water-cooled enclosure walls inside of which combustion takes place, and the combustion products are cooled to an appropriate furnace exit gas temperature (FEGT). The convection pass contains tube bundles, which compose the superheater, reheater, boiler bank, and economizer. The air heater usually follows the convection pass. The boiler, or evaporation-circulation system, typically includes one or more of the following subsystems: the furnace enclosure walls, mix headers, steam drum, steam-water separation equipment, boiler bank, and associated connecting piping (downcomers, supplies, and risers as discussed in Chapter 1). This chapter focuses on the boiler, superheater, and reheater components, plus systems for steam temperature control, steam bypass, and unit startup. Chapter 20 discusses economizers and air heaters.

Boilers

Boiler surface is defined as the tubes and drum shells that are part of the steam-water circulation system, and that are in contact with the hot gases (flue gas). Although the term *boiler* now frequently refers to the overall steam generating system, the term *boiler surface* excludes the economizer, superheater, reheater, or any component other than the steam-water circulation system itself.

While boilers can be broadly classified as fire tube and water tube types, as discussed in the *Introduction to Steam*, modern high capacity boilers are of the water tube type. In the water tube boiler, the water and steam flow inside the tubes and the hot gases flow over the outside surfaces. The boiler circulation system is constructed of tubes, headers, and drums joined in an arrangement that provides water flow to generate steam while cooling all parts. The water tube construction allows greater boiler capacity and higher pressure than shell or fire tube designs. The water tube boiler also offers greater versatility in arrangement; this permits the most efficient use of the furnace, superheater, reheater, and other heat recovery components.

Boiler configurations

Modern high capacity boilers come in a variety of designs, sizes, and configurations to suit a broad range of applications. Sizes range from 1000 to 10,000,000 lb/h (0.13 to 1260 kg/s) and pressures range from one atmosphere to above the critical pressure.

The combustion system, fuel, ash characteristics, operating pressure, and total capacity largely determine the boiler configuration. Figs. 1 through 5 illustrate the diversity in configurations.

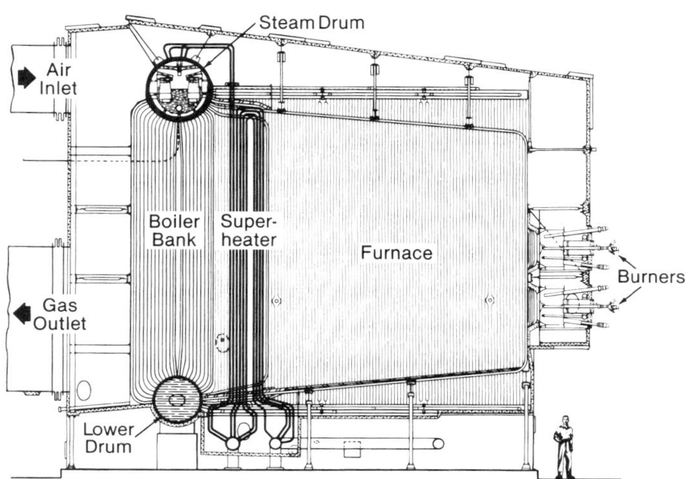

Fig. 1 Integral Furnace industrial boiler for oil and gas firing.

Typical industrial and small power boilers The Integral Furnace boiler shown in Fig. 1 is an oil- and gas-fired, low pressure two drum package boiler. In small capacities, it can be entirely shop assembled and shipped to the site. (See Chapter 1, Fig. 1.) Because it burns a clean fuel, there is no provision made for flyash collection or surface cleaning, and a small furnace volume can be used. A boiler bank (closely spaced tubes between a steam drum and lower drum) provides the heat transfer surface necessary for the rated steaming capacity. Fig. 2 illustrates a two drum Stirling® power boiler (SPB) designed for effective firing of high moisture wood and biomass. (See also Chapter 30.) A stoker (see Chapter 16) firing system is supplied and a boiler bank is provided for sufficient steam generating surface.

Additional unique designs include fluidized-bed boilers, process recovery boilers, and waste-to-energy boilers discussed in Chapters 17, 28 and 29, respectively.

Large utility boiler designs As discussed in Chapter 26, all utility boilers feature gas-tight, fully water-cooled furnace enclosure walls and floors made of all-welded membrane panel construction. Each design normally includes a single reheat section, although the supercritical once-through boiler has also been supplied as a double reheat unit. Fig. 3 is a supercritical 750 MW pulverized coal unit for variable pressure operation. Sloped tubes, commonly angled at 25 degrees from horizontal, wrap around the lower furnace in a single pass (spiral wound construction). Each tube passes through the various heat flux patterns typically encountered in the lower furnace, creating an essentially uniform heat absorption profile. At a point below the furnace nose, the furnace tubes transition to the vertical tube arrangement for the lower heat flux zone of the upper furnace. The consistency of the heat absorption profiles and outlet enthalpy patterns of the furnace afforded by the spiral wound construction allows for true variable pressure operation and on/off cycling by reducing thermal upsets. The inclined tube arrangement also reduces the number of flow paths through the furnace. Other Universal Pressure (UP®) designs are available for supercritical pressure operation with vertical tubing in both the lower and upper furnace. (See Chapter 26.)

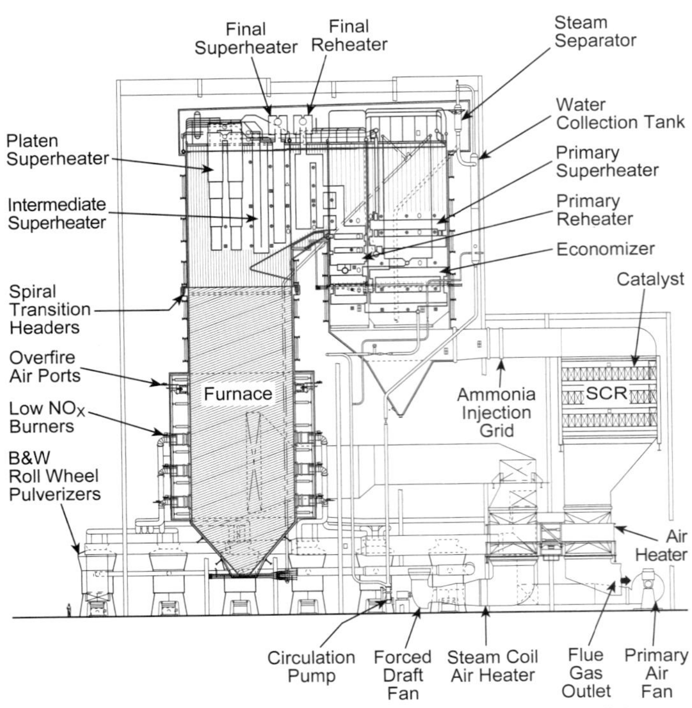

Fig. 3 Variable pressure 750 MW boiler for pulverized coal firing. Design pressure 3825 psig (26.4 MPa); 1054F (568C) superheat and 1105F (596C) reheat steam temperatures; capacity 5,000,000 lb steam/h (630 kg steam/s).

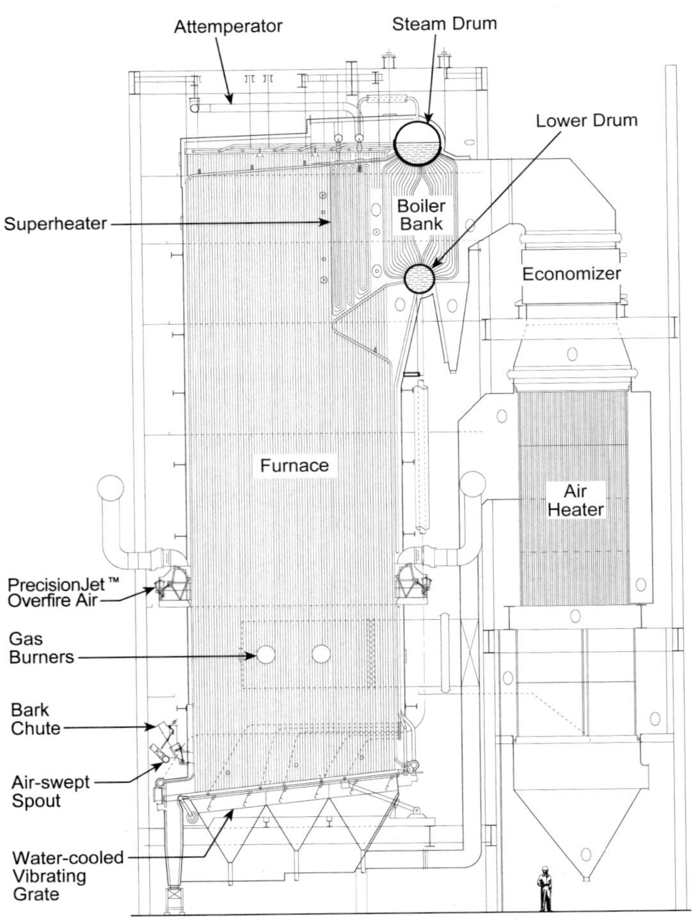

Fig. 2 Two drum Stirling® power boiler for bark firing.

Figs. 4 and 5 illustrate two variations of The Babcock & Wilcox Company's (B&W) Radiant boiler (RB) for natural circulation drum type steam generating systems. Fig. 4 shows an RB unit for coal firing. This is the Carolina-type design (RBC) with downflow convection backpass, which minimizes overall steam generator height. Provision is made for sootblower surface cleaning and for flyash collection. A Tower-type (RBT) configuration is also available that features fully drainable convection pass surfaces and a minimum plan area. (See Chapter 26.) A special type of radiant boiler, the RBW, is available for anthracite combustion. The RBW, named for the W shape its flame forms during combustion, has a similar design to the RBC, with the exception of the lower furnace and combustion system. (See Chapter 26.) Fig. 5 is a Radiant boiler of the El Paso (RBE) configuration for

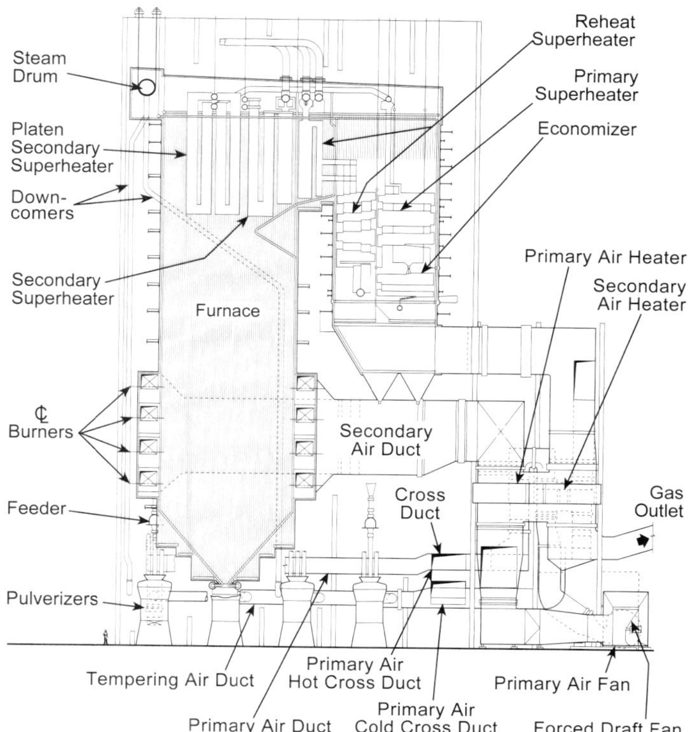

Fig. 4 Carolina-type Radiant boiler for pulverized coal firing. Design pressure 2975 psig (20.61 MPa); superheat and reheat steam temperatures 1005F (541C); capacity 6,600,000 lb steam/h (832 kg steam/s).

oil and gas firing. The relatively clean liquid or gaseous fuels normally burned in this unit allow for a very compact design that minimizes the boiler footprint (plan area) and support steel. The RBE design excludes ash hoppers and ash pits, but provision may be made for cleaning equipment.

Boiler design

Regardless of the size or configuration, modern boiler design remains driven by four key factors: 1) efficiency (boiler and cycle), 2) reliability, 3) capital and operating cost, and 4) environmental protection. These factors, combined with specific applications, produce the diversity of designs presented above and discussed at length in Chapters 26 through 31. However, all of these units share a number of fundamental elements upon which the site- and application-specific design is based.

The boiler evaluation begins by identifying the overall application requirements specified in Table 1. These are generally selected in an iterative process balancing initial capital cost, operating costs (especially fuel), steam process needs, and operating experience. The selection of these parameters can dramatically affect cost and thermal efficiency. Chapters 1 and 37 address these further.

From a boiler evaluation perspective, the temperature-enthalpy diagram shown in Fig. 6 (for a typical high pressure, single reheat unit) provides important design information about the unit configuration. In this example, the relative heat absorption for water preheating, evaporation, and superheating are 30%, 32%, and 38%, respectively. Reheating the steam increases the total heat absorption by approximately 20%. For cycles at supercritical operating pressures, a second stage of reheat may be added. For process applications, only the preheating and evaporation steps may be required.

Boilers can be designed for subcritical or supercritical pressure operation. At subcritical pressures, constant temperature boiling water cools the furnace enclosure, and the flow circuits must be designed to accommodate the two-phase steam-water flow and boiling phenomena addressed in Chapter 5. At supercritical pressures, the water acts as a single-phase fluid with a continuous increase in temperature as it passes through the boiler. These designs require special consideration to avoid excessive unbalances in tube metal temperatures, caused by variations in heat pickup across different flow circuits. In addition, special heat transfer phenomena must also be addressed. (See Chapter 5.)

Two basic fluid circulation systems are used, natural circulation and once-through. In natural circulation systems, which operate at subcritical pressures, water only partially evaporates in the boiler circuits producing a steam-water mixture at the tube outlets. Steam-water separation equipment is provided to separate the steam and water, supply saturated (dry) steam to the superheater, and recirculate water back to the boiler circuits. Natural circulation is the result of the density difference between the colder, denser fluid in the downcomers and the hotter, less dense fluid in the upflow portion of the boiler. In B&W UP once-through designs, the steam drum and internal steam separation equipment are eliminated and a separate startup system is added. UP boilers have been designed for both subcritical and supercritical operation. Supercritical pressures increase overall power cycle efficiency, but at a higher initial capital cost.

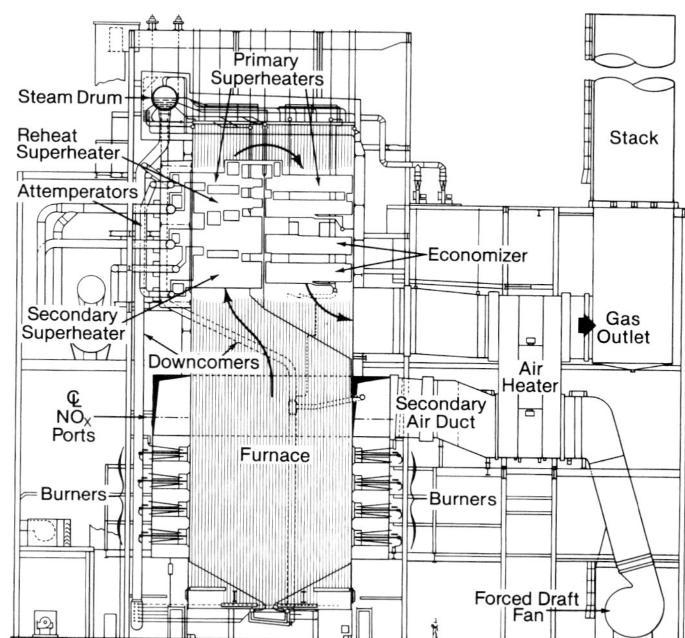

Fig. 5 El Paso-type Radiant boiler for gas firing. Design pressure 2550 psig (17.58 MPa); superheat and reheat steam temperatures 955F (513C); capacity 3,825,000 lb steam/h (482 kg steam/s).

Table 1
Use-Derived Specifications for Boiler Design

Specified Parameter	Comments
Steam use	Flow rates, pressures, temperatures — for utility boilers, the particular power cycle and turbine heat balance.
Fuel type and analysis	Combustion characteristics, fouling and slagging characteristics, ash analysis, etc.
Feedwater supply	Water source, analysis and economizer inlet temperatures.
Pressure drop limits	Gas side and steam side.
Government regulations	Including emission control requirements.
Site-specific factors	Geographical and seasonal characteristics.
Steam generator use	Base load, cycling, etc.
Customer preferences	Specific design guidelines such as flow conditions, design margins, equipment preferences and steam generator efficiency.
Economic evaluation criteria	Efficiency, fan power, reheat pressure drop and spray flows.

Design criteria Within the preceding framework, the important items in boiler design are the following:

1. Define the energy input based upon the steam flow requirements, feedwater temperature, and an assumed or specified boiler thermal efficiency.
2. Evaluate the energy absorption needed in the boiler and other heat transfer components.
3. Perform combustion calculations to establish fuel, air, and gas flow requirements. (See Chapter 10.)
4. Determine the size and shape of the furnace, considering the location and space requirements of the burners or other combustion system, and incorporating sufficient furnace volume for complete combustion and low emissions. Provision must be made for handling the ash contained in the fuel and cooling the flue gas so that the furnace exit gas temperature (FEGT) meets design requirements.
5. Determine the placement and configuration of convection heating surfaces. For utility applications, the superheater and reheater, when provided, must be placed where the gas temperature is high enough to produce effective heat transfer, yet not so high as to cause excessive tube metal temperatures or ash fouling. All convection surfaces must be designed to minimize the impact of slag or ash buildup and permit surface cleaning without erosion of the pressure parts.
6. For industrial applications, provide sufficient saturated boiler surface to generate the remainder of the steam not generated in the furnace walls. This can be accomplished, for example, with the addition of saturated enclosure surface, a boiler (generating) bank, an economizer, or a combination of these items.
7. Design pressure parts in accordance with applicable codes using approved materials.
8. Provide a gas-tight boiler setting or enclosure around the furnace, boiler, superheater, reheater, and economizer.
9. Design pressure part supports and the setting for expansion and local conditions, including wind and earthquake loading.

Fuel selection and specification are particularly important. Boiler systems are designed for specific fuels and frequently encounter combustion, slagging, fouling, or ash-handling problems if a fuel with characteristics other than those originally specified is fired. All potential boiler fuels must be assessed to determine the most demanding fuel.

Chapter 37 outlines procedures for optimizing the steam cycle and for evaluating the value of specific auxiliary equipment in a given application.

As discussed in Chapter 10, the boiler or combustion efficiency is usually evaluated as 100 minus the sum of the heat losses and credits, expressed as a percentage. Chapter 10 provides procedures for calculating the corresponding fuel, air, and gas flow rates.

Enclosure surface design

The furnace of a large pulverized coal-, oil-, or gas-fired boiler is essentially a large enclosed volume where fuel combustion and cooling of the combustion products take place prior to their entry into the convection pass. Excessive gas temperatures entering the convection pass tube banks could lead to elevated tube metal temperatures or unacceptable fouling and slagging. Radiation basically controls heat transfer to the furnace enclosure walls. These walls are typically cooled with boiling water (subcritical pressure) or high

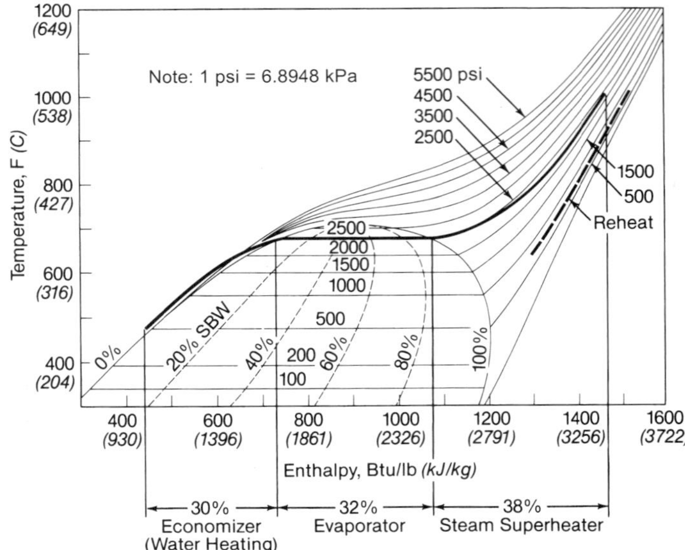

Fig. 6 Temperature-enthalpy diagram for subcritical pressure boiler absorption – one reheat section.

velocity supercritical pressure water. Occasionally, superheat or reheat steam is used.

The convection pass enclosure contains most of the superheater, reheater, and economizer surfaces. (See Fig. 4.) These enclosure surfaces can be water or steam cooled.

The furnace enclosures and convection pass enclosures are usually made with all-welded membrane construction. These enclosures have also been made from tangent tube construction or closely spaced tubes with an exterior gas-tight seal. For a membrane construction, the tube wall and membrane surfaces are exposed to the combustion process on the furnace side, while insulation and lagging (sheet metal) on the outside protect the boiler, minimize heat loss, and protect operating personnel. (See Chapter 23.)

Furnace size versus cycle requirements Besides providing the volume necessary for complete combustion and a means to cool the gas to an acceptable FEGT, the furnace enclosure also provides much of the steam generating surface in a boiler. These roles may not perfectly match one another. In coal-fired units, the minimum furnace volume is usually set to provide the fuel ash specific FEGT. Frequently this results in too much evaporator surface in high pressure boilers and too little surface in low pressure units to meet the thermodynamic requirements for the desired steam exit temperature.

Fig. 7 illustrates the effect of the steam cycle and the operating pressure and temperature on the relative energy absorption between the boiler/economizer and superheater/reheater. As the pressure and temperature increase, the total unit absorption for a given power production progressively declines because of increased cycle efficiency. The boiler and economizer absorption represents the relative amount of heat added to the entering feedwater to produce saturated steam or reach the critical point in a supercritical pressure UP boiler. As the operating pressure increases, the amount of heat required to produce saturated steam declines. Conversely, the amount of heat required for the superheater and reheater increases. The change in required boiler/economizer absorption may not seem significant. However, a 1% shift in absorption is equivalent to approximately 10F (6C) of superheat or reheat temperature.

On low pressure units, the furnace and economizer surfaces are typically not adequate to produce all of the required saturated steam. As a result, a boiler bank with water cooled walls is usually installed after the superheater. (See Figs. 1 and 2.) On a high-pressure unit, the heat absorbed by the furnace and economizer is adequate to produce all the saturated steam required.

Furnace design criteria The furnace is a large open volume enclosed by water-cooled walls for combustion. Fuel and combustion system selection establish the shape and volume of the furnace. For wall firing using circular burners (see Chapter 14), minimum clearances between individual burners, between burners and walls, as well as between burners and the furnace floor, are established based upon physical clearances and functional criteria for complete combustion.

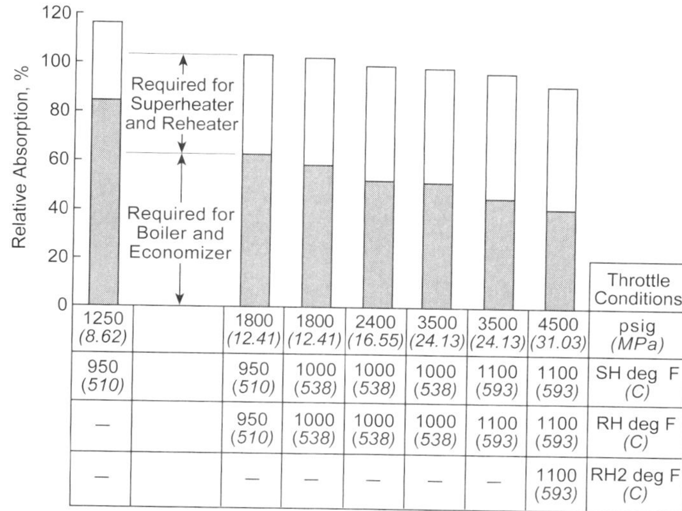

								Throttle Conditions
1250 (8.62)		1800 (12.41)	1800 (12.41)	2400 (16.55)	3500 (24.13)	3500 (24.13)	4500 (31.03)	psig (MPa)
950 (510)		950 (510)	1000 (538)	1000 (538)	1000 (538)	1100 (593)	1100 (593)	SH deg F (C)
—		950 (510)	1000 (538)	1000 (538)	1000 (538)	1100 (593)	1100 (593)	RH deg F (C)
—		—	—	—	—	—	1100 (593)	RH2 deg F (C)

Fig. 7 Relative heat absorption for selected power cycle operating conditions normalized to absorption per kWh in the 2400 psig cycle.

These clearances prevent fuel stream and flame interaction, assure complete combustion, avoid unacceptable flame impingement on the walls (which could lead to tube overheating or excessive deposits), and minimize the formation of nitrogen oxides (NO_x). The maximum fuel input rate, number of burners, and the associated clearances establish: 1) the furnace cross-sectional area, 2) the height of the combustion zone, 3) the height of the overfire air injection zone (if used), and 4) the distance between the burner zone and the furnace hopper or floor. Where the fuel is burned on stokers, the furnace cross-sectional area is established by a specified heat release rate per unit bed area. (See Chapter 16.)

As discussed in detail in Chapter 14, combustion system design and its impact on furnace volume and shape have become complex and critical as emissions limits have been reduced. Use of low NO_x burners, such as the B&W DRB-4Z®, reduce NO_x emissions. Other techniques, such as in-furnace staging with overfire air (NO_x ports), fuel reburning, and reactant injection are also successful in further reducing NO_x emissions. Some techniques for furnace sorbent injection to reduce sulfur dioxide (SO_2) emissions have also been developed. (See Chapter 35.) Each technique can have an impact on the furnace size and configuration.

Several criteria establish the overall furnace height and volume. For clean fuels, such as natural gas, the furnace volume and height are generally set to cool the combustion products to an FEGT that will avoid superheater tube overheating. For fuels such as coal and some oils that contain significant levels of ash, the furnace volume and height are established to cool the products of combustion to an FEGT that will prevent excessive fouling of the convection surfaces. Chapter 4 explores the relationships between furnace volume and FEGT. The specific effects of ash on furnace design are discussed below and in Chapter 21. The furnace height must also be set to provide at least the minimum time to complete combustion and to meet minimum clearance requirements from the burners and NO_x ports to the arch and convective surface.

Ash effects In the case of coal and to a lesser extent with oil, an extremely important consideration is the ash in the fuel. If this ash is not properly considered in the unit design and operation, it can deposit on the furnace walls, sloping surfaces, and throughout the convection pass tube banks. Ash not only reduces the heat absorbed by the unit, but it increases gas-side resistance (draft loss), erodes pressure parts, and eventually can result in unit outages for cleaning and repairs. (See Chapter 21.)

Ash problems are the most severe in coal-fired furnaces. There are two general approaches to ash handling: the dry-ash, or dry-bottom, furnace and the slag-tap, or wet-bottom, furnace.

In the dry-ash furnace, which is particularly applicable to coals with high ash fusion temperatures, a hopper bottom (Figs. 3 and 4) and sufficient cooling surface are provided so that the ash impinging on furnace walls or the hopper bottom is solid and dry; it can be removed essentially as dry particles. When pulverized coal is burned in a dry-ash furnace, about 80% of the ash is carried through the convection banks. The chemistry of the ash can have a dramatic impact on the furnace volume necessary for satisfactory dry-bottom unit operation. Fig. 8, which compares the furnace volume required for a nominal 500 MW boiler burning a low slagging bituminous or subbituminous coal to that for a high slagging lignite coal, illustrates this. Chapter 21 discusses the relationships defining the furnace size requirements in detail.

With coals having low ash fusion temperatures, it can be difficult to use a dry-bottom furnace because

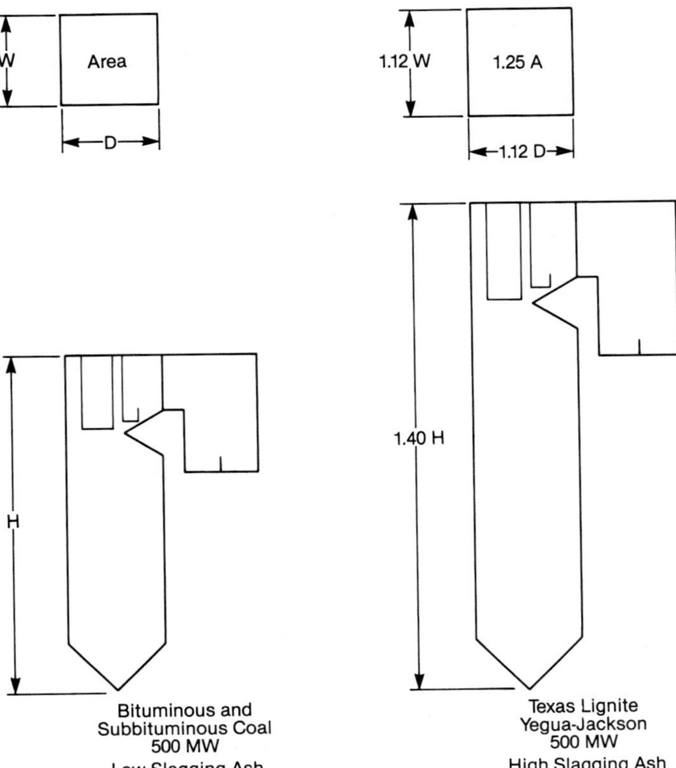

Fig. 8 Boiler size comparison for alternative coal types.

the slag is molten or sticky; it clings and builds up on the furnace walls and hopper bottom. The slag-tap or wet-bottom furnace was developed to handle these coals. The most successful form of the slag-tap furnace is that used with Cyclone furnace firing. (See Chapter 15, Figs. 7 and 8.) This furnace comprises a two-stage arrangement. In the lower part of the furnace, sufficient gas temperature is maintained so that the slag drops onto the floor in liquid form. Here, a pool of liquid slag is maintained and tapped into a slag tank containing water. In the upper part of the furnace, the gases are cooled below the ash fusion point, so ash carried over into the convection banks is dry and does not cause excessive fouling. Because of high NO_x emissions, slag-tap designs became less popular for new boiler designs. Recently, however, considerable reductions in NO_x levels have been demonstrated in Cyclone units firing western fuels and employing staged combustion.

Water-cooled walls Most boiler furnaces have all water-cooled membrane walls. In this type of construction, furnace wall tubes are spaced on close centers to keep membrane temperatures and thermal stresses within limits. The membrane panels (see Fig. 9) are composed of tube rows spaced on centers wider than a tube diameter and joined by a membrane bar securely welded to the adjacent tubes. This results in a continuous wall surface of rugged, pressure-tight construction capable of transferring the radiated heat from the furnace gas to the water or steam-water mixture in the tubes. The width and length of individual panels are suitable for economical manufacture and assembly, with bottom and top headers shop-attached prior to shipment for field assembly. Shipping clearances and erection constraints frequently govern size limitations. The lower furnace walls of Cyclone-fired, refuse, and fluidized-bed units use membrane construction with refractory lining.

Convection boiler surface

Some designs include boiler tubes as the first few rows of tubes in the convection bank. The tubes are spaced to provide gas lanes wide enough to prevent ash and slag pluggage and to facilitate cleaning for dirty fuels. These widely spaced boiler tubes, known as the *slag screen* or *boiler screen*, receive heat by radiation from the furnace and by both radiation and convection from the combustion gases passing through them. Another option is the use of wide spaced water- or steam-cooled surface (see Fig. 4) in the upper furnace. This provides additional radiant heat transfer surface in the furnace while allowing the furnace size to be optimized.

In the larger high pressure units, superheater surface generally forms the furnace outlet plane. The gas temperature entering the superheater must be high enough to give the desired superheat temperature with a reasonable amount of heating surface and the use of economical materials. The arrangements in Figs. 1 through 5 illustrate various configurations of superheater surface at the furnace outlet. To optimize superheater design, the furnace may also incorporate widely spaced steam-cooled platens, as shown in Figs. 3 and 4, or wingwalls in the upper furnace. Fig. 10

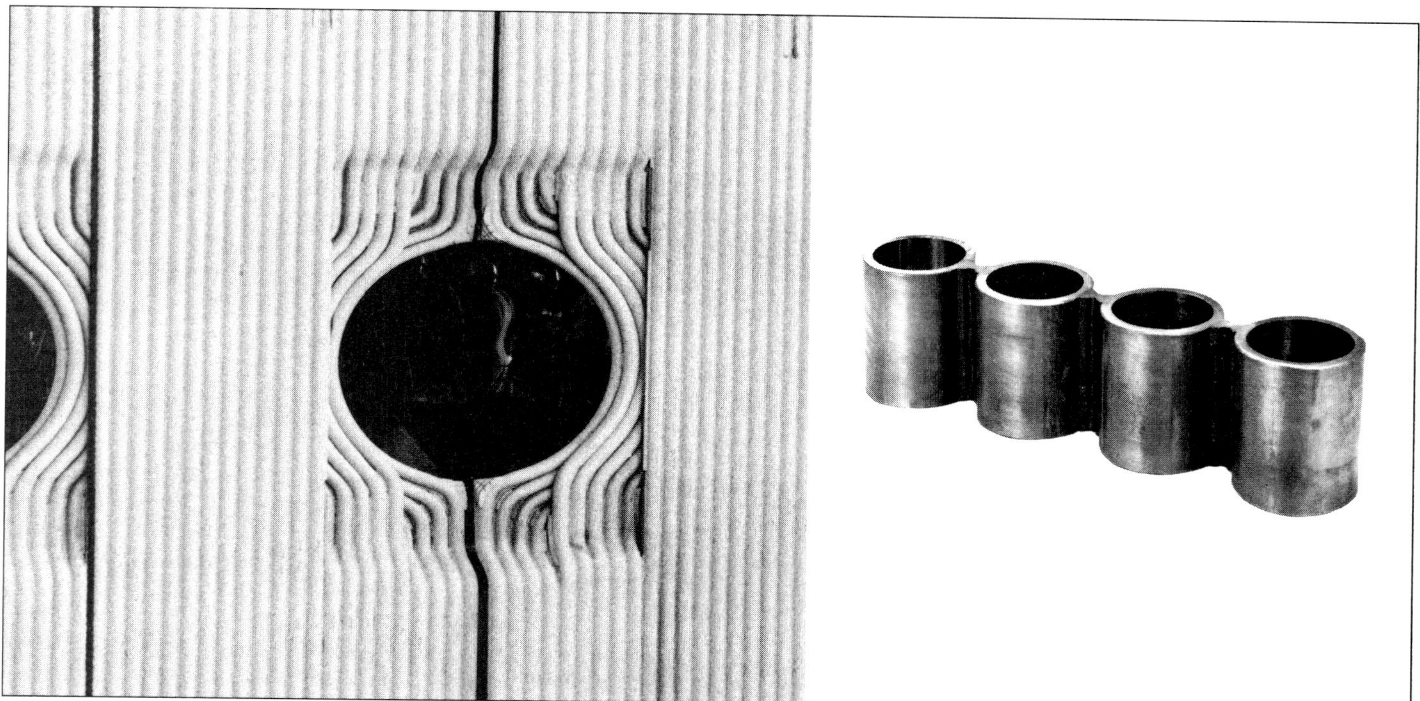

Fig. 9 Membrane wall construction at burner openings.

shows a plan arrangement of convection surface and the change in average gas temperature.

Design of boiler surface after the superheater depends on the type of unit (industrial or utility), desired flue gas temperature drop, and acceptable draft loss through the boiler surface. The preceding *Boiler design* section of this chapter illustrates typical arrangements of boiler and superheater surface for various types of boilers. The objective of convection heating surface design is to establish the proper combination of diameter, length, spacing, number, and orientation of tubes, as well as gas baffling, to provide the desired gas temperature drop with allowable flue gas resistance.

The quantity of heating surface (ft^2 or m^2) needed in the gas-side design of convection surfaces for a given load or duty (Btu/h or W) often inversely relates to gas side flow resistance or pressure loss. Design changes that increase resistance, such as tightening tube spacing perpendicular to flow, result in higher heat transfer rates (Btu/h ft^2 or W/m^2). This in turn reduces the amount of heating surface needed to carry the desired total thermal load. An optimal gas mass flux and design result from balancing the capital cost of the heating surface against the operating cost of fan power and the maintenance costs due to flyash erosion.

For a given flue gas flow rate, a considerably higher gas film heat transfer coefficient, heat absorption, and draft loss result when the gases flow at right angles to the tubes (cross flow) compared to flow parallel to the tubes (long flow). Gas turns between tube banks generally add draft loss and poor distribution with little benefit to heat absorption. Turns leaving the upstream bank and entering the downstream bank of tubes should then be designed for minimum resistance and optimum distribution.

Chapter 22 illustrates the determination of the amount of radiation and convection boiler surface required for a specified heat transfer.

Numerical analysis

When one-dimensional analysis and experimental data are insufficient for design alone, three-dimensional numerical computer modeling can be used to determine local heat transfer and flue gas conditions in boilers, superheaters, and reheaters. These numerical tools are complex, multi-dimensional computer

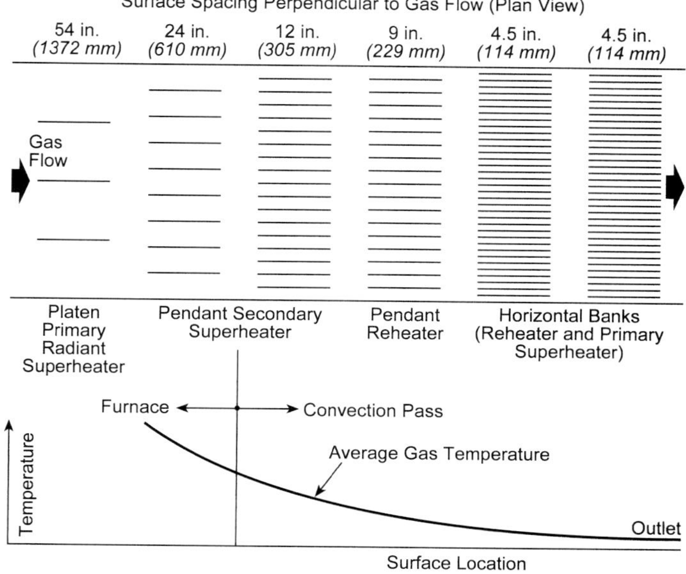

Fig. 10 Schematic plan arrangement of convection surface and change in average gas temperature.

codes that solve conservation equations for mass, momentum, energy, and species.

A numerical model of the superheater and/or reheater regions generally includes the boiler enclosure, burners, and any other geometric aspects which may affect the flow. Required numerical inputs include:

1. geometric data – boiler and heat exchanger surface configurations,
2. performance data – heat exchanger pressure drops and heat absorptions,
3. operating data – stoichiometry and furnace exit gas temperature,
4. burner setup – vane angles and turbulence intensity, and
5. fuel properties – fuel and flue gas quantities, composition and coal fineness, if applicable.

Numerical analyses fall into three general categories: 1) verification of proposed designs, 2) evaluation of design modifications, and 3) investigation of localized problems. Fig. 11 illustrates the use of numerical analysis to characterize the temperature profiles that exist at the furnace exit of a typical coal-fired utility boiler.

Accurate modeling of boilers and boiler components depends heavily upon the availability of information to determine the proper inlet conditions, boundary conditions, and fluid properties. Careful qualification of the model is required and is augmented by known global characteristics such as pressure drop and heat absorption. Correct application of accurate information to develop numerical models will improve designs and determine the cause of localized problems. See Chapter 6 for an in depth discussion of three-dimensional numerical modeling.

Design of pressure parts

Boilers have achieved today's level of safety and reliability through the use of sound materials and safe practices for determining acceptable stresses in drums, headers, tubes, and other pressure parts. Boilers must be designed to applicable codes. Stationary boilers in the United States (U.S.) and many parts of the world are designed to the American Society of Mechanical Engineers (ASME) Boiler and Pressure Vessel Code. (See Chapter 8 and Appendix 2.) The allowable design stress depends on the maximum temperature to which the part is subjected; therefore, it is important that pressure part design temperatures are known and not exceeded in operation. Drum boiler enclosure material temperatures are a function of upset spot heat flux, design pressure, metal conductivity, and the saturation temperature corresponding to the maximum boiler operating pressure. These parameters also determine each tube's outside diameter and thickness. For boiler tubes, temperatures are maintained at known levels by providing a sufficient flow of water to prevent the occurrence of CHF, or critical heat flux phenomena. (See Chapter 5.) An adequate saturated water velocity must exist for each tube, and particular attention must be given to high heat flux zones and sloped tubes with heat on top, such as furnace floor tubes.

Because steam drums have thick walls, it is necessary to limit the heat flow through them to avoid excessive thermal gradients during startup, shutdown, and normal operation. This is particularly important where the drum is exposed to flue gas. A number of tube holes penetrate the drum, and the flow of water through these holes serves to cool the drum wall. Where the heat input through a drum would be too high because of high gas temperature or velocity, insulation may be provided on the outside of the drum, or the drum could be relocated out of the heat.

The required steam purity is dependant on the intended use of the steam being produced. The most stringent requirements are typically associated with power generation applications. Fundamental to maintaining steam purity is maintaining the feedwater quality. (See Chapter 42.) In a drum boiler it is also important to properly design the steam separation equipment because entrained moisture will contain dissolved solids. (See Chapter 5.) In a once-through unit all of the moisture evaporates in the tubes, and steam separation equipment is not a factor in steam purity.

Boiler safety valves constitute very important protection items to ensure the safe design and operation of pressurized boiler components. (See Chapter 25.) Before drum safety valve set pressures are established, steam drum design pressure is determined by the pressure required at the point of use and the intervening pressure drop. As an example, where the steam is used in a turbine, the boiler operating pressure is determined by adding the turbine throttle pressure to the pressure drop through the steam piping, nonreturn valve, superheater, and drum internals at maximum unit steam flow. As a practical matter, to avoid unnecessary losses and maintenance from frequent popping of the safety valves, the first drum safety valve should then be set to relieve at least 5% higher than drum operating pressure. The ASME Code stipulates that the boiler design pressure must not be less than the low-set safety valve relief pressure.

Boiler enclosure

Chapter 23 describes the methods used to provide a tight boiler setting and a tight enclosure around the superheater, reheater, economizer, and air heater.

Boiler supports

Furnace wall tubes are usually supported by the headers to which they are attached, and generating bank tubes and screens are supported by the drum or headers to which they are connected. As discussed in Chapter 8, the following considerations for proper support design are important:

1. The tubes must be arranged and aligned so that they are not subjected to excessive bending-moment stresses in supporting the weight of the tubes, headers, drums, attachments, and fluid within. When the unit is bottom supported, the tubes must satisfy column buckling requirements.
2. The holding strength of the tube seats must not be exceeded.
3. Provision must be made to accommodate the expansion of the pressure parts. For a top supported unit, the hanger rods which support the pressure

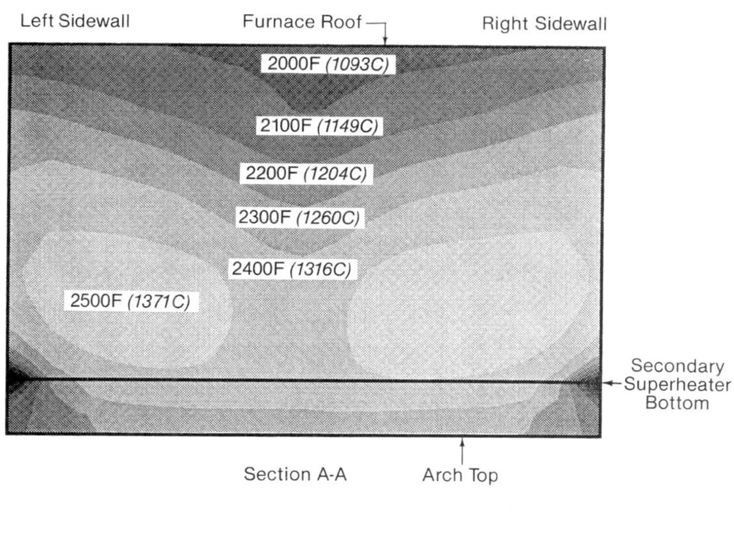

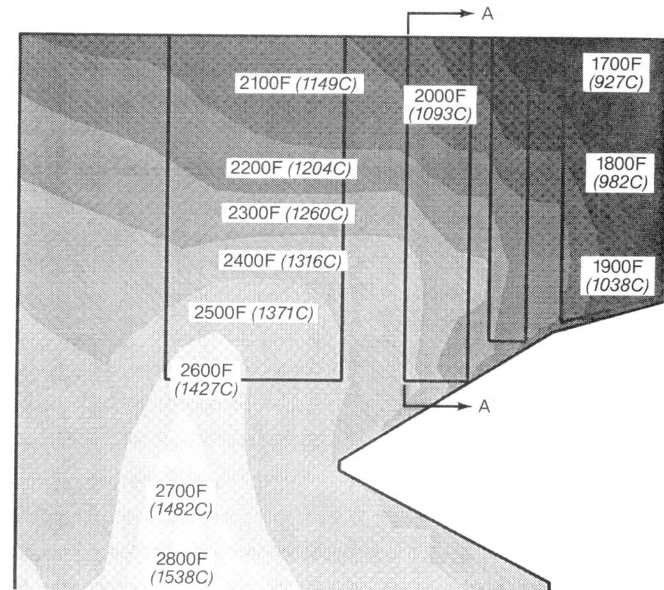

Fig. 11 Numerical modeling results – flue gas temperatures at furnace exit.

parts from the structural steel must be designed to swing at the proper angle. They must be long enough to withstand the movement without excessive stresses in the rods or the pressure parts. Bottom supported boilers should be anchored only at one point, guided along one line, and allowed to expand freely in all other directions. To reduce the frictional forces and resultant stresses in the pressure parts, roller saddles or mountings are desirable for bottom supported heavy loads.

Superheaters and reheaters

Advantages of superheat and reheat

When saturated steam is used in a turbine, the work done results in a loss of energy by the steam and subsequent condensation of a portion of the steam, even though there is a drop in pressure. The amount of moisture the turbine can handle without excessive turbine blade wear limits the amount of work the turbine can do. This is normally between 10% and 15% moisture. It is possible to increase the amount of work done with moisture separation between turbine stages, but this is economical only in special cases. Even with moisture separation, the total energy that can be transformed to work in the turbine is small compared to the amount of heat required to raise the water from feedwater temperature to saturation and then evaporate it. Therefore, moisture content constitutes a basic limitation in turbine design.

Because a turbine generally transforms the energy of superheat into work without forming moisture, this energy is essentially all recoverable in the turbine. The temperature-entropy diagram of the ideal Rankine cycle (shown in Chapter 2, Fig. 6) illustrates this. While this is not always entirely correct, the Rankine cycle diagrams in Chapter 2 indicate that this is essentially true in practical cycles.

The foregoing discussion is not specifically applicable at steam pressures at or above the critical pressure. In fact, the term *superheat* is not truly accurate in defining the temperature of the working fluid in this region. However, even at pressures exceeding 3200 psi (22.1 MPa), heat added at temperatures above 705F (374C) is essentially all recoverable in a turbine.

The reduction of cycle heat rate when steam temperatures entering the turbine are raised is an indication of the benefit of superheat. For example, in a simple calculation of a 2400 psig (16.5 MPa) ideal Rankine cycle with a single reheat stage, an increase in superheat temperature from 900 to 1100F (482 to 593C) reduces the gross heat rate from approximately 7550 to 7200 Btu/kWh. This is more than a 4.5% efficiency improvement attributable to the superheat.

Superheater types

Two basic types of superheaters are available depending upon the mode of heat transfer from the flue gas. One type is the *convection superheater*, for gas temperatures where the portion of heat transfer by radiation from the flue gas is small. Convective heat transfer is highly dependent on the characteristics of gas flow (i.e., quantity and velocity). At maximum load, or 100% steam output, the gas weight is highest; therefore, heat transfer is also high. The higher heat absorption results in higher steam temperatures leaving the superheater and a higher boiler output, or higher total absorption in the superheater per pound of steam. (See Fig. 12.)

A *radiant superheater* receives energy primarily by thermal radiation from the furnace with little energy from convective heat transfer. Radiant heat transfer is highly dependent on the high flue gas temperatures produced by the combustion of fuel. The steam temperature leaving the radiant superheater is lowest at higher loads for two reasons. Radiant absorption does not increase as rapidly as boiler output, and at higher loads there is more steam to cool the effects of the radiant absorption. (See Fig. 12.) The radiant superheater usually takes the form of widely spaced [24 in. (609.6 mm) or larger side spacing] steam-cooled wingwalls or pendant superheat platens located in the furnace. It is sometimes incorporated into the furnace enclosure as curtain walls.

In certain cases, the series combination of radiant and convection superheaters coordinates the two opposite sloping curves to give a flat superheat curve over a wide load range, as indicated in Fig. 12. A separately fired superheater can also be used to produce a flat superheat curve.

The design of radiant and convective superheaters requires extra care to avoid steam and flue gas distribution differences that could lead to tube overheating. Superheaters generally have steam mass fluxes of 100,000 to 1,000,000 lb/h ft^2 (136 to 1356 kg/m^2 s) or higher. These are set to provide adequate tube cooling while meeting allowable pressure drop limits. The mass flux selected depends upon the steam pressure and temperature as well as superheater thermal duty. In addition, the higher pressure loss associated with higher velocities improves the steam-side flow distribution.

Tube geometry

Cylindrical tubes of 1.75 to 2.5 in. (44.5 to 63.5 mm) outside diameter are typical in current superheaters and reheaters. Steam pressure drop is higher and alignment is more difficult with the smaller diameters, while larger diameters result in thicker tubes.

Superheaters and reheaters are almost exclusively made of bare tubes (see Fig. 13). Extended surface on superheater tubes in the form of fins, rings, or studs can increase metal temperature and thermal stress beyond tolerable limits.

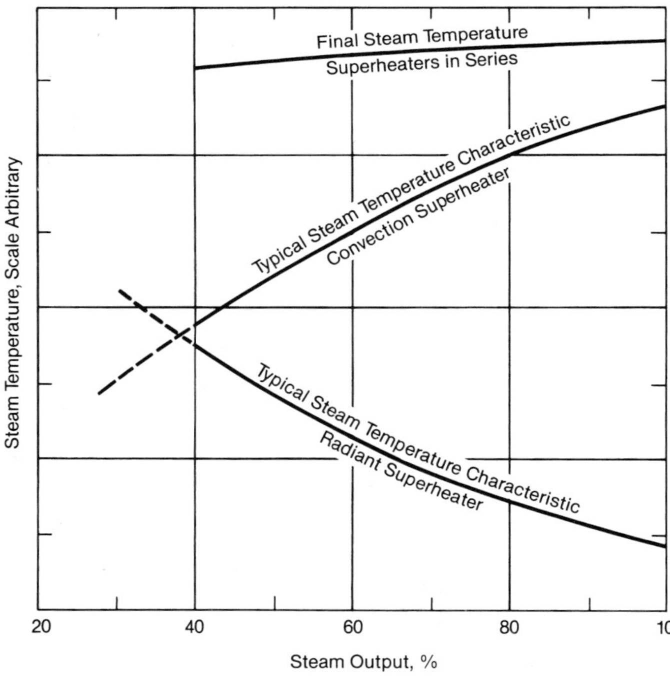

Fig. 12 A substantially uniform final steam temperature over a range of output can be attained by a series arrangement of radiant and convection superheater components.

Relationships in superheater design

Effective superheater design must consider several parameters including:

1. the outlet steam temperature specified,
2. the range of boiler load over which steam temperature is to be controlled,
3. the superheater surface required to give the specified steam temperature,
4. the gas temperature zone in which the surface is to be located,
5. the type of steel, alloy or other material best suited for the surface and supports,
6. the rate of steam flow through the tubes (mass flux or velocity), which is limited by the permissible steam pressure drop but that, in turn, exerts a dominant control over tube metal temperatures,
7. the arrangement of surface to meet the characteristics of the anticipated fuels, with particular reference to the spacing of the tubes to prevent accumulations of ash and slag or to provide for easy removal of these formations in their early stages, and
8. the physical design and type of superheater as a structure or component.

A change in any of these items may require a counterbalancing change in some or all of the other items.

The steam temperature desired in advanced power station design is typically the maximum for which the superheater designer can produce an economical component. Economics in this case requires the assessment of two interrelated costs – initial investment and the subsequent cost of upkeep to minimize operating problems, outages, and replacements. The steam temperature desired is based upon an iterative evaluation of variations in parameters 3 through 7, past operating experience, and the requirements of the particular project. Operating experience has resulted in the use

Fig. 13 Typical superheater elements for a replacement project.

of 1000F (538C) to 1050F (566C) steam temperatures for the superheat and reheat in many large utility boilers. However, advances in tube materials have allowed designers to increase these temperatures without compromising the reliability or economics of the units.

After the steam temperature is specified, the amount of surface necessary to provide this superheat must be established. This is dependent on parameters 5 through 8. Because there is no single correlation, this quantity must be determined by iteration, locating the superheater in a zone of gas temperature that satisfies the design criteria. In standard boilers, unit arrangement and the space designated for superheater surface basically establishes the zone.

After the amount of surface is established for the optimum location and tube spacing, the steam mass flux or velocity, steam pressure drop, and superheater tube metal temperatures are calculated considering material options and thickness requirements. The material determination is then made based on economic and functional optimization of the tubes, headers, and other components. It may be necessary to compare several arrangements to obtain an optimum combination that:

1. requires an alloy of less cost,
2. gives a more reasonable steam pressure drop without jeopardizing the tube metal temperatures,
3. gives a higher steam mass flux or velocity to lower tube temperatures,
4. gives the tube spacings that minimize ash accumulations and flyash erosion potential with various types of fuel,
5. permits closer spacing of the tubes, thereby making a more economical arrangement for a favorable fuel supply,
6. gives an arrangement of tubes that reduces the draft loss for an installation where this parameter evaluation is crucial, and
7. permits the superheater surface to be located in a zone of higher gas temperature, with a subsequent reduction of required surface.

It is possible to achieve a practical design with optimum economic and operational characteristics and with all criteria reasonably satisfied, but a large measure of experience and the application of sound physical principles are required for best results. Chapter 22 gives the calculation methods for superheater performance.

Relationships in reheater design

The fundamental considerations governing superheater design also apply to reheater design. However, the pressure drop in reheaters is critical because too much pressure loss through the reheater system can nullify the gain in heat rate attained with the reheat cycle.

Steam mass fluxes or velocities in reheater tubes should be sufficient to keep the difference between the bulk steam and metal surface temperature below 150F (83C). As a result of the high velocities required, the pressure drop through the reheater tubes is ordinarily 4 to 5% of reheater inlet pressure. This allows another 4 to 5% pressure drop for the reheat piping and valves without exceeding the usual 8 to 10% total allowable pressure loss. For economic reasons, the pressure drop allocated for piping is usually distributed with one-third to the cold (inlet) reheat piping and two-thirds to the hot (outlet) reheat piping.

Tube materials

Oxidation resistance, allowable stress, and economics determine the materials used for superheater and reheater tubes. The design process requires careful attention to all of these criteria to develop an economical and reliable configuration. Chapter 7 gives additional information on these materials, including maximum metal temperatures.

Fig. 14 shows variations in basic superheater heat transfer surface arrangements. These variations in surface arrangement permit the economic tradeoff between material unit costs and differences in the quantity of surface required due to thermal-hydraulic considerations. The comparison of counter flow versus parallel flow designs defines the basic relationships between heat transfer surface requirements and the corresponding tube metal temperatures and requirements.

Supports for superheaters and reheaters

Because superheaters and reheaters are located in zones of relatively high gas temperature, it is preferable to have the major support loads carried by the tubes themselves. For pendant superheaters, the major support points are located outside of the gas stream, with the pendant loops supporting themselves in simple tension. Fig. 15 illustrates standard support arrangements for a pendant superheater outlet section with major section supports above the roof line.

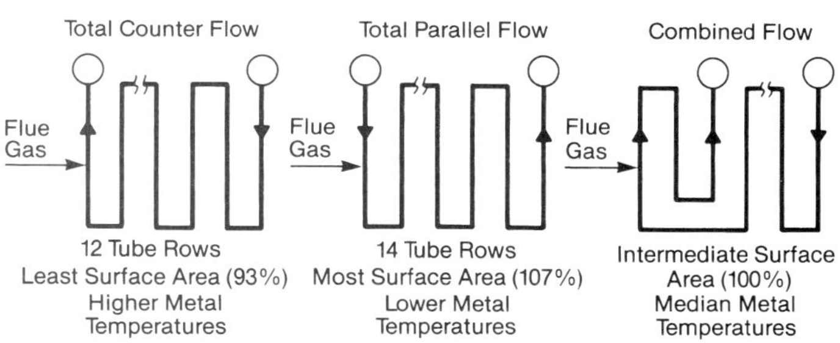

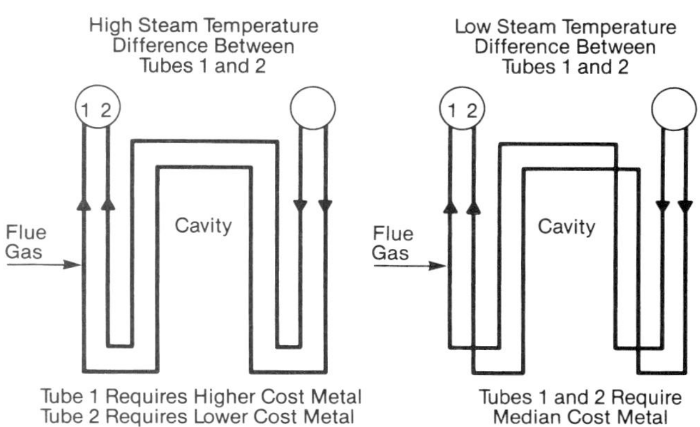

Fig. 14 Typical superheater heat transfer surface arrangements.

Where adequate side spacing is available, high chromium-nickel alloy ring-type guides are used. In addition, in the higher gas temperature zones, steam-cooled side-to-side ties are used to maintain side spacings. For closer side spaced elements, steam-cooled wraparounds are not practical and mechanical ties, such as D-links, are used to maintain alignment as shown in Fig. 16. In this case, the clear backspacing between tubes and the size of attachments have been kept to a minimum. This serves to reduce the thermal stresses imposed on the tube wall. Fig. 16 also shows a typical arrangement of a pendant reheater section and illustrates the support of a separated bank by a special loop of reheater, which permits all major supports to remain above the roof and out of the gas stream.

In horizontal superheaters, the support load is usually transferred to boiler or steam-cooled enclosure tubes or economizer stringer tubes by lugs, one welded to the support tubes and the other to the superheater tubes as indicated in Fig. 17. These lugs are made of carbon steel through high chromium-nickel alloy. Depending upon design considerations, they must slide on one another to provide relative movement between the boiler tubes and the superheater tubes. Saddle-type supports, also shown in Fig. 17, provide for relative movement between adjacent tubes of the superheater.

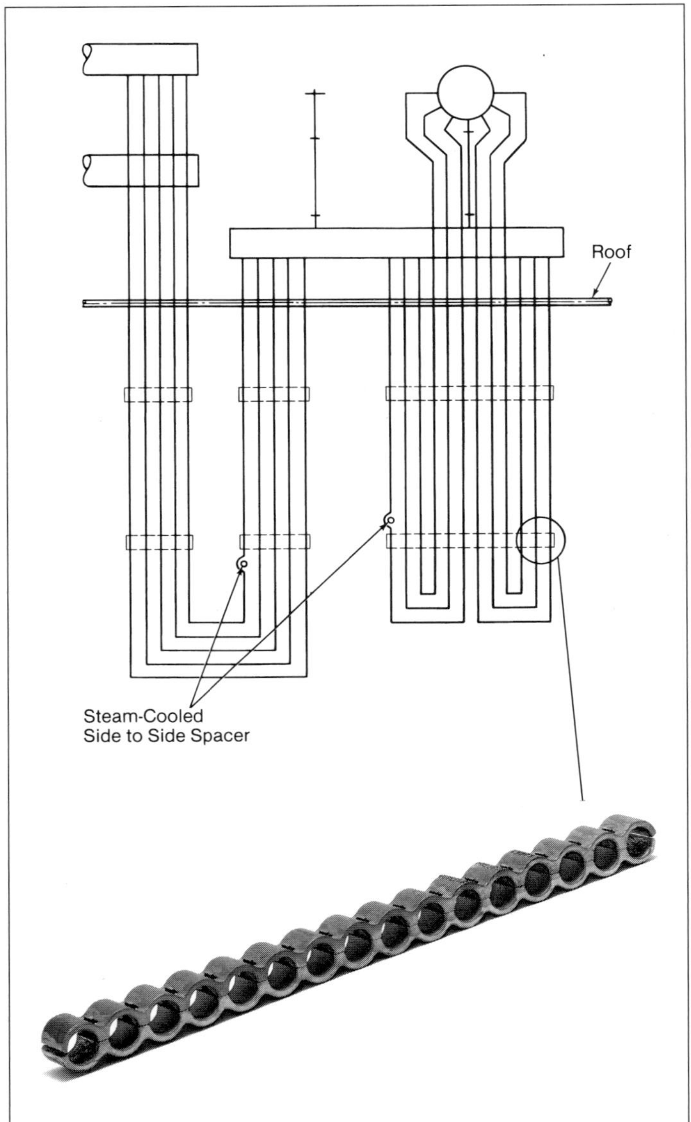

Fig. 15 Pendant superheater section with split ring casting supports.

As units grow in size, the span of the superheater tubes may become so great that it is impossible to end-support these tubes. Most of the larger units use stringer tubes (see Chapter 20 frontispiece), generally hung from the economizer outlet, to support the superheater tubes. Saddle-type supports maintain tube spacing within the section.

Internal cleaning

The internal cleaning of superheater and/or reheater surfaces is not normally required; however, under certain circumstances these surfaces have been chemically cleaned. Chapter 42 discusses this in depth. During startup, steam line blowing is used to remove residual scale, oils, and residual debris.

External cleaning and surface spacing

Most customers today are striving for long periods of continuous operation, in some cases as long as 18 to 24 months between outages, so gas side cleanability is critical. To enhance cleanability, the superheater sections (platens) of modern utility boilers are spaced according to the gas temperature and the fuel fired. Fig. 10 illustrates the spacing for pulverized coal-fired units. The backspacing in the direction of gas flow is usually set at 0.50 to 0.75 in. (12.7 to 19.1 mm) clear space between tubes. These spacings are empirical; they are based on tube fouling and erosion experience and on manufacturing requirements. Chapter 24 discusses surface arrangement and cleanability under *Sootblowers*.

Steam temperature adjustment and control

Improvement in the heat rate of modern boiler units and turbines results mostly from the high cycle efficiency possible with high steam temperatures.

Important reasons for accurate steam temperature regulation are to prevent failures due to excessive metal temperatures in the superheater, reheater, or turbine; to prevent thermal expansion from dangerously reducing turbine clearances; and to avoid erosion from excessive moisture in the last stages of the turbine.

The control of temperature fluctuations from variables of operation, such as slag or ash accumulation, is also important.

With drum-type boilers, steam output is maintained by firing rate, while the resulting superheat and reheat steam temperatures depend on basic design factors, such as total surface quantity and the ratio of convection to radiant heat absorbing surface. Steam temperatures are also affected by other important operating variables such as excess air, feedwater temperature, changes in fuel that affect burning characteristics, ash deposits on the heating surfaces, and the specific burner combination in service. In the once-

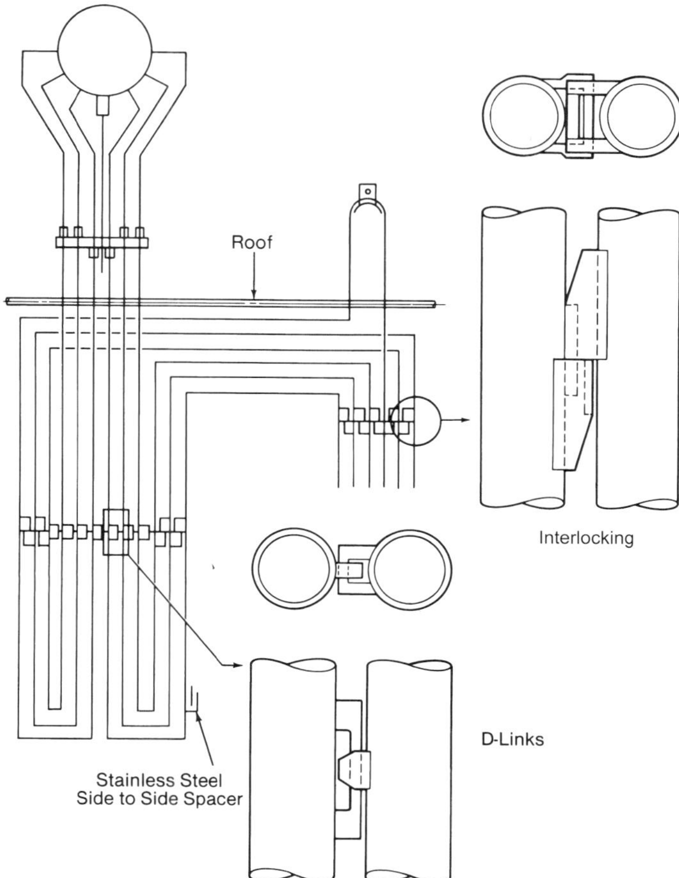

Fig. 16 Pendant reheater section with supports.

through boiler, which has a variable steam-water transition zone, coordination of the firing rate and the boiler feedwater flow rate control steam output pressure and temperature, leaving reheat steam temperature as a dependent variable. (See Chapter 26.) Standard performance practice for steam generating equipment usually permits a tolerance of ±10F (6C) in a specific steam outlet temperature.

Definitions of terms

Adjustment is a physical change in the arrangement of heating surface that affects steam temperature, but can not be used to vary steam temperature during operation. This could be the addition or deletion of component surface or boiler surface ahead of the superheater and/or reheater.

Control is the regulation of steam temperature during operation without changing the arrangement of surface. Examples are operation of attemperators, gas proportioning dampers, and gas recirculation fans.

The *attemperator* is an apparatus for reducing and controlling the temperature of a superheated fluid passing through it. This is accomplished by spraying high purity water into an interconnecting steam pipe usually between superheater stages or upstream of a reheater inlet.

The *gas proportioning damper* is used to control reheat steam temperatures. The damper opens to split the flow of hot flue gas between the superheater and reheater.

The *gas recirculation fan* is used for controlling reheat steam temperature by extracting flue gas from the economizer and injecting it into the furnace.

Effect of operating variables

Many operating variables affect steam temperatures in drum-type units. To maintain constant steam temperature, means must be provided to compensate for the effect of these variables.

Load As load increases, the quantity and temperature of the combustion gases increase. In a convection superheater (see *Superheater types*), steam temperature increases with load with the rate of increase being less for superheater surfaces located closer to the furnace. In a radiant superheater (see *Superheater types*), steam temperature declines as load increases. Normally a proportioned combination of radiant and convection superheater surface is installed in series in a steam generating unit to maintain steam temperature as constant as possible over the load range. (See Fig. 12.)

Excess air For a change in the amount of excess air entering the burner zone there is a corresponding change in the quantity of gas flowing over the convection superheater; therefore, an increase in excess air generally raises the steam temperature. However, the degradation of boiler efficiency due to the addition of excess air usually more than offsets the improvement in heat rate, making this technique uneconomical other than in special circumstances.

Feedwater temperature Decreasing the feedwater temperature by taking a feedwater heater out of service will increase the fuel input required to generate a given amount of steam. This will result in high gas weights and temperatures, which will yield higher steam temperatures in the superheater.

Heating surface cleanliness Removal of ash deposits from heat absorbing surfaces ahead of the superheater reduces the temperature of the gas entering the superheater and, subsequently, the steam temperature. Removal of deposits from the superheater surface increases superheater absorption and raises steam temperature.

Use of saturated steam If saturated steam from the boiler is used for sootblowers or auxiliaries, such as pump or fan drives, an increased firing rate is required to maintain constant main steam output. This raises the steam temperature.

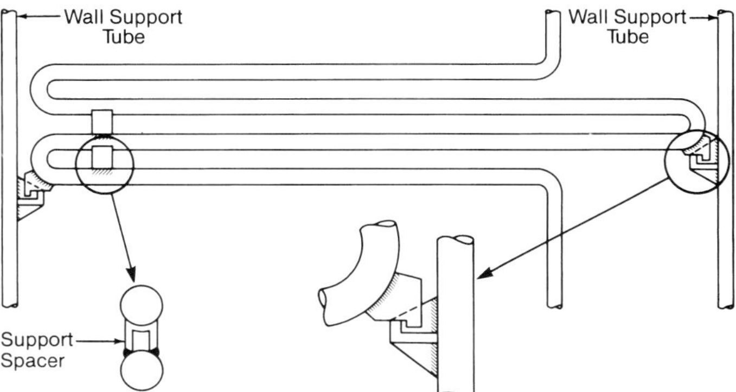

Fig. 17 Horizontal superheater section end supports on walls.

Blowdown The effect of blowdown (water bled from the boiler drum or steam supply system to reduce contaminants) is similar to the use of saturated steam, but to a lesser degree because of the low enthalpy of water as compared to steam.

Burner operation The distribution of heat input among burners at different locations or a change in burner adjustment usually has an effect on steam temperature due to changes in furnace heat absorption rate.

Fuel Variations in steam temperature may result from changing the type of fuel burned or from day to day changes in the characteristics of a given fuel.

Adjustment

Steam generating equipment represents a large, long-term capital investment. Over the life of the equipment, it is not unusual for the quality of the fuel or even the steam requirements to change. Ideally, provisions would be made in the original design to accommodate the modifications required by such changes. However, economics and space limitations often limit the designers ability to incorporate these provisions. As a result, if such changes do occur during the life of the equipment, it will usually be necessary to have the new conditions and/or fuel reviewed to determine what design changes can be made within the constraints imposed by the existing equipment.

Control

Control is necessary to regulate steam temperature within required limits in order to correct fluctuations caused by operating variables – particularly boiler load. Besides boiler load variation, ash deposition on heat transfer surfaces and fuel variations are the most frequent cause of steam temperature fluctuations. Changes in the sequence or frequency of sootblower operation can usually correct this condition. Selective operation of furnace wall blowers or implementation of unit load reductions to induce slag shedding from the furnace walls can reduce gas temperatures entering the superheater surfaces. The removal of feedwater heaters or pulverizers from service can also impact steam temperature and require steam temperature control.

The turbine manufacturer in accordance with a safe steam temperature-time curve establishes the time in which a turbine may be brought to full load. Because the temperature of the steam is directly related to the degree of expansion of the turbine elements and, consequently the maintenance of safe clearances, this temperature must be regulated within permissible limits by an accurate control device.

Among the means of control for regulating steam temperature are: attemperation, gas proportioning dampers, gas recirculation, excess air, burner selection, movable burners, divided furnace with differential firing, and separately fired superheaters. With attemperation, diluting high temperature steam with low temperature water or removing heat from the steam regulates steam temperature. By comparison, the other methods of control are based on varying the amount of heat absorbed by the steam superheating surfaces.

Attemperation Attemperators may be classified as two types – direct contact and surface. The spray type exemplifies the direct contact design, where the steam and the cooling medium (water and saturated steam) are mixed. In the surface design, which includes the shell type and the drum type, the steam is isolated from the cooling medium by the heat exchanger surface. Spray attemperators are usually preferred on all units that have attemperation requirements; however, surface type attemperators are occasionally used on some industrial units.

The superheater attemperator may be located either at some intermediate point between two sections of the superheater, or at the superheater outlet. The ideal location for the superheater attemperator for the purpose of *process control* is at the superheater outlet. Control would be direct and there would be no time lag. However, problems with this location include possible water carryover into the turbine and overheating of superheater tubes. Placing the superheater attemperator between superheater stages addresses these concerns, and is generally preferred. In addition, the attemperation fluid so thoroughly mixes with steam from the first stage superheater that it enters the second stage superheater at a uniform temperature. For reheat superheater applications, the attemperator is typically located at the reheater inlet.

The superheater spray attemperator, illustrated in Fig. 18, has proven most satisfactory for regulating steam temperature. High purity water is introduced into the superheated steam line through a spray nozzle at the throat of a venturi section within the line. Because of the spray action at the nozzle and the high velocity of the steam passing through the venturi throat, the water vaporizes and mixes with the superheated steam reducing the steam temperature. An important construction feature is the continuation of the venturi section into a thermal sleeve downstream from the spray nozzle. This protects the high temperature piping from thermal shock. This shock could result from non-evaporated water droplets striking the hot surface of the piping. The reheat attemperator is similar but does not have a venturi section.

The spray attemperator provides a quick acting and sensitive control for regulating steam temperature. It is important that the spray water be of highest purity, because solids entrained in the water enter the

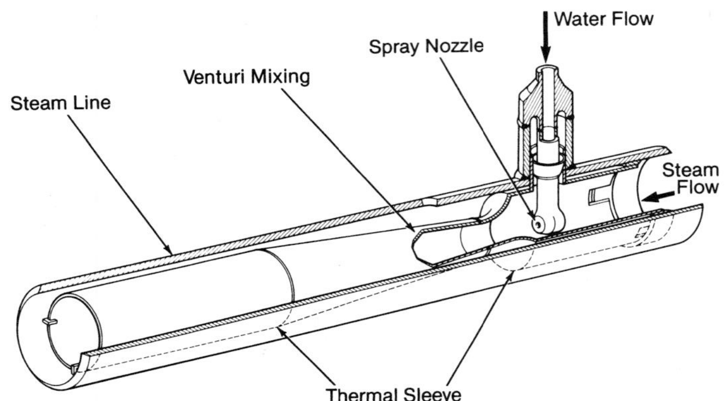

Fig. 18 Spray attemperator showing thermal sleeve.

steam and may cause troublesome deposits on superheater tubes, piping, or turbine blades. High-pressure heater drains are a source of extremely pure water, but require a separate high pressure corrosion resistant pump if used for attemperator supply. Normally, boiler feedwater is satisfactory, provided condenser leakage and makeup do not introduce too much contamination.

Three attemperator arrangements are possible, depending on the boiler performance requirements:

1. *Single-stage attemperator* A single attemperator may be installed in each of the connecting pipes between two stages of superheat.
2. *Tandem attemperator* A single-stage attemperator with two spray water nozzles may be installed in series in the connecting pipe between two stages of superheat. This arrangement is used where the spray quantity exceeds the capacity of a single nozzle or where the required turndown can not be achieved with a single spray water nozzle. The usual application requires a spray control valve for each spray nozzle. The operation is sequential with the control valve for the downstream spray nozzle opening first and closing last.
3. *Multiple-stage attemperator* Multiple single-stage attemperators are used. A spray control valve is required for each stage of spray. The first stage spray attemperator is used first, with the maximum spray flow based on a minimum allowable difference between the temperature of the steam leaving the attemperator and the saturation temperature. When the flow limit is reached on the first stage, the subsequent stages are activated sequentially.

The attemperation system branches from the feedwater piping. The boiler feed pump provides water pressure for all of the piping to the steam drum and the line to the attemperation system. In most instances, the pressure loss from the boiler feed pump through the feedwater heaters, piping, economizer and primary superheater to the attemperator location inside the connecting piping results in a boiler feed pump discharge pressure high enough to provide the required pressure differential in the attemperator system. The attemperator system pressure must be higher than the superheater steam pressure for proper operation. In certain cases, the feed pump discharge pressure is not sufficient due to low pressure loss in the boiler and feedwater system, or due to the high pressure differential the spray systems require to suit particular boiler characteristics. For these situations, a booster stage in the boiler feed pump is desirable to raise the spray water to the required pressure. Other options are a separate spray water booster pump or an additional feed line valve to increase boiler side resistance at the required loads.

In the past, there were problems in industrial steam power cycles involving deposits in the superheater and on turbine blading, which resulted in failures. Many of these problems were due to impurities in the attemperator spray water. To assure high quality spray water, additional cleanup equipment can be installed on the condensate return and feedwater makeup.

For low-pressure units with feedwater purity less than that required for spray attemperation, the condenser attemperator system is an economical and reliable system used to produce high quality spray water. (See Chapter 27.)

Drum and once-through boilers may use steam attemperation at very low loads, as described in the *Bypass and startup* section of this chapter.

UP boiler attemperator applications UP boilers are supplied with superheater spray attemperators to reduce variations in superheater outlet temperatures during transients. These attemperators may be single or multiple stage. Spray water is supplied from the UP boiler economizer inlet or directly from the boiler feed pump. Spray attemperation corrects main steam temperature deviations only on a temporary basis and is not the primary means of steam temperature control. Under steady-state conditions, the ratio of firing rate to feedwater flow determines steam temperature.

Gas proportioning dampers As shown in Fig. 4, the horizontal convection tube banks in the back end of a boiler can be divided into two or more separate flue gas passes separated by a baffle wall. The use of dampers in these gas passes then permits proportioning of the gas over the heat transfer surfaces and the control of reheat and superheat temperatures.

Design considerations for such systems include the following:

1. Dampers must be placed in a cool gas zone to assure maximum reliability (typically downstream of all boiler heat transfer surfaces).
2. Draft loss through the unit could increase for some designs, particularly with alternate fuels, so this parameter must be optimized.
3. Control system design and tuning are critical because damper control response is slower than with spray attemperators. Therefore, spray attemperators are used for transient control.
4. Under maximum bias conditions, when one damper is almost completely closed and the other is open, the gas temperatures at the dampers and the heat transfer surfaces nearest the dampers will be at their highest. These temperatures then set the tube metal design requirements.

Gas proportioning dampers are combined with spray attemperation for overall optimal steam temperature control systems. Spray attemperators provide for short-term transient temperature control. The gas proportioning dampers provide longer-term control and adjustment between superheat and reheat temperatures with a minimum impact on overall unit efficiency.

Gas recirculation Another method of controlling superheat or reheat is flue gas recirculation. As the name implies, gas from the boiler, economizer, or air heater outlet is reintroduced to the furnace by fans and flues. For the sake of clarity, recirculated gas introduced in the immediate vicinity of the initial burning zone of the furnace and used for steam temperature control is referred to as *gas recirculation*, and recirculated gas introduced near the furnace outlet and used for control of gas temperature is referred to as *gas tempering*. Fig. 19 shows an application of gas recirculation through the hopper bottom and gas tempering in the upper furnace of a Radiant boiler. In most instances, the gas is obtained from the economizer outlet. The recirculated gas must be introduced into the furnace in a manner that avoids interference with the fuel combustion. The amount of recirculated gas is expressed as a percentage of the sum of combustion gas weight and setting infiltration.

While recirculated gas may be used for several purposes, its basic function is to alter the heat absorption pattern within a steam generating unit. Recirculated gas has the special advantage of providing heat absorption adjustment for use as a design factor in initial surface arrangement, and as a method of controlling the heat absorption pattern under varying operating conditions.

Excess air An increase in the amount of excess combustion air in a drum type unit decreases the furnace heat absorption. At fractional loads, the increase in excess air and the decrease in absorption work together to increase the steam outlet temperature of a convection superheater. The resulting greater weight of gas sent to the stack increases the stack loss; however, the increase in turbine efficiency can offset some of this loss.

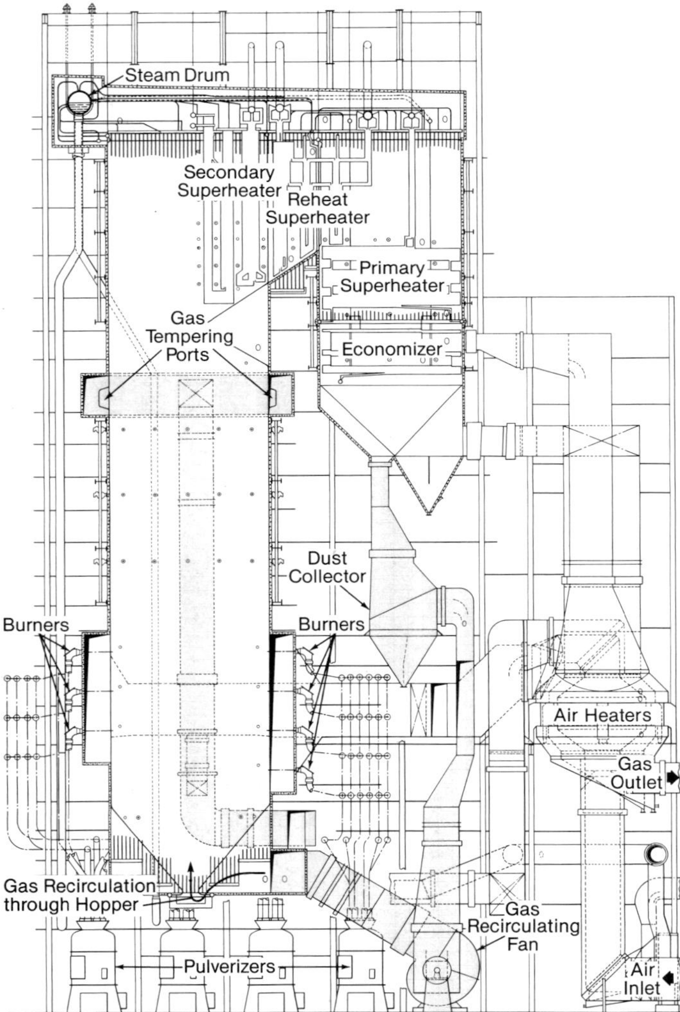

Fig. 19 Radiant boiler with gas tempering for gas temperature control and gas recirculation for control of furnace absorption and reheat temperature.

Burner selection It is often possible to regulate steam temperature by selective burner operation. Higher steam temperatures may be obtained at less than full load by operating only the burners giving the highest furnace outlet temperature. When steam temperature reduction is required, firing may be shifted to the lower burners. Distributing the burners over an extended height of the burner wall can improve this method of control.

Movable burners Using movable burners to raise or lower the main combustion zone in the furnace can also affect the regulation of steam temperature by changing the furnace absorption pattern. Tilting burners are used for this purpose.

Differentially-fired divided furnaces In some divided furnaces, the superheater receives heat from one section of the furnace only, while the other section of the furnace generates only saturated steam or may include a reheater. Changing the proportion of fuel input between the two furnace sections regulates the steam temperature. This arrangement, similar in principle to the separately-fired superheater, was formerly widely used in marine practice. The differentially-fired divided furnace method of steam temperature control is no longer in use today.

Separately-fired superheaters A superheater that is completely separate from the steam generating unit and independently fired may serve one or more saturated steam boilers. This arrangement is not economical for power generation, where a large quantity of high temperature steam is needed.

Bypass and startup systems

High pressure drum and once-through boilers must be capable of rapid, frequent, and reliable unit startups, and rapid load changes, to provide the most economical electrical production. During startup or low load (less than 20%) conditions in which outlet steam temperature tends to follow flue gas temperature, regular water attemperator steam temperature control systems are not sufficiently sensitive to cool the steam, as only a small amount of attemperation is required at low steam loads. However, provisions must still be made to match the differing flow, pressure, and temperature needs of the steam turbine and boiler. To address these special requirements, a variety of bypass and startup systems have been developed for drum and once-through boilers.

UP boiler startup systems

A key requirement of UP startup and bypass systems is the need for minimum design circulation flows in high heat absorbing circuits for cooling before the unit can be fired. Additional important features include providing a turbine bypass until steam pressures and temperatures are matched, reducing the bypass flow pressure and temperature before condenser and auxiliary equipment steam admission, recovering heat during startup, providing clean water for full startup, accelerating the startup processes, and providing greater unit flexibility through dual pressure boiler operation. UP boilers, while originally designed for base load operation, have been adapted to daily load and on/off cycling. The variable furnace pressure system is capable of on/off cycling, while the constant furnace pressure system is capable of daily low load operation on the bypass system.

Three different UP systems are available – one for constant furnace pressure operation over the load range and two for variable furnace pressure operation over the load range; however, only one variable furnace pressure system will be discussed in detail.

Spiral wound variable pressure startup system For variable furnace pressure systems, the startup system is equipped with pipes, valves, steam-water separators, and a water collection tank. Fig. 20 shows the general arrangement. The startup system controls operation during startup, shutdown, and minimum load operation. The spiral wound universal pressure (SWUP™) once-through boiler requires a minimum flow in the furnace circuits before the boiler can be fired. The flow must remain above this minimum flow during boiler operation. For loads where the steam flow falls below the minimum, the startup system maintains the necessary minimum steam flow through the boiler circuitry and supplies steam to the turbine that was provided to the superheater from the vertical steam separators.

Typical operation Using the boiler feed pump (BFP), water fills the economizer, furnace, vertical steam separator (VSS), and water collecting tank (WCT) until the WCT reaches normal level. The economizer vent valve (302) opens to vent air from the economizer to the separators. If the WCT water level rises above normal, the separator (WCT) high level control valves (341) will open to control the WCT water level.

During cold startup, the water must circulate through the water cleanup system until feedwater impurities reach an acceptable level before the boiler can be fired. Water flows through the condenser, polishers, and heaters to the boiler, and returns to the condenser by way of the 341 valves. Flow rate is approximately 30% maximum continuous rating (MCR) from the BFP to the condenser at this stage.

After water quality is established, the feedwater flow from the BFP to the boiler is reduced to about 7% MCR flow, where about 4% MCR flows into the economizer to avoid steaming economizer conditions, and about 3% MCR flows to the water collecting tank (WCT) through the subcooling control valve (383). The 383 valve controls the amount of subcooled water to the WCT, protecting the boiler circulation pump (BCP). Provided that the pump cooling system is in service, temperatures are within limits, the BCP inlet shutoff valve (380), the BCP recirculation valve (382), and the 383 valves are all open, the BCP is started, supplying flow to the economizer. The flow from the BFP combined with flow from the BCP provides minimum flow to the boiler. The WCT level control valve (381), which ensures flow to the furnace is over the mini-

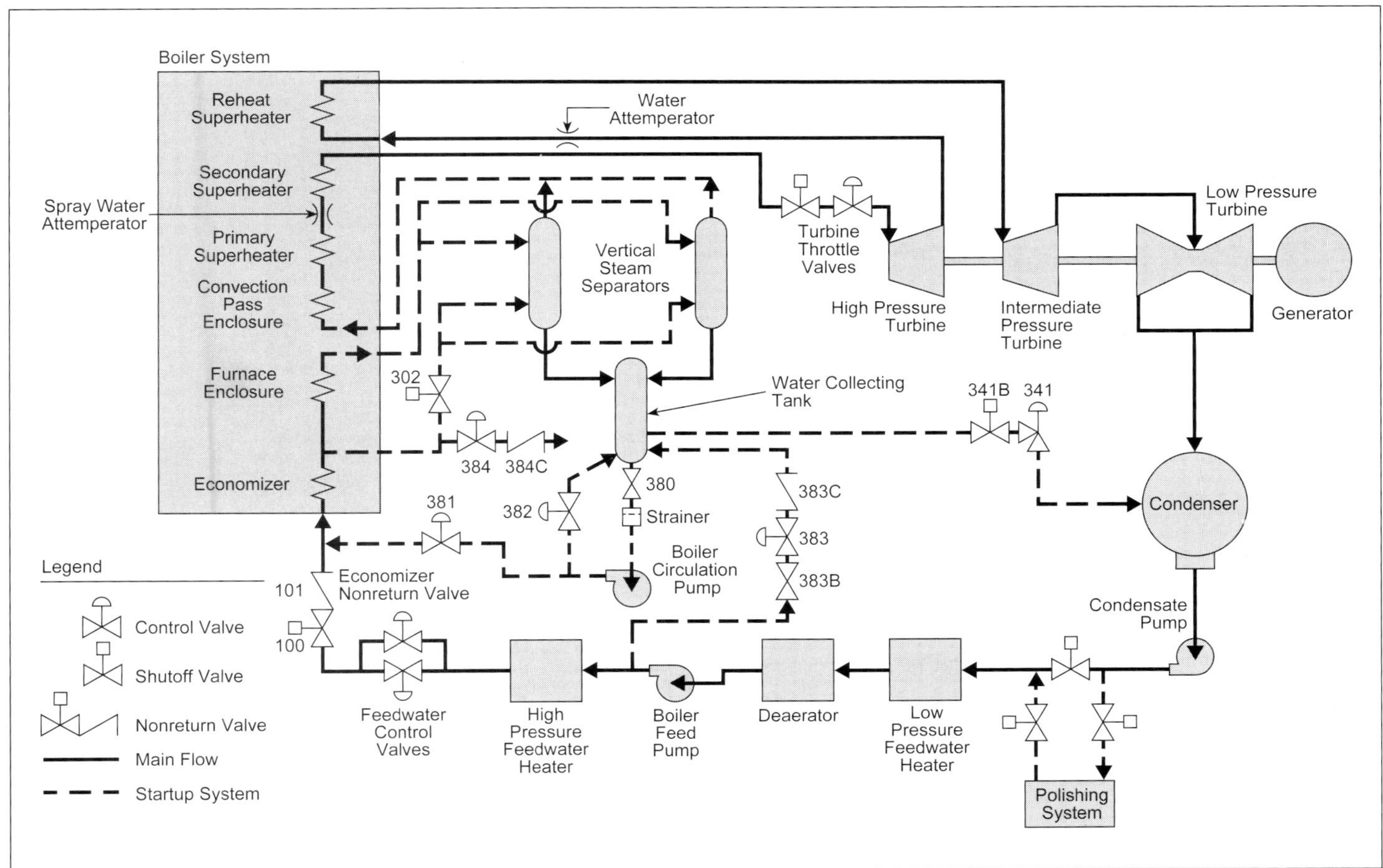

Fig. 20 Spiral wound universal pressure boiler startup system – variable pressure furnace operation.

mum flow required, opens, and tries to maintain the WCT level. With no steam flow, the 381 valve will not be able to regulate the fluid level of the WCT; therefore, the 341 valve controls the level until steam flow exceeds feedwater flow. When the BCP flow is above minimum, the 382 valve will close.

Once the circulation is established, the boiler is fired. A short time after firing, steam bubbles start forming in the furnace, causing a rapid expulsion of excess water to the vertical steam separator and WCT. The 341 valves will open quickly to control the WCT level.

As steam begins to form, some is contained for pressure, some is sent to boiler drains and vents, and the balance is used for steam line warming and eventually flows to the turbine. Once the steam production exceeds the total feedwater flow from the BFP, about 7% MCR, the WCT level will fall causing the 341 valves to close and the 381 valves to take control of the WCT level. The separator high level control block valves (341) are open in the event that an upset causes a high fluid level in the WCT and the 341 valves are forced to open. The feedwater pump system maintains the furnace flow at a minimum.

As steam production continues to increase, the 381 valve closes to control the WCT level. The BCP recirculation valve (382) opens to protect the BCP by providing sufficient flow, while the 383 valve begins to close. The BCP flow to the economizer inlet decreases, requiring the flow from the BFP to increase. When steam load reaches approximately 15%, the spray water attemperators release to control steam temperature for the cold turbine. Firing rate and steam temperature control operate in a manner similar to that of a steam drum boiler until the boiler transition point, the load at which the boiler becomes once-through and the startup system is no longer required.

At the transition to once-through, the fluid to the VSS is all steam, causing the WCT level to drop. As the WCT goes dry, the 381 valve closes to maintain WCT level. The BCP shuts down and the 383 valve closes. The 341 opens if the WCT level briefly becomes high. All steam is sent to the VSS and the BFP provides feedwater flow equal to the steam flow to the VSS; thus, the unit is in once-through operation.

Beyond the boiler transition point the startup system is no longer necessary, and the startup system warming system is placed into service. At 35% load the separator high level control block valve (341-B) closes. The warming system control valve (384) opens fully at 40% load to keep the startup system warm, and

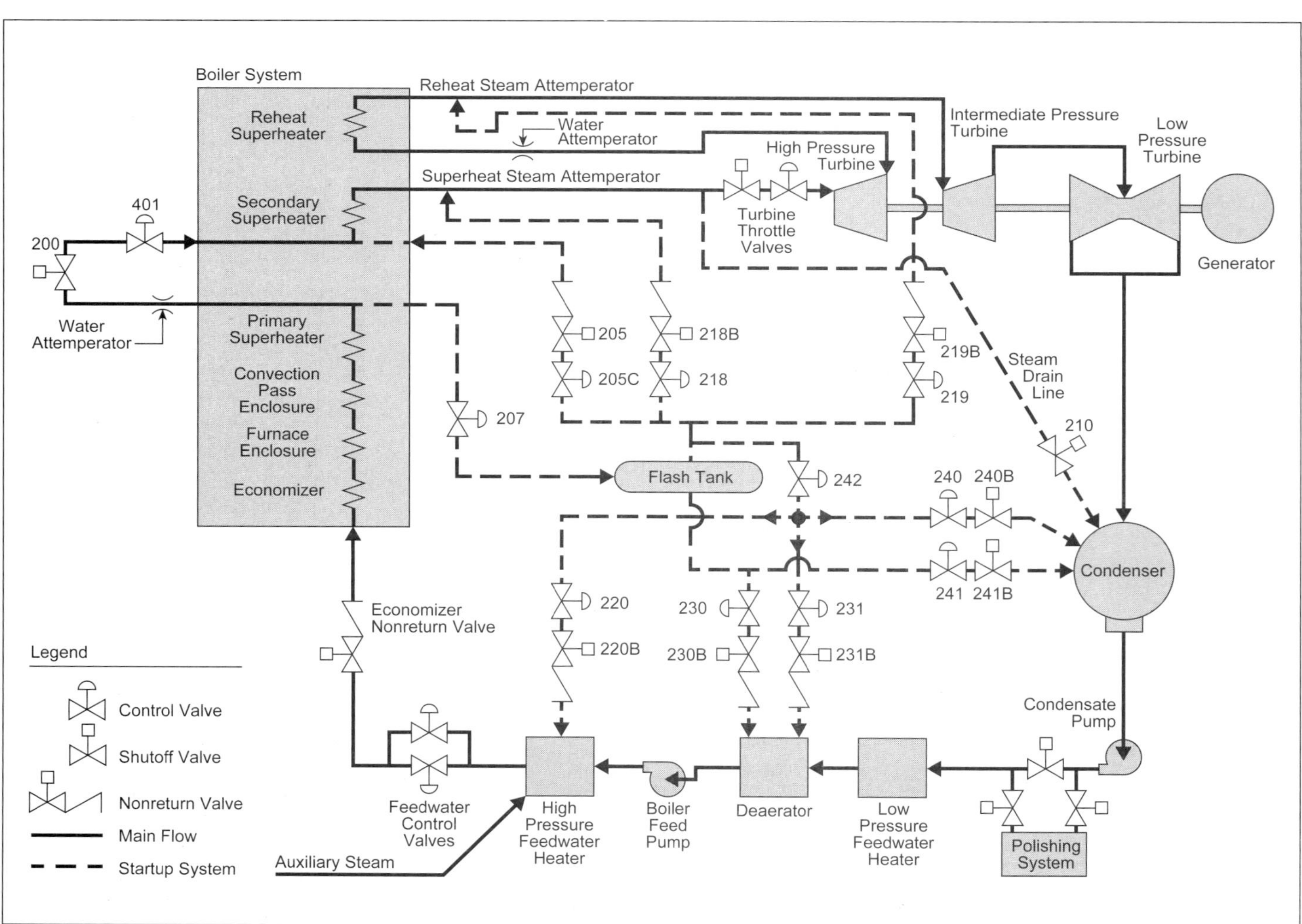

Fig. 21 Universal pressure boiler startup system – constant pressure furnace operation.

closes with increasing load, as more pressure differential is available to drive the warming flow. The 384 valve provides warm flow to the BCP, the 380 line, the 382 line, and the 341 line.

Constant furnace pressure startup system For constant furnace pressure the startup system has the steam separator (*flash tank*) located in a bypass that can be isolated from the boiler during normal operation. Fig. 21 shows the general arrangement.

Typical operation The boiler feed pump supplies the minimum required flow of feedwater during startup and low load operation to protect the furnace circuitry. Included in the startup circuitry are the economizer, furnace, and convection pass enclosures, and primary superheater.

The fluid leaving the primary superheater, at full pressure, is diverted through the pressure reducing valve (207) to the flash tank, where the steam-water mixture is separated during startup.

Drain valves 230 and 241 control the water level in the flash tank, with the 230 valve controlling flow to the deaerator for maximum heat recovery. Excess water above the capability of the deaerator is discharged to the condenser through valve 241. If the drains are not within water quality limits (see Chapter 42), all of the flow is through valve 241 to the condenser and polishing system. When water quality is acceptable, the 230 valve provides flash tank water to the deaerator for maximum heat recovery.

The 242 block valve remains closed until a level is established in the flash tank to assure that water does not enter the steam lines. Once a level is established, the 242 valve is opened and flash tank steam is provided to the deaerator for heat recovery and replacing deaerator auxiliary steam. The deaerator pressure is controlled with the 231 valve. Any excess steam not needed by the deaerator is first sent to the high pressure heater (220 valve), then excess steam not needed by the high pressure heater is sent to the condenser (240 valve).

Flash tank steam is provided to the secondary superheater inlet by the low pressure superheater nonreturn valve (205), secondary superheater low pressure control valve (205C), and associated piping. The 205 valves normally open at 300 psig (2.07 MPa), sending dry steam to the secondary superheater for warming steam lines. The turbine bypass valve (210) normally opens at 300 psig (2.07 MPa) main steam pressure to assist with warming and boiling out the superheater during the initial stage of startup. When sufficient steam is available, the turbine is rolled and placed on line.

The 240 valve relieves excess steam separated in the flash tank to the condenser. This valve also acts as an overpressure relief valve to avoid tripping spring loaded safety valves on the flash tank. The 240 valve has an adjustable set point, which can be set to hold various flash tank pressures at particular load points during startup.

The entire bypass system is sized to handle minimum required flow during startup and to permit operating up to the once-through load on the flash tank. Turbine load is increased from synchronization to once-through by increasing the boiler firing rate. As load is increased, the 220, 230, 231, 240, and 241 valves will close to provide a higher and higher percentage of the minimum boiler flow as steam from the flash tank to the turbine.

The transition from bypass system to once-through operation is made at the load corresponding to the minimum required flow, typically 25%. At this load, the steam entering and leaving the flash tank is slightly superheated, and all steam entering the flash tank goes to the superheater and turbine. By this load, the 220, 230, 231, 240, and 241 valves are all closed. The transition is made, at constant load, by slowly opening the superheater division valve (401). Opening the 401 causes the 200 block valve to go fully open, and the 207 valve to close to maintain boiler pressure. When the 207 valve is fully closed, the boiler is off the bypass system and the load can be increased. This method provides a very smooth and controlled transition from bypass to once-through operation.

Variable throttle pressure Above the minimum once-through load, the unit can be operated at constant or variable throttle pressure. To operate with variable throttle pressure, the 401 valve is used to vary throttle pressure across the load range, while maintaining full boiler pressure in the upstream circuits. This is called *dual or split pressure operation*. To operate with constant throttle pressure, the 401 valve is fully opened, with all throttling done by the turbine control valves.

The variable throttle pressure feature permits operating the unit with the throttle valves essentially wide open, and has three main advantages. First, this eliminates turbine metal temperature changes resulting from valve throttling and permits rapid load changes without being limited by turbine heating or cooling rates. Second, there can be a low load plant heat rate benefit. Third, shutdown with variable pressure maintains high temperatures in the turbine metals at low loads for reduced turbine stress from load cycling.

Steam temperature control The means for controlling main steam and reheat temperatures at normal operating loads are not effective during startup or very low loads. The startup system shown in Fig. 21 includes provision for steam attemperation from the flash tank to the main and reheat steam outlet headers. Steam attemperation precisely controls steam conditions during startup to meet the turbine metal temperature requirements.

The superheater outlet steam attemperator valve (218) is used at loads less than 20% to introduce saturated steam from the flash tank to the superheater outlet header. Initial rolling of the turbine, for a cold start, may be done with saturated steam passing from the flash tank through the 218 valve. This steam may be mixed with a limited quantity of steam passing through the 205 valves and the secondary superheater to control the high pressure turbine inlet temperature down to about 550F (288C). The 205C control valve achieves the necessary pressure drop between the flash tank and the secondary outlet header to allow steam attemperation to work.

The reheat outlet steam attemperator valve (219) is used at loads below about 20% of full load to introduce flash tank steam to the reheat outlet header. The ratio of flow through the attemperator valve to that through the high pressure turbine is limited mainly for turbine and turbine control considerations.

Overpressure relief The bypass system is also used to relieve excessive pressure in the boiler during a load trip. The use of the 207 valve, which sends excess steam to the flash tank, accomplishes this.

Bypass retrofits The above bypass system would be provided with any new UP constant furnace pressure boiler. Currently in operation are a variety of UP bypass systems that range from similar to vastly different than those described above. Many of these older units had their bypass systems retrofitted to incorporate some or all of the above features to improve operational flexibility and ease of startup.

Drum boiler bypass system

The B&W Radiant boiler bypass system minimizes startup time, controls shutdowns in anticipation of restarts, provides control of steam temperature to match turbine metal temperature, and allows dual pressure operation of the boiler and turbine for better load response. These features reduce stresses in the turbine for improved turbine availability and reduced maintenance costs. The drum boiler bypass system consists of an engineered system of steam valves and piping, as shown in Fig. 22.

This flexible bypass system can conform to most operating changes. For example, it can adapt to a turbine malfunction by adjusting the steam temperature and flow through any load range from synchronization to high load points. Although the bypass system enhances unit operation, the boiler can be operated as a conventional non-bypass unit at any time.

Operation

Dual pressure The drum boiler bypass system has a combination of valves consisting of a superheater stop valve (500) and a superheater stop valve bypass valve (501), which allows dual pressure operation. During dual pressure operation, the 501 valve controls the throttle pressure independently from the drum pressure. This permits constant pressure operation of the major boiler components and variable pressure operation of the turbine. A dual pressure shutdown keeps the boiler near full pressure and the turbine metal near maximum temperature in preparation for a quick restart, thus minimizing thermal stresses in the boiler and turbine. In addition, it allows more rapid load changes than full variable boiler pressure operation.

The 501 valve is a control valve typically sized for about 70% of full load steam flow. This permits dual pressure operation up to approximately 70% load. This percentage will vary depending on the specifics of the

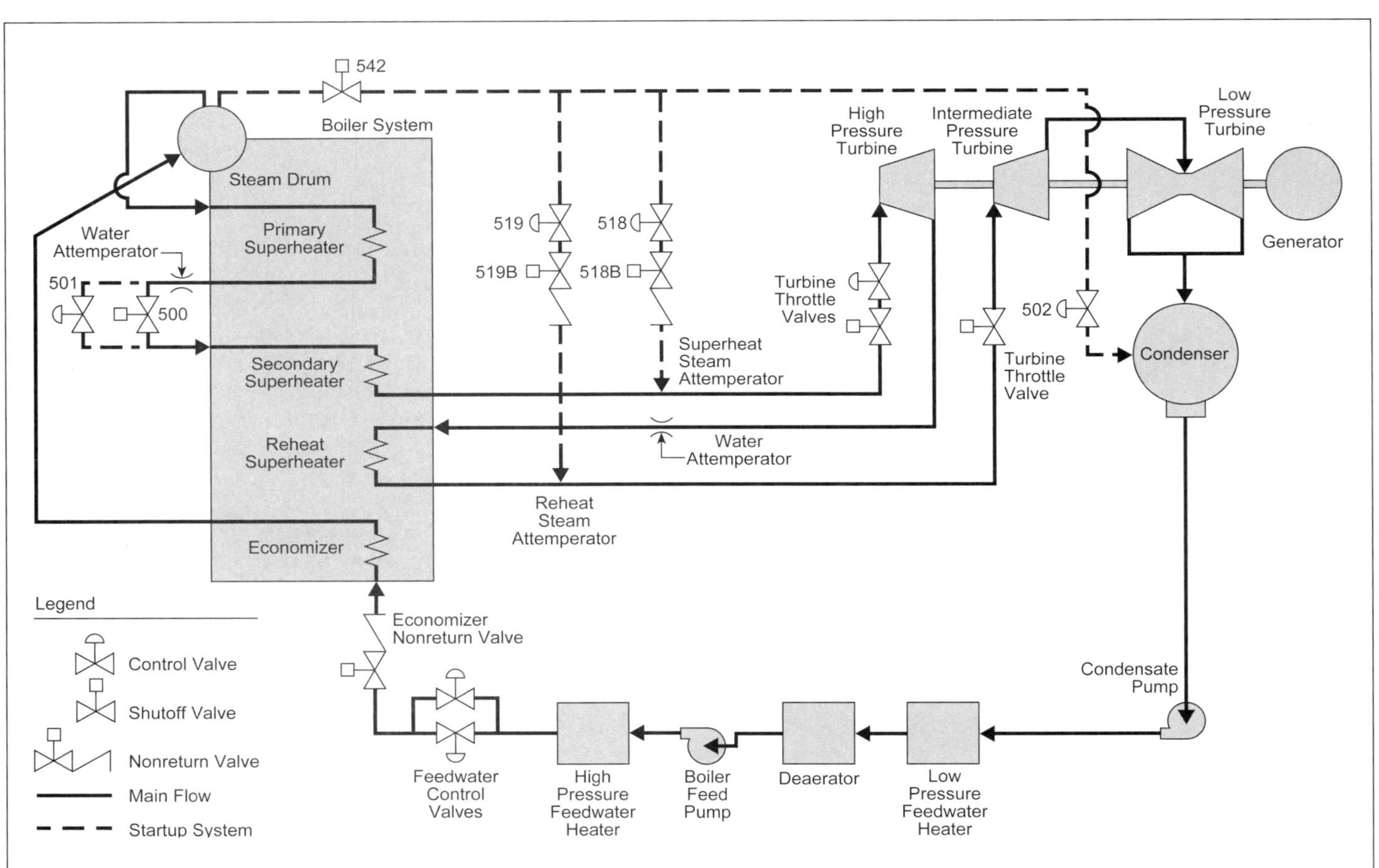

Fig. 22 Drum boiler bypass system schematic.

steam turbine. The 500 valve is a stop valve, which is pulsed open after the 501 is open. Once the 500 valve is completely open, the unit will operate in a constant pressure mode, with the turbine control valves opening to provide the increase from 70 to 100% load.

Steam attemperation The main steam outlet attemperator valve (518) and reheat outlet attemperator valve (519) are used for steam attemperation at low loads, when the normal water attemperation is not effective. Slightly superheated steam is mixed in the outlet headers, via special header internal nozzles, to produce a lower main and reheat steam temperature. Thus, the boiler can be fired to produce drum pressure and steam quickly without concern for the steam temperature to the turbine, reducing startup time and thermal stress in the turbine. The 518 and 519 valves are used for cold and warm startups, when the turbine metal temperature is low.

Drum pressure control The primary superheater bypass valve (502) controls drum pressure independent of firing rate. This allows the boiler to be fired to maximum furnace exit gas temperature producing the high steam temperature needed by the turbine for a hot startup, without concern for the drum pressure rising above maximum allowable pressure. The excess steam bypasses the primary and secondary superheater, to help provide maximum steam temperature. A primary superheater bypass shutoff valve (542) assures tight shutoff when the 502, 518, and 519 valves are closed.

Startup without bypass system The unit can be started and operated without use of the bypass system. Fully opening the 500 valves allows a conventional drum boiler startup.

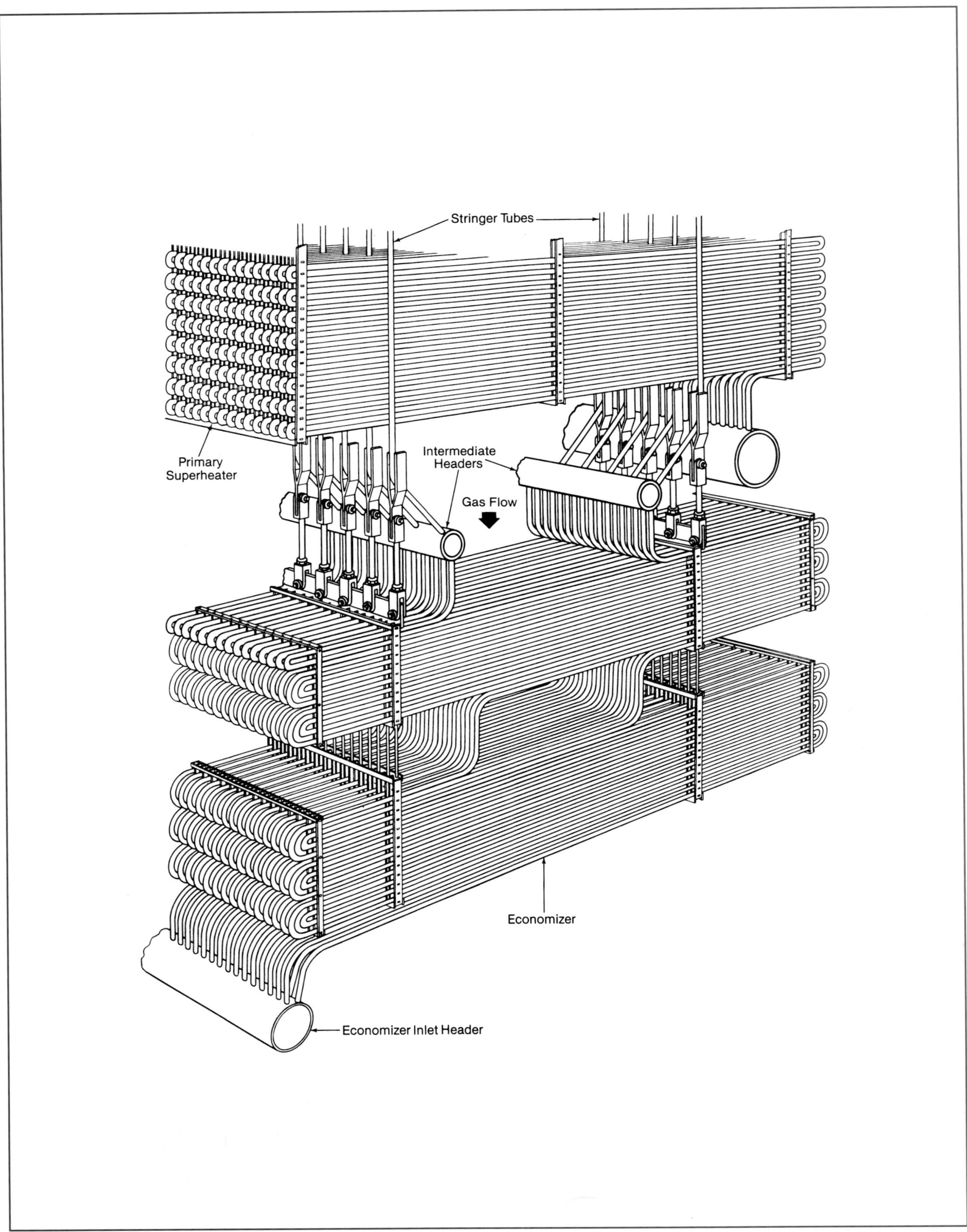

Typical utility boiler economizer.

Chapter 20

Economizers and Air Heaters

Economizers and air heaters perform a key function in providing high overall boiler thermal efficiency by recovering the low level (i.e., low temperature) energy from the flue gas before it is exhausted to the atmosphere. For each 40F (22C) that the flue gas is cooled by an economizer or air heater in a conventional boiler, the overall boiler efficiency increases by approximately 1% (Fig. 1). Economizers recover the energy by heating the boiler feedwater; air heaters heat the combustion air. Air heating also enhances the combustion of many fuels and is critical for pulverized coal firing to dry the fuel and ensure stable ignition.

In contrast to the furnace waterwalls, superheater and reheater, economizers and air heaters require a large amount of heat transfer surface per unit of heat recovered. This is because of the relatively small difference between the flue gas temperature and the temperature of either the feedwater or the combustion air. Use and arrangement of the economizer and/or air heater depend upon the particular fuel, application, boiler operating pressure, power cycle, and overall minimum cost configuration.

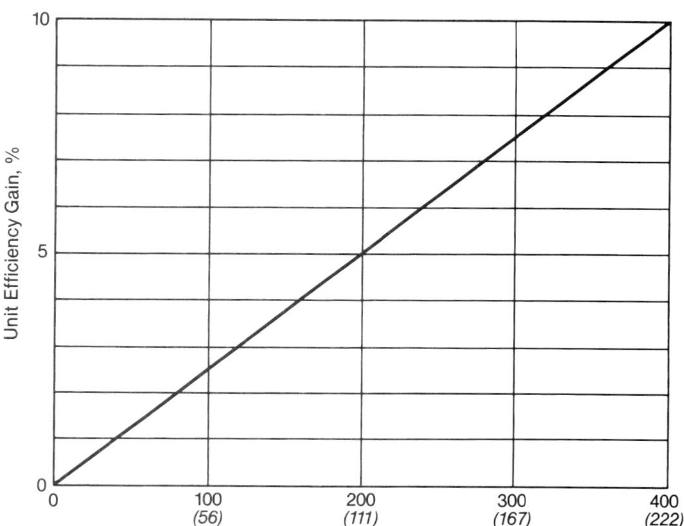

Fig. 1 Approximate unit efficiency increase due to an economizer and air heater in conventional boilers.

Increasingly stringent environmental regulations limiting nitrogen oxides (NO_x) and sulfur dioxide (SO_2) emissions can also affect economizer and air heater design. Selective catalytic reduction (SCR) systems for NO_x control operate within an optimal flue gas temperature range, and the arrangement and/or operation of the economizer may need to be modified accordingly. In the United States (U.S.), there has been a substantial increase in the use of low-sulfur Powder River Basin (PRB) coal to reduce SO_2 emissions. However, PRB coal ash has high fouling and plugging characteristics that prohibit extended surface economizers and require the use of sootblowers throughout the boiler.

Economizers

Economizers are basically tubular heat transfer surfaces used to preheat boiler feedwater before it enters the steam drum (recirculating units) or furnace surfaces (once-through units). The term economizer comes from early use of such heat exchangers to reduce operating costs or economize on fuel by recovering extra energy from the flue gas. Economizers also reduce the potential of thermal shock and strong water temperature fluctuations as the feedwater enters the drum or waterwalls. Fig. 2 shows an economizer location on a coal-fired boiler. The economizer is typically the last water-cooled heat transfer surface upstream of the air heater. (See chapter frontispiece.)

Economizer surface types

Bare tube

The most common and reliable economizer design is the bare tube, in-line, crossflow type. (See Fig. 3a.) When coal is fired, the flyash creates a high fouling and erosive environment. The bare tube, in-line arrangement minimizes the likelihood of erosion and trapping of ash as compared to a staggered arrangement shown in Fig. 3b. It is also the easiest geometry to be kept clean by sootblowers. However, these benefits must be evaluated against the possible larger weight, volume and cost of this arrangement.

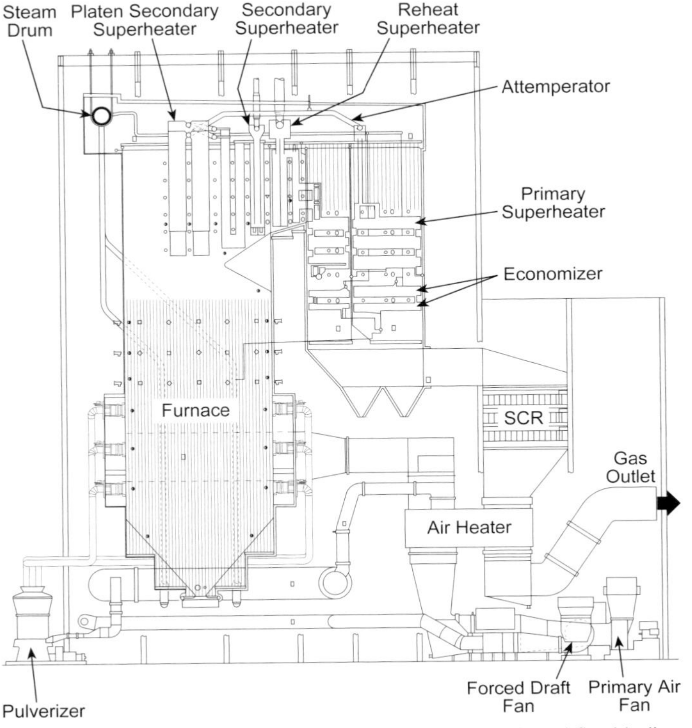

Fig. 2 Economizer and air heater locations in a typical coal-fired boiler.

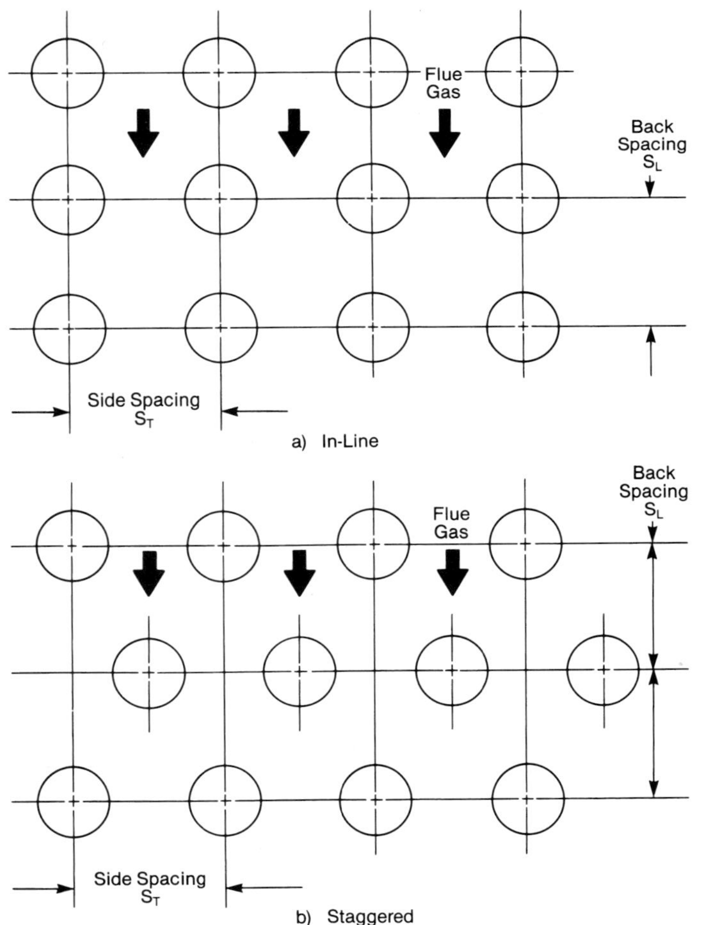

Fig. 3 Bare tube economizer arrangements.

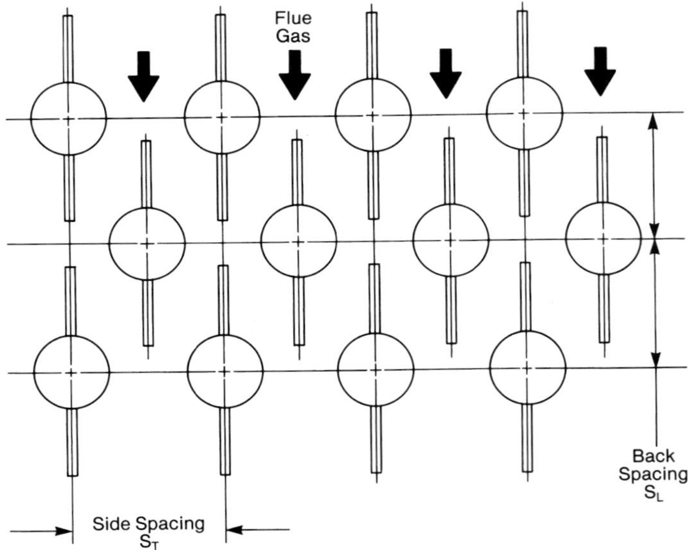

Fig. 4 Longitudinal fins, staggered tube arrangement. (Fin width exaggerated for clarity.)

Extended surfaces

To reduce capital costs, most boiler manufacturers have built economizers with a variety of fin types to enhance the controlling gas-side heat transfer rate. Fins are inexpensive parts which can reduce the overall size and cost of an economizer. However, successful application is very sensitive to the flue gas environment. Surface cleanability is a key concern. In selected boilers, such as PRB coal-fired units, extended surface economizers are not recommended because of the peculiar flyash characteristics.

Stud fins Stud fins have worked reasonably well in gas-fired boilers. However, stud finned economizers can have higher gas-side pressure drop than a comparable unit with helically-finned tubes. Studded fins have performed poorly in coal-fired boilers because of high erosion, loss of heat transfer, increased pressure loss and plugging resulting from flyash deposits.

Longitudinal fins Longitudinally-finned tubes in staggered crossflow arrangements, shown in Fig. 4, have also not performed well over long operating periods. Excessive plugging and erosion in coal-fired boilers have resulted in the replacement of many of these economizers. In oil- and gas-fired boilers, cracks have occurred at the points where the fins terminate. These cracks have propagated into the tube wall and caused tube failures in some applications. Plugging with flyash can also be a problem (tight spaces).

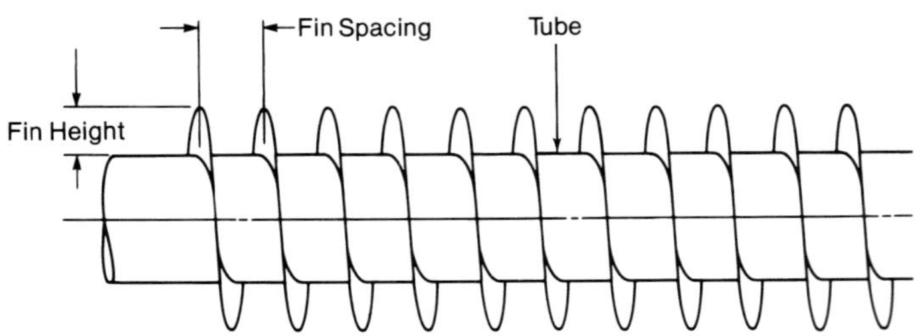

Fig. 5 Helically-finned tube.

Helical fins Helically-finned tubes (Fig. 5) have been successfully applied to some coal-, oil- and gas-fired units. The fins can be tightly spaced in the case of gas firing due to the absence of coal flyash or oil ash. Four fins per inch (1 fin per 6.4 mm), a fin thickness of 0.06 to 0.075 in. (1.5 to 1.9 mm) and a height of 0.75 in. (19.1 mm) are typical. For 2 in. (51 mm) outside diameter tubes, these fins provide ten times the effective area of bare tubes per unit tube length. If heavy fuel oil or coal is fired, a wider fin spacing must be used and adequate measures taken to keep the heating surface as clean as possible. Economizers in units fired with heavy fuel oil can be designed with helical fins, spaced at 0.5 in. (13 mm) intervals. Smaller fin spacings promote plugging with oil ash, while greater spacings reduce the amount of heating surface per unit length. Sootblowers are required and the maximum bank height should not exceed 4 to 5 ft (1.2 to 1.5 m) to assure reasonable cleanability of the heating surface. An in-line arrangement also facilitates cleaning and provides a lower gas-side resistance.

Rectangular fins The square or rectangular fins, arranged perpendicular to the tube axis on in-line tubes as shown in Fig. 6, have been used occasionally in retrofits. The fin spacing typically varies between 0.5 and 1 in. (13 and 25 mm) and the fins are usually 0.125 in. (3.18 mm) thick. There is a vertical slot down the middle because the two halves of the fin are welded to either side of the tube. Most designs are for gas velocities below 50 ft/s (15.2 m/s). However, because of the narrow, deep spaces, plugging with flyash is a danger with such designs.

Baffles The tube ends should be fully baffled (Fig. 7) to minimize flue gas bypass around finned bundles. Such bypass flow can reduce heat transfer, produce excessive casing temperatures, and with coal firing can lead to tube bend erosion because of very high gas velocities. Baffling is also used with bare tube bundles but is not as important as for finned tube bundles. Tube bend erosion can be alleviated by shielding the bends.

Velocity limits

The ultimate goal of economizer design is to achieve the necessary heat transfer at minimum cost. A key design criterion for economizers is the maximum allowable flue gas velocity (defined at the minimum cross-sectional free flow area in the tube bundle). Higher velocities provide better heat transfer and reduce capital cost. For clean burning fuels, such as gas and low ash oil, velocities are typically set by the maximum economical pressure loss. For high ash oil and coal, gas-side velocities are limited by the erosion potential of the flyash. This erosion potential is primarily determined by the percentage of Al_2O_3, and SiO_2 in the ash, the total ash in the fuel and the gas maximum velocity. Experience dictates acceptable flue gas velocities. Fig. 8 provides sample base velocity limits as a function of ash characteristics. Note: PRB coal contains ash with low erosion potential. Velocities above 70 ft/s (21.3 m/s) can typically be allowed for such coals.

Further criteria may also be needed. For example, a 5 ft/s (1.5 m/s) reduction in the base velocity limit is recommended when firing coals with less than 20% volatile matter. In other cases, such as Cyclone boilers, high flue gas velocities can be used because much

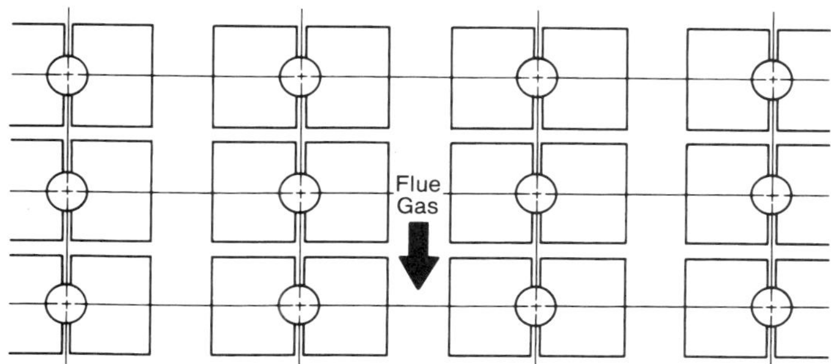

Fig. 6 Rectangular fins, in-line tube arrangement.

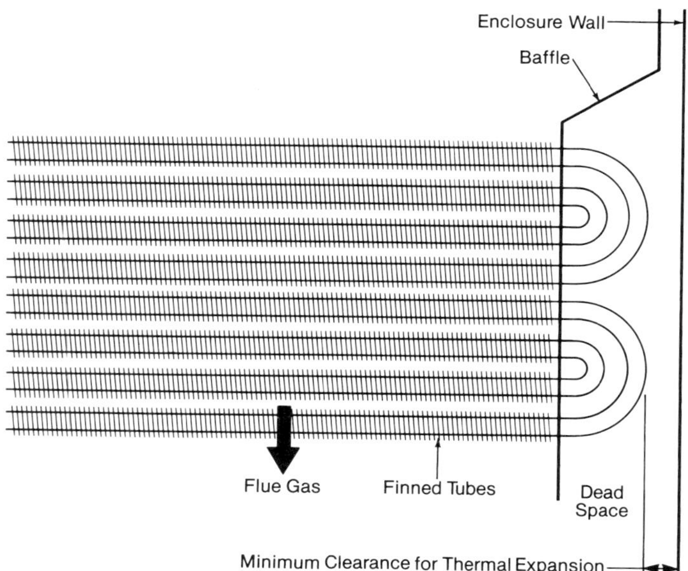

Fig. 7 Baffling of bare tube return bends for finned tube bundles.

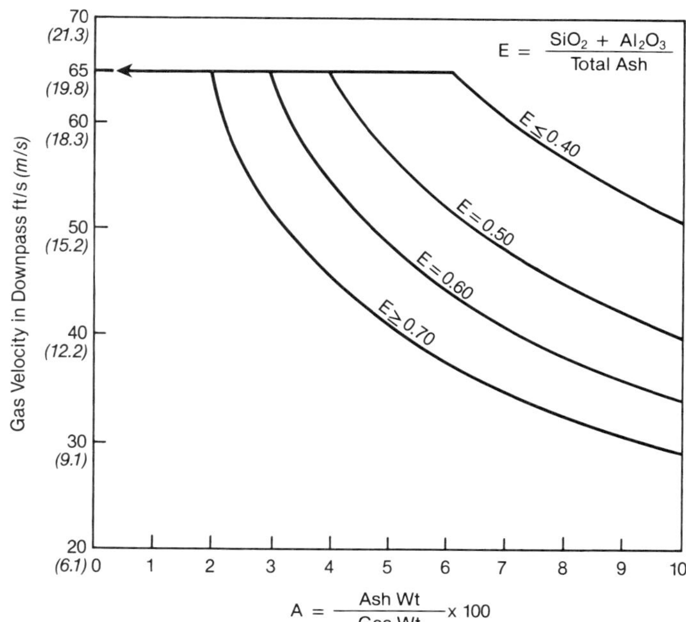

Fig. 8 Base maximum allowable velocity for pulverized coal-fired boiler economizers.

less flyash is carried into the convection pass, as much of the ash (> 50%) is collected in the bottom of the boiler as slag. Particles that enter the boiler furnace are also less erosive. (See Chapter 15.)

For a given tube arrangement and boiler load, the gas velocity depends on the specific volume of flue gas which falls as the flue gas is cooled in the economizer. To maintain the gas velocity, it can be economical to decrease the free flow gas-side cross-section by selecting a larger tube size in the lower bank of a multiple bank design. This achieves better heat transfer and reduces the total heating surface.

Other types of economizers

Fig. 9 depicts an industrial boiler with a longflow economizer, often used in chemical recovery boilers. Such heating surfaces consist of vertical, longitudinally-finned (membraned) tubes through which the feedwater flows upward. The gas flows downward in pure counterflow, outside the tubes and fins. While the heat transfer is less efficient than crossflow banks of tubes, there is minimal gas-side resistance and fouling products are removed through hoppers at the bottom of the enclosure.

Steaming economizers

Steaming economizers are defined as meeting the following enthalpy relationships:

$$H_2 - H_1 \geq \tfrac{2}{3}\left(H_f - H_1\right) \qquad (1)$$

where

H_2 = enthalpy of fluid leaving economizer (to drum)
H_1 = enthalpy of fluid (water) entering economizer

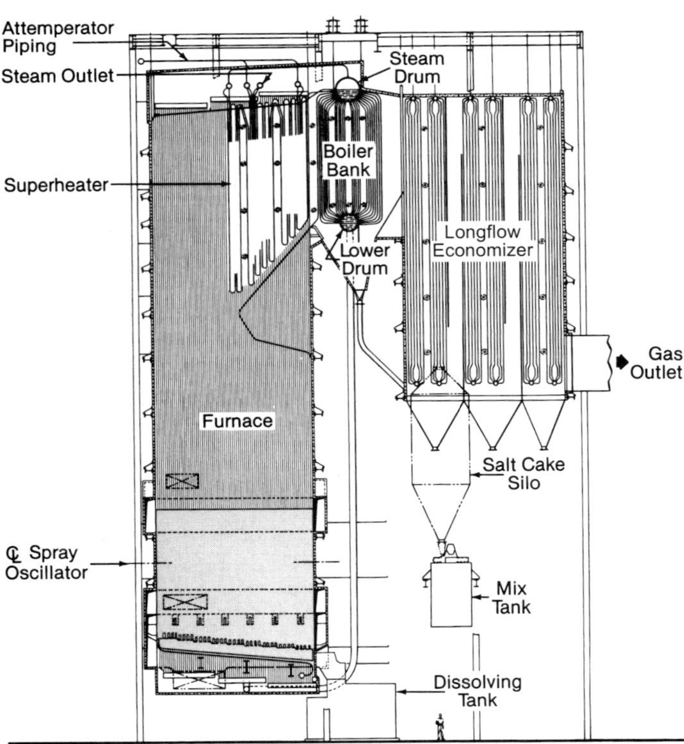

Fig. 9 Longflow economizer for a chemical recovery boiler.

H_f = enthalpy of saturated water at economizer outlet pressure

These economizers can be economical in certain boilers. They require careful design and must be oriented so the water flows upward and the outlet is below drum level. This avoids water hammer and excessive flow instabilities. The enthalpy equation accounts for possible steaming due to flow imbalances and differences in individual circuit heat absorptions.

High-pressure, drum-type units are usually sensitive to feedwater temperatures close to saturation. To enhance circulation, feedwater temperature to the drum should normally be at least 50F (28C) below saturation temperature. However, for some applications, due to specific design conditions, use of Equation 1 may result in a feedwater temperature slightly greater than 50F (28C) below saturation temperature.

Performance

Heat transfer

Bare tubes The equations discussed in Chapter 4 can be used to evaluate the quantity of surface for an economizer. For the economizer shown in Fig. 2 with the upflow of water and downflow of gas and nonsteaming conditions, the bundle can be treated as an ideal counterflow heat exchanger with the following characteristics:

1. bundle log mean temperature difference correction factor = 1.0,
2. heat absorbed by the tube wall enclosures and heat radiated into the tube banks from various cavities can generally be neglected,
3. all of the energy lost by the flue gas is absorbed by the water, i.e., no casing heat loss,
4. the water-side heat transfer coefficient is typically in the range of 2000 Btu/h ft² F (11,357 W/m² K) and has only a small overall impact on the economizer performance, and
5. the effect of gas-side ash deposition can be accounted for by a cleanliness factor based upon experience.

In general the heat transfer rate is primarily limited by the gas-side heat transfer for in-line bare tube bundles. In this case, the overall heat transfer coefficient (flue gas to feedwater) used in the heat exchanger calculation can be approximated by the following relationship:

$$U = 0.98\left(h_c + h_r\right) k_f \qquad (2)$$

where

U = overall heat transfer coefficient, Btu/h ft² F (W/m² K)
h_c = gas-side heat transfer coefficient for a bare tube bundle, Btu/h ft² F (W/m² K) (Chapter 4, Equations 60 and 61)
h_r = inter-tube radiation heat transfer coefficient, Btu/h ft² F (W/m² K) ≈ 1.0 for coal firing (for finned tubes h_r, is very small and is assumed to be 0)
k_f = surface effectiveness factor = 0.7 for coal, 0.8 for oil and 1.0 for gas

Finned tubes Heat transfer performance of finned tube economizers can be evaluated in a similar fashion except that appropriate relationships for extended surfaces should be used. In addition, because the gas-side heat transfer has been enhanced, the water-side heat transfer coefficient and tube wall thermal resistance are more significant and must be included in the evaluation. (See Chapter 4.) As a general guideline, the overall heat transfer coefficient can be approximated by the following relationship for most types of economizer fins:

$$U = 0.95\left(h_g k_f\right) \qquad (3)$$

where h_g is the gas-side heat transfer coefficient for the heat transfer across finned tube bundles evaluated with the procedures defined in Chapter 4. A calculated example is provided in Chapter 22 for a bare tube economizer tube bundle.

Gas-side resistance

The gas-side pressure loss across the economizer tube bank can be evaluated using the crossflow correlations presented in Chapter 3. The pressure loss should be adjusted for the number of tube rows using the correction factors provided. The gas-side resistance across the in-line finned tube banks is approximately 1.5 times the resistance of the underlying bare tubes.

Water-side pressure drop

The water-side pressure loss can be evaluated using the procedures in Chapter 3 where the total pressure loss ΔP_T is calculated:

$$\Delta P_T = \Delta P_f + \Delta P_l + \Delta P_z \qquad (4)$$

where

ΔP_f =	friction pressure loss	Ch. 3, Eq. 47
ΔP_l =	sum of the local losses (entrance, bends and exits)	Ch. 3, Eq. 52
ΔP_z =	static head loss	Ch. 5, Eq. 10

The design pressure is then evaluated by the sum of the drum design pressure and the total pressure loss ΔP_T rounded up to the nearest 25 psig (172 kPa).

If the calculated water-side pressure drop is excessive, the number of parallel flow paths must be increased. If the flue gas velocity can be increased, the water-side pressure drop can also be reduced by increasing the tube size, usually in increments of 0.125 in. (3 mm). As indicated in Chapter 3, the dynamic pressure drop is inversely proportional to the fifth power of the inside tube diameter.

Economizer support systems

Economizers are located within tube wall enclosures or within casing walls, depending on gas temperatures. In general, casing enclosures are used at or below 850F (454C) and inexpensive carbon steel can be used. If a casing enclosure is used, it must not support the economizer. However, tube wall enclosures may be used as supports.

The number of support points is determined by analyzing the allowable deflection in the tubes and tube assemblies. Deflection is important for tube drainability. Figs. 10 through 12 show typical support arrangements for bare tube economizers.

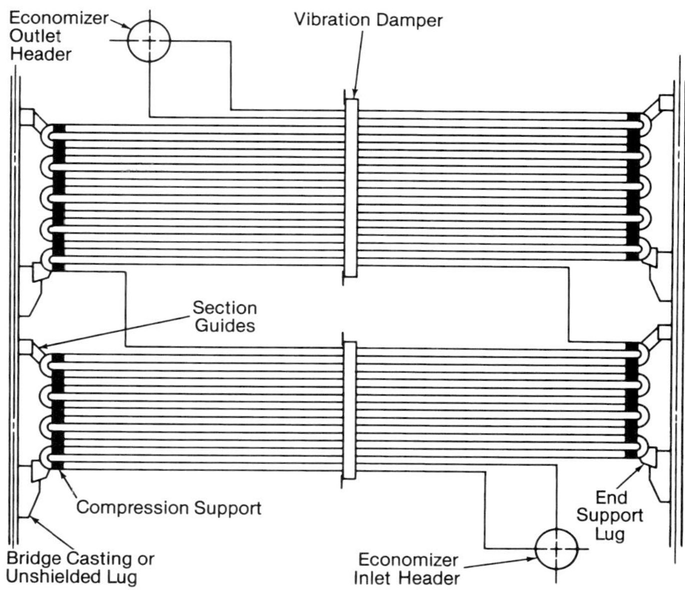

Fig. 10 Economizer supports – sample waterwall support arrangement.

Wall or end supports are usually chosen for relatively short spans and require bridge castings or individual lugs welded or attached to the tube wall enclosures. (See Fig. 10.) Another possibility exists if enclosure wall (usually primary superheater circuitry) headers are present above the economizer (for example, Fig. 11).

Quarter point stringer supports are used for spans exceeding the limits for end supports (Fig. 12). The stringers are mechanically connected to the economizer sections, which are held up by ladder type supports. The supports exposed to hot inlet gases may be made

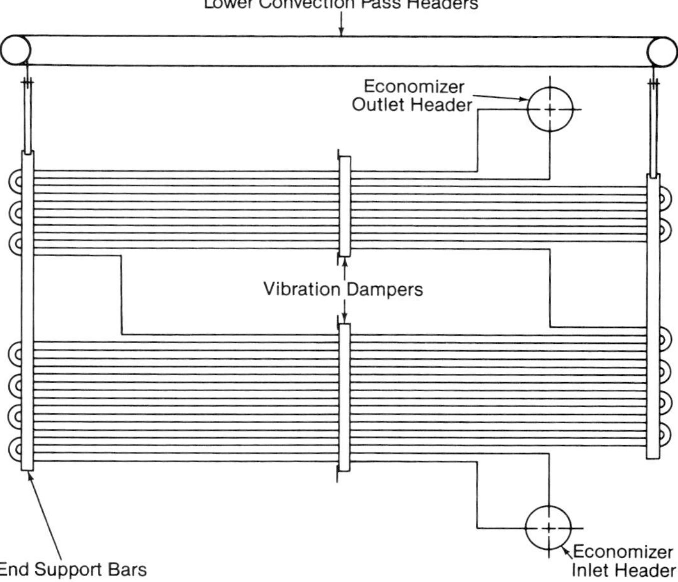

Fig. 11 Economizer supports – sample lower waterwall header arrangement.

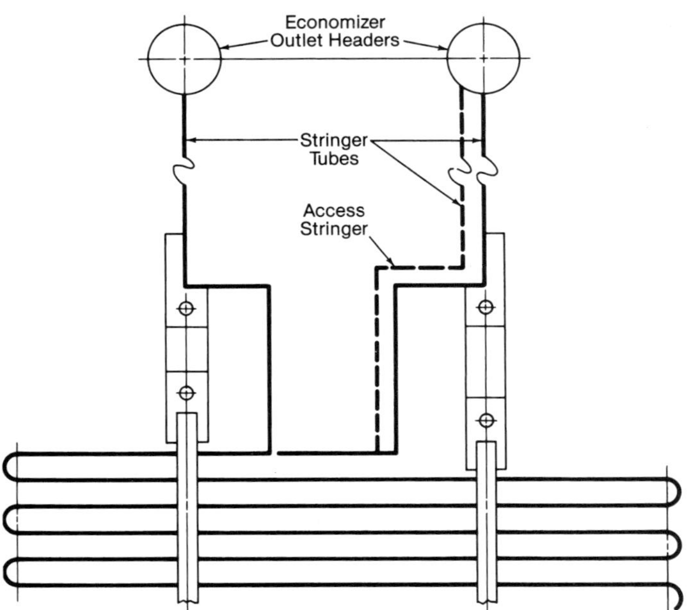

Fig. 12 Economizer supports – sample stringer support arrangement.

of stainless steel, while lower grade material is normally used to support the lower bank which is exposed to reduced gas temperatures. In The Babcock & Wilcox Company (B&W) designs, stringer tubes also usually support other horizontal convection surfaces above the economizer. (See frontispiece.) Bottom support is sometimes used if the gas temperature leaving the lowest economizer bank is low enough.

Bank size

The bank size is limited by the following constraints:

1. type of fuel,
2. fabrication limits,
3. sootblower range,
4. maximum shipping dimensions,
5. construction considerations, especially for retrofits, and
6. maintenance.

Bank depths greater than 6 ft (1.83 m) are rare in new boilers, while carefully designed larger banks can be tolerated in retrofits.

Access requirements

Cavities around the banks are needed for field welding, tube leg maintenance and sootblower clearance. A sufficient number of access doors must be arranged in the enclosure walls to access these cavities. Cavity access can be provided from the outside through individual doors or from the inside through special openings across stringers or collector frames. The minimum cavity height should be 2 ft (0.6 m) of crawl space.

Headers

B&W economizer header designs are typically based on American Society of Mechanical Engineers (ASME) Code requirements. Inlet headers are frequently located inside the gas stream and may receive feedwater through one or both ends. Regardless of design, it is necessary to properly seal the inlet pipe where it penetrates the enclosure by using brackets and flexible seals. The seal becomes especially important in pressurized (forced draft) units. Other important considerations are tube leg flexibility, differential expansion and potential gas temperature imbalances and upsets.

The outlet headers, not to be confused with intermediate or transition headers seen in larger boilers and stringer supports, receive the heated feedwater and convey it to the drum or, in the case of once-through boilers, to the downcomer(s) supplying the furnace circuitry. Inlet and outlet headers must be large enough to assure reasonable water flow distribution in the economizer banks. Flow velocities in headers are typically less than 20 ft/s (6.1 m/s).

Vibration ties

Vibration ties or tube guides are required on some end-supported tube sections. These ties may be needed if the natural frequencies within the boiler load range are in or near resonance with the vortex shedding frequency.

Stringer tubes are also subject to vibration. This vibration is magnified by long unsupported stringer tube lengths near the large cavity below the convection pass roof.

Tube geometry, materials and code requirements

Economizer tube diameters typically range between 1.75 and 2.5 in. (44.5 and 63.5 mm). Tubes outside this range are sometimes used in retrofits. Smaller tubes are normally used in once-through, supercritical boilers where water-side pressure drop is less of a consideration. In these units, tube wall thicknesses are minimized.

The ASME Code requires that the design temperature for internal boiler pressure parts is at least 700F (371C). The calculated mean tube wall temperature in economizers seldom reaches this temperature. It usually lies 10 to 20F (6 to 11C) above the fluid temperature, which seldom exceeds 650F (343C) along any economizer circuit.

The minimum tube wall thickness is determined in accordance with procedures outlined in Chapter 8.

In coal-fired boilers, the side spacing is usually determined by the maximum allowable gas velocity and gas-side resistance, which are functions of a given tube size. If fins are used, the side and back tube spacings should permit the fin tips to be at least 0.5 in. (13 mm) apart. For bare tubes, a minimum clear spacing of 0.75 in. (19 mm) is desirable.

PRB coal firing usually calls for increased clear side spacing to avoid plugging and bridging with flyash in the horizontal spaces.

The minimum back (vertical) spacing of the tubes should be no less than 1.25 times the tube outside diameter. Smaller ratios can reduce heat transfer by as much as 30%. Ratios larger than 1.25 have relatively little effect on heat transfer but increase the gas-side resistance and bank depth.

SCR considerations

SCR systems are normally located between the economizer and the air heater. NO_x reduction reactions take place within an optimal temperature range as the flue gas passes through the catalyst reactor (see Chapter 34). Therefore, flue gas temperature entering the SCR is critical to system performance.

Typical temperature control methods include:

1. *Gas bypass* The goal is to maintain a minimum gas temperature to the SCR, typically 600F (316C) to 650F (343C) for coal-fired applications, over the boiler load range in which the SCR will be operated. This can be achieved by extracting flue gas from a cavity above the economizer or between economizer banks, and routing the higher temperature gas to a point upstream of the SCR where it is mixed with lower temperature gas from the economizer outlet. The gas flow through the bypass flue is controlled by a damper.

   In retrofits, the bypass gas flow is usually limited by the available gas-side pressure differential and maximum allowable gas velocity through the bypass flue. Existing physical constraints limit the maximum flue size. The increase in flue gas temperature entering the SCR reduces boiler efficiency due to an increase in flue gas temperature leaving the unit. To reduce this loss in boiler efficiency, a separate economizer bank can be installed downstream of the SCR or additional air heater surface can be added if space is available.
2. *Water-side bypass* Water-side parameters (i.e., mass flow) indirectly control the flue gas temperature exiting the economizer by reducing the water flow through some of the economizer surface, thus reducing economizer heat absorption. (Instead of a gas-side damper controlling the gas bypass flow, several control valves regulate water flow to the various economizer banks.) Economizer water outlet temperature should not exceed certain limits dictated by furnace circulation.
3. *Duct burners* Special duct burners (usually natural gas fired) are sometimes used to raise the gas temperature entering the SCR. However, the economics of natural gas firing must be carefully evaluated.

Air heaters

Air heaters are used in most steam generating plants to heat the combustion air and enhance the combustion process. Most frequently, the flue gas is the source of energy and the air heater serves as a heat trap to collect and use waste heat from the flue gas stream. This can increase the overall boiler efficiency by 5 to 10%. Air heaters can also use extraction steam or other sources of energy depending upon the particular application. These units are usually employed to control air and gas temperatures by preheating air entering the main gas-air heaters.

Air heaters are typically located directly behind the boiler, as depicted in Fig. 2, where they receive hot flue gas from the economizer and cold combustion air from the forced draft fan(s). The hot air produced by air heaters enhances combustion of all fuels and is needed for drying and transporting the fuel in pulverized coal-fired units.

Classification of air heaters

Air heaters are classified according to their principle of operation as recuperative or regenerative.

Recuperative

In a recuperative heat exchanger, heat is transferred continuously and directly through stationary, solid heat transfer surfaces which separate the hot flow stream from the cold flow stream. The most common heat transfer surfaces are tubes and parallel plates. Recuperative heat exchangers function with little cross-contamination, or leakage, between streams.

Tubular air heaters In a typical tubular air heater, energy is transferred from the hot flue gas flowing inside many thin walled tubes to the cold combustion air flowing outside the tubes. The unit consists of a nest of straight tubes that are roll expanded or welded into tubesheets and enclosed in a steel casing. The casing serves as the enclosure for the air or gas passing outside of the tubes and has both air and gas inlet and outlet openings. In the vertical type (Fig. 13), tubes are supported from either the upper or lower tubesheet while the other (floating) tubesheet is free to move as tubes expand within the casing. An expansion joint between the floating tubesheet and casing provides an air/gas seal. Intermediate baffle plates parallel to the tubesheets are frequently used to separate the flow paths and eliminate tube damaging flow induced vibration.

Carbon steel or low alloy corrosion resistant tube materials are used in the tubes which range from 1.5 to 4 in. (38 to 102 mm) in diameter and have wall

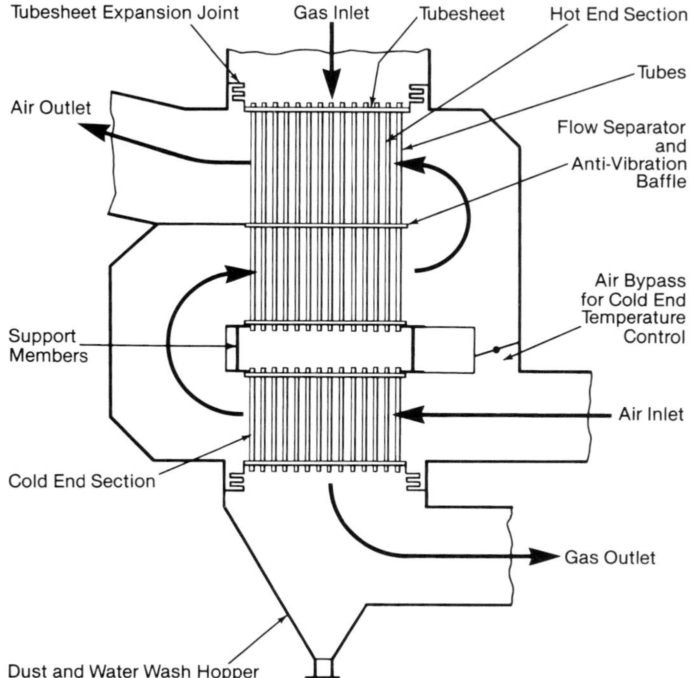

Fig. 13 Vertical type tubular air heater.

thicknesses of 18 to 11 gauge [0.049 to 0.120 in. (1.24 to 3.05 mm)]. Larger diameter, heavier gauge tubes are used when the potential for tube plugging and corrosion exists. Tube arrangement may be in-line or staggered with the latter being more thermally efficient.

Tubular air heaters may be fitted with steam or air sootblowers to remove ash accumulations from the gas-exposed side of tubing during operation. Permanent water wash piping above gas-side tube banks may also be used off line to soak and wash tube internal or external surfaces.

The most common flow arrangement is counterflow with gas passing vertically through the tubes and air passing horizontally in one or more passes outside the tubes. A variety of single and multiple gas and air path arrangements are used to accommodate plant layouts. Designs frequently include provisions for cold air bypass or hot air recirculation to control cold end corrosion and ash fouling. Modern tubular air heaters are shop assembled into large, transportable modules. Several arrangements are shown in Fig. 14.

Cast iron air heaters Cast iron tubular air heaters are heavy, large and durable. Their use is mainly limited to the petrochemical industry, but some are used on electric utility units. Cast iron is used because of its superior corrosion resistance. Rectangular, longitudinally split tubes are assembled from two cast iron plates and individual tubes are assembled into air heater sections. Air heaters are usually arranged for a single gas pass and multiple air passes with air flow inside the tubes. Heat transfer is maximized by fins cast into inside and outside tube surfaces.

Plate air heaters Plate air heaters transfer heat from hot gas flowing on one side of a plate to cold air flowing on the opposite side, usually in crossflow. Heaters consist of stacks of parallel plates. Sealing between air and gas streams at plate edges is accomplished by welding or by a combination of gaskets, springs and external compression of the plate stack. Plate materials and spacing can be varied to accommodate operating requirements and fuel types.

Steel plate air heaters were some of the earliest types

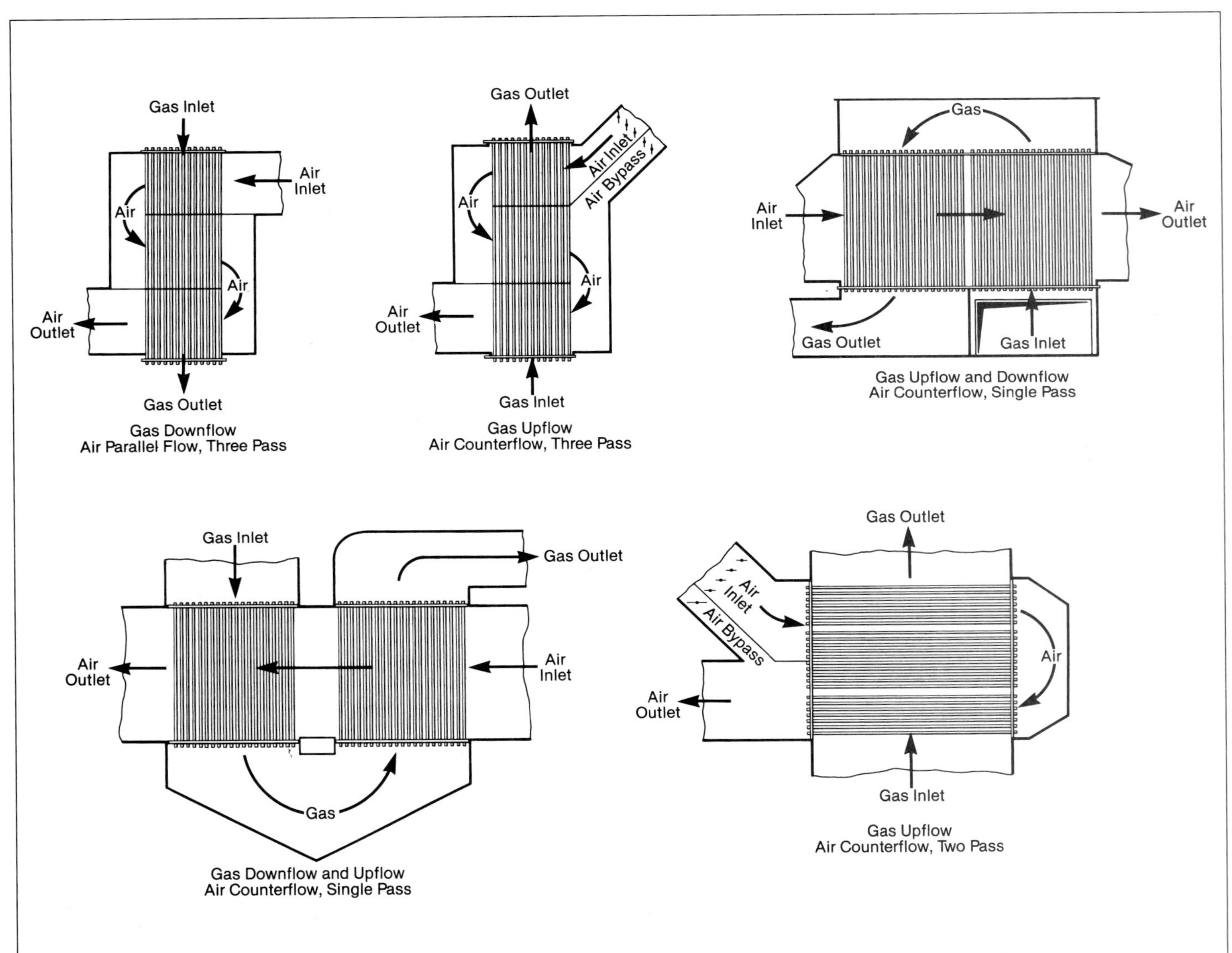

Fig. 14 Various tubular air heater arrangements.

used, but their use declined due to plate to plate sealing problems. However, sealing developments have prompted increased use in industrial and small utility applications. Plate modules may be combined to make different size air heaters with a variety of flow path arrangements. A single gas pass, two air pass plate air heater is shown in Fig. 15. Modern plate units are somewhat smaller than tubular units for a given capacity and exhibit minimal air to gas leakage.

Steam coil air heaters Steam coil and water coil recuperative air heaters are widely used in utility steam generating plants to preheat combustion air. Air preheating reduces the corrosion and plugging potential in the cold end of the main air heater. Occasionally, they serve as the only source of preheated combustion air. These heaters consist of banks of small diameter, externally finned tubes arranged horizontally or vertically in ducts between the combustion air fan and main air heater. Combustion air, passing in crossflow outside the tubes, is heated by turbine extraction steam or feedwater flowing inside the tubes. Ethylene glycol is sometimes used as the hot fluid to prevent out of service freezing damage.

Regenerative

Regenerative air heaters transfer heat indirectly by convection as a heat storage medium is periodically exposed to hot and cold flow streams. A variety of materials can be used as the medium and periodic exposure to hot and cold flow streams can be accomplished by rotary or valve switching devices. In steam generating plants, tightly packed bundles of corrugated steel plates serve as the storage medium. In these units either the steel plates, or surface elements, rotate through air and gas streams, or rotating ducts direct air and gas streams through stationary surface elements.

Regenerative air heaters are relatively compact and are the most widely used type for combustion air preheating in electric utility steam generating plants. Their most notable operating characteristic is that a small but significant amount of air leaks into the gas stream due to the rotary operation.

Ljungström The most common regenerative air heater is the Ljungström® type (Figs. 16 and 17), which features a cylindrical shell plus a rotor which is packed with bundles of heating surface elements and is rotated through counterflowing air and gas streams. The rotor is enclosed by a stationary housing which has ducts at both ends. Air flows through one half of the rotor and gas flows through the other half. Metallic leaf-type seals minimize air to gas leakage and flow bypass around the rotor. See Fig. 18. Bearings in upper and lower beam assemblies support and guide the rotor at the central shaft. A rotor speed of one to three rpm is provided by a motor driven pinion engaging a rotor encircling pinrack. Both vertical and horizontal shaft designs are used to accommodate various plant air and gas flow schemes. The vertical shaft design is more common.

The most prevalent flow arrangement has the hot gas entering the top of the rotor as cold air enters the bottom in counterflow fashion, as shown in Fig. 16. Heaters employing this flow scheme are identified as hot end on top and cold end on bottom. In operation, the rotor is subjected to a temperature differential, hot top and cold bottom, causing the rotor to expand and bow (or distort) upward. This rotor distortion opens gaps between the rotor and stationary parts through which objectionable air-side to gas-side leakage occurs.

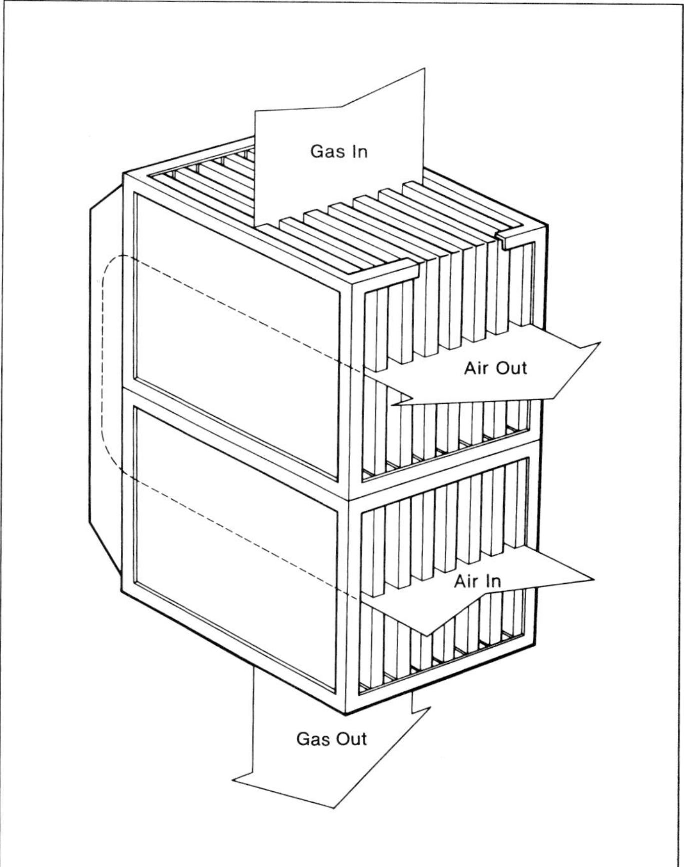

Fig. 15 Single gas pass, two air pass plate air heater.

Air to gas leakage is controlled by cold-presetting axial and radial seal plates to minimize gaps at the hot operating condition (Fig. 19). Note that a significant radial gap remains at the hot end which is tolerated in industrial and small utility boiler air heaters. However, for controlling leakage in larger units the hot end radial sealing plate is designed to adjust automatically during operation to follow rotor distortion and maintain a minimum seal gap.

Sootblowing devices in the gas outlet are used to direct superheated steam or dry air via nozzles into the rotor heating surface elements to periodically clean them of fuel residue accumulations during operation. For especially heavy or tenacious deposits, sootblowers may also be used at the air heater gas inlet.

Air heaters operating in dirty flue gas streams (coal, oil, waste, biomass, etc.) are also fitted with permanent water wash piping for soaking and cleaning heavy heating surface ash accumulations. The wash piping is located in the top of the heater above the rotor with nozzles arranged such that all heating surface elements are exposed to the nozzle spray pattern as the rotor turns. Air heaters are normally washed off-

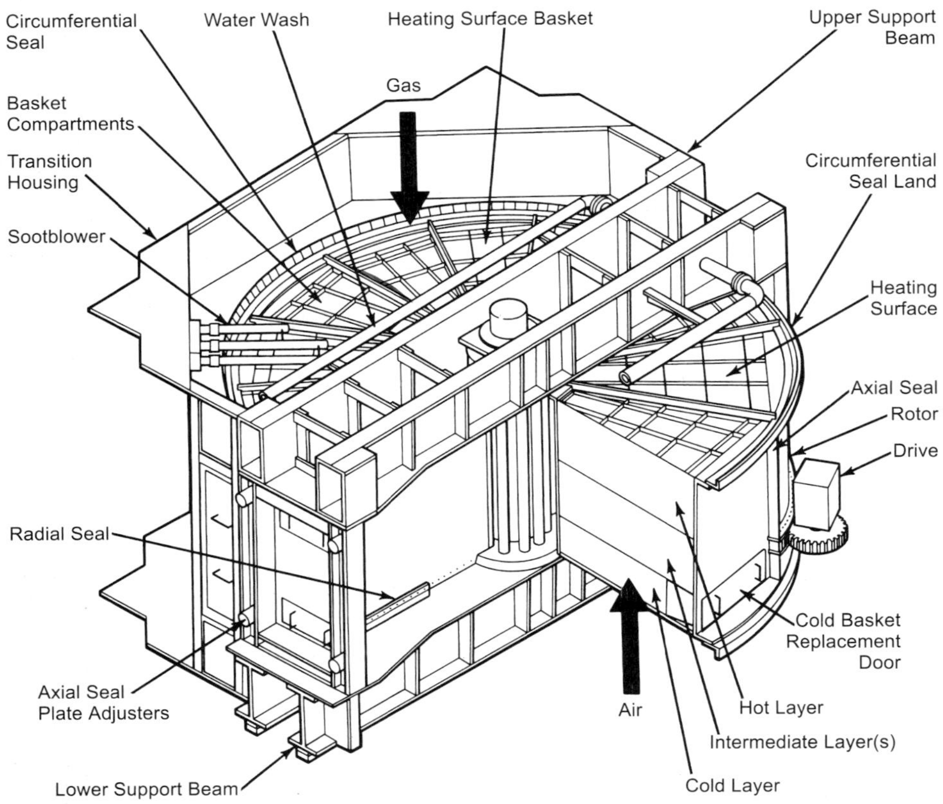

Fig. 16 Ljungström-type air heater.

Fig. 17 Air heater service project at a midwest U.S. power plant.

line as the rotor is turned via an auxiliary drive. Air heater washing through sootblowers may also be done with appropriate piping and valves.

Rothemühle The Rothemühle®-type regenerative air heater uses stationary surface elements and rotating ducts (Fig. 20). The surface elements are supported and contained within a stationary cylindrical shell called the stator. On both sides of the stator, a double wing symmetrical hood rotates synchronously on a common vertical shaft. The central shaft is supported by bearings within the stator and the hoods are driven slowly by a pinion which engages a pinrack encircling the lower hood. Stationary housings surround the hoods. Heat is transferred as flow streams are directed through the heating surface in counterflow fashion, one flow stream inside the hoods and the other outside. Either air or gas may pass through the hoods. However, air is more common because it requires less fan power.

Rothemühle air heater stators distort in a manner similar to Ljungström rotors. Special sealing systems mounted to rotating hoods at the interface with the stator are used to control leakage. Both preset and automated seal systems are used. Sealing at the rotating hood to stationary duct interface is maintained by a ring of spring-backed cast iron seal shoes and a rotating race.

Sootblowers mounted to the outside of rotating hoods are employed on cold and hot ends as fuels dictate. Water wash piping both inside the hoods (rotat-

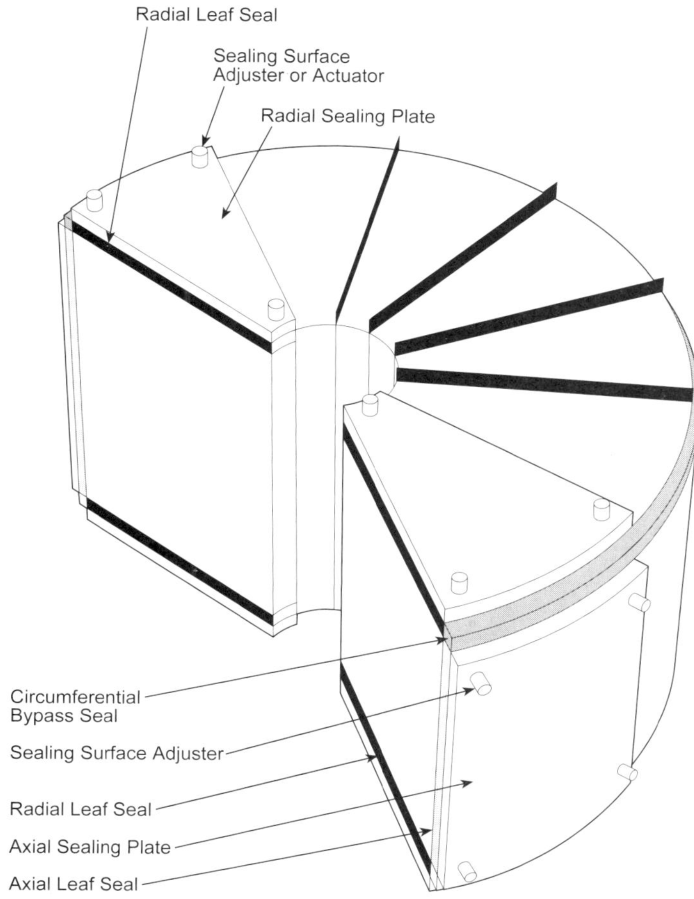

Fig. 18 Ljungström-type air heater seals.

ing piping) and outside the hoods (stationary piping) assures complete coverage of the heating surface. Special sootblower and rotating wash piping seals are used at the rotating-stationary interface.

Regenerative heating surface Regenerative air heater surface elements are a compact arrangement of two specially formed metal plates. Each element pair consists of a combination of flat, corrugated or undulated plate profiles. The roll formed corrugations and undulations serve to separate the plates to maintain flow paths, increase heating surface area and maximize heat transfer by creating flow turbulence. The steel plates, 26 to 18 gauge thick, are typically spaced 0.2 to 0.4 in. (5 to 10 mm) apart. Closely spaced, highly profiled element pairs exhibit a high heat transfer rate, pressure drop and fouling potential while widely spaced element combinations, where one plate is flat, exhibit a low heat transfer rate, low pressure drop and reduced fouling potential. The combination of plate profile, material and thickness is selected for maximum heat transfer, minimum pressure drop, good cleanability and high corrosion resistance.

Surface elements are stacked and bundled into self-contained baskets and are installed into air heater rotors and stators in two or more layers. The surface layer at the air inlet side, designated the cold layer, is distinguished from other layers by design. Cold layers, which are subject to corrosion and ash fouling, are typically 12 in. (305 mm) deep for economical replacement. Heavy gauge, open profile elements are used for corrosion resistance and cleanability. Practically all cold layer elements are low alloy corrosion resistant steel or, when high corrosion potential exists, porcelain enamel coated steel. Hot and intermediate surface layers are more compact than cold layers and use thinner plates. Fig. 21 illustrates several heating surface element profiles and air heater surface arrangements.

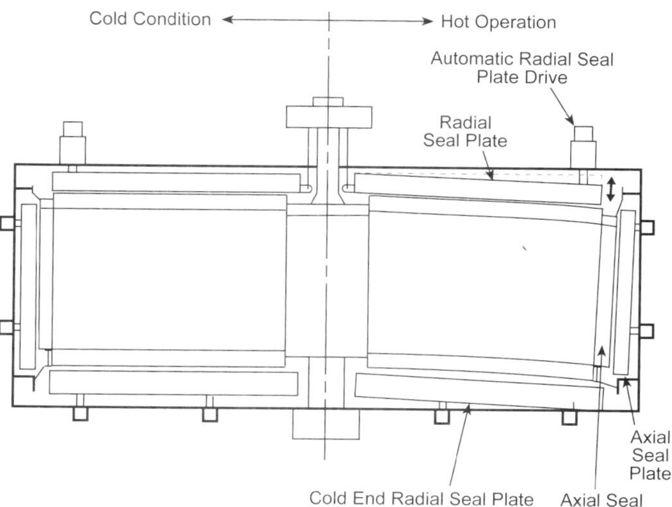

Fig. 19 Cross-section through air heater rotor and seal plates.

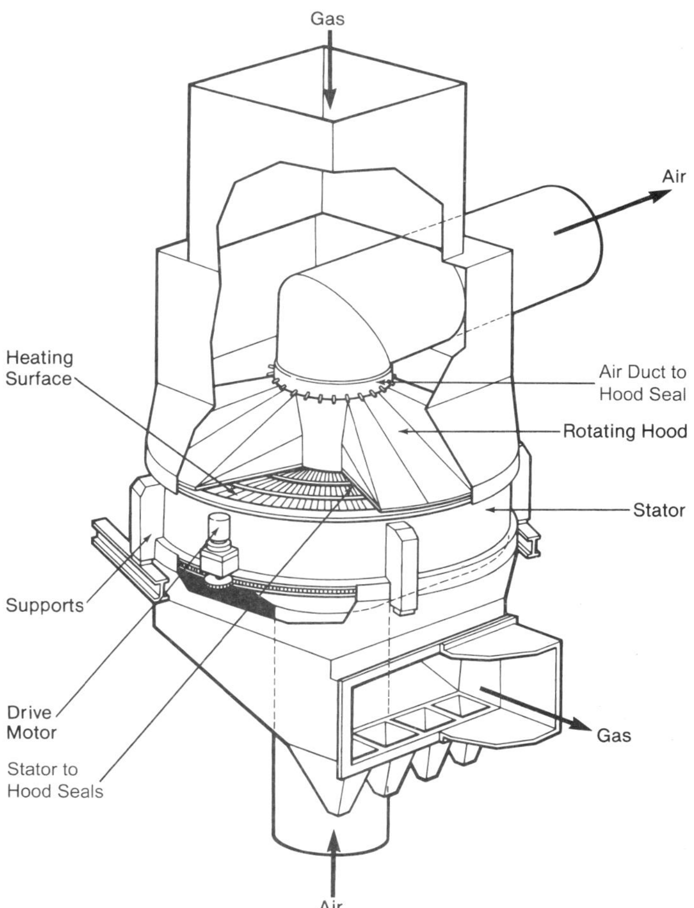

Fig. 20 Rothemühle-type air heater.

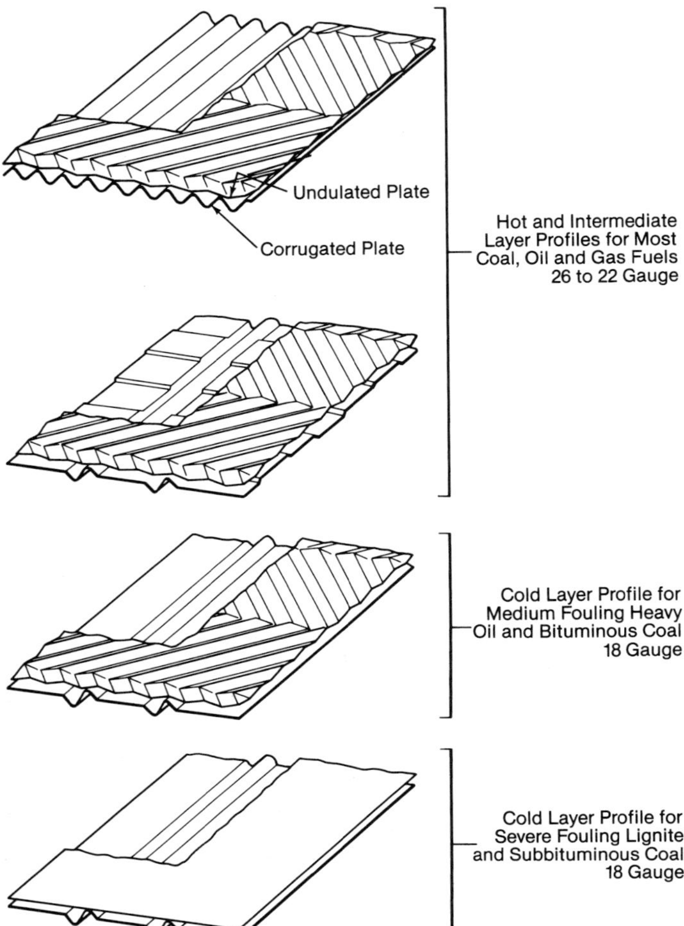

Fig. 21 Regenerative air heater surface element profiles.

Advantages and disadvantages

Many subtle differences exist between air heater designs within a particular type. However, there are some general advantages and disadvantages associated with each type which are listed in Table 1.

Performance and testing

Air heaters are designed to meet performance requirements in three areas: thermal, leakage and pressure drop. Low performance in any area increases boiler operating costs and may cause unit load curtailment.

Thermal performance

The thermal performance and surface area (A) of a recuperative air heater can be evaluated by:

$$A = q / (U \ LMTD \ F) \tag{5}$$

where q is the rate of heat transfer [Btu/h (W)], U is the overall heat transfer coefficient, $LMTD$ is the log mean temperature difference between the hot and cold fluids, and F is the heat exchanger arrangement factor. The performance, U, $LMTD$ and F can be evaluated using the correlations and methodology presented in Chapter 4. The overall U should include convection and radiation components as well as the appropriate gas- and air-side fouling factors. U typically ranges from 3 to 10 Btu/h ft^2 F (17 to 57 W/m^2 K).

Performance verification Thermal performance is measured by comparing the test gas outlet temperature to its design value. The true outlet temperature is obtained by correcting the measured temperature for air heater leakage and deviations from design conditions.

The ASME Performance Test Code, Section 4.3 (PTC 4.3), provides the following equation which is based on an air heater mass flow heat balance and assumes that the source of all leakage is from the entering air:

$$T_2 = T_{2m} + \left(\frac{\% \, lkg}{100}\right) \frac{c_{pa}}{c_{pg}} \left(T_{2m} - T_1\right) \tag{6}$$

where

T_2 = air heater gas outlet temperature corrected for leakage, F (C)
T_{2m} = measured gas temperature leaving air heater, F (C)
$\% \, lkg$ = percent air leakage with respect to inlet gas flow
c_{pa}, c_{pg} = specific heat of air and gas respectively, Btu/lb F (J/kg C)
T_1 = air inlet temperature, F (C)

The measured gas outlet temperature must also be corrected for deviations in various operating parameters such as mass flow rates and operating temperatures in order to accurately assess performance. Suppliers and the ASME Performance Test Codes provide various correction curves and factors for this purpose.

Leakage

Air flow passing from the air side to the gas side is called leakage. It is quantified in pounds per hour (kg/s) but is frequently expressed as a percentage of the gas inlet flow. Leakage is undesirable primarily because it represents fan power wasted in conveying air which bypasses the boiler combustion zone. Leakage can also reduce an air heater's thermal performance.

Recuperative units may begin operation with essentially zero leakage, but leakage occurs as time and thermal cycles accumulate. With regular maintenance, leakage can be kept below 3%. Air heater leakage is inherent with the rotary regenerative design. There are two types of leakage, gap and carryover.

Table 1
Advantages and Disadvantages of Air Heater Types

Type	Advantage	Disadvantage
Recuperative	Low leakage No moving parts	Large and heavy Difficult to replace surface
Regenerative	Compact Easy to replace surface	Leakage High maintenance Fire potential

Gap leakage occurs as higher pressure air passes to the lower pressure gas side through gaps between rotating and stationary parts. Its rate is given by the following general expression:

$$w_l = KA\left(2g_c \, \Delta P \rho\right)^{1/2} \tag{7}$$

where

w_l = leakage flow rate, lb/h (kg/s)
K = discharge coefficient, dimensionless (generally 0.4 to 1.0)
A = flow area, ft^2 (m^2)
g_c = 32.17 lbm ft/lbf $s^2 \times (3600 \text{ s/h})^2$ = 4.17×10^8 lbm ft/lbf h^2 (1 kg m/N s^2)
ΔP = pressure differential across gap, lb/ft^2 (kg/m^2)
ρ = density of leaking air, lb/ft^3 (kg/m^3)

Carryover leakage is the air carried into the gas stream from each rotor (stator) heating surface compartment as the surface passes from the air stream to the gas stream. This leakage is directly proportional to the void volume of the rotor and the rotation speed.

Regenerative air heater design leakage ranges from 5 to 15% but increases over time as seals wear. Effective automatic sealing systems, which nearly eliminate leakage rise due to seal wear, have been successfully applied. These systems monitor and adjust rotating to stationary seals on-line.

Another source of boiler air to gas flow leakage, which appears as air heater leakage, is outside air infiltration into lower pressure gas streams. Infiltration may occur at casing cracks or holes, flue expansion joints and access doors or gaskets. This sometimes neglected source can be significant and difficult to detect if leaks occur under lagging and insulation.

Air heater leakage can be obtained directly as the difference between air- or gas-side inlet and outlet flows based on velocity measurements. However, because velocity measurements are difficult to obtain accurately in large duct cross-sections, air heater leakage is more accurately based on calculated gas weights using gas analysis, boiler efficiency and fuel analysis data. (See Chapter 10.) Approximate air heater leakage can be determined by the following formula based on gas inlet and outlet oxygen (O_2) analysis (dry basis).

$$\%\text{ Leakage} = \frac{\%\,O_2 \text{ Leaving} - \%\,O_2 \text{ Entering}}{21 - \%\,O_2 \text{ Leaving}} \times 90 \tag{8}$$

Test air heater leakage should be corrected for deviations from design cold end air to gas differential pressure and inlet air temperature before comparison to design leakage.

Pressure drop

In recuperative air heaters, gas- or air-side pressure drop arises from frictional resistance to flow, inlet and exit shock losses and losses in return bends between flow passes. In regenerative air heaters, the main cause is heating surface frictional flow resistance. In both cases, pressure drop is proportional to the square of the mass flow rate. Typical values at full load flows are 2 to 7 in. wg (0.5 to 1.7 kPa).

Air- and gas-side pressure drop values are the differences between terminal inlet and outlet static gauge pressures. Correction of measured pressure drops for deviations from design flows and temperatures is necessary before comparison to design values.

Operational concerns

There are several operating conditions and maintenance concerns common to most air heaters. These include corrosion, plugging and cleaning, leakage performance degradation, erosion and fires. Air heaters used with high ash and/or high sulfur content fuels require more attention and maintenance than those firing clean fuels such as natural gas.

Corrosion

Air heaters used on units firing sulfur bearing fuels are subject to cold end corrosion of heating elements and nearby structures. In a boiler, a portion of the sulfur dioxide (SO_2) produced is converted to sulfur trioxide (SO_3) which combines with moisture to form sulfuric acid vapor. This vapor condenses on surfaces at temperatures below its dew point of 250 to 300F (121 to 149C). Because normal air heater cold end metal temperatures are frequently as low as 200F (93C), acid dew point corrosion potential exists. The obvious solution would be to operate at metal temperatures above the acid dew point but this results in unacceptable overall boiler heat losses. Most air heaters are designed to operate at minimum metal temperatures (MMTs) somewhat below the acid dew point, where the efficiency gained more than balances the additional maintenance costs. B&W recommends limiting MMTs to the values in Figs. 22 and 23 when burning sulfur bearing fuels.

When fuel sulfur levels are high, or ambient temperatures or operating loads are low, MMTs may be unacceptably low. These situations dictate the use of active or passive cold end corrosion control methods. Active systems used to raise MMT include: 1) steam- or water-coil air heaters to preheat inlet air, 2) cold air bypass, in which a portion of the inlet air is ducted around the air heater, and 3) hot air recirculation, in which a portion of the hot outlet air is ducted to combustion air fan inlets.

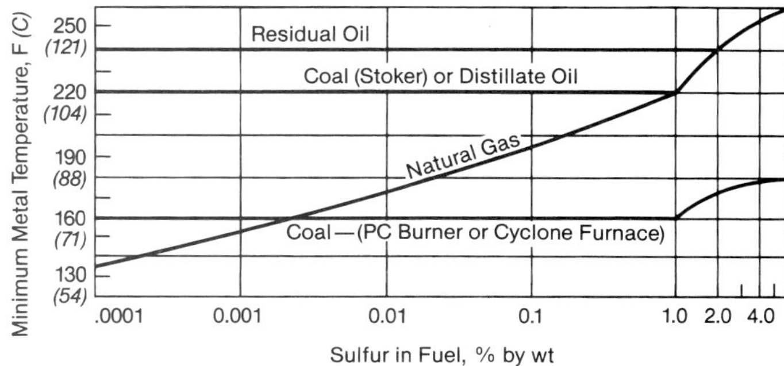

Fig. 22 Recuperative air heater cold end MMT limits when burning sulfur-bearing fuels.

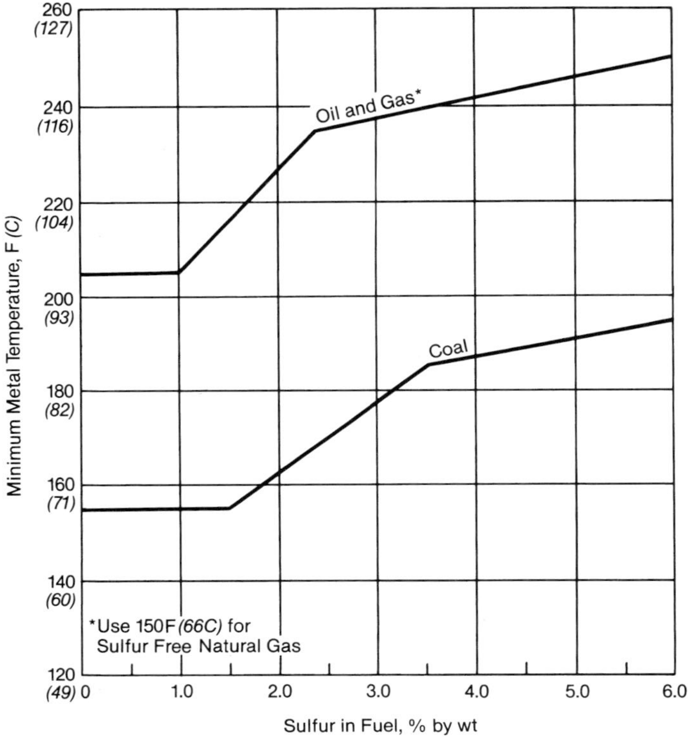

Fig. 23 Regenerative air heater cold end MMT limits when burning sulfur-bearing fuels.

Passive corrosion control methods incorporated in air heater design include: 1) thicker cold end materials, such as 11 or 14 gauge (3 or 2 mm) tubes and 18 gauge (1 mm) regenerative surface elements, 2) low or high alloy cold end surface materials which have at least twice the corrosion life of carbon steel, 3) nonmetallic coating, such as porcelain enamel, Teflon, or epoxies on cold elements, 4) nonmetallic cold end surface materials such as extruded ceramic in regeneratives and borosilicate glass tubes in tubulars, and 5) tubular air heater cold end tube arrangements which maximize MMT by providing higher gas flow and lower air flow velocities.

Plugging and cleaning

Plugging is the fouling and eventual closing of heat transfer flow passages by gas-entrained ash and corrosion products. It can occur at the air heater hot end but is most common at the cold end where ash particles adhere to acid moistened surfaces. Plugging increases air heater pressure drop and can limit unit load when fan capacity is reached at less than full load.

Air heater deposits are controlled and removed by periodic on-line sootblowing, and by various methods of cold end temperature control as described above to control corrosion. Deposits may also be controlled by employing regenerative heating surface designs with easier-to-clean, less tortuous flow paths, and by using smooth enamel coated regenerative surface elements that resist particle adherence.

When on-line cleaning methods are no longer able to stop or reverse fouling, off-line cleaning is necessary. Regenerative and recuperative air heater surface is normally washed off-line with permanent, low pressure wash piping. The surface is deluged (washed) continuously until the surface is deemed clean. Cleanliness may be proven by observed clarity of wash outlet water or by the ability to see light directed through the surface bank.

In cases where low pressure washing is not effective, high pressure washing is done. Power plant operators employ air heater washing specialists and equipment with water jets at nozzle pressure above 5000 psi (34.5 MPa). Special care must be taken to avoid breaking and/or flattening regenerative element plates.

Erosion

Heat transfer surfaces and other air heater parts can suffer erosion damage through impact of high velocity, gas-entrained ash particles. Erosion usually occurs near gas inlets where velocities are highest. However, areas near seals in regenerative air heaters can also be damaged as ash is accelerated through seal gaps. The undesirable effects of erosion are structural weakening, loss of heat transfer surface area and perforation of components which can cause air to gas or infiltration leakage. Erosion rate is a function of velocity, gas stream ash loading, physical nature of ash particles and angle of particle impact. It is controlled by reducing velocities, removing erosive elements from the gas stream, or using sacrificial material.

In the design stage, air heaters used with fuels containing highly erosive ash can be sized to limit gas inlet velocities to 50 ft/s (15 m/s). Inlet flues can also be designed to evenly distribute gas over the air heater inlet to eliminate local high velocity areas. Dust collectors, or strategically located screens and hoppers, may be used ahead of air heaters to remove some of the ash. In existing problem air heaters, flow distribution baffles may be installed to eliminate local high velocities, sacrificial materials such as abrasion resistant steel or ceramics may be placed over critical areas, or parts can be replaced with thicker materials for longer life.

Erosion in tubular air heaters frequently occurs within about 1 ft (0.3 m) of the gas inlet end due to turbulence as gas enters the tubes. Replaceable sacrificial sleeves may be installed in tube ends or egg crate-type flow straightening grids can be installed at tubular air heater inlets to reduce erosion.

Fires

Air heater fires are rare but do occur, particularly in regenerative units, and may be severe enough to completely destroy an air heater. Fires are detected by thermocouples in gas and air outlet ducts as well as with special early warning systems. Fires usually start near the cold end, which can be fouled with unburned combustible materials. Most fires occur during startup as unburned fuel oil deposited on ash fouled heating surfaces is ignited. Leaking bearing lubrication equipment and heavy accumulations of flyash are also fire hazards. Fires can be avoided by maintaining a clean air heater and proper tuning of boiler firing equipment. Frequent sootblowing during startup and just before shutdown is a strongly recommended fire prevention practice.

After a fire is confirmed, typical practice is to trip the boiler to stop air flow through the heater, to keep the air heater rotating, and to introduce as much water as possible through the permanent water wash system. Drains in air- and gas-side duct hoppers below the air heater must be opened to remove water and ash.

Utility applications

Gas to air recuperative and regenerative air heaters are usually used in utility units, primarily to enhance unit efficiency. Small increments of increased efficiency in large units amount to substantial fuel savings. Utility units generally use multiple air heaters for plant arrangement convenience, type of firing and maximum unit availability.

Pulverized coal-fired units require two streams of hot combustion air, i.e., primary air supplied at high pressure to pulverizers and secondary air supplied at lower pressure directly to burners. Two basic air flow systems are used, hot primary air and cold primary air. Each system uses air heaters. In the hot primary air scheme, used for smaller units, about one third of the combustion air heated in a secondary heater is ducted to hot primary air fans, where it is boosted in pressure and passed to the pulverizers; the remaining two thirds is ducted to the burners. The cold primary air system uses separate air heaters supplied by separate primary and secondary (forced draft) fans. In some units, both primary and secondary air are heated in a single regenerative unit which is more cost effective than separate air heaters. Fig. 24 shows a schematic Ljungström-type air heater for primary and secondary air, referred to as a tri-sector.

If separate regenerative primary and secondary air heaters are used, the primary air heaters, which operate at high air to gas pressure differentials, exhibit twice as much leakage as the secondary units. For this reason, low leakage recuperative air heaters may be used for primary air heating and regeneratives may be used for secondary air heating.

Recuperative and regenerative air heaters are used on oil- and gas-fired units. In general, regardless of fuel type, larger units use regenerative air heaters because of their smaller size and lower initial cost. However, for air to gas pressure differentials above 40 in. wg (10 kPa), in fluidized-bed applications for example, recuperative air heaters are usually preferred.

Industrial applications

Industrial units fire a variety of fuels such as wood, municipal refuse, sewage sludge and industrial waste gases as well as coal, oil and natural gas. As a result, many air heater types are used. In the small units, tubular, plate and cast iron heaters are widely used. Fuels fired on stoker grates, such as bituminous coal, wood and refuse, do not require high air temperatures, therefore water- or steam-coil air heaters can be used.

Environmental air heater application

For environmental reasons, emission of certain fossil-fired combustion products may be limited by law. (See Chapter 32.) Systems developed to limit emissions of two

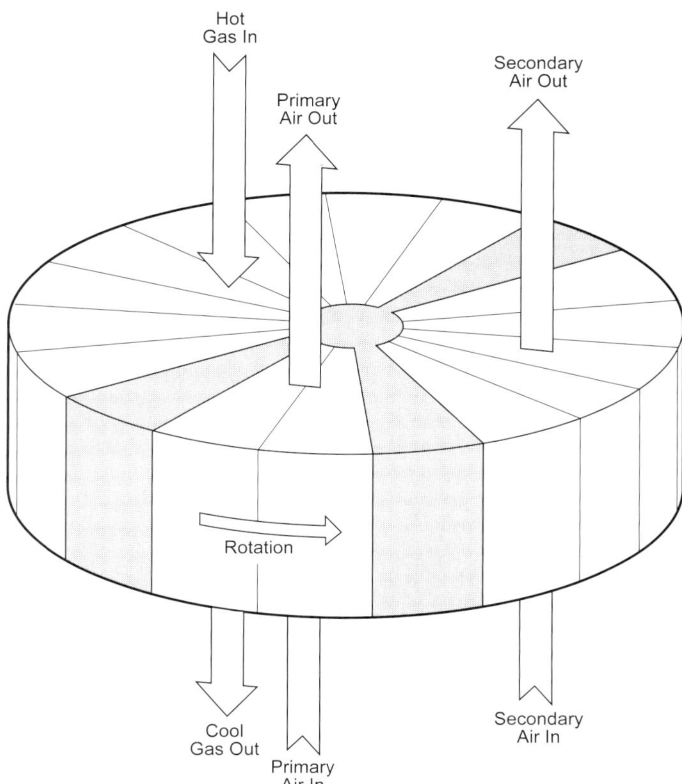

Fig. 24 Ljungström-type tri-sector air heater.

objectionable flue gas constituents, NO_x and SO_2, may require the use of specially modified heat exchangers.

NO_x removal

When selective catalytic reduction (SCR) technology is used on a boiler to control NO_x emissions, air heaters operating downstream of the SCR require special consideration. A typical, popular arrangement is to place the SCR reactor in the boiler flue gas stream between the economizer outlet and air heater inlet (high dust arrangement).

The SCR reduces NO_x by injecting ammonia (NH_3) into the flue gas just upstream of a catalyst which converts the NO_x to water and elemental nitrogen (N_2). However, the catalyst also converts some of the sulfur dioxide (SO_2) in the flue gas to sulfur trioxide (SO_3) which combines with unreacted NH_3 to form ammonium bisulfate (NH_4HSO_4). This condenses on air heater surface elements in the 510 to 340F (266 to 171C) temperature range. The objectionable NH_4HSO_4 fouling and corrosion in regenerative air heaters can be controlled to manageable levels by design features including:

1. Arranging heating surface basket layers so that NH_4HSO_4 deposition occurs within a layer rather than between layers. Deposition in the space between basket layers is more difficult to remove and keep clean. It is typical to provide two rather than three air heater basket layers.
2. Employing a closed profile heating element which allows deeper and more effective sootblower cleaning. Closed profile elements are designs in which

individual flow passages are closed through the depth of the element layer and do not allow sootblowing media to quickly dissipate.

3. Using porcelain enamel coated bottom layer heating surface which resists corrosion and deposition, and is easier to clean because of the smooth surface.
4. Fitting the heater with hot and cold end sootblowers which can be used to water wash heating surface off-line.

Heat exchangers with these features can typically operate reliably without off-line water washing for a year or more. As discussed in Chapter 34, some regenerative air heaters have been modified to simultaneously serve as selective catalytic NO_x reduction systems.

SO_2 reduction

When sulfur emission reduction is required, flue gas desulfurization (FGD) systems are frequently used. These systems remove SO_2 from the flue gas by reaction with injected compounds such as limestone. In most cases, the scrubbed flue gas exits the FGD system at a saturation temperature of 120 to 130F (49 to 54C) before entering the stack. In cases where acid dew point corrosion of flues and stack liners is a concern or increased gas buoyancy is needed to improve stack plume dispersal, gas exiting the FGD system is reheated to 180F (82C) or higher.

Regenerative heat exchangers similar to those for combustion air heating are used in FGD systems. The heat exchangers, referred to as gas-to-gas heaters (GGH), extract heat from untreated warm flue gas to heat treated (scrubbed) flue gas leaving an FGD tower before it enters the stack. A typical arrangement with gas temperatures is shown in Fig. 25.

Several necessary GGH design features distinguish them from air heaters and permit them to operate in a very corrosive environment without compromising FGD system SO_2 removal performance. Measures taken to protect a GGH from the corrosive flue gas include:

1. Fabricating rotors of low alloy corrosion-resistant steel and adding a corrosion margin to all rotor plate thicknesses.
2. Using porcelain enamel coated heating surface element plates.
3. Using stainless and/or non-metallic circumferential, radial and axial leaf seal materials.
4. Applying a layer of vinyl ester flake glass on internal surfaces of the untreated gas outlet duct, treated gas inlet duct, treated gas outlet duct, and rotor housing.
5. Employing hot and cold end air sootblowers for on-line cleaning, fitted with water wash nozzles for off-line washing.

Leakage of the untreated gas into the treated gas reduces FGD SO_2 reduction performance and must be controlled to low levels. The following features incorporated into GGH design can reduce the leakage to less than 1% of the untreated inlet gas flow:

1. Designing the rotor with radial heating surface division plates so that, during operation, there are always two under the radial and axial seal plates. Leakage flow, in this case, is forced to pass two seals rather than one seal producing a labyrinth effect which lowers leakage flow.
2. Employing an automatic sealing system to hot end radial seal plates that follows stator distortion to minimize seal leakage at all loads.
3. Purging stator compartments of untreated gas with treated gas just before they pass under radial seals and into the treated gas stream. Purging minimizes carryover leakage which is a component of total leakage. A purge system consists of a dedicated purge fan and associated ducting and controls.
4. Reducing rotor speed to the lowest possible level without significantly compromising thermal performance. Operation at low speed reduces carryover leakage.

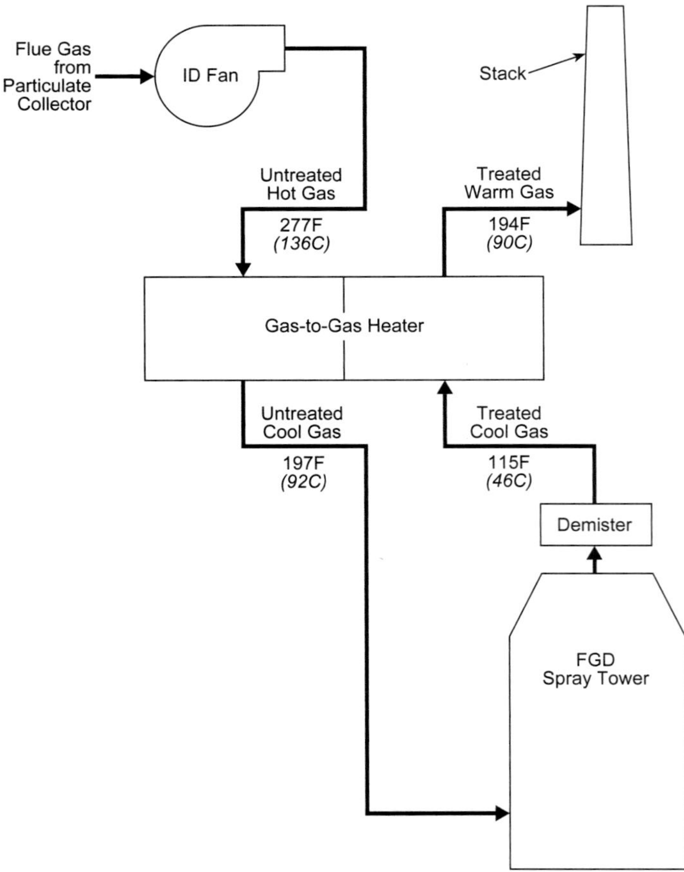

Fig. 25 Gas-to-gas heater in an FGD system.

Bibliography

Dubbel, H., *Taschenbuch für den Maschinenbau*, 11th Ed., Springer Verlag GmbH & Company, Berlin, Germany, 1958.

Ledinegg, M., *Dampferzeugung,* Dampfkessel, Feuerungen Springer Verlag-Wien, 1952.

Mayer, E.H., and McCarver, G.M., *Economizer Technology for Utility and Industrial Boilers, Course for the Center for Professional Advancement,* East Brunswick, New Jersey, 1991.

McAdams, W.H., *Heat Transmission,* Third Ed., McGraw-Hill Company, New York, New York, 1954.

Two coal-fired boilers: one 685 MW pulverized coal unit and one 844 MW unit with Cyclone™ furnaces.

Chapter 21

Fuel Ash Effects on Boiler Design and Operation

The effective utilization of fossil fuels for power generation depends to a great extent on the capability of the steam generating equipment to accommodate the inert residuals of combustion, commonly known as *ash*. The quantity and characteristics of the ash inherent to a particular fuel are major concerns to both the designer and the operator of the equipment.

With few exceptions, most commercial fuels contain sufficient ash to warrant specific design and operating considerations. The following focuses on these design and operating considerations, primarily as they relate to pulverized coal firing. Fuel ash characteristics relating to petroleum fuels are also discussed.

Ash dilutes the heating value of fuel, placing additional burdens on fuel storage, handling and preparation equipment. Extensive facilities are also needed to collect, remove and dispose of the ash. These material handling requirements represent significant costs in terms of equipment and real estate which are directly proportional to the amount of ash in the fuel. In the case of coal, ash quantities can be substantial. Consider, for example, a 650 MW utility steam generator firing a coal with a heating value of 10,000 Btu/lb (23,250 kJ/kg) containing 10% ash by weight. The unit would burn approximately 300 tons per hour (272 t_m/h) of coal, generating more than 700 tons per day (635 t_m/d) of ash.

In pulverized coal-fired boilers, most of this ash is carried out of the furnace by the gaseous products of combustion (flue gas). Abrasive ash particles suspended in the gas stream can cause erosion problems on convection pass heating surfaces. However, the most significant ash-related problem is deposition. During the combustion process, the mineral matter that forms ash is released from the coal at temperatures in the range of 3000F (1649C), well above the melting temperature of most mineral matter compounds. Ash can be released in a molten fluid or sticky plastic state. A portion of the ash, which is not cooled quickly to a dry solid state, impacts on and adheres to the furnace walls and other heating surfaces. Because such large total quantities of ash are involved, even a small fraction of the total can seriously interfere with boiler operation. Accumulation of ash deposits on furnace walls impedes heat transfer, delaying cooling of the flue gas and increasing the flue gas temperature leaving the furnace. Elevated temperatures at the furnace exit raise steam temperature and can extend deposition problems to pendant superheaters and other heat absorbing surfaces in the convection pass. In extreme cases, uncontrolled ash deposits can develop to the point where flow passages in tube banks are blocked, impeding gas flow and ultimately requiring the unit to be shut down for manual removal. Large deposits in the upper furnace or radiant superheater can become dislodged and fall, damaging pressure parts in the lower furnace. Under certain conditions, ash deposits can also cause fireside corrosion on tube surfaces.

Minimizing the potential for these ash-related problems is a primary goal of both the designers and operators of coal-fired boilers. The extent to which coal ash characteristics affect boiler design is illustrated in Fig. 1 which compares the relative size of a gas-fired and coal-fired boiler. Both are sized for the same steam generating capacity and similar steam conditions. While the combustion characteristics of coal play a role in sizing the furnace, the deposition and erosion potential of the ash are the primary design considerations driving the overall size and arrangement.

The variability of ash behavior is one of the biggest problems for boiler designers and operators. Although boilers are often designed to burn a wide range of coals satisfactorily, no unit can perform equally well with all types of coal.

Ash content of coal

The ash content of coal varies over a wide range. This variation occurs not only in coals from different geographical areas or from different seams in the same region, but also from different parts of the same mine. These variations result primarily from the wide range of conditions that introduced foreign material during or following the formation of the coal. (See Chapter 9.) Ash content can also be influenced by extraneous mineral matter introduced during the mining operation. Before being sold, some commercial coals are cleaned or washed to remove a portion of what would be labeled ash in the laboratory. However, the ash content of significance to the user is the content at the point of use. The values noted below are on that basis.

Most of the coal used for power generation in the United States (U.S.) has an ash content between 6 and 20%. Low values of 3 to 4% in bituminous coals are rare and these coals find other commercial uses, particularly in the metallurgical field. On the other hand,

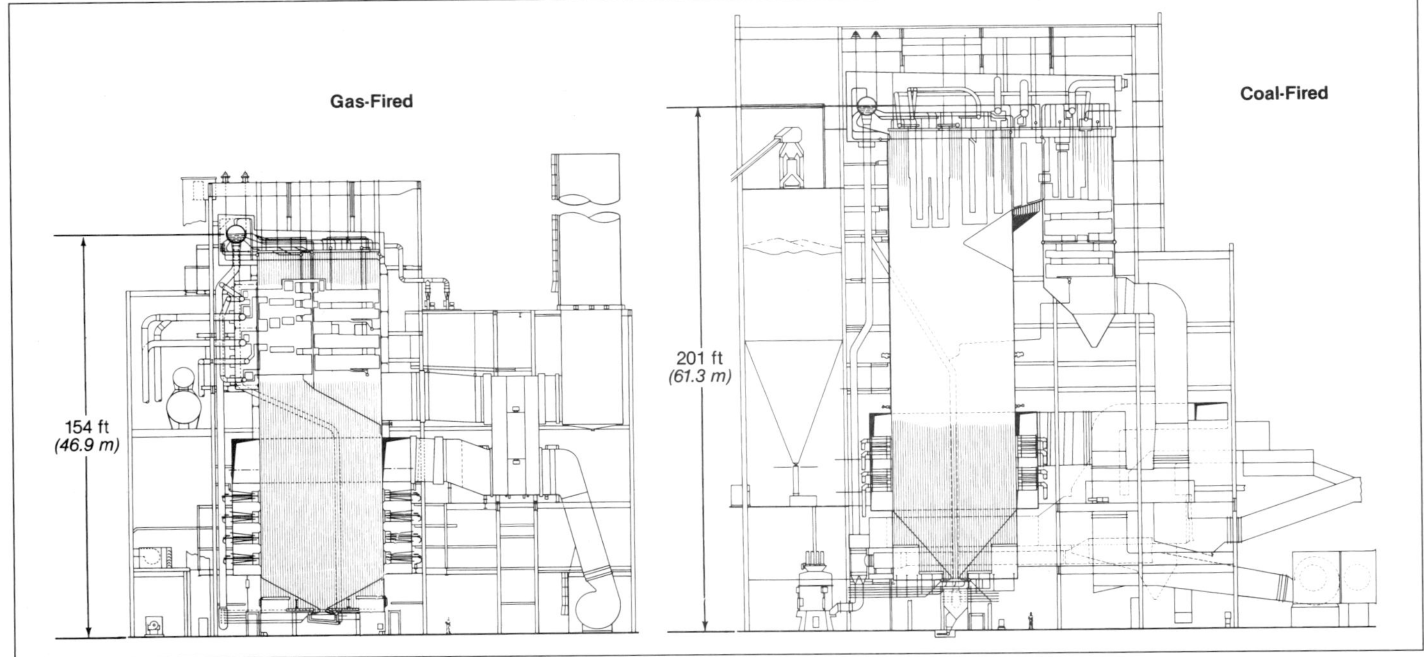

Fig. 1 Size comparison of gas-fired and coal-fired utility boilers.

some coals may have ash contents as high as 40%. Many high ash fuels can be successfully burned in utility (electric power generation) boilers. Their use has increased in areas where they offer an economic advantage.

Evaluation of ash content on a weight percentage basis alone does not take into account the heat input associated with the coal, which is also related to moisture content. It is common, for design and fuel evaluation purposes, to consider ash content on the basis of weight per unit of heat input, generally expressed as pounds of ash per million Btu. This factor is calculated as follows:

$$\frac{\text{Ash (\% by weight)}}{\text{HHV(Btu/lb)}} \times 10^4 = \text{lb ash}/10^6 \text{ Btu}$$

or (1)

$$\frac{\text{Ash (\% by weight)}}{\text{HHV (kJ/kg)}} \times 10^3 = \text{kg ash/MJ}$$

where HHV is the higher heating value of the fuel.

The relevance of this factor is illustrated in Table 1, which provides proximate analyses for three selected coals. Each coal has a moderate ash content of 9 to 10% by weight. However, on a heat input basis, ash quantities vary significantly. The lignite in this example would introduce almost three times as much ash as the high volatile bituminous coal at an equivalent heat input.

Furnace design for ash removal

Historically, two distinctly different types of furnace design were used to handle the ash from coal firing in large utility boilers. These are commonly referred to as the *dry-ash* or *dry-bottom* furnace and the *slag-tap* or *wet-bottom* furnace.

All modern pulverized coal-fired boilers use the dry-bottom arrangement. The coal-fired boiler in Fig. 1 is typical of this design. In a dry-bottom unit most of the ash, typically 70 to 80%, is entrained in the flue gas and carried out of the furnace. This portion of the ash is commonly known as *flyash*. Some of the flyash is collected in hoppers arranged under the economizer and air heaters, where coarse particles drop out of suspension when gas flow direction changes. The finer ash particles remain in suspension and are carried out of the unit for collection by particulate control equipment. (See Chapter 33.) The remaining 20 to 30% of the ash that settles in the furnace, or is dislodged from the furnace walls, is collected in a hopper formed by the frontwall and rearwall tube panels at the bottom of the furnace. This *bottom ash* is discharged through a 3 to 4 ft (0.9 to 1.2 m) wide opening that spans the entire width of the hopper.

Slag-tap furnaces were originally developed to resolve ash deposition and removal problems when firing coals with low ash fusion temperatures in dry-bottom furnaces. These units are intentionally de-

Table 1
Proximate Analyses of Three Selected Coals — Ash Content as Weight Per Unit of Heat Input

Rank	High Volatile Bituminous	Subbituminous	Lignite
Moisture, %	3.1	23.8	45.9
Volatile matter, %	42.2	36.9	22.7
Fixed carbon, %	45.4	29.5	21.8
Ash, %	9.4	9.8	9.6
Heating value, Btu/lb	12,770	8683	4469
lb Ash/10^6 Btu	7.4	11.3	21.5

signed to maintain ash in a fluid state in the lower furnace. Molten ash is collected on the furnace walls and other surfaces in the lower furnace and drained continuously to openings called *slag-taps* in the furnace floor. Water tanks positioned beneath the slag-taps solidify the liquid ash for disposal.

Slag-tap furnaces have been used with both pulverized coal and Cyclone™ furnace firing systems. (See Chapter 15.) Application is limited to coals having ash viscosity characteristics which would ensure that ash fluidity could be maintained over a reasonable boiler load range. Much of the coal ash research conducted by The Babcock & Wilcox Company (B&W) concerning the viscosity-temperature relationship of coal ash was initially directed at defining coal ash suitability limits for wet-bottom and Cyclone furnace applications. A minimum coal ash content was also specified to ensure sufficient ash quantities to maintain the required slag coating. One benefit of wet-bottom firing was a significant reduction in flyash quantity. In pulverized coal wet-bottom applications, as much as 50% of the total ash was collected in the furnace. Units equipped with Cyclone furnaces could retain up to 80% of the ash in the furnace.

The application of slag-tap units for pulverized coal firing began to decline in the late 1940s, primarily due to design improvements in dry-bottom units that minimized ash deposition problems. Slag-tap units equipped with Cyclone furnaces continued to be applied until the early 1970s when the federal Clean Air Act mandated control of nitrogen oxides (NO_x) emissions. The high furnace temperatures required for wet-bottom operation were highly conducive to NO_x formation.

Ash deposition

Regardless of the firing method, when coal is burned, a relatively small portion of the ash will cause deposition problems. Ash passing through the boiler is subject to various chemical reactions and physical forces which lead to deposition on heat absorbing surface. The process of deposition and the structure of deposits are variable due to a number of factors. Particle composition, particle size and shape, particle and surface temperatures, gas velocity, flow pattern and other factors influence the extent and nature of ash deposition.

Due primarily to the differences in deposition mechanisms involved, two general types of high temperature ash deposition have been defined as *slagging* and *fouling*.

Slagging is the formation of molten, partially fused or resolidified deposits on furnace walls and other surfaces exposed to radiant heat. Slagging can also extend into convective surface if gas temperatures are not sufficiently reduced.

Most ash particles melt or soften at combustion temperatures. The time-temperature history or cooling rate of the particle determines its physical state (solid, plastic or liquid) at a given location in the furnace. Generally, in order to adhere to a clean surface and form a deposit, the particle must have a viscosity low enough to wet the surface.

Slag deposits seldom form on clean tube surfaces. A conditioning period is required before significant deposition occurs. Assuming there is no direct flame impingement, as ash particles approach a clean tube, most tend to be resolidified due to the relatively lower temperature at the tube surface. The particles fracture on impact and partially disperse back into the flue gas stream. Over a period of time, however, a base deposit begins to form on the tube. The base deposit may be initiated by the settling of fine ash particles or the gradual accumulation of particles with very low melting point constituents. As the base deposit thickens, the temperature at its outside face increases significantly above the tube surface temperature. Eventually, the melting point of more of the ash constituents is exceeded and the deposit surface becomes molten. The process then becomes self-accelerating with the plastic slag trapping essentially all of the impinging ash particles. Ultimately, the deposit thickness reaches an equilibrium state as the slag begins to flow, or the deposit becomes so heavy that it falls away from the tubes. Depending on the strength and physical characteristics of the deposit, sootblowers using steam, compressed air or water as cleaning media (see Chapter 24) may be able to control or remove most of the deposit. However, the base deposit can remain attached to the tube, allowing subsequent deposits to accumulate much more rapidly.

Fouling is defined as the formation of high temperature bonded deposits on convection heat absorbing surfaces, such as superheaters and reheaters, that are not exposed to radiant heat. In general, fouling is caused by the vaporization of volatile inorganic elements in the coal during combustion. As heat is absorbed and temperatures are lowered in the convective section of the boiler, compounds formed by these elements condense on ash particles and heating surface, forming a glue which initiates deposition.

Areas where slagging and fouling can occur are shown in Fig. 2. Figs. 3 and 4 show heavily slagged and fouled surfaces. The characteristics of coal ash and their influence on slagging and fouling are discussed in the following sections.

Characteristics of coal ash

Sources of coal ash

Mineral matter is always present in coal and forms ash when the coal is burned. This mineral matter is usually classified as either inherent or extraneous. (See Chapter 9.) Inherent mineral matter is organically combined with the coal. This portion came from the chemical elements existing in the vegetation from which the coal was formed and from elements chemically bonded to the coal during its formation. Extraneous mineral matter is material that is foreign to the organic structure of the coal. This includes airborne and waterborne material that settled into the coal deposit during or after formation. It usually consists of mineral forms associated with clay, slate, shale, sandstone or limestone and includes pieces ranging from microscopic size to thick layers. Other extraneous material may be introduced through the mining process.

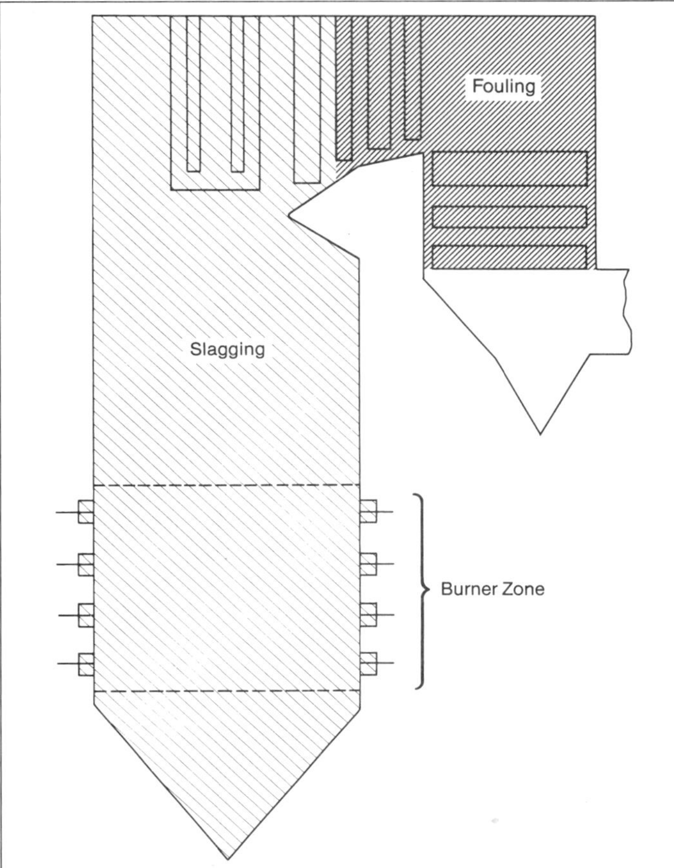

Fig. 2 Deposition zones in a coal-fired boiler.

Mineralogical composition

There are no standardized methods that are used routinely for determining the specific mineral constituents of coal. Mineralogical analysis requires the use of a low temperature ashing technique to separate the mineral matter from the organic portion of the coal. Standard high temperature ashing procedures would significantly alter the mineral forms. However, a number of researchers, using a variety of low temperature ashing methods and sophisticated analytical techniques, have identified an enormous variety of mineral species in coal, encompassing the entire spectrum of major mineral forms found in the earth's crust. Most of these minerals fall into one of several groups: clay minerals (aluminosilicates), sulfides/sulfates, carbonates, chlorides, silica/silicates and oxides. Some of the more common minerals in these groups are shown in Table 2.

Chemical composition

Because both quantitative and qualitative evaluation of mineral matter forms are extremely difficult, relatively simple chemical analyses are commonly used to determine the percentages of the major elements in the ash. Elemental ash analysis is performed on a coal ash sample produced in accordance with the American Society for Testing and Materials (ASTM) D 3174 ashing procedure. Pulverized coal is burned in a furnace with an oxidizing atmosphere at 1292 to 1382F (700 to 750C). The elements present in the ash are quantitatively measured using a combination of emission spectroscopy and flame photometry and are reported as weight percents of their oxides. Coal ash is consistently found to be composed mainly of silicon, aluminum, iron and calcium with smaller amounts of magnesium, titanium, sodium and potassium. The elemental analysis also identifies phosphorus as P_2O_5 and sulfur as sulfur trioxide (SO_3). Phosphorus is usually present in very small quantities and is sometimes omitted. Sulfur is reported as SO_3 because it is normally present as the sulfate form of one of the metals.

Percentages of the individual elements vary over a wide range for different coals; however, characteristic differences are evident between the older, high rank coals common in the Eastern U.S. and the younger, low rank Western coals. Bituminous coals typically have higher levels of silica, aluminum and iron, while the lower rank subbituminous coals and lignites generally have higher levels of the alkaline earth metals, calcium and magnesium, and the alkali metal sodium. These trends are evident in the ash analyses shown in Table 3.

Although the ash constituents are reported as oxides, they actually occur in the ash predominately as

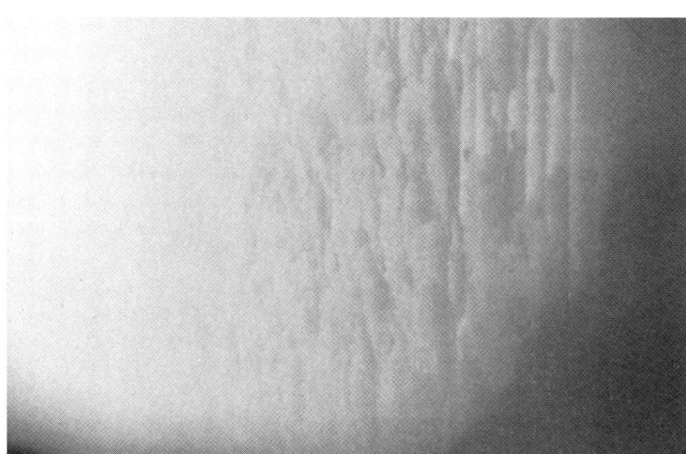

Fig. 3 Heavily slagged surface.

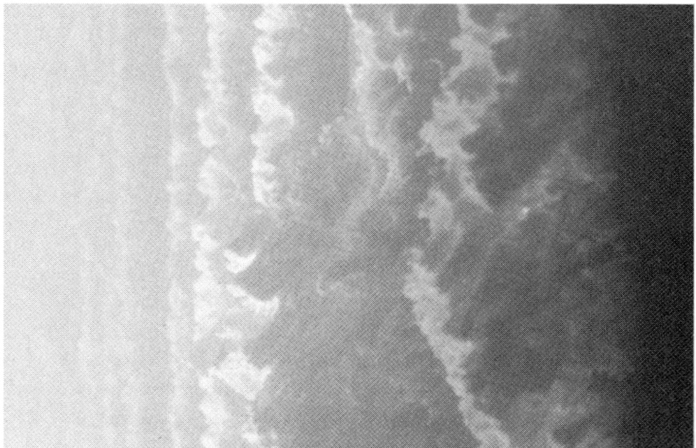

Fig. 4 Heavily fouled surface.

Table 2
Common Minerals Found in Coal

Mineral	Formula
Clay minerals:	
Montmorillonite	$Al_2Si_4O_{10}(OH)_2 \cdot H_2O$
Illite	$KAl_2(AlSi_3O_{10})(OH_2)$
Kaolinite	$Al_4Si_4O_{10}(OH)_8$
Sulfide minerals:	
Pyrite	FeS_2
Marcasite	FeS_2
Sulfate minerals:	
Gypsum	$CaSO_4 \cdot 2H_2O$
Anhydrite	$CaSO_4$
Jarosite	$(Na,K)Fe_3(SO_4)_2(OH)_6$
Carbonate minerals:	
Calcite	$CaCO_3$
Dolomite	$(Ca,Mg)CO_3$
Siderite	$FeCO_3$
Ankerite	$(Ca,Fe,Mg)CO_3$
Chloride minerals:	
Halite	NaCl
Sylvite	KCl
Silicate minerals:	
Quartz	SiO_2
Albite	$NaAlSi_3O_8$
Orthoclase	$KAlSi_3O_8$
Oxide minerals:	
Hematite	Fe_2O_3
Magnetite	Fe_3O_4
Rutile	TiO_2

a mixture of silicates, oxides and sulfates, with smaller quantities of other compounds. The silicates originate mainly from quartz and the clay minerals which contribute silicon, aluminum, sodium and much of the potassium. A principal source of iron oxide is pyrite (FeS_2) which is oxidized to form Fe_2O_3 and sulfur oxides. Part of the organic and pyritic sulfur that is oxidized combines with calcium and magnesium to form sulfates. Calcium and magnesium oxides result from the loss of carbon dioxide from carbonate minerals such as calcite ($CaCO_3$) and dolomite [$(Ca, Mg)(CO_3)$]. In low rank coals, a major portion of the sodium, calcium and magnesium oxides can originate from organically bound elements in the coal.

Laboratory ash is prepared from a coal sample in a controlled atmosphere at controlled temperatures to provide a reproducible and uniform ash. The actual ashing process during combustion in a pulverized coal-fired furnace is a much more complex process. In a boiler furnace, pulverized coal is burned in suspension as discrete particles. If all of the mineral matter were evenly distributed through the coal, the composition of each resulting ash particle would be the same as the bulk ash composition determined by the analysis of ASTM ash. A coal with no extraneous mineral matter might approach this hypothetical case, because organically combined inherent material would be expected to be evenly distributed. In reality, however, all coals contain non-uniformly distributed extraneous mineral matter in some of the wide variety of mineral forms shown on Table 2. When the coal is pulverized, some of the particles will be mostly coal with only inherent mineral matter, some will be pure mineral matter, and others will be combinations of both. Because the coal particles are burned discretely in suspension, the composition of an individual ash particle will depend on the specific mineral form or forms that were included in the coal particle. As a result, individual particle composition can vary significantly from the bulk ash composition.

During combustion, ash particles are exposed to temperatures as high as 3000F (1649C) and a variety of heating and cooling rates. The atmosphere in the burner zone can range from highly oxidizing to highly reducing. Depending on the composition of the specific particle, mineral forms in the ash can react with each other, with the organic and inorganic constituents of the coal, and with gaseous elements, such as sulfur dioxide (SO_2), in the flue gas. The compounds that are ultimately formed by these interactions are the materials that cause deposition problems. The compounds can have a wide variety of melting temperatures and viscosity-temperature characteristics. Some compounds combine to form eutectic mixtures that have melting temperatures lower than either of the original compounds. Particles that melt at lower temperatures and stay sticky long enough to reach a furnace wall become slag deposits. Volatile compounds that vaporize in the furnace tend to condense on and foul cooler convective heating surfaces.

Elemental ash analyses do not directly identify the compounds that cause deposition, or directly identify the mechanisms of deposit formation. Despite these limitations, no other data pertaining to coal ash composition are as widely available as the chemical analyses of ASTM ash. A large part of the coal ash research that has been conducted over the last sixty years has been directed at correlating analysis data and other characteristics of ASTM ash to observed ash behavior both in full scale boilers and in test facilities that closely simulate full scale conditions. Various evaluation methods have been developed based on these correlations to characterize ash behavior and predict deposition potential.

Ash fusibility

The measurement of ash fusibility temperatures is by far the most widely used method for predicting ash behavior at elevated temperatures. The preferred procedure in the U.S. is outlined in ASTM Standard D 1857, *Fusibility of Coal and Coke Ash*. An ash sample is prepared by burning coal under oxidizing conditions at temperatures of 1470 to 1650F (799 to 899C). The ash is pressed in a mold to form a triangular pyramid (cone) 0.75 in. (19 mm) in height with a 0.25 in. (6.35 mm) triangular base. The cone is heated in a furnace at a controlled rate to provide a temperature increase of 15F (8C) per minute. The atmosphere in the furnace is regulated to provide either oxidizing or reducing conditions. As the sample is heated, the temperatures at which the cone fuses and deforms to specific

Table 3
Ash Content and Ash Fusion Temperatures of Some U.S. Coals and Lignite

Rank:	Low Volatile Bituminous	High Volatile Bituminous				Sub-bituminous	Lignite
Seam	Pocahontas No. 3	No. 9	No.6	Pittsburgh		Antelope	
Location	West Virginia	Ohio	Illinois	West Virginia	Utah	Wyoming	Texas
Ash, dry basis,%	12.3	14.1	17.4	10.9	17.1	6.6	12.8
Sulfur, dry basis, %	0.7	3.3	4.2	3.5	0.8	0.4	1.1
Analysis of ash, % by wt							
SiO_2	60.0	47.3	47.5	37.6	61.1	28.6	41.8
Al_2O_3	30.0	23.0	17.9	20.1	21.6	11.7	13.6
TiO_2	1.6	1.0	0.8	0.8	1.1	0.9	1.5
Fe_2O_3	4.0	22.8	20.1	29.3	4.6	6.9	6.6
CaO	0.6	1.3	5.8	4.3	4.6	27.4	17.6
MgO	0.6	0.9	1.0	1.3	1.0	4.5	2.5
Na_2O	0.5	0.3	0.4	0.8	1.0	2.7	0.6
K_2O	1.5	2.0	1.8	1.6	1.2	0.5	0.1
SO_3	1.1	1.2	4.6	4.0	2.9	14.2	14.6
P_2O_5	0.1	0.2	0.1	0.2	0.4	2.3	0.1
Ash fusibility							
Initial deformation temp, F							
Reducing	2900 +	2030	2000	2030	2180	2280	1975
Oxidizing	2900 +	2420	2300	2265	2240	2275	2070
Softening temp, F							
Reducing		2450	2160	2175	2215	2290	2130
Oxidizing		2605	2430	2385	2300	2285	2190
Hemispherical temp, F							
Reducing		2480	2180	2225	2245	2295	2150
Oxidizing		2620	2450	2450	2325	2290	2210
Fluid temp, F							
Reducing		2620	2320	2370	2330	2315	2240
Oxidizing		2670	2610	2540	2410	2300	2290

shapes, as shown in Fig. 5, are recorded. Four deformation temperatures are reported as follows:

1. *Initial deformation temperature* (IT or ID) – the temperature at which the tip of the pyramid begins to fuse or show signs of deformation.
2. *Softening temperature* (ST) – the temperature at which the sample has deformed to a spherical shape where the height of the cone is equal to the width at the base (H = W). The softening temperature is commonly referred to as the fusion temperature.
3. *Hemispherical temperature* (HT) – the temperature at which the cone has fused down to a hemispherical lump and the height equals one half the width of the base (H = 1/2 W).
4. *Fluid temperature* (FT) – the temperature at which the ash cone has melted to a nearly flat layer with a maximum height of 0.0625 in. (1.59 mm).

The determination of ash fusion temperatures is strictly an empirical procedure, developed in standardized form, which can be duplicated with some degree of accuracy. Strict observance of test conditions is required to assure reproducible results. ASTM specified tolerances on reproducibility of the individual temperature measurements range from 100 to 150F (56 to 83C) when the test is performed by different operators and apparatus.

An earlier version of the ASTM D 1857 procedure specified the use of only a reducing atmosphere and had loosely defined criteria for identifying the softening and fluid points. When the atmosphere is not specified, it is generally assumed to be reducing. Reported softening temperatures are assumed to be the ST (H = W) point unless otherwise specified. Methods for determining fusibility of ash used by other countries are similar to the ASTM procedure but results may vary considerably due to differences in procedures or the definition of terms.

The gradual deformation of the ash cone is generally considered to result from differences in melting characteristics of the various ash constituents. As the temperature of the sample is increased, compounds with the lowest melting temperatures begin to melt, causing the initial deformation. As the temperature continues to increase, more of the compounds melt and the degree of deformation proceeds to the softening and hemispherical stages. The process continues until the

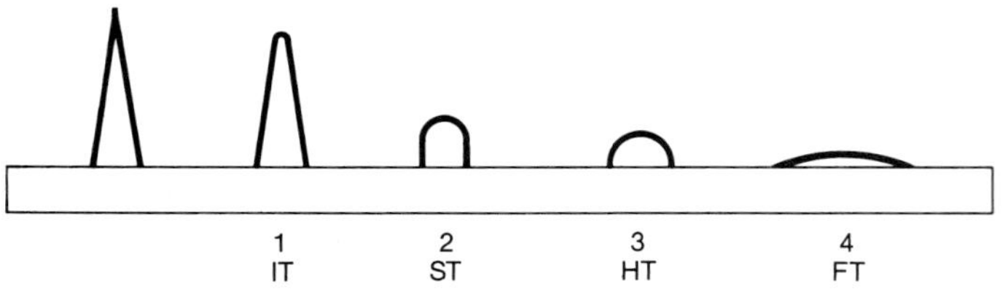

Fig. 5 Specific shapes as ash fuses and deforms with temperature.

temperature is higher than the melting point of most of the ash constituents and the fluid stage is reached.

Fusibility testing was originally developed to evaluate the clinkering (agglomerating) tendency of coal ash produced by combustion on a grate. In several respects, the test method is a somewhat better simulation of stoker firing than suspension burning of pulverized coal. During the fusion test, at a heating rate of 15F (8C) per minute, the transition from the IT to the FT stage may take up to two hours or more for a high fusion ash. Rather than slow heating and gradual melting of the ash, the process in a pulverized coal furnace is essentially reversed. Ash particles are rapidly heated, and then cooled at a relatively slow rate, as they pass through the furnace. During combustion, coal particles are heated almost instantaneously to temperatures ranging up to 3000F (1649C). As heat is removed from the flue gas, the ash is cooled over a period of less than two seconds to temperatures around 1900 to 2200F (1038 to 1204C) at the furnace exit.

In practical terms, for dry-bottom furnaces, fusion temperatures provide an indication of the temperature range over which portions of the ash will be in a molten fluid or semi-molten, plastic state. High fusion temperatures indicate that ash released in the furnace will cool quickly to a nonsticky state resulting in minimal potential for slagging. Conversely, low fusion temperatures indicate that ash will remain in a molten or plastic state longer, exposing more of the furnace surface or convective surface to potential deposition.

When temperatures in the furnace are below the measured initial deformation temperature, the majority of the ash particles are expected to be in a dry solid state. In this form, particles impacting on heating surface will bounce off and be re-entrained in the gas stream, or, at worst, settle on the surface as a dusty deposit which can be readily removed by sootblowers. At temperatures above the IT, the ash becomes increasingly more plastic in nature and impacting particles have a greater potential to stick to heating surfaces.

Fusibility temperatures also provide an indication of deposit characteristics as they relate to control and cleanability. When the temperature at a deposit surface is at or above the fluid temperature of the ash, slag will tend to flow or drip from the surface. While fluid slag can not be controlled with sootblowers, the deposits tend to be self-limiting in thickness and do not interfere significantly with heat transfer effectiveness. However, if the deposit surface temperature is in the plastic range, between the initial deformation and hemispherical temperatures, the slag will be too viscous to flow and will continue to build in thickness. Wide IT to HT differentials can result in deposits that build quickly to large proportions and are difficult to control, because sootblowers can be ineffective in penetrating the plastic shell that forms on the deposit surface.

In practice, very high and very low fusion values are relatively easy to interpret as being troublesome or non-troublesome with respect to slagging. Unfortunately, however, most coals fall in an intermediate range where evaluations can be much more difficult. Fusion temperatures have their most valid significance when used on a comparative basis against corresponding data from other fuels of known full-scale performance. Even comparisons can be misleading, however, when differences in data are within the range of reproducibility of the test. Actual ash viscosity measurements (described later) provide a much more accurate and less subjective definition of the viscosity/temperature relationship and are considered by B&W to provide a better assessment of slagging potential.

Influence of ash elements

Ash classification

Coal ash is classified into two categories based on its chemical composition. *Lignitic* ash is defined as having more ($CaO + MgO$) than Fe_2O_3. *Bituminous* ash is defined as having more Fe_2O_3 than the sum of CaO and MgO. Bituminous ash is generally characteristic of higher rank coals from the eastern U.S. Lower rank western coals typically have lignitic ash. As a result, bituminous ash is sometimes referred to as eastern ash and lignitic ash is sometimes referred to as western ash. However, ash classification is not specific to ASTM rank or geographical origin. In rare cases, lignites and subbituminous coals can have bituminous ash and bituminous coals can have lignitic ash. For example, the Utah coal shown in Table 3 is classified a bituminous, but has lignitic ash.

Effect of iron

Iron has a dominating influence on the slagging characteristics of coals with bituminous type ash. As shown in Table 2, iron can be present in coal in several mineral forms. These include pyrite (FeS_2), siderite ($FeCO_3$), hematite (Fe_2O_3), magnetite (Fe_3O_4) and ankerite [($Ca, Fe, Mg)CO_3$]. Pyrite is the major form of iron in most Eastern coals. In areas of the furnace where there is sufficient oxygen, pyrite is converted to Fe_2O_3 and SO_2. If the local atmosphere is reducing, however, pyrrhotite (FeS) is formed along with the lesser-oxidized iron forms such as FeO and metallic iron, Fe. The reduced forms have significantly lower melting temperatures than the oxidized forms. When completely oxidized to Fe_2O_3 iron tends to raise all four values of ash fusion temperatures: initial deformation, softening, hemispherical and fluid. In the lesser oxidized form (FeO) it tends to lower all of these values. The effect of iron in each of these forms is indicated in Fig. 6, plotted for a large number of ash samples from U.S. coals. The data show that as the amount of iron in the ash increases, there is a greater difference in ash fusibility between oxidizing and reducing conditions.

These effects may be negligible with coal ash containing small amounts of iron. Coals with lignitic ash generally have small amounts of iron and the ash fusion temperatures are affected very little by the state of iron oxidation. In fact, lignitic ash containing high levels of calcium and magnesium may have ash fusion temperatures that are lower on an oxidizing basis than on a reducing basis. The ash analysis and fusion temperatures shown for the subbituminous coal in Table 3 illustrate this effect.

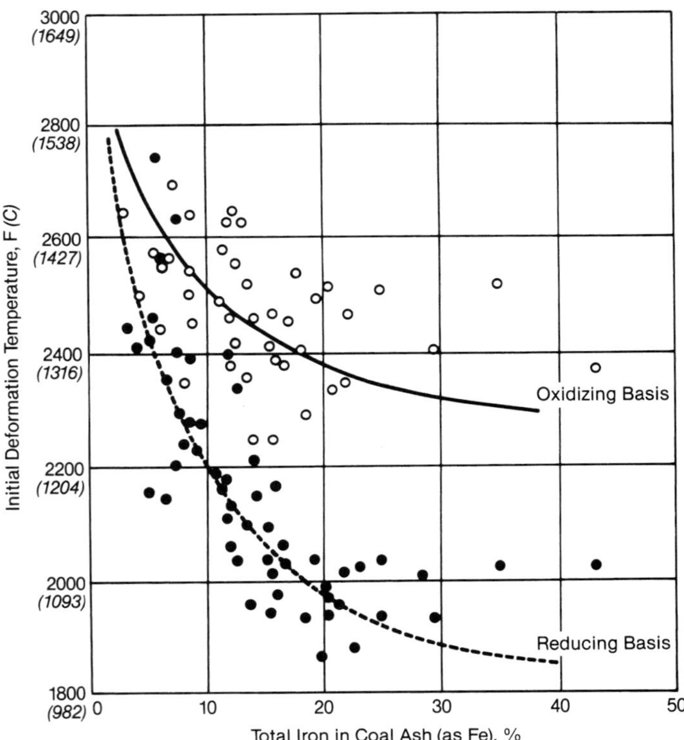

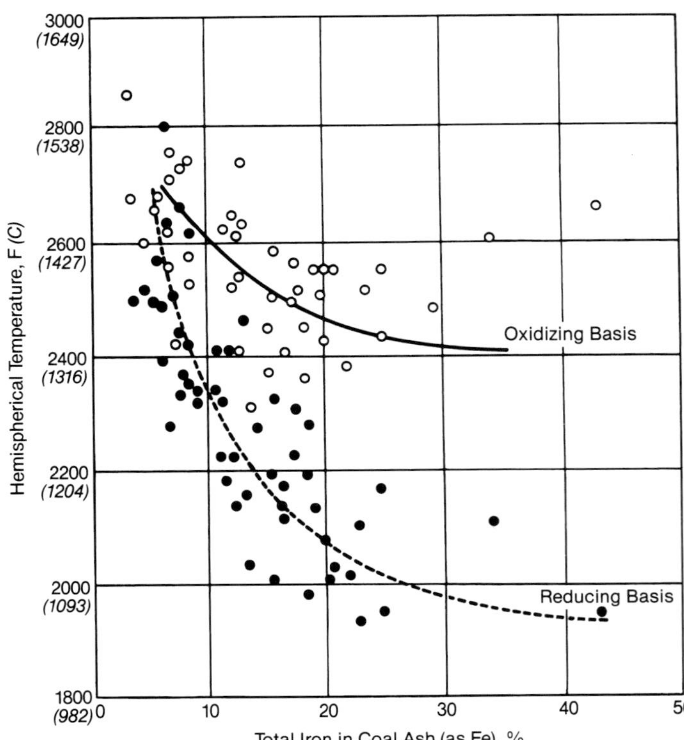

Fig. 6 Influence of iron on coal ash fusion temperatures.

Base to acid ratio

The constituents of coal ash can be classified as either basic or acidic. The basic constituents are iron, the alkaline earth metals calcium and magnesium, and the alkali metals sodium and potassium. Acidic constituents are silicon, aluminum and titanium. Bases and acids tend to combine to form compounds with lower melting temperatures. Experience has shown that the relative proportions of basic and acidic constituents provide an indication of the melting behavior and viscosity characteristics of coal ash.

The elemental analysis is used to calculate the percent base, percent acid and the base to acid ratio as follows:

$$\text{Percent base} = \frac{(Fe_2O_3 + CaO + MgO + Na_2O + K_2O) \times 100}{SiO_2 + Al_2O_3 + TiO_2 + Fe_2O_3 + CaO + MgO + Na_2O + K_2O} \quad (2)$$

$$\text{Percent acid} = \frac{(SiO_2 + Al_2O_3 + TiO_2) \times 100}{SiO_2 + Al_2O_3 + TiO_2 + Fe_2O_3 + CaO + MgO + Na_2O + K_2O} \quad (3)$$

$$\text{Base/acid ratio} = \frac{Fe_2O_3 + CaO + MgO + Na_2O + K_2O}{SiO_2 + Al_2O_3 + TiO_2} \quad (4)$$

The range of base to acid ratio extends from approximately 0.1 for highly acidic ash to 9.0 for ash that is high in base content.

Ash that is either highly acidic or highly basic generally has high ash fusion and melting temperatures. However, the presence of basic constituents in an acidic ash tends to flux or reduce the melting temperature and viscosity of the mixture. Conversely, the melting temperature and viscosity of a basic ash are reduced by relative proportions of acidic constituents. When the percent base and percent acid are nearly equal, fusion temperatures and ash viscosity tend to be reduced to minimum levels. The general trend is shown in Fig. 7. Minimum fusion temperatures typically occur at approximately 40 to 45% base which equates to base to acid ratios in the range of 0.7 to 0.8. Ratios in the range of 0.5 to 1.2 are generally considered to indicate high slagging potential.

The base to acid ratio considers all of the basic and acidic constituents to have equal effects on ash melting characteristics. However, research has shown that the various acids and bases have different fluxing strengths which must also be considered.

Studies conducted by B&W on the relationship of ash composition to ash viscosity have provided additional factors which improve the simple base to acid relationship. Ash viscosity is an important criterion for determining the suitability of a coal ash for use in a slag-tap furnace. Experience has shown that slag will flow readily at or below a viscosity of 250 poise. The temperature at which this viscosity occurs is called the T_{250} temperature of the ash. The preferred maximum T_{250} for wet-bottom applications is 2450F (1343C). Trends in T_{250} temperatures have been shown to correlate with ash fusion temperatures. Low T_{250} temperatures indicate low fusion temperatures and increased slagging potential.

Ash viscosity can be measured directly in a high temperature viscometer. Because viscosity measurements

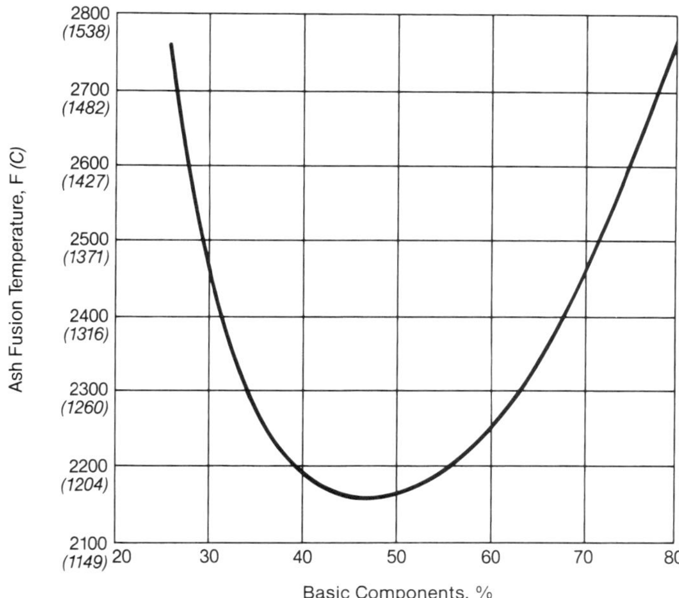

Fig. 7 Fusion temperatures and viscosities versus acidic constituents.

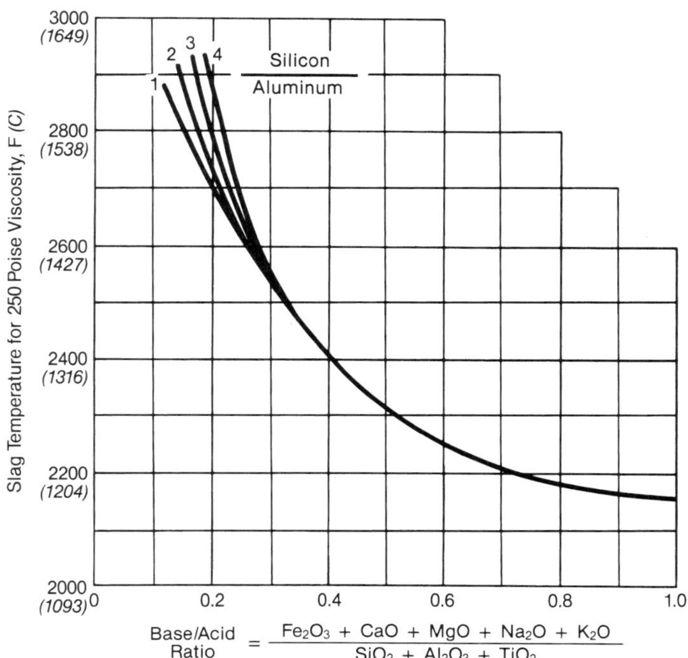

Fig. 8 Plot of temperature for 250 poise viscosity versus base to acid ratio – based on ferric percentage of 20.

require a considerable amount of coal ash that may not be readily available and are costly and time consuming, methods were developed to determine viscosity from chemical analysis of the coal ash. Based on a large number of direct viscosity measurements of bituminous and lignitic ash samples, T_{250} temperatures were related to ash composition as shown in Figs. 8 and 9. Fig. 8 is for bituminous ash and lignitic ash with an acidic content above 60%. At base to acid ratios less than 0.3, the silicon (SiO_2)/aluminum (Al_2O_3) ratio is taken into account. Silicon and aluminum are both acidic constituents; however, higher percentages of silicon tend to raise the T_{250} and the melting temperature.

Fig. 9 is for lignitic ash with an acidic content less than 60%. T_{250} is a function of both the percent base and the dolomite percentage which is defined as:

$$\text{Dolomite percentage} = \frac{(CaO + MgO) \times 100}{Fe_2O_3 + CaO + MgO + Na_2O + K_2O} \quad (5)$$

At a given percent base, higher dolomite percentages increase the T_{250} temperature, indicating that calcium and magnesium tend to raise ash viscosity and fusion temperature. Increasing amounts of the other base constituents (iron, sodium and potassium) tend to lower the T_{250} temperature.

Taken together, these trends indicate higher melting temperatures and higher viscosities at a given temperature for ash that is predominately composed of either silicon and aluminum or calcium and magnesium. Lower melting temperatures result from intermediate mixtures of these elements. However, in all combinations, iron, sodium and potassium act to flux the ash and increase the slagging potential.

As previously noted, the fluxing strength of iron is related to its state of oxidation. Metallic iron (Fe) and ferrous iron (FeO) are stronger fluxes than Fe_2O_3 and tend to reduce fusion temperatures and slag viscosity at a given temperature. The degree of iron oxidation is normally expressed as the ferric percentage where:

$$\text{Ferric percentage} = \frac{Fe_2O_3 \times 100}{Fe_2O_3 + 1.11FeO + 1.43Fe} \quad (6)$$

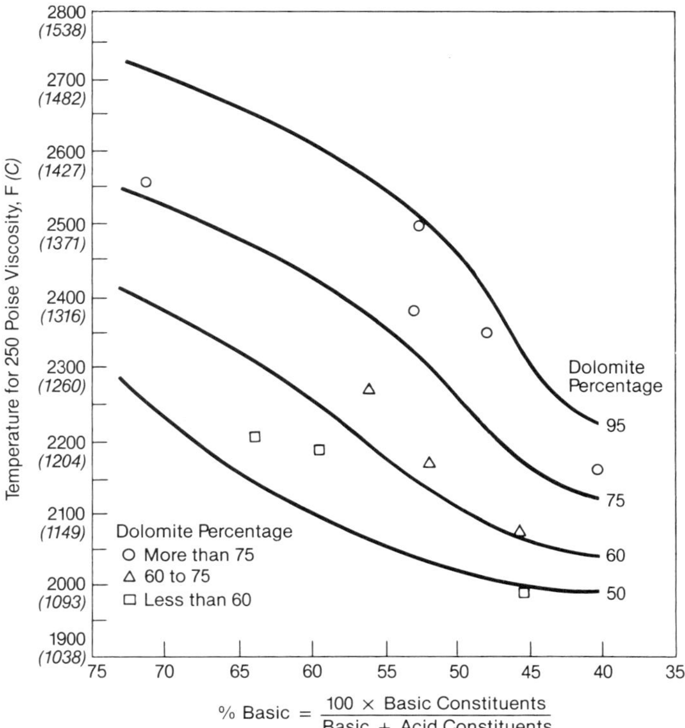

Fig. 9 Basic content and dolomite percentage of ash versus temperature for 250 poise viscosity.

The effect of ferric percentage on slag viscosity for a typical bituminous ash is shown in Fig. 10. Note that the T_{250} temperature can vary over a wide range depending on the degree of iron oxidation. Experience has shown that slag from boiler furnaces operating under normal conditions with 15 to 20% excess air has a ferric percentage of approximately 20%. The curves in Fig. 8 are based on this value.

Influence of alkalies on fouling

The alkali metals, sodium and potassium, have long been associated with the fouling tendencies of coal ash. Volatile forms of these elements are vaporized in the furnace at combustion temperatures. Subsequent reactions with sulfur in the flue gas and other elements in the ash form compounds that contribute to the formation of bonded deposits on convection heating surface.

Research conducted by B&W dating back to the 1950s identified a relationship between the total alkali content in bituminous coals and fouling potential. The specific laboratory procedure developed to establish this relationship, called the sintering strength test, is described in detail later in this chapter. Basically, the test involves measuring the compressive strength of flyash pellets heated in air for a period of time at temperatures of 1500 to 1800F (816 to 982C). The application of this method, combined with observations of fouling conditions in operating boilers, showed that high fouling coals produced flyash with high sintered strength. Conversely, low strength flyash was associated with low fouling coals. Correlation of standard ASTM ash analysis data with the sintering test results indicated a significant relationship (Fig. 11) between total alkali content (Na_2O and K_2O, expressed as equivalent total Na_2O) and flyash sintered strength. These correlations formed the basis for the first *fouling index* for bituminous coals which used the total alkali content in the coal to predict fouling potential.

Because ASTM ash produced in the laboratory could not be expected to represent the physical and chemical properties of flyash produced by full scale combustion, sintering strength testing required actual flyash samples aspirated from the flue gas in operating boilers. This meant full scale tests under steady-state conditions with a consistent coal supply, which became increasingly more difficult as unit size increased. To improve the efficiency and accuracy of obtaining data, a small laboratory ashing furnace (LAF) was constructed to burn pulverized coal at controlled conditions similar to those in a commercial boiler.

Subsequent tests on flyash produced in the LAF from a wide variety of bituminous coals demonstrated that sodium was the most important single factor affecting ash fouling. Potassium, which had been included in the previous alkali fouling indices, was found to make no significant contribution to sintering strength. Additionally, it was found that water soluble sodium, which was related to the more readily vaporized forms of sodium, had a major effect on sintered strength. This result was obtained by washing coals with hot condensate in the laboratory to remove the water soluble sodium. The washed coals were ashed in the LAF and sintered at various temperatures. Results for a high fouling Illinois coal are shown in Table 4. Water washing decreased the sodium content in the ash by approximately 70%, while the potassium content, which was initially higher than the sodium content, decreased by only 4%. Removing the soluble sodium resulted in a reduction in sintering strength at 1700F (927C) from 17,300 psi (119.3 MPa) for the raw coal to 550 psi (3.8 MPa) for the washed coal. Because the coal had a high chlorine content, it was concluded that most of the volatile sodium was probably in the form of NaCl. The insoluble potassium was likely associated with clay minerals or feldspar which would not readily decompose and vaporize during combustion.

The relationship of sintering strength to the percentage of soluble sodium in the ash was also found to be a function of the base to acid ratio, as shown in

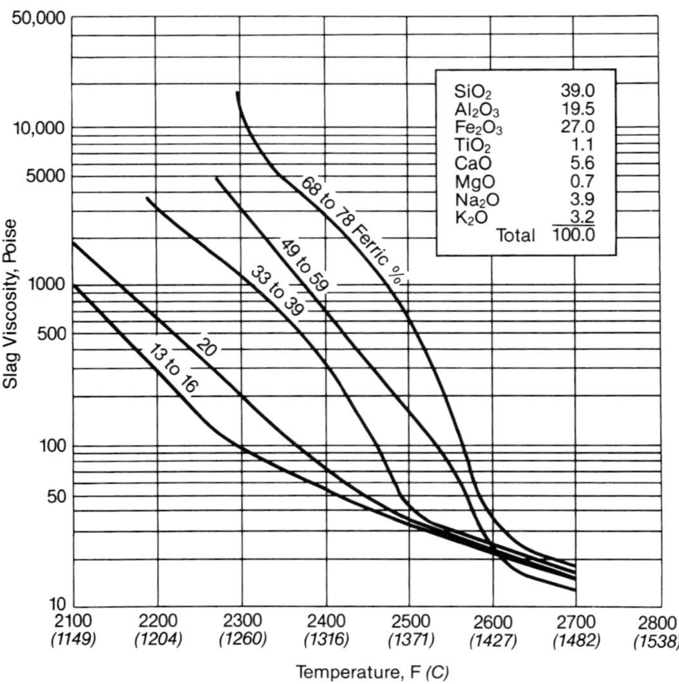

Fig. 10 Viscosity-temperature plots of a typical slag showing effect of ferric percentage.

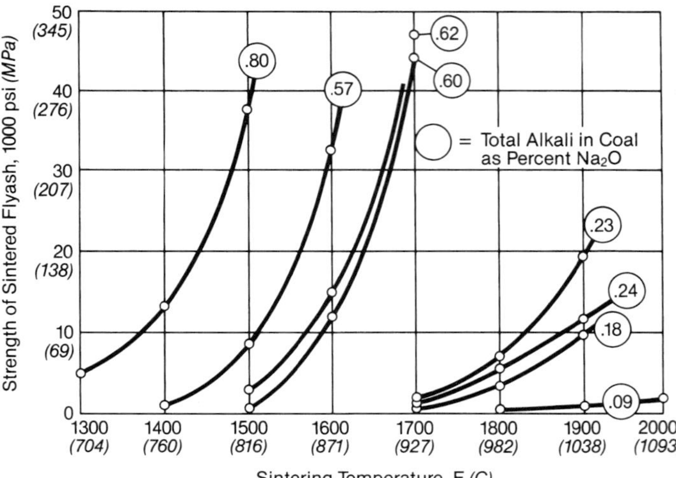

Fig. 11 Effect of alkali content in coal.

Table 4
Effect of Soluble Sodium on Sintered Strength

Ash Analysis	Raw Coal	Washed Coal
SiO_2	45.0	49.8
Al_2O_3	18.0	20.9
Fe_2O_3	21.0	22.9
TiO_2	0.8	1.0
CaO	8.8	1.6
MgO	0.9	1.0
Na_2O	1.6	0.5
K_2O	2.4	2.3
Ash sintered strength, psi	17,300	550
(MPa)	(119.3)	(3.8)

Fig. 12. The combination of high sodium and high base to acid ratios resulted in the highest sintering strengths. Low ratios and sodium contents resulted in reduced flyash strength at the same sintering temperature. Similar trends were noted for variations in sintering strength as a function of base to acid ratio and total Na_2O in the ash. Statistical evaluations of these relationships were used to develop the fouling index currently used for coals with bituminous ash.

Similar tests on the sintering characteristics of lignitic ash indicated that the sintering criteria associated with fouling for bituminous ash did not apply to lignitic ash with high alkaline (CaO, MgO) contents. However, sintering strength was found to be directly proportional to the total sodium content in the ash shown in Fig. 13. Full scale and pilot scale tests conducted by the U.S. Bureau of Mines at the Grand Forks Coal Research Laboratory in North Dakota also established a correlation between fouling rate and sodium content for coals with lignitic ash. As shown in Fig. 14, deposition rates were found to increase sharply as the Na_2O content increased up to approximately 6% and then level off at higher percentages of sodium.

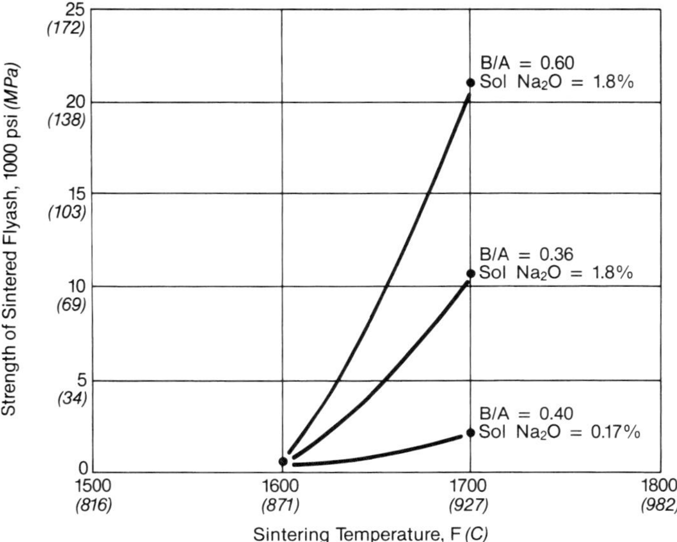

Fig. 12 Bituminous (Eastern) ash fouling effect of base to acid ratio and soluble sodium on sintered strength.

As previously noted, in low rank coals, a major portion of the alkali and alkaline earth metals can be organically bound in the coal. Because they are intimately mixed with the coal, it is believed that alkalies in this form are readily vaporized during combustion and play a dominating role in fouling. The organically associated elements occur in the form of cations chemically bonded to the organic structure of the coal. Ion exchange techniques have been developed to remove the cations from the coal for measurement. The method employed by B&W uses an ammonium acetate solution to provide a source of NH_4^+ ions which extract the ion-exchangeable cations. The laboratory procedure is described later in this chapter. Ion exchange data for a high fouling North Dakota lignite and a severe fouling Montana subbituminous coal are shown in Table 5. The data show that essentially all of the sodium in both coals is organically bound. In the lignite, the ion-exchangeable sodium actually exceeded the total sodium measured in ASTM ash. The difference most likely results from a loss of sodium due to vaporization during the high temperature ashing procedure. The relatively low percentages of ion-exchangeable K_2O indicate that most of the potassium exists in stable mineral forms.

Viscosity-temperature relationship of coal ash

The characteristics of slag deposits which form on furnace walls and other radiant surface are a function of deposit temperature and deposit composition.

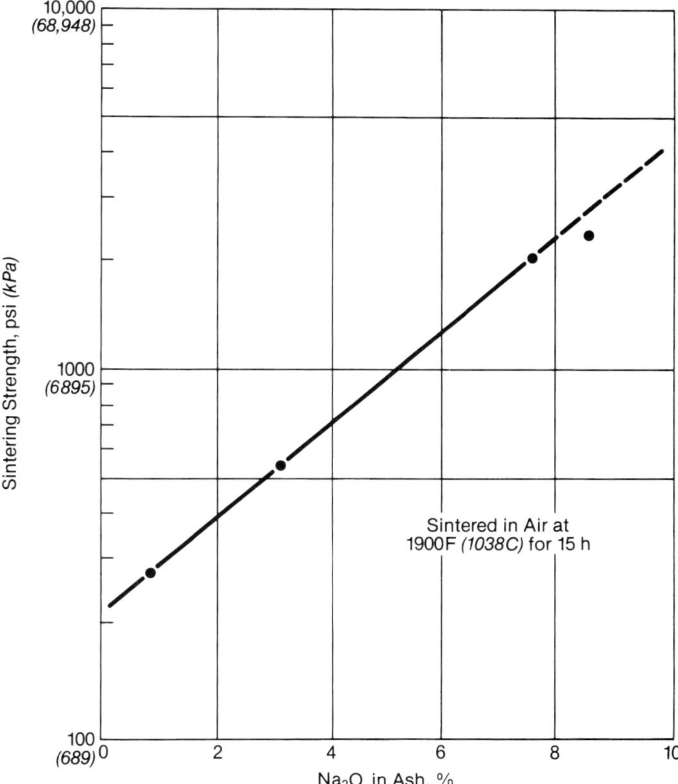

Fig. 13 Effect of Na_2O on sintering strength (North Dakota lignite ash).

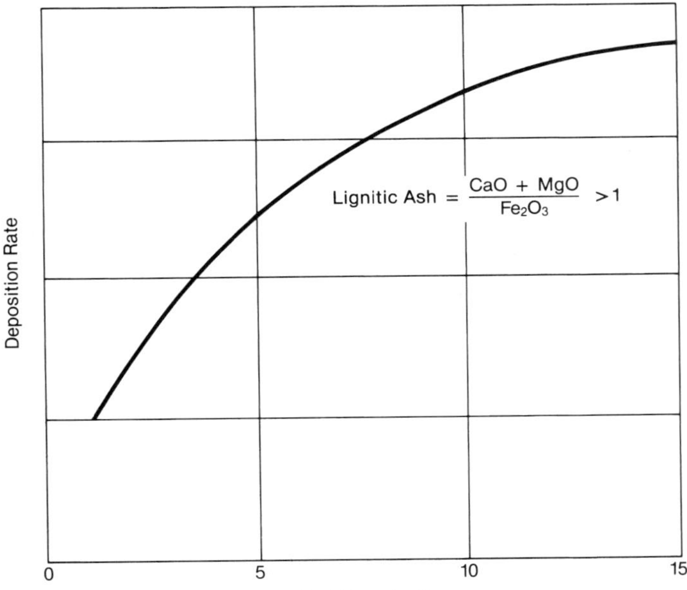

Fig. 14 Effect of Na_2O on deposition rate.

Deposit composition, in turn, is a function of the local atmosphere, particularly for ash with a significant iron content. Relationships between these factors determine the physical state of the deposit, which can range from a dry solid to plastic or even a viscous liquid if temperatures are sufficiently high. Dry deposits are usually not troublesome; they tend to be loosely bonded to the tube surface and relatively easy to remove by sootblowing. If deposits are allowed to build in thickness, the temperature increases and the surface of the deposit can become semi-molten or plastic. The plastic slag traps other transient ash particles and continues to build more and more rapidly as the surface temperature continues to increase. Ultimately, the deposit reaches an equilibrium state as the slag begins to flow.

Field experience has shown that plastic slag tends to form large deposits that are highly resistant to removal by conventional ash cleaning equipment. This observation led to an extensive study of the relationship between ash viscosity and potential slagging tendency. Viscosity measurements that had previously been used to determine flow characteristics for wet-bottom furnace applications were extended to higher viscosity ranges to define the temperature range where a given ash would exhibit plastic characteristics.

As liquid ash is cooled, the logarithm of its viscosity increases linearly with decreasing temperature as shown in Fig. 15. At some point, the progression deviates from the linear relationship, and viscosity begins to increase more rapidly as the temperature continues to decrease. This transition into the plastic region is caused by the selective separation of solid material from the liquid, resulting from crystallization of the higher melting point constituents of the ash. The temperature at which this deviation takes place is called the temperature of critical viscosity (T_{cv}). T_{cv} varies depending on ash composition but normally occurs in a range between 100 and 500 poise. The end of the plastic region is the point of solidification, or freeze point, of the slag. The freeze point typically occurs at a viscosity of approximately 10,000 poise. For convenience in comparing the viscosity-temperature relationship of various ashes, the viscosity range of 250 to 10,000 poise has been defined as the plastic region.

The temperature at which the plastic region begins and the range of temperature over which the ash is plastic provide an indication of the slagging tendency. The lower the temperature within this range and the wider the range, the greater the potential for slagging. Viscosity-temperature curves for a high slagging Illinois coal and a low slagging east Kentucky coal, shown in Fig. 16, illustrate this effect. The plastic range for the Illinois coal begins at a relatively low temperature and extends over a wide temperature range. In contrast, the east Kentucky coal has a very narrow plastic range which begins at a much higher temperature. In comparison to the Illinois coal, the Kentucky coal ash would be expected to cool quickly below the temperature where the ash is plastic, exposing much less of the furnace to potential deposition.

As previously noted, the iron content of coal ash and its degree of oxidation have a significant influence on

Table 5
Ion Exchange Data — High and Severe Fouling Coals

	Source:	North Dakota	Montana
	Rank:	Lignite	Subbituminous
Ash, dry basis, %		11.2	5.4
Total alkali, dry coal basis, %	Na_2O	4.25	6.74
	K_2O	0.37	0.65
Ion exchangeable alkali, dry coal basis, %	Na_2O	4.52	6.37
	K_2O	0.10	0.13
Relative ion exchange alkali, %	Na_2O	106%	95%
	K_2O	27%	20%

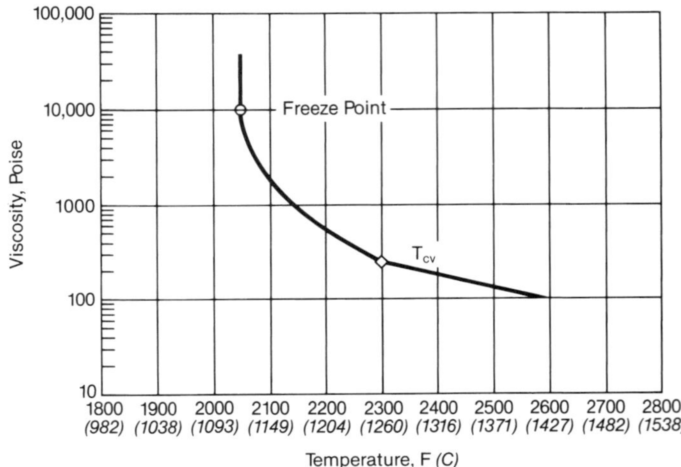

Fig. 15 Viscosity increase with decreasing temperature.

the viscosity of the ash. This effect is illustrated in Fig. 17 which shows the viscosity-temperature relationship for the high iron Illinois coal under both oxidizing and reducing conditions. Under reducing conditions, the viscosity at a given temperature is significantly lower and the ash remains plastic over a much wider temperature range.

Ash reflectivity

Ash from certain coals produces furnace deposits that have reflective rather than insulating properties. This is particularly true of low sulfur, low sodium coals found in the western U.S., from the Powder River Basin in Wyoming and Montana. Reflective deposits can significantly reduce furnace heat absorption and increase furnace exit gas temperature even when only a very thin deposit is present. This can result in excessive radiant superheater slagging and fouling of convection surfaces. Experience has shown that reflective ash deposits can be difficult to remove and require special considerations in selection of ash cleaning equipment and media. (See Chapter 24.) Proprietary methods based upon field experience and laboratory studies are used to evaluate the potential for reflective ash formation, and to address the impact on furnace design and boiler performance.

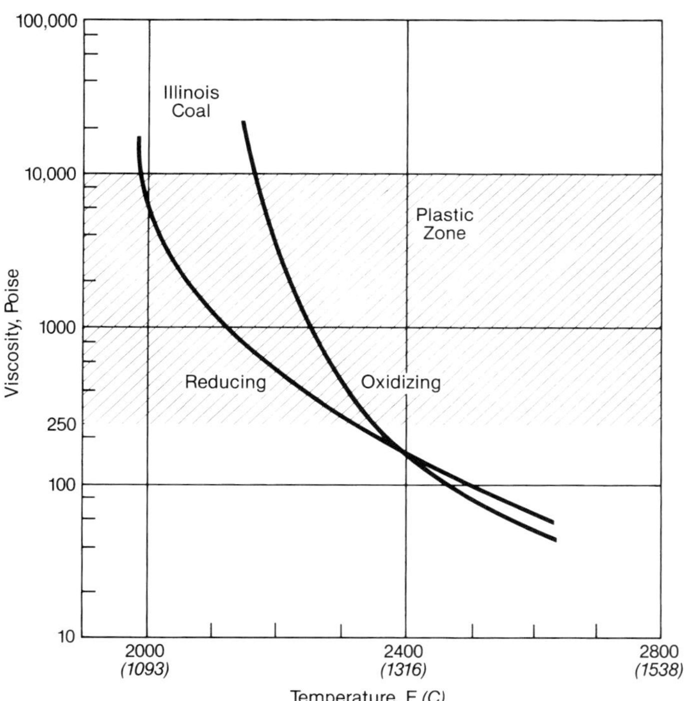

Fig. 17 Ash viscosity comparison – oxidizing and reducing conditions.

Ash characterization methods

Several slagging and fouling indices have been developed by B&W to provide criteria for various aspects of boiler design. Slagging indices establish design criteria for the furnace and other radiant surface while fouling indices establish design criteria for convective surface. Deposition characteristics are generally classified into four categories: low, medium, high and severe.

For the most part, the indices described below are based on readily available ASTM ash analysis and fusibility data. In actual practice, when evaluating coals, designers take into account full scale experience on similar fuels and results of non-routine testing which can, in some cases, modify the classification. These indices can also be used on a comparative basis to rank coals with respect to their slagging and fouling potential when evaluating a new coal supply for an existing unit.

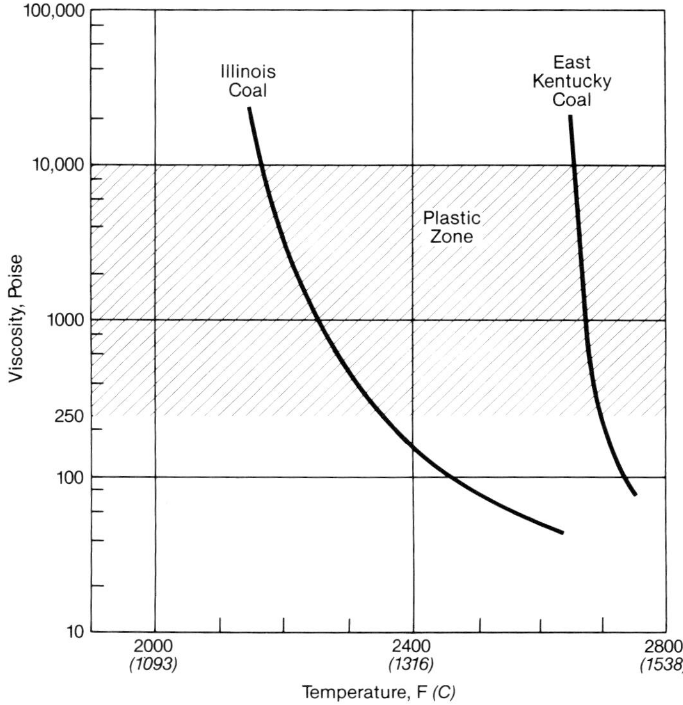

Fig. 16 Ash viscosity comparison for a high slagging and low slagging coal (oxidizing atmosphere).

Ash classification

Because the characteristics of bituminous and lignitic ash vary significantly, the first step in calculating slagging and fouling indices is the determination of ash type. In accordance with the criteria previously described, ash is classified as bituminous when:

$$Fe_2O_3 > CaO + MgO \qquad (7)$$

Ash is classified as lignitic when:

$$Fe_2O_3 < CaO + MgO \qquad (8)$$

Slagging index – bituminous ash (R_s) Calculation of the slagging index (R_s) for bituminous ash takes into account the base to acid ratio and the weight percent, on a dry basis, of the sulfur in the coal. The base to acid ratio indicates the tendency of the ash to form compounds with low melting temperatures. The sulfur content provides an indication of the amount of iron that is present as pyrite. The calculation is as follows:

$$R_s = \frac{B}{A} \times S \qquad (9)$$

where

B = $CaO + MgO + Fe_2O_3 + Na_2O + K_2O$
A = $SiO_2 + Al_2O_3 + TiO_2$
S = weight % sulfur, on a dry coal basis

Classification of slagging potential using R_s is as follows:

$R_s < 0.6$ = low
$0.6 < R_s < 2.0$ = medium
$2.0 < R_s < 2.6$ = high
$2.6 < R_s$ = severe

Slagging index – lignitic ash (R_s*) The slagging index for lignitic ash (R_s*) is based on ASTM ash fusibility temperatures. As previously noted, fusibility temperatures indicate the temperature range where plastic slag is likely to exist. The index is a weighted average of the maximum hemispherical temperature (HT) and the minimum initial deformation temperature (IT) as follows:

$$R_s^* = \frac{(\text{Max HT}) + 4\,(\text{Min IT})}{5} \quad \textbf{(10)}$$

where

Max HT = higher of the reducing or oxidizing hemispherical softening temperatures, F
Min IT = lower of the reducing or oxidizing initial deformation temperatures, F

Classification of slagging potential using R_s* is as follows:

$2450 < R_s^*$ = low
$2250 < R_s^* < 2450$ = medium
$2100 < R_s^* < 2250$ = high
$R_s^* < 2100$ = severe

Slagging index – viscosity (R_{vs}) As previously noted, B&W's most accurate method for predicting slagging potential is based on the viscosity-temperature relationship of the coal ash. This index (R_{vs}) is applicable to both bituminous and lignitic ash coals; however, measured ash viscosities are required.

$$R_{vs} = \frac{\left(T_{250\,oxid}\right) - \left(T_{10{,}000\,red}\right)}{97.5\left(fs\right)} \quad \textbf{(11)}$$

where

$T_{250\,oxid}$ = temperature, F, corresponding to a viscosity of 250 poise in an oxidizing atmosphere
$T_{10{,}000\,red}$ = temperature, F, corresponding to a viscosity of 10,000 poise in a reducing atmosphere

and fs is a correlation factor based on the average of the oxidizing and reducing temperatures (T_{fs}) corresponding to a viscosity of 2000 poise. Values for fs as a function of T_{fs} are provided in Fig. 18.

Classification of slagging potential using R_{vs} is as follows:

$R_{vs} < 0.5$ = low
$0.5 < R_{vs} < 1.0$ = medium
$1.0 < R_{vs} < 2.0$ = high
$2.0 < R_{vs}$ = severe

Fouling index – bituminous ash (R_f) The fouling index for bituminous ash is derived from sintering strength characteristics using the sodium content of the coal ash and the base to acid ratio as follows:

$$R_f = \frac{B}{A} \times Na_2O \quad \textbf{(12)}$$

where

B = $CaO + MgO + Fe_2O_3 + Na_2O + K_2O$
A = $SiO_2 + Al_2O_3 + TiO_2$
Na_2O = weight % from analysis of coal ash

Classification of fouling potential using R_f is as follows:

$R_f < 0.2$ = low
$0.2 < R_f < 0.5$ = medium
$0.5 < R_f < 1.0$ = high
$1.0 < R_f$ = severe

Fouling index – lignitic ash The fouling classification for lignitic ash coals is based on the sodium content in the ash as follows:

When $CaO + MgO + Fe_2O_3 > 20\%$ by weight of coal ash

$Na_2O < 3$ = low to medium
$3.0 < Na_2O < 6$ = high
$Na_2O > 6$ = severe

When $CaO + MgO + Fe_2O_3 < 20\%$ by weight of coal ash

$Na_2O < 1.2$ = low to medium
$1.2 < Na_2O < 3$ = high
$Na_2O > 3$ = severe

Coal ash effects on boiler design

Furnace design

The key to a successful overall gas-side design is proper sizing and arrangement of the furnace. As a first priority, the furnace must be designed to minimize slagging and to provide effective control of slag where and when it does form.

Ash deposition in the furnace can cause a number of problems. Slag deposits reduce furnace heat absorption and raise gas temperature levels at the furnace exit. This, in turn, can cause slagging and can aggravate fouling in the convection banks where ash deposits become increasingly more difficult to control as gas temperatures increase. The shift in heat absorption from the furnace to the superheater and reheater results in increased attemperator spray flow for control of steam temperatures, reducing cycle efficiency. Slag buildup at the top of a tall furnace is dangerous. Large deposits can become dislodged and fall, causing failures of furnace hopper tubes and loss of availability. Excessive slagging in the lower furnace can interfere with ash removal.

Experience has shown that several interrelated furnace design parameters are critical for slagging control. These parameters focus on keeping ash particles in suspension and away from furnace surfaces, distributing heat evenly to avoid high localized temperatures, and removing enough heat to achieve tempera-

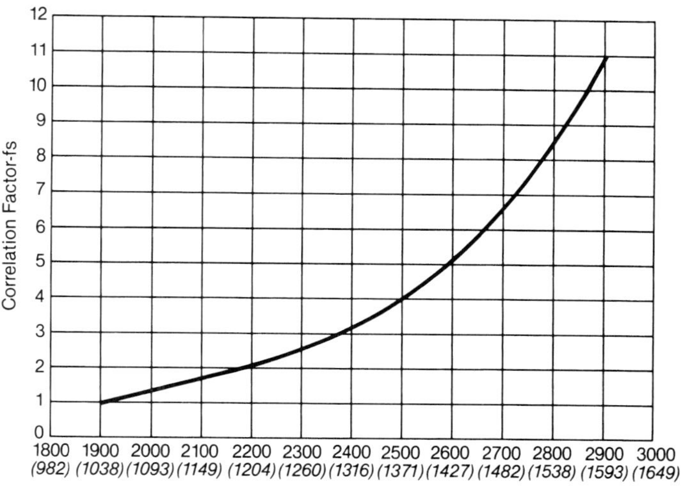

Fig. 18 Slagging index correction factor fs.

tures at the furnace exit that will minimize deposition on convection surface.

In the context of gas-side design, the furnace basically serves three functions. It must provide sufficient volume to completely burn the fuel, provide sufficient heat transfer surface to cool the flue gas and ash particles to a temperature suitable for admission to the convection surface, and minimize the formation of NO_x emissions (see Chapter 34). In general, for a coal-fired unit, it is the second criterion that determines the minimum furnace size.

The slagging classification of the coal establishes the upper limit on furnace exit gas temperature (FEGT) required to minimize the potential for slagging both in the radiant superheater and the close-spaced convection surface. As described in Chapter 22, furnace exit gas temperature is a function of furnace heat release rate. Limiting the FEGT, therefore, limits the heat release rate, resulting in lower average temperatures in the furnace. FEGT limits and corresponding heat release rates have been established by experience for different types of coal. In general, units using coals with low or medium slagging tendencies can have higher heat release rates and higher FEGTs. Units firing coals with high or severe slagging potential require lower heat release rates and lower FEGTs.

Ideally, the furnace would be an open box, sized with sufficient wall surface to cool the furnace gas and ash particles to the desired temperature before they reached any superheater surface. However, thermodynamic considerations in modern high pressure and high temperature cycles require that a significant portion of the total heat absorption be accomplished in the superheater and reheater. This requirement places a practical limit on the amount of furnace wall surface which, in a drum boiler, is dedicated to generating saturated steam. In order to achieve the required FEGT it becomes necessary to replace water-cooled furnace wall surface with steam-cooled superheater surface. These surfaces are generally in the form of widely spaced platens (see Chapter 19) located in the upper radiant zone of the furnace. Because platen surface is located in a relatively high gas temperature zone and subject to ash particle impaction, the side spacing must be sufficient to limit the potential for bridging and provide a degree of self-cleaning. Typical side spacing between platen sections is 4 to 5 ft (1.2 to 1.5 m). When platen superheater surface is used, the slagging classification of the coal establishes the upper limit on platen inlet gas temperature, in addition to limiting the FEGT.

An alternate method of controlling furnace exit gas temperature that has been widely used is gas tempering by flue gas recirculation. In this method, relatively cool gas from the economizer outlet is mixed with hot furnace gas near the furnace exit. Gas tempering offers a number of advantages. The FEGT can be limited with less furnace surface while the increased gas weight improves the thermal head for heat transfer, reducing the surface requirements in the convection pass. Proper introduction of the tempering flue gas provides a flat temperature profile at the furnace exit, reducing the possibility of localized slagging and fouling. Once the choice of gas recirculation is made, the system can also be used to control reheat steam temperature at partial loads. For this purpose, flue gas from the economizer outlet is introduced into the furnace through the furnace hopper opening. The cool gas reduces furnace heat absorption and makes more heat available to the reheater which offsets its natural characteristic of decreasing outlet steam temperature at partial loads.

The major disadvantages of gas recirculation are fan maintenance and power requirements. Fan erosion can be minimized to some extent by proper design and operation of a mechanical dust collector ahead of the fan. Extracting the recirculated gas after a hot precipitator offers the best potential for a relatively clean recirculated gas source.

In addition to having sufficient volume and heating surface, the furnace also must be correctly proportioned with respect to width, depth and height to minimize slagging. A significant design parameter in this regard is heat input from fuel to the furnace per unit of furnace plan area at the burners. Maximum limits on plan area heat release rate are a function of the slagging potential of the coal. Limits typically range from 1.5 to 1.8×10^6 Btu/h ft^2 (4.7 to 5.7 MW$_t$/m^2) for severe slagging and low slagging coals respectively.

The furnace must also be designed to limit the potential for ash particle impaction on furnace surfaces. Ample clearance must be provided between the burners and furnace walls as well as the furnace hopper and arch. These critical dimensions have been established by operating experience and keyed to the slagging classification of the coal.

The slagging classification also determines the locations, quantity and spacing of furnace wall blowers and long retractable sootblowers in the pendant radiant surface. (See Chapter 24.) These allow control of the deposition that inevitably occurs and are essential for maintaining furnace surface effectiveness and furnace exit gas temperature within the range provided for in the design. Some degree of slagging may be permitted above the burner zone but only to the extent that it can be controlled by selective operation of the wall blowers. The control of these deposits can help maintain steam temperature at reduced loads. Slag deposits on furnace walls must be avoided below and between burners, however, where they can not be controlled by sootblowers.

Effect of slagging potential on furnace sizing

Referring to Fig. 19, three large utility boilers are shown sized for 660 MW at maximum continuous load. The boilers are assumed to have the same width for purposes of illustration, with the boiler setting height and furnace depth varied to accommodate the slagging characteristics of the different fuels. Boiler (a) is designed to fire a bituminous coal having a low to medium slagging potential. The slightly larger boiler (b) is designed to fire a subbituminous coal classified as having a high slagging potential. The difference in size can be attributed primarily to the difference in slagging potential. The furnace (b) depth has been increased to control slagging by reducing the input per plan area. The input and gas weight are higher for the subbituminous coal due to its higher moisture content and resulting lower boiler efficiency. This increases the required furnace surface and the furnace exit area to maintain acceptable gas velocities entering the convection pass. Comparing boiler (c), firing a severe slagging lignite, to boiler (b), the furnace depth has again been increased due to the increased slagging potential. The furnace surface has also been increased to reduce the gas temperature leaving the furnace. The size differential of the three units is quantified in Table 6. This table shows the proportionate differences or increases using boiler (a) as a base. Boiler (a) is assigned a size factor of 1.0 for the various parameters shown.

Convection pass design

The key to successfully preparing a design that will control convection pass fouling reverts back to a furnace design that will maintain the furnace exit gas temperature at predicted levels. Temperature excursions at the furnace exit result in corresponding higher temperature levels throughout the convection pass which can cause deposition problems even with coals which normally would be considered to have a low or moderate fouling tendency.

In general, convective heating surface, both pendant and horizontal, is arranged to minimize the potential for bridging and obstruction of the gas lanes between adjacent sections. The minimum clear side spacing (measured perpendicular to the gas flow) between sections in a bank varies as a function of the average flue gas temperature entering the bank. The widest spacing is required in the superheater banks which are in close proximity to the furnace exit, where the gas temperature and fouling potential are high. As the flue gas temperature is reduced, the side spacing in succeeding banks can also be reduced. The specific side space dimensions at a given temperature entering the bank depend on the fouling classification of the coal. Severe fouling coals require the widest spacing. Adequate side spacing must be maintained even in low temperature horizontal banks such as economizers. While these surfaces are not normally subject to bonded deposits, sufficient clear space must be maintained between sections to ensure that accumulations of ash dislodged from upstream surfaces will not bridge and plug the gas lanes. (See Chapter 20.)

Bank depths (measured parallel to the direction of gas flow) are established as a function of fouling potential, clear side spacing and the temperature enter-

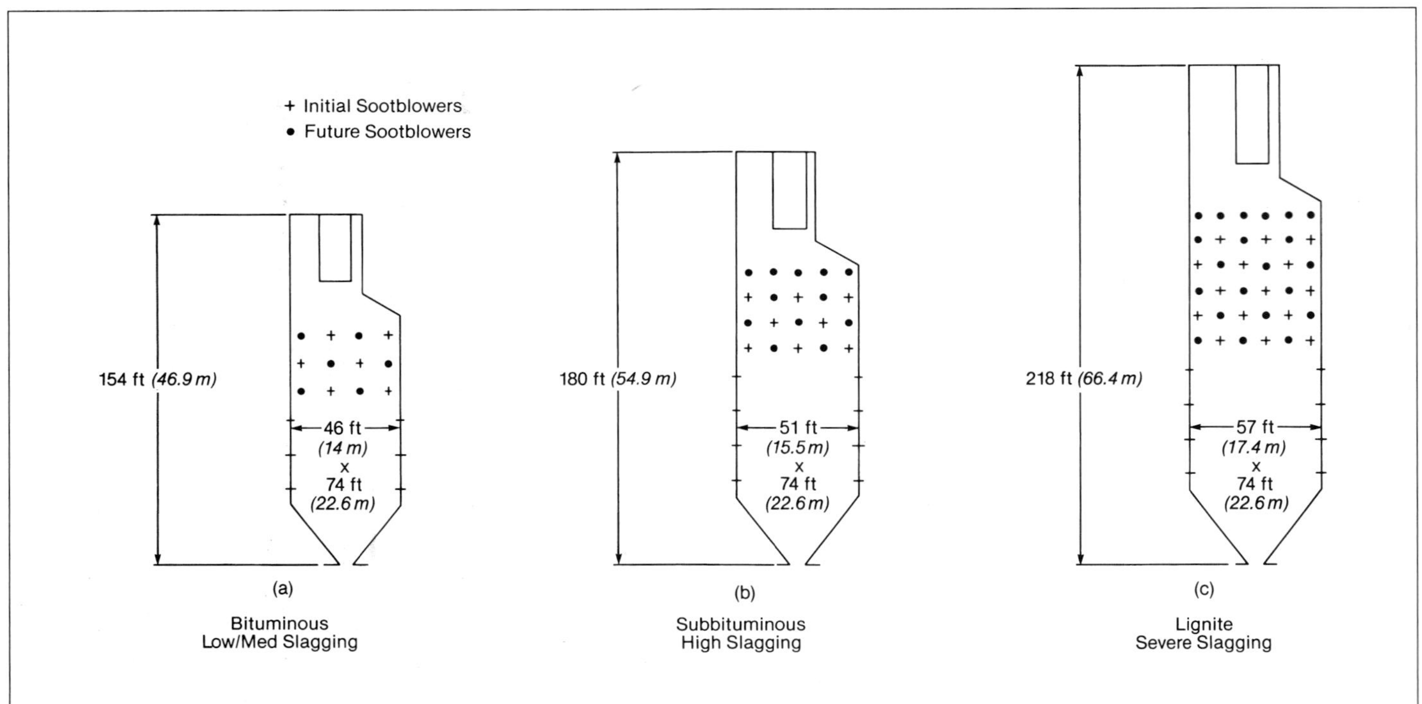

Fig. 19 Influence of slagging potential on furnace size. (See Table 6.)

Table 6
Boiler Size Versus Slagging Classification

	Boiler		
	(a)	(b)	(c)
Coal Rank	Bituminous	Subbituminous	Lignite
Slagging	Low/Med	High	Severe
Furnace plan area	1.0	1.11	1.24
Furnace surface	1.0	1.18	1.50
No. of furnce wall blowers	30	36	70

ing the bank. Cavities between the banks provide locations for long retractable sootblowers. At high gas temperatures, shallow bank depths are required to ensure adequate sootblower effectiveness. Sootblower jet penetration increases as temperatures are reduced and bank depths can be increased incrementally in cooler areas.

Flyash erosion

The metal loss on convection pass tubes due to flyash erosion is proportional to the total ash quantity passing through the boiler and is an exponential function of flue gas velocity. While with a given fuel there is no control of the ash quantity, erosion problems can be eased by reducing flue gas velocities. Velocity limits are determined based on the ash quantity on a pounds per million Btu (kg/MW_t) basis and the relative proportion of abrasive constituents in the ash. Typical limits range from 65 ft/s (19.8 m/s) for relatively non-abrasive low ash coals to 45 ft/s (13.7 m/s) or less for coals with high ash quantities and/or abrasive ash.

Effect of operating variables

Although the predominant factors affecting deposition are ash characteristics and boiler design, operating variables can also have a significant impact on slagging and fouling.

In general, operating variables associated with combustion optimization (see Chapter 14) tend to reduce the potential for deposition problems. These variables include air distribution, fuel distribution, coal fineness and excess air.

Air and fuel imbalances can result in high excess air at some burners while others operate with less than theoretical air. This, in turn, results in localized reducing conditions in the burner zone which can aggravate slagging, especially with coals having high iron content. High coal/air ratios can also delay combustion and upset heat distribution, resulting in elevated temperatures in the upper furnace and at the furnace exit. Long burnout times also increase the potential for burning particles to contact furnace walls and other heat transfer surfaces.

Secondary air imbalances can be minimized by adjusting individual burner flows to provide a flat O_2 profile at the economizer outlet. Care must be exercised to avoid burner adjustments that cause flame impingement on furnace walls. On the fuel side, burner line resistances should be balanced to maintain uniform coal flow to each burner. Coal feeders should be calibrated and adjusted to provide uniform coal flow to each pulverizer.

Low pulverizer fineness (see Chapter 13) can also cause problems associated with delayed combustion. Coarse particles require longer residence times for burnout and can cause slagging in the lower furnace.

Excess air has a tempering effect on average temperatures within the furnace and on furnace exit temperature. Excess air also reduces the potential for localized reducing conditions in the furnace when it is introduced through the burners. Air infiltration into the furnace or convection pass is far less beneficial and should be corrected or taken into account when establishing excess air requirements. While there is an associated efficiency loss, raising excess air above normal design levels is usually an effective tool for controlling deposition problems. In some cases, high excess air may also upset superheater/reheater absorption and steam temperatures.

Sootblowers (see Chapter 24) are the primary means of dealing directly with furnace wall slagging and convection pass fouling. The most important fundamental requirement is to use this equipment in a preventive, rather than corrective, manner. Sootblowers are most effective in controlling dry, loosely bonded deposits which typically occur in the early stages of deposition. If furnace slag is allowed to accumulate to the point that it becomes plastic or wet, or if convection pass deposits are allowed to build and sinter for long periods of time, removal becomes much more difficult. Sootblower sequencing requirements must be established by initial operating experience and updated when required, especially when fuel characteristics change. Boiler diagnostic systems, which are discussed in the following section, can assist in optimizing sootblower operation.

The least desirable operating technique for controlling deposition problems is load reduction. The most severe situations may require a permanent derate. However, in many marginal situations, temporary load reductions during off peak periods may provide sufficient cooling to shed slag and allow sootblowers to regain effectiveness.

Application of advanced diagnostic and control systems

Awareness of slagging and fouling conditions is critical to achieving reliability and availability on a coal-fired utility boiler. However, boiler surface cleanliness has been, traditionally, one of the most difficult operating variables to quantify. Typical indications of surface fouling appear to the operator indirectly in the form of steam temperatures, spray attemperation flows and draft losses (gas resistance). In some cases, experienced operators who are familiar with the operating characteristics of a unit can make judgments on slagging and fouling conditions based on operating conditions, but these secondary indications can be misleading. For example, the furnace can be slagged, causing undesirably high gas temperatures entering the convection surface. However, the steam temperatures and spray attemperation may be normal if the convection surfaces are also fouled.

Another indication of surface cleanliness is draft loss. By watching draft loss across a bank, an alert operator can determine that sootblowing is probably required. Usually, however, by the time a change in draft loss is detected across widely spaced pendant sections, the banks are already bridged and it may be too late for removal by the sootblowers.

Visual observation is frequently used to further quantify cleanliness conditions. In many instances, however, access is limited and subjective evaluations can leave considerable room for error. Advanced methods have been developed to overcome these shortcomings and to improve upon traditional time-based sootblowing control.

Computer based performance monitoring systems can provide a direct and quantitative assessment of furnace and convective surface cleanliness. B&W's Heat Transfer Manager™ (HTM) program is based on the heat transfer analysis program developed over many years for boiler design and validated by extensive empirical data. The HTM program is configured on a boiler-specific basis, taking into account the arrangement of the furnace and all convective surface. Measurements of temperatures, pressures, flows, and gas analysis data are used to perform heat transfer analysis in the furnace and convective section on a bank by bank basis.

Advanced intelligent sootblowing systems have also been developed to combine this real time assessment of furnace and convective surface cleanliness with closed loop control of the cleaning equipment. B&W's Powerclean™ system automatically determines where and when sootblowing should occur in the furnace and convection pass. Powerclean uses cleanliness data from the HTM program in an expert decision making structure that dictates when blowers should be cycled.

Intelligent systems such as Powerclean recognize problem areas early in their development, so that selective sootblowing can be directed at a specific problem area and ash cleaning equipment is operated based on need. Intelligent sootblowing systems can optimize blowing medium use and improve performance while reducing tube damage and providing consistency to boiler operations.

Slagging can be particularly troublesome in localized areas of the furnace. To help optimize wall cleaning, heat flux sensors can be used. These sensors are installed in the waterwalls of the furnace and provide a differential temperature across the wall which changes in proportion to the amount of deposition. Sensor data is integrated into the overall intelligent sootblowing system so that cleanliness can be optimized in the furnace region. If an array of sensors is installed, the furnace can be broken into regions for better control of wall cleaning equipment and to optimize operation.

Additional discussion of the application of control systems to local sootblower cleaning requirements is provided in Chapter 24.

Non-routine ash evaluation methods

The following describes the laboratory equipment and test procedures, referenced earlier, that are used to supplement the standard ASTM coal ash characterization methods.

Laboratory ashing furnace

As noted, the ASTM ashing procedure does not duplicate the ashing process that actually occurs in a boiler. A laboratory ashing furnace (LAF) provides a means to obtain flyash and deposit samples that are comparable to those obtained from full scale installations operating under similar conditions.

B&W's LAF, shown in Fig. 20, is designed to fire pulverized coal at rates typically between 5 and 10 lb/h (2.3 and 4.5 kg/h). The facility consists of a fuel feed system, pulverized coal burner and a refractory lined chamber. The combustion chamber is surrounded by an electrically heated guard furnace which controls the rate of heat removal from the chamber to simulate full scale furnace temperatures. The firing rate is established to approximate full scale furnace residence time. A deposition section located at the furnace exit contains air- or water-cooled probes. The surface temperature of the probes can be adjusted to simulate furnace and superheater tube operating temperatures.

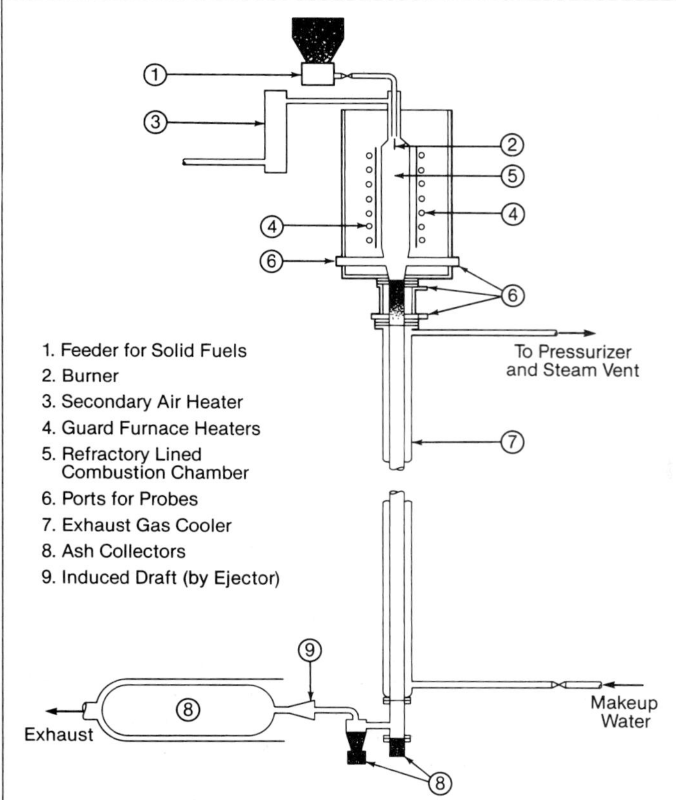

Fig. 20 Schematic of laboratory ashing furnace (LAF).

The probes are instrumented to allow measurement of metal temperatures, cooling fluid flow rates, and cooling fluid inlet and outlet temperatures. These data permit calculation of the total heat flux from the flue gas through the deposit and into the probe. The deposition section is also fitted with sootblowers to evaluate the effectiveness of ash removal equipment. Fig. 21 shows ash particles impacting a simulated superheater tube during a deposition test. The B&W LAF was used to develop extensive data which has now been correlated to the ash characteristics. These correlations have largely replaced the need for routine laboratory testing.

Measurement of ash viscosity

Viscosity of coal ash is measured in a high temperature rotating-bob viscometer (Fig. 22). The ash under study is contained in a cylindrical platinum-rhodium crucible, and a cylindrical bob is rotated in the liquid at a constant speed through a calibrated suspension wire. The torque or amount of twist produced in the suspension wire is proportional to the viscosity. The amount of twist is measured and recorded as the interval between impulses from light beams reflected from mirrors attached to the ends of the wire. The suspension wires are calibrated against viscosity standard oils obtained from the Bureau of Standards.

The electrically heated furnace is of the Globar tube type with temperature regulation provided through a controlling type potentiometer actuated by a thermocouple located in the furnace adjacent to the sample crucible. A thermocouple imbedded in the ash crucible support indicates sample temperature. Provision is made for controlling the atmosphere within the furnace. Ash is introduced into the crucible at an elevated temperature [2600 to 2800F (1427 to 1538C)] and held at that temperature until it becomes uniformly fluid. The temperature is then decreased in predetermined steps and the viscosity of the ash is measured at each temperature.

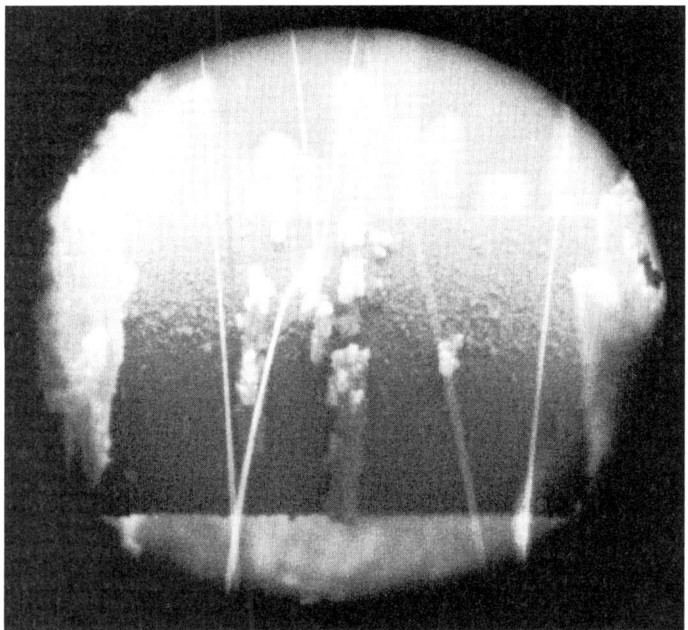

Fig. 21 Deposit formation on simulated superheater tube.

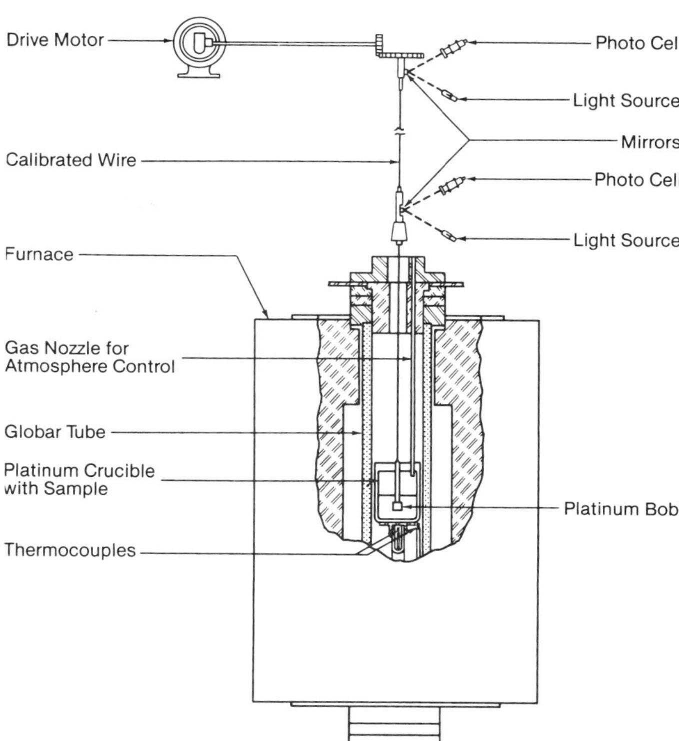

Fig. 22 Section through furnace of high temperature viscometer.

Ash sintering strength

The sintering strength test is performed on a flyash sample prepared in the LAF under a standard set of firing conditions. The flyash is passed through a 60 mesh (U.S. standard) (250 micron) screen to remove any particles of slag and then ignited to constant weight at 900F (482C) to remove any carbon that might be present. The ignited ash is then reduced to a minus 100 mesh size and at least 24 cylindrical specimens [0.6 in. (15.2 mm) diameter by 0.85 in. (21.6 mm) long] are formed in a hand press at a pressure of 150 psi (1034 kPa). At least six specimens are heated in air, usually at each of four temperature levels [1500, 1600, 1700 and 1800F (816, 871, 927 and 982C)] for 15 hours.

After the specimens have cooled slowly in the furnace, they are removed, measured and then crushed in a standard metallurgical testing machine. The sintered or compression strength is then computed from the applied force and the cross-sectional area of the sintered specimen. The average strength of six specimens is used as the strength of the sintered flyash at a particular sintering temperature.

Measurement of ion exchangeable cations in coal

Twenty grams of an air-dried minus 60 mesh coal sample are mixed with 100 ml of 1 N ammonium acetate in a 300 ml three-neck round bottom flask. A thermometer is inserted into the slurry. The slurry is stirred constantly and heated to 60 ±5C. The coal slurry sample is refluxed for 18 hours. The sample is filtered through a cellulosic filter media with 0.45 μ average pore size and washed twice with 25 ml of 1 N ammonium acetate solution.

The above procedure is repeated on the filtered coal except that the time is shortened to three hours. The combined filtrates are acidified by adding 2% by volume of glacial acetic acid and stored for inductive coupled plasma atomic emission spectrometric (ICPAES) analysis of Na, K, Ca and Mg.

Coal ash corrosion

Serious external wastage or corrosion of high temperature superheater and reheater tubes was first encountered in coal-fired boilers in 1955. Tube failures resulting from excessive thinning of the tube walls, as shown in Fig. 23, occurred almost simultaneously in the reheater of a dry ash furnace boiler and the secondary superheater of a slag-tap furnace unit. Corrosion was confined to the outlet tube sections of the reheater and the secondary superheater, which were made from chromium ferritic and stainless steel alloys, respectively.

Significantly, these boilers were among the first to be designed for 1050F (566C) main and reheat steam temperatures; also, both units burned high sulfur, high alkali Central and Southern Illinois coals, which were causing chronic ash fouling problems at the time.

Early investigations showed that corrosion occurred where complex alkali sulfates concentrated on tube surfaces beneath bulky layers of ash and slag. When dry, the complex sulfates were relatively innocuous; but when semi-molten [1100 to 1350F (593 to 732C)], they corroded the alloy steels used in superheater construction, and also other normally corrosion resistant alloys.

At first, it appeared that coal ash corrosion might be confined to boilers burning high alkali coals, but complex sulfate corrosion was soon found on superheaters and reheaters of several boilers burning low to medium alkali coals. Where there was no corrosion, the complex sulfates were either absent or the tube metal temperatures were moderate [less than 1100F (593C)]. The general conclusions drawn from this survey of corrosion were:

1. All bituminous coals contain enough sulfur and alkali metals to produce corrosive ash deposits on superheaters and reheaters, and those containing more than 3.5% sulfur and 0.25% chlorine may be particularly troublesome.
2. Experience has shown that corrosion rate is affected by both tube metal temperature and gas temperature. Fig. 24, which is used as a guide in design, indicates stable and corrosive zones of fuel ash corrosion as a function of gas and metal temperatures.

Based on this information, B&W modified the design of its boilers to greatly reduce the corrosion of superheaters and reheaters. These modifications included changes in furnace geometry, burner configuration, superheater arrangement and the use of gas tempering, all of which reduced metal and gas temperatures and reduced temperature imbalances. Experience from these installations has shown that it is possible to operate boilers with main and reheat steam temperatures up to 1050F (566C) with little, if any, corrosion from most coals.

Meanwhile, there was a gradual return to the 1000F (538C) steam conditions, due primarily to economic factors and secondarily to coal ash corrosion. This temperature level has permitted the use of lower cost alloys in the boiler, steam piping and turbine, with substantial savings in investment costs; it also has provided a greater margin of safety to avoid corrosion. Steam temperatures remained on the 1000F (538C) plateau for several decades. However, improved alloy creep and corrosion resistance and the potential for

Fig. 23 Typical corroded 18Cr-8Ni tube from secondary superheater.

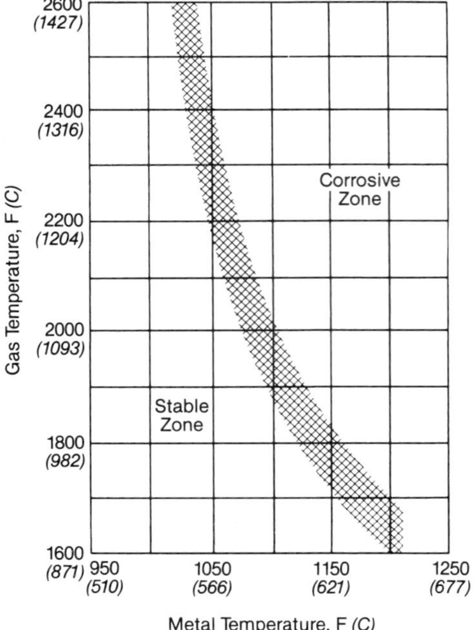

Fig. 24 Coal ash corrosion – stable and corrosive zones.

cycle efficiency improvement have led to increased steam temperatures in the newest generation of utility boilers. Some recent boilers have been designed with main and reheat steam temperatures at and above 1100F (593C), and temperatures of 1200F (640C) and above are envisioned.

General characteristics of corrosion

External corrosion of superheaters and reheaters is concentrated on the upstream side of the tube, as shown in Fig. 25. The greatest metal loss usually occurs on the 10 and 2 o'clock sectors of the tubes, and it tapers off to little or none on the back side of the tubes. The corroded surface of the tube is highly sculptured by a shallow macropitting type of attack. The amount of corrosion, as measured by reduction in tube wall thickness, varies considerably along the length of the tube, depending on local conditions, i.e., the position of the tube in the bank or platen, the proximity of sootblowers, the composition of ash deposits and, most importantly, the gas and metal temperatures.

The corrosion rate is a nonlinear function of metal temperature (Fig. 26). Typically, the corrosion of both chromium ferritic and 18Cr-8Ni stainless steels increases sharply above a temperature of 1150F (621C), passes through a broad maximum between 1250 and 1350F (677 and 732C) and then decreases rapidly at still higher temperatures. However, the corrosion behavior varies depending on gas and ash chemistries and other factors, as well as temperatures.

The highest corrosion rates are generally found on the outlet tubes of radiant superheater or reheater platens opposite retractable sootblowers. Values ranging from 50 to 250 mils/yr (1.27 to 6.35 mm/yr) have been observed on 18Cr-8Ni stainless steel tubes under these adverse conditions. When similar high temperature surfaces [1100 to 1175F (593 to 635C)] are arranged in convection tube banks so they are shielded from direct furnace radiation and sootblower action, corrosion rates are much lower, ranging between 5 and 20 mils/yr (0.13 to 0.51 mm/yr).

Corrosive ash deposits

Corrosion is rarely found on superheater or reheater tubes having only dusty deposits. It is nearly always associated with sintered or slag type deposits that are strongly bonded to the tubes. Such deposits consist of at least three distinct layers. The outer layer, shown diagrammatically in Fig. 27, constitutes the bulk of the deposit and has an elemental composition similar

←Direction of Gas Flow

Fig. 25 Transverse sections of corroded tubes from secondary superheater platens.

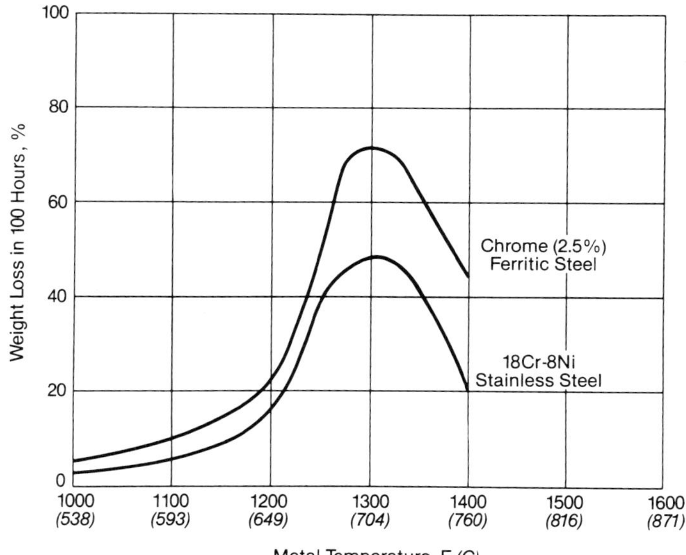

Fig. 26 Effect of temperature on corrosion rate.

to that of flyash. Though often hard and brittle, this layer is a porous structure through which gases may diffuse. Innocuous by itself, it plays an important part in the formation of an intermediate layer that contains the corrosive agents.

The intermediate layer, frequently called the white layer, is a white to yellow colored material which varies in thickness from 0.03 to 0.25 in. (0.76 to 6.35 mm). It usually has a chalky texture where corrosion is mild or nonexistent, but is fused and semi-glossy where corrosion is severe. In the latter condition this layer is difficult to remove as it is so firmly bonded to the corroded surface beneath.

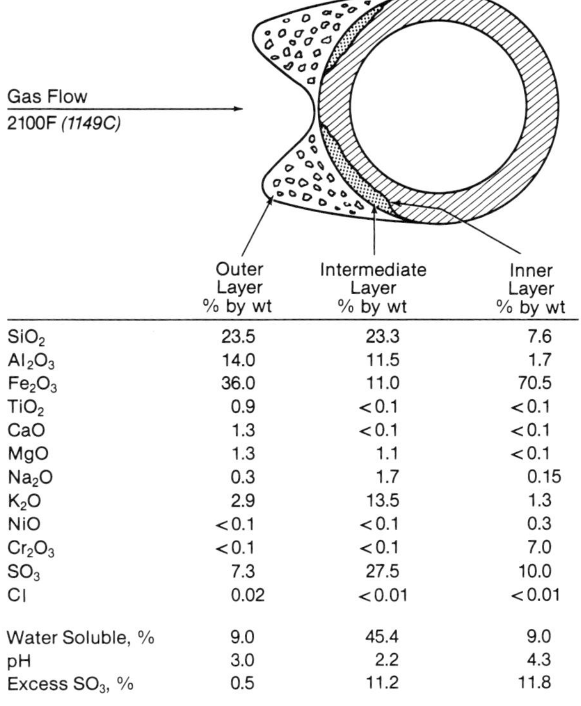

	Outer Layer % by wt	Intermediate Layer % by wt	Inner Layer % by wt
SiO_2	23.5	23.3	7.6
Al_2O_3	14.0	11.5	1.7
Fe_2O_3	36.0	11.0	70.5
TiO_2	0.9	<0.1	<0.1
CaO	1.3	<0.1	<0.1
MgO	1.3	1.1	<0.1
Na_2O	0.3	1.7	0.15
K_2O	2.9	13.5	1.3
NiO	<0.1	<0.1	0.3
Cr_2O_3	<0.1	<0.1	7.0
SO_3	7.3	27.5	10.0
Cl	0.02	<0.01	<0.01
Water Soluble, %	9.0	45.4	9.0
pH	3.0	2.2	4.3
Excess SO_3, %	0.5	11.2	11.8

Fig. 27 Analyses of typical ash deposit from 18Cr-8Ni superheater tube.

Upon heating in the air, the intermediate layer melts around 1000F (538C) and slowly discolors and hardens into a hard mass resembling rust. Chemical analyses of this layer show that it contains higher concentrations of potassium, sodium and sulfur than does the parent coal ash. A large part of this deposit is water soluble and the water soluble fraction is always acidic. The most common compounds found are $Na_3Fe(SO_4)_3$ and $KAl(SO_4)_2$.

Complex alkali sulfates, when molten, rapidly corrode most, if not all, superheater alloys. Corrosion begins between 1000 and 1150F (538 and 621C), depending on the relative amounts of complex sodium and potassium sulfates present and whether these are predominantly iron or aluminum base compounds. Corrosion usually begins at the lower temperature where the sodium-iron-sulfate system is the major part of the intermediate layer, but corrosion is more severe and persists into a higher temperature range when the potassium-aluminum-sulfate system dominates.

If the intermediate layer is carefully removed, a black, glassy inner layer is revealed, which appears to have replaced the normally protective oxide on the tube. This layer is composed primarily of corrosion products, i.e., oxides, sulfides, and sulfates of iron and other alloying constituents in the tube metal. It seldom exceeds 0.063 in. (1.59 mm) thickness on corroded 18Cr-8Ni stainless steel tubes, probably because of its strong tendency to spall when the tubes cool. The layer containing corrosion products from chromium ferritic steels often reaches 0.125 in. (3.18 mm) thickness and exhibits little tendency to spall as the tube cools.

Corrosion mechanisms

The elements in coal ash corrosion (sodium, potassium, aluminum, sulfur and iron) are derived from the mineral matter in coal. The minerals supplying these elements include shales, clays and pyrites which are commonly found in all coals.

During the combustion of coal, these minerals are exposed to high temperatures and strongly reducing effects of carbon for very short periods of time. Although comparatively stable, the mineral matter undergoes rapid decomposition under these conditions. Some of the alkalies are released or volatilized as relatively simple compounds, which have dew points in the 1000 to 1300F (538 to 704C) range. Furthermore, sulfur in the coal is oxidized, releasing SO_2 with the formation of a small amount of SO_3, leaving a residue of iron oxide (Fe_2O_3).

By far the largest portion of the mineral matter or its derived species react to form the glassy particulates of flyash. The flyash and volatile species in the flue gases deposit on the tube surfaces. Slowly, over a period of weeks, the alkalis and the sulfur oxides diffuse through the layer of flyash toward the tube surface. In the lower temperature zone of the ash deposit, chemical reactions between the alkalis, the sulfur oxides, and the iron and aluminum components of the flyash form complex alkali sulfates as follows:

$$3K_2SO_4 + Fe_2O_3 + 3SO_3 \rightarrow 2K_3Fe(SO_4)_3$$

and

$$K_2SO_4 + Al_2O_3 + 3SO_3 \rightarrow 2KAl(SO_4)_2 \qquad \textbf{(13)}$$

Similar reactions occur with sodium sulfate (Na_2SO_4), although the complex sodium sulfates are less apt to form at high temperatures because of their lower stability.

Work at B&W has shown that SO_3 concentrations in ash deposits must be very high (1000 to 1500 ppm) compared to the level in the flue gas (10 to 25 ppm) in order to form the complex alkali sulfates in the intermediate layer. Therefore, the bulk of the SO_3 must come from the catalytic oxidation of SO_2 in the outer layer of the deposit.

When the SO_3 produced in the outer deposit exceeds the partial pressure of SO_3 necessary for stability, the complex sulfates form through the above reactions. When the opposite is true, the complex sulfates begin to decompose according to the reverse of these reactions until a new equilibrium is reached. Because the formation of SO_3 is temperature dependent, the reversibility of these reactions is also temperature dependent. As shown in Fig. 26 the corrosion rate increases with temperature, passes through a maximum between 1250 and 1350F (677 and 732C), and then falls to a comparatively low level at higher temperatures.

The temperature range of this rapid liquid-phase attack is bracketed by: 1) the melting temperature of the mixture of complex alkali sulfates present, and 2) their thermal stability limits. The extreme width of this temperature band is approximately 400F (222C); corrosion due to the complex alkali sulfates may range from as low as 1000F (538C) to a maximum of 1400F (760C), depending on the species present in the intermediate layer.

Corrective measures

Various methods of combating corrosion of superheater and reheater tubes have been used or suggested, including the following:

1. the use of stainless steel shields to protect the most vulnerable tubes,
2. coal selection,
3. improvement of combustion conditions, i.e., providing proper coal fineness, fast ignition, good mixing and proper excess air, and
4. the use of more corrosion resistant alloys and alloy cladding on the most vulnerable superheater and reheater tubes. (See Chapter 19.)

New high chromium content stainless steels are generally more resistant to coal ash corrosion, but added chromium tends to degrade other metallurgical properties. New modified 25% chromium 20% nickel stainless steels offer improved performance for moderate conditions. Higher chromium nickel-based alloys and cladding offer maximum resistance for the most severe conditions.

Fuel oil ash

The ash content of residual fuel oil seldom exceeds 0.2%, an exceedingly small amount compared to that in coal. Nevertheless, even this small quantity of ash is capable of causing severe problems of deposition and corrosion in boilers. Of the many elements that may appear in oil ash deposits, the most important are vanadium, sodium and sulfur. Compounds of these elements are found in almost every deposit in boilers fired by residual fuel oil and often constitute the major portion of these deposits.

Origin of ash

As with coal, some of the ash-forming constituents in the crude oil had their origin in animal and vegetable matter from which the oil was derived. The remainder is extraneous material, resulting from contact of the crude oil with rock structures and salt brines, or is picked up during refining processes, storage and transportation. (See Chapter 9.)

In general, the ash content increases with increasing asphaltic constituents in which the sulfur acts largely as a bridge between aromatic rings. Elemental sulfur and hydrogen sulfide have been identified in crude oil, and simpler sulfur compounds are found in the distillates of crude oil including thioesters, disulfides, thiophenes and mercaptans.

Vanadium, iron, sodium, nickel and calcium in the fuel oil were probably derived from the rock strata but some elements, such as vanadium, nickel, zinc and copper, probably came from organic matter from which the petroleum was derived. Vanadium and nickel especially are known to be present in organo-metallic compounds known as porphyrins which are characteristic of certain forms of animal life. Table 7 indicates the amounts of vanadium, nickel and sodium present in residual fuel oils from various crudes.

Crude oil as such is not normally used as a fuel but is further processed to yield a wide range of more valuable products. For example, in a modern U.S. refinery, 92.4% of the crude is converted to lighter fraction fuels and products such as gasoline, leaving 7.6% of residual fuel oil or distillate residue. Virtually all metallic compounds and a large part of the sulfur compounds are concentrated in the distillation residue. Where low sulfur residual fuel oils are required, they are obtained by blending with suitable stocks, including both heavy distillates and distillation from low sulfur crudes. This procedure is also used occasionally if a residual fuel oil must meet specifications such as vanadium or ash content.

Table 7
Vanadium, Nickel and Sodium Content of Residual Fuel Oils (ppm by wt)

Source of Crude Oil	Vanadium	Nickel	Sodium
Africa:			
1	5.5	5	22
2	1	5	–
Middle East:			
3	7	–	1
4	173	51	–
5	47	10	8
U.S.:			
6	13	–	350
7	6	2.5	120
8	11	–	84
Venezuela:			
9	–	6	480
10	57	13	72
11	380	60	70
12	113	21	49
13	93	–	38

Release of ash during combustion

Residual fuel oil is preheated and atomized to provide enough reactive surface so that it will burn completely within the boiler furnace. (See Chapter 11.) The atomized fuel oil burns in two stages. In the first stage the volatile portion burns and leaves a porous coke residue and in the second stage the coke residue burns. In general, the rate of combustion of the coke residue is inversely proportional to the square of its diameter, which in turn is related to the droplet diameter. Therefore, small fuel droplets give rise to coke residues that burn very rapidly, and the ash forming constituents are exposed to the highest temperatures in the flame envelope. The ash forming constituents in the larger coke residues from the larger fuel droplets are heated more slowly, partly in association with carbon. Release of the ash from these residues is determined by the rate of oxidation of the carbon.

During combustion, the organic vanadium compounds in the residual fuel oil thermally decompose and oxidize in the gas stream to V_2O_3, V_2O_4 and finally V_2O_5. Although complete oxidation may not occur and there may be some dissociation, a large part of the vanadium originally present in the oil exists as vapor phase V_2O_5 in the flue gas. The sodium, usually present as chloride in the oil, vaporizes and reacts with sulfur oxides either in the gas stream or after deposition on tube surfaces.

Subsequently, reactions take place between the vanadium and sodium compounds, with the formation of complex vanadates having melting points lower than those of the parent compounds; for example:

$$Na_2SO_4 + V_2O_5 \rightarrow 2NaVO_3 + SO_3 \uparrow \quad (14)$$

Melting points: 1625F 1275F 1165F

Excess sodium or vanadium in the ash deposit, above that necessary for the formation of the sodium vanadates (or vanadylvanadates), may be present as Na_2SO_4 and V_2O_5, respectively.

The sulfur in residual fuel oil is progressively released during combustion and is promptly oxidized to SO_2. A small amount of SO_2 is further oxidized to SO_3 by a small amount of atomic oxygen present in the hottest part of the flame. Also, catalytic oxidation of SO_2 to SO_3 may occur as the flue gases pass over vanadium rich ash deposits on high temperature superheater tubes and refractories. (See Chapter 35.)

Oil slag formation and deposits

The deposition of oil ash constituents on the furnace walls and superheater surfaces can be a serious problem. This deposition, coupled with corrosion of superheater and reheater tubes by deposits, was largely responsible for the break in the trend towards higher steam temperatures that occurred in the early 1960s.

Practically all boiler installations are typically designed for steam temperatures in the 1000 to 1015F (538 to 546C) range to minimize those problems and to avoid the higher capital costs of the more expensive alloys required in the tubes, steam piping and turbine for 1050 to 1100F (566 to 593C) steam conditions.

There are many factors affecting oil ash deposition on boiler heat absorbing surfaces. These factors may be grouped into the following interrelated categories: characteristics of the fuel oil, design of the boiler, and operation of the boiler.

Characteristics of fuel oil ash

Sodium, sulfur and vanadium are the most significant elements in the fuel oil because they can form complex compounds having low melting temperatures, 480 to 1250F (249 to 677C), as shown in Table 8. Such temperatures fall within the range of tube metal temperatures generally encountered in furnace and superheater tube banks of many oil-fired boilers. However, because of its complex chemical composition, fuel oil ash seldom has a single sharp melting point, but rather softens and melts over a wide temperature range.

An ash particle that is in a sticky, semi-molten state at the tube surface temperature may adhere to the tube if it is brought into contact by the gas flow over the tube. Even a dry ash particle may adhere due to mutual attraction or surface roughness. Such an initial deposit layer will be at a higher temperature than that of the tube surface because of its relatively low thermal conductivity. This increased temperature promotes the formation of adherent deposits. Therefore, fouling will continue until the deposit surface temperature reaches a level at which all of the ash in the gas stream is in a molten state, so that the surface is merely washed by the liquid without freezing and continued buildup.

In experimental furnaces, it has been found that the initial rate of ash buildup was greatest when the sodium-vanadium ratio in the fuel oil was 1:6, but an equilibrium thickness of deposit [0.125 to 0.25 in. (3.175 to 6.35 mm)] was reached in approximately 100 hours of operation. When the fuel oil contained more refractory constituents, such as silica, alumina and iron oxide, in addition to sodium and vanadium, an equilibrium condition was not reached and the tube banks ultimately plugged with ash deposits. However, these ash deposits were less dense, i.e., more friable, than the glassy slags encountered with a 1:6 sodium-vanadium fuel oil. Both the rate of ash buildup and the ultimate thickness of the deposits are also influenced by physical factors such as the velocity and temperature of the flue gases and particularly the tube metal temperature.

In predicting the behavior of a residual oil insofar as slagging and tube bank fouling are concerned, several fuel variables are considered including: 1) ash content, 2) ash analysis, particularly the sodium and vanadium levels and the concentration of major constituents, 3) melting and freezing temperatures of the ash, and 4) the total sulfur content of the oil. Applying this information in boiler design is largely a matter of experience.

Boiler design

Generally speaking, progressive fouling of furnaces and superheaters should not occur as long as the ash characteristics are not severe compared to the tube metal temperatures. If such trouble is encountered, the solution can usually be found in improving combustion conditions in the furnace and/or modifying the sootblowing procedures.

Studies on both laboratory and field installations have shown that the rate of ash deposition is a function of the velocity and temperature of the flue gases and the concentration of oil ash constituents in the flue gases. The geometry of the furnace and the spacing of tubes in the convection banks are selected in the design of a boiler to minimize the rate of deposition. It is common practice to use in-line tube arrangements with wider lateral spacings for tubes located in higher gas temperature zones. This makes bridging of ash deposits between tubes less likely and facilitates cleaning of tube banks by the sootblowers.

Boiler operation

Poor atomization of the fuel oil results in longer flames and frequently increases the rate of slag buildup on furnace walls which, in turn, makes it more difficult to keep the convection sections of the boiler clean. Completing combustion before the gases pass over the first row of tubes is especially important.

Table 8
Melting Points of Some Oil Ash Constituents

Compound	Melting Point, F	(C)
Aluminum oxide, Al_2O_3	3720	(2049)
Aluminum sulfate, $Al_2(SO_4)_3$	1420*	(771)
Calcium oxide, CaO	4662	(2572)
Calcium sulfate, $CaSO_4$	2640	(1449)
Ferric oxide, Fe_2O_3	2850	(1566)
Ferric sulfate, $Fe_2(SO_4)_3$	895*	(479)
Nickel oxide, NiO	3795	(2091)
Nickel sulfate, $NiSO_4$	1545*	(841)
Silicon dioxide, SiO_2	3130	(1721)
Sodium sulfate, Na_2SO_4	1625	(885)
Sodium bisulfate, $NaHSO_4$	480*	(249)
Sodium pyrosulfate, $Na_2S_2O_7$	750*	(399)
Sodium ferric sulfate, $Na_3Fe(SO_4)_3$	1000	(538)
Vanadium trioxide, V_2O_3	3580	(1971)
Vanadium tetroxide, V_2O_4	3580	(1971)
Vanadium pentoxide, V_2O_5	1275	(691)
Sodium metavanadate,		
$Na_2O \cdot V_2O_5(NaVO_3)$	1165	(629)
Sodium pyrovanadate, $2Na_2O \cdot V_2O_5$	1185	(641)
Sodium orthovanadate, $3Na_2O \cdot V_2O_5$	1560	(849)
Sodium vanadylvanadates,		
$Na_2O \cdot V_2O_4 \cdot V_2O_5$	1160	(627)
$5Na_2O \cdot V_2O_4 \cdot 1V_2O_5$	995	(535)

* Decomposes at a temperature around the melting point.

Relatively large carbonaceous particles have a far greater tendency to impinge on the tubes than do the smaller ash particles. If these larger particles are in a sticky state, they will adhere to the tubes where oxidation will proceed at a slow rate with consequent formation of ash. Fouling from this cause is difficult to detect by inspection during boiler outages because the carbonaceous material has usually disappeared completely. It can generally be detected during operation because flames are usually long and smoky, and *sparklers* may be carried along in the flue gases.

Regular and thorough sootblowing can have a decisive effect on superheater and reheater fouling. (See Chapter 24.) To be fully effective, however, sootblowing cycles should be frequent enough so that ash deposits can not build to a thickness where their surfaces become semi-molten and difficult to remove. In instances of extreme slagging, it is sometimes necessary to relocate sootblowers, to install additional sootblowers to control deposition in a critical zone, or to use additives.

The boiler load cycle can also have a significant effect on the severity of slagging and superheater fouling. A unit that is base loaded for long periods is more apt to have fouling problems on a borderline fuel oil than a unit that takes daily swings in load. In the latter instance, the furnace generally remains cleaner due to periodic shedding of slag, with the result that the gas temperatures through the superheaters are appreciably lower. This eases the burden on the sootblowers and substantially controls ash deposit formation in the superheater-reheater tube banks. Overloading the boiler, even for an hour or two a day, should be avoided, especially if excess air has to be lowered to the point where some of the burners are starved of air. The furnace is apt to become slagged and ash deposition can creep into the superheater and reheater tube banks.

Oil ash corrosion

High temperature corrosion

The sodium-vanadium complexes, usually found in oil ash deposits, are corrosive when molten. A measurable corrosion rate can be observed over a wide range of metal and gas temperatures, depending on the amount and composition of the oil ash deposit. Fig. 28 shows the combined gas and metal temperature effects on corrosion for a specific fuel oil composition of 150 ppm vanadium, 70 ppm sodium and 2.5% sulfur. As the vanadium concentration of the fuel oil varies, the amount of corrosion, compared to a 150 ppm vanadium fuel, will increase or decrease according to the curve shown in Fig. 29. The effect of the sodium level in the fuel oil is not as clear. The sodium content does, however, definitely affect the minimum metal temperature at which corrosion will be significant.

At the present time there appears to be no alloy that is immune to oil ash corrosion. In general, the higher the chromium content of the alloy, the more resistant it is to attack. This is the main reason for the use of 18Cr-8Ni alloys for high temperature superheater tubes. High chromium contents, greater than 30%, give added corrosion resistance but at the expense of physical properties; 25Cr-20Ni has been used as a tube cladding but even this alloy has not provided complete protection. High-nickel high-chromium alloys may be more resistant to oil ash attack under oxidizing conditions, but the higher material cost must be justified by longer life, which is not always predictable.

Low temperature corrosion

In oil-fired boilers, the problem of low temperature corrosion resulting from the formation and condensation of sulfuric acid from the flue gases is similar to that previously described for coal firing.

Oil-fired boilers are more susceptible to low temperature corrosion than are most coal-fired units for two reasons: 1) the vanadium in the oil ash deposits is a good catalyst for the conversion of SO_2 to SO_3, and 2) there is a smaller quantity of ash in the flue gases. Ash particles in the flue gas react with and reduce the amount of SO_3 vapor in the gas, and oil has considerably less ash than coal. Furthermore, coal ash is more basic than oil ash and more effectively neutralizes acid.

Methods of control

The methods of control that have been used or proposed to control fouling and corrosion in oil-fired boilers are summarized in Table 9, but in every instance economics govern their applicability. There is no doubt that reducing the amount of ash and sulfur entering

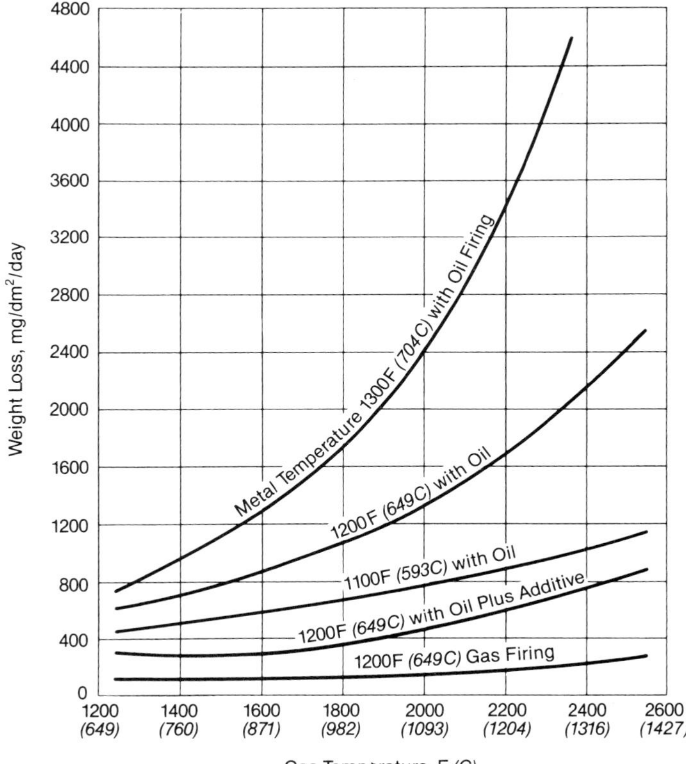

Fig. 28 Effect of gas and metal temperatures on corrosion of 304, 316 and 321 alloys in a unit fired with oil containing 150 ppm vanadium, 70 ppm sodium and 2.5% sulfur. Test duration 100 hours.

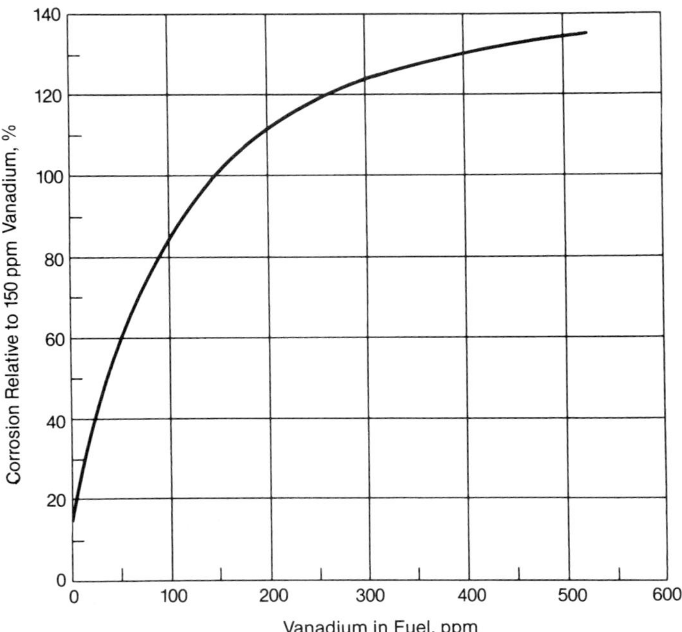

Fig. 29 Effect of vanadium concentration on oil ash corrosion.

the furnace is the surest means of control and that minimizing the effects of the ash constituents, once they have deposited on the tubes, is the least reliable. Because the severity of fouling and corrosion depends not only on the fuel oil characteristics but also on boiler design and operating variables, a generalized solution to these problems can not be prescribed.

Fuel oil supply

Although fuel selection and blending are practiced to some extent in the U.S., the common purpose is to provide safe and reliable handling and storage at the power plant rather than to avoid fouling difficulties. Because the threshold limits of sodium, sulfur and vanadium are not well defined for either fouling or corrosion, use of these means of control can not be fully exploited.

Fuel oil additives

An approach that is effective where the fuel oil ash is most troublesome involves adding, to the fuel or furnace, small amounts of materials that change the character of the ash sufficiently to permit its removal by steam or air sootblowers or air lances.

Additives are effective in reducing the problems associated with superheater fouling, high temperature ash corrosion and low temperature sulfuric acid corrosion. Most effective are alumina, dolomite and magnesia. Kaolin is also a source of alumina. Analyses of typical superheater deposits from a troublesome fuel oil, before and after treating it with alumina or dolomite, are shown in three bar graphs on the left of Fig. 30. The results for a different oil treated with magnesia are shown in the bar graph on the right.

The reduction of fouling and high temperature corrosion is accomplished basically by producing a high melting point ash deposit that is powdery or friable and easily removed by sootblowers or lances. When the ash is dry, corrosion is considerably reduced.

Low temperature sulfuric acid corrosion is reduced by the formation of refractory sulfates by reaction with the SO_3 gas in the flue gas stream. By removing the SO_3 gas, the dew point of the flue gases is sufficiently reduced to protect the metal surfaces. The sulfate compounds formed are relatively dry and easily removed by the normal cleaning equipment.

In general, the amount of additive used should be about equal to the ash content of the fuel oil. In some instances, slightly different proportions may be required for best results, especially for a high temperature corrosion reduction, in which it is generally accepted that the additive should be used in weight ratios of 2:1 or 3:1 (additive/ash), based on the vanadium content of the oil.

Several methods have been successfully used to introduce the additive materials into the furnace. The one in general use consists of metering a controlled amount of an additive oil slurry in the burner supply line. The additive material should be pulverized to 100% through a 325 mesh (44 micron) screen for good dispersion and minimum atomizer wear.

For a boiler fired by a high pressure return flow oil system (Chapter 11), it has been found advantageous to introduce the additive powders by blowing them into the furnace at the desired locations. The powder has to be reduced in size to 100% through a 325 mesh (44 micron) screen for good dispersion.

The choice of a particular additive depends on its availability and cost to the individual plant and the method of application chosen. For example, alumina causes greater sprayer plate wear than the other materials when used in an oil slurry.

The quantity of deposit formed is, of course, an important consideration for each unit from the aspect of cleaning. A comparison of the amounts of deposit

Table 9
Classification of Methods for Controlling Fouling and Corrosion in Oil-Fired Boilers

	Fuel Oil Supply
Reduce amount of fuel ash constituents to the furnace	Selection Blending Purification
	Design
Minimize amounts of fuel ash constituents reaching heat transfer surfaces	Furnace geometry Tube bank arrangement Metal temperature Gas temperature Sootblower arrangement
	Operation
Minimize effects of bonding and corrosive compounds in ash deposits	Load cycle Sootblowing schedule Combustion — excess air Additives Water washing

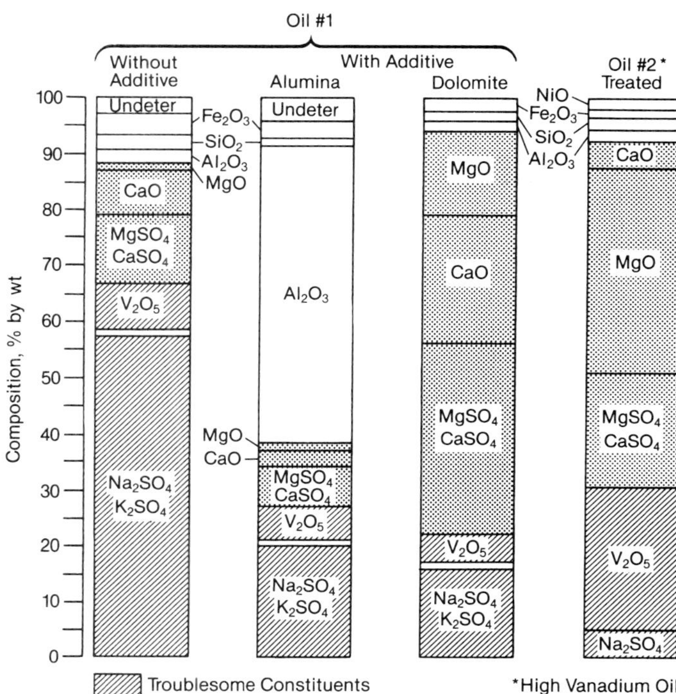

Fig. 30 Effect of fuel oil additives on composition of oil ash deposit.

formed with different additives shows that dolomite produces the greatest quantity because of its sulfating ability, magnesia is intermediate, and alumina and kaolin form the least. However, when adequate cleaning facilities are available, the deposits are easily removed and the quantities formed should not be a problem.

Excess air control

As mentioned previously the problems encountered in the combustion of residual fuel oils – high temperature deposits (fouling), high temperature corrosion and low temperature sulfuric acid corrosion – all arise from the presence of vanadium and sulfur in their highest states of oxidation. By reducing the excess air from 7% to 1 or 2%, it is possible to avoid the formation of fully oxidized vanadium and sulfur compounds and, thereby, reduce boiler fouling and corrosion problems.

In a series of tests on an experimental boiler, it was found that the maximum corrosion rate of type 304 stainless steel superheater alloy held at 1250F (677C) in 2100F (1149C) flue gas was reduced more than 75% (Fig. 31) when the excess air was reduced from an average of 7% to a level of 1 to 2%. Moreover, the ash deposits that formed on the superheater bank were soft and powdery, in contrast to hard, dense deposits that adhered tenaciously to the tubes when the excess air was around 7%. Also, the rate of ash buildup was only half as great. Operation at the 1 to 2% excess air level practically eliminated low temperature corrosion of carbon steel at all metal temperatures above the dew point of the flue gases (Fig. 32). However, much of the beneficial effects of low excess air combustion are lost if the excess air at the burner fluctuates even for short periods of time to a level of about 5%. Carbon loss values for low excess air were approximately 0.5%, which is generally acceptable for electric utility and industrial practice.

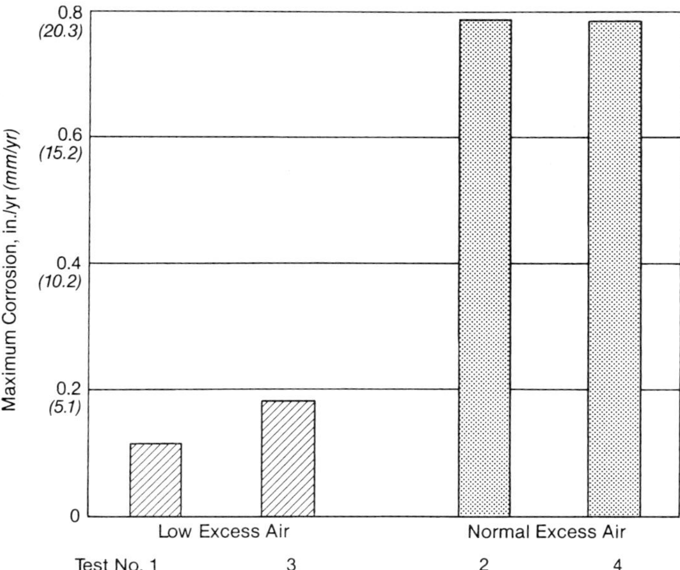

Fig. 31 Effect of low excess air combustion on high temperature oil ash corrosion.

A number of large industrial boilers both in the U.S. and in Europe have been operating with low excess air for several years. As a result, the benefits of reducing low temperature corrosion are well established. However, the benefits on high temperature slagging and corrosion are not wholly conclusive. In any event, great care must be exercised to distribute the air and fuel oil equally to the burners, and combustion conditions must be continuously monitored to assure that combustion of the fuel is complete before the combustion gases enter the convection tube banks.

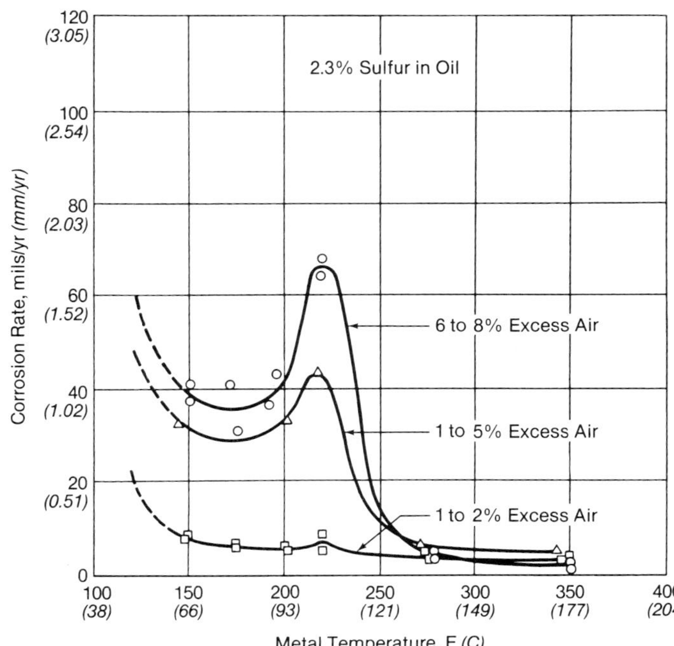

Fig. 32 Effect of excess air on low temperature corrosion of carbon steel.

Midwest power station burning bituminous coal.

Chapter 22

Performance Calculations

The evaluation of boiler performance involves many complex factors as indicated in preceding chapters. Only a few of these factors are subject to precise analysis; many others are the result of data taken from operating units. Ash in the fuel has perhaps the most dramatic impact on boiler performance as discussed in Chapter 21. In spite of the large number of variables, boilers are designed, built and operated in conformance with design specifications.

A well-designed and operated boiler completes combustion within the furnace. Flue gas temperatures leaving the furnace can be predicted by the methods presented in Chapter 4. Beyond the furnace, heat transfer surface arrangements represent a balance of temperature difference, space, pressure drop and draft losses. The final selection of these surface arrangements represents a compromise on the designer's part in meeting performance requirements while controlling ash deposition, corrosion, and erosion.

This chapter introduces the basic principles of boiler performance calculations. It illustrates the use of heat transfer, thermodynamics and fluid mechanics to determine heat and material balances for given boiler heat transfer components through practical application. These principles, as well as fundamental relationships, experimental data, operational experience and designer knowledge, are being incorporated into advanced numerical computational models for boiler evaluation. As discussed in Chapter 6 and elsewhere in *Steam*, these models are becoming increasingly useful in the design of boilers.

For the fossil fuel-fired steam generators being considered, the hotter heat transfer medium consists of the products of combustion, or flue gases. The cooler medium is superheated steam, steam-water mixtures at saturation, water or air depending upon the heat transfer component under consideration. Heat transfer surfaces can be categorized into one of four cases according to relative hot and cold medium flow direction and temperature as shown in Fig. 1. Typically, boiler banks or screens are Case I, superheaters and reheaters are either Case II or III, economizers are Case II or III, and air heaters are generally Case IV.

Performance calculations are typically used to establish one of three parameters: temperature, heat transfer surface area, or surface cleanliness. As in most thermal analysis problems, the evaluation of boiler performance is an iterative process. To evaluate flue gas and steam temperatures for a known boiler design arrangement, the surface area and surface cleanliness are normally known while the temperatures are assumed. The outlet temperature calculation updates subsequent iterations until convergence between assumed and calculated temperatures is achieved.

Heat transfer surface area or sizing can be determined for given fluid temperatures and surface cleanliness by assuming an initial surface arrangement and then confirming the desired thermal performance by calculation. High calculated outlet flue gas temperatures indicate the need for additional surface whereas low calculated outlet flue gas temperatures indicate the need to remove surface. Surface area is adjusted until calculated and specified temperatures converge.

Finally, for a given boiler configuration, measured temperatures can be used to assess surface cleanliness. An initially assumed cleanliness is used with the measured temperatures and known surface to verify the temperature data. The cleanliness factors are varied until temperature convergence is achieved.

When calculating the thermal performance of heat transfer equipment, initial temperatures are selected in part based upon past experience. These are used to establish the thermophysical properties, calculate mean temperature differences and solve the problem.

Note: To avoid the confusion of dual units, this chapter is provided in English units only. See Appendix 1 for SI conversion factors.

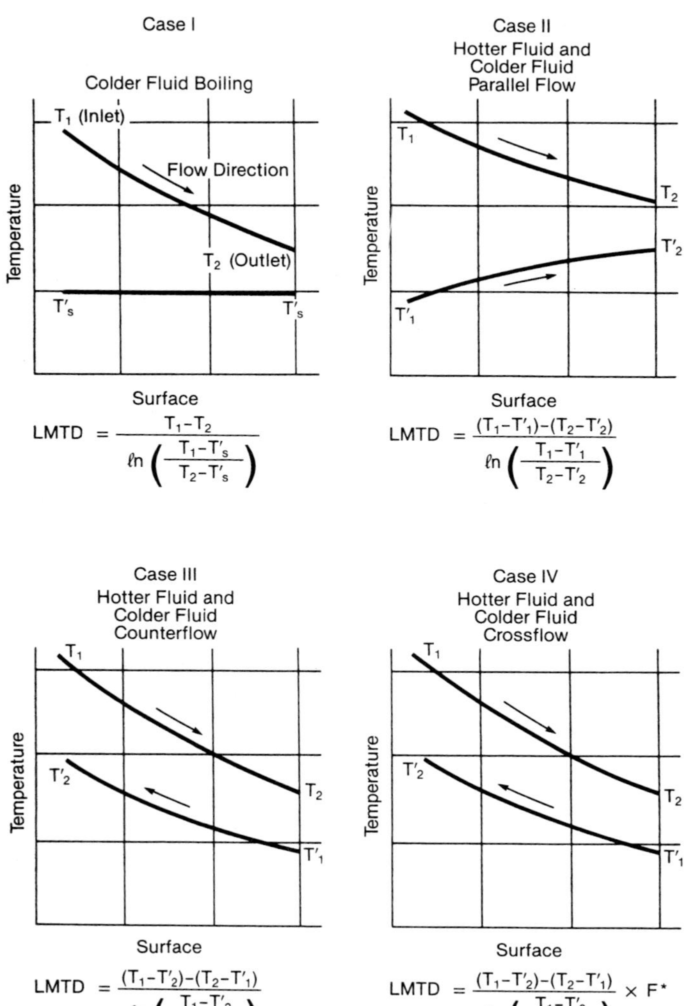

Fig. 1 Log mean temperature difference (LMTD) for selected heat exchanger configurations – single phase except for Case I colder fluid. (*As discussed in Chapter 4, a configuration factor *F* is needed for the case of crossflow.)

The resulting final temperatures are compared to the original assumptions. Generally, if an error of 5F or less is observed between iteration steps, the solution is considered complete. Otherwise, the new temperature is used to repeat the analysis.

To illustrate performance calculations, a small boiler (Fig. 2) with a simplified furnace and heating surface arrangement will serve as an example. The objective is to evaluate the outlet flue gas temperatures for the given surface area and cleanliness. The procedures developed here are fundamental and as such can be used on a wide range of boiler applications.

Operating conditions

Boiler performance specifications are defined by the customer. These specifications normally include: steam output conditions – pressure, temperature and flow; feedwater conditions; fuel and ash analysis; load range; capability; and efficiency. The final boiler design efficiently meets the specifications with a minimum of surface, materials and flow losses.

The specific analysis procedure is started at slightly different steps depending upon whether an existing piece of equipment is being analyzed or a new plant is being designed. For an existing installation, boiler performance calculations begin by confirming heat transfer equipment geometry and by defining the required operating conditions. Heat and material balances, including combustion calculations for the steam generator as a whole, are then determined. These calculations provide the information required to analyze each heat recovery component. The performance of each component is often intimately tied to other components within the boiler. As discussed in Chapter 1, the calculation process normally follows the direction of flue gas flow from furnace to stack and that is the approach used in the following example. Predictions of boiler performance are only complete when the performance of all devices within the boiler envelope agrees with the overall unit heat and material balances.

If a new plant is being analyzed, the process begins with the heat and material balances to establish air, fuel, feedwater and flue gas handling requirements. The process then continues with the iterative determination of heat transfer surface and equipment sizing as discussed above.

Because air resistance and draft loss calculations, including stack effects, are dependent on established air and gas temperature profiles through the boiler and accessories, they are determined after any thermal analysis. Pressure drop calculations are included in each component's evaluation. Flue and duct resistance calculations and the evaluation of stack effects are the final steps in the performance calculations.

For the example considered here, dimensional pa-

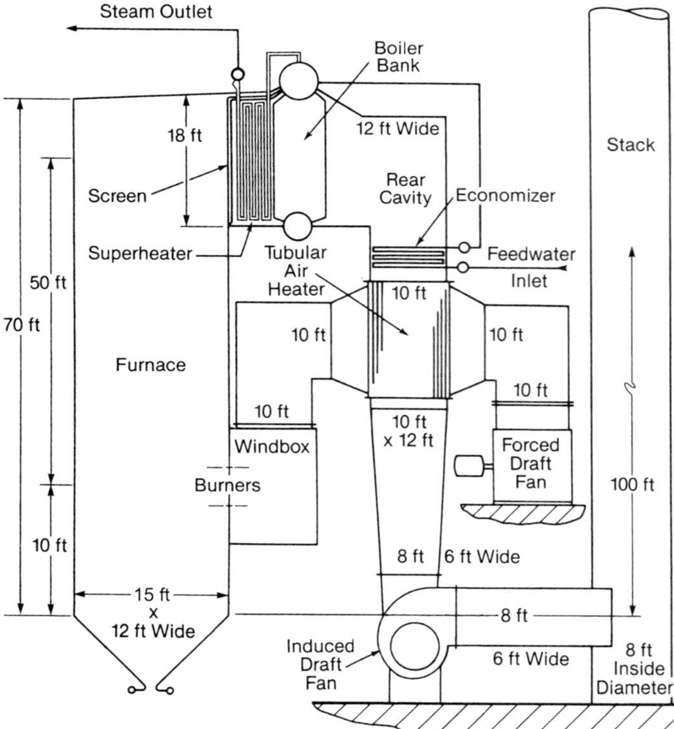

Fig. 2 Example of a coal-fired industrial boiler, sectional side view.

rameters are contained in Table 1, while Table 2 contains the specified operating conditions. The remainder of this chapter focuses on predicting performance for the hypothetical boiler.

Heat and material balances

Heat and material balances begin with combustion calculations. For this example, combustion calculations are determined by the *Btu method* as presented in Chapter 10. For the fuel analysis and losses specified in Table 2, the combustion calculations are summarized in Table 3.

The unit is expected to produce 250,000 lb/h superheated steam at 450 psig and 650F with feedwater conditions of 470 psig and 220F entering the economizer. The energy output, or heat in steam, is calculated to be 285.6×10^6 Btu/h (See Table 3, line 10).

The combustion calculations of Table 3 also define key parameters which guide much of the subsequent equipment design – heat input (328.6×10^6 Btu/h, Table 3, line 54), gas weight (324,100 lb/h, Table 3, line 56), air weight (302,500 lb/h, Table 3, line 58) and heat available (1034 Btu/lb Table 3, line 60). These values are noted for future reference.

This completes the heat and material balances and combustion calculations for the defined boiler envelope. As shown in Fig. 3, all pressures, temperatures and flows crossing the unit boundaries are established and calculations can now proceed on each component. A more detailed analysis would also account for items such as air infiltration, continuous boiler water discharge (blowdown), saturated steam extractions and steam reheaters, if applicable.

Component performance calculations

Furnace

Furnace exit gas temperatures must be determined to design downstream heat transfer components. Through testing and correlation of gas temperature data, furnace exit gas temperature has been found to have a relationship to the heat input of the fuel and to the effectiveness of the furnace walls. In Chapter 4, furnace exit gas temperature curves for various fuels are approximated. This extended family of curves represents the accumulation of extensive field experience and analytical evaluation. They are dependent on fuel and furnace geometry.

The heat which can be absorbed by the furnace was determined from the combustion calculations to be 1034 Btu/lb of flue gas. The layout of the furnace (Table 1) provides a flat projected surface of 4100 ft^2. For 2.5 in. outside diameter (OD) furnace tubes on 3 in. centers, Chapter 4, Fig. 33 indicates an effectiveness factor of 1.0. The heat release rate to the furnace is then:

$$\text{Heat release rate} = \frac{\text{Heat available} \times \text{Gas mass flow rate}}{\text{Flat projected area} \times \text{Effectiveness factor}} \quad (1)$$

$$= 81.7 \times 10^3 \text{ Btu/h ft}^2$$

Table 1
Physical Arrangement - Furnace

Construction: 2.5 in. outside diameter (OD) tubes on 3 in. centers with membrane construction. (See Chapter 23 for description.)
Width: 12 ft — Volume: 13,275 ft^3 to superheater entrance plane
Depth: 15 ft — Surface: 4100 ft^2 flat projected area, not including superheater exit plane

Physical Arrangement - Components

Parameter	Units	Screen	Superheater (Note 1)	Boiler Bank	Economizer	Air Heater
Tube OD	in.	2.5	2.5	2.5	2	2.25
Backspacing (centerline)	in.	6	3.25	4	3	4
Sidespacing (centerline)	in.	6	6	4	3	3
Rows deep		2	12	28	10	29
Rows wide		23	23	35	47	47
Tube length	ft	18	18	18 (Note 2)	10	19.5
Heating surface (Note 3)	ft^2	542	3169	11,545	2456	15,656
Free flow area (Note 4)	ft^2					
Gas		130	130	85	42	32
Air						62

Notes:
1. The superheater is a counterflow configuration; however, the steam flows in two parallel paths from steam drum to superheater outlet header. This is also called a two-flow arrangement.
2. Boiler bank tubes vary in length; the value listed represents an average length.
3. Heating surface is the external surface area of the tubes exposed to the flue gas except for the air heater where the flue gas flows through rather than over the tubes.
4. Free flow area is the minimum clear area between tubes, perpendicular to the direction of gas or air flow except for air heaters. The gas side free flow area for air heaters is the area defined by the inside diameter (ID) of the air heater tubes.

Table 2
Operating Conditions

Fuel: Bituminous Coal, Virginia
Analysis As-Fired

Ultimate, % by wt		Proximate, % by wt	
C	80.31	Moisture	2.90
H_2	4.47	Volatiles	22.05
S	1.54	Fixed carbon	68.50
O_2	2.85	Ash	6.55
N_2	1.38		100.00
H_2O	2.90		
Ash	6.55		
	100.00		

Higher heating value (HHV), as-fired:	14,100 Btu/lb
Excess air	20.0% by wt
Unburned carbon loss	0.4% by wt
Unaccounted loss (Table 3, Line 45)	1.5% by wt
ABMA radiation loss (see Chapter 23)	0.40% by wt
Furnace exit gas temperature	2000F
Superheater outlet:	
Steam flow	250,000 lb/h
Steam temperature	650 F
Steam pressure	450 psig
Steam enthalpy	1331 Btu/lb
Economizer inlet:	
Water flow	250,000 lb/h
Water temperature	220 F
Water pressure	470 psig
Water enthalpy	189 Btu/lb
Air heater:	
Air temperature entering	80 F
Barometric pressure	30 in. Hg
Gas temperature leaving	390 F

where

Heat available = 1034 Btu/lb
Gas mass flow rate = 324.1×10^3 lb/h
Flat projected area = 4100 ft^2
Effectiveness factor = 1.0

Furnace exit gas temperature is 2000F from Chapter 4, Fig. 34. Sufficient information is now known to begin analysis of the convection pass, i.e., screen, superheater, boiler bank, economizer and air heater.

The performance of the convection pass, however, is a function of the cleanliness, or effectiveness, of the heat transfer surfaces. The values of the heat transfer coefficient, as described in Chapter 4, apply to heat transfer surfaces free from ash and slag deposits. For calculation purposes, the effect of ash or other deposits on the heat transfer surfaces can be accounted for with a cleanliness factor:

$$\text{Cleanliness factor} = \frac{\text{Actual heat transfer rate}}{\text{Clean surface heat transfer rate}} \quad \textbf{(2)}$$

where gas side heat transfer coefficients are multiplied by the cleanliness factor to reproduce a given fuel's fouling characteristics. Unless specifically noted, the calculations that follow account for bituminous coal ash deposition and cleanliness factors have been omitted for simplicity. If the calculation procedures developed here are used with other dirty fuels, then commensurate cleanliness factors are required to accurately predict thermal performance.

Screen

Heat transfer

In this boiler design, the gases leaving the furnace first pass across the screen tubes. In this case, the tubes contain boiling water. These tubes control the amount of furnace radiation reaching the superheater surface. The heat transfer relationship for this surface is as follows:

$$q = U\,A\,(LMTD) = \dot{m}_g\, c_p\, \Delta T_g = \dot{m}_g\, c_p \left(T_1 - T_2\right) \quad \textbf{(3)}$$

where

q = heat transfer rate, Btu/h
U = h_g = combined heat transfer coefficient or overall heat transfer coefficient, Btu/h ft^2 F where the boiling water film and tube wall resistances are assumed negligible
h_g = $h_{rg} + h_{cg}$ = overall gas side heat transfer coefficient, Btu/h ft^2 F
h_{rg} = radiation heat transfer coefficient (gas side), Btu/h ft^2 F
h_{cg} = convection heat transfer coefficient (gas side), Btu/h ft^2 F
A = total surface area, ft^2
$LMTD$ = log mean temperature difference, gas and saturated water (T_s'), F
$\dot{m}_g$ = gas mass flow rate, lb/h
c_p = mean specific heat of gas, Btu/lb F
ΔT_g = $T_1 - T_2$
T_1 = gas temperature entering tube bank, F
T_2 = gas temperature leaving tube bank, F
T_s' = saturation temperature of boiler water, F

Using the nomenclature listed above, the log mean temperature difference for Case I of Fig. 1 is defined by:

$$LMTD = \frac{(T_1 - T_2)}{\ell n \dfrac{T_1 - T_s'}{T_2 - T_s'}} \quad \textbf{(4)}$$

Heat is transferred to the screen by direct furnace radiation, intertube radiation and convection. Furnace radiation to the screen is considered first. Radiation from the furnace to the screen is calculated from relationships developed in Chapter 4 using an effectiveness of 0.3 which approximates flue gas and screen surface emissivities. The temperature differential ($T_1^4 - T_2^4$) is defined by the furnace exit gas temperature and the screen tube temperature. Due to the large difference between gas and screen tube temperatures and the exponents in the formulation, it is sufficiently accurate to assume that the screen tube temperature is equal to the saturation temperature.

Table 3
Combustion Calculations – Btu Method

	INPUT CONDITIONS – BY TEST OR SPECIFICATION		
1	Excess air: at burner/leaving boiler/econ, % by weight		*20/20/20*
2	Entering air temperature, F		*80*
3	Reference temperature, F		*80*
4	Fuel temperature, F		*80*
5	Air temperature leaving air heater, F		*366*
6	Flue gas temperature leaving (excluding leakage), F		*390*
7	Moisture in air, lb/lb dry air		*0.013*
8	Additional moisture, lb/100 lb fuel		*0*
9	Residue leaving boiler/economizer, % Total		*85*
10	Output, 1,000,000 Btu/h		*285.6*
	Corrections for sorbent (from Chapter 10, Table 14 if used)		
11	Additional theoretical air, lb/10,000 Btu	Table 14, Item [21]	*0*
12	CO_2 from sorbent, lb/10,000 Btu	Table 14, Item [19]	*0*
13	H_2O from sorbent, lb/10,000 Btu	Table 14, Item [20]	*0*
14	Spent sorbent, lb/10,000 Btu	Table 14, Item [24]	*0*

	FUEL – *Bituminous coal, Virginia*					
	15 Ultimate Analysis		16 Theo Air, lb/100 lb fuel		17 H_2O, lb/100 lb fuel	
	Constituent	% by weight	K1	[15] x K1	K2	[15] x K2
A	C	*80.31*	11.51	*924.4*		
B	S	*1.54*	4.32	*6.7*		
C	H_2	*4.47*	34.29	*153.3*	8.94	*39.96*
D	H_2O	*2.90*			1.00	*2.90*
E	N_2	*1.38*				
F	O_2	*2.85*	– 4.32	*–12.3*		
G	Ash	*6.55*				
H	Total	*100.00*	Air	*1072.1*	H_2O	*42.86*

18	Higher heating value (HHV), Btu/lb fuel		*14,100*
19	Unburned carbon loss, % fuel input		*0.40*
20	Theoretical air, lb/10,000 Btu	[16H] x 100 / [18]	*7.603*
21	Unburned carbon, % of fuel	[19] x [18] / 14,500	*0.39*

COMBUSTION GAS CALCULATIONS, Quantity/10,000 Btu Fuel Input

22	Theoretical air (corrected), lb/10,000 Btu	[20] – [21] x 1151 / [18] + [11]	*7.572*
23	Residue from fuel, lb/10,000 Btu	([15G] + [21]) x 100 / [18]	*0.049*
24	Total residue, lb/10,000 Btu	[23] + [14]	*0.049*

			A At Burners	B Infiltration	C Leaving Furnace		D Leaving Blr/Econ	
25	Excess air, % by weight		*20.00*	*0.0*		*20.0*		*20.0*
26	Dry air, lb/10,000 Btu	(1 + [25] / 100) x [22]				*9.086*		*9.086*
27	H_2O from air, lb/10,000 Btu	[26] x [7]			*0.118*	*0.118*	*0.118*	*0.118*
28	Additional moisture, lb/10,000 Btu	[8] x 100 / [18]			*0.000*	*0.000*	*0.000*	*0.000*
29	H_2O from fuel, lb/10,000 Btu	[17H] x 100 / [18]			*0.304*		*0.304*	
30	Wet gas from fuel, lb/10,000 Btu	(100 – [15G] – [21]) x 100 / [18]				*0.660*		*0.660*
31	CO_2 from sorbent, lb/10,000 Btu	[12]				*0.000*		*0.000*
32	H_2O from sorbent, lb/10,000 Btu	[13]			*0.000*	*0.000*	*0.000*	*0.000*
33	Total wet gas, lb/10,000 Btu	Summation [26] through [32]				*9.864*		*9.864*
34	Water in wet gas, lb/10,000 Btu	Summation [27] + [28] + [29] + [32]			*0.422*	*0.422*	*0.422*	*0.422*
35	Dry gas, lb/10,000 Btu	[33] – [34]				*9.442*		*9.442*
36	H_2O in gas, % by weight	[100] x [34] / [33]				*4.28*		*4.28*
37	Residue, % by weight	[9] x [24] / [33]				*0.42*		*0.42*

EFFICIENCY CALCULATIONS, % Input from Fuel

	Losses			
38	Dry gas, %	0.0024 x [35D] x ([6] – [3])		*7.02*
39	Water from fuel, as fired % — Enthalpy of steam at 1 psi, T = [6]	H_1 = (3.958E – 5 x T + 0.4329) x T + 1062.2	*1237.1*	
40	Enthalpy of water at T = [3]	H_2 = [3] – 32	*48.0*	
41		[29] x ([39] – [40]) / 100		*3.61*
42	Moisture in air, %	0.0045 x [27D] x ([6] – [3])		*0.16*
43	Unburned carbon, %	[19] or [21] x 14,500 / [18]		*0.40*
44	Radiation and convection, %	ABMA curve, Chapter 23		*0.40*
45	Other, % (include manufacturers margin if applicable)			*1.50*
46	Sorbent net losses, % if sorbent is used	From Chapter 10, Table 14, Item [41]		*0.00*
47	Summation of losses, %	Summation [38] through [46]		*13.09*
	Credits			
48	Heat in dry air, %	0.0024 x [26D] x ([2] – [3])		*0.00*
49	Heat in moisture in air, %	0.0045 x [27D] x ([2] – [3])		*0.00*
50	Sensible heat in fuel, %	(H at T[4] – H at T[3]) x 100 / [18]	*0.0*	*0.00*
51	Other, %			*0.00*
52	Summation of credits, %	Summation [48] through [51]		*0.00*
53	**Efficiency**, %	100 – [47] + [52]		*86.91*

	KEY PERFORMANCE PARAMETERS		Leaving Furnace	Leaving Blr/Econ
54	Input from fuel, 1,000,000 Btu/h	100 x [10] / [53]		*328.6*
55	Fuel rate, 1000 lb/h	1000 x [54] / [18]		*23.3*
56	Wet gas weight, 1000 lb/h	[54] x [33] / 10	*324.1*	*324.1*
57	Air to burners (wet), lb/10,000 Btu	(1 + [7]) x (1 + [25A] / 100) x [22]	*9.204*	
58	Air to burners (wet), 1000 lb/h	[54] x [57] / 10	*302.5*	
59	Heat available, 1,000,000 Btu/h	[54] x {([18] – 10.30 x [17H]) / [18] – 0.005		
	Ha = *66.0* Btu/lb	x ([44] + [45]) + Ha at T[5] x [57] / 10,000}	*335.2*	
60	Heat available/lb wet gas, Btu/lb	1000 x [59] / [56]	*1034.0*	
61	Adiabatic flame temperature, F	From Chapter 10, Fig. 3 at H = [60], % H_2O = [36]	*3560*	

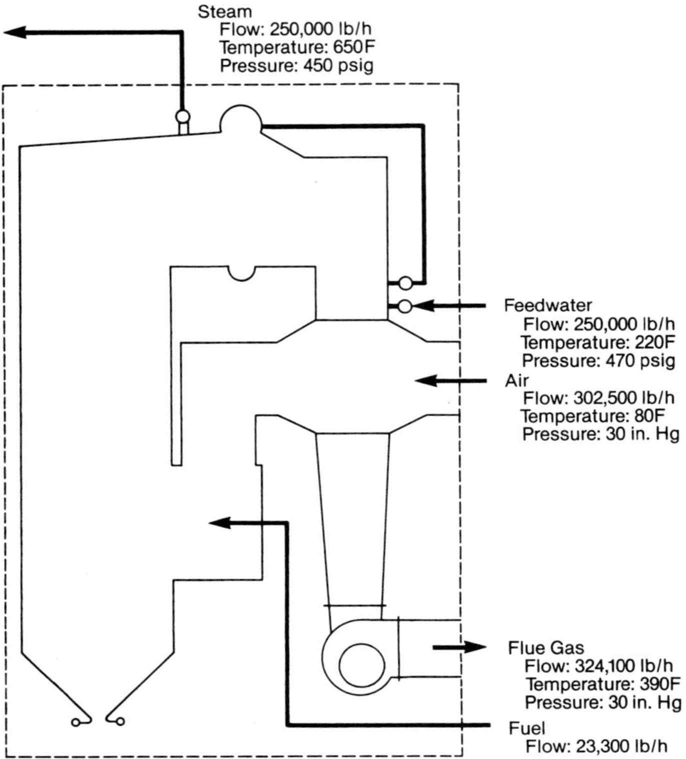

Fig. 3 Example of boiler fuel, air, gas, water and steam flow streams.

$$q'' = \sigma\, F_e \left(T_1^4 - T_2^4\right) = 18{,}420 \text{ Btu/h ft}^2 \tag{5}$$

where

q'' = heat flux, Btu/h ft²
σ = Stefan-Boltzmann constant = 1.71×10^{-9} Btu/h ft² R⁴
F_e = effectiveness factor = 0.3
T_1 = furnace exit gas temperature = 2000F = 2460R
T_2 = saturation temperature = 462F = 922R

The entrance to the screen is 18 ft high and 12 ft wide for a flat projected area of 216 ft². The heat transferred to the screen by furnace radiation is 216 ft² × 18,420 Btu/h ft² or 3.98×10^6 Btu/h.

Based on the configuration of the screen (two rows deep on 6 in. side- and back-spacing), some of the radiant heat is absorbed by the screen and the remainder is absorbed by the superheater. From Chapter 4, Fig. 33, Curve 1, an effectiveness factor of 0.55 (i.e., 55% of the radiant energy entering any row is absorbed) is used to determine radiant screen absorption. On a row by row basis, the radiation from the furnace is distributed as follows:

Furnace radiation to first screen row:	3.98×10^6 Btu/h
First screen row absorption: $0.55 \times 3.98 \times 10^6$ Btu/h =	$\underline{2.19} \times 10^6$ Btu/h
Furnace radiation to second screen row:	1.79×10^6 Btu/h
Second screen row absorption: $0.55 \times 1.79 \times 10^6$ Btu/h =	$\underline{0.99} \times 10^6$ Btu/h
Furnace radiation to superheater:	0.81×10^6 Btu/h

Furnace radiation does not affect the flue gas temperature drop across the screen; however, the furnace radiation absorbed by the screen is taken into account when determining screen steam generation rate. Furnace radiation passing through the screen will be absorbed by the superheater. The evaluation of furnace radiation on superheater performance will be addressed in the superheater section.

The heat transfer to the screen tubes is by convection and intertube radiation, and is calculated from the formulas provided below. Calculations start by assuming a gas temperature leaving the screen. This assumption is verified later. In this case, 1920F is assumed. Log mean temperature difference:

$$LMTD = \frac{(T_1 - T_2)}{\ell n \dfrac{T_1 - T_s'}{T_2 - T_s'}} = 1498\text{F} \tag{6}$$

where

T_1 = furnace exit gas temperature = 2000F
T_2 = gas temperature leaving screen = 1920F
T_s' = saturation temperature = 462F

Gas mass flux:

$$G_g = \dot{m}_g / A_g = 2498 \text{ lb/h ft}^2 \tag{7}$$

where

$\dot{m}_g$ = gas mass flow rate = 324,100 lb/h — Table 3
A_g = minimum gas free flow area = 130 ft² — Table 1

Gas film temperature:

$$T_f = T_s' + (LMTD/2) = 1211\text{F} \tag{8}$$

where

T_s' = 462F
$LMTD$ = 1498F

Gas Reynolds number:

$$\text{Re} = K_{\text{Re}}\, G_g = 5745 \tag{9}$$

where

K_{Re} = gas properties factor = 2.3 h ft²/lb — Fig. 4
G_g = 2498 lb/h ft²

Gas film convection heat transfer coefficient from Chapter 4, Equation 61:

$$h_{cg} = h_c'\, F_{pp}\, F_a\, F_d = 7.53 \text{ Btu/h ft}^2 \text{ F} \tag{10}$$

where

h_c' = basic convection crossflow geometry and velocity factor = 69.9 Btu/h ft² F — Ch. 4, Fig. 18

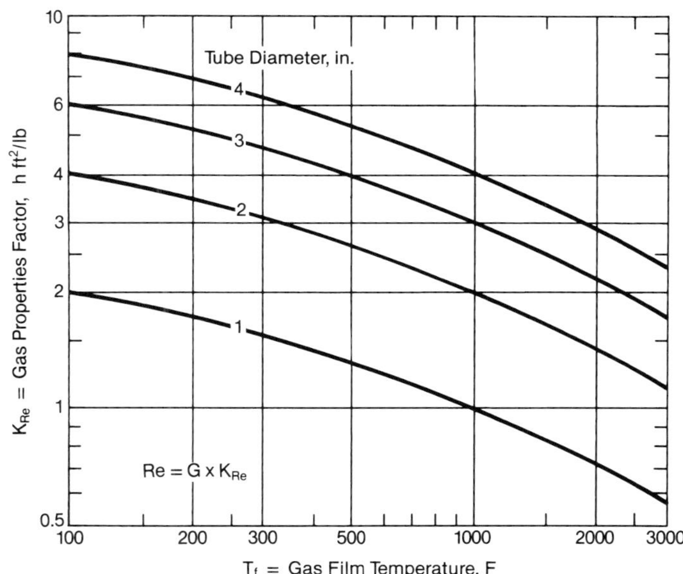

Fig. 4 Determination of Reynolds number, Re.

F_{pp} = physical properties factor = 0.133 — Ch. 4, Fig. 19
F_a = arrangement factor = 0.81 — Ch. 4, Fig. 22
F_d = heat transfer depth factor = 1.0 — Ch. 4, Fig. 23

In this calculation, F_d is 1.0 because the furnace flue gases turn prior to entering the screen.

The total radiation absorption in the screen is comprised of direct furnace radiation and intertube radiation. Direct furnace radiation affects screen steam generation rate, but does not affect the gas temperature leaving the screen. The intertube radiation, however, is a direct function of flue gas temperature leaving the bank. Also, direct furnace radiation is proportional to the planar area it crosses while intertube radiation is proportional to the total bank heating surface.

To determine intertube radiation, the radiation heat transfer coefficient must be adjusted to eliminate direct furnace radiation through the use of an effectiveness factor based on areas.

$$F_s = \frac{\text{Effective surface}}{\text{Total surface}} = \frac{A - A_p}{A} = 0.681 \quad \textbf{(11)}$$

where

A = total bank heating surface = 542 ft² — Table 1
A_p = planar area of the bank credited with radiation absorption. In this example, 80% of the direct furnace radiation was absorbed by the screen.
= $0.8 \times 12 \times 18 = 172.8$ ft²

Because this calculation subtracts the effect of direct furnace radiation, it will be added back in when the total screen absorption is finally determined.

Gas side radiation heat transfer coefficient adjusting for effective surface:

$$h_{rg} = h_r' \, K \, F_s = 2.35 \text{ Btu/h ft}^2 \text{ F} \quad \textbf{(12)}$$

where

h_r' = 8.2 Btu/h ft² F — Fig. 5
p_r = partial pressure = 0.19 atm — Fig. 6
L = mean radiating length = 1.42 ft — Fig. 7
K = fuel factor = 0.42 — Fig. 8
F_s = 0.681

Combined heat transfer coefficient:

$$h_g = h_{cg} + h_{rg} = 9.88 \text{ Btu/h ft}^2 \text{ F} \quad \textbf{(13)}$$

where

h_{cg} = 7.53 Btu/h ft² F
h_{rg} = 2.35 Btu/h ft² F

Overall heat transfer rate:

$$q = U \, A \left(LMTD\right) = 8.02 \times 10^6 \text{ Btu/h} \quad \textbf{(14)}$$

where

U = h_g = 9.88 Btu/h ft² F
A = 542 ft²
$LMTD$ = 1498F

To verify the flue gas exit temperature assumption of T_2 = 1920F for the screen tubes, the gas temperature leaving the screen can be calculated by an en-

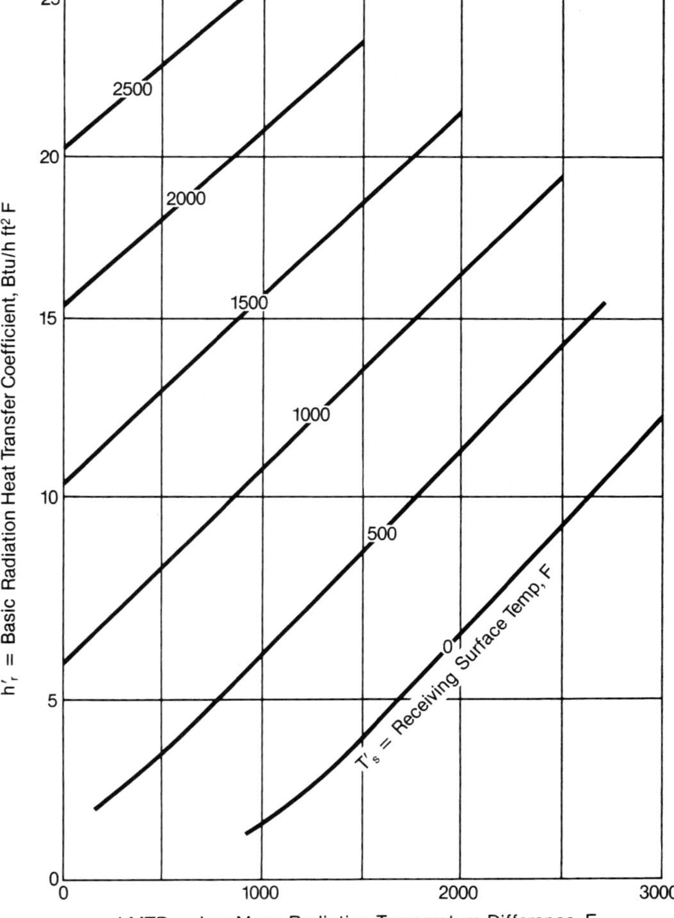

Fig. 5 Basic radiation heat transfer coefficient, h_r', for Equations 12, 32, and 44.

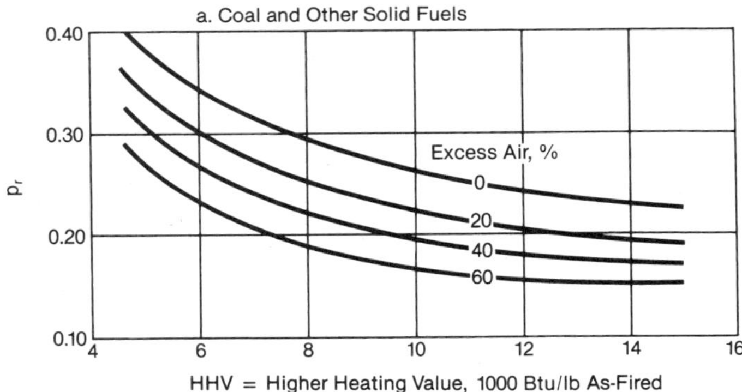

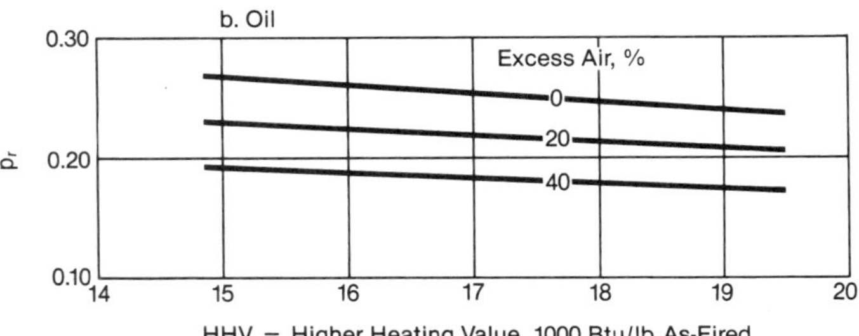

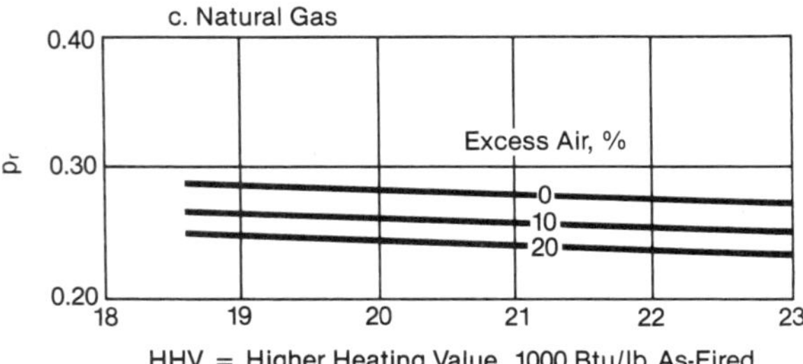

Fig. 6 Partial pressure, p_r, of principal radiating constituents (CO_2 + H_2O) of combustion gases in atmosphere for various fuels, heat values and excess air.

ergy balance between the energy absorbed by the screen tubes (excluding furnace radiation) and the energy lost by the flue gas.

$$T_2 = T_1 - q / \dot{m}_g \, c_p = 1920\text{F} \qquad \textbf{(15)}$$

where

T_1	= 2000F	
q	= 8.02 × 10⁶ Btu/h	
$\dot{m}_g$	= 324,100 lb/h	Eq. 3
c_p	= 0.31 Btu/lb F	Fig. 9

Solving the heat balance for T_2 indicates agreement with the earlier assumption of 1920F, and no recalculation is required. Otherwise, T_2 from Equation 15 would be used in Equation 6 and the calculation repeated.

To complete the screen heat transfer analysis, the total screen absorption is determined. Screen absorption is the sum of the convection and intertube radiation heat transfer rates (8.02 × 10^6 Btu/h) and direct furnace radiation rate (3.18 × 10^6 Btu/h) for a total screen absorption of 11.20 × 10^6 Btu/h.

Draft loss

Screen gas side draft loss is calculated from Chapter 3, Equations 52 and 56:

$$\Delta P = (f \; N \; F_d)\left(\frac{30}{B}\right)\left(\frac{T+460}{1.73\times10^5}\right)\left(\frac{G}{10^3}\right)^2 = 0.04 \text{ in. wg} \qquad \textbf{(16)}$$

where

f	= friction factor = 0.23	Ch. 3, Fig. 15
N	= number of tube rows = 2	Table 1
F_d	= tube bundle correction factor = 1.12	Ch. 3, Fig. 14
B	= barometric pressure = 30 in. Hg	Table 2
T	= 0.95 (T_1 + T_2)/2 = 1862F	(Note 1 below)
G	= 2498 lb/h ft²	Eq. 7

Pressure drop

The screen tubes are part of the furnace circuitry and tube side pressure drop calculations for this surface are included in a circulation analysis. Refer to Chapter 5 for circulation information.

Superheater

Heat transfer

The next component in the direction of flue gas flow is the superheater. Saturated steam from the steam drum is heated to the outlet conditions shown in Table 2.

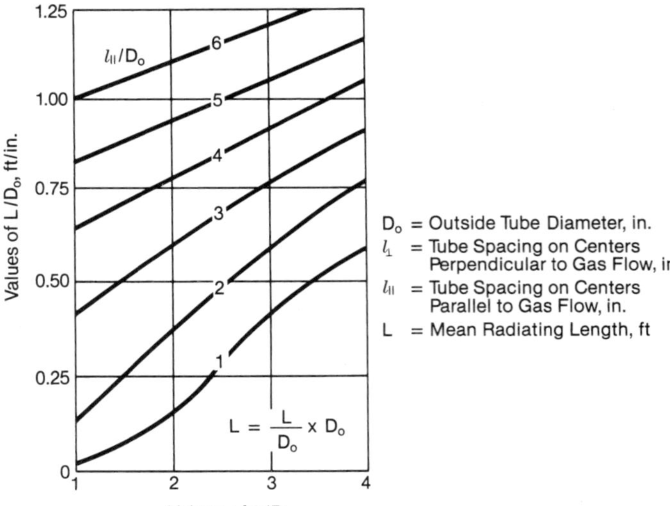

Fig. 7 Mean radiating length, *L*, for various tube diameters and arrangements or pitches – in-line tubes.

Note 1: Under most conditions, this equation is a good approximation of effective flue gas temperature.

The governing heat transfer equations for superheater surfaces are:

$$q = U\,A\,(LMTD) \quad \textbf{(17)}$$

$$LMTD = \frac{(T_1 - T_2') - (T_2 - T_1')}{\ell n \dfrac{(T_1 - T_2')}{(T_2 - T_1')}} \quad \textbf{(18)}$$

Fig. 1, Case III

$$q = \dot{m}_g\, c_p\, \Delta T_g = \dot{m}_g\, c_p\, (T_1 - T_2) \quad \textbf{(19)}$$

$$q = \dot{m}_s\, \Delta H \quad \textbf{(20)}$$

where

q = heat transfer rate, Btu/h
U = $(h_g\, h_s)/(h_g + h_s)$ = combined heat transfer coefficient, Btu/h ft^2 F (assuming negligible wall resistance)
h_g = $h_{rg} + h_{cg}$ = overall gas side heat transfer coefficient, Btu/h ft^2 F
h_{rg} = radiation heat transfer coefficient (gas side), Btu/h ft^2 F
h_{cg} = convection heat transfer coefficient (gas side), Btu/h ft^2 F
h_s = convection heat transfer coefficient (steam side), Btu/h ft^2 F
A = total surface area, ft^2
$LMTD$ = counterflow log mean temperature difference, gas and steam, F
T_1 = gas temperature entering superheater, F
T_2 = gas temperature leaving superheater, F
T_1' = steam temperature entering superheater, F
T_2' = steam temperature leaving superheater, F
$\dot{m}_g$ = mass flow of gas, lb/h
c_p = mean specific heat of gas, Btu/lb F
ΔT_g = $T_1 - T_2$, gas temperature differential, F
$\dot{m}_s$ = mass flow of steam, lb/h
ΔH = steam enthalpy difference, Btu/lb

Superheater steam side design conditions are:

Outlet:

$T_2' = 650$F, $P_2' = 450$ psig, $H_2' = 1331$ Btu/lb

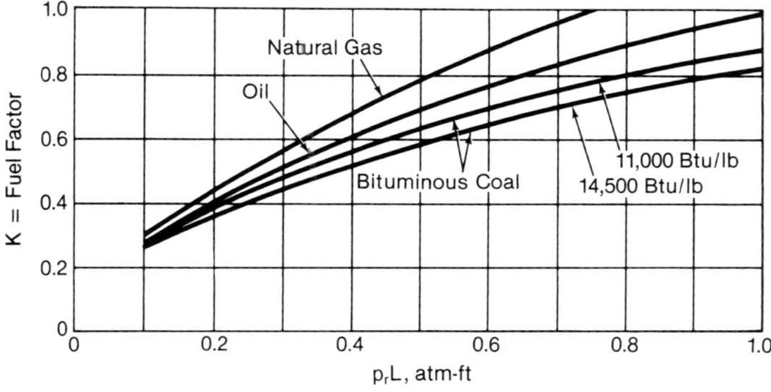

Fig. 8 Effect of fuel, partial pressure (H_2O and CO_2) and mean radiating length on radiation heat transfer coefficient.

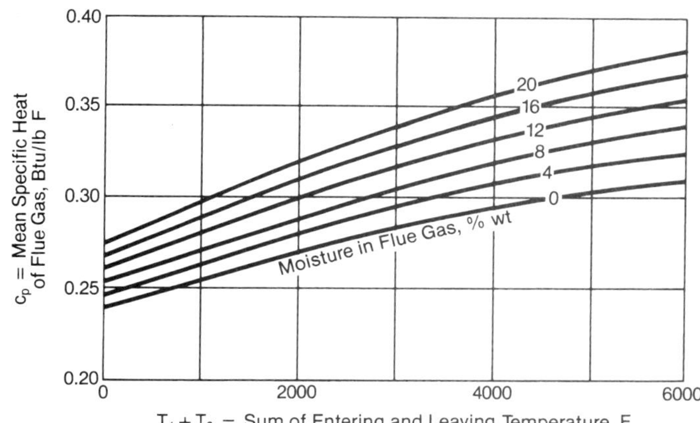

Fig. 9 Approximate mean specific heat, c_p, of flue gas.

Inlet:

$T_1' = 462$F, $P_1' = 460$ psig, $H_1' = \underline{1205}$ Btu/lb

$\Delta H = \;126$ Btu/lb

The outlet conditions are specified in Table 2 while the inlet conditions are assumed to be saturated steam at the drum pressure. Drum pressure is determined by superheater pressure drop and is assumed here based upon experience and verified later.

The heat transfer rate to the superheater is calculated as follows:

$$q = \dot{m}_s\, \Delta H = 31.50 \times 10^6 \text{ Btu/h} \quad \textbf{(21)}$$

where

$\dot{m}_s$ = 250,000 lb/h — Table 2
ΔH = 126 Btu/lb

Previous calculations determined that the superheater will receive 0.81×10^6 Btu/h furnace radiation. Therefore, the heat transferred by convection and intertube radiation is:

$$q_{ci} = q - q_r = 30.69 \times 10^6 \text{ Btu/h} \quad \textbf{(22)}$$

where

q = 31.50×10^6 Btu/h
q_r = 0.81×10^6 Btu/h

By rearranging Equation 19, the gas temperature leaving the superheater can be determined:

$$T_2 = T_1 - q_{ci} / (\dot{m}_g\, c_p) = 1608 \text{ F} \quad \textbf{(23)}$$

where

q_{ci} = 30.69×10^6 Btu/h
T_1 = gas temperature entering superheater
= gas temperature leaving screen =1920F
$\dot{m}_g$ = 324,100 lb/h
c_p = 0.303 Btu/lb F — Fig. 9

Superheater log mean temperature difference is:

$$LMTD = \frac{(T_1 - T_2') - (T_2 - T_1')}{\ell n \dfrac{(T_1 - T_2')}{(T_2 - T_1')}} = 1207\text{F} \quad \textbf{(24)}$$

where

T_1 = 1920F
T_2 = 1608F
T'_1 = 462F
T'_2 = 650F

The average gas side film temperature is approximated by (see Note 2 below):

$$T_f = (T'_1 + T'_2)/2 + LMTD/2 = 1160\text{F} \quad \textbf{(25)}$$

where

T'_1 = 462F
T'_2 = 650F
$LMTD$ = 1207F

Superheater tube material and thickness are selected according to the American Society of Mechanical Engineers (ASME) Code and manufacturing capabilities which are discussed in Chapters 7 and 8. For this example, a 2.50 in. OD carbon steel seamless tube with 0.220 in. wall thickness has been selected. Thickness is set by tube bending limitations and is normally greater than that required by Code. Allowing for manufacturing tolerances (+15% for pressure tubing), the average inside diameter (ID) of the tube is calculated to be 1.99 in. The flow area corresponding to this diameter is 3.12 in.^{2}. The total steam flow area is 1.0 ft^2 (2 flow × 23 rows × 3.12 in.^{2} × ft^2 /144 in.^{2}).

Steam mass flux:

$$G_s = \dot{m}_s / A_s = 250{,}600 \text{ lb/h ft}^2 \quad \textbf{(26)}$$

where

$\dot{m}_s$ = 250,000 lb/h
A_s = 0.998 ft^2

Steam Reynolds number:

$$\text{Re} = \frac{G_s D_e}{\mu} = 867{,}600 \quad \textbf{(27)}$$

where

D_e = 1.99 in. = 0.166 ft
μ = steam absolute viscosity
= 0.048 lb/h ft — Ch. 3, Fig. 5
G_s = 250,600 lb/h ft^2

Gas mass flux:

$$G_g = \dot{m}_g / A_g = 2498 \text{ lb/h ft}^2 \quad \textbf{(28)}$$

where

$\dot{m}_g$ = 324,100 lb/h
A_g = 130 ft^2 — Table 1

Gas side Reynolds number:

$$\text{Re} = K_{\text{Re}} \, G_g = 6495 \quad \textbf{(29)}$$

where

K_{Re} = 2.6 h ft^2/lb — Fig. 4
G_g = 2498 lb/h ft^2

Gas film convection heat transfer coefficient from Equation 61 in Chapter 4:

$$h_{cg} = h'_c F_{pp} F_a F_d = 5.91 \text{ Btu/h ft}^2 \text{ F} \quad \textbf{(30)}$$

where

h'_c = 69.9 Btu/h ft^2 F — Ch. 4, Fig. 18
F_{pp} = 0.133 — Ch. 4, Fig. 19
F_a = 0.635 — Ch. 4, Fig. 22
F_d = 1.0 — Ch. 4, Fig. 23

To obtain the gas side radiation heat transfer coefficient (h_{rg}), a factor, F_s, must be included to account for the furnace radiation absorbed in the superheater. In the screen calculations, it was shown that 80% of the furnace radiation was absorbed in the screen while 20% passed through the screen and was absorbed by the superheater. Similar to the screen calculations, superheater intertube radiation will be determined by eliminating direct furnace radiation from the radiation heat transfer coefficient through the use of an effectiveness factor:

$$F_s = \frac{A - A_p}{A} = 0.986 \quad \textbf{(31)}$$

where

A = 3169 ft^2
A_p = 0.2 (12 × 18) = 43.2 ft^2 — Fig. 2

Gas side radiation heat transfer coefficient:

$$h_{rg} = h'_r K F_s = 2.81 \text{ Btu/h ft}^2 \text{ F} \quad \textbf{(32)}$$

where

h'_r = 9.5 Btu/h ft^2 F — Fig. 5
p_r = 0.19 atm — Fig. 6
L = 0.85 ft — Fig. 7
K = 0.30 — Fig. 8
F_s = 0.986

In a superheater, the resistance to heat transfer through the steam film inside the tubes can not be assumed to be negligible as was done with the screen. Steam film convection heat transfer coefficient from Chapter 4, Equation 58, corrected to the OD surface area:

$$h_s = h'_l F_{pp} F_T D_i / D_o = 148 \text{ Btu/h ft}^2 \text{ F} \quad \textbf{(33)}$$

Note 2: This approximation applies when one working fluid controls the overall heat transfer coefficient.

where

h'_l = 687 Btu/h ft² F — Ch. 4, Fig. 13
F_{pp} = 0.33 — Ch. 4, Fig. 16
F_T = temperature factor = 0.82 — Ch. 4, Fig. 17
D_i = 1.99 in.
D_o = 2.50 in.

Overall heat transfer coefficient:

$$U = \frac{h_g\, h_s}{h_g + h_s} = \frac{\left(h_{rg} + h_{cg}\right) h_s}{h_{rg} + h_{cg} + h_s} \qquad \textbf{(34)}$$
$$= 8.23 \text{ Btu/h ft}^2 \text{ F}$$

where

h_{rg} = 2.81 Btu/h ft² F
h_{cg} = 5.91 Btu/h ft² F
h_s = 148 Btu/h ft² F

Overall heat transfer rate:

$$q = U\,A\left(LMTD\right) = 31.48 \times 10^6 \text{ Btu/h} \qquad \textbf{(35)}$$

where

U = 8.23 Btu/h ft² F
A = 3169 ft² — Table 1
$LMTD$ = 1207 F

Because this agrees with 31.50 × 10⁶ Btu/h from Equation 21, no iteration is required. If these heat transfer rate calculations do not agree, then the steam outlet temperature or the superheater surface must be re-estimated and the calculations repeated until agreement is achieved.

Draft loss

Superheater gas side draft loss is determined by combining Equations 52 and 56 in Chapter 3.

$$\Delta P = \left(f\ N\ F_d\right)\left(\frac{30}{B}\right)\left(\frac{T + 460}{1.73 \times 10^5}\right)\left(\frac{G}{10^3}\right)^2 \qquad \textbf{(36)}$$
$$= 0.06 \text{ in. wg}$$

where

f = 0.06 — Ch. 3, Fig. 15
N = 12 — Table 1
F_d = 1.0 — Ch. 3, Fig. 14
B = 30 in. Hg
T = 0.95 $(T_1 + T_2)/2$ = 1676F
G = 2498 lb/h ft²

Steam pressure drop

Superheater steam side pressure drop is the sum of friction or straight-flow losses, entrance and exit losses, and bend losses. Equations 47 and 51 in Chapter 3 can be combined as follows:

$$\Delta P = \Delta P_f + \Delta P_{e+e} + \Delta P_b$$
$$= \left(\frac{fL}{D_i} + \frac{1.5}{12} + \frac{N_b}{12}\right)\upsilon\left(\frac{G}{10^5}\right)^2 = 8.24 \text{ psi} \qquad \textbf{(37)}$$

where

ΔP_f = frictional pressure drop (fL/D_i)
ΔP_{e+e} = entrance (1/12) and exit (0.5/12) pressure drop
ΔP_b = bend loss $(N_b/12)$
G = steam mass flux = 250,600 lb/h ft²
f = 0.012 — Ch. 3, Fig. 1
L = length of one continuous superheater tube from steam drum to superheater outlet header = 130 ft
D_i = 1.99 in.
υ = average steam specific volume = 1.15 ft³/lb
N_b = bend loss factor = 2.8

The superheater is a two-flow design. Steam side pressure drop will be determined for the steam path with the highest bend loss factors which, in this example, is the path with three short radius 180 degree bends. A composite bend loss factor for this path is determined as follows:

N_b:		
Three 180 degree bends, R/D = 0.81, 3 × 0.6	= 1.80	Ch. 3, Fig. 9
Two 180 degree bends, R/D = 2.44, 2 × 0.28	= 0.56	Ch. 3, Fig. 9
Two 90 degree bends, R/D = 2.44, 2 × 0.22	= 0.44	Ch. 3, Fig. 9
Composite N_b	= 2.80	

Drum pressure and saturation temperature can now be determined and verified. The steam drum pressure is equal to the outlet steam pressure plus the superheater pressure loss calculated above, plus the steam separation equipment loss. Note, the steam separation equipment loss is manufacturer specific.

Superheater outlet pressure	450 psig	Table 2
Superheater-to-terminal pipe loss	+ 0 psi	(assumed)
Superheater pressure drop	+ 8 psi	
Steam separation equipment	+ 2 psi	(assumed)
Drum pressure	460 psig	
Sat. temperature (at 460 psig)	462 F	

This saturation pressure is in agreement with that originally assumed.

Boiler bank

Heat transfer

The function of the boiler bank, like the screen tubes, is to boil water, and the governing heat transfer equations defined for the screen are also applicable. Heat is transferred by convection, intertube radiation, and radiation from the rear cavity. In this example there are no cavities in the screen or superheater; however, in many applications cavities exist to accommodate sootblowers. Whenever cavities surround a bank of tubes, whether front or rear, the impact of the cavities on heat transfer must be considered (see discussion on cavity heat transfer).

Calculations begin by assuming the gas temperature leaving the boiler bank to be 818F based upon prior experience. The log mean temperature difference is calculated from Fig. 1, Case I:

$$LMTD = \frac{(T_1 - T_2)}{\ell n \dfrac{T_1 - T_s'}{T_2 - T_s'}} = 676\text{F} \tag{38}$$

where

T_1 = gas temperature entering the boiler bank
= gas temperature leaving the superheater
= 1608F
T_2 = assumed bank exit temperature = 818F
T_s' = saturation temperature = 462F

Gas mass flux:

$$G_g = \dot{m}_g / A_g = 3813 \text{ lb/h ft}^2 \tag{39}$$

where

$\dot{m}_g$ = 324,100 lb/h
A_g = 85 ft^2 — Table 1

Gas film temperature:

$$T_f = T_s' + LMTD/2 = 800\text{F} \tag{40}$$

where

T_s' = 462F
$LMTD$ = 676F

Gas Reynolds number:

$$\text{Re} = K_{\text{Re}}\, G_g = 10{,}700 \tag{41}$$

where

K_{Re} = 2.8 h ft^2/lb — Fig. 4
G_g = 3813 lb/h ft^2

Gas film heat transfer coefficient:

$$h_{cg} = h_c' \, F_{pp} \, F_a \, F_d = 8.45 \text{ Btu/h ft}^2 \text{ F} \tag{42}$$

where

h_c' = 90.7 Btu/h ft^2 F — Ch. 4, Fig. 18
F_{pp} = 0.118 — Ch. 4, Fig. 19
F_a = 0.79 — Ch. 4, Fig. 22
F_d = 1.0 — Ch. 4, Fig. 23

For the screen, it was determined that 80% of the direct furnace radiation is absorbed in a two-row bank. It can be shown that all rear cavity radiation is absorbed in the boiler bank because it is 28 rows deep. To calculate intertube radiation, an effectiveness factor must again be determined:

$$F_s = \frac{A - A_p}{A} = 0.98 \tag{43}$$

where

A = 11,545 ft^2 — Table 1
A_p at 100% = 12 × 18 = 216 ft^2 — Fig. 2

Gas side radiation heat transfer coefficient:

$$h_{rg} = h_r' \, K \, F_s = 0.83 \text{ Btu/h ft}^2 \text{ F} \tag{44}$$

where

h_r' = 4.0 Btu/h ft^2 F — Fig. 5
p_r = 0.19 — Fig. 6
L = 0.47 ft — Fig. 7
K = 0.212 — Fig. 8
F_s = 0.98

Combined heat transfer coefficient:

$$h_g = h_{cg} + h_{rg} = 9.29 \text{ Btu/h ft}^2 \text{ F} \tag{45}$$

where

h_{cg} = 8.45 Btu/h ft^2 F
h_{rg} = 0.83 Btu/h ft^2 F

Overall heat transfer rate assuming negligible wall and boiling resistances:

$$q = U\, A(LMTD) = 72.4 \times 10^6 \text{ Btu/h} \tag{46}$$

where

U = h_g = 9.29 Btu/h ft^2 F
A = 11,545 ft^2 — Table 1
$LMTD$ = 676F

Checking the gas temperature leaving the boiler bank:

$$T_2 = T_1 - q / (\dot{m}_g \, c_p) = 818\text{F} \tag{47}$$

where

T_1 = 1608F
q = 72.4 × 10^6 Btu/h
$\dot{m}_g$ = 324,100 lb/h
c_p = 0.283 Btu/lb F — Fig. 9

This agrees with the original outlet temperature assumption and therefore no recalculation is required. Otherwise, T_2 would be used in Equation 38 and the calculations repeated.

Draft loss

The governing equation for the gas side draft loss in the boiler bank is:

$$\Delta P = (f \, N \, F_d)\left(\frac{30}{B}\right)\left(\frac{T + 460}{1.73 \times 10^5}\right)\left(\frac{G}{10^3}\right)^2 = 1.03 \text{ in. wg} \tag{48}$$

where

f = 0.27 — Ch. 3, Fig. 15
N = 28 — Table 1
F_d = 1.0 — Ch. 3, Fig. 14
B = 30 in. Hg — Table 2
T = 0.95 $(T_1 + T_2)/2$ = 1152F
G = 3813 lb/h ft^2

Pressure drop

As with the screen tubes, the boiler bank is integral to the furnace circuitry. Pressure loss to the steam-water flow is discussed in Chapter 5.

Cavity between boiler bank and economizer

Heat transfer

Heat is transferred from each cavity to the cooler banks which form its boundaries. Cavity radiation is most significant at higher gas temperatures. In this example, cavity radiation has little impact on the overall results, but is included to illustrate the procedure for evaluating other configurations.

Assume the gas temperature leaving the cavity, entering the economizer to be 815F and the water temperature leaving the economizer to be 280F. Both assumptions are verified later. Consider cavity radiation to the boiler bank first.

Log mean temperature difference is approximated by:

$$LMTD = (T_1 + T_2)/2 - T_s' = 355\text{F} \tag{49}$$

where

T_1 = gas temperature entering the cavity
= gas temperature leaving the boiler bank
= 818F
T_2 = 815F (assumed)
T_s' = saturation temperature = 462F

By definition, mean radiating length:

$$L = 3.4\frac{V_L}{A} = 7.1\text{ ft} \tag{50}$$

where

V_L = volume of the cavity
= 12 ft × 18 ft × 10 ft = 2160 ft^3 — Fig. 2
A = area = 2 (12 × 18 + 12 × 10 + 10 × 18)
= 1032 ft^2 — Fig. 2

Gas side radiation heat transfer coefficient:

$$h_{rg} = h_r' K = 2.28\text{ Btu/h ft}^2\text{ F} \tag{51}$$

where

h_r' = 2.4 Btu/h ft^2 F — Fig. 5
p_r = 0.19 atm — Fig. 6
L = 7.1 ft
K = 0.95 — Fig. 8 (Extrapolated)

Because radiation is the only significant mode of heat transfer in the cavity, the overall heat transfer coefficient $U = h_{rg}$. The overall heat transfer rate to the boiler bank is then:

$$q = U\,A(LMTD) = 174{,}600\text{ Btu/h} \tag{52}$$

where

U = h_{rg} = 2.28 Btu/h ft^2 F
A = 12 × 18 = 216 ft^2 — Fig. 2
$LMTD$ = 355F

Radiation to the economizer follows the same logic.

$$LMTD = (T_1 + T_2)/2 - T_2' = 535\text{F} \tag{53}$$

where

T_1 = gas temperature entering the cavity
= 818F
T_2 = gas temperature leaving the cavity
= 815F
T_2' = assumed economizer water outlet temperature
= 282F

$$h_{rg} = h_r' K = 1.90\text{ Btu/h ft}^2\text{ F} \tag{54}$$

where

h_r' = 2.0 Btu/h ft^2 F — Fig. 5
p_r = 0.19 atm — Fig. 6
L = 7.1 ft
K = 0.95 — Fig. 8

$$q = U\,A(LMTD) = 121{,}900\text{ Btu/h} \tag{55}$$

where

U = h_{rg} = 1.90 Btu/h ft^2 F
A = 12 × 10 = 120 ft^2 — Fig. 2
$LMTD$ = 535F

Total heat transfer is the sum of the rates to the boiler bank and economizer, or 296,500 Btu/h. Checking the gas temperature leaving the cavity:

$$T_2 = T_1 - q/(\dot{m}_g\, c_p) = 815\text{F} \tag{56}$$

where

T_1 = 818F
q = 296,500 Btu/h
$\dot{m}_g$ = 324,100 lb/h
c_p = 0.27 Btu/lb F — Fig. 9

This is the same as the assumed value and further iteration is not needed. Otherwise, T_2 from Equation 56 would be used in Equation 49 and the calculations repeated. Verification of the economizer water outlet temperature is shown in the next section.

The total boiler bank absorption can now be determined. Absorption due to convection and intertube radiation is 72.4 × 10^6 Btu/h (Equation 46) while that from cavity radiation is 0.17 × 10^6 Btu/h (Equation 52) for a total boiler bank absorption of 72.6 × 10^6 Btu/h.

Economizer

The next component in the direction of flue gas flow is the economizer. The economizer heats the feedwater before it enters the steam drum.

Heat transfer

Economizer heat transfer follows the same formulations established for the superheater:

$$q = U\,A(LMTD) \quad \textbf{(57)}$$

$$q = \dot{m}_g\,c_p\,\Delta T_g = \dot{m}_g\,c_p(T_2 - T_1) \quad \textbf{(58)}$$

$$q = \dot{m}\,\Delta H \quad \textbf{(59)}$$

where the terms are generally defined after Equations 17 to 20 except the tube side fluid is water instead of steam.

Economizer outlet water temperature was previously assumed to be 282F. It was also established that 121,900 Btu/h is transferred from the cavity preceding the economizer. The total heat transfer rate to the economizer is calculated to be:

$$q = \dot{m}\,\Delta H = 15.70 \times 10^6 \text{ Btu/h} \quad \textbf{(60)}$$

where

$\dot{m}$ = 250,000 lb/h — Table 2
T_2' = 282F
H_2 = 252 Btu/lb — Ch. 2, Table 3
T_1' = 220F — Table 2
H_1 = 189 Btu/lb — Table 2
ΔH = $H_2 - H_1$ = 63.0 Btu/lb

The heat transfer by convection and intertube radiation is:

$$q_{ci} = q - q_r = 15.58 \times 10^6 \text{ Btu/h} \quad \textbf{(61)}$$

where

q = total heat transfer rate = 15.70 × 10[6] Btu/h
q_r = cavity radiation = 121,900 Btu/h

As can be seen from this result, cavity thermal radiation at low temperatures is not usually significant.

By rearranging Equation 58, the gas temperature leaving the economizer can be determined:

$$T_2 = T_1 - q_{ci}/(\dot{m}_g\,c_p) = 635\text{F} \quad \textbf{(62)}$$

where

q_{ci} = 15.58 × 10[6] Btu/h
$\dot{m}_g$ = 324,100 lb/h — Table 3
c_p = 0.268 Btu/lb F — Fig. 9
T_1 = 815F

From Fig. 1, Case III, for counterflow heat exchanger, the log mean temperature difference is:

$$LMTD = \frac{(T_1 - T_2') - (T_2 - T_1')}{\ell n \dfrac{(T_1 - T_2')}{(T_2 - T_1')}} = 472\text{F} \quad \textbf{(63)}$$

where

T_1 = economizer inlet gas temperature = 815F
T_2 = economizer outlet gas temperature = 635F
T_1' = economizer water inlet temperature = 220F — Table 2
T_2' = economizer water outlet temperature = 282F

Average gas film temperature is approximated by (See previous Note 2):

$$T_f = (T_1' + T_2')/2 + LMTD/2 = 487\text{F} \quad \textbf{(64)}$$

where

T_1' = 220F
T_2' = 282F
$LMTD$ = 472F

Gas mass flux:

$$G_g = \dot{m}_g / A_g = 7780 \text{ lb/h ft}^2 \quad \textbf{(65)}$$

where

$\dot{m}_g$ = 324,100 lb/h — Table 3
A_g = 42 ft[2] — Table 1

Reynolds number:

$$\text{Re} = K_{\text{Re}}\,G_g = 23{,}300 \quad \textbf{(66)}$$

where

K_{Re} = 3.0 h ft[2] /lb — Fig. 4
G_g = 7780 lb/h ft[2]

Gas film heat transfer coefficient:

$$h_{cg} = h_c'\,F_{pp}\,F_a\,F_d = 13.14 \text{ Btu/h ft}^2\text{ F} \quad \textbf{(67)}$$

where

h_c' = 152.6 Btu/h ft[2] F — Ch. 4, Fig. 18
F_{pp} = 0.105 — Ch. 4, Fig. 19
F_a = 0.82 — Ch. 4, Fig. 22
F_d = 1.0 — Ch. 4, Fig. 23

As for the preceding components, an effectiveness factor based on the total economizer surface area is determined:

$$F_s = \frac{A - A_p}{A} = 0.951 \quad \textbf{(68)}$$

where

A = 2456 ft[2] — Table 1
A_p at 100% = (12 ft) (10 ft) = 120 ft[2] — Fig. 2

Gas side radiation heat transfer coefficient:

$$h_{rg} = h_r'\,K\,F_s = 0.30 \text{ Btu/h ft}^2\text{ F} \quad \textbf{(69)}$$

where

h'_r = 1.6 Btu/h ft² F — Fig. 5
p_r = 0.19 — Fig. 6
L = 0.30 ft — Fig. 7
K = 0.20 — Fig. 8
F_s = 0.951

The water film and tube wall resistances are negligible so the total heat transfer coefficient is:

$$U = h_g = h_{cg} + h_{rg} = 13.44 \text{ Btu/h ft}^2 \text{ F} \qquad \textbf{(70)}$$

where

h_{cg} = 13.14 Btu/h ft² F
h_{rg} = 0.30 Btu/h ft² F

Overall heat transfer rate:

$$q = U\,A(LMTD) = 15.57 \times 10^6 \text{ Btu/h} \qquad \textbf{(71)}$$

where

U = 13.44 Btu/h ft² F
A = 2456 ft² — Table 1
$LMTD$ = 472F

After adding cavity radiation (121,900 Btu/h), the total heat transfer rate to the economizer is 15.69 × 10⁶ Btu/h.

Verifying the water outlet temperature assumption, the outlet enthalpy is evaluated from:

$$H_2 = H_1 + q/\dot{m} = 252 \text{ Btu/lb or } T'_2 = 282\text{F} \qquad \textbf{(72)}$$

where

q = 15.69 × 10^6 Btu/h
$\dot{m}$ = 250,000 lb/h
H_1 = 189 Btu/lb

From steam tables at an enthalpy of 252 Btu/lb, the water temperature is verified to be 282F. Therefore, the total economizer absorption is 15.69 × 10^6 Btu/h.

Draft loss

Economizer gas side draft loss is calculated from Chapter 3, Equations 52 and 56.

$$\Delta P = (f\,N\,F_d)\left(\frac{30}{B}\right)\left(\frac{T+460}{1.73\times10^5}\right)\left(\frac{G}{10^3}\right)^2 = 1.57 \text{ in. wg} \qquad \textbf{(73)}$$

where

f = 0.39 — Ch. 3, Fig. 15
N = 10 — Table 1
F_d = 1.0 — Ch. 3, Fig. 14
B = 30 in. H_2O
T_1 = 815F
T_2 = 635F
T = 0.95 $(T_1 + T_2)/2$ = 689F
G = 7780 lb/h ft²

Water side pressure drop

In this example, the economizer tubes are 2 in. OD with 0.148 in. wall thickness. Considering the manufacturing tolerance of plus 15% in the pressure tubing wall thickness, the tube inside diameter is 1.66 in. Water flow area is 2.16 in.²/tube and the total flow area is 0.706 ft² (2.16 in.²/tube × 47 tubes × 1.0 ft²/144 in.²). Water mass flux is:

$$G = \dot{m}/A = 354{,}100 \text{ lb/h ft}^2 \qquad \textbf{(74)}$$

where

$\dot{m}$ = 250,000 lb/h
A = 0.706 ft²

Water side Reynolds number:

$$\text{Re} = \frac{GD_i}{\mu} = 98{,}000 \qquad \textbf{(75)}$$

where

G = 354,100 lb/h ft²
D_i = 1.66 in. = 0.138 ft
μ = 0.5 lb/h ft — Ch. 3, Fig. 3

Economizer pressure drop is the sum of friction losses, entrance and exit losses, and bend losses.

$$\Delta P = \left(\frac{fL}{D_i} + \frac{1.5}{12} + \frac{N_b}{12}\right)\upsilon\left(\frac{G}{10^5}\right)^2 = 0.36 \text{ psi} \qquad \textbf{(76)}$$

where

f = 0.018 — Ch. 3, Fig. 1
L = 101 ft
D_i = 1.66 in.
N_b = nine 180 degree 0.79 R/D bends
= 9 × 0.61 = 5.49 — Ch. 3, Fig. 9
υ = specific volume = 0.017 ft³/lb — Ch. 2, Table 3
G = 354,100 lb/h ft²

The total pressure drop (economizer inlet to the steam drum) must include the static head of water for that elevation difference (25 ft) plus fittings and friction losses. Assuming the feedwater pipe diameter is large and fitting losses are negligible ($\Delta P_{piping} = 0$), then the static head is calculated by:

$$\Delta P_{static} = \frac{\Delta Z}{144\upsilon} = 10.0 \text{ psi} \qquad \textbf{(77)}$$

where

ΔZ = elevation between drum centerline and economizer inlet header
= 25 ft
υ = 0.0173 ft³/lb — Ch. 2, Table 3

Total pressure drop from economizer inlet to drum is:

$$\Delta P = \Delta P_{economizer} + \Delta P_{static} + \Delta P_{piping} = 10.4 \text{ psi} \qquad \textbf{(78)}$$

where

$\Delta P_{economizer}$ = 0.36 psi
ΔP_{static} = 10.0 psi
ΔP_{piping} = 0 (assumed)

Air heater

Heat transfer

The air heater is the last heat transfer component before the stack. In the overall heat and material balances, the air heater exit gas temperature is assumed to be 390F. The air heater, when sized properly, will have sufficient surface to provide the required air temperature to the fuel equipment (burners, pulverizers, etc.) and lower the gas temperature to that assumed in the combustion calculations. For the air heater, the heat transfer rate is determined as follows:

$$q = \dot{m}_g \, c_p \left(T_1 - T_2\right) = 21.09 \times 10^6 \text{ Btu/h} \quad \textbf{(79)}$$

where

$\dot{m}_g$ = 324,100 lb/h — Table 3
c_p = 0.265 Btu/lb F — Fig. 9
T_1 = gas temperature entering the air heater
= gas temperature leaving the economizer
= 635F
T_2 = assumed air heater exit temperature
= 390F

For the air side, the temperature rise is:

$$T_2' = T_1' + q / \left(\dot{m}_a \, c_p\right) = 366\text{F} \quad \textbf{(80)}$$

where

T_1' = 80F — Table 2
q = 21.09 × 10^6 Btu/h
$\dot{m}_a$ = 302,500 lb/h — Table 3
c_p = 0.244 Btu/lb F — Fig. 10

The tubular air heater in this example is a crossflow design. The log mean temperature difference is determined from Fig. 1, Case IV. This requires not only the two inlet and two outlet temperatures, but also a crossflow correction factor, F:

$$LMTD = \frac{\left(T_1 - T_2'\right) - \left(T_2 - T_1'\right)}{\ell n \dfrac{\left(T_1 - T_2'\right)}{\left(T_2 - T_1'\right)}} \times F = 260\text{F} \quad \textbf{(81)}$$

where

T_1 = 635F
T_2 = 390F — Fig. 3
T_1' = 80F
T_2' = 366F
F = crossflow correction factor
= 0.90 — Ch. 4, Fig. 26

In an air heater, gas and air film heat transfer coefficients are approximately equal. For this example, film temperatures are approximated by the following calculations.

$$\text{Gas:} \quad T_f = \left(T_1 + T_2\right)/2 - LMTD/4 = 448\text{F} \quad \textbf{(82)}$$

where

T_1 = 635F
T_2 = 390F
$LMTD$ = 260F

$$\text{Air:} \quad T_f = \left(T_1' + T_2'\right)/2 + LMTD/4 = 288\text{F} \quad \textbf{(83)}$$

where

T_1' = 80F
T_2' = 366F
$LMTD$ = 260F

The flue gas flows through 1363 tubes (29 rows of 47 tubes per row). The air heater tubes are electric resistance welded (ERW), 2.25 in. OD with a 0.083 in. wall thickness. Allowing for manufacturing tolerances (9% for nonpressure tubing), the average tube inside diameter is 2.07 in. Gas flow area is 3.36 in.^{2}/tube for a total of 31.8 ft^2. The gas mass flux is:

$$G_g = \dot{m}_g / A_g = 10{,}185 \text{ lb/h ft}^2 \quad \textbf{(84)}$$

where

$\dot{m}_g$ = 324,100 lb/h — Table 3
A_g = 31.8 ft^2

Gas Reynolds number:

$$\text{Re} = K_{\text{Re}} \, G_g = 30{,}600 \quad \textbf{(85)}$$

where

K_{Re} = 3.0 h ft^2/lb — Fig. 4
G_g = 10,185 lb/h ft^2

Gas film heat transfer coefficient is the sum of the convection heat transfer coefficient from the longitu-

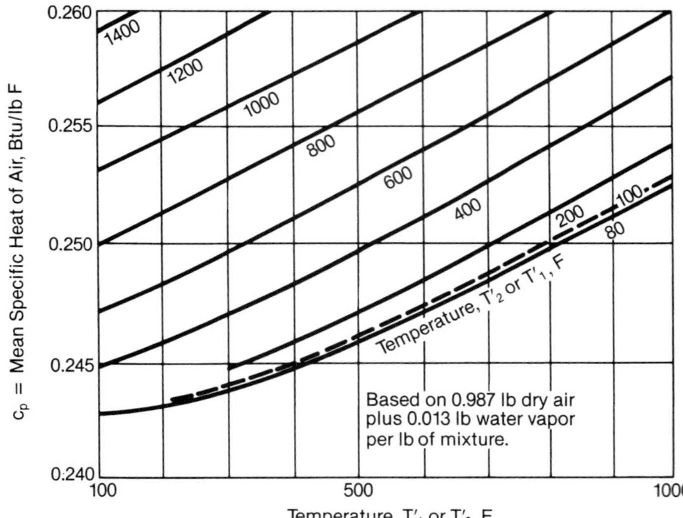

Fig. 10 Mean specific heat, c_p, of air at one atmosphere.

dinal gas flow inside the air heater tubes and a small gaseous radiation component from within the tube. The gas convection heat transfer coefficient h_{cg} is calculated from Equation 58 in Chapter 4. However, recognize that these coefficients do not account for fouling. To properly account for the ash layer inside the tubes, the gas side heat transfer coefficients will be multiplied by a cleanliness factor in the overall heat transfer calculation, Equation 91.

$$h_{cg} = h_l' F_{pp} F_T D_i / D_o = 9.91 \text{ Btu/h ft}^2 \quad \textbf{(86)}$$

where

h_l' = 52.6 Btu/h ft² F	Ch. 4, Fig. 13
F_{pp} = 0.188	Ch. 4, Fig. 14
F_T = 1.09	Ch. 4, Fig. 17
D_i = 2.07 in.	
D_o = 2.25 in.	Table 1

Gas side radiation heat transfer coefficient:

$$h_{rg} = h_r' K = 0.18 \text{ Btu/h ft}^2 \text{ F} \quad \textbf{(87)}$$

where

h_r' = 1.0 Btu/h ft² F	Fig. 5
p_r = 0.19	Fig. 6
L = 0.17 ft	Tube ID
K = 0.18	Fig. 8

From Table 1, the air side free flow area is 62 ft², hence the air mass flux is calculated to be:

$$G_a = \dot{m}_a / A_a = 4866 \text{ lb/h ft}^2 \quad \textbf{(88)}$$

where

$\dot{m}_a$ = 302,500 lb/h	Table 3
A_a = 62 ft²	

Air Reynolds number:

$$\text{Re} = K_{\text{Re}} G_a = 17{,}000 \quad \textbf{(89)}$$

where

K_{Re} = 3.5	Fig. 4
G_a = 4866 lb/h ft²	

The crossflow convection heat transfer coefficient for air is obtained from Chapter 4, Equation 61.

$$h_{ca} = h_c' F_{pp} F_a F_d = 11.50 \text{ Btu/h ft}^2 \text{ F} \quad \textbf{(90)}$$

where

h_c' = 109.4 Btu/h ft² F	Ch. 4, Fig. 18
F_{pp} = 0.104	Ch. 4, Fig. 20
F_a = 1.01	Ch. 4, Fig. 21
F_d = 1.0	Ch. 4, Fig. 23

Assuming negligible wall resistance, the overall heat transfer coefficient is:

$$U = \frac{F_{FR}\left(h_{cg} + h_{rg}\right) h_{ca}}{F_{FR}\left(h_{cg} + h_{rg}\right) + h_{ca}} = 5.07 \text{ Btu/h ft}^2 \text{ F} \quad \textbf{(91)}$$

where

F_{FR} = 0.90
h_{cg} = 9.91 Btu/h ft² F
h_{rg} = 0.18 Btu/h ft² F
h_{ca} = 11.50 Btu/h ft² F

The cleanliness or fouling resistance factor, F_{FR}, is empirically derived from field test data; 0.9 is representative of bituminous coal. The total heat transfer rate for the air heater is:

$$q = U A \left(LMTD\right) = 20.69 \times 10^6 \text{ Btu/h} \quad \textbf{(92)}$$

where

U = 5.07 Btu/h ft² F	
A = 15,656 ft²	Table 1
$LMTD$ = 260F	

Air heater exit gas temperature is calculated to be:

$$T_2 = T_1 - q / \left(\dot{m}_g c_p\right) = 390\text{F} \quad \textbf{(93)}$$

where

T_1 = 635F	
q = 20.69 × 10⁶ Btu/h	
$\dot{m}_g$ = 324,100 lb/h	
c_p = 0.26 Btu/lb F	Fig. 9

which is in agreement with the temperature assumed. Otherwise, T_2 would be substituted into Equation 81 and the calculations repeated.

Draft loss (gas inside tubes)

Air heater draft loss is comprised of friction (ΔP_f) plus entrance and exit losses (ΔP_{e+e}). Modifying and combining Chapter 3, Equations 47 and 52 for average gas temperatures, draft loss can be expressed as follows:

$$\Delta P = \Delta P_f + \Delta P_{e+e} = \left(12\frac{fL}{D_i} + 1.5\right)\left(\frac{30}{B}\right) \times \left(\frac{T + 460}{1.73 \times 10^5}\right)\left(\frac{G_g}{10^3}\right)^2 = 2.35 \text{ in. wg} \quad \textbf{(94)}$$

where

f = 0.024	Ch. 3, Fig. 1
L = 19.5 ft	Table 1
D_i = 2.07 in.	
T_1 = 635F	
T_2 = 390F	
T = $(T_1 + 2T_2)/3$ = 472F (Note 3 below)	
G_g = 10,185 lb/h ft²	
B = 30 in. Hg	

Note 3: An approximation used for mean gas temperature.

Air resistance (air crossflow over tubes)

The draft loss due to air flow across the air heater tubes is calculated from Chapter 3, Equations 52 and 56.

$$\Delta P = (f\,N\,F_d)\left(\frac{30}{B}\right)\left(\frac{T+460}{1.73\times10^5}\right)\left(\frac{G_a}{10^3}\right)^2 \quad \textbf{(95)}$$
$$= 1.52 \text{ in. wg}$$

where

f = 0.55 — Ch. 3, Fig. 15
N = 29 — Table 1
F_d = 1.0 — Ch. 3, Fig. 14
B = 30 in. Hg
T_1' = 80F
T_2' = 366F
T = 1/0.95 $(T_1 + T_2)/2$ = 235F (Note 4 below)
G_a = 4866 lb/h ft^2

Boiler thermal performance summary

Fig. 11 summarizes thermal performance. This figure illustrates the relative component absorption and temperature profiles. The furnace and boiler bank absorb the most heat. The furnace, screen and boiler bank are utilizing the heat available to generate saturated steam. The superheater raises the steam temperature to the desired outlet condition. The economizer is controlling gas temperature. The air heater is providing hot air for combustion and together with the economizer, is minimizing the flue gas temperature to the stack, reducing the sensible heat loss and increasing overall boiler efficiency.

Flues, ducts and stack

Performance calculations are not yet complete. Component draft loss and air resistance calculations have been made, but air and gas side performance still requires the evaluation of flues, ducts and stack effects. Once these are established, the designer can evaluate forced draft and induced draft fan conditions.

Air side loss – forced draft fan outlet to furnace

Air resistance calculations for forced draft fan outlet to windbox inlet are considered first. Static pressure at the windbox for this example is set at 5 in. wg. Windbox pressure is normally a function of the burner or fuel equipment design and is specified to assure proper operation. Starting at the windbox and working toward the forced draft fan:

Air mass flux:

$$G_a = \dot{m}_a / A_a = 2520 \text{ lb/h ft}^2 \quad \textbf{(96)}$$

where

$\dot{m}_a$ = 302,500 lb/h — Table 3
A_a = 10 × 12 = 120 ft^2 — Fig. 1

Reynolds number:

$$\text{Re} = \frac{G_a D_e}{\mu} = 4.4 \times 10^5 \quad \textbf{(97)}$$

where

G_a = 2520 lb/h ft^2
D_e = hydraulic diameter
= 4 × area/perimeter
= 4 × 120/44 = 10.9 ft
μ = 0.062 lb/h ft — Ch. 3, Fig. 4

The air resistance from windbox inlet to air heater outlet is:

$$\Delta P = \left(\frac{fL}{D_e}+N\right)\left(\frac{30}{B}\right)\left(\frac{T+460}{1.73\times10^5}\right)\left(\frac{G_a}{10^3}\right)^2 \quad \textbf{(98)}$$
$$= 0.04 \text{ in. wg}$$

where

f = 0.0134 — Ch. 3, Fig. 1
L = 25 ft
D_e = 10.9 ft
B = 30 in. Hg
T = 366F
G_a = 2520 lb/h ft^2

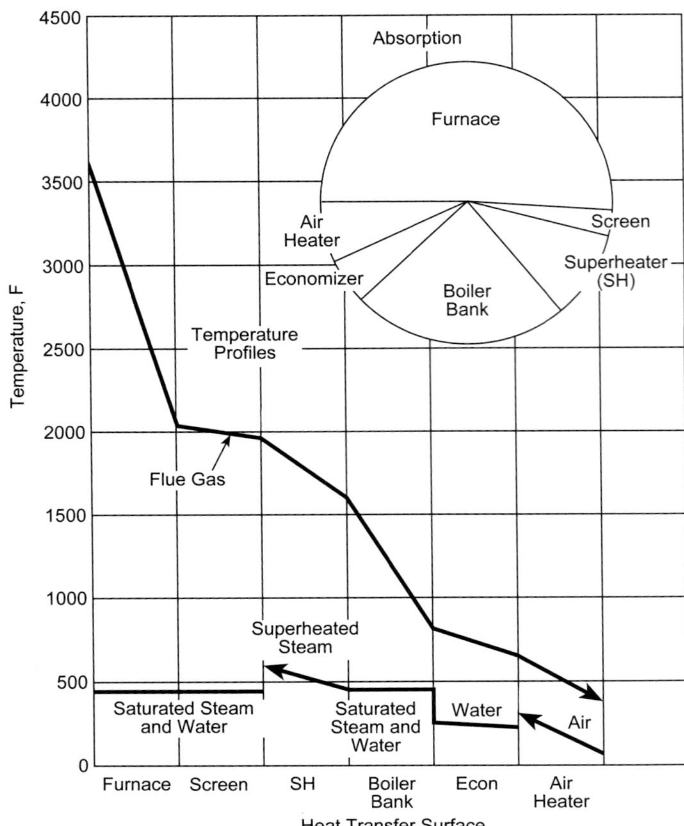

Fig. 11 Example boiler thermal performance summary.

Note 4: Under most conditions, this equation is a good approximation for effective air temperature.

$N = N_{bend} + N_{expansion} = 1.41$
$N_{bend} = 1.32$ Ch. 3, Fig. 10
$N_{expansion} = 0.09$ Ch. 3, Fig. 7

The frictional component of this equation (fL/D_e) is negligible. The draft loss from the air heater inlet to the forced draft fan transition outlet, neglecting friction, is:

$$\Delta P = N\left(\frac{30}{B}\right)\left(\frac{T+460}{1.73\times10^5}\right)\left(\frac{G_a}{10^3}\right)^2 = 0.03 \text{ in. wg} \quad \textbf{(99)}$$

where

$N = N_{contraction} + N_{bend} = 1.54$
$N_{contraction} = 0.22$ Ch. 3, Fig. 6
$N_{bend} = 1.32$ Ch. 3, Fig. 10
$B = 30$ in. Hg
$T = T_1'$ for the air heater = 80F Table 2
$G_a = 2520$ lb/h ft^2

The net static pressure at the forced draft (FD) fan outlet transition is then:

$$P_{total} = P_{windbox} + \Delta P_{windbox\ to\ air\ heater} + \Delta P_{air\ heater\ to\ FD\ fan} + \Delta P_{air\ heater} = 6.59 \text{ in. wg} \quad \textbf{(100)}$$

where

$P_{windbox}$ = 5.0 in. wg (set by burners)
$\Delta P_{windbox\ to\ air\ heater}$ = 0.04 in. wg
$\Delta P_{air\ heater\ to\ FD\ fan}$ = 0.03 in. wg
$\Delta P_{air\ heater}$ = 1.52 in. wg

Gas side loss – furnace to stack

The draft loss calculations begin at the furnace exit and work their way back to the induced draft fan. However, the net ID fan conditions must also include the draft loss for the downstream flues and the stack effect. The static pressure at the furnace exit in balanced draft boilers is controlled to be slightly negative; a value of – 0.1 in. wg is used in this example.

Although it doesn't affect the ID fan size evaluation, there may be a need to determine the furnace pressure at the burner level. The calculation that follows is provided for information and illustrates stack effect. Stack effect for the furnace is determined from methods developed in Chapter 25 as follows:

$$\Delta P_{SE} = SE \times Z = -0.58 \text{ in. wg} \quad \textbf{(101)}$$

where

Z = centerline of furnace exit to centerline of windbox = –50 ft Fig. 2
SE = stack effect
= 0.0116 in. wg/ft Ch. 25, Eq. 3 and Ch. 25, Table 3
T_1 = adiabatic temperature
= 3560F Table 3
T_2 = furnace exit gas temperature = 2000F

By controlling the furnace outlet to –0.1 in. wg, the net static pressure in the furnace at the burner elevation is approximately –0.68 in. wg.

From screen, superheater, and boiler bank component calculations, gas side draft losses were determined to be 0.04, 0.06 and 1.03 in. wg, respectively. The net static pressure at the boiler bank outlet is:

$$P_{boiler\ outlet\ bank} = P_{furnace} - \Delta P_{screen} - \Delta P_{superheater} - \Delta P_{boiler\ bank} = -1.23 \text{ in. wg} \quad \textbf{(102)}$$

where

$P_{furnace}$ = –0.10 in. wg
ΔP_{screen} = 0.04 in. wg
$\Delta P_{superheater}$ = 0.06 in. wg
$\Delta P_{boiler\ bank}$ = 1.03 in. wg

The calculations for the boiler bank outlet flue to the economizer inlet are handled the same as those for the air resistance calculations.

Gas mass flux:

$$G_g = \dot{m}_g / A_g = 2700 \text{ lb/h ft}^2 \quad \textbf{(103)}$$

where

$\dot{m}_g$ = 324,100 lb/h
A_g = 120 ft^2 Fig. 2

Reynolds number:

$$\text{Re} = \frac{G_g D_e}{\mu} = 368{,}000 \quad \textbf{(104)}$$

where

G_g = 2700 lb/h ft^2
D_e = hydraulic diameter
= 4 × area/perimeter
= 4 × 120/44 = 10.9 ft
μ = 0.08 lb/h ft Ch. 3, Fig. 4

Draft loss:

$$\Delta P = \left(\frac{fL}{D_e}+N\right)\left(\frac{30}{B}\right)\left(\frac{T+460}{1.73\times10^5}\right)\left(\frac{G_g}{10^3}\right)^2 = 0.07 \text{ in. wg} \quad \textbf{(105)}$$

where

f = 0.014 Ch. 3, Fig. 1
L = 15 ft
D_e = 10.9 ft
N = 1.35 Ch. 3, Fig. 10
T = 1/2 (818 + 815) = 817F
G_g = 2700 lb/h ft^2
B = 30 in. Hg

From the economizer component calculations, draft loss was calculated to be 1.57 in. wg. Stack effect from the boiler bank outlet to economizer outlet is:

$$\Delta P_{SE} = SE \times Z = 0.11 \text{ in. wg} \quad \textbf{(106)}$$

where

SE = 0.0071 in. wg/ft Ch. 25, Eq. 3 and Ch. 25, Table 3
Z = 15 ft

The net static pressure at the economizer outlet is then calculated:

$$P_{economizer\ outlet} = P_{boiler\ bank\ outlet} - \Delta P_{boiler\ bank\ to\ economizer} - \Delta P_{economizer} - \Delta P_{SE} = -2.98 \text{ in. wg} \quad \textbf{(107)}$$

where

$P_{boiler\ bank\ outlet}$ = −1.23 in. wg
$\Delta P_{boiler\ bank\ to\ economizer}$ = 0.07 in. wg
$\Delta P_{economizer}$ = 1.57 in. wg
ΔP_{SE} = 0.11 in. wg

The flue gas resistance from the economizer outlet to the air heater is due to friction only. However (fL/D_e) for this flue run is small, hence the resistance is negligible. Air heater gas side draft loss was previously calculated to be 2.35 in. wg. Referring to Fig. 2, the flue cross-section from the air heater outlet to the induced fan inlet decreases from 120 ft² to 48 ft². Again, as discussed above, frictional losses are negligible.

The mass flux is:

$$G_g = \dot{m}_g / A_g = 6750 \text{ lb/h ft}^2 \quad \textbf{(108)}$$

where

$\dot{m}_g$ = 324,100 lb/h — Table 3
A_g = 6 ft × 8 ft = 48 ft² — Fig. 2

Draft loss:

$$\Delta P = N\left(\frac{30}{B}\right)\left(\frac{T+460}{1.73\times 10^5}\right)\left(\frac{G_g}{10^3}\right)^2 = 0.07 \text{ in. wg} \quad \textbf{(109)}$$

where

N = 0.30 — Ch. 3, Fig. 6
B = 30 in. Hg
T = T_2 for the air heater = 390F
G_g = 6750 lb/h ft²

The stack effect from the economizer outlet to the induced draft (ID) fan inlet is determined to be:

$$\Delta P_{SE} = SE \times Z = 0.30 \text{ in. wg} \quad \textbf{(110)}$$

where

SE = 0.0059 in. wg/ft — Ch. 25, Eq. 3
Z = 50 ft — Fig. 2

The net static pressure at the induced draft fan inlet is calculated:

$$P_{ID\ fan\ inlet} = P_{economizer\ outlet} - \Delta P_{economizer\ to\ air\ heater} - \Delta P_{air\ heater} - \Delta P_{air\ heater\ to\ ID\ fan\ inlet} - \Delta P_{SE} = -5.69 \text{ in. wg} \quad \textbf{(111)}$$

where

$P_{economizer\ outlet}$ = −2.98 in. wg
$\Delta P_{economizer\ to\ air\ heater}$ = 0.0 in. wg
$\Delta P_{air\ heater}$ = 2.35 in. wg
$\Delta P_{air\ heater\ to\ ID\ fan\ inlet}$ = 0.07 in. wg
ΔP_{SE} = 0.30 in. wg

A straight flue runs from the induced draft fan outlet to the stack breaching. Friction again is negligible; however, an expansion loss at the stack breaching is included:

$$\Delta P = N\left(\frac{30}{B}\right)\left(\frac{T+460}{1.73\times 10^5}\right)\left(\frac{G_g}{10^3}\right)^2 = 0.22 \text{ in.wg} \quad \textbf{(112)}$$

where

N = 1.0 — Ch. 3, Fig. 7
B = 30 in. Hg
T = 390F
G_g = 6750 lb/h ft²

For the stack, two components must be determined: stack draft and stack resistance. For standard air with 0.013 lb wg/lb dry air (v = 13.70 ft³/lb at 80F and 30 in. Hg) and a typical flue gas (v_g = 13.23 ft³/lb at 80F and 30 in. Hg), Chapter 25, Equation 3 and the ideal gas law (Chapter 3, Equation 16a) can be combined to calculate the stack draft, ΔP_{SD}:

$$\Delta P_{SD} = 7.84 Z(0.00179 - 1/T)(B/30) = 0.45 \text{ in. wg} \quad \textbf{(113)}$$

where

D_i = stack diameter = 8 ft — Fig. 2
Z = stack height = 100 ft — Fig. 2
T_1 = stack inlet gas temperature = 390F
T_2 = stack exit gas temperature = 340F — Ch. 25, Fig. 11
T = (390 + 340)/2 = 365F = 825R
B = 30 in. Hg

The stack resistance is calculated from Chapter 25, Equation 7.

Gas mass flux:

$$G_g = \dot{m}_g / A_g = 6449 \text{ lb/h ft}^2 \quad \textbf{(114)}$$

where

$\dot{m}_g$ = 324,100 lb/h — Table 3
$A_g = \pi D_i^2/4$ = 50.3 ft²

Reynolds number:

$$\text{Re} = \frac{G_g D_i}{\mu} = 8.9 \times 10^5 \quad \textbf{(115)}$$

where

G_g = 6449 lb/h ft²
D_i = 8 ft
μ = 0.06 lb/h ft — Ch. 3, Fig. 4

Stack resistance:

$$\Delta P_{SR} = \frac{2.76}{B}\frac{T}{D_i^4}\left(\frac{\dot{m}_g}{10^5}\right)^2\left(\frac{fL}{D} + N_e\right) = 0.22 \text{ in. wg} \quad \textbf{(116)}$$

where

B = 30 in. wg
T = (390 + 340)/2 = 365F = 825R
D_i = 8 ft
$\dot{m}_g$ = 324,100 lb/h — Table 3
f = 0.012 — Ch. 3, Fig. 1
L = 100 ft — Fig. 2
N_e = stack exit loss = 1.0

The net static pressure at the induced draft fan outlet is calculated:

$$P_{ID\,fan\,outlet} = \Delta P_{SD} - \Delta P_{ID\,fan\,outlet\,to\,breaching} - \Delta P_{SR} = 0.01 \text{ in. wg} \quad \textbf{(117)}$$

where

ΔP_{SD} = 0.45 in. wg
$\Delta P_{ID\,fan\,outlet\,to\,breaching}$ = 0.22 in. wg
ΔP_{SR} = 0.22 in. wg

The net operating conditions for the fans are summarized in Table 4.

Fan purchasing specifications add test block factors to each net condition to accommodate deviations from design. See Chapter 25 for further discussion.

Table 4
Fan Operating Conditions

Net design conditions	Units	Fans	
		Forced draft	Induced draft
Flow	lb/h	302,500	324,100
Static pressure rise	in. wg	6.59	5.70
Inlet temperature	F	80	390

Summary

This chapter is intended to be an introduction. The approach is to give the reader a realistic yet basic overview of boiler performance calculations. Although there are many variables in these calculations, the designer must pay particular attention to fuel ash characteristics. Ash laden fuels will degrade heat transfer, increase draft loss and promote erosion. The principles presented in this chapter are sound but the impact of ash can dramatically change a given result. Whether analyzing existing equipment or sizing new, be sure to understand the slagging and fouling characteristics of the specified fuel. (See Chapter 21.)

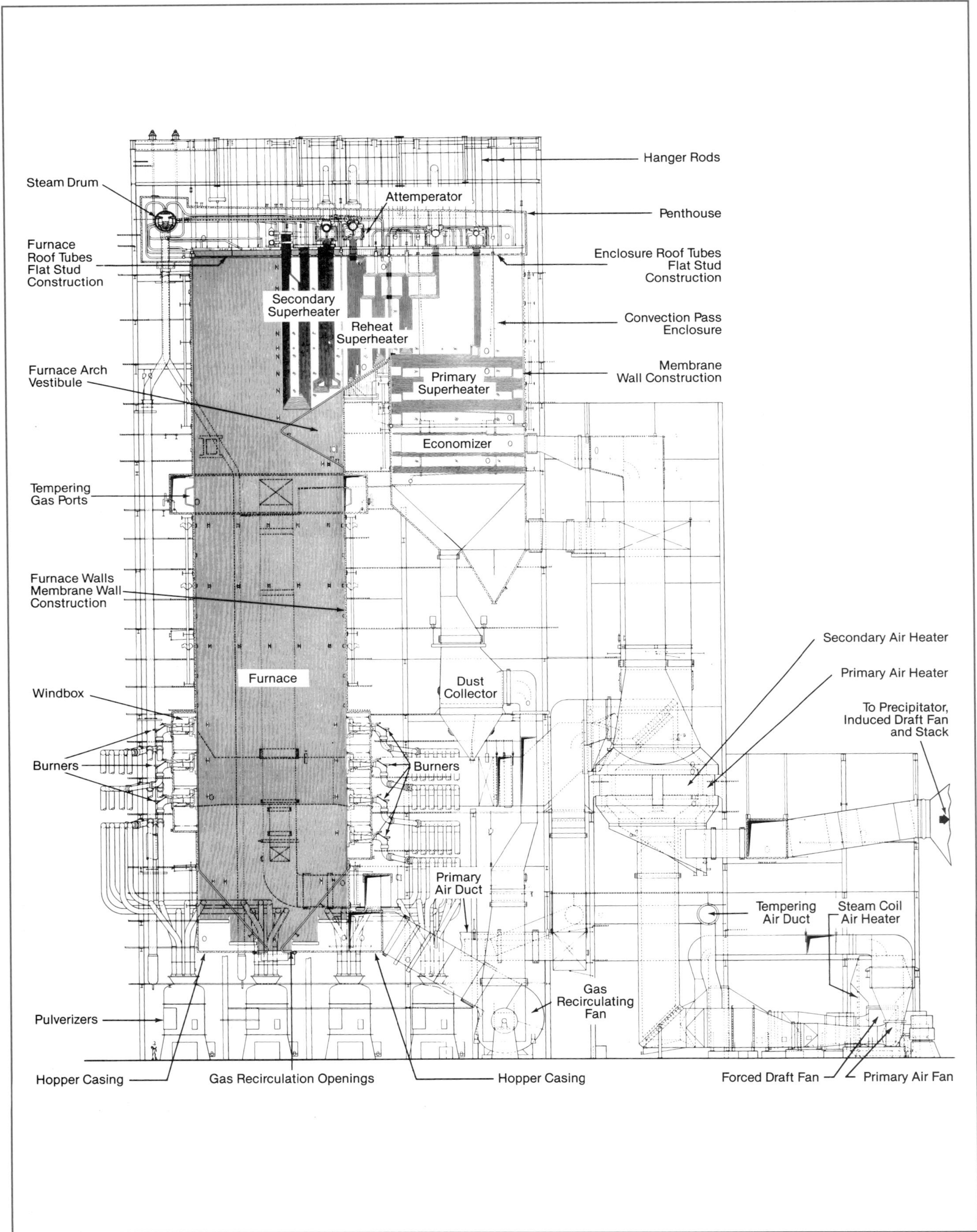

Typical coal-fired utility boiler setting and enclosure.

Chapter 23

Boiler Enclosures, Casing and Insulation

Boiler setting

The term *boiler setting* originally applied to the brick walls enclosing the furnace and heat transfer surfaces of the boiler. Today, boiler setting comprises all the water-cooled walls, casing, insulation, outer covering and reinforcement steel that form the outside envelope of the boiler and furnace enclosure. The term *enclosure* may refer to either the entire setting or to a part of it.

As larger capacity steam generating units were demanded, boiler settings underwent a long evolution from uncooled brick surfaces to today's water-cooled walls. Water-cooled walls began as widely spaced tubes exposed to the furnace and covered with insulating block. These progressed to tangent tubes covered with refractory. They gradually evolved to the present day construction of membrane tubes.

Design requirements

The boiler settings must safely contain high temperature pressurized gases and air. Leakage, heat loss and maintenance must be reduced to acceptable values. The following factors must be considered in the setting design:

1. Enclosures must withstand the effects of temperatures up to 3500F (1927C).
2. The effects of ash and slag, or molten ash, must be considered because:
   a. destructive chemical reactions between slag and metal or refractory can occur under certain conditions,
   b. accumulation of ash on the waterwalls can significantly reduce heat absorption,
   c. ash accumulations can fall causing injury to personnel or damage to the boiler, and
   d. high velocity ash particles can erode the pressure parts and refractory.
3. Provisions must be made for the thermal expansion of the enclosure and for differential expansion of attached components.
4. The buckstay system must accommodate the effects of thermal expansion, temperature and pressure stresses, as well as wind and earthquake loading appropriate to the plant site.
5. The effect of explosions and implosions must be considered to lessen the risk of injury to personnel and damage to equipment.
6. Vibrations caused by combustion pulsations and the flow characteristics of flue gas and air must be limited to acceptable values.
7. Insulation of the enclosures should limit the heat loss to an economical minimum.
8. Neither the exterior surface temperature nor the ambient air temperature should cause discomfort or hazard to operating personnel.
9. Enclosures must be gas-tight to minimize leakage into or out of the setting.
10. Settings of outdoor and indoor units that require periodic washdown must be weatherproof.
11. Settings must be designed for economic fabrication, erection and service life.
12. Serviceability, including access for inspection and maintenance, is essential.
13. Good appearance, in conjunction with cost and maintenance requirements, is desirable.

Tube wall enclosures

In today's units, water- or steam-cooled tubes, or both, are used as the basic structure of the enclosure in high temperature areas of the setting. Important types of water-cooled enclosures are membrane tubes, membrane tubes with refractory lining, flat stud tubes and tangent tubes. The facing page illustrates present day construction for tube wall enclosures.

Membrane tubes

Fig. 1 illustrates a typical furnace wall with membrane construction. These walls are water-cooled and constructed of bare tubes joined by thin membrane bars. The walls are gas-tight and do not require an exterior casing to contain the products of combustion. Insulation is placed on the outside of the wall and sheet metal or lagging is installed over the insulation to protect it.

Membrane tubes with refractory lining

There are several locations in selected types of boilers that require refractory lining on the furnace side of the tubes to protect the tubes from either erosion or

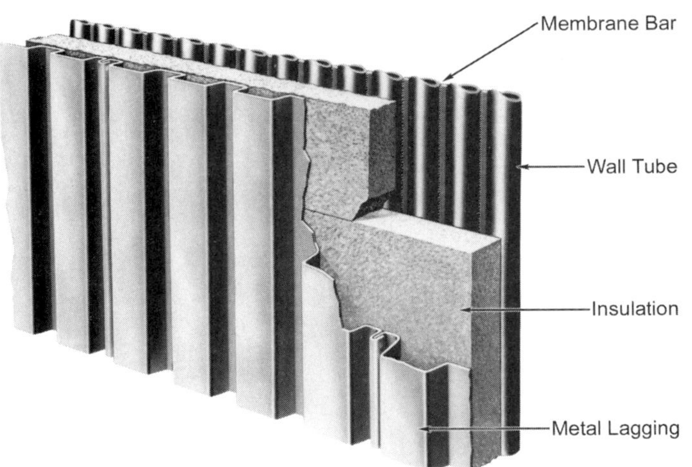

Fig. 1 Membrane wall construction.

corrosion from the products of combustion. Some of the most common applications are:

1. Cyclone-fired units: lower furnace and cyclone burner walls. (See Chapter 15.)
2. Circulating fluidized-bed boilers: lower furnace. (See Chapter 17.)
3. Refuse boilers: lower furnace. (See Chapter 29.)
4. Pulverized coal-fired boilers: burner throats. (See Chapter 26.)

Cylindrical pin studs, welded on the hot side of the tubes at close intervals, hold the refractory in place (Fig. 2). Lining the wall with refractory can also increase furnace temperature by reducing heat absorption where this is desired. The increase in temperature helps to maintain the coal, peat, or lignite ash in a liquid state, thereby preventing large ash buildup and allowing better removal of slag. These issues are discussed in more depth in Chapter 21. However, because of maintenance problems it is usually desirable to avoid refractory where technically acceptable.

Flat stud tube walls

These walls consist of tubes with small, flat bar studs welded at the sides (Fig. 3). These walls are typically backed by one of two construction methods which are usually found in the convection pass enclosure.

In the current method, the flat studded tubes are backed with refractory covered with a welded inner hot casing that is insulated and covered with metal lagging for protection. The casing is supported from channel tie bars welded to the tubes at each buckstay row. The walls are reinforced with buckstays and the inner casing is reinforced with stiffeners. Stiffener spacing and size are set by the design pressure of the walls between buckstays. This system provides a better gas-tight enclosure than the former method.

In former practice, as shown in Fig. 4, the tubes are backed with refractory, followed by a dense insulation and an outer cold casing. The casing is supported from the buckstays with expansion folds at the attachments. These folds minimize stresses in the casing caused by differential expansions between the hot tube wall and cold casing. This method is now obsolete, but found on many old boilers that are still in service.

While the construction of the casings described in the preceding paragraphs applies to areas of horizontal buckstay reinforcement, some industrial boiler designs require a vertical casing that is welded vertically to a bar located between two tubes (Fig. 5).

Tangent tube walls

These walls are constructed of bare tubes placed next to each other with a typical gap of 0.03125 in. (0.7937 mm). The refractory backup, casing and insulation system design, similar to that described for flat stud tube walls, has also been used with tangent tube walls. These walls are typically found in the furnace area of older boiler designs (Fig. 4).

Flat stud and tangent tube wall upgrades

In recent years, two methods have been used to provide a better enclosure seal on units with either inner (hot) or outer (cold) casing as the gas seal. In one method, on boiler enclosure areas with tangent tubes, a round bar is seal welded between each tube for the full length (Fig. 6). In the other method, where boiler enclosure areas have widely spaced tubes with flat studs, a flat bar is seal welded between the tubes just behind the flat studs over the full length (Fig. 7).

These methods have been effective on many boilers, providing an improved gas seal with considerably less maintenance and longer life than the casing seal

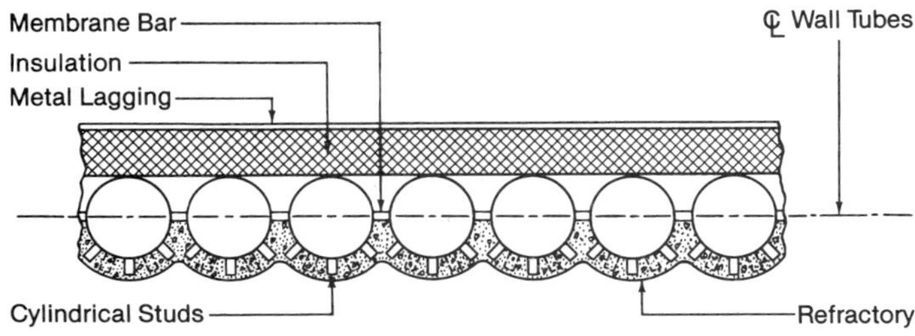

Fig. 2 Fully studded membrane walls.

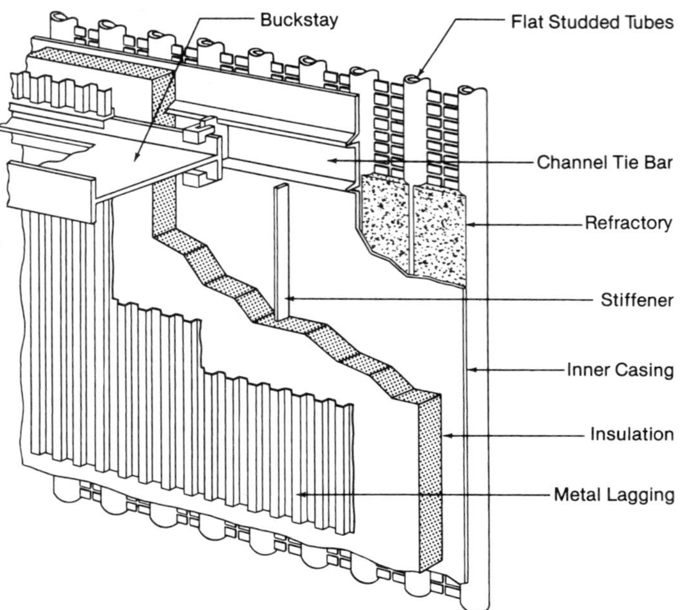

Fig. 3 Flat stud tube wall construction with inner casing shown.

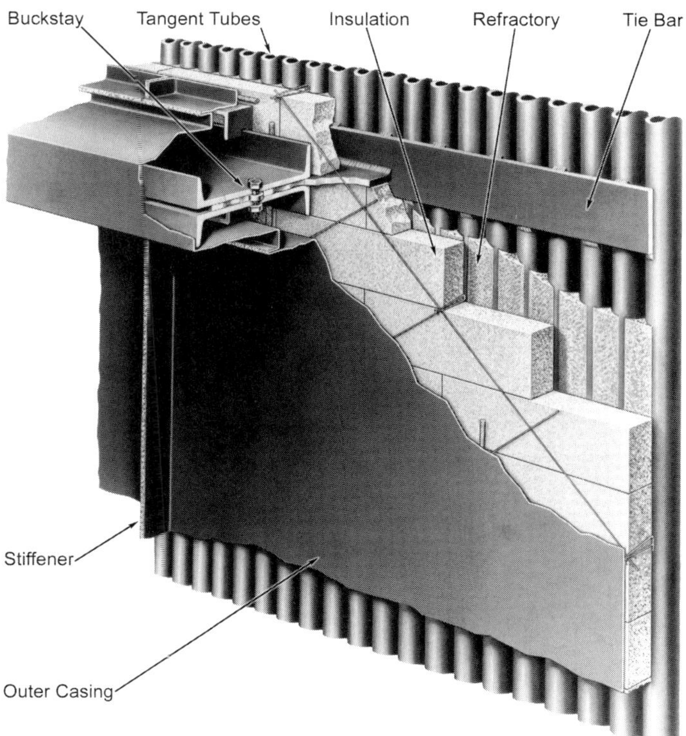

Fig. 4 Tangent tube wall construction with outer casing shown.

they replaced. Their biggest drawback is high installation costs because the entire boiler must be stripped of its existing casing and insulation, then a new insulation and lagging system must be installed.

Casing enclosures

The casing is the sheet or plate attached to pressure parts for supporting, insulating, or forming a gas-tight enclosure.

A boiler unit contains many cased enclosures that are not water-cooled. These enclosures must be designed to withstand relatively high temperatures while having external walls that minimize heat loss and protect operating personnel.

Casings are constructed of sheet or plate reinforced with stiffeners to withstand the design pressures and temperatures. When the casing is directly attached to the furnace walls, expansion elements are added to allow for differential thermal expansion of the tubes and casing. The frontispiece shows typical enclosures including the hopper casing, windbox, tempering gas plenum and penthouse casing.

Hopper

Hopper enclosures are used in various areas of the boiler setting that may include the economizer hopper, furnace hopper enclosure and the wash hopper for dry bottom units.

The enclosure provided by the hopper casing may also serve as a plenum for the recirculating gas which leaves the economizer hopper through ports and enters the furnace through openings between tubes in the furnace hopper.

Windbox

The windbox is a reinforced, metal-cased enclosure that attaches to the furnace wall, houses the burners, and distributes the combustion air. It may be located on one furnace wall, on two opposite furnace walls, or on all furnace walls using a wraparound configuration. The attachments to the furnace walls must be gas-tight and permit differential thermal expansion between the tubes and casing.

For large capacity boilers, the windbox may be compartmented and placed only on the front and rear furnace walls. The windboxes are compartmented (internally separated with horizontal division plates) for better combustion air control.

Tempering gas plenum

This enclosure, located above the windbox, provides for the distribution and injection of flue gas which is used to temper the furnace gases and control the ash fouling of heating surfaces. It is constructed similarly to the windbox, but is normally protected on the inside by a combination of refractory and stainless steel shields opposite the gas ports.

Penthouse

The penthouse casing forms the enclosure for all miscellaneous pressure parts located above the furnace and convection pass roofs. It is a series of reinforced flat plate panels welded together and to the top perimeter of the furnace pressure parts. Various seals are used at the penetrations through the penthouse walls, roof and roof tubes. Some examples are cylindrical bellows or flexible cans sealing the suspension hangers, large fold (pagoda) seals around the steam piping and refractory or casing seals around heating surface tube penetrations through the roof tubes. On many utility and some industrial boilers a gas-tight roof casing is used on top of the roof tubes as the pri-

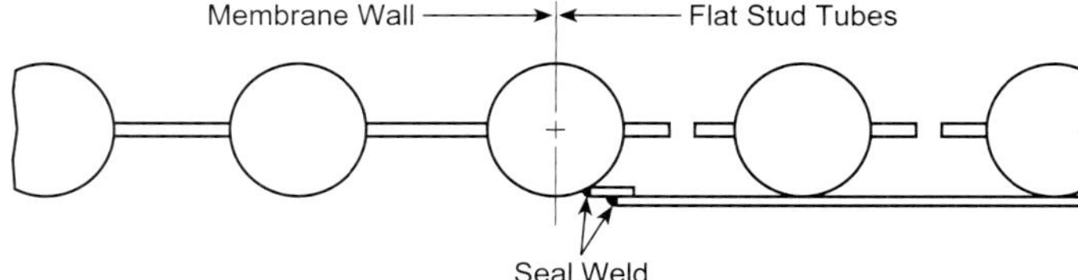

Fig. 5 Casing attachment to membrane wall.

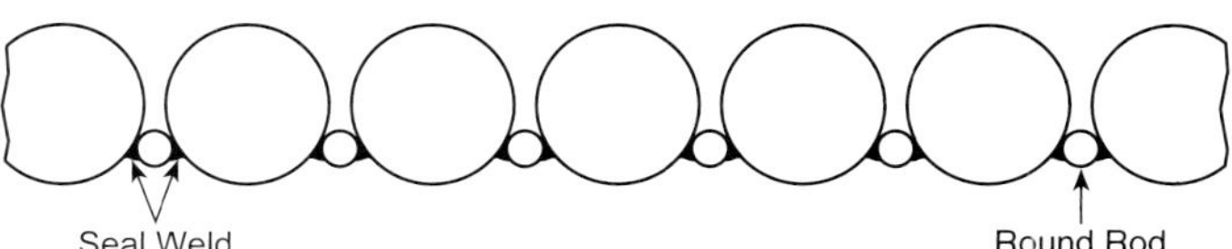

Fig. 6 Tangent tubes with closure rods.

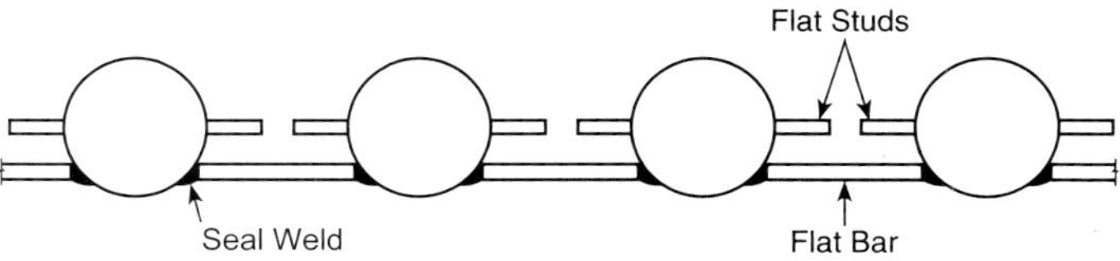

Fig. 7 Widely spaced tubes with flat studs and closure bars.

mary gas seal (pressure boundary). Penthouses may or may not be designed as pressure-tight enclosures with seal air. It depends upon whether the boiler is a pressure fired or a balanced draft unit and whether the roof seals are seal welded gas-tight or are the refractory type.

Design considerations

Resistance to ash, slag and erosion

Ash has a tendency to shed from a water- or steam-cooled metal surface, particularly when its temperature is well below the ash softening point. Wall type sootblowers remove ash in high temperature areas where it tends to adhere to the walls. (See Chapter 24.)

Extensive areas of exposed refractory must be avoided because large accumulations of ash could fall into the furnace damaging equipment or causing injury to personnel. Also, crotches in the tube wall should be designed to prevent ash and slag accumulation.

Pressure part erosion is reduced to acceptable levels by limiting the gas velocity through the unit. However, local high velocities can still occur in areas where gas bypasses baffles or heating surfaces. These high velocity lanes are best eliminated by proper design, baffle installation and routine maintenance.

Some unit designs such as process recovery, refuse and circulating fluidized-bed boilers require large areas of refractory on tube surfaces. In these units, the refractory system is designed to eliminate the effects of corrosion or erosion, or both, on pressure parts while minimizing the reduction in heat absorption by pressure parts.

Expansion

With the inner cased unit, Fig. 3, temperature differentials can occur between the casing and the tubes during startup. Expansion of the wall in the horizontal direction is governed by the temperature of the channel tie bar. Because the casing and the channels are at the same temperature, they can be welded together. Vertical expansion differences are accommodated by flexing of the casing flanges at the top and bottom of each casing section.

With a bottom supported unit, such as the PFI integral furnace boiler (Fig. 8) which is designed for pressure firing, the structure is fixed at a point at one end of the lower drum. Clearances, seals and supports are designed for known expansions in all directions from the fixed point.

With a top supported unit (see frontispiece), the expansion occurs downward from one elevation. The horizontal expansion of a boiler in both the fore and aft as well as in the side to side direction is determined by the location and method in which the hot boiler is tied to cold building steel. Typically, a combination of link ties and bumper ties is used to attach the hot boiler walls, via the buckstay system, to cold building steel.

Flues and ducts, piping, ash tanks and burner lines must be designed with expansion joints or seals to accommodate movement. Flexible metal bellows, metallic box fold joints, or non-metallic belts are used in flues and ducts while metal hoses and sliding or toggled gasketed couplings are used in piping. Water seals are generally used between ash hoppers or slag tanks and the associated furnace. With large units, the expansion may be as great as 12 in. (305 mm) between adjacent parts yet the joints must remain pressure-tight.

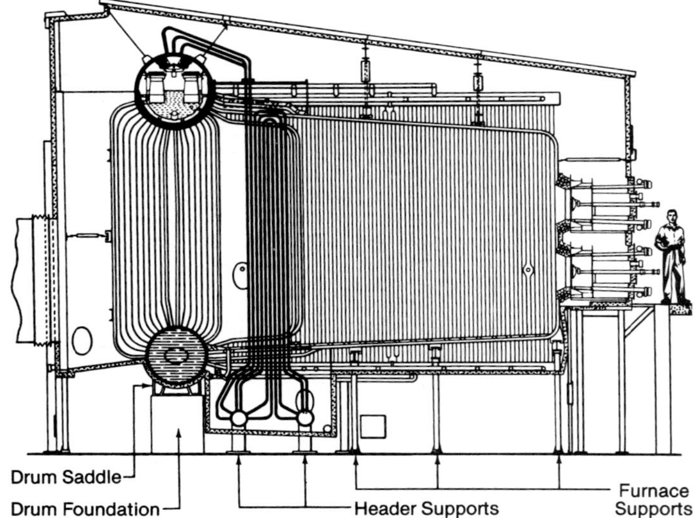

Fig. 8 Bottom supported unit.

Support

The support of boilers is discussed in Chapter 39. It is generally more economical to support the smaller units from the bottom and the larger units from the top. In either case, the boiler setting is formed by the waterwalls when these are available.

For bottom supported units, the enclosure is usually supported from a common foundation with the boiler as shown in Fig. 8. For top supported units, cased enclosures are supported from the pressure parts with the exception of the penthouse enclosure which is supported directly from the structural steel by the hanger rods.

Explosions

In the design of settings, the effect of possible explosions must be considered to minimize the possibilities of personnel injury and serious equipment damage. It is imperative that all types of boilers be designed to minimize the risk and effect of explosion. This requires that all new boilers and boilers undergoing major alterations be designed and evaluated so that they are in compliance with the National Fire Protection Association (NFPA) 85 Standard, Boiler Combustion Systems Hazards Code. On units with fluid or fluidized fuels, care must be taken to avoid puffs that can occur from improper fuel and air mixture during startup. (See Chapter 11.) A better understanding of the technical problems and the development of adequate design and operating codes have eliminated most explosions.

The enclosure is designed to withstand common puffs and large transient gas-side pressure excursions. A design can be provided that results in the failure of studs, stud attachments and welds rather than failure of tube walls in the event of a major furnace explosion. This minimizes the risk of release of large quantities of high pressure steam. Unfortunately, this design may also result in extremely hot gases being directed at platforms and steel in areas that were not designed to accommodate high temperature gases.

Explosion doors were once used on small furnaces to relieve excessive internal furnace pressure. These doors are no longer used because the rapid internal pressure increase from a fuel explosion is not significantly relieved by opening one or more doors. Explosion doors may also be more of a hazard than a safety margin because, in the event of a puff, they may discharge hot gases that would otherwise be completely contained within the setting.

The forces from normal operating negative or positive furnace gas-side pressures and from transient negative or positive furnace gas-side pressures, as defined by NFPA 85, are contained by rectangular bars called tie bars and/or channel tie bars attached to wall tubes to form continuous bands around the setting. Cold beams (*buckstays*), which are attached to the tie bars with slip connections, accommodate the gas-side pressure loadings and limit the inward and outward deflection of the wall tubes.

Because the buckstays are outside of the insulation, special corner connections are required that allow the walls to expand (Fig. 9). Forces generated by furnace gas-side pressures concentrate at the corner connections. These connections must be tight during startup when the walls have not fully expanded and during the normal operating fully expanded position.

The vertical tubes that span between the buckstays act as a beam to resist the internal furnace pressure. The larger the tube diameter and the heavier the tube wall, the farther apart the buckstays may be spaced, provided that allowable wall tube vibration limits are not exceeded. Permissible deflections and/or combinations of the positive or negative pressure loadings with wind or seismic loadings determine the size of the buckstay beam.

Implosions

Implosions are usually caused by an extremely rapid decay of furnace pressure due to sudden loss of fuel supply or by the improper operation of dampers on units with high static pressure induced draft fans. The risk of furnace implosions exists whenever a fan is located between the furnace and the stack. This risk exists even if the furnace is not normally being operated at a negative pressure since the rapid furnace temperature decay occurring on a master fuel trip (MFT) results in the furnace being exposed to the maximum head capacity of the fan on a transient basis during the fuel trip. This risk is also increased where axial flow forced draft (FD) fans are used since they can go into stall on the negative transient, blocking air flow into the furnace which is needed to restore the furnace pressure. The rules for determining minimum continuous and transient design pressures for the furnace enclosures can be found in the NFPA 85 Standard. In addition, induced draft fan controls are specified in NFPA 85 to minimize possible operating or control errors and to reduce the degree of furnace draft excursion following a fuel trip.

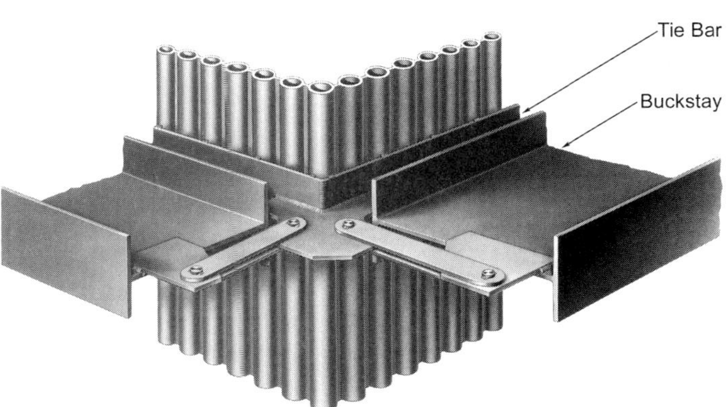

Fig. 9 Tie bar and buckstay arrangement at corner of furnace.

Vibration

Excessive vibration can cause failures of the tubes, insulation, casing and supports. These vibrations can be produced by external rotating equipment, furnace pulsations from the uneven combustion of the fuel, or turbulence in the flowing streams of air or gas in flues, ducts and tube banks.

Tube walls, flues and ducts are designed to limit vibration during normal operating conditions. In regard to wall tube vibrations, buckstays are typically spaced to ensure that the natural frequency of the wall tubes is greater than or equal to 6 hertz. The moment of inertia of a buckstay must be chosen to ensure that the buckstay natural frequency is greater than or equal to 3 hertz, based on a simply loaded uniform beam. Flues, ducts and casings are similarly stiffened by bars or structural shapes to limit vibration. This stiffening is particularly necessary in sections of flues and ducts where the flow is highly turbulent, as in the fan discharge connecting piece. Every effort should be made to eliminate the sources of severe vibration, such as unbalanced rotating equipment, poor combustion and highly turbulent or unbalanced air or gas flow.

Heat loss

Heat loss from a boiler setting is reduced by insulation, usually as an integral part of the boiler enclosure. (See Figs. 1 through 4.) There is an economical balance between the value of the heat loss and the cost of the insulation and installation.

The insulation system is designed to provide both safety for personnel and minimal heat loss. In addition, indoor units require ventilation for both operator comfort and room air change.

The materials used most frequently for heat insulation are listed below.

Mineral wool

This material is comprised of molten slag, glass or rock, blown into fibers by steam or an air jet or spun by high speed wheels.

Mineral wool base block Mineral wool fibers and clay, molded under heat and pressure to form blocks, are used to insulate membrane tube walls and boiler casings up to temperatures of 850, 1200 or 1900F (454, 649 or 1038C) depending upon the grade.

Mineral wool blanket Mineral wool fibers, compressed into blanket form and held in shape between hexagonal wire mesh or expanded metal lath, are used on all types of enclosures with external metal lagging or casing and for piping inside cased enclosures. The temperature limit is normally 1200F (649C).

Calcium silicate block

Reacted hydrous calcium silicate block is used on enclosures and piping, generally below 1200F (649C).

High temperature plastic

Insulating cement made of mineral wool fibers processed into nodules and dry mixed with clay forms a tough, fibrous monolithic insulation in final dried condition. Drying shrinkage is as much as 40%, but there is a tendency to crack upon drying. This material is used principally on irregularly shaped valves and fittings and to fill gaps between block insulation. Insulating cement is available in grades useable up to 1900F (1038C).

Ceramic fiber

High purity ceramic fibers, with melting points above 3000F (1649C), are occasionally used for tube enclosure seals where resiliency or high temperature insulation is required.

Heat loss calculations

Calculations of the heat flow through a composite wall are discussed in Chapter 4. Thermal conductivities of a wide range of commercial refractory and insulating materials, at the temperatures for which they are suitable, are given in Fig. 10. Combined heat losses (radiation plus convection) per square foot (square meter) of outer wall surface are given in Fig. 11 for various ambient air velocities and surface air temperature differences. The American Boiler Manufacturers Association (ABMA) Radiation Loss Chart provides a quick approximation for radiation loss, expressed as a percentage of gross heat input (Fig. 12).

Ventilation, surface temperature, conditions

To maintain satisfactory working conditions for personnel around a boiler, a cold face temperature of 135F (57C) or less is considered satisfactory. Heat losses, corresponding to these surface temperatures, range from 90 to 130 Btu/h ft^2 (284 to 410.1 W/m^2), which can be readily absorbed by the air circulation generally provided in present day boiler rooms.

Insulating a boiler to reduce heat loss to a value that can readily be absorbed by the total volume of room air does not in itself assure comfortable working conditions. Proper air circulation around all parts of the boiler is also necessary to prevent the accumulation of heat in the areas frequented by the operating personnel. This can be helped by using grating rather than solid floors, by ample aisle space between adjacent boilers, by the location of fans to assist the circulation of air around the boiler and by installing ventilating equipment to assure adequate air exchange.

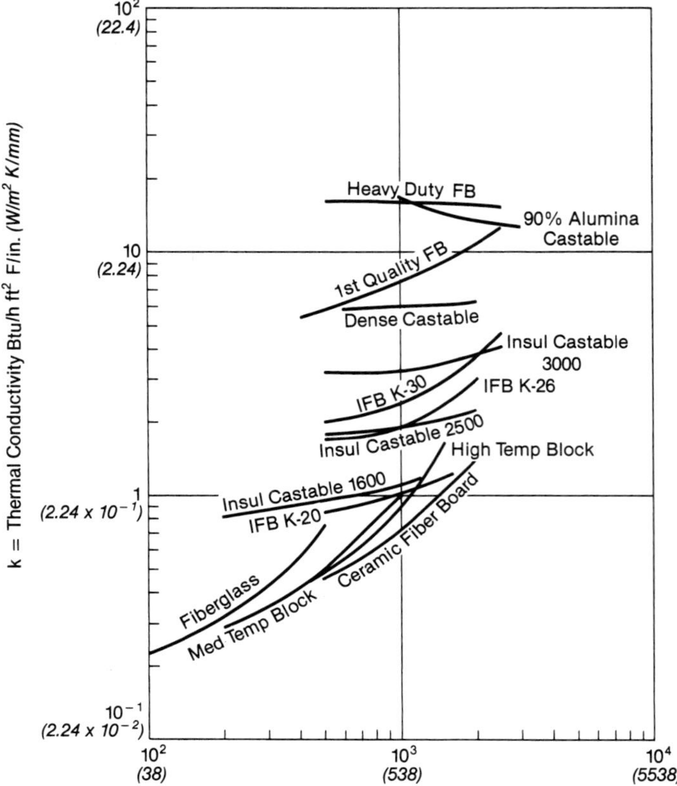

Fig. 10 Thermal conductivity of various refractory materials.

Good ventilation does not greatly increase overall heat loss. Air velocity affects the surface heat transfer coefficient. This can be verified by data from Fig. 11. However, surface conductance is only a small part of the total resistance to heat flow. For example, an increase in air velocity from 1 to 10 ft/s (0.3 to 3 m/s), for the conditions given in Fig. 13, will increase the heat loss rate through the wall by only 2%.

Unlike heat loss, outer surface temperature is considerably affected by the surrounding conditions. In the situation shown in Fig. 14, where two walls of similar temperatures are close together, the radiant heat transfer from either wall is negligible. The natural circulation of air through such a cavity is inadequate to cool the walls to a temperature suitable for personnel working in the vicinity. From Fig. 15 it can be seen that a considerable change in surface film resistance will cause an appreciable change in lagging or surface temperature while not affecting the heat loss through the wall to any extent.

Increased insulation thickness would not significantly reduce the surface temperature in the cavity shown in Fig. 14. Cavities should therefore be avoided in areas where operators work. Ventilating ducts to reduce the air temperature in such a cavity can be installed if necessary.

Heat Loss Rate, Btu/h ft^2 (W/m^2)
1600 (5047.3)
1400 (4416.4)
1200 (3785.5)
1000 (3154.6)
800 (2523.7)
600 (1892.8)
400 (1261.8)
200 (630.9)
0
Air Velocity/s — 30 ft (9.1 m)
25 ft (7.6 m)
20 ft (6.1 m)
15 ft (4.6 m)
10 ft (3.0 m)
5 ft (1.5 m)
0
50 (28)
100 (56)
150 (83)
200 (111)
250 (139)
300 (167)
Temperature Difference, Surface to Air, F (C)

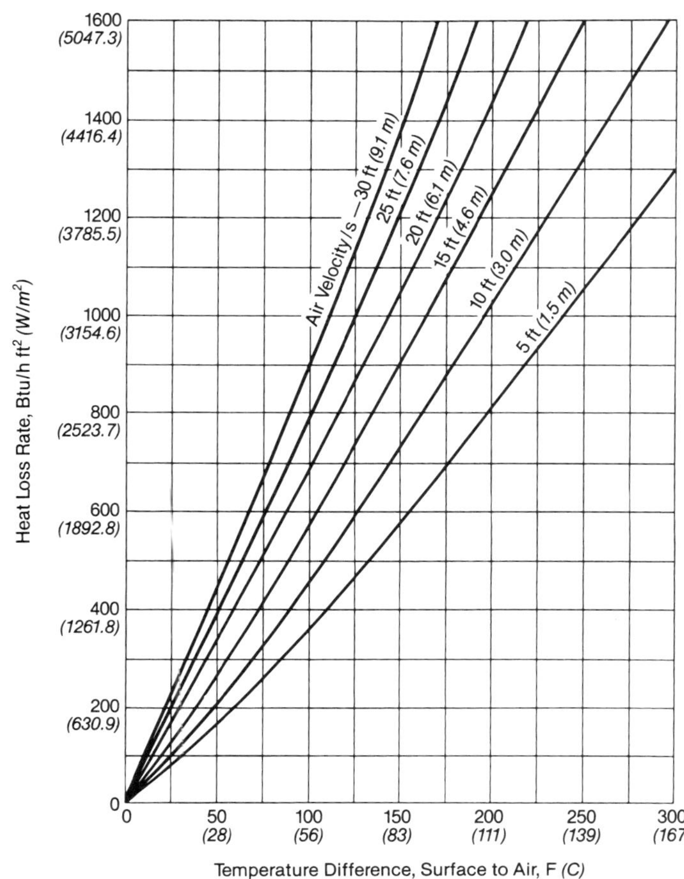

Fig. 11 Heat loss from wall surfaces (radiation + convection) (Source – *ASTM Standards, Part 13, 1969*).

Leakage

Continuing efforts have been made over the years to reduce air infiltration into boiler settings. Such in-leakage increases gas flow and the heat rejected to the stack, thereby lowering boiler efficiency and increasing the amount of induced draft fan power. (See Chapter 10.)

Corrosion

One of the advantages of membrane walls compared to cased walls is that they eliminate flue gas corrosion on the cold face of the enclosure walls. Most flue gases contain sulfur; therefore, metal parts of the setting must either be kept above the dew point of the gases or out of contact with the gases. (See Chapter 21.) The dew point generally ranges between 150 and 250F (66 and 121C) and is dependent on the fuel, its sulfur content and the firing method.

Flues carrying low temperature spent gases should be insulated on the outside to inhibit corrosion. This is particularly necessary on outdoor units. Water-cooled doors and slag tap coils require water temperatures above 150F (66C) to keep the cooling coils above the dew point of the gases.

When casing is located outside of insulation or refractory, it is still subject to the action of the flue gases. When this type of casing is subjected to a temperature below the dew point, an asphalt mastic or other type of coating is needed to protect it from corrosion on the inside. This problem requires special attention in the design of outdoor installations where temperatures may, at times, be below flue gas dew point temperatures.

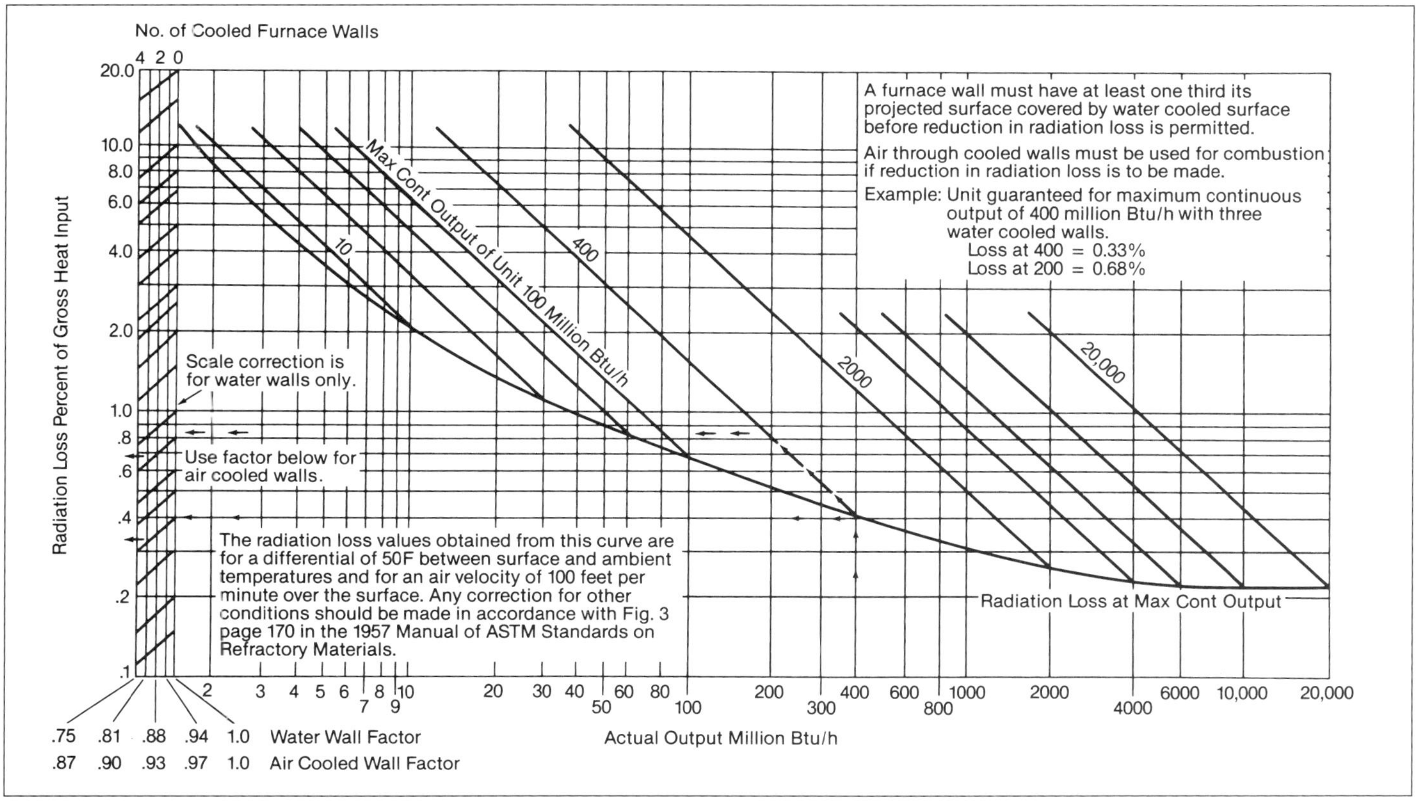

No. of Cooled Furnace Walls
4 2 0
Radiation Loss Percent of Gross Heat Input
20.0
10.0
8.0
6.0
4.0
2.0
1.0
.8
.6
.4
.2
.1
Max Cont Output of Unit 100 Million Btu/h
10
400
2000
20,000
A furnace wall must have at least one third its projected surface covered by water cooled surface before reduction in radiation loss is permitted.
Air through cooled walls must be used for combustion if reduction in radiation loss is to be made.
Example: Unit guaranteed for maximum continuous output of 400 million Btu/h with three water cooled walls.
Loss at 400 = 0.33%
Loss at 200 = 0.68%
Scale correction is for water walls only.
Use factor below for air cooled walls.
The radiation loss values obtained from this curve are for a differential of 50F between surface and ambient temperatures and for an air velocity of 100 feet per minute over the surface. Any correction for other conditions should be made in accordance with Fig. 3 page 170 in the 1957 Manual of ASTM Standards on Refractory Materials.
Radiation Loss at Max Cont Output
2 3 4 5 6 7 8 9 10 20 30 40 50 60 80 100 200 300 400 600 800 1000 2000 4000 6000 10,000 20,000
Actual Output Million Btu/h
.75 .81 .88 .94 1.0 Water Wall Factor
.87 .90 .93 .97 1.0 Air Cooled Wall Factor

Fig. 12 Radiation loss in percent of gross heat input (American Boiler Manufacturers Association).

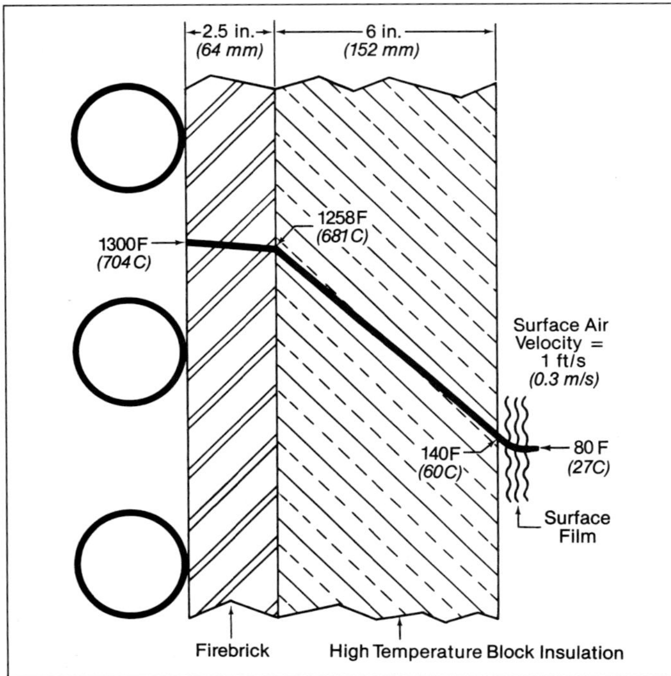

Fig. 13 Temperature gradients through tube and brick wall.

With the use of externally insulated casing, corrosion problems are greatly reduced because the flue gases are completely contained by a metal skin, which is well above the dew point temperature. However, even with the inner casing, seals and expansion joints must be insulated properly to avoid cold spots and consequent corrosion.

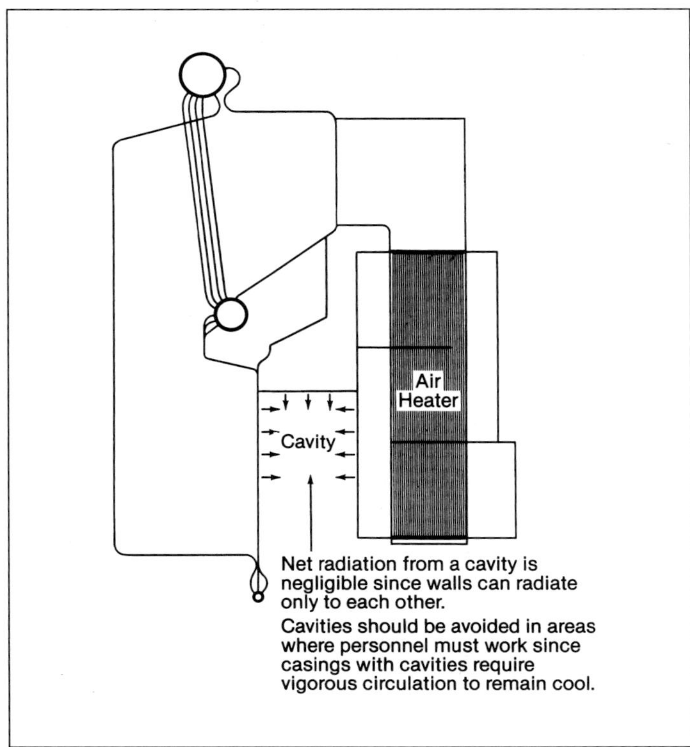

Fig. 14 Cavities tend to raise wall surface temperatures.

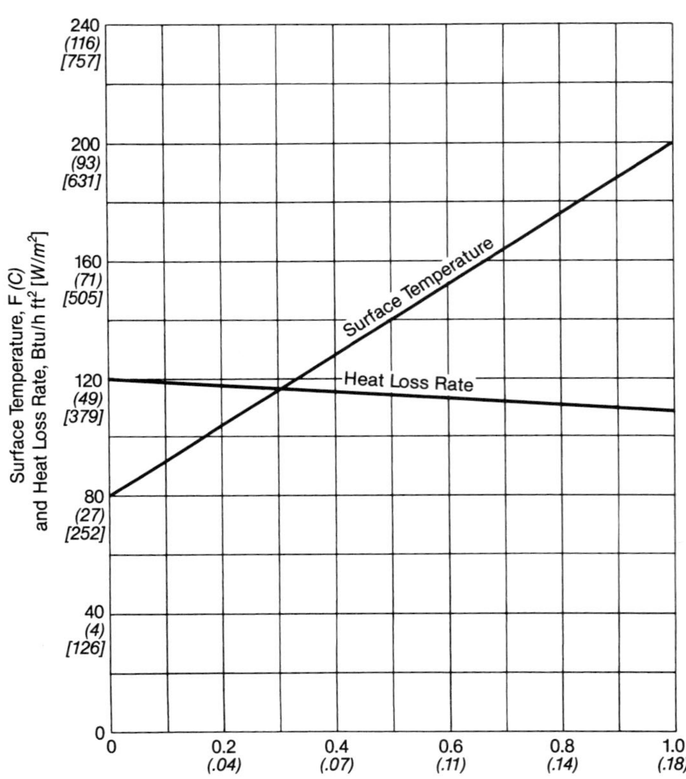

Fig. 15 Effect of surface film resistance on surface temperature and heat loss rate.

Resistance to weather

Outdoor boiler installations are possible in mild climates. While the initial cost of the plant is reduced, maintenance of the boiler and auxiliary equipment must be considered. Severe weather can extend outage time and increase maintenance expenses. These units must also have sufficient reinforcement to withstand the pressure and suction forces of the wind.

Lagging is an outer covering over a wall used for protecting the insulation from water or mechanical damage. It is relatively simple to make a metal lagged unit rainproof. Joints and flange connections are overlapped and flashings are used around openings. Welding of joints or the use of mastic compounds is necessary in areas which are difficult to seal.

Sloping roofs are required and are particularly important on aluminum lagging where pockets of water would eventually stain the surface. Direct contact between aluminum and steel must be avoided to prevent galvanic corrosion of the aluminum in the presence of moisture. Roof flashings should be designed so that water runoff does not wet the aluminum.

Weather hoods should be used to keep rain, snow and ice from contacting outdoor safety valves. Nozzle and valve necks must be insulated and protected with sheet metal or outer waterproof covering. Outdoor control lines containing air or flue gas, drain and sampling lines and intermittently operated steam and water lines should be insulated and protected by electric resistance heating tape. Steam pipe tracer lines may also be used in some cases. Dry air should be supplied for control lines and sootblowers. Steam and water lines outside the setting must be completely drainable.

Fabrication and assembly

The setting must be designed for economical fabrication and assembly. This requires integration of all shop and field methods and practices. Small units can be completely shop assembled. For larger units the trend has been toward shop subassembly of large components.

Shipping clearances usually limit the size of shop-assembled wall panels to 16 ft (4.9 m) in width or 105 ft (32 m) in length. These size criteria can not be used simultaneously. That is, as the panels are made longer, the maximum width will be less and as the width is increased the maximum length decreases. Shop assembly of components permits better quality control of the more complicated parts.

Casing enclosures, tube connections to headers, tie bars, doors and other attachments are normally of welded construction. New and improved materials and attachment methods reduce the manhour requirements for insulating boilers/auxiliary equipment and installing metal lagging.

Serviceability

Many setting details must be designed to simplify operation and maintenance. Working areas around the unit should have adequate lighting and comfortable temperatures. Clearances for servicing and removing parts should be provided. Access through the setting is necessary for inspection of boiler internals. Suitable platforms for access doors, sootblowers, instruments and controls are essential.

Inspection doors allow observation of combustion conditions and the cleanliness of heat absorbing surfaces. Figs. 16 and 17 illustrate inspection doors for balanced draft and pressurized settings. Safety is provided by two types of interlocks which assure that compressed air is properly aspirating the aperture before the door is opened. A feature of this door is that the aspirating jet does not restrict the comparatively wide view angle.

The tube bends that form openings must have the smallest possible radius. The length of the stud plate closures around the opening is minimized so that the plates can be adequately cooled by welded contact to the tubes, thereby preventing burn back and subsequent overheat of seals.

Appearance

The setting should present a good appearance and be designed so that it can be retained indefinitely with a minimum of housekeeping. The outer surface should be easy to clean. Equipment handling flue gas, coal, ash or oil should be designed to minimize leakage.

Light gauge metal lagging is generally used as an outer covering. This is particularly true for outdoor units where it is relatively simple to make the metal lagging water-tight. Many types of covering are found in older installations including plastic insulation, insulating cement, canvas and welded steel casing.

Figs. 1, 2 and 3 show metal lagging. Light gauge galvanized steel or clad aluminum sheets are commonly used. Galvanized steel is generally less expensive than aluminum, but for outdoor units it may be necessary to paint the galvanized steel after weathering, unless the climate is dry. The clad aluminum may be preferable because it only requires painting under more severe conditions.

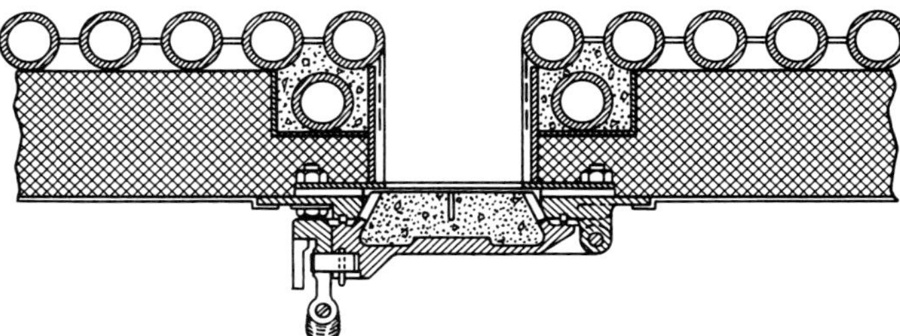

Fig. 16 Inspection door for balanced draft furnace.

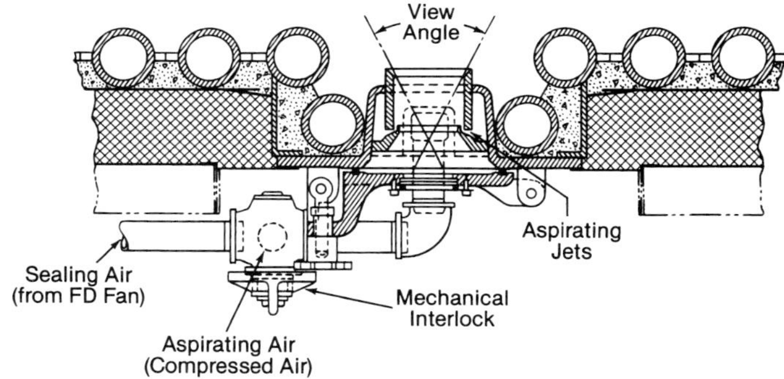

Fig. 17 Inspection door for pressurized furnace.

Hydrobin® dewatering storage bins.

Chapter 24

Boiler Cleaning and Ash Handling Systems

The combustion of virtually all fuels generates ash, or inert residuals of combustion. As discussed in Chapter 21, the ash can accumulate on surfaces exposed to the flue gas stream causing reduced performance, flow obstruction, accelerated corrosion and increased mechanical loadings. Specialized *boiler cleaning systems* are installed in boilers and environmental control systems to permit on-line deposit removal and control for optimized overall system operation. In addition, specialized *ash handling systems* are incorporated into the power plant designs to collect and remove the ash from the boiler and emissions control systems for safe and environmentally acceptable disposal.

Boiler cleaning systems

Steam generating plants are cleaned on-line by regularly removing ash deposits that accumulate on heat transfer surfaces. This maintains maximum thermal efficiency of the boiler (optimizes heat rate) and maintains flue gas temperatures within design conditions at key locations. This, in turn, supports the optimum operation of air pollution control equipment. On-line cleaning also prevents the blockage and plugging of gas passages in the boiler.

This chapter discusses the methods and types of equipment used for on-line cleaning (see Fig. 1). For this purpose, it is convenient to separate the discussion by regions of the boiler and downstream equipment, since these generally require different cleaning methods and devices.

The following terminology and definitions are useful to the discussions of regional cleaning.

Sootblower terminology

Cleaning medium

An effective method of on-line boiler cleaning is to direct a concentrated jet of cleaning medium against the soot or ash buildup. The cleaning medium may be saturated steam, superheated steam, compressed air or water. Combinations of water with other media, such as steam or air, have also proven effective but can cause boiler tube damage.

Superheated steam is the most widely used medium for several reasons. Boiler-generated steam avoids the cost and maintenance of a compressor. Because it is generally extracted from a high pressure steam outlet and its pressure reduced by regulating valves, steam provides flexibility in case more aggressive cleaning is required. Superheated steam is generally recommended over saturated steam to avoid condensation in the cleaning flow, which can accelerate boiler tube erosion or result in thermal cracking.

Compressed air is often used as the cleaning medium on larger boilers. The source may be high pressure 250 to 500 psig (1.7 to 3.4 MPa) reciprocating compressors or high flow rate centrifugal compressors discharging at pressures from 150 to 225 psig (1.03 to 1.55 MPa).

Water is used in the most aggressive cleaning situations, such as removing more tenacious slag from furnace walls, or cement-like deposits in air heater baskets.

Fig. 1 Long retractable IK sootblower used to remove ash deposits in utility boilers.

Peak impact pressure

Peak impact pressure (PIP) is the stagnation pressure measured directly along the centerline of the sootblower nozzle at distances downstream of the nozzle outlet. Peak impact pressure is maximized by designing the nozzle to achieve a highly concentrated jet at the greatest possible distance from the nozzle outlet. Because most sootblowers are operated with supersonic flow exiting the nozzles, ideal nozzle design involves the avoidance of shock waves at the nozzle outlet and minimum shear interaction with the flue gas environment.

Jet progression velocity

Jet progression velocity is the linear velocity at which the jet passes over a surface, directing a concentrated jet of cleaning medium. This is a function of the nozzle rotational speed and distance from the surface.

Cleaning radius and tube bank penetration

The cleaning radius refers to the distance to which a particular sootblower can clean a surface, when the path between the sootblower nozzle outlet and the surface is unobstructed. Although commonly used, this term is misleading because the area cleaned is not often circular. Tube bank penetration is the maximum effective cleaning distance when directed into a bank of tubes.

Wall box

A wall box attaches to the boiler side wall at the opening made for the cleaning device. The wall box provides structural support for the front end of long retractable sootblowers, or the entire weight of furnace wall blowers. The wall box provides an opening for the lance tube to pass into the boiler, and a sleeve and seal plate to minimize flue gas escape to the atmosphere when the lance is retracted from the boiler.

Lance tube

The lance tube is the tubular component of a retractable-type sootblower that travels into and out of the combustion zone and flue gas passages. The lance tube is fitted with nozzles at the tip and possibly along its length to accelerate and direct cleaning flow onto the surface to be cleaned.

Feed tube

The feed tube is the stationary tube that feeds cleaning medium to the moving lance tube or rotating element. The outer surface of the feed tube serves as a sealing surface for the packing assembly that slides and/or rotates with the lance tube.

Element

An element serves the same function as the lance tube, but remains inside the flue gas passage permanently.

Furnace cleaning

The furnace is a large unobstructed chamber where fuel and air are mixed and nearly all of the combustion occurs. The walls, slopes and floor tend to collect the non-combustible fuel components. In most power boilers (particularly pulverized coal-fired units), efficient heat transfer is necessary for effective boiler circulation, an efficient heat balance, and to minimize pollutant gas species. However, in some boiler types, such as black liquor recovery boilers and fluidized beds, little or no on-line furnace cleaning is attempted because of adverse impact on the combustion process.

Principles

Ash deposits on the side walls and hopper slopes are generally heavier than those found elsewhere in the boiler. When removed, these deposits fall and contribute to bottom ash.

Because these ash deposits collect on surfaces that are exposed to intense heat, they may melt at their surface after building to a sufficient thickness. Depending on chemical composition, the ash may collect in a loosely bound, porous layer that can be relatively easy to remove by on-line steam cleaning, or the ash may fuse into a dense sheet-like deposit known as *slag*. Slag removal often requires on-line water cleaning. The rapid quenching creates cracks and fissures, dislodging the slag in pieces or patches as the water jet continues its travel across the surface. The water jet also provides significant mechanical impact energy. This technique is not used on boiler surfaces lined with refractory.

Cleaning devices

Short retractable furnace wall sootblower When the furnace wall ash deposits are loosely adhered, a short travel retractable-type sootblower (called an *IR* or *wall blower*) directs steam or compressed air to remove the deposits. This design (see Fig. 2) consists of a poppet valve, feed tube, lance tube, and a drive mechanism for inserting the sootblower tube and rotating it in the fully inserted position. The lance tube has one nozzle near the tip. When at rest, the nozzle is located outside of the boiler and within a sleeve (a short segment of pipe welded to a wall box) to protect it from the furnace radiated heat. When the sootblower is activated, the lance tube is advanced to where the nozzle is about 1.5 in. (38 mm) from the face of the boiler tubes. The feed tube remains stationary while the lance tube telescopes to extend the nozzle.

The poppet valve is mechanically opened by the advancing lance tube drive mechanism and cleaning flow begins. Steam flows through the poppet valve, into the feed tube, to the lance tube and out through the nozzle. Once fully inserted, the lance tube is rotated one, two, or three revolutions while blowing a high pressure cleaning medium aimed nearly parallel to the boiler side wall. The cleaning radius is a function of the cleaning medium's pressure, the type of nozzle, the nature of the deposit, and the boiler surface. In general, on bituminous coal deposits, the normal effective cleaning area of an IR sootblower is elliptical with a vertical axis of 12 ft (3.66 m) and a horizontal axis of 10 ft (3.05 m). This area is typical when supplying either 150 psig (1034 kPa) air or 200 psig (1379 kPa) steam. For subbituminous or lignite deposits, the effective cleaning area may be much smaller.

Waterlance When ash deposits consolidate into a dense slag layer, mechanical impact energy alone may

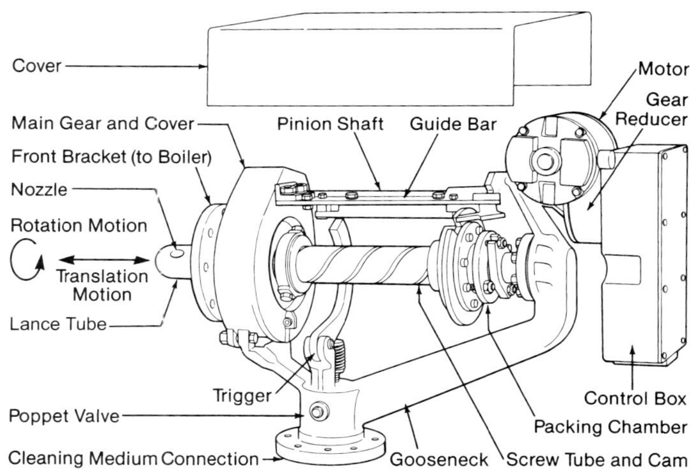

Fig. 2 Model IR wall blower (on newer designs, the control box is mounted separately).

not be sufficient to clean effectively. A waterlance removes slag by directing a concentrated water jet and precisely controlling the pattern and progression velocity of the water.

Shown in Fig. 3, the waterlance consists of a lance tube with a single nozzle, a drive assembly for inserting the lance tube into the boiler, and valves for turning a high pressure water source on and off. At rest, the nozzle is located inside a wall box sleeve. When activated, the lance tube is advanced along a helical path. Water flow is delayed until the nozzle is about 19 in. (48 cm) from the rest position. The nozzle is installed at a backrake angle to direct the water jet back toward the wall. Once water flow is established, the helical path creates a spiral cleaning pattern on the furnace wall.

The total travel length is about 3 ft (0.91 m). The most common backrake angles are 15 and 20 deg (0.26 and 0.35 rad) as measured from the perpendicular to the lance tube axis. These nozzle orientations achieve a typical cleaned area of 175 and 350 ft^2 (16.3 and 32.5 m^2), respectively.

Once the lance reaches its fully inserted position, water flow is shut off, compressed air is delivered to the lance tube to purge water from the lance, and the lance is removed as quickly as possible.

A partial arc waterlance operates similarly, except that the water is turned on and off during each rotation so that the water jet cleans a partial circle. Examples of cleaning patterns achieved by a partial arc waterlance compared to the standard waterlance are shown in Fig. 4. The partial arc waterlance avoids water jet impingement on corner wall tubes, around burners, or in other sensitive regions.

Selective pattern waterlance Because of the relatively low flow rate of its water jet, waterlance cleaning is generally effective up to 30 ft (9.14 m) from the nozzle. The selective pattern waterlance is designed to accommodate this distance limitation (between the nozzle and cleaning surface) by using a much longer lance tube and cleaning surfaces other than the wall through which the waterlance was inserted. This approach is particularly well-suited for the underside of the furnace's nose arch and hopper slopes. It has also been used effectively to clean wing walls and surfaces between wing walls, division panels, and the leading edge of tube pendants at the top of the furnace.

To maintain a constant jet progression velocity on the surfaces, drive motor control programming may be extensive. In addition, the water may be turned on and off one or more times within each revolution

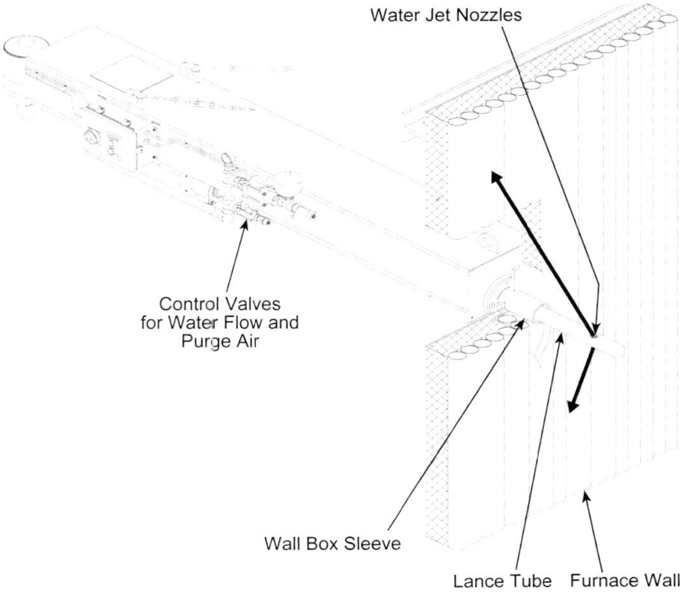

Water Jet Nozzles
Control Valves for Water Flow and Purge Air
Wall Box Sleeve
Lance Tube
Furnace Wall

Fig. 3 Waterlance.

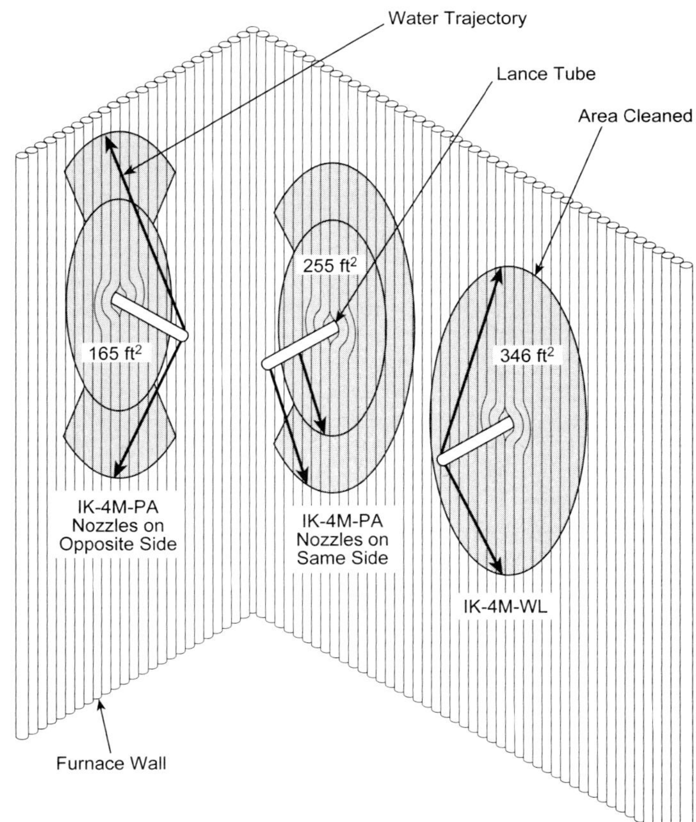

Fig. 4 Partial arc waterlance cleaning patterns.

to achieve a desired cleaning pattern. The selective pattern waterlance frequently uses more than a single nozzle in the lance tube to achieve close side-to-side spacing of the cleaning jet patterns, and for lance tube cooling. During the retraction of the lance tube, the water flow may be pulsed and is eventually terminated.

HydroJet® A more recently adopted technology is the HydroJet® boiler cleaning system shown in Fig. 5. The HydroJet directs a jet of water through a side wall opening to clean surfaces on the opposite side of the furnace. The lance tube on a HydroJet remains outside the boiler. Its nozzle is installed in the leading tip of the lance tube and the lance tube is articulated by a two-axis drive system to aim the nozzle at the surface to be cleaned.

The water flow required to deliver a concentrated jet across the width of the furnace and still achieve effective cleaning is greater than that of the waterlance. Overall, this injection of a sufficient volume of cold water into the flue gas significantly affects the boiler heat rate. However, the HydroJet offers several advantages that make it an attractive cleaning option. With a range of motion of ±45 deg (0.79 rad) in both the horizontal and vertical directions, a single HydroJet can provide the cleaning coverage equivalent to many waterlances, depending on the size and layout of the furnace. Without a traversing lance tube, this device requires much less outboard space on the side of the boiler and is tolerant of low-quality water.

The heat rate penalty of either a waterlance or Hydrojet can be estimated by calculating the heat energy required to evaporate the water jet. See Equation 1.

$$Q_{loss} = \dot{m}_w (H_1 - H_2) \qquad \textbf{(1)}$$

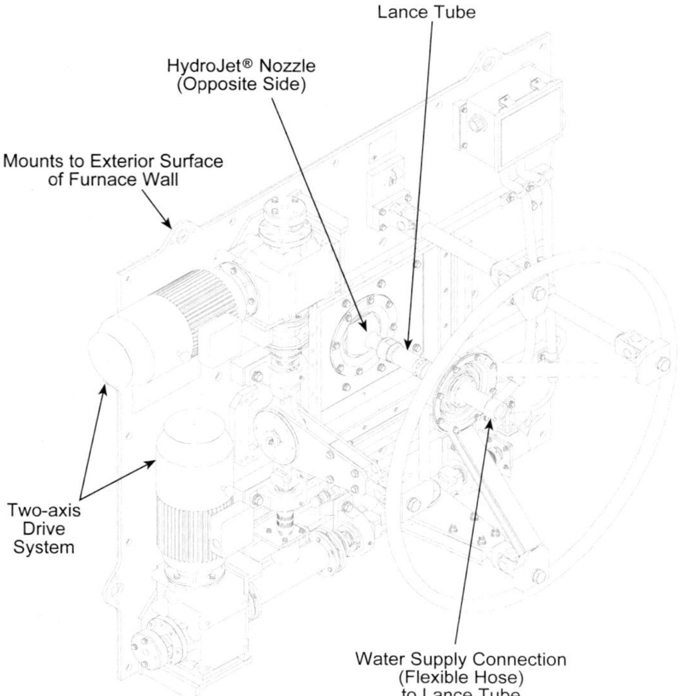

Fig. 5 HydroJet® boiler cleaning system.

where

Q_{loss} = heat penalty, Btu/h (kw)
$\dot{m}_w$ = mass flow rate of the water jet, lb/h (kg/s)
H_1 = enthalpy of the water leaving the HydroJet nozzle, Btu/lbm (kJ/kg)
H_2 = enthalpy of steam at the pressure and temperature of the flue gas exiting the air heater, Btu/lbm (kJ/kg)

Generally, atmospheric pressure and the flue gas exit temperature from the air heater are a good approximation for determining H_2 for this purpose. Although the evaporated water is heated to superheated temperatures in the furnace, this steam is cooled along with the flue gas through the remainder of the flue gas path and useful energy is extracted.

The boiler operator must also consider the impact on superheat temperatures, water wall circulation, furnace pressure, and burner tilts if the boiler is fitted with this feature. Each time the HydroJet is operated, excursions of these parameters are possible. This is managed by minimizing required flows through advanced nozzle designs, operating the HydroJet for short periods of time, cleaning small areas, and pausing between areas.

Control considerations

Permanently installed sensors have improved substantially and are successfully used in feedback control algorithms. Heat flux sensors may be installed in furnace wall tubes and monitored to detect heat transfer reduction associated with ash or slag buildup and the thermal shock incurred during cleaning. Emissivity probes can also be used to detect highly reflective deposits that will degrade radiative heat transfer to the furnace wall tubes. Optical pyrometers for measuring flue gas temperature can assess local or overall gas temperatures.

Control systems address furnace cleaning by dividing the surfaces that can be reached into relatively small segments or zones. In the case of a waterlance or an IR wall blower, a zone is the entire area that can be cleaned by one device. In the case of a HydroJet, a zone may be 5 to 10% of the total area that is within the reach of a single device. The division of the furnace area is made by observations to identify areas that generally accumulate deposits at about the same rate. Judgment is applied to ensure that cleaning a zone will not create an upset large enough to destabilize the boiler's overall operation.

In addition to using sensors to signal the control system when a particular area in the furnace requires cleaning, the control logic will employ permissives. Permissives are additional conditions that must be satisfied before the boiler cleaning device is activated. Permissives that are included in a furnace cleaning control system are specific to each boiler, and may include the following:

1. minimum elapsed time since the last cleaning of a particular zone,
2. furnace exit gas temperature, above the minimum recommended point for furnace cleaning,
3. superheat temperature, above the minimum recommended operating point,

4. reheat temperature, above the minimum recommended operating point,
5. superheater spray flows, above the minimum recommended operating point,
6. boiler load, above the minimum recommended load for furnace cleaning, and
7. burner tilts, lower than the recommended degree for furnace cleaning.

Unique applications

Hopper slopes High ash content fuels, in particular brown coal, tend to accumulate ash at such a high rate on the hopper slopes that it can bridge over the bottom opening in a matter of hours. This has been effectively avoided by installing selective pattern waterlances directly above the bottom ash opening and controlling the water jets to sweep the hopper slope on both sides.

Wing walls Tube sections that extend from the front wall of the furnace to the roof, referred to as wing walls, are cleaned by selective pattern waterlances. Slag accumulation on wing walls not only renders needed heat transfer surface ineffective, but is also a common source of major slag falls leading to boiler hopper damage. The face of these tube walls is only accessible by inserting a lance tube from the front wall in between adjacent wing walls. The selective pattern waterlance is often configured to continue past the wing walls to clean superheater pendants or division walls, maintaining the required jet progression velocity control despite the complex geometry changes between wing walls and other surfaces.

Overfire air ports When using overfire air ports to stage combustion and reduce nitrogen oxides (NO_x), a heavy slag buildup could accumulate above and around each air port. This slag accumulation can affect the flow distribution and penetration of overfire air into the furnace. This accumulation also presents a personnel hazard during an outage.

A water cleaning device, designed specifically for overfire air ports, has been used to remove this slag while the boiler is on-line. This special waterlance is referred to as an *eyebrow cleaner* because of the physical appearance of the slag accumulation pattern. The unit has a retractable lance tube that sits at rest inside a sleeve passing through the air duct to the overfire air ports. A high velocity water jet is used as the cleaning medium, and safeguards are included to ensure that refractory installed in the air ports is not damaged by the waterlance's operation.

Convection pass cleaning

Downstream of the furnace, heat is removed from the flue gas by steam-cooled or boiling-water cooled wall tubes and tube banks contained in the convection pass. (See Chapter 19.) In these tube banks, heat absorption is dominated by convection heat transfer from the flue gas to the outer surface of the tubes. Individual tube banks in this region may include primary and secondary superheaters or reheater.

The first segment of the convection pass is a critical region for on-line boiler cleaning because this is the first point at which the flue gas encounters narrow passageways between the tubes. At this point, the flue gas often contains a substantial concentration of ash particles, and the gas temperature may be high enough that the particles are in a semi-molten state. Under these conditions, deposits tend to collect on the leading edge of each tube bank and between tubes that are lined up behind one another. Deposits that collect between the tubes are referred to as *platenized buildup*, and are difficult to remove with on-line cleaning. Fortunately, it is usually leading edge buildup that has the more pronounced effect on gas path blockage.

Principles

Deposits in the convection pass are removed primarily by the mechanical impact energy delivered by a retractable sootblower. At the highest temperature locations, the outermost surface of the deposits may be in a semi-molten state, for which mechanical impact alone may not be sufficient. This condition is addressed in the section that describes Precision Clean® sootblower control technology. Throughout most of the convection pass, gas temperatures are lower and the deposits are relatively brittle if removed periodically to preclude sintering. In this case, the impact energy of a sootblower nozzle creates cracks and fissures in the deposits. Deposits break off in relatively large pieces, often from their own weight, once deep cracks have been created by the impact energy.

At even lower temperatures, the deposits are much more porous and are only loosely adhered to the tube surface, and to one another. In this case, cleaning is accomplished by the abrasion-like action of the cleaning medium. Ash particles are removed as small particles being swept off the surface, for as long as the nozzle is aimed at the surface.

The challenge in the convection section is to properly sequence the delivery of cleaning media at sufficiently high velocity to tube surfaces located within an array of tightly spaced rows and columns. The impact energy required to break up the deposits generally requires nozzle exit velocities well above the speed of sound. Immediately after exiting the nozzle, the cleaning media begins to decelerate. Depending on the strength of the deposits, this can limit the effective range of sootblower nozzles to between 2 and 13 ft. (0.6 to 4 m).

Cleaning devices

Long retractable sootblower Two common long retractable sootblowers, the IK-500 and IK-600®, are shown in Figs. 6 a and b. Each has a lance tube with nozzles near the front tip, a carriage for inserting the lance tube into the boiler, and a poppet valve for turning the cleaning medium on and off. A stationary feed tube delivers cleaning flow from the poppet valve to the lance tube as it slides over the feed tube. The carriage contains a gear set that drives the lance into and out of the boiler along a helical path.

The nozzles are commonly installed perpendicular to the longitudinal axis of the lance tube. As the cleaning media jet leaves the nozzle, it spreads outward and

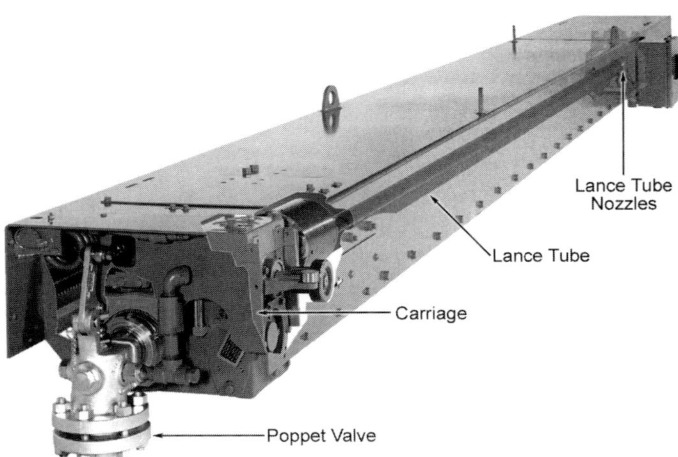

Fig. 6a Model IK-525 sootblower.

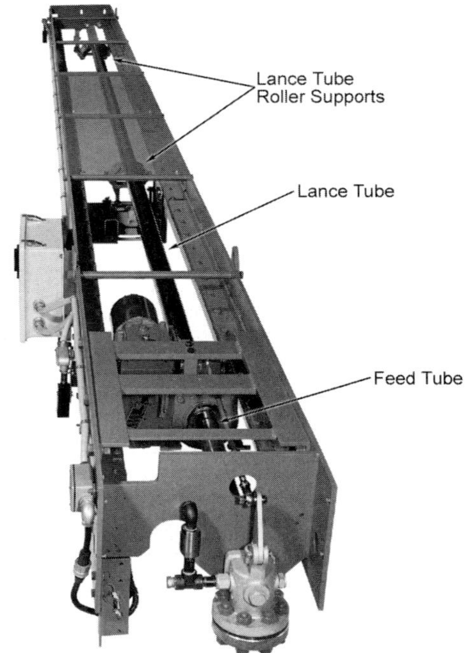

Fig. 6b Model IK-600® sootblower.

is effective at removing some of the platenized buildup in addition to deposits in the direct line of sight of the nozzle centerline. This perpendicular nozzle orientation is preferred when the side-to-side spacing between tube rows is less than 18 in. (46 cm). With smaller spacing between tubes, deposits can build up and bridge the gap to close off the gas passageway. The perpendicular nozzles provide the deepest penetration to ensure that the gas path remains unobstructed.

When the side-to-side spacing between tube rows is greater, the nozzles may be oriented 5 or 10 deg (0.088 to 0.175 rad) from the perpendicular axis. This configuration is referred to as *lead-lag* nozzles, since one will be oriented forward of the perpendicular and the other slightly backward. This design has greater potential for removing the platenized deposit buildup between tubes within a row, and is used when there is little or no potential for deposits to bridge from row to row. Lead-lag nozzles have been particularly effective at cleaning division panels and pendant sections at the entrance to the convection pass.

Sootblower nozzle performance In long retractable sootblowers, nozzles are installed in the side wall of the lance tube. Within the lance tube, the compressible cleaning media travels at subsonic speed, generally Mach 0.3 or lower. Most sootblower nozzles are designed to accelerate the flow to supersonic conditions.

Nozzle performance is assessed by measuring the peak impact pressure of the cleaning media jet at points downstream of the nozzle outlet. Within limits, the higher the impact energy, the more effective the nozzle will be at fracturing and removing ash deposits.

The effect that nozzle design has on peak impact pressure is illustrated by the comparison of three sootblower nozzles in Fig. 7. All three nozzles have the same throat diameter, and because all three accelerate the fluid to the speed of sound at the throat, the flow rates are identical at any pressure shown on the graph. In this figure, the peak impact pressure measured at a distance of 30 in. (76 cm) from the nozzle outlet is plotted as a function of supply pressure for each nozzle. The supply pressure at the nozzle inlet can be altered by adjusting the pressure drop in the sootblower's poppet valve.

The development of high performance nozzles has changed the previous operating approach. When using nozzles that were only slightly supersonic (Ground Flush and Hi-PIP, for example), it was generally true that the higher the density of the cleaning media, the better the cleaning. For a system using steam, this meant that less superheat was better for cleaning, and saturated steam was better than superheated steam. However, the performance of the newer class of supersonic nozzles depends on having no water droplets in the divergent section of the nozzle body to avoid the flowfield disturbances and to achieve full expansion of the steam jet. To avoid water droplets, it is generally necessary to have 25 to 50F (14 to 28C) of superheat at the inlet to the sootblower. The presence of condensate in the steam supply, or the formation of condensation in the sootblower lance tube, has a significant degrading effect on high performance nozzles. It can be severe enough to negate the benefit of a nozzle replacement project.

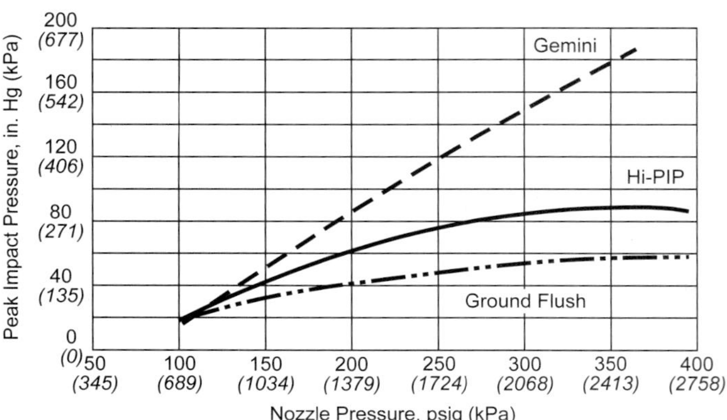

Fig. 7 Sootblower nozzle performance comparison at 30 in. (76.2 cm) from outlet.

Precision Clean® sootblower control A standard sootblower translates and rotates the lance tube at a constant speed throughout its travel. As the lance tube rotates, the nozzles sweep along the face of the tube bank. With constant speed rotation, the jet progression velocity (inverse of dwell time) varies significantly as the nozzle sweeps along the face of the convection tube bank, from a point close to a point far away.

The required dwell time varies with deposit types and deposit surface condition. Surface condition becomes particularly important when the deposit is semi-molten. In this situation, dwell time becomes one of the most significant factors in deposit removal.

A field-proven technique to address the dwell time issue is to control the rate of lance tube rotation to achieve a constant jet progression velocity within a defined arc. The arc includes surfaces that are within the effective range of the sootblower nozzle's impact energy. Outside this arc, the lance tube is simply rotated as quickly as possible until the next control segment (arc) for constant jet progression velocity is reached. This technique, referred to as Precision Clean sootblower control, significantly improves and extends the cleaning effectiveness of a sootblower and can be retrofit to existing sootblowers.

One way cleaning At the cooler end of the convection pass, sootblowers often clean adequately during a single pass of the lance tube alone. This is almost always the case in the economizer, and often in the primary superheater of a power boiler. In an industrial boiler, this may apply to the entire convection pass.

With one way cleaning, flow is most often low during lance insertion and high during lance withdrawal from the boiler. This patented technique most often involves a remotely activated isolation valve in the cleaning media supply line, with a bypass containing a flow orifice. The orifice provides a minimum amount of cleaning media flow when the isolation valve is closed, sufficient to keep the lance tube from overheating. The poppet valve remains unchanged. A transient heat transfer calculation is performed to determine the bypass cooling flow rate.

Oscillator sootblowers Sootblowers are often installed in locations where a high energy cleaning jet can cause damage. This might include the attachment point of generating bank tubes to the steam drum, locations near the face of a tube bank, and areas prone to sootblower-induced tube erosion or tube vibration.

Oscillator sootblowers have a mechanical linkage on the carriage that causes the lance tube to sweep a limited arc back and forth as it travels into and out of the boiler. The arc can be altered by gear set changes. Generally, the lance is fitted with pairs of nozzles so that cleaning is two-sided and the thrust of one nozzle jet is balanced by the thrust of the other. In some applications the lance tube may be fitted with a nozzle on only one side and thrust-balancing orifices on the other. A typical cleaning pattern is shown in Fig. 8.

Extended lance sootblowers As the flue gas temperature is lowered in the convection pass, additional cleaning options become available. Generally, when the gas temperature is below 1100F (593C), extended lance or fixed position sootblowers can be used. Use

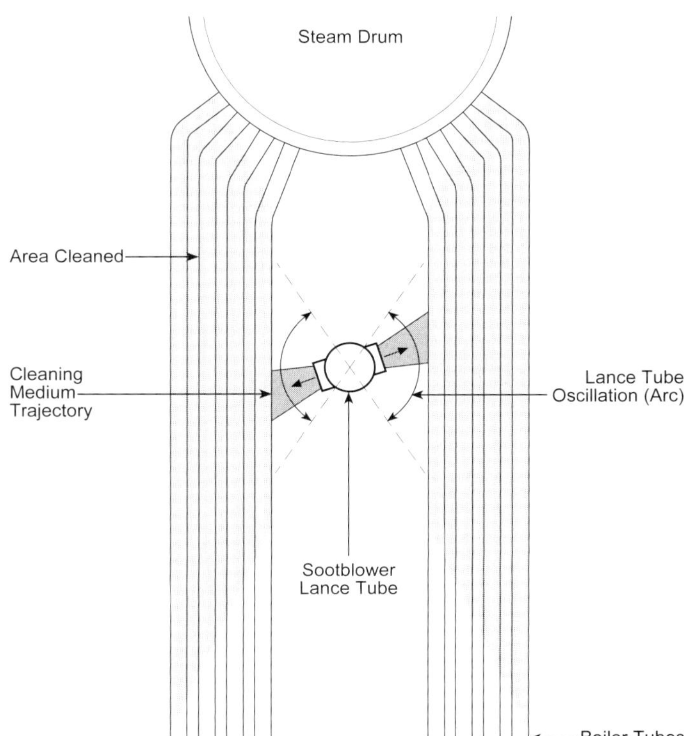

Fig. 8 Cleaning pattern of oscillator sootblower.

of specialized materials can permit the use of extended lance or fixed position sootblowers at higher temperatures. Both are distinguished by their lance tube design, that continuously remains inside the flue gas pathway. Fixed position sootblowers are discussed under *Back pass/economizer cleaning*.

An extended lance sootblower is still a retractable sootblower, although the beam and drive assembly are only long enough to translate the lance tube a fraction of its length. The extended lance tube is fitted with more than one nozzle assembly. A half-travel version, for example, has two nozzle assemblies: one at the front tip of the lance tube and the second at mid-length. Each assembly may consist of a pair of identical nozzles diametrically opposed to balance the thrust, or can have a single nozzle with or without thrust balancing orifices on the opposite side. Extended lance sootblowers fitted with 2, 3 and 4 nozzle assemblies are common.

Bearing plates support the lance tube inside the boiler. These plates are mounted to structural members inside the boiler, and the lance passes through an oversized opening in the plate to limit tube deflection. The most distant bearing is installed so that the lance tip is supported in the rest position (when the lance is retracted to the rear stop of the carriage travel). Additional bearings may be used if the gas temperature is high enough to induce stress relaxation in the lance tube material (sagging).

The extended lance sootblower is generally used to accommodate exterior space constraints. This sootblower is also used to address unique design challenges posed when the convection section is partitioned, and it is not permissible to expel a cleaning jet into a cavity or sleeve between the partitioned chambers. Extended lance sootblowers are used when a concentrated jet of supersonic cleaning media is needed for effective cleaning.

Diagnostic and sootblower control techniques

Modern boiler cleaning systems often incorporate on-line diagnostic sensors and advanced control algorithms to increase cleaning effectiveness and boiler efficiency while reducing the consumption of sootblower media and reducing potential damage to the boiler

Flue gas temperature is commonly monitored at the furnace exit to detect general fouling on the furnace walls. Load (fuel firing rate) and burner tilt position must be considered. Flue gas temperature is also useful for detecting fouling in segments of the convection pass. This is done with heat balance and heat transfer calculations for a particular tube bank to determine a fouling factor. Deposit accumulations are detected by tracking the fouling factor over time.

Pressure drop of the flue gas across tube banks is an indication of deposit accumulation. Unfortunately, by the time the deposit accumulation results in a measurable pressue drop increase, the deposits are already heavy and are likely difficult or impossible to remove with on-line cleaning alone. In recovery boilers, an operation called chill-and-blow can often disrupt the deposits enough that sootblowers can remove them, or can at least extend the time until the boiler must be shut down for a water wash.

The motor current on the forced or induced draft fan can also be monitored to detect deposit accumulation. Similar to flue gas pressure drop, by the time this parameter shows a measurable change the boiler deposits may be too extensive to remove with on-line cleaning.

Centralized and distributed computerized control systems can monitor fouling, verify that permissives have been satisfied, and activate sootblowers one by one through a prescribed sequence. Sequencing in the convection pass is situation-specific. Generally, the final sequence is selected by the boiler operators based on observed fouling patterns, accumulation rate, degree of difficulty in removing deposits in the various regions and temperature zones, and sootblowing system constraints. Constraints might include steam capacity (number of sootblowers that can be operated simultaneously) and the time it takes for each sootblower to run. These constraints may lengthen the elapsed time before critical sootblowers can be repeated.

The following guidelines are recommended as a starting point.

Convection pass deposits are easier to remove when they are fully solid and brittle. Even a thin layer of semi-molten deposits coating a thick and otherwise solid base increases the amount of impact energy required. For this reason, it is generally recommended to sequence sootblowers from front to back of the boiler. Starting at the positions nearest the furnace, each sootblower operation improves heat transfer at that point and makes it easier to remove deposits at points downstream, by lowering the gas temperature at these positions.

Among positions that are essentially at the same point (temperature) in the gas stream, the choices are bottom to top, or top to bottom. In convection pass sections where the gas flow is predominantly upward (such as tube pendants suspended over the furnace and before the arch), progressing from lower to upper elevations is recommended. However, when deposits weighing several hundred pounds could be dislodged, the recommendation would be top to bottom. This reduces the chance that a slag fall from an upper location would land on a lower sootblower's lance tube.

Back pass/economizer cleaning

Later segments of the convection pass generally require a denser heat transfer surface arrangement due to the lower gas temperature. Tube spacing is generally smaller as there is a lower likelihood of deposit bridging. For many fuels, tubes may be finned and positioned in a staggered arrangement rather than simple rows and columns. Deposit characteristics are clearly different in this region, dusty and non-sintered for some fuels, and sometimes accompanied by lightweight clumps of ash (commonly referred to as *popcorn ash*). However, the denser tube arrangement can be problematic for many units firing western United States (U.S.) coals. This back pass region includes the economizer.

Principles

Ash deposits in this region often begin as a thin layer of dust on the tubes, which may eventually accumulate as unstable flags (wedge-shaped ash buildup) built up on the leading edge of the tubes. Fixed position sootblowers may be used where the deposits are dusty or loosely adhered. Only a portion of the cleaning is actually accomplished by impact energy of the nozzle jet. Most cleaning is accomplished by the momentary increase in gas flow past the deposits when the cleaning media combines with the flue gas flow to disrupt the flow pattern. This is referred to as *mass blowing*.

When a greater degree of energy is required, for example due to ash compacted between radial fins on the boiler tubes, rake-type sootblowers may be used.

When ash deposits are loose or accumulated in unstable buildups, sonic horns operated at optimal frequencies may be effective.

Cleaning devices

Fixed position sootblowers This type of sootblower is characterized by a tubular element fitted with a number of nozzles along its length, usually positioned along one side of the element. The nozzles are generally spaced to correspond to the boiler tube spacing and aligned so that one nozzle is aimed between each row of tubes. The element serves as a lance tube, but remains permanently inside the boiler flue gas. The fixed position sootblower with a rotating element is also called a rotary sootblower. One common rotary sootblower, the model G9B, is shown in Fig. 9.

The element is supported along its length with bearings, similar to the extended lance sootblower. When the sootblower is activated, the element is rotated. The rotating motion operates a cam which mechanically opens the poppet valve, and cleaning medium flows

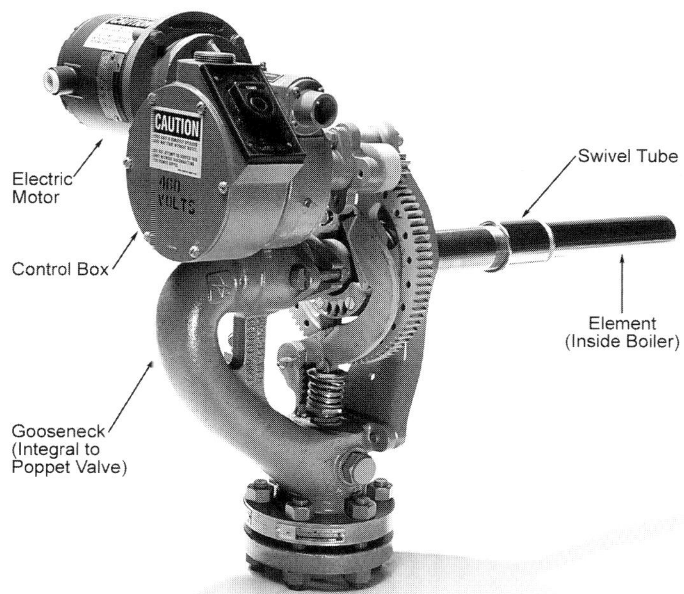

Fig. 9 Model G9B electric motor driven rotary sootblower.

into a short feed tube and then into one end of the element to distribute among the nozzles. The cam can be configured to close the poppet valve after a prescribed arc or after a full revolution of the element.

To clean effectively, the rotary sootblower must provide sufficient flow to all nozzles in the element. The nozzles nearest the feed tube entrance receive a slightly higher portion of the flow. Design guidelines are established to limit the flow maldistribution to less than approximately 10% between the closest and farthest nozzle. However, to accomplish this, the size of the nozzles must be restricted and thereby tube bank penetration is limited. Nozzles are a convergent/divergent design with throat diameters ranging from 0.25 to 0.375 in. (6.35 to 9.53 mm).

A non-rotating version of the fixed position sootblower omits the feed tube and gear drive assembly. An isolation valve simply turns flow on and off, and the isolation valve must be opened and closed with an electric or pneumatic actuator.

Rake-type sootblowers The rake-type sootblower is particularly effective at removing deposits from tightly spaced tubes with radial fins. This is a retractable sootblower with no lance rotation; the lance tube design is T-shaped as shown in Fig. 10. The nozzles are located in the rake segment of the lance tube after the tee, which is the section oriented perpendicular to the direction that the lance travels. This lance is never fully retracted from the flue gas. Also, the lance tube may have more than one rake. The rake provides the ability to traverse a relatively low number of nozzles across a large area of tube surface. The nozzles will always be pointed directly into the tube bank and can not contribute to cleaning the tube bank upstream of the sootblower. They can, however, deliver more cleaning energy than a rotary sootblower.

Traveling crosswise to the economizer tubes is preferable, since this provides the maximum opportunity for cleaning flow to attack tube deposit buildup from both sides (because of the diverging trajectory of cleaning flow) in addition to head on. However, the rake-type sootblower is still an effective approach to cleaning difficult ash buildups even if space constraints outside of the boiler dictate that the rake travel must be parallel to the economizer tubes. Nozzles range from 0.25 in. to 0.625 in. (6.35 to 15.9 mm).

Sonic horns The use of acoustic energy to propagate sound waves within the boiler enclosure can also dislodge deposits in the back pass regions of the boiler. Sonic horns produce acoustic waves at frequencies ranging from 20 to 200 hertz. Their use is generally limited to locations where the deposits are dry and loosely adhered. Example locations include a finned tube economizer with dry deposits trapped between the fins, and plates of an electrostatic precipitator. Fog horns have been used to clean deposits in fabric filter baghouses.

At present, most sonic horns operate at mid-range frequencies of about 75 hertz. At this frequency, a number of relatively closely spaced horns must be installed around the periphery of the boiler enclosure; each is operated intermittently, and operation alternates from one to the next. Their effectiveness is limited by the tendency for the sound pressure level to attenuate a short distance into the flue gas. A smaller number of horns is required when the operating frequency is lower, around 20 hertz, because of the ability of the long wavelength unit to establish acoustic resonance if the sound wave generator is carefully installed and adjusted. Lower frequencies are also more effective at extending the cleaning range deep into a staggered tube array.

SCR catalyst cleaning

The more common and reliable cleaning method for selective catalytic reduction (SCR) system catalyst is by sootblowing with steam or air. Catalyst plates have been cleaned by acoustic devices where site-specific conditions allow.

Principles

Sootblowers clean ash and ammonia salt particulate from the catalyst by providing a relatively low velocity of steam to each channel of the catalyst assem-

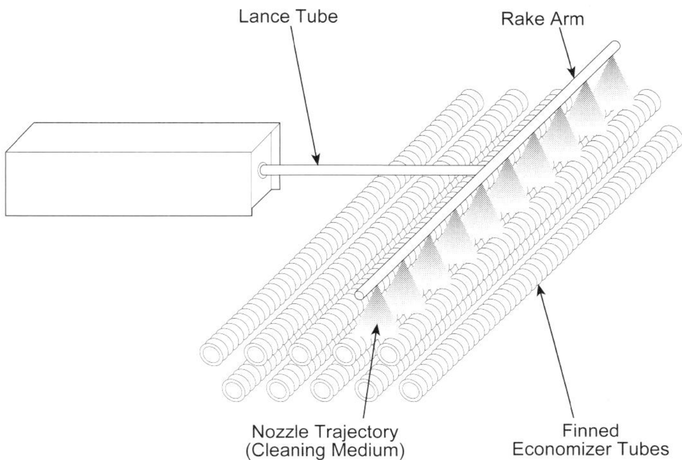

Fig. 10 Rake-type sootblower.

bly. The cleaning medium is almost always steam, but air can be used as an alternative. The steam velocity is low compared to sootblowers used for boiler cleaning to avoid mechanical or erosive damage to the catalyst substrate, yet high enough to dislodge loose particles in the catalyst channels and re-entrain them in the flue gas flow. Sootblowers cleaning the catalyst are designed to deliver steam in the range of 80 to 120 ft/s (24.4 to 36.6 m/s) to the catalyst surface. However, even at this relatively low velocity, it is critical to avoid moisture in the steam flow. Water droplets are erosive, can cause thermal shock on some catalyst designs, and could dissolve the thin vanadium pentoxide coating on the ceramic substrate.

Cleaning devices

Straight line sootblower The IK-525 Straight Line Cleaner is a retractable sootblower designed for applications where the lance tube is inserted and retracted without rotation. It is used with rake-type lance tube assemblies that reside permanently inside the flue gas. This sootblower cleans SCR catalyst assemblies, economizers, and air heaters, and is occasionally used for applications where lance tube rotation is undesirable or unnecessary, such as air duct cleaning.

SCR catalyst assemblies and air heaters are cleaned by directing steam or air flow directly into the narrow passages. An array of nozzles is positioned directly above or in front of the leading edge of the assembly. The nozzles are positioned in rake arms attached at right angles to a central lance tube. A single lance tube may be fitted with one to twelve rake arms. The lance tube is supported inside the flue gas chamber such that it holds the rake arm nozzles at a fixed distance from the leading edge of the catalyst or air heater assembly. The lance tube is inserted and retracted such that each rake arm advances to the starting position of the rake arm ahead of it. Cleaning can be during both the insert and retract pass of the rake arm, or during only the retract pass.

When the lance tube has a number of rake arms, it may be necessary to minimize the difference in cleaning medium flow as it is distributed to the arms. Therefore, the lance tube itself may consist of an upper and lower lance tube. The upper lance tube is aligned with the side wall penetration, and is the tube that penetrates the wall and telescopes over the stationary feed tube. Flow distribution pipes deliver cleaning medium flow to the lower lance tube. The rake arms are attached to the lower tube.

Flow distribution among the rake arms of each sootblower and among the nozzles is a primary consideration in the design of an SCR sootblower. The goal is to provide a uniform curtain of steam flow from each rake arm.

Various methods help avoid moisture, beginning with using a sufficiently superheated steam source. To avoid moisture from the enthalpy decrease associated with pressure drop and velocity increase between steam source and sootblower nozzles, it is generally recommended that the steam source for SCR applications be on the order of 50 to 100F (28 to 56C) superheated. The goal is to have the steam remain above saturation throughout the sootblower to the rake arm nozzles, while satisfying thermal shock criteria imposed by the catalyst design. A common requirement is that the cleaning medium temperature at the point of contact with the catalyst surface must be within 100F (56C) of the catalyst (i.e., flue gas) temperature to avoid thermal shock.

A superheated steam source can still produce condensate if precautions are not taken to drain condensate collected in the sootblower supply piping, and preheat the steam header piping before operating the sootblowers. Condensate drain control systems are recommended in the sootblower piping system to ensure that the sootblower piping is maintained free of condensate between sootblower operations. In addition, the control system should include steam header warm-up before the sootblowers are operated. Generally this would consist of opening the steam isolation and control valves and a sootblower piping drain valve for about five minutes before permitting the first sootblower in a sequence to operate.

Catalyst manufacturers have established maximum sootblower pressures for each of their designs. Generally, these fall into two groups: one recommends supply pressures from 80 to 100 psig (552 to 689 kPa), and the other prefers supply pressures of 10 to 20 psig (69 to 138 kPa). The latter has become more common.

Popcorn ash

Some of the ash that makes its way past the boiler arch is the result of small agglomerations referred to as popcorn or large particle ash. Most of this is removed from the hopper underneath the economizer, but some could be carried to the SCR. SCR designs include a screen positioned immediately above the inlet face of the catalyst bank or at the outlet of the economizer. The screens protect the catalyst from this ash. Because most of these particles will not pass through the screen, they must be removed by manual vacuuming during an outage. Aerodynamic devices have also been used at the outlet of the economizer to trap these large particles.

Air heater cleaning

Regenerative air heaters

Regenerative air heaters, common in large electric utility steam generating plants, are susceptible to ash particles collecting between the corrugated plates that form the heat storage elements or baskets (see Chapter 20). Some flow path blockage occurs at the leading edge of the element. Some particles collect loosely on the upper surface, and smaller particles become wedged between the plates. Most of the pluggage that occurs in the baskets, however, is an accumulation of fine particles that adhere to moistened surfaces near the cold end of the channels. The moisture is generally a result of acid precipitation from the flue gas. Ash particles may also attach to ammonia salt deposition associated with an SCR upstream of the air heater.

Regenerative air heater elements are made of thin, corrugated steel plates. Deposits are removed by means similar to that described in SCR cleaning. Cleaning media must be aimed directly at the top surface at relatively low velocities, to prevent the thin metal plates from being deformed or fatigued by flow-induced vibration. Unlike the SCR catalyst, the regenerative air heater is in motion.

Cleaning devices

Straight line sootblower The version of the straight line sootblower used in air heater cleaning applications is the model IKAH. The lance tube is shown in Fig. 11. It contains one small rake, so it is not fully retractable, although the components outside the boiler are similar to the fully retractable sootblower. The six nozzles located along the length of the lance tube provide five cleaning bands as the air heater rotates. When activated, the lance is slowly inserted and retracted while each nozzle traces a spiral cleaning path along the face of the air heater baskets. Generally, cleaning flow is only provided during the insertion pass. A pneumatic actuator opens and closes the poppet valve, directed by the sootblower control system.

Swing arm sootblower A swing arm sootblower was often included as part of the original equipment by the air heater manufacturer. The arm consists of a pivoting pipe extending into the space upstream of the air heater baskets with the tip turned to aim directly into the face of the baskets. A nozzle is installed in the tip to direct cleaning media into the baskets with sufficient flow to flush ash particles that have collected between the corrugated plates. The arm is slowly swiveled as the baskets are rotated past the nozzle. The cleaning path is spiral. The gear driven assembly that controls the arm movement is external to the air heater, and is located in the stationary part of the air heater element enclosure, ahead of the rotating basket segment.

Dual media air heater cleaner With increased use of low sulfur western North American coals and SCR systems, fouling of air heaters has become more difficult to manage. The high calcium content of western coals can lead to cementacious ash deposits if moisture is present, especially at the cold end of the air heater. Ammonia slip from the SCR can lead to tenacious salt deposits.

The dual media air heater cleaner has been used in recent years because of a superior cleaning capability with compressible media (steam or air), and a built-in capability to switch to high pressure water cleaning with highly precise nozzle position control. Dual media air heater cleaners have been installed at both the inlet and outlet face of the baskets on Ljüngstrom-type air heaters.

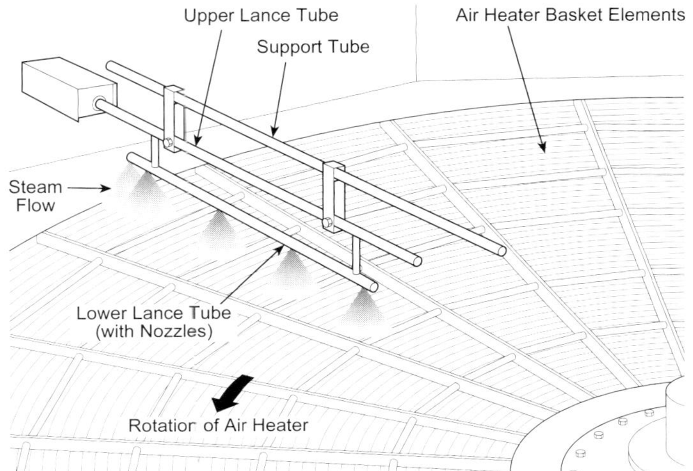

Fig. 11 Model IKAH sootblower for air heater cleaning.

The lance tube of the dual media cleaner is fully retractable, or if external clearance is not available, the lance tube can be configured with multiple nozzle assemblies and partial travel, making the sootblower beam much shorter. The nozzle assembly provides an array of converging/diverging nozzles for directing a jet of compressible cleaning media to the face of the baskets, and a second array of water jet nozzles. The lance tube does not rotate and is installed so that the nozzles are between 12 and 30 in. (30 to 76 cm) from the basket inlet. Cleaning is carried out during the retract pass of the lance, and the lance is controlled in a stepping mode; the lance tube advances one step and holds position for one, two, or three revolutions of the air heater before moving to the next position. The number of passes that the nozzle is held stationary is a function of radial position and operator selection.

The compressible flow nozzles are generally arranged in one of two possible arrays. The positioning of these nozzles provides a nearly uniform flow field between them and therefore thorough cleaning of the air heater basket channels that pass underneath the nozzle assembly. This allows for a substantial step size because of the wide cleaning band provided. Still, in this mode, the dual media sootblower typically requires 45 minutes to cover a 22 ft. (6.7 m) travel of the nozzle head. The compressible flow nozzle jets are typically supplied with steam or air at pressures ranging from 90 to 150 psig (621 to 1034 kPa).

The nozzle assembly also has a water jet nozzle array generally installed in pairs. The spacing between the pairs of water jet nozzles and the step size used during high pressure water wash ensure that every channel in the basket is cleaned by a water jet aimed directly at that single channel.

The compressible flow nozzles are occasionally supplied with low pressure, 100 psig (689 kPa) water to provide a soaking-type cleaning action. This is particularly effective when the material being removed is primarily salt deposits, and the dual media cleaner is positioned on the cold side (downstream) of the air heater baskets. With this flexibility, the dual media cleaner accommodates three modes of operation: high pressure steam cleaning for day-to-day cleaning maintenance, low pressure water wash for dissolving deposits, and high pressure water for semi-annual wash down. The high pressure water capability minimizes boiler down time and costly off-line water wash service contracts.

Tubular air heaters

Tubular air heaters are generally not as heavily fouled as regenerative air heaters, because their design affords less opportunity for deposits to collect. Most commonly, the flue gas passes through the tube side of this air heater design. Ash particles collect loosely on the inside walls of these tubes although there is the potential for moisture from acid precipitation to act as a wetting agent.

Cleaning devices

Rake-type sootblower for tubular air heaters The IK-525 straight line sootblower is used to pass a rake-type lance tube over the face of the tubesheet that forms the tubular air heater inlet. Nozzles direct cleaning media (steam or compressed air) directly into the air heater tubes as the lance advances and retracts. The lance tube may have multiple rakes to shorten the length of the sootblower needed on the outside of the air heater. The lance tube travel achieves complete cleaning coverage of the air heater tubes, with only minimal overlap at the start and stop positions of the adjacent rake arms.

Air puff sootblowers Fixed position sootblowers with a pulsating output are occasionally used to clean tubular air heaters when external clearance is not available for a rake-type sootblower. The model A2E air puff sootblower includes a gear driven mechanism that abruptly turns the flow on and off 20 times during each main drive rotation. A fixed position tubular element installed above the face of the tubesheet has nozzles installed along its length directed at the tube inlets. An abrupt and repetitive disruption of the flue gas flow into the tubes, along with modest impact energy, is accomplished by pulsing the cleaning medium flow. The most aggressive cleaning is achieved by a non-rotating element, but the number of A2E sootblowers can be reduced by rotating the element to direct cleaning flow into tube rows on either side of the element. Supply pressures of 125 to 150 psig (862 to 1034 kPa) are typical for coal-fired boilers.

The air puff sootblower is also used to clean the back pass regions of oil-fired boilers, including the economizer. Supply pressures up to 200 psig (1379 kPa) are generally required.

Other cleaning applications

This chapter focuses primarily on cleaning equipment found on large coal-fired utility boilers. While the same cleaning equipment is generally applicable to all boiler designs, a few additional special application features include:

Coal-fired (and biomass-fired) industrial boilers These units also usually have a boiler or generating bank located in the convection pass. (See Chapters 27 and 30.) Boiler banks are generally cleaned with retractable IK-type sootblowers.

Oil-fired boilers As discussed in Chapter 21, oil firing produces significantly less ash than coal firing though the ash is often more corrosive. Generally, fewer sootblowers are needed for oil versus coal firing. In these units, G9B sootblowers may replace some of the IK retractable blowers in the convection section upstream of the economizer.

Process recovery (PR) boilers As discussed in Chapter 28, deposits are often more extensive and harder to remove in PR boilers. As a result, IK sootblowers are used in larger quantities throughout all of the boiler surfaces and are cycled more frequently.

Waste-to-energy (WTE) boilers Because of the very corrosive nature of the WTE boiler environment, special care in cleaning of the surfaces is needed as outlined in Chapter 29. It is very desirable to not clean the superheater tubes completely to expose bare metal as this will accelerate corrosion. To address this, a *rapping* cleaning system is employed in the superheater along with carefully applied sootblowers in a range of units. Where sootblowers are used, tube shields are frequently employed to provide additional sacrificial metal on the tubes closest to the sootblower installation. More detail is provided in Chapter 29.

Fluidized-bed boilers Lower operating temperatures generally make cleaning these units easier than pulverized coal-fired units, reducing the sootblower requirements. In circulating fluidized-bed units, sootblowers are not installed in the vertical furnace shaft. (See Chapter 17.)

Ash handling systems

The ash handling systems of a steam generating plant collect the ash and residue from the combustion of solid fuels from different points along the boiler flue gas stream (see Fig. 12), transport it to storage bins or silos, and prepare the ash for transport or disposal. Because the characteristics of ash are very different from the front to the back of the boiler, the collection, transport, and storage systems are usually separate for the furnace and collection points downstream.

In a coal-fired boiler, the general categories of ash are:

1. Bottom ash – the material that collects at the bottom of the furnace, possibly a heavy slag.
2. Mill rejects – the heavy pieces of stone, slate and iron pyrite that are discharged from the coal pulverizer.
3. Economizer ash or popcorn ash – the course and comparatively dense particles that drop out of the flue gas when the gas changes direction abruptly, such as in the back pass and air heater ducts.
4. Flyash – the fine ash particles that are collected in the dust collection equipment.

Although bottom ash and mill rejects are unique to pulverized coal-fired boilers, the above classification is generally useful for describing the different types of ash handling equipment. Common terms and definitions are provided in Table 1.

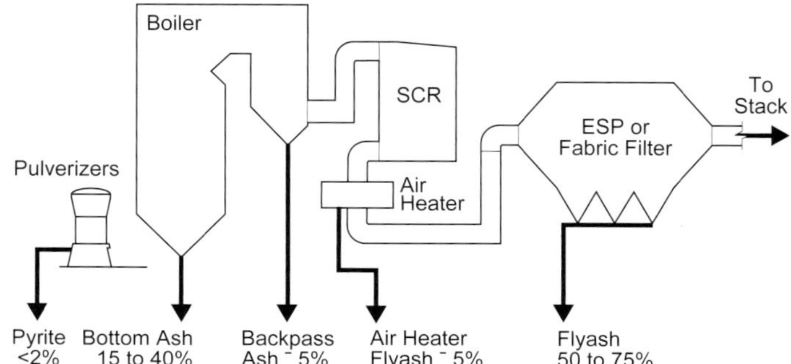

Fig. 12 Typical pulverized coal-fired boiler ash discharge locations and approximate amounts.

Bottom ash systems

Bottom ash is most often conveyed in a hydraulic system in which the ash is entrained in a high flow, circulating water system and delivered to either an ash pond or to dewatering storage bins. Alternatively, mechanical drag systems have been used to convey bottom ash to the dewatering storage bin because they use less water and usually have a lower initial cost. Both systems are designed to handle ash dropping into the conveyance system at temperatures as high as 2400F (1316C), requiring each system to have a quenching volume of water at the initial collection point. It is common for the bottom ash system to store as much as three days of ash output from the boiler.

Hydraulic bottom ash systems

The first component of the hydraulic conveying system is the water-filled hopper shown in Fig. 13. The bottom of the furnace terminates in a rectangular opening running across the width of the boiler (the floor of the boiler itself being a hopper). The bottom ash hopper fits up to this with its own sloped wall chamber so that ash and slag feed by gravity to a single outlet. On larger boilers, the bottom ash hopper may be divided into parallel chambers, referred to as *pant legs*, each having an outlet at its bottom as shown in Fig. 13. The outlet is sealed with a hydraulically driven gate controlled to open when the hopper is full, and coordinated with other hoppers so that only one hopper at a time empties into the conveying line.

The bottom ash hopper is filled with water to quench the hot ash as it enters, and is usually lined with refractory. The connection between the ash hopper and the bottom of the boiler seals the boiler gases and accommodates thermal expansion of the boiler. This is accomplished by a water trough design around the top of the hopper that keeps a steel plate skirt immersed in water. The skirt is attached to and moves up and down with the boiler, while the bottom ash hopper rests on the ground. Exposed refractory is kept cool by continuous water flow over the refractory surface. Cooling water supply is furnished either as overflow from the seal water trough or through embedded refractory cooling water distribution piping. The cooling method selected is determined by the boiler operating pressure.

When the boiler is a continuously slagging wet bottom design, the ash hopper is replaced with a slag tank. The wet bottom boiler terminates in a cooling trough for the flowing slag. To break up strings of slag forming at this point, the cooling trough has a slag swiper that swings a water-cooled arm back and forth across the inlet opening. The slag tank may include agitat-

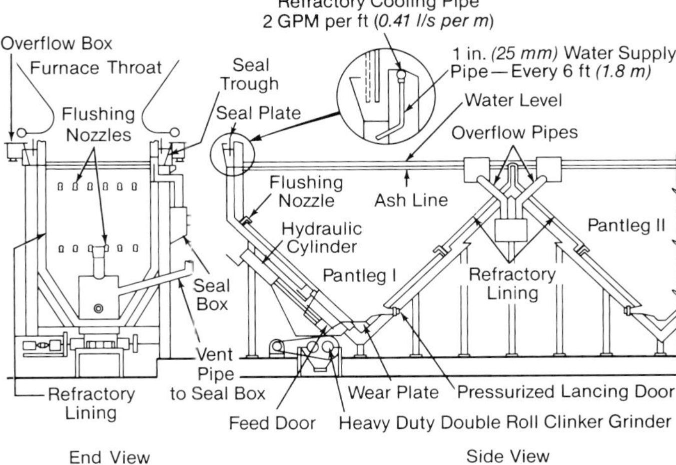

Fig. 13 Typical gravity feed bottom ash hopper.

Table 1
Ash Handling Terminology

Slag	Characteristic ash removed from furnace walls
Bottom ash	Ash removed from the bottom of the furnace
Clinker	Large piece of bottom ash that must be broken up before transport
Flyash	Ash carried by the flue gas into the backpass region and beyond
Popcorn ash	Characteristic ash collected in economizer and air heater hoppers
Dust	Fine ash particulate, typically 50 to 75% of all ash from boiler
Hydroejector	Venturi device that pulls bottom ash into hydraulic transport system
Hydrobin®	Dewatering bin for removing water from bottom ash
Hydrovactor	Venturi device that provides vacuum source for pneumatic transport
Airlock valve	Two chamber device for feeding ash into higher pressure transport line
Materials handling valve	Valve that admits flyash and air flow into vacuum transport line
Collector	Ash/air separator and transfer tank from vacuum system to silo
Transporter	Flyash collection and aeration vessel for dense phase pneumatic transport
Vacuum/pressure transfer station	Collection and transfer vessel from vacuum line to pressurized line in combination vacuum/pressure pneumatic system
Drag chain	Primary mover in mechanical drag conveyor system
Flights	Attached to drag chain to push ash along stationary platform
Unloader	Device to unload dry ash storage silo into transport vehicles
Conditioner	Adds water to flyash during unloading to minimize airborne dust
Pugmill	One version of combined silo unloader/conditioner

ing jets in the water pool to granulate the slag as it solidifies. The remaining features of the slag tank are generally the same as a bottom ash hopper.

The bottom ash hopper or slag tank is sized to hold about eight hours of ash produced by the boiler. To empty the hopper, the discharge gate opens and the contents enter a gate housing, which directs the ash into a clinker grinder where the slag is crushed. Clinker grinders have a variety of designs classified in terms of the maximum particle size that remains after passing through the grinder. Specified particle sizes range from cubes measuring 0.75 in. (1.9 cm) on all sides, to pieces 2.5 in. (6.4 cm) in diameter. Smaller particles require a lower water flow rate to transport them and have a lower tendency to plug the line, particularly in the valves and fittings of the transport piping. However, smaller particle sizes also tend to make it more difficult to separate the water at the end of the conveyance line.

Clinker grinders are either single or double roll design. The rolls are the elements that rotate and grind the slag with protruding teeth. A double roll design rotates the shafts counter to one another so that particles are pulled in and crushed as they pass through the grinder. A single roll design uses a stationary breaker bar for the opposing force.

The crushed particles are fed into the conveying line both by gravity and pump suction. The pump may be a centrifugal slurry pump or an ejector (jet pump). Slurry pumps are used to transport the ash longer distances, in particular when the frictional and elevation head loss exceeds about 150 ft. (45.7 m). Ejectors are often chosen for their ease of maintenance because they have no moving parts. A centrifugal pump is still required to produce the high pressure water that supplies the ejector and the energy efficiency of the system is lower, but by using the ejector, the course ash and slag particles do not pass through the pump itself.

In both cases the ash particles are diluted into a slurry that is about 20% solids by weight entering the piping system. The goal of the piping layout is to use the least number of bends possible because of the extremely erosive nature of this mixture, and the need to minimize pressure drop. The piping is either hard iron alloy or lined pipe. Lining materials include basalt and ceramics. Elbows and lateral fittings (used to merge two lines) are usually made of harder material, and are uniquely made with extra material thickness at points where solid particles tend to impact the pipe. When additional material is cast into the design, it is referred to as integral wear back fitting.

The piping system delivers the mixture to an ash pond or a Hydrobin® dewatering storage bin (Fig. 14 and chapter frontispiece). Two dewatering bins are used in each installation. One receives ash from the boiler while the other is draining or decanting water from the collected solid material.

A Hydrobin is initially filled partially with water. Incoming ash slurry is dropped into the center onto a sloped bar screen that acts as a partial classifier. Coarser material is forced to the sides of the cylindrical bin while the finer particles drop into the center. Later, while draining, the coarser particles act as a filter to trap fines before they reach the decanting elements along the wall.

The bin quickly fills with water from the incoming slurry, although it may take a day or more to fill to capacity with solid material. Until the bin is filled with solids, loading continues and excess water is extracted from the top and returned for reuse in the conveying lines. At the top of the Hydrobin, an underflow baffle directs all incoming material downward and helps prevent the fines from simply floating across the surface to the overflow. After the water flows under the baffle, it flows over a serrated weir as overflow. The concentration of suspended solids contained in the overflow return line is dependent on ash particle size distribution in the slurry, ash particle density, and the water flow rate going over the weir. The larger or denser the particles, the greater their tendency to sink rapidly to the bottom of the bin.

Mechanical drag systems for bottom ash

The first element of a mechanical drag system is a transition chute that mates up to the bottom opening of the boiler. The transition chute uses a water-filled trough to seal boiler gases by immersing the seal plate at the bottom of the boiler in the trough. The remainder of the transition chute provides storage volume for bottom ash, mainly for periods when the drag system is unavailable. The bottom of the chute funnels down with its sloped side walls to drop ash onto the center of a conveyor. For bottom ash, the conveyor itself is always submerged in water, but the transition chute may be dry or filled with water up to the top of the sloping side walls. These two types of transition chutes are shown in Fig. 15. When it is water-filled, the transition chute is referred to as a hopper and the overall system is referred to as a submerged chain conveyor (SCC).

The drag conveyor consists of two chains traveling horizontally underneath the chute, and once past the chute, turning upward along a dewatering incline. Flight bars are attached to the chains at regular intervals. The flight bars drag across a stationary plate while the bottom ash drops from the transition chute. The ash lands between the flight bars and gets pushed from underneath the chute, then up the incline to lift the ash from the quenching water pool. At the end of the incline, the residual moisture in the ash is generally about 20 to 30% by weight.

At the top of the incline, the ash is dumped into either a bunker or a discharge chute and onto another conveyor or series of conveyors. If necessary, a clinker grinder may be installed at the discharge end of the first conveyor. Ultimately, the bottom ash is dumped into a storage bin or dewatering bin. Normally, ash removal from the boiler is continuous, but the storage or dewatering bins may provide up to three days storage to accommodate pickup schedules. If a dewatering bin is used, a settling tank and surge/storage tank may be used to clarify the drained water for reuse.

To maintain the water trough seal at the boiler hopper skirt, water must be continuously circulated. Because of the low water inventory, a heat exchanger is commonly needed in the recirculation line to keep the water below about 140F (60C).

The submerged mechanical drag system requires

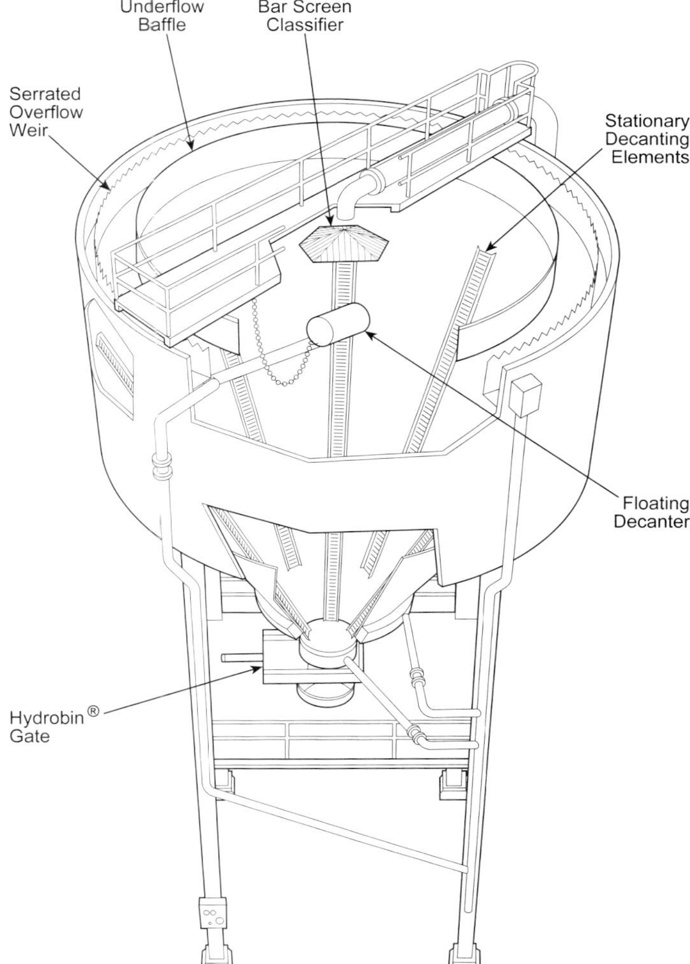

Fig. 14 Hydrobin® dewatering bin.

approximately 12 to 15 ft (3.7 to 4.6 m) height under the furnace throat opening on large utility boilers; smaller boilers require less distance. This allows the boiler to be located closer to the ground and reduces the cost of structural steel and other supporting components.

The drag chain system is typically installed on wheels in a rail or track system underneath the boiler, and one or both of the side plates at the bottom of the boiler are hinged. This allows the conveyor system to be removed for service (such as chain or flight replacement) and provides open access to the bottom of the boiler during outages.

Plunger ash extractor

The ash discharged from the stoker on mass burn municipal solid waste (MSW) units consists of ash and slag similar to that derived from other combusted fuels, except that the MSW fuel has a high ash content and includes oversize bulky wastes (OBW). This system is discussed in more detail in Chapter 29.

Mill rejects

Pulverizer rejects (often called pyrites) are commonly dumped into the bottom ash conveyance system. A mill storage hopper is installed at each pulverizer to collect the rejects. This hopper is sealed to the mill outlet spouts with seals that can accommodate positive pressure and pressure surges that occur when the mill is purged with inert gas.

When the bottom ash is conveyed with a hydraulic system, ejectors are used at each mill storage hopper to pull the dry mill rejects into the high velocity water flow for transport. The resulting slurry from each mill is piped into a common pyrites holding bin as shown in Fig. 16. The common holding bin is generally emptied at the same time that the boiler's bottom ash hoppers are emptied, using the same transport piping to the dewatering bin or pond.

Although it is common for mill rejects to be conveyed in the bottom ash hydraulic transport system, the much heavier pyrites impose a substantial burden on the hydraulic system. Often, the water flow must be increased to entrain these heavier particles. An economic tradeoff is made between the cost of an independent mill reject transport or a shared system. The shared system must include pumps selected for transporting the most difficult material, and the dewatering bins must be sized for higher water flow when mill rejects are transported with the bottom ash.

Economizer ash systems

The flyash removed from hoppers under the economizer, SCR, and air heater tends to contain larger particles along with the fine dust-like ash collected in air cleaning devices. The larger particles (popcorn ash) may contain concentrations of unburned carbon. Because of the high gas temperatures at these points, the ash may tend to sinter and become difficult to remove if permitted to sit for long periods. For this reason, it is common to provide crushers to reduce the particle sizes to about 0.25 in. (0.6 cm) in diameter for pneumatic transport. Alternatively, small, dry secondary hoppers can be installed under each economizer

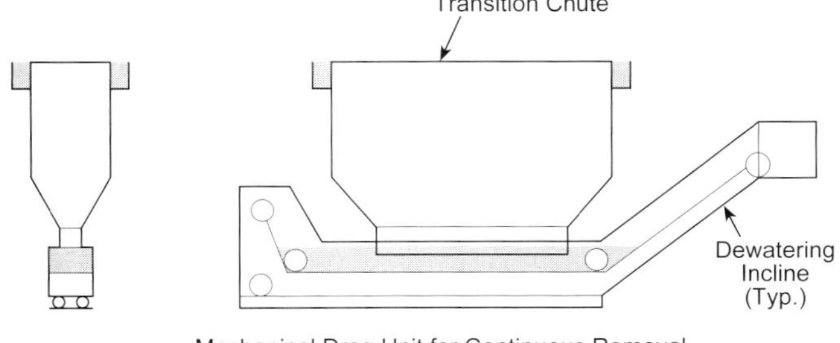

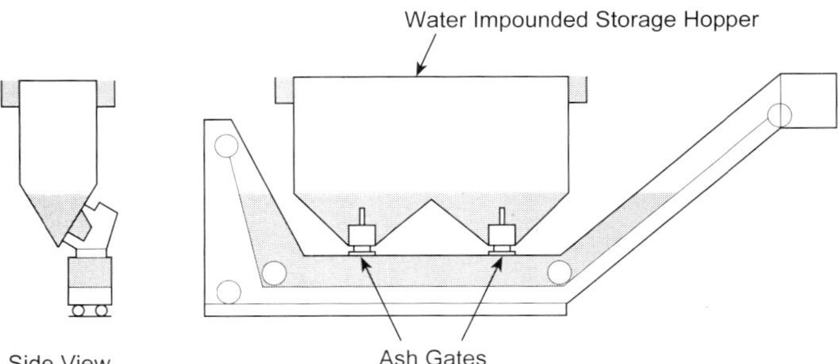

Fig. 15 Transition chutes for mechanical drag systems.

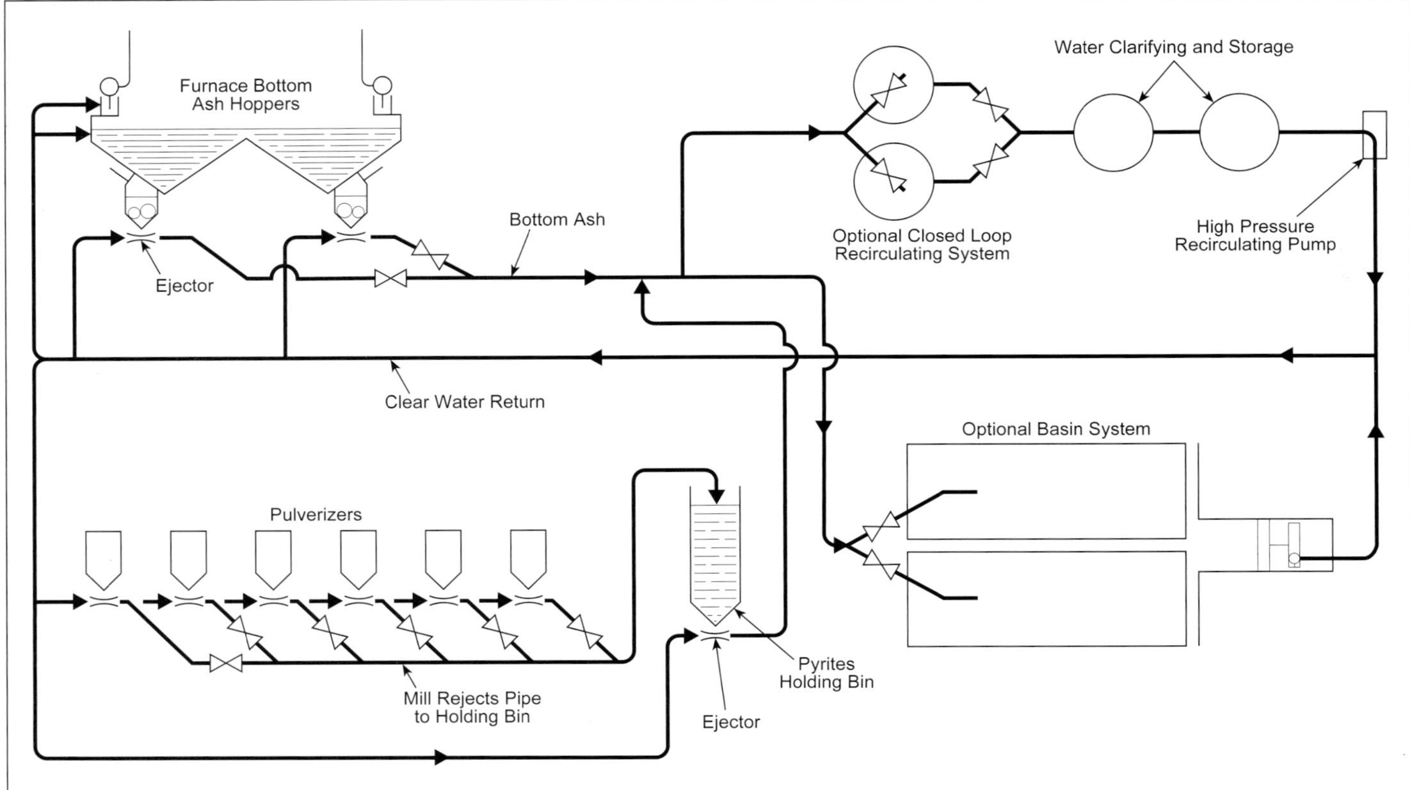

Fig. 16 Basic ash system including mill rejects.

or air heater hopper to continuously receive the ash. By removing the ash from the hot gas stream, the tendency to sinter is eliminated and the ash remains as primarily a fine dust with some popcorn ash particles.

This ash is generally combined with the flyash transport system, whether vacuum or pressurized pneumatic. A pressurized system requires an airlock valve to discharge the ash into the conveyance piping. A materials handling valve is used to dump ash into a vacuum system.

Economizer ash is occasionally mixed with a hydraulic bottom ash conveyance system. Because water is very difficult to remove from this ash, this is generally done only when the bottom ash is being disposed of in a fill area. Dewatering bins are not well suited to handling economizer ash.

Mechanical drag systems may also be used for economizer ash, and are particularly suitable when the ash is to be reinjected into the furnace. In this application, the drag unit is dry and sealed. A single drag unit will normally collect from all economizer hoppers. The ash is then discharged through a rotary feeder into a storage tank. If the ash is to be combined with the flyash transport, the storage tank discharges into the pneumatic system with an airlock valve or materials handling valve.

Flyash systems

Most of the ash from a pulverized coal-fired boiler is carried through the boiler and air heater by the flue gas. As much as 50 to 70% of the ash generated by combustion of pulverized coal is removed from the flue gas by an air cleaning device, such as a filter baghouse or electrostatic precipitator. (See Chapter 33.) Baghouses and precipitators have rows of collection hoppers that are emptied regularly by the flyash transport system.

Flyash consists of fine particles with low density. These particles are rarely transported in a water slurry because they can not be easily separated from the water at the end of the transport process. For this reason, flyash is almost always transported pneumatically. Pneumatic transport systems generally fall into three general types:

1. vacuum systems,
2. pressure systems, or
3. combination vacuum/pressure systems.

Vacuum systems

A vacuum system uses air, below atmospheric pressure, to entrain and convey ash particles. For systems with short transport distances, this often provides the lowest initial cost. A vacuum system uses fewer components and requires less clearance under the precipitator or baghouse. A vacuum system operates cleaner because any leaks merely pull ambient air into the system. Because of the low pressure of air in the system, transport distances are limited and larger line sizes may be necessary to avoid excessive piping wear.

The first point in the vacuum transport system is a material handling valve, attached to the outlet of each hopper. These valves empty the hoppers into the conveying system and in their full open position regulate the flow of ash to provide a quick but steady dispensing of the collected ash. A single pipe runs underneath each row of hoppers collecting the flow from the material handling valves. At one end, an air inlet valve opens to supply air for transporting the ash when a vacuum is applied at the other end of the pipe. Then, one at a time, the material handling valves are opened and ash flows into the pipe. Each material handling valve may use integrally mounted air inlet check valves. These are located just behind the gate so that they are opened by vacuum after the material handling valve itself is opened. These air inlet valves improve ash flow from the hopper into the conveying line.

The material handling valve can serve as a tee connection to the conveyance pipe, or may be connected to the conveying line with a short length of pipe. Integral wear back fittings or fittings with a replaceable wear back are commonly used in vacuum systems.

At the silo, the ash is separated from the conveying air in a two-chamber collection vessel. The collection vessel allows continuous collection of ash from the conveying line at vacuum pressures, and facilitates the gravity feed transfer of this ash into a silo at atmospheric pressure.

The vacuum source is located at the discharge end of the piping system, downstream of the separation equipment on the silo, to minimize the amount of abrasive ash that is drawn through the vacuum producer. The vacuum may be produced hydraulically or mechanically.

Pressure systems

Pressurized conveying of flyash offers some advantages over vacuum systems. Because they are not limited by one atmosphere of pressure differential from pickup point to the silo and vacuum producer, these systems can convey over longer distances. Also, the ash/air separation equipment can be simplified at the storage silo. However, because the feeding device at each pickup point is complicated by the need to transfer ash from the low pressure hoppers into the high pressure conveying line, the feeding device is larger. This requires more space under the precipitator or baghouse, and has a higher initial cost because the transfer equipment must be provided at each of the numerous pickup points. This offsets the cost savings of the simpler silo separation equipment. Pressurized systems are generally chosen with conveying distances greater than 750 ft (228 m), and considering the lower operating costs of a pressurized system. Pressurized systems may also be required at higher altitudes, because of the lower differential pressure available for a vacuum system.

The starting point for a pressurized system is a positive displacement blower which provides air flow and pressure for conveying the flyash. The blower feeds a piping system that branches out to collect ash from the numerous hoppers under the precipitator or baghouse.

An airlock valve dumps ash from each hopper into the conveying line (see Fig. 17). This valve consists of two chambers, upper and lower, with three gates: one to close off the opening to the hopper, one between the upper and lower chamber, and a cut-off gate to empty the lower chamber of each airlock into the conveying pipe one at a time.

An equalizer valve is required for each airlock valve. The equalizer valve has two positions: one to equalize the pressure between the hopper and upper chamber of the airlock, and the other to equalize pressure between the conveying pipe and the upper chamber. It is inevitable that some of the abrasive ash flows through the equalizer valve each time it changes position, so it is designed to be self-cleaning and includes wear resistant surfaces at key points.

To operate the airlock valve, all three gates are initially closed. The pressure of the upper (primary) chamber is equalized with that of the overhead hopper. The gate valve above the primary chamber is then opened for a preset period of time. This dwell time allows a measured amount of flyash to flow by gravity into the primary chamber. At the end of the loading cycle the upper gate is closed to isolate the chamber from the hopper. The upper chamber is then pressurized to a level slightly greater than the conveying line pressure. Finally, the gate between the upper and lower chamber is opened and ash is forced down into the conveying line through an intake tee. The intake tee imparts swirl within the lower chamber for better ash pickup performance. The cut-off gate below the lower chamber is a manual gate used to isolate the airlock for maintenance.

In some cases, it is necessary to include an aeration device to counteract the tendency of the ash to compact in the hopper and the upper chamber of the airlock valve. The design of the airlock is modified slightly as shown in Fig. 18.

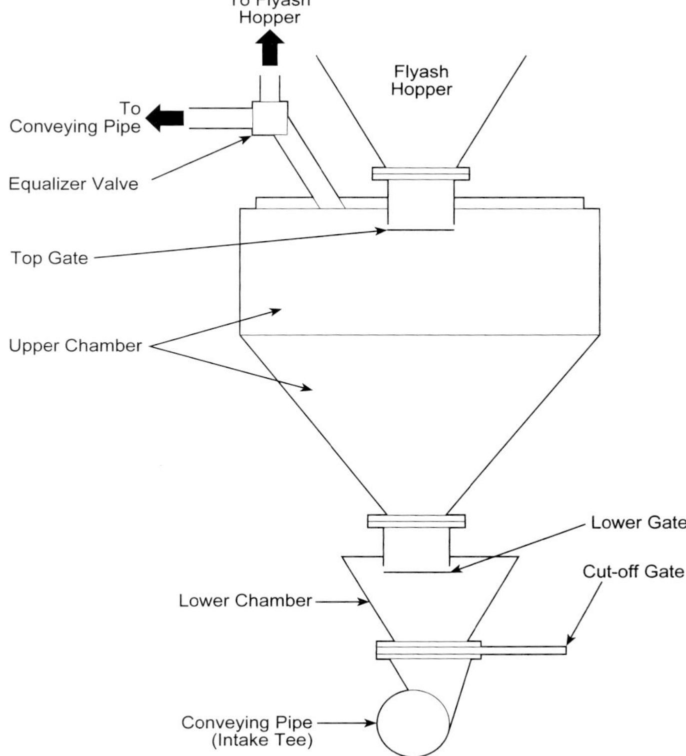

Fig. 17 Airlock valve (without aeration).

When the airlock valve is attached to an economizer hopper, the top gate is commonly left open until the upper chamber is ready to be emptied. This allows the much hotter economizer ash to collect in the upper chamber of the airlock valve to get the ash out of the hot flue gases and reduce the tendency of the ash to sinter and form large chunks. If this occurs despite the use of the airlock as a collection device, then a rotary feeder may be used to break the ash into smaller, transportable pieces.

Automatic butterfly valves at the heads of the branch lines provide branch isolation when ash is being removed from the other branches. Manually operated knife gate valves are provided at the ends of each branch line for maintenance purposes. An overall schematic of a pressurized conveyance system is shown in Fig. 19.

With a pressurized system, the conveying air and entrained ash are simply dumped into a dry storage silo. Ash settles in the silo and air is removed from the top through a self-cleaning bag vent filter. In some cases, the air flow is routed through a duct to the precipitator inlet.

Combination vacuum/pressure systems

Combination systems are used to transport ash longer distances than a vacuum system alone can accommodate, while keeping some of the advantages of the vacuum system, such as its lower space requirement. In the combination system, the initial withdrawal of ash from the hoppers is by a vacuum system. Ash is delivered to an intermediate collection vessel where it is transferred to a pressure conveying system for final delivery to a distant dry storage silo.

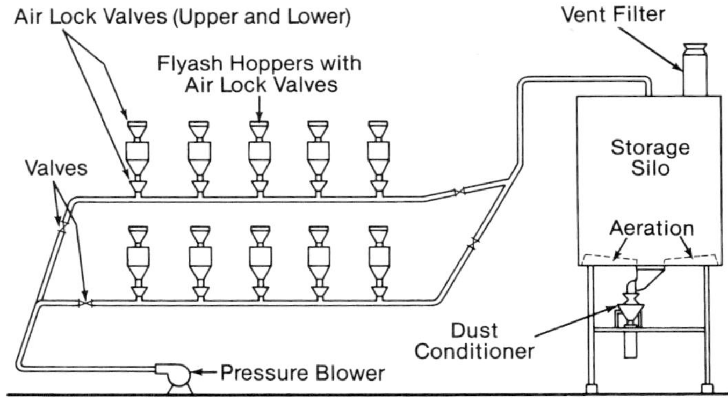

Fig. 19 Pressure flyash system.

The vacuum system is the same as that described from hopper pickup to the intermediate collection vessel. The intermediate collection vessel can be either a transfer tank or a surge bin. In principle, the surge bin is identical to a dry storage silo, including the primary and secondary separation equipment on the roof. The surge bin is smaller than a storage silo, but provides an intermediate storage volume for ash. The surge bin dumps into airlock valves underneath the bin, which feed ash into the higher pressure line of the pressure conveying system. The pressurized conveying system is the same as described earlier, except that the airlock valves collect from a surge bin.

A vacuum-pressure transfer tank is often used as the intermediate collection vessel. As shown in Fig. 20, the transfer tank is a vessel with three chambers: upper, middle, and lower. The upper chamber is always at the vacuum pressure of the vacuum conveying line, and the lower chamber is always at the higher pressure of the pressurized conveying line. Dump valves are located between the chambers. Ash continuously collects in the upper chamber, which also serves as the primary stage of air/ash separation with the centrifugal separator design of the upper chamber. After a preset period of time, the pressure in the upper and middle chamber is equalized, and the dump valve between the upper and middle chamber is opened. When the middle chamber has emptied the upper chamber, the upper dump valve closes. Another valve equalizes the pressure between the middle and lower chamber, and the lower dump valve is opened.

Air that is separated from the ash in the upper chamber is sent to a secondary collector, which includes a fabric bag filter, and then to the inlet of the vacuum producer.

A variation of the vacuum-pressure transfer tank is available for high temperature material separation and transfer. The high temperature tank has a single chamber that provides a centrifugal separator incorporated in the top of the tank, and a bottom section. The tank is constructed with materials and wear surfaces suitable for handling material up to 750F (399C). This tank dumps into airlock valves as shown in Fig. 21. The airlock valves are used as pickup points for the pressurized conveying system.

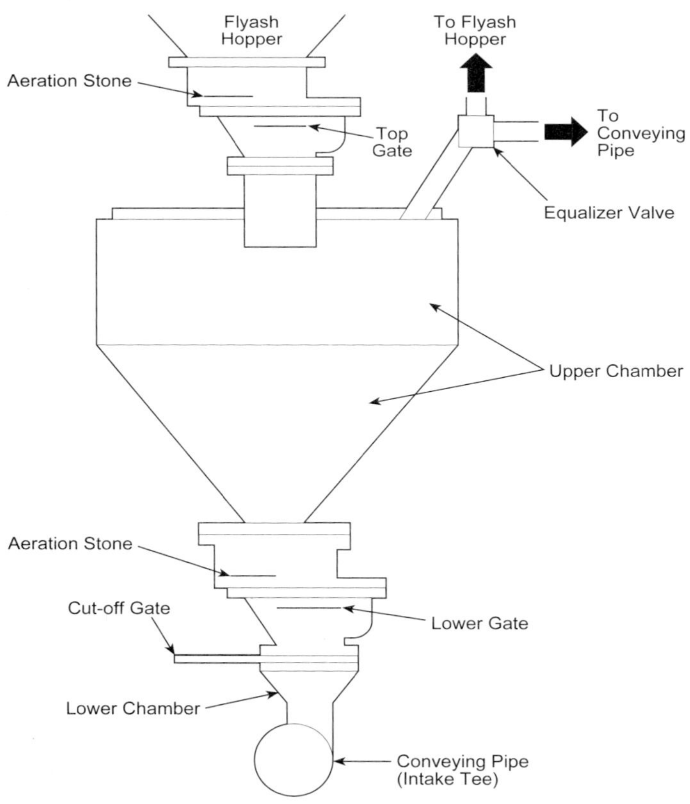

Fig. 18 Airlock valve (with aeration).

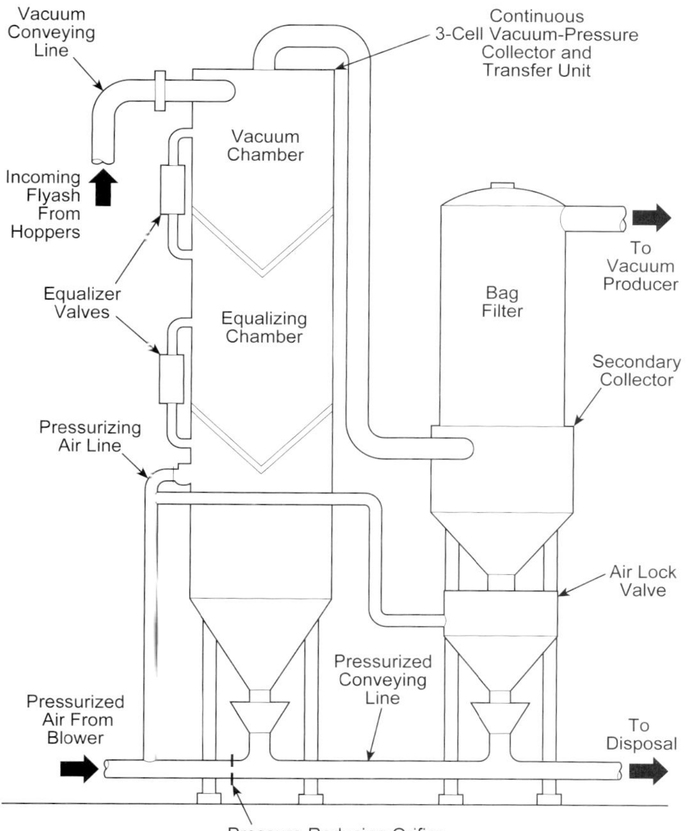

Fig. 20 Collection equipment for vacuum-pressure flyash system.

Dense phase systems

An option available for pressurized transport systems is to operate at higher ash to air ratios, which uses less air, smaller transport pipe sizes, and potentially causes less wear of system components. When the weight ratio of flyash to air exceeds about 15 to 1, the transport is referred to as dense phase.

In the systems described up to this point (dilute phase), the ash feed regulating valves were designed to deliver ash into the conveying line as individual particles that would become thoroughly entrained in the air flow. The air flow velocity would generally be maintained above the point where particles would drop out of the flow and drag along the bottom of the pipe.

In dense phase transport, the ash is initially fluidized and transported at lower velocities. The ash is transported through the line in a dense fluid-like condition, often dropping from suspension and resembling moving and shifting sand dunes as the ash moves along the pipe. The key to dense phase transport is that the ash must have a consistency that lends itself to fluidization. Different ash types have different tendencies.

The facilitating element of a dense phase system is the fluidizing vessel commonly called a *transporter*. Ash is delivered to this vessel in batches from the hoppers. When filled with ash, the inlet valve is closed and compressed air is pumped into the vessel.

The transporter vessel can be a top or bottom discharge design. A top discharge, fluidizing transporter vessel is shown in Fig. 22. Compressed air is pumped into the vessel through an aeration device (stone, ceramic, or metal element) in the bottom of the vessel. The ash becomes fluidized and the pressure in the transporter vessel builds. The aerated ash begins to flow into the conveying line soon after the compressed air is pumped into the transporter vessel. Elastomer-lined pinch valves (not shown) admit ash to the conveying line from one transporter vessel at a time.

A bottom discharge transporter vessel is shown in Fig. 23. In this design, compressed air is also pumped into the vessel, forcing ash into the conveying line. Transport is enhanced by pulsing the air flow into the vessel, and alternating between an injection point in the side of the transporter vessel and an injection point in the conveying line itself. The ash is also fluidized in this design before entering the conveying line, but not in the transporter vessel. Instead, the fluidization with compressed air is done in a holding bin above the transporter vessel so that the ash is already fluidized when it enters the transporter vessel.

A primary advantage of dense phase transport is lower air velocity. This allows use of less expensive, unlined steel pipe. However, much longer radius turns must be used in place of elbows or standard fittings to avoid plugging.

Separation and storage

At the end of the transport piping, ash is separated from the conveying air flow and collected in a dry storage silo. In a pressurized system, the air and ash flow empties into the silo and the air flow is vented from the top of the silo. A fabric (bag) filter may be used to capture any fine dust that is not separated by gravity in the silo, or a vent fan may be used to draw the air from the roof of the silo and pump it back into the

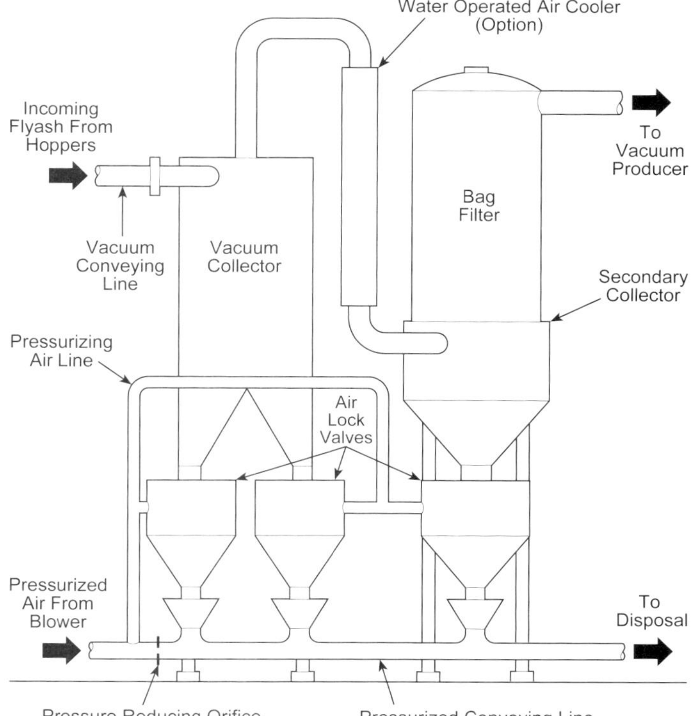

Fig. 21 Vacuum-pressure system for hot flyash.

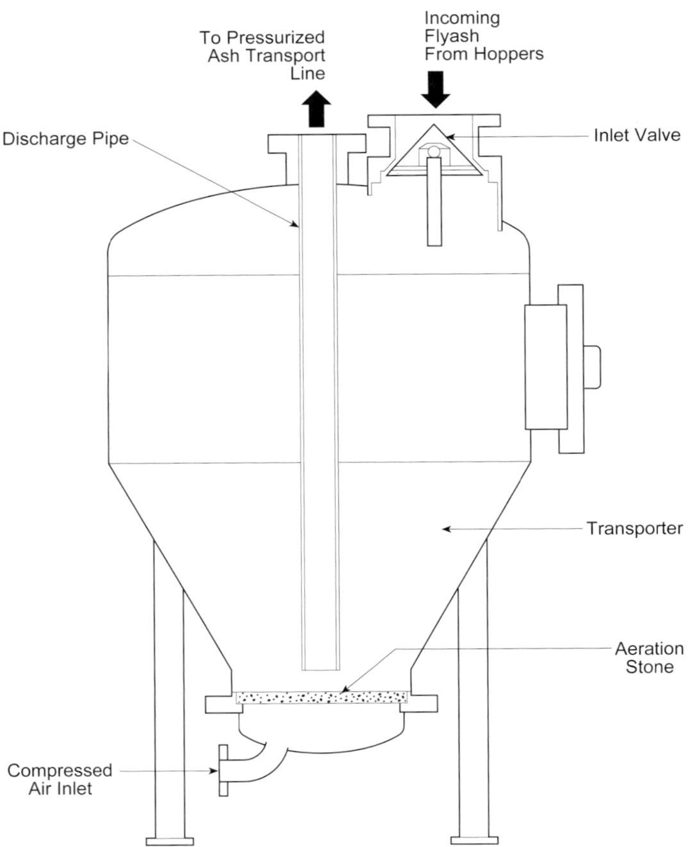

Fig. 22 Top discharge transporter vessel for dense phase system.

precipitator or baghouse. This venting system must accommodate both the conveying air flow rate and compressed air that is pumped into the silo to aerate the stored ash.

In a vacuum system, a primary collector transfers the ash from the low pressure of the vacuum transport line to the silo, which is at or slightly above atmospheric pressure. The collector is a two chamber device installed on the roof of the silo (Fig. 24). The upper chamber is a centrifugal separator and storage reservoir. A gate between the upper and lower chambers opens periodically to empty ash to the lower chamber, after the chamber pressures have been equalized. Then, the gate is closed, the pressure in the lower chamber is raised to that of the silo, and a gate valve between the lower chamber and the silo is opened. The collected ash flows into the silo by gravity. Therefore, the upper chamber can continuously collect ash from the transport line.

An alternate primary ash collector design replaces the centrifugal separator section shown in Fig. 24 with a stone box inlet pipe that directs the ash downward into the storage section of the upper chamber. Mounted directly above the upper chamber is the pulse-jet style bag filter that removes any fine dust from the transport air stream as the air travels to the vacuum source.

Air removed by the centrifugal separator of the primary collector is sent to a bag filter to remove any fine dust. A secondary collector is sometimes installed upstream of the bag filter. The vacuum source draws air from the bag filter, which by connection back through the secondary and primary collectors serves as the motive force for the vacuum transfer system.

Storage silos are usually sized to store three days of ash produced by the boiler at full load. Larger boilers have ash silos installed in pairs, so that one can be unloaded or serviced while the other collects ash.

Silos may be made of carbon steel or reinforced concrete construction. The silo has either a cone-shaped or flat bottom, with aeration devices installed inside along the bottom to ensure that the ash will flow freely when the silo is unloaded. Silos may be as tall as 75 ft (23 m). This height tends to compress the ash if well-distributed aeration is not maintained. Heaters may be used to heat the aeration air flow and prevent moisture.

The silo itself is elevated to provide clearance for rail cars or disposal trucks. When the ash is loaded into a closed transport, the silo is fitted with an unloader consisting of a retractable chute. The central telescoping pipe directs ash down into the vehicle's storage tank, while the chute provides a vent path for displaced air from the tank. An exhaust fan pumps the displaced air and dust back into the silo.

Silo unloading equipment

When flyash is to be hauled away by open trucks, dust conditioning is added to the silo unloading equipment. Most commonly, this is simply the addition of water during silo unloading to moisten the ash and minimize the fine particulate that otherwise becomes airborne. The challenge is to add and mix sufficient

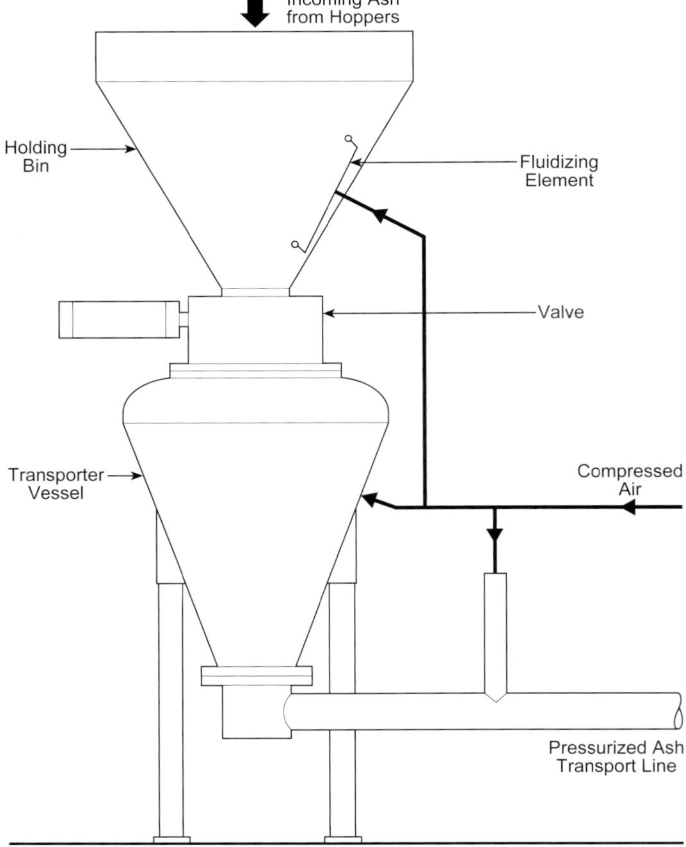

Fig. 23 Bottom discharge transporter vessel for dense phase system.

water to keep the dust down, while avoiding the potential to make cement by adding too much water to the calcium compounds. The appropriate amount of moisture is highly dependent on ash composition.

Dust conditioning equipment falls into two general types: batch mixers and continuous mixers. Batch mixers are vessels that collect a load of ash based on weight, stir the ash with mixing paddles and wall scrapers, and slowly add water until a well mixed batch of moistened ash is ready for unloading. Because the ash is weighed as it is loaded into the mixer, the moisture content can be carefully controlled to avoid problems such as cementing, dry pockets of ash, or free-standing water.

Batch mixers may not be rapid enough to support transport schedules. Continuous mixers are smaller and less expensive, and have capacities that allow trucks to be filled in a matter of minutes, one after another.

When the first several truckloads of ash are unloaded from the silo, the head of ash in the silo can be on the order of 40 to 50 ft (12 to 15 m). Simply opening a valve at the bottom of the silo would discharge at too high of a rate to be practical for an operator to control or condition. Continuous mixers use a motor-driven rotary feeder or metering feed gate at the inlet, or an inlet design that limits the input rate.

In high capacity, rotating drum continuous unloaders, incoming ash is aerated in the feed chute before entering the drum. The rotating drum tumbles the ash past a series of scrapers and a spray nozzle manifold. This type of continuous unloader can condition up to 8000 ft^3 (226 m^3) of flyash per hour, but can be more susceptible to dust bypass.

A lower cost option for continuous conditioning and unloading is referred to as a pugmill. A metering feed gate admits ash into the pugmill's mixing chamber. The mixing chamber is a U-shaped channel with two motor driven shafts with paddles for mixing and propelling the ash from inlet to exit. Spray nozzles are distributed within the chamber for thorough wetting. The chamber is oriented with a slight downward slope to assist ash movement and to drain the pugmill when it is manually cleaned. The top cover of the pugmill is hinged so that the entire chamber is accessible when out of service. For ash that has a high tendency to become cement-like, the mixing chamber can be lined and the paddles can be constructed of a high strength, non-stick polymer material.

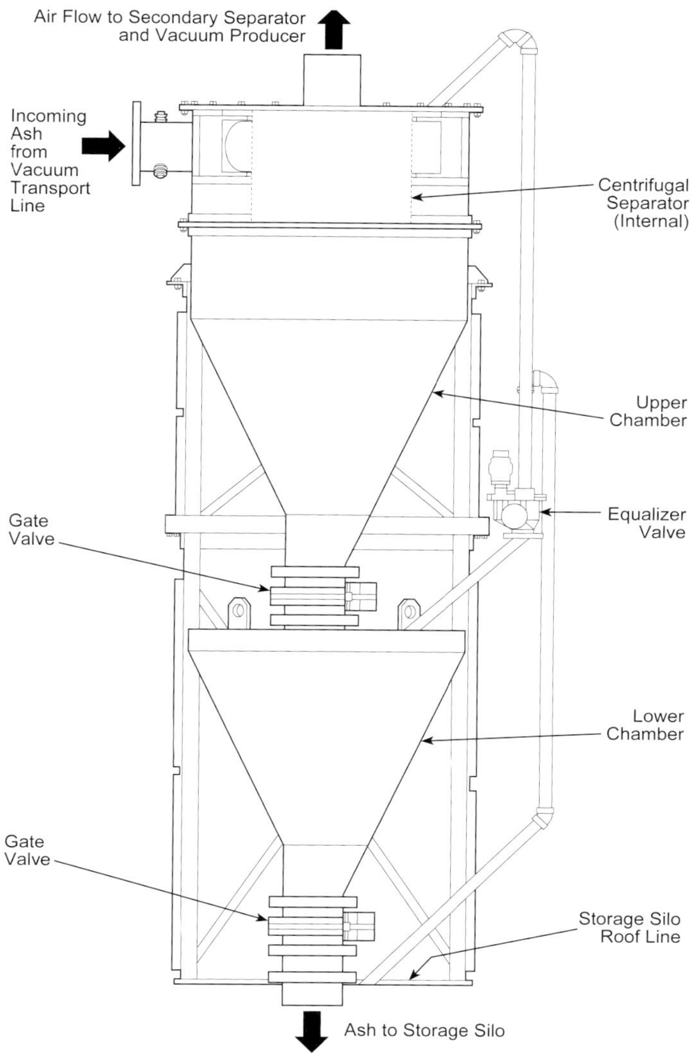

Fig. 24 Primary ash collector.

HydroJet, IK-600, Precision Clean and Hydrobin are trademarks of Diamond Power International, Inc.

Fan housing being prepared for installation (*courtesy of TLT-Babcock*).

Chapter 25

Boiler Auxiliaries

A variety of auxiliary components are needed for the modern steam generating system to function effectively and efficiently. While space prevents an in-depth review of all items, several deserve special attention. Safety and relief valves are critical to assure the continued safe operation of the boiler. The interplay of dampers, stack design, and fans ensures proper air and gas flow for optimum combustion. Finally, a specialty condenser is used to supply high purity water for attemperator spray in industrial boilers where such water is not typically available.

Safety and relief valves

The most critical valve on a boiler is the safety valve. Its purpose is to limit the internal boiler pressure to a point below its safe operating level. To accomplish this goal, one or more safety valves must be installed in an approved manner on the boiler pressure parts so that they can not be isolated from the steam space. The valves must be set to activate at approved set point pressures (discussed below) and then close when the pressure drops below the set point. When open, the set of safety valves must be capable of carrying all of the steam which the boiler is capable of generating without exceeding the specified pressure rise.

The American Society of Mechanical Engineers (ASME) Boiler and Pressure Vessel Code, Section I, outlines the minimum requirements for safety and safety relief valves applying to new stationary water tube power boilers. The Code also covers requirements for safety and safety relief valves for other applications beyond the scope of this text. By Code definition, a *safety valve* is used for gas or vapor service, a *relief valve* is used primarily for liquid service, and a *safety relief valve* may be suitable for use as either a safety or a relief valve.[1]

Fig. 1 shows a typical Code approved spring loaded safety valve for steam service. No other valves may be installed between the pressure vessel and the safety valve, nor on the discharge side of the safety valve. The inlet nozzle opening must not be less than the area of the valve inlet, and unnecessary pipe fittings must not be installed. These valves are designed for large initial opening at the upstream static pressure set point and for maximum discharge capacity at 3% above set point pressure.[1]

Fig. 2 shows a typical power actuated type safety valve that may be used in some Code approved applications. Power actuated valves are fully opened at the set point pressure by a controller with a source of power such as air, electricity, hydraulic fluid, or steam.

Fig. 3 shows a typical spring loaded relief valve for liquid service designed for a small initial opening at the upstream static pressure set point. The valve will

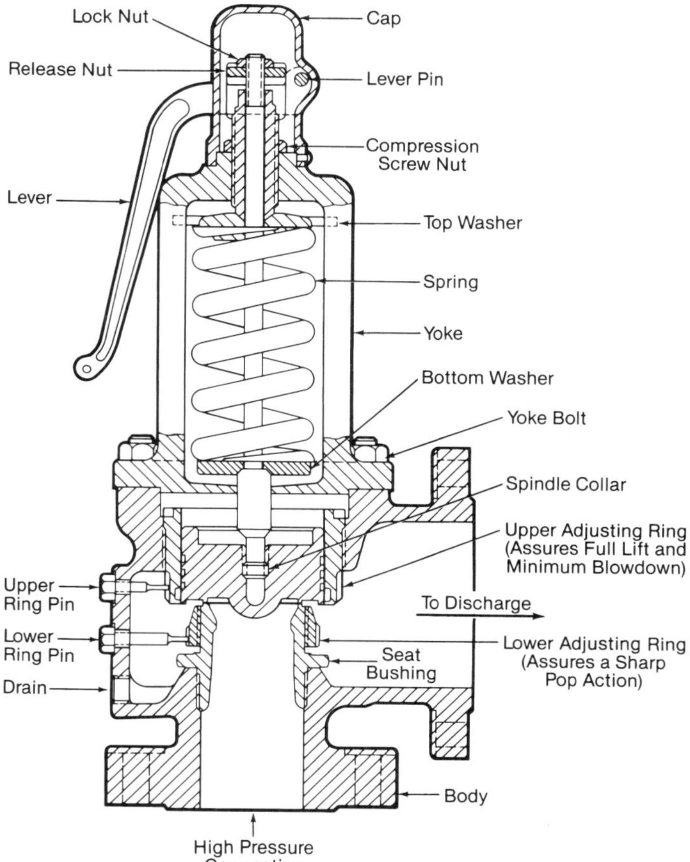

Fig. 1 Spring loaded safety valve (*courtesy of Dresser Industries, Inc.*).

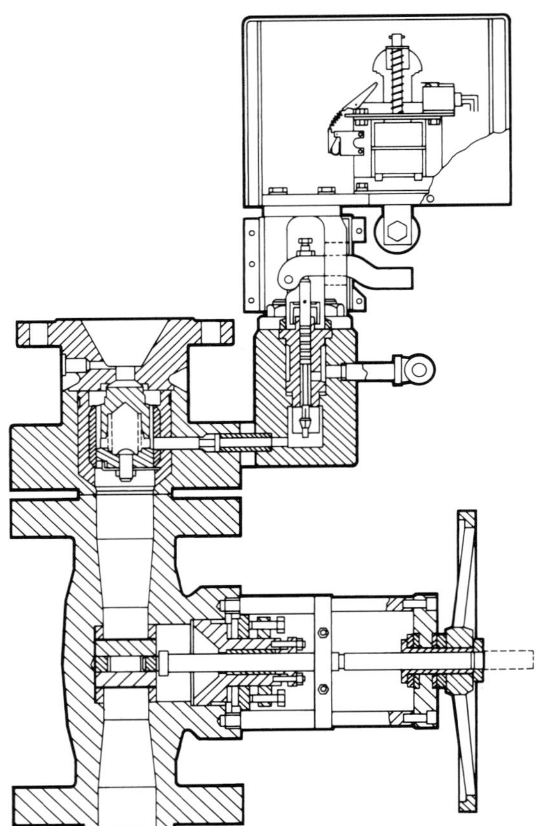

Fig. 2 Typical power actuated safety valve (*courtesy of Dresser Industries, Inc.*).

continue to open as pressure increases above set point pressure to prevent additional pressure rise.

Because of variations in power boiler designs, Code interpretations are sometimes necessary. It is also necessary to comply with local ordinances. The owner's approval must be obtained for all safety valve settings.

For drum boilers with superheaters, The Babcock & Wilcox Company (B&W) follows the Code procedure in which the safety valves are set such that the superheater valve(s) lift first at all loads. This maintains a flow of steam through the superheater(s) to provide a measure of overheat protection. This method permits the piping and valves downstream of the superheater to be designed for a lower pressure than other methods. This method is required on hand controlled units, stoker or other fuel-bed-fired units, and brick set units. Another method may be used for all other types of boilers that permits the drum safety valves to lift first. This method could allow a reduced flow condition to occur in the superheater while the boiler is still at a high heat input level. As a result, some superheater materials can exceed temperature limits.

The required valve relieving capacities for waste heat boiler applications are determined by the manufacturer. Auxiliary firing must be considered in the selection of safety or safety relief valves. The Code required relieving capacity must be based on the maximum boiler output capabilities by waste heat recovery, auxiliary firing, or the combination of waste heat recovery with auxiliary firing.[1]

Additional Code requirements are applicable for modified existing boilers or new boilers installed in parallel with old boilers, or boilers operated at an initial low pressure but designed for future high pressure (increases or decreases in operating pressure). Additional requirements, including mounting, operation, mechanical, material, inspection and testing of safety and safety relief valves are specified in the Code.

Code requirements for once-through boilers

For safety valve requirements on once-through boilers, the Code allows a choice of using the rules for a drum boiler or special rules for once-through boilers. Power operated valves may be used as Code required valves to account for up to 30% of the total required relief valve capacity. If the power relief valve discharges to an intermediate pressure (not atmospheric), the valve does not have to be capacity certified, but it must be marked with the design capacity at the specified relieving conditions. However, it is important to remember that, in order for a power relief valve discharging to an intermediate pressure to be counted as part of the required safety valve relieving capacity, the intermediate pressure zone must be fully protected with its own relieving capacity.

The power relief valve must be in direct communication with the boiler, and its controls must be part of the plant's essential service network, including required pressure recording instruments. A special iso-

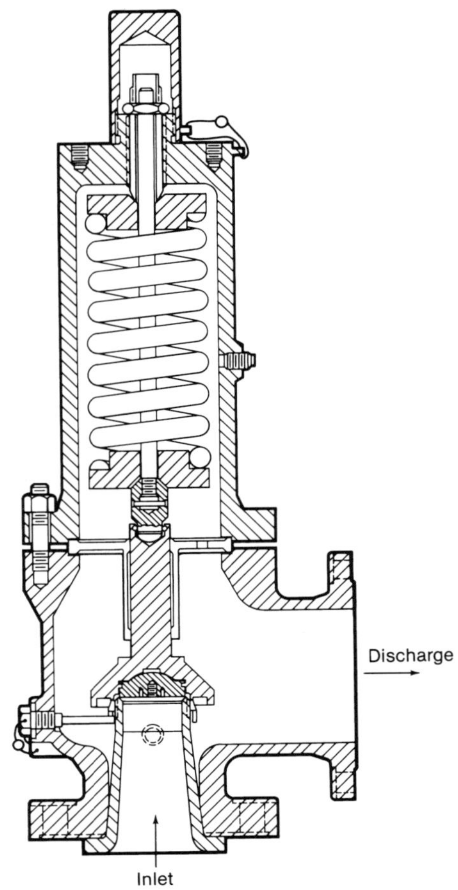

Fig. 3 Spring loaded pressure relief valve (*courtesy of Crosby Valve & Gage Company*).

lating stop valve may be installed for power relief valve maintenance provided redundant relieving capacity is installed. Provided all the Code requirements are met or exceeded, the remaining required relief capacity is met with spring loaded valves set at 17% above master stamping pressure. The superheater division valves (Chapter 19) are part of the power relief valve system so that credit may be taken for the spring loaded superheater outlet valve(s) relieving capacity as part of the total required relieving capacity. With a superheater division valve, the code requires 6 lb/h relieving capacity per each square foot of superheater heating surface (0.008 kg/s per m^2). The blowdown of the spring loaded valves shall not be less than 8% nor more than 10% of set pressure. (See Code, Section 1.)[1]

Air and flue gas dampers

Dampers are used to control flow and temperature of air and flue gas. They can also be used to isolate equipment in the air or flue gas streams when such equipment is out of service or requires maintenance.

The Air Movement and Control Association in Publication 850-02, *Application and Specification Guide for Flue Gas Isolation and Control Dampers*, defines both the isolation and control functions for the selection of dampers.[2]

Isolation dampers may be the nominal shutoff or zero leakage type. Shutoff dampers are used in applications where some limited leakage is tolerable. All types of dampers, with appropriate blade seals, can be used. Zero leakage dampers are designed to prevent any flow media leakage past the closed damper. This is accomplished by overpressurizing the blade seal periphery with seal air, which leaks back into the system. Guillotine-type dampers are best suited for this service. A pair of dampers or pairs of blades within one damper with an overpressurized air block between them can also be used. In heat recovery steam generator (HRSG) applications, diverter dampers are often used along with stack cap dampers to assist in maintaining boiler temperature during overnight shutdowns.

Control dampers are capable of providing a variable restriction to flow and may be of several varieties.[2]

Balancing dampers are used to balance flow in two or more ducts. *Preset position* dampers are normally open or normally closed dampers which move to an adjustable preset position on a signal. *Modulating* dampers are designed to assume any position between fully open and fully closed in response to a varied signal (either pneumatic or electric). A positioner with feedback indication is normally required.[2]

Dampers may also be classified by shape or configuration. Damper shape classifications that are commonly used in the fossil fuel-fired steam generation industry are *louver*, *round* (or wafer) and *guillotine* (or slidegate).

Louver dampers

A louver damper, as shown in Fig. 4, is characterized by one or more blades that mount into bearings located in a rigid frame. They are generally rectangular in shape. One blade shaft end extends far enough beyond the frame so that a drive can be mounted for damper operation. Ideally, blade shape is determined by the amount of pressure drop that can be tolerated across the open damper. In reality, it is usually a compromise among the following design goals: the ideal shape required for minimizing pressure drop, the ideal shape for minimizing initial cost, and the ideal shape to attain the required structural strength to withstand pressure and temperature conditions. An unreinforced flat plate is the simplest shape but is relatively weak and offers the greatest pressure drop. Air foil shaped blades have lower pressure losses and have somewhat greater strength. If uniform downstream flow distribution is required, opposed blade rotation is used; otherwise, parallel blade rotation may be used. (See Fig. 5.)

If the louver damper is to be used for isolation purposes, thin, flexible metallic seal strips are mounted on the blades to minimize leakage around the closed blades. Seal strips are not required in applications where louver dampers are used for flow control.

Depending on the application, louver dampers may have internal or external bearings. External bearings are of the self-aligning and self-lubricating ball or sleeve type, which require a mounting block and packing gland. In extreme conditions, packing glands may have a seal air provision. Internal bearings are machined castings with a self-cleaning feature. These types of bearings do not require lubrication and elimi-

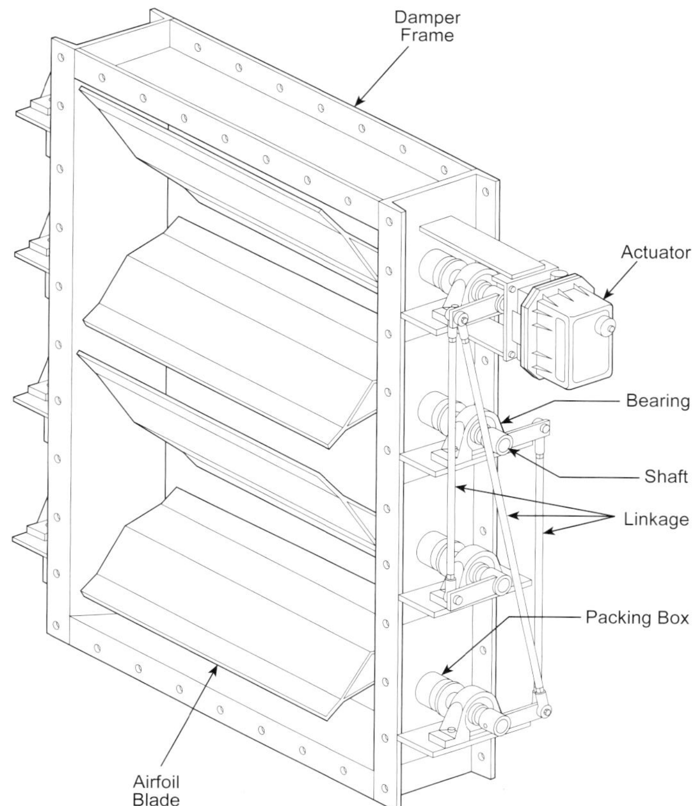

Fig. 4 Opposed-blade louver damper (*adapted from Mader Damper Co.*).

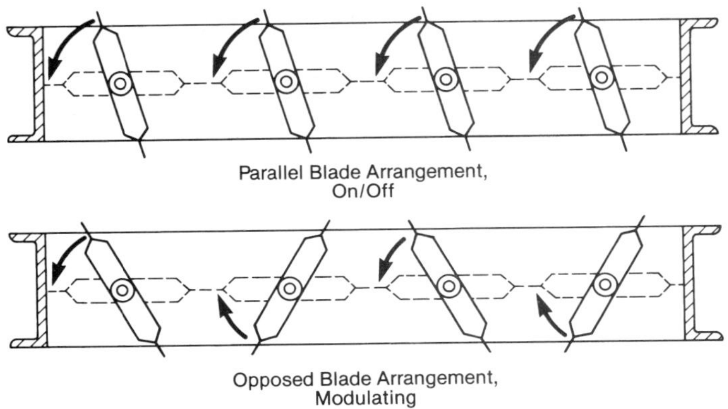

Fig. 5 Louver damper blade arrangement options (*courtesy of Air Movement and Control Association, Inc.*).

nate the need for the damper shaft to penetrate the frame, eliminating the maintenance associated with shaft seals.

Round or wafer dampers

Round dampers, as shown in Fig. 6, can be used for control or shutoff service. Due to higher allowable velocity limits, they are usually smaller than louver dampers and have a smaller seal edge to area ratio, which makes them more efficient for shutoff applications. They can also be configured in a two or more blade configuration with opposed operation for flow control as shown in Fig. 7, although this arrangement tends to reduce sealing efficiency.

Guillotine or slidegate dampers

Guillotine dampers have an external frame and drive system that can insert and withdraw the blade which acts as a blanking plate in the full cross-section of the duct. This minimizes the sealing edge required for any given duct size. The blade periphery is surrounded by, and forced between, flexible metal sealing strips. On zero leakage dampers, such as shown in Fig. 8, the blade edge extends through the frame into a seal chamber that is pressurized by a seal air blower. Seals for this type of damper are normally designed to eliminate reverse flexing. The blade is actuated by a chain and sprocket, pneumatic cylinder, hydraulic cylinder, or screw-type mechanism, connected to a drive which is mounted on the frame. Guillotines are typically large and used for isolation service, preferably in horizontal ducts. It is not uncommon for blade thickness to reach 0.75 in. (19 mm), and blade thicknesses exceeding 1 in. (25.4 mm) have been used.

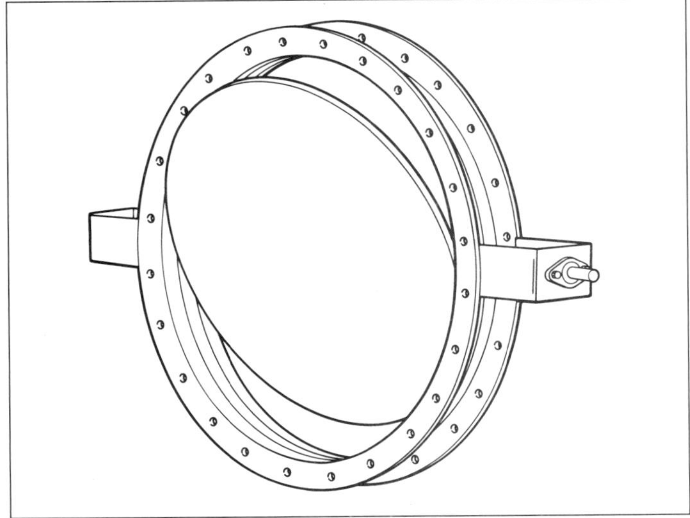

Fig. 6 Wedge seat round damper (*courtesy of Damper Design, Inc.*).

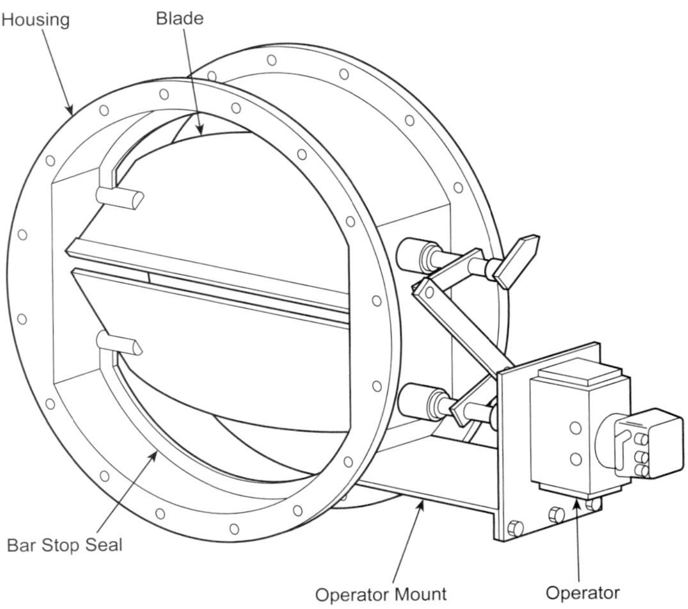

Fig. 7 Two-bladed wafer damper.

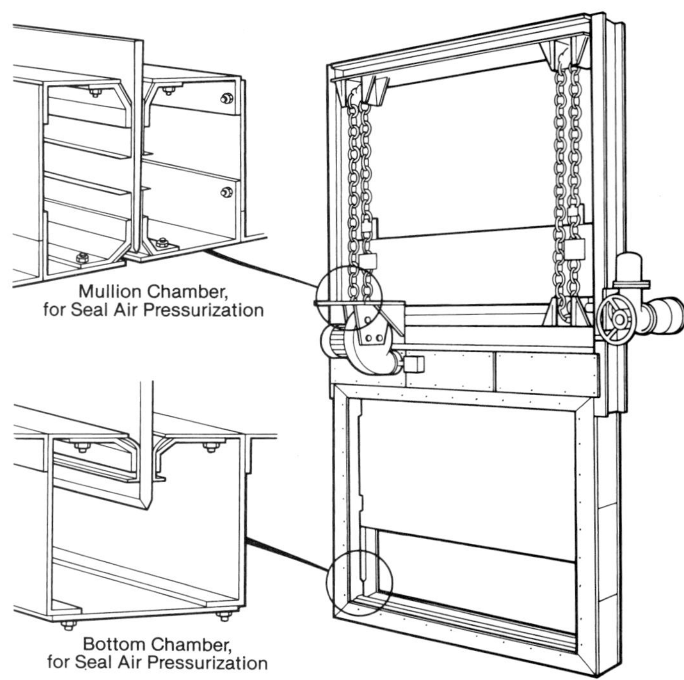

Fig. 8 Guillotine damper (*courtesy of Effox, Inc.*).

Expansion joints

Expansion joints are used in duct systems, windboxes, and other areas of the boiler and air quality control system islands to allow for relative movement between components of a system. This relative movement is usually caused by thermal expansion and contraction. Expansion joints are available in various configurations and can feature flow liners and insulation pillows to allow for a wide range of temperature and dust conditions. They are generally of metal construction or constructed of elastomers and composite materials, and are designed for maximum flexibility under the required service conditions.

The most common configuration for a metallic expansion joint is the multi-leaf or box fold joint. (See Fig. 9.) The box fold is normally external but can be internal if space considerations require. However, if an internal fold is used, the additional pressure drop caused by the flow area restriction must be accounted for when sizing the draft equipment. Also, internal folds could be subject to erosion damage in a dirty gas environment. If the external fold expansion joint is placed in a dirty environment, the folds are packed (intermediate temperature fiberglass insulating block is a typical packing) and the packed folds are shielded, typically with 10 gauge sheet of the same class material as the duct. Hinged shields are often used to permit field packing of the joint folds. The extent and location of packing and shielding are dictated by the joint designer's field experience. Nonmetallic joints also require packing and shielding as dictated by the designer's field experience.

A metallic expansion joint can accept toggle and axial motion but is unable to accept significant shear motion. For that reason, whenever two components must be connected that are displacing relative to one another in a shear plane, two metallic expansion joints must be employed by placing them at opposite ends of a duct *toggle section* as shown in Fig. 9. A metallic joint has a pinned support connection which defines its toggle axis. Side pins are used to transmit gravity loads across a joint and may pass through either round or slotted holes. Center pins may also be used to transmit loads of pressure and to actuate another expansion joint in series.

Nonmetallic joints (Fig. 10) are able to accept shear (three-way) motion, which is their primary advantage over metallic joints. For this reason, nonmetallic joints do not require a toggle section and only one joint is required for most applications. Nonmetallic joints are made with reinforced elastomers or Teflon coated woven fiberglass, and are normally the plain flat belt type. Flanged or U-type joints are also available. If design temperatures are above the use limit of the belt material, composite belts with an insulation liner are used and may include an insulation pillow for the higher temperatures.

It is essential that all relative motions between components be completely accounted for when designing

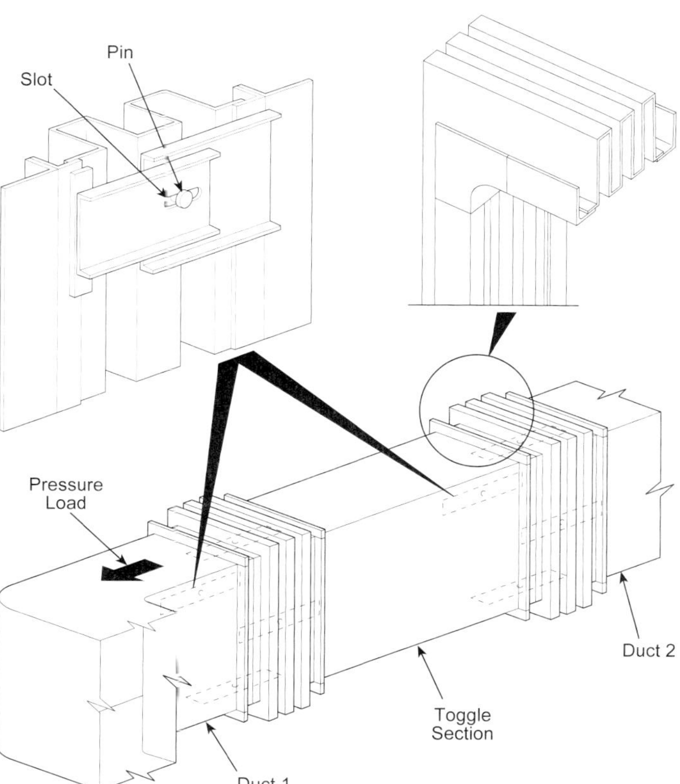

Fig. 9 Multi-leaf metallic expansion joint with pinned support connection.

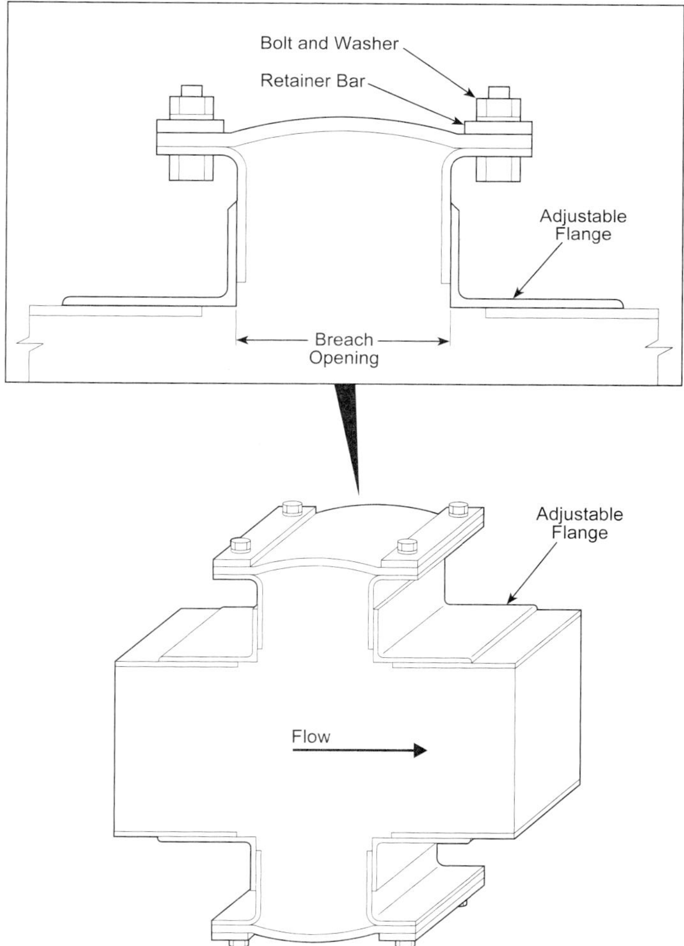

Fig. 10 Nonmetallic expansion joint.

the connecting expansion joint, and that design pressures and temperatures appropriate for the type of service be used for material and thickness selection. Failure to do so can result in premature failure of the expansion joint.

Stacks and draft

An adequate flow of air and combustion gases is required for the complete and effective combustion of fossil and chemical fuels. Flow is created and sustained by stacks and fans. Either the stack alone, or a combination of stack and fans, produces the required pressure head to generate the required flow.

Draft is the difference between atmospheric pressure and the static pressure of combustion gases in a furnace, gas passage, flue or stack. The flow of gases through the boiler can be achieved by four methods of creating draft, referred to as forced draft, induced draft, balanced draft and natural draft.

Forced draft boilers operate with the air and combustion products (flue gas) maintained above atmospheric pressure. Fans at the inlet to the boiler system provide sufficient pressure to force the air and flue gas through the system. Any openings in the boiler settings, such as opened doors, allow air or flue gas to escape unless the opening is also pressurized.

Induced draft boilers operate with air and gas static pressure below atmospheric. The static pressure is progressively lower as the gas travels from the air inlet to the induced draft fan. The required flow through the boiler can be achieved by the stack alone when the system pressure loss is low or when the stack is tall; this is called *natural draft*. For most modern boilers, a fan at the boiler system outlet is needed to draw flow through the boiler. Unlike forced draft units, air from the boiler surroundings enters through (infiltrates) any openings in the boiler setting.

Balanced draft boilers have a forced draft air fan at the system inlet and an induced draft fan near the system outlet. The static pressure is above atmospheric at the forced draft fan outlet and decreases to atmospheric pressure at some point within the system (typically the lower furnace). The static pressure is subatmospheric and progressively decreases as the gas travels from the balance point to the induced draft fan. This scheme reduces both flue gas pressure and the tendency of hot gases to escape. There are also power savings for this method because forced draft air fans require smaller volumetric flow rates and therefore less energy for a given mass flow. Most modern boilers are balanced draft for these reasons.

Draft loss is the reduction in static pressure of a gas caused by friction and other nonrecoverable pressure losses associated with the gas flow under real conditions. As discussed in Chapter 3, static pressure is related to the total pressure at a location by the addition of the velocity or dynamic pressure.

$$P_{total} = P_s + P_v = P_s + \frac{V^2}{2g_c v} \qquad \textbf{(1)}$$

where

P_{total} = total pressure, lb/ft^2 (N/m^2)
P_s = static pressure, lb/ft^2 (N/m^2)
P_v = velocity pressure, lb/ft^2 (N/m^2) = $V^2/2g_c v$
V = average gas velocity
v = specific volume of the gas
g_c = conversion constant
= 32.17 lbm ft/lbf s^2 = 1 kgm/Ns^2

Stack effect

Stack effect (SE), or chimney action, is the difference in pressure caused by the difference in elevation between two locations in vertical ducts or passages conveying heated gases at zero gas flow. It is caused by the difference in density between air and heated gases, whether air or flue gas. The stack effect is independent of gas flow and can not be measured with draft gauges. Draft gauges combine stack effect and flow losses. The intensity and distribution of this pressure difference depends on the height, the arrangement of ducts, and the average gas temperature in the duct and ambient air temperature.

Based upon this definition, the overall stack effect in its most general form is defined as:

$$\Delta P_{SE} = \frac{g}{g_c} Z \left(\rho_a - \rho_g\right) = \frac{g}{g_c} Z \left(\frac{1}{v_a} - \frac{1}{v_g}\right) \qquad \textbf{(2)}$$

where

ΔP_{SE} = stack draft effect driving pressure, lb/ft^2 (N/m^2)
g = acceleration of gravity = 32.17 ft/s^2 (9.8 m/s^2)
g_c = 32.17 lbm ft/lbf s^2 (1 kgm/Ns^2)
Z = elevation between points 1 and 2, ft (m)
ρ_a = density of air at ambient conditions, lb/ft^3 (kg/m^3)
ρ_g = average density of flue gas, lb/ft^3 (kg/m^3)
v_a = specific volume of air at ambient conditions, ft^3/lb (m^3/kg)
v_g = average specific volume of flue gas, ft^3/lb (m^3/kg)

The customary English units used for draft calculations are inches of water for draft pressure loss and feet for stack height or elevation. Using this system of units, the incremental stack effect (inches of pressure loss per foot of stack height) can be evaluated from:

$$SE = \left(\frac{1}{v_a} - \frac{1}{v_g}\right)\left(\frac{1}{5.2}\right) \qquad \textbf{(3)}$$

where

SE = stack effect, in./ft

For convenience, Table 1 provides the specific volume of air and flue gas at one atmosphere and 1000R (556K). Assuming air and flue gas can be treated as ideal gases, the ideal gas law in Chapter 3 permits calculation of the specific volume at other conditions:

Table 1
Sample Specific Volumes at 1000R (556K) and One Atmosphere

Gas	v_b (ft^3/lb)	v_b (m^3/kg)
Dry air	25.2	1.57
Combustion air (0.013 lb water/lb dry air)	25.4	1.58
Flue gas (3% water by wt)	24.3	1.5
Flue gas (5% water by wt)	24.7	1.54
Flue gas (10% water by wt)	25.7	1.60

$$v = v_R \left(\frac{T_f}{T_R}\right)\left(\frac{B_R}{B}\right) \tag{4}$$

where

v = specific volume at T_f and B, ft^3/lb (m^3/kg)
v_R = specific volume at T_R and B_R, ft^3/lb (m^3/kg)
T_f = average fluid temperature, R (K)
= T_f (F) + 460 (T_f (C) + 256)
B = barometric pressure (see Table 2), in. Hg (kPa)
T_R = 1000R (556K)
B_R = 30 in. Hg (101.6 kPa)

Table 3 provides a reference set of values for SE at one atmosphere. The total theoretical draft effect of a stack or duct at a given elevation above sea level can be calculated from Equation 1 or from:

$$\text{Stack draft} = Z\,(SE)\left(\frac{B_{\text{elevation}}}{B_{\text{sea level}}}\right) \tag{5}$$

where

Z = stack height, ft (m)
SE = stack effect, in./ft (Table 3 or Equation 2)
$B_{elevation}$ = barometric pressure at elevation (Table 2)
$B_{sea\ level}$ = barometric pressure at sea level (Table 2)

The average gas temperature in these calculations is assumed to be the arithmetic average temperature entering and leaving the stack or duct section. For gases flowing through an actual stack, there is some heat loss to the ambient air through the stack structure. There is also some infiltration of cold air. The total loss in temperature in a stack depends upon the type of stack, stack diameter, stack height, gas velocity, and a number of variables influencing the outside stack surface temperature. Fig. 11 indicates an approximate stack exit temperature relative to height, diameter, and inlet gas temperature.

Table 2
Barometric Pressure, B — Effect of Altitude

Ft Above Sea Level	Pressure in. Hg	Pressure kPa	Ft Above Sea Level	Pressure in. Hg	Pressure kPa
0	29.92	760	6000	23.98	609
1000	28.86	733	7000	23.09	586
2000	27.82	707	8000	22.22	564
3000	26.82	681	9000	21.39	543
4000	25.84	656	10,000	20.58	523
5000	24.90	632	15,000	16.89	429

Values from Publication 99, Air Moving and Conditioning Association, Inc., 1967.

Sample stack effect calculation

Fig. 12 illustrates the procedure used in calculating stack effect. The stack effect can either assist or resist the gas flow through the unit. The three gas passages are at different temperatures and the example is at sea level. For illustrative purposes, assume atmospheric pressure, i.e., (draft = 0) at point D.

Stack effect always assists up-flowing gas and resists down-flowing gas. Plus signs are assigned to up flows and minus signs to down flows. Using values from Table 3 for an ambient air temperature of 80F (27C), the stack effect in inches of water for each passage is:

Stack effect C to D = + (110 × 0.0030) = +0.33 in. H_2O
Stack effect B to C = − (100 × 0.0087) = −0.87 in. H_2O
Stack effect A to B = + (50 × 0.0100) = +0.50 in. H_2O

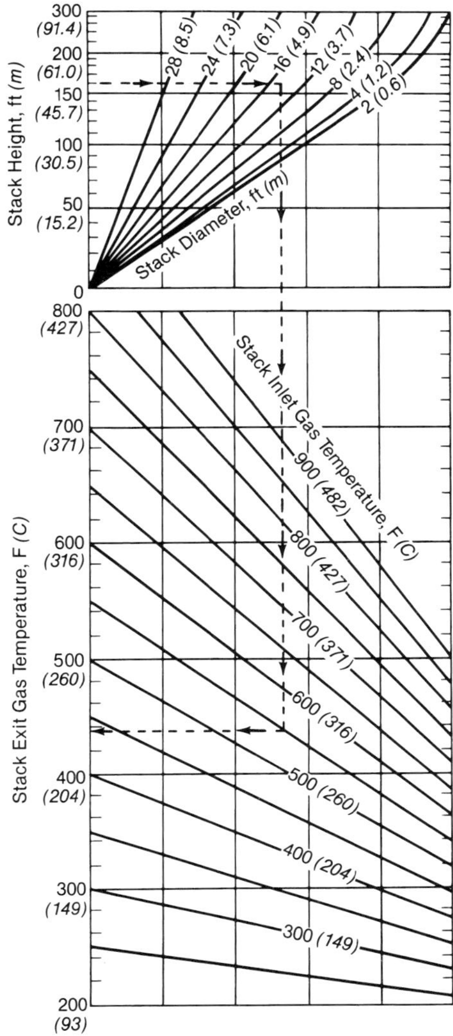

Fig. 11 Approximate relationship between stack exit gas temperature and stack dimensions.

Table 3
Reference Set of Stack Effect (SE) Values in. of H_2O/ft of Stack Height — English Units Only

Reference Conditions:
Air — 0.013 lb H_2O/lb dry air: 13.7 ft^3/lb, 80F, 30 in. Hg
Gas — 0.04 lb H_2O/lb dry gas: 13.23 ft^3/lb, 80F, 30 in. Hg
Barometric pressure 30 in. Hg

Avg. Temp in Flue or Stack, T_g	Ambient Air Temperature, T_a, F			
F	40	60	80	100
250	0.0041	0.0035	0.0030	0.0025
500	0.0070	0.0064	0.0059	0.0054
1000	0.0098	0.0092	0.0087	0.0082
1500	0.0112	0.0106	0.0100	0.0095
2000	0.0120	0.0114	0.0108	0.0103
2500	0.0125	0.0119	0.0114	0.0109

If draft gauges are placed with one end open to the atmosphere at locations A, B, C and D of Fig. 12, the theoretical zero flow draft readings are:

Draft at D = 0 in. H_2O
Draft at C = draft at D minus stack effect C to D
= 0 – (+ 0.33) = – 0.33 in. H_2O
Draft at B = draft at C minus stack effect B to C
= – 0.33 – (– 0.86) = + 0.53 in. H_2O
Draft at A = draft at B minus stack effect A to B
= + 0.53 – (+ 0.50) = + 0.03 in. H_2O

Note that because the calculation of stack effect in this example is opposite to the gas flow, stack effects are subtracted in calculating static pressures or drafts. If the stack effect is in the direction of gas flow, stack effects should be added.

The net stack effect from A to D in Fig. 12 is the sum of all three stack effects and is –0.03 in. For this reason, fans or stack height must be selected not only to provide the necessary draft to overcome flow loss through the unit, but also to allow for the net stack effect of the system.

In some boiler settings, gases leak from the upper portions when the unit is operating at very low loads or when it is taken out of service. The leakage can occur even though the outlet flue may show a substantial negative draft. The preceding example illustrates this type of low to no flow condition with a suction or negative pressure at the bottom of the uptake flue C-D and positive pressures at both points A and B in Fig. 12.

Chimney or stack

Early boilers operated with natural draft caused by the stack effect alone. However, for large units equipped with superheaters, economizers and especially air heaters, it is not practical or economical to operate the entire unit from stack induced draft alone. These units require fans to supplement the stack induced draft. The entire unit might be pressurized by a forced draft fan or the unit might use both induced and forced draft fans for balanced draft operation. The combination of only an induced draft fan and stack is not commonly used.

The required height and diameter of stacks for natural draft units depend upon:

1. draft loss through the boiler from the point of balanced draft to the stack entrance,
2. average temperature of the gases passing up the stack and the temperature of the surrounding air,
3. required gas flow from the stack, and
4. barometric pressure.

No single formula satisfactorily covers all of the factors involved in determining stack height and diameter. The most important points to consider are: 1) temperature of the surrounding atmosphere and temperature of the gases entering the stack, 2) drop in temperature of the gases within the stack because of heat loss to the atmosphere and air infiltration, and 3) stack draft losses associated with gas flow rate (due to fluid friction within the stack and the kinetic energy of gases leaving the stack).

Stack flow loss

The net stack draft, or available induced draft at the stack entrance, is the difference between the theoretical draft calculated by Equations 2, 3 and 4 and the pressure loss due to gas flow through the stack.

From Equation 46 in Chapter 3 for friction loss, plus one velocity head exit loss ($G^2v/2g_c$):

$$\Delta P_l = f \frac{L}{D} \frac{G^2}{2g_c} v + \frac{G^2}{2g_c} v \tag{6}$$

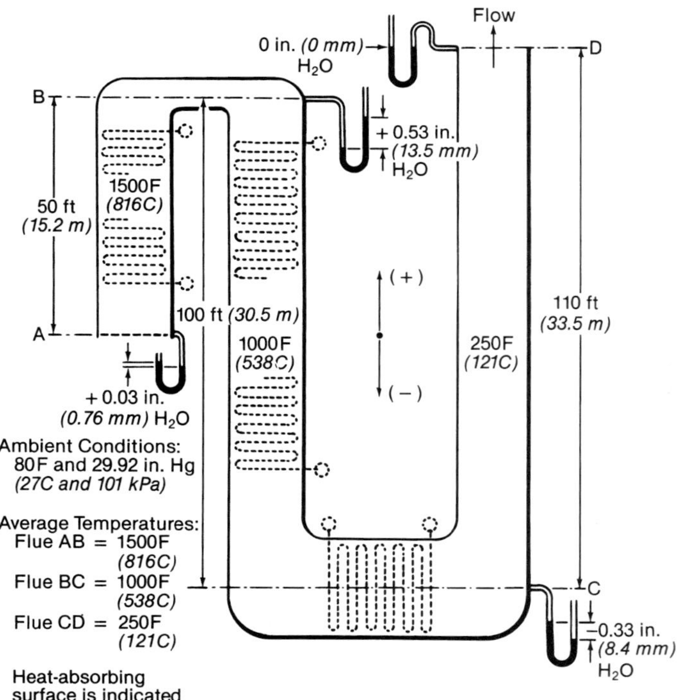

Fig. 12 Diagram illustrating stack effect, or chimney action, in three vertical gas passes arranged in series.

where

ΔP_l = stack flow loss, lb/ft^2 (N/m^2)
f = friction factor, dimensionless (Chapter 3, Fig. 1 ≈ 0.014 to 0.017)
L = length of stack = Z, ft (m)
D = stack diameter, ft (m)
G = mass flux $= \dot{m}/A$, = lb/h ft^2 (kg/m^2 s)
$\dot{m}$ = mass flow rate, lb/h (kg/s)
A = stack cross-sectional area, ft^2 (m^2)
v = specific volume at average temperature, ft^3/lb (m^3/kg)
g_c = 32.17 lbm ft/lbf s^2 (1 kgm/Ns^2)

For English units this reduces to:

Stack flow loss =

$$\Delta P_l = \frac{2.76}{B}\frac{T_g}{D_i^4}\left(\frac{\dot{m}}{10^5}\right)^2\left(\frac{fL}{D_i}+1\right) \qquad (7)$$

where

ΔP_l = stack flow loss, in. of water
$\dot{m}$ = mass flow rate, lb/h
B = barometric pressure, in. Hg
T_g = average absolute gas temperature, R
D_i = internal stack diameter, ft
f = friction factor from Chapter 3, Fig. 1, dimensionless
L = stack height above gas entrance, ft

The equation user is reminded that there is a gas-flue-to-stack-entrance exit loss that is not included in the above equation, which must be accounted for when determining overall system pressure losses.

Stack flow losses for natural draft units are typically less than 5% of the theoretical stack draft. Also, that part of the loss due to unrecoverable kinetic energy of flow (exit loss) is from three to seven times greater than the friction loss, depending on stack height and diameter.

Sample stack size selection

Tentative stack diameter and height for a given draft requirement can be calculated using English units with Figs. 11, 13 and 14 and an assumed stack exit gas temperature. Adjustments to these values are then made as required, by verification of the assumed stack exit temperature, a flow loss check and altitude correction, if necessary. The following example illustrates this sizing procedure:

Unit Specifications:	
Fuel	Pulverized coal
Steam generated, lb/h	360,000
Stack gas flow, lb/h	450,000
Stack inlet gas temp., F	550
Required stack draft (from point of balanced draft to stack gas entrance), in. H_2O	1.0
Plant altitude	Sea level
Initial Assumption:	
Stack exit gas temperature, F	450

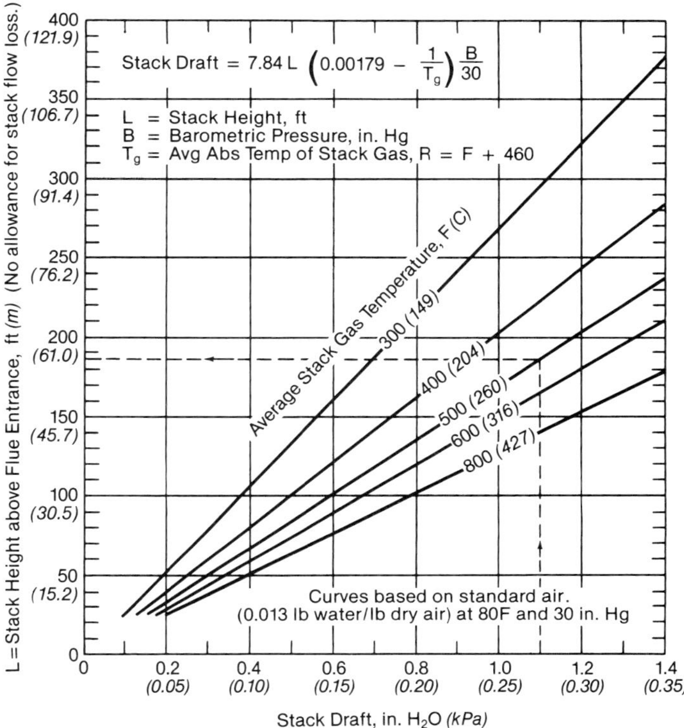

Fig. 13 Stack height required for a range of stack drafts and average stack gas temperatures.

If the stack gas flow is not specified, the following approximate ratios may be used:

Type of Firing	Gas Weight/Steam Flow Ratio
Oil or gas	1.15
Pulverized coal	1.25
Stoker	1.50

The stack diameter to the nearest 6 in. increment from Fig. 14 for 450,000 lb/h stack gas flow is 14 ft 6 in. For the required stack draft of 1.0 in. (increased to 1.1 in. for safety) and an average stack gas temperature of 500F, based on the specified inlet temperature of 550F and assumed exit temperature of 450F, Fig. 13 gives an approximate height of 187 ft. A check of the assumed stack exit temperature is obtained from Fig. 11, with the tentative height of 187 ft, diameter of 14 ft 6 in. and inlet temperature of 550F. This result is 430F, or an average stack temperature of 490F and draft of 1.1 in. H_2O. Fig. 13 is again used to establish a stack height neglecting stack flow losses. This height is 190 ft.

Assuming a stack flow loss of 5%, the final required stack height is 200 ft (190/0.95). This represents the active height of the stack. The height of any inactive section from foundation to stack entrance must also be included.

The stack flow loss is checked using the above values for diameter, height, average gas temperature and gas flow in Equation 6. A check of available net draft, using Equation 2, indicates that the 1.0 in. draft requirement is amply covered.

If the plant is not located at sea level, the draft re-

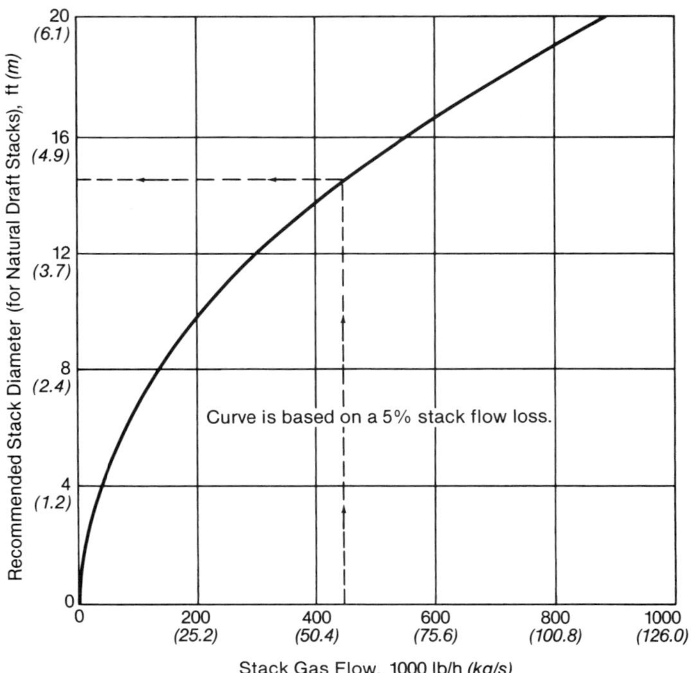

Fig. 14 Recommended stack diameter for a range of gas flows.

quirement of the unit must be increased by multiplying the draft by the altitude factor 30/B and the theoretical stack draft decreased by multiplying the theoretical draft by B/30, where B is the normal barometric pressure, inches of mercury, at the boiler site (Table 2).

External factors affecting stack height

The stack also functions to disperse combustion gases. Increasing stack height enlarges the area of dispersion. In narrow valleys or locations where there is a concentration of industry, it may be necessary to provide increased stack height.

Some power plants located near airports are prohibited from using stacks high enough to provide adequate dispersion. In such cases, the stack may be reduced in diameter at the top to increase the discharge velocity, simulating the effect of a higher stack. However, necking down the stack adds an appreciable amount of flow resistance which can only be accommodated by a mechanical draft system.

Stack design

After the correct stack height and diameter are established, there are economic and structural factors to consider in designing the stack. Stack material selection is influenced by material and erection costs, stack height, means of support (i.e., whether the stack is supported from a steel structure or a foundation), and erosive and corrosive constituents in the flue gas. After selecting the material, the stack is checked for structural adequacy, making both a static and a dynamic analysis of the loads.

Stack operation and maintenance

All connections to the stack should be air tight and sealed with dampers when not in use. Cold air leaks into the stack during operation, reducing the average gas temperature and the stack effect. Leaks also increase the gas flow and erosion potential in the stack.

A stack is subjected to the erosive action of particulate, acid corrosion from sulfur products, and weathering. Erosion is most common at the stack entrance, throats, necked down (reduced diameter) sections, and locations where the direction or velocity of gas changes. Abrasion resistant materials or erosion shields at these locations can reduce stack maintenance.

Fans

A fan moves a quantity of air or gas by adding sufficient energy to the stream to initiate motion and overcome resistance to flow. The fan consists of a bladed rotor, or impeller, which does the actual work, and usually a housing to collect and direct the air or gas discharged by the impeller. The power required depends upon the volume of air or gas moved per unit of time, the pressure difference across the fan, and the efficiency of the fan and its drives.

Power

Power may be expressed as shaft horsepower, input horsepower to motor terminals if motor driven, or theoretical horsepower which is computed by thermodynamic methods.

Fan power consumption can be expressed as:

$$\text{Power} = k\frac{\Delta P\dot{V}}{\eta_f C} \tag{8}$$

where

Power = shaft power input, hp (kW)
ΔP = pressure rise across fan, in. wg (kPa)
$\dot{V}$ = inlet volume flow rate, ft^3/min (m^3/s)
η_f = fan mechanical efficiency, 100% = 100
k = compressibility factor, dimensionless (see equation 9 below and Table 4)
C = constant of 6354 (1.00 for SI)

The compressibility factor k is calculated using the following formula:

$$k = \frac{\dfrac{\gamma}{\gamma-1}\left[\left(\dfrac{P_2}{P_1}\right)^{\frac{\gamma-1}{\gamma}}-1\right]}{\dfrac{P_2}{P_1}-1} \tag{9}$$

where

γ = specific heat ratio (1.4 for air)
P_1 = absolute inlet pressure (any unit)
P_2 = absolute outlet pressure (any unit, $P_2 = P_1 + P_T$)

An approximate value for k can be calculated by using:

$$k_{\text{approx}} = 1 - \frac{P_T}{C_{EST}}$$

where

P_T = fan total pressure, in. wg
C_{EST} = 1150 for $P_T < 10$ in. wg
1200 for $P_T < 40$ in. wg
1250 for $P_T < 70$ in. wg

All fluids and especially gases are compressible. The effects of compressibility are accounted for in the fan laws by including a compressibility coefficient.

Approximate ranges of fan efficiencies and compressibility factors for use in Equation 8 are provided in Table 4. The term fan efficiency can be misleading because there are a number of ways it can be defined. Fan efficiency can be calculated across the fan rotor only, across the fan housing (inlet to outlet) with no allowance for efficiency losses caused by inlet or outlet duct configuration, or across the housing with losses induced by inlet and outlet ducting included. The fan vendor can usually recommend the best duct arrangement at the fan inlet and outlet to minimize these losses. To select the proper fan motor, shaft input power must be calculated using the efficiency that accounts for all of the losses associated with the fan type, including losses caused by inlet and outlet duct arrangement.

In order to use the above formula, the volume flow must be known. Since the engineer often calculates boiler air and flue gas flows in mass units (lb/h, kg/s, etc.) by using the various heat and material balances published in performance test codes or other standards, it is advantageous to have a quick conversion method for calculating volume flow (cfm, m³/s). The conversion formula for arriving at volume flow when mass flow is known is:

$$\dot{V}(\text{cfm}) = \dot{m} \times \rho_{ref} \times \left(\frac{T_a}{T_{ref}}\right) \times \left(\frac{P_{ref}}{P_a}\right) \times \text{F} \qquad \textbf{(10)}$$

where

$\dot{V}$ = volumetric flow rate, cfm or ft³/min (m³/s)
$\dot{m}$ = gas flow, lb/h (kg/s)
ρ_{ref} = gas density at reference temperature, lb/ft³ (kg/m³)
T_a = actual temperature, R (K)
T_{ref} = reference temperature, R (K)
P_{ref} = reference pressure, psi (Pa)
P_a = actual pressure, psi (Pa)
F = time unit correction factor
= 1/60 English Units (1.00 for SI)

Another method used to calculate power consumption involves the concept of adiabatic head. If total pressure rise is known, adiabatic head can be calculated using the following formula:

$$Hd = \frac{(k)(\Delta P)(C)}{\rho} \qquad \textbf{(11)}$$

where

Hd = developed adiabatic head of gas column, ft (m)
k = compressibility factor

Table 4
Mechanical Efficiency — Approximate Ranges (η_f)

Centrifugal fan	
Paddle blade	45 to 60%
Foward curved blade	45 to 60%
Backward curved blades	75 to 85%
Radial tipped blades	60 to 70%
Air foil	80 to 90%
Axial flow fan	85 to 90%

Approximate compressibility factors (air)

ΔP/P	0	0.03	0.06	0.09	0.12	0.15	0.18
k	1	0.99	0.98	0.97	0.96	0.95	0.94

ΔP = total pressure increase, in. wg (kPa)
ρ = actual density, lb/ft³ (kg/m³)
C = constant of 5.20 (1.00 for SI)

Using the adiabatic head concept as defined by Equation 11, fan shaft input power can be calculated as:

$$\text{Power} = \frac{(\dot{m})(Hd)(C)}{\eta_f} \qquad \textbf{(12)}$$

where

$\dot{m}$ = gas flow, lb/h (kg/h)
C = constant of 0.505×10^{-6} (2.724×10^{-6} for SI)

Fan performance

Stacks seldom provide sufficient natural draft to cover the requirements of modern boiler units. These higher draft loss systems require the use of mechanical draft equipment, and a wide variety of fan designs and types is available to meet this need.

Fan performance is best expressed in graphic form as fan curves (Fig. 15) which provide static pressure (head), shaft horsepower and static efficiency as functions of capacity or volumetric flow rate. Because fan operation for a given capacity must match single values of head and horsepower on the characteristic curves, a balance between fan static pressure and system resistance is required.

Varying the operating speed (rpm) to yield a family of curves, as shown in Fig.16, will change the numerical performance values of the curve characteristics. However, the shape of the curves remains substantially unaltered. Changes in operation of fans can generally be predicted from the *Laws of Fan Performance*:

1. *Fan speed variation* (for constant fan size, density and system resistance)
   a) Capacity [ft³/min (m³/min)] varies directly with speed.
   b) Pressure varies as the square of the speed.
   c) Power varies as the cube of the speed.
2. *Fan size variation* (geometrically similar fans, constant pressure, density and rating)
   a) Capacity varies as the square of wheel diameter.
   b) Power varies as the square of wheel diameter.
   c) Rpm varies inversely as wheel diameter.

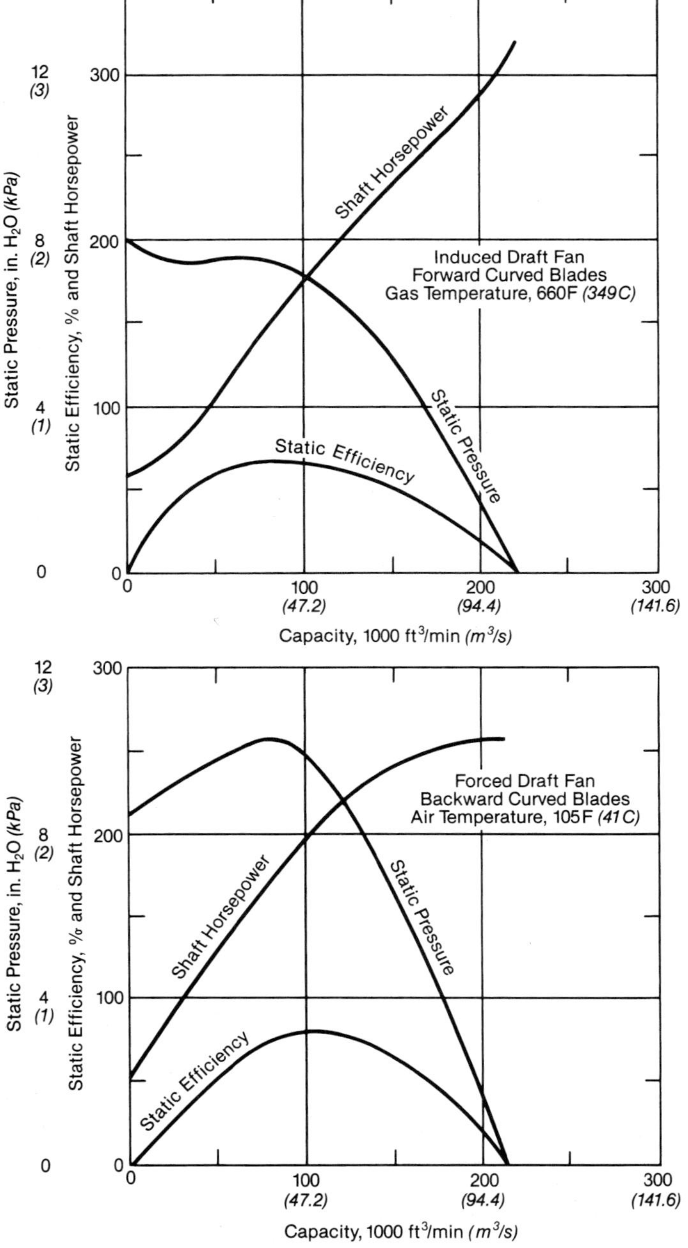

Fig. 15 Characteristic curves for two types of centrifugal fans operating at 5500 ft (1676 m) elevation and 965 rpm.

3. *Gas density variation* (constant size and speed plus constant system resistance or point of rating).
   a) Capacity remains constant.
   b) Pressure varies directly as gas density.
   c) Power varies directly as gas density.

Operating a fan at a temperature below design temperature will cause the shaft horsepower requirement to increase as a function of density ratio or absolute temperature ratio. In addition, if a different gas of a higher density is handled, the shaft horsepower will increase with the ratio of the densities. Because such operating conditions typically occur at startup, the inlet vanes or inlet louvers should be closed before the fan is started up.

The pressure increase of the fan also will rise as a function of the higher gas densities. This condition should be taken into account when designing flues, ducts, expansion joints, dampers, and other components.

In summary, as temperature decreases, density increases, and pressure rise of the fan increases. As an example:

Temperature	Density (lb/ft^3)	Static pressure rise (in. wg)
300F	0.0504	30.0
250F	0.0538	32.1
100F	0.0684	40.8

Geometrically similar fans have similar operating characteristics. Therefore, the performance of one fan can be predicted by knowing how a smaller or larger fan operates. The two main performance factors (speed and head) are linked in the concept of specific speed and specific diameter.

Specific speed is the rpm at which a fan would operate if reduced proportionately in size so that it delivers 1 ft^3/min of air at standard conditions, against a 1 in. wg static pressure.

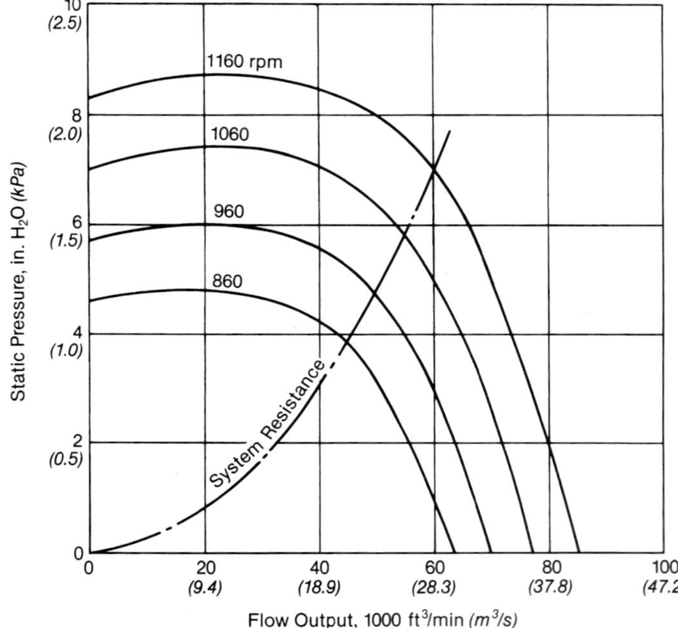

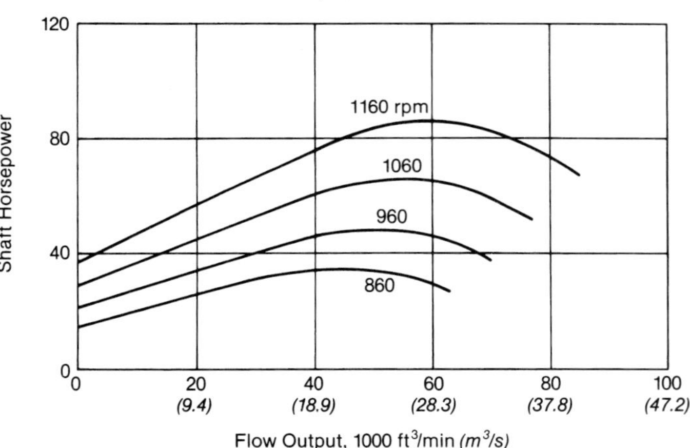

Fig. 16 Graph showing how desired output and static pressure can be obtained economically by varying fan speed to avoid large throttling losses.

Specific diameter is fan diameter required to deliver 1 ft^3/min standard air against a 1 in. wg static pressure at a given specific speed. The fan laws lead to these equations:

$$\text{Specific speed}\,(N) = \frac{\text{rpm}\left(\text{ft}^3/\text{min}\right)^{\frac{1}{2}}}{(\text{SP})^{\frac{3}{4}}} \quad \textbf{(13)}$$

$$\text{Specific diameter}\,(D) = \frac{D(SP)^{\frac{1}{4}}}{\left(\text{ft}^3/\text{min}\right)^{\frac{1}{2}}} \quad \textbf{(14)}$$

where flow in ft^3/min is at standard conditions, SP is static pressure (in. wg) and D is fan diameter (in.).

Because there is only one value of specific speed at the point of maximum efficiency for any fan design, that value serves to identify the particular design. The same is true for specific size. If either specific speed or specific size can be established from the requirements of an application, only those designs with corresponding identifying values need to be considered.

Fan capacity margins

To make sure that the fans will not limit boiler performance, margins of safety are added to the calculated or net fan requirements to arrive at a satisfactory test block specification. These margins are intended to cover conditions encountered in operation that can not be specifically evaluated. For example, variation in fuel ash characteristics or unusual operating conditions slag or foul heating surfaces. The unit then requires additional draft. Air heater leakage can increase to higher than expected levels because of incorrect seal adjustment or seal wear. Stoker-fired boilers, burning improperly sized coal, may require more than normal pressure to force air through the fuel bed. A need for rapid load increase or a short emergency overload often calls for overcapacity of the fans. The customary margins to allow for such conditions include: 1) 15 to 20% increase in the net weight flow of air or gas, 2) 20 to 30% increase in the net head, and 3) 25F (14C) increase in the air or gas temperature at the fan inlet.

System effects and fan margin

System effects is a label given to adverse conditions at the fan inlet and outlet that can cause increased static pressure losses. These conditions may not have been foreseen by the fan purchaser or the fan supplier. It is important to account for all system effects and to ensure that they are included in the fan's net static pressure requirements. If system effects are not fully considered, fan margins will be less than anticipated. System effects may include:

1. elbows located close to fan discharge,
2. insufficient duct length at the fan discharge,
3. improper inlet conditions, e.g. elbow located too close to the fan inlet, or
4. elbows without turning vanes or splitters.

The fan manufacturer can review the proposed ductwork layout and offer suggestions to minimize these effects. Of particular importance are effects caused by improper design of the transition duct from fan outlet(s) to final duct cross-section downstream of fan outlet. The ideal evase shape for low pressure drop may not be attainable in the space available.

Typical fan applications

Forced draft fan

Boilers operating with both forced and induced draft use the forced draft fan to push air through the combustion air supply system into the furnace. (See Fig. 17.) The fan must have a discharge pressure high enough to equal the total resistance of air ducts, air heater, burners or fuel bed, and any other resistance between the fan discharge and the furnace. This makes the furnace the point of balanced draft or zero pressure. Volume output of the forced draft fan must equal the total quantity of air required for combustion plus air heater leakage. In many installations, greater reliability is obtained by dividing the total fan capacity between two fans operating in parallel. If one fan is out of service, the other usually can carry 60% or more of full boiler load, depending on how the fans are sized.

To establish the required characteristics of the forced draft fan, the system resistance from fan to furnace is calculated for the actual weight of air required for combustion plus the expected leakage from

Fig. 17 Forced draft centrifugal fan for a 364 MW outdoor unit.

the air side of the air heater. It is design practice to base all calculations on 80F (27C) air temperature entering the fan. The results are then adjusted to test block specifications by the margin factors previously discussed.

Forced draft fan selection should consider the following general requirements:

Reliability Boilers must operate continuously for long periods (up to 18 months in some instances) without shutdown for repairs or maintenance. Therefore, the fan must have a rugged rotor and housing and conservatively loaded bearings. The fan must also be well balanced and the blades shaped so that they will not collect dirt and disturb this balance.

Efficiency High efficiency over a wide range of output is necessary because boilers operate under varying load conditions.

Stability Fan pressure should vary uniformly with volumetric flow rate over the capacity range. This facilitates boiler control and assures minimum disturbance of air flow when minor adjustments to the fuel burning equipment change the system resistance. When two or more fans operate in parallel, the pressure output curves should have characteristics similar to the radial tip or backward curved blade fans in order to share the load equally near the shutoff point.

Overloading It is desirable for motor driven fans to have self-limiting horsepower characteristics, so that the driving motor can not overload. This means that the horsepower should reach a peak and then drop off near the full load fan output (Fig. 16).

Induced draft fan

Units designed to operate with balanced furnace draft or without a forced draft fan require induced draft to move the gaseous products of combustion.

The gas weight used to calculate net induced draft requirements is the weight of combustion product gas at maximum boiler load, plus any air leakage into the boiler setting from the surroundings and from the air side to the gas side of the air heater. Net gas temperatures are based on the calculated unit performance at maximum load. Induced draft fan test block specifications of gas weight, negative static pressure and gas temperature are obtained by adjusting from net values by margins similar to those used for forced draft fans.

An induced draft fan has the same basic requirements as a forced draft fan except that it handles higher temperature gas which may contain erosive ash. Excessive maintenance from erosion can be avoided by protecting casing and blades with replaceable wear strips. Because of their lower resistance to erosion, air foil blades should be treated with caution when considering an induced draft application. Air foil blades are very susceptible to dust erosion and, if hollow, they can fill with dust and cause rotor imbalance should the blade surface wear through. Bearings, usually water-cooled, have radiation shields on the shaft between the rotor and bearings to avoid overheating.

Gas recirculation fans

As discussed in Chapter 19, gas recirculation fans are used in various boiler arrangements for controlling steam temperature, furnace heat absorption, slagging of heating surfaces, and control of NO_x emissions in oil- or gas-fired burners. They are generally located at the economizer outlet to extract gas and inject it into the furnace at locations dependent on the intended function. These multiple purposes are also an important consideration in properly sizing and specifying gas recirculation fans. Selection may be dictated by the high static pressure required for tempering furnace temperatures at full load on the boiler unit, or by the high volume requirement at partial loads for steam temperature control.

Even though gas recirculation fans have the same basic requirements as induced draft fans, there are additional factors to consider, such as being subjected to large amounts of abrasive ash. The gas recirculation fan typically operates at higher gas temperatures, so intermittent service may cause thermal shock or unbalance. When the fan is not in service, tight shutoff dampers and sealing air must be provided to prevent the back flow of hot furnace gas, and a turning gear is often used on large fans to turn the rotor slowly to avoid distortion.

Primary air fans

Primary air fans on pulverized coal-fired boilers supply pulverizers with the air needed to dry the coal and transport it to the boiler. Cold primary air fans should be designed for duty similar to forced draft fans. Primary air fans may be located before the air heater (cold primary air system) or downstream of the air heater (hot primary air system). The cold primary air system has the advantage of working with a smaller volumetric flow rate for a given mass flow rate. This method will pressurize the air side of the air heater and encourage leakage to the gas side. The hot primary air system avoids primary air heater leakage, but requires a higher fan design temperature and larger volumetric flow rate.

Fan maintenance

Fans require frequent inspection to detect and correct irregularities that might cause problems. However, they should also have long periods of continuous operation compared with other power plant equipment. This can be assured by proper lubrication and cooling of fan shafts, couplings and bearings.

A fan should be properly balanced, both statically and dynamically, to assure smooth and long-term service. This balance should be checked after each maintenance shutdown by running the fan at full speed, first with no air flow and second with full air flow.

Fans handling gases with entrained abrasive dust particles are subject to erosion. Abrasion resistant materials and liners can be used to reduce such wear. In some cases, beads of weld metal are applied to build up eroded surfaces.

Fan testing

It is difficult to obtain consistent data from a field test of fans installed in flue and duct systems because it is seldom possible to eliminate flow disturbances from such things as bends, change in flow area, and dampers. Structural arrangements at the fan entrance and discharge also materially affect field performance results. The consistent way to verify fan performance is on a test stand. (See Fig. 18.)

Fan types

There are essentially two different kinds of fans: centrifugal and axial flow. In a centrifugal fan, the air or gas enters the impeller axially and is changed by the impeller blades to flow radially at the impeller's discharge. The impeller is typically contained in a volute-type housing. In an axial flow fan, the air or gas is accelerated parallel to the fan axis.

Centrifugal fans

Centrifugal fans can utilize several types of impellers: airfoil, backward inclined, radial tipped, and forward curved. Fans with forward curved impellers are typically used for low-capacity applications and are seldom found in industrial applications. Characteristic traits of centrifugal fan impellers are:

Airfoil This impeller is capable of attaining the highest efficiency. It has the highest specific speed (it must run faster to develop a given pressure at a given diameter). Sound levels tend to be lower than the other types. It is not satisfactory for high dust loads; the blades can not easily be made wear resistant and it is prone to dust buildup on the backs of blades. The hollow blades have potential for water and dust ingress. Tip speed limits increase with reduced tip width.

Backward inclined

Curved backward inclined (CBI) offers high efficiency (although not as high as airfoil impellers). This design is satisfactory for dust loads and can accept removable wear plates. This design is prone to buildup on the backs of blades. It is economical to manufacture and sound levels are marginally lower than airfoil impellers. Specific speed is high.

Flat backward inclined is excellent for high dust loads and can easily accept removable wear plates. Efficiency is lower than for a curved backward design. It is economical to manufacture, has a high specific speed, and is prone to dust buildup on the backs of blades.

Radial tipped This design is excellent for dust loading, can accommodate removable wear plates satisfactorily, and has very few dust buildup problems. However, it has the lowest efficiency of the four types. This impeller has tip speed limitations and therefore is not suitable for very high pressure rise applications.

A further summary of these impeller types is given in Table 5.

Control of centrifugal fan output

Very few applications permit fans to always operate at the same pressure and volume discharge rate. Therefore, to meet requirements of the system, some means of varying the fan output are required such as damper control, inlet vane control, and variable speed control.

Damper control introduces sufficient variable resistance in the system to alter the fan output as required. The advantages are:

1. lowest initial capital cost of all control types,
2. ease of operation or adaptation to automatic control,
3. least expensive type of fan drive; a constant speed induction type AC motor may be used, and
4. continuous rather than a step-type control making this method effective throughout the entire fan operating range.

The primary disadvantage of damper control is wasted power. Excess pressure energy must be dissipated by throttling. Outlet damper control (usually with opposed blade type dampers) is inefficient and is not often used. Inlet box damper control with parallel blades pre-spins the entering air or gas in the same direction as impeller rotation, making this method more efficient than outlet damper control.

The most economical control is accomplished with variable inlet vanes (VIVs), designed for use with both dirty air and clean air. These vanes change position with flow and produce a more efficient pre-spin than inlet dampers. VIVs are the most common type of flow control for constant speed fans. The relative location and types of flow control devices are shown in Fig. 19.

Operating experience on forced draft, primary air and induced draft fans has proven that inlet vane control is reliable and reduces operating cost. It also controls stability, controls accuracy, and minimizes hysteresis. Inlet vane control (Fig. 20) regulates air flow entering the fan and requires less horsepower than outlet damper control at fractional loads. The inlet vanes give the air a varying degree of spin in the direction of wheel rotation, enabling the fan to produce the required head at proportionally lower power. Although vane control offers considerable savings in efficiency at any reduced load, it is most effective for moderate changes close to full load operation. The initial cost is greater than damper control and less than variable speed control.

Fig. 18 Full-scale testing of variable-pitch axial flow fan.

Table 5
Centrifugal Fans[3]

Type	Impeller Design	Housing Design	Performance Curves	Performance Characteristics	Applications
Airfoil	Highest efficiency of all centrifugal fan designs. 9 to 16 blades of airfoil contour curved away from the direction of rotation. Air leaves the impeller at a velocity less than its tip speed and relatively deep blades provide for efficient expansion within the blade passages. For given duty, this will be the highest speed of the centrifugal fan designs.	Scroll-type, usually designed to permit efficient conversion of velocity pressure to static pressure, thus permitting a high static efficiency; essential that clearance and alignment between wheel and inlet bell be very close to reach the maximum efficiency capability. Concentric housings can also be used as in power roof ventilators, because there is efficient pressure conversion in the wheel.	Pressure - HP; Efficiency; SP; ME; HP; Volume Flow Rate	Highest efficiencies occur 50 to 65% of wide open volume. This is also the area of good pressure characteristics; the horsepower curve reaches a maximum near the peak efficiency area and becomes lower toward free delivery, a self-limiting power characteristic as shown.	General heating, ventilating and air-conditioning systems. Used in large sizes for clean air industrial applications where power savings are significant.
Backward-Inclined Backward-Curved	Efficiency is only slightly less than that of airfoil fans. Backward-inclined or backward-curved blades are single thickness. 9 to 16 blades curved or inclined away from the direction of rotation. Efficient for the same reasons given for the airfoil fan above.	Utilizes the same housing configuration as the airfoil design.	Pressure - HP; Efficiency; SP; ME; HP; Volume Flow Rate	Operating characteristics of this fan are similar to the airfoil fan mentioned above. Peak efficiency for this fan is slightly lower than the airfoil fan. Normally unstable left of peak pressure.	Same heating, ventilating, and air-conditioning applications as the airfoil fan. Also used in some industrial applications where the airfoil blade is not acceptable because of corrosive and/or erosive environment.
Radial	Simplest of all centrifugal fans and least efficient. Has high mechanical strength and the wheel is easily repaired. For a given point of rating, this fan requires medium speed. This classification includes radial blades and modified radial blades, usually 6 to 10 in number.	Scroll-type, usually the narrowest design of all centrifugal fan designs described here because of required high velocity discharge. Dimensional requirements of this housing are more critical than for airfoil and backward-inclined blades.	Pressure - HP; Efficiency; SP; ME; HP; Volume Flow Rate	Higher pressure characteristics than the above mentioned fans. Power rises continually to free delivery.	Used primarily for material handling applications in industrial plants. Wheel can be of rugged construction and is simple to repair in the field. Wheel is sometimes coated with special material. This design also used for high pressure industrial requirements. Not commonly found in HVAC applications.
Forward-Curved	Efficiency is less than airfoil and backward-curved bladed fans. Usually fabricated of lightweight and low cost construction. Has 24 to 64 shallow blades with both the heel and tip curved forward. Air leaves wheel at velocity greater than wheel. Tip speed and primary energy transferred to the air is by use of high velocity in the wheel. For given duty, wheel is the smallest of all centrifugal types and operates at lowest speed.	Scroll is similar to other centrifugal fan designs. The fit between the wheel and inlet is not as critical as on airfoil and backward-inclined bladed fans. Uses large cut-off sheet in housing.	Pressure - HP; Efficiency; SP; ME; HP; Volume Flow Rate	Pressure curve is less steep than that of backward-curved bladed fans. There is a dip in the pressure curve left of the peak pressure point and highest efficiency occurs to the right of peak pressure, 40 to 50% of wide open volume. Fan should be rated to the right of peak pressure. Power curve rises continually toward free delivery and this must be taken into account when motor is selected.	Used primarily in low-pressure heating, ventilating, and air-conditioning applications such as domestic furnaces, central station units, and packaged air-conditioning equipment from room air-conditioning units to roof top units.

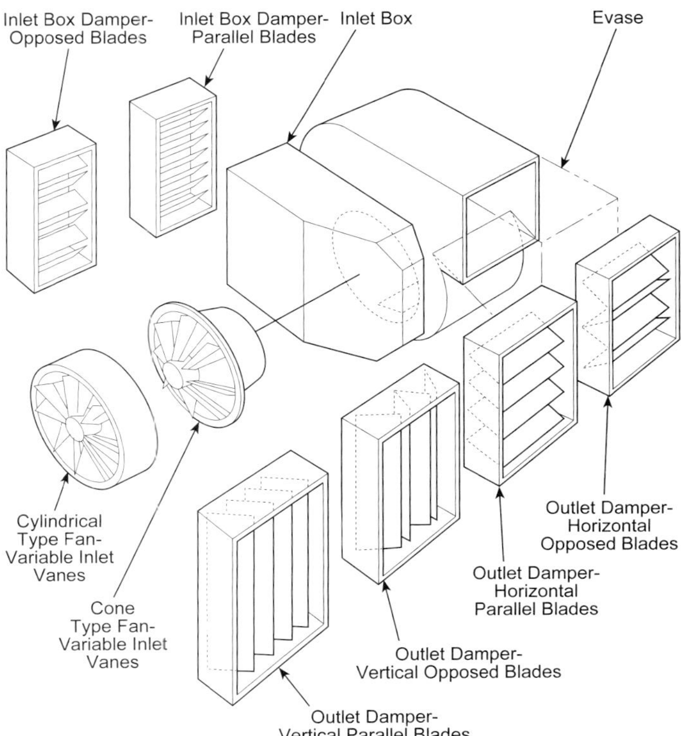

Fig. 19 Centrifugal fan flow control devices.[3]

Electric motors are normally used for fan drives because they are less expensive and more efficient than other types of drives. For fans of more than a few horsepower, squirrel cage induction motors predominate. This type of motor is relatively inexpensive, reliable, and highly efficient over a wide load range. It is frequently used in large sizes and can be used with a magnetic or hydraulic coupling for variable speed installations.

Variable speed motors are attractive because they reduce shaft speed (saving operating power) at reduced flow rates. However, variable speed motors require a higher initial cost which may not be offset by the lower power requirements at reduced fan load conditions. Speed control also results in some loss in motor efficiency because no variable speed driver works as efficiently throughout the entire fan load range as a direct connected, constant speed AC motor. The efficiency loss depends upon the type of speed variation equipment being used.

Variable speed control for centrifugal fans can be achieved by use of hydraulic couplings, variable speed DC motors, two-speed motors, and variable speed steam turbines. Hydraulic coupling losses can become quite severe at low loads, to where a thorough evaluation of power savings over the boiler operating load range does not produce enough monetary savings in power consumption to justify the high initial cost.

For some variable speed installations, particularly in the smaller sizes, wound rotor (slip-ring) induction motors are used. If a DC motor is required, the compound type is usually selected. The steam turbine drive costs more than a squirrel cage motor, but is less expensive than any of the variable speed electric motor arrangements in sizes above 50 hp (37 kW). A steam turbine may be more economical than the electric motor drive in plants where exhaust steam is needed for process, or on large utility units using the extraction steam for feedwater heating.

Two-speed motor drives can be a reasonable alternative because initial cost is more attractive when compared to the variable speed alternatives. Advantages offered by a two-speed motor drive over a single speed drive are:

1. sound level is lower at maximum continuous rating (MCR),
2. the rate of wear on impeller and other fan components is substantially reduced,
3. bearing life is increased,
4. impeller stresses are reduced and fatigue life is extended, and
5. power consumption at net and reduced load conditions is reduced.

Disadvantages of the two-speed motor drives are:

1. larger fans have higher initial cost,
2. higher motor cost,
3. plant operators are sometimes uncomfortable with shifting to alternate speeds,
4. if reserve margins on flow and pressure at net conditions are not sufficient, the fan may shift to high speed too often or at too low an operating load; anticipated power savings at high boiler loads may not be realized, and
5. if reserve margins at the low speed are too high, power savings at the normal operating condition are not enough to offset the increased initial cost of the two-speed system.

Fan curves for a unit set for both high and low speed operation are shown in Fig. 21.

Axial flow fans

In an axial flow fan, the impeller is predominantly parallel to the axis of rotation. The impeller is contained in a cylindrical housing. Axial fans for indus-

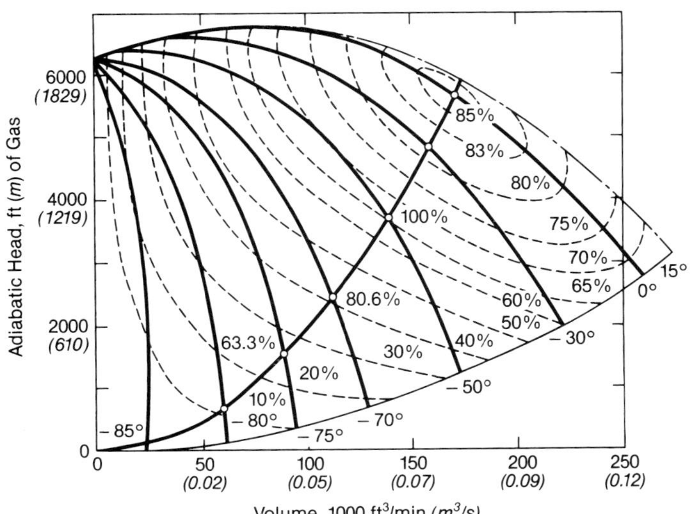

Fig. 20 Inlet vane control.

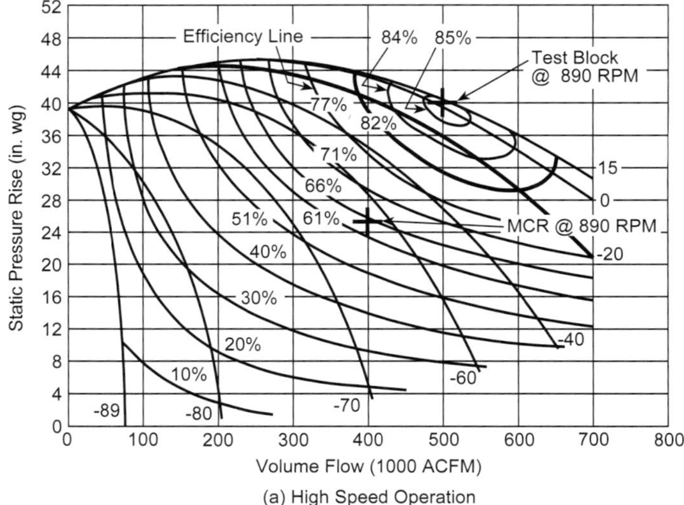

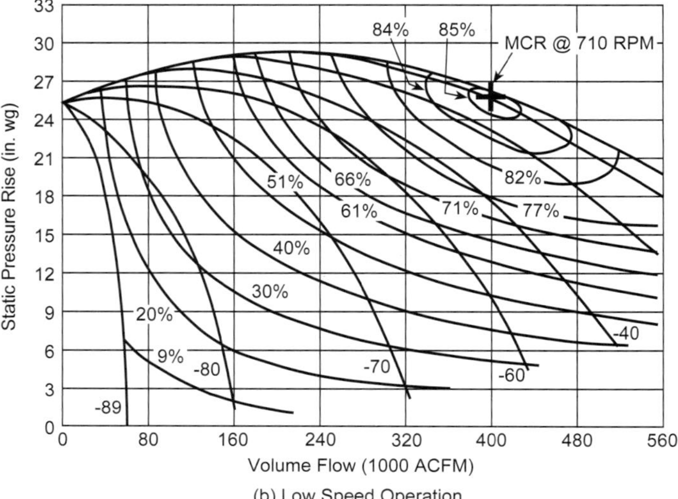

Fig. 21 Fan curves for high and low speed operation.

trial use are normally of the vane axial type, meaning they have either inlet or discharge guide vanes, or both. There are three types of vane axial fans:

1. Fixed pitch: Impeller blades are permanently secured at a given angle on the impeller hub.
2. Adjustable pitch: Impeller blade angle can by mechanically altered when the impeller is at a standstill.
3. Variable pitch: Impeller blade angle can be mechanically or hydraulically altered while the impeller is rotating.

Variable pitch axial flow fans used in fossil power generating systems can be more energy efficient than equivalent centrifugal type fans. Fig. 22 compares the power consumption of the primary air, forced draft and induced draft fans for a typical 500 MW coal-fired unit using variable pitch axial flow fans with the same unit using backward curved, inlet vane controlled centrifugal fans. At 100% unit load, auxiliary power savings using a variable pitch axial flow fan will be 4000 kW, or about 7% of the total auxiliary power consumption.

Performance and control characteristics

Fig. 23 shows the characteristic performance field of a variable pitch axial flow fan. Several major benefits observed from this figure for axial flow fans include:

1. The areas of constant efficiency run parallel to the boiler resistance line resulting in high efficiency over a wide boiler load range.
2. There is also a large control range above, as well as below, the area of maximum efficiency permitting the fan to be designed for net boiler conditions, while the test block point remains within the control range.
3. The lines of constant blade angle are actually individual fan curves for a given blade setting. Because the curves are very steep, a change in resistance produces very little volume change.
4. As the blade angle is adjusted from minimum to maximum position, the flow change is nearly linear, as shown in Fig. 24.

These last two characteristics provide stable fan and boiler control.

Parallel operation

Variable pitch axial flow fans can be operated in parallel provided care is taken to avoid operating either fan in the stall area discussed below. With two fans in operation, the resistance line for one fan is influenced by the other fan as well as by the boiler conditions. Two fans together will develop the pressure required to overcome the boiler resistance, but individual volume flows need not be equal. However, to obtain the most efficient fan operation and to avoid operation in a range close to the stall line, it is best to keep both fans operating in parallel in their design condition.

The suggested control logic for starting, stopping and supervising the operation of variable pitch axial flow forced draft and induced draft fans is very similar to that of centrifugal fans. An additional requirement of the logic is to prevent damage to the fan while

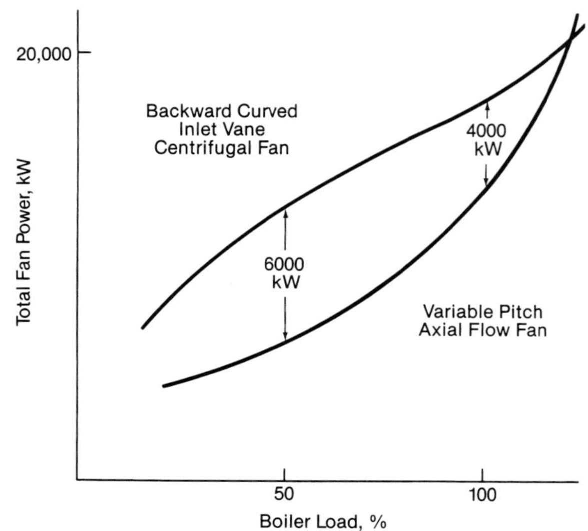

Fig. 22 Power savings.

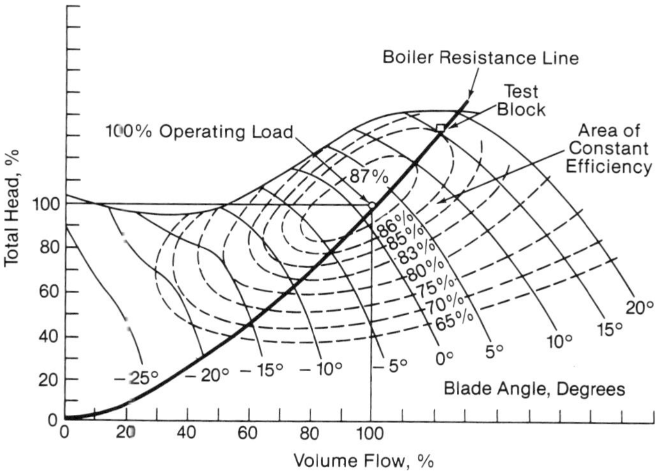

Fig. 23 Performance field for variable pitch axial flow fan.

maintaining an open flow path through the boiler to prevent furnace pressure excursions.

Stall characteristics

Axial flow fans have a unique characteristic called *stall*. Stall is the aerodynamic phenomenon which occurs when a fan operates beyond its performance limits and flow separation occurs around the blade. If this happens, the fan becomes unstable and no longer operates on its normal performance curve. Extended operation in the stall region should be avoided. Unpredictable flow vibrations occur which can damage the rotating blades.

The curves in Fig. 25, marked A, are the normal fan performance curves for a constant blade angle. Each blade angle curve has an individual stall point, identified as S on the diagram. The curve C connects all the stall points S and is generally referred to as the stall line.

The dashed curves B are the characteristic stall curves for three different blade angles. The curves show the path that the fan will follow when operating in a stalled condition.

Fig. 26 explains the stall phenomenon in relation to the fan and boiler system. If the normal boiler system resistance (curve B) increases for any reason (for instance, a furnace pressure excursion caused by a main fuel trip), the normal operating point X will change to meet a new higher system resistance (curve B_1) by traveling along the fan performance curve A. If the operating point arrives at point S, the fan will stall. Because of the relationship between the fan performance curve D in the stall area and the upset system resistance (curve B_1), a new operating point X_1 will be found where the system resistance (curve B_1) and the stall curve E intersect. When the system resistance is reduced to curve B, the fan will recover from the stall and return to its normal performance curve A.

In the case of an upset as described above, the blade angle can be reduced until the fan regains stability. The fan will be stable when the new performance curve A_1 provides a stall point S, which is higher than the system resistance (curve B_1).

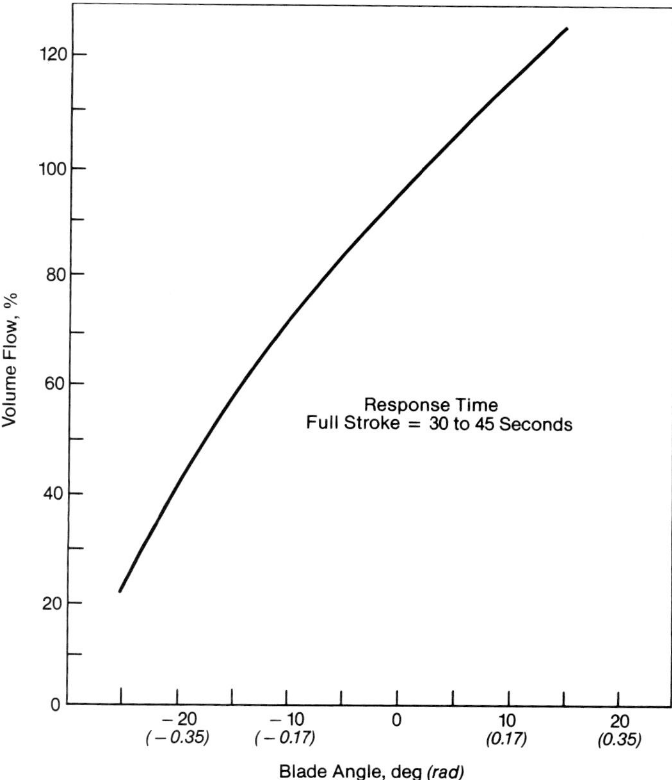

Fig. 24 Variable pitch blade control characteristics.

Stall prevention

When axial fans are sized properly and the system resistance line is parabolic in shape, the probability of experiencing stall is low. The possibility of a stall increases if a fan is oversized in regard to volume capacity, if the system resistance increases significantly, or if the fans are operated improperly.

The degree of stall protection desired in the control system depends on individual owner philosophy. A visual indication of head and volume (or blade angle) in the control room is a minimum requirement for satisfactory operation.

Fan arrangement

Axial flow fans designed for today's large fossil fueled steam generating systems are compact and rela-

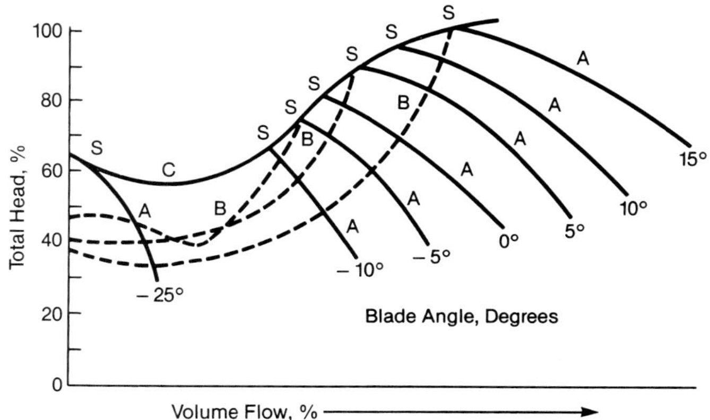

Fig. 25 Actual stall curves.

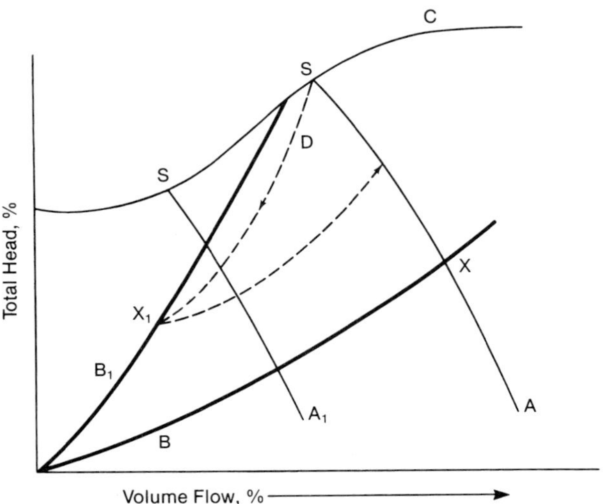

Fig. 26 Stall as related to the boiler.

tively light. By concentrating the load carrying parts of the rotor on a small radius, a low flywheel effect or moment of inertia (Wk^2) is obtained. The low weight, unbalanced forces and inertia of the axial flow fan permit either horizontal or vertical arrangement on steel construction using a dampening frame. This provides greater arrangement flexibility. Axial fans installed at grade require less of a concrete foundation than centrifugal fans.

A horizontal fan arrangement is generally the economic preference for forced draft and primary air fans. Induced draft fans can be arranged horizontally, or vertically inside the stack should available plant space be at a premium.

Induced draft fan blade wear

Fan blade wear for induced draft fans is a function of dust loading (including excursions due to particulate collection equipment malfunction), particle hardness, particle size distribution, the relative blade to particle velocity, the blade material, and the angle of attack. The fan designer has little control over the first three and must therefore look to other areas to address blade wear issues. Wear is inherently more uniform over variable pitch axial flow fan blades. By selecting a two-stage fan at a lower speed (rpm) instead of a faster one-stage fan, the tip velocity can be reduced. Abrasion resistant material coatings can also be added to the blade wear areas. Additionally, the blades can be made easily removable for fast replacement.

Aerodynamic characteristics

Fig. 27 shows a fan selection diagram used to determine the most economical fan for a given set of performance conditions. The performance requirements for a 500 MW coal-fired boiler using pairs of induced, forced and primary fans are superimposed on the fan-type selection diagram.

When specific speed increases, the diameter ratio of a centrifugal fan increases toward unity. As unity is approached, the design becomes less practical in the high specific speed range because the diameter ratios decrease.

System resistance is plotted in Figs. 16 and 20 along with the fan characteristic of static pressures at various speeds. If the fan operates at constant speed, any flow rate less than that shown at the intersection of the system resistance and specified rpm curves must be obtained by adding resistance through throttling the excess fan head.

Plotting characteristics on a percentage basis shows some of the many variations available in different fan designs. Fig. 28 is such a plot with 100% rated capacity selected at the point of maximum efficiency. Some fans give steep head characteristics, while others give flat head characteristics. Some horsepower characteristics are concave upward, others are concave downward. The latter have the advantage of being self-limiting, so that there is little danger of burning out the driver or little need for oversizing the motor.

Centrifugal fans have a steep head characteristic and are attractive for high static, low flow rate applications. Fans with steep head characteristics also require high tip speeds for the specified head. Therefore, centrifugal fans are usually equipped with wear liners when there is heavy dust loading in the gas. In contrast, axial flow fans are more suitable for use where static pressure requirements tend to decrease as flow requirements decrease. Application of axial flow fans to systems where pressure requirements may suddenly increase at reduced flow conditions must be carefully reviewed to ensure that the change in static requirements does not cause the fan to stall.

Types of fan shaft bearings

The types of bearings used in fans are anti-friction (grease lubricated roller or oil lubricated roller) and sleeve.

The choice in a given fan application depends largely on the end user. In most centrifugal fan applications, the industry prefers sleeve bearings for their reliability and for the ease of replacing worn liners. In some instances, anti-friction bearings, either grease or oil lubricated, may be chosen. Smaller higher speed fans and axial flow fans usually employ anti-friction bearings. Larger lower speed fans and fans with high moments of inertia will more often use sleeve bearings, as will centrifugal fans with large diameter impellers.

Sleeve bearings rely on a continuously maintained oil film to provide a very low friction protective layer between shaft and bearing surface. When the fan is at rest, there is contact between the shaft and bearing surface. Shortly after shaft rotation begins, an oil film is established (by the lubrication system) that prevents contact between metal surfaces.

Acoustic noise

Fans produce two distinct types of noise as follows:

1. *Single tone noise* is generated when the concentrated fluid flow channels leaving the rotating blades pass a stationary object (straightener vanes or nose). The distance from the rotating blades to stationary objects affects the sound intensity, with the blade-passing frequency and its first harmonic being most dominant.

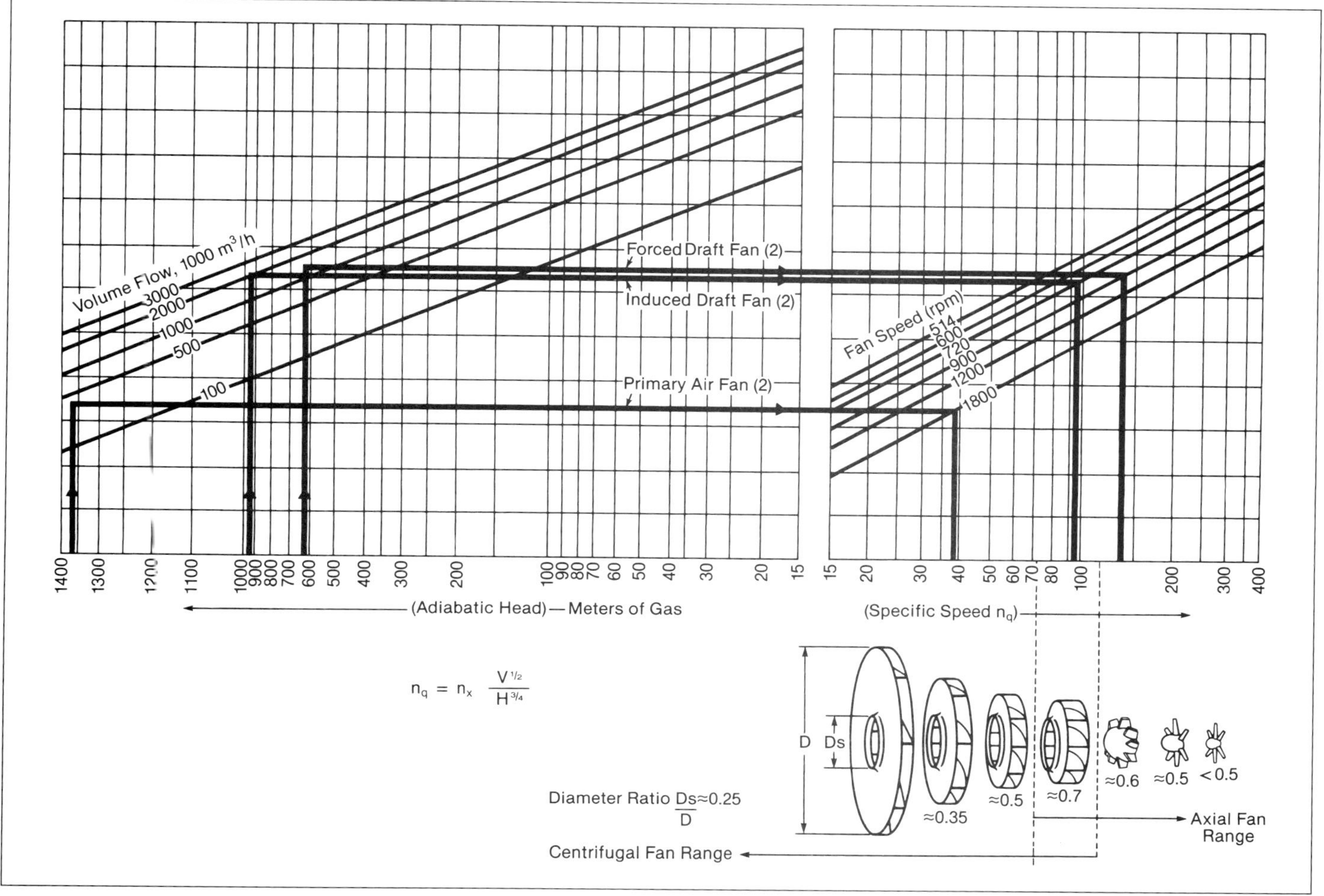

Fig. 27 Fan type selection diagram.

2. *Broad band noise* is produced by high velocity fluid rushing through the fan housing and, as the name implies, covers a wide frequency range.

From the outside, the apparent source of both types of fan noise is the fan housing. The sound travels out of the inlet box opening, through the discharge ductwork and through the casing of the fan. All three areas must be acoustically analyzed and individually treated to achieve an acceptable installation. Fig. 29 illustrates noise distribution characteristics around a fan.

Inlet sound levels from forced draft and primary air fans are most commonly controlled by the installation of an absorption-type silencer. Fan casing noise can usually be effectively controlled by the use of mineral wool insulation and acoustic lagging. Fan discharge noise, however, requires a more detailed evaluation to determine the most cost-effective method of control. For forced draft and primary air fans, insulating the outlet ducts or installing an absorption-type discharge silencer can be effective. For induced draft fans, installation of thermal insulation and lagging on the outlet flues will generally be sufficient. For situations where stack outlet noise must be reduced, an absorption-type discharge silencer can be used. However, this will not work on coal-fired units as the sound absorbing panels will become plugged with flyash. In this instance, a resonant-type discharge silencer with a self-cleaning design must be used. Typical noise attenuation devices are shown in Fig. 30.

Condensing attemperator system

Superheater spray water attemperation is the primary method of steam temperature control on most boilers. This spray water is introduced between stages of the superheater or, in some cases, immediately downstream of the superheater outlet. If feedwater is used as the source for this attemperation, it must have very low solids content in order to avoid introducing potentially damaging deposits in the superheater or in the turbine.

Many industrial plants do not have the water treatment facilities capable of meeting the low solids feedwater requirements for spray attemperation. If feedwater quality does not meet the solids criteria established for spray water, a condensing attemperator system can be used to provide a low solids spray water source.

A schematic diagram of a typical condensing attemperation system is shown in Fig. 31. The system typically consists of a condensing heat exchanger (condenser), spray water flow control valve, spray water

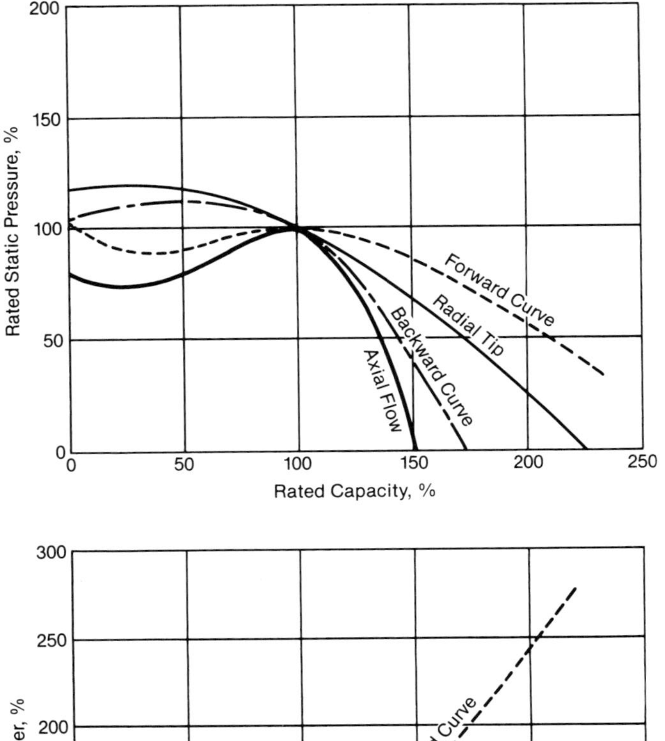

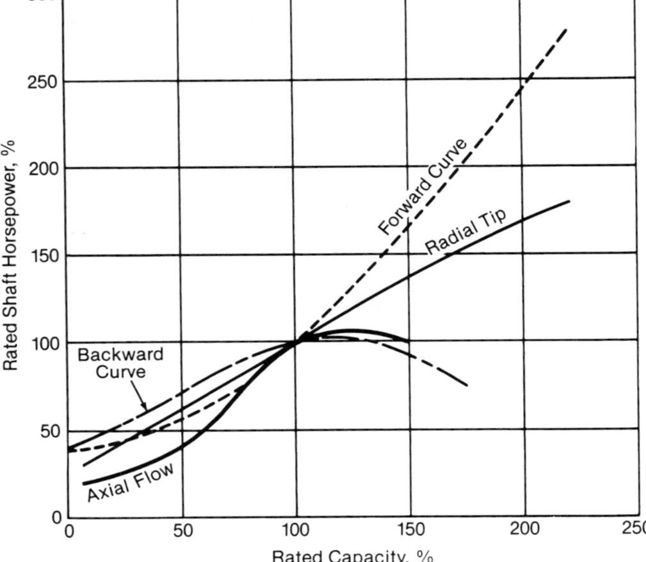

Fig. 28 Selected centrifugal and axial flow fan characteristic curves.

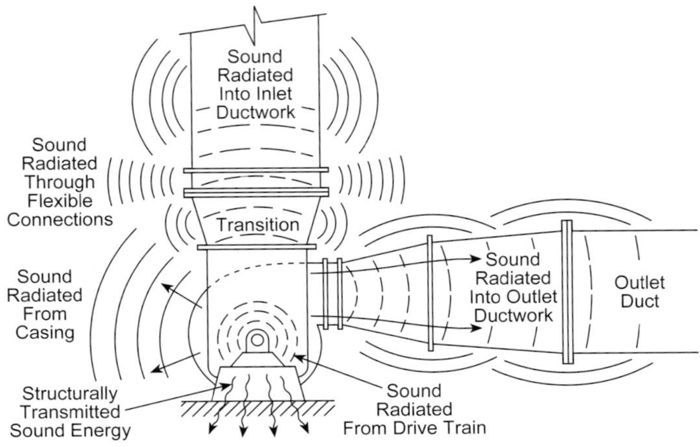

Fig. 29 Fan sound patterns.[4]

attemperator and occasionally, a condensate storage tank and a condensate pump. The basic design philosophy of this system is that saturated steam from the boiler steam drum is routed to the shell side of a vertical condenser, where it is condensed and subcooled. This condensate then flows through the spray water piping and control valve to an interstage attemperator or to a terminal attemperator.

Most condensing attemperation systems are driven by the pressure differential between the drum and the point in the steam path where the spray water is introduced. Condensers are normally located above the steam drum in order to increase the hydrostatic head available to the attemperator. If the available pressure differential is not sufficient to overcome the system resistance developed at the design spray water flow, then a condensate pump must be used. In order to avoid the need for a condensate pump, system resistance losses are usually minimized by using a low pressure drop attemperator if system spray characteristics over the boiler operating range permit its use.

Boiler feedwater is the cooling medium that flows through the tube side of the condenser. Fig. 31 shows the feedwater flowing from the economizer outlet to the condenser and from the condenser to the drum. Alternate locations may be upstream of the economizer inlet or an intermediate header in a multi-bank economizer.

The spray water condenser is a vertical, head down shell and tube heat exchanger arrangement (Fig. 32). The feedwater tube bundle is of the inverted U-tube type construction. Feedwater enters and leaves the tube bundle at the bottom of the inverted U-tubes. The cylindrical shell enclosure features a drum steam inlet near the top of the shell and a condensate outlet close to the bottom of the shell.

During normal operation, the condenser system will experience cyclic operation. At times there will be no spray flow demand and the condensate level on the shell side of the condenser will rise to the top of the tube bundle. This is a desirable operating characteristic of the inverted U-tube arrangement. It prevents

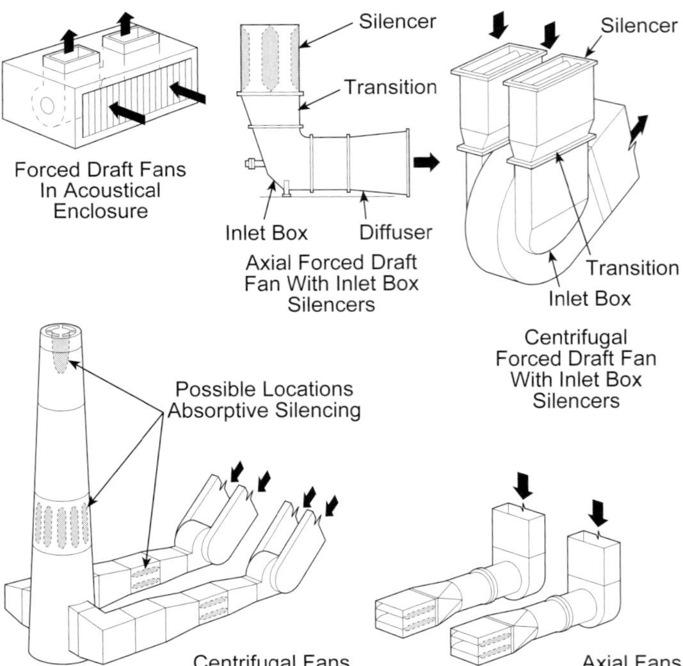

Fig. 30 Fan noise attenuation devices.[4]

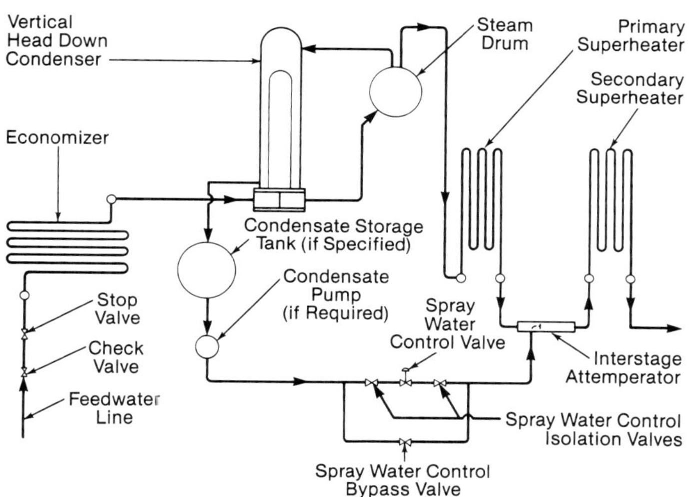

Fig. 31 Schematic arrangement of vertical condenser system.

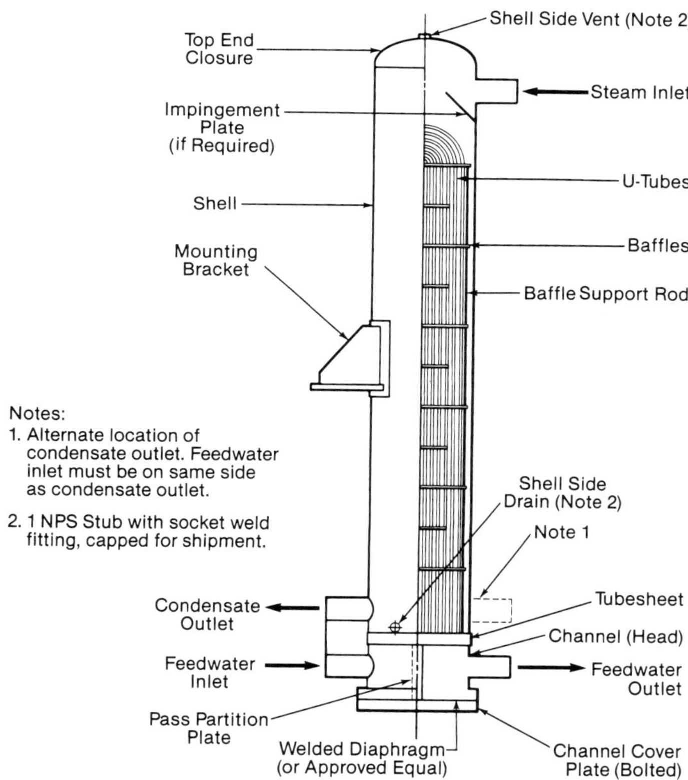

Fig. 32 General arrangement of vertical head down condenser.

steam pockets from being trapped at the top of the condenser shell which could cause water hammer and subsequent cracking of the shell.

The spray water condenser must be designed to withstand significant temperature transients and gradients. At times the condenser will be almost flooded with water and the whole assembly will be close to feedwater temperature. A sudden requirement for spray water can drop this water level quickly, exposing significant lengths of shell and internals to saturated steam temperature. In some cases this can mean a sudden rise in shell temperature of as much as 300F (167C).

References

1. *2004 ASME Boiler and Pressure Vessel Code, Section I: Power Boilers*, American Society of Mechanical Engineers, New York, New York, September 1, 2004.

2. AMCA Publication 850-02, "Industrial Process/Power Generation Heavy Duty Dampers for Isolation and Control," Air Movement and Control Association International, Inc., Arlington Heights, Illinois, 2004.

3. AMCA Publication 201-02, "Fans and Systems," Air Movement and Control Association International, Inc., Arlington Heights, Illinois, 2004.

4. AMCA Publication 801-01, "Industrial Process/Power Generation Fans: Specification Guidelines," Air Movement and Control Association International, Inc., Arlington Heights, Illinois, 2004.

Section III
Applications of Steam

The six chapters in this section illustrate how the subsystems described in Section II are combined to produce modern steam generating systems for specific applications. A number of steam generating unit and system designs for various applications are described and illustrated.

Chapter 26 begins the section with a discussion of large fossil fuel-fired equipment used to generate electric power. Both large and small industrial units, as well as those for small electric power applications, are then described in Chapter 27. The next four chapters address specialized equipment for specific applications. Unique designs for steam producing systems are used in pulp and paper mills, waste-to-energy plants, biomass-fired units, and marine applications. Biomass-fired systems in particular are receiving increased interest as renewable energy resources grow in importance.

This power plant features two B&W 420 MW variable pressure, supercritical boilers burning pulverized coal.

Chapter 26

Fossil Fuel Boilers for Electric Power

Most of the electric power generated in the United States (U.S.) is produced in steam plants using fossil fuels and high speed turbines. These plants deliver a kilowatt hour of electricity for each 8500 to 9500 Btu (8968 to 10,023 kJ) supplied from the fuel, for a net thermal efficiency of 36 to 40%. They use steam driven turbine-generators of up to 1300 MW capacity with boilers generating from one million to ten million pounds of steam per hour. A typical coal-fired facility is shown on the facing page.

Modern fossil fuel steam plants use reheat cycles with nominal steam conditions of 3500 psi/1050F/1050F (24.1 MPa/566C/566C) for supercritical pressure systems, and 2400 psi (16.5 MPa) with superheat and reheat steam temperatures ranging from 1000 to 1050F (538 to 566C) for subcritical pressure systems. For some very small units, lower steam conditions may be applied. In selected global locations where higher cycle efficiencies are required, supercritical pressure steam conditions on the order of 4300 psi/1075F/1110F (29.6 MPa/579C/599C) and 3626 psi/1112F/1130F (25.0 MPa/600C/610C) have been used.

Most power plants in the U.S. and around the world are owned and operated by: 1) investor owned electric companies, 2) federal, state or local governments, or 3) finance companies.

These owners, whether public or private, have been generally known as *utilities*. During the 1980s and 1990s, there was a trend toward new types of companies supplying significant portions of new generation. However, the fundamental approach to selecting power generation equipment will remain unchanged.

Selection of steam generating equipment

The owner has several technologies to choose from based upon fuel availability, emissions requirements, reliability, and project timing. One of the common choices for modern electric power generation is the high pressure, high temperature steam cycle with a fossil fuel-fired boiler.

Each new electric generating unit must satisfy the user's specific needs in the most economical manner. Achieving this requires close cooperation between the equipment designers and the owner's engineering staff or consultants. The designers, owner and engineering group must identify those equipment features and characteristics that will reliably produce low cost electricity. The primary costs of electricity include: 1) capital equipment, 2) financing charges, 3) fuel, and 4) operation and maintenance. The owner, prior to issuing equipment specifications, reviews and surveys all cost factors. (See Chapter 37.)

The capital cost survey must include all direct costs such as the boiler, steam turbine and electric generator, emissions control equipment, condenser, feedwater heaters and pumps, fuel handling facilities, buildings, and real estate. In addition, finance charges, including interest rates, loan periods, source of funds and tax considerations must be added. Fuel and emission control reagent costs need to be evaluated based on the initial costs, plant capacity variations expected during the life of the plant, and forecasts of cost changes during plant lifetime. The operation and maintenance costs should be estimated based on other current plants with similar equipment, fuels and operating characteristics. Operating and maintenance costs are heavily affected by personnel requirements, and consideration should be given to the availability of skilled labor as well as to the cost of retaining the skilled staff during the plant lifetime.

Plant efficiency, fuel use and capital cost are critically related. Higher plant thermal efficiency obviously reduces annual fuel costs; however, fuel savings are partially offset by the associated higher capital costs. Therefore, selection of the desired plant efficiency should carefully consider the economic tradeoffs between capital and operating costs.

Other important criteria are the location of the electric generating plant with respect to fuel supply and the areas where electricity is used. In some cases, it is more economical to transport electricity than fuel. Some large steam generating stations have been built at the coal mine mouth to generate electricity which is then used several hundred miles away. If the user is a member of a broader grid of interconnected utility companies, the future requirements of other system members may also be an important factor.

In power plant planning, substantial time and effort are required to establish accurate basic plant assessment data with comprehensive consideration of engineering factors and plans for future expansion or changes. The accuracy of these data is critical if the experience and craftsmanship of the boiler manufacturer and other suppliers are to fully benefit the plant designer and owner. The owner should, at the outset, decide who is to prepare these data. If the owner lacks personnel with the necessary qualifications, consulting engineers should be used. A thorough discussion with the boiler manufacturer will provide many details to help the owner make correct decisions.

Before the boiler and other major equipment items can be selected, the basis of operation and arrangement of the entire steam plant must be planned. Ultimately the available data must be translated into the form of equipment specifications so that the manufacturers of various components can provide apparatus in accordance with the user's requirements. After equipment selection, construction drawings must be prepared for the foundations, building, piping and walkways. The construction work must be coordinated utilizing modern schedule and control techniques for effective management and completion of erection. Ideally, the construction planning begins at the start of the conceptual engineering so that the site arrangement and project schedule facilitate efficient construction practices. Modularization of major components can be incorporated into the design where site conditions permit.

Boiler designer's requirements

The most important factors to the boiler designer are the steam conditions, fuel, project load service, and environmental constraints. Steam conditions identify the amount of steam required, the temperature and pressure of the main and reheat steam, and the feedwater temperature. Of equal importance is the type and chemical composition of the fuel, the ash characteristics, and the emissions control requirements for the specified plant site.

The boiler designer needs a complete set of data pertinent to steam generation to produce the most economical steam generating equipment and satisfy the needs of the user.

The requirements and conditions that form the basis for the designer's equipment selection can be outlined as follows:

1. Fuels – sources presently available or planned for future use; proximate, ultimate and ash analyses of each fuel as well as costs and future trends.
2. Steam requirements – ideally steam turbine heat balances should be provided for a more comprehensive understanding of the cycle design requirements. At a minimum the flow, pressure and temperature at each terminal location (main steam, feedwater, cold reheat, hot reheat, auxiliary steam, etc.) must be given for the various loads where performance is to be evaluated.
3. Boiler feedwater – source, chemical analysis and temperature of the feedwater entering the steam generating unit.
4. Environmental requirements – site limitations; permissible levels for nitrogen oxides (NO_x), sulfur dioxide (SO_2), particulate and other atmospheric emissions; solid waste specifications; and all other regulatory and commercial requirements.
5. Dispatch requirements – turndown, rate of load change, and on/off cycling.
6. Space and geographical considerations – space limitations, relation of new equipment to existing equipment, earthquake and wind resistance requirements, elevation above sea level, foundation conditions, climate, and accessibility for service and construction.
7. Auxiliary power – medium used and evaluated cost of energy.
8. Operating personnel – experience level of operating and maintenance personnel as well as the cost of labor.
9. Guarantees.
10. Evaluation basis – for unit efficiency, auxiliary power required, building volume, and various fixed charges.

With this information, the boiler designer is able to analyze the user's specific needs. As a result, an economical, reliable steam generator can be built and initial costs can be balanced with long-term savings. Setting these requirements based on anticipated future needs will provide a flexible boiler design within the constraints of cost and technical criteria.

Design practice

Utility boiler design is a custom design process due to the uniqueness of each project. Variations in projects are a result of the following fundamental differences: steam cycles, types and range of fuels, site conditions, emissions requirements, and the user's operating plans. Such variations require custom design in the overall arrangement of boiler components. The boiler designer works with standardized component concepts which are assembled to provide the optimum design arrangement to satisfy the unique requirements of the project.

While the overall design is customized, component design concepts are highly standardized and make use of pre-engineered elements where possible (attachment clips, access doors, etc.). In addition, some components, such as pulverizers, are largely pre-engineered.

At the start of a project, the designer reviews the available database and past experience to determine which large shop-fabricated components, called *modules*, may be applied to the design. Reviewing shipping sizes, traffic methods and routing, and the general economic tradeoff of greater shop assembly versus reduced field erection time are standard project practices. Modularization is now common practice but needs to be assessed on each project to determine the optimum cost and schedule impact.

Fuels

The basic types of fuels used for U.S. electricity generation are shown in Fig. 1. The dominant fuel in modern U.S. central stations is coal, either bituminous, subbituminous or lignite. A similar picture emerges for worldwide electric power production. (See Fig. 2.) While natural gas, and possibly fuel oil, used with gas turbine and combined cycle technology may be the fuel of choice for many new fossil fuel power plants, coal is expected to continue its important role in supplying energy to new, base-loaded utility power station boilers. The following material focuses primarily on coal-fired utility boilers but includes selected comments and discussion of other fossil fuels. General discussion of steam production as part of combined cycle and waste heat recovery applications is provided in Chapter 27. Specialized designs for unique, smaller power production facilities are also discussed.

Coal Coal is the most abundant fuel in the U.S. and in many parts of the world. The U.S. leads the world in coal production followed by China, India, South Africa and Australia. However, of the major fossil fuels, coal is also the most complicated and troublesome to burn. In the U.S., Europe, Japan, and increasingly in other parts of the world, there are also societal and political difficulties in locating, siting and permitting coal-fired power plants, especially considering the airborne, liquid and solid emissions compared to other fossil fuels. Coal use involves unloading, storage and handling facilities; preparation before firing using crushers and pulverizers; ash disposal equipment; emissions control equipment; and sootblowers. There is a wide variation in the properties of coal and its ash. As a result, while the design of the steam-producing unit must provide optimum performance when firing the intended coals, it must also accommodate reasonable alternate coals if necessary.

The coals selected for a prospective installation should be tabulated to indicate chemical analysis, heating value and grindability. In addition, each coal's ash analysis and ash temperature profile should be listed. The ash composition has a marked effect on a coal's slagging and fouling characteristics; the analysis must be provided to the boiler supplier for proper selection of unit geometry, features and surfaces. (See Chapter 21.)

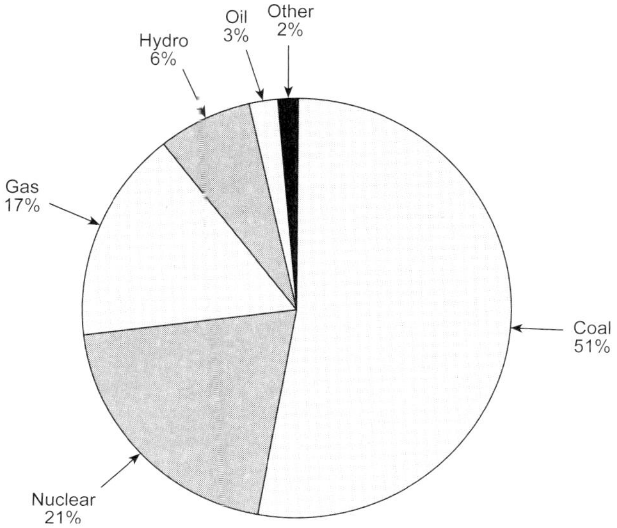

Oil 3% · Other 2% · Hydro 6% · Gas 17% · Coal 51% · Nuclear 21%

Fig. 1 U.S. electricity generation in 2001, by fuel.

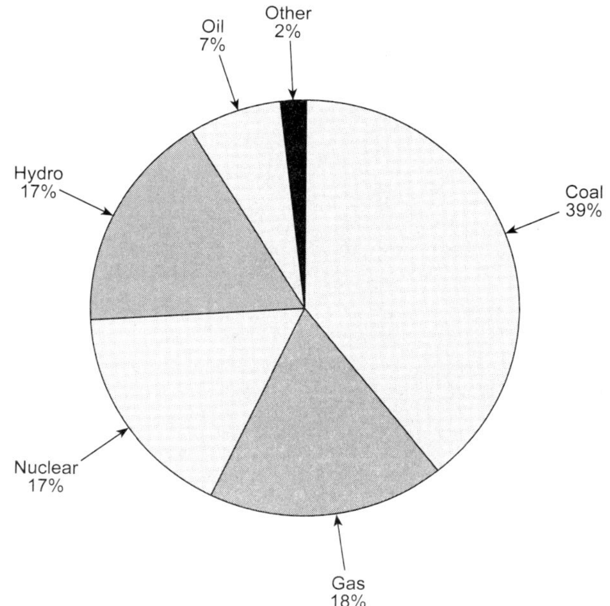

Fig. 2 Worldwide electricity generation in 2001, by fuel.

Steam requirements

Modern turbines are designed for nominal conditions of 3500 psi/1050F/1100F (24.1 MPa/566C/593C) for supercritical systems and 2400 psi/1000F/1000F (16.5 MPa/538C/538C) for subcritical systems. For some very small units, lower steam conditions such as 1800 psi (12.4 MPa) pressure may be selected. As noted earlier, some advanced steam pressure and temperature conditions may also be applied.

The steam temperature specified at the superheater outlet is generally considered to be equal to that required at the turbine plus 5F (3C). While there is very little heat loss in the well-insulated steam lines, the pressure loss in the piping results in a temperature change at constant enthalpy. For optimum performance and turbine maintenance, the steam temperature should remain constant over a broad load range. Means for controlling steam temperatures are required in utility practice. Steam temperature variation and control are discussed in Chapters 19 and 41.

The steam pressure at the superheater outlet is normally specified by the plant designer after evaluating the economics of pressure drop versus pipe sizing. Typically, the allowance for the main steam line pressure loss is 5% of the throttle pressure and the total allowance for the hot and cold reheat line loss is 5% of cold reheat pressure.

It has been customary stationary-boiler practice in the U.S. to hold the main steam pressure constant for all loads, on the premise that this condition satisfies all pressure and quantity requirements of the steam-using equipment (constant pressure). This is particularly the case where the unit is base loaded and significant cycling operation is not anticipated. In addition, constant pressure operation permits more rapid load transients while minimizing transient stresses in the boiler. However, constant pressure operation assesses some heat rate penalty at partial load conditions and results in some additional stresses in the turbine during rapid load changes.

Alternately, the system can be operated so that steam pressure varies with boiler load (variable pressure). This is more energy efficient at reduced load conditions and permits better temperature control to the turbine under load transients. However, the system may be less responsive due to the need to change the thermal inventory of the boiler during load changing, and additional stress is placed on the unit under transient conditions due to the saturated water temperature change. The thermal transient response may be improved by operating with throttle valve reserve where, for partial arc turbines, the turbine is controlled with constant throttle pressure until the first valve is closed and then operated with variable pressure. This allows the turbine valve to be opened for rapid load increase, essentially borrowing energy from the boiler system. This mode of operation is sometimes called turbine hybrid variable pressure operation.

Another mode of operation is boiler hybrid variable pressure or split pressure operation. In this mode, the turbine is operated variable pressure while the boiler through the primary superheater is operated at constant pressure. Pressure is controlled by a reducing valve located between the primary and secondary superheaters. The boiler can be maintained at a constant pressure, thereby minimizing stress and maximizing response to load changes. The temperature of the steam entering the turbine can be separately maintained, minimizing stress and maximizing load response of the steam turbine. This mode of operation provides the maximum responsiveness to load transients. While this mode of operation has a thermodynamic advantage at reduced loads compared to constant pressure operation, it is not as efficient as variable pressure operation.

These control approaches and their implications are addressed in Chapters 19 and 41.

Central station units use the reheat cycle for increased station thermal efficiency. The reheater, located within the boiler enclosure, receives steam exhausted from an intermediate pressure stage of the turbine. The steam temperature is raised and the steam is returned to the low pressure section of the turbine. The cold reheat enthalpy is impacted by the mode of turbine operation (constant vs. variable pressure) and this must be considered in the boiler design. The cold reheat enthalpy is greater under variable pressure operation (due to the increased main steam enthalpy at constant temperature but lower pressure at the throttle and reduced throttling loss) and therefore the heat absorption required of the reheater is reduced with variable pressure operation.

The double reheat cycle, where steam is returned to the boiler twice for reheating and then delivered to low pressure sections of the turbine, results in even higher station thermal efficiency than the single reheat cycle. However, the gain in plant efficiency must be weighed against the cost of additional piping systems and turbine and boiler equipment used to accommodate the flow of low pressure steam.

Steam flow requirement

Steam producing equipment must be of sufficient capacity, range of output and responsiveness to ensure prompt response to the turbine steam demand. The demand may be steady, as in base loaded systems, or it may fluctuate widely and rapidly, as in cycling units. The steam flow requirements should therefore be accurately established for peak flow, maximum continuous flow (usually steady maximum flow), minimum flow, and rate of change in flow. The peak load establishes the capacity of the steam producing equipment and all its auxiliaries.

Often the turbine is designed for 5% overpressure operation (operation at a pressure 5% greater than the nominal throttle pressure). This permits a 5% higher steam flow through the turbine and results in greater unit output. This design concept will determine both the boiler maximum capacity and the boiler maximum operating and design pressures.

Occasionally, the specification will require that the system be capable of peaking by removing one or more of the top feedwater heaters from service. The steam normally supplied to this feedwater heater then flows through the turbine, thereby increasing its output. The source of that steam is from extraction on the high pressure turbine and potentially the intermediate pressure turbine depending on the heaters removed from service. This method results in higher than normal heat rates (i.e. lower efficiencies) while overpressure operation reduces heat rate. This condition impacts the boiler design in determining the maximum output and because the change in turbine extraction impacts both the feedwater temperature and the reheat flow, thereby changing the required heat absorption in various boiler components.

The range from minimum to maximum output is an important factor in selecting the boiler's firing equipment. To maintain stable ignition and optimal combustion, this range must also be considered in designing the furnace.

The required rate of change in turbine load (steam flow) may affect the entire design of the steam producing equipment. Within a specified load range, good response is easily obtained with well-designed firing equipment for pulverized coal, oil and natural gas. A multiplicity of burners (and pulverizers, where used) widens the available range, but if there is a frequent change in load, remote manual on/off operation of burners and pulverizers may be objectionable. It is therefore important that the designer, manufacturer and operator understand what can be done by full automatic firing and what is required beyond that.

In selecting equipment to produce steam at minimum total cost, it is also necessary to establish the probable capacity factor – the ratio of the average output to the rated output. This capacity factor serves as a basis for establishing the value of incremental increases in steam generating efficiency. For example, the initial cost necessary to achieve high efficiency in a plant operating continuously at high output may not be justified for a plant operating as a cycling unit where the average capacity factor is relatively low.

Boiler feedwater

The steam passing through the turbine is condensed in the condenser. This condensate is the primary source of feedwater for the boiler. Makeup water is also required to be added to the system due to cycle losses. High purity makeup water is supplied by a comprehensive water treatment system which removes troublesome compounds, and oxygen normally found in raw water from surface or subsurface sources.

Dissolved oxygen will attack steel and the rate of attack increases sharply with a rise in temperature. High chemical concentrations in the boiler water and feedwater cause furnace tube deposition and allow solids carryover into the superheater and turbine, resulting in tube failures and turbine blade deposition or erosion.

As steam plant operating pressures have increased, the water treatment system has become more critical. This has led to the installation of more complete and refined water treatment facilities. (See Chapter 42.)

Environmental considerations

Environmental protection is an integral part of the boiler and power system design process. As outlined in Chapters 1 and 32, boilers emit varying levels of NO_x, SO_2, particulate, and other compounds into the air and they discharge ash. Government regulations tightly control these emissions in most parts of the world. The specific local regulations will govern the extent of back-end cleanup equipment required and many of the boiler design parameters. A new unit burning coal in the U.S. today might look like the unit shown in Fig. 3. The boiler is designed with low NO_x burners, an enlarged furnace, staged combustion (with overfire air or NO_x ports) to minimize NO_x emissions, and a selective catalytic reduction (SCR) system for further NO_x control. (See Chapters 14 and 34.) An electrostatic precipitator or fabric filter collects particulate (Chapter 33), while either a wet flue gas desulfurization system (wet scrubber) or a spray dryer absorber (dry scrubber) removes most of the SO_2 (Chapter 35). Careful review and economic evaluation of all local regulations and permitting requirements are needed to achieve the required environmental protection in the most cost effective manner.

Space and geographical considerations

Space and geographical considerations are extremely site specific and can dictate the equipment arrangement and even the construction sequence. In some cases, a new boiler is added to an existing plant to maximize the use of existing equipment and plant infrastructure. The most extreme case involves *repowering* an existing plant, where the old boiler is removed and a newer unit is installed in the same space.

The boiler and its support structures must be designed to accommodate expected seismic forces and wind loadings, in accordance with local structural laws and regulations. The site elevation above sea level significantly affects the design and sizing of ducts, fans and the stack due to the impact of site barometric pressure on the air and flue gas specific volume. Fan power consumption is also influenced by site elevation. Finally, boilers can be designed for indoor or outdoor installation, depending upon the prevailing climate.

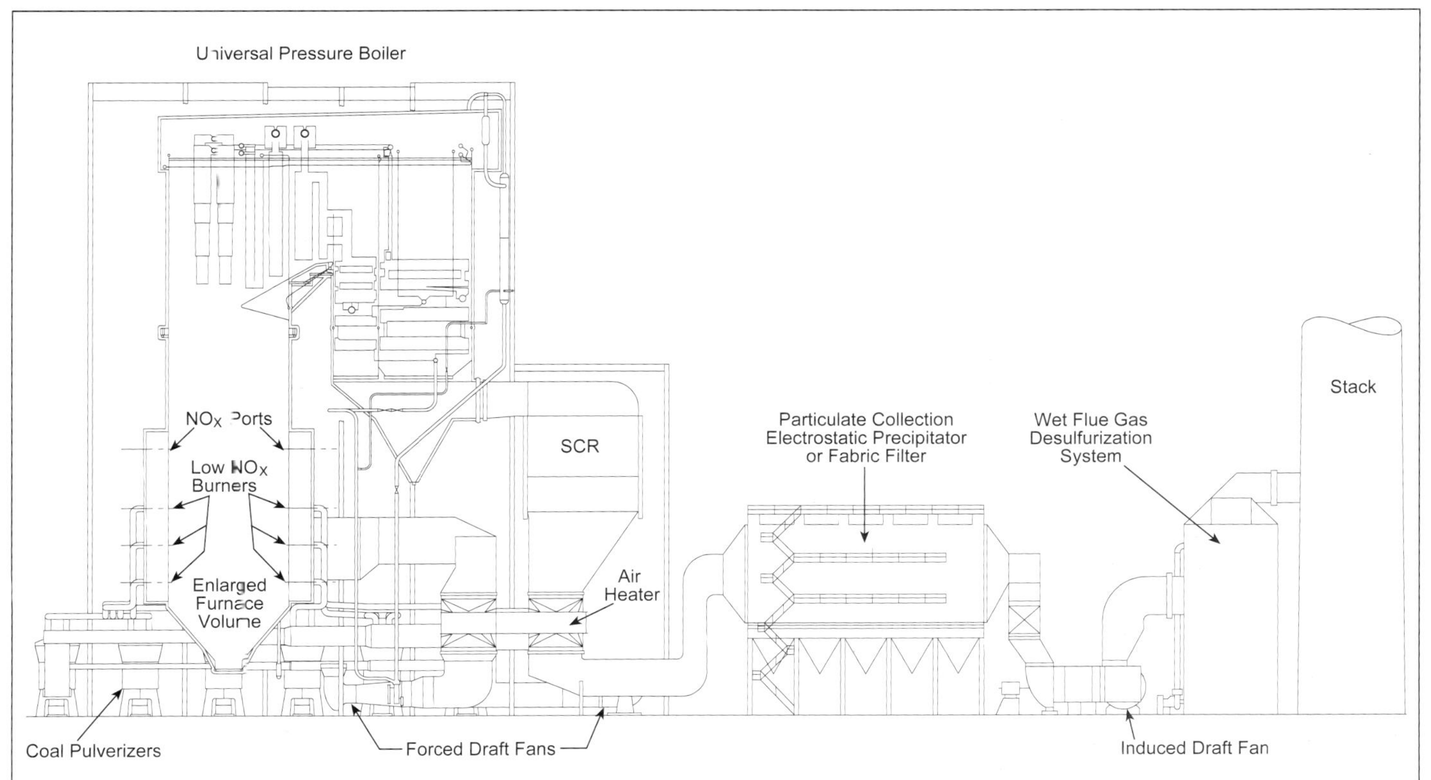

Fig. 3 Integrated pulverized coal-fired boiler power system with emissions control.

Power for driving auxiliaries

Modern practice in central stations generally calls for electric motor drives for rotating auxiliaries such as pumps, fans, pulverizers and crushers (although the boiler feed pump is usually turbine driven). The convenience and cost of the electrical drive substantiate this preference. The available plant voltage levels can determine the maximum motor size and may impact the number of forced draft and/or induced draft fans required.

Guarantees

It is common practice to obtain performance guarantees from component suppliers to provide a means to evaluate offerings and to validate the actual performance of the unit during operation.

The boiler manufacturer is usually requested to provide the following guarantees, depending on the arrangement, type of fuel and type of boiler.

1. For a given load point and fuel:
   a. efficiency,
   b. superheater steam temperature,
   c. reheater steam temperature,
   d. pressure drop from feedwater inlet to superheater outlet and from reheater inlet to reheater outlet,
   f. solids in steam (for drum boilers),
   g. auxiliary power consumption (fans, pulverizers and drives), and
   h. air heater leakage.
2. Unit maximum capacity (which is often greater than the turbine maximum capacity).
3. Superheater and reheater temperature control ranges.
4. Pulverizer capacity and fineness (where applicable).
5. NO_x and other emissions depending on project requirements.

Pulverized coal firing

The size of large pulverized coal-fired boilers and turbine-generators has peaked at 1300 MW. The equipment can be designed to burn practically any bituminous coal, subbituminous coal or lignite commercially available. Anthracite can be successfully burned in pulverized form, but requires a specialized boiler arrangement. While the special attention and additional expenses associated with plants designed for this fuel preclude its use in North America, anthracite is used today in parts of China, Vietnam and Europe.

The overall aspects of pulverized coal firing, as applied to boiler units, are as follows:

1. is suitable for almost any coal mined throughout the world,
2. is economically suitable for a very wide range of boiler capacities,
3. provides wide flexibility in operation and high thermal efficiency,
4. must have proper coal preparation and handling equipment, including moisture removal,
5. must have proper means of handling the ash refuse, and
6. must have controls for atmospheric emissions arising from elements in the coal and from the combustion process.

Babcock & Wilcox pulverized coal-fired boiler types for electric power

Prior chapters provide information on steam generation and power plant design and include fundamentals that are applied by designers and manufacturers worldwide. In particular, Chapter 19 discusses general boiler design, while Chapters 13 and 14 address the preparation and combustion of pulverized coal. This chapter concentrates on The Babcock & Wilcox Company (B&W) design philosophy which integrates these technologies into large reheat steam generators for electric power generation.

B&W has a broad base of experience that can be applied to most owner needs. B&W's basic design philosophy focuses on the key operating issues of availability and reliability, and on incorporating technology advances into each new unit design. Examples of the company's strengths include rapid startup and load shedding features, extended control ranges for superheater and reheater outlet temperatures, and experience in burning a wide variety of coals.

Supercritical (once-through) boilers

The B&W boiler for supercritical applications is the Universal Pressure (UP®) boiler. Originally designed for both supercritical and subcritical applications, this design is now used for supercritical applications; the drum boiler is used for subcritical applications. The supercritical application is usually applied to systems with a capacity of 300 MW or larger due to steam turbine considerations.

The original UP boiler design dates back to the mid-1950s when the 125 MW advanced supercritical boiler was provided for American Electric Power's Philo station. This pioneering boiler/turbine system, with advanced steam conditions of 4500 psi/1150F/1050F/1000F (31 MPa/621C/566C/538C), was the first supercritical system. The boiler design rapidly evolved to units as large as 1300 MW in the late 1960s. These boilers, nine of which are in operation, are among the largest capacity fossil fuel boilers in the world.

In modern units, the furnace is completely fluid-cooled, designed for balanced draft operation, and usually features dry ash removal. Superheater and reheater components are of the vertical pendant and/or horizontal design. Superheat temperature is controlled by the firing rate to feedwater flow ratio with typically one or more stages of attemperation for transient control. Reheat temperature control is by gas proportioning dampers.

Range in capacity, steam output – from 2,000,000 lb/h (252 kg/s) steam output to more than 10,000,000 lb/h (1260 kg/s).

Operating pressure – usually at 3500 psi (24.1 MPa) throttle pressure with 5% overpressure; higher pressures available.

Superheater steam temperatures – as required, usually 1050F (566C).

The principle of operation is that of the once-through or Benson cycle. The water, pumped into the unit as a subcooled liquid, passes sequentially through all the pressure part heating surfaces where it is converted to superheated steam as it absorbs heat; it leaves as steam at the desired temperature. There is no recirculation of water within the unit and, for this reason, a conventional drum is not required to separate water from steam.

The furnace is completely fluid-cooled and is usually designed for balanced draft operation. Heat transfer surface for single or two-stage reheat may be incorporated in the design. (See Chapter 2.)

Firing rate, feedwater flow, and turbine throttle valves are coordinated to control steam flow, pressure, and superheater steam temperature. Reheater steam temperature is controlled by gas proportioning dampers at the outlet of the steam and water heating surfaces.

The UP boiler is designed to maintain a minimum flow inside the furnace circuits to prevent furnace tube overheating during all operating conditions. This flow must be established before boiler startup. A startup system (boiler bypass), integral with the boiler, turbine, condensate and feedwater system, is provided. This system assures that the minimum design flow is maintained through pressure parts that are exposed to high temperature combustion gases during the startup operations and at other times when the required minimum flow exceeds the turbine steam demand. (See Chapter 19.)

Two types of UP boilers are available. The original design (UP) features the vertical tube furnace arrangement with high mass flux within the furnace tubes. This boiler is designed for load cycling and base load operation; fluid pressure in the furnace is at supercritical pressure at all loads. More recently, there has been demand for supercritical boilers capable of variable pressure operation and on-off cycling, as well as load cycling and base load operation. This boiler is the SWUP™, or spiral wound tube geometry UP boiler.

Spiral wound UP boiler – variable pressure operation with pulverized coal

The steam generating unit shown in Fig. 4 is a balanced draft coal-fired B&W Spiral Wound Universal Pressure (SWUP™) boiler, comprising a water-cooled dry-bottom furnace, superheater, reheater, economizer, and air heater components. The unit is designed to fire coal usually pulverized to a fineness of at least 70% through a 200 mesh (75 micron) screen. The B&W SWUP unit is designed for both base load and variable pressure load cycling operation as well as on-off cycling operation.

The unique feature of this boiler, compared to other boilers, is that the tubes in the furnace, from the lower furnace inlet headers to a location near the furnace arch, are wound around the furnace circumference rather than being vertical (see Fig. 5). With this arrangement, each tube in the furnace passes through similar heat absorbing areas so that the heat absorption from tube to tube is reasonably uniform. In addition, because the tubes are at an angle (typically 10 to 25 degrees from horizontal), the number of individual flow paths is reduced, compared to a vertical tube geometry. A high mass flux within the tube is obtained to maintain nucleate boiling during subcritical operation (Chapter 5).

Therefore, the water introduced from the economizer piping is heated at essentially the same rate to the same temperature, minimizing thermal upsets which restrain rapid load change. The capability of this design to operate at variable pressure is further enhanced by using a startup and bypass system specifically designed for rapid load change. (See Chapter 19.)

Fuel flow Raw coal is discharged from the feeders to the pulverizers. Pulverized coal is transported by the primary air to the burners through a system of pressurized fuel and air piping. The burners are located on the furnace walls with opposed firing (burners on the front and rear walls).

Air and gas flow Air from the forced draft fans is heated in the air heaters, then routed to the windbox where it is distributed to the burners as secondary air. In the arrangement shown in Fig. 4, high pressure fans provide air from the atmosphere to a separate section of the air heater known as the primary section. A portion of the air from the primary fans is passed unheated around the primary air heater as tempering primary air. Controlled quantities of preheated and tempering primary air are mixed before entering each pulverizer to obtain the desired pulverizer fuel-air mixture outlet temperature. The primary air is used for drying and transporting fuel from the pulverizer through the burners to the furnace.

Hot flue gas from the furnace passes successively across the finishing banks of the superheater and reheater. Before exiting the boiler, the gas stream is divided into two parallel paths, one gas stream passing over a portion of the reheater and the other stream passing over a portion of the superheater. Proportioning the gas flow between these two paths as unit load changes provides a tool for reheat steam temperature control. The flow quantities are adjusted by a set of dampers at the boiler exit. A controlled amount of gas passes over a portion of the reheater to obtain the reheater steam temperature set point, and the remaining gas travels a parallel path across the superheater surface. Attemperators are provided in the reheater and superheater systems. The reheater attemperator is used during transient loads and upset conditions, with reheater spray quantities held to a minimum to maximize cycle efficiency. The superheat attemperator spray is used to help control main steam temperatures during transient operation. This arrangement of convection surface and damper system provides extended capability to obtain design steam temperatures in the superheater and reheater over a broad load range.

The gases leaving the superheater and reheater sections of the convection pass cross the economizer, pass to the air heater(s) and then travel to the appropriate environmental control equipment. In areas where tight NO_x emissions requirements have been established and post-combustion control is needed, selective catalytic reduction (SCR) NO_x removal systems are frequently installed between the economizer and air heater (see Chapter 34).

Water and steam flow Feedwater (Fig. 6) enters the bottom header of the economizer and passes upward through the economizer tube bank into support tubing located between tube rows of the primary superheater. The heated feedwater is collected in outlet headers at the top of the unit. It is then piped to the lower furnace area from which multiple connecting pipes (supplies) are routed to the lower furnace headers.

From the lower furnace wall headers the fluid passes upward through the spiral furnace tubes to a transition section located below the furnace arch. From the transition section the tubes are routed vertically up the front and side walls and up the rear wall and furnace arch.

After discharging to the upper furnace wall headers, all of the fluid is piped to a fluid mix bottle, then to the front roof header, then through the roof tubes to the rear roof header. Pipes then convey the fluid to the vertical steam separators. The vertical steam separators are a part of the boiler startup system (see Chapter 19).

As illustrated in Fig. 7, steam from the vertical

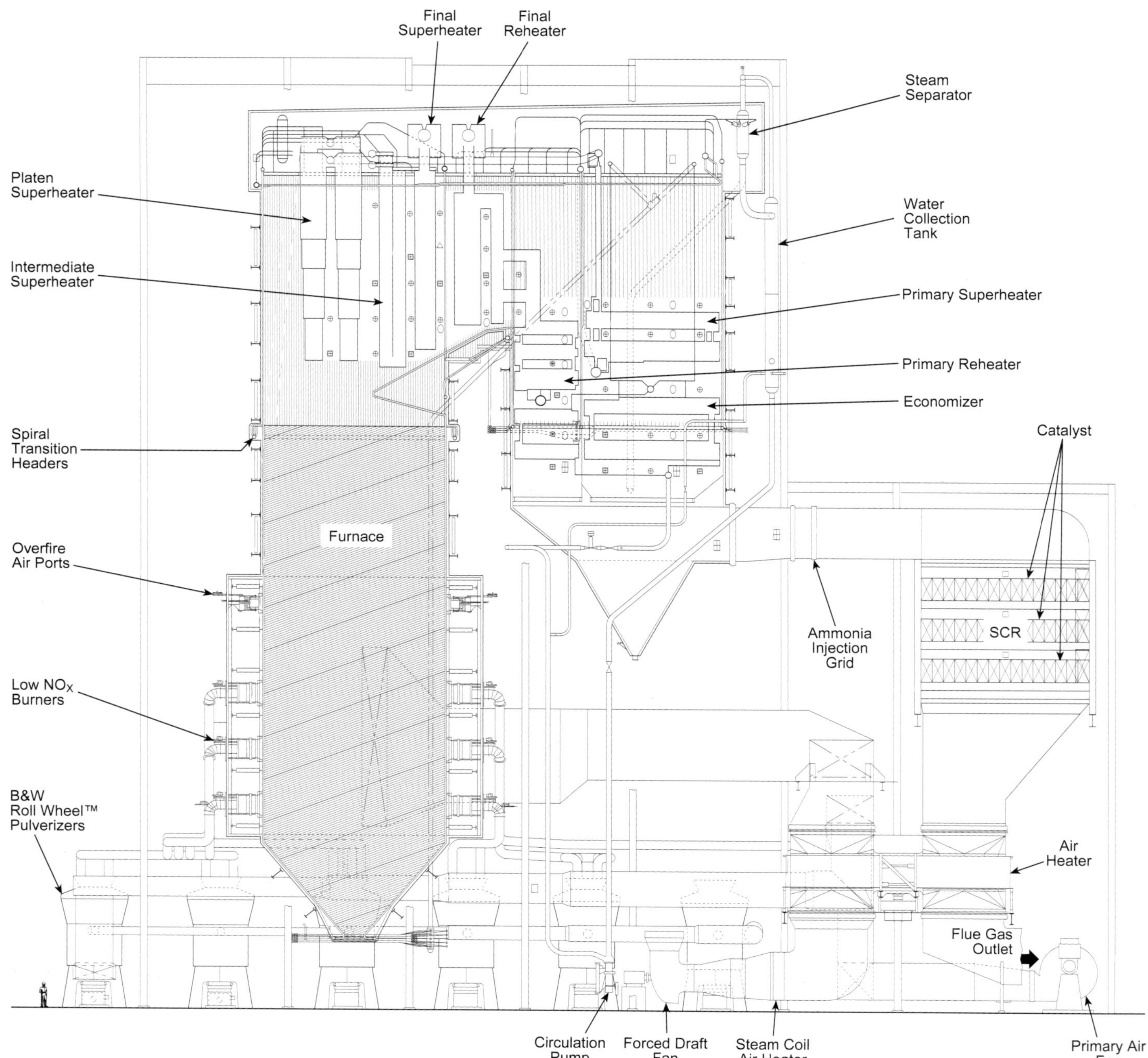

Fig. 4 750 MW once-through spiral wound universal pressure (SWUP™) boiler for pulverized coal firing.

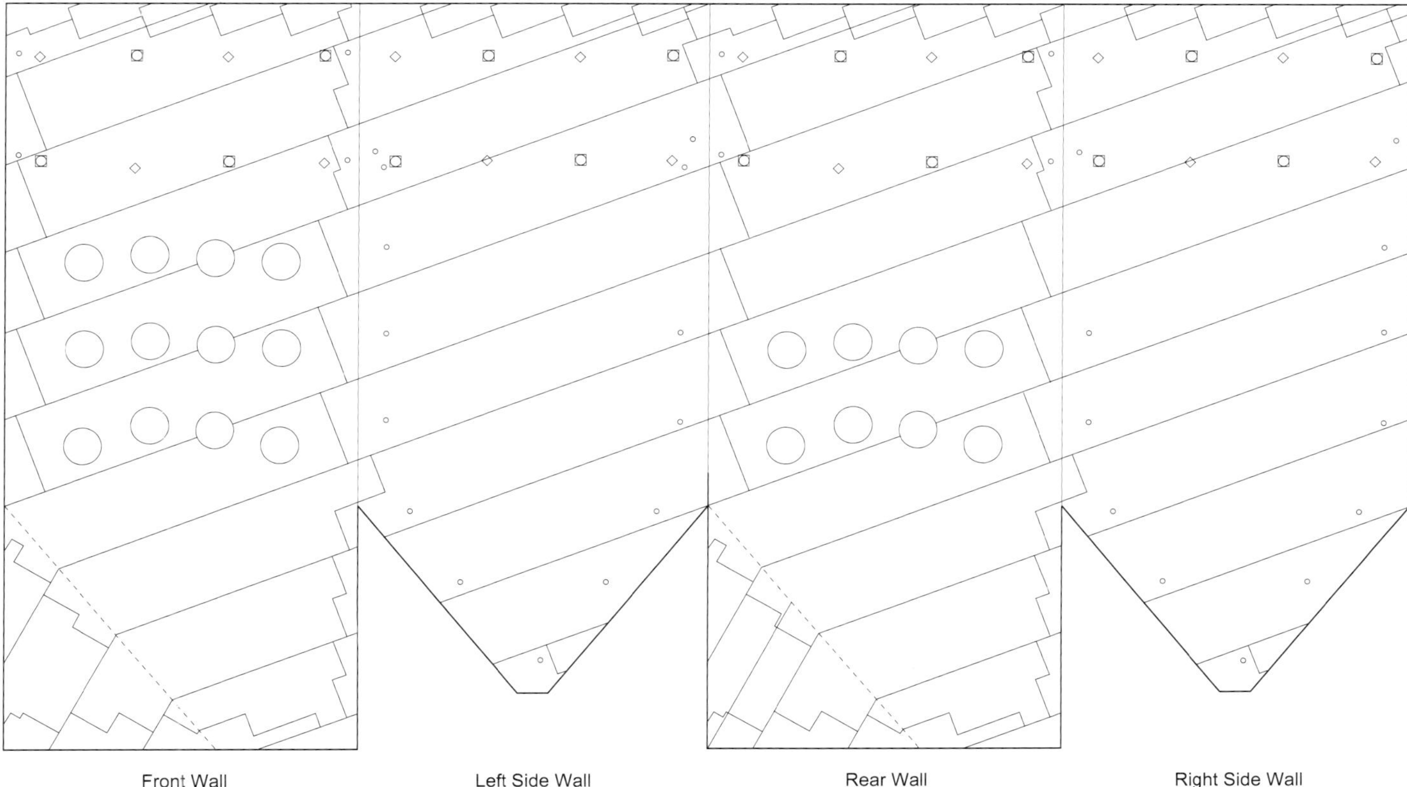

Fig. 5 Typical spiral wound tube arrangement for a 420 MW coal-fired boiler.

steam separator passes to the convection pass enclosure wall lower headers. The fluid flows up through the wall tubes and the baffle wall and is collected into the downcomer supplying the primary superheater inlet header. The steam rises through the primary superheater, discharges to its outlet header and flows through connecting piping equipped with a spray attemperator. The partially superheated steam then enters the platen secondary superheater and flows through the various superheater sections to its outlet headers. The steam flows through connecting pipes and the second set of spray attemperators to the finishing portion of the secondary superheater and finally to the outlet header and discharge pipes, which terminate at points outside of the unit penthouse. The superheated steam is directed to the high pressure section of the steam turbine. After partial expansion in the steam turbine (see Chapter 2), the low pressure steam is returned to the boiler for reheating.

The low pressure steam is reintroduced to the boiler at the reheater inlet header (RHSH inlet) and flows through the reheater tube bank to the reheater outlet header (RHSH outlet). Reheated steam is then routed to the intermediate pressure and then low pressure sections of the steam turbine-generator set.

Universal Pressure boiler for load cycling operation with pulverized coal

The steam generating unit shown in Fig. 8 is a balanced draft B&W Universal Pressure coal-fired Carolina-type boiler (UPC), comprising a water-cooled dry-bottom furnace and superheater, reheater, economizer and air heater components. The unit is designed to fire coal, usually pulverized to a fineness of at least 70% through a 200 mesh (75 micron) screen. The B&W UPC unit is particularly suited for base load duty and constant boiler pressure, load cycling operation.

Fuel flow Raw coal is discharged from the feeders to the pulverizers, which can be located at the front or sides of the unit. Pulverized coal is transported by the primary air to the burners through a system of pressurized fuel and air piping.

Air and gas flow The air flow arrangement and routing are similar to that used on the SWUP design previously discussed.

As indicated in Fig. 8, hot flue gases leaving the furnace pass successively across the fluid-cooled surface at the top of the furnace (wing walls), the secondary superheater and the pendant reheater, which are located in the convection pass out of the high radiant heat transfer zone of the furnace. The gas turns downward (convection pass) and crosses the horizontal primary superheater, horizontal reheater and economizer before passing to the air heaters. In areas where tight NO_x emissions requirements have been established and post-combustion control is needed, SCR NO_x control systems are frequently installed between the economizer and the air heater (see Chapter 34).

Water and steam flow Feedwater enters the bottom header of the economizer and passes upward through the economizer to the outlet header. It is then piped to the lower furnace area from which multiple connecting pipes (*supplies*) are routed to the lower furnace headers.

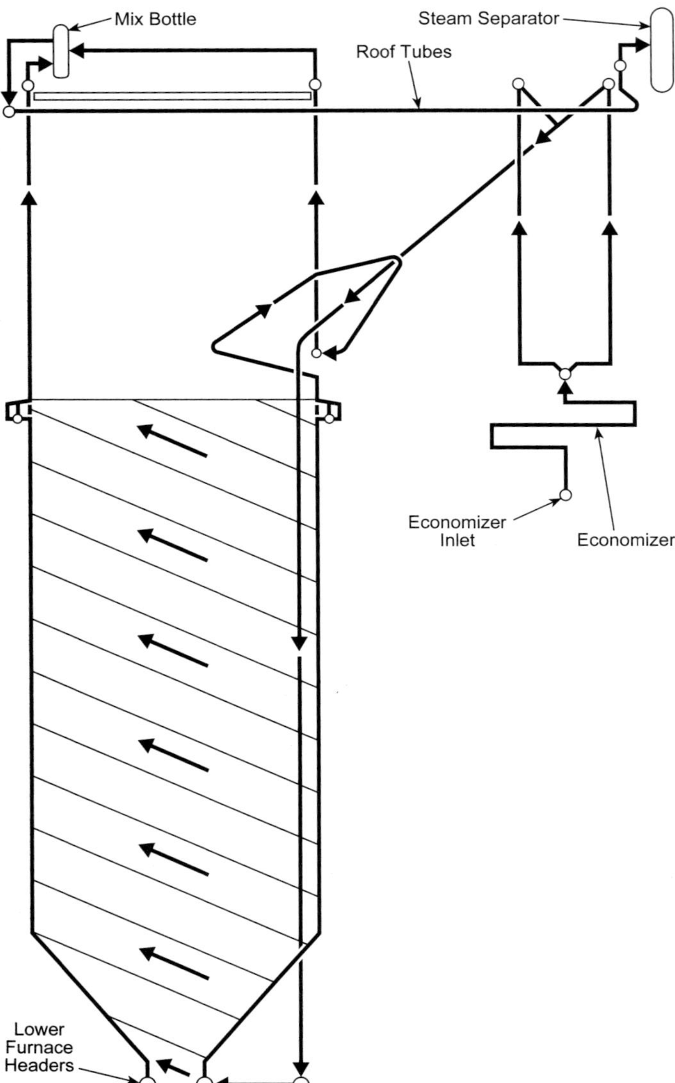

Fig. 6 Typical fluid flow for a SWUP™ boiler.

From the furnace wall headers the fluid is then passed upward through the first pass of vertical furnace tubes (Fig. 8). A transition section (Fig. 9) is provided some distance below the furnace arch and the first pass tubes exit the furnace enclosure and connect to headers at this location. The fluid from the various headers around the furnace circumference is mixed together in common piping. This equalizes the temperature of the fluid before it passes down to the second set of lower furnace inlet headers and flows upward through the second pass of vertical furnace tubes. The second pass tubes exit the furnace enclosure at the transition section and connect to another set of headers. The first and second pass are alternated around the furnace circumference to maintain uniform wall temperatures.

From the second pass outlet headers, the fluid flows through mix piping and on to the third pass inlet headers. The third pass furnace tubes form the furnace enclosure from the transition section to the furnace roof. In addition, furnace wing walls may be in a parallel flow circuit with the third pass tubes. This completes the furnace tube pass and the fluid exits to the upper furnace wall headers.

After discharging to the upper furnace wall headers, the fluid is piped to the front roof header, then through the roof tubes to the rear roof headers where mixing again takes place. It is then passed through a pipe distribution system to the convection pass enclosure wall lower headers. The fluid flows up through the wall tubes and the superheater screen. Pipes then convey the fluid to a common header and then to the primary superheater inlet header.

The fluid is collected and partially mixed before entering the primary superheater, then partially mixed again as it flows from the primary superheater through connecting piping to the secondary superheater. The furnace pressure control valves and connections to the boiler startup system (Chapter 19) and the spray attemperator are contained in this connecting piping. The steam flows through connecting piping to the secondary superheater and finally to the outlet header and discharge pipes, which terminate at points outside of the unit penthouse. The superheated steam is directed to the high pressure section of the steam turbine. After partial expansion in the steam turbine (see Chapter 2), the low pressure steam is returned to the boiler for reheating.

Advanced supercritical boiler designs

Vertical tube variable pressure boiler While the spiral wound furnace (SWUP) meets today's market needs for variable pressure operation, B&W is continuing the research and development of advanced supercritical boiler designs.

The ideal furnace design would have vertical tubes and be capable of variable pressure operation over the load range while exhibiting natural circulation flow characteristics (flow increasing as heat absorption increases), thus protecting the tubes from overheating. The high mass fluxes required by the ribbed tubes used

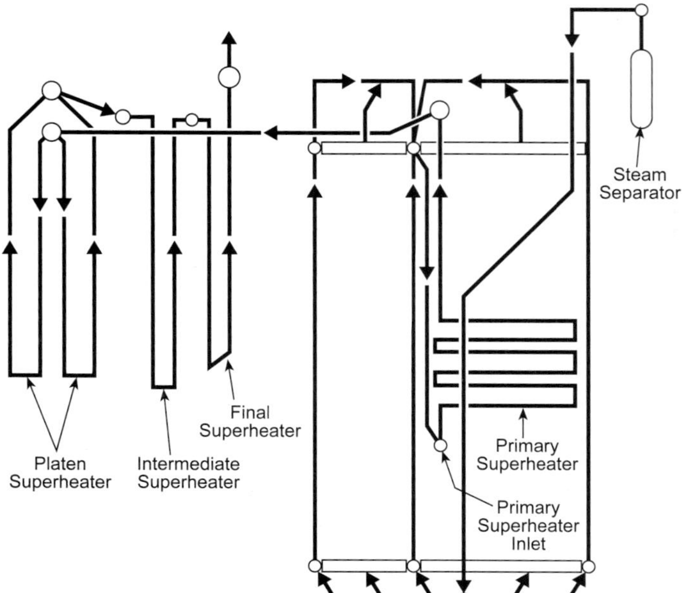

Fig. 7 Typical steam flow for a SWUP™ boiler.

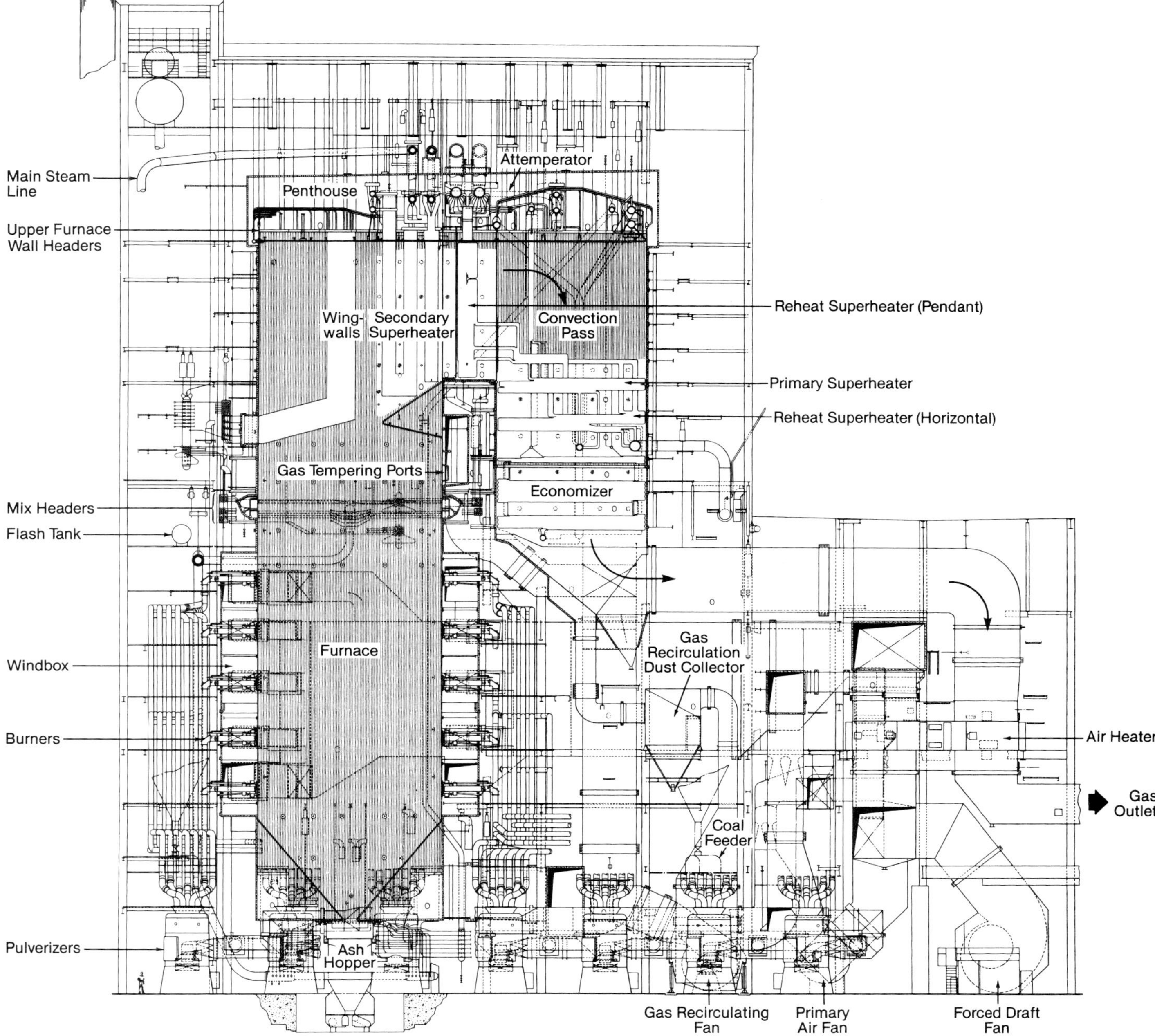

Fig. 8 1300 MW Universal Pressure (UPC) boiler for pulverized coal firing.

in the current UP boilers do not provide this characteristic. Instead, a ribbed tube design (Fig. 10) capable of preserving the desirable departure from nucleate boiling (DNB) characteristics (Chapter 5) at low mass fluxes in the high heat flux zones is necessary. The low mass flux is required to achieve the low dynamic pressure loss necessary to achieve the self-compensating, natural circulation characteristic. A secondary benefit is a reduction in pressure loss through the boiler so that the feed pump power is reduced and cycle efficiency is slightly increased. Research continues in ribbed tube development and the application of the advanced ribbed tube designs to the boiler. (See Chapter 19.)

Higher temperature and pressure The supercritical steam cycle provides an improvement in heat rate (efficiency) as compared to the subcritical steam cycle. Additional gains in efficiency are possible as the cycle temperature and pressure are further increased. Research continues in tube materials to develop higher efficiency boiler and turbine systems.

Subcritical (drum) boilers

The B&W boiler for subcritical applications is the Radiant boiler (RB), so named because the steam generation is by radiant heat transfer to the furnace enclosure tubes. Its components are pre-engineered with sufficient flexibility to adapt the design to various fuels and a broad range of steam conditions.

In modern units, the furnace is a natural circulation, water-cooled, balanced draft design and usually features dry ash removal. Superheater and reheater surfaces are of the vertical pendant and/or horizon-

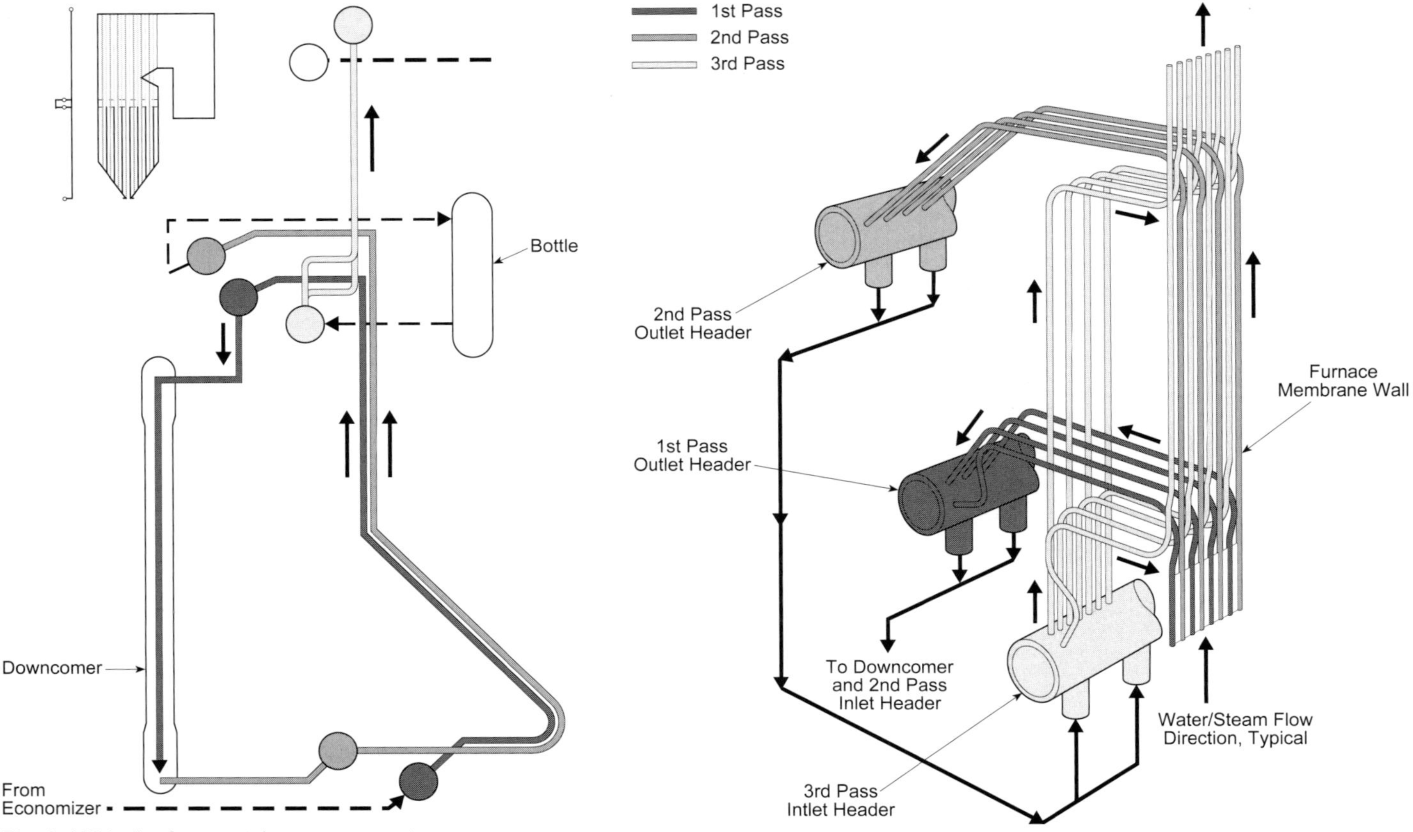

Fig. 9 UP boiler furnace tube arrangement.

tal designs. Superheat temperature control is accomplished by an attemperator. Reheat temperature control is by gas proportioning dampers.

Range in capacity, steam output – about 700,000 lb/h (90 kg/s) to a maximum that may exceed 7,000,000 lb/h (880 kg/s).

Pressure – subcritical, usually 1800 to 2400 psi (12.4 to 16.5 MPa) throttle pressure with 5% overpressure capability.

Superheater and reheater outlet temperatures – as required, usually in the range of 1000 to 1050F (538 to 566C).

Radiant boiler for pulverized coal – Carolina

The steam generating unit shown in Fig. 11 is a balanced-draft Carolina-type Radiant boiler (RBC). It is arranged with a natural circulation, water-cooled dry-bottom furnace, and superheater, reheater, economizer and air heater components. The unit is designed to burn coal, usually pulverized to a fineness of which at least 70% passes through a 200 mesh (75 micron) screen. The horizontal convection pass and vertical pendant heat transfer surfaces provide the benefits of: 1) high temperature supports outside of the flue gas stream, 2) minimal motion between the boiler roof penetrations, and 3) more control over pendant section spacing in the design. The lower furnace height compared to other pulverized coal designs also reduces structural steel and erection costs.

Fuel flow Raw coal is discharged from the feeders to the pulverizers. The pulverized coal is transported by the primary air to the burners through a system of pressurized fuel and air piping. The burners are located on the furnace walls with opposed firing (burners on the front and rear walls) being used for all but the smallest units.

Air and gas flow Air from the forced draft fans is heated in the air heaters, then routed to the windbox where it is distributed to the burners as secondary air. In the arrangement shown in Fig. 11, high pressure fans provide air from the atmosphere to a separate section of the air heater known as the primary section. A portion of the air from the primary fans is passed unheated around the primary air heater as tempering primary air. Controlled quantities of preheated and tempering primary air are mixed before entering each pulverizer to obtain the desired pulverizer fuel-air mixture outlet temperature. The primary air is used for drying and transporting fuel from the pulverizer through the burners to the furnace.

Hot gases from the furnace pass successively across the finishing banks of the superheater and reheater. Before exiting the boiler, the flue gas stream is divided into two parallel paths – one gas stream passing over a portion of the reheater and the other stream passing over a portion of the superheater. Proportioning the gas flow between these two paths as unit load changes provides a tool for reheat steam temperature control. The flow quantities are adjusted by a set of dampers at the boiler exit. A controlled amount of gas passes over a portion of the reheater to obtain the reheater steam temperature set point and the remain-

Fig. 10a Single-lead ribbed tube.

Fig. 10b Multi-lead ribbed tube.

Fig. 10c Optimized multi-lead ribbed tube.

ing gas travels a parallel path across the superheater surface. Attemperators are provided in the reheater and superheater systems. The reheater attemperator is used during transient loads and upset conditions, with reheater spray quantities held to a minimum to maximize cycle efficiency. Superheater spray is used to maintain main steam temperatures. This arrangement of convection surface and damper system provides extended capability to obtain design steam temperatures in the superheater and reheater over a broad load range.

The flue gases leaving the superheater and reheater sections of the convection pass cross the economizer. In new or recently retrofitted units, the gas leaves the economizer and passes through an SCR system to reduce NO_x emissions before the gas travels to the air heater. After the air heater, the gas passes to the appropriate environmental control equipment.

Water and steam flow Feedwater enters the bottom header of the economizer. The water passes upward through the economizer tube bank into support tubing that is located between tube rows of the primary superheater. The heated feedwater is collected in outlet headers at the top of the unit, then routed into the steam drum. As shown in Fig. 12, natural circulation provides water flow from the steam drum down through large diameter downcomer pipes to multiple supply distributor tubes or *supplies* connecting to the individual lower furnace headers. The fluid rises as it is heated in the furnace tubes and passes through riser tubes back into the steam drum.

The mixture entering the steam drum is separated into steam and water flows by cyclone separators. These provide essentially steam-free water to the downcomer inlets. The steam is further purified by passing it through the primary and secondary steam scrubbers as discussed in Chapter 5.

As illustrated in Fig. 13, steam from the steam drum passes through multiple connections to a header supplying the furnace roof and convection pass rear

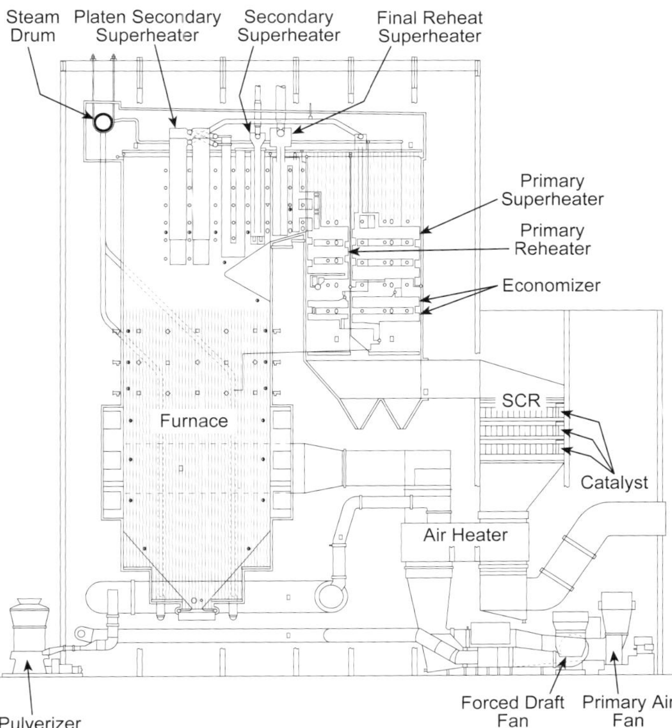

Fig. 11 Carolina-type Radiant boiler (RBC) for pulverized coal firing.

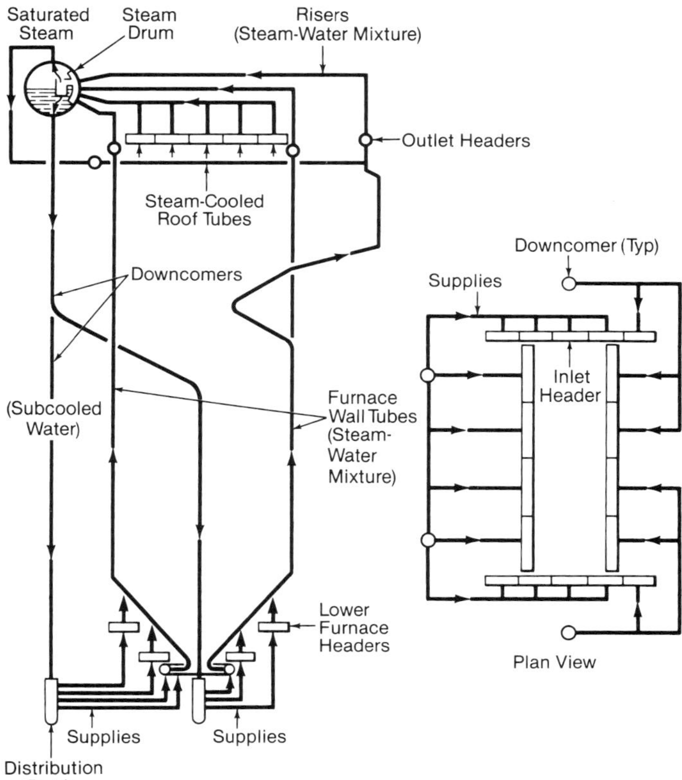

Fig. 12 Typical furnace circulation diagram for a Radiant boiler (RB).

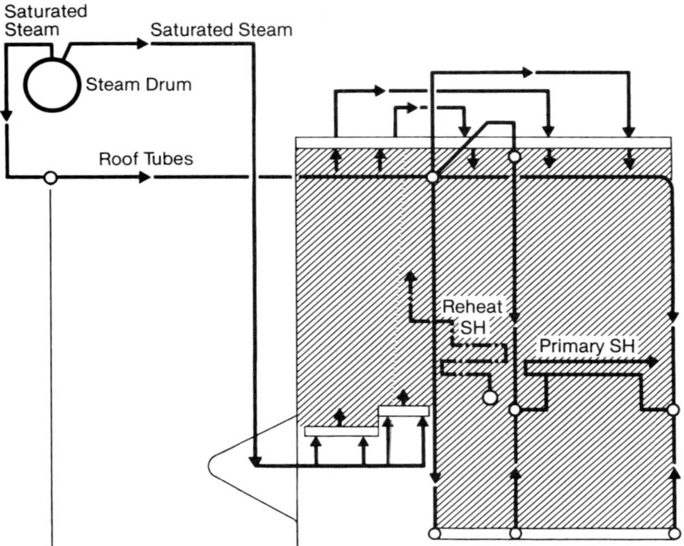

Fig. 13 Typical steam-cooled roof and convection pass enclosure wall circuits for a Radiant boiler (RB).

wall tubes, which directly connect to the primary superheater inlet header. Steam leaving the drum is also routed to a series of headers, which provide cooling steam for the convection pass side walls, before being directed to the primary superheater inlet header.

As shown in Fig. 14, the steam rises through the primary superheater, discharges to its outlet header and flows through connecting piping equipped with a spray attemperator. The partially superheated steam then enters the secondary superheater and flows through the various superheater sections to its outlet headers and discharge pipes, which terminate at points outside of the unit penthouse. The superheated steam is directed to the high pressure section of the steam turbine. After partial expansion in the steam turbine (see Chapter 2), the low pressure steam is returned to the boiler for reheating.

The low pressure steam is re-introduced to the boiler at the reheater inlet header (RHSH inlet) and flows through the reheater tube bank to the reheater outlet header (RHSH outlet). Reheated steam is then routed to the intermediate pressure and then low pressure sections of the steam turbine-generator set.

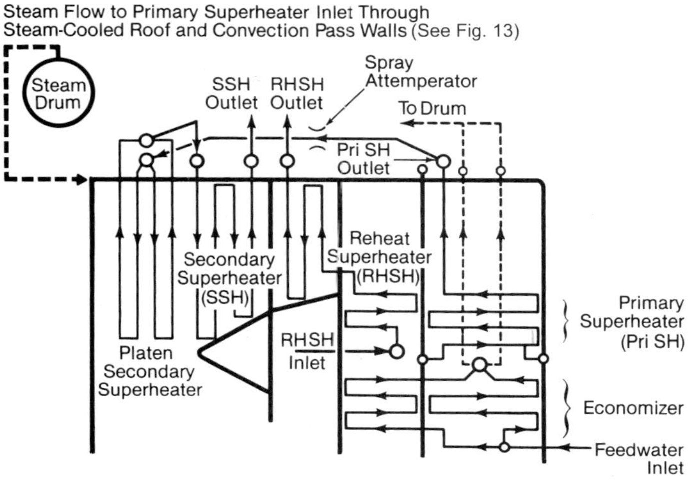

Steam Flow to Primary Superheater Inlet Through Steam-Cooled Roof and Convection Pass Walls (See Fig. 13)

Fig. 14 Superheater, reheater and economizer circuits for a Radiant boiler (RB).

Radiant boiler for pulverized coal – Tower

An alternate configuration of the Radiant boiler is the Tower design (RBT) shown in Figs. 15 and 16. A special feature of this design is that all surfaces are drainable. This feature was initially designed for service in northern latitudes to provide freeze protection by arranging the superheater, reheater and economizer sections so that they are drainable. The Tower design is a single gas path, as compared to the parallel gas path of the RBC, and therefore, has no inherent provision for steam temperature control other than spray attemperation. The Tower design may be equipped with gas recirculation and tempering for extended superheater and reheater temperature control range. Additional advantages of the Tower design include more uniform gas flow entering all convection sections, minimum tube bend erosion potential because of no convection pass turns, and removal of most of the ash and slag through the main furnace hopper. However, the Tower design requires a much taller structure and generally results in a longer erection schedule due to the single erection path as compared to the RBC design.

General characteristics of the design are water-cooled furnace, balanced draft, natural circulation, top supported and drainable. The draining feature is accomplished by placing all steam- and water-contained

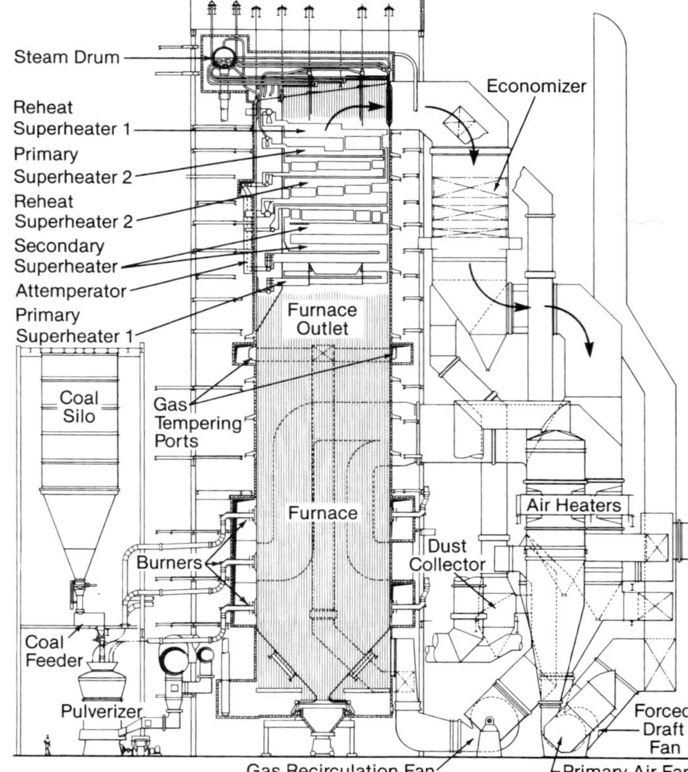

Fig. 15 400 MW Tower-type Radiant boiler (RBT) for pulverized coal firing.

Fig. 16 Four 400 MW RBT boilers and three 600 MW RBC boilers in Southeast Asia.

surfaces in the horizontal position so that they may be completely drained when the unit is taken out of service.

Fuel flow Similar to the RBC, the RBT is arranged so that raw coal is discharged from the feeders to the pulverizers. Primary air is used to transport the pulverized coal from the pulverizers to the burners through a system of pressurized fuel and air piping.

Air and gas flow Air is introduced to the RBT in the same pattern as the RBC. The geometric differences of the two units create varying paths for gas flow. The hot gases from the furnace pass vertically up and over the superheater and reheater banks located in the upper section of the boiler. The flue gas is then directed down to the economizer located outside of the boiler setting, finally passing to the SCR (if required) and air heater system, and then to the environmental control equipment for cleanup.

Water and steam flow Feedwater is introduced to the boiler system through the economizer where it is heated and discharged through a header/piping system to the steam drum. With this natural circulation system, the furnace circuitry follows the same general flow patterns as the RBC described above.

Saturated steam from the steam drum is directed through a series of pipes down to the outlet of the furnace, where a small bank of primary superheater tubes (primary superheater 1) is placed with its outlet tubing forming the internal steam-cooled support structure for the other banks of surface located farther up in the gas stream.

Steam from the support tubing is collected and directed to the finishing banks of primary superheater 2. The primary and secondary superheaters are connected with piping that contains a spray attemperator for superheater steam temperature control. Steam enters the secondary superheater and flows through the tube sections to the outlet header, which discharges to a terminal located outside of the unit. The superheated steam is directed to the high pressure section of the steam turbine. After partial expansion, low pressure steam is then returned to the boiler for reheating.

Steam returned to the boiler is introduced at the reheater inlet header. The steam flows through the reheater tube bank and is then routed to the outlet header and steam piping connection before being sent to the intermediate and low pressure sections of the steam turbine-generator.

Radiant boiler for anthracite – Downshot

The use of anthracite is limited to a relatively few locations in the world, mainly Spain and parts of Asia. Anthracite presents a special challenge for combustion due to its very low volatile matter content. As a result, a special furnace configuration is used to improve combustion. The furnace is arranged so that the burners fire downward into the lower furnace. The flame shape in this opposed-firing arrangement is in the form of a W, hence the name of the boiler style, the RBW. A typical steam generating unit is shown in Fig. 17.

Except for the lower furnace and the combustion system, this design is very similar to the RBC. The downshot firing arrangement provides increased residence time in the lower furnace. The lower furnace is refractory lined to minimize heat loss and to maintain peak flame temperature for high combustion efficiency. In addition, the anthracite fineness is increased, compared to normal coal, to 90% passing through a 200 mesh (75 micron) screen, to further enhance combustion efficiency.

The boiler is arranged with a water-cooled dry-bottom furnace, and superheater, reheater, economizer

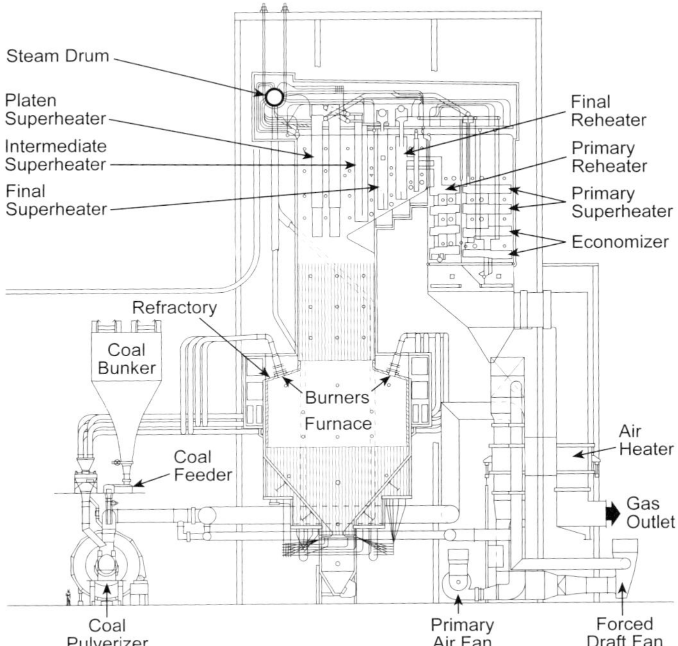

Fig. 17 300 MW RBW boiler for pulverized anthracite (downshot).

and air heater components. The horizontal convection pass and vertical pendant heat transfer surfaces provide the benefits of: 1) high temperature supports outside of the flue gas stream, 2) minimal motion between the boiler roof penetrations, and 3) more control over pendant section spacing in the design. The lower furnace height compared to other pulverized coal designs also reduces structural steel and erection costs.

Fuel flow Raw coal is discharged from the feeders to the pulverizers. The pulverized coal is transported by the primary air to the burners through a system of pressurized fuel and air piping. The burners are located on the shelves of the furnace front and rear wall so that they fire downward giving a flame pattern in the shape of a W.

Air and gas flow Air from the forced draft fans is heated in the air heaters, then routed to the windbox where it is distributed to the burners as secondary air. In addition, a portion of the air is introduced to the lower portion of the lower furnace as staging air. The staging air both increases the peak flame temperature near the burners by decreasing the amount of excess air and helps to draw the flame path down into the furnace to increase residence time.

In the arrangement shown in Fig. 17, high pressure fans provide air from the atmosphere to a separate section of the air heater known as the primary section. A portion of the air from the primary fans is passed unheated around the primary air heater as tempering primary air. The primary air is used for drying and transporting fuel from the pulverizer through the burners to the furnace. Generally, it is desirable to achieve the maximum pulverizer fuel-air mixture temperature possible. However, provisions are included for mixing the preheated and tempering primary air before entering each pulverizer to provide control of pulverizer fuel-air mixture temperature during pulverizer start up and shut down. In addition, special provisions may be included for further increasing the temperature of the fuel/air mixture to further enhance combustion.

Hot gases from the furnace pass successively across the finishing banks of the superheater and reheater in the same manner as described for the RBC boiler.

Water and steam flow The water and steam flow arrangement is the same as for the RBC boiler discussed earlier.

Oil- and gas-fired utility boilers

The use of oil and gas as fuels for new utility boilers has declined except for selected locations and conditions, primarily the oil producing nations. This is due in part to the price of these premium fuels and their availability. At the same time, advancements in gas turbine and gas turbine combined cycle systems (Chapter 27) have made the use of oil and gas in these systems more cost effective.

However, in selected cases, the proposed boiler location, fuel availability, local price or existing facility constraints may make oil or gas the preferred fuel. In addition, low sulfur oil- and natural gas-fired new utility boilers have the distinct advantage of low environmental emissions; NO_x, SO_2 and particulate emissions are less in these units.

Both the SWUP (and UP) and the RBC boiler designs can be used for oil and gas firing. However, for subcritical applications a design different than the RBC is normally used.

Fuels

Natural gas Natural gas (see Chapter 11) has the fewest design restrictions of the major fuels because it is relatively clean and easy to burn. If only natural gas is burned, fuel storage facilities, ash hoppers, ash pits and ash handling equipment are unnecessary. Sootblowers can be omitted and dust collectors are not needed. Control of heat input to the boiler furnace is simplified. Heating surfaces can be arranged for optimum heat transfer and gas resistance (draft loss) without consideration of ash deposits and erosion. The total enclosure volume is at a minimum and the adaptability for outdoor service is increased.

Fuel oil Fuel oil (see Chapter 11) has many of the desirable features of natural gas, including ease of handling and elimination of ash hoppers and ash pits. However, it requires storage, heating and pumping facilities. Oils with high sulfur and vanadium contents can cause troublesome deposits on surfaces throughout the unit. (See Chapter 21.) These deposits can be minimized by arranging the heating surfaces for optimum cleaning by sootblowing equipment. Provision should be made for water washing the furnace and all convection surfaces when the unit is shut down for maintenance. Air heater protection devices (a steam coil or hot water coil) should be installed to prevent gas condensation and acid attack on the gas side of the air heater surfaces (see Chapter 20) and in the equipment downstream of the air heater.

B&W Radiant boiler for natural gas and oil

The steam generator shown in Figs. 18 and 19 is an El Paso-type Radiant boiler (RBE) unit. It is arranged with a water-cooled hopper-bottom furnace and superheater, reheater, economizer and air heater components. The unit is designed to use natural gas and oil separately or in combination. The general characteristics of the Radiant boiler are discussed in *Pulverized coal firing*. The unique aspects of the El Paso design are the elimination of the pendant convection pass and the inclusion of the upflow and downflow convection passes within the footprint occupied by the boiler furnace. The flow rates and pressure ranges for the coal-fired Radiant boiler are also applicable here.

Air and gas flow Air from the forced draft fan is heated in the air heater and is distributed to the burner windbox. Hot gases from the furnace pass successively over the horizontal secondary superheater and reheater sections and one bank of the primary superheater. The gas then turns, flows downward and crosses the remainder of the primary superheater and economizer sections. It then travels out of the water-cooled enclosure to the air heater. In new or recently retrofitted units, the gas leaves the economizer and passes through an SCR system to reduce NO_x emissions before the gas travels to the air heater. After the

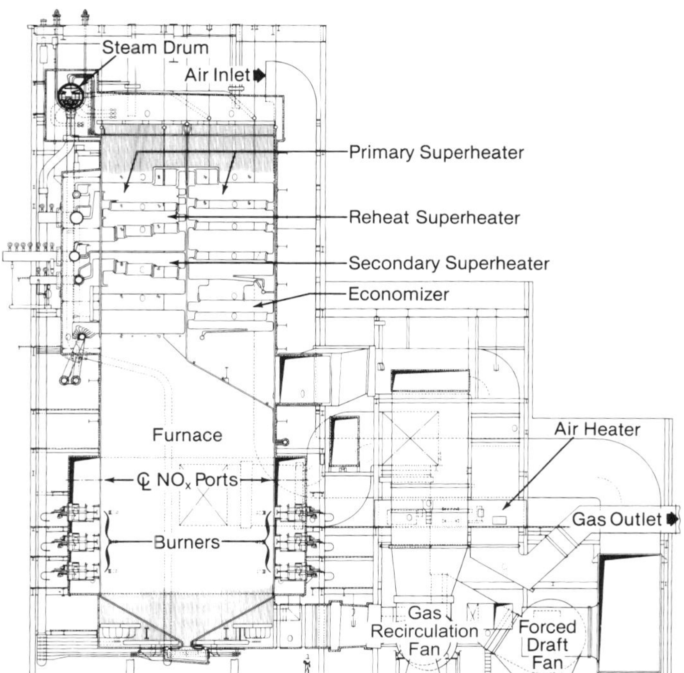

Fig. 18 El Paso-type Radiant (RBE) boiler for oil and gas firing.

Fig. 19 Two 300 MW RBE boilers in Egypt.

air heater, the gas passes to the appropriate environmental control equipment.

Water and steam flow Feedwater enters the bottom header of the economizer. The water flows upward through the economizer and discharges through the outlet header into piping that directs it to the steam drum. By means of natural circulation, water flows downward through downcomer pipes, then through supply distributor tubes to the lower furnace headers. It then rises through the furnace tubes (which enclose the convection area) and flows to the upper headers and through connecting riser tubes into the steam drum.

The steam-water mixture in the steam drum is passed through cyclone separators, which provide essentially steam-free water for the downcomers. The steam is further purified by passing it through the primary and secondary steam scrubbers within the drum.

Steam from the drum passes through multiple connections at the rear of the convection enclosure into the primary superheater inlet header.

The steam rises through part of the primary superheater and, by means of connecting tubes, is directed into the remainder of the primary superheater in the first gas pass. The flow discharges through the primary superheater outlet header into connecting piping equipped with a spray attemperator. The steam enters the secondary superheater inlet header and flows through the secondary superheater sections to the outlet header and to a discharge pipe, which terminates outside the casing at the front of the unit.

Low pressure steam, introduced to the reheater inlet header, flows through the reheater sections and out through the reheater outlet header to a connection terminating outside the casing, also at the front of the unit.

Type PFI Integral Furnace boiler installed in an industrial plant.

Chapter 27

Boilers for Industry and Small Power

Most manufacturing industries require steam for a variety of uses. Basic plant heating and air conditioning, prime movers such as turbine drives for blowers and compressors, drying, constant temperature reaction processes, large presses, soaking pits, water heating, cooking and cleaning are all examples of how steam is used.

Steam produced by industrial boilers can also be used to generate electricity in a cogeneration mode which uses a conventional steam turbine for electric power generation and low pressure extraction steam for the process. The electricity is then used by the plant or sold to a local electric utility company. As an alternate cogenerating system, a gas turbine can be used for power generation with a heat recovery steam generator for steam.

Thousands of boilers are installed in industrial and municipal plants, providing lower pressure and temperature steam than utility boilers dedicated to large, central station electric power generation. In an industrial plant, the dependability of steam generating equipment is critical. Most often, the industrial operation has a single steam plant with one or more boilers. If the steam flow is interrupted, production can be seriously impacted. Accordingly, industrial boilers must be very reliable because plant productivity relies so heavily on their availability. Loss of a boiler for a short time can stop production for days if, for example, materials cool and solidify in process lines. For this reason, some industries prefer multiple smaller units.

The principles governing the selection of boilers and related equipment are discussed in Chapter 37. Proper equipment selection can be accomplished only in the framework of a sound technical and cost evaluation. This requires a working knowledge and understanding of the performance of the different steam generating unit components under various conditions, including the significance of the many different arrangements of heat absorbing surfaces, the characteristics of available fuels, combustion methods and ash handling. The owner must also establish the present and future steam conditions and requirements. All pertinent environmental regulations must also be considered. A brief summary of boiler specifications is provided in Table 1.

Industrial boiler design

Industrial boilers generally have different performance characteristics than utility boilers. These are most apparent in steam pressures and temperatures as well as the fuel burning equipment.

Industrial units are built in a wide range of sizes, pressures and temperatures – from 2 psig (13.8 kPa) and 218F (103C) saturated steam for heating to 1800 psig (12.4 MPa) and 1000F (538C) steam for plant power production.

In addition, industrial units often supply steam for more than one application. For some applications, steam demand may be cyclic or fluctuating, thereby complicating unit operation and control of the equipment.

The Babcock & Wilcox Company (B&W) industrial boilers, such as the unit shown in Fig. 1, are water tube design and generally rely upon natural circulation for steam-water circulation.

Most utility boilers are designed to burn pulverized or crushed coal, oil, gas, or a combination of oil or gas with a solid fuel. Industrial boilers can be designed for the above fuels as well as coarsely crushed coal for stoker firing and a wide range of biomass or byproduct fuels.

Table 1
Typical Industrial Boiler Specification Factors

1. Steam pressure
2. Steam temperature and control range
3. Steam flow
   - Peak
   - Minimum
   - Load patterns
4. Feedwater temperature and quality
5. Standby capacity and number of units
6. Fuels and their properties
7. Ash properties
8. Firing method preferences
9. Environmental emission limitations — sulfur dioxide (SO_2), nitrogen oxides (NO_x), particulate, other compounds
10. Site space and access limitations
11. Auxiliaries
12. Operator requirements
13. Evaluation basis

Many industrial processes generate byproducts which can serve as boiler fuels, significantly contributing to the plant operating efficiency and effectively reducing product cost. Examples of these are gas products from the steel industry (blast furnace and coke oven gas), products from the petroleum industry (carbon monoxide (CO), refinery gas, petroleum coke), products from agriculture (sugar mill bagasse, peanut hulls, coffee grounds), waste from the pulp and paper industry (wood, bark, process chemicals, sludge) and municipal solid waste. Steam generation and fuel handling for some of these fuels and applications have become quite specialized. This chapter and the following chapters are devoted to these units:

Chapter 28 – Chemical and Heat Recovery in the Paper Industry
Chapter 29 – Waste-to-Energy Installations
Chapter 30 – Wood and Biomass Installations
Chapter 31 – Marine Applications

One of the distinguishable features of most industrial boilers is a large saturated water *boiler bank* between the steam drum and lower drum. (See Figs. 1 and 2.) The boiler bank serves the purpose of preheating the inlet feedwater to the saturation temperature and then evaporating the water (generating steam) while cooling the flue gas to a cost effective exit temperature.

In lower pressure boilers, insufficient heating surface is available in the furnace enclosure to absorb all of the energy needed to accomplish this function. Therefore, a boiler bank located downstream of the furnace and superheater (if included) in the flue gas stream provides the rest of the necessary heat transfer surface.

As shown in Fig. 3, as pressure increases, the amount of heat absorption required to evaporate water declines rapidly, and the absorption required for superheating and water preheating increases. In some modern very high pressure industrial units, a smaller boiler module separate from the steam drum provides the same function as the traditional boiler bank (see Fig. 4), but at a lower cost.

A separate economizer and/or air heater can be used downstream of the boiler bank to further reduce the flue gas exit temperature to a more economic value.

Steam requirements

To assure prompt fulfillment of all steam demands, i.e., the delivery of heat to all points of use at the required rates, it is necessary to select steam producing equipment of sufficient capacity, range of output and responsiveness. The demand may be steady, as in most space heating systems, or it may fluctuate widely and rapidly, as in a heavy forging plant. Many steam heated processes, as in the initial heating of a liquid batch, require high peak flows of short duration. Rapidly changing rates of steam flow characterize requirements to produce the electrical power to drive a steel rolling mill. Combined cycle systems often require rapid steam load responses during cycle transients. The steam flow requirements should, therefore, be accurately established for a number of conditions to ensure that the boiler system selected will meet all of

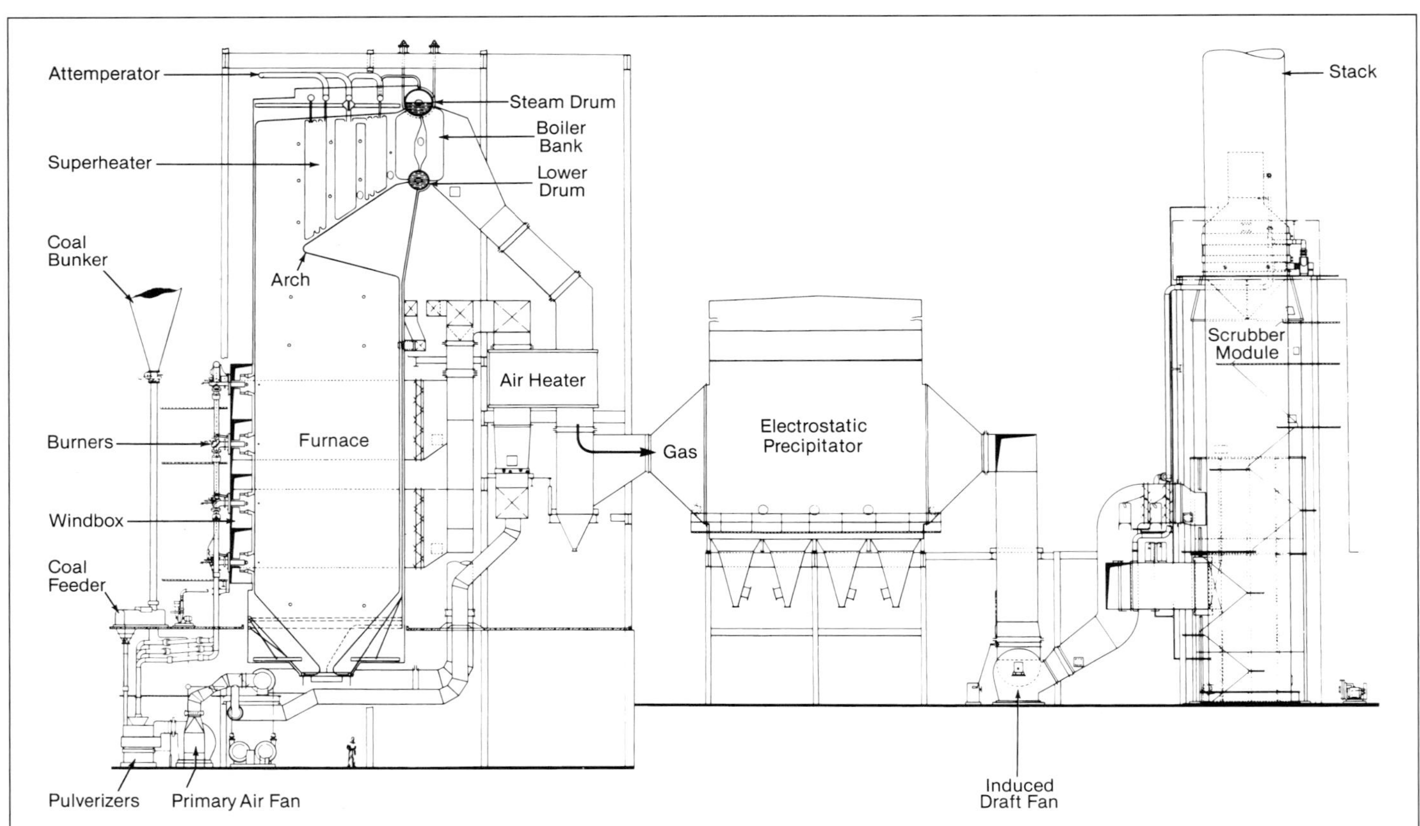

Fig. 1 Two drum Stirling® power boiler system for pulverized coal with environmental control equipment.

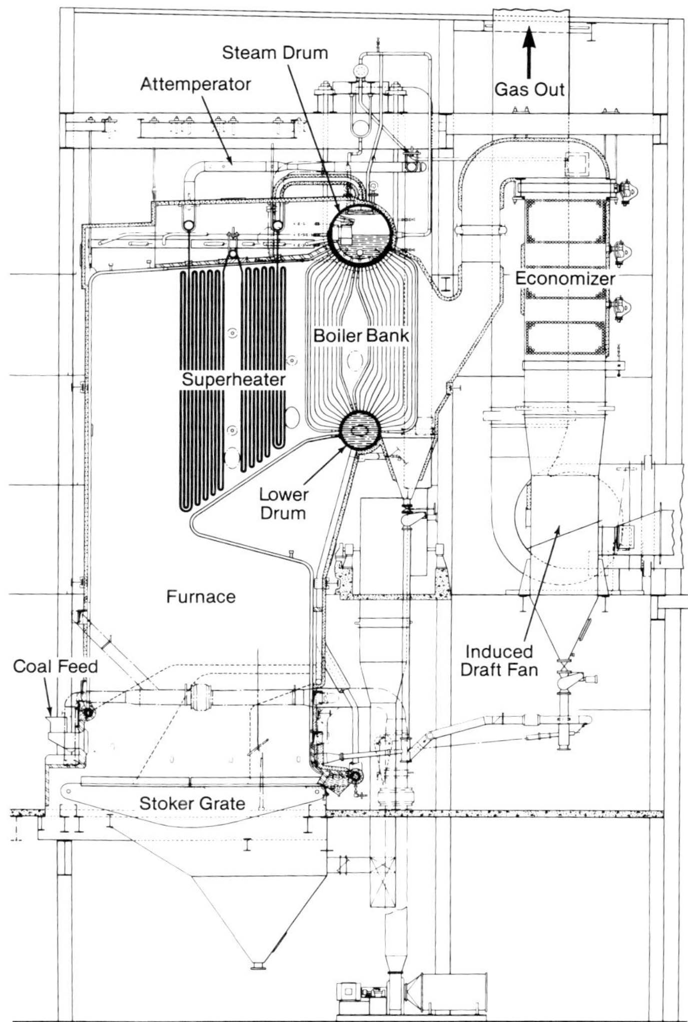

Fig. 2 Two drum Stirling® boiler for spreader-stoker firing.

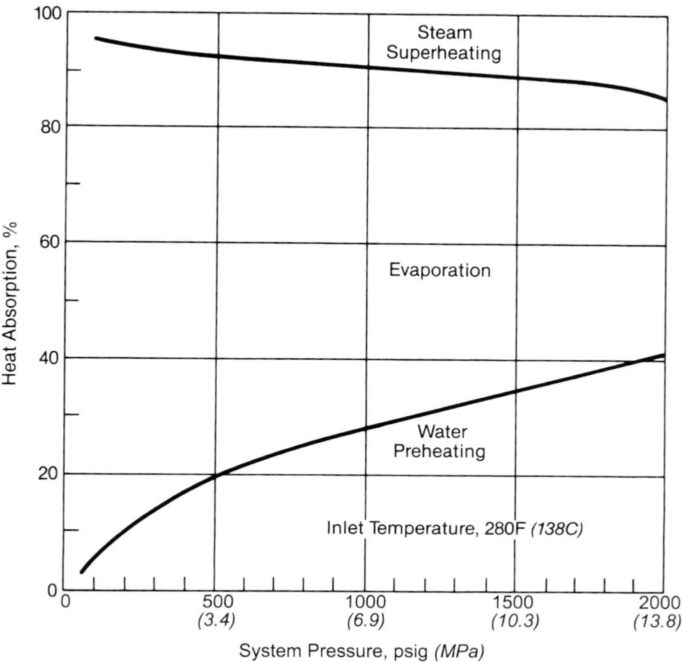

Fig. 3 Effect of steam pressure on evaporation in industrial boilers – 100F (56C) constant superheat.

its demand conditions, i.e., peak flow, maximum continuous flow (the usual steady maximum flow), minimum flow, and rate of change in flow.

The peak load will, of course, establish the top capacity for the steam producing equipment and all of its auxiliaries. For widely fluctuating loads, it is advisable to establish the 15 minute peak. In most systems, peaks of shorter duration can be met by the storage of heat inherent in the steam generating equipment.

Steam for process and heating The pressure of saturated steam (no superheat) used for process heating is such that the corresponding condensing steam temperature is somewhat above the required temperature of the materials to be heated. Generally, superheat is of no value for this kind of service and is often undesirable because of its interference with temperature control. Reclaiming or devulcanizing rubber, where the rubber in a caustic solution is heated to 400F (204C) by condensing saturated steam at 250 psig (1724 kPa) and 407F (208C) in the jacket of the devulcanizer, is a typical example of process heating with steam.

Pressures of saturated steam for comfort heating of buildings range from 2 psig (13.8 kPa) to as high as 80 psig (0.55 MPa) in the case of space heaters. It is seldom economical to distribute steam through long lines at pressures below 150 psig (1.0 MPa) because of piping costs. Furthermore, the usual requirements for steam within the boiler house for sootblowers, feed pumps and other auxiliaries make it desirable to operate boilers at a minimum of 125 psig (0.86 MPa). Consequently, few steam plants of any size are operated below this pressure. If the pressure required at points of use is lower, it is common practice to use pressure reducing stations at or near these locations.

The pressure required at the outlet of steam produc-

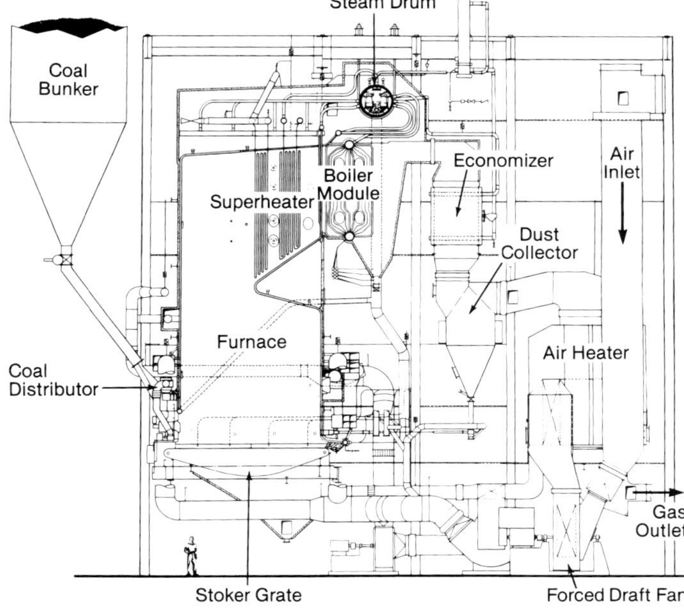

Fig. 4 Single drum Stirling® boiler for spreader-stoker firing.

ing equipment for process heating service usually ranges from 125 to 250 psig (0.86 to 1.7 MPa) and superheat is usually not required. For this service, boiler manufacturers have generally standardized on a pressure of 250 psig (1.7 MPa) for small water-tube boilers.

It is customary American stationary boiler practice to hold the main steam line pressure practically constant for all loads, on the premise that this condition satisfies all pressure and quantity requirements of the steam-using equipment. Automatic combustion control apparatuses are accordingly designed to function on this basis.

Cogeneration Many manufacturing operations, such as in paper and textile mills, in the production of chemicals and in processing rubber, require mechanical or electrical power as well as steam for process heating. For such cogeneration applications, studies are made of the relative merits and costs of: 1) a plant where the power is purchased and steam is generated to supply the heating requirements only, and 2) a plant where steam and power are generated in the same system. A sound appraisal of the relative merits of the two alternatives requires knowledge of the steam and power requirements, ability to correlate these requirements, economic studies and good judgement. The following general summary may be of assistance.

1. The basic economic advantage in generating steam and power in the same system arises from the use of a much larger portion of the heat supplied in the fuel. When generating electricity alone, as much as 60% of the heat supplied in the fuel is lost to the condensing system, even in a modern central station. (See Chapter 2.)
2. Despite this fundamental thermodynamic advantage, it is frequently more economical to purchase power when it is available at reasonable rates from a dependable source, except where:
   a. waste fuels and waste heat, such as bagasse, blast furnace gas, sawdust or hogged wood, and hot gases are available at low cost from the plant process, and
   b. the steam heating and power demands are reasonably parallel and relatively large, i.e., 50,000 lb/h (6.3 kg/s) of steam or more.
3. Two approaches are used for cogeneration. Where natural gas is available on site, a gas turbine can be used to generate power, with the waste heat in the turbine exhaust gas used to produce steam in a heat recovery steam generator (HRSG). HRSGs are discussed in more detail later in this chapter. Where a waste fuel such as petroleum coke, bark, biomass, tires or other fossil fuel such as coal is the economic fuel of choice, a steam turbine topping cycle is used. Here, high pressure high temperature steam is produced in the boiler system and is first passed through a steam turbine to generate power. The exhaust steam is then conditioned (brought to appropriate temperature and pressure) and sent to the process.
4. Variations in process heat and power demands usually do not coincide. To compensate for the differences, a variety of options can be used depending upon the economic evaluation:
   a. for a gas turbine-based system
      1. If steam demand typically exceeds power demand, auxiliary burners are supplied with the HRSG.
      2. If electrical load demand typically exceeds the steam demand requirements, the remainder of the power may be purchased outside.
   b. for a steam-based power system
      1. If steam demand typically exceeds power demand, the turbine exhaust steam flow can be supplemented with boiler auxiliary firing, and then the additional steam can be passed through a pressure reducing and attemperating system.
      2. If power demand is typically higher, either an extraction condensing turbine system can be used or the incremental power can be purchased outside.

If the process requirements for steam and power are reasonably parallel and steady, cogeneration, including capital, operating and maintenance costs, can be beneficial. Where discontinuous service, low capacity factor or significantly different steam and power requirements exist, electrical supply from the local power grid and on-site steam generation is frequently more cost effective.

Power generation Except for small isolated installations, the high speed turbine is the prime mover of choice for steam power generation because of its efficiency, compactness and low cost. Continued improvement in reliability, reduction in cost, and availability of packaged systems have made on-site power generation more popular. Where natural gas is available and cost effective, simple gas turbines, especially package units, have tended to dominate on-site power production needs. Where a waste fuel (or low cost coal) is available, a boiler and steam turbine system can prove to be the most economic system to supply on-site power. The selection of steam pressure and temperature for such systems depends upon an economic evaluation along the guidelines presented in Chapter 37. Steam temperature control using one or more of the methods outlined in Chapter 19 is usually provided when electrical output exceeds 25 MW. Steam temperature control is also very important where variations in the flow of the fuel and the fuel quality would otherwise lead to wide swings in power output.

Combined cycles

The concept of using waste energy for increased steam generation in industry has been around for many years. The progressive increase in fuel costs, the need to capture heat from various industrial processes, and the increasingly stringent environmental regulations have created the need for using waste heat to its fullest potential.

In the power industry, the waste heat from one power system such as a gas turbine can serve as the heat source for a steam turbine cycle. Such *combined cycles* can push overall electrical power cycle efficiency to nearly 50%. Overall energy use can substantially exceed even this level when electrical generation is combined with process steam use.

Industries such as steel making, oil refining, pulp and paper, and food processing have used many unique steam generating systems to get the most out of their waste heat. These systems allow reduced consumption of traditional fuels, recovery of waste heat for safety and economy, and elimination of process byproducts.

In its broadest terms, a *combined cycle* plant consists of the integration of two or more thermodynamic power cycles to more fully and efficiently convert input energy into work or power. With the advancements in reliability and availability of gas turbines, the term combined cycle plant today usually refers to a system composed of a gas turbine, heat recovery steam generator and a steam turbine. Thermodynamically, this implies the joining of a high temperature Brayton gas turbine cycle with a moderate and low temperature Rankine cycle – the waste heat from the Brayton cycle exhaust serves as the heat input to the Rankine cycle. The challenge in such systems is the degree of integration needed to maximize efficiency at an economic cost. Chapter 2 discusses the thermodynamic processes and benefits of this application.

The following brief discussion focuses on the combination of a gas turbine with a heat recovery steam generator and steam turbine.

Simple combined cycle system

As shown schematically in Figs. 5a and 5b, the simple combined cycle system can consist of a single gas turbine-generator, HRSG, single steam turbine-generator, condenser, and auxiliary systems. In addition, if the environmental regulations require, a selective catalytic reduction (SCR) system to control emissions of nitrogen oxides (NO_x) can be directly integrated within the steam generator. (See Chapter 34.) This is particularly attractive because the SCR catalyst can be positioned in an optimal temperature window within the HRSG. The gas temperature leaving the gas turbine is usually in the range of 950 to 1050F (510 to 566C) while the optimal SCR catalyst temperature is 650 to 750F (343 to 399C).

A variety of more complex configurations are possible. A key improvement in the steam cycle efficiency can be obtained by adding multiple separate pressure circuits to the HRSG to supply low pressure steam for deaeration and feedwater heating. This replaces the steam extraction regenerative feedwater heating used in conventional steam power cycles.

Commercial combined cycle systems

Actual configurations are typically more complex because of application requirements and the degree of integration. The gas turbine-generators and steam turbine-generators are commercially available in a number of specific sizes and arrangements. HRSGs are built from standardized components to suit a wide variety of steam uses and turbine exhaust conditions. Frequently, multiple gas turbines with HRSGs may feed a single steam turbine system. A gas bypass stack and silencer are typically installed downstream of the gas turbine so that it can be operated independently of the steam cycle. With the high levels of oxygen remaining in the gas turbine exhaust, supplemental firing systems can be installed upstream of the HRSG. This permits greater operating flexibility, improved steam temperature control, and higher overall power capacity. The HRSG can be designed with one to four separate operating pressure circuits to optimize heat recovery and cycle efficiency. In selected cases, further cycle efficiency gains are possible with the addition of steam reheat.

A range of cycle efficiencies is possible depending upon the complexity of the system and components. Sample overall electrical power generation cycle efficiencies for a system using a gas turbine with a 2200F (1204C) turbine inlet temperature are provided in Table 2. These are based upon the fuel higher heating value.

The environmental emissions from combined cycle systems are generally low. If natural gas is fired, sul-

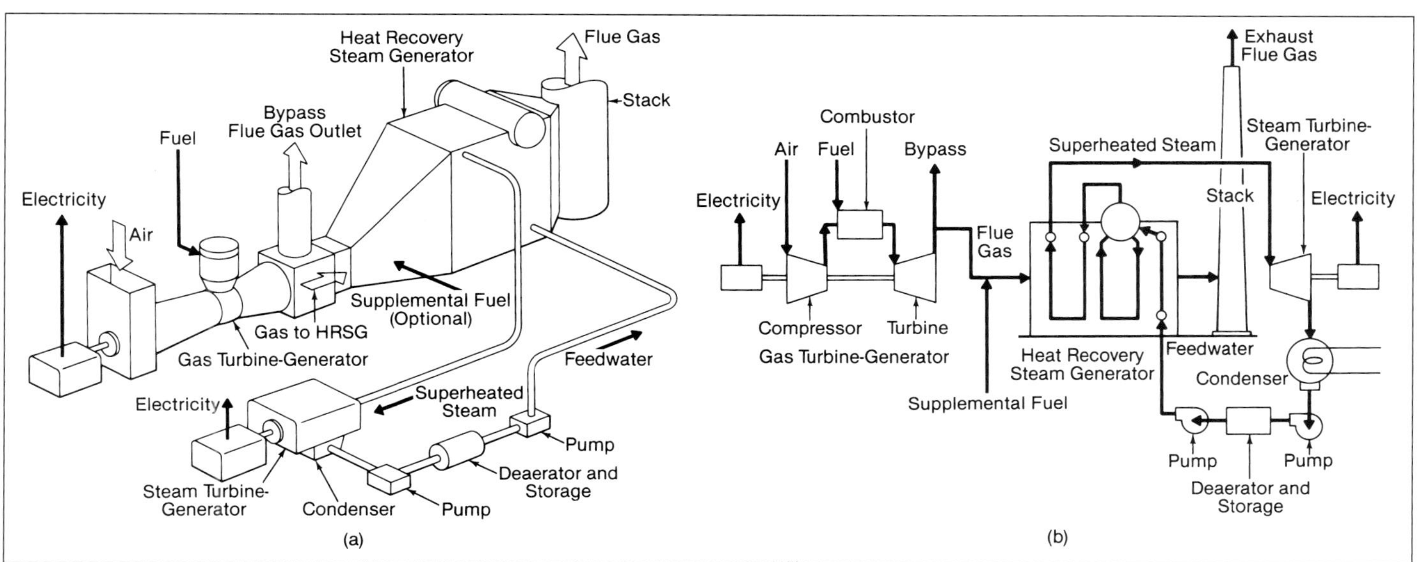

Fig. 5 Simplified combined cycle system schematics.

Table 2
Sample Cycle Efficiencies and Heat Rates

System	Efficiency (%)	Heat Rate (Btu/kWh)
Simple gas turbine	32	10,700
Gas turbine plus unfired single pressure steam cycle	42	8200
Advanced gas turbine plus unfired multiple pressure steam system	48	7100
Gas turbine plus dual pressure steam system plus process steam use (cogeneration)	61	—

Note: All values are calculated using the higher heating value (HHV) of the fuel. Use of the lower heating value (LHV) of the fuel would increase the efficiencies listed.

fur dioxide (SO_2) and particulate emissions will be negligible. With current gas turbine combustor designs, NO_x emissions from the gas turbine will be low, ranging generally between < 9 and 70 ppm. The final NO_x emissions then depend upon the supplemental firing system used (if any) and the potential incorporation of an SCR NO_x control system.

Beyond thermal efficiency and low environmental emissions, potential benefits of a gas turbine combined cycle plant include:

Schedule — Depending upon equipment size and complexity, delivery and construction of the gas turbine can take about one year. The steam turbine cycle system delivery and erection can frequently be completed in an additional year. (See Fig. 6.)

Flexibility — The gas turbine can be used alone for rapid startup and peaking service. The HRSG boiler system can usually be brought from a cold start to full load steam generation in about 60 minutes.

Capital — The system capital cost is typically low as a result of smaller standardized components, modular construction, prompt delivery, rapid erection, and minimum support system costs.

In the system selection, these benefits must be weighed against the usually higher cost of the cleaner premium fuels needed for the gas turbine, potential maintenance and availability issues, and load requirements.

Water treatment

Industry today is intensely competitive, and a company's survival may often depend on uninterrupted production. When industrial plants generate their own power or when steam is required for the manufacturing process, the availability of the boiler becomes a very important consideration.

Boiler reliability is directly related to water quality, as discussed in Chapter 42. The lack of proper water quality is frequently the cause of failure and lack of availability in industrial boilers. As pressures and temperatures have increased, so has the required level of water quality. However, the need for water quality is often underestimated and the result is expensive downtime for pressure part replacement. Higher water quality translates to higher steam quality, which is needed to protect superheaters and turbines.

The use of a spray water condenser (see Chapter 25) was a step taken to produce higher quality spray water to minimize the steam contamination that is often prevalent when using feedwater for spray attemperation. A sample condenser installation is shown in Fig. 7.

The maximum allowable boiler-water concentration (total solids in boiler water) in relation to the pressure at the outlet of a steam generating unit (applicable to the boiler types in the following descriptive outlines) is given in Table 3.

It is evident that good water treatment practice requires the proper attention be given to water quality throughout the boiler system. This includes makeup water, steam condensate, feedwater and boiler water. One of the major concerns in boiler operation is the prevention of fouling, regardless of operating pressure. This becomes more severe as operating pressures increase; conditions which are not regarded as troublesome at lower pressures may be entirely unsatisfactory at higher pressures.

Other design requirements

The other factors affecting industrial boiler evaluation and selection as outlined in Table 1 are addressed in depth in Chapter 37. Key among these are fuel and ash identification and evaluation. Chapters 9 and 21 address the characterization of fuels and the effect of the noncombustible residue or ash on boiler design. Chapters 11 through 17 discuss the various fuel preparation and combustion systems which may be used. The designs of the boiler, superheater, economizer and air heater systems are covered in Chapters 19 and 20 while auxiliaries are addressed in Chapters 24 and 25.

Fig. 6 Rapid installation of modular components minimizes schedules.

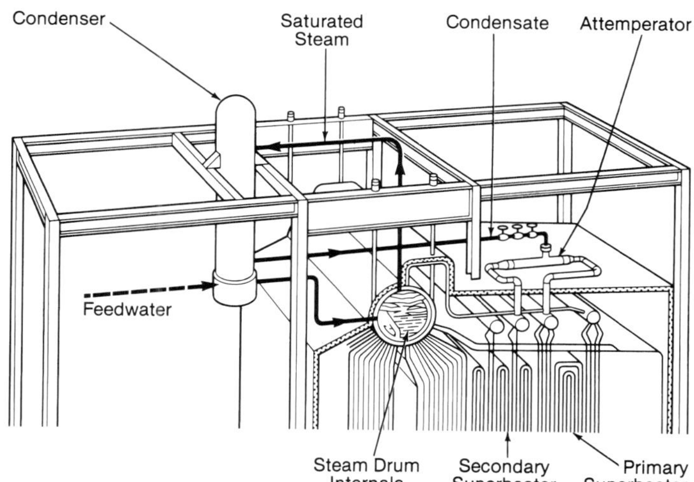

Fig. 7 B&W condenser-attemperator system supplies pure attemperator spray water by condensing steam from the steam drum with incoming boiler feedwater.

Environmental control

Atmospheric emissions from industrial boilers and other combustion sources have come under increasingly stringent regulation over the past 30 years. Pollutants emitted from fossil fuel fired industrial boilers can include particulate matter (PM), nitrogen oxides (NO_x), sulfur oxides (SO_x), carbon monoxide (CO), volatile organic compounds (VOCs), and other compounds commonly referred to as HAPs (hazardous air pollutants). These include hydrogen chloride and hydrogen fluorides (HCl and HF), mercury and other trace metals, and trace organic compounds. The emissions characteristics of industrial boilers are largely dependent on the type of fuel(s) combusted and the use of selected air pollution control systems and technologies.

The United States (U.S.) Environmental Protection Agency (EPA) has promulgated various environmental performance standards that apply to different sizes and classes of boilers firing both fossil and non-fossil fuels. The Federal New Source Performance Standards (NSPS), codified under Title 40, Part 60 of the Code of Federal Regulations (CFR), establish minimum requirements for many air emission sources throughout the U.S. In 40 CFR Part 60, Subparts Db and Dc specifically apply to fossil-fuel fired industrial steam generating units with heat input capacities greater than 100 million Btu per hour and greater than 10 million Btu per hour respectively (2.9 and 29.3 MW_t). The industrial boiler NSPS set minimum requirements governing allowable emissions of PM, SO_2, and NO_x from units constructed or substantially modified after June 19, 1984. Additional subparts under 40 CFR 60 set emissions standards for other classes of boilers including, but not limited to, electric steam generating units (40 CFR 60, Subpart Da), municipal waste combustors (40 CFR 60 Subpart(s) Eb, Cb, AAAA and BBBB), and chemical recovery boilers used in the pulp and paper industry (40 CFR 60, Subpart BB).

Table 3
Limits for Solids Content of Boiler Water in Drum Boilers, ppm

Pressure at Outlet of Steam Generating Unit, psi	Total Solids	Total Alkalinity	Suspended Solids
51 to 325	3500	700	15
326 to 450	3000	600	10
451 to 600	2500	500	8
601 to 750	2000	400	6
750 to 900	1500	300	4
900 to 1000	1250	250	2
1000 to 1500	100	20	1
1501 to 2000	50	10	1

A large industrial power boiler with environmental control equipment is shown in Fig. 1. This pulverized coal-fired Stirling® power boiler (SPB), discussed below, is equipped with low NO_x burners to reduce NO_x emissions, an electrostatic precipitator (ESP) to control particulate emissions, and a wet flue gas desulfurization (FGD) system to control emissions of SO_2.

The most common methods employed to control NO_x emissions from industrial boilers involve combustion modifications using low NO_x burner designs, staged combustion with overfire air ports, and flue gas recirculation techniques (see Chapters 11 and 14). Fluidized bed boilers, discussed in Chapter 17, can achieve very low NO_x emission rates due, in part, to the inherently low furnace combustion temperatures. If needed, post-combustion control technologies, including selective noncatalytic reduction (SNCR) and selective catalytic reduction (SCR) systems (see Chapter 34), can be used to achieve additional NO_x reductions on many industrial boiler applications.

CO and VOC emissions are best controlled by employing prudent combustion system design and operating practices The combustion system should facilitate good air/fuel mixing and allow for adequate residence time at required temperatures. Traditionally this has been referred to as the three Ts of combustion: time, temperature, and turbulence. Any combustion technique used to achieve low NO_x levels must be balanced to account for associated (negative) impacts on CO and VOC emissions.

To achieve compliance with current day particulate emission standards, most industrial boilers firing fuels other than a *clean* gas must utilize high efficiency particulate control equipment; typically a fabric filter (baghouse) or an electrostatic precipitator (ESP). (See Chapter 33.) Baghouses and ESPs can also effectively control HAPs (trace organic and metal compounds) that are affixed to the particulate matter.

The required levels and methods of SO_2 control are primarily driven by fuel choice. SO_2 controls are generally not needed if an industrial unit burns a clean fuel such as natural gas or a low sulfur fuel oil. While the use of low sulfur coal is still feasible for many older or smaller industrial boilers, the current NSPS regulation applicable to large industrial coal-fired boilers mandate the use of SO_2 control technologies capable of achieving 90% removal efficiency. Available post-combustion SO_2 controls include both dry and wet

FGD systems. (See Chapter 35.) Fluidized-bed boilers, discussed in Chapter 17, offer an alternative to FGD systems. These boilers are capable of achieving very high SO_2 removal rates via limestone addition directly into the combustion process (bed).

The combination of FGD systems with efficient particulate control has shown the potential for effective control of many HAPs. It is anticipated that the EPA's future HAP regulations may require retrofits of these combined controls on most large (new and existing) coal-fired industrial boilers.

B&W boilers for solid and multi-fuel applications

Brief descriptions of the boiler types that follow are intended to introduce the types of steam generating units that meet the wide range of fuel and performance requirements of the industrial sector.

Stirling® power boilers

Description The Stirling Power Boiler (SPB) is a top supported, two drum, single gas pass unit. (See Figs. 1 and 2.) In some cases, it is cost effective to replace the two drum design with a single drum and a smaller shop-assembled boiler module. (See Fig. 4.)

The furnace is completely water cooled using membrane wall construction [normally 3 in. (76.2 mm) tubes on 4 in. (101.6 mm) centers] and is satisfactory for either pressurized or balanced draft operation. Shop assembly of wall panels is maximized to facilitate field erection. Wall panels always come with the headers attached, regardless of the number of shipping pieces.

Cyclone steam-water separators, discussed in Chapter 5, along with primary and secondary steam scrubbers, are included in the steam drum to provide the high quality dry steam needed for present day superheater and turbine designs.

Furnaces include a nose arch which serves to direct gas flow over the superheater section and to shield the superheater.

SPBs are capable of firing solid, liquid or gaseous fuels. There are several furnace configurations to complement the type of fuel fired, as shown in Fig. 8. A hopper-bottom furnace is used for pulverized coal firing; a flat floor for gas or oil firing; and an open bottom to receive a stoker for stoker coal, wood, bagasse, biomass, refuse-derived fuel (RDF) and as-received municipal solid waste (MSW).

Furnaces for fuels having a significant amount of fines and/or high moisture such as wood, biomass, bagasse and RDF are arranged with an overfire air system consisting of multiple rows of nozzles. These nozzles inject air at various locations above the grate to provide the turbulence necessary for combustion.

The SPB is equipped with an economizer and/or air heater to provide economical heat recovery. For many fuels, air heating is important for combustion. Pulverized coal requires hot air to dry the fuel, and hot air is required to promote combustion of moist fuels such as wood, bagasse and biomass.

Because of the design features, the SPB is the preferred industrial boiler for many applications. There are some other designs that serve special fuels, capacities, pressures and temperatures, making them a good alternative under special conditions.

Pre-engineering While the SPB is custom designed to meet specific steam and fuel conditions, the design is done within a framework of pre-engineered components to minimize engineering costs and delivery time.

The furnace width and depth are pre-engineered in 1 ft (0.3 m) increments so that all of the closures at the corners are established. Drum centerlines, in 2 ft (0.6 m) increments between 16 and 32 ft (4.9 and 9.8 m), have been pre-engineered to locate all of the access doors, sootblower openings, buckstays and platforms. Combinations of steam and lower drum sizes are designed so that all of the bend angles for drum entry are established.

These pre-engineered increments allow flexibility in the design to satisfy the job-specific requirements of furnace exit gas temperature, burner clearances, residence time, grate size, gas velocity, convection surface spacing, etc.

SPB design range

Steam capacity	
pulverized coal, oil, gas	up to 1,200,000 lb/h (151.2 kg/s)
stoker coal	150,000 to 400,000 lb/h (18.9 to 50.4 kg/s)
stoker wood, bagasse, and biomass	180,000 to 600,000 lb/h (22.7 to 75.6 kg/s)
Steam pressure to	2200 psig (15.2 MPa) design
Steam temperature to	1000F (538C)

Towerpak® boiler

Description The Towerpak is a version of the SPB designed for lower capacities that are often required by smaller industrial plants. (See Fig. 9.) It incorporates many of the features of the SPB including membrane walls, cyclone steam-water separators, and stokers for wood or biomass combustion.

Towerpak boilers are one drum or two drum bottom supported units. For the smaller sizes, they can be shipped in a single unit or in modules for ease of field assembly (see Fig. 10). Larger units follow the SPB format of maximum subassembly of wall panels, again for ease of field assembly. This unit is a preferred design at low steam capacity for hard to burn solid fuels such as wood, biomass and stoker coal.

Pre-engineering Like the SPB, these units are custom designed to meet specific conditions for each application, but within the framework of pre-engineered components.

Towerpak design range

Steam capacity	40,000 to 300,000 lb/h (5.0 to 37.8 kg/s)
Steam pressure to	1800 psig (12.4 MPa) design
Steam temperature to	1000F (538C)

Circulating fluidized-bed boiler

Description Fluidized-bed boilers feature a unique concept of burning fuel in a bed of particles to control the combustion process and, when required, control

SO_2 and NO_x emissions. Two options are offered – the circulating fluidized-bed boiler (CFB) and the bubbling fluidized-bed boiler (BFB). (See Figs. 11 and 12.) Both options can be used in new as well as in retrofit applications. Fluidized-bed technologies are discussed in depth in Chapter 17.

The CFB is a top supported boiler. (See Fig. 11.) One or two drums are used depending on the need for a generating bank to absorb heat. Fuel is admitted to the lower part of the furnace by screws, air-assisted gravity chutes or pneumatic feed, depending on fuels fired and plant design.

The bed medium is typically composed of fuel ash and sulfated limestone particles (limestone is used as a sorbent for SO_2 removal). When firing low-ash and low-sulfur fuel, additional inert bed material (typically sand) may be introduced into the furnace. Compared to the fuel quantity present in the unit, the circulating bed material is many times greater. Total solids in the flue gas passing upwards through the furnace are a function of how much heat must be absorbed by the waterwalls.

Varying the bed density maintains the desired constant temperature necessary for maximum SO_2 removal [about 1550F (843C)]. The B&W Internal Recirculation CFB (IR-CFB) design features a two-stage

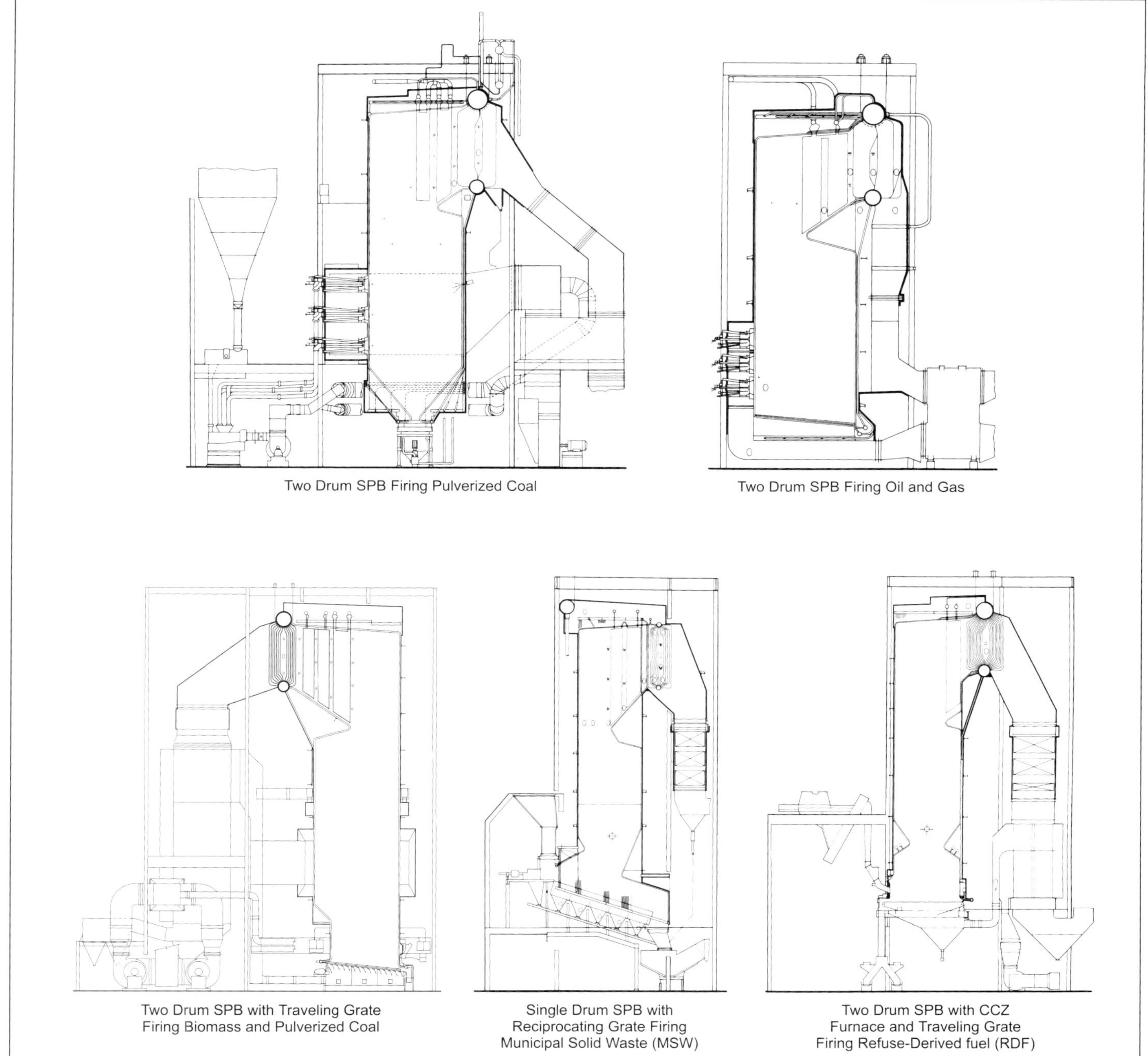

Fig. 8 Typical Stirling® power boiler furnace configurations.

solids collection system comprising U-beam particle separators (U-beams) at the furnace exit and a multi-cyclone dust collector (MDC) at the top of the convection downpass. Solids laden flue gas exits the furnace to U-beams where 90 to 95% of the solids are collected and directly (internally) returned to the furnace. Flue gas exiting the U-beams proceeds over convection surfaces, similar to other boiler designs. About 90% of the solids passing the U-beams are further collected by the MDC. The recirculation rate from the MDC is controlled to provide the flow necessary to maintain the required bed temperature and density.

The CFB has been selected for application with high-sulfur, high-ash and various waste fuels such as petroleum coke, waste coal, sludge, oil pitches, etc. It is also used for wood and other biomass fuels, however the bubbling fluidized bed is often a more economical choice for these fuels. The CFB, because it operates at a reduced combustion temperature, inherently generates about one half the NO_x as the other solid fuel-fired industrial boilers previously described.

Fig. 10 Field assembly of a Towerpak® boiler.

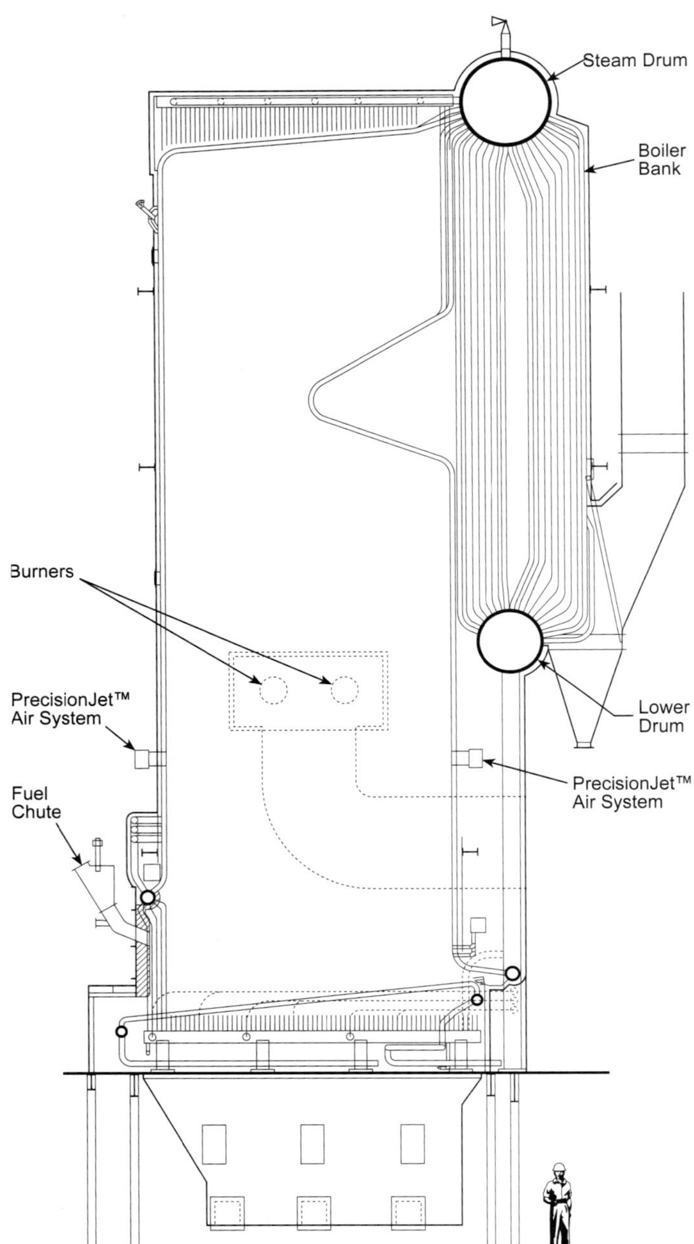

Fig. 9 Towerpak® boiler.

The CFB is an alternative to the pulverized coal or stoker coal-fired SPB which frequently must be equipped with a wet or dry scrubber (for SO_2 removal) and ammonia injection, catalytic or non-catalytic reduction (NO_x removal) equipment. The choice of technologies requires in depth evaluation of a number of factors including required amount of emissions removal, fuel cost, reagent cost and capital cost.

Pre-engineering The CFB is custom designed to meet each specific application, but like the SPB, it is designed within a framework of pre-engineered components to minimize engineering costs and delivery time.

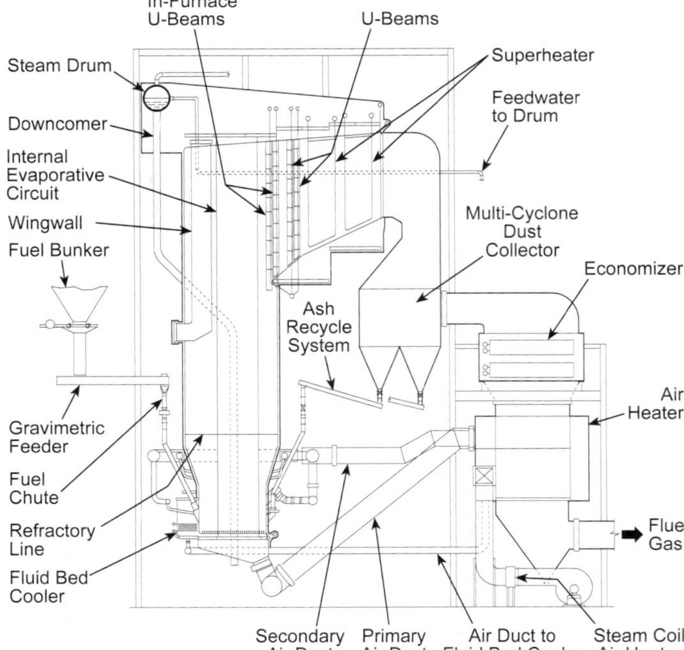

Fig. 11 Top supported circulating fluidized-bed boiler.

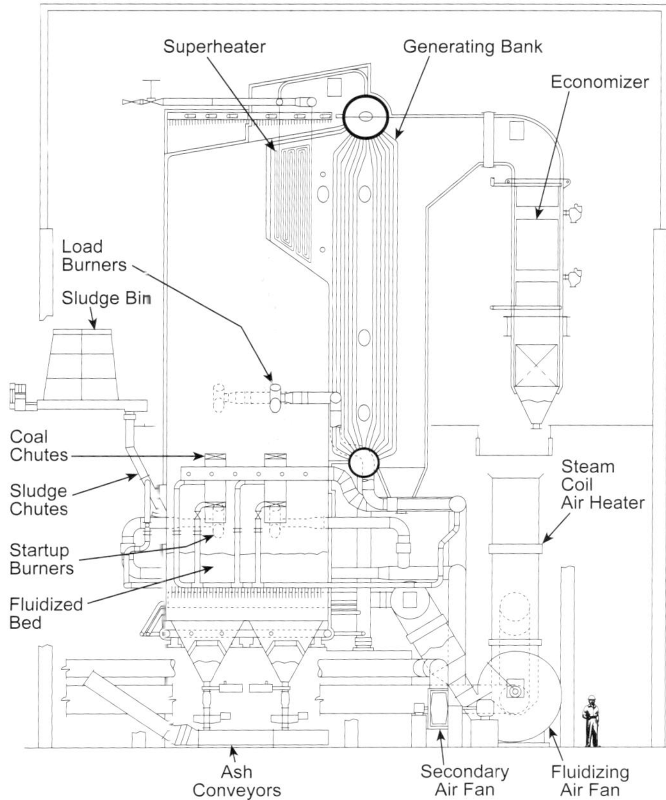

Fig. 12 Bottom supported bubbling fluidized-bed boiler.

CFB design range

Steam capacity	up to 1,500,000 lb/h (189 kg/s) or greater
Steam pressure to	2850 psig (19.7 MPa) design
Steam temperature to	1040F (560C)

Bubbling fluidized-bed boiler

Description The bubbling fluidized-bed (BFB) boiler is also a version of the Stirling power boiler in many ways. The unit can be top supported or bottom supported, can be a one or two drum design, and can burn a wide variety of fuels cleanly and efficiently. A bottom supported design similar to a Towerpak boiler (see Fig. 12) can be supplied for steam flow capacities up to 300,000 lb/h (37.8 kg/s). BFB technology is ideal for repowering an existing facility, recovery boiler conversion (Fig. 13), or when changing fuel or firing techniques of an existing boiler. It is particularly attractive in retrofit applications where the bottom of an existing furnace (including a stoker) can be removed and replaced with a bubbling fluidized-bed without major modifications to the balance of the furnace, convection pass enclosures and heat transfer surfaces. Such conversions have been effective in regaining boiler capacity lost because of a fuel change or a change in ash characteristics which are not compatible with the original boiler furnace design. Such retrofits also provide one option to reduce sulfur dioxide and nitrogen oxides emissions from industrial and small utility boilers.

The combustion process uses an inert material, typically sand, and fuel to form a bed in the bottom of the furnace. This bed is suspended by a stream of upwardly flowing fluidizing air. Fuel introduced to the bed is quickly volatized. A significant amount of volatiles escape the bed and are burned in the freeboard area above the bed using overfire air. The remaining volatiles and fixed carbon burn in the bed. To achieve efficient combustion and low emissions, the bed temperature is controlled in the range of 1350 to 1650F (732 to 899C).

The BFB is particularly well suited for high moisture waste fuels such as sewage sludge, and various sludges produced in pulp and paper mills and recycle paper plants, but can also burn a variety of other fuels including wood wastes, bark, coal, tire derived fuel, oil and natural gas.

Pre-engineering The BFB is custom designed to meet each specific application, but like the SPB and Towerpak, this is done within a framework of pre-engineered components.

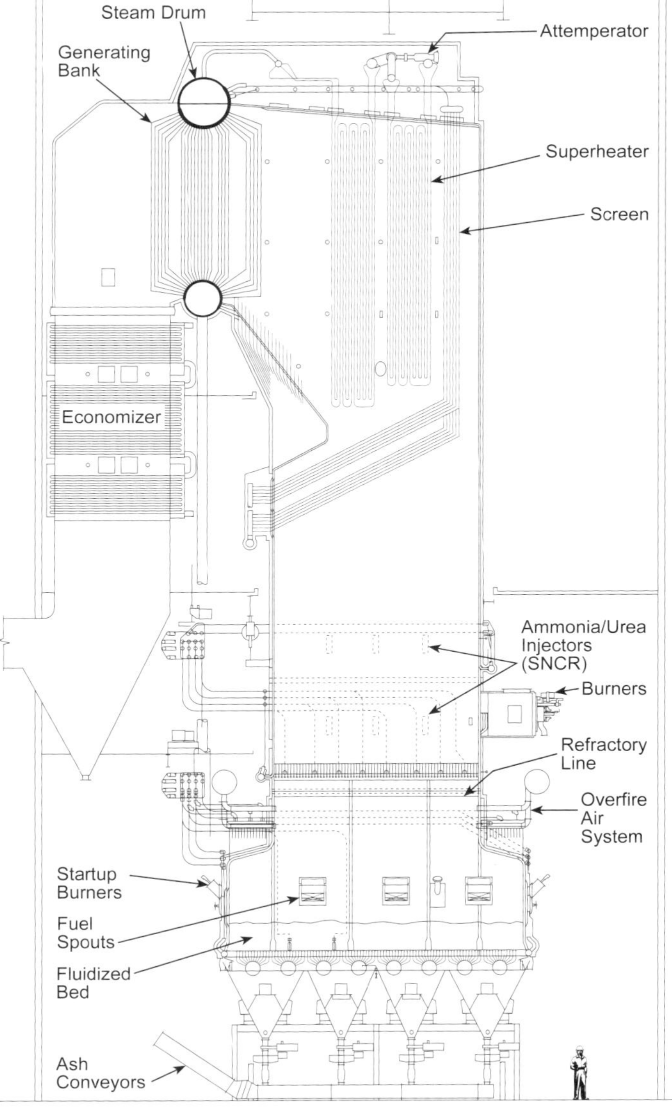

Fig. 13 Bubbling fluidized-bed retrofit of a recovery boiler.

BFB design range

Steam capacity	60,000 to 700,000 lb/h (7.5 to 88.2 kg/s)
Steam pressure to	2400 psig (16.5 MPa)
Steam temperature to	950F (510C)

B&W boilers for oil and gas applications

Boilers in this category have several similarities attributable to the fuel characteristics of oil and natural gas. Unlike coal-fired boilers, distillate oil and gas units do not have to consider slagging and fouling issues. Furnaces and convection banks can be arranged for optimum heat transfer. Some residual oils can be erosive and can contribute to fouling which is addressed in unit selection. The boilers in this section are all bottom supported.

FM boiler

Description The FM is a shop-assembled (package), two drum, bottom supported boiler. (See Fig. 14.) Package boilers can be shipped by rail, truck or barge. Most FM units are a D-type boiler design and have the furnace on one side and boiler bank on the other, separated by a baffle wall. In a D-type unit, firing is parallel to the drums toward the rear wall where the gas turns 180 degrees and flows frontward to the gas outlet.

B&W also provides an O-type FM boiler design where the furnace is located between the two drums. (See Fig. 15.) The flue gas splits at the rear of the furnace and flows forward through generating tube sections on each side of the furnace. Compared to the D-type boiler layout, the O-type boiler has a center of gravity that is on the unit centerline, which provides advantages for shipping.

Gas-tight furnaces, using membrane construction, are features of both the O and D boilers for pressure-fired (forced draft fan only) operation. With today's advanced low NO_x burner designs and steam separation technology, B&W offers rail-shippable units to 230,000 lb/h (29 kg/s).

In addition to burning oil and natural gas, these units can burn a variety of gaseous and liquid fuels such as landfill gas, refinery gas, and various waste liquid fuels. Emissions on natural gas are typically 30 ppm NO_x (0.036 lb/10^6 Btu), and 80 ppm NO_x (0.1 lb/10^6 Btu) on distillate oil. When required, 9 ppm NO_x is achievable on gas with advanced burners.

Fig. 14 Type FM integral furnace boiler – membrane wall construction.

Fig. 15 FM O-type boiler.

Pre-engineering The FM is totally pre-engineered in frame sizes to satisfy the capacity range of this package design. Series available are FM9, FM10, FM103, FM106, FM117 and FM120. Pre-engineered set furnace depths are the only variables in this series.

FM design range

Steam capacity	10,000 to 230,000 lb/h (1.3 to 29.0 kg/s)
Steam pressure to	1250 psig (8.62 MPa)
Steam temperature to	800F (427C) on oil 850F (454C) on natural gas (higher pressures and temperatures available)

High Capacity FM boiler

Description The High Capacity FM (HCFM) boilers are extensions of the FM D-type boiler design. (See Fig. 16.) High capacity boilers can be assembled in the shop, at a barge dock or in the field. Shipping dimensions of these units require that they be shipped by barge or ocean vessel.

The design is for oil and gas firing and uses membrane furnace walls for pressurized operation. Multiple burners are used for the increased capacity.

Pre-engineering Like the FM, these boilers are completely pre-engineered and set furnace depths are the only variable.

HCFM design range

Steam capacity	
HCFM	200,000 to 350,000 lb/h (25.2 to 44.1 kg/s)
Steam pressure to	1050 psig (7.24 MPa) design
Steam temperature to	825F (441C)

PFM boiler

Description Another higher capacity and design pressure D-type boiler, the Power for Manufacturing (PFM), is also designed for dock or field assembly. If

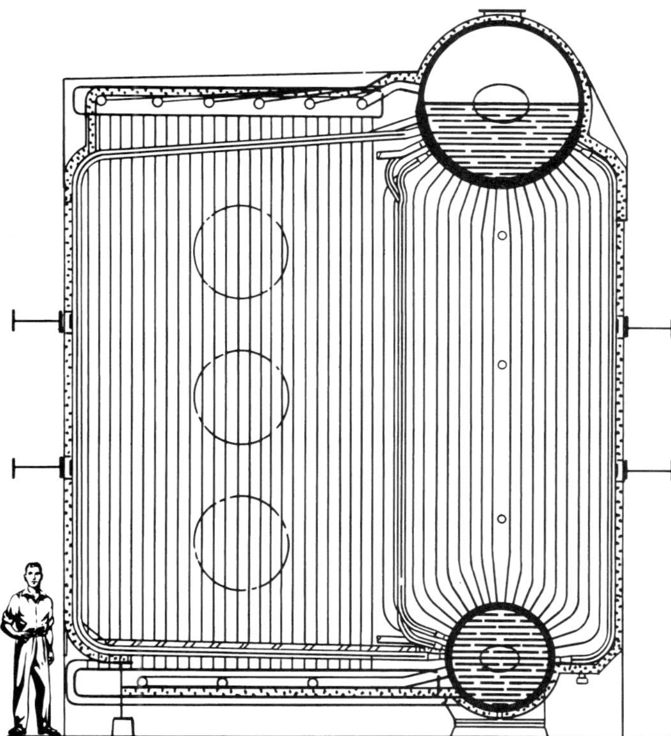

Fig. 16 High Capacity FM boiler.

shipped assembled, the size of these units requires special handling at the installation site. (See Fig. 17.) This design is also for gas and oil firing and uses a membrane furnace for pressure operation. Again, multiple burners are used for the increased capacity.

Pre-engineering Like the FM and HCFM, this boiler series, which includes PFM 140, 180, 220, 250 and 280, is completely pre-engineered and set furnace depths are the only variable.

PFM design range	
Steam capacity	200,000 to 600,000 lb/h (25.2 to 75.6 kg/s)
Steam pressure to	1800 psig (12.4 MPa) design
Steam temperature to	1000F (538C)

PFI boiler

Description The Power for Industry (PFI) boiler is a two drum, bottom supported, multiple gas pass unit designed specifically to burn liquid and/or gaseous fuels. (See Fig. 18.) Due to the large furnace enclosure, the PFI is an excellent choice for harder to burn byproduct fuels such as blast furnace gas and refinery catalytic cracker CO gas.

The furnace is completely water cooled using membrane wall construction [2.5 in. (63.5 mm) tubes on 3 in. (76.2 mm) centers].

The PFI was developed for maximum shop assembly of components. For example, the furnace is shipped in as few as ten membrane wall panels with headers attached at the top and bottom. The panels include two for each side wall and two each for roof, front wall and floor. The burner throats are integral with the front wall panel. The unit is bottom supported on simple concrete piers and incorporates a drainable superheater. A unique plenum encloses the upper front wall, roof and rear of the unit. This plenum serves as an integral air duct to the windbox and, with a division plate in the rear, a flue gas outlet. Either an air heater or economizer is used to reduce exit gas temperature and heat recovery.

This design features a gas pass extending the full length of the boiler bank. Flue gas flows horizontally and parallel to the drums through the bank. A gas baffle is used to direct the gas across the tubes in multiple passes to maximize heat transfer.

An inverted loop, drainable superheater is located behind a screen at the furnace outlet to protect it from direct furnace radiation. This superheater location gives a semi-radiant heat transfer characteristic that produces a relatively flat temperature curve across the load range. This temperature profile minimizes attemperation for steam temperature control.

Cyclone steam separators with primary and secondary scrubbers are included in the steam drum to produce the high quality steam needed for present day superheaters and turbines.

Pre-engineering The PFI unit is totally pre-engineered in a number of frame sizes to satisfy the capacity range of this boiler design. Several superheater arrangements are also pre-engineered for each frame size.

Units come in three different drum centerlines. Each centerline has a specific furnace depth and three or four furnace widths.

PFI design range	
Steam capacity	100,000 to 700,000 lb/h (12.6 to 88.2 kg/s)
Steam pressure to	1150 psig (7.9 MPa) design
Steam temperature to	960F (516C)

PFT boiler

Description The Power for Turbine (PFT) boiler was developed as an extension of the PFI design to accommodate the development of higher pressure and temperature turbine cycles. This unit incorporates many of the PFI features including: two drum, bottom supported, modularized furnace membrane walls [3 in.

Fig. 17 PFM boiler in transit.

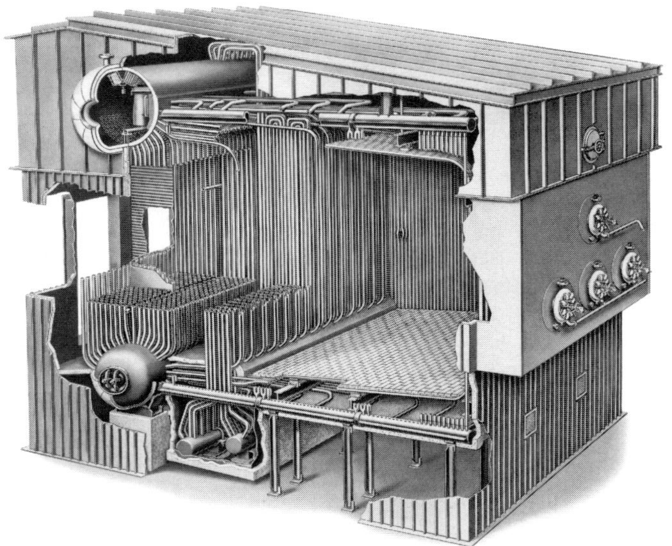

Fig. 18 Type PFI integral furnace boiler.

(76.2 mm) tubes on 3.75 in. (92.25 mm) centers], drum cyclones and a drainable superheater. (See Fig. 19.)

Some of the differences include an alternate pendant superheater (as shown in Fig. 19) and gas flow path the full width of the boiler bank with flow in vertical directions.

PFT units are particularly well suited to burn high ash liquid fuels, blast furnace gas and CO gas because cavities provide space for retractable sootblowers for cleaning requirements.

Pre-engineering The PFT, like the PFI, is totally pre-engineered in a number of frame sizes. The unit is designed with two drum centerlines and several furnace depths and widths to satisfy the capacity range of this design.

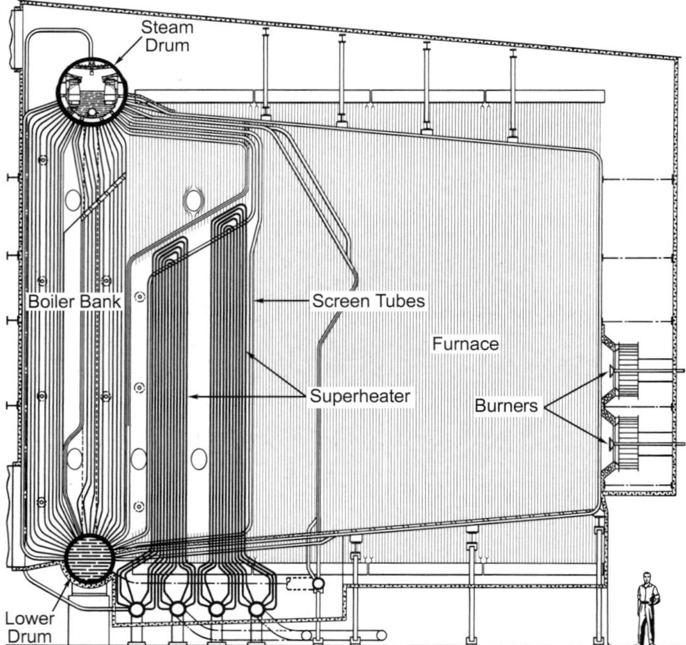

Fig. 19 PFT integral furnace boiler.

PFT design range

Steam capacity	300,000 to 800,000 lb/h (37.8 to 100.8 kg/s)
Steam pressure to	1800 psig (12.4 MPa) design
Steam temperature to	1000F (538C)

Enhanced Oil Recovery steam generator

Description The Enhanced Oil Recovery (EOR) steam generator (Fig. 20) was developed to meet a market need, as the name implies. High pressure, wet steam (approximately 80% quality) is produced by the unit and then injected into underground strata containing heavy oils. The steam enhances the recovery of oil by heating the heavy oil which reduces its viscosity and thereby aids in moving the oil to the producing wells.

A once-through steam water circuitry featuring all series flow coils is used for smaller units. Feedwater flows continuously in a single tube circuit through the convective (economizer) section to the radiant (furnace) section where the water is boiled to 80% (by weight) steam quality level. A multiple pass parallel flow design is used on larger units.

By maintaining wide flame clearances from the radiant walls (i.e., no direct flame impingement) and low heat releases, relatively poor feedwater (5000 ppm solids) can be tolerated. This permits minimal feedwater treatment and allows the water that is separated from the recovered oil to be recycled to the boiler with minimum cleanup. As stated, high solids content water is tolerable, however, near zero hardness is required to minimize the probability of solids depositing in the tubes. Only about 80% of the feedwater is vaporized to provide sufficient liquid at the steam outlet to keep water-soluble solids in solution.

Control of the process is achieved by pumping the required amount of feedwater at the specified pressure [up to 2500 psig (17.2 MPa)] into the convective section and by regulating the burner firing rate to maintain measured outlet steam quality.

Units are typically fired with natural gas or heavy oil.

Pre-engineering and design range Units come in pre-engineered sizes from 5 to 50 × 10^6 Btu/h (1.5 to 14.7 MW_t) output. The units are shop-assembled with units through 40 × 10^6 Btu/h (11.7 MW_t) trailer

Fig. 20 Enhanced oil recovery boiler on location.

mounted in one piece for shipment. Larger sizes are shop-assembled in several sections for final field assembly. Unit capacities are limited by shipping constraints and not because of design limitations.

EOR design range (oil and gas)

Steam capacity	up to 250,000 lb/h (31.5 kg/s)
Steam pressure to	2500 psig (17.2 MPa) design

B&W boilers for specialty and waste heat applications

Heat recovery steam generator

The heat recovery steam generator (HRSG) is sometimes referred to as a waste heat recovery boiler (WHRB) or turbine exhaust gas (TEG) boiler. The latter designation indicates the main application for these units today – waste heat recovery and steam generation from gas turbine exhaust gas. The HRSG is a key element in a combined cycle plant affecting initial capital cost, operating cost, and overall cycle efficiency.

HRSGs are flexible in design depending upon the specific application. The gas flow through the unit can either be horizontal or vertical with a tradeoff being made between the cost of floor space for the horizontal flow unit and structural steel requirements of the vertical flow unit. The horizontal flow case is the most common in North America, while the vertical design is popular in Europe. The HRSG may be designed for operation with multiple, separate pressure steam-water loops to meet application requirements and maximize heat recovery. Superheater or reheater surface may also be employed to further increase cycle efficiency. Circulation may be forced or natural (see Chapter 5) with most horizontal gas flow units using natural circulation. HRSGs may be unfired, i.e., only use the sensible heat of the gas as supplied, or may include supplemental fuel firing to raise the gas temperature to reduce heat transfer surface requirements, increase steam production, control superheat steam temperature, or meet process steam temperature requirements. Finally, HRSGs may be designed to incorporate an SCR or CO catalyst. (SCR catalysts are discussed in Chapter 34.)

HRSGs are suitable for use with gas turbine sizes from a small 1 MW unit up to modern machines exceeding 250 MW. See Figs. 21 and 22 for typical arrangements.

Heat recovery steam generators are designed to handle the unique requirements of a variety of combined cycle systems. Depending on the cycle configuration, the HRSG may consist of one to four separate boiler circuits (high pressure, two types of intermediate pressure and low pressure) within the same casing. The high pressure boiler is used for power generation [up to 1005F (541C) superheat]. The intermediate pressure boilers can be used for power generation, steam injection for NO_x control (water or steam injected into the gas turbine combustor to limit NO_x formation) and/or process steam. The low pressure boiler is normally used for feedwater heating and/or deaeration. Operating pressures can range from 300 to 1800 psig (2068 to 12,410 kPa) for the high pressure and intermediate pressure sections and 30 to 100 psig (206.8 to 689.5 kPa) for the low pressure section. To accommodate rapid startup, the design includes provisions for differential expansion between pressure parts and nonpressure parts. Expansion joints are used where a gas-tight seal is required.

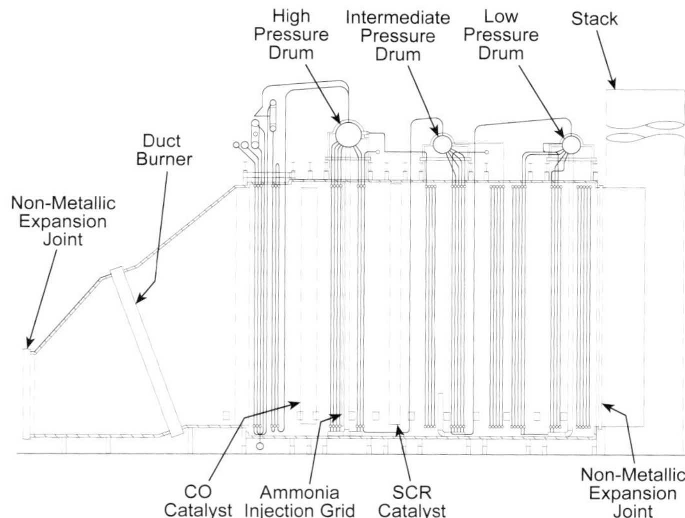

Fig. 21 Heat recovery steam generator (HRSG) arrangement.

HRSGs are designed to handle large gas flows with minimal pressure drop, resulting in greater gas turbine net electrical output. Special consideration is also given to the configuration of the interconnecting flue gas ductwork, transitions, and diverter valves to minimize pressure losses caused by changes in flow patterns or excessively high velocities. Heat losses through the boiler casing and ductwork are minimized through the use of an insulation system.

In the natural circulation design, the vertical tubes provide the pumping head necessary to develop sufficient circulation ratios for stable circulation. Circulation pumps are eliminated, therefore allowing plant efficiency to be maximized. The forced circulation design consists of horizontal tubes with gas flowing vertically. Floor space requirements are minimized. However, the additional load imposed by the circulation pumps reduces cycle efficiency.

Fig. 22 Large HRSG for enhanced oil recovery/cogeneration.

Once-through HRSG designs are also available. In a once-through design, feedwater is converted directly to superheated steam without the need for a drum. As a result, there is no distinction between economizer, evaporator or superheater surface. The water-to-steam transition point is variable within the unit and is dependant upon the heat input and mass flow of the water.

Most heat recovery steam generators are too large to permit shop assembly of an entire unit. Therefore, most HRSG components are shop fabricated to the maximum extent possible and delivered to the site as modules. Typically, each module is structurally independent and includes lifting lugs and temporary steel so that field transport and setting is readily accomplished. Some of the smaller HRSG units can be entirely shop-assembled and skid mounted.

Typical parameter ranges for HRSG units are summarized in Table 4.

Special HRSG designs are also available for enhanced oil recovery applications. The gas turbine exhaust gas is used in the HRSG to generate wet steam (approximately 80% quality) at pressures up to 2500 psig (17.2 MPa). The steam is injected into wells to enhance heavy oil recovery. A unique feature is that relatively dirty (up to 10,000 ppm dissolved solids) feedwater is used in a once-through steam generator design.

Technical considerations The HRSG is basically a counterflow heat exchanger composed of a series of superheater, reheater, boiler (or evaporator), and economizer sections positioned from gas inlet to gas outlet to maximize heat recovery and supply the rated steam flow at proper temperature and pressure. To provide the most economical and reliable design, it is necessary to evaluate the following:

1. Allowable back-pressure – Back-pressure is significantly influenced by the HRSG cross-sectional flow area. Higher back-pressures reduce HRSG cost but also reduce gas turbine efficiency. Back-pressures are typically 10 to 15 in. wg (2.5 to 3.7 kPa) in most units.

Table 4
HRSG Parameters

Turbine application:	
Gas turbine sizes	1 MW to 220 MW
Gas flow	25,000 to 5,000,000 lb/h (0.32 to 630 kg/s)
Gas turbine outlet temperature	≤ 1200F (≤ 649C)
Supplemental firing temperature	≤ 1650F (≤ 899C)
Steam flow:	15,000 to 600,000 lb/h (1.9 to 76 kg/s)
Operating pressures:	
High	≥400 psig (2.76 MPa)
Intermediate	50 to 400 psig (0.34 to 2.76 MPa)
Low	15 to 50 psig (0.10 to 0.34 MPa)
Steam temperature	up to 1005F (541C)
Supplemental fuels	#2 oil, natural gas

2. Steam pressure and temperature – The steam pressure and temperature are selected to provide an economical design. In general, higher steam pressures increase system efficiency but can limit total heat recovery from the flue gas in single pressure HRSGs due to the higher saturation temperature. To overcome this problem, multiple pressure HRSGs are offered. One to four separate pressure sections may be used. The superheater, reheater, boiler and economizer sections at each pressure are arranged to reduce overall cost and increase heat recovery.
3. Pinch point and approach temperatures – The pinch point temperature and approach temperatures have a significant impact on overall unit size. They are illustrated in Fig. 23 for a single pressure HRSG. Small pinch point and superheater approach temperatures result in larger heat transfer surfaces and higher capital costs, while the economizer approach temperature is typically set to avoid economizer steaming at the design point. Experience has generally indicated that the following ranges provide economical and technically satisfactory designs, although lower values may be appropriate in specific applications:

   Pinch point = ΔT_p = 20 to 50F (11 to 28C)
   Superheater approach = ΔT_{SH} = 40 to 60F (22 to 33C)
   Economizer approach = ΔT_E = 10 to 30F (6 to 17C)

4. Stack outlet temperature – As with feedwater temperature, the minimum flue gas exit or stack temperature needs to be controlled to avoid corrosion due to acid condensation. Typical values are identified in Chapter 20.
5. Load response and cycling requirements – Many HRSGs today are cycled. That is, they are brought up and down in load on a daily basis, thus requiring frequent warm-ups and cool-downs. As a re-

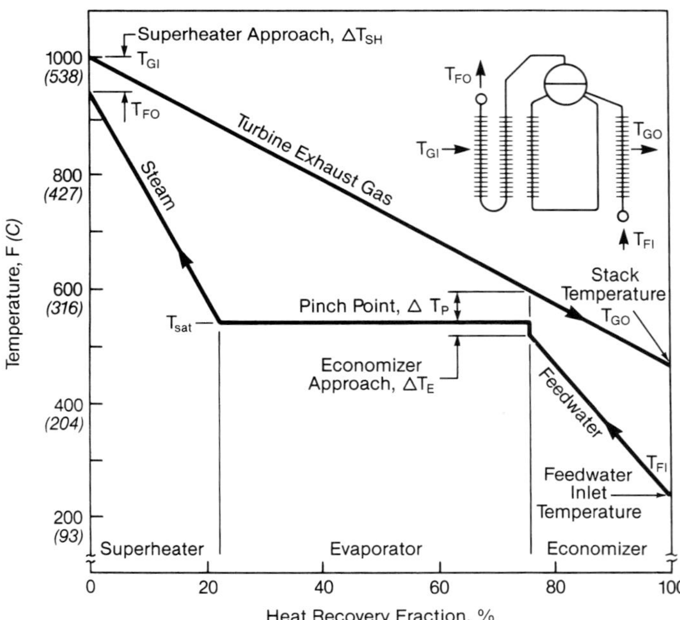

Fig. 23 Temperature profile in single pressure HRSG.

sult, special consideration must be given to the thermal and mechanical stresses associated with this type of operation. Pressure parts and their attachments must be designed in a manner which allows them to expand freely. Steaming in the economizer is inevitable at off-design points. As a result, economizers must incorporate up-flow tube circuitry, recirculation lines, steam-water separation equipment in the drum, or gas bypasses. Stack dampers may also be necessary to keep the HRSG warm if it is to be brought off line for short periods of time.

6. Emissions – Emission requirements can have a physical impact on the design of an HRSG. Depending on permit requirements and the type of fuels fired in the duct burner, an SCR or CO catalyst may be required. In some instances, it may be necessary to have both. The location of either is driven by temperature (see Chapter 34). Therefore, tube banks must be arranged to accommodate the SCR and/or CO catalyst blocks which are sized to meet the removal efficiency required at the design conditions specified. It may also be necessary to install a distributor grid upstream of the catalyst section to distribute the flue gas evenly across the catalyst.

CO boiler

In the hydrocarbon processing industry, the operation of a fluid catalytic cracking (FCC) unit, depending on the FCC arrangement, produces gases rich in carbon monoxide (CO). To reclaim the thermal energy in these gases, the FCC unit can be designed to include a CO boiler to generate steam.

For refineries generating large quantities of CO, field-erected boilers such as the PFI boiler are used. There are also many smaller refineries that have cracking units in the general range of 12,000 barrels (1908 m^3) per day or less which produce from 75,000 to 175,000 lb/h (9.5 to 22.1 kg/s) of CO gas. CO boilers for this capacity are often small enough to be shop assembled. A shop-assembled boiler modified for CO firing is shown in Fig. 24. CO gas is admitted through ports in the side walls and front wall to promote mixing and rapid combustion. The burners, for firing supplementary fuel, are located in a refractory front wall and fire into a horizontal furnace.

The maximum steam requirements of the cracking unit may occur at normal, full load operation or during startup of the cracking unit, depending on the plant steam cycle. The supply of CO is normally not sufficient to generate the maximum amount of steam required; supplementary fuel is then needed.

Supplementary fuel is also required to raise the temperature of the CO gases to the ignition point and to assure complete burning of the combustibles in the CO gas stream. The following design criteria have been established:

1. The basic firing rate should produce a temperature of 1800F (982C) in the furnace for a suitable residence to provide safe and stable combustion of the fuels.
2. Air is supplied by the forced draft fan to provide 2% oxygen leaving the unit when burning CO gases and supplementary fuel.
3. Supplementary firing equipment is provided which is capable of raising the temperature of the CO gases to 1450F (788C), the temperature needed for ignition of the combustibles.

Because of possible variations in the combustible and oxygen content of the CO gases, the sensible heat of the CO gases and the amount of supplementary fir-

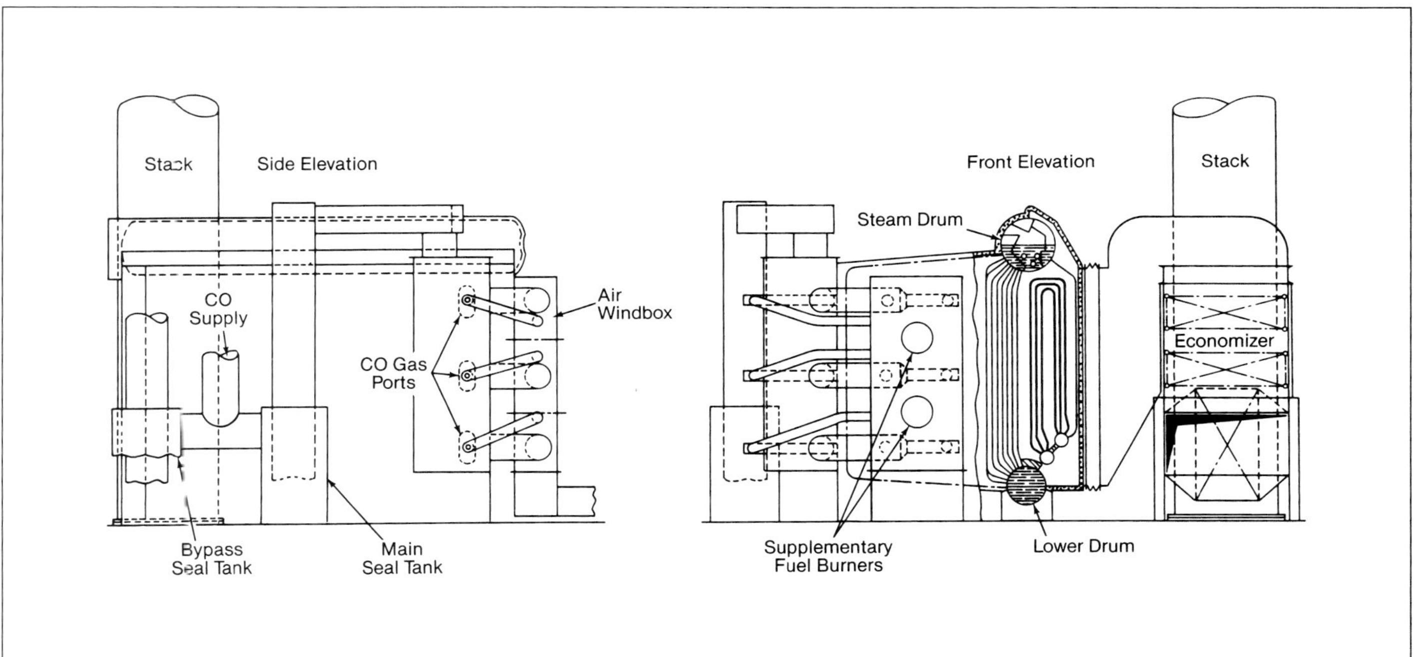

Fig. 24 Shop-assembled CO boiler.

ing, it is impractical to set up a fuel-air relationship. Consequently, it is necessary to directly determine the amount of excess oxygen leaving the unit. This may be determined intermittently by a portable oxygen analyzer, or continuously by an oxygen analyzer or combination oxygen-combustible recorder.

Water seal tanks are installed upstream of the CO boiler to act as shutoff valves in the large gas lines so that the CO gases from the catalyst regenerator may be passed through the boiler or sent directly to the stack. This permits independent operation of the CO boiler without interfering with the operation of the regenerator. Water seal tanks are preferred to mechanical shutoff dampers because of the high gas temperature, the size of the CO ducts, and the need for leak proof construction.

The operation of the CO boiler is coordinated with that of the catalytic cracker. Normally, the boiler will be required to supply steam for the operation of the catalytic cracking unit and will be started using supplementary fuel. The CO boiler should always be started using only the supplementary fuel burners and bypassing the gases from the regenerator to the atmosphere. CO gases should not be introduced into the boiler until it is brought up to temperature because these gases usually are at or below 1000F (538C) and, consequently, tend to cool the furnace. They ignite quite readily and burn with a non-luminous flame. As the CO is introduced to the boiler, it is necessary to reduce the supplementary fuel and the combustion air. This readjustment in the air requirement is determined from the oxygen recorder reading.

Because there are only slight variations in the operation of the catalytic cracking unit, the CO boiler is normally base loaded. It handles all the gases from the regenerator regardless of the carbon dioxide (CO_2) to CO ratio. A change in this ratio merely affects the quantity of supplementary fuel necessary to maintain the required furnace temperature of 1800F (982C). This temperature provides a reasonable operating margin for possible variation in the operation of the regenerator or the boiler. Stable operation can be maintained at a furnace temperature as low as 1500F (816C), but the margin above the ignition temperature of the CO gas is considerably reduced.

The economics of the CO boiler depend on the amount of available heat in the regenerator exhaust compared with an equivalent amount of heat from an alternate fuel. The heat from the CO gases is calculated by taking the sensible heat above an assumed boiler stack temperature plus all of the heat from the combustibles. The additional steam generated in the CO boiler by the supplementary fuel is comparable with the steam generated in a conventional power boiler. Normally, the supplementary fuel requirement will account for one fourth to one third of the output when the temperature of the entering CO gas is maintained at 1000F (538C).

Changes in FCC catalysts and in process conditions have permitted reductions in the CO content of the gases leaving the unit. These changes also result in the gas temperature to the CO boiler increasing from the 1000F (538C) level to as high as 1450F (788C). Many existing CO boilers have been upgraded to handle the new conditions. Changes have included elimination of combustion zone refractory, use of membraned water walls, and resizing of heat transfer surface. (See Fig. 25.) New heat recovery boilers on FCC units are designed for these new process conditions.

General waste heat boiler

The progressive increase in the cost of fuel has fostered technical progress in the utilization of waste energy, including specialized designs and applications of boilers. There are many industries or processes that generate large quantities of high temperature gases from which the sensible heat may be extracted for steam generation. Such gases are produced in catalytic regenerators; blast furnaces; copper reverberatory furnaces; annealing, forge and billet heating furnaces; and fired kilns of many types.

The heat contained in these exhaust gases can often generate all the steam required for an industrial process via properly designed boiler equipment. Where the waste gases carry some of the noncombustible process material in suspension, suitable hoppers will collect a portion of the material, and the cooled gases leaving the boiler may be passed through dust collectors to recover the remaining particulate. Many types of boilers are necessary to meet the wide range of requirements in this field. Boiler design depends on the chemical nature of the gases and their temperature, pressure, quantity and dust loading.

Heat transfer from waste gases

The rate of heat transfer from the gas to the boiler water depends on the temperature and thermophysical properties of the gases, velocity and direction of flow over the absorbing surfaces, and the surface cleanliness, as discussed in Chapter 4. Temperatures of many process gases are relatively low as shown in Table 5.

To obtain the proper velocity of the gases over the surfaces, a sufficient pressure difference or draft must be provided, either by a stack or a fan. The draft must overcome the pressure losses caused by the flow of gases through the unit, with adequate allowance for normal heating surface fouling.

The radiation heat transfer component is low and the tendency is to design many waste heat boilers for higher gas velocities than prevail on fuel-fired units. However, high velocities with dust-laden gases can cause tube erosion, particularly where there are changes in gas flow direction. Therefore, process-specific velocity limits must be met.

Diagrams A and B, Fig. 26, show the approximate convection heating surface required for usual conditions in waste heat boilers.

A water-cooled furnace is a feature of some waste heat boiler applications where it is necessary to cool the furnace gases to the temperature required to prevent slagging in the following tightly spaced convection surfaces. The approximate amount of surface required for such furnace applications is given in Diagram C of Fig. 26.

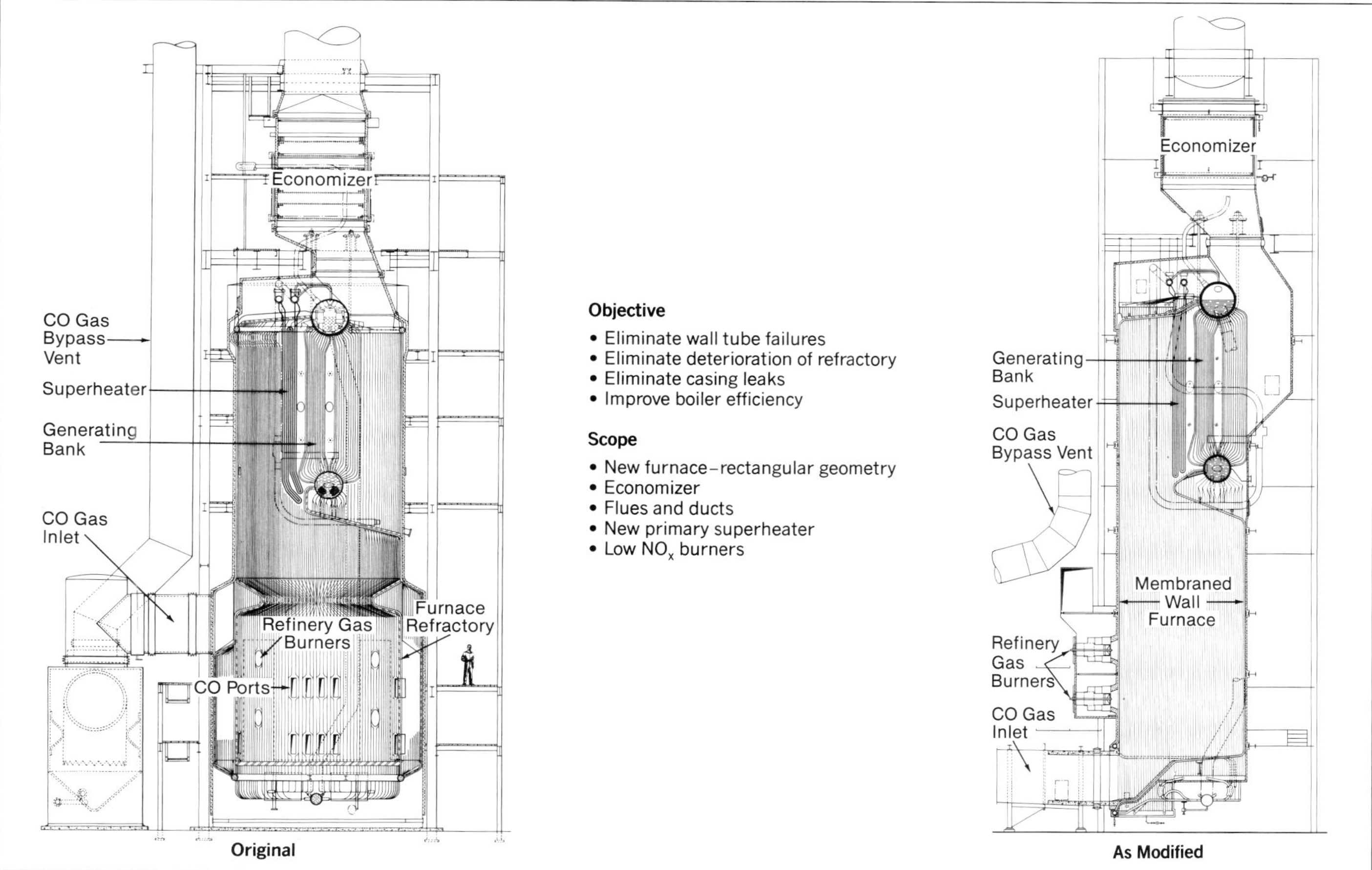

Fig. 25 Circular CO boiler modernization.

Application factors

The design of a boiler for a particular application depends on many factors, most of which vary from process to process and even within an industry. The cost of equipment, auxiliary power and maintenance must be compared with the expected savings. The boiler design depends somewhat on the cost of power at the plant. A smaller unit with closely spaced tubes will cost less but will require more fan power, because of high draft loss. A larger unit will be more expensive but have a lower draft loss and therefore a lower fan power requirement. Other important factors are the available space, locations for the proper flue connections, the corrosive nature of the gases, the effect of dust loading on erosion, and the process operating pressure conditions, i.e., pressurized or induced draft.

If the gases carry dust, attention must be given to tube spacing and to removal of dust dislodged from the heating surfaces. The tubes must be spaced reasonably close for good heat transfer yet far enough apart to prevent bridging of deposits which can lead to excessive gas-side pressure loss. Often the boiler is arranged for wide tube spacing where the gases are hottest, with spacing reduced where the gases are cooler to maintain gas velocities and good heat transfer.

Sometimes the particulate or dust carried into the boiler from the process can be removed by mechanical cleaners or sootblowers (see Chapter 24). In other cases, deposits from processes may require periodic manual cleaning with high pressure air, steam or water to keep the boiler passes open. In either case, suitable hoppers should be provided to collect the deposit material removed from the tubes.

Table 5
Temperature of Waste Heat Gases

Source of Gas	Temperature F	Temperature C
Ammonia oxidation process	1350 to 1475	732 to 802
Annealing furnace	1100 to 2000	593 to 1093
Cement kiln (dry process)	1150 to 1500	621 to 816
Cement kiln (wet process)	800 to 1100	427 to 593
Copper reverberatory furnace	2000 to 2500	1093 to 1371
Diesel engine exhaust	1000 to 1200	538 to 649
Forge and billet heating furnaces	1700 to 2200	927 to 1204
Open hearth steel furnace, air blown	1000 to 1300	538 to 704
Open hearth steel furnace, oxygen blown	1300 to 2100	704 to 1149
Basic oxygen furnace	3000 to 3500	1649 to 1927
Petroleum refinery	1000 to 1400	538 to 760
Sulfur ore processing	1600 to 1900	871 to 1038

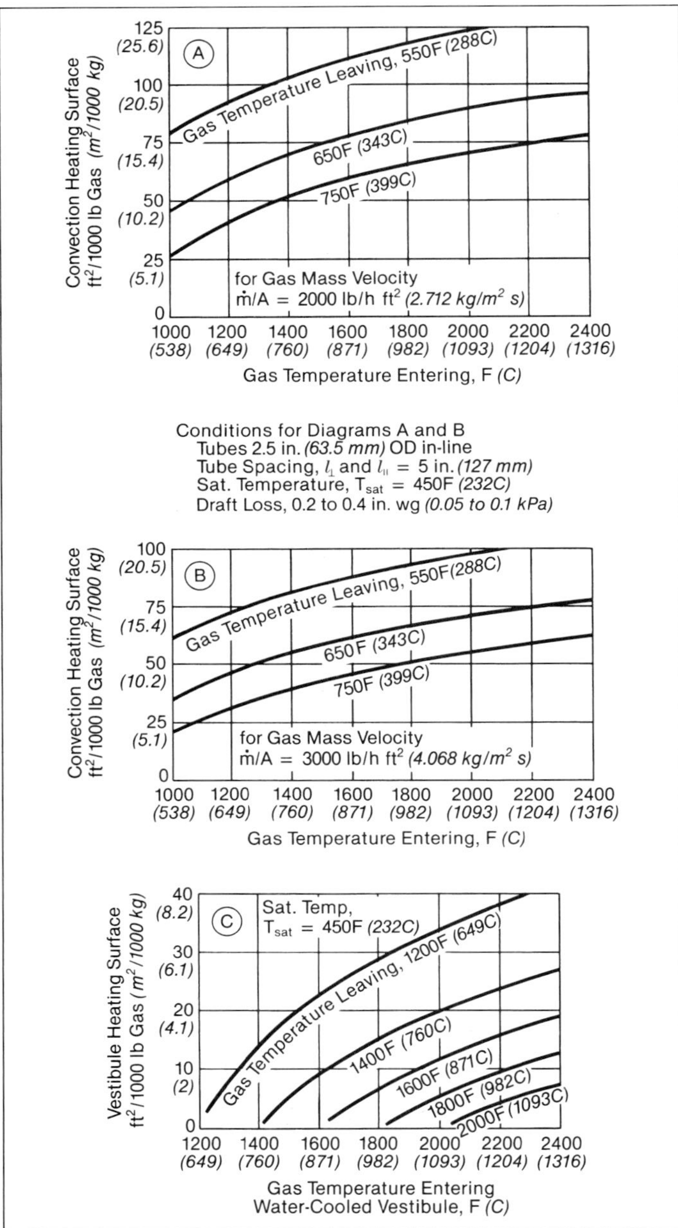

Fig. 26 Approximate surface required in convection tube bank and vestibule for various entering and leaving waste gas temperatures.

Gases from oil- or gas-fired process furnaces are relatively clean and, therefore, can be used in units with 1.0 in. (25.4 mm) clear spacing between bare tubes with little likelihood of deposit bridging and plugging.

Three-drum waste heat boiler An example of a simple three-drum waste heat boiler, specially designed to operate with dust-laden gases, is shown in Fig. 27. This type of unit is particularly well suited for use with high solids content waste gases from cement kilns. Maximum precipitation of solids is assured by the horizontal flow of the gases through the vertical tube banks and by the effective arrangement of baffles. With this tube arrangement, hand lancing is possible from both sides of the unit. Every space in the full width of the boiler can be lanced. Hand lancing can also be done through the roof and above the lower drums, making all absorbing surfaces accessible with a short hand lance.

With high solids content gases, it is often possible to reduce the amount of hand lancing by using long retractable sootblowers. These are located at one or sometimes two elevations along the depth of tube banks at gaps formed by eliminating a single row of tubes.

To maintain optimum heat transfer conditions without changing the direction of gas flow, the tubes in the rear sections of the boiler are more closely spaced than those at the inlet.

Circulation in this boiler is simple, with the boiler tubes in the hot gas end acting as risers and the tubes in the cooler gas zones acting as downcomers or supply tubes. The boiler has a relatively long drum in which steam separation takes place without the use of baffles. The steam is collected in a dry pipe located in the quiet, cool gas end of the drum. Feedwater is thoroughly mixed with the boiler water rising into the steam drum.

Expansion and contraction of the drums and tubes have no effect on the steel casing, firebrick, or insulation. The most common source of brickwork trouble is therefore eliminated and air infiltration is reduced to a minimum. All pressure parts rest on supports located below the lower drums.

The location of the superheater can be altered for superheat temperature requirements. To increase heat absorption further, an economizer can be installed in the flue leaving the boiler. The economizer is arranged for downward flow of gases to aid in the collection of solids.

Solids collected in the hoppers under the boiler and economizer can be readily removed while the boiler is in service. In a single boiler behind a cement kiln, from 20 to 40 t (18.14 to 36.3 t_m) of cement dust may be recovered each day from these hoppers.

Basic oxygen furnace/oxygen converter hood

The basic oxygen furnace (BOF) used in the steel industry is blown with pure oxygen through a retractable water-cooled lance mounted vertically above the

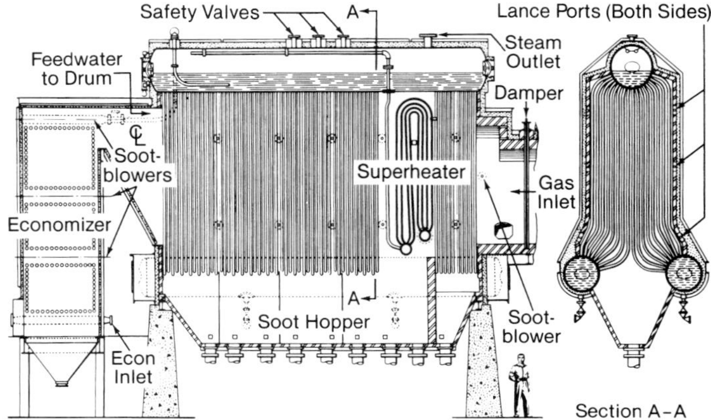

Fig. 27 Three-drum waste heat boiler with lance ports and sootblowers.

furnace. After each charge of molten iron, scrap steel and fluxing material are loaded into the furnace, the oxygen lance is lowered into position above the charged material, and a blowing period of 15 to 20 minutes begins.

During the blowing time, the oxygen starts a chemical reaction which brings the charge up to temperature by burning out the silicon and phosphorous impurities and reducing the carbon content, as required, for high grade steel. Large amounts of carbon monoxide gas, at 3000 to 3500F (1649 to 1927C), are released in this conversion process. The gas is collected in a water-cooled hood and burned with air introduced at the mouth of the hood. The products of combustion are then cooled by adding excess air, injecting spray water, or water cooling of the hood. Combinations of these cooling methods may be used.

The high temperature of the gases discharged from the BOF and their high carbon monoxide content (about 70% by volume) make them ideal for burning in the water-cooled hood located above the BOF. While there are basic similarities to usual boiler service, there are also significant differences, particularly the carryover of iron-bearing slag from the BOF and the short, intermittent operating periods.

The basic oxygen furnace (BOF) hoods are fabricated from membrane wall panels (Fig. 28). The membrane wall can be formed into a variety of hood configurations depending on plant layout. The hood may be a long flue type used to transport the gases to an evaporation or quench chamber, or it may be a bonnet type that collects the gases and immediately discharges them into a spark box where the temperature is sufficiently reduced with spray water for use in the gas cleanup system.

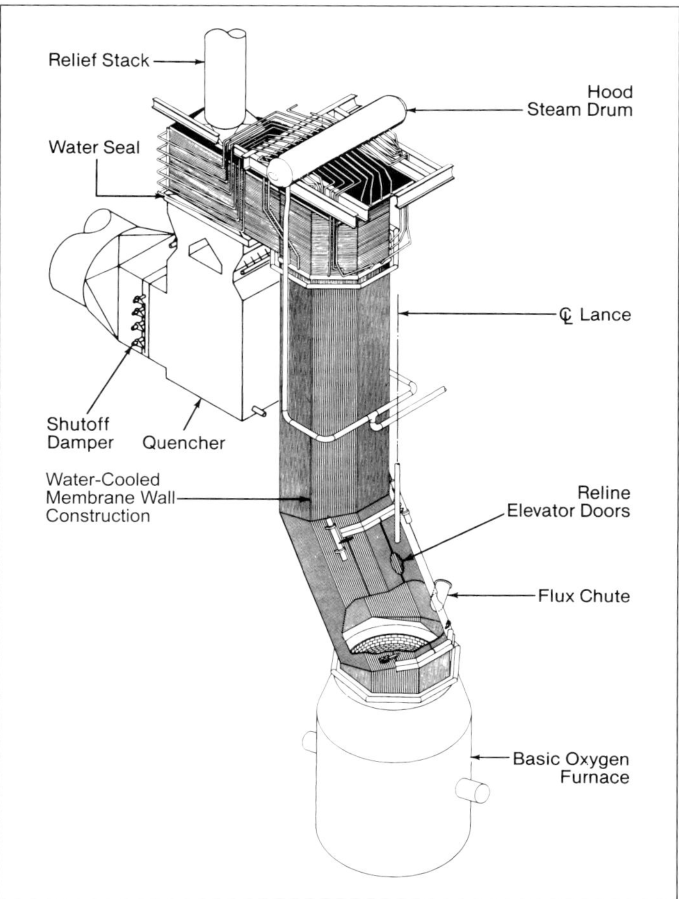

Fig. 28 Oxygen converter hood arrangement.

Steam generator hood The oxygen converter hood, when equipped with a steam drum, boiler circulation pumps, boiler mountings and controls, is an effective steam generator during the oxygen blowing portion of the converter cycle. Steam generation is limited to the time of the oxygen blowing period, which is intermittent or cyclical in operation because of the steelmaking process. The rate of steam generation varies from zero to a maximum and back to zero during this period which is normally about 20 minutes for a converter cycle of 40 to 45 minutes. This cyclic operation, coupled with the outage time required for relining the converter vessel every few weeks, limits the steam production of a single hood to 12 to 15% of the life of the lining.

Pulp and paper facility.

Chapter 28

Chemical and Heat Recovery in the Paper Industry

In the United States (U.S.), the forest products industry is the third largest industrial consumer of energy, accounting for more than 11% of the total U.S. manufacturing energy expenditures. In 2002, 57% of the pulp and paper industry relied on cogeneration for their electric power requirements.

Approximately one-half of the steam and power consumed by this industry is generated from fuels that are byproducts of the pulping process. The main source of self-generated fuel is the spent pulping liquor, followed by wood and bark. The energy required to produce pulp and paper products has been significantly reduced. Process improvements have allowed U.S. pulp and paper manufacturers to reduce energy consumption to 2.66×10^{12} Btu (2806.5×10^{12} J), a significant reduction.

Pulp and paper mill electric power requirements have increased disproportionately to process steam requirements. This factor, coupled with steadily rising fuel costs, has led to the greater cycle efficiencies afforded by higher steam pressures and temperatures in paper mill boilers. The increased value of steam has produced a demand for more reliable and efficient heat and chemical recovery boilers.

The heat value of the spent pulping liquor solids is a reliable fuel source for producing steam for power generation and process use. A large portion of the steam required for the pulp mills is produced in highly specialized heat and chemical recovery boilers. The balance of the steam demand is supplied by boilers designed to burn coal, oil, natural gas and biomass.

Major pulping processes

The U.S. and Canada have the highest combined consumption of paper and paperboard in the world (Fig. 1), consuming 105.6 million tons each year. With a base of more than 800 pulp, paper and paperboard mills, the U.S. and Canada are also the leader in the production of paper and paperboard. North America accounts for 32% of the total world output; pulp production is nearly 43%.

Total pulp production in the U.S. is divided among the following principal processes: 85% chemical, groundwood and thermomechanical; 6% semi-chemical; and 9% mechanical pulping. The dominant North America pulping process is the sulfate process, deriving its name from the use of sodium sulfate (Na_2SO_4) as the makeup chemical. The paper produced from this process was originally so strong in comparison with alternative processes that it was given the name *kraft*, which is the Swedish and German translation for strong. Kraft is an alkaline pulping process, as is the soda process which derives its name from the use of sodium carbonate, Na_2CO_3 (soda ash), as the makeup chemical. The soda process has limited use in the U.S. and is more prominent in countries pulping nonwood fiber. Recovery of chemicals and the production of steam from waste liquor are well established in the kraft and soda processes. The soda process accounts for less than 1% of alkaline pulp production and its importance is now largely historic.

Kraft pulping and recovery process

Kraft process

The kraft process flow diagram (Fig. 2) shows the typical relationship of the recovery boiler to the overall pulp and paper mill.[1] The kraft process starts with feeding wood chips, or alternatively a nonwood fibrous material, to the digester. Chips are cooked under pressure in a steam heated aqueous solution of

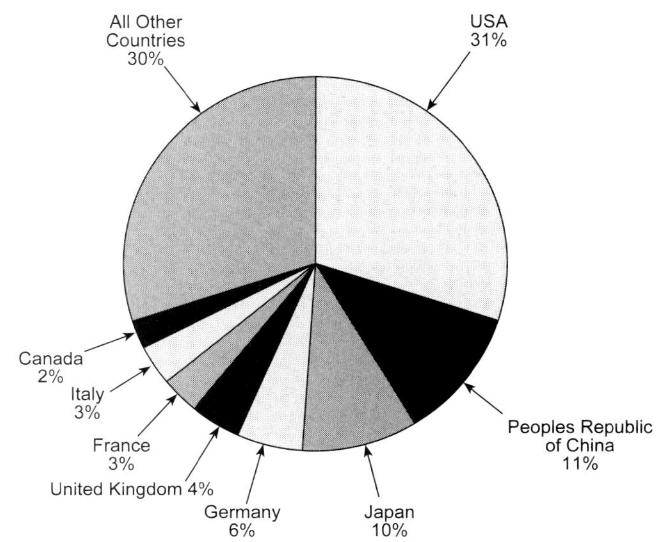

Fig. 1 World paper and board consumption by country, 2000.

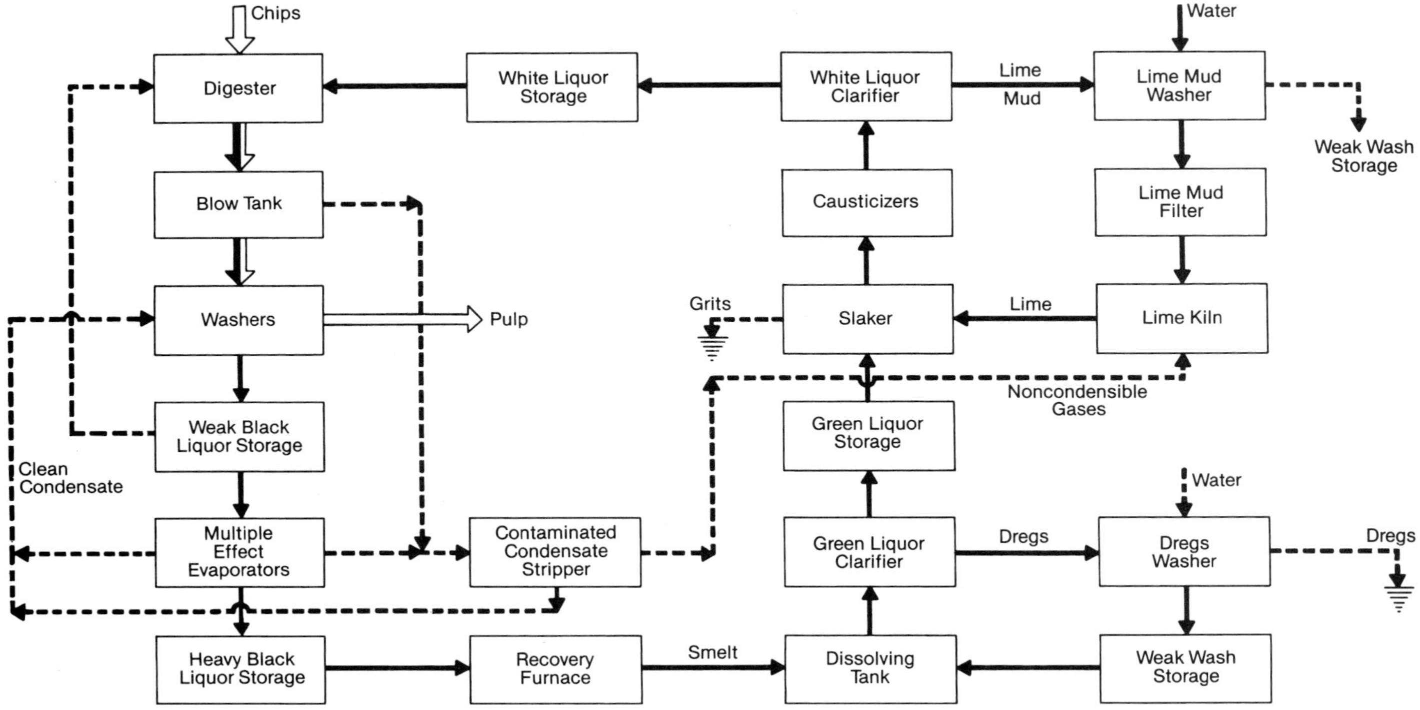

Fig. 2 Kraft process diagram.

sodium hydroxide (NaOH) and sodium sulfide (Na_2S) known as *white liquor* or cooking liquor. Cooking can take place in continuous or batch digesters.

After cooking, pulp is separated from the residual liquor in a process known as brown stock washing. The most common method features a countercurrent series of vacuum drum washers which displace the liquor with minimum dilution. Following washing, the pulp is screened and cleaned to remove knots and shives and to produce fiber for use in the final pulp and paper products.

The *black liquor* rinsed from the pulp in the washers is an aqueous solution containing wood lignins, organic material, and inorganic compounds oxidized in the cooking process. Typically, the combined organic and inorganic mixture is present at a 13 to 17% concentration of solids in weak black liquor. The kraft cycle processes this black liquor through a series of operations, including evaporation, combustion of organic materials, reduction of the spent inorganic compounds, and reconstitution of the white liquor. The physical and chemical changes in the unit operations are shown in Fig. 3.[2]

The unique recovery boiler furnace was developed for combusting the black liquor organic material while reducing the oxidized inorganic material in a pile, or bed, supported by the furnace floor. The molten inorganic chemicals or *smelt* in the bed are discharged to a tank and dissolved to form *green liquor*. Green liquor active chemicals are Na_2CO_3 and Na_2S.

Green liquor contains unburned carbon and inorganic impurities from the smelt, mostly calcium and iron compounds, and this insoluble material, or *dregs*, must be removed through clarification. This operation is basically settling of sediment and decantation of

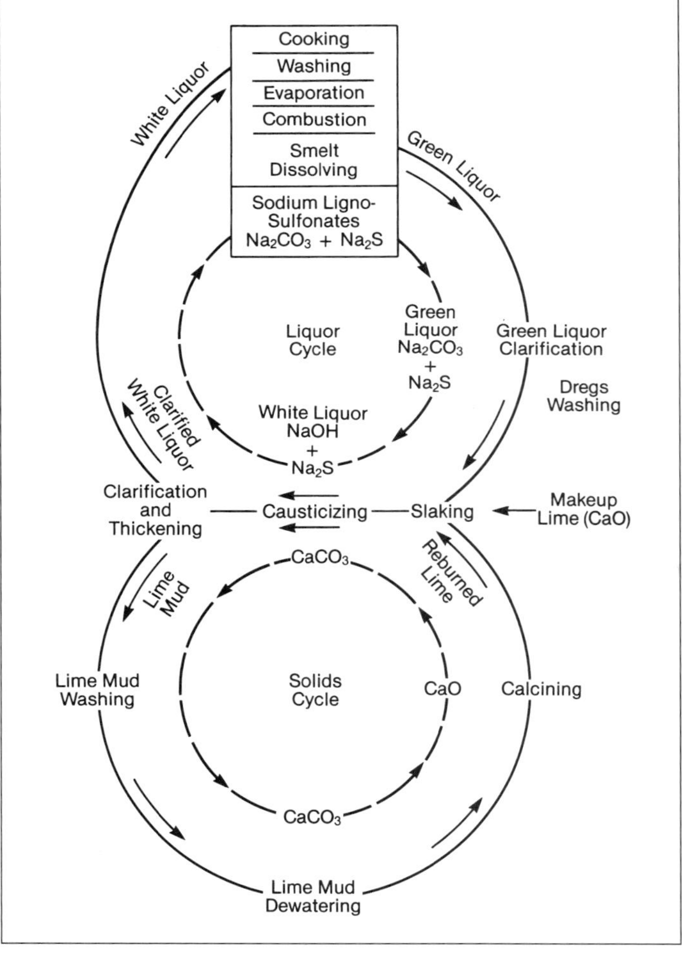

Fig. 3 Kraft process cycle.

clear green liquor that can be pumped to the slaker. The dregs are pumped out of the clarifier as a concentrated slurry. Normal operation is to water wash the dregs before landfill disposal. The water wash liquid containing the recovered sodium chemical is known as *weak wash*. The sodium chemicals are recovered by using the weak wash to dissolve the smelt in the dissolving tank.

Clarified green liquor and lime (CaO) are continuously fed to a slaker where high temperature and agitation promote rapid slaking of the CaO into calcium hydroxide ($Ca(OH)_2$). The liquor from the slaker flows to a series of agitated tanks that allow the relatively slow causticizing reaction to be carried to completion. The function of the causticizing plant is to convert sodium carbonate into active NaOH. The calcium carbonate ($CaCO_3$) formed in the conversion reaction precipitates in the causticizing operation to form a suspended *lime mud*.

The causticizing product must be clarified to remove the $CaCO_3$ precipitate and produce a clear white liquor for cooking. This clarification is carried out by either settling and decanting, in a manner similar to green liquor clarification, or by using pressure filters. In pressure filtration, the white liquor is filtered through a medium to provide a separation of clear white liquor from the lime mud. The lime mud is then washed to remove sodium chemicals that can lead to increased kiln emissions and clinkering, and further filtered to obtain the desired consistency for feed to the kiln.

The lime kiln calcines the washed lime mud feed into reburned lime. Calcination is the chemical breakdown with heat of the $CaCO_3$ into active lime and carbon dioxide (CO_2). The calcined lime is then slaked as previously described.

The reactions occurring in the solids cycle operations are as follows:

Slaking: $CaO + water = Ca(OH)_2 + heat$

Causticizing: $Ca(OH)_2 + Na_2CO_3 = CaCO_3 + 2NaOH$

Calcination: $CaCO_3 + heat = CaO + CO_2$

The combination of these process steps is referred to as *recausticization*.

In parallel with the reduction of sulfur compounds to form smelt, energy is released in the recovery furnace as the black liquor organic compounds are combusted. This combustion energy is used in the process recovery boiler to produce steam from feedwater. The steam can be introduced to a turbine generator to supply a large portion of the energy demand of the pulp and paper mill. Steam extracted from the turbine at low pressure is used for process requirements such as cooking wood chips, evaporation, recovery furnace air heating, and drying the pulp or paper products.

Rated capacity of a recovery unit

The capacity of a pulp mill is based on the daily tons of pulp produced. The primary objectives of a recovery boiler are to reclaim chemicals for reuse and to generate steam by burning the black liquor residue. Accordingly, the capacity of the recovery boiler should be based on its ability to burn or process the dry solids contained in the recovered liquor. Because the proper measure of recovery boiler capacity is the heat input to the furnace, The Babcock & Wilcox Company (B&W) has established a 24 hour heat input unit of 19,800,000 Btu (20,890 MJ). This unit, known as a B&W-Btu ton, corresponds to the heat input from 3000 lb (1361 kg) of solids (approximately equivalent to one ton of pulp produced) having a heating value of 6600 Btu/lb (15,352 kJ/kg) of solids. These were averages for the typical black liquor solids generated from a ton of kraft pulp production when this unit was originally defined. The black liquor solids produced from modern operations generally are characterized by considerable variation in the quantity of the solids per ton of pulp product and somewhat lower heating values. For this reason, a more common rating term applied to current recovery boilers is the amount of dry solids processed over a given period of time (hour or day). The recovery boiler is a heat input machine, and as such, The B&W-Btu rating recognizes the true indication of the unit's design capacity to process that energy input.

The nominal size of a B&W kraft recovery boiler can be determined by application of a simple formula as follows:

$$\text{Nominal size} = \frac{A \times B \times C}{19{,}800{,}000}, \text{ B\&W-Btu tons} \quad \textbf{(1)}$$

where

A = dry solids recovered, lb/t of pulp
B = pulp output of mill, t/24 h
C = heating value of dry solids, Btu/lb

and 19,800,000 is the product of 3000 lb/t and 6600 Btu/lb.

Today's recovery furnace is conservatively designed for a heat release rate (heat input rate divided by furnace plan area) of approximately 0.90×10^6 Btu/h ft^2 (2.84 MW/m^2). This heat release has increased over time, but was always kept below 1.0×10^6 Btu/h ft^2 (3.15 MW/m^2). Although new recovery boiler furnaces are sized for a low heat release rate, there are many recovery boilers that have experienced successful capacity increases at much higher heat release rates. Mills historically have increased pulp production making the recovery boiler a limiting factor at the mill. In order to maintain increased pulp production, the recovery boiler has been called upon to process an ever increasing amount of black liquor solids. Today it is common for a recovery boiler to successfully process solids that result in heat release rates up to 1.25×10^6 Btu/h ft^2 (3.94 MW/m^2). This has been achieved through improvements to the combustion air and liquor delivery systems, sootblower systems, reduction of chlorine in the as-fired liquor, and changes to convection pass arrangements.

There are several criteria commonly used to evaluate the potential success of capacity increases for recovery boilers. In addition to the heat release rate, the furnace exit gas temperature, superheater exit gas temperature, flue gas velocity, and furnace volume are important criteria. Depending on the extent of the capacity increase, most of these criteria can be met with changes to the original design. For example, flue gas temperatures can be reduced with furnace screen and superheater changes. Furnace volume has been modified by expanding the furnace forward, relocating the furnace front wall forward and making the side walls wider thus providing a greater furnace volume for higher rates of liquor solids.

The criteria B&W uses to predict recovery boiler performance is based on operating experience. Some of this has changed over time as equipment improvements have affected operation.

Process flows through the recovery boiler

The kraft process recovery boiler is similar in many respects to a conventional fossil fuel-fired boiler. The concentrated black liquor fuel is introduced into the furnace along with combustion air. Inside the furnace, the residual water is evaporated and the organic material is combusted. The inorganic portion of the black liquor solids is recovered as molten smelt. Most of the sulfur is in the reduced form of Na_2S and most of the remaining sodium is Na_2CO_3. The requirement to recover sulfur in a reduced state is the most unique aspect of recovery boiler design. Fig. 4 illustrates a typical modern recovery boiler.

Combustion air is introduced into the furnace at staged elevations: primary, secondary, tertiary, and at times, quaternary. One-fourth to one-half of the air enters at the primary level near the furnace floor. The balance is staged at the secondary, tertiary and quaternary levels. Heavy black liquor (solids greater than 60%) is fed to the furnace through multiple burners between the secondary and tertiary air levels.

The gases generated by the black liquor combustion rise out of the furnace and flow across convection heat transfer surface. Superheater surface is arranged at the entrance to the convection pass, followed by steam generating surface and finally the economizer. In designs featuring direct contact evaporators, the flue gas may flow from the boiler bank to the evaporator with no economizer surface provided, or a relatively small economizer may be required.

Feedwater enters the recovery boiler at the bottom of the first pass economizer. Heated water from the second pass economizer is discharged into the steam drum. From the drum, saturated water is routed through pipe downcomers to lower furnace enclosure wall headers and the boiler bank. From these steam generating circuits, the steam-water mixture is returned by natural circulation to the steam drum where the mixture is separated. From the drum, steam-free water is again returned to the furnace and boiler bank circuits, and water-free steam is directed to the superheater. After flowing through the superheater sections, the steam leaves the recovery boiler and is typically piped to a turbine-generator.

Boiler thermal performance

The thermal efficiency of a recovery boiler is defined as the ratio of energy output to energy input. The boiler output is a measure of the energy transferred to the feedwater in generating steam and can be expressed as:

$$\text{Output} = \left(H_s - H_{fw}\right) \dot{m}, \text{ Btu/h } \left(J/s\right) \tag{2}$$

where

H_s = enthalpy of steam leaving superheater, Btu/lb (J/kg)
H_{fw} = enthalpy of entering feedwater, Btu/lb (J/kg)
$\dot{m}$ = steam or water flow rate, lb/h (kg/s)

Boiler water is frequently withdrawn from the steam drum as blowdown to maintain steam purity. Steam may also be withdrawn prior to the final superheater stage for use in sootblowing. In these instances, the output expression must be corrected to account for the energy leaving the boiler prior to the superheater outlet.

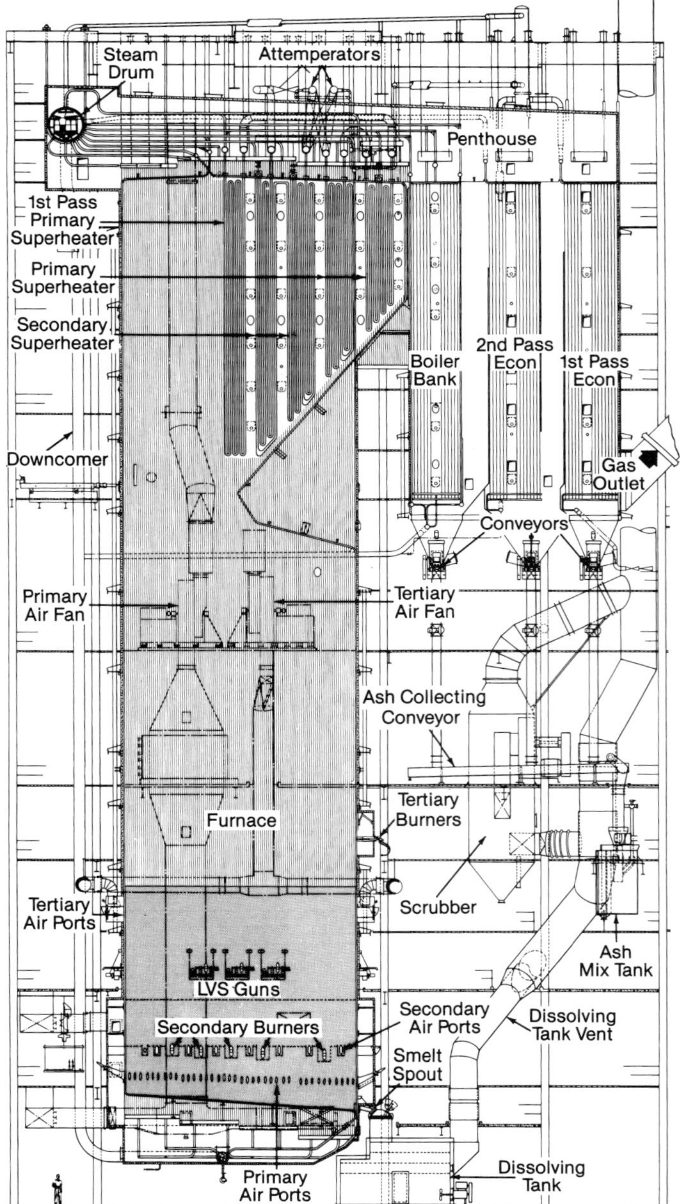

Fig. 4 Typical modern recovery boiler.

The portion of the input energy available to generate steam can be determined by calculating a steady-state heat and material balance around the boiler. Because steady-state output must equal the input less energy losses, boiler efficiency can also be expressed as:

$$\text{Boiler efficiency} = \frac{\text{Output}}{\text{Input}} = \frac{\text{Input} - \text{Losses}}{\text{Input}} \qquad (3)$$

Fig. 5 illustrates the major streams crossing the heat and material balance boundaries. The total heat input can be calculated by summing the chemical and thermal energy contained in the streams entering the boundary. The total losses are then calculated by summing the heat losses due to endothermic reactions occurring within the boiler and the thermal energy losses of the exiting streams.

In practice, it is not feasible to precisely measure all streams entering and leaving the system boundaries. An unaccounted for heat loss and a manufacturer's margin are added to the total losses to correct the calculated efficiency for these limitations.

The gross heating value or chemical energy of black liquor is determined by combusting a black liquor sample with an excess of oxidant, under pressure, in a bomb calorimeter. Under these laboratory conditions, the combustion products predominantly exist as CO_2, H_2O, Na_2CO_3, Na_2SO_4 and sodium chloride (NaCl). A key process in black liquor combustion is the reclamation of sodium compounds in a reduced state. The reduction reactions occurring in the recovery furnace result in different combustion products than those resulting from the bomb calorimeter procedure. These endothermic reactions account for a portion of the black liquor heating value that is not available in the recovery furnace to generate steam. To accurately determine recovery boiler efficiency, the bomb calorimeter gross heating value must be corrected for the heats of reaction of these different combustion products.

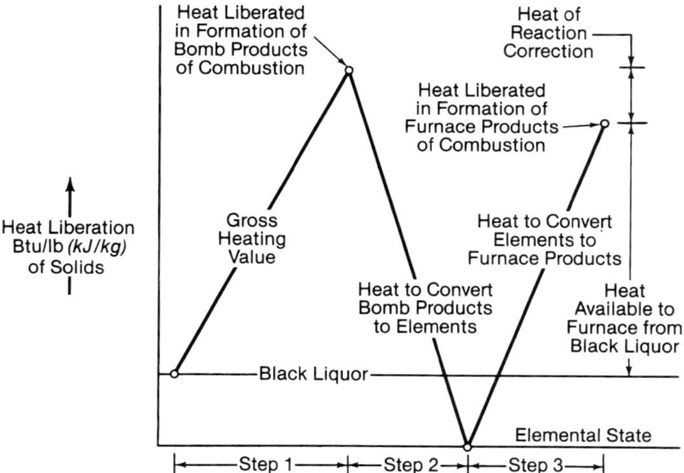

Fig. 6 Determination of black liquor heat of reaction correction.

The heat of reaction correction is the difference between the standard heat of formation of the bomb products and the heat of formation of the furnace products. Application of the heats of formation to determine a reaction correction is illustrated for kraft liquor in Fig. 6.

Step 1 This is the gross heating value of the black liquor sample determined in the bomb calorimeter.

Step 2 From the quantitative analysis of the fully oxidized bomb calorimeter compounds, the heat required to convert these products to their elemental state can be calculated from the standard heats of formation of the compounds from their elements.

Step 3 Similarly, from the quantity of each chemical compound present in the furnace combustion products, the heat of formation for the actual furnace products can be calculated.

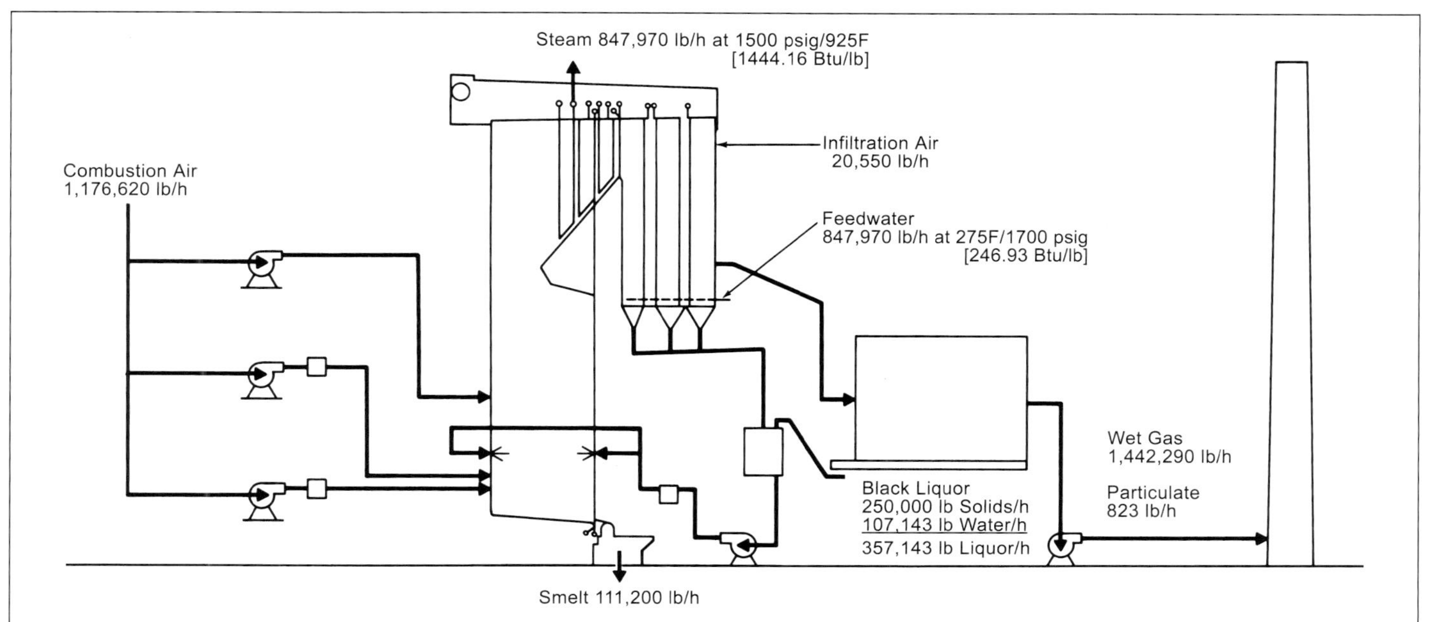

Fig. 5 Plant heat balance diagram. (See also Table 2.)

The difference between Step 2 and Step 3 is the heat of reaction correction.

Sulfur dioxide (SO_2) and Na_2S are the most significant recovery furnace combustion products that differ from those formed under bomb calorimeter conditions. The heat of reaction correction for Na_2S is calculated as follows:

$$\begin{aligned} Na_2SO_4 &= 2Na + S + 2O_2 \quad \text{(Step 2)} \\ 2Na + S &= Na_2S \quad \text{(Step 3)} \end{aligned} \tag{4}$$

The calculation is simplified by combining Steps 2 and 3 and using standard heats of formation:

$$Na_2SO_4 = Na_2S + 2O_2 \tag{5}$$

ΔH_f^0 (Na_2S)	=	89.2 kcal/gmole
ΔH_f^0 (O_2)	=	0.0
ΔH_f^0 (Na_2SO_4)	=	−330.9
Heat of reaction correction	=	−241.7 kcal/gmole
	=	−5550.0 Btu/lb Na_2S (−12,909.0 kJ/kg)

Similarly, the heat of reaction correction for sulfur dioxide can be determined from standard heats of formation from the bomb calorimeter combustion products:

$$Na_2SO_4 + CO_2 = SO_2 + Na_2CO_3 + \tfrac{1}{2}O_2 \tag{6}$$

ΔH_f^0 (SO_2)	=	71.0 kcal/gmole
ΔH_f^0 (Na_2CO_3)	=	270.3
ΔH_f^0 (O_2)	=	0.0
ΔH_f^0 (Na_2SO_4)	=	−330.9
ΔH_f^0 (CO_2)	=	−94.1
Heat of reaction correction	=	−83.7 kcal/gmole
	=	−2360.0 Btu/lb SO_2 (−5489.0 kJ/kg)

In actual furnace operations, there is a variety of partially reduced, partially oxidized combustion products. However, accounting only for the presence of SO_2 and Na_2S in correcting the bomb calorimeter gross heating value closely approximates black liquor combustion in a recovery furnace.

Salt cake makeup and other additives to the black liquor are treated in a manner similar to the heat of reaction correction in calculating recovery boiler efficiency. The heat of formation or gross heating value of the Na_2SO_4 salt cake is accounted for as a contribution to the total system energy input. The subsequent reduction of Na_2SO_4 to Na_2S and O_2 is then taken as a heat loss.

The black liquor elemental analysis and gross heating value are used to determine the chemical and thermal performance of the recovery boiler. A typical black liquor analysis is presented in Table 1.

Table 2 lists the inputs and losses for a recovery unit firing 250,000 lb/h (31.5 kg/s) dry solids at 70% black liquor concentration based on the composition and heating value given in Table 1. Industry practice is to express the various heat losses as a percentage of the total heat input, also shown in Table 2. The system boundaries for the heat and material balance are shown diagrammatically in Fig. 5.

Table 1
Black Liquor Analysis

	Dry solids, % by wt
Nitrogen (N)	0.10
Sodium (Na)	19.77
Sulfur (S)	4.36
Hydrogen (H_2)	3.86
Carbon (C)	35.14
Oxygen (O_2)	34.74
Inerts	0.30
Potassium (K)	1.31
Chlorine (Cl)	0.42
Total solids	100.00

Solids gross heating value = 5985 Btu/lb (13,921 kJ/kg)

The black liquor gross heating value is the predominant energy input to the recovery boiler system. The balance of the input is the sum of the sensible heats contributed by those process streams entering the boiler above a base reference temperature. The black liquor is typically preheated to 230 to 270F (110 to 132C) prior to firing. The majority of the combustion air (primary and secondary) is also generally preheated to promote stable furnace conditions.

The heat of reaction correction is expressed as a heat loss due to the endothermic reduction reactions in calculating recovery boiler efficiency. To determine this heat loss, the fraction of sodium and sulfur converted to Na_2S, Na_2SO_4 and SO_2 must be calculated from the chemical analysis of the smelt and flue gas leaving the recovery boiler. For the example presented in Table 2, 0.099 lb of Na_2S is formed for each pound of black liquor solids entering the recovery boiler. The heat of reaction correction or heat loss associated with the formation of Na_2S is calculated as follows:

$$\frac{0.099 \text{ lb } Na_2S}{\text{lb solids}} \times \frac{5550 \text{ Btu}}{\text{lb } Na_2S} \times \frac{250{,}000 \text{ lb solids}}{\text{h}} = 137.36 \times 10^6 \text{ Btu/h} \tag{7}$$

In addition, 0.0017 lb of SO_2 is formed in the reduction of Na_2SO_4 to Na_2CO_3:

$$\frac{0.0017 \text{ lb } SO_2}{\text{lb solids}} \times \frac{2360 \text{ Btu}}{\text{lb } SO_2} \times \frac{250{,}000 \text{ lb solids}}{\text{h}} = 1.0 \times 10^6 \text{ Btu/h} \tag{8}$$

The heat loss due to the reduction reactions is the sum of these heat of reaction corrections:

$$\begin{aligned} \text{Reduction reaction heat loss} &= 137.36 \times 10^6 + 1.0 \times 10^6 \\ &= 138.36 \times 10^6 \text{ Btu/h} \end{aligned}$$

In addition to the heat of reaction correction and the heat loss attributed to reducing salt cake makeup, energy is lost from the boiler in the form of sensible heat. Heat is also lost through water vaporization and through the molten smelt. Typically, smelt leaving the recovery furnace represents 532 Btu/lb (1237 kJ/kg) of heat consumed to melt the smelt and raise its tem-

Table 2
Material and Energy Balances for a Recovery Boiler Firing 250,000 lb/h Dry Solids at 70% Liquor Concentration

Material balance:			
Entering combustion air	=	1,176,620 lb/h	
Entering infiltration air	=	20,550 lb/h	
Entering black liquor	=	357,143 lb/h	
Total in		1,554,313 lb/h	
Smelt leaving	=	111,200 lb/h	
Wet gas leaving	=	1,442,290 lb/h	
Particulate leaving	=	823 lb/h	
Total out		1,554,313 lb/h	
Energy balance:		10^6 Btu/h	% Total
Chemical heat in liquor	=	1496.25	94.50
Sensible heat in liquor	=	37.47	2.37
Sensible heat in air	=	49.59	3.13
Input		1583.31	100.00
Sensible heat in dry gas	=	101.49	6.41
Moisture from air	=	2.37	0.15
Moisture from hydrogen	=	104.82	6.62
Moisture from liquor	=	129.20	8.16
Reduction reactions	=	138.36	8.74
Heat in smelt	=	71.25	4.50
Radiation	=	4.75	0.30
Unaccounted for and manufacturer's margin	=	15.83	1.00
Losses		568.07	35.88

Boiler efficiency =

$$\frac{\text{Input} - \text{Losses}}{\text{Input}} = \frac{1583.31 - 568.04}{1583.31} = 64.12\%$$

Output = Efficiency x Input =

$$\frac{64.12}{100} \times 1583.31 \times 10^6 = 1015.22 \times 10^6 \text{ Btu/h}$$

Steam flow =

$$\frac{\text{Output}}{H_s - H_{fw}} = \frac{1015.22 \times 10^6}{(1444.16 - 246.93)} = 847{,}970 \text{ lb/h}$$

perature to a nominal 1550F (843C). The balance of the heat losses are determined in a manner similar to those for conventional power boilers. (See Chapter 22.)

The contribution to boiler efficiency offered by the remaining streams crossing the recovery unit heat and material balance is primarily established by their sensible heat content at the temperature at which they cross the system boundary. The minimum temperature of the flue gas leaving the boiler is selected to minimize corrosion. The heat transfer surface arrangement and thermodynamic considerations then dictate the economic limit for the flue gas exit temperature, typically 350 to 400F (177 to 204C).

The gross heating value of a given black liquor sample is strongly influenced by its carbon content. As this content increases, the heating value typically increases, as illustrated in Fig. 7. An increased liquor heating value also generally corresponds to an increased hydrogen content, with a corresponding decrease in inorganic sodium and sulfur contents. These factors result in an increased quantity of theoretical air (see Chapter 10) required to combust the black liquor. The overall trends can be summarized as follows:

1. Carbon (and hydrogen) content increases with increasing heating value.
2. Inorganic sodium and sulfur contents decrease with increasing heating value.
3. Theoretical air increases with increasing carbon and hydrogen contents.
4. Theoretical air increases with increasing heating value.

These trends can be used as quick checks on laboratory results for a given black liquor sample chemical analysis and gross heating value.

Black liquor as a fuel

Black liquor

Black liquor is a complex mixture of inorganic and organic solids partially dissolved in an aqueous solution. Heavy or strong black liquor introduced to the recovery furnace ranges from 60 to 80% solids by weight. The organic fraction of the solids is principally derived from the hemicellulose and the lignin removed from the cellulose strands of the wood chips. The solids' inorganic fraction is primarily Na_2CO_3, sodium hydrosulfide (NaHS) and oxidized sulfur compounds. Black liquor also contains various chemical elements which enter the process with the wood, as impurities in makeup limestone and salt cake, and as contaminants in makeup water. These elements include potassium, chlorine, aluminum, iron, silicon, manganese, magnesium and phosphorous. The waste stream from a chlorine dioxide generator in a bleached mill can also contribute NaCl. Potassium and chlorine directly impact the recovery boiler design and operation if they are present in the black liquor in sufficient quantities.

It is not uncommon for many other waste streams and effluents from the mill to be added to the black liquor. This may include soap, various brines or other effluents from the bleach plant, or a variety of other streams. These may significantly affect the chemical composition or heating value of the black liquor fuel, which can have major impacts on performance, operations, emissions and/or cleanability of the recovery boiler.

Black liquor is sprayed into the furnace as coarse droplets which fall to the floor in a dry and partially combusted state to form a char bed. The mounded bed

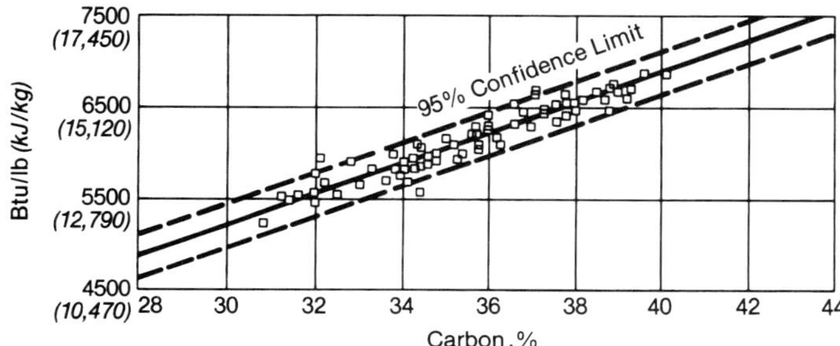

Fig. 7 Black liquor high heating value as a function of carbon content in dry solids.

consists of a matrix of carbon and inorganic sodium chemicals rising 3 to 6 ft (0.9 to 1.8 m). The black liquor droplets sprayed into the furnace must be large enough to minimize droplet entrainment in the rising combustion gases, yet small enough so that they fall to the bed nearly dry. Wet liquor droplets reaching the bed can quench the burning char and cause a bed blackout, or can result in high sulfur emissions.

The design of the recovery furnace must promote combustion of the black liquor in parallel with the efficient reduction of sodium compounds. The absolute reduction efficiency is determined by the degree, on a weight percent, to which sulfur is present in the smelt in a reduced state, such as Na_2S and NaHS.

$$\text{Reduction efficiency} = \frac{Na_2S + NaHS}{\text{Total sodium sulfur compounds}} \times 100\%,\ Na_2O \quad (9)$$

A common industry simplification is as follows:

$$\text{Reduction efficiency} = \frac{Na_2S}{Na_2S + Na_2SO_4} \times 100\%,\ Na_2O \quad (10)$$

The industry practice is to express the compounds in the equations as the equivalent weight of Na_2O.

Emissions

The black liquor combustion process is never theoretically complete. This results in small concentrations of unburned combustibles, typically carbon monoxide (CO), organic and sulfur compounds, and hydrogen sulfide (H_2S), being discharged to the atmosphere. The volatile organic compounds, or VOC, are generally expressed in terms of equivalent methane (CH_4) and are sometimes more specifically referred to as nonmethane volatile organic compounds (NMVOC). H_2S and sulfur-bearing organic compounds such as mercaptans are grouped together as total reduced sulfur (TRS). Trace amounts of SO_2 also exist in addition to TRS-bound sulfur. As in most combustion processes, nitrogen oxides (NO_x) are present and are expressed in terms of equivalent nitrogen oxide (NO_2). Black liquor combustion also creates particulate matter.

The modern recovery boiler achieves effective NO_x control by staged air combustion, control of excess air, and a uniform distribution of the black liquor through multiple burners. A recovery furnace inherently produces lower NO_x emissions compared to fossil fuel boilers. Burning 68 to 75% solids concentration black liquor, NO_x levels would normally be expected to be below 100 ppm.

Recovery boilers can often be upgraded in capacity with little or no increase in NO_x emissions, as air port sizes and locations can be modified to allow adequate control of operations, and emissions, when increasing solids input. Numerical modeling of the existing air system using actual operating data provides the engineer with accurate predictions of NO_x and particulate emissions before hardware is installed.

SO_2 emissions are a function of the sulfidity of the smelt. Fig. 8 shows the general variation of SO_2 with sulfidity. An environmental benefit of increased black liquor concentrations fired in recovery boilers is a reduction in SO_2 emissions.

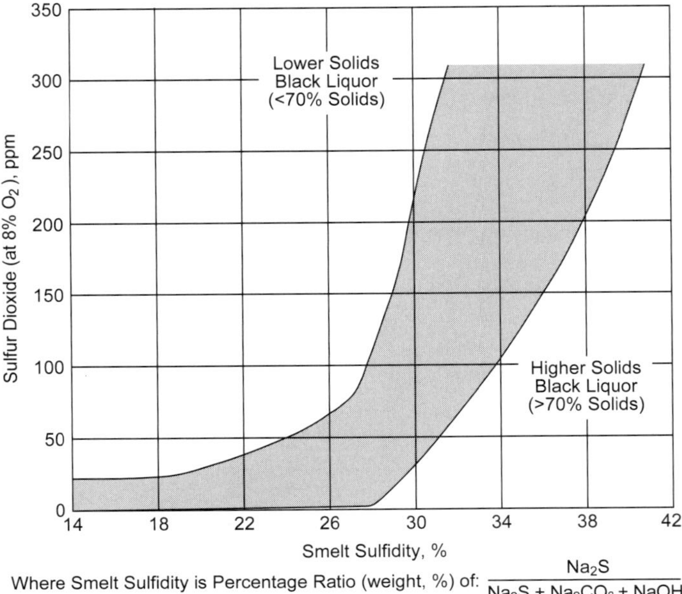

Fig. 8 Sulfur dioxide emissions.

Concentrations of TRS in the combustion gases leaving a modern boiler are readily controlled below 5 ppm, as the H_2S and volatile organic sulfide compounds are oxidized in the high temperature furnace. VOC and CO emissions can be controlled by proper furnace design and operation. VOC can be maintained below approximately 80 ppm (current Environmental Protection Agency/EPA limit) while CO emissions are controllable to less than 500 ppm. A hot furnace and thorough mixing of combustion air with the generated volatiles are essential in minimizing VOC, TRS and CO emissions. Particulate is removed from the combustion gases in a high efficiency electrostatic precipitator, which can control the stack discharge particulate to the current EPA limit of 0.044 gr/dscf.

Ash

Ash accumulations on heat transfer surfaces create an insulating barrier that reduces heat transfer to the boiler tubes. Consequently, as the ash deposits build and heat transfer decreases, steam outlet temperatures decay and the flue gases retain higher temperatures as they pass through the boiler surfaces. Higher flue gas temperatures lead to more ash in its molten phase being carried further back in the boiler convection pass where it adds to already present accumulations. As ash deposits grow, they also begin restricting gas flow through the boiler, plugging the gas passes, and eventually increasing the furnace draft loss to inoperable levels.

The characteristics of ash from black liquor combustion impact the design of the process recovery boiler. Approximately 45% by weight of the dry, as-fired solids is inorganic ash. The majority of these inorganics are removed from the furnace as Na_2S and Na_2CO_3 in the molten smelt. A significant amount of ash is present as particulate entrained in the existing flue gases. Generally, about 8% by weight of the entering black liquor dry solids leaves the furnace as ash.

Ash is generally categorized as *fume* or *carryover*. Carryover consists of char particles and black liquor droplets that are swept away from the char bed and liquor spray by the upward flue gas flow. Entrainment occurs when small particles caught in the furnace gases are not of sufficient size, shape or density to fall back into the furnace. Entrainment results in combustion of black liquor in the upper furnace which affects temperature and ash deposit properties. Entrainment of smelt and char materials is a major cause of convection surface plugging.

Once entrained, the black liquor carryover droplet follows the gas flow. When complete particle burnout occurs, the entrained droplet can settle out of the gas flow as a smelt bead. Otherwise, the partially combusted particles form sparklers that deposit on tubes as char, then continue to burn and yield a smelt deposit. At low loads with lower furnace gas flow rates, entrained droplets have time to burn out, and only small smelt droplets show up as carryover. As load is increased, larger drops can be entrained by the correspondingly increased gas flow. The particles can then include small smelt droplets and large char particles. Carryover can be controlled by furnace size and by proper design and operation of the firing and combustion air systems.

Fume consists of volatile sodium compounds and potassium compounds rising into the convection sections of recovery boilers. These volatiles condense into submicron particles that deposit onto the superheater, boiler bank and economizer surfaces. Fume particles in kraft recovery boilers are usually 0.25 to 1.0 microns in diameter and consist primarily of Na_2SO_4 and a lower content of Na_2CO_3. Fume also contains potassium and chloride salts.

The much larger carryover particles, typically greater than 100 microns, are easily distinguishable from the submicron fume particles. Fume and carryover ash are also different in their chemical analyses. Carryover is similar in composition to the smelt (see comments above). Fume is mostly Na_2SO_4 and is enriched in potassium and chloride relative to their concentration in the smelt.

Fume can contribute to deposit formation and plugging in the convection heat transfer sections of the boiler, particularly if allowed to sinter and harden. Fume particles are also the predominant source of particulate emissions from recovery boiler stacks.

A third category of ash is represented by intermediate size particles (ISP), a class of particles between carryover (> 100 microns) and submicron fume, which are produced during combustion of black liquor droplets and the char bed. While ISPs are abundant, exist everywhere, and are potentially important to fouling in recovery boilers, the mechanisms by which these particles are formed are not yet well understood.

Char bed temperature controls the fuming rate. A rate just sufficient to capture the sulfur released during combustion should be established. This minimizes the dust load to the precipitator and the SO_2 to the stack. More fuming than that sufficient to capture the sulfur causes the excess alkali to be converted to carbonate in the ash. Less fuming than that sufficient to capture sulfur causes excess sulfur to be released as SO_2, and chlorine to be released as HCl in the stack gas, rather than being converted to chloride in the ash.

As potassium (K) has a higher vapor pressure than sodium (Na), the fume contains a higher ratio of K/Na than that found in smelt or black liquor. This is referred to as potassium enrichment. Chlorides are also found at higher concentrations in fume. Potassium and chlorides can contribute to severe plugging in the recovery boiler convection surfaces.

Ash fouling and gas path plugging within recovery boilers are directly related to the melting properties of the ash in the boiler. Chloride and potassium concentrations within the liquor cycle are the most significant factors affecting ash melting points. High chloride concentrations are the result of high chloride sources such as in the wood supply or makeup chemicals. Environmental improvements within the last decade have reduced chemical losses throughout the system, and increased various dead-load chemicals, including chlorides, within the liquor stream.

One of the main factors in determining ash deposition rates is the ash stickiness.[3] Ash stickiness is a function of the amount of liquid phase present in the ash, that is dependent upon ash chloride content and temperature.

A simplified relationship between the ash chloride content and the sticky temperature is shown in Fig. 9. This graph may be used to illustrate the anticipated effects of elevated ash chloride levels in general terms. The conditions of this particular graph are at a 5% potassium molar ratio. Notice that the sticky band on this graph covers the widest temperature range from 5 to 10% molar chloride levels and would be in a narrow range at higher and lower chloride levels.

Ash enrichment by chloride reduces the ash melting point and increases the deposit sintering rate. The characteristic sticky temperature has been defined as the temperature where 15% of the alkali salt mixture is liquid. The presence of K in combination with chlorides further reduces the sticky temperature of depos-

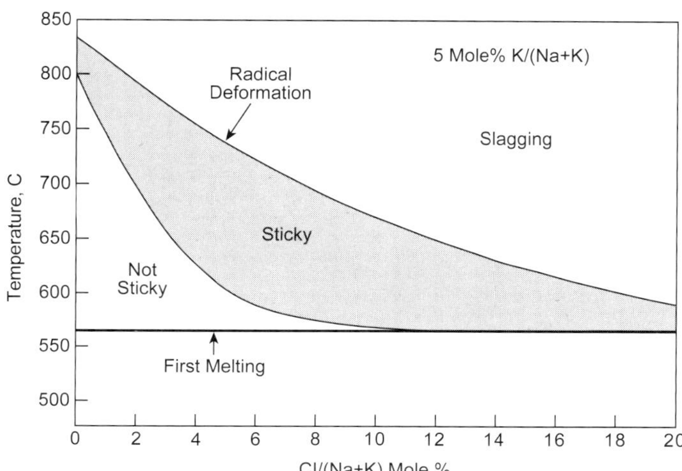

Fig. 9 Sticky temperature.[4]

its. The recovery boiler should be designed to reduce the gas temperature entering the boiler bank to below the ash sticky temperature to avoid bank plugging. Decreasing the Cl level in the black liquor can also decrease the plugging tendency of the resulting ash. The successful reduction of chloride levels has been achieved by ash purging. When purging ash (see *Ash system*), ash is removed periodically to maintain chloride levels at a specified target. For the most part, maintaining a chloride level in the precipitator ash of 1.5% or less has proven to be successful in reducing fouling and plugging within recovery boilers.

Recovery boiler design evolution

The kraft recovery process evolved in Danzing, Germany approximately 25 years after the soda process was developed in the United Kingdom in 1853. In 1907, the kraft recovery process was introduced in North America. From its inception, a variety of furnace types competed for a successful commercial design, including rotary and stationary furnaces. During the late 1920s and early 1930s, significant design developments were achieved by G.H. Tomlinson, working in conjunction with B&W engineers.

The first Tomlinson recovery boiler was supplied by B&W Canada in 1929, at the Canada Paper Company's Windsor Mills, Quebec plant (Fig. 10). This black liquor recovery boiler had refractory furnace walls that proved costly to maintain. The steam generated with the refractory furnace was also much less than that theoretically possible. Tomlinson decided that the black liquor recovery furnace should be completely water-cooled, with tube sections forming an integral part of the furnace. This new concept boiler, designed in cooperation with B&W, was installed at Windsor Mills in 1934. The water-cooled design was a complete success, and the boiler operated until 1988. The first Tomlinson recovery boilers in the U.S. were two 90 B&W-Btu t/day units sold to the Southern Kraft Corporation in Panama City, Florida in 1935.

The Tomlinson design evolved with a technique of spraying black liquor onto the furnace walls. The liquor is dehydrated in flight and on the furnace walls, where pyrolysis begins with the release of volatile combustibles and organically bound sodium and sulfur. As the liquor mass builds on the furnace walls, its weight eventually causes it to break off and fall to the hearth. There, pyrolysis is completed and the char is burned, providing the heat and carbon required in the reduction reaction.

By the end of World War II, the recovery boiler design (Fig. 11) had evolved to the general two-drum arrangement that represented B&W's standard product until the mid-1980s. Retractable sootblowers using steam as a medium eliminated hand lancing in the 1940s; this significant development made large recovery boiler designs practical.

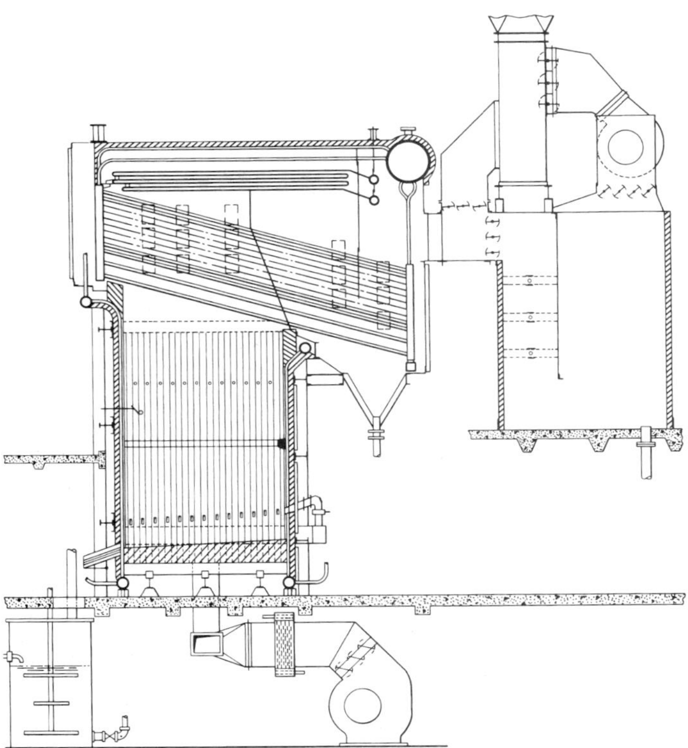

Fig. 10 First Tomlinson recovery boiler.

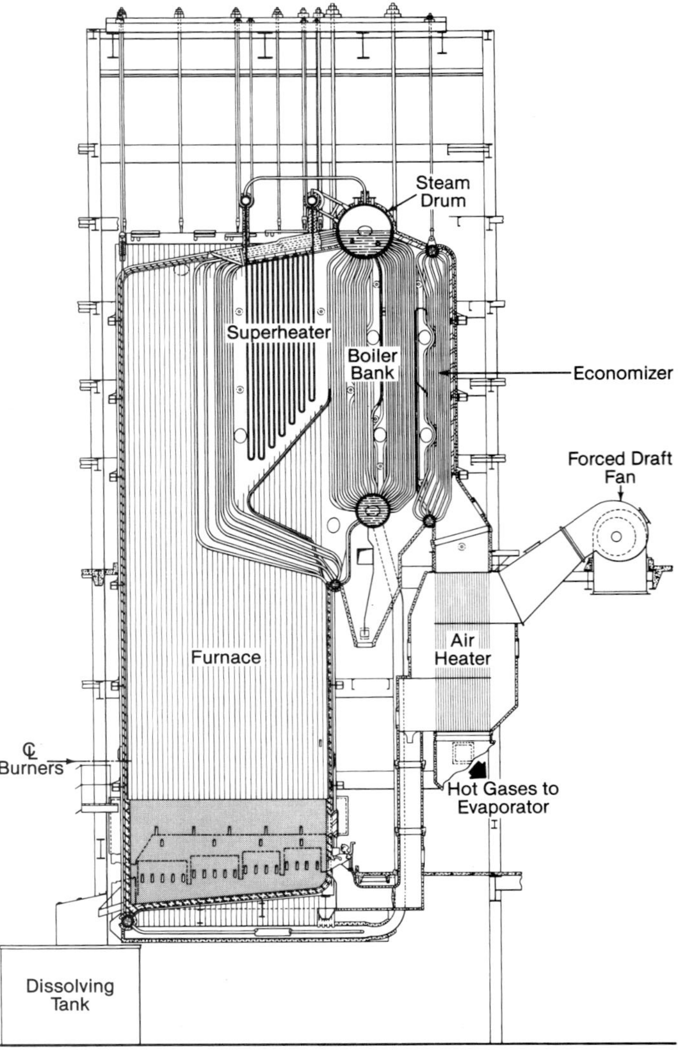

Fig. 11 General two-drum arrangement of the 1940s.

Wall and floor construction

By 1946, furnace wall construction had evolved from tube and refractory designs to a completely water-cooled furnace enclosure, using flat plate studs to close the space between tubes and to minimize smelt corrosion and the resultant smelt leaks. The flat stud design was superseded in 1963, with membrane tube construction where the gas-tight seal is along the plane of the wall rather than formed by casing behind the wall.

The 1963 furnace wall construction had 3 in. (76 mm) outside diameter (OD) tubes on 4 in. (102 mm) centers. The advantages of this construction included less air infiltration, reduced refractory maintenance, and a completely gas-tight unit. The design used cylindrical pin studs for corrosion protection of the tubes in the reducing zone of the lower furnace. The pin studs held solidified smelt, forming a barrier to the corrosive furnace environment. The current construction calls for 64 half-inch (13 mm) diameter studs per linear foot (0.3 m) of tubing.

The lower furnace design continued to evolve in the 1980s from the traditional pin stud arrangement to the use of composite or bimetallic tubes. The composite tubes are comprised of an outer protective layer of AISI 304L stainless steel and an inner core layer of standard American Society of Testing and Materials (ASTM) A 210 Grade A1 carbon steel. The composite tube inner and outer components are metallurgically bonded. The outer layer of austenitic stainless steel, which is also used to cover the furnace side of the carbon steel membrane bar, protects the core carbon steel material from furnace corrosion.

Not long after the introduction of 304L composite tubes, issues with sodium hydroxide attack of the stainless layer near air ports, and cracking of the 304L layer and tube-to-membrane weld, were discovered. The cracking and corrosion have typically been specific to the floor and primary air ports. Considerable investigation has occurred with other materials identified as probable substitutes for 304L. Differential expansion of the two layers has been identified as an issue. Table 3 shows the coefficients of expansion and tensile strength for different materials commonly used to protect recovery boiler lower furnace tube surfaces.

Several methods are used for lower furnace corrosion protection. Lower furnace protection should be determined from criteria such as steam pressure, the history of furnace corrosion, capital budget, and maintenance expectations. The list of common means of protection includes chromized carbon steel tubes, chromized pin studs, carbon steel pin studs, metallic spray coatings, high density pin studs, 304L, Alloy 825 and Alloy 625 composite tubes, and weld overlay of carbon steel tubes. Normally, carbon steel tubes are used below 900 psig (6.2 MPa) and tubes with an alloy outer surface are used above this pressure.

A new or rebuilt high pressure recovery boiler may have several different tubes in the lower furnace (see Fig. 12). The floor may be carbon steel with pin studs. To provide additional circulation margin in the floor, and protect against extreme heat absorption upset conditions, the use of multi-lead ribbed (MLR) tubing has become common, and has become the norm in carbon steel floors regardless of whether the floor is sloped or flat. It is also common for carbon steel floors to use tubes with an alloy outer surface for the first 3 ft (0.9 m) from each side wall, as these tubes can have accelerated corrosion rates over the remainder of the floor due to the continued exposure to molten smelt. The vertical walls from the floor to an elevation above the primary air port elevation would normally be 825 composite or 625 weld overlay. Above this elevation, up to approximately 3 to 15 ft (0.9 to 5 m) above the tertiary air ports (or quaternary zone), 304L can be used to reduce costs.

Materials continue to be studied and evaluated in the laboratory and in operating recovery boilers for corrosion protection.

Table 3
Properties of Materials Used to Protect Recovery Boiler Lower Furnace Tube Surfaces

Property	Carbon Steel	Type 304L	Incoloy 825	Inconel 625
Expansion coefficient Mean to 700F (371C) x 10^{-6} (in./in. F)	7.59	9.69	8.3	7.5
Thermal Conductivity at 700F (371C) (Btu/h ft² F)	320	142	115	117
Ultimate Tensile Strength ksi at 1000F (538C)		56	86	132

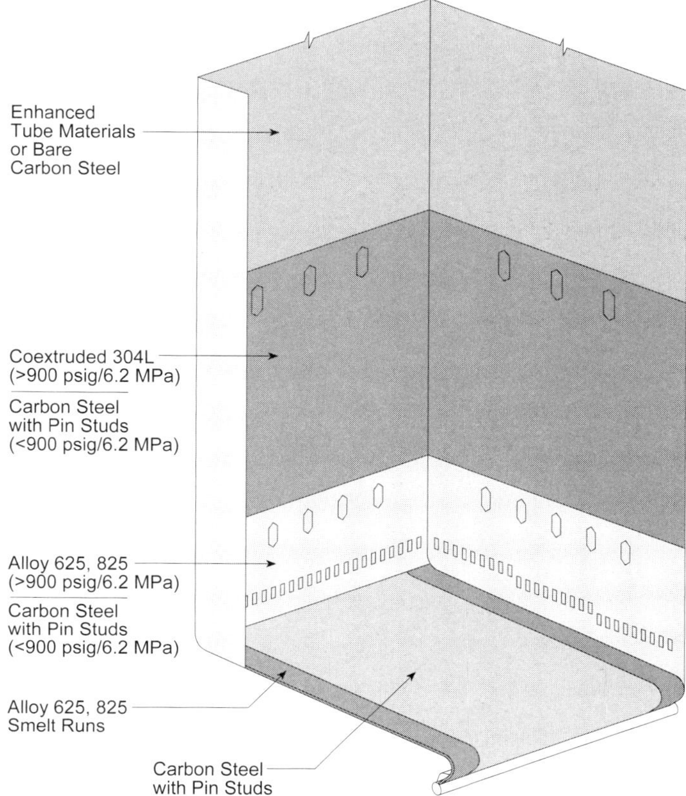

Fig. 12 Material zones in lower furnace.

Increased industry emphasis on high pressure and temperature operation, along with the higher availability demanded of large boilers and the trend of many mills being dependent upon a single recovery boiler, have required the decreased maintenance afforded by modern composite tubes. The single-drum boiler designed in 1987 featured readily available 2.5 in. (64 mm) OD composite tubes with 0.5 in. (13 mm) wide membrane bars. Fig. 13 chronicles the evolution of furnace wall construction with the decreasing width between tube seal bars.

Two-drum generating bank

The two-drum generating bank has evolved from a multi-pass design to a single pass design, common since the early 1960s. The multi-pass design generally had two flue gas passes, with the gases entering the bank near the top, being directed downward with a tile baffle wall integral to the bank, then turning and flowing upward, exiting the bank near the top and flowing out to the economizer.

In the 1960s, the generating bank became a single gas pass bank. The flue gas enters and traverses the entire generating bank height in one horizontal crossflow pass to the outlet flue. The tube spacing is 5 in. (127 mm) side spacing using 2.5 in. (64 mm) tubes. The generating bank screen, tubes between the superheater and generating bank, was also originally on 5 in. (127 mm) side spacing and became an area for pluggage, especially when the original capacity of the recovery boiler was exceeded. Today, an existing generating bank screen can be rearranged to provide a wider 10 in. (254 mm) side spacing to alleviate this pluggage potential.

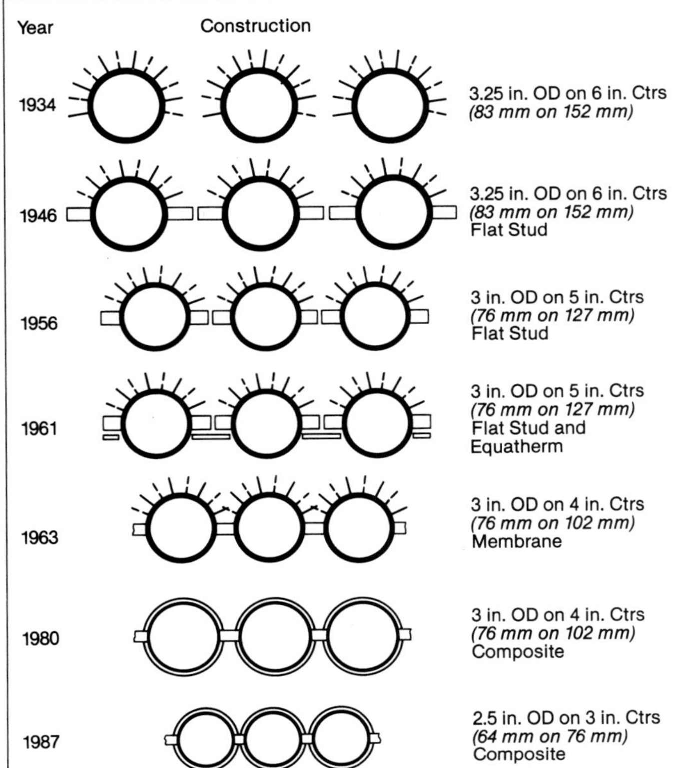

Fig. 13 Evolution of wall construction.

In the 1980s, a corrosion phenomenon was discovered in the lower portion of the generating bank just outside the lower drum. Near-drum corrosion could result in rapid corrosion of the generating bank tubes immediately above the lower drum surface. The corroded area is typically 0.25 to 0.50 in. (6 to 12 mm) in diameter and can be located 180 degrees apart on the sides of the tubes. The corrosion is normally within 1 in. (25 mm) of the lower drum top surface. The corrosion rate can be very high and is identified through ultrasonic testing of the tubes from inside the lower drum. In 2001-2002, B&W developed and patented a new product called GenClad® to protect the carbon steel tube surface immediately above the lower drum (see Fig. 14). The generating bank tube is coated with 309H stainless steel, approximately 0.050 in. (1.27 mm) thick, using a laser fusion method. The tube is then swaged to reduce the tube outside diameter to match the drum hole and the cladding is smoothed in the process, ready for rolling into the lower drum. The cladding is put on the tube in an area that is swaged and is extended into the gas stream several inches to provide protection from near-drum corrosion.

Evolution of the modern, single-drum design

The 1980s saw an increase in the pulp and paper industry's need for high pressure and temperature steam generation from the recovery boiler. This trend was paralleled by the demand for large, conservatively sized furnaces and general acceptance of the single-drum, all welded boiler design. B&W commissioned its first modern single-drum boiler in 1989, at Gaylord Container Corporation in Bogalusa, Louisiana.

In the two-drum design arrangement, the drums are exposed to combustion gases which limit the drum length that can be effectively supported. This maximum length established the design capacity of the two-drum arrangement at about 5×10^6 lb (2.3×10^6 kg) of dry solids per day. In the single-drum arrangement, the steam drum is moved out of the gas flow path, thereby removing this limitation and allowing recovery boilers designed to process daily solids rates of up to 9 to 10×10^6 lb (4.1 to 4.5×10^6 kg) of dry solids per day.

Black liquor solids concentration

The air pollution legislation of the mid-1960s forced major changes in recovery boiler design. To reduce malodorous emissions, the direct contact evaporator was replaced by additional multiple effect evaporator capacity to obtain the optimum liquor concentration. Economizer surface was added for flue gas cooling that was previously accomplished in the direct contact evaporator.

Floor design

There are two generally accepted floor arrangements for recovery boilers: a full sloped floor design and a flat or decanting style. The full sloped design slopes the entire floor at 4 to 6 degrees, and is intended to produce an arrangement that results in a complete drainage of the smelt. The flat or decanting style has

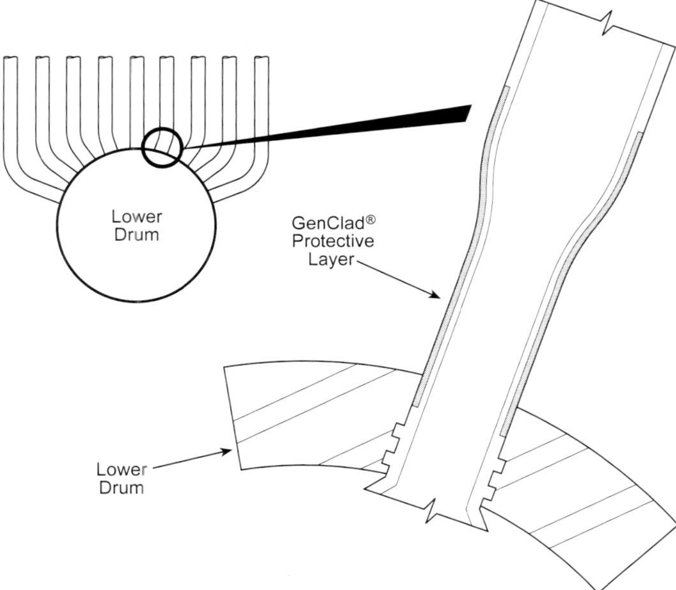

Fig. 14 Carbon steel tube surface protected with GenClad®.

little or no slope, and is intended to maintain a pool of smelt in the lower furnace. B&W traditionally provided only the sloped floor for all two-drum units and all single-drum units. Sloped tubes provide positive circulation and greater margin for waterside operational upsets in this high heat input area of the furnace. Floor arrangements are shown in Fig. 15.

Flat or decanting floors may be modified to incorporate some sloping capabilities, primarily at the front and rear of the furnace and along the side walls. B&W has a dual-sloped dihedral design, sloped at the sides, which can increase flow and circulation margin when conditions require. From an operational standpoint, the sloped or decanting styles allow some inventory of smelt in the lower furnace. Some operators feel this improves operation and reduces smelt spout and opening issues. Existing fully-sloped floors may be modified to achieve a partial decanting effect by dual sloping the floor from the front and rear. This provides the desired smelt inventory while maintaining sloped tubes for circulation. Smelt spouts may also be raised slightly above the floor level to provide a pool of molten smelt immediately in front of the smelt openings.

Lower furnace floor support design has evolved over the years. Floor beams under the floor tubes have increased in number to help strengthen the floor and reduce the potential for damage that can result from heavy salt cake falls from the upper furnace elevations. Floor beam end connections at the side walls have been redesigned for additional strength, to hold side walls against the floor in the event of a major furnace upset or explosion. These improvements have led to a more rigid design able to withstand a variety of operational conditions.

Superheater design

B&W's first recovery boiler designed for elevated pressure and temperature was placed in operation in 1957, at Continental Can Company (now Smurfit-Stone Container Corporation) in Hodge, Louisiana; it generated steam at 1250 psig (8.6 MPa) and 900F (482C).

B&W's high steam temperature design philosophy is reflected in the arrangement of superheater surface. The inlet primary superheater bank is placed following the furnace cavity, with steam flowing through the bank parallel to the gas flow. This results in the coolest available steam flowing through the superheater tubes exposed to the hottest gas temperatures and the radiant heat from the furnace. This arrangement minimizes the metal temperature of the superheater tubes.

Early recovery boilers had front superheater sections spaced on 10 in. (254 mm) side spacing and rear sections spaced on 5 in. (127 mm) side spacing. This arrangement was used when the furnace construction

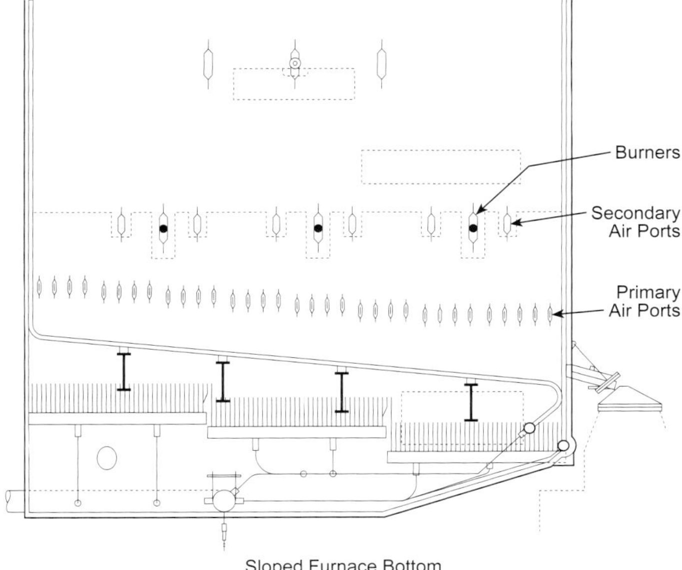

Sloped Furnace Bottom

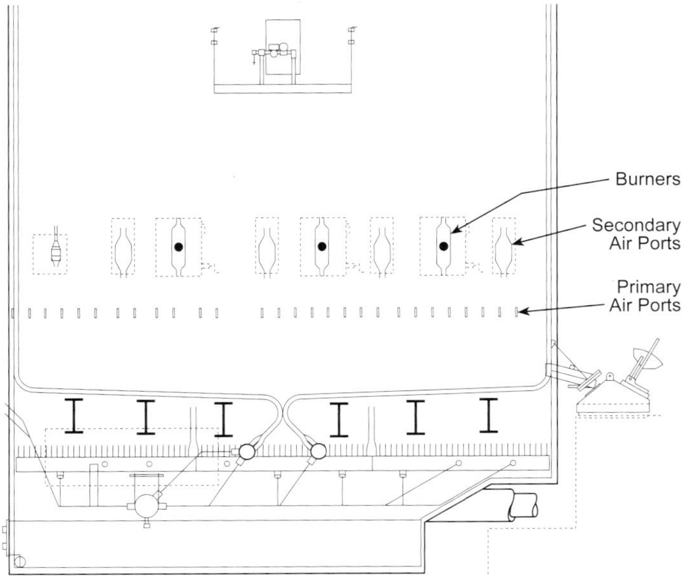

Decanting Floor

Fig. 15 Recovery boiler floor arrangements.

was 3 in. (76 mm) tubes on 5 in. (127 mm) centers. The closer spaced 5 in. (127 mm) side spaced sections may be modified to 10 in. (254 mm) side spacing to alleviate a potential pluggage area, particularly at higher loads. When these superheater components are replaced for any reason, consideration should be given to make the banks consistent with all 10 in. (254 mm) side spacing. This may require a change in superheater surface area, but will result in a superheater that offers improved cleanliness.

Ash buildup and superheater surface plugging have frequently limited availability of the recovery boiler. To avoid these conditions, in 1968 B&W established a 12 in. (305 mm) side spacing for the entire superheater bank, abandoning the conventional 5 and 6 in. (127 and 152 mm) spacing in the secondary superheater construction. This increased the clear side spacing in the superheater from 3.5 to 9.5 in. (89 to 241 mm).

Superheater tie arrangements have evolved as tube backspacing has changed and improved surface configurations have been adopted. Pendant superheaters (see Fig. 18) have utilized D-links or tongue and groove ties in the fore and aft direction. The tongue and groove tie allows for greater tube differential expansion and improved welding capabilities. Platen superheaters (see Fig. 18) have used pin and pipe ties, or a slip spacer type of tie, to restrain the tubes.

Economizer design

The recovery boiler economizer has evolved several times over the years in response to the change to low odor operation, cleanliness, increased heat utilization (efficiency), and reliability. There are several current designs which meet the needs of the industry (see Fig. 16).

The close side-spaced, horizontal, continuous-tube economizer was an early design used to reduce the temperature going to the direct contact evaporator or air heater. To keep gas temperatures high for the direct contact evaporator or air heater (600F/316C or greater), the economizer was small. This economizer was always a crossflow arrangement with the gases flowing perpendicular to the tube surface. This was an efficient method of heat absorption, but presented a cleaning problem. Instead of sootblowers, B&W developed and used shot systems for cleaning gas side deposits. This system took small pieces of metal, such as blanks from the manufacture of nuts and bolts, and dropped them across the economizer surface. The shot was collected in a hopper and returned to drop on the economizer surface. This was done repeatedly to clean any salt cake deposits from the outer surface of the tube. This method of cleaning was not as effective as sootblowing and presented problems with both the return system and tube damage over time.

When direct contact evaporators were discontinued, the economizer size needed to increase. The gas temperature exiting the direct contact evaporator is normally around 400F (204C) and a new, larger economizer was designed to provide the same exit temperature. Gas flow across the tubes (perpendicular) was maintained due to its efficient heat transfer, but to accommodate the increased economizer size necessary, the economizer arrangement was changed to vertical tubes with multiple (three or five) crossflow gas paths.

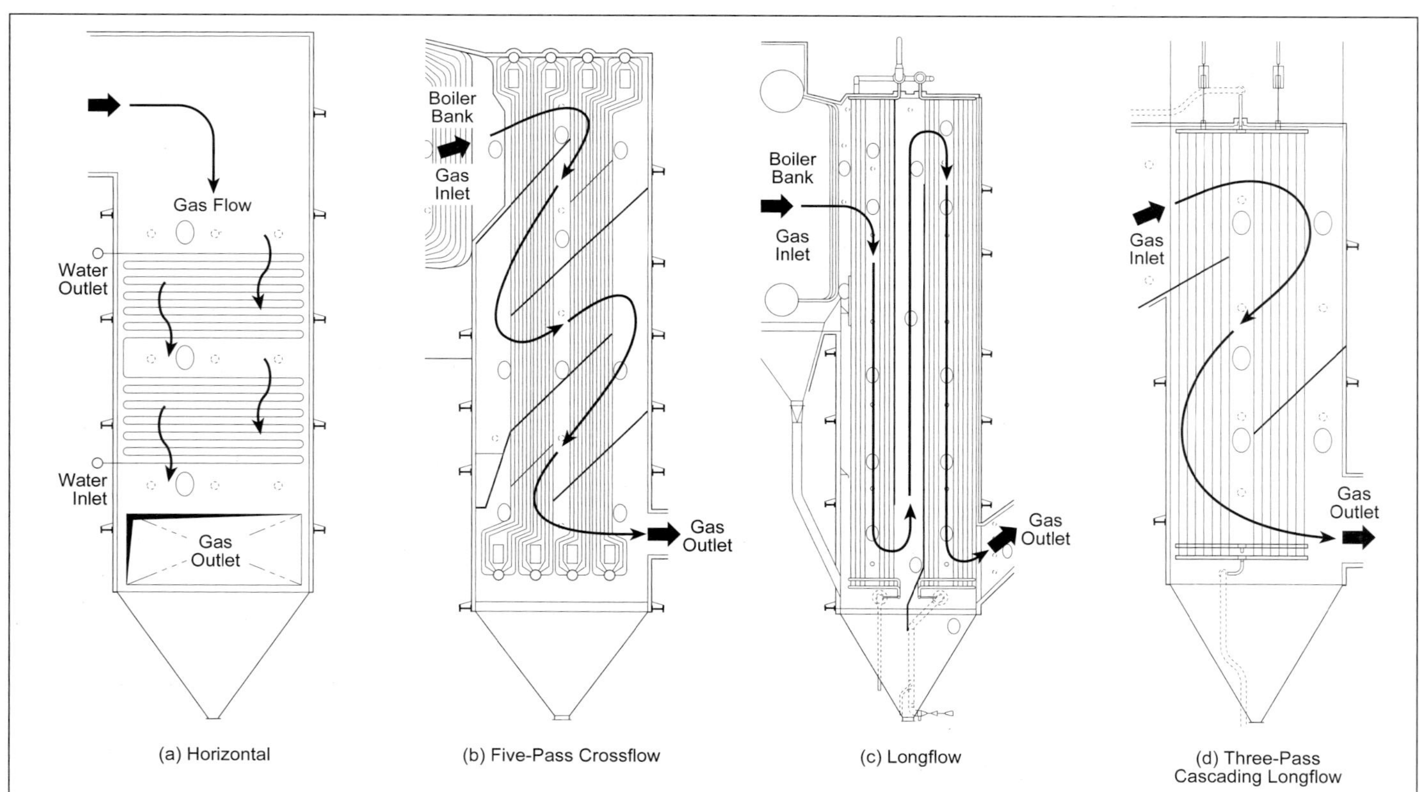

Fig. 16 Recovery boiler economizer arrangements.

The five-pass bare tube economizer design contained much greater heating surfaces than would have been available with the close side-spaced horizontal surface, and was designed to improve gas-side cleanliness.

As new recovery boilers became larger in capacity or existing boilers underwent capacity increases, the higher flue gas quantities, velocities and carryover made cleanliness an issue. This promoted the development of the longflow economizer (Figs. 4 and 20). The longflow economizer is a two gas/two water pass arrangement that has some limited amount of crossflow surface at entrances and exits of the banks, but primarily relies on the parallel flow of gases across the tubes for heat transfer. Because the flow of gases parallel to tube surfaces is less efficient than gas flow across tubes, longitudinal fins were added to the tube to enhance the effective tube surface area and improve heat transfer. Minor revisions have occurred to the longflow economizer over the years to make all tubes straight along with improving the tie design and arrangement. Water flow is always from bottom to top. The economizer bank closest to the gas outlet is the first water pass. The feedwater enters the bottom of this bank, exits the top, and is piped to the bottom of the bank closest to the generating bank. The water exits the top of this second water pass and is directed to the steam drum.

Space may sometimes be limited for a longflow economizer retrofit without major building modifications. An alternative finned tube design was developed by B&W for this condition, the three-pass cascading longflow arrangement. This design incorporates considerable crossflow effectiveness while opening up the bank to reduce gas-side velocity and improving the overall cleanliness.

The horizontal continuous tube economizer was reintroduced in a version that has increased the tube spacing and provided for effective sootblowing. This economizer can be supplied when space limitations dictate a more compact design to reduce building modifications. It is used primarily when large regenerative air heaters are replaced with an economizer.

The choice of economizers for retrofitting existing recovery boilers should be made by the boiler designer, working with the mill and the existing arrangement. New recovery boilers will be arranged with straight tube longflow designs, while retrofits may use any of the designs identified in the evolution discussion. Regardless of arrangement, the economizer must have good sootblower coverage and personnel access for inspection.

Combustion air system

The first recovery boiler to introduce air at three furnace levels – primary, secondary and tertiary – was built by B&W in the late 1940s. In 1956, a concentrated development effort was successful in providing fuller utilization of tertiary air, which had been largely ineffective in earlier designs. Today's advanced air management system is a result of extensive laboratory scale and computer flow modeling, theoretical consideration for the air penetration across the furnace at the secondary and tertiary air levels, and testing of different air system configurations on operating boilers. Several features resulting from this development program include the use of variable velocity control dampers on the secondary and tertiary air ports to better regulate air penetration across the furnace area, an interlaced port arrangement to provide better gas mixing within the furnace, and optimization of port location.

Design considerations for B&W recovery boiler

Furnace design

The design of the modern recovery boiler and its associated equipment systems must first consider efficient black liquor combustion.

The overall black liquor composition (chemistry, percent solids, and heating value) determines how the liquor dries in the furnace. Some compositions of liquor characteristics (higher solids and heating values) are more conducive to in-flight, suspension drying of the droplets, while other characteristics (lower heating values or solids) may require additional droplet drying time. In these instances, spraying liquor on the walls between the secondary and tertiary air port levels can provide greater dehydration prior to falling onto the hearth char bed. Flexibility in operating parameters (combustion air quantities, distribution and temperatures, liquor spray methods and characteristics) is required to accommodate potentially varying conditions.

Primary air enters the furnace around the perimeter of the hearth bed. The controlled reducing atmosphere at the hearth burns the char at the bed surface to maximize reduction of Na_2SO_4 to Na_2S in the smelt. The remaining air is admitted at the secondary and tertiary air zones. The total air admitted through the primary and secondary air ports is approximately the stoichiometric requirement for black liquor combustion. High pressure air entering through large secondary ports penetrates across the furnace to assure mixing with the volatile gases rising from the char bed. Combustion at the secondary air level achieves a maximum furnace temperature zone below the liquor spray for drying the liquor. Secondary air also limits the height of the char bed by providing air for combustion across the bed surface.

Further turbulence and mixing are created by admission of tertiary air, which assures complete combustion of unburned gases rising from the secondary zone and of volatiles escaping from the sprayed liquor. Tertiary air mass penetration also provides a uniform temperature and velocity profile of combustion gases entering the convection surface. Quaternary air is a fourth level of air which is sometimes applied above the tertiary zone. Quaternary air has been shown to lower NO_x and particulate emissions, and is being applied to recovery boilers in original design or as retrofits and upgrades.

The recovery boiler furnace must also be designed for efficient removal of the molten inorganic chemicals as smelt. Finally, the combustion gases and particulate carryover must be adequately cooled in the furnace to minimize deposition on convection surfaces.

The first step in the design of a new recovery boiler is the selection of furnace plan area, defined as the furnace width times depth. The plan area is generally set to achieve a black liquor solids heat input of 900,000 to 950,000 Btu/h ft^2 (2.83 to 2.99 MW/m^2). As the solids heating value decreases, a larger furnace plan area is desirable. However, an oversized plan area can lead to local cold spots on the smelt bed, which in turn limit reduction efficiency. Cold spots can also lead to unstable furnace blackouts. A large plan area further constrains load turndown with stable combustion. An undersized plan area generally leads to increased fume and particle carryover.

Once the plan area is established, the width and depth dimensions are selected. Maintaining a depth to width ratio between 1.0 and 1.15 allows an effective arrangement of the combustion air ports and generally permits an economical arrangement of convection pass heat transfer surfaces. In smaller recovery furnaces designed for high steam temperature, a higher aspect ratio permits increasing the depth to accommodate the large superheater surface. Furnace height is then determined by the radiant furnace heat transfer surface required to cool the combustion gases below 1700F (927C).

The surface of the floor and wall tubes in the lower furnace must be protected against the corrosiveness of the smelt and partially combusted gases. The most widely accepted approach today is to build the furnace and floor of composite tubes (alloy tube material over a base carbon steel tube). This tube construction should extend to 3 ft (0.9 m) above the tertiary air ports. Above this elevation, carbon steel tube and membrane construction is adequate.

Wide closure plates attached to tubes bent to form air ports and other openings can result in potentially high localized stress areas. Wide closure plates suffer from corrosion, burn back and cracking. This adversely affects the air flow area of the port opening and has the potential for closure plate cracking that propagates into the furnace wall tubes. The modern furnace openings are designed without closure plates to minimize this potential (Fig. 17).

Structural attachments must also minimize tube stresses. In high stress areas, a plate stamping is welded to the tubes, and the structural member is attached to this plate. This plate stamping is contoured to provide additional weld carrying load into the tube wall, permit flexibility for the absorption of thermally induced loads, and lower the potential for fatigue and stress corrosion related failures. The windbox attachment to the tube wall, shown in Fig. 17, uses plate stampings that are shop-attached, which permits field erection without having to make any field attachment welds directly to the pressure part. This design and construction philosophy is applied throughout the pressure part enclosure for all attachment welds.

Upper furnace and arch arrangement

Combustion is completed in the tertiary zone. The water-cooled furnace walls and volume above this zone provide the necessary surface and retention time to cool the gas to temperatures where sootblowers can effectively remove the chemical ash from convection surfaces.

Fig. 17 Burner and secondary air port windbox attachments.

The furnace arch, or *nose*, serves several important functions. The arch shields the superheater from the radiant heat of the furnace. The high temperature steam loops of the superheater are completely protected. Penetration of the arch into the furnace uniformly distributes the gas entering the superheater. An eddy above and behind the arch tip causes the gas to recirculate in the superheater tube bank, with a reverse gas flow between the superheater and the upper arch face preventing hot gas from bypassing the superheater surfaces. The angle of the arch is set to minimize the repose of deposited ash on its surface.

Furnace screen

In some recovery boilers, the superheater surface is insufficient to adequately cool the combustion gases before they enter the boiler bank. This is common in boilers designed for low steam temperature or when a recovery boiler's liquor burning capacity is significantly increased resulting in an increase in furnace exit gas temperature. A furnace screen can be used to absorb the additional heat, thereby maintaining an acceptable temperature of the gas entering the boiler bank section. When designing a new recovery boiler, the required heat absorption can often be accomplished through added furnace surface (height) and/ or an oversized superheater. Larger furnaces can impact the overall building size and therefore increase the cost of the project. The furnace screen represents an economical alternative to a larger furnace or superheater component.

When a furnace screen is required, the screen tubes are in line with the superheater sections to reduce pluggage potential. The sloped section of the screen originates inside the furnace arch to protect the tubes from falling salt cake deposits and to add strength to the sloped tubes. Tubes in the sloped section of the screen were originally attached to each other using round bars on either side of the tubes, welded parallel to the tubes to provide structural integrity. Current retrofit designs use membrane bar similar to furnace wall construction (see Chapter 23). The top most tube currently has a half-wide membrane bar on the top surface to improve structural integrity and further reduce potential tube damage caused by salt cake falls.

Convection surface

After leaving the furnace, the flue gases pass across the steam-cooled superheater banks to the longflow boiler bank and finally to the economizer sections. (See Fig. 4.) As the gas is cooled, entrained ash becomes less sticky and adheres less to the tube surfaces. As a result, it is possible to space the tubes in the convection banks progressively closer together. The closer spacing results in higher gas velocities and improved convection heat transfer rates, which in turn permit a more economical design as less heat transfer surface area is required.

Superheater

The superheater surfaces are exposed to the highest gas temperatures and, consequently, are arranged on a 12 in. (305 mm) side spacing. This results in very low gas velocities and prevents the bridging of deposits.

In arranging superheater surface, it is desirable to maintain low tube temperatures. Lower temperatures reduce the potential for high temperature corrosion and allow the use of less expensive low alloy steel. Temperatures are reduced by establishing a high steam flow through each tube, by arranging the coolest steam to flow through the superheater tubes exposed to the hottest gas temperatures, and by locating the majority of the superheater tube banks behind the furnace arch tip, shielded from furnace radiation. Superheater surface behind the furnace arch tip is considered convective surface and does not experience furnace radiation. Superheater surface types and arrangements are shown in Fig. 18.

From the drum, saturated steam enters the front tube row in the first or primary inlet superheater bank and flows through successive tube loops in parallel with the flue gas flow. (See Fig. 18.) The secondary superheater is located in the cooler gas region between the primary banks or behind the primary banks, depending upon final steam temperature. Steam flow in this secondary superheater is generally opposite to the gas flow to achieve a higher steam temperature. The superheater surface, steam direction and bank placement are determined by engineering and balancing steam temperature and flue gas temperature. Using these design guidelines for superheater design, arrangement, and materials selection permits final steam temperatures up to 950F (510C).

The superheater banks are top supported with the tube elements expanding downward. The tubes are interconnected with flexible support ties which allow independent tube expansion. Tube movement is critical to effective sootblowing. However, lack of tube restraints within a bank can lead to failure at the bank's top supports.

The superheater can be arranged as pendant surface (clear backspace between tubes) or platen surface (near-tangent backspace between tubes). Different heat transfer methods are used when calculating pendant and platen arrangements. The pendant arrangement has greater effective surface area due to a greater amount of the tube circumference available to absorb heat from the flue gases. This results in fewer superheater rows in depth for the same steam temperature rise. Superheater banks can be radiant, convective, or a combination (see Chapter 19). Radiant surface is in front of the furnace arch tip. This surface is the first steam pass after leaving the steam drum in order to utilize non-stainless steel tubing.

Boiler bank

Today's kraft recovery boiler generally incorporates a single steam drum, with a longflow boiler bank arranged downstream of the superheater. In passing across the superheater and rear wall screen tubes, the flue gas should be cooled below the ash sticky temperature prior to entering the boiler bank. For ashes with extremely low sticky temperatures, particular attention must be given to sootblower locations.

The boiler bank is constructed of shop-assembled tube sections arranged as modules inside a water-cooled enclosure (Fig. 19). Tubes in each section are connected to headers, with water entering the lower header and the steam-water mixture exiting the upper headers. As the flue gas enters the bank, it turns downward and flows parallel to the tube length, providing easy cleanability. Within the bank, a central cavity accommodates fully retractable sootblowers. The cavity permits personnel access for visual tube inspection adjacent to sootblower lance entry. Ash deposits dislodged during sootblowing are collected in

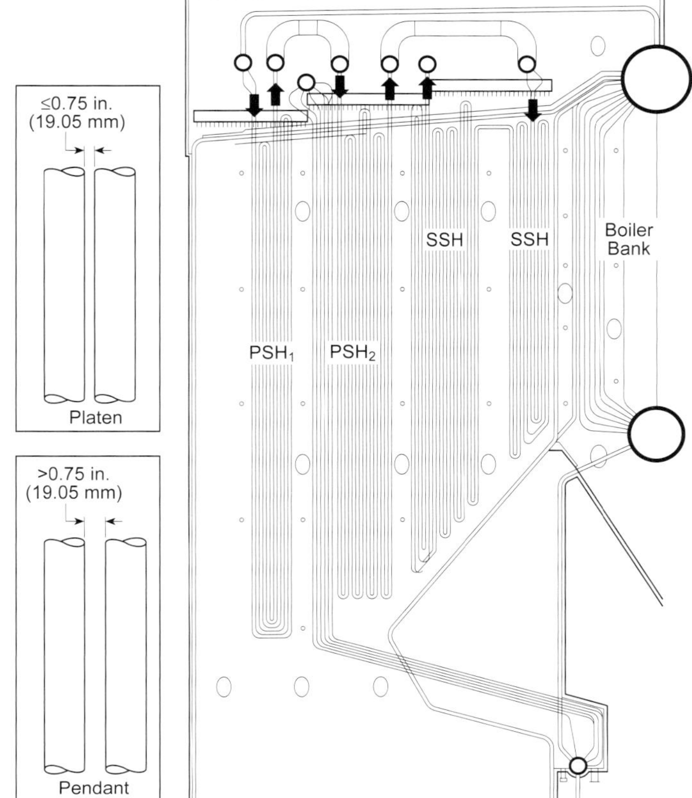

Fig. 18 Superheater arrangements.

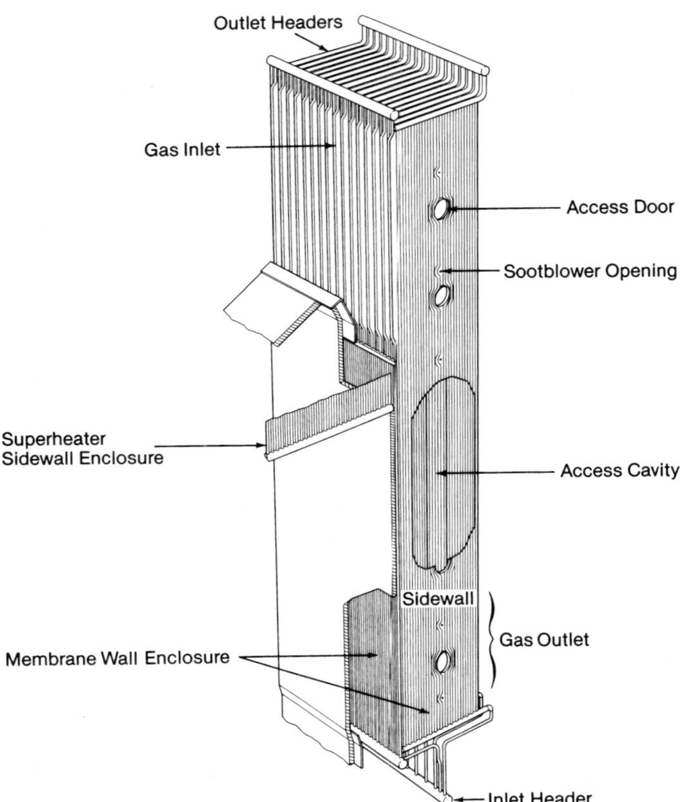

Fig. 19 Boiler bank isometric.

a trough hopper connected to the bank enclosure. Tube section inlet headers are widely spaced and vertically staggered to facilitate ash dropping into the hopper.

Impact-type particle deposition on the boiler bank tubes is less likely to occur with the gas longflow orientation. As a result, the allowable gas velocity in the downflow portion of the bank can be increased.

To improve heat transfer, longitudinal fins are welded to the front and back of each tube. Fins are tapered at the ends and welded to the tube on both sides. The welds are terminated by wrapping around the end of the fin. This combination of welding technique and tapered ends assures minimal stress concentration at fin termination for fins as large as 1.5 in. (38 mm).

Economizer

The boiler bank surface area is typically set to achieve a nominal exit gas temperature of about 800F (427C). This temperature maintains a reasonable differential with the saturated steam temperature [610F (321C) for a 1650 psig (11.4 MPa) drum pressure] and allows the use of carbon steel casing to enclose the downstream economizer banks. The modular economizer has vertical finned tubes arranged in multiple sections with upward water flow and downward longflow of gas (Fig. 20). The common arrangement features two banks. The flue gas enters at the upper end and discharges at the lower end of each bank. Gas flows down the length of the bank to provide good cleanability. As in the boiler bank, a central cavity dimensioned for personnel access accommodates fully retractable sootblowers. Trough hoppers are attached to the economizer casing to collect dislodged ash deposits.

The economizer surface area is set to achieve a final gas outlet temperature approximately 100F (56C) higher than the feedwater temperature. Although it is possible to achieve an exit gas temperature closer to that of the feedwater, the decreased temperature differential results in substantially increased surface requirements for small improvements in end temperature. In addition to this thermodynamic limitation, concern for cold end corrosion generally establishes a minimum gas exit temperature around 350F (177C). The minimum recommended temperature of the feedwater entering the economizer is 275F (135C) for corrosion protection of the tube surface. With special considerations, the feedwater entering the economizer can be designed for as low as 250F (121C).

Emergency shutdown system

An emergency shutdown procedure for black liquor recovery boilers has been adopted by the Black Liquor Recovery Boiler Advisory Committee (BLRBAC) in the U.S. An immediate emergency shutdown must be performed whenever water enters the furnace and can not be stopped immediately, or when there is evidence of a leak in the furnace setting pressure parts. The boiler must be drained as rapidly as possible to a level 8 ft (2.4 m) above the mid point of the furnace floor.

An auxiliary fuel explosion can occur when an accumulated combustible mixture is ignited within the confined spaces of the furnace and/or the associated boiler passes, duct work and fans which convey the combustion gases to the stack. A furnace explosion will result from ignition of this accumulation if the quantity of the combustion mixture and the proportion of air to fuel are within the explosive limit of the fuel involved. The magnitude and intensity of the explosion will depend upon both the quantity of accumulated combustibles and the proportion of air in the mixture at the moment of ignition.

Contacting molten smelt with water can also result in a very powerful explosion. The mechanism for a smelt-water explosion is keyed to the contact of water with hot liquid smelt. Rapid water vaporization causes the propagation of a physical detonation or shock wave.

In the design and operation of black liquor recovery boilers, every effort is made to exclude water from any source from getting to the furnace, or introducing liquor at less than 58% solids. For example, furnace attachment details are designed to prevent external tube loads, which can lead to stress assisted corrosion.

Recovery boiler auxiliary systems

Black liquor evaporation

The high black liquor solids concentration required for efficient burning is achieved by evaporating water from the weak black liquor. Large amounts of water can be economically evaporated by multiple effect evaporation. A multiple effect evaporator consists of a series of evaporator bodies, or effects, operating at different pressures.[5] Typically, low or medium pres-

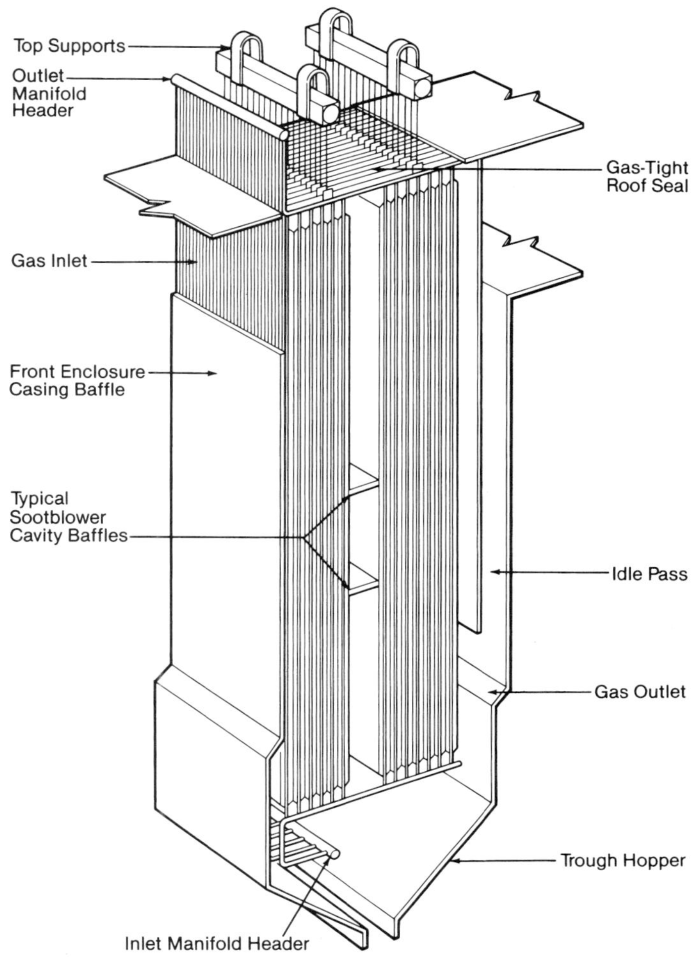

Fig. 20 Economizer isometric.

sure steam is utilized in the first evaporator effect and then the vapor from one body becomes the steam supply to the next, operating at a lower pressure. Modern evaporator systems integrate a concentrator into the flow sequence to achieve the final liquor concentration. As a general rule, each pound of water evaporated from the weak liquor results in one additional pound of high pressure steam generation in the recovery boiler.

For many years, direct contact evaporation was the technology used to achieve firing solids to the recovery boiler, and may still be necessary to evaporate liquor from some special fiber sources. In the direct contact evaporator, liquor and flue gas are brought together, heat is transferred from flue gases to the liquor, and mass transfer of liquor water vapor to the gas occurs across the liquor-gas interface. Adequate liquor surface must be provided for the heat and mass transfer. The gas contact acidifies the liquor by absorbing CO_2 and SO_2, which decreases the solubility of the dissolved solids and requires continuous agitation. The acidification also results in the release of malodorous compounds into the flue gas, a negative aspect of direct contact evaporation.

There are two types of direct contact evaporators used in the recovery unit, cyclone and cascade. The cyclone evaporator (Fig. 21) is a vertical, cylindrical vessel with the flue gas admitted through a tangential inlet near the conical bottom. The gas flows in a whirling helical path to the cylinder's top and leaves through a concentric re-entrant outlet. Black liquor is sprayed across the gas inlet to obtain contact with the gas. The liquor droplets mix intimately with the high velocity gas and are centrifugally forced to the cylinder wall. Recirculated liquor flowing down the cylinder wall carries the droplets and any dust or fumes from the gas to the conical bottom, out through the drain, and into an integral sump tank. Sufficient liquor from the sump tank is recirculated to the nozzles at the top of the evaporator to keep the interior wall wet, preventing ash accumulation or localized drying.

In the cascade evaporator, horizontally spaced tubular elements are supported between two circular side plates to form a wheel that is partially submerged in a liquor pool contained in the lower evaporator housing. The wetted tubes are slowly rotated into the gas stream. As the tubes rise above the liquor bath, the surface coated with black liquor contacts the gas stream flowing through the wheel.

Black liquor oxidation

When a direct contact evaporator is used, odor can be reduced by oxidation of sulfur compounds in the liquor before introduction to the evaporator. The oxidation stabilizes the sulfide compounds to preclude their reaction with flue gas in the evaporator and the consequent release of reduced sulfur gases. Oxidation can effectively reduce, but does not eliminate, discharge of these reduced sulfur gases. The direct contact evaporator is the prime source of odor.

Odor is generated in direct contact evaporators when the hot combustion gases strip hydrogen sulfide gas from the black liquor:

$$2NaHS + CO_2 + H_2O = Na_2CO_3 + 2H_2S \qquad \textbf{(11)}$$

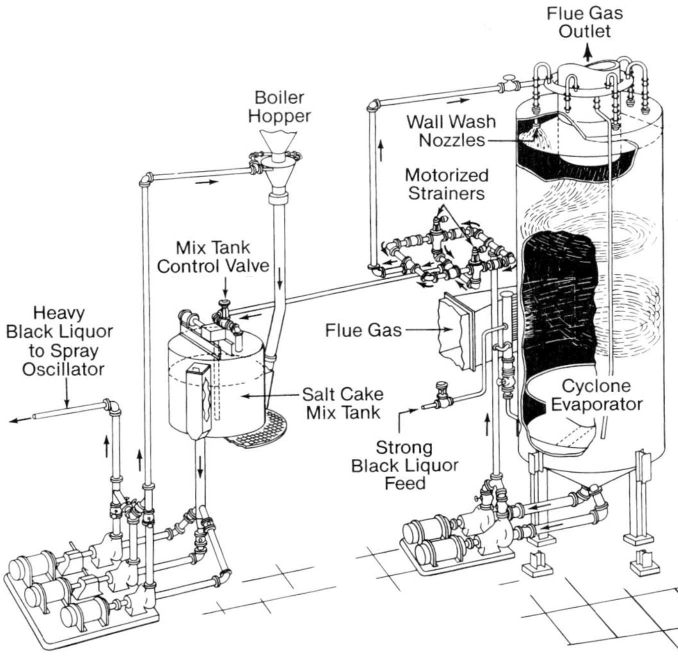

Fig. 21 Cyclone evaporator.

Oxidation stabilizes the black liquor sulfur by converting it to thiosulfate:

$$2NaHS + 2O_2 = Na_2S_2O_3 + H_2O \qquad \textbf{(12)}$$

Black liquor oxidation involves high capital and operating costs; the oxidation step also robs the liquor of heating value. Modern recovery facilities incorporate multiple effect evaporators, and the low odor design, which eliminates the need for a direct contact evaporator.

Black liquor system

Recovery boilers are operated primarily to recover pulping chemicals. This objective is best realized by maintaining steady-state operation. Recovery boilers are base loaded at a selected black liquor feed flow or heat input, in contrast to power boiler applications where fuel flow is varied in response to demand for steam generation.

The black liquor solids concentration can vary with the rate that recirculated ash sheds from heat transfer surfaces, the rate it is collected in the precipitator, and the rate it is returned to the black liquor stream system. Considerable fluctuations can occur, depending on which surface is being cleaned by sootblowers, the frequency of sootblower operation, and the rapping sequence of precipitator collection surfaces. The black liquor system design should provide uniform dispersion of the ash into the liquor to minimize the fluctuation of solids at the burner. This is increasingly important as the liquor solids concentration is increased into the regime where burners are maintained in a fixed position and liquor drying is in-flight (not on the walls). A fluctuation in ash flow would change the solids concentration by several percent, significantly impacting the characteristics of the liquor.

A heater is used to adjust the black liquor temperature to that required for optimum combustion and minimum liquor droplet entrainment in the gas stream. The black liquor is typically heated in a tube-and-shell heat exchanger, using low pressure steam on the shell side and black liquor on the tube side. There are two general categories of heater designs that can be operated with minimum scaling of the heat transfer surface. The first uses a conventional heat exchanger with once-through flow of liquor at high velocity in the tubes. The tubes have polished surfaces to inhibit scale formation. This heat exchanger is satisfactory for long periods of operation without cleaning.

The second approach is to recirculate liquor through a standard heat exchanger with stainless steel tube surface to maintain high velocities that inhibit scale formation. This approach also permits long periods of operation without cleaning.

As the liquor solids concentration increases above 70%, storage can become a problem. Moreover, it becomes increasingly difficult to blend recycled ash into the highly concentrated liquor. Recycled ash can be returned to an intermediate concentration liquor stream (of about 65% concentration) prior to final evaporation in a concentrator. In this type of arrangement, liquor is routed to the recovery furnace from an evaporator system product flash tank. The as-fired liquor temperature is established by controlling the flash tank operating pressure.

There are two designs of liquor burners, the oscillator and the limited vertical sweep (LVS). Oscillators are generally applied to liquors of lower heating values or percent solids where longer droplet drying time is necessary. LVS burners are best used for higher percent solids and heating value liquors where in-flight, suspension drying of the droplets works well. Both types of burners utilize a nozzle splash plate to produce a sheet spray of coarse droplets.

The oscillator sprays the black liquor on the furnace walls, where it is dehydrated and falls to the char bed. The oscillator burners, located in the center of the furnace wall between the secondary and tertiary air ports, are continuously rotated and oscillated, spraying liquor in a figure eight pattern to cover a wide band of the walls above the hearth. Oscillators work well on smaller units, or with a wide range of liquor characteristics.

With LVS burners (Fig. 22), black liquor is sprayed into the furnace for in-flight drying and devolatization of the combustible gas stream rising from the char bed. The objective of the LVS burner is to minimize the liquor on the wall. The LVS gun is normally used in a fixed position, but can also sweep vertically to burn low solids liquor or those with poor burning characteristics. LVS burners are normally applied to larger sized, heavier loaded units.

The temperature and pressure of atomized liquor directly impact recovery furnace operations. Lower temperature and pressure generally create a larger

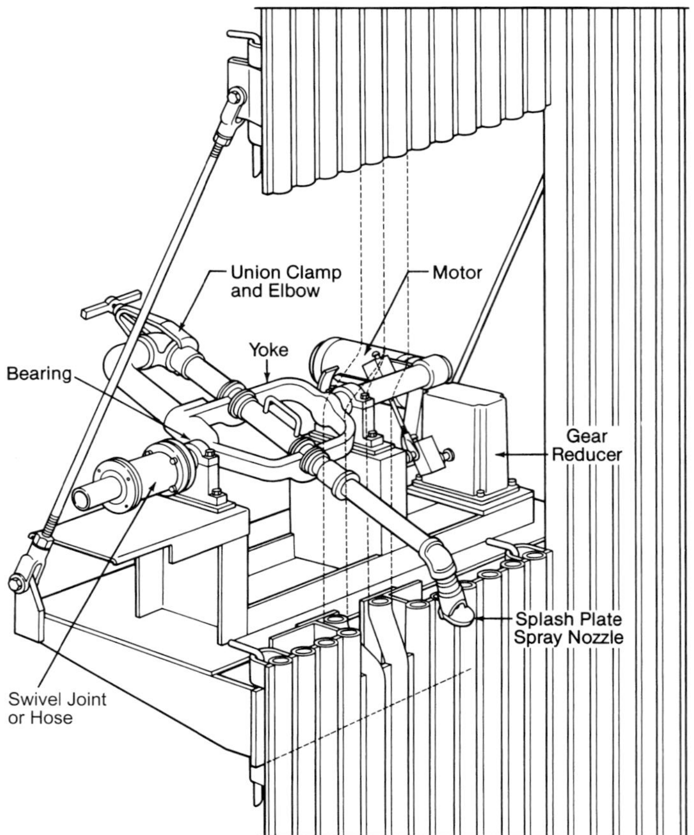

Fig. 22 Limited vertical sweep burner.

particle or droplet of atomized liquor. This minimizes the entrainment of liquor in the combustion gases passing to the heat absorbing surfaces. Where wall drying is carried out, large liquor droplets maximize the liquor sprayed on the wall and minimize in-flight drying. For oscillator firing, liquor at approximately 230F (110C) and 30 psig (207 kPa) generally provides the most satisfactory operation. For LVS burners, liquor with a slightly lower pressure of 25 psig (172 kPa) and higher temperature of 250F (121C) provides acceptable results.

As the liquor sprayed on the walls builds, it eventually falls to the char hearth. The majority of the char falling from the wall is deposited in front of the primary air ports, requiring 30 to 50% of the total air to be introduced through the primary ports.

In-flight drying deposits a minimum of char in the primary air zone around the periphery of the unit, as all the drying and a majority of the devolatization are in-flight over the furnace area. Consequently, less primary air flow is required to keep the char from in front of the primary air ports. Because the higher liquor solids concentration dictated by in-flight drying translates into less water evaporated in the furnace, a greater fraction of the combustion heat released is available to maintain bed temperatures. Less primary air is therefore required when burning liquor in-flight in contrast to the lower concentration liquors encountered in oscillator burner applications.

Ash system

The sodium compounds entrained in the flue gas originate from fume generation and liquor droplet carryover from the lower furnace. The resultant ash drops out of the flue gas stream and is collected in trough hoppers located below the boiler and economizer modules. The electrostatic precipitator removes nearly all the remaining ash.

The majority of the entrained ash is Na_2SO_4, and is commonly referred to as salt cake ash, or simply, salt cake. The ash collected in the trough hoppers and precipitator must be returned to the black liquor to recover its significant sodium and sulfur contents. This is accomplished by mixing the ash into the liquor in a specially designed tank. The salt cake ash is transferred to the mix tank through a wet ash sluice system or a drag chain ash conveyor system.

With a dry ash system (Fig. 23), mechanical drag chain conveyors are bolted to the bottom of the boiler and economizer trough hoppers, which extend across the full unit width. The drag chain conveyors are equipped with heat treated, high alloy forged link chains to support and convey the flights. The conveyors discharge through rotary seal valves into a collection conveyor. The collection conveyor discharges to the mix tank where the ash is uniformly mixed with the liquor.

Drag chain conveyors are also provided across the floor of the electrostatic precipitator, beneath the collecting surfaces. The conveyors discharge to mechanical combining conveyors, which in turn discharge the salt cake ash through rotary valves into the mix tank. Frequently, the dry ash system is arranged with two

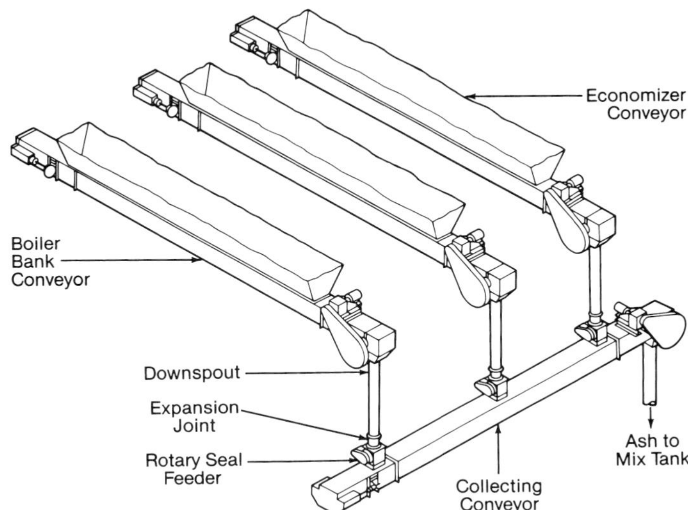

Fig. 23 Dry ash system.

mix tanks, one serving the boiler and economizer hopper ash system and one dedicated to the precipitator.

As discussed previously, successful reductions of chloride and potassium levels in the ash and black liquor have been achieved through precipitator ash purging. Ash can be removed periodically to maintain chloride or potassium concentrations at a specified level. An ash chute with a rotary valve is added to the back end of an existing precipitator ash transfer conveyer. When ash purging is desired, the ash conveyer is reversed and the ash falls through the chute into a sluice tank.

In a wet ash removal system, black liquor is circulated through the hoppers to sluice the collected ash. The sluice discharges directly from the hoppers through large pipes to the mix tank. However, these pipes can become plugged with ash and overflow liquor which create safety and cleanliness problems. This has led to wider acceptance of the dry ash mechanical conveyor system. However, the mechanical conveyors generally require more maintenance than the wet ash sluice designs.

The mix tank (Fig. 24) includes a mechanically scraped screen to assure that all material which passes to the fuel pumps is small enough to readily pass through the burner nozzles. Scraping is provided by flights on a low horsepower, slow moving agitator. The tank is designed to receive salt cake from the generating bank, economizer or precipitator hoppers, mixing it with the incoming black liquor.

Combustion air system

B&W's advanced air management system provides combustion air at three, and sometimes four, elevations of the furnace: primary, secondary, tertiary, and potentially quaternary (Fig. 25). The use of at least three levels allows performance optimization of the respective furnace zones – the lower furnace reducing zone, the intermediate liquor drying zone, and the upper furnace burnout or combustion completion zone. A quaternary air level may also be applied above the tertiary zone for additional NO_x or particulate control.

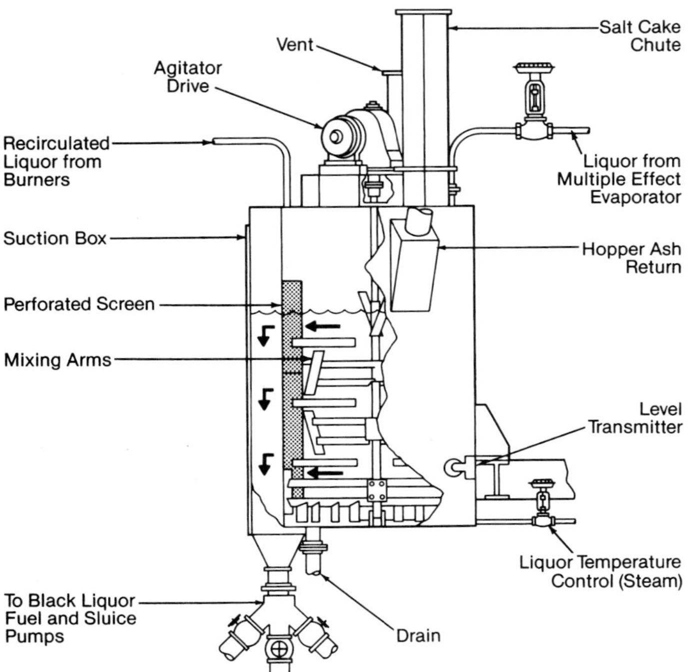

Fig. 24 Salt cake mix tank.

The primary air flow quantity must be sufficient to produce stable combustion and to provide the hot reducing zone for the molten smelt. Increasing primary air flow beyond the level required to achieve these objectives increases the amount of Na_2S re-oxidized to Na_2SO_4. The balance of the air at or less than the stoichiometric requirement is introduced above the char bed at the secondary level to control the rate at which the liquor dries and the volatiles combust, and to minimize the formation of NO_x. The additional air required to complete combustion is introduced at the tertiary level (and quaternary if supplied).

The air system may be arranged as a single-fan, two-fan (one primary and one secondary/tertiary/quaternary) or three-fan system. On larger systems, additional provisions are made to bias the air flow between respective air levels, from side to side on the primary and secondary air, or from front to rear on the tertiary air. This gives additional flexibility to the operator.

The primary air ports are arranged on all four furnace walls about 3 ft (0.9 m) above the floor. Air is introduced at a low velocity and 3 to 4 in. wg (0.75 to 1.0 kPa) static pressure which prevents it from penetrating the bed. The air lifts the carbon char in front of the port back onto the bed and maintains ignition.

Approximately 40% of the total air is admitted at the secondary zone. The air is introduced at the pressure and velocity needed to penetrate the furnace 4 to 6 ft (1.2 to 1.8 m) above the primary air ports. A velocity damper is used on each port.

Proper secondary air mixing with the gases rising from the char bed results in volatile combustion generating the heat required for in-flight liquor drying. This secondary zone typically exhibits the highest temperatures in the recovery furnace. The quantity of secondary air is dictated by the amount of burning required to dry liquor and control bed height, and decreases as the black liquor solids concentration increases.

The balance of the air is admitted at ambient temperature through the tertiary air ports located above the liquor guns, and through quaternary ports, if supplied. A velocity damper is again used on each port. The tertiary (and quaternary) flow increases proportionately as the solids increase.

The secondary air ports are normally arranged on the longest furnace wall, generally the side walls, as the furnace is deeper than it is wide. In contrast, the tertiary air ports are arranged on the front and rear walls.

Buildup of smelt and char that restricts the air port openings can cause a performance deterioration. A reduction in port area affects air pressure at the port or air flow through the port, depending on which is being controlled. When air flow is controlled, the effect of plugged ports is to increase air pressure and push the bed farther from the wall. If pressure is controlled, the effect is a resultant decrease in air flow. This results in less effective burning of the primary zone with a decrease in furnace temperature and increased emissions.

Primary air port openings are generally cleaned (rodded) every two hours to maintain proper air distribution and bed height. Automatic port rodders provide continuous cleaning to maintain a constant flow area. The automatic port rodding system stabilizes lower furnace combustion for maximum thermal efficiency and is vital in achieving low emissions.

Secondary and tertiary air ports also require periodic rodding, although the plugging that occurs at these ports is generally less severe than at the primary air ports. In the secondary and tertiary ports, the rodding equipment can be integrated with the velocity control damper. This is necessary to provide synchronization of the damper and rodder drives.

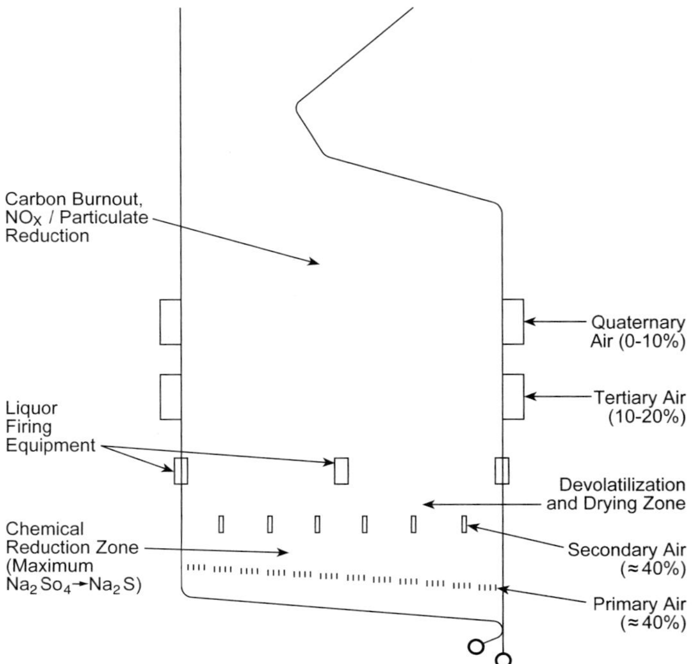

Fig. 25 Advanced air management system.

To further enhance stability in the lower furnace, the primary and secondary air is preheated in a steam coil air heater. Low pressure steam, normally 50 to 60 psig (345 to 414 kPa) is used for preheating. Additional preheating is provided by steam at 150 to 165 psig (1034 to 1138 kPa), typically achieving a 300F (149C) combustion air temperature. When burning liquors with a low heating value or from a nonwood fiber, air should be preheated to about 400F (204C). Water coil air heaters, sometimes used within the feedwater circuit around the economizer, may also be used to provide these air temperatures.

Flue gas system

Combustion gas exiting the economizer is routed through flues to the electrostatic precipitator and induced draft fan before discharge through a stack to the atmosphere.

The induced draft fan speed controls the pressure inside the furnace. The fan is normally located after the precipitator, allowing the fan to operate in the cleaner gas.

A two-chamber precipitator is normally used. Each chamber is equipped with isolation gates or dampers and a dedicated induced draft fan discharging to a common stack. Flue gas exiting the economizer is divided into two flues, which route the gases to the precipitator's two chambers. Each chamber typically has sufficient capacity to operate the recovery boiler at 70% load, corresponding to a stable black liquor firing mode. When designing the precipitator chamber, the expected gas flow rate should include that of the sootblower steam and the increased excess air under which the boiler would operate at a reduced rating. The maximum permissible particulate discharge rate then establishes the electrostatic precipitator size.

Cleaning system

Ash entrainment in furnace gases is affected by gas velocity, air distribution and liquor properties. The design of all recovery unit heating surfaces should include sootblowers using steam as the cleaning medium. Gas temperatures must be calculated to make certain that velocities and tube spacings are compatible with the sootblowers.

High levels of chloride and potassium in the black liquor may require greater cleaning frequency. As unit overload is increased, additional entrainment of ash and sublimated sodium compounds also leads to more frequent water washing. This term describes an off-line cleaning procedure which uses high pressure water to eliminate excessive deposits or buildups on the boiler's convection surfaces. In addition to excess quantities of flue gas ash, velocities and temperatures throughout the unit are increased, making ash deposits more difficult to remove.

Auxiliary fuel system

The primary objective of the recovery boiler is to process black liquor. However, the unit can fire auxiliary fuel, usually natural gas and/or fuel oil. This fuel is fired through specially designed burners, arranged at the secondary air level. These burners raise steam pressure during startup, sustain ignition while building a char bed, stabilize the furnace during upset conditions, carry load while operating as a power boiler, and burn out the char bed when shutting down.

In some installations, the recovery boiler must be able to generate full load steam flow and temperature on auxiliary fuel. These applications typically arise in mills that cogenerate electricity and/or have limited power boiler steam generating capability. Increased auxiliary fuel capacity can be accommodated by adding auxiliary burners above the secondary air level (typically at or above the tertiary air level).

Upper level burners allow combination black liquor and auxiliary fuel firing with minimal interference with lower furnace operations. This is not possible when operating the secondary level auxiliary burners. Also, the upper level burners can provide higher steam temperatures at lower loads during startup, which can be important in mills operating high steam temperature turbine-generators.

The auxiliary fuel burners are in a very hostile environment that compromises burner reliability and increases burner maintenance. B&W has always utilized rectangular shaped bent tube openings for burners to reduce accumulations of salt cake in the burner windbox and to keep the damaging furnace radiant heat away from vulnerable burner and lighter components. The oil atomizer or gas pipe elements are manually retracted when not in use. They would then be manually inserted when needed to start the burners.

The LM2100® auxiliary burner (Fig. 26) is designed to retract the burner and lighter elements into the windbox and insert them again to the same location using locally activated air cylinders. In addition to protection from radiant heat, the use of air cylinders also keeps the fuel elements at the correct position when inserted. The LM2100 improves burner reliability and maintenance. The LM2100 is standard design for new recovery boilers and can be retrofit to existing recovery boilers of any manufacturer.

Non-condensable gases/waste streams

Within the typical pulp and paper mill, there are numerous liquid and gaseous waste streams that are subject to stringent environmental disposal regulations. These waste streams are commonly incinerated on site. One simple and effective incineration method is to introduce these streams into an operating boiler.

Gaseous streams are typically characterized as non-condensable gas (NCG) or stripper off gas (SOG). NCG is a mixture of sulfur compounds and hydrocarbons mixed with air and is either dilute (DNCG) or concentrated (CNCG). These terms refer to the concentration of *active* compounds within the NCG. If the percentage of sulfur compounds and hydrocarbons is below the lower explosion limit (LEL) in air, the NCG is considered dilute, is essentially foul air, and is relatively easy to handle. It does, however, present minor challenges. It can be corrosive and is unpleasant to smell. If the percentage of active compounds is above the upper explosion limit (UEL) in air, the NCG is considered concentrated. CNCG should be treated as a corrosive fuel stream. These streams require added precautions.

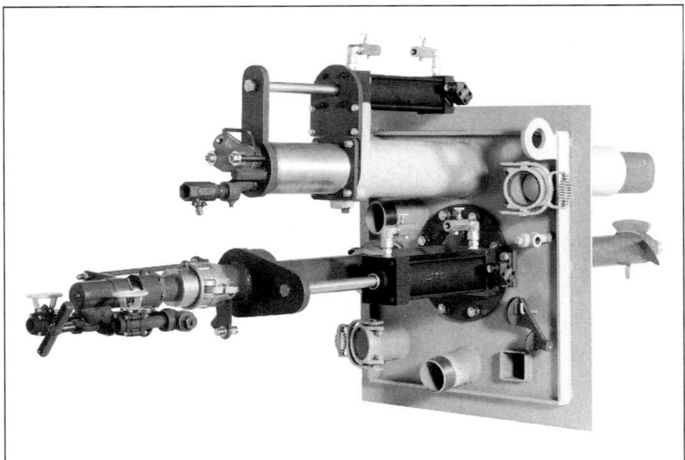

Fig. 26 B&W's LM2100® burner.

The explosion limits for some representative gases found in NCG are shown in Table 4. The generally accepted LEL for mixtures of gases in air is 2% by volume, and the generally accepted UEL is 50%. SOG is usually a mixture of methanol in air, and is typically handled like CNCG.

These gases require special safety precautions for collection and handling. Their escape must be prevented in and around the boiler building for personnel and equipment protection, and condensate must not be allowed to enter the furnace. For CNCG gases, additional precautions are required to avoid fire and explosion in equipment and transport systems. Positive ignition systems are required for gases entering the furnace.

It is generally preferred that these non-condensable gases be burned in mill power boilers, as these units can more easily include the appropriate safety equipment. One disadvantage to using power boilers is the requirement to capture the SO_2 generated by combustion of the sulfur compounds, unless the boiler is equipped with a scrubber. For recovery boilers, the explosion potential due to tube failures does not make this type of boiler a good application to introduce foreign streams. BLRBAC does not encourage the introduction of waste streams into recovery boilers. It has, however, become more commonplace in recent years to utilize recovery boilers for this disposal, as mills are required to meet more stringent regulations. Recovery boilers can have the ability to capture the SO_2 from NCG combustion. Significant caution and special system design are necessary to safely burn NCG in a recovery boiler.

Table 4
NCG Gases Explosion Limits

Constituent Gas	Lower Explosion Limit (LEL) % Volume in Air	Upper Explosion Limit (UEL) % Volume in Air
Turpentine	0.8	Not established
Methanol	7.3	36.0
Hydrogen Sulfide	4.3	45.0
Methyl Mercaptain	3.9	21.8
Dimethyl Sulfide	2.2	19.7
Dimethyl Disulfide	1.1	16.1

DNCGs can often be mixed with the combustion air for the tertiary air port level on a recovery boiler, or for the stoker on a power boiler. The primary concern is to assure adequate main fuel, usually equal to or above 50% of load, to assure burnout of the gases. To avoid odor problems, the ducts, windboxes and other air transport components must be air tight. Corrosion in the feed ducts can be reduced by mixing the gases with hot air or heating the mixture. For power boilers, if the sulfur input is high, boiler exit corrosion and emissions can increase. The corrosion rate can be reduced by upgrading alloys and/or coatings or by increasing the gas temperature to stay above the dew point.

For CNCG gases, transport must avoid air leakage that could move the mixture into the explosive range. Typically, these gases are fired with a support combustion source, such as an igniter in close proximity, to assure burnout. Stainless steel piping is used to avoid corrosion. One stream explicitly excluded from any boiler incineration consideration is terpenes (turpentine vapor). The UEL of these streams is not defined, and terpenes have been the most prevalent causes of explosions in non-condensable gas collection, transport and combustion systems.

For either gas, provision must be made to avoid slugs of liquid entering the furnace. System design uses flame propagation arresters to avoid explosions, and safety interlocks must be integrated into the existing controls.

For any type of waste gas, it is important that a complete analysis of the fuel parameters (constituents, amounts, moisture, etc.) be understood so that the appropriate systems and safeguards can be designed. It is then possible to determine the proper furnace temperatures and residence times required to assure complete combustion.

There are also several liquid waste streams with disposal requirements. These include soap, tall oil, methanol, spent acid and secondary sludge. These are normally blended and combusted with the black liquor fuel. Under specific circumstances, they may be fired in a separate, dedicated burner. Turpentine, another liquid stream, is extremely unpredictable and generally restricted from either of the above methods. Handling and combusting recommendations and cautions can be found in BLRBAC guidelines.

Smelt spout system

Smelt exits the furnace through specially designed openings at the low point of the floor and is conveyed to the dissolving tank in a sloped water-cooled trough, or smelt spout (Fig. 27). The spout is bolted to the furnace wall mounting box. Most original spout designs had the machined face of the spout positioned against the outside of the furnace wall tubes surrounding the opening. While many of this arrangement are still in operation, the fully insertable spout has become common. This type of spout inserts directly into the spout opening, where the spout taps the smelt directly from

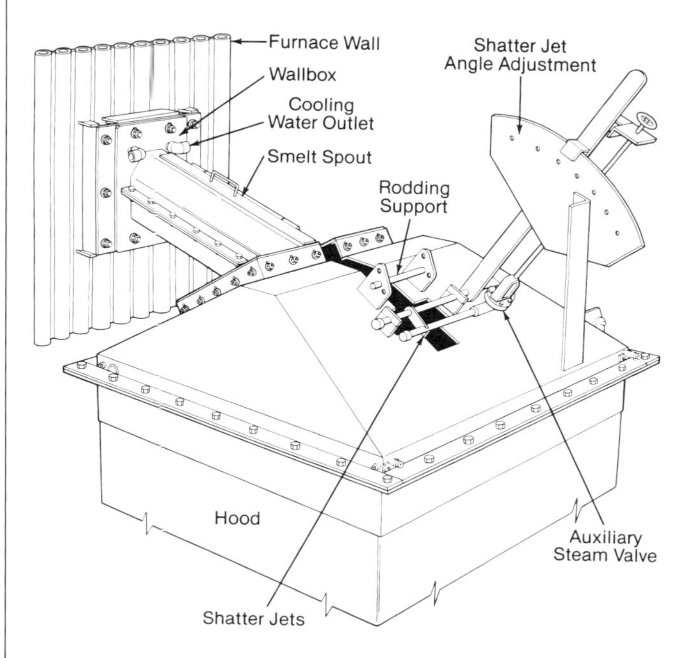

Fig. 27 Integral smelt spout hood with shatter jet assembly.

the furnace and no smelt contacts the spout opening tube surface. This has helped alleviate significant spout opening wear and deterioration.

Insertable spouts use a round-shaped trough. For noninserted spouts, the trough is V-shaped to correspond to the V-shaped bottom of the wall opening. Either spout design is constructed of a double wall carbon steel trough, with a continuous flow of cooling water passing between the inner and outer walls. Given the explosive nature of smelt-water reactions and the extreme temperature of the molten smelt, the spout must receive adequate cooling water. A dedicated cooling system with built in redundancy and multiple sources of backup water assures system reliability (Fig. 28). This system may either be a pressurized cooling system, or more commonly a gravity flow, induced vacuum system where the water in the smelt spout is at less than atmospheric pressure. The cooling water is treated in a dedicated, closed cycle to minimize scale-forming contaminants.

Green liquor system

The smelt is dissolved in green liquor in an agitated tank. Green liquor is withdrawn from the tank at a controlled density and the volume replaced with weak wash. The entering smelt stream must be finely dispersed, or shattered, to control the smelt-water reaction by rapid dissolution of the smelt into the green liquor. Excessively large smelt particles entering the tank can lead to explosions. The outer surface of a large smelt particle cools quickly as it contacts the green liquor, forming an outer shell around its hot core. As the shrinking forces build, the particle shell explodes, exposing the hot core to water and resulting in the sudden release of steam. To eliminate large smelt particles from reaching the dissolving tank contents, steam shatter jets are located above the smelt stream to disperse the flow into the tank.

The smelt spout discharge is enclosed by a hood (Fig. 27). The steam shatter jets are mounted on the top of the hood, with the nozzle angle adjusted to direct the jet at the smelt stream cascading off the spout end. The periphery of the hood is equipped with wash headers to flush the hood walls and prevent smelt buildup. A door is provided in the hood for operator access to inspect and to rod the spout and opening.

The dissolving tank is of heavy construction and is equipped for agitation. One or more agitators, either side or top entering, are generally used, including an emergency backup system, such as steam nozzles. A stainless steel band is commonly used in the carbon steel tank at the liquor level for corrosion protection, while the floor is protected with steel grating or poured refractory.

As the smelt is cooled and dissolved into the green liquor, large quantities of steam are released. The steam vapors are pulled through an oversized atmospheric vent on the tank to quickly relieve pressure in the event of a surge or explosion. Smelt particles and green liquor droplets are entrained in the steam-air mixture vented from the tank. These vented gases also contain H_2S that must be removed prior to discharge to the atmosphere. Typically, a vent stack scrubber (Fig. 29) is used to reduce H_2S emissions and to trap entrained particulates. Weak wash is used as the scrubbing medium to take advantage of its residual NaOH, which absorbs the malodorous gases.

Alternate processes and liquors

Soda process

With few exceptions, the requirements and features of the kraft recovery boiler apply to the soda recovery boiler. Because sulfur is not present in the soda pro-

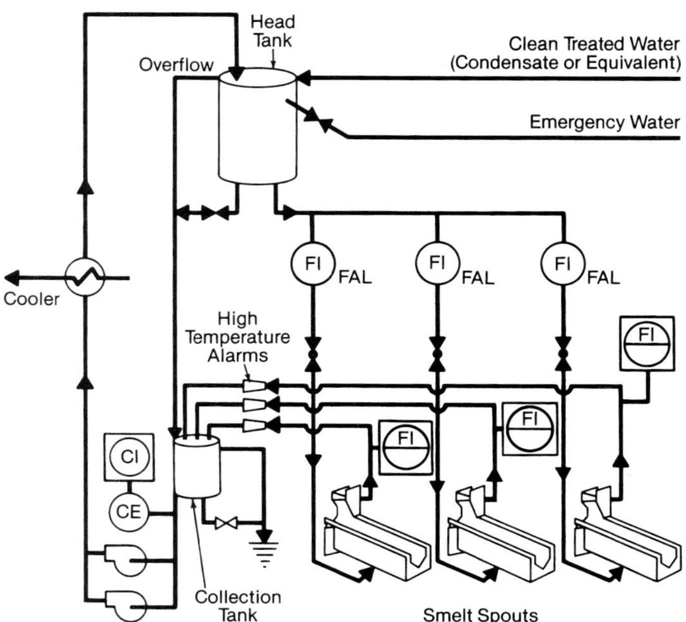

Fig. 28 Smelt spout cooling system.

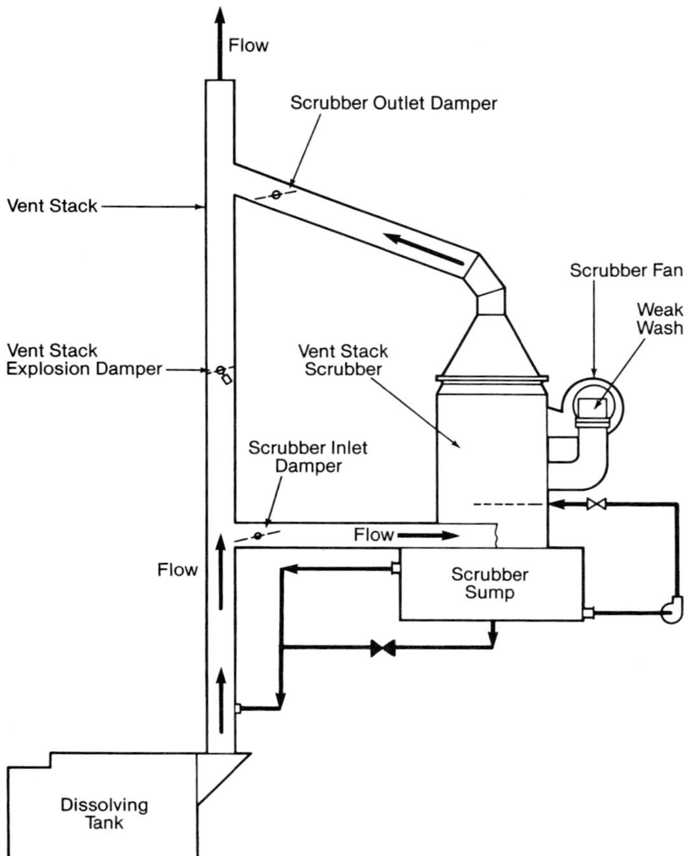

Fig. 29 Vent stack scrubber.

cess, there is no sulfur reduction in the recovery furnace. Soda ash (Na_2CO_3) is added to the recovered green liquor. The ash collected from the boiler hoppers and the electrostatic precipitator is in the form of Na_2CO_3 and can be added directly to the green liquor in the dissolving tank. A salt cake mix tank is not required.

In the furnace, the soda liquor does not form a suitably reactive char for burning in a bed. It is finely atomized and sprayed into the furnace by multiple soda liquor burners. The fine spray dehydrates in flight and combustion takes place largely in suspension in an oxidizing atmosphere. Combustion air is admitted through primary and secondary air ports around the furnace periphery, with the hottest part of the furnace just above the hearth. The Na_2CO_3 collects in molten form on the hearth and discharges through the smelt spouts.

The Na_2CO_3 smelt has a higher melting point than the kraft process smelt. This makes the soda smelt more difficult to tap from the furnace and shatter. Auxiliary fuel burners can be located low in the side walls, close to the spout wall, to keep the smelt hot for easier tapping.

Nonwood fiber liquor

Many countries do not have adequate forest resources for pulp production. As a result, alternative fiber sources are used, such as bamboo, sugar cane bagasse, reeds and straw. Black liquor from these fibrous materials requires special consideration in recovery system design.

The liquor is generally characterized by high viscosity and high silica content. The high viscosity limits the level at which liquor can be concentrated in multiple effect evaporators, and direct contact evaporators are generally required. Higher combustion air temperatures are used to compensate for the low liquor concentrations while allowing flexibility in adjusting the firing conditions.

Sulfite process

Pulp produced by the sulfite process is divided into two broad categories, semi-chemical and chemical. Semi-chemical pulp requires mechanical fiberizing of the wood chips after cooking. Chemical pulp is largely manufactured by the acid sulfite and bisulfite processes, which differ from the alkaline processes in that an acid liquor is used to cook the wood chips. The acid sulfite process is characterized by an initial cooking liquor pH of 1 to 2. The bisulfite process operates with a pH of 2 to 6. Cooking liquor used in a neutral sulfite process for manufacture of semi-chemical pulps has an initial pH of 6 to 10. Spent sulfite liquor, separated from chemical and semi-chemical pulp and containing the residual cooking chemicals and dissolved constituents of the wood, is evaporated and burned, and the chemicals are recovered in a system particular to each base.

Sulfite process pulp mills normally use one of four basic chemicals for digestion of wood chips to derive a large spectrum of pulp products: sodium, calcium, magnesium or ammonium. The principal differences in the sulfite waste liquor of the four bases are the physical properties of the base and the products of combustion.

Sodium This system reconstitutes the base in a form suitable for reuse in the pulping cycle. The sodium-base liquor may be burned alone or in combination with black liquor in a kraft recovery unit. The base chemical is recovered as a smelt of Na_2S with some Na_2CO_3 which would be a suitable makeup chemical for a nearby kraft mill. The combination of sulfite liquor and kraft liquor, referred to as *cross recovery*, is common to pulp mill operations. The proportion of sulfite liquor in cross recovery is generally the equivalent of the sodium makeup requirement of the kraft mill cycle.

Calcium Calcium-base liquor is concentrated and burned in specially designed furnaces. The furnace and boiler design applicable to magnesium-base liquor combustion can also be applied to calcium. This process is largely historic today due to its unacceptable effluents.

Magnesium With the magnesium base, a simple system is available for recovery of heat and total chemicals. This is due to the chemical and physical properties of the base. The spent liquor is burned at elevated temperatures in a controlled oxidizing atmosphere, and the base is recovered in the form of an active magnesium oxide (MgO) ash. The oxide can be readily recombined in a simple secondary system with the SO_2 produced in combustion, thereby reconstituting the cooking acid for pulping.

Industrial interest in the improved pulp from a variety of wood species stimulated the development of pulping techniques using a magnesium base. The two major pulping procedures, whereby a variety of pulps can be produced, are magnesium acid sulfite and bisulfite, or magnefite. The basic magnesium recovery system is appropriate for each of the pulping processes.

The heavy liquor is fired with steam-atomizing burners located in opposite furnace walls (see Fig. 30). The combustion products of the liquor's sulfur and magnesium are discharged from the furnace in the gas stream as sulfur dioxide and solid particles of MgO ash. Most of the MgO is removed from the gas stream in a mechanical collector or electrostatic precipitator and is then slaked to magnesium hydroxide, or $Mg(OH)_2$. In the complex secondary recovery system, the SO_2 is recovered by reaction with the $Mg(OH)_2$ to produce a magnesium bisulfite acid in an absorption system. This acid is passed through a fortification or bisulfiting system and is fortified with makeup SO_2. The finished cooking acid is filtered and placed in storage for reuse in the digester.

Ammonium Ammonium-base liquor is the ideal fuel for producing a low ash combustion product. Burning can be accomplished in a simple recovery boiler. The ammonia decomposes on burning to nitrogen and hydrogen, the latter oxidizing to water vapor and thus destroying the base. Concentrated ammonium sulfite liquor is burned in a furnace and recovery system similar to magnesium-base liquor. SO_2 produced in combustion can be absorbed in a secondary system to yield cooking acid for pulping. The SO_2 reacts in an absorption system with an anhydrous or aqueous ammonium makeup chemical to produce ammonium bisulfite acid. In a neutral sulfite semi-chemical plant, the absorption system produces cooking liquor consisting essentially of ammonium sulfite.

Black liquor gasification

Advanced technologies – gasifier development

A new technology being investigated for the combustion/reduction of black liquor (and thus replacing the process recovery boiler) is black liquor gasification (BLG). This technology involves the partial oxidation of the black liquor to produce a fuel gas that can be burned in a gas turbine/electric generator or fired in a power boiler. In the process, sulfur and sodium compounds are recovered from the fuel gas, before it is sent to the turbine or boiler, for regeneration of the pulping chemicals.

Potential BLG advantages

The pulp and paper process requires high capital investment in plant equipment, high internally generated energy consumption, and high purchased power consumption. BLG may have the potential to favorably impact mill energy efficiency, environmental performance, operating costs and pulp yields.

Energy efficiency The amount of electricity pulp mills need to import varies considerably from mill to mill and is dependant on the power/steam cycle used and on the type of pulp produced. Electrical loads have increased relative to steam loads as more electricity is needed to operate recycling, mechanical pulping, and pollution control equipment. Most mills import a significant portion of the mill's electricity demand, and BLG, integrated with combined cycle power production, may offer increased cycle efficiencies and electricity production.

Process flexibility Gasification technology can return sulfur to the pulping process in a variety of forms (sulfide, elemental sulfur, polysulfide, and/or sulfite). This provides opportunities to produce pulp with substantially improved yields and properties.

Availability The fouling of convection pass surfaces can reduce recovery boiler availability. BLG may provide a greater availability in that heat transfer surface fouling in either a power boiler or gas turbine heat recovery steam generator might be lower. Intermediate cleanup equipment removes sulfur compounds and alkali fume from the product gas.

Safety Smelt-water reactions are largely unpredictable and continue to pose a risk when operating kraft recovery boilers. Low-temperature BLG produces no molten smelt, eliminating the potential of smelt-water reactions, and high-temperature BLG at elevated pressure is expected to significantly reduce the likelihood and resultant force of an explosion.

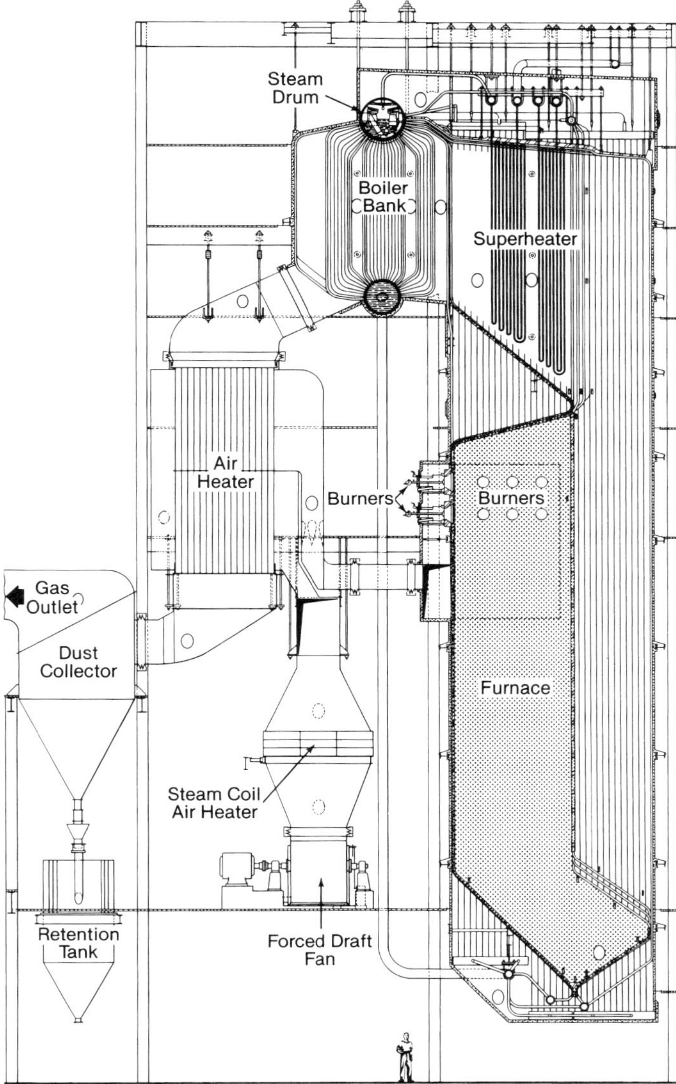

Fig. 30 MgO recovery boiler.

Emissions BLG processes require that sulfur and particulate emissions be removed from the fuel gas. As a result, overall emissions from a BLG-based process are expected to be low.

BLG processes

Black liquor gasification technologies can be classified by operating temperature. Low-temperature BLG is operated below the first melting temperature of the inorganic salts, typically 1112 to 1292F (600 to 700C), to produce a solid inorganic ash. High-temperature BLG is operated above the final melting temperature of the inorganic salts, typically 1652 to 1832F (900 to 1000C), to produce a molten smelt. Black liquor gasifiers have utilized either the fluidized-bed (for low-temperature BLG) or entrained-flow designs (for high-temperature, molten-phase designs), as described in Chapter 18. Typical BLG processes are shown in Fig. 31.

Low temperature gasifiers operate at 1292F (700C) or lower so that the organics leave as dry solids. This type of gasification process is normally conducted at atmospheric pressure and produces a product gas of approximately 200 to 300 Btu/DSCF (11.8 MJ/Nm^3). Minimal gas cleanup is required, and the product gas is generally fired to offset auxiliary fuel in a power boiler. This allows additional black liquor processing capacity, provides a useful product gas for auxiliary fuel use, and is well suited for incremental capacity additions at existing mills.

High temperature gasifiers generally operate at 1742F (950C) or higher and produce a product gas of approximately 85 to 100 Btu/DSCF (3.9 MJ/Nm^3) and a molten smelt of inorganic chemicals. High-temperature BLG can be pressurized and integrated with a combined cycle system for power generation. Product gas cleanup systems are required for the removal/reduction of sulfur and particulate emissions before the gas turbine. These systems can also separate the sulfur compounds from the product gas and return them to the pulping process.

Status and development needs

Atmospheric pressure BLG has had the most development to date. Various concepts, both high and low temperature, have been in design and pilot stages since the early 1990s. The first fully commercial process is an air-blown, high-temperature entrained flow gasifier with a processing capacity of 15 t/h of dry solids. This process provides incremental recovery capacity with product gas burned in a boiler.

High-temperature BLG with combined-cycle power generation is in development. Various pilot work is being done for pressurized, oxygen-blown entrained-flow reactors and large-scale commercial units are contemplated for the near future.

Black liquor gasification is a promising technology, but key issues must be resolved before full commercialization. While current Tomlinson technology is capital intensive, early BLG installations are expected to be quite expensive. Capital costs must be reduced, positive economics must result and the overall economic case must be strengthened for BLG to provide superior returns.

Specific gasifier design and process integration issues exist, including gasifier materials selection, refractory corrosion, product gas cleanup equipment, tar formation, sulfur removal, and an increased causticizing load. Many activities are underway within the industry to investigate and resolve these and other issues.

An important aspect of the overall process and economics is the effect of integrating higher yield pulping technologies into the BLG cycle (split sulfidity, polysulfide, etc.), as these can have significant positive impact on overall economics. They, too, are capital intensive and will take some development, but integrating them into the process is the key to providing acceptable, positive economic returns.

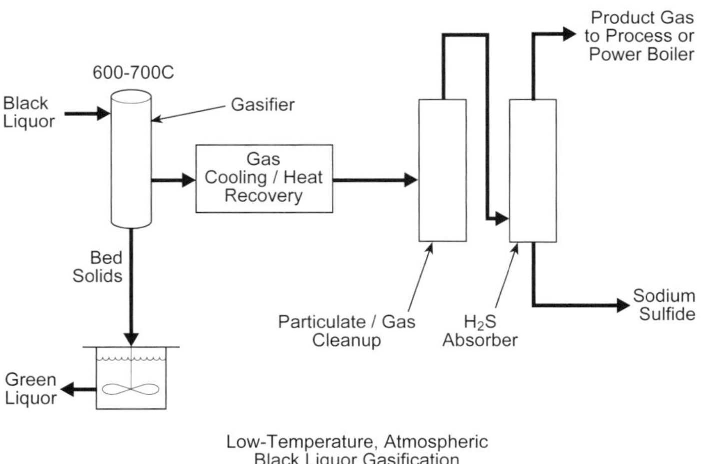

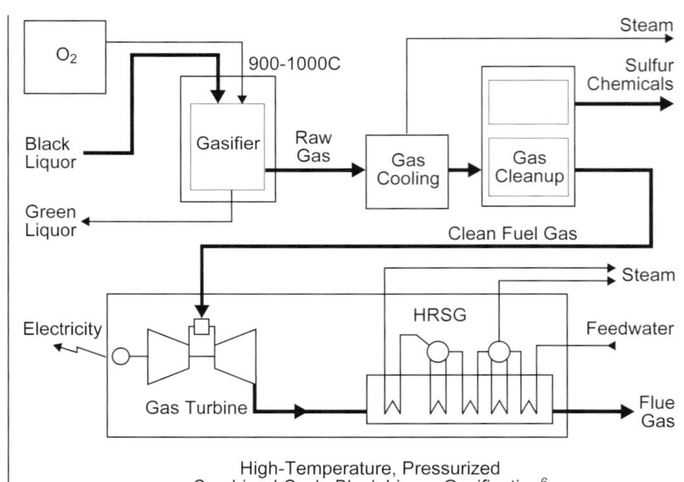

Fig. 31 Black liquor gasification processes.

References

1. Adapted from Hough, G.W., "Chemical Recovery in the Alkaline Pulping Processes," Technical Association of the Pulp and Paper Industry, Inc., (TAPPI), p. 197, 1985.

2. Adapted from Smook, G.A., *Handbook for Pulp and Paper Technologists,* Joint Textbook Committee of the Paper Industry TAPPI/CPPA, p. 69, 1986.

3. Wiggins, D., et al., "Liquor Cycle Chloride Control Restores Recovery Boiler Availability," TAPPI Engineering Conference, Atlanta, Georgia, Sept. 17-21, 2000.

4. Tran, H.N., "Kraft Recovery Boiler Plugging and Prevention," Notes from the Tappi Kraft Recovery Short Course, pp. 209-218, Orlando, Florida, January, 1992.

5. Adapted from Whitney, R.P., *Chemical Recovery in Alkaline Pulping Processes,* TAPPI, Monograph Series No. 32, p. 40, 1968.

6. Taken from Air Products & Chemicals, Inc. presentation at the American Institute of Chemical Engineers (AIChE) Annual Meeting, Miami Beach, Florida, November 15-20, 1998.

This facility features three 750 ton per day mass-fired units by B&W.

Chapter 29

Waste-to-Energy Installations

The disposal of garbage is a problem that has existed since civilization began. At various times throughout history, composting, animal feed, landfill and incineration have all been popular disposal methods. Today, refuse disposal methods are determined by cost and the effect on our environment.

The most common means of refuse disposal is still landfilling. Even in the late 1970s, nearly all of the refuse generated in North America was landfilled. Incineration with no heat recovery was a popular option that became economically unacceptable with the advent of environmentally responsible air pollution regulations and inexpensive landfill alternatives.

In Europe and Japan, where new landfill sites were less available, incineration continued as a viable option and those plants became the predecessors of today's refuse-to-energy plants. Heat recovery was added in the form of waste heat boilers which were originally hot water boilers and later low pressure and temperature steam boilers. These incinerators with waste heat boilers then evolved into waterwall boilers with integral stokers.

Refuse-fired boiler design parameters and operating characteristics are strongly affected by the components of the refuse, which change with time. The components also vary greatly by location. In North America, typical municipal solid waste (MSW) is high in paper and plastics content (Fig. 1) and typically has a lesser moisture content and greater heating value than that found worldwide. In a less industrialized country the refuse tends to have a greater moisture and ash content and lesser heating value. Table 1 shows representative refuse analyses ranging from 3000 to 6000 Btu/lb (6978 to 13,956 kJ/kg) higher heating value (HHV) basis.*

In the United States (U.S.) and North America, the refuse characteristics have changed dramatically in a short period of time. With more and more convenience foods, plastics, packaging, containers, and less food scraps due to home garbage disposals, the average refuse heating value has increased and the moisture content has decreased (Table 2). As more recycling programs are implemented, the analysis will continue to change. As glass, aluminum and other metals are recycled, the refuse heating value will increase; as paper and plastics are recycled the heating value will decrease.

In 2000, approximately 232 million tons (210 million t_m) of MSW were generated in the U.S. About 30% of that total was recycled (including composting), 15% sent to refuse-to-energy facilities, and the balance landfilled. The growth of refuse-to-energy facilities in the U.S. accelerated in the 1980s and early 1990s (Fig. 2) due to the growing disposal costs for landfills and a government-created market for the sale of electric power.

As old landfills closed, new landfills became more difficult and costly to open and tended to be located farther from the source of the refuse, increasing transportation costs. Concerns about ground water contamination resulted in more expensive landfill designs with several containment layers and leachate monitoring and control systems. The passage of the Public Utility

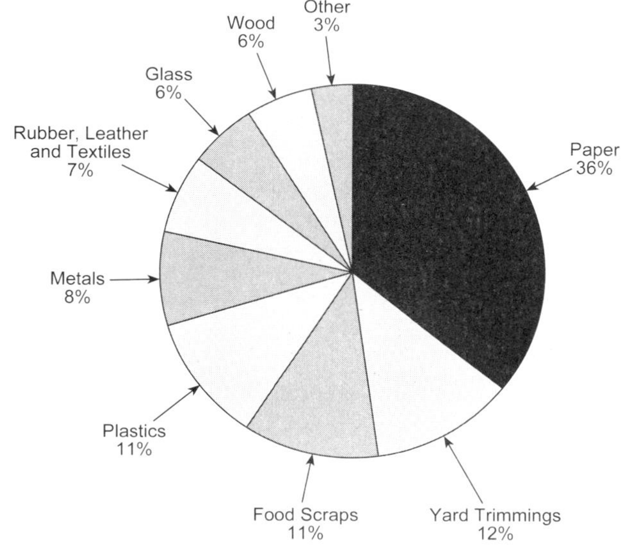

Fig. 1 U.S. municipal solid waste generation – 2001.

* Throughout this chapter, fuel energy content, or heating value, is stated as the higher (or gross) heating value which includes the total energy in the fuel without correcting for losses due to water evaporation.

Table 1
Range Of As-Received Refuse Fuel Analysis

	Weight Percent As-Received						
HHV, Btu/lb (kJ/kg)	3,000 (6,978)	3,500 (8,141)	4,000 (9,304)	4,500 (10,467)	5,000 (11,630)	5,500 (12,793)	6,000 (13,956)
Carbon	16.88	19.69	22.50	25.32	28.13	30.94	33.76
Hydrogen	2.33	2.72	3.10	3.49	3.88	4.27	4.66
Oxygen	12.36	14.42	16.49	18.55	20.62	22.68	24.75
Nitrogen	0.22	0.26	0.30	0.34	0.38	0.42	0.46
Sulfur	0.15	0.18	0.21	0.24	0.27	0.30	0.33
Chlorine	0.34	0.38	0.42	0.46	0.50	0.54	0.58
Moisture	35.72	32.35	29.98	28.60	25.22	22.85	21.46
Ash	32.00	30.00	27.00	23.00	21.00	18.00	14.00
Total	100.00	100.00	100.00	100.00	100.00	100.00	100.00

Regulatory Policies Act of 1978 (PURPA) required public utilities to purchase the electric power generated by refuse-to-energy plants. This created a revenue flow that helped offset the inherent high capital cost of these plants. These market forces resulted in a proliferation of refuse-to-energy facilities in the northeast U.S. where the costs of landfill and other disposal options were the highest, and selectively throughout North America in response to local environmental or economic factors.

Refuse disposal is a major problem worldwide and there is no single solution. An environmentally sound refuse disposal program includes generating less refuse, recycling components that can be economically reused, combustion of the balance of the refuse (including the efficient generation of electric power), and the landfill of the resulting ash.

Refuse combustion alternatives

Two main techniques are used for burning municipal refuse, distinguished by the degree of fuel preparation. The first technique, known as mass burning, uses the refuse in its as-received, unprepared state (Fig. 3). Only large or non-combustible items such as tree stumps, discarded appliances, and other bulky items are removed. Refuse collection vehicles dump the refuse directly into storage pits. Overhead cranes equipped with grapples move the refuse from the pit to the stoker charging hopper. Hydraulic rams located in the charging hopper move the refuse onto the stoker grates. The combustible portion of the refuse is burned off and the non-combustible portion passes through and drops into the ash pit for reclamation or disposal.

The second burning technique uses prepared refuse, or refuse-derived fuel (RDF), where the as-received refuse is first separated, classified, and reclaimed in various ways to yield salable or otherwise recyclable products (Fig. 4). The remaining material is then shredded and fed into the furnace through multiple feeders onto a traveling grate stoker. The RDF is burned, part in suspension and part on a stoker. The RDF can also be used as a fuel in circulating fluidized-bed (CFB) boilers and in bubbling fluidized-bed (BFB) boilers. More finely shredded RDF can be fired in suspension to supplement conventional fuels in large boilers used for power generation.

Corrosion

Combustion products from municipal refuse are very corrosive. The components that are present in coal, oil and other fuels that contribute to corrosion, as well as to high slagging and high fouling, are all present in refuse (Table 3). Corrosion in refuse-fired boilers is usually caused by the chloride compounds which deposit on the furnace, superheater and boiler tubes. Several modes of chloride corrosion may occur:

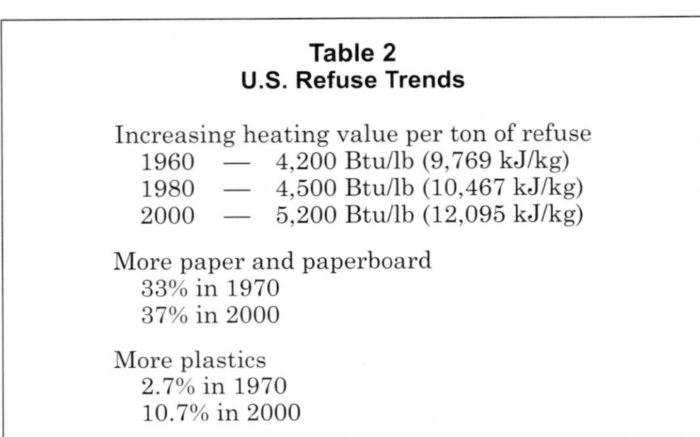

Table 2
U.S. Refuse Trends

Increasing heating value per ton of refuse
1960 — 4,200 Btu/lb (9,769 kJ/kg)
1980 — 4,500 Btu/lb (10,467 kJ/kg)
2000 — 5,200 Btu/lb (12,095 kJ/kg)

More paper and paperboard
33% in 1970
37% in 2000

More plastics
2.7% in 1970
10.7% in 2000

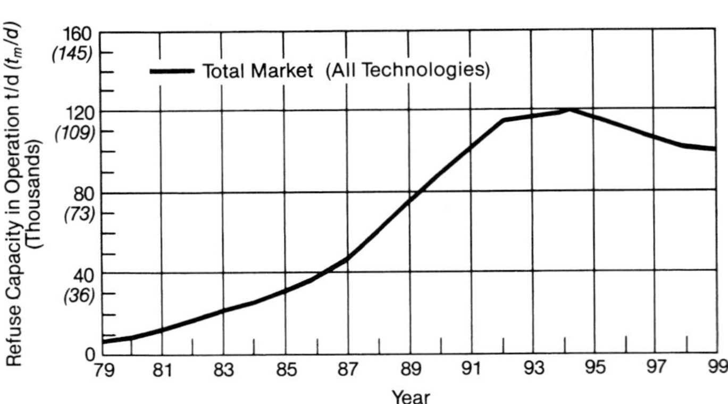

Fig. 2 U.S. refuse-to-energy market.

1. corrosion by hydrochlorides (HCl) in the combustion gas,
2. corrosion by NaCl and KCl deposits on tube surfaces,
3. corrosion by low melting point metal chlorides (mainly $ZnCl_2$ and $PbCl_2$), and
4. out-of-service corrosion by wet salts on the tube surface.

The rate of tube metal loss due to corrosion is temperature dependent with high metal temperatures correlating with high rates of metal loss (Fig. 5). Refuse boilers operating at higher steam pressures have higher temperature saturated water in the furnace tubes and, therefore, these furnace tubes have higher metal temperatures. Superheater tube metal temperatures are directly related to the steam temperature inside the tubes. In both cases, it is the temperature of the water or steam inside the tube that largely controls the tube metal temperature, rather than the temperature of the flue gas outside of the tube.

Furnace-side corrosion can be aggravated by poor water chemistry control. If water-side deposits are permitted to form, tube wall metal temperatures will rise and furnace corrosion will be accelerated. Standards for feedwater and boiler water quality are based on boiler operating pressures. These standards are no more stringent for refuse-fired units than for other fuels. However, the maintenance of feedwater and boiler water quality, and adherence to those standards, is more critical on refuse-fired boilers due to the highly corrosive nature of the fuel.

Lower furnace corrosion

The lower furnace environment of both mass-fired and RDF-fired units is constantly changing between an oxidizing atmosphere (an excess of O_2 beyond that needed for combustion) and a reducing atmosphere (a deficiency of O_2 below that needed for combustion) which can rapidly accelerate corrosion. Therefore, some form of corrosion protection is needed. Typically, the area of protection will encompass all four walls up to 30 ft (9.1 m) above the grate where there is reasonable assurance that an oxidizing atmosphere is prevalent and that combustion has been substantially completed.

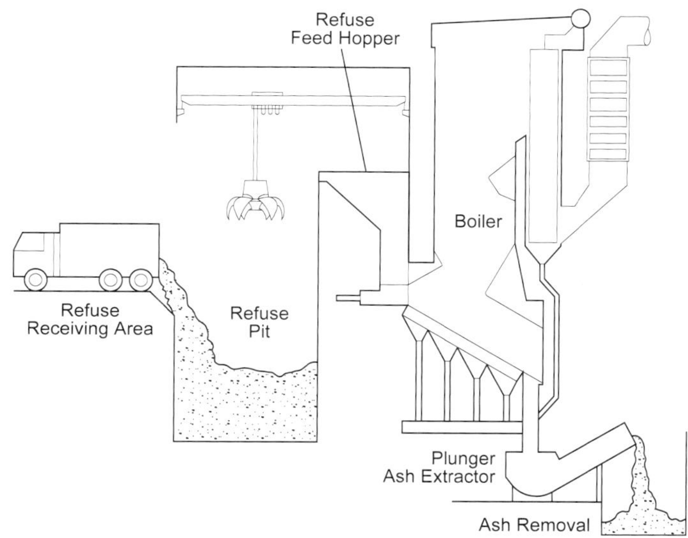

Fig. 3 Mass burning schematic.

Mass-fired units

A variety of solutions have been developed and used to address the issue of corrosion in the lower furnace. The lower furnace walls of virtually all mass-fired units have been protected by one or more of the following three systems: 1) pin studs with silicon carbide (SiC) refractory, 2) ceramic tiles manufactured from SiC (see Fig. 6), or 3) Inconel® weld cladding of the membrane tube panels.

The quality and physical characteristics of the sili-

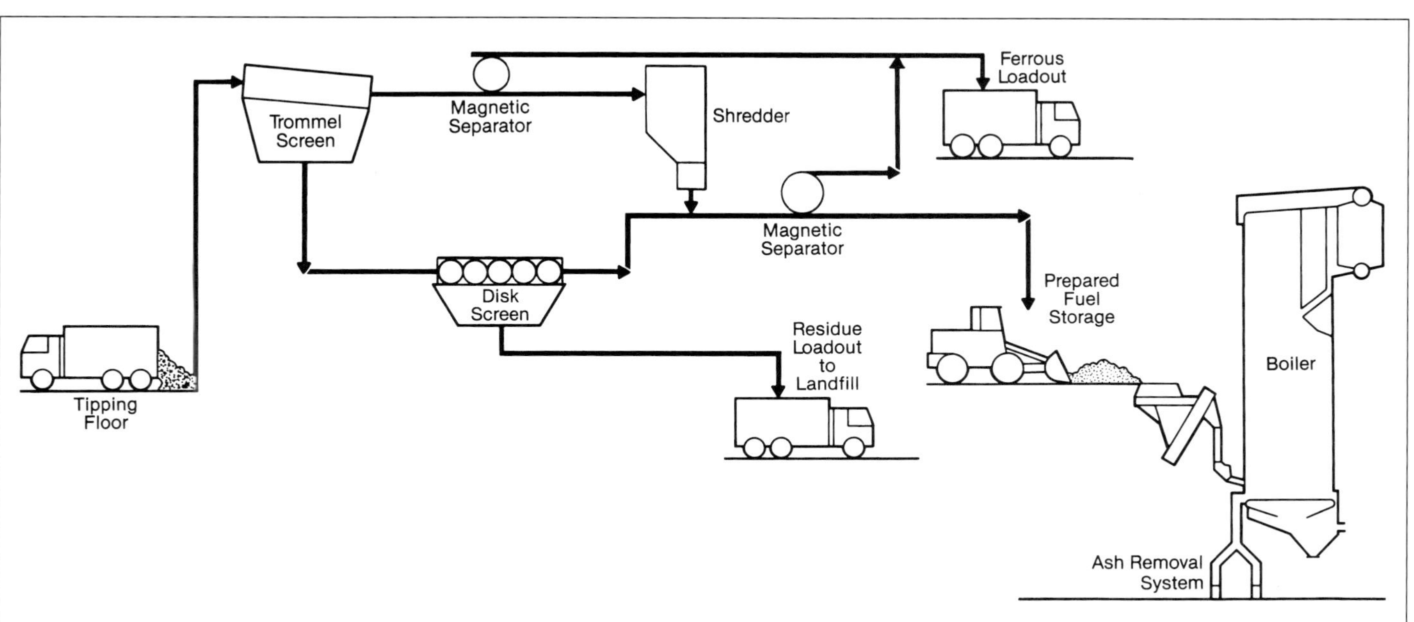

Fig. 4 RDF burning schematic.

Table 3
Corrosive Constituents in Fuels

Coal	Oil	Refuse	
Sodium	Sodium	Sodium	Chloride
Sulfur	Sulfur	Sulfur	Lead
Potassium	Vanadium	Potassium	Zinc
		Vanadium	

con carbide refractory placed over the pin studs must be maintained through proper application and curing. Lack of control during installation will result in refractory spalling, deterioration, and increased maintenance. The refractory material should have a high rate of thermal conductivity to maximize the effectiveness of the water-cooled surface it is protecting. The pin stud pattern, pin stud length and pin stud diameter must be carefully chosen for its ability to hold the refractory in place and to maximize the heat transfer through the stud to the furnace wall tubes. This, in turn, serves two purposes. One is to provide maximum cooling to keep as low a refractory temperature as possible. Maintaining a low refractory surface temperature has a dramatic effect on refractory life, furnace wall fouling and maintenance costs. Second, with more heat removed in this lower furnace area, less heating surface is required in the upper furnace to achieve the desired flue gas temperature leaving the furnace.

SiC tiles have a much harder and smoother surface than that produced with the pin studs and refractory; therefore, the tiles are more resistant to spalling and less prone to slag buildup. The attachment system used to anchor the ceramic tiles to the furnace walls must keep the tiles in contact with the furnace walls in order to cool the tiles. The anchors must also permit movement between the tiles to minimize expansion stresses that may damage the tiles.

Inconel weld overlay has gained general acceptance since the year 2000 for application in the lower furnace walls of large mass-fired boilers, as it had earlier for RDF-fired boilers. The good corrosion resistance, high thermal conductivity, metallurgical bond to the base tube and membrane bar metals, and wear resistance have made Inconel overlay the primary approach for lower furnace corrosion protection in new European designs. In such units, high temperature gunned-on and cast refractory systems also continue to be used in selected areas of the furnace where heat absorption must be limited or heat must be radiated back into the combustion zone while still providing corrosion protection. Typically, refractory is used on: 1) furnace arch areas over the grate feed and burnout zones, 2) on the intermediate elevation furnace walls where heat absorption must be minimized to maintain adequate furnace temperatures during continuous low-load firing, and 3) on the furnace walls in smaller boilers where the ratio of furnace wall area to furnace volume is so high that the gas temperature within the furnace is suppressed below the point required for optimal combustion (see Fig. 7).

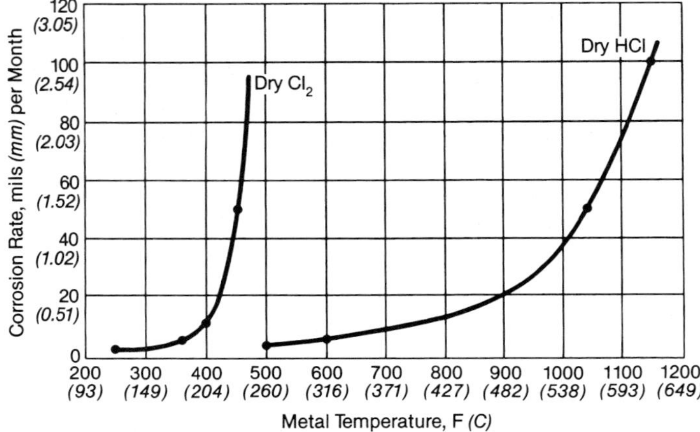

Fig. 5 Corrosion rate of carbon steel in chlorine and hydrogen chloride.

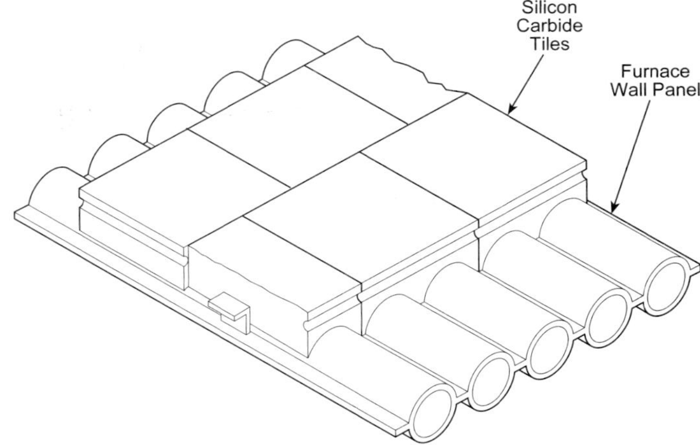

Fig. 6 Silicon carbide tiles for lower furnace wall corrosion protection.

The furnace side walls along the grate line generally experience the most erosion due to the scrubbing action of the refuse fuel and ash as it moves along the grate to the ash discharge. Increased erosion-resistant SiC materials are available for these zones. They do, however, have lower thermal conductivities.

An alternative near the grate line is the use of armour blocks or refractory blocks, rigidly attached to the furnace walls. These blocks extend up the full height of the charging hopper opening which is about 4 ft (1.2 m) high at the front of the furnace and tapers off to about 1 ft (0.3 m) high at the ash discharge end (Fig. 8).

Another method used for protection of the furnace walls immediately adjacent to the grate is the use of a separate water-cooled *wear zone panel* placed horizontally along the grate line and in front of the lower furnace walls as shown in Fig. 7. A flexible tight seal is created between the wear zone panel and the vertical furnace wall to accommodate expansion between the top-supported boiler and the bottom-supported grate structure. Depending upon the cooling water temperature and corrosion potential, the water-cooled wear zone panel may be protected from corrosion by Inconel weld overlay similar to the boiler walls. This low maintenance design provides a relatively cool surface which resists the buildup of significant slag while also providing corrosion protection and wear resistance with little demand for maintenance.

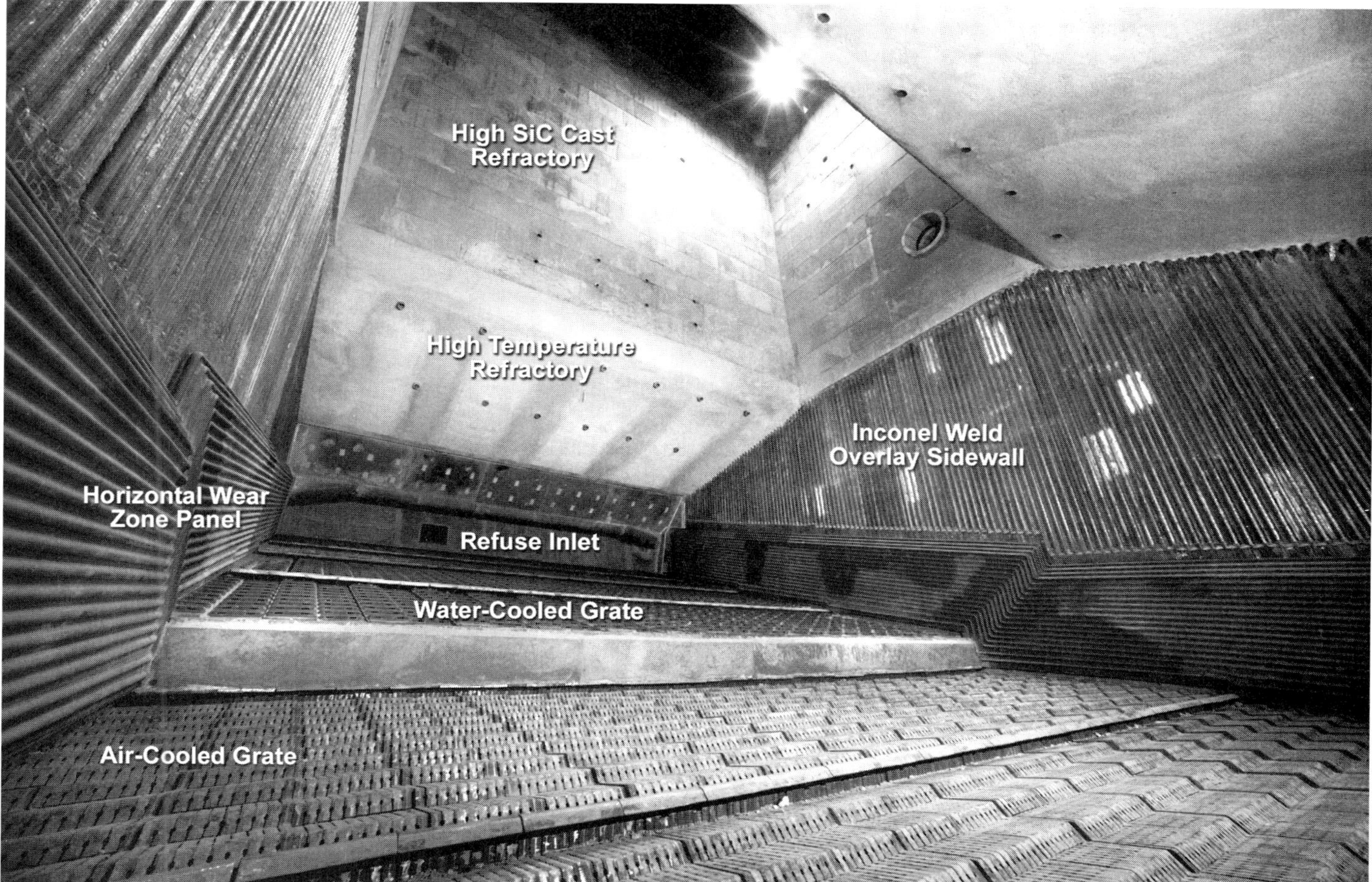

Fig. 7 Corrosion protection and design of a 26 t/h mass-fired unit.

RDF-fired units

Prior to the late 1980s, RDF boilers were installed with bare carbon steel tubes in the lower furnace and no corrosion protection. It was thought that with the more even combustion with a processed fuel, corrosion would not be a concern in the lower furnace. Early units, operating at low steam pressures and temperatures, did not experience corrosion problems. However, as higher pressure and temperature units went into operation, corrosion increased and lower furnace protection was needed.

The same pin stud and refractory design used on mass-fired units has also been tried on RDF units. This solved the corrosion problem but created another. Inherent in the RDF combustion process is a high degree of suspension firing and high flame temperatures in the lower furnace. When pin studs and refractory are applied, the lower furnace tubes are insulated, resulting in less heat transfer and hotter flue gas temperatures in the lower furnace. This, in turn, can result in significant slagging on the refractory wall surface. Pin studs and refractory were tried at two RDF facilities. However, increased furnace slagging resulted, and eventually the pin studs and refractory were removed.

The industry needed a material that was resistant to the chloride corrosion found in refuse boilers while not insulating the lower furnace tubes. The Babcock & Wilcox Company (B&W) pioneered the use of Inconel material as a solution to this lower furnace corrosion problem. In 1986, following rapid corrosion of the bare carbon steel tubes, the lower furnace of the Lawrence, Massachusetts unit was covered with a weld overlay of Inconel material. This overlay proved to be effective in minimizing corrosion in the lower furnace. Based on this early experience, the industry followed B&W's lead and Inconel weld overlay was subsequently field applied to the lower furnace of a number of operating boilers.

For the RDF boilers supplied as part of the refuse-to-energy plant in Palm Beach County, Florida, the decision was made to add the Inconel protection prior to manufacture. A bimetallic tube construction was used consisting of a carbon steel inner tube co-extruded with an Inconel outer tube (Fig. 9). These boilers went into operation in 1989 and the Inconel bimetallic tubes have proven to be an effective means of reducing lower furnace corrosion. Today, Inconel bimetallic tubing and carbon steel tubing with an Inconel weld overlay are the industry standard for lower furnace corrosion protection in RDF-fired boilers.

Mass burning

Mass burning is the most common refuse combustion technology worldwide. When the market for refuse-to-energy facilities expanded rapidly in the

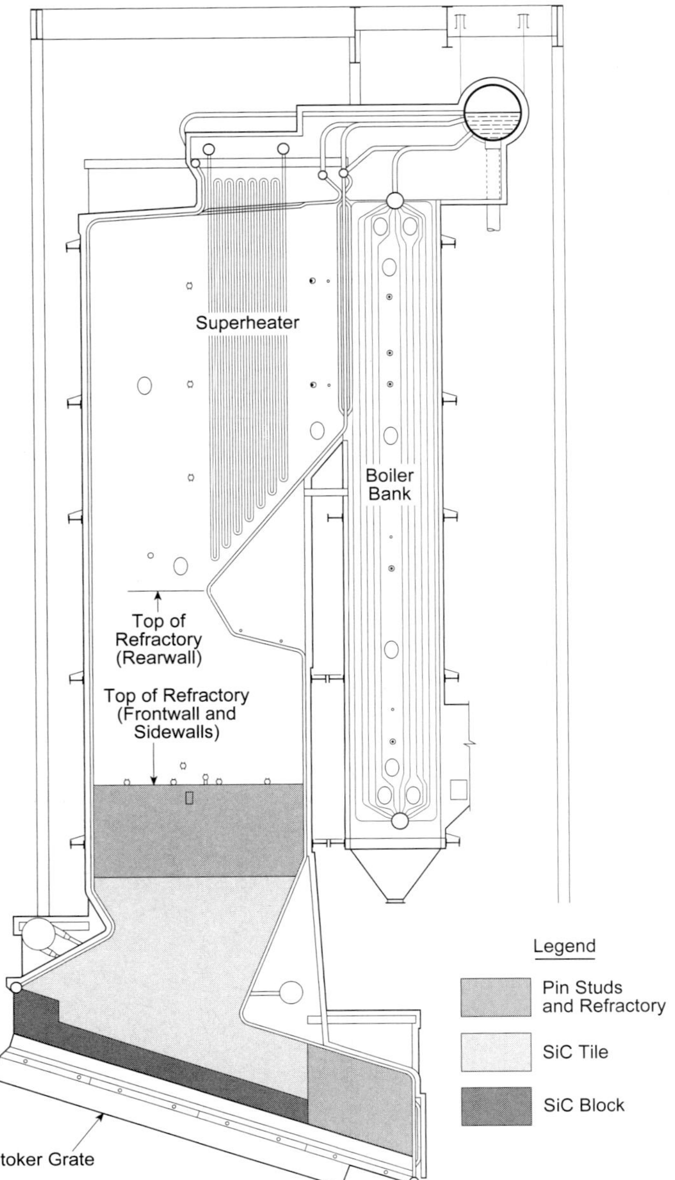

Fig. 8 Refractory type and location, mass-fired unit.

U.S. in the early 1980s, many of the U.S. refuse plants (Fig. 10) adopted the well-proven mass burning technology that had been developed in Europe. A modern B&W European mass burn power plant is shown in Fig. 11. The boiler is a multi-pass, top-supported design.

Some major differences in the application of the European technology in the U.S. resulted in several operational problems. U.S. applications tend to require large units to accommodate larger regional facilities than European and Japanese applications. U.S. plants were designed to operate at significantly higher operating pressures and temperatures to take advantage of the economics of production and sale of electric power while typical non-U.S. applications produced hot water and low pressure steam for heating applications. U.S. refuse fuel typically had a higher heating value and lower moisture content. Finally, these units were first installed at a time when environmental concerns were increasing.

The net result of these characteristics of the market resulted in many early refuse units experiencing operating problems related to:

1. high rates of slagging in the furnace,
2. higher gas temperature leaving the furnace resulting in overheating of superheaters and excessive fouling in the convection section,
3. tube failures from accelerated corrosion that were metal temperature related, and
4. concerns about the creation of dioxins (polychlorinated dibenzoparadioxin, PCDD) and furans (polychlorinated dibenzofurans, PCDF) during the combustion process that were related to less than optimum combustion systems, particularly less than optimum turbulence and mixing of fuel and air in the lower furnace.

Each of these problems has been taken into consideration as the design of mass-fired boilers has evolved.

Overall boiler design and gas flow path

B&W offers two basic designs for large mass-fired boilers. Fig. 12 shows a typical mass burn unit supplied in the U. S. while Fig. 13 illustrates a horizontal convection pass unit supplied in Europe. Both are single drum, top-supported, natural circulation boilers capable of supplying superheated steam at a range of pressures and temperatures to meet individual site requirements. The steam drums are supplied with the appropriate steam separation equipment to supply high purity dry steam to the superheater while spray attemperators closely control the steam temperature. The furnaces are fully water-cooled with membrane wall construction using the appropriate corrosion protection system (see earlier discussion). Cleaning of boiler surfaces is accomplished by a combination of mechanical rapping and sootblowers as discussed later in this chapter. Both designs utilize a reciprocating combustion grate and similar lower furnace configurations to provide efficient burnout of the fuel before the ash is discharged. The gas flow paths of the two units differ to meet market requirements and preferences.

As shown in Fig. 12 for the U.S. market design, the combustion products from the grate flow upward through the single furnace pass to provide sufficient residence time and temperature to burn the fuel con-

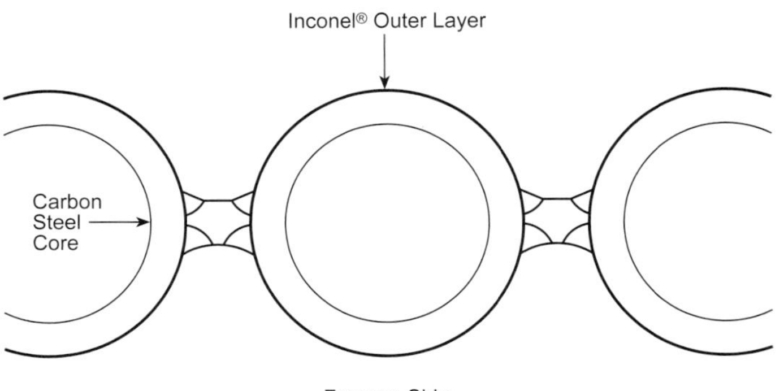

Fig. 9 Bimetallic tube.

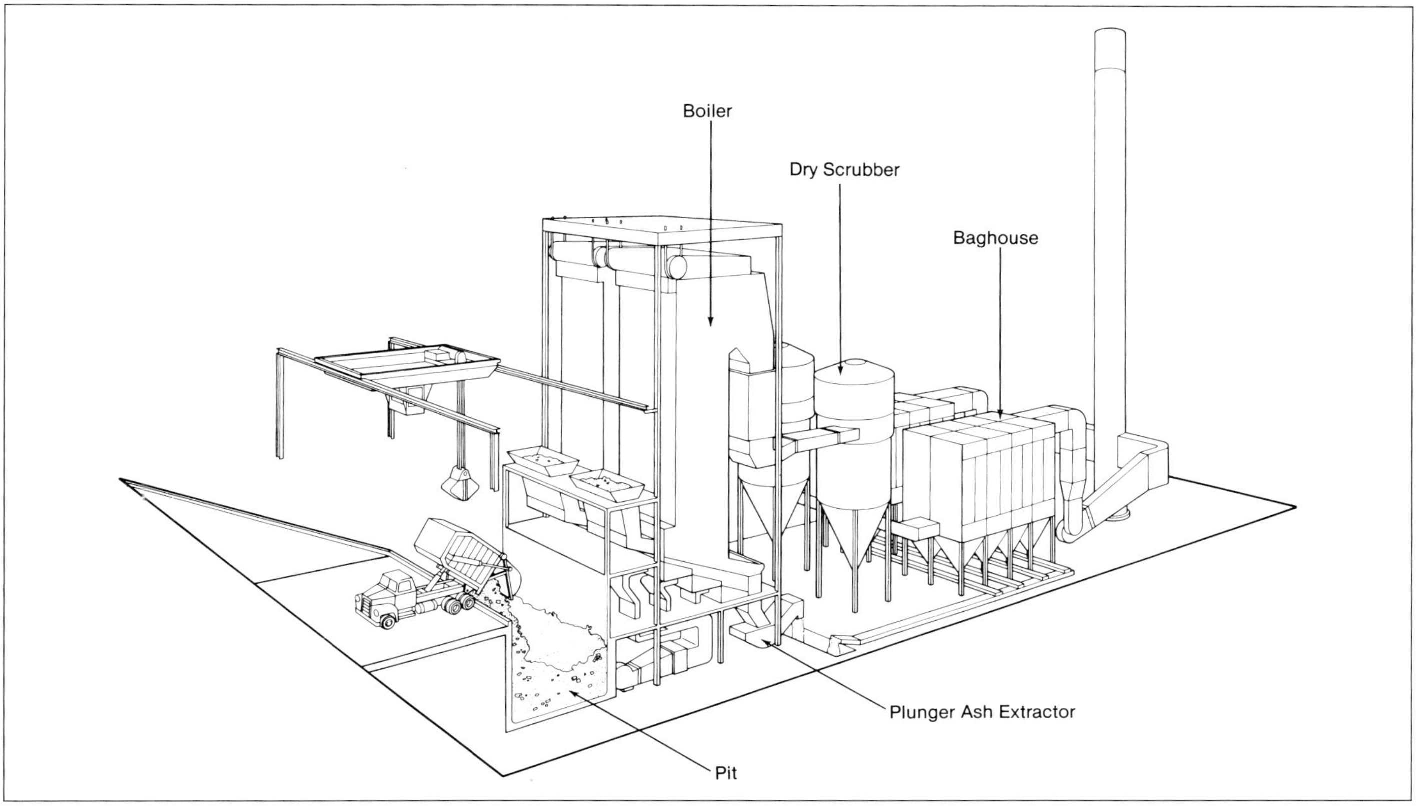

Fig. 10 Typical U.S. mass burning refuse-to-energy system.

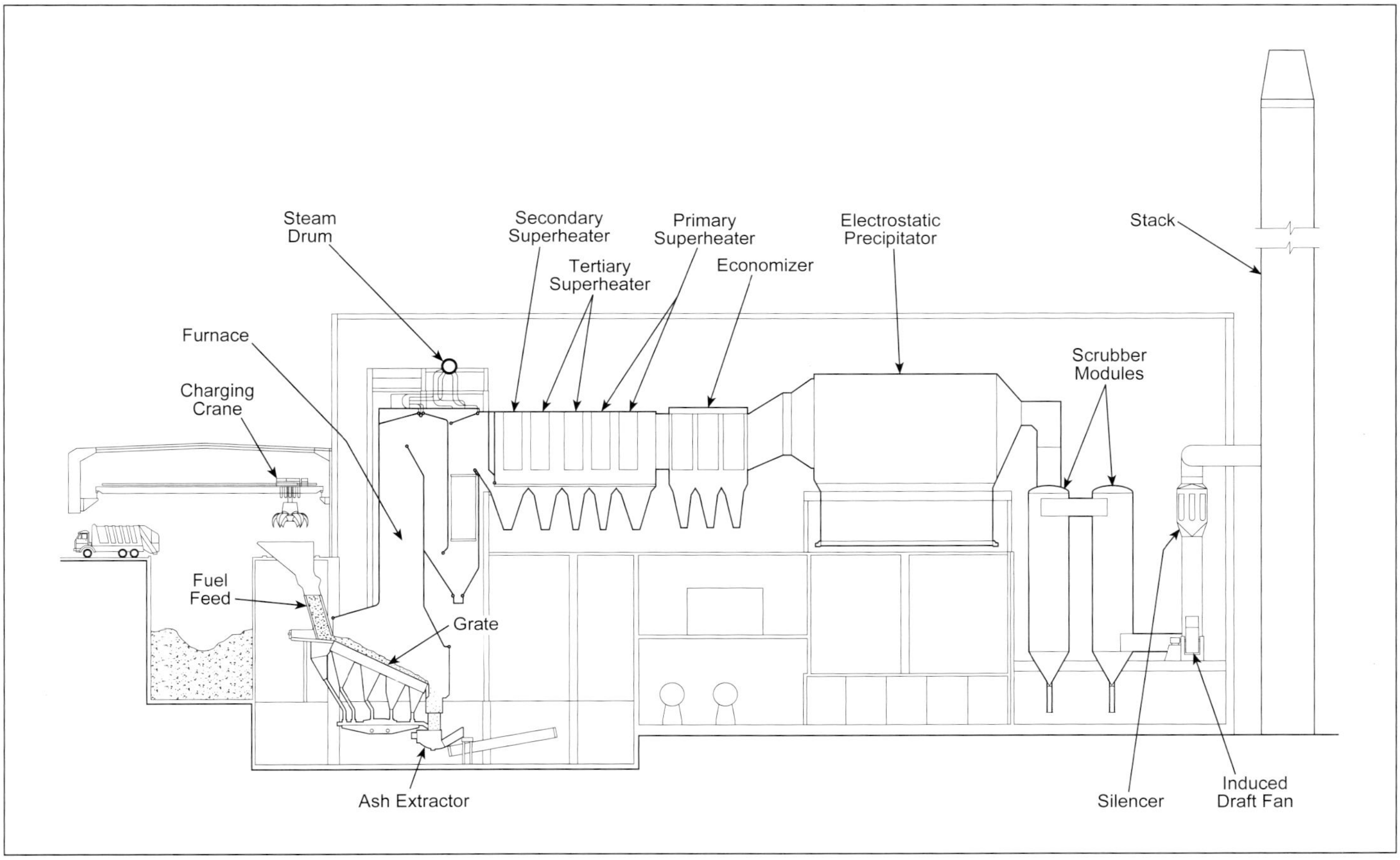

Fig. 11 B&W European mass burning power plant.

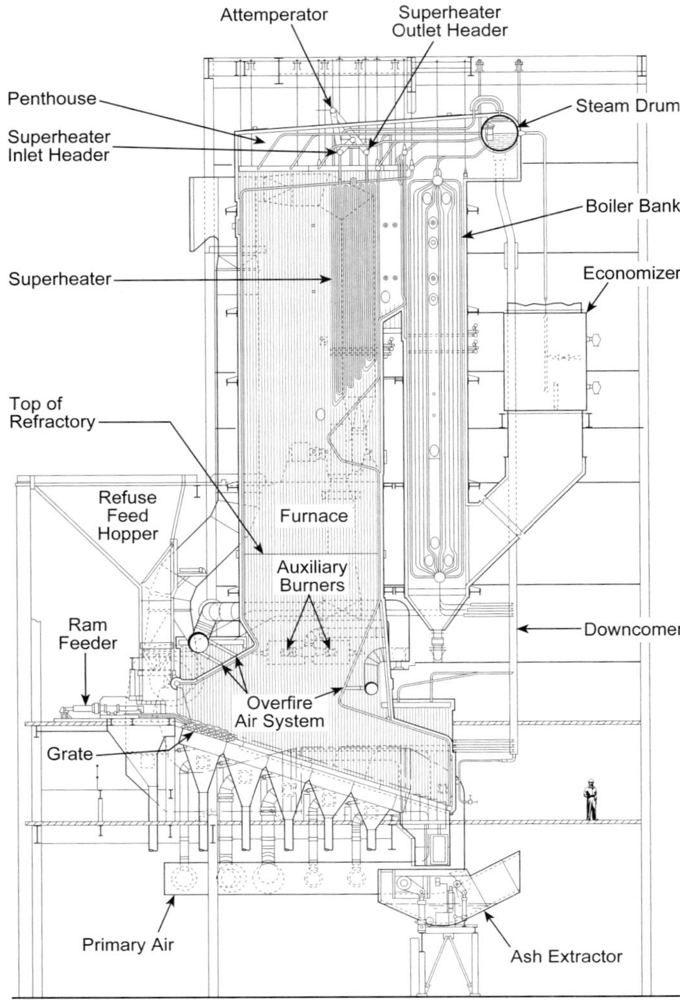

Fig. 12 Typical U.S. mass burning unit.

stituents and other hydrocarbons, and cool the flue gas to the required furnace exit gas temperature before entering the superheater. Where required, the design provides the necessary reaction time for a selective non-catalytic reduction (SNCR) process to control nitrogen oxides (NO_x). The flue gas then turns 90 degrees around a large furnace arch to pass through the crossflow superheater section before turning downward to flow parallel through the longflow boiler steam generating bank. The flue gas then flows upward through the crossflow economizer section and exits the boiler enclosure.

As shown in Fig. 13 for the European market design, a multiple pass furnace is used to cool the flue gas. The combustion products leave the grate and lower furnace to pass upward through the first open furnace pass and then downward through the second open pass. As noted earlier, a portion of the first pass may be covered with high SiC cast refractory to reduce heat absorption and permit the flue gas to maintain a suitable high temperature for a specified period of time for complete combustion of the fuel constituents and other hydrocarbons in smaller boilers and at lower load in larger units. At the bottom of the second open pass, the flue gas turns 180 degrees to pass upward through the third pass to cool the flue gas to the required gas temperature before entering the horizontal convection pass. Additional parallel flow boiler circuits or baffles in the form of widely spaced tube panels are typically used in the third pass to increase the heat absorption. The flue gas flow path is designed to provide reasonably uniform cross-section flow and temperature distributions throughout by the use of physical and numerical computational fluid dynamics (CFD) modeling (see Chapter 6). Noses or arches (not shown) are used at the first pass exit, second pass inlet and third pass inlet, while inclined walls and roofs are used in each pass to achieve the desired flue gas distribution.

The bottom-supported convection pass contains a series of convection heat transfer surface sections in the direction of gas flow: steam generating surface, secondary superheater, tertiary superheater, primary superheater, more steam generating surface in some cases, and finally the economizer before the gas exits the boiler enclosure. The convection pass surfaces are arranged to minimize corrosion, and are widely spaced to prevent pluggage and allow adequate space for mechanical sootblowing equipment. The first steam generating surface protects the superheater from high temperature peaks. The secondary and tertiary superheaters are designed for parallel steam and flue gas flow to minimize metal temperatures and corrosion. The second optional steam generating surface may be used to cool the flue gas to avoid steaming in the economizer.

Boiler plant sizing

A refuse plant must be sized to handle the physical amount of refuse that is delivered to that plant, regardless of the refuse heating value. A refuse boiler, on the other hand, is a heat input device and must be sized for the maximum heat input expected. When designing a refuse boiler, consideration must be given to both the design tons per day of refuse to be combusted and the typical range of heating values that is expected for the refuse in that location.

The boiler is typically designed for the maximum ton per day input at the maximum refuse heating value. A 1000 t/d (907 t_m/d) refuse plant is actually not the same size plant in all locations. As an example, a plant in an industrialized country might be designed to handle refuse with a heating value of approximately 5500 Btu/lb (12,793 kJ/kg), or 458 × 10^6 Btu/h (134 MW_t) total heat input to the boilers. At the other extreme, a plant in a less industrialized country might be designed to handle refuse with a heating value in the range of 3500 Btu/lb (8141 kJ/kg), or 292 × 10^6 Btu/h (85.6 MW_t) total heat input. Both are 1000 t/d (907 t_m/d) plants, but one has refuse boilers that have a 50% larger capacity in terms of heat input.

For many of the early U.S. plants, good data were not available on the true range of heating values of the refuse. Refuse boilers were sized for typical heating values of 4500 Btu/lb (10,467 kJ/kg). When the actual heating values were found to be as high as 5200 to 5500 Btu/lb (12,095 to 12,793 kJ/kg), the boilers were actually undersized and could not process the available refuse on a ton per day basis.

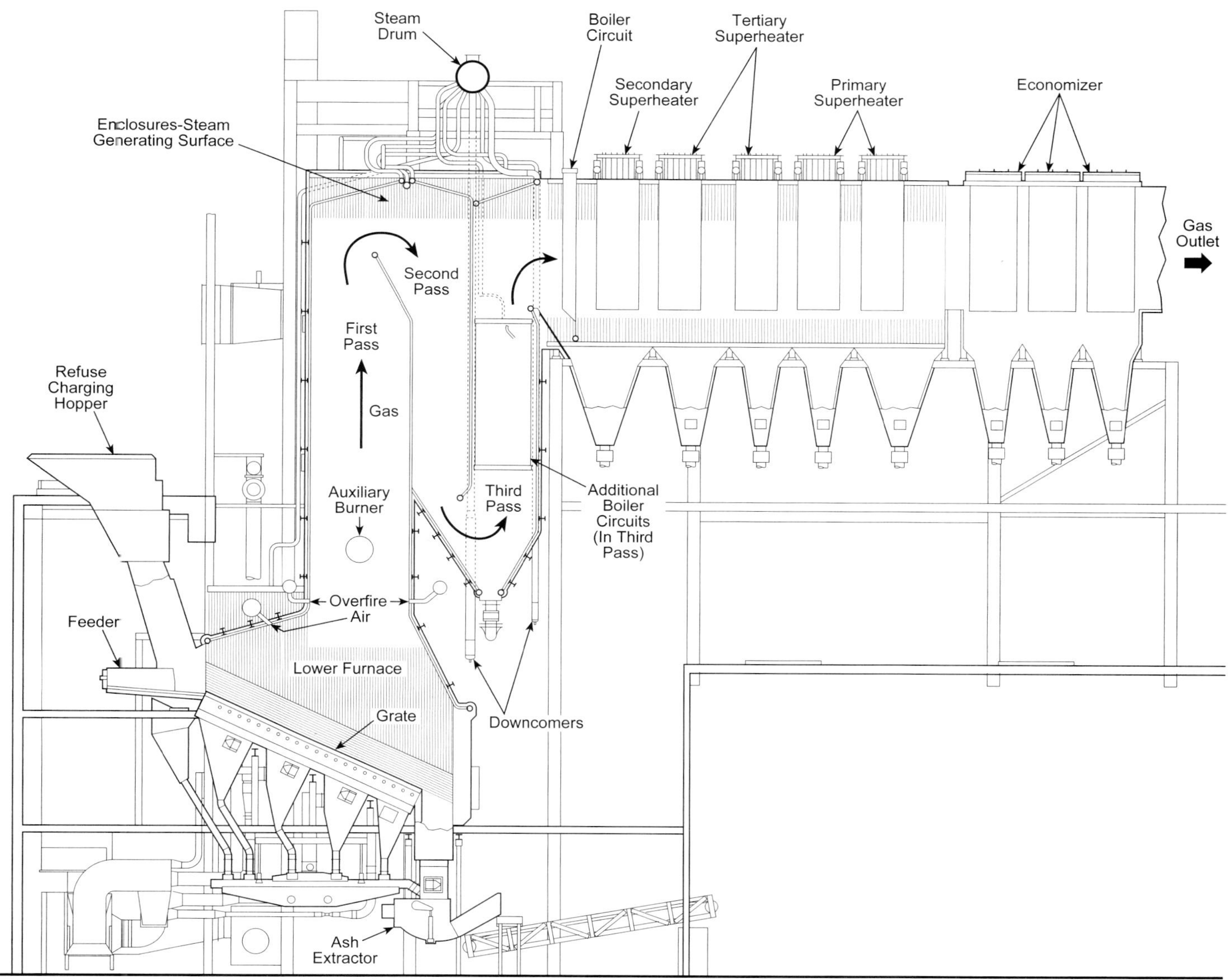

Fig. 13 Typica European mass burning unit.

Stoker capacity

A refuse stoker has both a heat input limit and a ton per day refuse throughput limit. If a typical 1000 t/d (907 t_m/d) refuse plant has two 500 t/d (454 t_m/d) boilers, and the design refuse heating value is 5000 Btu/lb (11,630 kJ/kg), each boiler would have a maximum heat input limit of 208.3×10^6 Btu/h (61.1 MW_t). If the actual refuse heating value is above 5000 Btu/lb (11,630 kJ/kg), the unit's ton per day capacity would be reduced below 500 t/d (454 t_m/d) so not to exceed the maximum heat input limit. On the other hand, if the actual refuse heating value is below 5000 Btu/lb (11,630 kJ/kg), then the unit could actually process more than 500 t/d (454 t_m/d) of refuse, up to the maximum ton per day limit of that stoker.

The ton per day limit is usually set by a refuse capacity per unit of stoker width, a limit for optimum fuel feed and distribution, or a structural limit based on refuse weight per square foot (m^2). These limits are in the range of 30 t/d (27 t_m/d) per front foot (0.3 m) of width and 65 lb/h ft^2 (2.74 kg/h m^2) of grate area. The grate surface area is set by a grate release rate generally in the range of 270,000 to 350,000 Btu/h ft^2 (852 to 1104 kW/m^2), but may be lower for low heating value, high moisture fuels. The design limits stated above represent typical values for equipment installed in North America. Stokers supplied for European and Asian refuse installations have typically used lower design limits, especially with respect to the grate heat release rate.

The stoker width and depth are also related to the specific fuel. A high heating value, low moisture fuel would require a wider, less deep stoker because the fuel will tend to burn more rapidly. A low heating value, high moisture fuel would require a narrow, deeper stoker because more residence time on the stoker is usually needed. The combination of all these criteria will set the maximum ton per day rating of the stoker.

There is also a minimum load that can be effectively handled on a given stoker. This load is also set by both a ton per day limit (minimum fuel inventory on the grate) and a heat input limit (minimum heat input for good combustion).

All of these limits can be incorporated into a capacity diagram, which provides the operator with the boundary limitations around a family of heating value curves. Fig. 14 is such a diagram for a typical 500 t/d (454 t_m/d) boiler burning 5000 Btu/lb (11,630 kJ/kg) refuse.

Stoker design

The combustion of MSW requires a rugged, reliable stoker to successfully convey and burn unsorted refuse. Most stokers use some variation of a reciprocating grate action, with either forward moving or reverse acting grate movement. Some arrangement of moving and stationary grates is used to move the refuse through the furnace and allow time for complete combustion.

A reciprocating grate stoker is typically designed with alternating moving and stationary rows of grate bars in a stairstep construction with a downward slope to help move the refuse along the length of the grate. Each row of grate bars overlaps the row beneath it and the alternate rows are supported from a moving frame driven by hydraulic cylinders. The moving rows of grate bars push the refuse over the stationary bars where it is picked up by the next row of moving bars. Fig. 15 shows a pivoting grate design where each row of grate bars is mounted on individual transverse shafts moving slowly 60 deg forward and backward so that the entire grate surface is in constant motion. The reciprocating grate shown in Fig. 16 has longitudinal grate girder assemblies filled with grate bars. Every other grate girder assembly is resting on a transverse shaft at the upper end, moving slowly 30 deg forward and backward so that the grate surface moves like a walking beam floor. The remaining assemblies are stationary. The typical grate drive units are equipped with hydraulic cylinders with a piston moving forward and backward and thereby creating the grate movement. The motion of these grates rolls and mixes the refuse, constantly exposing new material to the high temperatures in the fuel bed and allowing the combustion air to contact all of the burning refuse.

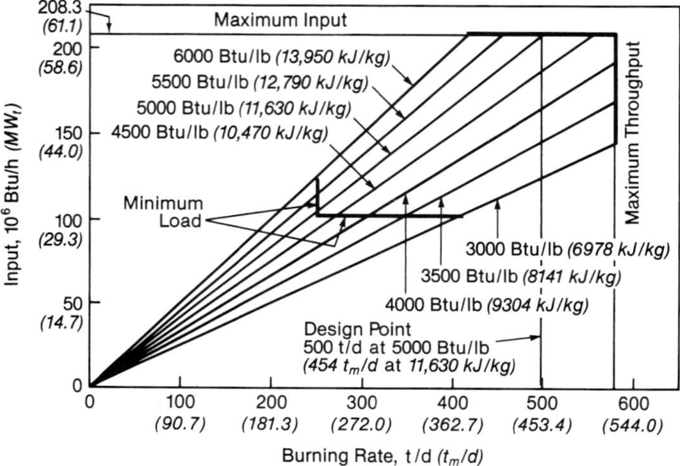

Fig. 14 Stoker capacity diagram.

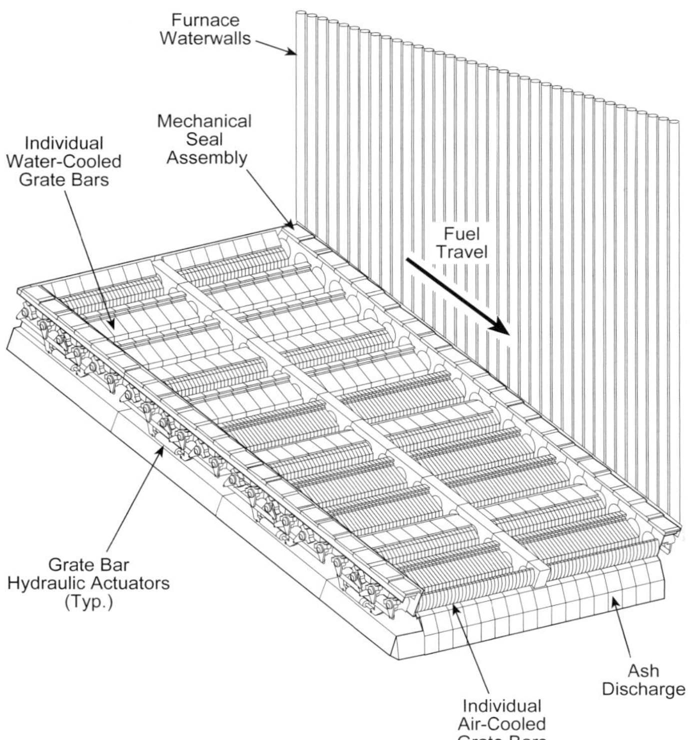

Fig. 15 Pivoting refuse grate.

More recently, MSW heating values have increased to levels where water cooling of the grate has become more common for new mass-fired units. Typically, the upper half of the grate, nearest to the fuel feed point and most intense combustion zone, is water-cooled while the balance of the grate is air-cooled. The grates in Figs. 15 and 16 can be supplied with either water- or air-cooled versions. Water cooling also has the operational advantage that combustion air temperature can be preheated as required for combustion without consideration for grate cooling, which is controlled by the water flow.

For lower heating value, high moisture refuse, drop-off steps are often incorporated into the stoker design. The steps are located at the end of each grate module resulting in one to three steps depending on the overall stoker length. These steps promote a tumbling and rolling action as the burning refuse falls off the step. This type of design was used on a number of the early refuse units in the U.S. but was found to be unnecessary with the higher heating value and lower moisture refuse. In fact, it can be a detriment as the tumbling can also result in excursions of high carbon monoxide (CO) emissions.

The grates are usually constructed in a series of standard modules with independent drives and air plenums. This allows the individual grate modules to be factory-assembled to limit field construction time and provide complete duplication of parts for easy maintenance and repair. This modular construction allows any size stoker to be constructed from a small number of standard modules. A typical stoker is usually from two to four modules in length and one to four modules in width.

Modular construction provides the operator with complete control over the amount of undergrate air introduced through each module. This ability to control the undergrate air in multiple air zones along the width and depth of the stoker is an important factor in minimizing CO and nitrogen oxides (NO_x) emissions. This method of construction also provides complete freedom in the operating speed of the individual grate modules to provide the required feed rate along the grate for complete burnout of the fuel.

Fuel handling

The MSW delivered to the refuse plant is generally dumped directly into the storage pit. This large pit also provides a place to mix the fuel. This mixing is done using the crane and grapple to move and restack the refuse as it is dumped into the pit. This produces a fuel, in both composition and heating value, that is as consistent as possible for the boilers. This is an essential job for the crane operator and any time not required to feed the furnaces is used to mix the fuel. It is not uncommon for the crane operator to mix four grapple loads back into the pit for every one load that goes to a charging hopper.

Fuel feed system

Controlled feed of the fuel is necessary for good combustion to minimize CO and NO_x emissions and to maintain constant steam output. At the bottom of the charging hopper feed chute, a hydraulic ram pushes the fuel into the furnace and onto the stoker grates at a controlled rate. On larger units, multiple charging rams are used across the width of the unit to provide a continuous fuel feed with optimum side to side distribution. The hydraulic rams stroke forward slowly and then retract quickly to provide the positive continuous fuel feed. These rams are simple to control with feed rate adjustments made by either the speed of travel or the number of strokes per hour.

Combustion air system

The primary combustion air, or undergrate air, is fed to the individual air plenums beneath each grate module. A control damper at the entrance to each air plenum controls the undergrate air to each section of the grate (Fig. 17). The grate surface is designed to meter the primary combustion air to the burning refuse uniformly over the entire grate area. This is accomplished by providing small air ports or tuyères in the surface of the individual grate bars. These air ports provide openings equal to approximately 3% of the grate area which results in sufficient pressure drop of air resistance across the grate to assure good distribution of the air flow through the grate, regardless of the depth of refuse on the grate. Undergrate air systems are generally designed for 70% of the total combustion air flow with expected normal operation at 60% of total air flow.

Because refuse contains a high percentage of volatiles, a large portion of the total combustion air should enter the furnace as secondary, or overfire, air through ports in the furnace walls. These secondary air ports are located only in the front and rear furnace walls so that the air flow parallels the normal flow pattern through the unit. Older design units generally provided 25 to 30% of the total air as overfire air. With today's emphasis on better combustion and lower emissions, the overfire air systems are designed for 50% of the total air to be overfire air with expected normal operation at 40%.

The basic function of the overfire air system is to provide the quantity of air and the turbulence necessary to mix the furnace gases with the combustion air and to provide the oxygen necessary for complete combustion of the volatiles in the lower furnace. Excess air in the furnace is usually maintained in the range

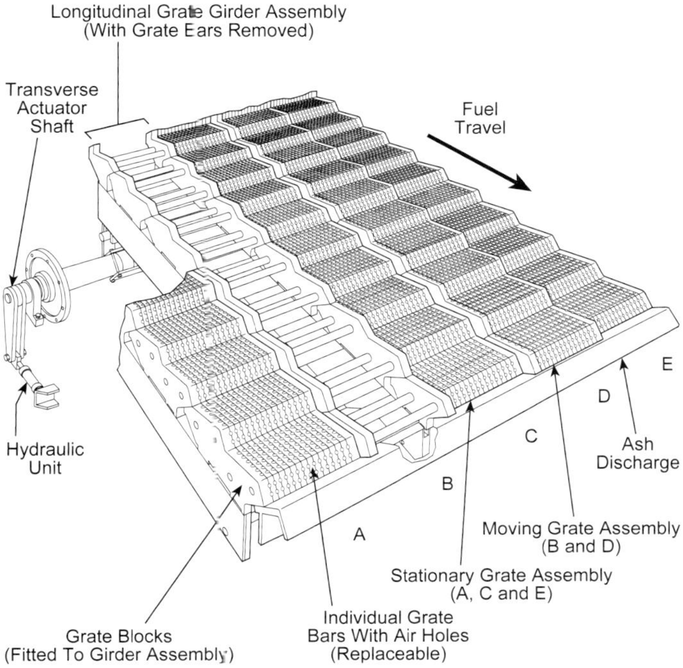

Fig. 16 Reciprocating refuse grate.

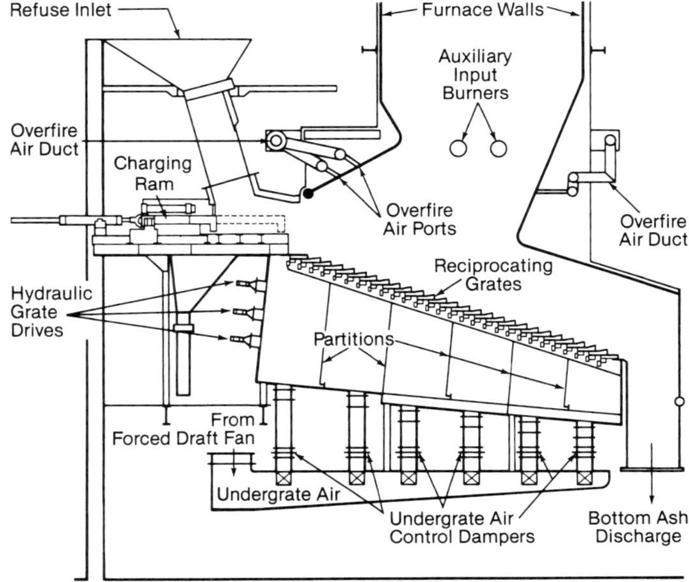

Fig. 17 Combustion air system.

of 80 to 100% and complete combustion is generally indicated by a CO value in the range of 10 to 100 ppm.

The high percentage of excess air needed to ensure proper combustion can be reduced through flue gas recirculation. Flue gas recirculation involves the use of cooled flue gas, generally extracted from the system down stream of the baghouse, to offset a certain portion of the fresh air used in the overfire air system. The recirculated flue gas maintains the total flow volume in the combustion air system while reducing the percentage of excess O_2, which improves boiler efficiency. Flue gas recirculation also reduces the production of NO_x by limiting the amount of O_2 present in the high temperature region of the furnace where NO_x is formed. The advantages of using flue gas recirculation can be offset by its disadvantages which include the capital cost of the system, power consumption of the recirculation fan, and overall system maintenance.

To aid in the combustion of wet fuels during extended periods of rainy weather, the air system includes steam coil air heaters designed to provide air temperatures in the range of 300 to 350F (149 to 177C) to help dry these fuels and maintain furnace temperature. These steam coil air heaters are commonly used only for the undergrate air because this is the air flow which directly aids in drying wet fuel. These air heaters must be conservatively designed with fin spacing not exceeding 4 to 5 fins/in. (1 fin/ 6.4 to 5.1 mm). Most plants take combustion air from the storage pit area to help minimize odors. This air is normally contaminated with dust and lint which could plug the steam coil. Some type of cleaning arrangement or filters must therefore be included to keep the steam coil clean.

When high moisture fuels are encountered, the first action by the operator is to use the steam coil air heater to provide hot air. However, for very high moisture fuels it may also be necessary to use the auxiliary fuel burners to stabilize combustion in the furnace. These cases are the exception, and in normal operation neither the auxiliary burners nor the steam coil air heater are needed for good combustion.

Ash handling systems

When refuse is burned, the ash takes the form of either light ash, called flyash, or coarse ash, which comes off the stoker. The flyash is entrained in the gas stream until it is removed in the particulate collection device or falls out into the boiler, economizer, or air heater hoppers. The stoker ash consists of ash from the fuel, slag deposits on the grate, ash from the furnace walls and, in some designs, ash from the superheater. The stoker ash is discharged through the stoker discharge chute and from the stoker siftings hoppers.

Ash extractor

The ash from the stoker discharge on mass-fired units may contain large pieces of non-combustible material, in addition to the normal ash from combustion. Ash consistency can vary from fine particles to large and heavy non-combustible objects in the fuel. The ash from the stoker discharge chute falls into a water bath in the ash extractor that quenches the ash and controls dusting. In the plunger ash extractor (Fig. 18), the ash is quenched, a slow moving, hydraulically operated ram cycles forward and back to push and squeeze the accumulated ash up an inclined dewatering section to the discharge of the extractor. The ram cycles continuously at a slow speed to push the ash out of the extractor. The dewatered ash has a moisture content of 15 to 20% as a result of the squeezing process on the incline. The lower moisture content can have an economic advantage as the cost of landfilling ash is based on total weight, which includes the weight of water in the ash. Another option is a wet conveyor ash extractor. This is a water-filled conveyor trough where noncombustible items, ash and grate siftings are transported on a steel apron belt submerged in water. As with the chain conveyor discussed later in the chapter, the belt carries the noncombustibles, ash and siftings up an incline, effectively dewatering the material before it is discharged from the belt to a bin or hopper, and the belt cycles around to the water-filled trough to collect more material.

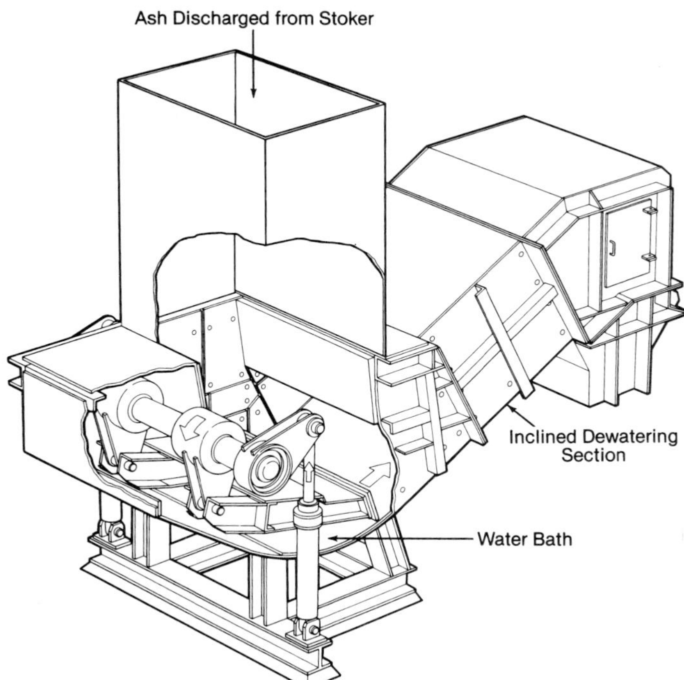

Fig. 18 Plunger ash extractor.

To keep the ash system simple and to minimize costs, ash from the extractor can discharge directly into a truck or bin for final disposal. To move the ash away from the vicinity of the stoker discharge, vibrating and belt-type conveyors are used. A short vibrating conveyor is placed at the discharge of the ash extractor; its metal trough can withstand the impact of the oversized, noncombustible material falling from the extractor. The vibrating conveyor then transfers the ash to a belt conveyor, minimizing the wear on the belt conveyor that would occur if the extractor discharged directly onto it.

Scrubber, precipitator, and baghouse flyash

The flyash collected in the scrubber, precipitator, or baghouse hoppers can be handled by dry mechanical screw or chain-type conveyors. These conveyors operate continuously to minimize hopper pluggage problems and they discharge onto a collecting conveyor, which is usually a dry chain type. Because the mechanical conveyors are dust-tight, but not designed to be gas-tight, separate sealing devices such as rotary seals or double flop valves are used. The collecting conveyor will collect the flyash discharged by all the conveyors under the rows of hoppers, and move it to a single collection point for ultimate disposal.

RDF firing

RDF technology was developed in North America as an alternative to the mass burning method. Initially, RDF was used as a supplementary fuel for large, usually coal-fired, utility boilers. For this application, the RDF is finely processed and sized to 1.5 in. (38.1 mm) maximum size. The resulting RDF is nearly all light plastics and paper.

For supplemental firing, B&W recommends a maximum RDF input of 20% on a heat input basis, and no RDF input until the boiler is operating above 50% load. In most cases the RDF is blown into the furnace at the pulverized coal burner elevation through an RDF burner with a fuel distribution impeller. Most of the RDF burns in suspension in the high heat input zone of the pulverized coal fire while the heavier fuel fraction falls out in the lower furnace. A dump grate stoker located in the neck of the ash hopper allows more complete burnout of the heavier fuel particles before they are discharged into the ash system (Fig. 19). RDF has been successfully co-fired in B&W boilers at Lakeland, Florida; Ames, Iowa; and Madison, Wisconsin.

RDF has also been successfully co-fired in B&W Cyclone furnaces where the finely processed and sized RDF is injected into the Cyclone secondary air stream moving tangentially inside of the Cyclone barrel. (See Chapter 15.) This method of RDF combustion has been used in Baltimore, Maryland.

Dedicated RDF-fired boilers

From this supplemental fuel experience, RDF then became the main fuel for boilers specifically designed to generate full load steam flow when burning RDF (Fig. 20). In some cases where steam flow was required even when refuse was not available, the boiler was

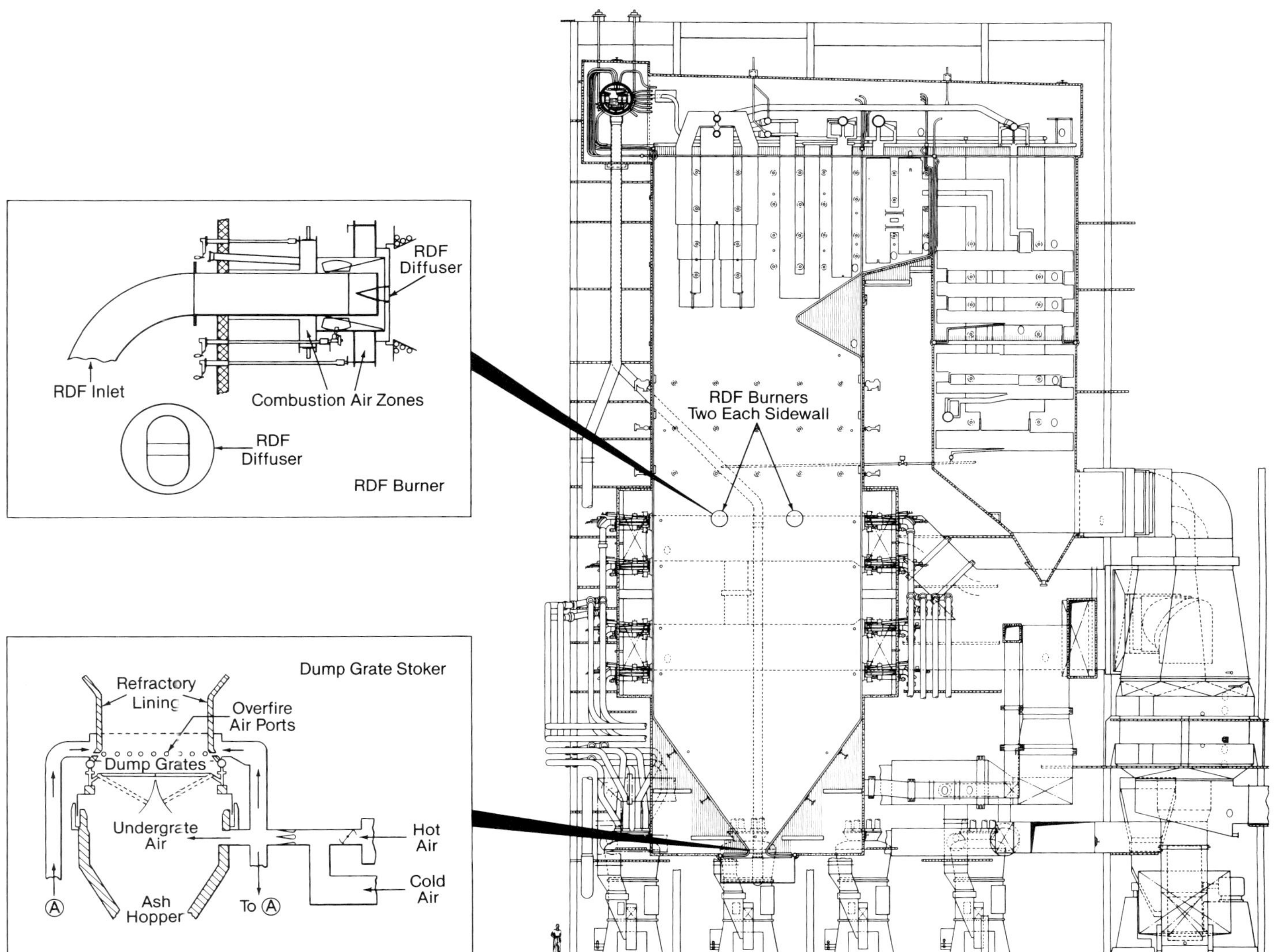

Fig. 19 Typical B&W RB-type utility boiler firing RDF as a supplementary fuel.

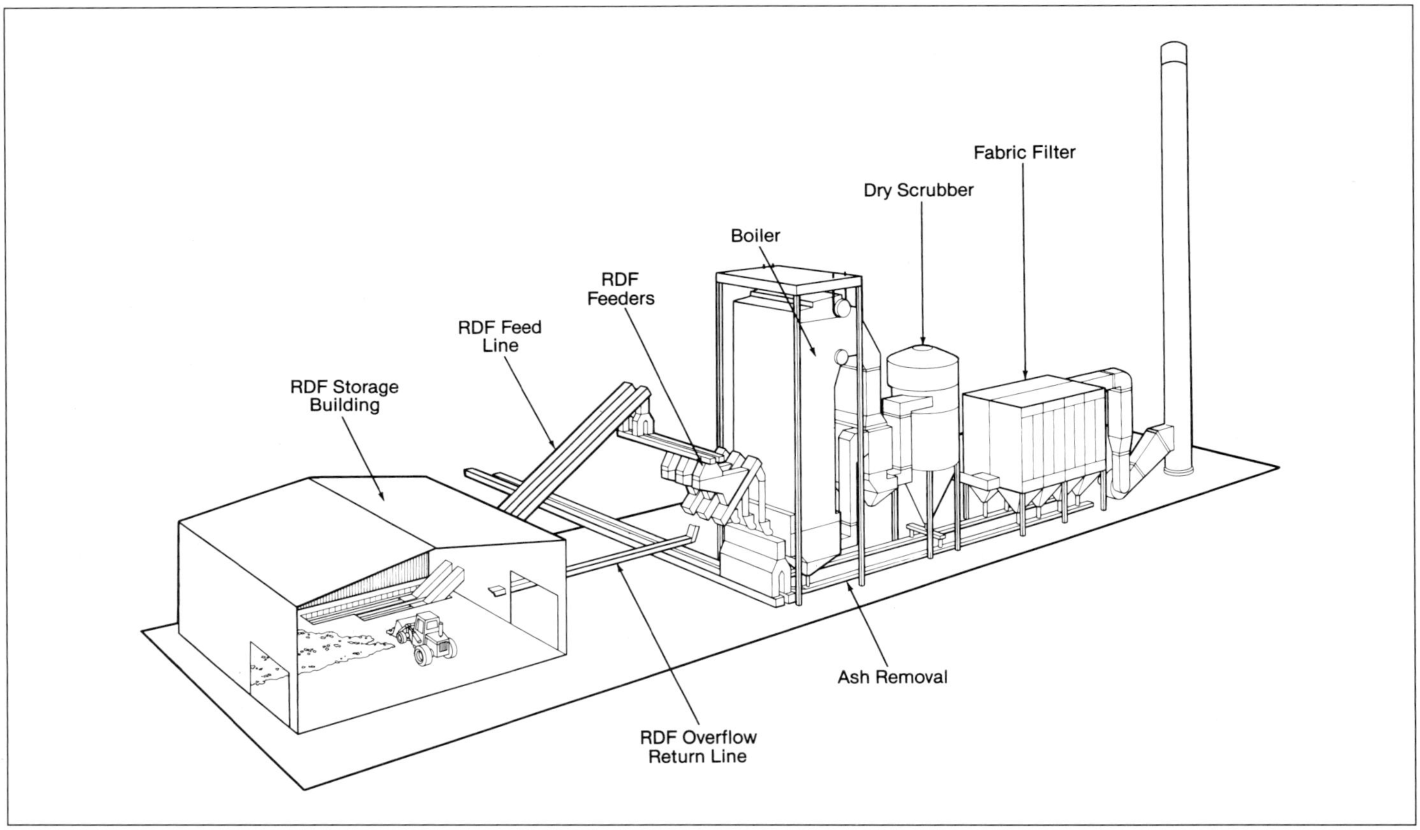

Fig. 20 Typical RDF refuse-to-energy system.

designed so that it could also reach full load on wood, coal or natural gas. More commonly there would only be auxiliary gas or oil burners for startup and shutdown.

The first boilers in the world to fire RDF as a dedicated fuel were B&W units which began operation in 1972 in Hamilton, Ontario, Canada. The first of such boilers in the U.S. went into operation in 1979 in Akron, Ohio. The boiler design was highly influenced by the proven technology of wood-fired boilers with respect to their fuel feed system, stoker design, furnace sizing, and overfire air system. The transfer of this technology from wood firing to RDF firing was successful in many areas, but in other areas design adjustments were needed to accommodate the unique aspects of RDF.

The operating experience from the first generation designs at Hamilton, Akron and other plants led to second generation designs with improved RDF processing systems, fuel feed systems and boiler design. Specific improvements included the first fuel feeder designed specifically for RDF and the use of alloy weld overlay in the lower furnace for corrosion protection.

The third generation of facilities is essentially today's state-of-the-art design. (See Fig. 21.) This boiler design has an enhanced overfire air system to significantly improve combustion efficiency. These third generation designs also incorporate dramatically improved fuel processing systems.

RDF processing

The first generation RDF processing systems were referred to as crunch and burn systems. The incom-

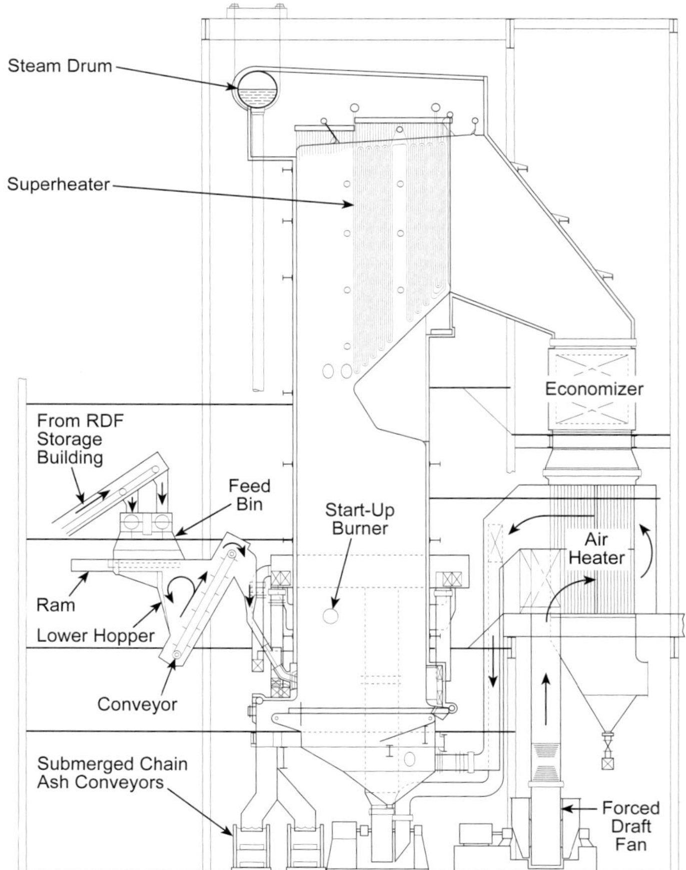

Fig. 21 Third generation RDF unit and fuel feeding system.

ing refuse first went to a hammer mill type shredder that produced an RDF with 6 by 6 in. (152 by 152 mm) top size. Ferrous metal was removed by magnetic separators. There was no other material separation and many undesirable components entered the boiler such as shredded particles of glass that became embedded in wood and paper resulting in a very abrasive fuel entering the boiler in suspension. Following the practice of wood-fired boilers, the RDF was generally stored in a hopper or bin. RDF, however, is much more compactible than wood and in nearly every case, significant problems were encountered getting the RDF out of the storage bins.

Second generation RDF processing systems recognized and corrected some of the problems. The shredder for final fuel sizing was moved to the back of the processing system and a rough sizing shredder was used as the first piece of equipment in the system. This reduced, but did not eliminate, the problem of abrasive particles embedding in the fuel. Size separation equipment was introduced, generally removing the small size fraction which is less than 1.5 in. (38.1 mm) composed mostly of broken glass, ceramics and dirt, which was sent to landfill. RDF was stored on the floor rather than in bins or hoppers and was moved by front-end loaders to conveyor belts. This greatly improved the reliability of fuel flow to the boiler.

In third generation RDF processing systems (Fig. 22) the first piece of equipment became a flail mill or similar equipment whose main function was to break open the garbage bags. The refuse is still size-separated using a trommel disk screen with the minus 1.5 in. (38.1 mm) size destined for landfill. Generally, a device such as an air density separator is added to collect the light fraction (paper, plastics, etc.) from this stream to achieve maximum heat recovery. Where it is economically attractive, separation of aluminum and other non-ferrous metals is added to the plus 1.5 in. (38.1 mm.) minus 6 in. (152 mm) stream.

RDF yield

RDF yield is defined as the percentage of RDF produced from a given quantity of MSW. For instance, a 70% yield means that 70 t (63 t_m) of RDF are produced for every 100 t (90 t_m) of incoming MSW. The ash content of RDF is directly related to the yield of the processing system. In a processing system with a lower yield, the portion of the MSW that is rejected is generally high in ash and inerts content and, therefore, the resulting RDF is low in ash content. When an RDF processing system is designed to obtain a higher yield, more of the ash is carried over into the RDF fuel fraction and the ash content is increased.

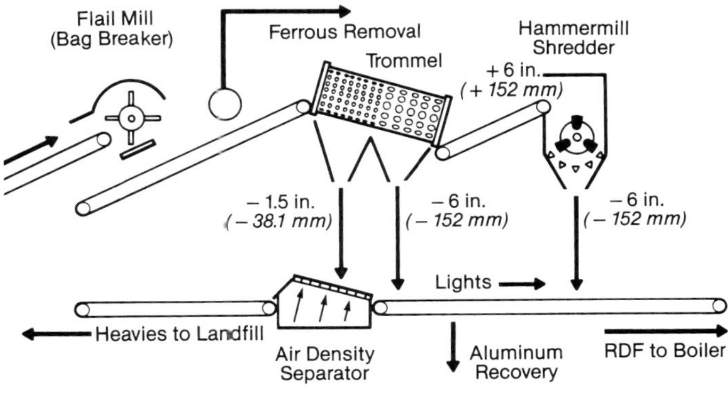

Fig. 22 Third generation RDF processing system.

The RDF heating value is inversely related to the yield; the lower the yield, the higher the heating value. A sophisticated RDF processing system will have multiple stages of material separation that remove a very large percentage of the metals, inert material, and high moisture organic material. Since a relatively large portion of the incoming MSW is removed by this processing system, the RDF yield is significantly lower than the yield produced by a less sophisticated crunch and burn system. The resultant RDF, however, will have a high heating value corresponding to its low percentage of non-combustible material (ash) and high-moisture organic material.

A typical MSW might have a composition comparable to the reference waste shown in Table 4. The majority of the waste is combustible materials, which have ash contents ranging from approximately 4% for wood to 12% for glossy magazine paper. The glass frac-

Table 4
Typical Reference Refuse

Component Analysis	Reference MSW (% by wt)	RDF (% by wt)
Corrugated board	5.53	—
Newspapers	17.39	—
Magazines	3.49	—
Other paper	19.72	—
Plastics	7.34	—
Rubber, leather	1.97	—
Wood	0.84	—
Textiles	3.11	—
Yard waste	1.12	—
Food waste	3.76	—
Mixed combustibles	17.75	—
Ferrous	5.50	—
Aluminum	0.50	—
Other nonferrous	0.32	—
Glass	11.66	—
Total	100.00	
Ultimate Analysis		
Carbon	26.65	31.00
Hydrogen	3.61	4.17
Sulfur	0.17	0.19
	(max. 0.30)	(max. 0.36)
Nitrogen	0.46	0.49
Oxygen	19.61	22.72
Chlorine	0.55	0.66
	(max. 1.00)	(max. 1.20)
Water	25.30	27.14
Ash	23.65	13.63
Total	100.00	100.00
Heating value	4,720 Btu/lb (10,979 kJ/kg)	5,500 Btu/lb (12,793 kJ/kg)
Fuel value recovery, % MSW	96	
Mass yield, % RDF/MSW	83	

tion, yard waste and mixed combustibles may also contain varying quantities of sand, grit and dirt. The predicted composition of the RDF will vary depending on the type of processing system and the resulting yield. Table 5 shows how the ash content in the fuel and the heating value of the fuel will vary as the RDF yield varies with different processing systems. Two cases are considered, one which assumes no front end recycling (curbside recycling or separate recycling facility), and a second case which assumes that such a system is in place in the community.

RDF quality

RDF used to supplement pulverized coal in utility boilers should be very low in ash; contain the least possible amount of ferrous metal, aluminum, and other nonferrous metal; and be small enough in particle size to be fed pneumatically to the boiler. The processing system for such a fuel would generally be a very low yield system, between 40 and 60%.

RDF for dedicated traveling grate stoker boilers should be low in ash; as free as possible of ferrous metal, aluminum and other nonferrous metals; and have a particle size distribution that is considerably larger than the particle size of RDF for use in Cyclone or pulverized coal boilers. The processing system for such a fuel will have a higher yield, around 70 to 85%.

RDF produced in a crunch and burn system, in which solid waste is shredded and only the ferrous metal removed, has a yield of about 93%; an inherently high ash content; and contains 100% of the aluminum, other non-ferrous metals, glass, stones and ceramics in the original MSW. The large quantities of aluminum, glass, and other inerts that remain in the RDF will result in higher wear on the stoker and lower furnace. The economic advantage of installing a less sophisticated crunch and burn processing system is generally offset by the need to provide a more conservatively designed stoker and furnace as well as a larger ash handling system.

RDF processing systems

An optimum RDF processing system for a dedicated boiler application achieves the highest yield with the highest heat recovery, while removing ferrous, aluminum and other non-ferrous materials, and glass. Such a system (Fig. 23) includes the following:

In-feed conveyors From the tipping floor, the solid waste is fed by front-end loaders or excavators to steel pan apron conveyors which feed the flail mill in-feed conveyors.

Initial size reduction The flail mill tears open the plastic garbage bags, coarsely shreds the refuse, and also breaks glass bottles to a size of approximately 1.5 in. (38.1 mm) or less.

Ferrous metal recovery Ferrous metal is extracted from the coarsely shredded MSW in each line by a single-stage overhead magnet. Recovered ferrous metal is moved to a ferrous air classifier where tramp materials such as paper, plastics or textiles are removed, thereby providing a clean ferrous product. A ferrous recovery of 90% is possible.

Size classification and final size reduction After ferrous removal, shredded waste is fed into a rotating trommel screen, a size separating device about 10 ft (3.0 m) in diameter by 60 ft (18.3 m) long. The trommel performs the following functions:

1. removes glass, sand, grit and nonferrous metal less than 1.5 in. (38.1 mm) in size, and

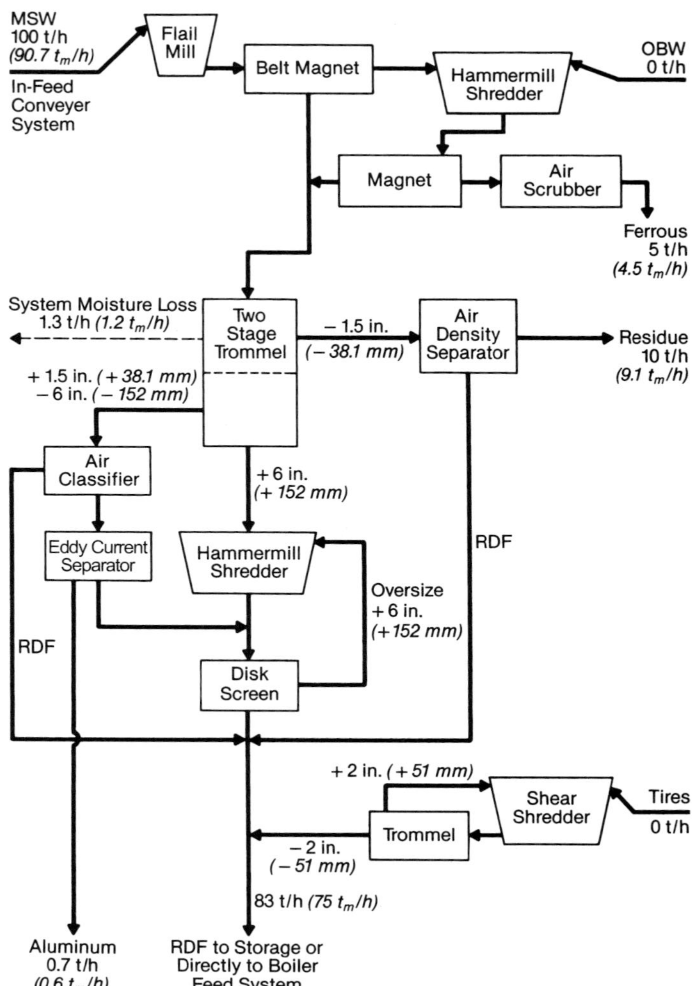

Fig. 23 Components of a complete RDF processing system.

Table 5
RDF Yield Versus Ash Content and Fuel Heating Value

Mode	RDF Yield %	Ash	Btu/lb	(kJ/kg)
Without front-end recycling:				
Mass burn	100%	23.64	4814	(11,197)
Crunch and burn	93%	19.87	5146	(11,970)
RDF	83	11.72	5641	(13,121)
	to 70%	to 8.87	to 5834	(13,570)
With front-end recycling:				
Mass burn	100%	19.58	5513	(12,823)
Crunch and burn	93%	17.16	5898	(13,714)
RDF	85	9.91	6328	(14,719)
	to 71%	to 6.59	to 6491	(15,098)

2. removes the minus 6 in. (152 mm), plus 1.5 in. (38.1 mm) fraction and contains the bulk of the aluminum cans.

The trommel oversize material, plus 6 in. (152 mm), is then shredded in a horizontal secondary shredder. Because the secondary shredder is a major consumer of energy and has high hammer maintenance costs, the RDF process is specifically designed to reduce the secondary shredder's load by shredding only those combustibles too large for the boiler. Particle size is primarily controlled by the design of the secondary shredder grate openings. Additional particle size control may be achieved by adding a disc screen downstream of the secondary shredder to recycle any oversized material back to the secondary shredder.

Separation of glass, stones, grit and dirt Trommel undersize material, minus 1.5 in. (38.1 mm), passes over an air density separator (ADS) designed to remove dense particles from less dense materials through vibration and air sweeping. This device can efficiently remove glass, stones, grit and dirt, as well as nonferrous metals. The light fraction, which can range from approximately 50 to 90% of the ADS feed, consists essentially of combustibles with high fuel value which are recovered and blended into the main fuel stream.

Aluminum can recovery To optimize aluminum can recovery, an air classifier is provided for the plus 1.5 in. (38.1 mm), minus 6 in. (152 mm) undersize fraction. The air classifier removes the light organic portion of the stream, allowing aluminum cans to be more visible for hand pickers. The air classifier heavy fraction drops onto a conveyor moving at approximately 2.5 ft/s (0.76 m/s) with numerous hand-picking stations on either side of the belt. Cans go into hoppers and, by conveyor, to a can flattener. A pneumatic conveyor then transfers the flattened cans into a trailer. An eddy current separator, for the removal of aluminum cans, can replace hand picking if the expected amount of cans is high enough to justify the additional capital cost. Aluminum recovery of 60% is possible with hand picking, while up to 90% recovery is possible with the eddy current separator.

Oversized bulky waste (OBW) The OBW shredder is generally a horizontal hammer mill used to shred ferrous metal recovered by the RDF processing lines and pre-separated oversized material which includes white goods such as refrigerators, washing machines, furniture and tree limbs. The ferrous metal is magnetically recovered and given a final cleaning by an air scrubber to remove tramp materials. The nonferrous material is integrated into the RDF stream.

Tire shredding line If there is a sufficient supply of tires, a separate tire shredding line can be included. A shear shredder, used specifically for shredding tires, can shred 500 passenger car tires per hour. The shredder includes a rotary screen classifier (trommel) for returning shredded tire chips above 2 in. (51 mm) back to the shredder. A tire chip 2 by 2 in. (51 by 51 mm) or less is the final product which is then blended in with the RDF stream.

RDF storage building RDF from each processing line is conveyed to an RDF storage building. From there, it is either fed directly to the boiler or fed directly to a shuttle conveyor and storage pile. When RDF feed is direct to the boiler, excess RDF from the boiler feed system is returned to the RDF storage building. RDF not being fed directly to the boiler is retrieved from the storage pile by a front-end loader and loaded onto inclined conveyors which transport the RDF to the boiler feed system.

Fuel feed system: metering feeders

A successful RDF metering feeder must meet the following design criteria:

1. controlled fuel metering to meet heat input demand,
2. homogenization of material to produce even density,
3. adequate access to deal with oversized material,
4. maintainability, in place, and
5. fire detection and suppression devices.

A reliable RDF metering feeder (see Fig. 21) is a key feature of the second generation RDF boiler design. One feeder is used for each air-swept fuel distributor spout. Each feeder has an upper feed bin which is kept full at all times by an over-running conveyor to ensure a continuous fuel supply. The fuel in this hopper is transferred to a lower hopper by a hydraulic ram. The ram feed from the upper hopper is controlled by level control switches in the lower hopper. The RDF is fluffed into a uniform density by a variable speed inclined pan conveyor which sets up a churning motion in the lower hopper. The pan conveyor delivers a constant volume of RDF per flight which is carried up the pan conveyor and deposited into the air-swept spout. The rate at which the fuel is deposited into the spout is based on fuel demand.

Air-swept distributor spouts

Air-swept fuel spouts, used extensively in the pulp and paper industry, proved to be equally effective for RDF firing. (See Chapter 28.) Lateral fuel distribution on the grate is achieved by locating multiple spouts across the width of the furnace. Longitudinal distribution is achieved by continuously varying the pressure of the air sweeping the spout floor. A major feature of this design is its simplicity.

Traveling grate stoker

To date, only traveling grates have been used for spreader-stoker firing of RDF. These grates move from the rear of the furnace to the front, into the direction of fuel distribution. A single undergrate air plenum is used. There is a wealth of experience worldwide with traveling grate stokers burning a myriad of waste and hard to burn fuels. The parameters for unit design shown in Table 6 were developed from this experience and the uniqueness of the RDF. (See also Chapter 16.)

On mass-fired stokers, a large volume of fuel at the front slowly burns down to a small volume of ash at the back. For an RDF stoker the key is to maintain an even 8 to 10 in. (203 to 254 mm) bed of fuel and ash over the entire stoker area. Grate problems are usually due to a shallow ash bed or localized piling of fuel. Operator tendency, when confronted with poor metering and/or fuel distribution, is to run the grates

Table 6
Stoker Design Criteria (English units)

Parameters	RDF	Wood
Grate heat release, 10^6 Btu/h ft^2	0.750	1.100
Input per ft of grate width, 10^6 Btu/h	15.5	29
Fuel per in. of distributor width, lb/h	450	1000
Feeding width as % of grate width	45 to 50	45 to 50
Grate speed, ft/h	25	N/A

faster. While this technique can minimize bed upset, it will shorten grate life due to higher wear rates and the overheating of the grate bars. With the recommended ash bed thickness, tramp material is minimized, grate temperatures are lowered, wear is reduced, and grate life is increased. To achieve this optimum ash bed requires controlled metering of the RDF and proper distribution of the fuel to the grates, as previously described.

A second problem is the accumulation of melted aluminum and lead on the grate. The best solution is total removal of these metals from the fuel stream. If this is not practical, experience has shown that maintaining proper ash bed thickness will cause the aluminum and lead to solidify in the ash bed rather than on the grate.

Fluidized-bed combustion

Beginning in the 1980s, fluidized-bed combustion technologies, both bubbling bed and circulating bed, had proven themselves to be efficient and cost-effective technologies for the combustion of waste fuels such as biomass, sludge, and waste coals. (See Chapters 17 and 30.) In the mid-1990s, there was growing interest in applying these fluidized-bed technologies to the combustion of RDF based upon higher efficiency, reduced emissions, and lower capital costs. While a number of waste-to-energy facilities were installed that utilized fluidized-bed technology, few of these facilities were able to achieve these high expectations due to several unique characteristics of RDF:

1. The quality and particle size of RDF is more critical for fluidized-bed combustion than for stoker firing to ensure fluidization of the fuel within the bed and to avoid plugging of the bed drain system. The fuel processing system needed to produce RDF suitable for fluidized bed combustion has a higher capital cost plus higher operating and maintenance costs.
2. A coal-fired fluidized-bed power generation facility can have a lower capital cost than a pulverized coal facility due to the in-bed capture of sulfur dioxide. This eliminates the need for a separate flue gas scrubbing system. The combustion of RDF, however, produces HCl as well as sulfur dioxide (SO_2) and a fluidized-bed boiler can not capture a sufficient amount of the HCl to meet current emissions limits. Therefore, a separate flue gas scrubbing system is required with significant capital and operation/maintenance costs.
3. Furnace and superheater tube corrosion is a major concern for any refuse-fired boiler; erosion is a major concern for a fluidized-bed boiler. One form of protection against corrosion in a refuse-fired boiler is the formation of an oxide layer on the surface of the tubes. This oxide layer acts as a barrier against continual contact between the corrosive flue gas and the tube material. The erosive action of the bed material in a fluidized-bed boiler will continuously remove this protective oxide layer as it is being formed, thereby accelerating the corrosion rate.

Fluidized-bed combustion technology is capable of firing RDF. However, the advantages of the technology must be carefully weighed against its disadvantages.

Lower furnace design configuration

The lower furnace designs of early RDF boilers were largely based upon technology used for wood-fired boilers. This included modest overfire air (OFA) systems with multiple small diameter nozzles designed for 25 to 30% of the total air supply, straight wall furnaces, and carbon reinjection systems. The result was less than desired combustion performance due to inadequate turbulent mixing in the furnace. Today's RDF units are designed with fewer large diameter OFA nozzles designed for 50% of the total air supply with nominal operation at 40%. This design has proven itself capable of injecting OFA deep into the combustion chamber for improved turbulent mixing of air with combustibles released from the grate. The Controlled Combustion Zone (CCZ) lower furnace design with twin arches for wood and biomass firing (see Chapters 16 and 30) has also been applied to some limited RDF applications.

Furnace exit gas temperature

Gas temperatures leaving the furnaces of first and second generation designs were higher than anticipated. There were not enough data on RDF firing to accurately predict the relationship between furnace surface area and furnace exit gas temperature. Compounding this problem was a continual increase in the heating value of the RDF due to changes in the composition of the raw refuse and the development of more efficient processing equipment. To achieve the desired furnace exit gas temperatures, the size of the third generation furnace has increased significantly. The furnace width and depth are set by the size of the stoker, therefore the furnace height has increased to achieve the required furnace exit gas temperatures.

Ash handling systems

Much of the non-combustible material in the RDF system is removed before it is fed to the boiler. Although systems vary, there is generally some effort made to remove ferrous metals and aluminum, both of which can be troublesome once they reach the stoker grates. Non-combustibles and most of the ash from combustion collect on the traveling grate stoker and discharge off the front into a submerged chain conveyor system.

The submerged chain conveyor (Fig. 24) is a mechanical conveyor that consists of a water-filled trough and a dry return trough, with two endless chain

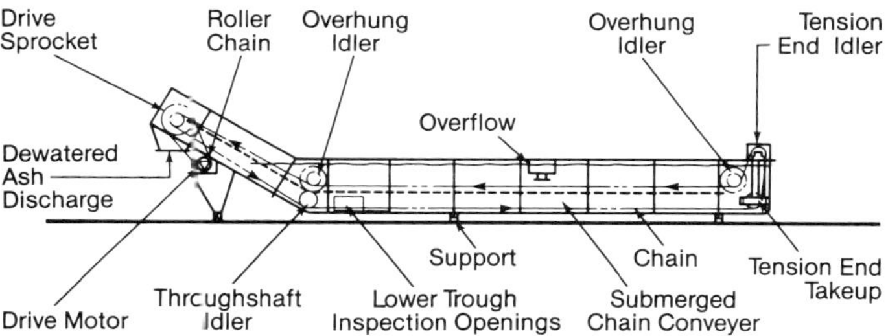

Fig. 24 Submerged chain conveyor for RDF stoker ash.

strands with flights connected between the strands. The return trough can be either above or below the water-filled trough. Ash from the stoker discharge chute drops into the water-filled trough. The water absorbs the impact of any larger ash pieces, quenches the ash, and provides a gas-tight seal with the stoker discharge chute.

The chains are usually driven by a variable speed drive to handle varying ash rates. The ash residue is conveyed from the bottom of a water-filled trough up an incline section where the ash dewaters and discharges directly into a truck, a storage bin, or onto another type of conveyor for final transport and disposal. Because this conveyor uses a dragging action to convey the ash, it is not used on mass-fired units where it can have problems in dragging the large noncombustible items up the incline section.

Fine ash from the boiler siftings hopper, flyash from the boiler, economizer and air heater hoppers, and flyash from the scrubber, baghouse or precipitator hoppers are all handled the same way as previously discussed for mass-fired boilers.

Retrofits to RDF

Most dedicated RDF boilers are new installations. However, it is possible to retrofit existing boilers to become dedicated RDF boilers. To be candidates, the existing boilers must be conservative designs for solid fuels, such as wood or coal. Typically, these are older units which are underutilized or used as standby units. These plants are often located near large metropolitan areas, sources of large quantities of refuse. The conversion of such an older power plant could represent a cost-effective solution to a community's refuse disposal problems.

B&W has converted several such boilers from coal-fired to dedicated RDF. Each retrofit is unique in that each of the coal-fired boilers was of different design and originally supplied by different manufacturers. Each retrofit was also the same in that all were designed to the same standards as new RDF units.

The principal modification involves enlarging the furnace to obtain the proper furnace volume for combustion. Although coal-fired boilers have conservatively sized furnaces, refuse firing requires even larger volumes. This is achieved by removing the existing stoker and lower furnace and installing new membrane furnace wall panel extensions (Fig. 25). The new lower furnace is protected from corrosion using either Inconel weld overlay or Inconel bimetallic tubes.

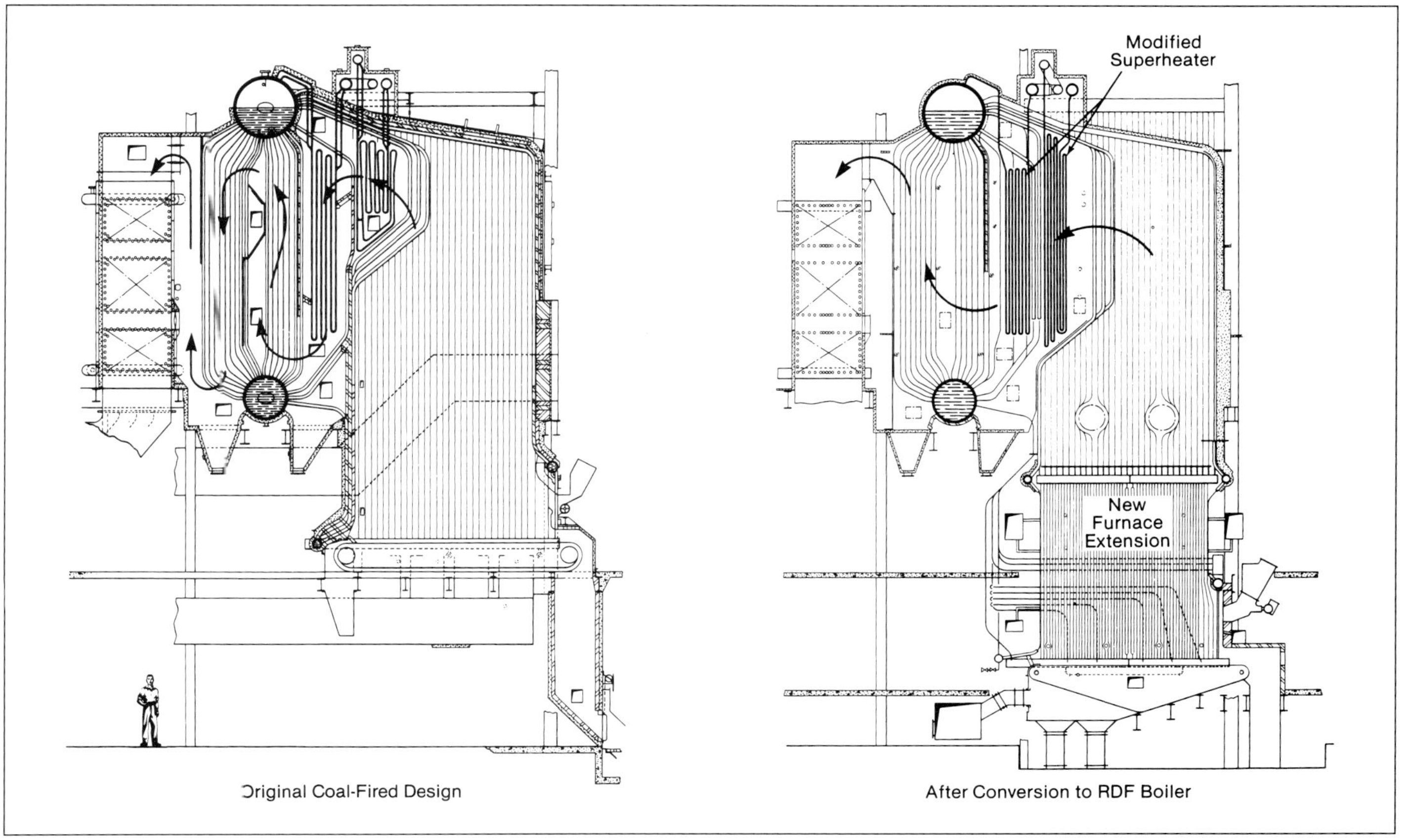

Original Coal-Fired Design

Modified Superheater

New Furnace Extension

After Conversion to RDF Boiler

Fig. 25 Coal-fired unit converted to RDF.

Other pressure part modifications could include:

1. converting the superheater to a parallel flow design while adding the proper metals for corrosion protection,
2. modifying the boiler, economizer and air heater surface for the proper heating surface distribution and to meet refuse standards for velocities, tube spacing, etc., and
3. possibly adding screen surface to lower the flue gas temperature entering the superheater.

In some cases, the coal stoker can be reused for RDF firing; in other cases it must be replaced. In either case the grate release rate, and other design criteria, must be set to the same design standards as new refuse boilers. Properly executed, the retrofit of an existing boiler to RDF firing will result in an RDF boiler as conservative as a new RDF boiler and capable of operating equally well.

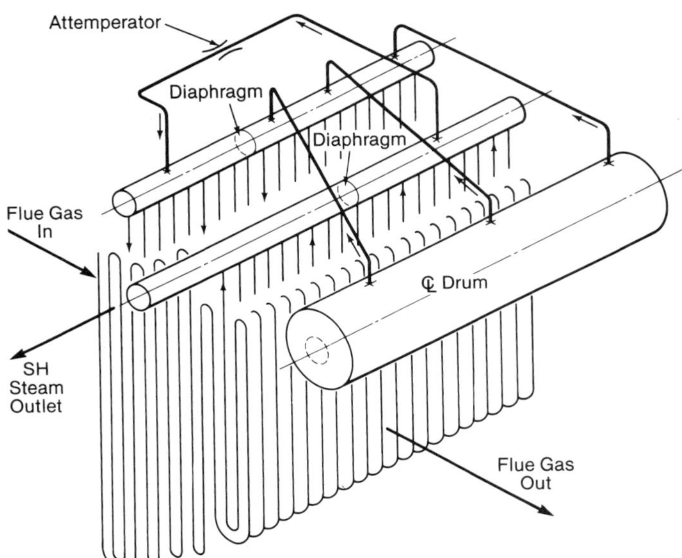

Fig. 26 Parallel flow superheater (SH).

Boiler components

Superheater

Superheater design is critical in both mass burn and RDF-fired refuse boilers because of the highly corrosive nature of the products of combustion. This is compounded in the U.S. by the desire for the highest possible steam temperature and pressure to maximize income from power production sales while still disposing of refuse in an environmentally safe manner. B&W has pioneered both the 900 psig (6.2 MPa) 830F (443C) and 1300 psig (8.96 MPa) 930F (499C) high pressure, high temperature steam cycles for refuse boiler application. This noteworthy increase in pressure and temperature over the more conventional refuse boiler steam conditions of 600 psig (4.14 MPa) and 700F (371C) has resulted in a significant improvement in cycle efficiency.

To accomplish this improvement, the superheater must be specifically designed for corrosion protection. Superheater corrosion is a function of many variables including flue gas temperature, flue gas velocity, tube spacing, tube metal temperature, tube metallurgy, and ash cleaning equipment. Each of these variables must be taken into consideration to avoid rapid superheater corrosion, even if the superheater steam outlet temperature is relatively low. For example, specifying only a low furnace exit gas temperature will not assure long superheater life.

Of these criteria, tube metal temperature and tube material are most critical. To obtain satisfactory refuse boiler superheater performance, two key design features are needed:

1. a parallel flow superheater design as shown in Fig. 26 where the coolest steam conditions are exposed to the hottest gas temperatures and the hottest steam temperatures are matched with the coolest gas temperatures. The result is a design with the lowest maximum superheater metal temperatures.
2. use of Incoloy tube material in the highest tube metal temperature sections of the superheater. Carbon steel is still used in the superheater sections with lower superheater metal temperatures.

B&W refuse boilers were the first to demonstrate successful commercial operation at the 900 psig (6.2 MPa), 830F (443C) steam cycle with the first three units going into operation at the Westchester County, New York refuse-to-energy facility in 1984. As of 2004, there were 21 B&W refuse boilers in operation at these steam conditions with more than 320 cumulative years of operating experience.

B&W extended this leadership position in superheater design in 1992 with the development of the first refuse boiler design to use the 1300 psig (8.96 MPa) and 930F (499C) steam cycle. This design was based on laboratory corrosion research and full scale superheater test sections installed in operating refuse boilers. In one test, two full scale superheater sections, composed of a variety of tube metallurgies, were installed in an operating 900 psig (6.2 MPa), 830F (443C) design refuse boiler with steam flow through the sections controlled to simulate 1000F (538C) operation. One section was removed after one year of operation and the second after 26 months. This test provided the basis for tube metallurgy selection for the 1300 psig (8.96 MPa), 930F (499C) design. In 1994, two mass-fired refuse boilers located in Falls Township, Pennsylvania, were placed into operation generating superheated steam at 1300 psig (8.96 MPa) 930F (499C). These boilers continue to operate successfully.

In addition to corrosion concerns, the superheater must be designed to minimize fouling and the potential for erosion due to excessively high flue gas velocities. Maximum design velocity is 30 ft/s (9.1 m/s), but in practice it is usually in the 10 to 15 ft/s (3 to 4.6 m/s) range. Minimum superheater side spacing is 6 in. (152 mm).

Boiler design

The lower furnace design, refuse stokers and refuse feed systems are markedly different for mass-fired and RDF boilers. However, the design requirements for the upper furnace, generating surface and economizer are the same. This is also true for auxiliary equipment such as burners and ash cleaning equipment.

Upper furnace design The upper furnace must be sized to provide adequate heat transfer surface to reduce the flue gas temperature entering the superheater to an acceptable level. This helps minimize fouling in the superheater and maintain low superheater tube metal temperatures to minimize corrosion. A certain amount of furnace volume is required for complete burnout of the fuel in the furnace and minimum CO emissions. The required volume should be measured from the point where all the combustion air has entered the furnace (the highest level of overfire air ports) to the point where the flue gas enters the first convective heating surface (at the tip of the furnace arch at the bottom of the superheater). Measured in this manner, the required furnace volume per unit of heat input is the same for both mass-fired and RDF boilers.

The furnace must also contain sufficient heating surface to lower the flue gas temperature to help reduce fouling and corrosion in the superheater and boiler bank. The gas temperature should be limited to 1600F (871C) entering the superheater and 1400F (760C) entering the boiler bank. Lower temperature values may be appropriate in certain geographic areas where the refuse composition is known to be highly corrosive. As a general rule, the furnace size is set by volumetric requirements in smaller capacity boilers and by maximum gas temperature limits in larger capacity boilers.

Boiler generating bank Refuse boilers in operation use both the one-drum and two-drum design. In the two-drum design there is both a steam drum (upper drum) and a lower drum, interconnected by the boiler generating bank tubes.

In the one-drum design the steam drum is located outside of the flue gas stream; there is no lower drum. The steam generating bank tubes are shop-assembled modules. These modules may be of either the vertical longflow (see Fig. 12) or a vertical crossflow design. Minimum side spacing in the two-drum design and for the generating bank modules used with the one-drum design is 5 in. (127 mm). Maximum design flue gas velocity is set at 30 ft/s (9.1 m/s).

Economizer The economizers can be either vertical longflow or horizontal crossflow. Economizer side spacing should be no less than 4 in. (102 mm) with a maximum flue gas velocity of 45 ft/s (13.7 m/s).

Air heater

Air heaters may be used for two reasons: 1) to supply preheated air to help dry and ignite the refuse on the stoker, and/or 2) to increase thermal efficiency where high feedwater temperatures preclude designing to lower exit gas temperatures with economizers. RDF-fired units have typically used air heaters to preheat the combustion air to the 300 to 350F (149 to 177C) range. Both tubular and regenerative air heaters have been used successfully. Due to air leakage into the air heater and the potential for fouling, regenerative types have been limited to the outlet side of hot electrostatic precipitators where the flue gases are relatively clean.

When either tubular or regenerative air heaters are used, the design and arrangement should minimize the potential for low-end temperature corrosion. To some extent, the surface arrangement in a tubular air heater will maintain adequate protection. Steam coil air heaters are required at the air inlet, on either type, to preheat the incoming ambient air and maintain the average metal temperature above acid dew points. (See also Chapter 20.)

Ash cleaning equipment

To maintain the effectiveness of all convective heating surfaces and to prevent pluggage of gas passages, it is necessary to remove ash and slag deposits from external tube surfaces. Steam or air sootblowers are most commonly used. Saturated steam is preferred for its higher density and better cleaning ability. One disadvantage of sootblowing is that localized erosion and corrosion can occur in areas swept too clean by the blowing medium. This problem can be addressed by installing tube shields on all tubes adjacent to each sootblower for localized protection.

A mechanical rapping system (Fig. 27) can be used to complement the sootblowers. In this system, a number of anvils strike designated pins to impart an acceleration through the superheater tube assembly. The purpose is to remove the bulk of the ash while leaving a light layer of ash on the tubes for corrosion protection. Mechanical rapping systems will not eliminate the need for sootblowers, but will reduce the number of sootblower cleaning cycles required.

Auxiliary input burners

Auxiliary fuel burners are used to maintain furnace temperature during startup, shutdown, and upset conditions to minimize the release of unburned hydrocarbons. In most cases, the auxiliary fuel (oil or gas) burners are designed for only 25 to 30% of the boiler's maximum heat input.

When not in service, the typical gas- or oil-fired burner requires some amount of air flow through the idle burner for protection against overheating. Because this air leakage represents an efficiency loss,

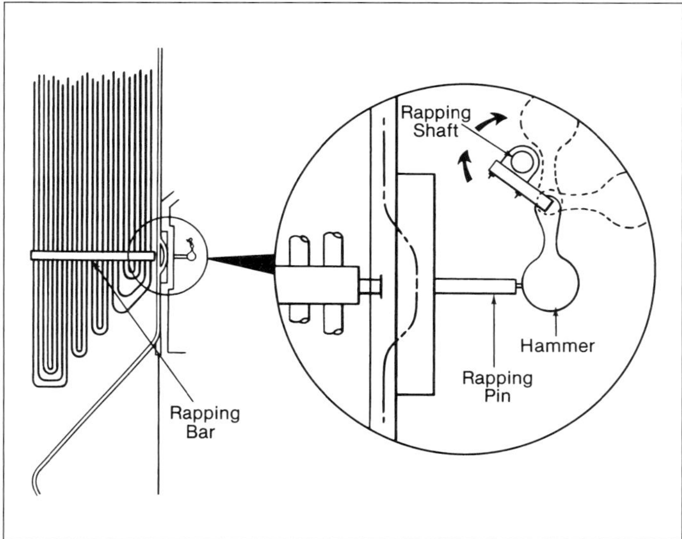

Fig. 27 B&W mechanical rapping system for cleaning superheaters.

and because these burners are used infrequently, a special design auxiliary input burner (AIB) is used for refuse boilers. The AIB is designed with a retractable burner element which is inserted toward the furnace when in use, and retracted when out of service. There is also a movable ceramic shutoff damper which provides protection against furnace radiation and debris when the burner is out of service. When the burner is put into service, this ceramic damper is retracted to one side and the burner element is inserted through an opening in the damper (Fig. 28).

Upper furnace maintenance platforms

Because refuse is a high fouling fuel, it is necessary to have good access to the convection sections. Maintenance platforms (Fig. 29) are often used to allow access to the superheater area for inspection and maintenance. Either retractable or light weight aluminum support beams are inserted into the furnace from access doors in the front wall to the superheater arch, where the beams are locked in place. Corrugated decking material is then inserted into the furnace through special side wall access doors, and arranged on top of the support beams. This system provides both a platform for working in the superheater and upper furnace and some overhead protection to those working in the lower furnace.

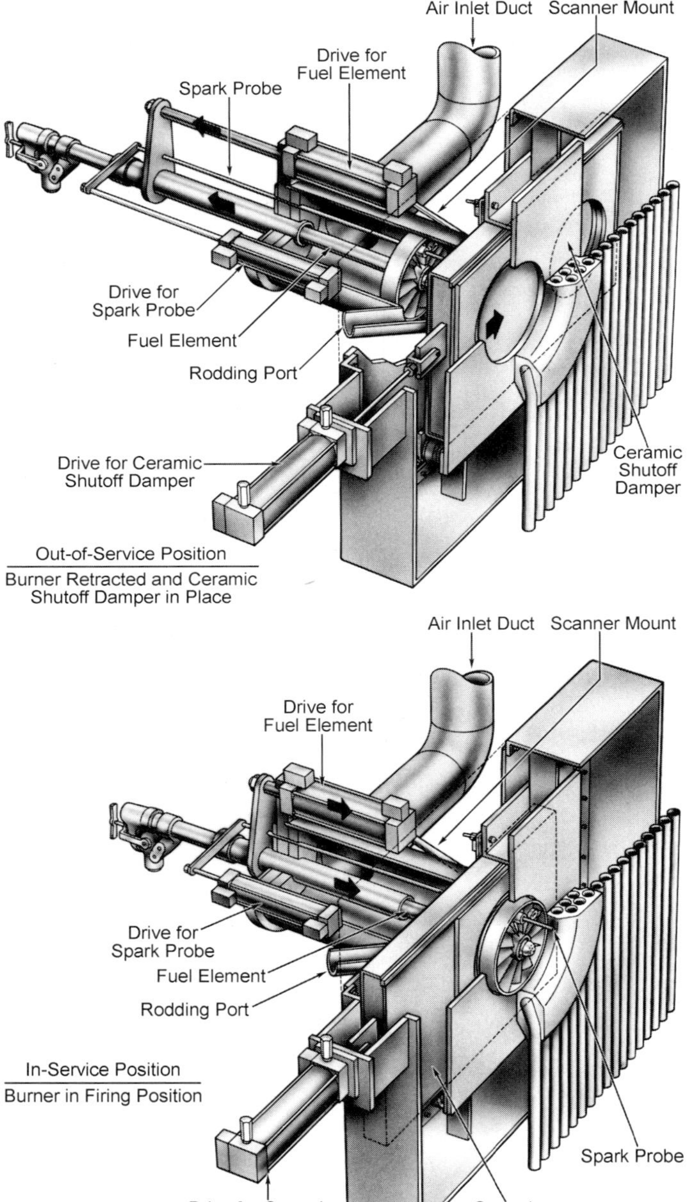

Fig. 28 Auxiliary input burner (out-of-service and in-service positions).

Air pollution control equipment

Various boiler fuels have specific components unique to that fuel. Some of these components, such as sulfur, create specific air pollution emissions that require unique boiler designs or specific air pollution control equipment. These fuels, such as high sulfur coal, are homogenous. This means the fuel will be the same in the future as it is today, and will be the same from one day to the next.

Refuse is a nonhomogenous fuel. It not only changes over the long term, but can change from day to day. Nearly every component of a fuel that can result in an unwanted air pollutant is present in refuse. However, in the early 1980s when the number of refuse-fired boilers began to rapidly grow in the U.S., the only emissions requirements were on particulates, NO_x and SO_x. Refuse boilers, due to their relatively cool burning systems and the generally low level of fuel bound nitrogen, are low NO_x generators. There are also very low levels of sulfur in refuse. Therefore, early boilers were generally equipped only with an electrostatic precipitator (ESP) for particulate control. As more boilers went into operation and further air emissions data were obtained, additional emissions requirements were applied. Initially, hydrochlorides were targeted for control. Soon the various state air pollution agencies set regulations for the control of dioxins and furans as well as a long list of heavy metals.

Dry systems

Dry scrubbers, used for years to control SO_2 emissions from coal-fired units, were found to be equally effective in controlling HCl emissions from refuse units.

With the initial use of dry scrubbers, there was a split in the preferred particulate collection system between the ESP and baghouse. ESPs were used in earlier applications due to their more extensive history of proven performance. However, it has been fairly well documented that the layer of ash and lime that collects on the bags themselves improves sorbent utilization and increases the removal of SO_2 and HCl in a baghouse. This allows better capture of pollutants for the same lime slurry rates, or the same level of pollutant capture at slightly reduced lime slurry rates. Dry scrubbers and baghouses were also found to be very effective in controlling dioxin, furan and heavy metal emissions. Today, the preferred system for nearly all refuse boilers in North America is the dry scrubber/baghouse combination. (See Chapters 33 and 35.)

To enhance the capture of mercury contained in the flue gas, refuse-fired boilers are typically equipped

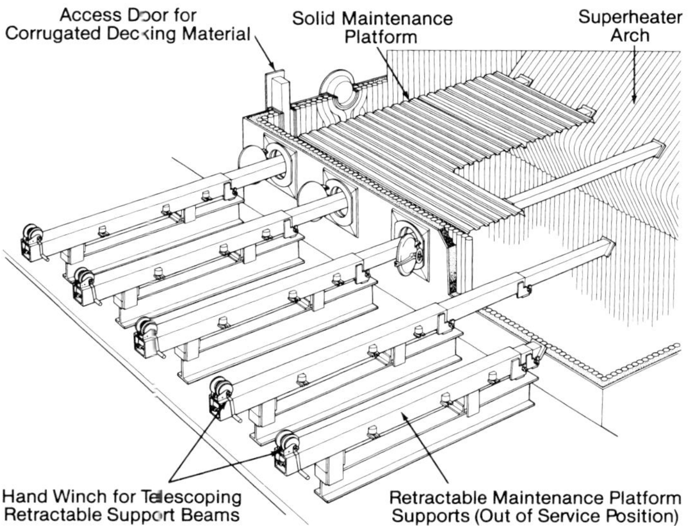

Fig. 29 Upper furnace maintenance platform.

with an activated carbon injection system that works in conjunction with a dry scrubber and baghouse. The process involves the injection of powdered activated carbon into the flue gas upstream of the dry scrubber. As the activated carbon mixes with the flue gas, it captures the mercury through adsorption. The activated carbon is then collected in the baghouse and removed along with the flyash.

NO_x and CO control

The permissible level of NO_x emissions has also been significantly reduced since the late 1980s. The lower NO_x limits can generally be achieved by installing an in-furnace ammonia or urea injection system. The ammonia injected into the furnace reacts with the NO_x contained in the flue gas to produce nitrogen and water. This form of NO_x control technology is known as selective non-catalytic reduction (SNCR) and can reduce the flue gas NO_x concentration by as much as 60%. In certain geographical nonattainment areas for NO_x emissions, selective catalytic reduction (SCR) systems may be required for even greater NO_x control. The SCR systems have been demonstrated to achieve up to 90% NO_x reduction on fossil fuel boilers. However, the catalyst itself is fairly easily poisoned, and therefore rendered less effective, by a multitude of substances, all of which are found in refuse to various degrees. At this time it is not clear what the long-term life of the SCR catalyst would be on a refuse-fired boiler. (See Chapters 33 through 35.)

During this same time period, the emission requirements for CO have also been driven to lower levels. These requirements have been met by a combination of:

1. better overfire air system,
2. more control of undergrate air (more compartments with individual air control),
3. better combustion control systems,
4. larger furnace volumes, and
5. operator training.

Wet systems

Wet emissions control systems are also used. A typical wet gas cleaning system is designed for the removal of HCl, SO_2, hydrogen fluoride, mercury, dioxins and solid particle pollutants from MSW-fired systems. A sample process used in Europe includes six major steps:

1. an ESP collecting most of the particles,
2. a heat exchanger that cools the gas to recover energy before the first scrubber,
3. a first spray scrubber to quench the gas further to saturated conditions and remove HCl and mercury,
4. a second spray scrubber to remove SO_2,
5. a third spray scrubber to allow removal of the dioxins, and
6. final filtering modules to remove excess dust particles coming from the ESP and other scrubbers.

An optional condensing heat exchanger can be used for additional heat removal if a hot water heating system is associated with the plant. Water treatment units process the liquid bleeds and overflows to yield a cake containing the solids for disposal.

Inconel is a trademark of the Special Metals Corporation group of companies.

Combined heat and power plant with straw-fired ultra-supercritical boiler (*courtesy of Kastrup Luftfoto A/S, Denmark*).

Chapter 30

Wood and Biomass Installations

The category of wood and biomass covers a wide range of material that can be used as a source of chemical energy. Wood encompasses a number of sources such as bark, wood sticks, sawdust, sander dust, over- and under-sized wood chips rejected from the pulping process, whole tree chips, and scrap shipping pallets. Biomass is anything that is or recently was alive such as straw, vine clippings, leaves, grasses, bamboo, sugar cane (called bagasse after the sugar has been extracted), palm oil, coffee grounds, and rice hulls from the food processing industry. All of these materials can be used as a fuel source to generate steam.

While wood and other biomass fuels were some of the first materials used as energy sources, their use declined around the early 1900s, when more consistent and easily transportable fossil fuels such as coal, oil and natural gas became more available. Nevertheless, there have always been applications where wood and biomass have been the preferred fuels. There has therefore been slow but steady progress in the development of equipment for firing these fuels.

Many factors have led to a recent, rapid development in this area. Some of these factors include the rising cost of certain fossil fuels, renewable energy options, development of new technology that allows better use of industrial byproducts, and the trend toward cogeneration in many industries. This includes industries that produce both electric power and process steam. Wood and biomass are used in many municipal combined heat and power (CHP) installations, particularly in Europe.

Biomass gasification is a relatively new technology being explored to supply fuel gas for power production by gas engines or gas turbines.

Equipment for chemical and heat recovery in the paper industry is discussed in Chapter 28. Equipment for using municipal solid waste as a fuel is discussed in Chapter 29. Stoker-related information is discussed in Chapter 16.

Steam supply for power production

Cogeneration in industry

Most heating requirements can be met with saturated steam at 150 psig (1.03 MPa) or less. Cogeneration, or the simultaneous production of electrical or mechanical energy and heat energy to a process, has very high thermal efficiency. Thus, waste heat from electricity production becomes part of a usable process rather than being rejected into the atmosphere.

In a cogeneration facility, relatively high pressure superheated steam passes through a steam turbine or steam engine where energy is extracted. The exhaust steam is then used as a heat source in a process. The conversion efficiency, or the amount of heat absorbed in the steam turbine plus the heat absorbed in the process, versus the amount contained in the steam, approaches 100%. The overall thermal efficiency of a cogeneration process is closely approximated by the thermal efficiency of the boiler alone.

As well as providing high energy efficiency, there is a second and often more compelling reason for industries, especially the pulp and paper industry, to practice cogeneration. These industrial plants are often located far from economical and reliable sources of electricity and conventional fossil fuels in order to be close to their source of fiber, the forests. Cogeneration facilities equipped with wood-fired power boilers, for example, are therefore often justified.

Biomass-fired utilities

Changing economic conditions and environmental regulations have made utility plants fired by biomass fuels a practical source of electrical energy, in spite of their relatively high capital costs. Sometimes the plant is built with a condensing turbine and other times it is built beside or within a plant that can use exhaust steam. Economic conditions favoring such facilities include the unpredictable and sometimes high cost of conventional fossil fuels, the relatively low cost of wood waste and other biomass fuels, and the high cost to transport and dispose of biomass in a landfill.

Occasionally an installation is justified with regard to the high cost of emissions control equipment. Most wood and biomass fuels generate lower nitrogen oxides (NO_x) and sulfur dioxide (SO_2) emissions than conventional fossil fuels. In one installation in California, for example, it was found that the amount of NO_x produced by incinerating grapevine clippings and other biomass could be substantially reduced by burning the material in a wood-fired utility installation in a controlled manner. This produced environmental benefits while generating electricity that would otherwise be produced by burning traditional fossil fuels.

The Babcock & Wilcox Company (B&W) operation in Europe completed a significant biomass-fired installation that burns straw in an ultra-supercritical power plant application. (See Fig. 1.) This straw-fired boiler consumes 105 MW_t of fuel producing 35 MW of electricity and 50 MW_t of heat. In this advanced design, carded (loosened by mechanical raking) straw is fed by screw feeders onto a water-cooled gasification grate where a high percentage of the energy content is released by pyrolysis and gasification. The remaining straw/carbon burns on a water-cooled vibrating grate. A bag filter removes more than 99% of particulates from the flue gas, and the plant operates within strict European emissions regulations.

Steam supply to process

Pulp and paper

The pulp and paper industry is the major consumer of biomass fuels because wastes such as bark, sawdust, shavings, lumber rejects and clarifier sludge are byproducts of the pulping, paper-making and lumber manufacturing processes. There is a great amount of heat energy available in these products, making them useful energy sources. (See Figs. 2 and 3.)

The production of pulp and paper requires large quantities of mechanical energy for grinding, chipping, cooking and refining. To produce a marketable product, the pulp must be dried using steam or heated surfaces such as paper machine dryer rolls. These energy needs are met by using steam in a variety of equipment (steam engines or turbines, steam coil air heaters, dryer rolls, and/or indirect heaters) and by direct steam injection.

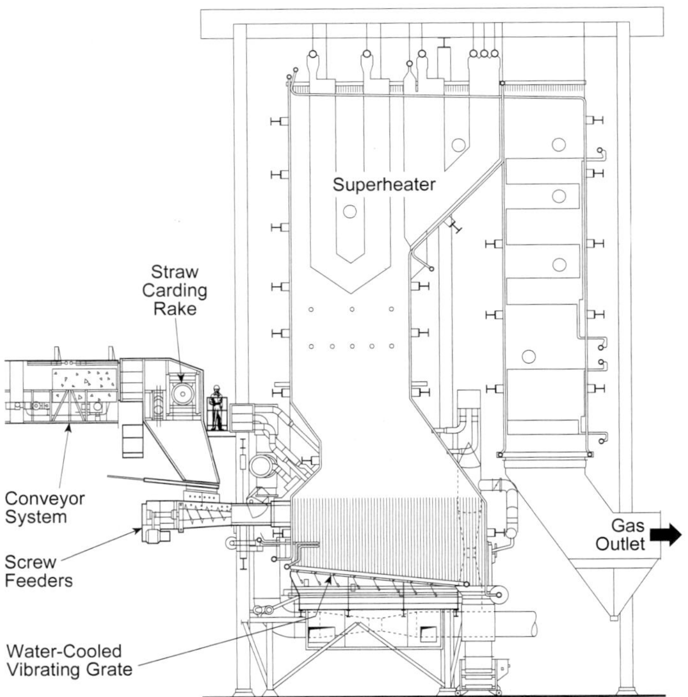

Fig. 1 Straw-fired ultra-supercritical boiler (*courtesy Energi E2 A/S*).

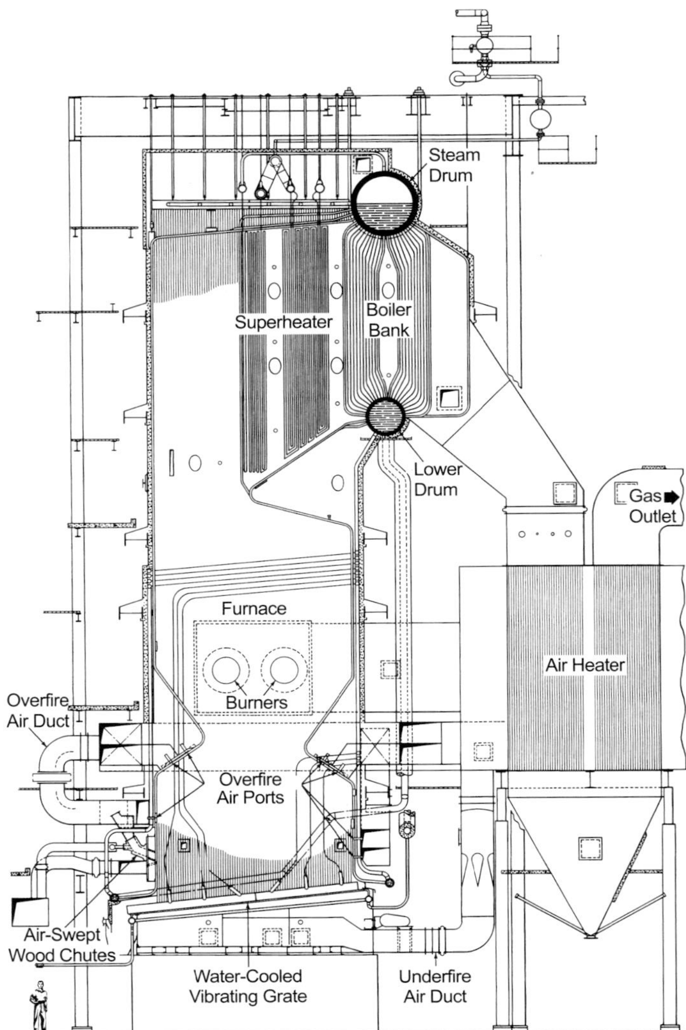

Fig. 2 Wood-fired two-drum Stirling® power boiler.

These requirements, coupled with the availability of the waste products, make boilers fired by wood and wood waste a logical choice for the pulp and paper industry. Most of the developments and improvements in equipment for these boilers have been driven by this industry's needs.

Food processing

The food processing industry also has energy needs that are provided by steam. Mechanical preparation, cooking, drying and canning all require a source of energy. Many foods leave behind waste byproducts rich in cellulose or other organic (hydrocarbon) material. Instant coffee production generates coffee grounds, sugar-making leaves bagasse from sugar cane, coconut preparation discards husks, rice has its hulls removed before packaging, and many types of nuts are sold roasted with their shells removed.

Many producers have installed boilers that burn such biomass materials, usually based very closely on equipment originally designed for the pulp and paper industry. These boilers produce steam that is then used as an energy source for the plant.

Fuels

Constituents

Wood and most biomass fuels are composed predominantly of cellulose and moisture. The high proportion of moisture is significant because it acts as a heat sink during the combustion process. The latent heat of evaporation (H_{fg}) depresses the flame temperature, contributing to the difficulty of efficiently burning biomass fuels.

Cellulose, as well as containing the chemical energy released during combustion, contributes fuel-bound oxygen. This oxygen decreases the theoretical air required for combustion and therefore the amount of nitrogen included in the products of combustion.

Most natural biomass fuels contain little ash. However, some byproducts, such as de-inking sludges, do contain a great deal of ash, in some cases up to 50% ash on a dry basis. De-inking sludges are particularly difficult to burn because they usually have high moisture and ash contents and low fuel-bound oxygen content.

Burning wood and biomass

The following general guidelines for wood and biomass combustion have been developed from experience:

1. Stable combustion can be maintained in most water-cooled furnaces at fuel moisture contents as high as 65% by weight, as-received.
2. The use of preheated combustion air reduces the time required for fuel drying prior to ignition and is essential to spreader-stoker combustion systems. The design air temperature will typically vary directly with fuel moisture content.

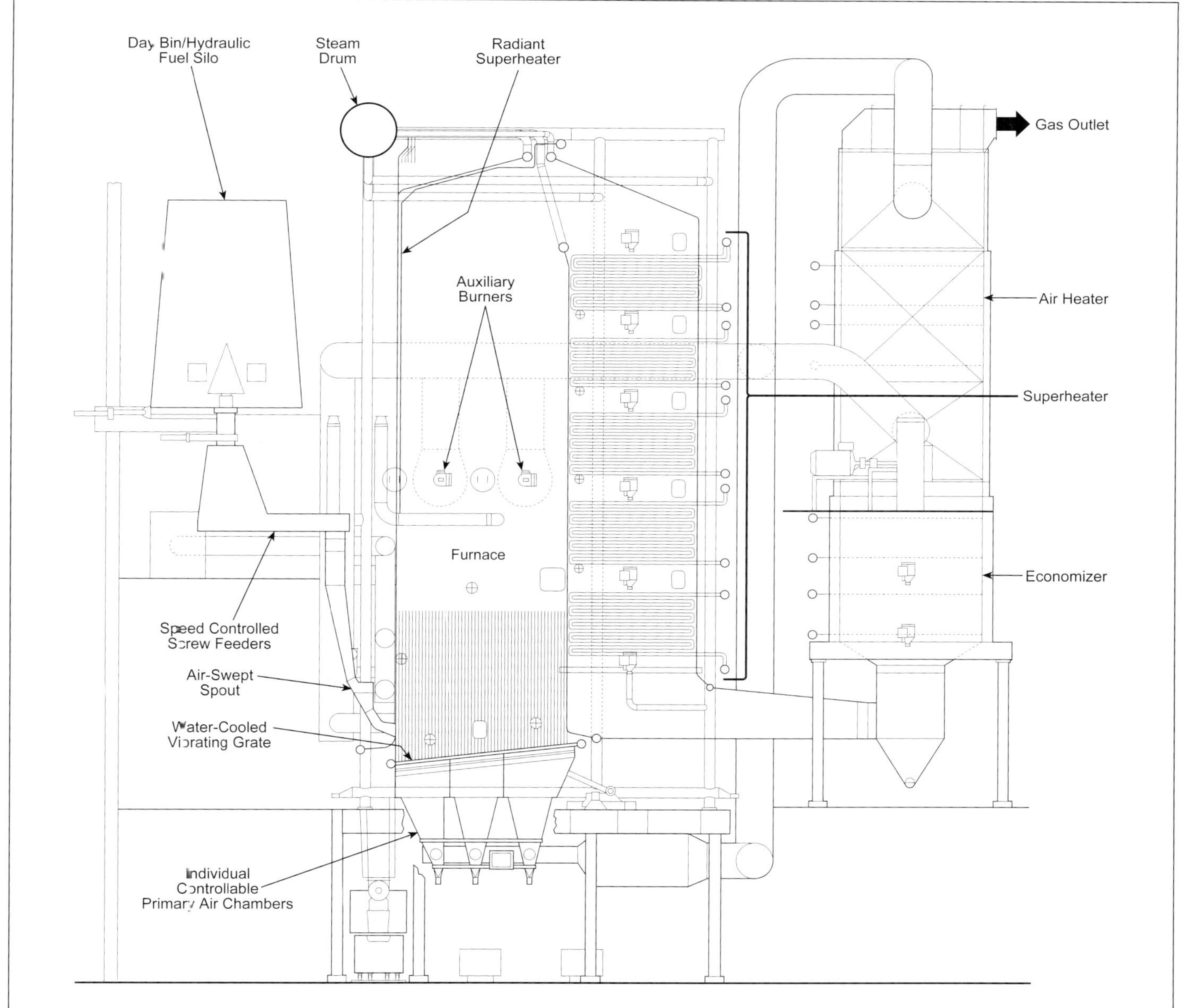

Fig. 3 Multifuel single-drum, 141,000 lb/h (17.8 kg/s) steam flow, bottom-supported boiler with vibrating grate stoker.

3. A high proportion of the combustible content of wood and biomass fuels burns in the form of volatile compounds. A large portion of the combustion air requirement is therefore added above the fuel, as overfire air (OFA).
4. Solid chars produced in the initial stages of combustion of these fuels are of a very low density. Conservative selection of furnace size and careful placement of the OFA system are used to reduce gas velocity and keep char entrainment at an acceptable level. Typical furnace selection criteria include a grate heat release rate of 0.47 to 1.1 $\times 10^6$ Btu/h ft^2 (1.5 to 3.5 MW$_t$/m^2) of grate surface area, a furnace liberation rate of 17,000 Btu/h ft^3 (176 kW/m^3) of furnace volume, and an upward gas velocity of 20 ft/s (6.1 m/s). This results in furnace residence times of approximately three seconds for larger units to enhance particulate burnout and minimize emissions.

Burning in combination with traditional fuels

Biomass can be burned on a traveling grate with stoker coal. The biomass is introduced to the furnace through a separate conveying system and either a separate windswept spout below the coal feeder or through a combination feeder. (See Chapter 16.)

Biomass can be burned with pulverized coal, oil or natural gas, using dedicated burners for the latter. In this case, the grate is selected for wood firing.

When burning biomass with substantial quantities of stoker or pulverized coal, the amount of ash from the coal is greater than the amount from the biomass. Therefore, the design parameters for the coal (slagging and fouling index) will govern the design.

When burning biomass with heavy fuel oil having high sulfur and vanadium content, the ash that forms on the convective surfaces can be tenacious and, once removed, can be very abrasive. It is preferred to design for low flue gas temperatures and low flue gas velocities, regardless of the specified contaminants in the heavy oil and in the biomass, as both flue gas parameters can vary widely over the typical range of boiler operating conditions. (See Chapter 10.)

Sludge burning

As mentioned above, paper mill sludges, especially de-inking sludges, are difficult fuels to burn. Their high moisture and ash content and low fuel-bound oxygen may limit their allowable proportion of the total heat input to the furnace.

Specific limits will vary with the sludge composition and combustion system. Higher sludge inputs for a given system can normally be achieved by combined firing with better quality fuels.

Wood waste mixed with sludge can prove difficult for the fuel handling system. Frequently, the sludge segregates from the other fuels (at transfer points on belt conveyors, for example) and can be fed preferentially to one feeder.

In spreader-stoker applications, the wet, dense sludge can pile up in one place on the grate very quickly. Therefore, when a portion of the fuel is sludge, the boiler operator must continually inspect the grate to determine whether adjustments are required.

Combustion systems

Many methods have been developed to burn wood and other biomass fuels. The best known and most successful methods use the following equipment.

Dutch oven

A dutch oven is a refractory-walled cell connected to a conventional boiler setting. Water cooling is sometimes provided to protect and extend the life of the refractory walls. Wood and other biomass fuels are introduced through an opening in the roof of the dutch oven and burn in a pile on its floor. OFA is introduced around the periphery of the cell through rows of holes or nozzles in the refractory walls.

The principal advantage of the dutch oven is that only a small portion of the energy released in combustion is absorbed because of its high percentage of refractory surface. Therefore, it is able to burn high moisture fuels (up to 60% moisture). In addition, as the fuel is pile-burned, there is a high thermal inventory in the cell that makes the unit less sensitive to interruptions in the fuel supply.

The dutch oven has distinct disadvantages when compared to more modern methods. The unit operates best when at a steady load and when burning a consistent fuel. It does not respond quickly to load demand. The refractory is subject to damage from the following sources: spalling and erosion caused by rocks or tramp metal introduced with the fuel, rapid cooling from contact with very wet fuels, and overheating when very dry wood is fired.

The dutch oven cell must be shut down regularly to allow manual removal, or rake out, of the accumulated ashes. During this period, either load must be drastically reduced as is done with multiple cell units, or auxiliary fuel must be fired as is required for units equipped with a single dutch oven.

Pinhole grate

The pinhole grate (Fig. 4) is a water-cooled grate formed by cast iron grate blocks, sometimes referred to as Bailey blocks, that are clamped and bonded to the spaced floor tubes of a water-cooled furnace. The grate blocks have venturi-type air holes to admit undergrate air to the fuel on the grate. This grate is used in conjunction with either mechanical fuel distributors or air-swept fuel spouts. Both produce a semi-suspension mode of burning, wherein the finer portion of the fuel is burned in suspension and the heavier fraction accumulates and burns on the grate. The ash and foreign material that stay on the grate are removed by raking. Typically, 25% of the air for combustion of wood is introduced as OFA through nozzles in the lower furnace walls and 75% is undergrate air.

This combustion system can follow minor load swings by varying both the fuel flow and air flow, and is suitable for biomass fuels containing up to approximately 55% moisture. A small amount of refractory is used and maintenance requirements are low.

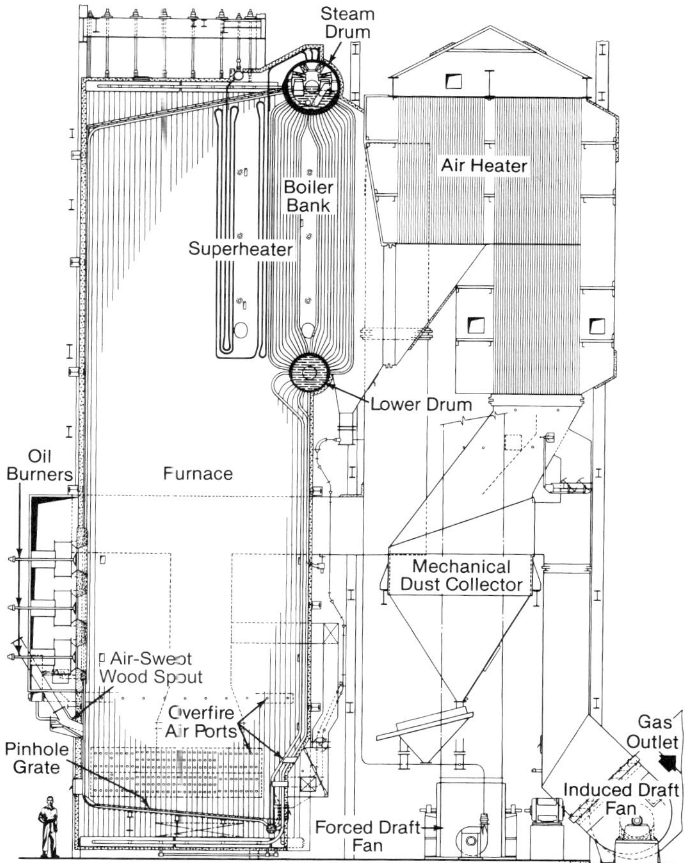

Fig. 4 Wood-fired boiler with pinhole grate.

The main disadvantage of this method is that the grate requires manual raking of the ashes. Therefore, biomass firing must be stopped on a regular basis. Manual raking also limits the depth of the furnace and therefore the steam capacity that can be economically built. Mechanical raking machines have been developed to increase the allowable furnace depth and to speed the raking process, while providing a certain degree of protection to the operator from hot gas, flames and hot ash.

Traveling grate

The traveling grate was introduced as an improvement over the pinhole grate. (See Fig. 5.) It is a moving grate that allows continuous automatic ash discharge and consists of cast iron or ductile iron grate bars attached to chains that are driven by a slow moving sprocket drive system. The grate bars have holes in them to admit undergrate air that is also used to cool the grate bar castings. The split between undergrate and overfire air flows depends largely upon fuel size, volatility and moisture content, while also meeting the grate cooling requirements. For coal firing, 60 to 85% of the air is typically supplied as undergrate with the balance used for OFA. For biomass firing, 25 to 50% of the air is typically supplied as undergrate. Rows of nozzles in the furnace front wall and rear wall are used for OFA. Fuel spreaders and burning mode are identical to those used with the pinhole grate. The main advantages of this grate are its ability to follow load swings and the automatic ash discharge that permits continuous operation on biomass fuels.

The traveling grate was originally developed for spreader-stoker firing of bituminous and subbituminous coals. With coal, the quantity of OFA required for efficient combustion can be as low as 15% of the total air. The quantity of ash in coal is also much higher than in wood. It is therefore possible to develop a relatively large bed of ash on the grate in order to protect the grate from high temperatures, and to help distribute the undergrate air flow. Lower moisture coal can often be burned without air preheating and rarely requires air temperatures in excess of 350F (177C).

The traveling grate, while an improvement over the pinhole grate, must be considered a compromise design for the burning of biomass because the use of preheated air and the usually low ash content of biomass reduce the cooling available to the grate. This grate has many moving parts that are subjected to the furnace heat, resulting in higher maintenance costs.

Vibrating grate

The vibrating grate allows intermittent, automatic ash discharge. The grate consists of cast iron grate bars or bare tube panels attached to a frame that vibrates on an intermittent basis, controlled by an adjustable timer.

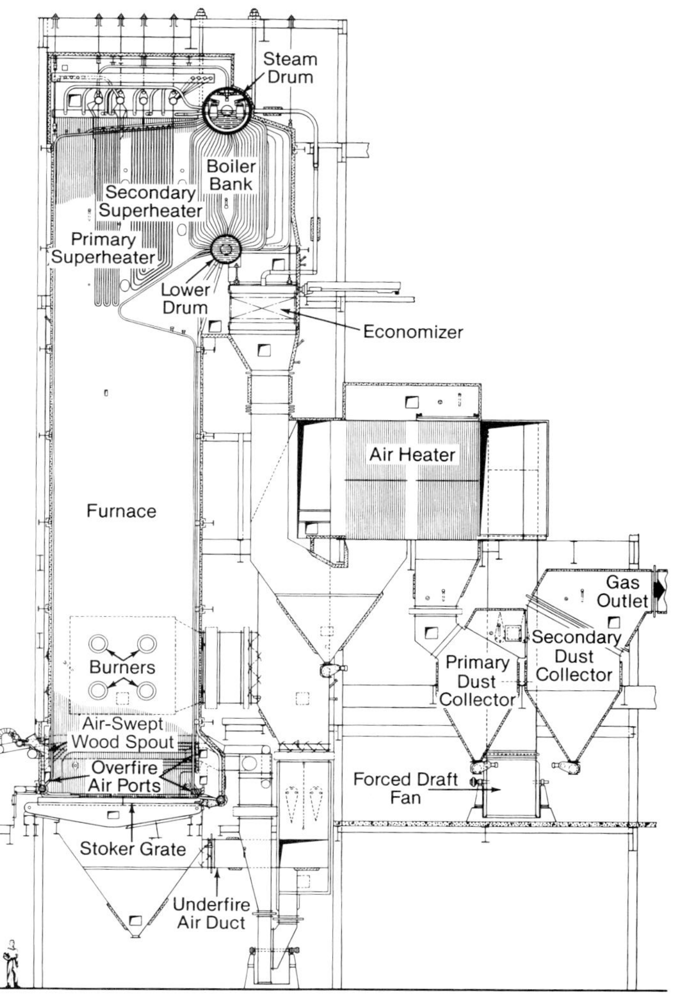

Fig. 5 Wood-fired boiler with traveling grate.

There are two major types that have been used for biomass fuels, one water-cooled and the other air-cooled.

The water-cooled vibrating grate (see Fig. 6) is used in conjunction with the semi-suspension firing mode. Because the grate is water-cooled, high temperature undergrate air, up to 650F (343C), can be used along with very high percentages of OFA and a relatively thin fuel bed. A key advantage of the vibrating grate is the low number of parts that are highly stressed, moving or in sliding contact. This results in reduced maintenance requirements. The vibration is intermittent and several vibration sequences (vibration frequency, duration, and time span between vibrating periods) are used. A common vibration sequence includes six cycles per second for a short duration of about two seconds every two minutes. The vibration time and the dwell time can be adjusted to suit the fuel characteristics. The vibrating action can help improve fuel distribution on the grate by causing fuel piles to collapse. This style of biomass combustion system is being used very successfully at a number of installations for a wide range of wood fuel moisture contents.

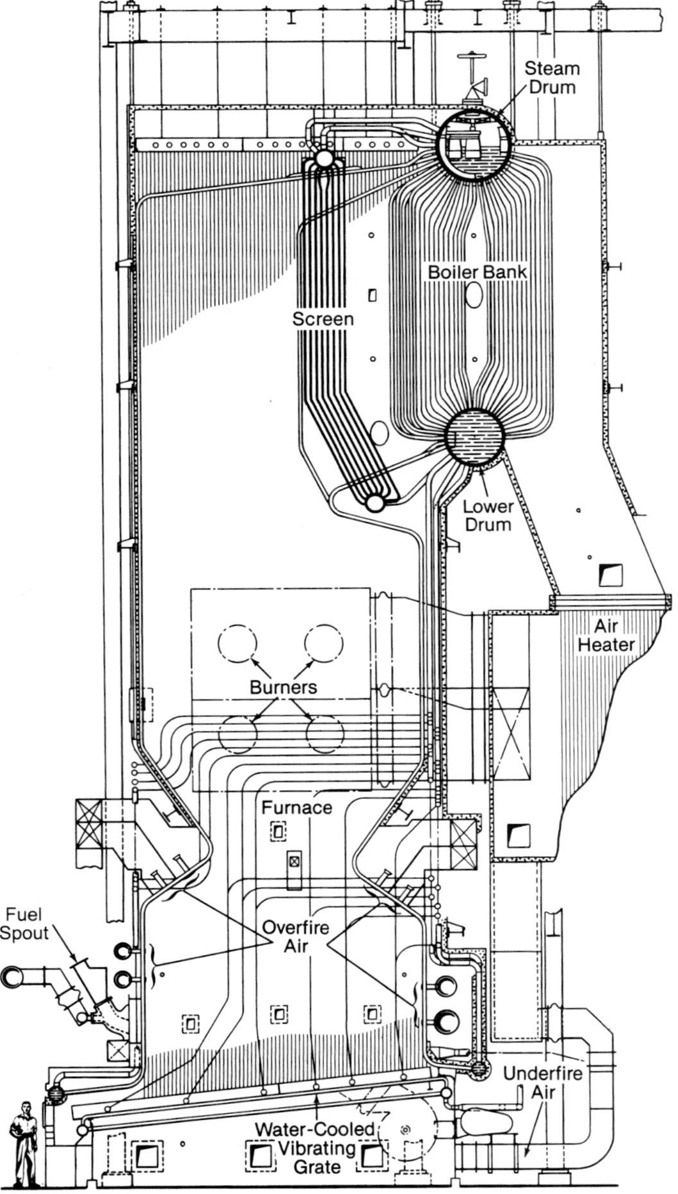

Fig. 6 Water-cooled vibrating grate unit with CCZ furnace.

The air-cooled vibrating grate can be used for applications similar to those suitable for the water-cooled vibrating grate, but the maximum allowable undergrate air temperature is 550F (288C), even with the use of stainless steel components. Because it can be installed with a horizontal grate surface, the air-cooled vibrating grate can be a very effective replacement for a traveling grate, when the traveling grate has been deemed unacceptable due to high maintenance or repair costs.

Furnace configurations

A variety of furnace configurations can be found in wood and wood-waste boilers. In the late 1970s, the controlled combustion zone or CCZ furnace was developed by B&W specifically for biomass combustion. As shown in Figs. 2 and 6, the CCZ design uses arches in the front and rear walls of the furnace to create a lower furnace zone in which combustion of biomass can be confined. OFA from nozzles located above and below the arches penetrates and burns off the volatile fuel released from the bed, as well as particles entrained in the upward flow. This furnace design incorporates a range of OFA systems that had multiple elevations of large ports and increased the OFA capacity to 40 to 50% of the total combustion air flow.

Tighter emission control regulations and advancing computational technology have stimulated the development and deployment of a straight-wall furnace design with fewer, larger OFA ports that can be more precisely located and operated. As shown in Fig. 7, the arches and multiple levels of OFA ports for the CCZ can be replaced with straight furnace walls and one level (sometimes two) of OFA ports. The B&W PrecisionJet™ OFA system has high velocity ports that include velocity dampers to fine-tune the combustion process. Computational models (see Chapter 6) permit the positioning and sizing of the higher velocity OFA ports to better complete the combustion process in a controlled fashion, thereby reducing NO_x, CO and particulate. OFA capacity of up to 60% of the combustion air flow can be used. As discussed in Chapter 16, a second option is a horizontal rotary overfire air system designed to create a double rotating circulation zone in the straight-wall furnace. (See Chapter 16, Fig. 13.) This can also increase turbulence and mixing of air and gaseous combustibles for enhanced performance. With a straight-wall design and fewer ports, these furnace configurations usually have lower initial capital costs.

Process recovery (PR) boilers (see Chapter 28) and coal-fired power boilers at pulp and paper mills have been retrofitted for wood and wood-waste firing capability as mill power and PR needs have evolved. Rising energy prices have also increased the benefit of higher internal power generation from waste product fuels. To accommodate the heat release rates discussed earlier, the furnace plan area of the original boiler must typically be enlarged to accommodate the grate area for the desired heat input. The new enlarged lower furnace is then tapered to connect with the origi-

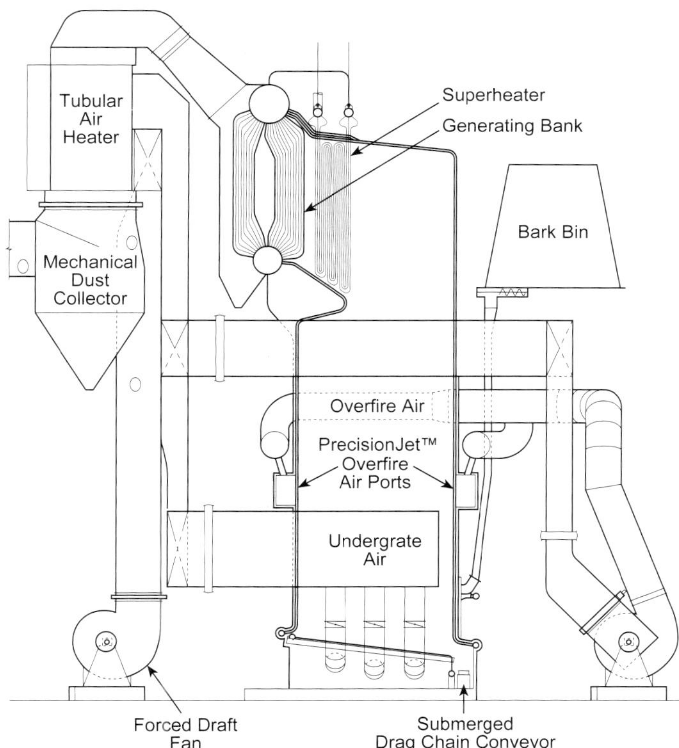

Fig. 7 Wood-fired boiler with straight furnace walls and overfire air.

nal upper boiler furnace to take full advantage of the original heating surfaces and boiler capacity. The same type of tapered furnace design may also be found where bubbling fluidized-bed combustion systems are retrofitted to PR or coal-fired boilers (see Fig. 8).

Dryers and pulverizers

If the biomass fuel available is particularly high in moisture, or if the capacity of an existing installation is to be increased, it is sometimes more economical to dry the fuel with boiler flue gas before firing it in the boiler furnace, rather than pressing the fuel to remove moisture or modifying the boiler for increased capacity. Dryers of the rotating drum type and cascading fuel type are available from several manufacturers. The dried fuel is then fired using one of the previously mentioned combustion systems.

Another means of burning biomass is by pulverizing/drying. This can be accomplished by mixing the biomass with hot gases removed from the boiler exit, pulverizing it in a fan/beater mill, and returning the mixture to the furnace through a burner. Milled peat is fired in several installations in the Nordic countries using this process. The milled peat, except for larger pieces, does not undergo much size reduction. Most other types of pulverizing systems have proven to require very high maintenance and to have poor availability and significant power consumption in biomass applications.

Fluidized-bed combustion

Fluidized-bed combustion has been successfully applied to a range of wood and biomass fuels and offers a number of features that may be advantageous in specific applications.

Only 2 to 3% of the bed is carbon; the remainder is comprised of inert material (sand). This inert material provides a large inventory of heat in the furnace, thereby dampening the effect of brief fluctuations in fuel heating value on steam generation.

Fluidized beds typically operate at 1350 to 1650F (732 to 899C), a range considerably lower than combustion temperatures for spreader stoker units [2200F (1204C)]. The lower temperatures produce less NO_x and would be expected to provide the most benefit on high nitrogen wood and biomass fuels.

SO_2 emissions from wood and biomass firing are generally considered insignificant, but where sulfur contamination of the fuel stream is a problem, limestone can be added to the fluid bed to achieve a high degree of sulfur capture. Fuels that are typically sulfur contaminated include tire-derived fuel and coal.

The type of fluidized bed selected will be a function of the as-fired calorific value of the wood waste or biomass fuel. Bubbling bed technology is generally selected for fuels of lower calorific value. For fuels of higher calorific value, the circulating fluid bed may be suitable.

Designs for fluidized-bed combustion are discussed in Chapter 17. Bubbling fluidized beds have been applied to retrofit (Fig. 8) and new installations (see Chapter 17, Fig. 10).

Boiler component design for wood and biomass burning

Grate

The grate forms the furnace floor and provides a surface on which the larger fuel particles burn. The grate may be air-cooled or water-cooled and stationary or arranged for automatic, continuous ash removal. Most grates consist of some form of cast iron or cast alloy grate bars. (See Chapter 16.)

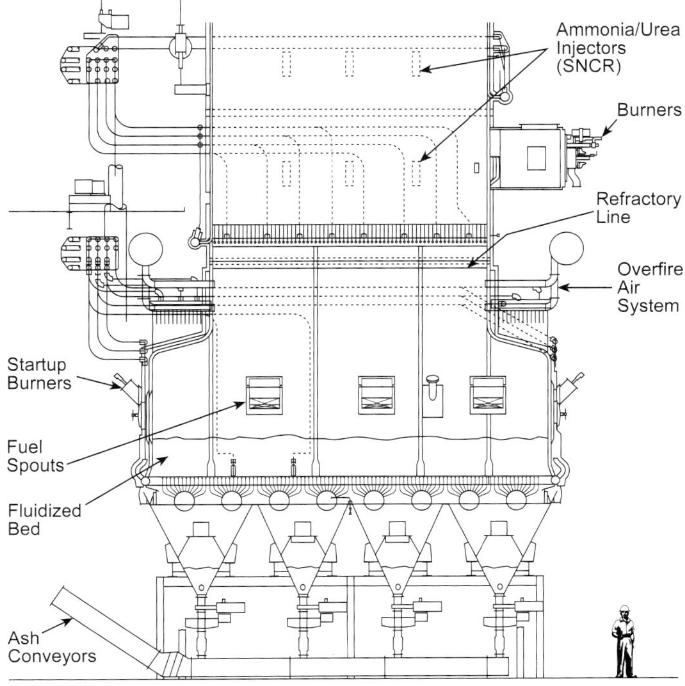

Fig. 8 Retrofit of process recovery boiler for wood-firing capability.

Fuel distributor

The two most common devices for introducing fuel into the furnace for semi-suspension firing are mechanical distributors and air-swept spouts. They are both designed to distribute fuel as evenly as possible over the grate surface.

The mechanical distributor uses a rotating paddle wheel to distribute the fuel. The speed of the wheel is varied to suit the fuel characteristics. In some installations, a continually varying speed is used to ensure good fuel distribution.

The air-swept spout uses high pressure air that is continuously varied by a rotary damper to distribute the fuel. Adjustments to the supply air pressure are made by the operator as the characteristics of the fuel change. The trajectory of the fuel leaving the spout is altered by an adjustable ramp at the bottom of the spout.

Burners

Burners are sometimes used to burn all or a portion of the biomass fuel. The burners are somewhat similar in design to a pulverized coal burner. (See Chapter 14.) Fuels that can be fired in a burner include sander dust, sawdust of less than 35% moisture content, and the fine material collected from a fuel dryer. Due to the possibility of inconsistent fuel flow and quality, a continuously operated auxiliary fuel pilot flame is recommended.

Furnace

A properly designed furnace has two main functions. The first is to provide a volume in which the fuel can be burned completely. The second is to absorb sufficient heat to cool the flue gas to a temperature at which the entrained flyash will not foul the convective surfaces. This must be accomplished while matching the dimensions of the grate and while providing sufficient clearance dimensions from auxiliary burners to prevent flame impingement on furnace walls.

Modern boilers are typically of membrane wall construction, but in certain limited cases, such as installations in developing countries, reverting to a tube and tile type of furnace construction may be appropriate. In such circumstances, the frequent maintenance requirement is outweighed by the reduction in first cost and the ease of operation due to the higher temperatures in the furnace.

Superheater

The sizing of the superheater on a wood-fired boiler can be complicated by several factors. For a given fuel being fired, the setting of the surface depends on the final steam temperature and on the control range required. The side and back spacing are selected to minimize fouling and erosion potential. (See Chapters 19 and 21.)

However, a wood-fired boiler rarely burns a consistent fuel. Variable moisture content and fuel analyses affect the steam-to-flue gas ratio, and a variety of auxiliary fuels, e.g., oil, gas or coal, may be available. Therefore, when designing a superheater, the full range of operating conditions must be completely understood.

Constituents in the ash can affect superheater design. For example, high levels of chloride are often found in bark from logs floated in sea water and can require the use of high alloy materials such as 310 stainless steel to minimize the corrosion rate of the superheater tubes in the high temperature zones.

Boiler bank

Due to the relatively high ratio of flue gas flow to steam flow and the relatively low pressures and temperatures at which most wood-fired boilers are operated, a large amount of saturated (boiling) surface is required. Furthermore, due to relatively low adiabatic flame temperatures, the amount of heat absorption in the furnace is usually low compared to fuels such as oil or natural gas. Therefore, a large portion of the total heating surface in a wood-fired boiler is usually provided as boiler bank.

In some cases, the amount of furnace surface is augmented by water-cooled screens in front of superheaters to lower flue gas temperatures entering the superheater and to protect it from thermal radiation or from the active burning areas in the furnace.

The amount of boiler bank surface is usually very substantial. This surface can be arranged as cross flow surface as normally found in a two-drum Stirling boiler (see Fig. 2), or may be arranged for longitudinal flow as found in a smaller bottom-supported Towerpak® boiler. (See Fig. 9.)

Because wood fuels frequently contain sand or other mineral matter in addition to the ash, flue gas velocities in the convection pass or boiler bank must be kept low, typically below 60 ft/s (18.3 m/s).

Economizer

In most cases, when an economizer is required to reduce the back-end temperature to a specified level, it is located between the boiler bank and the tubular air heater. The economizer is designed to reduce the flue gas temperature to that required at the air heater gas inlet.

Occasionally, the positions of the economizer and air heater need to be reversed. For example, it may be necessary to provide a higher gas temperature to the air heater as part of an installation that includes a fuel dryer. The dryer needs the hotter gas to remove moisture from the fuel. This same system may also be required to maintain a specified exit gas temperature entering the stack even when the dryer (and its thermal load) is out of service for maintenance or repair. For such a system to work effectively under both operating conditions, the economizer needs to be the final heat trap. In addition, a bypass, preferably on the economizer gas side, is needed to permit exit gas temperature control.

In the above example, it should be remembered that the economizer gas exit temperature would be lower when burning dry fuel from the dryer and would be higher with the dryer out of service, if no control method was provided.

There are no special mechanical design considerations for economizers on wood-fired units, other than to limit the flue gas velocity. In virtually all cases, a continuous bare tube economizer is used. (See also Chapter 20.)

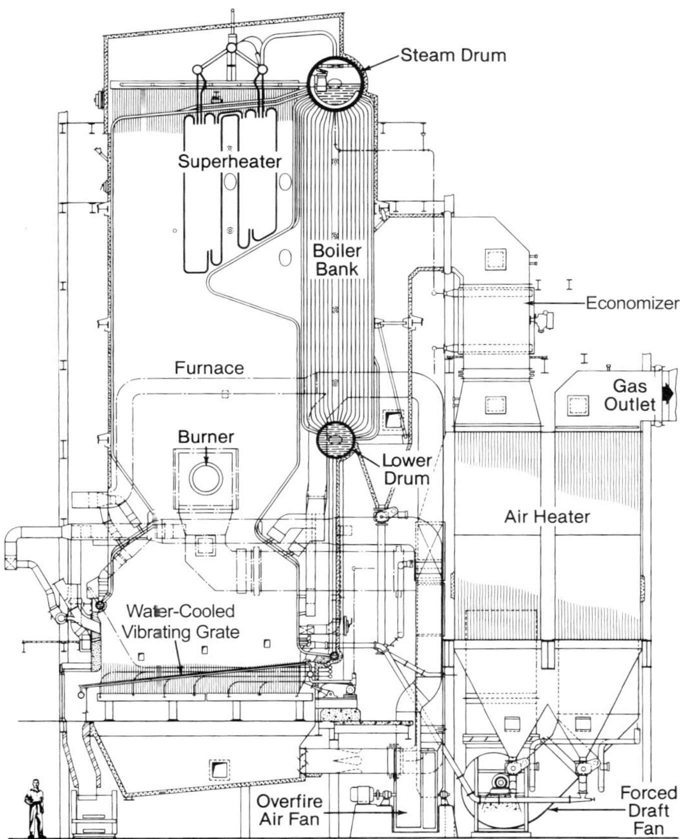

Fig. 9 Towerpak© boiler with longitudinal flow boiler bank.

Air heater

Due to the requirement to provide hot air to burn all but the driest of wood fuels, wood-fired boilers are usually equipped with air heaters. Because of the ash, sand and char in the flue gas, a recuperative type is normally selected, usually tubular. (See Chapter 20.) It is B&W practice to provide the tubes on square or rectangular pitch, rather than on triangular pitch, to lower air-side resistance and allow easier maintenance.

The common arrangement is for the flue gas to pass through the tubes and for the air to flow around the outside of the tubes. A two-gas pass design is favored for economic reasons. The low flue gas velocity in the hopper and the 180 deg change in direction also promote the separation of large, heavier particles of char and sand. The tubular air heater, when arranged for two gas passes, can act as a low efficiency (approximately 50%) mechanical dust collector.

Experience has shown the use of 2.5 in. (63.5 mm) outside diameter (OD) tubes to usually be the most cost effective. Available space limitations may require the use of smaller 2 in. (50.8 mm) OD tubes. In cases with a history of plugging by contaminants in the flue gas, it may be necessary to use 3 in. (76.2 mm) OD tubes.

Auxiliary equipment

Fans Wood-fired boilers require forced draft (FD), induced draft (ID) and usually overfire air (OFA) fans. (See Chapter 25.)

Forced draft fans require no special design considerations other than determining the required capacity and static pressure. The design may be determined by wood firing alone, wood firing in combination with auxiliary fuel, or auxiliary fuel alone, depending on the quantity and pressure of air required for each of these conditions. Normal test block margins for FD fans are sufficient. Usually these fans are controlled to maintain a constant pressure at the tubular air heater air outlet plenum.

ID fans must be designed to take into account the abrasiveness of the flue gas, the quantity of flue gas to be handled, the draft losses to be overcome, and the temperature of the flue gas. The abrasiveness of the flue gas depends on the type and efficiency of the dust collection equipment installed and whether the fan is located before or after this equipment. The ID fans are used to control the furnace pressure to a set point, usually at –0.1 to –0.5 in. wg (–0.025 to –0.12 kPa).

The quantity of gas and the draft loss used to specify the ID fan must take into account not only the expected operation of the boiler, but also possible wide variations in heating value, theoretical air, and moisture content of the fuel.

OFA fans are often exposed to severe service, as typically the air to be brought to 30 in. wg (7.5 kPa) has already been heated in the tubular air heater to as high as 650F (343C) before reaching the fan inlet. When specifying an OFA fan, it is particularly important to specify the maximum air temperature that it must be capable of handling.

Sootblowers Because biomass-fired boilers are susceptible to ash and carbon carryover, the convective heat transfer surfaces must be designed to accommodate sootblowers. Retractable sootblowers must be used for the superheater and high temperature boiler bank surfaces. Rotary sootblowers can be used for the low temperature boiler bank areas and the economizer, but retractable sootblowers are preferred, provided there is sufficient space. Traveling rake type sootblowers are typically used above the tubular air heater tubesheets on the flue gas side. (See Chapter 24.)

These sootblowers use either saturated or superheated steam to clean the gas lanes between the tubes. Blowing pressures are typically 150 to 250 psi (1034 to 1724 kPa) and the blowing sequence is usually initiated once per working shift. The high pressure jet of steam from the sootblower nozzle cleans the tube surfaces with a set radius around the sootblower lance depending on sootblower design [usually about 5 ft (1.52 m)].

The ash is typically nonsticky and is relatively easy to remove by sootblowing. If allowed to accumulate, the ash can plug gas passages, cause flow unbalances, affect boiler circulation and heat transfer, and ultimately lead to a forced outage of the boiler. If unburned carbon particles are allowed to accumulate, they can create a fire potential, particularly in the back-end equipment.

Fuel handling systems Biomass fuel systems can be quite complicated and maintenance-intensive due to the varying characteristics of the fuel. Typically, the fuel is continuously conveyed from storage to small surge bins at each fuel distributor that are kept full, with any overfeed returned to storage. The surge bins are equipped with variable speed screw feeders or chain feeders to control the rate of biomass fuel fed into the furnace.

These variable speed feeders must be capable of operation over a turndown range of four-to-one on automatic control. They must also be able to operate at very low speeds during startup conditions to build a fuel bed on the grate. The feeder drives must be of sufficient horsepower to allow the feeder to be started when the surge bin is full of fuel.

Upstream of the surge bins, it is common to have a large live-bottom storage bin with four to eight hours of biomass fuel inventory. This is to avoid interruptions in the fuel feed to the boiler when there are problems with the outside fuel handling equipment. (See Chapter 12.)

For many biomass fuels, special fuel preparation equipment may be needed to provide effective overall combustion. As an example, for the straw-fired unit shown in Fig. 1, the straw bale is carded or raked to loosen the straw before it is fed to the grate by screw feeders.

Ash handling Ash handling systems on biomass-fired boilers can be divided into two main areas, bottom ash and flyash.

Bottom ash is the ash that is raked or conveyed off the grate, plus the ash that falls through the grate bar holes into the undergrate hopper, called a riddlings or siftings hopper. Bottom ash consists mainly of sand and stones. The ash at the grate discharge is typically collected using a submerged drag chain conveyor with a dewatering incline at the discharge end. The siftings can be collected with drag chain or screw conveyors.

Flyash is the fine ash and unburned carbon that is collected from all the boiler bank, economizer, air heater and emissions control equipment hoppers. The ash handling equipment can be drag chains, screw conveyors or wet sluicing systems. Because the flyash contains a high percentage of hot carbon, it is important that rotary seal valves be used at each hopper discharge to prevent air infiltration that could create a fire in the hopper. For the same reason, all ash conveyors should be sealed.

In some instances, the flyash from the boiler bank and air heater hoppers is reinjected into the furnace to lower the unburned carbon loss and to reduce the quantity of material that must be disposed. However, the high maintenance requirements of these systems have limited their use.

Air systems

Air systems can be categorized as undergrate or underfire air and overfire air.

Undergrate air is typically low pressure [3 in. wg (0.75 kPa)] and, depending on the type of grate used, can be anywhere between 40 and 60% of the total air required for combustion. The purpose of undergrate air is to help dry the fuel, promote the release of the volatiles, provide the oxygen necessary for the combustion of the devolatilized char resting on the grate and, in the case of an air-cooled grate, cool the grate bars. The pinhole grate and vibrating grate can be provided with multiple undergrate air compartments with separate dampers for the operator to bias the undergrate air to the area of the furnace where the fuel is concentrated. Traveling grates can be provided with only one compartment per drive section.

OFA system capacities are varied and can range from 25 to 60% of the total air. Varying air port or nozzle sizes and air pressures are used to obtain adequate penetration of the air into the rising stream of volatiles from the grate. Typically, modern OFA nozzles are 3 to 6 in. (76.2 to 152.4 mm) in diameter or rectangular with velocity dampers (PrecisionJet™) and use air pressures up to 20 in. wg (4.98 kPa). Levels of nozzles are controlled independently such that the OFA can be varied with load and fuel characteristics. Where very high temperature OFA is used and it constitutes more than 40% of the total air flow, it is usually economical and energy efficient to provide a high pressure FD fan, rather than a low pressure FD fan plus a large high pressure OFA fan. The OFA ports can be designed to provide a double rotating circulating zone to further enhance combustion. (See Chapter 16.)

Emissions control equipment

Dust collector Mechanical dust collectors are used after the last heat trap on the boiler to collect the larger size flyash particulate, sometimes as protection for the ID fan. They typically consist of multi-cyclone tubes enclosed in a casing structure. The tubes consist of outer inlet tubes with spin vanes and inner tubes used without recovery vanes. The dust collector efficiency is in the range of 65 to 75% at an optimum draft loss of 2.5 to 3.0 in. wg (0.62 to 0.75 kPa). Due to the abrasive nature of the flyash, the outer collection tubes and cones are made of high hardness (450 Brinell) abrasion resistant material. (See Chapter 33.)

Precipitator Electrostatic precipitators are typically used after the mechanical collector to reduce the particulate concentration in the flue gas and to meet environmental requirements. Due to the high carbon content in the flyash, it is important to reduce the fire potential in the precipitator. It is necessary to ensure no tramp air enters the precipitator and that the flyash is continuously removed from the hoppers. Hopper level detectors and temperature detectors alert the operator.

Installations can be equipped with fire fighting apparatus such as steam inerting. Other suppliers recommend de-energizing the precipitator if a predetermined oxygen content in the flue gas is exceeded.

Fabric filter or baghouse Due to the fire potential, baghouse collectors have been rarely used for biomass fuels to date.

Wet scrubbers Wet scrubbers have also been used to control particulate emissions on biomass-fired boilers. Their main disadvantages are the high flue gas pressure drop, which increases the ID fan horsepower requirements, and high water consumption. Also, there is the need for a wet ash collection system and a water separation and clarification system. Wet scrubbers have given way to electrostatic precipitators as the preferred means of final flue gas cleanup, provided there is no need for a scrubber to reduce SO_2 emissions from auxiliary fuels.

Wet scrubbers with numerous small spray nozzles in a chamber can be used where low pressure drop and low water consumption are required. This is particularly well suited to retrofit applications where the scrubber can replace the mechanical dust collector for improved collection efficiency.

Environmental impact

Particulate emissions

About 80 to 95% of the total ash residue produced by a bark- and wood-fired spreader stoker is in the form of gas-borne particulate. This particulate is composed of a number of materials, including ash, sand contaminants introduced during fuel handling, unburned char from the furnace, and salt fume (usually present only where logs are seawater flumed).

The ash content of wood and bark fuels is low (0.2 to 5.3%, dry basis). Therefore, if fuel ash contaminants are not appreciable, the particulate will usually contain high percentages of unburned char.

The particulate loading in the flue gas exiting the boiler is influenced by factors related to both combustion and aerodynamics. Combustion-related factors affect particulates by determining the degree of burnout for the entrained char. They include plan and volumetric heat release rate (two design parameters that affect furnace temperature), residence time, and consequently, char burnout.

The importance of aerodynamic factors is based on the fact that bark- and wood-fired spreader-stoker units are designed to operate with some degree of suspension burning. Variables that would tend to increase the ratio of furnace velocity to mean char particle size would therefore tend to increase flue gas particulate loading.

Some of these factors include fuel moisture and fines content, boiler plan area, air staging, and excess air level.

For wood and bark fuels not containing appreciable quantities of sand, particulate loading at the air heater exit of a modern spreader-stoker unit would typically be in the range of 1 to 3 grains/DSCF (2.4 to 7.2 g/Nm^3).

Nitrogen oxides

NO_x emissions from wood and bark firing are low compared with those from traditional fossil fuels. Combustion temperatures in wood firing are sufficiently low, and little thermal NO_x is formed from the nitrogen in the combustion air. Corresponding NO_x emissions are therefore predominantly a function of the fuel nitrogen content. (See also Chapter 34.)

Conversion of fuel bound nitrogen to NO_x is dependent on a number of operating conditions including excess air, air staging, heat release rate, and fuel sizing and moisture content. Empirical studies have also found NO_x to vary inversely with fuel moisture content, although the magnitude of this correlation is less significant.

All factors considered, NO_x emissions from stoker firing of most wood and bark vary between 0.1 and 0.35 lb/10^6 Btu (0.04 to 0.15 g/MJ) heat input, expressed as nitrogen dioxide. Fuel contaminants that may introduce nitrogen compounds (glues and chemicals, for example) should receive special consideration.

Sulfur dioxide

Wood and bark typically contain 0.0 to 0.1% elemental sulfur on a dry basis. During the combustion process some of this sulfur can be converted to flue gas SO_2, but the conversion ratio is typically low (10 to 30%). Because the quantities of both wood sulfur and flue gas SO_2 are near the low end detection limit of corresponding analytical instruments, correlation of the two is not practical.

Typically, SO_2 emissions for stoker-fired wood and bark fuels do not exceed 0.03 lb/10^6 Btu (0.01 g/MJ) heat input. Special consideration should be given to fuels where sulfur bearing contaminants may be present.

Carbon monoxide

Of all emissions commonly associated with wood and bark firing, carbon monoxide (CO) is usually the most variable. As a gaseous product of incomplete combustion, CO is dependent on time, temperature and turbulence considerations.

At normal excess air levels, consistency of both fuel heating value and fuel distribution are considered the most important determinants of CO emissions. Typically, test data showing the highest standard deviation in CO correspond to the highest mean CO.

Conditions of high excess air, low excess air, high fuel moisture and reduced load (<70% of the maximum continuous rating) have all been demonstrated to increase flue gas CO concentration.

Modern spreader-stoker units firing wood and bark only, and operating at steady-state conditions, typically emit CO in the range of 0.05 to 0.5 lb/10^6 Btu (0.04 to 0.22 g/MJ) heat input.

Volatile organic compounds

Volatile organic compounds (VOCs) are also gaseous products of incomplete combustion. As such, the emission of VOCs during wood firing is influenced by the same factors affecting CO.

Typically, VOC emissions while stoker-firing wood and bark fuels do not exceed 0.05 lb/10^6 Btu (0.02 g/MJ) heat input, expressed as methane.

Naval steam power then and now: U.S.S. *Missouri* (top) and U.S.S. *John F. Kennedy*.

Chapter 31

Marine Applications

The ingenuity of many inventors and engineers has been devoted to the application of fossil fuel-fired steam boilers and engines to ship propulsion. The need for higher power, lower weight and smaller volume designs for ships drove many of the boiler system advances that ultimately appeared in stationary boilers designed and used today.

The primary propulsion systems in ships today, however, have progressed to diesel engines, gas turbines and even nuclear power systems in naval applications. (See Chapter 46.) Nevertheless, there remain requirements aboard new ships to produce steam for a variety of needs. In addition, a large existing population of contemporary ships, such as the LASH (lighter aboard ship) class container ships (Fig. 1), continues to use steam boilers as their primary source of power. Finally, increasing fuel costs have led to investigations of burning coal-water mixtures, coal-oil mixtures and other fuels in steam-based systems for use in the future.

The Babcock & Wilcox Company (B&W) marine boilers have established a reputation for dependability and efficiency. The history of these boilers reflects sound principles of design and fabrication. Since the first installation in the S.S. *Monroe* in 1875, B&W boilers have been installed in more than 4000 ships of the United States (U.S.) Navy and the Merchant Marine.

During World War I, B&W boilers were installed in 50 mine sweepers, 100 destroyers, and 500 Shipping Board vessels. During World War II, B&W boilers were furnished to the Navy for all but two battleships (such as the U.S.S. *Missouri*, shown in the chapter frontispiece), all the cruisers, all aircraft carriers (see U.S.S. *John F. Kennedy* in the chapter frontispiece), 90% of the destroyers, 33% of the destroyer escorts and numerous miscellaneous small craft.

General design considerations

Steam generating equipment for marine service is designed in accordance with the principles and considerations that apply to stationary units with modifications to meet specific requirements for operation at sea. The boiler must fit within a minimum engine room space, yet be accessible for operation, inspection and maintenance. Although lightweight, it must be sufficiently rugged to operate dependably under adverse sea conditions and to absorb vibration and forces resulting from rolling and pitching in heavy seas or from shocks that may result from accidental causes such as groundings and collisions. Special steam drum design requirements must be met to accommodate *roll* of the ship from side to side, *pitch* variations from bow up to bow down, permanent *list* to either side and permanent *trim* to either bow or stern. In the case of naval vessels, the shock effects of the detonation of explosives must also be considered. Operation over a wide load range with operating characteristics compatible with a high degree of automation is also required. Finally, the factors used in both the thermal and structural design must be conservative so that continuous operation over extended periods of time will be provided with minimum maintenance. These factors combined with the use of several different fuels have led to a variety of B&W marine boiler designs.

Integral Furnace naval boiler

The Integral Furnace naval type boiler (Fig. 2) is fitted with welded high strength, low weight stainless or low-alloy steel casings. Furnace roof, side and rear walls are water-cooled by 2 in. (50.8 mm) outside diameter (OD) close-spaced tubes. Front (burner) wall and floor are usually refractory, but some recent boilers have water-cooled front walls. Steam drum diameters range from 54 to 60 in. (1.37 to 1.52 m) and water drums vary from 27 to 36 in. (0.69 to 0.91 m). Two to four rows of 2 in. (50.8 mm) OD tubes form the superheater screen. The superheater has 1.00 or 1.25 in. (25.4 or 31.8 mm) OD tubes arranged in a maximum of 8 rows deep with provision for complete drainability. The boiler bank is usually inclined and

Fig. 1 S.S. *President Tyler*, LASH Class container ship (*courtesy of American President Companies, Ltd.*).

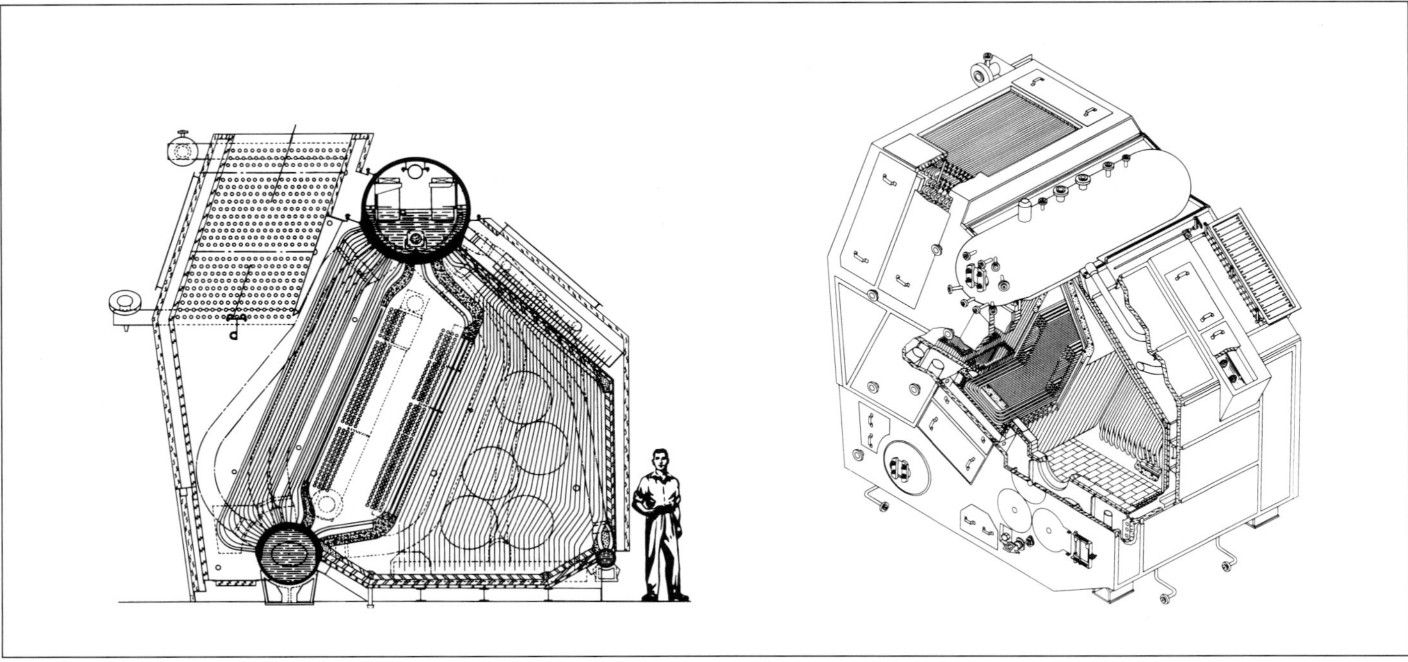

Fig. 2 Two-drum boiler, naval type.

has from 18 to 23 rows of 1.00 in. (25.4 mm) OD tubes. For maximum efficiency, the boiler is fitted with a stud-tube economizer and in some cases, steam air heaters.

Range in size, steam output:
 To 350,000 lb/h (44.1 kg/s) in no fixed increments
Operating pressure:
 Up to 1200 psig (8.27 MPa)
Steam temperature:
 Saturation to 1000F (538C)
Fuel:
 Navy special fuel oil (residual)
 Multipurpose fuel oil
Operational control:
 Manual to complete automatic, combustion and feedwater regulation.
Draft loss at maximum output:
 Up to about 75 in. wg (18.7 kPa) total through all components
Dimensions outside setting:
 Smallest – 14 ft 2 in. high × 15 ft 1 in. wide × 10 ft 6 in. deep (4.32 m × 4.60 m × 3.20 m)
 Largest – 21 ft 5 in. high × 18 ft 11 in. wide × 16 ft deep (6.53 m × 5.77 m × 4.88 m)

Indicated field of application

The primary use of these boilers is for combat or auxiliary naval vessels or special installations.

General comments

The nature of the construction and the rating at which these boilers operate limit their use to naval or high speed naval auxiliary vessels requiring maximum power in minimum space. They are designed for maximum efficiency at cruising speed, and some attainable efficiency is sacrificed to develop high power-to-weight and power-to-boiler volume ratios. The construction includes lightweight tubes, high-tensile drum plates, and other special features to minimize weight.

Integral Furnace merchant boiler

The Integral Furnace merchant boiler became the marine industry standard after World War II. The boilers were originally designed as tangent tube (Figs. 3 and 4) and refractory construction (water-cooled furnace) and more recently, membrane furnace construction (water-cooled, welded-wall furnace). Both types have a single gas pass with a 2 row, 2 in. (50.8 mm) OD screen before the superheater, steam drum diameters ranging from 54 to 72 in. (1.37 to 1.83 m) and water drums from 30 to 36 in. (0.76 to 0.91 m).

The tangent tube furnace is water-cooled by closely spaced 2 in. (50.8 mm) OD tubes on the side, roof, front and rear walls. Recent designs have sloped and bare furnace floor tubes with consequent reduction in exposed refractory. In designs with refractory floors, the tubes are usually buried in the floor to reduce refractory temperature and increase refractory life.

The membrane furnace is water-cooled by welded-wall construction consisting of 2.75 in. (69.9 mm) OD tubes on 3-9/16 in. (90.5 mm) centers. This applies to the furnace floor, front, rear and side walls. The floor is covered with firebrick to protect the tubes from overheating.

In either style boiler, an inclined or vertical boiler tube bank, composed of 2 to 4 rows of 2 in. (50.8 mm) OD screen tubes and 17 to 24 rows of 1.25 in. (31.8 mm) OD generating tubes, may be used. The superheater consists of either horizontal or vertical 1.25 in. (31.8 mm) OD tubes and has one or two access cavities to facilitate water washing, cleaning, inspection, and maintenance of the superheater and boiler bank. A single cavity is normally provided when the fouling characteristics of Bunker C oil are average, and two cavities are used when fouling characteristics are more severe. The boiler is generally equipped with steam atomizing burners, retractable sootblowers, and an air heater or an economizer.

Indicated field of application

These designs are primarily used for propulsion power, although they are also used without a superheater for auxiliary or heating service in the smaller sizes where space is at a premium, a compact lightweight design is needed, and feedwater is of good quality.

Range in size, steam output:
  To 400,000 lb/h (50.4 kg/s) in no fixed increments
Operating pressure:
  up to 1200 psig (0.69 to 8.27 MPa)
Steam temperature:
  Saturation to 1000F (538C)
Fuel:
  Oil (light fractions to heavy residuals) and liquified natural gas (LNG)
Operational control:
  Manual or completely automatic, including feedwater flow, combustion and steam temperature controls
Draft loss at maximum output:
  From 15 to 30 in. wg (3.7 to 7.5 kPa) total through all components
Dimensions outside setting:
  Smallest – 7 ft 8 in. high × 5 ft 5 in. wide × 5 ft 6 in. deep (2.34 m × 1.65 m × 1.68 m)
  Largest – 41 ft 6 in. high × 34 ft 3 in. wide × 24 ft 1 in. deep (12.65 m × 10.44 m × 7.34 m)

General comments

Designs of this type are suitable for vessels where large power plants must be installed in a minimum of space, and where weight saving is a vital consideration. Many possible variations in configuration permit application under limited space conditions. Mass action retractable sootblowers are recommended for the superheater zones.

Reheat boilers

Although reheat boilers have not been extensively used in the marine industry, there have been several applications. In considering a reheat boiler for shipboard applications, one design feature must be considered. That feature is the protection of the reheater tubes when the main propulsion turbine is in astern operation or stopped. B&W has designed the boiler with damper and bypass systems, as well as with extensive screen tube protection systems in front of the reheater.

The marine reheat boiler is of the two-drum divided inclined boiler bank type, capable of burning fuel oil or natural gas. Other than the reheater, reheat screen, baffle wall and damper system, the reheat boiler is similar to the tangent tube boiler.

Coal-fired marine boilers

Up until the late 1970s, coal-fired marine boilers put into service were for vessels on the Great Lakes.

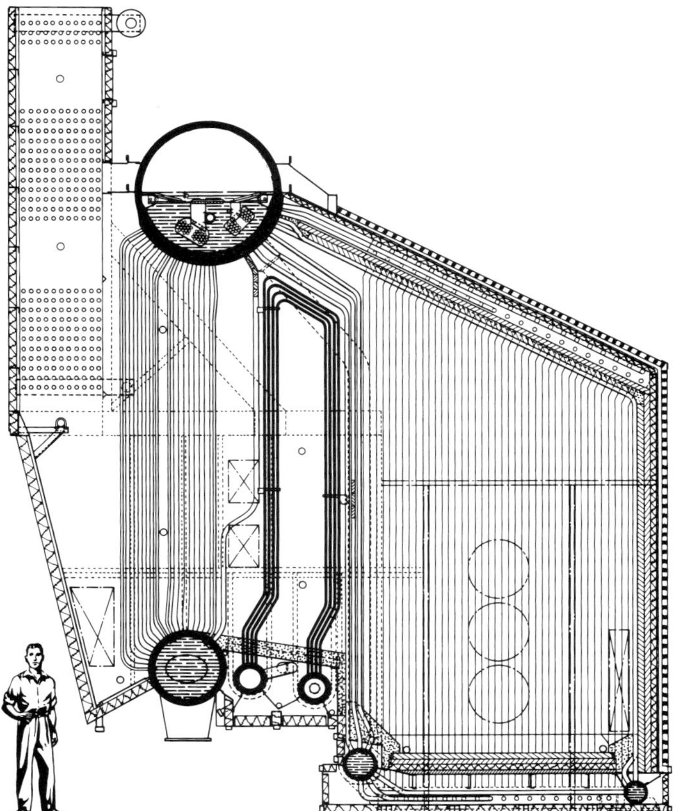

Fig. 3 Two-drum Integral Furnace boiler with vertical superheater.

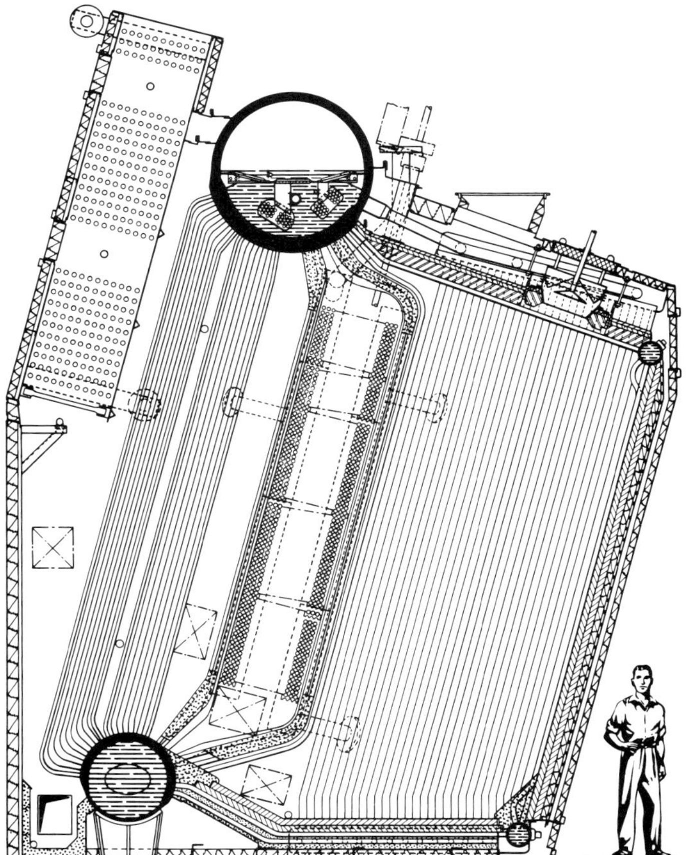

Fig. 4 Two-drum Integral Furnace boiler with horizontal superheater.

These units were built in the late 1950s, and were for relatively low-pressure and low-temperature conditions. The boiler designs were either the sectional header type or the two-drum type. This two-drum boiler had the same physical appearance as many of the newer oil-fired two-drum boilers. However, because coal firing requires a much lower furnace heat release rate than oil firing for proper combustion, the result was less steam generating capacity for the same shipboard space availability.

As with a land-based unit, boiler design begins with the stoker. The maximum stoker grate release rate for good operation and for prevention of grate overloading is 750,000 Btu/h ft^2 (2.37 MW$_t$ /m^2) of grate surface. Another requirement for good stoker operation, particularly when flyash and unburned carbon are re-injected into the furnace, is a limit of input to the stoker of 13 × 10^6 Btu/h ft (12.5 MW$_t$ /m) of stoker width. Exceeding this limit can cause poor fuel distribution on the grate, resulting in uneven burning.

The furnace volume should be set so that the furnace liberation does not exceed 30,000 Btu/h ft^3 (0.31 MW$_t$ /m^3). Adherence to this value will ensure sufficient residence time in the furnace to properly burn the fuel and minimize slagging and fouling.

The coal-fired marine boiler has a totally different appearance than that of the customary oil-fired marine boiler. The furnace extends considerably below the centerline of the lower drum. This dimension, commonly called the setting height of the furnace, is the dimension from the top of the stoker to the centerline of the water drum. Generally, on units of this capacity, setting heights range from 16 to 20 ft (4.9 to 6.1 m).

This boiler has been specifically designed to make full power by either burning 100% coal or 100% oil. When burning coal and with a grate release rate of 750,000 Btu/h ft^2 (2.37 MW$_t$ /m^2), the approximate turndown ratio of the stoker will be 3:1. However, with the oil burners, the turndown ratio can be as high as 16:1. Therefore, if required during maneuvering modes of operation, the boiler can be readily fired by oil and have the full flexibility and response time needed for this condition. The design operating conditions are 875 psig (6.03 MPa) and 900F (482C) with steam capacity ranging from 60,000 to 150,000 lb/h (7.6 to 18.9 kg/s).

Additional features of this design address other impacts of coal and its associated ash loading. Retractable sootblowers, hoppers, dust collection equipment and flyash re-injection are all included.

Auxiliary package marine boilers

As the marine industry has shifted its emphasis from main steam propulsion systems to diesel propulsion, the need developed for a simple, low-cost, self-contained auxiliary boiler.

B&W developed an Integral Furnace, two-drum D-type package boiler (FMB) that could operate as a stand-alone unit or in conjunction with a diesel or gas turbine exhaust waste heat boiler.

The boiler in Fig. 5 represents the marine package design conforming to the requirements of U.S. Coast Guard (USCG) and American Bureau of Shipping (ABS) Rules. The unit is shop-assembled, including furnace and firing equipment. It is not intended for ship propulsion, but as a means of supplying auxiliary steam.

The boiler incorporates a compact vertical bank of 2 in. (50.8 mm) OD tubes arranged between and terminating in an upper and lower drum. The tube ends are expanded into tube seats that are grooved to obtain maximum tightness.

A 54 in. (1.37 m) nominal diameter steam drum is required to promote water level stability even during wide variations in load, and helps to ensure dry, saturated steam. The requirements for this drum include satisfactory operation under the following conditions:

30 degree roll to each side for a 15 second period,
10 degree pitch from bow up to bow down for a 6 second period,
15 degree permanent list to either side, and
5 degree permanent trim to either the bow or stern.

Under any of these adverse conditions, the downcomers and boiler bank generating tubes will not uncover.

The furnace and boiler are covered by a 20 gauge (0.9 mm) nonpressurized galvanized steel ribbed lagging except in areas such as the steam drum joints and bent tube portions of the walls, which are covered by a 12 gauge (2.7 mm) carbon steel casing.

The boiler furnace unit is arranged for upward expansion and is mounted on a welded structural base frame, ready for positioning on the ship's foundations. The unit has a furnace with a refractory covered flat floor, and a boiler bank. The furnace side wall, roof, floor and rear wall are completely water-cooled and form an integral part of the boiler circuitry. One burner is used in the refractory front wall.

Water-cooled furnace wall tubes, comprising the side wall, roof and floor, are of membrane construction. The rear furnace wall is 2 in. (50.8 mm) OD flat-studded tubes on 4 in. (101.6 mm) centers. A 10 gauge (3.4 mm) steel inner casing backs the flat-studded wall. The furnace wall tubes receive their flow from the lower drum

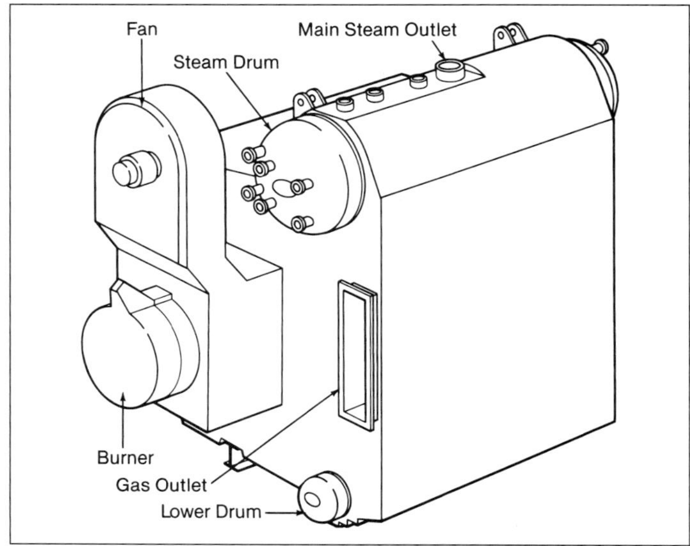

Fig. 5 Auxiliary package marine boiler.

and discharge the steam-water mixture into the steam drum. The 2.75 in. (69.9 mm) OD furnace, roof and floor tubes are a part of the furnace side wall circuitry. The furnace front wall (burner wall) is of refractory anchored construction and is backed by insulation and a gas-tight casing.

These marine auxiliary package boilers are designed to provide unattended operation with controls that are fully automatic after the boiler is on line.

Range in size, steam output:
To 100,000 lb/h (12.6 kg/s) in no fixed increments
Operating pressure:
225 psig (1.55 MPa)
Steam temperature:
Saturation only
Fuel:
Oil, light fractions to heavy residuals
Dimensions outside setting:
Overall width is 12 ft 6.25 in. (3.82 m), overall height (base to face of steam outlet flange) is 16 ft 9.75 in. (5.12 m) with drum centers at 12 ft 10 in. (3.91 m). These dimensions are constant for all boiler sizes. The length is variable to establish the various boiler sizes and ranges from about 12 to 18 ft (3.66 to 5.49 m).

Waste heat marine boilers

With the use of diesel or gas turbine propulsion systems, such as aboard the LaJolla class ship, shown in Fig. 6, it is often advantageous to incorporate a waste heat (exhaust gas) boiler system that, in addition to an auxiliary boiler, supplies saturated steam and improves overall plant efficiency.

The waste heat boiler system is composed of a forced circulation, spiral wound, finned tube or studded tube boiler (Fig. 7); auxiliary steam drum; and control equipment.

The waste heat boiler unit is arranged for upward expansion and is ready for positioning on the ship's foundation. It is designed to operate under either a wet or dry condition and can serve as a muffler or sound reducer whether filled with water or in a dry state.

The horizontal waste heat boiler uses the B&W continuous stud tube extended surface or a spiral wound, finned type surface. The elements including the inlet and outlet headers are located inside a welded gas-tight casing, arranged for bottom support. This boiler consists of 1.5 in. (38.1 mm) OD steel tubes and studs or fins.

The boiler elements, headers, casing, integral support steel and framing angles are shop-assembled. The tubes are supported by tube support plates. The unit is single-cased and of welded construction. The casing provides a gas-tight seal and is fitted with access panels for inspection purposes. Support of the exhaust gas boiler is on the ship's structural members.

A 36 or 42 in. (0.91 or 1.07 m) nominal diameter auxiliary steam drum is provided to meet anticipated list and trip operating conditions as well as to promote water level stability. The effect of shrink and swell is controlled by forced circulation with a minimum 4:1 circulation ratio, cyclone steam separators and the large diameter steam drum.

Fig. 6 M.V. *Potomac Trader*, LaJolla Class vessel with diesel propulsion plus waste heat and auxiliary steam boilers (*courtesy Penn-Attransco Corp.*).

One mode of operation has an auxiliary boiler operating in conjunction with the waste heat boiler system. (See Fig. 8.) The main diesel engine exhaust gas is directed to the waste heat boiler. The gas then enters the boiler and crossflows over the in-line extended surface to the waste heat boiler exit where it is then sent to the stack.

The auxiliary boiler feed pump supplies condensate to the auxiliary boiler steam drum. The circulating pumps take suction from the auxiliary boiler lower drum and feed the waste heat boiler. Flow is upward through the waste heat boiler to its outlet where the resulting steam-water mixture is conducted through piping to the auxiliary boiler's steam drum. All steam for ship services is drawn through the auxiliary boiler's cyclone steam separators so that the auxiliary boiler acts as the steam drum of the exhaust gas boiler. The auxiliary boiler will fire automatically to maintain the steam pressure within an adjustable band.

The alternate mode of operation for this system, with the auxiliary boiler, has the boiler feed pump supply condensate to the auxiliary steam drum. The circulating pumps take suction from the auxiliary steam

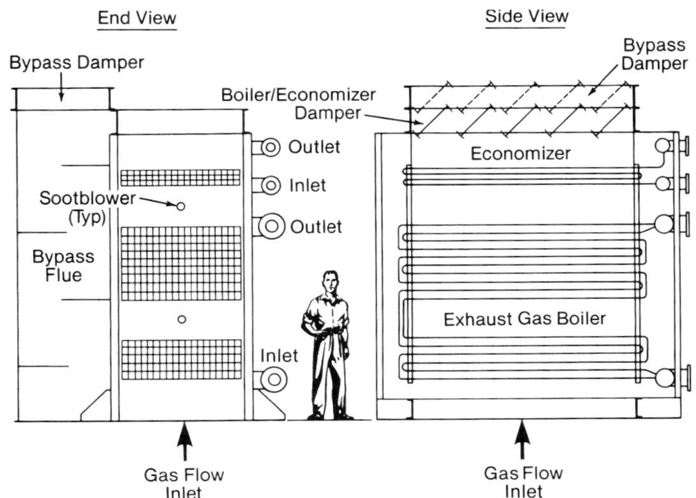

Fig. 7 Waste heat marine boiler.

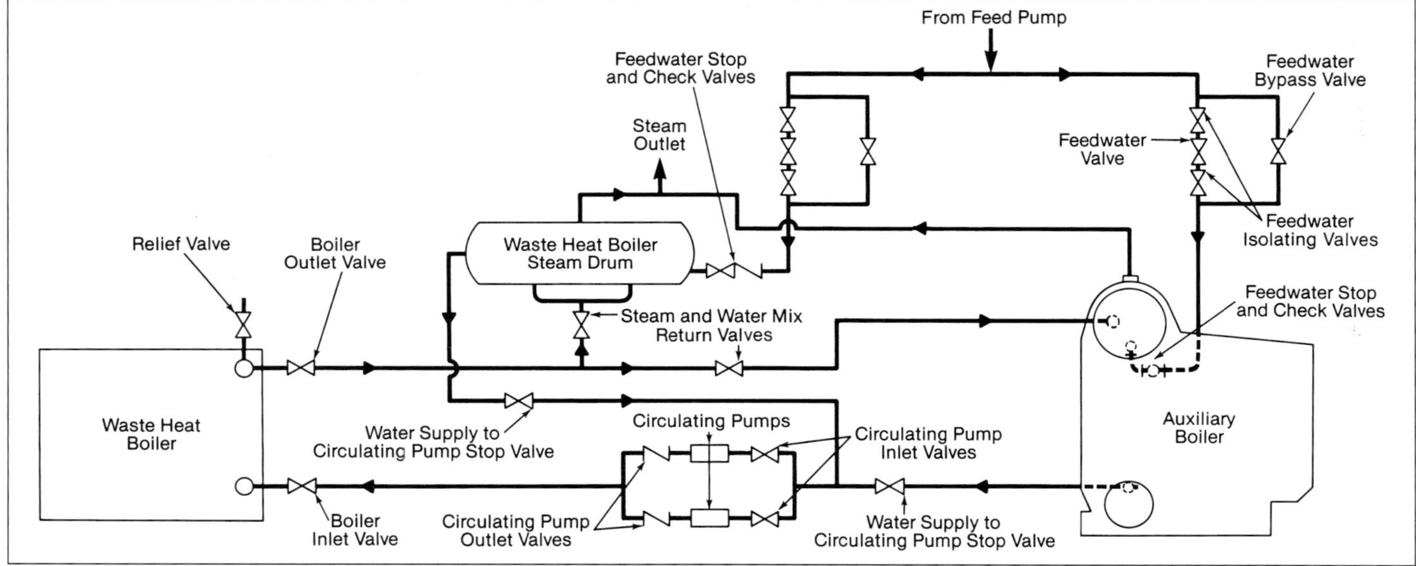

Fig. 8 Shipboard steam system combining auxiliary boiler and waste heat boiler.

drum and feed the waste heat boiler. Flow is upward through the boiler to its outlet, where the resulting steam-water mixture is conducted through piping to the auxiliary steam drum. The auxiliary boiler may be isolated and unfired in this condition for maintenance or to accommodate steam demand. It may also be fired automatically to maintain the steam pressure within an adjustable band.

Marine boiler design

The techniques used in the design of marine boilers are similar to those applied to stationary units (Chapters 19, 20 and 21). However, the design of marine boilers is also subject to constraints and requirements that are specific to marine plants. These include space conditions, fireroom configuration, list and trim, pitch and roll, and ship maneuverability. In addition to the codes and standards used for stationary boilers (i.e., ASME, ANSI), other codes apply. Examples include the United States Coast Guard Marine Engineering Regulations, subchapter F and other subchapters, American Bureau of Shipping (ABS) Rules, and Lloyd's Register of Shipping Rules.

Design procedures and guidelines for marine boilers are generally set in the Society of Naval Architects and Marine Engineers (SNAME) Technical Bulletin No. 3-32 (*Furnace Performance Criteria for Gas, Oil and Coal Fired Boilers*, 1981) and No. 3-11 (*Marine Steam Plant Heat Balance Practices*).

Fuel characteristics can have considerable bearing on the design. Regardless of whether oil or coal is used, the chemical analysis and ash characteristics are particularly important and should be specified so that heat transfer surface can be arranged to minimize corrosion and deposition from ash and slag.

Boiler design for a particular application starts by establishing the dimensions and arrangement of the furnace. Within this volume the desired fuel-burning equipment can be installed and the fuel burned efficiently, without exceeding allowable furnace heat absorption rates. This is a major concern where small furnaces are needed to fit into restricted shipboard engine room space.

The furnace arrangement and dimensions generally establish the overall size of the boiler bank. Therefore, the first step in designing a new boiler is to prepare a preliminary layout of a suitable furnace. To do this, the amount of fuel to be fired is calculated based on the required steam flow, pressure and temperature, as well as on feedwater temperature and efficiency. It is then possible to estimate the type and number of burners to be used, the allowable pressure drop through the burners for good combustion conditions, the size and shape of the furnace to properly accommodate the burners for good combustion, and the allowable furnace heat absorption rate.

Oil burners

The type and number of oil burners selected depend on the fuel rate and the allowable air resistance. The number is usually not less than two, so that at least one burner operates at all times, even when changing sprayer plates or cleaning atomizers of the idle burners. Many types of oil burners are available. The burners described in Chapter 11 are for stationary boilers and are for burning natural gas and oil, as well as other gaseous type fuels or mixtures.

Marine burners are generally designed solely for burning oil. Since World War II, three types or sizes of registers were applied to most merchant marine and naval vessels. They were the Iowa, Saratoga and Progress types and covered oil flow rates up to 7000 lb/h (0.9 kg/s) or 130×10^6 Btu/h (38.1 MW_t) input per register with excess air of 10 to 20%.

With the pursuit of more efficient steam propulsion systems in the 1970s, B&W developed a low excess air burner (5 to 10%) to be used on merchant ships. These burners were designed to burn natural gas or oil up to 130×10^6 Btu/h (38.1 MW_t) input. The burners are available in throat sizes from 15 to 30 in. (381 to 762 mm) (Fig. 9) and have pressure drops of 8 in. wg (2 kPa) at normal operating conditions.

Prior to 1970, most steam ships used a mechanical atomizer. However, B&W developed and introduced the wide range Racer® steam atomizers or sprayer plate. This steam atomizing sprayer plate provided coverage over the full range of operation. The properly sized and selected sprayer plate will cover the range from minimum to the designed maximum overload firing rate as long as atomizing steam is available. This steam atomizer and burner are capable of operating at extremely low firing rates without loss of good combustion or ignition as long as proper fuel oil temperature and proper air flow (air/oil ratio) are maintained.

When maneuvering, or at any time during port operation or normal operation at sea, it should not be necessary to secure burners unless the steam demand is less than that produced with all burners in service at the minimum fuel oil pressure. During normal operation, all burners should be kept in use for all rates of operation.

For all fuel oil pressures, the atomizing steam pressure should be constant and should be maintained at 135 psig (0.93 MPa) at the burner. Due to the pressure drop between the atomizing steam header and the burner, this is equivalent to a steam pressure of approximately 150 psig (1.03 MPa) at the atomizing steam header.

When burning heavy residual fuel oil, certain minimum clearances are required around the burners to prevent flame impingement and carbon deposits. To assure complete combustion of the fuel, the furnace depth is usually limited to a minimum of 6 ft (1.83 m), although both diesel and bunker oils have been burned successfully in furnaces 5 ft (1.52 m) or less in depth.

Furnace heat absorption

In the marine field, the heat release rate per unit furnace volume is frequently used for comparing boilers without regard to similarity of design. This ratio is not an important design criterion as it provides only an approximation of the time required for the products of combustion to pass through the furnace. However, it may be used to indicate the operating range for which the firing equipment is to maintain satisfactory combustion conditions. In naval vessels, the availability of suitable oil burners has permitted the installation of high capacity lightweight boiler units with heat released at a rate of 200,000 Btu/h ft^3 (2.07 MW$_t$ /m^3) of furnace volume at cruising conditions and more than 1,000,000 Btu/h ft^3 (10.3 MW$_t$ /m^3) at the maximum evaporative condition with satisfactory results. For merchant ships, arbitrary limits of 75,000 to 90,000 Btu/h ft^3 (0.77 to 0.93 MW$_t$ /m^3) at the normal rate of operation are commonly specified.

A more meaningful criterion of furnace design is the heat absorbed by the cold surfaces of the furnace, expressed as the amount of heat absorbed per square feet of radiant heat-absorbing surface as discussed in Chapter 19. The considerations described also apply to marine units, with particular emphasis on the determination of the furnace exit gas temperature to assure satisfactory superheater design and to establish the heat absorption by the furnace wall tubes for adequate circulation margins.

Marine boilers operate with higher heat input rates per square foot of boiler and superheater surfaces than stationary boilers designed for the same steam outputs. They have less heating surface and smaller overall dimensions. Flame temperatures more closely approach the adiabatic temperature, and furnace exit gas temperatures are considerably higher than those encountered in stationary boilers. Fig. 10a indicates the effect of excess air on furnace temperature.

The effectiveness of the water-cooled surface in the furnace is determined by applying the factors shown in Chapter 4 to the flat projected areas of the furnace walls and tube banks facing the furnace. Bare tangent tubes or membrane tubes are used wherever possible in furnace waterwalls. Where tubes must be spaced to facilitate replacement, it is preferable to limit the width of the exposed refractory areas between tubes to 1 in. (25.4 mm) or less to avoid wastage resulting from sodium compounds in the oil ash. The effect of wall tube spacing, or the ratio of water-cooled surface to refractory surface, on furnace heat absorption is shown in Fig. 10b. The heat absorption rate in the furnace increases with increased firing rate. However, furnace absorption as a percentage of the total boiler absorption decreases with increased firing rate as indicated in Fig. 10c.

Furnace tube temperatures

In all boiler tubes and particularly in tubes exposed to the high heat absorption rates of marine furnaces, adequate circulation must be provided to avoid critical heat flux (CHF) or departure from nucleate boiling (DNB) as discussed in Chapter 5. This is generally accomplished by empirical data and methods based on tests and operating experience. Tube wall thickness is usually set close to the minimum required for the design pressure, because weight is a primary consideration and excess thickness increases external tube temperature. The heat transfer coefficient across the boiling water film in the furnace steam generating tubes can be as high as 20,000 Btu/h ft^2 F (114 kW/m^2 K). However, in estimating tube temperature, a conservative value may be used, e.g., 2000 Btu/h ft^2 F (11.4 kW/m^2 K), to obtain a somewhat higher estimated tube temperature, resulting in lower allowable stress in the tube metal and therefore a somewhat more conservative tube thickness.

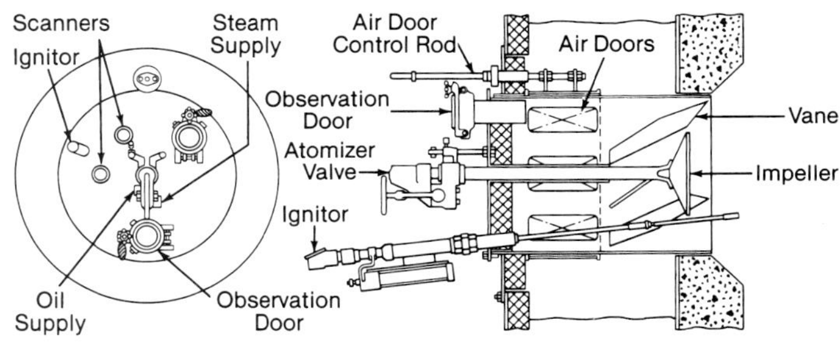

Fig. 9 B&W Progress-type oil burner with Racer® steam atomizer.

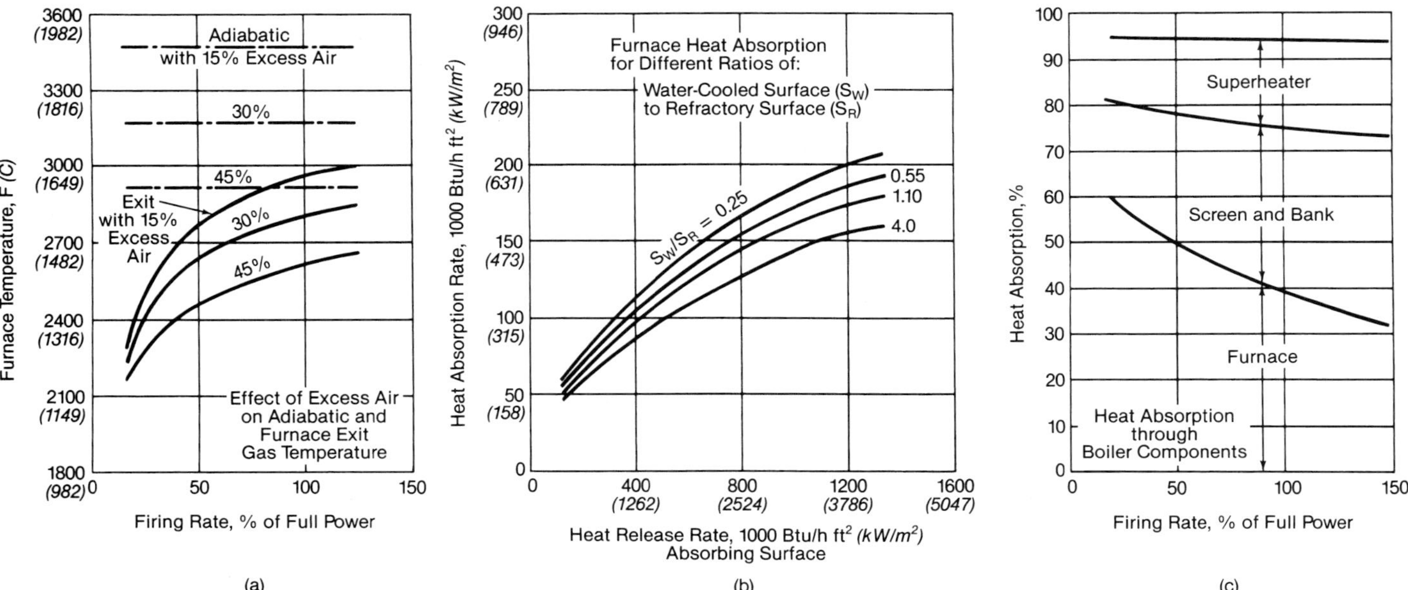

Fig. 10 General effect of excess air (a), ratio of heat-absorbing to refractory surface (b), and firing rate (c) on furnace heat absorption and temperature based on 18,500 Btu/lb (43 MJ/kg) fuel oil.

Scale deposits on the water side of boiler tubes are a long recognized cause of tube failure. These deposits can be particularly serious in the furnace tubes of marine units because of the high heat absorption rates. As an example, a calcium sulfate scale deposit on the inside of a tube with a thickness of only 0.024 in. (0.61 mm) and a thermal conductivity of 0.83 Btu/h ft F (1.44 W/m K) results in a 362F (201C) temperature differential across the scale. For a boiler operating at 665 psig (4.59 MPa), this scale would increase the tube outside metal temperature to 1004F (540C). This temperature exceeds the oxidation limit for steel, and oxidation and ultimate tube burnout will likely occur, even though boiler circulation may be adequate. Scale must be avoided by proper water conditioning. (See Chapter 42.)

Boiler tube banks

The boiler bank, composed of multiple rows of tubes where steam is generated, generally consists of a screen ahead of the superheater and a convection bank behind it. Both sections generate saturated steam but they are considered separately during the design stage. The screen tubes, by virtue of their location, absorb heat at a considerably higher rate than the main bank tubes. Consequently, to assure an adequate flow of water, the furnace screen tubes should be larger in diameter. For this reason, marine boilers are designed with two sizes of steam generating tubes. The diameters of screen tubes are generally 2 in. (50.8 mm) while those of the other generating tubes are 1, 1.25 or 2 in. (25.4, 31.8 or 50.8 mm), swaged to 1.5 in. (38.1 mm) at the drums. Circulation and the amount of heating surface required to obtain the desired gas temperature leaving the tube bank are the major factors in determining the tube size and the number of tube rows to be installed, although resistance to gas flow is also a factor.

In the design of the superheater screen, consideration must be given to the effect of tube spacing and the number of rows, the desired superheater outlet temperature and the maximum allowable superheater tube metal temperatures. The screen must be designed so that the gas temperature entering the superheater and the radiant heat penetration from the furnace will provide the desired steam temperature with a superheater of reasonable size and arrangement. Fig. 11 shows the effect of radiant heat penetration on the performance of the superheater over the designed load range of the boiler. The number of rows in the boiler bank is usually established by analyzing the economic advantages of economizers and air heaters as compared with boiler surface. The proper proportions of boiler surface and additional heat-absorbing surface beyond the boiler bank are based on the efficiency desired.

Superheaters

Superheaters of the convection type are generally used in marine boilers. While some radiant superheaters have been used, the difficulty of providing adequate cooling during fast startups and under maneuvering conditions has severely limited their application.

In the convection superheater, there is usually enough absorption by radiant heat penetration of the screen to give a flatter steam temperature characteristic than would be obtainable by convection alone. The characteristic steam temperature curves of radiant and convection superheaters are shown in Chapter 19.

With a properly designed screen between the convection superheater and the furnace, it is possible to maintain a high temperature differential between the gas and the steam. This minimizes the amount of heating surface necessary to obtain the desired steam temperature and reduces the size and weight of the superheater. Superheater heating surface usually consists of U-shaped tubes connected to headers at each end, although continuous tube superheaters are used in special designs.

Either a vertical or a horizontal arrangement may be used in drum-type boilers. The horizontal arrangement can be vented, drained, and readily cleaned by mechanical or chemical means. However, complete draining and venting is possible only with the vessel on an even keel. Vertical superheaters have inverted loops and, although drainable at all times, they can not be vented. Typical superheater arrangements for a two-drum boiler are shown in Figs. 3 and 4.

Superheater surface is arranged in loops of in-line tubes (usually one to four loops), and the most commonly used tube size is 1.25 in. (31.8 mm) OD. Because the tubes are relatively short, adequate strength is available in the usual range of thicknesses to maintain proper alignment with a minimum of supports. The number of steam passes is selected to provide sufficient pressure drop to obtain proper steam distribution and to assure satisfactory tube metal temperatures.

Good practice requires that maldistribution of gas and steam be kept to a minimum to maintain proper superheater tube temperatures. However, with various fuels and operating demands it is not possible to maintain the conditions necessary for perfect distribution. To provide a satisfactory margin when calculating steam and tube temperatures, the average gas side heat transfer rate is increased and the steam side rate is reduced.

For steam temperatures below 850F (454C), tubes are usually expanded into headers to minimize leakage and to reduce maintenance costs. For temperatures above 850F (454C), rolled and seal welded joints, or joints of the stub welded type, are necessary for satisfactory service.

In addition to being supported by the headers, superheater tubes may also be supported at one or more

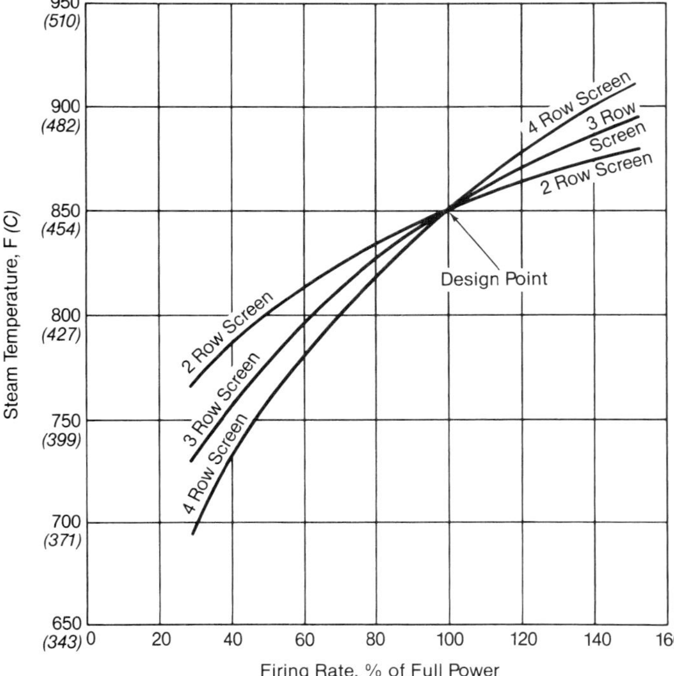

Fig. 11 Typical curves showing the effect of superheater location and firing rate on superheat.

points by alloy castings dovetailed into brackets welded to water-cooled support tubes. With these designs, the superheater tubes can be replaced without removing the supports.

Superheater fouling and high temperature corrosion

The use of higher steam temperatures has resulted in rapid fouling of superheater surfaces and corrosion and wastage of superheater supports. The heavy ash deposits come from burning oils with 0.05 to 0.20% or more ash by weight. The most significant constituents of the ash are vanadium, sodium and sulfur. Experience indicates that heavy deposits occur with a combination of high gas and heat-receiving surface temperatures. The heaviest ash accumulations occur on the furnace side of a superheater, where the gas temperatures are relatively high, while less ash forms at the back side, where the gas temperatures are much lower. Because of the potential of high corrosion rates, superheater metal temperatures should not exceed 1100F (593C). For further discussion of high temperature corrosion in oil-fired boilers, see *Oil ash corrosion* in Chapter 21.

Steam temperature control

In the marine power plant, just as in the stationary power plant, accurate control of steam temperature is required. In most cases steam temperature controls are used to assure that the allowable metal temperatures of the main steam piping and the main turbine are not exceeded. Close control of the maximum steam temperature permits the use of less expensive alloys in the superheater and piping. Several methods are used for superheat control, as described in Chapter 19. One or more methods may be used to alter the characteristic steam temperature curve of a particular unit design.

One method used in marine boilers is the dual or two-furnace boiler design. As shown in Fig. 12, the furnace is divided into two sections separated by a superheater and screen tube section. The flue gas flows successively through the superheat furnace (left hand furnace), the superheater heat transfer surfaces, the second furnace (where fuel and air are burned) and, finally, the boiler bank before being exhausted to the stack. By varying the oil firing rate in the superheater furnace, the quantity and temperature of the gases flowing across the superheater are controlled to obtain the desired steam temperature. At the same time, the firing rate in the saturated furnace, which contains no superheater surface, is adjusted to hold the desired steam pressure.

In most modern two-drum boilers, steam temperature is controlled, where required, through the use of a drum-type surface attemperator. With this type of control, all of the flue gas passes over the superheater, but a portion of the steam is passed through the attemperator, which is a steam-to-water heat exchanger submerged in boiler water in the steam or water drum. Fig. 13 indicates two piping arrangements for control attemperators. In a submerged attemperator where the inlet steam temperature exceeds 850F (454C), 16Cr-1Ni alloy should be considered to protect against methane gas corrosion.

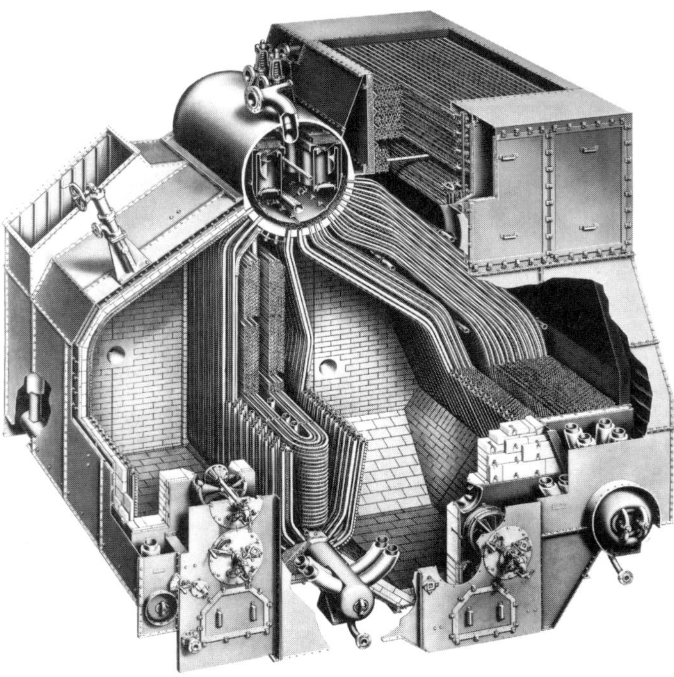

Fig. 12 Twin furnace design used for marine boiler superheat temperature control.

Auxiliary desuperheaters

Steam for auxiliary use may be required at a lower temperature than that delivered by the superheater to the main engine. To satisfy these demands, auxiliary desuperheaters are used, often in conjunction with pressure reducing stations. These desuperheaters may be inside the drum as discussed above or separate external spray types.

Reheaters

The use of steam reheat in a marine plant is more attractive as the speed and power requirements of the ship increase. Although reheaters increase the complexity of the boiler design and the first cost of the plant, fuel savings in high horsepower, high utilization plants can be significant. Design considerations are similar to those for superheaters but must be augmented by the requirements to protect the reheater from overheating during periods when there is no steam flow during maneuvering, running astern or at a stop bell.

Economizer and air heater cycles

Economizers or air heaters, and in some instances both, are required if a high boiler efficiency is to be obtained. The temperature of flue gas leaving the boiler bank at full power (the design rate) is a function of the saturation temperature corresponding to the drum pressure at which the unit is operating. Space, weight and economic considerations usually result in a boiler bank sized to reduce the exit gas temperature to within 50 to 100F (28 to 56C) of the saturation temperature. Gas temperatures in the range of 550 to 650F (288 to 343C) leaving the boiler bank are typical of merchant ship boilers. To reduce this flue gas temperature sufficiently in order to obtain acceptable efficiency, economizers and/or air heaters may be added. The choice of which to use depends on the design of the power plant and the desired performance characteristics of the unit.

Where the design includes a deaerating feedwater heater and a single stage of feedwater heating to supply water at 240 to 280F (116 to 138C), an economizer can be used to provide reasonably high boiler efficiency. The economizer can be used either alone or in conjunction with a steam air heater. In this range of feedwater temperature, an economizer can be economically designed to reduce the products of combustion to within 30F (17C) of the inlet water temperature at the normal rating. Therefore, with 280F (138C) feedwater, a boiler efficiency of 88.6% can be attained with an economizer alone and about 88.8% if a steam air heater is added.

When the feedwater temperature to the economizer is higher, the efficiency is limited, because the temperature of the gas leaving the economizer can not be lower than the inlet feedwater temperature. Consequently, when additional stages of regenerative feedwater heating are used, the inlet feedwater to the economizer may be at a temperature of 300 to 450F (149 to 232C), and it may not be economical to use an economizer unless it is followed by an air heater. Cycle efficiency is increased approximately 1% for each 100F (56C) rise in feedwater temperature through regenerative feedwater heating.

An air heater contributes additional efficiency even when the maximum practical amount of regenerative feedwater heating is used. At normal operating rates and an inlet air temperature of 100F (38C), exit gas temperatures of 300 to 320F (149 to 160C) are readily

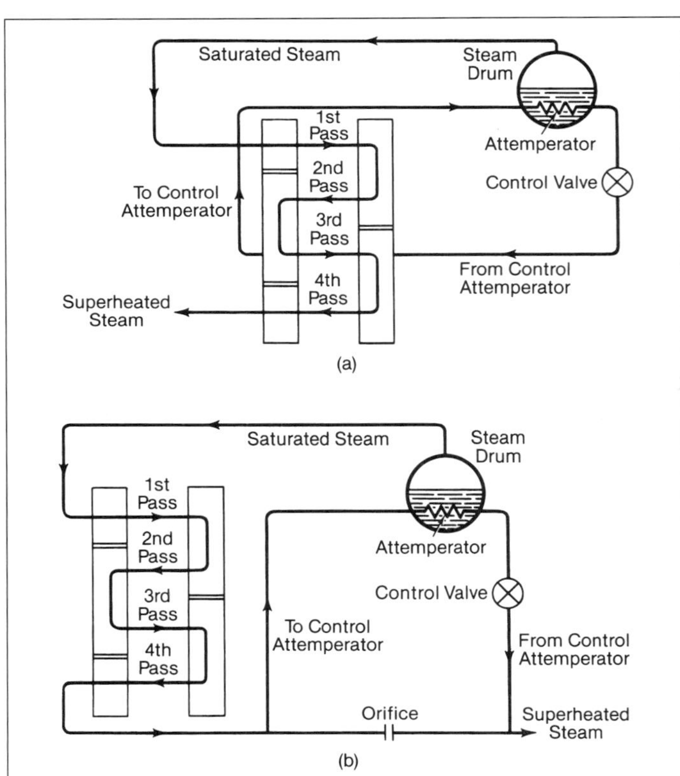

Fig. 13 Internal attemperators – (a) inter-pass, (b) after-pass.

obtained, corresponding to 88.5 to 88.0% efficiency. Air temperatures leaving air heaters are normally in the range of 300 to 450F (149 to 232C).

As an alternate to the standard economizer/air heater cycle, a dual economizer steam air heater cycle (DESAH), using a primary and secondary economizer, can be used. This arrangement will improve the cycle efficiency and reduce the fuel rate by utilizing turbine bleed steam to a high pressure feedwater heater. The actual cycle improvement will depend on the turbine characteristics with the additional bleed for the high pressure heater. The overall boiler efficiency will remain essentially unchanged over that with the regular economizer cycle; however, the fuel oil consumed will decrease because of additional heat added to the feedwater.

The cycle would use a feedwater temperature leaving the deaerator feedwater heater of about 285 to 300F (141 to 149C), presently being used on single economizer, steam air heater cycles. A feedwater temperature lower than this could result in economizer corrosion and plugging problems. If higher feedwater temperatures are desired in order to more efficiently use turbine bleeds, it should be pointed out that the boiler efficiency will correspondingly decrease. This is due to the higher gas temperatures occurring at the secondary economizer outlet because a terminal end temperature difference of 25F (14C) is required.

The feedwater would pass through the secondary economizer and be heated to about 335F (168C). It would then pass through a high pressure feedwater heater, that would raise the feedwater temperature about 50F (28C) before returning the water to the primary economizer and the steam drum (Fig. 14). This cycle shows a reduction in all purpose fuel use over that associated with the single economizer steam air heater cycle while being only slightly more complicated. Equipment costs rise somewhat due to the addition of a high pressure feedwater heater and its control. In addition, in order to realize the same exit gas temperature, the economizer heating surface would have to be increased 1.5 times the surface amount used with the single economizer cycle.

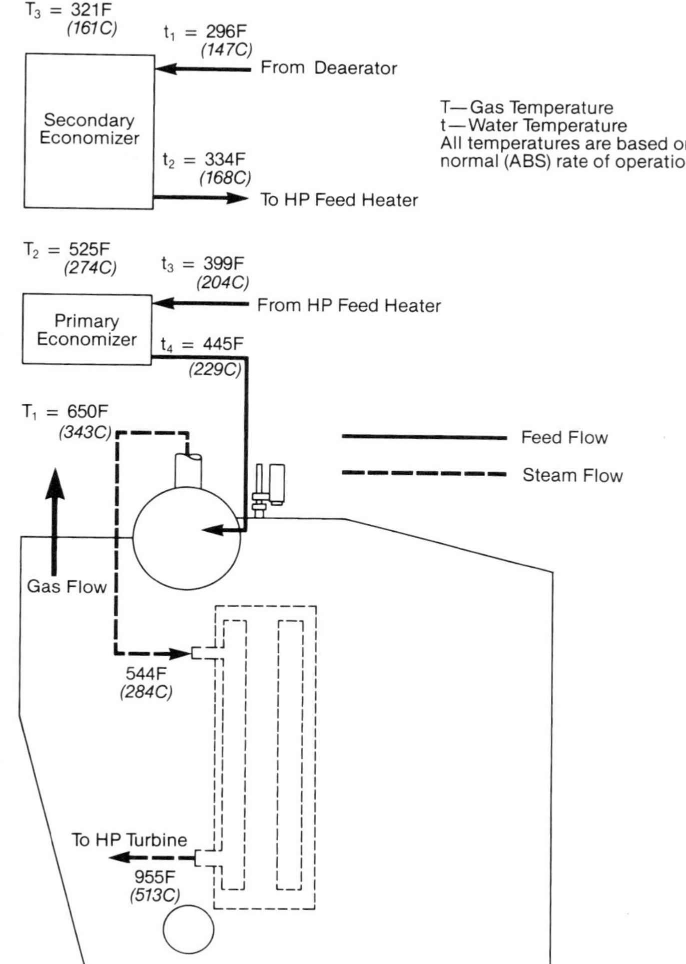

Fig. 14 Dual economizer steam air heater (DESAH) cycle schematic.

Economizers Two general types of economizers, the bare tube and extended surface types, are used in marine service. Both are nonsteaming and are almost always arranged for counterflow of water and flue gas to obtain the best possible heat transfer characteristics. (See Chapter 20.) The bare tube type is used where high feedwater temperature makes the application of both an economizer and air heater desirable. These economizers are designed to reduce the gas temperature about 100F (56C), with the remainder of the required temperature drop obtained with an air heater. All are drainable and ventable.

The second and more common type uses extended surface to reduce the size of the economizer. There are many types of extended surface, including cast iron or aluminum gill rings, spiral fins and small metal studs welded to the tubes. B&W has chosen the latter for its marine economizers due to the shape and size of the stud and because the method of attachment eliminates soot-collecting crevices where corrosion usually begins.

Extended surface economizers are usually constructed of 1.5 in. (38.1 mm) OD tubes formed into continuous loops. This construction requires only two headers. Tube-to-header joints are welded, or expanded and seal welded, to eliminate seat leakage. In the most recent designs, handhole fittings, except those required for inspection, are eliminated by externally welding the tubes in sockets or to stubs on the headers. A typical arrangement of an extended surface economizer with studded tubes is shown in Fig. 15.

When used alone, economizers are generally designed to reduce gas temperatures by 200 to 300F (111 to 167C), with a corresponding increase in the water temperature of 70 to 100F (39 to 56C). If the inlet water temperature is 280F (138C), the exit water temperature normally will range from 350 to 365F (177 to 185C), far enough below saturation temperature at a pressure of 600 psig (4.14 MPa) to prevent steaming.

In designing an economizer, the possibility of sulfuric acid corrosion of the economizer tubes must be considered. (See Chapter 20.) The tube metal temperature of bare tube economizers is essentially the same as that of the water within the tube. This is also true for the tube metal of extended surface elements, but the tip temperature of the studs is considerably higher.

In order to minimize gas side corrosion of carbon steel economizer tubes, B&W recommends that feed

Fig. 15 Extended surface stud tube economizer.

temperature be kept above 246F (119C) when burning Navy Special Fuel Oil (NSFO) and above 270F (132C) when burning Bunker C oil. Where these temperatures are not obtainable, the use of a corrosion resistant alloy tube is advised. The use of cast iron is not recommended as the corrosion resistance of cast iron is less than that of standard carbon steel, and any improvement in tube life with cast iron cladding is only due to its greater mass.

A bypass line is usually provided for operation of the boiler with the economizer out of service. During such operation, it is necessary to fire more fuel to maintain the required evaporative rating because of the decreased efficiency. This increases the steam temperature and requires either more attemperation or a reduction in rating to prevent overheating of superheater tubes or turbine. Normally there is no danger of metal oxidation in the economizer during bypass operation, because the gas temperatures entering the economizer are usually less than 850F (454C).

When economizers are included in the design, deaerating feedwater heaters should be installed to remove all traces of oxygen from the feedwater and to prevent internal corrosion of tubes and headers.

Air heaters (gas to air) Tubular and regenerative air heaters (see Chapter 20) have both been used for marine boiler applications. However, the tubular air heaters have become virtually obsolete due primarily to their extremely large space requirements and their relatively poor metal temperature characteristics.

Air heater corrosion can be reduced or practically eliminated by several means. For a given exit gas temperature, metal temperatures are somewhat higher with a regenerative air heater than with a recuperative type. Therefore, for a given satisfactory metal temperature, a lower exit gas temperature and a higher unit efficiency can be obtained with a tubular type air heater. With a regenerative air heater, air temperatures of 550F (288C) or higher can be produced with a resulting exit gas temperature of 235F (113C) or less. This in turn can result in a boiler efficiency of about 90.3% when using low excess air values.

As with the tubular air heater, corrosion can be a problem when the regenerative air heater is designed for low exit gas temperature and the corresponding high air temperature. This can result in an average cold-end metal temperature well below the acid dew point of the combustion gas. Corrosion can be reduced by the use of porcelain-enameled cold-end surface.

Another alternative available when using a rotary regenerative air heater is to place a small economizer in the gas stream and/or a steam air heater on the air side ahead of the regenerative air heater. Without the economizer, the regenerative air heater becomes large due to the relatively low heat transfer rates inherent in a gas-to-air heat exchanger. Often space and weight limitations, along with the increased cost associated with a large air preheater, dictate a reduction in its size. The addition of an economizer designed to reduce the gas temperature entering the air heater by 100 to 125F (56 to 69C) reduces the size of the air heater; the air temperature to the furnace is correspondingly reduced. This can frequently provide a better arrangement and at a lower overall cost.

Boiler casing

In marine service, the walls that form the boiler and furnace enclosure, together with those surrounding the economizer and air heater, are normally made gas-tight by using metal casing. Therefore, the term *casing* is generally used for the walls surrounding a marine unit.

To provide a comfortable fireroom, an outer casing temperature of 130F (54C) is usually specified. In the double case arrangement used for tangent tube boiler construction, the space between the inner and outer casing is pressurized with air at or near the forced draft fan discharge pressure to prevent leakage of combustion gases into the fireroom and to help maintain the desired outer casing temperature. With welded wall construction the boiler is, for the most part, single-cased and pressure-sealed to prevent leakage of combustion gases into the fireroom.

Circulation and steam separation

Satisfactory circulation characteristics and efficient steam separation are of prime importance in the successful operation of any boiler. These factors take on increased importance in a marine boiler that is rolling and pitching in a seaway or undergoing rapid load changes. In analyzing circulation, the procedures outlined in Chapter 5 are applied to the design at the maximum anticipated rate of evaporation.

Generally, steam drum diameters range from 54 to 72 in. (1.37 to 1.83 m) for merchant vessels and from 54 to 60 in. (1.37 to 1.52 m) for naval units. The drum must accommodate the desired number of tube rows and provide sufficient space for steam separation, feedwater introduction, distribution and treatment, and water level fluctuations caused by sudden load variations.

Marine steam drum baffling for lower rated boilers is usually simple in construction and arrangement. A V-type baffle using a triple layer of perforated steel plates is used in moderately-rated drum boilers. With relatively small clearances between the plates, it is installed (Fig. 15) just below the normal water level to break up steam and water jets issuing from the riser circuits. The duplex compartment baffle is more effective in higher-rated two-drum boilers. In current B&W marine boiler designs, horizontal steam separators and scrubbers are used (Fig. 16). Cyclone separators and scrubbers are used to ensure that dry steam is sent to the superheater where moisture and dissolved boiler water containment carryover would not be acceptable. The principles of these cyclone separators are similar to those of the vertical cyclones used in stationary units. (See Chapter 5.)

Drum boilers are generally rated conservatively and do not require external downcomers. Adequate circulation can be maintained by the rear rows of boiler bank tubes as downcomers, where the gas temperature is 850F (454C) or less. At low loads, the first several rows of tubes act as risers, and the remaining tubes serve as downcomers. As the firing rate increases, the high temperature gas zone moves deeper into the tube bank, the number of tubes acting as risers increases, and the number of downcomers declines. Excessive firing rates and rates beyond those contemplated in the design can reduce the number of downcomers below the minimum requirements. For highly rated units, it is necessary to install unheated downcomers to supplement the number of boiler bank tubes acting as downcomers. Circulation is benefited by interposing a convection superheater in the boiler bank because the heat absorbed by the superheater reduces the temperature of the gas flowing over the boiler tubes beyond it. More boiler tubes consequently act as downcomers over a wide operating range. In vertical superheater units, external downcomers are also required because of physical arrangement constraints.

In most horizontal superheater drum boilers (Fig. 4), furnace waterwall supply headers receive water from the lower drum through supply tubes located below the furnace floor. These tubes are spaced along the length of the waterwall headers to assure even distribution of water to the high duty furnace tubes and provide cooling for the furnace floor refractory. Naval and other high-rated designs have external downcomer tubes from the steam drum to the lower wall header because the supply of water from the boiler bank alone would be inadequate. In vertical superheater drum boilers (Fig. 3), furnace waterwall supply headers receive water from the steam or lower drum through external downcomers or feeder tubes.

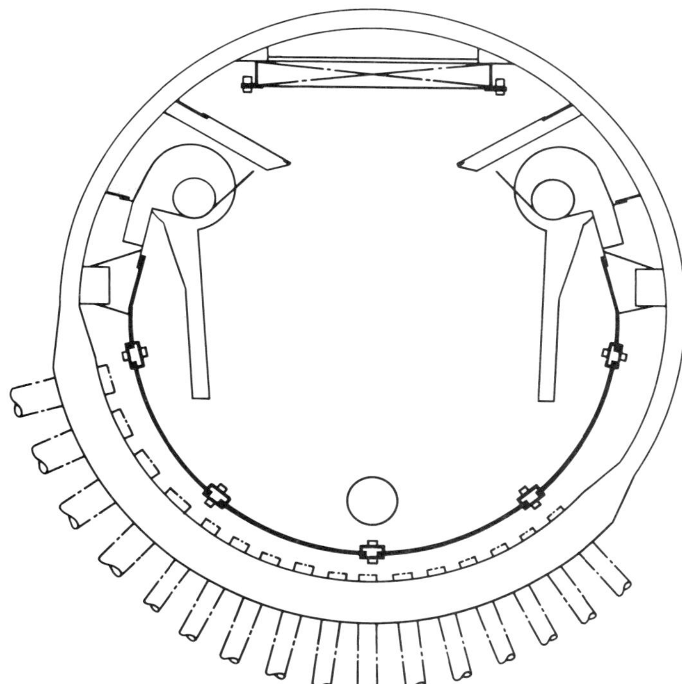

Fig. 16 Arrangement of horizontal cyclone separators in a marine boiler drum.

Drum boilers should not be located with the drum longitudinal axis athwart ship. In such a design, rolling motion may stop circulation in some riser tubes located in the sides of the boiler. This can result in tube overheating and failures. While this difficulty is alleviated by locating the drums fore and aft, ship rolling can still affect boiler circulation. During rolling, some of the rear boiler tubes in the drum may become uncovered and the overall downcomer water flow to the lower drum reduced. However, marine boilers operating at conservative load ratings are usually able to accommodate such conditions.

Section IV
Environmental Protection

Environmental protection and the control of solid, liquid and gaseous effluents or emissions are key elements in the design of all steam generating systems. The emissions from combustion systems are tightly regulated by local and federal governments, and specific rules and requirements are constantly changing. At present, the most significant of these emissions are sulfur dioxide (SO_2), oxides of nitrogen (NO_x), and fine airborne particulate. All of these require specialized equipment for control.

Chapter 32 begins this section with an overview of current regulatory requirements and overall emission control technologies. The chapter concludes with a discussion of mercury emissions control which is expected to become an integral part of overall plant multi-pollutant control strategy. Following this overview, Chapters 33, 34 and 35 discuss specific equipment to control atmospheric emissions of particulate, NO_x and SO_2 respectively. The NO_x discussion focuses on post-combustion technologies; combustion-related control options are addressed in Chapter 11 and Chapters 14 through 18.

Finally, a key element in a successful emissions control program is measurement and monitoring. Chapter 36 addresses a variety of issues and outlines a number of technologies for flue gas monitoring.

Integrated emission control systems on this western U.S. power plant include low NO_x burners, SCR, dry scrubber, and baghouse.

Chapter 32

Environmental Considerations

Since the early 1960s, there has been an increasing worldwide awareness that industrial growth and energy production from fossil fuels are accompanied by the release of potentially harmful pollutants into the environment. Studies to characterize emissions, sources and effects of various pollutants on human health and the environment have led to increasingly stringent legislation to control air emissions, waterway discharges and solids disposal.

Comparable concern for environmental quality has been manifest worldwide. Since the 1970s, countries of the Organization for Economic Cooperation and Development have reduced sulfur dioxide (SO_2) and nitrogen oxides (NO_x) emissions from power plants in relation to energy consumption. In at least the foreseeable future, emission trends are expected to continue downward due to a combination of factors: change in fuel mix to less polluting fuels, use of advanced technologies, and new and more strict regulations. In Japan, the reductions in SO_2 emissions were particularly pronounced due to strong environmental measures taken in the 1970s. As an example, in the United States (U.S.) between 1980 and 2001, electricity generation increased by 56%, while SO_2 emissions declined 38%.

Environmental control is primarily driven by government legislation and the resulting regulations at the local, national and international levels. These have evolved out of a public consensus that the real costs of environmental protection are worth the tangible and intangible benefits now and in the future. To address this growing awareness, the design philosophy of energy conversion systems such as steam generators has evolved from providing the lowest cost energy to providing low cost energy with an acceptable impact on the environment. Air pollution control with emphasis on particulate, NO_x, SO_2, and mercury emissions is perhaps the most significant environmental concern for fired systems and is the subject of Chapters 33 through 36. However, minimizing aqueous discharges and safely disposing of solid byproducts are also key issues for modern power systems.

Sources of plant emissions and discharges

Fig. 1 identifies most of the significant waste streams from a modern coal-fired power plant. Typical discharge rates for the primary emissions from a new, modern 615 MW coal-fired supercritical pressure boiler are summarized in Table 1.

Atmospheric emissions arise primarily from the byproducts of the combustion process (SO_2, NO_x, particulate flyash, and some trace quantities of other materials) and are exhausted from the stack. A second source of particulate is fugitive dust from coal piles and related fuel handling equipment. This is especially significant for highly dusting western U.S. subbituminous coals. Some low temperature devolatilization of the coal can also emit other organic compounds. A final source of air emissions is the cooling tower and the associated thermal rise plume which contains water vapor.

Solid wastes arise primarily from collection of the coal ash from the bottom of the boiler, economizer and air heater hoppers, as well as from the electrostatic precipitators and fabric filters. Pyrite collected in the pulverizers (see Chapter 13) is usually also included. Most of the ash is either transported wet to an ash settling pond where it settles out or is transported dry to silos from which it is taken by truck for beneficial use (e.g., cement additive). The chemical composition and characteristics of various ashes are discussed in Chapter 21.

The second major source of solids is the byproduct from the flue gas desulfurization (FGD) scrubbing process. Most frequently, this is a mixture containing primarily calcium sulfate for wet systems and calcium sulfite for dry systems. After dewatering, the wet system byproduct may be sold as gypsum or landfilled. Additional sources of solids include the sludge from cooling tower basins, wastes from the water treatment system and wastes from periodic boiler chemical cleaning.

Aqueous discharges arise from a number of sources. These include once-through cooling water (if used), cooling tower blowdown (if used), sluice water from the ash handling system (via the settling pond), FGD waste water (frequently minimal), coal pile runoff from rainfall, boiler chemical cleaning solutions, gas-side water washing waste solutions, as well as a variety of low volume wastes including ion exchange regeneration effluent, evaporator blowdown (if used), boiler blowdown and power plant floor drains. Many of these streams are chemically characterized in Chapter 42. Additional discussions of these systems as well as the controlling regulations are provided in References 1 and 2.

Air pollution control

U.S. legislation – Clean Air Act

The Federal Clean Air Act (CAA) is the core driving force for all air pollution control legislation in the United States (U.S.). The original CAA was first en-

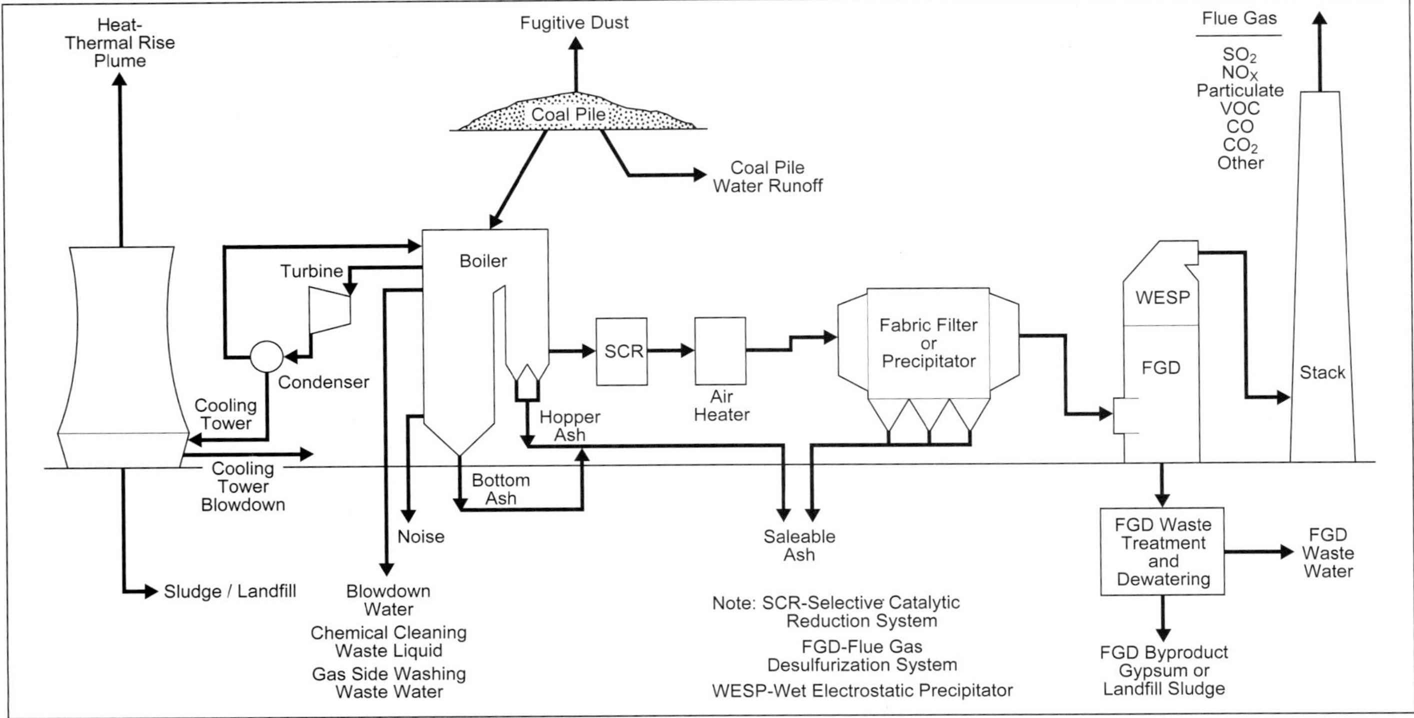

Fig. 1 Typical bituminous coal-fired power plant effluents and emissions.

acted in 1963, and since that time the Act has evolved through five significant amendment cycles in 1965, 1967, 1970, 1977, and 1990.

The primary objective of the CAA is to protect and enhance the quality of the nation's air resources to promote the public health and welfare and the productive capacity of its population.[3] The legislation generally provides for the U.S. Environmental Protection Agency (EPA) to set national air quality standards and other minimum regulatory requirements through federal regulations and guidance to state and local regulatory agencies. The individual states are required to develop state implementation plans (SIPs) to define how they will meet the minimum federal requirements. However, state and local government agencies may also develop and implement more stringent air pollution control requirements. The CAA as amended prior to 1990 included the following regulatory elements of potential interest to boiler owners and operators.

National Ambient Air Quality Standards (NAAQS) Federal standards were developed to define acceptable air quality levels necessary to protect public health and welfare. The EPA promulgated National Ambient Air Quality Standards for six *Criteria Pollutants*: sulfur dioxide (SO_2), nitrogen dioxide (NO_2), carbon monoxide (CO), ozone (O_3), particulate matter and lead. Two levels of standards have been established: primary standards aimed at prevention of adverse impacts on human health and secondary standards to prevent damage to property and the environment. All geographic areas of the country are divided into a number of identifiable areas known as air quality control regions which are classified according to their air quality. Air quality control regions that meet or better the NAAQS for a designated pollutant are classified as attainment areas for that pollutant, and regions that fail to meet the NAAQS are classified as nonattainment areas for that pollutant.

New Source Performance Standards (NSPS) Federal New Source Performance Standards were established for more than 70 categories of industrial processes and/or stationary sources. The NSPS rules set source-specific emission limitations and corresponding monitoring, recordkeeping and reporting requirements that must be met by new sources constructed on or after the effective date of an applicable standard. Sources constructed prior to the promulgation of an applicable NSPS are generally *grandfathered* and are not subject to the standards until such time that the source undergoes major modification or reconstruction. The EPA's NSPS regulations are published under Title 40, Part 60 of the Code of Federal Regulations.[4] Table 2 provides reference to select Subparts of 40 CFR 60 applicable to a variety of industrial and utility boilers. The various NSPS rules governing fossil fuel-fired boilers include emission limitations for NO_x, SO_2, particulate and opacity. The NSPS emission limits are based on the EPA's evaluation of best demonstrated technology, and these limits are subject to periodic review and revision. Finally, the NSPS rules generally establish the least stringent emission limitation a new source would have to meet. Typically, more stringent emission limitations are necessary to meet other federal, state or local permitting requirements. For example, any significant new source or major modification to an existing source of emissions may be subject to the federal *New Source Review* rules discussed below.

New Source Review (NSR) New Source Review regulations were established to: 1) preserve existing air quality in areas of the U.S. that are in compliance with

Table 1
Modern 615 MW Supercritical Coal-Fired Steam Generator Emissions and Byproducts*
(2.3% Sulfur, 7.7% Ash, 13,100 Btu/lb Fuel)

Emission	Typical Control Equipment**	Discharge Rate — t/h (t_m/h) Uncontrolled		Controlled	
SO_2	WFGD	9.1	(8.3)	0.27	(0.25)
SO_3	ESP or BH in conjunction with WFGD and WESP	0.2	(0.2)	0.03	(0.03)
NO_x as NO_2	SCR (with LNB)	0.8	(0.7)	0.08	(0.07)
CO_2	None (except cycle efficiency improvement)	536	(487)	—	—
Thermal discharge to water sources	Natural draft cooling tower	2.6 x 10^9 Btu/h	(750 MW_t)	~0	(~0)
Flyash to air***	ESP or BH in conjunction with WFGD and WESP	13.0	(11.8)	0.03	(0.02)
Ash to landfill	Landfill or saleable product	2.6	(2.3)	16	(14)
WFGD gypsum byproduct	Landfill or saleable product	0.0	(0.00)	24	(22)

* See Table 1 in Chapter 1 for a typical 500 MW subcritical coal-fired steam generator.
** Definitions:
WFGD – Wet flue gas desulfurization (limestone reagent with forced oxidation)
ESP – Electrostatic precipitator
BH – Baghouse fabric filter
WESP – Wet electrostatic precipitator
SCR – Selective catalytic reduction
LNB – Low NO_x burners with overfire air
*** As flyash emissions to the air decline, ash shipped to landfills or alternate uses increases.

the NAAQS, and 2) avoid further degradation and improve air quality in those areas of the U.S. that do not meet the NAAQS. NSR as discussed herein refers to two separate and distinct regulatory programs: *Prevention of Significant Deterioration (PSD)* and *nonattainment NSR*. In general, the NSR process requires any major new source or major modification to an existing source, which exceeds certain specified emission thresholds defined in the regulations,[5] to obtain permits and undertake other obligations prior to construction. A new or modified source subject to NSR generally must: 1) evaluate its potential impact on existing air quality, and 2) evaluate and utilize state-of-the-art air pollution control technologies. Under NSR, an affected project must conduct a pollutant-by-pollutant emissions evaluation. If a source is located in (or is impacting) an area which is in attainment with the NAAQS, the PSD rules apply. If the source is located in (or is impacting) a nonattainment area, the nonattainment NSR rules apply. Because a source may be located in an attainment area for one or more designated pollutants and in a nonattainment area for other designated pollutants, both the PSD and nonattainment NSR rules can apply simultaneously.

The PSD permitting process typically requires an air quality modeling analysis (which may require more than a year's worth of ambient air monitoring data) to demonstrate that new emissions will not cause a violation of the NAAQS or result in a significant deterioration of the existing air quality. Numerical limits (air quality increments) have been established by regulation for NO_x, SO_2 and fine particulate matter that restrict how much of the existing ambient air quality ($\mu g/m^3$) can be consumed by a new or modified emission source. All areas in the U.S. that meet the NAAQS are ranked into one of three classifications which determine the allowable increment consumption. Class I areas (national parks and designated wilderness areas) are to be kept in a pristine condition, whereas Class II or III areas allow for some further industrial growth. As might be expected, Class I areas allow almost no degradation of existing air quality. In addition to air quality impact analyses, the PSD process requires the use of *Best Available Control Technology* (BACT). A permit applicant must conduct a thorough evaluation of available control technologies (considering energy, environmental and economic impacts) for each pollutant subject to PSD review. The permitting authority then evaluates, accepts or rejects the proposed control technology. Once the control technology is agreed upon, the applicant and permitting authority determine a final permit limit that represents BACT. As part of this process, BACT limits contained in other PSD permits issued to similar sources throughout the U.S. must be considered and final BACT limits can not be less stringent than any applicable NSPS limit.

Unlike the PSD permitting process, nonattainment NSR does not require ambient air modeling or monitoring; however, a new or modified source subject to nonattainment NSR must offset the impact of new emissions by securing emission reductions from other sources in the area. The amount of reduction (offset) must be as great as, or greater than, the new increase in emissions, and the required offset is based on the severity of the area's air quality issues. The more polluted the air, the greater the required emissions offset. Offsets must be real reductions in existing emission rates, not otherwise required by regulations under the Clean Air Act, must be enforceable by the EPA, and must result in a positive net air quality benefit.

Table 2
Selected Categories – New Source Performance Standards (40 CFR Part 60)[4]

40 CFR Part 60 Subpart	Title/Description
D	Performance Standards for Fossil Fuel Fired Steam Generators (≥250 x 10^6 Btu/h) constructed after 08/17/71 (Note: generally superceded by Subparts Da or Db)
Da	Performance Standards for Electric Utility Steam Generating Units constructed after 09/18/78
Db	Performance Standards for Industrial, Commercial, Institutional Steam Generating Units (≥100 x 10^6 Btu/h) constructed after 06/19/84
Dc	Performance Standards for Small (<100 x 10^6 Btu/h) Industrial, Commercial, Institutional Steam Generating Units constructed after 06/09/89
Cb	Emission Guidelines for Large Municipal Waste Combustors constructed on or before 09/26/94
Ea	Performance Standards for Municipal Waste Combustors constructed after 12/20/89 and on or before 09/20/94
Eb	Performance Standards for Large Municipal Waste Combustors constructed after 09/20/94 or modified after 06/19/96
Y	Performance Standards for Coal Preparation Plants
BB	Performance Standards for Kraft Pulp Mills (Note: includes kraft recovery boilers)
AAAA	Performance Standards for Small Municipal Waste Combustors constructed after 08/30/99 or modified after 06/06/01
BBBB	Emission Guidelines for Small Municipal Waste Combusion Units constructed on or before 08/30/99

In general, offsets must be secured for the life of the source. In addition to achieving a net improvement in air quality through offsets, an affected source must utilize state-of-the-art controls to achieve the *Lowest Achievable Emission Rate* (LAER). LAER is based on the most stringent emission limitation contained in any SIP, contained in an existing permit, or achieved in practice by a similar source, regardless of cost or other economic consideration. Final LAER emission limits are established as an integral part of the NSR permitting process.

For further details and current information regarding the NSR rules, including NSR reforms that may impact the utility industry, see the following EPA Web page: *www.epa.gov/nsr/*

The 1990 Clean Air Act Amendments

The 1990 Clean Air Act Amendments added many new provisions to the existing Act, several of which have significantly impacted the electric utility industry. Under Title I of the Act, areas of the country that do not meet the National Ambient Air Quality Standards (NAAQS) were given new classifications and deadlines to achieve compliance. This provision of the Act has forced a dramatic reduction in NO_x emissions from utility and large industrial boilers located in the eastern U.S. because large portions of the northeast are ozone nonattainment areas and NO_x is a primary precursor to the formation of ozone. Title III of the Act authorized a new regulatory scenario for controlling 188 *hazardous air pollutants* (HAPs) from a wide range of industrial and commercial sources. Title IV of the Act established a new *Acid Deposition Control* program principally aimed at reducing SO_2 and NO_x emissions from older electric utility plants.

SO_2 and NO_x control under the 1990 Clean Air Act Prior to the 1990 CAA, a large population of existing electric utility boilers, generally built before 1971, were grandfathered and did not have to meet NSPS emission limits or comply with NSR requirements unless they were significantly modified or upgraded. Coal-fired utility boilers built between 1971 and 1978 were required (by promulgation of New Source Performance Standards) to limit SO_2 emissions to 1.2 lb/10^6 Btu heat input and compliance could be achieved by utilizing low-sulfur coals. However, coal-fired utility boilers built after 1978 were further required (by promulgation of additional New Source Performance Standards) to install scrubbers to achieve between 70 and 90% SO_2 removal efficiency.

Under the 1990 Amendments, Title IV of the Act (Acid Deposition Control) established a new SO_2 control program aimed at reducing emissions from all existing electric utility boilers through an innovative cap-and-trade program. The Acid Deposition Control program (commonly referred to as the Acid Rain program) set a goal of reducing annual SO_2 emissions by 10 million tons in a two-phased process. Phase I of the SO_2 program began in 1995 and initially affected 263 units at 110 mostly coal-burning electric utility plants located in 21 eastern and mid-western states. Phase II, which began in 2000, tightened the annual emissions limits imposed on the Phase I units and set limitations on smaller, cleaner plants fired by coal, oil, and natural gas, ultimately impacting more than 2,000 total units.

Under the SO_2 cap-and-trade program, an affected unit must hold (own) one *SO_2 Allowance* to cover each ton of SO_2 emitted in a given year. Existing units were initially given a pre-determined allocation of yearly allowances based on historical fuel consumption rates (heat input) and a mandated SO_2 emission rate (lb per million Btu heat input). A source owner has various options available to demonstrate compliance including: 1) reducing emissions through fuel switching, 2) adding more control equipment or improving the performance of existing control equipment, or 3) acquiring (purchasing) SO_2 allowances from other units. The bottom line compliance measure is straight forward; total SO_2 emissions from a unit in any given year must be covered by an equivalent number of SO_2 allowances. Once an allowance is used it is retired and any unused allowances may be sold or held (*banked*) with certain restrictions for future use. The total number of yearly SO_2 allowances allocated nationwide under Phase II of the program is fixed at 8.95 million tons and new power plants do not receive allowances. This effectively caps yearly SO_2 emissions from utility boilers at 8.95 million tons and ensures that the mandated reductions are maintained over time.

The Acid Rain program also mandated NO_x emission reductions from all existing utility boilers through a more traditional command-and-control approach where different types of boilers or combustion equipment were required to meet specified performance levels (i.e., a NO_x limit in pounds per million Btu of heat input). The NO_x command-and-control requirements were also implemented in a two-phased process with a goal of reducing utility boiler NO_x emissions by 2 million tons from 1980 levels.

The EPA's rules governing the Acid Rain program are published in the Code of Federal Regulations, under Title 40, Part(s) 72 through 78.

Further to the NO_x reductions mandated under the Acid Rain program, the EPA also implemented additional control measures (as authorized under Title I of the Act) to reduce interstate transport of NO_x emissions that contribute to ozone nonattainment problems in the eastern part of the U.S. These additional NO_x control measures (referred to herein as the *NO_x Budget* programs) are structured similarly to the SO_2 cap-and-trade program discussed above. However, unlike the SO_2 program which applies to all utility boilers year round, the NO_x Budget programs only apply to specified sources in the eastern part of the U.S. during the five month ozone season extending from May through September. As of this writing, the EPA has implemented the broadest of these NO_x Budget programs through a SIP Call (request for State Implementation Plans) mandating further NO_x reductions from electric utility boilers and other large industrial sources located in 21 eastern states and the District of Columbia (see Fig. 2).

For further details and current information regarding the SO_2 and NO_x trading programs see the following EPA Web page: *www.epa.gov/airmarkets/*

Control of air toxics under the 1990 Clean Air Act The 1990 CAA Amendments significantly changed the preexisting regulatory structure for addressing air toxics. Prior to the 1990 amendments, the Act required the EPA to identify and regulate, on a pollutant-by-pollutant basis, each non-criteria air pollutant that in its judgment caused significant health risks. This regulatory approach was largely ineffective as the EPA became involved in many legal, scientific, and policy debates over which pollutants to regulate and how stringently to regulate them. As a result, only seven pollutants had been regulated by 1990.

The 1990 CAA redirected the EPA to impose tighter controls for a broad range of air toxics through a two-phase approach. The first phase requires the EPA to develop technology-based standards for categories of sources that have the potential to emit significant quantities of 188 (originally 189) HAPs which were specifically identified in the Act. The second phase requires the EPA to impose further controls, beyond the initial technology-based standards, as necessary to address any remaining health risk concerns. Under the first phase, the EPA must develop *Maximum Achievable Control Technology* (MACT) standards for different classes or categories of sources that could be considered *major* emitters of HAP as well as other smaller *area* HAP sources. A major HAP source is defined as any type of stationary source with the potential to emit 10 tons per year of a single HAP or 25 tons per year of a combination of HAPs. An area source is defined to include smaller source categories (such as dry cleaners, gas stations, etc.) that are not by themselves a major source, but may collectively emit HAPs in sufficient quantities to warrant regulation. The applicable MACT standard for a given type or category of source must reflect emission performance levels that are already being achieved in practice by the better controlled, lower emitting units within a common source category. As of this writing, the EPA has proposed or promulgated MACT standards for more than 80 different source categories including but not limited to industrial, commercial and institutional boilers and process heaters; combustion sources at kraft, soda, and sulfite pulp and paper mills; and fossil fuel-fired electric utility boilers. The EPA's final MACT rules

Fig. 2 Area covered by NO_x SIP Call is shown in gray.

for various categories of sources are published in the Code of Federal Regulations, under Title 40, Part 63.

The 1990 CAA specifically directed the EPA to study hazardous air emissions, including mercury, from electric utility plants and to regulate such emissions as necessary to protect public health. In December 2000, the EPA issued a formal determination that regulation of HAP emissions, principally mercury from coal-fired electric utility boilers and nickel from oil-fired utility boilers, was appropriate and necessary. As discussed later in this chapter in the mercury emissions and control technologies section, several regulatory options are being proposed to address mercury control as of the date of this publication.

For further details and current information regarding the HAP regulations, including source-specific MACT standards and proposed alternatives for regulating mercury from electric utility boilers, see the following EPA Web page(s): *www.epa.gov/ttn/atw/index.html* and *www.epa.gov/mercury/*

Pending legislation/regulation As of the date of this publication, a number of U.S. Federal and State legislative and regulatory initiatives are underway to further reduce air emissions from fossil fuel-fired power plants. The most significant of the near-term initiatives are addressed in an inter-related suite of proposed rules collectively known as the *Clean Air Rules.* This suite of rules includes the proposed mercury cap-and-trade program discussed later; proposed rules to achieve further reductions in emissions of SO_2 and NO_x from power plants; and additional measures to address ozone and fine particle pollution throughout the U.S. While the final form of these regulatory initiatives can not be determined at the time of this writing, the direction of future regulatory actions will no doubt focus on achieving further improvements in air quality in the most effective way possible. For further details and current information regarding these initiatives and others, see the following EPA Web page: *www.epa.gov.*

International regulations – air pollution control

The passage of the Clean Air Act in the U.S. in 1963, marked the first enactment of air pollution control legislation by a major industrial nation. Since that time, air pollution control regulations have become more widespread in industrial and developing nations, particularly in Japan, Canada, and the European Union.[6,7] Many of the rapidly developing nations such as Korea, Taiwan, China, and the countries of former Eastern Europe are also aggressively controlling power plant emissions as their rapidly growing economies strain local ecosystems. As in the U.S., steam generating plants have been one focus of these regulatory measures, and the most common emissions of concern from combustion processes are SO_2, NO_x, particulates, and air toxics. The detailed regulations continue to evolve rapidly and are quite country specific. However, two trends are widespread:

1. Allowable emission limits for controlled pollutants will continue to decline with time.
2. A wider array of species will be considered for control.

Without attempting to be comprehensive, the following items provide a brief overview of worldwide SO_2 and NO_x regulatory efforts. These items provide a general indication of the range and application of control measures.

Control approaches One or more of the following measures have typically been adopted to control emissions:

1. *Emission standards* These limit the mass of SO_2, NO_x, or other pollutant emitted by volume, by heat input, by electric energy output, or by unit of time (hourly, daily, annually).
2. *Percent removal requirements* These specify the portion of the uncontrolled emissions that must be removed from the flue gas.
3. *Fuel requirements* Primarily aimed at SO_2 control, these either limit the type of fuel that can be burned or the fuel sulfur content.
4. *Technology requirements* These typically indicate the type of control technology specifically required or indicate the use of the best available control technology or reasonably available control technology at the time of installation. These requirements depend in many cases on some level of economic feasibility.

The most widely used control approach is emission standards, although this is usually combined with one or more of the other approaches. Emissions from new plants are usually more tightly controlled than emissions from existing capacity. Occasionally, older plants are not controlled, although this is changing, especially with the application of national cap-and-trade policies in the U.S. There have also been significant discussions on the applicability of international cap-and-trade programs that will link two or more national cap-and-trade programs together. The international cap-and-trade programs are proposed in the European Union (EU) Directive.

Emission control legislation and regulations throughout the world are currently in a state of flux as a variety of new and increasingly stringent regulations are phased in. Reference 6 provides a comprehensive and detailed summary of regulatory measures and current trends as of the end of 2003. The resulting compilation is quite complex as countries and areas consider the most effective alternatives. Reference 7 provides more detail for NO_x control.

Emissions standards worldwide are stated in a variety of units. See Appendix 1, Conversion Formulae for Emission Units.

Kinds of pollutants, sources and impacts

Air pollutants are contaminants in the atmosphere which, because of their quantity or characteristics, have deleterious effects on human health and/or the environment. The sources of these pollutants are classified as stationary, mobile or fugitive. Stationary sources generally include large individual point sources of emissions such as electric utility power plants and industrial furnaces where emissions are discharged through a stack. Mobile sources are those associated with transportation activities. Fugitive emissions generally include discharges to the atmosphere from conveyors, pumps, valves, seals and other process points not vented through a stack. They also include emissions from area sources such as coal piles, landfills, ponds and lagoons. They most often consist of particulates and occur in industry-related activities in which the emissions are not collected.

The focus of this chapter is stationary emission sources, particularly fired utility and industrial boiler systems. Key pollutants from these sources are SO_2, SO_3, NO_x, CO and particulate matter. Another class of emissions is called air toxics. These are potentially hazardous pollutants that generally occur in only trace quantities in the effluents from fired processes. However, they are undergoing more intense examination because of their potential health effects.

Sulfur dioxide Most of the sulfur in fuel converts to SO_2 with small quantities of sulfur trioxide (SO_3). The main source of sulfur oxides is from the combustion of coal, with lesser amounts from other fuels such as residual fuel oil. Based upon the August 2003 revision of the U.S. EPA National Emissions Inventory (NEI) program, the utility and industrial sectors (smelters, iron and steel mills, refineries) remain the largest emitters of sulfur oxides (see Table 3).[8] Sulfur oxides have been related to irritation of the human respiratory system, reduced visibility, materials corrosion and varying effects on vegetation. The reaction of sulfur oxides with moisture in the atmosphere has been identified as contributing to acid rain.

Sulfur trioxide Some of the sulfur dioxide that forms converts to sulfur trioxide (SO_3). The typical conversion rate is 1% or less in the boiler. However, the catalytic process that is frequently used to control NO_x levels has the undesirable side effect of converting additional SO_2 to SO_3, which can range from 0.5 to 2% additional conversion. The SO_3 readily combines with water to form sulfuric acid (H_2SO_4) at flue gas temperatures less than 500F (260C). This acid can create extremely corrosive conditions. The sulfuric acid condenses to form a fine mist when the flue gas passes through a wet flue gas desulfurization system that is used to remove sulfur dioxide (SO_2). This sulfuric acid mist contributes to the total stack particulate loading. Such mist is extremely fine, less than 0.5 micron, and very small amounts of this mist (5 ppm or less) can cause visible, bluish stack plumes, even in the absence of solid particulate.

Nitrogen oxides (NO_x) This category includes numerous species comprised of nitrogen and oxygen, although nitric oxide (NO) and nitrogen dioxide (NO_2) are the most significant in terms of quantity released to the atmosphere. NO is the primary nitrogen compound formed in high temperature combustion processes where nitrogen present in the fuel and/or combustion air combines with oxygen. The quantity of NO_x formed during combustion depends on the quantity of nitrogen and oxygen available, the temperature, the intensity of mixing and the time for reaction. Control of these parameters has formed the basis for a number of control strategies involving combustion process control and burner design. Based on the most recent EPA emissions inventory, utilities account for 22% of NO_x emitted in the U.S., with the transportation sector emitting 56%. Of the total utility NO_x emissions, approximately 90% comes from coal-fired boilers. The most deleterious effects come from NO_2 which forms from the reaction of NO and oxygen. NO_2 also absorbs the full visible spectrum and can reduce visibility. NO_x has been associated with respiratory disorders, corrosion and degradation of materials, and damage to vegetation. NO_x has also been identified as a precursor to ozone and smog formation.

Carbon monoxide This colorless, odorless gas is formed from incomplete combustion of carbonaceous fuels. CO emissions from properly designed and operated utility boilers are a relatively small percentage of total U.S. combustion source CO emissions, most of which come from the internal combustion engine in the transportation sector. The primary environmental significance of CO is its effect on human and animal health. It is absorbed by the lungs and reduces the oxygen carrying capacity of the blood. Depending on the concentration and exposure time, it can cause impaired motor skills and physiological stress.

Particulate matter Solid and liquid matter of organic or inorganic composition which is suspended in flue gas or the atmosphere is generally referred to as particulate. Particle sizes from combustion sources are in the 1 to 100 μm range, although particles smaller than 1 μm can occur through condensation processes. Among the effects of particulate emissions are impaired visibility, soiling of surrounding areas, aggravation of adverse effects of SO_2, and human respiratory problems.

PM_{10} and $PM_{2.5}$ Subsets of particulate matter, PM_{10} is particulate matter 10 μm and finer and $PM_{2.5}$ is particulate matter 2.5 μm and finer. Fine particles are emitted from industrial and residential oil combustion and from vehicle exhaust. Fine particles are also formed in the atmosphere when gases such as SO_2, NO_x and VOCs, emitted by combustion processes, are transformed into fine particulate by chemical reactions in the air (i.e., sulfuric acid, nitric acid, and photochemical smog). $PM_{2.5}$ is considered to have more deleterious health affects than coarser particulate.

VOC Volatile organic compounds represent a wide range of organic substances. These compounds con-

Table 3
2002 U.S. Emissions
U.S. EPA National Emission Inventory[8]

Category	Emissions (10^3 short tons) SO_2	NO_x	VOC	TSP
Utility combustion	10,293	4,699	52	695
Industrial combustion	2,299	2,870	170	269
Other combustion	575	725	790	405
Industrial process	1,399	1,000	7,418	1,127
Transportation	696	11,452	7,231	515
Other	91	356	883	19,142
Total	15,353	21,102	16,544	22,153

Notes: SO_2 — Sulfur dioxide
NO_x — Nitrogen oxides as NO_2
VOC — Volatile organic compounds
TSP — Total solid particulate (PM-10)

sist of molecules containing carbon and hydrogen, and include aromatics, olefins and paraffins. A major source is the refining and use of petroleum products. Also included among VOCs are compounds derived from primary hydrocarbons including aldehydes, ketones and halogenated hydrocarbons. The major source of these compounds is the transportation and the commercial/residential combustion sectors. VOCs are environmentally significant because of their role in the formation of photochemical smog through photochemical reactions with NO_x. Control of VOCs has been a primary means of addressing areas of ozone nonattainment. Smog arising from VOC emissions can cause respiratory problems, eye irritation, damage to vegetation and reduced visibility.

Toxic air pollutants This is a large category of air pollutants that could have hazardous effects.[9] The EPA had only promulgated standards for arsenic, asbestos, benzene, beryllium, mercury, radionuclides and vinyl chlorides for certain defined industries before the passage of the 1990 Amendments to the Clean Air Act. The 1990 Act first identified 189 toxic pollutants for which emissions are to be regulated. The list includes a wide range of simple and complex industrial organic chemicals and a small number of inorganics, particularly heavy metals. The EPA has identified hundreds of categories of air toxics sources, among which are municipal solid waste combustors, industrial boilers, and electric utility boilers. For these combustion sources, mercury has been the primary focus.

Mercury Present in only trace amounts in coal, mercury is released during the combustion process as elemental mercury, and is predominantly in the vapor phase at the exit of the furnace. Emissions from utility plants are extremely low. While other pollutants are recorded in pounds per million Btu (mg/Nm^3), mercury emissions are typically six orders of magnitude less and are frequently expressed in units of pounds per trillion Btu or µg/Nm. Fig. 3 indicates that the utility emissions account for approximately one-third of the U.S. manmade emissions of mercury or 48 tons/year (43.5t_m/year). (See Reference 10.)

Mercury in some chemical forms is very toxic. From whatever source, mercury can find its way into water sources where it can be converted into water soluble species such as methyl-mercury by microorganisms and accumulate in the fatty tissues of fish. Consumption of contaminated fish is the main identified risk to humans.

Carbon dioxide During the 1980s, concern increased about the potential impact of carbon dioxide (CO_2) emissions from manmade sources. CO_2 is one of several so-called *greenhouse* gases which may impact the climate and contribute to global warming. CO_2 is emitted from a variety of naturally occurring and manmade sources including the combustion of all fossil and hydrocarbon based fuels.

Greenhouse gas emissions in the U.S. totaled 1906 million metric tons in 2000, of which 83% or 1583 million metric tons was CO_2. CO_2 from the electric power sector totaled 642 million metric tons or 33.6% of total greenhouse gas emissions.[11]

Improving the power cycle efficiency (more power from less fuel) and the use of fuels with less carbon content are potential methods to address CO_2 emissions from any combustion source. Another option is separation and capture followed by sequestration. However, technology similar to SO_2 scrubbers or particulate collectors does not exist for carbon dioxide. Although CO_2 separation technologies exist, they are at present not economically viable for the large volumes of flue gas produced by electric power generation. Geologic, terrestrial, or ocean sequestration have their own technical and political challenges. Geologic sequestration holds considerable potential, as the storage capacity is estimated to be centuries of emissions, and this route may be the most environmentally acceptable.

Air pollution control technologies

The strategies for control of all emissions from a utility or industrial boiler are formulated by considering design fuels, kind and extent of emission reduction mandated, and economic factors such as boiler design, location, new or existing equipment, age and remaining life.

SO_2 control strategies and technologies SO_2 emissions from coal-fired boilers can be reduced using pre-combustion techniques, combustion modifications and post-combustion methods.

Pre-combustion These techniques include the use of natural gas or low sulfur oil in new units or the use of cleaned (beneficiated) coal or fuel switching in existing units. By using natural gas, sulfur emissions can be reduced to almost zero while the use of low sulfur oil will minimize SO_2 emissions. While the low sulfur content of oil and gas is advantageous, the price volatility and availability of these fuels make them less attractive. Switching to oil and gas in existing boilers requires attention be given to receiving equipment, storage facilities, combustion equipment including safety systems and boiler design. In the case of new systems, oil or gas firing can significantly reduce steam system capital costs. Even switching from one coal to another lower sulfur coal can have a dramatic impact on fuel handling, combustion and particulate collection equipment. These effects are explored in more detail in Chapters 21 and 44.

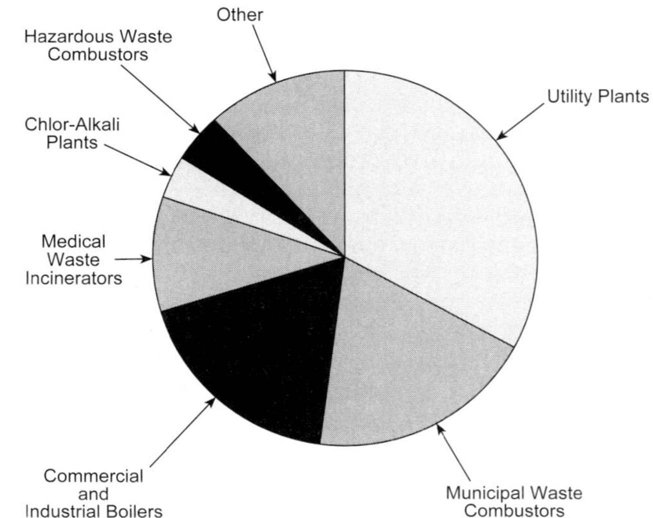

Fig. 3 Anthropogenic (manmade) emissions of mercury in the U.S.

Combustion modifications These techniques are primarily used to reduce NO_x emissions but can also be used to control SO_2 emissions in fluidized-bed combustion where limestone is used as the bed material. The limestone can absorb up to 90% of the sulfur released during the combustion process. (See Chapter 17.)

Sorbent injection technologies Sorbent injection, while not involving modification of the combustion process, is applied in temperature regions ranging from those just outside the combustion zone in the upper furnace to those at the economizer and flue work following the air heater. Sorbent injection involves adding an alkali compound to the coal combustion gases for reaction with the SO_2. Typical calcium sorbents include limestone [calcium carbonate ($CaCO_3$)], lime (CaO), hydrated lime [$Ca(OH)_2$] and modifications of these compounds with special additives. Sodium or magnesium based compounds are also used.

Wet and dry scrubbing technology Worldwide, wet and dry scrubbing or flue gas desulfurization (FGD) systems are the most commonly used technologies in the coal-fired electric utility industry. 2003 data indicate that in the U.S. more than 30% of the coal-fired utility capacity have SO_2 emission control. In contrast, more than 90% of coal-fired utility boilers in Germany and Japan have SO_2 control. Both wet and dry scrubbing use slurries of sorbent and water to react with SO_2 in flue gas, producing wet and dry waste products. (See Chapter 35.)

In the wet scrubbing process, a sorbent slurry consisting of water mixed with limestone, lime, magnesium promoted lime or sodium carbonate (Na_2CO_3) is contacted with flue gas in a reactor vessel. Wet scrubbing is a highly efficient (> 97% removal at calcium/sulfur molar ratios close to 1.0), well established technology which can produce usable byproducts.

Dry scrubbing involves spraying an aqueous sorbent slurry into a reactor vessel so that the slurry droplets dry as they contact the hot flue gas [~300F (~149C)]. The SO_2 reaction occurs during the drying process and results in a dry particulate containing reaction products and unreacted sorbent entrained in the flue gas along with flyash. These materials are captured downstream in the particulate control equipment. Dry scrubbing is a well established technology with considerable operational flexibility. The waste residue is dry.

NO_x control technologies NO_x emissions from fossil fuel-fired industrial and utility boilers arise from the nitrogen compounds in the fuel and molecular nitrogen in the air supplied for combustion. Conversion of molecular and fuel nitrogen into NO_x is promoted by high temperatures and high volumetric heat release rates found in boilers. The main strategies for reducing NO_x emissions take two forms: 1) modification of the combustion process to control fuel and air mixing, and reduce flame temperatures, and 2) post-combustion treatment of the flue gas to remove NO_x. (See Chapters 11, 14, 16, and 34.)

Combustion modification This approach to NO_x reduction can include the use of low NO_x burners, combustion staging, gas recirculation or reburn technology.

Low NO_x burners slow and control the rate of fuel and air mixing, thereby reducing oxygen availability in the ignition and main combustion zones. Low NO_x burners can reduce NO_x emissions by 50% or more, depending upon the initial conditions, are relatively low cost, and are applicable to new plants as well as retrofits.

Staged combustion uses low excess air levels in the primary combustion zone with the remaining (overfire) air added higher in the furnace to complete combustion. In some cases, the primary combustion zone may be operated substoichiometrically. Significant NO_x reductions are possible with staged combustion, although reducing zones and potential for corrosion and slagging exist.

Flue gas recirculation reduces oxygen concentration and combustion temperatures by recirculating some of the flue gas to the furnace without increasing total net gas mass flow. Large NO_x reductions are possible with oil and gas firing while only small reductions at best are possible with coal firing. Modifications to the boiler in the form of ducting and an efficiency penalty due to power requirements of the recirculation fans can make the cost of this option higher than some of the other in-furnace NO_x control methods.

Reburning is a technology used to reduce NO_x emissions in selected applications. In reburning, 75 to 80% of the furnace fuel input is burned in the primary combustion zone with minimum excess air. The remaining fuel (gas, oil or coal) is added to the furnace above the primary combustion zone. This secondary combustion zone is operated substoichiometrically to generate hydrocarbon radicals which reduce NO formed in the initial combustion to N_2. The combustion process is then completed by adding the balance of the combustion air through overfire air ports in a final burnout zone further up the furnace.

Post-combustion The two main post-combustion techniques for NO_x control are selective noncatalytic reduction (SNCR) and selective catalytic reduction (SCR). In SNCR, ammonia or other compounds such as urea (which thermally decomposes to produce ammonia) are injected downstream of the combustion zone in a temperature region of 1400 to 2000F (760 to 1093C). SCR systems remove NO_x from flue gases by reaction with ammonia in the presence of a catalyst (see Chapter 34). SCR is being used worldwide where high NO_x removal efficiencies are required in gas-, oil- or coal-fired industrial and utility boilers.

Particulate control technologies Particulate emissions from boilers arise from the noncombustible, ash forming mineral matter in the fuel that is released during the combustion process and is carried by the flue gas. Another source of particulate is the incomplete combustion of the fuel which results in unburned carbon particles. A brief description of the principal options for particulate emissions control in industrial and utility boilers follows while Chapter 33 provides an in-depth discussion.

Coal cleaning Historically, physical coal cleaning has been applied to reduce mineral matter, increase energy content and provide a more uniform boiler feed. Although reduction in flue gas particulate loading is one of the potential benefits, coal cleaning has been driven by the many other boiler performance benefits related to improved boiler maintenance and availability and, more recently, the reduction in SO_2 emissions.

Mechanical collectors These are generally cyclone collectors and have been widely used on small boilers when less stringent particulate emission limits apply. Cyclones are low-cost, simple, compact and rugged devices. However, conventional cyclones are limited to collection efficiencies of about 90% and are poor at collecting the smallest particles. Improvements in small particle collection are accompanied by high pressure drops.

Fabric filters These filters, also commonly referred to as baghouses, are available in a number of designs (reverse air, pulse jet, and shake/deflate), each having advantages and disadvantages in various applications. Applications include industrial and utility power plants firing coal or solid wastes, plants using sorbent injection and spray dryer FGD, and fluidized-bed combustors. Collection efficiency can be expected to be at least 99.9% or greater. Fabric filters have the potential for enhancing SO_2 capture in installations downstream of sorbent injection and dry scrubbing systems as discussed in Chapter 35.

Electrostatic precipitators (ESP) ESPs are available in a broad range of sizes for utility and industrial applications. Collecting efficiency can be expected to be 99.8% or greater of the inlet dust loading. ESPs are considered to be less sensitive to plant upsets than fabric filters because their materials are not as sensitive to maximum temperatures. They also have a very low pressure drop. ESP performance is sensitive to flyash loading, ash resistivity and coal sulfur content. Lower sulfur concentrations in the flue gas can lead to lower ESP collection efficiency. ESPs tend to collect coarser particulate more easily, whereas a fabric filter tends to have a more uniform collection efficiency across the particle size range. Therefore, a fabric filter has higher collection efficiency of fine particulate than an ESP. The desire to further control sulfuric acid mist emissions and very fine flyash has led to the utilization of wet ESPs downstream of wet flue gas desulfurization systems.

Mercury emissions and control technologies

Regulatory considerations

As discussed earlier in this chapter, mercury emissions are regulated under the hazardous air pollutants (HAPs) sector of the Clean Air Act. Mercury emission regulations are already in place for municipal solid waste (MSW) combustors, medical waste incinerators, and cement plants with the following levels as of this writing:

Source	Regulation
MSW combustion	0.08 mg/DSCM or 85% reduction
Medical waste incinerators	0.55 mg/DSCM or 85% reduction
Cement plants	0.12 mg/DSCM (existing plants), 0.056 mg/DSCM (new plants)

Note: DSCM = dry standard cubic meter (at 32F/0C)

The EPA has also determined that it is both appropriate and necessary to regulate utility units for emissions of mercury. As of this writing, the EPA has proposed several regulatory alternatives for controlling mercury emissions from new and existing electric utility boilers which generally fall into two categories: 1) MACT standards as discussed above, or 2) alternate performance standards coupled with a market based cap-and-trade program. EPA has promulgated rules for a cap-and-trade program where mercury reductions would occur over a two-phase process. When fully implemented by 2018, this would result in estimated mercury emission reductions of 33 tons (30.0 t_m) per year or approximately 70%. As of this writing, the pros and cons associated with the regulatory alternatives are the subject of extensive debate and the final outcome and timing are uncertain. However, mercury will be controlled and emissions reduced over time.

Mercury in U.S. coals

Coals vary in mercury content by rank, seam, mine, and mine internal location. Table 4 summarizes the mercury content data obtained for U.S. bituminous, subbituminous and lignite coals fired at U.S. utilities as part of an EPA study (see Reference 12). There are only modest differences between the average mercury contents of the three ranks of coal. However, the variation in mercury content among coals within a given rank is much larger.

Mercury in coal combustion flue gas

Mercury appears in coal combustion flue gas in both the solid and gas phases (*particulate-bound* mercury and *vapor-phase* mercury, respectively). Due to the high volatility of mercury and many of its compounds, most of the mercury found in flue gas is vapor-phase. Vapor-phase mercury, in turn, can appear as *elemental* mercury (elemental, metallic mercury vapor) or as *oxidized* mercury (vapor-phase species of various compounds of mercury). The form of mercury present, commonly referred to as *speciation*, is a key factor in the development of emissions control strategies.

Mercury speciation and emissions data have been gathered under an EPA study, wherein a subset of the U.S. boiler population was selected for flue gas emissions sampling on the basis of fuel type, boiler configuration, particulate control device, and SO_2 control technology (see Reference 12). This EPA study is known as the Information Collection Request (ICR). Flue gas sampling, in triplicate, was performed at approximately 84 U.S. plants to quantify total mercury and mercury speciation at the inlet and outlet of the last emission control device. The EPA Preliminary Test Method 003 [PRE-003: Standard Test Method for Elemental, Oxidized, Particle-Bound, and Total Mercury in Flue Gas Generated from Coal-Fired Stationary Sources (Ontario Hydro Method)] was used to determine the particulate-bound and vapor-phase mercury emissions. In the Ontario Hydro Method, flue gas samples are withdrawn isokinetically from the flue. The mercury species are collected on a quartz fiber filter, in potassium chloride (KCl) solutions, and in acidic peroxide and acidic potassium permangan-

Table 4
Mercury in Selected U.S. Coals[12]

Coal Type	Number of Analyses	lb Hg/trillion Btu (HHV) Low	High	Median
Bituminous	27,884	0.04	103.81	8.59
Subbituminous	8,193	0.39	71.08	5.74
Lignite	1,047	0.93	75.06	10.54

ate ($KMnO_4$) solutions. Mercury speciation is reported on the basis that the oxidized form of mercury is collected in the KCl impingers, and the elemental form of mercury is collected in the peroxide and potassium permanganate impingers.

Results of these extensive studies indicated that the speciation of vapor-phase mercury depends on coal type and other factors. Eastern U.S. bituminous coals tend to produce a higher percentage of oxidized mercury than do western subbituminous and lignite coals. Western coals have low chloride content compared to typical eastern bituminous coals. It has been recognized for several years that an empirical relationship holds between the chloride content of coal and the extent to which mercury appears in the oxidized form. Fig. 4 illustrates the relationship between coal chlorine content and vapor-phase mercury speciation for the plants and coals tested in the EPA study. An important reason for the significant uncertainty (scatter) is that mercury oxidation proceeds by both homogeneous (gas-to-gas) and heterogeneous (gas-to-solid) reaction mechanisms. Boiler convection pass and air heater temperature profiles, flue gas composition, flyash characteristics, and the presence of unburned carbon have all been shown to affect the conversion of elemental mercury to oxidized mercury species. Table 5 provides a summary of general mercury speciation by coal type.

Mercury control technologies

For industrial sources, such as MSW combustors, the typical mercury control technology is activated carbon injection in conjunction with a spray dryer FGD system and fabric filter. Activated carbon injection is used to reduce both mercury and dioxin emissions. The system is comprised of a series of injection nozzles located in the flue gas stream between the last boiler heat trap (economizer or air heater) and the entrance to the spray dryer absorber (SDA). At this location, the injected carbon particles adsorb both mercury and dioxin onto their surfaces by direct contact. When the carbon particles become trapped in the downstream fabric filter, additional adsorption takes place. A typical carbon injection rate is 0.2 lb (0.09 kg) of carbon per ton of MSW combusted. Thus, for a 750 t/d (680 t_m/d) refuse boiler, approximately 150 lb (68 kg)/day of activated carbon would be used. Long-term experience reveals that mercury reduction without carbon injection averages around 70% for a system with only a spray dryer FGD with fabric filter and about 90% with carbon injection added. Most of the mercury produced by these MSW combustors is oxidized mercury.

For utility coal-fired boilers, the development of mercury emissions control technology in the U.S. over the past decade has progressed from a very limited database to one in which utilities, various government agencies, and system suppliers have increasingly quantified various parameters that influence mercury emissions control. This work is ongoing and reflects the complexity of the measurement and control issues. There are three basic methodologies for controlling mercury stack emissions from coal-fired power plants:

1. Choice of a coal with as low a mercury concentration as possible. The degree to which the coal has been characterized and represents the inlet conditions for the emissions control devices is the first step in setting the expectations for equipment performance. The potential effectiveness of emission control devices is influenced most directly by the form of the mercury, either solid or vapor phase, and the vapor phase mercury speciation, elemental mercury versus oxidized mercury. The presence of a selective catalytic reduction system (SCR) for NO_x emission control upstream of the particulate control device tends to increase the oxidized mercury present, which is subsequently easier to control in a downstream FGD system.
2. Use of a suitably injected sorbent for mercury capture, such that the sorbent particulate matter is captured with the flyash. Powdered activated carbon is the most likely sorbent and can be effective for both elemental and oxidized mercury capture, though the costs can be significant. Typically, the amount of activated carbon needed with downstream fabric filtration is less than that required when using electrostatic precipitation. Flyash from coal-fired utility boilers frequently is sold for byproduct uses such as cement manufacturing. However, flyash contaminated with activated carbon is generally not marketable. One solution is to inject the powder activated carbon downstream of the primary dust collector and add a separate dust collector to trap the injected activated carbon.

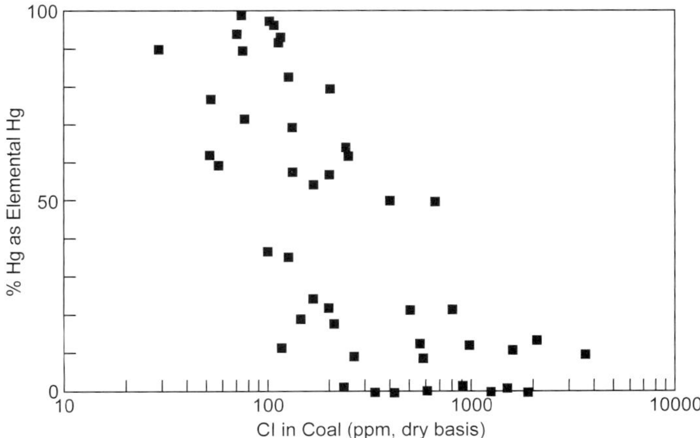

Fig. 4 Relationship between coal mercury content and mercury speciation for U.S. EPA Information Collection Request.

Table 5
Typical Mercury Speciation by Coal Type

Coal Type	Relative Content Mercury	Relative Content Chlorine	Primary Mercury Form in Flue Gas
Bituminous (Eastern U.S.)	Intermediate	High	Oxidized
Bituminous (Western U.S.)	Low	Intermediate	Mixture
Subbituminous (Powder River Basin)	Low	Low	Mixture, more elemental
Lignite	High	Low	Primarily elemental

3. Use of the dry or wet FGD system to provide substantial mercury removal from the gas stream. This approach is particularly suitable if the vapor phase mercury contains a relatively high proportion of oxidized mercury species.

Mercury capture in conventional coal-fired systems

Mercury removal efficiency across air pollution control devices at existing U.S. coal-fired plants was also quantified under EPA's ICR study.[12] Average mercury removal for bituminous coals was higher than that achieved for the subbituminous coals and lignites. This is due to the higher percentage of oxidized mercury produced by eastern coals, but also may result, in part, from differences in flyash properties and the tendency of bituminous coals to produce higher levels of unburned carbon.

FGD systems Both wet and spray dryer FGD systems perform better at removing mercury with bituminous coal than with the low rank fuels. The difference is due almost entirely to the higher percentage of oxidized mercury produced by the bituminous coals. Oxidized mercury, typically mercuric chloride, is soluble in water, making it amenable to removal in SO_2 scrubbers. Elemental mercury, insoluble in water, passes through most scrubbers. In the scrubber, the soluble mercury species react with the slurry to produce insoluble mercuric sulfide that is collected as a solid waste byproduct. There is a growing body of evidence that a phenomenon known as mercury re-emission can exist in a wet FGD system. Some of the captured oxidized mercury is actually reduced back to elemental mercury and released into the outlet flue gas. The Babcock & Wilcox Company (B&W) has developed a patented additive for wet FGD systems to minimize re-emission.

SCR systems The same catalysts that reduce NO_x to N_2 in SCR systems also help oxidize elemental mercury. As the flue gas passes through the SCR reactor and over the catalyst, a portion of the elemental mercury is converted to mercury oxidized species, for example mercuric chloride and mercury oxides, which can then be more readily removed in the FGD system. The current data suggests that the presence of HCl in the flue gas is needed for mercury oxidation to occur across the SCR.

Fabric filter systems Fabric filters tend to remove significantly more mercury than do ESPs. Both systems are capable of high-efficiency removal of particulate-bound mercury. Fabric filters may perform better on the same flue gas by providing more intimate contact between the flue gases and flyash as the flue gases pass through the filter cake on the filter bags. This intimate contact may promote the adsorption of mercury species onto the flyash or unburned carbon particles in the filter cake. While providing many insights into the behavior of mercury species in coal-fired systems, it should be noted that the EPA ICR data are subject to a variety of limitations. The individual plant data represent a brief period in time and give little insight into the variability of mercury emissions due to variability in coal properties, impacts of plant operating conditions, or mercury sampling uncertainties.

Enhanced mercury capture

Numerous other studies have been and continue to be conducted by B&W and others to shed additional light on the behavior of mercury in coal-fired systems, and to develop cost-effective approaches to its control.

Powdered activated carbon (PAC) Many studies have focused on the injection of some form of PAC. Adsorption is a technique that has often been successfully applied for the separation and removal of trace quantities of undesirable components. PAC injection is used to remove mercury in municipal waste combustor exhaust gases, as discussed above. Although this approach appeared attractive for coal-fired boilers in early work, the economics of high injection rates can be prohibitive when applied to these plants. More refined studies are now in progress to define more precisely what can and can not be achieved with PAC. Other studies seek to enhance PAC technology by impregnation with additives such as halides to yield improved chemisorption of the mercury species that may be present.

Enhanced scrubbing As noted above, conventional emissions control system components are capable of removing oxidized mercury due to its water solubility. A variety of advanced technologies are currently under development to enhance the performance of these systems. Some of these increase the effectiveness of the wet scrubbing systems in capturing the mercury once it is in the soluble state. Others introduce oxidizing agents to the flue gas to increase the conversion of elemental mercury to soluble oxidized species.

Commercial demonstration Efforts to develop mercury emission control technology have progressed to full-scale limited-term demonstration tests. During 2001, full-scale PAC injection tests were run at Alabama Power's Plant Gaston and at We Energies' Pleasant Prairie Power Plant. At about the same time, B&W conducted full-scale tests of its enhanced wet FGD process at both Michigan South Central Power Agency's Endicott Station and at Cinergy's Zimmer Station. Such demonstrations are an important step in gaining commercial acceptance of the technologies. As of this writing, additional testing is ongoing.

Water pollution control

U.S. legislation – Clean Water Act

In 1972, the U.S. Congress enacted landmark legislation that greatly expanded existing laws designed to control water pollution. Commonly referred to as the Clean Water Act (CWA),[13] the Federal Water Pollution Control Act of 1972 and successive amendments, the Clean Water Act of 1977, and the Water Quality Act of 1987 provide the basic framework for protecting the quality of surface waters in the United States. The stated objective of the CWA is to restore and maintain the chemical, physical and biological integrity of the nation's waters.

Two significant elements of the 1972 Act were the creation of the National Pollution Discharge Elimination System (NPDES), a permitting program to regulate pollutant discharges from industrial and municipal sources, and the establishment of technology-based *end of pipe* effluent standards for specified industries and processes.

Requirements under the Act The NPDES permit program requires that every industrial or municipal facility or other *point source* discharging into public waters obtain a NPDES permit. The NPDES program also regulates wet weather discharges including storm water runoff associated with many industrial activities.

NPDES permits include specific numerical limits for a wide range of conventional, toxic and other nonconventional pollutants to ensure that discharges to receiving waters are protective of human health and the environment. NPDES permits may contain technology-based effluent limits (generally industry/process specific standards), water quality-based effluent limits (as may be necessary to protect the quality of the receiving waters according to designated use classifications), or a combination of both.

The intent of technology-based effluent limits is to establish a minimum level of treatment or control of pollutants in point source discharges using available treatment technologies. In 1982, the EPA promulgated technology-based effluent limits and pretreatment standards for 21 major industrial categories including new and existing steam power facilities [Note: pretreatment standards apply to any indirect discharges to publicly owned treatment works (POTWs) that are not otherwise covered under a facility's NPDES discharge permit]. The key effluent limits applicable to steam power plants are summarized in Table 6. The detailed requirements are set forth in the Code of Federal Regulations – 40 CFR Part 423.

Another provision of the CWA, initially focused on large steam electric facilities, required the EPA to develop regulations governing the capacity, location, design and construction of water intake structures used to draw large volumes of cooling water from lakes, rivers, estuaries and oceans. This CWA requirement is unique in that it applies to the intake and not the discharge of water. The major goal of the requirement is to protect aquatic life by minimizing the impingement and entrainment of fish and other organisms into cooling water systems. Phase I rules, applicable to new utility facilities with a water intake flow of greater than two million gallons per day and using 25% or more for cooling purposes, were promulgated in December 2001 (Re: 40 CFR Part 125). Phase II rules governing cooling water intake structures at existing electric utility plants withdrawing 50 million gallons or more of water per day were promulgated in July of 2004. Phase III rules governing water intake structures at certain other existing utility and industrial facilities (not covered under the Phase II rules) were proposed in November of 2004 and are due to be finalized by June of 2006.

Visit the EPA's home page on the World Wide Web at *www.epa.gov* for further information regarding the CWA and related regulatory programs.

Power plant discharge sources

The following describes the principal aqueous discharge streams from utility power plants.

Once-through cooling water Water from rivers, lakes or oceans is used to absorb heat from the steam condenser. The cooling water exiting the steam condenser is at an elevated temperature and can be returned to the source or pumped to a cooling tower for evaporative cooling before being returned to the steam condenser. In the former case, the cooling water contains significant concentrations of only one principal regulated pollutant, total residual chlorine (TRC), which arises out of chlorine addition for condenser fouling control. The duration of each chlorination event is limited. The concerns over TRC discharge include toxicity to living organisms and the generation of halogenated hydrocarbons.

Cooling tower blowdown When the heated cooling water from the steam condenser is cooled in an evaporative cooling tower, a buildup of dissolved solids and suspended matter occurs. Most of this buildup is removed from the system by cooling tower blowdown. Some of the suspended matter can settle out in the cooling tower basin and is removed at infrequent intervals. All of the dissolved solids and the remaining suspended solids are removed largely by cooling tower blowdown. Blowdown flow is adjusted to keep the concentration of dissolved and suspended solids below the limits required to control condenser tube fouling and corrosion. Other sources of chemical pollutants in blowdown include chlorine and organic chemicals for control of biofouling, corrosion inhibitors (consisting of chromate, zinc, polyphosphates, etc.), chemicals for scale control and products of corrosion. Some of these maintenance chemicals appear on the EPA's regulated toxic pollutants list and none are permitted to be present in detectable levels in cooling tower blowdown after treatment (except for chromium and zinc, which are separately regulated).

Table 6
Aqueous Discharge Limits for New Steam Power Generating Systems (Note 4)

Source and Pollutant (Note 1)	NSPS Effluent Limits, mg/l (Notes 2 and 3) Maximum	Average
All discharges		
pH (unitless)	6 to 9	—
PCBs	No discharge	—
Low volume waste (Note 5):		
TSS	100	30
OG	20	15
Chemical metal cleaning wastes:		
TSS	100	30
OG	20	15
Copper	1.0	1.0
Iron	1.0	1.0
Bottom ash transport water:		
TSS	100.0	30.0
OG	20.0	15.0
Once-through cooling water:		
Total residual chlorine (TRC)	0.2	—
Free available chlorine	0.5	0.2
Cooling tower blowdown:		
Free available chlorine	0.5	0.2
Zinc	1.0	1.0
Chromium	0.2	0.2
Other 126 priority pollutants	No detectable amount	
Coal pile runoff:		
TSS (1980)	50	—

Notes:
1. Nomenclature: TSS – total suspended solids; OG – oil and grease.
2. New Source Performance Standards.
3. 30 day rolling daily average (Average); maximum any one day (Maximum).
4. Adapted from Reference 1.
5. Low volume wastes include ion exchange, water treatment, evaporator blowdown, boiler blowdown, lab and floor drains, plus FGD waste water.

Ash handling water waste Ash produced from the combustion of fuel, whether oil or coal, is collected at different points in the combustion process. Flyash is the finer size ash collected by particulate collection systems and bottom ash is removed from hoppers at the furnace bottom. Additional hoppers at intermediate points also accumulate ash. In many cases, ash is moved from these points with sluice water, which then goes to a settling pond and can typically contain 5% suspended solids by weight.

The ash settling pond overflow contains dissolved and suspended solids, the quantities of which will depend on the source of the ash, the type of combustion process and the point from which it is extracted from the combustion process. Coal ash contains, in addition to the eight or nine major elemental constituents, a number of trace elements that can appear in pond overflow and which may need to be treated.

Coal pile runoff Open storage of large quantities of coal is required for an uninterrupted fuel supply to utility plants [on the average, 800 to 2400 yd^3 (611 to 1834 m^3) per megawatt of rated capacity is kept on hand]. The water and oxygen from the air react with the minerals in bituminous coal to produce a leachate contaminated with ferrous sulfate and sulfuric acid. The low pH from the acid accelerates dissolution of many of the metals present in the coal minerals.

FGD blowdown In wet FGD systems, a portion of the absorber slurry that is sprayed into the flue gas stream to remove SO_2 is removed from the absorber tank for dewatering. In the dewatering process, the solid reaction products are separated from the liquor. The liquor is recycled to the absorber tank where additional sorbent is added. Recycling of the liquor can result in chloride buildup which, in turn, can cause increased corrosion to the alloys in the system. This buildup can be controlled by the loss of liquor retained in the dewatered sludge or by a blowdown. An aqueous blowdown discharge would typically contain calcium sulfate, calcium sulfite and calcium chloride. Also, depending on flyash carryover, traces of metal ions could also be present. In setting effluent limitations in 1982, the EPA reserved regulating FGD aqueous discharge to a future date.

Metal cleaning wastes These aqueous wastes can arise from either *chemical* or *nonchemical* cleaning of metal heat transfer surfaces in the boiler.

Chemical metal cleaning uses chemical solvents for water-side cleaning of boiler system components to remove corrosion products. Cleaning intervals are measured in years for large utility boilers, and produce three to four boiler water volumes [20,000 to 100,000 gal (75,707 to 378,533 l)] of waste water per cleaning. The composition of the waste solvents depends on the construction material of the boiler system, but largely consists of iron with lesser amounts of copper, nickel, zinc, chromium, calcium and magnesium. The disposal method for the spent solvent depends on the type of chemical cleaning solvents used. When hydrochloric acid based solvents are used, spent solvent is treated on-site by neutralization and is discharged subject to the effluent limits in Table 6 or more stringent water quality standards. With approval from appropriate regulating bodies, organic-based solvent wastes are often incinerated in other operating boilers at the site. The metals in the chemical cleaning wastes are retained with the normal boiler ash.

Nonchemical water cleaning is used to remove fireside deposits by means of high pressure jets of water. The waste water can contain the same metals and pollutants contained in the ash deposits being removed. Because the deposit composition varies with location in the boiler, the wash water composition will depend on the location of the area being cleaned. These waste waters may be classified as either low volume wastes or metal cleaning wastes and are treated according to the corresponding effluent limits.

Low volume wastes These include discharges from ion exchange water treatment, evaporator blowdown, boiler blowdown, cooling tower basin cleaning, laboratory and floor drains, and drains and losses from house service water systems. FGD blowdown is also included until the EPA develops specific regulations for this stream. By EPA definition, low volume wastes are those from all sources taken collectively as if they were from one source. Excluded are those wastes for which specific effluent limits are established.

Water pollution control technologies

The technologies for waste water treatment used to meet limits for discharge include clarification and filtration.

Clarification This process is used to settle out larger suspended particles and condition smaller colloidal particles to make them settle and allow filtration for removal. A pond, reservoir or tank is used to allow larger particles to settle in a matter of hours. The finer particles overflow and are made to settle more quickly by the addition of chemical agents, coagulants and polymers that cause agglomeration to sizes large enough to settle out of suspension.

Filtration This uses a porous barrier across flowing liquid to remove suspended materials. Filtration can be used to supplement clarification and permits reducing suspended solids to the parts per million level. Filter types include sand filters that are generally slow and do not handle fine clay well. Preconditioning with coagulants can improve filtration rates. Dual media filters improve on sand filters by superimposing a coarse, granular material over the fine bed. This allows more of the filter bed to be used, reduces head loss, and provides higher flow rates and longer operating cycles before cleaning.

As required, and with approvals from appropriate regulating bodies, final waste stream pH is controlled by combining various plant streams to provide a neutral pH product. Where needed, acid or alkali addition can be used to achieve the final pH. Other treatments are also available to address other criteria pollutants where concentration warrants.

In selected cases, zero discharge water management is provided which does not return any waste water to water sources. Effectively all water brought into the plant is evaporated through cooling towers, ponds, evaporators, or the stack. Residual solids are then sent to disposal.

Solid waste disposal

The Resource Conservation and Recovery Act

The rapid growth of industrial activity and use of consumer goods by society have resulted in an explosive growth in the generation of solid wastes. The EPA has estimated that over 100 million tons of fossil fuel combustion wastes are generated annually in the U.S., principally from coal ash residues generated at utility power plants.

In 1976, the U.S. Congress passed the Resource Conservation and Recovery Act (RCRA)[14] as an amendment to the Solid Waste Disposal act of 1965. The RCRA establishes the regulatory framework to ensure that solid and hazardous wastes are properly managed to protect human health and the environment. Subtitle C of the Act establishes a comprehensive management system to regulate hazardous waste from the time it is generated until its final disposal, commonly referred to as a *cradle to grave* approach. Subtitle D of the Act directs states to develop and implement solid waste management plans and regulations to promote environmentally sound management practices for non-hazardous wastes which address landfill performance standards and recycling/beneficial use programs. Subtitle I of the Act regulates petroleum products and other hazardous substances that are stored in underground tanks.

As originally drafted, RCRA did not clearly address how fossil fuel combustion wastes should be regulated and in 1980, the U.S. Congress passed the Solid Waste Disposal Act Amendments which included what is commonly referred to as the *Bevill Exemption.* The Bevill Exemption excluded, among other things, fossil fuel combustion wastes from being regulated as hazardous waste under RCRA Subtitle C, pending completion of a report to Congress and a formal determination by the EPA as to whether regulation under Subtitle C is warranted.

In August 1993, the EPA published the first of two regulatory determinations concerning fossil fuel combustion wastes. This first determination applied only to separately managed, *large volume* wastes including flyash, bottom ash, boiler slag and flue gas emission control waste generated at coal-fired electric utilities and independent power facilities. The EPA determined that if managed properly, these materials would remain exempt from regulation as a hazardous waste under Subtitle C of the RCRA. The EPA's second ruling, published in May 2000, generally extended the hazardous waste exemption to other defined categories of *low volume* wastes generated at electric utilities and industrial facilities that combust coal, oil, natural gas, petroleum coke or mixtures of coal and other fuels. These low volume wastes, when properly co-managed with large volume combustion wastes, would typically include (but are not limited to) coal pile runoff, coal mill rejects/pyrite, water treatment sludge, and boiler chemical cleaning waste. However, if certain wastes are managed independently and exhibit the characteristics of a hazardous waste, they are subject to the hazardous waste regulations. For more detailed information refer to the EPA's final regulatory determination published in the Federal Register on May 22, 2000 (65 FR 32213).

While the EPA has concluded that fossil fuel combustion wastes do not warrant regulation as hazardous waste, the agency further determined that national non-hazardous waste standards under RCRA Subtitle D are needed for coal combustion wastes disposed in landfills and surface impoundments, or used in minefill applications. However, national regulations have yet to be developed under Subtitle D; therefore, the management and regulation of combustion waste products are currently handled by the individual states.

Finally, the EPA has concluded that beneficial uses of coal combustion waste (other than for minefill) pose no significant risk and no additional national regulations are needed. The EPA has stated they do not want to add any unnecessary barriers to the beneficial uses of combustion waste products because such uses conserve natural resources, reduce disposal costs and reduce the total amount of waste ultimately destined for disposal.

Solid combustion wastes

The principal solid waste streams in coal- and oil-fired utility boilers include the following:

1. *Bottom ash* is the portion of fuel ash that falls to the bottom of the furnace or from the stoker discharge. In coal-fired Cyclone furnace boilers, the bottom ash consists of slag that drops from the bottom of the furnace into a slag tank for solidification.
2. *Flyash* is the finer ash material that is borne by the flue gas from the furnace to the back end of the boiler; it drops out in the economizer and air heater hoppers or is collected by particulate control equipment.
3. *Pyrite* is iron sulfide, an impurity which is separated from coal in the pulverizer and which is combined with bottom ash for disposal.

FGD waste characteristics depend on the particular technology used:

1. *Wet scrubbing (calcium-based system)* A natural oxidation system produces a wet sludge containing a mixture of calcium sulfite and calcium sulfate reaction products, trace amounts of flyash and unreacted limestone. In a forced oxidation system, the principal difference is that the reaction product is almost totally in the form of calcium sulfate or gypsum, which is more easily dewatered to a filter cake for wallboard, landfill or other use.
2. *Dry scrubbing* Waste is dry and contains calcium sulfite, calcium sulfate, flyash and unreacted sorbent (hydrated lime). The potential uses for this material are more limited.
3. *Dry lime injection* Waste is dry and contains calcium sulfite, calcium sulfate, flyash and a large proportion of calcium oxide (CaO).

Solid waste treatment methods

To dispose of waste materials from wet FGD systems, treatment methods are applied to ultimately produce a solid. These methods include dewatering, stabilization and fixation, and are designed to achieve waste volume reduction, stability and better handling, and liquid recovery for reuse.

Dewatering This process is used to physically separate water from solids to increase the solids content of the product and recover water for reuse and further treatment.

A settling pond is the simplest method for dewatering, is not sensitive to inlet solids content, requires low maintenance and is highly reliable. Ponds are often used for ash or wet FGD scrubber slurries. Sizing provides low flow velocity so that solids can settle undisturbed by gravity. Settling ponds are unpopular with regulatory agencies, require substantial acreage and must be shut down for solids removal.

Thickeners are large cylindrical tanks with radial rakes at the bottom. The rakes carry plows to push material on the bottom that slopes toward the center. The plows push settled material toward the underflow discharge. Thickeners rely on gravity to separate high specific gravity solids. Although thickeners are complicated and have high capital and maintenance costs, they have high throughput rates and require less land area than settling ponds.

Liquid cyclones or hydroclones are now commonly used in place of thickeners to remove solids from slurries by centrifugal and liquid shear effects. Hydroclones separate and collect particles down to a particular size, with finer particles staying with the liquid overflow. Hydroclones do not separate material less than 5 μm effectively and are not efficient with slurries containing more than 20% solids. They are low in cost, have low space requirements, and produce low solids content in the overflow and a high solids fraction in the underflow.

Vacuum filters, either of the drum or belt type, are generally used for second stage dewatering of wet scrubber slurries. They take little space and produce a high solids content product, up to 95% for FGD slurry. Centrifuges are also an option for second stage dewatering.

Stabilization This process increases the solids content of scrubber waste by adding dry solids such as flyash. Stabilization is applied to impart greater physical stability to the waste, making it easier to place in the landfill and making it less susceptible to future problems. Stabilization and fixation are generally applied to scrubber wastes as the final treatment step after dewatering. Bottom ash and flyash, because of their more granular nature, generally dewater easily and do not require stabilization for disposal. For stabilization, a dry solid such as soil or flyash is mixed with the waste slurry, spreading the water in the waste over a larger mass of solids. Also, there is improvement in particle size distribution that leads to closer packing, lower permeability and lower combined volume. Stabilization can be reversible and if the waste is rewetted, it may fluidize and fail structurally.

Fixation This process involves the addition of an agent such as lime to produce a chemical reaction to bind free water and produce a dry product. Fixation includes a number of processes. Mixing suitable proportions of scrubber slurry with alkaline flyash containing sufficient CaO produces chemical reactions that result in a material with compressive strength comparable to low-strength concrete and with very low permeability. Both characteristics contribute to ease of placement and minimal leaching problems.

When the flyash does not have sufficient alkalinity, lime may be added to the flyash and scrubber slurry mixture to produce the cementitious reaction. Four percent addition of lime has produced material with the necessary physical properties for disposal or use. The cured material is suitable for structural fill, providing a site that can be used for building construction after completion of the landfill. Comparable fixation reactions with scrubber sludge have been obtained with additions of 5 to 10% blast furnace slag.

Disposal and utilization methods and requirements

Ultimate disposition of utility plant wastes (ashes and FGD residues) is by disposal (in landfills or impoundments) or by utilization. Where disposal is used, the waste stream is analyzed, and the site is permitted and approved by the appropriate regulatory agencies.

Disposal methods These can be either wet or dry, depending on the physical condition of the material. Wet disposal requires construction of a pond which may be below or above grade with impermeable barriers or dikes. Below grade construction may be considered and depends on suitable geology and hydrology at the site. With wet disposal, the waste is placed in slurry or liquid form. After settling, the liquid that has separated is collected, treated and either released or recycled. Dry disposal can use a simple method of landfill construction in which the waste is placed and compacted to form an artificial hill. The trend is toward dry disposal because of smaller volumes and more options for site or material reclamation.

Utilization methods These become more attractive as waste management costs increase. Bottom ash, flyash and boiler slag are used in applications where they can be substituted for sand or gravel. The characteristics of boiler slag and bottom ash also make these materials useful for blasting grit, roofing granules and controlled fills. Flyash, because of its chemistry and physical properties, is applicable in the manufacture of Portland cement and concrete mixes. The value of these materials is so low that the cost of transportation severely limits their use to applications close to the producing power plant.

With the increased use of ammonia reagents for NO_x emissions control, flyash (particularly from bituminous coal) can adsorb excess ammonia present in the flue gas and this adsorption may hinder the beneficial use or end disposal of flyash or FGD byproduct streams. Ammonia absorbed by flyash can be released (off-gassed) when coming in contact with water. Currently, there are various processes in development to remove residual ammonia from flyash.

FGD byproduct use is potentially in the areas of agriculture, sulfur recovery and gypsum. Agricultural use is limited. Trace elements from flyash contamination could have an unacceptable impact and make wide use doubtful. Use for sulfur recovery is limited by incomplete technology development, high capital cost, and the low market price of sulfur.

FGD byproduct from forced oxidation wet scrubbing systems, primarily gypsum, has seen extensive commercial use in wallboard production. However, gypsum byproduct specifications may vary significantly by end user and must be established and confirmed prior to FGD system design.

References

1. Elliot, T.C., Ed., Chen, K., Swanekamp, R.C., *Standard Hankbook of Power Plant Engineering*, Second Ed., McGraw-Hill Company, New York, New York, 1998. See chapter entitled, "Legislation and pollution sources," by D.A. Kellermeyer.

2. Corbitt, R.A., Ed., *Standard Handbook of Environmental Engineering*, Second Ed., McGraw-Hill Company, New York, New York, 1998.

3. Clean Air Act, 42 USCA S7401 *et seq.*, Sec. 101 (b) (1).

4. Title 40, Code of Federal Regulations, Part 60, United States (U.S.) Government Printing Office, Washington, D.C., July, 2004.

5. Title 40, Code of Federal Regulations, Parts 51 and 52 (51.165, 51.166, and 52.21), United States (U.S.) Government Printing Office, Washington, D.C., July, 2004.

6. Lesley, S., "Trends in Emissions Standards," Report CCC/77, International Energy Agency (IEA), Clean Coal Centre (CCC), London, England, United Kingdom, November, 2003.

7. Wu, Z., "NO_x Control for Pulverized Coal-Fired Power Stations," Report CCC/69, International Energy Agency, Clean Coal Centre, London, England, United Kingdom, December, 2002.

8. United States (U.S.) Environmental Protection Agency (EPA), National Emissions Inventory Criteria Pollutant Data: Current Emissions Trends Summary 1970-2002, Accessed August 28, 2003. Available online at: www.epa.gov/ttn/chief/trends/index.html

9. Patrick, D.R., *Toxic Air Pollution Handbook*, Van Nostrand Reinhold, New York, New York, 1994.

10. "Mercury Study Report to Congress," United States (U.S.) Environmental Protection Agency (EPA) Report EPA-452/R-97-2003, December, 1997.

11. "Emissions of Greenhouse Gases in the United States (U.S.), 2000 Summary," Report DOE/EIA-0573, Department of Energy (DOE) Energy Information Agency (EIA), U.S. Department of Energy, February 22, 2002.

12. Senior, C.L., "Behavior of Mercury in Air Pollution Control Devices on Coal-Fired Utility Boilers," Power Production in the 21st Century: Impacts of Fuel Quality and Operations Conference, Engineering Foundation, New York, New York, 2001.

13. Clean Water Act, 33 U.S.C., S1251 et seq., Section 101(a), United States (U.S.) Environmental Protection Agency (EPA).

14. The Resource Conservation and Recovery Act (RCRA), 42 Act, 42 U.S.C., S9601 et seq., Section 1004(5).

Pulse jet fabric filter under construction at a 550 MW power plant.

Chapter 33

Particulate Control

As steam is widely generated by the combustion of most fossil fuels, the flue gas carries particulate matter, principally ash, from the furnace. Except for natural gas, all other fossil fuels contain some quantity of non-combustibles which form the majority of the particulate. Unburned carbon also appears as particulate. Particulate control is needed to collect this material and to limit its release to the atmosphere.

All coals contain some amount of ash. The amount of ash varies depending on the type of coal, location, depth of mine and mining method. In the United States (U.S.), eastern bituminous coals typically contain 5 to 15% ash while the western subbituminous coal ash content may range from 5 to 30% ash by weight. Texas lignites also contain up to 30% ash. Mining methods on thin seams of coal may also contribute to higher ash quantities (see Chapter 9).

When coal is burned in conventional boilers, a portion of the ash drops out of the bottom of the furnace (*bottom ash*) while the remainder of the ash is carried out of the furnace in the flue gas. It is this remaining ash (*flyash*) that must be collected after the furnace and before exhausting the flue gas to the atmosphere.

Different combustion methods contribute different proportions of the total coal ash content to the flue gas. With pulverized coal firing, 70 to 90% of the ash is carried out of the boiler with the flue gas. A stoker-fired unit will emit about 40% of its ash in the flue gas along with some amount of unburned carbon. With Cyclone firing, only 15 to 40% of the ash is normally carried by the flue gas. On circulating fluidized-bed boilers, all of the ash, along with the fluidized-bed material, is carried by the flue gas. Therefore, the selection and design of particulate control equipment are closely tied to the type of firing system.

An American Society for Testing and Materials (ASTM) ash composition analysis test (ASTM D-3682-01) of a coal ash sample reveals the major ash components. Proximate and ultimate coal analyses commonly offer additional insight and also provide total ash content. Ash components are typically reported in the oxide form and include silicon dioxide, titanium dioxide, iron oxide, aluminum dioxide, calcium oxide, magnesium oxide, sodium oxide, potassium oxide, sulfur trioxide and diphosphorous pentoxide. Trace quantities of many more elements are also found in ash. The proportion of the major constituents varies significantly between coal type and mine location. The analysis and composition of flyash are discussed in greater detail in Chapters 9 and 21.

Other significant coal ash properties are particle size distribution and shape, both of which are dependent on the type of firing method. Stoker-fired units generally produce the largest particles. Pulverized coal-fired boilers produce smaller, spherical shaped particles of 7 to 12 microns (Fig. 1). Particles from Cyclone-fired units, also mostly spherical, are among the smallest. Fluidized-bed units produce a wide range of particles that are generally less spherical and are shaped more like crystals. Knowledge of ash properties is important in the selection of the correct particulate control equipment.

Regulation of particulate emissions

Particulate control equipment was first used by utilities in the 1920s and before that time in some industrial applications.[1] Prior to 1971, however, controls were installed largely on a best effort basis. In 1971, the first Environmental Protection Agency (EPA) performance standard limited outlet particulate emissions to 0.1 lb/10^6 Btu (123 mg/Nm^3 at 6% O_2) heat input

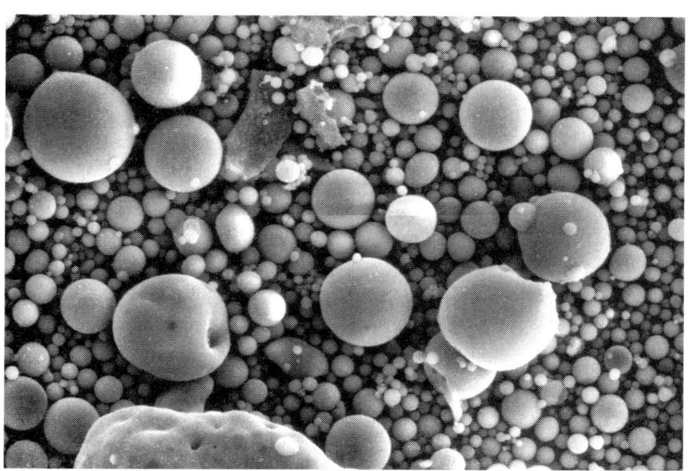

Fig. 1 Flyash from pulverized coal (magnified X 1000).

and stack opacity to 20% for those units larger than 250×10^6 Btu/h (73.3 MW_t) heat input. *Opacity,* measured by an instrument called a transmissometer, is the portion of light which is scattered or absorbed by particulate as the source of light passes across a flue gas stream. Therefore, both the amount and appearance of the stack emissions are regulated. Since 1979, the EPA New Source Performance Standards (NSPS) for particulate control permit a maximum of 0.03 lb/10^6 Btu (36.9 mg/Nm^3 at 6% O_2) heat input for these units. A 20% opacity is still permissible.

Federal and state EPA regulations set the primary guidelines for particulate emissions. In addition, many local regulatory bodies have generally stricter regulations than those set by the federal EPA. There are separate emissions standards for a variety of combustion processes including steam generators firing coal, oil, refuse and biomass. Currently there are three major classification levels for steam generating units: one for units greater than 250×10^6 Btu/h (73.3 MW_t), one for the 100 to 250×10^6 Btu/h (29.3 to 73.3 MW_t) units, and a third for those units less than 100×10^6 Btu/h (29.3 MW_t) heat input.[2,3] Finally, if a new plant is in a nonattainment area, the permissible particulate emissions and opacity may be significantly reduced from nominal control levels.

Particulate control technologies

Particulate emissions from the combustion process are collected by particulate control equipment (Fig. 2). This equipment must remove the particulate from the flue gas, keep the particulate from re-entering the gas, and discharge the collected material. There are several major types of equipment available including electrostatic precipitators, fabric filters (baghouses), mechanical collectors and venturi scrubbers. Each of these uses a different collection process with different factors affecting the collection performance. Prior to the 1990s, the technology of choice for large coal-fired utility plants was the dry electrostatic precipitator, with fabric filters a distant second. Today, preference is given to the fabric filter (pulse jet type) for reasons described below.

Dry electrostatic precipitators

A dry electrostatic precipitator (ESP) electrically charges the ash particles in the flue gas to collect and remove them. The unit is normally comprised of a series of parallel, vertical metallic plates (collecting electrodes or CEs) forming ducts or lanes through which the flue gas passes. Centered between the CEs are discharge electrodes (DEs) which provide the particle charging and electric field. Fig. 3 is a plan view of a typical ESP section which indicates the process arrangement.

Charging The CEs are typically electrically grounded and connected to the positive polarity of the high voltage power supply. The DEs are suspended off of electrical insulators in the flue gas stream and are connected to the output (negative polarity) of a high voltage power source, typically 55 to 85 average kV DC. An electric field is established between the DEs and the CEs, and the DEs will exhibit an active glow, or *corona*. As the flue gas passes through the electric field, the particulate takes on a negative charge which, depending on particle size, is accomplished by field charging or diffusion.

Collection The negatively charged particles are attracted toward the grounded CEs and migrate across the gas flow. Some particles are difficult to charge, requiring a strong electric field. Other particles are charged easily and driven toward the plates, but also may lose the charge easily, requiring recharging and recollection. Gas velocity between the plates is also an important factor in the collection process since lower velocities permit more time for the charged particles to move to the CEs and reduce the likelihood of loss back into the gas stream (re-entrainment). In addi-

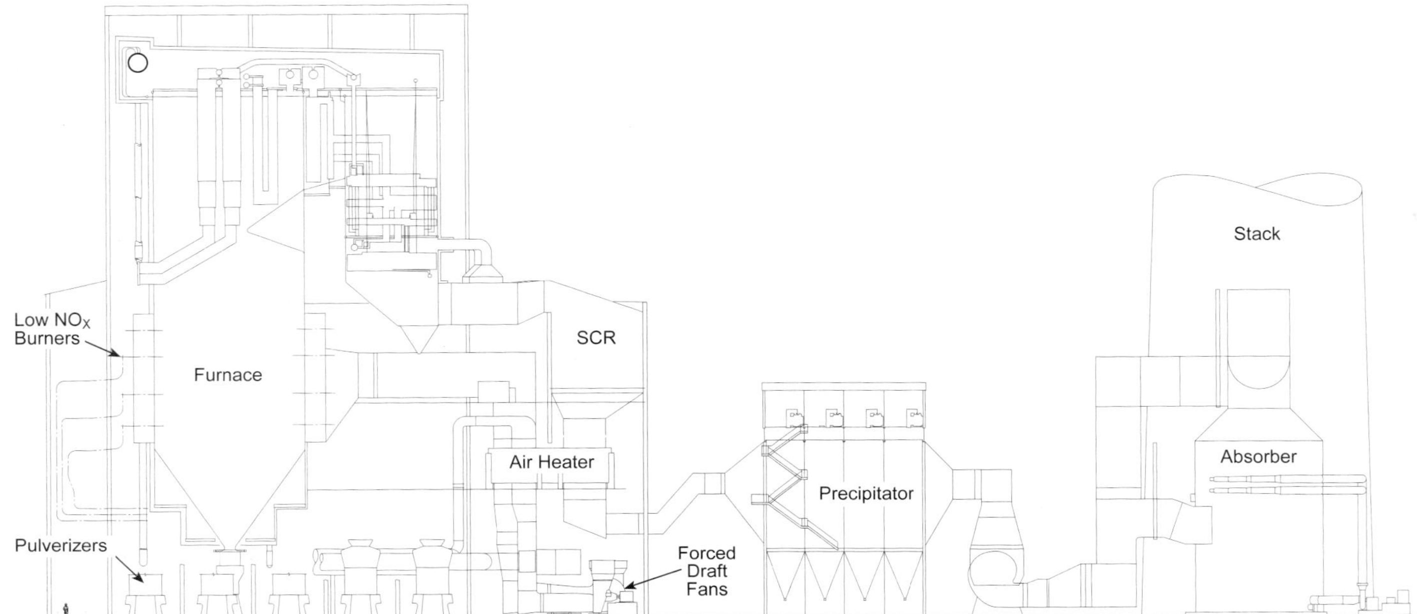

Fig. 2 Particulate control equipment – plant side view.

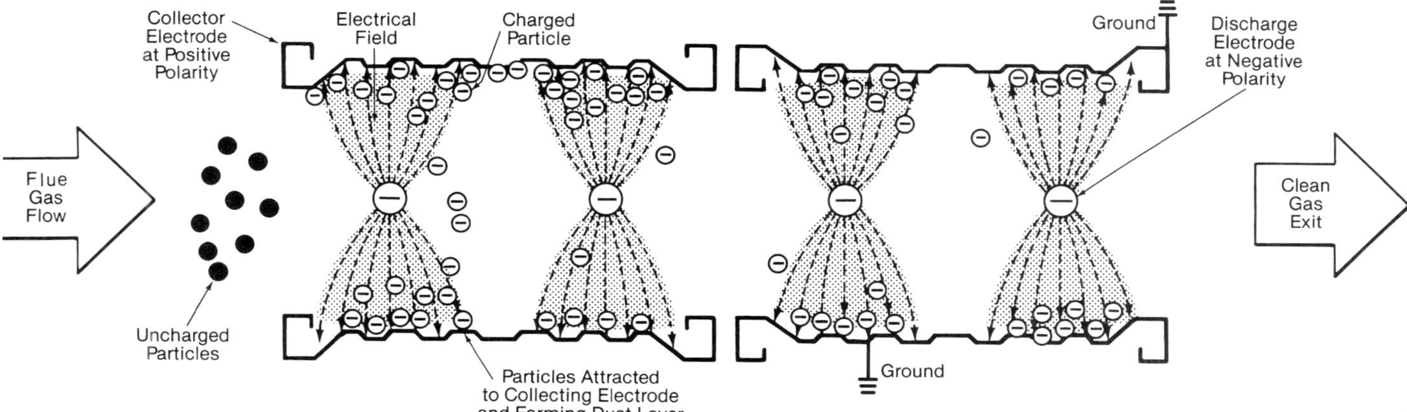

Fig. 3 Particle charging and collection within an ESP.

tion, a series of CE and DE sections is generally necessary to achieve overall particulate collection requirements. In modern ESPs designed for utility flyash, four or more sections in series are required to achieve design collection levels.

The ash particles form an ash layer as they accumulate on the collection plates. The particles remain on the collection surface due to the forces from the electric field as well as the molecular and mechanical cohesive forces between particles. These forces also tend to make the individual particles agglomerate, or cling together.

Rapping The ash layer must be periodically removed. The most common removal method is rapping which consists of mechanically striking the collection surface; this rapping force dislodges the ash. Because particulate tends to agglomerate, the ash layer is removed in sheets. This sheeting is important to prevent the re-entrainment of individual particles into the flue gas stream, requiring recharging and recollection downstream.

While most of the particles are driven to the CEs, some positively charged particles attach to the DEs. A separate rapping system is therefore used to remove deposits from these electrodes and maintain proper operation.

Ash removal The dislodged particulate falls from the collection surface into hoppers. Once the particulate has reached the hopper it is important to ensure, by proper design, that it remains there in bulk form with minimal re-entrainment until the hopper is emptied. (See Chapter 24 for hopper ash removal methods and equipment.)

Dry ESP applications

Utility Because coal is the most common fuel for steam generation, collection of the coal ash particles is the greatest use of a particulate collector. The electrostatic precipitator has been the most commonly used collector. To meet the particulate control regulations for utility units and considering the resulting high collection efficiency, special attention must be given to details of precipitator sizing, rapping, flow distribution and gas bypass around the collector plates. The result will then be a collector that can be confidently and consistently designed and operated to meet the outlet emissions requirements. Operating collection efficiencies which exceed 99.9% are common on the medium and higher ash coals with outlet emissions levels of 0.01 to 0.03 lb/10^6 Btu (12.3 to 36.9 mg/Nm^3 at 6% O_2) heat input common on all coals. ESPs are also widely installed on utility boilers that fire oil as their principal fuel.

Industrial Other common noncoal-fired industrial units where ESPs are successfully being applied include municipal refuse incinerators and wood, bark, and oil-fired boilers. For these, the ash in the flue gas is typically more easily collected than coal flyash so an ESP of modest size will easily collect the particulate. The moisture content in the refuse, wood and bark is the major contributor to the low resistivity. The carbon content of the residue, ash and unburned combustibles also contributes to low resistivity. (See note below.)

Pulp and paper In the pulp and paper industry, precipitators are used on power boilers and chemical recovery boilers. The power boiler particulate emissions requirements are the same as those for the industrial units using the same fuels. For the recovery boilers, precipitators are used to collect the residual salt cake in the flue gas. Chapter 28 contains further information on the recovery boiler processes and the reuse of the collected material.

A recovery boiler is a unique application for a precipitator due to the small particulate size and the tendency for the cohesive ash particles to stick together. The resistivity of the particulate is low so it is collected easily in the ESP. However, the fine particulate can also cause problems with the generation of effective corona by the DEs due to an effect called space charge. Because the particulate is so small, gas bypass around collector plates and re-entrainment of rapped particulates in the flue gas are more of a design concern. Re-entrainment is minimized by proper gas flow control and by lower gas velocities. Precipitator collection efficiencies are 99.7 to 99.8% to meet the 20% opacity and the local emissions requirements. Due to the characteristics of the salt cake particulates, a drag chain conveyor across a precipitator floor, rather than a

---

Note: Resistivity is an inverse measure of a particle's ability to accept and hold a charge. Lower resistivity indicates improved ability to accept a charge and be collected in an ESP.

normal hopper, is used for salt cake removal. In addition, casing corrosion is a more significant concern and as a result more insulation is required to reduce casing heat loss. Finally, in order to improve system reliability, two precipitator chambers are commonly used, each capable of handling 70% of the gas flow and each equipped with separate isolation capabilities.

Precipitators have also been applied in the steel industry to collect and recover the fine dust given off by some processes.

Dry precipitator components

All ESPs have several components in common (Fig. 4) although there is some shape and size variation between units.

Casing As shown in Fig. 4, the structure forming the sides and roof of an ESP is a gas-tight metal cased enclosure. The structure rests on a lower grid, which serves as a base and is free to move as needed to accommodate thermal expansion. All of the collecting plates and the discharge electrode system are top supported from the plate girder assemblies. The entire enclosure is covered with insulation and lagging. Access doors in the casing and adequately sized walkways between the fields assist in maintenance access for the internals.

Materials for the precipitator enclosure and internals are normally carbon steel, ASTM A-36 or equivalent, because gas constituents are noncorrosive at normal operating gas and casing temperatures. Projects with special conditions may warrant an upgrade in some component materials.

Hoppers For utility flyash applications, metal inverted pyramid-shaped hoppers are supported from the lower grid and are made of externally stiffened casing. The hoppers provide the lower portion of the overall enclosure and complete the gas seal. Their sides are designed with an inclination angle of at least 60 deg from horizontal. Hoppers are generally designed as particulate collection devices that can store

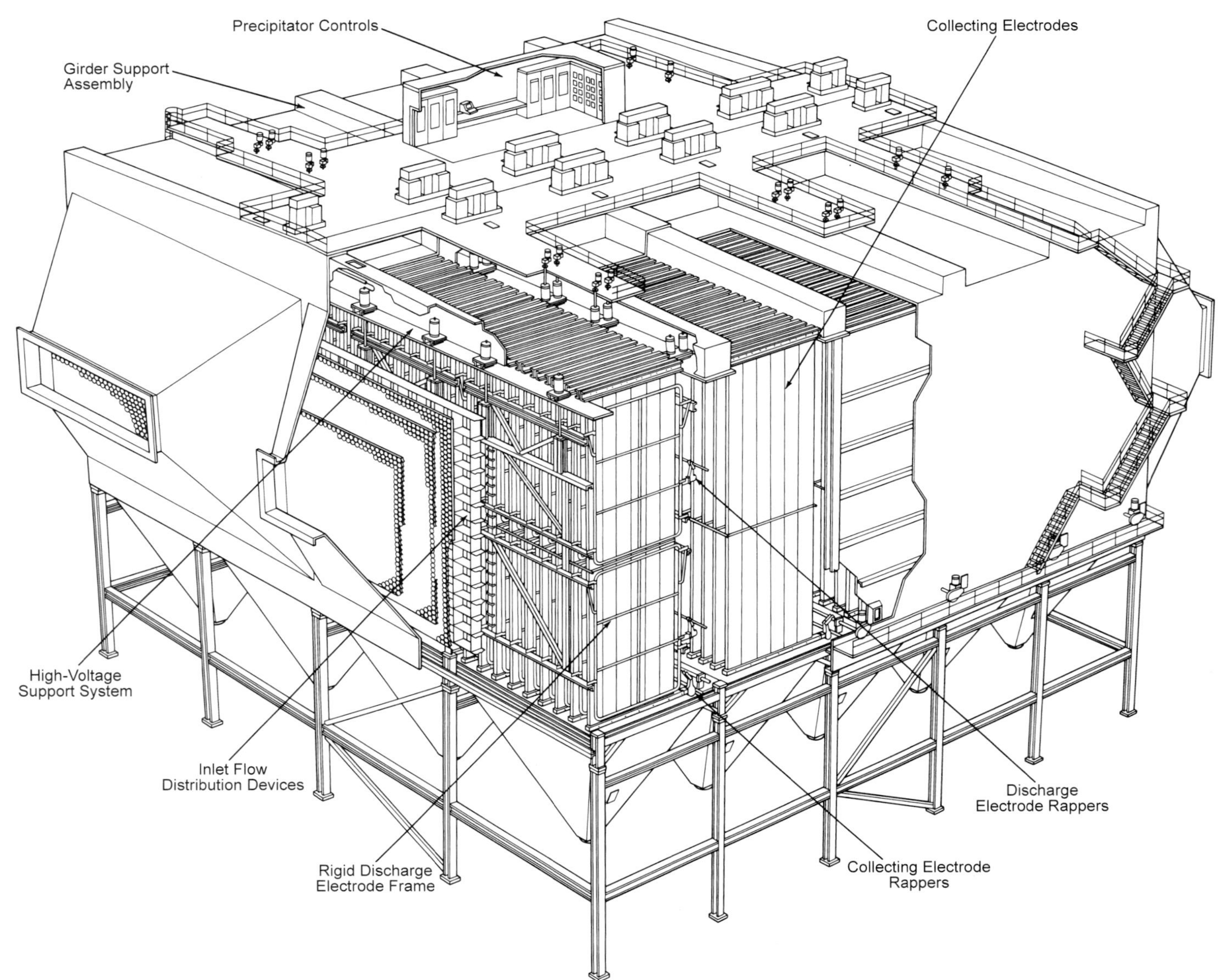

Fig. 4 B&W rigid frame electrostatic precipitator.

ash for short periods of time when the ash removal system is out of service.

Because many ash removal systems are noncontinuous, the following items are normally supplied with the precipitator hoppers to ensure good particulate removal: hopper heaters, electromagnetic vibrators, poke holes, anvil bars and level detectors. Hot air fluidizing systems are also sometimes supplied to assist in ash removal.

For non-utility flyash applications, the use of alternative ash holding/removal designs are common. These include flat-bottom with drag scrapers, wet-bottom, and trough-type hoppers.

Collecting electrodes As shown in Fig. 4, the CE surface typically consists of a series of roll-formed collector plates assembled into a CE and supported from the top. CEs are also widely referred to as plates or curtains. The CEs are spaced in rows across the width of the precipitator, typically on 12 or 16 in. (305 or 406 mm) centers in modern designs that also use rigid DEs. In older designs, CE spacings of 9 or 10 in. (229 or 254 mm) were generally seen when used with wire (0.109 in./2.8 mm diameter typical) DEs. In the direction of gas flow, the CE sections are arranged into fields which are normally powered by separate and dedicated power supplies. The collection surface area in the Deutsch-Anderson equation (see Equation 1) is the total CE plate area required for particulate collection. For calculating surface area, the CE assembly is treated as a plane and includes both sides of the CE where exposed to gas flow. The rolled plates can be up to 50 ft (15.2 m) in length with a shop straightness tolerance of 0.5 in. (12.7 mm). CEs may also be large flat plates with stiffener bars added to maintain straightness. For optimum performance with a uniform electric field and with minimal inter-electrode sparkover or electrical arcing (high current spark), the alignment of collection surface and electrodes must be maintained within tight tolerances.

Discharge electrodes and insulators As described in the section on charging, the DEs, connected to the high voltage power source, are located in the gas stream and serve as the source of the corona discharge. These electrodes are the central components of the discharge system which is electrically isolated from the grounded portions of the ESP. An electrical clearance gap of 6 to 8 in. (152 to 203 mm), depending on CE spacing, must be maintained throughout the ESP between the DE system and any grounded components.

Discharge electrodes are found in several shapes. Common types include the rigid frame, rigid electrode and weighted wire. The rigid frame, shown in Fig. 5, consists of strips of electrode supported between sections of frame tubing. Each frame is attached to a structural carrier, both front and rear. This assembly is supported by insulators forming a four point suspension system. The rigid electrode consists of a member with proprietary shape that is top supported and hangs the full height of the precipitator. The typical rigid electrode top support is also a frame hanging from insulators. The lower ends of the rigid electrode have a guide bar and side to side spacers. The third type of discharge electrode, weighted wire, consists of a round or barbed wire supported at the top and held straight and in tension with a weight at the bottom. The upper frame is supported from insulators and there is a lower steadying frame to guide and space the electrodes.

For highest equipment reliability, either the rigid frame or rigid electrode is the most common configuration. Discharge electrode failure in the form of broken wires has been a recurring problem with the weighted wire electrodes, particularly with lengths of 30 ft (9.1 m) or more, resulting in performance deterioration.

Gas flow control devices ESPs operate most efficiently when the gas flow is distributed evenly across the ESP cross section. Flow control devices such as turning vanes, flow straighteners and perforated plates are frequently installed in the inlet and outlet flues, and flow transition pieces, to provide the desired degree of flow uniformity while optimizing resultant pressure drop.

Rapping systems As shown in Fig. 4, the most effective method of cleaning the collector curtains is to rap each one separately and in the direction of gas flow. This method assures that each curtain receives a rapping force. The rapping system shown is a tumbling hammer type, where the hammer assemblies are mounted on a shaft extending across the ESP in a staggered arrangement. The shaft is turned slowly by an external drive controlled by timers for rapping fre-

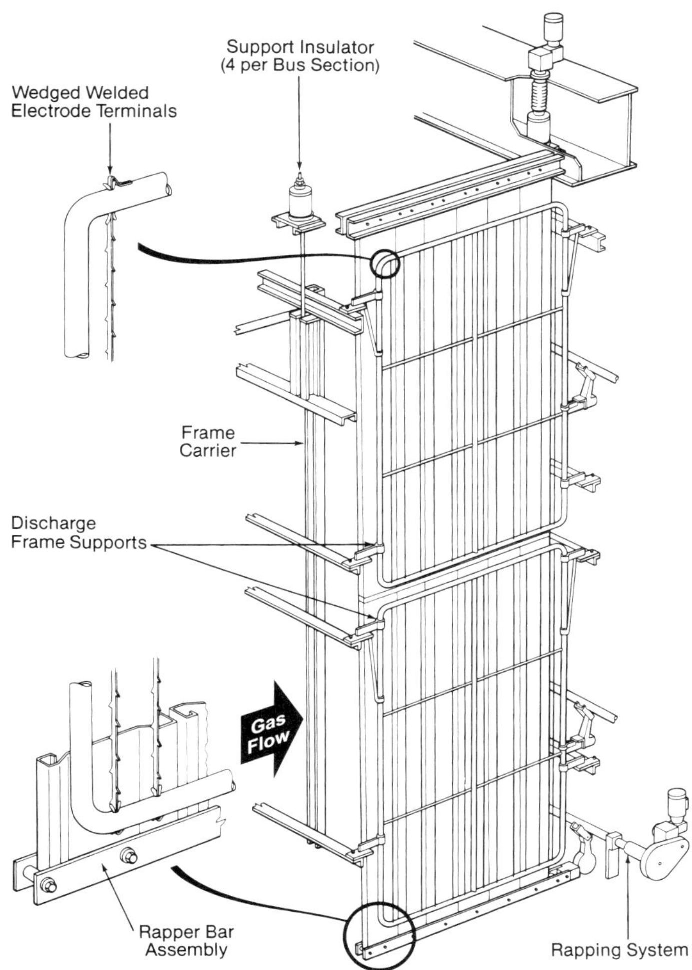

Fig. 5 Rigid frame discharge electrode and rapping system for an ESP.

quency and optimum cleaning. Hammer size is selected to match the application and size of the collector curtain. External, top-mounted rapping technology has also been widely used, principally in the U.S. Typically, more than one collector curtain is rapped at a time with this method, and the rapping force is in the downward direction on the top edge of the curtains. Both a drop rod and magnetic impulse are the drive mechanisms used.

Due to the difficulty of cleaning high resistivity flyash from the collection surface, considerable tests have been performed to ensure that adequate rapping forces are transmitted across the entire collection surface. A minimum acceleration of 100 g applied at the farthest point from impact has been established as an industry standard.

Typically, rapping of the rigid frame discharge electrodes is accomplished using a tumbling hammer system as shown in Fig. 5. The hammer assemblies are mounted in a staggered arrangement on a shaft across the width of an electrical section. Note that a smaller hammer than that used for the collector system is required for proper cleaning of the discharge electrodes. An external drive unit mounted on the precipitator roof is used to slowly turn the rapper shaft and, because it is attached directly to the carrier frame, the drive shaft must also be electrically isolated with an insulator. As with the collector system, top rapping of the discharge electrodes is another method of cleaning that is typically used with the rigid electrode designs.

A rapping system is sometimes used on the flow distribution devices at the precipitator inlet.

Power supplies and controls A transformer rectifier (TR) set along with a controller supply the high voltage power to the discharge electrode system. Several TR sets are normally needed to power a precipitator. With this combination of electrical components, the single-phase 480 V AC line voltage is regulated in the controller and then transformed into a nominal 55,000 to 75,000 V before being rectified to a negative DC output for the discharge system. Electrically, a precipitator most closely resembles a capacitive load. Due to the capacitive load and the nature of the precipitator internals, the TR set must be designed to handle the current surges caused by arcs between the discharge electrodes and the grounded collection surface. A current-limiting reactor in series with the TR set primary also helps to temporarily limit the current surges.

Traditionally, a voltage controller tries to maximize the voltage input to the precipitator. To achieve this input and when operating as designed, the controller must periodically raise the voltage to the point of sparking between the discharge electrode and the collection surface. The controller must then also detect the sparks and reduce the voltage to avoid an arc.

Today's microprocessor TR controls with quick response times, interface advantages and programming capabilities provide many functions to optimize particulate collection.

Dry precipitator sizing factors

An ESP is sized to meet a required performance or particulate collection efficiency. The sizing procedure determines the amount of collection surface to meet the specified performance. An equation which relates the collection efficiency (E) to the unit size, the particle charging and the collection surface is the Deutsch-Anderson equation:[1]

$$E = 100\% \times \left(\frac{\text{Inlet dust loading} - \text{Outlet dust loading}}{\text{Inlet dust loading}} \right) \quad (1)$$

$$1 - \frac{E}{100} = e^{\frac{-wA}{V}} \quad (2)$$

or

$$A = \left[\ell n \left(\frac{1}{1-(E/100)} \right) \right] \frac{V}{w} \quad (3)$$

where

E = ESP removal efficiency, % of inlet particulate removed
w = migration velocity, ft/min (m/s)
A = collection surface area, ft^2 (m^2)
V = gas flow, ft^3/min (m^3/s)

Migration velocity is the theoretical average velocity at which the charged particles travel toward the collection surface. This velocity is dependent upon how easily the particulate is charged, and the value is normally selected by empirical means based on experience. The factors which affect migration velocity are the fuel and ash characteristics, the operating conditions, and the effects of gas flow distribution.

These factors also have an effect on the ability of the particulate to accept a charge. A commonly used indication of this effect is resistivity, measured in ohm-cm. Fig. 6 illustrates typical resistivity curves for two fuel ashes. High resistivity ashes result in low migration velocities and large collection surface areas while average resistivity ashes result in moderately sized surface areas, i.e., lower resistivity indicates improved collection.

As previously discussed, the flue gas will pass through a series of collection fields. ESPs collect particulate in geometric fashion meaning each field will tend to collect very roughly 80% of the particulate entering that particular field. Therefore, the inlet field will collect the vast majority of the total mass particulate with each successive field collecting geometrically less of the total. For example, the first field could collect 80% of the total particulate loading while the second field would collect 80% of the remaining 20% or 16% of the total particulate loading, etc.

Fuel and ash characteristics The fuel and ash constituents which reduce resistivity or that are favorable to ash collection in an ESP include moisture, sulfur, sodium and potassium. Applications with sufficient quantities of these components usually result in moderately sized precipitators. Constituents which hamper ash collection and increase outlet emissions include calcium and magnesium. High percentage concentrations of these items without offsetting quantities of the favorable constituents result in poorer collection and larger precipitators. The fuel and ash constituents and their relative quantities must be reviewed in the sizing process to determine the overall effect on migration velocity. The migration velocity/resistivity can then be altered to some extent by controlling the content of the critical constituents.

Gas temperature, volume and distribution As indicated in Fig. 6, gas temperature has a direct effect on resistivity and on the gas volume passing through the ESP. Gas flow from the boiler also has an effect on sizing as indicated by the Deutsch-Anderson equation. There is an optimum gas velocity range within an ESP for maximum performance which must be considered as part of the design selection. Maximum ESP efficiency is achieved when the gas flow is distributed evenly across the unit cross section. Uniform flow is assumed in the ESP sizing calculations and should be verified during the design stage by using a flow model. These models should include the precipitator as well as the inlet and outlet flues. Flow uniformity is typically achieved by installing distribution devices in the flue transition sections immediately upstream and downstream of the ESP. Hopper design must also prevent high velocity areas to avoid flyash re-entrainment. The industry standard for flow distribution and modeling is the Institute of Clean Air Companies EP-7.[5]

Particle size distribution Particle size also affects ESP design and performance. A particle size distribution versus collection efficiency curve (Fig. 7) indicates that an ESP is less efficient for smaller particles (less than 2 microns) than for larger ones. Therefore, ESP applications with a high percentage of particles less than 2 microns will require more collection surface and/or lower gas velocities.

Sectionalization Sectionalization refers to how many or how few independent high voltage power supplies are installed on an ESP. ESP performance improves with additional, independently energized sections, but improving performance must be weighed against additional capital and operating costs. Proper sectionalization can overcome the effects of a poorly operating section(s) due to adverse process or mechanical/electrical conditions.

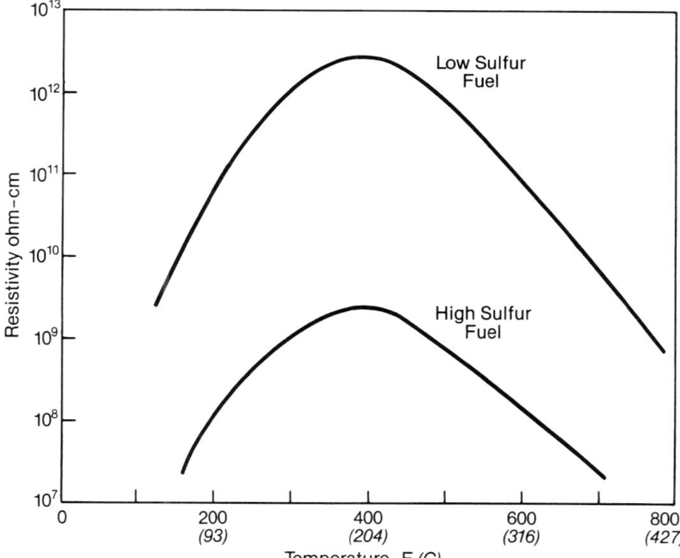

Fig. 6 Typical ash resistivity.

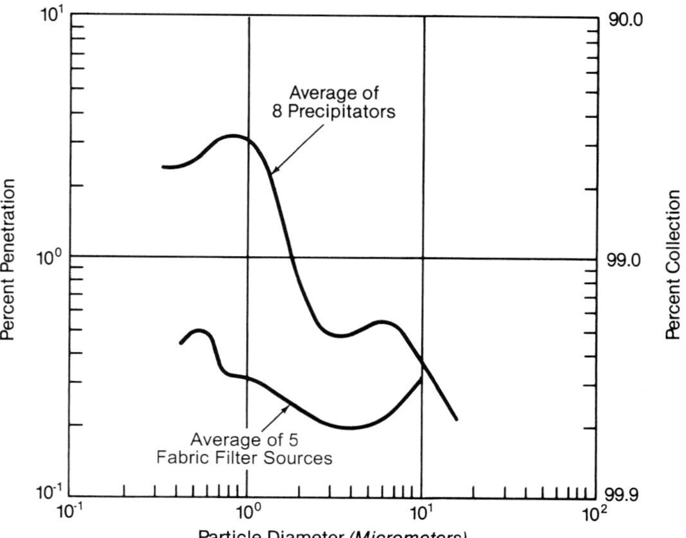

Fig. 7 Summary of fine particulate collection (adapted from Reference 4).

Rapping effects Rapping causes some amount of particulate re-entrainment in the flue gas flow. Given the nature of the precipitation process, adjustment of rapping (how often and how intense) can greatly impact outlet emissions, either positively or negatively. Proper rapping adjustment in the outlet sections is especially important because there are no downstream sections to handle re-entrainment due to over-rapping. There are no set industry standards for rapping adjustments although many suppliers begin with proprietary guidelines. Effective fine-tuning is usually based upon the supplier's field experience or plant personnel familiar with the specific ESP.

Ash removal Proper ESP design will prevent operational issues associated with re-entrainment from the hoppers. However, a malfunctioning ash removal system can also cause problems. High ash levels can permit excessive re-entrainment, and can cause electrical malfunction if the ash pile contacts the discharge electrode system (high resistance grounding). If not corrected, ash removal malfunction can also cause the ash pile to float upward and distort the DEs.

Performance enhancements

A change in fuel, a boiler upgrade, a change in regulation, or performance deterioration may call for a precipitator performance enhancement. Enhancement techniques include additional collection surface, flue gas conditioning, improved/modified gas flow distribution, additional sectionalization, additional rapping, control upgrades and internals replacement. Gas conditioning alters resistivity by adding sulfur trioxide (SO_3), ammonia, moisture, or sodium compounds while the other modifications involve only mechanical and/or electrical hardware changes.

After identifying the causes of current or anticipated performance deterioration, the equipment is surveyed to determine the need for replacement or upgrade. Additional collection surface, in series or in parallel with existing surface, may be needed to meet improved particulate collection needs. Gas conditioning may be used to offset some collection surface deficiency or to enhance the performance of a marginal precipitator. Large dust accumulations near the precipitator entrance, flow patterns on the collection surface, and a velocity traverse across the precipitator face indicate possible flow maldistribution. In addition, TR set controllers made before the mid-1980s can potentially benefit from an upgrade to improve performance. Finally, a detailed internal inspection will determine a possible need for replacement of collection surface and discharge electrodes and the need to upgrade the rapping system. A combination of enhancement techniques may be needed.

Wet electrostatic precipitators[6]

The collection of acid mists consisting of fine particulate has been accomplished with wet electrostatic precipitators (WESPs) for nearly 100 years, principally on industrial processes. These WESPs differ from the dry ESPs in both materials and configuration; however, the collection mechanism is basically the same. Typical operation is at or below the flue gas acid moisture dew point temperature for the gas being filtered, and particulate loading is low compared to normal coal-fired dry ESP applications. Collection efficiencies of 99% have been reported with wet ESPs when precipitator sections or modules are placed in series. (See Chapter 35, Fig. 15.)

Boiler applications Wet ESPs have not historically been used for utility or industrial boiler emissions control when firing coal, oil or gas. However, with the emergence of expanded emission control requirements, reduced emission limits, use of non-traditional fuels and the interactions of other emissions control equipment, there has been renewed interest in the use of WESPs to control selected emissions, especially of sulfuric acid mist and fine dry particulates. Firing of petroleum coke and Orimulsion® fuel can result in elevated levels of SO_3 in the flue gas. In addition, as noted in Chapter 34, SCRs on coal-fired systems have a tendency to increase the concentration of SO_3 in the flue gas due to the oxidation action of the catalyst. Depending upon the catalyst selection, this oxidation can increase the SO_3 significantly on the higher sulfur fuels. As the SO_3 is carried through the cooler backend equipment, it condenses and becomes an acid mist (H_2SO_4). When wet flue gas desulfurization systems are used for sulfur control, significant levels of SO_3/acid mist tend to pass through the system and result in opacity/visibility problems at the stack. (See Chapter 35.) Acid mist has also been noted on some units without the wet FGD systems. WESPs have demonstrated the ability to effectively collect the H_2SO_4 and other aerosols in such environments.

Wet precipitator design The WESP design generally follows that of dry ESPs. However, WESPs differ from dry ESPs in three key areas. First, the physical arrangement of WESPs can be non-standard compared to dry ESPs. The WESP configuration and shapes of the key components can be adapted to integrate the WESP with other system equipment, such as the case of the integration of a WESP in a wet flue gas desulfurization system discussed in Chapter 35. The components can also be configured in a more conventional stand-alone arrangement. Second, the moist corrosive atmosphere present requires careful selection of material in critical areas. Third, instead of a rapping system to remove the collected particulate, water spray or a water film removes the material deposited on the collection surfaces and discharge system. The physical arrangement of the bottom of the WESP is then adapted to collect the water film discharge.

Fabric filters

A fabric filter, or baghouse, collects the dry particulate matter as the cooled flue gas passes through the filter material. The fabric filter is comprised of a multiple compartment enclosure (see Fig. 8 and chapter frontispiece) with each compartment containing up to several thousand long, vertically supported, small diameter fabric bags. The gas passes through the porous bag material which separates the particulate from the flue gas.

Operating fundamentals

With the typical coal-fired boiler, the particle laden flue gas leaves the boiler and air heater and enters the filter inlet plenum which in turn distributes the gas to each of the compartments for cleaning. An outlet plenum collects the cleaned flue gas from each compartment and directs it toward the induced draft fan and the stack. Inlet and outlet dampers then allow isolation of each compartment for bag cleaning and maintenance. Each compartment has a hopper for inlet gas flow as well as for particulate collection and removal by conventional equipment, as discussed further in Chapter 24. The individual bags are closed at one end and connected to a tubesheet at the other end to permit the gas to pass through the bag assembly. The layer of dust accumulating on the bag is usually referred to as the *dustcake.*

Collection of the particulate on the bag fabric is the heart of the filtering process. The major forces causing this collection include impingement by either direct contact or impaction and dustcake sieving. Minor forces which assist in the collection are diffusion, electrostatic forces, London-Van der Waal's forces and gravity.[7] The dustcake is formed by the accumulation of particulate on the bags over an operating period. Once formed, the dustcake, and not the filter bag material, provides most of the filtration. Some filter media such as membrane and tight needle felts provide higher removal efficiency than normally achieved by filter cake alone. Although impingement collection is most effective on the larger particles and the sieving process collects all particle sizes, a dustcake must form to maximize overall collection.

As the dustcake builds and the flue gas pressure drop across the fabric filter increases, the bags must be cleaned. This occurs after a predetermined operating period or when the pressure drop reaches a set point. Each compartment is then sequentially cleaned to remove the excess dustcake and to reduce the pressure drop. A residual dust coating is preferred to enhance further collection. In pulse jet fabric filters (discussed below), it is common to perform cleaning online which maintains all compartments in operation.

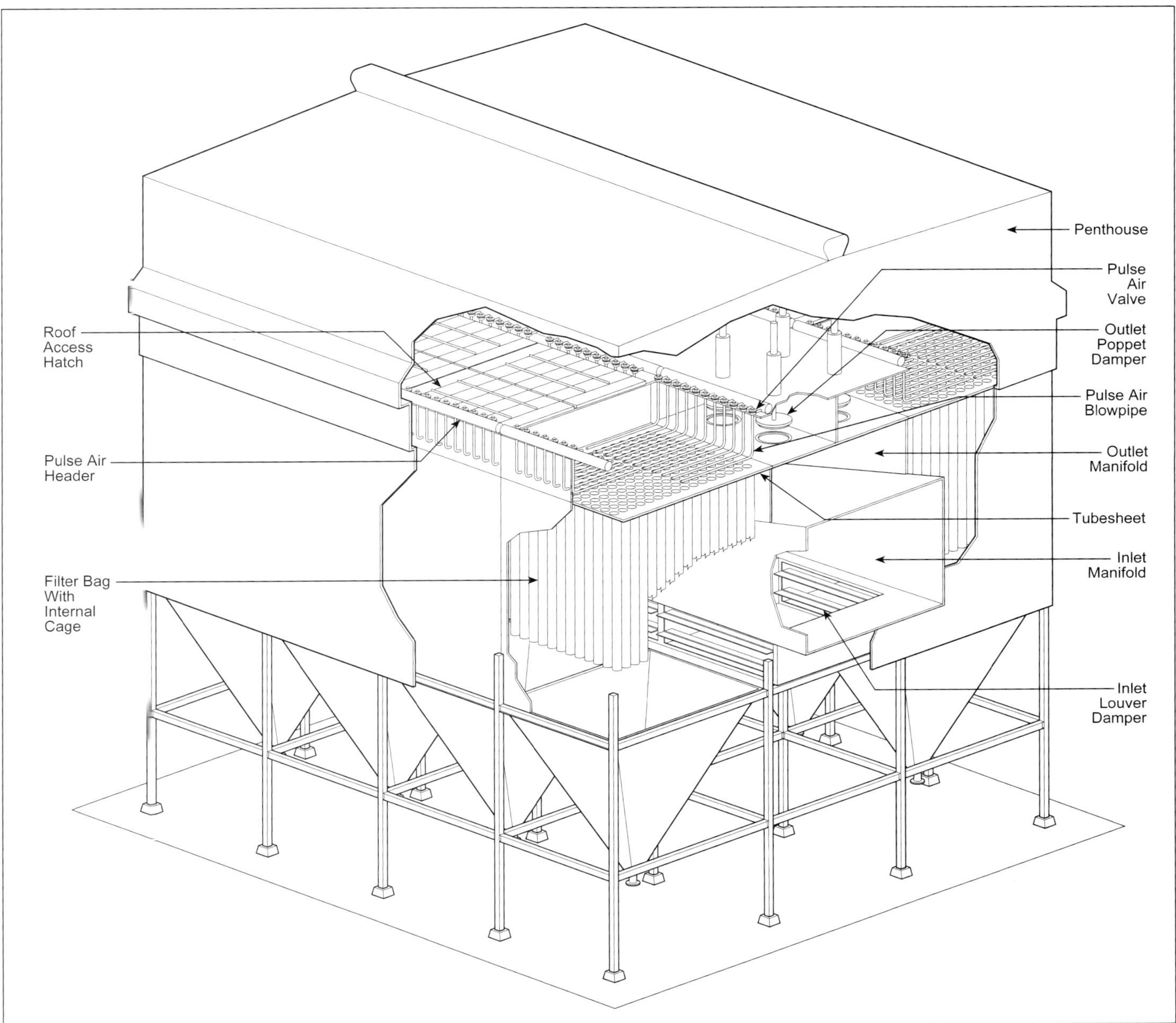

Fig. 8 Pulse jet fabric filter.

Initial filter bags have a pre-coat filter media. The pre-coat is an optimum particle size mixture which aids in developing the first filter cake and tends to stay within the outer surface of the filter media.

Types of fabric filters

Bag cleaning methods distinguish the types of fabric filters, with the three most common types being reverse air, shake deflate and pulse jet. The cleaning method also determines the relative size by the air/cloth (A/C) ratio and the filtering side of the bag. Both the reverse air and the shake deflate are inside-the-bag filters with gas flow from inside the bag to outside; the pulse jet is an outside-the-bag filter with the flow from outside to inside (Fig. 9). Note that the tubesheet on the inside-bag filtering is located below the bags; for the pulse jet the tubesheet is above the bags.

A reverse air, more correctly termed *reverse gas*, filter reverses the flow of clean gas from the outlet plenum back into the bag compartment to collapse the bags in an isolated compartment and dislodge the dustcake. This is a gentle cleaning motion. Once the dislodged particulate falls to the hopper, the bags are gently reinflated before full gas flow is allowed for filtering. This system requires a reverse gas fan to supply the cleaning gas flow along with additional dampers for flow control. This type of filter system has been used in most large utility power plant fabric filters in the U.S. to date. Experience with this type of fabric filter on some coal flyash applications has demonstrated that reverse gas cleaning alone does not provide an acceptable operating pressure drop. Therefore, some units have added sonic air horns to each compartment to assist in the cleaning.

Shake deflate (*shaker*) filters are similar to reverse air units in that the cleaning occurs in an isolated compartment and a small amount of cleaned flue gas is used to slightly deflate the bags. In addition, a mechanical motion is used to shake the bags and dislodge the accumulated dustcake.

Pulse jet technology is a more rigorous cleaning method and can be used when the compartment is either isolated or in service. A pulse of compressed air is directed into the bag from the open top which causes a shock wave to travel down its length, dislodging the dustcake from the outside surface of the bag. A unique aspect of the pulse jet system is the use of a wire cage in each bag to keep it from collapsing during normal filtration. The bag hangs from the tubesheet. A series of parallel pulse jet pipes are located above the bags with each pipe row having a solenoid valve. This permits the bags to be pulsed clean one row at a time. Use of pulse jet units with utility boilers is increasing and they are expected to be the major contributor for retrofit or new units.

A variant of the pulse jet fabric filter called COHPAC® (compact hybrid particulate collector) is the unique use of a high velocity fabric filter commonly installed after a malperforming electrostatic precipitator. This arrangement reduces particulate emissions and may prove to be a good solution in special applications.

Design parameters

Three key fabric filter design parameters are air/cloth (A/C) ratio, emission rate and drag. A/C ratio is the gas volumetric flow rate divided by the exposed bag surface area and is commonly referred to as the filtration velocity. Industry standards, along with operating experience, establish the design A/C ratios. The A/C has a significant impact on useful bag life. Lowering the A/C ratio for a specific fabric filter design tends to increase bag life and conversely increasing A/C ratio decreases bag life. A/C ratios are typically stated with one compartment out of service for cleaning (net condition).

A/C ratios commonly applied to coal-fired boilers may range from 1.5 to 2.3 ft/min (0.45 to 0.7 m/min) with reverse air fabric filters and 3.0 to 4.0 ft/min (0.9 to 1.2 m/min) for pulse jet fabric filters. The type of firing and ranges in fuel significantly affect the appropriate filtration velocity.

The emission rates are generally established based upon historical data. Several prediction models have been developed but they are of limited use. Bench tests using collected particulate samples have proven beneficial as an indicator for media selection.

The pressure drop includes the drop across the bags, the dustcake and the attachment of the bag to the tubesheet. The calculation of drag is useful in evaluating performance.

The general equation for drag can be expressed as follows:

$$\Delta P = a \times V + b \times c \times t \times V^2 \quad \textbf{(4)}$$

where

ΔP = pressure drop, in. wg (kPa)
a = constant with units of in. wg-min/ft (kPa-min/m) (This value is determined based upon actual operating data. This constant increases as the fabric filter media ages. The product of $a \times V$ is referred to as the effective residual pressure drop.)
V = the filtration velocity, ft/min (m/min)
b = the cake coefficient with units of in. wg-min-ft/lb. Values range significantly from 0.1 to 700. Flyash from coal ranges from 5 to 20. (kPa-min-m/kg)
c = the inlet dust concentration with units of lb/ft^3 (kg/m^3)
t = time, min.

Applications

The Babcock & Wilcox Company (B&W) participated in the first U.S. utility fabric filter installation in the 1960s, on an oil-fired boiler in southern California.[8] Several utilities followed by the late 1970s. Interest in these systems continues to grow due to the high particu-

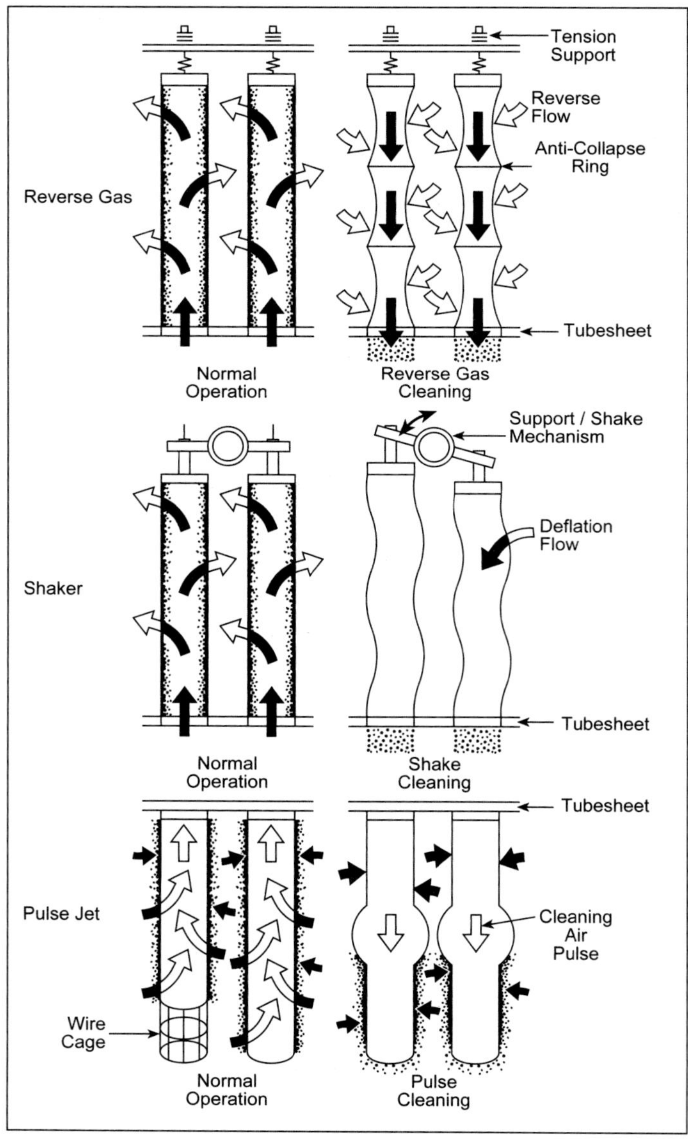

Fig. 9 Fabric filter types.

late removal efficiency. Well designed filters routinely achieve greater than 99.9% particulate removal, meeting all current U.S. EPA and local regulations.

Besides standard utility coal-fired applications, fabric filters are used on circulating fluidized-bed boilers, industrial pulverized and stoker coal units, refuse-fired units in combination with a dry flue gas scrubber, and in the cement and steel industries. Fabric filters are not currently used on oil-fired units due to the sticky nature of the ash.

A unique advantage with fabric filters is that all of the flue gas passes through the dustcake as it is cleaned. When the dustcake has high alkalinity, it can be used to remove other flue gas constituents and acid gases, such as sulfur dioxide (SO_2). (See Chapter 35.) Addition of activated carbon into the flue gas has been used to control mercury emissions. Fabric filters generally are better collectors of the carbon and mercury components than are electrostatic precipitators.

Fabric filter components

Configurations Pulse jet fabric filters have two general arrangements. The first is called *hatch style*, that has large horizontal covers used to access the area above the tubesheet for bag replacement. The most common arrangement includes a penthouse above the hatch covers. The second type, called *walk-in*, has a gas-tight compartment above the tubesheet with vertical doors to access the area above the tubesheet for bag replacement. Walk-in arrangements are more common for industrial sized units with shorter bags.

Reverse air and shaker units generally have access through vertical doors located at the tubesheet level and also located at a level near the top of the bags.

All styles of fabric filters can be provided with integral compartments sharing common division walls or with separate walls forming the compartments. The latter style is commonly referred to as modular and generally used on small industrial units.

The gas conveying flues of all styles of fabric filters can be of a design integral to the compartment casing walls or totally separated. Most utility size fabric filters utilize common wall construction.

Casing, hoppers and dampers

Enclosure or casing The fabric filter is a metal encased structure with individual bag compartments. The inlet and outlet plenums are typically located between two rows of compartments to provide short inlet and outlet flue connections (see Fig. 8). This enclosure rests on a support steel structure. For reverse air units, interior access is required at both the lower tubesheet and bag support elevations. In a pulse jet filter, access is required above the tubesheet for bag cage removal. This is provided by large roof access doors or by a top plenum and a side manway. Typical enclosure materials are carbon steel ASTM A-36 or equivalent under normal coal-fired boiler conditions. The entire enclosure is covered with insulation and lagging to keep metal temperatures high and to minimize corrosion potential.

Hoppers Each filter compartment has a hopper to collect the dislodged particulate and to channel its flow to the ash removal system. Most filters also use the hopper as part of the flue gas inlet to each compartment. Therefore, the hopper is designed with steep sides for ash removal along with considerations for proper gas flow distribution. Hopper heaters, level detectors, poke holes and an access door are common hopper features.

Dampers Each filter compartment may be provided with both inlet, outlet and on reverse air filters, deflation dampers. These dampers may be manual or automated. Louver and poppet dampers are normally used. Dampers allow isolation of compartments for offline cleaning and online service entry. Some units include bypass dampers.

Bag materials and supports

Substantial research and development on bags and their materials have taken place to lengthen their life and to select bags for various applications. (See Table 1.) The flexing action during cleaning is the major factor affecting bag life. Bag blinding, which occurs when small particulate becomes trapped in the fabric interstices, limits bag life by causing excessive pressure drop in the flue gas. Finishes on the bag surface are also used to make some bags more acid resistant and to enhance cleaning.

The most common bag material in coal-fired utility units with reverse fabric filters is woven fiberglass. Typical bag size is 12 in. (305 mm) diameter with a length of 30 to 36 ft (9.1 to 11.0 m). Bag life of three to five years is common. The shake deflate filters also use mostly fiberglass bags. On both of these units, the fiberglass bag is fastened at the bottom to a thimble in the tubesheet. At the top, a metal cap is fitted into the bag and the bag has a spring loaded support for the reverse air filters. The bags are attached to the tubesheet commonly using a thimble and clampless designed bag. The upper operating temperature limit is 500F (260C) for most fiberglass bags.

The most common bag material in coal-fired utility units with pulse jet fabric filters is polyphenylene sulfide (PPS) needled felt. In addition to polyphenylene sulfide, fiberglass, acrylic, polyester, polypropylene, Nomex®, P84®, special high temperature fiberglass media, membrane covered media, and ceramic are used in various applications. For the pulse jet filters, the typical bag size is 5 or 6 in. (127 or 152 mm) diameter round or oval with a length of 10 to 26 ft (3 to 8 m). Advances in cleaning technology are increasing the ability to provide longer bags. Pulse jet bags are commonly sealed to the tubesheet using a snap-band seal. The bag is supported internally from a metal wire cage to prevent bag collapse during operation. Cages are normally carbon steel and may include a variety of coatings from pre-galvanized to coated wire. Some applications use a stainless steel cage.

Filter media cleaning

Reverse air fabric filters employ the most gentle form of cleaning. The method dislodges collected particulate by closing compartment dampers and, with the use of a reverse air fan, reversing the direction of gas flow. During reverse flow, gas passes from the outside of the bag to the inside, dislodging the collected ash from the inside surface of the bag. Sonic horns have proven successful is certain applications to assist in reverse air cleaning.

The shaker style of cleaning begins with the isolation of a compartment. Next, mechanical energy in the form of sinusoidal acceleration is applied to the top of the bag and travels along the bag to dislodge the material caked on the filter.

Pulse jet cleaning can be accomplished with all compartments operating, which is more normal, or by removing a compartment from service. Pulse air varies from low pressure, high volume to high pressure, low volume. The components of most pulse cleaning systems are similar. Blowers or air compressors are used to supply the cleaning air. Pulse valves are rapid opening and closing valves. Air is discharged from the outlet of the pulse valve along a supply pipe to nozzles or holes for discharge into the top of the bag. This air inflates the bag, dislodging collected filter cake from the outside surface of the bag. The more common style of pulse system has fixed supply pipes called blowpipes. An alternate type of pulse system has a rotating blowpipe.

Mechanical collectors

Mechanical dust collectors, often called cyclones or multiclones, have been used extensively to separate large particles from a flue gas stream. The cyclonic flow of gas within the collector and the centrifugal force on the particulate drive the particulate out of the flue gas (Fig. 10). Hoppers below the cyclones collect the particulate and feed an ash removal system. The mechanical collector is most effective on particles larger than 10 microns. For smaller particles, the collection efficiency drops considerably below 90%.

Mechanical collectors were adequate when the emissions regulations were less stringent and when popular firing techniques produced larger particles. These collectors were frequently used for reinjection to improve unit efficiency on stoker firing of coal and biomass. With stricter emissions regulations, mechanical collectors can no longer be used as the primary control device. However, with the onset of fluidized-bed boilers, there has been a resurgence of mechanical collectors for recirculating the bed material. A high efficiency collector is then used in series with the mechanical one to meet particulate emissions requirements. (See Chapter 17 for more information on fluidized-bed combustion.)

Wet particulate scrubbers

A wet scrubber can be used to collect particulate from a flue gas stream with the intimate contact between a gas stream and the scrubber liquid. The venturi-type wet scrubber (Fig. 11) is used to transfer the suspended particulate from the gas to the liquid. Collection efficiency, dust particle size and gas pressure drop are closely related in the operation of a wet scrubber. The required operating pressure drop varies inversely with the dust particle size for a given collection efficiency; or, for a given dust particle size, collection efficiency increases as operating pressure drop increases.

Due to the excessive pressure drop and the stringent particulate regulations, wet particulate scrubbers are now infrequently used as a primary collection device. However, on most coal-fired applications where wet FGD scrubbers are required in series with a high efficiency collector for control of acid gas emissions, the extra particulate removal of the FGD scrubber is an added benefit.

Equipment selection

Major evaluation factors to consider when selecting particulate control equipment include emissions requirements, boiler operating conditions with resulting particulate quantity and sizing, allowable pressure drop/power consumption, combined pollution control requirements, capital cost, operating cost, and maintenance cost. For new units that must meet the stringent federal, state and local regulations, the selection is reduced to a comparison of electrostatic precipitators and fabric filters because these are the only high efficiency, high reliability choices. For retrofits on operating units, the performance of existing control equipment as well as unique flue gas conditions may require specialized equipment.

The advantages of a well designed ESP are high total collection efficiency, high reliability, low flue gas pressure loss, resistance to moisture and temperature upsets, and low maintenance. Advantages of a fabric filter include high collection efficiency throughout the particle size range, high reliability, resistance to flow upsets, little impact of ash chemical constituents on performance, and good dustcake characteristics for combination with dry acid gas removal equipment.

Table 1
Fabric Filter Typical Media Applications

Fabric Filter Type	Construction	Materials	Typical Maximum Operating Temperature	Chemical Resistance
Reverse gas Shake deflate	Woven	Treated fiberglass	500F (260C) ±	Acid resistant coatings available
Pulse jet	Felted	PPS Nomex® P84® Acrylic Polyester	Ambient to 350F (177C) ± depending on specific media selected	Highly variable — consult manufacturer

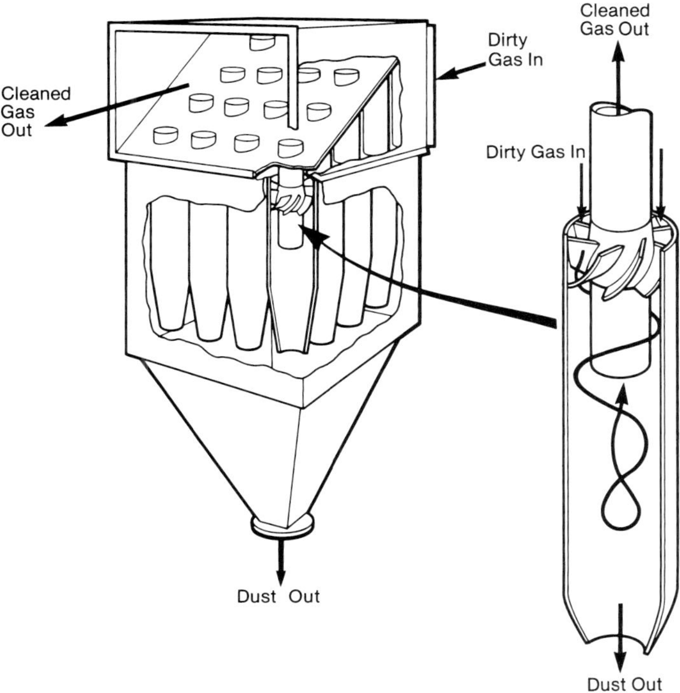

Fig. 10 Mechanical collector.

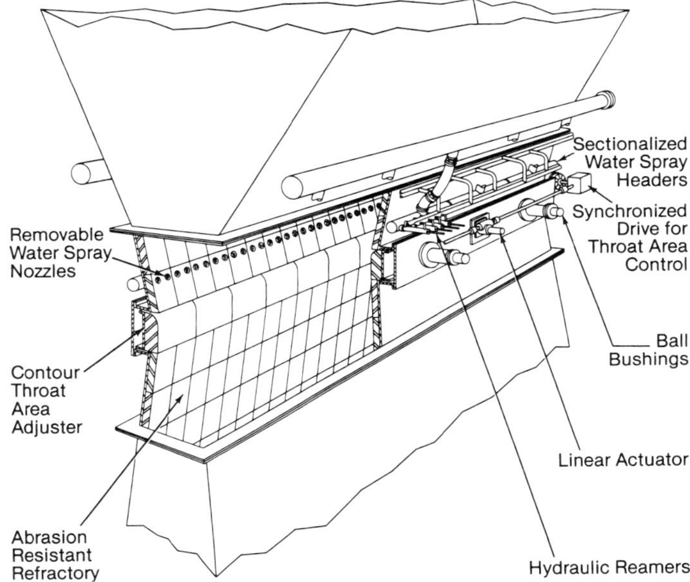

Fig. 11 Venturi-type wet scrubber.

Plants equipped with dry scrubbing systems will typically require a fabric filter if high SO_2 collection efficiencies (greater than 90%) are required due to the secondary scrubbing effect of the fabric filter. A comparison of overall and particle size collection efficiencies for precipitators and fabric filters is shown in Fig. 7. An application where small particulate dominates would favor a fabric filter for maximum control as long as bag blinding does not occur.

For those applications where an ESP or fabric filter is technically acceptable and high collection efficiencies are required, some general guidelines on initial capital costs are: 1) on small units, a pulse jet fabric filter is generally more economical, 2) on units with medium or high sulfur coal, an ESP is economical, and 3) on low sulfur coal-fired large units, a pulse-jet fabric filter may again be more economical. However, when operating and maintenance costs are also considered, the lowest capital cost may not provide the lowest overall cost. Therefore, it is important to perform a detailed engineering study to quantify all of the variables for a specific site to obtain a true assessment of the real cost.

References

1. White, H.J., *Industrial Electrostatic Precipitation,* Addison-Wesley Publishing Company, Reading, Massachusetts, 1963.

2. United States Code of Federal Regulations, Title 40, Vol. 6, Part 60, Subparts D, Da, Db and Dc, Revised July 1, 2003.

3. Federal Register, Environmental Protection Agency, 40 CFR Part 60, Vol. 55, No. 177, September 12, 1990.

4. Lane, W.R., Khosla, A., "Comparison of Baghouse and Electrostatic Precipitator Fine Particulate, Trace Element and Total Emissions," ASME-IEEE Joint Power Conference, Indianapolis, Indiana, September 27, 1983.

5. *Electrostatic Precipitator Gas Flow Model Studies,* Publication No. EP-7, Institute of Clean Air Companies, Washington, D.C., Revised 1997.

6. Staehle, R.C., Triscori, R.J., Kumar, K.S., et al., "The Past, Present and Future of Wet Electrostatic Precipitators in Power Plant Applications," Paper 207 presented to the *Combined Power Plant Pollution Control Mega Symposium,* Washington, D.C., May, 2003.

7. Bustard, C.J., et al., *Fabric Filters for the Electric Utility Industry,* Vol. 2, EPRI CS-5161, Electric Power Research Institute, Palo Alto, California, 1988.

8. Bagwell, F.A., Cox, L.F., and Pirsh, E.A., *Design and Operating Experience With a Filterhouse Installed on an Oil-Fired Boiler,* Air Pollution Control Association, June, 1969.

COHPAC is a trademark of Electric Power Research Institute, Inc.

Nomex is a trademark of E.I. du Pont de Nemours Company.

Orimulsion is a trademark of Bitumenes Orinoco, S.A.

P84 is a trademark of Inspec Fibres GmbH.

Coal-fired utility boiler selective catalytic reduction system retrofit.

Chapter 34

Nitrogen Oxides Control

Nitrogen oxides (NO_x) are one of the primary pollutants emitted during combustion processes. Along with sulfur oxides (SO_x) and particulate matter, NO_x emissions contribute to acid rain and ozone formation, visibility degradation, and human health concerns. As a result, NO_x emissions from most combustion sources are regulated, monitored, and require some level of control. This control is closely tied to the combustion process and to various boiler system components.

NO_x formation mechanisms

NO_x refers to the cumulative emissions of nitric oxide (NO), nitrogen dioxide (NO_2), and trace quantities of other nitrogen-bearing species generated during combustion. Combustion of any fossil fuel generates some level of NO_x due to high temperatures and the availability of oxygen and nitrogen from both air and fuel.

NO_x emissions from fired processes are typically 90 to 95% NO with the balance NO_2. However, once the flue gas leaves the stack, most of the NO is eventually oxidized in the atmosphere to NO_2. It is the NO_2 in the flue gas which contributes to the brownish plume sometimes seen in a power plant stack discharge.

There are two principal mechanisms of NO_x formation in steam generation: *thermal* NO_x and *fuel* NO_x which are explored more in depth below. An additional source is *prompt* NO_x. Prompt NO_x is produced in fuel-rich conditions by a complex series of reactions between hydrocarbon radicals and molecular nitrogen in the flame zone, resulting in the formation of amines and cyano compounds and their subsequent oxidation to NO.[1] It is a relatively small contributor to total NO_x formation, especially in the combustion of coal and other nitrogen-containing fuels.

Thermal NO_x

Thermal NO_x refers to the NO_x formed through high temperature oxidation of the nitrogen found in the combustion air. The formation rate is a function of temperature as well as residence time at temperature. Significant levels of NO_x are usually formed above 2200F (1204C) under oxidizing conditions, with exponential increases as the temperature is increased. At these high temperatures, molecular nitrogen (N_2) and oxygen (O_2) dissociate into their atomic states and participate in a series of reactions.[2,3] One product of these reactions is NO. Since the traditional factors leading to complete combustion (high temperatures, long residence time, and high turbulence or mixing) all tend to increase the rate of thermal NO_x formation, some compromise between effective combustion and controlled NO_x formation is needed.

Thermal NO_x formation is typically controlled by reducing peak and average flame temperatures. Controlled mixing burners can be used to slow the combustion process. A second approach is staged combustion, where only part of the combustion air is initially added to burn the fuel; the remaining air is added separately to complete the combustion process. A third method is to mix some of the flue gas with the combustion air at the burner, referred to as flue gas recirculation (FGR). This increases the gas weight which must be heated by the chemical energy in the fuel, thereby reducing the flame temperature. Refer to Chapters 10 through 17 for more detail on the combustion processes and the fuel effects on NO_x formation.

Fuel NO_x

The major source of NO_x emissions from nitrogen-bearing fuels such as coal and oil is the conversion of fuel-bound nitrogen to NO_x during combustion. Laboratory studies indicate that fuel NO_x contributes approximately 50% of the total uncontrolled emissions when firing residual oil and more than 80% when firing coal. Nitrogen found in these fuels is typically bound to the fuel as part of organic compounds. During combustion, the nitrogen is released as a free radical to ultimately form NO or N_2. Although it is a major factor in NO_x emissions, only 20 to 30% of the fuel-bound nitrogen is converted to NO.

The majority of NO_x formation from fuel-bound nitrogen occurs through a series of reactions that are not fully understood. However, it appears that this conversion occurs by two separate paths (see Fig. 1).

The first path involves the oxidation of volatile nitrogen species during the initial phase of combustion. During the release and prior to the oxidation of the volatile compounds, nitrogen reacts to form several intermediate compounds in the fuel-rich flame re-

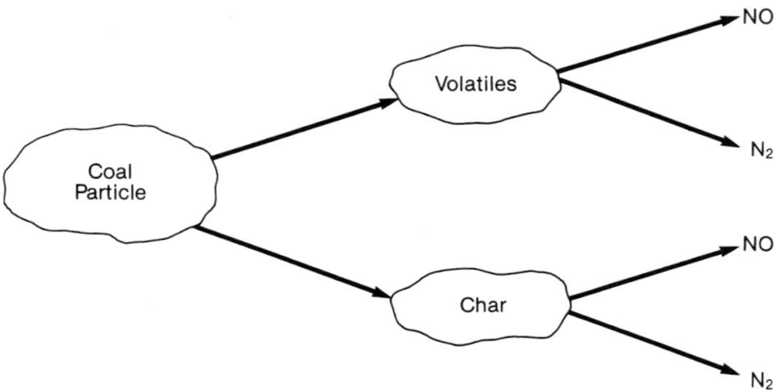

Fig. 1 NO_x formation from nitrogen contained in coal during combustion.

gions. These intermediate compounds are then oxidized to NO or reduced to N_2 in the post-combustion zone. The formation of either NO or N_2 is strongly dependent on the local fuel/air stoichiometric ratio. It is estimated that this volatile release mechanism accounts for 60 to 90% of the fuel NO_x contribution.

The second path involves the release of nitrogen radicals during combustion of the char fraction of the fuel. These reactions occur much more slowly than the reactions involving the volatile species.

Conversion of fuel-bound nitrogen to NO_x is strongly dependent on the fuel/air stoichiometry but is relatively independent of variations in combustion zone temperature. Therefore, this conversion can be controlled by reducing oxygen availability during the initial stages of combustion. Techniques such as controlled fuel-air mixing and staged combustion provide a significant reduction in NO_x emissions by controlling stoichiometry in the initial devolatilization zone.

Global effects of NO_x in atmosphere

Once in the atmosphere, the NO_2 is involved in a series of reactions that form secondary pollutants. The NO_2 can react with sunlight and hydrocarbon radicals to produce ground level ozone/photochemical (urban) smog, acid rain constituents, and particulate matter.[4] Each of these formations can have a significant effect on human health and on the environment. Automobiles and other vehicles are the major source of NO_x emissions with greater emissions concentration in the large metropolitan areas (see Fig. 2). Utility power plants and other fossil fuel burning steam generation sources also contribute to global NO_x emissions.

Acid rain

NO_x and sulfur dioxide (SO_2) contribute to the formation of *acid rain*, which includes a dilute solution of sulfuric and nitric acids with small amounts of carbonic and other organic acids. The NO_x and SO_2 react with water vapor to form acidic compounds; these acids account for more than 90% of acid rain. Nitric acid is formed through the reaction:

$$NO_2 + OH \rightarrow HNO_3 \qquad (1)$$

Most acid rain control has concentrated on SO_2 contributions instead of those due to NO_x because it is estimated that NO_x contributes less than one-third of the acid rain generated.

Particulate matter

NO_x emissions can also contribute to fine, ambient particulate matter. In the atmosphere, NO_x reacts with other airborne chemicals to form nitrates, which contribute to the less than 2.5 micron particulate (inhalable portion of the ambient pollutants). NO_x also promotes the transformation of SO_2 into sulfate particulate compounds that can also be less than 2.5 microns in size.

NO_x control technologies in steam generation

Combustion technology-based NO_x control (see Chapters 11 through 16) is typically the lowest cost solution, provided it meets local and federal emissions requirements. Where further control is needed, fuel switching is used and/or post-combustion control is added.

Pre-combustion fuel switching

Unlike sulfur and particulate constituents in coal, nitrogen species contained in the fuel can not be easily reduced or removed. Therefore, the most common pre-combustion option for reducing NO_x levels is to switch to a fuel with inherently lower nitrogen content. For boilers capable of firing multiple fuels with minimal impact on the steam cycle, this may be the most cost-effective solution.

Coal combustion generally produces the highest NO_x emissions. Oil combustion generates less NO_x while gas firing produces even less. When firing oil fuels, a reduction in fuel nitrogen content results in reduced NO_x emissions. However, there does not appear to be a similar correlation between coal nitrogen content and NO_x formation. Other factors in coal chemistry, including volatile species, oxygen, and moisture content, appear to dominate the formation of NO_x during coal combustion. Therefore, reducing coal nitrogen content may not provide a corresponding NO_x reduction.

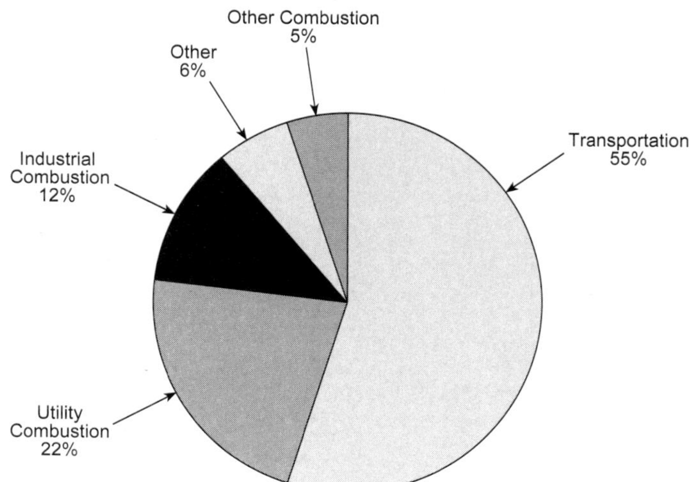

Fig. 2 Sources of United States NO_x emissions – 2001.

Combustion control

NO_x formation is promoted by rapid fuel-air mixing. This produces high peak flame temperatures and excess available oxygen which, in turn, promotes NO_x emissions. Combustion system developments responsible for reducing NO_x formation include low NO_x burners (see Fig. 3), staged burning techniques, and FGR. The specific NO_x reduction mechanisms include controlling the rate of fuel-air mixing, reducing oxygen availability in the initial combustion zone, and reducing peak flame temperatures. These are explored further in Chapters 11 through 16.

Post-combustion reduction

With current and proposed regulations, it is economically logical to consider NO_x controls that achieve the lowest emission levels possible. Additional NO_x control techniques can be applied downstream of the combustion zone to achieve the further reductions. These post-combustion control systems are referred to as selective catalytic reduction (SCR) and selective noncatalytic reduction (SNCR). In either technology, NO_x is reduced to nitrogen (N_2) and water (H_2O) through a series of reactions with a reagent (or reagents) injected into the flue gas. The most common reagents used in commercial applications are ammonia and urea for both SCR and SNCR systems.

SCR can effectively reduce NO_x levels leaving the boiler by 90% or greater depending on the specific project and space availability. SNCR typically is limited to lower NO_x reduction levels but may be the more economical choice depending on the required NO_x reduction or the unique project requirements. Both technologies have been successfully applied for NO_x reduction on multiple fuels and boiler types.

A combination of combustion and post-combustion NO_x control is frequently the most economical approach for existing installations and is the favored approach for installations that can not achieve emissions by combustion alone.

Selective catalytic reduction

Selective catalytic reduction technology is currently the most effective method of post-combustion NO_x reduction. SCR technology has been applied to a variety of fuels including natural gas, refinery gas, coals (bituminous, subbituminous, and lignite), fuel oils, petroleum coke, Orimulsion®, biomass, and refuse. It has been applied with conventional boilers as well as fluidized-bed boilers, process heaters, steel mill furnaces, and combined cycle gas turbines. SCRs have also been successfully installed on gas turbine based systems, reciprocating engines, and nitric acid plants. The technology and performance are sensitive to temperature and flue gas constituents and therefore are more closely related to boiler operation than are other types of environmental control equipment.

Initial development of SCR technology began in Japan around 1963. Early laboratory tests included the evaluation of many catalyst formulations and their life expectancies, as well as process design optimizations.

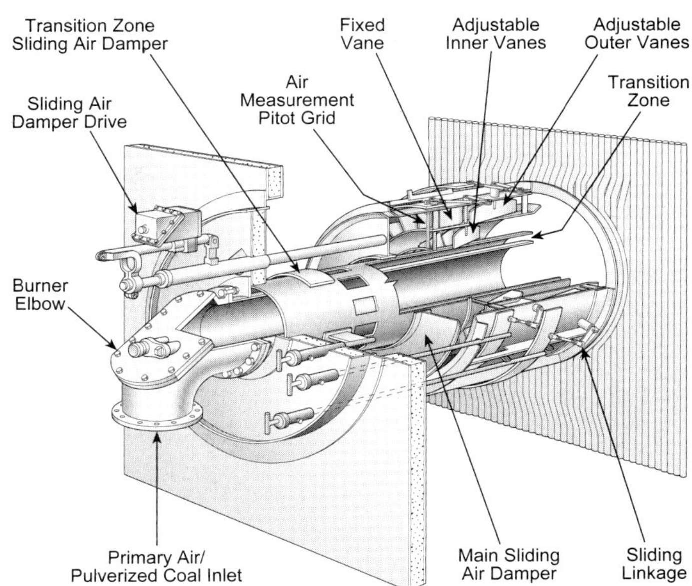

Fig. 3 DRB-4Z® low NO_x burner for coal firing.

Development tests were followed by several catalytic pilot plant installations during the early 1970s; the first commercial installations occurred in 1978.

NO_x reduction process

In the SCR process, a catalyst facilitates a chemical reaction between NO_x and a reagent (usually ammonia) to produce nitrogen and water vapor. The most common and predominant reactions are represented by the following equations. While the reactions are exothermic, the temperature rise across the catalyst is typically negligible due to the low NO_x concentrations.

$$4NO + 4NH_3 + O_2 \rightarrow 4N_2 + 6H_2O \quad \textbf{(2)}$$

$$NO + NO_2 + 2NH_3 \rightarrow 2N_2 + 3H_2O \quad \textbf{(3)}$$

As shown in Fig. 4, the ammonia (NH_3) is injected into the hot flue gas through an injection grid, the flue gas with the ammonia passes across the catalyst surface, and the NO_x reactions occur within the micropores of the catalyst. Nitrogen oxides are present in flue gas in the form of both NO and NO_2 with NO being the greater; however, NO_2 is preferentially reduced in this process.

SCR system description

While the main component of the SCR process is the catalyst, the full system contains a variety of components and subsystems. The catalyst is housed in a

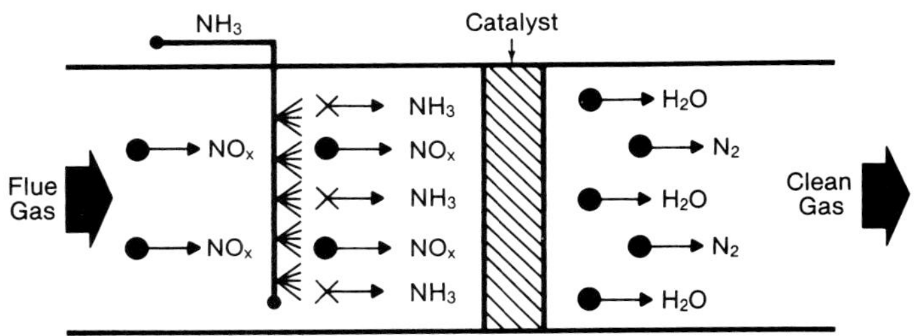

Fig. 4 Principles of NO_x removal process for SCR.

large reactor. Flue gas is directed to and from the reactor by large flues that may contain control and shutoff dampers as well as flow distribution devices and static mixers, depending upon project-specific requirements. The reactor is a gas-tight steel enclosure that permits the flue gas and reagent to contact the catalyst for the required residence time to achieve design performance. The reactor structure includes internal stiffeners and a catalyst support system that also permits rapid loading and unloading of catalyst materials. The design includes a sealing system to avoid flue gas bypass, and may include catalyst cleaning devices (sootblowers and/or sonic horns). The reactor can be configured for vertical or horizontal flue gas flow depending upon the fuel, space availability, and upstream and downstream equipment. Separate economizer and SCR flue gas bypass flues are frequently included.

The reagent (usually ammonia) subsystems include truck or rail unloading, storage, vaporization or hydrolysis equipment, flow control, dilution air supply and mixing, and reagent distribution and injection components. Each subsystem contains piping valves, instrumentation, controls and safety equipment for appropriate ammonia handling and material compatibility. Further details are provided in subsequent sections of this chapter.

A critical part of SCR system design is integration into the overall boiler system. This is particularly important because the SCR system is normally integrated into the heat recovery portion of the boiler system between the close-coupled economizer and air heater. As a result, SCR system design must accommodate swings in flue gas temperature while maintaining the desired performance, without affecting boiler operation.

Hot side/high dust The ideal gas temperature for NO_x reduction ranges from 700 to 750F (371 to 399C). In utility boilers, this places the typical SCR system at the economizer outlet, preceding the air heater. (See Fig. 5.) This is the hot side location for all fuel types and the high dust location for ash-laden fuels. Units with hot side electrostatic precipitators (ESPs) can locate the SCR between the ESP and air heater to take advantage of the ideal temperature and to avoid ash effects on the catalyst. This has been labeled hot side/low dust. For some boiler applications, the ideal temperature range does not occur at a convenient location, and system design therefore requires special accommodations, such as a split economizer.

Cold side/low dust An alternative SCR location is after the air heater and particulate collection or flue gas desulfurization (FGD) systems, as shown in Fig. 6. SCRs installed in these arrangements are designated as cold side/low dust. These systems include all of the components referenced previously, but must also include a method to increase the flue gas temperature. On utility boiler applications, the cold side location has been reserved for retrofit installations where lack of space and outage duration have driven the decision. The additional equipment required for this design may include gas-to-gas air heaters, high pressure steam heat exchangers, duct burners, or a combination of

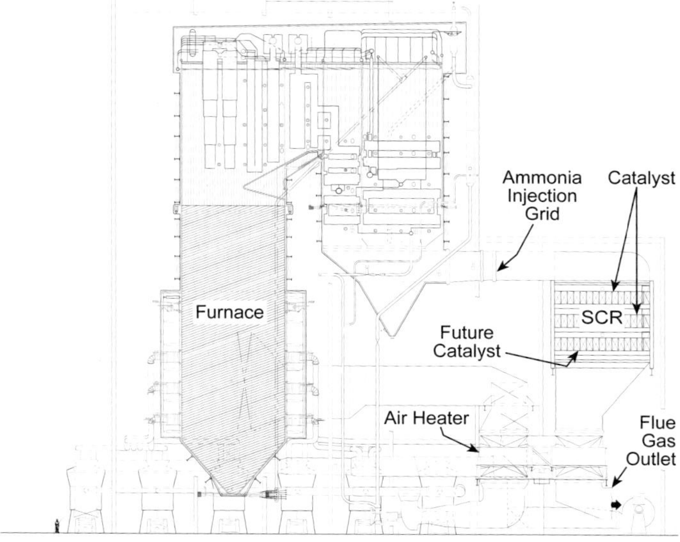

Fig. 5 New utility boiler with SCR.

these components. When installed after an FGD system or on a low sulfur fuel and after the particulate collector, this typically results in a design gas temperature of 600 to 650F (316 to 343C). This arrangement also allows a constant operating gas temperature throughout the boiler load range.

Industrial boiler applications may have a greater need for this arrangement. Package boilers may require a duct burner to achieve the required gas temperature. Boiler systems for biomass or refuse/waste fuels that contain rapid catalyst deactivation components should also consider the cold side/low dust arrangement.

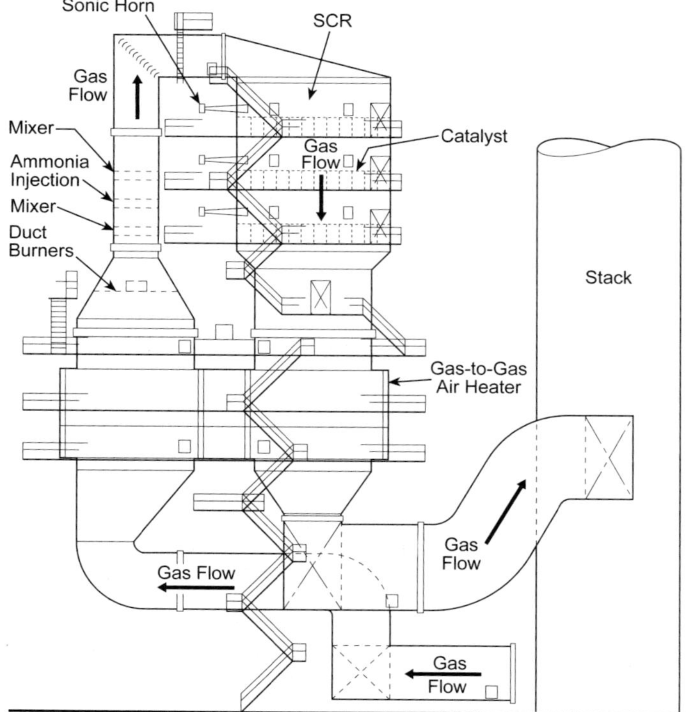

Fig. 6 SCR located after air heater.

Catalyst

Catalyst formulation, pitch and sizing are the keys to successful SCR system application and performance. Although catalyst formulations are proprietary, the most common components in steam generation applications use titanium dioxide with small amounts of vanadium, molybdenum, tungsten, or a combination of other active reagents. Formulation is matched to the flue gas conditions based on ash-free flue gas, low-ash flue gas, high-dust flue gas and the ash constituents.

Catalyst types Catalyst types and materials typically fall into one of three categories: base metal, zeolite, and precious metal.

Most steam generation experience has been with base metal catalysts. These are typically described as titania-vanadia and can be provided as plate type, honeycomb type, or corrugated fiber type (see below). All three are similar in performance characteristics and can be supplied in a homogeneous construction or as a coated active layer on a base material.

The plate type catalyst (Fig. 7) is manufactured from a roll of stainless steel expanded metal substrate. The homogeneous catalyst is paste-pressed into and around the substrate to form a catalyst plate of 0.9 to 1.0 mm thickness. After cutting to length, correctly sized undulations, to set the catalyst pitch, are formed onto each plate prior to final processing and drying. The completed plates are then assembled into encased blocks; multiple blocks are packaged into a metal-frame shipping module. (See Fig. 8.)

Homogeneous honeycomb catalyst (see Fig. 9) is formed in an extrusion process that produces a catalyst log approximately 6 in. × 6 in. (152 × 152 mm) in cross section with the selected honeycomb cell pitch, and cut to proper length. After final processing and drying, the logs are assembled and packaged into metal-frame shipping modules.

Corrugated catalyst (Fig. 10) consists of a corrugated glass fiber substrate coated with titanium dioxide. After further processing, the base is impregnated with the active components. This catalyst is more porous and lighter than either the honeycomb or plate type. The resulting catalyst is cut to size and assembled into encased blocks and metal-frame shipping modules.

Fig. 8 Installation of catalyst blocks.

Zeolite catalysts are less common than the base metal catalysts but are effective in certain applications. These are aluminosilicate materials which function similarly to base metal catalysts. One advantage is the higher allowable operating temperature of greater than 1000F (538C). This catalyst will also oxidize SO_2 to SO_3, is more sensitive to ash-laden flue gas, and is best applied to clean flue gas projects. Zeolite catalysts offer advantages for simple cycle gas turbines.

Most precious metal catalysts are manufactured from platinum and rhodium. There is little operating experience with these catalysts in steam generation systems. They also require careful consideration of flue gas constituents and operating temperatures. While effective in reducing NO_x at gas temperatures below

Fig. 7 Plate-type catalyst block.

Fig. 9 Honeycomb catalyst geometry.

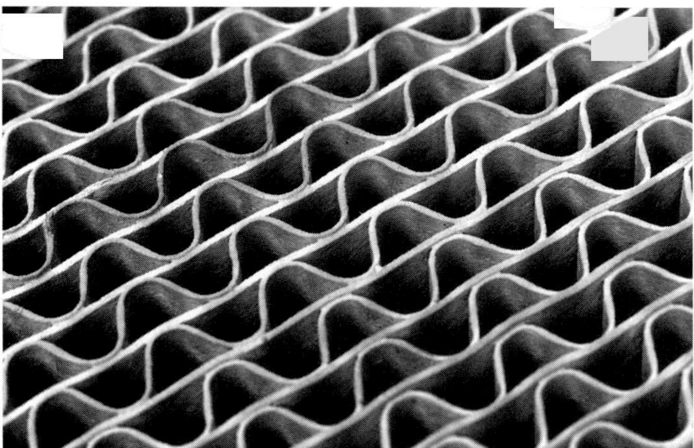

Fig. 10 Corrugated catalyst geometry (*courtesy of Haldor Topsoe, Inc.*).

550F (288C) without reagent, these catalysts can also be used as oxidizing catalysts, converting CO to CO_2 under proper temperature conditions. However, SO_2 oxidation to SO_3 and high material costs often make these catalysts less attractive and limit the application to clean flue gas. Some catalyst manufacturers have combined the precious metal and base metal catalysts to allow both NO_x reduction and CO oxidation within one catalyst chamber.

Sizing criteria Catalyst performance depends on the surface area velocity, enabling the NO_x reduction reactions to occur. A common term that relates catalyst volume with a specific surface area is space velocity (S_v), expressed as Equation 4.

$$S_v = \text{flue gas flow } (Nm^3/\text{hr}) \; / \text{ catalyst volume } (m^3) \tag{4}$$

Thus, the greater the space velocity the less catalyst required. Each catalyst configuration and formulation has a unique base space velocity. Catalyst pitch (plate centerline or cell spacing dimension) and catalyst type contribute to this space velocity.

A project's design space velocity is dependent on many factors. Primary factors include required NO_x reduction, flue gas flow, ammonia slip, gas velocity, flue gas constituents, required catalyst life, and distribution profiles (gas temperature, velocity, NO_x and NH_3/NO_x). An iterative process optimizes the selected catalyst volume and arrangement, accounting for gas velocity, pressure drop and fan adequacy. Typical performance guarantees include NO_x reduction/NO_x emissions, ammonia slip, catalyst life, pressure drop, ammonia consumption and SO_2 oxidation.

Operating issues Catalyst is not consumed during operation but does deteriorate. Deterioration is typically caused by poisoning, masking or blinding, and is most pronounced in the high dust arrangement. Deterioration occurs when the active sites within the catalyst micro-pore structure become contaminated and inactive. Moisture in liquid form can accelerate the deterioration. Operation at gas temperatures greater than 800F (427C) can also accelerate deterioration due to sintering of the pore structure. In some cases, poisoning from coal constituents like arsenic can be eliminated with limestone addition to the coal. Catalyst management recommendations to address the deterioration issues are provided in a later section.

Catalyst protection is needed for extended boiler outage periods and when the catalyst is isolated during the periods when high NO_x control is not required. Humidity control and ash removal are necessary.

On projects with SO_3 in the flue gas, ammonium sulfates can deposit on the catalyst surface when gas temperature is below the precipitation temperature, as determined by capillary condensation at the ammonia concentration entering the catalyst. Therefore, a minimum continuous operating temperature is required. At the catalyst outlet there is a concern for deposits formed by the combination of ammonia slip and available SO_3 in the flue gas. This precipitation temperature typically occurs within the air heater. Remedies are provided later in this chapter. SO_2 oxidation to SO_3 across the catalyst may also increase the possibility of ammonia bisulfate formation and can affect downstream corrosion concerns; however, the amount of deposit is ammonia-limited.

Ammonia reagents and control

Reagents most commonly used today are anhydrous ammonia, aqueous ammonia, and urea. When urea is used, it is typically decomposed on site to ammonia, CO_2 and water. Regardless of the source, once the ammonia is in vapor form, it is common to dilute the ammonia stream with air to assist with transport, flow control and injection functions.

A typical ammonia control and supply system is shown in Fig. 11.

Anhydrous ammonia reagent systems Various grades of anhydrous ammonia are available. Commercial grade ammonia is the most common grade used for SCR systems. It is approximately 99.5 to 99.7% pure NH_3. The remaining fraction is water with a small amount of oil. The water component helps combat stress corrosion cracking, which is more likely to occur when the highest metallurgical grade ammonia at 99.995% pure NH_3 is used with standard carbon steel storage tanks.

Anhydrous ammonia is the purest form of ammonia available and allows the greatest storage volume efficiency. The storage tanks are typically horizontal, consisting of a cylindrical section with elliptical ends. These tanks are designed for the high pressures exhibited by the ammonia vapor at normal ambient conditions. Tank design pressure ratings are usually in the range of 250 to 265 psi (1724 to 1827 kPa) while normal operating pressures are generally between 80 and 120 psi (552 to 827 kPa). Storage and use of anhydrous ammonia are considered more hazardous compared to other reagent systems. This is primarily due to high concentrations, pressures, volatility, and explosiveness once it is evaporated to gaseous form. The typical tank capacity ranges from 10,000 to 100,000 gallons. In the United States (U.S.), federal and potentially state and local regulations may affect storage system design. To supply ammonia from the storage tank to the ammonia injection grid (AIG), the ammonia is typically vaporized and diluted with the

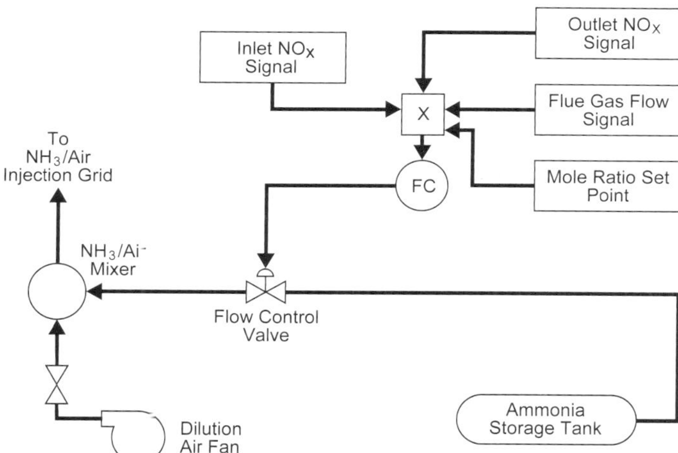

Fig. 11 NH_3/dilution air supply and control system.

air transport media. Vaporization can be integral to the storage tank or separate and closer to the SCR system. Sufficient dilution air is added to produce an ammonia-air mixture that is not within the combustible limits of approximately 15 to 30% NH_3 in air and to provide sufficient transport media volume for the injection process. The typical dilution air quantity will yield an approximate concentration of 5% NH_3 in air by volume.

In most cases, the dilution air flow is constant and the ammonia flow is adjusted to satisfy SCR system demand based on operating load, flue gas flow, inlet NO_x concentration, and the design or set point NO_x removal or SCR outlet concentration. An in-line static mixer is typically provided immediately downstream of the point where ammonia and dilution air combine. This ensures the mixture is adequately blended prior to its arrival at the manifold valve station. The manifold valve station controls the flow of this uniform ammonia-air mixture to individual ammonia injection locations within the SCR system inlet flue.

Aqueous ammonia reagent systems Aqueous ammonia is a diluted form of ammonia with NH_3 dissolved in water to form NH_4OH. Aqueous ammonia is typically available at concentrations up to a maximum of 29.4% [solubility limit at ambient conditions of 86F (30C) and 1 atm] by weight of NH_3 equivalent. This concentration is also frequently described as 26°Bé, where degrees baumé is an alternate measurement of density.

In the past, the 29.4% solution was typically preferred to minimize storage volume, cost associated with transportation weight of water, and cost of water vaporization. However, a legal classification change in the U.S. now lists ammonia-water solution concentrations below 20% as non-hazardous. This has encouraged the industry to consider a 19% concentration. This, however, increases storage tank volume and vaporization energy requirements due to the increased water content.

The storage tanks are typically horizontal, similar to those for anhydrous ammonia. Where anhydrous ammonia vessels are typically designed for approximately 250 psi (1724 kPa), aqueous tanks are normally rated at 20 to 50 psi (138 to 345 kPa). Tank material is typically carbon steel. Some installations utilize stainless steel to avoid startup problems associated with rust. Most storage system designs will also incorporate a containment basin in case of a ruptured line or vessel.

The delivery method and unloading equipment are typically the same as for anhydrous systems. However, while the vapor pressure of ammonia in the water solution can present similar vapor lock or cavitation problems when using pumps, the problem potential is less severe. Therefore, pumped delivery is more common.

Various vaporization methods exist for aqueous ammonia. Many utilize a spray nozzle to atomize the aqueous ammonia into a hot dilution air stream in an evaporation chamber. The vaporized mixture is then transported to the manifold valve station which in turn supplies the AIG. Kettle-type systems that utilize an alternate heat source independent of the dilution air have also been used. These systems essentially boil off the ammonia, which is subsequently mixed with hot dilution air at a downstream location. The dilution air is hot in all cases, as it is important to maintain a stream temperature above the water saturation point during transport to the AIG. Transport lines are often heat traced and/or heavily insulated to prevent heat loss along the travel length. When the dilution air is the source of the heat of vaporization, the supply temperature must be even higher to account for vaporization and heat loss along the travel distance. Hot combustion air or flue gas is often used as the source of dilution medium on clean-fuel applications. Even so, it is sometimes necessary to incorporate electric or steam heaters to maintain minimum temperature requirements. When dilution air is pulled from ambient, heater cost and power consumption can become significant.

Similar to anhydrous systems, the dilution air quantity is typically that sufficient to provide an approximate 5% by volume NH_3 in air. If the design air temperature is lower than desired and/or the design ammonia flow demand is high, it can become necessary to dilute to lower NH_3 concentrations to maintain minimum transport temperatures. As with anhydrous systems, the dilution air flow is typically constant and the ammonia flow is modulated to satisfy the SCR system ammonia demand.

Ammonia from urea reagent systems Due to reduced handling concerns, urea has gained popularity. These systems utilize the relatively benign urea as the source of ammonia. Urea is commonly available in either liquid solution or solid form. The systems attempt to produce ammonia at a rate as close to demand as possible to essentially eliminate storage issues associated with large quantities of ammonia.

Most of these systems are designed to produce a gaseous mixture of ammonia, carbon dioxide and water by the hydrolysis of a urea-in-water solution. Often, the urea solution is produced at the site by blending deionized water and solid urea. Solutions are typically maintained in the 40 to 50% by weight range to avoid crystallization. Urea has an ammonia equivalence of approximately 53% by weight. The hydrolysis reaction is also endothermic and thus requires heat input to maintain the reaction. The basic hydrolysis reaction is as follows:

$$NH_2CONH_2 + H_2O \rightarrow 2NH_3 + CO_2 \quad (5)$$

Other processes utilizing either variations on the hydrolysis process, high temperature flash vaporization, or catalysts are in initial commercial use. The primary goal among all of these systems is improved safety in transportation, storage, and handling of the required bulk reagent.

While solid urea is less costly to transport, it does present material handling challenges. The material tends to melt and break, leading to compaction and cohesive/adhesive handling and storage problems. These processes typically produce a moist gaseous product of approximately 20 to 30% ammonia. Similar to aqueous systems, the presence of water vapor requires a hot dilution air stream to prevent water condensation during transport to the AIG. Depending on system design, the dilution air may also be used to maintain line temperatures above those of undesirable corrosive substances. System capabilities to produce ammonia at changing demand rates can impact plant operation. The ramp rates of ammonia production may not closely follow the plant boiler load and subsequent ammonia demand. Due to these variations in system designs, operation and production capabilities, a thorough review is typically needed on a case-by-case basis.

Ammonia flow controls

As indicated in the equations for the ammonia reaction with the NO_x, one mole of ammonia is required to reduce one mole of NO_x and thus the required injection mole ratio of ammonia to NO_x is equal to the required NO_x reduction plus the ammonia slip value (ppm) divided by the inlet NO_x concentration (ppm). The control system logic uses a feed forward and feed back input with this equation, along with a flue gas flow input, to provide the ammonia flow control demand. The feed forward input is determined by either an inlet NO_x analyzer or a curve-generated value for repeatable boiler conditions. The feed back input can use the output of an SCR outlet analyzer or in some cases the stack continuous emissions monitor (CEM). Because of the hazardous nature of the ammonia, typically there are several interlocks that will interrupt ammonia flow upon upset conditions.

Distribution requirements

SCR reactor performance is influenced by the distribution of velocity, temperature and reacting gas components through the catalyst bed (see Fig. 12). The degree of a given profile uniformity can be generally described by the coefficient of variation (Cv), commonly reported as the percentage of the profile standard deviation from the arithmetic mean value.

$$Cv = \frac{\sigma}{\overline{x}} 100\% \quad (6)$$

$$\sigma = \sqrt{\frac{1}{(n-1)} \sum_{i=1}^{n} (x_i - \overline{x})^2} \quad (7)$$

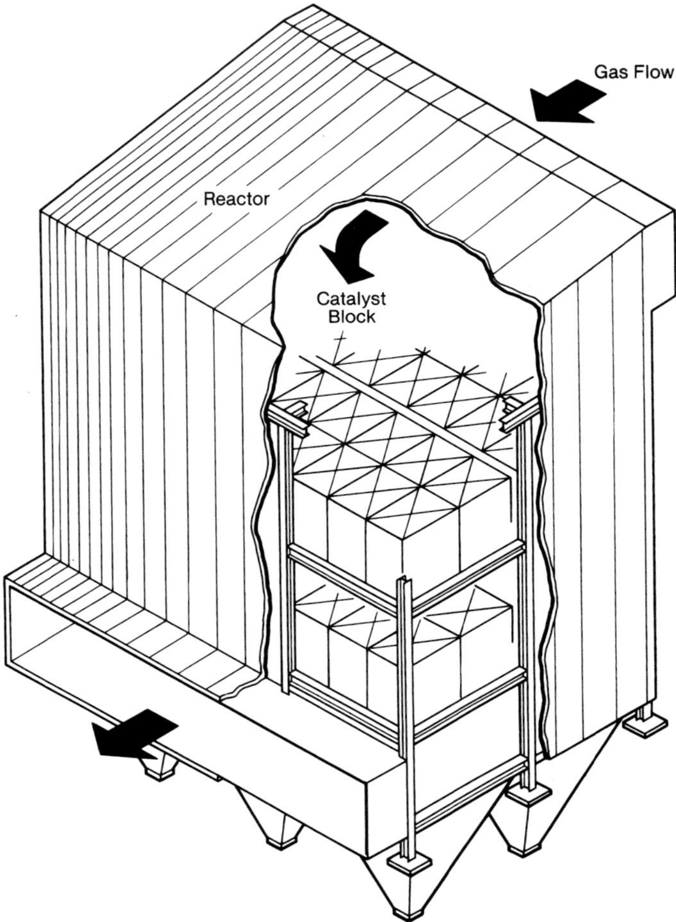

Fig. 12 Typical downflow SCR reactor module.

$$\overline{x} = \frac{1}{n} \sum_{i=1}^{n} x_i \quad (8)$$

where

σ = standard deviation
$\overline{x}$ = arithmetic mean

SCR reactor distribution goals for flow and molar ratio are commonly stipulated in this coefficient of variation format. Temperature distribution criteria is the exception by typically being expressed as a minimum and maximum (±) deviation about the arithmetic mean. The goal is to achieve the most uniform profile possible at the SCR catalyst inlet for NO_x concentration, ammonia to NO_x molar ratio, velocity and temperature. The degree of uniformity required for each parameter, coupled with the physical space and arrangement limitations presented, drives the level of complexity and intricacy of a successful design. Numerical or physical flow model studies are almost always used to determine the necessary distribution devices to achieve the desired uniformities.

Velocity distributions Velocity distributions are typically reviewed at two primary locations, at the inlet to the AIG and at the inlet to the catalyst. The velocity distribution at the AIG can impact the injection and subsequent or dosing concentration profile of ammonia into the flue gas. This is where the initial NH_3/NO_x molar ratio variance is produced. Independently controllable AIG injection zones are often incorporated to allow some compensation for imperfect flow profiles. These designs offer the ability to improve the ammonia concentration profile by reducing flow to low gas velocity areas and increasing flow to high velocity regions.

Variations in flow across the reactor cross section produce variations in local space velocities. The effect of this on the overall NO_x removal and NH_3 slip is negligible if the Cv is less than 15%.

The high flow zones increase the localized catalyst operating space velocity. This reduces the residence time in that region of the catalyst bed. The result is a localized reduction in reagent utilization and efficiency while producing a corresponding increase in the local ammonia slip concentration. The low velocity areas exhibit the opposite. These areas allow excess ammonia to be more effectively utilized. The result is increased local efficiency and reduced local ammonia slip. The ultimate amount of excess ammonia in any zone is, however, limited to that of the local NO_x concentration.

When systems process dust-laden gases, the concern for velocity maldistribution extends to the potential for particulate deposition and particle impact induced erosion. The pressure drop across the catalyst bed can act to rebalance the flow distribution approaching the catalyst bed inlet. Thus, the actual flow through the catalyst may improve from that measured upstream. However, when gas approaches with a poor velocity distribution, the redistribution effect can aggravate erosion and deposition potential. The typical value for the catalyst inlet velocity coefficient of variation is 15% or less.

Temperature distributions Temperature distribution concerns are usually associated with the extremes in the profile. Concerns for reduced catalyst activity and/or ammonium bisulfate precipitation are associated with the low temperature zones, while catalyst degradation is a concern in the high temperature areas. Due to this interest in the extremes, the temperature distribution requirements are typically expressed as an allowable range and are often specified as a specific minimum and maximum (±) deviation about the arithmetic mean.

A commonly specified limit for profile extremes is ±20F (11C). This range is fairly strict when compared to typical mean values of 550 to 750F (288 to 399C) and at times can be difficult to satisfy. This is especially true for sulfur-laden gases, which during low-load conditions require a portion of the flue gas to bypass the economizer to maintain a minimum operating reactor temperature. A review of the specific catalyst temperature requirements against the expected mean operating temperature can potentially allow for an increased deviation range.

Ammonia-to-NO_x molar ratio distribution Multiple factors determine the allowable NH_3/NO_x variation at the catalyst. These factors include the design NO_x removal efficiency, allowable ammonia slip, inlet NO_x concentration, and amount of excess catalyst.

NO_x removal efficiency is a significant factor. The ability to satisfy low average and peak ammonia slip concentrations by the addition of extra catalyst volume becomes exceedingly difficult at high removal efficiencies. As 100% NO_x removal is approached, the problem is exemplified by the potential for a given deviation in the mole ratio distribution to produce localized zones in the reactor where the ammonia concentration exceeds that of the NO_x concentration. At these locations, once all of the NO_x is removed, no further ammonia can be utilized; the overall average removal efficiency becomes limited and average slip is increased. This is particularly problematic for coal or other sulfur-containing fuel systems where the average slip is more limited.

While there are no specific break points, systems designed for NO_x removal efficiencies below 80% can often allow a molar ratio coefficient of variation of 10% for both coal- and gas-fired projects. In the range of approximately 80 to 90% NO_x removal, the assessment becomes more complex. Coal or sulfur-laden gas projects typically require a molar ratio Cv of 5%, where gas-fired projects may still allow a Cv of 10%. The difference is due to the level of average excess ammonia reagent. Gas-fired designs usually have higher allowable ammonia slip values, typically on the order of 10 ppm. The resulting increase in excess reagent makes operation at higher inlet molar ratio maldistributions more practical. Above 90% removal, the performance becomes very sensitive to the precise design efficiency, design slip and inlet NO_x level, and a molar ratio Cv of 5% is typically required. For lower ammonia slip designs and/or when approaching 95% removal, the molar ratio Cv may need to move toward values of 3% or better. Such uniformity levels can require significant mixing and/or flue length.

Injection design and mixing

Two principal factors affect ammonia distributions prior to the catalyst. One is the degree with which ammonia disperses or merges from one injection point to another. The other is the dispersed, but locally imbalanced, concentrations produced by the imperfect injection of local ammonia flux to local NO_x flux within the duct. An additional factor, to be discussed later, is the degree of turbulent mixing occurring in the downstream flue, either naturally or by a forced action (i.e., mixing devices). (See Fig. 13.)

The first factor of dispersing ammonia from point-to-point can be improved by increasing the injection point area concentration or the number of injection points per square foot (meter). The quantity of injection points required considers the distance from grid to catalyst and the degree of mixing along that path. Obviously, there is a point of diminishing returns. The second factor can be improved by locally adjusting the ammonia flow to more uniformly inject ammonia flow into the local NO_x mass flux across the injection plane. Local NO_x mass flux is determined by the local volumetric gas flow times the local NO_x concentration. Designing the injection grid with independently controllable zones allows adjustment of the ammonia flow based on the gas flow variance and NO_x concentration variance across this injection plane. This can lead

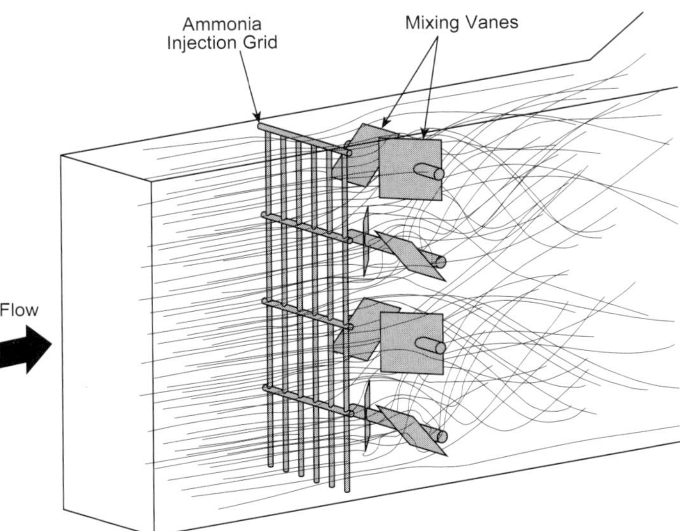

Fig. 13 Turbulent flow model.

to improved final mixtures through better injection, and a reduced initial molar ratio variance.

Imbalances due to flow and NO_x concentration variations across the plane of ammonia injection are best addressed by upstream flue gas flow correction, independently controllable injection zones, and/or downstream mixing.

Static mixing (blending of multiple fluids across fixed vanes and/or obstructions) is often incorporated for high uniformity levels. Static mixing devices (see Fig. 14) vary in shape, size and functionality. In many cases, a coordinated approach between the ammonia injection grid design and the static mixers is required. The use of static mixers is not without consequences. The general pros and cons are:

Pros:

1. higher level of achievable uniformity in temperature and/or chemical component distributions,
2. increased stability in the degree of uniformity achieved, and
3. simpler ammonia injection grid design and/or operation.

Cons:

1. a need for sufficient downstream flue length to ensure mixer performance,
2. additional pressure drop attributed to the mixer(s),
3. reduced traceability of the injected ammonia path from a particular injection grid zone to the catalyst, and
4. production of velocity disturbances downstream of the mixer(s).

Operating issues

In addition to the catalyst operating issues discussed previously, general operating issues and system effects must also be considered.

Ash accumulation in horizontal flues during periods of reduced gas flow can cause buildup to form, move, and accumulate above the catalyst. Larger ash agglomerate particles, often referred to as *popcorn ash* or *large particle ash* (LPA), can rapidly cover and block the inlet catalyst surface if allowed to reach the reactor. Flow model studies and arrangement design reviews are required to address these issues.

Sulfur-containing gases will experience a partial oxidation of SO_2 to SO_3 across the catalyst surface. While typically low, this oxidation level can influence air heater plugging when ammonium bisulfate is formed. It is also important when downstream scrubber systems are used and acid mist plume generation is of concern. Accurate ammonia flow control is required to minimize ammonia slip, and air heater design should consider the recommended features for reducing sulfate compound deposits.

Retrofit and general equipment issues

When an SCR system is retrofit on an existing plant, boiler modifications may be necessary for optimum performance. (See Figs. 15, 16 and 17.) Burner alterations/replacements to reduce furnace outlet NO_x can reduce reactor requirements. However, these same burner changes may increase furnace outlet NO_x maldistribution, and thus impact ammonia injection and/or mixing.

Physical space requirements of the reactor are often significant, and can require relocation of existing plant equipment. Air heaters are sometimes moved to allow proper flues between the upstream economizer outlets, the reactor, and the downstream air heaters. Arrangements become further complicated if economizer flue gas bypass is needed to sustain low load temperatures.

Air heater cold end surfaces can be enameled for ease of cleaning and to reduce corrosion. Changes in air heater surface design, as well as sootblower design and application, are often considered. These changes relate to ammonium bisulfate deposition and clean-

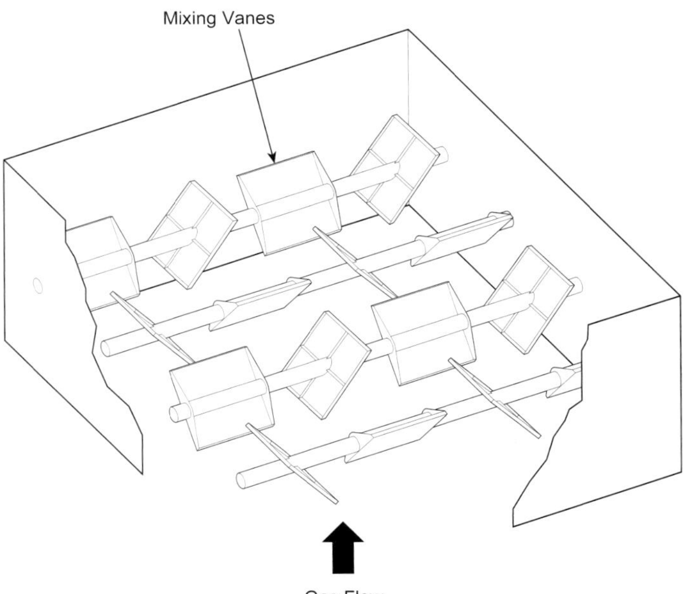

Fig. 14 Static mixing device for utility boiler SCR.

ing concerns when handling sulfur-laden gases and increased SO_3 concentrations.

The source of ammonia dilution air will vary depending on the ammonia system chosen and the nature of the boiler fuel being fired. Gas-fired systems produce a hot secondary air source that is essentially free of particulate. The coal-fired systems can exhibit sufficient entrainment of ash and traces of SO_3 into the secondary air stream of regenerative type air heaters; ammonia vaporization and injection system fouling can preclude using this source of heated air. Ambient air can be used in combination with steam or electric heaters. Designing for SCR reactor bypass can present additional arrangement challenges and costs. A catalyst protection system is also required to maintain dry catalyst conditions.

Pressure drop increase associated with the addition of flues, ammonia injection systems, mixers and catalyst can exceed existing fan limitations. Higher upstream static pressures in the boiler may also require costly furnace modifications to raise the design pressure. When new induced draft (ID) fans are added, other equipment such as precipitators may require a design pressure review.

The existing sale and/or disposal of flyash can be influenced by the retrofit. Flyash can absorb a portion of the ammonia, influencing its value as a saleable byproduct. Ammonia off-gassing can also occur when ammoniated high alkaline ash is wetted, as in the cement industry.

Whether a retrofit or a new system installation, structural design costs can be considerable. Catalyst weights and reactor spans are significant and the reuse and attachment to existing steel can complicate the analysis and the final design.

Operating system service needs

Due to the effects of flue gas (especially ash-laden), changes in control valve settings, or subtle changes in distribution, performance will change with operation. An increase in ammonia slip is the major indicator of this change. Therefore, catalyst activity, AIG valve settings, ash accumulation, and reactor pressure loss must be closely monitored.

The catalyst bed will typically contain samples which can be removed and tested during a short unit outage. Yearly sample analysis will establish an activity deterioration curve to establish a catalyst man-

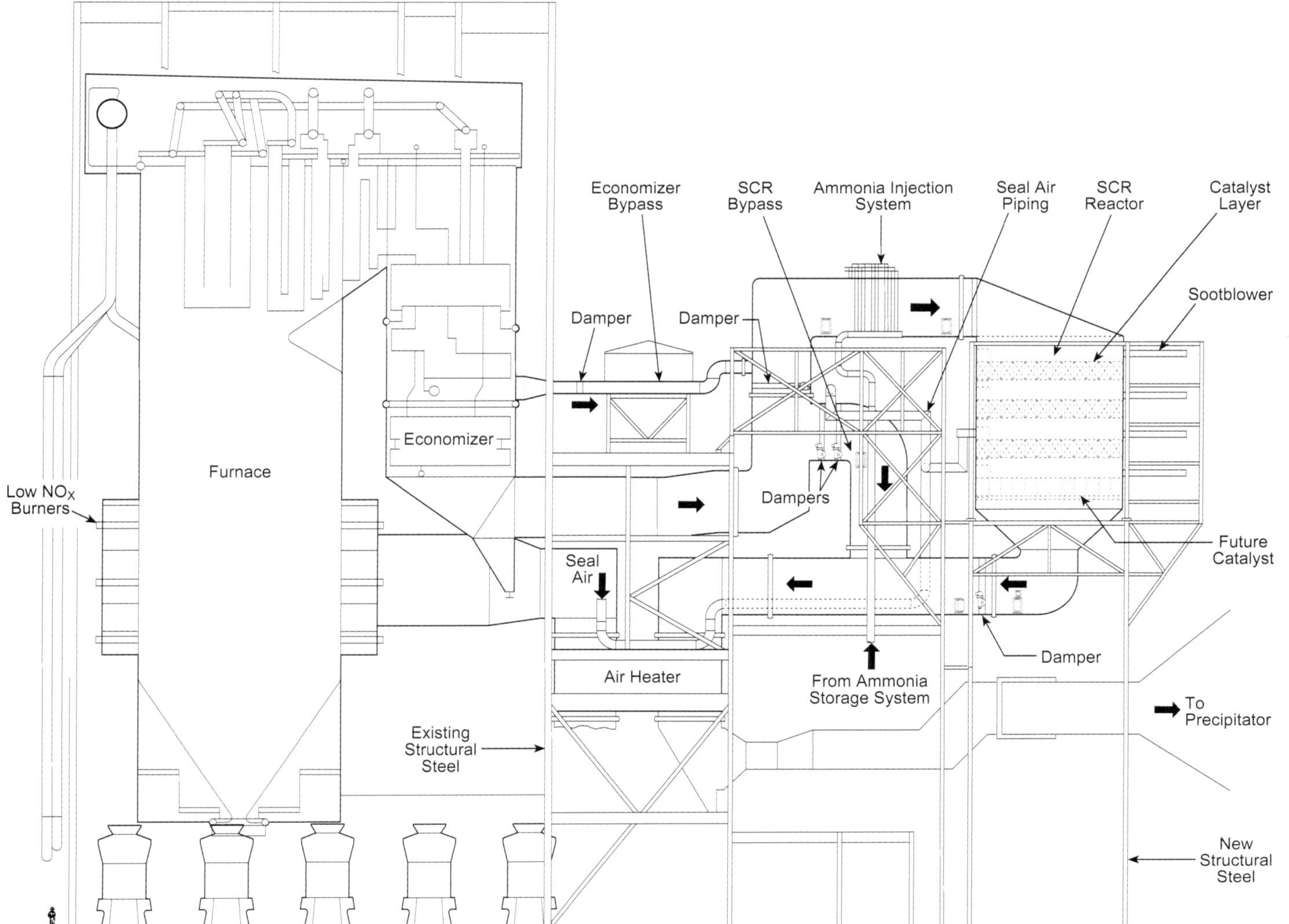

Fig. 15 SCR retrofit on a 655 MW coal-fired boiler.

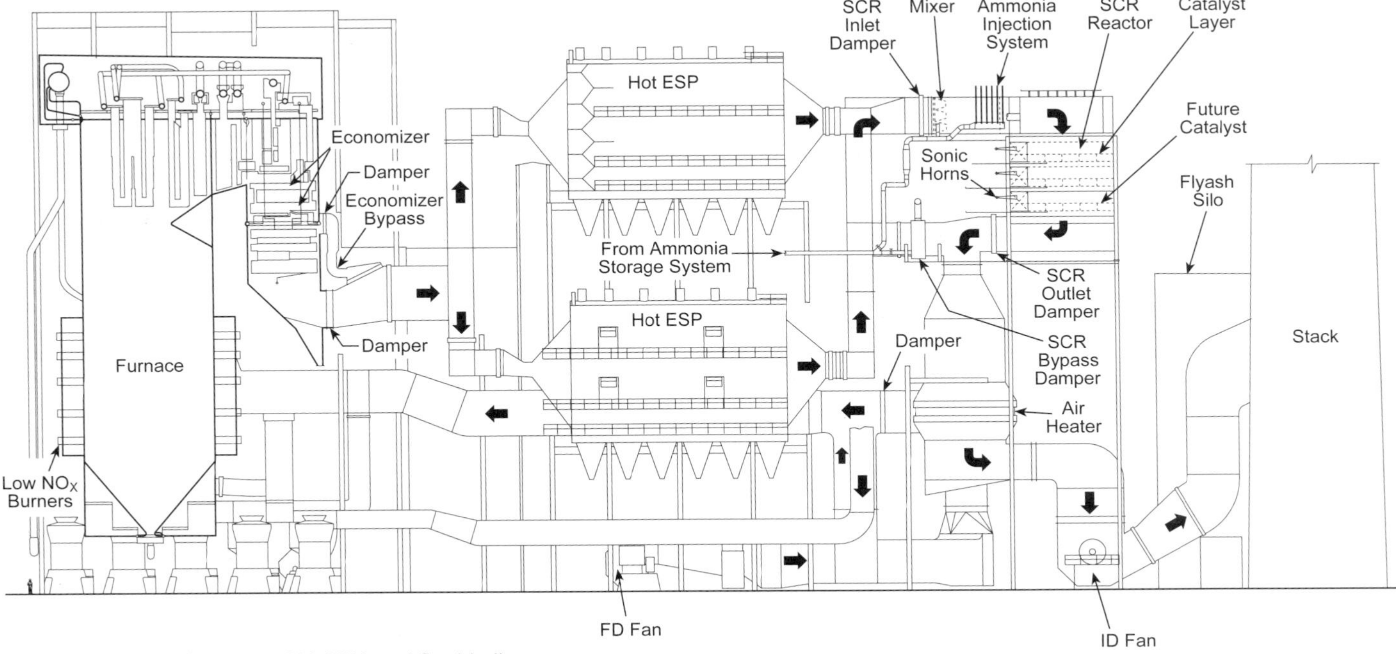

Fig. 16 SCR retrofit on two 600 MW coal-fired boilers.

agement plan. Frequent measurement and tracking of ammonia in most fly ashes can also be used to monitor reactor and catalyst performance.

Catalyst management activities include catalyst addition or replacement, depending on reactor design, and catalyst regeneration. Regeneration techniques include in-situ washing and catalyst layer removal for washing by specialists. Regeneration involves the removal of ash from the catalyst surface and, in some cases, addition of active catalyst components. Some management plans will require a combination of methods. The cost of regeneration is generally less than the cost of new catalyst.

Eventually, the catalyst or a potion of the modules will require disposal due to erosion or unrecoverable poisoning. Since the catalyst is contaminated with the

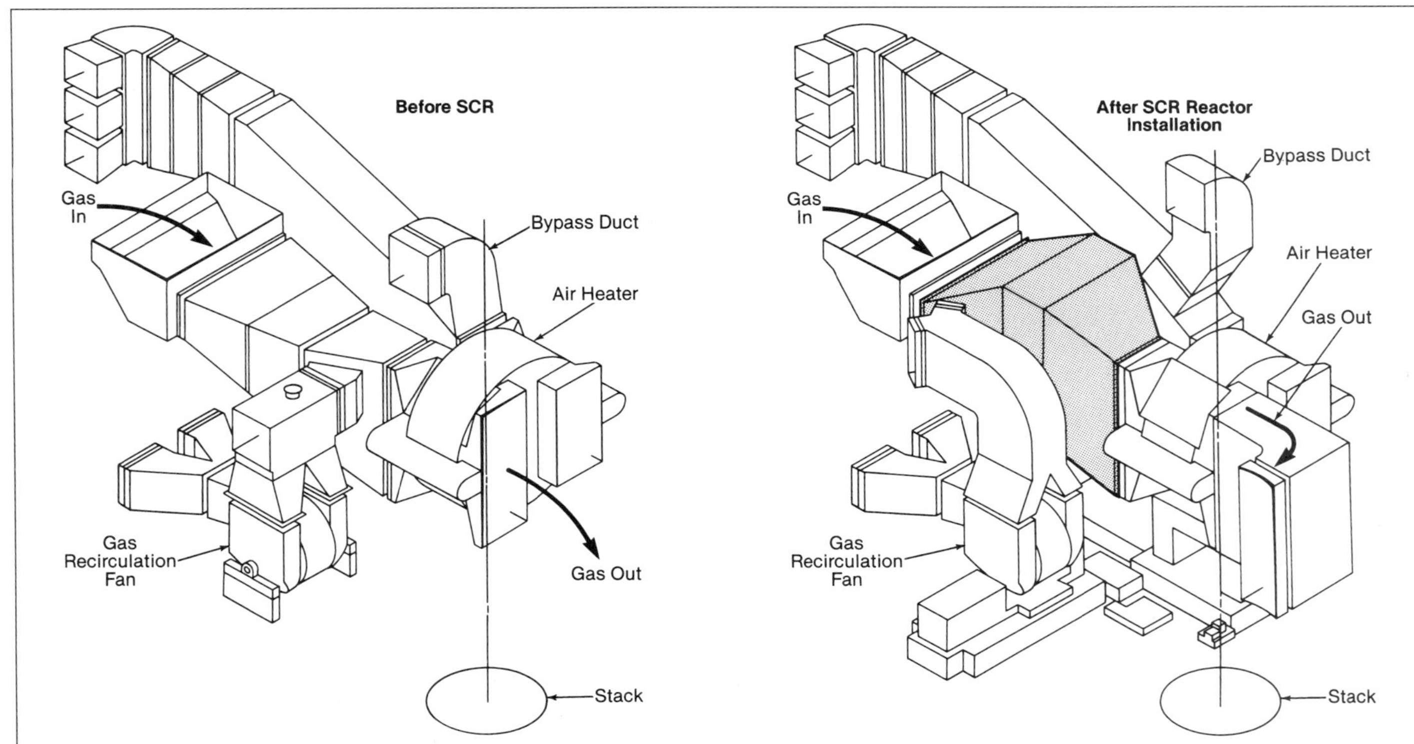

Fig. 17 Oil- and gas-fired utility boiler SCR retrofit.

constituents from some fuels, the spent catalyst may require special landfill disposal procedures. Crushing of the honeycomb modules to reduce volume, sale of plate catalyst to steel mills, or negotiating disposal as a condition of new catalyst purchase are common solutions.

Re-optimization of the AIG control valve settings may also be required. A potential cause is a change in flow characteristics throughout the dilution medium and ammonia systems. Changes in the uniformity of gas flow, NO_x and gas temperature may also lead to re-optimization.

On high dust applications, cleaning during outage periods is recommended. Other periodic checks include removal of water accumulations from anhydrous ammonia systems, instrument and NO_x analyzer maintenance, and cleaning of impurities from aqueous ammonia and urea systems.

Technology enhancements

SCR advancements have centered on catalyst development for improved erosion resistance, resistance to chemical deactivation, lower SO_2 to SO_3 conversion, greater activity configurations, and an expanded temperature performance range. Catalyst materials such as zeolite and precious metals are more recent innovations.

Application of neural network control technology offers the potential to further optimize NO_x reduction for load-following operation.

Selective noncatalytic reduction technology

There are currently two basic selective noncatalytic reduction (SNCR) processes available. An ammonia-based system and a urea-based technology have been developed and commercially operated for many years. Although there are distinct differences to each technology, the overall processes are similar and will be presented as the same in the following sections. While the current emphasis in NO_x control has centered on the SCR technology, SNCR is an effective control technology for certain applications or in combination with combustion and control system techniques.

NO_x reduction reaction

SNCR technologies inject a reducing agent into NO_x-laden flue gas within a specific temperature zone. In addition, it is important to properly mix the reagent with the flue gas. Finally, the mixture must have adequate residence time at temperature for the reduction reactions to occur. The chemical reactions for the processes are represented by:

Ammonia
$$4NO + 4NH_3 + O_2 \rightarrow 4N_2 + 6H_2O \tag{9}$$

Urea
$$2NO + (NH_2)_2CO + 1/2\,O_2 \rightarrow 2N_2 + 2H_2O + CO_2 \tag{10}$$

The acceptable temperature range for either reaction is 1400 to 2000F (760 to 1093C), although temperatures above 1700F (927C) are preferred. Below 1600F (871C), chemical enhancers, such as hydrogen, are needed to assist the reactions. As the temperature increases within this range, the ammonia or urea may react with available oxygen to form NO_x. This reaction becomes significant at temperatures above 2000F (1093C) and may become dominant as temperatures approach 2200F (1204C).

The application of SNCR technology to boilers which by design have the suitable gas temperatures in the upper furnace is an attractive use of the technology. Examples of this boiler type are those applied to municipal solid waste and biomass firing. (See Fig. 18.) Multiple injection levels are used to maintain NO_x reduction efficiencies at acceptable levels as boiler load changes. Based on operating experience, four or more injection levels may be required for larger units. Multiple levels are required because the flue gas temperature profile changes with boiler load and the reagent injection point must be adjusted accordingly. To determine the most appropriate and effective injection locations, it is common practice to perform both numerical and chemical kinetics modeling of the furnace and combustion.

When using ammonia reagents with SNCR, it is important to control the excess unreacted ammonia. As flue gas temperatures are reduced, the excess ammonia can react with other combustion species, primarily sulfur trioxide (SO_3), to form ammonium salts. The major ammonium products formed are ammonium sulfate [$(NH_4)_2SO_4$] and ammonium bisulfate

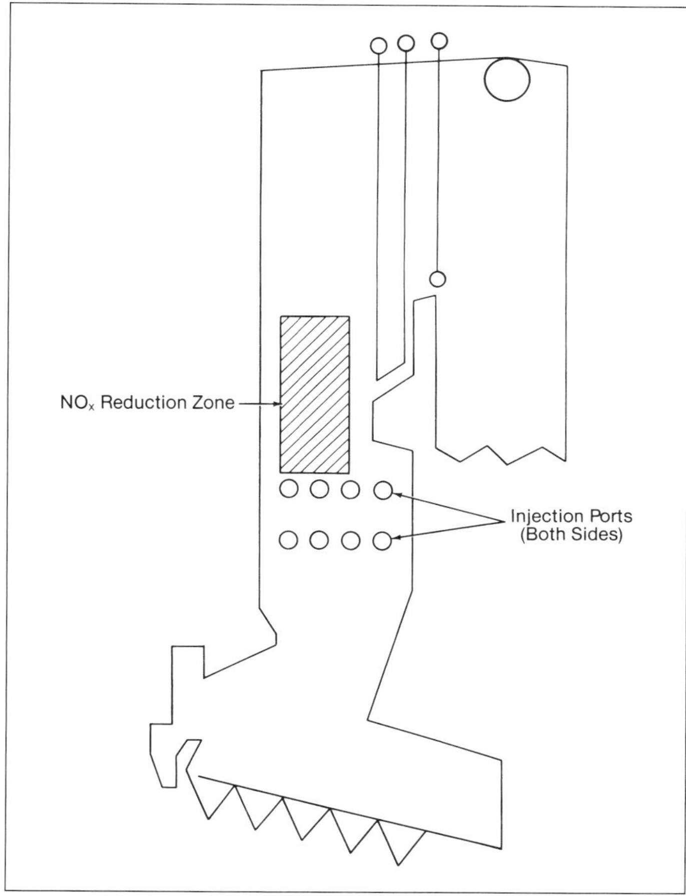

Fig. 18 Typical SNCR application for a municipal solid waste boiler.

(NH_4HSO_4). Ammonium sulfate, typically formed when ammonia concentration exceeds SO_3 concentrations, is a dry, fine particulate (1 to 3 microns in diameter) that may contribute to plume formation. The ammonium bisulfate, typically formed when SO_3 concentration exceeds the ammonia concentration, is a highly acidic and sticky compound which, when deposited on downstream equipment such as economizers and air heaters, contributes to significant fouling and corrosion. The ammonia may also attach to the ash and affect its reuse or market value.

While NO_x reduction levels of 70% are possible under carefully controlled conditions, 30 to 50% reductions in NO_x emissions are more typically used in practice to maintain acceptable levels of reagent consumption and ammonia slip.

Most of the current applications for these technologies have been on stoker-fired and fluidized-bed boilers, where the appropriate temperature range and residence time are available. (See also Chapters 16 and 17.) In large utility units (see Fig. 19), the proper temperature range occurs in the convection pass cavities, making application, especially on retrofits, more challenging. For these applications, overall control may be limited to the 20 to 30% range. New boiler applications may require special design of the boiler convection pass with a dedicated open cavity in an optimal temperature window for the reactions to occur, although with current NO_x emission requirements this technology may be limited to lower NO_x fuels and processes. Other flue gas constituents, such as CO, may also affect the potential reduction achievable with SNCR.

Most SNCR process development has centered on optimization of the basic process. The use of natural gas or other injection agents above the reagent injection zone have been studied and installed with some success to extend the useful NO_x reduction temperature range and achieve greater NO_x reduction. A combination of SNCR and SCR catalyst to control ammonia slip has also been studied.

Reagent equipment and injection

The SNCR system consists of storage and handling equipment for the ammonia or urea, equipment for mixing the chemical with the carrier (compressed air, steam or water), and the injection equipment. Much of the equipment is similar to the SCR storage and handling equipment. The key component, the injection system, consists of nozzles generally located at various elevations on the furnace walls to match the expected flue gas operating temperature. The number and location of the nozzles are established by the supplier and are based on obtaining good reagent distribution within the flue gas.

One major difference between the ammonia and urea based processes is that the ammonia is usually injected into the gas stream in a gaseous state, whereas the urea is injected as an aqueous solution. The urea technology requires a longer residence time for reactions due to the time required to vaporize the liquid droplets once they are in the gas stream. Aqueous ammonia systems may also be injected as a liquid with additional residence time required for vaporization.

Emerging technologies

A variety of advanced post-combustion NO_x control systems are at various stages of development and demonstration. Many of these are combined NO_x and SO_2 control systems. While the SNCR and SCR systems discussed here basically feature dry reduction of NO_x by ammonia or urea, advanced systems under development offer a variety of options, for example wet (aqueous) or dry absorption by solids, absorption plus oxidation by a liquid, and absorption plus reduction by a liquid. Technologies include low temperature NO_x oxidation with injection of an oxidizing agent like ozone, where NO_x is converted to soluble species that are easily removed with a water wash. Another example is the injection of a chelating agent to flue gas desulfurization (FGD) slurry to enable NO_x scrubbing. Although high NO_x reduction is claimed, further development and demonstration of these systems are required.

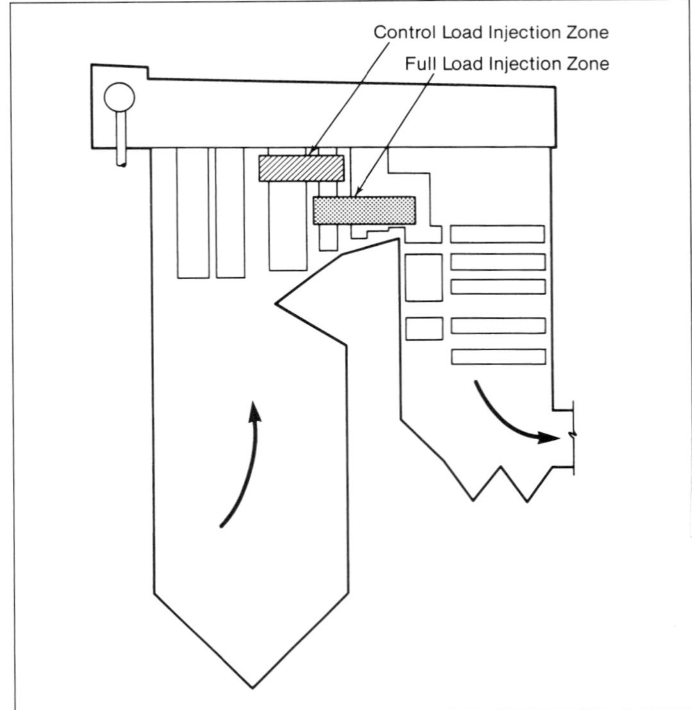

Fig. 19 Coal-fired utility boiler with SNCR temperature window location at full and part (control) load.

References

1. Miller, J.A., and Bowman, C.T., "Mechanism and Modeling of Nitrogen Chemistry in Combustion," Presented at the 1988 Fall Meeting of the Western States Section of the Combustion Institute, Dana Point, California, October 17-18, 1988.

2. Zeldovich, Y.B., "Oxidation of Nitrogen in Combustion and Explosion," *Academic des Sciences de L'URSS Comptes Rendus (Doklady)*, Vol. 51, No. 3, pp. 217-220, January 30, 1946.

3. Sarofim, A.F., and Pohl, J.H., "Kinetics of Nitric Oxide Formation in Premixed Laminar Flames," Proceedings of the 14th International Symposium on Combustion, pp. 739-754, The Combustion Institute, Pittsburgh, Pennsylvania, 1973.

4. "Why NO_x? The Costs, Consequences and Control of Nitrogen Oxides in the Human and Natural Environment," American Lung Association Report, Washington, D.C., April, 1989.

Bibliography

Campbell, L.M., Stone, D.K., and Shareef, G.S., Sourcebook: NO_x Control Technology Data, Publication No. EPA-600/2-91-029, Prepared for the U. S. Environmental Protection Agency (EPA), Research Triangle Park, North Carolina, July, 1991.

Gouker, T.R., and Brundrett, C.P., "SCR Catalyst Developments for the U.S. Market," Presented at the Joint EPA/EPRI $DeNO_x$ Symposium, Washington, D.C., March 25-28, 1991.

Hurst, B.E., and White, C.M., "Thermal $DeNO_x$: A Commercial Selective Non-Catalytic NO_x Reduction Process for Waste-to-Energy Applications," proceedings of the ASME Twelfth Biennial National Waste Processing Conference, Denver, Colorado, June 2, 1986.

Lim, K.J., et al., "Environmental Assessment of Utility Boiler Combustion Modification NO_x Controls," U.S. Environmental Protection Agency Report, Contract No. 68-02-2160, April, 1980.

Perhac, R.M., "Environmental Effects of Nitrogen Oxides," 1989 Symposium on Stationary Combustion Nitrogen Oxide Control, Report GS-6423, Vol. 1, Electric Power Research Institute, Palo Alto, California, July, 1989.

Pohl, J.H., and Sarofim, A.F., "Devolitilization and Oxidation of Coal Nitrogen," Sixteenth Symposium on Combustion, Massachusetts Institute of Technology, Cambridge, Massachusetts, pp. 491-501, August 15, 1976.

"Selective Catalytic Reduction (SCR) Controls to Abate NO_x Emissions," A white paper prepared by the SCR Committee of the Institute of Clean Air Companies, Inc. (ICAC), Washington, D.C., November, 1997.

"Selective Non-Catalytic Reduction (SNCR) for Controlling NO_x Emissions," A white paper prepared by the SNCR Committee, of the Institute of Clean Air Companies, Inc. (ICAC), Washington, D.C., May, 2000.

"Standard on the Storage and Handling of Anhydrous Ammonia: Extension of the Office of Management and Budget's Approval of Information Collection (Paperwork) Requirements," 69:42781-42782, OSHA Standards 29 CFR, 1910.111, Occupational Safety and Health Association, United States (U.S.) Department of Labor, Washington, D.C., July 16, 2004.

Orimulsion is a trademark of Bitumenes Orinoco, S.A.

Wet flue gas desulfurization module at a utility installation.

Chapter 35

Sulfur Dioxide Control

Sulfur appears in the life cycle of most plants and animals. Most sulfur emitted to the atmosphere originates in the form of hydrogen sulfide from the decay of organic matter. These emissions slowly oxidize to sulfur dioxide (SO_2). Under atmospheric conditions, SO_2 is a reactive, acrid gas that can be rapidly assimilated back to the environment. However, the combustion of fossil fuels, in which large quantities of SO_2 are emitted to relatively small portions of the atmosphere, can stress the ecosystem in the path of these emissions.

Man is responsible for the majority of the SO_2 emitted to the atmosphere. Annual worldwide emissions are approximately 160 million tons, nearly half of which are from industrial sources. The two principal industrial sources are fossil fuel combustion and metallurgical ore refining.

When gaseous SO_2 combines with liquid (ℓ) water, it forms a dilute aqueous solution of sulfurous acid (H_2SO_3). Sulfurous acid can easily oxidize in the atmosphere to form sulfuric acid (H_2SO_4). Dilute sulfuric acid is a major constituent of *acid rain*. (Nitric acid is the other major acidic constituent of acid rain.) The respective reactions are written:

$$SO_2\,(g) + H_2O(\ell) \rightarrow H_2SO_3\,(aq) \tag{1}$$

$$O_2\,(g) + 2H_2SO_3\,(aq) \rightarrow 2H_2SO_4\,(aq) \tag{2}$$

SO_2 can also oxidize in the atmosphere to produce gaseous sulfur trioxide (SO_3). Sulfur trioxide reactions are written:

$$2SO_2\,(g) + O_2\,(g) \rightarrow 2SO_3\,(g) \tag{3}$$

$$SO_3\,(g) + H_2O(g) \rightarrow H_2SO_4\,(\ell) \tag{4}$$

While Equations 1 and 2 describe the mechanism by which SO_2 is converted to sulfuric acid in acid rain, Equations 3 and 4 characterize dry deposition of acidified dust particles and aerosols.

The pH scale, a measure of the degree of acidity or alkalinity, is the method used to quantify the acidity of acid rain.

Pure water has a pH of 7 and is defined as neutral, while lower values are defined as acidic and higher values as alkaline. If rainwater contained no sulfuric or nitric acid, its pH would be approximately 5.7 due to absorption of carbon dioxide (CO_2) from the atmosphere. The contributions of man-made SO_2 and nitrogen oxides (NO_x) further reduce the pH of rainwater. No uniformly accepted definition exists as to what pH constitutes acid rain. Some authorities believe that a pH of about 4.6 is sufficient to cause sustained damage to lakes and forests in the northeastern portion of North America and in the Black Forest region of Europe.

SO_2 emissions regulations

Legislative action has been responsible for most industrial SO_2 controls. Major landmark regulations include the Clean Air Act Amendments of 1970, 1977 and 1990 in the United States (U.S.), the Stationary Emissions Standards of 1970 in Japan, and the 1983 SO_2 Emissions Regulations of the Federal Republic of Germany. Since the mid-1980s, SO_2 emissions regulations have been implemented in most other industrialized nations and many developing nations.

SO_2 control

Most utilities have adopted one of two strategies for SO_2 control, either switching to low sulfur coal or installing scrubbers.

A variety of SO_2 control processes and technologies are in use and others are in various stages of development. Commercialized processes include wet, semi-dry (slurry spray with drying) and completely dry processes. The wet flue gas desulfurization (WFGD) scrubber is the dominant worldwide technology for the control of SO_2 from utility power plants, with approximately 85% of the installed capacity, although the dry flue gas desulfurization (DFGD) systems are also used for selected lower sulfur applications.

Total annual SO_2 emissions in the U.S., including electric utility SO_2 emissions, have declined since 1970 as various regulations have been adopted. During the same period, electricity generation from coal has almost tripled (see Table 1).[1,2]

A significant portion of this emissions reduction has been the result of switching to low sulfur coal, predominantly from the western U.S. In 1970 virtually all of the utility coal came from the eastern, higher sulfur coal fields, while by 2000 approximately half of the coal came from western low sulfur sources. Slightly less than two-thirds of SO_2 emission reductions have been attributed to fuel switching while over a third has been through the installation of flue gas desulfurization systems, predominantly wet scrubbers. More than 30% of the U.S. coal-fired capacity already has FGD systems installed and operating. This may come close to doubling over the next decade and a half as existing regulations are implemented and proposed regulations are adopted.

Wet scrubbers

Reagents

Wet scrubbing processes are often categorized by reagent and other process parameters. The primary reagent used in wet scrubbers is limestone. However, any alkaline reagent can be used, especially where site-specific economics provide an advantage. Other common reagents are lime (CaO), magnesium enhanced lime (MgO and CaO), ammonia (NH_3), and sodium carbonate (Na_2CO_3). The first part of this chapter concentrates on limestone-based wet scrubbing systems with forced oxidation to produce gypsum. The next major section focuses on lime-based semi-dry or dry systems.

Reagent regeneration A number of the wet processes are also classified as either non-regenerable or regenerable systems. In non-regenerable systems, the reagent in the scrubber is consumed to directly generate a byproduct containing the sulfur, such as gypsum. In regenerable systems, the spent reagent is regenerated in a separate step to renew the reagent material for further use and to produce a separate byproduct, such as elemental sulfur. The dominant limestone and lime reagent systems used today are non-regenerable. In many cases the regenerable systems have been retrofitted with non-regenerable limestone or lime reagent systems to reduce costs and improve unit availability.

Absorber design

Evolution The first WFGD scrubbers installed in the U.S. were combined particulate collectors and SO_2 absorbers. However, the energy requirements for the venturi scrubbers used for particulate collection proved to be excessive. High efficiency dust collectors, usually electrostatic precipitators (ESPs), replaced venturi scrubbers, and separate, much lower pressure drop absorber towers were used to absorb the SO_2.

The WFGD systems installed in the 1970s and early 1980s were typically sized with multiple modules with spare spray headers and sometimes spare scrubber (absorber) modules. Prior to about 1978, most of the lime and limestone WFGD systems were not designed with forced oxidation. These absorber internals were subject to severe scaling and plugging of internal components, such as spray nozzles. With the development of forced oxidation, the scaling has effectively been eliminated. This, along with improvements in accessory equipment such as pumps, has eliminated the need for spare scrubber modules. Most wet scrubbers include a spare recirculation pump and spray header. However, the pump and header reliability are such that even spare pumps are not required in some designs.

Since the mid-1990s the use of a single absorber module in WFGD systems has become an accepted design. Single absorber modules have been supplied to handle 1300 MW boilers. Some designs use a single absorber to treat the flue gas from multiple boilers. The availability of most of these systems has been approaching 100%. In most cases, forced outages or load reductions have been due to problems with support systems such as reagent preparation, dewatering and gypsum handling rather than the absorber module.

General arrangement and design Fig. 1 provides a side view of a 660 MW modern coal-fired power system showing the location of the SO_2 scrubber absorber module. Absorber modules are located on either side of the stack though other arrangements are possible. The dust collector, in this case an ESP, is placed upstream of the wet scrubber. In this arrangement, the induced draft (ID) fan is between the dust collector and the scrubber, permitting the fan to operate in a particulate-free, dry flue gas. In cases of retrofit to existing units, it is sometimes necessary to add a booster fan between the ID fan and the scrubber. In some systems, the booster fan is installed after the wet scrubber in what is referred to as the wet position. This is not the preferred method as wet fans require more maintenance and are more expensive.

The most common WFGD absorber module is the spray tower design shown in Fig. 2. The flue gas enters the side of the spray tower at approximately its midpoint and exits through a transition at the top. The upper portion of the module (absorption zone) provides for the scrubbing of the flue gas to remove the SO_2 while the lower portion of the module serves as an integral slurry reaction tank (also frequently referred to as the recirculation tank and oxidation zone) to complete the chemical reactions to produce gypsum. The self-supporting absorber towers typically range in diameter from 20 to 80 ft (6 to 24 m) and can reach 150 ft (46 m) in height. In some designs, the lower reaction tank is flared downward to provide a larger di-

Table 1
U.S. SO_2 Emissions and Coal-Fired Power Generation

	Total U.S. SO_2 10^6 t/yr	Utility SO_2 10^6 t/yr	Coal Fired Utility Generation 10^{12} kWh
1970	31	17	0.7
1980	26	17	1.2
1990	23	16	1.6
2000	16	11	2.0

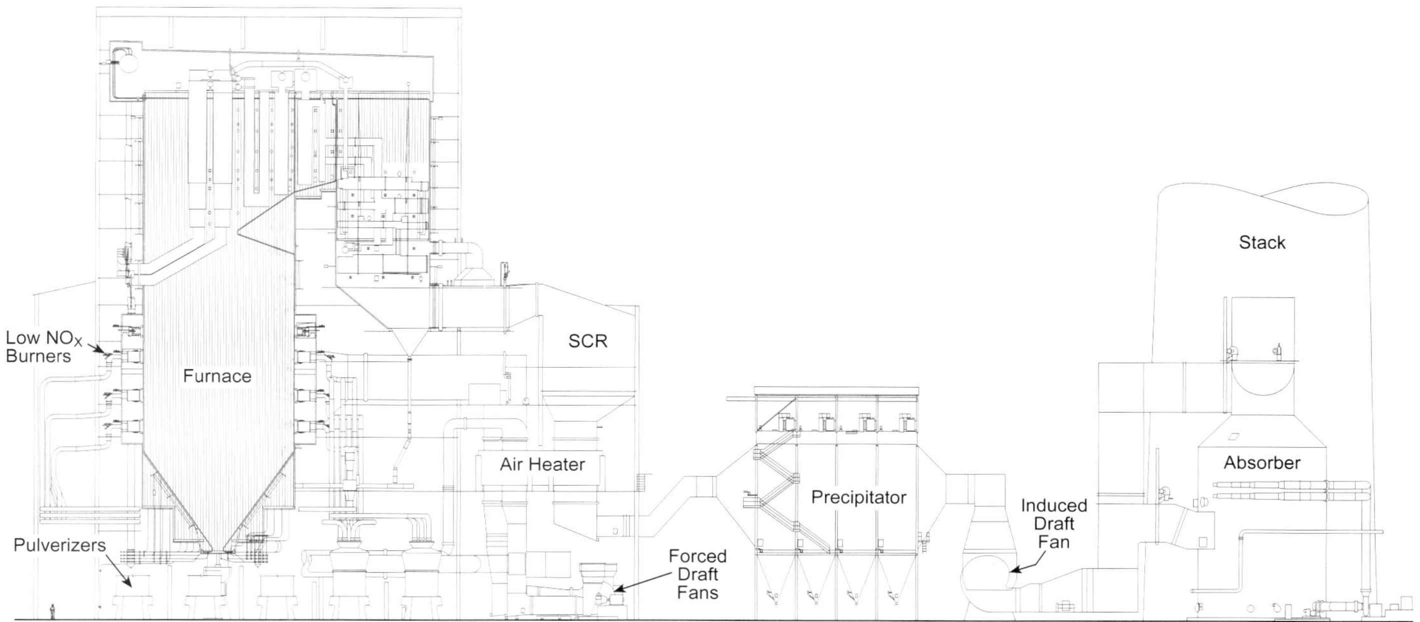

Fig. 1 Typical emission control components for a coal-fired utility boiler.

ameter tank for larger slurry inventory and longer retention time. Other key components shown include the slurry recirculation pumps, interspatial spray headers and nozzles for slurry injection, moisture separators to minimize moisture carryover, oxidizing air injection system, slurry reaction tank agitators to prevent settling, and the perforated tray to enhance SO_2 removal performance. The following sections discuss the gas flow path, slurry flow path, and the materials used in the tower shell design to accommodate the highly corrosive environment.

Gas flow path As noted above, the flue gas enters the absorber at about the midpoint and turns and flows up through an absorption tray, a spray zone, and moisture separators before exiting the absorber and flowing to the stack. Because the flue gas enters the absorber from the side, gas flow non-uniformity in the tower is a potential issue. This non-uniformity reduces overall SO_2 removal performance and aggravates moisture separator carryover. The absorber design depicted in Fig. 2 incorporates a perforated plate tray that reduces flue gas flow maldistribution. The pressure drop across the tray is usually between 1 and 3 in. wg (0.2 and 0.7 kPa). The tray provides intimate gas/liquid contacting and increases the slurry residence time in the absorption zone. Some absorbers have two trays providing multiple contact zones for SO_2 removal. Absorber modules that do not use a tray are referred to as open spray towers. Flue gas enters the scrubber module at a temperature of 250 to 350F (121 to 177C) and is evaporatively cooled to its adiabatic saturation temperature by a slurry cascading from the absorber tray. A wet-dry interface exists at the scrubber inlet where the inlet fluework at the gas temperature of approximately 300F (149C) merges with the scrubber shell which is at saturation temperature of around 125F (52C). Reagent slurry droplets impinging and drying out on the relatively hot surfaces can create growing deposits which can affect scrubber operation. Deposits are minimized by a combination of features that effectively keep the slurry away from the wet-dry interface region.

The fluework from the exit of the wet scrubber to the stack is an important facet of the system design. The potential for severe corrosion and deposition in these flues is well documented. This potential for severe corrosion arises from many factors. The flue gas leaves the moisture separator saturated with water vapor. Some carryover of slurry droplets smaller than 20 microns is inevitable. These droplets will usually be slightly acidic and may contain high concentrations of dissolved chlorides. The flue gases will contain some residual SO_2 and ample oxygen to oxidize the SO_2 to SO_3. Because the flue gas is saturated with water vapor, surface condensation is inevitable. This condensate can become severely acidic (pH less than 1) leading to the formation and accumulation of acidic deposits of sulfuric acid and calcium salts on the walls and floor of the flue.

Two approaches are used to minimize these effects: flue gas reheat and flue/stack lining. The former option involves reheating the flue gas so that no droplets remain. Reheating the flue gas that is leaving the scrubber has been accomplished by various means:

1. steam coil heaters,
2. mixing with some hot flue gas which is bypassed around the scrubber,
3. mixing with hot air, and
4. regenerative heat exchangers that transfer heat from the hot flue gas inlet to the cooler flue gas outlet.

Several problems are associated with each of the reheat methods, primarily corrosion and deposition. As a result, operation without flue gas reheat, i.e., with a wet stack, has become popular in the U.S. Under these conditions, the fluework from the scrubber to the stack is made of high alloy materials or lined with corrosion resistant materials, and the stack is lined with acid resistant linings or is made of alloys. Moisture collection devices are installed in the outlet flues and stack, and drains are provided to capture the moisture carryover and condensate collected in these devices. The use of regenerative gas-gas heaters is prevalent internationally because of restrictions on stack plumes.

Slurry flow path The bottom of the absorber is an integral reaction or recirculation tank. Slurry-containing reaction products and unreacted fresh reagent are pumped from the recirculation tank to the spray headers and fall through the spray zone, tray and gas inlet zone to the tank. The slurry is continuously recirculated in this loop. Fresh reagent such as limestone is added to replenish the alkalinity required to remove SO_2. Reaction products are pumped from the absorber to the dewatering system. Spray nozzles are used in wet scrubbers to atomize the slurry into fine droplets and provide contact surface area for the slurry and flue gas. The operating pressures typically vary between about 5 and 20 psi (34 and 138 kPa). Spray nozzles without internal obstructions are favored to minimize plugging by tramp debris. The arrangement of the spray headers and nozzles is designed to ensure complete coverage of the absorber cross-section and prevent gas bypassing through areas of low slurry spray flux, which reduces overall SO_2 removal efficiency.

A portion of the slurry droplets is entrained by the gas and carried up to the moisture separators. The

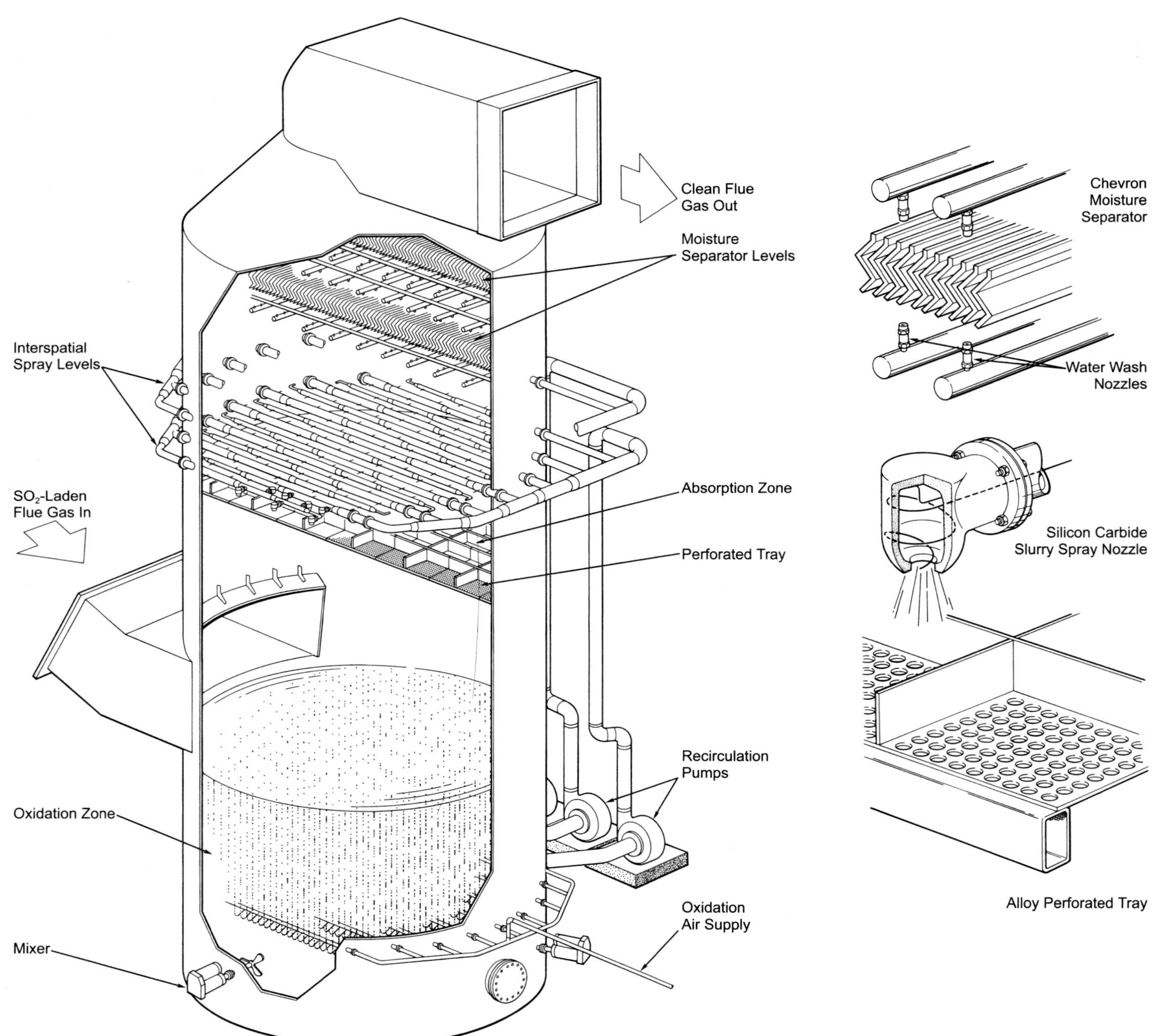

Fig. 2 Wet flue gas desulfurization absorber module.

moisture separators collect and coalesce these slurry droplets so that they will drain back down into the absorber. Otherwise, slurry droplet carryover would result in the buildup of slurry in outlet flues, fallout in the area around the power plant, and excessive particulate emissions. The moisture separators in most wet scrubbers are of the *chevron* design (see Fig. 2). Chevrons are closely spaced corrugated plates that collect slurry droplets by impaction. They efficiently collect droplets larger than about 20 microns in diameter.

The recirculation tank size is set: 1) to meet the requirements of the recirculation pump suction head, 2) to allow time for the chemical reactions of sulfite oxidation and limestone dissolution, and 3) to provide time and surface area for gypsum crystal growth development. Some of these rate processes are described in more detail later.

Materials of construction The slurry pH of the recirculated slurry is typically between 5 and 6 in the recirculation tank of a limestone system. The pH in the spray zone and on the tray can be as low as 3.5 to 4. Chloride concentrations (controlled by blowdown) are usually designed at 20,000 ppm by weight. However, some systems run up to 50,000 ppm. Systems that use seawater as makeup water can run as high as 100,000 ppm. Alloy construction has been the most popular selection for absorber materials in the U.S. (see Table 2). Rubber-lined, carbon steel scrubber modules are popular in Europe. In addition, there are a number of quality lining systems available.

The industry has used a number of different alloys for the absorber shell, tray and internal supports. Rubber linings, flake-glass linings and ceramic tile systems have also been used. The material selection on any project is dependent on the process chemistry and the cost-benefit analysis of the material from a lifecycle perspective.

Flake-glass lined carbon steel is the least expensive and provides the lowest installed capital cost, particularly in areas with low field labor costs. However, there are risks to using flake-glass linings and they require the most maintenance. It is expected that these linings would require annual inspections and repair of about 3% of the surface area. This is, therefore, not a good selection for absorbers requiring two to three years between outages. Flake-glass or epoxy resin lining systems are frequently suggested for slurry tanks, process trenches and sumps. The lining in tanks and sumps is less subject to problems because there is less erosion, no internals, a limited number of connections, and no direct contact with the flue gas.

Rubber linings have provided a range of operating issues in the U.S. because of the many failures in the early use of scrubbers. The failures were due to incorrect rubber selection and poor application. The technology has advanced and rubber lining can provide good corrosion and erosion resistance over the life of the plant. With rubber linings, inspection intervals of 12 to 18 months are recommended. Typically, minimal repair is required in the first four years. After that it is expected that 3% surface repair will be required each year.

Tile-lined concrete is a good selection for absorber modules. The tile provides excellent erosion resistance in the impingement areas and under the agitators. It also has high chloride resistance. The tile does require some grout repair. The recommended time between outages is 18 to 24 months.

Alloy construction is frequently selected for the absorber vessel. Areas of wear can be addressed with a limited amount of erosion resistant coatings/linings. The outage time required to inspect and repair erosion barriers (linings) is reduced compared to coated or lined designs.

Absorption of SO_2

There are a number of parameters that control the SO_2 removal capability of a wet absorber as listed in Table 3. The effects of these parameters can be better understood when the two primary functions of the

Table 2
Some Alloy Choices for Wet FGD Slurry Service

Alloy	UNS Number	Cr Wt %	Mo Wt %	Ni Wt %	N Wt %	Maximum Cl ppm
316L	S31603	16.0	2.0	10.0	0	10,000
20	N08020	19.0	2.0	32.0	0	12,000
317L	S31703	18.0	3.0	11.0	0	15,000
825	N08825	19.5	2.5	38.0	0	15,000
317LM	S31725	18.0	4.0	13.5	0	18,000
317LMN	S31726	17.0	4.0	13.5	0.1	20,000
904L	N08904	19.0	4.0	23.0	0	20,000
2205	S32205	22.0	3.0	4.5	0.14	30,000
255	S32550	24.0	2.9	4.5	0.1	45,000
G	N06007	21.0	5.5	36.0	0	50,000
254-SMO	S31254	19.5	6.0	17.5	0.18	55,000
AL-6XN	N08367	20.0	6.0	23.5	0.18	55,000
625	N06625	20.0	8.0	58.0	0	55,000
C-22	N06022	20.0	12.5	50.0	0	100,000
C-276	N10276	14.5	15.0	51.0	0	100,000

Note: Compositions are ASTM minimums.

absorber, physical and chemical, are considered. The physical parameters set the gas/slurry contacting surface created in the absorber and the chemical factors control the rate that absorbed SO_2 is removed from the liquid phase of the slurry.

The SO_2 removal efficiency of a scrubber is determined by the amount of SO_2 that can be absorbed by a unit volume of slurry. Limestone based systems are reaction-rate limited. That is, the SO_2 absorption is controlled by how fast the absorbed SO_2 can be reacted with limestone in the scrubber. The system must provide sufficient alkalinity to react with the SO_2. The alkalinity can be present as Ca and CO_3 ions or as solid $CaCO_3$. There are three primary methods of increasing the alkalinity in the absorption zone.

The first method is to increase the liquid to gas ratio (L/G) in the absorber. By increasing the flow of slurry, more liquid contact is provided. The increased flow does not change the effective surface in terms of area per unit volume of slurry. It does provide more total surface area and alkalinity to the absorption zone. This does not increase the absorption per unit volume of slurry. However, it increases the total volume of slurry in the spray zone.

The second method is to increase the total and dissolved alkalinity per unit volume of slurry. This is accomplished by increasing the limestone to SO_2 ratio in the slurry and/or by increasing the absorber reaction tank solids residence time. Increasing the stoichiometry increases both the dissolved alkalinity and the solid phase alkalinity. Increasing the tank residence time increases the dissolved alkalinity because there is more time for limestone to dissolve.

The third method is to use one or more trays. A tray is a much more efficient contact device than a slurry spray. This is because a tray creates more surface area between the slurry and the gas. Therefore, the tray generates more surface area per unit volume of slurry. The tray also provides significant holdup time for the slurry. This increases the limestone dissolution in the absorption zone and increases the absorption per unit volume. The limestone dissolution on the tray can be as much as 50% of the dissolution in the entire absorber.

The benefits of the absorber tray compared to an open spray tower include:

1. reduced liquid to gas ratios
2. increased absorption for the same L/G
3. more uniform gas distribution
4. fewer recirculation pumps
5. fewer spray headers
6. reduced pump maintenance
7. acts as a maintenance platform

There are absorber designs based on reagents such as sodium carbonate which are dissolved chemistry absorbers. In these designs, there is sufficient liquid phase alkalinity available so that the SO_2 removal becomes gas phase limited. For these systems, the L/G and tray pressure drop are much lower. Such systems can achieve removal efficiencies as high as 99% at low tray pressure drops and an L/G of about 40 gpm/1000 acfm. Magnesium enhanced lime systems are solid chemistry systems that run like liquid systems because the magnesium will be a dissolved species. Organic acid can be added to limestone systems to achieve the removal of liquid based (dissolved species) systems.

Table 3
Effect of Various Design Parameters on SO_2 Removal

Parameter	Type	When Parameter Increases
Inlet SO_2	Chemical	Removal decreases
L/G	Physical	Removal increases
Tray ΔP	Physical	Removal increases
Stoichiometry	Chemical	Removal increases
pH	Chemical	Removal increases
Nozzle pressure	Physical	Removal increases
Cl-concentration	Chemical	Removal decreases

Limestone forced oxidized (LSFO) process description

Limestone based wet FGD systems are classified as non-regenerable. This means that the reagent is consumed by the process and must therefore be continually replenished. A generic diagram of non-regenerative processes is shown in Fig. 3. Each consists of four process steps: reagent preparation, SO_2 absorption, slurry dewatering, and final disposal. Within each of these process steps many variations exist. Essentially all wet FGD installations have some unique aspects. The following discussion will consider the limestone forced oxidized system, which is the most common system in use today. The LSFO system produces a gypsum byproduct that is typically sold for use in the manufacture of wallboard. It can also be sold to the cement industry, used as fertilizer, or sent to a landfill.

Reagent preparation

The primary advantages of limestone are its wide availability and cost effectiveness; roughly 5 to 6% of the earth's crust consists of calcium and magnesium carbonates and silicates. Limestone, consisting mostly of calcium carbonate ($CaCO_3$), is easily mined, transported and stored. Its storage and conveying at the plant site are similar to coal handling. A typical analysis of a limestone suitable for use in a wet scrubber is listed in Table 4.

Limestone can be ground dry in an air-swept mill, or wet in an overflow ball mill and can be ground onsite at the power plant, or offsite at a separate location. In North America, most limestone wet FGD systems feature onsite wet grinding for slurry preparation. In most cases, the system of choice is a closed loop ball mill circuit. A typical ball mill circuit is shown in Fig. 4. The energy required to achieve a given grind size is estimated by the Bond relationship:

$$W = \frac{10 W_i}{\sqrt{D_P}} - \frac{10 W_i}{\sqrt{D_F}} \qquad (5)$$

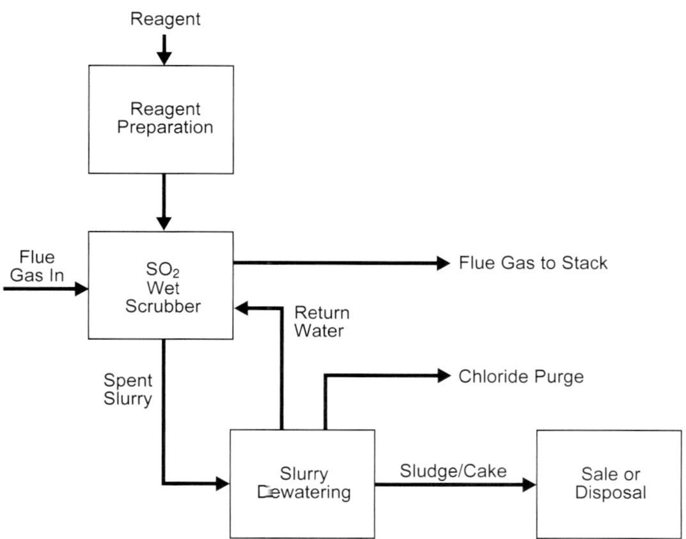

Fig. 3 Wet FGD system diagram.

where

W = power, kWh/t of product
W_i = Bond work index, kWh (micron)$^{1/2}$/t
D_P = diameter in microns for which 80% of product is finer
D_F = diameter in microns for which 80% of feed is finer

The Bond work index for calcitic limestones ranges from about 8 to 15 kWh (micron)$^{1/2}$/t; 10 to 12 is typical.

For onsite grinding systems following Fig. 4, the limestone is received with a maximum diameter of about 0.75 in. (19 mm) or less and is fed through a weigh belt feeder to the ball mill. Either fresh or recycled water is added at the ball mill feed chute in proportion to the feed rate of the limestone. The output from the mill overflows to the mill product tank where it is pumped to a set of hydroclones to separate the fines from the coarse fraction. The coarse fraction (underflow) is returned to the ball mill for further grinding while the fine material is sent to the limestone feed tank. The water balance is maintained to provide 25 to 35% suspended limestone solids in the feed tank.

The limestone grind is usually expressed as a percent passing a certain sieve size. The typical grind is 90 to 95% passing 325 mesh (44 microns). Fine grinding requires a larger ball mill system and higher operating power consumption. Finer material provides greater limestone utilization in a smaller reaction tank and higher dissolution rate due to the higher surface area of finer limestone particles. Stoichiometry, or reagent ratio, is defined as the molar ratio of the reactant, $CaCO_3$ for limestone systems, to the SO_2 removed.

The limestone grinding mill shown in Fig. 4 is a horizontal ball mill. The mill consists of a cylindrical shell containing steel balls ranging in size from 1 to 4 in. (25 to 102 mm). The cylinder is rotated and the limestone is ground by the tumbling of the balls and limestone in the mill. Other types of wet mills include vertical tower mills and attrition mills.

Table 4
Typical Limestone Composition

Component	%
$CaCO_3$	93 – 97
Available $CaCO_3$	92 – 96
$MgCO_3$	0.5 – 2
Total inerts	3 – 7
SiO_2	0.5 – 3
Fe_2O_3	0.5 – 3
R_2O_3 (other metal oxides)	0.5 – 4

For systems making marketable gypsum, limestone purity is a primary factor (see Table 4). The concentration of inerts affects the gypsum purity, defined as the percent pure gypsum ($CaSO_4 \cdot 2H_2O$). In addition there are limits to the amount of SiO_2, Fe_2O_3 and total metal oxides (R_2O_3) in the gypsum.

Limestone can also be dry ground at a central location and transported by truck to the FGD system. At the site, the pulverized limestone is pneumatically conveyed to a storage bunker. It is then fed to water-filled slurry preparation tanks. This system requires less space than onsite grinding facilities and reduces some of the FGD operating burden. However, offsite grinding is typically more expensive than onsite preparation.

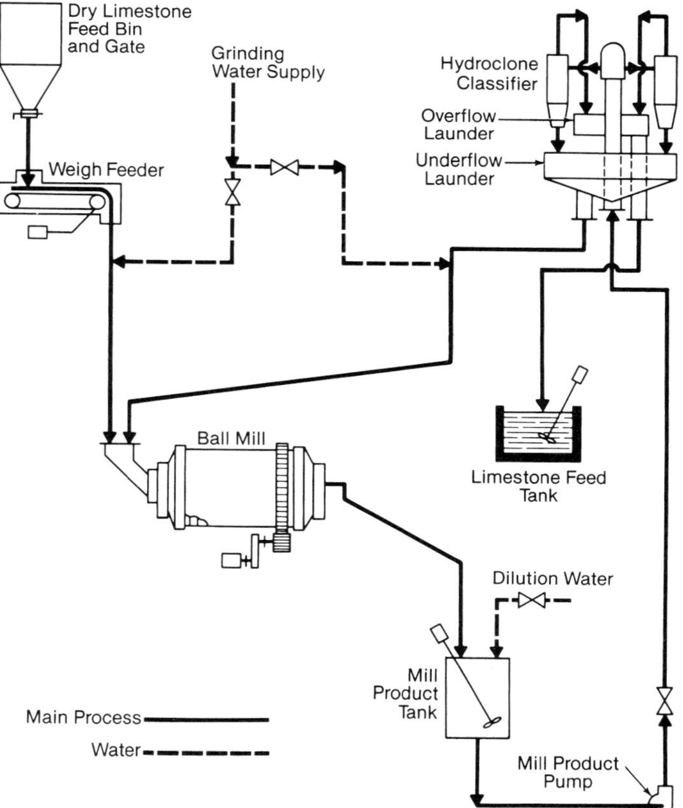

Fig. 4 Limestone reagent preparation system.

Slurry dewatering and disposal

The disposal of reaction products from limestone wet scrubbers includes ponding, landfilling, and the production of high grade gypsum for wallboard, cement and fertilizer. In the production of high grade gypsum, the spent slurry consisting of the reaction products, inerts from the limestone, excess reagent, and flyash is dewatered in two stages (primary and secondary dewatering). In addition, a purge stream is used to remove chlorides and fine particles from the WFGD system. A typical system is shown in Fig. 5.

Hydroclones are used for primary dewatering of gypsum. Multiple hydroclones are mounted in a cluster and fed by slurry pumped from the absorber reaction tank to a radial distributor and then to the individual hydroclones. The hydroclone classifies the solids in the feed slurry by particle size. A dilute slurry of fine particles leaves through the *hydroclone overflow* and a concentrated slurry of coarse particles is discharged from the hydroclone underflow. Typically a feed slurry containing 15 to 20% suspended solids will produce an overflow stream containing 3 to 4% suspended solids and an underflow stream containing 50 to 55% suspended solids from the primary cyclone. Most of the hydroclone overflow is returned to the absorber. Recycling of this finer fraction of the solids allows for crystal growth of the gypsum and better utilization of unreacted limestone. A portion of the overflow is sent to a purge system for removal of fine particles and chlorides.

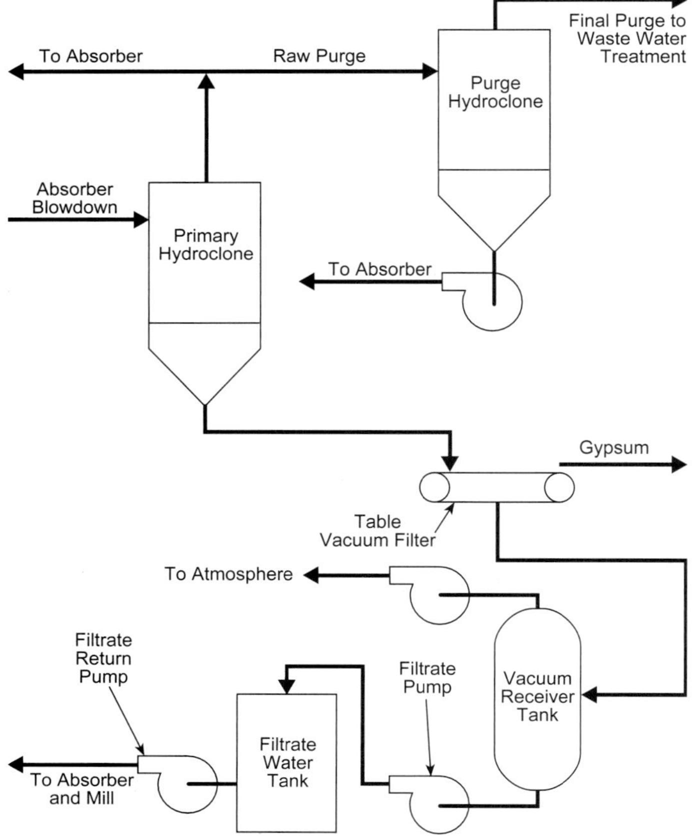

Fig. 5 Slurry dewatering system.

Secondary dewatering of the gypsum is typically accomplished using either horizontal belt or drum vacuum filters. Horizontal belt filters are typically used for wallboard grade gypsum production because the final cake must contain no more than 10% moisture. Drum filters are prevalent when higher moisture fractions are acceptable (20%), such as for the cement industry and fertilizer. However, with recent design improvements, some drum filters can achieve 10% cake moisture. Centrifuges have also been used in secondary dewatering.

The production of wallboard grade gypsum requires that the slurry on the vacuum filter be washed with fresh water to remove total dissolved solids (TDS) and specific ions such as chloride, Cl^-. The actual specifications will vary based on the wallboard manufacturer's requirements.

Purge system

The hydroclone overflow contains a higher percentage of fine particles of inerts and flyash than the slurry fed to the hydroclones. This stream is the source for the purge stream. The purge stream flow rate can be set by the amount needed to purge dissolved solids from the system or by the need to remove impurities and fines. Typically, the raw purge stream is sent to a second set of (purge) hydroclones to further concentrate fines in the overflow and reduce the final purge stream total suspended solids (TSS) to about 1.5%. The underflow is returned to the absorber. Concentrating the fines in the overflow helps reduce the amount of pure gypsum in the final purge stream.

The purge stream flow is usually set to control chloride concentration, typically to a design concentration of 20,000 ppm or less.

The final purge stream is typically ponded or sent to a waste water treatment system for removal of other chemical species and suspended solids. The treated water can then be returned to a river or other body of water. The solids removed in the treatment system are concentrated or sent to a landfill.

WFGD system water balance

The wet FGD system loses water to evaporation when the incoming gas is quenched. There is also some loss to the gypsum cake. In addition, there is a loss of water in the purge stream. Within the system, water is used for preparing the limestone slurry, washing the moisture separators, cake wash, and as seal water for the vacuum filter seal pumps. Reclaim water (filtrate) from the vacuum filters is collected and used for limestone preparation and level control in the absorber reaction tank.

Cake wash water and vacuum pump seal water must be fresh water from a well, river or lake. For all other uses including moisture separator wash water, the requirements are less stringent. Sources for the remaining water can be cooling tower blowdown, secondary sewage treatment plant effluent, and seawater, among others.

Process flow diagram and mass balance

A typical limestone based wet scrubbing process using in-situ forced oxidation is shown in Fig. 6. In this example, the reagent preparation system includes a closed circuit ball mill system using water that is recycled from the slurry dewatering system. The feed slurry is pumped to the absorber reaction tank as required to control the pH in the tank or to achieve the desired SO_2 removal efficiency or emission concentration. The use of pH control can result in excessive limestone consumption at low boiler loads or when firing lower than design sulfur coals without operator intervention. Excess limestone can adversely affect gypsum purity. By contrast, limestone consumption can be minimized by controlling to a removal efficiency or SO_2 emission set point.

Air is also pumped to this reaction tank and distributed by a sparge grid or agitator. Oxidation air can also be introduced into the tank by air lances and dispersed with specially designed agitators. The oxygen in the air reacts with sulfite present in the slurry to produce gypsum ($CaSO_4 \cdot 2H_2O$).

The slurry is pumped from the reaction tank to the spray headers shown in Fig. 2. The slurry is sprayed countercurrently into the flue gas where it absorbs the SO_2. From there, the slurry falls to the sieve tray where additional SO_2 is absorbed into the froth created by the interaction of the flue gas and slurry on the tray. The slurry then drains to the reaction tank.

The reaction products, primarily gypsum, are continuously withdrawn from the wet scrubber. This spent slurry is pumped via gypsum blowdown pumps to the dewatering system. The spent slurry typically contains about 15 to 20% suspended solids. A hydroclone is used to concentrate the slurry. The underflow from the hydroclone is concentrated to 50% solids. The overflow containing 4% suspended solids is sent back to the absorber and to the purge stream.

The underflow from the hydroclone is directed to a vacuum filter where the filtered solids are washed with fresh water and dewatered to form a filter cake containing about 10% free moisture. The cake is then sent by truck to a wallboard manufacturer. A mass balance for this example is presented in Table 5.

A summary of the major power requirements for this limestone to gypsum process is shown in Table 6. The operating power requirement for the example displayed in this table is 1.9% of gross unit output. This percentage will vary with coal sulfur content and the required removal efficiency. Higher overall sulfur removal or high sulfur loading requires higher L/G, more pressure drop, more oxidation air, etc.

Wet scrubber chemistry

SO_2 absorption in a wet scrubber and its subsequent reaction with alkaline earth materials such as limestone is an elementary acid-base reaction which takes place in an aqueous environment. However, the chemical processes involved are complex. SO_2 is a relatively insoluble gas in water. Calcium carbonate ($CaCO_3$) also has a low solubility in water. The principal reaction products are calcium sulfite hemi-hydrate ($CaSO_3 \cdot \frac{1}{2} H_2O$) and calcium sulfate dihydrate ($CaSO_4 \cdot 2H_2O$), or gypsum. These two salts also have low solubility in water.

In a limestone system with forced oxidation, the following reaction model can be used to describe the process using the chemical species in Table 7.

In the gas/liquid contact zone:

$$SO_2\,(g) \rightleftharpoons SO_2\,(aq) \tag{6}$$

Dissolving gaseous SO_2

$$SO_2\,(aq) + H_2O \rightleftharpoons HSO_3^- + H^+ \tag{7}$$

Hydrolysis of SO_2

$$CaCO_3\,(s) + H^+ \rightleftharpoons Ca^{++} + HCO_3^- \tag{8}$$

Dissolution of limestone

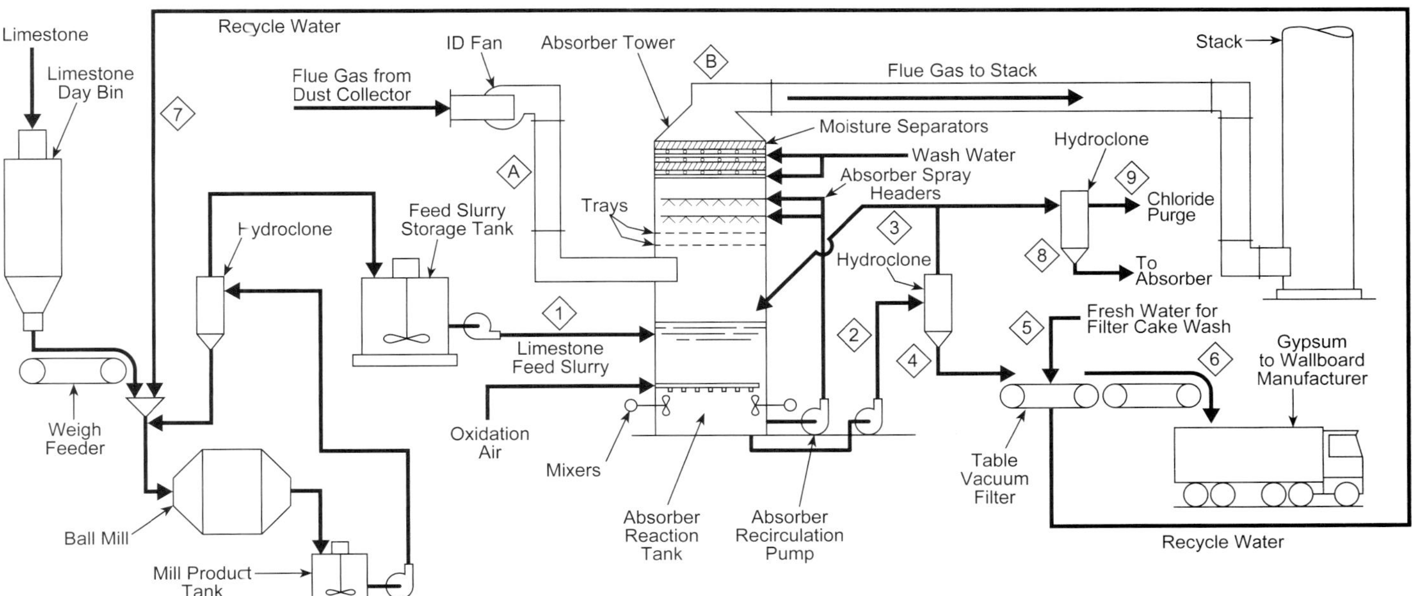

Fig. 6 Wet scrubber FGD system flow diagram (see Table 5 for mass balance).

Table 5
Mass Balance for the Limestone Forced Oxidation System Shown in Fig. 6

Gas side	Inlet	Exit
Stream designation	A	B
Flow rate, ACFM	1,710,000	1,462,000
Total mass flow rate, lb/h	5,521,000	5,924,175
Mass flow rate of H_2O, lb/h	240,000	550,000
Mass flow rate of SO_2, lb/h	34,900	698
Mass flow rate of HCl, lb/h	458	5
Static pressure, in. wg	12	1
Temperature, F	300	129

500 MW boiler burning coal with 4.1% sulfur
Limestone stoichiometry = 1.03
SO_2 efficiency = 98%

Liquid side	Feed Slurry	To Hydroclone	Hydroclone Overflow	Hydroclone Underflow	Filter Wash Water	Filter Cake	Recycle Water	Hydroclone Underflow	Chloride Purge
Stream designation	1	2	3	4	5	6	7	8	9
Flow rate, GPM	330	961	648	271	53	22	71	11	41
Total flow rate, lb/h	204,225	573,964	354,000	197,435	26,500	107,493	390,338	6,296	20,500
Flow rate of sus. solids, lb/h	61,268	114,793	15,000	98,718	0	96,743	0	944	0
% suspended solids	30	20	4.3	50	0	90	0	15	0
Chloride conc., ppm in liquid	8,000	20,000	20,000	20,000	50	1,000	8,000	20,000	20,000
pH	7.8	5.5	5.5	5.5	7	7	6	5.5	5.5
Temperature, F	100	129	129	129	120	120	120	129	129

$$HCO_3^- + H^+ \rightleftharpoons CO_2(aq) + H_2O \tag{9}$$
Acid-base neutralization

$$CO_2(aq) \rightleftharpoons CO_2(g) \tag{10}$$
CO_2 stripping

In the reaction tank:

$$CaCO_3(s) + H^+ \rightleftharpoons Ca^{++} + HCO_3^- \tag{11}$$
Dissolution of limestone

$$HCO_3^- + H^+ \rightleftharpoons CO_2(aq) + H_2O \tag{12}$$
Acid-base neutralization

$$CO_2(aq) \rightleftharpoons CO_2(g) \tag{13}$$
CO_2 stripping

$$O_2(g) + 2HSO_3^- \rightarrow 2SO_4^= + 2H^+ \tag{14}$$
Sulfite oxidation

$$Ca^{++} + SO_4^= + 2H_2O \rightleftharpoons CaSO_4 \cdot 2H_2O \tag{15}$$
Precipitation of gypsum

Reaction 6 expresses the mass transfer rate of SO_2 from the gas phase to the liquid or aqueous phase. Its mass transfer rate can be expressed by:

$$\frac{d(Gy)}{dV} = k_g a\,(y - y^*) \tag{16a}$$

or

$$N_g = \int \frac{dy}{y - y^*} = \int \frac{k_g a}{G}\, dV \tag{16b}$$

where

G = molar gas flow rate, moles/s
y = mole fraction of SO_2 in flue gas
k_g = gas film mass transfer coefficient, moles/m^2 s
a = interfacial surface area, m^2/m^3
y^* = equilibrium SO_2 concentration at the gas/liquid interface
V = volume of the gas/liquid regime, m^3
N_g = number of gas phase transfer units, dimensionless

Although k_g can be approximated, the interfacial surface area can not. The gas phase mass transfer rate must be determined experimentally. This involves operating the scrubber under conditions in which $y^* \rightarrow 0$. Equation 16 can then be integrated to:

$$N_g = -\ell n\,(1 - E) = k_g a\; V/G \tag{17}$$

where

N_g = overall number of gas phase transfer units, dimensionless
E = overall SO_2 fractional efficiency

Many factors determine the number of gas phase transfer units (N_g). These include the slurry spray rate, the droplet size and spatial distributions, the gas phase residence time (height of spray zone), the liquid residence time, wall effects, and the gas flow distribution.

In a limestone based wet scrubber, the rate-limiting reactions in the gas/liquid contact zone are believed to be Reaction 8. The reaction rate for limestone dissolution can be expressed by:

$$\frac{d\,[CaCO_3]}{dt} = k_c \left([H^+] - [H^+]_{eq}\right) Sp_c\,[CaCO_3] \quad \textbf{(18)}$$

where

$[CaCO_3]$ = calcium carbonate concentration in the slurry, moles/l
k_c = reaction rate constant
$[H^+]$ = hydrogen ion concentration, moles/l
$[H^+]_{eq}$ = equilibrium (H^+) at the limestone surface, moles/l
Sp_c = specific surface area of limestone in slurry

The reaction tank permits Reactions 12 through 16 to approach completion. In a limestone scrubber, limestone is added directly to the reaction tank. The pH of the slurry returning from the gas/liquid contact zone to the reaction tank can be as low as 3.5. The pH in the reaction tank is usually 5.2 to 6.2. Therefore, the overall reaction in the reaction tank is:

$$CaCO_3(s) + H^+ + HSO_3^- + \tfrac{1}{2}\,O_2 + H_2O \rightarrow CaSO_4 \cdot 2H_2O(s) + CO_2(g) \quad \textbf{(19)}$$

The rate of gypsum crystallization in the reaction tank can be expressed by:

$$\frac{d[CaSO_4 \cdot 2H_2O]}{dt} = k(R-1)Sp_g\,[CaSO_4 \cdot 2H_2O] \quad \textbf{(20)}$$

where

k = crystallization rate constant
R = $A_{Ca++}\,A_{SO_4^=}\,/\,K_{Sp}$
A_{Ca++} = activity of Ca^{++} ion
$A_{SO_4^=}$ = activity of $SO_4^=$ ion
K_{sp} = solubility product of gypsum
Sp_g = specific surface area of gypsum

R is a measure of the level of supersaturation. If R is greater than 1, the solution is supersaturated with gypsum. If R is less than 1, the solution is subsaturated in gypsum.

The reaction tank serves a second important function in lime/limestone scrubber systems. The tank is sized to provide sufficient time for the dissolved gypsum to crystallize and precipitate. Typically, the reaction tank is designed for three to five minutes of residence time based upon the recirculation rate. An additional consideration in sizing the reaction tank is to provide sufficient solids retention time, based on gypsum produced, to allow the crystals to grow to a suitable particle size.

Table 6
Typical Power Requirements for Limestone Scrubber with Forced Oxidation

Absorber System	Avg. Power (kW)
Oxidation air blower	1300
Absorber recirc. pumps	3250
Absorber recirc. tank agitators	250
Moisture separator wash water pump	50
Misc. pumps and agitators	50
Subtotal	4900
Dewatering Area	**Avg. Power (kW)**
Vacuum pump for filter	330
Reclaim water pump	50
Filter drive	15
Misc. pumps and agitators	30
Subtotal	425
Reagent Preparation	**Avg. Power (kW)**
Ball mill	1200
Mill product tank pump	30
Limestone feed tank agitator	80
Misc. pumps and agitators	30
Subtotal	1340
Other Systems	**Avg. Power (kW)**
General-instrument air	80
Booster fan power	2830
Subtotal	2910
Total	9575

Notes:
500 MW boiler burning coal with 4.1% sulfur
Heating value of coal = 10,950 Btu/lb
Absorber L/G = 130 gal/1000 ACF
Total pressure drop = 12 in. wg
Parasitic power = 9575 kW/500 MW
= 1.9%

Table 7
Chemical Species Found in Scrubber Reaction Model

SO_2	sulfur dioxide
H_2O	water
HSO_3^-	bisulfite ion
H^+	hydrogen ion
$CaCO_3$	calcium carbonate (main limestone constituent)
Ca^{++}	calcium ion
CO_2	carbon dioxide
HCO_3^-	bicarbonate ion
$CaSO_3 \cdot 1/2\,H_2O$	calcium sulfite hemihydrate
$CaSO_4 \cdot 2\,H_2O$	calcium sulfate dihydrate (gypsum)
(l)	denotes liquid phase
(g)	denotes gaseous phase
(aq)	denotes dissolved specie in water
(s)	denotes solid phase

Performance enhancing additives

The steady-state hydrogen ion concentration of the slurry in the gas/liquid contact zone is determined by the balance between the rate of H^+ generation in Reaction 7 and the rate of H^+ consumption by Reactions 8 and 9. As the hydrogen ion concentration increases, i.e., as the pH drops, the equilibrium SO_2 vapor pressure increases and SO_2 removal efficiency is reduced. This equilibrium relationship can be expressed by:

$$y^* = \frac{C' \times 10^3 \left[H^+\right]}{k_1 \left(k_2 + \left[H^+\right]\right)} \quad \textbf{(21)}$$

where

k_1 = the equilibrium constant for Reaction 7, moles/l atm
k_2 = the equilibrium constant for Reaction 8, moles/l
C' = total concentration of dissolved SO_2, millimoles/l
= $[SO_2]_{aq} + [HSO_3^-]$
$[H^+]$ = steady-state hydrogen ion concentration
y^* = SO_2 vapor pressure expressed as ppm (assuming barometric pressure = 1 atm)

At 122F (50C):

k_1 = 0.4643 moles/l atm
k_2 = 7.162×10^{-3} moles/l

As the SO_2 vapor pressure rises, the rate of SO_2 absorption diminishes and approaches zero as $y^* \rightarrow y$. (See Equation 16.) If a simple means existed to reduce the hydrogen ion concentration in the gas/liquid contact zone, then the SO_2 vapor pressure could be controlled and the SO_2 absorption rate would be maximized. A buffer is a chemical specie which performs this function. A class of weak organic acids has been found suitable for use in limestone scrubbers to control pH and improve overall SO_2 removal. These buffers can be described by:

$$H^+ + A^- \rightleftharpoons AH \quad \textbf{(22)}$$

where AH is the generalized acid group. When the pH falls in the gas/liquid contact zone, Reaction 22 is driven to the right. Some organic acids which buffer in this range include adipic, formic and succinic acid. Because these are water soluble, the pH buffering is nearly instantaneous compared to the lime or limestone dissolution. Because the SO_2 vapor pressure, i.e., the equilibrium SO_2 concentration of the gas/liquid interface, is proportional to the hydrogen ion concentration, buffers minimize the rise in SO_2 vapor pressure. The buffer concentration required to achieve a given absorption depends upon SO_2 concentration, L/G ratio (the ratio of slurry flow to gas flow), and contact time. Typically, the concentration of these additives which is required to achieve adequate control ranges from 3 to 30 millimoles/l.

A second additive used in wet FGD systems is magnesium oxide, which reacts with SO_2 to form magnesium sulfite. Because magnesium sulfite is highly soluble, the sulfite ion is the primary reactant in the gas/liquid interface as follows:

$$SO_3^= + H^+ \rightleftharpoons HSO_3^- \quad \textbf{(23)}$$

Sodium carbonate can also be added to the lime/limestone system for a similar benefit.

The total concentration of dissolved alkaline species such as $CO_3^=$, HCO_3^-, $SO_3^=$ and OH^- in the slurry is referred to as *dissolved alkalinity*. If the dissolved alkalinity is sufficiently high, the scrubber may become gas-phase diffusion controlled. This is illustrated in Fig. 7.[3] Under these conditions the rate of SO_2 absorption is dependent only upon the amount of interfacial surface area, i.e., the spray droplet surface area plus the tray froth surface area.

Dry scrubbers

Dry scrubbing is more correctly referred to as spray dryer absorption (SDA) to reflect the primary reaction mechanisms involved in the process: drying alkaline reagent slurry atomized into fine drops in the hot flue gas stream and absorption of SO_2 and other acid gases from the gas stream. The process is also called semi-dry scrubbing to distinguish it from injection of a dry solid reagent into the flue gas.

Dry scrubbing is the principal alternative to wet scrubbing for SO_2 control on utility boilers. Since the initial installation in 1980, more than 13,000 MW of dry scrubbers have been installed at U.S. electric utilities. The allowable SO_2 emission permit limits have steadily decreased over time to levels requiring 90 to 95% removal, and the technology has proven to operate reliably at these levels. The application of dry scrubbers to large electric utility boilers is generally limited to those burning low sulfur coals. This is due primarily to the higher unit reagent costs for dry scrubbing. However, for smaller utility and industrial boiler applications, the simplicity and lower capital costs of dry scrubbing make it an attractive alternative for higher sulfur coals. Spray dryer systems are also considered maximum achievable control technology (MACT) for combined HCl and SO_2 control on waste-to-energy units. In the U.S. utility market, dry scrubbers have mainly been applied to units west of the Mississippi River, burning low sulfur fuels, with generating capacities of 90 to 900 MW. However, of the more than 50 current U.S. utility and large coal-fired industrial cogeneration installations, over 40% of the installations are located in the east.

The advantages of dry scrubbing over wet scrubbing include:

1. less costly construction materials (carbon steel),
2. fewer process unit operations,
3. simplicity of control and operation,
4. lower water consumption,
5. lower auxiliary power consumption,
6. use of available alkalinity in the flyash for SO_2 absorption,
7. integral SO_3 emissions control, and
8. production of dry solid byproducts without the need for dewatering.

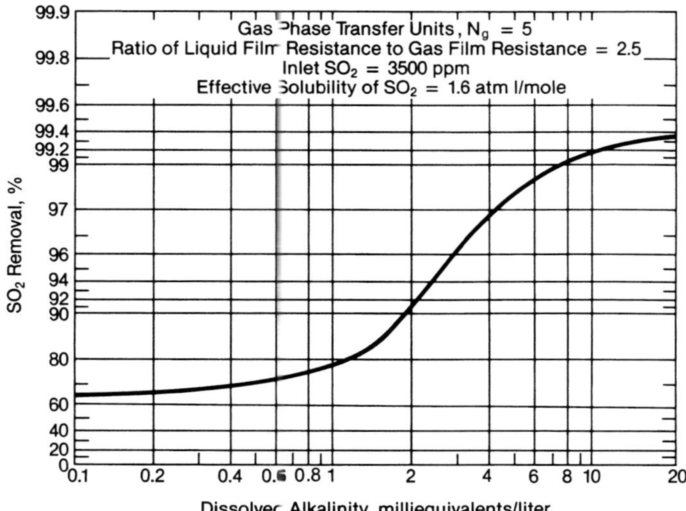

Fig. 7 Influence of dissolved alkalinity on SO_2 performance (1 milliequivalent = 50 ppm $CaCO_3$).

System design

Fig. 8 depicts the equipment orientation for a typical utility SDA installation coupled with a baghouse (BH). Unlike a wet scrubber installation, the SDA is positioned before the dust collector. Flue gases leaving the air heater at a temperature of 250 to 350F (121 to 177C) enter the spray chamber where the reagent slurry is sprayed into the gas stream, cooling the gas to 150 to 170F (66 to 77C). An electrostatic precipitator (ESP) or fabric filter (baghouse) may be used to collect the reagent, flyash and reaction products. Baghouses are the dominant selection for U.S. SDA installations (over 90%) and provide for lower reagent consumption to achieve similar overall system SO_2 emissions reductions Both reverse-air and pulse-jet baghouse designs are in use (see Chapter 33). The baghouse is an integral part of the process and acts as a second stage SO_2 removal reactor that can contribute a significant fraction of the overall performance. Operating data from an SDA/baghouse system achieving 90% overall SO_2 emissions reduction has shown that the SDA accounted for 79% of the total and the remaining 21% was obtained across the baghouse. The particulate filter cake on the bags was actually removing 63% of the SO_2 entering the baghouse.

For large utility applications, the flue gases are introduced to the spray chamber through a compound gas disperser (upper and lower) as shown in Fig. 9 for a vertical flow design. For lower gas flow applications, all of the gas flow enters through the roof gas disperser. The gas disperser is designed to assure good mixing of the flue gases with the reagent spray, and to make use of the full volume of the spray chamber. The vertical gas flow configuration dominates utility boiler SDA applications although there are a few horizontal flow units in operation. The SDA chamber is sized to provide sufficient gas-phase residence time for drying of the reagent slurry to produce free-flowing solids in the particulate collector. The required residence time depends on the inlet flue gas temperature, feed slurry solids loading, degree of atomization, and exit gas temperature. In general, for atomization systems producing a Sauter mean drop diameter of 50 to 60 microns, 10 to 12 seconds is sufficient. This residence time is calculated based on the spray chamber volume and the outlet flue gas temperature and pressure. Field studies have shown that the flue gas temperature is within ±10 to 20F (6 to 11C) of the SDA outlet temperature throughout the entire spray chamber.

The reagent slurry is introduced to the reaction chamber as a fine mist of droplets using a single, high-capacity rotary atomizer or multiple dual-fluid (high pressure air/slurry) atomizers. A rotary atomizer assembly for a large utility application is shown in Fig. 10. The assembly shown includes the 900 hp drive motor, a speed-increasing gearbox, the atomizer

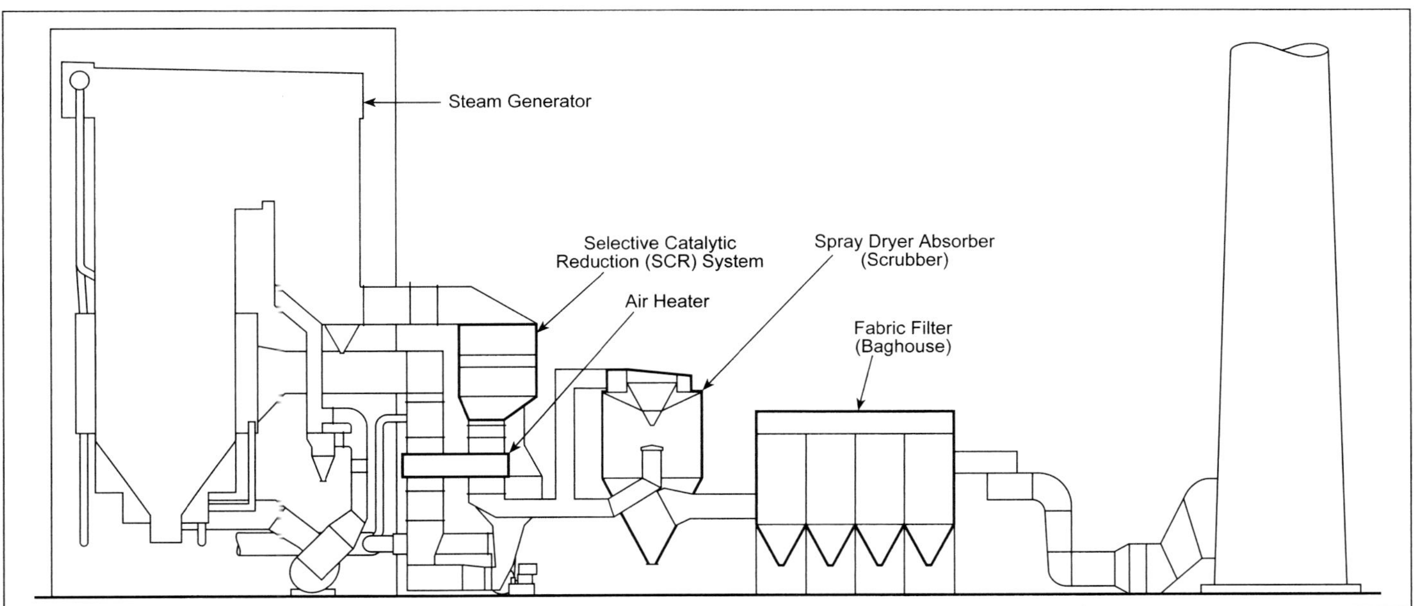

Fig. 8 Utility spray dryer absorption (SDA) system.

wheel, and associated hardware. Rotary atomization far dominates (90+%) the utility and large industrial installed technology base. This is primarily because the rotary design provides higher slurry flow capacity per atomizer and better performance with higher solids slurries. Both systems have proven reliable and capable of consistently achieving low SO_2 emissions. Key performance requirements for SDA atomizers include:

1. ability to handle variable slurry flow rates,
2. rapid response to flow rate changes,
3. uniform spray coverage of the SDA chamber,
4. low susceptibility to pluggage and build-up, and
5. ease of removal for servicing.

For rotary atomizers, the spray droplet size distribution is a function of the speed of rotation, wheel diameter and wheel design as well as, to a lesser extent, reagent slurry feed properties including the flow rate, viscosity, density, and surface tension.[4] Increasing the wheel speed reduces the droplet size at the cost of higher power consumption. Note that for a rotary unit, atomization quality is not sensitive to the diameter of the exit nozzles and so does not degrade as the nozzles wear from erosion by the reagent slurry. In addition to the slurry properties, key design criteria for dual-fluid nozzle atomizers include the nozzle exit orifice dimensions and orientation, air pressure, and air-to-slurry ratio. Water spray from a typical dual-fluid atomizer with multiple exit orifices is shown in Fig. 11. For this system, minimizing formation of large (+100 micron) drops from primary atomization or the coalescence of sprays from multiple atomizers is a key criteria for successful operation.

Fig. 12 shows a compact selective catalytic reduction/SDA/BH installation for a 550 MW power plant. At this site, the two SDA chambers, each with a single rotary atomizer, treat approximately 1,800,000 ACFM (850 m^3/s) of flue gases from a pulverized coal-fired boiler burning Powder River Basin coal to reduce SO_2 emissions by 94%.

Process description

The quantity of water in the reagent slurry introduced into the SDA is controlled so that it almost completely evaporates in suspension, leaving the solids exiting the spray chamber with a free moisture content of 1 to 2%. SO_2 absorption takes place primarily while the water is evaporating and the flue gas is adiabatically cooled by the spray. The difference between the temperature of flue gas leaving the dry scrubber and the adiabatic saturation temperature is known as the *approach temperature*. Flue gas saturation tem-

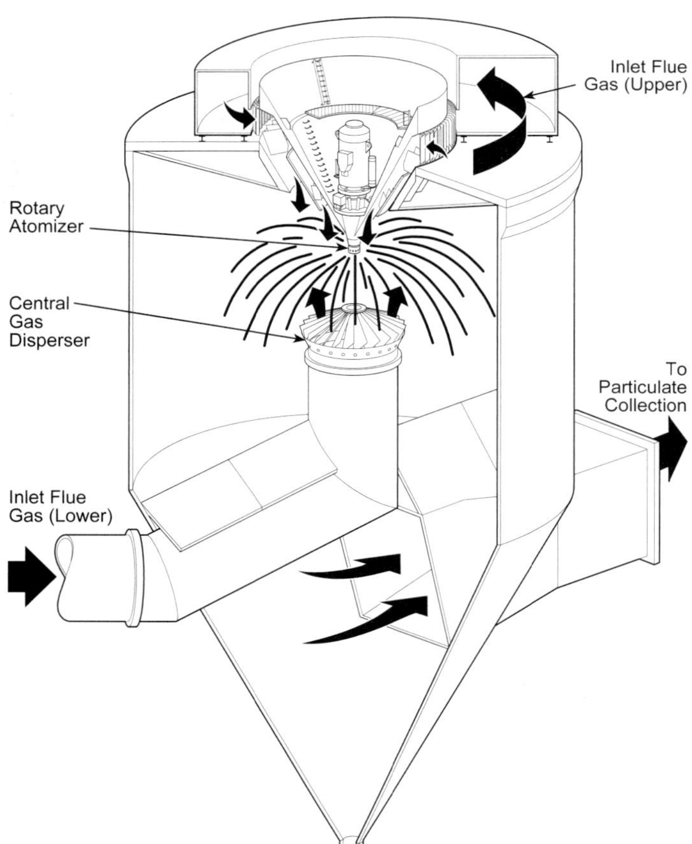

Fig. 9 Spray dryer absorber cutaway.

Fig. 10 Rotary atomizer assembly.

peratures are typically in the range of 115 to 125F (46 to 52C) for low moisture bituminous coals and 125 to 135F (52 to 57C) for high moisture subbituminous coals. The *stoichiometry* is the molar ratio of the reagent consumed to either the inlet SO_2 or the quantity of SO_2 removed in the process. Both definitions are in common use. Reagent stoichiometry and approach temperature are the two primary variables that control the scrubber's SO_2 removal efficiency. The optimal conditions for SO_2 absorption must be balanced with practical drying considerations. The approach temperature selected must be consistent with the atomizer feed slurry solids loading for efficient drying of the reagent solids. The *spraydown*, or difference between the inlet and outlet flue gas temperatures, and the solids loading of the atomizer feed slurry, determine how much reagent may be introduced to the process.

The predominant reagent used in dry scrubbers is lime slurry produced by slaking a high-calcium pebble lime. The slaking process can use a ball mill or a simple detention slaker. SDA systems that use only lime slurry as the reagent are known as *single pass* systems. Some of the lime remains unreacted following an initial pass through the spray chamber and is potentially available for further SO_2 collection. Solids collected in the ESP or baghouse may be mixed with water and reinjected in the spray chamber. This solids *recycle system* is used at most utility SDA installations to make use of the available alkalinity inherent in the flyash and to minimize reagent use. Some of the recycled

Fig. 11 Dual-fluid atomizer.

Fig. 12 B&W's SDAs and fabric filter baghouse provide an integrated SO_2 and particulate removal process.

solids form an inert core supporting the finer lime solids on the surface, thereby exposing more lime surface area for SO_2 absorption. The flyash may be removed from the flue gas before the SDA by a pre-collector. This is common practice in Europe where marketing of the flyash from coal-fired utility boilers is a well-established industry practice. Even with flyash pre-collection, a recycle system results in more efficient lime utilization.

The process is controlled by monitoring the flue gas temperature at the SDA outlet to control the amount of water or recycle slurry added to the atomizer feed. SO_2 emissions at the stack are compared to a desired operating set point and the lime slurry flow to the atomizers is adjusted to reduce the outlet SO_2 as necessary. For a rotary atomizer, the atomizer spindle, which connects the motor to the atomizer wheel, is monitored continuously for vibrations to provide an early indication of unbalance or pluggage of the wheel.

Process flow diagram and mass balance

A typical SDA process flow diagram is shown in Fig. 13. The flue gases, with or without pre-collection of the flyash, enter the spray dryer absorber where the gas stream is cooled by the reagent slurry spray. The mixture then passes on to the baghouse for removal of particulate before entering the ID fan and passing up the stack. Pebble lime (CaO) is mixed with water at a controlled rate to maintain a high slaking temperature that helps generate fine hydrated lime ($Ca(OH)_2$) particles with high surface area in the hydrated lime slurry (18 to 25% solids). A portion of the flyash, unreacted lime and reaction products collected in the baghouse is mixed with water and returned to the process as a high solids (35 to 45% typical) slurry. The remaining solids are directed to a storage silo for byproduct utilization or disposal. The fresh lime and recycle slurries are combined just prior to the atomizer(s) to enable fast response to changes in gas flow, inlet SO_2 concentrations, and SO_2 emissions as well as to minimize the potential for scaling.

Process mass balances for recycle and single pass SDA system designs for a typical application are presented in Table 8 while the power requirements are provided in Table 9. In this example, use of the recycle system reduces lime consumption by 3141 lb/h (1424.7 kg/h), or 42% over that expected with single-pass operation. The stoichiometry for the single pass system is 1.78 on an inlet SO_2 basis compared to 1.04 for the recycle system. Note the difference in acceptable approach temperatures for long-term operation of the two systems. The 7% solids feed slurry with the single pass system requires operation at a 45F (25C) approach temperature or above for adequate drying while a 30F (17C) approach is used with the recycle system.

Dry scrubber chemistry

The mechanism of SO_2 absorption in an SDA is similar to that attained by wet scrubbing when viewed on the level of what takes place in the individual slurry droplets. Most of the reactions take place in the aqueous phase; the SO_2 and the alkaline constituents dissolve into the liquid phase where ionic reactions pro-

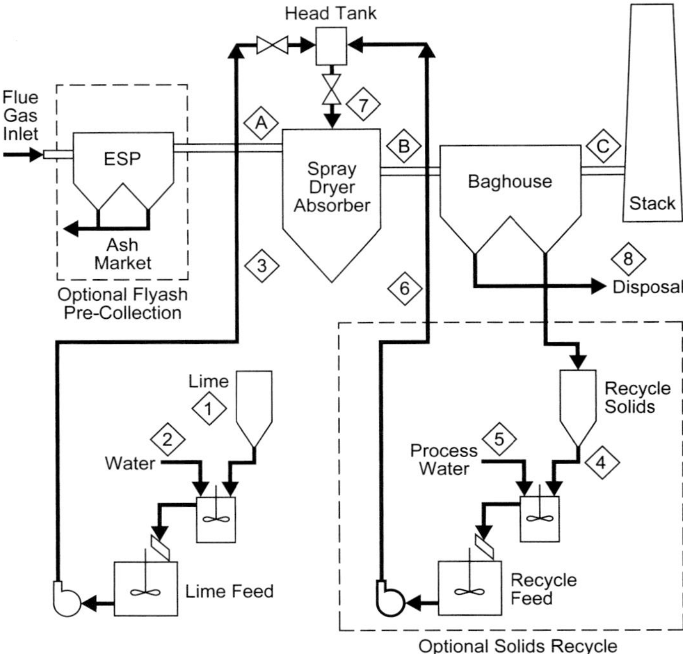

Fig. 13 Dry scrubber FGD system flow diagram (see Table 8 for mass balance).

duce relatively insoluble products. The reaction path can be described as follows:

$$SO_2(g) \rightleftharpoons SO_2(aq) \quad \text{Dissolving gaseous } SO_2 \tag{24}$$

$$Ca(OH)_2(s) \rightarrow Ca^{++} + 2OH^- \quad \text{Dissolution of lime} \tag{25}$$

$$SO_2(aq) + H_2O \rightleftharpoons HSO_3^- + H^+ \quad \text{Hydrolysis of } SO_2 \tag{26}$$

$$SO_2(aq) + OH^- \rightleftharpoons HSO_3^- \tag{27}$$

$$OH^- + H^+ \rightleftharpoons H_2O \tag{28}$$

$$HSO_3^- + OH^- \rightleftharpoons SO_3^= + H_2O \quad \text{Neutralization} \tag{29}$$

$$Ca^{++} + SO_3^= + \tfrac{1}{2} H_2O \rightarrow CaSO_3 \cdot \tfrac{1}{2} H_2O(s) \quad \text{Precipitation} \tag{30}$$

These reactions generally describe activity that takes place as heat transfer from the flue gas to the slurry droplet is evaporating water from the droplet. The rate-determining reaction may vary at different stages of the drying and absorption process. The high pH in the droplet environment (10 to 12.5) helps maintain a low concentration of acid in the liquid phase that enhances SO_2 absorption from the gas stream. Rapid SO_2 absorption occurs when liquid water is present. The drying rate can be slowed down to prolong this period of efficient SO_2 removal by adding deliquescent salts to the reagent feed slurry. Salts such

Table 8
Mass Balance for Process Flowsheet Shown in Fig. 13

Recycle System Design

Flue gas stream	SDA Inlet	BH Inlet	ID Fan Inlet
Stream designation	A	B	C
Flow rate, ACFM	1,729,848	1,571,675	1,609,250
Flow rate, lb/h	5,244,973	5,454,664	5,508,400
SO_2, ppmv	380	89	21
SO_2, lb/h	4,381	1,086	263
Particulate, lb/h	26,835	148,511	76
Temperature, F	275	163	159
Pressure, in. wg	–16.0	–19.5	–27.0

- 500 MW boiler burning 0.44% sulfur coal with high calcium flyash
- Available flyash alkalinity = 0.40 milliequivalents/gram
- Flue gas saturation temperature = 133F
- SO_2 emissions reduction = 94%
- SO_2 emissions rate = 0.05 lb/10^6 Btu
- Particulate emissions rate = 0.015 lb/10^6 Btu

Solids/slurry streams	Pebble Lime	Slaking Water	Lime Slurry	Recycle Solids	Process Water	Recycle Slurry	Atomizer Feed	Byproduct Solids
Stream designation	1	2	3	4	5	6	7	8
Total flow rate, GPM		48	50		269	363	413	
Total flow rate, lb/h	4,394	24,069	28,463	114,384	134,747	249,132	277,594	34,861
Solids flow, lb/h	4,394	0	5,693	112,096	0	112,096	117,789	34,167
Flyash, lb/h								24,273
Byproducts, lb/h								9,894
Water flow, lb/h		24,069	22,770	2,288	134,747	137,036	159,805	694
Solids loading, wt. %	100	0	20	98	0	45	42	98

Single Pass System Design

Flue gas stream	SDA Inlet	BH Inlet	ID Fan Inlet
Stream designation	A	B	C
Flow rate, ACFM	1,729,848	1,594,410	1,632,843
Flow rate, lb/h	5,244,973	5,424,712	5,478,350
SO_2, ppmv	380	73	21
SO_2, lb/h	4,381	880	263
Particulate, lb/h	26,835	40,936	76
Temperature, F	275	177	173
Pressure, in. wg	–16.0	–19.5	–27.0

Solids/slurry streams	Pebble Lime	Slaking Water	Lime Slurry	Recycle Solids	Process Water*	Recycle Slurry	Atomizer Feed	Byproduct Solids
Stream designation	1	2	3	4	5	6	7	8
Total flow rate, GPM		83	86		183	0	269	
Total flow rate, lb/h	7,535	41,278	48,813	0	91,254	0	140,067	41,467
Solids flow, lb/h	7,535	0	9,763	0	0	0	9,763	40,638
Flyash, lb/h								26,787
Byproducts, lb/h								13,851
Water flow, lb/h		41,278	39,050	0	91,254	0	130,304	829
Solids loading, wt. %	100	0	20		0	0	7	98

*Process water addition to lime slurry for atomizer feed to achieve desired SDA outlet temperature.

as calcium chloride also increase the equilibrium moisture content of the end product. However, since the use of these additives alters the drying performance of the system, the operating conditions must be adjusted (generally increasing the approach temperature) to provide for good long-term operability of the SDA and the ash handling system. Ammonia injection upstream of a dry scrubber also increases SO_2 removal performance. SO_2 absorption continues at a slower rate by reaction with the solids in the downstream particulate collector.

An SDA/baghouse combination also provides efficient control of HCl, HF and SO_3 emissions by the summary reactions of:

$$Ca(OH)_2 + 2HCl \rightarrow CaCl_2 + 2H_2O \qquad \textbf{(31)}$$

Table 9
Typical Power Requirements for 500 MW SDA/Baghouse System in Table 8

	Recycle System Design	Single Pass System Design
Atomizers and accessories, kW	1190	640
Reagent preparation, kW	230	125
Baghouse accessories, kW	445	445
ID fan power	6780	6880
Total	8645	8090
% of net 500 MW output	1.73	1.62

$$Ca(OH)_2 + 2HF \rightarrow CaF_2 + 2H_2O \quad \textbf{(32)}$$

$$Ca(OH)_2 + SO_3 \rightarrow CaSO_4 + H_2O \quad \textbf{(33)}$$

Proper accounting of the reagent consumption must include these side reactions, in addition to the SO_2 removed in the process.

The SDA process can make use of a variety of water sources for preparation of the reagent slurry and cooling of the flue gas. River water, well water, cooling tower blowdown, municipal waste water treatment effluent, and seawater have all been used successfully in SDA applications. In general, lime slaking requires good quality water with limitations on the sulfite, sulfate and TDS concentrations being the primary concerns for producing good quality fresh reagent slurry. Preparation of the recycle solids slurry and humidification of the flue gas to the desired operating temperature can be achieved with lower quality water sources. Selection of the process operating conditions must also consider the chloride content of the water sources to avoid problems with drying the reagent slurry and potential corrosion.

Sulfur trioxide (SO_3)

SO_3 and acid mist formation

Sulfur trioxide (SO_3) is formed directly in the combustion of sulfur-containing fuels, and indirectly by the conversion of small quantities of flue gas SO_2 to SO_3 in the presence of iron, some ash constituents such as vanadium, and some selective catalytic reduction (SCR) catalysts. Under normal operating conditions in coal-fired boilers without SCR systems, 0.75% of the SO_2 typically is converted to SO_3. High vanadium containing ashes can increase this range to 2%. Conversion of SO_2 to SO_3 by SCR catalysts is dependent upon SO_2 concentration, gas temperature, and catalyst formulation with a typical range of an additional 0.7 to 1.5%. Catalysts have been formulated with relatively low SO_2 to SO_3 oxidation potential.

As the flue gas temperature falls below 1000F (538C), SO_3 begins to react with water in the flue gas to form sulfuric acid vapor, H_2SO_4. By the time the flue gas exits the air heater, virtually all of the SO_3 has reacted to form H_2SO_4.

Sulfuric acid condenses on metal surfaces beginning in the temperature range of 200 to 300F (93 to 149C) and results in corrosion of metal surfaces in flues, air heaters, dry ESPs and fans. Acid dew point curves are presented in Fig. 14. In addition, in units with wet scrubbers, any significant residual sulfuric acid mist can cause visible plume or opacity problems (blue plume), visible several miles from the power plant. Wet scrubbers are not effective in removing acid mists because the rapid flue gas quenching in the scrubber produces submicron aerosol mists which are not easily collected (20 to 30%). Dry scrubbers in combination with downstream baghouses are effective in removing H_2SO_4 (> 95% typical). Finally, in units equipped with SCR systems, excess ammonia (NH_3) leaving the SCR can react with SO_3 to form sticky, acidic particles of ammonium bisulfate (NH_4HSO_4) which can foul air heaters.

Control methods

Sorbent injection The injection of wet or dry alkali compounds (usually calcium or magnesium oxide compounds), typically in the furnace, after the SCR, or upstream of the particulate collection device has been shown to be an effective means of controlling SO_3/H_2SO_4. Sorbent injection has the advantage of low capital cost, but has very high reagent cost due to the high stoichiometric ratio (4 to 8) that is needed to achieve 80 to 90% SO_3/H_2SO_4 control. Dry ESP performance may be adversely affected by removing the SO_3 from the flue gas (see Chapter 33).

Humidification Cooling the flue gas by humidification upstream of an existing dry ESP will allow the dry ESP to collect some sulfuric acid. It is possible to operate a dry ESP at 20F (11C) below the acid dew point without encountering excessive corrosion. 20 to 40% sulfuric acid removal can be achieved in some instances. This is possible because the acid is absorbed on the flyash rather than on the metal surfaces of the dry ESP. Careful attention must be paid to thermal insulation integrity and in-leakage of seal air streams.

Ammonia injection Ammonia injection upstream of the dry ESP has been used to control acid mist. However, there can be problems with ash handling, since the ash tends to be more cohesive. It is also possible that excess ammonia will react with the SO_3 to form

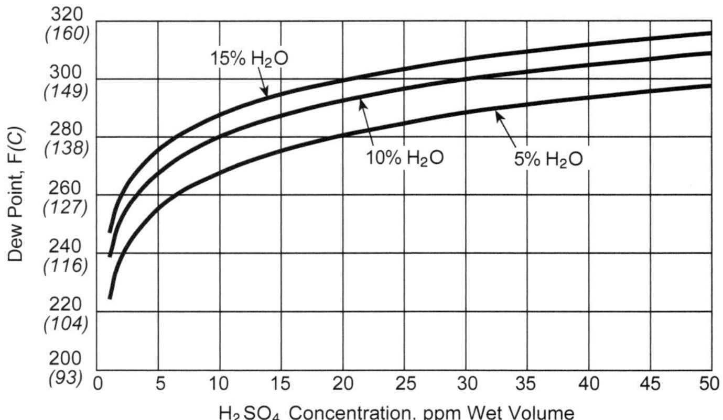

Fig. 14 Dew point versus concentration.

submicron aerosols that are difficult to collect. Capital costs are low, but reagent costs are high.

Wet precipitator To achieve very high removals of acid mist, a wet ESP can be used. For a utility boiler, the wet ESP would be used downstream of the wet FGD system.

Either a stand-alone arrangement or an integrated close-coupled wet ESP can be used to control SO_3; however, the standalone arrangement is typically more costly than the integrated design and consumes considerable additional space. Fig. 15 shows the integrated design.

The submicron nature of the acid mist can only be treated in a multi-field unit. Acid mist removal efficiencies of 80 to 90% can be achieved. For example, if the inlet is 34 ppmdv, less than 5 ppmdv at the stack can be achieved. Wet ESP can also remove 80 to 90% of any solid particulate that is present, such as residual flyash from the dry ESP and the small amounts of gypsum that result from the wet FGD moisture separator wash system. Even if a wet ESP is not included in a wet FGD project today, careful thought should be given to future needs. The addition of a wet ESP to an existing wet FGD system may not be possible if provisions are not made before the initial wet FGD system installation.

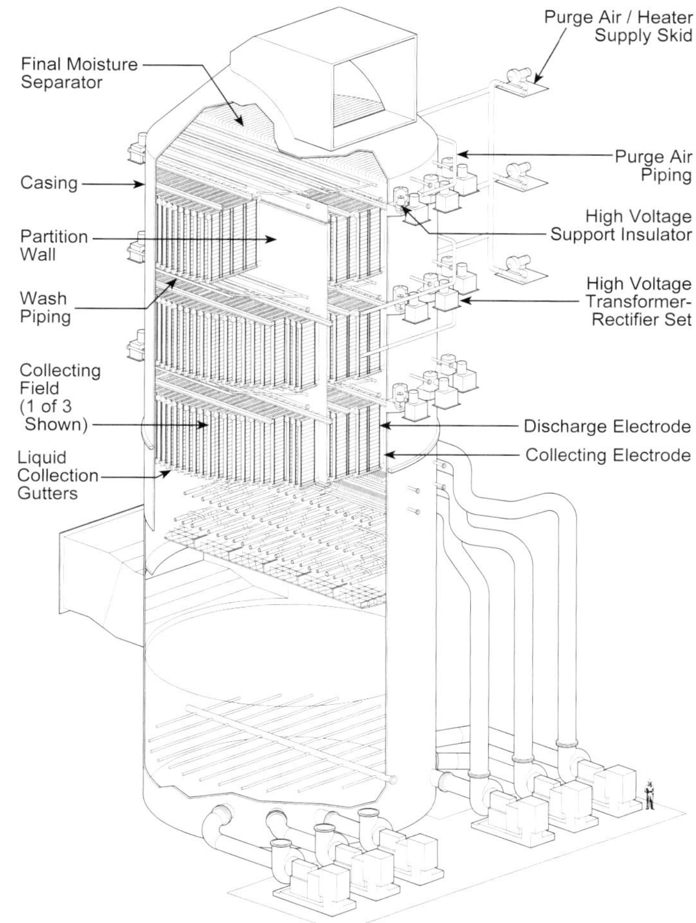

Fig. 15 Integrated wet ESP and wet FGD system.

References

1. "United States (U.S.) Environmental Protection Agency (EPA), National Emissions Inventory Criteria Pollutant Data: Current Emissions Trends Summary 1970-2002," www.epa.gov/ttn/chief/trends/index.html, accessed August 28, 2003.

2. *Annual Energy Review 2002*, U.S. Department of Energy, Energy Information Agency, Washington, D.C., November, 2003.

3. Plummer, L.N., Wigley, T.M.L., and Parkhurst, D.L., "The Kinetics of Calcite Dissolution in CO_2-Water Systems at 5C to 60C and 0.0 to 1.0 atm CO_2," *American Journal of Science*, Vol. 278, pp. 179-216, 1978.

4 Master, K., *Spray Drying in Practice*, SprayDryConsult International, Denmark, 2002.

Bibliography

Hardman, R., Stacy, R., and Dismukes, E., "Estimating Sulfuric Acid Aerosol Emissions from Coal-Fired Power Plants," Taken from the Conference on Formation, Distribution, Impact and Fate of Sulfur Trioxide in Utility Flue Gas Streams, United States Department of Energy-FETC, March, 1998.

Lesley, S., "Trends in Emissions Standards," CCC/77 Report, IEA Clean Coal Centre, International Energy Agency, London, England, United Kingdom, November, 2003.

Dry flue gas desulfurization system installation in western U.S.

Chapter 36

Environmental Measurement

Environment related measurements at modern power plants are primarily performed to determine compliance with government regulations. However, measurements are also necessary to monitor and optimize process performance, assess guarantees, calibrate continuous emission monitoring systems (CEMS) and verify design information. As discussed in Chapter 32, fossil fuel power plants emit air, water and solid pollutants.

This chapter focuses on measurements for monitoring air pollution control systems and, in particular, equipment for controlling sulfur dioxide (SO_2), nitrogen oxides (NO_x) and particulate emissions. Special emphasis is placed on measurements associated with wet SO_2 scrubbing systems which involve perhaps the most complex process.

Post-combustion emissions can be controlled by the use of one or more of the following systems: 1) particulate – electrostatic precipitators (ESP) or fabric filters (baghouses), 2) NO_x – selective catalytic reduction (SCR) systems or selective noncatalytic reduction (SNCR) systems, and 3) SO_2 – wet or dry flue gas desulfurization (FGD) scrubbing systems or other forms of sorbent injection. These systems are discussed in Chapters 33, 34 and 35. In each case, the flow rate and composition must be measured for the associated gas, liquid and solid streams. In addition, the overall stack emissions are monitored and recorded.

Instrumentation

Primary measurements taken on environmental equipment include pH, liquid level, solids level, solids concentration (density), continuous emission monitoring for regulated pollutants, static and differential pressure, temperature and flow of liquid, flue gas, air and steam.

Chapter 40 discusses the measurement of pressure, temperature and fluid flow.

pH measurement

The acidic or alkaline nature of a liquid stream is monitored by measuring pH or the potential of hydrogen which is expressed as the negative log of the hydrogen ion concentration ($-\log_{10}[H^+]$). This is the most important parameter in controlling wet FGD systems. pH measurement is also important in water softening, acid neutralization, and other auxiliary processes. The alkalinity, or lack of acidity, is generally proportional to the reagent use in the FGD system. Solutions with a pH less than 7 are considered to be acidic; those having a pH greater than 7 are alkaline.

pH measurements are used to control the feed of fresh reagent slurry to the wet FGD system. (See Chapter 35.) A bleed stream of recirculation slurry is passed through one or two pH sensors. The outputs are typically combined with signals for boiler load, SO_2 level, fresh reagent slurry density and flow to establish the flow of fresh reagent to the FGD absorber module. An increasing pH reduces the fresh reagent feed flow and a decreasing pH increases the flow.

pH is measured by means of electrodes and a voltmeter. The electrode (Fig. 1) is an electrochemical device comparable to a battery whose voltage changes with pH. In combination electrodes, one of the half-cells is the measurement electrode or pH sensing bulb and the other is the reference electrode or reference junction. Because the measurement electrode reacts to the pH of the solution, its potential is variable though repeatable. Theoretical electrode output varies from +414 mV at a pH of 0 to −414 mV at a pH of 14. The actual output is sensitive to the solution concentration at a reference temperature of 77F (25C).

The reference electrode completes the electrical circuit and produces a comparative voltage, so changes in the overall sensor are only a function of changes in the measuring electrode. The reference electrode is comprised of a potassium chloride (KCl) salt solution with a controlled amount of silver chloride (AgCl) dissolved in it. An Ag/AgCl electrode is placed into this electrolyte. The reference potential is a function of the KCl and AgCl concentrations. For example, a 1.0 M (molar) KCl electrolyte produces an offset (E_x), or voltage difference from theoretical, of −8 mV, while 3.3 M KCl produces a −45 mV offset (E_x). The difference, which is consistent across the entire range, is compensated for by standardization or zero adjustment.

The pH sensor is also affected by temperature. The slope (mV per pH unit) changes 1 mV per 5C (see Fig. 2). Temperature is compensated for by slope or temperature adjustment or by automatic temperature compensation.

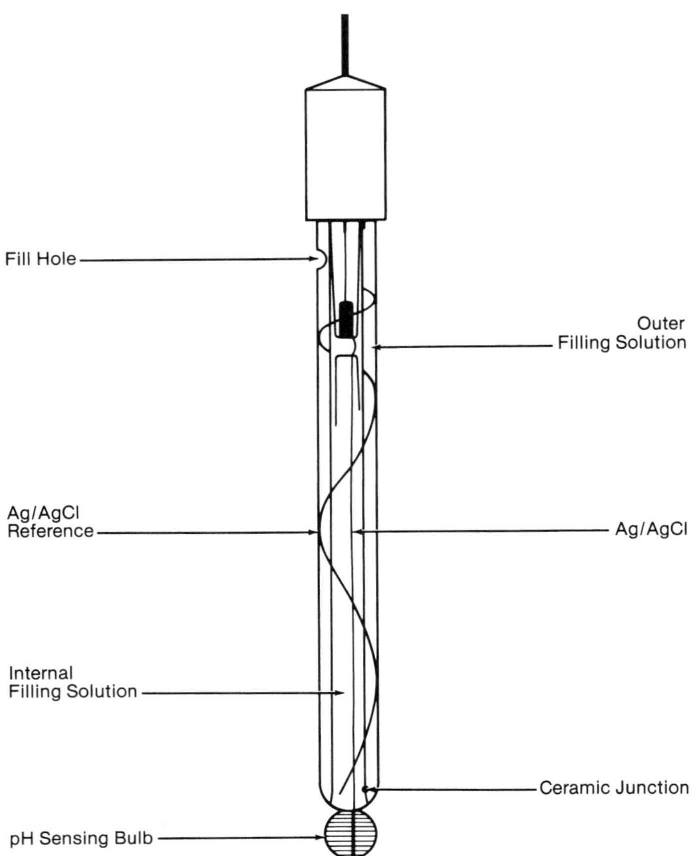

Fig. 1 Combination pH electrode (laboratory version).

The pH electrode used in wet FGD processes may be subject to fouling and aging. This is compounded by the slurry's abrasive or scaling nature which can cause plugging of the liquid junction. This junction is the point where the KCl reference solution meets the solution being measured. The liquid junction can be a small hole or may be made of fiber, wood, porous ceramic, Teflon or kynar. It holds the electrolyte in the probe and enables the electrolyte to form a salt bridge with the measured solution.

There are three types of pH sensors: dip, flow-through and insertion. The dip sensor is inserted into the tank and can be removed for maintenance and calibration. The flow-through sensor is placed in a flowing stream of process solution. The insertion sensor is a variation of the flow-through sensor, but does not require sample lines. It is inserted through packing and a valve into the flowing stream. To maintain or calibrate the insertion sensor, it is retracted through the valve, which is then closed to isolate the process.

All sensor types have advantages and disadvantages. The dip sensor is easy to maintain but prone to leakage from scrubber vessels that operate at positive pressure. Sample lines for flow-through sensors can plug. Both flow-through and insertion types are subject to high rates of erosion. All types are subject to scaling, although ultrasonic cleaning devices that are designed to remove brittle, insoluble coatings have been used with some success. Ultrasonic cleaners work best if used intermittently, but they can cause probe breakage.

The pH meter is a voltmeter that measures the electrode potential, converts the potential at a given temperature into pH terms, and corrects for nonideal behavior. pH meter operation depends upon proper standardization and calibration. This is best done with stable pH solutions known as buffers. Four of the best buffers, measured at 77F (25C), are pH 4.01, 6.87, 9.18 and 12.45 because they are traceable to the National Institute of Science and Technology (NIST).

Standardization (offset or buffer adjustment) is best done over the pH range anticipated. Standardization adjusts the offset so the meter reads 0.0 mV at pH 7.0. Calibration, or slope adjustment, compensates for alterations in an electrode's response to pH changes. To perform the calibration, the buffer should span the pH of the measured solution. For example, if the expected process pH is 5.4, then a buffer near pH 4 is preferred. The standardization/calibration procedure is summarized below:

1. standardize using a pH 6.87 buffer,
2 calibrate with a buffer that spans the process pH using a different adjustment than that used for standardization (calibration or slope), and
3. use the correct temperature setting, use an automatic temperature compensator or use the same temperature for the two buffers and the process.

Liquid level

Liquid level measurement is needed in any liquid storage tank that is not of the overflow design to control flow into and out of the tank. Examples in a wet FGD system (Chapter 35) include the tanks for absorber slurry recirculation, reagent storage, thickener underflow and quench recycle. In a typical dry FGD system, slurry levels in the lime feed slurry tank, recycle slurry tank and atomizer head tank are monitored. Common liquid level measurement devices are

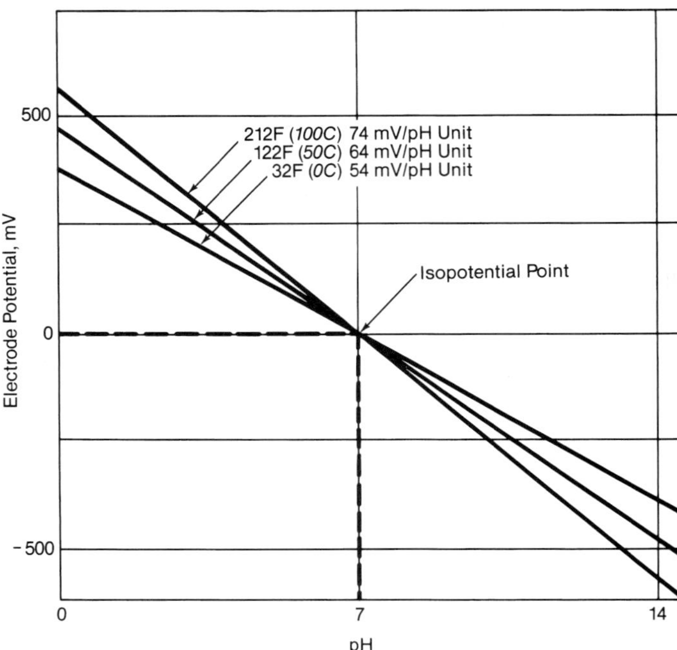

Fig. 2 Typical pH electrode response as a function of temperature.

Table 1
Common Liquid Level Measurement Devices for Environmental Control Systems

Sensor Type	Theory of Operation	Environmental Control System Applications	Comments
Differential pressure	Measures differential static pressure with sensors or pressure taps mounted on the wall of the vessel. The sensor contacts the fluid.	Slurry tank level. This technique is often used with a ram valve or a flush system.	Reliable, subject to plugging in slurry application. Two types are used — flange mounted and extended diaphragm.
Ultrasonic	Emits ultrasonic pulse to the liquid surface. The transit time of the return pulse reflected from surface is converted to the liquid level.	Open tanks containing most fluids or slurries.	Sensors do not contact the liquid. Some problems result from waves, dust, vapor, foam, etc. Background noise can often be filtered electronically.
Electrical capacitance	Detects level by measuring the capacitance between the probe and the tank wall.	High/low level alarm. Discrete level capacitance probes can detect liquid surface.	Some problems with solids buildup or scale.
Float	Float moves up and down with level. The float is on an arm inserted into the vessel or within a housing.	Clean water, mill product tanks (nonscaling): high/low level alarm or constant tank volume.	Magnetic float types are available but are not considered good for scaling applications.
Radio frequency (phase tracking)	Detects an electrical signal which passes through a sensor and is reflected back at both the liquid surface and the bottom of the sensor. The signal phase shift provides a continuous indication of level.	Liquid and solids level.	Relatively new, promising technology. The electric field is much larger than the conductors.
Optical	Detects changes in light intensity.	Sludge level in thickeners.	For sludge level, the change in light intensity is measured at a fixed distance from the surface.

listed in Table 1. In SCR NO_x reduction systems, aqueous and anhydrous ammonia storage tank levels are monitored.

Differential pressure level transmitters are the most common for absorber recirculation tank level control because of their proven reliability. However, they are not the most accurate because of errors due to changes in slurry density or errors due to flue gas pressure above the liquid level

Other liquid level devices have varying degrees of accuracy due to the tank conditions. For example, most tanks used in FGD systems contain slurry and are therefore agitated, causing waves or a vortex on the surface. The liquid surface level in the absorber tower recirculation tank is also disrupted by: 1) the very high spray flow rates impinging on the surface, 2) foaming due to gas released during the reactions or entrained by the spray, and 3) oxidation air sparged into the tank. Sensors immersed in the slurry also tend to scale. Some level devices use stilling wells to reduce the effect of agitation and to shield the level probe from spray.

Solids level

Many types of sensors used for measuring liquid level are also used for solids level measurement. However, solids level measurement presents some unique problems due to the often dusty environments, the uneven level caused by bridging or normal angles of repose characteristic of the solids, or the changes in density or properties due to compaction or aeration.

For environmental equipment, solids level measurement is used mainly for storage silos (ash, lime, limestone, gypsum) but is also used for hoppers on precipitators or fabric filters. Types of silo measurement devices include electrical capacitance, ultrasonic, and radio frequency (phase tracking) sensors. In addition, load cells or strain gauges that measure the weight change of the silo as it fills or empties are common. Another type of solids level device is a cable measurement system in which a weight or float is lowered automatically from the top of the bin or silo. When the float reaches the solids surface, loss of cable tension is detected. The device then measures the distance traveled by counting electronic pulses as the float retracts. The measured level is therefore the silo maximum level less the distance traveled by the probe. This device should not be used when the silo is being filled.

A typical device used to determine high hopper level is a nuclear transmitter and detector. When the ash level rises above the sensor, the strength of the signal from the nuclear source to the receiver is decreased. This signal is converted to a level indication. There have been objections to the use of this type of device due to the nuclear material present and licensing requirements.

Another high/low solids level device is the paddle wheel, which includes a motor that turns a paddle. When material is present, rotation of the paddle stops. The increased torque of the motor as it attempts to turn the stopped paddle triggers a microswitch alarm.

Finally, vibrating probes are sometimes used for single point solids level detection. The vibration of a piezoelectric transducer-driven probe inserted into the vessel is dampened by contact with material. This is detected as a voltage change, which is electronically amplified to yield a level indication. The capacitance probe is a similar device.

Solids concentration (density)

The most commonly used devices for density measurement are nuclear absorption meters (see Fig. 3). These devices measure the absorption of gamma rays from a radioactive source, and the degree of absorption is proportional to the solids content. Nuclear density meters do not contact the slurry; instead, they are installed on the outside of a slurry pipe. Some disadvantages of nuclear density meters are:

1. The signal is not linear with solids content unless a linearizer is used. If accurate chemical analysis of the slurry is available, the device can be pre-calibrated.
2. For some models, a nuclear license is required in the United States (U.S.) and in most other countries.
3. The device does not distinguish between suspended and dissolved solids.
4. Errors can result due to the buildup of solids or scale in the pipe.

Nuclear density meters are typically low maintenance devices. The inaccuracies mentioned above are routinely verified by manual sampling and measurement. Low level sources which do not require a site license have been used successfully in wet and dry FGD applications.

Other devices used for density measurement include differential pressure and ultrasonic devices and vibrating reed instruments. The reed instruments electronically convert the slurry dampening effect on electrically-driven coil vibrations to a density measurement.

Pressure, flow and temperature

Pressure, pressure differential, flow and temperature measurements are common in environmental system applications. They indicate process performance, energy consumption, operational problems and ability to meet design or operating requirements. Theory, instrumentation and application for these devices are discussed in detail in Chapter 40. Selected uses in environmental systems are discussed in Chapters 33 through 35. Comments with particular importance in environmental systems include:

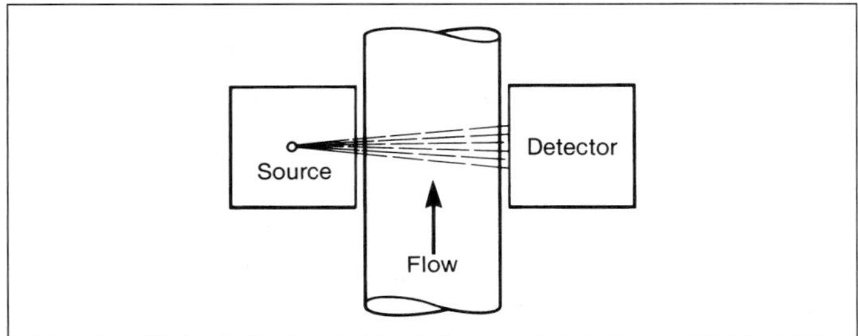

Fig. 3 Nuclear absorption meter for density measurement.

1. Slurries tend to be highly erosive and potentially corrosive, and provision must be made for appropriate environments.
2. The presence of solids in liquid or gas flows can lead to plugging of measuring devices. Continuous purging systems are frequently included to ensure long-term operation.
3. Temperature monitoring can be particularly important for freeze protection, crystallization prevention, thermal gradient prevention (thickeners and clarifiers), corrosion prevention (precipitators, fabric filters, dry FGD systems and flues/ducts) and general process control. Elastomeric linings and fiberglass or plastic components must be protected from high temperatures. Unlined steel stacks must be protected from low temperatures.

Continuous emission monitoring systems

Early power plant emission measurements focused on providing information for combustion control. The manual Orsat gas analyzer, which sampled oxygen (O_2), carbon monoxide (CO) and carbon dioxide (CO_2) flue gas concentrations, was used extensively. However, this analyzer is not a continuous monitor and requires a trained operator.

Power plants are required to continuously monitor key emissions to maintain compliance with operating permits. In addition, continuous emission monitoring systems (CEMS) permit process control of the emissions control equipment and the combustion system.

During the 1980s, CEMS were used primarily to measure particulate (opacity), O_2, SO_2, NO_x and CO_2. By the early 1990s, CEMS to monitor CO, hydrogen chloride (HCl), ammonia and volatile organic compounds (VOC) were introduced for commercial combustion applications.

The common methods for continuously measuring flue gas constituents are described in Table 2 and are further detailed in Table 3. All of the analyzers are similar in that they use a measuring cell and a reference cell. They are set to zero with air or a calibration standard, the span is set with a calibration standard, and a sample is taken. The result is obtained by comparing the voltages detected across the sample, the zero standard and span standard.

There are three types of analyzers – extraction, dilution-extraction, and in situ.

In the extraction method (Fig. 4), a sample of flue gas is withdrawn from the process stream and directed to an external analyzer for the constituent concentration determination. This type of analyzer is typically used where there are short distances between the sample point and the analyzer. The probe and sample line are often heated to prevent condensation from interfering with the measurement. Because the gas sample is typically drawn by vacuum to the external analyzer, sample line leaks can compromise the integrity of the sample.

The dilution-extraction method uses a carrier/dilution media, typically instrument quality air, which is supplied to the extraction probe in the flue gas process stream. Problems associated with dirty, wet and corrosive flue gases are drastically reduced by the 50

Table 2
CEM Technologies

Technology	Operating Characteristics
Infrared radiation (IR)	An infrared beam passes through a measurement filter and is absorbed by the constituent gas. A light detector creates a signal which is used to monitor concentrations.
Ultraviolet absorption (UV)	A split beam, with optical filters, phototubes and amplifiers, measures the difference in light beam absorption between the reference and sample.
Chemiluminescence	NO_x is measured by first converting any NO_2 to NO, then reacting with ozone to convert all the NO to NO_2 and its characteristic light generating effect. Ozone does not react with NO_2.
Flame ionization detection	Hydrocarbons are ionized with strong light. The signals are received by the flame ionization detector.
Transmissometer	Light is passed through the stack where it is reflected by a mirror on the opposite side. The quantity of light returning is proportional to particulate matter and aerosols in the flue gas. (See Fig. 5.) Alternatively, the receiver may be placed on the opposite wall.
Electrochemical cells	The voltage measured when a gas sample is injected into a solution with a strong base is compared to a reference voltage.
Chromatography	A sample, zero and calibration gas are eluted through a column. The output as measured by flame photometric or thermal conductivity detectors is compared.

to 200 dilution ratio provided by the sample probe. Precise dilution is assured by accurately controlling the dilution air pressure which yields a constant air flow. The air flow creates a vacuum on the downstream side of a critical sample orifice. The gas sample rate, limited by sonic flow through the orifice, is therefore held constant. Furthermore, because the sample lines from the probe to the analyzer are under positive pressure, minor sample line leakage does not affect the integrity of the sample concentration as it would in a vacuum system. This method has gained wide acceptance in utility and other systems where long sample lines are required.

In situ analyzers, such as that shown in Fig. 5, provide measurement of flue gas concentrations directly in the process stream. Several types of detection devices are available, and all are based on absorption spectroscopy. Since gas stream particulates can reduce the level of light transmission, ceramic filters are used to exclude these particles. A single in situ probe can measure several gas species and opacity. Problems with in situ analyzers generally occur in the optic or electronic components. Because a separate analyzer is required for each sample point, installations that require multiple sample points frequently find the extractive systems more cost effective.

Continuous analyzers generally require flue gas conditioning. Sample conditioning includes filtration to remove particulate (see Fig. 6), chilling to remove water, heating to maintain process temperature, and dilution (used with extractive monitor and in situ probe).

The need for conditioning coupled with zero and span calibration make extractive systems complicated,

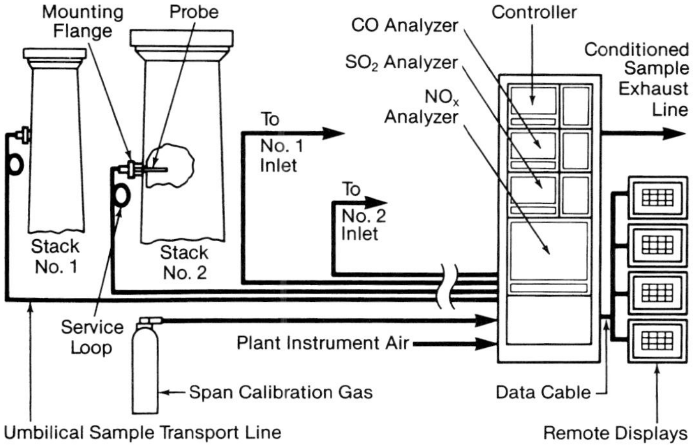

Fig. 4 Extraction stack/inlet CEM system.

Table 3
Continuous Emission Monitoring
(Reference Table 2 for Technologies)

Constituent	CEM Technology
Particulate matter (opacity)	Transmissometer, beta ray absorption
Sulfur dioxide (SO_2)	UV absorption, IR pulsed fluorescence
Nitrogen oxides (NO_x)	Chemiluminescence, UV spectroscopy, IR
Hydrogen chloride (HCl)	IR with gas filter
Carbon monoxide (CO)	IR
Carbon dioxide (CO_2)	IR
Oxygen (O_2)	Electrochemical cell
Volatile organic compounds (VOC)	Flame ionization detection
Other organic air toxics	Chromatography
Ammonia (NH_3)	Same as NO_x*

*NH_3 is converted to NO_x in one of two split streams. Both streams are analyzed for NO_x. NH_3 is determined as the difference.

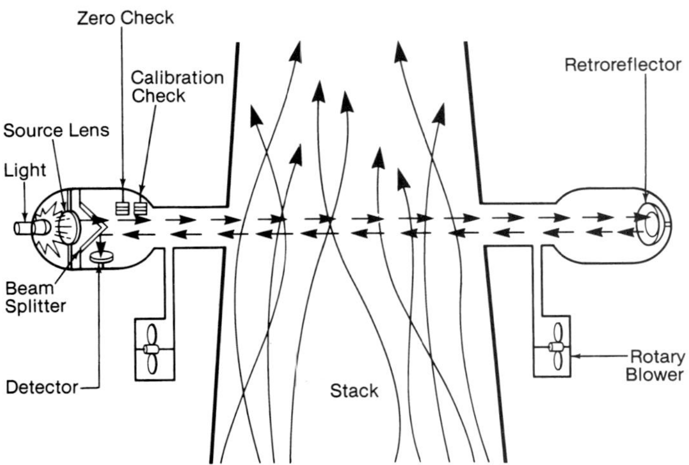

Fig. 5 In situ transmissometer system.

costly, and vulnerable to plugging, corrosion and leaking. These systems are also slow to respond due to the time lapse between sampling and analysis. In situ systems use conditioning but do not rely on long sample lines. These systems have faster response, fewer components and better availability than extractive systems. However, they are prone to interference and fouling and are less accessible for service. Extractive systems can be expanded by adding sample points as shown in Fig. 4. It is important to consider the location and ambient conditions in which CEMS equipment is placed. Many of the problems associated with CEMS can be traced to dirty conditions, excessive heat or cold, vibration, humidity, etc.

Monitoring process performance

Environmental control systems involve physical and chemical processes. Chemical methods are used to maintain desired removal or emission levels, avoid operating problems, improve reagent utilization, improve process efficiency, troubleshoot equipment, and monitor process changes.

In a wet FGD system, the quality of the reagent and the recirculating slurry stream are monitored. Analysis is more critical for processes that produce a usable byproduct such as gypsum. In this case, the analysis extends into product purity and feed stream impurities. The quality of the reagent is normally measured extensively when the reagent source is selected. Later it is evaluated if the source is changed or process problems occur.

Tables 4 through 7 list the important analyses for common wet FGD systems. Additional analyses are required for makeup water streams, sodium-based FGD systems, dual loop systems, regenerative systems, ex situ oxidation processes, water treatment, and sulfuric acid or SO_2 recovery processes. However, the major species and methods would be similar to those presented in Tables 4 through 7; only the concentrations would vary. The analyses common for spray dry FGD

Table 4
Limestone Quality Analysis (Wet Limestone FGD Systems)

Parameter	Suggested Method*	Suggested Frequency	Use of Data
Reactivity	pH stat test followed by particle size measurement	Duplicate analysis of candidate limestones	Compared to reference to determine acceptability.
Grindability (Bond Work Index)	Laboratory ball mill test	Duplicate analysis of candidate limestones	Limited by specification of wet milling system design.
Particle size of limestone as delivered	Sieve method	Duplicate analysis of candidate limestones then monitor monthly	Limited by specification of wet milling system design.
Inerts	Acid dissolution	Duplicate analysis of candidate limestones then monitor weekly for gypsum systems	Quality indicator; important for gypsum systems.
Carbonate	CO_2 evolution, $Ba(OH)_2$ absorption, alkalinity titration	Duplicate analysis of candidate limestones then monitor weekly for gypsum systems	Check reagent quality; important for gypsum systems.
Calcium and magnesium	AA, IR, EDTA titration	Duplicate analysis of candidate limestones then monitor weekly for gypsum systems	Check reagent quality.
Limestone slurry solids content, % by wt	Gravimetric	Daily	Check density meter and indicate problems in milling circuit.
Limestone slurry particle size	Wet sieve	Daily	Troubleshooting problems in milling circuit or process such as low utilization or poor gypsum quality.

* Many methods have been standardized through the American Society for Testing and Materials, Philadelphia, Pennsylvania.

Table 5
Lime Quality Analysis (Wet Lime and Spray Dry FGD Systems)

Parameter	Suggested Method*	Suggested Frequency	Use of Data
Slaking rate	ASTM Method C-110-76a	Duplicate analysis on all candidate limes	Temperature rise rate is related to reactivity and lime quality.
Grit	Acid dissolution	Duplicate analysis on all candidate limes	High grit levels indicate poor lime quality.
Weight loss on ignition	Gravimetric	Duplicate analysis on all candidate limes	Weight loss is an indirect measure of lime quality via carbonate and hydroxide content.
Available lime index	Rapid sugar test	Duplicate analysis on all candidate limes	Lime quality.
Calcium/ magnesium	Atomic absorption (AA) EDTA titration or Ion Chromatograph (IC)	Duplicate analysis on all candidate limes then as needed to check reagent quality	Analysis is required for processes producing a usable byproduct or high magnesium lime systems.
Lime slurry solids content, % by wt	Gravimetric	Daily	Check density meter calibration and indicate slaker performance.

* Many methods have been standardized through the American Society for Testing and Materials, Philadelphia, Pennsylvania.

systems are presented in Table 8. The lime quality analysis in Table 5 is applicable for wet lime or spray dry FGD systems. In general, there is significantly less process chemistry monitoring for a spray dry FGD system.

Performance testing

As with any system, performance tests are conducted on environmental control systems to determine if the units are operating properly. A typical new environmental control system might include low NO_x burners, SCR, ESP, wet FGD scrubber modules, and gypsum processing equipment. The ten most common parameters measured to determine performance for such a system include:

1. system pressure drop,
2. particulate emissions,
3. stack opacity,
4. NO_x emission concentration,
5. SO_2 removal efficiency and SO_2 emissions,
6. makeup water requirements,
7. reagent consumption,
8. gypsum purity,
9. mist eliminator carryover, and
10. electric power consumption.

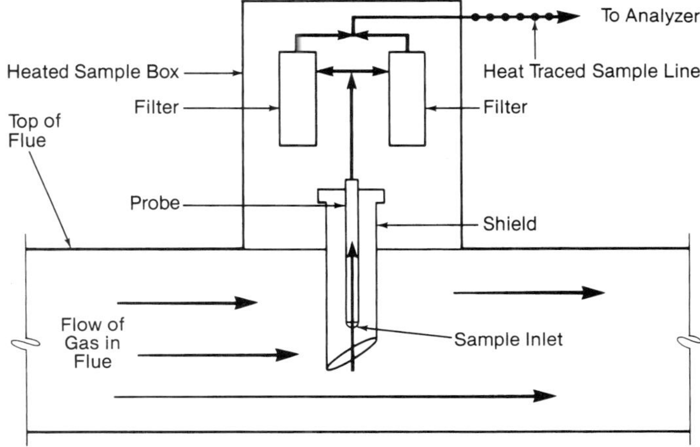

Fig. 6 Extractive sample probe with dust conditioning.

Other performance parameters that would be measured depend on the type of system. Performance testing is usually extensive and expensive. Multiple performance tests may be completed for a new environmental control system depending on specific contractual requirements. An initial test may be followed by additional testing after a defined period of operation.

Sample location selection

Selecting the sample locations is the first step in the performance test and is often incorporated in the design of the environmental control system. Outlet sampling is not difficult if it is done in the stack. However, the inlet and intermediate samples are more difficult to obtain. In Fig. 7, the system inlet at the far left is a section of flue followed by an immediate upward bend. The dry precipitator outlet is at the induced draft (ID) fan inlet. Fig. 7 also shows test ports in the downward flue section after the absorber tower and at the stack entrance. Test port locations are selected to give the best flow profile within the plant layout limitations.

The number and location of traverse points are selected using U.S. Environmental Protection Agency (EPA) Method 1. Application of Method 1 is illustrated in Fig. 8. The distance downstream from the flow disturbance, which in this case is the stack inlet, is 138 ft (42.1 m). This is more than 8 diameters (138/16 = 8.6). The upstream distance to the stack exit is 5 diameters. Therefore, 12 traverse points are required by Method 1. For rectangular flues, Method 1 is applied by first determining an effective or equivalent diameter which is calculated as follows:

Table 6
Recirculation Slurry/Blowdown Analysis (Wet FGD Systems)

Parameter	Suggested Method*	Suggested Frequency	Use of Data
pH	2-buffer method calibration or grab sample	Once per shift by grab sample, daily by buffer method	Troubleshoot pH meter and maintain process efficiency.
Slurry suspended solids content	Gravimetric	Weekly	Calibrate density meter and maintain process efficiency.
Cations: Ca^{++} and Mg^{++} primary; K^+ and Na^+ secondary	IC, AA, ICP or wet chemistry methods	Daily for Ca^{++} and Mg^{++} in calcium-based systems	To calculate stoichiometry and monitor process chemistry.
Anions: $SO_3^=$, $CO_3^=$, $SO_4^=$ and Cl^- primary; F^- and NO_3^- secondary	IC or wet chemistry methods	Daily for $SO_3^=$, $CO_3^=$, $SO_4^=$, Cl^-	To calculate stoichiometry, measure chloride concentration, and monitor process performance and gypsum production.
Dissolved metals: Fe, Mn, Al, etc.	AA	Infrequently	Process troubleshooting.
Additives: adipic acid, dibasic acid, formate, thiosulfate, scale inhibitors, etc.	IC or wet chemistry methods	Daily	Determine or correct dosage.
TDS (total dissolved solids)	Gravimetric	Weekly	Calibrate density meter for suspended solids; important for mag-lime, high chloride or suspended solids.
Dissolved alkalinity	Acid-base titration	Weekly	Measure alkali loss in filter cake of mag-lime process; important for mag-lime or sodium systems.
Settling tests	Graduated cylinder	Weekly	Predict thickener performance and assist process troubleshooting.
Filter leaf test	Filter and weight	During startup	Predict filter cake properties.
Particle size distribution	Coulter counter or laser diffraction	During startup	Determine settling or dewatering characteristics; important for gypsum systems.

* Many methods have been standardized through the American Society for Testing and Materials, Philadelphia, Pennsylvania, and the Electric Power Research Institute, Palo Alto, California.

$$DC = \frac{2LW}{L + W} \tag{1}$$

where

DC = equivalent diameter
L = length
W = width

Then a matrix layout is used for sample location selection. The cross-section is divided into as many equal rectangular areas as there are traverse points, and the sample is taken at the centroid of each area.

Gas-side measurements

For a performance test, a combination of EPA Methods 5 and 8 is the most often used procedure to simultaneously measure particulate, SO_2, and SO_3 or acid mist (H_2SO_4) levels. The Method 5 sampling train, shown in Fig. 9, is used with Method 8 impingers shown in this figure. The data obtained from this traverse also represent EPA Methods 1 through 4. Results include gas velocity by S-type pitot tube (Fig. 10) and CO_2, O_2, moisture, SO_2, particulate, and SO_3 or H_2SO_4 levels. The following calculations are used:

Determination of moisture content in stack gas

$$B_w = \frac{(MWC)(1.34)}{\left[(MWC)(1.34)\right] + \frac{(DGV)(P_m)}{T_m}(500)} \tag{2}$$

where

B_w = moisture fraction
MWC = impinger weight gain, g
DGV = sampled dry gas volume at meter conditions, ft^3
P_m = pressure at meter, in. Hg
T_m = temperature at meter, R
1.34 = constant for converting grams of H_2O to liters of H_2O vapor at standard conditions, l/g
500 = conversion of cubic feet at meter conditions to liters at standard conditions, l R/ft^3/in. Hg

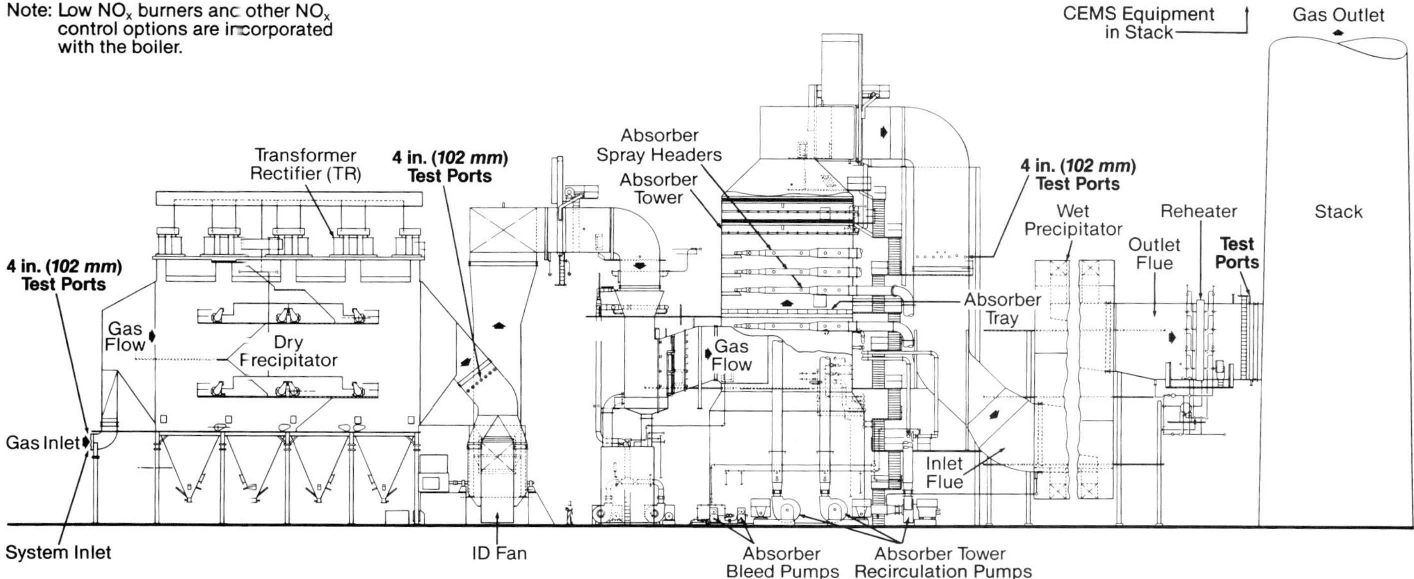

Fig. 7 Typical test port locations indicating compromises made in measurement due to equipment arrangement.

Determination of flue gas molecular weight – dry basis

$$M_d = \begin{bmatrix} (44)(\%\,CO_2\,dry) + (32)(\%\,O_2\,dry) \\ + (28)(\%\,N_2\,dry) \end{bmatrix} \times 0.01 \qquad (3)$$

where

M_d = molecular weight of gas on dry basis, lb/lb-mole
44 = molecular weight of CO_2, lb/lb-mole
32 = molecular weight of O_2, lb/lb-mole
28 = molecular weight of N_2, lb/lb-mole

Determination of flue gas molecular weight – wet basis

$$M_w = M_d\left(1 - B_w\right) + 18B_w \qquad (4)$$

where

M_w = molecular weight of gas on wet basis, lb/lb-mole
M_d = molecular weight of gas on dry basis, lb/lb-mole
B_w = moisture fraction, lb-mole H_2O/lb-mole wet gas
18 = molecular weight of H_2O, lb/lb-mole

Determination of gas humidity

$$H = \frac{18B_w}{M_d\left(1 - B_w\right)} \qquad (5)$$

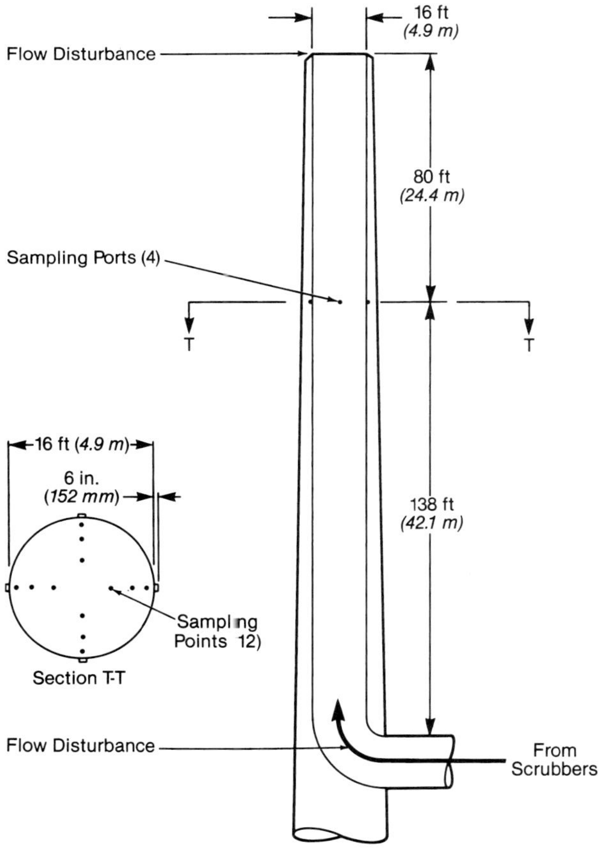

Fig. 8 Stack dimensions and sampling port locations for EPA Method 1.

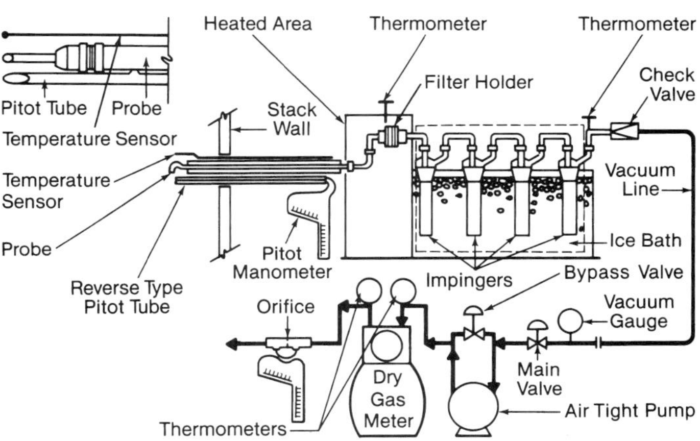

Fig. 9 EPA Method 5 particulates and SO_2 emissions sampling train.

Table 7
Dewatering System Analysis (Wet FGD Systems)

Parameter	Suggested Method*	Suggested Frequency	Use of Data
Thickener or hydroclone underflow solids, %	Gravimetric	Daily	Troubleshoot process and calibrated lime density meters
Clarified water, water analysis (anions, cations, suspended solids)	See Table 6	Weekly	Monitor dewatering system performance
Supernatent, filtrant solids content, % by wt	Gravimetric	Daily	Monitor filter cloth status
Gypsum Quality Analysis			
CA^{++}, Mg^{++}, Na^{+}, K^{+}, Fe_2O_3, R_2O_3, SiO_2	AA or ICP	Weekly	Product purity and properties
Cl^{-}, $SO_4^{=}$	IC	Daily	Product purity and properties
$SO_3^{=}$, $CaSO_3 \cdot 1/2\ H_2O$	Iodimetric titration	Daily	Product purity and properties
Moisture content	Gravimetric	Daily	Important for marketable gypsum and monitoring dewatering system performance
Surface area	BET analysis (subsieve sizer)	Startup or process change	N_2 gas is condensed on surface and absorbed; N_2 corresponds to surface area
Combined water	Heat to 482F (250C)	Weekly	Gypsum purity
Mean particle size	Coulter counter, Sedigraph plus sieve analysis or laser diffraction	Startup or process change	Product purity and properties
Aspect ratio	Scanning electron microscope	Startup or process change as required by end users	Product purity and properties
Total water soluble salts	Sum ($Na^{+} + K^{+} + Mg^{++}$)	As needed	Product purity and properties
Flyash	Determined as acid insoluble	Daily	Determined as acid insolubles along with silica (SiO_2) and other impurities
pH	Electrode	Daily	Product purity and properties

* Many methods have been standardized through the American Society for Testing and Materials, Philadelphia, Pennsylvania, and the Electric Power Research Institute, Palo Alto, California.

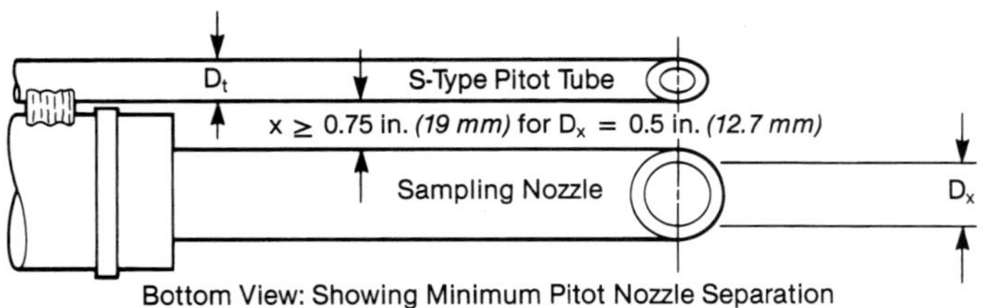

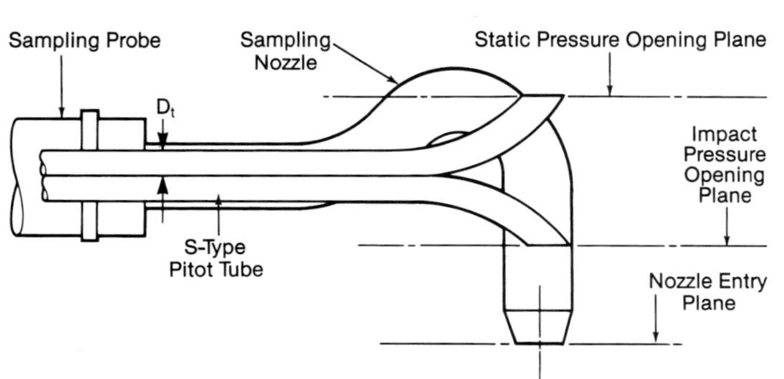

Fig. 10 S-type pitot tube with sampling probe.

where

H = gas humidity, lb H_2O/lb dry gas
B_w = moisture fraction, lb-mole H_2O/lb-mole wet gas
M_d = molecular weight of dry gas, lb/lb-mole
18 = molecular weight of H_2O, lb/lb-mole

Determination of stack gas velocity

$$V_s = 85.49\, C_p \left(\frac{T_s\, \Delta P}{P_s\, M_w} \right)^{1/2} \tag{6}$$

where

85.49 = pitot tube constant in the following units:

$$\frac{\text{ft}}{\text{s}} \left(\frac{(\text{lb / lb-mole})(\text{in. Hg})}{(\text{R})(\text{in. H}_2\text{O})} \right)^{1/2}$$

V_s = velocity of the flue gas, ft/s

ΔP = average differential pressure measured by S-type pitot tube, in. H_2O
T_s = average gas temperature in duct, R
P_s = absolute duct pressure, in. Hg
M_w = wet gas molecular weight, lb/lb-mole
C_p = S-type pitot tube correction factor (normally 0.84)

Volumetric flow rate – actual conditions

$$Q_{ac} = 60 V_s A \tag{7}$$

where

Q_{ac} = actual volumetric flow rate, ACFM
V_s = velocity of the flue gas, ft/s
A = cross-sectional area of the duct, ft^2
60 = conversion factor, s/min

Volumetric flow rate – dry standard conditions

$$Q_{sd} = \frac{528 Q_{ac} P_s (1 - B_w)}{29.92\, T_s} \tag{8}$$

where

Q_{sd} = dry volumetric flow rate at standard conditions, DSCFM
Q_{ac} = actual volumetric flow rate, ACFM
B_w = moisture fraction, lb-mole H_2O/lb-mole wet gas
T_s = average gas temperature in duct, R
P_s = absolute duct pressure, in. Hg
528 = standard temperature, R
29.92 = standard pressure, in. Hg

Total particulate and H_2SO_4 – grains/ACF

$$C_a = \frac{(0.001)(15.43)\, M_n}{V_m} \tag{9}$$

where

C_a = concentration of specie in the flue gas at duct conditions, grains/ACF
M_n = mass of specie collected, mg
V_m = dry gas volume sampled as measured by dry gas meter, ACF
0.001 = conversion of milligrams to grams, g/mg
15.43 = conversion of grams to grains, grains/g

Total particulate, H_2SO_4, condensible total particulate and condensible H_2SO_4

$$C_s = \frac{(0.001)(15.43)\, M_n\, (29.92)\, T_s}{V_m\, Y\, (528)\, P_s} \tag{10}$$

where

C_s = concentration of specie in the flue gas at dry conditions, grains/DSCF
M_n = mass of specie collected, mg
T_s = average gas temperature in duct, R
V_m = dry gas volume sampled at meter, DCF
Y = meter correction factor
P_s = average gas pressure in duct, in. Hg
0.001 = conversion of milligrams to grams, g/mg
15.43 = conversion of grams to grains, grains/g
29.92 = standard pressure, in. Hg
528 = standard temperature, R

Total particulate, H_2SO_4, NO_x and SO_2 – lb/h

$$C_m = C_s (1.428 \times 10^{-4})(60)(Q_{sd}) \tag{11}$$

where

C_m = mass flow rate of specie in the flue gas, × lb/h
C_s = concentration of specie in the flue gas at dry standard conditions, grains/DSCF
Q_{sd} = dry volumetric flow rate at standard conditions, DSCFM
60 = conversion of minutes to hours, min/h
1.428×10^{-4} = conversion of grains to pounds, lb/grain

Total particulate, H_2SO_4, NO_x and SO_2

$$E = \frac{20.9\, C_m\, F_d}{60\, Q_{sd}\, (20.9 - \%\, O_{2ds})} \tag{12}$$

Table 8
Spray Dryer FGD System Process Analysis

Parameter	Suggested Method	Suggested Frequency	Use of Data
Lime slurry solids content	Gravimetric	Once per shift	Confirm process meter
Recycle slurry solids content	Gravimetric	Once per shift	Confirm process meter
Flue gas wet bulb temperature	Manual	Startup, fuel change	Process troubleshooting
Flyash alkalinity	Acid titration	Startup, fuel change	Process troubleshooting
Baghouse ash free moisture	Gravimetric	Startup, process change	Process troubleshooting
Process water TDS (total dissolved solids)	Gravimetric	Startup, water source change	Confirm operations
Slaking water anions	Wet chemistry	Startup, water source change	Process troubleshooting

where

E = emission rate on a dry basis at 0% O_2, lb/10^6 Btu
C_m = mass flow rate of specie in the flue gas, lb/h
F_d = dry F-factor at 0% O_2, determined from fuel analysis, DSCF/10^6 Btu (see Equation 17)
Q_{sd} = dry volumetric gas flow rate at standard conditions, DSCFM
% O_{2ds} = O_2 in the gas on a dry basis, %
20.9 = O_2 in air, used with % O_2 in the gas to correct to 0% O_2, %
60 = conversion of minutes to hours, min/h

SO_2 – parts per million (ppm) wet

$$C_{ppm,w} = \frac{(32.03)\ \mathrm{N}\ (V_t - V_{tb})\ (V_{soln} / V_a)\ (1 - B_w)}{V_d \times 2.66} \tag{13}$$

where

$C_{ppm,w}$ = gas SO_2 concentration on a wet basis, ppm wet
N = barium standard titrant normality, meq/ml
V_t = volume of barium titrant used for sample, ml
V_{tb} = volume of barium titrant used for blank, ml
V_{soln} = total volume of impinger solution, ml
V_a = volume of sample aliquot used, ml
B_w = gas moisture fraction, g-mole H_2O/g-mole wet gas
V_d = gas volume sampled at dry standard conditions, DSCF
32.03 = number of milligrams of SO_2 per milliequivalents, mg SO_2/meq
2.66 = conversion of mg/SCM to parts per million, mg/SCM / ppm

SO_2 – ppm dry

$$C_{ppm,d} = \frac{20.9\ C_{ppm,w}}{(1 - B_w)(20.9 - \%\ O_{2ds})} \tag{14}$$

where

$C_{ppm,d}$ = gas specie (SO_2) concentration on a dry basis, corrected to 0% O_2, ppm dry
$C_{ppm,w}$ = gas specie (SO_2) concentration on a wet basis, ppm wet
B_w = moisture fraction of the gas, lb-mole H_2O/lb-mole wet gas
% O_{2ds} = O_2 in the gas on a dry basis, %
20.9 = O_2 in air, used with % O_2 in the gas to correct to 0% O_2, %

NO_x – ppm dry and ppm wet

NO_x is evaluated directly by monitoring systems as parts per million, dry basis. It can be converted to ppm wet if needed by solving Equation 14 for $C_{ppm,w}$.

Collection/removal efficiency – total particulate, H_2SO_4 and SO_2

$$C_{eff} = \left[(C_{in} - C_{out})/(C_{in})\right] \times 100 \tag{15}$$

where

C_{eff} = collection/removal efficiency, %
C_{in} = concentration of matter at inlet, lb/10^6 Btu
C_{out} = concentration of matter at stack, lb/10^6 Btu

Percent isokinetic

$$\%\ I = (V_n / V_s) \times 100 \tag{16}$$

where

$$V_n = \frac{V_d\ T_s\ 29.92}{60\ tA_n\ 528\ P_s (1 - B_w)}$$

and where

% I = % isokinetic
V_s = average flue gas velocity in duct during sampling, ft/s
V_n = average sample gas velocity through the nozzle during sampling, ft/s
V_d = dry gas volume sampled at standard conditions, ft^3
t = elapsed sampling time, min
A_n = cross-sectional area of nozzles, ft^2
B_w = gas moisture fraction, lb-mole H_2O/lb-mole wet gas
T_s = flue gas temperature in the duct, R
P_s = absolute pressure in the duct, in. Hg
60 = conversion from minutes to seconds, min/s
528 = standard temperature, R
29.92 = standard pressure, in. Hg

The % isokinetic value is used as a measure of how representative the sample is of the actual flue gas. Generally, if the % isokinetic is between 95 and 105%, the sample is considered acceptable.

F-factor

The F-factor used in making emissions rate calculations was developed from chemical and combustion analyses of the fuel (reference EPA Method 19). This factor is the ratio of the theoretical volume of dry gases at 0% excess air (O_2) given off by complete combustion of a known amount of fuel, to the gross caloric value of the burned fuel. The data generated from the ultimate analysis of the fuel are used to calculate the F-factor as follows:

$$F_d = (10^6)\frac{3.64(\mathrm{H}) + 1.53(\mathrm{C}) + 0.57(\mathrm{S}) + 0.14(\mathrm{N}) - 0.46(\mathrm{O})}{\mathrm{HHV}} \tag{17}$$

where

F_d = dry F-factor at 0% O_2, DSCF/10^6 Btu
HHV = higher heating value, Btu/lb
H = hydrogen in the fuel, % by wt
C = carbon in the fuel, % by wt
S = sulfur in the fuel, % by wt
N = nitrogen in the fuel, % by wt
O = oxygen in the fuel, % by wt

F-factors are constant for a given fuel category (within ±3%). Standard average values approved by the EPA are:

Fuel	DSCF/10^6 Btu
Bituminous coal	9780
Oil	9190
Natural gas	8710
Wood bark	9600
Wood chips	9240
Municipal solid waste	9570

The F_d factor is used to convert the measured pollutant concentration to an emissions rate in lb/10^6 Btu by:

$$\text{lb}/10^6\ \text{Btu} = F_d\ C_{ppm,d}\ K\ 20.9/(20.9 - \%\ O_{2,d})$$

Where K is the conversion factor from ppmdv to lb/scf.

$K\ NO_X = 1.194 \times 10^{-7}$
$K\ SO_2 = 1.660 \times 10^{-7}$

Particulate and SO_2 removal efficiencies

The SO_2 and particulate removal efficiencies are determined by inlet and outlet sampling using Equations 12 and 15.

Stack opacity

Stack opacity is determined by the installed transmissometer or by EPA Method 9 which is a visual method by a certified observer.

NO_x concentration

NO_x concentration is determined by EPA Method 7. This measurement is taken in the stack on a composite sample. NO_x concentration may be adjusted to design conditions from actual test conditions.

NH_3 concentration

NH_3 concentration is determined by EPA Method CTM-27. This measurement is taken at the SCR outlet on a composite sample before the air heater. NH_3 concentration may be adjusted to design conditions from actual test conditions.

Condensible particulate matter

New regulations limit the amount of total particulate matter finer than 10 microns (PM_{10}) which includes both solids and vapors which may condense in the atmosphere to form particulates after the flue gas exits the stack. These condensibles may include H_2SO_4 and VOCs. EPA Methods 201 and 202 are designed to measure the condensible fraction of particulate emissions. There is some concern that Method 202 may overstate the condensibles by inclusion of SO_2 which is not purged from the impingers before the analysis.

Makeup water requirements

Makeup water requirements are measured at the appropriate terminal point with a certified flow meter. Adjustments from test conditions may be necessary due to variations in gas volume or temperature or other process deviations from design conditions.

Reagent consumption

There are two common ways to measure reagent consumption. The first method is to measure the solid reagent flow (for lime/limestone systems) using a calibrated weight belt feeder. The measurement is time averaged. This mass flow method is cumbersome because the initial and final storage tank volumes and densities must also be measured. This method is best done over a period of at least 24 hours. The second and preferred method is chemical analysis. For a wet FGD system, the reagent consumption is calculated from the stoichiometric ratio, limestone (or reagent) analysis, and gas side measurements by the following equation:

$$LS = \frac{100\,(SR)\,(SO_2)}{64\,(CaCO_3)} \tag{18}$$

where

LS	=	limestone consumption, lb/h
SR	=	stoichiometric ratio (see Chapter 35)
SO_2	=	SO_2 removed, lb/h
64	=	molecular weight of SO_2, lb/lb-mole
100	=	molecular weight of $CaCO_3$, lb/lb-mole
$CaCO_3$	=	available $CaCO_3$ in limestone (wt fraction: lb available $CaCO_3$/lb limestone)

Reagent consumption may be adjusted for deviations from SO_2 removal rate and mass flow (or concentration) design conditions.

Gypsum purity

A composite sample of gypsum, taken over the test period, is analyzed as outlined in Table 7. Adjustments may be necessary for flyash contamination, chloride concentration, and inerts if the inlet levels to the FGD system substantially deviate from design.

Other test methods

There are numerous other procedures for measuring system parameters. The example in this chapter is a formal performance test; other methods could be used. SO_2 can be measured with a CEMS or by other temporary instrument methods following EPA Method 6. Particulate level can be measured by EPA Method 5 or 17 or alternate methods such as the probe shown in Fig. 11.

Mercury emissions measurement

Continuous or semi-continuous measurement of mercury emissions is a developing technology that will be used to assure compliance with future regulation of mercury emissions from coal-fired boilers. Various monitors under development share common features of sample pre-treatment, adsorption of mercury onto a gold trap and then periodically purging the mercury from the trap and quantitatively measuring the mercury using cold vapor atomic spectroscopy (CVASS), atomic fluorescence spectroscopy (AFS) or Zeeman-modulated atomic spectroscopy (Pavlish, 2003). The measurement challenges include low concentrations (typically less than 10μg/Nm^3), fouling with particulates, and interference from common flue gas constituents such as SO_2, HCl and NO_x. An alternative method accepted by the EPA involves drawing a measured flue gas sample volume through a carbon trap for a specified time period (1 month for example) and then measuring the mercury in the carbon trap to calculate an

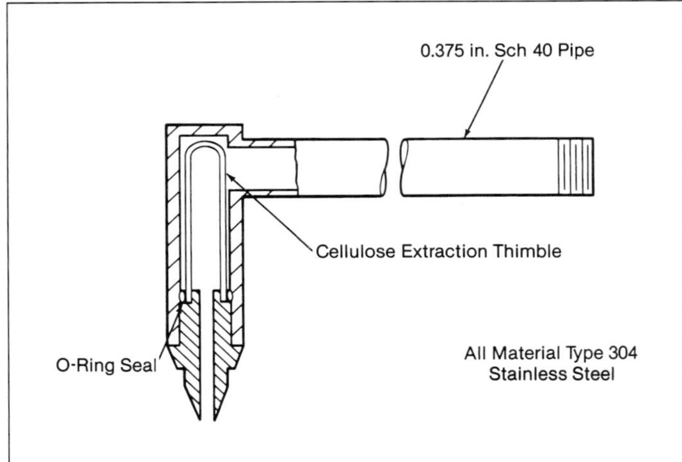

Fig. 11 Probe detail, isokinetic dust sampling.

emission rate in lb/TBtu. ASTM Method D6784-02 is the validated and accepted stack sampling test method for measuring mercury speciation and total mercury in coal-fired flue gas (see Fig. 12).

The future of environmental measurement

Advances in environmental measurement techniques will certainly continue. Technologies such as fiber optics, video imaging and Fourier transform infrared analysis (FTIR), which have been used successfully in the laboratory, are expected to move into the field. Measurement will be extended to include other air and water pollutants, and air toxics monitoring will become increasingly important.

Fig. 12 Mercury testing at research and development facility.

Bibliography

American Society of Mechanical Engineers (ASME) Performance Test Codes: PTC 1 on General Instructions, 1999; PTC 2 on Definitions and Values, 1980 (R1985); and PTC 3.2 on Solid Fuels.

ASME PTC Supplements on Instruments and Apparatus: PTC 19.2 on Pressure Measurement, 1987 (R1998); PTC 19.3 on Temperature Measurement, 1974 (R1998); PTC 19.6 on Electrical Measurements (IEEE Standard 120), and PTC 19.10 on Flue and Exhaust Gas Analyses, 1981.

United States Code of Federal Regulations, CFR 40, Protection of the Environment, Parts 53 to 80.

United States Code of Federal Regulations, (e-CFR), Website www.gpoaccess.gov/cfr/index.html, February 7, 2005.

Continuous Emission Monitoring Guidelines: Update, Report CS-5998, Electric Power Research Institute (EPRI), Palo Alto, California, September, 1988.

United States Code of Federal Regulations, CFR 40, Part 60 Environmental Protection Agency (EPA), Method 19: Determination of Sulfur Dioxide Removal Efficiency and Particulate Matter, Sulfur Dioxide and Nitrogen Oxide Emission Rates.

FGD Chemistry and Analytical Methods Handbook, Report CS-3612, Vols. 1 and 2, Electric Power Research Institute (EPRI), Palo Alto, California, July, 1984.

Handbook of Electrode Technology, Orion Research, Cambridge, Massachusetts, 1982.

Liptak, B.G., Ed., *Instrument Engineers' Handbook: Process Measurement and Analysis,* Vol. 1, Fourth Ed., CRC Press, Boca Raton, Florida, June 27, 2003.

Performance Test Code Committee No. 40, Flue Gas Desulfurization Units, Draft VII, The American Society of Mechanical Engineers, New York, New York, October, 1989.

Elliot, T.C., "Monitoring Pollution," *Power,* pp. 51-56, September, 1986.

Elliot, T.C., "Level Monitoring," *Power*, pp. 41-58, September, 1990.

Gilbert, G., "Three Ways To Measure Flow in Power Plant Ducts," *Power,* pp. 63-72, June, 1987.

Kiser, J.V.L., "Continuous Emissions Monitoring: A Primer," *Waste Age,* pp. 64-68, May, 1988.

Kiser, J.V.L., "More on Continuous Emissions Monitoring," *Waste Age,* pp. 119-122, June, 1988.

FGD Manual, The McIlvaine Company, Northbrook, Illinois, August, 1989.

Pavlish, J.H., et al., "Status Review of Mercury Control Options for Coal-Fired Power Plants," Fuel Processing Technology, 82: 89-165, pp 98-99, 2003.

Robertson, R.L., "Continuous Emission Monitoring and Quality Assurance Requirements for New Power Plants," appearing in *Continuous Emission Monitoring Systems for Power Plants: The State-of-the-Art,* American Society of Mechanical Engineers (ASME), pp. 13-15, New York, New York, 1988.

Rothstein, F., and Fisher, J.E., "pH Measurement: The Meter," *American Laboratory,* September, 1985.

Spriggs, D., "Calibration and Troubleshooting of pH Loops," presented to the American Water Works Association Ohio Section Northeast District Meeting, Painesville, Ohio, July 20, 1989.

Section V
Specification, Manufacturing and Construction

This section begins with an in-depth discussion of the specification, evaluation and procurement process for large capital expense items. This includes discussions about project scope, terms and conditions, and general bid evaluation. Also discussed are power system economics, and procedures for the evaluation of equipment characteristics in terms of justifiable expenditures. Examples and calculations are included for both utility and industrial units.

This is followed by a discussion of the manufacturing processes for fossil fuel-fired equipment. Welding and metal removal techniques, as well as fabrication of the various components and component parts, are covered. Examination and quality control are also discussed. The section ends with a discussion of various construction techniques, labor requirements, on-site considerations, safety issues, and post-construction testing prior to unit startup.

Many factors are involved in specifying and evaluating the economics of a steam generating or emissions control equipment project.

Chapter 37

Equipment Specification, Economics and Evaluation

Accurate and complete technical specifications are critical to the successful development of any new project. Steam generating and environmental emissions control systems are complex and typically require the evaluation of many competing factors to select the equipment and potential options that best meet project needs.

Each application of new or upgraded equipment has a number of unique factors and requirements that need to be identified and communicated to the equipment or system designer/supplier. Specifications should be developed to document these needs. However, the developer of those specifications should also allow the equipment designers flexibility to adapt their products to best meet the true requirements of the project. There are often multiple options or alternative solutions that can be considered. The solution should be selected after evaluating the initial capital investment in conjunction with several variables including the relative longer-term operating and maintenance cost benefits, feasibility of obtaining required regulatory construction and environmental permits, ability to meet varying operating conditions during the life of the equipment, and costs associated with the disruption or outage of existing equipment during construction or tie-in.

The following discusses some of the primary factors that should be considered when developing and planning a project requiring new or upgraded steam generating or environmental control systems. It describes key technical and commercial criteria that should be developed and communicated to the equipment suppliers.

Project feasibility studies and technology selection

Before detailed equipment specifications can be developed, groundwork is required to establish the basic project needs and evaluate major alternatives. Some aspects that should be evaluated during the early feasibility stages include:

1. steam capacity requirements,
2. range of plant operating conditions,
3. steam cycle conditions (pressure/temperature),
4. fuel type(s),
5. combustion technology,
6. site location and arrangement logistics, including construction feasibility,
7. stack emissions and air quality control system technology,
8. reagent type(s),
9. solid waste disposal logistics, and
10. water quality and treatment.

For a utility power plant, the need for new generating capacity is identified from the forecast of both peak and average long-term electrical demand (load). The overall power and steam generation plan must also recognize unit retirements within the system, spinning reserve, reserve for scheduled and unscheduled outages, and current generating equipment mix. The specific power plant type selected is based upon an evaluation of the amount and type of demand, assessment of current boiler types and sizes within the system, long-term fuel availability and cost, condensing water supply, new equipment types available, environmental impact and, in many cases, public opinion. Extensive feasibility studies are typically conducted to provide such data.[1,2]

For an industrial installation, the needs are usually tied to process steam or heating requirements. The timing of equipment addition, the steam flow rate, and the frequency of operation are governed by the application. In addition, with the growth of cogeneration and independent power production, steam supply needs have been increasingly influenced by the demands of electrical power production. In all cases, the steam flow requirements will determine the type, size and number of steam generating units. Particular attention should be given to the variety, cost and availability of fuel sources. Process byproducts, waste materials and waste heat offer potential low-cost sources of energy that can be used to displace the use of higher-cost traditional fuels to meet industrial steam generation needs.

For projects involving addition of emissions control systems to existing boilers, many of the design conditions for the system are typically established by the current boiler operations. However, issues such as technology selection, reagent type(s), solid waste disposal, and water treatment requirements should be considered and evaluated during early stages of project development. It may also be important to evaluate potential future changes in environmental permitting requirements and operating conditions, such as the use of an alternate fuel type, that could affect equipment sizing or even technology selection.

For repair and retrofit projects, the evaluation is focused on the scope being considered. However, appropriate additional consideration should be given to other factors potentially being affected by the repair or retrofit. Examples include delivery schedule to meet outage requirements, changes in emissions rates, and effects on auxiliaries such as fans and downstream equipment.

During these early stages, it is often helpful to obtain input from the equipment suppliers. On large, complex projects, some plant owners and project developers have found it valuable to form a strategic alliance or partnership with a qualified equipment supplier. As the project develops, such a business model can help to evolve the design of a unique or customized new system. This alternative business structure will be discussed later in this chapter.

Specifications

Once the major feasibility issues and alternatives have been evaluated and established, specifications need to be developed. These specifications provide a *definition* from which the primary equipment or system will be designed and by which that equipment will be integrated into the overall project.

Specifications can vary in detail but as a minimum, performance and functional requirements need to be defined. The basic design is derived from the functional requirements of the application. The equipment designer needs to know the key characteristics of all major system inputs and all performance output requirements. This information is essential for setting vital system design parameters such as size, capacity, materials, equipment redundancy level, etc.

The design specification for new or replacement steam generating equipment should include at least the following information:

1. steam flow characteristics
2. steam temperature and pressure conditions
3. feedwater supply
4. fuel type(s) and characteristics, including ash properties
5. boiler efficiency requirements
6. emissions requirements
7. performance guarantees and test conditions

The design specification for new or replacement environmental control systems should include at least the following information:

1. conditions and composition of flue gas and flyash entering the system
2. emissions performance requirements
3. reagent type(s) and characteristics
4. solid byproduct quality requirements
5. wastewater quality limitations
6. makeup water composition
7. performance guarantees and test conditions

The specification should also define how the equipment or system design will be integrated into the overall project. The following is a list of typical project definition information that needs to be included in a specification to an equipment supplier:

1. complete scope of work, including the definition and location of all terminal and interface points within the overall plant or existing facilities;
2. site-specific requirements, design conditions, layout/arrangement limitations, delivery logistics, and access restrictions, among others;
3. plant standardization requirements (if applicable);
4. project schedule, including equipment delivery, construction and startup/commissioning, consistent with the scope of work definition;
5. commercial contract terms and conditions; and
6. evaluation criteria/factors to be used in assessing design and operating tradeoffs.

Beyond these items, it is usually appropriate to only specify material and workmanship quality requirements, allowing potential suppliers to apply their experience and knowledge to propose the most dependable and cost-effective design.

Performance – steam generating equipment

Performance specifications, defining the functional requirements for utility, industrial and environmental equipment, are summarized in the following section. For steam generating equipment, this generally consists of steam flow, steam temperature and pressure, feedwater temperature and quality, fuel, boiler system efficiency, emissions requirements, and performance guarantees.

Steam flow Primary performance specifications include the steam flow requirements. For utility and other units that produce steam for electric power generation, these usually include:

1. maximum continuous rating (MCR),
2. performance load steam flow rate (to match optimal turbine heat rate load),
3. minimum steam flow rate,
4. type of service (base load, cycling or peaking), and
5. expected cycle of operation.

In power station practice, typically one boiler is matched to an individual turbine-generator set ranging from 25 to 1300 MW. Depending on the load requirements, one or more steam generating units are purchased. In general, economies of scale dictate the purchase of the fewest number of units possible to meet overall load and system requirements.

For some industrial units, specified steam flow requirements include:

1. maximum steam flow capacity,
2. turndown (ratio of maximum to minimum load), and
3. expected steam load pattern variations (daily, weekly or seasonal).

However, in some industrial applications the primary function of the steam generator is to consume or process a quantity of fuel. In such cases, the fuel input is specified and steam capacity is determined indirectly from the maximum fuel input rating.

Industrial unit operators frequently prefer multiple smaller boilers rather than a single large unit, giving them more operating flexibility while maximizing steam system availability. It is normally critical that the industrial process be available at all times and not be limited by the steam generating system. Typically, the number of steam generating units is established by assessing the load pattern, load duration, availability requirements, maintenance periods, types of boilers in the system, sources of backup steam or auxiliary power, and possible future expansion plans. Minimum standby requirements are set by the minimum required steam flow plus an allowance for routine maintenance and forced outages. Where the load is seasonal and steam is not a critical process requirement, standby may not be required. Where steam is required throughout the year, some level of economical excess capacity will be needed (see Table 1).

The number of units to be installed needs to be evaluated on the basis of total cost, including initial capital cost, when compared with tradeoffs in thermal efficiency and the true value of availability or cost penalty for unavailability. The cost for more, smaller units will be greater than the cost for fewer, larger units. However, if the larger units will be operated for a significant amount of time at reduced load, the effective thermal efficiency will be lower and operating fuel-related costs will be higher. In addition, installing a number of smaller units over time can defer capital costs until the additional capacity is required.

Steam temperature and pressure For utility applications, a turbine-generator set is generally specified, and therefore the turbine cycle selected will determine the pressure and temperature relationships. As part of the specifications, the complete turbine heat balance at various operating conditions such as minimum load, control load, and full load should be provided to give the designers more insight into system requirements. The turbine specifications will typically establish acceptable deviation limits in outlet steam temperature as well as the maximum rate of change. Constant pressure operation or variable pressure operation, each with advantages and disadvantages, may be selected to meet operator needs. (See Chapter 26.) Constant pressure versus variable pressure operation also has a significant impact on supercritical pressure boilers as referenced in other chapters. When the application calls for superheat and reheat steam conditions, the load range and load pattern will help determine the steam temperature control ranges. The benefits of higher steam temperatures at lower loads must be compared to the higher costs of providing an extended range. The evaluation penalties and allowable variations in steam temperature will also affect the type of temperature control system included. Allowance must also be made for pressure and heat loss from piping between the boiler outlet and the steam use location. Typically, the heat loss is negligible with today's insulation products and application techniques. It is normally left to the supplier to determine the appropriate design margin necessary to achieve the specified steam pressure and temperature conditions at the terminals.

In industrial applications where there may be some flexibility in the selection of steam pressure and temperature, interaction with the steam system supplier may permit selection of a more economical set of operating conditions.

Feedwater The temperature and chemical analysis of the feedwater must be specified. The minimum required feedwater temperature is one of the components that establishes boiler output (firing rate) and the water chemistry will affect water treatment needs. Recommended water chemistry specifications are provided in Chapter 42. For small units, a deaerator treats the feedwater before it enters the boiler. The resulting feedwater temperature is typically 220 to 240F (104 to 116C). Oxygen in the feedwater must be minimized to prevent internal boiler corrosion.

In larger high temperature, high pressure boilers, feedwater heaters are used to improve cycle efficiency as discussed in Chapter 2. If increased output (MWs) is desired when feedwater heaters are out of service, then this design requirement needs to be conveyed along with the applicable feedwater temperature and steam flow conditions. If turbine heat balances are provided for all operating conditions as part of the specification, feedwater temperature requirements will inherently be included. In any case, the feedwater temperature needs to be specified for all boiler conditions over the load range.

Fuel The selected fuel has a major impact on steam generating equipment design and capital cost due to the type of combustion equipment, fuel delivery/processing equipment, fouling and erosion characteristics of the flyash, and heat transfer properties of the flue gas (including moisture content). Refuse fuel, biomass and most byproduct or waste fuels require significantly different combustion systems than pulverized coal, oil or natural gas and can also result in different boiler corrosion characteristics. Boilers designed to fire multiple fuel types require a compromise between the various fuels, particularly for steam temperature control. They therefore may be more expensive than boilers designed to burn a single fuel type. Furthermore, boilers designed to accommodate multiple classifications of a fuel type, such as bituminous and subbituminous coals, will require the application of some tradeoffs in the design of the boiler, fuel preparation equipment, and combustion system. Designing for multiple fuel types and classifications is especially challenging for supercritical boilers by further complicating the design requirements for the circulation system to accommodate a wider variation in heat transfer rates through the critical furnace steam/water circuits. Environmental equipment design require-

Table 1
Boiler Capacity with Standby Units

Total Number of Units	Number of Standby Units	Per Unit Capacity, %	Total Capacity, %
2	1	100	200
3	1	50	150
4	1	33	133
5	1	25	125

ments can also be significantly different when designed for multiple fuel types and classifications.

Complete analyses should be provided for all fuels being considered. For each fuel, ultimate analysis, proximate analysis, heating value and ash analysis should be provided in accordance with applicable American Society for Testing and Materials (ASTM) or equivalent international standard requirements. The buyer normally provides a range analysis that indicates the extreme values (low and high) for each constituent. However, a fuel analysis derived statistically from such constituents may represent an erroneous and unrealistic combination. It is best to provide a set of actual fuel analyses, which comprise the range, to enable suppliers to appropriately design and size the equipment. The analyses should be provided as moisture and ash free, moisture free and/or *as-fired*. Additional information needed includes as-received solid fuel sizing, liquid fuel viscosities, and liquid/gaseous fuel pressures. The effect of the fuel and ash on unit design is discussed in Chapter 21.

Boiler system efficiency Maximizing boiler operating efficiency minimizes fuel usage and reduces costs. To achieve the highest efficiency, heat traps are used to extract the last heat from the flue gas. Heat traps include air heaters and economizers. (See Chapter 20.) The temperature of the flue gas exiting the final trap should be low to maximize boiler efficiency. However, it should normally exceed the acid dew point of the gas to avoid corrosion and other operating problems in the last heat trap and in any downstream equipment. In addition, some environmental control equipment such as dry flue gas desulfurization (DFGD) systems requires the entering gas temperature to be somewhat higher for proper operation.

Utility applications commonly use both an economizer and an air heater. On other steam generating applications when a single heat trap is required, an air heater is usually chosen to aid in combustion of pulverized coal, wood or biomass. For other fuels, an economizer is frequently selected because of lower cost, possible increased efficiency, and reduced nitrogen oxides (NO_x) levels compared to an air heater installation (increased air temperatures can promote NO_x formation in the furnace).

If the boiler supplier is to design the heat traps effectively, the specification should include evaluation factors such as fuel cost and capitalization period. For competitive bids, it is also prudent for the buyer to specify the design exit gas temperature so that the impact on boiler efficiency between competing designs is normalized. Minimum exit gas temperature is typically set by corrosion and operating concerns of downstream equipment and should be consistent between boiler suppliers.

Environmental limits Meeting appropriate environmental emissions limits is essential. Joint discussions between the buyer, equipment suppliers and regulating agencies are recommended at early phases of project development to ensure a clear understanding of permissible and achievable emissions limits. Emissions controlled by the steam generating equipment include NO_x, carbon monoxide (CO), and volatile organic compounds (VOC). These emissions have a significant impact on design requirements of the combustion system, boiler and NO_x control systems and therefore need to be defined. Further, test methods such as applicable Environmental Protection Agency (EPA) test methods, test duration, e.g., three-hour test period, and test conditions, e.g., boiler load range, should be documented to ensure each party has a common understanding of all requirements and conditions. Other emissions, captured by downstream environmental equipment, are discussed later in this section.

Performance guarantees Specific guarantees are normally a function of the scope structure discussed later in this chapter. They normally include items that can be controlled by a specific piece of equipment and items that support the overall plant performance. Ultimately these are capacity, net plant heat rate (NPHR), stack emissions, reagent use, and other operating factors such as water usage and waste disposal. Specific guarantees relating to the steam generator scope that impact NPHR and stack emissions usually include MCR capacity, boiler efficiency, auxiliary power consumption, water and steam side pressure drop, superheater and reheater steam temperature control range, spray water quantities, and NO_x, CO, and VOC emissions. If the spray water source is downstream of the feedwater heaters, then heat rate is not impacted by superheater spray and is therefore not considered. For subcritical drum type boilers, steam purity is commonly included.

Performance – environmental control systems

Flue gas conditions and design heat input Environmental control systems are designed to reduce the level of regulated pollutants such as NO_x, sulfur dioxide (SO_2) and particulate matter present in the combustion product gases (flue gas) that leave the steam generator. The size, function and design of these systems are dependent on the chemical composition, temperature and volume of the entering flue gas. If the environmental control system is to be supplied with the boiler as an integrated system, the boiler design conditions for steam requirements, feedwater, fuel (including the range of the chemical composition of the fuel and ash) plus the ambient conditions will provide the input required to define the range of flue gas operating conditions relevant to integrated system design.

If the environmental control system is to be supplied separately, the following information for the flue gas entering the system should be defined in the specification:

1. chemical analysis of the flue gas, including concentration of the pollutant(s) to be controlled,
2. temperature of the flue gas,
3. chemical analysis and type of flyash, and
4. flow rate of the flue gas (on a mass and volumetric flow rate basis).

It is important to provide the design range of each of these items at maximum and minimum operating loads to ensure that the system can accommodate normally expected operating variations. A chemical analysis of the design fuels and the boiler design heat input rate are also helpful when correlating the specified flue gas conditions with the expected fuel(s) and establishing a consistent basis for emissions calculations reported per unit of heat input.

Emissions performance The environmental control system is typically designed to achieve a maximum allowable outlet emissions rate that is driven by the project air permits, other environmental regulations or performance objectives. Alternatively, in some cases, a minimum percent reduction performance rate may be required. The specification should define the required performance objective including the guaranteed emission rate at the stack for each controlled pollutant and specific test conditions. The specific test conditions include the test method to be used, test duration, measurement frequency and period over which the measurements will be averaged.

Reagents Many environmental control systems require a consumable reagent to create the chemical reactions required to control the target pollutant.

Selective catalytic reduction (SCR) and selective non-catalytic reduction (SNCR) type NO_x control systems use ammonia or urea based reagents. Various economic, plant logistics and permitting requirements affect reagent selection. Each type will require different equipment for storage, handling and use. Wet flue gas desulfurization (WFGD) systems have used a variety of reagent types including limestone, magnesium-enhanced lime and sodium based upon relative economics and specific performance requirements for each application. Dry flue gas desulfurization (DFGD) systems typically use lime.

Other additives and regents can also be used for various performance enhancements and specialty processes.

The chemical composition of the reagent and the extent that composition is expected to vary should be specified.

Quality requirements for solid wastes and byproducts The cost to landfill solid waste materials can vary significantly from project to project. Some projects depend on producing solid byproduct in a form that can be sold for commercial application. Coal flyash, for example, can be sold if the carbon concentration is sufficiently low. Such a limitation on carbon content may affect the design of the boiler's combustion system and/or coal selection, or may dictate the use of multiple stage particulate collection to capture most of the flyash prior to a dry FGD system.

Wet FGD systems can be designed to produce a grade of gypsum used in wallboard production. The specific requirements will affect the selection of dewatering equipment and may affect the design of the purge and wastewater treatment systems. Any requirements on the quality, purity or composition of a solid waste byproduct should be specified to the equipment supplier.

Process makeup water (FGD systems) Potential sources of makeup water should be identified and the analysis and range of contaminants defined. The analysis can affect the selection of system materials as well as the ability to achieve a specified byproduct quality requirement. Various plant wastewater streams may be desirable to use as makeup water in a dry or wet FGD system, but the composition and quantity of each potential source should be reviewed and analyzed on a case-by-case basis.

Wastewater requirements Typically, a wet FGD system will require a portion of the filtrate water to be purged from the system to avoid excessive buildup of chlorides and/or fines. The purge flow rate is typically set to control to a maximum chloride concentration in the recycled slurry system which is also a function of the chloride concentration in the fuel. Higher design chloride levels will increase the corrosiveness of the slurry and more expensive materials will be needed for the absorber tower and outlet flue.

Performance guarantees Specific guarantees for environmental control systems are customized to fit the requirements and objectives of each project. If the environmental control system is provided as a separate, stand-alone system, performance guarantees must be tied to flue gas conditions that are specified at the system inlet.

Typical performance items guaranteed over the operating load range include:

1. emission levels for pollutants being controlled by the system being installed, and
2. gypsum quality (for wet FGD systems only).

Performance items guaranteed at a specific load condition include:

1. auxiliary power consumption,
2. reagent flow rate (limestone, lime, ammonia, other),
3. flue gas pressure drop across the system (if supply of induced draft or booster fans is not within the supplier's scope),
4. makeup water consumption rate, and
5. wastewater purge flow rate.

Project requirements

In addition to critical inputs required to size and select equipment, specifications should provide definition and direction for how the equipment supplier's work will integrate into the overall project.

Scope of work

Determining scope structure is a critical decision. The buyer may specify whether or not bids should be for a complete *turnkey* project, in which all components including erection are within the supplier's scope. Other approaches include *system island* scope and *equipment packages*.

Regardless of how the scope is structured, suppliers must understand the scope responsibility to ensure that the buyer's needs are met. If the project is executed without addressing ambiguous terminal points, scope can be missed and the project can face problems of schedule delay and unexpected costs. Scope definition is an evolutionary process that occurs during project development. Several types of documents are needed to define a project and ensure a common understanding of scope. The following should be included as part of a contract:

1. equipment listing and/or system descriptions,
2. division of work, including work scope provided by other parties,
3. terminal points listing,
4. process and instrumentation diagrams (P&IDs),
5. equipment arrangement drawing(s), and
6. definition of the extent of shop fabrication.

Degrees of scope breakdown structure There are a variety of general breakdown structures which are used in power plant steam generation and environmental system projects, each with its strengths, weaknesses, costs and risks. Key examples include:

1. Turnkey approach for the material supply and erection of a complete facility.
2. System island approach focusing on a major subsystem such as the boiler island, turbine island, and environmental equipment island, each including structural steel, architectural materials, process piping, auxiliary components, etc.
3. Equipment package approach focusing on smaller scopes such as individual components and related auxiliaries.

For larger scope approaches, operation and maintenance agreements or equity/financing arrangements may be included.

Considerations for scope breakdown The scope structure is dependent on the buyer's ability to manage the project and the selected commercial contracting structure. Different scope structures result in different impacts to the steam generating system. Listed below are a few items to consider:

1. If a turnkey project is pursued, some suppliers may be eliminated because of a lack of experience, inadequate financial backing to obtain required project security instruments, or their inability to support the magnitude of financial risk associated with large projects. As a result, the buyer may lose or disqualify potential suppliers who, otherwise, could provide good technology and services at a competitive price.
2. In a turnkey project, the selection of sub-suppliers may be affected by the turnkey contractor's evaluation process and work experience/relationships.
3. Physical interfaces must be considered to minimize the risk of missed scope relative to material, engineering, and construction. As the degree of scope breakdown increases, risks related to physical interfaces increase. Also, some scope is integral to equipment, such as the structural steel used to support a baghouse or dry scrubber. Piping and other equipment terminals need to be coordinated to properly meet in space (type, size and location) and so that the system is properly engineered from a stress analysis and performance perspective.
4. Environmental equipment sizing is dependent on the inlet gas conditions as well as the effects of potential variations in the operation of upstream systems (boiler, air heater, etc.). Proper margins should be considered to ensure that emissions limits are continuously met without adding overlapping layers of conservatism. If the equipment is oversized, a higher cost could be incurred unnecessarily. Equipment that is too small may under-perform or fail due to overloading caused by the effects of upstream systems. Purchasing a complete, integrated system from a single supplier allows the downstream systems to be designed using the supplier's experience and knowledge of the operating conditions of the upstream equipment. Purchasing as a complete package also eliminates the complications of performance risk responsibility that exist with a process interface point such as between the boiler and environmental equipment systems.
5. Performance interface risks also increase as scope is divided and performance guarantees are required for individual components. This risk can cause equipment within a particular system to be over-designed. It may also cause the operation of the facility to be less efficient because of margin added by suppliers to manage guarantee risk. For turnkey projects, overall unit performance, such as capacity, NPHR and emissions are critical and become contractual obligations. Other performance elements can be guaranteed separately as scope becomes further broken down to support the overall NPHR. The same philosophy could be applied to emissions. Guarantees for a piece of equipment may depend on the performance of other equipment. This can cause each supplier to assess their individual risk and apply appropriate margins. As a result, the overall system may be over-designed with margin applied on top of margin. Guarantee conditions would be based on the scope boundary and the more boundaries that exist, the more measurements are required to prove individual sub-system performance.
6. When a specification is prepared, consolidation of construction subcontracts should be considered. It may be more economical to consolidate electrical, civil, insulation and lagging, architectural, and related scope across the entire project rather than having multiple contractors responsible for the same work but in different areas. This will reduce interface risks and bring consistency to the project.

Site conditions and requirements There are numerous requirements specific to each project site and plant that should be communicated to the equipment supplier. If the supplier's work includes construction and extended balance of plant scope, site-specific requirements are especially critical. During a bidding process, it is often valuable for the equipment supplier and construction contractor to visit and investigate the site thoroughly to understand all site logistics.

Equipment arrangement The specification should clearly describe the site and include dimensioned plot plans showing the location of existing equipment, the expected location of new equipment and major locations for tie-in connections to existing plant services, among other items. The specification should also define any limitations or restrictions to be imposed on the equipment supplier. The supplier should be encouraged to arrange the equipment within these limitations for optimal overall cost.

Site access Rail, truck, barge and personnel access to the site should be defined. Sufficient lay down areas adjacent to the site generally reduce handling costs. However, provisions can often be made for off-site fabrication and extended modularization if space is limited and there is sufficient delivery access. It is important to state which existing equipment and facilities will remain operational or on-line during construction.

Site design criteria and applicable codes The specification should also include information on climate conditions and design codes that apply at the site, including:

1. seismic zone criteria,
2. wind load criteria,
3. snow load criteria,
4. range of ambient conditions for temperature and relative humidity,
5. soil design conditions, and
6. site altitude.

Also, the specification should identify if the new equipment will be located in an enclosed building or outdoors. Even in colder climates, some owners prefer to use as few building enclosures as possible.

Standardization requirements and design criteria It is important to define any plant standardization requirements that may have been developed to make operation and maintenance easier and more consistent throughout the plant. Such items could include common suppliers for valves, instrumentation, motors, etc. This can be done by providing a list of acceptable suppliers for certain equipment. However, caution should be taken not to over-specify equipment requirements. Different technologies can require different criteria and while clear performance expectations should be outlined, excessive detail may add unnecessary cost.

Project schedule The specification should accurately state project schedule requirements for engineering work and document submittals consistent with the scope and needs of multiple suppliers, equipment deliveries to site, construction and commissioning, startup and testing. The project schedule needs to be developed backwards from the required on-line date, allowing adequate time for equipment to be checked and commissioned after installation. Equipment deliveries should be scheduled to support the expected erection sequence, but also need to recognize lead-times required for engineering, procurement of raw materials, and fabrication. For projects involving the modification of or addition to existing equipment, the project construction schedule may need to be carefully coordinated around normally planned maintenance outages or lower demand periods. If the required outage span is of critical economic importance to the project, the incremental value should be defined to the bidders so that added construction costs for multiple shifts and peak labor can be evaluated and optimized. Typically, the value to the plant owner of having the equipment on-line sooner easily justifies the additional cost required to minimize an outage span. However, these tradeoffs need to be properly considered.

Commercial terms and conditions

A bid specification should identify the expected commercial terms that would apply to a resulting contract. Major clauses that are typically part of a contract include liability terms and limitations thereon, warranty, force majeure, contract completion dates, title and risk of loss, changes to the work, proprietary data, choice of law, cancellation, payment terms, and insurance. In many instances these are prepared and negotiated for specific projects to meet both supplier and buyer needs. In other cases, the United States (U.S.) Uniform Commercial Code, as adopted in the various states, or other laws, rules or regulations provide protection for the buyer and supplier in the absence of specific contract provisions.

Payment terms Matching payment terms to a supplier's cash flow needs minimizes project cost by reducing working capital requirements. *Milestone payments* provide a payment upon receipt of order, with additional payments following drawing submittal, major material orders, and completion of major component fabrication. *Progress payments* are made at given dates following project initiation.

Invariably, payment terms must balance two competing interests: the owner's desire to assure that the project will be timely completed in accordance with the contract, and the contractor's interest in assuring a positive cash flow. For this reason, many projects provide for some retained amount from each invoice, often 5%, which can be reduced or liquidated once the contract is completed. The contractor may be allowed to substitute a letter of credit or bond for the retention once performance guarantees are met.

Evaluation criteria

The design of steam generating units and environmental control systems involves complex tradeoffs between hardware options and operating costs. For the supplier to provide the best possible offering to meet the buyer's needs, it is important to include the general criteria and quantitative factors the buyer will use to evaluate competing proposals. The following is a list of potential key performance items for which quantitative evaluation factors should be provided, depending on project application:

1. boiler fuel efficiency,
2. main steam and reheat steam pressure drop,
3. auxiliary power consumption,
4. auxiliary steam flow,
5. reagent flow rate (ammonia, lime/limestone, etc.),
6. SCR catalyst life, and
7. flue gas pressure drop.

Other evaluation factors can include ash disposal costs, revenue from byproduct (e.g., gypsum) sale, and extended warranties, etc.

General bid evaluation

Once the specification has been prepared and proposals have been submitted, bids are evaluated. A consulting engineer may be used to help with the evaluation. Although the best bid is usually competitive, it is not always the lowest price. A thorough review must consider whether the supplier has offered the intended scope, and any exceptions to the specification must be evaluated. Bid evaluation generally includes a review of scope, operating ease, maintenance and operating costs, service, design and construction features, hardware, schedule, experience, commercial terms, and price.

Scope

Scope of supply typically differs between suppliers. It is important to verify what items are explicitly covered in each bid. Inconsistencies between bids can be resolved by discussions with the suppliers and by assessing a cost for discrepancies in the bid evaluation.

Operating ease

Design differences will appear among the product offerings. Seemingly minor design features can significantly alter costs. It can be valuable to have boiler operators evaluate the benefit(s) of such design features.

Maintenance costs

Designs that require minimal maintenance provide ongoing cost savings. Areas that previously demanded high maintenance, such as refractory and brickwork, have generally been eliminated from modern boiler designs. Current gas-tight water-cooled furnace construction has replaced the tube and tile design. However, some corner seals, hangers, supports and spacers may still be high maintenance items if not properly designed. Pulverizer designs can vary widely resulting in significantly different annual maintenance costs. These factors can be incorporated into the final evaluation discussed later.

Operating costs

Annual boiler operating costs can approach the initial unit investment. Therefore, fuel and auxiliary power costs should be quantitatively evaluated. Based on estimated fuel cost, discount rate and the predicted unit life, present values can be assigned to calculate annual fuel and power costs. Other operating costs include reagent usage for emissions control systems, wastewater flow rate (water treatment system cost), and solid waste disposal costs, among others.

Service organization

The quality and capability of a supplier's after-market service department are important. An experienced staff can troubleshoot problems quickly and minimize costly downtime. Factors that should be considered when evaluating supplier alternatives include the proximity of the supplier's office or representative to the plant, the number of available personnel, and the training and experience of the service personnel.

Design features

A thorough bid evaluation must review the equipment's design and construction features and determine if certain technologies favor one supplier. Although some designs may present a low initial cost, accompanying higher maintenance requirements may quickly offset these savings.

Areas that should be reviewed include, but are not limited to:

1. furnace design and arrangement to determine how the supplier has accommodated the specified fuel characteristics in the proposed design to ensure that the effects of slagging, fouling, erosion and corrosion are minimized, while achieving low NO_x, CO and unburned carbon performance;
2. low NO_x burners to determine if they have been effective on a similar installation;
3. pulverizer design regarding maintenance requirements and degradation between overhauls; and
4. superheater supports.

If special design requirements were specified, the evaluator needs to ensure that each supplier accommodated the requirement. For example, if very rapid load change requirements are specified for a supercritical boiler, additional hardware may be required for the steam generator to comply.

Construction features

If the scope of work is limited to material supply only, the degree of shop fabrication and the extent of design for constructability offered by different bidders can vary and have a significant effect on the material installation cost. The degree of shop fabrication can range from supply of many small components to large prefabricated modules. Even seemingly minor differences can result in significantly different field labor requirements that need to be evaluated. The degree of shop fabrication on a given project is typically driven by transportation logistics and site access limitations. Plants located with good access to water transport can benefit from increased shop fabrication efficiencies, reduced construction risks, and reduced peak labor requirements at the site.

Schedule

As-bid schedules should be compared to one another and to the specification. A shorter overall project schedule can reduce project financing costs or increase revenue through early completion and startup. However, early delivery of some equipment does not necessarily reduce the overall project span if that equipment does not fall on the project's critical path. If the scope being proposed does bring the project economic benefit from a shorter span, then differences can be subject to a quantitative evaluation credit or penalty. However, if that credit is to be realized, sufficient commercial motivation (such as a bonus for early completion or liquidated damages for late performance) should be established in the contract to ensure that a bidder offers a realistic commitment.

Project management and project team

An equipment supplier's project manager and key project personnel can greatly influence project execution performance. It is the responsibility of this project team to ensure that the specified product is provided in accordance with the contract schedule and within project cost constraints. Sub-suppliers and vendors must be managed effectively, the team needs to respond to buyer changes in scope and design, and communication with the buyer and other project entities must be maintained. The ability of the supplier's personnel to manage these critical functions can have a significant effect on project execution. It is in the buyer's best interest to review and evaluate the experience and capability of the personnel that a supplier will assign to a project, especially if the project is large in scope, complex, or has extraordinary requirements.

Experience

Demonstrated success in executing projects using similar equipment with similar scope and complexity is an important factor to ensuring success of using a selected supplier. If the technology being selected has limited commercial experience, the buyer must depend on the supplier's experience with similar technologies and ability to support performance guarantees. The buyer should contact organizations that have purchased similar units. Visits to operating installations can provide important insights.

Terms and price

Proposed payment terms and schedule should also be evaluated carefully. Differences in payment terms can represent a significant value to the project. However, deferred payments can also be expensive. Commercial terms and conditions essentially define the level of risk that a supplier will share with the buyer. The lowest cost can be achieved when the buyer and supplier share a reasonable amount of the project risk. Also, performance guarantees rely on a supplier's ability to back them up. This ability should be evaluated before awarding a contract.

Commercial structure

Another major decision is the commercial contracting structure. Traditionally, the developer of a capital equipment project will issue a tender for major equipment packages in which the owner or developer will manage the risks of integrating these packages.

Some projects such as those financed off-balance sheet (or *project financed*) need to have single-point guarantees of plant performance and construction schedule to address financial risks. An owner or developer may therefore need to employ a general contractor or possibly a consortium of companies that will manage the project integration risk and provide overall guarantees.

Another concept is to establish a strategic project alliance or partnership with one or several major equipment or system suppliers to jointly develop, manage, and share the risk of project integration and execution.

Traditional commercial structures

A company that is planning a major capital equipment project, such as a new coal-fired power plant, must work through many of the following development process steps to obtain corporate management approvals, project financing, local community support and any required government permits:

1. Define basic needs, objectives and goals.
2. Explore critical logistics and feasibility.
3. Identify permitting requirements.
4. Assess potential alternate approaches or technologies.
5. Optimize design requirements to meet the goals and objectives.
6. Evolve a plan and timetable for development and implementation.
7. Secure capital financing.
8. Identify internal and external resources needed to develop and execute the project.
9. Develop specifications for the equipment and system packages that will be purchased.
10. Evaluate bids and select supplier(s).

Traditionally, the project developer will work through these steps using in-house resources or by hiring a consulting engineering company to support the project feasibility and development activities. The equipment supplier can also be a key resource in a number of these areas. However, because the traditional procurement philosophy depends on maintaining competitive pressure on the qualified equipment suppliers, the opportunity to involve an equipment supplier in the developmental stages can be limited.

Traditional procurement methods require the equipment supplier to bid competitively to a fixed specification after most of these steps have been completed. The success of this approach depends on the expertise, oversight, direction and integration capabilities of the owner and its hired consultant to develop and define the specific scope and design requirements.

Selection of equipment suppliers is primarily based on low evaluated bid price. With emphasis on low capital cost and the need to compare technically equivalent offers, long-term value relating to operating costs, maintenance, and availability can be lost in the process. In addition, a low price may motivate the selected supplier to work in a manner that may not be in the interest of the owner during the execution of the contract in an attempt to avoid losses on a project with low profitability.

An owner also can find that project costs increase with changes in scope or design that evolve during the project. Changes are more costly when they occur later in the process as they will affect work already completed, and may impact the work and scope of other contractors/suppliers.

Collaborative partnerships/alliances

Recently, collaborative alliance-type business structures have been used to integrate the equipment supplier into the project development process. Such an approach allows the equipment supplier to contribute resources and expertise when the project is taking shape and project direction is still flexible. The equipment supplier can function as a team member or partner, rather than simply a service provider addressing a fixed specification.

In such an alliance, all parties can contribute to the final solution and can benefit from the result. This approach involves the major supplier(s) in the development phase of the project, a time when specific product knowledge can best be applied. This establishes project design requirements that recognize plant operational needs, establishes appropriate levels of redundancy, and evaluates the layout of equipment to best facilitate erection and maintenance access. The ideal time to evaluate and optimize a project is during these early stages.

The commercial framework of an effective alliance should include a system of risk and reward incentives that will keep all partners motivated to work towards the common goals of project cost, project schedule execution, and equipment performance. Partners should be motivated to earn their profit by performing at or better than a set of measurable, mutually beneficial goals established for the project.

Benefits of a project alliance Large complex projects are well suited for collaborative alliances. Involving an experienced equipment supplier in the development of the plant design, definition of scope, and estimation of capital costs can bring significant value to the process and can avoid fatal flaws of assumptions that can be made during the initial conceptual stages of a project.

Once an alliance is established, the owner and the supplier can develop a relationship that capitalizes on the strengths of each organization as they work through the issues and move forward to creatively address challenges of operations, maintenance and project execution. Some of the benefits that can be realized from such an alliance program structure include the following:

1. Many aspects of the bid specification development, proposal development, proposal evaluation, supplier selection and contract negotiation processes are eliminated, saving time and money.
2. The project team can employ a cooperative approach to focus on optimizing the owner's true costs. Decision-making can focus on the best long-term value rather than on the lowest near-term capital cost.
3. The project specification can be prepared interactively, allowing innovative solutions to be identified, more thorough project planning to be implemented, and false starts to be minimized.
4. The legal and financial framework is typically established at the beginning of the alliance and usually requires little adjustment during the life of the alliance.
5. Project resource assignment and loading are planned to achieve the best utilization of personnel from all participant organizations, eliminating redundancy and extra costs.
6. Open communication can be used in all aspects of planning and job tracking, which builds trust and minimizes conflict.
7. For multiple unit or multiple plant projects, lessons learned on earlier project phases can benefit later stages of the project. Implementing lessons learned is a key factor to achieving economic benefit over the duration of the alliance.

Factors to consider When considering the alliance option, the development process is inherently different. The equipment supplier needs to be selected at a much earlier point in the process. Because this selection can not be based on competitively bid prices, other factors need to be evaluated to determine the relative degree of confidence that the candidate partner will contribute. Such factors include:

1. experience and success record for the technology required,
2. experience with other projects performed under an alliance agreement,
3. experience performing a project of similar scope, design and complexity,
4. qualifications, experience and capabilities of the project team that will be assigned to work on the project,
5. capabilities to manage cost and project risk to competitive levels through a combination of materials sourcing capability, systems project management and control, labor charge rates, and markups (overhead and profit) that would be applied against actual costs incurred,
6. degree that the partner will share and limit project cost overruns relative to a mutually developed target cost,
7. degree that the partner will share the financial risk associated with the operating performance of the equipment and systems to be supplied, and
8. degree that the partner will share the risk of project schedule performance to meet targeted commercial operation.

Power system economics

The relative economics of the options must be compared as part of the evaluation of alternate power systems or different product offerings. Unfortunately, this is not a straightforward, simple task. Large projects involve a complex series of time dependent cash flows which usually include an initial near-term investment in plant and equipment, and a long-term, uneven series of operating costs, maintenance costs and revenues from the sale of power or end product. This is further complicated by the time value of money and the effects of inflation. Money today is generally worth more than the same amount of money a year from now because of: 1) the potential interest which could be earned during the next year, and 2) the erosion in the future buying power due to inflation. The following provides a very brief overview of several methodologies used to place the total evaluated costs of different projects on the same financial basis so they can be directly compared. More comprehensive discussions of such evaluations are provided in References 3 to 6. This discussion is broken down into the following subsections:

Constant versus current cost basis
Definitions
System cost evaluation
Evaluation parameters
Methods for evaluating operating costs
Other considerations
Examples

Constant versus current cost basis

Analyses are conducted on either a current or constant cost basis. The current cost basis presents costs as they would be expected to appear on a company balance sheet in the future while the constant cost basis does not include inflation.

The selection of the constant or current cost approach usually does not change the economic choice between similar projects. For nearer term projects, such as plant upgrades or the addition of small components, current costs are frequently used because of near-term impact. For longer term projects such as new plants, constant costs are frequently used as at least one of the evaluation tools so that the effect of inflation does not distort real cost trends. In either case, it is important to use the same basis for evaluating all alternatives. The examples provided below are evaluated on a constant cost basis.

Definitions

The following parameters provide the basis for the economic evaluation:

Discount rate (k) This is the cost of money for a power system owner and includes the weighted cost of capital for each class of debt and equity. Establishing the appropriate cost of capital or discount rate is increasingly complex. References 4 and 7 provide detailed discussion on the sources of funds and the evaluation of their costs. A simplified discount rate (k_c) estimate can be obtained from the following equation which takes into account that interest payments on debt are generally tax deductible expenses:

$$k_c = (FD)\, k_d\,(1 - TR) + (FE)\, k_e \qquad (1)$$

where

FD = fraction of capital consisting of debt
FE = fraction of capital consisting of equity
k_d = annual debt interest rate
k_e = annual stockholder return
TR = fractional income tax rate

The annual debt interest rate and stockholder return reflect the anticipated impact of inflation and are used in current cost basis analyses. If a constant cost based analysis is desired, the *current cost* discount rate (k_c) can be related to the *constant cost* discount rate (k_f) through the escalation or inflation rate (e) in the following equation:

$$k_f = \left[(1 + k_c) / (1 + e)\right] - 1 \qquad (2)$$

The escalation rate may have to be treated as a time dependent variable if it is expected to change significantly over the period of analysis.

Fixed charges Also referred to as carrying charges, these are the fixed annual costs associated with the initial capital investment and are generally incurred whether or not the plant is operated. The components include the return on investment, capital recovery (or depreciation), property taxes and insurance as well as federal and state taxes. Over the life of a new plant the annual revenue requirement declines because of income tax effects and depreciation of the initial capital investment. However, a weighted average or levelized value is frequently used for evaluations. To obtain a levelized value the actual annual fixed charges are discounted to the present and then levelized using the equations discussed below. Annual fixed charge rates vary from 10 to 21% of the initial total capital investment.

Interest during construction This refers to the cost of money during construction prior to initial plant operation. It is not included if all project costs are incurred in a single year.[3]

Capacity factor This is the total power produced in a given period (typically one year) divided by the product of the net unit capacity and the given time period. It accounts for part load operation and outage times. The capacity factor varies over the plant life, typically climbing to a maximum in the first few years, remaining relatively constant for a period of several years and then falling as the unit begins to age.

Load factor This is the fraction of the full load or maximum continuous rating (MCR) at which the steam generating system is normally expected to operate for extended periods of time.

Net plant heat rate (NPHR) This is the total fuel heat input expressed in Btu divided by the net busbar power leaving the power plant expressed in kWh. Plant total energy efficiency (E) can be obtained from the following relationship:

$$E = \frac{3412 \text{ Btu/kWh}}{NPHR} \times 100\% \qquad (3)$$

The $NPHR$ typically varies with plant load.

Demand and auxiliary power charges These are the charges placed against the internal use of energy for auxiliary equipment. Separate rates are established by electricity and steam usage. The size of the charges depends upon the cost of fuel and other operating costs.

System cost evaluation

In the economic evaluation of steam generating systems, it is important to include all appropriate costs and expenses. Table 2 provides a list of typical capital, variable and fixed operating and maintenance costs which could be associated with a steam generating system installation. Capital costs basically include all expenditures for the purchase, installation and startup of the steam generating system and its auxiliaries. Care must be taken to ensure that different product offerings contain the same scope of work. To obtain the total plant investment, interest during construction, initial inventory items (such as fuel) and contingency are added. Variable operating costs include delivered fuel cost, other consumable material purchases, power and steam demand charges, and waste disposal or byproduct use.

Fixed operating and maintenance costs cover operating expenses which are likely to be incurred whether or not the plant is in operation. These change from year to year and generally increase as the plant ages. They also tend to vary with the unit load pattern, plant size, and primary fuel.

Table 2
Typical Steam Generating System Cost Items

Capital Costs
Boiler (nominal scope including burner system and air heater)
Fans and drives
Appropriate flues and ducts
Coal crushers
Pulverizers
Pyrite removal (pulverized coal units)
Dust collectors
Ash handling
Steam coil air heaters
Steam piping
Injection water piping
Startup pump and drive
Feedwater heaters and piping (if used)
Feedwater treatment
Selective catalytic reduction (SCR) system
Flue gas desulfurization (FGD) system (dry or wet)
Baghouse or electrostatic precipitator (ESP)
Wet ESP (if required)
Reagent storage and preparation systems
Reagent slurry pumps and piping
Gypsum dewatering system (if wet FGD system is used)
Tanks and sumps
Instrumentation
Controls
Foundations and electrical connections
Structural steel
Building
Platforms
Freight
Sales taxes
Site preparation
Erection/insulation/painting
Startup
Performance tests
Miscellaneous engineering and supervision services
Variable Operating Costs (Credits)
Primary fuel cost
Electrical and steam power demand:
Fan power
Pulverizer power
Recirculating pump power
Boiler feedpump power
Crusher power
Makeup water and chemicals
Wastewater treatment
Steam demand:
Sootblowers
Flue gas reheat
Air heating
Compressed air demand
Ash removal and disposal
Auxiliary fuel for low load stability
Other consumables (ammonia, limestone, etc.)
Credits for any byproduct sales
Fixed Operating and Maintenance Costs
Plant operators
Maintenance labor
Maintenance material
Chemical cleaning
Replacement material
Supervision
Other overhead items

In the financial evaluations, selected penalties or operating costs can be associated with superheater, reheat superheater and economizer pressure drops (steam-water side and gas side). Depending upon the buyer's perspective, these can be explicitly listed as separate operating costs or they may be indirectly accounted for as part of the general electrical and steam power requirements listed above. Reheat superheater steam side pressure loss is of particular interest because of its strong connection with thermal cycle efficiency.

Evaluation parameters

Present value With an economic life stretching many decades, power plants experience operating expenses over extended periods of time. It is necessary to account for the time value of money in evaluating projects which have differing initial capital investments and future expenses over several years. This is usually accomplished by using a *present value* (PV) calculation for each annual cash flow. The PV of a *future value* or payment (FV) in year n is defined as:

$$PV = FV(1+k)^{-n} = FV(PVF) \tag{4}$$

where

$PVF = (1 + k)^{-n}$ = present value factor
k = interest or discount rate
n = number of years or periods at rate k

An alternate use of the PVF is to evaluate the future value of a present value:

$$FV = PV/PVF \tag{5}$$

Where a series of equal future annual payments, A, will be made for n years, the PV is defined as:

$$PV = A\left(\frac{(1+k)^n - 1}{k(1+k)^n}\right) = A\,(SPVF) \tag{6}$$

where

$$SPVF = \text{uniform annual } \textit{series present value factor}$$
$$= \frac{(1+k)^n - 1}{k(1+k)^n}$$

As with the single payment present value factor, $SPVF$ can be used to convert a PV into a series of future equal payments for n years:

$$A = PV/SPVF \tag{7}$$

For illustration purposes, Table 3 provides values of PVF and $SPVF$ for a discount rate of 6% from 1 to 30 years.

Levelization In evaluating various power system options, it is frequently desirable to compare weighted average annual values of future costs such as fuel instead of the lumped sum present values. Such a parameter would be a levelized cost evaluated using the following equation:

$$\text{Levelized value} = \frac{\sum_{j=1}^{n} (FV_j)(PVF_j)}{SPVF} \tag{8}$$

where

PVF_j = present value factor for year j and rate k
FV_j = future annual payment in year j
$SPVF$ = uniform annual series present value factor in year n for discount rate k
j = individual years up to n years
k = annual discount rate

For the special case where the annual payment escalates or is inflated from an initial value of S_1 with a constant rate, e, the future value of the annual payment in year j is:

$$S_j = (1+e)^j S_1 \tag{9}$$

The resulting levelization equation then becomes:

Levelized value of S

$$= S_1 \left(\frac{1 - \left[(1+e)/(1+k) \right]^n}{k - e} \right) / \; SPVF \tag{10}$$

Methods for evaluating operating costs

There are several approaches to the evaluation of operating costs. The revenue requirements method and the capitalized cost method are reviewed here and examples are given. While each method used involves a large number of inherent assumptions, these are the dominant methods used to evaluate steam supply and emissions control equipment. See References 3 to 7 for more discussion.

Revenue requirements method This method determines the minimum gross receipts which are required to make each alternative pay its way. The alternative that requires the lowest receipt of revenues is the economic choice as it will permit profitable operation with the smallest revenues. This method takes into account the present value of all future expenditures and costs.

The capital (first cost) expenditure which is justified in order to put two alternatives on an equal basis is less than the difference in future revenue requirements. The future revenue requirements are converted to current dollars by multiplying each future expenditure by the PVF for each year in which the expenditure will occur. By summing the present value of the future revenue requirements, a total present value is determined. To convert this sum into a justifiable capital expense, it must be levelized, or converted to an equivalent equal annual cost. This is done by dividing the total present value by the uniform annual series present value factor (SPVF) (which is 13.765 for a 30 year period and 6% interest rate, used in the examples that follow). A tabulation of present value factors (PVF) for 6% interest rate is shown in Table 3. Present value factors for other interest rates and evaluation periods can be evaluated from Equations 4 and 6. After converting the total into an equal annual cost, the justifiable capital expenditure is determined by dividing the annual cost by the fixed charge rate.

Capitalized cost method This method assumes that the equipment being operated has an infinite lifetime and that the costs of operating the unit will continue indefinitely. It is a simpler evaluation procedure to use than the revenue requirements method described above, but is not as exact or flexible in accounting for future changes in such items as fuel and maintenance costs which will change during the life of the unit.

The amount calculated using this method is the justifiable capital expense which can be spent to purchase the additional equipment and still give an operating advantage.

For example, if it is found that additional air heater surface will achieve higher unit efficiency, i.e., fuel savings over the lifetime of the unit, the additional air heater surface is justified if its capitalized cost is less than the cost of the fuel which can be saved over the remaining unit life net of any tax impact.

Other considerations

Risk All new projects involve some level of risk or uncertainty. Short-term projects using established technology typically are considered low risk while long-term projects using new, untried technology are generally considered higher risk. A number of approaches are used to account for the risk or uncertainty during project evaluations. One approach is to establish special guarantee provisions in the specifications and contract. This will typically increase the price from the suppliers since they assume more of the risk. A second approach is to increase the buyer's in-

Table 3
6% Compound Interest Factors

Year	Single Payment Present Value Factor (PVF)	Uniform Annual Series Present Value Factor (SPVF)	Year	Single Payment Present Value Factor (PVF)	Uniform Annual Series Present Value Factor (SPVF)
1	0.9434	0.943	16	0.3936	10.106
2	0.8900	1.833	17	0.3714	10.477
3	0.8396	2.673	18	0.3503	10.828
4	0.7921	3.465	19	0.3305	11.158
5	0.7473	4.212	20	0.3118	11.470
6	0.7050	4.917	21	0.2942	11.764
7	0.6651	5.582	22	0.2775	12.042
8	0.6274	6.210	23	0.2618	12.303
9	0.5919	6.802	24	0.2470	12.550
10	0.5584	7.360	25	0.2330	12.783
11	0.5268	7.887	26	0.2198	13.003
12	0.4970	8.384	27	0.2074	13.211
13	0.4688	8.853	28	0.1956	13.406
14	0.4423	9.295	29	0.1846	13.591
15	0.4173	9.712	30	0.1741	13.765

ternal contingency account in the project cost evaluation (see Reference 3). A third approach involves making adjustments to the discount rate to accommodate the level of risk or uncertainty with higher rates applied to riskier projects (see Reference 7). One or more of these options, as well as other approaches, may be used in the ranking of project options depending upon the buyer's internal needs.

Unequal periods When comparing project alternatives of equal life, the various evaluation techniques generally provide similar rankings. However, when project lives are significantly different, additional care must be exercised in performing the analysis and interpreting the results. As an example, the costs of additional future equipment replacement may need to be considered when short-term and long-term repairs to a particular equipment problem are considered. In addition, when interpreting the evaluation results, each financial model uses different assumptions about capital reinvestment which could result in different project rankings.

Examples

Four examples follow which illustrate the use of the evaluation methodologies. The first two examples compare the use of the revenue requirement method and the capitalization method in evaluating the choice between a subcritical pressure 2400 psi utility steam cycle and a supercritical pressure 3500 psi utility steam cycle assuming that a new power plant is needed. Example 3 provides a comparison between two industrial boiler bids where one bidder has a higher price but higher boiler efficiency. Example 4 evaluates the lifetime auxiliary power difference between two industrial boiler bids. All of these examples are evaluated on a constant U.S. dollar basis, i.e., excluding inflation.

References

1. Aschner, F.S., *Planning Fundamentals of Thermal Power Plants*, John Wiley & Sons, New York, New York, April, 1978.

2. Li, K.W., and Priddy, A.P., *Power Plant System Design*, Wiley Text Books, New York, New York, March, 1985.

3. Electric Power Research Institute, *Technical Assessment Guide (TAG): Electrical Supply, 1989*, Vol. 1, Revision 6, Report EPRI P-6587-L, Palo Alto, California, September, 1989.

4. Brigham, E.F., and Ehrhardt, M.C., *Financial Management: Theory and Practice*, Eleventh Ed., South-Western College Publishing, Mason, Ohio, 2004.

5. Grant, E.L., and Ireson, W.G., Leavenworth, R.S., *Principles of Engineering Economy*, Eighth Ed., Ronald Press, New York, New York, February, 1990.

6. Leung, P., and Durning, R.F., "Power system economics: On selection of economic alternatives," *Journal of Engineering for Power*, Vol. 100, pp. 333-346, April, 1978.

7. Copeland, T.E., Weston, J.F., and Shastri, K., *Financial Theory and Corporate Policy*, Fourth Ed., Pearson Addison-Wesley, Reading, Massachusetts, December 31, 2003.

Example 1
Revenue Requirements Method Evaluation of 2400 psi and 3500 psi Steam Cycles

A) Problem: Determine the economic choice between a 2400 psi cycle and a 3500 psi cycle with 1000F superheat (SH) and reheat superheat (RH) temperatures using a constant cost basis.

B) Given:

	Unit 1	Unit 2
1. Throttle pressure, psi	2,400	3,500
2. Temperature, SH/RH,F	1,000/1,000	1,000/1,000
3. Plant size, kW	600,000	600,000
4. Net plant heat rate (NPHR), Btu/kWh: 100% Load	9,000	8,860
80% Load	9,000	8,860
60% Load	9,180	9,040
35% Load	10,080	9,920
5. Fuel cost, $/million Btu	$2.00	$2.00
6. Evaluation period, yrs	30	30
7. Interest rate or discount rate, %	6	6
8. Maintenance costs	Same	Same
9. Availability	Same	Same
10. Operating costs (excluding fuel)	Same	Same
11. Fixed charge rate, % (Interest, taxes, insurance, debt return)	14.0	14.0

12. Plant capacity factors (Units 1 and 2)

Years	1 to 10	11 to 20	21 to 30
100% Load	44%	23%	6%
80% Load	28%	23%	23%
60% Load	13%	34%	34%
35% Load	6%	11%	23%
10% Load	9%	9%	14%
	100%	100%	100%
Average capacity factor for period	76.30%	65.65%	52.85%
Average capacity factor for lifetime		64.93%	

C) Solution:

1. Cost of fuel consumed per year — 2400 psi unit
(h/yr) x capacity factor x kW load x Btu/kWh = Btu/yr
(Btu/yr) x ($/Btu) = $/yr

(a) Years 1 to 10 — Btu/yr

8,760 x 0.44 x 1.00 x 600,000 x 9,000 = 20,814 x 10^9
8,760 x 0.28 x 0.80 x 600,000 x 9,000 = 10,596 x 10^9
8,760 x 0.13 x 0.60 x 600,000 x 9,180 = 3,764 x 10^9
8,760 x 0.06 x 0.35 x 600,000 x 10,080 = 1,113 x 10^9
36,287 x 10^9

36,287 x 10^9 x $2.00/$10^6$ = $72,574,000/yr

(b) Years 11 to 20

8,760 x 0.23 x 1.00 x 600,000 x 9,000 = 10,880 x 10^9
8,760 x 0.23 x 0.80 x 600,000 x 9,000 = 8,704 x 10^9
8,760 x 0.34 x 0.60 x 600,000 x 9,180 = 9,843 x 10^9
8,760 x 0.11 x 0.35 x 600,000 x 10,080 = 2,040 x 10^9
31,467 x 10^9

31,467 x 10^9 x $2.00/$10^6$ = $62,934,000/yr

(c) Years 21 to 30

8,760 x 0.06 x 1.00 x 600,000 x 9,000 = 2,838 x 10^9
8,760 x 0.23 x 0.80 x 600,000 x 9,000 = 8,704 x 10^9
8,760 x 0.34 x 0.60 x 600,000 x 9,180 = 9,843 x 10^9
8,760 x 0.23 x 0.35 x 600,000 x 10,080 = 4,265 x 10^9
25,650 x 10^9

25,650 x 10^9 x $2.00/$10^6$ = $51,300,000/yr

2. Present value of fuel costs — 2400 psi unit
Fuel costs for period x PVF for period = PV of fuel costs for period

(a) Years 1 to 10
$72,574,000/yr x 7.36 = $534,145,000
(b) Years 11 to 20
$62,934,000/yr x (11.47 – 7.36) = $258,659,000
(c) Years 21 to 30
$51,300,000/yr x (13.76 – 11.47) = $117,477,000
$910,281,000

3. Cost of fuel consumed per year — 3500 psi unit
(h/yr) x capacity factor x kW load x Btu/kWh = Btu/yr
(Btu/yr) x ($/Btu) = $/yr

(a) Years 1 to 10 — Btu/yr

8,760 x 0.44 x 1.00 x 600,000 x 8,860 = 20,490 x 10^9
8,760 x 0.28 x 0.80 x 600,000 x 8,860 = 10,431 x 10^9
8,760 x 0.13 x 0.60 x 600,000 x 9,040 = 3,706 x 10^9
8,760 x 0.06 x 0.35 x 600,000 x 9,920 = 1,095 x 10^9
35,722 x 10^9

35,722 x 10^9 x $2.00/$10^6$ = $71,444,000/yr

(b) Years 11 to 20

8,760 x 0.23 x 1.00 x 600,000 x 8,860 = 10,711 x 10^9
8,760 x 0.23 x 0.80 x 600,000 x 8,860 = 8,569 x 10^9
8,760 x 0.34 x 0.60 x 600,000 x 9,040 = 9,693 x 10^9
8,760 x 0.11 x 0.35 x 600,000 x 9,920 = 2,007 x 10^9
30,980 x 10^9

30,980 x 10^9 x $2.00/$10^6$ = $61,960,000/yr

(c) Years 21 to 30

8,760 x 0.06 x 1.00 x 600,000 x 8,860 = 2,794 x 10^9
8,760 x 0.23 x 0.80 x 600,000 x 8,860 = 8,569 x 10^9
8,760 x 0.34 x 0.60 x 600,000 x 9,040 = 9,693 x 10^9
8,760 x 0.23 x 0.35 x 600,000 x 9,920 = 4,197 x 10^9
25,253 x 10^9

25,253 x 10^9 x $2.00/$10^6$ = $50,506,000/yr

4. PV of fuel costs — 3500 psi unit
(a) Years 1 to 10
$71,444,000/yr x 7.36 = $525,828,000
(b) Years 11 to 20
$61,960,000/yr x (11.47 – 7.36) = $254,656,000
(c) Years 21 to 30
$50,506,000/yr — (13.76 – 11.47) = $115,659,000
$896,143,000

5. Saving in fuel costs — 3500 psi versus 2400 psi
PV of fuel costs
2400 psi unit $910,281,000
3500 psi unit $896,143,000
PV of savings $ 14,138,000

$14,138,000 / 600,000 kW = $23.56/kW

6. Justifiable capital expenditure — 3500 psi versus 2400 psi

Justifiable additional expenditure =

$$\frac{\text{PV of fuel cost saving}}{\text{Fixed charge rate x PVF}}$$

PVF for 30 years at 6% interest = 13.76

Fixed charge rate = 14% = 0.14

Justifiable additional expenditure =

$$\frac{\$14{,}138{,}000}{0.14 \times 13.76} = \$7{,}339{,}000$$

Justifiable expenditure/kW =

$7,339,000 / 600,000 kW = $12.23/kW

D) Result:
1. In terms of today's worth, the 3500 psi cycle will save $14,138,000 or 23.56/kW in fuel costs over its life as compared with the 2400 psi cycle.
2. The saving will justify the expenditure of $7,339,000 or $12.23/kW more for the 3500 psi cycle than for the 2400 psi cycle, all other factors being equal.

Essentially any auxiliary power usage, difference in boiler efficiency, or difference in net plant heat rate can be treated in the same manner as this example. This method basically converts the differences between designs into fuel costs. The fuel costs, which account for future load factor, cycle heat rates and future fuel costs, are converted into present day dollars using present value mathematics for the projected interest rate over the life of the unit. The justified capital expenditure can then be calculated.

Example 2
Capitalized Cost Method Evaluation of 2400 psi and 3500 psi Steam Cycles

A) Problem: Compare the economics of a 2400 psi cycle and a 3500 psi cycle using a constant cost basis.

B) Given: Same information as Example 1.

C) Solution:

1. Average NPHR — 2400 psi cycle — Btu/kWh

100% Load:	(0.44 + 0.23 + 0.06) x	9,000 =	6,570
80% Load:	(0.28 + 0.23 + 0.23) x	9,000 =	6,660
60% Load:	(0.13 + 0.34 + 0.34) x	9,180 =	7,436
35% Load:	(0.06 + 0.11 + 0.23) x	10,080 =	4,032
			24,698

Average NPHR = 24,698/2.68* = 9,216

* This is the sum of all capacity factors in Item B12, Example 1, except the factors for zero load.

2. Average NPHR — 3500 psi cycle — Btu/kWh

100% Load:	(0.44 + 0.23 + 0.06) x	8,860 =	6,468
80% Load:	(0.28 + 0.23 + 0.23) x	8,860 =	6,556
60% Load:	(0.13 + 0.34 + 0.34) x	9,040 =	7,322
35% Load:	(0.06 + 0.11 + 0.23) x	9,920 =	3,968
			24,314

Average NPHR = 24,314/2.68 = 9,072

3. Capitalized value of heat rate difference

Difference = 9,216 – 9,072 = 144 Btu/kWh

Average capacity factor for lifetime = 64.93% = 0.6493

Capitalized value =

$$\frac{\text{NPHR x kWh generation / yr x Fuel cost}}{\text{Fixed charge rate}}$$

kWh generation/yr = Capacity factor x Capacity x h/yr

Capitalized value =

$$\frac{144 \times 0.6493 \times 600{,}000 \times 8{,}760 \times \$2.00 / 10^6 \text{ Btu}}{0.14}$$

= \$7,020,000

\$7,020,000 / 600,000 kW = \$11.70/kW

D) Result:
For these conditions the expenditure of up to \$7,020,000 or \$11.70/kW is economically justified for a 3500 psi unit versus a 2400 psi unit. (Note that this is quite close to the \$12.23/kW answer reached by the revenue requirements method in Example 1.)

The capitalized cost method of evaluation can be adapted to evaluate auxiliary power, boiler efficiency, or other operating costs by converting them to fuel cost.

Example 3
Comparing Industrial Boiler Bids with Differing Boiler Efficiencies and Capital Cost

A) Problem: Determine which of two boiler bids provides the most economical system offering.

B) Given:
1. Boiler produces 600,000 lb/h superheated steam at 1550 psi and 955F (H_g = 1460 Btu/lb from Mollier diagram, Chapter 2)
2. Boiler feedwater is 400F (H_f = 375 Btu/lb, Chapter 2)
3. Fuel cost = \$5.00/10^6 Btu (gas)
4. Load factor (average annual fraction of MCR) = 0.9
5. Use factor (fraction of in service time) = 0.9
6. Discount rate = 10%
7. Constant dollar basis
8. Expected unit life = 30 years
9. SPVF = 9.43 (30 years; 10% discount rate)
10. Bid No. 1 guarantees 86.5% efficiency with a price of \$10,400,000
11. Bid No. 2 guarantees 87.5% efficiency with a price of \$11,200,000
12. Income tax rate = 38% (composite)
13. Depreciation tax factor = 0.17

C) Solution:
Equations:

Unit input = (steam flow rate) (ΔH) (load factor)/efficiency

Annual fuel cost = (per Btu cost) (annual unit output) (use factor)

Net PV of fuel savings = Δ fuel cost x SPVF

For bid No. 1:

Unit input = (600,000 lb/h) (1460 – 375 Btu/lb) (0.9) / 0.865
= 677.3 x 10^6 Btu/h

Annual fuel cost = (\$5.00/10^6 Btu) (677.3 x 10^6 Btu/h) x (8760 h/yr) (0.9) = \$26,701,000/yr

For bid No. 2:

Unit input = (600,000 lb/h) (1460 – 375 Btu/lb) (0.9) / 0.875
= 669.6 x 10^6 Btu/h

Annual fuel cost = (\$5.00/10^6 Btu) (669.6 x 10^6 Btu/h) x (8760 h/yr) (0.9) = \$26,396,000/yr

Difference between bids (savings before tax) =
\$26,701,000 – \$26,396,000 = \$305,000/yr

Corporate income tax on savings = \$305,000 x 0.38
= \$115,900/yr

After tax savings = \$305,000 – \$115,900 = \$189,100/yr
Net PV of savings = \$189,100/yr x 9.43 = \$1,783,200

Difference in price = \$11,200,000 – \$10,400,000
= \$800,000

Tax savings on depreciation= Δ price x depreciation factor
= \$800,000 x 0.17 = \$136,000
(dependent upon government depreciation schedule and discount rate)

Difference in price net of tax savings = \$800,000 – \$136,000
= \$664,000

D) Result:
Because the net present value of the energy savings (\$1,783,200) is greater than the net price for the more efficient unit after depreciation tax savings (\$664,000), bid No. 2 provides the more economical offering.

Example 4
Evaluating the Cost Impact of Different Auxiliary Power Requirements of Two Industrial Boiler Bids

A) Problem: Determine the economic impact of differing auxiliary power requirements of two boiler bids.

B) Given:

1. Power cost = $0.025/kWh
2. Load factor = 0.9
3. Use factor = 0.9
4. Auxiliary power requirements

Component	Bid No. 1 kW at MCR	Bid No. 2 kW at MCR
Pulverizer	430	320
Coal feeder	3	3
Primary air fans	200	175
Forced draft fans	450	425
Auxiliary fans and blowers	30	30
Regenerative air heater	10	10
Total maximum auxiliary power	1,123	963

5. Assume the same discount rate, unit life, and SPVF as in Example 3.

C) Solution:
Total maximum auxiliary power corrected for capacity and load factors:

Bid No. 1 = (1,123 kW) (0.9) (0.9) = 910 kW
Bid No. 2 = (963 kW) (0.9) (0.9) = 780 kW

Annual power costs:

Bid No. 1	= (910 kW) ($0.025/kWh) x (8760 h/yr) =	$199,290
Bid No. 2	= (780 kW) ($0.025/kWh) x (8760 h/yr) =	$170,820
	Difference =	$28,470

Tax on difference = $28,470 x 0.38 = $10,820

Difference net of tax = $28,470 – $10,820 = $17,650

Net PV of difference = $17,650 x 9.43 = $166,440

D) Result:
The power system supplied as bid No. 1 will cost approximately $166,440 more to operate than the system supplied under bid No. 2 on a present value basis.

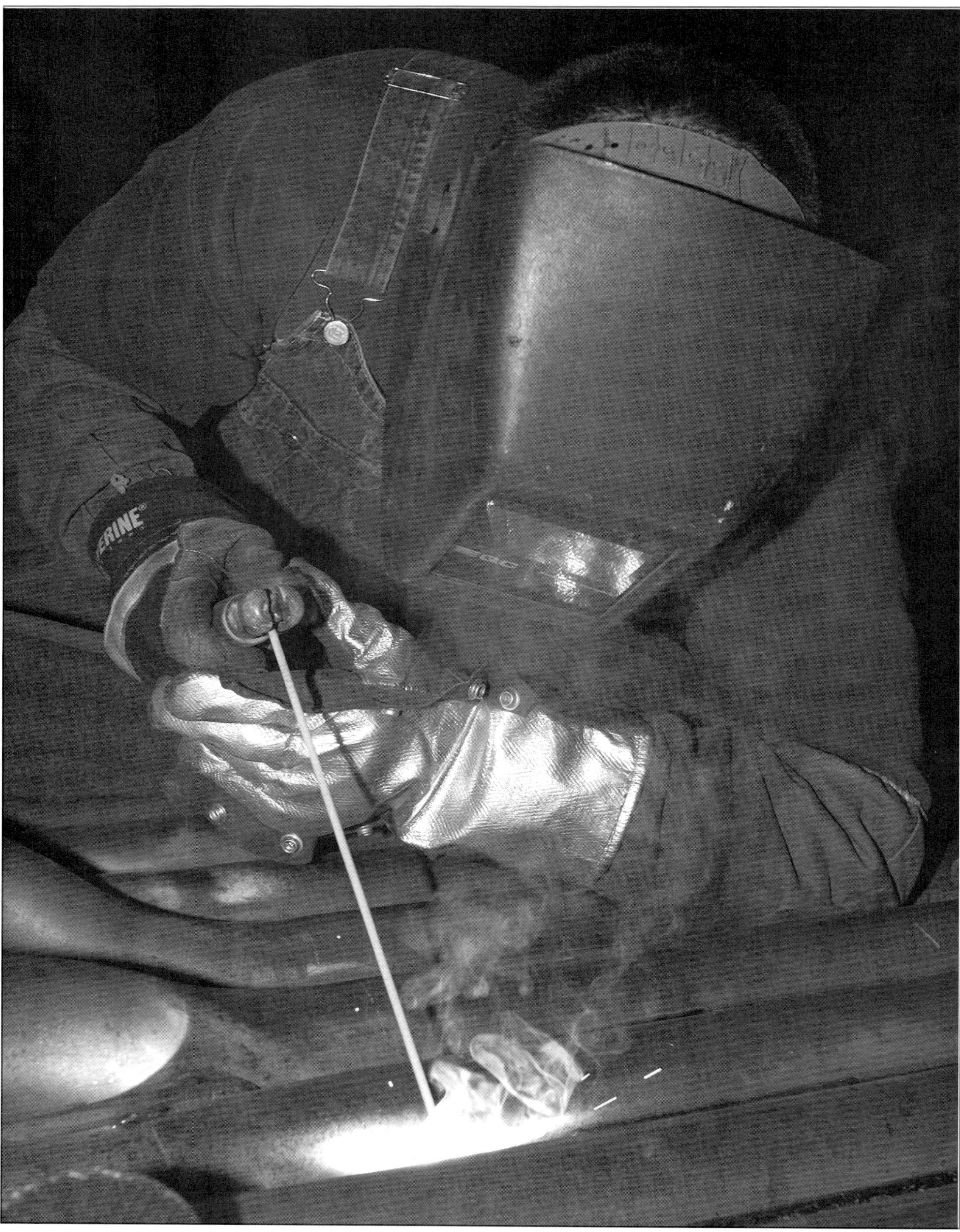

Wall panel welding.

Chapter 38

Manufacturing

The installation of a steam generating unit involves the design and manufacture of the components and final erection of the unit. The power generation industry has achieved unprecedented levels of quality and reliability in producing modern steam generating equipment. State-of-the-art design and manufacturing techniques as well as the use of advanced materials are key reasons for these gains.

Component manufacturing involves a variety of skills and unique manufacturing methods and requires facilities for virtually all welding, metal removal, metal forming and casting processes. Manufacturing operations are continuously evolving to meet the increased quality and productivity demanded by today's markets. These manufacturing and quality evolutions have enabled the fabrication of large, complex boilers of sizes and technical competence once thought to be unachievable.

Gaining functionality at low cost is an ever present goal in the mature fossil power equipment industry. Achieving high quality levels in manufactured components requires that the manufacturing facility focus on small details. It is, therefore, important for Manufacturing to participate early in design discussions. The resultant design-for-manufacturability improvements add value to the product. *Value engineering*, a structured approach that enhances the manufacturing process, is used by The Babcock & Wilcox Company (B&W) to maintain its leadership position in the industry.

Once a design approach is determined, communication of the design information to the manufacturing facilities must be accurate and timely. Today's computers, integrated systems, and communication technology provide unprecedented capability to handle these tasks.

Traditional techniques as well as innovations implemented by B&W to manufacture steam generating equipment are described in the following sections. For information specific to nuclear equipment, see Chapter 50.

Welding

Welding and welding related processes are central to boiler fabrication. These processes are used in manufacturing a large percentage of components produced by B&W. In addition, many of today's basic welding technologies were pioneered by B&W.

A *weld* can basically be defined as a coalescence of one or more materials by the application of heat localized in the region of the intended joint. For most welding processes, the induced heat raises the temperature of the material above the melting point to create a weld pool. The heat source is then removed, allowing the pool to solidify and creating a metallurgical bond between the joined surfaces. Depending upon the application, filler material may be added to the weld pool to fill a cavity between the parts being joined.

There are many welding processes from which to select. The choice is primarily based on the best method of generating and conveying the energy needed to create the weld. Electric arc welding is by far the most common group of processes used in fabricating fossil fueled steam generating equipment, although other processes also apply. The most widely used arc welding processes are gas tungsten arc welding (GTAW), gas metal arc welding (GMAW), submerged arc welding (SAW), shielded metal arc welding (SMAW) and plasma arc welding (PAW). Of the non-arc welding processes, high frequency resistance welding (HFRW) and laser beam welding (LBW) are the most common. Table 1 summarizes the advantages and disadvantages of each process and provides examples of steam generator applications. The *Welding Handbook* and other sources provide specific information on the distinguishing features, attributes and limitations of each process.[1]

Selecting a welding process requires consideration of several factors. The composition of the materials and the anticipated service conditions dictate the appropriate filler material. Base material composition also affects the use of preheat and post weld heat treatment to control the resultant hardness of the material and to minimize the susceptibility of the weld to cracking. Also, the structural mass of the component and the orientation and design of the weld joints influence the welding processes selected and the process parameters used. The quality and productivity associated with each welding process must also be factored into the selection.

Welding automation

A given welding process can be performed in a number of ways. Historically, the most common approach was manual welding, in which the operator moves the welding torch along the joint to be welded and makes the

Table 1
Comparison of Welding Processes

Process	Advantages	Disadvantages	Applications
LBW	Very deep penetration Very high quality Little distortion Good mechanical properties	High capital expense Equipment difficult to maintain Not portable Very good fit-up needed	Weld overlay of tubes and panels
GTAW	High quality Low capital expense Readily automated	Shallow penetration High operator skill required Low productivity Arc radiation hazard	Tube-to-tube weld root pass Join superheater and economizer tube sections
GMAW	Moderate productivity Moderate quality Readily automated	Shallow penetration Contact tip wear requires periodic stoppages Arc radiation hazard	Panel wall repair and attachments Tube-to-tube welds
PAW	Deep penetration High productivity High quality	Complex process Difficult to maintain High capital expense Arc radiation hazard	Sootblower tube-to-tube welds
SAW	Very high productivity High quality Operator protected from arc radiation	Limited in welding position	Header fabrication Water wall fabrication
SMAW	Very low capital expense Very portable	Can not be automated High operator skill required Lower weld quality Low productivity	Repair Tube-to-header welds
HFRW	Very high productivity High quality	High capital expense Very limited in application	Tube manufacturing

necessary adjustments to the welding equipment. However, automation of welding processes has become very common because it provides more consistent quality and lower costs as compared to non-automated processes.

One form of automation is *orbital* welding, used with the GTAW process to join superheater and economizer tube sections. In this process, the welding head is driven around the outside diameter (OD) of the tube to produce a weld. A computer coordinates the motion of the welding head, the weld filler wire feed rate and the process parameters to maintain high weld quality.

Robotic welding and cutting are also commonly used. Robots are used in a number of boiler-related welding applications including seal welding of tangent tubes in a Cyclone furnace, fabricating coal pulverizer components, applying nonferrous overlay cladding to steam generator tubesheets as well as nozzle bores, and making complex tube-to-tubesheet welds inside small diameter bores. The flexibility of robots also promotes their use in plasma cutting of nozzle openings in shells and to contour burn nozzle mating surfaces. A number of robot types, including rectilinear, articulated arm, cylindrical coordinate and gantry style units are used in fabricating boiler components (see Fig. 1).

Automated determination of weld joint positions has also been implemented. Electromechanical devices, called *probes*, are used to track the weld groove to ensure proper placement of the weld head during submerged arc welding. An alternative guidance method, which has been used with the GMAW process, is *through the arc* tracking. This method senses the welding arc's electrical characteristics to guide the weld head. A more recent development for GTAW, GMAW and PAW is laser-based vision tracking, in which a noncontact sensor determines the torch position within the groove and adjusts the torch as necessary. This type of sensor also allows automated adaptive control of the welding process parameters by real time response to changes in weld joint topography.

Metal removal

Metal removal is performed either to prepare a component for another manufacturing operation, such as in machining weld joint preparations, or to produce a finished shape. Traditional machining processes are considered to be either chip removal or abrasive removal. The chip removal operations used in steam generation equipment production include turning, mill-

Fig. 1 Robotic welding.

ing, drilling, boring and broaching, while grinding, sanding and grit blasting are examples of abrasive operations. Nontraditional metal removal operations, such as electro discharge machining (EDM) and arc gouging, are also used.

B&W's fabricating shops have a variety of manufacturing equipment ranging in size from small hand tools to machines with 100 t (91 t_m) programmable rotary tables. Larger pressure vessels require larger machine tools. For example, the vertical boring mill shown in Fig. 2 can accommodate a part 16 ft (4.9 m) in diameter and 10 ft (3.0 m) high, while the horizontal boring mill in Fig. 3 has a rotary table capable of supporting 100 t (91 t_m).

As in other manufacturing areas, the most apparent innovation in machining is computer control of the equipment. Part configurations, programmed off line, are transmitted by distributed numerical control (DNC) to many machines on the shop floor. (See Fig. 4.) More

Fig. 2 Vertical boring mill.

Fig. 3 Horizontal boring mill.

recent user-friendly computer numerical control equipment also permits sequences to be programmed by the machine operator. Computer numerical control (CNC) machining centers with automatic tool changing are common. In addition, multi-spindle computer numerical control drills are often used to prepare steam generator tube bundle support plates. (See Chapter 50.) These and other refinements permit some machine tools to be accurate within 0.00005 in. (0.00127 mm).

CNC thermal cutting machines, oxy-fuel and plasma, make use of downloaded computer generated nesting programs to maximize material use. The addition of a multi-station CNC punch, featuring an integral plasma burning head, complements the thermal cutting tables.

Electrical discharge machining (EDM) removes metal by spark erosion between an electrode and the work piece. In this process, intricate shapes matching those of the solid carbon electrode can be burned into the work piece. The solid electrodes are being replaced in many cases by CNC wire electrode machines.

Automation is increasing as computers often control entire shop sections. Material, automatically moved into storage buffers, then to CNC machines, is scheduled and routed using computerized identification systems. Probes are often used to check the final machined dimensions.

Forming

Forming is the permanent plastic deformation of a work piece accomplished by applying mechanical forces, through tooling, to the work piece. The primary objective of forming is the production of a desired shape.

Basic forming elements include the material to be formed, the tooling to change the shape, the lubrica-

Fig. 4 Computer numerical control machining.

tion between the tools and the work piece, and the machine used to hold the tooling and to apply the forces needed to form the material.

In sheet forming operations, the forces are primarily applied in the plane of the sheet. The thickness is not constrained between opposing tools but may change due to in-plane strains. Sheet forming operations also include tube forming processes using an internal or external tool and sometimes both. Most sheet forming operations are performed cold.

B&W's principal forming applications are in the areas of plate and tube bending. Plate is bent to form large cylindrical shells, while tubes are bent to form flow loops or offsets. Two methods are used to bend plate for steam drum fabrication. Press die forming is used to make half shell courses (two seam welds) while roll bending is used to form full round shell courses (one seam weld). (See Chapter 50.)

Tube forming is done primarily on draw benders. The clamp die, bend die and pressure die are always used while the wiper die and mandrel are only required for tight radius bends with thin wall tubing. Forming dimensions are typically expressed relative to the diameter of the tubing being bent. For example, a 1D bend has a centerline radius equal to the tube diameter. Tubes with wall thickness greater than 10% of the OD do not require a mandrel or wiper die for forming.

A variation of draw bending is *boost bending*. Here the tube is clamped ahead of the bend die and an axial force is applied to push the tube into the bend region. This shifts the neutral axis toward the outside of the bend to reduce thinning on the outside and increase thickening on the inside.

The application of computer numerical control to bending, as shown in Fig. 5, is highly advantageous for tubes containing multiple bends of the same radius. The tube is fed into the bender, gripped by the axial transporter and moved to a reference location. All bends are then made automatically as programmed. The transporter rotates the tube for out-of-plane bends and feeds it axially to the next tangent point.

Castings

Castings are used throughout modern boilers and steam generating equipment. The choice of castings over other manufacturing processes, such as fabricated or machined components, is driven by functional design requirements and by final part configuration. Geometry often dictates the casting process as the least costly production method.

Castings are produced from an array of materials, from simple grades of irons and steels to complex alloys and proprietary metal chemistries and heat treatments. Material choice for particular applications is determined by tradeoffs between the function of the part

Fig. 5 Tube bending by computer numerical control.

and the design parameters, compared to the ultimate cost of the part.

Castings are produced using a multitude of processes including sand casting, permanent mold, die casting, investment casting, lost foam casting, and vacuum casting. The process selected is established by the material being poured, the part configuration and size, and the required tolerances. The weights of individual castings in modern steam generating equipment can range from a few ounces to many tons.

Process control is critical in producing quality castings. Process is even more important when castings are subjected to severe operating conditions (e.g., wear metals in coal pulverizers). Because of this control requirement, B&W operates its own foundry, producing pulverizer wear metal components from proprietary high chrome irons. (See Fig. 6.)

Fig. 7 Utility steam drum ready for shipment.

Fabrication of fossil fuel equipment

Drum fabrication

A large steam drum, complete and ready for shipment, is shown in Fig. 7. The drum consists of multiple cylindrical sections, or *courses,* and two hemispherical heads. Large stubs and nozzles along its length serve as flow connections for the steam generating and superheating surfaces of the boiler. The large nozzles along the bottom and on the hemispherical heads are connections for the downcomers, which carry water to the various generating system circuits.

Drum fabrication begins by pressing flat plate into half cylinders or by rolling the plate into full cylindrical shells. When short shell sections must be used because of material dimensions or heat treatment requirements, the plate is rolled into a cylinder. For large drums, the plate is pressed into half cylinders then welded longitudinally, forming a course. The drum length is achieved by circumferentially welding the required number of shell sections. Longitudinal and circumferential welds are made using the automatic submerged arc process.

Fig. 6 Casting process at foundry.

Drum heads may be hot formed by pressing or spinning flat plate in suitable dies. The heads are attached to the completed cylinders by circumferential welds.

Automatic welding techniques are often used to join the various shell courses and heads. Fig. 8 shows a two wire submerged arc welding machine on a longitudinal seam. Filler wire is continuously fed to the weld area, which is completely covered by a granular flux to prevent oxidation and enhance metal properties. The electrode carriage, moving along the weld joint, is guided by a seam tracker. Depending on the plate material, the material thickness and the configuration of the weld, multiple passes may be required.

For submerged arc circumferential welds, the drum is rotated while the welding head and flux applying

Fig. 8 Welding longitudinal drum seam with double submerged arc welding machine.

equipment are stationary. An electromechanical guidance device may also be used to keep the arc tracking the weld groove.

Prior to and throughout all welding operations, preheat may be applied to minimize stresses and metallurgical transformations. In some cases, heating is maintained after welding until the vessel has been stress relieved.

Engineering advances in designs and the use of high strength, high temperature alloys have required the development of new weld techniques. This is especially true for stainless steels, Inconel® and other high nickel alloys. The tungsten inert gas (TIG) process is often specified to attain full root penetration. These processes also assure a uniform inside surface contour that does not interfere with fluid flow.

Nozzles and stubs are machined from hot forged billets, pipe or tubing. Larger connections may be integral with the heads, formed from rolled plate, or made from heavy wall pipe as shown along the bottom of the drum in Fig. 7. Attachments are manually welded to the drum with the shielded metal arc process using coated electrodes, by automatic satellite welding with the submerged arc process, or by semiautomatic welding with the metal inert gas (MIG) process. Other techniques are being developed that will significantly contribute to welded joint quality and increase the deposition rate of weld metal. Narrow weld grooves in thick plates are used to reduce weld volume in the joint.

Every carbon steel drum in which the material thickness is greater than 0.75 in. (19.1 mm) must be subjected to a final post weld heat treatment. In some cases, several short cycle stress relief operations may be applied to reduce preheating. For applications requiring high impact strength, formed heavy plate is water quenched to cool it rapidly . This heat treatment is followed by a normal stress relief.

All longitudinal and circumferential drum welds and the adjacent base material must be subjected to radiographic examination. For this purpose extensive radiography (x-ray) equipment is available. (See Fig. 9.) Ultrasonic evaluations are used where x-ray examinations are difficult or impractical. Surface and subsurface defects may be located using magnetic particle examination. Liquid penetrant examination also reveals surface flaws.

After all pressure parts have been welded, the vessel is ready for the finishing operations, including the installation of attachments and internals.

Headers as terminal pressure vessels

Headers are used in steam boilers to connect two or more tubes to a circuit. Headers can be fabricated from seamless tubing, pipe or hollow forgings. Access to their interiors is gained through the use of handholes. Headers vary in diameter from 3.0 in. to 48 in. (76.23 to 1,219.68 mm.) OD and can reach lengths in excess of 100 ft (30 m). Fig. 10 shows superheater headers nearing completion. Headers can be either shop installed to components such as panels and sections or shipped to the job site for field installation into the boiler.

Fig. 9 Drum weld testing with linear accelerator.

To fabricate a header, several pipes or hollow forgings are first joined by submerged arc welding. The ends of the header stock may have spherical or elliptical forged heads welded in place, or may be machined to accept a welded-in flat closure plate.

Completion of the header includes drilling and machining the tube and nozzle openings, and fitting and welding the nozzles and stubs (Figs. 11 and 12). Other attachments are added by manual or semiautomatic welding followed by heat treating, cleaning and finishing steps. These operations are conducted following strict quality control procedures.

Tubes as heat transfer surfaces

Tubes, in a variety of arrangements, are the heat absorbing surfaces and flow circuits in today's steam generating units (see Figs. 14 and 15). Tubes range in size from 0.875 to 5.0 in. (22.23 to 127.0 mm) OD and may be plain, finned or studded. Finned and studded tubes are used to create extended surface in some furnace wall

Fig. 10 Superheater headers nearing completion.

Fig. 11 Header drilling on vertical boring mill.

panels or sections to aid in heat transfer. Fig. 13 shows an opening in a wall panel surrounded by studded tubes. Materials used include carbon steel, chromium molybdenum alloys, composites with a metallurgically bonded stainless steel, or Inconel cladding over carbon steel.

Tubes can be shipped loose to the job site for field installation, or shop assembled into panels. Fabrication of a membrane panel begins by placing a metal strip between each row of tubes. This strip is then welded to the tubes on automatic equipment, which can perform numerous welds simultaneously (Fig. 15). A gas-tight metal wall surrounding the enclosed volume is created in the boiler by the membraned wall panels. The membraned wall panel design eliminates the need for a pressure-tight casing and reduces field erection costs.

Walls are typically assembled in panels up to 10 ft (3.0 m) wide and 90 ft (27.4 m) long, although larger panels have been fabricated. The panels may be formed into various configurations by using one of two bending machines (Fig. 16). To further reduce field erection costs, some panels and sections are preassembled into modules. Headers are often welded to these modules to complete the assembly (see Fig. 17).

Convection superheater surface uses continuous tube sections. A typical horizontal superheater section, shown in Fig. 18, is made from relatively short lengths

Fig. 12 Orbital welding.

of straight tube that are first bent, then welded together to give the proper configuration.

Close return bends are made by first hot upsetting a section of straight tube to increase the wall thickness at the bend location. Bending and hot sizing then form the tube to the desired configuration. The prepared bends are joined by computer controlled orbital GTAW to form a section. Upsetting and hot bending operations are often replaced by CNC booster bending. All other bends in a section are typically cold formed.

To complete a wall panel or section, support lugs, brackets and, in the case of panels, wall boxes are welded into position. Sections are hydro tested at 1.5 times design pressure, as required by the American

Fig. 13 Studded wall panel manufacturing.

Fig. 14 Hopper panel fabrication for a spiral wound universal pressure boiler.

Society of Mechanical Engineers (ASME) Code. Sections and panels are generally painted prior to shipment. Wall panels are often painted white to reflect light and thereby enhance worker vision within the boiler during erection.

Fig. 15 Wall panel membrane operation.

Longflow economizers

A *longflow economizer* is an assembly of finned tubes and headers installed within a process recovery boiler (see Chapter 28). The economizer absorbs much of the residual heat from the low temperature flue gases to improve boiler efficiency.

Value engineering has been beneficial in redesigning a longflow economizer, reducing manufacturing costs, lead time and field erection costs. One economizer design change eliminates bending the tube ends before fitting them into the inlet and outlet headers. The ends may be swaged to a smaller diameter in an automated process where the outside diameter and inside diameter (ID) of the swaged section remain concentric. Fins, formed from cold rolled flat stock and sheared to size, are automatically submerged arc welded on both sides to a shot blasted tube. All welds are then inspected. The tube ends are welded to inlet and outlet headers in jigs to form platens, which are furnace stress relieved. Finally the platens are assembled into an economizer module in a fixture. This fixture, with stiffeners added, is also used as a shipping frame for the unit. The frame is then used at the job site to lift the economizer module into the boiler (Fig. 19).

Structural steel and casings

As much as 11,000 t (10,000 t_m) of structural steel may be required to support a modern utility boiler. Most of this steel is available in standard mill shapes, however some members are fabricated from plate material into T- and H-shaped sections using submerged arc welding.

In modern units, water or steam cooled tubes constitute the basic structure of the boiler enclosure. Most boilers also contain non-water-cooled or cased enclosures. (See Chapter 23.)

Fuel preparation and burning equipment

B&W manufactures a great variety of burners for many fuels including oil, gas, coal, refuse, black liquor, wood bark, combinations of fuels, and specialty fuels

Fig. 16 Wall panel bending.

Fig. 17 Overhead view of wall panel construction.

(see Fig. 20). Because many similar or duplicate burners are built, product standardization and volume manufacturing techniques have evolved. This includes the manufacture of Cyclone™ furnaces (Fig. 21).

Cyclone barrels, ranging from 5 to 10 ft (1.5 to 3.0 m) in diameter, are fabricated using many duplicate tubes which are bent to a semicircular shape and joined to form a cylinder. Jigs, fixtures and power tools are used for these repetitive operations.

Cyclone furnaces are completely shop assembled in erection frames which are also used for stress relieving the final product. This permits close quality control to assure that ASME Code and B&W standards are met. Before shipment, a refractory coating is applied to the studded inside surfaces and the outside is covered with a protective casing. See Chapter 15 for further details on Cyclone furnaces.

The B&W coal pulverizer is another product that benefits from standardization, although some variations are required due to installation configurations. B&W's EL pulverizer series, for smaller boiler applications, has been supplemented by the B&W Roll-Wheel™ series. A large B&W Roll Wheel pulverizer has the capacity to grind 105 t (95 t_m) of coal per hour. (See Chapter 13.)

The top, intermediate and lower pulverizer housings are fabricated from plate material. Other components such as tires and grinding ring segments for B&W's Roll Wheel pulverizers, and balls and rings for B&W's EL pulverizers, are cast from wear resistant parts.

Fig. 18 Superheater sections.

Fig. 19 Longflow economizer prepared for shipment.

Fig. 20 Fully assembled pulverized coal burner.

The pulverizer gearbox which is a welded and machined fabrication encloses precision bevel and helical gears. Each pulverizer gearbox assembly is run tested before being shipped.

Tube manufacturing

Thousands of feet (meters) of tubes made from a variety of steels and alloys are required for a large steam generating unit. Tubes are used for heat transfer and steam generating surfaces, including furnace walls, floors, roofs, and in superheaters, reheaters, economizers and air heaters.

Two types of tubes are used: seamless and welded. Each has applications within the boiler based on user and ASME Code requirements that define diameter, wall thickness and chemistry. (See Chapter 7.)

Seamless tubes Seamless tubes are pierced from solid rounds of steel, which are produced in electric arc furnaces. The process is carefully monitored to produce quality steel consistent with application requirements.

Tubes for boiler application are furnished hot finished or cold drawn, depending on size, tolerance and finish desired. Metallurgically, the two are similar; the differences are surface finish and manufacturing tolerances. For certain applications, the surface finish of the tubes is critical to reduce the pressure drop. To obtain a smooth, even surface and close tolerances, stainless tubing is generally finished by cold drawing. This process also permits grain size control in final heat treatment, which is important in high temperature applications of austenitic stainless steels. The production of seamless tubes requires special tools and careful control of heating and manufacturing procedures.

Electric resistance welded (ERW) tubes The ERW tube making process begins with a coil of high quality, fully killed steel strip that has been shot blasted for surface cleanliness. The strip is uncoiled and leveled and its edges are skived to remove defects. Passing through sequential forming rolls, the tube edges are fused without filler metal by an electrical current of 450 kHz. An exclusive feature of the process is online electronic weld imaging, which determines weld integrity. This is typically joined at the welding station by ultrasonic testing.

ERW tubing is typically produced in sizes from 1.0 to 4.5 in. (25 to 114 mm) OD with walls from 0.095 to 0.650 in. (2.413 to 16.51 mm) in carbon, carbon-molybdenum and alloy steel grades. Tubes can be delivered as cold drawn and normalized.

Engineering-manufacturing interface

Communications between the key areas of manufacturing and engineering are essential in achieving state-of-the-art manufacturing. As in other manufacturing areas, computers are largely responsible for many of the improvements. Information such as design data and drawings can be electronically transmitted to the manufacturing facilities thereby reducing total product lead times.

Value engineering, a logical planning tool, has also streamlined projects from the engineering analysis through the manufacturing steps. The reduction of component complexity and improvements in product quality and delivery are results of this technique. Equally important as computerization and value engineering is the early participation of manufacturing personnel in the product design.

B&W's manufacturing emphasis continues to be on high quality, competitively priced boiler components and complete factory-assembled boilers as described in Chapter 27 (see Fig. 22). State-of-the-art engineering designs, innovative manufacturing processes and the highest levels of quality control assure that components meet all codes and customer specifications.

Quality assurance and quality control

The quality assurance department establishes procedures and controls to assure that all products conform to product technical requirements. Quality con-

Fig. 21 Cyclone furnace manufacturing.

Fig. 22 Factory-assembled boilers in various stages of completion.

trol personnel in the manufacturing plants work to ensure that all products meet applicable codes and specifications.

The ASME Boiler and Pressure Vessel Code (see Appendix 2, Codes and Standards Overview) provides rules for the safe construction of boilers and other pressure vessels. In the United States, these rules are also usually part of state laws. The ASME Code is written by industry members, inspection agencies and users of boilers and pressure vessels. It strikes an acceptable balance between safety, functionality and manufacturability.

Section I of the Code, *Power Boilers*, requires a boiler manufacturer to have the fabrication procedures, as well as the design and material selection, monitored by a third party inspector. This authorized inspector is usually employed by an insurance agency and is licensed by the National Board of Boiler and Pressure Vessel Inspectors. In plants that produce a large volume of boiler components, this inspector may be a full-time resident. The inspector checks drawings, reviews material certifications, inspects welds, reviews radiographs, witnesses hydrostatic tests and monitors the general workmanship achieved by the fabricator. At the completion of a job, when the inspector is satisfied that the parts comply with all aspects of the Code, the inspector signs the ASME Code data sheet.

All welding performed on power boilers and pressure vessels must be done by Code-qualified personnel. The personnel must be trained to use the correct filler material, welding positions and processes. In addition, they must be trained to apply any required preheat and post weld heat treatments. All welding procedures must be qualified to assure consistent, high quality welds. To maintain the qualifications of the welding personnel and assure appropriate implementation of procedures, B&W operates welding training centers and employs trained instructors at its manufacturing facilities.

Nondestructive examination

Nondestructive examination (NDE) provides the mechanism for quality control of base metals and weldments. There are four predominant methods of nondestructive examination that pertain to the pressure vessel industry: radiography (RT), ultrasonics (UT), magnetic particle inspection (MT), and dye penetrant inspection (PT). Radiography and ultrasonics are used for volumetric examination, while magnetic particle and dye penetrant inspection techniques provide for surface examination.

Radiography (RT) Radiography (x-ray) has been a primary weld quality control technique for many years. The two types of equipment commonly used are x-ray machines and gamma ray equipment.

The choice of using x-ray machines or gamma ray equipment is based on the mobility of the part to be examined and the availability of an x-ray facility. X-ray units can best be used in shops where fixed position equipment is available. Gamma ray equipment may be used to reduce the time needed to examine a circumferential weld seam. This examination is done by locating the isotope, usually cobalt-60, on the centerline of the unit, wrapping the outside diameter of the weld seam with x-ray film and exposing the film through remote control. Gamma ray equipment is usually preferred for field use or when the component can not be readily moved.

To assure correct exposure time, a small three hole penetrometer is placed on the object being x-rayed which causes an image to form on the film. The penetrometer thickness and hole size are specified in applicable regulatory documents. When properly used, an image of the holes appears on the film. Flaws of equal or larger size are expected to appear on the film as well. The ASME Code specifies acceptance standards with respect to the size of porosity and slag inclusions. It also specifies those welds that must be radiographed in pressure vessels and pressure part fabrications.

Real-time radiography permits viewing the weld seam on a video screen while the radiographic process is occurring. The results may also be recorded on video tape.

A significant advantage of radiography over other nondestructive methods is that a permanent film record is made of each inspection. These records are retained for at least five years or as required by a contract.

Ultrasonics (UT) UT examination has proven to be a valuable tool in evaluating welds, plate and forgings of pressure vessels. This examination can be applied to most sizes and configurations of materials.

In generating the sound for ultrasonic testing, piezoelectric materials are used. When pulsed by an electric current, these materials vibrate and generate a sound wave. In addition, when the materials receive sound wave vibrations, they also generate electronic impulses. These impulses may be viewed on an oscilloscope to locate a defect in the material or weld. Sound is generated in frequencies typically ranging from 200 kHz (1 Hz = 1 cycle/s) to 25 MHz.

Electrical pulses are directed from an oscillator to a transducer that is applied to the material being examined. A liquid couplant is used for good contact between the transducer and the material being examined. The transducer generates high frequency sound that is directed into the material. If a flaw is encountered, some of the sound is reflected to the transducer, which in turn generates an electrical impulse. This impulse is then displayed on an oscilloscope, showing the size and location of the defect. With proper manipulation of the transducer, the entire sample can be examined. Flaw acceptance standards are given in the ASME Code.

In addition to flaw detection, ultrasonic techniques are used to measure material thicknesses.

Magnetic particle examination (MT) Magnetic particle examination can only be used on ferrous materials. This nondestructive examination technique will involve either the dry method, using alternating current (AC) or direct current (DC), or the wet method, using normal or ultraviolet light for examination. All procedures involve generating a magnetic field in the part being examined. A flaw on or near the surface becomes discernible by the accumulation of magnetic particles. The method used depends on the size, shape and surface condition of the material being tested.

Magnetic particle examination provides an inexpensive method of determining the surface and near surface condition of the materials being tested. The process can be used in the manufacturing facilities or in the field.

Liquid penetrant examination (PT) In contrast to magnetic particle examination, liquid penetrant examination can be used on ferrous or non-ferrous materials, however the process only reveals surface flaws. All examinations begin by coating the surface with a red dye. The bulk of the dye is removed by wiping the surface with a cloth, leaving dye in any small discontinuities that exist on the surface of the material being tested. Dye remaining in surface flaws of the material is made visible by applying a white developer.

Fabricators use liquid penetrant examination for welds or base metals of non-ferrous materials, such as stainless steel or Inconel. This examination can be performed during manufacturing or in the field. The equipment is inexpensive and portable and personnel can be easily trained to perform the inspection.

Quality control of fossil fuel equipment

A quality control system must be maintained throughout all manufacturing operations. The initial step is the inspection of incoming material to assure that it conforms to the purchase order requirements and the appropriate ASME Codes.

In addition to examinations stipulated by the Code, B&W performs other nondestructive tests as process quality control measures. Shop butt welds in boiler wall panels and superheater, reheater and economizer sections are radiographed. In addition, stub-to-header welds are examined by magnetic particle examination. These components may also be hydrostatically tested.

Internal and external cleanliness of pressure parts is an important part of quality control. Tube IDs are cleaned with sponges while ODs are shot blasted and coated with a rust inhibitor. Steel balls are blown through sections and panels with compressed air to ensure that the components are free of internal obstructions.

In-process and final inspections, which consist of checking critical dimensions and verifying that all required nondestructive examinations have been completed, are part of the quality program. A final inspection of truck and rail car loading is made to assure safe delivery of components to the job site.

Technologies

The development and use of state-of-the-art manufacturing technologies have continued to be important to B&W. The company maintains an entire group which is dedicated to manufacturing technology research and development.

Work has been conducted to improve all aspects of the welding operations, including planning, materials, processes, new applications and automation. In the area of weld planning, computer based models have been generated to assist in weld procedure optimization. When commercially available alloys are found to be lacking in a given operation, new materials which can provide the required mechanical and corrosion properties are developed.

Improved welding procedures are continuously being developed for the common processes as well as for the more exotic processes, such as explosion welding, laser welding, electron beam welding, plasma MIG and variable polarity plasma arc welding. In the area of welding automation, B&W has advanced off-line programming and sensor technologies to improve the flexibility of automated systems.

Dramatic changes in manufacturing are occurring with the continuous development of advanced computer technologies. Newer machine tools will provide greater accuracy in motion control, repeatability and in process measurement. Bends can be made in sequence rather than in a batch mode, thereby reducing part inventories. Heat treating has improved with computer control of furnace temperatures and atmospheres. Inspection is being automated through the use of computer controlled coordinate measuring equipment, vision systems and real time radiography. Furthermore, manufacturing operations are being integrated with computers on a plant wide basis to facilitate material flow and to apply statistical process control tools.

Reference

1. Connor, L.P., *Welding Handbook,* Vols. 1 and 2, Eighth Ed., American Welding Society, Miami, Florida, 1987.

Inconel is a trademark of the Special Metals Corporation group of companies.

Central station utility steam generating unit under construction.

Chapter 39

Construction

Construction plays a major role in the success of any project. The construction organization must employ erection techniques that complement the designers' skill and the fabricators' craftsmanship. Sound field assembly is essential for correct functional performance of the completed installation. Efficient and safe execution of any project requires that components be on the job site at the right time along with the necessary tools, equipment, labor and supervision. In addition to properly erecting the components, a successful construction organization possesses experience in estimating, planning, safety, quality control, cost analysis, labor relations, tool and equipment design, technical services, and finance and accounting. A competitive marketplace requires continuous improvements in quality, safety and cost, together with reduced construction schedules, as contractors and owners strive to minimize project costs and schedules. The most successful projects maintain communication between engineering and manufacturing operations, ensuring that field erection is a continuation of the shop fabrication process and not a completely independent event.

During initial project development, the construction plan and work performance logistics must be evaluated, then incorporated into the overall project design. Schedule requirements, together with safety and site-specific conditions, will affect the labor and erection methods that will be used.

A number of boiler types are small enough to permit shipment completely assembled. However, larger industrial boilers, central station boilers, and environmental systems are shipped to job sites in various stages of fabrication and subassembly. A central station boiler with its associated heat transfer equipment and auxiliaries may weigh more than 12,000 t (10,885 t_m), requiring the equivalent of 500 railroad cars to ship the material to the site over a period of several months. The modularized construction practices in use today result in some boiler components weighing 400 to 1000 t (363 to 907 t_m). Flue gas desulfurization (FGD or scrubber) and selective catalytic reduction (SCR) system modules for environmental retrofit work may weigh between 50 to 150 tons (45 to 136 t_m). In these cases, overland and barge shipments can become unique operations.

The field assembly of power generation facilities and the associated environmental equipment requires efficient, well engineered, well organized erection methods to permit installation within a reasonable time and at minimum cost without sacrificing quality and safety. Modularization of complex components together with continuous advances in computerized drafting, modeling, scheduling and tracking have become the norm for these complex or fast track projects. The cost of erecting a steam generating unit, such as that shown in Fig. 1 represents a significant part of the total investment in the plant.

Retrofitting an existing plant with environmental equipment presents a unique set of challenges requiring extensive planning and coordination between owners, engineers, manufacturers and constructors. (See Fig. 2.)

Fig. 1 Modern power plant construction includes appropriate environmental equipment.

Fig. 2 SCR retrofit construction.

Estimating and cost development

The estimating process begins with a complete understanding of the project work scope and specifications for performance. The estimator prepares a detailed list of work activities or tasks for the entire project utilizing specifications and drawings obtained from the owner and the component supplier's engineering function. Direct work hours for each labor category are then assigned to each activity based upon historical data and general experience. For example, insulation is estimated based upon the number of hours per unit area [ft^2 (m^2)] times the insulated area of the proposed equipment. The total hours for each craft and for the total project are then calculated. The totals are frequently corrected for a variety of factors, e.g., local labor productivity, overtime and site factors. To the direct labor total, indirect labor such as supervisors, foremen and housekeeping, etc., are added to produce a total project estimate.

Scope

Historical performance, accumulated through ongoing experience and applied through an estimating database, provides the basis for the initial estimate. The estimating process utilizes a variety of sources to develop a detailed understanding of a project's specific work scope. These sources include product specifications, equipment and layout drawings, site visits, customer interaction, engineering, and internal data developed from experience on similar projects.

Direct costs

The direct costs associated with field construction are developed and tied to a detailed task list by applying historical work hour rates associated with a specific quantity of material. These quantities may include tonnage, linear feet or square feet (meters).

Installation factors

Following the direct cost estimate, several specific factors are considered to finalize the direct work hour requirements for each activity. These factors include local labor productivity, overtime and multiple shift work inefficiencies, repair difficulty factors, and site-specific conditions.

Indirect labor and support requirements

Once the direct work hour costs have been developed, indirect labor support requirements can be applied. These are driven by man days and individuals available for the project timeline. Examples include the general foreman, laborers, operating engineers, unloading crews, and any other individuals needed to support the installation labor activities.

Other factors

While detailed estimates are required for specific projects, total estimates for nominal scope jobs frequently fall into general ranges. For coal-fired utility boilers of nominal scope, erection work hours usually range from 1.25 to 2.0 work hours per kilowatt. In other cases such as retrofit SCRs or FGD systems, the estimates do not fit such ranges well because of the dramatically different equipment scopes and configurations possible for a given size unit.

The estimated time required to complete a project is related to the number of estimated work hours. There is a limit, however, for any given unit, to the number of workers that can be efficiently employed. If more than an optimum number are employed, the lack of adequate work space results in diminishing returns and excessive costs. Fig. 3 shows the typical number of weeks required for various work hour expenditures for a single boiler unit, and the maximum number of people utilized during peak operations.

Erection time span can also be affected by performing on a multiple shift or overtime basis. With these alternatives, significant additional costs can be incurred due to the inefficiency resulting from working longer hours, hours not worked, fatigue, or shift turnover inefficiencies.

Schedule and project plan development

The construction schedule is a key input to the integrated project schedules that focus on all of the disciplines needed to complete a given project.

Upon contract award, the construction project team

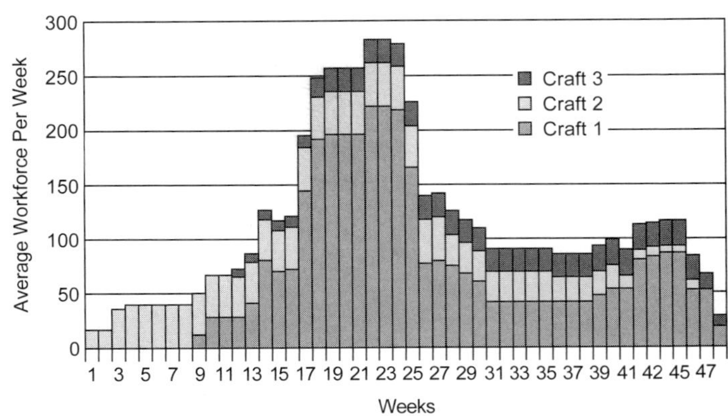

Fig. 3 Typical labor requirements for a single boiler unit.

conducts a detailed planning session. Typical participants include the project manager, scheduler/planner, project superintendent(s), and other contributors as selected by the project manager. The basis for this planning session is the scope of work as originally defined and estimated, and the existing conditions of the construction site. During the session, missed scope items or recent customer-initiated changes to the as-bid project are evaluated. The primary objective of this session is to document a detailed construction plan. The information inputs needed and the outputs from a successful planning session are shown in Table 1.

Projects are defined by a series of work activities arranged in a logical sequence. Detailed schedules are then developed by sequencing all activities according to their predecessor/successor relationships. These relationships are used to assign activity durations, and to allocate resources to complete the activity within the desired time period. Resource allocations typically include labor plus heavy and specialized equipment.

Work activities are arranged into a work breakdown structure and assigned activity codes. The activity codes allow the work to be sorted by contract, area of work, responsibility, component, or other required factors. These activities can then be used to monitor and track progress throughout the course of the project.

Schedules are established using computer-based scheduling systems. A broad range of detailed graphic schedules and reports, including time scale logic diagrams, bar charts, precedence diagrams, labor graphs, earned value curves and job completion percentages are derived from such systems. These outputs permit the overall coordination of several items: the work to be performed, labor by craft or specialty, erection equipment assignments, material deliveries, and critical resource requirements. The goal of the planning session is to initially define the project in enough detail to ensure that subsequent tracking and monitoring activities can effectively control all resources and job progress, identify positive and negative deviations, and prescribe necessary corrective actions.

The level of detail developed in the work breakdown structure is dependent on project complexity. Straightforward, short duration projects may be broken down into less than fifty activities, whereas more complex, long-term projects may be defined by hundreds.

Construction project management

The successful completion of a large utility construction project requires a disciplined and proven approach for executing design, engineering, procurement, expediting, erection, inspection and commissioning. This approach is required whether the project involves new boiler erection, significant modifications to an existing unit, or the retrofit of environmental equipment. Detailed project execution plans are developed to ensure proactive control of scope, safety, quality, cost and schedule; report progress; identify deviations; forecast trends; take corrective actions when necessary; promote communications; and coordinate the activities of all participants. The project team is responsible to ensure that timely constructability reviews are performed to integrate design and construction considerations, assuring a technically complete and constructable design.

An initial project kickoff period provides a detailed project definition before any engineering releases are authorized. This definition period permits all involved to clarify details and better define owner and engineering requirements. This phase is used to:

1. document owner design, operating and maintenance philosophy,
2. jointly review the specifications,
3. review subcontract vendor specifications,
4. allow the owner to review acceptable vendors,
5. establish procedures for project execution, and
6. firmly establish the project's schedule.

A project team is utilized to effectively manage and coordinate the multiple interfaces required for a large project. This team will establish all procedures and

Table 1
Construction Planning Session Inputs and Outputs

Session Inputs

- Integrated overall project schedule
- Contract documents such as commercial terms and conditions and technical scope
- Final estimate including worksheets and specified material quantities
- Subcontract scope and commitments
- Material and equipment delivery schedules and configurations
- Prior site reviews
- Customer/plant knowledge
- Labor agreements and availability

Session Outputs

- Detailed construction project management (CPM) schedule and associated reports
- Project safety plan
- Manpower loading and leveling
- Equipment and critical resource loading and leveling
- Identification of subcontract work scope and schedule requirements
- Project work assignments for support disciplines
- Site organization chart
- Earned value curves (measurement of performance against a detailed plan to predict final costs and schedule)*
- Financial reporting
- Material delivery schedule
- Preferred material configuration
    - Modularization
    - Field weld locations
    - Maximum component rigging weight
    - Maximum component size
- Baseline weekly progress report format

* Reference – Fleming, Quentin W. and Koppelman, Joel M., *Earned Value Project Management*, Project Management Institute, Inc., Pennsylvania, September, 2000.

controls required for the coordination and routing of documentation, design and engineering information, technical interfaces, and correspondence.

Once the construction phase of a project begins, the site organization serves as the primary conduit for communication. Regular project meetings keep all participants informed.

The key element of the project management approach is the scope-of-work document. Each working group involved in the project will develop a definition of objectives, scope of work for their activities, data collection needs, methodology or approach to the work, end products, responsibilities for all involved, time durations and schedules.

An integrated project schedule and report is developed to cover all activities by involved participants. Monthly progress reporting and performance measurements are completed using the earned value concept. Schedule trends/projections are evaluated with action plans formulated to regain or advance the schedule targets included in each progress report. Outstanding or new actions are included. Potential threats to the project's goals are identified with alternative suggestions to reduce or eliminate any negative impact.

Design for constructability

The Construction Industry Institute (CII) defines constructability as a process of optimum integration of construction knowledge and experience into project planning, design, procurement, and field operations to achieve overall project objectives.[1] CII recognizes that maximum benefits are achieved when people with construction knowledge and experience are involved with the project from its inception, and continue throughout its duration. The significance of this philosophy is best illustrated by Fig. 4. The ability to influence overall project costs is greatest at the beginning of the project and diminishes asymptotically with the progression of time.

In the past, the term *constructability* was generally associated with the quality of the various engineering groups' approaches to erecting components. Specifically, it referred to issues or potential assumptions made in association with the designer's intent that could preclude a component from being satisfactorily built or inspected (including errors and omissions). Information that lacked clarity was typically resolved through various means including drawing revisions, requests for information, engineering memoranda, and nonconformance reports. Resolution almost always increased cost as a result of rework and/or field delays while awaiting the design engineers' disposition. Many long-term projects now implement constructability reviews of these design outputs prior to the start of construction.

The endorsement of CII's philosophy and recognition of its inherent benefits provides the foundation for a successful project. The common goal is to ensure the timely delivery of quality steam generation and environmental control equipment that is erected safely, efficiently, and cost effectively, subject to specific project constraints. This is accomplished when the project managers, superintendents, and construction engineers work hand-in-hand with the engineering groups early in a project's development cycle.

Since construction costs for some projects can exceed 50% of a total project, the cost of labor presents both the greatest opportunity and greatest risk for almost any project. A design for constructability seeks to minimize impacts and improve productivity through the elimination of rework and/or corrective action, improve material deliveries, and incorporate some type of modularization.

Implementation

Typical barriers to constructability implementation, as identified by CII, include lack of disciplined coordination, failure to dedicate an individual administrator, and improper funding for the overall program. For a constructability program to be effective, these barriers must be recognized and overcome.

Evaluation

To help evaluate constructability efforts, feedback is needed throughout the project. During construction, it is important to obtain an objective assessment of the design documents from contractors and subcontractors.

Coordination and distribution of design drawings will be monitored and reviewed prior to issuance for construction. Innovative approaches, erection sequences, and cost saving measures will be reviewed to assure that the final product exhibits construction-friendly components. Constructability will also contribute to minimizing interference with existing plant facilities, thereby minimizing the effects on the operating facility.

The benefits are most evident during the construction phase of a project once engineering and procurement are essentially complete. If flexibility in the project schedule has already been consumed by engineering and procurement, the cost impact on the project becomes significant. The cost of a day's delay early in a project is minimal, whereas, once construction has started, costs could exceed 100,000 U.S. dollars for each day of delay. Savings and benefits lie in the reliability and predictability provided to the construction plan through the efficient use of labor and equipment. Every job will have its share of delays due to late deliveries, equipment failure, and acts of nature. Therefore, the goal of constructability is to select those attributes of the project that can be controlled and then exercise maximum influence over those offering the largest payback. The proper utilization of constructability minimizes project risks such as labor availability and maximizes opportunities through productivity improvements.

Integrated design process for retrofits

The integrated design process requires close coordination of design, manufacture, shipping and construction processes. Retrofit projects for complex utility systems are unique and therefore require a significant amount of design integration. Several key steps are critical when these retrofit projects are considered:

1. Determine the detailed scope early in the process.
2. Define the general arrangement of all components in concert with existing structures and equipment.
3. Define the construction approach.
4. Perform joint project management, engineering and construction reviews of the design.
5. Iterate the second, third, and fourth steps to advance and evolve the design.

The design/construction team must share a common and detailed understanding of the defined scope as early as possible. This scope description and the resulting assignment of responsibilities establish the direction of the project team.

Project definition for all members of the design and construction team is not complete until participants have a clear spatial understanding of new equipment location within the existing facilities. Interferences are relatively inexpensive to resolve at the design stage, but can be devastating to the project budget and schedule if recognized during the construction process.

Large retrofit projects typically span 15 to 30 months from award to startup. This span is dependent upon the number of units to be retrofit, the complexity of the installation, and the unit's planned outage schedule. These projects are impacted by numerous site-specific constraints that may lengthen the overall project.

When the total project and schedule integration process is not followed, project management and engineering will typically initiate the design process and begin the fabrication of materials well before the involvement of the construction manager and/or superintendent. This approach generally results in project delays and increased costs after mobilization of the construction team. Joint planning and an ongoing review of the design by team members from project management, engineering and construction have frequently been shown to result in a high quality, safe, economical, and on-schedule project.

Integrating the different perspectives, knowledge and concerns of these three key project disciplines yields significant benefits. The first joint review only begins the process. Ongoing reviews generate the greatest long-term benefits. Significant design and construction decisions must be made in ways that balance the scope, schedule, and cost triangle to benefit the overall project.

Some of the key decisions that need to be addressed by the team are:

1. the best physical arrangement of the equipment,
2. supports for the new equipment,
3. utilization of existing or new steel,
4. defined and scheduled pre-outage, outage, and post-outage activities,
5. planned outage schedule,
6. definition of the erection sequence,
7. existing access limitations and additional requirements,
8. shop or field modularization, and
9. field labor availability.

A critical period occurs early in a project when scope, general arrangement, construction approach and schedule decisions are being defined and key subcontractors are being integrated into the project. Additional resources are needed to focus on technical, commercial and interface issues associated with subcontractors, architect-engineers, structural steel fabricators, and flue and duct fabricators. This support early in the project provides the necessary attention to effectively address schedule-critical activities during a time frame when resources are typically overloaded. Once these specialized functions are executed, these resources are dismissed and project team members then consult as required.

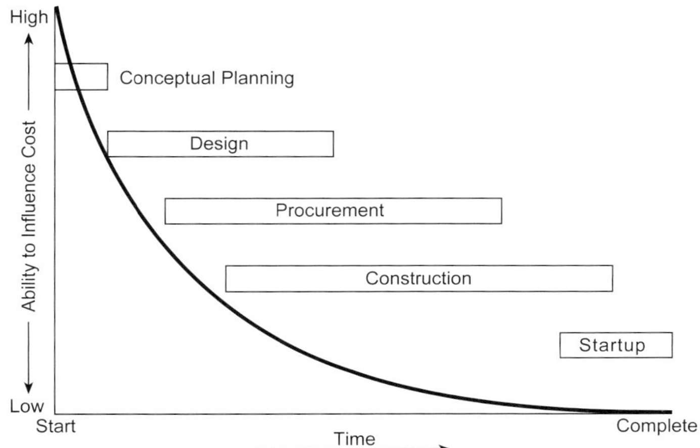

Fig. 4 The ability to influence cost diminishes as a construction project progresses.

Construction technology

The construction of steam and power generation facilities, and all associated equipment, is a complex and challenging undertaking. Each project presents a unique set of objectives, subject to specific constraints and opportunities. Successful projects are the result of detailed planning and forethought, as well as the creative application of construction means and methods to deal with the constraints, and to take advantage of the opportunities. Critical decisions regarding product configuration, constructability, construction equipment selection and labor issues must be made early and followed throughout the project.

Traditionally, the construction site supervisor has acquired the necessary skills through experience, and therefore is well-qualified. Technological advances including computer-aided design, global positioning systems, and more powerful lift cranes have enabled construction engineers and technicians to support supervision by refining the planning process and expediting the work in the field.

Aggressive schedules, improved financial performance, increased scrutiny, and regulations regarding safety and quality require that projects are executed with increasingly refined margins for error, both technically and economically. The goals established for these projects include: 1) reduced field operations costs, 2) a minimal project time span, 3) the construction of a quality product, and 4) provision for a safe working environment. The integration of technology into the construction process increases the opportunity to accomplish these goals through the proper, effective and innovative application of means, methods, techniques, tooling and equipment.

A construction technology group normally supports the field construction program. These engineers provide construction engineering services, construction logistics, and constructability reviews for each specific product and configuration. Typical outputs include recommended shipping sizes and configurations, rigging weight and center of gravity calculations, crane placement drawings, heavy lift and hauling plans, and structural analyses of existing facilities for imposed construction loads.

Approaches to construction – erection methodology

There are three basic approaches that may be used in a construction project: 1) *knocked down*, the traditional method, 2) *field modularization*, and 3) *shop modularization*. The knocked down approach is based on the erection of shop fabricated, standard-sized shipping units that readily fit on a truck/trailer or rail car. The standard shipping units are erected one piece at a time in *stick* fashion. Field modularization combines these traditional ship units into larger components, or modules, at the construction site for final installation. Off-site assembly, shop modularization, or prefabrication combine the traditional ship units into subassemblies at the manufacturer's plant.

Modularization is an integrated approach to power plant construction whose concepts must be incorporated at the earliest possible stages of a project's planning. The most successful projects have continued commitment to this approach by the customer, the equipment supplier, and the erecting contractor.

While modularization may be cost effective, the extent of effective modularization is site specific. Modularization divides plant materials into subsets, or modules, that are assembled before being brought to the construction site. (See Fig. 5.) The degree and extent of modularization can vary considerably. In its simplest form, a structural steel beam with shop-attached clip angles is a module. At the opposite end of the scale, a shop-assembled boiler complete with all trim piping, instrumentation, and insulation and lagging, or a 50-ton section of ductwork insulated and lagged is also considered a module.

Boiler components have been modularized for some time. Fig. 6 depicts furnace wall panels in which loose boiler tubing and membrane stock are welded to form the panels. Large modularized scrubber components are shown in Fig. 7.

A modularized project requires more engineering and design time than a conventional project. Planning must consider the water and land routes used for shipping the modules. Construction sites located on navigable waterways have the greatest potential for shipment of large modularized components. (See Fig. 8.)

Module unloading, site movements and erection procedures must be carefully considered. Special consideration must also be given to foundations, backfilling and compacting at the site. Other concurrent site activities must be evaluated when planning for modularization.

Construction participation by all involved parties is required in planning any project to enhance constructability and identify intended erection methods. The erection method is influenced by the following factors.

Retrofit versus new plant installation The construction logistics associated with retrofitting, repairing or altering an existing unit are much different from those required for constructing a new unit. Logistics for new plant construction are further diversified, depending on whether the unit is to be built on an existing plant site, adjacent to existing units, or on a new site. Retrofit operations are generally performed in a pre-outage/outage/post-outage mode.

In planning retrofit work, an erection method that permits performing many tasks concurrently is chosen. This optimizes equipment and labor and minimizes outage times. When constructing on an existing plant site, the erection method must maintain a productive interface between construction activities and ongoing plant operations. Construction on a new site offers a contractor the greatest amount of freedom in selecting the most cost-effective, efficient method for executing the work; however, potential obstructions include those areas of work occupied by other contractors.

Plant arrangement The overall plant arrangement plays a key role in establishing the sequence for erecting components and equipment. Plant arrangement defines the available access routes for material flow from the laydown and storage areas to the final installation points.

Scope The scope of work directly affects the construction plan and completion of the construction. Tools and equipment needed for one phase of the project may also be influenced by the requirements of other phases.

Parameters Product and component parameters, including size, weight and shipping configurations, dictate the type and size of material handling equip-

Fig. 5 Large scrubber modular components on a barge.

Fig. 6 Raising furnace wall section into position.

ment and the need for special handling procedures, methods, jigs and fixtures. Material type, thickness, and required erection tolerances are considered in establishing methods for component alignment, welding, post weld heat treatment, stress relieving, and nondestructive examination.

Site conditions Existing site conditions and accessibility play major roles in determining the size, weight and configuration of the components to be shipped. Water accessibility represents the greatest opportunity for shipping large shop-assembled modules; otherwise, rail or truck shipments are required. If multiple transportation modes are used, the size of the component is limited by the most restrictive shipping method.

Shipping accessibility only partially controls the extent of material and equipment shipments to the job site. Upon arrival, the material must be off-loaded and transported to its storage area or final installation point. Material off-loading schemes depend upon the mode of shipment and the site conditions at the off-loading area. Transportation of the material is then limited by the width, overhead clearances and load limitations of existing roadways.

Contractor interface Interface with other contractors is an important part of a construction project. Once portions of the project have been completed by individual contractors, effective interfacing between contractors is required to achieve certain location tolerances and to complete the erection.

Schedule The project schedule, including customer-specified start, finish and milestone dates, dictates the required erection sequence, labor needs and timing for project completion. The quality and skills of the local labor force play major roles in the erection plan and overall sequencing of the work.

Because cost is an important criterion governing selection of the erection method, the economic advantages and disadvantages of the other selection criteria must also be evaluated. Costs of construction operations versus component shipping configurations are often arranged in a matrix. The shipping configuration reflecting the lowest total cost is considered to be the most effective approach and becomes the preferred choice for erection. Tabulated construction costs are coupled with corresponding costs for component engineering, shop fabrication and assembly, and shipment to the job site to identify the most effective overall approach for a specific project.

Extent and location of field welds Overall product quality and constructability are enhanced by minimizing the number of field welds. The required welds must be readily accessible to allow the use of state-of-the-art equipment and processes.

Fig. 7 Modular components for an environmental control installation.

Fig. 8 A barge carrying SCR components is positioned for off-loading.

The scope of work to be performed and the component configuration influence the extent and location of field welds. The construction engineer must provide input regarding preferred field weld locations, suggested component configurations for these weld locations, and intended welding processes.

Construction engineering and rigging design

Background

Babcock & Wilcox Construction Co., Inc. (BWCC) takes an engineered approach to the execution of heavy lifts. Specialized tools, equipment, rigs, jigs and fixtures are provided for each project.

Modern construction includes modularized components that are erected on short timetables. To reduce construction and shipping costs, the weight of support steel components has been reduced. As a result, the calculation of imposed erection loads is critical.

In recent years, customer specifications have required the submittal of engineered rigging drawings and procedures prior to construction. These submittals often require the stamp of a licensed professional engineer.

Engineering principles

The construction engineer involved with product erection must be familiar with the product's construction details and function as well as responsible for the engineering, design and specification of all systems, tools, equipment, devices and mechanisms required for erection. This requires a command of engineering fundamentals and knowledge of principles taken from civil, structural, mechanical and electrical engineering disciplines.

Engineered systems

Engineered systems required to support construction operations may generally be identified by:

1. material handling (discussed below),
2. rigs, jigs and fixtures,
3. temporary supports, shoring and reinforcement of existing structural components,
4. access and protection structures,
5. temporary structures and construction facilities,
6. civil/site work, and
7. specialized tooling, machining and equipment.

Material handling and setting considerations

Like any phase of a construction project, material handling operations must be planned and integrated into the total construction plan. The size, weight and shipping schedules of material and components to be received and installed must be known in order to plan for storage area needs and to schedule crews and equipment. Special handling provisions and special inspection and storage requirements must be known and accommodated. In addition, the plant structure must be designed to support the loads involved during construction and the clearances required to get equipment in place.

Material handling plans must reflect actual site conditions including rail and truck access, locations and condition of haul roads, and availability and location of laydown areas and warehouses. Although most material shipped to a construction site can be handled with conventional construction equipment, material handling plans must identify any special equipment or fixtures required to handle large or unusual components.

Equipment used for material handling operations may generally be grouped by function into two categories: 1) transporting, and 2) lifting.

Equipment for transporting material is limited to that intended for on-site application, and does not include trucks, railroad cars or barges listed for material delivery to the job site.

A wide range of equipment is available for material transport and lifting on the job site. Selection and use depend on the configuration and weight of the material to be handled and the relative distance between transport points and available site clearances. Selected techniques and types of equipment that can be utilized for material transport and lifting are shown in Figs. 9 through 11.

While this listing is intended to provide an idea of the range of tools, methods and equipment available for material handling, it is not all inclusive. Various suppliers and manufacturers maintain and supply their own version of the generic equipment identified. The selection, specification and use of this equipment require a thorough construction engineering effort from the planning phase of the project through the execution of the work.

Fig. 9 Hydraulic platform trailer.

Depending on the equipment selected or available, various operations are involved to get the components to their final in-place position. These operations, represented in Fig. 12, may be accomplished by use of a single hydraulic crane or may require a sophisticated handling system custom designed for the application. (See Fig. 13.)

Serviceability

The serviceability of a material handling system is a measure of how well the system performs its intended function subject to conditions produced in its work environment.

Imposed loads and forces

The weight of the component being handled is a major consideration in the design and selection of the material handling system; however, component weight is not the only load imposed. All systems operate dynamically while simultaneously generating and being acted upon by forces that include steady-state and time-varying components. Forces are introduced by the operating environment, by the work being done, by the inertia of the masses put into motion, and by the weights of the load and the system components.[2]

Imposed loads are generally categorized as gravity or nongravity. Furthermore, both loadings may be static or dynamic. While dynamic loading must be recognized in designing the material handling system, the magnitude of acceleration/deceleration loads is small in most cases. As a result, static loading generally governs the system design.

Static gravity loading conditions are categorized as lifted load, dead load, live load, friction or impact.

The *lifted load* is the calculated rigging weight of the component being handled. This calculation must account for manufacturing variability and normally includes a safety factor. Shapiro suggests multiplying the calculated weight by 1.03 for mass produced items and by 1.07 for singular devices.[2]

Dead load is defined as the weight of the material handling system components themselves, independent of the component being lifted. In calculating this weight, the same considerations must be made with regard to potential dimensional variations in components. The suggested weight parameters act as multipliers as they relate to imposed loading or induced stress and as divisors as they provide stability or uplift resistance. The *imposed dead load* is the cumulative weight of all other components supported from the component being designed.

The *live load* is an allowance that accounts for the weight of construction personnel, scaffolding, ash (slag or dust accumulations), and snow or ice accumulations on the handled component. Because these loads are frequently unknown when the material handling system is designed, they are applied on the basis of a specified unit weight per surface area or a specified density per volume.

Friction loads are those associated with a body's resistance to movement. Friction takes on various forms depending on the system arrangement and load handling application

Fig. 10 Hydraulic gantry system.

The coefficients of static, rolling and sliding friction define friction loads. Depending on the arrangement of load handling components, friction may become a significant part of the total design load.

Impact loads are associated with sudden shocks that occur during a load handling operation. An impact allowance (25% of the handled load) is typically applied in the direction of load movement. Minimal impact loading is achieved by moving components at constant, slow speeds.

Nongravity loading conditions include wind loading, seismic loading and stabilization. Most applicable codes and specifications have reduced these dynamic conditions to static load equivalents. For the purpose of material handling system design, these conditions are therefore treated as such.

Wind loading, when imposed on material handling system components, is evaluated and applied in accordance with Uniform Building Code (UBC) or American National Standards Institute (ANSI) guidelines.

Seismic loading is not a major design consideration for most material handling systems. Nevertheless, seismic forces may be evaluated with UBC or ANSI specifications.

Stabilization forces are those forces required to maintain lateral stability of compression and flexural structural members. This force is arbitrarily consid-

Fig. 11 Manitowoc crane.

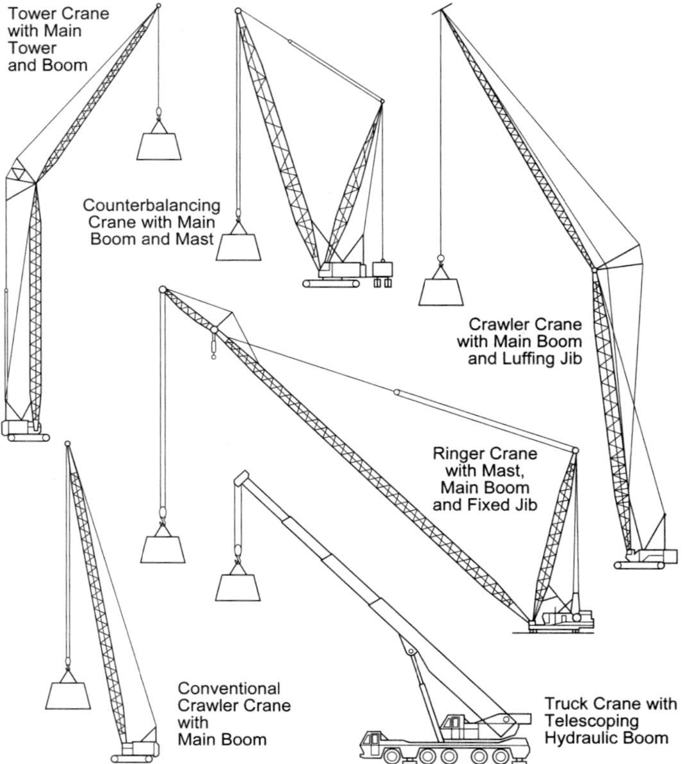

Fig. 12 Configurations of typical lift cranes.

ered to be 2% of the total force developed in the compression zone of the member being stabilized. In the case of compression members, the compression zone is the total cross section; for flexural members, this zone is limited to the compression flange.

Load stability

Load stability is the inherent tendency of a load in equilibrium, either static or dynamic, to remain in equilibrium and to minimize any translational or rotational effects produced via the imposition of unforeseen forces and/or moments throughout the course of load handling operations. A rigging system (see Fig. 14) must be configured to ensure that load stability is maintained throughout the course of load handling operations. The specific conditions of equilibrium are stable, unstable, and neutral.

Fig. 13 Site-specific handling system for an environmental control installation.

Load control

Load control is an assurance that the load will assume a predetermined, expected orientation without deviation throughout the course of handling operations or at any point in those operations.

Load distribution

Throughout the load handling operation, the component being handled seeks a position of static equilibrium. A lifted load always vertically aligns its center of gravity with the centerline of the lifting hook. Prudent arrangement and design of the material handling system is based on knowledge of the component's center of gravity as well as the magnitude of the resultant loads and forces acting on the system.

Depending on the handling operations involved, the relative position of a component's center of gravity with respect to its supports can vary with time. This variation produces a time-dependent variation in the magnitude of the support reactions, and thereby, the load imposed on the material handling system. Consider the steam drum illustrated in Fig. 15. Two separate sets of load blocks and rigging are used for raising the drum to its final position in the boiler support steel. Throughout the course of the raising operation, it may be necessary to incline the drum to clear local interferences with boiler support steel. As the drum is inclined at some angle θ from the horizontal, the load imposed on each set of blocks varies in accordance with the following expressions:

$$W_{L1} = \frac{1}{2}\left(1 + \frac{Y}{L}\tan\theta\right)W_L \tag{1}$$

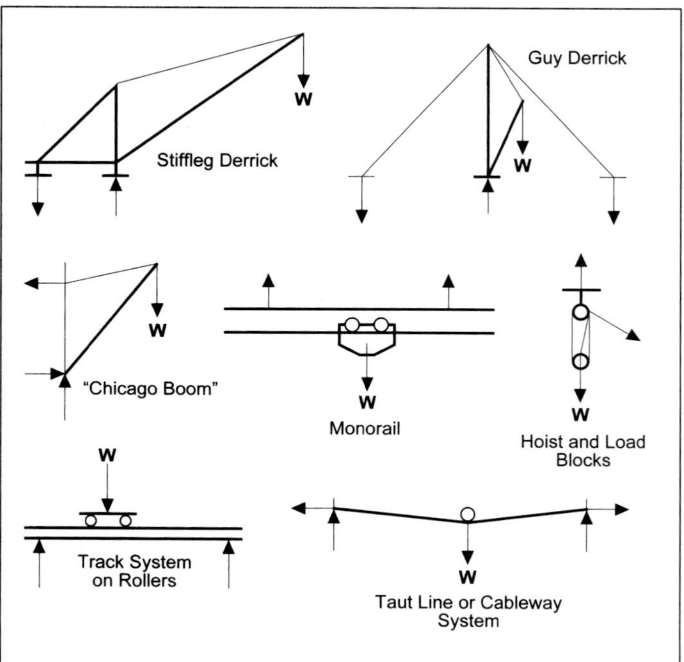

Fig. 14 Typical rigging configurations.

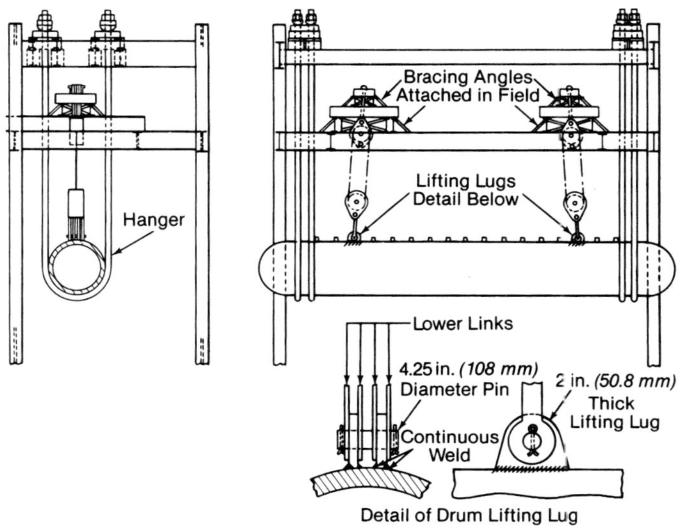

Fig. 15 Steam drum lift and support arrangement.

$$W_{L2} = \frac{1}{2}\left(1 - \frac{Y}{L}\tan\theta\right)W_L \qquad \textbf{(2)}$$

where

W_{L1} = portion of W_L supported by high end blocks
W_{L2} = portion of W_L supported by low end blocks
W_L = weight of component being raised
Y = 2 × (transverse distance between longitudinal centerline of drum and centerline of lifting lugs)
L = center-to-center spacing between lifting lugs

Per these expressions, when the drum is horizontal, i.e., θ = 0 deg:

$$W_{L1} = W_{L2} = \frac{W_L}{2} \qquad \textbf{(3)}$$

That is, the rigging weight of the drum is shared equally by each set of load blocks. However, as the angle θ is increased, the load transferred to the blocks on the upper end of the drum is increased. Therefore, each set of load blocks, rigging and drum lifting lugs are conservatively sized to accommodate the total rigging weight of the drum (Fig. 16).

Design loading

The material handling system design loading is established by identifying the maximum of several critical load combinations and then multiplying this value by a safety factor of 1.10. Typical combinations include:

lifted load + dead load + live load + impact,
lifted load + dead load + live load + friction, and
lifted load + dead load + live load + impact + greatest of wind seismic or stabilization load.

Friction and impact loading are mutually exclusive events, and therefore are never considered to be a viable combination.

Other site considerations – storage

Storage is also a critical issue. Storage areas must be laid out and material placed in storage to provide ready access to support the construction effort. Generally, a numbered grid system is used in laydown yards and warehouses to identify specific storage locations. Fig. 17 shows a typical storage layout.

To support the planned construction activities, material must not only be on site and available, but must be in good condition. In addition to inspection of the material upon receipt, the material and the storage areas are inspected periodically to satisfy specific storage requirements and to ensure proper protection of the material.

With the amount of material involved in most construction projects, it is essential that an effective documentation system be used. A complete listing of items to be received is needed to ensure that a complete inventory is maintained. For each item, material receiving and storage records must effectively document receipt and disbursement status, condition of the material, and current storage location.

Construction of a fossil fuel unit

The installation of a top-supported steam generating unit is unique in that the heaviest components are at or near the top and most other major components are hung from steel rods. Field assembly of the boiler begins by positioning the major upper components in a manner that provides a design deflection after completing the boiler installation. While the following discussion is primarily concerned with top-supported units, many of the basic procedures apply to all units.

Structural supports

The design of the supporting structural steel must consider the loading and stresses encountered during the installation of the boiler. The major construction-

Fig. 16 Typical utility boiler steam drum lift.

Fig. 17 Site laydown area.

imposed loads must be evaluated. Because certain structural members may be temporarily omitted to permit access for large subassemblies, the associated additional structural loads must also be considered.

Before erection is started, all foundations must be completed and checked for specification compliance. Because most large units are top-supported, as discussed in Chapter 26, erection of the structural supporting steel is the first step in building the unit. This is done by conventional methods, using cranes and derricks. Stairways and walkways are completed to the fullest possible extent to provide safe access to all parts of the unit.

Because many of the structural members govern the location of boiler and auxiliary components, supporting steel must be aligned with established building centerlines. After the structure has been aligned and plumbed, the members are rigidly connected with high tensile bolts or by welding before supporting any structural loads.

Assembly of pressure components

The construction schedule and plan include a timely sequence for installation of boiler components. The steam drum is usually the first major assembly placed in the structure. All large or heavy components, superheater modules, top headers, interconnecting pipes and wall tube panels are positioned, following the plan, in the boiler cavity while unrestricted space is available to access and hoist them.

Drums Drums for the largest boilers may weigh more than 400 t (363 t_m); they may be 100 ft (30.5 m) in length and may be located more than 200 ft (61 m) above the ground. To clear supporting steel, the drum may be inclined by as much as 60 deg (1.05 rad) during the lifting operation. Capacity and safety of the drum rigging are of primary importance in selecting the equipment and methods.

Fig. 16 shows a typical drum lifting arrangement using lugs that are shop-welded to the drum. Attachment pieces for the top of tackle blocks are located on temporary steel members above the boiler structure. These blocks should permit simple linkage connections at the top support and drum attachments. The capacity of the hoisting engine and the sizes of tackle blocks and wire rope are determined with ample safety factors.

When the drum is in its final position, it is supported from the top structural steel by U-bolts that permit linear movement with temperature variations. Fig. 18 shows a typical drum and support arrangement. The drum is a major anchor point for other boiler components and, for this reason, accurate location is important. The structural members, platforms and stairways that were omitted for drum erection are then installed.

Downcomers, headers and large pipes Plant access consideration is important and the sequential plan should be carefully followed in the erection of downcomers, loose headers and large pipes. These components are shipped in the longest lengths possible to avoid costly field welds. Large downcomers are often placed in the boiler cavity while steel is being erected and are moved to their final location when convenient. Without proper integration into the planning sequence, these components must be shipped in shorter lengths or additional structural steel must be omitted to provide clearances during erection.

Tube walls To reduce cost and the time required for field erection, a large percentage of furnace enclosure tubes are shop-assembled into membrane panels. B&W's preference is to furnish wide, long panels. However, as with any modularization approach, shipping and field conditions determine the panel sizes that can be provided. The wall panels are erected as-received from the supplier or, if conditions permit, the construction plan may include on-site assembly into larger wall sections. Panels may even be ground-assembled into complete walls with headers, casing, buckstays and doors attached.

Superheater, reheater and economizer These components can be furnished with loose headers and individual elements or in shop-assembled modules with or without headers. Again, the design features of the particular unit and project determine the degree of field or shop modularization. When large, heavy surface modules are to be erected, the construction plan normally includes early installation while access and hoisting clearance are available.

Tube connections Tubes are attached to drums or headers by welding, expanding, or a combination of the two (see Fig. 19). Generally, tubes that withstand pressures greater than 1500 psi (10,342 kPa) are expanded and seal-welded or welded to shop-attached stubs. Shop-assembled tube-to-header connections, such

Fig. 18 Radiant boiler steam drum in position.

as those in wall panels and superheater modules, are usually welded directly into the headers. (See Fig. 20.)

Design standards permit expanded tubes for low-pressure boilers. Tube expanding, or rolling, is a process of cold-working the end of a tube into contact with the metal of the drum or header tube hole or seat. The end of the tube is inserted into the hole and then plastically expanded by internal pressure; relieving of the pressure leaves the tube tightly seated in the drum or header.

A typical roller expander is shown in Fig. 21. This tool contains rolls set at a slight angle to the body of the expander, causing the tapered mandrel to feed inward when it is turned. For conditions of widely fluctuating temperatures and bending loads, the expanded joint must be seal-welded or replaced by a shop-attached tube stub.

Field welding

The welds that are made in the erection of a steam generating unit can be divided into two classes – those for assembly of plate and structural materials to support the unit and those for assembly of tubes, pipes and other pressure parts making up the steam generating unit. Welding on plate and structural materials is performed in accordance with the Structural Welding Code AWS D 1.1.

Welding that involves boiler pressure parts must conform to the requirements of the American Society of Mechanical Engineers (ASME) Boiler and Pressure Vessel Code, The fabrication Codes reference ASME IX for Welding and Brazing Qualification requirements. ASME IX provides the requirements for qualification of procedures and welding personnel.

Fig. 19 Expanding tube connections.

Various conditions make field welding different in many respects from welding in the shop (discussed in Chapter 38). A typical field welding environment includes dust, wind, variable temperature, rain, high scaffolding, locations remote from supply and maintenance facilities, and a mobile labor force. Under these conditions, it is important to consider the selection of the welding processes and equipment. Key factors include simplicity, reliability, portability, ease of maintenance, and availability of spare parts. Equally important are the skills and availability of the workforce.

Manual and semiautomatic welding methods are therefore preferred for field erection of standard boiler components. While there is a continuing search for and evaluation of reliable automated welding equipment for field erection and repair, it is seldom used. For example, use of orbital gas tungsten arc welding (GTAW) or gas metal arc welding (GMAW) have seen limited applications due to the precise weld joint fitup requirements, development and tight control of welding parameters and technique, limited space restriction for the equipment, lack of highly trained operators and the associated cost and maintenance of the welding systems.

Fig. 20 Headers with shop-assembled stubs.

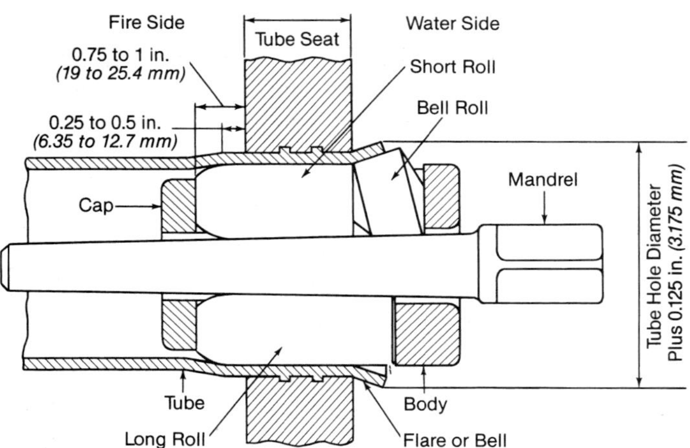

Fig. 21 Position of expander and mandrel after tube is expanded and flared.

Processes and equipment

Manual GTAW, shielded metal arc welding (SMAW), and flux core arc welding (FCAW) are the most common processes used in the field. (See Chapter 38, Table 1.) These processes are versatile and can be used on all field-weld materials utilizing equipment that is simple and easily maintained. In addition, most field welders are qualified for the GTAW and SMAW processes.

For welding non-pressure parts or making non-pressure attachments to pressure parts, there is increasing use of semiautomatic welding. Welders trained and qualified in GMAW or FCAW processes can make quality welds at significantly higher deposition rates compared to the GTAW or SMAW process.

As with the GTAW process, adequate precautions must be taken to protect the weld area from the elements, especially excessive wind. When a flux core welding electrode that is self-shielded, i.e., does not require external gas for protection of the molten weld puddle, is used for erecting boiler casing and other structural components, excessive wind must still be avoided.

Applications

Pressure welds Welds joining dissimilar metals are typically made in the shop because they are often difficult to make under field conditions and require special qualification. Field butt welds in tubes and pipes are made using pre-machined weld joints that have been covered with a protective cap. Thorough inspection and cleaning are still required before fitup and welding to avoid weld defects.

Butt welds may be made with an open groove joint or with a backing ring that is left in place. When an open groove joint is used, the root pass is typically made with the GTAW process because of the inherent better control of the inside weld surface contour. The groove is completed using the SMAW process.

When the weld joint contains a backing ring, the entire weld is usually made with the SMAW process. This method is normally used for the original erection of riser and downcomer piping. Rings are never used for furnace walls. However, they may be used for other pressure boundary tube welds depending upon engineering or customer requirements, and, if permitted, is the preferred method of fabrication.

Seal welds Seal welds are used to make mechanical joints fluid-tight. Joint strength is developed by pipe threads, tubes rolled into drum holes containing grooves, or header master handholes fitted with specially-designed plugs (see Fig. 22). The maximum size of seal welds made without post weld heat treatment is 0.375 in. (9.5 mm).

Special precautions are required to obtain sound seal welds; welding over pipe threads is not permitted. After seal welding, tubes should be lightly rerolled. Handhole plug seal welds should be thoroughly cleaned and are preheated to avoid cracking.

Qualification of welding procedures and personnel

Welding qualification requirements are contained in the ASME Code, Section IX. The ASME fabrication Codes and the National Board Inspection Code (NBIC), published by the National Board of Boiler and Pressure Vessel Inspectors, references this Code for qualification of repairs and alterations to in-service boilers.

A welding procedure qualification test verifies that required mechanical properties are obtained when designated materials are welded by following a specific technique. BWCC has established many procedural qualification records based upon the ASME Code. These procedure qualifications support hundreds of welding procedure specifications.

A welder qualification test is required to verify that an individual has the skill to deposit sound weld metal in a given position while following an established welding procedure. Welders are qualified in accordance with ASME IX standards either at the job site or at simultaneous testing sites by participating organizations.

Thermal treatment

Preheat Preheating the weld metal is the most effective way of preventing cracks in low alloy steel welds because it reduces the cooling rate of the weld and base metal. The required minimum preheat temperature is specified in the field weld schedule and on BWCC's welding procedure specifications (WPS). Preheating methods include the use of electric-resistance blankets and oxy-gas or oxy-fuel torches. Special temperature-indicating crayons are used to assure that the proper preheat temperature has been attained. (See also Chapter 38.)

Post weld heat treatment Requirements for post weld stress relief are also specified in the field weld schedule. Electric resistance heaters are commonly used for field post weld heat. These heaters offer the advantages of ease of application and precise temperature control.

Codes, inspections and examinations

The ASME Code requires a valid ASME Certification of Authorization to install power boilers. To obtain an ASME Certificate of Authorization, an organization is required to maintain a Quality Control System that addresses the Code requirements. A review is performed by ASME to evaluate the Quality Control System and verify implementation. Once issued, the Certificate of Authorization is renewed triannually by review. A boiler installed in accordance with the ASME Code must be inspected by an authorized inspector at various stages of assembly to assure that all Code requirements are met.

Section I of the ASME Code contains requirements for the installation of power boilers. ASME B31.1 contains requirements for the installation of boiler and non-boiler external piping. The state or jurisdictional authority maintains rules and regulations that govern the installation of boilers and establishes the applicable governing Codes. Section I of the ASME Code has been adopted by law in most states or jurisdictions.

Whereas ASME governs new construction, the NBIC governs repairs and alterations to boilers and other pressure-retaining items. The NBIC requires the repairs and/or alterations to comply, to the extent possible, with the original Code of construction. As with the ASME Code, the NBIC is mandatory only when adopted by a state or other jurisdictional authority.

Section I of the ASME Code specifies the nondestructive examination (NDE) requirements for new construction of power boilers. ASME B31.1 specifies the NDE requirements for boiler and non-boiler external piping. NDE requirements for repairs and alterations are specified in the NBIC. The type and extent of NDE must be considered during the proposal, planning and scheduling processes. For all construction, the applicable regulations and any additional NDE requirements exceeding the Code should clearly be specified in contract documents. The nondestructive technologies are discussed in more depth in Chapters 38 and 45.

Construction of environmental equipment

Regulatory requirements associated with the Clean Air Act Amendments of 1990 caused many fossil fuel-fired electric generating plants in the United States (U.S.) to install, or plan for the installation of, emission reduction equipment. Many stations installed sulfur dioxide (SO_2) scrubbers in the early 1990s which were then followed in the early 2000s by the installation of selective catalytic reduction (SCR) systems for nitrogen oxides (NO_x) removal.

While erection of environmental equipment is significantly different from boiler erection, many of the same principles and attributes of boiler erection apply. A modularized approach generally pays significant dividends over stick-built construction at the job site. The project benefits in both labor savings and in the potential for a reduced site completion schedule when high-cost field labor is replaced with lower-cost shop labor.

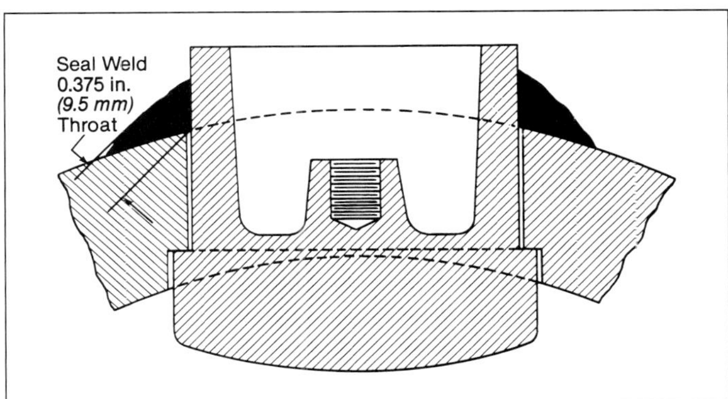

Fig. 22 Welded handhole fitting for higher pressure boilers.

The construction principles utilized for environmental equipment are the same. However, the method of construction differs significantly. SCR projects predominantly consist of structural steel and fabricated plate, while FGD system projects consist of fabricated plate, structural steel, process equipment and piping. There is little in the way of operating equipment with the exception of dampers, sonic horns or sootblowers, ammonia vaporization equipment, and some piping and instrumentation. In these installations, the chemical process required to support the operation utilizes additional equipment to handle the lime or limestone reagents and gypsum removal and drying.

Most SCR projects are designed with the reactor and flues bottom-supported off structural steel. The primary component of FGD systems is the absorber tower which is independently bottom-supported on a separate foundation. For SCR systems, close coordination is required between steel erection activities and the installation of the interconnecting flues and reactor. Complicating matters on a retrofit project is the space available for the SCR. The process temperature for most SCR systems requires the boiler flue gas to be taken from an area immediately downstream of the economizer, and before the air heater inlet. On large boilers, this tie-in point can be more than one hundred feet (tens of meters) above ground level. Also, since a reactor can contain nearly a million pounds (450,000 kg) of catalyst at a significant elevation above grade, concerns for overturn and uplift forces may require a complex foundation design.

At many sites constructed decades ago, little consideration was given to future expansion and therefore, current space limitations are a serious concern. Many projects are faced with a significant rerouting of the boiler exit gas to a point where space is available for either an SCR or FGD system. The flue gas must then be routed back to continue its path through the air heater for SCR applications or to the stack for FGD systems. Each project poses a unique set of circumstances.

Sequencing and modularization

Early in the project, specific information about the weight of ductwork sections and equipment must be estimated. Many times the module boundaries are easily identified, due to equipment terminal points or expansion joint locations. Other times, sound judgments must be made, based on experience. The erector is the prime contributor to determining splice locations that are best from a variety of standpoints, including access, crane capacity and geometry.

The resulting erection sequence can then be used in establishing schedule priorities for detail design, material procurement and fabrication. To allow maximum flexibility for installation, schedule priorities should be developed in accordance with erection plans. Receiving materials far in advance of the time required at the job, or in a sequence that does not support planned erection, results in increased handling costs. Without the proper materials, the constructor is forced to work on activities that deviate from the intended plan, increasing overall project cost.

The erector should complete a constructability review prior to the issuance of key design outputs. This review often uncovers missing or incorrect information. Early identification of potential problems can alleviate costly field rework or production delays due to additional engineering evaluations and the resulting drawing revisions.

Once modules are identified and an erection sequence is established, the project team interacts with plate fabricators for delivery options. A modular construction approach provides the project team with the opportunity to review and approve the manufacturer's plan for component fabrication. This process may result in on-site productivity improvements as the constructor advises on alternative shipping configurations and rigging requirements, and gains valuable information about the material to be received, thus helping to identify the extent of preparatory work that will be required on site prior to erection.

Fig. 23 is a typical arrangement for an SCR retrofit project. Identified in the figure are module boundaries utilized for design, procurement, planning and erection activities. The individual modules are installed in a defined sequence to maximize efficiency and minimize erection time. This sequence, which usually starts from the lowest to the highest point, builds from the reactor back to the boiler and involves the erection of sufficient steel to set the outlet or bypass flues. Once the initial flues are set, steel progresses again to the point where additional flue sections can be set and the reactor built. Ideally, horizontal runs of ductwork are bottom-supported, and require that the support steel is first installed beneath the duct. Structural engineers should be aware of the construction/erection plan so that the steel erection sequence and requirements support the equipment's erection sequence. These activities require close coordination with the design engineers to ensure that the uncompleted steel structure is adequate to support the initial duct sections as they are installed. Cover-plating the existing steel is frequently required prior to imposing additional loads from the new structure and can be a time-consuming and costly effort. If present, lead-based paint must be abated from existing steel prior to any welding or grinding operations. The use of bolt interface connections rather than welded connections could minimize lead abatement requirements. Close coordination with plant operations is also required as a result of the numerous interferences with plant utilities supported from the columns or beams being modified.

Tie-in to the unit is the last activity phase, accomplished during a planned unit outage. Depending upon planned unit outages, however, there is flexibility in the erection sequence. Many times a tie-in outage occurs before construction of the SCR is initiated, especially if the plant has a major outage scheduled well in advance of the SCR completion. In these instances, a gas bypass flue is installed which allows the construction of the SCR during unit operation. This requires dampers and blanking plates to allow the balance of construction to take place subsequent to the tie-in. The blanking plates allow safe entry to the flues during plant operation. However, this approach generally impacts the erection sequence and potentially impacts the total construction cost. A second, shorter outage is then required to make the final tie-in of fluework.

Construction safety

A comprehensive safety program must be implemented at each construction site. The objective is to promote an awareness of safety among personnel and create a safe working environment to prevent injuries. BWCC has developed a Safety Program Manual in support of a construction site safety program. The manual has been specifically designed for construction and includes the requirements of 29 CFR 1926 Safety and Health Regulations for Construction by the Occupational Safety and Health Administration (OSHA). The contents of the Safety Program Manual include, but are not limited to, the safety procedures outlined in Table 2.

Internal cleanliness of pressure parts

Manufacturing and shipping

Many problems can develop during operations because of dirt or other foreign matter in the circulating system of a boiler. During the fabrication of drums, tubes, headers, pipes or other pressure parts, precautionary measures must be taken to assure internal cleanliness. For boilers that use welded stub construction, the ends of all tubes and pipes, as well as the stubs in the drums and headers, are closed with metal caps and sealed with plastic tape. This usually makes it unnecessary to clean these components in the field. Inevitably there are a few closure caps displaced during shipping, unloading and handling. In such instances, the construction crew must examine the components involved, perform any necessary cleaning, and replace the caps as soon as possible.

Field assembly

Boiler pressure parts should be searched before handhole and manhole closure fittings are installed. Reasonably straight tubes between headers or drums may be probed by passing a ball through them. Bent tubes that will not clear themselves of foreign material by gravity may be probed by blowing a tight-fitting sponge through them. The butt welds in small diameter tubes are examined radiographically for misalignment or obstructions that may cause flow restrictions. Headers are checked with a light and mirror immediately before the end handhole fittings are installed.

No foreign material may enter the pressure part system during assembly. Closures are left in place until removal for welding or positioning equipment. If shop-installed closures are removed to permit installation, temporary closures are placed over the openings to prevent debris from entering the equipment. Prior to flushing or final closing of equipment, a visual inspection is made and foreign objects are removed from the equipment.

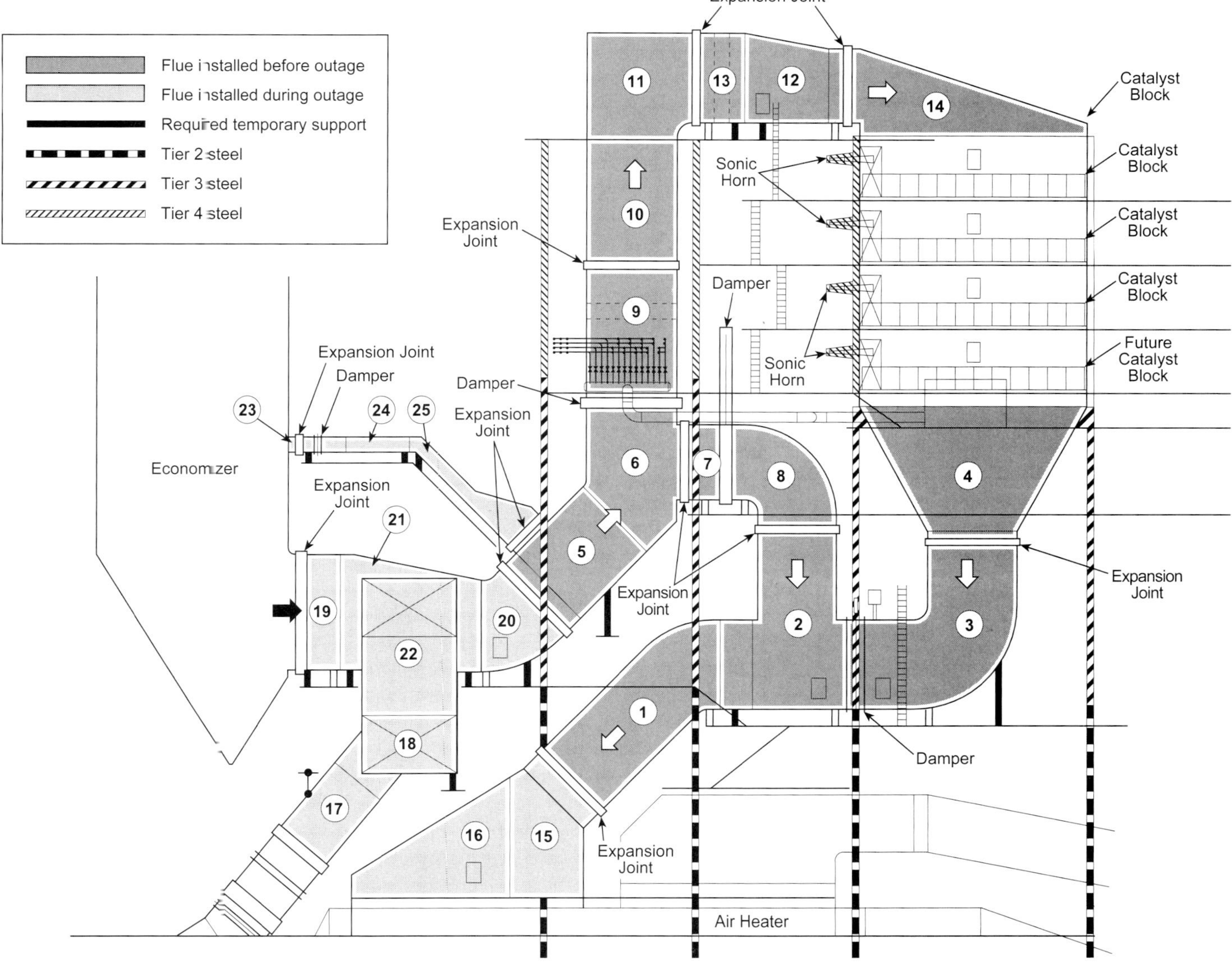

Fig. 23 Typical arrangement for an SCR retrofit project.

Removal of tube sections during repair or alteration must be performed in a manner that prevents foreign material from entering the tubes during the cutting operation. Ends of existing tubes are plugged before machining new weld preparations.

Post-construction testing

Hydrostatic testing

When all pressure part connections have been completed, the steam generating unit is tested at a pressure specified by the applicable code; this is usually 1.5 times the maximum allowable working pressure (MAWP). Before the test is applied, a final inspection is made of all welding. External connections, including all fittings, are completed within Code requirements. Connections for flange safety valves are usually blanked. Welded safety valves are closed with an internal plug.

Hydrostatic testing water must be clean and warm enough to bring pressure part metal temperatures to at least 70F (21C). Nondrainable superheaters and reheaters are filled with demineralized water or condensate, if available. As a boiler is being filled, it is vented at every available connection to allow water to reach every circuit.

Hydrostatic pressure is applied after the boiler is filled. If a boiler feed pump or other high capacity pump is used, precautions must be taken to control the pressure at all times. While water is being pumped into the boiler, continuous inspection is made to detect any leakage. During hydrostatic testing, safety precautions must be observed, and nonessential personnel must be kept away from the test area.

The test pressure must be held long enough to allow minute leaks to be detected. Following this holding period, pressure is reduced to MAWP and retained for an inspection period. During final inspection, every area of the pressure system is viewed for leakage and repaired if necessary.

After inspection, the unit is drained and all nondrainable parts are protected from freezing, if necessary. The refractory, insulation and casing work are then completed in areas left open for inspection.

Air testing

Gas-bearing components, such as the boiler, flues and ducts are field-tested for leakage and tightness. All boilers are tested by one of two procedures. Boilers designed to operate with a positive pressure, wherein furnace leaks would emit gases from the boiler setting, are tested following a pressure-fired procedure. Units designed for suction or balanced draft are tested following the balanced draft procedure.

The pressure-fired air test procedure is a pressure drop test. A test is considered satisfactory when the pressure is raised to 15 in. wg (3.74 kPa) maximum and does not drop more than 5 in. wg (1.25 kPa) in ten minutes. If pressure can not be maintained, leaks are sought and corrected and the test is rerun until satisfactory tightness is achieved. On large units, it is common to install temporary bulkheads in flue and duct runs for individual tests.

The balanced draft testing procedure is commonly performed by pressurizing the system with the forced draft fan while throttling the dampers at the downstream test terminal. With the system pressurized, all field welds are inspected by sight and sound to detect leaks.

Assembly of nonpressure parts and auxiliaries

While pressure parts are being assembled and tested, work is in progress on nonpressure parts and auxiliary equipment. These components, including air heaters, fuel equipment, fans, duct work, refractory, insulation and casing, require a large portion of the construction labor. They must be scheduled with consideration for items that must be assembled in a prescribed sequence. For example, air heaters are finished prior to erection of flues and ducts. Pulverizers are placed early enough for piping, drives and other auxiliaries to be completed.

Table 2
Selected BWCC Safety Procedures for Construction

Posting requirements	Welding, thermal cutting and compressed gas cylinders
Medical facilities and treatment	Fire protection
Safety training	Steel erection
Hazard communication	Confined space entry
Accident and personal injury reporting	Lockout/tagout
Personal protective equipment	Excavation and trenching
Job safety analysis	Radiation protection
Housekeeping	Documentation and record-keeping
Operator pre-qualification	Project site safety inspections
Lifting equipment	Subcontractor safety
Hand tools	Material safety data sheets
Electrical	Fall protection
Scaffolds and ladders	Floor, wall openings and stairways

Refractory materials

The furnace and other wall areas of modern fossil fuel boilers are made almost entirely of water-cooled tubes. The increased use of membrane walls has reduced the use of refractory in these areas. However, castable and plastic refractories may still be used to seal flat studded areas, wall penetrations, and door and wall box seals. (See Chapter 23.)

Other than membrane wall construction, when tubes are tangent or flat studded, several types of plastic refractory materials are applied to the outside of tubes for insulation or sealing purposes. When an inner casing is to be applied directly against the tubes, the refractory serves principally as an inert filler material for the lanes between tubes. It has a binder that cements it to the tubes and it is troweled flush with the surface of the wall.

Smaller boilers may have furnaces constructed of tubes on wide-spaced centers backed with a layer of brick or tile. Brickwork of this type is supported by the pressure parts and held in place by studs. The brick is insulated and made air-tight by various combinations of plastic refractory, plastic or block insulation and casing.

High quality workmanship is mandatory in the application of refractory material. Construction details are clearly outlined on drawings and instructions that also designate the materials to be used. These materials must be applied to correct contour and thickness without voids or excessive cracking. Skilled mechanics and close supervision are essential.

Casing

In general, inner or outer boiler casing is used in areas that are not membraned. Outer casing is applied outside of insulating materials and has, for the most part, been replaced by sheet metal lagging. Inner casing is positioned inside of insulation and may be pressure or nonpressure casing.

The primary purpose of pressure inner casing is to prevent air and gas leakage from a steam generating unit. The inner casing is seal welded to prevent leakage from the boiler setting and also serves as a base for the application of insulation and lagging.

In most cases, modern boiler casings feature welded construction. For pressure casing, it is essential that the strength of the weld equal the strength of the plate, that the welds be free of cracks and pinholes, and that they present a good appearance. For pressure casing, tightness is checked by a pressure drop air test. Nonpressure casings are given a thorough visual inspection.

Insulation

Selecting the type and thickness of insulation applied to boiler surfaces is an essential part of the design work. Specifications are written and drawings are made to indicate the applied materials and the method of attachment.

Typical standards specify insulation blocks or insulation blankets. Some plastic insulation is used for filling voids. The thickness of the insulation for all surfaces is selected to give a specific surface temperature, e.g., 130F (54C) with 80F (27C) ambient air temperature and 50 ft/s (15.2 m/s) surface velocity. The prime requisites for installing boiler insulation are that the material be tight, free of voids, well anchored, and reinforced where necessary.

Ductwork with heavy stiffeners at frequent intervals presents a difficult insulating job. With the stiffeners outside the flue and duct plate, the insulation thickness is often increased over minimum requirements to assure adequate coverage of the stiffeners. If the stiffeners are placed in a manner where standard insulation coverage is too costly, the insulation may be placed over the stiffeners. As an alternative, inner lagging may be applied over the stiffeners and the minimum required insulation is placed over the lagging. When insulation is applied at an appreciable distance from a hot surface with a long vertical run, air may circulate in the void. This increases heat transfer into the insulation, which is undesirable. Where vertical runs exceed 10 ft (3 m), horizontal barriers are installed between the duct and/or tube wall and insulation.

Sheet metal lagging

Outdoor boilers require the use of a durable waterproof lagging. A light gauge metal lagging (galvanized steel or aluminum) or outer casing is used on walls, flues, ducts, air heaters, vestibules, windboxes, downcomers, steam/water lines and other exposed components. The metal covering does not seal against air infiltration but serves as a barrier to protect the insulation against water and physical damage. The cost is reasonable and it is extensively used in both indoor and outdoor installations.

Pre-operational inspection

A final inspection is made after erection work is completed and the unit is ready for operation. This is the combined responsibility of the erection, service and operating organizations.

Externally, all components of the unit are checked for expansion clearances. Obscure corners are examined for any construction blocking or bracing that might have been left in place. Points reviewed for expansion clearances are platforms and walkways constructed adjacent to external members that move with the unit.

Of special importance in fossil fuel units is the removal of combustible materials that may present an explosion hazard during initial firing. Internal cavities are checked to ensure that all debris has been removed. Tubes are given a final inspection for alignment, particularly those that might interfere with sootblower operations. Movable connections between tubes are examined for expansion clearances and to see that all attachments are properly anchored.

On most units, construction schedules do not permit completion of insulation before the unit is fired. It is important that this work be completed in areas that will be inaccessible or cause a safety hazard when the unit is in operation.

References

1. *Constructability: A Primer*, Publication 3-1, the Construction Industry Institute, Constructability Task Force, Bureau of Engineering Research, The University of Texas at Austin, Austin, Texas, Third Printing, April, 1990.

2. Shapiro, H.I., Shapiro, J.P., and Shapiro, L.K., Ed., *Cranes & Derricks,* McGraw-Hill Education, November, 1980.

Section VI
Operations

With proper design, manufacture and construction, modern steam generating systems are capable of operating efficiently for long periods of service. However, successful operation requires adherence to basic operating principles. These principles begin with the careful monitoring of operating conditions so that a system functions within design limits. Chapter 40 describes the instrumentation for monitoring pressures, temperatures and flows – the key process parameters. (Specialized instrumentation for environmental equipment is covered in Section IV, Chapter 36.) These operating parameters then serve as the inputs to the control system. The fundamentals of control theory and modern integrated control systems are reviewed in Chapter 41. These systems have advanced rapidly during the past few decades to provide greater operator knowledge and flexibility to optimize plant performance.

Successful long-term operation of steam producing systems requires careful attention to water treatment and water chemistry control. Chapter 42 provides a discussion of water treatment practices from startup through operation and chemical cleaning. Drum and once-through boilers have different requirements, and each boiler requires individual consideration.

General operating principles and guidelines outlined in Chapter 43 conclude this section. Each steam generating system is unique and requires specific operating guides. However, a number of general principles covering initial operations serve as a basis.

Virtually all areas of a power plant are continually monitored to ensure efficient and reliable daily operation.

Chapter 40

Pressure, Temperature, Quality and Flow Measurement

Instruments for measuring pressure, temperature, fluid flow, and the quality and purity of steam are essential in the operation of a power plant. The measurements obtained permit safe, economical and reliable operation of the steam generating equipment. Instrumentation may range from a simple indicating device to a complex automatic measuring device. The fundamental purpose of the instrument is to convert some physical property of the material being measured into useful information. The methods used to obtain the measurement depend on available technology, economics, and the purpose for which the information is being obtained.

Test instrumentation, often portable, may be used to determine flow, pressures, and temperatures required to satisfy the user and the boiler equipment supplier that the design operating conditions have been achieved. Requirements for these instruments are summarized in the American Society of Mechanical Engineers (ASME) Performance Test Codes. Many test instruments require skilled technical operators, careful handling, and frequent calibration. Some instruments and measurement methods may not be suitable for long-term continuous operation because of a harsh environment, material limitations, or excessive cost. Instruments used for continuous measurement may require compromises in accuracy for long-term, dependable, reliable operation.

Pressure measurement

The pressure gauge is probably the earliest instrument used in boiler operation. Today, with complex control systems in operation and more than one hundred thirty years after the first water tube safety boiler, a pressure gauge is still used to determine steam drum pressure. The Bourdon tube pressure gauge is shown in Fig. 1. Although improvements have been made in construction and accuracy, its basic principle of operation remains unchanged. A closed end oval tube in a semicircular shape straightens with internal pressure. The movement of the closed end is converted to an indication (needle position).

Pressure measuring instruments take various forms, depending on the magnitude of the pressure, the accuracy desired and the application.

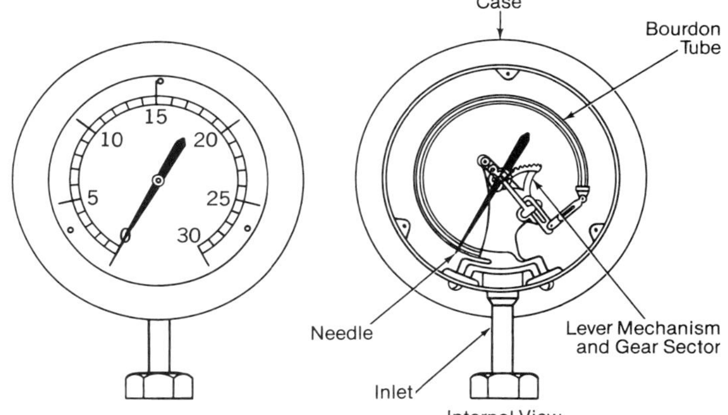

Fig. 1 Bourdon gauge.

Manometers are considered an accurate means of pressure or pressure differential measurement. These instruments, containing a wide variety of fluids depending on the pressure, are capable of high accuracy with careful use. The fluids used vary from those lighter than water for low pressures to mercury for relatively high pressures. Fig. 2 illustrates an inclined manometer used to measure low pressure differentials

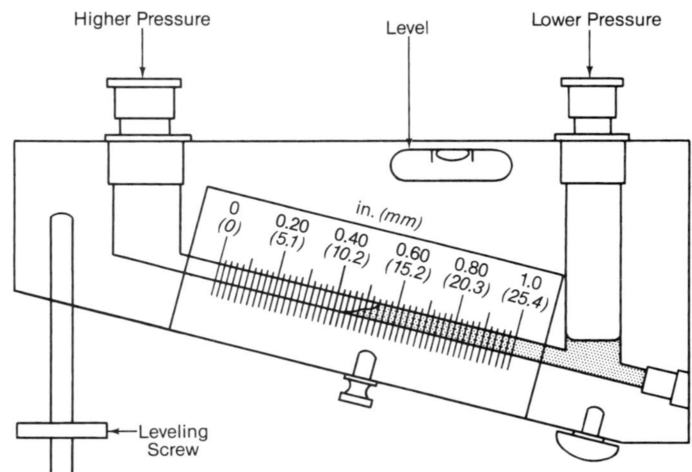

Fig. 2 Inclined differential manometer (*courtesy of Dwyer Instruments, Inc.*).

at low static pressure. Differential diaphragm gauges with a magnetic linkage are also used for low pressure measurement. Fig. 3 shows a high pressure mercury manometer. For greater precision in measuring small pressure differentials, such as the measurement of flow orifice differentials, hook gauges or micromanometers may be used.

Bourdon tube pressure gauges are available for measuring a wide range of static pressures in varying degrees of precision and accuracy. Pressure gauges used as operating guides need not be of high precision and normally have scale subdivisions of about 1% of the full scale range. For certain tests, such as hydrostatic testing of pressure parts and boiler efficiency, more precision is required. Gauges with 0.1% full scale subdivisions are used. For performance calculations, high precision temperature and pressure measurements are required to accurately determine steam and water enthalpies. Dead weight gauges or calibrated pressure transmitters are preferable to Bourdon gauges for high precision pressure measurement.

Diaphragm type gauges are used for measuring differential pressures. A slack diaphragm pressure gauge can be used to measure small differentials where total pressure does not exceed about 1.0 psig (6.9 kPa). For high static pressures, opposed bellows gauges (Fig. 4) read a wide range of differential pressures. They are suitable for measuring differential pressure across boiler circuits and can be used to measure differentials from 2 to 100 psi (0.01 to 0.69 MPa) at static pressures up to 6000 psi (41.37 MPa).

Incorporated into many pressure measurement devices is the capability of producing an output signal proportional to the measurement. The output signal may be transmitted to a central measurement or control system. Pneumatic transmitted signals are

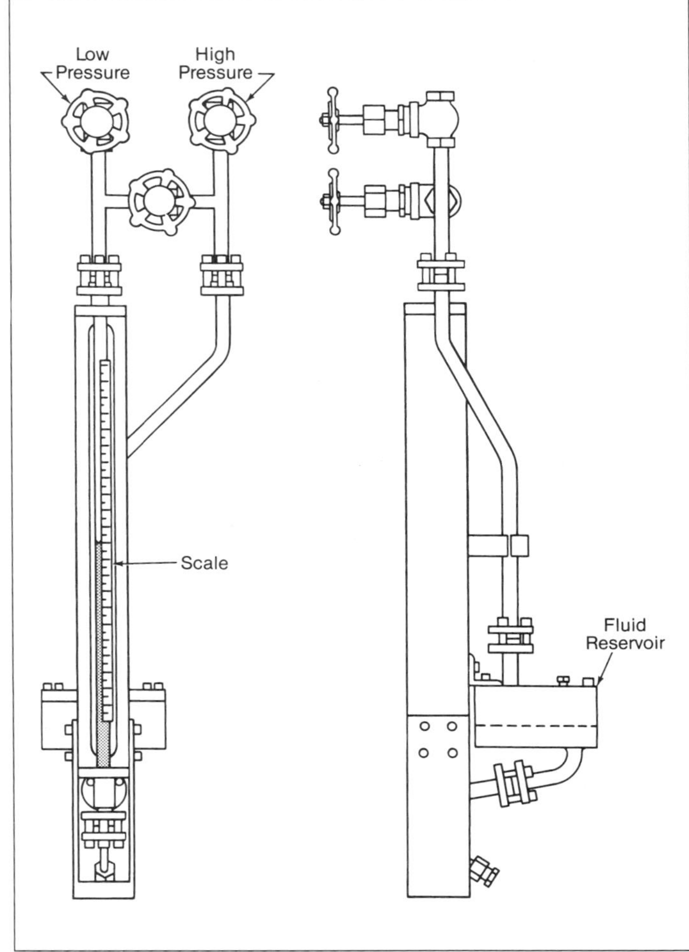

Fig. 3 High pressure mercury manometer (*courtesy of Meriam Process Technologies*).

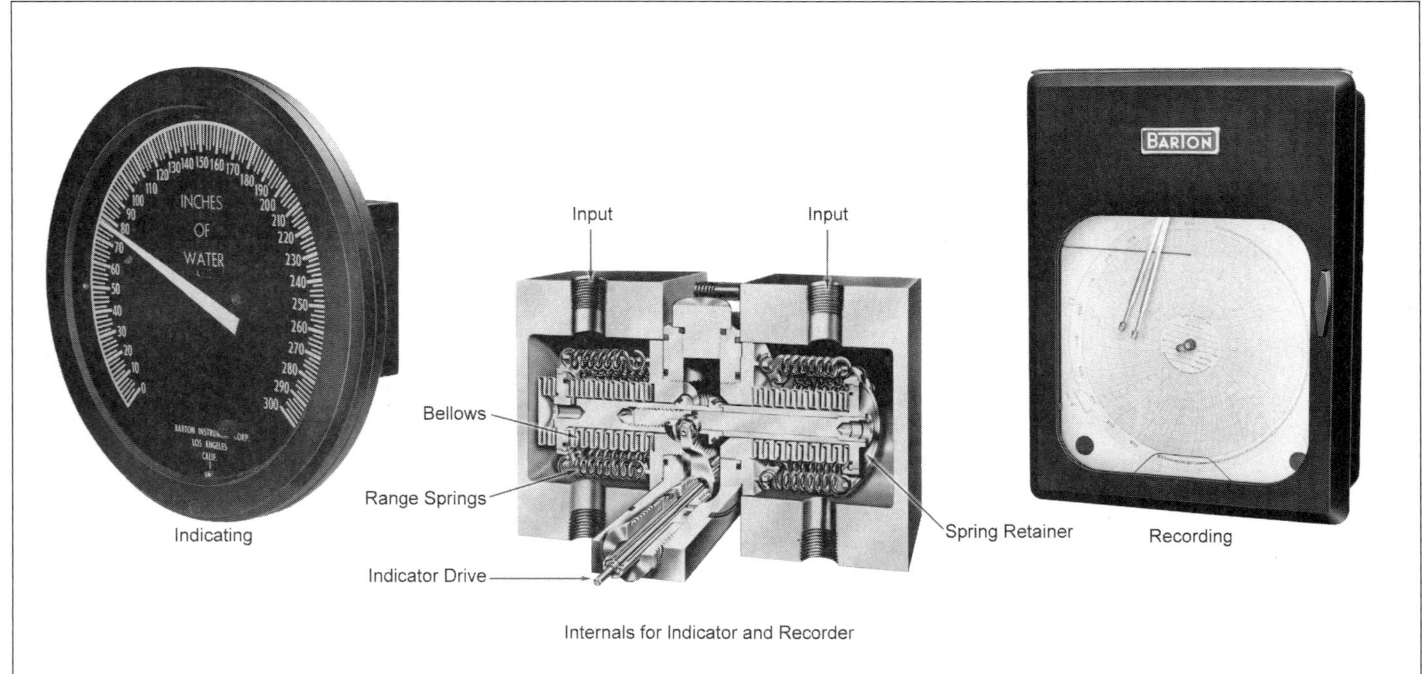

Fig. 4 Opposed bellows gauge (*courtesy of ITT Barton, a unit of ITT Industries, Inc.*).

used in control systems, but the more common designs use electrical circuitry to sense the changing position of basic mechanical motion and to produce an electrical output signal to be transmitted. These are readily adaptable to computer based systems. Some transducers use a piezoelectric crystal which changes electrical resistance as the element is deformed under pressure. Other pressure transmitter designs use a diaphragm with strain gauges attached; minor changes in diaphragm strain are transferred to the strain gauges, which change resistance. Electrical circuits recognize the resistance change in the crystal or strain gauges and produce an indication of pressure. Advanced applications have used an optic glass fiber embedded in a metal diaphragm, with pressure measurement determined as a change in the light beam traversing the fiber. Pressure transducers incorporating electrical components have the advantage of fewer mechanical parts for improved durability. However, this may restrict the installation location due to temperature limitations of the electrical components. Instruments configured to generate electric signals have been adapted to incorporate computing capability to assist in calibration. Pressure transmitters and transducers have been readily adapted for power plant computer data acquisition and control systems.

Instrument corrections

In recording and reporting pressure readings, corrections may be necessary for water leg and for conversion to absolute pressure by adding atmospheric pressure. Effective water leg is the added pressure imposed on the gauge by the leg of water standing above the gauge. Fig. 5 illustrates a water leg correction to a pressure gauge reading. The density of the water in the leg may change with ambient temperature and, for precise measurement, the water leg temperature should be monitored. On some gauges it may be possible to zero the gauge at zero operating pressure with the water leg completely filled to compensate for the water leg static head. Care must be taken in performing water leg adjustments as this may affect the instrument calibration over the operating range. Differential pressure measurements require a water leg correction for both sides of the measuring instrument. For some liquid level devices, the differential pressure measurements from two water legs of different elevations are used to provide the indication of liquid level. Differential pressure measurements across orifices, nozzles or pitot tubes to measure flow are described below.

Location of pressure measurement connections

The criteria for selecting a connection location to the pressure source for the measuring device are the same regardless of the magnitude of the pressure, the type of measuring device or the fluid being measured. Pressure connections, or *taps*, in piping, flues or ducts should be located to avoid impacts or eddies; this assures an accurate static pressure measurement. Under certain circumstances a through the wall pressure connection may not be representative. It may be necessary to use a probe incorporating a tip configuration that minimizes the effects of the flowing fluid to obtain a representative static pressure measurement.

Connecting lines to instruments should be as short and direct as possible and must be leak-free. For differential pressure readings, it is preferable to use a differential pressure measuring device rather than to calculate the difference between the readings on two instruments. Particular attention should be paid to the placement of the instrument, the potential for condensation in the impulse lines, and the accumulation of debris or non-condensable gases in the completed installation.

Temperature measurement

Today, the Fahrenheit and Celsius (formerly centigrade) scales are the most common and firmly established temperature scales.

The Fahrenheit scale is fixed at the freezing point (32F) and the boiling point (212F) of pure water at atmospheric pressure, with 180 equal degrees between the two points. The centigrade scale was originally based on the same freezing (0C) and boiling (100C) points of water at standard atmospheric pressure, using the convenient 100 degree interval of the decimal system.

In 1960, the General Conference of Weights and Measures changed the defining fixed point from the freezing point to the triple point (0.0C) of water. The triple point is the condition under which the three phases of matter (solid, liquid and vapor) coexist in equilibrium. This point is more easily and accurately reproduced than the freezing point. At standard atmospheric pressure the interval between the triple point for water and steam on the centigrade scale became 99.99C instead of 100C.

In 1990, temperature standards were further refined by the International Temperature Scale 1990

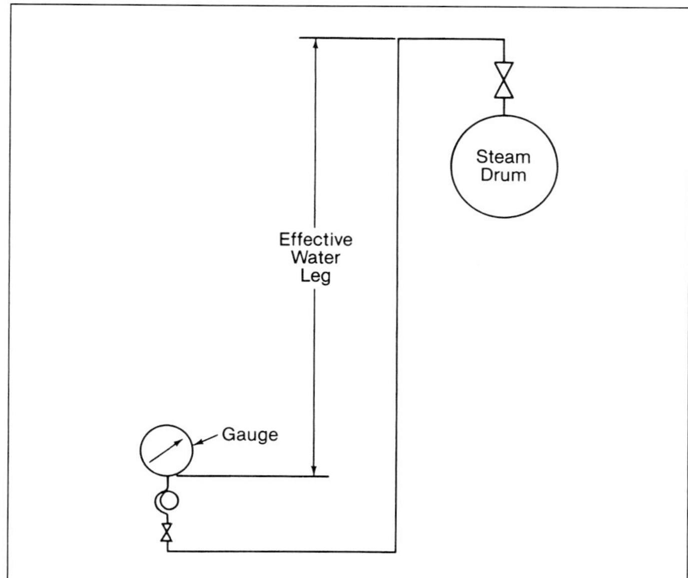

Fig. 5 Application of water leg correction to pressure gauge reading (Correction $\Delta P = g\rho L/g_c$).

(ITS-90). These changes, however small, reflect the continued review of temperature measurement. As measured by the current standard, water at standard conditions boils at 211.953F (99.974C).

In thermometer calibrations, the fusion and vaporization of various pure substances have been carefully determined and are known as fixed points. Two additional scales in scientific work use the absolute zero of temperature, i.e., –273.16C or –459.69R. The absolute scale using Celsius degrees as the temperature interval is called the Kelvin (K) scale; the absolute scale using Fahrenheit degrees is called the Rankine (R) scale.

Methods of measuring temperature

Heat affected properties of substances, such as thermal expansion, radiation and electrical effects, are incorporated into commercial temperature measuring instruments. These instruments vary in their precision depending on the property measured, the substance used and the design of the instrument. The care taken in selecting the correct instrument largely determines the accuracy of the results. It is the engineer's responsibility to make certain that the application is correct and that the results are not affected by extraneous factors.

Changes of state

Fusion For a pure chemical element or compound such as mercury or water, fusion, or change of state from solid to liquid, occurs at a fixed temperature. The melting points of these materials are suitable fixed reference points for temperature scales.

The fusion of pyrometric cones is widely used in the ceramic industry as a method of measuring high temperatures in refractory heating furnaces. These cones, small pyramids about 2 in. (50.8 mm) high, are made of oxide and glass mixtures that soften and melt at established temperatures. The pyrometric cone is suitable for temperatures ranging from 1100 to 3600F (593 to 1982C). Its use in the power industry is generally limited to the laboratory. An adaptation of this method of temperature determination is used when observing ash and slag accumulations soften and flow in a coal-fired furnace or superheater.

Fusion pyrometers are also made in the form of crayons, paint and pellets. The crayons and paint, applied to a metal surface, produce a flat-finish mark. A change in the finish from flat to glossy indicates that the surface temperature has reached or exceeded a selected value. The pellets melt at specified temperatures when in contact with a hot surface and may be seen more easily than crayon marks.

Vaporization The vapor pressure of a liquid depends on its temperature. When the liquid is heated to its boiling temperature, the vapor pressure is equal to the total pressure above the liquid surface. The boiling points of various pure chemical elements or compounds at standard atmospheric pressure [29.92 in. Hg (101.33 kPa)] can be used as thermometric fixed points. If the liquid and vapor are confined, the increase in vapor pressure may be used to measure temperature using a calibrated pressure gauge. (See Fig. 6.)

Expansion properties

Most substances expand when heated and, for many materials, the amount of expansion is almost directly proportional to the change in temperature. This effect is the basis of various thermometer types using gases, liquids or solids.

Gases The expansion of gases follows the relationship, in English units:

$$Pv_M = \mathbf{R}T \tag{1}$$

where

P = absolute pressure, lb/ft^2
v_M = volume, ft^3/mole of gas
$\mathbf{R}$ = universal gas constant, 1545 ft lb/mole R
T = absolute temperature, R = F + 460

At high pressures or near the condensation point, gases deviate considerably from this relationship. Under other conditions, however, the deviation is small.

Two types of gas thermometers are based on this relationship. In one, a constant gas volume is maintained and pressure changes are used to measure changes in temperature. Very accurate instruments of this type have been developed for laboratory work. The constant volume type thermometer is widely used commercially.

Nitrogen is commonly used for the gas-filled thermometer in industrial applications. It is suitable for a temperature range of –200 to 1000F (–129 to 538C). The construction is similar to the vapor pressure thermometer with nitrogen gas replacing the liquid and vapor. Expansion of the heated nitrogen in the bulb increases the pressure in the system and actuates a temperature indicator.

Liquids Liquid expansion properties may be incorporated into a thermometer by using a bulb and capillary to confine the liquid. In this thermometer, the bulb and capillary tube are completely filled with liquid and a calibrated pressure gauge is used to measure temperature. Mercury has been used for a temperature range of –40 to 1000F (–40 to 538C). Instrument readings can be in error if the capillary tubing is subjected to temperature changes.

The liquid-in-glass thermometer is a simple, direct

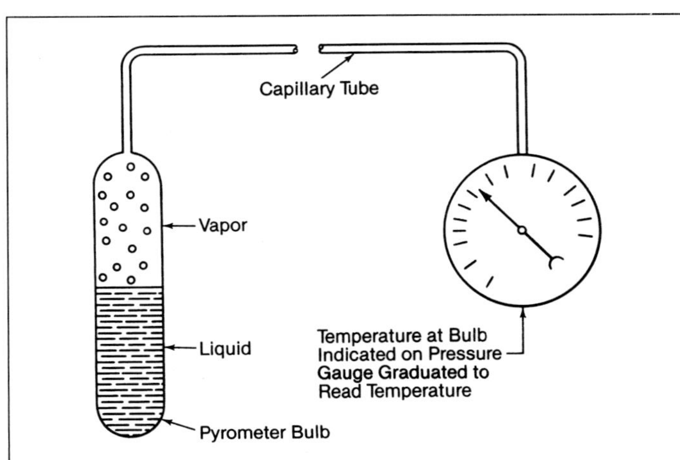

Fig. 6 Schematic assembly of vapor pressure thermometer.

reading, portable instrument. Low precision thermometers are inexpensive and instruments of moderate precision are available for laboratory use. This thermometer features a reservoir of liquid in a glass bulb that is connected directly to a glass capillary tube with graduated markings. Mercury, the most common liquid, is satisfactory from –40F (–40C), just above its freezing point, to about 600F (316C) if the capillary space above the mercury is evacuated. The upper temperature limit may be 900F (482C) or higher if this space is filled with pressurized nitrogen or carbon dioxide (CO_2).

The use of unprotected glass thermometers is restricted to laboratory or research applications. For more severe service there are various designs of industrial thermometers with the bulb and stem protected by a metal casing and usually arranged in a thermometer well. Response to rapid temperature changes is slower than with the unprotected laboratory type instrument.

Solids The expansion of solids when heated is recognized in thermometers using a bimetallic strip. Flat ribbons of two metals with different coefficients of thermal expansion are joined face to face by riveting or welding to form a bimetallic strip. When the strip is heated, the expansion is greater for one metal than for the other. A flat strip bends, or changes curvature if it was initially in a spiral form. Bimetallic strips are seldom used in power plant thermometers but are widely used in inexpensive thermometers, household thermostats and other temperature control and regulating equipment. They are particularly useful for automatic temperature compensation in other instrument mechanisms.

Radiation properties

All solid bodies emit radiation. The amount is very small at low temperatures and larger at high temperatures. The quantity of radiation may be calculated by the Stefan-Boltzman formula:

$$q/S = \sigma \varepsilon T^4 \quad (2)$$

where

q = radiant energy per unit time, Btu/h
S = surface area, ft^2
σ = Stefan-Boltzman constant, 1.71×10^{-9} Btu/h ft^2 R^4
ε = emissivity of the surface, dimensionless (usually between 0.80 and 0.95 for boiler materials)
T = absolute temperature, R = F + 460

At low temperatures the radiation is primarily in the invisible infrared range. As the temperature rises, an increasing proportion of the radiation is in shorter wavelengths, becoming visible as a dull red glow at about 1000F (538C) and passing through yellow toward white at higher temperatures. The temperature of hot metals [above 1000F (538C)] can be estimated by color. For iron or steel, the approximate color scale is:

dark red	1000F (538C)
medium cherry red	1250F (677C)
orange	1650F (899C)
yellow	1850F (1010C)
white	2200F (1204C)

Within the visible range, two types of temperature measuring instruments, the optical pyrometer and the radiation pyrometer, sense the emitted radiation to indicate temperature.

Optical pyrometers The optical pyrometer visually compares the brightness of an object to a reference source of radiation. The typical internal calibration source is an electrically heated tungsten filament. A red filter may also be used to restrict this visual comparison to a particular wavelength. This instrument measures the temperature of surfaces with an emissivity of 1.0, which is equivalent to that of a blackbody. By definition, a blackbody absorbs all radiation incident upon it, reflecting and transmitting none. When calibrated and used properly, the pyrometer yields excellent results above 1500F (816C). Temperature measurement of the interior surface of a uniformly heated enclosure, such as a muffle furnace, is a typical application. When used to measure the temperature of an object outside a furnace, the optical pyrometer always reads low. The error is small [20F (11C)] for high emissivity bodies, such as steel ingots, and large [200 to 300F (111 to 167C)] for unoxidized liquid steel or iron surfaces.

The optical pyrometer is widely used for measuring objects in furnaces at steel mills and iron foundries. It is not applicable for measuring gas temperature because clean gases do not radiate in the visible range.

Radiation pyrometers In one type of radiation pyrometer, all radiation from the hot body, regardless of wavelength, is absorbed by the instrument. The heat absorption is measured by the temperature rise of a delicate thermocouple (TC) within the instrument. The thermocouple is calibrated to indicate the temperature of the hot surface at which the pyrometer is sighted, on the assumption that the surface emissivity equals 1.0. The hot surface must fill the instrument's entire field of view.

This type of radiation pyrometer has high sensitivity and precision over a wide range of temperatures. The instrument gives good results above 1000F (538C) when used to measure temperatures of high emissivity bodies, such as the interiors of uniformly heated enclosures. Because operation of the instrument does not require visual comparison, radiation pyrometers may be used as remotely operated temperature indicators. Errors in temperature measurements of hot bodies with emissivities of less than 1.0, especially if they are in the open, are extremely large.

Radiation pyrometers sensitive to selective wavelengths in the infrared band give good results when measuring temperatures of bodies or flames. One design uses a system of lenses, a lead sulfide cell and electronic circuitry to produce a measurement.

An advancement of the single band pyrometer is the two color pyrometer, which measures the intensities of two selected wave bands of the visible spectrum emitted by a heated object. This unit computes the total of the emitted energies and converts it into a temperature indication. As with the optical pyrometer, indications depend on sighting visible rays and therefore its minimum temperature measurement is 1000F (538C).

Infrared radiation is produced by all matter at temperatures above absolute zero and detectors sensitive to the infrared band may be used to measure objects at low temperatures where the radiation is not in the visible band. The development of metal-silicide detectors permits sensitivities in the 1 to 5 μm band. Metal-silicides are formed by the reaction of metals, such as platinum or palladium, with p-type silicon. When coupled with electronic circuitry for producing an image, the detectors are sophisticated devices for low temperature measurement.[1]

Infrared thermal imaging equipment is available for quantitative measurement or qualitative imaging. The instruments for making quantitative measurements and providing target temperature indications are called *thermographic imagers* or *imaging radiometers*. Qualitative imaging instruments are called *thermal viewers*. These viewers are mechanically or electronically scanned; the electronically scanned units are called *pyroelectric vidicons* or *pyrovidicons*.[2]

Infrared temperature measurement is applicable to surveys of surfaces, such as boiler casings and insulated steam pipes, for locating temperature variations. These instruments may also assist in determining electrical component temperatures in operating circuits. Radiation pyrometers are not capable of determining gas temperatures.

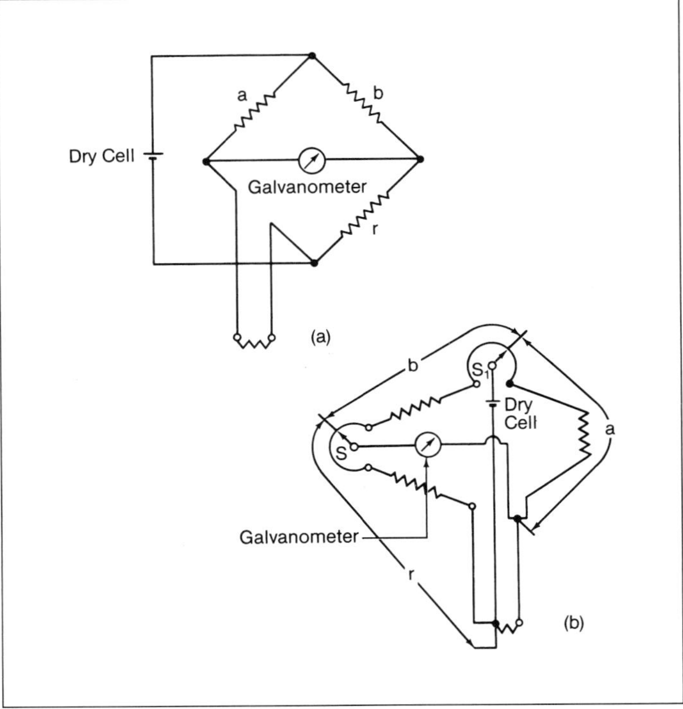

Fig. 7 Diagrams of electrical circuits for resistance thermometers (*courtesy of Leeds and Northrup Company*).

Optical properties

Optical fibers have been used in fixed (end) and distributed (average temperature along length) temperature measurement. Development is continuing in the application of this method to temperatures above 450F (232C).[3]

Electrical properties

Two classes of widely used temperature measuring instruments, the thermocouple and the electrical resistance thermometer, are based on the relation of temperature to the electrical properties of metals.

Because of its versatility, convenience and durability, the thermocouple is of particular importance in power plant temperature measurements. A separate section later in this chapter is devoted to the thermocouple and its applications.

Resistance thermometer The electrical resistance thermometer, used over a temperature range of –400 to 1800F (–240 to 982C), depends on the increase in electrical resistance of metals with increasing temperature; this relationship is nearly directly proportional. If the electrical resistance of a calibrated wire material is measured (by a Wheatstone bridge or other device), the temperature of the wire can be determined.

In the simplest form of the Wheatstone bridge system, shown in Fig. 7a, the reading would be the sum of the resistances of the calibrated wire and the leads connecting this wire to the Wheatstone bridge. However, this value would be subject to error from temperature changes in the leads. By using a more refined circuit shown in Fig. 7b, the resistance of the leads can be eliminated from the instrument reading. To localize the point of temperature measurement, the resistance wire may be made in the form of a small coil. From room temperature to 250F (121C), commercial instruments usually have resistance coils of nickel or copper. Platinum is used for higher temperatures and in many high precision laboratory instruments to cover a wide range of temperatures.

The electrical resistance thermometer can be used for the remote operations of indicating, recording or automatic controlling. If proper precautions are taken, it is stable and accurate, but it is less rugged and less versatile than a thermocouple.

Acoustic properties

Acoustic properties of gases vary with temperature and are used as the basis for temperature measuring instruments. The velocity of sound in a gas varies directly with the square root of its absolute temperature, as indicated in the following formula:

$$c = \sqrt{g_c kRT / M} \tag{3}$$

where

c = speed of sound, ft/s
g_c = proportionality constant, 32.17 lbm ft/lbf s^2
k = specific heat ratio, dimensionless
R = universal gas constant, 1545 ft lb/mole R
T = absolute temperature, R = F + 460
M = molecular weight, lb/mole

By measuring the time required for sound to traverse a specific distance, the average gas temperature can be calculated. This phenomenon has been commercially applied to measuring furnace gas temperatures. By developing a number of independent temperature paths a thermal contour may be developed.

Other properties

Other material properties that vary with temperature are also used as the basis for temperature measuring instruments. By flowing gas through two orifices in series, a method has been developed for measuring gas temperatures up to 2000F (1093C) commercially and up to 4500F (2482C) in the laboratory. Temperatures ranging from absolute zero to 5000F (2760C) have been accurately determined by using a laboratory instrument that measures electron motions. High temperature research has been stimulated by gas turbine and jet propulsion applications, and further development of new and improved temperature measuring instruments is expected.

Thermocouples

A thermocouple consists of two electrical conductors of dissimilar materials joined at the ends to form a circuit. If one of the junctions is maintained at a temperature higher than the other, an electromotive force (emf) is generated, producing current flow through the circuit, as shown in Fig. 8.

The magnitude of the net emf depends on the temperature difference between the two junctions and the materials used for the conductors. No unbalance, or net emf, is generated if the two junctions are at the same temperature or if the conductors are of the same material at different temperatures. If one junction of a calibrated thermocouple is maintained at a known temperature, the temperature of the other junction can be determined by measuring the net emf produced; this emf is almost directly proportional to the temperature difference of the two junctions. The relationship between emf and the corresponding temperature difference has been established by laboratory tests throughout the temperature ranges for common thermocouple materials. These values are plotted in Fig. 9. The thermocouple is a low cost, versatile, durable, simple device providing fast response for accurate temperature measurement. It also allows convenient centralized reading.

In the potentiometer circuit, copper leads may connect the reference junction terminals to the measuring instrument without affecting the net emf of the thermocouple. If the instrument is at a uniform temperature, no emf is set up between the copper conductors and slide wire materials within the potentiometer. However, if temperature differences exist within the instrument, there will be disturbing emf values in the circuit and the readings will be affected.

Multiple circuits

If two or more thermocouples are connected in series, the net emf at the outside terminals is equal to the sum of the emf values developed by the individual thermocouples. When all of the individual hot and cold junctions are maintained at the same respective temperatures, as in the device known as the thermopile, this multiplied emf value makes it possible to detect and measure extremely small temperature variations.

Two or more thermocouples may be connected in parallel to obtain a single reading of average temperature. In this case, the resistance of each thermocouple must be the same. The emf across the terminals of such a circuit is the average of all the individual emf values and may be read on a potentiometer normally used for a single thermocouple.

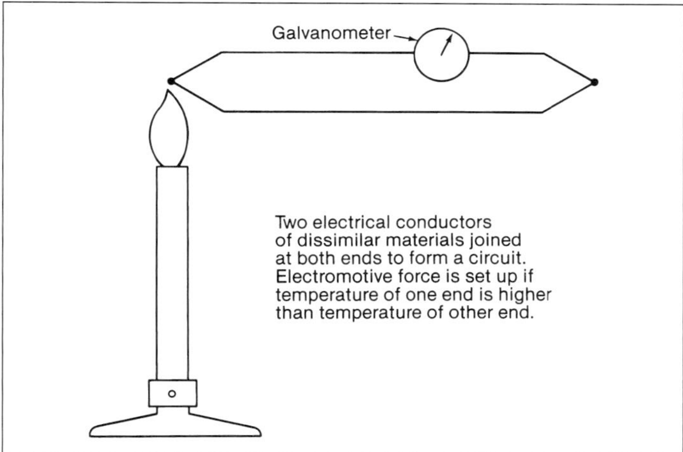

Fig. 8 Principle of the thermocouple illustrated.

Practical application and multiple circuits

The limitation of the potentiometer circuitry to measure small emf values has led to various circuits of multiple thermocouples connected in series or parallel. These circuits make it possible to measure extremely small temperature variations or average temperatures. Thermocouple circuit considerations of wire diameter and length, while less important for steady-

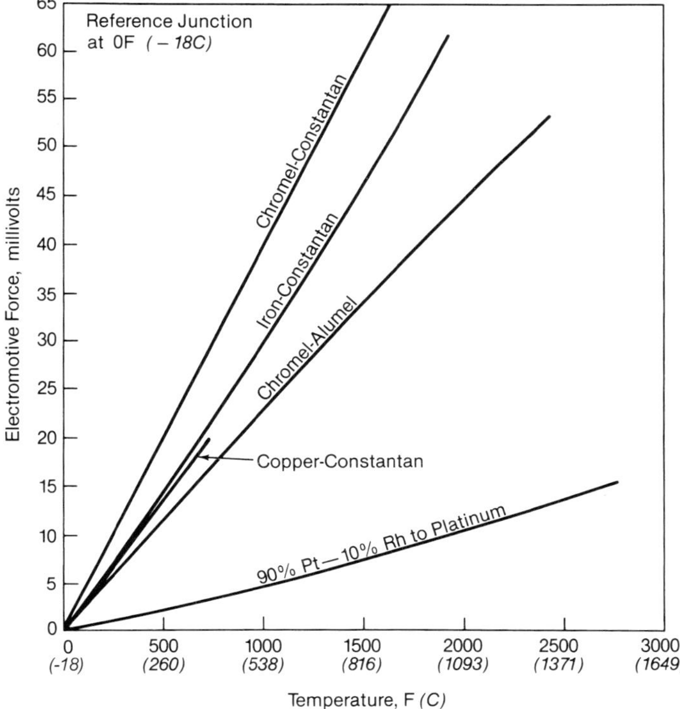

Fig. 9 Relation of temperature to electromotive force produced by several commonly used thermocouples.

state measurement, will affect thermocouple readings during dynamic temperature conditions. These applications require detailed circuit design investigation.

The application of thermocouples requires attention to the insulation of the lead wire to prevent erroneous circuits. Parasitic couples may be introduced by mechanical, soldered, or brazed connections if temperature variations exist within different metals of the connections.

Selection of materials The combinations of metals and alloys most frequently used for thermocouples are listed in Table 1 with their general characteristics and useful temperature ranges. Selection depends largely on the ability to withstand oxidation attack at the maximum expected service temperature. Durability depends on wire size, use or omission of protection tubes, and nature of the surrounding atmosphere.

All thermocouple materials deteriorate when exposed at the upper portion of their temperature range to air or flue gases and when in contact with other materials. Platinum in particular is affected by metallic oxides and by carbon and hydrocarbon gases when used at temperatures above 1000F (538C) and is subject to calibration drift. The upper temperature limits shown in Table 1 are well below the temperatures at which rapid deterioration occurs.

For high temperature duty in a permanent installation or where destructive contact is likely, service life may be extended, at some sacrifice in response speed, by using closed-end protection tubes of alloy or ceramic material. The arrangement should permit removal of the thermocouple element from the protection tube for calibration or replacement. Omission of the protection tube may require frequent calibration and corrections for calibration drift. For short periods of service within the useful range of the thermocouple selected, the correction for calibration change is usually negligible.

Sheathed-type thermocouple Sheathed-type magnesium oxide insulated thermocouples have been in use for some time. The thermocouple wires are insulated with inert magnesium oxide insulation that protects the wires from the deteriorating effects of the environment. Sheaths can be made of stainless steel and other resistant materials, ensuring relatively long life

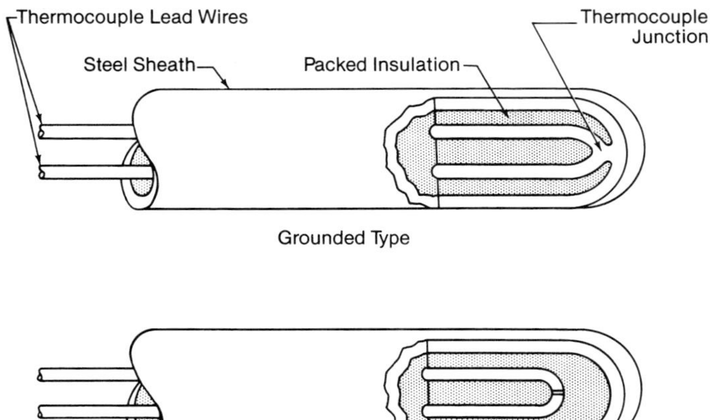

Fig. 10 Sheathed-type thermocouple.

and resistance to oxidizing, reducing, or otherwise corrosive atmospheres. Sheathed thermocouples are available as grounded or nongrounded types, as illustrated in Fig. 10. The grounded type has a more rapid response to temperature change but can not be used for connection in series or parallel, because it is grounded to the sheath. For these applications, the nongrounded type should be used. The grounded type is susceptible to separation or parting of the thermocouple wire where long leads at high temperatures are used, whereas the nongrounded type appears satisfactory for this service.

Sheathed thermocouples may be equipped with pads for welding to tubes, as illustrated in Fig. 11. Pad-type thermocouples should not be used when the pad is exposed to temperatures that are markedly different from that of the body being measured, because this added metal surface can radiate or absorb heat. Typical applications of pad-type thermocouples are drum, superheater header, or tube surface temperature measurements. (See Fig. 12.)

Thermocouple and lead wire There are two classes of wire for thermocouples, the closely standardized and matched thermocouple wire and the less accurate

Table 1
Types of Thermocouples in General Use

Type of Thermocouple (Note 1)	Useful Temperature Range, F (C)		Maximum Temperature, F (C)	Millivolts at 500F (260C) (Note 2)	Magnetic Wire
(+) Copper to constantan (–)	–300 to 650	(–184 to 343)	1100 (593)	13.24	–
(+) Iron to constantan (–)	0 to 1400	(–18 to 760)	1800 (982)	15.01	Iron (+)
(+) Chromel to constantan (–)	–300 to 1600	(–184 to 871)	1800 (982)	17.94	–
(+) Chromel to alumel (–)	0 to 2300	(–18 to 1260)	2500 (1371)	11.24	Alumel (–)
(+) 90% Pt — 10% Rh to platinum (–)	900 to 2600	(482 to 1427)	3190 (1754) (Note 3)	2.048	–

Notes:
1. Nominal composition: constantan, 55% Cu, 45% Ni; chromel, 90% Ni, 10% Cr; alumel, 95% Ni, 5% Al, Si and Mn combined.
2. Reference junction 0F (–18.0C).
3. Melting point.

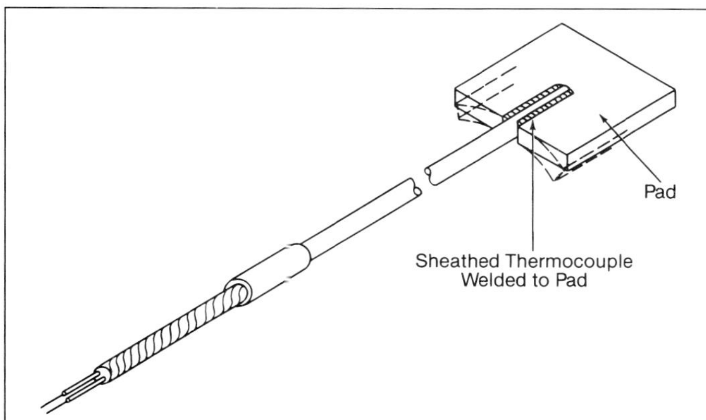

Fig. 11 Sheathed thermocouple with pad.

compensating lead wire. For thermocouples of noble metal, extension leads of copper and copper-nickel alloy, which have an emf characteristic approximating the noble metal pair, are used to reduce cost. For thermocouples of base metal, the extension lead wire is the same composition as the thermocouple wires, but it is less expensive because the control in manufacture and calibration is less rigorous.

For accuracy, the matched thermocouple wire should be used at the hot junction and continued through the zone of greatest temperature gradient to a point close to room temperature, where a compensating lead wire may be spliced in for connection to the reference junction. The reference junction is usually located at the recorder or central observation point. Where a number of thermocouples are used, it may be economical to establish a zone box (see Fig. 13) from which copper conductors are extended to the measuring instrument.

Splicing of extension lead wires may be done by twisting the wire ends together, by using screw clamp connectors, or by fusion welding, soldering or brazing.

Care should be taken to maintain correct polarity

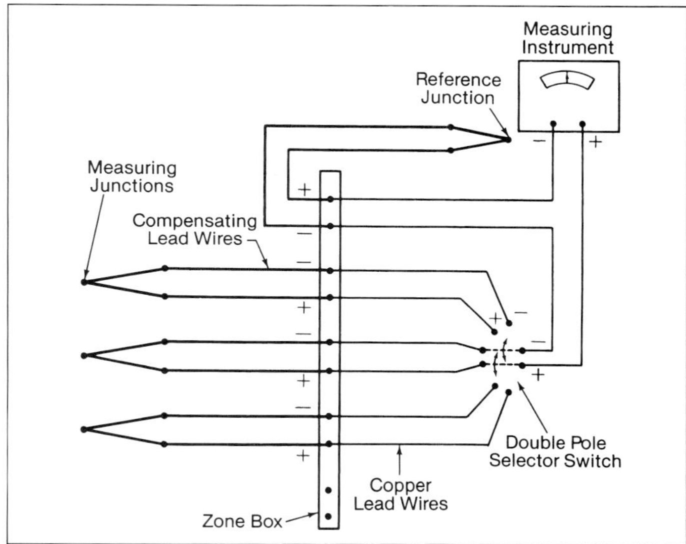

Fig. 13 Thermocouple and lead wire arrangement to measuring instrument through selector switch.

by joining wires of the same composition. Polarity is usually identified by color code or tracers in the wire covering.

Hot junction Wire and pipe-type elements for hot junctions may be purchased or may be made from stock thermocouple wire, as shown in Fig. 14. These types are useful for direct immersion in a gas stream, for insertion into thermometer wells, or for thermal contact with solid surfaces.

Where the temperature of a metallic surface is to be measured, a versatile hot junction may be formed by peening the wires separately into holes drilled in the surface of the metal. Steps for making this type of junction are illustrated in Fig. 15.

The peened junction has the advantage of being completely mechanical, with a tight and remarkably strong attachment and minimum interference with the temperature of the object. The temperature indicated is essentially that of the metallic surface, which is the first point of contact with the conductors. The depth of the drilled hole has no temperature effect; it only provides mechanical strength. However, the junc-

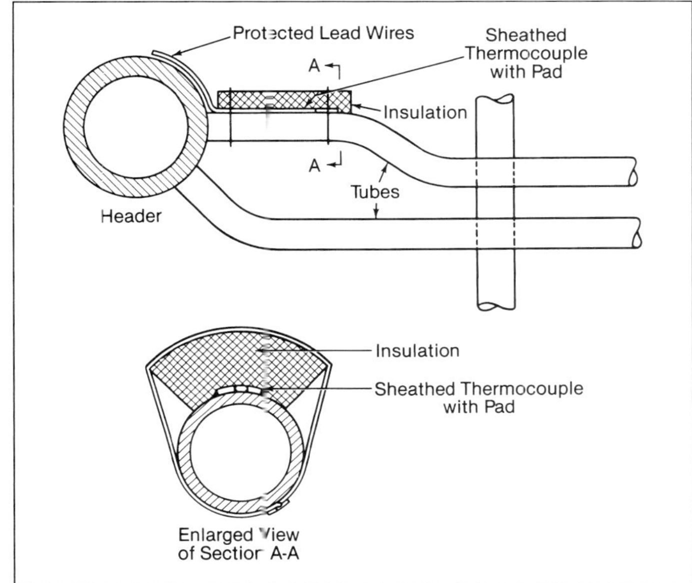

Fig. 12 Application of pad-type thermocouple.

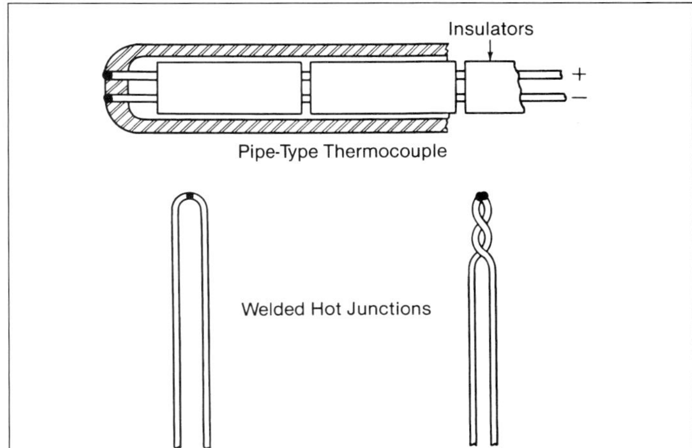

Fig. 14 Thermocouple elements and hot junctions.

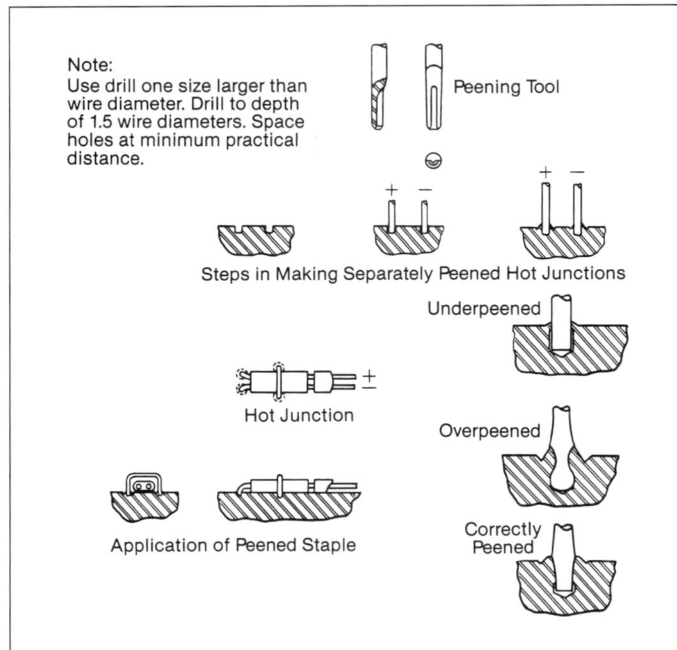

Fig. 15 Procedure for installing a peened-junction thermocouple to measure surface temperature of metal.

tion contact points are subject to thermal conduction in the wires. Where this type of error is significant, precautions should be taken to prevent temperature gradients in the wires. Drilled holes for the wire ends should allow a snug fit before peening. For installations requiring a significant number of thermocouples, there are portable spot welding machines available to join the ends of the wire directly to a steel surface.

Millivolt potentiometer

The temperature-emf relationships of standardized thermocouple wires, as established and published by the manufacturer, should be used to convert the potentiometer readings to equivalent temperature values. These tabulated values represent the net emf impressed in the potentiometer terminals when the reference junction is at 0F (–18C) and the measuring junction is at the temperature listed. The millivolt potentiometer may be used with any type of thermocouple. It is frequently used when several different thermocouple types are installed.

It is usually impractical to maintain the reference junction at 0F (–18C) while taking readings. Correction for any reference junction temperature can be made by adding to each observed emf reading the emf value corresponding to the reference junction temperature. The temperature of the measuring junction can then be determined. Most millivolt potentiometers are equipped with a compensator, which should be used to correct for the reference junction temperature. Direct millivolt readings then correspond to the actual thermocouple temperature.

Direct reading potentiometer

A potentiometer may be graduated to read the hot junction temperature directly instead of in millivolts. When calibrated for use with a specific type of thermocouple, the reference junction temperature is usually automatically compensated for by an internal resistor circuit. This circuit compensates for deviations from standard room temperature. The resistor circuit characteristics are specific to the type of thermocouple material for which the potentiometer is calibrated. Some potentiometers provide a selector switch to directly read different types of thermocouples.

Temperature recorders

In temperature recorders, a power operated mechanism automatically adjusts the slide wire of a potentiometer circuit to balance the emf received from the thermocouple. The action of the slide wire is coordinated with the movement of a recording pen or print wheel that is drawn over, or impressed on, a moving temperature graduated chart.

As a result, an essentially continuous record may be traced for one thermocouple, or a rapid scanning of many thermocouples may be recorded.

Checking recorder calibration Compared with the manually operated potentiometer, all recorders are inherently less accurate because of wear and lost motion in the actuating mechanism, shrinkage, or printing inaccuracies of paper and chart scales.

Temperature of fluids inside pipes

The temperature of a fluid (liquid, gas or vapor) flowing under pressure through a pipe is usually measured by a glass thermometer, an electrical resistance thermometer, or a thermocouple. Each thermometer is inserted into a well, or thermowell, projecting into the fluid. The thermowell is preferred but a thermocouple properly attached to the outside of a pipe wall can provide good results. The thermowell provides

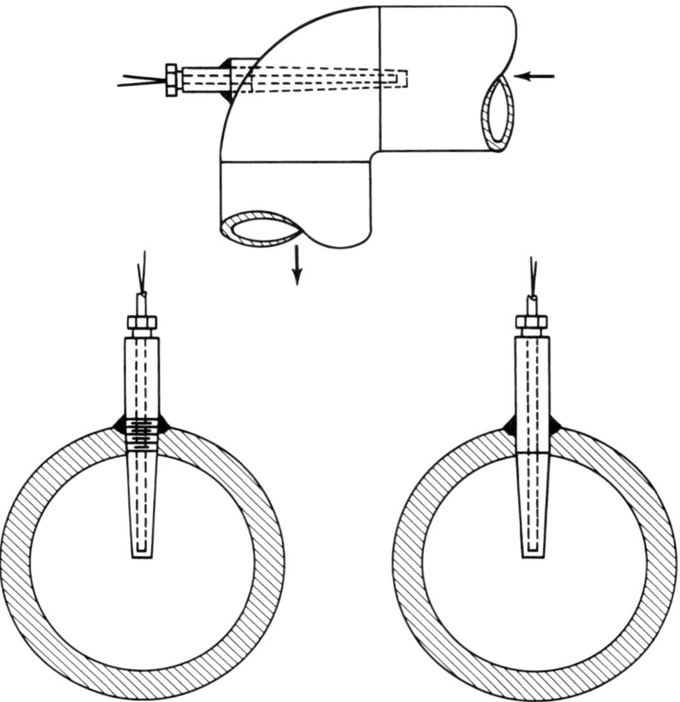

Fig. 16 Thermocouple well installation.

an average temperature if temperature stratification exists in the fluid, because thermal conduction along the well equalizes the temperature.

A metal tube closed at one end, screwed into or welded to a pipe wall, and projecting into the fluid, serves as a thermowell (Fig. 16). It must have strength and rigidity to withstand hydrostatic pressure, resistance to bending and vibration caused by its resistance to fluid flow, and material compatibility for a welded attachment. To minimize heat conduction to the surroundings and to give rapid response to temperature changes, the well should be as small as possible. The portion outside the pipe should also be small. The material of the well must resist the erosive or corrosive action of the fluid. Because the temperature may be locally depressed by acceleration of a compressible fluid through a constriction in a pipe, the well should be carefully positioned, taking these factors into account. Care should also be observed when locating the well for measuring steam temperature downstream of spray water attemperator injection points, as inadequate mixing of the steam and spray water may result in fluid temperature measurement errors. Projecting parts of the thermowell and the pipe wall should be thoroughly insulated to prevent heat loss. Substantial error can result from air circulation within the thermowell and should be minimized by packing with insulating material.

Thermowell designs vary. If the fluid is a liquid or a saturated vapor, good heat transfer is assured and a plain well is satisfactory. If the fluid is a gas or superheated vapor, a finned well is sometimes used. The method of attachment, a mechanical thread or welded attachment to the pipe wall, is optional provided safety code requirements are met (see Fig. 16).

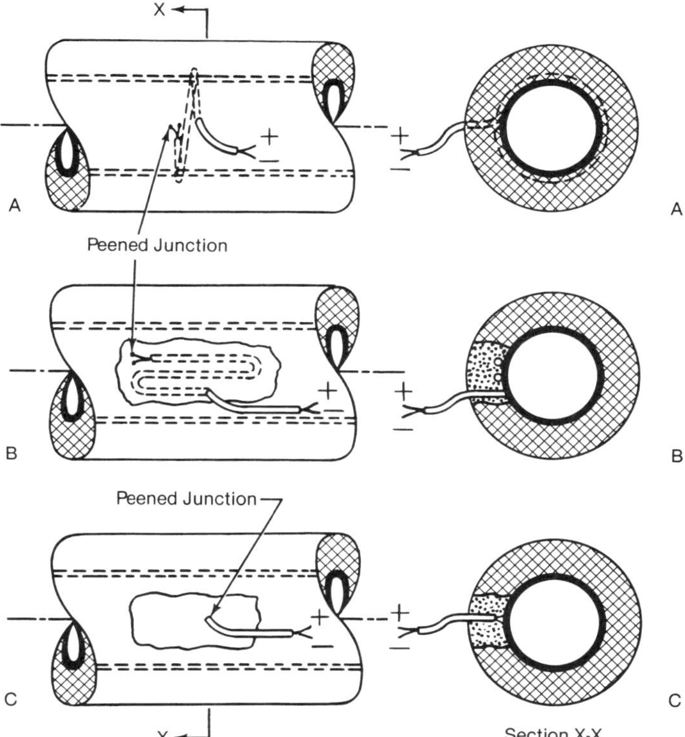

Fig. 17 Thermocouple wires extending from hot junction disposed on pipe wall before leading outside.

Tube temperature measurement

It is frequently desirable to know the metal temperature of tubes in different classes of service. These classes include furnace wall or boiler bank tubes that are cooled by water and steam at saturation temperature, economizer tubes cooled by water below the temperature of saturation, and superheater and reheater tubes cooled by steam above saturation temperature. These temperature measurements may be used to determine the safety of pressure parts, the uniformity among tubes in parallel flow circuits, or the fluid temperature increase between inlet and outlet conditions. The peened-type hot junction, illustrated in Fig. 17, is satisfactory in many cases and is probably the simplest form of thermocouple application. Sheathed-type thermocouples with pads are generally used to permanently monitor temperatures of tubes not exposed to external heat. (See Fig. 11.) Properly applied insulation of sufficient thickness in the vicinity of the thermocouple attachment will reduce the effects of thermal conduction along the tube. The thermocouple wire wrapped around the tube will minimize the conduction along the thermocouple wire. These techniques are illustrated in Fig. 17. When properly installed the surface thermocouple may be used to measure both metal and fluid temperatures.

Furnace wall tubes

In measuring water-cooled furnace wall tube temperatures, special protection for the thermocouple and its lead wires must be provided because of the destructive high temperature furnace atmosphere and, in some cases, the accumulation and shedding of ash and slag deposits. If a simple peened thermocouple is used, the results may lack accuracy because of errors from conduction; in most cases, the service life would be short because of physical damage or wire deterioration from overheating. Cover plates welded to the tubes have been used to protect the wires, but these plates interfere with the normal heat transfer and cause local ash deposits to create abnormal conditions at the temperature measurement point.

The use of chord-drilled holes through which the thermocouple wires are laced (Fig. 18) was developed by The Babcock & Wilcox Company (B&W); this is a satisfactory temperature measurement method for furnace wall tubes. The tube surface is free of projections, the wires are protected, and the thermal conduction effect at the hot junction is minimized because the wires pass through an essentially isothermal zone before emerging into cooler surroundings at the rear of the tube. To minimize the effect on the heat flow pattern within the tube metal, the chord-drilled holes should

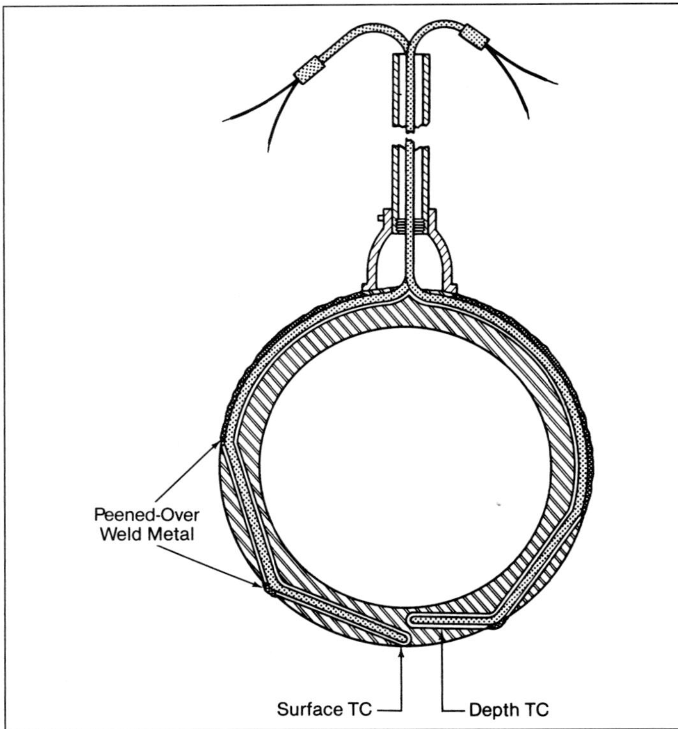

Fig. 18 Chordal thermocouple.

be as small as possible. The effect of these holes on the strength of the tube is small. This effect is minimal in the critical direction of hoop stress and is readily tolerated in the direction of longitudinal stress. The use of chord-drilled holes in the thicker, high pressure wall tubes is a practical method of obtaining surface metal temperatures.

Gradient thermocouple

Measuring the temperature gradient through a tube wall is a means of determining the heat flow rate through the wall and of detecting the accumulation of certain types of internal deposits.

To obtain the best results, a large temperature gradient through the wall is desirable. However, ideal conditions may be compromised to permit an arrangement that provides measurable metal temperatures and is practical to fabricate. Fig. 18 shows a section through a tube which illustrates the drilling and installation of typical surface and depth thermocouples.

Calibration Sheathed thermocouples are suitable for this temperature gradient application. For greater accuracy the thermocouples may be laboratory calibrated. No procedure is available to accurately or economically calibrate the thermocouple tube assembly in the field. The most satisfactory method is as follows.

Following installation, obtain a series of temperature readings on all thermocouples at two or three different ratings on the boiler, from a low rate to maximum rate, while the tube is known to be internally clean; read all temperatures as simultaneously as possible. Several series of readings should be taken on the thermocouples at each rating. An average should be calculated for each thermocouple to correct for minor differences in temperatures resulting from heat input variations. This also compensates for the time lag between surface and depth thermocouples. By using the surface and depth dimensions, the equivalent lengths can be determined and the temperature gradients can be plotted. In practice, the heat flows through a tube wall in a path of decreasing sectional area and the temperature gradient through the metal is not linear with respect to thickness. The gradient, however, may be drawn as a straight line if temperatures are plotted against an equivalent length of flow path (l_e). This represents an equivalent flat plate having a uniform flow path with an equivalent but greater thickness. This is illustrated in Fig. 19.

The equivalent plate thickness is calculated in the following manner:

$$l_e = R \, \ell n \left(R / r \right) \qquad \textbf{(4)}$$

where

R = outside radius of tube, in.
r = inside radius of tube, in.
l_e = equivalent flat plate thickness, in.

In many cases, thermocouple error may represent a significant portion of the gradient being measured. Ideally, lines drawn through the surface and depth temperature (Fig. 20) should intersect the inner surface equivalent thickness line slightly above the fluid temperature as measured by a thermocouple peened on the outside surface of the tube opposite the furnace. The amount above the fluid temperature represents the temperature drop across the fluid film and this drop increases with increased heat input. It is difficult to obtain consistently drilled thermocouple holes that reflect the pattern shown in Fig. 20. If, at low inputs, the line drawn through the surface and depth temperature intersects the inner surface equivalent thickness line at fluid temperature or slightly above, it may be considered as satisfactory. If it falls below or well above this point, a correction or calibration may be made by ignoring the measured equivalent depth dimension, plotting a line through the outer and inner surface temperatures, then adjusting the equivalent depth to a value represented by the intersection of the adjusted gradient and depth thermocouple temperature (Fig. 21).

Heat flux measurement Once a series of temperature plots shows consistent results, heat inputs may be calculated:

$$q'' = \frac{k \left(T_1 - T_2 \right)}{l_e} \qquad \textbf{(5)}$$

where

q'' = heat flux, Btu/h ft^2
k = thermal conductivity, Btu/h ft F
T_1 = outside surface thermocouple temperature, F
T_2 = depth thermocouple temperature, F
l_e = equivalent plate thickness (surface to depth), ft (Fig. 19)

Scale detection If the temperature gradients are to be used as a scale detector, then thermocouples should be read periodically, plotted, and checked for a change in the temperature above saturation.

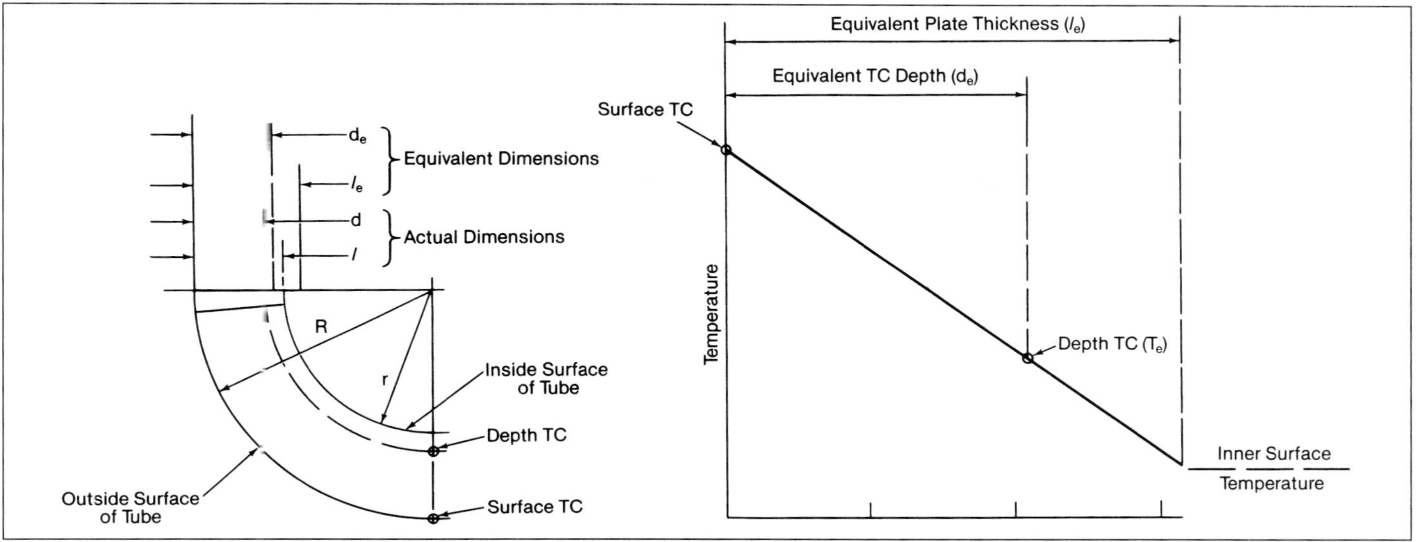

Fig. 19 Thermocouple installation.

To assure a close comparison, the thermocouple temperatures should always be read with the calibrated equipment and under operating conditions close to those existing during the original setup. An accumulation of internal deposits in the thermocouple zone is marked by an increased temperature difference from inner surface to saturation.

When the average of the surface and depth temperatures approaches the limit for the tube metal, the boiler should be shut down and the inside tube surfaces inspected. Sufficient allowance should be made for tube to tube temperature variations.

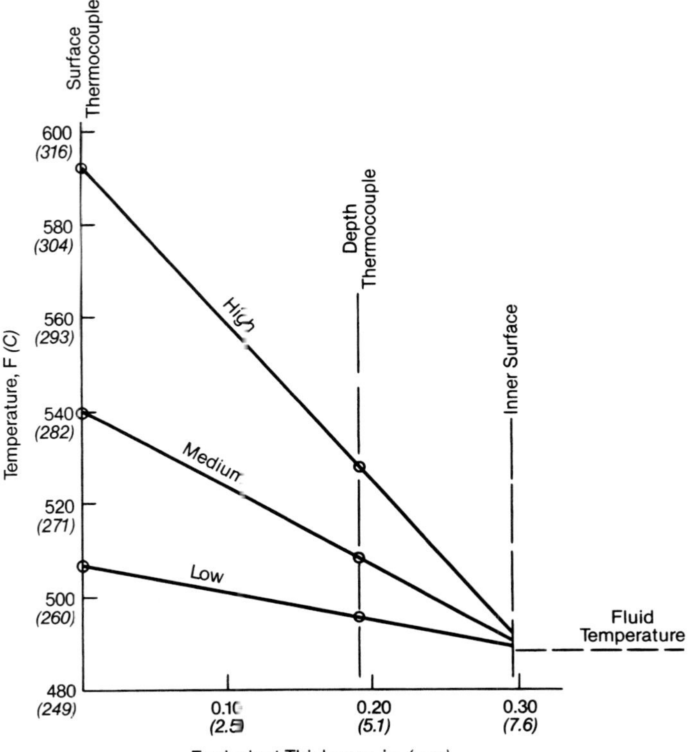

Fig. 20 Sample plot of temperature gradients for low, medium and high heat rates.

Limitations of gradient thermocouples The gradient thermocouple method of measuring heat flow through tube walls is a guide rather than an absolute measure. Relatively small dimensions and small differential temperatures exist between surface thermocouple and depth thermocouple locations. Also, any small error in determining temperatures or in metal thickness between thermocouples usually represents a large percentage of this small difference.

The effectiveness of the gradient temperature plot as an internal scale detector is also dependent on the nature of the scale. Certain types of scale, such as carbonate or silica deposits, accumulate uniformly, while iron oxide deposits can accumulate irregularly. The location of nonuniform deposits is uncertain and the gradient thermocouple method is not reliable for detecting such accumulations.

For a quick determination of internal tube deposit changes, the difference between surface and depth temperature readings can be plotted against the difference between surface and fluid temperature readings. With a clean tube, points should fall along a straight line over a range of inputs. As deposits form, the difference between surface and fluid temperature increases for an unchanging surface and depth temperature difference. Fig. 22 shows a typical plot of such conditions.

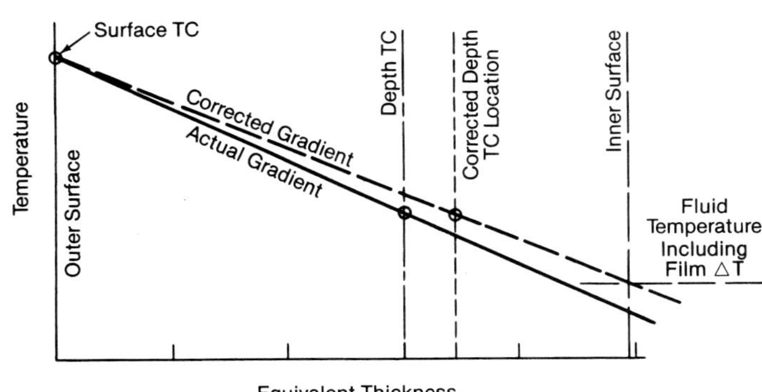

Fig. 21 Adjustment of temperature gradient.

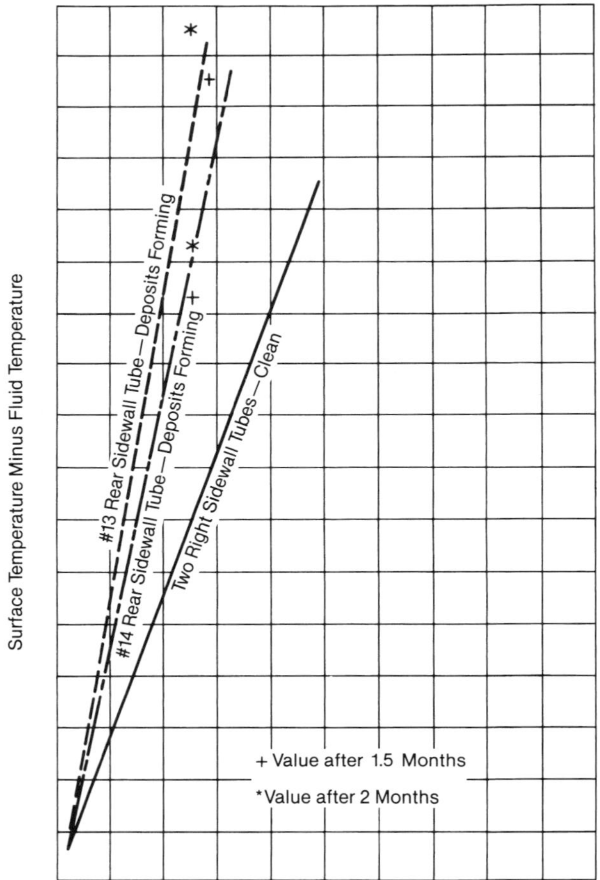

Fig. 22 Plot of chordal thermocouple temperatures showing effect of internal deposits.

Superheater and reheater applications

Chordal surface thermocouples can be used to measure metal temperatures of superheater or reheater tubes in the gas stream, using the principles previously described. However, special provision must be made to protect the wires between the point of measurement and the exit from the boiler setting. This can be done by containing the sheathed thermocouple in stainless steel tubing that is welded to the superheater or reheater tube. This maintains the sheath and protection tubing at a temperature which is approximately that of the superheater or reheater tube. Additional cooling of the stainless steel protection tube is frequently required.

Gas temperature measurement

Temperature measurement of combustion gas at different locations within the boiler is important to the boiler design engineer and the operating plant engineer. Accurate temperature measurements of gas entering and leaving the heat absorbing components can confirm design predictions and operating performance.

In all cases of gas temperature measurement, the temperature sensitive element approaches a temperature that is in equilibrium with the conditions of its environment. While it receives heat primarily by convection transfer from the hot gases in which it is immersed, it is also subject to heat exchange by radiation to and from the surrounding surfaces and by conduction through the instrument itself. If the temperature of the surrounding surfaces does not differ from that of the gas (gas flowing through an insulated duct), the temperature indicated by the instrument should accurately represent the gas temperature. If the temperature of the surrounding surfaces is higher or lower than that of the gas, the indicated temperature will be correspondingly higher or lower than the gas temperature.

The variation from the true gas temperature depends on the temperature and velocity of the gas, the temperature of the surroundings, the size of the temperature measuring element, and the construction of the element and its supports. To correct for the errors in temperature measurement caused by the surroundings, it is best to calibrate the instrument to a known and reliable source.

As an example, consider a 22 gauge (0.7112 mm) bare thermocouple. When it is used to measure the gas temperature in boiler, economizer, or air heater cavities with surrounding walls cooler than the gas, the error in the observed readings may be found from Fig. 23 (line for bare TC).

High velocity thermocouple

The design and operation of steam generating units depend on the evaluation of gas temperatures in the furnace and superheater sections of the equipment. Boiler design to achieve successful thermal performance must take into account the limitations imposed by the allowable metal temperatures of superheater

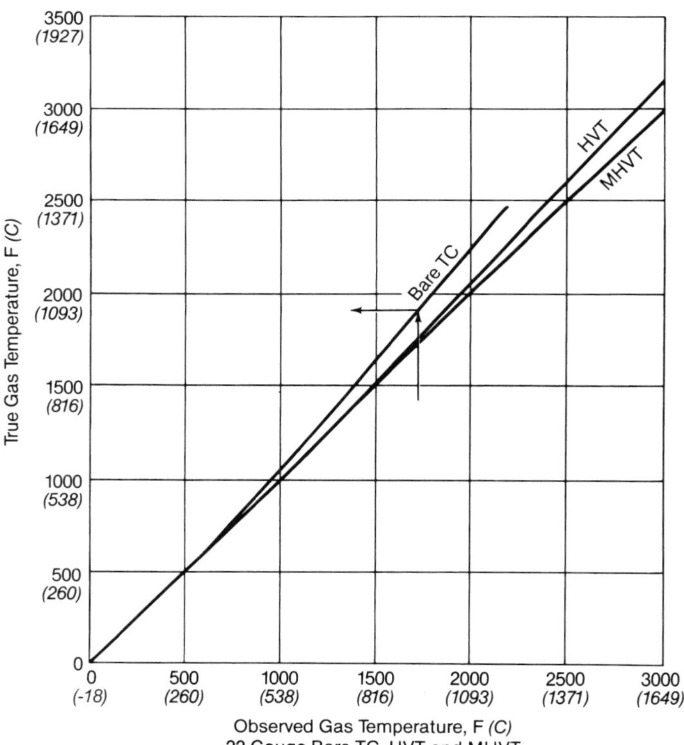

Fig. 23 General magnitude of error in observed readings when measuring temperature in boiler cavities with thermocouples.

tubes and by the fusing characteristics of ash and slag from the current or expected fuel. The overall complexity of combustion and heat transfer relationships prevents exact calculation and accurate measurement of the gas temperatures provides information to confirm calculation methods.

The optical pyrometer and radiation pyrometer are not designed to measure gas temperature. Excessive error is also encountered when gas temperatures in the furnace and superheater areas are measured with a bare thermocouple.

The high velocity and multiple-shield high velocity thermocouples (HVT and MHVT), developed to correct for radiation effects, are the best instruments available for measuring high gas temperatures in cooler surroundings or low gas temperatures in hotter surroundings. Cross-sections through single and multiple-shield high velocity thermocouples developed for use in boiler testing are shown in Fig. 24.

The surfaces (water-cooled walls, or superheater or boiler tube banks) surrounding the usual location of a gas traverse are cooler than the gases. Consequently, the readings from a bare unshielded thermocouple indicate lower temperatures than those obtained with an HVT. For the same reason, an HVT generally indicates lower values than an MHVT. A comparison between bare thermocouple, HVT and MHVT results in typical boiler furnaces and cavities is given in Fig. 23.

MHVT measurements closely approach true gas temperatures. In this design, the thermocouple junction is surrounded by multiple shields, all of which receive heat by convection induced by the high gas flow rate. In this manner, the heat transfer by radiation is so reduced that there is virtually no heat exchange between the junction and the innermost shield. Because of the small flow areas that rapidly become clogged by ash, use of the MHVT is limited to clean gas conditions. Where traverses are taken in dust- or slag-laden gases, it is usually necessary to use the HVT. The readings are corrected by comparison with results obtained under clean gas conditions from an MHVT.

For temperatures exceeding 2200F (1204C), noble metal thermocouples are required and it is important to protect the thermocouple from contamination by the gases or entrained ash. Various coverings, shown in Fig. 24, provide some protection for the wires, especially when fouling occurs from molten slag at temperatures above 2400F (1316C). When platinum thermocouples are used in gas above 2600F (1427C), appreciable calibration drift may occur even while taking measurements requiring only several minutes of exposure time. The thermocouple elements should be checked before and after use with corrections applied to the observed readings. When the error (ΔT) reaches 40 to 60F (22 to 33C), the contaminated end of the thermocouple should be removed and a new hot junction should be made using the sound portion of the wire.

Because heat transfer by radiation is proportional to the surface area, emissivity, and difference of the source and receiver absolute temperatures, the effects of radiation increase as the temperature difference between the thermocouple hot junction and the surrounding surfaces increases. Also, because heat transfer to the thermocouple by convection is proportional to the gas mass velocity and the temperature difference between gas and thermocouple, the junction temperature may be brought closer to the true temperature of the gas by increasing the rate of mass velocity and convection heat transfer at the thermocouple while shielding the junction from radiation.

A special high velocity thermocouple probe for measuring high gas temperatures in boilers is illustrated in Fig. 25. This portable assembly is primarily used for making test traverses in high duty zones by insertion through inspection doors or other test openings in the setting.

This thermocouple is supported by a water-cooled probe. The measuring junction is surrounded by a tubular porcelain radiation shield through which gas flow is induced at high velocity by an attached aspirator. The gas aspiration rate over the thermocouple can be checked by an orifice incorporated with the aspirator and connected to the probe by a flexible hose. The gas mass velocity (or mass flux) over the thermocouple junction should be at least 15,000 lb/h ft^2 (20.34 kg/m^2 s). Convection heat transfer to the junction and shield is simultaneous and both approach the temperature of the gas stream. Radiation transfer at the junction is diminished by the shield. Because the shield is exposed to the radiation effect of the surroundings, it may gain or lose heat and its temperature may be slightly different from the junction temperature.

With the increasing size of steam generators, handling long HVT probes has become more difficult but,

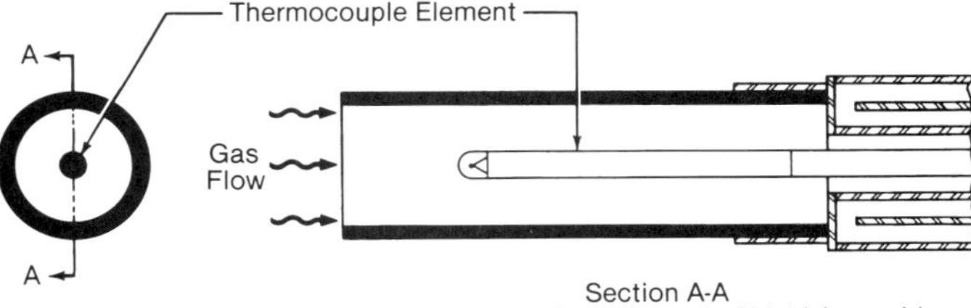

Section A-A
High Velocity Thermocouple Shield Assembly

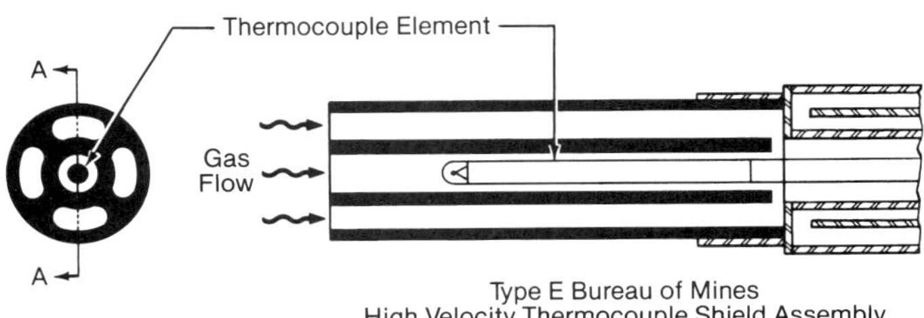

Type E Bureau of Mines
High Velocity Thermocouple Shield Assembly

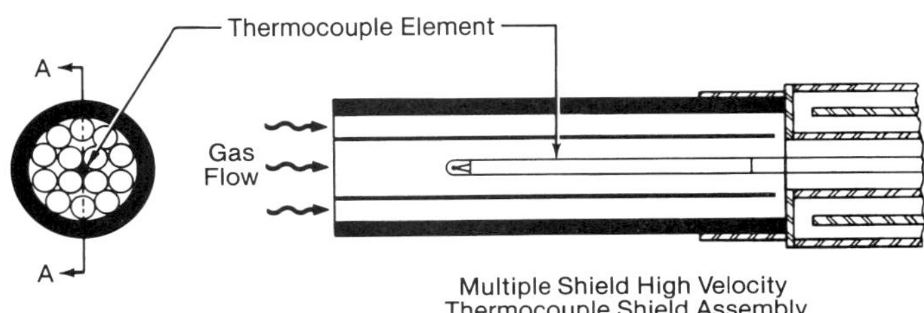

Multiple Shield High Velocity
Thermocouple Shield Assembly

Fig. 24 Shield assemblies for high velocity thermocouple (HVT) and multiple-shield, high velocity thermocouple (MHVT).

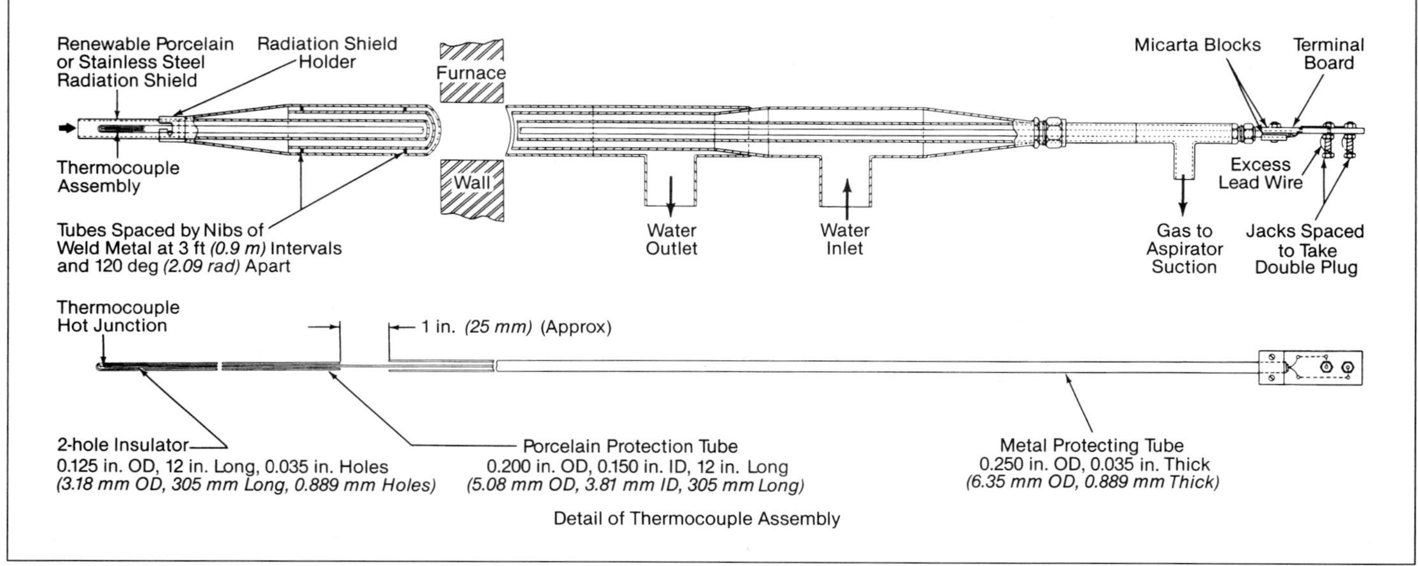

Fig. 25 Rugged, water-cooled, high velocity thermocouple (HVT) for determining high gas temperatures.

to date, no acceptable point by point measurement of high temperature gases has been fully developed.

Measurement evaluations

Large boilers may have significant variation in actual measurements at a particular measurement plane. Where significant data variation exists, data reduction may require mathematical methods to weight-average individual measured values during the data reduction process.

Flue gas temperature measurements at the economizer outlet during performance tests are an example of a large number of data points which require reduction to a single representative temperature.

A weighted average temperature becomes time consuming and the need for such accuracy must be justified. An average of the individual temperature measurements is usually acceptable if a sufficient number of points are obtained. The average gas temperature may be approached by increasing the number of points or by instrument design to satisfy a particular requirement. For example, in a heat loss efficiency test, multiple measuring locations are used to indicate flue gas exit temperatures for determining dry gas loss. The number of points required by the ASME Performance Test Code permits efficiency accuracies within 0.05%.

Under certain circumstances, the maximum, rather than average, gas temperature may be required for equipment protection. A moveable probe with one or two thermocouples may be used to determine the location of this high temperature. For instance, during boiler pressurization and before steam is flowing through the superheater or reheater tubes, a bare thermocouple temporarily installed in the gas stream immediately before the tubes may be used to indicate the highest temperature to prevent tube overheating. These thermocouples are normally removed after steam flow is adequate for cooling. On units designed for remote operation, it is possible to remotely retract and insert these thermocouples using a sootblower carriage. The remotely insertable thermocouple is called a thermoprobe.

Insulation and casing temperature

Outer surfaces

The measurement method should be carefully selected to avoid significant errors in measuring uncased insulation surface temperature. Portable contact thermocouple instruments designed to be pressed against a surface are unsatisfactory on insulated surfaces because the instrument cools the surface at the point of contact, and the low rate of heat transfer through the insulating material prevents adequate heat flow from surrounding areas to the contact point.

Thermocouple attachment to the insulation surface must not appreciably alter the normal rate of heat transmission through the insulation and from the surface to the surroundings. Fine wires can be attached and maintained at the surface temperature more easily than heavy wires. If the insulation is plastic at the time of application, press the thermocouple junction and several feet of lead wire into the surface of the insulation; the thermocouple will adhere when the insulation hardens. If the insulation is hard and dry, the junction and lead wires may be cemented to the surface using a minimum of cement. Fastening the wires to the surface with staples introduces conduction errors. Covering the wires with tape changes the heat transfer characteristics of the surface and imposes an undesirable insulation layer between the wire and the ambient air.

Steel casings

The temperature of steel boiler casings may be accurately measured with portable contact thermocouple instruments, because lateral heat flow from adjoining metal areas quickly compensates for the small quantity of heat drawn by the instrument at the contact point. Thermocouple wires may be peened into or fused onto the metal surface to form the hot junction. Thermal contact between the lead wires and the surface should be maintained for several feet. The wire installation should minimize any disturbance of heat transfer from the surface to the surrounding air. Approximate surface temperatures may be conveniently measured with fusion paints or crayons.

Thermometers are sometimes fastened to metal surfaces with putty. This method gives an approximate surface temperature indication if the metal is massive and near ambient temperature. It is not recommended for boiler casing temperature measurement and is completely unsatisfactory for measuring insulation surface temperature.

Through-steel components, such as metal ribs imbedded in insulation and studs or door frames extending through insulation, cause considerable local upsets in surface temperatures and their influence may spread laterally along a metal casing. These effects must be considered in planning and interpreting surface temperature measurement.

Infrared cameras may be used to measure surface temperatures. These instruments can minimize the need for complex thermocouple installations, especially if large surface areas are to be measured. Improved equipment permits recording for further review.

Measurement of steam quality and purity

The most common methods of field testing for steam quality or purity are:

1. sodium tracer (flame photometry),
2. electrical conductivity (for dissolved solids),
3. throttling calorimeter (for direct determination of steam quality), and
4. gravimetric (for total solids).

Each of these methods is described in ASME Performance Test Code 19.11, *Water and Steam in the Power Cycle*. The throttling calorimeter determines steam quality directly, whereas the other methods determine the steam solids content.

Most of the solids content of steam comes from the boiler water, largely in the carryover of water droplets. (See also Chapter 5.) It is customary to relate steam quality and solids content of steam by the following equation:

$$x = 100 - \left[\frac{\text{solids in steam} \times 100}{\text{solids in boiler water}}\right] \qquad \textbf{(6)}$$

where x equals % steam quality in percent by weight and solids are expressed as equivalent parts per million (ppm) (by weight) of steam or water.

By the use of Equation 6, steam quality can be determined if the solids content is known. This relationship is subject to error resulting from carryover of solids in solution in the steam or in vaporized form. This effect occurs principally with silica at pressures above 2000 psi (13.79 MPa).

The four steam quality measurement methods may be summarized as follows:

1. The sodium tracer method, with continuous recording of total dissolved solids in steam, is used for the highest accuracy.
2. The conductivity method is useful within certain limitations but is less accurate.
3. The calorimeter method is not suitable for measuring extremely small quantities of carryover and at pressures exceeding 600 psi (4.14 MPa).
4. Gravimetric analyses require large samples and do not detect carryover peaks.

Obtaining the steam sample

If the steam quality results are to be accurate, the instruments must be supplied with a representative steam sample. The method of obtaining a steam sample and the operation of the boiler during testing are fundamentally the same for each testing method.

The sampling nozzle design should be as recommended by ASME Performance Test Code 19.11. It should be located after a run of straight pipe equal to at least ten diameters. Locations in the order of preference are:

1. vertical pipe, downward flow,
2. vertical pipe, upward flow,
3. horizontal pipe, vertical insertion, and
4. horizontal pipe, horizontal insertion.

The nozzle should be installed in the saturated steam pipe in the plane of a preceding bend and in a position that ensures the sampling ports directly face the steam flow. On boilers with multiple superheater supply tubes, sampling nozzles should be located in tubes spaced across the width of the drum. These sampling points should be at no greater than 5 ft (1.52 m) intervals.

When a calorimeter is used the connection from the sampling nozzle to the calorimeter must be short and well insulated to minimize radiation losses. Connections must be steam-tight so the insulation remains dry.

When steam purity is tested by conductivity or sodium tracer methods, the tubing from the sampling nozzles to the condenser should be steel (preferably stainless), with an inside diameter not exceeding 0.25 in. (6.4 mm). It should be of minimal length to reduce the storage capacity of the line. Multiple connections can be run to a common line and then to the condenser. However, these connections should be valved so that each one can be sampled individually to isolate possible selective carryover. Cooling coils, or condensers, should be located close to the sampling nozzles to minimize settling of solids in the sample line.

Sodium tracer method

The sodium tracer technique permits measuring dissolved solids impurities (carryover) in steam condensate to as low as 0.001 ppm.

Sodium is present in all boiler water where chemical treatment is in the form of solids. The ratio of the total dissolved solids in the steam condensate to the total dissolved solids in the boiler water is proportional to the ratio of the sodium in the steam condensate to the sodium in the boiler water. By determining the sodium content of the steam condensate and boiler water and the total dissolved solids in the boiler water, the solids content of the steam and the percent moisture carryover may be calculated. The sodium content of steam condensate and water is usually determined by using a flame photometer. The effect of any change in boiler operating conditions on carryover is promptly indicated, facilitating problem analysis.

The operation of a flame photometer is illustrated in Fig. 26. The condensed steam sample is aspirated through a small tube in the burner into the oxygen-hydrogen flame. The flame, at 3000 to 3500F (1649 to 1927C), vaporizes the water and excites the sodium atoms, which emit a characteristic yellow light having a definite wavelength. The intensity of the emitted yellow light is a measure of the sodium in the sample. The intensity of the light is measured with a spectrophotometer equipped with a photomultiplier attachment.

The light from the flame is focused by the condensing mirror and is directed to the diagonal entrance mirror. The entrance mirror deflects the light through the entrance slit and into the monochromator to the plane mirror. Light striking the plane mirror is reflected to the fiery prism where it is dispersed into its component wavelengths. The desired light wavelength is obtained by rotating a wavelength selector which adjusts the position of the prism. The selected wavelength is directed back to the plane mirror where it is reflected through the adjustable exit slit and lens. The light impinges on the photomultiplier tube, causing a current gain which registers on the meter. The amount of sodium in the sample is obtained by comparing the emission from the water sample to emissions obtained from solutions of known sodium concentration.

Conductivity method

Electric conductivity can be used to determine steam purity in certain boilers. This method is applicable to units operating with significant solids concentration in the boiler water and with a total steam solids content in excess of 0.5 ppm.

The conductivity method is based on the fact that dissolved solids, whether acids, bases or salts, are completely ionized in dilute solution and, therefore, conduct electricity in proportion to the total solids dissolved. On the basis of solids normally present in boiler water, the solids content in parts per million equals the electrical conductivity of the sample in micromho per mm times 0.055.

The condensed sample should ideally be free from dissolved gases, especially ammonia (NH_3) and CO_2, which contribute nothing to the solids content of the steam but have a significant effect on conductivity.

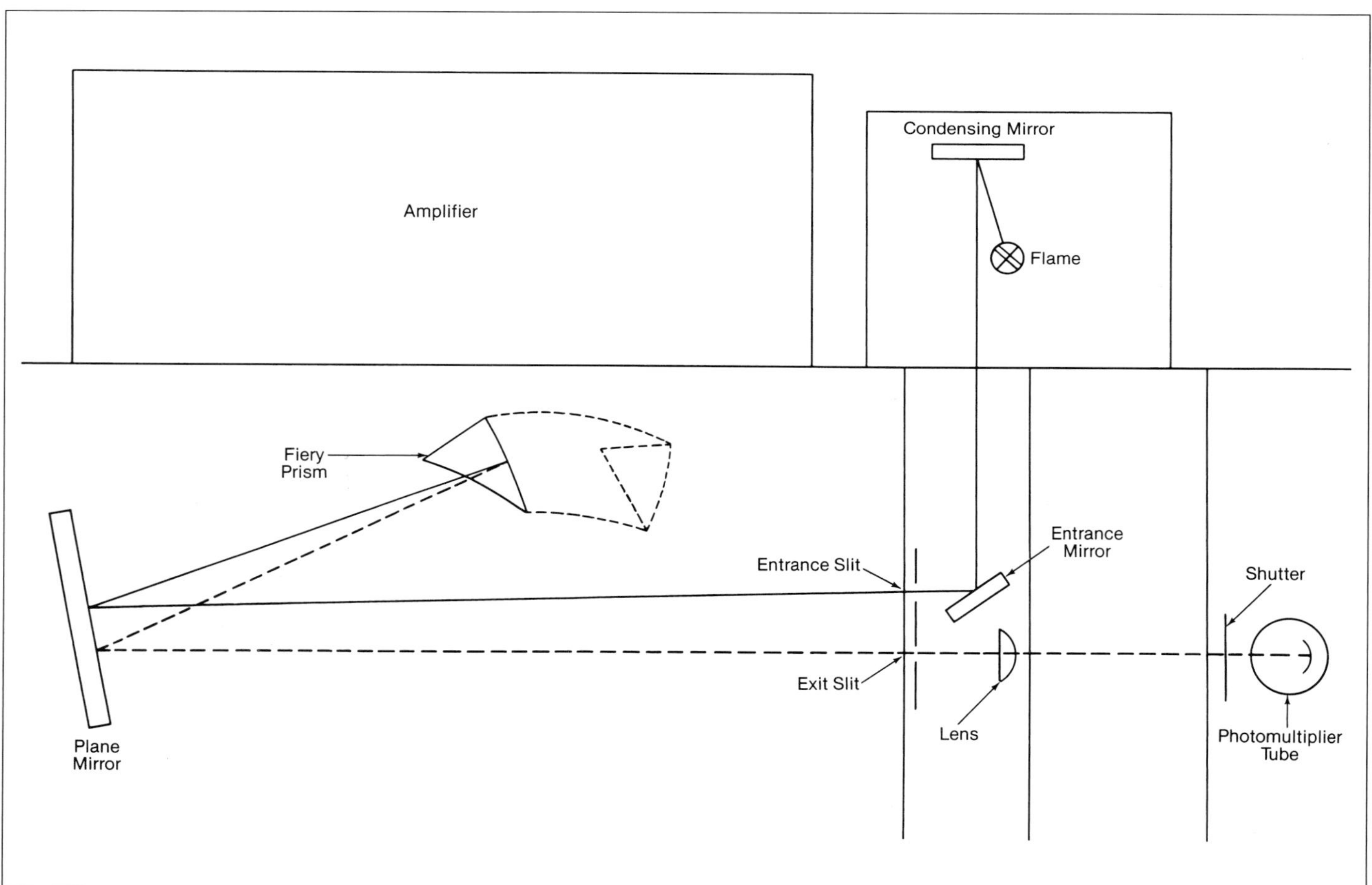

Fig. 26 Schematic arrangement of flame photometer.

Throttling calorimeter

When steam expands adiabatically without doing work, as through an orifice, the enthalpies of the high and low pressure steam are equal, provided there is no net change in steam velocity. Such an expansion is termed throttling. As can be seen from a Mollier chart (see Chapter 2), wet steam with an enthalpy exceeding 1150 Btu/lb (2675 kJ/kg) becomes superheated when throttled to atmospheric pressure. The temperature of the expanded steam determines the enthalpy, which may be used with the pressure of the wet steam to determine the percent moisture in the wet steam sample.

The foregoing principle is used in various forms of throttling calorimeters. Each incorporates a small orifice for expansion of the steam sample into an exhaust chamber, where the temperature of the expanded steam is measured at atmospheric pressure. Velocity changes in properly designed calorimeters are negligible. The variations in the different designs are chiefly in the means of shielding the unit from external temperature influences. At pressures below 600 psi (4.14 MPa), the calorimeter shown in Fig. 27 is used. This is a small, low cost, simple, accurate instrument which gives good results for steam having appreciable quantities of moisture: up to 4.3% at 100 psi (0.69 MPa), 5.6% at 200 psi (1.38 MPa) and 7.0% at 400 psi (2.76 MPa).

Fig. 27 shows the calorimeter installed and ready for use. The connection should be short; the connection and the calorimeter should be well insulated. It is essential that the calorimeter discharge be completely unobstructed, so no backpressure can build in the exhaust chamber. The orifices must be clear, full opening and of the correct diameter: 0.125 in. (3.175 mm) for pressures from atmospheric to 450 psi (3.10 MPa) and 0.0625 in. (1.588 mm) from 451 to 600 psi (3.11 to 4.14 MPa). The thermometer should be immersed in oil of suitably high flashpoint. The calorimeter is operated by fully opening the shutoff valve and letting the steam discharge through the unit to the atmosphere. The calorimeter must be thoroughly warmed up and in temperature equilibrium before normal temperature or test readings are taken.

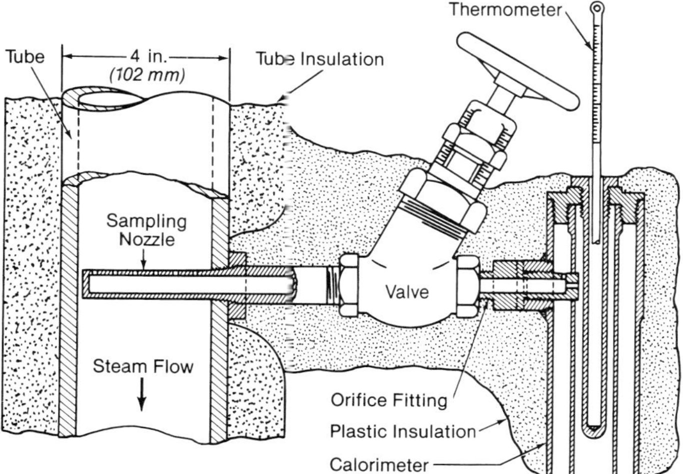

Fig. 27 Throttling calorimeter showing sampling tube in steam pipe.

In obtaining calorimeter readings, the temperature of the expanded and superheated steam is measured by a thermometer inserted in the thermometer well. Due to radiation from the calorimeter installation, thermometer corrections and orifice irregularities, the observed temperature is generally lower than the actual temperature. To determine a suitable correction, the as-installed temperature of the calorimeter should be determined. This can be done by taking readings when the boiler is known to be delivering dry saturated steam. This is the case when the boiler output is steady at about 20% of rated capacity with low water concentration and steady water level. The normal correction is obtained by subtracting the as-installed temperature from the theoretical temperature, read from the zero moisture curve of Fig. 28. The normal correction should not exceed 5F (2.8C). If it does, the calorimeter orifice is clogged, the insulation is faulty, or some other test feature is incorrect. When the unit is used to determine the quality of a wet steam sample, the percent moisture can be found from the curves of Fig. 28 using measured drum pressure and corrected calorimeter temperature. When properly installed, insulated and operated, the calorimeter can give accurate results for steam moisture contents as low as 0.25% for low pressure boilers. For pressures above 600 psi (4.14 MPa) and for greater accuracy of steam purity measurement in the low ppm range, other methods of measurement should be used.

Gravimetric analysis

Gravimetric analysis is used to determine solids levels, particularly total solids, in a condensed steam sample. It consists of evaporating a known quantity

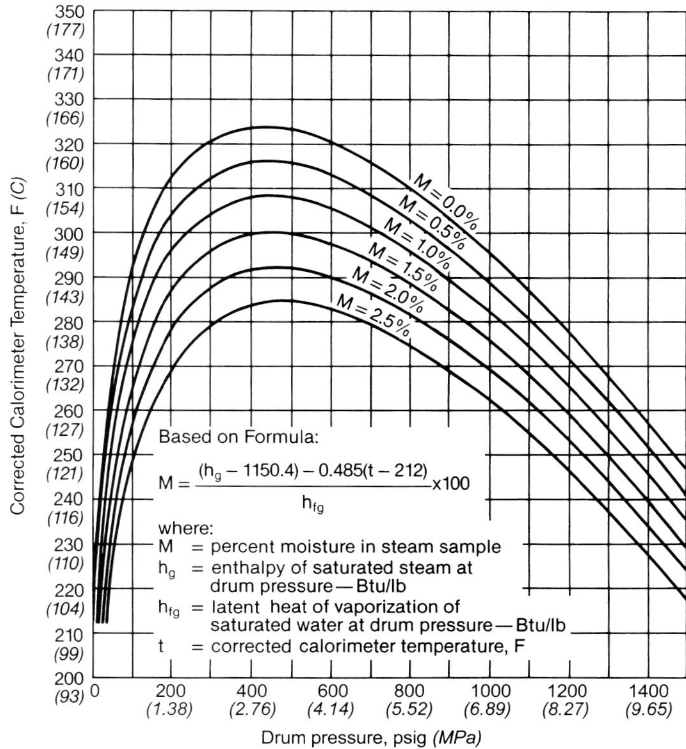

Fig. 28 Percent moisture in steam versus calorimeter temperature.

of condensed steam to complete dryness and weighing the residual. This can be done on a batch basis or by a continuous sample system.

While a gravimetric analysis provides an accurate measure of total solids in water, its main disadvantage is that it requires a relatively large quantity of water obtained over an extended period of time. It therefore does not localize carryover peaks when they occur. Time lag in other methods of steam purity measurement is usually small enough to permit attributing peaks of carryover to periods of high water level, high boiler water solids, or other operating upsets.

Flow measurement

Fluid flow includes water, steam, air and gas flow. While there are many means of measuring flow, the basic methods for the accuracy required by the ASME Performance Test Code use the orifice, flow nozzle (flow tube), or venturi tubes as primary elements. The pressure drop or differential pressure created by these restrictions can be converted into a flow rate.

The flow of any fluid through an orifice, nozzle or venturi tube may be determined by the equation:

$$\dot{m} = C_q Y A \sqrt{\frac{2 g_c \rho_1 (P_1 - P_2)}{1 - \beta^4}} \quad (7)$$

where

$\dot{m}$ = flow rate, lb/s
C_q = coefficient of discharge, dimensionless, dependent on the device used, its dimensions and installation
Y = compressibility factor, dimensionless, equals 1.0 for most liquids and for gases where the pressure drop across the device is less than 20% of the initial pressure
A = cross-sectional area of throat, ft^2
g_c = proportionality constant, 32.17 lbm ft/lbf s^2
P_1 = upstream static pressure, lb/ft^2
P_2 = downstream static pressure, lb/ft^2
β = ratio of throat diameter to pipe diameter, dimensionless
ρ_1 = density at upstream temperature and pressure, lb/ft^3

Details of primary element sizing, fabrication and flow calculations can be obtained from the ASME publication *Fluid Meters, Their Theory and Application.*

Table 2 is extracted from the ASME Performance Test Code 19.5, Section 4, *Flow Measurement,* to show the advantages and disadvantages of the three primary element types. The throat tap nozzle has the highest accuracy of the primary flow measuring devices.

The primary elements listed should be fabricated of erosion and corrosion resistant materials. The orifice and flow nozzles are shown in Fig. 29; the venturi tube is shown in Fig. 30.

Even though these elements can be sized with considerable accuracy by calculation, they should be laboratory calibrated prior to precision testing. Calibration is usually by weighed water tests using scales. For commercial use, calculated flow rates or rates based on prototype testing or calibration are adequate.

Certain important factors of the primary element installation should be considered for accurate measurement. These include:

Table 2
Advantages and Disadvantages of Various Types of Primary Elements

Advantages	Disadvantages
Orifice	
1. Lowest cost	1. High nonrecoverable head loss
2. Easily installed and/or replaced	2. Suspended matter may build up at the inlet side of horizontally installed pipe unless eccentric or segmental types of orifices are used with the hole flush with the bottom of the pipe
3. Well established coefficient of discharge	3. Low capacity
4. Will not wiredraw or wear in service during test period	4. Requires pipeline flanges, unless of special construction
5. Sharp edge will not foul with scale or other suspended matter	
Flow Nozzle	
1. Can be used where no pipeline flanges exist	1. Higher cost than orifice
2. Costs less than venturi tubes and capable of handling same capacities	2. Same head loss as orifice for same capacity
	3. Inlet pressure connections and throat taps when used must be made very carefully
Venturi Tube	
1. Lowest head loss	1. Highest cost
2. Has integral pressure connections	2. Greatest weight and largest size for a given size line
3. Requires shortest length of straight pipe on inlet side	
4. Will not obstruct flow of suspended matter	
5. Can be used where no pipeline flanges exist	
6. Coefficient of discharge well established	

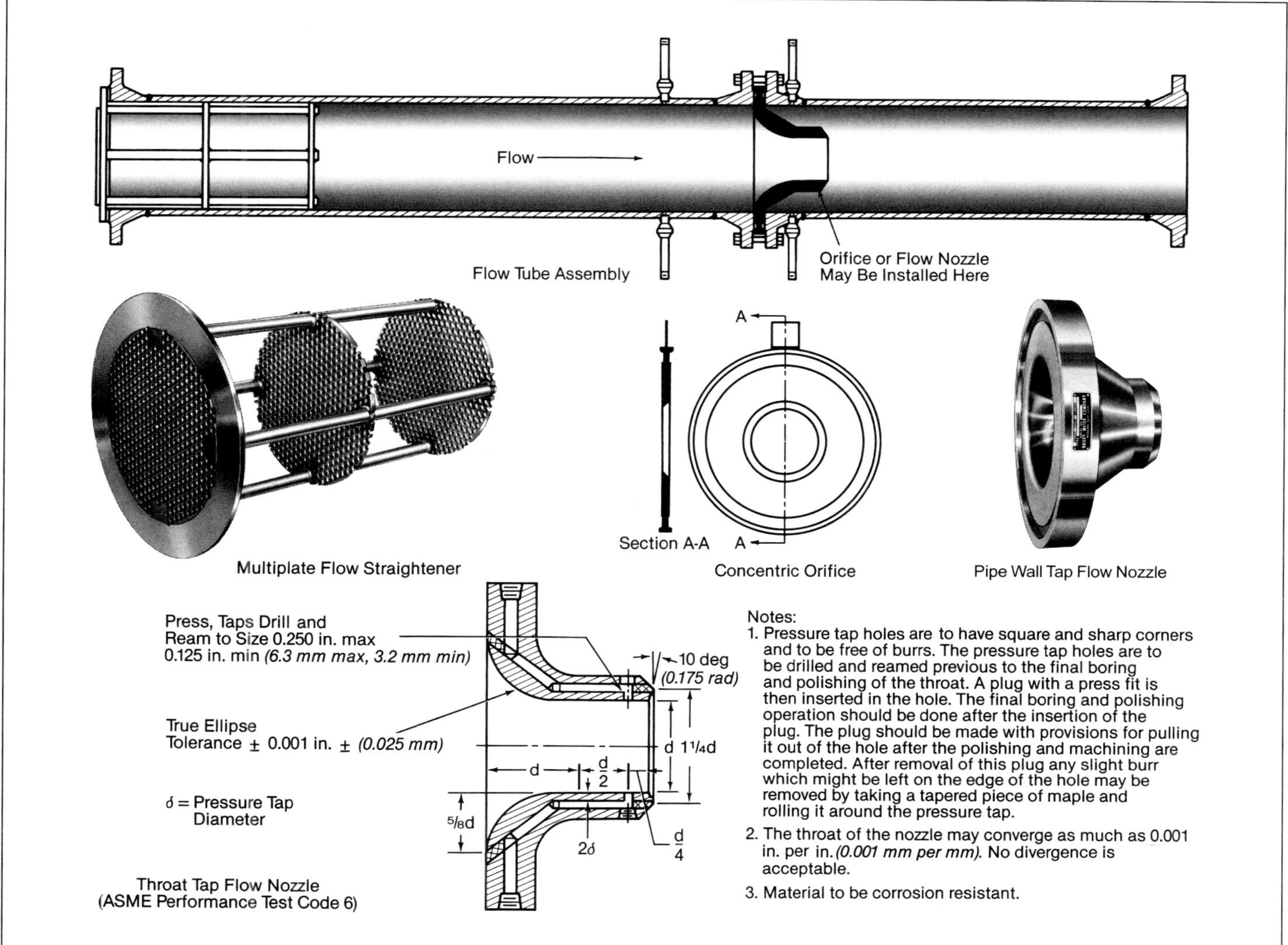

Fig. 29 Orifice and flow nozzles.

1. location in the piping in relation to bends or changes in cross-section,
2. possible need for approach straightening vanes,
3. location and type of pressure taps,
4. dimensions and condition of surface or piping before and after the element,
5. position of the element relative to direction of fluid flow, and
6. type and arrangement of piping from primary element to differential pressure measuring instrument.

Details of these requirements for precision testing are also covered in the ASME publication *Fluid Meters, Their Theory and Application*. These devices produce a differential pressure across the element and a differential pressure gauge should be used for the flow measurement.

Flow measurement of combustion air or flue gas generally does not require a high degree of precision. Orifices, flow nozzles, or venturis are also used, but the rigid requirements of construction and location are usually not possible because of space limitations. As a substitute, a component of the steam generator, located in the flow path, can create a suitable pressure differential, which is used to obtain a flow measurement. For example, the draft loss across the gas side of an air heater may be used as an index of the flue gas flow. Because pressure differentials vary with a change in surface cleanliness, these are not highly reliable methods. The use of orifices, flow nozzles, venturi air foils, or impact suction tubes in areas free of entrained dust and at relatively stable temperatures prove more satisfactory. Fig. 31 illustrates two types of venturi sections used in air ducts for measuring combustion air flow. Fig. 32 illustrates an impact suction pitot tube arrangement for metering primary air flow to a pulverizer. For a dependable determination of air flow, the primary element should be calibrated at normal operating pressure and temperature. Test connections, or sampling points, for this purpose are located at a zone in the duct or pipe where good flow characteristics are obtainable, and the calibrating flow measurements are made using a hand held pitot tube or equivalent probe. The pitot tube, when inserted facing the air or gas flow stream, measures velocity pressure, that is, the difference between total and static pressure. The velocity pressure measurement can be converted into a velocity reading for English units only by:

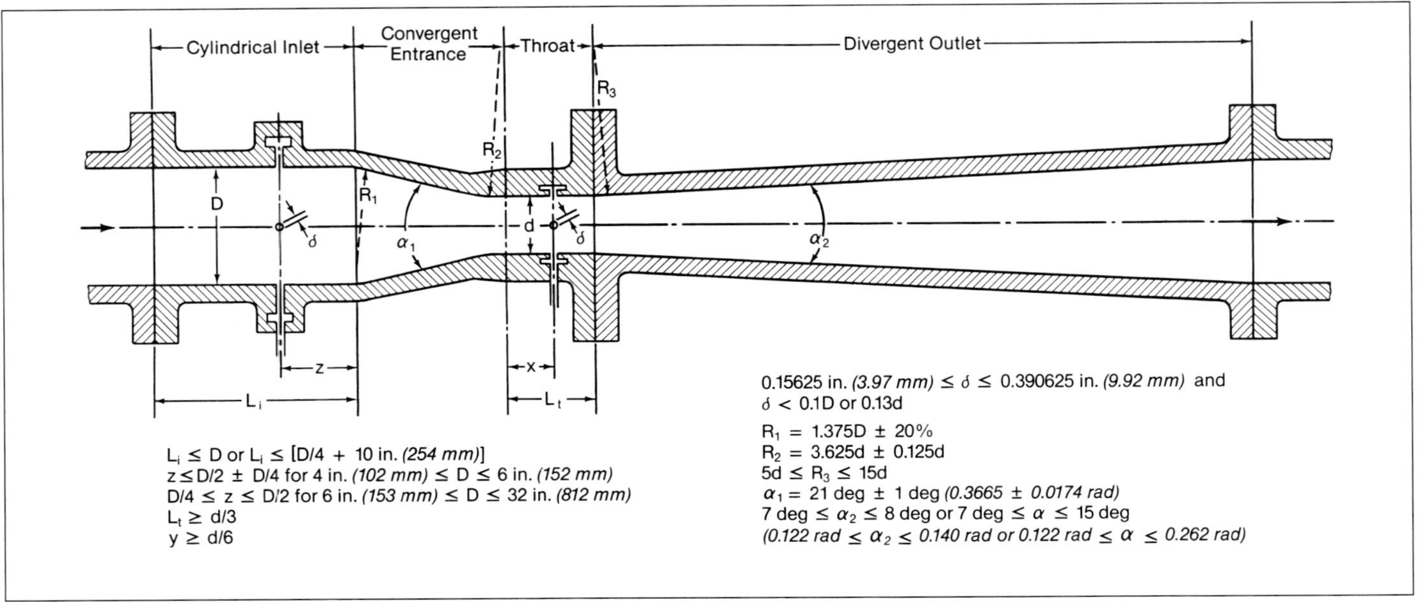

Fig. 30 Dimensional proportions of classical (Herschel) venturi tubes with a rough-cast, convergent inlet cone. (Source: ASME, *Fluid Meters, Their Theory and Application*, Sixth Edition, 1971.)

$$V = 1097\sqrt{\frac{h_V}{\rho}} \qquad \textbf{(8)}$$

where

V = velocity, ft/min
h_V = velocity head as indicated by pitot tube differential, in. wg
ρ = density at the temperature of the sampling location, lb/ft^3

A typical pitot tube and measuring manometer are illustrated in Fig. 33. The pitot tube has a coefficient of 1.0, which is applied to velocity head readings, eliminating the need for corrections. Fig. 34 shows a method of traversing a round duct using a pitot tube.

An additional portable flow measuring device that can be used for calibration purposes is the Fechheimer probe. It also has a coefficient of 1.0 and incorporates a null balance feature that permits determining when the probe faces the direction of gas flow. This probe is illustrated in Fig. 35. In this schematic, the outer holes in the probe are each at exactly 39.25 deg from the centerline hole, placing them at a point of zero impact pressure and providing a reading of true static pres-

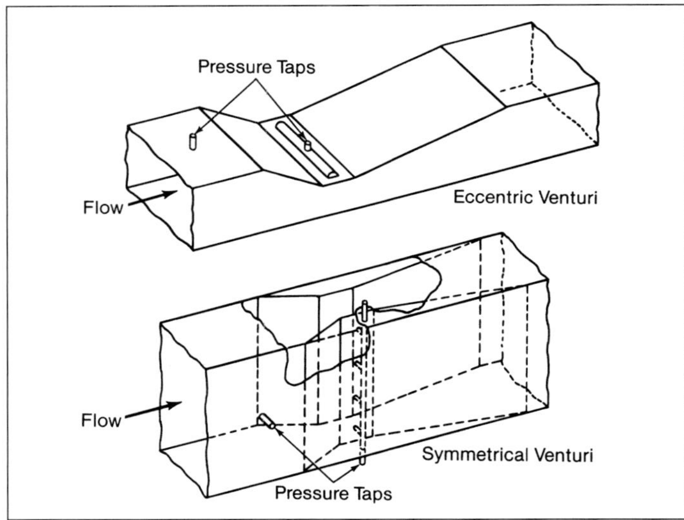

Fig. 31 Venturi sections for air ducts.

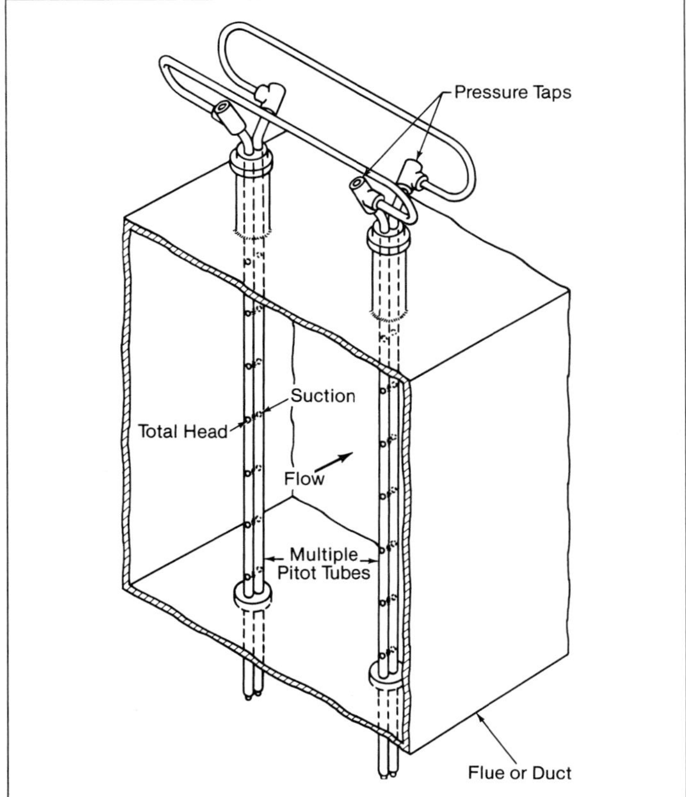

Fig. 32 Averaging pitot tubes.

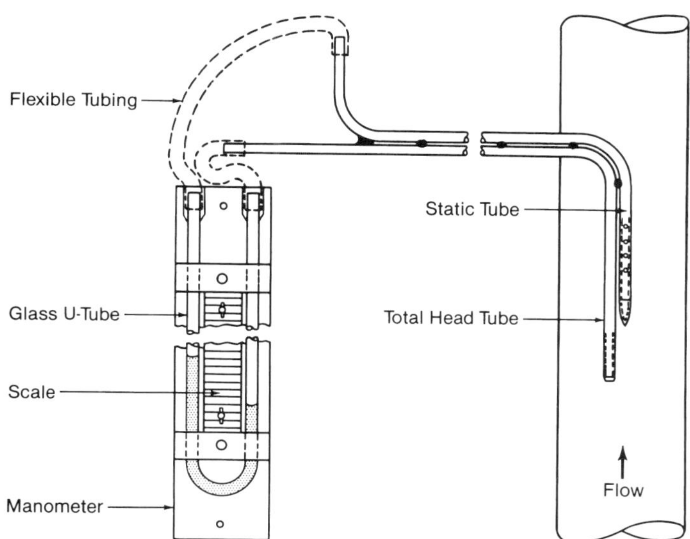

Fig. 33 Arrangement of pitot tube and manometer.

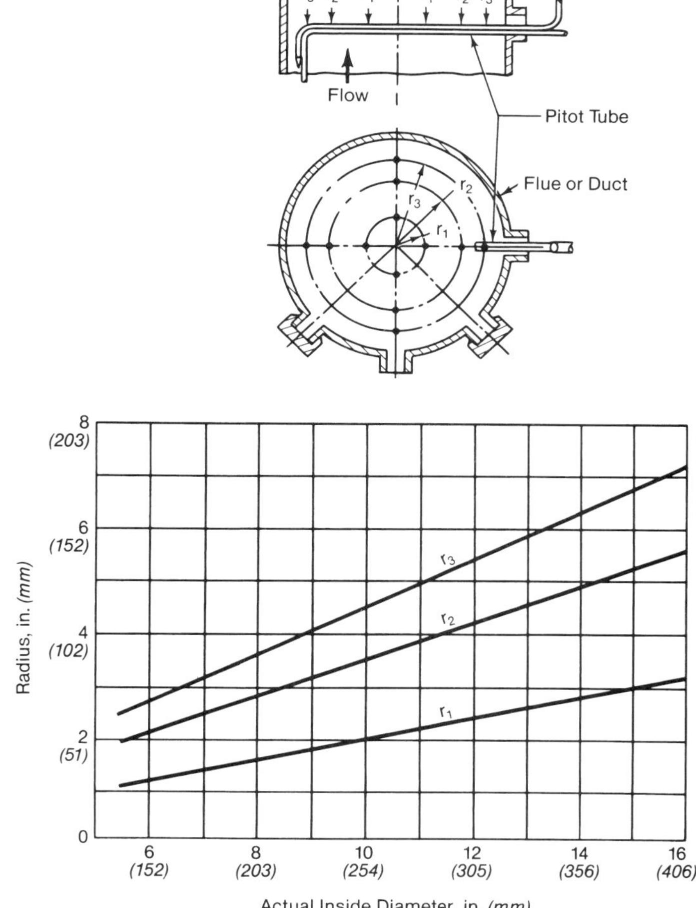

Fig. 34 Pitot tube traverse.

sure. When the probe faces the gas flow, a manometer connected across the two outer holes shows a zero, or null, balance. Because the centerline hole receives full impact or total pressure when facing into the gas stream, a manometer connected across one outer hole and the centerline hole gives true velocity pressure, that is, impact pressure less static pressure.

The impact suction or reversed type pitot tube, illustrated in Fig. 36, produces a differential pressure greater than one velocity head, which gives a magnified reading. This type of pitot tube measures flow in either direction. To use the reversed type pitot tube for flow measurement, calibration is required for the Reynolds number range as well as for the geometry of the flow channel and its orientation in the stream.

Permanently installed pitot tubes are used where relatively high pressure losses due to the primary measuring element are undesirable. Flow tubes are also used in these cases. These tubes differ from most other primary elements in that the flow channel cross-section varies symmetrically in the inlet and exit directions from the contoured throat section.

The throat section is equipped with two sets of pressure taps, one set pointing upstream and one downstream, similar to impact suction pitot tubes. Each set of taps is interconnected and the two sets in turn are connected to a differential pressure gauge. The pressure taps are located at cross-sections of equal area so that the differential developed is a result of impact and suction pressure differences only and is, therefore, a function of the velocity head alone.

Many other types of flow measuring devices exist. The pitot tube grid incorporates an array of pitot tubes in a flow path. The primary signal is a pressure differential. The array of pitot tubes measures a number of flows and minimizes measurement errors caused by flow unbalance and upstream conditions. The hot wire anemometer measures flow using a probe with a small wire at the tip. Flow over the wire removes heat, and flow may be measured when the hot wire is in equilibrium with electrical power input and heat lost. Doppler phenomena may be applied to flow measure-

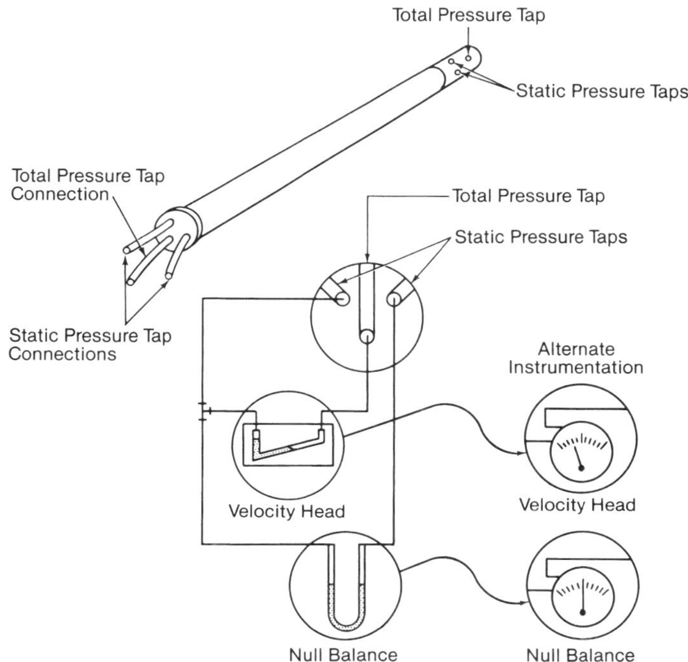

Fig. 35 Fechheimer probe.

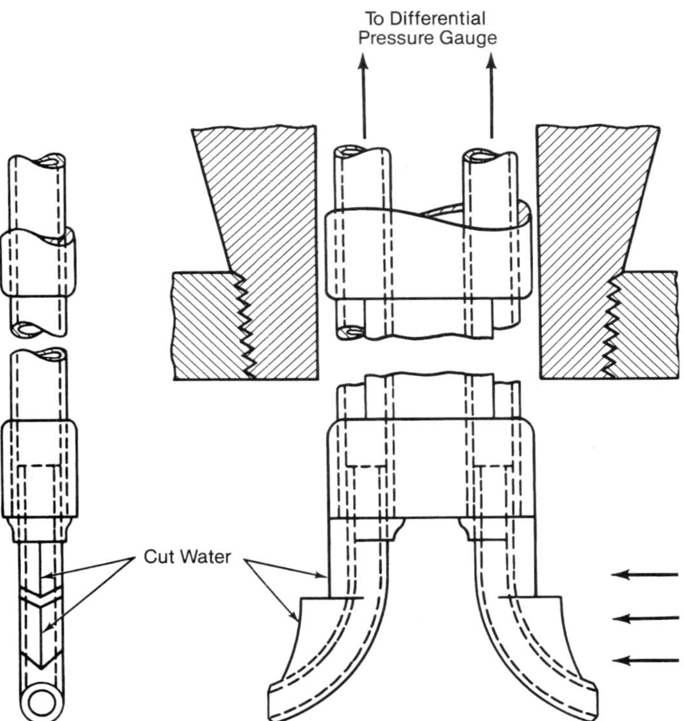

Fig. 36 Combined-type pitot tube with reversed static tube. (Source: ASME, *Fluid Meters, Their Theory and Application*, Sixth Edition, 1971.)

ment with laser or acoustic transmissions. The transmission signal and type of fluid make these methods very application specific.

Computer application to measurement

The application of computer technology to instrumentation provides the capability to further understand the operating characteristics of many interrelated power plant components and processes at a much faster rate than was previously possible. Measurements have progressed from readings recorded at local gauges, strip charts, or manually recorded test gauges to significant quantities of information acquired in seconds.

The instrument readings are not made directly at the instrument, but rather transmitted signals from the instrument are entered into a data acquisition system and the reading is indicated at a computer. The computer gathers, stores, interprets and displays the information. The display appears as an extension of the instrument. The information gathered may be further manipulated for process control or equipment performance computations.

Performance computation

Computer computation permits calculations in near real time for boiler efficiency and other plant performance indicators. These performance indicators provide rapid updates to operating information and provide a check on measurement errors. Measurements of temperature, pressure and flow that might be in error because of instrument calibration are identified rapidly.

Process control

A computer based control system receives its basic information from a data acquisition system which accumulates information from the plant instrumentation. Process control is achieved according to specific mathematical algorithms. The control system, updated by current data or operator interface, produces output signals to modulate components based on these algorithms.

Data acquisition system

A typical data acquisition system consists of computer hardware and related software. The basic hardware components are the computer, a signal multiplexer and the measuring instrument. The computer program (software) controls the hardware operation and data recording. In operation, the program causes the signal multiplexer to read a specific instrument analog output signal. The signal is converted from an analog to a digital signal, which is conveyed to the computer and stored in memory. The first processing of the data is the storage of the digitized signal in a data array. Data acquisition, incorporating many instruments, other computer outputs and manual inputs, requires that each of the signals be assigned a specific location in the array. The stored signal may be converted to engineering units and stored in another array. Furthermore, the data may be averaged over time and then stored. Performance calculations are generally performed using data converted to engineering units and averaged over time. Portable data acquisition systems are generally used for specific test purposes and may use high precision transmitters and thermocouples.

Signal multiplexer

The signal multiplexer in a data acquisition system provides an interface between the transmitted instrument signal and the computer. The typical multiplexer provides signal conditioning, analog to digital signal conversion and switching capability for multiple inputs. It produces a digital output signal for computer processing and assignment to a position in the data array. Significant engineering design is required to accommodate the variety of signals entering the multiplexer.

Data acquisition computer program

The data acquisition software may be program oriented or menu oriented but should readily permit changes to the quantity of signals, the rate at which the computer reads each signal and the computations performed with each signal. Data storage and retrieval are but two of many design considerations for computer program configurations.

References

1. Tower, J.R., "Staring PtSi IR cameras: More diversity, more applications," *Photonics Spectra,* (USPS 448870), Laurin Publishing Company, Pittsfield, Massachusetts, February, 1991.

2. Kaplan, H., "What's new in IR thermal imagers," *Photonics Spectra,* (USPS 448870), Laurin Publishing Company, Pittsfield, Massachusetts, February, 1991.

3. Weiss, J., and Esselman, W., Electric Power Research Institute and R. Lee Foster-Miller Inc., "Assess fiberoptic sensors for key power plant measurements," *Power,* p. 55, October, 1990.

Bibliography

1986 Digital Systems Techniques, Instruments and Apparatus, Supplement to ASME Performance Test Codes, ANSI/ASME PTC 19.22, American National Standards Institute/American Society of Mechanical Engineers, New York, New York, 1986.

Manual on the Use of Thermocouples in Temperature Measurement, ASTM Special Technical Publication 470B, American Society for Testing and Materials, Philadelphia, Pennsylvania, 1981.

The OMEGA Temperature Handbook and Encyclopedia, Vol. MMV, Fifth Ed., Omega Engineering, Inc., 2004.

Temperature Measurement Designer's Guide, Thermo Electric Company Inc., Saddle Brook, New Jersey, 1986. Also available online at www.thermo-electric-direct.com.

Typical utility power plant control console.

Chapter 41

Controls for Fossil Fuel-Fired Steam Generating Plants

Instruments and controls are essential to all steam generating installations, promoting safe, economic and reliable operation of the equipment. They range from the simplest manual devices to complete systems that automatically control the boiler and all associated equipment.

Boiler control systems, particularly those applied to boiler turbine-generator units, are of several types that perform varied control functions. In the past, the accepted practice has often been to identify these functions as separate and independent systems, often applied to the boiler by different suppliers. Today, a power plant control system is most often an integrated package with demand requirements applied simultaneously to the boiler, turbine and major auxiliary equipment. This minimizes the number of complex interactions between subsystems.

Various types of boiler control systems for fossil fuel boilers include:

1. combustion (fuel and air) control,
2. steam temperature control for superheater and reheater outlet,
3. drum level and feedwater flow control,
4. burner sequence control and management systems,
5. bypass and startup,
6. systems to integrate all of the above with the turbine and electric generator control,
7. data processing, event recording, trend recording and display,
8. performance calculation and analysis,
9. alarm annunciation system,
10. management information system, and
11. unit trip system.

An overall integrated control system for a fossil fuel-fired drum boiler is described later in this chapter.

Basic control theory

Boiler control is the regulation of the boiler outlet conditions of steam flow, pressure and temperature to their desired values. In control terminology, the boiler outlet steam conditions are called the outputs or controlled variables. The desired values of the outlet conditions are the set points or input demand signals. The quantities of fuel, air, and water are adjusted to obtain the desired outlet steam conditions and are called the manipulated or controlled variables.

Disruptive influences on the boiler, both internal and external, such as fuel heating value variations, unit load change or change in cycle efficiency, are called disturbance inputs. The controller or control system looks at the desired (set points) and actual (output variables) values of the outlet steam conditions and adjusts the amounts of fuel, air and water (manipulated variables) to make the outlet conditions match their desired values. The controller can be manual, with an operator making the adjustments, or automatic, with a pneumatic system, electronic analog computer, or a digital computer making the adjustments. A block diagram of a conceptual boiler control system is shown in Fig. 1.

While it is theoretically possible to operate a boiler with manual control, the operator must maintain a constant watch for a disturbance. Time is needed for the boiler to respond to a correction and this can lead to overcorrection with further upset to the boiler. An automatic controller, on the other hand, once properly

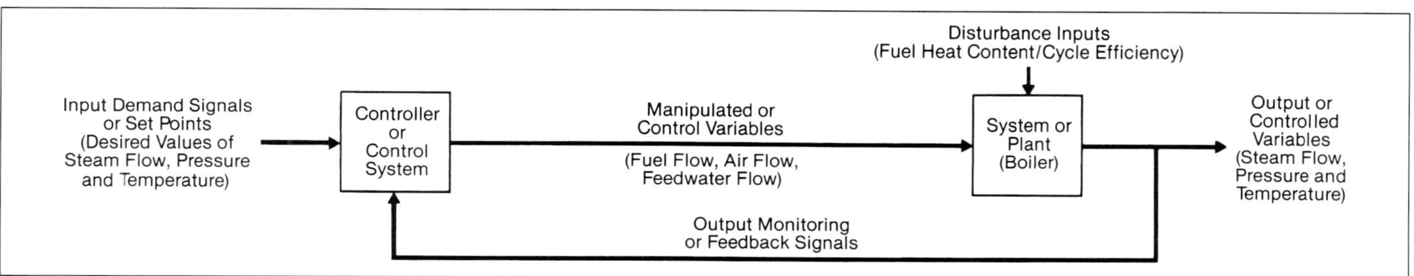

Fig. 1 Block diagram of conceptual boiler control system.

tuned will make the proper adjustment quickly to minimize upsets and will control the system more accurately and reliably.

The various types of automatic control can be illustrated by Fig. 2, which represents the control of one output variable, steam pressure, in a hypothetical boiler. The set points for steam pressure and steam flow will be used as reference values, and the manipulated variable is the fuel valve position. Disturbance inputs act on the boiler, which is the system to be controlled.

Open loop control

The simplest control mode is an open loop, feedforward, or non-feedback control. The manipulated variables of fuel, air and water are adjusted only from the input demand signals without monitoring the outlet conditions or output variables. As an example, Fig. 3 illustrates an open loop control system used to accomplish the control function illustrated by Fig. 2. Fig. 3a is a block diagram representing the action to be taken, which is feedforward only. Fig. 3b is a calibration curve expressing fuel valve position as f(x), a function of steam flow demand. This curve, established by calibrating or programming the fuel valve position required to maintain the desired steam flow with a constant steam pressure, is entered into the controller. The response of this open loop control is fast and depends only on the accuracy of the calibration curve.

One difficulty with open loop control is that the calibration curve is accurate only as long as the boiler conditions are the same as when the calibration was set. Another problem is that the output changes as the load changes on the output signal (the fuel valve in Fig. 3). For example, a change in friction on the valve stem will cause a variation in valve position in response to a given input demand signal. This introduces an error into the performance of the open loop control. Such disadvantages normally outweigh the open loop advantages of stability and simplicity.

Closed loop control

Usually the process requirements can not be met by an open loop control, and a closed loop or feedback control must be used. In closed loop control, the actual output of the system is measured and compared to the input demand signal with the difference between the two signals (the error signal) used to eliminate this difference.

The open loop system of Fig. 3 could be closed by having an operator observe the measured steam pressure and then manually adjust the fuel valve position to obtain the fuel flow necessary to maintain the desired steam pressure and flow. To avoid the disadvantages of manual control, an automatic controller can be provided in a closed loop or feedback system.

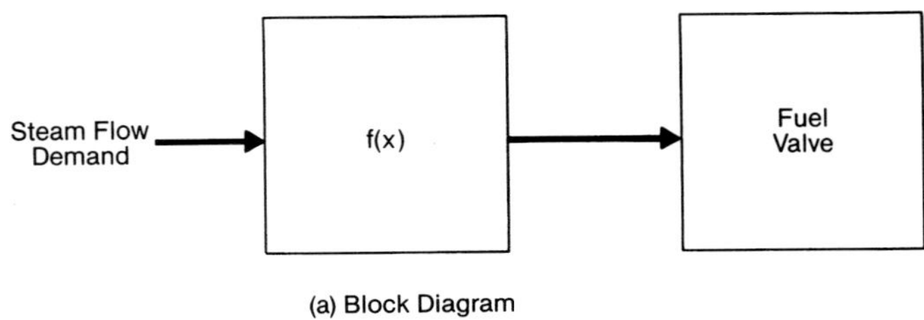

(a) Block Diagram

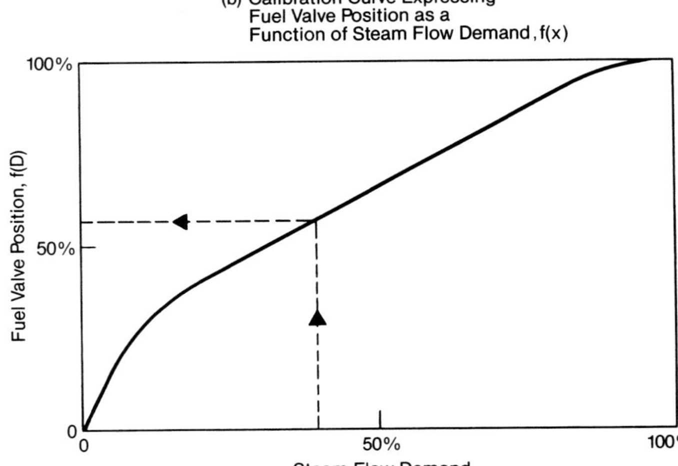

Fig. 3 Open loop mode using steam flow demand to vary fuel input.

Proportional control

The simplest type of closed loop system is proportional control. The manipulated variable or controller output is proportional to the deviation of the controlled variable from its desired value. The deviation is called the error signal. Depending upon controller arrangement, the output signal of a proportional control system will always be either directly or inversely proportional to the controlled variable.

Fig. 4 illustrates the application of proportional control to the control function of Fig. 2. Table 1 explains symbols in Fig. 4 and subsequent figures. In this system, for every deviation of steam pressure from its set point, the fuel valve will be moved to a specific position, as shown in Fig. 5, for three values of relative

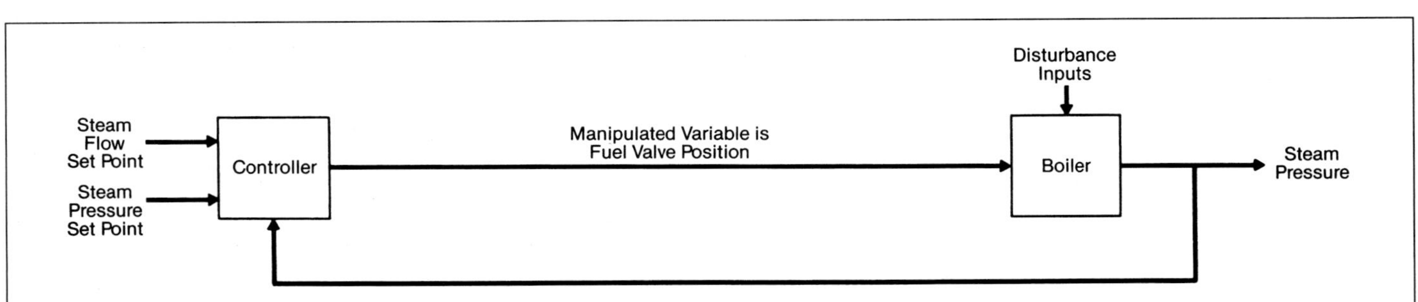

Fig. 2 Simplified block diagram of a boiler control function.

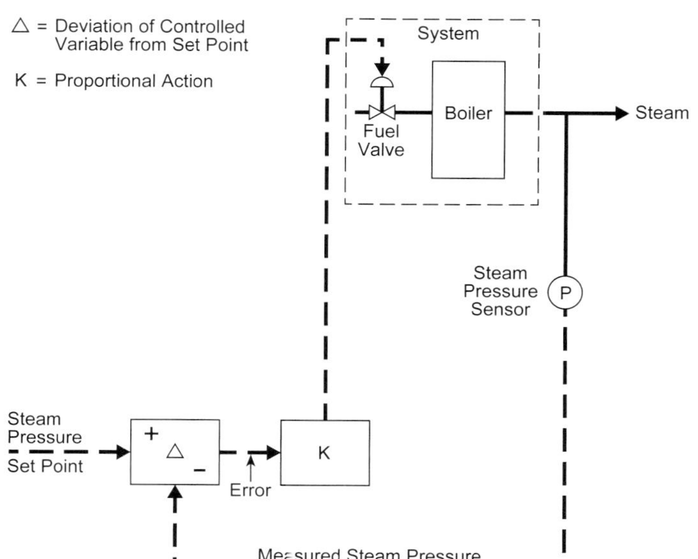

Fig. 4 Proportional control.

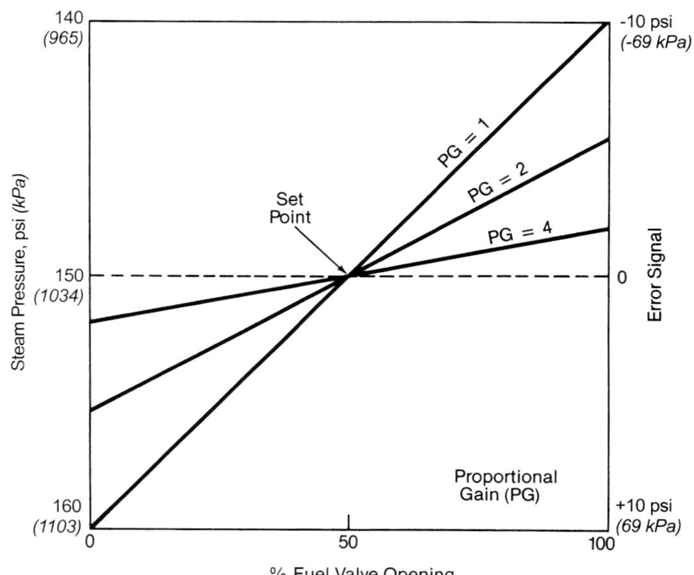

Fig. 5 Fuel valve opening for various pressure deviations in proportional control system.

proportional gain. In Fig. 5, the relative proportional gain could be set so that a pressure error of 10 psi (68 kPa) initiates a 50% change in position of the fuel valve which in the system calibration would be called a gain of 1.0. If the control were set to be two or four times as sensitive, the relative proportional gain would be said to be 2.0 or 4.0 respectively. In the latter case, a change of 2.5 psi (17.2 kPa) in steam pressure initiates a 50% change in fuel valve opening.

The response of steam pressure to step increases in load on the boiler is shown in Fig. 6 for two values of proportional gain on the controller. A step increase in load or steam flow immediately decreases the steam pressure which, through the proportional controller, causes the fuel valve to open by a proportional amount. The additional heat input due to the increased fuel flow causes the steam pressure to return toward its set point, which in turn reduces the fuel valve opening.

There may be a certain amount of cycling or searching, depending on the proportional gain of the controller, before the steam pressure stabilizes. It is noted in Fig. 6 that the steam pressure does not stabilize at its set point, but is offset to a lower value. A characteristic of proportional only control is that an error or

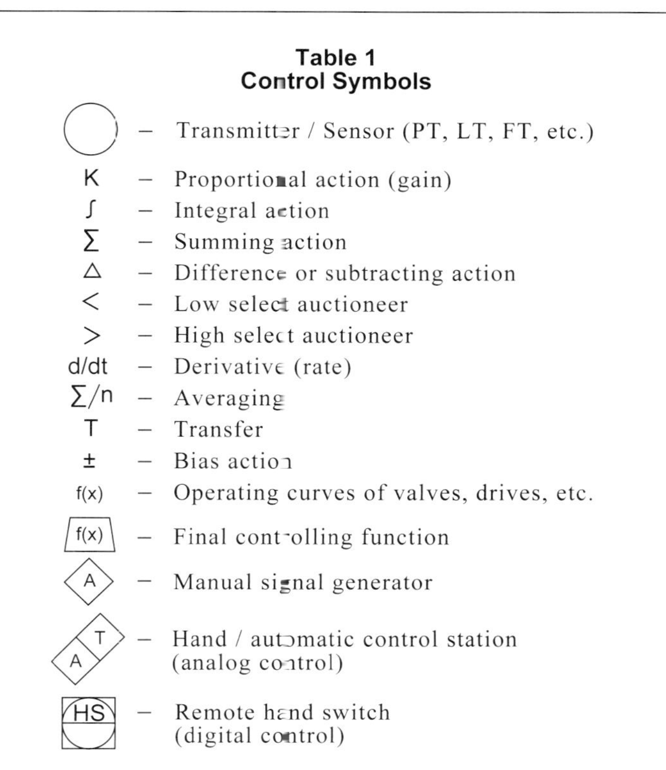

Table 1
Control Symbols

◯	–	Transmitter / Sensor (PT, LT, FT, etc.)
K	–	Proportional action (gain)
∫	–	Integral action
Σ	–	Summing action
△	–	Difference or subtracting action
<	–	Low select auctioneer
>	–	High select auctioneer
d/dt	–	Derivative (rate)
Σ/n	–	Averaging
T	–	Transfer
±	–	Bias action
f(x)	–	Operating curves of valves, drives, etc.
f(x) (boxed)	–	Final controlling function
A (diamond)	–	Manual signal generator
T / A (diamonds)	–	Hand / automatic control station (analog control)
HS	–	Remote hand switch (digital control)

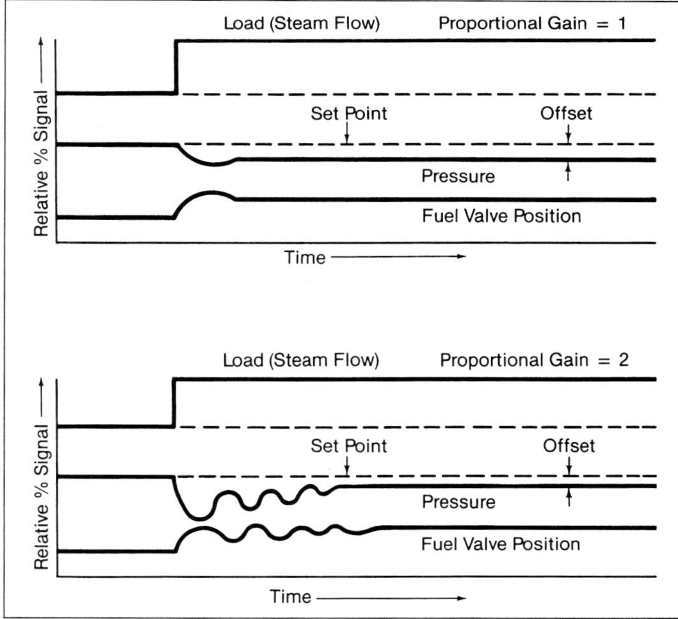

Fig. 6 Response of steam pressure to step increase of load in a proportional control system.

offset is necessary to provide a steady-state fuel valve opening which will support the desired load, except for a single load condition.

An increase in the relative proportional gain or sensitivity will reduce the final offset in the steam pressure as shown in Fig. 6. An increase in proportional gain may increase the time required for the system to reach a steady-state condition. This offset is always present with proportional control and, if the gain is increased even more to reduce the offset, undamped oscillations will result.

Proportional plus integral control

The offset may be eliminated by adding integral or reset control to the proportional control system (Fig. 7). The response of the boiler to a step increase in load, when using a proportional plus integral control system, is shown in Fig. 8. The steam pressure is returned to its set point without the offset. However, the system may be less stable as it takes a longer period of time for the steam pressure to stabilize with the offset completely eliminated.

Integral control, as its name implies, is based on the repetitive integration of the difference between the controlled variable and its set point over the time the deviation occurs. Integral control is also referred to as reset control because the band of proportional action is shifted or reset so that the manipulated variable operates about a new base point.

Proportional plus integral plus derivative control

The stability and response of the system can be improved further by adding a third mode of action called derivative or rate control. Derivative control action is a function of the rate of controlled variable change from its set point, as shown in Fig. 9. The addition of this control function is shown in Fig. 10. As soon as the step change is made, the pressure starts to drop and the proportional mode begins to open the fuel valve. The derivative mode will also open the fuel valve further, as a function of the rate at which the pressure is changing, anticipating where the valve should be positioned. When the rate of steam pressure change decreases, the derivative control has less effect and the proportional and integral action do the final valve positioning.

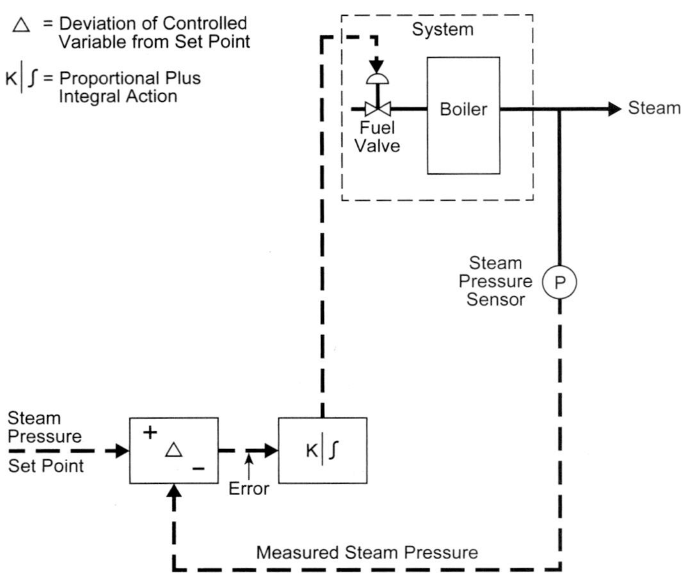

Fig. 7 Proportional plus integral control.

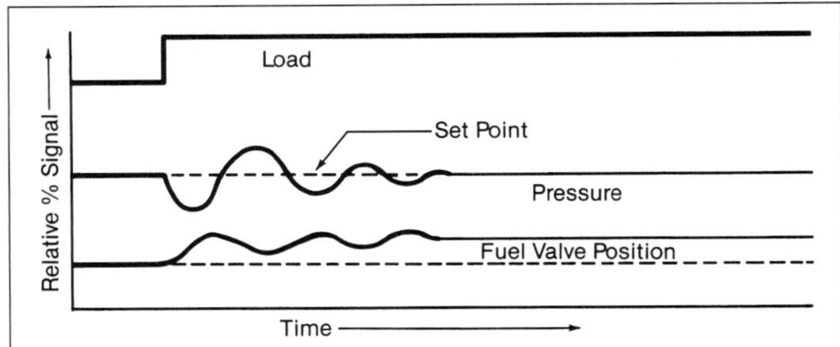

Fig. 8 Response of steam pressure to step increase of load in proportional plus integral control system.

Feedforward/feedback control

In a closed loop system, the controlled variable must deviate from its set point before the controller initiates a corrective action. The open loop system therefore has a faster response because it takes corrective action before the controlled variable starts to change. When the open loop or feedforward system is combined with the closed loop or feedback system, the result is a fast response system that can compensate for changes in the calibration curve. Fig. 11 represents a feedforward/feedback control system for the control function of Fig. 2.

As the step change in load occurs, the feedforward signal immediately positions the fuel valve to meet the calibration curve requirements. If this curve is exact, no error develops and the feedback loop has no work to do. If the calibration curve is in error, the feedback loop readjusts the fuel valve position to eliminate any pressure error which may develop due to shifts in the calibration. However, the response will be faster and the magnitude of the system upset will be smaller because the feedforward signal moves the valve near its final steady-state position leaving a smaller range of action for the feedback loop.

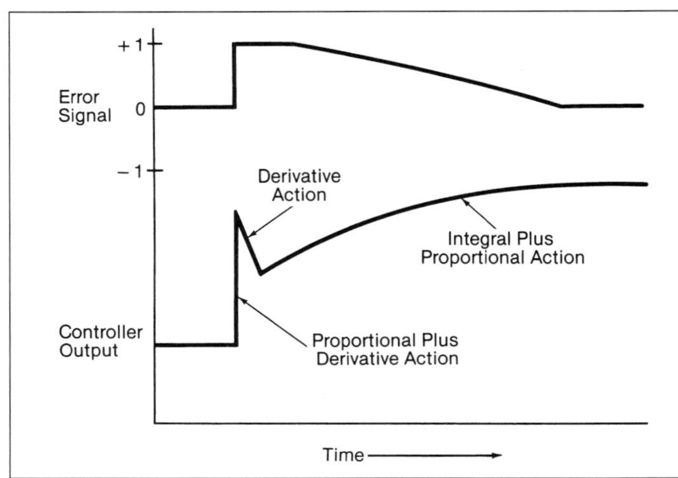

Fig. 9 Proportional plus integral plus derivative action.

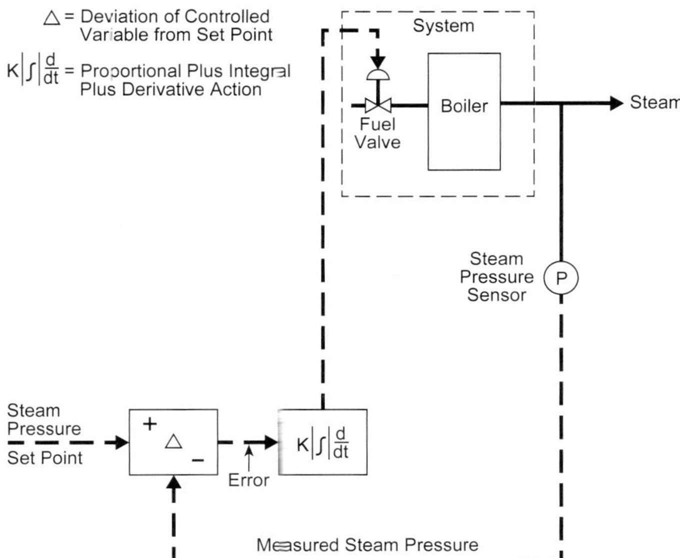

Fig. 10 Proportional plus integral plus derivative control.

Control system design philosophy for a utility boiler

The effectiveness of a boiler turbine-generator unit, as a power producing system, depends on how well it can respond to demands. To achieve optimum response from the unit while maintaining stable pressure and temperature, the control system must coordinate the boiler and turbine. Because the boiler can not produce rapid changes in steam generation by changes in throttle pressure, the turbine valves open and close for initial load response. This initial change permits the new steam flow requirements to be met at the existing throttle pressure. The control system then acts to restore the proper amount of stored energy to the boiler by coordinating the boiler and turbine. This must be done without exceeding the operating limits of the boiler or turbine-generator.

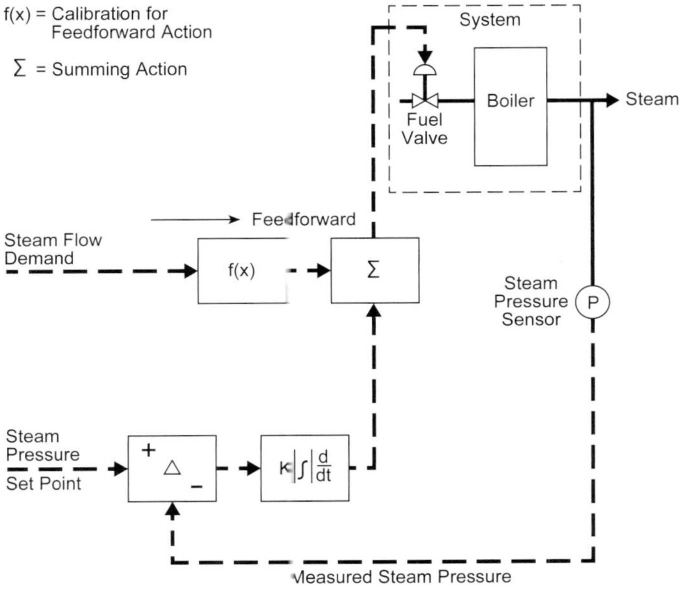

Fig. 11 Feedforward/feedback control.

Characteristics of different control modes

Boiler-following control This mode of operation is designed for the boiler response to follow turbine response. Megawatt load control is the responsibility of the turbine-generator. The boiler is assigned secondary responsibility for throttle pressure control. The demand for a load change repositions the turbine control valves. Following a load change, the boiler control modifies the firing rate to reach the new load level and to restore throttle pressure to its normal operating value.

Load response with this type of system is rapid because the stored energy in the boiler provides the initial change in load. The fast load response is obtained at the expense of less stable throttle pressure control. A boiler-following control system is illustrated in Fig. 12.

Turbine-following control In this operating mode, turbine response follows boiler response. Megawatt load control is the responsibility of the boiler while the turbine-generator is assigned secondary responsibility for throttle pressure control. With increased load demand, the boiler control increases the firing rate which, in turn, raises throttle pressure. To maintain a constant throttle pressure, the turbine control valves open, increasing megawatt output. When a decrease in load is demanded, this process is reversed. Load response with this type of system is rather slow because the turbine-generator must wait for the boiler to change its energy output before repositioning the turbine control valves to change load. However, this mode of operation will provide minimal steam pressure and temperature fluctuation during load change. A conventional turbine-following control system is shown in Fig. 13.

Coordinated boiler turbine control While both of the above systems can provide satisfactory control, each

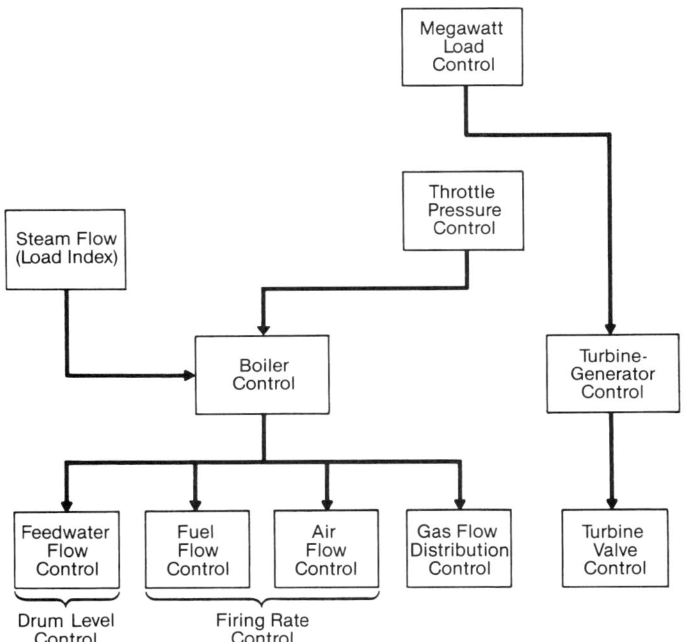

Fig. 12 Boiler-following control system.

has inherent disadvantages and neither fully exploits the capabilities of both the boiler and turbine-generator. The turbine-following and boiler-following systems may be combined into a coordinated control system giving the advantages of both systems and minimizing the disadvantages. The coordinated boiler turbine-generator control system is shown in Fig. 14. Megawatt load control and throttle pressure control are the responsibility of both the boiler and the turbine-generator. One type of a coordinated system assigns the responsibility of throttle pressure control to the turbine-generator, taking advantage of the stability of a turbine-following system. In addition, the system uses the stored energy in the boiler, taking advantage of the fast load response of a boiler-following system.

Because the boiler is not capable of producing rapid changes in steam generation at constant pressure, the turbine is used to provide the initial load response. When a change in load is demanded, the throttle pressure set point is modified using megawatt error (the difference between the actual load and the load demand), and the turbine control valves respond to the change in set point to give the new load level quickly. As the boiler modifies the firing rate to reach the new load and restore throttle pressure, the throttle pressure set point returns to its normal value. The result is a fast and efficient production of electric power through proper coordination of the boiler and turbine-generator. The response is much faster than the turbine-following system, (Fig. 13). However, it is not as fast as the boiler-following system because the effect of megawatt error is limited to maintain a balance between boiler response and stability. Wider limits would provide more rapid response while narrower limits would provide more stability.

A comparison of response characteristics of various arrangements of a boiler turbine-generator control system for a large change in load demand is shown in Fig. 15.

Development of the coordinated control system received added impetus when electric power generating companies introduced the wide area economic load dispatch system. This system requires precise load control. Each controlled unit must produce a specific level of megawatts and the area load control system adjusts the turbine control valves until this level is reached. Even relatively minor interactions between steam flow and throttle pressure can create a continuous interaction between the boiler, the turbine-generator, and the area load control. The conventional boiler-following system, which is subject to such interactions between steam flow and throttle pressure, does not lend itself to area load control because two highly responsive systems would be adversely interacting with each other.

Wide area economic load dispatch systems require controlling the boiler and turbine-generator together in a coordinated system. This achieves automatic control over a wide operating load range.

Integrated boiler turbine-generator control The integrated control system, in its basic form, consists of ratio controls that monitor pairs of controlled inputs, including:

1. boiler energy input to generator energy output,
2. superheater spray water flow to feedwater flow,
3. fuel flow to feedwater flow,
4. fuel flow to air flow,
5. recirculated gas to air flow; this, in effect, is a ratio of reheater absorption to absorption in primary water and steam, and
6. fuel to primary air flow for pulverized coal-fired units.

Fig. 16 is a simplified diagram showing all system inputs controlled in a parallel relationship by ratio controls.

The integrated control system, shown in Fig. 17, coordinates the boiler and turbine-generator for fast

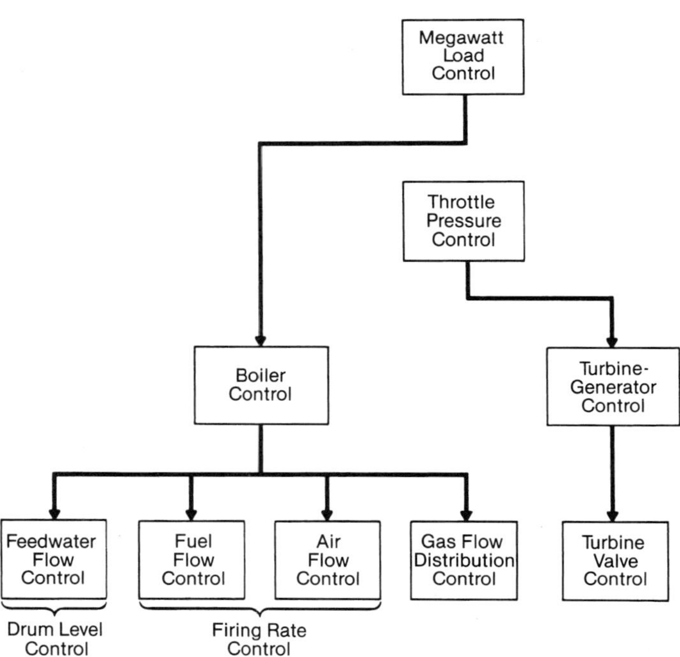

Fig. 13 Turbine-following control system.

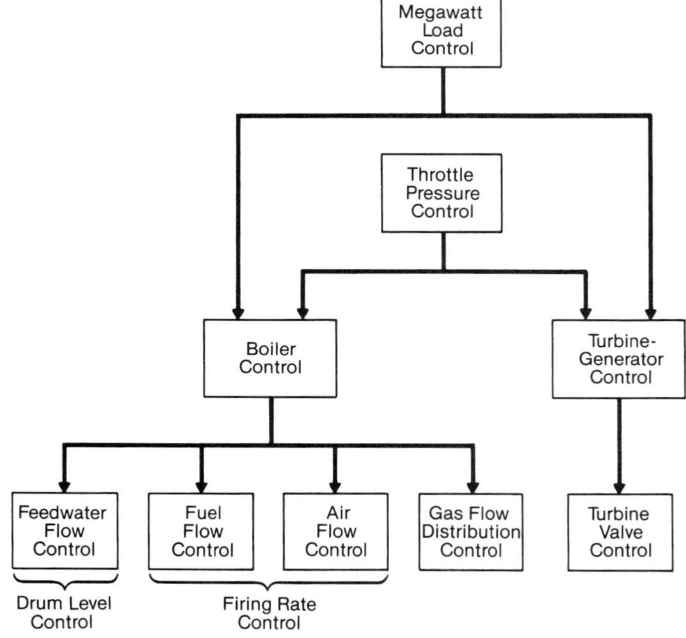

Fig. 14 Coordinated boiler turbine-generator control system.

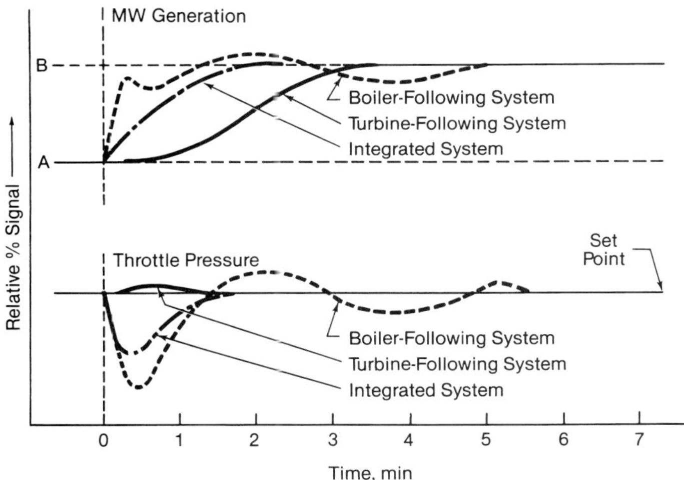

Fig. 15 Megawatt load change and throttle pressure deviation.

and efficient response to load demand initiated by the automatic load dispatch system. See Table 1 for control symbols.

Variable pressure operation

A variable pressure boiler operates with the throttle pressure varying with load. In its purest form, the throttle valves on the turbine are left wide open and throttle pressure varies directly with load. There are recognized advantages of variable pressure operation at lower loads. These include minimum change in turbine metal temperature due to the elimination of the adiabatic throttling effect of turbine valves. Improved cycle efficiency at reduced loads results from reduced boiler feed pump power and higher turbine inlet steam temperatures. However, there are disadvantages including sluggish response to load change demand because of little energy storage during operation. Typically, the load demand change response rate for variable pressure operation is approximately 50% of the response rate for constant pressure operation.

The Babcock & Wilcox Company (B&W) drum boiler bypass and startup system adds another dimension to variable pressure operation. The system, shown in Fig. 18, can generate and deliver steam to the turbine with a wide range of controlled throttle pressure and temperature to assure the least thermal stress to the turbine and consequently the most rapid startups.

This system combines the advantages of constant pressure operation of the boiler with variable pressure operation of the superheater and turbine. The throttle pressure is controlled by modulating the superheater division valves with the drum pressure held at or near full load pressure. The control strategy and operating procedure are equivalent to variable throttle pressure operation with load or steam flow controlled by the turbine control valves.

Cycling operation

Most of the fossil fuel-fired drum boilers being installed today will encounter cycling operation. Cycling involves taking a unit off-line on nights and weekends and reloading it on weekday mornings.

This analysis of cycling units assumes that the daily load cycle starts at full load, full temperature operation and passes successively through a load reduction, an off period and an idle period. This is followed by a restart and a reloading back to full load, full temperature. The basic difficulty is that steam temperatures must be controlled to limit cyclic thermal stresses on the turbine. The boiler lacks capability for temperature control during startup and low load. Therefore, the turbine requirements and the boiler capability are not compatible unless an interfacing system is provided. The startup system described above (Fig. 18) provides this interface.

A more aggressive integrated control strategy is needed to take full advantage of the boiler turbine system capabilities while recognizing the limitations of each. The principles applied to such a control system include:

1. The fuel heat content effectively released in the boiler furnace will establish and maintain the megawatt output from the unit on a steady-state basis. Therefore, the only valid boiler load index for fuel and air input is megawatt demand and/or generation.

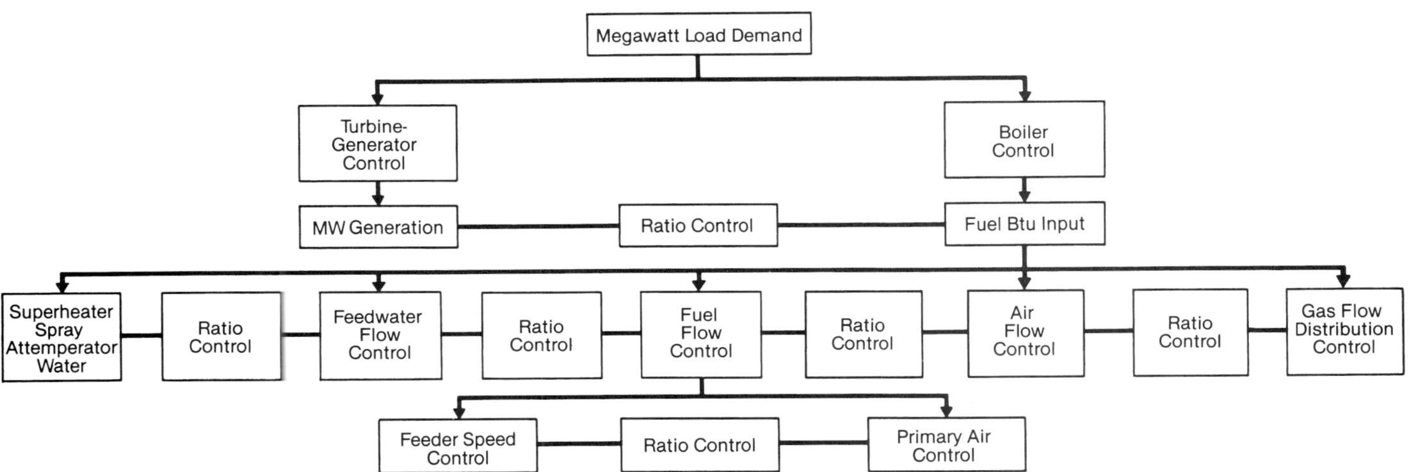

Fig. 16 Simplified diagram of ratio control.

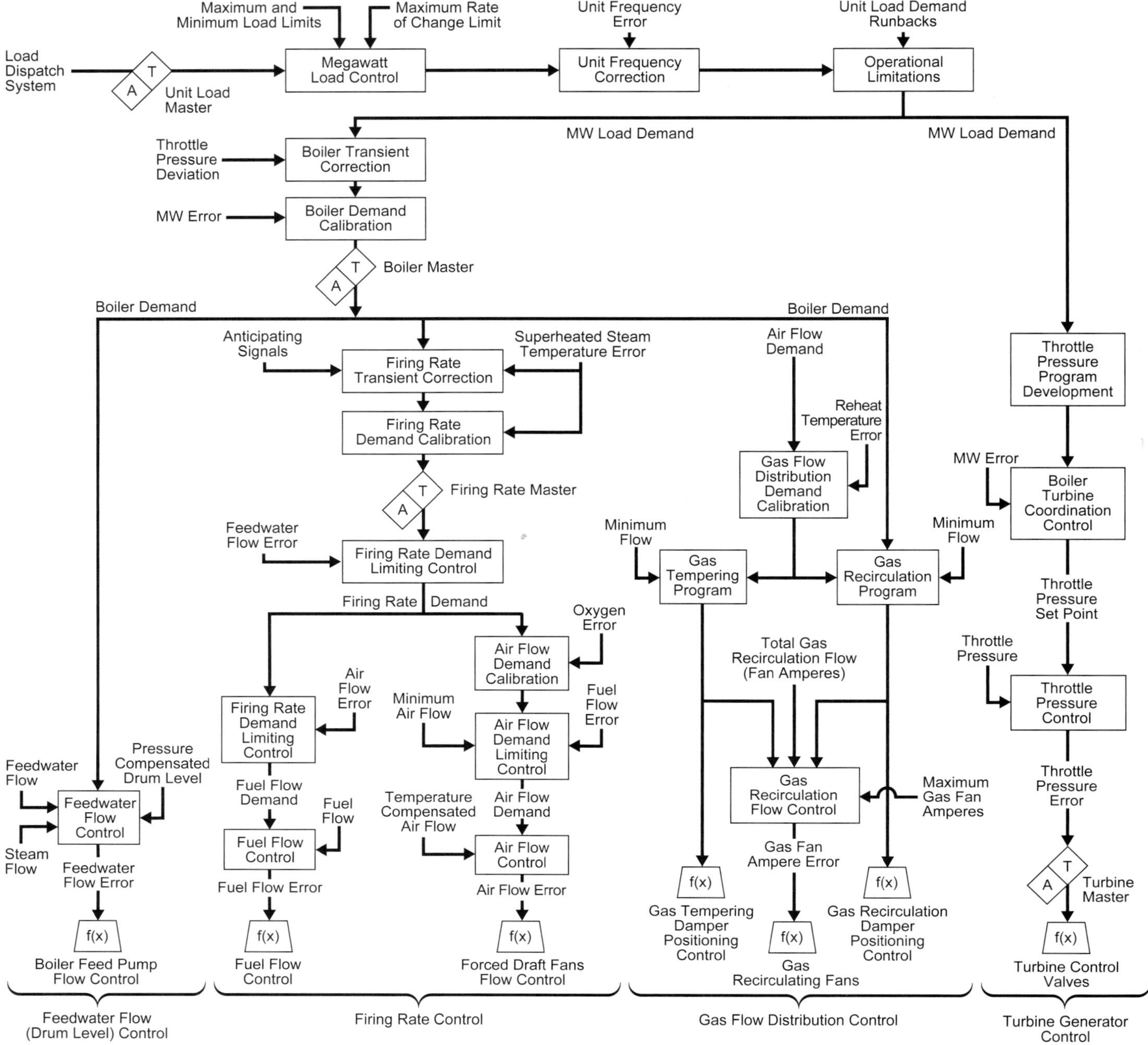

Fig. 17 Integrated boiler turbine-generator control system.

2. The turbine valve control will control steam flow and megawatt generation but only on a short term transient basis.
3. The relationship between steam flow and megawatt generation is not dependent on the turbine valve position.

Feedwater control systems

Feedwater control regulates the flow of water to a drum-type boiler to maintain the level in the drum within the desired limits. The control system will vary with the type and capacity of the boiler as well as with load characteristics.

The control strategy of the feedwater control system can be one of three types. They are classified as one-, two- or three-element feedwater control systems.

In single element feedwater control (Fig. 19) the water in the drum is at the desired level when the signal from the level transmitter equals its set point. If a deviation of water level exists, the controller applies proportional plus integral action to the difference between the drum level and set point signals to change the position of the regulating valve. A hand-automatic station gives the operator complete control over the valve. A valve positioner can be included in the control valve assembly to match the valve characteristics to the individual requirements of the system.

Single element control will maintain a constant drum level for slow changes in load, steam pressure

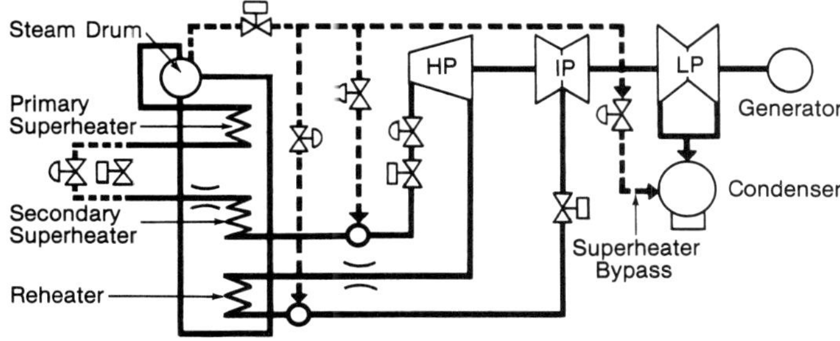

Fig. 18 Superheater bypass and startup system for drum boiler.

or feedwater pressure. However, because the control signal satisfies the requirements of drum level only, excessive swell (increase in steam bubble volume due to an increase in load) or shrink (decrease in steam bubble volume due to a decrease in load) effects will result in wider drum level variations and a longer time for restoring drum level to set points following a load change with single element control than with two- or three-element control

Two-element control comprises a feedforward control loop which utilizes steam-flow measurement to control feedwater input, with level measurement assuring correct drum level. This control strategy is primarily used in industrial applications and seldom used in power generating plants.

The most widely used feedwater control system, especially in the utility industry, is the three-element feedwater control.

Three-element control is a cascaded-feedforward control loop which maintains water flow input equal to feedwater demand. Drum level measurement keeps the level from changing due to flow meter errors, blowdown, or other causes.

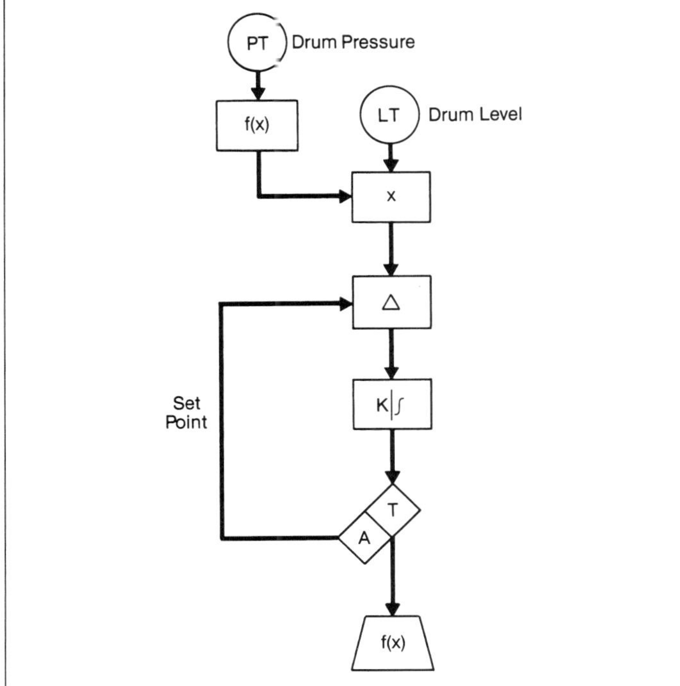

Fig. 19 Single element feedwater control.

The control applies proportional action to the error between the drum level signal and its set point. The sum of the drum level error signal and the steam flow signal is the feedwater demand signal. This is the output of the summer. The feedwater demand signal is compared with the water flow input and the difference is the combined output of the controller. Proportional plus integral action is added to provide a feedwater correction signal for valve regulation or pump speed control. A hand-automatic station gives the operator complete control over the valve. In single boiler turbine units, the turbine first stage shell pressure can be used as a measure of steam flow. The details of this system are illustrated in Fig. 20, using the symbols given in Table 1.

Three-element feedwater control systems can be adjusted to restore a predetermined drum level at different loads. In boilers with severely fluctuating loads, the system can be modified to permit drum water level to vary with boiler load to compensate for swell and shrink effects. If the drum level is allowed to vary in this manner, a nearly constant inventory of water, as opposed to a constant level, is maintained.

During low load operation (0 to 20%), it is difficult to obtain steam and water flows accurately because flow transmitters are usually calibrated for high load operation. It is advisable to transfer the feedwater control system to single element control. Drum level is the only variable used in the control scheme. Automatic transfer as a function of load between three element and single element control is often provided.

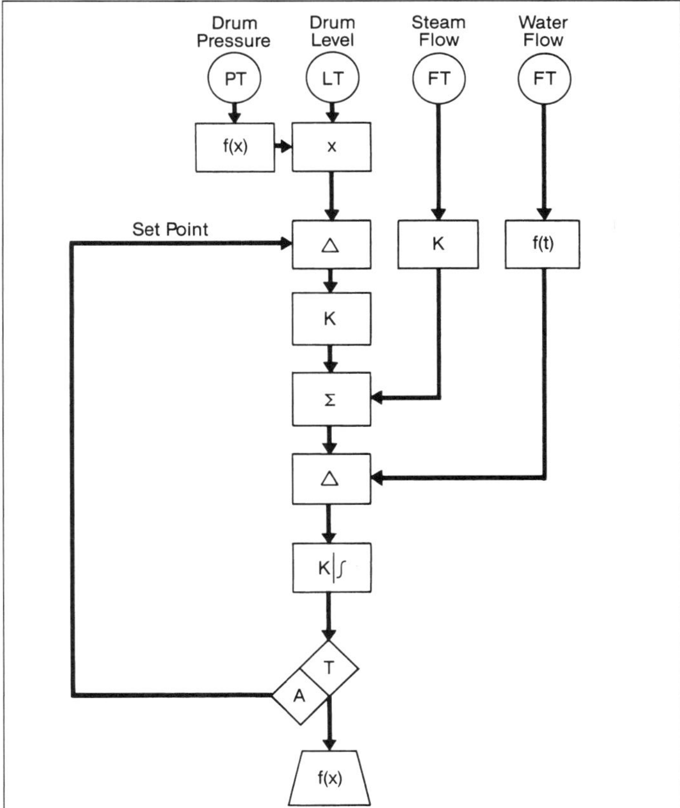

Fig. 20 Three-element cascade feedforward control loop for feedwater flow control.

Drum water level is one of the most important measurements for safe and reliable boiler operation. If the level is too high, water can flow into the superheater with droplets carried into the turbine. This will leave deposits in the superheater and turbine, and in the extreme, cause superheater tube failure and turbine water damage. Consequences of low water level are even more severe. An insufficient head of water may cause a reduction in furnace circulation, causing tube overheating and failure. To provide a more reliable system B&W recommends the use of three independent drum level transmitters as inputs to an auctioneered remote level signal system for control, interlocking, alarming and tripping functions. To provide an additional safeguard, the control logic for drum level should be independent of the boiler tripping logic. It is also important to install and calibrate drum level instrumentation with care. Instrumentation to be provided must be in accordance with The American Society of Mechanical Engineers (ASME) Boiler and Pressure Vessel Code. Minimum drum water level instrumentation used on utility size power boilers should be one water gauge glass and two indirect water level indicators with alarm and trip points clearly marked.

Furnace draft control

The furnace draft control regulates the induced draft (ID) fans to provide the proper exhausting force for flue gas flow through the boiler. When the ID fans are properly balanced with the forced draft (FD) fans, the furnace pressure will be at the desired operating level. To improve reliability and meet National Fire Protection Association NFPA 85 requirements, furnace pressure is measured with three independent transmitters connected to independent connections. A median select auctioneering system selects the median or middle of the three transmitter signals for control. The ID fans are regulated by a feedforward program based on air flow demand which is then corrected by any deviation of furnace pressure from its desired set point. This control strategy must also include furnace overpressure and implosion protection overrides. The most common and most dangerous disturbance is furnace implosion.

Furnace implosion is usually the result of a combination of a master fuel trip (flame collapse) and an equipment malfunction resulting in a vacuum condition in the furnace. The control system must include two furnace override circuits to minimize any negative pressure excursions. The first is a low furnace pressure override which will proportionally drive the inlet dampers or vanes closed when the median furnace pressure signal shows a large negative furnace pressure excursion. The second is a master fuel trip (MFT) kicker (or rundown) circuit which will temporarily start driving the ID fan inlet damper or vanes in the closed direction as soon as a MFT occurs without waiting for a negative furnace excursion to develop. The magnitude of the closing of the ID fan inlet dampers or vanes is dependent on the existing unit load, as represented by steam flow. These override circuits are required by NFPA 85 to be downstream of the hand/automatic control stations of the ID fans so that operator action is not permitted. Fig. 21 illustrates the furnace draft control strategy.

Superheat and reheat temperature control

From the viewpoint of control system designers, there are inherent advantages to considering boiler outlet steam temperature control as an independent temperature control system. A greater understanding is gained if the temperature control systems are recognized as a system for distributing the boiler heat input between steam generation, steam superheating and steam reheating. Starting from that viewpoint, some of the inherent limitations of boilers to control temperatures and the difficulties in reaching design solutions can be identified. For the various methods of steam temperature control, refer to Chapter 19.

Combustion control systems (air and fuel flow control)

A combustion control system regulates the fuel and air input, or firing rate, to the furnace in response to a load index. The demand for firing rate is, therefore, a demand for energy input into the system to match a withdrawal of energy at some point in the cycle. For boiler operation and control systems, variations in the boiler outlet steam pressure are often used as an index of an unbalance between fuel-energy input and energy withdrawal in the output steam.

A great variety of combustion control systems have been developed over the years to fit the needs of particular applications. Load demands, operating philosophy, plant layout, and types of firing must be considered. The following examples show some of the systems for various types of fuel firing. The control symbols shown in these figures are listed in Table 1.

Oil- and gas-fired boilers Fig. 22 illustrates a system for burning oil and gas, separately or together. The fuel and air flows are controlled from steam pressure through the boiler master with the fuel readjusted from fuel flow and air flow. The oil or gas header pressure may be used as an index of fuel flow and the windbox to furnace differential as an index of air flow on a per burner basis. Such indices are common for single burner boilers. For multiple burner boilers, a pitot tube or other flow measuring instrument is used to provide more accurate air flow measurement.

Pulverized coal-fired boilers Fig. 23 illustrates a more sophisticated combustion control that would be used on larger boilers having several pulverizers. Each pulverizer supplies a group of burners that are located in an open or compartmented windbox. Primary and secondary air are admitted and controlled on a per-pulverizer basis.

To compensate for boiler and pulverizer response characteristics, a firing rate kicker (derivative action) is usually required on the pulverizer system. The boiler firing rate demand is then compared to the total measured fuel flow (summation of all feeders in service delivering coal) to develop the demand to the fuel-air master. The fuel-air master demand signal is then applied in parallel to all operating pulverizers.

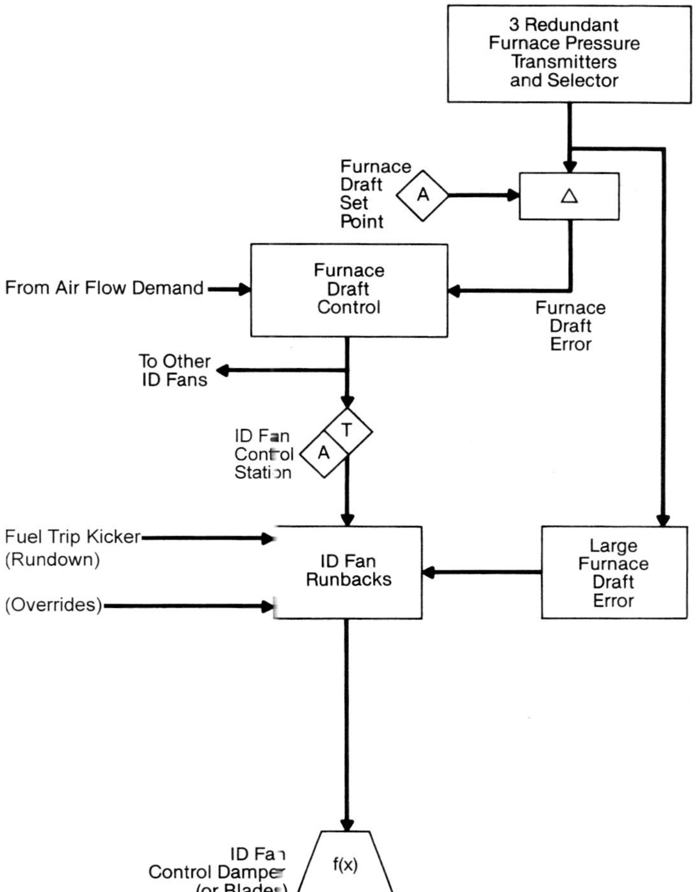

Fig. 21 Furnace draft control.

The individually biased pulverizer demand signal is then applied in parallel to meet demands for coal flow, primary air flow, and total pulverizer group air flow (primary and secondary air flows). When an error develops between demanded and measured primary air flow or total air flow, proportional plus integral action will be applied, adjusting the primary or secondary air dampers to reduce the error to zero. A low primary air flow or total air flow cross-limit is applied in the individual pulverizer control. If either measured primary air flow or total air flow is low relative to coal rate demand, this condition is recognized in the air flow cross limits in the coal feeder demand to reduce the demand to that equivalent to the measured air flows. A minimum pulverizer loading limit, a minimum primary air flow limit, and a minimum total air flow limit are applied to the respective demands to keep the pulverizers above their minimum safe operating load. (See Chapter 13.) A minimum load limit maintains sufficient burner nozzle velocities at all times, and maintains primary air/fuel and total air/fuel ratios above prescribed levels.

High capacity vertical shaft pulverizers, such as the B&W Roll Wheel™ pulverizer (Chapter 13), are being supplied on most large boilers today to achieve the required steam flow capacities with coal having low heating values. Pulverizer response can be improved by overshooting both primary air flow and feeder speed. This overfeed is obtained with present control systems by the use of kickers on primary air flow and coal flow developed from a change in pulverizer demand.

Cyclone-fired boilers The Cyclone™ boiler (see Chapter 15) controls shown in Fig. 24 are similar to those described for pulverized coal-fired units. Although the Cyclone furnace functions as an individual furnace, the principles of combustion control are the same.

On multiple Cyclone installations, the feeder drives are calibrated so that all feeders run at the same speed for the same master signal. The total air flow is controlled by the velocity damper in each Cyclone furnace to maintain the proper air/fuel relationship. This air flow is automatically temperature compensated to provide the correct amount of air under all boiler loads. The total air flow to the Cyclone furnace is controlled by the windbox to furnace differential pressure, which varies with load to increase or decrease the forced draft fan output.

Automatic compensation for the number of Cyclone furnaces in service has been incorporated, along with an oxygen analyzer. This gas analyzer helps the operator monitor excess air for optimum firing conditions.

Atmospheric pressure fluidized-bed boilers The circulating fluidized bed thermal performance (load, power consumption, furnace temperature, combustion efficiency, gaseous emissions) is strongly dependent on the solids mass or inventory distribution within the system, solids size, and solids flow within and through the system. (See Chapter 17.) Specific examples of the special role of solids in a circulating fluidized-bed process include the following:

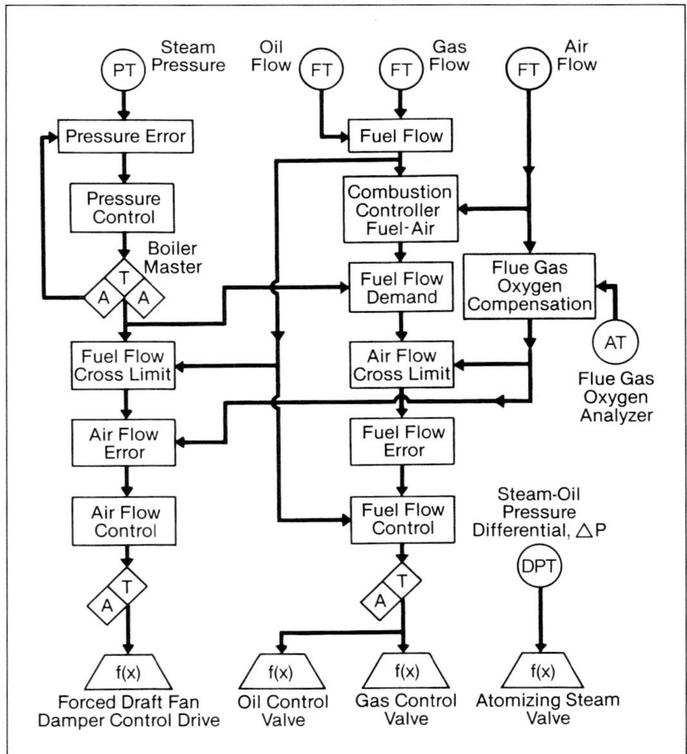

Fig. 22 Diagram of combustion control for an oil- and gas-fired boiler.

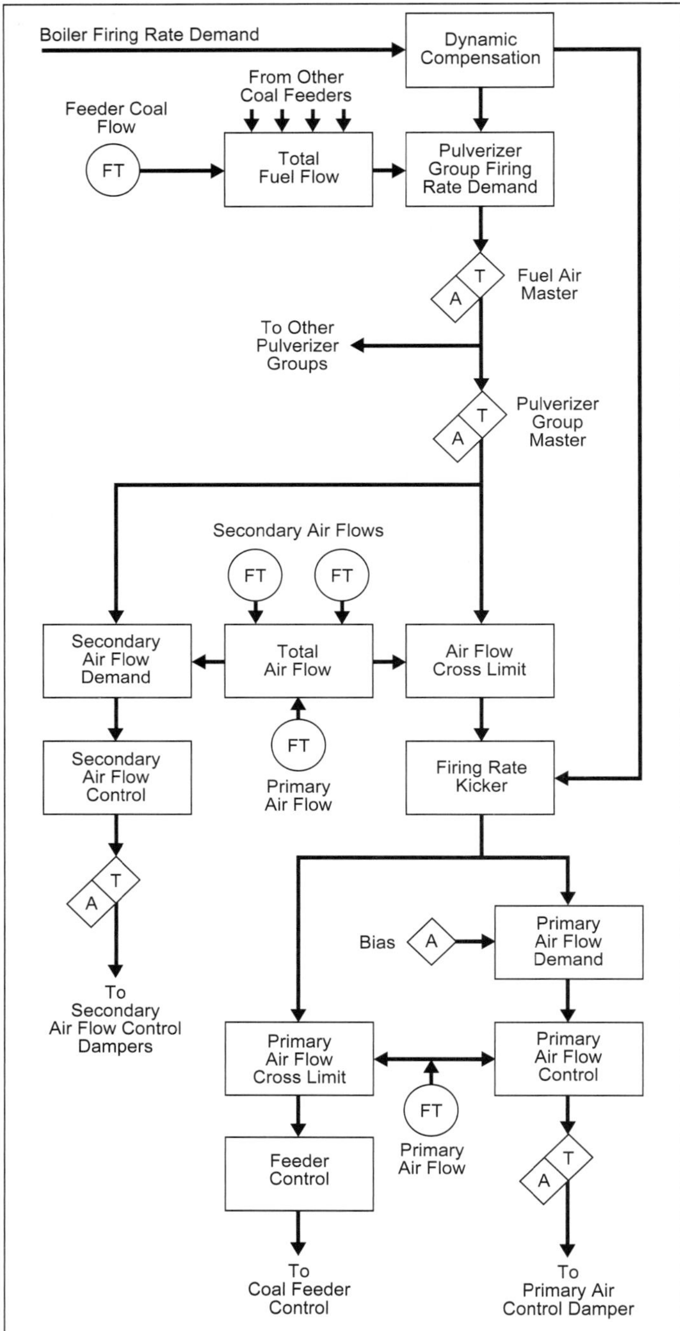

Fig. 23 Diagram of combustion control for a pulverized coal-fired boiler.

1. furnace heat absorption is controlled through changes in solids inventory,
2. fuel combustion is affected by the solids inventory distribution in the furnace,
3. solids size strongly affects furnace heat transfer and inventory distribution, U-beam and multiclone dust collector efficiencies, solids flow in the system, and solids losses,
4. carbon conversion efficiency and sorbent consumption depend on amount and distribution of solid effluents,
5. U-beam collection efficiency depends on the solids level in the particle storage hopper, and
6. forced draft fan operating point, power consumption and air flow depend on the furnace solids inventory.

Different areas of circulating fluidized-bed process control are interrelated, with the solids inventory control function playing a central role. The complexity of process control makes it difficult to analyze and predict system performance for multiple combinations of the process variables. Additional problems have been caused by continuous variation of material and energy inputs and widely different speeds of the system dynamic responses to these variations. Much is to be learned from operation of commercial and laboratory circulating fluidized-bed units to define more exactly the optimal control strategy.

Control system design philosophy for Universal Pressure (UP®) once-through boilers

The use of the once-through boiler design in the United States (U.S.) stems from continuing efforts to reduce electric power generating costs. The once-through boiler can operate at full steam temperature over a wider load range, and can operate at supercritical pressures to increase cycle efficiency. The B&W Universal Pressure UP® boiler is one type of a once-through boiler.

The once-through concept can best be visualized as a single tube as discussed in Chapters 1 and 26. Feed-

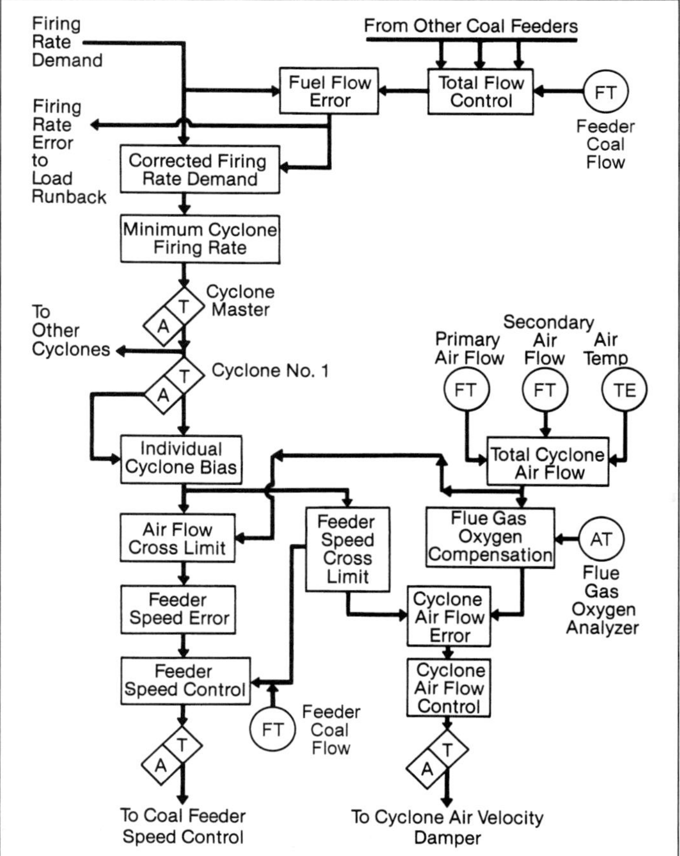

Fig. 24 Diagram of combustion control for a Cyclone-fired boiler.

water is pumped into one end, heat is applied along the length of the tube, and steam flows out of the other end. The output is a function of the feedwater flow and the amount of heat supplied. The outlet fluid enthalpy, or heat content, depends only on the ratio of heat input to feedwater flow. A valve at the outlet provides a means of varying the pressure level. When the pressure level is maintained constant, the outlet steam temperature is also dependent only on the ratio of heat input to feedwater flow.

At steady-state conditions, feedwater flow equals steam flow. The pressure level will be influenced by the valve restriction at the outlet and by the density of the fluid throughout the system. Therefore, a change in heat input will influence both pressure and temperature. It is possible to vary the flow and pressure at the outlet by changing the amount of valve restriction, or at the inlet by changing both the feedwater flow and the heat input. It is important to note that a change in feedwater flow without a corresponding change in the heat input will result in a change in the outlet steam temperature. During transient conditions, other factions, such as the fluid and energy storage requirements which change with load, must be considered because they influence the feedwater flow and heat inputs.

Combustion guides

With the increase in complexity and automation of steam generating plants, it is necessary to provide the operating personnel with information about significant operating events to assist the operator to make rapid evaluations of the operating situation. The NPFA 85 Boiler and Combustion Systems Hazards Code requires that the following information, at a minimum, be provided to the operating personnel in a continuous trend display:

1. Steam flow
2. Feedwater flow rate
3. Total fuel flow rate (% of maximum unit load)
4. Total air flow rate (% of maximum unit load)
5. Drum level
6. Final steam temperature
7. Main steam pressure
8. Furnace pressure/draft

Excess Air

The level of excess air is one index commonly used to determine the performance of the unit and to guide its operation. Excess air is the amount of air supplied above the theoretical air needed for combustion. (See Chapter 10.) It is always necessary to supply some excess air to assure complete combustion of the fuel. Any excess air not used constitutes a loss in boiler efficiency by exhausting excess heat up the stack.

Steam flow-air flow

The comparison of steam flow to air flow can give an immediate indication of fuel consumption efficiency. Because boiler efficiency at a specific load remains essentially constant for a given amount of excess air, fuel consumption can be determined from steam flow, which is established by standard flow measuring devices. Steam flow is used as an index of heat absorption. The air flow indication or index can be established by primary elements located on either the air side or the flue gas side of the unit.

Air flow measurement

From a control standpoint, the preferred location for air flow measurement is in the clean air duct between the air heater and the burners. Available air measuring devices are discussed in Chapter 40.

The alternate location for air flow measurement is on the flue gas side, with connections to measure pressure differences resulting from the flow of combustion products through a section of the main boiler or air heater. The products of combustion at a given flue gas temperature are proportional to the amount of air supplied; therefore, the flue gas flow can be an index of air flow.

The measurement of combustion air supplied to a boiler is made difficult by several factors:

1. The air measuring device is typically of the restrictive type, similar to the primary elements used in other flow measurements. The resulting unrestored head loss means added fan power.
2. On large and complicated units it is usually impossible to find one duct through which all of the combustion air passes. Metering equipment capable of accurately totaling several flows must then be used.
3. Air density must be considered in both the calibration and operation of any air flow meter. Either manual or automatic temperature compensation must be applied to the air flow indication for it to reflect the mass rate of air flow.

Fuel flow-air flow

The rates of admission of fuel and air flow before combustion provide an important indication of the fuel flow to air flow ratio. With reference to NFPA 85, under no circumstances is the air flow demand to be less than the boiler purge rate and shall not be reduced below the amount required by the actual fuel inputs. It is important that the fuel and air flow control subsystem be designed to keep the fuel and air in proper calibration and enable the operator to adjust the fuel-air ratio within a span consistent with the fuel being fired.

Fuel flow measurement

The measurement of gaseous fuels must consider the physical properties of the gas, particularly temperature, pressure and specific gravity. Where widely changing properties are encountered, automatic compensation should be used. A primary element installed in the main header will produce a differential pressure to actuate the recording. On less complex installations, particularly those with single burners, the pressure at the burner can be used to indicate flow.

With liquid fuels, various types of flow meters can be installed directly in the supply header to indicate and record fuel flow. As with gaseous fuels, oil-header pressure can be used to indicate flow.

The measurement of solid fuel flow can be accomplished through weight scales or feeder speeds. The use of mass flow, however, is more accurate. The feeder can be calibrated in units of fuel flow per feeder revolution to produce a continuous flow rate. Gravimetric feeders (see Chapter 12) provide controlled weight flow rates to the pulverizers.

Gas analysis

Gas analyzers are used to determine correct fuel-air relationships. Representative flue gas samples are continuously analyzed either chemically or electronically to record the oxygen and combustibles present in the products of combustion. Because there is direct correlation between oxygen content in the flue gas and the quantity of excess air supplied to the combustion zone, the operator receives a continuous and direct reading of combustion efficiency. The indicated or recorded oxygen signal is used in adjusting the total air flow.

Comparison of combustion guides

Each combustion guide has its particular merits; none is infallible. The advantages of one may be used to overcome the disadvantages of another.

The fuel flow to air flow ratio index will illustrate how the control system proportions fuel and air continuously during severe load swings. Because coal has wide variations in fuel heating value and moisture content, the coal flow as measured by the feeders only provides an approximation of the fuel flow to the boiler. When the fuel heating value changes, the calibration of the correct fuel-air relationship also changes. Therefore, to maintain a given excess air, the proportions of air and fuel must be corrected. To compensate for this, the energy per unit mass can be calculated using the boiler as a calorimeter.

On major load changes, the steam flow to air flow is temporarily unbalanced because of the overfiring and underfiring necessary for responsive load change. The fuel consumption is normally higher or lower than indicated by the steam flow load index. This guide is also affected by changing feedwater or steam temperatures, because a variation of either temperature demands more or less transfer of heat per unit of steam produced at the specified pressure and temperature. These deviations are inherently minimized where megawatt generation is used as the index of heat absorption. It is necessary, when substantial changes may occur in steam or feedwater temperature, to provide manual or automatic compensation to the steam flow-air flow circulation.

A gas analyzer gives true excess air determinations. There is some delay because combustion must be completed before a representative sample can be obtained. If the sample is taken near the combustion zone, this delay is not usually significant. Considerable study may be needed to locate sample points that will give correct average excess air values. The dirty, hot, and corrosive conditions of the flue gases at these points make periodic maintenance necessary for continuous, dependable sampling. The gas analyzer is an accurate index for feedback control.

Burner management systems

Burner management systems (BMS) are applied to all boilers to prevent continued operation during hazardous furnace conditions, and to assist the operator in starting and stopping burners and fuel equipment.

The most important burner control function is to prevent furnace or pulverizer explosions which could threaten the safety of operating personnel and damage the boiler and auxiliary equipment. The control system must also prevent damage to burners and fuel equipment from maloperation while avoiding false trips of fuel equipment when a truly unsafe condition does not exist.

Other important factors in the design of the burner management system are the method and location to be used in the startup, shutdown, operation and control of the fuel equipment. These factors must be understood before purchase, design, or application of a burner control system. The minimum requirements for a burner management system are specified in NFPA 85, Boiler and Combustion Systems Hazard Code.

The different categories of burner control are summarized below.

Supervised manual control

Supervised manual control (Fig. 25) is a system in which a trained operator has primary responsibility to operate the burner equipment at the burner platform. Checks on the existing conditions are dependent on observation and evaluation by the local operator. However, this system must include certain interlocks for preventing improper operator action, certain safety trips and flame detection. Good communication between the local operator and the control room is needed to coordinate the startup and shutdown of burner equipment. Supervised manual controls may be found on older boilers but are not permitted by NFPA 85 to be used on new boilers.

Remote manual sequence control

Illustrated in Fig. 26, this system represents a major improvement over manual control, permitting remote manual operation using instrumentation systems and position indications in the control room for intelligence. This system requires various burner permissives (prequalifying conditions), interlocks, and trips to account for the position of fuel valves and air registers in addition to main flame detection. With this system, the operator participates in the operation of the fuel equipment. Each sequence of the burner operating procedure is controlled from the control room and no steps are taken except by operator command.

Automatic sequence control

The next logical step (Fig. 27) is to automate the sequence control to permit startup of burner equipment from a single pushbutton or switch control. Automation, then, replaces the operator in controlling the operating sequences. Because the operator initiates

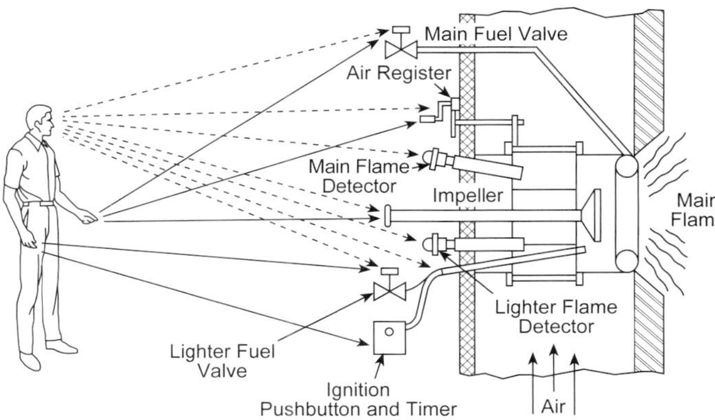

Fig. 25 Fuel burner supervised manual control.

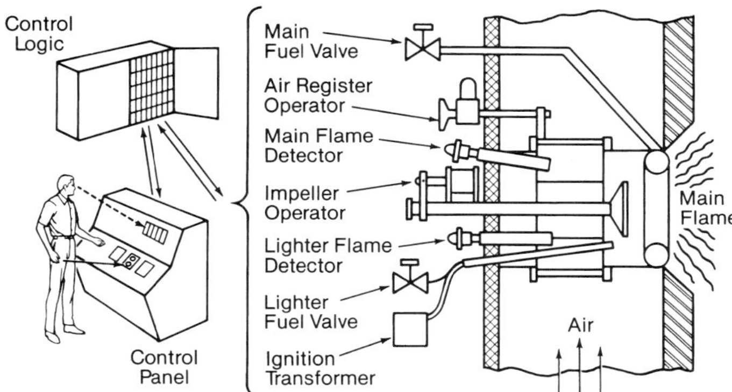

Fig. 27 Automatic sequence control.

the start command, it is expected that the operator participates in, or at least monitors each step of the operating sequence as indicated by signal lights and instrumentation signals, as the startup process proceeds to completion with the fuel utilization unit in service on automatic control. This type of system is the most common system applied to gas, oil, and coal burning systems.

Fuel management

A degree of fuel system automation can be achieved that will permit fuel equipment to be placed in service without operator supervision (Fig. 28). A fuel management system can be applied that will recognize the level of fuel demand to the boiler. This system will know the operating range of the fuel equipment in service, will reach a decision concerning the need for starting up or shutting down the next increment of fuel equipment, and will select the next increment based on the firing pattern of burners in service. Such demands for the startup or shutdown of fuel preparation and burning equipment can be initiated by the management system without the immediate knowledge of the operator.

The degree of operating flexibility allowed with a burner control system is closely related to the degree of operator participation. A higher level of automation reduces the flexibility of the operator in handling situations where a piece of equipment fails to perform as expected. One method used to provide more operating flexibility is to allow operator participation at two or more levels. Increased flexibility is also obtained by the careful grouping of equipment so that a fault anywhere in the system will affect a limited amount of equipment. An example of this is the grouping of burners on a pulverizer so that the failure of a piece of hardware on a burner will only affect the failed burner, and not the pulverizer and its remaining burners.

Fig. 29 illustrates the provision for manual intervention in a fuel management type system. In this figure, functions are detailed at four levels of a pulverizer/burner group control. Manual intervention is provided at level 2 to permit remote manual operation with interlock protection features.

Flame detection

One of the most important items required for a burner management system is individual burner flame detection, regardless of fuel.

Ultraviolet (UV) flame detectors have been successfully applied to natural gas-fired burners due to the abundance of ultraviolet radiation produced by the combustion of the hydrogen in natural gas. Flicker or infrared detectors are better suited for coal- or oil-fired burners. A flicker detector uses the high frequency dynamic flickering of the primary combustion process. An infrared detector uses the infrared radiation produced in the primary combustion of the flame.

The proper location of the flame detectors on a burner is dependent on many factors, including

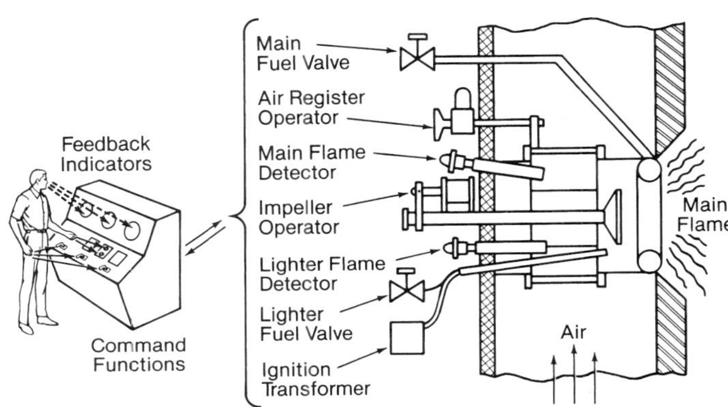

Fig. 26 Remote manual sequence control.

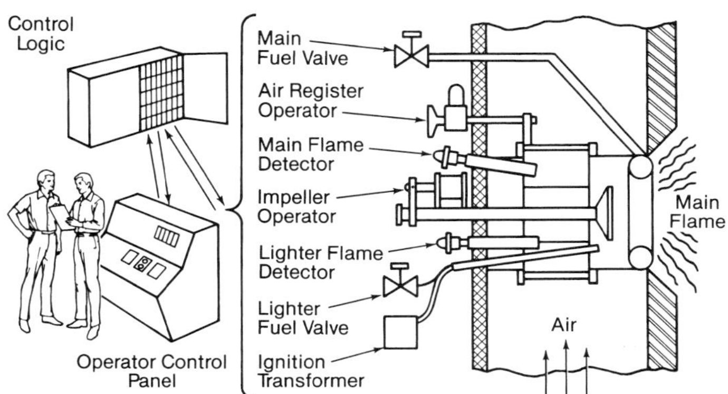

Fig. 28 Fuel management.

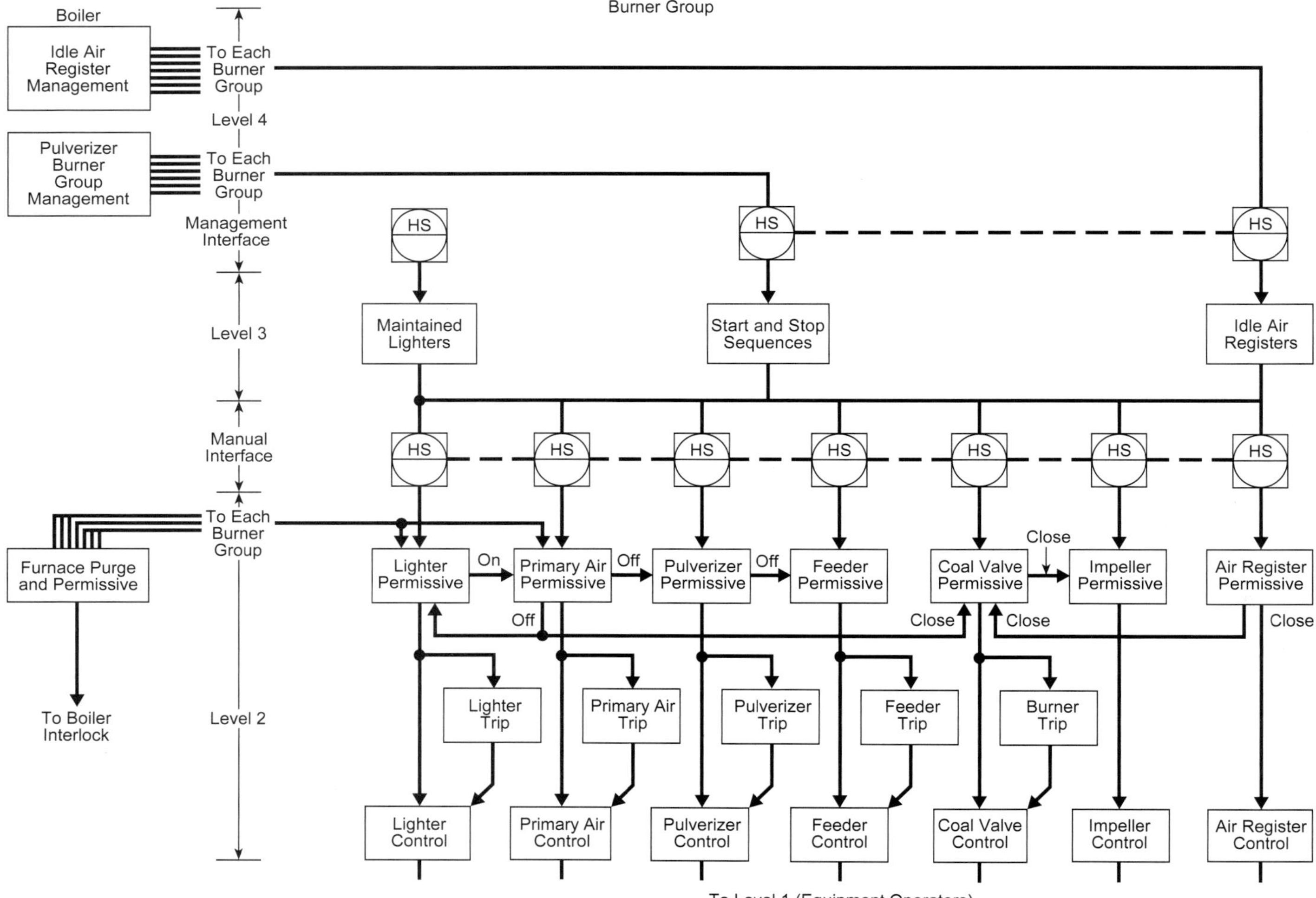

Fig. 29 Pulverizer/burner group control.

burner management system design. It should, therefore, be determined by the boiler manufacturer in conjunction with the BMS manufacturer as an integral part of the control system design.

Flame monitoring devices in current use are designed for on/off operation based on the presence or absence of flame. Although the basic detector units provide a variable analog output signal which can be read on a meter, there is no direct claim to an established relationship between this relative output signal and the flame quality. Therefore, the analog signals should never be relied upon as flame quality indicators. They should be used only to provide helpful information for initial setup adjustments and continuous on-line observance.

Some power plants use closed circuit television for continuous furnace observation by the operator from the control room.

Classification of lighters/igniters

The combustion of solid, liquid and gaseous fuels is hazardous because upset conditions may lead to an air-fuel mixture that could cause furnace explosion. These hazardous conditions are especially prevalent during startup and shutdown of the boiler. Therefore, it is important to have reliable instrumentation, well designed safety interlocks, and clear understanding of operating sequences by the operator. However, a reliable and noninterruptible ignition source is probably the single most important factor. The National Fire Protection Association (NFPA) provides guidelines on the application of igniters. The following definitions are extracted from the NFPA 85 Boiler and Combustion Systems Hazards Code.

Class 1 Igniter. An igniter that is applied to ignite the fuel input through the burner and to support ignition under any burner lightoff or operating conditions. Its location and capacity are such that it will provide sufficient ignition energy, generally in excess of 10% of full load burner input, at its associated burner to raise any credible combination of burner inputs of both fuel and air above the minimum ignition temperature.

Class 2 Igniter. An igniter that is applied to ignite the fuel input through the burner under prescribed lightoff conditions. It is also used to support ignition under low load or certain adverse operating conditions. The range of capacity of such igniters is generally 4 to 10% of full load burner fuel input.

Class 3 Igniter. A small igniter applied particularly to gas and oil burners to ignite the fuel input to the burner under prescribed lightoff conditions. The ca-

pacity of such igniters generally does not exceed 4% of the full load burner fuel input.

Class 3 Special Igniter. A special Class 3 high energy electrical igniter capable of directly igniting the main burner fuel.

Failure mode analysis

NFPA 85 states that the burner management system must be totally independent of all other control systems. Therefore, the control system designer must consider carefully every aspect of the operation and equipment limitations. A great number of failure modes must be analyzed and addressed. Due to space limitation, the following are only a few of the issues:

1. The typical safe way to implement the master fuel trip (MFT) in the burner management system is *de-energize to trip* the equipment. As the description indicates, a circuit should be energized continuously. De-energizing of the circuit represents an abnormal condition. Therefore, the equipment must be removed from service. For an *energize to trip* system, in the event of specific power failure, the equipment can not be stopped.
2. Redundancy of control hardware may be necessary or desirable in many areas. However, too many backups may render the system too cumbersome for practical use.
3. System design must not prevent the operator from shutting down any equipment desired. The system will function to complement the operator's action so that he or she will not generate a hazardous operating condition.

Post-combustion NO_x emissions control

Various nitrogen oxides (NO_x) control technologies are discussed in Chapters 32 and 34. The following focuses on the control strategy associated with post-combustion selective catalytic reduction (SCR) technology. The B&W SCR NO_x removal system is a dry process in which ammonia (NH_3) is used as a reducing agent, and the NO_x contained in the flue gas is reduced to N_2 and H_2O.

As explained in Chapter 34, ammonia (anhydrous or aqueous) is injected into the flue gas upstream of the SCR reactor, through an injection grid. The grid is designed to ensure an even NH_3 distribution and allow for fine tuning of ammonia distribution. The flue gas then passes to the reactor and through the catalyst layers. As this occurs the inlet flue gas is constantly monitored to ensure that the NO_x is being reduced to an acceptable level without overspraying ammonia.

There are two primary control loops for an SCR system: ammonia flow and flue gas inlet temperature.

Ammonia flow control

Ammonia flow demand is calculated based primarily on flue gas flow and the outlet NO_x deviation from set point. This demand signal is limited on anhydrous systems to ensure that the ammonia to dilution air ratio is always below the hazardous buildup limit of ammonia in the system If this ratio exceeds the limit, the ammonia block valve is interlocked to close (Fig. 30).

Flue gas inlet temperature

Economizer outlet and bypass dampers are modulated to maintain a minimum allowable flue gas temperature at the SCR inlet. This prevents the formation of ammonium bisulfate upstream of the catalyst. This salt can mask the catalyst pores and cause a loss of activity. The gas temperature demand is developed as a function of boiler load. This gas temperature demand signal is compared to a measured SCR system inlet temperature. The resulting error signal is conditioned by the firing rate demand to generate the demand for the economizer flue gas dampers (Fig. 31).

SO_2 emissions control

Various sulfur dioxide (SO_2) control strategies and technologies are discussed in Chapters 32 and 35. The following focuses on the main control loops associated with wet and dry scrubbing technologies.

Wet scrubbers

The wet scrubbing process typically includes reagent preparation, absorber area, primary dewatering, and secondary dewatering systems. Sulfur dioxide laden flue gas is contacted with a sorbent liquid which saturates the flue gas and produces a wet byproduct.

Sulfur dioxide removal is accomplished by a countercurrent spray absorption process within the absorber. By spraying limestone slurry into the flue gas, the sul-

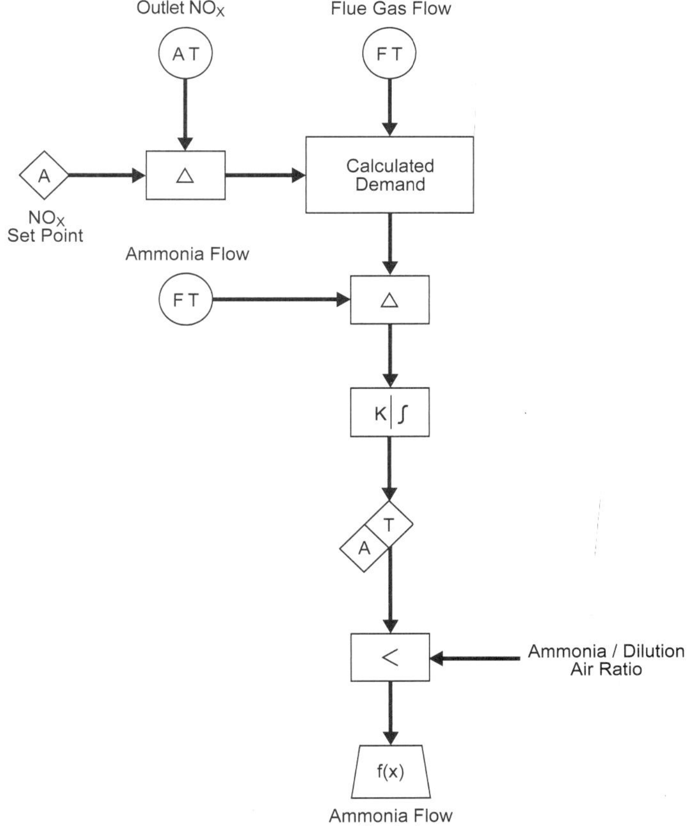

Fig. 30 SCR ammonia flow control.

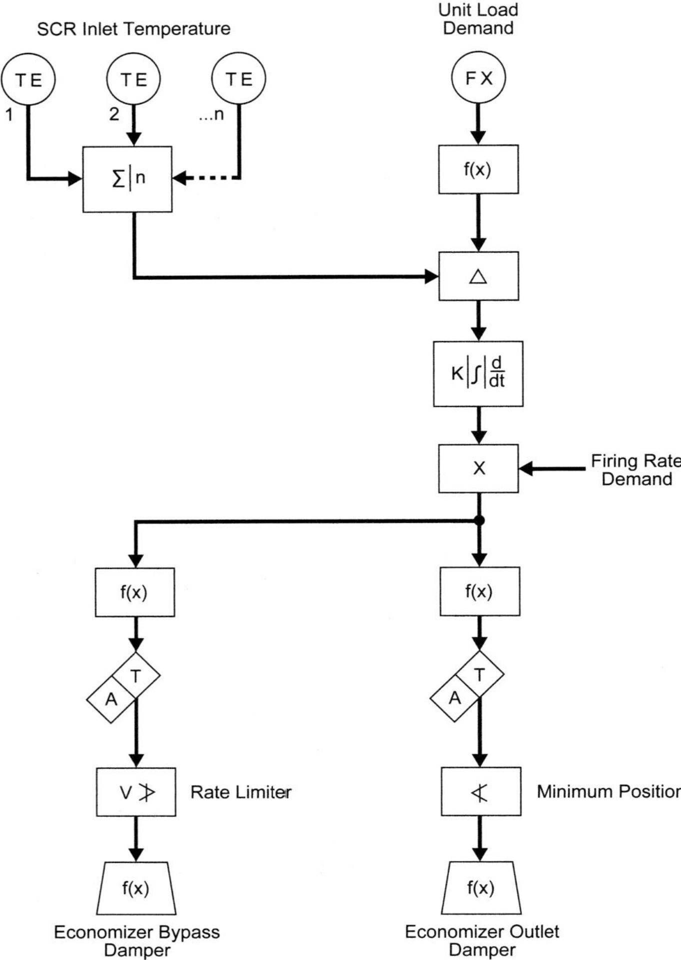

Fig. 31 SCR temperature control.

fur dioxide is converted to hydrates of calcium sulfite ($CaSO_3{\cdot}H_2O$) and calcium sulfate ($CaSO_4{\cdot}H_2O$).

As the hot flue gas enters the absorber, the flue gas is quenched to saturation by the recycle slurry sprayed from the first absorber spray level, and the initial removal of sulfur dioxide occurs. The flue gas then passes upward through the perforated absorber tray which uniformly distributes the flue gas and recycle slurry from the second and third absorber spray levels across the absorber cross-sectional area. The flue gas then passes through the absorber spray zone above the tray where recycle slurry is sprayed downward, countercurrent to the flue gas flow, from the second and third absorber spray levels, completing the SO_2 removal process.

The flue gas then flows upward to the first stage mist eliminator located near the top of the absorber. The first stage mist eliminator is a three pass vertical chevron, designed to collect carryover (mist) by inertial contact. Above and below the mist eliminators are a series of headers with nozzles that spray fresh or blended water onto the first stage mist eliminators. This removes any solids that have collected on the upstream and downstream faces of the chevrons. After passing through the first stage mist eliminator, the flue gas enters the second stage mist eliminator where the process is repeated. The flue gas then exits the absorber module.

The recycle slurry and the mist eliminator wash water pass down through the absorber and are collected in the lower section of the absorber module, referred to as the absorber reaction tank. Fresh limestone slurry is added to control the pH, and reclaim water is added to control the recycle slurry density. Side agitators keep the recycle slurry solids in suspension.

There are three critical control loops associated with the wet scrubber: pH, density, and reaction tank level.

pH control The recycle slurry pH is controlled by adding fresh limestone slurry to the absorber. The amount added is a function of expected boiler load and SO_2 loading as shown in Fig. 32. Due to the large volume of slurry in the reaction tank, a change in flue gas flow or flue gas SO_2 concentration will not show up as a change in recycle slurry pH until after a relatively long delay. To compensate for this delay, feedforward signals from flue gas SO_2 concentration and boiler load are used to adjust the demand for fresh limestone slurry feed (Fig. 32).

Level control Absorber level control compares the level measurement of the absorber reaction tank to an operator-adjusted set point. Proportional plus integral action is applied to any error signal. The resultant demand signal positions the reclaim water control valve (Fig. 33). A level measurement greater than the set point will decrease reclaim water flow, while a level measurement less than set point will increase reclaim water flow.

Density control The absorber reaction tank density

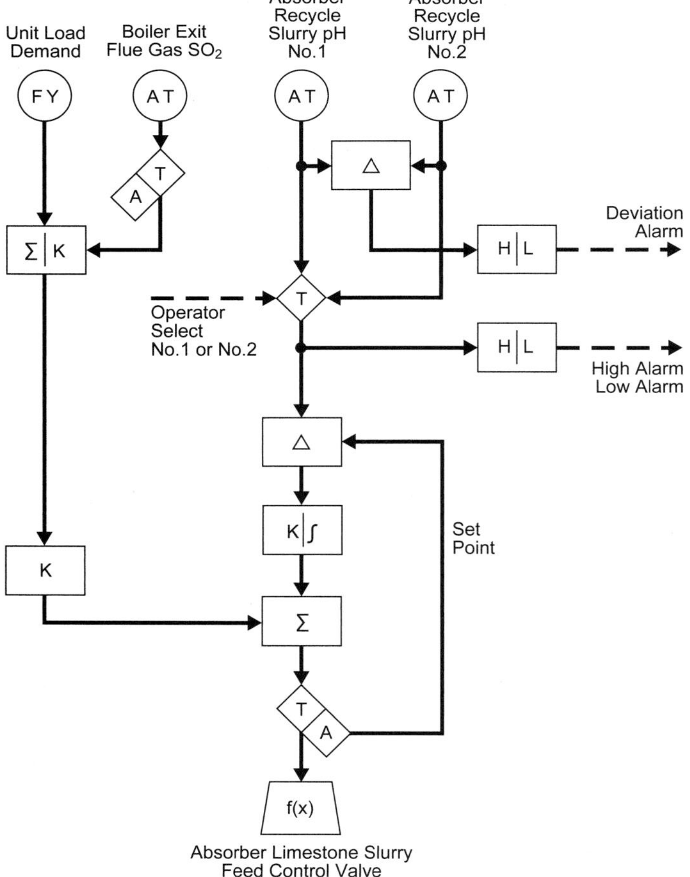

Fig. 32 Absorber slurry feed control.

is affected by the balance between water addition (fresh limestone slurry, mist eliminator wash, level control) and water removal (evaporation by the flue gas, slurry removal during dewatering). The balance between water removal and water addition, in conjunction with the increase in the density of the solids (as the fresh limestone reacts with absorbed SO_2), causes the absorber reaction tank density to rise. When the density reaches a high set point, the density control loop begins the dewatering process (removing solids). Once the density has reached a low set point, the slurry is recirculated back to the absorber reaction tank, thus completing the density control cycle.

Dry scrubbers

A dry scrubber is commonly referred to as a spray dryer absorber (SDA) module. Boiler flue gas exiting the air heater is directed to the SDA module which includes a rotary atomizer unit. Atomizer feed slurry, a mixture of hydrated lime slurry and recycle ash slurry, is supplied to the atomizer at a controlled rate and atomized within the SDA module into a cloud of extremely fine mist. Heat from the flue gas evaporates the moisture in the slurry cloud while the alkaline slurry simultaneously absorbs the SO_2 contained in the flue gas. The result is the conversion of the calcium hydroxide slurry component into a fine powder of calcium/sulfur compounds, and the lowering of flue gas temperature. The ratio of fresh lime slurry to recycle ash slurry is controlled to maintain the required alkalinity in the atomizer feed slurry, to achieve the desired level of SO_2 removal. The feed slurry to the atomizer is controlled to maintain the desired degree of evaporative cooling in the module. Module outlet flue gas temperature is regulated by the atomizer feed slurry control valve to be above the adiabatic saturation temperature (dew point), to ensure the spray-dried calcium/sulfur product remains dry and free flowing.

SO_2 control The reagent preparation system provides a mixture of hydrated lime slurry and recycle ash slurry to the atomizer head tank to obtain the desired feed slurry blend for proper SO_2 absorption. The atomizer head tank lime slurry feed flow control valve regulates the flow of lime slurry to the atomizer head tank to maintain the required percent SO_2 removal and/or stack SO_2 concentration. The basic design of this control loop is illustrated in Fig. 34.

Developments in instrumentation and controls

Power plant automation

Many power plants today are essentially automatic when operating between one third and full load using the conventional systems discussed earlier in this chapter. To achieve complete automation of a plant, it is first necessary to add control subloops for the auxiliaries which were previously considered to be the exclusive responsibility of the operator. All of these subloops must then be integrated through unit management control to provide complete plant automation. Fig. 35 shows the arrangement of such a control system.

The basic concept of full automation is to provide the maximum amount of control and monitoring at the lowest possible level. The organization of the different levels, as shown in Fig. 36 for a fossil fuel plant, is summarized as follows:

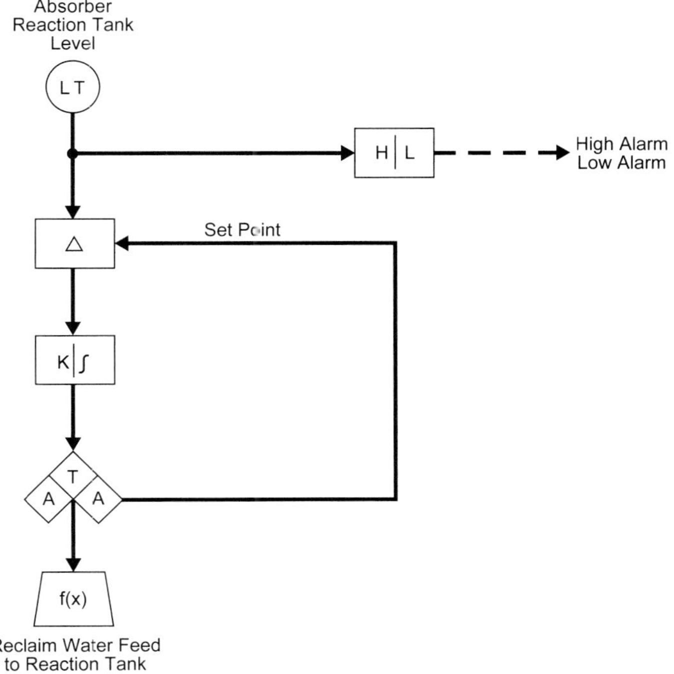

Fig. 33 Absorber reaction tank level control.

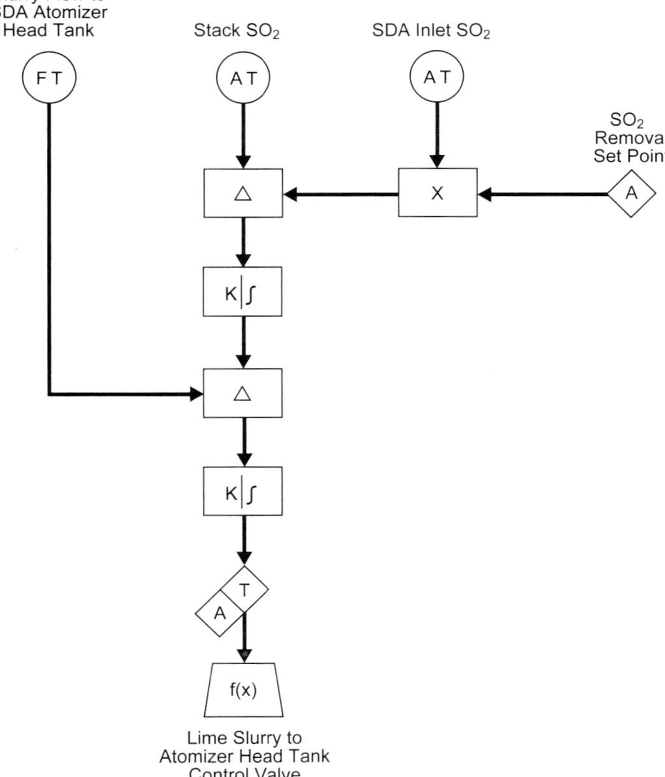

Fig. 34 Spray dryer absorber control loop.

Level 1 – equipment hardware being controlled
Level 2 – internal control of the individual equipment items or group of functionally identical items of equipment always operated in parallel
Level 3 – coordination of the sequence of actions and interlocking with respect to more than one equipment group
Level 4 – modulation and sequencing of major cycle components involving the entire unit
Level 5 – generation of the major load demand signals to coordinate the operation of the overall plant

Advantages obtained with the hierarchical control concept include self-regulation of equipment groups. This allows the equipment group to protect itself through low level interlocking, which provides greater reliability and control decentralization and permits continued plant operation with some portions of the control system out of service.

Some of the advantages of overall unit automation are:

1. Improved protection of personnel and equipment through more complete instrumentation, simplified information display, and more extensive and thorough supervisory control.
2. Reduced outages, reduced maintenance, and longer equipment life through more uniform and complete control procedures of startup, on-line operation, and shutdown.
3. Better plant efficiency through continuous and automatic adjustments to the controls to optimize plant operation.
4. More efficient use of personnel during startup and on-line operation.

The trend of boiler automation is clear and its future scope can be predicted with considerable confidence. The achievement of some goals will be difficult. There are operating problems to be solved, traditions to be overcome, and new philosophies and approaches to be accepted. Reliable and adequate equipment is already available and its application, through the cooperation of manufacturers and users, will make the automatic power plant a reality.

Human factor engineering

Until the 1990s, most control rooms were designed with only one thing in mind, i.e., how to group the indicators, recorders, and operator stations within reach and sight of an operator who stands in front of the panel and operates the unit. With a state-of-the-art control room, the operators are surrounded by display screens and keyboards, which make human factor engineering a very important part of the design. (See chapter frontispiece.) It will affect the physical well being of operators as well as his or her ability to operate the unit effectively. The following are just a few examples of why human factor engineering is important.

1. With microprocessor and computer based distributed control systems, the amount of information available to the operator has increased dramatically. The system has to be designed so that important information is presented and instantly understood by the operator. For example, an alarm list that is not recognized properly will cause a great deal of confusion. On the other hand, several levels of alarms using different color codes will help the operator make correct decisions.
2. To be able to read display screens for long periods of time, lighting in the control room must be designed to avoid eye strain of the operators.
3. Furnishings to accommodate display screens and keyboards to maximize operator alertness must be provided.

Smart transmitters

In the past, a transmitter was used to measure pressure, temperature, and flow (differential pressure) from remote locations. A signal representing the calibrated value of that measurement was sent to the control system located in or near the central control room. The signal was then conditioned to provide a usable

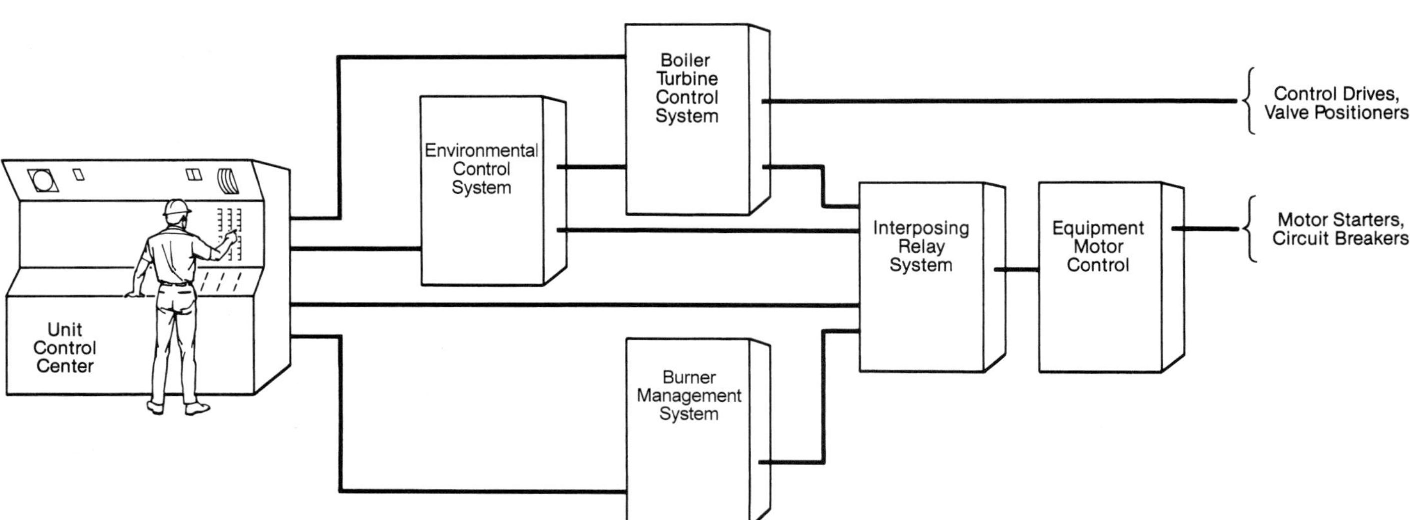

Fig. 35 Composite of automation components.

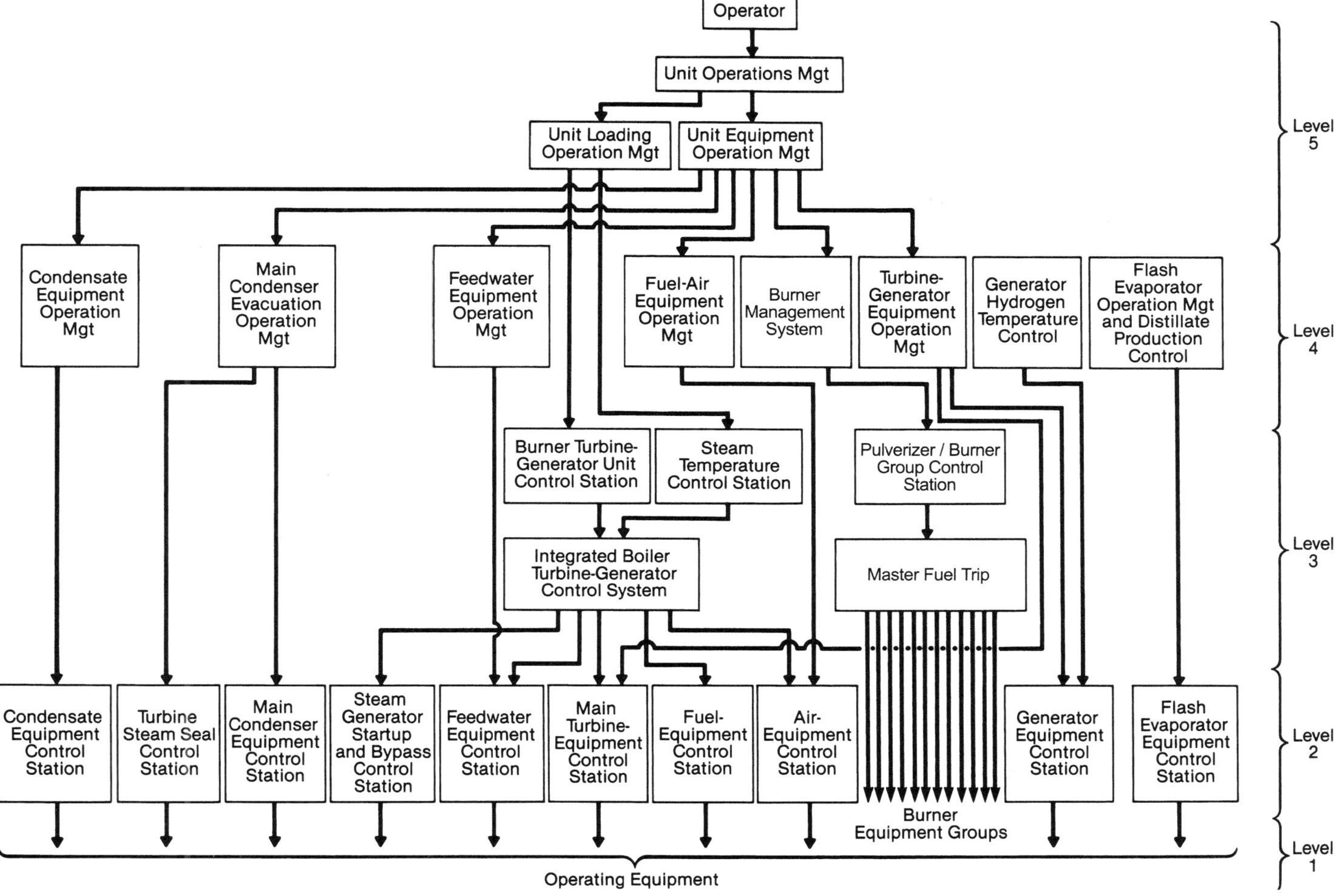

Fig. 36 Multi-level approach to power plant operation.

signal to the control system, indicator, or recorder. The smart transmitters available today can condition the signal locally – for example, square root extraction to convert differential pressure to flow and temperature compensation for flow measurements. The smart transmitter can also provide remote configuration, reranging, rezeroing, calibration and diagnostics features to provide direct access to the field device from the control room, virtually eliminating periodic maintenance visits to the instruments.

Installation and service requirements

It is important to make certain that instrumentation and control equipment are properly installed and serviced. Supervision of installation and the required calibration adjustments should be performed by qualified and experienced personnel. Adequate time to tune the controls should be provided during initial boiler operation.

Maintenance can be performed by trained plant personnel. The personnel who will be responsible for this maintenance should be available to observe and assist the control manufacturer's representatives in the installation and calibration of the control system. Most control manufacturers offer training programs to familiarize customer personnel with the control equipment. A planned program of preventive maintenance should be developed for the control system.

Performance monitoring

The continuing growth in size and complexity of units has added importance to the frequent analysis of performance. Many operating companies are analyzing their performance evaluation programs and developing monitoring techniques that make possible the detection of malfunctions before substantial losses have occurred.

A number of performance evaluation procedures are practiced today. In some power companies, the performance of each unit is checked periodically by extensive heat rate tests. Techniques and procedures for performing such evaluations are covered under the ASME Performance Test Codes.

The computer is also being used to conduct on-line performance calculations periodically, hourly and daily in most systems. This information can be compared with expected performance data stored in the computer and any deviation recorded for future correction. This could be especially useful for monitoring and controlling emissions.

Degassed conductivity analyzer for water quality analysis.

Chapter 42

Water and Steam Chemistry, Deposits and Corrosion

Steam generation and use involve thermal and physical processes of heat transfer, fluid flow, evaporation, and condensation. However, steam and water are not chemically inert physical media. Pure water dissociates to form low concentrations of hydrogen and hydroxide ions, H^+ and OH^-, and both water and steam dissolve some amount of each material that they contact. They also chemically react with materials to form oxides, hydroxides, hydrates, and hydrogen. As temperatures and velocities of water and steam vary, materials may dissolve in some areas and redeposit in others. Such changes are especially prevalent where water evaporates to form steam or steam condenses back to water, but they also occur where the only change is temperature, pressure, or velocity. In addition, chemical impurities in water and steam can form harmful deposits and facilitate dissolution (corrosion) of boiler structural materials. Therefore, to protect vessels, tubing, and other components used to contain and control these working fluids, water and steam chemistry must be controlled.

Water used in boilers must be purified and treated to inhibit scale formation, corrosion, and impurity contamination of steam. Two general approaches are used to optimize boiler water chemistry. First, impurities in the water are minimized by purification of makeup water, condensate polishing, deaeration and blowdown. Second, chemicals are added to control pH, electrochemical potential, and oxygen concentration. Chemicals may also be added to otherwise inhibit scale formation and corrosion. Proper water chemistry control improves boiler efficiency and reduces maintenance and component replacement costs. It also improves performance and life of heat exchangers, pumps, turbines, and piping throughout the steam generation, use and condensation cycle.

The primary goals of boiler water chemistry treatment and control are acceptable steam purity and acceptably low corrosion and deposition rates. In addition to customized boiler-specific guidelines and procedures, qualified operators are essential to achieving these goals, and vigilance is required to detect early signs of chemistry upsets. Operators responsible for plant cycle chemistry must understand boiler water chemistry guidelines and how they are derived and customized. They must also understand how water impurities, treatment chemicals, and boiler components interact. Training must therefore be an integral, ongoing part of operations and should include management, control room operators, chemists, and laboratory staff.

General water chemistry control limits and guidelines have been developed and issued by various groups of boiler owners and operators (e.g., ASME,[1,2,3] EPRI[4] and VGB[5]), water treatment specialists[6,7,8] utilities and industries. Also, manufacturers provide recommended chemistry control limits for each boiler and for other major cycle components. However, optimum water and steam chemistry limits for specific boilers, turbines, and other cycle components depend on equipment design and materials of construction for the combination of equipment employed. Hence, for each boiler system, boiler-specific water chemistry limits and treatment practices must be developed and tailored to changing conditions by competent specialists familiar with the specific boiler and its operating environment.

Chemistry-boiler interactions

To understand how water impurities, treatment chemicals and boiler components interact, one must first understand boiler circuitry, and steam generation and separation processes.

Boiler feed pumps provide feedwater pressure and flow for the boiler. From the pumps, feedwater often passes through external heaters and then through an economizer where it is further heated before entering the boiler. In a natural circulation drum-type unit, boiling occurs within steel tubes through which a water-steam mixture rises to a steam drum. Devices in the drum separate steam from water, and steam leaves through connections at the top of the drum. This steam is replaced by feedwater which is supplied by the feedwater pumps and injected into the drum just above the downcomers through a feedwater pipe where it mixes with recirculating boiler water which has been separated from steam. By way of downcomers, the water then flows back through the furnace and boiler tubes. *Boiler water* refers to the concentrated water circulating within the drum and steam generation circuits. Chapters 1 and 5 provide detailed descriptions of steam generation and boiler circulation.

Boiler feedwater always contains some dissolved solids, and evaporation of water leaves these dissolved impurities behind to concentrate in the steam generation circuits. If the concentration process is not limited, these solids can cause excessive deposition and corrosion within the boiler and excessive impurity carryover with the steam. To avoid this, some concentrated boiler water is discarded to drain by way of a blowdown line. Because the boiler water is concentrated, a little blowdown eliminates a large amount of the dissolved solids. Since steam carries very little dissolved solids from the boiler, dissolved and suspended solids entering in the feedwater concentrate in the boiler water until the solids removed in the blowdown (boiler water concentration times the blowdown rate lb/h or kg/s) equal the solids carried in with the feedwater (lb/h or kg/s).

A small amount of dissolved solids is carried from the drum by moisture (water) droplets with the steam. Because moisture separation from steam depends on the difference between their densities, moisture separation is less efficient at high pressures where there is less difference between the densities. Therefore, to attain the same steam purity at a higher pressure, the dissolved solids concentration in boiler water must generally be lower.

In a drum boiler, the amount of steam generated is small compared to the amount of water circulating through the boiler. However, circulation is also largely driven by the difference in densities between the two fluids, so as pressure increases the ratio of water flow to steam flow decreases. At 200 psi (1379 kPa), water flow through the boiler must be on the order of 25,000 pph (3 kg/s) to produce just 1000 pounds per hour of steam. Even at 2700 psi (18.6 MPa), 2500 to 4000 pounds of water circulates to produce 1000 pounds of steam. By contrast, most or all of the water entering a once-through boiler is converted to steam without recirculation.

Some boiler operators have asked why boiler water concentrations change so slowly once a source of contamination is eliminated and the continuous blowdown rate is increased. How quickly can excess chemical be purged from a boiler? How much impurity or additive is needed to upset boiler water chemistry? How quickly do chemical additions circulate through the boiler? To answer these questions and explore some other chemistry-boiler interactions, consider for example a typical 450 MW natural circulation boiler, generating 3,000,000 pounds of steam per hour. It has a room temperature water capacity of 240,000 pounds and an operating water capacity of 115,000 pounds. The furnace wall area is 33,000 square feet, about 5800 of which are in the maximum heat flux burner zone.

Impurities purge slowly from the boiler because the boiler volume is large compared to the blowdown rate. For example, at maximum steaming capacity with a blowdown rate 0.3% of the steam flow from the drum, 17 hours may be required to decrease the boiler water concentration of a non-volatile impurity by 50%. Almost two hours are required to effect a 50% reduction in the boiler water concentration even at a blowdown rate of 3%. Without blowdown, dissolved sodium with a fractional carryover factor of 0.1% would have a half life of 52 hours. While long periods of time are generally required to purge impurities, mixing within the boiler is rapid. For the boiler being used as an example, the internal recycle rate is about one boiler volume per minute, and steam is generated at a rate of one boiler volume every 5 minutes.

The rate of steam generation is such that replacement feedwater must be essentially free of hardness minerals and oxides that deposit in the boiler. For example, feedwater carrying only 1 ppm of hardness minerals and oxides could deposit up to 25,000 lb (11,340 kg) per year of solids in the boiler, so the boiler might require chemical cleaning as often as 3 or 4 times per year. Also, small chemical additions have a large effect on boiler water chemistry. For example, addition of 0.2 lb (0.09 kg) of sodium hydroxide to the boiler water increases the sodium concentration by 1 ppm, which can significantly affect the boiler water chemistry. Similarly, a small amount of chemical hideout can have a large effect on boiler water concentration. Hideout or hideout return of only 0.01 gram per square foot (0.1 g/m^2) in the burner zone can change the boiler water concentration by 1 ppm.

Control of deposition, corrosion, and steam purity

The potential for deposition and corrosion is inherent to boilers and increases with boiler operating pressure and temperature. Evaporation of water concentrates boiler water impurities and solid treatment chemicals at the heat transfer surfaces. During the normal nucleate boiling process in boiler tubes, small bubbles form on tube walls and are immediately swept away by the upward flow of water. As steam forms, dissolved solids in the boiler water concentrate along the tube wall. Additionally, the boundary layer of water along the wall is slightly superheated, and many dissolved minerals are less soluble at higher temperatures (common phenomenon referred to as *inverse temperature solubility)*. Both of these factors favor deposition of solids left behind by the evolution of steam in high heat flux areas, as illustrated in Fig. 1.

These deposits in turn provide a sheltered environment which can further increase chemical concentrations and deposition rates. In a relatively clean boiler tube, concentration of chemicals at the tube surface is limited by the free exchange of fluid between the surface and boiler water flowing through the tube. Wick boiling as illustrated in Fig. 2 generally produces sufficient flow within the deposits to limit the degree of concentration. However, as heavy deposits as illus-

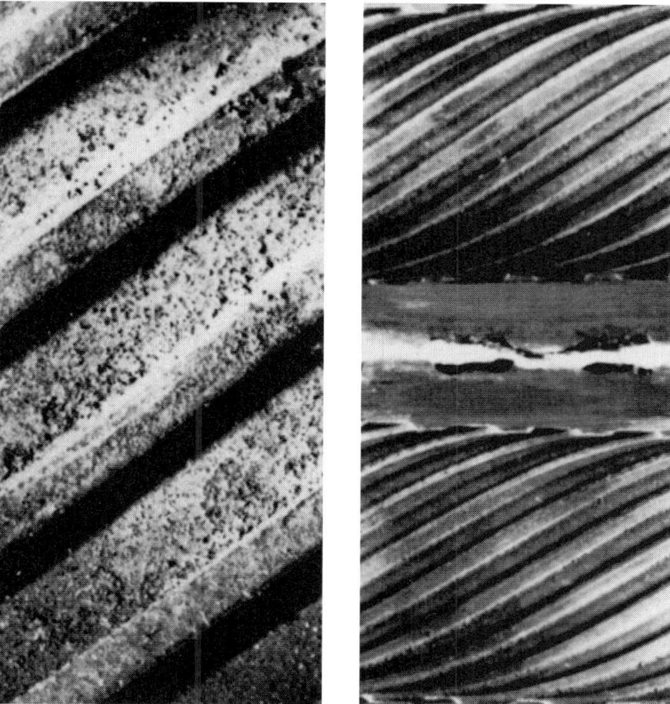

Fig. 1 Three years of operation resulted in light deposits because of good water treatment. The upper right figure is the heated side, and the lower right figure is the unheated side.

trated in Fig. 3 accumulate, they restrict flow to the surface. Some boiler water chemicals concentrate on tube walls during periods of high load and then return to the boiler water when the load is reduced. This is termed *hideout* and *hideout return*. This can greatly complicate efforts to control boiler water chemistry.

Typical boiler deposits are largely hardness precipitates and metal oxides. Hardness, easily precipitated minerals (mainly calcium and magnesium), enters the cycle as impurities in makeup water and in cooling water from condenser leaks. Metal oxides are largely from corrosion of pre-boiler cycle components. *Scaling* occurs when these minerals and oxides precipitate and adhere to boiler internal surfaces where they impede heat transfer. The result can be overheating of tubes, followed by failure and equipment damage. Deposits also increase circuitry pressure drop, especially detrimental in once-through boilers. Effective feedwater and boiler water purification and chemical treatment minimizes deposition by minimizing feedwater hardness and by minimizing corrosion and associated iron pickup from the condensate and feedwater systems. Also, phosphate and other water treatment chemicals are used in drum boilers to impede the formation of particularly adherent and low thermal conductivity deposits.

Some chemicals become corrosive as they concentrate. Corrosion can occur even in a clean boiler, but the likelihood of substantial corrosion is much greater beneath thick porous deposits that facilitate the concentration process. Concentration at the base of deposits can be more than 1000 times higher than that in the boiler water and the temperature at the base of these deposits can substantially exceed the saturation temperature. Hence, as deposits accumulate, control of boiler water chemistry to avoid the formation of corrosive concentrates becomes increasingly important. Since chemistry upsets do occur, operation of a boiler with excessively thick deposits should be avoided.

Because local concentration of boiler water impurities and treatment chemicals is inherent to steam generation, water chemistries must be controlled so the concentrates are not corrosive. On-line corrosion is often caused by concentration of sodium hydroxide, concentration of caustic-forming salts such as sodium carbonate, or concentration of acid-forming salts such as magnesium chloride or sulfate.[10] Effective feedwater and boiler water treatment minimizes corrosion by minimizing ingress of these impurities and by add-

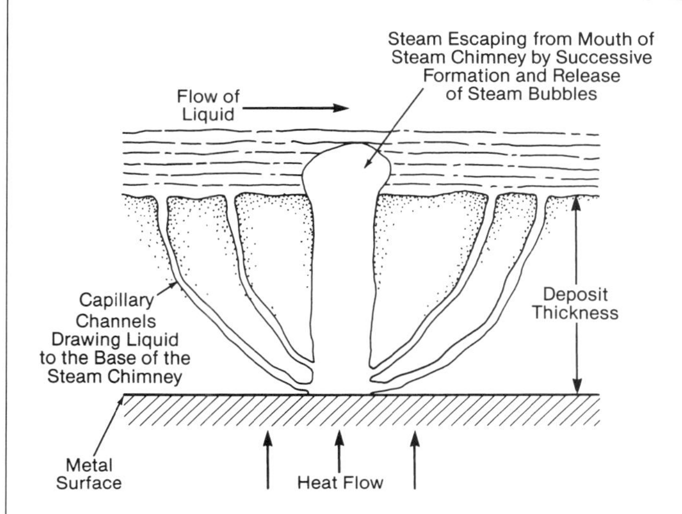

Fig. 2 Schematic of the wick boiling mechanism (*adapted from Reference 9*).

Fig. 3 An example of internal deposits resulting from poor boiler water treatment. These deposits, besides hindering heat transfer, allowed boiler water salts to concentrate, causing corrosion.

ing treatment chemicals (such as trisodium phosphate) that buffer against acid or caustic formation. However, corrosion and excessive precipitation can also be caused by improper use of buffering agents and other treatment chemicals. For example, underfeeding or overfeeding of treatment chemicals, out-of-specification sodium-to-phosphate ratios, or out-of-specification free-chelant concentrations can cause corrosion.

Dissolved carbon dioxide and oxygen can also be corrosive and must be eliminated from feedwater. Carbon dioxide from air in-leakage and from decomposition of carbonates and organic compounds tends to acidify feedwater and steam condensate. Oxygen is especially corrosive because it facilitates oxidation of iron, copper, and other metals to form soluble metal ions. At higher temperatures, oxygen is less soluble in water and the rate of chemical reaction is increased. As boiler feedwater is heated, oxygen is driven out of solution and rapidly corrodes heat transfer surfaces. The combination of oxygen and residual chloride is especially corrosive, as is the combination of oxygen and free chelant.

Carryover of impurities from boiler water to steam is also inherent to boiler operation. Though separation devices remove most water droplets carried by steam, some residual droplets containing small amounts of dissolved solids always carry through with the steam. Also, at higher pressures, there is some vaporous carryover. Excessive impurities can damage superheaters, steam turbines, or downstream process equipment.

Boiler feedwater

To maintain boiler integrity and performance and to provide steam of suitable turbine or process purity, boiler feedwater must be purified and chemically conditioned. The amount and nature of feedwater impurities that can be accommodated depend on boiler operating pressure, boiler design, steam purity requirements, type of boiler water internal treatment, blowdown rate, and whether the feedwater is used for steam attemperation. Feedwater chemistry parameters to be controlled include dissolved solids, pH, dissolved oxygen, hardness, suspended solids, total organic carbon (TOC), oil, chlorides, sulfides, alkalinity, and acid or base forming tendencies.

At a minimum, boiler feedwater must be softened water for low pressure boilers and demineralized water for high pressure boilers. It must be free of oxygen and essentially free of hardness constituents and suspended solids. Recommended feedwater limits are shown in Table 1. Use of high-purity feedwater minimizes blowdown requirements and minimizes the potential for carryover, deposition, and corrosion problems throughout the steam-water cycle.

Operation within these guidelines does not by itself ensure trouble-free operation. Some feedwater contaminants such as calcium, magnesium, organics, and carbonates can be problematic at concentrations below the detection limits of analytical methods commonly used

Table 1
Recommended Feedwater Limits

	Drum Boilers							Once-Through Boilers	
Pressure, psig (MPa)	15 to 300 (0.10 to 2.07)	301 to 600 (2.08 to 4.14)	601 to 900 (4.14 to 6.21)	901 to 1000 (6.21 to 6.90)	1001 to 1500 (6.90 to 10.34)	>1500 (>10.34)	with AVT* All	AVT All	Oxygen Treatment All
pH, all ferrous heaters	9.3 to 10.0	9.3 to 10.0	9.3 to 10.0	9.3 to 9.6	9.3 to 9.6	9.3 to 9.6	9.3 to 9.6	9.3 to 9.6	8.0 to 8.5
pH, copper-bearing heaters	8.8 to 9.2	8.8 to 9.2	8.8 to 9.2	8.8 to 9.2	8.8 to 9.2	8.8 to 9.2	8.8 to 9.2**	8.8 to 9.2	N/A
Total hardness, as ppm $CaCO_3$, maximum	0.3	0.2	0.1	0.05	0.003	0.003	0.003	0.003	0.001
Oxygen, ppm maximum***	0.007	0.007	0.007	0.007	0.007	0.007	0.007	0.007	0.030 to 0.150
Iron, ppm maximum	0.1	0.04	0.02	0.02	0.01	0.01	0.01	0.010	0.005
Copper, ppm maximum	0.05	0.02	0.01	0.01	0.005	0.002	0.005	0.002	0.001
Organic, ppm TOC max.	1.0	1.0	0.5	0.2	0.2	0.2	0.2	0.200	0.200
Cation conductivity, μS/cm max.	—	—	—	—	0.5	0.2	0.2	0.15	0.15

* All volatile treatment.
** AVT not recommended for copper-bearing cycles and associated low feedwater pH where the drum pressure is less than 400 psig.
*** By mechanical deaeration before chemical scavenging of residual.

Note:
ppm = mg/kg

for industrial boilers. Also, operators must be sensitive to changes in feedwater chemistry and boiler operating conditions, and must adapt accordingly.

Makeup water

Boiler feedwater is generally a mix of returned steam condensate and fresh makeup water. For utility boilers, most of the steam is usually returned as condensate, and only 1 to 2% makeup is necessary. However, for some industrial cycles, there is little or no returned condensate, so as much as 100% makeup may be necessary.

Chemistry requirements for makeup water depend on the amount and quality of returned steam condensate. Where a large portion of the feedwater is uncontaminated condensate, makeup water can generally be of lesser purity so long as the mixture of condensate and makeup meet boiler feedwater requirements. The feedwater concentration for each chemical species is the weighted average of the feedwater and makeup water concentrations:

Feedwater concentration = (condensate concentration × flow + makeup concentration × makeup flow) / total feedwater flow **(1)**

The selection of equipment for purification of makeup water must consider the water chemistry requirements, raw water composition, and quantity of makeup required. All natural waters contain dissolved and suspended matter. The type and amount of impurities vary with the source, such as lake, river, well or rain, and with the location of the source. Major dissolved chemical species in source water include sodium, calcium, and magnesium positive ions (cations) as well as bicarbonate, carbonate, sulfate, chloride, and silicate negative ions (anions). Organics are also abundant.

The first steps in water purification are coagulation and filtration of suspended materials. Natural settling in still water removes relatively coarse suspended solids. Required settling time depends on specific gravity, shape and size of particles, and currents within the settling basin. Settling and filtration can be expedited by coagulation (use of chemicals to cause agglomeration of small particles to form larger ones that settle more rapidly). Typical coagulation chemicals are alum and iron sulfate. Following coagulation and settling, water is normally passed through filters. The water is chlorinated to kill micro-organisms, then activated charcoal filters may be used to remove the final traces of organics and excess chlorine.

Subsequently, various processes may be used to remove dissolved scale-forming constituents (hardness minerals) from the water. For some low pressure boilers, removal of hardness minerals and scale-forming minerals is adequate. For other boilers, the concentration of *all* dissolved solids must be reduced or nearly eliminated. For low pressure boilers, the capital and operating cost for removal must be weighed against costs associated with residual dissolved solids and hardness. These include increased costs for boiler water treatment, more frequent chemical cleaning of the boiler, and possibly higher rates of boiler repair. Demineralized water nearly free of all dissolved solids is recommended for higher pressure boilers and especially for all boilers operating at pressures greater than 1000 psi (6.9 MPa).

Sodium cycle softening, often called *sodium zeolite softening*, replaces easily precipitated hardness minerals with sodium salts, which remain in solution as water is heated and concentrated. The major hardness ions are calcium and magnesium. However, zeolite ion-exchange softening also removes dissolved iron, manganese, and other divalent and trivalent cations. Sodium held by a bed of organic resin is exchanged for calcium and magnesium ions dissolved in the water. The process continues until the sodium ions in the resin are depleted and the resin can no longer absorb calcium and magnesium efficiently. The depleted resin is then regenerated by washing it with a high concentration sodium chloride solution. At the high sodium concentrations of this regeneration solution, the calcium and magnesium are displaced by sodium. Variations of the process, in combination with chemical pre-treatments and post-treatments, can substantially reduce hardness concentrations and can often reduce silica and carbonate concentrations.

For higher pressure boilers, evaporative or more complete ion-exchange demineralization of makeup water is recommended. Any of several processes may be used. *Evaporative distillation* forms a vapor which is recondensed as purified water. *Ion exchange demineralization* replaces cations (sodium, calcium, and magnesium in solution) with hydrogen ions and replaces anions (bicarbonate, sulfate, chloride, and silicate) with hydroxide ions. For makeup water treatment, two tanks are normally used in series in a cation-anion sequence. The anion resin is usually regenerated with a solution of sodium hydroxide, and the cation resin is regenerated with hydrochloric or sulfuric acid. *Reverse osmosis* purifies water by forcing it through a semi-permeable membrane or a series of such membranes. It is increasingly used to reduce total dissolved solids (TDS) in steam cycle makeup water. Where complete removal of hardness is necessary, reverse osmosis may be followed by a mixed-bed demineralizer. *Mixed-bed demineralization* uses simultaneous cation and anion exchange to remove residual impurities left by reverse osmosis, evaporator, or two-bed ion exchange systems. Mixed-bed demineralizers are also used for polishing (removing impurities from) returned steam condensate. Before regenerating mixed-bed demineralizers, the anion and cation resins must be hydraulically separated. **Caustic and acids used for regeneration of demineralizers and other water purification and treatment chemicals present serious safety, health, and environmental concerns. Material Safety Data Sheets must be obtained for each chemical and appropriate precautions for handling and use must be formulated and followed.**

Dissolved organic contaminants (carbon-based molecules) are problematic in that they are often detrimental to boilers but are not necessarily removed by deionization or evaporative distillation. Organic contamination of feedwater can cause boiler corrosion, furnace wall tube overheating, drum level instability, carryover, superheater tube failures, and turbine corrosion. The degree to which any of these difficulties occurs depends on the concentration and nature of the organic contaminant. Removal of organics may require activated carbon filters or other auxiliary purification equipment.

Returned condensate – condensate polishing

For many boilers, a large fraction of the feedwater is returned condensate. Condensate has been purified by prior evaporation, so uncontaminated condensate does not generally require purification. Makeup water can be mixed directly with the condensate to form boiler feedwater.

In some cases, however, steam condensate is contaminated by corrosion products or by in-leakage of cooling water. Where returned condensate is contaminated to the extent that it no longer meets feedwater purity requirements, mixed-bed ion-exchange purification systems are commonly used to remove the dissolved impurities and filter out suspended solids. Such demineralization is referred to as *condensate polishing*. This is essential for satisfactory operation of once-through utility boilers, for which feedwater purity requirements are especially stringent. While high pressure drum boilers can operate satisfactorily without condensate polishing, many utilities recognize the benefits in high pressure plants. These benefits include shorter unit startup time, protection from condenser leakage impurities, and longer intervals between acid cleanings. Condensate polishing is recommended for all boilers operating with all volatile treatment (AVT) and is essential for all boilers operating with all volatile treatment and seawater cooled condensers. Provisions for polishing vary from adequate capacity for 100% polishing of all returned condensate to polishing only a portion of the condensate. However, all must be adequate to meet feedwater requirements under all anticipated load and operating conditions.

Most of the pressure vessels that contain ion exchange resins have under-drain systems and downstream traps or strainers to prevent leakage of ion exchange resins into the cycle water. These resins can form harmful decomposition products if allowed to enter the high temperature portions of the cycle. Unfortunately, the under-drain systems and the traps and strainers are not designed to retain resin fragments that result from resin bead fracture. Also, the resin traps and strainers can fail, resulting in resin bursts. Resin intrusion can be minimized by controlling flow transients, reducing the strainer's screen size, increasing flow gradually during vessel cut-in, and returning the polisher vessel effluent to the condenser during the first few minutes of cut-in.

Feedwater pH control

Boiler feedwater pH is monitored at the condensate pump discharge and at the economizer inlet. When the pH is below the required minimum value, ammonium hydroxide or an alternate alkalizer is added. Chemicals for pH control are added either downstream of the condensate polishers or at the condensate pump discharge for plants without polishers. For high purity demineralized feedwater, ammonium hydroxide injection pumps or alternative feedwater pH control is achieved using a feedback signal from a specific conductivity monitor. Conductivity provides a good measure of ammonium hydroxide concentration, and automated conductivity measurement is more reliable than automated pH measurement. Also, the linear rather than logarithmic relationship of conductivity to ammonia concentration gives better control. Fig. 4 shows the relationship between ammonium concentration, pH, and conductivity of demineralized water.

While an equilibrium concentration of ammonium hydroxide remains in the boiler water, much of the ammonium hydroxide added to feedwater volatilizes with the steam. Conversely, the solubility of ammonium hydroxide is such that little ammonia is lost by deaeration. Hence, returned condensate often has a substantial concentration of ammonium hydroxide before further addition.

Common alternative pH control agents include neutralizing amines, such as cyclohexylamine and morpholine. For high pressure utility boilers with superheaters, the more complex amines are thermally unstable and the decomposition products can be problematic.

Deaeration and chemical oxygen scavengers

Oxygen and carbon dioxide enter the cycle with undeaerated makeup water, with cooling water which

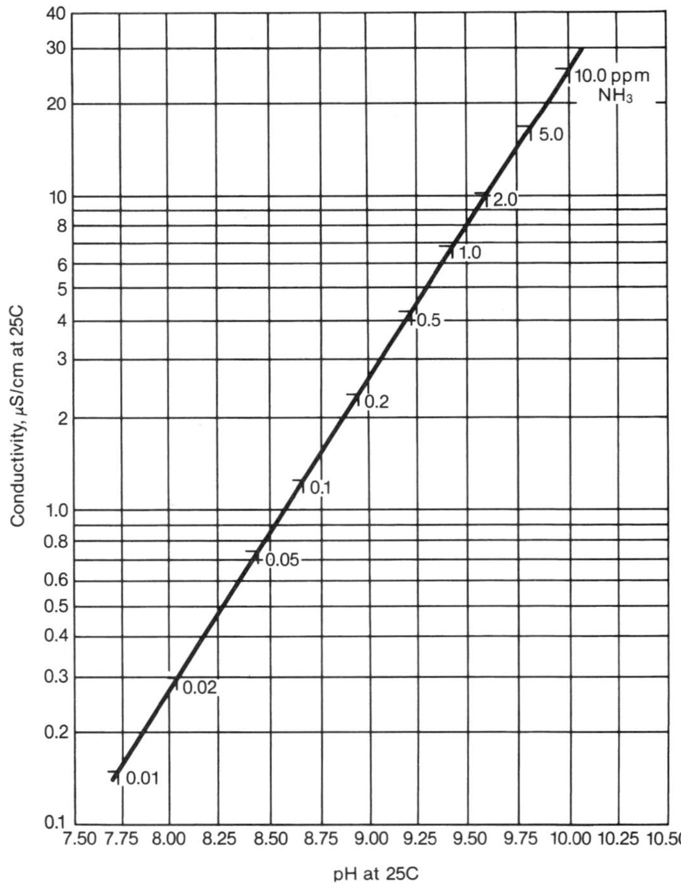

Fig. 4 Approximate relationship between conductivity and pH for ammonia solutions in demineralized water.

leaks into the condenser, and as air leaking into the vacuum portion of the cycle.

For turbine cycles, aeration of the feedwater is initially limited by use of air ejectors to remove air from the condenser. Utility industry standard practice is to limit total air in-leakage to less than one standard cubic foot of air per minute per 100 MW of generating capacity (approximately 0.027 Nm^3/100 MW), as measured at the condenser air ejectors. Final removal of oxygen and other dissolved gases adequate for boiler feedwater applications is generally accomplished by thermal deaeration of the water ahead of the boiler feed pumps. Thermal deaeration is accomplished by heating water to reduce gas solubility. Gases are then carried away by a counter flow of steam. The process is typically facilitated by the use of nozzles and trays which disperse water droplets to increase the steam-to-water interfacial area. Thermal deaeration can reduce feedwater oxygen concentration to less than 7 parts per billion (ppb). It also essentially eliminates dissolved carbon dioxide, nitrogen, and argon.

Chemical agents are generally used to scavenge residual oxygen not removed by thermal deaeration. Traditional oxygen scavengers have been sodium sulfite for low pressure boilers and hydrazine for high pressure boilers. Sulfite must not be used where the boiler pressure is greater than 900 psig (6.2 MPa). Other oxygen scavengers (erythorbic acid, diethylhydroxylamine, hydroquinone and carbohydrazide) are also used. Hydrazine has been identified as a carcinogen and this has increased the use of alternative scavengers. Scavengers are generally fed at the exit of the condensate polishing system and/or at the boiler feed pump suction.

Attemperation water

Water spray attemperation is used to control steam temperature. Attemperation for utility boilers is discussed in the *Steam temperature adjustment and control* section of Chapter 19, *Boilers, Superheaters and Reheaters*. The spray water is feedwater, polished feedwater, or steam condensate. As the spray water evaporates, all chemicals and contaminants in the water remain in the steam. This addition must not be excessive. It must not form deposits in the attemperator piping, and it must not excessively contaminate the steam. If a superheated steam purity limit is imposed, the steam purity after attemperation must not exceed this limit. To meet this requirement, the weighted average of the spray water total solids concentration and the saturated steam total solids concentration must not exceed the final steam total solids limit. Additionally, spray water attemperation must not increase the steam total solids concentration by more than 0.040 ppm. Independent of other considerations, the spray water solids concentration must never exceed 2.5 ppm. Ideally, the purity of attemperation water should equal the desired purity of the steam.

Drum boilers and internal boiler water

Boiler water that recirculates in drum and steam generation circuits has a relatively high concentration of dissolved solids that have been left behind by water evaporation. Water chemistry must be carefully controlled to assure that this concentrate does not precipitate solids or cause corrosion within the boiler circuitry. Boiler water chemistry must also be controlled to prevent excessive carryover of impurities or chemicals with the steam.

Customized chemistry limits and treatment practices must be established for each boiler. These limits depend on steam purity requirements, feedwater chemistry, and boiler design. They also depend on boiler owner/operator preferences regarding economic tradeoffs between feedwater purification, blowdown rate, frequency of chemical cleaning, and boiler maintenance and repair. Direct boiler water treatment (usually referred to as *internal treatment*) practices commonly used to control boiler water chemistry include all volatile treatment, coordinated phosphate treatments, high-alkalinity phosphate treatments, and high-alkalinity chelant and polymer treatments. In all cases, when treatment chemicals are mixed, the identity and purity of chemicals must be verified and water of hydration in the weight of chemicals must be taken into account. The specific treatment used must always be developed and managed by competent water chemistry specialists.

Feedwater is the primary source of solids that concentrate in boiler water, and feedwater purity defines the practical limit below which the boiler water solids concentration can not be reduced with an acceptable blowdown rate. Additionally, hardness and pre-boiler corrosion products carried by the feedwater play major roles in defining the type of boiler water treatment that must be employed. Where substantial hardness is present in feedwater, provision must be made to ensure that the hardness constituents remain in solution in the boiler water or to otherwise minimize the formation of adherent deposits. This is often accomplished by use of chelant, polymer, or high-alkalinity phosphate boiler water treatment. Where substantial hardness is not present, boiler water treatment can be optimized to minimize impurity carryover in the steam and to minimize the potential for boiler tube corrosion.

Because boiler water impurities and treatment chemicals carry over in the steam, steam purity requirements play a major role in defining boiler water chemistry limits. Boiler specifications normally include a list of boiler-specific water chemistry limits that must be imposed to attain a specified steam purity. Limits must always be placed on the maximum dissolved solids concentration. Limits must also be placed on impurities and conditions that cause foaming at the steam-water interface in the drum. These include limits on oil and other organic contaminants, suspended solids, and alkalinity. The *carryover factor* is the ratio of an impurity or chemical species in the steam to that in the boiler water.

Blowdown

The dissolved solids concentration of boiler water is intermittently or continuously reduced by blowing down some of the boiler water and replacing it with feedwater. Blowdown rate is generally expressed as a percent relative to the steam flow rate from the drum.

Blowdown is accomplished through a pressure letdown valve and flash tank. Heat loss is often minimized by use of a regenerative heat exchanger.

The ratio of the concentration of a feedwater impurity in the boiler water to its concentration in the feedwater is the *concentration factor*, which can be estimated by use of Equation 1. However, a more complex formula must be used where there is substantial carryover.

If there is no blowdown, solids concentrate until carryover with the steam is sufficient to carry away all of the solids that enter the boiler with the feedwater. For example, where the feedwater silica concentration is 0.01 ppm and 10% of the silica in the boiler water carries over with the steam, the equilibrium boiler water silica concentration is 0.1 ppm.

Traditional all volatile treatment

For all volatile treatment (AVT), no solid chemicals are added to the boiler or pre-boiler cycle. Boiler water chemistry control is by boiler feedwater treatment only. No chemical additions are made directly to the drum. Feedwater pH is controlled with ammonia or an alternate amine. Because ammonia carries away preferentially with the steam, the boiler water pH may be slightly lower (0.2 to 0.4 pH units) than the feedwater pH. For traditional all volatile treatment, as opposed to oxygen treatment, hydrazine or a suitable alternate is added to scavenge residual oxygen. Table 1 shows the recommended AVT feedwater control limits. Because all volatile treatment adds no solids to the boiler water, solids carryover is generally minimized.

All volatile treatment provides no chemical control for hardness deposition and provides no buffer against caustic or acid-forming impurities. Hence, feedwater must contain no hardness minerals from condenser leakage or other sources. It must be high-purity condensate or polished condensate with mixed-bed quality demineralized makeup water.

All volatile treatment can be, but rarely is, used below 1000 psig (6.9 MPa). Normally it is used only for boilers operating at or above 2000 psig (13.8 MPa) drum pressure. It is not recommended for lower pressure boilers where other options are feasible. While all volatile treatment is one of several options for drum boilers, it is the only option for once-through boilers.

Oxygen treatment

Even in the absence of dissolved oxygen, steel surfaces react with water to form some soluble Fe^{++} ions which may deposit in the boiler, superheater, turbine, or other downstream components. However, in the absence of impurities, oxygen can form an especially protective Fe^{+++} iron oxide that is less soluble than that formed under oxygen-free conditions. To take advantage of this, some copper-free boiler cycles operating with ultra pure feedwater maintain a controlled concentration of oxygen in the feedwater. Most of these are high pressure once-through utility boilers, but this approach is also used successfully in some drum boilers.

Oxygen treatment was developed in Europe, largely by Vereinigung der Grosskesselbetreiber (VGB),[11] and there is also extensive experience in the former Soviet Union (FSU). It can only be used where there is no copper in the pre-boiler components beyond the condensate polisher, and where feedwater is consistently of the highest purity, e.g., cation conductivity < 0.15 μS/cm at 77F (25C). A low concentration of oxygen is added to the condensate. The target oxygen concentration is 0.050 to 0.150 ppm for once-through boilers and 0.040 ppm for drum boilers. With oxygen treatment, the feedwater pH can be reduced, e.g., down to 8.0 to 8.5. An advantage of oxygen treatment is decreased chemical cleaning frequencies for the boiler. In addition, when oxygen treatment is used in combination with lower pH, the condensate polisher regeneration frequency is reduced.

Coordinated phosphate treatment

Coordinated phosphate-pH treatment, introduced by Whirl and Purcell of the Duquesne Light Company,[12] controls boiler water alkalinity with mixtures of disodium and trisodium phosphate added to the drum through a chemical feed pipe. The objective of this treatment is largely to keep the pH of boiler water and underdeposit boiler water concentrates within an acceptable range. Fig. 5 indicates the phosphate concentration range that is generally necessary and sufficient for this purpose. Phosphate treatment must not be used where the drum pressure exceeds 2800 psig (19.3 MPa). All volatile treatment is recommended at the higher pressures.

In sodium phosphate solutions, an $H^+ + PO4^{\equiv} \rightarrow HPO4^{=}$ balance buffers the pH (i.e., retards H^+ ion concentration changes). Solution pH depends on the phosphate concentration and the molar sodium-to-phosphate ratio. The relationship between pH, phosphate concentration, and molar sodium-to-phosphate ratio is shown in Fig. 6. Where solutions contain other dissolved salts (e.g., sodium and potassium chloride and sulfate), sodium phosphate can still be used to control pH, and the curves of Fig. 6 are still applicable. However, for such solutions, the sodium-to-phosphate ratio labels on these curves are only apparent values with reference to pure sodium phosphate solutions. Measured sodium concentrations can not be used in calculating sodium-to-phosphate ratios for control of

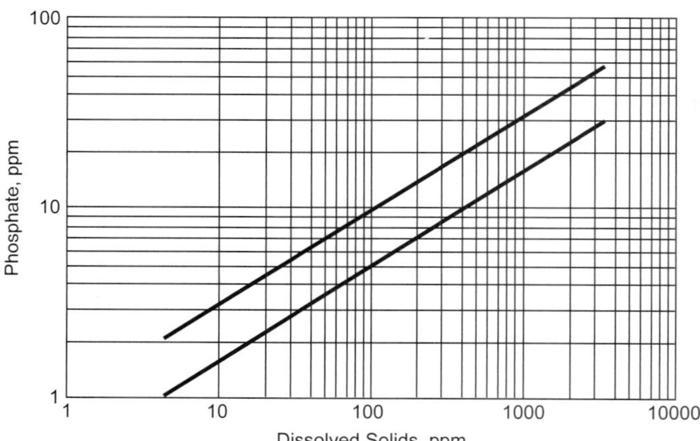

Fig. 5 Phosphate concentrations to control boiler water chemistry (little or no residual hardness in the feedwater). Indicates phosphate range at a given dissolved solids concentration.

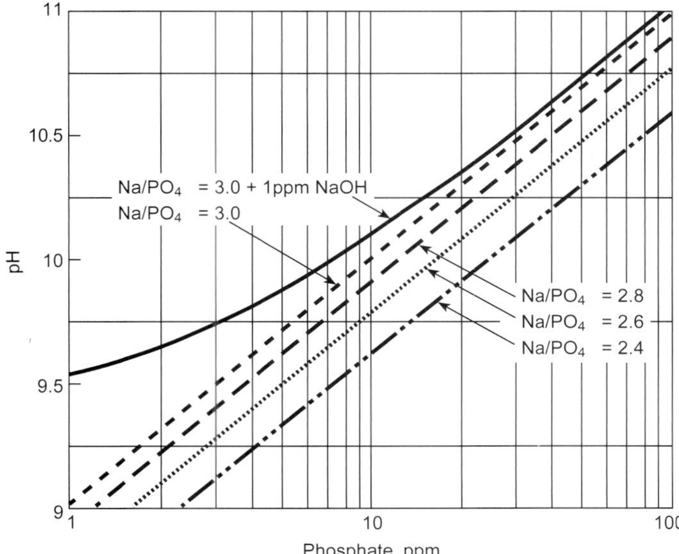

Fig. 6 Estimated pH of sodium phosphate solutions. Note: pH values can differ by up to 0.2 pH units, depending on the choice of chemical equilibrium constants used, but more often agree within 0.05 pH units.

boiler water pH because measured sodium concentrations include non-phosphate sodium salts. While dissolved sodium chloride and sulfate do not alter boiler water pH, ammonia does alter the pH. Hence, the presence of ammonia must be taken into account where ammonia concentrations are significant compared to phosphate concentrations.

Historically, the initial goal of coordinated pH-phosphate control was to keep the effective molar sodium-to-phosphate ratio just below 3, to prevent caustic stress corrosion cracking, acid corrosion, and hydrogen damage. This proved to be an effective method for control of deposition and corrosion in many boilers. However, caustic gouging of furnace wall tubes occurred in some boilers using coordinated pH-phosphate control, and laboratory tests indicated that solutions with molar sodium-to-phosphate ratios greater than about 2.85 can become caustic when highly concentrated. Subsequently, many boilers were operated under congruent control with a target effective sodium-to-phosphate ratio of less than 2.85, generally about 2.6, and often less than 2.6. Again, this proved to be an effective method of control for many boilers, but some of the boilers operating with low molar sodium-to-phosphate ratios experienced acid phosphate corrosion. Instances of boiler tube corrosion generally occurred in boilers that experienced substantial phosphate hideout and hide-out-return when the boiler load changed.

Phosphate hideout, hideout-return, and associated corrosion problems are now addressed by *equilibrium phosphate treatment.*[13] The concentration of phosphate in the boiler water is kept low enough to avoid hideout and hideout return associated with load changes, thus it is always in equilibrium with the boiler. The effective molar sodium-to-phosphate ratio is kept above 2.8. The free hydroxide, as depicted in Fig. 6, is not to exceed the equivalent of 1 ppm sodium hydroxide. Concern about caustic gouging at the higher ratios is largely reduced by experience with this treatment regime and by experience with caustic boiler water treatment.

Tables 2 and 3 show recommended boiler water chemistry limits. Customized limits for a specific boiler depend on the steam purity requirements for the boiler. Boiler and laboratory experience indicate that, under some conditions, phosphate-magnetite interactions can degrade protective oxide scale and corrode the underlying metal. To minimize these interactions, the pH must be greater than that corresponding to the 2.6 sodium-to-phosphate ratio curve of Fig. 6, and preferably greater than that corresponding to the 2.8 curve. The pH must always be above the 2.8 curve when the drum pressure is above 2600 psig (17.9 MPa). The maximum pH is that of trisodium phosphate plus 1 ppm sodium hydroxide. Additionally, the boiler water pH is not to be less than 9 nor greater than 10. As discussed below, it may be necessary to reduce the maximum boiler water phosphate concentration to avoid hideout and hideout return, and to avoid associated control and corrosion problems.

Table 2
Boiler Water Limits

Pressure Range, psig (MPa)	Maximum TDS, ppm	Maximum Suspended Solids, ppm	Maximum Silica, ppm
15 to 50* (0.10 to 0.35)	1250	15	30
15 to 50** (0.10 to 0.35)	3500	15	150
51 to 325 (0.35 to 2.24)	3500	10	150
326 to 450 (2.25 to 3.10)	3000	8	90
451 to 600 (3.11 to 4.14)	2500	6	40
601 to 750 (4.14 to 5.17)	2000	4	30
751 to 900 (5.18 to 6.21)	1500	2	20
901 to 1000 (6.21 to 6.90)	1250	1	8

* For natural separation with no diffuser baffles in the steam drum.
** Where the drum includes baffles that separate water droplets from steam.

Notes:
1. Operation outside the limits defined in this table is not recommended. Within these broad limits, more restrictive limits must be imposed, consistent with the type of boiler water treatment method chosen to control deposition, corrosion, and carryover. For example, the boiler water pH range of 8.5 to 9.0 is only acceptable with all volatile treatment (AVT).
2. ppm = mg/kg

Table 3
Boiler Water Limits for Coordinated Phosphate Boiler Water Treatment

	Pressure, psig (MPa)			
	15 to 1000 (0.10 to 6.90)	1001 to 1500 (6.90 to 10.34)	1501 to 2600 (10.35 to 17.93)	2601 to 2800 (17.93 to 19.31)
Maximum TDS, ppm	Defined by Table 2 or as necessary to attain required steam purity, whichever is less	100 ppm or as necessary to attain required steam purity, whichever is less	50 ppm or as necessary to attain required steam purity, whichever is less	15 ppm or as necessary to attain required steam purity, whichever is less
Maximum sodium, ppm	Maximum sodium concentration (if any) as necessary to attain required steam purity			
Maximum silica, ppm	Defined by Table 2 or as necessary to attain required steam purity, whichever is less	2 ppm or as necessary to attain required steam purity, whichever is less	0.5 ppm or as necessary to attain required steam purity, whichever is less	0.1 ppm or as necessary to attain required steam purity, whichever is less.
Phosphate as PO_4, ppm	See Fig. 5			
"Effective" Na/PO_4 molar ratio	2.6 to 3.0	2.6 to 3.0 + 1 ppm NaOH	2.6 to 3.0 + 1 ppm NaOH	2.8 to 3.0 + 1 ppm NaOH
pH	See Figs. 6 and 7	9.4 to 10.5 and as dictated by Figs. 6 and 7	9.0 to 10.0 and as dictated by Figs. 6 and 7	9.0 to 10.0 and as dictated by Figs. 6 and 7
Maximum specific conductivity, μS/cm	Twice the maximum TDS (ppm).			

Note:
ppm = mg/kg

Phosphate treatment chemicals may hide out during periods of high-load operation, then return to the boiler water when the load and pressure are reduced. This type of hideout makes control of boiler water chemistry difficult and can cause corrosion of furnace wall tubes. This hideout and return phenomena is caused by concentration of phosphate at the tube/water interface in high heat flux areas. In these areas, phosphates accumulate in the concentrated liquid. The concentrated phosphates then precipitate, or they adsorb on or react with surface deposits and scale.[13,14,15] Where excessive deposits are not present, this hideout and hideout return associated with load and pressure changes can be eliminated by decreasing the phosphate concentration in the boiler water or possibly by increasing the sodium-to-phosphate ratio. Where hideout and hideout-return are caused by excessive deposits, the boiler must be chemically cleaned. The amount of phosphate hideout or return accompanying load changes must not be more than 5 ppm. Corrective action is necessary if the amount of phosphate hideout or return accompanying load changes is more than 5 ppm and/or the boiler water pH change is more than 0.2 pH units, or where there are changes in the hideout/hideout-return behavior.

This phenomenon must be distinguished from loss of phosphate to passive film formation. As the passive oxide film reforms following a chemical cleaning of the boiler, some phosphate is irreversibly lost from the boiler water. This is minimized if chemical cleaning is followed by a phosphate boilout repassivation of the boiler.

Operators should not over-correct for deviations of pH and phosphate concentration from target values. Corrective action must be taken with an understanding of system response times, the amounts of impurities being neutralized, and the amount of treatment chemicals likely to be required.

Where phosphate treatment is used, pH is an especially critical parameter, so the accuracy of pH measuring devices and temperature corrections must be assured. The boiler water pH must also be corrected to discount the pH effect of residual ammonia in the boiler water. Fig. 7 shows the estimated effect of ammonia on boiler water pH. The figure indicates the expected pH for solutions with different concentrations of sodium phosphate and 0.2 ppm ammonia. Where these species dominate the solution chemistry, such figures may be used to estimate sodium-to-phosphate molar ratios.

With high purity feedwater, the recommended boiler water pH can be attained with appropriate additions of trisodium phosphate. If the recommended boiler water pH can not be maintained within the above limits using trisodium phosphate or a mixture of trisodium and disodium phosphate, this is indicative of alkaline or acid-forming impurities in the feedwater or excessive hideout, and the root cause must be addressed. An exception is low level equilibrium phosphate treatment, where the small amount of trisodium phosphate added to the boiler water may at times be insufficient to achieve the recommended pH. A small amount of sodium hydroxide may be added to attain the recommended pH, but the excess sodium hydroxide must not exceed 1.0 ppm.[13] Even 1.0 ppm sodium

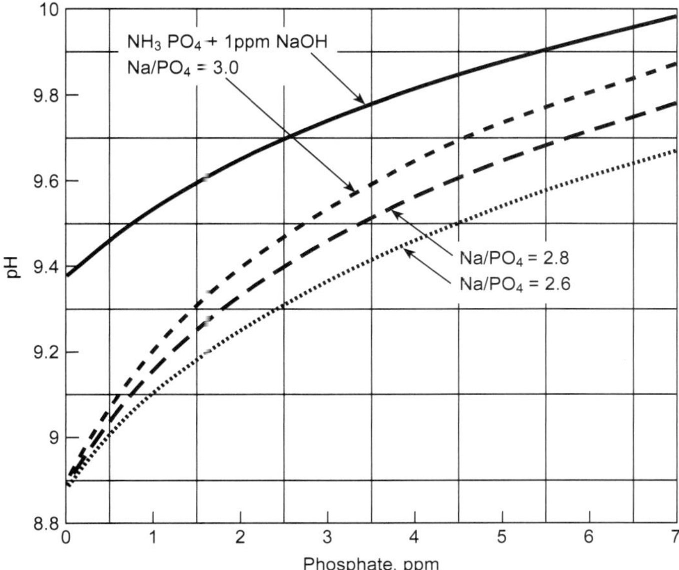

Fig. 7 Estimated pH of sodium phosphate solutions containing 0.2 ppm ammonium hydroxide. Note: pH values can differ by up to 0.2 pH units, depending on the choice of chemical equilibrium constants used, but more often agree within 0.05 pH units.

hydroxide may be excessive for some units, for example oil-fired boilers with especially high heat fluxes in some areas of the furnace.

When mixing boiler water treatment chemicals, operators should verify the identity and purity of the chemicals and take into account water of hydration in the weight of the chemicals. Neither phosphoric acid nor monosodium phosphate should be used for routine boiler water treatment. If monosodium phosphate is used to counter an isolated incident of alkali contamination of the boiler water, it must be used with caution, and at reduced load.

High-alkalinity phosphate treatment (low-pressure boilers only)

Minimal carryover and deposition are achieved with demineralized makeup water and minimal dissolved solids, but this is not necessarily cost-effective for all low pressure industrial boilers. Where softened water with 0.02 to 0.5 ppm residual hardness (as $CaCO_3$) is used as makeup water for low pressure industrial boilers, high alkalinity or *conventional* phosphate treatment may be used to control scale formation. This high alkalinity treatment must only be used for boilers operating below 1000 psig (6.9 MPa). The pH and phosphate concentrations are attained by addition of a trisodium phosphate and (if necessary) sodium hydroxide solution through a chemical feed line into the drum. With high-alkalinity phosphate treatment, the boiler water pH is maintained in the range of 10.8 to 11.4. This high pH precipitates hardness constituents that are less adherent than those formed at lower pH.

Where high alkalinity boiler water is excessively concentrated by evaporation, the concentrate can become sufficiently caustic to produce caustic gouging or stress corrosion cracking of carbon steel. Hence, high-alkalinity boiler water treatment must not be used where waterside deposits are excessively thick, where there is steam blanketing or critical heat flux (see Chapter 5), or where there is seepage (e.g., through rolled seals or cracks).

Fig. 8 shows phosphate concentration limits for high-alkalinity phosphate treatment. With some feedwaters (e.g., high-magnesium low-silica), lower phosphate concentrations may be advisable. The required pH is attained by adjusting the sodium hydroxide concentration in the chemical feed solution. The total (M alkalinity in calcium carbonate equivalents) must not exceed 20% of the actual boiler water solids concentration.

Dispersants, polymers, and chelants (low pressure boilers only)

Where substantial hardness (e.g., 0.1 ppm as $CaCO_3$) is present in feedwater, chelant treatment is often used to ensure that the hardness constituents remain in solution in the boiler water, or polymer treatment is used to keep precipitates in suspension. Blowdown of the dissolved contaminants and colloids is more effective than that of noncolloidal hardness precipitates and metal oxides.

While phosphate treatment precipitates residual calcium and magnesium in a less detrimental form than occurs in the absence of phosphate, chelants react with calcium and magnesium to form soluble compounds that remain in solution. Chelants commonly employed include ethylene-diaminetetraacetic acid (EDTA) and nitrilotriacetic acid (NTA). Because of concern about thermal stability, the use of chelants and polymers should be limited to boilers operating at less than 1000 psi (6.9 MPa).

To be most effective, chelant must mix with the feedwater and form thermally stable calcium and magnesium complexes before there is substantial residence time at high temperature, where free chelant is not thermally stable. Because the combination of free chelant and dissolved oxygen can be corrosive, chelant must be added only after completion of oxygen removal and scavenging. Also, there must be no copper-bearing components in the feedwater train beyond the chelant feed point.

Control limits depend on the feedwater chemistry, specific treatment chemicals used, and other factors. However, the boiler feedwater pH is generally be-

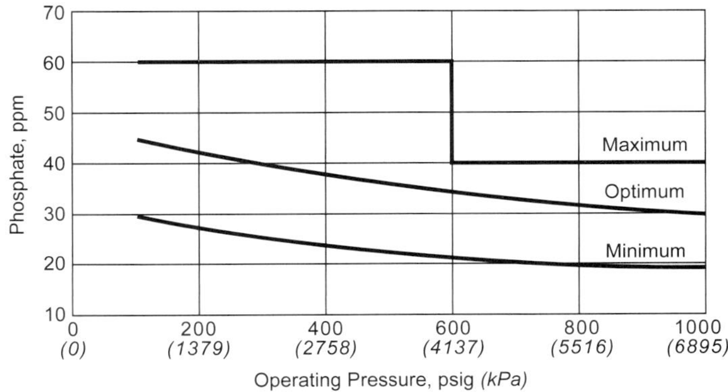

Fig. 8 Phosphate concentration limits for high-alkalinity phosphate treatment.

tween 9.0 and 9.6 and hardness as calcium carbonate is less than 0.5 ppm. The boiler water pH is generally maintained in the range of 10.0 to 11.4. The boiler water pH is attained by a combination of alkalinity derived from the chelant feed (e.g., as Na_4 EDTA), evolution of CO_2 from softened feedwater, and addition of sodium hydroxide. Polymeric dispersants are generally used to impede formation of scale by residual solids.

Once-through universal pressure boilers

In a subcritical once-through boiler, there is no steam drum. As water passes through boiler tubing, it evaporates entirely into steam. Because steam does not cool the tube as effectively as water, the tube temperature increases beyond this dry-out location. Subcritical once-through boilers are designed so this transition occurs in a lower heat flux region of the boiler where the temperature increase is not sufficient to cause a problem. However, because the water evaporates completely, it must be of exceptional purity to avoid corrosion and rapid deposition, and carryover of dissolved solids.

Similarly stringent water purity requirements must be imposed for supercritical boilers. While there is no distinction between water and steam in a supercritical boiler, the physical and chemical properties of the fluid change as it is heated, and there is a temperature about which dissolved solids precipitate much as they do in the dry-out zone of a subcritical once-through boiler. This is termed the *pseudo transition zone*. (See Chapter 5.)

Satisfactory operation of a once-through boiler and associated turbine requires that the total feedwater solids be less than 0.030 ppm total dissolved solids with cation conductivity less than 0.15 μS/cm. Table 1 lists recommended limits for other feedwater parameters. Feedwater purification must include condensate polishing, and water treatment chemicals must all be volatile. Ammonia is typically added to control pH. For traditional all volatile treatment, hydrazine or a suitable volatile substitute is used for oxygen scavenging. Iron pickup from pre-boiler components can be minimized by maintaining a feedwater pH of 9.3 to 9.6. Prior to plant startup, feedwater must be circulated through the condensate polishing system to remove dissolved and suspended solids. Temperatures should not exceed 550F (288C) at the convection pass outlet until the iron levels are less than 0.1 ppm at the economizer inlet.

Utility once-through boilers with copper-free cycle metallurgy commonly use oxygen treatment. Table 1 includes recommended limits for other feedwater chemical parameters for oxygen treatment. Startup is with increased pH and no oxygen feed. Oxygen addition to feedwater is initiated and pH is reduced only after feedwater cation conductivity is less than 0.15 μS/cm.

System transients and upsets inevitably cause excursions above recommend limits. Increased rates of deposition and corrosion are likely to be in proportion to the deviations. Small brief deviations may individually be of little consequence, but the extent, duration, and frequency of such deviations should be minimized. Otherwise, over a period of years the accumulative effects will be significant. Potential effects include increased deposition, pitting, pressure drop, and fatigue cracking. Particular care is required to minimize the extent and duration of chemistry deviations for cycling units where operational transients are frequent.

Steam purity

Purity or chemistry requirements for steam can be as simple as a specified maximum moisture content, or they can include maximum concentrations for a variety of chemical species. Often, for low-pressure building or process heater steam, only a maximum moisture content is specified. This may be as high as 0.5% or as low as 0.1%. Conversely, some turbine manufacturers specify steam condensate maximum cation conductivity, pH, and maximum concentrations for total dissolved solids, sodium and potassium, silica, iron, and copper. Turbine steam must generally have total dissolved solids less than 0.050 ppm, and in some cases less than 0.030 ppm. Individual species limits may be still lower. If steam is to be superheated, a maximum steam dissolved solids limit must be imposed to avoid excessive deposition and corrosion of the superheater. This limit is generally 0.100 ppm or less. Even where steam purity requirements are not imposed by the application, steam dissolved solids concentrations less than 1.0 ppm are recommended at pressures up to 600 psig (4.1 MPa), dissolved solids concentrations less than 0.5 ppm are recommended at 600 to 1000 psig (4.1 to 6.9 MPa), and dissolved solids concentrations less than 0.1 ppm are recommended above 1000 psig (6.9 MPa).

Up to 2000 psig (13.8 MPa), most non-volatile chemicals and impurities in the steam are carried by small water droplets entrained in the separated steam. Because these droplets contain dissolved solids in the same concentration as the boiler water, the amount of impurities in steam contributed by this mechanical carryover is the sum of the boiler water impurities concentration multiplied by the steam moisture content. Mechanical carryover is limited by moisture separation devices placed in the steam path, as described in Chapter 5.

High water levels in the drum and boiler water chemistries that cause foaming can cause excessive moisture carryover and therefore excessive steam impurity concentrations. *Foaming* is the formation of foam or excessive spray above the water line in the drum. Common causes of foaming are excessive solids or alkalinity, and the presence of organic matter such as oil. To keep dissolved solids below the concentration that causes foaming requires continuous or periodic blowdown of the boiler. High boiler water alkalinity increases the potential for foaming, particularly in the presence of suspended matter.

Where a chemical species is sufficiently volatile, it also carries over as a vapor in the steam. Total carryover is the sum of the mechanical and vaporous carryover. Vaporous carryover depends on solubility in steam and is different for each chemical species. For most dissolved solids found in boiler water, it is negligible by comparison to mechanical carryover at pressures less than 2000 psig (13.8 MPa). An exception is silica for which vaporous carryover can be substantial at lower pressures. Fig. 9 shows typical vaporous carryover fractions (distribution ratios) for common boiler water constituents under typical conditions over a wide range of boiler pressures. Fig. 10 shows expected total dissolved solids carryover for typical high-pressure boilers. Vaporous carryover depends on pressure and on boiler water chemistry. It is not affected by boiler design. Hence, if vaporous carryover for a species is excessive, the carryover can only be reduced by altering the boiler water chemistry. Only mechanical carryover is affected by boiler design. Non-interactive gases such as nitrogen, argon, and oxygen carry over almost entirely with the steam, having no relationship to moisture carryover.

Excessive steam impurity concentrations can also be caused by feedwater and boiler water chemistries that favor volatile species formation. Carryover of volatile silica can be problematic at pressures above 1000 psig (6.9 MPa). Fig. 11 shows boiler water silica concentration limits recommended to obtain steam silica concentrations less than 0.010 ppm at pressures up to 2900 psig (20.0 MPa) where the pH may be as low as 8.8. Vaporous silica carryover at a pH of 10.0 is 88% of that at a pH of 8.8. The vaporous silica carryover at a pH of 11.0 is 74% of that at 8.8. The only effective method for preventing excessive silica or other vaporous carryover is reduction of the boiler water concentrations. Another common source of excessive impurities in steam is inadequate attemperation spray water purity. All impurities in the spray water enter directly into the steam. Procedures for measuring steam quality and purity are discussed in Chapter 40.

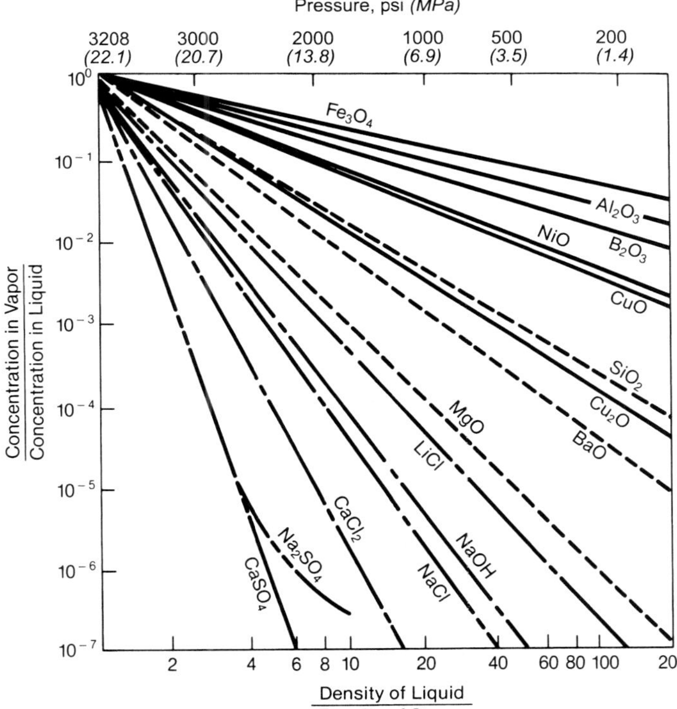

Fig. 9 Impurity carryover coefficients of salts and metal oxides in boiler water (*adapted from Reference 16*).

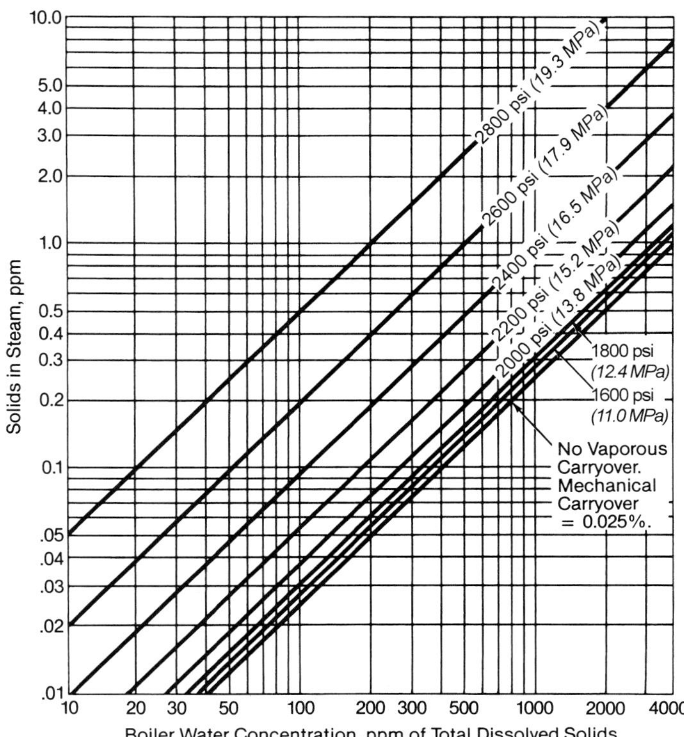

Fig. 10 Solids in steam versus dissolved solids in boiler water.

Water sampling and analysis

A key element in control of water and steam chemistry is effective sampling to obtain representative samples, prevent contamination of the samples, and prevent loss of the species to be measured.[17] References 18 and 19 provide detailed procedures. In general, sample lines should be as short as possible and made of stainless steel, except where conditions dictate otherwise. Samples should be obtained from a continuously flowing sample stream. The time between sampling and analysis should be as short as possible. Samples should be cooled quickly to 100F (38C) to avoid loss of the species of interest. Sample nozzles and lines should provide for isokinetic sample velocity and maintain constant high water velocities [minimum of 6 ft/s (1.8 m/s)] to avoid loss of materials. Sample points should be at least 10 diameters downstream of the last bend or flow disturbance.

Guidelines and techniques for chemical analysis of grab samples are listed in Table 4. The detailed methods are readily available from the American Society for Testing and Materials (ASTM) in Philadelphia, Pennsylvania, U.S. and the American Society of Mechanical Engineers (ASME) in New York, New York, U.S.

Wherever practical, on-line monitoring should be considered as an alternative to grab samples. This gives real-time data, enables trends to be followed, and provides historical data. However, on-line monitors require calibration, maintenance, and checks with grab samples or on-line synthesized standard samples to ensure reliability. Table 5 lists important on-line monitoring measurements and references to specific methods. In addition to the measurements listed, in-

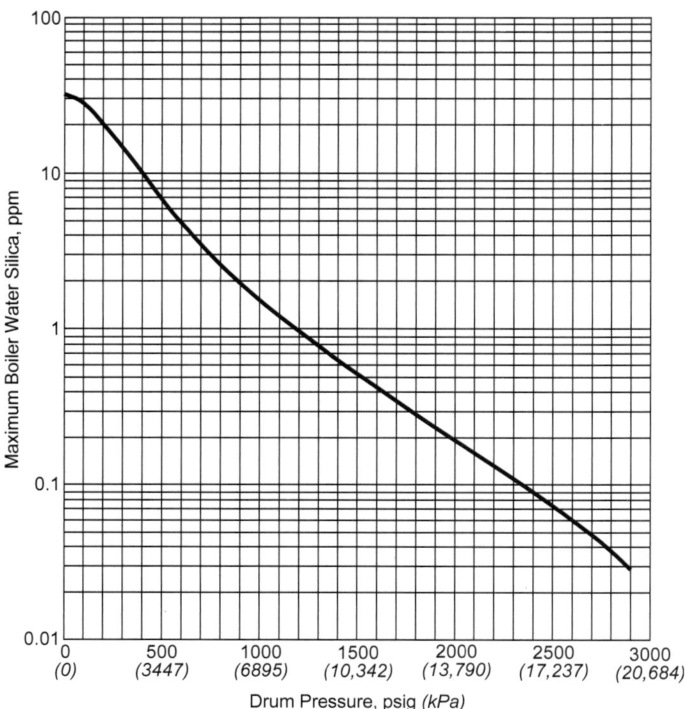

Fig. 11 Boiler water silica concentration limit, where maximum steam silica is 0.010 ppm and boiler water pH is greater than 8.8.

strumentation is commercially available to monitor chloride, dissolved oxygen, dissolved hydrogen, silica, phosphate, ammonia and hydrazine.

Adequate water chemistry control depends on the ability of boiler operators to consistently measure the specified parameters. Hence, formal quality assurance programs should be used to quantify and track the precision and bias of measurements. Detailed procedures should be in place to cover laboratory structure, training, standardization, calibration, sample collection/storage/analysis, reporting, maintenance records, and corrective action procedures. Further discussion is provided in Reference 20.

Common fluid-side corrosion problems

Water and steam react with most metals to form oxides or hydroxides. Formation of a protective oxide layer such as magnetite (Fe_3O_4) on the metal surface causes reaction rates to slow with time. Boiler cycle water treatment programs are designed to maintain such protective oxide films on internal surfaces and thus prevent corrosion in boilers and other cycle components. With adequate control of water and steam chemistry, internal corrosion of boiler circuitry can be minimized. Yet, chemistry upsets (transient losses of control) do occur. Vigilant monitoring of system chemistry permits quick detection of upsets and quick remedial action to prevent boiler damage. Where these measures fail and corrosion occurs, good monitoring and documentation of system chemistry can facilitate identification of the root cause, and identification of the cause can be an essential step toward avoiding further corrosion. Where corrosion occurs and the origin is unknown, the documented water chemistry, location of

Table 4
Guidelines for Measurements on Grab Samples

Measurement	Technique(s)	Reference/ Comment (Notes 1 and 2)
pH	Electrometric	ASTM D 1293 Method A
Conductivity	Dip or flow type conductivity cells energized with alternating current at a constant frequency (Wheatstone Bridge)	ASTM D 1125 Methods A or B
Dissolved oxygen	Colorimetric or titrimetric	ASTM D 888 Methods A, B, C
Suspended iron oxides	Membrane comparison charts	ASME PTC 31 Ion exchange equipment
Iron	Photometric (bathophenanthroline) or atomic absorption (graphite furnace)	ASTM D 1068 Method C, D
Copper	Atomic absorption (graphite furnace)	ASTM D 1688 Method C
Sodium	Atomic absorption or flame photometry	ASTM D 4191 or ASTM D 1428
Silica	Colorimetric or atomic absorption	ASTM D 859 or ASTM D 4517
Phosphate	Ion chromatography or photometric	ASTM D 4327 or ASTM D 515 Method A
Ammonia	Colorimetric (nesslerization) or ion-selective electrode	ASTM D 1426 Method A or B
Hydrazine	Colorimetric	ASTM D 1385
Chloride	Colorimetric, ion-selective electrode or ion chromatography	ASTM D 512 or ASTM D 4327
Sulfate	Turbidimetric or ion chromatography	ASTM D 516 or ASTM D 4327
Calcium and magnesium	Atomic absorption; gravimetric or titrimetric	ASTM D 511 or ASTM D 1126
Fluoride	Ion-selective electrode or ion chromatography	ASTM D 1179 or ASTM D 4327
Morpholine	Colorimetric	ASTM D 1942
Alkalinity	Color-change titration	ASTM D 1067 Method B
Hydroxide ion in water	Titrimetric	ASTM D 514
Total organic carbon	Instrumental (oxidation and infrared detection)	ASTM D 4779

Notes:
1. ASME PTC refers to Performance Test Codes of the American Society of Mechanical Engineers, New York, New York.
2. ASTM refers to testing procedures of the American Society for Testing and Materials, Philadelphia, Pennsylvania.

Table 5
On-line Monitoring Measurements

Measurement	Technique(s)	Reference
pH	Electrometric	ASTM D 5128
Conductivities (general, cation and degassed)	Electrical conductivity measurement before and after hydrogen cation exchanges and at atmospheric boiling water after acidic gas removal	ASTM D 4519
Sodium	Selective ion electrode flame photometry	ASTM D 2791
Total organic carbon	Instrumental (oxidation and measurement of carbon dioxide)	ASTM D 5173

the corrosion, appearance of the corrosion, and chemistry of localized deposits and corrosion products often suggest the cause. Common causes are flow accelerated corrosion, oxygen pitting, chelant corrosion, caustic corrosion, acid corrosion, organic corrosion, acid phosphate corrosion, hydrogen damage, and corrosion assisted cracking. Figs. 12 and 13 show typical locations of common fluid-side corrosion problems. Further discussion of corrosion and failure mechanisms is provided in References 21, 22, 23, and 24. For EPRI members, Boiler Tube Failures: Theory and Practice[25] provides an especially thorough description of utility boiler corrosion problems, causes, and remedial measures.

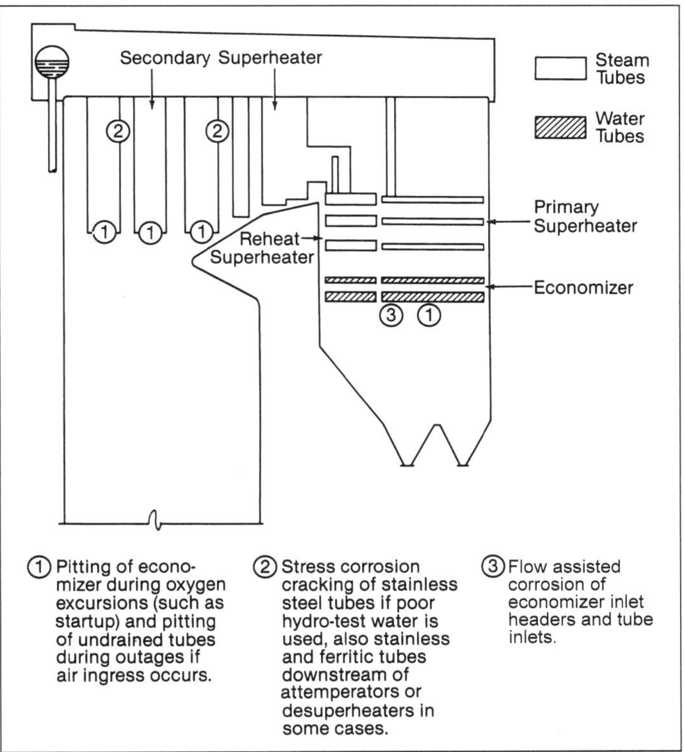

Fig. 13 Boiler convection pass showing typical locations of various types of water-side corrosion.

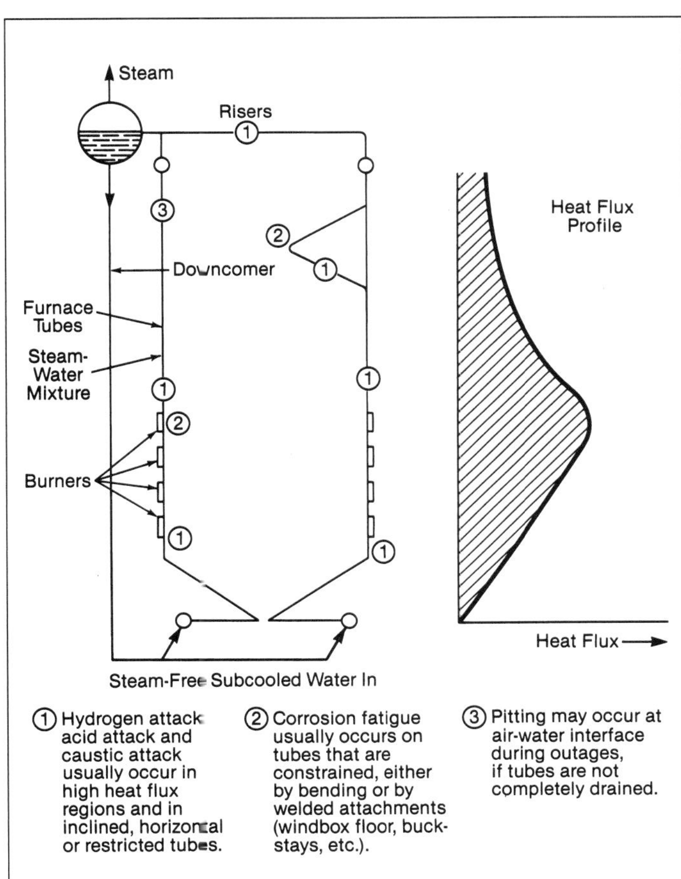

Fig. 12 Typical locations of various types of water-side corrosion in a boiler furnace water circuit.

One distinguishing feature of corrosion is its appearance. Metal loss may be uniform so the surface appears smooth. Conversely, the surface may be gouged, scalloped, or pitted. Other forms of corrosion are microscopic in breadth, and subsurface, so they are not initially discernible. Subsurface forms of corrosion include intergranular corrosion, corrosion fatigue, stress corrosion cracking, and hydrogen damage. Such corrosion can occur alone or in combination with surface wastage. In the absence of component failure, detection of subsurface corrosion often requires ultrasonic, dye penetrant, or magnetic particle inspection (Chapter 45). These forms of corrosion are best diagnosed with destructive cross-section metallography.

Another distinguishing feature is the chemical composition of associated surface deposits and corrosion products. Deposits may contain residual corrosives such as caustic or acid. Magnesium hydroxide in deposits can suggest the presence of an acid-forming precipitation process. Sodium ferrate (Na_2FeO_4) indicates caustic conditions. Sodium iron phosphate indicates acid phosphate wastage. Organic deposits suggest corrosion by organics, and excessive amounts of ferric oxide or hydroxide with pitting suggest oxygen attack.

Flow accelerated corrosion is the localized dissolution of feedwater piping in areas of flow impingement. It occurs where metal dissolution dominates over protective oxide scale formation. For example, localized conditions are sufficiently oxidizing to form soluble Fe^{++} ions but not sufficiently oxidizing to form Fe^{+++} ions needed for protective oxide formation. Conditions known to accelerate thinning include: flow impingement on pipe walls, low pH, excessive oxygen scavenger concentrations, temperatures in the range of 250 to 400F/121 to 204C (although thinning can occur at any feedwater temperature), chemicals (such as chelants) that increase iron solubility, and thermal degradation of organic chemicals. Thinned areas often have a scalloped or pitted appearance. Failures, such as that shown in Fig. 14, can occur unexpectedly and close to work areas and walkways. To assure continued integrity of boiler feedwater piping, it must be periodically inspected for internal corrosion and wall thinning. Any thinned areas must be identified and replaced before they become a safety hazard. The affected piping should be replaced with low-alloy chromium-bearing steel piping, and the water chemistry control should be appropriately altered.

Oxygen pitting and corrosion during boiler operation largely occur in pre-boiler feedwater heaters and economizers where oxygen from poorly deaerated feedwater is consumed by corrosion before it reaches the boiler. A typical area of oxygen pitting is shown in Fig. 15. Oxygen pitting within boilers occurs when poorly deaerated water is used for startup or for accelerated cooling of a boiler. It also occurs in feedwater piping, drums, and downcomers in some low pressure boilers which have no feedwater heaters or economizer. Because increasing scavenger concentrations to eliminate residual traces of oxygen can aggravate flow accelerated corrosion, care must be taken to distinguish between oxygen pitting and flow accelerated corrosion which generally occurs only where all traces of oxygen have been eliminated.

Fig. 14 Rupture of 6 in. (152 mm) feedwater pipe in an area thinned by internal corrosion.

Chelant corrosion occurs where appropriate feedwater and boiler water chemistries for chelant treatment are not maintained. Potentially corrosive conditions include excessive concentration of free chelant and low pH. (See prior discussion of boiler water treatment with dispersants, polymers, and chelants.) Especially susceptible surfaces include flow impingement areas of feedwater piping, riser tubes, and cyclone steam/water separators. Affected areas are often dark colored and have the appearance of uniform thinning or of flow accelerated corrosion.

Corrosion fatigue is cracking well below the yield strength of a material by the combined action of corrosion and alternating stresses. Cyclic stress may be of mechanical or thermal origin (Chapter 8). In boilers, corrosion fatigue is most common in water-wetted surfaces where there is a mechanical constraint on the tubing. For example, corrosion fatigue occurs in furnace wall tubes adjacent to windbox, buckstay, and other welded attachments. Failures are thick lipped. On examination of the internal tube surface, multiple initiation sites are evident. Cracking is transgranular. Environmental conditions facilitate fatigue cracking where it would not otherwise occur in a benign environment. Water chemistry factors that facilitate cracking include dissolved oxygen and low pH transients associated with, for example, cyclic operation, condenser leaks, and phosphate hideout and hideout-return.

Acid phosphate corrosion occurs on the inner steam-forming side of boiler tubes by reaction of the steel with phosphate to form maricite ($NaFePO_4$). Fig. 16 shows ribbed tubing that has suffered this type of wastage. The affected surface has a gouged appearance with maricite and magnetite deposits. Acid phosphate corrosion occurs where the boiler water effective sodium-to-phosphate ratio is less than 2.8, although ratios as low as 2.6 may be tolerated at lower pressures. Though not always apparent, common signs of acid phosphate corrosion include difficulty maintaining target phosphate concentrations, phosphate hideout and pH increase with increasing boiler load or pressure, phosphate hideout return and decreasing pH with decreasing load or pressure, and periods of high iron concentration in boiler water. The potential for acid phosphate corrosion increases with increasing internal deposit loading, decreasing effective sodium-to-phosphate molar ratio below 2.8, increasing phosphate concentration, inclusion of acid phosphates (disodium and especially monosodium phosphate) in phosphate feed solution, and increasing boiler pressure. To avoid acid phosphate corrosion, operators should monitor boiler water conditions closely, assure accuracy of pH and phosphate measurements, assure purity and reliability of chemical feed solutions, assure that target boiler water chemistry parameters are appropriate and are attained in practice, and watch for aforementioned signs of acid phosphate corrosion.

Under-deposit acid corrosion and *hydrogen damage* occur where boiler water acidifies as it concentrates beneath deposits on steam generating surfaces. Hydrogen from acid corrosion diffuses into the steel where

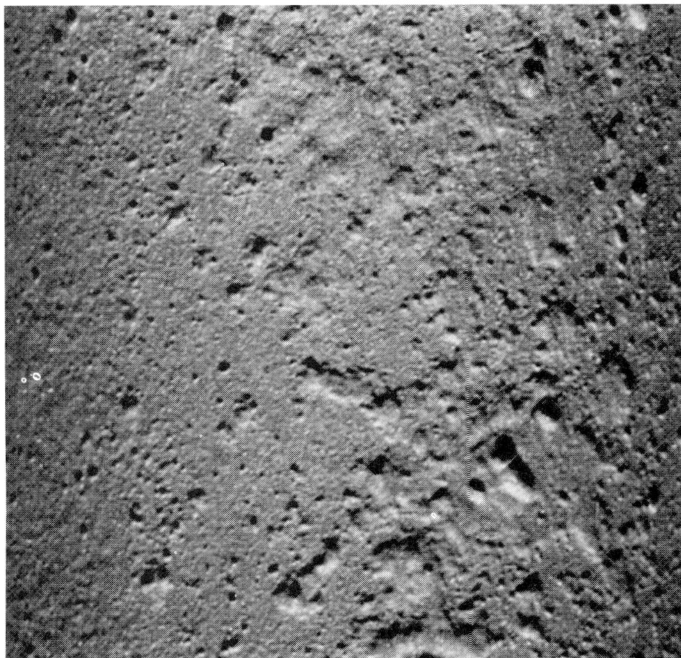

Fig. 15 Oxygen pitting of economizer feedwater inlet.

it reacts with carbon to form methane as depicted in Fig. 17. The resultant decarburization and methane formation weakens the steel and creates microfissures. Thick lipped failures like that shown in Fig. 18 occur when the degraded steel no longer has sufficient strength to hold the internal tube pressure. Signs of hydrogen damage include under deposit corrosion, thick lipped failure, and steel decarburization and microfissures. The corrosion product from acid corrosion is mostly magnetite. Affected tubing, which may extend far beyond the failure, must be replaced. The boiler must be chemically cleaned to remove internal tube deposits, and boiler water chemistry must be altered or better controlled to prevent acid-formation as the water concentrates. Operators should reduce acid-forming impurities by improving makeup water, reducing condenser leakage, or adding condensate polishing. For drum boilers, operators should use phosphate treatment with an effective sodium-to-phosphate molar ratio of 2.8 or greater.

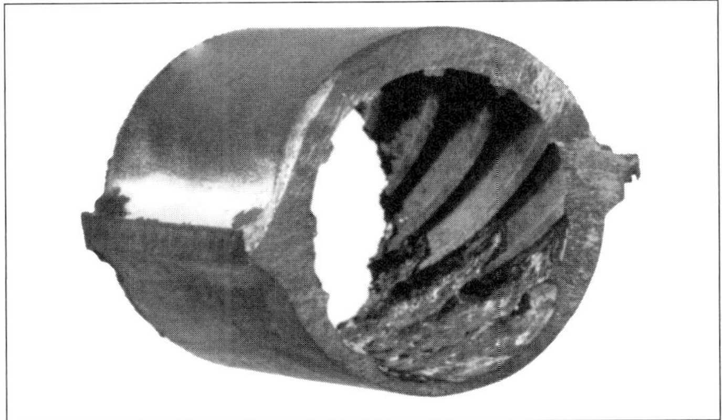

Fig. 16 Acid phosphate corrosion of ribbed tubing.

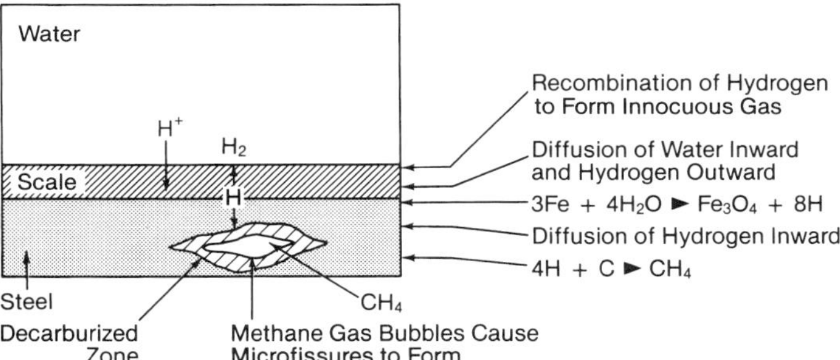

Fig. 17 Schematic of hydrogen attack, showing steps that occur and the final result. Hydrogen attack can occur in both carbon and low alloy steels in acidic or hydrogen environments.

Caustic corrosion, gouging and grooving occur where boiler water leaves a caustic residue as it evaporates. In vertical furnace wall tubes, this occurs beneath deposits that facilitate a high degree of concentration and the corroded surface has a gouged appearance as shown in Fig. 19. In inclined tubes where the heat flux is directed through the upper half of the tube, caustic concentrates by evaporation of boiler water in the steam space on the upper tube surface. Resulting corrosion is in the form of a wide smooth groove with the groove generally free of deposits and centered on the crown of the tube. Deposits associated with caustic gouging often include Na_2FeO_4. To prevent reoccurrence of caustic gouging, operators should prevent accumulation of excessive deposits and control water chemistry so boiler water does not form caustic as it concentrates. The latter can generally be achieved by assuring appropriate feedwater chemistry with coordinated phosphate boiler water treatment, taking care to control the effective sodium-to-phosphate molar ratio as appropriate for the specific boiler and the specific chemical and operating conditions. In some instances, where caustic grooving along the top of a sloped tube is associated with steam/water separation, such separation can be avoided by use of ribbed tubes which cause swirling motion that keeps water on the tube wall.

Caustic cracking can occur where caustic concentrates in contact with steel that is highly stressed, to or beyond the steel's yield strength. Caustic cracking is rare in boilers with all welded connections. This

Fig. 18 Brittle tube failure in hydrogen damaged area.

Fig. 19 Caustic gouging initiated along weld backing ring.

generally occurs in boilers using a high alkalinity caustic boiler water treatment, and it is normally associated with unwelded rolled joints and welds that are not stress relieved. On metallographic examination, caustic cracking is intergranular and has the branched appearance characteristic of stress corrosion cracking as illustrated in Fig. 20. It can generally be avoided by use of coordinated phosphate treatment. Where a high alkalinity caustic phosphate boiler water treatment is used for low pressure boilers, nitrate is often added to inhibit caustic cracking.

Overheat failures like that shown in Fig. 21 occur where deposits impede internal heat transfer to the extent that a tube no longer retains adequate strength and bulges or ruptures. Internal tube deposits generally cause moderate overheating for extended periods of time, causing *long-term overheat failures. Short-term overheat failures* generally occur only when there is gross interruption of internal flow to cool the tube, or grossly excessive heat input.

Out-of-service corrosion is predominantly oxygen pitting. Pitting attributed to out-of-service corrosion occurs during outages but also as aerated water is heated when boilers return to service. Especially common locations include the waterline in steam drums, areas where water stands along the bottom of horizontal pipe and tube runs, and lower bends of pendant superheaters and reheaters. Pinhole failures are more common in thinner walled reheater and economizer tubing. Such corrosion can be minimized by following appropriate layup procedures for boiler outages and by improving oxygen control during boiler startups.

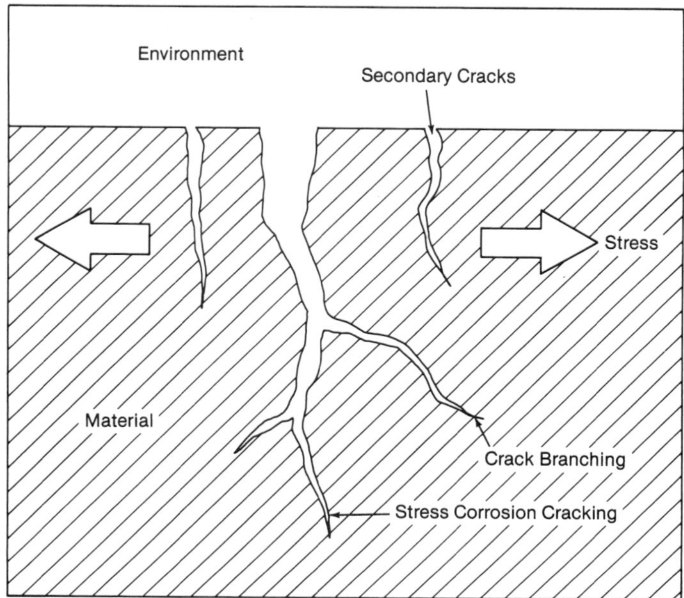

Fig. 20 Schematic of stress corrosion cracking.

Pre-operational cleaning

In general, all new boiler systems receive an alkaline boilout, i.e. hot circulation of an alkaline mixture with intermittent blowdown and final draining of the unit. Many systems also receive a pre-operational chemical cleaning. The superheater and reheater should receive a conventional steam blow (a period of high velocity steam flow which carries debris from the system). Chemical cleaning of superheater and reheat surfaces is effective in reducing the number of steam blows to obtain clean surfaces, but is not required to obtain a clean superheater and reheater.

Alkaline boilout

All new boilers should be flushed and given an alkaline boilout to remove debris, oil, grease and paint. This can be accomplished with a combination of trisodium phosphate (Na_3PO_4) and disodium phosphate (Na_2HPO_4), with a small amount of surfactant added as a wetting agent. The use of caustic NaOH and/or soda ash (Na_2CO_3) is not recommended. If either is used, special precautions are required to protect boiler components.

Chemical cleaning

After boilout and flushing are completed, corrosion products may remain in the feedwater system and boiler in the form of iron oxide and mill scale. Chemical cleaning should be delayed until full load operation has carried the loose scale and oxides from the feedwater system to the boiler. Some exceptions are units that incorporate a full flow condensate polishing system and boilers whose pre-boiler system has been chemically cleaned. In general, these units can be chemically cleaned immediately following pre-operational boilout.

Different solvents and cleaning processes are used for pre-operational chemical cleaning, usually determined by boiler type, metallic makeup of boiler components, and environmental concerns or restrictions. The four most frequently used are: 1) inhibited 5% hydrochloric acid with 0.25% ammonium bifluoride, 2) 2%

hydroxyacetic/1% formic acids with 0.25% ammonium bifluoride and a corrosion inhibitor, 3) 3% inhibited ammonium salts of ethylene-diaminetetraacetic acid (EDTA), and 4) 3% inhibited ammoniated citric acid.

Steam line blowing

The steam line blow procedure depends on unit design. Temporary piping to the atmosphere is required with all procedures. This piping must be anchored to resist high nozzle reaction force.

All normal startup precautions should be observed for steam line blowing. The unit should be filled with treated demineralized water. Sufficient feedwater pump capacity and condensate storage must be available to replace the water lost during the blowing period.

Numerous short blows are most effective. The color of the steam discharged to the atmosphere provides an indication as to the quantity of debris being removed from the piping. Coupons (targets) of polished steel attached to the end of the exhaust piping are typically used as final indicators.

Periodic chemical cleaning

Cleaning frequency

Internal surfaces of boiler water-side components (including supply tubes, headers and drums) accumulate deposits even though standard water treatment practices are followed. These deposits are generally classified as hardness-type scales or soft, porous-type deposits.

To determine the need for cleaning, tube samples containing internal deposits should be removed from high heat input zones of the furnace and/or areas where deposition problems have occurred. The deposit weight is first determined by visually selecting a heavily deposited section. After sectioning the tube (hot and cold sides), the water-formed deposit is removed by scraping from a measured area. The weight of the dry material is reported as weight per unit area: either grams of deposit per square foot of tube surface or mg/cm^2. Procedures for mechanical and chemical methods of deposit removal are provided in ASTM D3483.[26] General guidelines for determining when a boiler should be chemically cleaned are shown in Table 6. The deposit weights shown are based on the mechanical scraping method. This removes the porous deposit of external origin and most of the dense inner oxide scale. Values are slightly lower than those obtained from the chemical dissolution method.

Because of the corrosive nature of the fuel and its combustion products, furnace tubes in Kraft recovery and refuse-fired boilers are particularly susceptible to gas-side corrosion which can be aggravated by relatively modest elevated tube metal temperatures. (See Chapters 28 and 29.) Through-wall failures due to external metal corrosion can occur in these tubes at water-side deposit weights much less than 40 g/ft^2 (43 mg/cm^2). In addition, for Kraft recovery boilers there are significant safety concerns for water leakage in the lower furnace. (See Chapters 28 and 43.) For these units, a more conservative cleaning criterion is recommended for all operating pressures.

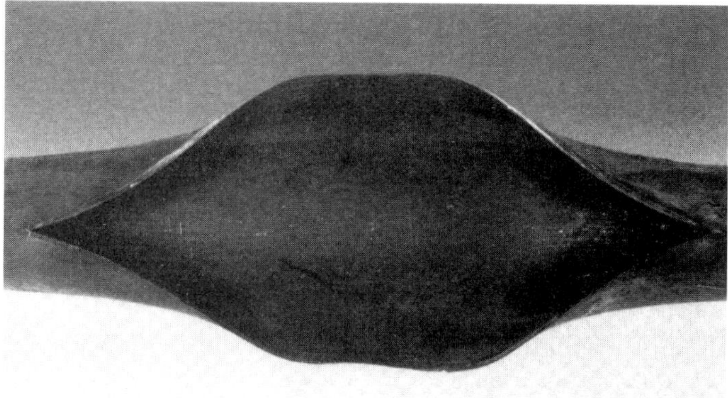

Fig. 21 Short-term overheat thin edged failure.

Chordal thermocouples

The chordal thermocouple (see Chapter 40) can be an effective diagnostic tool for evaluating deposits on operating boilers. Properly located thermocouples can indicate a tube metal temperature increase caused by excess internal deposits, and can alert the operator to conditions that may cause tube failures. Thermocouples are often located in furnace wall tubes adjacent to the combustion zone where the heat input is highest and the external tube temperatures are also high. (See Fig. 22.)

Deposition inside tubes can be detected by instrumenting key furnace tubes with chordal thermocouples. These thermocouples compare the surface temperature of the tube exposed to the combustion process with the temperature of saturated water. As deposits grow, they insulate the tube from the cooling water and cause tube metal temperature increases.

Beginning with a clean, deposit-free boiler, the instrumented tubes are monitored to establish the temperature differential at two or three boiler ratings; this establishes a base curve. At maximum load, with clean tubes, the surface thermocouple typically indicates metal temperatures 25 to 40F (14 to 22C) above saturation in low duty units and 80 to 100F (44 to 56C) in high duty units as shown in Fig. 23. The temperature variation for a typical clean instrumented tube is dependent upon the tube's location in the furnace, tube thickness, inside fluid pressure, and the depth of the

Table 6
Guidelines for Chemical Cleaning

Unit Operating Pressure, psig (MPa)	Water-side Deposit Weight* (g/ft^2)
Below 1000 (6.9)	20 to 40
1000 to 2000 (6.9 to 13.8) including all Kraft recovery and refuse-fired boilers	12 to 20
Above 2000 (13.8)	10 to 12

* Deposit removed from hot or furnace side of tube using the mechanical scraping method. (1 g/ft^2 = 1.07 mg/cm^2)

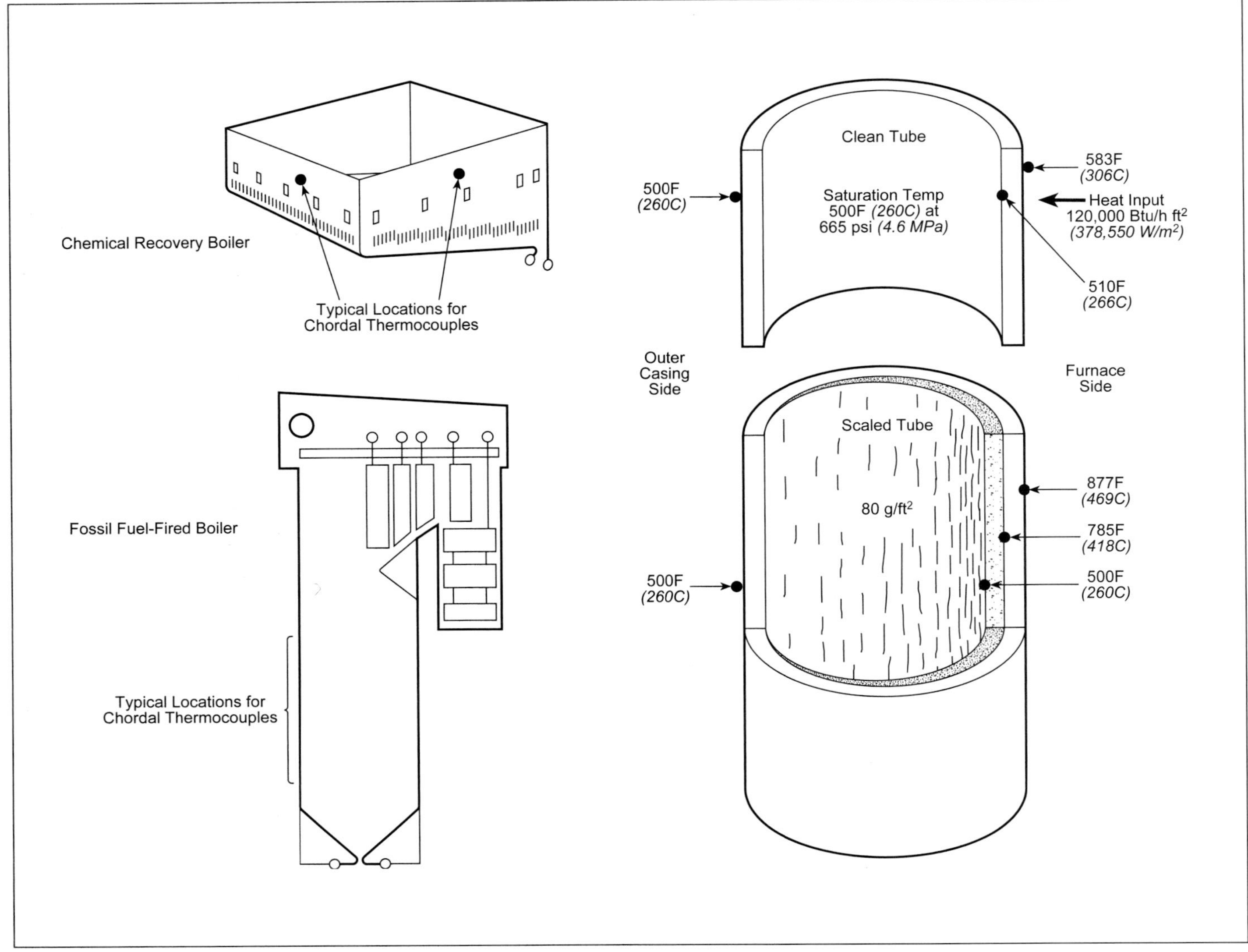

Fig. 22 Typical locations of chordal thermocouples.

surface thermocouple. Internal scale buildup is detected by an increase in temperature differential above the base curve. Chemical cleaning should normally be considered if the temperature differential at maximum boiler load reaches 100F (56C).

Initially, readings should be taken weekly, preferably using the same equipment and procedure as used for establishing the base curve. Under upset conditions, when deposits form rapidly, the checking frequency should be increased.

Chemical cleaning procedures and methods

In general, four steps are required in a complete chemical cleaning process:

1. The internal heating surfaces are washed with a solvent containing an inhibitor to dissolve or disintegrate the deposits.
2. Clean water is used to flush out loose deposits, solvent adhering to the surface, and soluble iron salts. Corrosive or explosive gases that may have formed in the unit are also displaced.
3. The unit is treated to neutralize and *passivate* the heating surfaces. This treatment produces a passive surface, i.e., it forms a very thin protective film on freshly cleaned ferrous surfaces.
4. The unit is flushed with clean water to remove any remaining loose deposits.

The two generally accepted chemical cleaning methods are: 1) continuous circulation of the solvent (Fig. 24), and 2) filling the unit with solvent, allowing it to soak, then flushing the unit (Fig. 25).

Circulation cleaning method

In the circulation (dynamic) cleaning method (Fig. 24), after filling the unit with demineralized water, the water is circulated and heated to the required cleaning temperature. At this time, the selected solvent is injected into the circulating water and recirculated until the cleaning is completed. Samples of the return solvent are periodically tested. Cleaning is considered complete when the acid strength and the iron content of the returned solvent reach equilibrium (Fig. 26), indicating that no further reaction with the deposits is taking place. In the circulation method, additional solvent can be injected if the dissipation of the solvent concentration drops below the recommended minimum concentration.

The circulation method is particularly suitable for cleaning once-through boilers, superheaters, and economizers with positive liquid flow paths to assure circulation of the solvent through all parts of the unit. Complete cleaning can not be assured by this method unless the solvent reaches and passes through every circuit of the unit.

Soaking method

The soaking (static) cleaning method (Fig. 25) involves preheating the unit to a specified temperature, filling the unit with the hot solvent, then allowing the unit to soak for a period of time, depending on deposit conditions. To assure complete deposit removal, the acid strength of the solvent must be somewhat greater than that required by the actual conditions; unlike the circulation method, control testing during the course of the cleaning is not conclusive, and samples of solvent drawn from convenient locations may not truly represent conditions in all parts of the unit.

The soaking method is preferable for cleaning units where definite liquid distribution to all circuits (by the circulation method) is not possible without the use of many chemical inlet connections. The soaking method is also preferred when deposits are extremely heavy, or if circulation through all circuits at an appreciable rate can not be assured without an impractically-sized circulating pump. These conditions may exist in large natural circulation units that have complex furnace wall cooling systems.

Advantages of this method are simplicity of piping connections and assurance that all parts are reached by a solvent of adequate acid strength.

Solvents

Many acids and alkaline compounds have been evaluated for removing boiler deposits. Hydrochloric acid (HCl) is the most practical cleaning solvent when using the soaking method on natural circulation boilers. Chelates and other acids have also been used.

An organic acid mixture such as hydroxyacetic-formic (HAF) is the safest chemical solvent when applying the circulation cleaning method to once-through boilers. These acids decompose into gases in the event of incomplete flushing.

For certain deposits, the solvent may require additional reagents, such as ammonium bifluoride, to promote deposit penetration. Alloy steel pressure parts, particularly those high in chromium, should generally not be cleaned with certain acid solvents. A general guideline for solvent selection can be found in Table 7.

Prior to chemically cleaning, it is strongly recommended that a representative tube section be removed and subjected to a laboratory cleaning test to determine and verify the proper solvent chemical, and concentrations of that solvent.

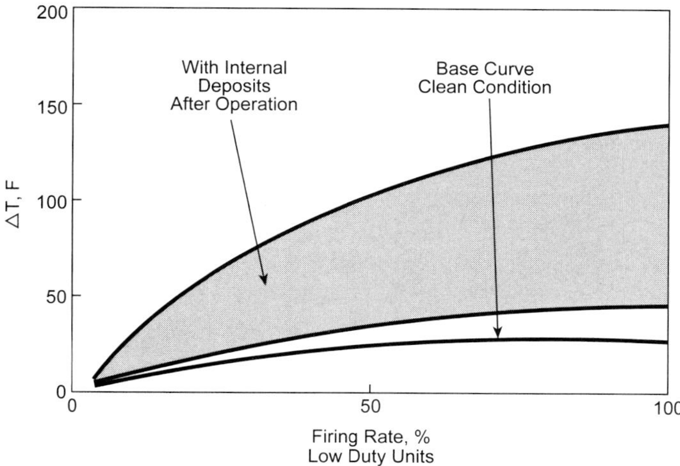

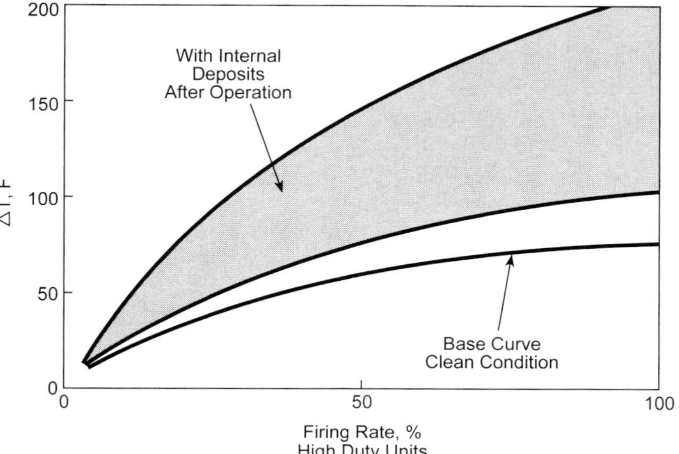

Fig. 23 Temperature difference between surface and saturation thermocouples.

Deposits

Scale deposits formed on the internal heating surfaces of a boiler generally come from the water. Most of the constituents belong to one or more of the following groups: iron oxides, metallic copper, carbonates, phosphates, calcium and magnesium sulfates, silica, and silicates. The deposits may also contain various amounts of oil.

Pre-cleaning procedures include analysis of the deposit and tests to determine solvent strength and contact time and temperature. The deposit analyses should include a deposit weight in grams per square foot (or milligrams per square centimeter) and a spectrographic analysis to detect the individual elements. X-ray diffraction identifying the major crystalline constituents is also used.

If the deposit analysis indicates the presence of copper (usually from corrosion of pre-boiler equipment, such as feedwater heaters and condensers), one of three procedures is commonly used: 1) a copper complexing agent is added directly to the acid solvent, 2) a separate cleaning step, featuring a copper solvent, is used followed by an acid solvent, and 3) a chelant-based solvent at high temperature is used to remove iron, followed by addition of an oxidizing agent at re-

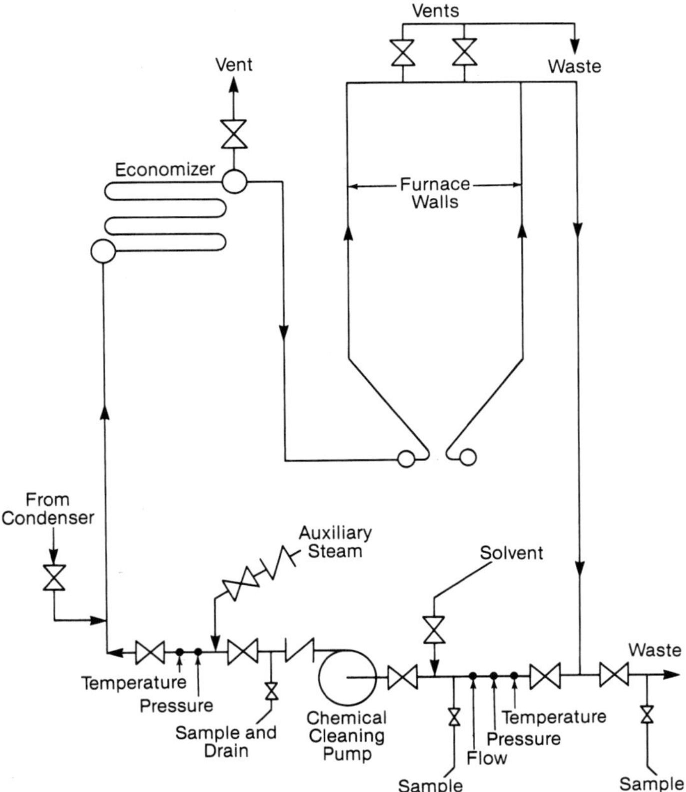

Fig. 24 Chemical cleaning by the circulation method – simplified arrangement of connections for once-through boilers.

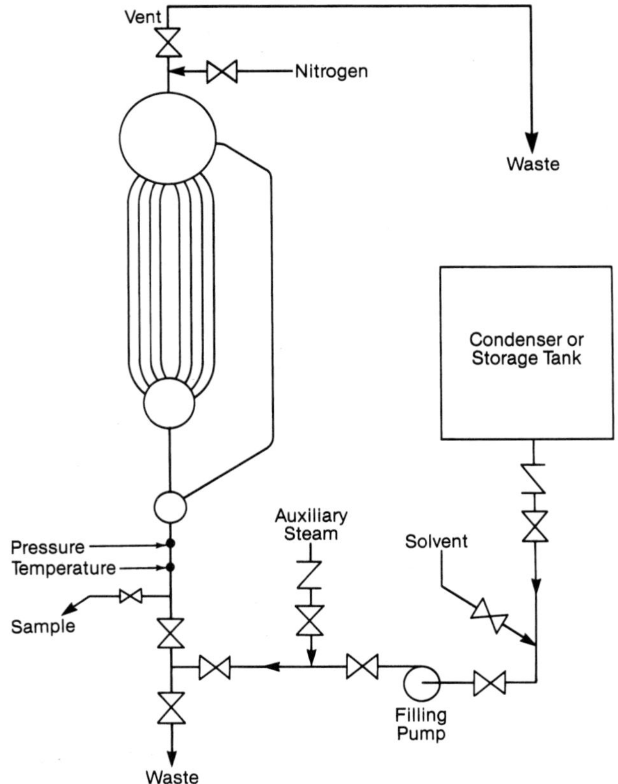

Fig. 25 Chemical cleaning by the soaking method – simplified arrangement of connections for drum-type boilers.

duced temperature for copper removal. The decision to use one of these methods depends on the estimated quantity of copper present in the deposit.

When deposits are dissolved and disintegrated, oil is removed simultaneously, provided it is present only in small amounts. For higher percentages of oil contamination, a wetting agent or surfactant may be added to the solvent to promote deposit penetration. If the deposit is predominantly oil or grease, boiling out with alkaline compounds must precede the acid cleaning.

Inhibitors

The following equations represent the reactions of hydrochloric acid with constituents of boiler deposits:

$$Fe_3O_4 + 8HCl \rightarrow 2FeCl_3 + FeCl_2 + 4H_2O \quad \textbf{(2)}$$

$$CaCO_3 + 2HCl \rightarrow CaCl_2 + H_2O + CO_2 \quad \textbf{(3)}$$

At the same time, however, the acid can also react with and thin the boiler metal, as represented by the equation:

$$Fe + 2HCl \rightarrow FeCl_2 + H_2 \quad \textbf{(4)}$$

unless means are provided to slow this reaction without affecting the deposit removal. A number of excellent commercial inhibitors are available to perform this function. The aggressiveness of acids toward boiler deposits and steel increases rapidly with temperature. However, the inhibitor effectiveness decreases as the temperature rises and, at a certain temperature, the inhibitor may decompose. Additionally, all inhibitors are not effective with all acids.

Determination of solvent conditions

Deposit samples The preferred type of deposit sample is a small section of tube with the adhering deposit, though sometimes tube samples are not easily obtained. Selection of the solvent system is made from the deposit analyses. After selection of the solvent system, it is necessary to determine the strength of the solvent, the solvent temperature, and the length of time required for the cleaning process.

Solvent strength The solvent strength should be proportional to the amount of deposit. Commonly used formulations are:

1. Natural circulation boilers (soaking method)
   (a) pre-operational – inhibited 5% hydrochloric acid + 0.25% ammonium bifluoride
   (b) operational – inhibited 5 to 7.5% hydrochloric acid and ammonium bifluoride based on deposit analysis
2 Once-through boilers (circulation method)
   (a) pre-operational – inhibited 2% hydroxyacetic-1% formic acids + 0.25% ammonium bifluoride
   (b) operational – inhibited 4% hydroxyacetic-2% formic acids + ammonium bifluoride based on deposit analysis

Solvent temperature The temperature of the solvent should be as high as possible without seriously reducing the effectiveness of the inhibitor. An inhibitor test should be performed prior to any chemical cleaning to determine the maximum permissible temperature for a given solvent.

When using hydrochloric acid, commercial inhibitors generally lose their effectiveness above 170F (77C) and corrosion rate increases rapidly. Therefore, the temperature of the solvent, as fed to the unit, should be 160 to 170F (71 to 77C). In using the circulation method with a hydroxyacetic-formic acid mixture, a temperature of 200F (93C) is necessary for adequate cleaning. Chelate-based solvents are generally applied at higher temperatures (about 275F/135C). In these cases, the boiler is fired to a specific temperature. The chelate chemicals are introduced and the boiler temperature is cycled by alternately firing and cooling to predetermined limits.

Steam must be supplied from an auxiliary source to heat the acid as it is fed to the unit. When using the circulation method, steam is also used to heat the circulating water to the predetermined and desired temperature before injecting the acid solution. Heat should be added by direct contact or closed cycle heat exchangers. The temperature of the solvent should never be raised by firing the unit when using an acid solvent.

Cleaning time When cleaning by the circulation method, process completion is determined by analyzing samples of the return solvent for iron concentration and acid strength. (See Fig. 26.) However, acid circulation for a minimum of six hours is recommended.

In using the soaking method, the cleaning time should be predetermined but is generally between six to eight hours in duration.

Preparation for cleaning

Heat transfer equipment All parts not to be cleaned should be isolated from the rest of the unit. To exclude acid, appropriate valves should be closed and checked for leaks. Where arrangements permit, parts of the unit such as the superheater can be isolated by filling with demineralized water. Temporary piping should be installed to flush dead legs after cleaning. In addition to filling the superheater with demineralized water, once-through type units should be pressurized with a pump or nitrogen. The pressure should exceed the chemical cleaning pump head.

Bronze or brass parts should be removed or temporarily replaced with steel. All valves should be steel or steel alloy. Galvanized piping or fittings should not be used. Gauge and meter connections should be closed or removed.

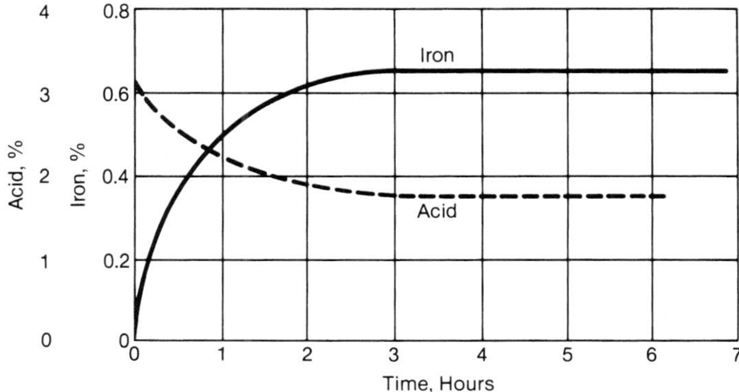

Fig. 26 Solvent conditions during cleaning by the circulation method.

Table 7
Comparative Cleaning Effectiveness

Type of Cleaning	Makeup of Deposit: Iron	Copper	Silica	Hardness (Ca/Mg)
HCl	Good	Medium	Medium	Good
HAF	Good	Poor	Medium	Medium/poor
EDTA	Good	Medium	Poor	Medium/poor
Citric	Good	Medium	Poor	Poor
Bromate	N/A	Good	N/A	N/A

All parts not otherwise protected by blanking off or by flooding with water will be exposed to the inhibited solvent. Vents to a safe discharge should be provided wherever vapors might accumulate, because acid vapors from the cleaning solution do not retain the inhibitor.

Cleaning equipment The cleaning equipment should be connected as shown in Fig. 24 if the continuous circulation method is used, or as shown in Fig. 25 if the soaking method is used. Continuous circulation requires an inlet connection to assure distribution. It also requires a return line to the chemical cleaning pump from the unit. The soaking method does not require a return line. The pump discharge should be connected to the lowermost unit inlet.

The filling or circulating pump should not be fitted with bronze or brass parts; a standby pump is recommended. A filling pump should have the capacity to deliver a volume of liquid equal to that of the vessel within two hours at 100 psi (0.7 MPa). A circulating pump should have sufficient capacity to meet recommended cleaning velocities. With modern once-through boilers, a capacity of 3600 GPM (227 l/s) at 300 psi (2.1 MPa) is common. A solvent pump, closed mixing tank and suitable thermometers, pressure gauges, and flow meters are required. An adequate supply of clean water and steam for heating the solvent should be provided. Provision should be made for adding the inhibited solvent to the suction side of the filling or recirculating pump.

Cleaning solutions Estimating the content of the vessel and adding 10% to allow for losses will determine the amount of solvent required. Sufficient commercial acid should then be obtained. An inhibitor qualified for use with the solvent also needs to be procured and added to the solvent.

Cleaning procedures

The chemical cleaning of steam generating equipment consists of a series of distinct steps which may include the following:

1. isolation of the system to be cleaned,
2. hydrostatic testing for leaks,
3. leak detection during each stage of the process,
4. back flushing of the superheater and forward flushing of the economizer,
5. preheating of the system and temperature control,

6. solvent injection/circulation (if circulation is used),
7. draining and/or displacement of the solvent,
8. neutralization of residual solvent,
9. passivation of cleaned surfaces,
10. flushing and inspection of cleaned surfaces, and
11. layup of the unit.

Every cleaning should be considered unique, and sound engineering judgment should be used throughout the process. The most important design and procedural considerations include reducing system leakage, controlling temperature, maintaining operational flexibility and redundancy, and ensuring personnel safety.

Precautions

Cleaning must not be considered a substitute for proper water treatment. Intervals between cleanings should be extended or reduced as conditions dictate. Every effort should be used to extend the time between chemical cleanings. Hazards related to chemical cleaning of power plant equipment are fairly well recognized and understood, and appropriate personnel safety steps must be instituted.[27]

Chemical cleaning of superheater, reheater and steam piping

In the past, chemical cleaning of superheaters and reheaters was not performed because it was considered unnecessary and expensive. With the use of higher steam temperatures, cleaning procedures for superheaters, reheaters and steam piping have gained importance and acceptance.

When chemically cleaning surfaces that have experienced severe high-temperature oxide exfoliation (spalling of hard oxide particles from surfaces), it is important to first remove a tube sample representing the worst condition. Oxidation progresses at about the same rate on the outside of the tubes as on the inside; exfoliation follows a similar pattern.

The tube sample should be tested in a facility capable of producing a flow rate similar to that used in the actual cleaning. This allows development of an appropriate solvent mixture.

To determine the circulating pump size and flows required, it is usually necessary to contact the boiler manufacturer.

Figs. 27 and 28 show possible superheater/reheater chemical cleaning piping schematics for drum boiler and once-through boiler systems, respectively.

If, in the case of a drum boiler, the unit is to be cleaned along with the superheater and reheater, it is usually necessary to orifice the downcomers to obtain the desired velocities through the furnace walls.

A steam blow to purge all air and to completely fill the system must precede cleaning in all systems containing pendant non-drainable surfaces. Most drainable systems also benefit from such a steam blow.

Presently, two solvent mixtures are available to clean superheater, reheater and steam piping. One is a combination of hydroxyacetic and formic acids containing ammonium bifluoride; the other is an EDTA (ethylenediaminetetraacetic acid)-based solvent.

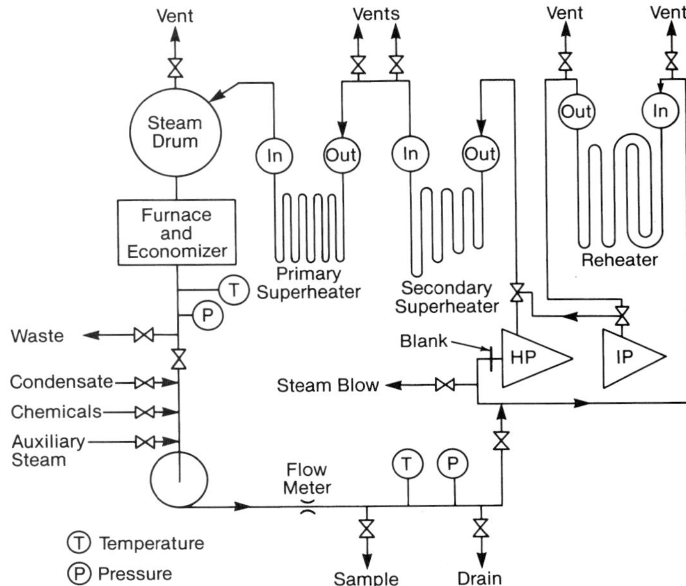

Fig. 27 Typical superheater/reheater chemical cleaning circuit for a drum-type boiler.

Solvent disposal

General considerations A boiler chemical cleaning is not complete until the resultant process waste water stream is disposed of. Selection of handling and disposal methods depends on whether the wastes are classified as hazardous or non-hazardous. Boiler chemical cleaning wastes (BCCW) are different in volume and frequency of generation and have different discharge regulations from other power plant waste streams. Of all power plant discharges, BCCWs are most likely to be classified as hazardous. Depending upon the cleaning process, the resultant BCCW may become one of the driving forces in solvent selection.

Under National Pollutant Discharge Elimination System (NPDES) requirements, boiler cleaning wastes are considered chemical metal cleaning wastes. The primary parameters of concern are iron, copper, chromium and pH. In all cases, waste management must be performed in accordance with current regulatory requirements.

Waste management options Table 8 lists the handling practices for BCCW. In *co-ponding*, the BCCW is mixed in an ash pond with other waste streams from the power plant. Acid wastes are neutralized by the alkaline ash, and the metals are precipitated as insoluble metal oxides and hydroxides, or absorbed on ash particles. Co-ponding is the least expensive and the easiest disposal option.

Incineration of organic-based cleaning wastes by direct injection into the firebox of the utility boiler is another common disposal practice. Potential emissions from the boiler must be carefully monitored to ensure regulatory compliance.

Large quantities of BCCW are often disposed of in a secure landfill. Evaporation can reduce waste volume and, thereby, reduce overall landfill disposal costs.

HCl cleaning wastes can be treated to NPDES standards using lime or caustic precipitation. It is more

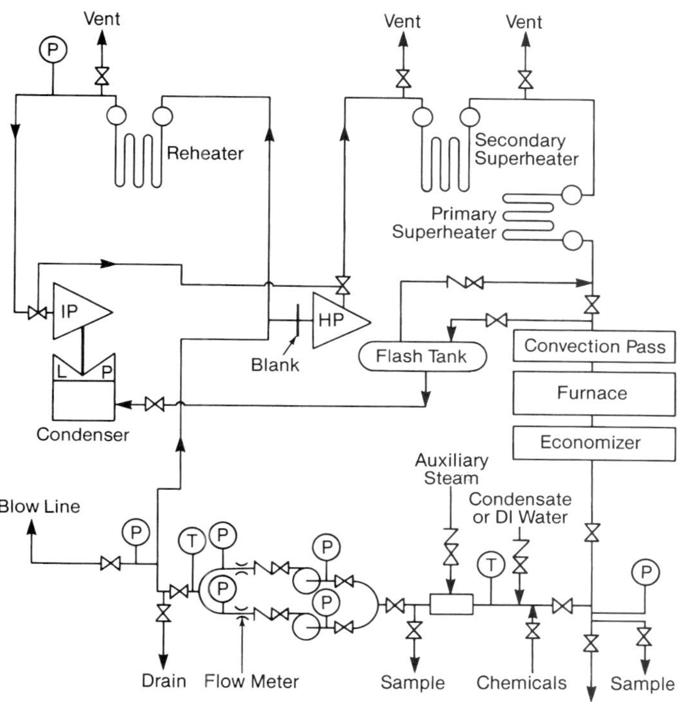

Fig. 28 Typical superheater/reheater chemical cleaning circuit for a once-through boiler.

difficult to treat the organic cleaning agents (such as EDTA) by current techniques. Treatment methods with permanganate, ultraviolet light, and hydrogen peroxide (wet oxidation) have been used with limited success. Several vendors have proprietary processes which claim to successfully treat chelated wastes.[28,29]

Table 8
Boiler Chemical Cleaning Wastes Practices/Options

Source Reduction
- Optimize cleaning frequency
- Reduce volume of cleaning solution
- Improve boiler water chemistry

Alternate Solutions
- Change the cleaning solvent

Disposal
- Evaporation
- Incineration
- Co-ponding
- Secured landfill

Treatment
- Neutralization
- Physical waste treatment
- Chemical waste treatment

Recycle and Reuse
- Recycle for metal recovery
- Reuse acid in alternate applications

Removing metal ions and reusing chemical cleaning waste are subjects receiving increased attention. As the regulatory environment continues to change, more emphasis will be placed on the treatment and reuse of BCCW.

Layup

During periods when boiler operation is interrupted, substantial pitting and general corrosion can occur within unprotected water-steam circuitry. When boilers return to operation, corrosion products migrate to high heat flux areas of the boiler or carryover to the turbine. Out-of-service corrosion can therefore impede boiler startup and lead to operational problems such as deposition, under-deposit corrosion, corrosion fatigue, and cycle efficiency loss.

Preservation methods inhibit out-of-service corrosion by eliminating or controlling moisture, oxygen, and chemical contaminants that cause corrosion. Table 9 provides a brief summary and comparison of common preservation methods.

These methods are designed to limit corrosion caused by the normal range of boiler and atmospheric contaminants. Gross contamination must be avoided and, if it occurs, the contaminants must be immediately neutralized and removed. Respective vendors should be contacted for specific recommendations for balance-of-plant equipment (turbine, condensate, feedwater, and atmospheric pollution control systems). Vendor procedures should also be followed for boiler auxiliary equipment such as pulverizer gearboxes, sootblowers, fans, and motors.

Boiler shutdown for layup

Appropriate shutdown procedures can facilitate preservation for subsequent idle periods. Reducing load dissipates fluid-side salts that concentrate on tube surfaces in high heat flux areas. As boiler load is reduced, feedwater and boiler water pH should be increased to the upper end of the target operating range, preferably 9.6 or higher. Where an oxygen scavenger/inhibitor is employed, concentration should increase to 0.050 ppm hydrazine* or equivalent. When boiler water is to remain in the boiler for a substantial cold layup period, the oxygen scavenger/inhibitor concentration in the water should be increased to 20 ppm hydrazine or equivalent after the boiler pressure has decayed to below 200 psig (1.4 MPa). At higher pressures and associated temperatures, the scavenger may rapidly decompose. An oxygen scavenger/inhibitor is not added to boilers that employ oxygen treatment. For boilers on oxygen treatment, the oxygen feed should be stopped at least an hour be-

*WARNING: HYDRAZINE IS A PROVEN GENERIC CHEMICAL FOR THIS APPLICATION. HOWEVER, HYDRAZINE IS A KNOWN CARCINOGEN, AND CAN BE REPLACED WITH OTHER PRODUCTS THAT HAVE EQUIVALENT ABILITY TO SCAVENGE OXYGEN AND INHIBIT CORROSION.

Table 9
Summary and Comparison of Boiler Lay-up Methods

Lay-up Method	Effectiveness	Costs	Safety and Environmental Concerns	Strengths	Weaknesses
Drained and dry for erection or maintenance	Poor	Minimal	Minimal	Allows full access to internal surfaces Quick and easy	Not effective
Vaporous corrosion inhibitors	Variable, generally fair	Chemicals Chemical application Chemical removal and disposal	Handling and disposal of chemicals	Minimal maintenance requirements	Remaining inhibitive capacity is difficult to monitor Chemicals must be replaced periodically Difficult to distribute through components
Nitrogen blanketing	Excellent	Nitrogen distribution system Nitrogen	Nitrogen suffocation	Consistently effective Easy to monitor	Safety concerns Nitrogen leakage
Hot standby	Good	Heat	Residual temperature and pressure	Fast restart	Not recommended for more than 3 days
Cold standby	Poor	Minimal	Minimal	Fast restart	Not recommended for more than 30 days
Wet, water-filled	Variable, generally good	Demineralized water Chemical treatment Disposal of treated water	Handling of chemicals Disposal of treated water	Easily applicable to non-drainable components Facilitates rapid return to service	Freeze damage Valve seepage and associated corrosion damage Difficult to monitor and inspect May corrode copper alloys beyond boiler
Dry, dehumidified air	Excellent	Dehumidifier and blower unit Air recirculation piping	Minimal	Consistently effective Safe No disposal problems Easy to monitor	Boiler must be totally drained Initial plumbing and equipment requirements

fore shutdown; there can be no oxygen leakage into the cycle. Where possible, as steam pressure decays to atmospheric pressure, nitrogen should be introduced through upper vents to keep the internal pressure positive and prevent air ingress. Boiler-specific operating instructions should be consulted for other important shutdown procedures and precautions.

Draining for boiler maintenance and inspection

Draining boilers without further preservation is often necessary for maintenance and inspection, and the unit should be drained and dried as thoroughly as possible. Draining the boiler hot (for example, at 20 psig/0.14 MPa) will facilitate drying, but may leave condensate in non-drainable superheater elements. Boiler components that need not be open for maintenance should, where possible, be isolated and protected. For example, while the lower furnace is open for maintenance, the superheater should, if possible, be protected with an appropriate wet or dry layup method. All openings should remain covered to prevent ingress of contaminants.

The unpreserved maintenance period should be as brief as possible, preferably less than 3 weeks. Where a long maintenance outage with minimal preservation is necessary, the boiler and open sections of the pre-boiler system should generally be chemically cleaned and repassivated to reform a protective oxide film following the outage.

Hot standby and hot layup

When a boiler returns to operation while steam pressure remains above atmospheric pressure, preservation requirements are minimal. If the shutdown period is less than 72 hours and there is no air in-leakage, ammonia and oxygen scavenger/inhibitor concentrations need only be raised to the high end of the normal operating range with an oxygen scavenger/inhibitor concentration equivalent to about 0.050 ppm hydrazine. Oxygen scavengers are not used for boilers that employ oxygen treatment. However, a low oxygen concentration must be maintained by effective deaeration, and water must remain at a high level of purity with cation conductivity less than 0.15 µS/cm.

Extended hot standby is not recommended. Hot layup can be extended by use of auxiliary heat, but non-corrosive and non-depositing conditions can be difficult to assure, especially in economizers, superheaters, and reheaters. If temperature is maintained by injecting blowdown from an adjacent boiler, excessive blowdown solids can accumulate under low-flow conditions in the standby boiler. When steam is injected from another boiler, deposition and corrosion problems can arise from temperature and chemistry differentials. Also, care must be taken to avoid cavitation and to control the chemistry of steam condensate. Water leakage across valves can cause corrosion in seepage areas, and the corrosion is aggravated by relatively high material and water temperatures.

Cold standby for up to thirty days

For short periods (not more than a week), boilers may remain on cold standby status with little or no chemical additions. Moderate ammonia and oxygen scavenger/inhibitor additions can extend this period up to thirty days. Treated demineralized water for short-term cold standby should have a minimum concentration of 10 ppm ammonia. The concentration of oxygen scavenger/inhibitor normally used in the boiler should be increased to the equivalent of 25 ppm of hydrazine. Nitrogen blanketing (see below) is recommended.

For boilers employing oxygen treatment, no scavenger should be used for short cold standby periods. However, oxygen ingress must be avoided and low oxygen concentrations must be maintained (for example, by use of nitrogen blanketing of boiler and deaeration of makeup water). Throughout the storage period, the pH must be 9.2 or higher, and cation conductivity throughout the boiler must be less than 0.15 μS/cm.

Boiler storage for more than seven days

Any of several layup practices are acceptable for storage periods up to six months. Alternatives for fluid-side layup include nitrogen blanketing, wet layup with treated demineralized water, and dry dehumidified layup. For extended outages, it is important that the pre-boiler feedwater train, balance of the water-steam cycle, auxiliary equipment, and fireside of the boiler are adequately preserved. For idle periods longer than six months, dry (dehumidified) storage is recommended. For boilers with non-drainable components, water removal issues must be weighed against the advantages of dry layup.

Reheaters – all periods

Reheaters are generally stored dry because they can not easily be isolated from the turbine. Wet storage requires installation of blanks or special valves in the connecting lines. Drying can be performed by exposing reheaters to condenser vacuum after the fires have been removed, the unit tripped, and the turbine seals maintained. With approximately 20 in. (5 kPa) of vacuum on the condenser, vents or drains on the reheater inlet should be opened to allow air to pass through and remove all moisture. Dehumidified dry storage is recommended where the storage period exceeds 30 days. Alternatively, the evacuated reheater may be isolated and nitrogen blanketed, or it may be nitrogen blanketed in conjunction with the turbine.

Nitrogen blanketing

Even where water is present, corrosion can be prevented by eliminating oxygen from the environment. Oxygen can be eliminated by sealing and pressurizing the entire boiler, or the space above water level, with nitrogen to prevent air in-leakage. In the absence of acids and other oxidants, eliminating air stops corrosion. Nitrogen blanketing is a highly effective method for preventing corrosion. It is easy to monitor and alarm, so effective preservation can be assured. However, the boiler must be well sealed to prevent excessive leakage.

It is absolutely imperative that working spaces around nitrogen blanketed equipment be well ventilated. Venting of nitrogen during purging or water filling operations can release large amounts of nitrogen into surrounding areas. Also, before entry, areas that have been nitrogen blanketed must be well ventilated and the air tested to confirm that all parts have adequate oxygen concentrations.

Wet (water-filled) layup

Wet layup in combination with nitrogen blanketing is often the most practical method of protection, especially for boilers that are not fully or easily drainable. However, for longer storage periods, advantages of wet layup are offset by accumulative corrosion in areas of valve seepage and by accumulative cost of replacement water, chemicals, treated water disposal, nitrogen cover gas, and heat in cold climates. Consequently, dry layup is generally recommended for storage periods longer than six months.

Before a boiler is flooded with layup water, provision must be made to support the additional weight when drum (if present), superheater, and steam piping are filled. If the boiler is to be completely filled with layup water, an expansion tank or surge tank above the highest vent is necessary to accommodate volume changes that are caused by normal temperature fluctuations. The expansion space at the top of the boiler (whether in a drum or in a surge tank) should be nitrogen blanketed to assure that there is no air ingress. Where freezing conditions are expected or possible, provision must be made for heating water-filled components.

Wet layup generally requires demineralized water having a specific conductivity less than 1.0 μS/cm before treatment chemicals are added. Use of demineralized water and all volatile treatment chemicals is essential where boilers include non-drainable or stainless steel components. Use ammonium hydroxide to raise water pH into the range of 10.0 to 10.4. Use an oxygen scavenger/inhibitor to further retard corrosion.

Dry (dehumidified) layup

Dry layup requires the removal of all water and the dehumidification of air to maintain a relative humidity less than 50%, and preferably less than 40%. This prevents corrosion by hygroscopic salts. Dry-air (dehumidified) storage is highly effective, and its continued effectiveness is easy to monitor. Dry layup allows easy and safe access for maintenance, with no potential for suffocation and no exposure to toxic chemicals.

It also eliminates the potential for biological activity, damage from freezing water, and corrosion by leaking water. However, implementation requires that the boiler be completely drained and dried, and this is a major problem for boilers with non-drainable superheaters or other non-drainable circuits. Also, mechanical circulation and dehumidification equipment and piping installation can be costly.

In preparation for dry storage, water must be drained completely from all boiler circuits, including feedwater piping, economizer, superheater, and reheater. All non-drainable boiler tubes and superheater tubes should be blown with pressurized air. Auxiliary sources of heat are used to dry fluid-side surfaces. Deposits of sufficient thickness to retain moisture must also be removed.

The preferred dry layup method is continuous recirculation of dehumidified air. Fans force air through the dehumidifier, boiler fluid-side circuitry, and back to the dehumidifier. The system must include instrumentation for measuring relative humidity. Recirculating dehumidification also requires external (usually flexible plastic) piping to complete the path. The system must be sized to handle the residual moisture and moist air in-leakage, and must be monitored to assure that relative humidity remains less than 50%.

An alternative, but inferior, dry layup method is the use of static desiccant to absorb moisture with no forced air circulation. This method is effective for boiler components and small (package) boilers, but not generally adequate for large complex boiler circuitry.

Termination of the storage period requires removal of the recirculation, dehumidification, and monitoring materials and equipment. Any loose desiccant particles or dust (which generally contain silica or sulfite chemicals) must be cleaned from the boiler.

Vaporous corrosion inhibitors (VCI)

Vaporous corrosion inhibitors retard corrosion by forming a thin protective film over metal surfaces. Where such chemicals are sealed into a closed space, they can retard corrosion even in the presence of both water and oxygen. These inhibitors are not generally used for completed boilers, where the size and complexity of boiler circuits precludes effective distribution of dry powders. However, they are often used to inhibit internal corrosion of boiler components, as supplements to dry storage for small boilers, and as an alternative treatment for hydrotest water.

References

1. Klein, H.A., and Rice, J.K., "A research study on internal corrosion in high pressure boilers," *Journal of Engineering for Power,* Vol. 88, No. 3, pp. 232-242, July, 1966.

2. Cohen, P., Ed., *The ASME Handbook on Water Technology in Thermal Power Systems,* The American Society of Mechanical Engineers, New York, New York, 1989.

3. "Consensus on Operating Practices for the Control of Feedwater and Boiler Water Quality in Modern Industrial Boilers," The American Society of Mechanical Engineers, New York, New York, 1994.

4. "Interim consensus guidelines on fossil plant cycle chemistry," Report CS-4629, Electric Power Research Institute, Palo Alto, California, June, 1986.

5. "Guideline for Boiler Feedwater, Boiler Water, and Steam of Steam Generators with a Permissible Operating Pressure > 68 bar," VGB PowerTech e.V., Essen, Germany, VGB-R 450 Le, (in German), 1988.

6. *Betz Handbook of Industrial Water Conditioning,* Ninth Ed., Betz Laboratories, Trevose, Pennsylvania, September, 1991.

7. *Drew Principles of Industrial Water Treatment,* 11th Ed., Drew Industrial Division, Ashland Chemical Co., Boonton, New Jersey, 1994.

8. Kemmer, F.N., Ed., *The Nalco Water Handbook,* Second Ed., McGraw-Hill, New York, New York, 1988.

9. Macbeth, R.V., et al., UKAEA Report No. AAEW-R711, Winfrith, Dorchester, United Kingdom, 1971.

10. Goldstein, P., and Burton, C.L., "A research study on internal corrosion of high-pressure boilers – final report," *Journal of Engineering for Power,* Vol. 91, pp. 75-101, April, 1969.

11. Freier, R.K., "Cover layer formation on steel by oxygen in neutral salt free water," *VGB Speiserwassertagung 1969 Sunderheft,* pp. 11-17 (in German), 1969.

12. Whirl, S.F., and Purcell, T.E., "Protection against caustic embrittlement by coordinated pH control," Third Annual Meeting of the Water Conference of the Engineers' Society of Western Pennsylvania, Pittsburgh, Pennsylvania, 1942.

13. Stodola, J., "Review of boiler water alkalinity control," Proceedings of the 47th Annual Meeting of The International Water Conference, Pittsburgh, Pennsylvania, pp. 235-242, October 27-29, 1986.

14. Economy, G., et al., "Sodium phosphate solutions at boiler conditions: solubility, phase equilibrium and interactions with magnetite," Proceedings of the International Water Technology Conference, Pittsburgh, Pennsylvania, pp. 161-173, 1975.

15. Tremaine, P., et al., "Interactions of sodium phosphate salts with transitional metal oxides at 360C," Proceedings of the International Conference on Interaction of Iron Based Materials with Water and Steam, Heidelberg, Germany, June 3-5, 1992.

16. Martynova, O.I., "Transport and concentration processes of steam and water impurities in steam generating systems," *Water and Steam: Their Properties and Current Industrial Applications,* J. Staub and K. Scheffler, Eds., Pergamon Press, Oxford, United Kingdom, pp. 547-562, 1980.

17. Nagda, N.L. and Harper, J.P., *Monitoring Water in the 1990s: Meeting New Challenges,* STP1102, American Society for Testing and Materials, Philadelphia, Pennsylvania, 1991.

18. *Annual Book of ASTM Standards,* Section 11, Water and Environmental Technology, Vol. 11.01 (D 1192 Specification for Equipment for Sampling Water and Steam; D 1066 Practice for Sampling Steam; D 3370 Practice for Sampling Water; and D 4453 Handling of Ultra-pure Water Samples), American Society for Testing and Materials, Philadelphia, Pennsylvania, 2004.

19. *Steam and Water Sampling, Conditioning, and Analysis in the Power Cycle, ASME Performance Test Code 19.11,* The American Society of Mechanical Engineers, New York, New York, 1997.

20. Rice, J., "Quality assurance for continuous cycle chemistry monitoring," Proceedings of the International Conference on Fossil Plant Cycle Chemistry, Report TR-100195, Electric Power Research Institute, Palo Alto, California, December, 1991.

21. Uhlig, H.H., and Revic, R.W., *Corrosion and Corrosion Control,* Third Ed., John Wiley & Sons, New York, New York, 1985.

22. Lamping, G.A., and Arrowood, Jr., R.M., *Manual for Investigation and Correction of Boiler Tube Failures,* Report CS-3945, Electric Power Research Institute, Palo Alto, California, 1985.

23. French, D.N., *Metallurgical Failures in Fossil Fired Boilers,* Second Ed., Wiley, New York, New York, 1993.

24. Port, R.D., and Herro, H.M., *The Nalco Guide to Boiler Failure Analysis,* McGraw-Hill Company, New York, New York, 1991.

25. Dooley, R.B., and McNaughton, W.P., "Boiler Tube Failures: Theory and Practice," Electric Power Research Institute, Palo Alto, California, 1996. LICENSED MATERIAL available to EPRI members.

26. *Annual Book of ASTM Standards,* Section 11, Water and Environmental Technology, Vols. 11.01 and 11.02, American Society for Testing and Materials, Philadelphia, Pennsylvania, 2003.

27. Wackenhuth, E.C., et al., "Manual on chemical cleaning of fossil-fuel steam generating equipment," Report CS-3289, Electric Power Research Institute, Palo Alto, California, 1984.

28. Samuelson, M.L., McConnell, S.B., and Hoy, E.F., "An on-site chemical treatment for removing iron and copper from chelant cleaning wastes," Proceedings of the 49th International Water Conference, Pittsburgh, Pennsylvania, p. 380,1988.

29. "Nalmet heavy metal removal program," Nalco Chemical Company, Naperville, Illinois, February, 1989.

Babcock & Wilcox built and operates this refuse-to-energy plant in the southern U.S.

Chapter 43

Boiler Operations

Effective operation of steam generating equipment has always been critical to maintaining system efficiency, reliability and availability. With advances in sensor technology, control software, and steam generator hardware, efficient operation today means balancing equipment performance with safety requirements and emissions mandates. The procedures used to run steam generating equipment vary widely depending upon the type of system, fuel and application. Systems can range from simple and fully automated requiring a minimum of attention, such as small gas-fired package boilers, to the very complex requiring constant operator attention and interaction, such as a large utility plant. There are, however, a set of relatively common fundamental operating guidelines that safeguard personnel and optimize equipment performance and reliability. When combined with equipment-specific procedures, these guidelines promote the best possible operations. The first half of this chapter focuses on general practices applicable to large multi-burner fossil-fuel fired natural circulation and once-through boilers, though these practices are also generally applicable to most other boiler designs. Starting with *Operation of Cyclone furnaces*, the remaining sections focus on the unique practices for several important special cases.

Because of the intimate relationship between equipment design and operation, additional operating guidelines may be found in other chapters of *Steam*. Selected areas of particular interest are listed in Table 1.

General boiler operations

Fundamental principles

Although boiler design and power production have become sophisticated, basic operating principles still apply. Combustion safety and proper steam/water cooling of boiler pressure parts are essential.

Combustion safety and steam/water cooling requirements Before firing a furnace, there must be no lingering combustible material inside the unit. *Purging*, or removal of this material, assures that the furnace is ready for firing. A standard operating rule for multiple burner boilers is to purge the unit at no less than 25% of the maximum continuous rating (MCR) mass air flow for the greater of at least five minutes or five volume changes.

Once combustion is established, the correct air/fuel ratio must be maintained. Insufficient air flow may permit the formation of combustible gas pockets and provide an explosion potential. Sufficient air flow to match the combustion requirements of the fuel should be maintained and a small amount of excess air should be admitted to cover imperfect mixing and to promote air and fuel distribution. For emissions reduction, many newer firing systems are staged with overfire air (OFA) ports installed in the upper combustion zones of the furnace. The controlled introduction of combustion air does not eliminate the need to provide adequate operating excess air for safety.

In addition to these combustion precautions, it is important to verify boiler water levels. Combustion should never be established until adequate cooling water is in the tubes and steam drum for natural circulation boilers and minimum flow rates are established for once-through boilers.

General safety considerations Pressure part failure remains a major concern in the boiler industry. The

Table 1
Index to Additional Operating Guidelines Found in *Steam*

System or Component	Chapter
Air heaters	20
Ash handling equipment	24
Auxiliary equipment	25
Burners, coal	14
Burners, oil and gas	11
Bypass systems	19
Chemical cleaning	42
Chemical recovery units	28
Cyclone furnaces	15
Fluidized-bed boilers	17
Particulate removal equipment	33
Pulverizers	13
Scrubbers (flue gas desulfurization)	35
Selective catalytic reduction systems	34
Sootblowers	24
Startup systems	19
Stokers	16
Waste-to-energy systems	29
Water chemistry	42
Water treatment	42

American Society of Mechanical Engineers (ASME), National Board of Boiler and Pressure Vessel Inspectors, and other organizations have issued extensive codes directed at minimizing these failures. The codes are continuously updated to support quality pressure part design. (See Appendix 2.)

Combustion products have also received considerable safety attention. Carbon dioxide (CO_2), carbon monoxide (CO), sulfur oxides and chlorine compounds must be considered. While CO_2 is not poisonous, it reduces oxygen availability in the flue gas. Sulfur oxides can form acids when breathed into the lungs. Chlorine compounds can form hydrochloric acids or carcinogens such as dioxin.

Safety of fuel ash must also be considered. While few ashes contain hazardous materials, heavy metals and arsenic can be present in dangerous levels. In addition, residual combustion may continue in ash collected in hoppers. This is especially true when new emissions reduction combustion systems are being installed. Carbon-in-ash characteristics may be altered under the new low nitrogen oxides (NO_x) firing configurations.

Finally, safeguards apply when performing maintenance and other work on out-of-service boilers. A confined space may have insufficient oxygen for personnel. Good lighting is important for a safe working environment. In addition, ash accumulations in out-of-service boiler areas may still be too hot for maintenance activities.

Initial commissioning operations

Operator requirements

Good operation begins before equipment installation is complete. It includes training of the operators as well as preparation of the equipment.

Every operator must be trained to understand and fulfill the responsibility assumed for the successful performance of the equipment and for the safety of all personnel involved. To be prepared for all situations that may arise, the operator must have a complete knowledge of all components: their designs, purposes, limitations and relationships to other components. This includes thoroughly inspecting the equipment and studying the drawings and instructions. The ideal time to become familiar with new equipment is during the pre-operational phase when the equipment is being installed.

A distributed control system (DCS) integrates the individual process controllers of a steam generation process into a coordinated, interactive system. It enables the operator to manage the process as a complete system, with control over the interrelationship of various subsystems.

Modern distributed control systems are extendable to provide operating personnel with advanced simulations of equipment behavior. Simulator training is typically in real-time, ensuring not only a working understanding of the various steam generator systems, but also the reaction time and rate of the equipment. Combined with simulated equipment failures and what-if events, the operating personnel can gain valuable experience prior to initial operation.

Preliminary operation should not be entrusted to inexperienced personnel who are not familiar with the equipment and the correct operating procedures. Considerable equipment damage and potential safety events can result from improper preparation of equipment or its misuse during preliminary checkout.

Operator training takes on special significance during the initial operating period required to prepare the unit for commercial operation. Knowledgeable and experienced operators are valuable during this period when the controls and interlocks are being adjusted, fuel burning equipment is being regulated, operating procedures are being perfected and preliminary tests are being conducted to demonstrate the performance and capabilities of the unit.

Preparations for startup

A systematic approach is required when a new boiler is being installed or when any boiler has undergone major repairs or alterations. The procedure varies with boiler design; however, certain steps are required for all boilers. The steps may be classified as inspection, cleaning, hydrostatic testing, pre-calibration of instruments and controls, auxiliary equipment preparation, refractory conditioning, chemical cleaning, steam line cleaning (blowing), safety valve testing and settings, and initial operations for adjustments and testing.

Inspection An inspection of the boiler and auxiliary equipment serves two purposes: 1) it familiarizes the operator with the equipment, and 2) it verifies the condition of the equipment. The inspection should begin some time during the construction phase and continue until all items are completed.

One item frequently overlooked during the inspection is the provision or lack of provision for expansion. The boiler expands as the temperature and pressure are increased, as do steam lines, flues and ducts, sootblower piping and drain piping. Before pressure is raised in the boiler, temporary braces, hangers or ties used during construction must be removed.

Cleaning Debris and foreign material that accumulate during shipment, storage, erection or repairs must be removed. Debris on the water side can restrict circulation or plug drain lines. Debris on the gas side can alter gas or air flows. Combustible material on the gas side can ignite and burn at uncontrollable rates and cause considerable damage. Glowing embers can be the source of ignition at times when ignition is not desired.

Fuel lines, especially oil and gas lines, should be cleaned to prevent subsequent damage to valves and the plugging of burner parts. Steam cleaning is recommended for all oil and gas lines. Atomizing steam and atomizing air lines should be cleaned.

Hydrostatic test After the pressure parts are assembled, but before the refractory and casing are installed, a hydrostatic test at 1.5 times the boiler design pressure is applied to all new boilers and maintained for a sufficient time to detect any leaks. Testing is equally important following pressure part replacements at pressures specified in local code requirements.

The hydrostatic test is normally the first time the new boiler is filled with water; therefore, this is the time to begin using high quality water for the prevention of internal fouling and corrosion. (See Chapter 42.) Demineralized water or condensate treated with 10 ppm of ammonia for pH control and 500 ppm of hydrazine for control of oxygen should be used for all nondrainable superheaters and reheaters. A clear filtered water is suitable for components that will be drained immediately after the hydrostatic test.

Temperature plays an important part in hydrostatic testing. The metal temperature and therefore the water temperature must be at or above applicable Code restrictions for hydrostatic testing. For example, these restrictions, as stated in the ASME Code, specify that the hydrostatic test temperature will not be below 70F (21C) to take advantage of the inherent toughness of carbon steel materials as related to temperature. The water temperature should be kept low enough so that the pressure parts can be touched and close inspections can be made. It should not be so high that water escaping from small leaks evaporates immediately or flashes to steam. Also, the water temperature should not be more than 100F (56C) above the metal temperature to avoid excessive metal stress transients.

Finally, no air should be trapped in the unit during the hydrostatic test. As the unit is being filled, each available vent should be open until water appears.

Instrumentation and controls Every natural circulation boiler has at least two indicating instruments: a water/steam pressure gauge and a water level gauge glass. If a superheater is involved, some type of steam temperature indicator is also used. Once-through type boilers have steam pressure gauges, flow meters and temperature indicators. These indicators are important in that the pressure, temperature levels and flows indicated by them must be controlled within design limits. Therefore, the indicators must be correct. Operation should not be attempted until these instruments are calibrated, and the calibrations should include corrections for actual operating conditions. Water-leg correction for the pressure gauge is one example. (See Chapter 40.)

Most modern boilers are controlled automatically once they are fully commissioned. However, before these boilers can be started up, the controls must be operated on manual until the automatic controls have been adjusted for site-specific conditions. Automatic control operations should not be attempted until the automatic functions have been calibrated and proven reliable over the load range.

Auxiliary equipment The auxiliary equipment must also be prepared for operation. This equipment includes fans to supply air for combustion and to transport fuel, feedwater pumps to supply water, fuel equipment to prepare and burn the fuel, and air heaters to heat the air for combustion. This equipment also includes an economizer to heat the water and cool the flue gas, ash removal equipment, a drain system to drain the boiler when required, a functioning sootblowing system to adequately clean the heat transfer surfaces, and post-combustion environmental control equipment.

Chemical cleaning Water-side cleanliness is important for all boilers because water-side impurities can lead to boiler tube failures. They can also lead to carryover of solids in the steam, resulting in superheater tube failure or turbine blade damage.

The flushing of all loose debris from the feedwater system and boiler, and the use of high quality water for the hydrostatic test, must be supplemented by proper water-side cleaning before startup. To remove accumulations of oil, grease and paint, the natural circulation boiler is given a caustic and phosphate boilout after the feedwater system has been given a phosphate flush. The once-through boiler and its associated pre-boiler equipment are given a similar flushing. This boiling out and/or flushing should be accomplished before operation.

After boiling out and flushing are completed, products of corrosion still remain in the feedwater system and boiler in the form of iron oxide and mill scale. It is recommended that acid cleaning for the removal of this mill scale and iron oxide be delayed until operations at fairly high capacities have carried loose scale and oxides from the feedwater system to the boiler. This results in a cleaner boiler for subsequent operations. Chemical cleaning of internal heating surfaces is described in Chapter 42.

Steam line blowing Fine mesh strainers are customarily installed in turbine inlet steam lines to protect turbine blades or valves against damage from scale or other solid material that may be carried by the initial flow of steam. In addition, many operators use high velocity steam to clean the superheater and steam lines of any loose scale or foreign material before coupling the steam line to the turbine. The actual procedure used depends on the design of the unit. Temporary piping for flow bypass to the atmosphere is required with all procedures. This piping and any noise reduction silencers must be securely anchored to resist the high nozzle reaction created during the high velocity blowing period.

Several methods are used for blowing steam lines, including particularly high pressure air and steam blowing. The recommended method and the one most used is steam blowing because experience has shown that temperature shock and high velocities are the most effective means of removing loose scale. Sufficient shock is obtained with a series of blows where the steam temperature changes during each blow. Lower pressure, high velocity steam blowing techniques are also available, including those with high velocity treated water injection to promote thermal shocking of the mill scale and debris. Designed and managed correctly, these techniques provide safe removal of scale and other solid particles detrimental to turbine components.

There are two basic methods of supplying steam for steam line blowing with a natural or forced circulation boiler. The first method is to use steam returning from the flash tank to the superheater on the forced circulation unit or steam flow from the drum on the natural circulation unit. The second method is to use high pressure steam directly from the boiler. The latter method supplies large quantities of steam at higher pressures. Temporary valving is required to control the flow rates during the blowing period.

Boiler pressure and temperature can be maintained during the blowing period by continuous firing. If firing is discontinued during the blowing period, it must be remembered that any change in boiler pressure changes the saturation temperature throughout the system. To avoid excessive thermal shock, changes in boiler pressure should be limited to those corresponding to 75F (42C) in saturation temperature during the relatively short blowing periods.

Safety valves Safety valves are essential to the safe operation of any pressure vessel, allowing adequate relief of excess pressure during abnormal operating conditions. The set point of each safety valve is normally immediately checked and adjusted if necessary after reaching full operating pressure for the first time with steam. Safety valve seats are susceptible to damage from wet steam or grit. This is an essential reason for cleaning the boiler and blowing out the superheater and steam line before testing safety valves.

Safety valves on drum-type boilers are normally tested both for set point pressure and for the closing pressures. This generally requires that the boiler pressure be raised until the valve opens and then reduced for the valves to close.

The testing of safety valves always requires caution. Safety valve exhaust piping and vent piping should not exert any excessive forces on the safety valve. Gags should always be used as a safety measure while making adjustments to the valves.

As an alternate, safety valves can be tested and set with the boiler pressure below the safety valve design pressure by supplementing the boiler pressure with a hydraulic lift, attached to the valve stem, in accordance with the manufacturer's instructions. This hydraulic assist method reduces the risk of damage to the valve seat from extended steam flow in a conventional test.

Startup Operating procedures vary with boiler design. There are, however, certain objectives that should be included in the operating procedures of every boiler. These objectives include:

1. protection of pressure parts against corrosion, overheating and thermal stresses,
2. prevention of furnace explosions,
3. production of steam at the desired temperature, pressure and purity, and
4. compliance with environmental regulations.

Filling In filling the boiler for startup, certain precautions should be taken to protect the pressure parts. First, high quality water should be used to minimize water-side corrosion and deposits. Second, the temperature of the water should be regulated to match the temperature of the boiler metal to prevent thermal stresses. High temperature differentials can cause thermal stresses in the pressure parts and, if severe, will adversely affect the life of the pressure parts. High temperature differentials can also distort the pressure parts enough to break studs, lugs and other attachments. Differential temperatures up to 100F (56C) are generally considered acceptable.

A third precaution taken during the filling operation is the use of vents to displace all air with water. This reduces oxygen corrosion and assures that all boiler tubes are filled with water.

A fourth precaution on drum-type boilers is to establish the correct water level before firing begins. The water level rises with temperature. Therefore, only 1 in. (25 mm) of water is typically required in the gauge glass, except with certain special designs that may require a higher starting level in order to fill all circulating tubes exposed to the hot flue gases.

Circulation Overheating of boiler tubes is prevented by the flow of fluid through the tubes. Flow is produced in the natural circulation-type boiler by the force of gravity acting on fluids of different densities. Flow starts when the density of the water in the heated tubes is less than that in the downcomers. This flow increases as firing rate is increased. (See Chapter 5.) Some drum-type boilers are designed for forced circulation and depend on a circulating pump to assist this flow.

The once-through type boiler depends on the boiler feedwater pump to produce the necessary flow. Whenever a once-through boiler is being fired, a minimum design flow must be maintained through the furnace circuits. With the use of the bypass system, the fluid can bypass the superheater and turbine to maintain this minimum design flow until saturated steam is available for admission to the superheater and until the turbine is using sufficient steam to maintain the design minimum furnace circuit flow. (See Chapter 19.)

Purging Considerable attention has been given to the prevention of furnace explosions, especially on units burning fuel in suspension. Most furnace explosions occur during startup and low load periods. Whenever the possibility exists for the accumulation of combustible gases or combustible dust in any part of the unit, no attempt should be made to light the burners until the unit has been thoroughly purged. The National Fire Protection Association Boiler and Combustion Systems Code (NFPA 85:2004) provides the consensus procedures for purging of various boiler and combustion systems. For multi-burner boilers, the purge rate from the forced draft fan through the stack is at least 25% of the full load mass air flow which must be maintained for the greater of five minutes or five volume changes of the boiler enclosure. A maximum of 40% of the full load mass air flow is specified for coal-fired units only, while no upper limit is set for oil- or gas-fired units.

Protection of economizer Very little water, if any, is added to the drum-type boiler during the pressure raising period; consequently, there is no feedwater flow through the economizer. Economizers are located in relatively low temperature zones. Nevertheless, some economizers generate steam during the pressure raising period. This steam remains trapped until feedwater is fed through the economizer. It not only makes the control of steam drum water level difficult; it causes water hammer. This difficulty is overcome if feedwater is supplied continuously by venting the economizer of steam or by recirculating boiler water through the economizer. If a recirculating line is available, the valve in this line must remain open until feedwater is being fed continuously through the economizer to the boiler.

Protection of primary, secondary and reheat superheater During normal operation, every superheater tube must have steam flow sufficient to prevent overheating. During startup and before there is steam flow through every tube, the combustion gas temperature entering the superheater section must be controlled to limit superheater metal temperatures to 900F (482C) for carbon steel tubes and 950 to 1000F (510 to 538C) for various alloy tubes. While firing rate is used primarily to control gas temperatures, other means are useful, e.g., excess air, gas recirculation and burner selection.

Gas temperature entering the superheater during startup is typically measured with retractable thermoprobes that are removed as soon as steam flow is established in every superheater tube. (See Chapter 40.) Other permanent gas temperature monitoring devices are available, but precautions must be taken so that their installed locations, usage and maintenance are adequate for the task of measuring gas temperatures entering the superheater. These devices are required for the first few startups to establish acceptable firing rates. They should always be used on the once-through type boiler.

The two prerequisites for steam flow through every superheater tube are: 1) removal of all water from each tube, and 2) a total steam flow equal to or greater than approximately 10% of rated steam flow. Water is removed from drainable superheaters by simply opening the header drains and vents. Nondrainable superheaters are not as simple, because the water must be boiled away. There will be no steam flow through a tube partially filled with water, and those portions of the tube not in contact with water will be subjected to excessive temperatures unless the gas temperature is limited.

Thermocouples attached to the outlet legs of nondrainable superheater tubes, where they pass into the unheated vestibule, will read saturation temperature at the existing pressure until a flow of steam is established through the tube. The temperature of these outlet legs rises sharply to significant increases above saturation immediately after flow is established. Superheater tubes adjacent to side walls and division walls are normally the last to boil clear. These should, therefore, have thermocouples.

Protection of drums and headers In most installations, the time required to place a boiler in service is limited to the time necessary for raising pressure and protecting the superheater and the reheater against overheating. In some cases, however, the time for both starting up and shutting down may be determined by the time required to limit the thermal stresses in the drums and headers. Protection of drums and headers can pose significant challenges for operators faced with the need to cycle the steam generator on/off for load demand.

Various rules, based on thermal stress analysis and supported by operating experience, have been formulated and accepted as general practice. The rules fall into three categories: one for drums and headers with rolled tube joints, a second set for headers with welded tube joints, and a third set for steam drums with welded tube joints.

On drums and headers with rolled tube joints, the relatively thin tubes contract and expand at a much faster rate than the thicker drum or header walls; therefore, tube seat leaks are likely to occur if heating and cooling rates are not controlled. Heating and cooling rates have, therefore, been established at 100F (56C) change in saturation temperature per hour.

Headers with welded tube connections present no problem with tube seat leaks or with header distortion because the tubes are welded to the headers and the headers are normally filled with fluid at a constant temperature. The concern is mainly with temperature differentials through the header wall and the resulting incipient cracking if excessive thermal stresses occur.

Steam drums with welded tube connections are not subject to tube seat leaks but, because they contain water in the bottom and steam in the top, the heating and cooling rates vary between the top and the bottom. This results in temperature differences between the top and the bottom. Stress analyses show that the principal criterion for reliable rates of heating and cooling should be based on the relationship between the temperature differential through the drum wall and the temperature differential between the top and bottom of the drum, both of which are measured in the same circumferential plane. Analysis also shows that the allowable temperature differentials are based on tensile strength, drum diameter, wall thickness and pressure. Therefore, each steam drum has its own characteristics.

To determine the temperature differentials during the periods of startup and shutdown, it is necessary to continuously and accurately obtain the outside and inside surface temperatures of the drum shell. Temperatures of the outside of the drum shell are best determined by thermocouples. At least six outside thermocouples are required: two on each end, one on the top and bottom (all outside the internal baffle and scrubber area) and two in the center, one top and one bottom. Temperatures of the inside of the drum shell are best determined by placing thermocouples on one of the riser tubes entering the bottom of the drum, one on each end, and one in the center.

There are two periods when the steam in the top of the drum is hotter than the water occupying the lower half. One period is soon after firing begins on boilers with large nondrainable superheaters. Steam forms first in the superheater and flows back to the steam drum where it heats the top of the drum. A second period is during cooling when cold air is pumped through the boiler by the forced draft fans. The boiler water cools, but the steam in the top of the steam drum remains at essentially the same temperature.

The temperatures of the top and bottom can be brought close together by flooding the steam drum with water. Firing must be stopped and water must not be allowed to spill over into the superheater. A high level gauge glass is needed for this operation. However, thermocouples attached to the superheater supply tubes can be used to indicate when the drum is flooded because there is a sharp change in temperature when water enters these supply tubes. Allowable temperature differentials for cooling are stringent, considerably more so than for heating. Fast cooling should, therefore, be supervised with the use of thermocouples and a high level gauge glass.

Cooling is particularly important when the boiler is to be drained for short outages. The steam drum must be cooled to permit filling with the available water without exceeding the allowable temperature rate of change. (See Fig. 1.) A wider latitude is permitted if the water used for filling is hotter than the steam drum.

Before removing the unit from service, a more uniform cooling can be achieved by reducing pressure to approximately two-thirds of the normal operating pressure. This reduces the temperature of both the water and steam in the steam drum.

Operating techniques for maximum efficiency

Fuel is a major cost in boiler operation. It is therefore important to minimize fuel consumption and maximize steam production. Although a boiler's efficiency is primarily a result of its design, the operator can maintain or significantly improve efficiency by controlling heat losses to the stack and losses to the ash pit.

Stack losses The total heat that exits the stack is controlled by the quantity and the temperature of the flue gas. The quantity of gas is dependent on the fuel being burned, but is also influenced by the amount of excess air supplied to the burners. While sufficient air must be provided to complete the combustion process, excessive quantities of air simply carry extra heat out of the stack.

The temperature of the flue gas is affected by the cleanliness of the boiler heat transfer surfaces. This in turn is dependent on sootblower and air heater operation. Optimal cleanliness of the heat transfer surfaces is achievable with commercial sensors and control systems that measure the heat transfer effectiveness of the boiler banks, evaluate against the design duty, and clean only on an as-needed basis. While high gas exit temperatures waste energy, excessively low temperatures may also be unacceptable. Corrosion can occur at the acid dew point of the gas where corrosive constituents condense on the cooler metal surfaces. Ash plugging of heat transfer surfaces can be aggravated by the presence of condensate. Equipment protection and performance of backend environmental components [selective catalytic reduction (SCR) system protection, for example] can also be constraints in desirable steam generator gas temperature.

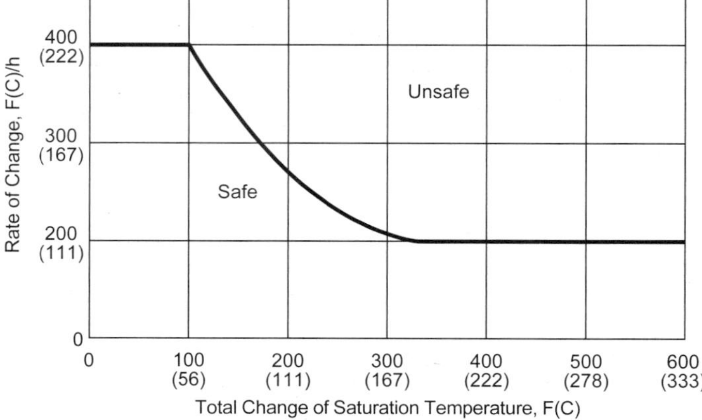

Fig. 1 Permissible rate of change for saturation temperature, drum and furnace.

Pressures and temperatures Most boilers supply steam to turbines or processes that require heat. These processes rely on design pressures and temperatures. Deviations from the set points may result in lower overall efficiency of the power cycle, loss of production, or damage to the product or process.

In units that produce saturated steam, boiler temperature is directly related to the operating pressure. For many industrial applications, the process requirements dictate the steam temperature and consequently the operating pressure.

For electric power producing steam systems, steam temperature has a significant impact on turbine efficiency. Modern controls can typically maintain this temperature within 10F (6C) of the desired setting. Each 50F (28C) reduction in steam temperature reduces cycle efficiency by approximately 1% on a supercritical pressure unit.

Variable pressure operation Historically, utility power boilers in the United States (U.S.) have been operated at a constant steam pressure and the turbine load has been controlled by varying the throttle valves at the turbine inlet. This causes an efficiency loss due to the temperature drop across the throttle valves.

In variable pressure operation, boiler pressure is varied to meet turbine requirements. This can significantly improve cycle efficiency when operating at low loads (15 to 40% of MCR). However, boiler response to turbine requirements is slower in this mode. When fast response is needed, pressure increments are used. At a given pressure, the throttle valves control the steam to the turbine and provide quick response. As turbine requirements increase and the valves approach full open position, the boiler pressure is increased to the next increment. This arrangement provides high efficiency while retaining quick response.

Emissions requirements The majority of the steam generators in operation today were not originally designed for today's environmental regulations. Hardware changes made to accomplish the emissions reduction are typically accompanied by new operating philosophies and guidelines for efficiency and equipment protection. To maintain optimal SCR efficiency and minimal ammonia usage/slip the combustion system is optimized to generate the lowest level of nitrogen oxides (NO_x) possible. Staging of combustion air is frequently used in firing system upgrades. Sufficient care must be taken to ensure that lower furnace corrosion associated with the reducing atmosphere is not excessive, especially with high sulfur content fuels. Depending on system design, there may be tradeoffs between lower NO_x generation and increases in the carbon content in the ash and carbon monoxide (CO) in the flue gas. To maintain ash quality suitable for disposal/sale, operational balance must be achieved within the combustion process. Furnace combustion is also usually controlled to achieve any specific CO emissions rate.

Combustion optimization A modern DCS is equipped

to offer significantly more information and analysis, assisting the operating personnel in optimizing the steam generator. Closed loop neural network systems and other advanced intelligent control systems have made strides in maintaining optimum operations consistently throughout the operating range. Variants of these systems use complex databases and algorithms of unit-specific operating scenarios to guide the controlling parameters during steady state and transient operation. Extensive parametric testing of the unit is conducted to define the characteristics and boundaries of the operating variables. As long as no subsequent mechanical modifications are added and proper equipment maintenance is available, these advanced control systems are capable of consistently returning the steam generator to its most optimum operating condition. A significant attribute of these systems is the ability to target specific results during optimization and change the targeted parameter as needed. As the result, NO_x can be minimized during selected periods and heat rate can be minimized the rest of the operating year.

Key operation functions

Burner adjustments The fuel and combustion air combine and release heat at the burners. To maintain even heat distribution across the width of the furnace, air and fuel flows must be evenly supplied to all burners. While individual burner adjustments affect a particular burner, they also impact adjacent burners. Most boilers have multiple burners in parallel flow paths on the front and/or rear furnace walls. Burners must be similar in design and must be adjusted in the same way to optimize air flow distribution. For a detailed discussion of burner features refer to Chapters 11 and 14.

Adjustments can typically vary the turbulence and flow rates in burners. Increased turbulence increases the air-fuel mixing. It also increases the combustion intensity, provides faster heat release, reduces unburned carbon (UBC) in the ash, permits operation with less excess air, and increases boiler efficiency. However, increased slagging, increased NO_x emissions and higher fan power consumption can also result.

Fuel adjustments on an individual burner basis are also possible, either through diverter mechanisms in the fuel preparation system or through flow balancing devices in the coal transport piping. While possible, the dynamic balancing of fuel flows is difficult. As with combustion air adjustments, potential impacts to adjacent burners will require reliable feedback on relative flow rates to be effective as a dynamic operational tool. Advances in flame characterization systems and sensors for post-combustion constituents are providing more precise feedback targets for on-line fuel balancing, as are in situ coal and combustion air flow measurement devices.

Overfire air port adjustments Burner systems intended for NO_x reduction are typically staged at the primary combustion elevation and usually operated with less than theoretical air for complete combustion. The remaining air needed to complete the burning process is introduced separately, above the highest burner zone. The overfire air (OFA) flows are typically varied as a function of load, based on parametric testing performed during commissioning. While normally satisfactory for ensuring complete combustion and emissions compliance, any changes to fuel type, fuel preparation quality, or the fuel/air transport systems will require close monitoring of OFA system performance. More advanced dynamic control schemes are available, dependent on measurement feedback of the combustion process in the upper furnace zones.

Excess air The total combustion air flow to a boiler is generally controlled by adjusting forced and induced draft fan dampers in relation to the fuel flow. *Excess air* is the amount of additional combustion air over that required to theoretically burn a given amount of fuel. The benefits of increasing excess air include increased combustion intensity, reduced carbon loss and/or CO formation, and reduced slagging conditions. Disadvantages include increased fan power consumption, increased heat loss up the stack, increased tube erosion, and possibly increased NO_x formation.

For most coal ashes, particularly those from eastern U.S. bituminous coals, the solid to liquid phase changes occur at lower temperatures if free oxygen is not present (reducing conditions) around the ash particles. As a result, more slagging occurs in a boiler operating with insufficient excess air where localized reducing conditions can occur.

Localized tube metal wastage may also occur in furnace walls under low excess air conditions but the impact is less clearly defined. The absence of free oxygen (a reducing atmosphere) and the presence of sulfur (from the fuel) are known causes of tube metal wastage. The sulfur combines with hydrogen from the fuel to form hydrogen sulfide (H_2S). The H_2S reacts with the iron in the tube metal and forms iron sulfide, which is subsequently swept away with the flue gas. High chlorine levels can also promote tube wastage. Although coals and most conventional fuels burned in the U.S. contain very little chlorine, it is a problem in refuse-derived fuels. (See Chapter 29.)

Fuel conditions Fuel conditions are temperature, pressure and, if solid, mean particle size and distribution. Cooler temperatures, lower pressure and larger particle size contribute to less complete combustion and increased unburned carbon in the ash. Conversely, if the fuel is hotter, finer and at a higher pressure, combustion is improved. Unfortunately NO_x emissions and slagging can also increase with these conditions. Current day combustion technologies employ multi-zone low NO_x burners to control the pace of combustion and are more demanding on mean particle size and distribution for control of NO_x and UBC production.

Effects of fuel preparation equipment on boiler performance Fuel preparation equipment readies the fuel for combustion and can have a significant impact on pollutant emissions. The preparation equipment may be the crushers, pulverizers and drying systems on coal-fired units; fuel oil heaters and pumps on oil-fired units; refuse handling, mixing or drying equipment on refuse-fired boilers; or fuel handling, blending, sizing and delivery equipment on stoker-fired units. If this equipment is not properly operated, the fuel may not be completely burned, leaving UBC in the ash or carbon monoxide (CO) in the flue gas.

The operator must monitor the fuel preparation equipment. Knowledge of the equipment, its maintenance record and operating characteristics is essential.

Feedwater and boiler water conditioning requirements The main function of a boiler is to transfer heat from combustion gases through tube walls to heat water and produce steam. Clean metal tubes are good conductors, but impurities in the water can collect on the inside surface of the tubes. These deposits reduce heat transfer, elevate tube temperatures, and can lead to tube failures. Water conditioning is essential to minimize deposits and maintain unit availability. (See Chapter 42.)

Sootblower operations As discussed in Chapter 24, a *sootblower* is an automated device that uses steam, compressed air, or high pressure water to remove ash deposits from tube surfaces. Sootblowing improves heat transfer by reducing fouling and plugging. However, excessive sootblowing can result in increased operating cost, tube erosion and increased sootblower maintenance. Conversely, infrequent sootblower operation can reduce boiler efficiency and capacity. Optimum sootblowing depends on load conditions, combustion quality and fuel. With the increased dependency in the U.S. on Powder River Basin (PRB) coal and the cost attractiveness of spot market fuels, advanced sootblowing control systems are being increasingly deployed to assist the operators. Unlike manual guidelines, these intelligent systems do not rely on fuel characteristics to influence cleaning patterns. Rather, they focus primarily on the heat transfer surfaces and monitor efficiency. Since cleaning decisions are based solely on an as-needed basis, the intelligent sootblowing systems are ideally suited for operations dealing with changing fuel blends. Furnace-based sootblowing equipment is increasingly being controlled with advanced control schemes that monitor the heat flux in select zones, supervise blower operation for effectiveness, and minimize thermal shock to the furnace wall tubes.

Personnel safety Operating instructions usually deal primarily with the protection of equipment. Rules and devices for personnel protection are also essential. The items listed here are based on actual operating experience and point out some personnel safety considerations.

1. When viewing flames or furnace conditions, always wear tinted goggles or a tinted shield to protect the eyes from harmful light intensity and flying ash or slag particles.
2. Do not stand directly in front of open ports or doors, especially when they are being opened. Furnace pulsations caused by firing conditions, sootblower operation, or tube failure can blow hot furnace gases out of open doors, even on suction-fired units. Aspirating air is used on inspection doors and ports of pressure-fired units to prevent the escape of hot furnace gases. The aspirating jets can become blocked, or the aspirating air supply can fail. In some cases, the entire observation port or door can be covered with slag, causing the aspirating air to blast slag and ash out into the boiler room.
3. Do not use open-ended pipes for rodding observation ports or slag on furnace walls. Hot gases can be discharged through the open ended pipe directly onto its handler. The pipe can also become excessively hot.
4. When handling any type of rod or probe in the furnace, especially in coal-fired furnaces, be prepared for falling slag striking the rod or probe. The fulcrum action can inflict severe injuries.
5. Be prepared for slag leaks. Iron oxides in coal can be reduced to molten iron or iron sulfides in a reducing atmosphere in the furnace resulting from combustion with insufficient air. This molten iron can wash away refractory, seals and tubes, and leak out onto equipment or personnel.
6. Never enter a vessel, especially a boiler drum, until all steam and water valves, including drain and blowdown valves, have been closed and locked or tagged. It is possible for steam and hot water to back up through drain and blowdown piping, especially when more than one boiler or vessel is connected to the same drain or blowdown tank.
7. Be prepared for hot water in drums and headers when removing manhole plates and handhole covers.
8. Do not enter a confined space until it has been cooled, purged of combustible and dangerous gases and properly ventilated with precautions taken to keep the entrance open. Station a worker at the entrance and notify the responsible person.
9. Be prepared for falling slag and dust when entering the boiler setting or ash pit.
10. Use low voltage extension cords or cords with ground fault interrupters. Bulbs on extension cords and flashlights should be explosion proof.
11. Never step into flyash. It can be cold on the surface yet remain hot and smoldering underneath for extended periods even after the pressure parts are cool.
12. Never use toxic or volatile fluids in confined spaces.
13. Never open or enter rotating equipment until it has come to a complete stop and its circuit breaker is locked open and any other drive devices are immobilized. Some types of rotating equipment can be set into motion with very little force. These types should be locked with a brake or other suitable device to prevent rotation.
14. Always secure the drive mechanism of dampers, gates and doors before passing through them.
15. Do not inspect for tube leak locations until metal and refractory surfaces are cool, and ash accumulations are removed.

Performance tests Many steam generating units are operating day after day with efficiencies at or near design values. Any unit can operate at these efficiencies with the proper instrumentation, a reasonable equipment and instrument maintenance program, and proper operating procedures.

Early in the life of the unit when the gas side is relatively clean, the casing is tight, the insulation is new, the fuel burning equipment has been adjusted for optimum performance and the fuel/air ratio has been set correctly, performance tests should be conducted to determine the major controllable heat losses. These losses are the dry gas to the stack and, on coal-fired units, combustible in the ash or slag.

During these tests, accurate data should be taken to serve as reference points for future operation. Sampling points that give representative indications of gas temperatures, excess air and combustibles in the ash should be established and data from these points recorded. Items related to the major controllable losses should also be recorded at this time, e.g., draft losses, air flows, burner settings, steam flow, steam and feedwater temperatures, fuel flow and air temperatures.

Procedures for performance tests are provided in the American Society of Mechanical Engineers (ASME) Performance Test Codes, PTC 4, *Steam Generating Units*; PTC 4.2, *Coal Pulverizers*; and PTC 4.3, *Air Heaters*.

Abnormal operation

Low water If water level in the drum drops below the minimum required (as determined by the manufacturer), fuel firing should be stopped. Caution should be exercised when adding water to restore the drum level due to the potential of temperature shock from the relatively cooler water coming in contact with hot drum metal. Thermocouples on the top and bottom of the drum will indicate if the bottom of the drum is being rapidly cooled by feedwater addition, which would result in unacceptable top-to-bottom temperature differentials. If water level indicators show there is still some water remaining in the drum, then feedwater may be slowly added using the thermocouples as a guide. If the drum is completely empty, then water may only be added periodically with soak times provided to allow drum temperature to equalize. (See *Protection of drums and headers*.)

Tube failures Operating a boiler with a known tube leak is not recommended. Steam or water escaping from a small leak can cut other tubes by impingement and set up a chain reaction of tube failures. By the loss of water or steam, a tube failure can alter boiler circulation or flow and result in other circuits being overheated. This is one reason why furnace risers on once-through type boilers should be continuously monitored. A tube failure can also cause loss of ignition and a furnace explosion if re-ignition occurs. As discussed later in the chapter, process recovery boilers are particularly sensitive to tube leaks because of the potential for smelt-water reactions, which can lead to boiler explosions.

Any unusual increase in furnace riser temperature on the once-through type boiler is an indication of furnace tube leakage. Small leaks can sometimes be detected by the loss of water from the system, the loss of chemicals from a drum-type boiler, or by the noise made by the leak. If a leak is suspected, the boiler should be shut down as soon as normal operating procedures permit. After the leak is then located by hydrostatic testing, it should be repaired.

Several items must be considered when a tube failure occurs. In some cases where the steam drum water level can not be maintained, the operator should shut off all fuel flow and completely shut off any output of steam from the boiler. When the fuel has been turned off, the furnace should be purged of any combustible gases and feedwater flow to the boiler should be stopped. The air flow should be reduced to a minimum as soon as the furnace purge is completed. This procedure reduces the loss of boiler pressure and the corresponding drop in water temperature within the boiler.

The firing rate or the flow of hot gases can not be stopped immediately on some waste heat boilers or on certain types of stoker-fired boilers. Several factors are involved in the decision to continue the flow of feedwater, even though the steam drum water level can not be maintained. In general, as long as the temperature of the combustion gases is hot enough to damage the unit, the feedwater flow should be continued. (See later discussion for chemical recovery units.) The thermal shock resulting from feeding relatively cold feedwater into an empty steam drum should also be considered. (See *Protection of drums and headers*.) Thermal shock is minimized if the feedwater is hot, the unit has an economizer, and the feedwater mixes with the existing boiler water.

After the unit has been cooled, personnel should make a complete inspection for evidence of overheating and for incipient cracks, especially to headers, drums, and welded attachments. (See *Personnel safety*, especially when the potential of hot materials, boiler fluids, or combustibles is present.)

An investigation of the tube failure is very important so that the condition(s) causing the tube failure can be eliminated and future failures prevented. This investigation should include a careful visual inspection of the failed tube. In some cases, a laboratory analysis or consideration of background information leading up to the tube failure is required. This information should include the location of the failure, the length of time the unit has been in operation, load conditions, startup and shutdown conditions, feedwater treatment and internal deposits.

Shutdown operations

Boiler shutdown is less complicated than startup. The emphasis again is on safety and protection of unit materials. Two shutdown situations may occur: a controlled shutdown, or one required in an emergency.

Under controlled conditions, the firing rate is gradually reduced. Once the combustion equipment is brought to its minimum capacity, the fuel is shut off and the boiler is purged with fresh air. If some pressure is to be maintained the fans are shut down and the dampers are closed. The drum pressure gradually lowers as heat is lost from the boiler setting; a minimal amount of air drifts out of the stack. If inspection and maintenance are required, the draft fans remain on to cool the boiler more quickly. If the unit is equipped with a regenerative air heater, it is shut down, allowing the boiler to cool faster. If a tubular air heater is present, this heat trap can only be bypassed to help cool the boiler. The cool down rate should not exceed 100F (56C) per hour of saturation temperature change to prevent damage due to thermal stress.

In an emergency shutdown, the fuel is immediately shut off and the boiler is purged of combustible gases. Additional procedures may apply to the fuel feed equipment. The boiler may be held at a reduced pressure or may be completely cooled as described above.

Operation of Cyclone furnaces

The Cyclone component of a Cyclone™ furnace is a cylindrical chamber designed to burn crushed coal. The basic design, sizing and general operation of Cyclone furnaces are covered in Chapter 15. The key feature that affects operations is the collection of most of the coal ash as a liquid slag in the Cyclone chamber. This slag is continuously tapped into the furnace through a hole at the discharge end of the Cyclone. The slag collects on the furnace floor and flows through a tap into a water filled tank where the chilling effect of the water leaves the slag in granular form. Unique operational issues center upon maintaining desired furnace slag tapping without excessive maintenance.

Fuels

A key operating parameter of Cyclone units is the selection of a proper coal. Fuels acceptable for Cyclone firing must generally meet the following specifications subject to site-specific conditions:

Bituminous coals:
1. maximum total moisture – 20%
2. ash content – 6 to 25%, dry basis
3. minimum volatile matter – 15%, dry basis
4. ash (slag) viscosity – refer to Table 2
5. ash iron ratio and sulfur content – See Chapter 15, Fig. 4

Subbituminous coals and lignite:

1a. maximum total moisture for a direct-fired system, but without a pre-dry system – 30%

1b. maximum total moisture with pre-dry system – 42%

2. minimum high heating value – 6000 Btu/lb (13,956 kJ/kg), as-fired
3. ash content – 5 to 25%, dry basis
4. ash (slag) viscosity – refer to Table 2

For proper combustion and efficient unit operation, properly sized crushed coal is required. This is especially important for subbituminous coals, where the higher moisture content requires a finer coal grind to maintain Cyclone temperatures and to minimize unburned carbon in the flyash.

Combustion air

All Cyclone furnaces use heated air at a high static pressure. The air temperature ranges from 500 to 750F (260 to 399C), depending on the unit design and the rank of the fuel. The static differential across the Cyclone ranges from 32 to 50 in. wg (8 to 12.5 kPa). This high pressure produces the very high velocities which, in turn, produce the scrubbing action required for complete combustion of the crushed coal.

A key variable in Cyclone operation is excess air. For bituminous coals, which are generally high in sulfur and iron content, 15 to 18% excess air is recommended at full load operation. Operation at lower excess air can promote an oxygen-deficient reducing atmosphere, which can significantly increase unit maintenance. Operation with molten slag and deficient air can result in iron sulfide attack (wastage) of the boiler tubes. When operated with insufficient excess air, fuels with high iron content can smelt the iron from the ash. A pig iron, which forms a strong bond to the boiler tubing, can form and may result in tube damage during deposit removal.

Operation with high levels of excess air leads to lower thermal efficiency and results in Cyclone cooling. This hinders slag tapping. On multiple Cyclone units, it is important to accurately measure the secondary air and coal flow to each Cyclone to ensure it is operating with the proper air/fuel ratio.

Because proper excess air is essential, theoretical air curves should be used to readjust the excess air levels when one unit of a multiple Cyclone furnace is removed from service.

Firing subbituminous coals is generally more difficult due to increased fuel moisture which must be evaporated during combustion. Units designed for subbituminous coal firing have provisions for higher air temperature and generally run with 10 to 12% excess air.

Due to environmental concerns, some bituminous coal-fired Cyclone furnaces have been converted to burn low sulfur subbituminous coal or coal blends. Subbituminous coal has a lower ash content than most bituminous coals. As a result, less slag is available to trap the raw coal particles. This, combined with the depressed flame temperature caused by the increased moisture content, can result in less coal being entrapped in the slag where the combustion is completed. Acting on the additional environmental concerns, utilities have installed OFA systems for combustion NO_x control. While furnace wall corrosion remains a concern, worthwhile NO_x reductions have been attained with fuel rich stoichiometry and lower sulfur fuels. These Cyclone operating techniques have changed the operating practices of these slagging combustors.

When firing higher moisture subbituminous coal, smaller crushed coal particle size is required and higher transport air temperature is desired.

Table 2
Ash Viscosity Requirements

Coal Rank	Maximum T_{250}* Ash Viscosity	As-Fired Total Moisture %
Bituminous	2450F (1343C)	0 to 20
Subbituminous – direct-fired	2300F (1260C)	21 to 30
Subbituminous/lignite**	2300F (1260C)	31 to 35
Lignite***	2300F (1260C)	36 to 42

* T_{250} is the temperature at which the ash viscosity is 250 poise.

** For lignite firing, a fuel pre-dry system is required.

*** For high moisture lignite firing, pre-dry and moisture separator systems are required.

As the moisture content reaches the 20% range at design Cyclone fuel inputs, clinkering and UBC in the Cyclone and on the furnace floor can increase dramatically. Levels of unburned carbon also increase in the ash in the economizer, air heater and precipitator hoppers. Supplemental firing with fuel oil or natural gas along with design changes may be required if the unit was not originally designed for this condition.

Several Cyclone units are operating on high moisture North Dakota lignite fuel. This is accomplished by the use of a pre-dry system, as illustrated in Chapter 15. In this system, hot air [750F (399C)] is introduced to the raw coal stream prior to crushing in a pressurized and heated conditioner. The hot air-coal mixture then travels to a cyclone separator, where the saturated air and some coal fines are separated from the coal stream and are injected into the lower furnace. The dried coal is then carried by a lift line to the Cyclone burner.

Operation of once-through boilers

Principles

In full once-through operation, high pressure water enters the economizer and high pressure steam leaves the superheater; there is no recirculation of steam or water within the unit. The path is through multiple parallel tube circuits arranged in series. Design and construction details are found in Chapters 19 and 26. This type of boiler can be conceptualized as a long heated pipe with continuous coolant flowing through it. As the fluid progresses through the components, heat is absorbed from the combustion process. The final steam temperature is dependent on the feedwater inlet temperature and ratio of the fluid flow to the heat available, which have important implications on unit operation.

Operating practice

The once-through steam generator, referred to by The Babcock & Wilcox Company (B&W) as the Universal Pressure (UP®) boiler, has operating characteristics not seen in drum units. The UP boiler is capable of operating above the critical pressure point [3200 psi (22.1 MPa)] and can deliver steam pressure and temperature conditions without a steam-water separation device after startup. The original UP boilers are typically designed and operated for base load operation. To address the increasing demand for operational flexibility, the B&W Spiral Wound Universal Pressure (SWUP™) boiler is capable of variable pressure operation and on/off cycling, as well as load cycling and base load operation. This spiral wound tube geometry UP boiler, including its startup and bypass systems, minimize the thermal upsets during transients, allowing rapid load changes.

To start up a UP boiler, a steam/water separation device (flash tank or vertical steam separator), with appropriate valving and piping to bypass and return fluid to the cycle, is supplied. The modern bypass systems are highly automated and very effective in achieving smooth operation. The bypass system is in service only during low load operation.

The major control functions for fuel, air, water and steam flow are highly interactive placing unique constraints on operation. Though today's integrated control systems manage many operating functions, it is still important for the operator to fully understand the relationships between short-term (transient) and long-term (steady-state) energy transfer effects and to carefully monitor and coordinate, as required, all control actions. Modern control systems are often referred to as coordinated systems. (See Chapter 41.) The combustion systems are conventional; they are subject to the normal concerns of fuel utilization, efficiency and safety.

The operator has three main controls for operating a UP boiler:

1. firing rate,
2. feedwater control, and
3. steam flow/steam pressure.

Firing rate The long-term demand for firing rate is directly proportional to load. The short-term effects of firing rate impact steam temperature and pressure. During increases in firing rate the steam temperature increases until the feedwater flow is increased to compensate and balance the new firing rate. Steam pressure increases in a similar fashion to that of a drum boiler. In a supercritical pressure application, there is no large rise in specific volume during the transition from liquid to vapor conditions. However, there is still a large increase in specific volume as a supercritical fluid is heated. With this expansion, the turbine control valves are opened and flow is increased to maintain pressure. With the many interactions that are a consequence of firing rate, the operator and automated control system should strive to position firing rate for the desired electrical load and then manipulate other variables to control temperature and pressure.

Feedwater flow The outlet steam temperature is dependent in steady-state conditions on the ratio of feedwater flow rate to firing rate. Because it is desirable to position firing rate based on electrical demand, feedwater flow control is based on outlet steam temperature. The short term effect of feedwater flow changes is to increase or decrease steam pressure and electrical load. The load change results because energy is placed into or brought out of storage through cooling or heating of the boiler metals. However, this load change is transitory. Pressure is a measure of the balance between steam flow and feedwater flow. The outlet pressure is constant if they are matched. The operator can, therefore, use feedwater to assist in recovery from transients and upsets, but ultimately must position feedwater flow for the appropriate outlet steam temperature.

Steam flow/steam pressure In steady-state conditions, steam flow is the same as feedwater flow, and boiler pressure is determined by turbine throttle valve position. In the short term, a change in steam flow impacts steam temperature and electrical load. This is because increasing steam flow at a constant firing rate affects the balance of the two and the steam temperature drops initially as a consequence to electrical load increase. This can not be a lasting effect, however, as steam and feedwater flows eventually match each other at a different pressure. Similarly, load may be changed by adjusting the turbine throttle valve which increases or decreases steam flow, but this load change comes from depleting or building stored energy in the form of pressure. As above, steam flow may be manipulated to help restore control of steam temperature and load, but it ultimately must be controlled to achieve the desired operating pressure.

Operating skills

Efficiency The supercritical boiler provides very high cycle efficiency. This inherent efficiency can be maintained throughout boiler life with proper instrumentation and maintenance. The operator must be thoroughly trained to know the optimum operating parameters. Deviations in these parameters may then be addressed with instrument maintenance, direct operator action (such as sootblowing) or an engineering investigation. Performance tests should be conducted on a regular basis. Besides the conventional need to minimize heat losses, these tests can be used to confirm control system calibration. Test results should be reviewed with the operator.

Startup systems Chapter 19 provides a comprehensive overview of the configuration and operation of state-of-the-art startup and bypass systems for variable and constant furnace pressure once-through boiler systems. Systems for UP boilers have evolved as the understanding and experience of the once-through concept have increased. A key requirement for the startup system is to maintain adequate flow in the furnace walls (30% for most variable pressure supercritical units, 25% for most constant pressure supercritical units, and 33% for most subcritical units) to protect them from overheating during startup and low load operation. Initial designs (first generation) simply bypassed any excess flow from the furnace, not required for the turbine power generation, directly to the condenser from the high pressure turbine inlet. Second generation startup systems added a steam-water separation device called a flash tank (including steam-water separation equipment) after the convection pass enclosure circuits, but upstream of the primary superheater. This permitted the generation of dry steam for turbine roll and synchronization. The flash tank was effectively a small drum including drum internals. (See Fig. 2.)

Subsequent constant pressure third generation designs moved the bypass line plus necessary valves (200 and 201) downstream of the primary superheater so that the primary superheater is always in the initial flow circuit. (See Fig. 3 for key elements.)

Steam is produced in the flash tank because the pressure reducing valve (207 in Fig. 3) drops the pressure from supercritical to approximately 1000 psig (6.89 MPa) in current designs. For a smooth transition or changeover from startup (flash tank) to once-through operation, the enthalpy of the flash tank steam must be matched with the enthalpy leaving the 201 valve. Because the flash tank generally operates at 1000 psig (6.89 MPa), the throttle pressure to the turbine is only ramped to full pressure after transition from the flash tank.

Due to the need for flexibility of the variable furnace pressure SWUP designs, startup and bypass systems have changed to incorporate features of the subcritical drum and supercritical once-through startup systems. As shown in Fig. 4 (key elements only), the SWUP system has evolved from the second generation system shown in Fig. 2 with the replacement of the valves and flask tank with vertical steam separators, a water collection tank and a boiler circulation pump. At startup and low load, the flow passes through the vertical steam separators where steam is removed and sent to the superheater and water is discharged to the water collection tank. Water then flows through the boiler circulation pump to maintain the minimum furnace wall circulation. Any excess water accumulated during startup is sent to the condenser through the 341 valves. When unit operation reaches the load where the sum of the furnace flow plus the attemperator spray flows is equal to the main steam flow rate, once-through operation is achieved, the 381 valves are closed, and the boiler circulation pump is shut down.

Limits and precautions The once-through nature of UP boilers results in special limits and precautions that must be observed for reliable and dependable operation:

1. Feedwater conductivity – unlike a drum boiler which can release suspended solids through blowdown, all hardness and other contaminants that enter the boiler in the feedwater are deposited on the water side of the heating surface or in the superheater. Deposition can lead to overheating and tube failures. Boiler firing must be stopped if conductivity exceeds 2.0 microsiemens for five minutes or if it exceeds 5.0 microsiemens for two minutes. The operator should be trained in water quality control requirements. (See Chapter 42.)
2. Minimum feedwater flow – firing is not permitted unless the boiler feedwater flow is above the specified minimum. The boiler is to be immediately shut down if flow falls below 85% of minimum for twenty seconds or 70% of minimum for one second.

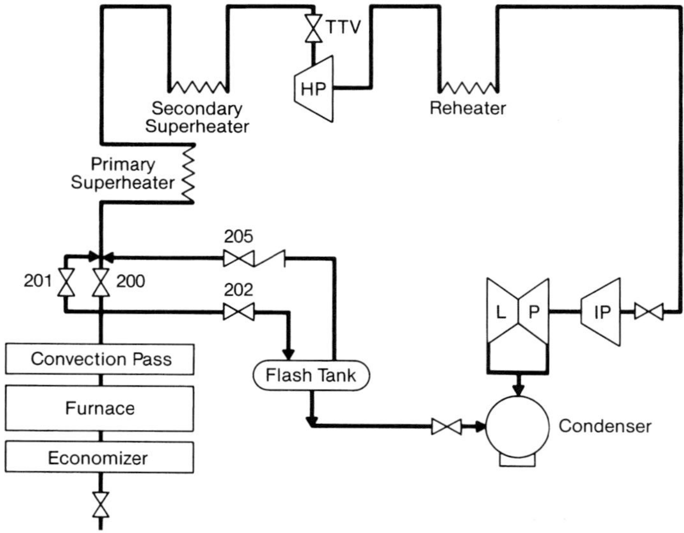

Fig. 2 Second generation UP startup system schematic.

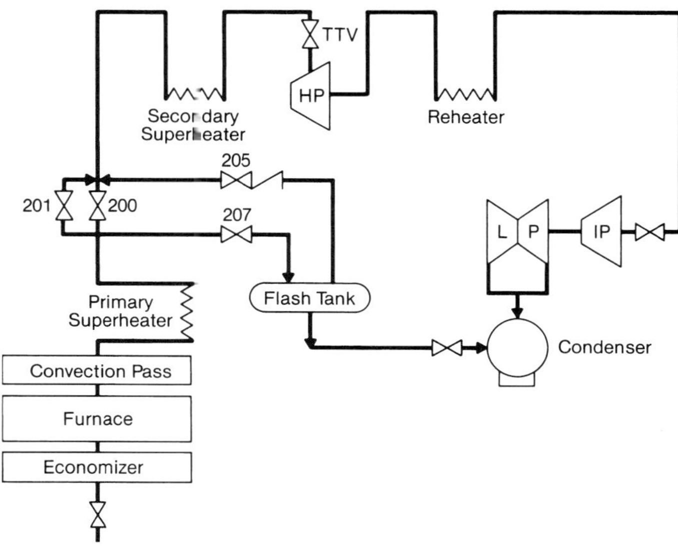

Fig. 3 Third generation UP startup system with primary superheater in initial flow circuit.

3. Feedwater temperature – the heat input required per unit of water flow is dependent on the difference between the feedwater inlet temperature and the controlled outlet steam temperature. If feedwater temperature is reduced while the outlet temperature is maintained (by overfiring), the heat input to the furnace can exceed design levels that may result in damage to the tube material. For units that were not designed for a feedwater heater out of service, removal of a feedwater heater will require that the final steam temperature be reduced to hold furnace heat input rates within design limits. Unless specific information is provided the general rule is to reduce steam temperature one degree F for every two degrees F that the feedwater temperature is low.

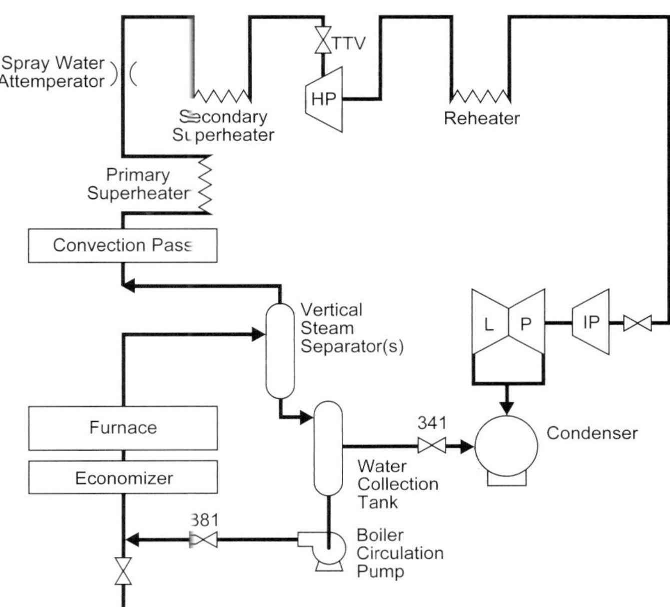

Fig. 4 Startup system for a SWUP boiler (key elements).

4. Overfiring – for supercritical units, the fluid in the furnace has no latent heat of vaporization as in a subcritical boiler. Consequently, as heat is absorbed, the fluid temperature is always increasing and the furnace tube metal temperatures increase. Firing in excess of design rates can result in excessive tube metal temperatures in selected furnace locations. The operator must therefore observe a strict limit of transient heat input. A maximum of 1.15 times the steady-state rate is a typical guideline. This rate, at various load conditions, should be clearly established by combustion tests.
5. Tube leaks – operation with a tube leak is not recommended on any boiler. For once-through units, the situation is more critical because the leakage can reduce cooling flow to subsequent tubes or circuits downstream.
6. Low pressure limit – a constant pressure supercritical UP boiler must be tripped if the fluid pressure in the furnace falls below 3200 psig (22.1 MPa) for fifteen seconds. Failure to do so can result in improper circuit flow, inadequate tube cooling and furnace tube failures. For the SWUP boilers the units are designed for specific variable pressure ramps. To avoid possible tube damage, the units must operate along design pressure ramp curves.

Operation of fluidized-bed boilers

Chapter 17 provides a detailed discussion of fluidized-bed combustion systems. The following sections provide selected remarks on general fluidized-bed boiler operation, both circulating fluidized-bed (CFB) systems and bubbling fluidized-bed (BFB) systems.

CFB systems overview

The CFB is constructed and behaves like a conventional drum boiler in many respects. However, thermal performance, combustion efficiency, furnace absorption pattern and sulfur dioxide (SO_2) control are strongly influenced by the mass of solids in the bed (inventory), the size distribution of the bed material (bed sizing), and the circulating rate of the bed (flow rate). Bed density can be described as bulk density which is a measure of the material weight per unit volume. There is a higher density and inventory in the lower furnace, made of coarser particles (average size 200 to 400 micron), and a lower density and inventory in the upper furnace (also called furnace shaft) made of finer particles (average size 100 to 200 micron). Fig. 5 shows the density of the bed with respect to furnace height and indicates the effect of load.

To integrate these influences into operating procedures, the CFB is equipped with additional instrumentation which provides operator input for bed management. These inputs are as follows:

1. Primary zone ΔP – The gas-side differential pressure is measured across the primary furnace zone [from the air distribution grid to an intermediate elevation of about 6 ft (1.8 m) above the grid] and is an indication of the coarser bed inventory. This variable is managed as a function of boiler load and desired furnace shaft ΔP, utilizing the bed drain flow for control.

2. Shaft ΔP – The furnace shaft ΔP, from the intermediate elevation to the furnace roof, indicates the shaft bed density and is managed to follow boiler load with corrections for primary zone temperature control. It is influenced by the multi-cyclone dust collector (MDC) recycle rate, the primary to secondary air split, and excess air setting.
3. Lower bed (or primary zone) temperature – The lower bed temperature is set by the operator as a primary control variable for optimal combustion and emission performance. The shaft ΔP, primary to secondary air split, and excess air controls are used to control this operating parameter.

The variables mentioned above are typically not available on conventional boilers. However, the overall operation is similar. Fuel feed is set by steam requirements (pressure), total air flow is adjusted to follow fuel flow, and excess air is set to optimize combustion and unit efficiency. Shaft bed density is then predominantly used to control the lower bed temperature, the most important variable on a CFB boiler. The lower bed temperature is usually controlled to a fuel-dependent set temperature [for example 1550F (843C) for lignite and 1620F (882C) for petroleum coke] which results in optimal combustion efficiency and SO_2 absorption. The optimum bed operating temperature from combustion and emission standpoints is from 1500 to 1650F (816 to 899C), depending on fuel.

Another variable that is controlled somewhat differently in a CFB is the air admission to the furnace. The total air flow is divided into primary and secondary flows. Because total air flow is based on fuel flow, the primary to secondary air split becomes another important parameter affecting emissions and bed temperature. The split affects the lower bed density, shaft density, and control of the primary zone temperature. The optimum air split is determined as a function of steam flow during commissioning tests. Once the optimum primary to secondary air split (as a function of steam flow) has been determined for automated control, the operator should not need to independently manipulate the split.

Solids management is also unique to CFBs. Once the bed material inventory has been established, bed inventory is maintained by the fuel ash and sorbent-derived solids flows in most cases. Any change in inventory necessary to accommodate load change comes from the MDC hoppers. In the case of firing low-ash and low-sulfur fuels, solids input with fuel ash and sorbent may not be adequate to provide enough bed material for furnace temperature control. This also may be the case when firing waste fuels requiring a high bed drain rate for removing oversized material, e.g., rocks. If a shortage of solids for maintaining bed inventory occurs, the first source to add bed material would be recycling a usable part (properly sized material) from the lower furnace bed drain. If this is not sufficient, inert bed material fed from an external source, e.g., sand, would be provided.

With the sometimes complex relationships where changes in one variable impact several temperatures and flows, control logics are provided to automatically control these inter-relating parameters with minimal, if any, operator actions needed. However, the operator must be trained to understand these relationships and to control upsets to establish and maintain stability if necessary.

Fuel sizing and fuel characteristics

Coal firing Fuel sizing is very important to successful boiler operation. It impacts the furnace heat release profile: fine particles tend to burn higher in the furnace shaft while coarse ones burn predominantly in the primary zone. For medium and high ash fuels, fuel sizing also impacts sizing of bed particles and, correspondingly, bottom ash to flyash split. When fuel variation occurs, the operator should be thoroughly trained on the required adjustments in case the variations are outside the range of the programmed control logics.

Biomass firing Biomass fuels provide a particular challenge in CFB operations because of continuous variations in size, shape, moisture content, and heating characteristics. To minimize operating problems, fuel sources should be continuously blended to achieve as uniform a consistency as possible. Modest inventories are needed – large enough to permit blending for uniformity of feed but not large enough to result in excessive inventory storage times. Bed material sizing should be checked on a regular basis to ensure that the bed material is not deteriorating. Care must be taken in such sampling procedures because of the high bed material temperature.

Startup and shutdown considerations Chapter 17 provides general guidelines for the startup of CFBs. Hot and cold startup conditions are distinguished by the temperature of the furnace solids inventory, either higher or lower than the auto-ignition temperature of the main fuel established at the end of the boiler purge sequence. As with any boiler, prior to startup and after shutdown the unit must be purged according to the latest applicable NFPA codes. An important part of a cold startup is the warming of the bed material with auxiliary fuel overbed burners designed for this purpose. During normal operation, a

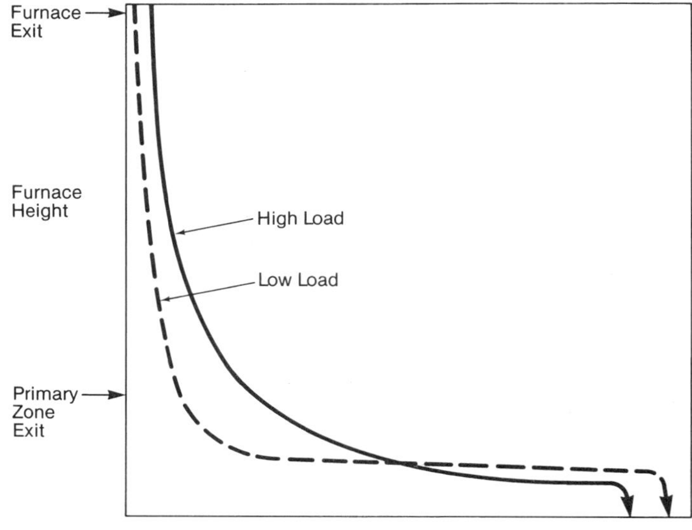

Fig. 5 Typical bed density profile for a circulating fluidized bed.

CFB is loaded with enough bed material to provide a 20 in. wg (50 kPa) pressure differential across the primary zone. During the warming process, bed materials are introduced to raise the bed pressure drop from 5 in. wg (1.25 kPa). Auxiliary fuel is used to increase and stabilize the primary zone temperature at about 1000F/538C (varies somewhat depending on main fuel) in preparation for main fuel introduction. Simultaneously, the boiler metal temperatures as indicated by steam pressure (saturation temperature) must be within the heatup rate allowed for the boiler pressure parts. Upon ignition of the auxiliary fuel, the flue gas temperature entering the U-beams and the steam-cooled surfaces must be monitored and maintained below 950F (510C) until 10% of the MCR steam flow is established through the steam-cooled circuits. Once the primary zone temperature is stabilized at the fuel auto-ignition point, the main fuel is introduced and the bed temperature can be gradually raised to about 1500F (816C) where the auxiliary burner may be shut off. Boiler load can be increased from this low load operation with circulating mode being established with increasing solids, fuel and air flow. The rate of increase must be controlled such that the bed temperature remains stable. Primary air to total air ratio is further adjusted to improve combustion and reduce emissions.

Overbed burners are used to assist with the burnup of unburned carbon particles in the primary zone of the furnace. If the boiler is being shut down for maintenance and personnel entry into the boiler setting is required, all solids must be completely removed from the boiler and hoppers, and the boiler temperature must be below 120F (49C) prior to entry. (See *personnel safety* discussed earlier.)

BFB systems

Overview As is the case with its CFB counterpart, bed management of a BFB is important. Bed inventory, bed sizing, makeup and drainage must be integrated into operating procedures from data collected during commissioning tests. The BFB is also equipped with additional instrumentation to assist the operator in developing the unit-specific operating guidelines.

1. Primary zone ΔP – the gas-side differential pressure is used as an indication of bed height. The primary zone may be compartmented on the air side for coal firing, and multiple ΔP cells are then displayed.
2. Bed temperature – the typical bed temperature range is 1350 to 1650F (732 to 899C) and is a primary control variable. Because control of this variable is critical, multiple thermocouples are installed in the bed.

The operation of a BFB is similar to that of other combustion technologies in that fuel and air demands are set by the required steaming conditions. Bed temperature is controlled within the desired range by the primary/overfire air split for biomass firing and by selected compartment slumping for coal firing (see Chapter 17). If an SO_2 sorbent is used, it is usually metered in proportion to the fuel flow. Various interactions can occur when manipulating control variables and the operator must be specifically trained to respond to upsets. These responses include:

1. Fuel flow control – fuel feed is increased or decreased based on steaming requirement. Changes in fuel feed have a short-term impact on bed temperature.
2. Bed height – height is controlled by adding material from the makeup system or by removing material through the drain system.
3. Air control – total air is primarily controlled based on steam flow. Air may be biased by the operator to change bed conditions. Increasing primary or bed air at a given load increases bed turbulence and burnout in the bed, while the secondary air changes opposite to the bed air change to control constant total air at a given load. Reducing bed air at a given load would have the opposite effect. The operator must, therefore, bias the set point based on combustion conditions. Fuel sizing variations, moisture content, and higher heating value are most likely to influence combustion conditions.
4. Bed material flow – material flow to the primary zone is through makeup and bed drains. Because these are generally intermittent devices, the operator must observe control variable trends, particularly height, and make appropriate adjustments.

Coal sizing Careful attention to coal sizing (minimum and maximum) is critical to successful unit operation for both under-bed and over-bed feed systems. Finer particles are elutriated too quickly and coarser particles tend to take too long to burn completely. In either case, UBC and SO_2 emissions (if being controlled) both increase. An optimum size range exists for each fluidized-bed system and fuel type. The proper size range is fine tuned through unit operating practice. In addition, for the over-bed feed system fuel size interacts with spreader speed, angle of injection and gas velocity.

Operation of Kraft recovery boilers

Overview

The Kraft recovery boiler has three purposes in today's pulp and paper industry:

1. recovery of sodium and sulfur compounds from the spent pulping liquor in forms suitable for regeneration,
2. efficient heat recovery from burning the liquor to generate steam for process use, and
3. operation in an environmentally responsible manner, cooling the combustion gases to allow back end particulate collection and minimizing the discharge of objectionable gases.

Steam flows can exceed 1,000,000 lb/h (126 kg/s) at superheater outlet pressures as high as 1500 psi (10.34 MPa) and final steam temperatures of up to 950F (510C). These units burn black liquor with solids contents up to 75 to 80% and with heating values decreased to about 5500 to 5600 Btu/lb (12,793 to 13,026 kJ/kg). A detailed review of designs and general operating practice is provided in Chapter 28. Selected operating issues are highlighted below:

1. black liquor combustion process and air flow,
2. auxiliary burners,
3. operating problems, and
4. smelt-water reactions.

Black liquor combustion process and air flow

Unlike the firing of conventional fossil fuels, recovery boiler combustion goes through several distinct stages. Black liquor firing is composed of drying, volatile burning, char burning and smelt coalescence as discussed in Chapter 28. This has a distinct impact on unit operation and air addition. All modern B&W recovery boilers are equipped with three levels of combustion air. These are known as the primary, secondary and tertiary air streams in order of increasing elevation in the furnace. A fourth level, called quaternary, can be added for additional NO_x and particulate control (see Chapter 28).

Primary air Primary air is in the lowest zone in a recovery boiler furnace. The air ports are located approximately 3 ft (0.9 m) above the furnace floor. Admitted on all four walls of the unit, the primary air provides perimeter air around the bed at low velocity and low penetration, so the boiler does not lose combustion. The primary air is critical to bed stability, bed temperature and reduction efficiency.

Secondary air High pressure secondary air penetrates the full cross-section of the unit and is admitted at the top of the bed providing the combustion air for ignition and the heat to dry black liquor droplets and support pyrolysis. This air also helps control SO_2, total reduced sulfur (TRS) and CO emissions. Finally, secondary air controls the shape of the top of the bed. Fig. 6 indicates the proper bed shape with the relative locations of the air ports and liquor nozzles.

Tertiary air Tertiary air is admitted at the upper elevation, above the liquor guns. Like secondary air, it is admitted at high pressure to provide penetration across the width of the furnace to complete the combustion process. This is an oxidizing environment to control CO and TRS emissions.

To maintain combustion in the proper location, the air flow splits and pressures must be properly maintained. Ranges of typical air flow distributions, inlet temperatures and pressure differentials are summarized in Table 3 for the two dominant firing systems, stationary and oscillating.

Successful stationary or oscillating firing depends on the correct liquor droplet size. This size must permit complete in-flight drying before reaching the furnace wall or bed. The variables in black liquor combustion include liquor percent solids, temperature and pressure at the spray nozzles, spray nozzle sizing and splash plate angle, and number and position of black liquor guns in service. Other variables include the vertical and angular movement of the oscillators or the off-horizontal angle of the liquor guns for stationary firing. Other factors on the air side include total excess air; air temperature(s); air splits between the primary, secondary and tertiary ports; and the static air pressure in each of these air zones.

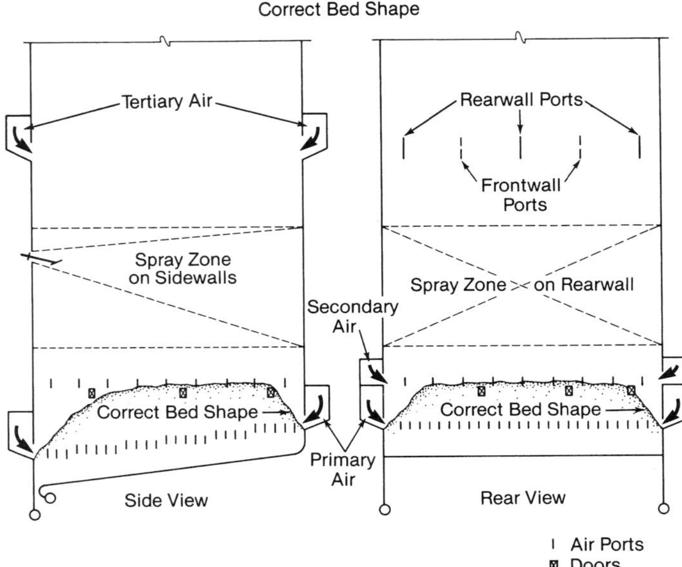

Fig. 6 Proper bed shape for a black liquor recovery boiler – three air level design.

The objective in the manipulation of all of these variables is to control the steps of the drying and combustion of the liquor, and control bed temperatures. The result is stability of operation, with minimum plugging and emissions.

A comprehensive discussion of recovery boiler emissions is provided in Chapter 28.

Auxiliary burners

Recovery boilers are equipped with oil- and/or natural gas-fired auxiliary burners located at the secondary and tertiary air port elevations. The burners in the secondary windboxes are used for unit startup, as black liquor can only be fired into a heated furnace with an auxiliary ignition source. The secondary burners are also used to stabilize the smelt bed during an upset condition and to burn the bed out of the unit on shutdown.

Operating problems

Two problems unique to recovery boiler operations are plugging and aggressive tube corrosion.

Plugging of the superheater, boiler bank or economizer is generally caused by two mechanisms. The first is condensation of the fume or normal gases given off by the black liquor combustion. The hot combustion gases that leave the lower furnace contain vaporized compounds that condense when the gas temperature is cooled in the upper zones of the boiler, superheater or boiler bank. This condensation is dependent on gas temperature. The material condenses on the cooled heat transfer surfaces such as furnace screen tubes or primary superheater tubes. It also precipitates out when the gas temperature falls below approximately 1100F (593C). In either case, this material is a source of fouling in convection surfaces. Condensation or precipitation depend upon the temperature regime and chemistry of the fume. The deposit can be in the plastic range and very difficult to remove. On a properly designed and operated unit, this transition of the fume

Table 3
Typical Air Flow Splits and Operating Conditions*

	Firing Technique Stationary Firing	 Oscillating Firing
Primary air	30 to 40% 300F (149C) 3 to 4 in. wg ΔP (0.7 to 1 kPa)	40 to 50% 300F (149C) 1 to 3 in. wg ΔP (0.2 to 0.7 kPa)
Secondary air	40 to 50% 300F (149C) 8 to 18 in. wg ΔP (2 to 4.5 kPa)	20 to 30% 300F (149C) 6 to 10 in. wg ΔP (1.5 to 2.5 kPa)
Tertiary air	10 to 20% 80F (27C) 10 to 20 in. wg ΔP (2.5 to 5 kPa)	20 to 30% 300F (149C) 8 to 12 in. wg ΔP (2 to 3 kPa)
Economizer outlet conditions	< 2.5% O_2 100 to 200 ppm CO	2.5 to 3% O_2 200 to 300 ppm CO

* Units with four air levels are discussed in Chapter 28.

from a dry gas to a sticky substance takes place in the upper furnace, wide spaced screen or superheater. Here the deposits can be controlled by sootblowers. On an overloaded or improperly operated unit, the gas temperature remains elevated farther back into the closer side-spaced boiler bank or economizer, where controlling the plugging with sootblowing equipment becomes difficult. The gas temperature at which the fume turns to a sticky liquid is also liquor chemistry dependent, with higher levels of chlorides or potassium compounds being detrimental to unit cleanability.

The second major cause of plugging is mechanical carryover of smelt, or unburned black liquor, into the convective heat transfer sections of the boiler. This material is predominantly made of sodium carbonate (Na_2CO_3), sodium sulfate (Na_2SO_4) and sodium sulfide (Na_2S) compounds. The cause of this carryover is related to operation of the liquor nozzles, oscillators and various air port settings. As previously noted, the black liquor firing equipment and air system settings must produce complete in-flight or wall drying of the liquor droplets. If this drying occurs too high above the smelt bed, the less dense droplets are entrained in the gas stream and carry over into the convection sections of the boiler. In practice, most recovery boiler plugging results from a combination of fume condensation and mechanical carryover.

Recovery boilers are also subject to aggressive corrosion compared to conventional fossil fuel-fired units due to the presence of corrosive sulfur, chloride and other trace compounds in an elevated temperature environment. The Kraft recovery boiler also operates in both an oxidizing and reducing atmosphere due to the combustion process. Because of these conditions, the floor and lower furnace walls are constructed using one or a combination of corrosion protection systems: metallic spray coatings or carbon steel tubes, high-density pin studs with refractory, 304L stainless steel tubes, Incoloy® alloy 825 and Inconel® alloy 625 composites tubes, and weld overlay of carbon steel tubes.

Smelt-water reactions

One unique and undesirable feature of a Kraft recovery boiler is the possibility of water entering an operating furnace through a tube leak or external source, resulting in a smelt-water reaction. A smelt-water reaction occurs when water combines with hot or molten smelt, and a violent explosion can result. This concern has resulted in years of study and testing and has prompted enhanced industry standards on unit design, operation and maintenance. Modern units are equipped with the provisions for an emergency shutdown procedure. This is initiated if water is suspected to have entered the furnace of an operating recovery boiler. Refer to Fig. 7 for procedural overview.

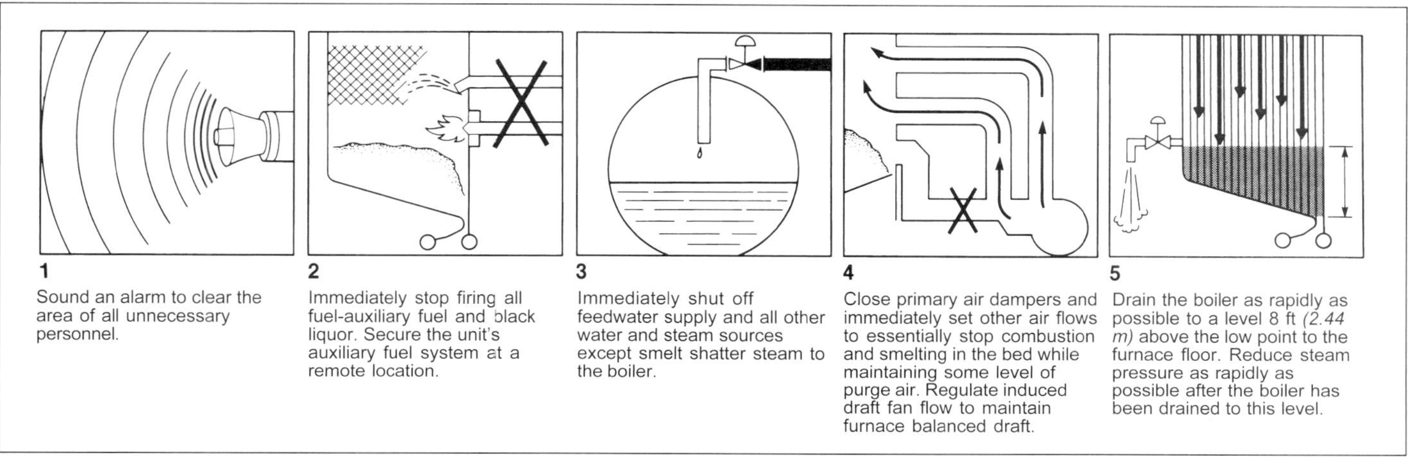

Fig. 7 Emergency shutdown procedure overview for an operating recovery boiler (*adapted from the Black Liquor Recovery Boiler Advisory Committee, October, 2003*).

Inconel and Incoloy are trademarks of the Special Metals Corporation group of companies.

Section VII
Service and Maintenance

This section describes the last element of a successful steam generating system life cycle plan – service and maintenance.

As owners and operators of steam plants search for optimum performance, efficiency, and life cycle for all equipment, issues of maintenance and availability have become increasingly important.

The section begins with a discussion of service and maintenance encountered with all plants, both utility and industrial. A well-crafted service and maintenance program is essential in sustaining the availability of critical steam generating assets and maximizing overall performance and output. Condition assessment is then addressed with detailed discussion about examination techniques, assessment of various components, and analysis techniques for determining remaining life. The effects of cycling operation are also addressed.

Upgrade work on lower hopper of once-through Universal Pressure boiler.

Chapter 44

Maintaining Availability

The design of boiler systems involves the balancing of near-term and long-term capital costs to maximize the availability and useful life of the equipment. Fossil fuel-fired boilers operate in a very aggressive environment where: 1) materials and technology are pushed to their economic limits to optimize efficiency and availability, and 2) the erosive and corrosive nature of the fuels and combustion products result in continuous and expected degradation of the boiler and fuel handling components over time. As a result, the original boiler design is optimized to balance the initial customer capital requirement and the long-term expected maintenance, component replacement, and service costs for a possible operating life of many decades.

When a new power plant is started up, there is a relatively short learning period when the operators and maintenance crews learn to work with the new system and resolve minor issues. This period may be marked by a high forced outage rate, but this quickly declines as the system is broken in and operating procedures are refined.

As the plant matures, the personnel adapt to the new system, and any limitations in the plant design are either overcome or better understood. During this phase, the forced outage rate remains low, availability is high, and the operating and maintenance costs are minimal. The power plant is usually operated near rated capacity with high availability.

As the plant continues to operate, a number of the major boiler pressure part components reach the point where they are expected to be replaced because of erosion, corrosion, creep, and fatigue. Without this planned replacement, increasingly frequent component failures occur resulting in reduced availability. In some instances such as waste-to-energy systems, this period can be as short as one to three years for superheaters because of the very corrosive flue gas composition. However, for most fossil fuel-fired utility boilers operating on their design fuels, major pressure part components are economically designed for more than two decades of operation before economic replacement. Failures of major components such as steam lines, steam headers and drums can cause major, prolonged forced outages. Significant capital expenditures are normally required to replace such components.

A strategic availability improvement program that includes capital expenditures to replace or repair this equipment before major forced outages occur can smooth out and raise the availability curve. Higher availabilities usually require higher maintenance, higher capital expenditures, and better strategic planning. The large expenditures needed for high availability in older plants require a strategic plan to yield the best balance of expenditures and availability.

Strategic plan for high availability

Mature boilers represent important resources in meeting energy production needs. A systematic strategic approach is required to assure that these units remain a viable and productive resource. The more efficient, but older boilers in the system can be the backbone of the commercially available power for a utility.

Emphasis on high availability

Today, the need for high commercial availability is of prime importance to the financial livelihood of a power supplier. This means that the low- cost units in a system must be available for full capacity power production during critical peak periods, such as hot summer days. Competition in the electrical supply industry requires that low-cost units be available so that the system can supply power to the grid at low overall costs. Usually, the large fossil powered units are the lowest cost units in the system. Lost revenue associated with having a large, low-cost unit out of service for repairs can be in excess of one million U.S. dollars per day. Owners are attempting to maintain availability levels of 90% or more on these large workhorses in the system.

The emphasis on maintaining or even improving availability means that a strategic plan must be put in place. Times between planned outages have been increased. Planned outage times have been decreased. Units are being run continuously for more than a year. In some instances, units are run up to four years without being shut down. This requires complete system and component reliability. Problems that were once tolerated can not be permitted to exist.

One example of a change in availability philosophy is illustrated by the technique used to maintain low water-side tube deposit levels in the furnace. It was once common to chemically clean the boiler furnace every four to seven years. Deposit buildup in the unit dictated the cleaning cycle. Chemical cleaning normally required extending an outage time by several days to a week. Today, many units have undergone extensive changes to the boiler water treatment system. Many high pressure boilers have been converted to oxygen water treatment (see Chapter 42). Reports have indicated that oxygen water treatment has drastically reduced the rate at which furnace wall deposits form, and as a result, chemically cleaning the furnace in these high pressure units has been dramatically reduced. The reduced furnace deposits resulting from oxygen water treatment and the associated reduced need for chemical cleaning are significant factors in improving unit availability. This is a good example of how new technology and a systems approach to availability can have dramatic effects on overall operating costs.

Increased capacity and operability

During the early phases of a boiler's life cycle, less maintenance is required to maintain high availability. However, as the unit matures and components wear, more significant steps become necessary to maintain the desired availability. As the plant matures, the need for component replacement becomes expected and routine. Often at this point in the life of a power plant, system demands, fuels and cost structures have changed. Units must operate in ways unforeseen when they were built. They must have greater operability, or the ability to effectively perform their role in the manner required at the time. A plant that meets environmental emissions limits by means of system changes and fuel switching may have introduced capacity problems and operability limits that can become intolerable. Strategic equipment plans must address this issue of operating under new circumstances.

Boiler changes should be considered that incorporate technology advances to increase unit capability, operability and availability. This also provides the opportunity to address current operating issues or changing operating needs. Design changes might be incorporated to accommodate new fuel sources or to more successfully withstand cycling service. The potential for improvement is extensive. However, without a strategic plan, the replacement and repair of boiler components might be unsystematic, and the results may not achieve the desired returns on investment.

Key issues to consider when optimizing a unit include understanding how boilers age, identifying the critical components, and determining the general types of changes and enhancements possible. These elements are then combined with operating experience and equipment condition assessment (see Chapter 45) to develop an integrated strategic plan for each boiler and each component in the boiler.

Strategic plans yield high returns on investment

While the driving force for the availability improvement plan may be to stretch present capacity resources to meet current demands as well as to improve operability and availability, a well conceived strategic plan can also achieve an attractive return on investment. The payback can be realized when the capital improvement results in higher availability, a lower heat rate, higher capacity, reduced routine maintenance, or lower forced outage rates. There have been numerous programs that have significantly improved plant capacity, operability and availability while achieving a payback in as little as one year.

Impact of environmental regulations

In most developed nations today, regulations at national and local levels control air emissions from power plants and other industrial boiler applications (see Chapter 32). Evolving emissions control limits and changing plant operating requirements add another dimension to the strategic plan. These interactions typically take the form of:

1. increasing plant capacity resulting in absolute emissions constraints,
2. increasing plant availability and operation (MWh) resulting in absolute emissions constraints,
3. economic fuel switching, changing absolute emissions and emissions rates,
4. increasingly stringent emissions requirements, reducing plant availability, and
5. reduced plant output due to parasitic load.

To take full advantage of the plant capacity as well as maintenance and upgrade capital expenditures, the addition of appropriate emissions control technology becomes an integral part of the strategic plan for all plants. In fact, combining the addition of emissions control equipment with plant upgrades can result in enhanced power plant economic performance that offsets part or all of the emissions equipment expenditures. Thus, the full strategic plan incorporates appropriate retrofits and upgrades of emissions control equipment.

Maintaining availability

The availability of a boiler is determined by the combined availability of its various critical components. If critical components become unreliable, boiler availability declines. These critical components include the pulverizers and burners, pressure retaining components such as economizers and superheater headers, and the various balance of plant components such as fans, controls, valves, etc.

The following section discusses the factors that affect the availability of key components of a typical boiler. The mechanisms that affect component degradation and reliability are revealed. Strategies are outlined in this section for improving availability through selective replacement upgrades of components that have become unreliable through normal degradation. Examples of these upgrades are detailed in this section.

Pulverizer availability improvement strategies – E and EL mills

Reliable pulverizer performance is an important element of the combustion system, and is essential for responsive power plant operation with good availability. Other subcomponents of the fuel preparation and delivery system such as burners, coal feeders, motors, dampers, coal pipes, combustion controls and combustion air fans must also perform reliably if the combustion system is to ensure good unit availability. See Chapters 12, 13, 14, 25 and 41 for more information about these components.

The first vertical air-swept coal pulverizer by The Babcock & Wilcox Company (B&W) was the E pulverizer introduced in 1937, as shown in Fig. 1a. The EL pulverizer (Fig. 1b) was developed in the 1950s as an upgrade to the E pulverizer. However, there are still a number of E pulverizers in operation today. Both the E and EL pulverizers are of the ball-and-ring (sometimes referred to as a ball-and-race) design.

Pulverizer degradation mechanisms

As discussed in Chapter 13, the E pulverizer is a machine that tends to wear from the inside due to the abrasive action of a continuous stream of fine coal particles. As the throat of the E pulverizer wears out, coal begins to fall past seals and into the lower parts of the pulverizer (*dribble*) causing severe wear to these parts. To prevent higher coal dribble rates as the throat wears, operators must increase air flow to maintain adequate velocity through the ever-widening throat gap. However, as air flow increases, fineness deteriorates, and erosion wear of other downstream pulverizer components increases. Upgrading the E pulverizer can help solve this problem and increase pulverizer availability.

Impact loading and thermal stresses may cause the E pulverizer rings to fail. Under certain operating conditions, there is considerable vibration generated and transmitted to the pulverizer components. As a result, shaft or spring fatigue failures can result which cause the loss of pulverizer availability.

E and EL type pulverizers can experience operating problems such as cracking of the upper or lower grinding rings. In addition, as grinding balls attempt to leave the ball race under the centripetal force of their rotation around the ring, they put asymmetrical loads on the rings and main shaft. Main shaft breakage can result.

E and EL upgrades

Since the introduction of the E and EL pulverizers, a significant number of functional and mechanical improvements have been made. Some of the major improvements leading to increased pulverizer availability and reliability are summarized below.

E to EL conversion The preferred improvement to an E pulverizer is to upgrade it to an EL mill. The EL pulverizer is much less prone to coal dribble problems due to improvements to the bottom grinding ring and

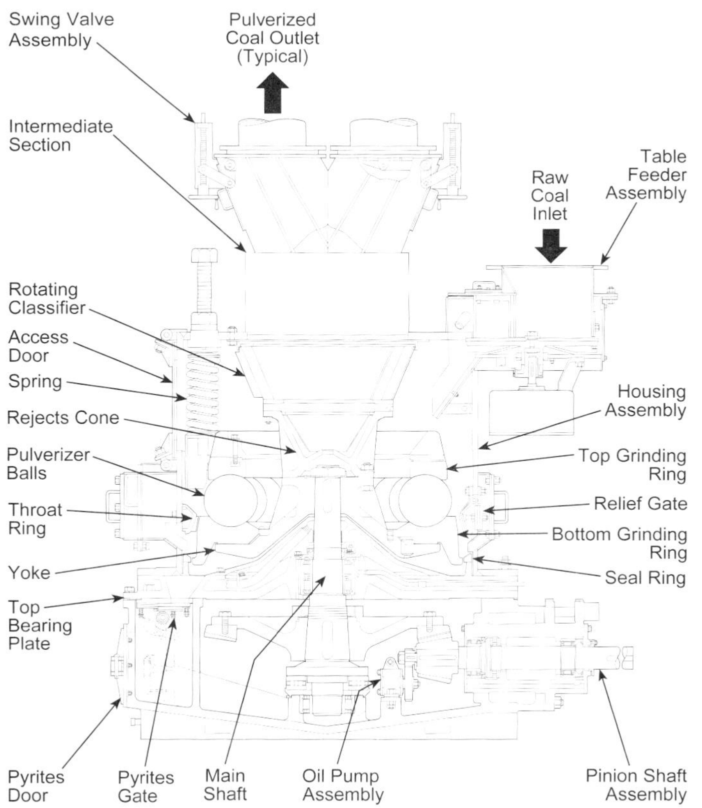

Fig. 1a Babcock & Wilcox E-type pulverizer.

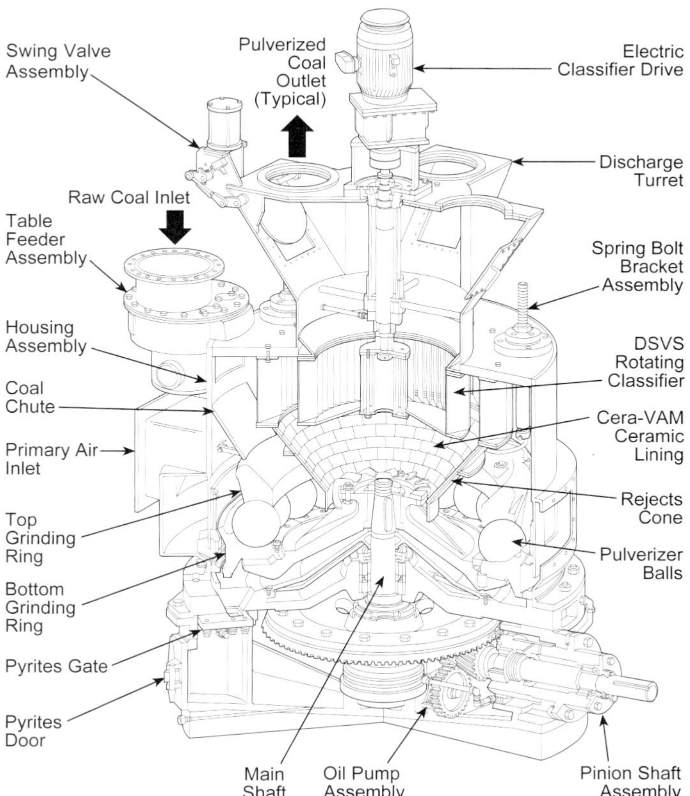

Fig. 1b Babcock & Wilcox EL-type pulverizer.

throat configuration. A complete E to EL conversion as shown in Fig. 2 includes new top and bottom grinding rings, springs, classifier, a more wear-resistant throat, relief gate, and housing units to protect the inside wall of the pulverizer just above the throat. Larger pulverizer motors and primary air (PA) fans may be needed to improve capacity. E to EL conversions have significantly improved the availability, capacity and reliability of E type pulverizers.

EL grinding zone improvements The B&W On-Track™ grinding zone retrofit package incorporates a number of design improvements that increase EL pulverizer availability and reliability by reducing grinding zone overhauls. Five key design changes are included:

1. *Deep dish grinding rings* The upper and lower grinding rings include a deeper and improved grinding track contour that limits radial movement of the balls as they travel around the grinding track. (See Fig. 3). This lowers the asymmetric loading and displacement of the upper ring and lowers the loading on the lower ring and main shaft, helping to prevent failures of the ring and shaft.
2. *Integrally cast metal snubbers* Snubbers (or bumpers) are cast into the upper ring. The snubbers help reduce the excessive side-to-side movement of the top grinding rings. Excessive side-to-side movement causes bending stresses and premature failure of the main shaft. With snubbers, spring and main shaft loading is reduced and breakage is less likely.
3. *Heavy duty upper ring flutes* Heavy duty outer flutes incorporated in the upper ring resist breaking of the fluted areas of the rings, enabling the upper rings to remain in service until they reach the end of their normal wear cycle. Fig. 4 shows the increased cross section of the heavy duty flutes compared to the narrower standard flutes. These larger flutes result in smaller gaps between the flutes and therefore, provide more control over the grinding balls as they orbit around the pulverizer track. This helps reduce stresses on the main pulverizer shaft.
4. *Larger diameter grinding balls* Larger diameter (13.63 in./34.61 cm) grinding balls increase the time between grinding zone overhauls because there is more wear metal on each ball compared to the smaller 12.5 in. (31.75 cm) diameter standard grinding balls. Longer wear cycles associated with these larger balls improve pulverizer availability and reduce pulverizer operating costs.
5. *Internally reinforced lower ring* The lower grinding ring of the EL pulverizer may experience thermal stress cracking over time. These cracks may require replacement of the lower ring, affecting mill availability. Efforts have been made to allow cracked rings to remain functional in the pulverizer without the need for replacement. One solution to bottom ring cracking is the internally reinforced ring. The grinding ring looks like the standard ring externally, but it is internally reinforced with steel. The harder, less ductile wear material of the ring propagates cracks much easier than tough mild steel. The mild steel reinforcement does not crack easily and holds the cracked segments of the ring together so the ring can remain in service.

Erosion protection for EL pulverizers Pulverizers experience internal wear from the erosive properties of coal circulating within the mill. High quantities of silica and alumina in the coal can reduce the life of most internal components. Fig. 5 shows the areas of an EL pulverizer that typically experience high wear from the erosive effects of coal. These areas can be protected by cladding with Cera-VAM® high density alumina ceramic. This ceramic cladding significantly reduces erosion rates and reduces maintenance work during overhauls. The application of ceramic lining materials to existing pulverizers must be done with care to minimize the impact on other operational areas.

Roll wheel pulverizer upgrades

Roll wheel pulverizers are described in detail in Chapter 13. These pulverizers are installed on many modern coal-fired boilers. As with the ball and race mill, wear affects the pulverizer availability. Upgrades can increase wear resistance.

Improved roll wheel design The typical pulverizer roll wheel wears in a pattern shown in Fig. 6. When localized wear reaches a point where the wheel must be replaced, there is considerable metal left in the remainder of the wheel.

To improve wear life, the standard *tires* may be

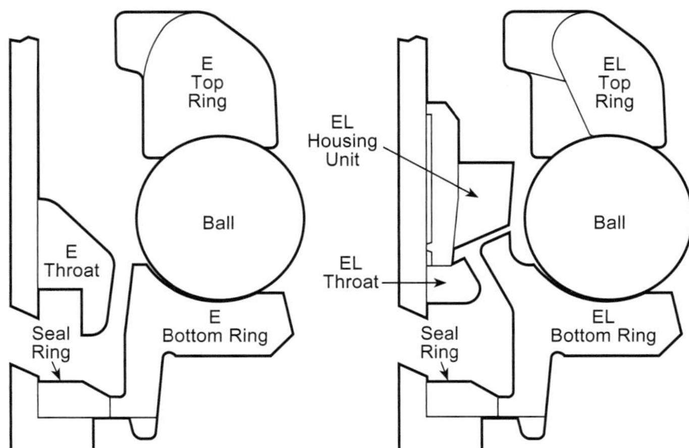

Fig. 2 E to El pulverizer upgrade.

Fig. 3 Improved upper grinding ring for an EL pulverizer.

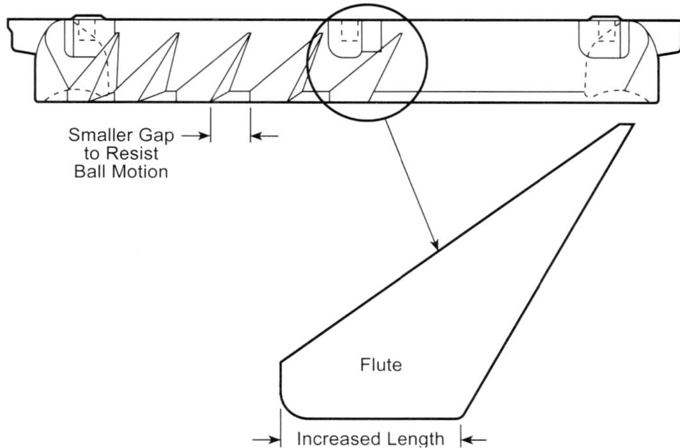

Fig. 4 Heavy duty flutes for upper grinding ring.

upgraded to an asymmetric design that places more material in the high wear area. As a further enhancement, the tires may incorporate a low profile. This lower profile enhances wear life and improves mill performance in certain circumstances. The upgraded tires reduce the frequency of mill overhauls and improve mill availability.

Rotating throat The throat area of a roll wheel pulverizer (see Chapter 13, Fig. 3) is critical to the distribution of air through the pulverizer. This area can experience high erosion rates from the fluidized air/coal particle mixture, which could lead to premature component failure or increased maintenance. High erosion rates can also lower component availability. High velocities and turbulent air flow lead to relatively high pressure drop through most standard stationary throat designs.

The B&W rotating throat (shown in Fig. 7) is designed to improve mill performance while dramatically reducing the maintenance costs associated with stationary throat wear. The wear resistance of the throat segment is increased by applying chromium carbide weld overlay to the upper area of the vanes. The B&W rotating throat incorporates a patented air foil vane design that promotes more uniform air flow within the port, and has resulted in significant reductions in pulverizer pressure drop. In addition, the uniform air flow combined with the carbide weld overlay significantly reduces throat segment wear.

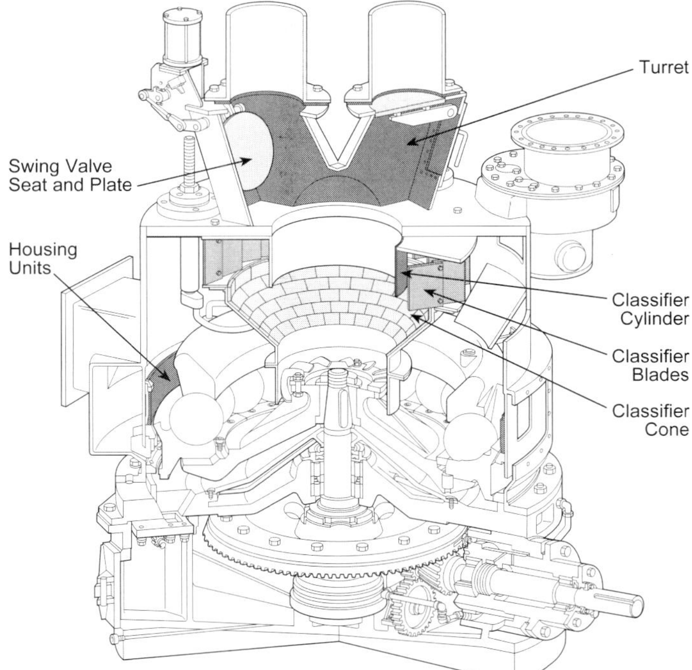

Fig. 5 EL pulverizer with Cera-VAM® ceramic cladding.

Low temperature pressure components

The low temperature pressure components are defined here as those below 900F (482C) metal temperature. These components normally are the economizer, furnace and convection pass walls, drums and flash tanks (steam separators), along with their downcomers and connecting piping and headers. These components suffer from corrosion, erosion, fatigue and overheating. Creep failure of these components only occurs when local overheating occurs.

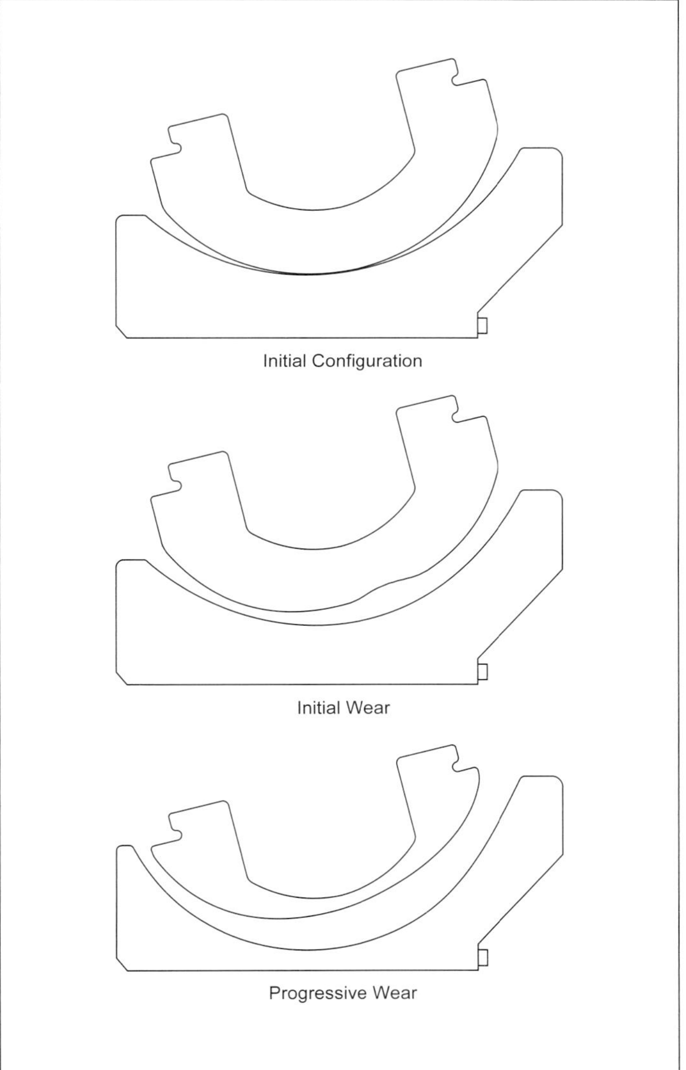

Fig. 6 Typical roll wheel wear pattern.

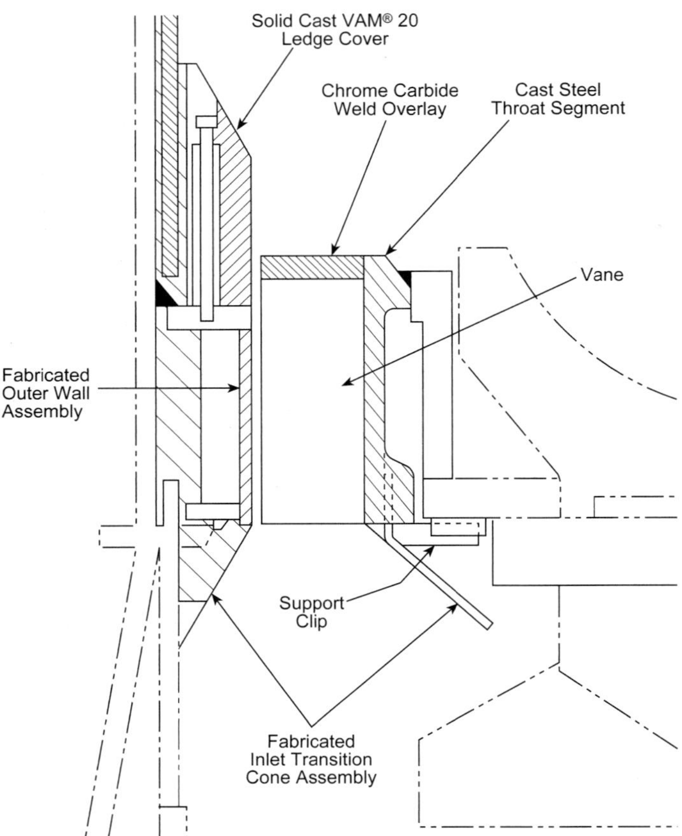

Fig. 7 B&W rotating throat for a roll wheel pulverizer.

Degradation mechanisms

Corrosion Corrosion occurs inside and outside the tubes, pipes, drums and headers of these lower temperature components. Internal corrosion is usually associated with the boiler water, contaminants in the water, and improper chemical cleaning or poor storage procedures (see Chapter 42). External corrosion can be caused by corrosive combustion products, a reducing atmosphere in the furnace, moisture between insulation and a component, and acid formed on components in the colder flue gas zones when the temperature reaches the acid dew point. Corrosion results in wall metal loss. This wall thinning raises the local stresses of the component and can lead to leaks or component failure.

Corrosion may also be accelerated by the thermal fatigue stresses associated with startup and shutdown cycles. Furnace wall tubes, in areas of high structural restraint or high heat flux, often contain internal longitudinal or external circumferential or longitudinal corrosion fatigue cracks in cycled units.

Corrosion fatigue can occur in the steam drum around rolled tube joints. The residual stresses from the tube rolling process are additive to the welding and operating pressure stresses. Corrosion from chemical cleaning and water chemistry upsets acts on this highly stressed area to produce cracking around the seal weld or the tube hole. Extensive cracking can require drum replacement. Corrosion problems are further discussed in Chapter 42.

Erosion Erosion of boiler components is a function of the percent ash in the fuel, ash composition, and local gas velocity or sootblower activity. Changing fuels to a high-ash western United States (U.S.) fuel may lead to more erosion, slagging and fouling problems. Changing fuels might also require a change in the lower temperature convection pass elements to accommodate higher fouling and erosion. The tube wall loss associated with erosion weakens the component and makes it more likely to fail under normal thermal and pressure stresses. Erosion is common near sootblowers; on the leading edges of economizers, superheaters and reheaters; and where there are vortices or around eddies in the flue gas at changes in gas velocity or direction. Such changes are caused by closely spaced tube surfaces, slag deposits, or other obstructions including extended surfaces and staggered tube arrangements.

Fatigue The thermal stresses from temperature differentials that develop between components during boiler startup and shutdown can lead to fatigue cracks. These cracks can develop at tube or pipe bends; at tube-to-header, pipe-to-drum, fitting-to-tube, and support attachment welds; and at other areas of stress concentration. Smaller, lower temperature boilers are less prone to fatigue failures because the thermal differentials are lower and operate over small distances in these units. As unit size and steam temperature increase, the potential for thermal stresses and the resulting fatigue cracking also rises.

Overheating Overheating is generally a problem that occurs early in the life of the plant and can often result in tube ruptures. The nature of failures attributable to overheating is discussed in Chapters 7, 42 and 45. These problems may go undetected until a tube failure occurs. Overheating attributable to operation is generally resolved during the early stages of boiler life. Other problems regarding overheating may be difficult to ascertain, and specialized boiler performance testing (see Chapter 40) is generally required to identify the source and determine corrective actions.

In spite of these aging mechanisms, low temperature components are normally expected to be replaced after more than two decades without major overhauls unless the unit burns a corrosive fuel, burns fuel in a reducing atmosphere, or is improperly operated. When erosion, corrosion, fatigue, or overheating lead to frequent leaking, failures, or the threat of a major safety related failure, then component repair, redesign, or replacement is appropriate.

Hydrogen damage Boilers operating at pressures above 1200 psi (8274 kPa) and 900F (482C) final steam temperature suffer from more complicated aging mechanisms than lower temperature units. These boilers are generally larger than the low pressure, low temperature units and this increases the likelihood of thermal fatigue from boiler cycling. The higher pressures and associated higher furnace wall temperatures make these units more susceptible to water-side corrosion. The high temperatures in combination with any furnace wall internal deposits may promote hydrogen damage of the furnace tubing in areas of high corrosion or heavy internal deposits. Chapter 42 discusses hydrogen damage in more detail. Severe cases of furnace wall hydrogen damage have forced the retirement of older units.

Furnace wall wastage mechanisms associated with staged low NO_x burners The advent of low NO_x (nitrogen oxides) burners with staged combustion for coal-fired units has increased the wastage rates of some boiler furnace walls near the burner zone. Corrosion rates as high as 0.040 in. (1.02 mm) per year have been reported for small local regions in furnaces of units with staged combustion. The pattern of high corrosion rates is not uniform throughout the furnace or from boiler to boiler. The pattern seems to be unit-specific and rather unpredictable as to its extent and rate. Normal combustion gases with unstaged combustion lead to an oxidizing flue gas that is not particularly corrosive. However, in staged combustion there is incomplete mixing of the air and fuel, and pockets of a reducing gas are formed in the furnace in the burner zone. The reducing gas is high in H_2S, CO and unburned fuel (carbon and iron pyrites), and is low in O_2. This gas mixture together with the sulfur and chlorides in the fuel does not allow the tubing to form a protective oxide outer covering. High corrosion rates can occur in areas in contact with the reducing gas and unburned fuel.

The corrosion rate in the high wastage areas of the boiler furnace is temperature dependant. As a result, the corrosion rate is higher for high pressure units with staged combustion than for lower pressure units with the same burner arrangement. Inside diameter (ID) deposits in the furnace wall tubing also raise the metal temperature and increase the outside diameter (OD) wastage rate. The highest corrosion rates typically occur in localized areas adjacent to and above the burner elevation up to just below the overfire air ports, where the reducing gas concentrations are the highest adjacent to the walls.

Availability improvement strategy examples

Economizer upgrades One example of an availability improvement strategy involves an upgraded bare tube economizer. A 600 MW coal-fired boiler was designed with a staggered, finned tube economizer. The economizer, shown in Fig. 8, was equipped with longitudinal fins. In clean condition, these fins increase heat transfer and reduce the amount of tubing required.

The economizer experienced severe flyash plugging. As the plugging spread through both banks, system gas flow resistance more than doubled. The increased flue gas resistance pushed the boiler fan to its limit and restricted boiler load. High velocity gas lanes were created around the plugged areas. Extremely high ash velocities caused tube erosion and resulted in numerous tube failures. Unit availability and reliability were greatly reduced.

An engineering evaluation determined that the existing finned economizer was not appropriate for the high ash coal used as the main fuel. The staggered tube pattern only increased the severe flyash plugging. An analysis showed that an in-line bare tube economizer with equivalent performance could be fitted into the existing space. Fig. 9 illustrates the upgraded in-line economizer design.

The upgraded economizer included the following features:

1. Vertical spacing between tubes was decreased to allow installation of more tube rows and to minimize gas-side resistance.
2. An in-line arrangement was used to reduce gas-side pressure drop and potential for ash erosion.
3. The economizer tube diameter was increased from 1.75 to 2 in. (44.5 to 50.8 mm) to achieve a higher flue gas velocity and better convection heat trans-

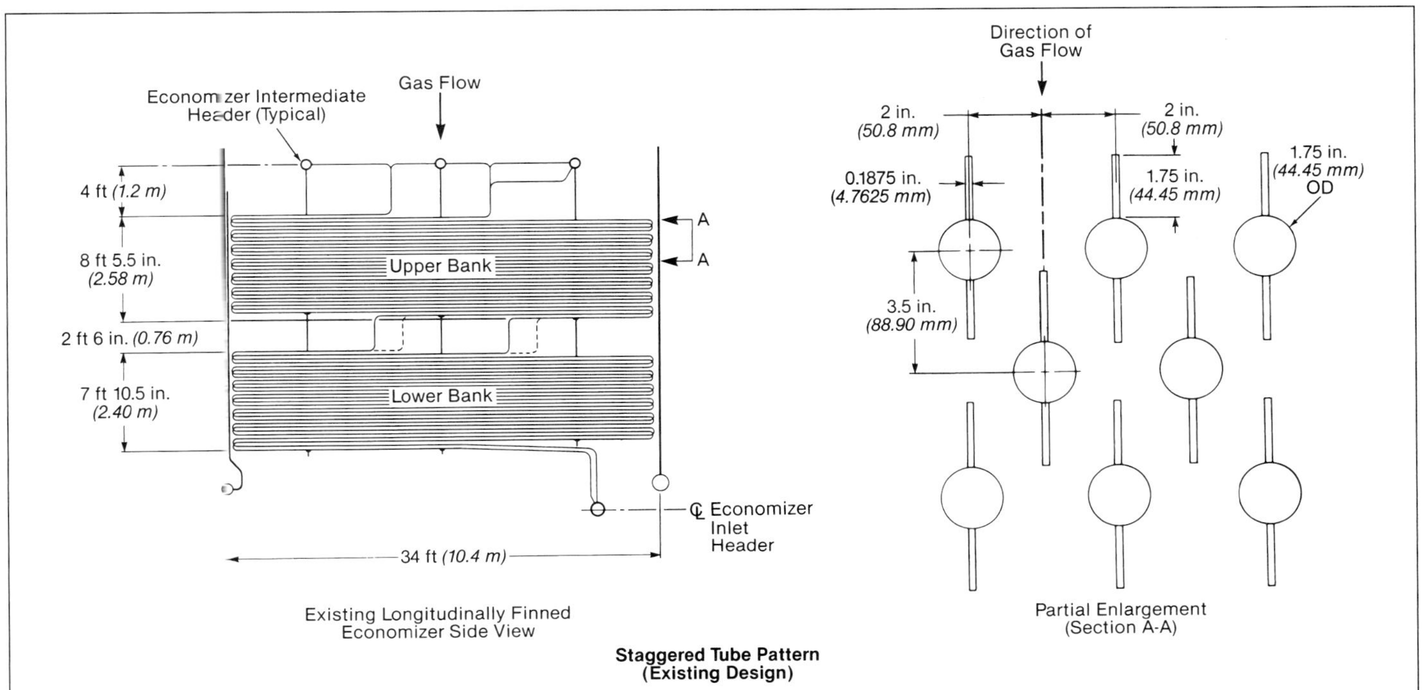

Fig. 8 Original staggered, finned tube economizer that experienced severe flyash plugging.

fer without exceeding the allowable velocity limit for the percent ash and percentage of abrasive silica and alumina constituents in the fuel. Bare tubes arranged in-line are most conservative in such hostile environments, are least likely to plug, and have the lowest gas-side resistance per unit of heat transfer. The weight of the new economizer did not exceed the original weight. Also, the new economizer required no more space than the original design.

4. The tube bends were protected by new erosion barriers on top and bottom of both banks. Properly designed perforated barriers do not noticeably reduce the effective heating surface, but they are very effective in throttling the flue gas flow across the tube bends along the enclosure walls. Without barriers, the open space between the return bends and the enclosure walls would be the path of least resistance, and flue gas would stream through these gaps at very high velocities. The effect would be excessive erosion of the bends, reduced overall heat transfer across the banks, and possible damage to uncooled casing. The existing inlet and intermediate headers were reused. Access doors and platforms did not have to be relocated or modified.

The upgraded economizer has experienced no flyash plugging. The recurring tube erosion problem has also been eliminated. As a result, unit reliability and availability have been significantly restored. The boiler operates at full load and has not been limited by economizer problems.

Furnace wall corrosion reduction Another example of an availability improvement strategy involved the reduction in furnace wall corrosion associated with staged combustion low NO_x burners burning higher sulfur coals. As discussed earlier, staged combustion can lead to localized corrosion of the furnace wall tubes in areas where reducing conditions are produced near the walls. In this example, the corrosion rates in some regions of the furnace approached 0.040 in. (1.02 mm) per year. The areas of high corrosion on the furnace side walls are shown in Fig. 10. The corrosion was highest on the side walls near the corners of the furnace at the burner elevation.

The high corrosion pattern shown in Fig. 10 was caused by pockets of high CO, unburned fuel and H_2S formed in areas adjacent to the furnace walls (see Fig. 11). These areas resulted in high corrosion rates that led to repeated tube leaks and boiler outages. Considerable outage time was required to repair leaks and replace tube wall panels on a recurring basis. The unit suffered from low availability due to furnace corrosion.

The current method of combating this high corrosion is to protect the tubing with a cladding that is more resistant to the corrosive atmosphere. High chromium nickel alloys similar to Inconel® alloy 622 (21% Cr, 13% Mo, 4% Fe, 3% W, and balance Ni) are being clad over the boiler tube material. Cladding can be applied onto the tubing or furnace panels, or can be formed as part of the tube during the production process (coextrusion). Other coatings have been tried over the years, such as high-velocity oxy-fuel (HVOF) spray coatings. These spray coatings are satisfactory where low corrosion rates [less than 0.010 in. (0.25 mm) per year] are involved. However, where more severe corrosion is encountered, the use of weld clad tubing or coextruded tubing for corrosion protection is most often chosen. Fig. 12 shows the laser cladding of a furnace wall panel with Inconel alloy 622. In this example, the panel was fabricated, laser clad, and then installed in the side walls to protect the walls from corrosion.

Fig. 13 shows a laser clad sootblower opening. This opening had experienced rapid corrosion rates due to

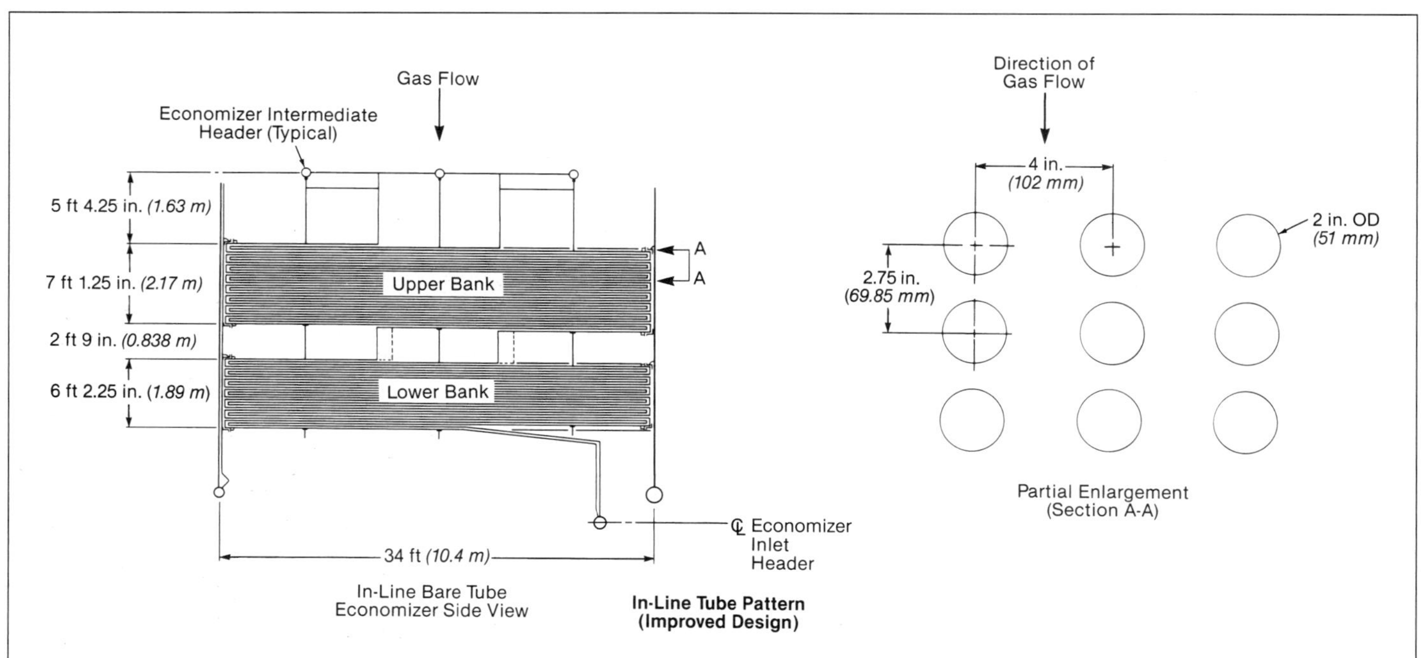

Fig. 9 Upgraded bare tube in-line economizer for high ash coal.

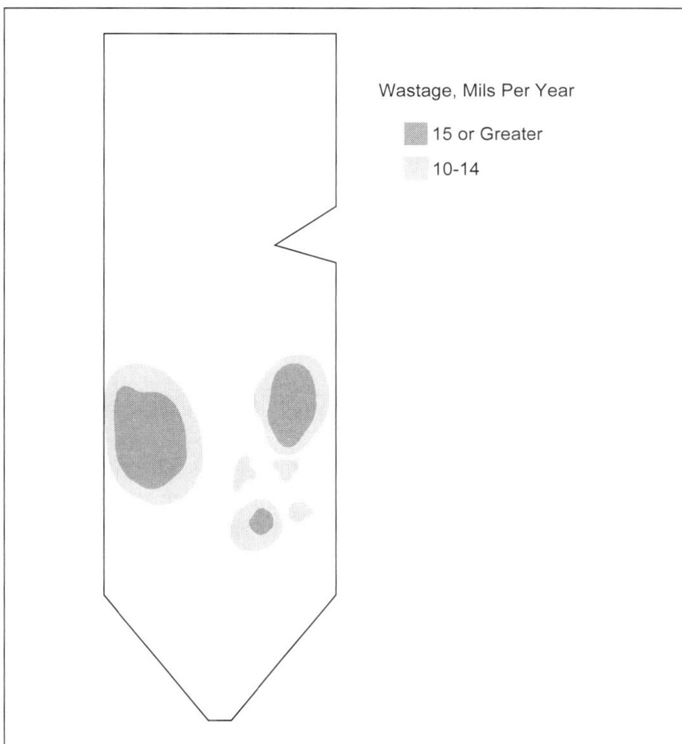

Fig. 10 High corrosion pattern on furnace side walls.

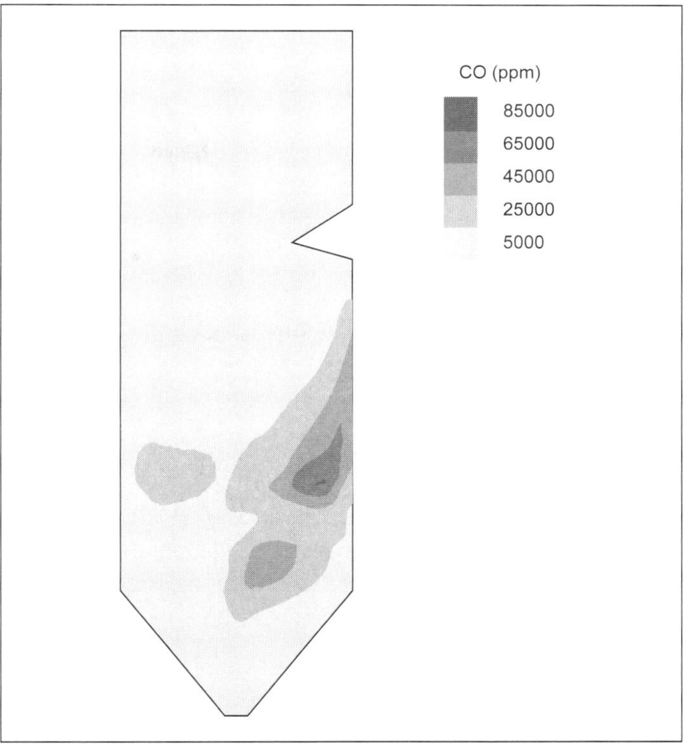

Fig. 11 Computer model showing areas of high CO near the side walls.

adjacent reducing conditions and the action of the sootblowers. Inconel alloy 622 has been shown to dramatically decrease corrosion (as well as erosion) in areas that have experienced high wastage.

Boilers that are experiencing corrosion from staged combustion conditions should be carefully monitored. Wall thickness measurements at suspected high corrosion areas should be taken at frequent intervals (e.g., yearly or every two years). Maps of the wall thickness changes (such as shown in Fig. 10) can help determine which areas to protect with cladded tubing. With the lower corrosion rates associated with cladded tubing, the frequency of weld repairs and wall panel replacements is dramatically reduced. The beneficial effects on unit availability can be dramatic.

Hopper panel impact protection Many units burning western U.S. coals and to a lesser degree eastern U.S. coals experience slagging problems. Large accumulated masses of slag can fall onto the hopper slopes of the lower hopper area of the boiler furnace (see Fig. 14 for a hopper design). Falling slag results in very high impact loads on the hopper panel tubes and the supporting structure. Tubes erode, deform, crack, and eventually leak. Hopper supports deform and lose their functionality. This situation may lead to unexpected major outages for lower hopper repairs or replacement.

An innovative hopper design has been patented (U.S. patent no. 5,692,457) that improves the ability of the hopper to withstand these massive slag falls. (See Fig. 14.) The heart of the system is the crush tube that is designed to absorb the impact load. The crush tube is replaceable and located under the hopper panel and is designed to deform preferentially (or crush on impact loading) compared to the pressure-containing tube of the hopper panel and the support truss members. In addition, the hopper panel tube is typically made thicker to improve its ability to withstand falling slag. The hopper tubes can also be weld clad to resist erosion and improve longevity in this harsh environment as the ash and slag slide down the hopper.

The combination of a reinforced hopper tube with the impact absorption capability of the replaceable crush tube improves the likelihood that the hopper can withstand a typical slag fall. The crush tubes that have been deformed are replaced during regular

Fig. 12 Laser cladding process on a furnace wall panel (*courtesy of Praxair Surface Technologies*).

Fig. 13 Laser clad sootblower opening.

maintenance outages. Hoppers utilizing this unique design concept have performed well in service for many years and have improved unit availability.

Wear resistant materials Most older boiler components do not contain highly wear resistant materials in high erosion areas. These areas, such as within coal pulverizers, require frequent and sometimes costly maintenance. Down time to repair and replace these eroded components reduces availability.

Advanced materials have been developed specifically for boiler fuel handling applications. Materials have evolved from basic low carbon steels to ceramic linings. These linings include high density, high alumina ceramics and silicon carbide ceramics. The alumina ceramics with 96% alpha alumina and a density of about 232 lb/ft^3 (3716 kg/m^3) are ideal for components exposed to high velocity coal particles. These alumina linings provide up to ten times the wear resistance of carbon steel. The nitride-bonded silicon carbide ceramics have good erosion and corrosion properties, and they provide high thermal conductivity and superior thermal shock resistance. These ceramics are used in the most severe environments.

Typically, ceramic linings are used in coal transport equipment, pulverizers, piping, exhaust fans and burner nozzles. The extent of application depends upon the relative erosive nature of the specific coal and the associated velocities. Table 1 offers guidelines for applying ceramic linings to coal piping.

Ash composition affects the erosiveness of the coal. Large percentages of clay, alumina, silica, pyrite and quartz also increase coal's abrasiveness. The guidelines given in Table 1 assume a silica composition of 40%. High alumina and silica content increases the abrasiveness of the coal. Following these guidelines can help improve unit availability by reducing repair frequency and the associated replacement time.

Additional availability improvement strategies are contained in the following chapters:

Chapter 7 discusses various materials that might be used in upgrades for new components.

Chapter 20 discusses economizer design.

Chapter 42 contains information about proper water chemistry and layup procedures.

Chapter 45 discusses methods for evaluating the condition of low temperature components.

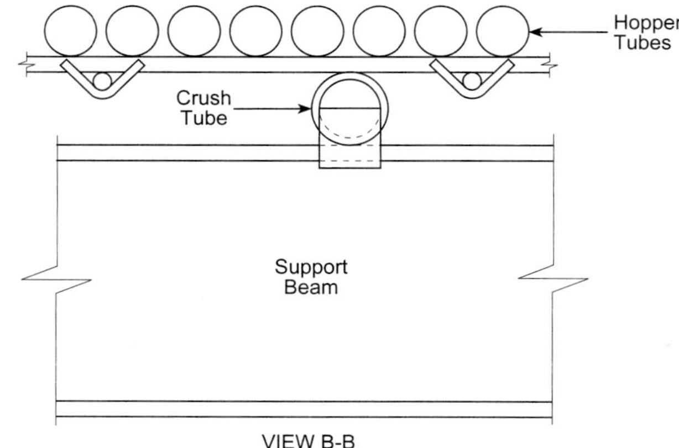

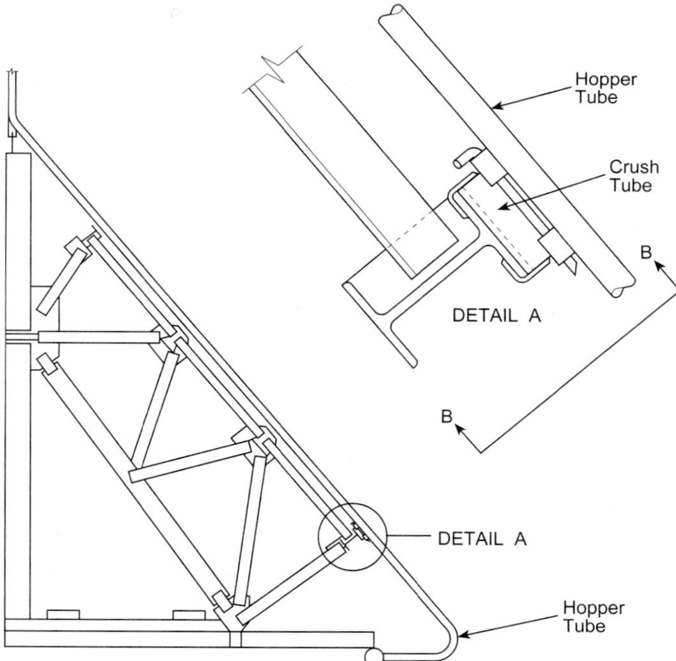

Fig. 14 Impact resistant hopper design.

Table 1
Recommendations for the Application of High Density Alumina Ceramic Linings in Coal Piping

Ash Content of Coal	Recommendations
Less than 6%	No ceramic lining required.
6% to 9%	Use ceramic lining from the pulverizers to the first major bend plus two pipe diameters beyond the bend. Line all short radius bends and two pipe diameters beyond these bends.
More than 9%	Same as above except, also line all bends and the pipe two diameters beyond these bends.

High temperature pressure components

The high temperature pressure components are defined here as those that operate at or above 900F (482C) metal temperature. The components normally at this higher temperature are the superheaters, reheaters and attemperators with their connecting tubing and piping. These components operate in a temperature range high enough where they can experience degradation from oxidation, fuel ash corrosion, carburization, graphitization and creep. These are the high temperature failure modes. Corrosion, erosion, fatigue, and overheating are also possible in a manner similar to that discussed in this chapter on the lower temperature components. (See Chapter 45 for information on evaluating the condition of these components and some of their common failure modes.)

High temperature creep rupture and creep fatigue failure (see Chapters 7 and 45) are the two primary aging mechanisms for the high temperature components. All components that operate at temperatures above 900F (482C) are subject to some degree of creep. Most of the tubes, piping and headers from the primary superheater to the turbine, including the superheater and reheater, are designed to operate in the creep regime. As a result, most of these high temperature components have a finite useful life and can fail by creep or creep fatigue after 20 to 40 years of operation. Replacement and redesign must be considered in any strategy for operating high temperature, high pressure boilers.

Chapter 45 discusses methods for predicting creep ruptures in high temperature superheaters and reheaters, and for assessing the condition of high temperature piping and headers. Availability of these high temperature units in the later years of operation generally falls, mainly because of creep ruptures and creep fatigue failures. It is common to replace a superheater after 25 years of service due to creep rupture incidents. Superheater outlet headers have also been replaced after 25 years due to creep fatigue cracking. Fig. 15 illustrates the typical replacement age for reheaters in power boilers.

The aging process and rate of component degradation vary from unit to unit. Table 2 presents the component replacement sequence for a typical high pressure, high temperature unit.

High temperature component degradation mechanisms

Oxidation The most common degradation mechanism for superheaters and reheaters is tube OD oxidation. At higher temperatures the oxidation rates from metals increase. Chapter 7 discusses the oxidation limits placed on tube metals to prevent rapid oxidation. Superheater and reheater tubes are designed with OD metal temperatures below the oxidation limits. However, some oxidation does take place. The rate of oxidation is typically 0.001 to 0.002 in. (0.025 to 0.051 mm) per year. Failures from oxidation do not usually occur to tubes unless the oxidation process is accelerated by overheating or sootblower activity.

When oxidation does occur it appears as wastage on the upstream side of the tube OD. Flat spots on the OD or thin spots in the tubing may be obvious where the wastage has occurred. The wastage is usually highest on the upstream side of the leading edge tubing near the sootblowers. Wastage from oxidation may also occur on the tubing OD near the outlets, or near changes to higher material grades (higher metal temperatures). When oxidation occurs, the ID oxide scale thickness will also usually be thicker than expected for the hours of operation. Fuel ash corrosion can accelerate oxidation wastage and is discussed below.

Changes in operating conditions that are associated with changes in fuel type or changes in burner designs may alter the operating temperature of a tube. A temperature increase above the oxidation limit can cause rapid wall loss from oxidation. When this occurs, tube failures can follow. The tube that is thinned by oxidation will ultimately fail by a creep mechanism due to the higher stresses in the thinner wall (see *Creep* section below).

Tubes that operate near sootblowers are periodically cleaned by the sootblower action. This action removes some of the protective oxide from the tube OD. The tube regenerates this protective layer, but in doing so, some wall is lost to form the protective oxide. Over time this repeated oxidation/cleaning/oxidation cycling can lead to wall loss that is sufficient for creep fail-

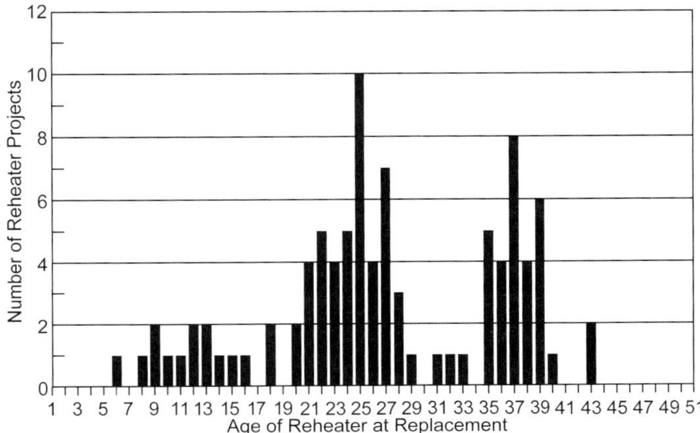

Fig. 15 Reheater replacement projects by age.

Table 2
Component Replacement Schedule for a Typical High Temperature, High Pressure Boiler

Typical Replacement Period (Years)	Component Replaced	Cause for Replacement
20	Miscellaneous tubing	Corrosion, erosion, overheating
	Attemperator	Fatigue
25	Secondary superheater (SSH)	Creep
	SH outlet header	Creep fatigue
	Burners and throats	Overheating, corrosion
30	Reheater	Creep
35	Primary superheater and economizer	Corrosion
40	Lower furnace	Overheating, corrosion

Note: The actual component replacement period is highly variable depending on the specific design, operation, maintenance and fuel.

ures to occur. Often, oxidation around the sootblower area is mistakenly labeled as coal ash corrosion. Coal ash can cause corrosion failures that appear to be similar to oxidation failures (see section below). Oxidation, however, does not rely on a corrosive ash.

Fuel ash corrosion Fuel ash corrosion is sometimes very aggressive, but usually does not occur or becomes only a minor problem. It is sometimes confused with oxidation because both occur in the same components and have similar appearance. Fuel ash corrosion is the wastage of tubing caused by a chemical reaction with the fuel ash. Elements in the fuel can react with the normally protective oxides. Chloride and sulfur are examples of elements that degrade the oxidation resistance of the metal tubing and markedly increase oxidation rates.

Usually, in fuel ash corrosion, the ash forms a liquid layer on the tubing during unit operation. Complex alkali metal salts form the liquid corrosive layer in some coal-fired boilers. $Na_3Fe(SO_4)_3$ is one member of the complex sodium or potassium metal (Fe or Al) sulfate family that forms a corrosive liquid film on superheaters and reheaters in some coal-fired boilers. The liquid layer is a low melting point mixture of compounds formed by the ash and it dissolves away the normally protective oxide scale. In other cases, corrosive liquid deposits can be formed by vanadium compounds or chlorides in combination with sulfates.

It is difficult to distinguish between oxidation and fuel ash corrosion. They may be similar in appearance. One way to distinguish if a superheater or reheater is suffering from oxidation or fuel ash corrosion is to analyze the OD scale on the tubing with x-ray diffraction and chemical analysis. Metallographic examination of the tube metal near the OD surface may also help show the advance of corrosive compounds along tube grain boundaries. If the tube scale is shown to have high percentages of elements or compounds [such as $Na_3Fe(SO_4)_3$] associated with fuel ash corrosion, then fuel ash corrosion is present, not just oxidation. Oxidation by itself does not show evidence of high percentages of these elements or compounds, although low concentrations may be present.

Carburization Although some carburization is often associated with fuel ash corrosion, carburization of superheaters and reheaters is rare. It has become more common with the advent of staged combustion for low NO_x burners. Carburization occurs when incomplete combustion of the coal occurs before the flue gas passes through the superheater or reheater. Incomplete combustion leads to increased amounts of CO and unburned fuel contacting the superheater tubes.

Metal tubes require an oxidizing atmosphere to form a protective oxide film on the surface of the tube. When localized reducing conditions exist around the tube at an elevated temperature, the tube can not protect itself from wastage from the sulfur and other corrosive species. In addition, both the unburned carbon in the fuel, and flue gas CO, react with the dissolved chromium in the tube to form chrome carbides.

Carbon diffuses into the surface of the tube, raises the carbon level, and forms chrome carbides. If this occurs, there is so much carbon in the metal that all of the available chromium reacts to form chromium carbides. When the chromium is tied up as chromium carbide it can not strengthen the metal and can not help form protective chromium-rich oxides. Essentially, the affected portion of the tube near the OD becomes a weak low-chromium steel material that is less corrosion-resistant at the elevated temperatures encountered in superheaters and reheaters.

Graphitization Carbon in both carbon steel and carbon-moly tubes (example: SA209-T1a) is free to diffuse through the metal. Carbon is most stable as graphite in the metal, so with time, the carbon may coalesce into graphite flakes. These flakes may line up in a linear and continuous spiral pattern along the tube. When they form this pattern, the graphite forms spiraling planes of weakness in the tube. Under the influence of operational stresses, cracks can develop along the spiraling paths of graphite. Sudden tube failures occur with no warning. Reliability suffers when this process nears its final phase, as tube failure rates increase.

Graphitization can occur when carbon steel or carbon-moly steel tubes are heated to operational temperatures above about 900F (482C) for long periods of time. Not all units experience graphitization of the carbon or carbon-moly tubing. It is currently not clear why some tubes undergo graphitization and others do not. When severe graphitization occurs, the tubes fail with little ductility and the fracture occurs in a spiral around the tube. Tubes with chromium do not suffer this problem because the carbon becomes associated with the chromium as chromium carbides. (See Chapter 7 for more information.)

Creep Chapter 7 discusses creep and creep rupture of tubing. Superheaters and reheaters will eventually fail by creep if other failure mechanisms are not present. Most of the reheaters replaced in Fig. 15 were replaced due to creep rupture failures. Overheating due to some form of internal obstruction can lead to creep failures in short periods of time. Normally, creep rupture failures take many years to occur. As the frequency of these failures increases, then component replacement is required in order to maintain unit availability.

Availability improvement strategy examples

Superheater upgrade to improve fuel ash corrosion resistance One example of an availability improvement strategy involves an upgraded superheater. This eastern U.S. utility had recurring tube failures in the secondary superheater of an oil-fired boiler, and the boiler suffered from low availability as a result of these failures. Fuel ash corrosion from the fuel oil caused failures in the tubes exposed to the higher operating temperatures. Fig. 16 illustrates the arrangement of the boiler and shows the problem superheater.

The furnace gas temperature at the leading edge of the superheater was nearly 2800F (1538C). High superheater tube temperatures resulted from this unusually high furnace exit gas temperature (FEGT). External tube corrosion was accelerated by this high temperature and by the high vanadium and sodium contents in the oil. Tube failures occurred so frequently that the bottom three rows of the horizontal superheater had to be replaced every 18 months.

An upgrade in superheater design usually requires field testing to define the operating conditions, followed by a detailed engineering study. To correctly design the upgraded superheater, boiler operating temperatures were obtained. A computerized boiler model was used for the performance testing. This system revealed that the actual FEGT was 125F (69C) higher than the design value. The high FEGT was a result of furnace design and burner arrangement. Because it was impractical to redesign the furnace or change to a less corrosive fuel, and the owner did not wish to lower the unit rating in order to lower the corrosion rate, a design study concentrated on a superheater upgrade.

A new superheater was designed using the actual FEGT that was obtained from the computer model of the unit. The new design accommodated the higher operating metal temperatures and allowed the tubes to continuously function at the higher temperatures without producing overheat or corrosion related failures. In the final upgrade arrangement, the three lowest tube rows that had been attacked by corrosion were replaced with Incoclad® 671/800HT®, a corrosion resistant coextruded high chromium nickel alloy material. The tubing was made by co-extruding an Inconel alloy 671 layer 0.070 in. (1.78 mm) thick on the outside of the inner core of Incoloy® alloy 800HT. The outer layer of Inconel alloy 671 provided the high corrosion resistance while the inner core of Incoloy alloy 800HT provided the high strength at the elevated tube temperatures. The remaining tubes in this bank were replaced with a stainless steel material to accommodate the higher metal temperatures. A new support method that allowed greater material expansion at the elevated operating temperatures was included.

The modified superheater allowed the unit to run with high availability and at full rated capacity. The fuel ash corrosion problem was greatly reduced.

Superheater upgrade to restore full load capability and improve availability This example of a superheater and reheater upgrade also involved the upgrade of the outlet headers to P91 (SA335P91) material. The boiler was a B&W coal-fired unit with single reheat. The radiant boiler provided 1,600,000 lb/h (201.6 kg/s) main steam at 1050F (566C) and 2203 psig (15.2 MPa), and 1,263,800 lb/h (159.2 kg/s) reheat steam at 1050F (566C) and 525 psig (3.6 MPa). The goals of the project were to restore unit capacity and improve availability.

Steam generator reliability had been high until the late 1980s, when superheater components began to suffer from the latter stages of normal aging processes, including outlet header maintenance. The unit had also lost 10 MW of capacity. Subsequent study suggested that an increase in throttle pressure of 70 psig (483 kPa) would restore the lost 10 MW. This meant that the boiler steam-side pressure drop needed to be reduced by approximately 70 psig (483 kPa).

The project involved an upgraded superheater, reheater and primary superheater together with the superheater and reheater outlet headers as shown in Fig. 17.

Selected features of the redesigned high temperature components include:

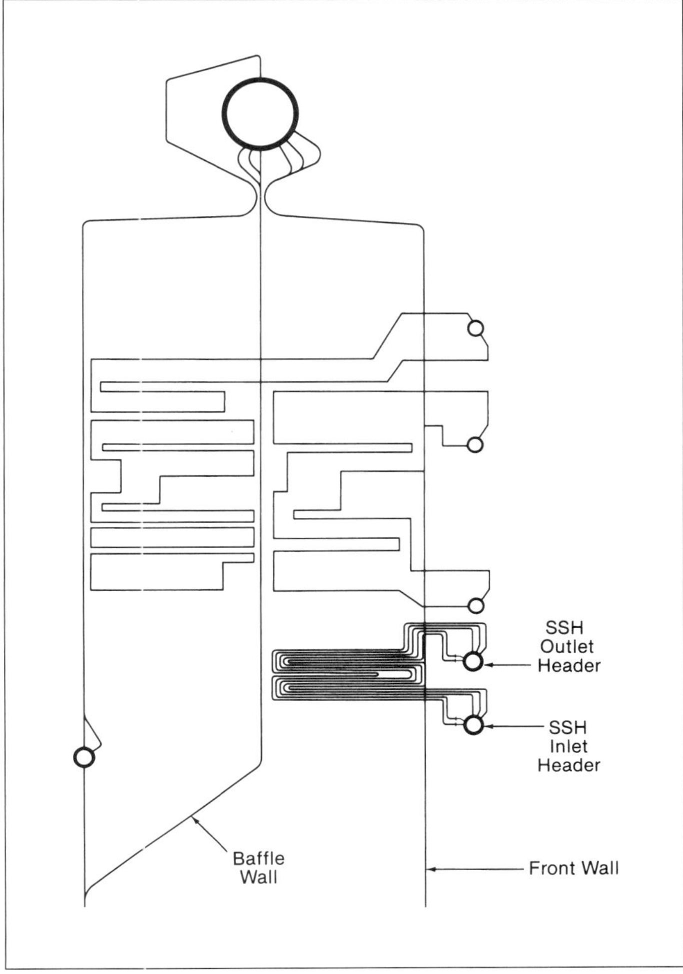

Fig. 16 Boiler arrangement showing secondary superheater (SSH) directly above furnace where gas temperatures reach 2800F (1538C).

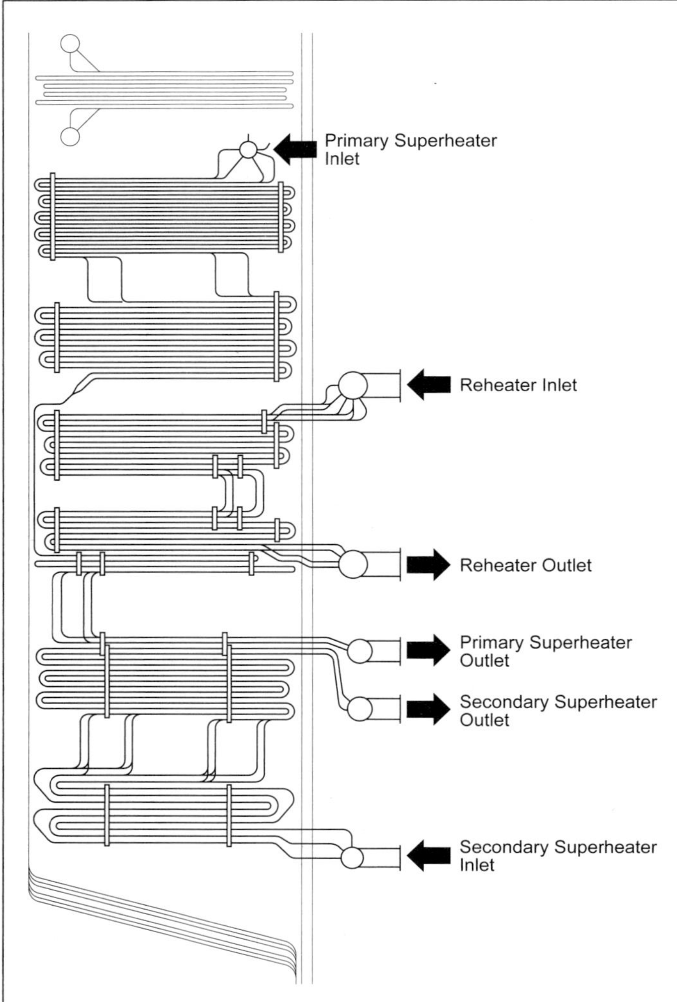

Fig. 17 Upgraded superheater and reheater sections and headers.

1. Tube material of T91 (SA213T91) was used in the secondary and reheat outlet banks to provide longer component life and reduce steam-side pressure drop.
2. Reduced diameter reheat outlet tube legs and header stubs [2 in. (5.08 cm) OD versus original 2.5 in. (6.35 cm) OD] were used to improve flexibility and reduce stress in an area where tube failures had been occurring.
3. Secondary superheater outlet header material of P91 (SA335P91) was used to reduce size and increase component life.
4. The number of parallel steam flow paths in the primary superheater (PSH) inlet bank, PSH intermediate bank and secondary superheater (SSH) outlet bank were increased to reduce steam-side pressure drop. A reduction in steam drum to SSH steam outlet pressure drop by 70 psig (483 kPa) was accomplished by increasing the number of parallel steam flow paths and employing upgraded tube alloys permitting thinner tube walls. The superheater pressure drop reduction permitted a corresponding increase in turbine throttle pressure to regain the original 10 MW higher generator output.

Superheater header upgrades Chapter 45 describes the mechanisms for reheater and superheater outlet header failures. The main failure modes are tube-to-header weld failures, ligament cracking and nozzle cracking. These failures are caused by the combined effects of thermal cycling and creep, often called *creep fatigue*. Advanced design standards for headers have been developed to help reduce these failures. When an outlet header must be replaced, a design upgrade should be considered. An upgraded header can provide long-term, reliable service even under cycling conditions.

Nozzle cracking is associated with the weld in a tee nozzle or in the internal sharp corners of a forged design. Any header upgrade could include a forged tee to eliminate the welded connection. The forging would include generous radii to reduce stress concentrations.

Under cycling service, the tube-to-header weld area may develop cracks due to inadequate tube leg flexibility or due to thermal stresses in the header. Cracking due to inadequate tube leg flexibility can be overcome by providing longer or more flexible tube legs between the furnace penetration and the header. However, this modification may require header relocation and/or relocation of the tube penetrations.

Relocating the tube penetrations around the header may also be necessary to avoid header ligament cracking. Many headers were designed with closely spaced or non-symmetrical tube penetrations, as shown in Fig. 18, that are more prone to creep fatigue ligament

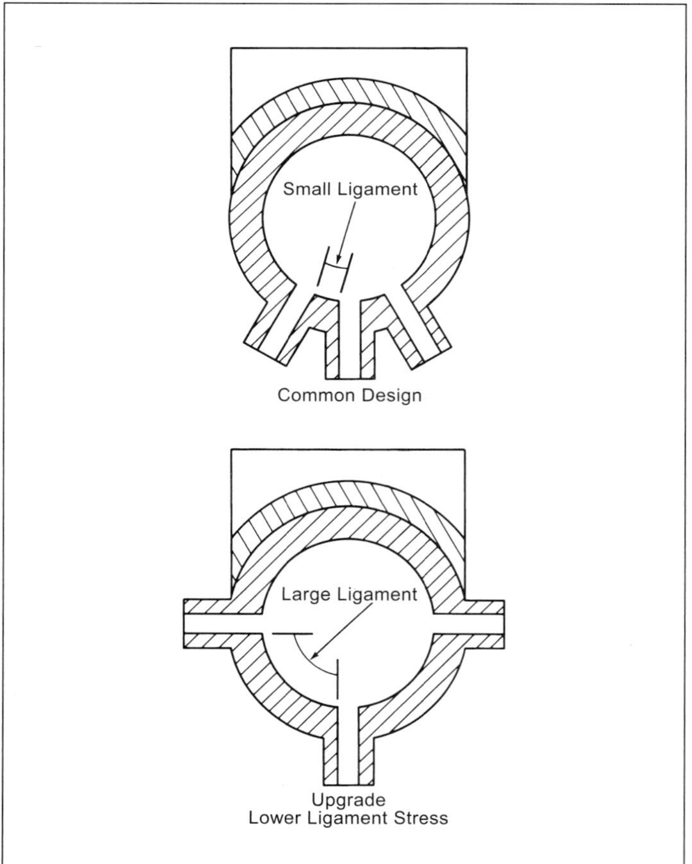

Fig. 18 Increased ligament size for longer design life.

cracking. Studies have shown that creep fatigue cracking can be decreased by using larger ligaments. Upgraded headers with widely spaced tube penetrations provide larger ligaments and lower ligament temperatures and stresses. As a result, header life is significantly increased.

Redesign of the tube-to-header weld configuration, as shown in Fig. 19, can also increase the life of a high temperature header. Chamfering the tube end and redesigning the weld preparation allows the weld to be made with a near full penetration tube-to-header weld. Reduction of the large partial penetration area at the end of the tube stub reduces the stresses around the weld joint and reduces the rate of crack propagation from this area of the weld joint.

The tube hole in the header creates a sharp discontinuity at the intersection of the hole and header internal surface. Very high stresses associated with this sharp corner cause creep fatigue cracks to initiate from the hole. A redesigned header may also be fabricated with a chamfer at the tube hole penetration as shown in Fig. 19. The chamfer lowers the stress around the hole and reduces creep fatigue cracking.

The most common upgrade for a high temperature header is an upgrade in material. American Society of Mechanical Engineers (ASME) SA335P11 headers may be upgraded to SA335P22 or P91 materials, and SA335P22 headers may be upgraded to SA335P91. The material upgrade generally results in a dramatic increase in header life, and in many cases, it is anticipated that these upgrades may nearly double the life of certain header components.

Additional availability improvement strategies are contained in the following chapters:

Chapter 7 discusses various materials that might be used in upgrades for new components.

Chapter 19 explains boiler, attemperator, superheater and reheater design.

Chapter 45 discusses methods for evaluating the condition of high temperature components.

Circulation upgrades

Cycling boiler upgrades

Many older and larger fossil utility boilers were not designed to accommodate frequent on/off cycles. Design criteria for cycling service boilers are limited due to the lack of long-term experience with large units. Cycling service can cause fatigue failures in the economizer tubing and inlet header, lower furnace wall tubes and headers, structural components such as buckstays, and some steam drum internals. This fatigue cracking can be caused by the sudden flow of cold water into hot boiler components. Thermal differentials of 200 to 400F (111 to 222C) can be created. Furnace subcooling, boiler-forced cooling during a shutdown, and intermittent cold feedwater flow into the boiler during startup are three sources of thermal differentials and cyclic cracking. The thermal stresses produced within the components may be sufficient to produce low cycle fatigue cracks. The solution in most cases is to modify the boiler and/or feedwater system to prevent the sudden entry of cold water into hot boiler components.

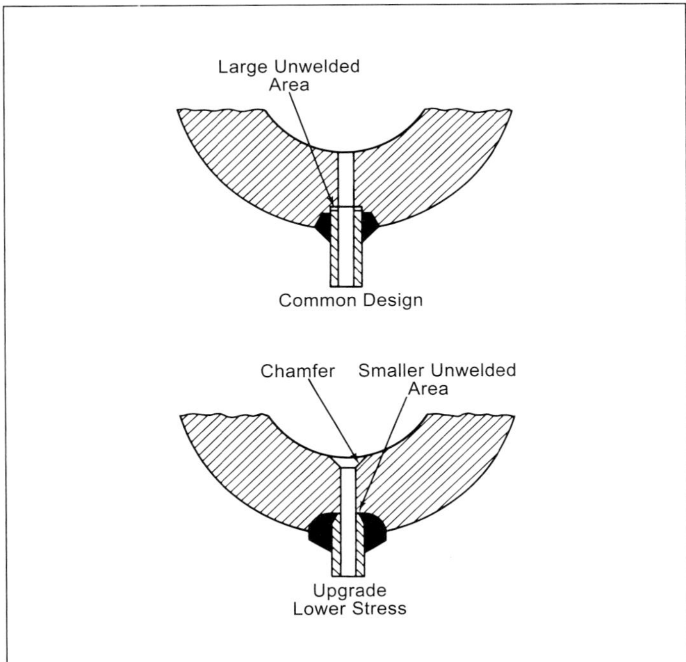

Fig. 19 Header upgrade with redesigned tube penetration.

When a boiler is experiencing cracking due to subcooling and cold feedwater flow upon startup, an off-line pump-assisted circulation system may be installed to reduce the thermal transients. The system, as shown in Fig. 20, consists of an off-line circulation pump, a thermal sleeved tee connection between the off-line pump and the feedwater line, a connection line from the boiler downcomer to the pump, a warming bypass system, various valves, and a control system.

The off-line pump is only operated when the boiler is shut down. Its purpose is to provide a small amount

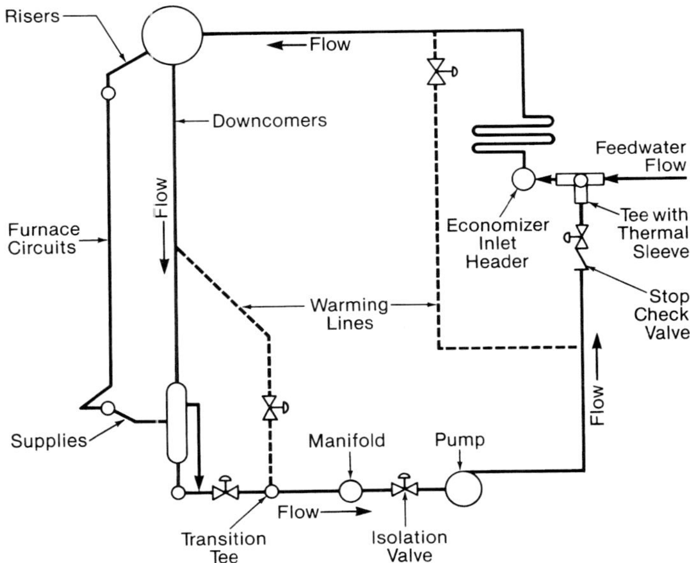

Fig. 20 Off-line recirculation system to reduce thermal shock.

of circulation within the furnace circuit and through the economizer to prevent temperature stratification in the water circuits. The tee connection permits the introduction of a small amount of hot water from the furnace into the feedwater stream when feedwater is intermittently supplied to the boiler and before a steady feedwater flow is established. The warm furnace water introduced at the tee connection raises the feedwater temperature enough to prevent thermal shock to the economizer. The connection contains an internal thermal sleeve that protects the tee from a thermal shock when cold feedwater is first fed to the economizer. A control system monitors the feedwater temperature and flow and controls the recirculation pump. When the boiler startup sequence is initiated, the off-line system is shut down and isolated. Warming lines permit natural circulation through the pump when it is shut down and the boiler is off-line.

Experience with off-line recirculation systems has shown that thermal shock differentials can be reduced to less than 100F (56C) from previous levels of 200 to 400F (111 to 222C). Such a reduction may eliminate the fatigue cracking that is associated with frequent unit cycling.

Drum boiler bypass system upgrade

Steam turbine transient stresses associated with on/off operation or load cycling shorten turbine life. Long startup times required to minimize these stresses lead to costly fuel consumption. Drum boilers can be upgraded with a superheater bypass system for improved cycling capability. The bypass system upgrade can be installed on a drum boiler to minimize startup time, provide control of steam temperature to match turbine metal temperature, and allow dual pressure operation of the boiler and turbine for better load response. These features reduce turbine stresses for improved availability and reduced maintenance costs.

The common means of steam temperature control in a drum boiler (water attemperation in conjunction with parallel convection surface, gas recirculation, excess air, or burner input adjustments) do not permit easy control of turbine temperatures during startup and low loads. This is because there is a large mismatch between flue gas flow and steam flow. The drum boiler bypass system, shown in Fig. 21, provides direct control of the steam temperature by saturated attemperation of the superheat and reheat outlet steam. This arrangement provides the desired steam temperature for the turbine without undue restrictions on the startup firing rate.

The drum boiler bypass system substantially reduces cold startup time because it controls the temperature differences of the saturated boiler surface, superheater surface, and turbine. The bypass system consists of a control system with steam piping and valves and is described in detail in Chapter 19.

The system has a set of superheater stop and bypass control valves downstream of the primary superheater to allow dual pressure operation; the turbine throttle pressure is controlled separately from drum pressure. This control system is designed to maintain constant pressure operation of the drum and boiler

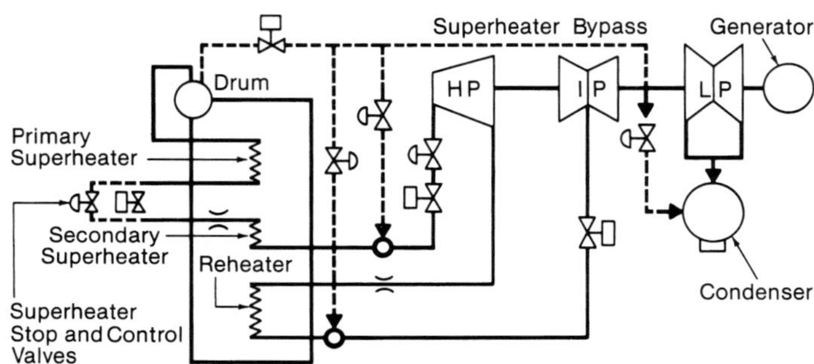

Fig. 21 Typical superheater bypass system for a drum boiler to minimize turbine stress.

furnace circuits and variable pressure operation of the turbine because the turbine throttle valve is open over most of the load range.

Dual pressure operation minimizes thermal stresses to the boiler and the turbine. A dual pressure shutdown keeps the boiler near full pressure and the turbine metal near full load temperature in preparation for a quick restart. In addition, it allows more rapid load changes than variable boiler pressure operation.

A superheater bypass diverts excess steam from the boiler to the condenser, thereby separating firing rate from drum pressure during shutdown and startup. Gas temperature probes, located near the superheater and reheater outlet tubes, monitor flue gas temperatures.

The superheater bypass system permits cycling the unit rapidly and often without accumulating major fatigue turbine damage.

Circulation improvements

Water circulation upgrades are often required to overcome operational and maintenance problems. Symptoms of a circulation problem include localized tube failures in particular boiler circuits, increases in water-side and steam-side deposits and pressure drops, fluctuations in feedwater control, and drum water level excursions and fluctuations. Occasionally, circulation problems may appear when supply tubes or headers for the furnace waterwall circuits crack and distort. Circulation problems can directly affect boiler performance and, if severe, can reduce capacity and availability.

Determining causes and solutions for circulation problems requires boiler performance testing, heat transfer and circulation computer modeling, and a thorough knowledge of boiler dynamics. Thorough discussions of heat transfer and circulation are covered in Chapters 4 and 5.

Consider an example boiler with circulation problems. This unit was experiencing an abnormally high forced outage rate due to tube failures in the platen wing walls as well as extensive failures of the furnace rear wall arch tubes. Fig. 22a illustrates the upper furnace configuration and arrangement of the wing walls.

The majority of tube failures in this unit were found in the lower 17 deg (0.3 rad) inclined portion of the wing walls. Some time later, the rear wall arch tubes began failing, especially in the lower inclined area. Poor water chemistry and resultant heavy internal tube deposits were originally blamed for the tube failures. However, water chemistry was monitored and found to be acceptable in this case.

A complete engineering study was required to review the plant operating procedures and the original boiler design. The study goals included:

1. analyzing the overall boiler circulation system,
2. reviewing the platen wing wall design,
3. examining the design and performance of steam drum internals,
4. investigating the rear wall arch and supply circuit design, and
5. providing operational and hardware modifications to eliminate the tube failures in the platen wing walls and rear wall arch tubes.

Field tests were conducted to verify the operating parameters. A computer simulation focused on the circulation of the steam drum, wing wall and rear wall arch tube circuits. Mass flow velocities in these tubes were found to be low. These low velocities contributed to flow imbalances between circuits as well as film boiling conditions inside the tubes. The steam film acted as an insulator, producing numerous overheat failures.

The existing drum internals were also found to be significant contributors to the circulation problem. The separators in the steam drum generated a high pressure drop, by reducing the pumping head and causing flow imbalances. The drum water level, 10 in. (254 mm) below the drum centerline, permitted steam to be drawn into the downcomers. This further reduced the effective water density (pumping head) and mass flow velocities in the furnace circuits.

The key modifications implemented in the circulation upgrade were as follows (See Fig. 22b):

1. New drum internals. Three rows of highly efficient, low pressure drop cyclone separators replaced the existing two rows of high pressure drop turboseparators. (See Fig. 23.) The drum water level was increased from −10 in. (−254 mm) to the centerline. These modifications increased the available head for all circuits.

2. Redesigned wing walls. The overall length of the wing walls was increased to raise the heat absorption in these vulnerable circuits, based on the natural circulation principle of more heat, more flow. This also lowered the furnace exit gas temperature by about 35F (19C) which reduced the tendency of slagging beyond the furnace. A third benefit was a slight reduction in superheat and, especially, reheat spray, both being excessive on the existing unit.

   The bottom slope was increased from 17 to 40 deg (0.3 to 0.7 rad) and the tube diameter was decreased. Multi-lead ribbed tubes were used instead of internally smooth tubes in the sloped portion.

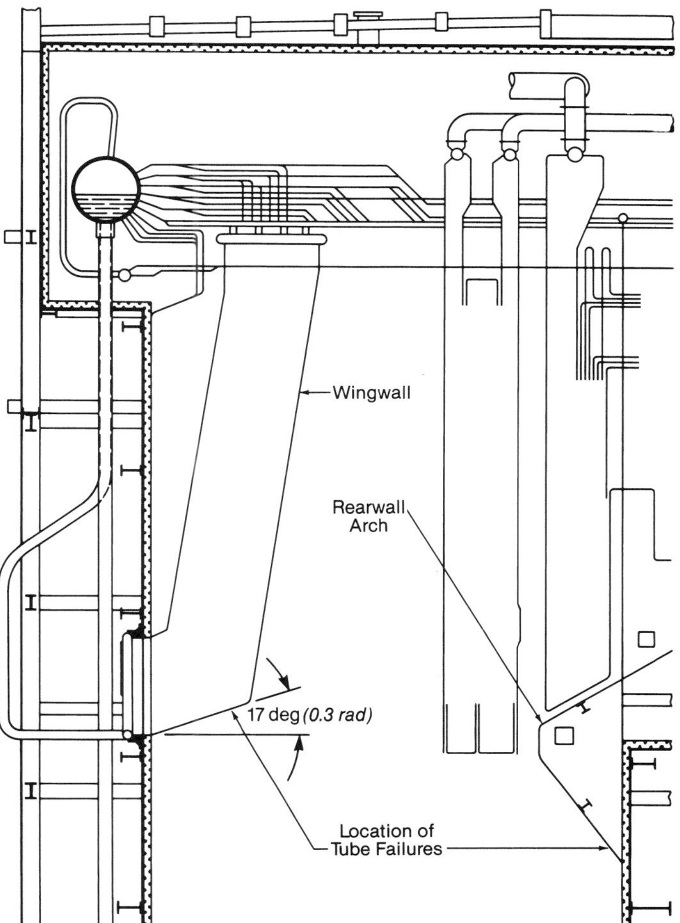

Fig. 22a Original upper furnace arrangement.

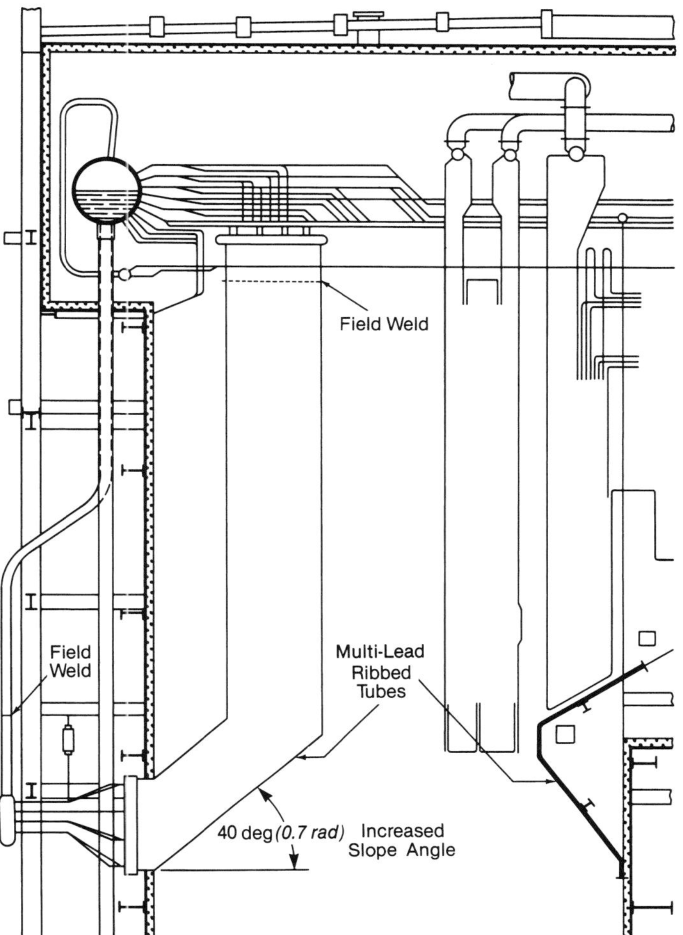

Fig. 22b Upper furnace arrangement for a circulation upgrade.

These measures were aimed at eliminating film boiling and flow instabilities, identified as the root cause of tube failures in these circuits. Film boiling in these circuits promoted formation of internal deposits and corrosion due to stagnant flow conditions. The increased flow resistance was more than offset by the greater pumping head generated by the new design.
3. Redesigned water supply system. The existing downcomers were lengthened and closed off by new downcomer bottles. Each bottle was equipped with eight supply pipes, four feeding each vertical wing wall inlet header.
4. Redesigned rear wall arch. Multi-lead ribbed tubes were installed in the rear wall arch in the form of shop-membraned tube panels. This eliminated the failures in the lower bend area. The original transition from the 2.5 in. (63.5 mm) OD rear wall tubes to the 3 in. (76.2 mm) arch tubes created a steam-water jet along the unheated side (inside of the extrados) of these bends and a corresponding void on the inside of the tube bends. This had caused the crucial bend sections facing the furnace heat to fail due to insufficient cooling. For additional margin, the arch tubes were also redesigned as multi-lead ribbed tubes.

The overall boiler circulation was significantly improved, and a high design margin prevented the recurrence of failures in the wing walls or rear wall tubes. The circulation upgrade restored 100% load capability and improved unit reliability and availability.

Strategies for fuel switching while maintaining high commercial availability

To help reduce sulfur dioxide (SO_2) emissions, many power plants are now burning a Powder River Basin (PRB) coal blend. The ash of some PRB blends forms a white deposit in the furnace and elsewhere in the boiler. The deposit tends to be reflective. This characteristic suppresses heat transfer, especially in the furnace, and leads to flyash plugging of the tightly-spaced convection heating surfaces. The inhibition of heat transfer raises flue gas temperatures, and often results in some loss of thermal efficiency.

The elevated gas temperatures can lead to long-term overheating of the non-cooled mechanical economizer support hangers, causing premature failure. Under some conditions, the temperature of the flue gas leaving the economizer can exceed the design temperature of the casing enclosure and flues between the economizer and the air heater downstream. The high gas temperature can also elevate economizer water outlet enthalpy, which causes cavitation problems in boiler circulation pumps in forced circulation boilers.

A boiler's economizer is, by definition, intended to reduce operating costs by recovering energy from the flue gas. This heat energy is transferred to water entering the boiler. Since overall boiler efficiency is increased 1% for every 40F (22C) reduction in flue gas stack temperature, the design of the economizer and its interaction with the boiler's fuel and flue gas are important concerns to power plant operators.

Fig. 23 Cyclone steam separator installation.

Many utility boilers were installed with continuous fin surface (CFS) economizers using a staggered tube arrangement. This surface used 2.0 in. (5.08 cm) high and 0.25 in. (0.635 cm) thick fins welded to the top and bottom of each tube [usually 2 in. (5.08 cm) outside diameter] parallel to the tube axis. CFS economizers originally appealed to many plant designers because the torturous path created for the flue gas enhanced heat absorption, and the fins could capture heat and transfer it to the tubing. This made the CFS economizer less costly and permitted installation in a relatively small area.

With the increasing use of western U.S. and lower quality fuels over the past few decades, however, the very factor that had been an advantage of the CFS economizer design became a disadvantage. The spacing proved more susceptible to plugging and flyash erosion and required more frequent cleaning.

Fuel switching can, therefore, lead to many related boiler problems that affect efficiency, availability and capacity. Power plant operators are faced with many interrelated problems when operating a plant that has changed from the design fuel to a low cost compliance fuel such as a PRB coal. Margins for acceptable operation are reduced and the power plant can suffer from high heat rates (i.e., lower efficiency), high operating costs, and frequent outages to remove ash buildup. A strategy is needed to deal with these problems so that the new low-cost compliance fuel can be used without the plant suffering from degradations in capacity, heat rate, operating cost and availability.

Fuel switching example

In one example of fuel switching, the economizer experienced ash plugging problems with PRB coal, making the unit unable to achieve the design main steam temperature at higher loads. The ash plugging resulted in periodic shutdowns for cleaning and maintenance. Because of the economizer's staggered, finned tube design, plant operators could not use sootblowers to restore heat transfer capabilities. At higher loads, the superheater temperature could not reach design levels, leading to a reduction in boiler efficiency. The resulting low main steam temperature to the turbine is equivalent to a reduction in cycle efficiency.

The example unit is a 750 MW (gross), pulverized coal-fired, balanced draft boiler. At maximum continuous rating (MCR), original design conditions were 4,985,000 lb/h (628 kg/s) main steam flow at 2400 psig (16.5 MPa) throttle pressure and 1007F (542C). The ash plugging resulted in periodic shutdowns for cleaning and maintenance.

The primary goal of the upgrade project was to replace the worn and increasingly unreliable economizer with a design that corrected these existing problems, and while doing so, return the main steam temperature to its original design level. Additional project tasks included complete boiler modeling, pressure part design and fabrication (including the complete redesign of the economizer support system), supply of sootblower equipment and hangers, and the fabrication and supply of steam-cooled wall openings for the sootblowers and access doors.

The existing economizer was replaced with a bare tube, in-line design. (See the example of a finned tube economizer replacement earlier in this chapter.) The bare tube, in-line economizer design minimized the erosion and ash trapping problems that are common to staggered arrangements. The in-line design also improved the effectiveness of sootblowers and resulted in the maintenance of adequate heat transfer. Fig. 24 illustrates the upgraded design and its redesigned economizer, primary superheater and sootblowers.

To maintain the necessary amount of heat transfer surface, the upgrade increased the number of economizer elements by about 25%. The installation of a new inlet header and a number of intermediate headers accommodated the increased number of water flow circuits. Tube shields to protect the tubes in the sootblower lane were added, in addition to erosion barriers along the walls of the unit.

To raise the superheat steam temperature, relatively small horizontal primary superheater loops above the economizer were added. The existing economizer support tubes were modified to provide water-cooled support for the small bank of new primary superheater surface.

New sootblowers were also incorporated into the retrofit design. Limited outboard space prohibited the installation of full travel, long retract sootblowers. Extended lance, half-track sootblowers were installed as part of the upgrade. Gas temperatures entering the economizer exceeded the traditional maximum temperature for these extended lances, so the lances were manufactured from a high temperature alloy material.

The eight sootblowers were supplied with the Progressive Helix Mechanism feature to reduce the potential for boiler tube erosion (see Chapter 24). This patented sootblower enhancement reduces tube erosion by shifting the nozzle cleaning path on every blowing cycle. More than 400 cycles occur before any nozzle path is repeated.

The previous economizer design had suffered from failures of the supports that were of a welded design. To simplify installation and to provide a support system that did not rely on tension welds, the economizer support system was designed to accommodate small banks of elements and to support the banks with pins and collector plates. Stainless steel and Incoloy alloy 800HT pins and plates were used to ensure high-temperature strength and oxidation resistance without relying on welds. Fig. 25 shows the collector plates and pin connections used in this upgrade.

The use of factory-installed pins and plates reduced on-site installation times. Construction crews removed the pins as the sections were brought up to elevation, and then reinstalled the pins after the collector plates had been mated to the support brackets. Because the process required no field welding of supports, the entire installation proceeded smoothly. After the sections had been hung in place, project personnel performed tube-to-tube welds between banks as time and space permitted.

The finned tube economizer replacement resulted in the following unit enhancements:

1. provided a solution to the problem of burning the PRB fuel blend,
2. eliminated plugging of the finned tube economizer,
3. reduced draft loss,

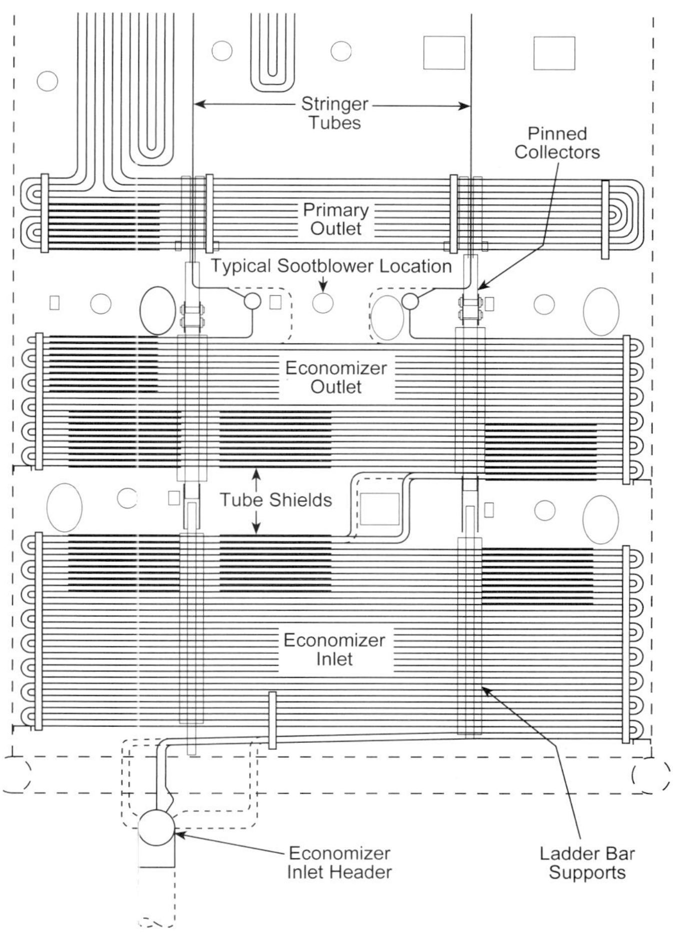

Fig. 24 Upgraded economizer and superheater arrangement to accommodate PRB fuel.

Fig. 25 The use of pins and collectors simplified construction and improved reliability by eliminating original welded supports.

4. reduced economizer exit gas temperatures to improve unit efficiency,
5. returned main steam temperatures to original design conditions to improve unit efficiency, and
6. improved unit availability through elimination of fouling problems and restoration of unit capacity and efficiency.

Capacity upgrades and environmental regulations

A capacity increase with higher efficiency and availability (sometimes called uprates), together with lower emissions, can lead to attractive returns on investment. Increases up to 10% in capacity and improvements in heat rate of 1 to 2% have been demonstrated.

With the current interpretation of the rules for environmental standards relative to capacity increases, the boiler operator must consider meeting, for example, U.S. New Source Performance Standards when a capacity upgrade is implemented. The addition of scrubbers and selective catalytic reduction (SCR) systems for the boiler flue gas must be considered when appropriate. This environmental equipment, combined with the capacity upgrade, may result in lower total regulated emissions when the project is properly configured. However, future changes in rules and policies may change the feasibility of this scenario.

Example strategy for capacity upgrade while maintaining high commercial availability

The typical capacity upgrade starts with an engineering study. To optimize the return on investment for the capacity increase, the project requires cooperation between the boiler manufacturer and the turbine vendor. During the engineering study, there is an interaction between the boiler and the turbine equipment suppliers such that the boiler and turbine uprate designs may be considered simultaneously. A series of design discussions must be held between the boiler and turbine designer to develop an upgrade strategy that can provide the optimum combination of performance and capital expenditure while considering the limits of the existing equipment and the potential design improvements. Often, new technology has been developed for the boiler and turbine generator set since the commissioning of the power station, and therefore, there is an opportunity to apply this new technology to enhance the efficiency and capacity of the unit.

The boiler manufacturer looks at key components that will be affected by the increased steam and water flow requirements. Operating pressures and heat transfer are affected by the increased flows. The designer must look at each of the key components versus the original design and possible design improvements to develop a plan for the capacity increase. As an example, a furnace circulation analysis must be run to determine if the furnace has the capacity to operate safely with the higher flows and heat input. Components that are commonly evaluated are shown in Table 3.

To properly evaluate the unit in the engineering study, a performance test of the unit under various conditions is required. This test will calibrate and validate the boiler computer model that is used to evaluate the boiler upgrade. The boiler model takes into account changes in boiler surface and fuels that have occurred since the boiler was first commissioned. The model will also consider changes to the unit, or operating procedures that were made for any pollution control equipment that may have been added or modified.

The increased firing rate will increase gas velocities in the convection pass and will increase the possibility for fouling and slag accumulation. The need for increased sootblowing and erosion protection is evaluated.

After the engineering study is completed and the equipment modifications are determined, costs are evaluated. A typical upgrade might include advanced turbine components to accommodate the higher steam flows and to achieve higher efficiency.

Table 3
Key Boiler Components Evaluated in a Capacity Upgrade Engineering Study

Pressure components	Economizer
	Furnace and convection pass water circuits, including drums and steam separators
	Superheaters and reheaters
	Attemperators
	Connection piping and headers
Nonpressure boiler components	Fuel preparation equipment
	Burners
	Air heaters
	Ductwork and supports
Balance of plant equipment	Fans
	Feed pumps
	Pollution control equipment
	Controls

Because turbine efficiency is determined by the design of the steam path components such as the nozzles, bucket profiles, as well as the leakage flows, these components may be upgraded with significant impact on overall plant performance. Much work has been done to improve the aerodynamics and reduce losses associated with buckets and nozzles. Most turbine upgrades involve a reduction in secondary flow losses. Designs have evolved that improve the resistance to solid particle erosion for longer efficient design life. There are numerous examples of the advantages of turbine upgrades. However, the best strategy is to coordinate the boiler and turbine upgrade to achieve the best return on investment.

Normally, a capacity uprate capital project improves heat rate and extends times between major turbine overhaul outages. These benefits are the result of replacing nozzles and final stage buckets with improved designs and improvements in steam losses together with higher steam flows.

Typical improvements to the boiler might include the replacement of the superheater or reheater outlet bank(s), a new attemperator, new burners, new higher efficiency separators in the steam drum, and the addition of air quality control equipment. The actual modifications are very specific to the original unit design, the fuel, and the environmental constraints.

Inconel, Incoclad, Incoloy and 671/800HT are trademarks of the Special Metals Corporation group of companies.

Electromagnetic acoustic transducer (EMAT) technology is one method used to determine remaining life of boiler tubes.

Chapter 45

Condition Assessment

The assessment of accumulated damage, or condition assessment, has a long history in the boiler industry. Whenever a component was found to contain damage or had failed, engineers asked what caused the damage and whether other components would fail. These questions typically pertained to tubing and headers, which caused the majority of downtime. As boiler cycling became more common, the need for more routine condition assessment increased to avoid component failure and unscheduled outages.

Condition assessment includes the use of tools or methods in the evaluation of specific components and then the interpretation of the results to identify 1) the component's remaining life and 2) areas requiring immediate attention.

A boiler component's damage assessment, typically compared to its design life, is based on accumulated damage, and can be performed in three phases. In Phase 1 of the assessment, design and overall operating records are reviewed and interviews are held with operating personnel. In Phase 2, nondestructive examinations, stress analysis, verification of dimensions, and operating parameters are undertaken. If required, the more complex Phase 3 includes finite element analysis, operational testing and evaluation, and material properties measurement.[1] (See Fig. 1.)

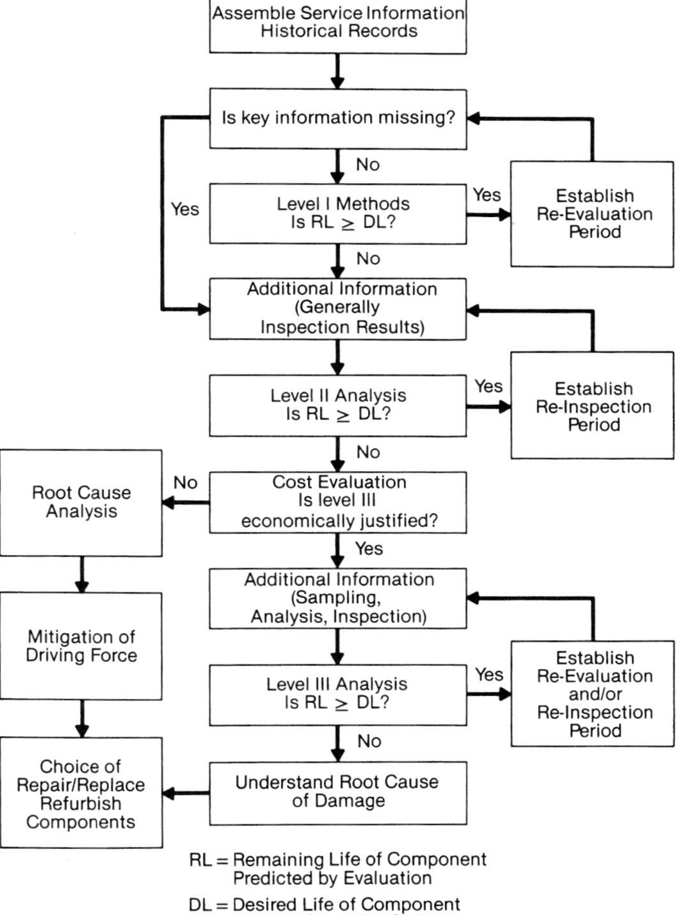

Fig. 1 Three phases (levels) of boiler damage assessment (*courtesy of the Electric Power Research Institute*).[1]

Condition assessment examination methods

The major boiler components must be examined by nondestructive and destructive tests.

Nondestructive examinations

Most nondestructive examination (NDE) methods for fossil fuel-fired plants have been in use for many years, although new methods are being developed for major components. Nondestructive testing does not damage the component.

The NDE methods used in evaluating electric utility power stations and industrial process plants include: visual, magnetic particle, liquid penetrant, ultrasonic, eddy current, radiography, nuclear fluorescence, electromagnetic acoustics, acoustic emissions, metallographic replication, strain measurement and temperature measurement.

Visual Whether the inspected component is subject to mechanical wear, chemical attack, or damage from thermal stress, visual examination can detect and identify some of the damage. Visual inspection is enhanced by lighting, magnification, mirrors, and optical equipment such as borescopes, fiberscopes and binoculars.

Magnetic particle Magnetic particle testing (MT) and wet fluorescent magnetic particle testing (WFMT) detect surface and near surface flaws. Because a magnetic field must be imparted to the test piece, these tests are only applicable to ferromagnetic materials.[2] The choice between these techniques generally depends on the geometry of the component and the required sensitivity. For typical power plant applications, one of two methods is used: 1) the component is indirectly magnetized using an electromagnetic yoke with alternating current (AC), or 2) the part is directly magnetized by prods driven by AC or direct current (DC).

In magnetic particle testing, any discontinuity disrupts the lines of magnetic force passing through the test area creating a leakage field. Iron particles applied to the area accumulate along the lines of magnetic force. Any leakage field created by a discontinuity is easily identified by the pattern of the iron particles. Dry magnetic particle testing is performed using a dry medium composed of colored iron particles that are dusted onto the magnetized area. In areas where a dry medium is ineffective, such as in testing overhead components or the inside surfaces of pressure vessels, the wet fluorescent method is more effective. With this method, fluorescent ferromagnetic particles are suspended in a liquid medium such as kerosene. The liquid-borne particles adhere to the test area. Because the particles are fluorescent, they are highly visible when viewed under an ultraviolet light.

Liquid penetrant Liquid penetrant testing (PT) detects surface cracking in a component. PT is not dependent on the magnetic properties of the material and is less dependent on component geometry.[2] It is used by The Babcock & Wilcox Company (B&W) in limited access areas such as tube stub welds on high temperature headers which are generally closely spaced. PT detects surface flaws by capillary action of the liquid dye penetrant and is only effective where the discontinuity is open to the component surface. Following proper surface cleaning the liquid dye is applied. The penetrant is left on the test area for about ten minutes to allow it to penetrate the discontinuity. A cleaner is used to remove excess penetrant and the area is allowed to dry. A developer is then sprayed onto the surface. Any dye that has been drawn into the surface at a crack bleeds into the developer by reverse capillary action and becomes highly visible.

Ultrasonic Ultrasonic testing (UT) is the fastest developing technology for nondestructive testing of pressure components. Numerous specialized UT methods have been developed. A piezoelectric transducer is placed in contact with the test material, causing disturbances in the interatomic spacings and inducing an elastic sound wave that moves through the material.[3] The ultrasonic wave is reflected by any discontinuity it encounters as it passes through the material. The reflected wave is received back at the transducer and is displayed on an oscilloscope.

Ultrasonic thickness testing Ultrasonic thickness testing (UTT) is the most basic ultrasonic technology. A common cause of pressure part failure is the loss of material due to oxidation, corrosion or erosion. UTT is relatively fast and is used extensively for measuring wall thicknesses of tubes or piping.

The surface of the component must first be thoroughly cleaned. Because ultrasonic waves do not pass through air, a couplant such as glycerine, a water soluble gel, is brushed onto the surface. The transducer is then positioned onto the component surface within the couplant. A high frequency (2 to 5 MHz) signal is transmitted by the transducer and passes through the metal. UTT is performed using a longitudinal wave which travels perpendicular to the contacted surface. Because the travel time for the reflected wave varies with distance, the metal thickness is determined by the signal displacement, as shown on the oscilloscope screen (Fig. 2).

Ultrasonic oxide measurement In the mid 1980s, B&W developed an ultrasonic technique specifically to evaluate high temperature tubing found in superheaters and reheaters. This NDE method, called the *Nondestructive Oxide Thickness Inspection Service* (NOTIS®), measures the oxide layer on the internal surfaces of high temperature tubes. The test is generally applicable to low alloy steels because these materials are commonly used in outlet sections of the superheater and reheater.

Low alloy steels grow an oxide layer on their internal surfaces when exposed to high temperatures for long time periods (Fig. 3). The NOTIS test is not applicable to stainless steels because they do not develop a measurable oxide layer.

The technique used for NOTIS testing is similar to UTT; the major difference between the two is the frequency range of the ultrasonic signal. A much higher frequency is necessary to differentiate the interface between the oxide layer and inside diameter (ID) surface of the tube. Using data obtained from this NOTIS testing, tube remaining creep life can also be calculated as discussed later in *Analysis techniques*. NOTIS and UTT are methods in which the transducer is placed in contact with the tube using a couplant gel. Because of the high sensitivity of the NOTIS method, it is less tolerant of rough tube surfaces or poor surface preparation.

Ultrasonic measurement of internal tube damage Several ultrasonic methods have been investigated for detecting damage within boiler tubes. All techniques

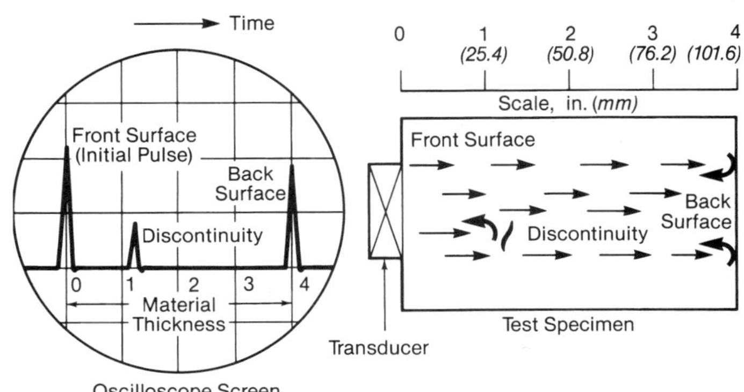

Fig. 2 Typical ultrasonic signal response.[4]

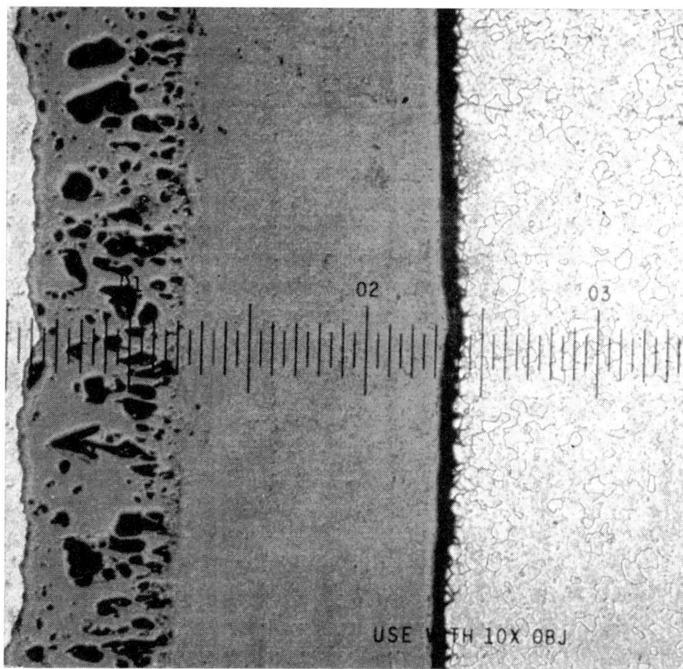

Fig. 3 Steam-side oxide scale on ID surface.

use contact UT where a transducer is placed on the outside diameter (OD) or tube surface using a couplant, and an ultrasonic signal is transmitted through the material. The techniques can be categorized by type of signal evaluation: backscatter, the evaluation of UT wave scatter when reflected by damaged material; attenuation, the evaluation of UT signal loss associated with transmission through damaged material; and velocity, the measurement and comparison of UT wave velocity through the tube material.[4]

When a longitudinal wave passes through a tube, part of the signal is not reflected to the receiver if it encounters damaged material. The damaged areas reflect part of the wave at various angles, backscattering the reflected signal. The loss of wave amplitude that is received back at the transducer is then used to evaluate the degree of damage.

Damage in the tube can also be assessed by evaluating the loss of signal amplitude (attenuation) as a shear wave is transmitted through the tube wall. The technique uses a fixture with two transducers mounted at angles to each other. One unit transmits a shear wave into the tube and the second transducer, the receiver, picks up the signal as the wave is reflected from the tube ID. A drop in signal amplitude indicates damage in the tube wall.

This technology is the basis of the B&W patented *Furnace wall Hydrogen damage Nondestructive Examination Service* (FHyNES®)test method (Fig. 4). The velocity test method uses either longitudinal or shear ultrasonic waves. As a wave passes through a chordal section of tube with hydrogen damage, there is a measurable decrease in velocity. Because the signal is not reflected from the tube inside surface, ultrasonic velocity measurement is not affected by damage to the inside of the tube and therefore specifically detects hydrogen damage.

Immersion ultrasonic testing In immersion ultrasonic testing, the part is placed in a water bath which acts as the couplant. B&W uses a form of immersion UT for tube wall thickness measurements. In two-drum industrial power boilers, process recovery boilers and some utility power generation boilers, most of the tubes in the convective bank between the drums are inaccessible for conventional contact UTT measurements. For these applications, an ultrasonic test probe was developed which is inserted into the tubes from the steam drum; it measures the wall thickness from inside the tubes. As the probe is withdrawn in measured increments, the transducers measure the tube wall thicknesses. A limitation of this technique is that the ID surface of the tubes must be relatively clean.

Shear wave ultrasonic testing This is a contact ultrasonic technique in which a shear wave is directed at an angle into the test material. Angles of 45 and 60 deg (0.79 and 1.05 rad) are typically used for defect detection and weld assessment. The entire weld must be inspected for a quality examination.

Time of flight defraction (TOFD) TOFD is an ultrasonic technique that relies on the diffraction of ultrasonic energies from defects in the component being tested. The primary application is weld inspection on piping, pressure vessels, and tanks. TOFD is an automated inspection that uses a pitch-catch arrangement with two probes, one on each side of the weld. The weld material is saturated with angled longitudinal waves to inspect for discontinuities. Because the time separation of the diffracted waves is directly related to flaw size (height), TOFD can detect both the flaw and allow estimation of the flaw size.

Eddy current Measuring the effects of induced eddy currents on the primary or driving electromagnetic field is the basis of eddy current testing. The electromagnetic induction needed for eddy current testing is created by using an alternating current. This develops the electromagnetic field necessary to produce eddy currents in a test piece.

Eddy current testing is applicable to any materials that conduct electricity and can be performed on magnetic and nonmagnetic materials. The test is therefore

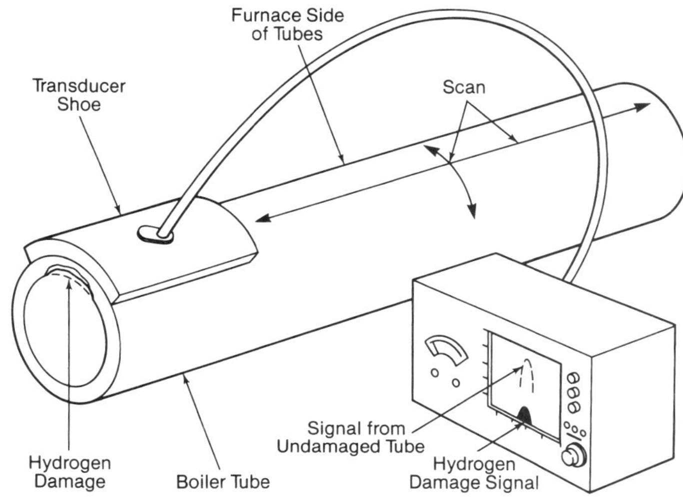

Fig. 4 Shear wave technique for detecting hydrogen damage.

applicable to all metals encountered in power station condition assessment work.

Parameters affecting eddy current testing include the resistivity, conductivity, and magnetic permeability of the test material; the frequency of the current producing the eddy currents; and the geometry and thickness of the component being tested.

Radiography Radiography testing (RT) is the most common NDE method used during field erection of a boiler. Radiography is also valuable in condition assessments of piping. As x-rays and gamma rays pass through a material, some of the rays are absorbed. Absorption depends upon material thickness and density. When the rays passing through an object are exposed to a special film, an image of the object is produced due to the partial absorption of the rays.

In practical terms, a radioactive source is placed on one side of a component such as a pipe, at a weld, and a film is placed on the opposite side. If x-rays are directed through the weld and there is a void within the weld, more rays pass through this void and reach the film, producing a darker image at that point. By examining the radiographic films, the weld integrity can be determined. During the field erection of a boiler and power station, thousands of tube and pipe welds are made and radiographed. (See also Chapters 38 and 39.)

The major disadvantage of radiography is the harmful effect of excessive exposure to the radioactive rays. RT is also limited in its ability to provide the orientation and depth of an indication.

Nuclear fluorescence The primary use of this testing in condition assessment is the verification of alloy materials in high temperature piping systems. When certain elements are exposed to an external source of x-rays they fluoresce (emit) additional x-rays that vary in energy level. This fluorescence is characteristic of the key alloys common to high temperature piping and headers. Chromium and molybdenum are the key elements measured. The nuclear alloy analyzer is a portable instrument that contains a low level source of x-rays. A point on the surface of the pipe is exposed to x-rays emitted from the analyzer. As the source x-rays interact with the atoms of the metal, the alloys emit x-rays back to the analyzer. Within the detector system of the analyzer, the fluoresced x-rays are separated into discrete energy regions. By measuring the x-ray intensity in each energy region, the elemental composition is also determined.

Electromagnetic acoustics Electromagnetic acoustics combine two nondestructive testing sciences, ultrasonics (UT) and electromagnetic induction. This technology uses an electromagnetic acoustic transducer (EMAT) to generate high frequency sound waves in materials, similar to conventional ultrasonics. Conventional UT transducers used for field testing convert electrical impulses to mechanical pulses by use of piezoelectric crystals. These crystals must be coupled to the test piece through a fluid couplant. For electrically conductive materials, ultrasonic waves can be produced by electromagnetic acoustic wave generation.[5] In contrast to conventional contact UT where a mechanical pulse is coupled to the material, the acoustic wave is produced by the interaction of two magnetic sources. The first magnetic source modulates a time-dependent magnetic field by electromagnetic induction as in eddy current testing. A second constant magnetic field provided by an AC or DC driven electromagnet or a permanent magnet is positioned near the first field. The interaction of these two fields generates a force, called the Lorentz force, in the direction perpendicular to the two other fields. This Lorentz force interacts with the material to produce a shock wave analogous to an ultrasonic pulse, eliminating the need for a couplant.

Fig. 5 shows the basic principles of EMAT operation. A strong magnetic field (B) is produced at the surface of the test piece by either a permanent magnet or electromagnet. Eddy currents (J) are induced in the test material surface. An alternating eddy flow in the presence of the magnetic field generates a Lorentz force (F) that produces an ultrasonic wave in the material. For boiler tubes that are electromagnetically conductive (including alloys such as SA-213T22), the EMAT technology is ideal.

B&W, working with the Electric Power Research Institute (EPRI), developed a nondestructive rapid scan system to inspect boiler tubes using EMAT technology. This EMAT based system is known as the Fast-Scanning Thickness Gage (FST-GAGE®) and was developed specifically to scan boiler tubes and provide a continuous measurement of tube wall thickness. (See Fig. 6.) The system conducts tests at exceptional speeds, allowing scanning of thousands of feet (m) of boiler tubing in a single shift. To perform an inspection, the FST-GAGE system is manually scanned along individual boiler tubes. System sampling rates greater than 65 samples per second supports rapid scanning of tubes. During a scan, the system provides an immediate display of both tube wall thickness and signal amplitude. At the conclusion of each tube scan, a complete record of the inspection is electronically stored and is traceable to each boiler tube and position.

As with conventional UT, the FST-GAGE system can assess internal tube damage by evaluating the loss of signal amplitude (attenuation) as a shear wave is transmitted through the tube wall. By monitoring and indicating signal amplitude, the system can also be used to detect tube damage such as hydrogen damage, similar to B&W's patented FHyNES technique. The FST-GAGE has also demonstrated the ability to detect internal tube pitting, caustic gouging, and under-deposit corrosion.

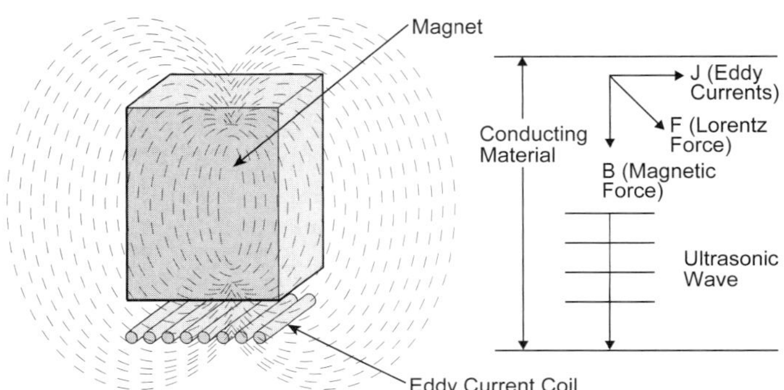

Fig. 5 Basic principles of EMAT operation.

Fig. 6 B&W's Fast-Scanning Thickness Gage (FST-GAGE®) EMAT based system can provide continuous measurement of tube wall thickness.

As with any NDE method, surface preparation is important for effective testing with EMATs. However, EMAT is not as sensitive to scale as conventional UT since it produces the ultrasonic wave within the material. Some scales, such as magnetite oxide of uniform thickness, have no detrimental effect on the signal generation of the EMAT probe. When the plant burns a clean fuel such as natural gas, testing may be conducted without any special surface cleaning. To protect the coil from damage, surface preparation will normally be required for boilers firing oil or solid fuels. Some gas-fired units may require surface cleaning if external buildup or corrosion is present. Grit blasting or water blasting are effective methods of cleaning larger areas. Smooth metal is the preferred surface to ensure rapid testing.

EMAT technology continues to be applied where its unique properties have advantages over conventional UT techniques. B&W and EPRI are developing a system for the detection of cracking in boiler tubes associated with corrosion fatigue. Waterside corrosion fatigue is a serious boiler tube failure mechanism. The failures usually occur close to attachments such as buckstay welds, windbox attachment welds, or membrane welds. The combination of thermal fatigue stresses and corrosion leads to ID-initiated cracking that is oriented along the tube axis. The EMAT system under development for corrosion fatigue has unique characteristics that enhance its ability to scan past welds and attachments and scan the full circumference of the boiler tube. The EMAT equipment uses a tone burst EMAT signal allowing the use of horizontally polarized shear wave (SH waves) to detect cracking adjacent to external tube attachment welds.

B&W has developed an EMAT application to inspect horizontal banks (i.e., economizer, reheater) of tubing within the boiler. Horizontal bank tubing may experience tube failures caused by out of service corrosion pitting forming aligned voids in the tube ID. The purpose of this EMAT test is to detect the internal aligned pitting at the lower portion of the horizontal tube internal surface. The test is accomplished by scanning along the outside of the horizontal tube at either the 3 or 9 o'clock positions with an EMAT transducer generating a Lamb wave (ultrasonic waves that travel at right angles to the tube surface) which is focused at the 6 o'clock position on the tube.

B&W has also developed a surface wave EMAT application to show surface indications including axially oriented cracks in boiler tubes. Conventional surface nondestructive test methods were unsatisfactory because they lacked adequate sensitivity and had slow production rates for testing on large areas. A tone burst EMAT technique was developed that uses a bidirectional focused surface wave EMAT that follows the tube surface circumferentially until the signal is reflected back from a longitudinally orientated OD crack.

Acoustics Acoustics refers to the use of transmitted sound waves for nondestructive testing. It is differentiated from ultrasonics and electromagnetic acoustics in that it features low frequency, audible sound. B&W uses acoustic technology in testing tubular air heaters. Because the sound waves are low frequency, they can only be transmitted through air. A pulse of sound is sent into the air heater tube. As the wave travels along the tube, it is reflected by holes, blockage or partial obstructions. By evaluating the reflected wave on an oscilloscope, the type of flaw and its location along the tube can be determined.

Acoustic emissions Acoustic emissions (AE) detect subsurface crack growth in pressure vessels. When a structure such as a pipe is pressurized and heated, the metal experiences mechanical and thermal stresses. Due to the stress concentration at a defect such as a crack, a small overall stress in the pipe can produce localized yield and fracture stresses resulting in plastic deformation. These localized yields release bursts of energy or stress wave emissions that are commonly called acoustic emissions. AE testing uses acoustic transducers that are positioned along the vessel being monitored. AE signals are received at various transducers on the vessel. By measuring the time required for the signal to reach each of the transducers, the data can be interpreted to identify the location of the defect.

Metallographic replication Metallographic replication is an in situ test method that enables an image of the metal grain structure to be nondestructively lifted from a component. Replication is important in evaluating high temperature headers and piping because it allows the structure to be examined for creep damage. Prior to the use of replication techniques, it was necessary to remove samples of the material for laboratory analysis. The replication process involves three steps: grinding, polishing and etching, and replicating. In the first step, the surface is rough ground then flapper wheel ground with finer grit paper. In the second step, the surface is polished using increasingly finer grades of diamond paste while intermittently applying a mixture of nitric acid and methanol in solution. The acid solution preferentially attacks the grain boundaries of the metal. In the final step, the replica, which is a plastic tape, is prepared by coating one face of the tape with acetone for softening. The tape is then firmly pressed onto the prepared surface. Following a suitable drying time, the tape is removed and mounted onto a glass slide for microscopic examination.

Strain measurement Strain measurements are obtained nondestructively by using strain gauges. Gauges used for piping measurements are characterized by an electrical resistance that varies as a function of the applied mechanical strain.[6] For high temperature components, the gauge is made of an alloy, such as platinum-tungsten, which can be used at temperatures up to 1200F (649C). The gauge is welded to the surface of the pipe and the strain is measured as the pipe ramps through a temperature-pressure cycle to operating temperature. Strain gauges used for lower temperature applications such as for analysis of hanger support rods are made of conventional copper-nickel alloy (constantan). These low temperature gauges are made of thin foil bonded to a flexible backing and are attached to the test surface by a special adhesive.

Temperature measurement Most temperature measurements can be obtained with sheathed thermocouples (TC). In special applications where temperature gradients are needed such as detailed stress analysis of header ligaments, special embedded TCs are used. The embedded unit is constructed by drilling a small hole into the header. A sheathed TC wire is then inserted and peened in place. (See Chapter 40.)

Destructive examinations

B&W tries to minimize the use of sample analysis because it is generally more expensive to perform destructive testing. However, for certain components, complete evaluation can only be done by removing and analyzing test samples. Destructive testing is described for two types of specimens, tube samples and boat samples.

Tube samples Tubes are the most common destructively tested components. Tube samples are generally removed from water- and steam-cooled circuits. A relatively large number of samples may be removed for visual inspections, from which a smaller number are selected for complete laboratory analysis. A tube analysis usually includes the following: 1) as-received sample photo documentation, 2) complete visual inspection under magnification, 3) dimensional evaluation of a ring section removed from the sample, 4) material verification by spectrographic analysis, 5) optical metallography, and 6) material hardness measurement. On waterwall tubes removed from the boiler furnace, the analysis includes a measurement of the internal deposit loading [g/ft^2 (g/m^2)] and elemental composition of the deposit. On steam-cooled superheater and reheater tubes, the thickness of the high temperature oxide layer is also provided.

Specialized tests are performed as required to provide more in-depth information. Failure analysis is a common example. When failures occur in which the root cause is not readily known from standard tests, fractography is performed. Fractography involves examination of the fracture surface using a scanning electron microscope.

Boat samples Boat samples are wedge shaped slices removed from larger components such as headers, piping and drums. The shape of the cut allows the material to be replaced by welding. Because the repairs usually require post weld heat treating, the use of boat samples is expensive. In most instances, replication is adequate for metallographic examination of these components and boat sample removal is not required.

Condition assessment of boiler components and auxiliaries

In Phase 1 of a condition assessment program, interviews of plant personnel and review of historical maintenance records help identify problem components. These components are targeted for a closer onsite examination during Phase 2 of the program. Nondestructive and destructive examination methods can then be used to evaluate the remaining life of the boiler components and its major auxiliaries.

Boiler drums

Steam drum The steam drum is the most expensive boiler component and must be included in any comprehensive condition assessment program. There are two types of steam drums, the all-welded design used predominantly in electric utilities where the operating pressures exceed 1800 psi (12.4 MPa), and drums with rolled tubes. The steam drum operates at saturation temperature [less than 700F (371C)]. Because of this relatively low operating temperature, the drum is made of carbon steel and is not subject to significant creep. Creep is defined as increasing strain at a constant stress over time.

Regardless of drum type, damage is primarily due to internal metal loss. The causes of metal loss include: corrosion and oxidation, which can occur during extended outages; acid attack; oxygen pitting; and chelant attack discussed in Chapter 42. Damage can also occur from mechanical and thermal stresses on the drum that concentrate at nozzle and attachment welds. These stresses, most often associated with boilers that are on/off cycled, can result in crack development. Cyclic operation can lead to drum distortion (humping) and can result in concentrated stresses at the major support welds, seam welds, and girth welds. The feedwater penetration area has the greatest thermal differential because incoming feedwater can be several hundred degrees below drum temperature.

A problem unique to steam drums with rolled tube seats is tube seat weepage (slight seeping of water through the rolled joint). If the leak is not stopped, the joint, with its high residual stresses from the tube rolling operation, can experience caustic embrittlement. (See Chapter 42.) In addition, the act of eliminating the tube seat leak by repeated tube rolling can overstress the drum shell between tube seats and lead to ligament cracking.

Condition assessment of the steam drum can include visual and fiber optic scope examination, MT, PT, WFMT, UT and replication.

Lower drum The lower or mud drum is most often found in industrial boilers. (See Chapter 27.) Part of the boiler's water circuit, the lower drum is not subject to large thermal differentials or mechanical stresses. However, as in steam drums with rolled tubes, seat weepage and excessive stresses from tube rolling can occur. In most cases, visual inspection, including fiber optic probe examination of selected tube penetrations, is sufficient. Kraft recovery boiler lower drums are subject to corrosion of the tube-drum interface on the OD. This area of the drum is inaccessible, therefore inspections are conducted from the ID using UTT and EMATs to check for cracking and wall thinning.

The downcomers carry water from the steam drum to the mud drum and the various wall circuits. Two areas on the downcomers that should be inspected are termination welds for cracks and horizontal runs of piping for internal corrosion pitting and thinning.

Boiler tubing

Steam-cooled Steam-cooled tubing is found in the superheater and reheat superheater. Both components have tubes subjected to the effects of metal creep. Creep is a function of temperature, stress and operating time. The creep life of the superheater tubes is reduced by higher than expected operating temperature, thermal cycling, and by other damage mechanisms, such as erosion and corrosion, causing tube wall thinning and increased stresses. Excessive stresses associated with thermal expansion and mechanical loading can also occur, leading to tube cracks and leaks independent of the predicted creep life.

As discussed in Chapters 19, 21, 29 and 45, superheater tubing can also experience erosion, corrosion, and interacting combinations of both.

Condition assessment of the superheater tubes includes visual inspection, NOTIS, UTT and tube sample analysis. Problems due to erosion, corrosion, expansion, or excessive temperature can generally be located by visual examination.

Water-cooled Water-cooled tubes include those of the economizer, boiler (generating) bank and furnace. The convection pass side wall and screen tubes may also be water-cooled as discussed in Chapter 19. These tubes operate at or below saturation temperature and are not subject to significant creep. Modern boilers in electric utilities and many industrial plants operate at high pressures. Because these boilers are not tolerant of waterside deposits, they must be chemically cleaned periodically, which results in some tube material loss. As discussed in Chapter 42, proper water chemistry control will limit tube inside surface material loss due to ongoing operations and cleaning.

With the exception of creep deformation, the factors that reduce steam-cooled tube life can also act upon water-cooled tubes. Erosion is most likely to occur on tube outside surfaces in the boiler or economizer bank from sootblowing or ash particle impingement. Corrosion of the water-cooled tubes is most common on internal tube surfaces and results from excessive waterside deposits. Deposit accumulations promote corrosion, caustic gouging or hydrogen damage.

Risers The riser tubes are generally found in the penthouse or over the roof of the boiler. They carry the saturated steam-water mixture exiting the upper waterwall headers to the steam drum. Condition assessment includes UTT measurements on nondrainable sections and on the extrados (outside surface) of bends. When access is available it is advantageous to perform internal visual inspection with a fiber optic or video probe.

Headers

Headers and their associated problems can be grouped according to operating temperature. High temperature steam-carrying headers are a major concern because they have a finite creep life and their replacement cost is high. Lower temperature water- and steam-cooled headers are not susceptible to creep but may be damaged by corrosion, erosion, or severe thermal stresses.

High temperature The high temperature headers are the superheater and reheater outlets that operate at a bulk temperature of 900F (482C) or higher. Headers operating at high temperature experience creep under normal conditions. The mechanics of creep crack initiation and crack growth are further discussed in the data analysis section of this chapter. Fig. 7 illustrates the locations where cracking is most likely to occur on high temperature headers. In addition to material degradation resulting from creep, high temperature headers can experience thermal and mechanical fatigue. Creep stresses in combination with thermal fatigue stress lead to failure much sooner than those resulting from creep alone.

There are three factors influencing creep fatigue in superheater high temperature headers: combustion, steam flow and boiler load. Heat distribution within the boiler is not uniform: burner inputs can vary, air distribution is not uniform, and slagging and fouling can occur. The net effect of these combustion parameters is variations in heat input to individual superheater and reheater tubes. When combined with steam flow differences between tubes within a bank, significant variations in steam temperature entering the header can occur. (See Fig. 8.) Changes in boiler load further aggravate the temperature difference between the individual tube legs and the bulk header. As boiler load increases, the firing rate must increase to maintain pressure. During this transient, the boiler is temporarily over fired to compensate for the increasing steam flow and decreasing pressure. During load decreases, the firing rate decreases slightly faster than steam flow in the superheater with a resulting decrease in tube outlet temperature relative to that of the bulk header (Fig. 9). As a consequence of these

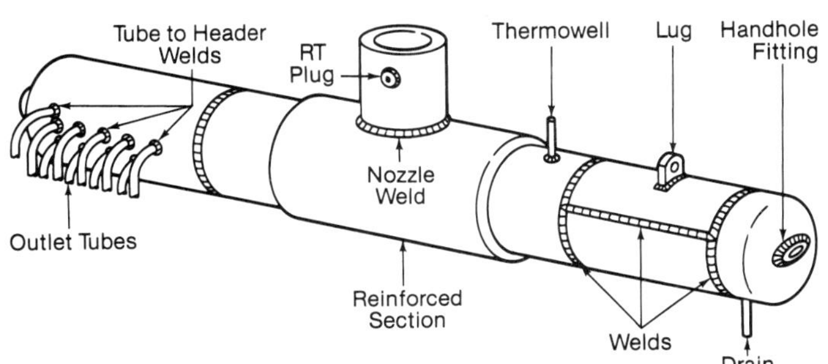

Fig. 7 Header locations susceptible to cracking.

The Babcock & Wilcox Company

tion can begin. This evaluation program should be as complete as economically possible.

Detailed evaluation program

Phase I To determine where physical testing is required, the following preliminary steps are part of a Phase I evaluation: plant personnel interviews, plant history review, walkdowns, stress analysis, and life fraction analysis.

Plant personnel interviews are conducted to gather information that is not readily available from plant records. Significant history may only be found in recollections of experienced personnel.

Operating history reviews complement personnel interviews. They can provide problem histories and design or operating solutions.

Piping system walkdowns serve three major functions: to evaluate pipe supports and hangers, to find major bending or warpage, and to verify changes.

Pipe hangers and supports should be carefully examined. This can be done by creating a baseline inspection record of all supports.

While the data are being taken on the piping walkdown, the general appearance should also be noted. In particular, inspections may reveal the following damages: necked-down rods or yokes, spring coil fractures, deterioration of the hanger can, and deterioration of tiebacks into building steel.

Many times a walkdown reveals that a modification was performed. If the entire system was not reviewed during the modification, other problems may result.

Stress analysis of the piping system can now be performed. Typically, a computer program is used to perform the stress calculations based on design and any abnormal conditions found during the walkdown. Once a piping system is modeled, the analysis allows the engineer to pinpoint high stress locations. The objective is to limit the nondestructive examination work to these high stress areas.

Life fraction analysis (LFA) of a pipe is done if the primary failure mode is creep due to operating temperatures above 900F (482C). The LFA is based on the unit's operating history and stress levels are calculated using design conditions and minimum wall thicknesses. This analysis is discussed at length later.

Phase II Phase II of the evaluation includes all physical testing of the piping system. The majority of the testing should be nondestructive; however, some destructive testing may be required. The results from Phase I testing provide test location priority. Specific test recommendations are shown in Table 1.

The test data generated from the inspections must be evaluated to determine the remaining component life. This is known as the Phase III evaluation and is covered under *Analysis techniques*.

Typical failures The most typical steam pipe failure is cracking of attachment welds (support welds or shear lugs). These cracks are caused by thermal fatigue, improper support, or improper welding.

Radiograph plugs often have cracked seal welds. Although the plug threads are the pressure bearing surfaces, they can become disengaged over time due to corrosion, creep swelling or oxidation.

Steam pipe warping is another serious problem. If the pipe has deformed, it has undoubtedly gone through a severe thermal shock. The high strain between the upper and lower sections of pipe can cause permanent deformation.

Two final common failure areas are the boiler outlet headers and turbine stop (throttle) valves. These areas should always be considered in any piping evaluation.

Table 1
Typical Piping System Tests

Test Area	Test Type	Optional*
Circumferential welds	A, B, D, E, F, G, H	C, I, J
Longitudinal welds	A, B, D, E, F, G, H	C, I, J
Wye blocks	A, B, E, F, G, H	C, D, I, J
Hanger shear lugs	A, B	C, F, G, J
Hanger bracket and supports	A, B	C, F, G, J
Branch connections	A, B, D, E, F, G, H	C, I, J
RT plugs	A, B	C, F, G, I
Misc. taps and drains	A, B	C, I
Elbows/bend	A, E, F	B, C, G

A - Visual
B - Magnetic particle
C - Liquid penetrant
D - Ultrasonic shear
E - Ultrasonic thickness
F - Replication
G - Material ID
H - Dimensional
I - Radiography
J - Metallography

* Optional tests should be used to gather more detailed information.

Low temperature piping

Low temperature piping operating at less than 800F (427C) is not damaged by creep. These systems typically fail due to fatigue, erosion or corrosion. The evaluation methods are the same as those for high temperature piping; however, a finite life is not predicted. Low temperature pipes, if maintained, last much longer than their high temperature counterparts. Typical systems are reheat inlet steam lines, extraction lines, feedwater lines and general service water lines.

Typical failures Many high temperature failure modes occur in low temperature pipes. Cold reheat lines experience thermal shock because the reheat temperature control is typically an in-line attemperator. The attemperator spray can shock the line if the liner is damaged or the nozzle is broken. Economizer discharge lines that run from the economizer outlet header to the boiler drum can be damaged during startup sequences. If the economizer is steaming and flow is initiated as a water slug, the line can experience severe shocks. This can cause line distortion and cracking at the end connections and support brackets. Other low temperature piping can be damaged by oxygen pitting caused by inadequate water treatment. Erosion due to flow cavitation around intrusion points can cause severe wall thinning. If solid particles are entrained in the fluid, erosion of pipe elbows results. General corrosion of the inside pipe surface can be caused by extended outage periods. Proper line draining is recommended unless protective materials are in place.

Tubular air heaters

Tubular air heaters are large heat exchangers that transfer heat from the boiler flue gas to the incoming combustion air, as discussed in Chapter 20. On large utility boilers, tubular air heaters can contain up to 90,000 tubes with lengths of 50 ft (15.2 m) each. These 2 in. (50.8 mm) OD tubes are densely grouped with spacings of 3 to 6 in. (76.2 to 152.4 mm) centers in two directions. Flue gas flow direction is typically opposite that of the combustion air to maximize thermal efficiency. Unfortunately, this promotes corrosion on the gas side cold end.

Condensate formation promotes acid corrosion from the flue gas which causes wall thinning. If left unchecked for several years the tubes eventually corrode through, causing air leakage from the air to gas side.

Because access to air heater tubes is limited, eddy current and acoustic technologies are used to test for blockage, holes and wall thinning. Eddy current technology is used to measure wall thicknesses of thin [< 0.065 in. (< 1.65 mm)] nonferrous heat exchanger tubing. Holes and partial and complete blockage are located using acoustic technology. When an audible sound is introduced into a tube it travels the length of the tube and exits the open end. If a hole exists in the tube, however, it changes the signal pitch in the same manner as a flutist changes a note pitch. In a like manner, partial or total tube blockage yields a pitch change. B&W uses The Acoustic Ranger® inspection probe for this test. (See Fig. 13.)

Boiler settings

The boiler components that are not part of the steam-water pressure boundary are general maintenance items that do not have a significant impact on remaining life of the unit. The nonpressure components include the penthouse, boiler casing, brickwork and refractories, and flues and ducts. Deterioration of these components results from mechanical and thermal fatigue, overheat, erosion and corrosion. In all cases, condition assessment is done by performing a detailed visual inspection. For flues, ducts and casing, it is of value to inspect the in-service boiler to detect hot spots, air leaks and flue gas leaks that can indicate a failed seal.

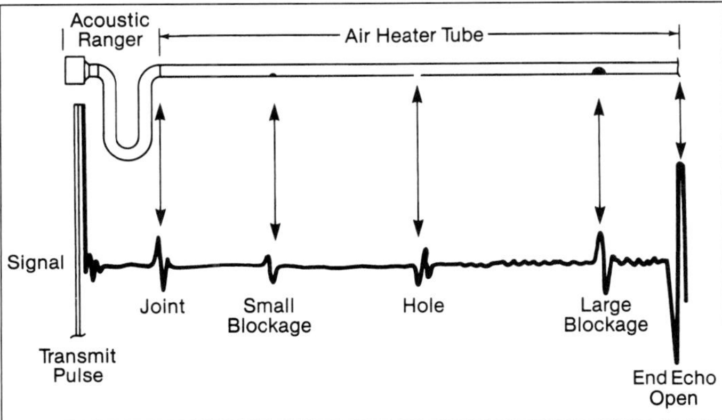

Fig. 13 Acoustic Ranger® schematic.

The structural members of the boiler must be reviewed during a condition assessment inspection. Normally these members, along with the support rods above the boiler and auxiliaries, last the life of the boiler. However, because nonuniform expansion can lead to boiler load movement, the support system should be examined during the boiler outage inspections. Particular attention should be given to header and drum supports that could be damaged if the vessel is distorted.

Analysis techniques

Once the testing is complete and the data are compiled, the next step in condition assessment is to decide whether to repair, replace or re-inspect certain components. For high temperature components with finite lives, this decision is aided by computers that predict failures by modeling analyses.

Component end of life is defined as the point at which failures occur frequently, the costs of inspection and repair exceed replacement cost, or personnel are at risk. Therefore, remaining life can be considered as the interval between the present time (t_p) with accumulated damage and the time at end of life (t_e). This can be written as:

$$R.L. = t_e - t_p \tag{1}$$

For waterwall tubing that is eroding at a linear rate, the remaining life is as follows:

$$R.L. = \frac{(t_c - t_r)}{e.r.} \tag{2}$$

where

$R.L.$ = remaining life
t_c = current wall thickness, in. (mm)
t_r = preset replacement wall thickness, in. (mm)
$e.r.$ = erosion rate, in./yr (mm/yr)

Remaining life of headers is calculated using modeling software due to the complexity of crack growth.

Steam-cooled tubes

Steam-carrying superheater and reheater tubes operating at temperatures above 900F (482C) are subject to failure by creep rupture. The creep life of a tube can be estimated from tabulated data, provided the applied (hoop) stress and the operating temperature are known.

When a tube is put in service, the metal contacting the steam begins to form a layer of oxide scale known as magnetite (Fe_3O_4). As the life progresses, the ID oxide layer grows at a rate that is dependent on temperature. This scale acts as a heat transfer barrier and causes an increase in the tube metal temperature as discussed in Chapter 4. The metal temperature, therefore, also gradually increases with time.

Internal oxide thickness measurements are necessary for estimating a tube's operating temperature and remaining creep life as well as for assessing the overall condition of the superheater. In the past, these measurements have been obtained by removing tube samples for laboratory examination. To avoid destructive tests, B&W developed the NOTIS NDE technique discussed earlier.

Life prediction methodology The prediction of tube creep life begins with creep rupture data taken in short-term laboratory studies. Creep specimens, similar to cylindrical tensile specimens, are machined from various tube steels. Each specimen is heated to a temperature (T) and is pulled uniaxially at a stress (S) until failure occurs; at this point, a time to failure (t) is measured. A matrix of stress and temperature values has been tested.

The Larson-Miller parameter (LMP) is a function relating T, S, and t. This parameter is defined as:

$$LMP = (T + 460)(20 + \log t)$$

where

T = constant temperature applied to the creep specimen, F
t = time at temperature T, h

Every tube in service has an associated LMP number that increases with time. These LMP data can be related to stress, as is illustrated in Fig. 14. This relationship between stress and LMP can then be used to predict a time to creep rupture for a superheater or reheater tube. Knowing two of the three factors affecting creep rupture, i.e., hours in service and hoop stress, the third factor, temperature history, can be estimated.

There are numerous mathematical models relating steam-side oxide thickness, time, and mean metal temperature for low chromium-molybdenum alloys. The following relationship is most widely used to calculate oxide thickness (x) in mils:

$$\log x = \left[0.00022\,(T + 460)(20 + \log t)\right] - 7.25 \qquad (3)$$

Although this formula works well with 1-1/4 Cr-Mo alloys, it must be modified for use with higher chromium alloys and carbon steels.

Creep life fraction analysis The life fraction is defined as the ratio of the time a tube withstands a given stress and temperature (t) to the time for creep rupture conditions (t_f).

Robinson's Rule of life fractions states that if the applied stress and temperature conditions are varied, the sum of the life fractions (or damage) associated with each set of conditions equals 1 at failure. It may also be written as follows:

$$(t/t_f)_1 + (t/t_f)_2 + \ldots + (t/t_f)_n = 1 \qquad (4)$$

where subscripts 1 through n indicate unique stress temperature conditions.

Example – Part I

Assume a tube operates at a hoop stress of 5 ksi (34,474 kPa) and a temperature of 1050F (566C). What is the predicted time to failure? Using these parameters and the stress-LMP curve presented in Fig. 14, the effective LMP at failure is 38,015.

From the LMP equation, the expected time to failure (t_f) can be calculated:

$$\begin{aligned} \text{LMP} &= \left[(T + 460)(20 + \log t_f)\right] \\ t_f &= 150{,}000 \text{ h} \end{aligned} \qquad (5)$$

If this tube has operated for 100,000 hours at these parameters, the creep life fraction expended is:

$$f_{\text{expended}} = t/t_f = \frac{100{,}000}{150{,}000} = 0.6667 \qquad (6)$$

The creep life fraction remaining is:

$$\begin{aligned} f_{\text{remaining}} = 1 - t_{\text{expended}} &= 1 - 0.6667 = 0.3334 \\ t_{\text{remaining}} &= f_{\text{remaining}} \times t_f \\ t_{\text{remaining}} &= 0.3334 \times 150{,}000 \\ t_{\text{remaining}} &= 50{,}000 \text{ h} \end{aligned} \qquad (7)$$

Example – Part II

Assume that, after operating at 1050F (566C) for 100,000 hours, tube temperature increases to 1065F (574C).

Using the same effective LMP, the LMP equation is used to calculate t_f at 1065F (574C) as follows:

$$\begin{aligned} \text{LMP} &= (T + 460)(20 + \log t_f) \\ t_f &= 85{,}000 \text{ h} \end{aligned} \qquad (8)$$

Therefore, a new tube operating at 1065F would have an expected life of 85,000 hours. However, from Part I, this tube has used up two-thirds of its life at 1050F and has a remaining life fraction of 0.33.

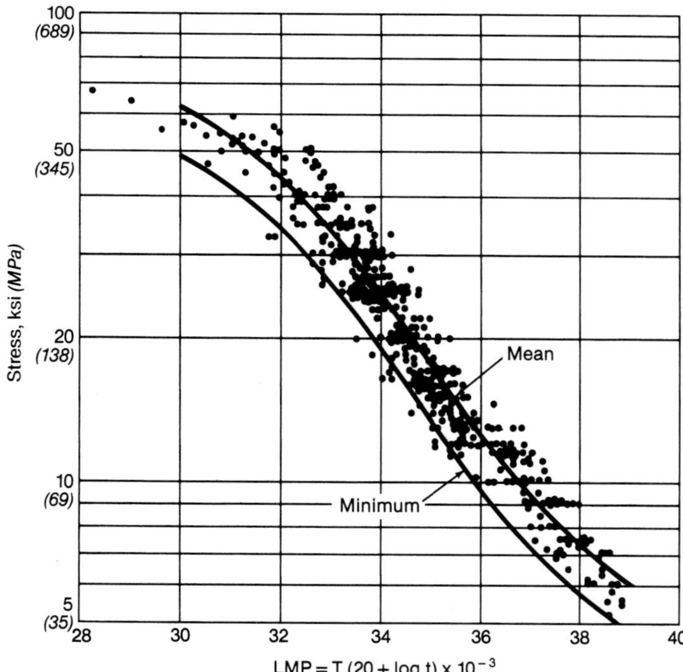

Fig. 14 Stress versus Larsen-Miller parameter (LMP).

Robinson's Rule can be applied to determine the higher temperature service life after the tube is exposed to 100,000 hours at 1050F.

Recall Robinson's Rule – the sum of the life fractions is equal to 1:

$$\left(t/t_f\right)_{1050} + \left(t/t_f\right)_{1065} = 1$$
$$100{,}000/150{,}000 + t/85{,}000 = 1 \qquad \textbf{(9)}$$
$$t = 28{,}000 \text{ h}$$

Note that in this example, the total tube life would be 100,000 + 28,000, or 128,000 hours.

Analysis procedure The following analysis procedure is used with tube wall oxide thickness measurements from the NOTIS system.

1. The life of the tube, past and future, is broken into time intervals, each of length t:

   An oxide growth rate is determined knowing the present oxide thickness and the time in service and assuming the initial oxide thickness was zero. Once a mathematical function describing oxide thickness with respect to time and temperature is defined, the thickness in each analysis interval can be calculated. The tube metal temperature in each interval, taking into account the insulating property of the oxide, is also calculated.

   A linear wall thinning rate is determined for the tube, knowing the present wall thickness as measured by NOTIS, the original wall thickness, and the service time of the tube. Once a function describing wall thickness with respect to time is defined, the wall thickness in each analysis interval can be calculated. The hoop stress is calculated using the ASME Boiler Code Section I tube formula in each interval.
2. The creep life fraction used up in each interval is determined:

   Given the stress, the LMP value may be found from the creep rupture database. Given the operating temperature and the LMP, the time a new tube would last at these conditions (t_f) is determined. The interval creep life fraction used up is then t/t_f.
3. Because the life fractions, summed over the analysis intervals, are equal to 1, the remaining life is obtained by subtracting the tube service time from this total life.

Accuracy of creep life prediction analysis Life fraction analysis is the most accurate and widely accepted method for estimating tube life. Although this method is straightforward and well documented, it is not precise.

Result inaccuracies are due to inherent material property variations. Additionally, during service, short excursions to higher temperatures lower the remaining life fraction.

Rather than attempting to determine the precise time of a creep rupture, the evaluation places each tube into a band of expected remaining lives. These bands take into account the shortcomings of the life fraction analysis, as well as the inaccuracy of the operating parameters for the unit being assessed. In effect, these bands are confidence limits.

Capabilities and limitations At elevated temperatures, the external and internal surfaces of boiler tubes slowly oxidize. The external scale is normally removed, whereas the internal scale usually remains intact. Typically, the multilaminated scale that is formed on the inner surface is characterized by an iron-rich inner layer and an oxygen-rich outer layer. The latter generally contains a large number of pores or voids. The ultrasonic response with the NOTIS system from the inner/outer layer interface is small compared to the signals associated with the metal/oxide and oxide/air interfaces. Therefore, a tightly adhering porous oxide does not affect the accuracy of the NOTIS system. However, if the two oxide layers become disbanded due to exfoliation processes, the NOTIS operator may only measure the oxide thickness to the disbanded area and therefore indicate that the oxide is exfoliating. Exfoliation is the flaking of scale particles off of the internal oxide layer.

Water-cooled tubes

Water-carrying tubes operate at or below saturation temperature and are not subject to creep damage. Therefore, these tubes have no defined design life as do high temperature components. However, erosion, corrosion, thermal expansion and mechanical stresses act on water carrying tubes. This limits useful practical life.

Outside tube wall thinning Erosion and corrosion are the most common causes of OD wall thinning. Erosion typically occurs on the tube outside diameter in the form of wall thickness loss. A tube's wall thickness is designed to the ASME Boiler and Pressure Vessel Code to withstand a given pressure, temperature, and mechanical load. An example is as follows:

Given:		
	Design pressure (P)	= 2400 psi (16.5 MPa)
	Design temperature (T)	= 700F (371C)
	Tube outside diameter (OD)	= 3 in. (76.2 mm)
	Tube material	= SA210 A1

From the ASME Code the wall thickness formula is:

$$t = \frac{PD}{2S + P} + 0.005D \qquad \textbf{(10)}$$

where

t = minimum wall thickness
P = design pressure
D = outside diameter
S = allowable stress

By knowing the material and temperature, the allowable stress can be found in the ASME Code. In this case, the allowable stress is 14,400 psi (99,285 kPa). Solving the equation gives t = 0.245 in. (6.22 mm).

In the case of industrial and utility boilers, the next higher standard tube thickness would be supplied. In the case of a chemical recovery boiler used in the pulp and paper industry a much higher thickness would be used. This allows for wall thickness reduction when operated in a reducing atmosphere. (See Chapter 28.)

As a guide, the following are B&W recommendations for utility and industrial boiler tubes:

1. Water-cooled tubes should be repaired to original wall thickness or replaced if reduced to 70% of original.
2. Steam-cooled tubes should be repaired to original wall thickness or replaced if reduced to 85% of original.

Inside tube wall corrosion Internal corrosion due to hydrogen damage, caustic gouging and under-deposit corrosion is more difficult to evaluate for remaining life because it is not easily quantified.

In addition to wall thickness, other factors, such as the effect of water quality, microstructure and damaged area size, must also be considered. Internal corrosion data can be mapped similarly to that for ultrasonic wall thickness.

Internal deposits Internal deposits lead to tube failures and provide initiation sites for hydrogen damage and under-deposit corrosion as discussed in Chapter 42. The type of corrosion depends upon the nature of the deposit (dense or porous) and the composition of the chemicals beneath the deposit. Chemical cleaning of the boiler removes these deposits. The cleaning frequency depends on the type of unit, operating pressure, fuel, and water treatment program. The hours of operation and data from tube samples also determine the cleaning frequency.

Tube samples should be taken from locations with the heaviest deposits which are usually located in the high heat input burner zone areas of the furnace.

Headers and piping

High temperature headers and piping typically fail due to creep, fatigue, or a combination of the two. Although a header is more complex in geometry, it is essentially a pipe with tubes welded to it. Therefore, many of the remaining life analyses are similar to those for tubes. However, it must be noted that the root cause of damage can be different in headers.

On thick walled sections, where thermal and stress gradients can occur, remaining life is based on crack initiation and propagation. Calculational methods, statistics, dimensional measurements, metallographic methods, and post service creep rupture testing are considered mainly with crack initiation and the events preceding it. When dealing with thick sections, these analytical techniques must be followed by crack growth analyses. Once a component is cracked, the pre-crack evaluation methods do not apply.

Crack growth analysis (CGA) There are three steps used in predicting crack propagation based on time dependent fracture mechanics. (See also Chapter 8.) The schematic shown in Fig. 15 illustrates the various steps.

Step 1 consists of identifying the creep growth and deformation behavior of the material. Step 2 consists of the expressions for the crack tip driving force for creep, C_t. The basic expression for C_t is:

$$\frac{da}{dt} = bC_t^m \quad \textbf{(11)}$$

where a is crack depth, t is time, and both b and m are material constants.

The final step is the development of results from Steps 1 and 2. These results are the remaining life or crack growth curves that are presented two ways. The first is a plot of crack versus time. This curve is exponential because the crack grows faster as it becomes larger. The second curve is the inverse of the first, plotting size versus remaining life. Remaining life is based on a critical predetermined through-wall crack size. Once the crack reaches this value, the remaining life is zero. From the previous equation, this can be written:

$$\text{R.L.} = \int_{a_i}^{a_c} \frac{da}{bC_t^m} \quad \textbf{(12)}$$

where a_c is the critical crack depth and a_i is the initial crack depth.

Leak before break analysis (LBBA) When evaluating cracks that are growing at a stable rate, even though they are through-wall, a condition occurs where fluid leaks through the crack but does not cause

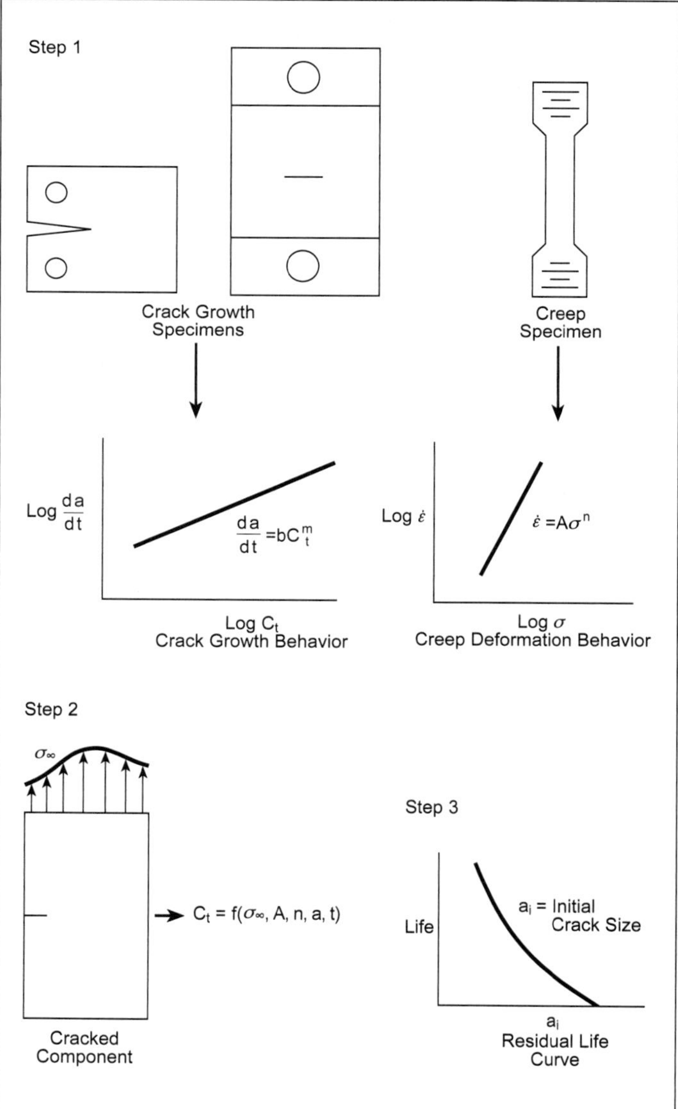

Fig. 15 Three-step methodology for crack growth analysis.

a catastrophic rupture. If the flaw is characterized by depth a and length $2C$, then the following expression applies to leak before break:

$$\text{R.L.} = \int_{a_i}^{a_c} \frac{da}{bC_t^m} = \int_{C_i}^{C_{cr}} \frac{dc}{bC_t^m} \tag{13}$$

where C_{cr} and C_i are the critical crack and initial crack half lengths respectively. Because all variables except C_i are known or determined, the equation can be solved for C_i. B&W has developed a software code, Failure Analysis Diagram (PCFAD®), that models many crack scenarios encountered in header and piping systems. The failure assessment procedure uses a safety/failure plane diagram. The plane is defined by the stress intensity factor/fracture toughness ratio (K_r') as the ordinate and the applied stress/net section plastic collapse stress ratio (S_r') as the abscissa. If an assessment point lies within the curve (Fig. 16), the structure is safe. The distance from a point to the curve indicates the margin of safety. Chapter 8 provides further discussion. These two approaches allow the engineer to determine 1) how long a flaw will take to reach a predetermined critical size (crack growth, PC CREEP), and 2) whether the flaw will cause a leak or catastrophic failure.

Life fraction analysis Life fraction analyses (LFA) are performed on piping at temperatures above 900F (482C) in which the primary failure mode is creep. Stress levels are calculated using design conditions. To determine the minimum creep rupture life, published creep rupture data (LMP) are used. The calculation is similar to that for steam-cooled tubes, however piping is not influenced by gas stream heat flux and, therefore, operates at a fairly consistent temperature. The life fraction expended (LFE) is expressed as:

$$\text{LFE} = \frac{t}{t_m} \tag{14}$$

By using Minor's sum, an expression can be created to represent the unit's history of operating at different conditions:

$$\text{LFE} = \sum_{i=1}^{z} \frac{t_i}{t_{mi}} \tag{15}$$

where

z = number of different conditions
i = ith temperature and stress level
t_i = time of operation at given conditions
t_{mi} = minimum time to creep rupture (LMP)

When considering only creep rupture stress, a long remaining life results. However, most piping systems are also subjected to fatigue stress. Fatigue damage can be calculated by using standard fatigue curves. By combining the creep rupture and fatigue components, the expression becomes:

$$\text{LFE} = \sum_{i=1}^{z} \frac{t_i}{t_{mi}} + \sum_{j=1}^{x} \frac{c_j}{C_j} \tag{16}$$

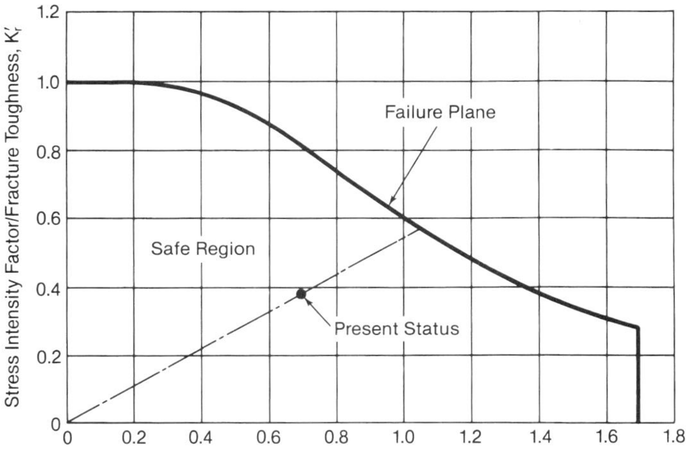

Fig. 16 Sample PCFAD® failure plant diagram.

where

x = number of different stress modes
j = jth stress level
c_j = number of cycles at stress level
C_j = number of cycles to failure at stress level

This simple approach requires detailed operating history. In addition, it only provides a gross estimate of expected life.

Stress calculation The calculation of stresses for headers is complex because of tube stub geometry, tube bank loading, differential tube temperatures, and piping stresses on the outlet nozzle. Using specially developed software codes, it is possible to perform a finite element analysis for the entire header. B&W has performed finite element analysis on the tube stub ligament region and found that this area contains very high stresses, especially during temperature transients.

The total stress (S_t) applied to the flaw is equal to the sum of the primary stresses (axial or hoop) (S_p), bending stresses (S_b) and pressure stresses on the crack face (S_c):

$$S_t = S_p + S_b + S_c \tag{17}$$

The primary stress is determined by the orientation of the flaw. If the flaw is located axially along the length of the component, then the hoop stress is primary. If it is located circumferentially, then the axial stress is primary. Bending stresses are caused by dead loads, hanger spacing, thermal differentials and restraints. The pressure stresses applied to the crack face are only considered when the crack is open to the pressure. When this is the case, the applied stress is equal to the internal pressure.

BLESS Code analysis BLESS is an acronym for Boiler Life Evaluation and Simulation System. The complete BLESS code considers crack initiation and growth in headers as well as crack growth in pipes. The BLESS Code can also perform leak before break analysis (LBBA) of axial cracks in pipes. The Code was developed for EPRI by B&W as a subcontractor to General Atomics. The BLESS Code was derived from B&W's previous software codes for failure analysis

(PCFAD) and crack growth analysis (PC CREEP). Because stress calculations in headers and piping require difficult finite analysis due to complex geometries, simplifying assumptions were made in the development of the BLESS Code to allow for analysis.

The BLESS Code greatly facilitates the life assessment of elevated temperature headers and piping by eliminating the need for finite element thermal and stress analysis and utilizing developments in nonlinear creep-fatigue crack growth. The evaluation includes both crack initiation and crack growth. The Code permits the evaluation of the effects of extremely detailed thermal and mechanical load histories on headers with very complicated geometric details. The estimated remaining life is calculated by BLESS either as a single value (when re-run in the deterministic mode) or a statistical distribution. This distribution is obtained when BLESS is run in the probabilistic mode and defines the probability of failure as a function of time. Such information can be useful in making run/repair/replace and re-inspection decisions for aging or cracked headers and piping.

Flaw characterization Flaws found in headers and piping must be characterized prior to crack growth, LBBA, or BLESS analysis. This characterization involves accurately determining the flaw's length and depth. Through-wall depth is considered the most critical. The most common characterization methods include standard NDE techniques such as MT, PT, RT and UT.

Destructive samples

Destructive sampling is frequently done when data from nondestructive evaluations are inconclusive. Material properties, damage and deposits can be quantified. Tube samples and boat samples are discussed earlier under *Destructive examinations*.

Leak detection

Leaks in boilers, piping and feedwater heaters are major contributors to power plant unavailability and performance losses. In their early stages, leaks are often undetected because they are inaudible and/or concealed by insulation.

In the early 1980s, acoustic monitoring equipment began to be used for leak detection. By using a piezoelectric pressure transducer to detect acoustic energy emitted by a leak, the detection of smaller leaks became possible. Leak noise is transmitted by air and by the structure. Boiler tube leaks are best detected through an airborne sensor because of the large boiler structure volume. Feedwater heaters, headers and piping leaks are best detected through a structural sensor. Because these components are small and self-contained, direct contact monitoring is used.

Leak noise is caused by a fluctuating pressure field associated with turbulence in the fluid. Turbulence is a condition of flow instability in which the inertial effects are highly dominant over the viscous drag effects. Once turbulence is established, the acoustic energy radiated from a leak increases strongly with pressure and flow rates.

Acoustic leak detection technology has been demonstrated through laboratory work, field testing and operational experience. The sensitivity of the sensor depends on three factors: sound radiated from the leak, attenuation of sound between the leak and the sensor, and background noise. Leak noise occurs in a broad band, ranging from below 1 kHz to above 20 kHz. Because of the low frequency background noise and the greater attenuation of high frequencies, most airborne systems operate in the range of 1 to 25 kHz. This is important because the acoustical signal diminishes in amplitude as it travels away from the source. Therefore, in designing a system, there is a tradeoff between sensor spacing and minimum detectable leak.

Table 2 shows typical leak monitor signals for boiler tubes, feedwater heaters, steam piping, and crack detection acoustic emissions.[7]

Boiler leak detection The background noise in a boiler is primarily due to combustion in the furnace. Direct measurement of background noise is needed to determine the spectral characteristics. In addition, the background noise level must be stable. The magnitudes of background noise from different parts of the boiler are similar; a large component is due to sootblower operation.

Using background noise and leak characterization data, full scale tests have been run to optimize sensor listening distances, sensor orientation, and signal processing equipment. A typical 500 MW utility boiler can

Table 2
Typical Acoustic Monitoring Signals[8]

	Leak Detection				Crack Detection
	Boiler Tubes	Feedwater Heaters		Steam Lines	Steam Lines/Headers
Acoustic path	Gasborne	Waterborne	Metalborne	Metalborne	Metalborne
Detection band	0.5 to 25 kHz	5 to 25 kHz	100 to 400 kHz	200 to 500 kHz	100 to 500 kHz
Background noise frequency	Below 2 kHz	Below 8 kHz	Below 125 kHz	Below 200 kHz	Below 200 kHz
Background noise amplitude*	20 to 30 dB	40 to 60 dB	30 to 50 dB	40 to 60 dB	40 to 60 dB
Total system gain	60 dB	50 to 60 dB	40 to 60 dB	40 dB	60 to 80 dB
Number of sensors	12 to 18	1 to 2 per heater	1 per heater	Every 20 ft (6.1 m)	Every 10 to 20 ft (3 to 6.1 m)

* Reference 1 μbar

be monitored using 16 to 24 sensor channels. Fig. 17 shows a typical airborne sensor and waveguide arrangement.

Waveguides are installed through the boiler enclosure at various locations. A typical sensor has a detection range of approximately 50 ft (15.2 m).

Header and piping leak detection Headers and piping are monitored with structural sensors similar to those of feedwater heaters. The sensors are placed approximately 15 to 20 ft (4.6 to 6.1 m) apart.

These sensors can also be used to detect crack growth. When a crack grows, it emits acoustic energy that can be detected. The processing of crack growth signals from leak detection sensors requires different, more powerful computing hardware than the systems required for leak detection. With structural leak detection sensors installed, and the proper fitting and signal processing hardware, periodic monitoring for crack growth can be performed.

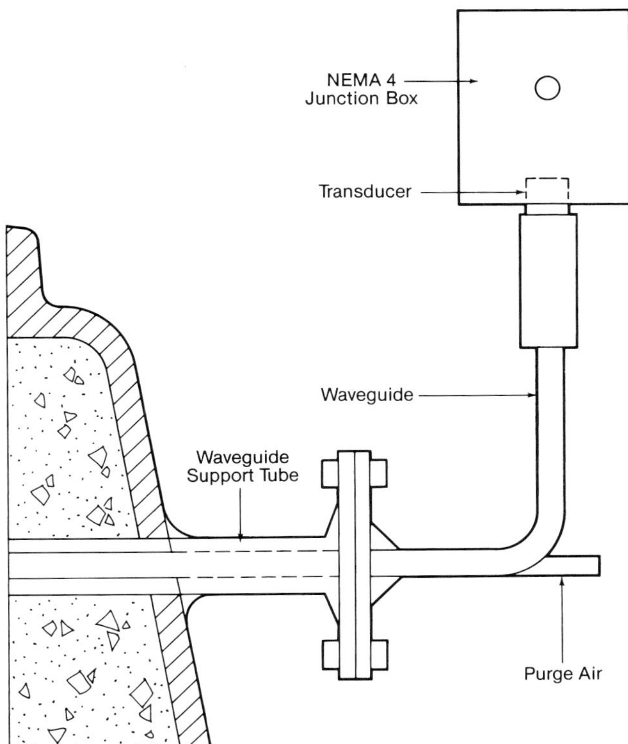

Fig. 17 Typical airborne noise sensors.

Cycling effects and solutions

In assessing a boiler's ability to withstand cycling, those components most vulnerable to cycling are reviewed first. These components are discussed from two standpoints: the operating methods which minimize cyclic damage, and design modifications which permit the component to better withstand cycling conditions.

Cycle definition

Two types of cycling service are usually considered: load cycling and on/off cycling. The on/off type has also been called *two shifting*.

A cycle is considered to start at full load, full temperature steady-state conditions. It goes through a load change, then returns to the initial conditions. A typical load cycle is then composed of three phases:

1) load reduction,
2) low load operation, and
3) reloading.

A typical on/off cycle has four phases:

1) load reduction,
2) idle,
3) restart, and
4) reload.

The phase that is often ignored, the idle period, can offer the greatest potential for reducing cyclic damage.

Economizer thermal shock

On boilers that are on/off cycled, economizers often show more cyclic damage than the other components. The economizer receives water from the extraction feedwater heater system, and the inner metal surfaces follow the feedwater temperature with practically no time delay. As a result, high rates of metal temperature change can occur with resultingly high local stresses.

Fig. 18 shows economizer inlet temperatures during an overnight shutdown cycle. The first two hours are for load reduction, followed by eight hours of idle or banked condition. Next, the boiler is fired in preparation for restart. The rates of temperature change during the load decrease and increase are usually not excessive, but would represent load cycling conditions for the economizer.

During the banked period, there is some air leakage through the boiler with a resulting decay in boiler pressure. As this happens, the drum water level decreases. At the same time, the leaking air passing through the boiler is heated to near saturation temperature, and that air then heats the economizer. An economizer metal temperature can increase at 30 to 50F/h (17 to 28C/h) during this period and can approach saturation temperatures. When the drum level drops, the operator usually refills the boiler so that it is ready for firing. Because there is no extraction steam available, the feedwater temperature is low. This slug of cold water quickly chills the economizer, causing thermal shock as indicated by the solid lines in Fig. 18. The inlet header and tubes receive the greatest shock.

When the boiler is fired in preparation for turbine restart, rollup, and synchronization, the economizer heats up rapidly, often nearing saturation temperature. Feedwater is started when the initial load is applied to the turbine. Because little extraction heating is available, feedwater temperature is low. A severe shock occurs at this point, as the temperature can increase 300F (167C) in a few minutes.

Typically, early damage consists of cracks initiating in the tube holes of the inlet header which are closest to the feedwater inlet connection (Fig. 19).

Other damage has also been seen from this cyclic service. Outlet headers have shown damage similar to inlet headers. Furthermore, some tube bank support systems can not accommodate the high temperature differences between rows.

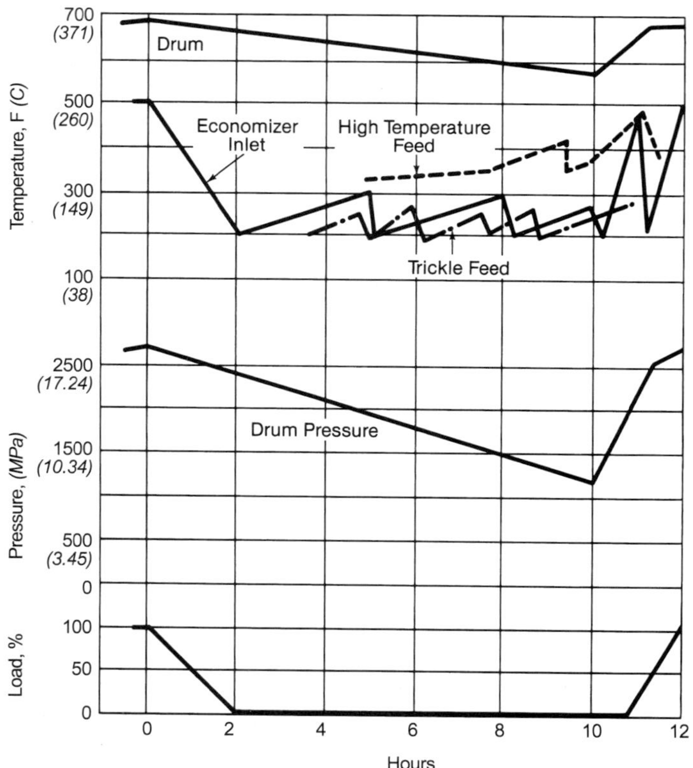

Fig. 18 Economizer temperatures during overnight shutdown cycles.

Solutions are available to reduce the frequency and magnitude of thermal shocks. These have taken two forms and address the out of service and restart conditions.

The first solution is called trickle feed cooling. Very small quantities of feedwater are frequently introduced during the shutdown and restart periods. This prevents the inlet header from reheating and reduces the cooling rate during feedwater introduction. Because feedwater introduction is controlled to limit economizer temperature rise, some drum blowdown may be necessary to prevent a high water level.

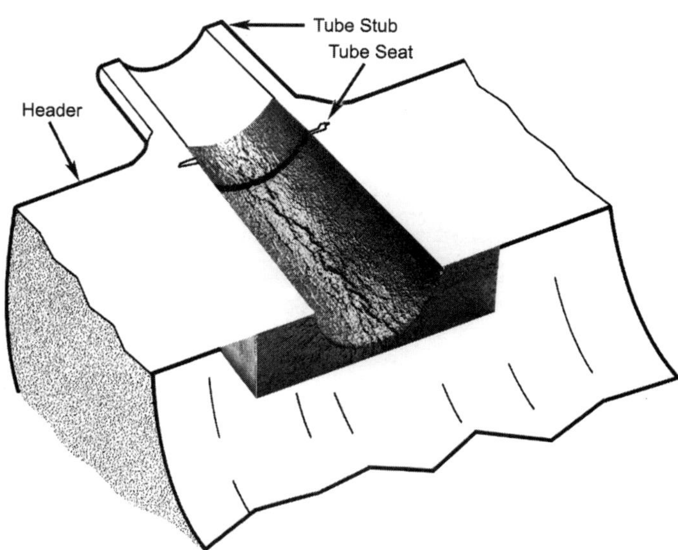

Fig. 19 Cracking in economizer inlet header occurs first in bore holes nearest water inlet.

A second method of reducing thermal shock is to permit the economizer to reheat during the idle period and then to provide higher temperature feedwater for restart. This can occur by pressurizing a high pressure heater with steam from an auxiliary source or from the drum of the unit. The quantity of steam required is low because it only heats the initial low flow of feedwater.

Furnace subcooling

Several drum boilers that have been subjected to on/off cycling have developed multiple cracks in lower furnace wall tubes where the tubes are restrained from diameter expansion or contraction. Typical cracking areas have been at the lower windbox attachment where filler bars or plates are welded to the tubes (Fig. 20).

Investigation of these failures indicated that, during the shutdown or idle period, relatively cold (cooled below saturation temperature) water settled in the lowest circuits of the furnace bottom. When circulation was started by initial firing or the circulating pump, the cold water interface moved upward through the walls, rapidly cooling the tubes. As the interface moved, its temperature gradient decreased and the rate of cooling decreased, therefore reducing damage higher in the furnace. Experience indicates that if the subcooling can be limited to 100F (56C), there is a low probability of damage.

An out-of-service circulating pump system may be used to limit the subcooling (Fig. 21). This is a low capacity pump that draws from the bottom of the downcomers and discharges water to the drum, therefore

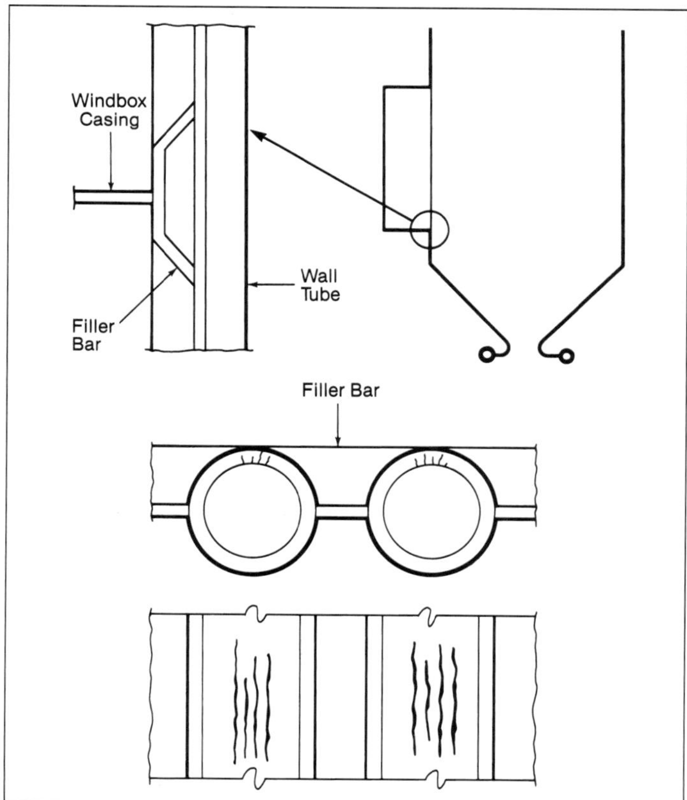

Fig. 20 Lower windbox attachment cracking due to subcooling.

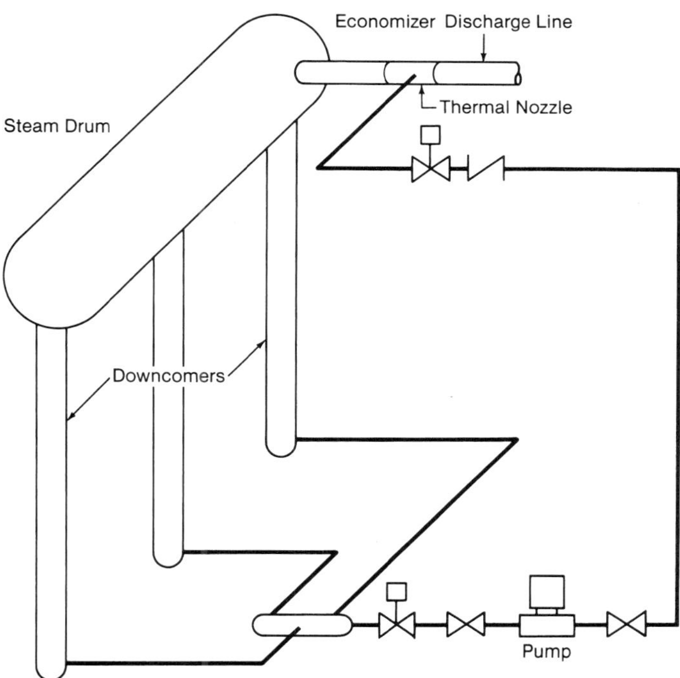

Fig. 21 Out-of-service circulating pump system.

preventing the stratification temperatures of water within the unit.

Tube leg flexibility

The enclosure walls of most boilers are water- or steam-cooled. The water-cooled circuits carry boiling water, and the steam-cooled circuits carry steam from the drum. As a result, they operate near the saturation temperature corresponding to the drum pressure. Whether the boiler is being fired or shut down, considerable heat absorption or loss is necessary to change the temperature of the walls. As a result, they change temperature more slowly than the other components.

The economizer, superheater and reheaters penetrate these walls; the penetrations are designed to be gas-tight. At the point of penetration the expansion then follows saturation temperature. However, the header that forms the inlet or outlet of the other circuits expands with the temperature of the steam or water that it is handling. Fig. 22 shows the motions of the header end and the outermost connecting leg for a superheater or reheater outlet. Note that the greatest deflection is when the header temperature is at a maximum.

In the case of economizers and the reheater inlet, the deflection is in the opposite direction because these headers operate below saturation temperature. For these components the greatest temperature difference, and therefore the greatest deflection, is at low loads.

Regardless of the direction, the greatest temperature difference produces the maximum differential expansion and the maximum bending stresses in the connecting legs. The stress range and amplitude dictate component fatigue life.

Consider the superheater outlet and the drum (penetration point) temperature differences at constant and variable drum pressures, as shown in Fig. 23. In this example, for constant drum pressure, the maximum differential occurs at 50% load (hour 1) and is:

1000F – 685F = 315F (538C – 363C = 175C)

The minimum difference is at the end of the idle period (hour 10) and is:

655F – 570F = 85F (346C – 299C = 47C)

This is a stress range proportional to:

315F – 85F = 230F (157C – 29C = 128C)

For variable drum pressure, the maximum difference occurs at 35% load and is:

1000F – 540F = 460F (538C – 282C = 256C)

The minimum temperature difference at the end of the idle period is:

655F – 400F = 255F (346C – 204C = 142C)

producing a stress range proportional to:

460F – 255F = 205F (238C – 124C = 114C)

The variable drum pressure mode of operation is then slightly less severe for the superheater and reheater outlets and much less severe for the reheater inlet.

In considering operational changes, it should be noted that the steam temperature characteristics shown are the maximum available at a given load.

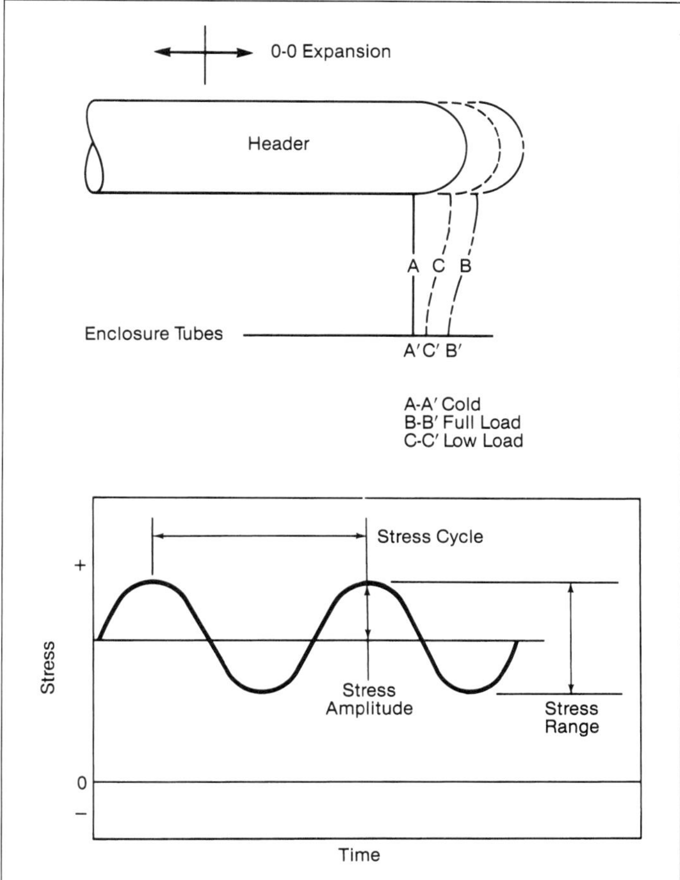

Fig. 22 Superheater tube leg flexibility.

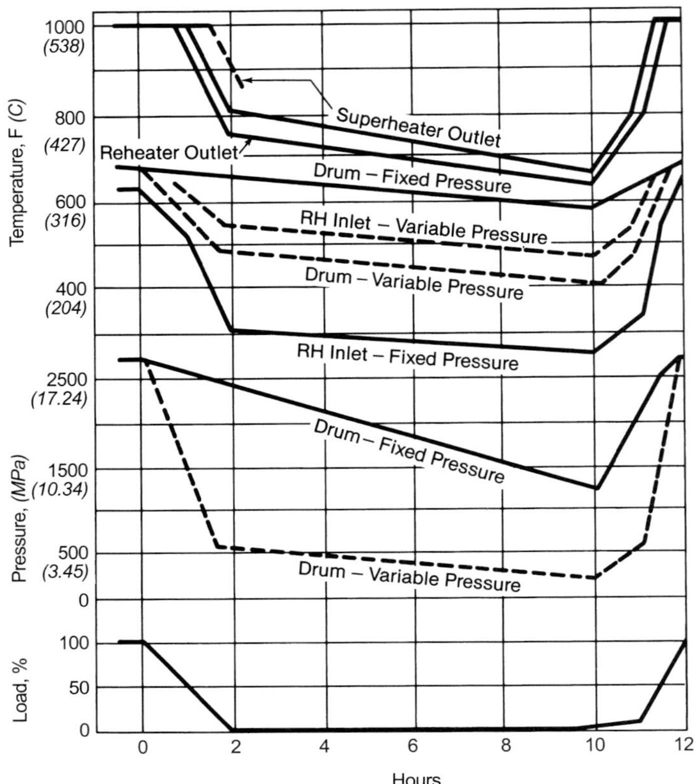

Fig. 23 Overnight shutdown temperatures.

Lower temperature differences of 50 to 75F (28 to 42C) may be obtained in practice.

The first indications of cyclic damage are external cracks on the tube to header welds or the stub to tube welds of the outermost header legs. This damage is relatively easy to inspect and repair. Successive damage is also usually limited to closely adjacent legs because they have experienced similar stress levels. Most sensitive are high temperature headers which are a short distance from the penetration seal on wide units.

Drums

It is important to limit the rate of saturation temperature change in a steam drum. When operating in a variable drum pressure mode, considerable overfiring or underfiring is necessary to quickly change the drum pressure. These firing effects on steam temperature control also prevent rapid drum pressure changes.

Top to bottom drum temperature differences must also be limited. Only small differences result when pressure changes are made under load. The greatest differences develop during pressure reduction, when little or no steam is being taken from the drum.

If drum pressure is rapidly lost during the idle period, top to bottom differentials develop and the drum *humps*. Because most drums have two point supports, the humping is unrestrained and causes little change in the drum or support stress levels. However, the drum acts as a stiff beam, and connected parts move with it (Fig. 24). If those parts do not have sufficient flexibility they can experience unacceptably high stresses.

While most recent units were designed with flexibility in the drum connections, there are some older units where the front furnace wall tubes are routed directly to the drum and are supported by the drum. In such cases, it is difficult to add flexibility by rerouting and the humping must be limited.

Developments

Symptoms of advanced creep and fatigue have been found in some older superheater and reheater outlet headers. Diametral swelling has been observed. Interior cracking around the header ID tube holes and along the length has also been present. To better understand the failure mechanism, finite element stress analyses have been performed. These analyses consider steady-state temperature differences between individual tubes and the bulk header and transient temperature changes of both.

Advances in nondestructive examinations

Innovative techniques are being developed to replace or enhance existing NDE methods. Some are becoming viable due to advancements in microprocessor technology. Others are relatively new and may replace current methods. Advanced techniques include: 1) infrared scanning,[8] 2) automated Phased Array UT, 3) pipe and wall scanners which automatically cover large areas, and 4) through-insulation radiography. EMAT technology is being refined and studied for further applications in the NDE field.

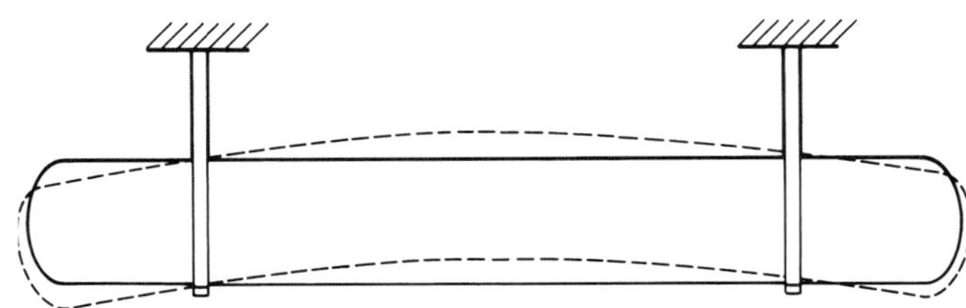

Fig. 24 Drum humping.

References

1. "Condition Assessment Guidelines for Fossil Fuel Power Plant Components," Report GS-6724, pp. l-l to 1-6, Electric Power Research Institute (EPRI), Palo Alto, California, March, 1990.

2. "Guide for the Nondestructive Inspection of Welds," American National Standards/American Welding Society (ANSI/AWS) B1.10:1999, American Welding Society, Miami, Florida, 1999.

3. Bar-Cohen, Y., and Mal, A.K., "Ultrasonic Inspection," *Metals Handbook,* Ninth Ed., Vol. 17, Nondestructive Evaluation and Quality Control, pp. 232-233, 254, ASM International, Metals Park, Ohio, 1989.

4. Alcazar, D.G., et al., "Ultrasonic Detection of Hydrogen Damage," *Materials Evaluation,* Vol. 47 (3), March, 1989.

5. Bar-Cohen, Y., and Mal, A.K., "Ultrasonic Inspection," *Metals Handbook*, Ninth Ed., Vol. 17, Nondestructive Evaluation and Quality Control, pp. 255-256, ASM International, Metals Park, Ohio, 1989.

6. Bar-Cohen, Y., and Mal, A.K., "Strain Measurement for Stress Analysis," *Metals Handbook,* Ninth Ed., Vol. 17, Nondestructive Evaluation and Quality Control, pp. 448-449, ASM International, Metals Park, Ohio, 1989.

7. "Acoustic Leak Detection," Technical Brief TB.CCS.32.9.87, Electric Power Research Institute (EPRI), Palo Alto, California, September, 1987.

8. "Atlantic Electric Demonstrates Infrared Inspection of Boiler Waterwalls," First Use, Electric Power Research Institute (EPRI), Palo Alto, California, December, 1989.

Section VIII
Steam Generation from Nuclear Energy

Nuclear power generation provides a critical element in the energy supply of virtually all developed nations today, and offers the promise to address growing power needs in an environmentally acceptable and safe manner in the future. This section describes the application of steam generation fundamentals to the design of nuclear steam supply systems (NSSS) in which steam is generated by heat released from nuclear fuels.

This section begins with an overview of nuclear installations, concentrating on the pressurized water reactor. Principles of nuclear reactions and the nuclear fuel cycle are then explored in Chapter 47. Chapter 48 is dedicated to nuclear steam generators. Operating experience indicates that this component is a particularly challenging and important part of the NSSS. As nuclear power plants age, the steam generators are increasingly being replaced to optimize plant performance and extend the operating plant life.

Chapter 49 explores the key service, maintenance and operating characteristic of a nuclear steam system that can optimize life and performance. The section concludes in Chapter 50 with an overview of the highly specialized manufacturing requirements and capabilities that are necessary for successful component fabrication.

Oconee Nuclear Power Station, Duke Power Company.

Chapter 46

Steam Generation from Nuclear Energy

Since the early 1950s, nuclear fission technology has been explored on a large scale for electric power generation and has evolved into the modern nuclear power plants. (See frontispiece and Fig. 1.) Many advantages of nuclear energy are not well understood by the general public, but this safe, environmentally benign source of electricity is still likely to play a major role in the future world energy picture. Nuclear electric power generation is ideally suited to provide large amounts of power while minimizing the overall environmental impact.

First generation power plants

The concept of an energy generating plant using nuclear fission was first considered by nuclear physicists in the 1930s. However, peaceful use of the atom was delayed until after World War II. The United States (U.S.) had a head start on nuclear technology because of its work in the atomic weapons program. The U.S. Atomic Energy Commission (AEC) took the lead in research and development for a controlled chain reaction application to energy generation. Many concepts were hypothesized and several promising paths were explored, but the real momentum developed when U.S. Navy Captain Hyman G. Rickover established a division in the AEC to develop a nuclear power plant for a submarine. This program, established in 1949, was to become the forerunner of commercial generating stations in the U.S. and the world. Rickover's design succeeded in 1953. Technology and materials developed by his team became the cornerstone of future U.S. nuclear plants. Concurrently, the AEC established a large testing site in Arco, Idaho where, in 1951, the fast neutron reactor produced the first electricity (100 kW) generated by controlled fission.

The world's first civil nuclear power station became operational in Obninsk in the former Soviet Union (FSU) in mid-1954, with a generating capability of 5 MW. This was about the same energy level produced in the U.S. submarine design.

In 1953, the Navy canceled Captain Rickover's plans to develop a larger nuclear power plant to be used in an aircraft carrier. However, he subsequently transformed this project into a design for the first U.S. civilian power stations. Duquesne Light Company of Pittsburgh, Pennsylvania agreed to build and operate the conventional portion of the plant and to buy steam from the nuclear facility to offset its cost of operation. On December 2, 1957, the Shippingport, Pennsylvania reactor plant was placed in service with a power output of 60 MW. This event marked the beginning of the first generation U.S. commercial nuclear plants.

Several basic concepts were being explored, developed and demonstrated throughout the world during this period. The U.S. submarine and Shippingport plants were pressurized water reactors (PWR) that used subcooled water as the fuel coolant and moderator. The FSU developed enriched uranium, graphite-

Fig. 1 Indian Point Station, New York.

moderated, water-cooled designs. British and French engineers explored natural uranium, graphite-moderated, carbon dioxide-cooled stations. In all these designs, the coolant (gas, liquid metal or pressurized water) transferred heat to a heat exchanger, where secondary water was vaporized to provide steam to drive a turbine-generator. The other major competing approach in the U.S. was the boiling water reactor (BWR). This was similar to the PWR except that the need for a heat exchanger to transfer heat from the coolant to the secondary steam cycle was eliminated by boiling the water in the reactor core and by using this slightly radioactive steam to drive the turbine. (See also Chapter 1.)

Two main classes of reactor fuel evolved: enriched and natural uranium. In the early stage of nuclear power development, only the two major nuclear powers, the U.S. and the FSU, had sufficient fuel production capacity for civilian power generation using enriched uranium. Therefore, natural uranium was chosen as the principal fuel in the United Kingdom (U.K.), France, Canada and Sweden. While the enriched uranium-fueled plants could be smaller and therefore required a lower initial investment, higher costs of the enriched fuel over the life of the plant made its use economically equivalent to that of natural uranium. All six countries launched programs to build civil, commercial nuclear power stations. Other countries collaborated with one of these six for construction technology.

In the U.S., the emerging technology focused on the enriched uranium light water reactor (LWR) that used regular water for the coolant and moderator. This had been the technology selected by the Navy programs and, as a result, the civilian sector gained from the experience of naval applications. The AEC also financed construction of the nuclear-powered merchant ship N.S. *Savannah*, which used this technology in a PWR designed and constructed by The Babcock & Wilcox Company (B&W).

In 1955, the AEC began the Power Reactor Demonstration Program to assist private industry. However, by 1963, only one demonstration plant had started up, and two others had been privately financed. Yankee Rowe in Massachusetts, Indian Point Unit 1 in New York, and Dresden 1 in Illinois operated successfully and demonstrated the viability of this technology for larger plants.

Second generation power plants

From these beginnings, the second generation of nuclear plants began in 1963, when a New Jersey public utility (Jersey Central Power & Light) invited bids for a nuclear power station to be built at Oyster Creek. A comprehensive study showed that it was economically attractive to operate, and the utility was ready to proceed. This bid was won with a turnkey proposal to deliver a plant at a fixed price. The suppliers were expected to absorb initial losses in expectation of profitably building a series of similar plants in the near future. From this study and resulting order came the realization that nuclear power stations could be competitive with other electricity sources. Enthusiasm spread and, in the following years, more new nuclear generating capacity than conventional fossil fuel capacity was ordered. By the mid-1970s, nuclear plants, including new orders, totaled a significant portion of the nation's electric generating capacity. The second generation nuclear plants were comparable in size or larger than contemporary fossil-fueled plants. While nuclear plant capital costs were generally higher than those of the current fossil-fueled plants, the lower fuel cost of the nuclear plant offered a reduction in total generating cost.

The U.S. designs were also attractive for export. U.S. PWR designs were ordered in Japan, South Korea, the Philippines, Spain, Sweden, Switzerland, Yugoslavia, Taiwan, Italy, Brazil, Belgium and Germany, while Italy, Mexico, Spain, India, the Netherlands, Switzerland, Taiwan and Japan ordered BWRs of U.S. design.

In parallel with the rapid U.S. growth in nuclear electric generation capacity, the rest of the economically developed countries also began development and construction. In France, early efforts using natural uranium and gas-cooled reactors could not compete with oil-fired plants and enriched uranium light water reactors. After France developed uranium enrichment capability, the country's national utility, Electricité de France, in conjunction with heavy component designer and manufacturer Framatome, ordered PWR and BWR designs based on U.S. technology. An innovative concept applied by the French was to build a series of nearly identical plants of each design, thereby achieving economies of scale. Subsequently, the PWR design was selected by the French for their standardized series of plants. The basic designs developed were also eventually exported to Belgium, South Korea and South Africa.

In the U.K., initial designs for weapons production reactors were applied to generate commercial power from natural uranium-fueled, graphite-moderated, gas-cooled reactors. In the early 1960s, an advanced gas-cooled reactor (AGR), which used enriched uranium, was designed and subsequently built. Italy and Japan also built versions of these British designs.

The FSU developed a graphite-moderated, boiling water-cooled, enriched uranium reactor from its early Obninsk power station design. A major difference in this reactor from similar Western designs was that it did not include a sturdy reactor containment building. The FSU also developed naval propulsion reactors based upon the PWR concept that were then applied to civilian use. This PWR design was exported to several former Soviet bloc countries and Finland. The FSU PWR was similar to the U.S. designs except it generally provided significantly lower electrical output.

The excellent moderator characteristics of deuterium oxide (heavy water) became the basis for the Canada Deuterium Uranium (CANDU) reactor systems. These natural uranium-fueled reactors were first operated in 1962 (25 MW Nuclear Power Demonstration reactor at Rolphton, Ontario), and subsequent designs increased output to 800 MW.

In Japan, electric utilities experimented with several types of imported nuclear generation plant designs. They ordered gas-cooled reactors from the U.K., and BWR and PWR designs from the U.S. The Japanese government also sponsored a breeder reactor design.

Germany was actively involved in designs primarily associated with high temperature gas reactor concepts. In 1969, the formation of Kraftwerk Union introduced German PWR and BWR designs based on U.S. technology. The intent was to construct a series of these plants to achieve economies of scale. However, unlike the French program, few plants were constructed. Several German PWRs were exported to Brazil, Spain, Switzerland and the Netherlands.

The energy crisis of the mid-1970s caused significant reductions in worldwide electric usage growth rates, and electric utilities began canceling and delaying new nuclear generating capacity construction. Cost increases caused by a number of factors also contributed to this construction decline. The delay of schedules due to slower load growth and/or regulatory hurdles greatly increased financing costs and drove unit costs higher. The high inflation of the period further increased construction costs. In addition, continually changing safety regulations increased costs as changes in design of plants already under construction were required.

PWR installations

By 1978, nine second generation nuclear units designed by B&W were placed into commercial operation, each generating 850 to 900 MW. The first unit was one of three built at Duke Power Company's Oconee Nuclear Station located near Seneca, South Carolina. The chapter frontispiece shows the three units of this station. The three vertical, cylindrical concrete structures are the reactor containment buildings. Although some equipment is shared among units, each is operationally independent. The rectangular building at the right contains the turbine-generators. The control room, auxiliary systems and fuel handling facilities for Units 1 and 2 are located in a structure between reactor buildings 1 and 2. Corresponding equipment for Unit 3 is located separately. The following general description applies to Oconee Unit 1, although it is generally applicable to all U.S. B&W plants; the concepts are applicable to all PWRs.

Containment building

Figs. 2 and 3 show a vertical section and plan of the reactor containment building. The structure is post-tensioned, reinforced concrete with a shallow domed roof and a flat foundation slab. The cylindrical portion is pre-stressed by a post-tensioning system consisting of horizontal and vertical tendons. The dome has a three way post-tensioning system. The foundation slab is conventionally reinforced with high strength steel. The entire structure is lined with 0.25 in. (6.3 mm) welded steel plate to provide a vapor seal.

The containment building dimensions are: inside diameter 116 ft (35.4 m); inside height 208 ft (63.4 m); wall thickness 3.75 ft (1.143 m); dome thickness 3.25 ft (0.99 m); and foundation slab thickness 8.5 ft (2.59

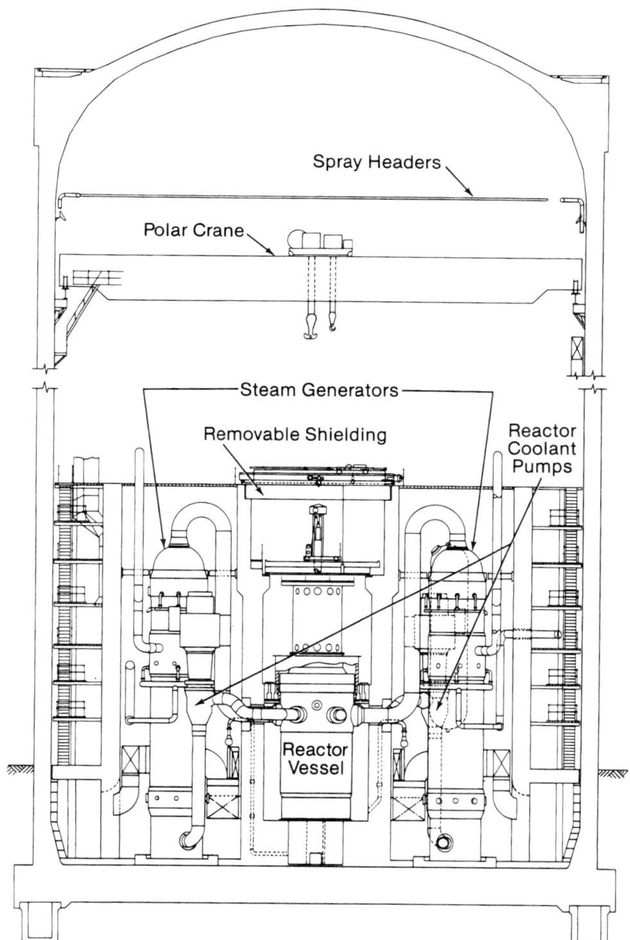

Fig. 2 PWR containment building, sectional view.

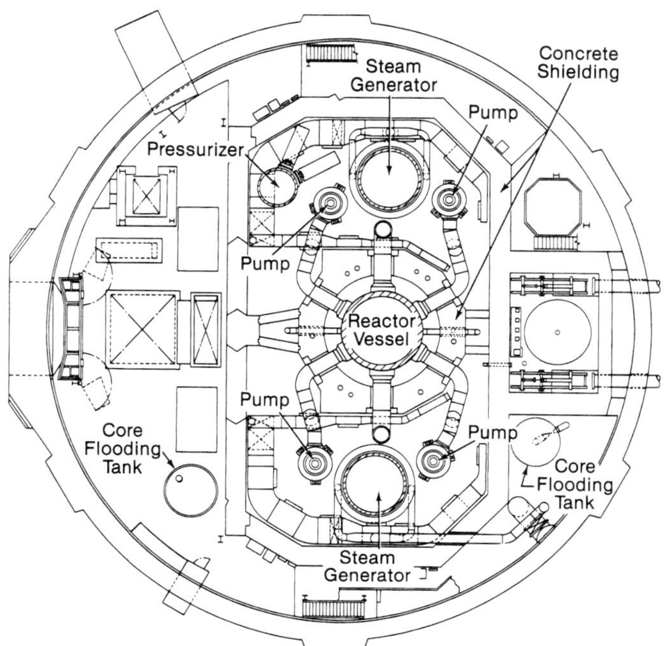

Fig. 3 Reactor containment building, ground floor plan view.

m). The building encloses the nuclear steam supply system and portions of the auxiliary and safeguard systems. The interior arrangement meets the requirements for all anticipated operating conditions and maintenance, including refueling.

The building is designed to sustain all internal and external loading conditions that may occur during its design life. In the event of a major loss-of-coolant accident, the building is designed to sustain the pressure caused by the release of the high-pressure water. To protect against external accidents, extensive tests and analyses have been conducted to show that airplane crashes or objects propelled by tornadoes will not penetrate the wall of the containment.

Nuclear steam supply system (NSSS)

B&W PWR system The major components of the B&W pressurized water reactor NSSS, shown in Fig. 4, include the reactor vessel, two once-through steam generators (OTSGs), pressurizer, primary reactor coolant pumps and piping. As shown in the simplified diagram in Fig. 5 (one of two steam generators shown), the NSSS is comprised of two flow circuits or loops.

The primary flow loop uses pressurized water to transfer heat from the reactor core to the steam generators. Flow is provided by the reactor coolant pump. The pressurizer maintains the primary loop pressure high enough to prevent steam generation in the re-

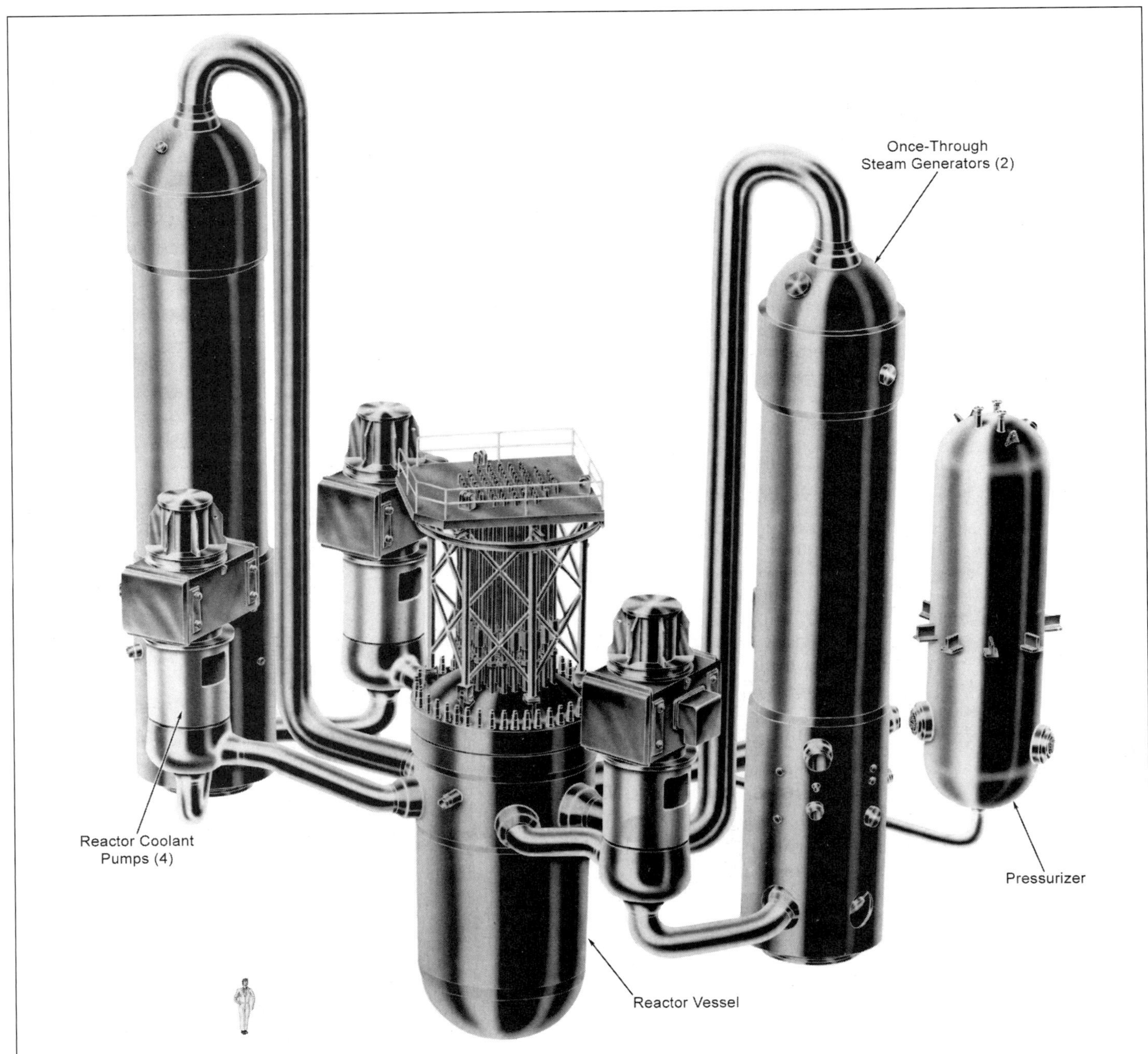

Fig. 4 B&W nuclear steam supply system.

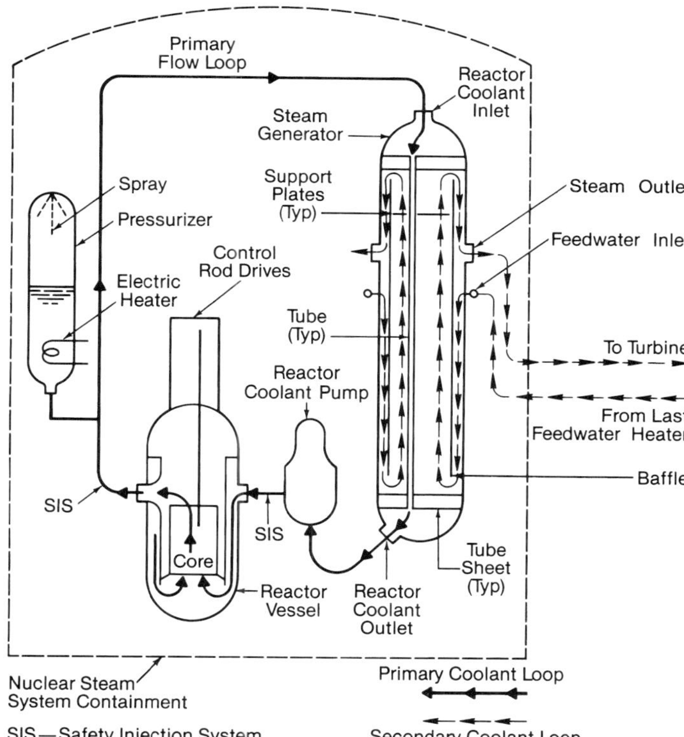

Fig. 5 Simplified schematic of primary and secondary loops.

actor core during normal operation. The steam generators provide the link between the primary coolant loops and the power producing secondary flow loop. In the B&W design, subcooled secondary side water enters the steam generator and emerges as superheated steam, which is sent to the steam turbine to produce power. The primary side water is cooled as it flows vertically downward through the straight tubes and supplies the energy to generate the steam. Table 1 in Chapter 47 of Reference 1 provides a listing of the important design parameters for the 900 MW Oconee type NSSS as well as for a larger 1300 MW system. The first 1300 MW design went into operation in 1987 at Muelheim Kaerlich power station in Germany.

Other PWR systems Fig. 6 shows a PWR system using the recirculating steam generator design. The key components remain the same as the B&W design except that two to four recirculating steam generators replace the OTSGs in providing the link between the primary coolant loop and the power producing secondary side flow loop. In the recirculating steam generator (RSG) design (Fig. 7), the primary coolant passes inside of the U-tubes while the secondary side water passes over the outside of the tubes where it is partially converted to steam. Steam separation equipment then removes residual water which is recirculated back to the bundle for further evaporation. The moisture-free steam is sent to the steam turbine to produce power.

CANDU The CANDU system uses natural uranium fuel and heavy water as the reactor coolant and moderator. The calandria (Fig. 8) is a low-pressure, thin-walled vessel containing the moderator and provides support for the horizontal fuel channel assemblies, shielding, and control mechanisms. In a CANDU 6 system, there are 380 fuel assemblies traversing the entire width of the calandria. Heavy-water coolant flows through the fuel assembly with a full-power flow rate of 1.98×10^6 lb/h (249.5 kg/s) at an operating pressure of 1450 psi (10 MPa) and a mean temperature of 553F (289C). From the calandria, reactor coolant is piped to four inverted U-tube recirculating steam generators with design and layout similar to that shown in Fig. 6. In the CANDU system, each of the fuel assemblies can be isolated from the feedwater system, thereby allowing online refueling at full-power operation.

The balance of the following discussion focuses on B&W system components as the key concepts are generally applicable to all PWRs.

PWR reactor and fuel

Reactor vessel and internals

The reactor vessel, which houses the core, is the central component of the reactor coolant system (RCS). The vessel has a cylindrical shell with a spherical bottom head and a ring flange at the top. A closure head is bolted to this flange.

The vessel is constructed of low alloy steel with an internal stainless steel cladding to protect the vessel from corrosion. The vessel with closure head is almost 41 ft (12.5 m) tall and has an inside diameter of 171 in. (4343 mm). The minimum thicknesses of the shell wall and inside cladding are 8.4 and 0.125 in. (213 and 3.18 mm), respectively. Reflective metal (mirror) insulation is installed on the exterior surfaces of the reactor vessel.

The vessel has six major nozzles for reactor coolant flow (two outlet hot leg and four inlet cold leg). The coolant water enters the vessel through the four cold leg nozzles located above the midpoint of the vessel. Water flows downward in the annular space between the reactor vessel and the internal thermal shield, up

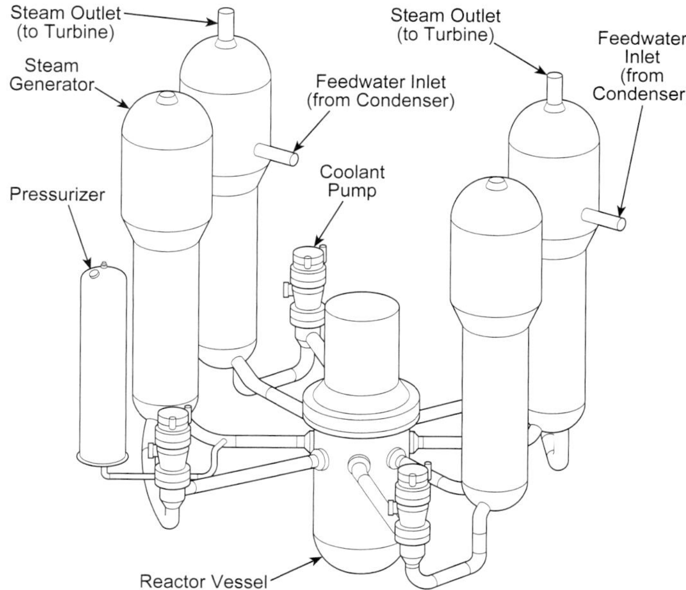

Fig. 6 Four steam generator loop configuration for a pressurized water reactor with recirculating steam generators.

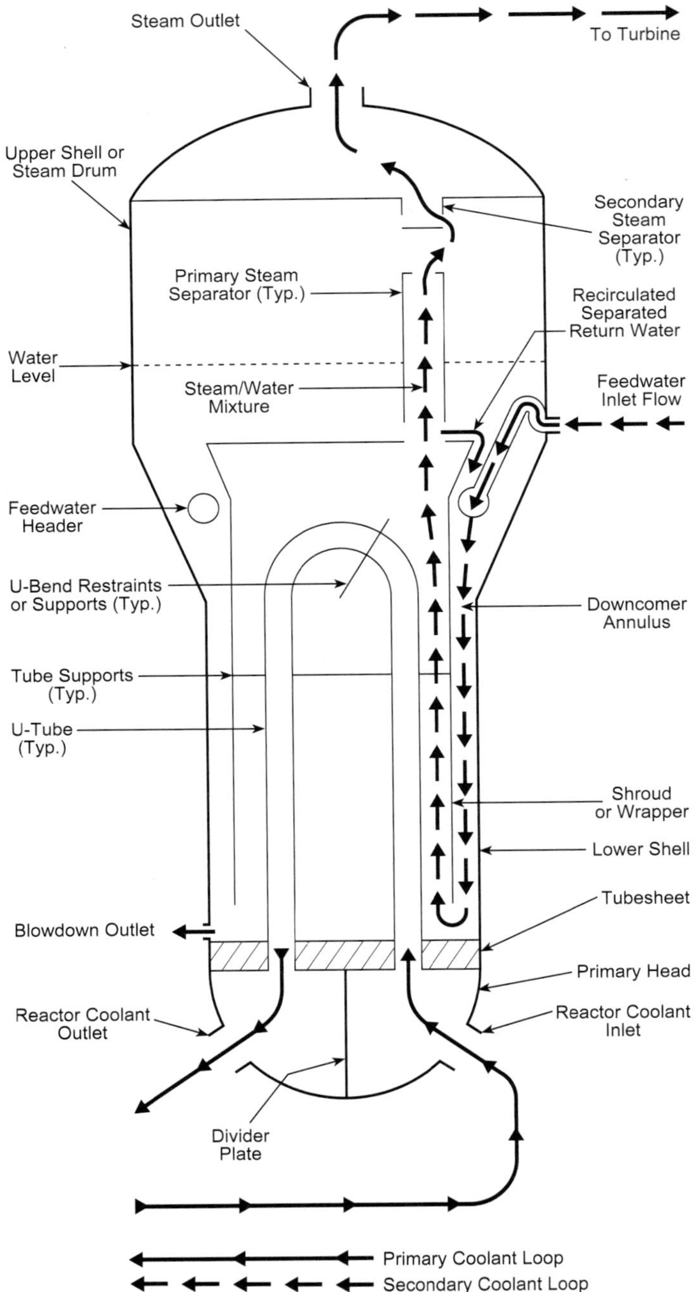

Fig. 7 Simplified recirculating steam generator (RSG) with water preheating configuration.

through the core and upper plenum, down around the inside of the upper portion of the vessel and out through the two hot leg nozzles. Fig. 9 shows a reactor vessel being installed at a nuclear power plant site.

The core, containing the fuel bundle assemblies, is the primary component of the reactor vessel. The remaining major components are the plenum and the core support assemblies (see Figs. 10 and 11).

Plenum assembly The plenum assembly, located directly above the core, includes the plenum cover, upper grid, control rod guide tube assemblies, and flanged plenum cylinder. The plenum cylinder has multiple openings for reactor coolant outlet flow. These flow openings are arranged to form the coolant profile leaving the core. The plenum cover is attached to the top flange of the plenum cylinder. It consists of a square lattice of parallel flat plates, a perforated top plate and a flange. The control rod guide tube assemblies are positioned by the upper grid. These assemblies shield the control rods from coolant crossflow and maintain the alignment of the rods.

Core support assembly The core support assembly consists of the following components:

1. core support shield,
2. core barrel,
3. lower grid assembly,
4. flow distributor,
5. thermal shield,
6. surveillance specimen holder tubes,
7. in-core instrument guide tubes, and
8. internals vent valves.

The core support shield is a flanged cylinder that mates with the reactor vessel opening. The forged top flange rests on the circumferential ledge in the reactor vessel top closure flange, while the core barrel is bolted to the vessel's lower flange. The cylindrical wall of the core support shield has two nozzle openings. These openings form a seal with the reactor vessel out-

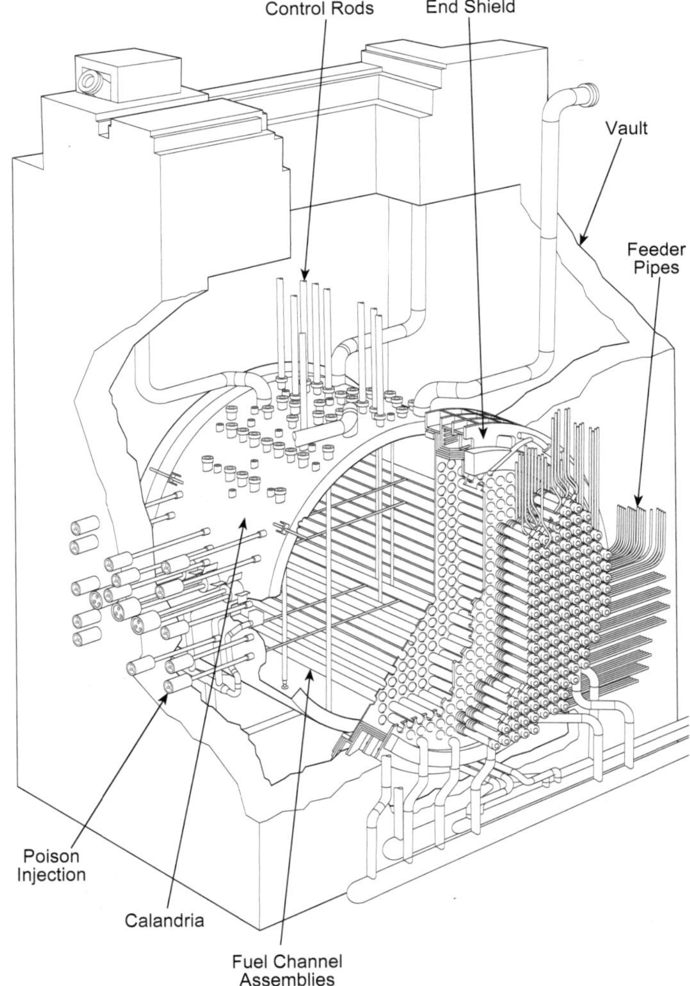

Fig. 8 CANDU 6 reactor assembly.

Fig. 9 Reactor vessel placed into position between steam generators.

let nozzles by the differential thermal expansion of stainless and carbon steel. The core support shield also has eight holes in which the internals vent valves are mounted.

The core barrel supports the fuel assemblies, lower grid, flow distributor and in-core instrument guide tubes. This cylinder is flanged at both ends. The upper flange is bolted to the core support shield assembly and the lower flange is bolted to the lower grid assembly. A series of horizontal spacers is bolted to the inside of the cylinder and a series of vertical plates is bolted to the inside surfaces of the horizontal spacers to form walls enclosing the fuel assemblies. Coolant flows downward along the outside of the core barrel cylinder and upward through the fuel assemblies. Approximately 90% of the coolant flow traverses the heat transfer surfaces of the core fuel assemblies. Of the remaining 10% flow, some is directed into the gap between the core barrel and thermal shield primarily to cool the thermal shield. The remainder of the flow bypasses the core through various leak paths in the internals.

Core guide lugs are welded to the inside wall of the reactor vessel. In the event of a major internal component failure, these lugs limit the drop of the core barrel to 0.5 in. (12.7 mm) and prevent its rotation about the vertical axis.

The lower grid assembly, consisting of two lattice structures surrounded by a forged flanged cylinder, supports the fuel assemblies, thermal shield and flow distributor. The top flange is bolted to the lower flange of the core barrel. Pads bolted to the top surface of the upper lattice structure provide fuel assembly alignment. A perforated plate midway between the lattice structures aids the uniform distribution of coolant flow to the core.

The flow distributor is a perforated, dished head that is bolted to the bottom flange of the lower grid. The distributor supports the in-core instrument guide tubes and shapes the inlet flow to the core.

The thermal shield is a stainless steel cylinder located in the annulus between the core barrel and the reactor vessel. It is supported and positioned by the

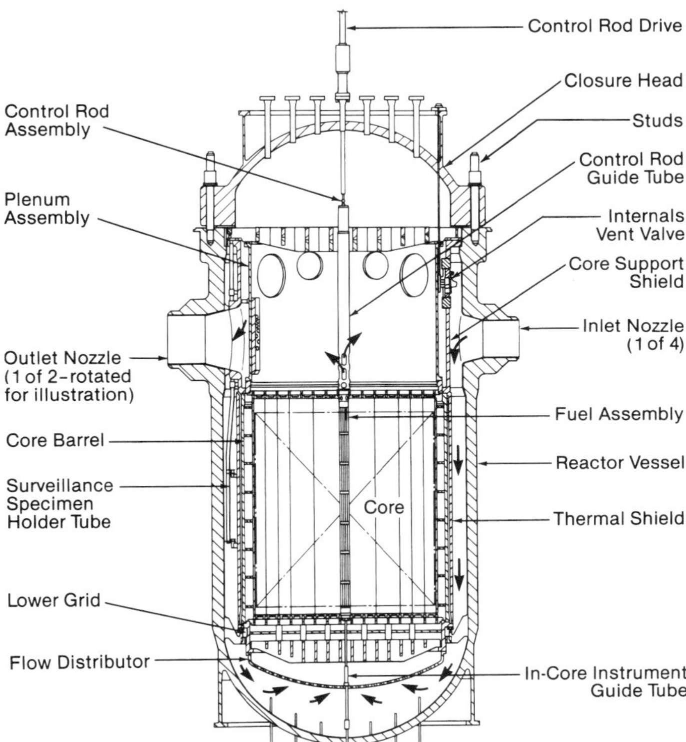

Fig. 10 Cross-sectional view of reactor vessel internals.

top flange of the lower grid. The thermal shield and intervening water annuli reduce the radiation exposure and internal heat generation of the reactor vessel wall by attenuating neutron and gamma radiation.

The surveillance specimen holder tubes are installed on the outer wall of the core support assembly, at approximately mid-height of the core.

The in-core instrument guide tube assemblies guide the in-core assemblies between the penetrations in the bottom head of the reactor vessel and the instrument tubes in the fuel assemblies.

Eight co-planar internals vent valves are installed on the core support shield; they are 42 in. (1067 mm) above the centerlines of the reactor vessel inlet and

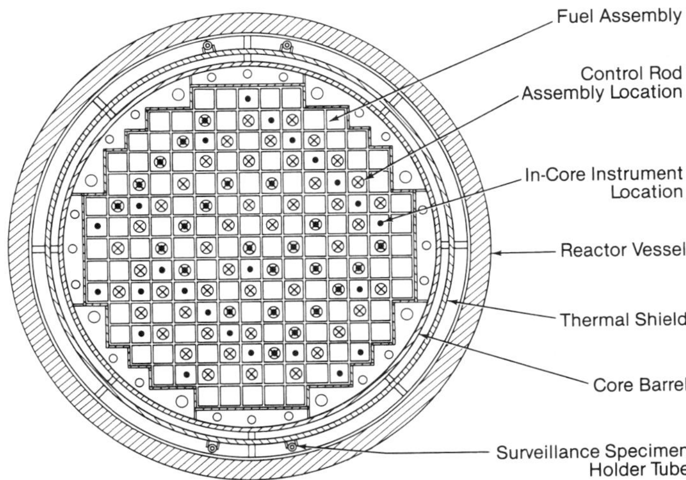

Fig. 11 Plan view of reactor vessel internals.

outlet nozzles. The valve seats are inclined 5 deg (0.087 rad) from vertical and the valve discs naturally hang closed. During normal operation, the valves are forced closed by the differential pressure, approximately 43 psi (296 kPa), between the outer annulus of the reactor vessel and the core outlet region. During a severe accident, such as a large break loss of coolant accident (LBLOCA), however, the differential pressure reverses and the internals vent valves open, permitting steam to be vented directly from the core region to the upper downcomer.

Reactor vessel closure head The closure head is an integral part of the reactor vessel pressure boundary. The head provides access for the replacement of spent fuel, and Alloy 600 penetration nozzles for control rod drive mechanisms and instrumentation. The closure head is typically made of low alloy steel and clad with stainless steel like the rest of the reactor vessel. Recently, many reactor closure heads in existing PWR systems have experienced corrosion damage and are being replaced (see summary in box on the following page.)

Fuel assemblies

In the reactor core, 177 fuel bundle assemblies rest on a lower grid attached to the core barrel. Each fuel bundle assembly contains 208 fuel rods (Fig. 12). Each fuel rod is made up of enriched uranium oxide pellets contained in zircaloy-4 tubing/cladding (Fig. 13). Fuel pellets (Fig. 14) are made up primarily from uranium dioxide (UO_2) powder. The powder is pressed and sintered in a dry hydrogen atmosphere to produce the required pellet size. Final machining of the pellet is done under water. Within the rods, the fuel pellets are spring loaded at both ends to ensure contact between the pellets. The rod is pressurized with dry helium to improve heat transfer across the gap between the pellets and cladding. End caps are laser welded to seal the rods and ensure integrity of the fuel/coolant boundary.

Spacing of the fuel rods in the assembly is maintained with spacer grids at several locations along the length of the bundle (Fig. 15). These grids are formed from strips (Inconel® for the end spacers and zircaloy for the intermediate spacers) that are stamped to form the detailed configurations necessary to ensure adequate contact points with the fuel rods. The strips are welded together to form the square-shaped spacer grid that holds the rods in correct spacing and alignment. Stainless steel end fittings complete the bundle assembly (Fig. 16).

Each fuel assembly is fitted with an instrumentation tube at the center and with 16 guide tubes, each accommodating 16 control rods. (See Fig. 17.) Each group of 16 control rods is coupled together to form a

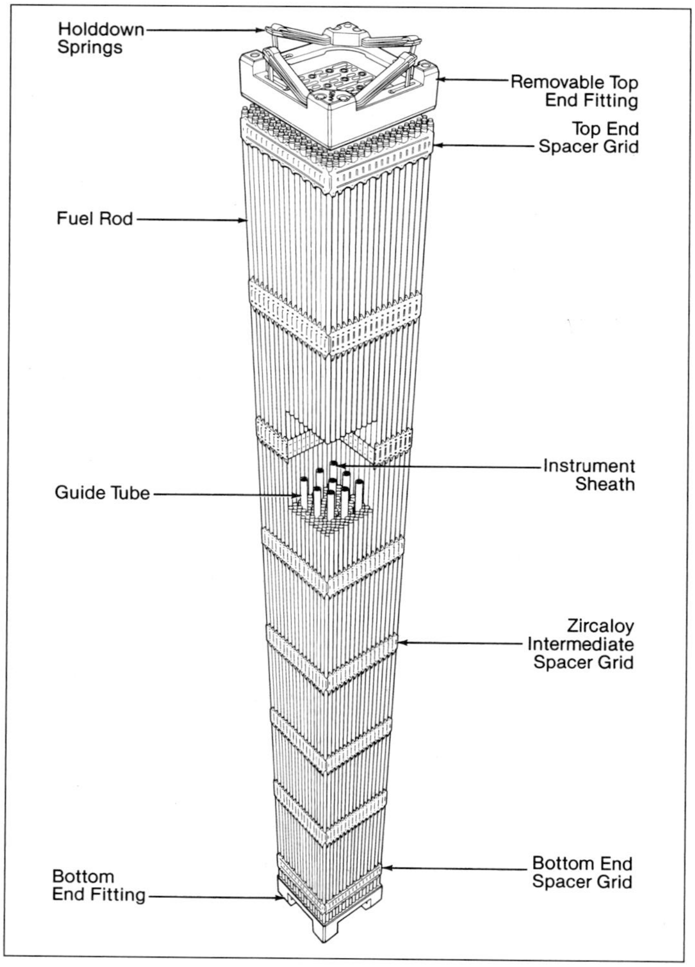

Holddown Springs
Removable Top End Fitting
Top End Spacer Grid
Fuel Rod
Guide Tube
Instrument Sheath
Zircaloy Intermediate Spacer Grid
Bottom End Spacer Grid
Bottom End Fitting

Fig. 12 Nuclear fuel assembly.

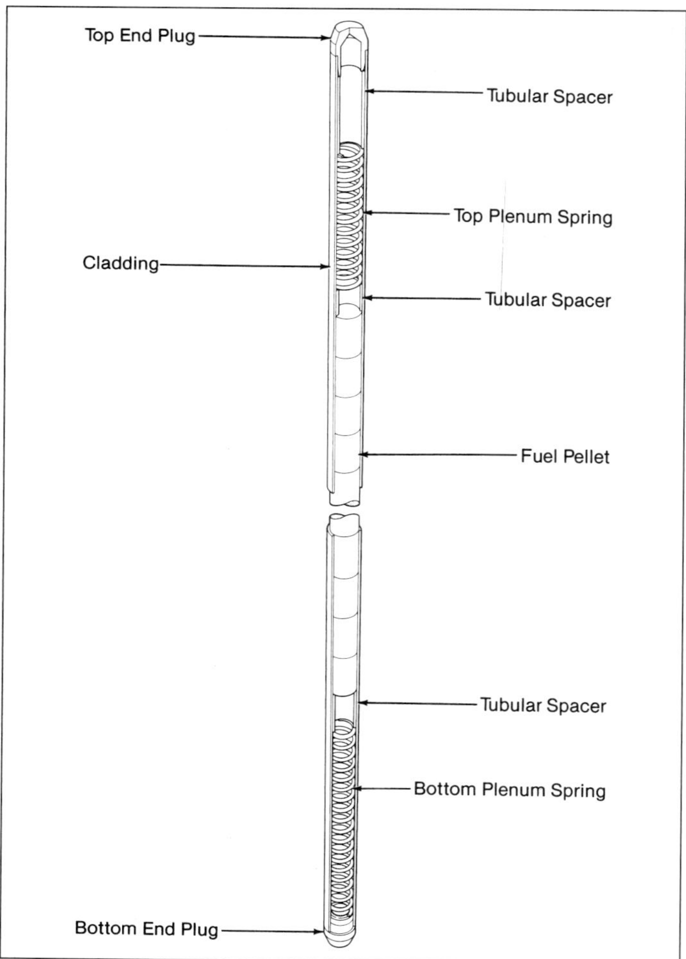

Fig. 13 Typical fuel rod assembly.

Reactor vessel closure heads

In recent years, discovery of cracked and leaking closure head Alloy 600 penetration nozzles, mainly the Control Rod Drive Mechanism (CRDM) nozzles, has raised concerns about the structural integrity of the closure heads. The CRDM nozzle is attached to the reactor vessel closure head and provides access to the core for the CRDMs.

Background

The problem was first discovered at the French Bugey Unit 3 power plant in 1991, when primary water leakage was observed on the vessel closure head at the CRDM nozzle region. Investigation of other French units showed cracking in the CRDM nozzles and the associated welds, and the cause was identified as primary water stress corrosion cracking (PWSCC).

This early French experience prompted examination of some PWR nozzle penetrations in the U.S. Detailed nondestructive examinations showed similar damage. Because the cracks were not through-wall and were within the approved acceptance criteria, the Nuclear Regulatory Commission (NRC) allowed continued plant operation contingent upon increased monitoring. However, in 2001, visual inspections of Oconee Unit 3 CRDM nozzles identified boric acid crystals at 9 of the 69 head penetrations, indicating that some primary water leakage had occurred. Further investigations revealed PWSCC initiated cracks that propagated from the weld radially and axially into the nozzle, allowing primary water leakage.

Evaluation of PWSCC effects has been completed at many of the PWR plants in the U.S. with subsequent decisions to replace some of the heads. These decisions to replace, rather than to follow a monitor and repair strategy, are strongly influenced by the expense of inspection and repair activities, which are challenging to perform in the hostile radioactive environment. It has proven desirable to avoid repairs and replace reactor vessel closure heads in their entirety, as quickly as possible, rather than bear the cost and schedule implications of difficult inspections and time-consuming repairs. In 2001, B&W received the first replacement head order in the U.S. to provide heads for Oconee Units 1, 2 and 3.

Closure head replacement requirements

The principal requirement for replacement closure heads is to provide a *form, fit and function* replacement that is compatible with the original equipment configuration and that provides enhanced reliability suitable for at least a 40 year design life. A form, fit and function replacement design ensures proper and efficient integration of a new closure head into the existing plant, thereby simplifying the safety assessments required to show compliance with NRC 10CFR 50.59 regulations. However, enhanced reliability can only be achieved by having a sound understanding of the cause of the degradation, and by optimizing all aspects of the design that are related to known degradation mechanisms.

Replacement head features

The conditions that induce PWSCC, and that are present in the current CRDM-to-head-attachment region, are:

1. the corrosive environment with primary borated water,
2. the presence of tensile stresses, and
3. the use of materials susceptible to corrosion.

Accepting the presence of the primary water environment, the major focus of the replacement head design has been to provide superior materials and to control the stresses induced at all stages of manufacture.

Materials The material for the CRDM nozzles has been changed from Alloy 600 to thermally treated Alloy 690TT. Similarly, the weld consumables for the CRDM nozzle-to-head weld have been changed from Alloy 182 to Alloy 52 and/or Alloy 152, both having material compositions compatible with the Alloy 690 base material. All of these replacement materials have demonstrated superior resistance to PWSCC if they are specified and procured under a carefully controlled and monitored program.

Fabrication In addition, optimum resistance to PWSCC requires control of mechanical factors such as cold work and residual stresses. Materials for fabrication will start in the annealed condition, essentially free of bulk cold work and residual stress. Some manufacturing steps can reintroduce cold work and stresses which can negatively affect PWSCC resistance (both internal and surface conditions are important). Fabrication sequence and welding methods are important in minimizing internal residual stresses; finishing methods for both individual parts and completed assemblies are important in controlling surface cold work and related residual stresses.

Nozzle-to-head welds The original equipment nozzle-to-head welds were made with a manual shielded metal arc welding (SMAW) process. The replacement units use an automated gas tungsten arc welding (GTAW) process which improves weld quality, and minimizes residual as-welded stresses. To address concerns regarding the effect of as-manufactured surface condition on PWSCC resistance, B&W has employed an electropolishing process for both the CRDM nozzle prior to installation and the finished weld after surface conditioning. This process, unlike abrasive honing or other surface treatments, entirely removes the cold worked surface layer associated with machining and grinding.

Fig. 14 Nuclear fuel pellets.

control rod assembly as shown in Fig. 18. At any given time, 69 of the 177 fuel assemblies contain a control rod assembly. Because all fuel assemblies are identical and can be used anywhere in the core, the 108 remaining fuel assemblies have their guide tubes partially filled at the top with an orifice rod assembly. This restricts excessive coolant flow through the unused control rod guide tubes and equalizes coolant flow among the fuel assemblies.

The control rod assemblies regulate the reactor reactivity and, therefore, the power output. Each control rod consists of an absorber section made from a neutron absorbing material such as stainless steel with silver-indium-cadmium cladding, and an upper and lower stainless steel end piece. The rods regulate relatively fast reactivity phenomena such as Doppler, xenon buildup and decay, and moderator temperature change effects. The slower reactivity effects, such as fuel burnup, fission product buildup, and hot-to-cold moderator reactivity deficit, are controlled by a soluble

Fig. 16 Final visual inspection of fuel assemblies.

neutron absorbent (usually boron) in the reactor coolant. The concentration of the absorbent is monitored and adjusted by the plant operators.

Refueling

Most U.S. power reactors operate for a planned period of between one and two years before the reactivity of the nuclear fuel is reduced and the unit must be reloaded with fresh fuel. Because only part of the fuel is replaced during the shutdown, this is referred to as *batch refueling*. In some reactors, such as the CANDU design, a continuous refueling mode that does not require reactor shutdown is used.

Most PWRs were designed for annual refueling, replacing approximately one-third of the core at each outage. As operating experience has been gained, the trend has shifted to 18-month refueling cycles. Although fuel costs are higher, the longer refueling cycle is cost-effec-

Fig. 15 Spacer grid coordinate measuring machine.

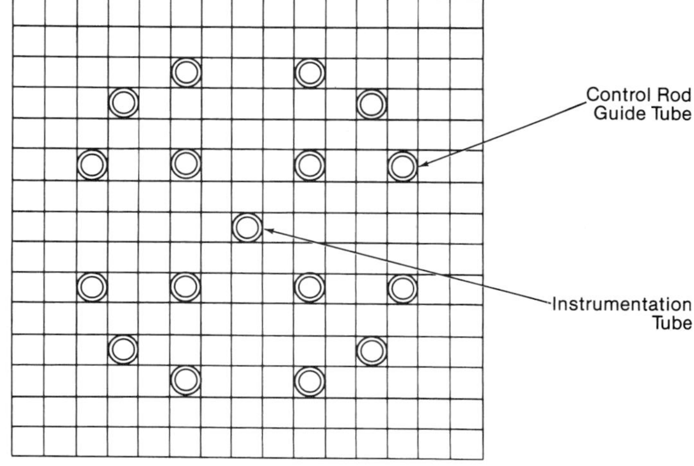

Fig. 17 Diagrammatic cross-section of fuel assemblies showing instrumentation tube and control rod guide tube locations.

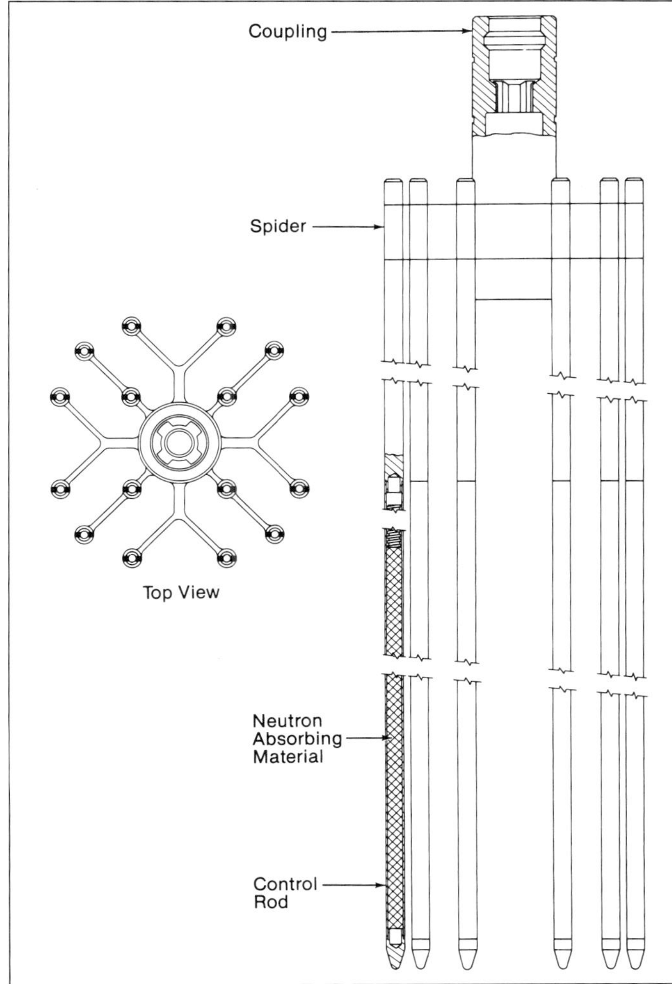

Fig. 18 Control rod assembly.

tive due to reductions in the number of refueling outages, licensing submittals, and replacement power cost. Furthermore, the burnup capability of the fuel has increased, permitting subsequent batch size reductions.

As nuclear fuel is continually consumed in the reactor, the fuel bundle assemblies are subjected to temperature and irradiation effects that change some of their characteristics. Dimensional and structural changes in the assemblies include growth of the fuel rods, spring relaxation, and fuel rod bow. Some corrosion of the cladding takes place on the coolant side. This corrosion effect liberates hydrogen from the coolant, some of which is absorbed by the cladding, and can subsequently decrease cladding mechanical properties. Fission gas release in the fuel rod can cause high internal pressure. The coolant is continually monitored during operation to detect any leakage of these fission gases. While these various effects have been considered and monitored over the years of reactor operation, they have not been found to limit operation of the nuclear systems.

Extreme care and detailed handling procedures are required to remove the spent fuel assemblies for disposition. During their removal from the reactor core and near-term storage in spent fuel pools, water serves as both a shielding and as a cooling medium. The spent fuel assemblies are kept in the spent fuel pool at the reactor site until the radioactive decay and heat generation have declined sufficiently for safe transportation and processing.

Energy transport

Each fission of uranium-235 (U-235) produces approximately 200 MeV, predominantly in the form of kinetic energy of the fission products. This energy dissipates to thermal energy of the fuel. The core heat transfer process involves the fuel, fuel-cladding gap, cladding and coolant. A typical temperature profile is shown in Fig. 19. The mode of heat transfer from cladding to coolant is subcooled forced convection. Film boiling is avoided by maintaining a DNBR above 1.3. DNBR is the ratio of local heat flux at the departure from nucleate boiling (DNB) to actual local heat flux. (See Chapter 5.) The coolant enters the core at approximately 557F (292C) and exits at 607F (319C), thereby establishing the sink temperature for fuel to coolant heat transfer. The cladding to coolant convective and cladding conductive heat transfer rates are relatively high, therefore the fuel temperatures primarily depend on the gap and fuel conductances and the rate of energy deposition. The heat transfer calculation methods are described in Chapters 4 and 5. The melting temperature of unirradiated uranium dioxide (UO_2) is 5080F (2804C). This temperature decreases with burnup at the rate of approximately 58F (32C) per 10,000 MWd/t_m U, due to the accumulation of fission products. A design overpower maximum fuel centerline temperature of 4700F (2593C) has been selected to preclude melting at the fuel centerline.

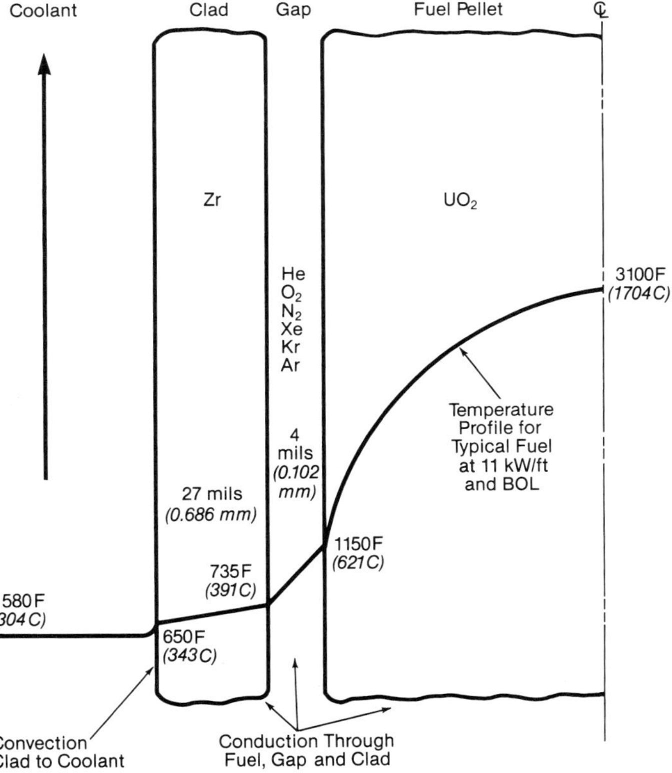

Fig. 19 Radial temperature profile for a fuel rod at 11 kW/ft.

Instrumentation

The reactor vessel instrumentation consists of the nuclear instrumentation and in-core monitoring system.

Nuclear instrumentation Neutron flux levels are monitored using a combination of source, intermediate and power range detectors. Each of these detectors is located at core mid-height, outside the reactor vessel but inside the primary shield. The distribution and types of detectors are:

Range	Detector Type	Number
Source	BF_3 Proportional	2
Intermediate	Compensated ion chamber	2
Power	Uncompensated ion chamber	4

These measurements are supplied to the reactor operator, safety parameter display system, reactor control portion of the integrated control system and reactor protection system. Additionally, the source and intermediate range readings are differentiated to provide startup rate, and the outputs of each pair of power range detectors are differenced to provide top to bottom flux imbalance.

In-core monitoring system Core performance is monitored using 52 in-core detector assemblies. These assemblies are arranged in a spiral fashion outward from the center of the core. Each assembly consists of seven local flux detectors, one background detector and one thermocouple, arranged as shown in Fig. 20. The seven local flux detectors, distributed over the axial core length, indicate the axial flux shape and provide fuel burnup information. The background detector indicates the integrated axial flux. These flux detectors are self-powered rhodium-103 instruments. The thermocouple indicates the exit coolant temperature.

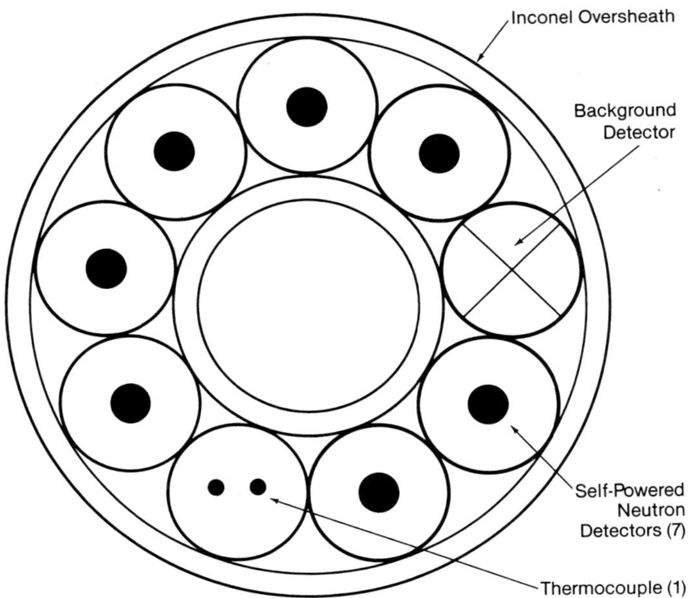

Fig. 20 Instrument tube cross-section.

Steam generators

Nuclear steam generators are particularly significant components in the NSSS and are discussed in more detail in Chapters 48 and 50.

The steam generators in PWR systems transfer heat from the primary coolant loop to the secondary flow loop, effectively functioning as heat sinks for the reactor core. The decrease of primary coolant temperature through the steam generators is virtually the same as the increase of primary coolant temperature through the core. Any difference is attributable to the fluid energy supplied by the reactor coolant pumps, less heat losses to ambient.

A variety of steam generator designs have been used in nuclear reactor systems.[2] However, two have been primarily used on second generation NSSS: the once-through steam generator (OTSG) and the more prevalent recirculating steam generator (RSG).

The OTSG is a straight tube counterflow heat exchanger where the primary coolant flows vertically downward inside the tubes and the secondary side water flows upward changing from slightly subcooled liquid to superheated steam at the outlet.

The RSG uses an inverted vertical U-tube bundle to transfer the energy and generate steam. The primary coolant passes through the U-tubes. The subcooled secondary side water mixes with water from the steam separators and passes up over the outside of the U-tube bundle as it is partially converted to steam. The steam-water mixture passes through multiple levels of steam separation equipment which returns the water to the U-tube bundle for further heating and evaporation and passes the saturated steam to the power-producing system.

Pressurizer

Description

The pressurizer (Fig. 21) is a tall, cylindrical tank connected at the bottom to a reactor coolant loop hot leg through 10 in. (254 mm) diameter surge line piping. Spray is introduced near the top of the pressurizer through a nozzle and 4 in. (102 mm) diameter line from a cold leg. Three replaceable heater bundles are installed over the lower portion of the pressurizer. Pressure relief devices are mounted at the top of the unit; these include two code safety valves and the power operated relief valve (PORV).

Function

The pressurizer controls primary system pressure. During normal operation, the pressurizer volume of 1500 ft^3 (42.5 m^3) contains equal portions of saturated liquid and saturated steam. Should the primary system liquid expand, such as through an increase of coolant temperature due to a load reduction, the excess volume displaces liquid into the surge line, raising the pressurizer level. This evolution is termed an *insurge*. The pressurizer steam is compressed, raising the primary system pressure and causing some of the steam to be condensed. The specific volume of liquid is much less than that of steam, therefore the steam condensation counteracts the ongoing pressure rise. The opposite occurs during an *outsurge*, the displacement of pressurizer liquid toward the hot leg through the surge line. In this case, the primary system pressure decreases, saturated pressurizer liquid flashes to steam and the net increase of specific volume through the change of phase suppresses the depressurization.

The pressurizer liquid volume is sized to maintain liquid above the pressurizer heaters and to maintain pressure above the HPI actuation point during the outsurge due to a reactor trip. The pressurizer steam volume is sized to retain a steam bubble during the insurge due to a turbine trip. Operation without a pressurizer steam bubble is termed solid plant operation. The pressurizer is sized to avoid this mode in which pressure control is encumbered.

Heaters

The pressurizer heaters maintain the pressurizer liquid at the saturation temperature, replacing the heat losses to ambient. They also raise and/or restore primary system pressure by elevating the saturation temperature of the pressurizer fluid. The three bundles of heaters are divided into five banks. There are two essential banks and three nonessential banks. The essential banks are energized automatically, based on primary system pressure. Essential bank no. 1 is energized below 2150 psig (14.82 MPa) and is subsequently de-energized above 2160 psig (14.89 MPa); the corresponding pressures for essential bank no. 2 are 2145 and 2155 psig (14.79 and 14.86 MPa). Nonessential bank no. 3 is sized to maintain the pressurizer fluid temperature during normal, steady-state operation. Banks no. 4 and no. 5 provide additional heater capacity that can be used during system startup or load changes.

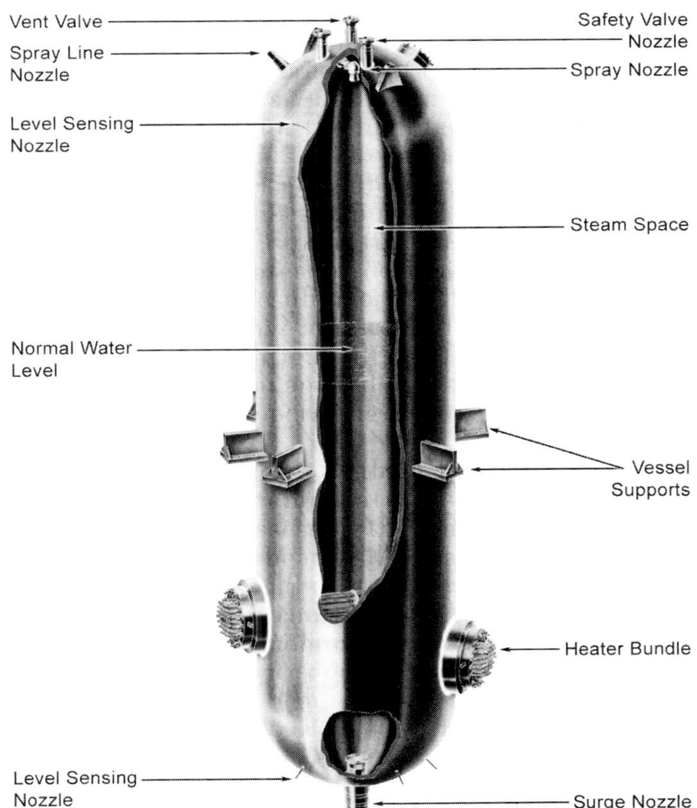

Fig. 21 Nuclear steam system pressurizer.

Spray

The pressurizer spray system lowers primary system pressure and counters an increasing pressure transient. The actuation of pressurizer spray introduces 550F (288C) cold leg fluid into the pressurizer steam space. The resulting condensation decreases the pressurizer fluid volume, thereby decreasing pressure.

The spray enters the pressurizer through the spray nozzle. This nozzle imparts rotational motion to the fluid, generating a downward directed, hollow spray cone. This pattern enhances steam condensation efficiency while minimizing contact of the cold spray with the pressurizer vessel walls. Between the pressurizer wall and the spray nozzle, the spray line makes a bend in the vertical plane. This loop seal line configuration is designed to maintain liquid in the spray line, thereby alleviating the temperature changes experienced by the spray nozzle.

The spray line is attached to the cold leg piping just downstream of the reactor coolant pump. The spray is driven by the RCS fluid pressure drop across the reactor vessel, and the full spray flow rate at normal operating conditions is approximately 170 GPM (10.7 l/s). The spray flow control valve is normally operated in a modulated rather than on/off mode. When in automatic mode, the spray flow control valve opens 40% when the primary system pressure increases to 2200 psig (15.17 MPa) and closes when pressure decreases to 2150 psig (14.82 MPa). The 40% open setting permits a spray flow rate of approximately 90 GPM (5.68 l/s).

A continuous flow rate of approximately 1 GPM (0.063 l/s) is maintained to minimize the thermal transients of the spray nozzle; this is accomplished by a bypass line around the spray flow control valve. An auxiliary system provides pressurizer spray when the reactor coolant pumps are inactive, such as during decay heat removal cooling of the RCS.

Pressure control devices

Pressure increases are countered by pressurizer steam condensation through compression and by pressurizer spray actuation. Should pressure continue to rise, the PORV opens at 2450 psig (16.89 MPa). A pressure actuated solenoid operates the pilot valve, which actuates the PORV. The PORV discharges into the pressurizer quench tank and can be isolated by closing the PORV block valve. A hypothetical continuous rod withdrawal accident from low power would cause core power generation to greatly exceed steam generator heat removal. The two code safety valves are sized to prevent RCS pressure from exceeding the 2750 psig (18.96 MPa) limit. These valves open at 2500 psig (17.24 MPa) and also discharge to the pressurizer quench tank.

Pumps and piping

Description and function

Coolant is transported from the reactor vessel to the two steam generators through hot leg piping. The coolant is returned to the reactor vessel from the generators through four cold legs, two per generator. Fluid circulation is provided by four reactor coolant pumps, one per cold leg. The arrangement of the piping and pumps is shown in Figs. 2, 3 and 4.

The hot leg and cold leg piping is fabricated from carbon steel and is clad on the inside with type 304 or 316 stainless steel. The inside diameters of the hot leg and cold leg piping are 36 and 28 in. (914 and 711 mm) and their nominal wall thicknesses are 2.75 and 2.5 in. (69.9 and 63.5 mm), respectively. The other piping attached to the RCS is either fabricated from stainless steel or clad with stainless steel.

Thermal sleeves are used to minimize the thermal stresses generated by rapid fluid temperature changes; they are installed on the four high pressure injection (HPI) nozzles and two core flood nozzles.

Pump characteristics

The reactor coolant pumps are vertical suction, horizontal discharge, centrifugal units. They are further characterized as diffused flow, single stage, single suction, constant speed, vertical centrifugal, controlled leakage pumps having five-vane impellers. The pumps are driven by constant speed, vertical squirrel cage induction motors. Pump sealing is accomplished using three-stage mechanical seals. The pump and motor are just over 27 ft (8.2 m) tall (see Fig. 22). The motor develops approximately 9000 hp (6714 kW) at cold conditions and 1185 rpm (124 rad/s) using a 6600 volt, three-phase power supply. The locked rotor starting current is approximately 3800 amperes. At normal operating conditions, the motor develops approximately 8300 hp (6192 kW), drawing a current of 685 amperes. Each unit's rotational moment of inertia is in excess of 70,000 lbf ft^2 (28,928 Nm^2). This rotational inertia is sufficient to sustain rotation for approximately 45 seconds following a power interruption. Each reactor coolant pump has a nominal capacity of 92,400 GPM (5829 l/s) with a discharge head of 403.5 ft (122.99 m) at operating RCS conditions.

Integrated control system

Function

The integrated control system (ICS) simultaneously controls the reactor, steam generators and turbine to obtain a smooth, rapid response to load changes. The ICS operates the control rods to regulate core power, adjusts the feedwater flow to control the rate of steam production, and operates the turbine throttle valves to control electrical power output. This control method combines the advantages of two alternative approaches.

The NSSS responds inherently to load changes without an ICS, but this response causes undesirable system variations. Consider the response to an increased load demand without integrated control. The increased electrical load causes the turbine throttle valves to open to maintain turbine speed. The increased steam demand lowers steam pressure at the turbine and within the steam lines and steam generators. Feed flow rate is increased to match the increased steam flow rate. The increased steam generator heat transfer lowers the temperature of the reactor coolant returning to the core. Core power gradually increases due to the reactivity gained from the negative temperature coefficient. This means that the increased reactor coolant density enhances the moderation of neutrons, thereby increasing the thermal neutrons available to cause thermal fissions. Also, the control rods are withdrawn to maintain the average primary fluid temperature. These primary system interactions would be reversed as the core and steam generator stabilized at the new, higher power level. This method of feedback control has the advantage of immediately supplying the required electrical load.

An extremely stable but slow method of NSSS control is termed the turbine-following method. The turbine output is varied only as the steam pressure is adjusted for the revised conditions. Again consider the response to an increased load demand. The control rods are withdrawn to increase reactor power and the feed flow rate is increased to support a higher rate of steam generation. As the turbine throttle pressure begins to increase, the throttle valves are gradually opened, allowing the turbine output to respond to the increased load demand.

The ICS operates all three systems: turbine, feedwater flow and control rods, to combine the rapid response of the unregulated system with the stability of the turbine-following system. The key to ICS operation is the turbine header pressure setpoint. The turbine governor valves maintain turbine speed and turbine header steam pressure. Because of the thermal inertia of the NSSS, the steam production rate can not be changed as rapidly as the steam demand varies in response to a turbine load change. This time delay is handled by temporarily changing the turbine

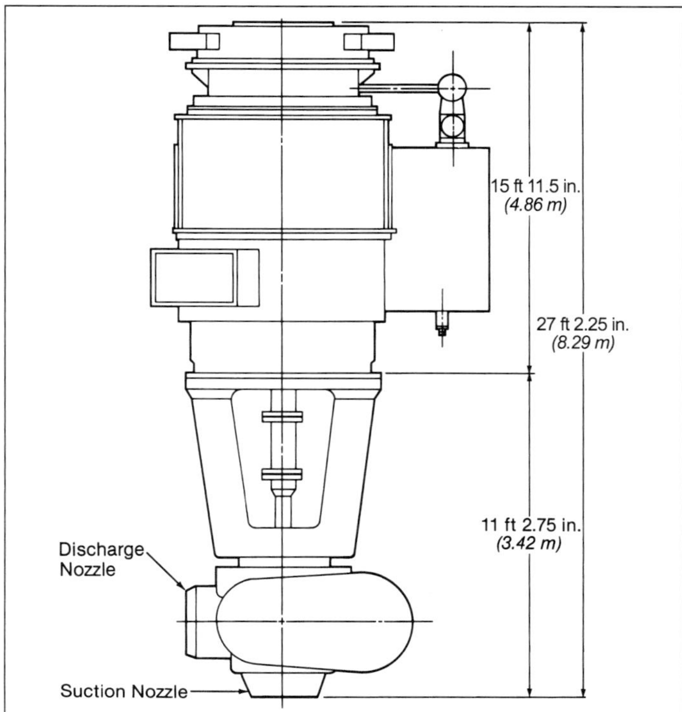

Fig. 22 Reactor coolant pump arrangement.

header pressure setpoint. For example, again consider a load increase. The ICS causes the control rods to be withdrawn and the feedwater flow rate to be increased. Simultaneously, the ICS temporarily reduces the turbine header pressure setpoint. The turbine governor valves open to maintain turbine speed and to reduce the header steam pressure to the new setpoint. As core power, primary to secondary heat transfer and steam flow rate gradually increase, the turbine header pressure recovers and its set point is returned to the steady-state value. In this fashion, the ICS obtains a rapid response and a smooth transition between turbine loads.

Limits and controls

The ICS recognizes a variety of limiting conditions, such as a reactor coolant pump trip, a feed pump trip or an asymmetric control rod fault. It imposes the appropriate load and load change limits corresponding to these conditions.

Special ICS signals are imposed below 15% of full power. During system startup, the reactor and steam generator power levels are increased to approximately 10% power by steaming through the turbine bypass valves. The turbine-generator is rolled, synchronized with the electrical distribution system and gradually loaded. The turbine control station is placed in automatic mode at approximately 15% turbine load. The reactor is then controlled to maintain a constant average coolant temperature of 582F (306C) above 15% power, using the RC subsystem. From 0 to 15% power, manual control is used to increase the average coolant temperature from 532F to 582F (278 to 306C).

Reactor protection system

Function

The reactor protection system (RPS) trips the reactor when limiting conditions are approached. These limits prevent boiling of the reactor coolant, limit the local core power generation rate (the linear heat rate) and minimize challenges to the PORV.

Multiple independent RPS channels monitor critical reactor trip functions. If any two of these channels detect single or multiple trip functions outside normal parameters, the control rod drive breakers are opened, tripping the reactor. Depending upon the NSSS design, the trip functions can include:

1. ARTS (anticipatory reactor trip system) main feedwater pump trip,
2. ARTS turbine trip,
3. overpower trip,
4. high outlet temperature trip,
5. high pressure trip,
6. reactor building high pressure trip,
7. pressure/temperature trip,
8. low pressure trip,
9. power/imbalance/flow trip, and
10. power/reactor coolant pumps trip.

The two ARTS trip functions trip the reactor before the RCS pressure rises to the reactor set point, thereby minimizing the activation of the PORV (power operated relief valve). The reactor can also be manually tripped from the control room.

Safety features actuation system

The safety features actuation system (SFAS) engages the emergency core cooling system (ECCS) in the event of a breach of the primary system boundary. The two SFAS signals are low RCS pressure and high reactor building pressure. A schematic of the various systems is shown in Fig. 23. The SFAS operates the following systems:

1. high pressure injection,
2. low pressure injection,
3. reactor building cooling, and
4. reactor building spray.

In addition, the SFAS activates the two emergency diesel generators.

Each of the three independent measurements of RCS pressure is fed to a trip bistable. This input is combined with the output of a reactor building pressure bistable; the trip of either pressure bistable trips the output. This output signal, combined with those of the other two pressure logic signals, engages the SFAS should two out of three trip. Reactor building spray is also initiated using two out of three logic, but is based only on reactor building pressure. In addition, the reactor building spray is delayed five minutes from the time of actuation.

Certain features are bypassed when the system is normally depressurized. These include high pressure injection, low pressure injection, reactor building isolation and reactor building cooling. Bypass must be initiated at an RCS pressure of less than 1850 psig (12.76 MPa); the system is automatically reinstated when the pressure exceeds 1850 psig (12.76 MPa).

High pressure injection system

The high pressure injection (HPI) system provides water to the RCS during a loss of coolant accident (LOCA). It is designed to prevent core uncovery in the event of a small RCS leak, during which the RCS pressure remains elevated, and to delay core uncovery in the event of an intermediate size break. The HPI system is activated by the SFAS signal when RCS pressure decreases to 1600 psig (11.03 MPa) or when reactor building pressure increases to 4 psig (27.6 kPa).

The HPI system consists of three pumps, a common discharge header and four lines that are equipped with motor operated isolation valves leading to the four cold legs. One HPI pump operates continuously to provide water for the reactor coolant pump seals. Two pumps are always available for automatic safety injection. The SFAS signals activate both pumps and their auxiliary equipment. The system also opens the four isolation valves to preset throttled positions to produce 125 GPM (7.89 l/s) per line using either of the two HPI pumps.

The HPI system is designed redundantly. Either pump is sufficient to meet the system requirements. The pumps are located in separate rooms and draw

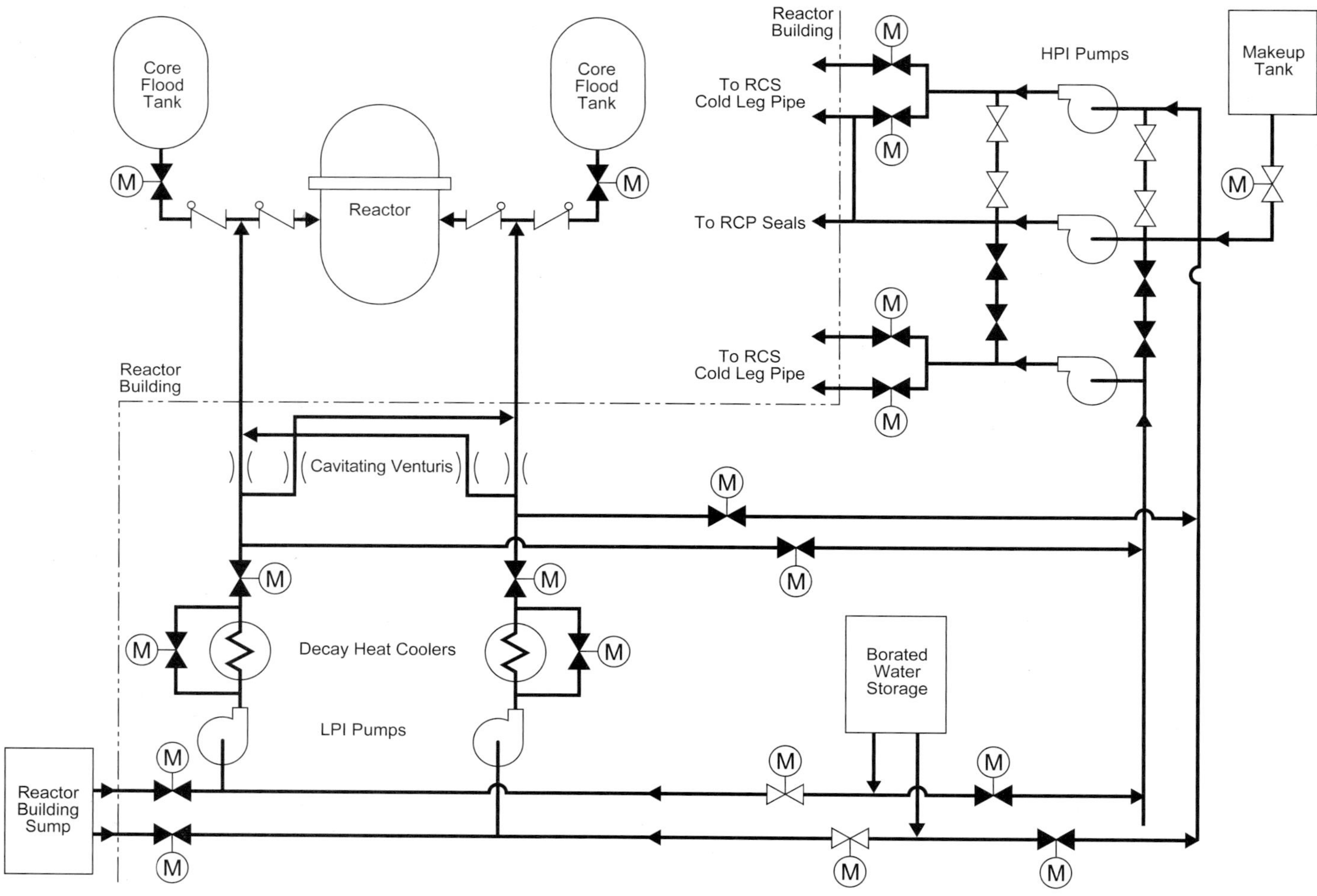

Fig. 23 Emergency core cooling/injection system.

water from the borated water storage tank (BWST) through separate lines. The SFAS channels are physically separated. The HPI pump and valve power supply lines are independent, physically separated and energized from independent power sources. The SFAS also activates the emergency diesel generators that provide backup power to the HPI system.

HPI fluid enters each RCS cold leg midway in the down-sloping piping run at the reactor coolant pump discharge. The fluid enters horizontally at the side of the cold leg pipe. The HPI nozzles are equipped with sleeves to minimize the thermal stress caused by the cold water injection.

The HPI system can be used during normal operation as part of the makeup and purification system. Either HPI pump can supply RCS makeup and seal injection flow to the reactor coolant pumps; these functions are normally performed by the makeup pump.

Low pressure injection system

The low pressure injection (LPI) system provides water to the core in the event of a large LOCA which depressurizes the RCS. The two LPI pumps discharge 3000 GPM (189.2 l/s) each at a rated head of 350 ft (106.7 m). [Shutoff head is 435 ft (132.6 m).] The borated water storage tank (BWST) supplies the LPI pumps. The pumps are actuated independently and automatically by the SFAS signal, as are the four motor operated isolation valves. They discharge to the reactor vessel upper downcomer through two core flood nozzles.

The two BWSTs have a capacity of 450,000 gal (1,703,430 l) each. The boron concentration is maintained above 1800 ppm at 80F (27C). The storage tank inventory provides at least 30 minutes of operation of all ECCS pumps. When the BWST inventory is depleted, the LPI pumps are manually transferred to the reactor building emergency sump, which collects water lost from the RCS. In the event that the HPI pumps are still required, such as with RCS pressure higher than the LPI pump shutoff head, the HPI pumps can be realigned to draw from the discharge of the LPI pumps. This is referred to as piggy-back operation.

Core flood system

The core flood system provides a passive water supply to the core in the event of a large break loss of coolant accident. The system consists of two tanks, associated piping and valves. Each tank has a volume of 1410 ft^3 (39.9 m^3). Approximately two-thirds of this volume, or 7500 gal (28,391 l), is borated water; the remainder is nitrogen gas pressurized to 600 psig (4.14 MPa). Should the RCS depressurize to less than

600 psig (4.14 MPa) in the event of an LBLOCA, the nitrogen cover gas forces the core flood tank (CFT) water into the RCS. The boron concentration of the CFT water is maintained between 2270 and 3490 ppm.

Each CFT discharges through 14 in. (356 mm) lines, dual check valves and a core flood nozzle into the upper downcomer of the reactor vessel. The two CFTs use separate piping and nozzles. Each tank is equipped with a motor operated isolation valve to prevent tank discharge when the RCS is normally depressurized. These valves are open when the RCS is pressurized.

Reactor building cooling and spray systems

The reactor building emergency cooling and spray systems are actuated in the event of a loss of coolant accident. The spray system also reduces the post accident level of fission products in the reactor building atmosphere through chemical reaction. The systems are designed such that either one can offset the heat released by the escaping reactor coolant.

The systems are activated by the SFAS signals. The emergency cooling system is actuated within 35 seconds after the RCS pressure decreases to 1600 psig (10.03 MPa) or the reactor building pressure increases to 4 psig (27.6 kPa). The emergency spray system is actuated 5 minutes after the reactor building pressure reaches 30 psig (206.8 kPa). Both systems can also be actuated manually.

The reactor building emergency cooling system includes four units, each consisting of an air circulator and a cooling coil. The 40,000 ft^3/min (18.9 m^3/s) circulators draw air from a point high within the reactor building dome and discharge downward toward the cooling units. Two of the four units are equipped with activated charcoal filters for removing fission products. The cooling units reject heat to the nuclear service cooling water system.

The reactor building emergency spray system includes two spray trains, each consisting of a pump, spray header, spray additive tank and eductor and associated isolation valves, piping and controls. The pumps are supplied from the BWST and, later in an accident, from the reactor building emergency sump.

The 300 hp (224 kW), 1500 GPM (94.6 l/s) spray pumps discharge through the spray additive eductors, drawing in sodium hydroxide. The spray additive concentration is sufficient to bring the entire post-accident inventory of reactor building water to a pH of 9.3. The spray discharge is distributed through 100 spray nozzles per header that are arranged to provide a uniform spray throughout the reactor building, above the operating floor.

Decay heat removal system

The RCS is periodically cooled and depressurized for maintenance and refueling. The first portion of the cooldown is accomplished by circulating the primary coolant using reactor coolant pumps and by removing heat using the steam generators. This method becomes impractical as the RCS pressure is reduced towards the net positive suction head (NPSH) of the reactor coolant pumps and as the primary to secondary system temperature difference diminishes.

The decay heat removal system (DHRS) is used to complete the RCS cooldown below 225 psig (1.55 MPa) and 290F (143C). The DHRS also performs the following functions:

1. purifies the reactor coolant during cold shutdown,
2. refills the RCS following maintenance,
3. cools and adds boron to the spent fuel pool, and
4. transfers water between the borated water storage tank and the fuel transfer canal during refueling.

The DHRS also functions as the low pressure injection system during a loss of coolant accident.

The DHRS is activated approximately six hours after reactor shutdown. It can reduce the RCS temperature from 280F to 140F (138 to 60C) within 14 hours. The DHRS pumps draw reactor coolant from the hot leg piping, just beyond the reactor vessel, through a 12 in. (305 mm) decay heat drop line. The DHRS fluid is discharged through coolers to the reactor vessel upper downcomer through the core flood nozzles. The injected DHRS fluid then follows the usual flow path down the reactor vessel downcomer, up through the core and out the hot leg to the DHRS inlet, thereby completing the flow circuit. The DHRS flow rate and rate of RCS cooling are controlled by bypassing a portion of the DHRS flow around coolers. The flow rate per cooler is limited to approximately 3000 GPM (189.2 l/s). The heat transferred in the coolers is removed by the nuclear service cooling water (NSCW) system, and the temperature difference between the RCS and the DHRS cooler outlet is maintained at approximately 30F (17C). The decay heat removal suction header temperature is measured to monitor RCS cooldown and the return header flow rates are also measured and remotely indicated. The DHRS suction block valve is interlocked to prevent inadvertent actuation at RCS pressures above 225 psig (1.55 MPa).

A portion of the DHRS flow rate can be diverted to the makeup and purification system for purifying the reactor coolant. This flow path is from the discharge of a DHRS pump; through the letdown filter, purification demineralizers and makeup filters; and back to a DHRS pump inlet header. The purification demineralizers are limited to a maximum fluid temperature of 135F (57C).

Makeup and purification system

The functions of the makeup and purification system are as follows:

1. control RCS inventory,
2. purify and degasify the reactor coolant,
3. maintain coolant boron concentration,
4. add chemicals to the reactor coolant for pH control,
5. supply seal injection flow to the reactor coolant pumps and handle seal return flow, and
6. add borated water to the core flood tanks.

The makeup and purification system can also function as part of the high pressure injection system.

The makeup and purification system consists of the letdown, purification and makeup portions. The letdown portion draws reactor coolant from the cold leg suction through a 2.5 in. (63.5 mm) line. The effluent is cooled using one of the three letdown coolers. Radioactive nitrogen-16 is allowed to decay in the letdown delay line, a 15 ft (4.6 m) length of 12 in. (305 mm) piping. Depressurization and flow control are provided by the letdown orifice; the nominal flow rate of 45 GPM (2.83 l/s) may be increased to 140 GPM (8.83 l/s) by bypassing the orifice.

The flow rate of 45 GPM (2.83 l/s) processes one RCS volume daily. Letdown flow rate, temperature and pressure are measured. Multiple motor operated valves are used to isolate the system in the event of SFAS operation.

The purification portion of the system consists primarily of the letdown filters, mixed bed demineralizers and makeup filters. The demineralizers are provided temperature protection by temperature actuated isolation valves. The purified effluent may be directed to the makeup tank or diverted to the bleed tanks when reducing the RCS fluid inventory.

The 4400 gal (16,657 l) makeup tank is central to the makeup and purification system. It receives purified letdown flow and seal return flow. It also serves as the point of boron and hydrogen addition to the RCS. Boron is used for reactivity control, lithium hydroxide provides pH control and hydrogen or hydrazine is used for oxygen control. The makeup tank acts as a surge tank to accommodate temporary changes of RCS inventory and provides water for the makeup pump. Finally, it can be used to degasify the reactor coolant.

A 4 in. (101.6 mm) line connects the makeup tank to its pump, with cross connects to the HPI pumps. The pump discharges are recombined and a 2.5 in. (63.5 mm) diameter flow line is routed back to the seal return coolers to cool the operating pumps. The makeup flow rate is measured and remotely indicated and is controlled by the pressurizer level control valve. Makeup is injected into the RCS through the cold leg HPI nozzle at the reactor coolant pump discharge, thereby completing the flow circuit.

During power operation, the makeup and purification system operates continuously to regulate RCS inventory. The makeup flow control valve is adjusted automatically to maintain pressurizer level. The flow rate of the makeup pump is the sum of the makeup flow rate (to the RCS), the seal injection flow rate to the reactor coolant pumps and the makeup pump recirculation flow rate. The net makeup system flow rate to the RCS is the sum of the makeup and the seal injection flow rates, less the seal return flow rate.

Abnormal transient operating guidelines

The abnormal transient operating guidelines (ATOG) represent a symptom-oriented response to plant transients. These guidelines provide the operator with a clear and effective method of correcting abnormal conditions. ATOG involves the identification and correction of key upset conditions, regardless of their cause. These actions alone are sufficient to ensure core covery and cooling and to ensure plant safety.

The three key symptoms of ATOG are loss of subcooling margin, inadequate heat transfer and excessive heat transfer.

The subcooling margin refers to the difference between the saturation temperature at RCS pressure and an RCS temperature. A positive subcooling margin ensures that the core is covered and therefore cooled. A loss of subcooling margin, on the other hand, indicates steam generation within the RCS and may indicate core uncovery. The operator must take action to restore the subcooling margin, such as by activating HPI pumps or a primary system heat removal process.

Subcooling margin is the key ATOG indicator. However, inadequate or excessive heat transfer indications must also be remedied. Inadequate heat transfer is countered by restoring primary to secondary system cooling and/or by initiating HPI-PORV cooling. In the latter method, core decay heat is removed by actuating full HPI and the PORV. The HPI fluid flows into the cold leg discharge piping leading to the reactor vessel. It flows through the core, out the hot leg, into the pressurizer through the surge line and out the PORV. Excessive heat transfer is remedied by reducing the rate of primary to secondary system heat transfer. This can be done by throttling the flow of auxiliary feedwater to the steam generators, restoring steam generator secondary pressure and reducing the primary to secondary temperature difference.

Exclusive among the events that can give rise to abnormal ATOG symptoms is the steam generator tube rupture (SGTR) accident. This event is readily identified using radiation indications and alarms and indications of steam generator conditions: pressure, temperature and level. Event oriented operator actions are taken to minimize radiation release and to ensure core cooling.

NSSS design: today and in the future

There are currently 103 operating, fully licensed nuclear power reactors in the United States representing 97.5 GW of capacity and generating 780 billion kWh in 2002, or 20% of the total U.S. generation.[2,3] These 103 NSSS units fall into one of three categories:

	No.	MW
Pressurized Water Reactor (PWR)		
Recirculating Steam Generator (RSG)	62	59,793
Once-Through Steam Generator (OTSG)	7	5,915
Boiling Water Reactor (BWR)	34	31,792
Total	103	97,500

As of May 2004 there were 440 operating nuclear power systems worldwide with a capacity of 362 GW producing approximately 16% of global electricity generation.[4,5] More than 30 new plants are under construction.[6] As discussed in the introduction to this chapter, commercial NSSS designs outside of the U.S. fall into one of six general categories:

1. PWR – RSG
2. PWR – OTSG
3. BWR
4. PHWR (pressurized heavy water reactor, including the Canadian CANDU design)
5. GCR (gas-cooled reactor)
6. FBR (fast breeder reactor)

As with the PWRs, the PHWR, GCR and FBR also include steam generators to provide a heat sink for the primary side reactor coolant system and to generate steam for the secondary side power cycle. The PHWR system steam generators are very similar in design to the PWR recirculating steam generators. (See Chapter 48.) In particular, 33 Canadian CANDU PHWR systems with 21 GW of capacity are in operation in Canada, India, Pakistan, China, Argentina, Romania and South Korea. An additional 18 units (48 GW) in India are based upon the CANDU technology.

The current (second) generation of nuclear power plants in the U.S. has demonstrated decades of safe and reliable performance. Continual improvements have been made in maintenance and operation. In the past 20 years, the average capacity factor has increased from about 60% to more than 90%. This significant increase translates into an additional 23,000 MW of power to the grid – the equivalent of 23 new plants. Production costs (fuel, operations and maintenance) of most plants are less than US$0.02/kWh, and in the best plants about $0.01/kWh. The proven reliability has resulted in 25 of the current operating plants being granted licenses to continue operations for another 20 years beyond their original license period.

Despite this excellent performance, no new nuclear plants have been ordered in the U.S. for the past 26 years, although design of a third generation of nuclear systems, commonly designated Generation III, has proceeded. The key attributes of these new designs are:

1. simple and more rugged making these more resistant to upsets and easier to operate,
2. standardized to reduce cost and schedule,
3. increased availability,
4. extended design life (approximately 60 years),
5. reduced possibility of core meltdown accidents,
6. minimized environmental impact,
7. enhanced fuel burn-up to reduce waste, and
8. extended fuel life.

The most significant change has been the incorporation of passive or inherent safety features which do not require operator or control system intervention to avoid accidents if equipment malfunctions. These systems may include gravity assisted flow, natural circulation within and between components, and resistance to deterioration at elevated temperatures. Most of these NSSS designs (including PWRs, BWRs, PHWR/PWR hybrid, high temperature GCR) are evolutionary from the existing second generation systems. These systems are being introduced in other countries with two in operation by 2003 and others being built or ordered.

Beyond the Generation III designs, a number of countries including the U.S. have formulated a general agreement to explore the potential of several more revolutionary designs, designated Generation IV. Introduction of these new designs may be feasible starting around 2020. In addition to further improving the key attributes of the Generation III designs, the Generation IV designs will try to offer some radical new approaches to solve future energy needs and resolve environmental concerns. The new designs will employ fast cores and coolants that allow a marked increase in operating temperatures. The high temperatures can be used for chemical processes, including production of low cost hydrogen. The advanced fuel cycles can be arranged to extract more energy from the spent fuel and also to drastically reduce the toxicity of high level waste.

Nuclear ship propulsion

Nuclear power for propulsion has been applied to both commercial and naval vessels beginning with the U.S.S. *Nautilus*, the world's first nuclear ship (see Fig. 24). The development of nuclear ship propulsion also formed the basis for commercial pressurized water reactors (PWR) for land-based electric power generation. For submarines, nuclear propulsion has proven to be the greatest single advancement in post-World War II technology. This dramatic achievement virtually abolished range limitations and enabled the submarine to become a true submersible, freed from the need to make regular forays to the surface to recharge batteries.

Nuclear power has also proven extremely valuable for naval surface ships such as aircraft carriers and escort vessels. While the advantages are not as readily apparent as for submarines, the unlimited, sustained power available with nuclear propulsion allows the ship commander to devote his full attention toward effectively executing his mission without the continuous logistical concern for fuel oil supplies.

While any type of reactor system may be used for ship applications, only the PWR and the sodium-cooled reactor have been used. Of these, the PWR is the predominant system. A typical PWR schematic arrangement for shipboard application is shown on Fig. 25.

Fig. 24 Launching of the U.S.S. *Nautilus*, the world's first nuclear powered ship.

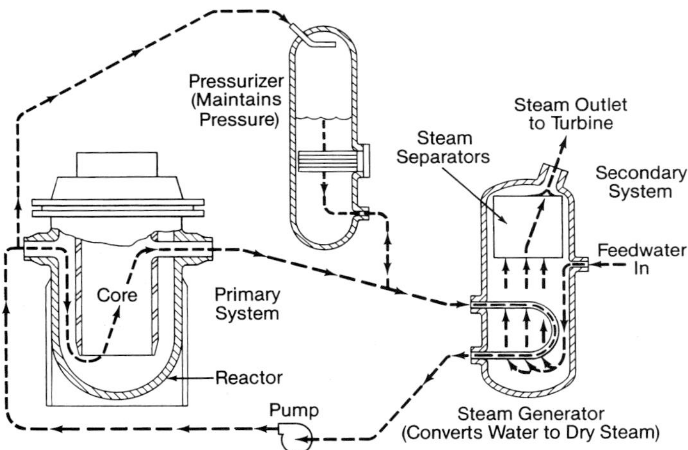

Fig. 25 Pressurized water reactor system – naval type.

There are important differences between shipboard reactors and land-based installations. These differences involve weight and space limitations, plant reliability and on-board maintenance, plant safety, and problems inherent with a moving platform. While the weight and size of the nuclear reactor itself does not present a problem, the weight and size of the radiation shielding around the NSSS is significant. These considerations usually dictate a more compact arrangement of the major components, as compared to land installations. However, the need to account for maintenance requirements forces the designer to balance compactness against access needs. Because access to ship units is more limited, reliability of shipboard components assumes even greater importance.

Safety is, of course, a major consideration. Shipboard nuclear plants are more subject to external hazards than land-based plants, particularly for naval units that may be exposed to extreme battle shock conditions. In the early stages of development, commercial ships showed promise. However, the public fear of accidents, especially with ships in harbor, essentially has prevented any further exploration of commercial ship installations. On the other hand, the U.S. Navy has maintained an excellent performance and safety record over the years, and U.S. naval ships are welcomed in major ports throughout the world.

Commercial nuclear ships

In the 1950s, several countries became interested in applying nuclear power to commercial shipping. These efforts resulted in the launching of the U.S. ship N. S. *Savannah* and the soviet ice-breaker *Lenin*. Later in the 1970s, Germany launched the N. S. *Otto Hahn* and Japan launched the cargo ship N. S. *Mutsu*. Fig. 26 is a schematic diagram of a typical marine propulsion plant.

Nuclear merchant ship *Savannah*

A nuclear merchant ship was first proposed by President Dwight D. Eisenhower in 1955, as evidence of the U.S.'s interest in promoting the peaceful use of atomic energy. After Congress approved the funding in 1956, the President directed the AEC and the Maritime Administration (MARAD) to design and construct the vessel that was subsequently named the *Savannah*.

The program's major objectives were to demonstrate the peaceful use of nuclear energy and resolve the problems of commercial marine reactor operation. Plant requirements included a conservative design with a long core life, use of commercially available materials and equipment wherever practical, and safety of operation.

The *Savannah* (Fig. 27) was a single-screw, geared turbine vessel, 595 ft (181.4 m) long, with a beam of 78 ft (23.8 m), draft of 29.5 ft (9 m), and a displacement of 21,900 t (19,867 t_m). The ship had a design speed of 22 knots (40.7 km/h) at 22,000 shaft hp (16,412 kW), accommodations for 60 passengers and 652,000 ft^3 (18,463 m^3) of cargo space.

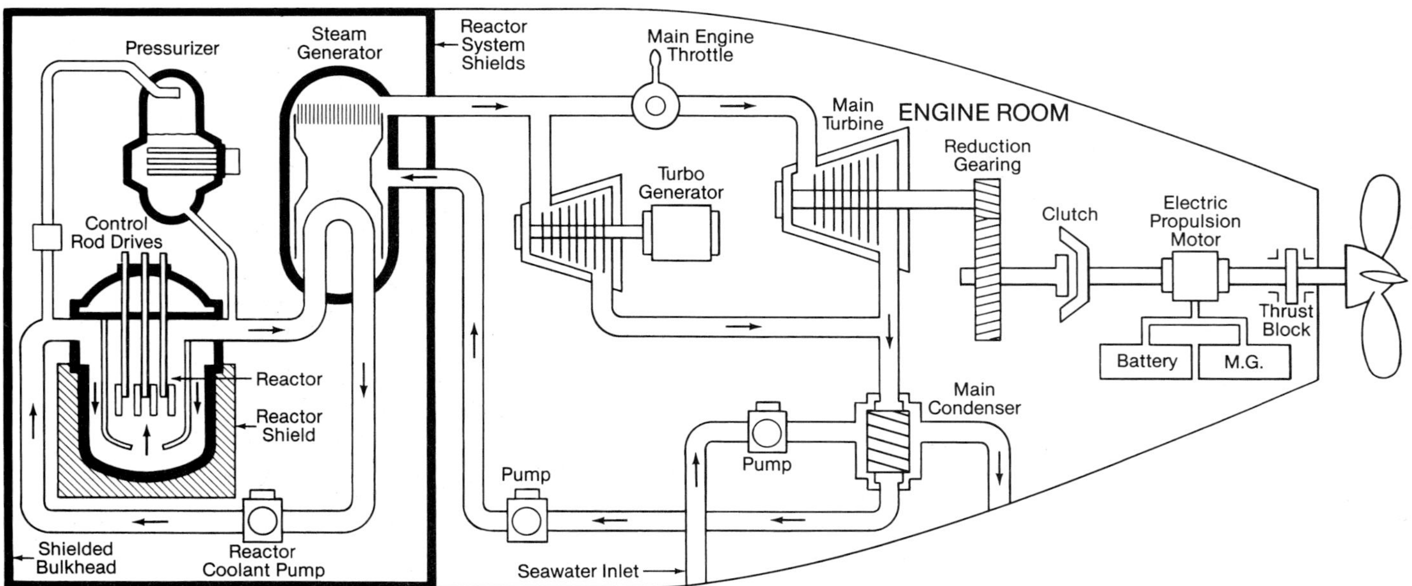

Fig. 26 Shipboard nuclear propulsion system.

Fig. 27 N.S. *Savannah* – first nuclear merchant ship.

B&W supplied the PWR nuclear propulsion plant and auxiliaries (see Fig. 28) for the *Savannah* and trained the operating crew. Pertinent design data for the power plant are given in Table 1.

The keel of the *Savannah* was laid in 1958, and the ship was launched 14 months later. In 1962, the ship was delivered to the operating agent, and port visitations began. The *Savannah's* reactor was fueled by 15,653 lb (7100 kg) of uranium enriched to an average of 4.4% uranium-235. On her first fuel core, the *Savannah* traveled approximately 330,000 mi (531,089 km) and developed 15,000 full power hours with no shuffling of fuel. By late 1970, the *Savannah* had traveled more than 450,000 mi (724,205 km), visited 32 different U.S. ports in 20 states and 45 different foreign ports in 27 countries, and had been visited by more than 1,500,000 people.

The operation of the *Savannah* provided technology for future development of nuclear ships, and established standards for the design of the ship and reactor, operating practices and safety. After completing her mission, the *Savannah* was retired and is now a museum ship located in Charleston, South Carolina.

Nuclear merchant ship *Otto Hahn*

After building the N.S. *Savannah* reactor, B&W developed an improved nuclear marine plant known as the Consolidated Nuclear Steam Generator (CNSG). The CNSG was designed to achieve more economic nuclear propulsion systems for merchant ships, and has potential application for small- to medium-sized land-based plants.

The CNSG design incorporates the reactor and once-through steam generators within a single pressure vessel, achieving a compact arrangement and eliminating some of the auxiliary equipment.

The CNSG design was used successfully for the nuclear plant of the German N.S. *Otto Hahn* (Fig. 29). This ship was a single-screw geared turbine ore carrier 565 ft (172.2 m) long, with a beam of 77 ft (23.5 m), a draft of 30 ft (9.1 m), dead weight of about 15,000 t (13,608 t_m) and a design speed of 16 knots (30 km/h) at 10,000 shaft hp (7460 kW).

The nuclear plant for this ship was designed and constructed by the German-Babcock Interatom consortium with the assistance of B&W. A cross-section of the pressure vessel is shown in Fig. 30. Pertinent design data for the power plant at normal load are given in Table 2. The *Otto Hahn* began commercial service in 1970, and was decommissioned in 1978.

Naval nuclear ships

U.S. naval nuclear propulsion program

Background In 1946, the U.S. Navy's Bureau of Ships recognized that atomic energy might ultimately be developed for ship propulsion. Under the direction of Captain (later Admiral) Hyman G. Rickover, a team was established to transform existing theory and concepts into practical engineering designs.

The basic requirements for applying nuclear power to shipboard propulsion were clear. The reactor would have to produce sufficient power so that the ship would have military usefulness, and it would have to produce that power safely and reliably. The reactor plant would have to be rugged enough to meet the stringent requirements of a combatant ship, and be

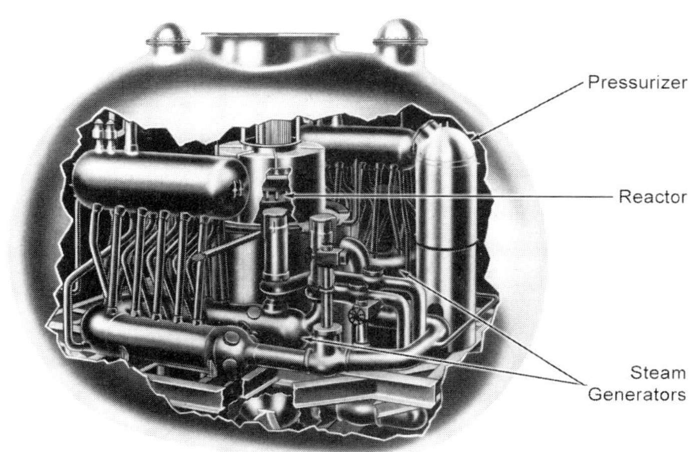

Fig. 28 N.S. *Savannah* nuclear steam supply system arrangement.

Table 1
N.S. *Savannah*

Power plant design data:	
Maximum shaft power	22,000 hp (16,412 kW)
Reactor power	70 MW
Total steam flow	226,000 lb/h (28.5 kg/s)
Turbine inlet pressure	430 psi (2.96 MPa)
Feedwater temperature	340 F (171C)
Reactor coolant data:	
Pressure in reactor	1735 psi (11.96 MPa)
Temperature in core	508 F (264C)
Flow	9,400,000 lb/h (1184.4 kg/s)

designed for operation by a Navy crew. A suitable reactor would require new corrosion resistant metals that could sustain prolonged periods of intense radiation, effective shielding to protect personnel from radiation, and the development of new components that would operate safely and reliably for prolonged time periods.

These problems were even more difficult for submarine applications because the reactor and associated steam plant had to fit within the confines of a small hull, and be able to withstand extreme battle shock. Although the application of nuclear power to submarines was a major challenge, success would revolutionize submarine warfare. No longer limited to submerged operation on battery power, a true submarine was possible, one that could travel submerged at high speed for long periods. Because of the immense challenge posed by this revolutionary approach, the Navy established extremely rigorous guidelines covering design, fabrication, quality, testing, and training for the suppliers and Navy personnel. These new guidelines demanded significant improvements in performance from all involved parties and established the design and engineering philosophy which underlies the basis of the commercial and Naval nuclear industry today.

Three reactor concepts were initially considered for naval nuclear propulsion. A study of a gas-cooled reactor showed that this concept was not then suitable. The pressurized water reactor and liquid metal-cooled reactor approaches were found promising and carried through to full scale prototype plants, and thereafter to shipboard applications. B&W became actively involved in both designs.

Fig. 29 N.S. *Otto Hahn (courtesy of Gesellschaft fur Kernenergieverwertung in Schiffbau und Schiffahrt mbH).*

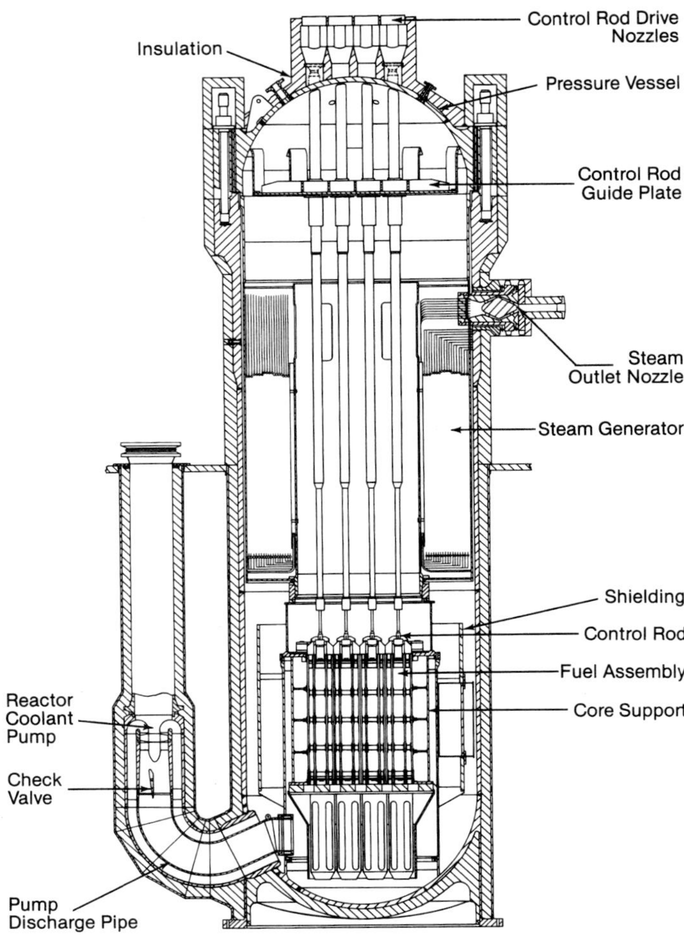

Fig. 30 Reactor vessel cutaway (N.S. *Otto Hahn*) (*courtesy of Gesellschaft fur Kernenergieverwertung in Schiffbau und Schiffahrt mbH*).

U.S.S. *Nautilus*

In 1949, the U.S. Navy's Chief of Naval Operations issued a formal requirement for the development of a nuclear powered submarine. The following year, the U.S. Congress authorized funds for a land-based prototype of the pressurized water reactor that would power the world's first nuclear powered ship. B&W was

Table 2
German N.S. *Otto Hahn*

Power plant design data:	
Shaft power	10,000 hp (7460 kW)
Reactor power	38 MW
Total steam flow	141,000 lb/h (17.8 kg/s)
Superheater outlet press.	440 psi (3.03 MPa)
Superheater outlet temp.	523 F (273C)
Feedwater temp.	365 F (185C)
Reactor coolant data:	
Pressure in reactor	918 psi (6.33 MPa)
Temp. in reactor core	523 F (273C)
Flow	5,280,000 lb/h (665.3 kg/s)

selected to provide major reactor system components. Just three years later, the prototype began operation and for the first time, a reactor produced sufficient energy to drive power machinery.

On January 17, 1955, the *Nautilus* (Fig. 31) put to sea for the first time and radioed her historic message: "underway on nuclear power." On her shakedown cruise, the *Nautilus* steamed submerged more than 1300 mi (2092 km) in 84 hours – a distance that was 10 times greater than had been traveled continuously by a submerged submarine.

Within months, the *Nautilus* would break virtually every submarine speed and endurance record. Her global odyssey carried her under the seven seas, and on the first voyage in history across the top of the world, passing submerged beneath the North Pole.

On her first fuel core the *Nautilus* steamed more than 62,500 mi (100,584 km), more than half of which were totally submerged. During 25 years of service, she traveled a total of 600,000 mi (965,606 km) on nuclear energy, making the dream of Jules Verne come alive. The *Nautilus* is now a historic museum ship and is located in New London, Connecticut.

The *Nautilus* was the first application of a reactor power plant using pressurized water both as the primary coolant and as the heating fluid for converting the secondary water into steam. The recirculating steam generator (Fig. 32) was comprised of a straight tube and shell heat exchanger with riser and downcomer pipes connected to a separate steam drum. B&W designed and fabricated the prototype test steam generator, and the reactor vessel and pressurizer for the *Nautilus*.

U.S.S. *Seawolf*

The U.S.S. *Seawolf*, the second U.S. nuclear submarine, was launched in 1955, and her liquid metal-cooled reactor attained initial criticality in 1956. To help ensure tube integrity, the *Seawolf's* sodium-heated steam generators (Fig. 33) utilized sodium-potassium as a third or *monitoring* fluid in the annulus of the double-tube design. Although the ship operated satisfactorily for almost two years on its sodium-cooled reactor, overriding technical and safety considerations (mainly the potential for violent reaction between sodium and water) led to the abandonment of this type of reactor for propelling U.S. naval ships.

Fig. 31 U.S.S. *Nautilus*.

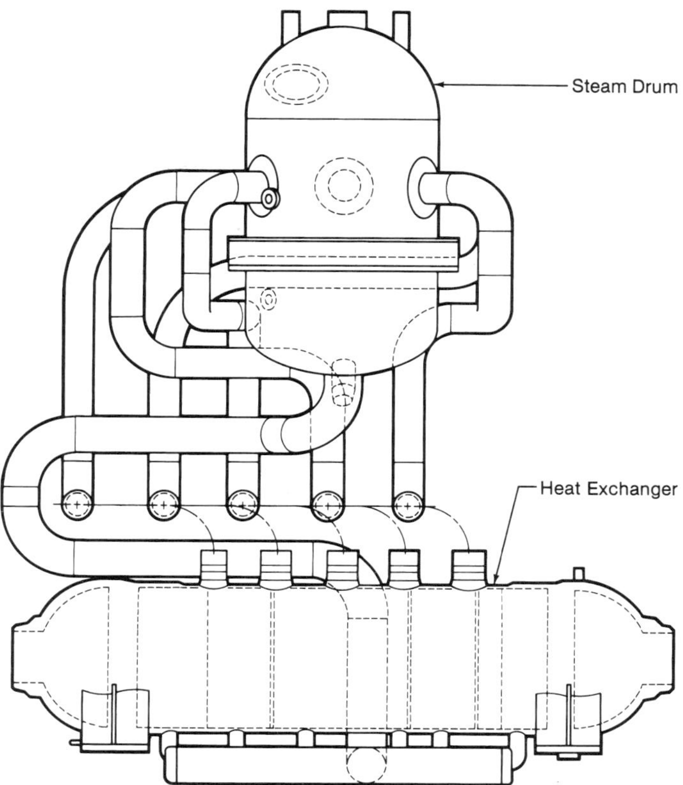

Fig. 32 U.S.S. *Nautilus* steam generator.

While liquid sodium is a much more efficient heat transfer medium than water, it can be very troublesome in service. Two problems are particularly noteworthy: the sodium has to be kept molten at all times or it will solidify and can damage primary system piping, and the sodium must be kept isolated from water to prevent a violent reaction.

In 1958, the *Seawolf* entered a shipyard where her sodium-cooled plant was replaced with a pressurized

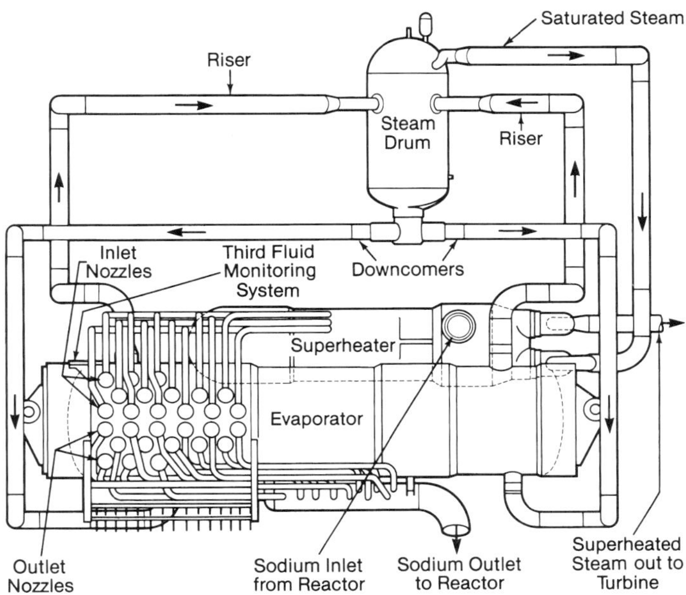

Fig. 33 Sodium-heated steam generator for U.S.S. *Seawolf*.

water reactor similar to that in the *Nautilus*. When her sodium plant was shut down for the last time, the *Seawolf* had steamed a total of 71,611 mi (115,247 km) of which 57,118 mi (91,923 km) were fully submerged.

U.S.S. *Skipjack*

The U.S.S. *Skipjack* attack submarine class combined the PWR plant with a streamlined *Albacore* hull shape to provide increased speed and reduced flow noise. In 1956, B&W received a contract for the design and fabrication of steam generators for the *Skipjack*. These were the first vertical recirculating PWR steam generators with integral steam separators and were the forerunner to the recirculating steam generator designs in current commercial PWR (non-B&W) plants and in CANDU reactor plants.

Technological advancements

Each new naval propulsion plant is a balance between the desire to make technological advances and a commitment to deliver a reliable fighting ship on schedule. Each new design can incorporate only a portion of all the potential improvements in technology. The ultimate success of any power plant design will depend on how well that balance is achieved.

In developing power plant components for modern naval vessels, improvements have been made in thermal hydraulic design, materials, structural design, and fabrication. Significant improvements in the art of steam-water separation have contributed greatly to the performance and compactness of naval steam generators.

Significant advances have also been made in reactor fuel technology. Reactor fuel lifetime has been extended from 2 years for the first Nautilus core, to more than 15 years for cores delivered in the 1980s. Efforts have been underway to develop reactor fuel that will last the life of the ship, making expensive and time consuming refueling unnecessary.

While much research has been done in the U.S. and in other countries on alternate forms of submarine power production, such as magnetohydrodynamics and fuel cells, nuclear power has proven to be highly reliable, safe and cost effective, and continues to be the system of choice for the U.S. Navy and other international fleets for the foreseeable future.

International naval nuclear programs

All of the principal maritime nations have studied the application of nuclear power to naval ship propulsion. The U.S., Great Britain, France, the People's Republic of China, and the FSU have built nuclear vessels. These nations all have naval nuclear fleets that rely primarily on pressurized water reactor technology. However, the Soviet Navy reportedly utilizes liquid metal cooled reactors in at least one attack submarine class for higher power output and greater operational speeds.

The Soviet *Typhoon*-class ballistic missile submarine (Fig. 34) is the largest submarine type ever built with a length of 563 ft (171.6 m) and a submerged displacement of 26,500 t (24,041 t_m). The Soviets' titanium-hulled *Alfa*-class attack submarines are reportedly the world's fastest and deepest diving with a speed of 45 knots (83 km/h) and an operational depth of 2500 ft (762 m).

Nuclear ship development is proceeding in other countries. The FSU is continuing development of new attack and missile submarines. The overall picture is somewhat unclear due to the economic conditions in the FSU and also because of continued safety issues, particularly in the aftermath of the sinking of the Kursk submarine in 2000. Both issues have led to a marked downturn in Soviet ship deployment. China has been developing new attack and ballistic missile nuclear submarines based on early Soviet designs. India is trying to develop a nuclear ballistic missile submarine.

Future U.S. Navy

The U.S. Navy has the largest nuclear fleet in the world, comprising 72 submarines and 10 aircraft carriers. In line with the ever changing situation in the world, the naval strategies and ship requirements are constantly evolving to meet the perceived global issues. Ship designs and power propulsion systems are constantly upgraded to meet the navy needs and to incorporate the newest technological advances.

Submarines

With the demise of the Cold War, the need for a large contingent of ballistic missile submarines has greatly diminished and strategy is now more focused on providing multi-purpose submarines with a broad range of missions. The U.S.S. *Virginia* (SSN774) is the first of a new class of attack submarines intended as a more cost-effective follow-on to the current Seawolf class. The ship is capable of both deep ocean

Fig. 34 Soviet *Typhoon*-class ballistic missile submarine.

warfare and shallow water operations of all types. Missions of the new ship include anti-submarine warfare, covert operations, personnel delivery, intelligence gathering, convert mine warfare, and Battle Group Support. The Virginia displaces 7800 tons and is 370 feet long. Fig. 35 shows a Virginia-class submarine under construction.

Aircraft carriers

In 2006, the U.S.S. *George W. Bush* will be the last of the Nimitz-class carriers to be built (see Fig. 36). Construction of a new class of carrier, the CVN 21 with 100,000 ton displacement, will begin at Newport News Shipbuilding in 2007 for delivery in 2014. The new ship will incorporate a new design of nuclear power plant, expanded flight deck, and a new electrical power distribution system. The new power plant will greatly increase the electrical power supply enabling the deployment of an electromagnetic aircraft launch system and offering scope for advancements in new electromagnetic weapons systems. The power plant and related equipment are designed to reduce maintenance and provide substantial reduction of personnel required to operate and staff the ship.

Fig. 35 Virginia-class submarine nears completion.

Fig. 36 U.S.S. *Nimitz* and U.S.S. *Ronald Reagan* are part of the current Nimitz-class aircraft carrier group.

References

1. Stultz, S.C., and Kitto, J.B., Eds., *Steam/its generation and use*, 40th Ed., The Babcock & Wilcox Company, Barberton, Ohio, 1992.

2. *Annual Energy Review 2002*, Report DOE/EIA-0384 (2002), Energy Information Agency, U.S. Department of Energy, Washington, D.C., October, 2003.

3. *Annual Energy Outlook 2004*, Report DOE/EIA-0383 (2004), Energy Information Agency, U.S. Department of Energy, Washington, D.C., January, 2004.

4. *International Energy Outlook 2004*, Report DOE/EIA-0484 (2004), Energy Information Agency, U.S. Department of Energy, Washington, D.C., April, 2004.

5. International Atomic Energy Agency, *Power Reactor Information*, Reference Data Series 2, website www.iaea.org/programmes/a2/, (2004).

6. *Nuclear News*, Vol. 47, No. 1, American Nuclear Society, LaGrange Park, Illinois, January, 2004.

Inconel is a trademark of the Special Metals Corporation group of companies.

Spent fuel storage at a nuclear power plant.

Chapter 47

Fundamentals of Nuclear Energy

Fundamental particles and structure of the atom[1]

In the fifth century B.C. Greek philosophers postulated that all matter is composed of indivisible particles called *atoms*. Over the next 24 centuries there were many speculations on the basic structure of matter but all theories lacked any experimental basis. It was not until late in the nineteenth and early twentieth centuries that the existence of fundamental particles, the *proton, neutron* and *electron*, was confirmed.

The discovery of nuclear fission, credited to Otto Hahn, Lise Meitner and Fritz Strassman, was made possible through an accumulation of knowledge on the structure of matter beginning with Becquerel's 1896 detection of radioactivity.

As shown in Fig. 1, the structure of an atom is pictured as a dense, positively charged nucleus surrounded by an array of negatively charged electrons. Each proton, the equivalent of a hydrogen atom nucleus, carries an elemental positive charge of electricity. The number of protons determines the type of chemical element of the atom. Each neutron is an electrically neutral particle with a mass slightly greater than the proton. Because of their association with the nucleus of the atom, protons and neutrons are also referred to as *nucleons*. Each electron shown orbiting around the nucleus has an elemental negative charge and a mass about 1/2000 that of a proton. The nucleus of an atom with a characteristic number of neutrons and protons is called a *nuclide*.

Despite the minute size of the atom, there is a relatively great distance between the nucleus and the orbiting electrons. This distance, approximately 10^5 times the dimension of the nucleus, accounts for the ability of various radiations to pass through apparently dense materials.

Fig. 1 also indicates the relationship of the positively charged protons in the nucleus and the negatively charged electrons. In the un-ionized state, the number of protons is balanced by an equal number of electrons. An atom becomes ionized by gaining or losing one or more electrons. A gain of electrons yields negative ions and a loss results in positive ions. An atom in an ionized state can interact with other elements to form various compounds.

When an atom has more than two electrons, their orbits are located in a series of separate and distinct groupings of energy levels or *shells*. Each shell is capable of containing a specific number of electrons. In general, an inner shell fills to its maximum number of electrons before electrons begin to form in the next shell. An x-ray is a quantum of electromagnetic energy that is emitted when an electron transitions from an outer shell to an inner shell or between energy levels within the same shell.

The number of electrons in the outermost shell determines certain chemical properties of the elements. The properties are similar for elements which have similar electron distributions in the outer shell regardless of the number of inner shells. This accounts for the repetition of chemical properties in the Periodic Table of the Elements. (See Appendix 1.)

Nuclides and isotopes

A nuclide is characterized by its atomic number Z (number of protons in the nucleus) and the mass number A (total number of protons plus neutrons). When nuclides are described by chemical symbol, the atomic

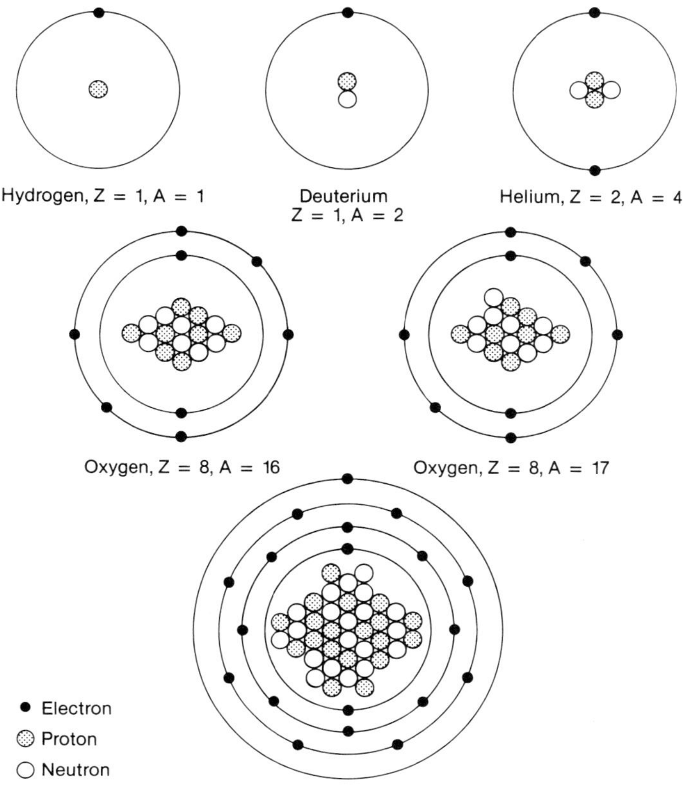

Fig. 1 Structure of the atom (Z = atomic number; A = mass number).

number is the left hand subscript and the mass number the right hand superscript. For example, the α particle can be represented as α, ${}_2\alpha^4$, or ${}_2He^4$. With most chemical elements there are several types of atoms having different mass numbers, i.e., different numbers of neutrons in the nucleus but the same number of protons. An isotope is one of two or more nuclides of the same chemical element having different mass numbers.

Two isotopes of hydrogen and two isotopes of oxygen are depicted in Fig. 1. An ordinary hydrogen atom ${}_1H^1$, contains one proton and no neutrons. Its atomic number Z and mass number A are 1. It combines with oxygen to form H_2O or regular or light water. The deuterium atom, ${}_1H^2$ or ${}_1D^2$, has one proton and one neutron; Z is 1 and A is 2. It combines with oxygen to form D_2O or heavy water. A third isotope of hydrogen, tritium, ${}_1H^3$ or ${}_1T^3$, has one proton and two neutrons.

Heavier nuclides are also identified by chemical name and mass number. Oxygen-16 and oxygen-17 (Fig. 1) have 16 and 17 nucleons respectively, although each has eight protons. For example, symbols for the two isotopes of oxygen described above are ${}_8O^{16}$ and ${}_8O^{17}$.

Although either the atomic number or the chemical symbol could identify the chemical element, subscripts sometimes are useful in accounting for the total number of charges in an equation.

Mass

It is customary to list the mass of atoms and fundamental particles in atomic mass units (AMU). This is a relative scale in which the nuclide ${}_6C^{12}$ (carbon-12) is assigned the exact mass of 12 AMU by agreement at the 1962 International Union of Chemists and Physicists. One AMU is the equivalent of approximately 1.66×10^{-24} grams, or the reciprocal of the presently accepted Avogadro number, 0.602214×10^{24} atoms per gram-atom. A gram-atom of an element is a quantity having a mass in grams numerically equal to the atomic weight of the element.

Table 1 lists the masses of the fundamental particles and the atoms of hydrogen, deuterium and helium in atomic mass units.

Mass defect

The mass of the hydrogen atom ${}_1H^1$ listed in Table 1 is almost but not quite equal to the sum of the masses of its individual particles, one proton and one electron. However, the mass of a deuterium atom ${}_1H^2$ is noticeably less than the sum of its constituents – a proton, neutron and electron. Measurements show that the mass of a nuclide is always less than the sum of the masses of its protons, neutrons and electrons. This difference, the *mass defect* (MD), is customarily calculated as:

$$MD = Zm_h + (A - Z)\,m_n - m_e \qquad (1)$$

where

MD = mass defect, AMU
Z = number of protons in the nucleus of the nuclide
m_h = mass of the hydrogen atom, AMU
$A - Z$ = number of neutrons in the nucleus

Table 1
Masses of Particles and Light Atoms

Isotope or Particle	Mass, AMU
Electron	0.000549
Proton, ${}_1p^1$	1.007277
Neutron, ${}_0n^1$	1.008665
Hydrogen, ${}_1H^1$	1.007825
Deuterium, ${}_1H^2$	2.01410
Helium, ${}_2He^4$	4.00260

m_n = mass of neutron, AMU
m_e = mass of the nuclide including its Z electrons, AMU

Binding energy

Although most nuclei contain a plurality of protons with mutually repulsive positive charges, the nucleus remains tightly bound together, and it takes considerable energy to cause disintegration. This energy, called *binding energy*, is equivalent to the mass defect. From the equivalence of mass and energy, as defined by Einstein's equation $E = mc^2$, one AMU equals 931 million electron volts (*MeV*). An electron volt is the energy gained by a unit electrical charge when it passes, without resistance, through a potential difference of 1 volt. Therefore,

$$\text{Binding energy (MeV)} = \left\{ Zm_h + (A - Z)\,m_n - m_e \right\} \times 931 \qquad (2)$$

This represents the amount of radiant or heat energy released when an atom is formed from neutrons and hydrogen atoms. It also represents the energy which must be added to fission an atom into its basic nucleons, i.e., neutrons and protons.

Dividing Equation 2 by A, the number of nucleons in the nucleus, yields the binding energy per nucleon. This in turn can be plotted as a function of A, the mass number as shown in Fig. 2. The result shows that the binding energy per nucleon rapidly increases for low mass numbers, reaches a maximum for mass numbers in the range of 40 to 80, and then drops off with an increasing slope. On the rising portion of this curve, *fusion* or joining of nucleons to atoms of higher mass number means that there is an increased binding energy per nucleon and consequently a release of energy. On the falling portion of this curve, the *fission* process or splitting of an atom results in nuclides of lesser mass numbers and greater binding energy per nucleon. Again, energy is released because of the increased mass defect.

Radioactivity and decay

Nuclear radiations

Nuclides that occur in nature are stable in most cases. However, a few are unstable, especially those of atomic number 84 and above. The unstable nuclides undergo spontaneous change at specific rates by radioactive disintegration or decay. Many nuclides decay into other unstable nuclides, resulting in a decay chain that continues until a stable isotope is formed.

There are generally three types of radiation commonly arising out of the decay of specific nuclides: alpha (α) particles, beta (β) particles and gamma (γ) rays.

Alpha particle The α particle is equivalent to the nucleus of a helium atom comprising two neutrons and two protons. It results from radioactive decay of an unstable nuclide and, with very few exceptions, is observed only in the decay of heavy nuclides.

Beta particle The β particle results from radioactive decay and has the same mass and charge as an electron. It is believed that the nucleus of an atom does not contain electrons and, in radioactive β decay, the β radiation arises from conversion of a neutron into a proton and β particle; therefore:

$$\text{neutron} \rightarrow \text{proton} + \beta + \text{energy}$$

In some instances a positive β particle (called a *positron*) is produced from the conversion of a proton to a neutron.

Gamma ray Gamma rays are electromagnetic radiation resulting from a nuclear reaction. They can be treated like particles in many nuclear reactions and are included among the fundamental particles. Although γ rays have physical characteristics similar to x-rays, their energy is greater and wave length shorter. The only other difference between the two rays is that γ rays originate from within the nucleus while x-rays originate from within the shell structure of the atom.

Biological effects of radiation

The organs of the human body are composed of tissue which is composed of atoms. When electrons are knocked out of or added to atoms (ionized), the chemical bonds which bind atoms together to make molecules are broken. This process is called *ionization*. The recombination of these broken molecules can result in changes in the molecular structure of a cell which may affect the way the cell functions, its growth characteristics and its interaction with other cells. In cases of high radiation exposure, cancer may result. The term used to describe the effects of radiation on humans is *biological damage*.

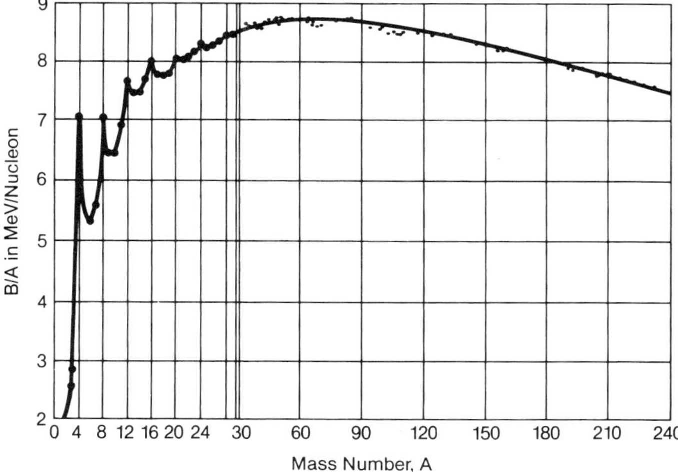

Fig. 2 Binding energy per nucleon versus mass number.

Ionization is a direct result of alpha and beta particles interacting with human tissue. These particles produce a continuous path of ionization as they travel throughout the body. Gamma radiation produces ionization indirectly in matter by the photoelectric effect, the Compton effect and pair production. All three processes yield electrons which in turn produce most of the ionization which occurs within the body.[2]

In the photoelectric effect, the γ ray transfers all its energy to the electron that it strikes. The electron then causes ionization in the medium. In the Compton effect, only part of the γ ray's energy is transferred to the electron as kinetic energy. The remaining energy gives rise to a lower energy γ ray. Pair production occurs when the γ ray has an energy greater than 1.02 MeV. All of its energy is given up and two particles, an electron and a positron, are produced. (All three processes produce electrons which then ionize the absorbing matter.)

Although ionization causes biological damage, the seriousness of the damage is determined by many factors such as the types of cells effected (how radiosensitive they are), the age of the person receiving the exposure (young cells are more radiosensitive) and whether the dose is received over a short (acute exposure) or long (chronic exposure) period of time.

Acute exposure is more detrimental because a large amount of damage is incurred rapidly and the cells do not have an adequate amount of time to repair themselves. Under chronic exposure, cells are very effective in repairing injury. The effects of chronic exposure and the amount required to produce damage have been extrapolated primarily from known cases of acute radiation exposure. More recent radiobiology laboratory and other studies have developed data which assess the effects of chronic exposure. As the nuclear industry continues, the database for low level occupational exposure grows and it becomes more evident that the associated risks are low. The data available suggest that the body can tolerate low doses of radiation received over long periods of time with little risk. This is supported by the exposure and health history of x-ray technicians, physicians and radiation workers. In many laboratory and historic exposure studies, beneficial effects of low level radiation are demonstrated. These studies certainly counter the argument that no level is low enough and that allowable levels must be endlessly reduced in the pursuit of safety.

Radiation protection

Alpha particles, due to their large mass and double positive charge, travel very short distances. Their range in air is only a few centimeters. Fig. 3 illustrates the penetration ranges of various types of radiation. Alpha particles are seldom harmful when the source is located outside the human body because an α particle of the highest energy will barely penetrate the outer layer of skin. If emitted inside the body, however, α radiation can be serious. It is important, therefore, to prevent the ingestion or inhalation of α emitting nuclides. Particular care is required to keep the air in working spaces free of dust containing α emitters. Fabrication of α emitters such as plutonium normally takes place inside gloveboxes that remain under a slight negative pressure. The boxes discharge air effluent through a filter designed to prevent α bearing dust from entering the working spaces and the atmosphere in general. The air is continuously monitored and analyzed.

Beta particles penetrate up to an inch of wood or plastic material and travel several yards in air. The skin and the lenses of the eye are most vulnerable to external β radiation. However, clothing and safety glasses provide adequate protection for external exposure to β radiation. The β particle is not as great an internal hazard as the α particle. The β particle, due to its smaller mass and lower charge, will travel farther than the α particle through tissue and will deposit less energy in a localized area.

Gamma rays penetrate deeply and deposit their energy throughout the entire body. They have great ranges in air and may present a hazard at large distances from the source; however, they do not present as large an internal hazard as α particles.

Neutrons, like gamma radiation, are an external hazard and their damage extends throughout the body. In addition, the effectiveness of neutrons to produce biological damage is 2.5 to 10 times greater than γ rays. The concrete and water shielding provide the primary protection against neutrons and gamma rays.

Decay rate

The rate at which a radioactive nuclide emits radiation is a characteristic of the nuclide and is unaffected by temperature, pressure or the presence of other elements that may dilute the radioactive substance. Each nucleus of a specific radioactive nuclide has the same probability of decaying in a definite period of time at a rate characterized by its radioactive decay constant. The rate of decay at any time, t, always remains proportional to the number of radioactive atoms existing at that particular instant. The decay is calculated by:

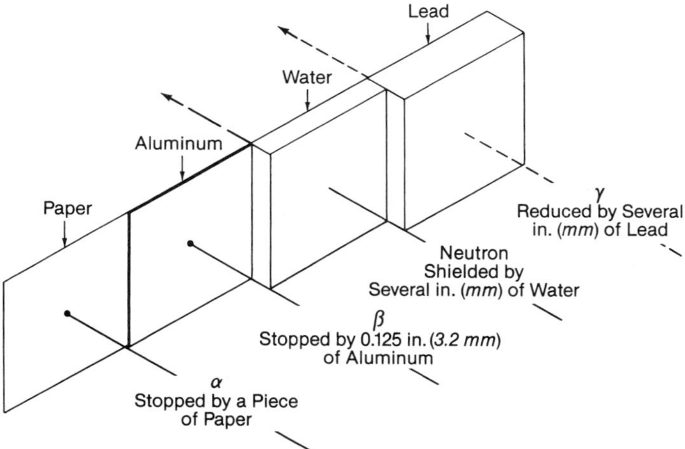

Fig. 3 Ranges of various types of radiation in materials.

$$N(t) = N(0)\,e^{-\lambda t} \quad \textbf{(3)}$$

where

$N(t)$ = the number of atoms per cm^3 at time t
$N(0)$ = the number of atoms per cm^3 at time zero
λ = radioactive decay constant

Half life

Decay is usually expressed in terms of a unit of time called radioactive half life, $T_{1/2}$. This represents a measurable period of time, the period it takes for a quantity of radioactive material to decay to one half of its original amount. The relationship between half life and decay constant can be determined by substituting ($T_{1/2}$) for t and $N(0)/2$ for $N(t)$ in Equation 3 and solving for λ. The result is:

$$\lambda = \frac{0.693}{T_{1/2}} \quad \textbf{(4)}$$

Decay constants for radioactivity isotopes are easily obtainable with a listing of measured half lives.[3]

If $N(t)$ represents the number of radioactive atoms present at time t, then $\lambda N(t)$ becomes the number of radioactive nuclei that decay per unit of time at time t. This is referred to as the *radioactivity*, or more simply the *activity*, of the atoms and is expressed in *curies*. A curie is 3.70×10^{10} disintegrations per second. $N(t)$ can be converted directly to curies by the relation $N(t)/(3.70 \times 10^{10})$. Fig. 4 shows on a relative scale how the activity decreases during several half lives. This curve applies to all radioactive substances.

It is sometimes difficult to distinguish between stable and radioactive nuclides. All nuclides heavier than bismuth-209 (Bi-209) are unstable. However, the specific nuclides thorium-232 (Th-232), uranium-235 (U-235) and U-238 have half lives of 10^8 to 10^{10} years and can be considered stable. In fact, they are generally referred to as the stable isotopes of the heavier chemical elements. Table 2 shows half lives of some nuclides of high atomic number.

Induced nuclear reactions

By providing sufficient energy, all of the fundamental particles can be made to react with various nuclei. In this procedure, the particle strikes or enters an atomic nucleus causing a transfiguration or change in structure of the nucleus and the release of a quantity of energy. Particles which activate these reactions include neutrons, deuterons, α, β, protons, electrons and γ. Many reactions produce artificial radioactive nuclides.

Charged particles such as alphas, betas and protons do not have great penetration power in matter because the interaction with the existing electrical field within the atoms either slows or stops them. Collisions between like charged particles require exceedingly high kinetic energy. Electrical fields, however, can not deflect electrically neutral neutrons; therefore, they collide with nuclei of the material on a statistical basis. The neutron is therefore the most effective particle for inducing nuclear reactions including fission.

Reactions of principal interest in the design of nuclear reactors are those that involve neutrons and those that involve the interaction of the particles pro-

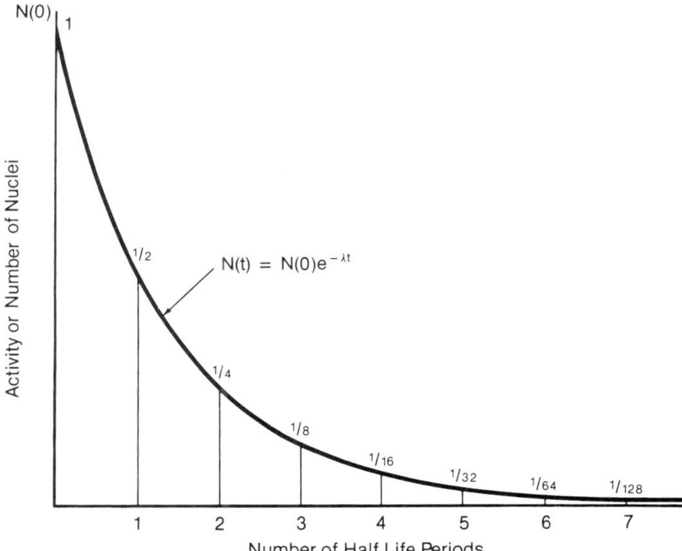

Fig. 4 Exponential decay of radioactive nuclides.

duced by fission and other nuclear processes within the reactor and surrounding materials.

The expression:

$$ {}_{Z1}X^{A1}\left(P_1\,P_2\right)_{Z2}X^{A2} \tag{5}$$

is generally used to denote a nuclear reaction and is the short form for:

$$ {}_{Z1}X^{A1} + P_1 \rightarrow {}_{Z}X^{A*} \rightarrow {}_{Z2}X^{A2} + P_2 $$

The X^* notation indicates formation of a compound nucleus which is generally unstable. Terms P_1 and P_2 represent incident and resultant particles respectively. The individual sum of the Zs (atomic numbers) and the sum of the As (mass numbers) on the left side of the equation always equals, respectively, the sum of the Zs and the sum of the As on the right side of the equation. (See Table 3).

Nuclear reactions always conserve total mass-energy on the two sides of the equation in accordance with Einstein's equation for the equivalence of mass and energy. However, there is usually an energy difference and a mass difference. A good statistical probability for the reaction to occur exists when energy is released as a result of reaction. The more probable reactions generally can be initiated by lower energy particles. The less probable ones, those that require significant energy addition, can be initiated only by high energy particles. Table 3 gives some typical induced nuclear reactions.

In a nuclear reactor, many heavy nuclides with mass numbers greater than 238 are produced by a series of neutron captures accompanied by β and/or α decay. These artificially produced nuclides have many uses – some are readily fissionable, some are fertile because they absorb a neutron and then transmute to a fissionable nuclide, some have high energy α decay modes that can be used in neutron sources, and some are used as sources in medical procedures.

Table 2
Half Lives of Heavy Elements

	Decay Mode	Half Life
Naturally occurring nuclides:		
Thorium-232	α	1.39×10^{10} yr
Uranium-238	α	4.51×10^{9} yr
Uranium-235	α	7.13×10^{8} yr
Artificial nuclides:		
Thorium-233	β	22.1 min
Protactinium-233	β	27.4 d
Uranium-233	α	1.62×10^{5} yr
Uranium-239	β	23.5 min
Neptunium-239	β	2.35 d
Plutonium-239	α	2.44×10^{4} yr

Probability of nuclear reactions – cross-sections

The probability of a nuclear reaction between a neutron, or other fundamental particle, and a particular nuclide is expressed as the *cross-section* for that reaction. The probability is dependent on the energy of the interacting particle. Because neutron interactions are most important in a nuclear reactor, the following discussion is directed toward neutron cross-sections; however, the concepts can be directly applied to all particle interactions.

Two types of nuclear cross-sections are defined:

1. The microscopic cross-section or interaction probability is an intrinsic characteristic of the nuclei of the material. It has dimensions of an area and is normally expressed in *barns* where one barn equals 10^{-24} cm^2.

Table 3
Typical Induced Nuclear Reactions

Short Form	Long Form
Alpha:	
${}_4Be^9(\alpha,n)_6C^{12}$	${}_4Be^9 + {}_2\alpha^4 \rightarrow {}_6C^{12} + {}_0n^1$
${}_7N^{14}(\alpha,p)_8O^{17}$	${}_7N^{14} + {}_2\alpha^4 \rightarrow {}_9F^{18*} \rightarrow {}_8O^{17} + {}_1p^1$
Deuteron:	
${}_{15}P^{31}(d,p)_{15}P^{32}$	${}_{15}P^{31} + {}_1d^2 \rightarrow {}_{16}S^{33*} \rightarrow {}_{15}P^{32} + {}_1p^1$
${}_4Be^9(d,n)_5B^{10}$	${}_4Be^9 + {}_1d^2 \rightarrow {}_5B^{11*} \rightarrow {}_5B^{10} + {}_0n^1$
Gamma:	
${}_4Be^9(\gamma,n)_4Be^8$	${}_4Be^9 + {}_0\gamma^0 \rightarrow {}_4Be^8 + {}_0n^1$
${}_1H^2(\gamma,n)_1H^1$	${}_1H^2 + {}_0\gamma^0 \rightarrow {}_1H^{2*} \rightarrow {}_1H^1 + {}_0n$
Neutron:	
${}_5B^{10}(n,\alpha)_3Li^7$	${}_5B^{10} + {}_0n^1 \rightarrow {}_5B^{11*} \rightarrow {}_3Li^7 + {}_2\alpha^4$
${}_{48}Cd^{113}(n,\gamma)_{48}Cd^{114}$	${}_{48}Cd^{113} + {}_0n^1 \rightarrow {}_{48}Cd^{114*} \rightarrow {}_{48}Cd^{114} + {}_0\gamma^0$
${}_1H^1(n,\gamma)_1H^2$	${}_1H^1 + {}_0n^1 \rightarrow {}_1H^{2*} \rightarrow {}_1H^2 + {}_0\gamma^0$
${}_8O^{16}(n,p)_7N^{16}$	${}_8O^{16} + {}_0n^1 \rightarrow {}_8O^{17*} \rightarrow {}_7N^{16} + {}_1p^1$
Proton:	
${}_6C^{12}(p,\gamma)_7N^{13}$	${}_6C^{12} + {}_1p^1 \rightarrow {}_7N^{13*} \rightarrow {}_7N^{13} + {}_0\gamma^0$
${}_4Be^9(p,d)_4Be^8$	${}_4Be^9 + {}_1p^1 \rightarrow {}_5B^{10*} \rightarrow {}_4Be^8 + {}_1d^2$

* Indicates formation of a compound nucleus which is generally unstable.

2. The macroscopic cross-section is a probability of interaction per centimeter of neutron path and takes into account the density of the material. It has the dimensions cm^{-1}.

The symbol σ represents the microscopic cross-section. In the case of a neutron approaching a fissionable atom it becomes possible to consider a total cross-section σ_T in which:

$$\sigma_T = \sigma_c + \sigma_s + \sigma_f = \sigma_a + \sigma_s \quad \textbf{(6)}$$

where

σ_c = the capture cross-section, a measure of the probability for absorption without fission
σ_s = the scattering cross-section, the probability that the nucleus will scatter the neutron
σ_f = the fission cross-section (present in only a few of the many nuclei), the probability for a neutron to strike and cause a fission to occur
σ_a = the absorption cross-section, the sum of the probabilities for capture and fission

The macroscopic cross-section can be obtained from:

$$\Sigma = N_0 \rho \sigma / M \quad \textbf{(7)}$$

where

Σ = macroscopic cross-section, cm^{-1}
N_0 = Avogadro's number, 0.602214×10^{24} atoms/gram-atom
ρ = density of the material, $grams/cm^3$
σ = microscopic cross-section, cm^2/atom
M = atomic weight, grams/gram-atom

Experimental measurements determine microscopic cross-sections for each element. Total cross-section measurements are generally made by transmission techniques. For example, placing a material of known density and thickness in front of a neutron source permits measuring the intensity of neutrons at a particular energy on each side of the material. The difference represents the loss or attenuation of neutrons by the material. The cross-section required to obtain this attenuation is derived by calculation.

Isotopes of the same element can have very different cross-sections. For example, the isotope xenon-134 (Xe-134) has a microscopic cross-section for neutron absorption of about 0.2 barn, yet Xe-135 has a microscopic cross-section of 2.7×10^6 barns. U-235 is fissionable at any energy, and U-238 only at high energy.

Cross-sections of some nuclides also contain abrupt peaks called *resonances* at certain energy bands (Fig. 5). Because a cross-section is really a function of relative energy of the neutron and the nucleus, an effective change in the cross-section results when the energies of the neutron and the nucleus increase or decrease as a result of temperature changes.

Where the curve of cross-section versus energy is fairly smooth, the effect of a temperature change of the target nucleus is relatively small. However, the effect of a temperature change is large and important in the vicinity of a resonance. An increase in temperature results in increased vibration of the nucleus with a corresponding increase in the number of probable collisions between the nucleus and neutron occurring at energies in the vicinity of the resonance. Therefore, an increase in the temperature of the nucleus results in an apparent broadening of the energy width of the resonance. This in turn results in a very effective increase in resonance neutron absorption, i.e., capture and fission. Conversely, a decrease in temperature of the nucleus results in narrowing the resonance width and decreasing resonance absorption. The change in resonance energy width with temperature, known as the *Doppler effect*, is important in reactor control.

Fig. 5 illustrates a typical cross-section curve showing the neutron capture cross-section of U-238 as a function of neutron energy. Below 10^2 eV the height of the resonance peaks is about 10^3 barns. Between 10^2 and 4×10^3 eV the great number of resonances makes it impractical to show the cross-section as a curve, although the resonance parameter data are available for use on a computer. Between 4×10^3 and 10^5 eV the curve represents a statistical average of the measurements which have been made.

The fission process

Nuclear fission is the splitting of a nucleus into two or more separate nuclei accompanied by release of a large amount of energy. In the fission of an atom due to a neutron, the mass of the neutron plus its energy must be equal to or greater than the mass defect associated with the two fission products. U-235 is the only naturally occurring nuclide that is capable of undergoing fission by interaction with low energy or slow neutrons. Some of the artificially produced heavy nuclides are also fissionable with slow neutrons, including plutonium-239 (Pu-239) and U-233. Other nuclides such as U-238 and Pu-242 require higher energy neutrons to cause fission.

This difference in fission capability occurs because the binding energy of a nucleus is not only determined by its mass number but also by whether the number of protons and neutrons is even or odd. A nuclide with an even number of neutrons and protons, such as U-238, has the highest binding energy per nucleon and requires the most added energy to fission. Fission only occurs with high energy or fast neutrons (> 1MeV). A nuclide with an odd number of protons and an even number of neutrons or an even number of protons and an odd number of neutrons, such as U-235, Pu-239 and U-233, has a lower binding energy per nucleon. Nucleons with both an odd number of protons and neutrons have the lowest binding energy per nucleon for essentially equivalent mass numbers. U-235 and Pu-239 are fissionable with slow or thermal neutrons (~0.025 eV). The term *thermal neutron* refers to a neutron energy distribution that is in thermal equilibrium with the temperature of the surrounding materials.

Fission occurs when the fissionable nucleus absorbs a neutron. In the case of U-235 the reaction is:

$$_{92}U^{235} + {}_0n^1 \rightarrow {}_{92}U^{236*}$$

$$_{92}U^{236*} \rightarrow {}_{Z1}X^{A1} + {}_{Z2}Y^{A2} + 2.43\ {}_0n^1 + \text{Energy}$$

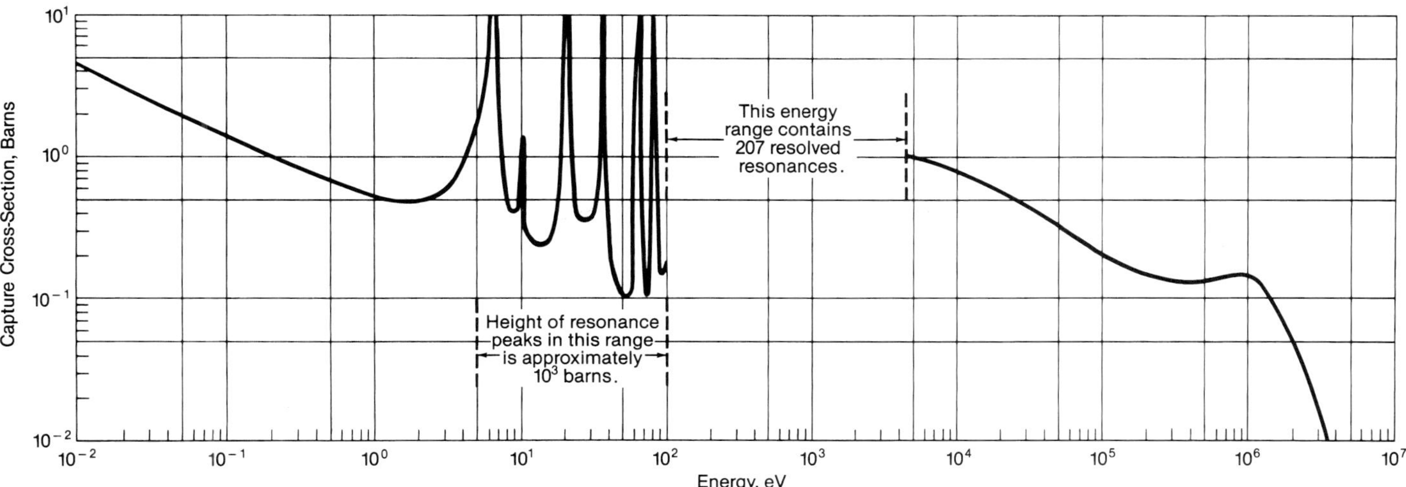

Fig. 5 Capture cross-section of uranium-238.

As previously noted, the asterisk in ${}_{92}U^{236*}$ indicates an unstable nuclide. The value 2.43 applies to U-235 fission by a thermal neutron and is the statistical average of the number of neutrons produced per reaction. X and Y represent the fission products which are distributed as shown in Fig. 6.

Energy from fission

Fission also produces γ rays, neutrons, β particles and other particles. The energy release per fission amounts to about 204 MeV for U-235 and is distributed as shown in Table 4. Release of approximately this amount of energy per fission can be predicted by examination of Fig. 2 for binding energy, considering that two fission products are formed as shown in Fig. 6. Table 4 also includes data for Pu-239.

Neutrons from fission

The fact that additional neutrons are born or generated by a fission event makes it possible to establish a *chain reaction*, as depicted in Fig. 7, that can sustain itself as long as sufficient fissionable material and neutrons are present. The neutrons produced per fission event, υ, and the average number of neutrons produced per neutron absorbed in the fuel, η, vary with the different fissionable isotopes and with the energy of the neutron producing the fission. Statistical averages for these quantities are given in Table 5. There are individual fission reactions that produce only one neutron, possibly none, or as many as five. Because some of the neutrons are absorbed without producing fission, η becomes a more meaningful quantity in reactor design than υ. The values in the table are for fission by low energy or thermal neutrons at room temperature [0.025 eV neutron energy or 7218 ft/s (220 m/s) neutron velocity]. To maintain a chain reaction, the average number of neutrons per absorption must be significantly greater than 1 because some of the neutrons will be lost to absorption in the moderator and structural materials, to absorption in control materials, and to leakage from the core. Control materials such as boron and silver-indium-cadmium are used to maintain a steady-state chain reaction by absorbing any additional neutrons over that needed for steady-state operation.

Fission neutron energy distribution

Neutrons released from fission vary in initial energy over a wide range up to 15 MeV and above. The distribution of neutrons produced by fission as a function of energy has been determined from several experimental measurements, and typical results for U-235 and Pu-239 are shown in Fig. 8. Based on these measurements, the average energy of fission neutrons is about 2 MeV; the peak of the energy distribution occurs at about 0.8 MeV.

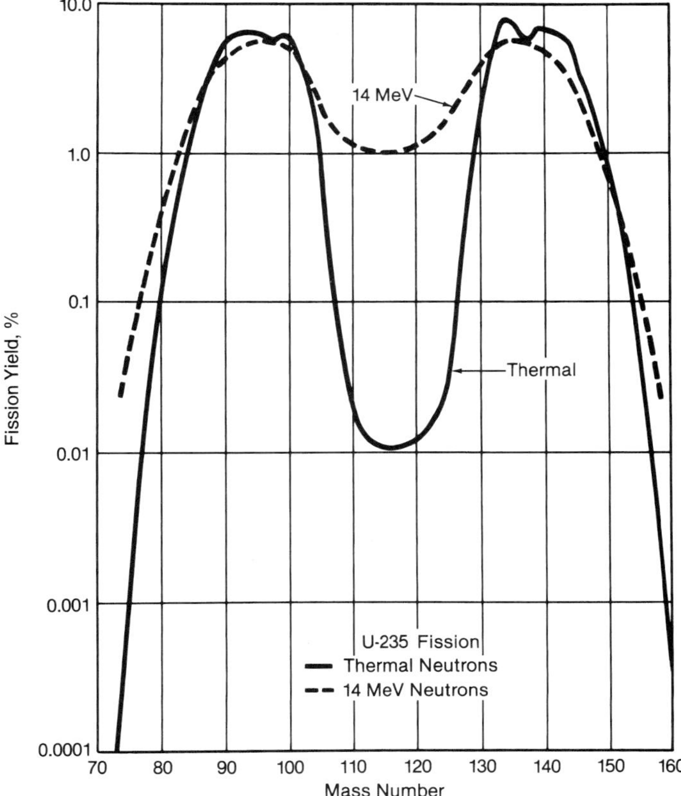

Fig. 6 Mass distribution of fission products from fission of U-235.

Table 4
Energy Produced in Fission (MeV/fission)

	U-235	Pu-239
Instantaneous		
Kinetic energy of fission products	169	175
γ ray energy	8	8
Kinetic energy of fission neutrons	5	6
	182	189
Delayed		
β particles from fission products	8	8
γ rays from fission products	7	6
Neutron-capture γs*	7	10
	22	24
Total	204	213

* Energy produced depends on reactor composition.

Burnup

As a mass of uranium undergoes fission it produces energy. The term *burnup* is used to represent the amount of energy produced per unit mass of the material. The units of burnup are megawatt days per metric ton (MWd/t_m) of initial heavy metal, i.e., uranium. One megawatt day represents about 2.6×10^{21} fissions. A nuclear fuel assembly is typically discharged from the reactor when it has achieved a burnup of about 50,000 MWd/t_m. In commercial power reactors that are fueled with uranium comprising mainly the isotope U-238, with 4% or less U-235, most of the energy produced comes from the fissioning of U-235. At burnups of 50,000 MWd/t_m only about 5% of the initial uranium content of the fuel assembly has fissioned. Even when the other nuclear reactions that cause loss of the original uranium atoms are considered, there is still more than 90% of the initial uranium remaining in the fuel assembly when it is discharged.

Fission products

When a nucleus undergoes fission, experimental measurements have shown that predominantly two fission products or fragments are generated. The distribution of these fission products for U-235 fissions has also been measured and is shown in Fig. 6. In fission, as in any nuclear reaction, there is always conservation of

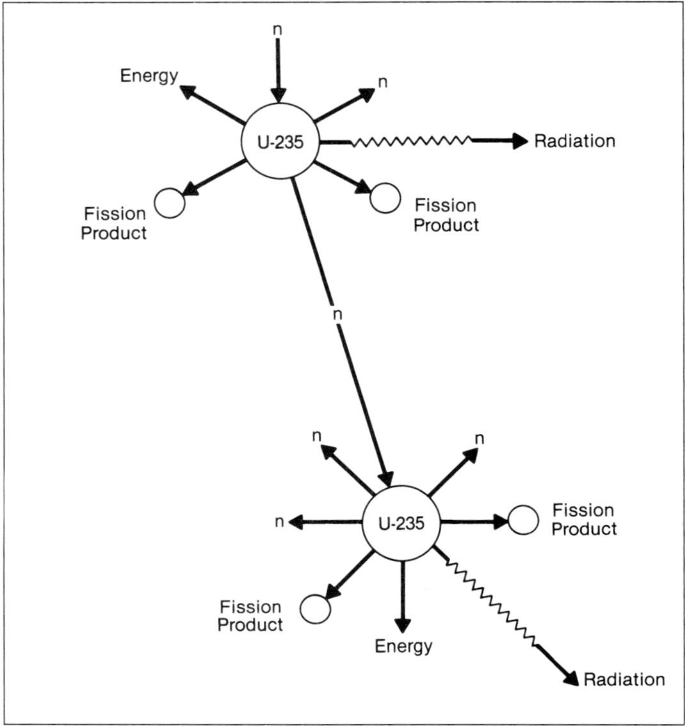

Fig. 7 Chain reaction.

total mass-energy so that one of the two fission products will come from each hump of the distribution.

Examination of Fig. 6 reveals that mass number of the fission products ranges from about 70 to 170 with two plateaus at approximately 95 and 135. Curves are shown for fission caused by thermal energy neutrons and by 14 MeV neutrons. The higher the neutron energy causing fission the more uniform the fission product distribution.

Many fission products interact with neutrons, absorbing them so that they are not available to the chain reaction. As these fission products build up (see Fig. 9), they act as absorbers to retard the chain reaction. All long lived fission products except samarium build up as the core is operated, reaching a maximum effect at the end of core life.

Table 5
Neutrons from Fission

Fuel	Avg. Neutrons per Fission, υ	Avg. Neutrons per Absorption in Fuel, η
Uranium-233	2.51	2.28
Uranium-235	2.43	2.07
Natural uranium	2.43	1.34
Plutonium-239	2.90	2.10
Plutonium-241	3.06	2.24

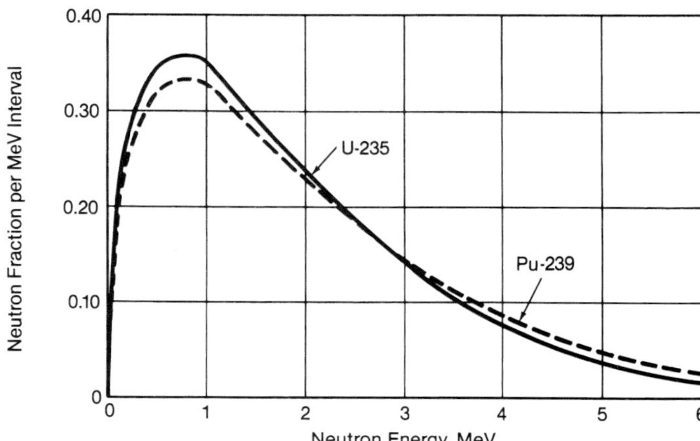

Fig. 8 Fission neutron energy distribution.

After discharge from the reactor and sitting in storage for a few years, most of the fission products will have decayed due to their relative short half lives. However, some of the long half life fission products will contribute significantly to the γ ray source that must be shielded against in handling the fuel during its ultimate disposal. These isotopes include strontium-90 (Sr-90), ruthenium-106 (Ru-106), cesium-134 (Cs-134) and 137 (Cs-137), cerium-144 (Ce-144) and europium-154 (Eu-154). These isotopes will either be bound in the fuel pellets or on the inside surface of the cladding. Other isotopes important to fuel handling are the artificially produced nuclides in the fuel pellet that decay by spontaneously fissioning. These isotopes include Pu-238, americium-241 (Am-241), and curium-242 (Cm-242) and 244 (Cm-244). In addition, the γ emitting isotope cobalt-60 (Co-60) is produced in the structural steels and Inconel® through neutron activation and must be shielded against.

Fission product behavior with time

Certain fission products, specifically Xe-135 and samarium-149 (Sm-149), both of which have very high cross-sections for absorption of thermal neutrons, are not only produced directly from fission but also are the decay products of other fission products.

Essentially all initial products of fission are highly radioactive and decay rapidly to less active isotopes with somewhat longer half lives. There are usually several isotopes in the chain before a stable end product is reached. The most significant such decay chain is the following:

$$\begin{array}{ccccccccc} & \beta & & \beta & & \beta & & \beta & \\ {}_{52}Te^{135} & \rightarrow & {}_{53}I^{135} & \rightarrow & {}_{54}Xe^{135} & \rightarrow & {}_{55}Cs^{135} & \rightarrow & {}_{56}Ba^{135} \\ <1.0\,\text{min.} & & 6.7\,\text{h} & & 9.2\,\text{h} & & 2\times10^{6}\,\text{yr} & & \end{array}$$

In this chain, tellurium-135 (Te-135) with a one minute half life decays to iodine-135 (I-135) with a 6.7 hour half life and then to Xe-135. This xenon isotope, which fortunately has only a 9.2 hour half life, has a macroscopic cross-section for thermal neutron absorption approximately 100,000 cm^{-1}, as great as all the long lived fission products together. Unfortunately, the nuclides of this chain occur abundantly as fission products – a predictable happening, because the mass number 135 occurs at a peak in the fission product distribution (Fig. 6). The Xe-135 absorbs an appreciable fraction of available neutrons as long as the reactor is operating. This changes some of the Xe-135 to Xe-136 (which has negligible neutron absorption). As a result, when a water reactor operates at constant power level, the Xe-135 builds up to its equilibrium value in 36 to 48 hours.

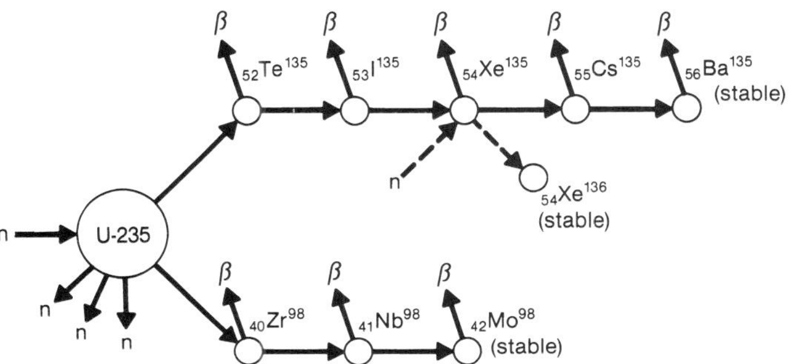

Fig. 9 Fission product chain.

When reactor power lessens and, particularly, when the reactor is shut down, the I-135 formed at the original power level continues for a time to generate Xe-135 at a rate corresponding to the original power level. Therefore, the Xe-135 builds up rapidly after shutdown because fewer neutrons are available for conversion to Xe-136. The buildup reaches a peak 4 to 12 hours after shutdown and then slowly decays. The time behavior of Xe-135 must be addressed in reactor control.

Sm-149, the second important fission product in the reactor core during operation, is generated as follows:

$${}_{60}Nd^{149} \rightarrow {}_{61}Pm^{149} \rightarrow {}_{62}Sm^{149} \text{ (stable)}$$

In the chain, neodymium-149 (Nd-149) decays (1.7 hour half life) into promethium-149 (Pm-149), which in turn decays (47 hour half life) into Sm-149. Although Sm-149 is a stable isotope it is destroyed so rapidly by neutron absorption that it reaches an equilibrium value when the reactor operates at constant power. Typically, Sm-149 reaches equilibrium in a pressurized water reactor after 50 to 100 days. Buildup after shutdown is slower and less extensive than for Xe-135.

Decay after shutdown has little consequence for all other fission fragments because of their long lives and comparable capture cross-sections between parent and daughter nuclide.

Nuclear reactor composition

A reactor core is composed of fuel, structural and control materials, and a moderator and/or coolant. Typical pressurized water reactor (PWR) fuel assembly and control component designs are shown in Fig. 10. A depiction of the interactions that take place within a fuel assembly is shown in Fig. 11. Neutrons react not only with uranium, but also with the nuclei of most other elements present. Therefore, many materials which have high neutron absorption properties should not be used for structural purposes. Fortunately, a few materials such as aluminum and zirconium have low neutron absorption cross-sections. Inconel, steel and stainless steel also can be used to a limited extent.

The choice of coolants includes water, helium and liquid sodium. Thermal reactor design, where the neutron energy distribution is essentially in thermal equilibrium with its surroundings, requires a moderator (a material containing atoms of a light element such as hydrogen, deuterium or carbon) in the core to reduce the kinetic energy of the neutrons. Hydrogen is the ideal moderator because its nucleus is as light as a neutron and, therefore, it can absorb the energy of the neutron in a single direct collision. Hydrogen is normally present in the form of water. The heavier the atom the less energy it absorbs per collision, and the less slowing effect on the neutron. Carbon, in the form of graphite, is about the heaviest atom that can be used practically as a moderator. Carbon has the additional advantage of absorbing very few neutrons, even less than hydrogen. Other low weight atoms can be used, but they are less practical because of excess neutron absorption or high material cost. Because the helium nucleus absorbs essentially no neutrons, it would be an ideal moderator except that helium is a gas at reactor operating temperatures. In this state it would be impossible to provide the core with a sufficient number of atoms to be an effective moderator.

All thermal reactors contain some fast (high energy) neutrons because only fast neutrons result from fission. The amount of moderator provided determines the degree of thermalization (energy reduction) or the percentage of thermal neutrons present in the reactor. Perhaps the greatest advantage of the thermal reactor is its compatibility with the water coolant. Water has proven to be the most practical and economical coolant.

The low neutrons per fission available with natural uranium makes it difficult to design a natural uranium (0.71% U-235) reactor. However, it can be accomplished if all materials are of especially high purity and if discrete sections of fuel are placed in a heterogeneous array such that the neutrons can slow down in the moderator and then re-enter the fuel material as slow neutrons. Under these conditions, there is a relative high probability of causing fission. The normal moderators used in natural uranium reactors are either graphite or deuterium oxide (heavy water). The absorption cross-section of normal hydrogen makes it less desirable than the other moderators. Also, minute quantities of any high neutron absorbers can prevent a chain reaction. Nevertheless, all early reactors used natural uranium and many still operate, notably the large plutonium producing reactors and the Canadian CANDU pressurized heavy water reactors. (See Chapter 46.)

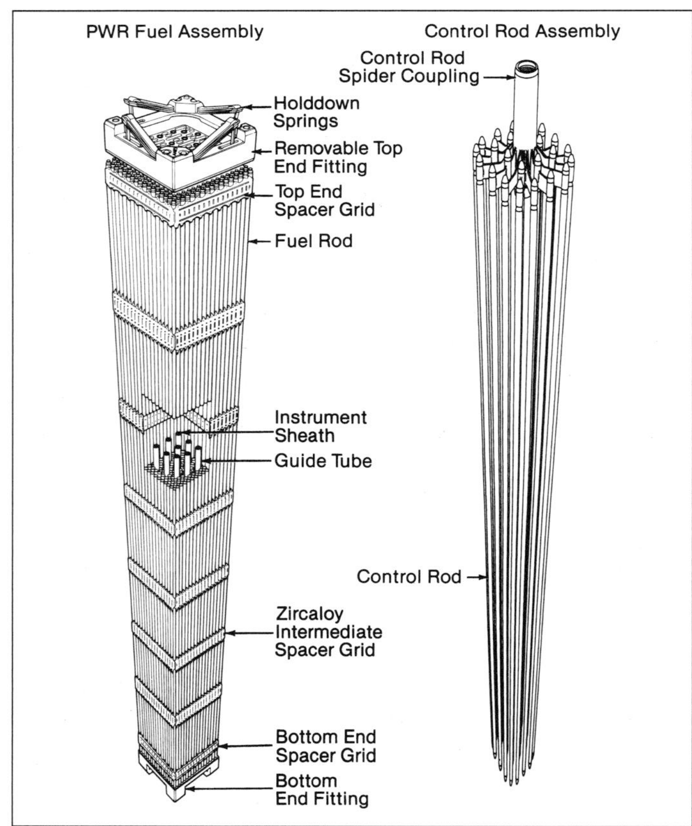

Fig. 10 Typical PWR fuel assembly and control components.

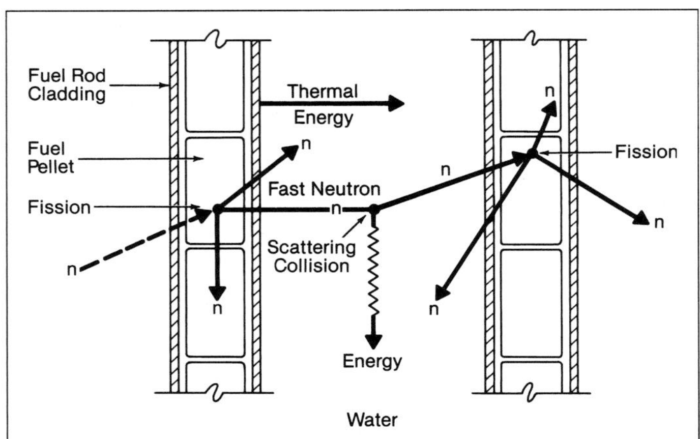

Fig. 11 PWR neutron moderator/coolant-fuel interactions.

Physical description of nuclear chain reactions[4-7]

To use fission as a continuous process for power production, it is necessary to initiate and maintain a fission chain reaction at a controlled rate or level which can be varied with power demand. Obtaining a steady-state chain reaction requires the availability of more than one neutron for each fission produced because non-fission reactions absorb some neutrons and others leak from the reactor. For a reactor as a whole, the neutrons born in each instant of time constitute one generation of neutrons. The term *effective multiplication factor*, k_{eff}, is defined as the ratio of the number of neutrons in one generation to that in the previous generation. With a multiplication factor of less than one, the system is a decaying one and will never be self-sustaining. With a multiplication factor greater than one, a nuclear system produces more neutrons than it uses, and power increases. For a steady-state chain reaction, $k_{eff} = 1$.

The steady-state equation for neutron balance in a chain reaction can also be written as:

$$\text{Production} = \text{Absorption} + \text{Leakage} \qquad (8)$$

When this condition is obtained, the reactor has gone *critical* meaning that the necessary amount of neutron production has been achieved to balance the leakage and the absorption of neutrons. The mass of fissionable material required to achieve this condition is called the *critical mass*.

Production, absorption and leakage all depend heavily on interactions of neutrons with nuclei of the various materials in the reactor. Production depends primarily on those interactions with U-235 nuclei which result in fission. Absorption depends on interactions of neutrons with any nuclei in the core that result in absorption of neutrons with or without fission. Leakage depends on the scattering effect of collisions between neutrons, nuclei and other particles, which results in transport of neutrons toward the boundaries of the chain reacting system and ultimate escape from the system.

Table 5 indicates that each fission of a U-235 atom by a thermal neutron makes available an average of 2.43 neutrons. However, only 2.07 neutrons are produced per thermal neutron absorbed in U-235 as also indicated in Table 5. The other 0.36 neutrons are absorbed in non-fission reactions. If natural uranium is used, the number of neutrons produced per thermal neutron absorbed is reduced to 1.34 because of the neutron absorption in U-238. Most of this absorption results in the ultimate production of fissionable Pu-239. However, neutrons used in this manner are no longer available to help maintain the chain reaction.

A neutron chain reaction is possible because each fission event on the average produces more than one neutron. If the process is controlled such that just one of the neutrons produced from fission causes another fission reaction then a steady state process is maintained and a constant power level is achieved. To best understand the process, a general description of the lifetime of a neutron follows. It must be realized that for a power reactor the number of neutrons crossing a 1 cm^2 surface (neutron flux) near the center of the reactor is on the order of 10^{13} neutrons/cm^2/s. The lifetime of a neutron is on the order of 10^{-9} seconds. Therefore, this description is only representative of what happens in an instant of time.

As described previously, a neutron produced by the fission process has an average energy of about 2 MeV and is termed a fast neutron. At this energy, elastic or inelastic scattering reaction with the moderator, the fuel or with the structural materials are most probable. However, about 2% of the fast neutrons cause a fission in U-238 and produce an even higher number of neutrons than produced from thermal fission.

In an elastic scattering reaction some of the neutron energy is transferred to the nucleus with which it collides in the form of kinetic energy; this can cause the nucleus to be displaced from its normal lattice position in the material. In an inelastic collision a compound nucleus is formed in an excited state and it reduces its energy by releasing a lower energy neutron and a γ ray. Some of the original neutron's energy is also deposited as the kinetic energy on the nucleus. The net impact of the scattering reactions is to cause the neutron to lose energy or slow down into the energy range where resonance interactions with various materials (especially U-238) can take place. In the resonance range, neutrons will have a high probability of being captured if their energy coincides with any of the resonances associated with the uranium fuel or the fission products and will not be available to react with the U-235 to help sustain the chain reaction. The resonance escape probability is the fraction of neutrons that escape capture while slowing down to the thermal energy range. The term *resonance escape* refers particularly to U-238 because in the intermediate range of neutron energy there are several resonance peaks of absorption cross-section for this isotope. These resonances are useful in the production of plutonium; however, to maintain criticality in a thermal reactor, sufficient neutrons must escape absorption in the resonance region.

In contrast, collisions with hydrogen atoms in the moderator can slow down the neutron very rapidly, and therefore the resonance escape probability is high for water moderated systems. Once past the resonance range the neutron is said to become thermalized and becomes part of a Maxwellian energy distribution with an average energy of about 0.2 eV. As the fission cross-sections for U-235 and Pu-239 are the highest in the thermal energy range, most fission reactions take place in this range.

The probability of having a fission reaction is referred to as the *thermal utilization* and is the ratio of the fission cross-section to the absorption cross-section (capture plus fission) averaged over the moderator plus the fuel and structural materials.

The other mechanism for loss of neutrons is leakage from the system. This is dependent on the size of the system and the length of the path neutrons travel between source and capture. For a thermal reactor this length is approximately 2.95 in. (75 mm). Today's power reactors are on the order of 13.1 ft (4 m) in diameter and therefore only neutrons born in the fuel assemblies on the periphery of the core have a significant probability of being lost from the system.

Control of the chain reaction

More than 99% of neutrons produced in fission are *prompt* neutrons; that is, they are produced almost instantaneously. About 0.73%, in the case of U-235, are released by the decay of fission products rather than directly from fission. The average half life for these delayed neutrons from fissioning of U-235 is about 13 seconds. This provides time for the reactor operator to respond to small changes in either power demand by the system or reactor system parameters.

The chain reaction can be regulated by placing materials with high neutron absorption capability in the reactor and providing a means of varying their amounts. These materials tend to stop the chain reaction and include boron, cadmium-silver-indium, hafnium and gadolinium. One or more of these materials usually goes into the reactor in the form of control rods that can be withdrawn to start up the reactor or to increase power level and reinserted to reduce power or to shut down the unit. To assure accurate control of power, at least some of the rods must be capable of fine regulation.

Reactivity

The first step in establishing a chain reaction is to bring the reactor to critical condition at essentially no thermal power or zero power level. The reactor power is gradually increased up to the desired level by the removal of a control material and maintained there. The objective in each step of this procedure is to obtain and hold a constant value of $k_{eff} = 1.0$. In view of the changes that continually occur in a nuclear system, it is never possible to keep $k_{eff} = 1.0$ for more than a short period of time without adjustments to compensate for variations. Operating a reactor at any constant power level at an effective multiplication factor of unity corresponds to steering a ship on a compass course. It takes a continual effort, either automatic or manual, to hold the ship on the exact course.

If a reactor operates at a specific power level with $k_{eff} = 1.0$, and if anything changes to increase or decrease the multiplication factor, a reactivity change is said to have occurred. It may be of positive or negative change depending upon the direction of change in k_{eff}. Reactivity, represented by the symbol ρ, is defined as the ratio:

$$\rho = \left(k_{eff} - 1\right) / k_{eff} \qquad \textbf{(9)}$$

Reactivity has been given units of dollars ($) and cents (¢) or inhours. It is usually expressed in the units of pcm (per cent milli) which corresponds to a ρ of 1×10^{-5}. As will be shown, most processes in the reactor, except for control rod insertion, generate reactivities of 1 to 10 pcm which constitute a very small change from a k_{eff} of 1.

Calculation of reactor physics parameters[4-7]

The calculation of a chain reaction (or the design of a nuclear reactor) requires solution of the steady-state equation:

$$\text{Production} = \text{Absorption} + \text{Leakage}$$

or its counterpart for nonsteady-state (transient) conditions:

$$\text{Production} - \text{Absorption} - \text{Leakage} = dn/dt \qquad \textbf{(10)}$$

where dn/dt is the variation of the neutron density with time.

The production rate is evaluated as:

$$\text{Production} = \upsilon \Sigma_f n v \qquad \textbf{(11)}$$

where

υ = neutrons per fission
Σ_f = macroscopic fission cross-sections, cm^{-1}
n = neutron density, neutrons/cm^3
v = neutron velocity, cm/s

If the neutron flux is defined as the product of the neutron density and neutron velocity, then the production rate can be defined as:

$$\text{Production} = \upsilon \Sigma_f \phi \qquad \textbf{(12)}$$

where

ϕ = nv = neutron flux

Similarly, the absorption rate is defined as:

$$\text{Absorption} = \Sigma_a \phi \qquad \textbf{(13)}$$

where

Σ_a = macroscopic absorption cross-section, cm^{-1}

Leakage is a complicated function of the gradients in the neutron flux at boundaries of the region under consideration.

In performing reactor calculations, the core is divided into discrete spatial regions or nodes, and the continuous neutron energy range is divided into a number of groups. The above equations then become a coupled set of partial differential equations that are solved in both the space and energy domains to find the neutron flux and reaction rates in each node.

In the design of the early reactors, most calculations of reactor physics parameters were done by hand using homogeneous models of the core components. Neutron cross-sections were obtained from experimental plots of total, absorption, scattering and fission cross-sections. Critical experiments were performed mocking up the fuel designs and modeling both the fuel and control rods. From these experiments, modeling adjustments were developed to ensure high accuracy predictions of reactor performance parameters. As with most other disciplines, the advent of more powerful computers has permitted the designer to perform detailed calculations in both space and energy. The following sections describe some techniques that have been used in calculations for commercial PWR fuel cycle design and licensing.

Cross-sections

Basic neutron and gamma cross-sections are compiled and verified as part of the industry effort to maintain the cross-section libraries. These libraries provide the basic data needed to calculate the cross-section of each isotope as a function of incident neutron energy. Typically two major energy groups, each containing subgroups, are used by the designer. The first, the fast and epithermal group, spans energies from 1.85 eV to 20 MeV and is composed of 40 or more subgroups. The second, the thermal energy group, covers energies from 0.00001 eV to 1.85 eV and is composed of 50 or more subgroups. In the fast and epithermal energy range, only scattering reactions that decrease neutron energy are used in the calculations. In the thermal energy range, the neutron energies are in thermal equilibrium with the moderator, and energy can be imparted to the neutron during scattering collisions. Therefore, scattering reactions, which can increase or decrease neutron energy, are permitted in the calculations for this group.

When designing a reactor core, specific cross-sections are calculated using the total multigroup cross-section set (40 epithermal plus 50 thermal groups) for each material present. Each cell type in the fuel assembly can then be modeled (Fig. 12). This calculation determines the neutron flux distribution in each energy group and cell. The resulting neutron flux is then used to average the cross-sections over the two major energy groups and over various regions of the fuel assembly. A full core representation, indicating the position of fuel assemblies and control components, is shown in Fig. 13.

The important quantities that describe the reactor core and its performance include critical boron concentration, core power generation distribution, control rod reactivity worths, and the reactivity coefficients associated with changes in reactor conditions.

Critical boron concentration

There is sufficient U-235 in a commercial power reactor to power the unit for 12 to 24 months before refueling. Without additional absorbers present in the system, the k_{eff} would be greater than 1. It is impractical to provide the absorption using control rods. Therefore, for long-term control $k_{eff} = 1$ is maintained by soluble boron in the moderator (boron dissolved in water) and/or burnable absorbers. Burnable absorbers contain a limited concentration of absorber atoms such that, as neutrons are absorbed, the effectiveness of absorption decreases and essentially burns out with time. Burnable absorbers may be in the fuel rods or in fixed absorber rods that are placed in fuel assembly guide tubes. The boron concentration in the coolant can be near 1800 ppm at the beginning of a fuel cycle. This concentration can decrease to near zero at the end of the cycle, when a significant portion of the uranium is depleted and fission products have built up. The critical boron concentration is the boron level required to maintain steady-state reactor power levels. The burnable absorber limits the amount of soluble boron that is used at the beginning of the cycle.

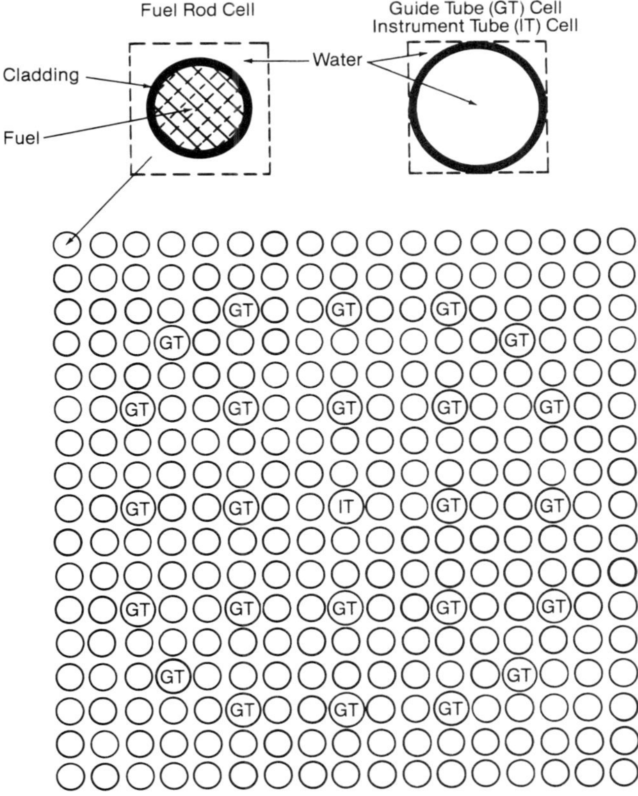

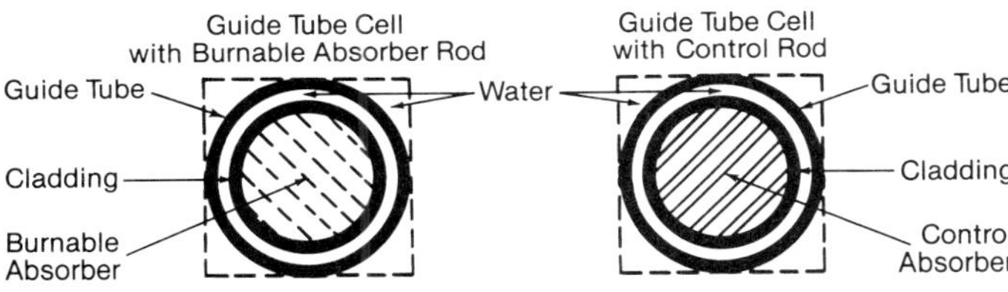

Fig. 12 Cross-sectional view of fuel assembly.

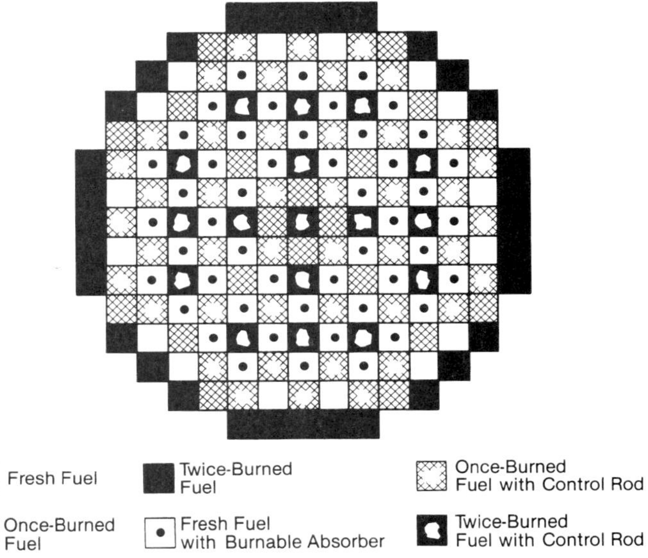

Fig. 13 Cross-sectional computer modeling of PWR core.

Power distribution

The primary factor governing acceptable reactor operation is the energy production rate at every point in each fuel rod. If the rate is too high, the fuel can melt and cause rod failure. Excessive energy rates can also cause coolant steaming, which results in poor heat transfer and can lead to cladding burnout.

The energy production rate, or power, at each fuel rod location is referred to as the core power distribution. This has typically been calculated by forming a three-dimensional model of the reactor and representing each fuel assembly as a homogenized region over an axial interval. A more recent method for power calculations involves reconstructing the power production in each fuel rod based on detailed fuel assembly calculations. This method is more accurate because it provides a better fuel assembly model that can include water, burnable absorbers or guide tube control rods. With proper fuel assembly placement, the core power distribution can be optimized to limit power peaks in fuel rod segments. As part of this primary analysis, initial enrichments of the fuel and burnable absorbers can be determined, as well as the necessary soluble boron concentration in the coolant over the cycle.

Control rod worths

Control rods are used in a pressurized water reactor to change power levels and to shut down the reactor. Approximately 48 rods, divided into banks of 8, are commonly used and core reactivity is changed by sequentially moving the banks. The control rod banks are grouped into shutdown and control (or regulating) banks. The shutdown banks provide negative reactivity to bring the reactor from hot to cold conditions. During a shutdown, the core temperature decreases and its reactivity increases. The shutdown control rod banks counteract this increase. The controlling banks are used in conjunction with variations in the soluble boron concentration to take the reactor from hot zero power to full power. These banks also provide the reactivity to handle rapid power changes. The negative reactivity caused by the buildup of xenon in the core following a shutdown is offset by a combination of control rod removal and decreases in the coolant boron concentration.

The reactivity *worth* of a control rod is a measure of its ability to reduce core reactivity. Control rod worths are calculated by first calculating the core with the control rod inserted to determine its k_{eff}. The core is then similarly calculated with the rod withdrawn. Control rod worth is defined as:

$$\frac{k_{eff(i)} - k_{eff(w)}}{k_{eff(w)}\, k_{eff(i)}} \tag{14}$$

where

$k_{eff(w)}$ = multiplication factor, control rod withdrawn
$k_{eff(i)}$ = multiplication factor, control rod inserted

Bank worths are similarly calculated.

Reactivity coefficients

Reactivity coefficients define the rate of core reactivity change associated with the rate of change in reactor conditions. These coefficients indicate the relative sensitivity of the core operation to changes in operating parameters.

In determining a coefficient, the core is first modeled with a given power and/or temperature distribution, and k_{eff} is calculated. The reactor is then similarly modeled at a different power or temperature distribution. The coefficient is defined as the reactivity change per unit of power or temperature. Fig. 14 shows the impact of various core parameter changes on core reactivity.

There are three basic reactivity coefficients that are important to reactor operation: 1) the moderator temperature, 2) Doppler, and 3) power coefficients. The moderator temperature coefficient is defined as the change in reactivity associated with a change in moderator temperature. It includes the effects of moderator density changes and the changes in nuclear cross-sections. The Doppler coefficient is defined as the change in reactivity associated with changes in fuel temperature that occurs primarily because of fuel nuclides with large resonances. The power coefficient is defined as the change in reactivity associated with a change in power level. It is a combination of the moderator and Doppler coefficients and is based on the change in temperature as power changes.

When the coolant temperature increases, the associated density reduction normally decreases its moderating capability. A net decline in core reactivity, i.e., a negative moderator temperature coefficient, occurs. Similarly, when the fuel temperature increases, core reactivity decreases. Both mechanisms provide an inherent safety response for the system. However, if the soluble boron concentration is too high, a positive moderator temperature coefficient can occur, and core

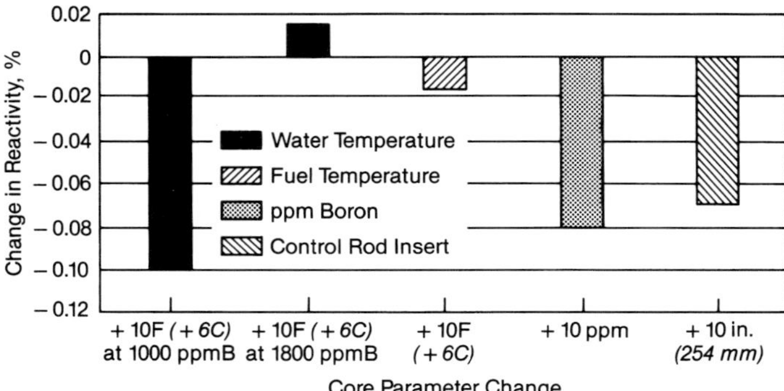

Fig. 14 Core reactivity as a function of operating parameters.

reactivity could increase. For this reason boron concentration is limited to ensure a negative moderator coefficient at full power.

Neutron detectors

When operating a nuclear chain reaction system, neutron detectors are used to measure the intensity of the neutron radiation (flux). This radiation is a direct indication of nuclear reaction intensity. Neutron detectors can be counters or ionization chambers. Counters, which detect neutrons by sensing the individual ionizations they produce, are most useful when the neutron flux is low. Ionization chambers are more useful at high neutron fluxes. They measure the electrical current that flows when neutrons ionize gas in a chamber. These two types of detectors, placed on the exterior of the pressure vessel, only measure neutrons leaking from the reactor.

Two other types of detectors, a self-powered neutron detector and a miniature fission chamber, can be used to measure the neutrons inside the instrument tube of operating fuel assemblies. Self-powered detectors are usually arranged in strings and are positioned to continuously measure the neutron reactions at up to seven axial locations in the fuel assembly. The miniature fission chamber contains uranium, and fission events are detected by the electronics. The fission chamber is moved into and out of the core on a periodic basis, normally monthly, and measures the neutron level along the entire length of the fuel assembly.

Neutron source

During the start of the chain reaction, it is essential that the operator monitor nuclear instrumentation that is counting neutrons and not gamma rays. The mass of fuel in the core is much greater than the critical mass required to sustain a chain reaction. Control rods and soluble boron dissolved in the reactor coolant keep the core subcritical when no power output is required.

When the core is to be brought to criticality, control rods are withdrawn to initiate the chain reaction. However, if the control rods are withdrawn before a measurable neutron flux is available, a reaction could be initiated by a stray neutron and build up to a high power level before the control rods could be reinserted to maintain the desired core output.

To control the rate of buildup of the reaction, it is necessary to have a neutron source present in the core for startup. With a neutron source available, it is possible to measure neutron flux before moving control rods and help ensure a safe reactor startup.

Several types of neutron sources are available. One is an americium-beryllium-curium source rod. The radioactive isotopes americium and curium emit α particles which react with the beryllium to produce neutrons.

Fast reactors

It is possible to design for a chain reaction to occur predominantly with either fast (high energy) neutrons or slow (thermal energy) neutrons. In fast reactors the chain reaction is maintained primarily by fast neutrons. Thermal reactors are those in which the chain reaction occurs primarily from thermal neutrons. Today's commercial water reactors are thermal reactors.

Fertile material such as U-238 can be used as a fuel by first converting it to plutonium in a reactor. Fast reactors and particularly fast breeder reactors accomplish this conversion most effectively. Liquid sodium, which has essentially no moderating effect, is the coolant most often considered for fast breeders. Helium is used as a coolant in high temperature gas-cooled reactor designs.

Conversion and breeding

Most important of the conversion reactions is the capture of a neutron by a U-238 nucleus, resulting in a nucleus of fissionable Pu-239:

$$_{92}U^{238} + {_0}n^1 \rightarrow {_{92}}U^{239} + {_0}\gamma^0$$

$$\beta \qquad \beta$$

$$_{92}U^{239} \rightarrow {_{93}}Np^{239} \rightarrow {_{94}}Pu^{239}$$

$$23 \text{ min.} \qquad 2.3 \text{ d}$$

The U-238 nucleus absorbs a neutron and becomes U-239, which decays with a half life of 23 minutes, into neptunium-239 (Np-239). Again this nuclide, with a half life of 2.3 days, is transmuted to Pu-239 by β decay. This is the nuclear process by which the most useful isotope of plutonium is formed.

There are other reactions during operation which form different isotopes of plutonium. Some of these isotopes, including Pu-241, are fissionable, and others, including Pu-240, are converted to fissionable isotopes by neutron absorption. These isotopes can not be separated economically from Pu-239, and therefore must be taken into account when plutonium is used in a reactor.

The large commercial PWRs in the United States operate on slightly enriched uranium fuel. These reactors convert U-238 to plutonium and produce about 50% as much plutonium as the U-235 consumed. A reactor is considered to be a converter when the amount of fissionable material produced, e.g., plutonium, is less than the amount of fissionable material consumed, e.g., U-235. A breeder reactor is one in which more fissionable material is produced than consumed.

By definition, a breeder reactor must have the value of η (neutrons produced per neutron absorbed in fissionable fuel, Table 5) greater than 2.0. One neutron is required to maintain the chain reaction, one or more for the breeding, and an additional fraction for absorption in nonfuel materials and leakage.

It is not possible to make a breeder reactor with natural uranium fuel, because η is less than 2.0. Table 5 shows that η does not greatly exceed 2.0 with any common fissionable isotopes for fissions produced by thermal neutrons. Consequently, it becomes difficult and impractical to make a breeder with a thermal reactor.

Fortunately, at high neutron energies, η has a greater value than at thermal energies, particularly with Pu-239. For this reason, and because the absorption cross-sections of most materials are less than at thermal energy, fast reactors breed plutonium more effectively than thermal reactors. In addition, fast reactors operate best with plutonium fuel.

References

1. Evans, R.D., *The Atomic Nucleus*, R.E. Krieger Publishing Company, New York, New York, 1982.

2. Moe, H.J., et al., *Radiation Safety Training Technician Training Course*, ANL-7291, Rev. 1, Argonne National Laboratory, Argonne, Illinois, 1972.

3. Firestone, R.B., and Shirley, V.S., Eds., *Table of Isotopes*, Sixth Ed., Wiley-Interscience, Hobocan, New Jersey, 1998.

4. Lamarsh, J.R., *Introduction to Nuclear Reactor Theory*, American Nuclear Society, LaGrange Park, Illinois, 2002.

5. Lamarsh, J.R. and Baratta, A., *Introduction to Nuclear Engineering*, Third Ed., Prentice-Hall, Reading, Massachusetts, 2001.

6. Duderstadt, J., and Hamilton, L., *Nuclear Reactor Analysis*, Wiley Publishers, Hobocan, New Jersey, 1976.

7. Bell, G., *Nuclear Reactor Theory*, R.E. Krieger Publishing Company, New York, New York, 1979.

Inconel is a trademark of the Special Metals Corporation group of companies.

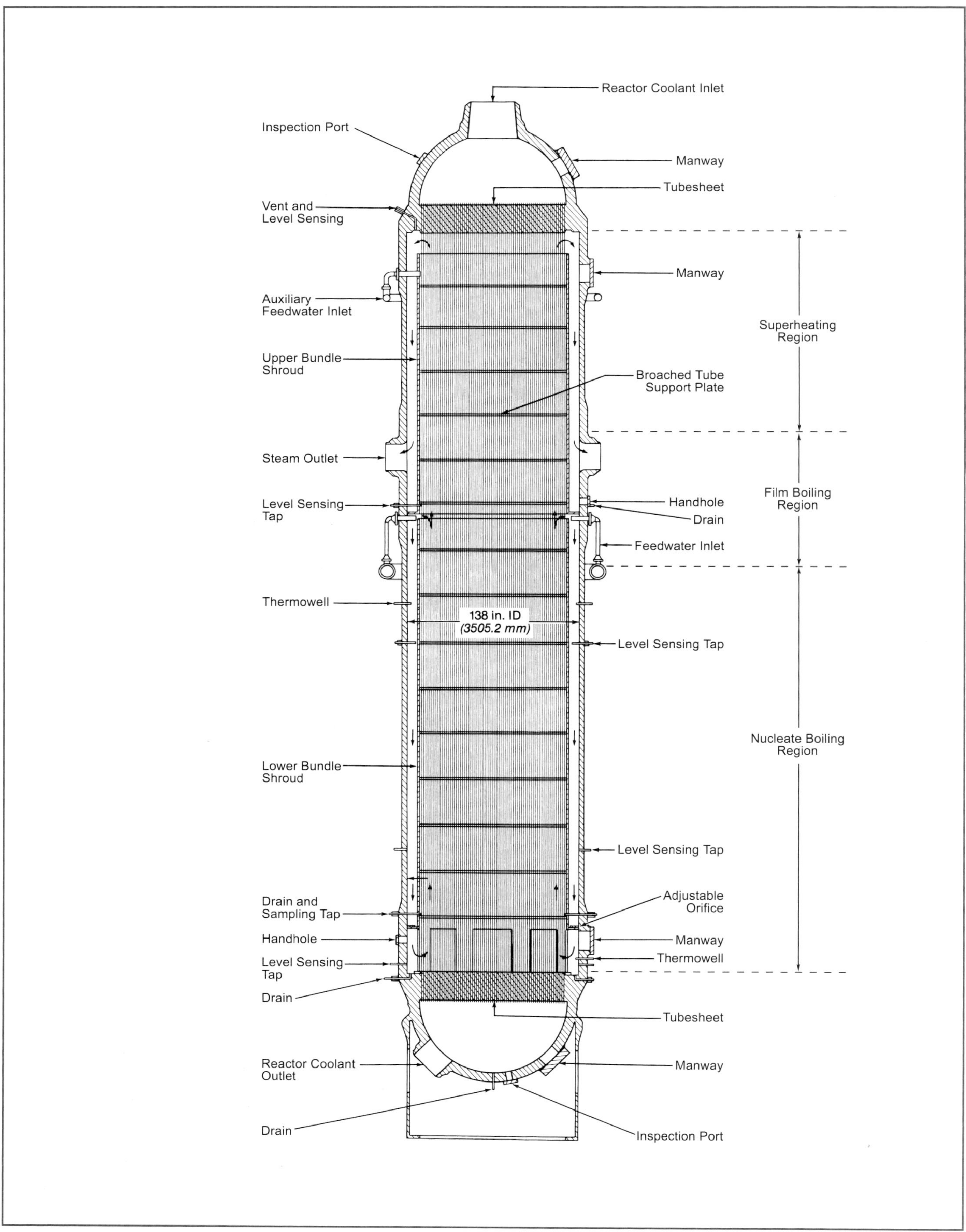

Fig. 1 B&W nuclear once-through steam generator (OTSG).

Chapter 48

Nuclear Steam Generators

Nuclear steam generators provide the principal isolation barrier between the primary coolant system in contact with the reactor core and the secondary power cycle system, as discussed in Chapter 46. Nuclear steam generators also serve two other functional requirements as they transfer energy from the primary reactor coolant system to the secondary-side power-producing system. The first functional requirement is as the heat sink for the reactor core. The decrease of primary coolant temperature through the steam generators is virtually the same as the increase of the primary coolant temperature through the reactor core. Any difference is attributable to the fluid energy supplied by the reactor pumps, less any heat loss to ambient. The second functional requirement is to generate the flow rate of steam from the feedwater supply at the temperature, pressure and enthalpy conditions necessary to efficiently drive the steam turbine/electric generating system.

Each commercial pressurized water reactor (PWR) power system combines two to four steam generators with a nuclear reactor, pressurizer, and multiple primary coolant pumps to form the nuclear steam supply system (NSSS). The 69 operating PWR systems in the United States (U.S.) use a total of 209 steam generators. In addition, the 22 Canada Deuterium Uranium (CANDU) pressurized heavy water reactor (PHWR) systems installed in Canada employ a total of 184 with as many as 12 steam generators per reactor, although 4 is more typical.

There are two fundamental steam generator designs used in commercial power systems in North America. The Babcock & Wilcox Company (B&W) once-through steam generator (OTSG) is a vertical shell counterflow straight-tube heat exchanger design which directly generates superheated steam as the feedwater flows through the steam generator in a single pass. The other fundamental design used by a number of suppliers and in the CANDU PHWR is a recirculating steam generator (RSG) where only part of the feedwater is converted to steam as the water passes through the unit. After the steam is separated from the water, the steam is sent to the turbine for power generation while the water is returned to the tube bundle for additional steam generation. These steam generator designs use a vertical shell and an inverted U-tube heat exchanger bundle, with steam separation equipment located inside the top shell of the RSG. With this design there are significant differences between suppliers and between different models regarding materials of construction, thermal hydraulic parameters, manufacturing techniques and tube supports, all of which can significantly influence performance.

A few other commercial PWR steam generator designs have been used, primarily in early units. The first commercial U.S. power reactor at Shippingport, Pennsylvania, used two different horizontal tube bundle steam generator designs. NPD (Nuclear Power Demonstration), the first CANDU power reactor, had one horizontal U-tube, U-shell steam generator. Current Russian PWR plants use horizontal bundle steam generators. No horizontal units are in commercial operation today in North America.

A variety of aging issues have been observed in the existing fleet of PWR steam generators. These factors have reduced the performance of some steam generators and thus effectively reduced their useful life. Intergranular attack, stress corrosion cracking, tube denting, tube wall thinning, fretting wear damage, foreign object damage and other issues have all had an impact on unit performance because of the need to plug significant numbers of steam generator tubes.[1] In some cases, the degradation has reached a stage where repair is no longer a viable option and steam generator replacement is needed. These replacement steam generators employ new technologies to address the aging issues and to extend their useful life.

The balance of this chapter focuses on the fundamentals of steam generator design and on the design improvements that are incorporated into the replacement OTSG and RSG units. Steam generator service issues are discussed in Chapter 49, *Nuclear Services and Operations*. Steam generator manufacturing is discussed in Chapter 50.

B&W once-through nuclear steam generator (OTSG)

Design configuration and flow

The B&W once-through nuclear steam generator (see Fig. 1) is a vertical, straight tube and shell, counterflow heat exchanger, approximately 73 ft (22.25 m) tall and 13 ft (3.96 m) in diameter, each weighing approximately 570 tons (517 t_m). Each steam generator produces approximately 5.4 million lb/h (680.4 kg/s) of steam at 925 psi (6.38 MPa) and 595F (313C). This provides 60F (33C) of superheat. The more than 15,000 Inconel® 600 tubes in each steam generator are positioned by broached tube support plates (see Figs. 1 and 2) which are set within the bundle shroud, which is in turn positioned to the vessel shell by shroud pins. These plates are spaced to minimize tube vibration. The distance between tubesheets is 52.1 ft (15.9 m).

As shown in Figs. 1 and 3, primary coolant enters at the top, flows downward through the tubes, and exits at the bottom through two primary outlets. The carbon steel shell surfaces exposed to the primary side coolant are clad with stainless steel to avoid corrosion in the primary side chemistry environment. The generator functions as a counterflow heat exchanger, i.e., the primary coolant flows downward within the tubes and the secondary fluid flows upward around the tubes. The steam generator is a once-through design in that the secondary fluid makes only one pass through the unit. Because the OTSG produces superheated steam, it does not require steam separators or dryers.

Referring to Figs. 1 and 3, the feedwater flows downward within the annular region formed by the lower shroud and the steam generator inside wall. Near the bottom of the downcomer the feedwater passes through the adjustable orifice plate, and then turns inward toward the bottom of the bundle. The fluid then flows upward around the tubes, passing

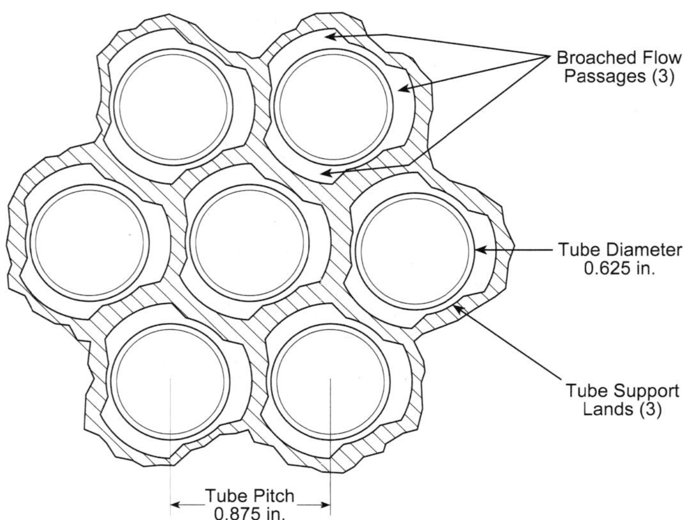

Fig. 2 OTSG broached support plate.

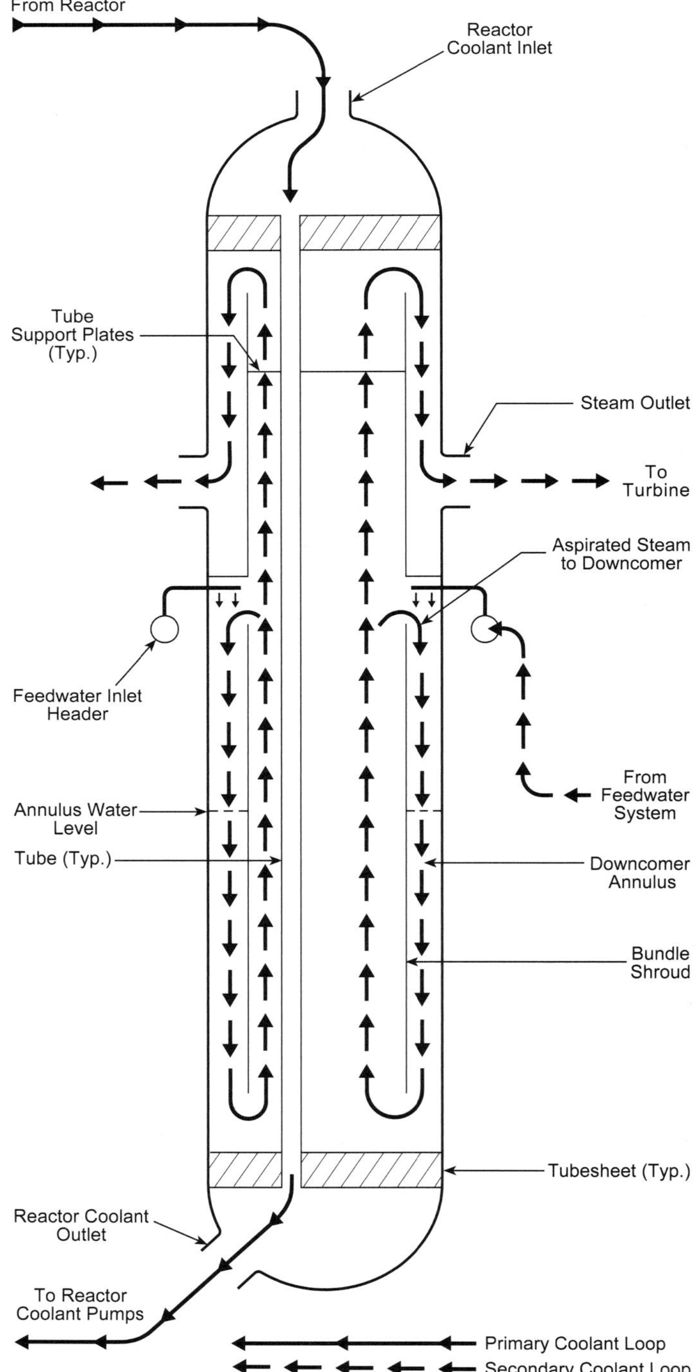

Fig. 3 Simplified OTSG flow circuits.

through broached holes in the 15 tube support plates. The steam is superheated prior to leaving the top of the tubed region. It turns outward and downward, flowing down the upper annular downcomer to the steam outlet nozzles.

The feedwater is introduced into the annular downcomer through thirty-two 3 in. (76 mm) connections with attached spray nozzles located 31.9 ft (9.7 m) above the lower tubesheet. The exit flow from these nozzles aspirates steam through a gap in the bundle shroud, thereby allowing the feedwater to reach saturation temperature before entering the tube bundle. This arrangement was designed to mitigate temperature mismatches between the shell and tube bundle. Later integral economizer once-through steam generator (IEOTSG) designs introduced feedwater through two nozzles lower in the shell where it directly enters an integral economizer section in the lower portion of the tube bundle region as shown in Fig. 4. With this design, external feedwater heating is used at some loads to adjust the feedwater temperature to approximate tube bundle and shell temperatures, thereby avoiding excessive temperature differentials between shell and tube bundle and within the lower tubesheet. This design is able to achieve more steam output at a higher pressure to match the output of the 1300 MW 205 fuel bundle core. Most of the discussion which follows relates to the standard OTSG design.

In the OTSG design, an adjustable orifice is located near the bottom of the downcomer, just above the ports into the tubed region. This device consists of a movable plate overlaying a fixed plate; both plates contain 24 identical flow passages. This adjustable orifice allows downcomer flow resistance to be adjusted to optimize flow stability and downcomer water level. These factors may change with operation as tube support plate deposits increase riser-side flow resistance.

Key information about the OTSG design is provided in Table 1.

Modes of secondary-side heat transfer

The secondary-side feed entering an OTSG downcomer annulus is quickly brought to saturation temperature by contact heating between the feedwater spray and aspirated steam drawn from the riser (in an IEOTSG it is heated by convective heat transfer as it begins to pass upward within the tube bundle). The remaining heat transfer is by nucleate boiling, film boiling and superheating. Nucleate boiling which begins at the bundle entrance in the OTSG (and at the top of the preheater zone in the IEOTSG) involves the formation of discrete vapor bubbles on the tubes and the release of these bubbles into the secondary flow. This combination of phase change and bubble motion creates very efficient heat transfer, in contrast to film boiling and superheating which both involve heat removal from the tube by steam convection and, consequently, are much less efficient. (See Chapter 5.) As a result, these latter modes occupy a large part of the bundle surface to transfer a relatively small portion of the heat load. The ratio of steam mass to total fluid mass (i.e., steam quality) increases as the secondary fluid flows upward. As the steam quality

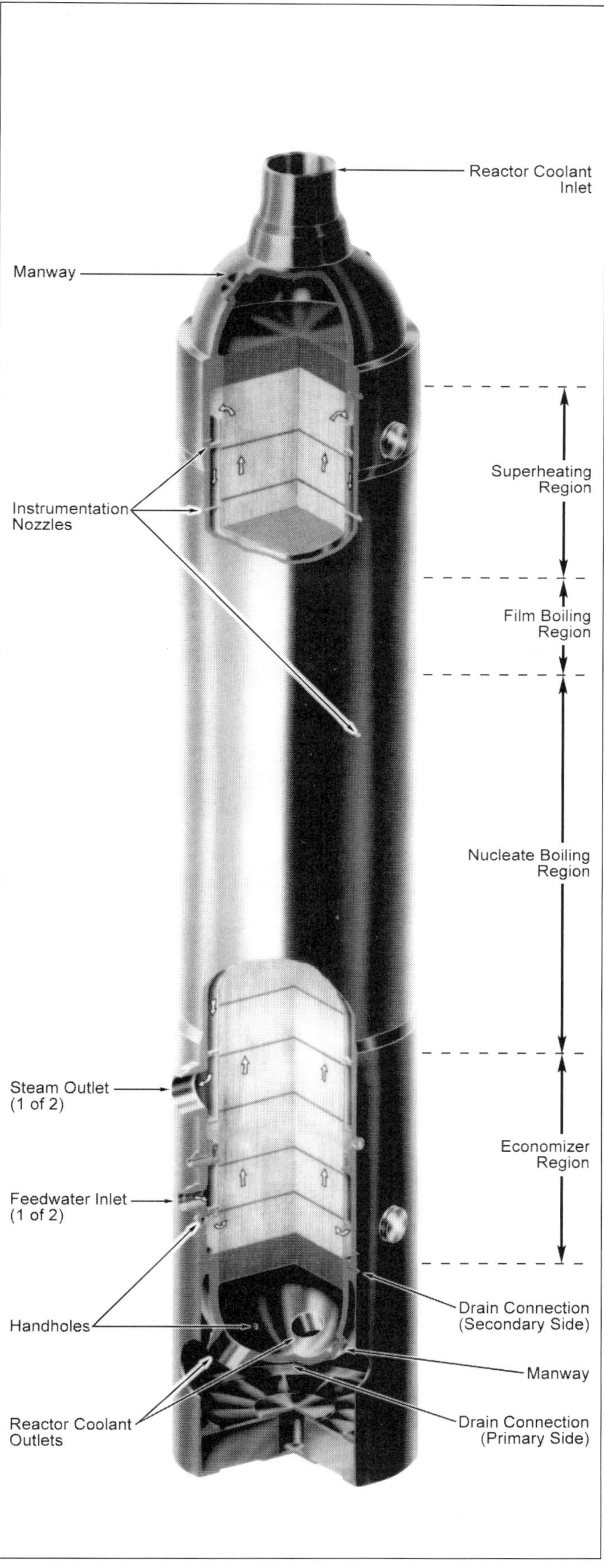

Fig. 4 Integral economizer once-through steam generator (IEOTSG).

Table 1
Once-Through Steam Generator (OTSG) Operating and Design Information

	OTSG Oconee (177)
Plant Information	
Power level, MW_t	2568
Power level, MW_e	846
No. steam generators	2
Steam Generator Design Conditions	
Primary T (Hot), F (C)	604 (317.7)
Primary T (Cold) F (C)	554 (290.0)
Primary flow, lb/h (kg/s)	65.66×10^6 (8273.16)
Primary operating pressure, psia (MPa)	2200 (15.17)
Secondary steam temperature, F (C)	595 (312.8)
Secondary feedwater temperature, F (C)	460 (237.8)
Steam flow, lb/h (kg/s)	5.4×10^6 (680.4)
Secondary operating pressure, psia (MPa)	925 (6.38)
Steam Generator Design Data	
No. tubes per generator	15,531
Tube OD, in. (mm)	0.625 (15.875)
Tube minimum wall thickness, in. (mm)	0.034 (0.864)
Tube material	Inconel® 600
Surface area, ft^2 (m^2)	133,000 (12,356)
Tube pitch, in. (mm)	0.875 (22.225)
No. support plates	15
Support plate type	Broached
Support plate material	Carbon steel
Integral economizer	No

reaches the high nineties, there is insufficient liquid to wet the tubes and nucleate boiling gives way to film boiling. Still farther up the generator where the moisture content of the fluid is depleted and the steam begins to superheat, heat transfer is by forced convection of vapor. The primary and secondary fluid temperatures and the modes of secondary heat transfer are indicated in Figs. 5 and 6. The secondary-side temperature remains at saturation through the nucleate and film boiling regions. The counterflowing primary fluid gradually cools as it flows through the steam generator. The drop in primary fluid temperature in the upper areas of the unit is relatively slow due to the less efficient heat transfer in the film boiling and superheater regions. In the lower regions of the unit, the primary fluid temperature change is much more rapid due to more efficient nucleate boiling heat transfer.

Instrumentation

The steam generator thermal performance measurements include fluid pressure, temperature, and level. Shell temperatures are measured at five equally spaced locations along the length of the generator. These measurements are used to determine tube to shell temperature differentials. A steam generator downcomer temperature measurement indicates the temperature of the incoming feed; this temperature is used in the temperature compensation of level. There are three ranges of steam generator level measurement: full, operate and startup. The operate range begins 8.5 ft (2.59 m) above the lower tubesheet. It is graduated in percentage of full (operating) range, from 0% at 102 in. (2.6 m) above the lower tubesheet to 100% at a height of 394 in. (10.0 m). Whereas the common upper tap of the startup and operate range levels extends into the tubed region, their lower taps are in the steam generator lower downcomer. Feed and steam system measurements also indicate steam generator status. These include steam pressure and the pressure drop across the feedwater control valves. The flow rate of main feedwater is indicated in two ranges: startup to 15% of full flow range, and operating range. The majority of these steam generator measurements are supplied to the plant computer and the integrated control system.

Recirculating steam generator (RSG)

Design configuration and flow

Recirculating steam generators are supplied by a number of manufacturers worldwide as part of PWR and PHWR (mainly CANDU) systems. Two, three or four steam generators are supplied for each PWR system and 4 to 12 steam generators are supplied for each CANDU system. They range in height from approximately 38 to 73 ft (11.6 to 22.3 m) weighing from approximately 50 to 790 tons (45 to 717 t_m) each. See Fig. 7 for selected comparisons of sample units.

Each RSG is a vertical shell, inverted U-tube heat exchanger with steam-water separation equipment located above the tube bundle inside the upper shell (or steam drum). A cylindrical shroud or bundle wrap-

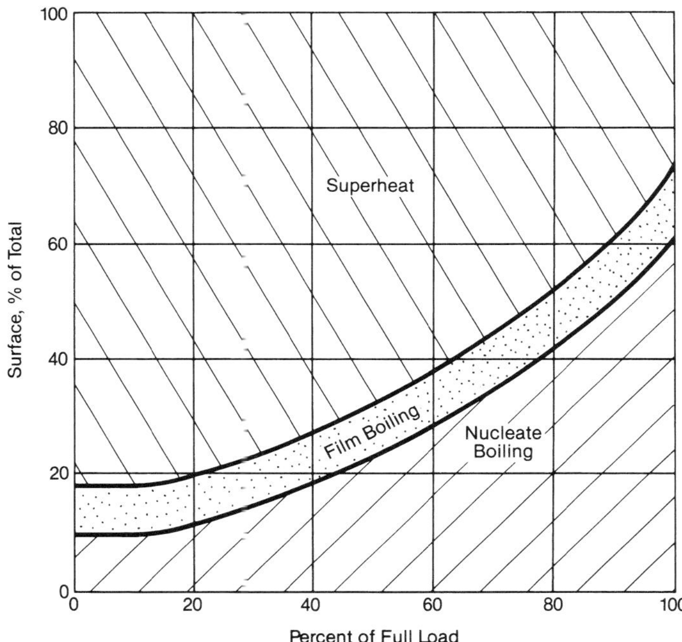

Fig. 5 Steam generator heat transfer for an OTSG.

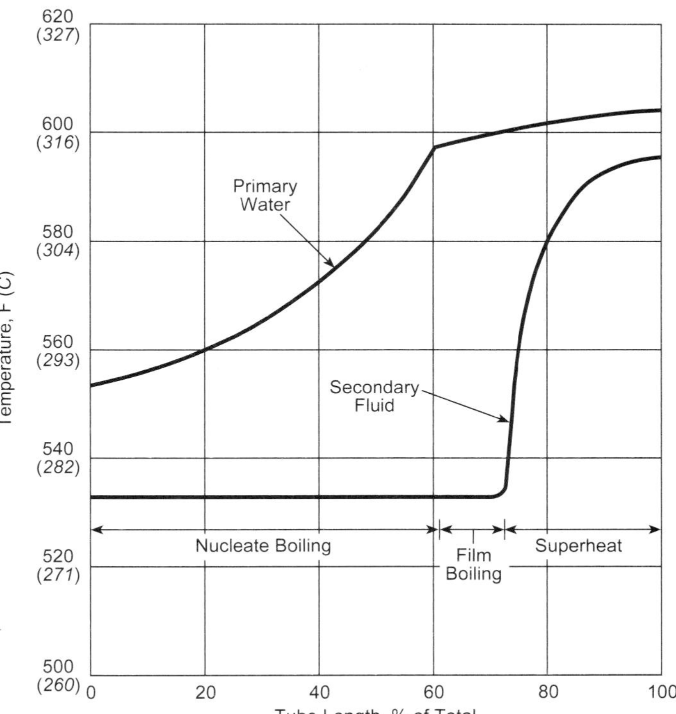

Fig. 6 Approximate primary and secondary temperature profiles at full load for an OTSG.

per surrounds the tube bundle separating it from the lower shell. This creates an annular region which serves as the downcomer to return the recirculated water from the steam separators to the tube bundle inlet at the bottom of the unit. In a feed ring type RSG, feedwater is introduced by a nozzle and header to the top of the downcomer, and flows with the separator return flow down and into the tube bundle. In a preheater type RSG, feed flow enters the steam generator through a nozzle and feedwater distribution box to the baffled section at the cold leg outlet end of the tube bundle where it is heated to saturation before joining with the hot leg riser flow within the tube bundle. A typical RSG supplied by B&W to replace an original equipment manufacture (OEM) unit is shown in Fig. 8.

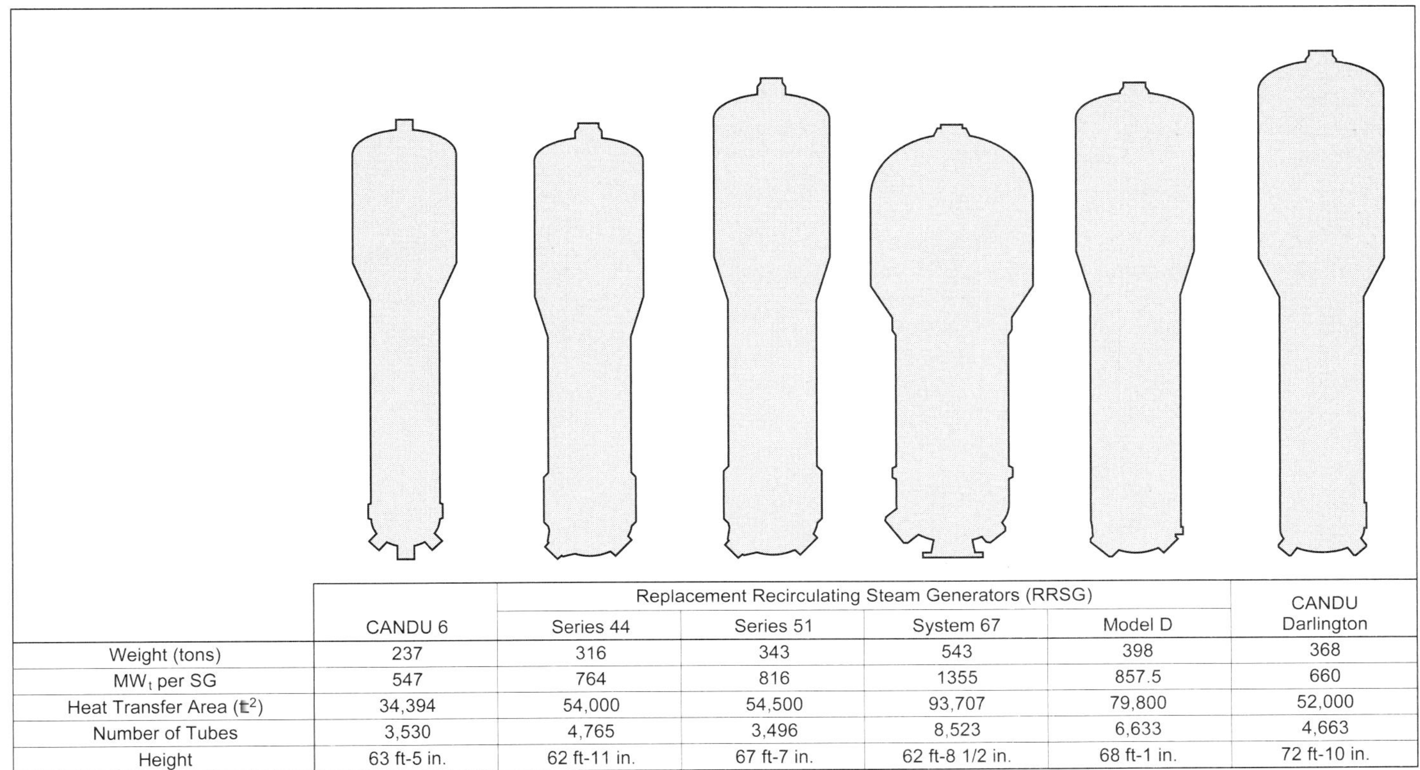

	CANDU 6	Replacement Recirculating Steam Generators (RRSG)				CANDU Darlington
		Series 44	Series 51	System 67	Model D	
Weight (tons)	237	316	343	543	398	368
MW_t per SG	547	764	816	1355	857.5	660
Heat Transfer Area (ft^2)	34,394	54,000	54,500	93,707	79,800	52,000
Number of Tubes	3,530	4,765	3,496	8,523	6,633	4,663
Height	63 ft-5 in.	62 ft-11 in.	67 ft-7 in.	62 ft-8 1/2 in.	68 ft-1 in.	72 ft-10 in.

Fig. 7 Relative sizes of recirculating steam generators (RSG).

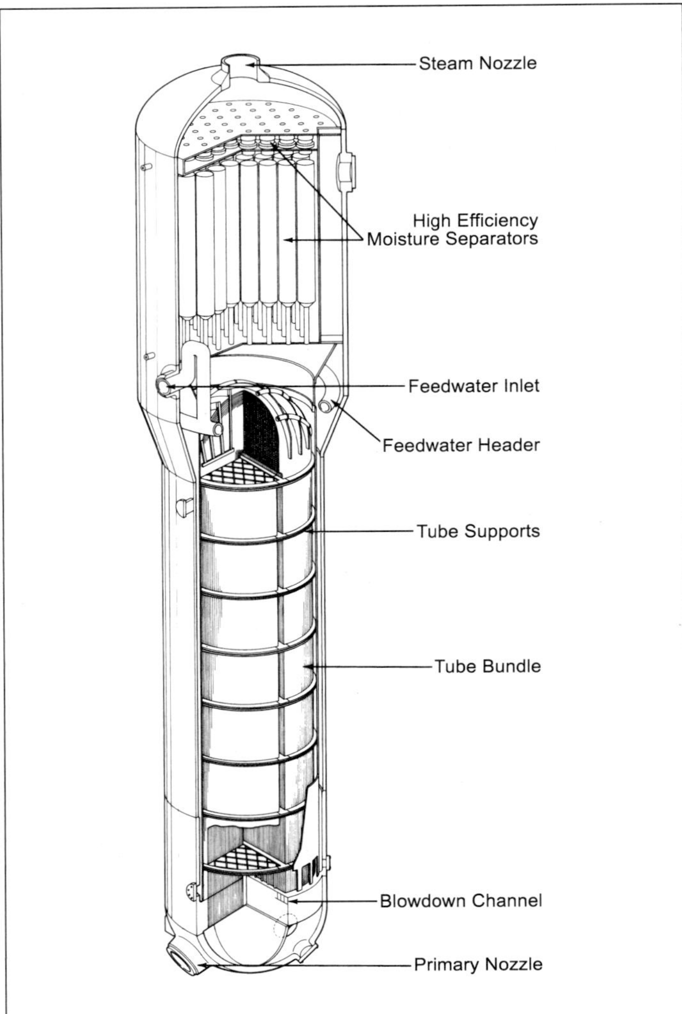

Fig. 8 B&W replacement recirculating steam generator.

A simplified schematic shown in Fig. 9 provides an overview of the flow configuration and the major design features of a typical feed ring type RSG. The hot primary coolant enters a portion of the vessel primary head which is separated into two plenums by a divider plate. The primary coolant flows through the inside of the U-tube bundle and exits the steam generator through the primary head outlet plenum. In most RSG designs, the U-tubes make a continuous 180 degree bend at the top of the tube bundle; one major OEM design uses two 90 degree bends separated by a horizontal section at the top of the tube bundle.

In the configuration shown, secondary-side feedwater enters the upper shell via a feed ring and is mixed with water returning from the steam-water separation equipment. The water flows down the downcomer annulus between the shroud and the shell to the tubesheet where it enters the tube bundle. The secondary-side water is heated as it passes up through the tube bundle generating steam through nucleate boiling heat transfer, creating a two-phase flow. Steam of 10 to 40% quality, depending on hot-side or cold-side U-tube bundle location, exits the tube bundle and is distributed to the primary and secondary steam separators in the upper shell to send effectively moisture-free (<0.25% water) steam to the secondary-side power cycle. Water leaving the steam separators is recirculated down the annulus where it mixes with the feedwater before being returned to the bundle inlet for further steam generation.

The natural recirculation within the RSG is driven by the density difference between the higher density water column in the downcomer annulus area and the lower density steam-water mixture in the tube bundle. The RSG recirculation rate is set by the balance between the pressure drop through the flow path (downcomer, tube bundle and steam separation equipment) with the driving head created by the density differential (see Chapter 5). To minimize the potential

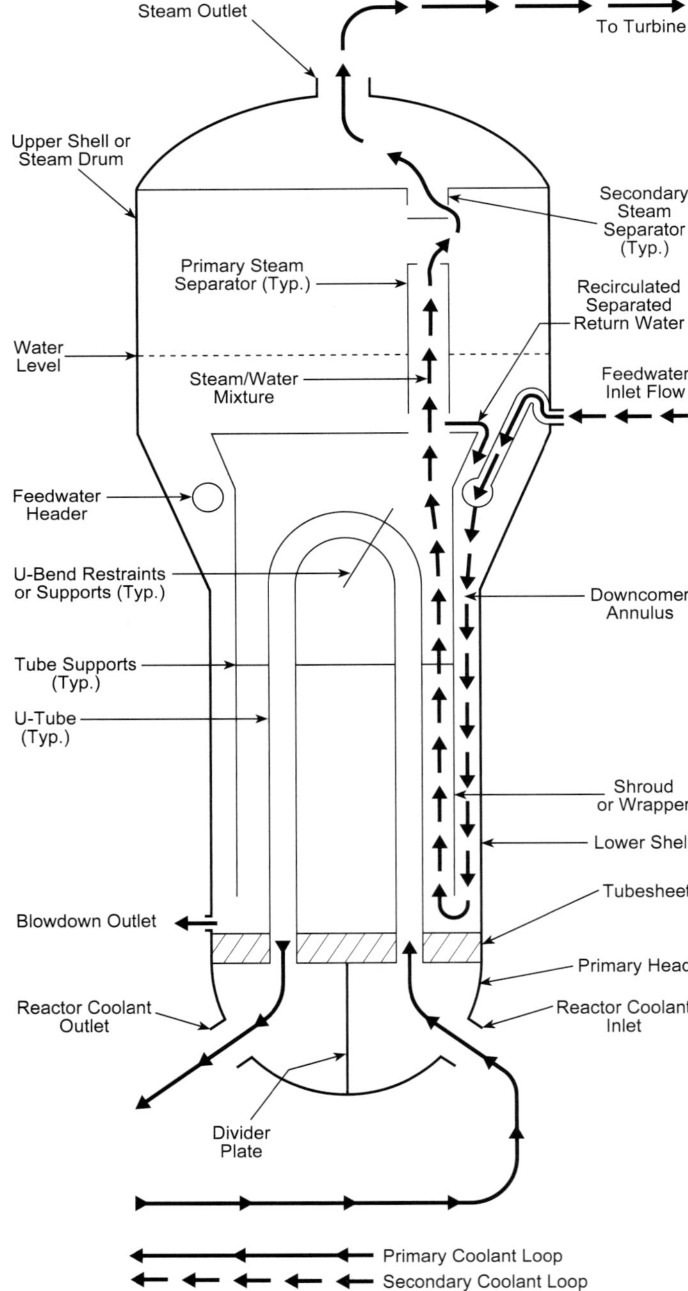

Fig. 9 Simplified recirculating steam generator (RSG) flow circuits.

buildup of any contaminants, a blowdown system is located in the bottom of the RSG near or as part of the tubesheet to remove a controlled quantity of water inventory. Blowdown is not very efficient at removal of many feedwater contaminants but it is the only way, other than moisture carryover, for feedwater contaminants to be removed from the steam generator during operation. Contaminants remaining in the steam generator can deposit onto the tubes and to a lesser degree build up on the tubesheet and tube supports.

Preheater (integral economizer) designs are also used for some RSGs. Here the feedwater is introduced near the tube bundle cold leg outlet to increase overall thermal efficiency. Several options have been used for this integral economizer section: counterflow, axial flow and split flow (cold leg side of the tube bundle). Simplified schematics are shown in Fig. 10.

Horizontal tube supports (support plates and/or lattice bar grids) are provided at various elevations along the vertical straight tube bundle section (see Fig. 8). A variety of supports have been used in the various original equipment designs (see Fig. 11). The tube supports provide lateral support to the tubes during steam generator manufacturing, shipment and operation and protect against seismic loads. They are spaced at intervals to minimize the likelihood of flow-induced vibration and are designed to provide adequate open area for the recirculating flow. Many of these designs were prone to accelerated corrosion, particularly if made of carbon steel (see Reference 1). B&W's advanced lattice grid tube supports (Figs. 12 and 13) have addressed many of these issues.

Additional tube supports or restraining systems are required in the upper bundle U-bend area (see Fig. 14). These maintain the position of the tubes and prevent excessive flow-induced vibration as the high velocity steam-water mixture flows out through the U-bends.

Key information on the various RSG designs is provided in Table 2.

Steam-water separation

High efficiency steam-water separation is a vital function in all recirculating steam generators, as discussed in Chapter 5. This is particularly true in nuclear RSGs. Low moisture carryover in the steam improves turbine efficiency and total power output, and minimizes the carryover of any contaminants into the turbine. The traditional limit for nuclear RSG moisture carryover at full power is less than 0.25% moisture by weight, while more recent requirements have been for 0.10% or less.

Low steam carryunder (steam entrainment) within the downcomer return water maximizes the downcomer annulus driving head, thereby maximizing the internal circulation rate. Low separator pressure drop also increases the natural circulation rate through the bundle by lowering the overall flow resistance.

The B&W steam separation system combines curved arm primary (CAP) separators (see Chapter 5, Fig. 23) with cyclone secondary separators (see Fig. 15) to accommodate the necessary steam throughput.[2] The actual moisture carryover as measured at steam generator startup has been well below 0.10% moisture by weight in all B&W replacement RSGs.

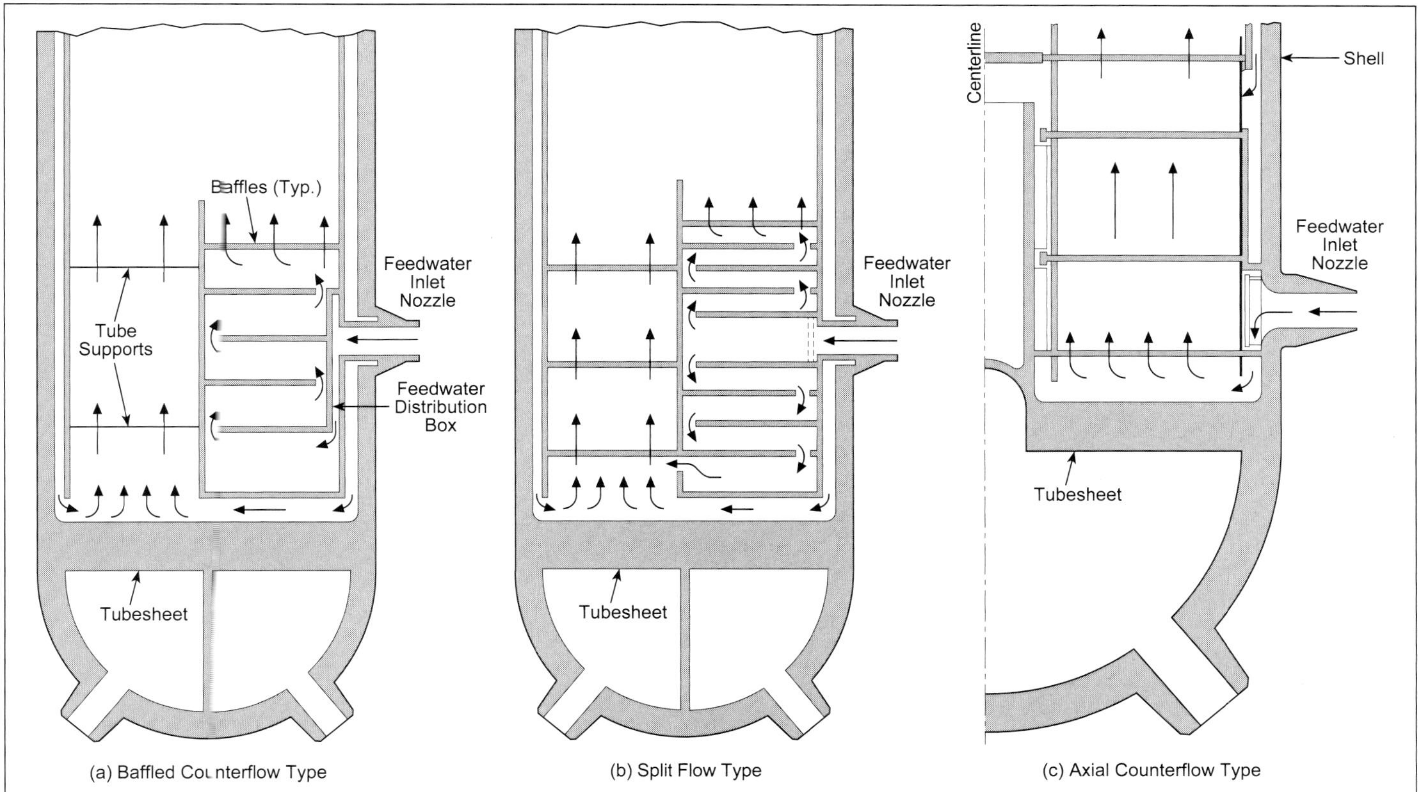

Fig. 10 Integral economizer designs used in RSGs.

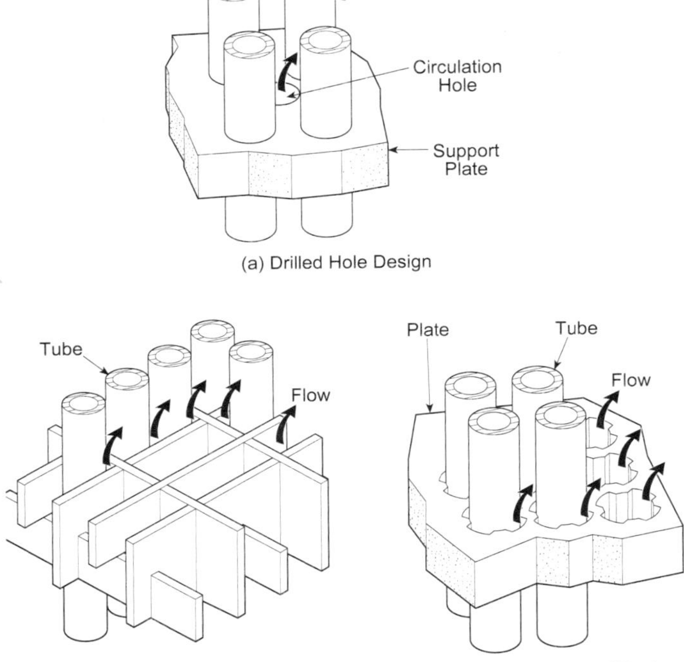

Fig. 11 Original RSG support plate designs by various manufacturers.

Circulation and hydraulic performance

Detailed thermal-hydraulic design analyses are conducted on all RSG designs and include, among others:

1. heat transfer analysis on a local basis,
2. three-dimensional analysis of flow (velocities, pressures, qualities, etc.),
3. dynamic response of the steam generator to normal and abnormal transients,
4. potential deposit accumulation on the secondary side of the RSG,
5. recirculation rate within the tube bundle, and
6. flow-induced loadings on internal components.

The circulation ratio (total flow rising through the tube bundle divided by the exiting steam flow) is a particularly important parameter. Higher circulation rates reduce the overall steam voidage in the tube bundle and increase the local velocities. Reduced voidage improves the steam drum water stability during transients. A higher circulation ratio also increases the margin for stability, and increases the margin for tube surface dryout and for critical heat flux in the tube support areas. It also reduces low velocity areas, especially near the tubesheet where deposits may accumulate.

Circulation rate is increased by low flow path pressure drop at the tube supports and the steam separators, and by low steam carryunder from the steam-water separators to the downcomer. High circulation

Fig. 12 B&W lattice grid tube support assembly.

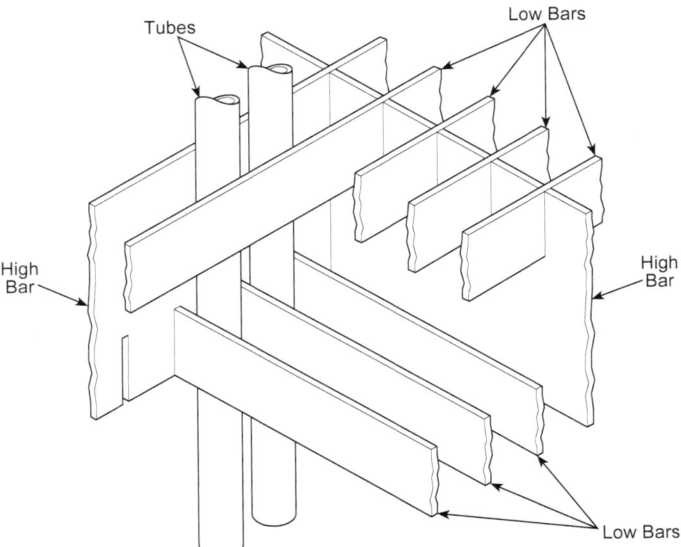

Fig. 13 B&W lattice grid tube support design details.

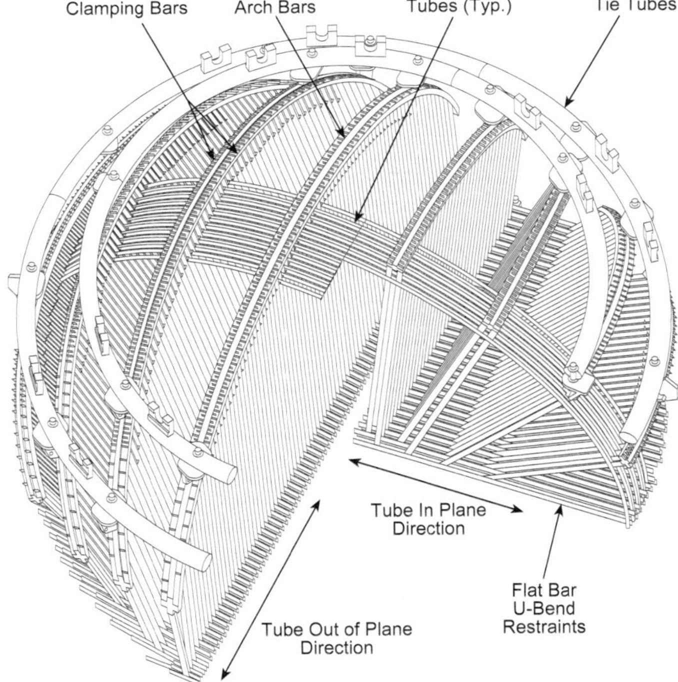

Fig. 14 B&W flat bar U-bend restraint system.

Table 2
Selected OEM Recirculating Steam Generator Operating and Design Information
(Reference 3, Supplemented)

RSG Series Model or Type		System 67	System 80	Series 44	Series 51	Model F	Model D2/D3
Plant Information							
Plant capacity	MW_t	2,700	3,817	1,520/2,200/3,000	1,650/2,660/3,300	3,425	2,800/3,411
Plant capacity	MW_e	865	1,270	470/700/1,000	560/833/1,100	1,150	930/1,180
No. steam generators		2	2	2/3/4	2/3/4	4	3/4
Steam Generator Operating Conditions (per steam generator)							
Primary T (Hot)	F	594.0	621.2	603.9	599.1	620.0	619.0
Primary T (Cold)	F	548.0	564.5	543.1	535.5	557.0	558.0
Primary flow	10^6 lb/h	74.00	82.00	32.95	34.1	37.60	35.10
Primary operating pressure	psig	2,235	2,235	2,235	2,235	2,235	2,235
Secondary steam temp.	F	530.0	554.6	511.7	513.1	541.0	546.5
Secondary feed temp.	F	435.0	450.0	425.0	432.0	440.0	
Steam flow	10^6 lb/h	5.90	8.59	3.30	3.58	3.95	
Secondary operating pressure	psig	865	1,070	740	750	955	1,000
Steam Generator Design Data (per steam generator)							
No. tubes		8,519	11,000	3,260	3,388	5,626	4,674
Tube OD	in.	0.750	0.750	0.875	0.875	0.6875	0.750
Tube wall	in.	0.048	0.042	0.050	0.050	0.040	0.043
Tube material		Inconel® 600	Inconel 600	Inconel 600	Inconel 600	Inconel 600	Inconel 600
Surface area (OD)	ft^2	90,675	124,800	44,430	51,000	55,000	48,300
Tube pitch	in.	1.000	1.000	1.234	1.280	0.980	1.0625
Tube pitch		Triangular	Triangular	Square	Square	Square	Square
No. of support plates		6	8	6	7	7	8
Support type		Egg crate	Egg crate	Drilled	Drilled	Quatrefoil	Quatrefoil
Support material		Carbon steel	Carbon steel	Carbon steel	Carbon steel	Ferritic Stainless steel	Carbon steel
Integral economizer		No	Yes	No	No	No	Yes

Note: SI Conversions: C = (F–32) x 5/9; mm = in. x 25.4; $m^2 = ft^2 \times 0.09290$; kg/s = lb/h x 0.000126; MPa = psi x 0.006895

ratios (up to 6.0) have been a significant feature in B&W designs.

Replacement steam generators

Of the 69 operating PWR nuclear power plants in the U.S., 56 had 133 individual steam generators replaced or awarded by 2005. Additional RSG replacements are anticipated as the remaining installations age.

The previously unanticipated aging issues experienced by nuclear steam generators generally involve complex interactions between the materials of construction, water chemistry operation and contaminant ingress, tube support arrangements, fabrication techniques, and local thermal-hydraulic conditions. These only became apparent with long-term operation of the equipment. Examples of this damage include:

1. intergranular attack,
2. stress corrosion cracking (SCC),
3. primary- and secondary-side cracking at the tubesheet expansion transitions,
4. primary-side cracking at the U-bends,
5. wall thinning/wastage/pitting involving sludge near the tubesheets,
6. denting or permanent deformation of the tubes at the carbon steel support plates due to corrosion product buildup and volume expansion,
7. fretting wear,
8. tube circumferential cracking due to high frequency fatigue,
9. erosion-corrosion in higher velocity regions, and
10. foreign object damage.

In many cases, the affected tubes are plugged when significant leakage is detected or when degradation reaches a critical level. However, with time the overall bundle degradation can reach a stage where repair is no longer a viable option and steam generator replacement is needed. References 1 and 3 explore this in more depth.

B&W's advanced series replacement steam generators are designed with the most current technology available to overcome the steam generator issues while still meeting or exceeding all of the steam generator functional requirements. Replacement units must fit within the

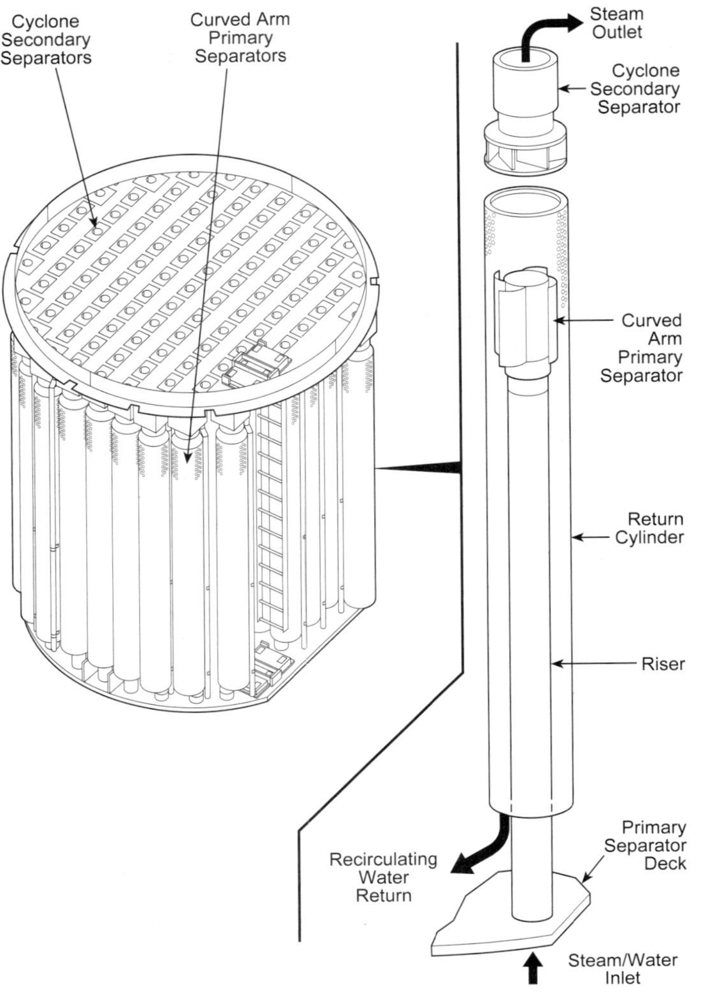

Fig. 15 Primary and secondary RSG steam separator system.

established plant configuration, and must meet all performance and operational requirements. In addition, these replacement units must be redesigned to minimize or eliminate the original problematic degradation.

Replacement steam generators are also designed to support extension of power plant operating life. In many cases, the plants with replacement units have received extended licenses to operate 20 years beyond the original plant life span and may eventually operate even further. To achieve these objectives, while combating the causes of component degradation, emphasis is placed on upgrading component materials, introducing advanced design features and manufacturing methods, and incorporating the latest technical developments.

Replacement recirculating steam generators (RRSG)

The contract for the first replacement nuclear steam generator supplied by B&W was awarded in 1988, for Northeast Utilities' Millstone Unit 2 nuclear power plant. Since then and through 2004, B&W has designed and manufactured a total of 6 once-through and 34 replacement recirculating nuclear steam generator units for the U.S. market.

All of the RSG units are similar but are adapted to the requirements of the specific power plant. Significant features are shown in Fig. 8. While the basic envelope of each RSG type remained the same as the respective OEM equipment, major changes were made to the tubing and tube support materials. Design features were included to enhance reliability and facilitate in-service inspection and maintenance (see Table 3).

Tubing material The majority of critical degradation mechanisms involved the tube bundle and the tube supports; therefore, the selection and quality of the tubing are the most important factors in steam generator design. The original equipment tube material was mill-annealed (MA) Alloy 600 (Inconel: Ni-Cr-Fe alloy). However, experience has shown this material to be susceptible to stress corrosion cracking (SCC), especially in areas of cold work such as at the original tubesheet roll expansion transition regions (see Fig. 16) and in areas of harsh environments such as the top of the tubesheet. The replacement units feature

Table 3
Replacement Recirculating Steam Generator Design Features

Features	Original Design	Replacement Design	Improvement
Tube material	Alloy 600MA	Alloy 690TT	Corrosion resistance
Surface area	Nominal	Increase	Accommodates reduced conductivity and elimination of preheater Provides power uprate capacity and/or steam pressure increase
Tube support plate material	Carbon steel	Type 410S	Corrosion resistance
Tube support plate design	Drilled/broached	Lattice bars	Improves flow area Minimizes crevices
Divider plate	Floating	All welded	Minimizes bundle bypass
Drum internals	Original	Upgraded by test and field experience	Improves water inventory and performance characteristics

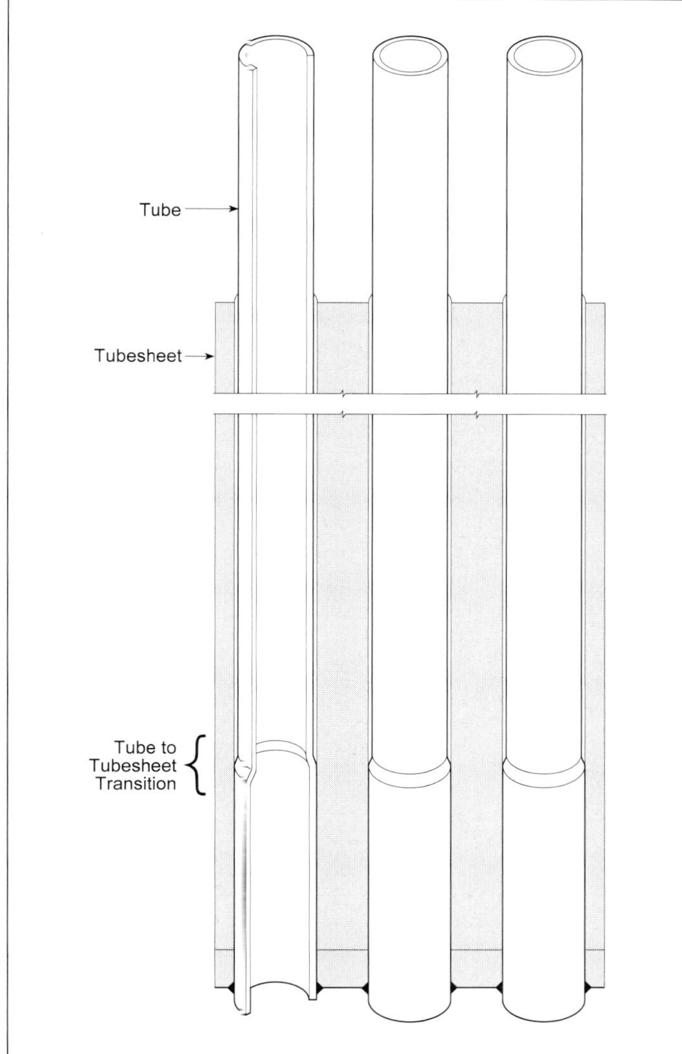

Fig. 16 Original equipment tube to tubesheet expansion transition.

Alloy 690TT (thermally treated) tubing which has a high resistance to corrosion. This replacement tubing is particularly resistant to SCC due to its higher chrome content and thermal treatment.

Tubing today is manufactured to very stringent specifications, resulting in a very high quality product. The new metallurgical treatment offers a material grain structure that is much more resistant to cracking. In addition, these tubes are metallurgically very clean and have a minimum of defects and inclusions. As a result, these tubes have very low eddy current noise, thus improving the resolution of in-service eddy current testing (ECT) inspections. Any small, tight cracks or other defects that may occur will be found more easily.

General design features The first requirement in the design of any replacement steam generator is the selection of a general architecture which will, as closely as possible, match that of the original unit. This is necessary to comply with Nuclear Regulatory Commission (NRC) requirements and ensure licensing. The pressure vessel envelope is designed to match the original in size, dimensions, and pressure rating. The heat transfer surface (external surface of the tubing) is sized to match the function of the original surface and/or any upgrades required by the plant owner.

In defining the heat transfer surface, certain effects must be taken into account including:

1. the net effect of the reduced thermal conductivity of A690TT relative to A600MA,
2. any tube wall thickness and/or diameter change,
3. preheater deletion if applicable, which requires the addition of a large amount of surface, and
4. any change of rated performance such as power increase, steam pressure increase or primary inlet temperature reduction.

Having defined the vessel envelope and heat transfer surface requirements, the design is refined to address other thermal-hydraulic parameters such as primary-side flow resistance (which must closely match the original) and the range of secondary-side thermal-hydraulic issues described above. Finally, structural and reliability design features are introduced to ensure prolonged and reliable operation.

Tube supports The original equipment incorporated a variety of tube supports including flat plates with drilled flow holes, broached plates, and lattice grids (see Fig. 11). These have been replaced with a refined lattice grid arrangement made of intersecting flat bars of Type 410S stainless steel. (See Figs. 12 and 13.) This design is based on the lattice bar arrangement used in the later CANDU steam generator designs. (See also Chapter 50.) These lattice bar tube supports provide anti-vibration restraint for the straight leg portions of the tubes and provide lateral support for the entire bundle. Lateral loads include handling and shipping loads plus seismic and other accident loads. Normally the lateral in-service loads from the U-bends are taken through the uppermost lattice grid tube support; these can be well in excess of handling loads.

The U-bends are supported against flow-induced vibration (FIV) and mechanical loads by flat bar U-bend supports (see Fig. 14). FIV restraint is provided by positioning sufficient flat bar supports within the U-bends to avoid excessive tube response to turbulent or fluid-elastic instability flow excitation. Support against mechanical loads, including seismic, is basically provided by coordination and transfer of the U-bend swaying forces to the uppermost lattice, designed to sustain the lateral load of the overhung U-bends. The flat bar configurations provide an open flow arrangement and line contact tube support to ensure free flowing recirculation conditions and to avoid tube-conforming crevices or trapped spaces that may accumulate vapor and deposits, which may encourage corrosion.

The Type 410S material selected for lattice bars and U-bend supports is a low carbon version of Type 410. This material has excellent strength in its quenched and tempered condition, is resistant to tube fretting, and has good resistance to local and general corrosion. Unlike carbon steel, any oxide that does form by corrosion has a volume not greater than that of the base metal. This avoids tube and support plate degradation caused by voluminous rapid linear oxide growth, usually referred to as denting corrosion, that plagued many earlier carbon steel support plates.

Welded divider plate All recirculating steam generators feature a flat plate in the primary head to separate the incoming primary flow from the primary outlet flow. In certain original designs, this was essentially a floating design where the divider plate was located by engaging seat bars that were welded to the head and tubesheet surfaces. Because of gaps around the plate, there was some primary flow bypass from the inlet to the outlet side of the primary head. Replacement steam generators typically have a fully welded divider plate that is joined around the full perimeter to the primary head and tubesheet, as shown in Fig. 17. This was not done in some original designs due to concerns about thermal stresses in the combined head/plate/tubesheet structure. However, three-dimensional finite element elastic-plastic analysis now allows a more accurate assessment of the operational thermal and mechanical loads and shows that the all-welded arrangement meets all design and pressure vessel code requirements.

Steam drum internals The steam drum internals for all B&W replacement RSGs incorporate the high effi-

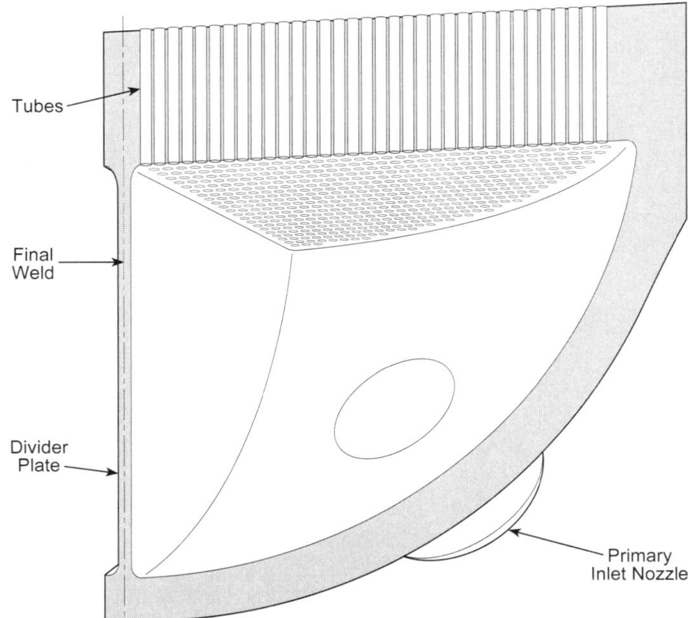

Fig. 17 Fully welded primary head divider plate.

Fig. 18 Replacement recirculating steam generators can be supplied that use the original upper shell, which remains in containment.

ciency CAP separators and cyclone secondary separators as described earlier. The drum internals have been replaced for all replacement units whether the steam drum pressure vessel itself was replaced or retained. In many cases the drum was too large to remove, so new separators were installed into the existing drums within the reactor building. Such replacement RSGs are shipped without the upper shell, or steam drum (see Fig. 18).

Replacement once-through steam generators (ROTSG)

The Oconee nuclear plant units in the U.S. were the original lead units of the B&W PWR program. Over the years, the once-through steam generators performed very well, but like the recirculating steam generators, they eventually showed corrosion problems with their Alloy 600 tubes. Steam generators for all three Oconee units (two steam generators per unit) have been replaced.[4] (See Fig. 19.)

A major difference between the OTSGs and the recirculating steam generators is that since there is no steam drum there is no large secondary water inventory in the once-through units. Because of such thermal-hydraulic characteristics and because the straight tubes and the straight shell are bound together at both ends, the design of these steam generators is very tightly integrated. Thermal characteristics of heat-up and cool-down and the thermal hydraulics of the downcomer flow, feedwater heating, level control, etc. are highly interdependent with each other and with the structural and hydraulic configuration of the equipment. The replacement units have been designed to retain, as closely as possible, the thermal and performance characteristics of the original units so that there is as little change as possible to this carefully integrated balance.

As with the recirculating design, the tubing material for the ROTSGs has been changed from A600MA to A690TT. The tube support configuration of the original OTSGs introduced the industry to the broached plate design. The broached plate design has been retained for the replacement units but with Type 410S material. In addition to the broaching, the flow passages use a patented hourglass configuration and the surface finish is enhanced by an electropolishing process. This minimizes deposition and reduces the potential for tube fretting and any support contact surface anomalies.

The original OTSGs were constructed with a radial open lane through the bundle to allow inspection of the bundle inner region. Long-term operation has shown this to be an undesirable feature. Secondary-side flow non-uniformity led to thermal tensile stress, localized high velocities, and localized corrosive environments, all of which contributed to fatigue in tubes adjacent to the lane.

In designing the steam generator shell, the original envelope is retained but the material is replaced with high strength steel to minimize weight, and with forgings to reduce the number of longitudinal welds. The original vessel cylindrical support skirt is replaced with a conical base to improve access for pipe work and

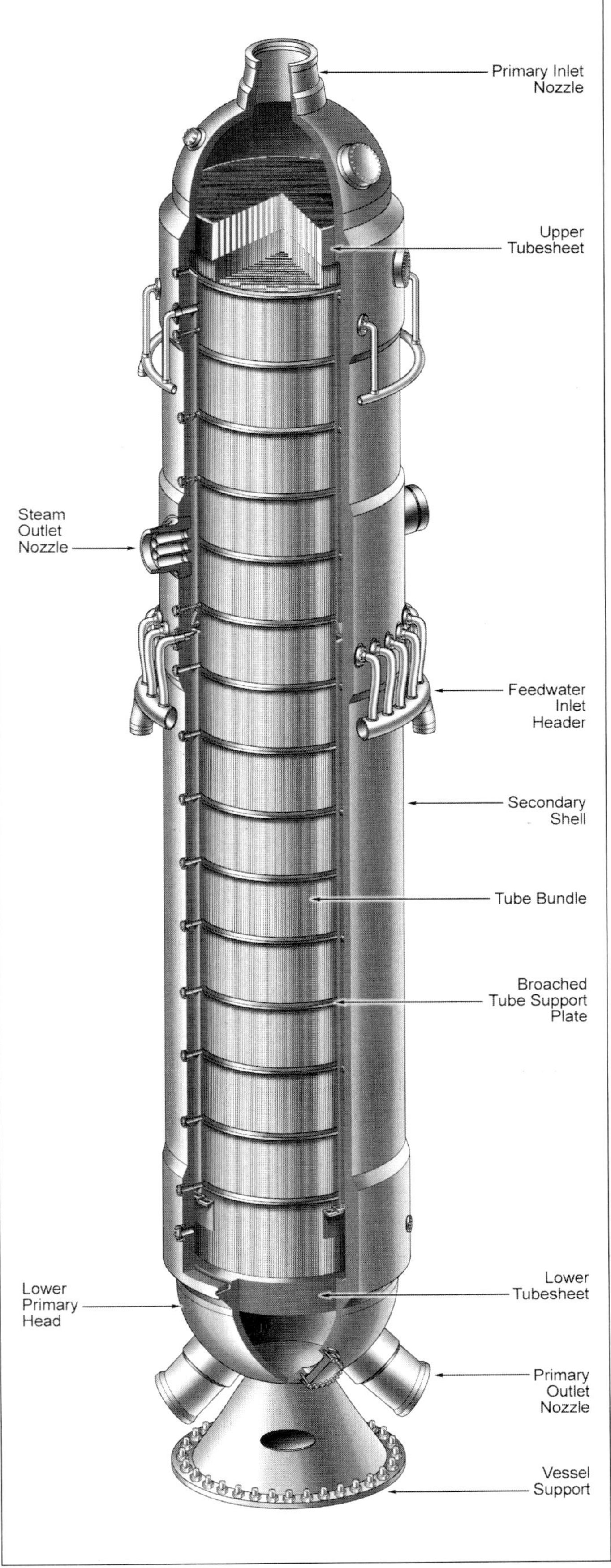

Fig. 19 Replacement once-through steam generator.

Table 4
Replacement Once-Through Steam Generator Design Features

Features	Original Design	Replacement Design	Improvement
Tube material	Alloy–600MA	Alloy–690TT	Corrosion resistance
Pressure vessel	Carbon steel	High strength steel	Reduces weight
Vessel support skirt	Cylindrical	Conical	Improves maintenance access
Tube support plate	Broached	Broached with added hour-glassing, enhanced surface finish, 410S material	Mitigates deposits Corrosion resistance Reduces pressure drop
Tube bundle layout	Tube-free lane	Open lane eliminated	Improves flow distribution

for manway access during manufacture, installation and service (see Fig. 19).

Other design features have been added such as feedwater header configuration improvements, addition of more inspection ports, and seal welding of infrequently used inspection ports. A summary of the major changes to the original design is given in Table 4.

Further development

Research and development (R&D) work defines the technology needs and establishes internal or external programs to achieve those objectives. The work includes development of new design configurations and features; development of manufacturing processes including specialized welding, cleaning, tube expansion and surface conditioning procedures; development of processes, tooling and robotics for field service and inspection processes including water lancing, visual inspection, cleaning and tube inspection; thermal-hydraulic, stress and flow-induced vibration design procedures; and design elements including steam separators and tube support structures. This R&D work includes the materials, chemistry and corrosion technology to support component and pressure vessel design and tube longevity. A summary of B&W's R&D activities is shown in Table 5.

References

1. *Steam Generator Reference Book*, Electric Power Research Institute, Palo Alto, California, 1985.

2. Parkinson, J. R., et al., "Steam Separation Uprate by Elimination of Capacity-Limiting Mechanisms," American Society of Mechanical Engineers (ASME) Winter Annual Meeting, Miami Beach, Florida, November 17-22, 1985.

3. Cohn, P. Ed., *The ASME Handbook on Water in Thermal Power Systems*, American Society of Mechanical Engineers, New York, New York, 1989.

4. Klarner, R., Boyd, J.T., and Keck, M., "Once Through Times Two," *Nuclear Engineering International*, April, 2004.

Inconel is a trademark of the Special Metals Corporation group of companies.

Table 5
Major Replacement Recirculating Steam Generator Development Activities

Activity	Description
Materials	Cooperative efforts with international R&D agencies to evaluate new materials such as Alloy 690TT
Weld metals	Corrosion resistance and mechanical properties of Alloy 52 and Alloy 152 and other materials
Water chemistry	Continued involvement with industry to establish operating guidelines Research on leading issues: tubing SCC, feedwater additives, and flow accelerated corrosion of SG internals
Thermal-hydraulic design	3-D design codes and CFD analysis Flow tests of various internal geometries
Mechanical design	Mechanical tests of tube supports, gasketed closures, tube-to-tubesheet joint
Vibration	Cooperation with external labs to develop tube supports and internal support structures Reduced U-bend tube/support clearance to reduce wear and fretting
Manufacturing	Development of pressure vessel welding and tube-to-tubesheet welding techniques
Steam separation	Testing of primary and secondary separators to improve overall performance
Maintenance services	Testing of new ECT, UT, and video probes for tube examination Tube plugging, sleeving, and waterlancing equipment development
Tubing	Passivating tube inner surfaces to reduce primary radiological activation

Bibliography

Kakaç, S., Ed., *Boilers, Evaporators and Condensers*, Wiley-Interscience, New York, New York, 1991. See Chapter 9, "Nuclear Steam Generators and Waste Heat Boilers" by Collier, J.G.

Klarner, R., Albert, J., and Schneider, W.G., "SG Replacement, Operation and Support," PLIM & PLEX 2003 Conference, New Orleans, Louisiana, October 13-14, 2003.

Klarner, R., Fluit, S., and Schneider, W. G., "Replacement Steam Generators for Calvert Cliffs, Oconee and Future Replacement Designs," Fourth CNS International Steam Generator Conference, Toronto, Canada, May, 2002.

Schneider, W. G., Klarner, R. and Smith, J., "Replacement Steam Generators," *Nuclear Engineering International*, January, 2002.

Tong, L. S. and Weisman, J., *Thermal Analysis of Pressurized Water Reactors*, Third Ed., American Nuclear Society, La Grange Park, Illinois, 1996.

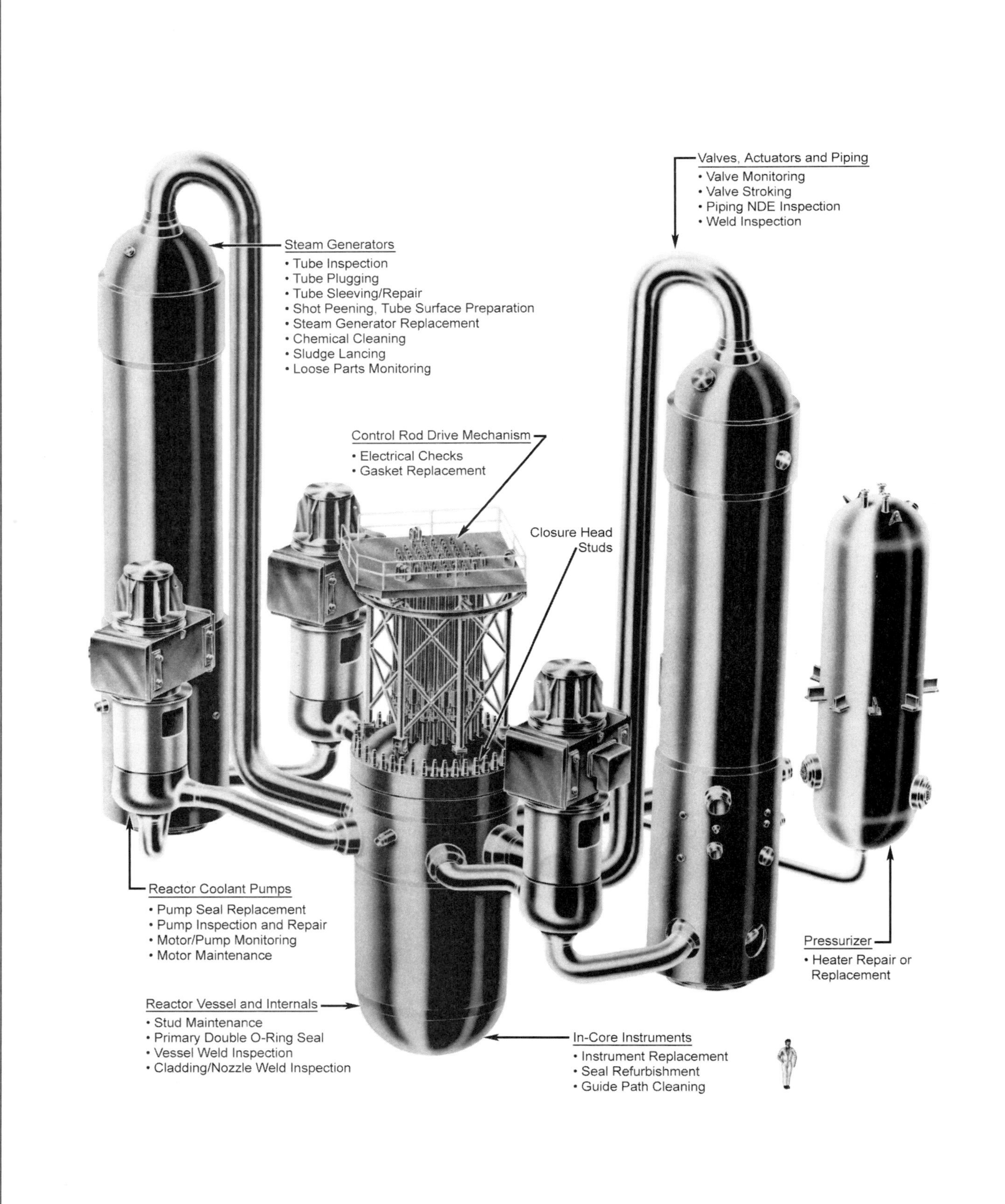

Fig. 1 Key areas of service and life extension in a pressurized water reactor (once-through steam generator reactor shown).

Chapter 49

Nuclear Services and Operations

Nuclear service activity includes all life management planning, inspection, analysis, assessment, maintenance and repair of equipment as required to support the plant equipment throughout its operating life. This life management activity must achieve continued high reliability and high availability for the entire life of the plant. Because most routine nuclear plant maintenance is done during scheduled refueling outages and because of the extended time between such outages (12 to 24 months), successful long range outage planning is essential. Ultimately, all of this planning effort is designed to obtain approval from the appropriate regulating authority to return the plant to full operation.

In some areas, the routine maintenance for nuclear plants is similar to that for fossil-fueled plants. This includes the turbine-generator systems, pumps, valves, fans, filters, electrical distribution systems, and some of the instrumentation and control systems. Nuclear plants, however, have many inspection and maintenance activities which are different in nature, precision, extent or available maintenance time period. Also, in parts of the plant subject to radiation exposure there are significant differences due to radiological considerations. Dealing with the radiological issues imposes unique requirements on plant maintenance, especially in personnel protection, handling of radioactive materials, disposal of waste items and the need for remote robotic equipment.

The following sections focus on the maintenance of the major components and the unique aspects of working in a nuclear facility. Most of the discussion is centered on the key areas of the reactor primary heat transport system as shown in Fig. 1. The reactor shown has once-through steam generators (OTSGs). Recirculating steam generators (RSGs) have similar tubing inspection and repair requirements but also have the steam drum/upper bundle area which requires periodic inspection.

Maintenance – key issues

Personnel protection

Although stringent precautions are taken, nuclear plant maintenance workers are exposed to low levels of radiation. In the United States (U.S.), allowable technician dose limits are regulated by the Nuclear Regulatory Commission (NRC). Other countries (e.g., Canada) set their own allowable technician dose limits. Some operating companies curtail exposure even further.

Workers entering a radioactive area must wear protective clothing. This typically consists of cotton overalls, paper and rubber shoe covers, cotton and rubber gloves, and a cotton hood or paper hat. Respiratory protection and/or plastic suits with air supply may also be required.

Limited accessibility

A major restraint to performing nuclear plant maintenance is the limited accessibility of much of the equipment. The primary reactor system is not accessible during power operation. As a result, this equipment must be maintained during outages. Repair, replacement or return-to-service of parts must be based upon maintaining fitness-for-service to the next point in time when that part will be inspected.

Most of the plant equipment is required by license to be operable during normal operation. If a safety-related system fails, the operator typically has some limited period of time for its repair and return-to-service, otherwise the entire unit must be shut down. Examples of such critical equipment include the high pressure injection system, low pressure/decay heat removal system, core flood system, building spray system, reactor protection system, and nuclear instrumentation.

Application of remote technology

Principles known as *time, distance and shielding* and ALARA (as low as reasonably achievable) are the cornerstones of performing maintenance in a radiation-exposure efficient manner. In applying remote technology, emphasis is placed on the time of personnel exposure as well as the distance between the personnel and the work piece.

When considering whether remote technology is applicable, the level of radiation, the time required for the maintenance and the work place environment are evaluated. Key areas where this technology has been applied include work in or around the reactor vessel and its internals and within the steam generators. Although the reactor vessel is highly radioactive, it is submerged in water during maintenance. While the water shields the workers from radiation, remote and/or fully robotic equipment is necessary for the underwater repairs.

While remote and robotic technology have improved the extent, quantity and refinement of inspection data being gathered, the cleaning and repair performed have increased enormously. The inspection of tens of thousands of steam generator tubes in a short outage is not unusual compared to hundreds in prior years. At the same time, efforts to optimize these activities have led to shorter outages. As a result of these shorter, more effective outages, the average U.S. reactor availability has climbed from 60% in the past 20 years to more than 90%.

Although remote technology often encompasses the use of robots, it can also be an extension of a hand tool or the development of a flexible probe and manipulator to increase distance from the work piece and to allow internal access to tight or complicated spaces. Each maintenance task requires a different degree of sophistication based on the environment and the complexity of the task. As remote technology advances, the amount of personnel radiation exposure declines even as the scope and sophistication of the work increases.

Component maintenance

Plant life management

The objective of plant life management is to ensure that the plant equipment achieves a continuing state of high availability throughout its operating life and the plant reaches at least its desired lifetime. Life management places emphasis on the assessment of component condition prior to scheduled outages and on the scheduling, planning and mobilization of workforce, equipment and methodology to achieve timely execution of the outage work. Outages may only end when all work is successfully completed and documented, and approval is granted by the appropriate regulatory authority to return to service.

Because experience has shown that much of the plant availability is dependent on the successful operation of the steam generators, the discussion and illustration of plant life management methods will concentrate on these components.

Long range life assessment Prior to development of a life management plan, a life assessment process is completed for each major component. This life assessment is necessary to establish the operating history of the component and to provide a basis for planning a multi-outage inspection and maintenance program. For steam generators, the life assessment and other work should individually address the three major areas of: 1) pressure vessel boundary, 2) the heat transfer tubing and 3) internals such as tube supports, steam separators and feedwater headers. Breakdown into these three areas is necessary because the regulatory requirements and the inspection and maintenance requirements for each are unique.

Pressure vessel boundary The design, manufacture, in-service inspection, and repair of all component pressure vessel boundaries is regulated by the American Society of Mechanical Engineers (ASME) Boiler and Pressure Vessel Code as augmented by other regulatory requirements. During manufacture, the Code-defined volumetric inspections ensure the components meet stringent standards and establish a baseline for in-service evaluations. In-service requirements specify examination of the pressure vessel boundary, including visual inspection and volumetric nondestructive examination (NDE) of the welds. An assessment of the steam generator must determine the condition of the equipment and any areas of non-compliance with the requirements and must identify any work needed to restore the unit to a compliant state.

Tubing The steam generator tubing is the boundary between the primary and the secondary heat transport systems and is very important to the integrity of the plant. Tube integrity is the primary concern when reviewing the steam generator. In The Babcock & Wilcox Company (B&W) plants, there are about 15,000 tubes in each of the two once-through steam generators (Fig. 1). In plants with recirculating steam generators, there are about 3000 to 11,000 tubes in each of two to four steam generators per plant. With a typical tube wall thickness of about 0.04 in. (1 mm) or less, the tubing is a vulnerable boundary and must be carefully maintained. Tube leaks are a major concern and primary to secondary leakage above certain prescribed limits requires shutdown of the plant.

Experience has shown the tubing to be a major cause of loss of plant availability, long outages, and expensive repair and replacement work. Common problems with steam generator tubing are generally related to corrosion involving a combination of materials and water chemistry issues. Some of the more common problems are stress corrosion cracking, pitting, wastage, erosion, fretting at tube supports, and denting due to corrosion of any carbon steel tube supports. Fretting due to flow-induced vibration and fretting due to foreign object ingress may also exist. In modern replacement steam generators with highly corrosion resistant tubing and good tube support, foreign object damage may be the most serious concern regarding tube integrity. Because of the magnitude and frequency of problems over the years, the tubing receives the majority of attention in the life management plan.

Internals Internals common to all types of steam generators include tube supports, shrouds (bundle wrapper) and feedwater and/or auxiliary water headers. For recirculating units, additional internals include U-bend tube supports, steam separators and various supports and baffles internal to the steam drum. In contrast to the tube bundle that contains many identical elements (tubes), the internals consist of a large number of parts that are dissimilar in terms of function, configuration and accessibility for service. The internals are subject to a variety of flow forces and flow-induced vibrations. Bundle supports in close contact with the tubing can cause tube degradation due to fretting, and tube denting due to corrosion of the supports. The life assessment process must consider the potential for a wide variety of potential problems with these internals.

Life management program plan Following the assessment of the component, a long range life management plan can be developed. Table 1 illustrates a small portion of such a plan for a set of steam generators. It shows the scope, timing, inspection, cleaning and repair activities for outages 11 through 14 of the total life

Table 1
Steam Generator Life Management Program Plan
Unit ABCD/SG 1, 2, 3, 4

Outage No. Activity	11 October 2003	12 April 2005	13 October 2006	14 April 2008
Pressure vessel NDE program				
Tube ECT		SG 2, 4 50% Bobbin 10% TS MRPC		SG 1, 3 50% Bobbin 10% TS MRPC
Tubesheet waterlancing		SG 1, 2, 3, 4		
Tubesheet FOSAR		SG 1, 2, 3, 4		SG 1, 3
Drum and upper bundle visual inspection		SG 2		
Tube plugging		If required		If required
Planned retrofit of bundle inspection ports				SG 1, 3 TSP 6, 8
Chemical cleaning		Assess need		

Tube ECT: Eddy current testing, with 50% of bundle using the standard Bobbin and 10% bundle using the MRPC coil.
Tubesheet FOSAR: Foreign object search and retrieval, visual examination.
Drum and upper bundle inspection: Visual/robotic inspection to meet requirements of NRC Generic letter 97-06(1977).
Tube plugging: Based on inspection results.
Planned retrofit of bundle inspection nozzles: See chapter 48.
Tubesheet water lancing: Cleaning sludge deposits on top of tubesheet surface.
Chemical cleaning: chemical dissolution of secondary-side tube deposits.

cycle. A full plan would cover many more outage cycles, more plant components, and many more activities.

Having identified the necessary inspection and maintenance tasks, significant effort is then employed to define work scope, the preparation of detailed plans, detailed procedures, and to obtain regulatory approvals to proceed. A typical outage would result in a large volume of tube eddy current testing (ECT), pressure vessel inspections, and visual and foreign object search and retrieval (FOSAR) inspections of the secondary and primary side internals. There would also be cleaning activities, repairs and possibly some component modifications that had been planned in advance. The assessment of this large database of information, which is done as soon as the inspection data is available, forms the basis for the next phase to determine fitness-for-service.

Fitness-for-service assessment The fitness-for-service assessment is a methodology for determining that a particular plant component is in an acceptable condition at the time of inspection (referred to as *condition monitoring*) and confirming that it will be in an acceptable condition at the time of its next planned inspection (*operational assessment*). The method is illustrated by Fig. 2 which shows the current as-measured size of a defect in a steam generator tube wall, the estimated growth for a period of operations and the estimated size at the time of its next inspection. Essentially, the initial as-measured size from tube inspection, with an added allowance for measurement error, is plotted at the time of the current inspection. A projection is made of the growth of this defect to the time of the next inspection. If the projected flaw size at the time of the next inspection is greater than the allowable flaw size, the tube is considered potentially unsafe and is most likely removed from service by plugging.

The probability of detecting a particular defect in tubing is dependent on the nature of the defect (pits, frets, cracks, etc.) and its location relative to other sources of electronic signals such as tube supports, tube wall expansion transitions and corrosion deposits. With current technology, pits measuring a small percentage of the wall thickness can be detected. By contrast, tight cracks are much more difficult to detect and may sometimes be missed, even at 30% of wall thickness. Also, allowance for measurement error is necessary to ensure that a high percentage of the actual defect sizes are within the bounds of these derived defect sizes.

Steam generator primary side maintenance

Maintenance of a steam generator primary side includes the major activities of:

1. inspection of the tubing,
2. plugging tubes that fail to meet the inspection criteria,
3. repair of tubes to keep them in service and to thus reduce performance degradation, and
4. stabilizing out-of-service tubes to keep them from damaging their neighbors.

Steam generator tube inspection Based on prior experience with the steam generators under examination, some fraction of the tubing will be required to be in-

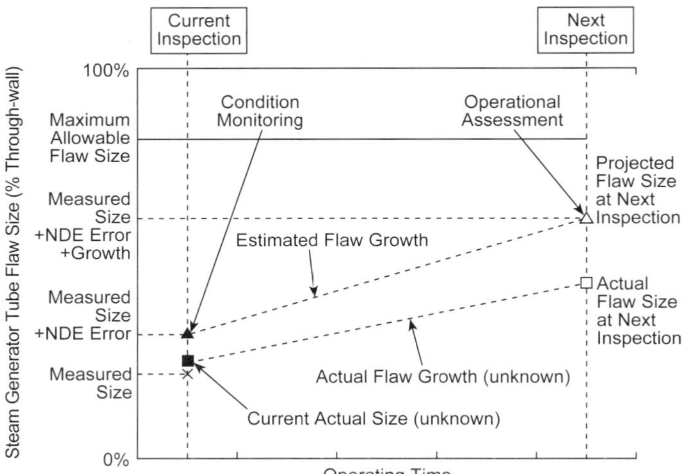

Fig. 2 Fitness-for-service assessment for a steam generator tube with defect indication.

spected on a periodic basis. For example, it may be determined that half of the steam generators in the plant will be inspected at every second outage (as in Table 1). There may also be commitments for specialized inspections or for other tube inspections as a condition for a return-to-service approval from prior outages.

Eddy current testing (ECT) ECT is the primary inspection method, although ultrasonic testing (UT) is used in special cases. The ECT is conducted by imposing an alternating current (AC) magnetic field. This field induces eddy currents in the material that creates a secondary magnetic field. If discontinuities are present, the eddy currents are altered and the apparent impedance of a detecting coil is changed. The size of the discontinuity can then be related to this change by prior calibration.

Tubes are inspected by inserting the probe and then inspecting the tube while withdrawing the probe at a controlled rate. RSG and OTSG tubes can be inspected full length except that some smaller radius RSG tubes may require access from both ends. A typical robotic positioner for the ECT probes is shown in Fig. 3.

Several different probes are used to characterize the tubing condition. A bobbin coil ECT probe basically consists of a bobbin wound with coils of wire. Eddy currents generated by the coil electromagnetically couple the probe and the test piece with the eddy currents flowing in a direction parallel to the coil winding. Eddy current signals from a bobbin coil are measured as changes in primary coil electrical impedance induced by the eddy current magnetic field. Bobbin coil ECT has been used throughout the industry for many years, although other probe types are essential to complement the basic information received. One limitation is that the signal is integrated 360 deg around the tube and is therefore unable to differentiate within that range. The bobbin coil is able to size many types of defects once the nature of such defects has been determined by laboratory analysis or by other probe types, but it is not good for exploring unknown situations such as detecting tight cracks of moderate size or circumferentially oriented cracks. It can, however, be used to see further in the radial direction than other ECT techniques. It is sensitive at low operating frequencies to conditions (supports, deposits) beyond the tube outer surface; at high frequencies it has a shorter range of vision.

Because of these limitations, a number of rotating and array probes have been developed. MRPC (motorized rotating pancake coil), +Point® (plus point) rotating probes and X-Probe® array probes are examples. Rotating probes like MRPC and +Point provide excellent and finely detailed data but have the disadvantage of very slow probe travel speed. The X-Probe offers the speed of the bobbin with the detail of rotating probes, although with the generation of a very large amount of data. Fig. 4 shows a typical arrangement of the bobbin coil and X-Probe; the data acquisition and analysis system is shown in Fig. 5.

Ultrasonic testing (UT) UT of tubing is a valuable tool, particularly in the early investigative phase where a newly discovered condition must be characterized. The process requires a rotating surface-riding UT probe where the signal path must be flooded by a couplant such as water. Because of the rotation and the couplant, the process is slow and somewhat awkward, but it does have excellent resolution. It has the advantage of seeing indications differently than ECT, thereby providing results to supplement the ECT information.

Destructive examination (DE) DE is a vital part of investigating any newly discovered degradation condition in tubing (or in other component areas). Except perhaps for phenomena like fretting, it is not possible to characterize a degradation condition and to determine its root cause without sample removal and laboratory examination. Until newly discovered defects are characterized in the lab, it is not possible to know with any certainty the relationship between the defect and the ECT signal it produced in the field. Because some

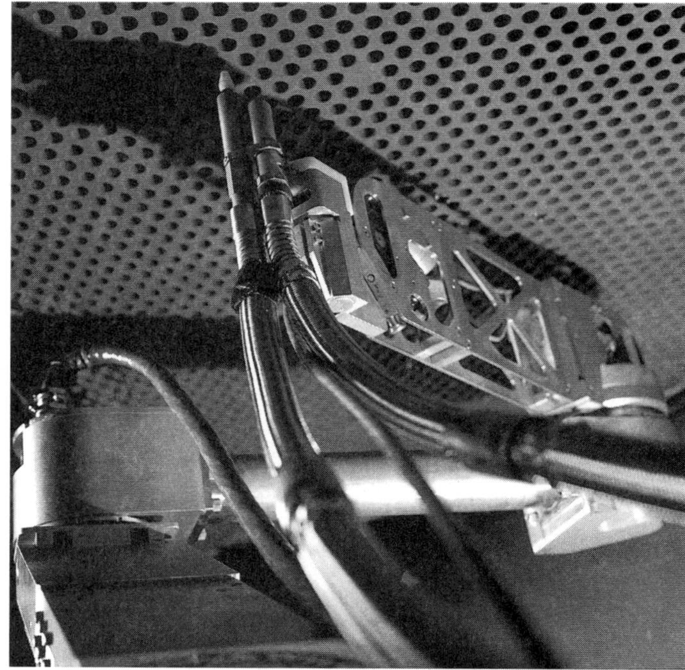

Fig. 3 Robotic eddy current inspection shown in a recirculating steam generator mockup.

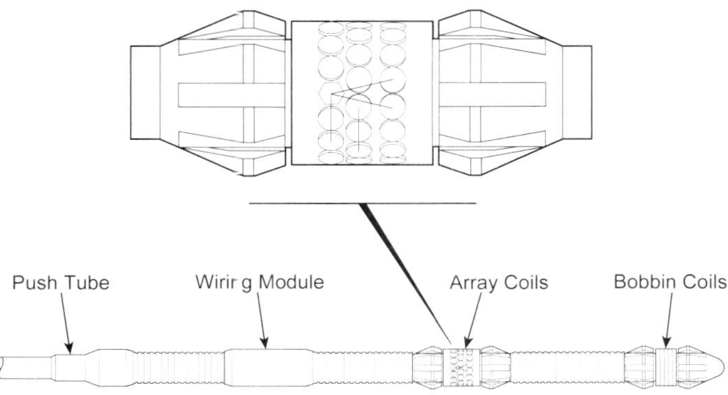

Fig. 4 Multi-coil array probe with bobbin coil.

of the more critical defects such as tight cracks are difficult to see at any time, even where they are relatively large, such characterization is vital.

Remote technology In the early years of nuclear power all of the maintenance work was performed manually by technicians who entered the steam generator head. As maintenance activities and radiation levels increased with plant aging, it became increasingly difficult to perform repairs manually. As inspecting ever larger numbers of tubes during refueling outages became mandatory, this work was a logical candidate for early remote technology. B&W's first manipulator used on OTSGs was positioned manually on the tubesheet, locked in place, then computer controlled from a remote location. For RSGs, an articulated arm manipulator was used. Although also manually installed, it was not computer controlled. Both manipulators performed well for inspection, but they could not support the heavy tools required for plugging and other repair operations. In addition, because of reduced radiation dose limits, the practice of manually installing the manipulators had to be eliminated.

Advanced manipulators Advanced manipulators have been developed to minimize human exposure, to perform multiple inspection and repair duties, and to be computer controlled.

An earlier design of a remote manipulator is shown in Fig. 6. The unit is installed in the steam generator head by a series of cables and pulleys as shown in Fig. 7. Although lightweight in construction, it has a load capacity of 200 lb (91 kg). Once calibrated, it can be directed to any tube in the steam generator. With the exception of tool head installations, all operations are remotely controlled and video equipment is used to monitor the work. The control center is located outside the containment building.

Fig. 8 shows an advanced manipulator that loads itself into the channel head while its arm exits the manway for attachment of tool heads. Being mounted on the manway opening, it provides unobstructed access to all parts of the tubesheet. It also reduces personnel radiation exposure.

Tube plugging Any tube not meeting the established criteria for acceptability must be taken out of service or repaired. While explosive plugs were used in the

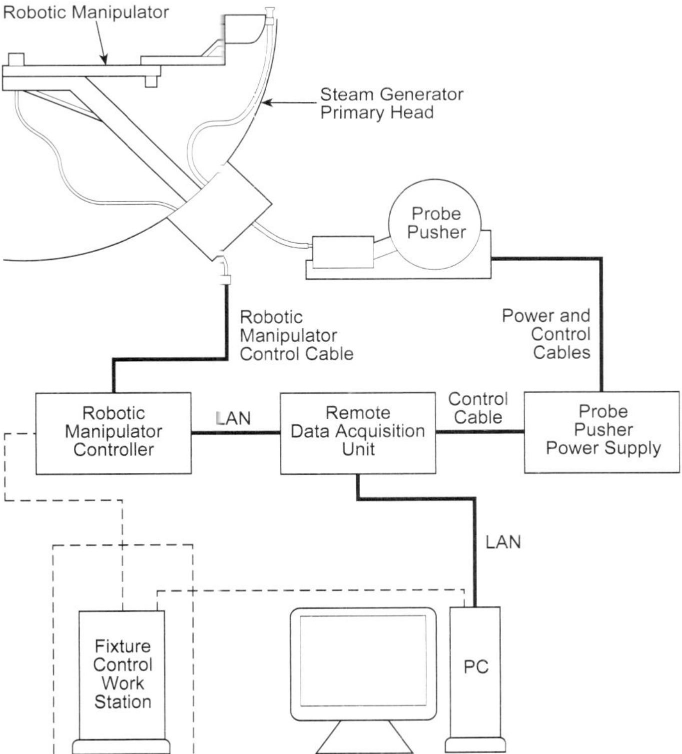

Fig. 5 Eddy current data acquisition and analysis system for multi-coil array probe.

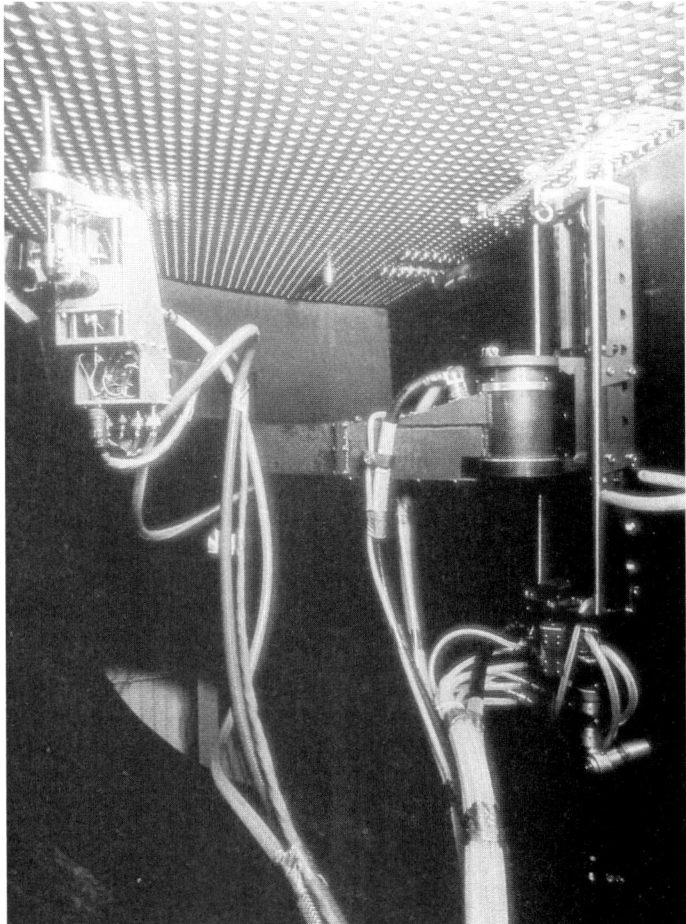

Fig. 6 Tubesheet-mounted manipulator inside a recirculating steam generator primary head.

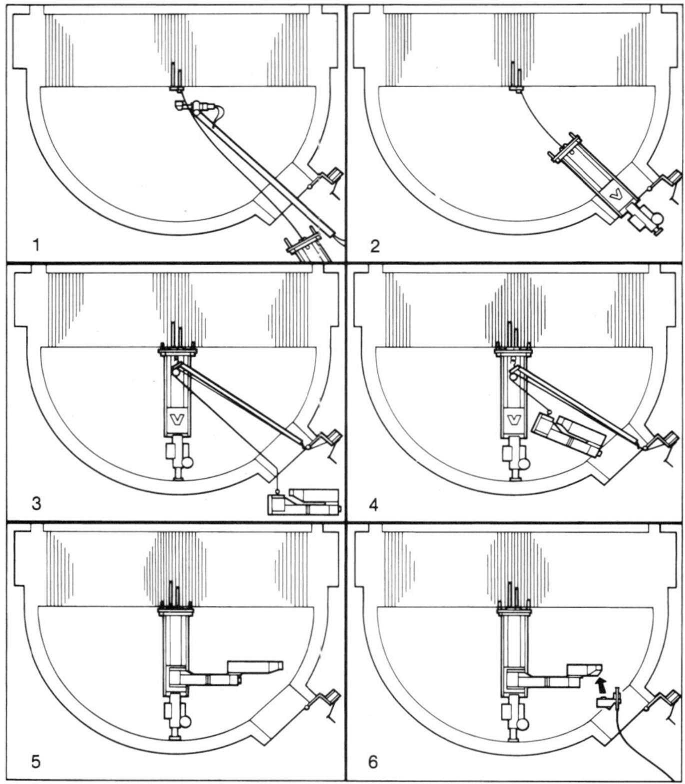

Fig. 7 Installation sequence for the tubesheet-mounted manipulator in Fig. 6.

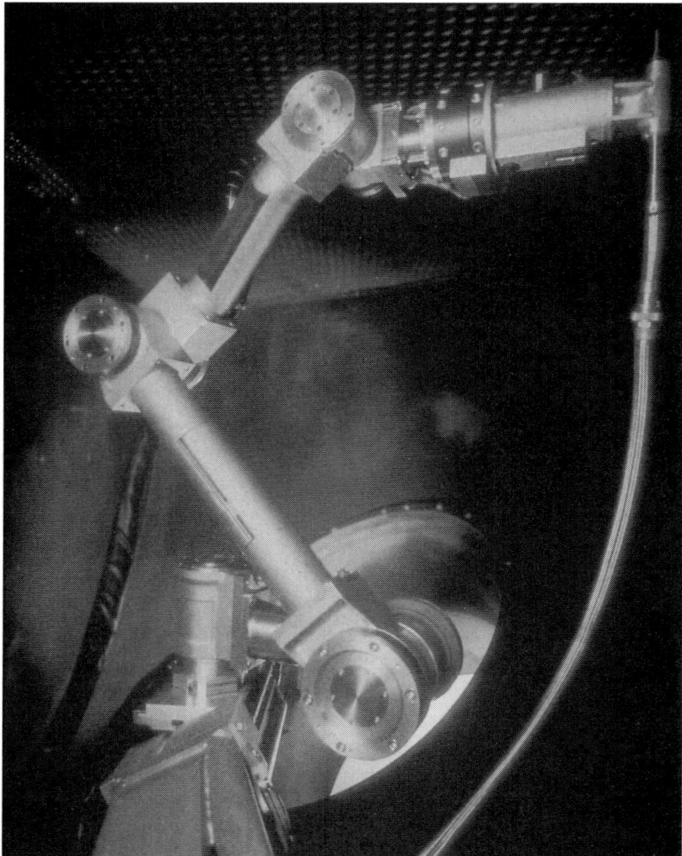

Fig. 8 Manway-mounted manipulator in an RSG head mockup.

1970s and 1980s, mechanical plugs are now more common. Fig. 9 shows typical mechanical plugs. Seal welded plugs are also used. The rolled plug is installed in one tube end with an air-powered roll expander, which forces the plug wall outward into the tube wall. Expansion continues until the plug and tube wall contact the tubesheet bore and the proper torque is reached by the tool. This process is repeated at the opposite end of the tube.

Tube repair A limited number of tubes may be plugged before unit efficiency is impaired. For nuclear steam generators, this limit is set by thermal performance parameters. One method of repairing damaged tubes and returning plugged tubes to service is *tube sleeving*. In this technique, a short tube (sleeve) is inserted into the base tube to bridge the degraded area. Should the defect continue to grow, the sleeve effectively seals the leak from the secondary side. However, this technique is generally limited to the portion of the tubes near the tubesheet. Fig. 10 shows a typical tube sleeved with roll expanded sleeves.

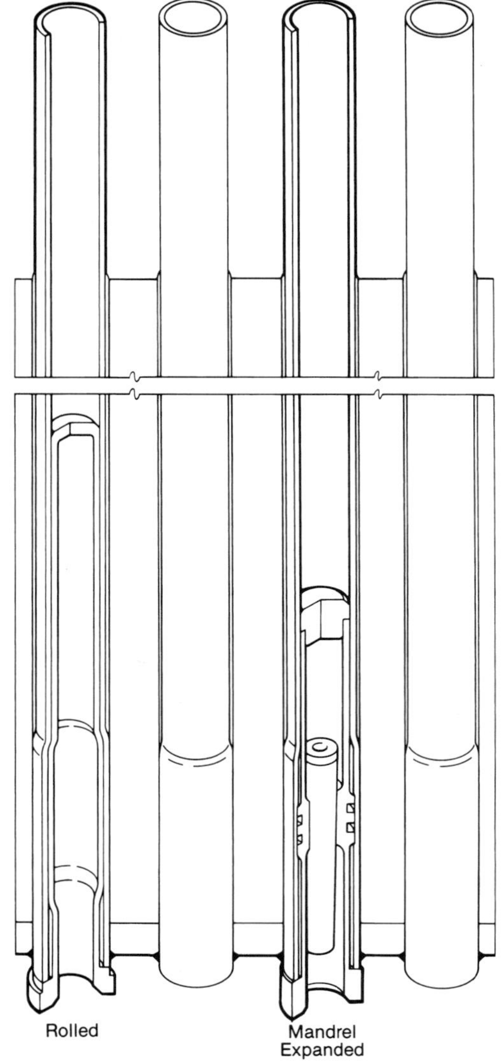

Fig. 9 Steam generator tube mechanical plugs – rolled (left) and expanding mandrel (right). Note: clearances and expansion transition profiles are greatly exaggerated radially for illustration.

Because of extensive steam generator replacement taking place, the need and practice of sleeving have been reduced.

Steam generator secondary side maintenance

While control of secondary side water chemistry helps to minimize repairs in this area, some maintenance is necessary. Two conditions, fouling and sludge buildup, must be eliminated. Fouling consists of deposition of solids, primarily iron, on the outside of tubes and at the tube support plates. Support plate deposits can increase pressure drop resulting in reduced steam generator performance. Tube bundle deposits can reduce heat transfer capability and also degrade performance. More important, buildup of these deposits can concentrate contaminants such as chlorides and fluorides that promote stress corrosion cracking – a problem which has destroyed tens of thousands of tubes and has led to the need for extensive steam generator replacement.

Sludge deposit buildup occurs at the lower tubesheet of the steam generator where solids from the feedwater are deposited. Again, other contaminants can concentrate in this sludge, degrading the tubes.

Different techniques are used to remove these sludge buildups. Chemical cleaning is used for the tube support plate and tube bundle. A chemical solution is introduced into the secondary side during a shutdown. Then, following a soaking period, the unit is flushed. Sludge piles are removed by water lancing. This process, also known as *sludge lancing*, uses high pressure water introduced robotically through a secondary side opening to loosen deposits. The loose deposits and water are then suctioned out of the steam generator and processed through a filter system.

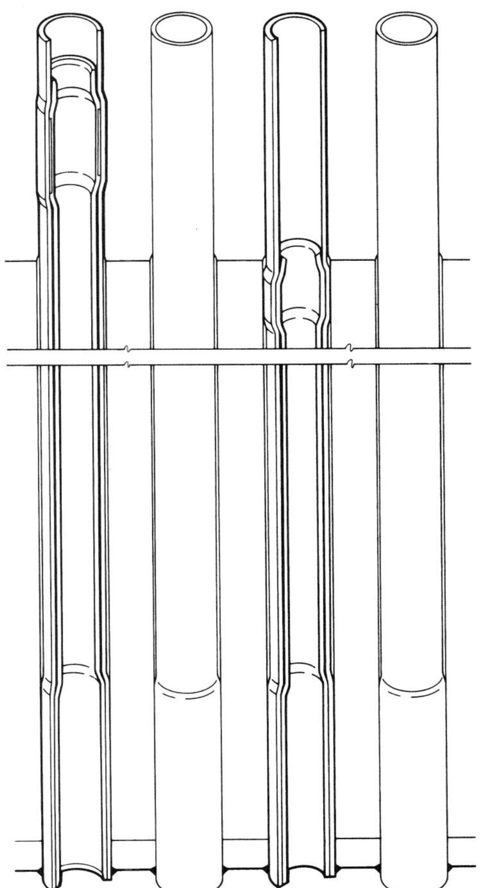

Fig. 10 Steam generator roller expanded tube sleeves – free span upper expansion (left) and tubesheet region upper expansion (right).

Fig. 11a shows the flexible inter-tube water lancing system used by B&W for steam generator cleaning. The system uses six or eight small-diameter pressure tubes to deliver high pressure water to the very small nozzles in its end manifold. These end manifolds direct intense water jets at the tubesheet face and upward toward the lowermost support. The flexible lance is fed into the steam generator through a secondary side inspection port just above the tubesheet. A semi-rigid guide directs the flexible lance along the steam generator's central tube-free lane, guides it through a tight right-angle turn, and projects it into the individual intertube lanes and ultimately to the bundle periphery. The jet action dislodges the sludge deposits from the tubesheet (and/or from tube supports) and permits the deposits to be flushed out of the bundle and removed by suctioning. This inter-tube system is particularly effective for hard deposits. Fig. 11b shows the water jet spraying from a flexible lance positioned within a small demonstration tube bundle.

The need for chemical cleaning or sludge lancing is determined by performance degradation, visual inspections using small fiberscopes and cameras, and by the removal of tube samples to detect degradation (or in some cases to directly measure the deposit loading).

Reactor vessel and internals

While the reactor vessel and its internals are static equipment, they are not free from either degradation or the need for maintenance. Maintenance ranges from care for the studs, gaskets and gasket surfaces during each refueling outage to repairs for various severe and costly degradation mechanisms.

Reactor vessel The vessel closure is the large diameter, high pressure bolted flange with its delicate gasket face which is the joint between vessel and head. During refueling, the reactor head must be removed from the vessel to gain access to the core. Each of the very large studs that retain the head is also removed from the vessel for cleaning. Special precautions are taken to contain the contaminated particulate that is released by stud cleaning equipment. Following cleaning, each stud is lubricated for reinsertion into the vessel flange. Likewise, each stud hole is cleaned prior to retensioning the studs.

The primary seal between the reactor vessel flange and head is a double O-ring configuration. Both mating surfaces must be thoroughly cleaned before installing the O-rings, which are replaced each time the head is removed. The seals are installed on the head and, just prior to setting the head, the vessel flange is cleaned and inspected for scratches that could cause leakage.

Reactor vessel heads worldwide have, of late, been subject to extensive degradation and some leakage due to stress corrosion cracking of the weld joints between the control rod drive mechanism (CRDM) guide tubes and the head. These guide tubes penetrate the

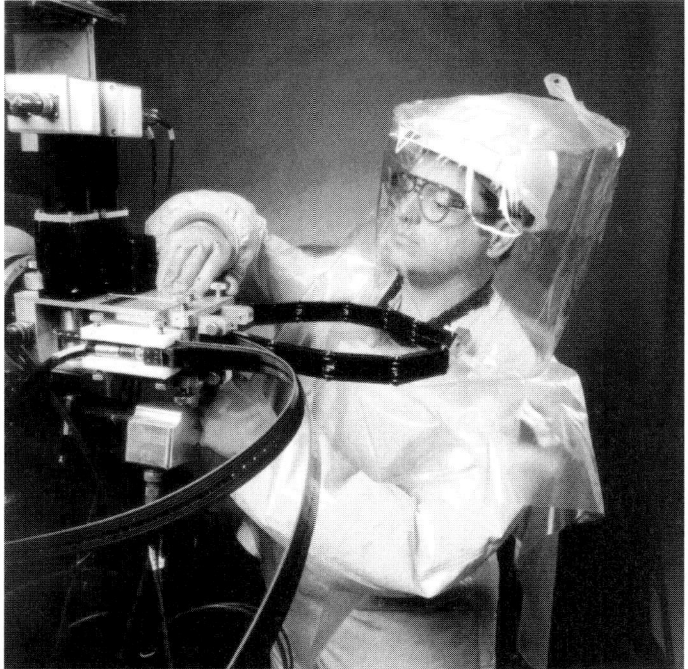

Fig. 11a Flexible inter-tube waterlance system – waterlance and drive unit mounted to a steam generator mockup.

Fig. 11b Waterlance jet spray pattern in a small demonstration tube bundle.

head and are welded to the cladding at the inner surface of the head. In at least one case, extensive corrosion of the head itself had occurred before this problem was discovered.

Because these welds are difficult and costly to inspect and repair, many utilities have chosen to replace the heads with components made with new, more corrosion resistant materials (Alloy 690, 52, 152 versus the prior Alloy 600, 82, 182 materials).

The small instrument guide tubes in the lower head of the reactor have also required attention at times but to a much lesser degree than the head penetrations.

In addition to the above more conventional forms of degradation, the effect of neutron radiation on reactor vessels must be considered. Neutrons can reduce the toughness of steels, increasing their susceptibility to brittle fracture. As a result, the welds in the reactor core region are subjected to extensive UT examination. Of particular interest is the surface just under the cladding. Special techniques are used to achieve maximum coverage and sensitivity.

To gain access to the inside surface of the reactor vessel, all the fuel and internals must be removed. Due to its highly irradiated condition, the vessel must remain filled with water to provide shielding. Each weld, including those between the nozzles and primary piping, must be examined. This can be accomplished using an automated reactor inspection system, shown in Fig. 12. A typical vessel inspection can take from 10 to 20 days, depending on the quantity of welds and the need for rescanning.

Internals While the reactor vessel internals require little routine maintenance, there have been various core barrel integrity problems, mostly related to vibration due to inlet flow buffeting. Because the internals are highly radioactive, any repairs must be done under water. The bolted joints of the reactor internals have required the most repair activity. Stress corrosion cracking of high strength steel or Inconel® bolts has been the most common cause for repair. Although the performance of A-286 stainless and X-750 Inconel bolts was expected to be good, pure water stress corrosion has been a problem.

In replacing these bolts, standard wrenches, drills and sockets have been adapted to be operated from an extension or underwater platform. Underwater cameras facilitate these remote repairs.

Control rod drive mechanisms (CRDM)

The CRDMs are mounted to the top of the reactor vessel head and control the movement of the reactivity control rods in the core. A central shaft known as either a lead screw or drive shaft attaches to the control rod and is moved by the mechanism. The CRDM is powered electrically, and position indication (PI) switches are used to let the reactor operator know the position of the control rod. During refueling outages, inspection, tests and routine maintenance are performed on the mechanisms to assure the motors and PI systems are functioning properly. Motors with unacceptable characteristics are replaced. PIs are also replaced if not functioning properly. Some CRDM designs attach to the reactor vessel head with bolts and use gaskets to provide the sealing. Occasionally these joints leak and require replacement of the gaskets and rework of the flange faces. This work is performed during the refueling outage while the reactor vessel head is on its storage stand. Remote tooling and appropriate shielding are used to minimize the radiation dose to the workers.

Reactor head service support structures are designed for several major functions. They provide cooling and lateral support for the CRDMs, support power and instrumentation cables, facilitate reactor head lifts, and act as an internal missile shield. On some reactors the original service structures were not fully optimized to simplify refueling activities. These original inefficiencies together with increased economic performance goals have resulted in the replacement of original service structures independent of or in conjunction with reactor vessel closure head (RVCH) replacement. The primary objective of an enhanced service structure (ESS) is to reduce the disassembly and reassembly times of the reactor head during refueling outages thereby shortening critical path activities as well as reducing personnel dose and improving safety. These objectives can be met by striving for a redesigned configuration which integrates all functions into a single assembly which can be removed with a single point pickup after disconnecting instrumentation, CRDM power cables and the head closure studs. Service structure enhancement can entail integrating the entire cooling system into the ESS, missile shield integration and building in a single point pickup tripod arrangement. Improvements to allow quick disconnects for cables and more efficient RVCH stud removal will also improve critical path time. Despite building more functionality into the ESS, overall access for reactor head inspections can be improved by providing handholes and by spacing the RVCH insulation away from the head surface, thereby allowing remote visual inspection of the head with the insulation still in place.

In-core instrumentation

Most pressurized water reactors (PWRs) utilize in-core instrumentation inserted through the bottom of the reactor vessel. During refueling, these very long, slender rods are withdrawn from the fuel assemblies. At that time, instruments that are not performing properly are replaced. Because the instrument (detector) is radioactively hot, the end that was in the core is pushed back into the reactor vessel and cut off using underwater remote tooling and is then disposed. The remainder of the instrument is pulled out the other end and also disposed. Additional inspections and maintenance to the sealing system can be performed. New detectors are inserted for use during the next operating cycle.

Reactor coolant pumps

Depending on the size of the plant, there are from one to four reactor coolant pumps in the primary system. Because the pumps and motors are constantly engaged, they require more maintenance than passive components. Pump seals are the items most frequently replaced. Nuclear plant motor maintenance is similar to that performed in other industries, except that the motors are very large and must operate with very little vibration for the sake of the seals and, of course, they must be maintained within the reactor building.

Pump repair While infrequent, pump repairs, other than seals, and general overhauls are important in a comprehensive maintenance program. Overhauls are conducted to minimize pump failures due to wear.

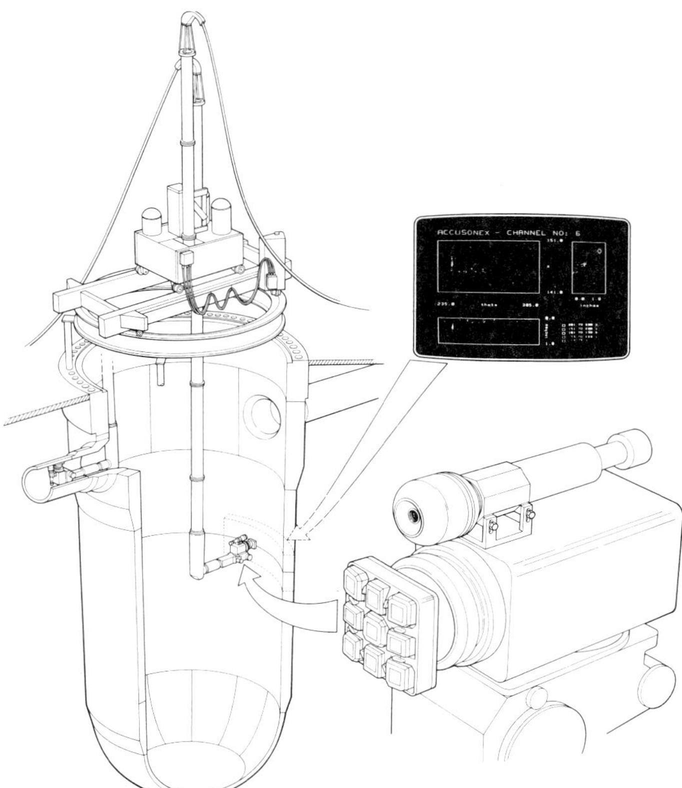

Fig. 12 Automated reactor inspection system.

Because of the contaminated environment in which a pump operates, special tooling is used when making repairs to minimize radiation exposure to technicians.

Mechanical shaft seal maintenance The mechanical shaft seal is a primary boundary for the reactor coolant. The seals are typically a three stage design, with each successive stage being subjected to a lower pressure drop. Sealing is provided by rotating and stationary surfaces that contact each other or are separated by a thin film. The seal faces are typically carbon or aluminum oxide. Because they must operate continuously in vibrating conditions with exceedingly low leakage, pump seals need regular attention and sometimes replacement during outage maintenance.

Pressurizer

The pressurizer serves to establish and control the operating pressure within the primary heat transport system. This is achieved by electrical heaters that raise the saturation pressure of the pressurizer's water and steam inventory to the appropriate value.

Heaters within the pressurizer are generally the only components that require maintenance. These heaters consist of side mounted bundles, used in B&W designs (Fig. 13), or bottom mounted individual heaters, used in other pressurized water reactors. Heater maintenance primarily consists of element/bundle replacement, power cable replacement, cable-to-element connector repair, and element recovery.

Moisture is often the cause of a defective heater element. Element recovery can be done by drawing a vacuum at the electrical connector and sucking out the

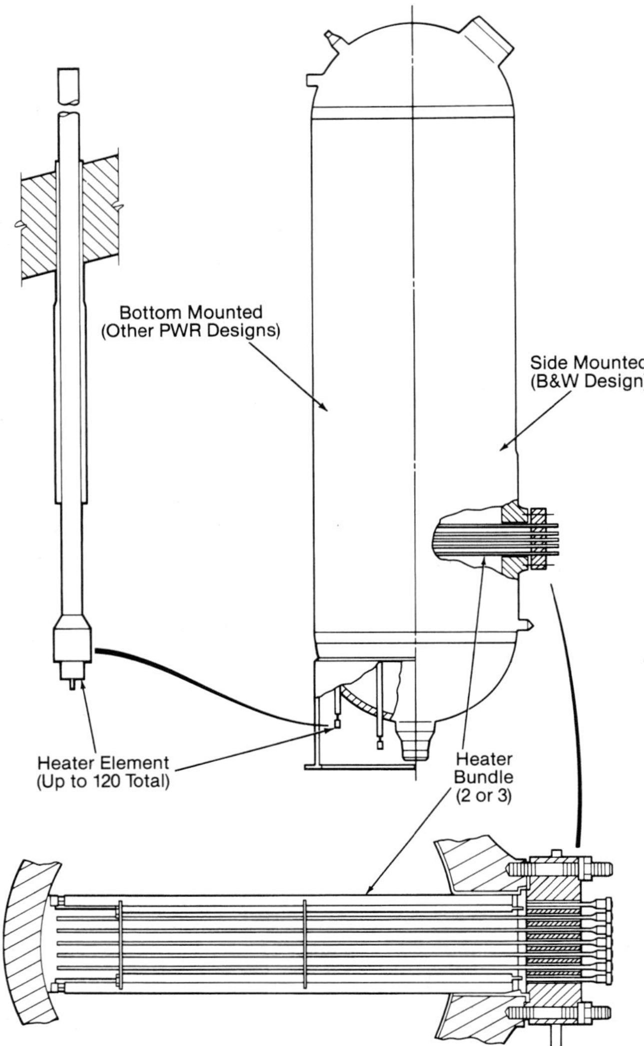

Fig. 13 Pressurizer heater bundles – bottom mounted (left) and side mounted (right).

moisture. If moisture is present in the connector area, the connector pins may be damaged by arcing and may break when removing the power cable from the heater. This damage may be repaired. However, if these intermediate repairs are not successful, the heater element/bundle must be replaced.

Instrumentation and controls (IC)

The critical nature of and general inaccessibility to nuclear plant IC dictate that they be extremely reliable. This reliability is achieved by stringent qualification and by incorporating multiple channels to measure the same parameters. Such *coincident logic* reduces the possibility that a single failure will trigger a safety action.

If a component fails suddenly or gradually degrades, this must be detectable. Detection is achieved by monitoring parameters with the plant computer and annunciator systems and by alarming abnormal conditions.

Checks are often made by comparing the output from redundant sensors. On critical components, these tests are run and documented at specified intervals. Channel calibrations are performed routinely to ensure that instrumentation and electrical devices meet specifications for accuracy and response time. Integrated system tests may also be performed. If a device fails a test, it is replaced or declared inoperable. The number of such failures permitted before output reduction or shutdown is very strictly regulated.

Valves and actuators

Valves in a nuclear plant serve two basic functions: control and isolation. Nevertheless, all valves have their individual requirements of opening and closing time, shutoff tightness, monitorability, etc. In addition, the number of valves of all types in a nuclear plant is very large – and each valve can change the configuration of the plant.

Therefore, in a nuclear plant, a systematic maintenance plan is mandatory. For example, a plan for motorized isolation valves should address packing and seat leakage, opening and closing stroke times, actuator output and lubrication, function of control switches, wear of internal actuator parts, and the reliability of its controller, etc.

Piping

Piping is subject to operating pressure and to imposed forces and motions due to its own thermal and pressure effects and to that of the equipment it serves. Piping is subject to stress, fatigue, cracking, corrosion and to flow accelerated corrosion (FAC) which is an ongoing industry challenge of pipe wall thinning and, in some cases, failure.

Piping and pipe welds are examined using ultrasonic testing (UT), liquid penetrant (PT) or magnetic particle testing (MT). Where UT is required, it may be performed manually or with automated scanners. Complex shapes, such as nozzle intersections, can be scanned using robotic manipulators as depicted in Fig. 14. Robots and scanners provide precise positioning and automatic data analysis with minimal personnel radiation exposure.

Equipment diagnostics

Loose parts monitoring Loose parts, or foreign objects, such as bolts and small metal pieces caught in the high velocity flow of the reactor coolant system can bounce around in the primary system and cause significant damage to areas such as steam generator tubesheets and seal welds. While loose parts are seldom encountered, the NRC requires a system to detect them. A typical system features acoustic sensors in the primary system to detect loose parts. These sensors, or accelerometers, detect the noise generated by impacting loose parts. The system records impact data and provides an audible alarm for impacts beyond a prescribed energy level.

Foreign object damage to steam generator tubing on the secondary side persists; sources include the windings of upstream valve gaskets which unravel in service or bits of metal, weld slag and tooling. In new replacement steam generators with high corrosion resistance, foreign object damage is often the most significant remaining failure risk. Such foreign object change is normally detected by eddy current and/or visual tube inspection during an outage.

Neutron noise monitoring The internals of the reactor vessel provide support for and alignment of the core. Maintenance of both is important to assure that the control rods can be inserted for reactivity control. The internals are subjected to high velocity reactor coolant flow, which causes them to vibrate. Although some vibration is normal, excessive amounts must be detected. During power operations, neutron flux is monitored by special instrumentation located outside the reactor vessel. Certain characteristics of this neutron signal are affected by vibration of the reactor internals. Analysis of this signal can detect abnormal vibration patterns. Measurements are typically made at the end of a fuel cycle and at the beginning of the next cycle to verify that the internals are aligned and secure.

Reactor coolant pump monitoring Two principal pump characteristics are monitored, shaft vibration and seal performance. Accelerometers and noncontacting displacement probes located on the pump provide continuous vibration monitoring. The displacement probes are mounted radially; the accelerometers are located on and near the pump bearing. Monitoring equipment in or near the control room records the sensor data and sounds an alarm if certain limits are exceeded.

Seal performance is also monitored. Parameters such as seal cavity pressures, seal injection flow and temperature indicate when seal replacement is needed.

Valve monitoring and diagnostics The valve population of a typical nuclear plant is very large and all valves are critical to plant operation. Nuclear plant valves control and direct flow, isolate systems and relieve pressure. Motor operated valves (MOVs) and check valves are especially important because they provide isolation for critical systems and are often required to operate under severe temperature and pressure conditions. To determine the condition of these valves, diagnostic systems have been developed.

MOVs provide isolation. They may remain open or closed during plant operation. The ability of the MOV to function is directly related to the thrust applied to the valve stem to overcome differential pressure and subsequent high friction. However, control switches limit the thrust of the motor actuator to avoid damaging the valves and internals. Special diagnostic systems determine the available thrust of the actuator and assure that the switches are properly set. Preventive maintenance programs use the data generated by these diagnostic systems.

Check valves are passive devices that only permit flow in one direction. Because the active parts of the valves are wholly contained in the fluid stream, it has been difficult to determine functionality without disassembly. However, diagnostic systems that evaluate in-service check valves have been developed. These systems use ultrasonics and acoustics to evaluate the disc position and the impact of the disc on the seat and back stop. They can detect wear at the hinge pin and flutter of the disc, which can lead to premature failure. As with MOV diagnostics, data from these systems are used in scheduling maintenance.

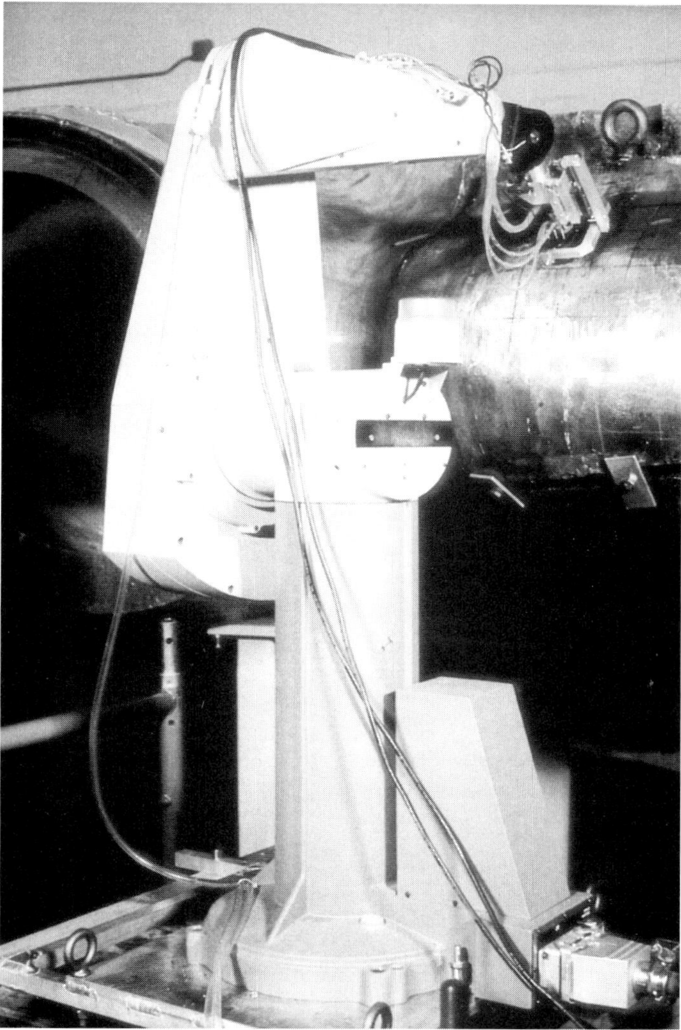

Fig. 14 Automated UT inspection of complex pipe shapes.

Refueling

Major activities

During refueling, approximately one-third of the fuel assemblies are removed and replaced and the remaining assemblies are usually moved inward within the core. Scheduled maintenance is performed during refueling outages. Depending on the extent of maintenance required, either that maintenance or the refueling operation determines the length of the outage.

Fuel movement activities are centered around the reactor vessel and the spent fuel building. To access the fuel, the reactor vessel must be opened. The following major steps are involved in this process:

1. remove the missile shields,
2. vent the primary system,
3. uncouple the control rods from their drives,
4. raise and park the drive lead screws,
5. disconnect the power to the control rod drives,
6. install the canal seal plate,
7. detension (loosen) and remove the reactor vessel studs,
8. install the guide studs,

9. install the lifting rig, lift the vessel head and place on a stand,
10. install the internals guide fixture,
11. raise the water level in the refueling canal,
12. install the lift fixture, remove the upper internals and place on a stand under water, and
13. retract and park in-core instruments (detectors).

The fuel can now be moved.

These operations are shown in Fig. 15. Fuel movement is performed under water with permanently installed equipment. When each fuel assembly is removed, it is transferred to the on-site spent fuel storage area. Here, control rod assembly changes and the appropriate inspections are made. Fuel that is to be reused is moved back to the reactor building and placed in its new core location; new fuel is also installed at this time. Final verification of fuel assemblies is recorded with underwater video equipment.

Fuel handling equipment

Fuel handling equipment can be grouped into three areas: 1) main fuel handling, 2) fuel transfer, and 3) spent fuel/new fuel handling. The main fuel handling system, used to remove and install fuel in the reactor vessel, consists of a motorized bridge spanning the canal or cavity and riding on rails. A handling mast, mounted to and extending below the carriage, is positioned over the fuel assembly to be moved. The grapple, or gripping device, is then lowered and engages the assembly. Once engaged, the grapple is raised and the fuel is drawn into the hollow mast. Sensors indicating load and fuel assembly position assist the operator. The bridge is then moved to the fuel transfer area, where the fuel is lowered into a vertical transfer basket and released from the grapple. At this point, the basket is lowered to the horizontal position and is moved through a transfer tube into the spent fuel area. Finally, the fuel is upended and moved to storage.

The main fuel handling and transfer systems are only accessible during a plant shutdown, because they are located inside the reactor building. As a result, maintenance on this equipment is scheduled at the beginning of an outage.

Reactor vessel disassembly and reassembly

To allow access to the fuel, the reactor vessel head and upper internals must be removed from the reac-

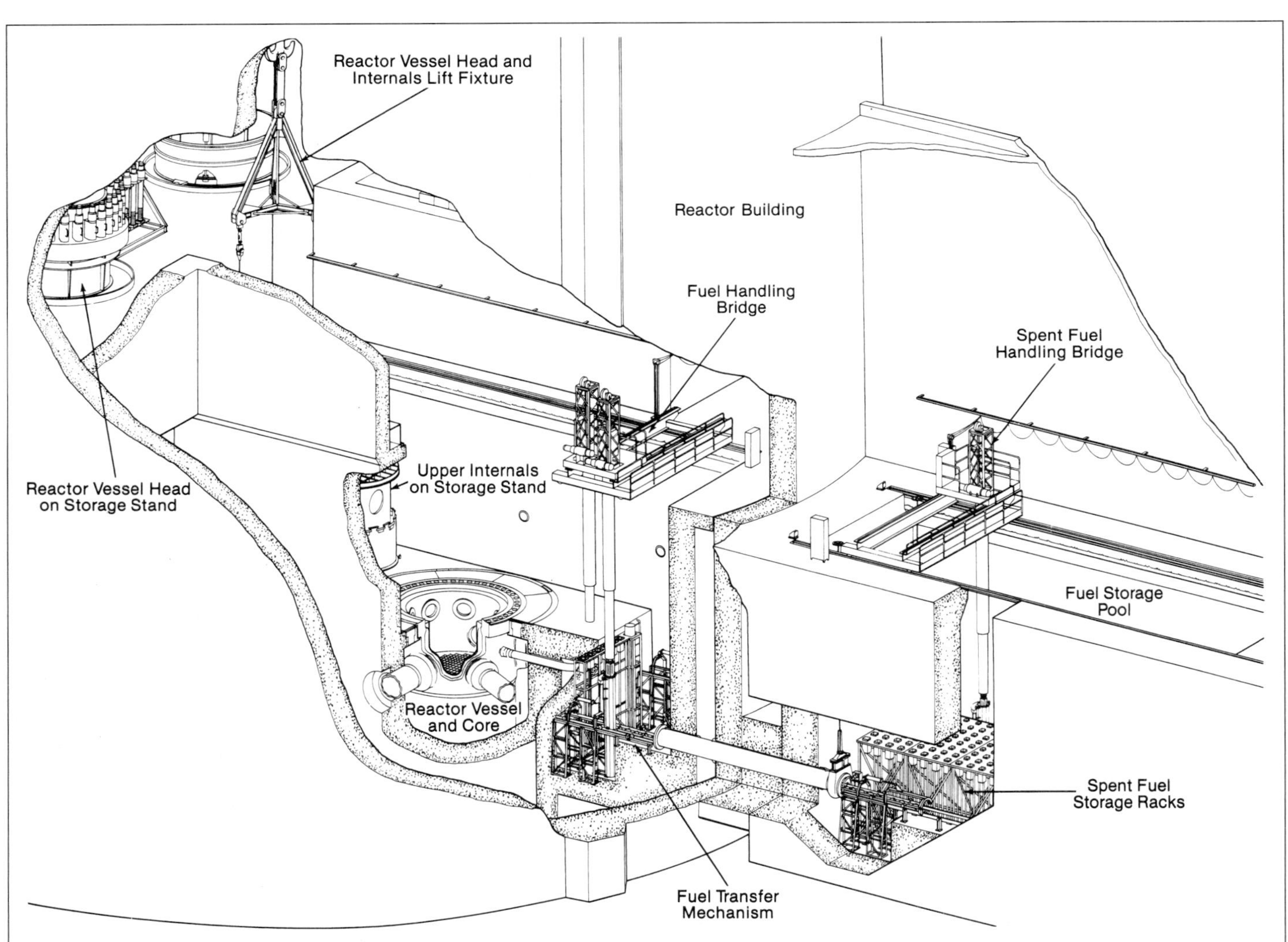

Fig. 15 On-site fuel movement system.

tor vessel. In the B&W nuclear steam supply system (NSSS), each control rod drive must be disconnected from its control rod before this operation can begin. The control rod remains with the fuel. Special tooling is used to reach down through the top of the drive, and to attach to and rotate the lead screw and disconnect it from the control rod. The lead screw is then raised and parked inside the drive assembly. Other NSSS designs require similar procedures; however, the sequence and exact techniques may differ.

The tensioned studs and nuts that secure the reactor vessel head to the reactor vessel must be detensioned and removed before fuel handling can continue. A typical hydraulic detensioning operation is shown in Fig. 16. Generally, three detensioners are used simultaneously and follow a prescribed pattern. Once the studs are detensioned, they are screwed out of the threaded holes in the reactor vessel flange and either removed or parked on the head. A lifting fixture is then attached to the head. The head is then raised using the reactor building crane and moved aside to its storage stand inside the reactor building.

The upper internals are lifted in a similar manner; however, due to their highly contaminated and activated nature, the lifting fixture is attached remotely. The upper internals are stored on a stand under water in the refueling cavity.

Once the refueling operations are completed, the upper internals are reinstalled, the head is placed back on the reactor vessel and the studs are tensioned. In preparation for these activities the flanges on both the reactor head and vessel are cleaned and new O-rings are installed on the head flange. The control rod drives are recoupled to the control rods, power cables and cooling water lines are reattached and, at the appropriate time, the system is filled and vented.

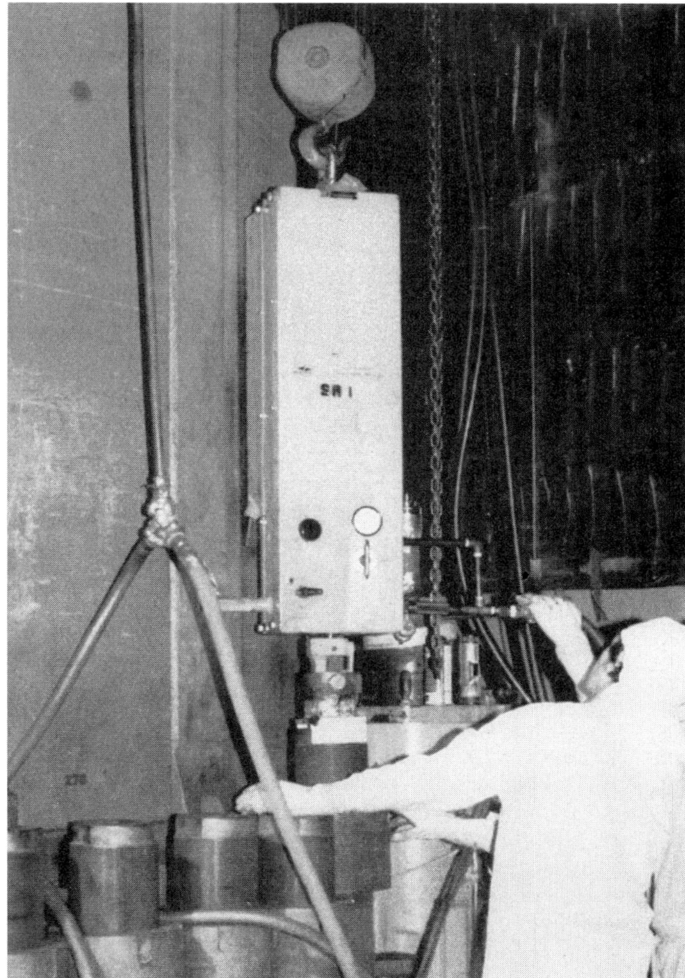

Fig. 16 Service technicians operating reactor closure head stud detensioning equipment during a plant service outage.

Radiation detection and monitoring

Because no human sense can detect nuclear radiations, special instruments and techniques are used for detecting and measuring radiation. In general, three types of checks are required as follows:

1. Measurement of beta and gamma radiations by instruments such as the Geiger counter and *ionization chamber* provide instantaneous indications of the total level of β and γ. Most instruments are read as dose rate, usually in Sieverts. All persons in areas where radiation may be encountered normally wear a film badge or a thermoluminescent dosimeter (TLD). These devices are used for official records for exposure control. However, one's exposure can not be determined without evaluating the device with special processes or instrumentation. These devices measure the amount of β and γ radiation a person absorbs in terms of the total dose of radiation for a day, week or month.
2. Protection against neutrons is provided, primarily, by the concrete and water shielding around nuclear reactors. When a reactor first goes into service, neutron counters as well as gamma counters are used to make a careful survey of the surrounding area. Should leakage exist, the shielding is modified. Thereafter, periodic checks are made for γ and neutron leakage. In experimental installations, or where neutron surveys indicate the need for continuous surveillance, special film badges and neutron dosimeters may also be used.
3. Keeping the concentration of alpha emitters in working spaces and in outgoing air below prescribed tolerances is sufficient to protect against α particles. Alpha counters monitor air quality in working spaces on a regular schedule. Samples of dust accumulating in various areas of a room are monitored periodically for alpha radiation. The air concentrations are then converted into internal exposures for the worker. Whole body counters and urinalysis can also help ensure that the workers' internal exposures are well within the regulatory limits. Portable alpha *friskers* are also used to monitor hands, shoes and clothing of personnel leaving risk areas.

When high radioactive air concentrations occur, other precautions such as respirators, self-contained breathing air and glove boxes are used to protect the workers from internal exposures.

Nuclear operations

Nuclear power plant operations and services described here are generally applicable to light water moderated and cooled pressurized water reactor (PWR) systems, as depicted in Fig. 17, and are modeled on plants with reactors designed by B&W. As design details vary among PWR types, operational differences are noted.

Major licensing requirements

In the U.S., civilian use of nuclear energy was authorized by the Atomic Energy Act of 1954. The act also established an Atomic Energy Commission (AEC) for developing and regulating nuclear energy. A major objective of the AEC was to protect the public health and safety. The Energy Reorganization Act of 1974 replaced the AEC with the Nuclear Regulatory Commission (NRC), established to regulate the construction and operation of nuclear power plants as well as research facilities. Public health and safety continue to be a major goal of the NRC.

The safety of nuclear power plants is based on:

1. *Defense in depth* This is a feature of PWR reactors based on three barriers between the nuclear fuel and the environment: 1) the fuel cladding, 2) the reactor coolant system (RCS) pressure boundary (i.e., the envelope created by reactor vessel, pumps, piping, pressurizer and the steam generators and their tubing), and 3) the reactor containment building (typically a pre-stressed concrete structure with an inner steel liner).
2. *Design approval* This includes NRC review and approval of the site and the design of safety systems and safety-related equipment.
3. *Quality assurance program* Construction of the station and erection of the equipment as well as maintenance and operation are guided by NRC approved quality assurance programs.
4. *Operating license review* Following construction, the NRC grants the utility (owner) an operating license. The license and the technical specifications define the operating bases of the plant.

The nuclear licensing steps, described as follows, are primarily designed to ensure public safety.

Plant licensing

In addition to licensing by the NRC, approvals for the plant site, construction and operating licenses must be obtained from the respective authorities. Additional regulating agencies in the U.S. include the Federal Energy Regulatory Commission (FERC), the Environmental Protection Agency (EPA) and the Corps of Engineers. Public Utilities Commission (PUC), state environmental control board and local governmental approvals are also often required for nuclear power plant siting, construction and operation.

NRC power plant licensing takes place in two major steps: the construction permit and the operating license. Completion of each step entails hearings conducted by an advisory safety licensing board.

In the construction permit stage, certain documentation must be prepared for the NRC. The proposed plant site is characterized, the preliminary design features are outlined, and conformance to the Code of Federal Regulations standards is presented.

A preliminary safety analysis report, which contains design and safety information, is prepared by the utility and its contractors, and is submitted to the NRC. The review includes questions by the NRC and subsequent responses from the plant owner. Along with the review, these become part of the public record and licensing process. The NRC staff prepares a safety evaluation report as a result of their review. This report leads to an Atomic Safety and Licensing Board hearing, which is open to the public. Upon approval by the board, a construction permit is issued by the NRC and plant construction can begin.

As the plant design becomes final and construction proceeds, the information contained in the review is revised and expanded. A complete description of the site characteristics, safety systems, quality assurance programs and technical specifications is included. This revised information becomes the final safety analysis report. This report is also submitted to the NRC for review and approval. As with the preliminary report, the review includes NRC questions and owner responses. Following review, the NRC again prepares a safety evaluation report of the site and plant. This provides the basis for a second public hearing to approve an operating license for the utility. During construction, on-site NRC inspectors are assigned to assure conformance to the final report.

Upon receipt of the operating license, the utility performs system pre-operational and hot functional testing and prepares to load fuel. Approval by the NRC is required for fuel loading and low power operation at about 5% of full power. Following successful fuel loading and low power operation, the NRC grants permission for power escalation.

In 1989, the nuclear power plant approval procedure in the U.S. was revised to allow one-step licensing. The objectives of this simplified procedure are to streamline the licensing process, to promote standardized vendor designs and to minimize NRC required design changes during and after construction. It also provides for combined licensing (construction and operation) which allows the owner and other stakeholders a much higher level of confidence in the viability of a project.

The simplified rule differs from the original licensing process in that: 1) a whole plant design certification by the NRC is required; 2) following design certification, a combined construction and operating permit may be obtained; and 3) site permits may be obtained prior to design certification.

Technical specifications

The technical specifications for a system or plant design are, in the broadest sense, the licensed limits for operation. These specifications are based on the analyses and evaluations of the final safety analysis report and include the following.

Safety limits Limits are imposed on important process variables to prevent an uncontrolled release of radioactivity. If any limit is exceeded the reactor must be shut down, the licensee must notify the commission, review the matter, and record the results of the review. The review includes the cause of the condition and the corrective action taken to prevent reoccurrence. Operation shall not be resumed until authorized by the commission.

Limiting conditions for operation These conditions are the lowest permissible equipment performance levels for safe operation of the facility. When a limiting condition for a nuclear reactor is exceeded, the licensee shall shut down the reactor or follow any remedial action permitted by the technical specifications until the condition can be met. In most cases, the licensee must notify the commission, review the incident, record the results of the review, and submit corrective actions.

Surveillance requirements Safety-related nuclear power plant equipment must be tested periodically; frequency depends on the specific equipment. If safety equipment fails a surveillance test, corrective action must be implemented within a specified time. If corrections can not be made within this time period, the plant must be shut down.

Design features Certain plant design features are included in the technical specifications. These features, such as construction materials and layout, would affect plant safety if altered.

Administrative controls Administrative controls are the provisions relating to organization, management, procedures, record keeping, review and audit, and reporting that are necessary to assure safe operation of the facility.

Pre-operational testing

System checkout During nuclear power plant construction, it is desirable to test each system (electrical, hydraulic, pneumatic, etc.) as soon as possible. Pre-operational testing is integrated into the plant erection schedule so that system testing can begin as soon as support systems are available. Test procedures are prepared for each component and system. The tests are performed and witnessed by designated engineers and reviewers.

Pre-critical testing is performed on all systems and components. These tests are conducted on the reactor coolant system, the feedwater system, the steam system, the reactor makeup and purification system, the emergency feedwater system, and the control and protective systems. In general, pre-critical testing consists of the following sequence of events:

1. *Cleaning and inspection* Fluid systems are flushed to remove construction debris and are inspected for cleanliness and proper installation of piping and components.
2. *Electrical tests* Alternating current (AC) and direct current (DC) power wiring and distributions are inspected and tested for proper hookup and circuit continuity. Rotating and moving equipment, such as pumps and valves, are given no-load tests to determine proper direction of motion.
3. *Instrumentation tests* Instrumentation and control wiring is inspected and tested for proper hookup and electrical continuity. The instrumentation (transmitters, temperature sensors, flow meters

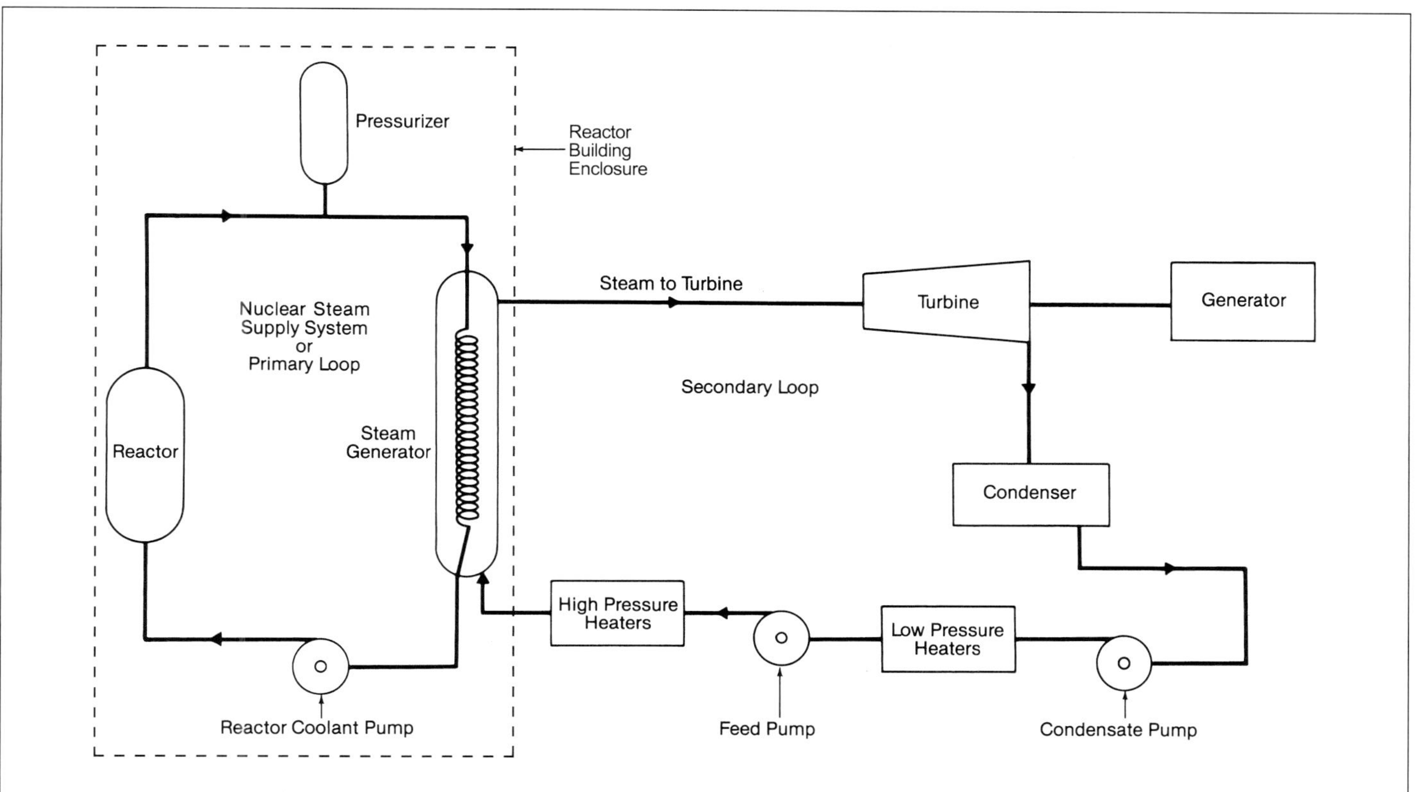

Fig. 17 Pressurized water reactor (PWR) primary and secondary heat transport and electricity generation systems.

and level indicators) is calibrated and the control systems are tested. A control simulator is used by some utilities for cold system testing.

4. *Hydrostatic tests* All fluid systems and components except the reactor coolant system (RCS) are cold tested hydrostatically to greater than design pressure. Because of the relatively high nil ductility temperatures of the reactor vessel welds, the RCS is hydrotested at a temperature slightly above ambient.
5. *Functional tests* Functional tests which demonstrate component performance are conducted. Typically, system flow rates, pump head capacity relationships, valve stroke times and controls are tested.
6. *Operational tests* Additional tests at near normal operating conditions are performed for low temperature and pressure systems. Instrument air systems, the spent fuel pool cooling system, component cooling water systems and the decay heat cooling systems are typical systems tested.

Performance of the high temperature and pressure systems, such as the RCS, the makeup and letdown system, the feedwater system, the steam system and the emergency feedwater system are verified during hot functional testing.

Water quality Many of the elements of water treatment, water chemistry and corrosion discussed in Chapter 42 for fossil fuel systems also apply to nuclear steam systems. Key differences exist between these treatments, however. In a pressurized water reactor (PWR) system, two independent flow circuits are used as discussed earlier: the primary or reactor coolant system and the secondary or steam generator condensate feedwater system. Additional differences include more extensive use of high alloy steel, nickel-based alloys and other non-steel materials. The ultimate goals of the water treatment system remain to minimize corrosion and deposition. The basic premise of the nuclear steam supply system chemistry program is that it is better to prevent contaminants (both ionic and particulate) from entering the cycle than to attempt to remove them later. This is the basis of the ALARA (As Low As Reasonably Achievable) chemistry philosophy employed at nuclear plants.

For their primary circuits, PWR plants use chemical shim reactivity control which utilizes a soluble neutron poison (boric acid) to regulate core reactivity. In addition, a pH chemical additive, lithium hydroxide, is used to adjust the pH toward the alkaline side. The industry has developed a set of guidelines for the proper concentrations of these chemicals, including the reduction of boric acid concentration as fuel burnup progresses.[1,2]

These guidelines attempt to minimize the impact of water chemistry on the integrity of the primary system internals, including the pressure boundary and fuel cladding, and to minimize the transport of radioactive corrosion particles throughout the system. In addition to lithium and boric acid concentration, primary system chemistry control parameters include chloride, fluoride, dissolved hydrogen and dissolved oxygen. Hydrogen is added to maintain reducing conditions in the primary coolant to minimize primary system corrosion.

For the secondary circuit, the major goals of the chemistry program are to minimize both corrosion product transport and the corrosion of steam generator tubes. Corrosion product transport and corrosion of the tube material is mainly affected by the water chemistry related factors of: 1) pH, 2) electrochemical potential (ECP) and 3) the particular species beyond those which affect pH (e.g., lead, copper, and sulfur). Corrosion is also affected by material susceptibility, tensile stress and operating temperature.

The pH of the secondary circuit is controlled by the use of volatile amines (e.g., ammonia, morpholine, and ethanolamine). Reducing conditions are maintained in the secondary circuit by the use of a volatile oxygen scavenger (e.g., hydrazine). Strict water purity requirements are imposed to minimize the transport of contaminants in the system. To achieve maximum system reliability, the industry has developed a set of secondary-side water chemistry guidelines for the proper control of operating chemistry conditions.[3]

Hot functional testing (HFT) During commissioning and prior to installation of the fuel and plant operation, a process called hot functional testing is implemented. This serves to check out the operation and performance of all of the functioning systems. It also serves to establish a protective oxide layer on all carbon steel and alloy materials in the primary and secondary heat transport systems which, in turn, minimizes corrosion during normal operation.

HFT occurs after all nuclear plant systems have been erected and have completed pre-operational testing. This testing provides warmup and operation at zero power operating conditions. The fuel is not loaded in the core for HFT. In B&W plants, a plate is placed in the core region to simulate the core pressure drop for the RCS pump and flow testing.

The HFT begins by establishing water levels in the RCS as indicated by the pressurizer level in the steam generators and in the condenser hot well.

The pressurizer heaters ( ) are energized to raise the RCS to a pressure that provides a net positive suction head of about 250 psig (1.72 MPa). This permits operation of the reactor coolant pumps. Upon reaching this pressure, the reactor coolant (RC) pumps are energized to heat the reactor coolant and steam systems at a rate of about 30F/h (17C/h) from about 70 to 530F (21 to 277C).

There are several temperature hold points during the HFT warmup at which specific tests are performed. Typical testing includes the pressurizer spray bypass flow calibration. This bypass has a trickle flow of about 1 GPM (0.06 l/s), which keeps the pressurizer spray line and the spray nozzle at a relatively uniform temperature.

At about 600 psig (4.24 MPa), the RCS high pressure safety injection pumps are tested to demonstrate flow capability and to adjust the safety injection flow distribution.

Following testing at the hold points, system heating continues to the hot zero power (HZP) level. Typically, in pressurized water reactors, the HZP temperature and pressure ranges are:

RC temperature – 500 to 550F (260 to 288C)
RC pressure – 2100 to 2200 psig (14.6 to 15.3 MPa)
Steam temperature – 525 to 570F (274 to 299C)
Steam pressure – 800 to 1200 psig (5.62 to 8.37 MPa)

Before fuel loading, the reactor protection system (RPS) and the control systems are calibrated and checked.

In addition, all protective system instrumentation, including the out-of-core neutron detectors (Fig. 18b), RCS pressure transmitters and temperature sensors, nuclear instrumentation and transmitters, RC pump monitors and the feedwater pump turbine hydraulic oil pressure switches, are calibrated and tested.

The reactor, coolant system pressure and inventory, feedwater, steam pressure and electrical generation are controlled by the integrated control system (ICS) (Figs. 18a and 18b) and the non-nuclear instrumentation (NNI). The NNI provides most of the instrumentation for the reactor, feed and steam systems, including signal inputs to the ICS, the control console indicators and recorders, and the plant computer as well as many of the test inputs to the reactimeter. The ICS is made up of unit load demand controls, integrated master controls, turbine and turbine bypass system controls, feedwater controls and reactor control subsystems. The ICS also controls the reactor, feedwater flow and the turbine-generator output to meet the electrical power demand.

The NNI also controls the RC system makeup flow, the pressurizer heaters, the pressurizer power operated relief valve (PORV) and the reactor coolant pump seal flow.

The feedwater and steam control subsections of the ICS provide automatic control over the full power range; manual reactor control is used up to 15% power and automatic control is available from 15% to full power. Manual control can be used in each subsection of the ICS or its actuated components such as control rods, feedwater valves, feed pump turbine and turbine bypass valves. Alternately, the turbine can be manually operated from the control room.

Below 15% power, the ICS maintains the steam generator water level at a constant set point and the turbine bypass system similarly controls the steam pressure. The RCS temperature rises with power. The temperature is typically 532F (278C) at zero load and increases to about 580F (304C) at 15% power. As the load increases to full power, the temperature remains at about 580F (304C). Above 15% power, the RCS temperature and steam pressure are controlled by the ICS.

PWRs designed and built by other suppliers use recirculating steam generators (RSG), which have different operating characteristics than the once-through steam generators (OTSG). In the RSG plants, the RCS average temperature rises with the power level, the steam generator operating pressure drops with increasing power (Fig. 19), and the steam generator water level is controlled to a constant RSG secondary inventory or to a relatively narrow variable RSG band.

An RSG plant's average RCS temperature is controlled to a variable set point; the feedwater flow is controlled by a multi-element unit that monitors steam flow, feedwater flow, steam generator water level and, for some plants, RC temperature and steam pressure.

The RSG units typically control electrical output by controlling the turbine to a variable first stage pres-

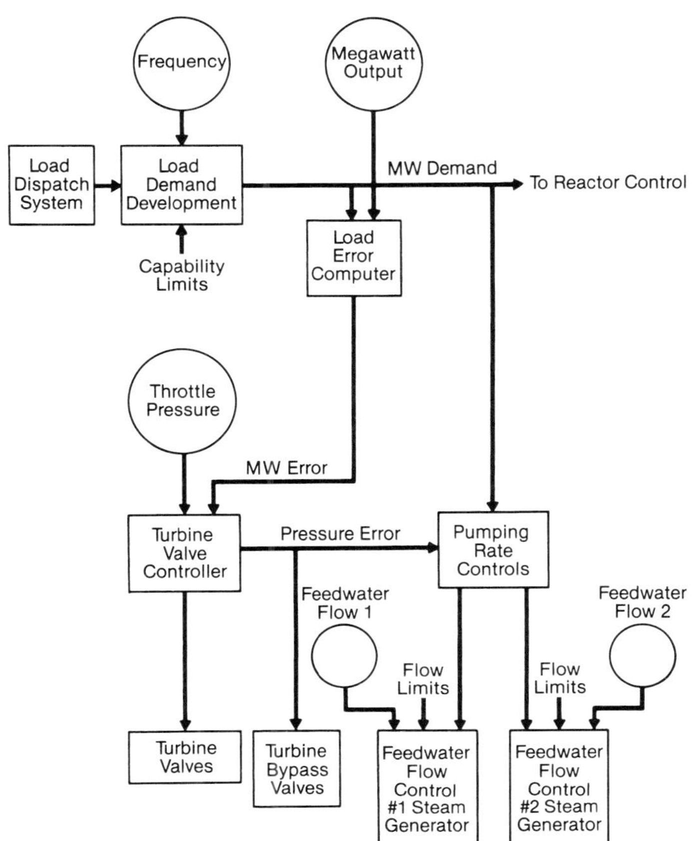

Fig. 18a Integrated control system.

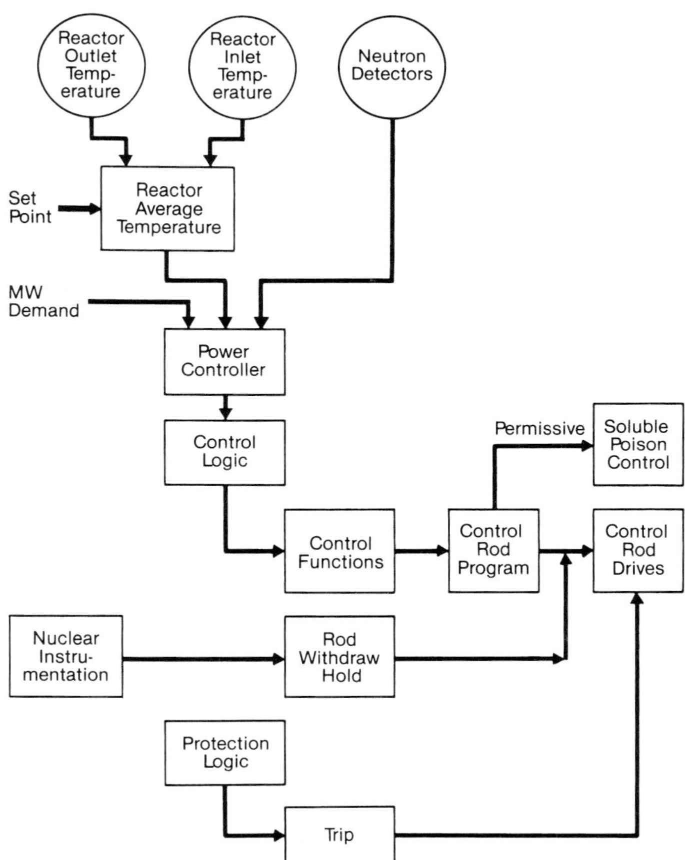

Fig. 18b Integrated control system – reactor controls.

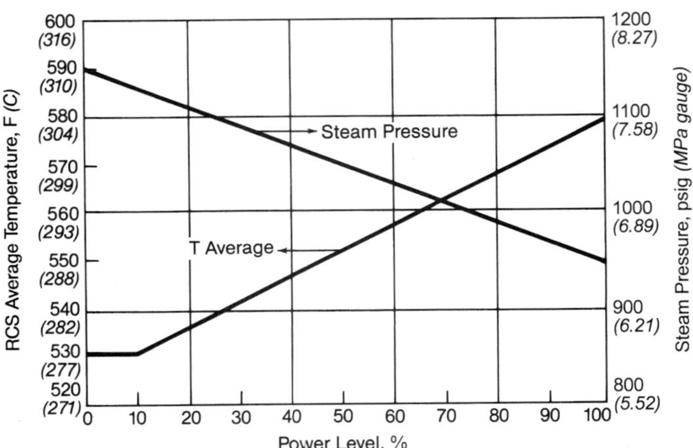

Fig. 19 Typical recirculating steam generator plant reactor coolant system (RCS) average temperature and steam pressure.

sure signal. This control allows the steam pressure to change with power level.

The duration of an HFT is usually 500 hours for a first-of-a-kind unit and about 250 hours for all other units.

A significant portion of the HFT period is allocated to operator training because the plant conditions are near those of normal operation. In performing the hot tests, the operators gain valuable experience in managing the systems under normal and abnormal conditions. This operating experience constitutes one of the final stages of operator training prior to the NRC operator's license examination.

After completion of the test, the RCS is cooled and depressurized and the reactor vessel head is removed. The steam, feedwater and fluid safety systems are inspected and modified as necessary. Following successful HFT, permission for fuel loading is granted by the NRC.

Operator training U.S. regulations require that the reactor operators and supervisory personnel be licensed. An individual who has the basic operational qualifications is called a reactor operator; a senior reactor operator has completed more intensive training. The utilities develop, manage and conduct NRC approved operator training and requalification programs to satisfy regulatory requirements and to maintain the required operating staff. The training programs cover all facets of nuclear operations, including equipment malfunctions, design faults and severe transients. An operator's license is issued for a period of two years and is renewable by requalification.

Fuel loading and zero power testing

The nuclear fuel is first loaded into the reactor vessel only after pre-loading tests and practice fuel handling operations are completed. Neutron detectors and sources placed in the reactor vessel are used to monitor neutron multiplication during the core loading process.

Fuel loading takes place with the reactor vessel and the fuel handling pool filled with ambient temperature demineralized water. The borated water is a neutron poison and assures a multiplication factor of 0.95 or less with a fully loaded core.

With the system prepared, fuel loading begins. Neutron multiplication measurements, slow fuel assembly insertion and frequent pauses to relocate sources and neutron monitors take place until all fuel assemblies are installed. After loading, the system is closed and prepared for low power. Testing following fuel loading includes calibrating neutron source range detectors, assuring sufficient source range signal for warmup to hot zero power, and verification that the reactor is sufficiently subcritical for the warmup.

Following completion of this testing, safety control rod groups are withdrawn to provide a subcritical shutdown margin should a reactor trip be required during the warmup.

After reaching about 250 psig (1.83 MPa) in the RCS and with one or more RCS pumps running, the system high points and control rod drives are vented to release noncondensable gases. Venting of the drives is particularly important because gas in the control drive path can interfere with the hydraulic snubbers which prevent impact of the rods upon trip.

Upon reaching HZP conditions, excess heat from the RCS pumps is removed by the turbine bypass system. This maintains constant RCS temperature and steam pressure conditions.

Initial reactor criticality is achieved in one of two ways. One method is to dilute the boric acid concentration in the RCS until a subcritical margin is achieved. Criticality is then achieved by slowly withdrawing the control rods. A second method is preferred by most B&W plant owners. All control rods are withdrawn while maintaining subcriticality by boric acid concentration. The concentration is then diluted until criticality is achieved. After initial criticality is attained, a series of zero power physics tests are performed:

1. *Control rod reactivity worth* This verifies that the individual rod worth, control rod group (banks) rod worth, and total rod worth are within acceptance criteria. The tests must show that the control rod worth is capable of power operation. Performance of the safety function, which brings the reactor to a subcritical condition following a trip, is also verified. Typical control rod reactivity worth for each of seven banks is shown in Table 2.
2. *Core reactivity coefficients* The reactivity coefficients are the moderator deficit (or coefficient), the fuel temperature (Doppler) coefficient, the RCS pressure coefficient and the power (overall) coefficient. This testing verifies that the reactivity coefficients are within acceptance criteria and that the power coefficient is negative at levels above about 20% power.
3. *Soluble poison (boric acid) reactivity worth* Tests are conducted by exchanging control rod position with soluble poison concentration to determine that the reactivity worth of the poison is within core acceptance criteria. Boric acid dissolved in the reactor coolant is used to limit the core's reactivity throughout the fuel cycle. The boron concentration at HZP is about 1500 ppm. As xenon and other reactivity poisons build into the fuel and the fuel is depleted, the boron concentration is reduced to maintain full power (see Fig. 20).

Table 2
Control Rod Group Reactivity Worths at Hot Zero Power, Beginning of Fuel Cycle

Group	Purpose	Reactivity Worth % Δ K/K
1	Safety	0.95
2	Safety	3.05
3	Safety	0.79
4	Safety	1.75
5	Control	1.15
6	Control	1.26
7	Control	1.17
Total		10.12

4. *Nuclear instrumentation range and overlap tests* At criticality, the three neutron detector ranges (startup, intermediate and power) must be on scale and must have the proper overlap. Testing is conducted at initial criticality to assure proper neutron detection signal strength.
5. *Pseudo-control rod ejection tests* In addition to the standard control rod testing, pseudo-control rod tests are performed. This testing assures that the control rod with the most reactivity worth does not exceed the licensed limits.

The zero power physics tests as well as testing of the protection systems are performed to:

1. measure operational parameters,
2. measure performance against design criteria,
3. provide compliance with NRC requirements, and
4. obtain data for design improvements and potential power upgrades.

Most initial operation parameters are recorded by special equipment and by the plant computer.

Power ascension tests

After completion of the zero power physics tests and low power operation (5% or less), the power is raised in stages. Typically, the power plateaus are 15, 45, and 75% and full power.

The objectives of power escalation include:

1. determining the plant operating characteristics,
2. familiarizing the operators with the plant,
3. verifying that the plant is capable of operating safely and within the design and technical specifications,
4. verifying that the operating procedures provide reliable operation, and
5. bringing the unit to rated capacity.

Testing at each stage includes operating the reactor systems and the entire power plant. The test time at each plateau is sufficient to provide adequate operator familiarization and training.

Operation to the first test plateau is by manual control of the reactor and with the turbine-generator off-line. Plant heat removal is by the turbine bypass system, which controls steam flow from the steam generators to the condenser. In this low power range, the operating performance of the turbine bypass system, feedwater system, and the feedwater and steam pressure controls is tested for the first time. Initial heat balances are performed by calculating the primary heat balances (RCS flow and temperature change across the reactor) and the secondary heat balances (feedwater flow and enthalpy rise across the steam generators).

The heat balances, in addition to RCS flow measurements, permit calibrating the nuclear instrumentation using the plant thermal power as a baseline. The secondary heat balances also provide an initial cross check of the RCS flow measurements.

Near the completion of 15% power level testing, a reactor shutdown from outside the control room is performed. Reactor control is from a remote panel and the reactor is brought to and maintained at HZP conditions. This test demonstrates safe control and shutdown of the unit should a fire or toxic gas require the operators to evacuate the main control room.

The performance at the next power plateau is extrapolated from the previous plateau. For example, operation at the 45% power plateau is estimated from the operating experience at the 15% plateau. Following satisfactory operation at 15% power, the system may be placed in automatic control and the power level raised to the 45% level. Power escalation, however, is typically performed with the reactor in manual control and the feedwater and steam systems in automatic control. As the power is raised above 15%, the turbine-generator is brought up to speed, loaded and synchronized to the electrical grid. At the 45% power level, heat balances are again calculated and the neutron power and RCS flow are calibrated. Control system

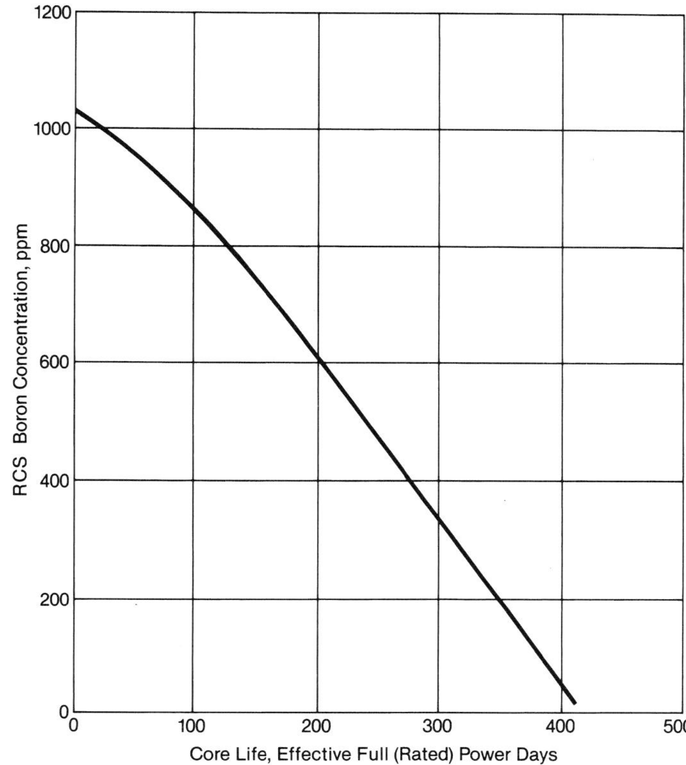

Fig. 20 Boron concentration in the RCS during a fuel cycle.

tests, consisting of decreasing and increasing power ramps, are also conducted. At the 45% power level, the in-core instrumentation (rhodium detectors) is on-line, and the core performance indicators can be investigated by mapping the core power distribution.

Following successful testing at the 45% power level, the power is raised to about 75%. During this increase, the second main feed pump is typically placed in operation at about 45 to 50% power. Heat balances at 75% power are again calculated for the reactor and the secondary systems, and the neutron detectors and RCS flow are calibrated.

Testing with three of the four reactor coolant pumps is conducted at 75% power. This provides the core and RCS temperature distribution characteristics. It also verifies proper feedwater flow to the steam generators.

Ramp power changes at rates up to 10% per minute are also conducted at the 75% power level. This tests the control system operation and permits control setting changes.

As testing is completed at the 75% power level, the unit's performance is reviewed and its full power performance is extrapolated. The reactor is then gradually raised to full power and further testing preparations are made.

At full power, the RCS and the secondary system heat balances are again performed, the neutron instrumentation is calibrated, and final reactor coolant flow calibrations are performed.

During steady-state, full power operation, power maps are obtained from the in-core neutron monitors and the core neutron flux distribution is evaluated against the acceptance criteria. Final tests of the reactivity coefficients, control rod worth and boron reactivity worth are also performed and compared with design values.

Several important safety tests are conducted at the 100% power level, including:

1. *RC flow coast down testing* For first-of-a-kind plants, the RCS pumps are tripped in various combinations to characterize the flow coast down performance and to test the trip functions related to loss of RCS flow or pumps.
2. *Reactor trip testing* A reactor trip test is performed from full power to verify that the control rod insertion rates, the pressurizer performance, and the control of steam pressure and feedwater flow are within acceptance criteria.
3. *Loss of off-site power* Testing of the unit's response to separation from the electrical grid is conducted to verify that performance during the subsequent reactor trip, startup of the emergency diesel generators, operation of the turbine driven emergency feedwater pumps, and the natural circulation of the RCS are within acceptance criteria.
4. *Control system tests* Steady-state and transient control systems are tested at full power to verify their performance.
5. *Turbine trip testing* The turbine-generator is tripped at full power to verify that the turbine, the main steam system, the RCS, and the feedwater system perform as designed to bring the unit to reduced power.

Acceptance testing

After testing at full power is complete, final acceptance testing is performed before turning the unit over to the utility. Acceptance testing demonstrates that the steam pressure, flow and temperatures meet specifications and that the electrical output of the turbine-generator is as designed. Additional tests, such as transient capability, operation with three of four RC pumps, and load rejection capability are often included in the acceptance testing.

Typically, steady-state acceptance testing to demonstrate that steam conditions and electric generation meet the warranties is carried out for about 100 hours. Parameters such as reactor power, RCS temperatures, feedwater flow and temperatures, steam temperature and pressure, and electrical generation are monitored to maintain a running account of the performance.

Transient tests are scheduled and are typically of short duration. A typical transient acceptance test is as follows. Power is reduced from 100 to 60% at the design rate. Power is held at 60% until peak xenon poison builds into the core, which is normally six to eight hours after power reduction. The unit is then returned to full power at the design rate. This test demonstrates the transient capability of the control systems, as well as the boron-handling system's capability to dilute and then increase the RCS boric acid concentration.

Certain plant parameters, and those necessary to demonstrate performance, are recorded. The data are used for performance evaluation and to demonstrate that the warranties are met.

Upon satisfactory completion of the acceptance tests, the plant is turned over to the utility and begins commercial operation.

Commercial operation

Most U.S. nuclear power plants are base loaded, primarily because of associated lower fuel costs. During base load operation, the plant operates at or near full power, regardless of demand.

Electrical power production

The main control room and its adjacent control equipment, offices and work areas form the center of nuclear plant operations. The operations personnel, including the reactor operators, auxiliary operators and supervisory personnel, perform their functions in and around the control room. All control stations, switches, controllers and indicators necessary to start up, operate and shut down the nuclear units are located here. Operations that maintain safe conditions after upsets are also coordinated from this central control room. Controls for certain auxiliary systems are located at remote stations.

The station operator has many duties including monitoring plant performance and making adjustments as necessary. The operator must periodically reduce the RCS boron concentration, as indicated in Fig. 20. This compensates for reactivity loss due to fuel burnup and maintains full power with the control rods within a specified operating band. The operator must also maintain a log of operations, review and approve maintenance activities, and supervise and perform surveillance tests.

One of the most time consuming functions of the plant operating staff is scheduling and performing mandatory surveillance testing of the safety equipment. There are about 400 surveillance tests required for each nuclear power plant. The emergency diesel generators, for example, are tested monthly. Most fluid systems, such as high pressure injection, emergency feedwater and decay heat removal, are also tested monthly. Finally, the reactor protection system and related safety systems are tested on a similar schedule.

During power production, the staff conducts maintenance on accessible systems and plans the inspection, maintenance and repair work scope for the next planned outage. Preventive maintenance and, more recently, reliability centered maintenance programs have been instituted and have contributed to improved station reliability.

Nuclear power plant operations are audited periodically by teams from the Institute for Nuclear Power Operations and the NRC.

References

1. *Pressurized Water Reactor Primary Water Chemistry Guidelines*, Vol. 1, Rev. 5, Electrical Power Research Institute (EPRI), 1002884, Palo Alto, California, September, 2003.

2. *Pressurized Water Reactor Primary Water Chemistry Guidelines*, Vol. 2, Rev. 5, EPRI, 1002884, Palo Alto, California, September, 2003.

3. *Pressurized Water Reactor Secondary Water Chemistry Guidelines*, Rev. 5, TR-102131-R5, EPRI, Palo Alto, California, 2000.

Inconel is a trademark of the Special Metals Corporation group of companies.

+Point and X-Probe are trademarks of Zetec, Incorporated.

B&W nuclear steam generator U-bend supports provide close-tolerance fit for improved vibration control.

Chapter 50

Nuclear Equipment Manufacture

Many manufacturing operations for nuclear steam supply system components are similar to those used for fabricating fossil fueled equipment. However, the unique requirements and geometry of nuclear components call for special manufacturing methods, equipment and facilities. Nuclear equipment must operate reliably with minimal maintenance because of its operating environment. For this reason there are rigorous equipment specifications by customers and regulatory authorities. Quality assurance requirements are stringent and special attention must be given to cleanliness and material control.

Selecting the manufacturing sequence and methods requires balancing design requirements with manufacturing capabilities while meeting commercial obligations. As a result, special equipment and fabrication sequences are usually required.

Typical component description

Commercial nuclear components are typically cylindrical pressure vessels. Some of the components are large and heavy and require customized shipping and handling arrangements. Reactor vessels and components for pressurized water reactor (PWR) systems up to 32 ft (9.75 m) in diameter, 125 ft (38.1 m) long and weighing 1000 t (907 t_m) have been built by The Babcock & Wilcox Company (B&W). Some nuclear components also have close assembly tolerances. This requires specialized equipment and extraordinary care during machining and welding operations.

Manufacture of reactor vessel components

Different reactor systems have differing reactor vessel designs. B&W has experience in several systems, including the PWR, which is the focus of this chapter.

Each reactor vessel is made up of a cylindrical main section and two hemispherically shaped heads. The reactor closure head is attached by a bolted flange and has several penetrations for the reactor control rods, venting and/or instrumentation (Fig. 1). Control rod drive mechanism (CRDM) tubes penetrate the hemispherical section and are welded at the inside surface. Similar tubes for venting and instrumentation may also be present, and follow the same general arrangement. In recent years, plant owners have replaced reactor closure heads because of stress corrosion cracking of the weld between the head and the reactor CRDM tubes, and boric acid wastage of the carbon steel base material. (See Chapter 46.) The inside of the reactor vessel includes many attachments for the fuel support system. (See Chapter 49.) Since there have been no reactor vessels manufactured recently by B&W, this section will only focus on reactor vessel components, such as the reactor closure head.

Fabrication of replacement reactor vessel closure heads

The closure head has a hemispherical dome with a bolting ring flange. The closure heads for typical B&W reactors measure 16.67 ft (5.08 m) outside diameter (OD), with a dome radius of approximately 7.25 ft (2.21 m) and a nominal dome thickness of 7 in. (178 mm). The bolting ring flange is 30 in. (762 mm) thick.

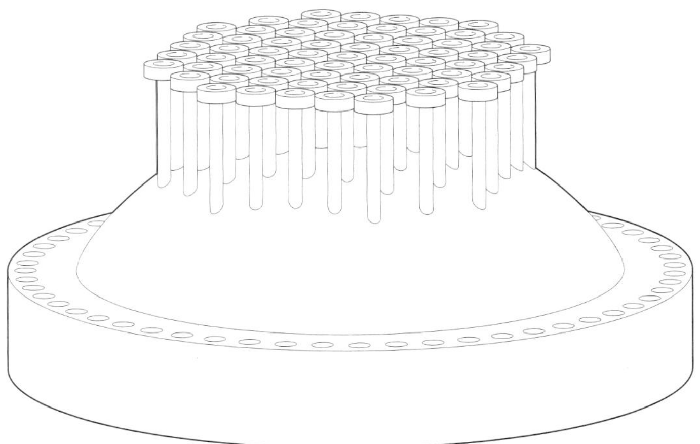

Fig. 1 Reactor vessel closure head.

Typical replacement closure heads are fabricated from a one-piece forging. This eliminates the periodic in-service inspection required for pressure boundary welds. Large diameter heads may have their hemispherical portion made from a single forged partial hemisphere with additional forged rings or segments that are welded together. The bolting ring is forged separately and welded to the hemisphere. All reactor components made of ferritic material have their internal surfaces clad with austenitic stainless steel or Inconel® to provide resistance to corrosion from the reactor coolant. The cladding can be applied using various welding methods such as submerged arc welding (SAW) or gas metal arc welding (GMAW) using wire or strip.

Critical features There are several critical features on the closure head that must be fabricated to close tolerances. The gasket surface on the bolting ring must be flat and have the required surface texture and lay to ensure a leak-tight joint. The CRDM center lines must be true to the axis of the head and the CRDM tube diameters must be held within tightly controlled tolerances to ensure that the operation of the control rods is not hampered in any way. The weld between the CRDM tube and the head is designed and fabricated to avoid residual stress and to maintain the position of the CRDM tube. It is critical to control residual stresses during the manufacturing process to avoid stress corrosion cracking of the closure head during operation. Residual stress is controlled during manufacture by limiting the cold work and by electropolishing the CRDM tubes and welds.

Machining The CRDM tube weld preparations (preps) are machined on a computer numerical control (CNC) horizontal boring mill. Machining the geometry of the weld prep requires careful three-dimensional modeling and programming of the tool path because the geometry varies depending upon the radial location of the hole on the head. The stud holes and CRDM penetrations are drilled using a trepanning cutter (Fig. 2).

Testing and inspection The closure head is tested and inspected to the requirements of the American Society of Mechanical Engineers (ASME) Boiler & Pressure Vessel Code Section III and V. Material testing includes chemistry and mechanical properties and may include other tests such as ultrasonic, hydrostatic and surface finish depending on the material form and application. Fabrication testing includes magnetic particle (MT), dye penetrant (PT), radiographic (RT) and ultrasonic (UT). Pressure testing includes hydrostatic testing of the tube material and the closure head assembly. This is done by using a thick circular disc that is bolted to the head using tooling studs (Fig. 3).

Electropolishing Electropolishing is an electrochemical technique that removes a very thin layer of material from a work piece. Electropolishing can be used to smooth a surface or to remove a cold worked layer of material to reduce residual stress. The work piece is connected electrically as an anode and, along with an adjacent cathode, simultaneously brought in contact with an electrolyte. The amount of material removed is a function of the electrolyte chemistry and flow, the electrical current and exposure time. Electropolishing is applied to CRDM nozzle welds to reduce residual stress and is applied to once-through steam generator (OTSG) tube support plates to smooth their surface.

Fig. 2 Horizontal boring mill used to drill holes in reactor pressure vessel head.

Manufacture of steam generators

B&W manufactures once-through steam generators (OTSG) and recirculating steam generators (RSG) for PWR and Canada Deuterium Uranium (CANDU) reactor systems. (See Chapter 48.) Both steam generators have many components and manufacturing methods in common. The following sections describe the fabrication of the common components and the manufacture of items that are unique to each design. Fig. 4 shows a simplified cross section of an RSG, and Fig. 5 shows a simplified cross section of an OTSG. An RSG consists of cylindrical and conical shells, hemispherical heads, nozzles, a tubesheet, U-tubes, lattice grid or broached-hole tube supports, baffle and divider plates, a shroud, and steam drum internals. A typical RSG is about 75 ft (22.9 m) long, 12 to 15 ft (3.7 to 4.6 m) in diameter and weighs up to 400 t (363 tm). An OTSG consists of cylindrical shells, hemispherical heads, nozzles, two tubesheets, broached-hole tube support plates, straight

Fig. 3 Hydrostatic testing of closure head assembly.

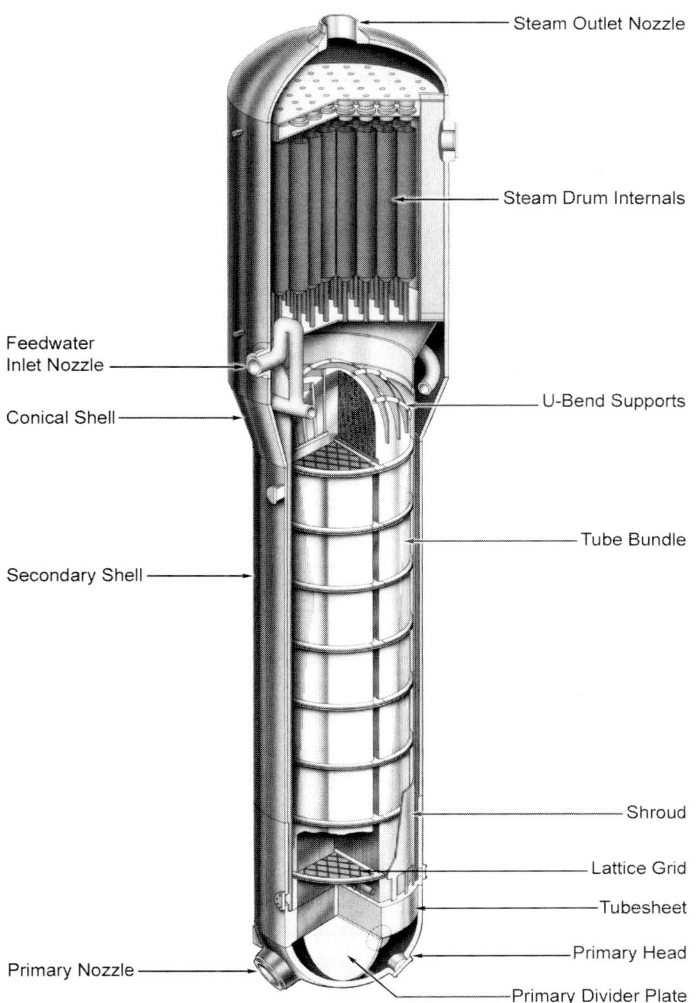

Fig. 4 Advanced series recirculating nuclear steam generator.

tubes, a shroud, and feedwater inlet header. A typical OTSG is about 73 ft (22.3 m) long, 12 ft (3.7 m) in diameter and weighs up to 470 t (426 tm).

Cylindrical shells

The shell *courses* used in fabricating the cylindrical portion of the steam generators may be made from forged ring sections or from plate formed by rolls as shown in Fig. 6. The rolling method depends upon the size, thickness and material. Cylindrical shells for steam generators are typically less than 4 in. (101.6 mm) thick and are cold rolled. Some designs require thicker sections near the tubesheet area or where reinforcement for nozzle openings is required. These shells are usually hot rolled.

For shells made from rolled plate, the longitudinal seam of each course is submerged arc welded using an automatic guidance system. In preparation for circumferentially welding several courses, each course edge is machined to improve weld quality. The sections are then aligned, tack welded and set up on drum rotators. A stationary submerged arc welder, automatically guided, is positioned over the seam and the circumferential welds are made while accurately rotating the drum.

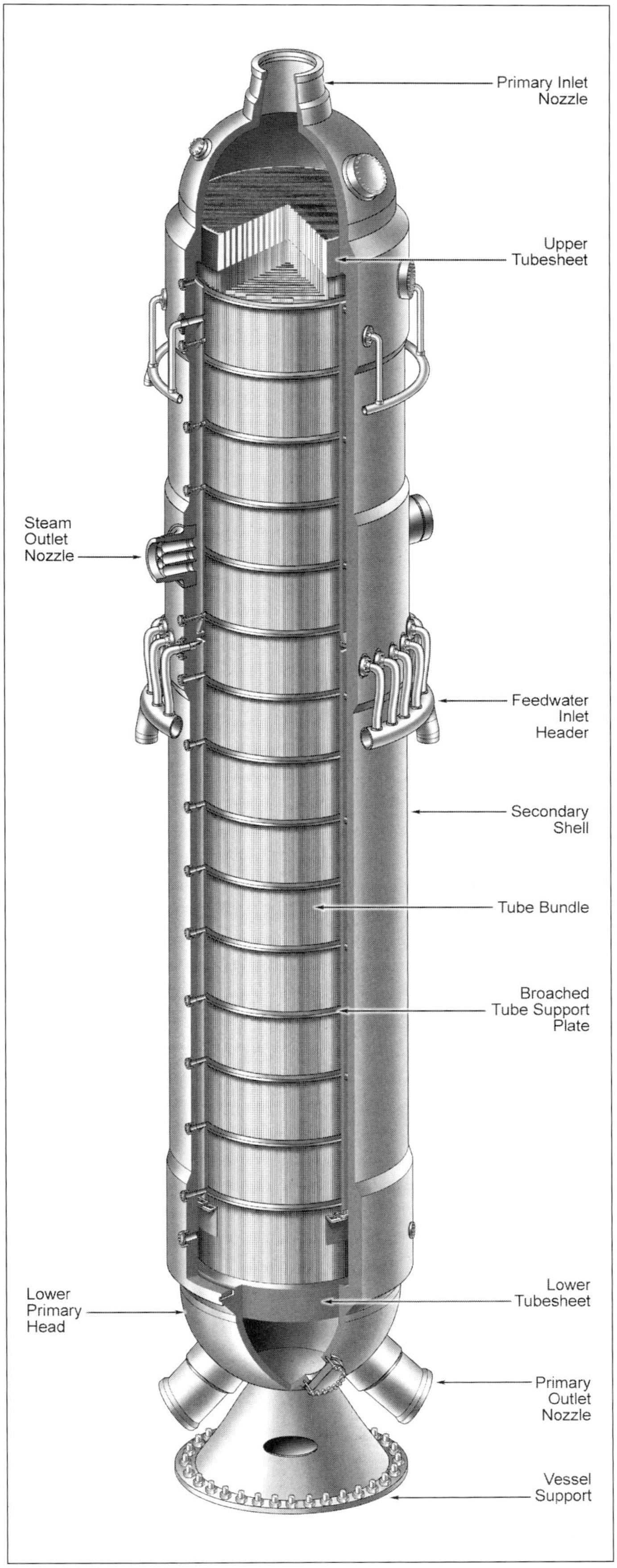

Fig. 5 Once-through nuclear steam generator.

Fig. 6 Cold rolling plate for steam generator cylindrical shell.

The advantage of using forgings for cylindrical shells is the elimination of the longitudinal seam. This eliminates the cost of making and examining the weld during manufacture, and eliminates the need to conduct in-service inspection of the weld. Forged cylinders can usually be made to tighter diameter tolerances.

Tubesheet manufacture

The tubesheet is made from a forging that can weigh up to 80 t (73 t_m). The first major manufacturing operation is cladding the primary side of the tubesheet with stainless steal or Inconel using GMAW welding methods, discussed in Chapter 38. The tubesheet is heated before, during and after welding using large electric heaters. Usually the welding head is moved over the stationary tubesheet, however, in some cases the tubesheet geometry requires it to be rotated under a stationary welding head. In these cases, speed control of the rotator is critical to maintaining weld integrity. The clad tubesheet is post weld heat treated to 1125F (607C). The cladding is then machined and examined using UT. Cladding thickness after machining is typically 0.3125 in. (7.9 mm).

Drilling the holes through which the tubes are inserted is the most critical operation for tubesheet manufacture. A typical tubesheet is 24 in. (610 mm) thick and has up to 15,000 holes or more ranging in diameter from 0.5 to 0.875 in. (12.7 to 22 mm), depending upon the tubing size. These holes must be located accurately, be precise in diameter, and be straight and perpendicular to the tubesheet surface. Variations from perpendicularity, often called drill drift, are typically limited to 0.015 in. (0.38 mm) through the thickness of the tubesheet. Extreme care is therefore given to gundrilling.

Gundrilling is done on a multi-spindle, computer-controlled horizontal drilling machine (Fig. 7). Gundrills are designed to drill holes with large depth-to-diameter ratios. Coolant travels through the center of the drill and removes cutting chips through a channel along its shank.

Regular dimensional and visual inspections of the drill tools and the hole diameters and locations are conducted. After gundrilling, the tubesheet has its circumferential weld preparations machined on a vertical or horizontal boring mill.

Steam generator heads

Recirculating steam generators The primary head of an RSG is typically 10 ft (3.1 m) in diameter, 6 in. (152.4 mm) thick and weighs up to 35 t (31 t_m). Primary heads used in PWR plant steam generators are clad on the inside surface. Similar ferritic material primary surfaces on CANDU RSGs do not require cladding. Two or three nozzles as large as 41 in. (1041 mm) in diameter are attached to the head. The head is mounted on a large capacity welding positioner so that it can be properly oriented for submerged arc welding of nozzles and attachments. Several attachments for internal components, such as the divider plate seat and manway hinge, are also welded to the inside surface of the head.

The steam drum head on an RSG is either semi-elliptical or hemispherical, is forged or formed from plate, and is up to 15 ft (4.6 m) in diameter. It also contains one large diameter 30 in. (762 mm) steam outlet nozzle. The nozzle to head weld is performed using the submerged arc welding process.

Once-through steam generators The primary head of an OTSG (Fig. 8) is similar in size to an RSG and is also clad on the inside surface. However, there is no divider plate in the primary head of an OTSG. The lower head is set up on a boring mill and the base support weld prep is machined. The manway gasket seat and stud holes are drilled and the threads milled.

Fig. 7 Gundrilling a tubesheet.

The base support is fitted and set up on a weld positioner for submerged arc welding. The primary nozzles are fit and welded using an orbital gas tungsten arc welder (GTAW) that travels around the weld seam utilizing specialized guidance systems, and video cameras to provide visibility for the operators. This equipment is necessary due to the restricted access around the weld seam created by the base support. All nozzle welds are backclad to ensure that there is a continuous corrosion resistant cladding on surfaces in contact with primary side water. The welds then receive a post weld heat treatment (PWHT) and volumetric examination by ultrasonic testing (UT) or radiographic testing (RT), and surface examination using magnetic particle (MT) or dye penetrant testing (PT). The head is then set up on a vertical boring mill and the base support and circumferential seam weld prep are machined.

Pressure boundary assembly

Pressure boundary components include the cylindrical shells, heads, the tubesheet, and external and internal supports.

Alignment, welding, nondestructive examination (NDE), PWHT and machining are critical operations when assembling these components. The sequence of adding nozzles and attachments depends on the accuracy requirements and anticipated weld distortion during assembly. If there are tight tolerance requirements on the location of nozzles and attachments, these features are machined after final assembly and PWHT of the pressure boundary components. However, this adds to the critical path schedule and, where it can be accommodated, these features are attached to the individual components and machined prior to component assembly.

Circumferential weld seams are typically made with submerged arc welding (SAW) (Fig. 9). The components are aligned and then rotated under a stationary submerged arc weld head. The welds are generally two-sided (first welded on the inside and then on the outside of the wall thickness) so that the root of the weld can be ground clear prior to welding the opposite side. If access to the inside is restricted, the welds are one-sided and are made from the outside. This requires precise alignment of the mating weld preps to avoid any defects at the root of the weld. This area may have restricted access for grinding and examining the surface. The welds receive surface NDE (MT, PT) and volumetric NDE (UT, RT). The ASME Code specifies when these examinations shall be made with respect to the PWHT.

The OTSG pressure boundary assembly has some unique differences from an RSG, mainly because the second tubesheet must be installed after the shroud and tube bundle supports are installed. The lower primary head is installed after the tubes are inserted. This sequence is explained in more detail later in this chapter.

Pressure boundary post weld heat treatment PWHT is an important consideration in planning the assembly of a steam generator. All carbon steel pressure boundary welds require PWHT. In the assembly sequence, these welds must be accessible for NDE after PWHT, as specified by the ASME Code (Fig. 10). PWHT is conducted on as large an assembly as possible to save on the cost of the PWHT furnace operation. Offsetting this are the risk and cost of having to re-PWHT this large assembly in the event of a major weld repair. Consequently, an assurance NDE is conducted prior to PWHT. If there are many complex welds in an assembly, a subassembly of these components may receive its own PWHT and NDE.

Fig. 8 Once-through steam generator primary head.

Fig. 9 Submerged arc welding.

Fig. 10 Radiographic examination of a steam generator weld.

The pressure boundary assembly is prepared for PWHT by sealing all openings. The tubesheet holes are cleaned to ensure that all contaminants that might adhere to the surface during PWHT are removed. Thermocouples are attached in strategic locations to monitor and control the heating and cooling operation and to ensure that the welds achieve their required temperature. The vessel is filled with argon, which is slowly circulated through the inside of the vessel during PWHT to eliminate oxidation. The pressure boundary is then heated in a computer-controlled gas-fired furnace.

PWHT of RSG pressure boundary PWHT can cause significant temperature differences and thermal stress in the pressure boundary assembly, especially if its geometry is complex. Therefore, the pressure boundary is assembled and post weld heat treated prior to installation of the tube bundle to avoid harmful differential thermal expansion effects. The primary head is welded to the tubesheet prior to the tube bundle assembly to avoid PWHT effects, although this means that the tube seal welding must be done in a confined area. After installing the tube bundle, the final closing seam is welded and locally heat treated. This is described later in this chapter. The welds are then examined according to ASME Code.

PWHT of OTSG pressure boundary The upper tubesheet and upper primary head of an OTSG are installed after the shroud and bundle supports are installed. These two circumferential seams receive a local PWHT. The straight tubes are inserted through the lower tubesheet and seal welded at both ends. Then the lower primary head is installed. The circumferential seam between the lower primary head and lower tubesheet requires local PWHT after welding. These operations are described later in this chapter.

Tube bundle components

Tube support assemblies The tube supports of modern RSGs are lattice grids. These grids consist of a ring assembly and a grid of flat bars that are assembled in the ring to form the lattice pattern. The flat bars intersecting each other on their edges form a diamond shape around each tube. This provides good vibration dampening yet allows the steam-water mixture to flow through the bundle with minimal pressure drop (Fig. 11).

Lattice grids (see also Chapter 48) must be accurately manufactured so each diamond is precisely aligned with the tubesheet drilling pattern. During tube bundle assembly, each tube is inserted through each grid then through the holes in the tubesheet. The accuracy of the lattice grids determines the ease with which the tube bundle is assembled.

The most critical operation in manufacturing lattice grids is cutting slots in the rings and bars where the bars intersect and fit together. Slotting is done using a special milling cutter setup driven by a boring mill. It is extremely important for the cutter setup to be accurate and the work piece to be set up true to the machine to control slot depth and angle. The mill must be calibrated to ensure that it is capable of precisely positioning the tool. CNC machines are essential to ensure consistent quality. After final assembly, the completed grid is inspected to verify that the diamonds are properly located.

OTSGs use broached tube support plates (TSP) to support the bundle. These TSPs are typically flat plates made of stainless steel that are drilled to the tube bundle pattern using a multi-spindle drill. Each hole is then machined by broaching so that the final hole shape resembles a trefoil pattern consisting of three contact areas that support the tube and three enlarged areas that allow flow through the plate (Fig. 12). Broaching is done on a multi-spindle CNC ma-

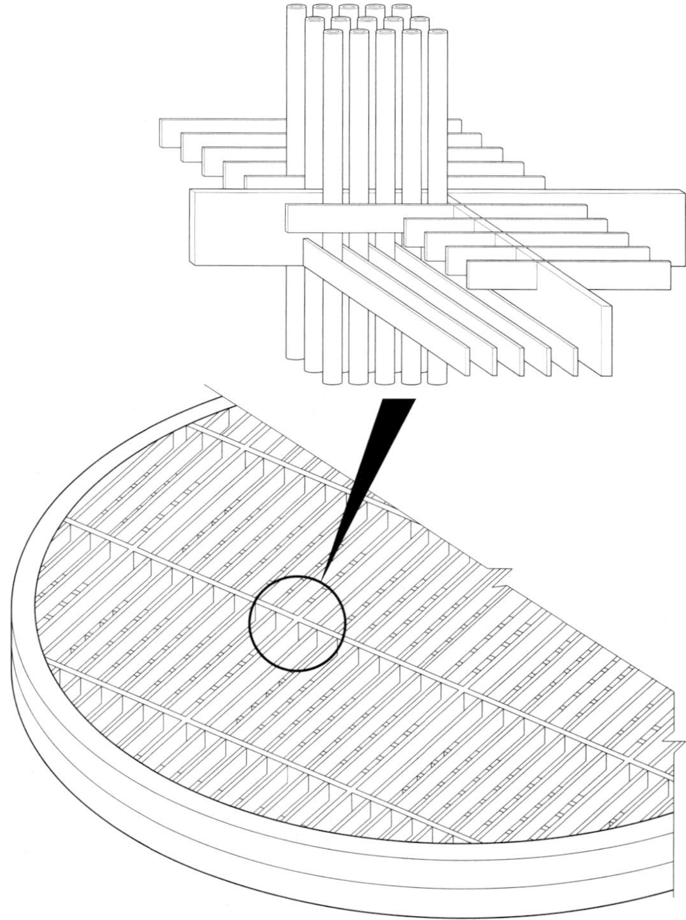

Fig. 11 Lattice-grid tube supports.

chine that pulls the broaches through the plate vertically downward (Fig. 13). Before broaching, the holes are tapered on one or both ends depending upon their location in the bundle to accommodate tube bundle assembly and to provide improved pressure drop for secondary side flow. After broaching, each plate is electropolished by immersing the plate into a tank containing electrolyte and passing appropriate current between the plate and a nearby cathode. Electropolishing removes burrs and smooths the surface to avoid damage to the tubes during insertion.

Baffle plates Baffle plates are installed in the preheater section of RSGs to direct water flow. Baffles are fabricated from machined plates. The plates are normally drilled by assembling them into a tightly clamped stack using bolts and welds and then gundrilling. Drilling a stack of plates requires the plates to be ground flat and to uniform thickness to ensure that the holes in individual plates have accurate perpendicularity to their individual plate surface. Generally, larger diameter baffle plates are best drilled individually using a gantry drill because of the difficulty of clamping, handling and accurately drilling a stack of large diameter plates.

U-bend supports The tube U-bends (Fig. 14) of RSGs are protected from flow-induced vibration by U-bend supports. These consist of an assembly of stainless steel flat bars that are GTAW welded to a collector bar to form a fan bar assembly. Some types of stainless steel require that these GTAW welds receive a PWHT in a controlled atmosphere. The angular separation of the fan bars depends upon the U-bend size and flow conditions, and the bars are located to minimize unsupported tube lengths. The fan bar assemblies are installed between the tube rows, and the outer ends of the fan bars are joined after assembly in the bundle with an arch bar. The U-bend support assembly is not connected to the shroud, thus allowing the assembly to float with the tube bundle during operation.

Tube bending Tubes for steam generators are typically 0.5 to 0.875 in. (12.7 to 22 mm) diameter with a typical nominal wall thickness of 0.045 in. (1.14 mm).

Fig. 12 Broached tube support plate.

Fig. 13 Multi-spindle CNC broaching machine.

In the OTSG design, the bundle consists of straight tubes. In the RSG, bundles are composed of U-tubes.

The U-tubes must have accurate bend radii, straight leg lengths, and must be within a strict ovality and wall thinning tolerance. These tolerances are required to ensure that tolerance stack-up does not cause interference in the U-bend region. To achieve these tolerances, the tubes are purchased to a length tolerance of 0.125 in. (3.18 mm). The tubes are formed on a draw bender over dies to a radius tolerance of ±0.03 in. (0.8 mm). The dies for the tight radius bends are

Fig. 14 U-bend tube supports.

designed to contain the tube and to minimize ovality during bending. After the tubes are bent, they are dimensionally checked to verify that all tolerances have been met.

Shrouds The shrouds are fabricated from steel plates rolled into cylinders and their longitudinal seams are joined by SAW. The sections of the shroud near the preheater section of the RSG may have their inside diameters machined round. This ensures that baffle plates can slide axially and can expand during various operating conditions. These shroud sections for RSGs may also contain manifolds to distribute the incoming feedwater and recirculating water.

It is critical to ensure that the inside radius of the shroud is accurate so that the required clearance exists to allow lattice grids and baffles to be aligned with the tubesheet pattern. The inside radius of the RSG shroud is measured by projecting a laser down its centerline and using a radial micrometer. Alternatively, an optical instrument can be used in a similar setup.

Tube bundle assembly

The tube bundle is assembled in a *clean room* (Fig. 15) to ensure that the components are not contaminated. Cleanliness and material control procedures are carefully monitored. All internal components are cleaned just prior to enclosing them.

Installation of shroud and bundle supports The shroud must be installed and aligned to ensure that the tube bundle supports can be aligned with the tubesheet drilling pattern. The shroud is installed horizontally by using wheels or bearing rollers bolted onto one end. This allows the tubesheet end of the shroud to roll into the shell as a crane supports the other end and provides a horizontal push. The shroud is then aligned using pins located around its circumference.

If the RSG design includes an integral preheater, a secondary divider plate is installed with a powered cart, which also adjusts the position of the divider plate. This plate fits into grooves in the shroud. The preheater baffle plates are then installed by mounting them on the end of a boom that is driven by a floor mounted power unit. The holes in the baffle plates must be aligned with those in the tubesheet.

Baffle plates are supported by tie rods machined to length that ensure parallel placement of the plates to the tubesheet. The baffle plates must have radial clearance to allow for thermal expansion. This clearance, however, creates an installation problem because the baffles must support the tube bundle and remain in alignment with the tubesheet while the tubes are being installed. This problem is overcome by installing a special shim that fits into the clearance between the baffle plates and the shroud. This shim dissolves during vessel operation, thereby maintaining the proper clearance.

The lattice grids for RSGs are installed by mounting them on a boom (Fig. 16). The grids are aligned with the tubesheet using optical scopes mounted on the primary side of the tubesheet. A small target is positioned in selected tube locations on the lattice grids, and the grid is adjusted using wedges around its circumference until the target aligns with the scope's line of sight. Alignment rods that are the same diameter as the tube are also used to check the alignment of the lattice grids. The grids are held in place by support blocks welded to the inside of the shroud.

To ensure that the tubes can be installed in an OTSG without misalignment or damage to their surface, it is critical that the TSPs are aligned with the hole pattern in the tubesheet and are parallel to the tubesheet face. The TSPs for OTSGs are connected together with tie rods. The first set of tie rods is installed into the tubesheet, and the ends are checked with a coordinate measuring machine to ensure they provide a flat plane for the first TSP to rest against. Support blocks are aligned and welded to the shroud, providing support to the circumferential edge of the TSP. The first support plate is installed and aligned using similar optical techniques to that used for aligning lattice grids. Then, wedges and keys are installed to firmly support the TSP and to keep it in alignment during rotation of the vessel for subsequent manufacturing operations. This sequence is repeated for each TSP. After all the TSPs are installed, the top tubesheet is set up on an alignment fixture in preparation for installation. Optical alignment scopes are set up, and the tubesheet is fit to the cylindrical shell. Ceramic backing bar is installed to the inside diameter of the circumferential seam along with alignment wedges.

Fig. 15 Nuclear clean room.

Fig. 16 Installation of lattice grid tube bundle support.

After fitup and alignment have been confirmed, the root passes of the weld are made while monitoring the alignment. The ceramic backing bar is removed, the optical instruments are removed, and the remainder of the weld is completed with SAW. The completed seam then receives MT, UT (Fig. 17) and RT.

Installation of tubes Tubes are installed with the vessel in a horizontal position. Each tube hole and tube are dry-swabbed clean just prior to assembly. Cleaning operations at this stage are done without liquids. Any moisture or contaminants trapped between the tube and tube hole can cause tube-to-tubesheet welding defects.

In an RSG, the U-tubes are fitted with a plastic tube end pointer and installed in layers from the smallest bend radius tubes to the largest bend radius tubes (Fig. 18) starting with the lowest layer. This tubing sequence simplifies the temporary supports that must be installed to keep the U-bend supported during tube installation. As each layer is installed, the appropriate U-bend support fan is installed. The partially completed bundle is temporarily supported to prevent sagging due to the weight of the tubes. Each U-tube is inserted until it protrudes the specified amount from the primary face of the tubesheet. This amount of protrusion depends upon whether the tube-to-tubesheet weld is a fillet weld or a flush weld. The installer also checks for proper clearance between each U-bend. The tubes are tack expanded by inserting an expander inside the tubes about 0.75 in. (19 mm). The expander consists of a split collet actuated by a hydraulically driven tapered mandrel. This yields the tube sufficiently to hold it in place and to close the gap for welding. This expansion method produces less residual stress in the tube than other methods, such as roller expansion.

In an OTSG, the straight tubes are fitted with plastic tube end pointers, cleaned, and installed in layers starting from the top layer and working down. This tubing sequence avoids any particulate matter that may be picked up by the tube during insertion from falling down onto tube ends that have already been installed. This sequence also simplifies retrieval of a tube end pointer in the unlikely event that one comes loose during tube insertion. Tubes are inserted through the bottom tubesheet, through the TSP assembly, and through the upper tubesheet. The tubes are inserted so that the correct tube projection is achieved at the upper tubesheet. The tubes are then tack expanded using the same tack expansion techniques that are used for RSGs. The tubes in the upper tubesheet are final tack expanded and seal welded. The tube ends in the lower tubesheet receive an initial expansion to provide some support to the tube while the tube ends are machined to the proper protrusion at the lower tubesheet. This is followed by a final tack expansion.

Tube-to-tubesheet welding The tubes and tubesheets are joined with automatic gas tungsten arc welding (GTAW). The tube-to-tubesheet welding heads are specially designed units to provide welds of consistent quality (Fig. 19). At the beginning of each shift, each operator makes a tube-to-tubesheet weld test

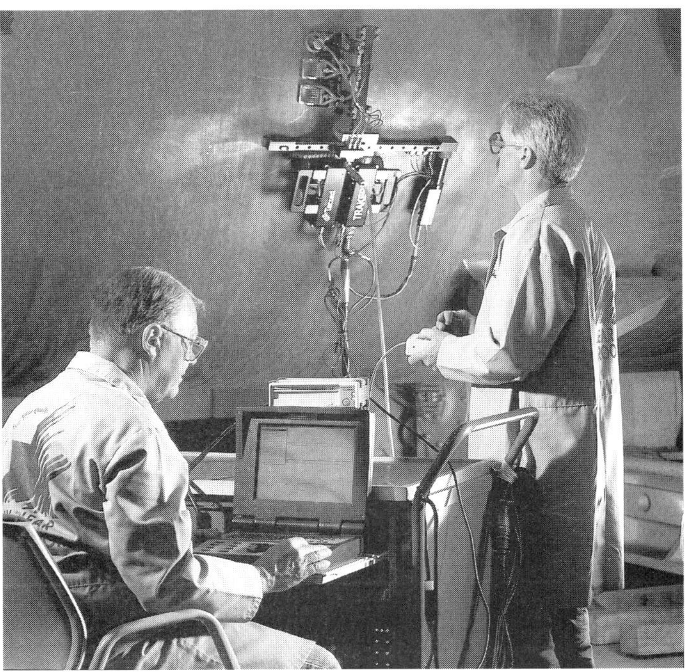

Fig. 17 Ultrasonic inspection of seam weld.

Fig. 18 Installation of tubing.

block. The block receives a visual and PT inspection, and is then sectioned for further metallurgical examination.

Each weld is visually compared to a workmanship sample. All welds not meeting the sample quality are repaired and then rewelded. This is followed by PT examination of all tube-to-tubesheet welds. Finally, a leak test is done by pressurizing the secondary side with a mixture of air and helium. The primary tubesheet face is then monitored for helium leaks at the welds.

Tube expansion Each tube is hydraulically expanded into its hole after tube-to-tubesheet welding. This expansion closes the crevice between the tube and the hole to avoid a potential corrosion site. The tube may be expanded near the secondary face of the tubesheet or it may be expanded full depth or at the primary face depending on customer specifications. Hydraulic expansion is the recommended method for nuclear steam generators because it produces less residual stress in the tube and reduces the potential for stress corrosion cracking compared to other expansion methods. Each tube is expanded by inserting a probe that has a seal positioned at each end of the expansion zone. Distilled water at approximately 35,000 psi (241.3 MPa) is pumped through the probe, expanding the tube and sealing it against the tube hole.

The expansion probe must be carefully positioned with respect to the secondary face of the tubesheet. If the probe is positioned beyond the face, then unacceptable tube deformation could occur. If the probe is too far inside the tube hole, an unacceptably long crevice could possibly result. Therefore, the tubesheet thickness variation is measured and the probe length is adjusted to ensure proper positioning.

Tubes located near the tubesheet periphery may require expansion using special probes because the curvature of the head encroaches on the probe insertion area. A flexible extension is attached to the probe, which allows it to be inserted in confined areas. In cases where the tube must be expanded full depth, the expansion is done in two or more steps. Following expansion, each tube is examined with eddy current testing (ECT) techniques.

Fig. 19 Tube-to-tubesheet welding.

Steam drum internals

RSGs require a steam drum and steam separators to remove moisture from the steam (Fig. 20). OTSGs provide steam at slightly superheated conditions and do not require steam separators. The steam separators are similar in design to those used on fossil-fuel boilers. Generally, the separators in nuclear steam generators are an axial flow type, although tangential flow separators have been used. There are typically more than 100 separators in a large RSG. Most designs also have smaller secondary separators located above the primary units.

The fabrication and assembly tolerances of steam separators are critical. The shape and assembly clearances of internal components have a significant effect on separator performance. Custom made assembly jigs and fixtures are used to ensure that the required tolerances are met. In addition, all internal welds are carefully ground to avoid discontinuities that affect separator performance.

The steam separators are installed in a deck structure made of steel plate. This deck support structure is welded to the inside of the drum, and the supports are designed to accommodate differential thermal expansion during steam generator operation as well as loads due to specified accidental pipe break and seismic events. The separator and deck structure is usually of modular construction to simplify installation into the steam drum.

Final assembly

Closing seam fitting and welding of RSGs The final assembly operation for RSG fabrication consists of fitting and welding the closing seam, which is usually one of the circumferential seams on the cone. The steam drum and the cylindrical shell are positioned on rotators and the circumferential seam is aligned. Previous machining of the component mating surfaces assures good alignment. The weld could be either a two-sided weld if there is sufficient access to the inside of the vessel, or a one-sided weld. If a one-sided weld is used, it is generally of a narrow groove design to limit the volume of weld deposit. The sequence of making a one-sided weld is described later.

Prior to starting work on the inside of the vessel, the tube bundle and the steam separator assembly are sealed off to prevent weld flux and other material from entering. The drum and cylindrical shell are aligned and the seam is tack welded to hold the components in position. The next several weld passes are completed using shielded metal arc welding (SMAW). This additional welding provides enough structural strength to allow the assembly to be moved to a submerged arc welding station where the outside of the weld is finished using SAW. As the vessel is rotated, the weld is completed.

Due to access limitations at some sites installing replacement RSGs, fitting and welding of the closing seam may not occur until after the RSG is inside the

Fig. 20 Fabrication of steam separator subassembly.

containment building. In these cases, the lower half of the steam generator is generally assembled to the refitted upper half of the original steam generator. Typically, the steam drum and head will be refit with new separators and steam flow control equipment before the closing seam is made with the steam generator vertical and in place.

Local PWHT of RSG closing seam The closing circumferential seam requires post weld heat treatment. Because temperature differences during this procedure can exceed those experienced during normal operation of the vessel, care must be taken to avoid excessive and harmful differential expansion of internal components. Temperatures can be controlled using an electric furnace designed to fit around the closing seam. Insulation is applied to the outside of the vessel shell and to the attachments in the vicinity of the seam. The tube bundle U-bend region is also insulated to reduce radiant heat transfer. Internal heat transfer is further reduced by drawing a vacuum on the secondary side of the vessel. During PWHT, air may be blown through the tube bundle primary side to limit tube temperature. Thermocouples are mounted in each critical area to monitor and control the process.

The electric furnace has several zones that can be independently controlled to keep the temperatures within prescribed limits. A computer is used to control the furnace and to collect data and provide a record of temperatures.

Final assembly and local PWHT of OTSG pressure boundary After the straight tubes are inserted through the lower tubesheet, seal welded and helium leak tested, the lower primary head is installed. The circumferential seam between the lower primary head and lower tubesheet requires local PWHT after welding. This operation is complex because the tube bundle will be partially heated during the operation. Temperature gradient limits are imposed to limit the stresses in each component and to limit the loads on the tubes due to the difference in coefficient of thermal expansion between the tubes and carbon steel pressure boundary. An electrically heated clamshell furnace is installed around each circumferential weld seam. Supplementary electric heaters may be placed on other components such as the tubesheet primary face to further control the temperature gradients. Heat transfer by convection to the secondary side may be controlled either by drawing a vacuum, or by providing air circulation, depending on the requirements during heat-up or cool-down.

Final inspection and testing

A hydrostatic test is done on the completed vessel; the primary and secondary sides are tested separately. The test consists of filling the vessel with treated water that is pressurized to 125% of the design pressure. The pressure is reduced and held at the design pressure while an inspector examines the vessel for leaks. The metal temperature must be held above 70F (21C) to keep it well above its brittle transition temperature.

After hydrostatic testing, the vessel is drained and dried. A vacuum is drawn on the secondary side to assist drying. Felt plugs are also blown through the tubes to dry them. All accessible pressure boundary welds are then examined by the MT method.

Nozzles and integral support surfaces, called terminal points, connect to mating components in the field and must be within specified tolerances to simplify field installation and minimize stresses caused by fitup. Special coordinate measuring machines (Fig. 21) are used to set up the tools used to machine the final terminal point geometry (Fig. 22). It is common practice to finish machine nozzle weld preps at this stage to bring them within tolerance.

Just prior to closing the vessel, the insides of the tubes are cleaned by blowing felt plugs through them. The secondary side of the steam generator is examined and cleaned, and final drum internal components are installed. All manway covers are bolted on. The primary and secondary sides of the vessel are evacuated and backfilled with nitrogen to reduce the formation of oxides. Steam generators are wrapped in shrink-wrap plastic to protect them during shipment.

Surface conditioning

Some steam generator owners specify that the inside surfaces of the primary side (excluding the tubing) are to be surface conditioned or electropolished. This process produces a very smooth surface finish,

Fig. 21 Set up of nozzle machining tools using coordinated measuring machine.

Fig. 22 Final nozzle end machining.

typically 10 microinches or better. This results in a significant reduction in true surface area of the work piece which in turn reduces radioactive isotope update and occupational exposure. Reduced occupational exposure makes these components more easily maintained in the field.

The first step in surface conditioning is to grind the surface with abrasive flap wheels. Several passes are required, each one made with progressively finer abrasives until a surface finish of 40 microinches is achieved. This operation can be done by machine on an individual component prior to assembly. However, this is usually impractical because the component must then be protected from surface damage for the remainder of the manufacturing operation. Usually, the grinding is done near the end of the assembly sequence with hand tools. This is a time consuming operation, especially if as-welded surfaces are involved.

The second step in surface conditioning is to impart further improvements in surface roughness on a microscopic scale by electropolishing. In electropolishing, an applied electric current flows from the metal surface (anode or work piece) through a conductive electrolytic solution to another conducting surface (cathode). This process smooths the micropeaks formed by mechanical grinding.

Shipping

Steam generators may be shipped by rail or by special road transport directly to the plant site, or may travel part of the way by barge or heavy lift ship. For rail transportation, the load must be centered on the car for even weight distribution. If the vessel is long, it can span two railway cars. If it is short, a single heavy-duty flat car is usually sufficient (Fig. 23). The vessel is oriented to minimize its width and height and is loaded onto bunks fastened to the car. Loads up to 16 ft (4.9 m) in diameter can be shipped along most routes in the United States and Canada. Shipping larger units requires measurement of bridge clearances and other potential obstructions. If the vessel spans two cars, the bunks must be designed to allow movement of the load while negotiating curves. Additional ballast may be attached to the rail car to lower the center of gravity.

Manufacture of pressurizers

Pressurizers are cylindrical vessels that help stabilize the pressure in the primary heat transport system. (See Chapter 46.) Each unit is fitted with special penetrations in which electrical heaters are installed.

Pressure boundary assembly

The pressure boundary consists of cylindrical shells and hemispherical heads. The pressure boundary contains a surge nozzle, spray inlet nozzle, several connections for electric heaters, and various small water level nozzles. The manufacturing methods and quality assurance requirements for these components are the same as those described previously for steam generators.

Heater connections and installation The heaters are the direct immersion type, sheathed in stainless steel or Inconel and assembled in bundles. In some designs, the heaters consist of a single straight element. Each of the elements is field assembled through penetrations in the vessel wall and is sealed by means of a bolted closure. The closure is sealed by gaskets or patented mechanical seals. An electrical connection is then made to the end of each heater using a special insulated fitting.

Post weld heat treatment

PWHT of pressurizers is simpler than that of steam generators because there are no complex geometries or internals that limit heating and cooling rates. PWHT is done by putting the completed pressurizer in a gas furnace and heating to 1125F (607C).

Some patented mechanical seals used for the heater connections must be installed prior to PWHT because of their required machined surface finish. This machining can not be done after the component is welded onto the vessel because of inadequate access for specialized equipment. As a result, the sealing surface is protected from the furnace environment by coating it with an antiscaling compound. Some minor surface dressing of the sealing surface may be required after PWHT.

Fig. 23 Steam generator awaiting offloading to heavy lift ship.

Manufacture of dry shielded canisters[1]

Dry shielded canisters are part of a system of containers used to store and ship spent nuclear fuel. The canisters are designed to safely store bundles of spent fuel at plant sites, and some designs allow transport to long-term storage sites. The design of the canisters and internal structures serves as the containment boundary to confine radioactive spent fuel and provide a leak-tight, inert atmosphere to ensure that the integrity of the fuel cladding is maintained. A typical canister consists of an outer cylindrical shell made of stainless steel, typically 5.5 ft (1.7 m) in diameter by 15.5 ft (4.7 m) in length. The internal assembly consists of an array of guide sleeves that are square in cross-section. Each one is designed to accept a bundle of spent fuel. The guide sleeves are made of stainless steel. The ends of the canisters are capped with shield plugs that include lead shielding. Although many individualized components are similar to other nuclear pressure vessels, there are several unique features that require specialized manufacturing operations. For example, for some designs, it is necessary to limit the canister weight due to specific lifting capabilities at the plant site. Therefore, material thickness must be checked frequently to allow the manufacturing tolerances to be worked toward their minimum material condition.

The cylindrical shells are made from stainless steel plate using manufacturing operations common to other components described earlier. However, their outside surface must meet tight tolerances on circularity, diameter and straightness because the cylinder must fit inside a shielded storage container. The minimum diameter is limited by the requirement to insert the basket assembly into the cylinder without restriction. The diameter must also provide sufficient allowance for differential expansion of shell and basket assembly components when the assembly is subject to elevated temperature during service.

The shield plugs consist of a stainless or carbon steel circular disk with ribs or stiffeners welded to it. The entire shield plug assembly is poured full of lead typically 5 in. (127 mm) thick. A gamma ray scan confirms the effective thickness of the lead shielding.

The top cover plates are welded to the canister in the field. Consequently, it is critical to maintain a specific radial gap between the cover plate and the inside diameter (ID) of the cylinder. One way of accomplishing this is by machining and hand working the OD of the cover plates to suit the as-built ID of the cylinder. The cover plate must also meet a specific thickness tolerance to ensure that the maximum amount of fuel can physically fit into the canister. This requires flattening operations of the plates during machining and frequent UT thickness checks. There are similar requirements to custom machine, flatten and hand work the top shield plug to achieve specified radial gap.

The guide sleeves are made to form a square cross-section and are sized to suit the dimensions of PWR and BWR fuel assemblies. The tight tolerances on the guide sleeves ensure that a spent fuel bundle can be inserted without binding. The guide sleeves are formed from stainless steel sheet, and the longitudinal seams are welded using a welding jig that maintains the sleeve geometry. The welding equipment utilizes cooling methods and heat input techniques that minimize distortion due to welding.

During manufacture and assembly of the basket assembly, it is critical to align the spacer discs that support the guide sleeve so the sleeves can be installed without distortion or binding. The spacer discs are set up in an alignment jig and welded to the support rods. The entire assembly must meet exacting geometric requirements of overall perpendicularity, cylindricity and height to accommodate thermal expansion during service and to resist potential impact loading during transport.

After the basket assembly is assembled into the cylinder, the final alignment of each guide sleeve is checked using a plug gauge. The plug gauge, representing the maximum size of fuel to be stored in the canister, must pass through the guide sleeve within a specified load limit.

Reference

1. Johnson, E.R., and Saverot, P.M., Eds., *Monograph on Spent Nuclear Fuel Storage Technologies*, Institute of Nuclear Materials Management, Northbrook, Illinois, 1997.

Appendices

Appendix 1

Conversion Factors, SI Steam Properties and Useful Tables

This appendix provides tables, charts and figures that can be useful in evaluating the generation and use of steam. They supplement the material presented throughout *Steam* and are divided into four subsections: selected conversion factors; SI versions of the Steam Tables and graphical data; useful charts and figures; and typical operating parameters for a 500 MW pulverized coal-fired power plant.

Part 1: Conversion Factors

Table 1.1 Length

Table 1.2 Length, Small Units

Table 1.3 Area

Table 1.4 Volume

Table 1.5 Mass

Table 1.6 Force

Table 1.7 Pressure or Force per Unit Area

Table 1.8 Energy, Work and Heat

Table 1.9 Power or Rate of Doing Work

Table 1.10 Density (Mass/Volume)

Table 1.11 Thermal, Heat Transfer and Flow Parameters

Table 1.12 Multiplying Prefixes

Table 1.13 Selected Additional Conversion Factors

Part 2: Systemè International d'Unitès (SI), Steam Tables Plus Selected Charts and Figures

Table 2.1 SI Properties of Saturated Steam and Saturated Water (Temperature)

Table 2.2 SI Properties of Saturated Steam and Saturated Water (Pressure)

Table 2.3 SI Properties of Superheated Steam and Compressed Water (Temperature and Pressure)

Fig. 2.1 Mollier diagram (*H-s*) for steam – SI units

Fig. 2.2 Pressure-enthalpy diagram for steam – SI units.

Fig. 2.3 Psychrometric chart – SI units

Part 3: Useful Charts and Tables

Fig. 3.1 Temperature-enthalpy diagram (MPa = 0.006895 × psi)

Fig. 3.2 Estimates of the weight of air required for selected fuels and excess air levels

Fig. 3.3 Estimates of the weight of the products of combustion for selected fuels and excess air levels

Fig. 3.4 Estimates of the percent total air for flue gas percent oxygen for selected fuels

Fig. 3.5 Estimates of efficiency loss due to unburned combustibles for selected fuels and ash contents

Fig. 3.6 Estimates of theoretical air required for the combustion of coal based upon volatile matter content

Table 3.1 SO_2, SO_3 Estimated Emissions from Boilers Firing Coal or Fuel Oil

Table 3.2 Conversion Formulae for Emission Rates

Table 3.3 Periodic Table of the Elements

Part 4: Typical 500 MW Pulverized Coal-Fired Power Plant Parameters

Table 4.1 Coal and Ash Analysis, an Example of a High Sulfur Bituminous Coal

Table 4.2 Mechanical Equipment – 500 MW PC-Fired Plant

Table 4.3 Environmental Equipment – 500 MW PC-Fired Plant

Table 4.4 Plant Performance Summary – 500 MW PC-Fired Plant

Table 1.1 Length

Starting With → Multiply By To Obtain ↓	Inch	Foot	Yard	Statute Mile	Millimeter	Centimeter	Meter	Kilometer
Inch	1	12	36	63.36×10^{3}	3.937×10^{-2}	0.3937	39.37	3.937×10^{4}
Foot	0.08333	1	3	5280	3.281×10^{-3}	3.281×10^{-2}	3.281	3281
Yard	0.02778	0.3333	1	1760	1.094×10^{-3}	1.094×10^{-2}	1.093	1094
Statute Mile	1.578×10^{-5}	1.894×10^{-4}	5.682×10^{-4}	1	6.214×10^{-7}	6.214×10^{-6}	6.214×10^{-4}	0.6214
Millimeter	25.40	304.8	914.4	1609×10^{6}	1	10.00	1000	1.000×10^{6}
Centimeter	2.540	30.48	91.44	1.609×10^{5}	0.1000	1	100	1.000×10^{5}
Meter	0.0254	0.3048	0.9144	1609	1.000×10^{-3}	0.010	1	1000
Kilometer	2.540×10^{-5}	3.048×10^{-4}	9.144×10^{-4}	1.609	1.000×10^{-6}	1×10^{-5}	0.0010	1

Table 1.2 Length, Small Units

Starting With → Multiply By To Obtain ↓	Mil	Inch	Angstrom	Nanometer	Micron	Millimeter	Centimeter
Mil	1	1000	3.937×10^{-6}	3.937×10^{-5}	3.937×10^{-2}	39.37	393.7
Inch	0.0010	1	3.937×10^{-9}	3.937×10^{-8}	3.937×10^{-5}	3.937×10^{-2}	0.3937
Angstrom	2.540×10^{5}	2.540×10^{8}	1	10.00	1.000×10^{4}	1.000×10^{7}	1.000×10^{8}
Nanometer	2.540×10^{4}	2.540×10^{7}	0.100	1	1000	1.000×10^{6}	1.000×10^{7}
Micron	25.40	2.540×10^{4}	1.000×10^{-4}	1.000×10^{-3}	1	1000	1.000×10^{4}
Millimeter	2.540×10^{-2}	25.40	1.000×10^{-7}	1.000×10^{-6}	1.000×10^{-3}	1	10
Centimeter	2.540×10^{-3}	2.54	1.000×10^{-8}	1.000×10^{-7}	1.000×10^{-4}	0.100	1

Table 1.3 Area

Starting With → Multiply By To Obtain ↓	Square Inch	Square Foot	Square Yard	Acre	Square Mile	Square Centimeter	Square Meter	Square Kilometer
Square Inch	1	144.0	1296	6.273×10^{6}	4.014×10^{9}	0.1550	1550	1.550×10^{9}
Square Foot	6.944×10^{-3}	1	9.000	4.356×10^{4}	2.788×10^{7}	1.076×10^{-3}	10.764	1.076×10^{7}
Sqaure Yard	7.716×10^{-4}	0.1111	1	4840	3.098×10^{6}	1.196×10^{-4}	1.196	1.196×10^{6}
Acre	1.594×10^{-7}	2.296×10^{-5}	2.066×10^{-4}	1	640	2.471×10^{-8}	2.471×10^{-4}	247.1
Square Mile	2.491×10^{-10}	3.587×10^{-8}	3.228×10^{-7}	1.563×10^{-3}	1	3.861×10^{-11}	3.861×10^{-7}	0.3861
Square Centimeter	6.452	929.0	8361	4.047×10^{7}	2.590×10^{10}	1	1.000×10^{4}	1.000×10^{10}
Square Meter	6.452×10^{-4}	0.09290	0.8361	4047	2.590×10^{6}	1.000×10^{-4}	1	1.000×10^{6}
Square Kilometer	6.452×10^{-10}	9.290×10^{-8}	8.361×10^{-7}	4.047×10^{-3}	2.590	1.000×10^{-10}	1.000×10^{-6}	1

Table 1.4 Volume

Starting With → Multiply By To Obtain ↓	Fluid Ounce	Quart	Gallon	Barrel (Petroleum)	Cubic Foot	Cubic Yard	Cubic Centimeter (Milliliter)	Liter	Cubic Meter
Fluid Ounce	1	32.00	128.0	5376	957.5	2.585×10^{4}	3.381×10^{-2}	33.82	3.382×10^{4}
Quart	0.03125	1	4.000	168.0	29.92	807.9	1.057×10^{-3}	1.057	1057
Gallon	7.813×10^{-3}	0.2500	1	42.00	7.481	202.0	2.642×10^{-4}	0.2642	264.2
Barrel (Petroleum)	1.860×10^{-4}	5.952×10^{-3}	0.0238	1	0.1781	4.809	6.290×10^{-6}	6.290×10^{-3}	6.290
Cubic Foot	1.044×10^{-3}	3.342×10^{-2}	0.1336	5.615	1	27.00	3.532×10^{-5}	3.532×10^{-2}	35.32
Cubic Yard	3.868×10^{-5}	1.238×10^{-3}	4.951×10^{-3}	0.2079	3.704×10^{-2}	1	1.308×10^{-6}	1.308×10^{-3}	1.308
Cubic Centimeter (Milliliter)	29.57	946.4	3785	1.590×10^{5}	2.832×10^{4}	7.646×10^{5}	1	1000	1.000×10^{6}
Liter	2.957×10^{-2}	0.9464	3.785	159.0	28.32	764.6	1.000×10^{-3}	1	1000
Cubic Meter	2.957×10^{-5}	9.464×10^{-4}	3.785×10^{-3}	0.1590	2.832×10^{-2}	0.7646	1.000×10^{-6}	1.000×10^{-3}	1

Table 1.5 Mass

Starting With → Multiply By To Obtain ↓	Pound	Short Ton	Long Ton	Gram	Kilogram	Metric Ton
Pound	1	2000	2240	2.205×10^{-3}	2.205	2205
Short Ton	5.000×10^{-4}	1	1.12	1.102×10^{-6}	1.102×10^{-3}	1.102
Long Ton	4.464×10^{-4}	0.8929	1	9.842×10^{-7}	9.842×10^{-4}	0.9842
Gram	453.6	9.072×10^{5}	1.016×10^{6}	1	1000	1.000×10^{6}
Kilogram	0.4536	907.2	1016	1.000×10^{-3}	1	1000
Metric Ton	4.536×10^{-4}	0.9072	1.016	1.000×10^{-6}	0.0010	1

Table 1.6 Force

Starting With → Multiply By To Obtain ↓	Newton	Grams	Kilograms	Pounds
Newton	1	9.807×10^{-3}	9.807	4.448
Grams (f)	102.0	1	1000	453.6
Kilograms (f)	0.1020	0.001	1	0.4536
Pounds (f)	0.2248	2.205×10^{-3}	2.205	1

Table 1.7
Pressure or Force per Unit Area

Starting With → Multiply By To Obtain ↓	Atmospheres	Cm of Mercury at 0C	Inch of Mercury at 0C	Inch of Water at 4C	Feet of Water at 4C	Kilograms per Square Meter	Pounds per Square Inch (abs)	Pascal	Bar
Atmospheres	1	1.316 X 10^{-2}	3.342 X 10^{-2}	2.458 X 10^{-3}	2.950 X 10^{-2}	9.678 X 10^{-5}	6.804 X 10^{-2}	9.869 X 10^{-6}	0.9869
Cm of Mercury at 0C	76.00	1	2.540	0.1868	2.242	7.356 X 10^{-3}	5.171	7.501 X 10^{-4}	75.01
Inch of Mercury at 0C	29.92	0.3937	1	7.355 X 10^{-2}	0.8826	2.896 X 10^{-3}	2.036	2.953 X 10^{-4}	29.53
Inch of Water at 4C	406.8	5.354	13.6	1	12	3.937 X 10^{-2}	27.68	4.015 X 10^{-3}	401.9
Feet of Water at 4C	33.90	0.4460	1.133	8.333 X 10^{-2}	1	3.281 X 10^{-3}	2.307	3.345 X 10^{-4}	33.49
Kilograms per Square Meter	1.033 X 10^{4}	136.0	345.3	25.40	304.8	1	703.1	1.0197 X 10^{-1}	1.0197 X 10^{4}
Pounds per Square Inch (abs)	14.70	0.1934	0.4912	3.613 X 10^{-2}	0.4335	1.422 X 10^{-3}	1	1.450 X 10^{-4}	14.50
Pascal	1.01325 X 10^{5}	1.333	3387.4	249.0	2989	9.80665	6895	1	1.000 X 10^{5}
Bar	1.013	1.333 X 10^{-2}	3.387 X 10^{-2}	2.49 X 10^{-3}	2.989 X 10^{-2}	9.80665 X 10^{-5}	6.895 X 10^{-2}	1 X 10^{-5}	1

PRESSURE is the force per unit area exerted on or by a fluid.
Absolute pressure is the sum of atmospheric and gauge pressure.
Standard atmospheric pressure is 14.696 lbf/in. or 29.92 mm Hg.

Table 1.8
Energy, Work and Heat

Starting With → Multiply By To Obtain ↓	British Thermal Units	Centimeter-grams	Foot-pounds	Horsepower-hours	Joules Watt-second	Kilowatt-hours	Kilogram-meters	Watt-hours
British Thermal Units	1	9.297 X 10^{-8}	1.285 X 10^{-3}	2545	9.480 X 10^{-4}	3413	9.297 X 10^{-3}	3.413
Centimeter-grams	1.076 X 10^{7}	1	1.383 X 10^{4}	2.737 X 10^{10}	1.020 X 10^{4}	3.671 X 10^{10}	10^{5}	3.671 X 10^{7}
Foot-pounds	778	7.233 X 10^{-5}	1	1.98 X 10^{6}	0.7376	2.655 X 10^{6}	7.233	2655
Horsepower-hours	3.929 X 10^{-4}	3.654 X 10^{-11}	5.050 X 10^{-7}	1	3.725 X 10^{-7}	1.341	3.653 X 10^{-6}	1.341 X 10^{-3}
Joules Watt-second	1054.8	9.807 X 10^{-5}	1.356	2.684 X 10^{6}	1	3.6 X 10^{6}	9.807	3600
Kilowatt-hours	2.93 X 10^{-4}	2.724 X 10^{-11}	3.766 X 10^{-7}	0.7457	2.778 X 10^{-7}	1	2.724 X 10^{-6}	0.001
Kilogram-meters	107.6	10^{-5}	0.1383	2.737 X 10^{5}	0.1020	3.671 X 10^{5}	1	367.1
Watt-hours	0.293	2.724 X 10^{-8}	3.766 X 10^{-4}	745.7	2.778 X 10^{-4}	1000	2.724 X 10^{-3}	1

Table 1.9
Power or Rate of Doing Work

Starting With → Multiply By To Obtain ↓	Btu per Minute	Foot-pounds per Minute	Foot-pounds per Second	Horse-power	Kilowatts	Watts
Btu per Minute	1	1.285 X 10^{-3}	7.716 X 10^{-2}	42.44	56.91	5.691 X 10^{-2}
Foot-pounds per minute	778	1	60	3.3 X 10^{4}	4.426 X 10^{4}	44.26
Foot-pounds per second	12.97	1.667 X 10^{-2}	1	550	737.6	0.7376
Horsepower	2.357 X 10^{-2}	3.030 X 10^{-5}	1.818 X 10^{-3}	1	1.341	1.341 X 10^{-3}
Kilowatts	1.757 X 10^{-2}	2.260 X 10^{-5}	1.356 X 10^{-3}	0.7457	1	10^{-3}
Watts	17.5725	2.260 X 10^{-2}	1.356	745.7	1000	1

Table 1.10
Density (Mass/Volume)

Starting With → Multiply By To Obtain ↓	Pounds per Cubic Foot	Pounds per Gallon (U.S. Liquid)	Grams per Cubic Centimeter	Kilograms per Liter	Kilograms per Cubic Meter
Pounds per Cubic Foot	1	7.481	62.43	62.43	6.243 X 10^{-2}
Pounds per Gallon (U.S. Liquid)	0.1337	1	8.345	8.345	8.345 X 10^{-3}
Grams per Cubic Centimeter	1.602 X 10^{-2}	0.1198	1	1.000	1.000 X 10^{-3}
Kilograms per Liter	1.602 X 10^{-2}	0.1198	1.000	1	1.000 X 10^{-3}
Kilograms per Cubic Meter	16.02	119.8	1000	1000	1

Table 1.11
Thermal, Heat Transfer and Flow Parameters

Thermal Conductivity

1 Btu/h ft F	= 1 Btu/h ft R
	= 12 Btu in./h ft^2 F
	= 1.7307 W/m K

Heat Content

1 Btu/lb	= 2.326 kJ/kg
1 Btu/Standard cubic foot	= 0.039338 MJ/Nm^3

Heat Flux, or Heat Transfer Rate

1 Btu/h ft^2	= 3.1546 W/m^2

Volumetric Heat Release Rate

1 Btu/h ft^3	= 10.3497 W/m^3

Heat Transfer Coefficient or Conductance

1 Btu/h ft^2 F	= 5.6784 W/m^{2}K

Mass Flow Rate

1 lb/h	= 0.000126 kg/s

Mass Flux

1 lb/h ft^2	= 0.001356 kg/m^{2}s

Volumetric Flow Rate

1 ft^3/min (ACFM)	= 0.000472 m^3/s
1 GPM	= 0.06308 l/s

Table 1.12
Multiplying Prefixes

Multiplication Factors	Prefix	SI Symbol
1 000 000 000 000 = 10^{12}	tera	T
1 000 000 000 = 10^{9}	giga	G
1 000 000 = 10^{6}	mega	M
1 000 = 10^{3}	kilo	k
100 = 10^{2}	hecto*	h
10 = 10^{1}	deka*	da
0.1 = 10^{-1}	deci*	d
0.01 = 10^{-2}	centi*	c
0.001 = 10^{-3}	milli	m
0.000 001 = 10^{-6}	micro	μ
0.000 000 001 = 10^{-9}	nano	n
0.000 000 000 001 = 10^{-12}	pico	p
0.000 000 000 000 001 = 10^{-15}	femto	f
0.000 000 000 000 000 001 = 10^{-18}	atto	a

* To be avoided where possible.

Table 1.13
Selected Additional Conversion Factors

Length and Area

100 feet (ft)/minute (min)	= 0.508 meter (m)/second(s)
1 square mile (mi^2)	= 640 acres
	= 259 hectares (ha)
1 m/s	= 196.9 ft/min
1 square kilometer (km^2)	= 100 ha
	= 0.3861 mi^2
1 ha	= 10,000 m^2
	= 2.471 acres
1 nautical mile	= 6080 ft
	= 1.151 statute mi
	= 1.853 km
1 nautical mile/hour (h)	= 1 knot

Weight

1 pound (lb)	= 16 ounces (oz)
	= 7000 grains (gr)
	= 0.454 kilogram (kg)
1 oz	= 0.0625 lb
	= 28.35 grams (g)
1 grain (gr)	= 64.8 milligrams (mg)
	= 0.0023 oz
1 lb/ft	= 1.488 kg/m
1 g	= 1000 mg
	= 0.03527 oz
	= 15.43 grains
1 kg/m	= 0.672 lb/ft

Volume

1 cubic foot (ft^3)	= 1728 cubic inches (in.^{3})
1 cubic inch (in.^{3})	= 16,390 cubic millimeters (mm^3)
1 imperial gallon	= 277.4 in.^{3}
	= 4.55 liters (l)
1 U.S. gallon (gal)	= 0.833 imperial gallon
	= 3.785 l
	= 231 in.^{3}
1 U.S. barrel (bbl) (petroleum)	= 42 U.S. gallons
	= 35 imperial gallons
1 liter (1)	= 10^6 mm^3
	= 0.220 imperial gallon
	= 0.2642 U.S. gallon
	= 61.0 in.^{3}
1 board ft	= 12 in. x 12 in. x 1 in. thick
	= 144 in.^{3}
1 ft^3/min	= 1.699 m^3/h
1 m^3/h	= 0.589 ft^3/min

Temperature, Measured

F	= (C x 9/5) + 32
	= R – 459.67
C	= (F – 32) x 5/9
	= K – 273.15

Density

1 ft^3/lb	= 0.0624 m^3/kg
1 grain/ft^3	= 2.288 g/m^3
1 grain/U.S. gallon	= 17.11 g/m^3
	= 17.11 mg/l
1 m^3/kg	= 16.02 ft^3/lb
1 g/m^3	= 0.437 grain/ft^3
	= 0.0584 grain/ U.S. gallon
1 g/l	= 58.4 grain/U.S. gallon

Water at 62F (16.7C)

1 ft^3	= 62.3 lb
1 lb	= 0.01604 ft^3
1 U.S. gallon	= 8.33 lb

Water at 39.2F (4C), maximum density

1 ft^3	= 62.4 lb
1 m^3	= 1000 kg
1 lb	= 0.01602 ft^3
1 l	= 1.0 kg
1 kg/m^3	= 1 g/l
	= 1 part per thousand
1 g/m^3	= 1 mg/l
	= 1 part per million (ppm)

Pressure

1 atmosphere (metric)	= 98,066.5 pascals (Pa)
	= 1 kg force (kgf)/square centimeter (cm^2)
	= 10 m head of water
	= 14.22 lb/in.^{2}
1 lb/ft^2	= 47.88 Pa
	= 0.1924 in. of water
	= 4.88 kg/m^2
1 ton (t)/in.^{2}	= 13,789 kilopascals (kPa)
	= 1.406 kg/mm^2
1 in. head of water	= 5.20 lb/ft^2
1 m head of water	= 9806 Pa
	= 0.1 kg/cm^2
1 m head of mercury (Hg)	= 133.3 kPa
	= 1.360 kg/cm^2
	= 1333 millibars
1 kg/m^2	= 9.806 Pa
	= 1 mm head of water
	= 0.2048 lb/ft^2
1 kg/cm^2	= 98.066 kPa
	= 735.5 mm Hg
	= 0.981 bar
	= 14.22 lb/in.^{2}
1 kg/mm^2	= 9.8066 megapascals (MPa)
	= 0.711 t/in^2

In these conversions, inches and feet of water are measured at 62F (16.7C), millimeters and meters of water at 39.2F (4C), and inches, millimeters, and meters of mercury at 32F (0C).

Note: The following steam tables and Fig. 2.2 have been abstracted from *ASME International Steam Tables for Industrial Use* (copyright 2000 by The American Society of Mechanical Engineers), based on the IAPWS industrial formulation 1997 for the Thermodynamic Properties of Water and Steam (IAPWS-IF97)

Table 2.1
SI Properties of Saturated Steam and Saturated Water (Temperature)[1]

Temp K	Press. MPa	Volume, m^3/kg			Enthalpy, kJ/kg			Entropy, kJ/(kg K)			Temp K
		Water v_f	Evap v_{fg}	Steam v_g	Water H_f	Evap H_{fg}	Steam H_g	Water s_f	Evap s_{fg}	Steam s_g	
273.16	0.00061166	0.001000	206.0	206.0	0	2500.9	2500.9	0	9.1555	9.1555	**273.16**
275	0.00069845	0.001000	181.6	181.6	7.8	2496.5	2504.3	0.0283	9.0783	9.1066	**275**
280	0.00099182	0.001000	130.2	130.2	28.8	2484.7	2513.5	0.1041	8.8738	8.9779	**280**
285	0.001389	0.001001	94.61	94.61	49.8	2472.8	2522.6	0.1784	8.6766	8.8550	**285**
290	0.001920	0.001001	69.63	69.63	70.7	2461.0	2531.7	0.2513	8.4862	8.7375	**290**
295	0.002621	0.001002	51.87	51.87	91.7	2449.2	2540.8	0.3228	8.3023	8.6251	**295**
300	0.003537	0.001003	39.08	39.08	112.6	2437.3	2549.9	0.3931	8.1244	8.5175	**300**
305	0.004719	0.001005	29.77	29.77	133.5	2425.4	2558.9	0.4622	7.9523	8.4145	**305**
310	0.006231	0.001007	22.90	22.91	154.4	2413.5	2567.9	0.5302	7.7856	8.3158	**310**
315	0.008145	0.001009	17.80	17.80	175.3	2401.6	2576.8	0.5970	7.6241	8.2211	**315**
320	0.01055	0.001011	13.95	13.96	196.2	2389.6	2585.7	0.6629	7.4674	8.1303	**320**
325	0.01353	0.001013	11.04	11.04	217.1	2377.5	2594.6	0.7277	7.3154	8.0431	**325**
330	0.01721	0.001015	8.805	8.806	238.0	2365.4	2603.3	0.7915	7.1678	7.9593	**330**
335	0.02172	0.001018	7.078	7.079	258.9	2353.2	2612.1	0.8544	7.0244	7.8787	**335**
340	0.02719	0.001021	5.733	5.734	279.8	2340.9	2620.7	0.9164	6.8849	7.8013	**340**
345	0.03378	0.001024	4.677	4.678	300.8	2328.5	2629.3	0.9775	6.7492	7.7267	**345**
350	0.04168	0.001027	3.841	3.842	321.7	2316.0	2637.7	1.0378	6.6171	7.6549	**350**
355	0.05108	0.001030	3.175	3.176	342.7	2303.4	2646.1	1.0973	6.4884	7.5857	**355**
360	0.06219	0.001034	2.641	2.642	363.7	2290.7	2654.4	1.1560	6.3629	7.5189	**360**
365	0.07526	0.001037	2.209	2.210	384.8	2277.8	2662.5	1.2140	6.2405	7.4545	**365**
370	0.09054	0.001041	1.858	1.859	405.8	2264.8	2670.6	1.2713	6.1209	7.3922	**370**
375	0.1083	0.001045	1.571	1.572	426.9	2251.6	2678.5	1.3279	6.0042	7.3320	**375**
380	0.1289	0.001049	1.335	1.336	448.0	2238.2	2686.3	1.3838	5.8900	7.2738	**380**
385	0.1525	0.001053	1.140	1.141	469.2	2224.7	2693.9	1.4390	5.7783	7.2174	**385**
390	0.1796	0.001058	0.9783	0.9794	490.4	2210.9	2701.3	1.4937	5.6690	7.1627	**390**
395	0.2106	0.001062	0.8429	0.8440	511.7	2197.0	2708.6	1.5477	5.5619	7.1096	**395**
400	0.2458	0.001067	0.7293	0.7303	532.9	2182.8	2715.7	1.6012	5.4569	7.0581	**400**
405	0.2856	0.001072	0.6334	0.6345	554.3	2168.3	2722.6	1.6542	5.3539	7.0080	**405**
410	0.3304	0.001077	0.5522	0.5533	575.7	2153.6	2729.3	1.7065	5.2528	6.9593	**410**
415	0.3808	0.001082	0.4832	0.4842	597.1	2138.7	2735.8	1.7584	5.1534	6.9119	**415**
420	0.4372	0.001087	0.4242	0.4253	618.7	2123.4	2742.1	1.8098	5.0558	6.8656	**420**
425	0.5002	0.001093	0.3736	0.3747	640.2	2107.9	2748.1	1.8607	4.9597	6.8205	**425**
430	0.5702	0.001098	0.3300	0.3311	661.9	2092.0	2753.9	1.9112	4.8652	6.7764	**430**
435	0.6478	0.001104	0.2924	0.2935	683.6	2075.8	2759.4	1.9612	4.7720	6.7333	**435**
440	0.7335	0.001110	0.2598	0.2609	705.4	2059.3	2764.7	2.0109	4.6802	6.6911	**440**
445	0.8281	0.001117	0.2315	0.2326	727.3	2042.4	2769.7	2.0601	4.5897	6.6498	**445**
450	0.9320	0.001123	0.2067	0.2078	749.3	2025.1	2774.4	2.1089	4.5003	6.6092	**450**
455	1.046	0.001130	0.1850	0.1862	771.4	2007.4	2778.8	2.1574	4.4120	6.5694	**455**
460	1.171	0.001137	0.1660	0.1672	793.5	1989.4	2782.9	2.2056	4.3247	6.5303	**460**
465	1.307	0.001144	0.1493	0.1504	815.8	1970.8	2786.7	2.2534	4.2384	6.4918	**465**
470	1.455	0.001152	0.1345	0.1356	838.2	1951.9	2790.1	2.3010	4.1529	6.4539	**470**
475	1.616	0.001159	0.1214	0.1226	860.7	1932.4	2793.2	2.3483	4.0683	6.4165	**475**
480	1.790	0.001167	0.1098	0.1110	883.4	1912.5	2795.9	2.3953	3.9843	6.3796	**480**
490	2.183	0.001185	0.09021	0.09140	929.1	1871.0	2800.1	2.4886	3.8183	6.3069	**490**
500	2.639	0.001203	0.07457	0.07577	975.5	1827.1	2802.6	2.5811	3.6543	6.2354	**500**
510	3.165	0.001223	0.06194	0.06317	1022.5	1780.7	2803.3	2.6731	3.4916	6.1647	**510**
520	3.769	0.001245	0.05167	0.05291	1070.4	1731.5	2801.9	2.7646	3.3298	6.0944	**520**
530	4.457	0.001268	0.04324	0.04451	1119.2	1679.1	2798.3	2.8559	3.1680	6.0240	**530**
540	5.237	0.001294	0.03626	0.03756	1169.1	1623.1	2792.2	2.9473	3.0056	5.9530	**540**
550	6.117	0.001323	0.03045	0.03177	1220.3	1563.0	2783.3	3.0391	2.8418	5.8809	**550**
560	7.106	0.001355	0.02557	0.02692	1272.9	1498.3	2771.2	3.1315	2.6755	5.8070	**560**
570	8.213	0.001392	0.02143	0.02282	1327.2	1428.2	2755.4	3.2250	2.5056	5.7305	**570**
580	9.448	0.001433	0.01790	0.01933	1383.7	1351.6	2735.3	3.3201	2.3303	5.6505	**580**
590	10.82	0.001482	0.01485	0.01633	1442.8	1267.1	2709.9	3.4177	2.1477	5.5654	**590**
600	12.34	0.001540	0.01219	0.01373	1505.2	1172.8	2678.0	3.5188	1.9546	5.4734	**600**
610	14.03	0.001611	0.00984	0.01145	1572.1	1065.1	2637.3	3.6250	1.7461	5.3711	**610**
620	15.90	0.001704	0.007703	0.009407	1645.7	938.2	2583.9	3.7396	1.5133	5.2528	**620**
630	17.97	0.001837	0.005688	0.007525	1730.7	780.1	2510.8	3.8697	1.2382	5.1079	**630**
640	20.27	0.002076	0.003561	0.005637	1842.0	552.4	2394.4	4.0378	0.8632	4.9010	**640**
647.0960	22.064	0.003106	0	0	2087.5	0	2087.5	4.4120	0	4.4120	**647.0960**

1. U.S. Customary steam tables are provided in Chapter 2, Table 1.
2. In the balance of *Steam*, enthalpy is denoted by H in place of h to avoid confusion with heat transfer coefficient.

Table 2.2
SI Properties of Saturated Steam and Saturated Water (Pressure)

Press. MPa	Temp K	Volume, m³/kg Water v_f	Evap v_{fg}	Steam v_g	Enthalpy, kJ/kg Water H_f	Evap H_{fg}	Steam H_g	Entropy, kJ/(kg K) Water s_f	Evap s_{fg}	Steam s_g
0.00080	276.911	0.001000	159.6	159.6	15.8	2492.0	2507.8	0.0575	8.9992	9.0567
0.0010	280.120	0.001000	129.2	129.2	29.3	2484.4	2513.7	0.1059	8.8690	8.9749
0.0012	282.804	0.001000	108.7	108.7	40.6	2478.0	2518.6	0.1460	8.7624	8.9083
0.0014	285.119	0.001001	93.90	93.90	50.3	2472.5	2522.8	0.1802	8.6720	8.8521
0.0016	287.160	0.001001	82.75	82.75	58.8	2467.7	2526.6	0.2101	8.5935	8.8036
0.0018	288.988	0.001001	74.01	74.01	66.5	2463.4	2529.9	0.2366	8.5242	8.7609
0.0020	290.645	0.001001	66.99	66.99	73.4	2459.5	2532.9	0.2606	8.4621	8.7227
0.0025	294.228	0.001002	54.24	54.24	88.4	2451.0	2539.4	0.3119	8.3303	8.6422
0.0030	297.23	0.001003	45.65	45.66	101.0	2443.9	2544.9	0.3543	8.2222	8.5766
0.0040	302.11	0.001004	34.79	34.79	121.4	2432.3	2553.7	0.4224	8.0510	8.4735
0.0050	306.03	0.001005	28.19	28.19	137.8	2423.0	2560.8	0.4763	7.9177	8.3939
0.0060	309.31	0.001006	23.73	23.73	151.5	2415.2	2566.7	0.5209	7.8083	8.3291
0.0080	314.66	0.001008	18.10	18.10	173.9	2402.4	2576.2	0.5925	7.6349	8.2274
0.010	318.96	0.001010	14.67	14.67	191.8	2392.1	2583.9	0.6492	7.4997	8.1489
0.012	322.57	0.001012	12.36	12.36	206.9	2383.4	2590.3	0.6963	7.3887	8.0850
0.014	325.70	0.001013	10.69	10.69	220.0	2375.8	2595.8	0.7366	7.2945	8.0312
0.016	328.46	0.001015	9.430	9.431	231.6	2369.1	2600.7	0.7720	7.2127	7.9847
0.018	330.95	0.001016	8.442	8.443	241.9	2363.1	2605.0	0.8035	7.1403	7.9437
0.020	333.21	0.001017	7.647	7.648	251.4	2357.5	2608.9	0.8320	7.0753	7.9072
0.025	338.11	0.001020	6.202	6.203	271.9	2345.5	2617.4	0.8931	6.9371	7.8302
0.030	342.25	0.001022	5.228	5.229	289.2	2335.3	2624.6	0.9439	6.8235	7.7675
0.040	349.01	0.001026	3.992	3.993	317.6	2318.5	2636.1	1.0259	6.6431	7.6690
0.050	354.47	0.001030	3.239	3.240	340.5	2304.7	2645.2	1.0910	6.5020	7.5930
0.060	359.08	0.001033	2.731	2.732	359.8	2293.0	2652.9	1.1452	6.3859	7.5311
0.080	366.64	0.001038	2.086	2.087	391.6	2273.5	2665.2	1.2328	6.2011	7.4339
0.10	372.76	0.001043	1.693	1.694	417.4	2257.5	2674.9	1.3026	6.0562	7.3588
0.101325	373.12	0.001043	1.672	1.673	419.0	2256.5	2675.5	1.3067	6.0477	7.3544
0.12	377.93	0.001047	1.427	1.428	439.3	2243.8	2683.1	1.3608	5.9369	7.2976
0.14	382.44	0.001051	1.236	1.237	458.4	2231.6	2690.0	1.4109	5.8352	7.2460
0.16	386.45	0.001054	1.090	1.091	475.3	2220.7	2696.0	1.4549	5.7464	7.2014
0.18	390.06	0.001058	0.97648	0.97753	490.7	2210.7	2701.4	1.4944	5.6677	7.1620
0.20	393.36	0.001061	0.88467	0.88574	504.7	2201.6	2706.2	1.5301	5.5968	7.1269
0.25	400.56	0.001067	0.71763	0.71870	535.4	2181.2	2716.5	1.6072	5.4452	7.0524
0.30	406.68	0.001073	0.60471	0.60579	561.5	2163.4	2724.9	1.6718	5.3198	6.9916
0.40	416.76	0.001084	0.46131	0.46239	604.7	2133.3	2738.1	1.7766	5.1188	6.8954
0.50	424.99	0.001093	0.37371	0.37480	640.2	2107.9	2748.1	1.8606	4.9600	6.8206
0.60	431.98	0.001101	0.31447	0.31558	670.5	2085.6	2756.1	1.9311	4.8281	6.7592
0.80	443.56	0.001115	0.23921	0.24033	721.0	2047.3	2768.3	2.0460	4.6156	6.6615
1.0	453.04	0.001127	0.19322	0.19435	762.7	2014.4	2777.1	2.1384	4.4465	6.5850
1.2	461.11	0.001139	0.16211	0.16325	798.5	1985.3	2783.8	2.2163	4.3054	6.5217
1.4	468.20	0.001149	0.13962	0.14077	830.1	1958.8	2788.9	2.2839	4.1836	6.4675
1.6	474.53	0.001159	0.12257	0.12373	858.6	1934.3	2792.9	2.3438	4.0762	6.4200
1.8	480.27	0.001168	0.10919	0.11036	884.6	1911.4	2796.0	2.3978	3.9798	6.3776
2.0	485.53	0.001177	0.09840	0.09958	908.6	1889.8	2798.4	2.4470	3.8921	6.3392
2.5	497.11	0.001197	0.07875	0.07995	962.0	1840.1	2802.0	2.5544	3.7015	6.2560
3.0	507.01	0.001217	0.06545	0.06666	1008.4	1794.9	2803.3	2.6456	3.5402	6.1858
4.0	523.51	0.001253	0.04852	0.04978	1087.4	1713.5	2800.9	2.7967	3.2731	6.0697
5.0	537.09	0.001286	0.03816	0.03945	1154.5	1639.7	2794.2	2.9207	3.0530	5.9737
6.0	548.74	0.001319	0.03113	0.03245	1213.7	1570.8	2784.6	3.0274	2.8626	5.8901
7.0	558.98	0.001352	0.02603	0.02738	1267.4	1505.1	2772.6	3.1220	2.6926	5.8146
8.0	568.16	0.001385	0.02214	0.02353	1317.1	1441.5	2758.6	3.2077	2.5372	5.7448
10	584.15	0.001453	0.01658	0.01803	1407.9	1317.6	2725.5	3.3603	2.2556	5.6159
11	591.23	0.001489	0.01451	0.01599	1450.3	1256.1	2706.4	3.4300	2.1246	5.5545
12	597.83	0.001526	0.01274	0.01427	1491.3	1194.3	2685.6	3.4965	1.9977	5.4941
13	604.01	0.001566	0.01122	0.01279	1531.4	1131.5	2662.9	3.5606	1.8733	5.4339
14	609.82	0.001610	0.00988	0.01149	1570.9	1067.2	2638.1	3.6230	1.7500	5.3730
16	620.51	0.001710	0.00760	0.00931	1649.7	931.1	2580.8	3.7457	1.5006	5.2463
18	630.14	0.001839	0.00566	0.00750	1732.0	777.5	2509.5	3.8717	1.2339	5.1055
20	638.90	0.002039	0.00382	0.00586	1827.1	584.3	2411.4	4.0154	0.9145	4.9299
22.064	647.096	0.003106	0.0	0.00311	2087.5	0.0	2087.5	4.4120	0.0	4.4120

1. See Note 1, Table 2.1
2. See Note 2, Table 2.1

Table 2.3 (Part 1 of 2)
SI Properties of Superheated Steam and Compressed Water (Temperature and Pressure)[1]

Pressure MPa (sat. temp.)		275	300	350	400	450	500	550	600	650	700	750	800	850	900	950	1000	1100	1200
		Temperature, K																	
0.001	v,m³/kg	0.001000	138.4	161.5	184.6	207.7	230.8	253.8	276.9	300.0	323.1	346.1	369.2	392.3	415.4	438.4	461.5	507.7	553.8
(280.12)	H,kJ/kg	7.760	2551.0	2644.9	2739.5	2835.2	2932.3	3030.8	3130.9	3232.5	3335.7	3440.6	3547.2	3655.6	3765.7	3877.6	3991.2	4223.9	4463.4
	s,kJ/(kg K)	0.0283	9.1037	9.3930	9.6456	9.8712	10.0757	10.2635	10.4376	10.6002	10.7532	10.8980	11.0356	11.1669	11.2928	11.4137	11.5303	11.7520	11.9604
0.002	v,m³/kg	0.001000	69.16	80.73	92.29	103.83	115.37	126.91	138.45	149.99	161.53	173.07	184.61	196.15	207.69	219.22	230.76	253.84	276.92
(290.65)	H,kJ/kg	7.701	2550.0	2611.7	2720.4	2835.2	2932.3	3030.8	3130.8	3232.5	3335.7	3440.6	3547.2	3655.6	3765.7	3877.6	3991.2	4223.8	4463.4
	s,kJ/(kg K)	0.0283	8.7826	9.0727	9.3256	9.5512	9.7557	9.9436	10.1176	10.2803	10.4333	10.5780	10.7156	10.8470	10.9729	11.0938	11.2104	11.4321	11.6405
0.004	v,m³/kg	0.001000	0.001003	40.351	46.133	51.909	57.681	63.453	69.223	74.994	80.763	86.533	92.303	98.072	103.842	109.611	115.380	126.919	138.457
(302.11)	H,kJ/kg	7.763	112.575	2644.3	2739.2	2835.1	2932.2	3030.7	3130.8	3232.4	3335.7	3440.6	3547.2	3655.6	3765.7	3877.5	3991.2	4223.8	4463.4
	s,kJ/(kg K)	0.0283	0.3931	8.7520	9.0053	9.2311	9.4357	9.6236	9.7977	9.9604	10.1134	10.2581	10.3957	10.5271	10.6529	10.7739	10.8905	11.1122	11.3206
0.01	v,m³/kg	0.001000	0.001003	16.121	18.442	20.756	23.067	25.377	27.686	29.995	32.303	34.611	36.920	39.228	41.536	43.844	46.151	50.767	55.383
(318.96)	H,kJ/kg	7.769	112.581	2643.3	2738.7	2834.7	2932.0	3030.6	3130.7	3232.3	3335.6	3440.5	3547.1	3655.5	3765.6	3877.5	3991.2	4223.8	4463.4
	s,kJ/(kg K)	0.0283	0.3931	8.3268	8.5814	8.8076	9.0125	9.2005	9.3746	9.5373	9.6904	9.8352	9.9728	10.1041	10.2300	10.3510	10.4676	10.6893	10.8977
0.02	v,m³/kg	0.001000	0.001003	8.044	9.211	10.371	11.529	12.685	13.840	14.995	16.150	17.304	18.459	19.613	20.767	21.921	23.075	25.383	27.691
(333.21)	H,kJ/kg	7.779	112.590	2641.6	2737.8	2834.2	2931.6	3030.3	3130.5	3232.2	3335.5	3440.4	3547.0	3655.4	3765.6	3877.4	3991.1	4223.8	4463.3
	s,kJ/(kg K)	0.0283	0.3931	8.0029	8.2597	8.4868	8.6920	8.8802	9.0545	9.2172	9.3703	9.5151	9.6528	9.7842	9.9101	10.0310	10.1476	10.3693	10.5777
0.04	v,m³/kg	0.001000	0.001003	4.005	4.596	5.179	5.760	6.339	6.917	7.495	8.073	8.651	9.228	9.805	10.383	10.960	11.537	12.691	13.845
(349.10)	H,kJ/kg	7.799	112.609	2638.0	2736.0	2833.1	2930.8	3029.8	3130.0	3231.8	3335.2	3440.2	3546.9	3655.3	3765.4	3877.3	3991.0	4223.7	4463.3
	s,kJ/(kg K)	0.0283	0.3931	7.6747	7.9363	8.1650	8.3710	8.5595	8.7340	8.8970	9.0501	9.1950	9.3327	9.4641	9.5900	9.7110	9.8276	10.0494	10.2578
0.07	v,m³/kg	0.001000	0.001003	0.001027	2.618	2.954	3.287	3.619	3.950	4.281	4.612	4.942	5.272	5.602	5.932	6.262	6.592	7.252	7.911
(363.08)	H,kJ/kg	7.830	112.636	321.755	2733.2	2831.4	2929.7	3028.9	3129.4	3231.3	3334.8	3439.8	3546.6	3655.0	3765.2	3877.1	3990.8	4223.6	4463.2
	s,kJ/(kg K)	0.0283	0.3931	1.0378	7.6726	7.9039	8.1111	8.3001	8.4750	8.6381	8.7914	8.9364	9.0741	9.2056	9.3316	9.4526	9.5692	9.7910	9.9994
0.101325	v,m³/kg	0.001000	0.001003	0.001027	1.802	2.037	2.268	2.498	2.727	2.956	3.185	3.413	3.641	3.870	4.098	4.326	4.554	5.010	5.465
(373.12)	H,kJ/kg	7.861	112.665	321.780	2730.3	2829.7	2928.5	3028.1	3128.7	3230.8	3334.3	3439.5	3546.3	3654.7	3765.0	3876.9	3990.7	4223.4	4463.0
	s,kJ/(kg K)	0.0283	0.3931	1.0378	7.4961	7.7303	7.9386	8.1283	8.3035	8.4668	8.6203	8.7653	8.9032	9.0347	9.1607	9.2818	9.3984	9.6202	9.8287
0.2	v,m³/kg	0.001000	0.001003	0.001027	0.903	1.025	1.144	1.262	1.379	1.495	1.612	1.728	1.844	1.959	2.075	2.191	2.306	2.538	2.769
(393.36)	H,kJ/kg	7.961	112.756	321.859	2720.5	2824.0	2924.8	3025.3	3126.6	3229.1	3332.9	3438.3	3545.3	3653.9	3764.3	3876.3	3990.1	4223.0	4462.7
	s,kJ/(kg K)	0.0283	0.3931	1.0377	7.1629	7.4068	7.6191	7.8107	7.9870	8.1510	8.3050	8.4503	8.5884	8.7201	8.8463	8.9674	9.0842	9.3061	9.5146
0.4	v,m³/kg	0.001000	0.001003	0.001027	0.001067	0.5054	0.5672	0.6274	0.6867	0.7455	0.8040	0.8624	0.9206	0.9787	1.0367	1.0947	1.1526	1.2684	1.3840
(416.76)	H,kJ/kg	8.163	112.940	322.020	533.052	2811.9	2916.9	3019.6	3122.3	3225.6	3330.1	3436.0	3543.3	3652.2	3762.8	3875.1	3989.0	4222.2	4462.0
	s,kJ/(kg K)	0.0283	0.3930	1.0376	1.6011	7.0659	7.2872	7.4832	7.6618	7.8272	7.9821	8.1281	8.2667	8.3987	8.5251	8.6465	8.7634	8.9855	9.1942
0.6	v,m³/kg	0.001000	0.001003	0.001027	0.001066	0.3320	0.3748	0.4158	0.4559	0.4955	0.5348	0.5739	0.6129	0.6518	0.6906	0.7293	0.7680	0.8453	0.9225
(431.98)	H,kJ/kg	8.364	113.124	322.180	533.189	2798.8	2908.7	3013.9	3117.9	3222.2	3327.3	3433.6	3541.4	3650.6	3761.4	3873.8	3988.0	4221.3	4461.3
	s,kJ/(kg K)	0.0283	0.3930	1.0375	1.6009	6.8560	7.0876	7.2881	7.4692	7.6361	7.7920	7.9387	8.0777	8.2101	8.3368	8.4584	8.5754	8.7978	9.0066
0.8	v,m³/kg	0.001000	0.001003	0.001027	0.001066	0.2450	0.2785	0.3099	0.3405	0.3705	0.4002	0.4297	0.4591	0.4883	0.5175	0.5466	0.5757	0.6338	0.6918
(443.56)	H,kJ/kg	8.566	113.308	322.341	533.326	2784.7	2900.1	3007.9	3113.5	3218.7	3324.5	3431.3	3539.4	3648.9	3759.9	3872.6	3986.9	4220.4	4460.7
	s,kJ/(kg K)	0.0283	0.3929	1.0373	1.6007	6.6982	6.9417	7.1473	7.3309	7.4994	7.6562	7.8036	7.9431	8.0758	8.2028	8.3246	8.4418	8.6644	8.8733
1	v,m³/kg	0.001000	0.001003	0.001027	0.001066	0.001123	0.220627	0.2464	0.2712	0.2955	0.3194	0.3432	0.3668	0.3903	0.4137	0.4370	0.4603	0.5069	0.5533
(453.04)	H,kJ/kg	8.768	113.492	322.501	533.463	749.328	2891.277	3001.9	3109.0	3215.2	3321.6	3428.9	3537.4	3647.2	3758.5	3871.3	3985.8	4219.6	4460.0
	s,kJ/(kg K)	0.0283	0.3928	1.0372	1.6005	2.1089	6.8251	7.0360	7.2224	7.3924	7.5502	7.6982	7.8382	7.9714	8.0986	8.2206	8.3380	8.5608	8.7699
1.5	v,m³/kg	0.000999	0.001003	0.001026	0.001066	0.001123	0.143300	0.1616	0.1788	0.1954	0.2117	0.2278	0.2437	0.2595	0.2752	0.2909	0.3065	0.3376	0.3687
(471.45)	H,kJ/kg	9.272	113.952	322.903	533.806	749.587	2867.592	2986.2	3097.5	3206.3	3314.5	3423.0	3532.5	3643.0	3754.9	3868.2	3983.0	4217.5	4458.3
	s,kJ/(kg K)	0.0284	0.3927	1.0369	1.6000	2.1082	6.6009	6.8272	7.0209	7.1950	7.3554	7.5052	7.6464	7.7805	7.9084	8.0309	8.1487	8.3721	8.5815
2	v,m³/kg	0.000999	0.001003	0.001026	0.001066	0.001122	0.104394	0.1192	0.1326	0.1454	0.1579	0.1701	0.1821	0.1941	0.2060	0.2178	0.2296	0.2530	0.2764
(485.53)	H,kJ/kg	9.776	114.412	323.304	534.149	749.847	2841.382	2969.7	3085.7	3197.2	3307.2	3417.1	3527.5	3638.8	3751.3	3865.1	3980.3	4215.3	4456.6
	s,kJ/(kg K)	0.0284	0.3926	1.0366	1.5996	2.1075	6.4265	6.6713	6.8733	7.0518	7.2149	7.3665	7.5090	7.6439	7.7725	7.8956	8.0138	8.2377	8.4476
2.5	v,m³/kg	0.000999	0.001002	0.001026	0.001065	0.001122	0.080796	0.0936	0.1048	0.1154	0.1255	0.1355	0.1452	0.1549	0.1644	0.1740	0.1834	0.2023	0.2210
(497.11)	H,kJ/kg	10.279	114.872	323.705	534.492	750.107	2811.705	2952.2	3073.5	3188.0	3299.9	3411.1	3522.5	3634.5	3747.6	3861.9	3977.6	4213.2	4454.8
	s,kJ/(kg K)	0.0284	0.3924	1.0362	1.5991	2.1068	6.2754	6.5436	6.7548	6.9382	7.1041	7.2575	7.4013	7.5372	7.6664	7.7900	7.9086	8.1331	8.3434
3	v,m³/kg	0.000999	0.001002	0.001026	0.001065	0.001121	0.001202	0.0765	0.0863	0.0953	0.1040	0.1124	0.1206	0.1287	0.1368	0.1447	0.1527	0.1684	0.1841
(507.01)	H,kJ/kg	10.782	115.331	324.107	534.835	750.367	975.542	2933.7	3060.9	3178.5	3292.5	3405.0	3517.4	3630.3	3744.0	3858.8	3974.8	4211.0	4453.1
	s,kJ/(kg K)	0.0284	0.3923	1.0359	1.5986	2.1062	2.5804	6.4331	6.6546	6.8431	7.0120	7.1673	7.3124	7.4492	7.5792	7.7033	7.8223	8.0474	8.2580
3.5	v,m³/kg	0.000998	0.001002	0.001025	0.001065	0.001121	0.001202	0.0642	0.0730	0.0810	0.0886	0.0959	0.1030	0.1100	0.1170	0.1238	0.1307	0.1442	0.1577
(515.71)	H,kJ/kg	11.284	115.791	324.508	535.179	750.629	975.651	2914.0	3047.8	3168.9	3284.9	3398.9	3512.4	3626.0	3740.3	3855.6	3972.1	4208.9	4451.4
	s,kJ/(kg K)	0.0284	0.3922	1.0356	1.5981	2.1055	2.5794	6.3337	6.5668	6.7608	6.9327	7.0900	7.2364	7.3742	7.5049	7.6296	7.7490	7.9747	8.1857
4	v,m³/kg	0.000998	0.001002	0.001025	0.001065	0.001121	0.001201	0.0549	0.0631	0.0703	0.0770	0.0835	0.0898	0.0960	0.1021	0.1082	0.1142	0.1261	0.1379
(523.51)	H,kJ/kg	11.786	116.250	324.910	535.523	750.890	975.761	2893.1	3034.3	3159.1	3277.3	3392.8	3507.3	3621.7	3736.7	3852.5	3969.3	4206.7	4449.7
	s,kJ/(kg K)	0.0284	0.3920	1.0353	1.5977	2.1048	2.5785	6.2417	6.4877	6.6877	6.8629	7.0222	7.1700	7.3088	7.4402	7.5654	7.6852	7.9115	8.1228

1. See Notes 1 and 2, Table 2.1

Table 2.3 (Part 2 of 2)
SI Properties of Superheated Steam and Compressed Water (Temperature and Pressure)[1]

Pressure MPa (sat. temp.)		Temperature, K 275	300	350	400	450	500	550	600	650	700	750	800	850	900	950	1000	1100	1200
5	v,m³/kg	0.0010	0.0010	0.0010	0.0011	0.0011	0.0012	0.0418	0.0490	0.0552	0.0608	0.0662	0.0713	0.0764	0.0814	0.0863	0.0911	0.1007	0.1102
	H,kJ/kg	12.79	117.17	325.71	536.21	751.42	975.99	2846.5	3005.8	3138.9	3261.8	3380.3	3497.0	3613.1	3729.3	3846.1	3963.8	4202.4	4446.2
(537.09)	s,kJ/(kg K)	0.0284	0.391737	1.0347	1.5967	2.1035	2.5765	6.0698	6.3477	6.5609	6.7431	6.9067	7.0573	7.1980	7.3309	7.4572	7.5779	7.8053	8.0174
8	v,m³/kg	0.0010	0.0010	0.0010	0.0011	0.0011	0.0012	0.0013	0.0276	0.0324	0.0365	0.0401	0.0436	0.0470	0.0502	0.0534	0.0565	0.0626	0.0687
	H,kJ/kg	15.79	119.92	328.12	538.28	753.00	976.70	1219.4	2905.5	3072.4	3212.6	3341.6	3465.4	3586.7	3706.9	3826.8	3947.0	4189.4	4435.9
(568.16)	s,kJ/(kg K)	0.0285	0.3909	1.0328	1.5939	2.0996	2.5708	3.0329	5.9969	6.2645	6.4724	6.6505	6.8103	6.9574	7.0948	7.2245	7.3478	7.5788	7.7932
10	v,m³/kg	0.0010	0.0010	0.0010	0.0011	0.0011	0.0012	0.0013	0.0201	0.0247	0.0283	0.0314	0.0344	0.0371	0.0398	0.0424	0.0450	0.0500	0.0548
	H,kJ/kg	17.78	121.75	329.73	539.66	754.07	977.21	1218.5	2819.8	3022.5	3177.3	3314.6	3443.7	3568.7	3691.7	3813.8	3935.8	4180.7	4429.0
(584.15)	s,kJ/(kg K)	0.0285	0.3903	1.0315	1.5921	2.0970	2.5670	3.0266	5.7754	6.1007	6.3304	6.5199	6.6866	6.8382	6.9788	7.1109	7.2359	7.4694	7.6854
15	v,m³/kg	0.0010	0.0010	0.0010	0.0011	0.0011	0.0012	0.0013	0.0015	0.0140	0.0173	0.0198	0.0220	0.0240	0.0259	0.0278	0.0296		
	H,kJ/kg	22.74	126.31	333.74	543.13	756.78	978.6	1216.9	1497.5	2868.6	3078.8	3242.2	3386.9	3522.4	3653.0	3780.8	3907.3		
(615.31)	s,kJ/(kg K)	0.0285	0.3889	1.0284	1.5875	2.0907	2.5578	3.0117	3.4992	5.7198	6.0322	6.2579	6.4447	6.6090	6.7583	6.8966	7.0263		
20	v,m³/kg	0.0010	0.0010	0.0010	0.0011	0.0011	0.0012	0.0013	0.0015	0.0079	0.0116	0.0139	0.0158	0.0175	0.0190	0.0205	0.0219		
	H,kJ/kg	27.67	130.85	337.75	546.61	759.54	980.1	1215.6	1486.3	2624.9	2961.6	3162.4	3326.5	3474.1	3613.1	3747.1	3878.4		
(638.9)	s,kJ/(kg K)	0.0283	0.3874	1.0253	1.5830	2.0844	2.5490	2.9977	3.4679	5.2618	5.7635	6.0410	6.2530	6.4321	6.5910	6.7360	6.8706		
22.064	v,m³/kg	0.0010	0.0010	0.0010	0.0011	0.0011	0.0012	0.0013	0.0015	0.0054	0.0099	0.0122	0.0140	0.0156	0.0171	0.0184	0.0197		
	H,kJ/kg	29.70	132.72	339.41	548.05	760.69	980.8	1215.3	1482.5	2418.6	2906.0	3127.0	3300.4	3453.6	3596.3	3733.0	3866.3		
(647.1)	s,kJ/(kg K)	0.0283	0.3868	1.0240	1.5811	2.0819	2.5455	2.9922	3.4566	4.9231	5.6524	5.9580	6.1820	6.3679	6.5310	6.6789	6.8157		
25	v,m³/kg	0.0010	0.0010	0.0010	0.0011	0.0011	0.0012	0.0013	0.0015	0.0020	0.0080	0.0103	0.0120	0.0135	0.0148	0.0161	0.0173		
	H,kJ/kg	32.57	135.38	341.76	550.11	762.33	981.7	1214.8	1477.8	1876.4	2817.3	3074.0	3262.2	3423.8	3572.1	3712.8	3849.1		
	s,kJ/(kg K)	0.0281	0.3860	1.0222	1.5785	2.0784	2.5405	2.9845	3.4416	4.0760	5.4884	5.8434	6.0867	6.2827	6.4523	6.6045	6.7443		
30	v,m³/kg	0.0010	0.0010	0.0010	0.0011	0.0011	0.0012	0.0013	0.0014	0.0018	0.0054	0.0079	0.0095	0.0109	0.0121	0.0132	0.0142		
	H,kJ/kg	37.43	139.89	345.77	553.62	765.17	983.5	1214.4	1471.2	1808.5	2631.5	2976.2	3194.4	3371.8	3530.2	3678.1	3819.7		
	s,kJ/(kg K)	0.0279	0.3845	1.0192	1.5742	2.0724	2.5322	2.9721	3.4186	3.9569	5.1754	5.6529	5.9350	6.1503	6.3315	6.4914	6.6366		
35	v,m³/kg	0.0010	0.0010	0.0010	0.0010	0.0011	0.0012	0.0013	0.0014	0.0017	0.0036	0.0061	0.0077	0.0090	0.0101	0.0111	0.0120		
	H,kJ/kg	42.26	144.38	349.78	557.14	768.05	985.30	1214.2	1466.0	1776.3	2402.0	2868.9	3123.2	3318.4	3487.6	3643.0	3790.1		
	s,kJ/(kg K)	0.0276	0.3830	1.0162	1.5699	2.0666	2.5242	2.9603	3.3980	3.8937	4.8159	5.4635	5.7924	6.0293	6.2229	6.3909	6.5418		
40	v,m³/kg	0.0010	0.0010	0.0010	0.0010	0.0011	0.0012	0.0013	0.0014	0.0017	0.0026	0.0048	0.0064	0.0076	0.0086	0.0095	0.0104		
	H,kJ/kg	47.06	148.86	353.79	560.67	770.96	987.22	1214.3	1461.7	1755.1	2222.5	2754.1	3049.3	3263.9	3444.6	3607.7	3760.4		
	s,kJ/(kg K)	0.0271	0.3814	1.0132	1.5656	2.0609	2.5164	2.9490	3.3793	3.8482	4.5379	5.2741	5.6560	5.9165	6.1232	6.2997	6.4564		
45	v,m³/kg	0.0010	0.0010	0.0010	0.0010	0.0011	0.0012	0.0012	0.0014	0.0016	0.0022	0.0039	0.0054	0.0065	0.0075	0.0083	0.0091		
	H,kJ/kg	51.83	153.32	357.80	564.22	773.90	989.23	1214.6	1458.3	1739.6	2129.0	2639.2	2973.9	3208.8	3401.4	3572.4	3730.9		
	s,kJ/(kg K)	0.0267	0.3799	1.0102	1.5614	2.0553	2.5089	2.9383	3.3621	3.8118	4.3873	5.0920	5.5251	5.8103	6.0306	6.2156	6.3782		
50	v,m³/kg	0.0010	0.0010	0.0010	0.0010	0.0011	0.0012	0.0012	0.0014	0.0016	0.0020	0.0032	0.0046	0.0057	0.0066	0.0074	0.0081		
	H,kJ/kg	56.57	157.77	361.80	567.77	776.88	991.32	1215.1	1455.6	1727.6	2075.5	2536.4	2898.8	3153.7	3358.2	3537.3	3701.5		
	s,kJ/(kg K)	0.0261	0.3783	1.0073	1.5573	2.0498	2.5015	2.9280	3.3461	3.7811	4.2956	4.9314	5.4003	5.7097	5.9437	6.1374	6.3059		
55	v,m³/kg	0.0010	0.0010	0.0010	0.0010	0.0011	0.0011	0.0012	0.0013	0.0015	0.0019	0.0028	0.0040	0.0050	0.0058	0.0066	0.0073		
	H,kJ/kg	61.29	162.20	365.80	571.34	779.88	993.48	1215.8	1453.3	1718.0	2039.9	2453.4	2826.6	3099.2	3315.5	3502.5	3672.4		
	s,kJ/(kg K)	0.0256	0.3767	1.0044	1.5532	2.0444	2.4944	2.9180	3.3311	3.7543	4.2308	4.8008	5.2833	5.6144	5.8618	6.0641	6.2385		
60	v,m³/kg	0.0010	0.0010	0.0010	0.0010	0.0011	0.0011	0.0012	0.0013	0.0015	0.0018	0.0025	0.0035	0.0044	0.0053	0.0060	0.0066		
	H,kJ/kg	65.98	166.61	369.80	574.91	782.91	995.70	1216.68	1451.6	1710.1	2014.0	2390.0	2759.7	3046.3	3273.6	3468.3	3643.8		
	s,kJ/(kg K)	0.0249	0.3751	1.0015	1.5492	2.0391	2.4874	2.9084	3.3170	3.7305	4.1804	4.6986	5.1764	5.5244	5.7845	5.9951	6.1752		
65	v,m³/kg	0.0010	0.0010	0.0010	0.0010	0.0011	0.0011	0.0012	0.0013	0.0015	0.0018	0.0023	0.0031	0.0040	0.0048	0.0054	0.0060		
	H,kJ/kg	70.64	171.02	373.80	578.50	785.97	997.99	1217.70	1450.2	1703.6	1994.1	2341.8	2700.4	2995.7	3232.8	3434.8	3615.6		
	s,kJ/(kg K)	0.0242	0.3735	0.9986	1.5452	2.0339	2.4805	2.8992	3.3036	3.7089	4.1391	4.6184	5.0815	5.4400	5.7114	5.9298	6.1155		
70	v,m³/kg	0.0010	0.0010	0.0010	0.0010	0.0011	0.0011	0.0012	0.0013	0.0015	0.0017	0.0022	0.0029	0.0036	0.0043	0.0050	0.0056		
	H,kJ/kg	75.28	175.41	377.80	582.09	789.05	1000.33	1218.84	1449.2	1698.1	1978.2	2304.5	2648.9	2948.2	3193.5	3402.1	3588.1		
	s,kJ/(kg K)	0.0234	0.3719	0.9958	1.5413	2.0288	2.4739	2.8903	3.2910	3.6892	4.1040	4.5539	4.9985	5.3617	5.6424	5.8681	6.0590		
75	v,m³/kg	0.0010	0.0010	0.0010	0.0010	0.0011	0.0011	0.0012	0.0013	0.0014	0.0017	0.0021	0.0026	0.0033	0.0040	0.0046	0.0051		
	H,kJ/kg	79.90	179.78	381.79	585.69	792.15	1002.72	1220.11	1448.5	1693.5	1965.2	2275.0	2605.0	2904.4	3155.8	3370.5	3561.3		
	s,kJ/(kg K)	0.0226	0.3702	0.9929	1.5374	2.0237	2.4673	2.8816	3.2789	3.6709	4.0732	4.5005	4.9265	5.2898	5.5774	5.8096	6.0055		
80	v,m³/kg	0.0010	0.0010	0.0010	0.0010	0.0011	0.0011	0.0012	0.0013	0.0014	0.0016	0.0020	0.0025	0.0031	0.0037	0.0043	0.0048		
	H,kJ/kg	84.49	184.14	385.78	589.29	795.28	1005.17	1221.50	1448.06	1689.7	1954.3	2251.1	2567.9	2864.7	3120.2	3340.0	3535.3		
	s,kJ/(kg K)	0.0217	0.3686	0.9901	1.5336	2.0187	2.4609	2.8732	3.2673	3.6539	4.0459	4.4552	4.8640	5.2241	5.5164	5.7543	5.9547		
100	v,m³/kg	0.0010	0.0010	0.0010	0.0010	0.0011	0.0011	0.0012	0.0013	0.0014	0.0015	0.0018	0.0021	0.0025	0.0029	0.0033	0.0038		
	H,kJ/kg	102.62	201.46	401.71	603.78	807.98	1015.39	1227.99	1448.46	1679.5	1924.9	2188.4	2466.7	2743.0	3000.5	3231.5	3440.4		
	s,kJ/(kg K)	0.0178	0.3618	0.9791	1.5187	1.9996	2.4366	2.8418	3.2253	3.5950	3.9586	4.3221	4.6813	5.0163	5.3109	5.5607	5.7751		

1. See Notes 1 and 2, Table 2.1

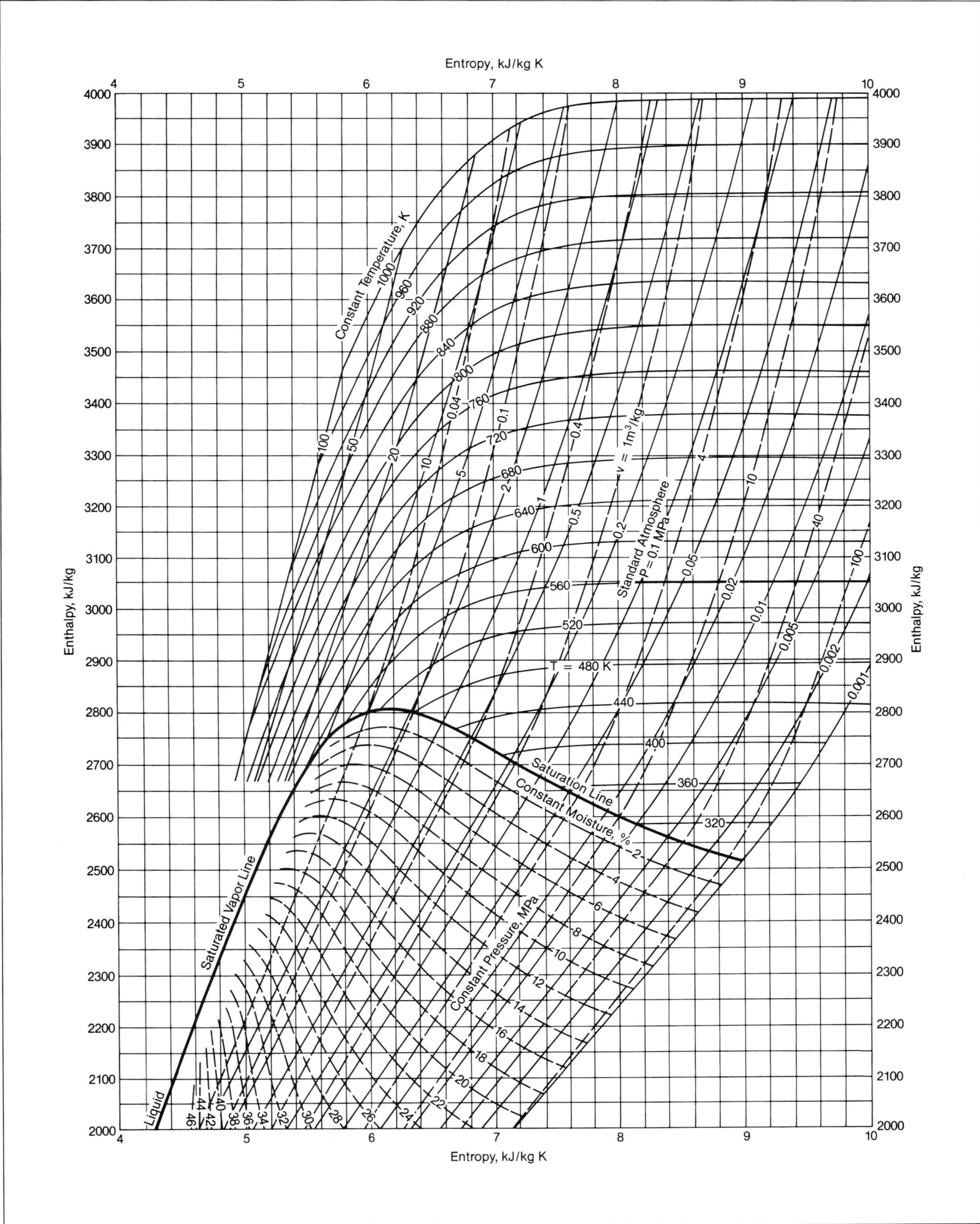

Fig. 2.1 Mollier diagram (*H-s*) for steam – SI units.

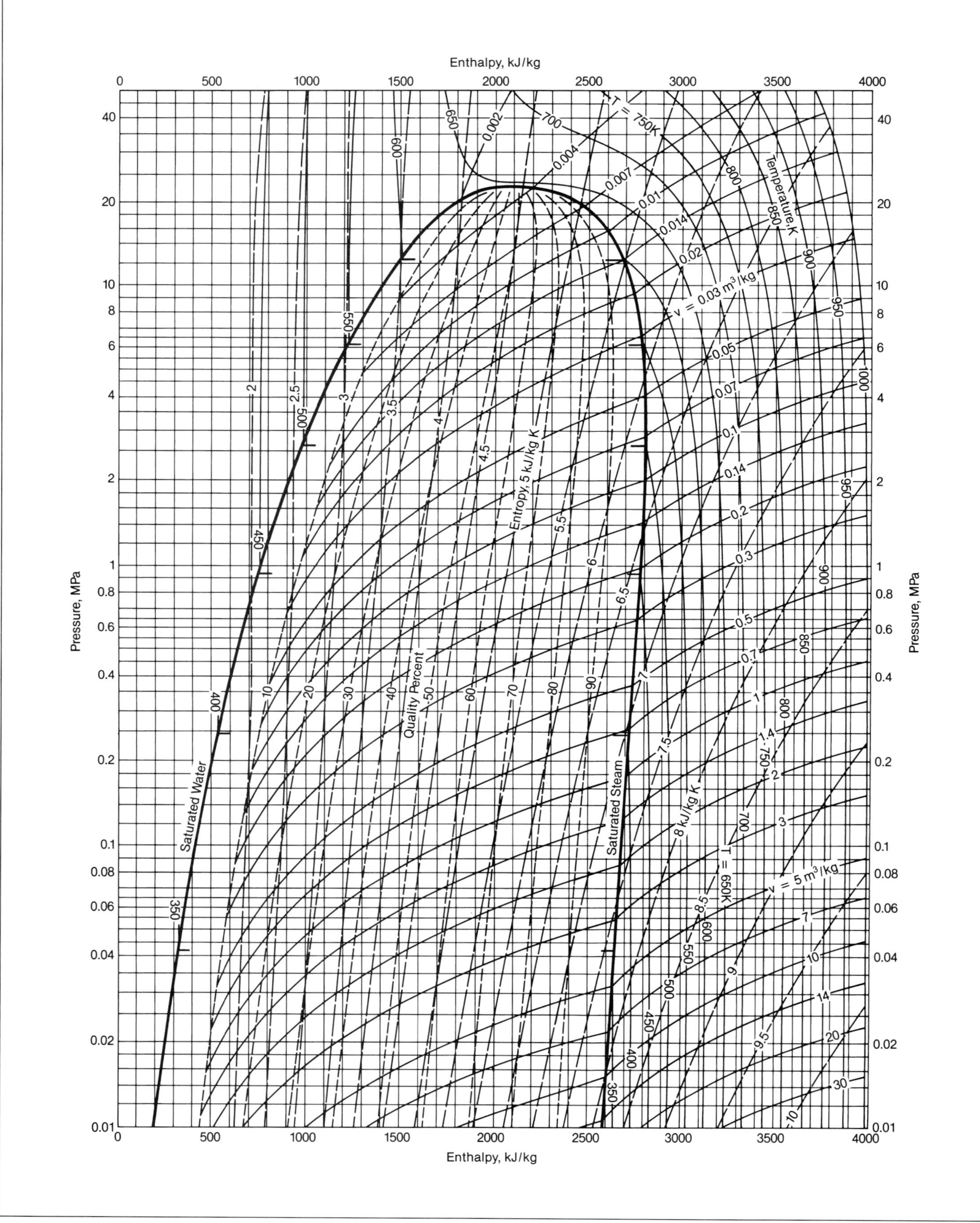

Fig. 2.2 Pressure-enthalpy diagram for steam – SI units.

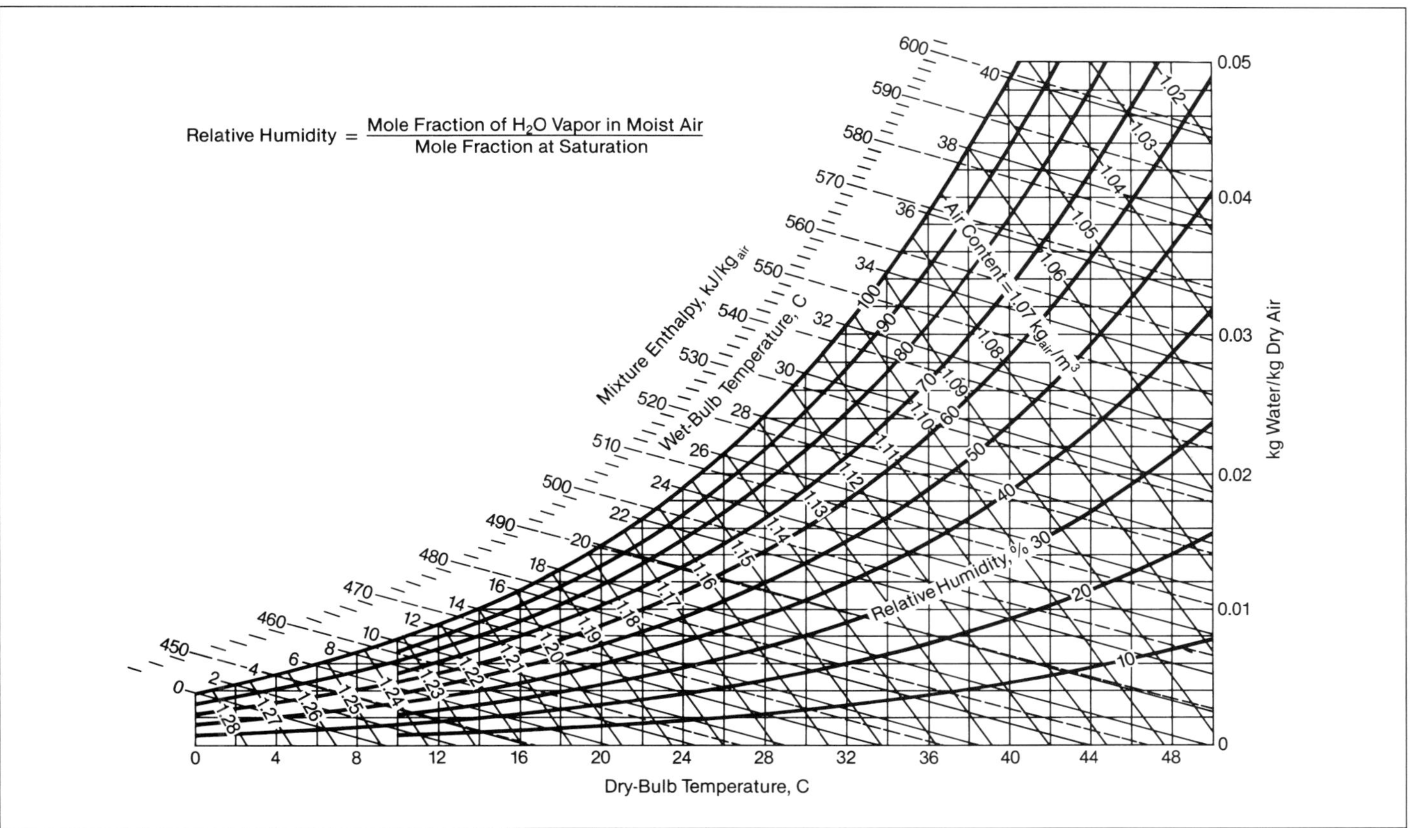

Fig. 2.3 Psychrometric chart – SI units.

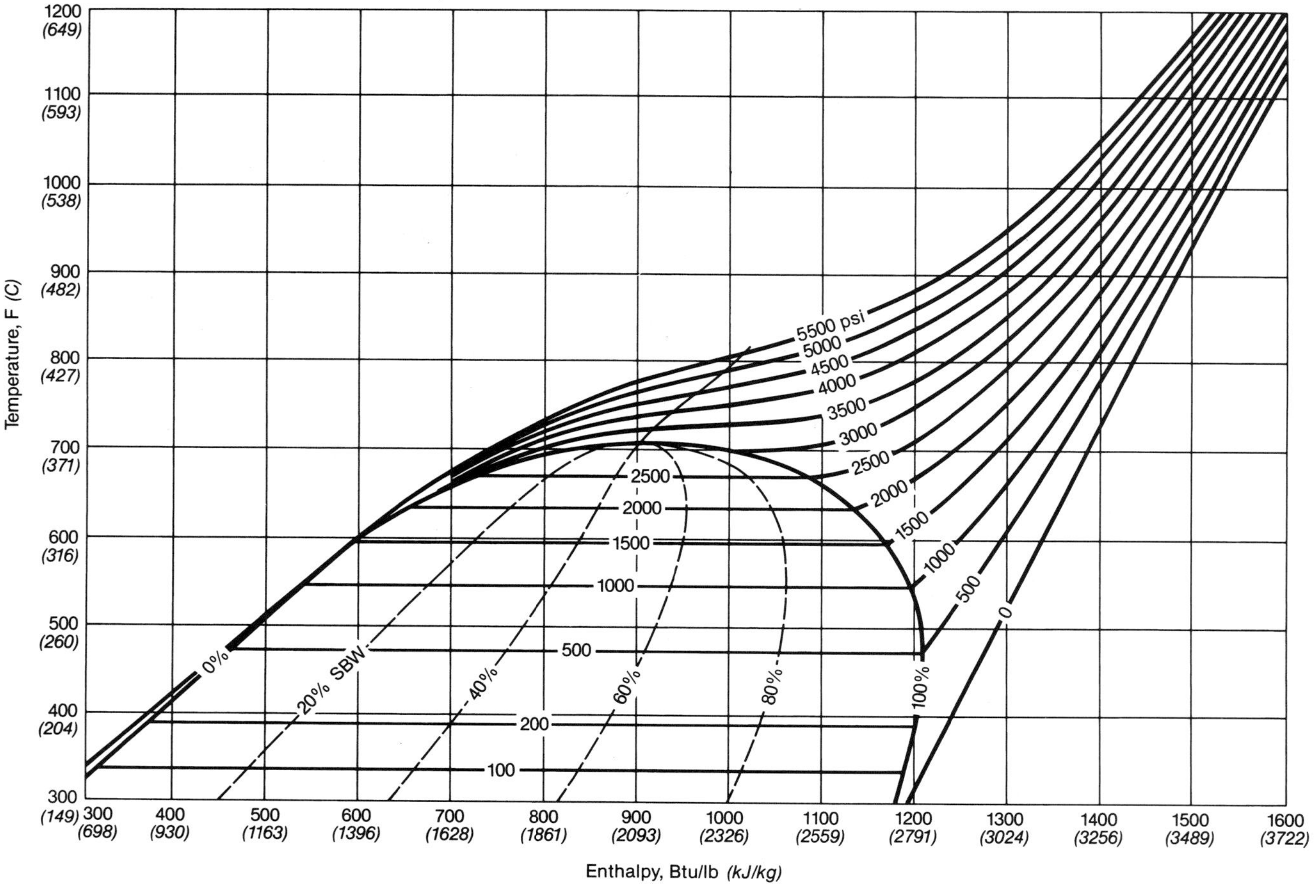

Fig. 3.1 Temperature-enthalpy diagram (MPa = 0.006895 × psi).

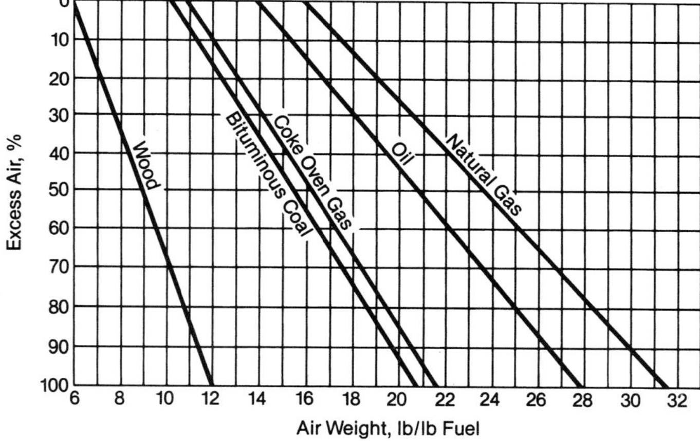

Fig. 3.2 Estimates of the weight of air required for selected fuels and excess air levels.

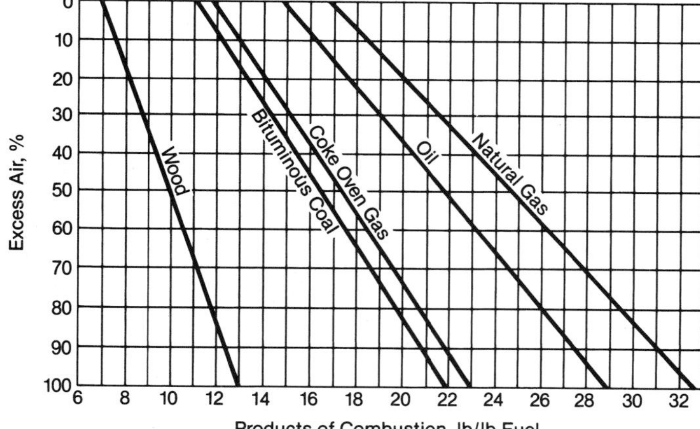

Fig. 3.3 Estimates of the weight of the products of combustion for selected fuels and excess air levels.

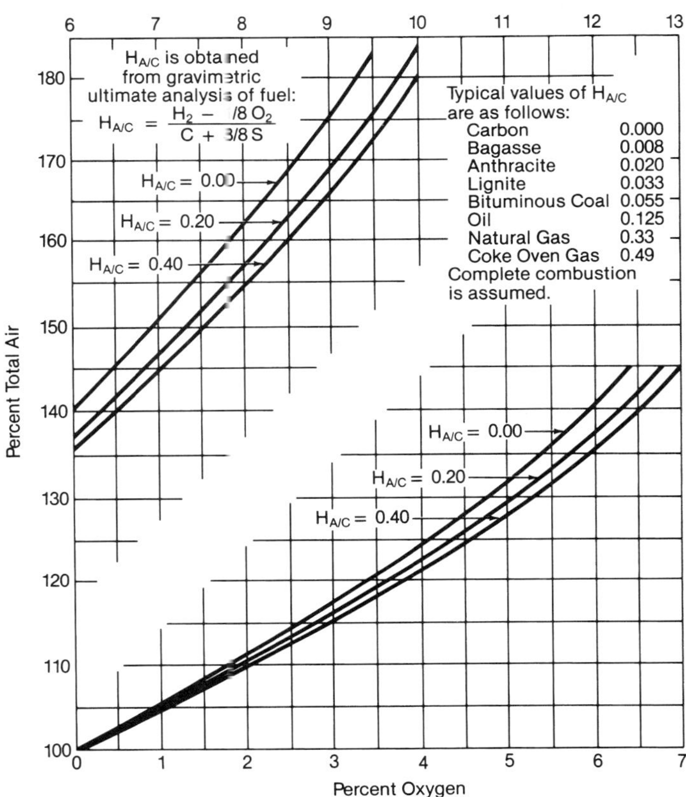

Fig. 3.4 Estimates of the percent total air for flue gas percent oxygen for selected fuels.

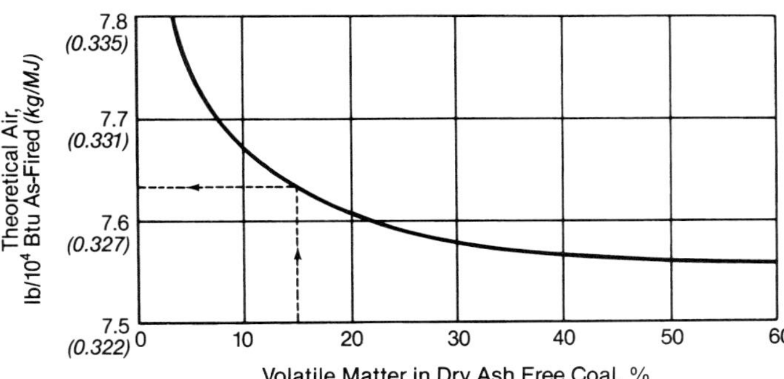

Fig. 3.6 Estimates of theoretical air required for the combustion of coal based on volatile matter contents.

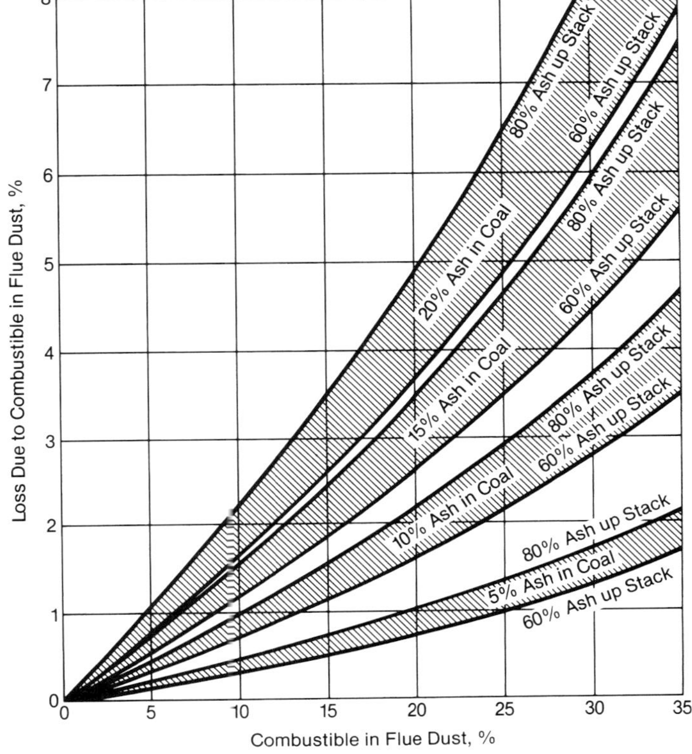

Fig. 3.5 Estimates of efficiency loss due to unburned combustibles for selected fuels and ash contents.

Table 3.1
SO_2, SO_3 Estimated Emissions from Boilers Firing Coal or Fuel Oil

SO_2:

$$\text{lb } SO_2 \text{ produced/lb fuel} = K\left(\frac{\text{lb sulfur}}{\text{lb fuel}}\right)\left(\frac{64 \text{ lb } SO_2\text{/mole}}{32 \text{ lb sulfur/mole}}\right)$$

$$\text{lb } SO_2 \text{ produced/h} = K\left(\frac{\text{lb fuel}}{\text{h}}\right)\left(\frac{\text{\% sulfur in fuel x 2}}{100}\right)$$

$$\text{\% } SO_2 \text{ by weight} = \frac{\text{lb } SO_2\text{/h x } 10^2}{\text{lb flue gas/h}}$$

$$\text{ppm } SO_2 \text{ by weight} = (\text{\% } SO_2 \text{ by weight}) \text{ x } 10^4$$

$$\text{\% } SO_2 \text{ by volume} = \frac{100\,(\text{\% } SO_2 \text{ by weight/64})}{\frac{(\text{\% } SO_2 \text{ by weight})}{64} + \frac{100-(\text{\% } SO_2 \text{ by weight})}{M}}$$

$$\text{ppm } SO_2 \text{ by volume} = \text{\% } SO_2 \text{ by volume x } 10^4$$

where K = 0.95 Cyclone furnace firing
0.97 dry bottom PC firing
0.99 oil firing

M = molecular weight of flue gas
30.2 coal firing
29.0 oil firing

SO_3:
SO_3 quantity = 1% of SO_2 quantity (coal firing)
= 2% of SO_2 quantity (oil firing)

Table 3.2
Conversion Formulae for Emission Rates

EPA F-Factor Method

A common method of reporting emissions is in pounds of pollutant per million Btu of fuel input. An EPA-approved conversion formula called the dry, oxygen based F-factor method can be used to convert ppm by dry volume of a pollutant to lb/10^6 Btu as follows:

$$\frac{\text{lb}}{10^6\ \text{Btu}} = \text{ppmdv} \times \frac{20.9}{(20.9 - \%O_2\text{dv})} \times F_d \times (2.59 \times 10^{-9}) \times M$$

where: F_d = a fuel factor representing the dry standard cubic feet (DSCF) of flue gas produced by burning a million Btu of fuel at theoretical air (0% excess air). Standard conditions are 68F and 14.7 psia. F_d can be calculated from an ultimate fuel analysis, or a fuel specific default value may be used as follows:

$$F_d = \frac{1{,}000{,}000\left[3.64(\%\ H) + 1.53(\%\ C) + 0.57(\%\ S) + 0.14(\%\ N) - 0.46(\%\ O)\right]}{\text{Higher heating value, Btu/lb}}$$

Note: The percentage by weight of each constituent in the above equation is acquired from an ultimate fuel analysis and must be on the same basis as the higher heating value (both wet or both dry).

Fuel Type	F_d Default Values
Refuse (Municipal solid waste)	9,570
Bituminous and subbituminous coal	9,780
Anthracite	10,100
Lignite	9,860
Natural gas	8,710
Fuel oils	9,190
Black liquor	8,800
Wood	9,400

2.59×10^{-9} = A constant based on the ideal gas law that equals 10^6 moles per DSCF. The ideal gas law, PV=nRT, says that 1 mole of ideal gas will occupy 385.5 ft^3 at 68F and 14.7 psia. 10^6 moles would, therefore, occupy 385.5 x 10^6 ft^3 and 10^6 moles per DSCF = 1/385.5 x 10^6 = 2.59 x 10^{-9}.

M = the molecular weight of the pollutant being measured (lb/lb-mole), where

Pollutant	Expressed as Equivalent	M
NO_x	NO_2	46.01
CO	CO	28.01
SO_2	SO_2	64.06
VOC	CH_4	16.04
SO_3	SO_3	80.06

ppmdv = the parts per million of pollutant in the flue gas measured on a dry volume basis.

$\frac{20.9}{(20.9 - \%O_2\text{dv})}$ = a correction factor for the amount of excess air as measured by percent O_2 on a dry volume basis in the flue gas.

Additional Conversion Methods

Instructions: Multiply the given pollutant value by the factor listed to convert to the units listed in the left-hand column.

Given Value →	ppm (dry vol.)	grains/DSCF	mg/DSCM	mg/Nm^3	lb/10^6 Btu	ng/Joule*
ppm (dry vol.)	1	$\frac{5.504 \times 10^4}{(M)}$	$\frac{24.056}{(M)}$	$\frac{22.417}{(M)}$	$\frac{1 \times 10^6\ (HI)}{(M)(N')}$	$\frac{2.321 \times 10^3\ (HI)}{(M)(N')}$
grains/DSCF	1.817×10^{-5} (M)	1	4.371×10^{-4}	4.074×10^{-4}	$\frac{18.162\ (HI)}{(N')}$	$\frac{4.217 \times 10^{-2}\ (HI)}{(N')}$
mg/DSCM	4.157×10^{-2} (M)	2.288×10^3	1	0.932	$\frac{4.156 \times 10^4\ (HI)}{(N')}$	$\frac{96.432\ (HI)}{(N')}$
mg/Nm^3	4.461×10^{-2} (M)	2.455×10^3	1.073	1	$\frac{4.458 \times 10^4\ (HI)}{(N')}$	$\frac{1.035 \times 10^2\ (HI)}{(N')}$
lb/10^6 Btu	$\frac{(M)(N')}{1 \times 10^6\ (HI)}$	$\frac{5.506 \times 10^{-2}\ (N')}{(HI)}$	$\frac{2.406 \times 10^{-5}\ (N')}{(HI)}$	$\frac{2.243 \times 10^{-5}\ (N')}{(HI)}$	1	2.321×10^{-3}
ng/Joule*	$\frac{4.308 \times 10^{-4}\ (M)(N')}{(HI)}$	$\frac{23.716\ (N')}{(HI)}$	$\frac{1.037 \times 10^{-2}\ (N')}{(HI)}$	$\frac{9.664 \times 10^{-3}\ (N')}{(HI)}$	430.8	1

grains/DSCF = grains per dry standard cubic foot (68F and 29.92 in. Hg)

mg/DSCM = milligrams per dry standard cubic meter (68F and 29.92 in. Hg)

mg/Nm^3 = milligrams per normal cubic meter (0C and 760 mm Hg)

lb/10^6 Btu = pounds per million Btu, fuel heat input (HHV basis)

ng/Joule = nanograms per Joule, fuel heat input (HHV basis)

* **Caution:** Should not be used for conversion calculations when ng/Joule is stated as an electrical output based standard.

M = molecular weight of pollutant (lb/lb-mole)

N′ = moles of dry flue gas per hour

HI = heat input in fuel, 10^6 Btu/h (HHV basis)

$\frac{N'}{HI}$ and its reciprocal occur in many of the above conversion formulae.

$\frac{N'}{HI}$ can be calculated using the EPA F-factor (F_d) discussed above as follows:

$$\frac{N'}{HI} = \frac{F_d}{385.5} \times \frac{20.9}{(20.9 - \%O_2\text{dv})} \quad \frac{\text{moles dry flue gas}}{10^6\ \text{Btu}}$$

Table 3.3
Periodic Table of the Elements

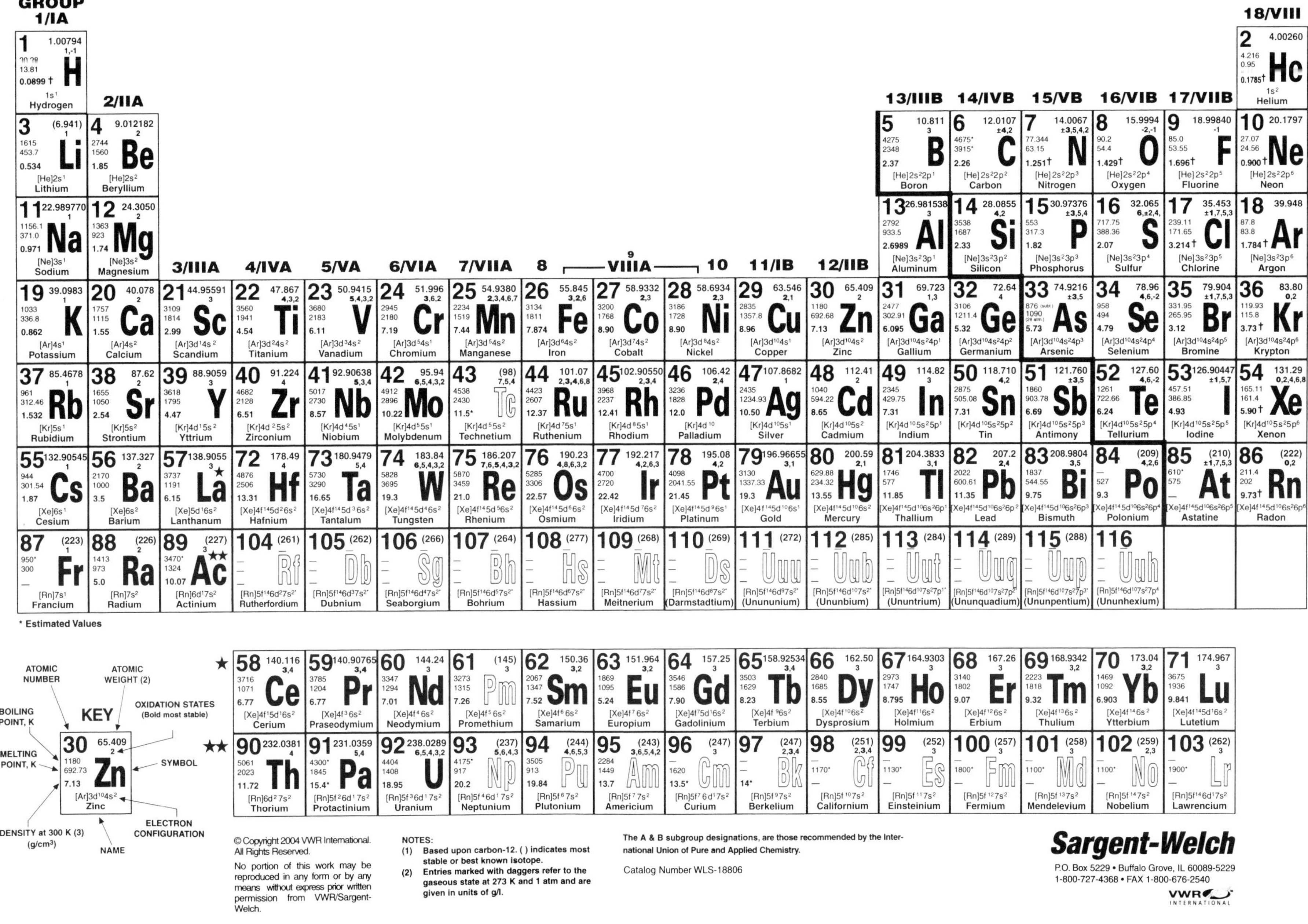

* Estimated Values

NOTES:
(1) Based upon carbon-12. ( ) indicates most stable or best known isotope.
(2) Entries marked with daggers refer to the gaseous state at 273 K and 1 atm and are given in units of g/l.

The A & B subgroup designations, are those recommended by the International Union of Pure and Applied Chemistry.

Catalog Number WLS-18806

Sargent-Welch
P.O. Box 5229 • Buffalo Grove, IL 60089-5229
1-800-727-4368 • FAX 1-800-676-2540

Table 4.1
Coal and Ash Analysis, an Example of a High Sulfur Bituminous Coal

Proximate Analysis, %		Ultimate Analysis, %		Sulfur Forms, %	
Moisture	12.0	Moisture	12.0	Pyritic	2.0
VM	35.7	Carbon	60.6	Organic	2.0
FC	40.3	Hydrogen	4.1	Sulfate	0.1
Ash	12.0	Nitrogen	1.2		4.1
		Sulfur	4.1		
		Oxygen	5.9		
		Ash	12.0		
		Chlorine	0.1		

Ash Fusion Temperatures, F (Reducing Atmosphere)	
Initial Deformation	1950
Softening	2030
Hemispherical	2100
Fluid	2150

Other

Btu/lb = 10,950 as received
Grindability = 56 HGI
Theoretical air = 7.6 lb/10,000 Btu

Ash Analysis, %	
Si as SiO_2	45.0
Al as Al_2O_3	20.0
Fe as Fe_2O_3	18.0
Ti as TiO_2	1.0
Ca as CaO	7.0
Mg as MgO	1.0
Na as Na_2O	0.6
K as K_2O	1.9
S as SO_3	3.5
P as P_2O_5	0.2

Table 4.2
Mechanical Equipment
Nominal 500 MW PC-Fired Drum Boiler Plant

Steam Generator
Balanced draft, direct-fired, pulverized coal

Pulverizers
Six, roll wheel mills, 50 t/h each

Main Steam		Reheat Steam	
Flow	3.625 x 10^6 lb/h	Flow	3.336 x 10^6 lb/h
Temp	1005F	Temp	1000F
Pressure	2415 psi	Pressure	532 psi

Turbine/Generator
Tandem-compound, four-flow, 3600 rpm, 640,000 kVa at 24 kV, 3 phase, 60 Hz.

Fans
FD: 2 motor-driven, total 4,000,000 lb/h
PA: 2 motor-driven, total 710,000 lb/h
ID: 4 motor-driven, total 5,174,000 lb/h

Fan Horsepower Requirements
FD: 2400 HP based on static rise from –0.5 to 14 in. wg
PA: 1700 HP based on static rise from –0.5 to 50 in. wg
ID: 9900 HP based on static rise from –23 to 10 in. wg
Fan data assumes 8% air heater leakage and 87% fan/motor efficiency.

Feedwater Pumps

Main:	2 turbine-driven, 50% capacity, 5000 GPM each, TDH: 7200 ft
Booster:	2 turbine-driven, 50% capacity, 4800 GPM each, TDH: 1280 ft
Condensate:	2 motor-driven, 100% capacity, 1400 hp, 7000 GPM each, TDH: 630 ft

Table 4.3
Environmental Equipment
Nominal 500 MW PC-Fired Drum Boiler Plant

Wet Limestone Scrubber (full load)

Limestone usage rate:	30.3 t/h
Gypsum production:	48.4 t/h
SO_2 at scrubber inlet:	36,568 lb/h
SO_2 at scrubber outlet:	731 lb/h
Required SO_2 removal:	98%

NO_x Control

Type:	SCR, OFA and low NO_x burners
Emissions:	0.07 lb/10^6 Btu

Particulate Control

Type:	Electrostatic precipitator
Emissions:	0.03 lb/10^6 Btu
Specific collector area:	350 ft^2/1000 SCFM

Table 4.4
Plant Performance Summary
Nominal 500 MW PC-Fired Drum Boiler Plant

Turbine Steam Conditions	
Pressure	2400psig
Temperature	1000F
Reheat temperature	1000F

Fuel Input

Coal feed 223 t/h
Btu input 4.88 x 10^9 Btu/h

Ash Flow Rates	**t/h**
Furnace bottom ash	5.1
Economizer ash	2.2
Flyash	20.7
Total through unit	28.0
Mill rejects (pyrites)	0.5

Heat Rates at Full Load	
Steam cycle	7880 Btu/kWh
Gross	8985 Btu/kWh
Net	9871 Btu/kWh

Rated Output Power Production	
Total gross	536 MW
Total Net	494 MW

Thermal Discharge	**10^6 Btu/h**
Cooling towers	2500
Stack	587

Controlled Emissions	**lb/h**
SO_2	731*
NO_x as NO_2	342*
Particulate	146

Efficiencies	
Boiler (coal to steam)	88%
Plant (coal to busbar)	35%

Untreated Flue Gas Characteristics

Component	**Volume % Dry**	**Volume % Wet**
Nitrogen	81.07	73.47
Carbon dioxide	15.28	13.84
Oxygen	3.26	2.95
Water	–	9.39
Sulfur dioxide	0.39	0.35
	100.00	100.00

Flue Gas Molecular Weight
30.85 (dry basis)
29.64 (wet basis)

Excess air leaving economizer 17.9%

Mass Flows	**lb/10,000 Btu**
Actual dry air	8.888
Wet gas from fuel	0.800
Moisture in air	0.116
Wet gas weight	9.804
Water from fuel	0.444
Water in wet gas	0.560
Dry gas weight	9.244

Water in wet gas (as%) 5.711%

Losses	**%**
Dry gas loss	4.659
Water from fuel	5.056
Moisture in air	0.116
Unburned combustibles	0.500
Radiation	0.170
Unaccounted for manufacturer's margin	1.500
	12.001

*Local requirements vary.

Appendix 2

Codes and Standards

There are many codes and specifications which control the design and construction of boilers, steam generators, pressure vessels and piping. Two of the most important are the American Society of Mechanical Engineers (ASME) Boiler and Pressure Vessel Code (BPVC) and the ASME B31 Pressure Piping Code. Pressure-containing components may be required by law to meet the requirements of these codes. In the United States (U.S.) and Canada (and recently in some other countries), state, city and provincial laws, enforced by local jurisdictions (see Note below), require new pressure parts to comply with the requirements of the ASME BPVC. The design and construction of pressure parts for repair or replacement are established by the local jurisdiction (see note below) and/or insurance carrier generally following the National Board Inspection Code (NBIC) in the U.S.

In addition to the BPVC and B31 Pressure Piping Codes, the ASME Performance Test Code provides uniform procedures for the testing and performance evaluation of power plant equipment. This code is frequently used to establish contract compliance of a particular component.

The ASME Boiler and Pressure Vessel Code

The ASME BPVC establishes rules of safety governing the design, fabrication and inspection during construction of boilers, pressure vessels and nuclear power plant components. The objectives of the rules are to assure reasonably certain protection of life and property and to provide a margin for deterioration in service. These rules do not provide criteria for thermal performance, but rather set minimum necessary guidelines for structural integrity to ensure safe operation during the expected component life. The BPVC provides a systematic approach to evaluating the stresses and applying material properties in a way which provides a safe pressure vessel design. The rules are established through a structured voluntary consensus code writing system and are implemented through specific contract terms and/or adoption by various jurisdictions (see Note below).

ASME BPVC history

To understand why the ASME Boiler and Pressure Vessel Committee was formed and how it operates, some historical background is necessary.

During the mid 1880s, explosions in fire tube boilers were common. Design knowledge and construction expertise generally were held by large companies, such as The Babcock & Wilcox Company (B&W), and were based essentially on experience.

In 1887, the American Boiler Manufacturers Association (ABMA) was organized to develop general rules for design and construction of boilers and pressure vessels. However, participants were reluctant to give up trade secrets and the effort to develop rules or standards failed. There was, however, some exchange of technical information such as materials data, riveting methods, safety factors, head and flange design rules and hydrostatic testing which helped manufacturers improve the safety of their products.

From 1898 to 1903, more than 1200 people were killed in the U.S. in 1900 separate boiler explosions. The catastrophic explosion of a fire tube boiler in a factory in Brockton, Massachusetts, in 1905 killed 58 people. This tragedy led the governor of Massachusetts to request and obtain the first legal set of rules for the design and construction of boilers. The next year a similar set of boiler design and construction rules was issued by the State of Ohio. Other jurisdictions followed with their own rules. This wide array of rules and regulations caused problems for manufacturers and users because equipment which was acceptable and met the rules in one state was frequently not acceptable in another.

Colonel E.D. Meier, a founder of the ABMA, campaigned to get acceptance of uniform boiler laws. In 1911, he was elected president of the ASME. Finally, on February 13, 1915, the first ASME Boiler and Pressure Vessel Code of 147 pages was issued. It was entitled *Boiler Construction Code, 1914 Edition*.

Note: *Local jurisdiction* refers to the municipal, state, provincial or federal authorities enforcing boiler and pressure vessel laws or regulations.

ASME BPVC additions

After the 1914 Edition, the following major sections were issued:

1921 Section III – Boilers for Locomotives. (With the 1962 Edition, this Section was integrated into Section I and the Section III designation was later reassigned.)
1922 Section V – Miniature Boilers. (With the 1962 Edition, this Section was integrated into Section I and the Section V designation was later reassigned.)
1923 Section IV – Low Pressure Heating Boilers.
1924 Section II – Material specifications. (Until 1924, materials were included as part of Section I.)
1925 Section VIII – Unfired Pressure Vessels. (With the 1968 Edition, this title was changed to Rules for Construction of Pressure Vessels, Division 1 and Division 2 was first issued.)
1926 Section VI – Rules for Inspection (of Power Boilers). (With the 1970 Addenda, this section was reassigned with a new title.)
1926 Section VII – Suggested Rules for Care of Power Boilers.
1937 Section IX – Welding Qualifications. (This section was originally a supplement to Section VIII, but with the 1941 Edition the section was published separately.)
1963 Section III – Nuclear Vessels. (Now Rules for Construction of Nuclear Facility Components.)
1968 Section VIII, Division 2 – Alternative Rules for Pressure Vessels.
1968 Section X – Fiber-Reinforced Plastic Pressure Vessels.
1970 Section XI – Rules for Inservice Inspection of Nuclear Power Plant Components.
1970 Section VI – Recommended Rules for Care and Operation of Heating Boilers.
1971 Section V – Nondestructive Examination.
1975 Section III, Division 2 – Code for Concrete Reactor Vessels and Containments. (Now Concrete Containments.)
1997 Section III, Division 3 – Containment Systems for Storage and Transport Packagings of Spent Nuclear Fuel and High Level Radioactive Material and Waste. (Now Containments for Transportation and Storage.)
1997 Section VIII, Division 3 – Alternative Rules for Construction of High Pressure Vessels.
2004 Section XII – Rules for Construction and Continued Service of Transport Tanks.
2004 Section II – Metric values of material properties were added as Part D.

As of 2004, the current sections of the BPVC are summarized in Table 1.

Table 1
ASME Boiler and Pressure Vessel Code Sections, 2004

I	Power Boilers
II	Materials
III-1	Division 1, Rules for Construction of Nuclear Facility Components
III-2	Division 2, Code for Concrete Containments
III-3	Containments for Transportation and Storage (of nuclear spent fuel or high level radioactive waste)
IV	Heating Boilers
V	Nondestructive Examination
VI	Recommended Rules for the Care and Operation of Heating Boilers
VII	Recommended Guidelines for the Care of Power Boilers
VIII-1	Pressure Vessels, Division 1
VIII-2	Pressure Vessels, Division 2 – Alternative Rules
VIII-3	Alternative Rules for Construction of High Pressure Vessels
IX	Welding and Brazing Qualifications
X	Fiber-Reinforced Plastic Pressure Vessels
XI	Rules for Inservice Inspection of Nuclear Power Plant Components
XII	Rules for Construction and Continued Service of Transport Tanks

Organization of the ASME Boiler and Pressure Vessel Committee

The first ASME Boiler and Pressure Vessel Committee in 1911 had seven members. By the early 1990s, the ASME Boiler and Pressure Vessel Committee had grown to about 800 people involved at all levels from working groups to the Boiler and Pressure Vessel Main Committee. The BPVC tries to maintain balance on the committees with representatives from manufacturers, users, insurance carriers, jurisdictions and other areas. However, the representatives from these groups may maintain their position within the Boiler and Pressure Vessel Committee even if they change their affiliation.

The administrative structure of the ASME Boiler and Pressure Vessel Committee consists of the Main Committee, the Executive Committee, the Conference Committee and the Headquarters Staff. Subcommittees report to the Main Committee.

In general, there are three different types of subcommittees reporting to the main committee. First, there are the component or *book* committees which are responsible for all the rules of design and construction for a particular type of component and for that Section of the BPVC in which the rules are given. Presently, the book subcommittees are:

Section I – Subcommittee on Power Boilers (SC I)
Section III – Subcommittee on Nuclear Power (SC III)
Section IV – Subcommittee on Heating Boilers (SC IV)
Section VIII – Subcommittee on Pressure Vessels (SC VIII)
Section X – Subcommittee on Reinforced Plastic Pressure Vessels (SC X)
Section XII – Subcommittee on Transport Tanks (SC XII)

Second, there are the service book subcommittees. These subcommittees are responsible for developing a specific set of rules which may be used by the component committees. Currently, these are:

Section II – Subcommittee on Materials (SC II)
Section V – Subcommittee on Nondestructive Examination (SC V)
Section IX – Subcommittee on Welding (SC IX)

Section XI – Subcommittee on Nuclear Inservice Inspection (SC XI)

Finally, there are the service subcommittees which are responsible for a specific area of technology or administration of the BPVC that may be applied to the book sections. These currently are:

Subcommittee on Design (SC D)
Subcommittee on Safety Valve Requirements (SC SVR)
Subcommittee on Boiler and Pressure Vessel Accreditation (SC BPVA)
Subcommittee on Nuclear Accreditation (SC NA)

Code Editions, Code Addenda and Code Cases

A new edition of the BPVC is issued every three years (...1998, 2001, 2004...). The code requirements become mandatory on the date of issue. No items are included in the new edition of the BPVC which have not been covered by the previous edition or its three Addenda.

Additions, revisions and deletions to the BPVC are contained in the Code Addenda, issued every year on December 31. The Code Addenda becomes mandatory six months after the date of issue although its contents are optional and may be used up to that date. This permits a thorough review by the public without having to comply with mandatory rules. It also gives a transition period for changing any affected company standards.

Except for Section III, the date of the appropriate edition and addenda of the BPVC for a particular contract is the date the contract is negotiated or signed. Most sections indicate that it is the date of the contract, although the actual signing of the contract may come very much later. Section III permits, for some applications, the use of an edition and addenda which are not the latest versions. In general, this is used for construction of a component which is used for repair or replacement. The other sections of the BPVC do not control the repair or replacement of components after the product has been installed; consequently, the decision of the date of the edition and addenda is the responsibility of the owner, the local jurisdiction and/or the insurance carrier.

Code Cases are contained in two special volumes independent from the BPVC Sections. Code Cases are issued four times a year after each Main Committee meeting and are subject to the same approval procedures as other changes and additions. Code Cases permit the use of a material or construction for which there are no current BPVC rules. The scope of Code Cases is purposely kept very specific to limit their usage. Each is reviewed for incorporation into the text of the appropriate BPVC Section. Code Cases are numbered sequentially, and nuclear Code Cases carry an N prefix. Code Case use is optional, not mandatory. When one is chosen, all parts are mandatory and the Code Case number shall be listed on the Manufacturer's Data Report form.

Technical Inquiries

When any user of the BPVC is unable to interpret a specific rule, a Technical Inquiry may be sent to the Boiler and Pressure Vessel Committee for an interpretation. Each Section of the BPVC includes an appendix that describes the preparation of Technical Inquiries. It is important that the inquiry meet all of the requirements so that action can be taken without delay. Inquiries are approved by two methods. The first is consideration by a special group of five people who must vote unanimously on the answer. If that method fails, the inquiry must be considered and approved by the entire subcommittee. Once the inquiry and reply are approved, a reply is sent to the inquirer. The question and reply are also published in the *Interpretations*, issued twice each year to BPVC subscribers.

There are several reasons that an inquiry will not be handled by the committee:

1. indefinite question – no reference to BPVC paragraph or rule,
2. semi-commercial question – doubt as to whether question is related to BPVC requirements or is asking for design approval,
3. approval of specific design – inquirer wants approval of specific design or construction as meeting BPVC requirements, and
4. basis or background of BPVC rules – inquirer wants rationale or basis of BPVC requirements. These are not given. Suggested revisions and additions are accepted for consideration.

General design philosophy and safety factors

The basic design philosophy of the BPVC from its beginning was one in which the primary membrane stress, or maximum direct stress, does not exceed the allowable design stress. (See Chapter 8 for an in-depth discussion of stress analysis and stress categorization, e.g., primary membrane stress.) The calculated stress is based on the maximum stress theory. This design philosophy is still used in Section I; Section VIII, Division 1; and Section IV. In most instances, these sections do not call for a detailed analysis to determine the local and secondary stresses. Instead, they rely on a design-by-formula approach.

The BPVC makes a significant distinction between allowable stress limits and the physical properties of the material such as ultimate tensile strength. The published allowable stresses incorporate a safety factor which reduces measured properties to account for: 1) the degree of complexity of the stress evaluation method, 2) certain levels and types of stress concentration, 3) certain nonuniformity in the materials, and 4) geometric factors. The safety factors are based upon experience, experimental evidence and theoretical evaluations. In general, where the BPVC allows a more detailed exact stress evaluation, a smaller safety factor is permitted in determining the allowable stress. The BPVC therefore specifies the evaluation procedures and the applicable allowable stresses (including safety factors) to provide a uniform, safe design procedure.

In the original BPVC in 1914, the allowable stress limits were based on a safety factor of five applied to the minimum specified ultimate tensile strength of the material. However, in December 1931, a joint American Petroleum Institute (API)-ASME Committee on Unfired Pressure Vessels was formed to develop a special code for Petroleum Liquids and Gases. The first edition was issued in September 1934. The API-ASME Code permitted a safety factor of four with only minor changes in the design equations used for establishing the minimum wall thickness of cylinders and heads, for reinforced openings, and for other parts. At that time, the ASME BPVC maintained a safety factor of five with some special provisions which permitted a safety factor of four. One of the reasons for the special provisions was to conserve metal during World War II.

The safety factor of five, except for the special provisions in Section VIII, was used by all sections of the BPVC through the 1949 Edition. Based upon the satisfactory experience in using a safety factor of four with the API-ASME Code and the special paragraphs of Section VIII, the 1950 editions of Section I and Section VIII were issued with a safety factor of four. The safety factor of Section IV remains at five.

During the 1950s, Admiral Hyman Rickover led the development of the first nuclear submarine, the *Nautilus*. Westinghouse's Bettis Atomic Power Laboratory was the prime contractor, and B&W designed and built the reactor vessel, pressurizer and steam generators. Bettis, General Electric's Knolls Atomic Power Laboratory and the U.S. Atomic Energy Commission developed and published rules for the stress analysis and design basis. This was entitled *Tentative Structural Design Basis for Reactor Vessels*. These rules permitted increased allowable stresses because a thorough and accurate stress analysis was required.

During the late 1950s, the ASME Code Committee and the U.S. government began to see the need for an ASME code which would contain design rules similar to the tentative basis. The first consideration was to do a major upgrading of Section VIII either by a supplemental document or major revision of Section VIII. After much discussion in various committees, it was decided to develop a new code section.

In 1958, the ASME established a Special Committee to Review Code Stress Basis. This special group met many times to develop and review all facets of nuclear vessel requirements – not only design rules, but also fabrication, inspection, materials and other kinds of rules. Fatigue analysis methods were introduced for explicit use for the first time.

In 1963, a public hearing was held in Baltimore, Maryland. As a result of the meeting, Section III, *Nuclear Vessels*, was issued which permitted a safety factor of three (on minimum specified ultimate tensile strength). However, in order to use this factor, many requirements had to be met. A very thorough stress analysis procedure, called design by analysis based on the maximum shear stress theory, had to be applied. Also, more rigid fabrication, examination and testing procedures were required to allow a reduction in the safety factor.

While the development of the Nuclear Code was being concluded in late 1961, a task group was formed to develop an extension of Section VIII. The original title was *Task Group on Code for Industrial Pressure Vessels of Superior Quality*. Later, this name was changed to *Alternative Rules for Pressure Vessels* and ultimately it became Section VIII, Division 2.

In 1968, the new division was issued as an alternative method for constructing pressure vessels, Section VIII, Division 2. It required a much more extensive structural investigation in order to reduce the safety factor to three.

In 1998, the safety factor for Section I, Section III Class 2 and 3, and Section VIII, Division 1 was reduced from 4.0 to 3.5 on minimum specified ultimate tensile strength, based mainly on acceptable experience, and improvements in materials achieved over the years.

Allowable stress values

The BPVC provides the basis for establishing allowable stress values for Section I, Section VIII, Division 1, and Section III, Division 1, Classes 2 and 3. (See Chapter 8.) For wrought or cast ferrous and nonferrous materials, the allowable stress values are based on the following criteria:

1. at temperatures below the creep range, the lowest of the following:

   1/3.5 of the specified minimum tensile strength at room temperature,

   (1.1)/3.5 of the tensile strength at temperature,

   2/3 of the specified minimum yield strength at room temperature, and

   2/3 of the yield strength at temperature.

   In addition, for austenitic stainless steels and certain nonferrous nickel alloys, excluding bolting, flanges and other strain sensitive usage where slightly greater deformation is objectionable, the factor on yield strength at temperature may be increased from 2/3 to 0.90.

2. at temperatures in the creep range, the lowest of the following applies:

   100% of the average stress to produce a creep rate of 0.01% in 1000 hours,

   80% of the minimum stress to cause rupture at the end of 100,000 hours, and

   67% of the average stress to cause rupture at the end of 100,000 hours.

Method for development of allowable stress values

Allowable stress values are developed by the Subcommittee on Materials (SC II) by applying the various safety factors to data which the committee has obtained from tests conducted for the committee or various other industrial sources. The published allowable stress values at any given temperature are then the lowest stresses evaluated using the criteria discussed above. (See also Chapter 8.)

The BPVC also provides the basis for establishing allowable stress values, called *design stress intensity values*, for Section VIII, Division 2, and Section III, Division 1, Class 1. No design stress intensity values have been established in the creep range for Section VIII, Division 2. For Section III, Division 1, Class 1 applications in the elevated temperature range, rules for some materials are given in Section III, Division 1, NH. For the various materials permitted, except for bolting materials, the design stress intensity values, i.e., allowable stress values, are based on the lowest value of the following criteria:

1/3 of the specified minimum tensile strength at room temperature,

(1.1)/3 of the tensile strength at temperature,

2/3 of the specified minimum yield strength at room temperature, and

2/3 of the yield strength at temperature except for most austenitic stainless steels and certain nonferrous materials where the factor may reach as high as 90% of the yield strength at temperature. This may result in a permanent strain of as much as 0.1%. When this amount of deformation is not acceptable, the designer must reduce the allowable stress to obtain an acceptable amount of deformation.

There are tables in Section VIII, Division 2 which list factors for limiting permanent strain in high alloy steels and in nickel and high-nickel alloys. In effect, the criteria permit the designer to choose a factor between 67 and 90% of the yield strength depending on the tolerable deformation. For example, suppose that by some limiting clearances the tolerable strain is limited to 0.05% in a shell. The designer can then refer to the table of limiting permanent strain for 18-8 stainless steel and see that the factor on yield strength should be limited to 0.80. The designer then goes to the yield strength values to establish his own allowable stress values.

Bolting materials have their design stress intensity values, i.e., allowable stresses, established by similar, but more restrictive, criteria. (See Section II, Part D, Table 3 and also Appendix 2 of the Standard.)

Strength theories

The three commonly used theories to predict failure are the maximum principal stress theory, the maximum shear stress theory, and the distortion energy theory as discussed in Chapter 8.

Test results show that for ductile materials, such as those ferrous and nonferrous materials permitted by the BPVC, either the maximum shear stress theory or the distortion energy theory predicts yield and fatigue failure better than the maximum stress theory. However, the maximum stress theory is easy to apply in boiler components and pressure vessels where the circumferential and the longitudinal stresses are calculated using simple formulas with an adequate safety factor to set the allowable stress values. Where a more exact analysis is desired and the safety factor may be less, the BPVC uses the maximum shear stress theory. Section I, Section IV, Section VIII, Division 1, and Section III, Division 1, Subsections NC, ND and NE use the maximum stress theory while Section III, Division 1, Subsection NB and Section VIII, Division 2, use the maximum shear stress theory. The former use a safety factor of 3.5 or 5 on tensile strength; the latter use a safety factor of three.

Design criteria

The design criteria for different sections of the BPVC are related to the theory of failure and safety factors which are applied to the allowable stress values. For both Section I, *Power Boilers,* and Section IV, *Heating Boilers*, the requirements call for the calculation of the required minimum wall thickness which will keep the basic circumferential stress at or below the allowable stress limit for that section. Additional rules, equations and charts are given to determine the minimum required thicknesses of components other than those with circular cross-sections. A more detailed stress analysis may be required for special designs and configurations for which there are no rules. Both Section I and Section IV recognize that high local stresses (discussed below) and secondary stresses may exist within a component, but those sections have established design rules that keep stresses at a safe level with a minimum of additional analysis.

Section VIII, Division 1, *Pressure Vessels,* and Section III, Division 1, Class 2 and 3, *Nuclear Facility Components*, have similar design criteria to Sections I and IV except Sections VIII, Division 1, and III, Division 1, Class 2 and 3 require the minimum thickness of cylindrical shells to be calculated in both the circumferential and the longitudinal directions. The actual minimum required thickness of the cylinder may be set by stress in either direction. In addition, Sections VIII, Division 1, and III, Division 1, Class 2 and 3 have rules which permit the combination of primary membrane stress and primary bending stress to be as high as 1.5 S where S is the basic allowable stress value. At elevated temperatures, consideration must be given to inelastic strain due to creep.

The design criteria for Section VIII, Division 2 provide rules for the more common shell and head geometries for design temperatures that are below the creep range. A detailed stress analysis is required for most complex geometries and loadings including internal pressure loading. For Section III, Division 1, Classes 2 and 3, the requirements are similar to those of Section VIII, Division 2. However, rules are provided requiring detailed analysis and there are fewer formulas.

Both Sections VIII, Division 2, and III, Division 1, Class 1 have adopted the maximum shear stress criteria. This method requires calculated stresses to be assigned to various categories and subcategories that permit different allowable stresses for various combinations of calculated stresses. The various categories and subcategories of stresses are:

1. primary stresses
   a. general primary membrane stress
   b. local primary membrane stress
   c. primary bending stress
2. secondary stresses
3. peak stresses

Primary stresses are caused by loadings which are necessary to satisfy the laws of equilibrium with applied pressure and other loads, and they are not self-limiting by deformation and redistribution. Secondary stresses are developed by self-constraint in the structure. A basic characteristic of secondary stresses is that they are self-limiting. That is, rotation, deflection or deformation takes place until the forces and moments are in balance even though some geometric change may have occurred. Peak stresses are highly localized stresses in a structure usually caused by an abrupt change in the geometry such as a notch or a nozzle fillet weld. Peak stresses are considered only during fatigue analysis of cyclic loadings.

Failure modes, stress limits and stress categories are related. One of the most important aspects of a stress analysis is to make sure that various stresses are assigned to the proper stress categories. Limits on primary stress are set to prevent plastic deformation and burst. Secondary stress limits are set to prevent excessive plastic deformation which may lead to incremental collapse and to ensure the validity of the use of an elastic analysis for making a fatigue analysis. Peak stress limits are set to prevent fatigue failure due to excessive cyclic loadings. For analysis according to Section VIII, Division 2, and Section III, Division 1, Class 1, thermal stresses are considered only in the secondary and peak categories.

The ASME Code for Pressure Piping, B31

The ASME Code for Pressure Piping is divided into many sections relative to different applications. Currently, these sections are:

B31.1	Power Piping
B31.3	Process Piping
B31.4	Pipeline Transportation Systems for Liquid Hydrocarbons and Other Liquids
B31.5	Refrigeration Piping and Heat Transfer Components
B31.8	Gas Transmission and Distribution Piping Systems
B31.9	Building Services Piping
B31.11	Slurry Transportation Piping Systems

The scope of jurisdiction of the ASME BPVC Section I applies to the boiler proper and to the boiler external piping. The boiler external piping is that piping connecting the boiler proper with the nonboiler external piping as defined in PG-58 of Section I. For boiler external piping, the materials, design, fabrication, installation and testing shall be done according to ASME B31.1, *Power Piping*. In addition, most of the connecting piping outside of the Section I jurisdiction is done according to ASME B31.1.

Editions, Addenda, Cases and Interpretations

With the responsibility of the B31 Pressure Piping Code Committee resting with ASME, procedures of releasing editions, addenda, cases and interpretations are the same for the Piping Code as for the BPVC.

Design philosophy, safety factors and strength theory

The design philosophy of the Piping Code is one where the primary membrane stress is not permitted to exceed the allowable stress. In the Piping Code, the longitudinal pressure stress is a major consideration because it combines with longitudinal bending stress developed from the supporting of dead loads and from flexibility stress due to the thermal expansion of the piping system. Although circumferential pressure stress is calculated and sets the minimum required thickness for internal pressure, the longitudinal stress often influences piping arrangement.

Two sections of the ASME B31 Code for Pressure Piping are primarily used in combination with the ASME BPVC sections. The two are: ASME B31.1 Power Piping, and ASME B31.3 Process Piping. B31.1 is closely associated with Section I while B31.3 is associated with Section VIII. However, there are no ASME requirements assigning one or the other. This is essentially done by the jurisdictions and the laws they enforce.

Both sections of the Piping Code use the maximum stress theory for combining and evaluating stresses. The allowable stress values for Section I and B31.1 are the same as are the safety factors. However, B31.3 is unique because it uses the maximum stress theory but permits higher allowable stresses which are based on a safety factor of three on ultimate tensile strength similar to Section VIII, Division 2 and Section III, Division 1, Class 1.

Bibliography

Cross, W., *The Code: An Authorized History of the ASME Boiler and Pressure Vessel Code*, American Society of Mechanical Engineers, New York, New York, 1990.

Greene, A.M., History of the ASME Boiler Code, American Society of Mechnanical Engineers, New York, New York, 1955.

Symbols, Acronyms and Abbreviations

Symbols

Symbols used in *Steam* are defined within the chapters. The following is a summary of those that are commonly used. Specialized symbols are not repeated here. Symbols for chemical elements appear in Appendix 1, Table 3.3. The frequently used units are listed for each parameter, for illustration purposes. Other units used are defined in the chapter text.

a	crack size, ft or in. (m or mm)
a	inside radius, ft or in. (m or mm)
a	major axis of an ellipsoid, ft or in. (m or mm)
a, b, c	general coefficients
A	area, cross-sectional area or surface area, ft^2 (m^2)
A	mass number (total number of protons plus neutrons)
A_n	cross-sectional area of nozzles, ft^2 (m^2)
b	minor axis of an ellipsoid, ft or in. (m or mm)
b	outside radius, ft or in. (m or mm)
B	barometric pressure, in. Hg
B	ratio of throat diameter to pipe diameter, dimensionless
B_w	moisture fraction, lb/lb (kg/kg)
c	distance, ft (m)
c	speed of light
c_p	specific heat at constant pressure, Btu/lb F (J/kg K)
c_{pa}	specific heat of air, Btu/lb F (J/kg K)
c_{pg}	specific heat of gas, Btu/lb F (J/kg K)
c_v	specific heat at constant volume, Btu/lb F (J/kg K)
C	crack geometry factor, dimensionless
C or c	coefficient
C_a	concentration of specie at duct conditions, lb/lb (kg/kg)
C_{eff}	collection/removal efficiency, dimensionless or percent
C_f	cleanliness factor, dimensionless
C_i	concentration of specie in the impinger, lb/lb (kg/kg)
C_{in}	concentration at inlet, lb/lb (kg/kg)
C_m	mass flow of specie, lb/h (kg/s)
C_{out}	concentration at outlet, lb/lb (kg/kg)
C_p	pitot tube correction factor, dimensionless
$C_{ppm,d}$	concentration of specie on a dry basis, ppm
$C_{ppm,w}$	concentration of specie on a wet basis, ppm
C_q	coefficient of discharge, dimensionless
C_s	concentration of specie at standard conditions, lb/lb (kg/kg)
C_t	energy release rate (power) parameter, in. lb/in.^{2} h (J/m^2 s)
C_t	thermal capacitance, Btu/ft^3 F (J/m^3 K)
C_V	valve flow coefficient
d	diameter, ft or in. (m or mm)
d_a	percent reduction in area
D	diameter, ft or in. (m or mm)
D or d	displacement, ft (m)
D_e or D_H	equivalent or hydraulic diameter = 4A/Per, ft or in. (m or mm)
D_i	internal stack diameter, ft
D_{sm}	diameter, Sauter mean
DL	dead load, psi (Pa)
e	available energy, Btu/lb (J/kg)
E	emission rate, dry, at 0% O_2, lb/h (kg/s)
E	energy, Btu (J)
E	weld joint efficiency or ligament efficiency, dimensionless
E	Young's modulus of elasticity, psi (Pa)
ΔE	change in energy
E_b	blackbody emissive power, Btu/h ft^2 (W/m^2)

EQ	seismic loading, psi (Pa)
Eu	Euler number = $\Delta P/\rho V^2$, dimensionless
f	friction factor, dimensionless
f_x	body force
F	geometry correction factor, dimensionless
F	heat exchanger arrangement factor, dimensionless
F	peak stresses
F	radiation configuration factor, dimensionless
$\mathcal{F}$	total radiation exchange factor, dimensionless
F_a	crossflow arrangement factor, dimensionless
F_A	allowable axial compressive stress, psi (Pa)
F_d	F-factor, dry, at 0% O_2, dimensionless
F_d	tube bundle depth factor, dimensionless
F_e	effectiveness factor, dimensionless
F_{pp}	fluid property factor
F_T	fluid temperature factor, dimensionless
FS	factor of safety, dimensionless
FV	future value
g	acceleration of gravity, 32.17 ft/s^2 (9.8 m/s^2)
g	Gibbs free energy, Btu/lb (J/kg)
g_c	proportionality factor, 32.17 lbm ft/lbf s^2 (1 kg m/N s^2)
G	incident thermal radiation, Btu/h ft^2 (W/m^2)
G	mass flux or mass velocity, lb/h ft^2 (kg/m^2 s)
G_{max}	critical flow mass flux, lb/h ft^2 (kg/m^2 s)
Gr	Grashof number $= \dfrac{g\beta\left(T_s - T_\infty\right)\rho^2 L^3}{\mu^2}$, dimensionless
h	heat transfer coefficient, Btu/h ft^2 F (W/m^2 K)
h	thickness, ft or in. (m or mm)
h_{ct}	contact coefficient, Btu/h ft^2 F (W/m^2 K)
h_c	crossflow heat transfer coefficient, Btu/h ft^2 F (W/m^2 K)
h_c'	crossflow velocity and geometry factor
h_l	longitudinal heat transfer coefficient, Btu/h ft^2 F (W/m^2 K)
h_l'	longitudinal flow velocity and geometry factor
H	enthalpy, Btu/lb (J/kg)
H_f	enthalpy of saturated water, Btu/lb (J/kg)
H_{fg}	enthalpy of vaporization, Btu/lb (J/kg)
H_{fw}	enthalpy of feedwater, Btu/lb (J/kg)
H_g	enthalpy of saturated steam, Btu/lb (J/kg)
I	electrical current, amperes
I	intensity
I	moment of inertia, in.^{4} (m^4)
J	mechanical equivalent of heat, 778.17 ft lbf/Btu (1 N m/J)
J	radiosity, Btu/h ft^2 (W/m^2)
J_I	J-integral, in. lb/in.^{2} (J/m^2)
J_{IC}	J-integral, critical, in. lb/in.^{2} (J/m^2)
J_R	material crack growth resistance, in. lb/in.^{2} (J/m^2)
k	coefficient or parameter
k	specific heat ratio compressibility factor, dimensionless
k	thermal conductivity, Btu/h ft F (W/m K)
k_c	insulation thermal conductivity, Btu/h ft F (W/m K)
k_T	isothermal compressibility
K	discharge coefficient, dimensionless
K	factor or coefficient
K	thermal conductance, Btu/h F (W/K)
K_{eg}	equilibrium constant, dimensionless
K_I	stress intensity factor, ksi in.$^{1/2}$ (MPa m$^{1/2}$)
K_{IC}	critical stress intensity factor or fracture toughness, ksi in.$^{1/2}$ (MPa m$^{1/2}$)
K_r	stress intensity factor/toughness ratio, dimensionless
K_y	mass transfer coefficient, lb/ft^2 s (kg/m^2 s)
ΔK	stress intensity range
$l_\parallel$	tube centerline spacing parallel to flow, ft (m)
$l_\perp$	tube centerline spacing transverse to flow, ft (m)
l_e	equivalent plate thickness, ft or in. (m or mm)
L or l	length or dimension, ft or in. (m or mm)
L_h	fin height, ft or in. (m or mm)
$LMTD$	log mean temperature difference, F (C)
LS	limestone consumption, lb/h (kg/s)
L_t	fin spacing, ft or in. (m or mm)
m	mass, lb (kg)
$\dot{m}$	mass flow rate, lb/h (kg/s)
m_s	mass of steam, lb (kg)
m_w	mass of water, lb (kg)
M	moisture in steam sample, percent
M	molecular weight, lb/lb-mole (kg/kg-mole)
M_d	molecular weight of gas, dry basis
M_o	redundant bending moment, lb ft (N m)
M_w	molecular weight of gas, wet basis
n	parameter, coefficient or number of cycles
N	coefficient, number of cycles or number of concentric thermocouple shields
Nu	Nusselt number = $h\,L/k$, dimensionless
N_v or N	number of velocity heads, dimensionless
p	particle size, in. (mm)
p_r	partial pressure, atm
P	parameter
P	pressure or partial pressure, psi (Pa)
P	primary stress
P	temperature ratio for surface arrangement factor, dimensionless
P_b	primary bending stress

Pe Peclet number = RePr, dimensionless
Per perimeter, ft or in. (m or mm)
P_L local primary membrane stress
PL furnace gas pressure loading, psi (Pa)
P_m pressure at meter, psi (Pa)
P_m general primary membrane stress
Pr Prandtl number = $\frac{c_p \mu}{k}$, dimensionless
P_s absolute pressure at duct, psi (Pa)
P_{total} total pressure, lb/ft² (N/m²)
P_v total vertical design load
PV present value
PVF present value factor
ΔP pressure differential, psi (Pa)
ΔP_l stack flow loss, in. of water
ΔP_{SE} stack effect driving pressure, lb/ft² (N/m²)

q heat flow rate, Btu/h (W)
q'' heat flux, Btu/h ft² (W/m²)
q''_{CHF} heat flux at CHF conditions, Btu/h ft² (W/m²)
q''' volumetric heat generation rate, Btu/h ft³ (W/m³)
q_{rel} heat release rate, Btu/h ft³ (W/m³)
q_{rev} reversible heat input per unit mass, Btu/lb (J/kg)
Q or q heat input, Btu (J), or heat input per unit mass, Btu/lb (J/kg)
Q gas flow rate, lb/h (kg/s)
Q heat absorption or heat added to system, Btu (J)
Q secondary stresses
Q_{ac} actual volumetric flow rate, ft³/h (m³/s)
Q_{sd} dry volumetric flow rate at standard conditions, ft³/h (m³/s)

r radius or inside radius, ft or in. (m or mm)
R constant or coefficient
R forcing function
R radius or outside radius, ft or in. (m or mm)
R specific gas constant – **R**/M, lbf ft/lbm R (J/kg K)
R temperature ratio for surface arrangement factor, dimensionless
R thermal resistance, h ft² F/Btu (m² K/W)
R universal gas constant, 1545 ft lb/lb-mole R (8.3143 kJ/kg-mole K)
Ra Rayleigh number = GrPr, dimensionless
Re Reynolds number = $\frac{\rho V L}{\mu} = \frac{G L}{\mu}$, dimensionless
R_f fouling index – bituminous ash
R_s slagging index – bituminous ash
R_s^* slagging index – lignitic ash
R_{vs} slagging index – viscosity

s specific entropy, Btu/lb F (J/kg K)
s_f entropy of water, Btu/lb F (J/kg K)
s_{fg} entropy of evaporation, Btu/lb F (J/kg K)
s_g entropy of steam, Btu/lb F (J/kg K)
S allowable stress
S entropy, Btu/F (J/K)
S surface area, or total exposed surface area for a finned surface, ft² (m²)
S general source term
S quantity of entropy, Btu/F (J/K)
S steam-water slip ratio, dimensionless
SE stack effect, in. wg/ft
S_f fin surface area; sides plus peripheral area, ft² (m²)
S_L longitudinal (parallel) spacing, ft or in. (m or mm)
S_m allowable stress intensity, psi (Pa)
S_r applied stress/net section plastic collapse stress ratio, dimensionless
S_T transverse (side) spacing, ft (m)
St Stanton number = Nu/(RePr), dimensionless
S_u material tensile strength, psi (Pa)
Sy material yield strength, psi (Pa)
$SPVF$ series present value factor

t thickness, ft or in. (m or mm)
t time, h or s
t_c current wall thickness, in. (mm)
t_e time at end of life
t_p present time
T temperature, F or R (C or K)
T_{cv} temperature at critical viscosity, F (C)
T_{sat} saturation temperature, F (C)
T_w wall temperature, F (C)
T_{250} temperature at 250 poise viscosity, F
T^o temperature at initial time, F (C)
T_o sink temperature, F or R (C or K)
TD terminal temperature difference, F (C)
TS tensile strength at temperature, psi (Pa)
Δt time interval, h or s
ΔT temperature difference, F (C)
ΔT_{LMTD} log mean temperature difference, F (C)
ΔT_{sat} $T_{sat} - T_w$, F (C)

u internal stored energy, Btu/lb (J/kg)
u, v, w velocity in x, y, z coordinates respectively, ft/s (m/s)
U overall heat transfer coefficient, Btu/h ft² F (W/m² K)
υ specific volume, ft³/lb (m³/kg)
υ_a specific volume, air
υ_f specific volume, water
υ_{fg} specific volume, evaporation
υ_g specific volume, steam

V electrical voltage, volts
V velocity, ft/s (m/s)
V volume, ft³ (m³)

V_d dry gas volume, ft^3 (m^3)
V_m dry gas volume at meter conditions, ft^3 (m^3)
V_n sample gas velocity through the nozzle, ft/s (m/s)
V_o redundant shear force, psi (Pa)
V_s velocity of the flue gas, ft/s (m/s)

w work per unit mass, ft-lb/lbm (J/kg)
w_{rev} reversible work per unit mass, ft-lb/lbm (J/kg)
W width, ft or in. (m or mm)
W work, ft-lb (J)
W_l leakage flow rate, lb/h (kg/s)
WL wind loading, psi (Pa)

x dimension, ft or in. (m or mm)
x steam quality, mass fraction steam
x_{CHF} steam quality at CHF conditions, mass fraction steam
x, y, z dimensions in Cartesian coordinate system, ft or in. (m or mm)
Δx change in length, ft or in. (m or mm)
X weight fraction, dimensionless

Y compressibility factor, dimensionless
Y Schmidt fin geometry factor, dimensionless
Y_g concentration in bulk fluid, lb/lb (kg/kg)
Y_i concentration at condensate interface, lb/lb (kg/kg)

Z or z elevation or length, ft or in. (m or mm)
Z atomic number (number of protons in the nucleus)
Z Schmidt fin geometry factor, dimensionless

Greek symbols

α absorptivity, dimensionless
α angle, deg (rad)
α coefficient of thermal expansion, ft/ft (m/m)
α thermal diffusivity = $k/\rho c_p$, ft^2/s (m^2/s)
α void fraction, dimensionless
β deflection or rotation influence coefficients
β volume coefficient of expansion, 1/R (1/K)
Γ effective diffusion coefficient
δ diameter or film thickness, ft or in. (m or mm)
ε emissivity, dimensionless
ε roughness, in. (mm)
ε strain, ft/ft (m/m)
η efficiency as a fraction or percentage
η Schmidt fin efficiency, dimensionless
η_{net} net efficiency, dimensionless
η_{th} thermal efficiency, dimensionless
θ angle, deg (rad)
μ dynamic viscosity, lbm/ft s (kg/m s)
μ Poisson's ratio, dimensionless
ν kinetic viscosity, ft^2/s (m^2/s)
π 3.1415926
ρ density, lb/ft^3 (kg/m^3)
ρ reflectivity, dimensionless
$\bar{\rho}_d$ average downcomer density, lb/ft^3 (kg/m^3)
ρ_{hom} homogenous density, lb/ft^3 (kg/m^3)
$\rho(z)$ local density, lb/ft^3 (kg/m^3)
σ atomic cross-section, barns
σ standard deviation
σ Stefan-Boltzmann constant, 0.1713 x 10^{-8} Btu/h ft^2 R^4 (5.669 x 10^{-8} W/m^2 K^4)
σ stress, psi (Pa)
σ_a allowable alternating stress, psi (Pa)
σ_r, σ_t, σ_z radial, tangential and axial stress, psi (Pa)
σ_t thermal stress, psi (Pa)
$\sigma_{y.p.}$ yield point stress, psi (Pa)
σ_1 longitudinal stress, psi (Pa)
σ_2 hoop stress, psi (Pa)
τ shear stress, psi (Pa)
τ transmissivity, dimensionless
ϕ^2_{LO} two-phase friction multiplier, all fluid as a liquid, dimensionless
ϕ general dependent variable
Φ two-phase multiplier for a steam-water separator, dimensionless

Subscripts

1 initial or inlet condition
2 final or outlet condition
a air
amb ambient
aux auxiliary component or system
b bulk fluid
c contact or conduction
cg convection, gas-side
cv convection
$clean$ unfouled or clean condition
e node point east
$econ$ economizer
eq or e equivalent
f fluid, fin or saturated water
g gas or saturated steam
i inside or ith parameter or initial condition
in entering a system
j jth parameter
l liquid
max maximum
min minimum
o outside or final condition
out leaving a system
p node point under evaluation
P products
r radiation
ref reference
rg radiation, gas-side
R reactants
s steam
sg surface-to-gas
SE stack effect

t	thermal
w	node point west
w	wall
δ	liquid film surface (gas liquid interface)

Acronyms and Abbreviations

AA atomic absorption
ABMA American Boiler Manufacturers Association
abs absolute
ABS American Bureau of Shipping
A/C air/cloth
AC alternating current
ACF actual cubic feet
ACFM actual cubic feet per minute
ADS air density separator
AE acoustic emissions
AEC Atomic Energy Commission, U.S.
AECL Atomic Energy of Canada Limited
AFBC atmospheric pressure fluidized-bed combustion
AFS atomic fluorescence spectroscopy
AFW auxiliary feedwater
AGR advanced gas-cooled reactor
AH air heater
AIB auxiliary input burner
AIG ammonia injection grid
AISC American Institute of Steel Construction
ALARA as low as reasonably achievable
amb ambient
AMU atomic mass units
ANSI American National Standards Institute
API American Petroleum Institute
approx approximate
ARIS Automated Reactor Inspection System
ARTS Anticipatory Reactor Trip System
ASD allowable stress design
ASME American Society of Mechanical Engineers
ASTM American Society for Testing and Materials
attemp attemperator
atm atmosphere
ATOG abnormal transient operating guidelines
avg average
AVT all-volatile treatment
AZS air zone swirler

B/A base-to-acid ratio
B&W The Babcock & Wilcox Company
BAC boiler as calorimeter
BACT best available control technology
BAT best available technology
bbl barrel
BCC body-centered cubic
BCCW boiler chemical cleaning waste
BCM boiler condenser mode
BCT body-centered tetragonal
BCTMP bleached chemical thermomechanical pulp
BFB bubbling fluidized bed
BFP boiler feed pump
BFW boiler feed water
BGC British Gas Corporation
BH baghouse
bit or bitum bituminous
BLESS Boiler Life Evaluation and Simulation System
BLG black liquor gasification
BLRBAC Black Liquor Recovery Boiler Advisory Committee
BOF basic oxygen furnace
BOL beginning of life
BOOS burners out of service
BPVC Boiler and Pressure Vessel Code (ASME)
BRC below regulatory concern
BSW bottom sediment and water
Btu British thermal unit
BWCC Babcock & Wilcox Construction Co., Inc.
BWR boiling water reactor
BWST borated water storage tank

C Celsius or centigrade degrees
CAA Clean Air Act, U.S.
CAD computer-assisted design/drafting
CANDU Canada Deuterium Uranium
CAP curved arm primary
CBI curved backward inclined
CBK carbon burnout kinetic
cc cubic centimeter
CC combined cycle
CCZ Controlled Combustion Zone
CE collecting electrode
CEMS continuous emissions monitoring system
CFB circulating fluidized bed
CFCC continuous fiber ceramic composite
CFD computational fluid dynamics
cfm cubic feet per minute
CFR Code of Federal Regulations
CGA crack growth analysis
CHF critical heat flux
CHP combined heat and power
CII Construction Industry Institute
C.I.S. Confederation of Independent States
cm centimeter
CNC computer numerical control
CNCG concentrated noncondensable gas
CNSG consolidated nuclear steam generator
COE Corps of Engineers, U.S. Army
coef coefficient
col column
cond condenser
COG coke oven gas
cP centipoise
CRDM control rod drive mechanism
CVD chemical vapor deposition
CWA Clean Water Act

d day
DAF dry ash-free
dB decibels
DC direct current
DC drain cooler
DCS distributed control system
DCS digital control system
DE destructive examination
DE discharge electrode
deg degree
DESAH dual economizer steam air heater
DGV sampled dry gas volume
DHRS decay heat removal system
DNB departure from nucleate boiling
DNBR DNB ratio
DNC distributed numerical control
DNCG dilute noncondensable gas
DOD Department of Defense, U.S.
DOE Department of Energy, U.S.
DPFAD deformation plasticity failure assessment diagram
DRB dual register burner
DRE destruction and removal efficiency
DSCF(M) dry standard cubic feet (per minute)
DSCM dry standard cubic meter

EAR estimated additional resources
EBW electron beam welding
EC European Community
ECCS emergency core cooling system
ECON economizer
ECP electrochemical potential
ECS environmental control systems
ECT eddy current testing
EDC eddy dissipation concept model
EDM eddy dissipation combustion model
EDM electrical discharge machining
EDTA ethylenediaminetetraacetic acid
eff efficiency
EI enhanced ignition (burner)
EIA Energy Information Administration
EMAT electromagnetic acoustic transducer
emf electromotive force
EMF electromagnetic filter
EOR enhanced oil recovery
EPA Environmental Protection Agency, U.S.
EPFM elastic-plastic fracture mechanics
EPRI Electric Power Research Institute
ERW electric resistance welded
ESP electrostatic precipitator
ESS enhanced service structure
ESW electroslag welding
EU European Union
evap evaporation

F Fahrenheit degrees
FAC flow accelerated corrosion
FB firebrick
FBC fluidized-bed combustion
FBR fast breeder reactor
FC feedwater control subsystem
FC fixed carbon
FC/VM fixed carbon/volatile matter ratio
FCAW flux core arc welding
FCC face-centered cubic
FCC fluid catalytic cracking
FD forced draft
FEA finite element analysis
FEGT furnace exit gas temperature
FERC Federal Energy Regulatory Commission
FES finite element system
FGD flue gas desulfurization
FGR flue gas recirculation
FHyNES furnace hydrogen damage nondestructive examination system
FIV flow-induced vibration
FMB marine package boilers
FMEC failure mode evaluation chart
FOSAR foreign object search and retrieval
FPS Fossil Power Systems
FRP fiberglass reinforced plastic
FS factor of safety
FSS flame safety system
FST-GAGE Fast-Scanning Thickness Gage
FSU former Soviet Union
ft foot/feet
FT fluid temperature
FURNIS furnace inspection system
FW field weld

g gram
Ga gauge
gal gallon
GCR gas-cooled reactor
gen generator
GGH gas-to-gas heaters
GJ giga joule
GMAW gas metal arc welding
GPM gallons per minute
gr grains
GR gas recirculation
GT gas turbine
GTAW gas tungsten arc welding
GTCC gas turbine combined cycle

h hour
HAP hazardous air pollutant
HAZ heat-affected zone
HCGE high capacity gas element
HFRW high frequency resistance welding
HFT hot function testing
HGI Hardgrove Grindability Index
HHV higher heating value
HLW high level waste
hp horsepower
HP high pressure (turbine)

HPI high pressure injection
HRSG heat recovery steam generator
HT hemispherical temperature
HTM Heat Transfer Manager
HTW high temperature Winkler
HV high volatile
HVOF high-velocity oxy-fuel
HVT high velocity thermocouple
HZP hot zero power

IAPS International Association for the Properties of Steam
IAWPS International Association of Wood Products Societies
IC Instrument and Controls
IC ion chromatography
ICPAES inductive coupled plasma atomic emission spectrometric (analysis)
ICR Information Collection Request
ICS integrated control system
ID induced draft
ID inside diameter
ID initial deformation (temperature)
IEOTSG integral economizer once-through steam generator (nuclear)
IFGR induced flue gas recirculation
IGA intergranular attack
IGCC integrated gasification combined cycle
IGSCC intergranular stress corrosion cracking
IMC integrated master control subsystem
in. inch
in. Hg inches of mercury
in. wg inches water gauge
INPO Institute of Nuclear Power Operations
IP intermediate pressure (turbine)
IPP independent power producer
IR infrared radiation
IR-CFB internal recirculation circulating fluidized-bed
ISO International Standards Organization
ISP intermediate size particles
IT initial deformation temperature

J joule

K Kelvin degrees
kcal kilocalories
kg kilogram
kHz kilohertz
kJ kilojoule
km kilometer
kmole or kg-mole kilogram-mole
kPa kilopascal
ksi 1000 pounds per square inch
kV kilovolt
kW kilowatt
kWh kilowatt hour

l liter
LAER lowest achievable emission rate
LAF laboratory ashing furnace
lb pound
LBBA leak before break analysis
lbf pound force
LBLOCA large break loss of coolant accident
lbm pound mass
lb-mole pound-mole
LBW laser beam welding
LEA low excess air
LEFM linear elastic fracture mechanics
LEL lower explosion limit
LFA life fraction analysis
LFE life fraction expended
L/G liquid to gas ratio (FGD)
LHV lower heating value
LLW low level waste
LMP Larson-Miller parameter
LMTD log mean temperature difference
LNB low NO_x burner
LNCB low NO_x cell burner
LNG liquefied natural gas
LOCA loss of coolant accident
LP low pressure (turbine)
LPG liquefied petroleum gas
LPI low pressure injection
LPMS loose parts monitoring system
LSD Lagrangian stochastic deterministic (model)
LVS limited vertical sweep
LWR light water reactor

m meter
MA mill-annealed
MACT maximum achievable control technology
MAF moisture and ash free
MARAD Maritime Administration, U.S.
MAWP maximum allowable working pressure
max maximum
MCR maximum continuous rating
MD mass defect
MDC multi-cyclone dust collector
med medium
MeV million electron volts
MFR mass flow rate
MFT master fuel trip
mg milligram
MHVT multiple shield high velocity thermocouple
mi mile
MIC microbial induced corrosion
MIG metal inert gas
min minimum or minute
MIST multiple loop integral systems test
MJ megajoule
ml milliliter
MLR multi-lead ribbed
mm millimeter
MMP managed maintenance program

MMT minimum metal temperature
mol molecular
MOV motor operated valve
mPa millipascal
MPa megapascal
mph miles per hour
MRPC motorized rotating pancake coil
MRS monitored retrievable storage
MSDS material safety data sheets
MSW municipal solid waste
MT magnetic particle testing
MTD mean temperature difference
mV millivolt
MV medium volatile
MW megawatt, megawatts-electric
MWd megawatt-day
MW_t megawatts-thermal
μS microSiemens

N Newton
N_{2a} atmospheric nitrogen
NAAQS National Ambient Air Quality Standards
NAPAP National Acid Precipitation Assessment Program
NBIC National Board Inspection Code
NC numerical control
NCG noncondensable gas
NDE nondestructive examination
NDTT nil-ductility transition temperature
NEI National Emissions Inventory
NFPA National Fire Protection Association
NIST National Institute of Science & Technology
Nm^3 Normal cubic meter
NMVOC nonmethane volatile organic compounds
NNI non-nuclear instrumentation
NOTIS nondestructive oxide thickness inspection system
NPD nuclear power demonstration
NPDES National Pollutant Discharge Elimination System
NPHR net plant heat rate
NPSH net positive suction head
NRC Nuclear Regulatory Commission, U.S.
N.S. nuclear ship
NSAC Nuclear Safety Analysis Center
NSCW nuclear service cooling water
NSFO Navy special fuel oil
NSPS New Source Performance Standards
NSR New Source Review
NSSS nuclear steam supply system
NTU net transfer unit
NUG nonutility generator
NWMS nuclear waste management system
NWPA Nuclear Waste Policy Act

O/G oil and gas
OBW oversize bulky waste
OD outside diameter
ODE ordinary differential equations
OEM original equipment manufacturer
OFA overfire air
OMLR optimized multi-lead ribbed
OPEC Organization of Petroleum Exporting Countries
OSHA Occupational Safety & Health Administration
OT oxygen treatment
OTSG once-through steam generator
oz ounce

Pa pascal
PA primary air
PA/PC primary air and pulverized coal mixture
PAC powdered activated carbon
PAW plasma arc welding
PAX primary air exchange (burner)
PC pulverized coal
PCB polychlorinated biphenyl
PCDD polychlorinated dibenzoparadioxin
PCDF polychlorinated dibenzofurans
PCFAD computer-based failure analysis diagram
PDE partial differential equations
PDVSA Petroleos de Venezuela South America
PFBC pressurized fluidized-bed combustion
PFI Power for Industry (boiler)
PFM Power for Manufacturing (boiler)
PFT Power for Turbine (boiler)
pH negative log of hydrogen ion concentration
PHWR pressurized heavy water reactor
P&ID process and instrumentation diagram
PIP peak impact pressure
PM particulate matter
POHC principal organic hazardous constituent
PORV power (pilot) operated relief valve
ppb parts per billion (mass)
ppm parts per million (mass)
ppmv parts per million by volume
PPS polyphenylene sulfide
PR process recovery
PRB Powder River Basin
press. pressure
pri primary
prox proximate
PRV pressure reducing valve
PSD prevention of significant deterioration
PSDF Power Systems Development Facility
PSH primary superheater
psi pounds per square inch (absolute or difference)
psia pounds per square inch absolute
psig pounds per square inch gauge
P/T pressure/temperature
PT penetrant testing (liquid/dye)
PTC Performance Test Code

PUC Public Utilities Commission
pulv pulverizer
PURPA Public Utility Regulatory Policies Act
PVC polyvinyl chloride
PWHT post weld heat treatment
PWR pressurized water reactor
PWSCC primary water stress corrosion cracking

R Rankine degrees
R&D research and development
rad radian
RAR reasonably assured resources
RB Radiant boiler
RBC Radiant boiler Carolina-type
RBE Radiant boiler El Paso-type
RBT Radiant boiler Tower-type
RBW Radiant boiler W-type
RC reactor coolant
RCM reliability centered maintenance
RCRA Resource Conservation and Recovery Act
RCS reactor coolant system
RDF refuse-derived fuel
recirc recirculation
red. reducing atmosphere
rem roentgen equivalent man
rev revolutions
RH reheater
RHAS radiant heat absorbing surface
RHSH reheat superheater
RH-TRU remote handled transuranic waste
RO reverse osmosis
ROTSG replacement once-through steam generator
rpm revolutions per minute
RPS reactor protection system
RRSG replacement recirculating steam generator
RSG recirculating steam generator
RT radiography testing
RVCH reactor vessel closure head
RWP radiation work permit

s second(s)
SA secondary air
SAE Society of Automotive Engineers
sat. saturated
SAW submerged arc welding
SBLOCA small break loss of coolant accident
SBW steam by weight
SCC stress corrosion cracking
SCC submerged chain conveyor
SCFM standard cubic feet per minute
SCR selective catalytic reduction
SD stack draft
SDA spray dryer absorber
SE stack effect
sec second(s)
SFAS safety features actuation system
SFS Saybolt Furol Seconds
SG steam generator
SGTR steam generator tube rupture
SH superheater
SI international system of units
SIP state implementation plan
SLR single-lead ribbed
SMAW shielded metal arc welding
SMD Sauter mean diameter
SNAME Society of Naval Architects and Marine Engineers
SNCR selective non-catalytic reduction
SNG synthetic natural gas
SOG stripper off gas
SPB Stirling power boiler
sp gr specific gravity
SR stoichiometric ratio
SS stainless steel
S.S. steam ship
SSH secondary superheater
ST softening temperature
subbitum subbituminous
SUS Saybolt Universal Seconds
SWUP spiral wound Universal Pressure (boiler)

t ton (English units)
t_m metric ton
T_{250} viscosity at 250 centipoise
TBS turbine bypass system
TC thermocouple
TD terminal temperature difference
TDF tire-derived fuel
TDFM time dependent fracture mechanics
TDS total dissolved solids
TEG turbine exhaust gas
temp temperature
theo theoretical
TIG tungsten inert gas
thk thick
TLD thermoluminescent dosimeter
TOC total organic carbon
TOFD time of flight defraction
TR set transformer rectifier set
TRC total residual chlorine
TRS total reduced sulfur
TRU transuranic
TS tensile strength
TSC two-stage combustion
TSP total suspended particulate
TSP tube support plates
TSS total suspended solids
TSV turbine stop valve
TTT time-temperature-transformation
TTV turbine throttle valve
typ typical

UBC Uniform Building Code
UBC unburned carbon
UBCL unburned carbon loss

UCL	unburned carbon loss
UEL	upper explosion limit
UGA	undergrate air
UIC	uncompensated ion chamber counter
U.K.	United Kingdom
ULD	unit load demand subsystem
UP	Universal Pressure (boiler)
UPC	Universal Pressure Carolina-type (boiler)
UPC	Universal Pressure coal-fired (boiler)
U.S.	United States
USCG	United States Coast Guard
USD	United States dollar
U.S.S.	United States ship
UT	ultrasonic testing
UTT	ultrasonic thickness testing
UV	ultraviolet
V	volts
VAC	volts alternating current
VIV	variable inlet vanes
VM	volatile matter
VOC	volatile organic compounds
vol	volume
vs	versus
VSS	vertical steam separator
W	watt
WCT	water collecting tank
WESP	wet electrostatic precipitator
WFGD	wet flue gas desulfurization
wg	water gauge
WHRB	waste heat recovery boiler
WPS	welding procedure specifications
WRC	Welding Research Council
wt	weight
WTE	waste-to-energy boiler
YP	yield point
yr	year
YS	yield strength

The Babcock & Wilcox Company Trademarks in Edition: 41

The following is a list of trademarks used in *Steam, Edition: 41* that are owned by The Babcock & Wilcox Company or one of its affiliated companies.

Acoustic Ranger
Auto Spring
B&W
B&W Roll Wheel
B&W-56
B&W-67
B&W-89
B&W-118
Babcock & Wilcox
Cera-VAM
Cyclone
DRB-4Z
DRB-XCL
DSVS
Elverite
FHyNES
FST-GAGE
GenClad
Heat Transfer Manager
Hydrobin
HydroJet
IK-600
LM2100
Multiclone
NOTIS
On-Track
PCFAD
PowerClean
Precision Clean
PrecisionJet
Racer
Stirling
SWUP
Towerpak
UP
VAM
XCL-S

Index

Abrasiveness index, 9-9
Absorber (FGD), 35-2
Absorption, 35-5
Absorptivity
  definition, 4-3
  of gases, 4-12, 4-31
Access doors, 20-6, 23-9
Acid mist, 35-18
Acid rain, 32-4, 34-2, 35-1
Acidity, 36-1
Acoustics
  electromagnetic, 45-4
  emissions, 45-5
  leak detection, 45-16
Adiabatic, 3-4
Adiabatic flame temperature, 2-26, 10-11
Adipic acid, 35-12
Aging (degradation) mechanisms
  corrosion, 44-6
  erosion, 44-6
  fatigue, 44-6
  stress, 44-6
Air
  combustion, 10-5, 10-16
  composition, 10-4
  control, 41-10
  distributor, fluidized-bed, 17-1, 17-3
  enthalpy, 10-19
  flow measurement, 40-20, 41-13
  infiltration, leakage, 10-16, 10-23, 23-7
  moisture, 10-5, 10-6
  properties of, 4-11
  theoretical, 10-5, 10-7, 10-9, 10-16
Air, quaternary, 28-4
Air flow-steam flow control, 41-13
Air heater, Chapter 20
  applications, 20-15
  calculations, 22-15
  cast iron, 20-8
  cold end minimum metal temperatures, 20-13
  condition assessment of, 45-5, 45-11
  corrosion, 20-13
  environmental, 20-15
  erosion, 20-3, 20-14
  fires, 20-14
  industrial, 20-15
  leakage, 20-12
  Ljungström, 20-9
  marine, 31-12
  operation, 20-13
  performance, 20-12
  plate, 20-8
  plugging, 20-14
  recuperative, 20-7
  regenerative, 20-9
  Rothemühle, 20-10
  seal(s), 20-10, 20-11
  steam
    coil, 20-9, 29-12, 29-21
    marine, 31-10
  testing, 20-12
  tubular, 20-7, 29-21
Air pollution, Chapter 32
  air toxics, 32-5, 32-8
  carbon dioxide, 32-8
  carbon monoxide, 32-2, 32-7
  greenhouse gas, 32-8
  international regulations, 32-6
  mercury, 32-8
  nitrogen oxides (NO_x), 32-1, 32-4, 32-7 (see also NO_x control)
  particulate matter, 32-7 (see also Particulate control)
  sources, 32-1
  sulfur oxides (SO_x), 32-1, 32-4, 32-7 (see also SO_2 control)
  technologies, 32-8
  U.S. legislation, 32-1
Air swept spout, 16-7, 29-17
Air testing, 39-18
Albacore hull, 46-24
Alkalinity, 35-1, 36-1
Allowable stress, 7-20, 8-3, C-4
Allowances, 32-4
Alloying elements, 7-5
  interstitial, 7-2
  substitutional, 7-2
Alpha
  emitters, 47-4
  particle, 47-3
Alumina ceramics, 13-11, 44-10
Aluminizing, 7-15
American Boiler Manufacturers Association (ABMA), 23-6, C-1
American Society for Testing and Materials (ASTM), 9-5, 9-7
American Society of Mechanical Engineers (ASME), 1-14, 2-1
  Boiler and Pressure Vessel Code, 8-1, 19-8, 20-6, C-1
    allowable stresses, 7-20, 8-3, C-4
    calculations, 8-5
    design criteria, 8-2, 8-14
    strength theories, 8-2
    stress classifications, 8-2, 8-9, 8-10
  Performance Test Code, 10-10, 10-18, 10-21, 40-1
  Pressure Piping (B31), C-1, C-6
Ammonia, 34-3, 35-2, 35-18
  anhydrous, 34-6
  aqueous, 34-7
  flow control, 34-8, 41-17
  injection, 34-6
  reagent systems, 34-6
Ammonia sulfates, 20-15, 34-6
Annealing, 7-8
Anticipatory reactor trip system (ARTS), 46-15
API gravity, fuel oil, 9-14, 10-20
Approach temperature
  flue gas desulfurization, 35-14
As-received, coal, 9-5, 9-7
Ash, black liquor, 28-8 (see also Recovery boiler, Kraft process)
  carryover, 28-9

chloride/potassium purging, 28-10, 28-21
fume, 28-9
stickiness, 28-9
Ash, coal, 9-4, 9-8, 21-1 (see also Slag)
characteristics and composition, 21-3, 21-6
characterization methods, 21-13
classification, 20-13
composition, 9-8
content, 21-1
corrosion, 15-4, 15-10, 21-20
deposition, deposits, slagging and fouling, 21-3, 44-7
effects of reducing atmosphere, 21-6, 43-7
effects on boiler design, 19-6, 21-14, 21-16, 43-7
convection pass design, 21-16
furnace design, 21-14
erosion, 21-17 (see also Erosion)
fusibility, fusion temperature(s), 15-1, 15-3, 21-5
fusion, 9-8
gasifier, Chapter 18
hoppers, 20-4
influence of elements, 21-7
numerical model, 6-12
operating variables, 21-17
particles, 20-3
pulverized coal combustion, 14-4
removal, dry-bottom, wet-bottom, flyash, 21-2
residue, 10-11, 10-14
sintering strength, 21-19
viscosity, 21-8
measurement, 21-9
temperature relationship, 21-10
Ash, fuel oil, 21-22
characteristics, 21-24
control, 21-25
corrosion, 21-25
high temperature, 21-25
low temperature, 21-25
deposition, 21-23
design considerations, 21-24
operating considerations, 21-24
origin, 21-23
Ash cleaning equipment (see also Sootblowers)
mechanical rapping system, 29-21
sootblowers, 24-1, 29-21
Ash deposits
emittance, 4-10
thermal conductance, 4-9
Ash handling systems, 24-12, 29-12, 29-18
air heater flyash, 24-12
bottom ash handling systems
clinker grinders, 24-14
dewatering storage bin, 24-14
hydraulic, 24-13
mechanical drag, 24-14
plunger ash extractor, 24-15, 29-12
slag tank, 24-14
submerged chain conveyor, 24-14, 29-18
water impounded hopper system, 24-15
dry flyash handling systems, pneumatic, 24-16
airlock, 24-17
pressure system, 24-17
vacuum system, 24-16
flyash handling system, 24-15, 29-12, 29-19
collector, 24-20
drag chain conveyor, 24-15, 29-12
screw conveyor, 29-12
silo, 24-20
silo unloading, 24-20
mechanical conveying, 24-14
pyrite, 24-15
Aspirator, 3-16
Atom, 47-1
Atomic Energy Act, 49-14
Atomic Energy Commission (AEC, now NRC), 46-1, 49-14
Atomic number, 47-1, T-15
Atomic Safety and Licensing Board, 49-14
Atomization
oil, 11-11
mechanical, 11-11
steam, 11-12
slurry (FGD)
dual fluid, 35-13
rotary, 35-13
Attemperation, 1-8, 19-14
marine, 31-9
water, 42-7
Attemperator, 19-13, 19-14
arrangement(s), 19-15
condition assessment, 45-9
control, 19-14
spray, 1-8, 19-14
Austenite, 7-3
Austenitic stainless steel, 7-13
Automatic sequence burner controls, 41-14
Automation
power plant, 41-19
welding, 38-1
Auxiliaries, 39-18
ash handling, Chapter 24
boiler, 1-9, Chapter 25
power requirements, 15-10, 26-6
sootblowers, Chapter 24
Auxiliary power, cost, 37-11, 37-17
Availability
maintaining, Chapter 44
Available energy, 2-17

Babcock, George Herman, Intro-5, Intro-6
Baffle(s)
air heater, 20-7
drum, steam, 5-14
economizer, 20-3
marine, 31-12, 31-13
numerical modeling, 6-13
Bagasse, analysis, 9-12
Baghouse, 29-22, Color Plate 7 (see also Fabric filter)
Bainite, 7-5
Balanced draft, 23-4
Ball or ball-and-tube mill, 13-5, 35-6
Banks of tubes
convection heat transfer in, 4-16
crossflow around, 4-16
effect of number of tube rows in, 4-18
longitudinal flow around, 4-18
Bare tube, arrangement
in-line, 20-1
staggered, 20-1
Bark fuel, 27-2, 30-1
Barns, 47-5
Basic oxygen furnace, 27-20
Batch refueling, 46-10
Bernoulli's equation, 3-4
Best Achievable Control Technology (BACT), 32-3
Beta particle, 47-3
Bid evaluation, 37-7
Bimetallic materials and tubes, 7-14, 28-11, 29-5
Binding energy, 47-2
Biological effects, radiation, 47-3
Biomass (see also Chapter 30)
as a fuel, 9-18, 30-3
fluidized-bed, 17-1, 17-3, 30-7
Black liquor – see Recovery boiler, Kraft process
Blackbody
definition of, 4-4
radiation from, 4-4, 4-7
Blast furnace, 27-2, 27-18
gas, 9-10, 27-4, 27-13
Blowdown
steam generator, fossil, 19-14, 42-7
steam generator, nuclear, 48-7
Blowing, steam line, 43-3
Boiler
bank, 1-8, 19-2, 22-11, 27-2, 28-4, 28-17, 29-21
chemistry, 42-1
circulation, general theory, 3-17, 5-17
cleaning, 43-2
codes, 8-5, C-1
configurations, 19-1
control, Chapter 41, 43-3
automatic sequence, 43-3
basic theory, 41-1
coal, pulverized, drum, 41-10
coal, pulverized, Universal Pressure, 41-12
combustion optimization, 43-6
Cyclone, 41-11
fluidized-bed, 41-11
key operating functions, 43-7
manual, 43-3
oil and gas, 41-10

system design, utility boiler, 41-5
temperature
reheat, 41-10
superheat, 41-10
definition, 1-1, 19-1
design, 1-14, 19-3
ASME Boiler and Pressure Vessel Code, 19-8, C-1
basic circulation systems, 19-3
calculations, performance, 22-1
combustion effects, 19-5
convection surface, 19-6
criteria, 19-4
critical heat flux, 5-2, 5-4, 19-8
heat absorption, 19-3
heat and chemical recovery boilers, paper industry, 28-1
heating surface, 19-4, 19-6
pressure parts, 19-8
recovery, chemical and heat, Chapter 28
safety valves, 19-8
specifications, 19-4
supports, 19-8
thermal gradients, 19-8
design practice, 26-2
development, history of, Intro-1
enclosure, 19-8, 23-1
energy equation, 2-9
expansion, 23-4
historical (see also Steam engine)
Blakey, William, Intro-4
centennial, Intro-5
shell, Intro-2
Stevens, John, Intro-4
straight tube, Intro-7
Trevithick, Richard, Intro-4
waggon, Intro-3
Watt, James, Intro-3
industrial, 27-1
industrial byproduct, 30-2
numerical model, 6-12
pulverized coal, 6-14
recovery, 6-14
waste-to-energy, 6-21
windbox, 6-16
once-through, 1-3, 1-12
operation, 43-1
abnormal, 43-9
operator requirements, 43-2
performance tests, 43-8
personnel safety, 43-8
shutdown, 43-9
startup, 43-2
screen, 19-6
selection, 26-1
setting, 23-1
supports, 19-8
surface, 19-1
types
anthracite combustion (RBW), 19-2, 26-6, 26-15
basic oxygen furnace (BOF), 27-20
bent tube, Intro-7
biomass-fired, Chapter 30
blast furnace gas, 27-4
bubbling fluidized-bed (BFB), 17-2, 27-11
carbon monoxide (CO), 27-17
Carolina (RBC), 19-2, 26-12, Color Plate 2
circulating fluidized-bed (CFB), 17-9, 27-8, Color Plate 5
El Paso (RBE), 19-2, 26-16
electric utility, Chapter 26
Enhanced Oil Recovery (EOR), 27-14
fire tube, Intro-4
FM, 27-12
FMB (marine package boiler), 31-4, 31-5
heat recovery steam generator (HRSG), 27-15
High Capacity FM (HCFM), 27-12
industrial, Chapter 27, 28-1, 30-1
Integral Furnace, Intro-9, 19-2
fluidized-bed, 17-1
large utility, 19-2, Chapter 26
marine, Intro-6, Chapter 31 (see also Marine boiler)
oxygen converter hood, 27-20
package, Intro-10, 27-12
PFI, 27-13
PFM, 27-12
PFT, 27-13
pressurized fluidized-bed, 17-13
process recovery (PR), 28-1, Color Plate 4
Radiant, Intro-10, 19-2, 26-11
shop-assembled, Intro-10, 27-12, 38-11
spiral circuitry or spiral wound, 19-2, 26-7, 26-10
Spiral Wound Universal Pressure (SWUP), 26-7, Color Plate 1
Stirling (SPB), Intro-7, 19-2, 27-8, 30-2, Color Plate 3
subcritical, 19-3
supercritical, 19-3
Tomlinson, 28-10
Tower (RBT), 19-2, 26-14
Towerpak, 27-8, 30-8
Universal Pressure (UP), Intro-10, 19-2, 26-6, 26-9
vertical tube, variable pressure, 26-10
waste heat, 27-18
water tube, Intro-4, 19-1
wood-fired, Chapter 30
Boiler as a calorimeter (BAC) calculation method, 10-23
Boiler-following control, 41-5
Boiler support steel, 39-11
Boiling, 1-2, 4-22, 5-1
convective, forced, 5-2, 5-3
departure from nucleate, 5-4, 46-11 (see also Critical heat flux)
film, 5-2
flow, 5-2, 5-3
heat transfer, 4-22, 5-3
incipient, 5-2
nucleate, 4-22, 5-2
subcooled, 4-23, 5-2
transition, 5-2
Boiling water reactor (BWR), 46-2
Bomb calorimeter, 10-9
Bond work index, 35-7
Boric acid, 49-16, 49-18
Boring mill, 38-3
Bottom ash, 21-2, 24-13, 32-16
Boundary layer, fluid flow, 4-3, 4-14
Bourdon tube pressure gauge, 40-1
Brayton cycle, 2-22
IGCC plants, 18-14
Breeder reactor, 47-15
Broaching, 50-6
Buckstays, 23-2, 23-5
Buffer, 35-12
Bulk fluid temperature, 4-15
Buoyancy effects, 4-3
Burnable absorbers, nuclear, 47-13
Burner
auxiliary, 28-23, 29-21
coal, pulverized, Chapter 14
air-fuel mixing, 14-7
boiler integration, 14-5
burner nozzle, 14-7
cell, 14-9
circular, 14-9
corner-firing system, 14-16
DRB-4Z, 14-13
DRB-XCL, 14-12
dual register, 14-11
enhanced ignition, 14-18
flame safety system, 14-20
flame stability, 14-2, 14-7
impeller, 14-8
low NO_x Cell, 14-14
movable, 19-16
numerical model, 6-21
oil or gas firing, 14-20
performance, 14-8
primary air exchange, 14-18
register, 14-7
S-type, 14-9
excess air, 10-15 (see also Excess air)
flame detection, 41-15
fuel management, 41-15
management system, 41-14
automatic sequence, 41-14
failure mode analysis, 41-17
remote manual sequence, 41-14
supervised manual control, 41-14
oil and gas, Chapter 11
circular, 11-9
corner-firing system, 11-10
gas elements, 11-13
igniters, 11-15
marine, 31-6
mechanical atomizers, 11-11

steam atomizers, 11-12
XCL-S type, 11-9
operation
adjustments, 43-7
emissions 43-6
excess air, 15-2, 43-7
fuel conditions, 43-7
optimization, 43-6
overfire air adjustments 43-7
recovery boiler
limited vertical sweep, 28-20
oscillator, 28-20
selection, effect on steam temperature, 19-16
Burner management systems, 41-14
Burning profiles, 9-8
Burnup, nuclear, 47-8
Bypass and startup systems
drum boilers, 19-20, 43-4
drum pressure control, 19-21, 44-16
dual pressure 19-20
operation, 19-20
steam pressure control, 19-21
upgrades, 44-16
UP boilers, 19-16, 43-12
constant furnace pressure startup system, 19-19
flash tank, 19-19
heat recovery, 19-19
minimum required flow, 19-19
overpressure relief, 19-20
steam temperature control, 19-19
variable throttle pressure, 19-19
variable furnace pressure startup system, 19-17
minimum flow, 19-17
vertical steam separator, 19-17
water collection tank, 19-17
Byproduct fuels, solid, 30-1

Calcination, 17-12
Calcium
carbonate, 35-9
sulfate, 35-9
sulfite, 35-9
Calculations, performance, 22-1
Calorimeter
boiler as (BAC), 10-23
bomb, 10-7
Calorimeter throttling, 40-17
CANDU, 46-2, 46-5, 48-1, 50-2
calandria, 46-6
Capacity
factor, 37-11
increase, 44-2, 44-20
Capitalized cost method, 37-13
Carbon dioxide in flue gas
from fuel analysis, 10-6, 10-21, 32-8
properties, 10-2
Carbon steel, 7-12
Carburization, 44-12
Carnot cycle, 2-12
Carrying charges, 37-11
Carryover, 5-13, 48-7
Carryover, impurities
mechanical, 42-12
moisture, 5-13
vaporous, 42-12
Carryunder, 5-13, 48-7
Casing, 23-3, 39-18
marine boilers, 31-12
Cast iron, 7-3, 7-14
air heater, 20-8
Cast steel, 7-15
Castings, 38-4
Catalyst – see NO_x control, catalyst
Catalytic cracker, 27-13
Cavity
heat transfer, 4-28, 23-6
radiation calculation, 22-13
Celsius, 40-4
Cementite, 7-3
Centigrade temperature scale, 40-3
Centrifuge, 35-8
Ceramic tile, 35-5
Ceramics, 7-15
Ceramics and refractories, 15-10, 29-3
Chain reaction, nuclear, 47-7, 47-10, 47-11
Change of phase, solid, 7-2
Char, 14-3, 18-2
Charpy test, 7-17
Chemical
analyses
dewatering system, 36-10
EPA method, 36-7, 36-8
gypsum quality, 36-10, 36-13
lime quality, 36-7
limestone quality, 36-6
recirculation slurry/blowdown, 36-8
cleaning, fossil – see Chapter 42
advantages, 43-3
alkaline boilout, 42-18
circulation cleaning method, 42-20
cleaning equipment, 42-23
cleaning frequency, 42-19
cleaning time, 42-23
deposit samples, 42-22
deposits, 42-21
inhibitors, 42-22
precautions, 42-24
preoperational, 42-18
preparation, 42-23
procedures, 42-20, 42-23
soaking (cleaning) method, 42-21
solvent strength, 42-22
solvent temperature, 42-22
solvents, 42-21
superheater/reheater, 42-24
cleaning, nuclear, 49-3, 49-7
chemical addition, 49-16
startup, 49-16
cleaning solvent disposal
considerations, 42-24
co-ponding, 42-24
evaporation, 42-24
incineration, 42-24
options, 42-24
recovery process, pulp, 28-1, 43-16
vapor deposition, 7-15
Chemistry, fuel and water, Intro-17
Chimney – see Stack and draft
Chloride, 35-5
Chromizing, 7-15
Circulation, 1-3, 3-17, 5-17, 19-3, 20-4, 48-8
design criteria, 5-19
forced, 1-3, 5-19
improvements, 44-16
marine, 31-12
natural, 1-3, 5-18, 19-3, 43-4
once-through, 5-17, 19-3
startup, 43-4
steam generator, nuclear, 48-8
Circulation or flow circuits
Radiant boiler (RB), 26-13, 26-14
Spiral Wound Universal Pressure (SWUP) boiler, 26-10
Cladding, 46-5, 46-8, 50-2, 50-4, 50-13 (see also Weld, cladding)
Classifier, coal, 13-2, 13-8
rotating, 13-13
Clean Air Act, U.S., 32-1, 35-1
1990 Amendments, 32-4
acid deposition control, 32-4
air quality control regions, 32-2
air toxics, 32-5
criteria pollutants, 32-2
evolution of legislation, 32-1
Maximum Achievable Control Technology, 32-5, 32-10
National Ambient Air Quality Standards, 32-2
New Source Performance Standards, 32-2
New Source Review, 32-2
NO_x SIP Call, 32-5
state implementation plans (SIPs), 32-2
Clean Air Rules, 32-6
Clean Water Act (CWA), 32-13
Cleaning
air heater, 24-10
back pass/economizer, 24-8
convection pass, 24-5
external, 19-12
furnace, 24-2
internal, 19-12, 43-3 (see also Chemical, cleaning, fossil)
SCR catalyst, 24-9
sonic horns, 24-9
Cleaning and inspection, nuclear, 49-15
Cleanliness, internal, 39-16
Cleanliness factor, 4-20, 22-4
Climate factor, fossil fuel units, 26-5
CO catalyst, 27-17
Coal
abrasiveness, Chapter 13
testing, 13-8
wear effects, 13-1, 13-4, 13-9
analysis
ash, 15-3
burning profile, 9-8

crushers, 15-6, 15-11
equilibrium moisture, 9-7
grindability, 9-8, 13-8
mineral-matter-free, 9-5, 9-7
moisture, 9-7, 15-3
moisture-free, 9-7
sizing, 15-6
sulfur, 9-4, 9-7, 15-12
angle of repose, 12-14
anthracite, 9-6
bituminous, 9-6
blending, 12-16
breakers, 12-3
bulk density, 12-14
caking, 9-8
char, 9-9
characterization (impurities), 9-7, 12-5
organic sulfur, 12-5
pyritic sulfur, 12-5
sulfate sulfur, 12-5
classification, 9-5
cleaning, 12-4, 32-8
dewatering, 12-8
dry processing, 12-8
flotation, 12-8
gravity concentration, 12-6
tramp metal, 12-3, 12-13
coalification, 9-3, 12-5
crushers, 12-3, 12-12
dry, mineral-matter-free, 9-5
drying, 13-2
dust suppression and collection, 12-17
elemental ash analysis, 9-8
enthalpy, 10-20
erosiveness, 9-9, 13-9
feeders, 12-15, 13-2, 15-7, 44-3
gravimetric, 12-15
volumetric, 12-15
fixed carbon, 9-5
fouling indices, 9-8, 21-13
free swelling index, 9-8
friability, 12-8
frozen coal, 12-17
gasification, 9-11, Chapter 18 (see also Gasification, coal)
grindability, 9-8, 13-8
grinding, 13-2
gross calorific value, 9-8
gross (higher) heating value (HHV), 9-8
Hardgrove Grindability Index (HGI), 9-8, 13-8
heating value, 9-8, 10-10
lignite, 9-6
lower (net) heating value (LHV), 9-8
mining, 12-2
moisture, 9-5, 9-7, 13-10
oxidation, 12-17
preparation, 12-4
processing and handling, Chapter 12
properties, effect on design, 26-3
proximate analysis, 9-5
pulverizer – see Pulverizer
pyritic sulfur, 9-4
rank, 9-5
reserves, 9-1
screening, 12-4
size reduction, 12-3
sizing, 12-3
slagging indices, 9-9, 21-13
slagging potentials, 9-9
spontaneous combustion, 12-13
stoker, 16-1, 16-4
storage, 12-10
bunker, 12-14
design, 12-14
downspout design, 12-15
seal height, 12-15
fires, 12-18
flow problems, 12-17
funnel flow, 12-15
mass flow, 12-15
shear cell, 12-15
silo, 12-14
stacker, 12-13
stockpile, 12-13
fires, 12-17
inspection and maintenance, 12-14
reclaim, 12-14
subbituminous, 9-1, 9-6
transportation, 12-8
barge, 12-9
continuous transport, 12-10
rail, 12-9
truck, 12-10
ultimate analysis, 9-8
U.S. availability, 9-2
volatile matter, 9-5
world availability, 9-1
Coalification, 9-3
Coatings, 7-15
Codes, 8-5, C-1
Cogeneration, 1-4, 27-4, 30-1
Coke, 9-6
breeze, 9-9
delayed coke, 9-17
fluid coke, 9-17
oven gas, 9-10
Cold
bending, 7-9
drawing tubes, 38-10
forging, 7-9
rolling, 7-9
work, 7-9
Combined cycle, 1-17, 27-4
Combined heat and power, 30-1
Combustion
biomass fluidized-bed, 17-3, 17-5, 30-7
Btu method, 10-6, 10-23, 10-26
calculation methods, 10-3, 10-23
calculations, 22-3, 22-5 (see also Chapter 10)
coal, fluidized-bed, 17-1 (see also Fluidized-bed boiler)
coal, pulverized, 14-1
char oxidation, burnout, 14-3
char reactivity, 14-3
combustion rate, 14-2
devolatilization, 14-3
mineral matter effect, 14-4
moisture effect, 14-3
unburned carbon loss, 14-8, 14-19
volatile matter effect, 14-3 (see also Volatile matter, coal)
constants, 10-2
control(s), 41-10
corner-firing, 11-10, 14-16
definition, 10-1
Dulong's formula, 10-10
equilibrium, 2-26
fundamental laws, 10-1
guides, operating, 41-13
heat of, 10-7
mole method, 10-3, 10-6, 10-8, 10-23
numerical model, 6-1 (see also Numerical modeling)
advanced burner, example, 6-21
combustion reactions
heterogeneous, 6-8
homogeneous, 6-7
mesh generation, 6-19
radiation, 6-8
stoker, example, 6-21
theory, 6-3
oil and gas
efficiency, 11-5
excess air, 11-4
performance requirements, 11-4
pulsation, 11-5
stability, 11-4
turndown, 11-4
pressurized fluidized-bed, 17-13
principles of, Chapter 10
pulverized coal – see Combustion, coal, pulverized
system, coal
applications for metals and cement industry, 14-21
compartmented windbox, 14-11
conventional, 14-4
corner-firing, 14-16
downshot firing, 14-5, 14-18
high moisture coal, 14-5
integration with boiler, 14-5
low volatile coal, 14-5
overfire air (OFA) ports, 14-10, 14-18
primary air, 14-2, 14-6
pulverized coal, 14-1
safety, 14-20
secondary air, 14-2, 14-7
research and development, Intro-14
sorbent, 10-11, 10-27
stoker – see Stoker
three Ts of, 10-1
use of mole, 10-3
Commercial operation, nuclear, 49-20
Commercial structure, 37-9
Compressed water, properties, 2-4, T-7

Compressible, 3-1
Compressor, energy equation, 2-9
Computational fluid dynamics (CFD), 6-1
Computer numerical control, 38-3
Condensate/condensate polishing, 42-6
Condensation, film, 4-22
Condensing attemperator system, 25-21, 27-7
Condition assessment, Chapter 45
  analysis techniques
    BLESS code, 45-15
    crack growth, 45-14
    cycling effects, 45-17
    leak before break, 45-14
    life fraction, 45-10, 45-12, 45-15
    remaining life, 45-11
  boiler components
    air heater, tubular, 45-11
    attemperators, 45-9
    drums, 45-6, 45-20
    economizer, 45-17
    headers
      high temperature, 45-7
      low temperature, 45-8
    settings, 45-11
    tubing
      steam-cooled, 45-7, 45-11
      water-cooled, 45-7, 45-13
Conductance, heat transfer
  for conduction, 4-1
  for convection, 4-3
  in thermal circuit, 4-6
  unit-thermal conductance, 4-3
Conduction, heat transfer
  cylindrical geometries, 4-6
  definition, 4-1
  Fourier's law, 4-1
Conductivity
  feedwater limits, 43-12
  method, steam purity, 40-17
  thermal, refractory and insulation, 23-6
Conservation
  of energy, 3-3, 6-4
  of mass, 3-2, 6-4
  of momentum, 3-2, 6-4
Consolidated nuclear steam generator, 46-21
Constructability, 39-4
Construction, Chapter 39
  approaches, 39-6
    field modularization, 39-6
    knocked down, 39-6
    shop modularization, 39-6
  cost, 39-2
  project management, 39-3
  safety, 39-16
  schedule, 39-2
  technology, 39-5
Contact resistance, heat transfer, 4-1
Containment building, 46-3
Continuity equation, 3-2, 6-5
Continuous emission monitoring (CEM), 34-8, 36-4
Contraction
  pressure loss, 3-11
  static pressure, difference, 3-13
Control (and protective systems), nuclear
  in-core instrumentation, 49-9
  integrated control system, 46-14, 49-17
  neutron detectors, 47-14, 49-17
  non-nuclear instrumentation, 49-17
  nuclear, 49-17
  reactor protection system, 49-17
  rod, 49-8, 49-18
    assemblies, 46-8
    drive mechanisms, 49-8
    worth, 47-13, 49-18
  room, 49-20
Control system design philosophy
  Universal Pressure once-through boiler, 41-12
  utility drum boiler, 41-5
Control volume
  fluid dynamics, 3-2
  for energy balance, 4-5, 4-23
  for numerical modeling, 6-7
Control(s) and control systems, Chapter 41
  atmospheric fluidized-bed boiler, 41-11
  auxiliary boiler (marine), 31-5
  basic theory, 41-1
  boiler-following, 41-5
  burner, 41-14
  burner management systems, 41-14
  closed loop, 41-2
  combustion, 41-10
  coordinated boiler-turbine, 41-6
  Cyclone-fired boiler, 41-11
  during startup, 43-3
  feedforward/feedback, 41-4
  fluidized-bed, 41-11
  furnace draft, 41-10
  human factor engineering, 41-20
  integrated boiler and turbine-generator, 41-6
  merchant marine boiler, 31-2, 31-3
  naval boiler, 31-1
  neural networks, 43-7
  nuclear plant, 49-10, 49-17
  oil- and gas-fired boiler, 41-10
  open loop, 41-2
  proportional, 41-2
  proportional plus integral, 41-4
  proportional plus integral plus derivative, 41-4
  pulverized coal-fired boiler, 41-10
  pulverizer, 13-2
  selective catalytic reduction systems, 41-17
  scrubbers, 41-17
  sootblowing, 43-8
  steam flow-air flow, 41-13
  steam temperature, 19-12, 31-9 (see also Steam, temperature)
  turbine-following, 41-5
  Universal Pressure boiler, 43-11
Convection, heat transfer, 4-3
  boiler surface, 19-6
  definition, 4-3
  fouling, 4-20
  heat flow, rate of heat transfer, 4-3
Convection pass, 1-7, 7-21, 19-1, 20-4
Convective heat-transfer coefficient, (see also Nusselt number)
  in flow over banks of tubes, 4-16, 4-18
  in flow through tubes (and ducts)
    laminar, 4-15
    turbulent, 4-16
  in free convection, 4-14
Conversion factors, Appendix 1
Conversion formulae for emission rates, T-14
Converter reactor, 47-15
Coordinate measuring machine, 50-8, 50-11
Coordinated boiler-turbine control, 41-5
Coordination curves, pulverizer, 13-7
Core – see Reactor
Core reactivity coefficients, 49-18
Corrosion, 23-7, 42-14 (see also Ash, coal, corrosion and Ash, fuel oil, corrosion)
  acid corrosion, 42-16
  acid dew point, 27-16
  acid phosphate corrosion, 42-16
  air heater, 20-13
  caustic corrosion, 42-17
  caustic cracking, 42-17
  caustic gouging, 42-17
  chelant corrosion, 42-16
  corrosion fatigue, 42-16
  decarburization, 42-17
  erosion-corrosion, 45-9 (see also flow accelerated corrosion)
  flow accelerated corrosion, 42-15
  fretting, 48-9, 49-2
  fuel ash, 7-22, 21-20
  general, 42-1
  hydrogen damage, 42-16
  intergranular, 7-6, 42-15, 42-18, 48-9
  locations, 42-15
  out-of-service corrosion, 42-18
  pitting, 42-16, 48-9, 49-2
  protection, process recovery boiler, 28-11
  stress corrosion, 48-9, 49-2, 49-7, 49-8
  stress corrosion cracking, 42-18, 48-9
  syngas cooler, gasification, 18-11
  under-deposit corrosion, 42-16
  wastage, 49-2
  waste-to-energy system, 29-2
    ash cleaning equipment, 29-21
    bimetallic tube, 29-5
    ceramic tiles, 29-3
    chloride, 29-2
    flue gas temperature, 29-3, 29-18
    hydrochlorides, 29-3

Inconel, 29-4, 29-19
Inconel bimetallic, 29-5, 29-19
lower furnace, 29-3
metal temperature, 29-3
oxidizing atmosphere, 29-3
parallel flow superheater, 29-20
pin stud, 29-3
protection, 29-3
reducing atmosphere, 29-3
refractory, 29-3, 29-8
silicon carbide, 29-3, 29-8
superheater, 29-20
wear zone panel, 29-4
weld overlay, 29-3
Cost(s)
constant, 37-11
current, 37-11
evaluation, 37-11
future value, 37-12
levelization, 37-12
present value, 37-12
Cranes, 39-9
Creep, 7-1, 7-19, 8-12, 44-12
crack growth, 8-12
detection, 45-7
flaw size, 8-10
headers, 45-7
life prediction, 45-12
piping, 45-9
replication, 45-5
rupture, 7-19
Creep-fatigue, 44-14
Criteria Pollutants, 32-2
Critical boron concentration, 47-13
Critical flow, 5-20
Critical heat flux (CHF), 5-2, 5-4, 5-19, 19-8
criteria, 5-6
definition, 5-4
evaluation, 5-6
marine boiler, 31-7
ribbed bore tube, 5-6
Critical mass, 47-10
Critical transformation temperature, 7-3
lower critical, 7-4
upper critical, 7-4
Cross-section, nuclear, 47-5, 47-12
absorption, 47-6
capture, 47-6
fission, 47-6
macroscopic, 47-6
microscopic, 47-6
resonance, 47-6
scattering, 47-6
Crossflow, friction factor, 3-15
Crushers, 12-3
double roll, 12-3
hammermill, 12-4
rotary breaker, 12-3
single roll, 12-3
Crystal structure, 7-1
body-centered cubic, 7-1
body-centered tetragonal, 7-4
face-centered cubic, 7-1
Curies, 47-4
Cycle
Brayton, 2-22
Carnot, 2-12
nuclear fuel, 47-12
Rankine, 2-13
for nuclear plant, 2-19
regenerative, 2-14
reheat, 2-15
steam, 2-1, 2-12
Cycling, 27-17, 44-15
definition, 45-17
drum, 45-20
economizer, thermal shock, 45-17
furnace, subcooling, 45-18
operation control, 41-7
superheater, tube leg flexibility, 45-19
Cyclone furnace, Chapter 15, 43-10
advancements, 15-12
advantages, 15-1
air pollution control, 15-12
ash, 15-3
burner, 15-2, 15-4
radial, 15-4
thin door radial, 15-4
scroll, 15-4
vortex, 15-4
capacities, 15-5, 15-9
coal
crushers, 15-6, 15-11
feeders, 15-7
sizing, 15-6, 43-10
suitability, 15-3, 15-5, 43-10
combustion air, 43-10
combustion control, 15-9
controls, 41-11
design features, 15-7
erosion, 15-11
fabrication, 38-9
firing arrangements, 15-5
flyash emission and recovery, 15-3, 15-12
fuels suitable for, 15-3, 15-8, 43-10
igniters, 15-10
oil and gas burners, 15-8
operation and maintenance, 15-10, 43-10
power requirements, 15-10
waste fuels, 15-3, 15-9

D-type boiler (FM), 27-12
Dampers
control, 25-3
guillotine, 25-4
isolation, 25-3
louver, 25-3
round, 25-4
Data acquisition, 40-3, 40-24
Dead weight gauge, 40-2
Decay
radioactive, 47-4
Degradation mechanisms, 44-6, 44-11
Demand charge, 37-11
Density
of flue gas, 4-11
of gases, 4-11
Departure from nucleate boiling, DNB – see Critical heat flux
Deposits
ash, 21-3, 24-1
scaling, 42-3
slag, 21-11, 24-2
water-side, 42-15, 42-19
Desuperheater (see also Attemperator)
marine, 31-10
Diffusion
atomic, 7-1
thermal, 4-7, 4-22
Dimensional analysis, heat transfer
correlation of data, 4-15
dimensionless numbers or groups, 4-15
Discount rate, 37-11
Discrete ordinates method, 6-9
Dislocations, 7-2
Dissimilar metal welds, 7-11
Dissolved alkalinity, 35-12
Distillate fuels, 9-12
Distributor grid, 27-17, 34-6
DNB – see Critical heat flux
Doppler
coefficient, nuclear fuel, 47-14, 49-18
effect, 47-6
Dose
equivalent, 49-13
radiation, 47-3
radiation absorbed, 49-13
Downcomer, 48-5
definition, 1-3, 1-9
Draft (see also Stack and draft)
calculations – see Chapter 22
loss, 19-7, 25-6
Draft effect, 3-15
Drainability, economizer, 20-5
Drum, steam, 1-3, 5-13, 39-10
allowable temperature differential, 43-5
baffle(s), 5-14
capacity, 5-17
condition assessment of, 45-6
cycling, 45-20
fabrication, 38-5
head fabrication, 38-5
lifting, 39-11
marine, 31-11
materials, 7-23
protection, startup and shutdown, 43-5, 45-20
support, 39-11
U-bolts, 39-12
Dry scrubbers – see SO_2 control
controls, 41-19
Dry shielded canister, 50-13
Dryout – see Critical heat flux
Dual economizer steam air heater (DESAH), 31-11
Dual furnace, 31-9
Duct burner, 27-17
Ductile iron, 7-15
Ductility, 7-17

Duplex alloys, 7-14
Dust collector, 33-12, 35-2

Earthquake requirements, 26-2
Economic evaluation, 37-1, 37-10
Economizer, 1-8, Chapter 20, 27-2
  calculations, 20-4, 22-13
  cascading longflow, 28-14
  cleanability, 20-2
  condition assessment of, 45-7
  continuous, 28-14
  crossflow, 28-14
  cycling, 45-17
  drainability, 20-5
  erosion, 20-3
  fabrication, 38-8
  fin, 20-2
  gas velocity limits, 20-3
  header, 20-6
  heat transfer, 20-4
  longflow, 20-4, 28-14
  marine, 31-11
  materials, 7-21
  mean temperature, 20-4
  plugging, 20-2, 20-14
  pressure drop, 20-5
  sootblowers, 20-3
  startup protection, 43-4
  steaming, 20-4
  stringer tubes, 20-6
  supports, 20-5
  tube
    baffles, 20-3
    diameter, 20-6
  tube deflection, 20-5
  upgrade, 44-7
  vibration, 20-6
  wall enclosures, 20-5
Eddies, 3-8
Eddy current testing – see Nondestructive examination
Edison, Thomas Alva, Intro-6, Intro-8
Efficiency, 44-18
  boiler or steam generating unit, 19-4
  combustible fuel, 10-17
  cycle – see Chapter 2
  gasification systems, 18-2, 18-14
  heat balance, 10-18
  marine boiler, 31-10
  operating techniques, 43-6
  particulate removal, 36-13
  recovery boiler, 28-4
    calculations, 28-6
  SO_2 removal, 35-6, 36-13
Effluents (aqueous), 32-13
Ejector, 3-16
Electric resistance welded (ERW) tubes, 7-21
Electric utilities
  history of, Intro-7
Electromagnetic acoustics, 45-4
Electron, 47-1
Electropolishing, 50-2, 50-11
Electrostatic precipitator, 29-22, 32-10, 33-2, 33-8, 35-2, 35-13
  applications, 33-3, 33-8
Electrostatic precipitator, dry, 33-2
  components, 33-4
    casing, 33-4
    collecting electrodes, 33-5
    discharge electrodes, 33-5
    enclosure, 33-4
    hoppers, 33-4
    insulators, 33-5
    power supplies and controls, 33-6
    rapping systems, 33-5
    transformer rectifier, 33-6
  particle size distribution, 33-7
  particulate control, 33-3
    ash removal, 33-3
    charging, 33-2
    collection, 33-2
    rapping, 33-3
  performance
    Deutsch-Anderson equation, 33-6
    enhancement, 33-7
    fuel and ash, 33-3
    gas flow control, 33-5
    sizing factors, 33-6
  reentrainment, 33-3
  resistivity, 33-3
  sectionalization, 33-7
Electrostatic precipitator, wet, 33-8, 35-19
Emergency core cooling system (ECCS), 46-15
Emergency feedwater system, 49-15
Emissions, air pollution (see also individual pollutants)
  cold gas, 18-2
  F-factor, 36-12
  gasification plants, 18-12
  industrial boilers
    carbon monoxides (CO), 27-7
    nitrogen oxides (NO_x), 27-7
    particulate matter (PM), 27-7
    sulfur oxides (SO_x), 27-7
    volatile organic compounds (VOC), 27-7
  oil and gas, 11-5
    opacity, 11-9
    oxides of nitrogen, 11-5
    oxides of sulfur, 11-8
    particular matter, 11-8
  stokers, 16-10
  utility boilers, 1-10, 26-5, 32-2
  waste-to-energy system
    activated carbon injection, 29-23
    baghouse, 29-22
    carbon monoxide, 29-23
    dry scrubbing system, 29-22
    electrostatic precipitator, 29-22, 29-23
    hydrochloric acid, 29-22
    mercury, 29-22
    nitrogen oxides, 29-23
    polychlorinated dibenzofurans, 29-6
    polychlorinated dibenzopara-dioxin, 29-6
    selective catalytic reduction, 29-23
    selective non-catalytic reduction, 29-23
    sulfur dioxide, 29-22
Emissive power, 4-4
Emissivity (and emittance), 4-4, 4-10
  coal-ash deposits, 4-10
  gases, 4-12
  gray surfaces, 4-4, 4-10
  non-gray surfaces, 4-4
  of various surfaces, 4-12
Enclosure, Chapter 23 (see also Setting, boiler)
Energy
  applications, 2-9
  available, 2-17
  consumption, 9-1
  equation, 2-7, 2-9, 3-4, 4-5, 4-7
  production, 9-1
Enhanced oil recovery, 27-14
Enthalpy, 2-1, 2-7, 5-1
  steam and water, 2-1, 2-2, T-5, T-6
Entrainment, 3-16
Entropy, 2-10
  increase, 2-12
  steam and water, 2-1, 2-2, T-5, T-6
Environmental
  gas-side measurements, 36-8
  legislation – major U.S., Chapter 32
    Clean Air Act, 32-1 (see also Clean Air Act, U.S.)
    Clean Water Act, 32-13
    research, development and evolution of, Intro-14
    The Resource Conservation and Recovery Act (RCRA), 32-15
  process monitoring, 36-6
Environmental control, 27-4, Chapter 32
Environmental equipment, construction 39-15
Environmental regulations, Chapter 32, 44-2, 44-20
Equilibrium diagram, 7-3
Equipment diagnostics, nuclear (see also Nondestructive examination)
  check valves, 49-11
  destructive examination, 49-4
  eddy current testing, 49-4, 50-10
  fitness-for-service assessment, 49-3
  foreign object search and retrieval (FOSAR), 49-3
  loose parts monitoring, 49-10
  motor operated valves, 49-11
  neutron noise monitoring, 49-11
  reactor coolant pump monitoring, 49-11
  valve monitoring and diagnostics, 49-11
Equipment selection, 26-1
Equipment specification, 37-2
Equivalent
  diameter, 3-7, 4-15, 4-18 (see also Hydraulic diameter)

Erection (see also Construction)
approaches, 39-3
auxiliaries, 39-18
cleanliness, 39-16
cost estimates, 39-2
field welding, 39-13
material handling, 39-8
nonpressure parts, 39-18
procedure, fossil fuel unit, 39-11
structural supports, 39-11
time estimates, 39-2
time span, 39-2
Erection methodology, 39-6
contractor interface, 39-7
new construction, 39-6
plant arrangement, 39-6
product parameters, 39-7
retrofit, 39-6
schedule, 39-2, 39-7
scope, 39-6
site conditions, 39-7
Erosion, 21-17, 23-4
air heater, 20-14
barriers, 44-8
protection, 44-4, 44-20
tube bends, economizer, 20-3
Estimating, construction, 39-2
Euler
equation of motion, 3-3
number, definition, 3-7
Evaluation criteria, 37-7
Evaporation, 4-22, 5-1, 42-2
Evaporator, chemical process recovery
cascade, 28-19
cyclone, 28-19
direct contact, 28-19
multiple effect, 28-19
Examination (see also Condition assessment)
destructive
boat samples, 45-6
tube samples, 45-6
methodology, 45-1
nondestructive (see also Nondestructive examination)
acoustic emissions, 45-5
eddy current, 45-3
electromagnetic acoustics, 45-4
liquid penetrant, 38-12, 45-2
magnetic particle, 38-12, 45-2
nuclear fluorescence, 45-4
radiography, 38-11, 45-4
replication, 45-5
strain, 45-6
temperature, 45-6
ultrasonic, 38-12, 45-2
visual, 45-1
Excess air, 10-15
effect on Cyclone corrosion, 15-4
effect on Cyclone operation, 43-10
effect on efficiency, 43-1
effect on steam temperature, 19-13
from flue gas analysis, 10-21, 10-23
loss, 43-6
measurement of, 10-21
Expansion, boiler, 23-4
Expansion joints, 25-5
configurations used in flues and ducts, 25-5
metallic, 25-5
nonmetallic, 25-5
types, 25-5
Extended surface, 4-20
Extruded tubing, 44-8

Fabric filter (and baghouse), 32-10, 33-8, 35-13, Color Plate 7
air/cloth ratio, 33-10
applications, 33-10
bag materials and supports, 33-11
cleaning, 33-11
configurations, 33-11
hatch style, 33-11
walk-in, 33-11
design parameters, 33-10
drag, 33-10
dustcake, 33-8
enclosure, 33-11
hoppers, 33-11
operating fundamentals, 33-8
tubesheet, 33-8
types, 33-9
pulse jet, 33-10
reverse gas, 33-9
shake deflate, 33-10
shaker, 33-10
Fahrenheit temperature scale, 40-4
Fans
acoustic noise, 25-20
aerodynamic characteristics, 25-20
axial flow, 25-17
centrifugal, 25-15
drives, 25-17
forced draft, 25-13
gas recirculating, 25-14
induced draft, 25-14
laws of performance, 25-11, 25-18
maintenance, 25-14
output control, 25-15
performance, 25-11
power, 25-10
primary air, 13-6, 25-14
specific diameter, 25-12
specific speed, 25-15
stall, 25-19
testing, 25-14
variable speed, 25-15
Fast reactors, 47-15
Fatigue
crack growth, 8-11, 8-13
damage evaluation, 8-6, 45-5
design curve, 8-6
maintaining availability, effect on, 44-6, 44-11, 44-14
Feasibility, project, 37-1
Fechheimer probe, 40-22
Feedwater
control, 41-8
design considerations, 26-5
economizer, 20-6
requirements, Universal Pressure boiler, 43-11
temperature, effect on final steam temperature, 19-13
treatment
effect on boiler availability, 43-8
Ferrite, 7-3
Ferritic stainless steels, 7-14
Filling boiler, 43-4
Film condensation, 4-22
Filter
sludge, vacuum, 35-8
Fine grinding, coal, 13-14
Fineness, coal, 13-8
capacity factors, 13-10
testing, 13-10
Finite element analysis (FEA), 8-8
application of, 8-8, 8-9
computer software, 8-9
limitations, 8-9
nonlinear, 8-9
stress analysis, 8-8
Finned tube economizer, 44-7, 44-18
Fins
circumferentially attached, 4-20
economizer, 20-2
efficiency, 4-21
Fire(s) and fire protection
air heater, 20-14
pulverizer, 13-13
Firing rate control, Universal Pressure boiler, 43-11
Fission, Intro-12, 1-11, 46-1 (see also Fuel, nuclear)
process, 47-2
product(s), 46-11
Fittings
flow resistance, 3-9
Fixation, 32-16
Fixed charges, 37-11
Flake glass, 35-5
Flame
coal, pulverized, 14-7
detection or scanners, 14-20, 41-15
stability, 14-7
photometer, 40-18
Flat stud tubes, 23-2
Floor design, process recovery boiler, 28-12
decanting, 28-12
sloped, 28-12
Flow
around tubes, 4-16, 4-18
boiling, 5-2
coast down, 49-20
in tubes (and ducts), 4-15, 4-16
instability, 5-11
laminar, 3-8, 4-15
over flat plate, 4-15
patterns, 5-8
turbulent, 3-8, 4-16
two-phase, steam-water, 5-8
Flue gas, 1-5, 1-10
analysis, 10-21
enthalpy, 10-19
excess air determination, 10-21

mass flow, 10-16
moisture, 10-6, 10-17
residue, 10-14
sampling, 10-21
specific heat, 22-9
volumetric combustion chart, 10-22
Flue gas desulfurization (FGD) (see also SO_2 control, scrubber (FGD))
construction, 39-15
dry scrubbing, 32-9, 35-12
sorbent injection, 32-9
wet scrubbing/scrubber, 32-9, 35-2, 36-1
Flue gas sampling
isokinetic, 36-12
location selection, 36-7
Flue gas velocity, 20-3
Fluid
dynamics, Chapter 3
research and development of, Intro-15
entrainment, 3-16
flow, energy equation, 3-4
flow measurement, 40-20
flow nozzle, 40-20
orifices, 40-20
venturi tube, 40-20
friction, pressure loss, 3-5
pressure loss, two-phase, 5-9
straightening vanes, 40-20
Fluidization, 17-2
Fluidized-bed boiler, Chapter 17, 27-8, Color Plate 5
air staging, 17-8
ash, 17-10, 17-12
attrition, 17-13
bubble caps, 17-2
bubbling bed, 17-1, 17-2, 27-11
calcination, 17-12
circulating bed, 17-1, 17-9, 27-8
control, 41-11
distributor, 17-3
fixed bed, 17-1
fuel feed systems, 17-7
mean diameter, particle, 17-2
minimum fluidization, 17-1
NO_x control, 17-8, 17-13
operations, 43-13
bubbling bed, 43-15
fuel sizing, 43-15
operating overview, 43-15
circulating bed, 43-13
biomass firing, 43-14
coal firing, 43-14
fuel sizing and characteristics, 43-14
operating overview, 43-13
operations, 43-13
startup and shutdown, 43-14
overfire air system, 17-6, 17-8
pressurized bed, 17-13
Sauter mean diameter, 17-2
slumped bed, 17-3
solids separation, 17-11
staged combustion, 17-13
superficial bed velocity, 17-3
temperature control, 17-4, 17-10
turbulent bed, 17-2
U-beams, 17-11
volume-surface mean diameter, 17-2
waste fuels, 17-1
Flyash, 32-1, 32-16 (see also Ash, coal)
Flyash plugging, 44-7
Forced circulation, 1-3, 5-19 (see also Circulation)
Forced convection, heat transfer
entrance effects, 4-14
heat transfer coefficient inside tubes (and ducts), 4-3, 4-15, 4-16
laminar flow, 4-15
turbulent flow, 4-16
Forced outage rate, 44-1
Formic acid, 35-12
Forming, 38-3
Fouling, 4-20, 9-8, 28-9, 44-6 (see also Ash, coal, deposition)
Fracture mechanics, 8-10
creep
crack growth, 8-12
flaw size, 8-12
elastic-plastic, 8-11
failure assessment
deformation plasticity, 8-11
diagrams, 8-11, 8-12
leak-before-break, 8-12
plastic instability, 8-12
safety margins, 8-12
fatigue crack growth, 8-12
linear elastic, 8-10
time dependent, 8-13
toughness, 7-18
Free convection, heat transfer, 4-14
Free surface, 3-1
Friction factor
crossflow gases, 3-15
flow in pipes and ducts, 3-7, 3-8
Friction loss, 3-5
two-phase flow, steam-water, 5-9
Fuel (see also specific fuel and Chapter 9)
biomass, 9-18, 16-1, 16-7, 30-1
boiler operations, 43-1
coal, 15-1, 26-3
flow control, 41-10
flow measurement, 41-13
natural gas, 15-8 (see also Natural gas)
firing application, 26-16
nuclear,
assembly, 46-8, 47-9
burnup, 47-8
fission, 47-6
chamber, 47-14
energy, 47-7
neutrons, 47-7
process, 47-6
products, 47-8
fissionable, 47-5
fusion, 47-2
handling, 49-12
loading, 49-18
refueling, 46-10
oil, 9-12, 11-1, 15-8 (see also Petroleum)
API gravity, 9-14, 11-2
asphaltene content, 11-2
auxiliary, marine, 31-5
burning equipment, 11-9
burning profile, 11-2
carbon residue, 11-2
combustion calculations, Chapter 10
distillation, 11-2
enthalpy, 10-20
fire point, 11-2
firing application, 26-16
flash point, 9-14, 11-2
heating value, 11-2
kinematic viscosity, 9-14
merchant marine, 31-3
naval, 31-2
No. 2, 9-12
No. 6, 9-12
Nos. 4 and 5, 9-13
pour point, 9-14, 11-2
preparation, 11-1
properties, 11-2
reserves, 9-12
residual fuel oils, 9-13
Saybolt Furol viscosity, 9-14
Saybolt Universal viscosity, 9-14
specific gravity, 11-2
transportation, storage and handling, 11-1
ultimate analysis, 11-2
viscosity, 11-2
water and sediment, 9-14, 11-2
Orimulsion, 11-3
other gases, 11-3
purging, 43-1
research and development of, Intro-14
sludge, 30-4
straw, 30-2
switching, 32-5, 32-8
switching, SO_2 control, 35-1, 44-18
wood, 30-2
Fuel ash corrosion, 44-12
Fuel NO_x, 34-1
Furnace, 1-7
Controlled Combustion Zone (CCZ), 16-7, 30-6
design, enclosure, 19-4
dry ash, 19-6
recovery boiler, 28-10
sizing, 19-5, 21-14
slag-tap, 19-6
water-cooled walls, 19-6
wet-bottom, 19-6
divided, 19-16
draft control, 41-10
dutch oven, 30-4
exit gas temperature, calculation, 22-4
heat transfer calculations, 22-3

heat transfer methods
empirical, 4-24
numerical, 4-26
marine, 31-1, 31-2
materials, 7-21
subcooling, 44-15
wall corrosion, 44-8
wall wastage, 44-7
Fused coatings, 7-16
Future value, 37-12

Galvanizing, 7-16
Gamma ray, 47-3, 49-13
Gas (see also Natural gas)
constant, 3-4, 40-4
enthalpy, 10-12, 10-19
equation, 40-4
expansion, 40-4
temperature evaluation, 40-16
temperature measurement, 40-14
flue gas, 10-16
forced convection around cylinders, 4-15
free convection, 4-14
laminar forced convection in tubes, 4-15
laws, 3-4, 10-1
physical properties, 3-10, 4-9, 4-11, 10-2
turbulent forced convection in tubes (and ducts), 4-16
viscosity, 3-11
Gas, flue (see also Flue gas)
analysis, 41-14
biasing effects, 19-15
proportioning damper, 19-15
recirculation, 19-15
tempering, 19-15
Gas disperser, 35-13
Gaseous
fuels from coal, 9-9, 18-1
radiation, 4-4, 4-8, 4-12
state, 3-1
Gasification, black liquor, 28-27
high temperature, 28-28
low temperature, 28-28
Gasification, coal, 9-11, Chapter 18
air-blown, 18-2
crossflow, 18-4
definition, 18-1
developmental issues, 18-11
developments, 18-7
down-flow, 18-4
entrained flow, 18-6
fixed (moving) bed, 18-4
fluidized bed, 18-5
gas compositions, 18-3
gas cooling, 18-11
oxygen blown, 18-2, 18-10
power generation, 18-14
processes, 18-2
product cleanup, 18-12
reactions, 18-1
transport, 18-7
upflow, 18-3
Geiger counter, 49-13
Generating bank (see also Boiler, bank)
crossflow, 28-12
longflow, 28-17
Geographical considerations in boiler design, 26-5
Global warming, 32-8
Grain boundaries, 7-2
Graphitization, 7-4, 7-11, 44-12
Grashof number, definition of, 4-2, 4-14
Grate
air-cooled, 16-6, 16-8
chain, 16-3
moving, 16-2
pinhole, 30-4
stationary, 16-2
traveling, 16-3, 16-5, 16-8, 30-5
vibrating, 16-3, 16-5, 16-9, 30-5
water-cooled, 16-3, 16-10
Gray iron, 7-14
Graybody, thermal radiation, 4-10
Green liquor, 28-25
Greenhouse gases, 32-8
Grid – see Mesh generation
Guarantees
equipment, 26-6
performance, 37-4, 37-5
Gundrilling, 50-4
Gypsum, 32-1, 32-16, 35-8

Half-life, 47-4
heavy elements, 47-5
radioactive, 47-4
Hardening, quenching, 7-9
Hardness, 7-17
Hazardous Air Pollutants (HAPs), 32-4
Hazardous waste, 32-15
Header(s)
analysis techniques, 45-14
condition assessment of, 45-7
creep, 45-7
fabrication, 38-6
leak detection, 45-17
ligament, 44-14
material, 7-23
protection, startup, 43-5
structural analysis – see Chapter 8
tube-to-header weld, 44-15
upgrade, 44-14
Heat
affected zone (HAZ), 7-10
available, 10-11, 10-23
credits, 10-18, 10-23
Dulong's formula, 10-10
exchanger
air heater(s), 20-7
effectiveness, 4-22
mean temperature difference in, 4-19, 4-21
nuclear, 48-1
types, arrangements, 4-20
losses, 10-15, 10-18, 10-19, 23-5, 43-6
mechanical equivalent, 2-8
of combustion, 10-7
of formation, 28-5
of reaction correction, 28-5
of vaporization, 5-1
rate, 19-9
reduction due to deposits, 43-6
transfer, Chapter 4
applications, 4-28
basic modes, 4-1
boiling, 4-21, 4-27, 5-3
calculations, Chapter 22
cleanliness factor, 4-20
coefficient of, 4-3
condensation, 4-22
conduction, 4-1
contact resistance, 4-1
convection, 4-3, 4-14
discrete ordinates, 6-9
electrical analogy, 4-6, 4-23
extended surface, 4-20
furnaces, 4-24
gas-side, 20-4
gas-to-gas, 20-16
insulation, 4-22, 4-29
Kirchoff's law, 4-4
modeling of, 6-5
NTU method, 4-21
Ohm's law, 4-6
porous materials, 4-22
radiation, 4-3, 4-7, 6-8
research and development of, Intro-15
steam generator, nuclear, 48-3, 48-6
supercritical (pressure water), 5-7
surface, 20-1
waste gases, 27-18
water-side, 20-4
treatment
post weld, 7-11, 50-5, 50-12
welding, 38-6
Heat flux measurement, 40-12
Heat rate, 37-11
Heat recovery steam generator (HRSG), 27-15
Heat transfer (see Heat, transfer)
Heating surface, regenerative air heater, 20-11
Heating values, high and low, 9-8, 10-10
Heavy lifting, 39-8
Heavy water, 46-2
Hero's engine, Intro-1
Hideout, chemical, 42-3
History of steam generation and use, Intro-1
Hook gauge, 40-2
Hooke's law, 7-16
Hopper, 23-3
numerical model, 6-21
Hot
functional testing, 49-16

rolling, 7-9
shortness, 7-6
work, 7-9
zero power, 49-18
Human factor engineering, control systems, 41-20
Hydraulic diameter, 3-5 (see also Equivalent, diameter)
Hydroclones, 35-8
Hydrogen damage, 42-17, 44-6
detection of, 45-3
Hydrolysis, 34-7
Hydrostatic test, 39-17, 42-23, 43-2, 50-2, 50-11

Ideal gas, 3-4
Igniter, 11-15, 14-20, 15-10
capacity, 15-10
classification of, 41-16
coal burner, 14-20, 15-10
oil and gas, 11-15, 15-10
operation, 15-10
Ignition
stability, flame, 14-7
temperatures, 10-10
Imaging radiometer, 40-6
Impact protection, 44-9
Impact suction pitot tube, 40-21
In-core instrumentation, 49-9
Incineration, 28-23, 29-1
Incompressible, 3-1
Induced nuclear reactions, 47-4
Industrial boiler design, 27-1
Injector, 3-16
Inspection (see also Equipment diagnostics, nuclear)
pre-operation, 39-19
startup, 43-2
Instability, flow, 5-11
Instrumentation, 40-1
corrections, 40-3
water leg, 40-3
environmental, 36-1
in-core, 49-9
nuclear, 49-10
startup, 43-3
Insulating material
properties of, 4-1, 4-9
thermal conductivity, 23-6
Insulation, 4-22, 4-29, 23-1, 23-5, 23-6, 39-18
ambient air conditions, 4-29
heat loss, 4-29
installation, 23-5
temperature limits and thermal conductivities, 4-29
Integral Furnace boiler, Intro-9, 19-2
Internal deposits, water-side, 43-9, 44-6
Internal recirculation, CFB, 27-10
International temperature scale, 40-3
Inverse temperature solubility, 42-2
Inviscid, 3-4
Ionization, 47-3
chamber, 47-14, 49-13
Iron, effects on coal ash, 21-7, 43-10
Irrecoverable, pressure losses, 3-6
Irreversible, 3-5
Isentropic, 3-4
Isolation valves, nuclear, 49-10
Isothermal transformation diagrams, 7-4
Isotope, 47-1

Jet pump, 3-16

Kelvin temperature scale, 40-4
Kinematic viscosity, 3-10
Kraft process – see Recovery boiler, Kraft process

Lagging, 23-2, 23-9
Lamellar tearing, 7-11
Laminar flow (see also Flow, laminar)
definition, 4-14
inside tubes (and ducts), 4-15
Larson-Miller parameter
creep life fraction, 45-12
life prediction, 45-12
Laser alignment, 50-8
Laws of thermodynamics, 2-10
Layup, 42-25
dry layup, 42- 27
hot standby, 42- 26
nitrogen blanketing, 42-27
wet layup, 42-27
Leak detection
acoustics, 45-16
headers, 45-17
piping, 45-17
tubes, 45-16
Leak test, 50-10, 50-11
Levelization, cost, 37-12
Licensing, nuclear, 49-14
Life management, nuclear, 49-2
Lighter – see Igniter
Lignite, 9-6
ash fusion temperature, 43-10
Lime, 35-2, 35-15
hydrated, 35-16
pebble, 35-16
Limestone, 35-2 (see also Sorbent)
forced oxidation, 35-6
Linear accelerator, 38-6
Linings, rubber, 35-5
Liquid
forced convection around tubes, 4-16
laminar forced convection in tubes (and ducts), 4-15
metal coolants, 46-2, 46-22
penetrant examination, 38-12
physical properties, 3-10
state, 3-1
turbulent forced convection in tubes (and ducts), 4-16
viscosity, 3-11
Ljungström air heater, 20-9
Load, 39-18
blocks, 39-11
dead, 39-9
distribution, 39-10
static equilibrium, 39-10
factor, 37-11
friction, 39-9
impact, 39-9
imposed, 39-9
lifted, 39-9
live, 39-9
seismic, 39-9
stabilization force, 39-9
wind, 39-9
Local heat transfer coefficient
definition, 4-3
in boiling, 5-4
in free convection, 4-14
in laminar flow, 4-15
in turbulent regions, 4-16
Longflow economizer, 20-4, 28-15
Low water, 43-9
Lowest Achievable Emission Rate, 32-4

Machine tools, 38-3
Machining, 38-2
electrical discharge, 38-3
Magnesium oxide, 35-12, 35-18
Magnetic particle examination, 38-12
Maintaining availability, Chapter 44
Maintenance
boiler, Chapter 44
economizer, 20-6
fans, 25-14
nuclear equipment, 49-2
pulverizer, 13-11
remote technology, 49-1, 49-5
stack, 25-10
strategic plan, 44-1
Makeup water and purification system, 42-5, 49-15
Malleable cast irons, 7-14
Manometer, 40-1
high pressure, 40-2
inclined, 40-1
micromanometer, 40-2
Manpower loading, construction, 39-3
Manufacturing
development, Intro-13, Intro-16
fossil fuel equipment, Chapter 38
methods, 38-1
nuclear equipment, Chapter 50
requirements, 38-1
tubes, 38-10
Marine boiler
circulation, 31-12
codes, 31-6
design, 31-6
development, 31-1
efficiency, 31-10
furnace design, 31-7
nuclear, 46-19
requirements, 31-1
steam separation, 31-12

steam temperature control, 31-9
superheater design, 31-8
tube bends, 31-8
types
coal, 31-3
merchant, 31-2
naval, 31-1
package, 31-4
reheat, 31-3
waste heat, 31-5
Martensite, 7-5
Mass
conservation of, 3-2
defect, 47-2
diffusion and transfer, 4-22
energy equivalence, 47-2
flow rate per unit area, 3-6
flux, 3-6
number, 47-2
transfer coefficients, 4-22
transfer rate, 35-10
velocity, 3-6
Mass burning, refuse, 29-3
air plenums, 29-11
combustion air system, 29-11
excess air, 29-11
fuel feed system, 29-11
overfire air, 29-11
plunger ash extractor, 29-12
rams, 29-11
reciprocating grate, 29-10
steam coil air heaters, 29-12
stoker capacity diagram, 29-10
stoker design, 29-10
undergrate air, 29-11
vibrating conveyor, 29-12
Master fuel trip, 23-5
Material handling and storage
equipment, 39-8
erection, 39-8
plan, 39-8
transportation, 39-8
Materials – see Chapter 7 and specific materials
Materials, bimetallic, 28-11
Maximum Achievable Control Technology (MACT), 32-5
Mean film temperature, 4-19
Measurement, instrument and methods (see also Chapters 36 and 40)
atomic absorption, 42-14
conductivity, 42-14
density, 36-4
flame photometer, 42-14
flow, 36-4, 40-20
gravimetric, 42-14
ion chromatography, 42-14
liquid level, 36-2
mercury emissions, 36-13
on grab samples, 42-14
on-line, 42-13, 42-15
opacity, 36-5, 36-13
pH, 36-1
pressure, 36-4, 40-1
purity, steam, 40-17
quality, steam, 40-17
solids level, 36-3
temperature, 36-4, 40-3
Mechanical equivalent of heat, 2-8, 3-3, 10-9
Membrane wall, 19-6, 23-1
construction, 39-18
fabrication, 38-7
gasifier design, 18-9
tubes, 23-1, 23-2
Mercury, 32-8
control technologies, 32-10, 32-11, 36-13, 36-14
in U.S. coals, 32-10
regulations, 32-6, 32-10
Mesh generation, 6-9
Metal lagging, 23-2
Metal removal, 38-2
Microalloyed steels, 7-12
Mineral wool, 23-5
Modeling (see also Numerical modeling)
ash, coal, 6-12
boiler, 6-14
burner, coal, pulverized, 6-21
combustion, 6-1, 6-7
large particle (popcorn) ash (LPA), 6-12
numerical, Chapter 6
recovery boiler, 6-14
stoker, 6-21
turbulent flow, 6-5
windbox, 6-16
Moderator, 46-1
temperature coefficient, 47-14, 49-18
Modular construction, 27-6, 39-6
field, 39-6
shop, 39-6
Moisture
air, 10-5
flue gas, 10-6, 10-17
gypsum, 36-10
in stack gas, 36-8
pulverization, 13-10
pulverized coal combustion, 14-3
Molar quantities
conversion to mass units, 10-4
Molar ratio, NO_x control, 34-8
Mole
concept of, 10-1
in combustion calculations, 10-3, 10-6
Mollier diagram
nuclear steam cycle, 2-21
steam, Chapter 2 Frontispiece, T-9
Momentum
conservation of, 3-2
Mud drum, 1-8
Multi-cyclone (multiclone) dust collector, 17-11, 27-10
Multi-lead ribbed tubes, 5-5, 26-13, 44-17
Municipal solid waste (MSW), 9-19, 12-18, 27-2, 29-1
National Ambient Air Quality Standards, 32-2
National Fire Protection Association (NFPA), 11-1, 13-12, 23-4
National Pollution Discharge Elimination System, 32-13
Natural circulation, 1-3, 5-18, 26-10, 26-11, 27-1, 27-15, 48-6
Natural gas – see Chapter 11
burner, 11-9
burning equipment, 15-10
consumption, 9-16
firing application, 26-16, 27-4
gas elements, 11-13
igniter, 11-15
preparation, 11-2
production, 9-16
properties, 11-3
reserves, 9-16
transportation, storage and handling, 11-2
Nautilus, 46-19
Navier-Stokes equation, 3-2
Navy special fuel oil (NSFO), 31-12
Net plant heat rate, 37-11
Neutron, 47-1
absorption, 47-6
delayed, 47-11
detector, 47-14, 49-17
energy distribution from fission, 47-8
flux, 47-11, 49-17
leakage, 47-11
production, 47-10
prompt, 47-11
reactions, 47-5
source, 47-14
thermal, 47-6
New Source Performance Standards (NSPS), 32-2
New Source Review (NSR), 32-2
Newton's law of cooling, 4-3
Newtonian fluid, 3-1
Nitrates, 34-2
Nitric acid, 34-2, 35-1
Nitrogen oxides (NO_x), 32-7, 34-1, 35-1 (see also NO_x and NO_x control)
combustion optimization, 43-6
emissions, 32-7, 34-1
formations, 34-1
nitric oxide (NO), 32-7, 34-1
nitrogen dioxide (NO_2), 32-7, 34-1
Non-condensable gases/waste stream, 28-23
Nonattainment Areas, 32-2
Nondestructive examination, 38-11, 45-1 (see also Examination, nondestructive)
eddy current testing (ECT), 45-3, 49-4, 50-10
probes, 49-4
electromagnetic acoustics, 45-4
leak testing, 50-10
dye penetrant (PT), 50-2
liquid penetrant (PT), 45-2, 49-10

magnetic particle (MT), 45-2, 49-10, 50-2
nuclear fluorescence, 45-4
radiographic (RT), 50-2
ultrasonic testing (UT), 45-2, 49-4, 50-2
Nonpressure parts, materials, 7-24
Nonrecoverable – see Irrecoverable
Normalizing, 7-8
NO_x
emissions control, post-combustion, 41-17
ammonia flow control, 41-17
flue gas inlet temperature, 41-17
NO_x control, 32-4, 32-9, Chapter 34
analyzer, 34-8
catalyst, 34-3
base metal, 34-5
corrugated fiber configuration, 34-5
honeycomb configuration, 34-5
operating issues, 34-6
parallel plate configuration, 34-5
precious metal, 34-5
sizing, 34-6
zeolite, 34-5
chemical reagents, 34-3
anhydrous ammonia (NH_3), 34-3
aqueous ammonia, 34-6
urea, 34-3
coal, and wood/bark stoker, 16-10
coal, Cyclones, 15-13
coal, fluidized-bed, 17-1, 17-8, 17-13
coal, gasification systems, 18-15
coal, pulverized, 14-9
air staging, 14-14
corner-fired, 14-16
wall-fired, 14-15
control by combustion, 14-10
fuel NO_x, 14-9
fuel ratio (FC/VM), 14-10, 14-18
fuel staging, 14-10
NO_x formation, 14-9
NO_x ports, 14-10
thermal NO_x, 14-10
combustion, 34-1
oil and gas
burners out-of-service, 11-6
flue gas recirculation, 11-7
low excess air, 11-6
overfire air ports, 11-6
oxides of nitrogen, 11-5
reburning, 11-7
two-stage combustion, 11-6
options, Chapter 34
post-combustion, 34-3
selective catalytic reduction, Chapter 34
numerical model, 6-18
NO_x formation, 34-1
NO_x ports, 14-10
Nozzle
energy equations, 2-9
fabrication, 38-6
spray, FGD, 35-3, 35-4
Nuclear (see also Fuel, nuclear, and Chapters 46 through 50)
binding energy, 47-2
chain reaction, 47-7
component maintenance, 49-2
cross-section, 47-12
emergency core cooling system (ECCS), 46-15
fission, 46-1, 47-1
nucleons, 47-1
nucleus, 47-1
nuclide, 47-1
fuel, 46-2, 46-8, 47-7
fuel assembly, 46-8
fuel cycle, 47-12
fundamentals, 47-1
icebreaker, *Lenin*, 46-20
merchant ship, *Otto Hahn*, 46-21
merchant ship, *Savannah*, 46-20
moderator temperature coefficient, 47-14
multiplication factor, effective, 47-10
operations, 49-14
power, Intro-12, Chapter 46
pumps and piping, 46-13
radiation, 47-2, 49-1
reactions, Chapter 47
reactor, 46-1, 46-5 (see also Reactor)
resonance, 47-6
safety systems, 1-13
service, Chapter 49
steam generator, 1-12, 46-12 (see also Steam generator, nuclear)
once-through, 46-12
recirculating, 46-12
steam supply system (NSSS), 1-11, 46-4
future of, 46-18
integrated control system, 46-14
Nuclear Regulatory Commission (NRC), 46-9, 49-14
Nuclear ships, 46-19
Nuclides, 47-1
Numerical analysis, 19-7
Numerical modeling, Chapter 6
benefits of, 6-2
carbon burnout kinetic (CBK) model, 6-8
chemical percolation devolatilization (CPD) model, 6-8
chemical reactions, 6-4, 6-7
combustion, 6-7
eddy dissipation combustion model (EDM), 6-7
eddy dissipation concept (EDC) model, 6-7
heterogeneous chemical reaction, 6-8
homogeneous chemical reaction, 6-7
computational fluid dynamics, 6-1
conduction heat transfer, 4-23
discrete ordinates, 6-9
discrete phase transport, 6-7
equations
discretization of, 6-8
fluid flow/heat transfer, 6-5
fundamental, 6-5
turbulence, 6-5
Eulerian reference frame, 6-7
examples of, 6-10
advanced burner, 6-21
Kraft recovery boilers, 6-14
popcorn (large particle) ash, 6-12
SCR systems, 6-18
wall-fired pulverized-coal boiler furnaces, 6-14
waste-to-energy system, 6-21
wet scrubbers, 6-10
windbox, 6-16
finite difference method, 6-9
finite volume approach, 6-9
fluid transport, 6-4
furnace heat transfer, 4-26
history of, 6-2
Lagrangian reference frame, 6-7
limitations of, 6-3
mathematical, 6-1
mesh generation, 6-9
particle, transport, 6-4, 6-7 (see also Numerical modeling, discrete phase transport)
process, 6-2
radiation heat transfer, 4-23
radiative heat transfer, 6-4, 6-8
theory of, 6-3
turbulence, 6-5
uses, 6-3
Nusselt number
definition, 4-2, 4-15
in long tubes or conduits, 4-16

O-type FM Boiler, 27-12
Oil – see Fuel, oil
Oil emulsions, 9-18
Once-through
boiler, 1-3, 26-9 (see also Universal Pressure (UP) boiler)
circulation, 1-3, 5-17
steam generator, nuclear, 1-12, 46-12, 48-2, 48-13, 49-1, 49-17, 50-2, 50-4
Ontario Hydro Method, 32-10
Opacity, 36-13
Openings, pressure vessels
ligaments between, 8-13
reinforcement, 8-13
Operating issues, 34-10
retrofits, 34-10
Operating license, nuclear, 49-14
Operation
fossil fuel-fired boilers, Chapter 43
abnormal, 43-9
combustion safety, 43-1
cooling requirements, 43-1
fundamental principles, 43-1

general, 43-1
general safety considerations, 43-1
low water, 43-9
personnel safety, 43-8
preparations for startup, 43-2
shutdown, 43-9
startup, 43-2, 43-4
tube failures, 43-9
water chemistry and treatment – see Chapter 42
nuclear steam supply systems, Chapter 49
acceptance tests, 49-20
operator requirements, 49-18
power ascension testing, 49-19
startup, nuclear, 49-18
zero power tests, 49-18
Operator
requirements, fossil fuel units, 43-2
training, nuclear, 49-18
Optical
fiber, 40-6
pyrometer, 15-3, 40-5
Orifice, 40-19
Orimulsion, 9-18
Orsat analysis, 10-21
Otto Hahn, 46-21
Overfire air (OFA), 11-6, 14-14, 15-13, 16-1, 16-3, 16-9, 17-4, 17-9
Overfire air ports – see NO_x ports
Overheat failures, 42-18
Overheating, 44-6
Oxidation, 44-11
FGD, forced, 35-3
resistance, 7-12
Oxygen
analyzer, 10-21
in feedwater, 26-5
required for combustion, 10-5
scavengers, 42-6
treatment, water, 42-8
Ozone, 32-2, 34-2

Package boiler, Intro-10, 27-9, 27-12
Paper industry – see Chapter 28
recovery boiler operation, 43-15
recovery processes and units, 28-1
steam requirements, 28-1
waste fuels, 28-1
Parr formulas, 9-5
Particle(s), atomic
alpha, 47-3
beta, 47-3
fundamental, 47-1
mass, 47-2
Particulate control, 32-9, Chapter 33
coal, gasification systems, 18-12
equipment, 33-2
electrostatic precipitators, 33-2 (see also Electrostatic precipitator)
fabric filters, 33-8 (see also Fabric filter)
mechanical collectors, 33-12
selection, 33-12
venturi scrubbers, 33-12
New Source Performance Standards, 32-2
opacity, 33-2
Particulate matter
condensable, 36-13
constituents, 33-1
Partnerships, 37-9
Pearlite, 7-4, 7-5
Peat, 9-6
Penthouse, 23-3
Performance
air heater, 20-12
calculations, Chapter 22
economizer, 20-4
environmental system, 37-4
monitoring, 41-21
steam generating system, 37-2
tests
environmental control systems, 36-7
fossil fuel-fired units, 43-8
nuclear units, 49-15
Periodic Table of the Elements, T-15
Personnel safety, rules, 43-8
Petroleum, 9-12 (see also Fuel, oil)
byproducts, 9-17
consumption, 9-14
pitch, 9-17
pH, 36-1
calibration, 36-2
control, 41-18
electrode, 36-1
feedwater, 42-6
for corrosion control, 42-6
fossil, drum, 42-6, 42-8
fossil, once-through, 42-12
sensors, 36-2
standardization, 36-2
Phase diagram, 7-2
iron carbon, 7-2
Phosphate hideout, 42-9
Photochemical smog, 34-2
Physical metallurgy of steel, 7-2
Pinch point, thermal, HRSG, 27-16
Piping
condition assessment of, 45-9, 45-14
creep, 45-9
leak detection, 45-17
materials, 7-23
nuclear, 49-10
Pitot tube, 3-4, 40-21 (see also S-type pitot tube)
Plant safety, nuclear, 46-15
Plate air heater, 20-8
Plate forming, 38-4
Plutonium, 47-4, 47-15
$PM_{2.5}$, 32-7
PM_{10}, 32-7
Pneumatic signals, 40-2
Pollution control – see Emissions, air pollution and individual pollutants)
Post weld heat treatment, 7-11, 50-5, 50-12
Potentiometer
direct reading, 40-10
millivolt, 40-10
Powder River Basin (PRB) coal, 9-2, 12-15, 21-13, 44-18
Powdered activated carbon (PAC), 32-12
Power
coefficient, nuclear, 47-14
costs, 26-1
cycle diagram
fossil fuel, 2-16, 2-18
nuclear fuel, 2-20
distribution, neutron, 47-7
for auxiliaries, fossil fuel equipment, 15-10, 26-5
Power plant
automation, 41-19
emissions, discharges
aqueous discharges, 32-13
atmospheric emissions, 32-1, 32-7
solid wastes, 32-1, 32-10
Prandtl number, 4-2, 4-14
Precipitation hardened stainless steels, 7-14
PrecisionJet, 16-9
Present value, present value factor, 37-12
Pressure
component assembly, 38-6, 39-12
difference, contraction and enlargement, 3-11
drop (or loss)
acceleration loss, 5-10
bend, 3-13
calculation, Chapter 22
contraction, 3-11
enlargement, 3-11
equation, closed channel, 3-7
fittings, 3-9
fluidized-bed, 43-14
friction loss, 3-5, 3-7, 5-10
gas flow across tube banks, 3-15
hydraulic or static loss, 5-10
local loss, 5-10
rectangular duct, 3-13
reheater relationship, 19-11
stack, 3-15
steam-water, 5-9
separator, 5-17
superheater, 19-11
two-phase (steam-water), 3-16, 5-9
valve, 3-9
gauge
dead weight, 40-2
opposed bellows, 40-2
slack diaphragm, 40-2
gradient, 3-1
loss – see Pressure, drop
manometer, 40-1
high pressure, 40-1
measurement
instrument connections, 40-3
part welds, 39-13

stress analysis, 8-3
taps, 40-3
transducer, piezoelectric, 40-3
transmitters, 40-2
vessel design, 8-1
Pressurized casing, marine, 31-12
Pressurized fluidized bed, 17-13
cycle, 17-13
pressure, 17-13
turbine, gas, 17-13
Pressurized water reactor (PWR), 1-12, Chapter 46, 48-1, 49-14
component, manufacturing, 50-1
components, 49-2
coolant pumps, 49-9
fuel handling, 49-11
naval type, 46-20
reactor vessel, 49-7
vessel internals, 49-8
Pressurizer, nuclear steam system, 46-12, 49-9, 50-12
heater, 49-9, 49-16, 50-12
Prevention of Significant Deterioration, 32-3
Primary air, 13-6
Process steam, 27-5, 27-15
Producer gas, 9-11
Prompt NO_x, 34-1
Properties (see also Chapters 2, 3, 4, 7 and 9)
acoustic, 40-6
boiling point, 40-3
coefficients of thermal, 40-5
electrical, 40-6
emissivity, 40-5
expansion, 40-4
freezing point, 40-3
gases, 40-4
liquids, 40-4
optical, 40-6
triple point, 40-3
Proportioning dampers, 19-15, 26-6
Propulsion, nuclear ship, 46-19
Proton, 47-1
Psychrometric chart, T-11
Pulping, 28-1
Pulverized coal (see also Combustion, coal, pulverized)
firing, application, 26-6, 27-2
introduction of, Intro-9
Pulverizer, Chapter 13
availability improvement, 44-3
ball-and-tube, 13-5
capacity correction factors, 13-11
capacity testing, 13-10
controls, 13-12
fabrication, 38-9
fluid energy, 13-15
high speed, 13-4
loading system, 13-14
maintenance, 13-11
principles of operation, 13-1
roll wheel, 13-1, 44-4
safety systems, 13-13
systems, 13-6
temperatures, 13-8, 13-12
throat, rotating, 44-5
types E and EL, 13-3, 44-3
upgrades, 44-3, 44-4
Pump, energy equations, 2-9
Pump seals, nuclear, 49-9
Purging, furnace, 43-4
Purification system, 49-15
Pyroelectric vidicons, 40-6
Pyrometer
Cyclone application, 15-3
fusion, 40-4
optical, 40-5
radiation, 40-5
Pyrometric cone, 40-4

Quality assurance, 38-10, 49-14, 50-1
Quality control, Intro-13, 38-10
Quaternary air, 28-21

Radiant boiler – see Boiler, types, Radiant
bypass systems, 19-20
Radiation, nuclear, 47-1, 49-1
absorbed, 47-3
detector, 47-3, 49-13
dose, 47-3
factor in maintenance, 49-1
protection, 47-3, 49-1
Radiation pyrometer, 40-5
Radiation, thermal, 4-3, 4-7
absorptivity, 4-3
between blackbody surfaces, 4-4
between gray surfaces, 4-4
blackbody, 4-4
carbon dioxide, from, 4-12
cavity, 4-28, 4-32, 20-4, 22-13
discrete ordinates, 6-9
emissive power of, 4-4
emissivity (emittance), 4-4
enclosures for, 4-7
gases, from, 4-8
geometry factor, 4-4
heat loss from setting, 23-6
heat transfer coefficient, 22-7
intertube, 20-4, 22-7
modeling of, 6-8
particles from, 4-8
reflection of, 4-3
reflectivity, 4-3
shield, 4-31
transmissivity, 4-3
water vapor, from, 4-12
Radioactive
decay constant, 47-4
half-life, 47-4
Radioactivity, 47-4
Radiographic examination, 50-2, 50-6
Radiography, 38-11
Rankine cycle, 2-13 (see also Cycle, Rankine)
in IGCC plants, 18-14
nuclear heat source, 2-19
Rankine temperature scale, 40-4
RDF
processing, 12-18 (see also Chapter 29)
RDF firing, refuse, 29-5, 29-13
combustion
air-swept distributor spouts, 29-17
Controlled Combustion Zone, 29-18
dump grate stokers, 29-13
fluidized-bed combustion, 29-18
fuel distribution, 29-17
metering feeder, 29-17
overfire air system, 29-18
refuse-derived fuel (RDF), 29-2
technology, 29-13
supplemental firing (co-firing), 29-13
traveling grate stoker, 29-2, 29-17
processing, 29-14
air density separator, 29-15, 29-17
aluminum separator, 29-17
disk screen, 29-15
eddy current separator, 29-17
flail mill, 29-15
magnetic separator, 29-15
oversized bulky waste, 29-17
quality, 29-16
shredder, 29-15
storage, 29-17
system, 29-14
tire shredding, 29-17
trommel, 29-15
yield, 29-15
RDF retrofit, 29-19
Reactivity, 47-11
coefficients, 47-14, 49-18
limestone, 36-6
Reactor
closure head, 46-9, 49-9, 50-1
replacement, 46-9, 50-1
containment building, 46-3
control rod drive mechanism (CRDM), 50-1
control rods, 46-8, 49-8
coolant pumps and motors, 46-13, 49-9
pump seals, 49-9
coolant system (RCS), 46-5, 49-14
core support, 46-6
decay heat removal system, 46-17
electropolishing, 50-2
emergency core cooling system, 46-15
hydrostatic testing, 50-2
inspection, 50-2
instrumentation, 46-11
machining, 50-2
operator, 49-18
plenum, 46-6
protection system, 46-15
refueling, 46-10
service support structure, 49-9
shielding, 46-5, 46-20
trip testing, 49-20

types
boiling water, 46-2
breeder, 47-15
converter, 47-15
fast, 47-15
pressurized heavy water (CANDU), 46-2, 46-5
water, Chapter 46, 47-9
naval, 46-20
thermal, 47-15
vessel, 46-5
vessel and internals, 49-7
Reactor vessel, 46-5, 50-1
Reburning, 32-9
Recirculating steam generator, nuclear, Intro-13, 1-12, 46-12, 48-4, 49-1, 49-17, 50-2, 50-4
Recirculation, gas, 19-15
Recovery boiler
economizer, 20-4
Kraft process, 28-1
black liquor, 28-2, 28-7
elemental analysis, 28-6
evaporation, 28-18
gross heating value, 28-5
limited vertical sweep burner, 28-20
mix tank, 28-21
oscillator burner, 28-20
oxidation, 28-19
splash plate spray nozzle, 28-20
combustion air system, 28-15
convection surface, 28-17
direct contact evaporator, 28-19
dissolving tank, 28-25
emergency shutdown system, 28-18
emissions, 28-8
evaporator, 28-18
feedwater temperature, 28-18
furnace, 28-3, 28-15
heat balance, 28-5
heat inputs, 28-5
heat losses, 28-5
numerical model, 6-14
operation, 43-15
black liquor combustion process, 43-16
primary air, 43-16
secondary air, 43-16
tertiary air, 43-16
overview, 43-15
pluggage, 43-16
tube corrosion, 43-16
rated capacity, 28-3
smelt spout system, 28-24
smelt water reactions, 43-17
soda process, 28-25
sootblowers, 28-23
streams entering and leaving, 28-5
sulfite process
magnesium, 28-26
thermal performance, 28-4
vent stack scrubber, 28-25
wall construction, 28-11
Rectangular ducts, flow, 3-13
Recuperative, air heater, 20-7
Red liquor – see Recovery boiler, sulfite process, magnesium
Refinery gas, 27-2
Reflooding, 5-4
Refractories, 7-15
Refractory, 23-1
materials, 23-6, 39-18
Refueling, nuclear, 49-11
fuel handling equipment, 49-12
Refuse, 10-14, 29-1 (see also Residue or Ash, coal)
Refuse-derived duel – see RDF firing
Regenerative
air heater, 20-9
heating surface, 20-11
Rankine cycle, 2-14
Reheat
cycle, 2-15
flue gas, FGD, 35-3
temperature adjustment and control, 19-15, 19-21
use of, 19-3
Reheater (or reheat superheater), 1-7, 19-11
attemperator, 19-14
cleaning, external, 19-12
cleaning, internal, 19-12
design, 19-11
marine, 31-10
materials, 7-21
Reinjection
stoker, carbon, 16-6
stoker, flyash, 16-9
Relative roughness, 3-9
Remote manual sequence burner controls, 41-14
Repairs, nuclear equipment, 49-1
Repowering, 26-5
Research and development, Intro-13, 48-14
Residual elements, 7-12
Residual stress, 50-2
Residue, 10-14
Resin fragments, 42-6
Resistance, heat transfer, 4-1
Resource Conservation and Recovery Act (RCRA), 32-15
Retort, 16-2
Revenue requirements method, economics, 37-13
Reverse osmosis, 42-5
Reynolds number, 4-2, 4-15
Reynolds number, friction factor relationship, 3-7
Ribbed tube, 5-5, 26-13, 44-17
Rickover, Admiral Hyman G., 46-1, 46-21
Rigging design, 39-8
Riser (tubes), 1-9
Risk, financial, 37-13
Rothemühle air heater, 20-10
Roughness, effect on friction factor, 3-9
S-type pitot tube, 36-8, 36-10
Safety
burner, coal, 14-20
burner, oil and gas, 11-16
combustion, 14-20, 43-1
construction, 39-16
factors, C-3
nuclear,
analysis report, 49-14
limits, 49-15
pulverizer, 13-13
relief valve, 25-1
valve, 19-8
valve, setting, 43-4
Sampling, water and steam, 42-13
Saturated steam, properties, 2-2, 2-3, T-5, T-6
Saturated water, properties, 2-2, 2-3, T-5, T-6
Savannah, 46-20
Scale, marine, 31-8
Scale detection, fossil fuel boiler, 40-12
Scaling, 35-2
Schedule, construction, 39-2
Scope of supply or work, 37-5
SCR – see Selective catalytic reduction
Screen
heat transfer and draft loss calculations, 22-4
Screw feeders, 16-7
Scrubber, dry, Color Plate 7 – see SO_2 control
Scrubber, wet, Color Plate 6 – see SO_2 control
Selective catalytic reduction (SCR), 27-5, 27-7, 29-23, 32-9, 32-12, 34-3, 39-15 (see also NO_x control)
applications, 34-3
arrangements, 34-4
catalyst – see NO_x control, catalyst
cold side/low dust, 34-4
distributions, 34-8
ammonia to NO_x, 34-9
temperature, 34-9
velocity, 34-8
hot side/high dust, 34-4
mixing, 34-10
retrofits, 34-10
systems, 34-4
Selective non-catalytic reduction (SNCR), 29-23, 32-9, 34-3, 34-13 (see also NO_x control)
Series present value factor, economics, 37-12
Setting, boiler
appearance, 23-9
cased enclosures and casing, 23-3, 39-18
construction, 23-1
corrosion, 23-7
definition, 23-1
enclosures, 23-1
expansion, 23-4
explosion effect, 23-4

fabrication and assembly, 23-9
heat loss, 23-5
implosion effect, 23-5
inner casing, 23-2
insulation, 23-5, 23-6, 39-18
leakage, 23-7
outer casing, 23-2
penthouse, 23-3
resistance to ash and slag, 23-4
resistance to weather, 23-8
serviceability, 23-9
tube wall, 23-1, 23-2
ventilation, 23-6
vibration, 23-5
windbox, 23-3
Shape or configuration factors, thermal
definition, 4-4
for radiation, 4-4, 4-7, 4-31
Sheet metal lagging, 23-8, 39-19
Shielding, nuclear, 46-5, 46-20, 47-4, 49-8, 49-13
Shipboard nuclear reactors, 46-20
Shipment and shipping, 39-6, 50-12
Shutdown
fossil fuel-fired units, 43-9
Signal multiplexer, 40-24
Slag, 21-3, 43-7, 43-10
entrained flow gasifier, 18-6
layer, 15-3
removal, 15-3, 15-8, 24-2, 24-13
screen, 19-6
viscosity, 9-8 (see also Ash, coal)
Slagging, 21-3, 44-9 (see also Ash, coal, deposition)
Slaker, FGD
ball mill, 35-15
detention, 35-15
Slip, ammonia (NH_3), 34-6
Slurry, dewatering, FGD, 35-7
Smart transmitters, 41-20
Smelt, Kraft process, 28-2, 43-19
reduction efficiency, 28-8
SO_2 control, Chapter 35
fuel switching, 32-5, 32-8, 35-1, 44-18
reagent(s)
lime, 35-2, 35-12, 35-16
limestone, 35-2, 35-6, 35-10
magnesium oxide, 35-12
scrubber (FGD)
chemical analysis, 36-6, 36-8
control, 41-17
density, 41-18
level, 41-18
pH, 41-18
dry, 35-12, Color Plate 7 (see also Spray dryer absorber (SDA))
lime, 35-2
limestone, 35-2, 35-6
liquid level, 36-2, 36-3
magnesium oxide, 35-12
numerical model, 6-10
performance testing, 36-7
pH measurement, 36-1
reagent consumption, 36-13
spray tower, 35-2
stoichiometry, 35-7
tray, 35-3, 35-4
wet, 35-2, Color Plate 6
technologies, 32-8
SO_3 control, 35-18
Society of Naval Architects and Marine Engineers (SNAME), 31-6
Soda process – see Recovery boiler, Kraft process
Sodium
carbonate, 35-2
cycle softener, 42-5
selective ion electrode, 42-15
tracer method, 40-17
zeolite softener, 42-5
Solid waste
disposal, 32-15
treatment, 32-16
utilization, 32-17
Solids
carryover, 5-13
concentration, 19-14
gasifier discharge, 18-12
gravimetric analysis, 40-17
in boiler water, 5-13
in steam, 5-13, 40-17
Solution annealing, 7-8
Sootblowers, 20-3, 20-14, 24-1, 44-19
application, 24-5, 28-23
cleaning media, 24-1
control(s), 24-4
intelligent controls, 43-8
marine, 31-3
operations, 43-8
RDF boiler, 29-21
recovery boiler, 28-23
terminology, 24-1
type
air puff, 24-12
extended lance, 24-7
fixed position, 24-8
G9B, 24-8
HydroJet, 24-3
IK, 24-5
IR, 24-2
long retractable, 24-5
one way, 24-7
oscillator, 24-7
Precision Clean, 24-7
rake, 24-9
rotary, 24-8
short retractable, 24-2
straight line, 24-10
swing arm, 24-11
waterlance, 24-2
Sorbent (see also SO_2 control, reagent)
calcination, 10-14, 10-28
calcium to sulfur molar ratio, 10-11, 10-14, 10-28
combustion calculations, 10-11, 10-28
injection, 35-18
spent sorbent, 10-14
sulfation, 10-14
Specific gravity, 9-14
Specific heat, 2-22, 4-9
ratio, 3-4
Specific volume
water and steam, 2-1, 2-2, T-5, T-6
Specifications, Chapter 37
Spent fuel, nuclear, 46-11, 50-13
Spheroidizing, 7-9
Spout – see Air swept spout
Spray dryer absorber (SDA), 35-12 (see also SO_2 control)
approach temperature, 35-14
atomizer, 35-13
gas disperser, 35-13
process, 35-16
reagent, 35-15
recycle, 35-15
single pass, 35-15
spray chamber, 35-13
spraydown, 35-15
stoichiometry, 35-15
Stability, 5-12
Stack and draft, 25-6
barometric pressure, 25-7
draft
balanced, 25-6
forced, 25-6
induced, 25-5
loss, 25-6
natural, 25-6
loss (or resistance), 43-6
friction, 25-8
velocity head, 25-8
specific volume, 25-6
air and flue gas, 25-6
stack
effect, 22-18, 25-6
flow, 3-15
flow loss, 25-8, 43-6
operation and maintenance, 25-10
Staged combustion, 14-11, 15-13, 16-6, 44-8
Startup
fossil fuel-fired units, 43-2, 43-4
State Implementation Plans (SIPs), 32-2
Steady-state
flowing systems, 4-6
heat conduction, 4-5
Steam
air heater, 31-10
as a thermodynamic system, 2-1
capacity, 27-8
coil air heater, 20-9
conditions, 37-2
cycle, 2-14
nuclear plant, 2-19
enthalpy, 2-1, 2-2, T-5, T-6
entropy, 2-1, 2-2, T-5, T-6
flow requirements, 27-2
generating bank, 1-8 (see also Boiler, bank)
generation fundamentals, 1-2
heat recovery steam generator,

gasification, 18-14
history of generation and use, Intro-1
impurities, 5-13
line blowing, 42-3
Mollier diagram, Chapter 2 Frontispiece, T-9
natural separation, 5-14
pressure-enthalpy diagram, 2-6, T-10
process and heating, 27-3
properties, Chapter 2, 4-9, 5-1, T-5, T-6
purity, 40-17, 42-2, 42-12
measurement, 40-17
quality, 5-1, 40-17
measurement, 40-17
requirements, 23-3, 26-4
samples, 42-13
sampling nozzle 40-17
installation location, 40-17
scrubber, secondary separator, 5-13
separation, 5-13
separators, 5-13, 31-13, 48-2, 48-7, 50-10
solids removal, washing, 5-16
specific volume, 2-1, 2-7, T-5, T-6
tables, 2-2, T-5
temperature
adjustment, 19-13
control, 1-8, 19-12, 41-10, 43-6
attemperation, 19-14
gas recirculation, 19-15
gas tempering, 19-15
temperature-enthalpy diagram, T-12
viscosity, 3-1, 3-9, 3-12
Steam engine (see also Boiler, historical)
aeolipile, Intro-1
beam, Intro-3
Branca, Giovanni, Intro-1
Corliss, Intro-5
Hero's, Intro-1
Newcomen, Thomas, Intro-3
Savery, Thomas, Intro-3
Steam flow-air flow control, 41-13
Steam generation, Chapter 1
history of, Intro-1
Steam generator
fossil fuel, 1-6 (see also Boiler)
nuclear, Intro-12, 1-12 (see also Chapter 48)
baffle plate, 50-7
blowdown, 48-7
CANDU, 48-1, 50-2
carryover, 48-7
carryunder, 48-7
circulation, 48-3
clean room, 50-8
cleaning, cleanliness, 50-8
closing seam, 50-10
cracking, 48-9
damage (mechanisms), 48-9
denting, 48-9
divider plate, 48-12, 50-4, 50-8
downcomer, 48-5
electropolishing, 50-11
erosion-corrosion, 48-9
feedwater, 48-2, 48-6
flow circuits, 48-2, 48-6
foreign object damage, 48-9
fretting, 48-9
heat transfer, 48-3, 48-6
instrumentation, 48-4
integral economizer (preheater), 48-3, 48-7, 50-8
intergranular attack, 48-9
lattice grid (tube support), 48-7, 48-11, 50-6, 50-8
manipulator, 49-5
manufacture, 50-2
natural circulation, 48-6
naval, 46-20
once-through (OTSG), 1-12, 46-4, 46-12, 48-2, 48-13, 49-1, 49-17, 50-2, 50-4
pitting, 48-9
post weld heat treatment, 50-5, 50-11
preheater, 48-7
pressure boundary, 50-5
primary head, 50-4, 50-6, 50-11
primary side maintenance, 49-3
tubing, 49-3
eddy current testing, 49-4
plugging, 49-5
repair, 49-6
recirculating (RSG), 46-5, 46-12, 48-4, 48-10, 49-1, 49-17, 50-2, 50-4
replacement, 48-9
secondary side maintenance, 49-7
chemical cleaning, 49-7
sludge lancing, 49-7
shell (course), 50-3
shroud (wrapper), 48-2, 48-6, 50-8
steam drum, 48-4, 48-12
steam drum head, 50-4
steam drum internals, 50-10
stress corrosion cracking (SCC), 49-2, 49-7, 48-9, 49-8
support plate, 48-2, 48-7
surface conditioning, 50-11
tack expansion, 50-9
terminal points, 50-11
tie rods, 50-8
tube
bending, 50-7
bundle, 50-6, 50-8
expansion, 50-10
installation, 50-9
ovality, 50-7
plugging, 49-5
repair, 49-6
sleeving, 49-6
support plate (TSP), 50-6, 50-8
tube material, 48-4, 48-10
tube to tubesheet welding, 50-9
tubes, 50-6
tubesheet, 48-2, 48-6, 50-4
U-bend support (restraint) 48-6, 48-8, 50-7
oxygen converter hood, 27-20
Steam line blowing, 42-19
Steam tables (SI) and selected charts, Appendix 1
Steaming economizers, 20-4
Steel, structural, 8-14, 38-8
Stefan-Boltzmann constant and law, 4-4, 40-5
Stoichiometry
NO_x control, 34-7
SO_2 control, 35-16
Stoker, Chapter 16
coal characteristics, 16-4, 16-6
industrial boiler applications, 16-1, 27-1, 29-9, 30-4
marine, 31-4
mass feed, 16-3
numerical model, 6-21
overfeed, 16-1
reciprocating grates, refuse, 29-10
spreader stoker, coal-fired, 16-4
air system, 16-7
ash removal, 16-7
carbon injection, 16-6
feeders, 16-4, 16-7
grates, 16-5
selection/sizing, 16-7
spreader stoker, wood-, bark-, biomass-fired, 16-7, 30-4
air system, 16-9
ash removal, 16-10
combustion air temperature, 16-10
feeders, distributors, 16-7
grates, 16-8, 30-4
selection/sizing, 16-10
traveling grate, refuse, 29-17
underfeed, 16-2
Storage areas, site, 39-11
Strain gauges, 40-3
Stress
allowable, ASME Code, 7-20, 8-3, C-4
alternating, 8-6
analysis methods, 8-3
classifications, 8-2, 8-9, 8-10
concentrations, 8-2, 8-6, 8-14
discontinuity, 8-7
drum, 43-5
failure theories, 8-2
hoop, 8-3, 8-4
intensity factor, 8-10
longitudinal, 8-3
peak, 8-3
pressure, 8-3
primary, 8-2
secondary, 8-3
significance, 8-1
steady-state, 8-2
strength theories, 8-2
thermal, 8-5
transient, 8-2
Stress corrosion cracking, 50-1, 50-10

(see also Corrosion, stress corrosion cracking)
Stress-strain diagrams, 7-16
Stringer tubes, 20-5
Structural analysis, Intro-16, Chapter 8
Structural support
  codes and standards, 8-15
  design loads, 8-14
  linear type, 8-16
  plate type, 8-15
  shell type, 8-15
Stud-tube walls
  pin stud, recovery furnace, 28-11
  sulfite process, 28-26
    ammonium-base, 28-27
    calcium-base, 28-26
    magnesium-base, 28-26
    sodium-base, 28-26
Studded surfaces, 15-10, 29-3
Subcooling, 44-15
Subcritical (pressure) drum boilers, 26-11
Sulfation, 17-8, 17-13
Sulfite process – see Recovery boiler, sulfite process
Sulfur, 9-3
  acid gas removal, 18-13
  capture, 10-16, 10-31
  Claus recovery process, 18-15
  coal, gasification systems, 18-13
  content – U.S. coals, 9-3, 12-5
    forms, 12-5
    organic, 12-5
    pyritic, 12-5
  dioxide (SO_2)
    control, Chapter 35 (see also SO_2 control)
    emissions in flue gas, 35-1
  emissions control, FGD, 35-1
  hot gas desulfurization, 18-13
  in syngas, 18-13
  trioxide (SO_3), 32-7, 35-18
Sulfuric acid, 35-1
Supercritical (pressure) boilers, 19-2, 20-6, 26-6
Supercritical pressure steam
  heat transfer, 5-7
  properties, 5-7
Superheat, 19-9 (see also Steam, temperature)
Superheated steam, properties, 2-4, T-7
Superheater, 1-7, 19-9
  cleaning, external, 19-12
  cleaning, internal, 19-12
  condition assessment of, 45-7
  convection, 19-9
  design, 19-10
  fabrication, 38-7
  heat transfer surface, 19-11
  horizontal, 19-12, 19-13
  ligament damage, 45-8
  marine, 31-8
  mass velocity or mass flux, 19-10
  materials, 7-21
  pendant, 19-11, 19-13, 26-6, 26-11, 28-17
  performance calculations, 22-8
  platen, 28-17
  radiant, 19-9, 28-17
  remaining life, 45-11
  separately fired, 19-16
  startup protection, 43-5
  supports, 19-11
  tube metals, 7-21, 19-11
  tube sizes, 19-10
  tube spacing, 19-7, 19-11
  types, 19-9
Supply tubes, 1-9
Supports
  boiler, 19-8
  reheater, 19-11
  superheater, 19-11
Surface tension
  water and vapor, 3-1
Surveillance
  requirements, 49-15
Syngas compositions, 18-11

Temperature
  adiabatic flame, 2-26, 10-11
  difference, log mean, 4-19
  entropy diagrams, 2-11
  exit gas, 19-4
  gradient, 40-12
  ignition, 10-10
  measurement, 40-3
    acoustic, 40-6
    bimetallic strip, 40-5
    casing, 40-16
    fluid, 40-10
    fusion, 40-4
    gas, 40-14
      evaluation, 40-16
      weighted average, 40-16
    gradient, 40-12
    infrared, 40-6
    insulation, 40-16
    methods, 40-4
    standards, 40-3
    tube, 40-11
  recorders, 40-10
  scales, 40-3
Tempering
  gas, 19-15
  gas plenum, 23-3
  materials, 7-9
Tensile test, 7-16
Terms and conditions, contract, 37-7
Testing, nuclear, 49-15
  acceptance, 49-20
  electrical, 49-15
  full load, 49-18
  functional, 49-16
  hot functional, 49-16
  hydrostatic, 49-16
  instrumentation, 49-16
  operational, 49-16
  preoperational, 49-15
  procedures, 49-15
  zero power, 49-18
Thermal
  analysis, 8-9
  circuit, 4-6, 4-30
  conductivity, 4-1, 4-9
    of gases, 4-9, 4-11
    of insulating materials, 4-9
    of liquids, 4-9
    of steels and alloys, 4-10
  efficiency, Carnot cycle, 2-12
  neutron, 47-6
  probe, 43-5
  resistance, 4-1, 4-6
    contact, 4-1
    in conduction, 4-3
  utilization, 47-11
  viewers, 40-6
Thermal conductivity
  of insulating materials, 23-6
Thermal NO_x, 34-1
Thermocouple, 40-7, 43-5
  chordal, 40-12, 42-19
  circuit arrangement, 40-7
  cold junction, 40-7
  correction for, 4-31, 40-8
  emf (chart), 40-7
  gas temperature error graph, 40-14
  gradient, 40-12
    limitations, 40-13
    scale detection, 40-12
  high velocity, 40-14
  hot junction, 40-7, 40-9
  lead wire, 40-8
  material selection, 40-8
  multiple circuits, 40-7
  multiple-shield high velocity, 40-15
  pad type, 40-8
  peened, 40-9
  principles of, 40-7
  sheathed type, 40-8
  types, 40-7
Thermodynamics – see Chapter 2
  cycles, 2-1, 2-12
  definitions, 2-1
  first law of, 4-5
  laws of, 2-10
  process, 2-7
    irreversible, 2-11
    reversible, 2-10
Thermographic imagers, 40-6
Thermoluminescent dosimeter, 49-13
Thermometer
  gas, 40-4
  liquid in glass, 40-4
  mercury, 40-4
  resistance, 40-6
  vapor pressure, 40-4
  well, 40-10
Thermopile, 40-7
Thermoprobe, 40-16
Thermowell, 40-10
Thorium, 47-4
Throat tap nozzle, 40-20
Throttle pressure control, 43-12
Throttling calorimeter, 40-19

Time-temperature-transformation (TTT) diagrams, 7-4
Tire shredding, 29-17
Title IV (Acid Deposition Control), 32-4
Total organic carbon (TOC), 42-4
Toughness, 7-17
Trademarks (B&W) appearing in this edition, TM-1
Transducer, 40-3
Transfer units, 35-10
Transformation diagrams, isothermal, 7-4
Transient or unsteady-state conduction, heat transfer, 4-5
Tray, perforated plate, 35-3, 35-6
Trepanning, 50-2
Trim, marine, 31-1
Tube
  banks, 19-7 (see also Banks of tubes)
  bending, 38-4, 50-7
  cold drawing, 38-10
  composite, 28-11
  crossflow pressure drop, 3-15
  electric resistance welding, 38-10
  expansion, 50-10
  flat stud, 23-2
  forming, 38-4
  hole drilling, 38-6
  leak detection, 45-16
  leg flexibility, 45-19
  manufacture, 38-10
  membrane, 23-1
  plugging, nuclear, 49-5
  refractory lining, 23-1
  remaining life, 45-11
  repair, nuclear, 49-6
  seamless, 38-10
  sizes, 19-10, 31-2, 31-5
  sleeving, 49-6
  spacing, 19-7, 19-11
  surface, 40-11
  tangent, 23-2
  temperature, 40-11
  upgrades, 23-2
  wall enclosures, 23-1
Tube banks, 4-27
  arrangement factors, 4-19, 4-20, 4-21
  crossflow over, 4-16
  longitudinal flow over, 4-18
  radiation, 4-28
Tube rolling and expanding, 39-13
Tubular air heater, 20-7
Turbine
  combustion in IGCC plants, 18-14
  energy equation, 2-9
  feed pump, nuclear, 49-17
  following control, steam, 41-5
  gas, 17-13
  nuclear, bypass valve, 49-17
  steam, 17-13
  trip, testing, 49-20
Turbine-generator control, 43-6
Turbulence fluctuations, 6-5
Turbulence model, 6-5
Turbulent flow, 4-14, 4-16 (see also Flow, turbulent)
  crossflow around tubes, 4-16
  inside tubes, 4-16
  longitudinal flow around tubes, 4-18
  structure of, 4-16
Turning vanes, flow through, 3-13
Turnkey, 37-5
Two-phase flow, steam-water, 5-8
  heat transfer, boiling, 5-3
  instability, 5-11
  pressure drop or loss, 5-9
  separation, 5-13
  stability, 5-12
  void fraction, 5-9

U-beams, 17-11, 27-10
Ultimate tensile strength, 7-17
Ultrasonic examination, 38-12, 45-2, 49-4, 49-10, 50-2, 50-5
  immersion testing, 45-3
  oxide measurement, 45-2
  shear wave, 45-3
  thickness testing, 45-2
  time of flight defraction, 45-3
Unburned carbon or unburned combustible, 10-14, 44-7
  pulverized coal, 14-8, 14-19
Undergrate air, 16-1, 16-6, 16-9
Universal gas constant, 3-4, 10-3, 10-9, 40-4
Universal Pressure (UP) boiler, Intro-10, 19-2, 26-6, 26-9
  attemperators, 19-15
  bypass and startup systems, 19-16, 19-20
  control system philosophy, 41-12
  controls, 41-12
  operation, 43-11
    bypass system, 43-11
    control functions, 43-11
    limits and precautions, 43-12
    practice, 43-11
    principles, 43-11
    skills, 43-12
    startup, 43-12
Upgrades, boiler
  bypass system, 44-16
  capacity, 44-20
  circulation, 44-16
  cycling operation, 44-15
  economizer, 44-7
  headers, 44-14
  superheater, 44-13
Uranium, 47-4
Uranium dioxide, 46-8
Urea, 34-3
Useful charts and tables, Appendix 1
U.S. Naval Nuclear Propulsion Program, 46-21
U.S.S. *Kennedy*, 31-1
U.S.S. *Missouri*, 31-1
U.S.S. *Nautilus*, Intro-12, 46-22
U.S.S. *Ronald Reagan*, 46-25
U.S.S. *Seawolf*, 46-23
U.S.S. *Skipjack*, 46-24
U.S.S. *Virginia*, 46-24
Utility power plant, Color Plate 8

Vacuum filter, 35-8
Valve
  flow, coefficient, 3-12
  flow resistance, 3-9
  nuclear, 49-10
  safety and relief – see Safety, relief valve
  Universal Pressure boiler, bypass, 43-12
Vapor, superheated, 3-1
Vaporization (ammonia), 34-7
Variable pressure operation, 19-2, 19-17, 19-20, 43-6
  operation control, 41-7
Vegetation wastes, 9-19
Velocity
  fluidizing, 17-2
  head, 3-6
  head, mass-velocity relationship for air, 3-15
  head loss for valves and fittings, 3-12
  minimum, 17-2
  ranges, 3-9
  steam generating system, 3-9
Ventilation, boiler room, 23-6
Venturi, 3-4
  air foil, 40-21
  tube, 40-20
Venturi scrubber, 35-2
Vibration
  buckstays, 23-5
  casing, 23-5
Viscosity, 3-1
  absolute, 9-14
  apparent, 3-9
  coal ash slag
    effect of iron, 15-4
  compared with kinematic, 3-10
  gases of, 4-11
  in common gases, 3-11
  in common liquids, 3-11
  oxidizing versus reducing atmosphere, 15-13
  requirements for Cyclone furnaces, 15-4
  steam, 3-12
  T_{250}, 15-4
Void fraction, 5-9
Volatile matter, coal
  ASTM test, 14-3
  fuel ratio (FC/VM), 14-10, 14-18
  NO_x influence, 14-5
Volatile organic compounds (VOCs), 32-7

Warm working, 7-9
Waste, 12-18
  energy, 27-4
  gas turbine exhaust, 18-14

heat
marine boiler, 31-5
steam generator, 27-15
Waste burning (see also Fuel)
fuel
application, 15-3, 15-9
Water
alkalinity, 42-8
analysis (see also Measurement), 42-13
attemperation, 42-7
boiler, 42-2
chemistry, Chapter 42
quality assurance, 42-14
chloride, 42-4
coil, air heater, 20-9
condensate, 42-6
conductivity, 42-14
demineralizer, 42-5
enthalpy, 2-2, T-5, T-6
entropy, 2-2, T-5, T-6
feedwater, 42-4
foaming, 42-12
for spray attemperators, 42-7
gas-shift reaction, 18-2
gasification system treatment, 18-15
hammer, 20-4
hardness, 42-2, 42-3
hydroxide, 42-1
iron, 42-4
makeup, 42-5
organics, in water, 42-5
oxygen, 42-4
pollution control, 32-13
technologies, 32-15
U.S. legislation, 32-13
quality, 42-5, 49-16
boric acid, 49-16
corrosion product transport, 49-16
pH, 49-16
requirements
fossil fuel boilers, Chapter 42, 43-8
nuclear units, 49-16
seal tanks, 27-18
silica, 42-9, 42-13, 42-14
sodium, 42-2, 42-5
specific volume, 2-1, 2-2, T-5, T-6
total dissolved solids, 42-9, 42-12
treatment, 32-15, 42-1
all volatile, 42-8
chelants, 42-11
conventional phosphate, 42-11
coordinated phosphate, 42-8
deaeration, 42-6
demineralization, 42-5
dispersants, 42-11
equilibrium phosphate, 42-9
evaporation, 42-2
for drum boilers, 42-7
high alkalinity phosphate, 42-11
hydroxide, 42-1
internal, 42-7
oxygen, 42-8
pH control, 42-6
phosphate, 42-8
polymers, 42-7, 42-11
raw water, 42-5
reverse osmosis, 42-5
softening, 42-5
Universal Pressure boilers, 42-12
waste water, 42-24
Water-cooled furnace
introduction of, Intro-9
Water-cooled walls, 19-6
Water leg correction, 40-3
Weak wash, pulp, 28-3, 28-5
Wear, pulverizer, 13-9
Wear metals
castings, 38-4
Wear resistant materials, 44-10
Weld
cladding, 7-14, 44-8
Welding, 7-10, 38-1
arc, 38-1
automation, 38-1
cladding/overlay, 50-2, 50-4, 50-13
development of, Intro-13
electrodes, 38-6
gas metal arc welding, 38-1, 50-2
gas shielded metal arc, 38-1
gas shielded tungsten arc, 38-1
gas tungsten, 38-1, 50-5, 50-9
heat treatment, 38-1
high frequency resistance, 38-1
laser beam, 38-1
orbital, 38-2, 50-5
plasma arc, 38-1
processes, 38-1
robotic, 38-2
shielded metal arc, 38-1
submerged arc, 38-5, 50-2, 50-5
tube to tubesheet welding, 50-9
Welds
applications, 39-14
codes and inspection, 39-14
field, 39-13
nondestructive examination, 38-11, 49-2
pressure, 39-14
processes and equipment, 39-14
qualification, 39-14
seal, 39-14, 50-6
thermal treatment, 39-14
Wet electrostatic precipitator, 33-8
Wet scrubbers – see SO_2 control
Wheatstone bridge, 40-6
White iron, 7-14
White liquor, pulp, 28-2
Wick boiling, 42-2
Wilcox, Stephen, Intro-5, Intro-7
Wind loading, 26-4, 39-9
Wind swept spout – see Air swept spout
Windbox, 23-3
Wood
as a fuel, 30-2
fuel properties, 9-18
Work breakdown structures, 39-3
Work-hardening, 7-9

Xenon, 47-6, 47-9, 49-20
X-ray, 47-1
machine, 38-6, 38-11

Yancey-Geer Price, 9-9, 13-9
Yield point, 7-17
Yield strength, 7-17
Young's modulus, 7-16